Geological Survey of Canada

Geology of Canada, no. 6

GEOLOGY OF THE
APPALACHIAN-CALEDONIAN OROGEN
IN CANADA AND GREENLAND

edited by

Harold Williams

1995

This is volume F-1 of the Geological Society of America's Geology of North America series
produced as part of the Decade of North American Geology project.

© Minister of Energy, Mines and Resources Canada 1995

Available in Canada through

Geological Survey of Canada offices:

601 Booth Street
Ottawa, Canada K1A 0E8

3303-33rd Street N.W.
Calgary, Alberta T2L 2A7

100 West Pender Street
Vancouver, B.C. V6B 1R8

or from

Canada Communications Group - Publishing
Ottawa, Canada   K1A 0S9

A deposit copy of this publication is also available for reference
in public libraries across Canada

Cat. No. M40-49/6E
ISBN 0-660-13134-X

Price subject to change without notice
Cette publication est aussi disponible en français

Design and layout
P.A. Melbourne
S. Leslie

Cover
*Gently dipping late Precambrian redbeds of the Signal Hill Group, Cape Spear,
Newfoundland looking north to Cape Spear. Cape Spear is the most easterly
point of North America. Photo by Harold Williams GSC 1994-800.*

Printed in Canada

PREFACE

The *Geology of North America* series has been prepared to mark the Centennial of The Geological Society of America. It represents the co-operative efforts of more than 1000 individuals from academia, state and federal agencies of many countries, and industry, to prepare syntheses that are as current and authoritative as possible about the geology of the North American continent and adjacent oceanic regions.

This series is part of the Decade of North American Geology (DNAG) Project which also includes eight wall maps at a scale of 1:5 000 000 that summarize the geology, tectonics, magnetic and gravity anomaly patterns, regional stress fields, thermal aspects, seismicity, and neotectonics of North America and its surroundings. Together the synthesis volumes and maps are the first co-ordinated effort to integrate all available knowledge about the geology and geophysics of a crustal plate on a regional scale.

The products of the DNAG Project present the state of knowledge of the geology and geophysics of North America in the 1980s, and they point the way toward work to be done in the decade ahead.

From time to time since its foundation in 1842 the Geological Survey of Canada has prepared and published overviews of the geology of Canada. This volume represents a part of the seventh such synthesis and besides forming part of the DNAG Project series is one of the nine volumes that make up the latest *Geology of Canada*.

J.O. Wheeler
General Editor for the volumes
published by the
Geological Survey of Canada

A.R. Palmer
General Editor for the volumes
published by the
Geological Society of America

ACKNOWLEDGMENTS

Although the *Geology of Canada* is produced and published by the Geological Survey of Canada, additional support from the following contributors through the Canadian Geological Foundation assisted in defraying special costs related to the volume on the Appalachian Orogen in Canada and Greenland.

Alberta Energy Co. Ltd.
Bow Valley Industries Ltd.
B.P. Canada Ltd.
Canterra Energy Ltd.
Norcen Energy Resources Ltd.
Petro-Canada
Shell Canada Ltd.
Westmin Resources Ltd.

J.J. Brummer
D.R. Derry (deceased)
R.E. Folinsbee

CONTENTS

FOREWORD

Most of this volume and its systematics focus on the Canadian Appalachian region. The chapter on the East Greenland Caledonides stands alone and there has been no attempt to integrate the geological accounts of the two far removed regions. The East Greenland Caledonides are an extension of the Appalachian miogeocline or Laurentian continental margin, and outboard terranes are absent.

Rocks of the Canadian Appalachian region are described under four broad temporal divisions: lower Paleozoic and older, middle Paleozoic, upper Paleozoic, and Mesozoic. The rocks of these temporal divisions define geographic zones, belts, basins, and graben, respectively. Zone names, such as Humber, Dunnage, Gander, Avalon, and Meguma are now just about household words among Canadian geologists and they are as well known and well-worn in eastern Canada as any structural or morphological subdivisions of mountain systems elsewhere. Some belts (such as Gaspé), basins (such as Magdalen), and graben (such as Fundy) are also well known and many more will become familiar from the accounts that follow.

The Appalachian region of North America is highly populated and the mountain belt is probably one of the best known and best mapped areas of its size in the world. Many of our concepts on mountain building for more than 100 years, from the geosynclinal theory of James Hall to the opening and closing Proto-Atlantic Ocean of Tuzo Wilson, were based on Appalachian rocks and structures. Canadian studies of the Appalachian Orogen had tremendous impetus from the work of Sir William Logan, first director of the Geological Survey of Canada. Logan came close to our present understanding of the Appalachian miogeocline as an ancient continental margin destroyed by structural imbrication, marked by his Great Dislocation, or Logan's Line.

We have made major advances since Sir William Logan's first edition of the Geology of Canada in 1863. The penultimate edition by R.J.W. Douglas was printed in 1970, although much of it was written long before at a time when geosynclinal theory was still in vogue. Between the time of its writing and printing, plate tectonics and continental drift became widely accepted and the conceptual aspects of the volume were immediately outdated. However, its usefulness remains as a report of progress and a factual account complete with geological and thematic maps. The aim of this volume is also a report of progress with an effort to separate facts and concepts, although this has not always been possible, given the present enthusiasm for plate models. The database grew considerably since the writing began about 10 years ago. Some manuscripts were updated more than once and recent references are added to brief addenda and to the Summary and Overview in Chapter 11. These changes are obvious in some cases and the insertions are not everywhere smooth.

Thanks to all those who contributed to this volume, those who provided reviews, and those who helped with editing. Their names and addresses are listed elsewhere. E.R.W. Neale played a major role as co-editor in its early stages and Wang Shao-Ming in its later stages. Thanks also to the Natural Sciences and Engineering Research Council of Canada, the Department of Natural Resources Canada, and the Canadian Geological Foundation for their support. The editorial, desktop publishing, and cartographic staff of the Geological Survey of Canada produced the volume. The 1:5 000 000 geological maps and figures for Chapter 1 were done by Charles Conway of Memorial University's Cartographic Laboratory. I wish to thank them all on behalf of myself and the other contributors.

Why would someone spend nights, weekends, and statutory holidays on a volume such as this for more than 10 years? There is much to be said and this is a welcome opportunity to summarize and publish the results of our efforts since the advent of plate tectonics. I thank John Wheeler for inviting me as editor and co-ordinator. Appalachian geology has always been fun and informative, and I sincerely hope that those coming behind will have as much enjoyment and satisfaction as those of us who went before.

We have advanced considerably in the last 25 years since the Douglas edition of the Geology of Canada. But in the words of D.M. Baird "how strange it is that the more we seem to find out....the horizon is still there, always inviting us to go closer." We have more problems now than Douglas and Logan. "And where will the horizon be teasing us to approach in the next..." 25 or 12 years, "and will we be then, as far away from where we stand now as our present position is.." from that of Douglas and Logan.

Harold Williams
Department of Earth Sciences
Memorial University of Newfoundland
St. John's, Newfoundland

Chapter 1

INTRODUCTION

Chapter 1

INTRODUCTION

Harold Williams

PREAMBLE AND DEFINITION

The Canadian Appalachian region includes the provinces of insular Newfoundland, Nova Scotia, New Brunswick, Prince Edward Island, and the southern part of Quebec along the south side of the St. Lawrence River (Fig. 1.1). It has an area of approximately 500 000 km^2 and it is widest (600 km) at the Canada-United States International Boundary in New Brunswick and Nova Scotia. A larger unexposed area of Appalachian rocks and structures extends across the Gulf of St. Lawrence and seaward to the Atlantic continental edge. Because of its coastal setting and insular makeup, the region offers tremendous shoreline exposures along marine passages.

The Appalachian region is a Paleozoic geological mountain belt or orogen. This means that its rocks have been affected by orogeny, the combined effects of folding, faulting, metamorphism, and plutonism. Paleozoic folds and faults of several generations trend northeastward. Regional metamorphic rocks occupy continuous belts in interior parts of the orogen, and granitic batholiths are common throughout its length (Maps 1 and 2).

The word "Appalachian" was first used in a geographic context for the morphological mountains in the southeast United States. It has displaced the word "Acadian" formerly applied to this region of eastern Canada. In the present context, the word "Appalachian" is used for the geological mountain belt without regard for its morphological expression.

Like the Cordilleran and Innuitian orogens, the Appalachian Orogen occupies a position peripheral to the stable interior craton of North America (Fig. 1.2). Undeformed Paleozoic rocks of the craton overlie a crystalline Precambrian basement. The exposed basement forms the Canadian Shield. The cover rocks define the Interior Platform. The Canadian Appalachian region is bordered to the west by that part of the Canadian Shield known as the Grenville Structural Province, and by that part of the Interior Platform known as the St. Lawrence Platform.

Williams, H.
1995: Introduction: Chapter 1 in Geology of the Appalachian-Caledonian Orogen in Canada and Greenland, (ed.) H. Williams; Geological Survey of Canada, Geology of Canada, no. 6, p. 1-19 (also Geological Society of America, The Geology of North America, v. F-1).

Rocks of the Appalachian Orogen are mainly of Paleozoic age, and they contrast with Paleozoic rocks of the St. Lawrence Platform. Apart from the obvious structural contrasts, the Paleozoic rocks of the orogen are thicker and contain discontinuous deep marine clastic and volcanic units that contrast with the sheet-like shallow water limestones and mature quartz sandstones of the platform. As well, there are major differences in contained Paleozoic faunas, metallogenic characteristics, and geophysical expression.

Along the west flank of the Appalachian Orogen, the Paleozoic rocks overlie a Grenville gneissic basement and are continuous and correlative with cover rocks of the St. Lawrence Platform, although they have slightly older units at the base of the section. This part of the orogen is known as the Appalachian miogeocline. East of the miogeocline, fault-bounded Zones of lower Paleozoic rocks exhibit sharp and rapid facies contrasts. Volcanic rocks are common and in places overlie an ophiolitic basement. In other places, mixed Paleozoic sedimentary and volcanic rocks overlie continental rocks that are unlike those of the Grenville Structural Province. These contrasts in lower Paleozoic rocks, coupled with contrasting basement relationships, allow the definition of a number of distinct geological zones or terranes that lie outboard of the miogeocline. The geological zonation in most common usage is that of the Humber (miogeocline), and successively outboard Dunnage, Gander, Avalon, and Meguma zones (Williams, 1976, 1978, 1979; Fig. 1.3 and Map 2).

For more than a century, North American geologists viewed the Appalachian region as a fixed and permanent "geosyncline", which through deformation, metamorphism, and plutonism was transformed into a geological mountain belt. Since the advent of plate tectonics and the wide acceptance of continental drift, the development of orogens such as the Appalachians is viewed as the result of rifting, ocean opening, subduction, accretion of terranes during ocean closing, and eventual continental collision. Accordingly, the Appalachian miogeocline (Humber Zone) is viewed now as the Paleozoic passive margin of eastern North America. Outboard zones (Dunnage, Gander, Avalon, and Meguma) are suspect terranes, or composite suspect terranes, accreted to North America during the closing of a Paleozoic ocean.

The boundary between the Appalachian Orogen and the St. Lawrence Platform is drawn at the structural front between the deformed rocks of the orogen and the undeformed rocks of the platform (Fig. 1.3). This boundary is

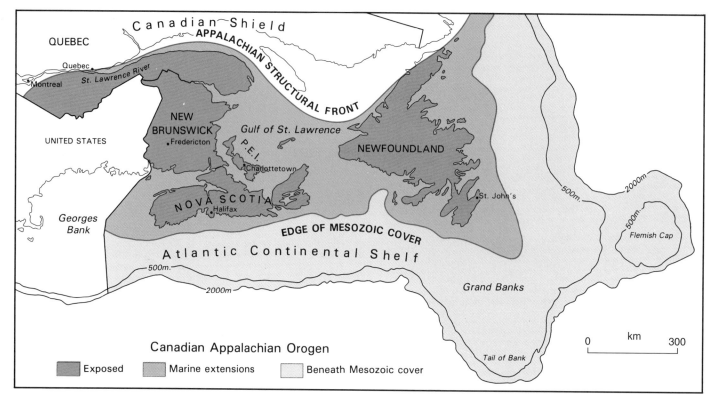

Figure 1.1. Canadian Appalachian region.

Figure 1.2. Phanerozoic orogens of North America.

coincident, or nearly so, with Logan's Line, a major structural junction between transported rocks from well within the orogen and structurally underlying rocks of the miogeocline. The exposed edge of the miogeocline occurs much farther east where it is marked by a narrow steep belt of ophiolite occurrences, the Baie Verte-Brompton Line (Williams and St-Julien, 1982).

Stratigraphic and sedimentological analyses of the Canadian and United States Appalachians indicate that terranes east of the miogeocline were accreted during three main events (Williams and Hatcher, 1983). The accretionary events are approximately coeval with three major orogenic episodes that occurred during the Early-Middle Ordovician (Taconic), Silurian-Devonian (Acadian), and Carboniferous-Permian (Alleghanian). Opening of the present North Atlantic Ocean was initiated well east of the accreted Appalachian terranes. Thus, a variety of Appalachian terranes were stranded at the margin of the North American craton. The sinuous form of the miogeocline in Canada, expressed in the Quebec Reentrant, St. Lawrence Promontory, and Newfoundland Reentrant (Fig. 1.3), probably reflects an orthogonal ancient continental margin bounded by rifts and transform faults analogous to the modern Grand Banks (Thomas, 1977; Williams and Doolan, 1979).

The closing of the Paleozoic Iapetus Ocean (Harland and Gayer, 1972) and opening of the North Atlantic explain why segments of the Paleozoic North American miogeocline are now found on the eastern side of the Atlantic as parts

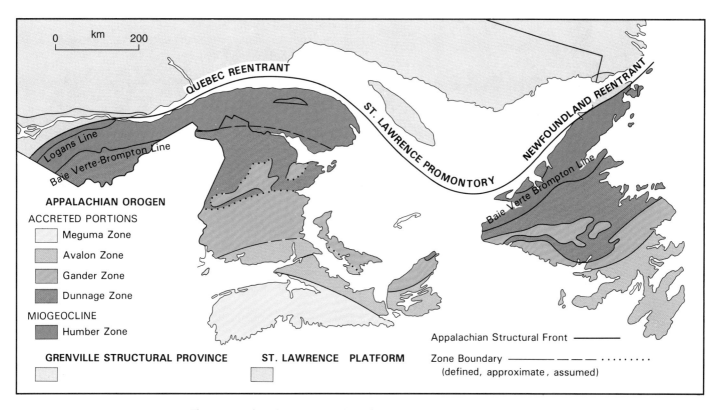

Figure 1.3. Simple zonation of the Canadian Appalachian region.

of the European plate, e.g. Hebredian foreland of the British Caledonides (Fig. 1.4). Other segments of the miogeocline, although not now part of the North American continent, remain within the American plate. The Greenland Caledonides represents such a segment and for that reason the geology of East Greenland is discussed in this volume (Chapter 12).

PHYSIOGRAPHY AND GLACIATION

The Canadian Appalachian region has a glaciated surface of highlands, uplands, lowlands, valleys, and fiords (Fig. 1.5). Its broken hummocky nature contrasts with the St. Lawrence Lowlands to the northwest and the smooth flat surface of the Atlantic Continental Shelf to the southeast. Highest elevations occur to the west and northwest where the rocks of the miogeocline form local highlands. From there, an undulating upland slopes gently southeastward to the coast, although it is dissected by valleys and interrupted by lowlands.

Lowlands of the Canadian Appalachian region occur in the vicinity of the Gulf of St. Lawrence and these are underlain by mainly subhorizontal Carboniferous rocks. Within the Gulf of St. Lawrence, there is no morphological distinction between the St. Lawrence and Appalachian lowland provinces. The absence of an elevated miogeocline makes this region unique for the length of the Appalachian mountain chain.

The Atlantic Continental Shelf is underlain by Mesozoic and Cenozoic strata (Fig. 1.1). These thicken profoundly toward the shelf edge and they form Canada's newest petroleum frontier. Offshore landforms were developed during periods of Late Cretaceous-Tertiary subareal erosion, producing mesas, cuestas, interfluves, and stream valleys (King, 1972; Grant, 1989).

Physiography of the Canadian Appalachian region probably relates to a long and continuous erosional cycle that had its beginning in the Jurassic Period, contemporaneous with continental breakup and the opening of the North Atlantic Ocean (King, 1972; Grant, 1989). The present setting of the Atlantic Continental Shelf, which is partly emerged as the Atlantic Coastal Plain farther south in the eastern United States, and the conspicuous drowned coastline of the Atlantic Provinces may reflect pre-Pleistocene tilting and tectonic subsidence (King, 1972; Grant, 1989).

Glacial erosion and deposition contributed to the landscape by surficial modification of former features. Pleistocene ice sheets advanced south and southeastward across the region, and the island of Newfoundland supported its own ice cap that flowed radially to the sea (Fig. 1.6). Fiord development is extensive around the Newfoundland coast but overdeepening is less evident elsewhere. Glacial retreat began about 20 000 years ago, and glacial rebound brought the surface of the Atlantic region to its present position (Grant, 1989; Occhietti, 1989).

ACCESS AND CULTURE

The Appalachian region is the oldest settled part of Canada. It is covered by a network of paved roads and secondary gravel roads in populated areas. Because forest industries are important, there are numerous logging roads

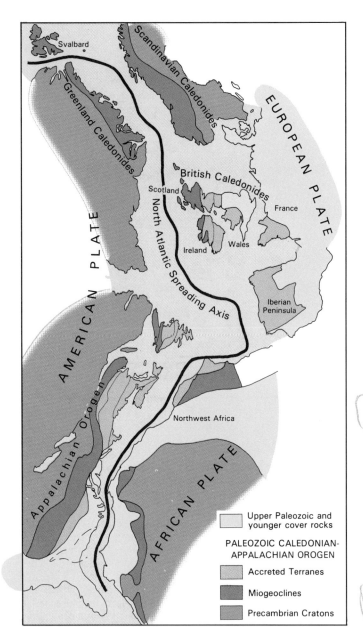

Figure 1.4. Tectonic elements of the restored North Atlantic region (after Williams, 1984).

rank among the nation's best, and ophiolite suites in western Newfoundland are as well exposed, widely studied, and intimately understood as any in the world. Gros Morne National Park in western Newfoundland is a UNESCO World Heritage Site, recognized for its variety of rocks and relationships, superbly exposed in a glaciated, rugged coastal setting. Maritime Canadians are noted for their friendliness, helpfulness, and hospitality so that anyone may visit, traverse the countryside, and inspect the outcrops at their leisure, free of harassment of any kind.

GEOPHYSICAL EXPRESSION AND OFFSHORE EXTENSIONS

The Appalachian Orogen has a geophysical expression that allows extension of its onland rocks and structures seaward to the submerged and covered edge of continental crust at the Atlantic margin.

The Bouguer anomaly field over the onland orogen, and the free-air anomaly field over marine areas, have a general level tens of milligals higher than that over the Grenville Province of the Canadian Shield (Map 3). Major anomalies and gradients trend northeast, parallel to structural trends and lithofacies belts, and are traceable to the continental edge. A positive gradient from west to east lies at or near the exposed edge of the Appalachian miogeocline. Ophiolite complexes and mafic volcanic belts in central parts of the orogen have a strong positive expression (Haworth and Jacobi, 1983; Williams and Haworth, 1984a; Shih et al., 1993a).

Similarly, aeromagnetic and sea magnetic anomalies follow major structures and are especially useful in extending ophiolitic and volcanic belts offshore (Map 4, Williams and Haworth, 1984b; Shih et al., 1993b). Where the Appalachian magnetic basement is deeply covered by younger sediments at the continental edge, the magnetic anomalies are expectedly broad and less distinct. A prominent positive anomaly, the East Coast Magnetic Anomaly, occurs at the morphological shelf edge east of Nova Scotia and southward, but it is absent along the torturous rifted margin of the Grand Banks in Newfoundland. One suggestion is that the East Coast Magnetic Anomaly is a Paleozoic collisional zone that was the locus for opening of the North Atlantic (Nelson et al., 1985). Part of this collisional zone occurs inland in the southeastern United States (Brunswick Magnetic Anomaly) and it may be truncated off Nova Scotia by the axis of Atlantic spreading, and therefore displaced to the African continental margin.

The Avalon-Meguma zone boundary is marked by the Collector Magnetic Anomaly extending offshore to the continental edge where it is collinear with the Newfoundland Seamounts (Williams and Haworth, 1984b; Shih et al., 1993b). Other features of the modern margin also reflect ancestral controls, e.g. the Charlie Gibbs Fracture Zone coincides with the extension of the Gander-Avalon zone boundary off northeast Newfoundland, and the Tail of the Bank mimics the St. Lawrence Promontory.

Seismic reflection studies indicate that a Grenville lower crustal block extends in subsurface well beyond the exposed edge of the miogeocline (Fig. 1.7). It meets a Central lower crustal block beneath the Dunnage Zone. A steep boundary between the Gander and Avalon zones extends

and trails. Navigable rivers and large lakes provide access to other parts of the region. Float planes or helicopters are available for charter near most major centres. Some shoreline sections can be walked, but a boat is usually necessary to negotiate headlands and cliffed shoreline. Boats are easily hired in the numerous fishing villages scattered along the coast.

Exposures across northeast Newfoundland provide an unsurpassed cross-section of the orogen, and many of the latest Appalachian concepts emanated from this area. Allochthonous terranes such as those along the St. Lawrence River in Quebec and western Newfoundland

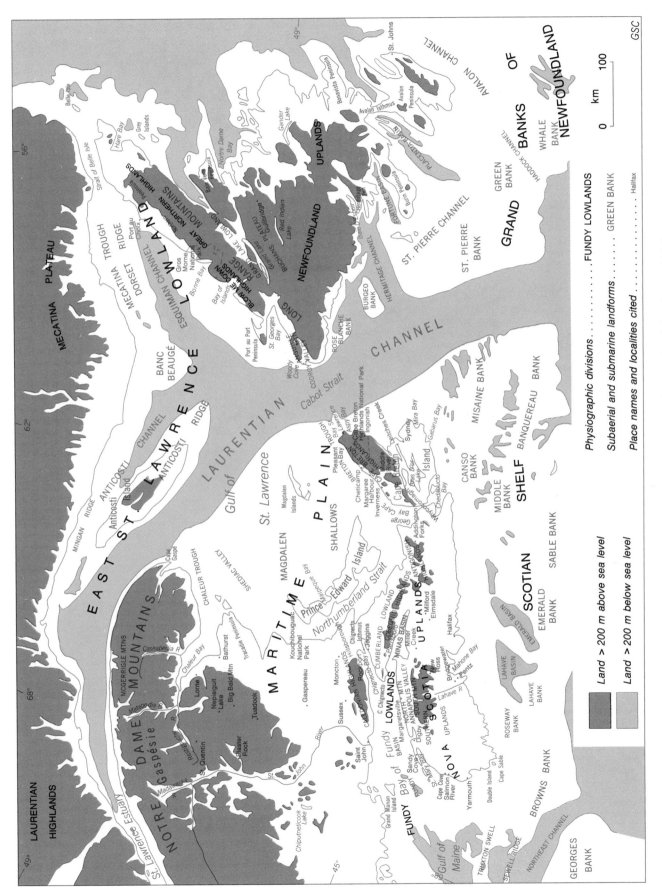

Figure 1.5. Physiography of the Canadian Appalachian region (after Grant, 1989).

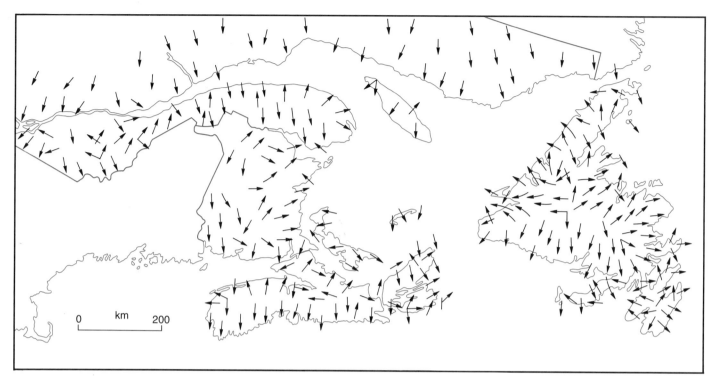

Figure 1.6. Glacial features of the Canadian Appalachian region (modified from Grant, 1989 and Occhietti, 1989). Arrows indicate direction of glacial ice movement.

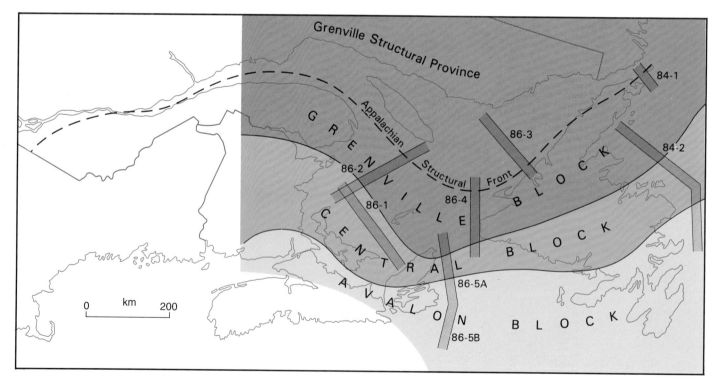

Figure 1.7. Lower crustal blocks of the Canadian Appalachian region (after Marillier et al., 1989), as determined along deep reflection lines, e.g. 86-1, etc.

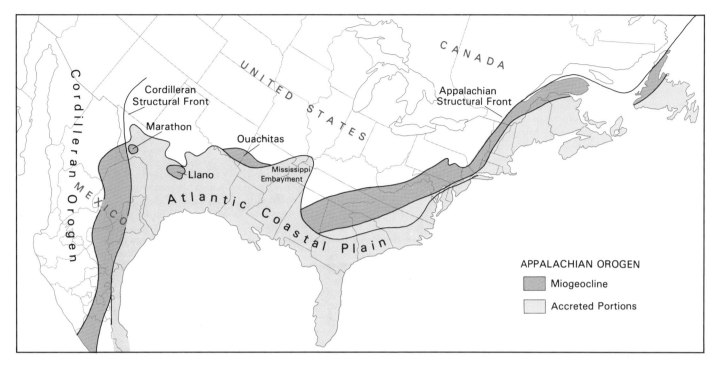

Figure 1.8. Appalachian Orogen, Newfoundland to Mexico. Approximate area of Paleozoic rocks in Mexico contained in Sierra Madre and Acatlan terranes (Ruiz et al., 1988).

to the mantle and separates Central and Avalon lower crustal blocks. The Humber Zone is the surface expression of the Grenville lower crustal block. The Dunnage Zone is allochthonous above the Grenville and Central lower crustal blocks. The Gander Zone may be the surface expression of the Central lower crustal block or it too may be allochthonous. The Avalon Zone and corresponding lower crustal block is a microplate rooted in the mantle (Keen et al., 1986; Marillier et al., 1989).

EXTENSIONS AND SETTING IN THE NORTH ATLANTIC

The Canadian Appalachians extend southwestward through the eastern United States to Alabama. There, Paleozoic rocks and structures are overlapped unconformably by Mesozoic and younger strata of the Atlantic Coastal Plain (Fig. 1.8). West of the Mississippi Embayment, deformed Paleozoic rocks reappear in the Ouachitas of Arkansas and Oklahoma and they occur still farther west in the Llano and Marathon uplifts of Texas. Paleozoic rocks and structures are known in Mexico where they are overprinted by younger structures of the Cordilleran Orogen (Ruiz et al., 1988). The Appalachian Orogen is therefore continuous to the Pacific margin of the continent.

The Appalachian miogeocline, everywhere developed upon Grenville basement, is the most continuous feature of the orogen. Basement inliers and their cover rocks form the Blue Ridge Province of the U.S. Appalachians and imbricated cover rocks farther west form the Valley and Ridge

Province. Fossiliferous and little metamorphosed Paleozoic rocks that make up the accreted parts of the Canadian Appalachians continue into New England but equivalents are largely absent in the southern U.S. Appalachians where the centrally located Inner Piedmont and Charlotte belts are composed entirely of crystalline rocks. Farther east however, the U.S. Carolina Slate Belt, which extends to the coastal plain onlap, is a natural continuation of the Canadian Avalon Zone. Detailed treatment of the U.S. Appalachians with maps and illustrations is contained in Hatcher et al. (1989). The Tectonic Lithofacies Map of the Appalachian Orogen (Williams, 1978a) in its largest format at 1:1 000 000 scale is also a useful illustration for Canadian-United States comparisons.

Restoration of the North Atlantic (Fig. 1.4) juxtaposes the Atlantic continental shelf with the marine shelves of Ireland, the Iberian Peninsula, and northwest Africa. A miogeocline, equivalent to that of the Appalachians, occurs in the northwest British Caledonides. In the British Isles, most of the miogeoclinal rocks are metamorphosed but locally in Scotland the Durness succession has an early Paleozoic stratigraphy and faunas identical to those of the Appalachian miogeocline. Reworked Grenville basement is recognized locally in the miogeocline of Ireland and Scotland but the undeformed foreland to the miogeocline has much older Precambrian rocks of the Laxfordian (1800 Ma) and Scourian (2500 Ma) provinces. Beyond the British Isles, rocks like those of the Appalachian miogeocline reappear in the Caledonides of East Greenland. Equivalents of the 1000 Ma Grenville Structural Province locally form the basement to the miogeoclinal cover rocks there but the

Greenland miogeocline, like that of Scotland, is developed mainly upon much older Precambrian rocks. Beyond Greenland, miogeoclinal rocks of the Hecla Hoek Group occur in Northern Svalbard (Williams, 1984).

The Appalachian miogeocline and its extensions therefore formed a remarkably continuous continental margin, the western margin of Iapetus, that can be traced for almost 10 000 km and is open ended.

Accreted terranes that locally include volcanic rocks on ophiolitic basement have equivalents in the British Isles, and ophiolitic complexes are common on the opposing margin of Iapetus in the Scandinavian Caledonides. Other easternmost Appalachian terranes have natural continuations in northwest Africa, the Iberian Peninsula, France, and Wales. Some of these are huge and rival the Appalachian miogeocline in width and continuity, e.g. Avalon Zone. The Tectonic Map of Pre-Mesozoic Terranes in Circum-Atlantic Phanerozoic Orogens (Keppie and Dallmeyer, 1989) is a useful illustration for North Atlantic connections. Collections of papers are found in Kerr and Fergusson (1981) and Dallmeyer (1989).

Parallelism of the Grenville Structural Province and the Appalachian miogeocline in eastern North America implies Precambrian ancestral control for the development of the Paleozoic margin. Similarly, parallelism of the Appalachian Orogen and the North Atlantic margin implies further mimicry and Paleozoic ancestral control of the modern continental margin (Fig. 1.9).

The North Atlantic Ocean and its continental margins provide an actualistic model for the Paleozoic Iapetus Ocean that led to the development of the Appalachian Orogen. The widths of the North Atlantic continental shelves and the thicknesses of sediments at the North Atlantic margins are comparable to palinspastically restored widths of Paleozoic margins and thicknesses of their miogeoclinal sections. The form of the North Atlantic margin mimics and provides an explanation for the sinuosity of the deformed Appalachian miogeocline. The crust and mantle beneath the North Atlantic is analogous to Paleozoic volcanic rocks and ophiolite suites, and marine microcontinents (e.g. Rockall Plateau) and oceanic volcanic

islands and seamounts (e.g. Iceland, the Faeroes, Newfoundland seamounts) are potential suspect terranes (Fig. 1.10).

HISTORY OF INVESTIGATION

Geological investigations in the Appalachian region began even before the creation of the Geological Survey of Canada in 1842. These include studies in Nova Scotia by Abraham Gesner and in Newfoundland by J.B. Jukes. With the creation of the Geological Survey of Canada, the newly appointed William E. Logan began systematic fieldwork in Gaspé Peninsula (now Gaspésie) and the Quebec Eastern Townships. He summarized fully his own work and that of his associates such as Alexander Murray, James Richardson, Robert Bell, and Elkanah Billings in the momentous 983-page volume published in 1863 and entitled "Geology of Canada". Logan's stratigraphy and his insight into such current concerns as ancient continental margins and allochthonous terranes are still respected (Stevens, 1974).

Systematic studies and reconnaissance mapping were carried out in the Appalachian region during the latter half of the 19th century. L.W. Bailey, G.F. Matthew, R.W. Ells, W. McInnes, and C. Robb together studied and mapped the geology of New Brunswick, outlining the major lithic units and general distribution of mineral deposits. Hugh Fletcher and E.R. Faribault produced a series of 1:63 360 geological maps covering the important coal producing and gold mining districts of Nova Scotia, the combined result of more than 86 years of fieldwork. Alexander Murray was appointed in 1864 to the newly created Geological Survey of Newfoundland. Murray's appointment coincided with the beginning of a 40-year boom in copper mining, and he and J.P. Howley published numerous reports and a generalized geological map of the island (Murray and Howley, 1881, 1918; Howley, 1925).

More detailed studies of regional geology and of mineral districts and mines were advancing in the early 1900s. Most of this work was carried out by the Geological Survey of Canada supplemented by that of provincial mines departments.

Geological mapping of the Quebec Appalachians, initiated by the Geological Survey of Canada, was expanded and eventually succeeded by work of the Quebec Department of Mines in the 1950s and 1960s. Classic stratigraphic and paleontological studies were made in Gaspésie and along the St. Lawrence River by J.M. Clarke and Charles Schuchert.

Geological mapping continued in New Brunswick, in large part under the guidance of W.J. Wright, up to 1953 and the staking rush that signalled the beginning of the Bathurst mining boom. The New Brunswick Research and Productivity Council was established in 1962, and in 1965 the Provincial Department of Mines embarked upon a program of expanded geological fieldwork with increased permanent staff.

In Nova Scotia, a program of modern-day 1:63 360 mapping coupled with biostratigraphic studies was begun by the Geological Survey of Canada with W.A. Bell and others. The program continued through the 1940s and 1950s to its completion. Onland geological investigations are continuing under the direction of the Provincial Mines

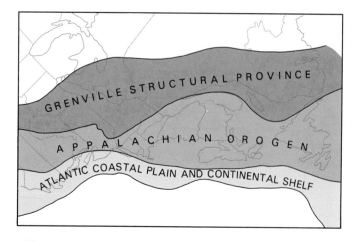

Figure 1.9. Ancestral controls and tectonic mimicry in eastern North America.

Branch, and offshore studies are being conducted by the Atlantic Geoscience Centre and the Nova Scotia Research Council.

Few geological studies were made in Newfoundland during the early 1900s, except for those of the Princeton University expeditions. The Geological Survey of Newfoundland was reactivated in 1933 under A.K. Snelgrove and later under C.K. Howse. In 1949, the year of confederation with Canada, responsibilities for regional geology were assumed by the Geological Survey of Canada and the Geological Survey of Newfoundland was disbanded. The Geological Survey of Newfoundland was reactivated in 1952 under the direction of D.M. Baird. Since 1949, the Geological Survey of Canada completed a program of 1:253 440 reconnaissance mapping. Most recent studies are carried out by a much expanded Geological Survey of Newfoundland that is mainly concentrating on 1:50 000 mapping and a study of mineral deposits.

As a result of previous and present geological activity, all of the Canadian Appalachian region is covered by 1:253 440 or 1:250 000 reconnaissance mapping and more than half is covered by modern 1:63 360 or 1:50 000 mapping. Mineral districts are covered in more detail. Contributions by university research teams have added much to our understanding of the orogen over the past 20 years, espec-ially by Memorial University in Newfoundland, Dalhousie University in Nova Scotia, the University of New Brunswick in New Brunswick, and Université Laval and Université de Montréal in Québec. Recent, or relatively recent geological map compilations are available for all provinces at various scales (Keppie, 1979; Potter et al., 1979; van de Poll, 1983, 1989; Avramtchev, 1985; Colman-Sadd et al., 1990).

Aeromagnetic maps at 1:63 360 or 1:50 000 scale are available for the Canadian Appalachian region through the Geological Survey of Canada, and gravity data are also available, although in less detail. Seismic refraction and reflection work has focused mainly on offshore regions, but several seismic reflection lines were completed across the miogeocline in Quebec (St-Julien et al., 1983). Deep reflection lines were done off northeast Newfoundland (Keen et al., 1986) and in the Gulf of St. Lawrence (Marillier et al., 1989) as part of the Frontier Geoscience Project. Deep reflection transects across insular Newfoundland were completed as part of the Canadian Lithoprobe Project in 1989 (Williams et al., 1989; Piasecki et al., 1990). Isotopic ages are available for most large plutons and ophiolite complexes, with precision ranging from that of early K-Ar ages to more reliable Rb-Sr, U-Pb, and Sm-Nd ages.

A number of syntheses, reviews and collections of papers on Canadian Appalachian geology, with rather complete reference lists, have been published over the last 25 years. These include Neale and Williams, 1967; Kay, 1969; Poole et al., 1970; Rodgers, 1970; Geological Association of Canada, 1971; Williams et al., 1972; Tozer and Schenk, 1978; Williams, 1979; Wones, 1980; St-Julien and Béland, 1982; and Hatcher et al., 1983.

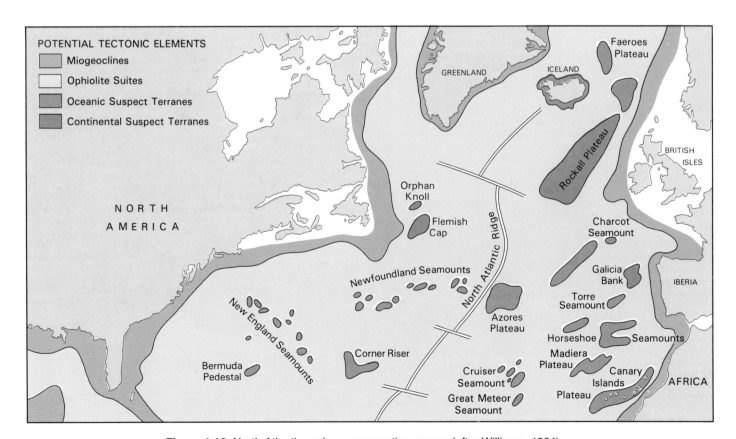

Figure 1.10. North Atlantic region – prospective orogen (after Williams, 1984).

The International Geological Correlation Program, Project 27, the Caledonian-Appalachian Orogen, has led to an inventory of Appalachian geology (Schenk, 1978). It terminated in 1984 and several 1:1 000 000 and/or 1:2 000 000 Appalachian compilation maps are published: tectonic lithofacies (Williams, 1978a,b), magnetic anomaly (Zietz et al., 1980a, b), Bouguer gravity anomaly (Haworth et al., 1980a, b), and a structural map (Keppie, 1982). Project 27 is replaced by the International Geological Correlation Program, Project 233, Terranes in Circum Atlantic Paleozoic Orogens.

The discovery of major offshore oil and gas deposits brought new interest to the geology of the North Atlantic continental margin and a much accelerated level of activity. Collections of papers are found in Yorath et al., 1975; Vogt and Tucholke, 1986; Sheridan and Grow, 1988; Tankard and Balkwill, 1989; and Keen and Williams, 1990. Transects of the Atlantic margin, including the onland Appalachian Orogen, are part of the Decade of North American Geology (DNAG) Project on North American continental margin transects, and this work is leading to worthwhile comparisons between the Appalachian miogeocline and the modern Atlantic margin (Haworth et al., 1985, in press). Perhaps the most exciting recent work is the deep reflection experiments that led to the definition of lower crustal blocks. Relating surface geology to deep structure is a new and exciting challenge in the study of the Canadian Appalachian region.

HISTORY OF IDEAS

Since the acceptance of plate tectonics and the realization that the earth's crust is a mosaic of moving plates, great strides have been made in understanding where geological mountain belts are sited and how they evolved. In view of our present streamlined theories it is interesting to reflect on what went before, and to look at earlier attempts to rationalize the development of the Appalachian mountain system from the viewpoint of fixed continents and permanent oceans. "No thorough grasp of a subject can be gained unless the history of its development is clearly appreciated" (Geikie, 1905). Since many previous ideas emanated from the Appalachian region of eastern North America, it is appropriate to review some of them here.

Before the advent of plate tectonics about 25 years ago, the protolith of the Appalachian Orogen was viewed as a geosyncline. The nature of geosynclines, their developmental patterns, and their position with respect to continents and ocean basins were controversial and enigmatic topics. Few agreed on such fundamental concepts as controls of initiation and siting, basement relationships, and causes of subsidence and ensuing mountain building.

The idea of a geosyncline is summarized by King (1959) and partly reproduced here. It began with James Hall in 1857 (published in 1883), a stratigrapher and paleontologist for the State of New York where rocks of the Interior Platform pass eastward into the Appalachian Orogen. Hall observed that undeformed sedimentary rocks of western New York are thin, whereas deformed rocks of the Appalachian region in eastern New York are thick; although both formed during the same Paleozoic time span and were deposited in shallow water. He reasoned that

there must be some relationship between the greater thickness of shallow water sedimentary rocks in the Appalachians and their deformed nature. The Appalachian region must have subsided more than the platformal region and Hall suggested that the greater subsidence resulted from gradual yielding of the crust beneath the weight of the sediments themselves. Along the axis of the subsiding area, the sedimentary rocks were folded and in places ruptured to localize igneous intrusions. This was followed by uplift and erosion to produce the geological mountain belt. Hall viewed mountain building as largely a matter of thick sedimentary accumulation, subsidence of the crust under the weight of the sediments, and subsequent uplift. In this model, deformation was incidental to the sediment accumulation and a consequence of the subsidence. Geological mountains were thus related to the slow surficial processes of erosion and sedimentation, and drainage patterns determined their siting.

James Dwight Dana, a well known contemporary of Hall, while agreeing with the facts was critical of the explanation for mountain building (Dana, 1873). Dana appraised Hall's explanation as "a theory of the origin of mountains with the origin of mountains left out". Dana envisaged fundamental differences between continents and ocean basins, and he believed in a cooling earth with a contracting interior that led to crustal compression. Greatest yielding to such compression was between the

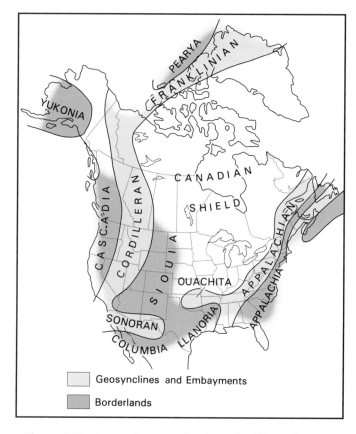

Figure 1.11. Geosynclines and borderlands of North America (modified after Schuchert, 1923).

continents and oceans along belts warped down to form the "geosynclinals" (later geosynclines) and warped up to form the "geanticlinals" (later geanticlines). The crust was therefore bent down by outside forces to form a sediment receptacle. Eventually the geosynclinals were destroyed by the compression to which they owed their beginnings through "a catastrophe of plications and solidifications" (Dana, 1873).

Dana was the first to use the term geosyncline and because the theories of both Hall and Dana were based on Appalachian examples, the Appalachian system is the type geosyncline. Dana envisaged a geanticline to the east of the geosyncline which assured a continual supply of detrital material. He also envisaged compression toward the continent, related to an oceanic crust that contracted more quickly than continental crust. This further suggested a position for geosynclines peripheral to continents.

Charles Schuchert, a ruling figure among North American stratigraphers, was a champion of Dana's views and for many years perpetuated the idea of troughs (geosynclines) and borderlands (geanticlines). He outlined the borderland and geosynclinal elements of the North American continent (Schuchert, 1923; Fig. 1.11), and further subdivided the Canadian Appalachians into two geosynclinal and two geanticlinal elements (Schuchert and Dunbar, 1934; Fig. 1.12). This model influenced thinking on the nature of the Canadian Appalachians for many years and was still in use, in modified form, up to the time of the fifth edition of the Geology and Economic Minerals of Canada (Douglas, 1970).

The standard work on North American geosynclines is that of Marshall Kay (1951). He named and classified geosynclines (partly after the German geologist Hans Stille) according to their position with respect to platforms and continental margins, and according to their shape and relative age. He further subdivided his real geosynclines or orthogeosynclines into two main parts, dependent upon rock types; the miogeosyncline with mainly shallow water sedimentary rocks that bordered the platform, and the eugeosyncline with deep marine sedimentary and volcanic rocks that lay farther offshore at the continental margin. Kay drew attention to the presence of volcanic rocks in North American Phanerozoic orogenic belts. He interpreted the volcanic rocks as products of island arcs that ringed the North American continent during its Paleozoic evolution (Fig. 1.13). Sedimentary studies (Krynine, 1948) supported Kay's model and suggested that volcanic islands and contemporary tectonic lands were a more likely provenance for mixed detritus of the geosynclines than the crystalline basement rocks of the borderlands of Dana and Schuchert. Furthermore, the borderland concept found little support in emerging geochronological and geophysical studies.

The notion that geosynclines were peripheral to continents was always favoured by North American geologists since the time of Dana, because of the symmetry of the North American continent and the annular arrangement of its Paleozoic mountain belts (Fig. 1.2). Almost any cross-section of the continent from the interior outward contains the same progression of morphological and tectonic elements; uplands of the ancient Canadian Shield, Phanerozoic lowlands of the Interior Platform, Phanerozoic mountains, modern continental margin, and ocean basin (Fig. 1.14). Kay's work and the substitution of contemporary volcanic arcs for ancient crystalline borderlands firmly established the hypothesis that the North American continent grew, much like a tree, by the addition of younger and outward geosynclinal belts. This was the fixist idea of continental accretion that persisted as a popular model in North America up to the time of plate tectonics, e.g. Wilson (1954), Engel (1963). Probably the most widely cited cross-section to confirm the validity of the concept is that from the Archean Superior Province of the Canadian Shield eastward across the Proterozoic Grenville Province, the Paleozoic Appalachian Orogen, and modern Atlantic margin (Fig. 1.15).

Even before the wide acceptance of continental drift and plate tectonics in North America, models for the evolution of the Appalachian geosyncline emphasized analogies with modern continental margins. Thus, Drake et al. (1959) noted the similarity between the paired inner trough and continental rise prism of the present North Atlantic margin and the Paleozoic miogeosynclinal and eugeosynclinal couple of Kay. Dietz (1963) and Dietz and Holden (1966) argued that Kay's miogeosynclinal zone represented an eastward thickening sedimentary wedge, "the miogeocline", that changed eastward at a bank edge into a continental rise clastic-volcanic facies, "the eugeocline". Rodgers (1968) strengthened the analogue between the Appalachian miogeocline and modern continental shelf by delineating a Paleozoic carbonate bank along the eastern margin of North America that was marked by a bank-edge limestone breccia facies. He furthermore compared the Appalachian Paleozoic bank and the modern Bahama Bank, and suggested that the Paleozoic bank was bounded

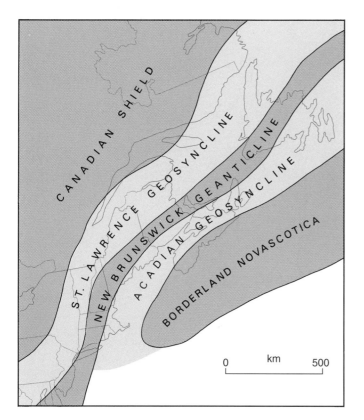

Figure 1.12. Geosynclines and borderlands of the Canadian Appalachian region (after Schuchert and Dunbar, 1934).

eastward by a drop-off in bathymetry or sharp declivity, which possibly coincided with the edge of North American Grenville basement.

Because most of these actualistic North American models failed to invoke some form of continental drift, or failed to explain orogens that occur in central parts of continents, they were not generally accepted. Even at home in the Canadian Appalachians, the fixist accretionary models did not adequately accommodate the tectonic relations across the well exposed and remapped northeast Newfoundland section (Williams, 1964). There, the Paleozoic miogeocline-eugeocline pair is bounded to the east, not by a contemporary oceanic domain, but by Precambrian rocks overlain

by a Cambrian shelf sequence. The outboard position of the eastern Newfoundland Precambrian terrane (Avalon Zone), and the two-sided nature of the system, refuted the fixist idea of Paleozoic accretion (Fig. 1.16).

These enigmas led Wilson (1966) to propose that the western and eastern parts of the Appalachian Orogen were separated in the early Paleozoic by a proto-Atlantic ocean (Fig. 1.17), which closed in the late Paleozoic to juxtapose the North American and African-European continents. Mesozoic opening of the modern North Atlantic occurred along a slightly more eastern axis so that the African-European Avalon terrane remained in North America. Wilson's model was viewed with scepticism and extreme

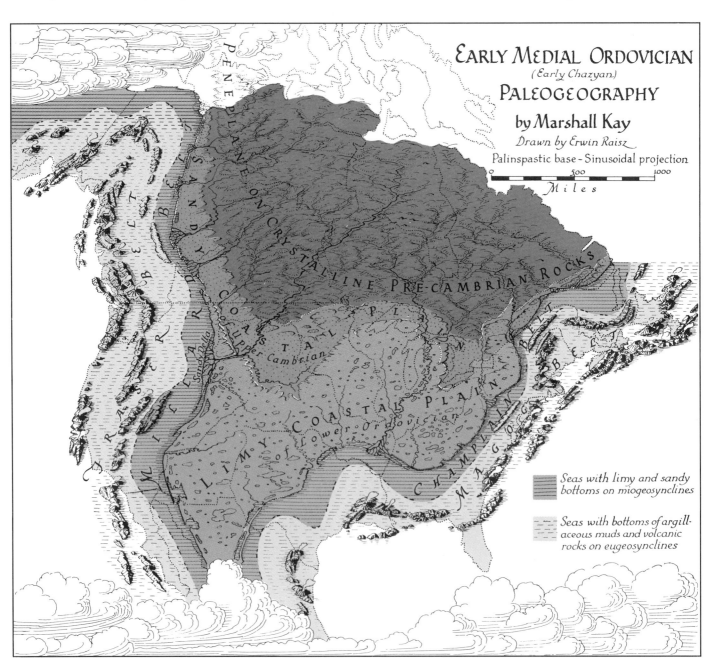

Figure 1.13. Ordovician paleogeography of North America (after Kay, 1951).

caution for the next few years but was endorsed enthusiastically after the emergence of plate tectonic theory (McKenzie and Parker, 1967). Today, the model of producing an orogen through a cycle of ocean opening and closing is known as the "Wilson cycle" in honour of its originator.

The first plate tectonic model for the Appalachian Orogen that incorporated rocks and structures into the scheme of a Wilson cycle is that of Dewey (1969) and Bird and Dewey (1970). It defined the margins of Iapetus and its oceanic tract, and traced the evolution of the Appalachian miogeocline from a rifted margin, through its passive development, to its collisional destruction (Fig. 1.18). The implications of onland ophiolite suites were worked out by Stevens (1970), Church and Stevens (1971), Dewey and Bird (1971), and Williams (1971). As in previous decades, many of these latest Appalachian concepts drew heavily upon relationships in Newfoundland.

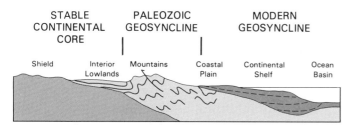

Figure 1.14. Cross-section of North America from the centre of the continent to its eastern margin showing the ideal arrangement of morphological and tectonic elements (after King, 1959).

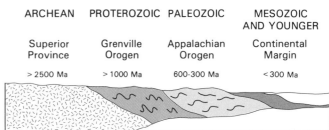

Figure 1.15. Generalized cross-section southeastward from the Superior Province of the Canadian Shield to the Atlantic continental margin (after King, 1959).

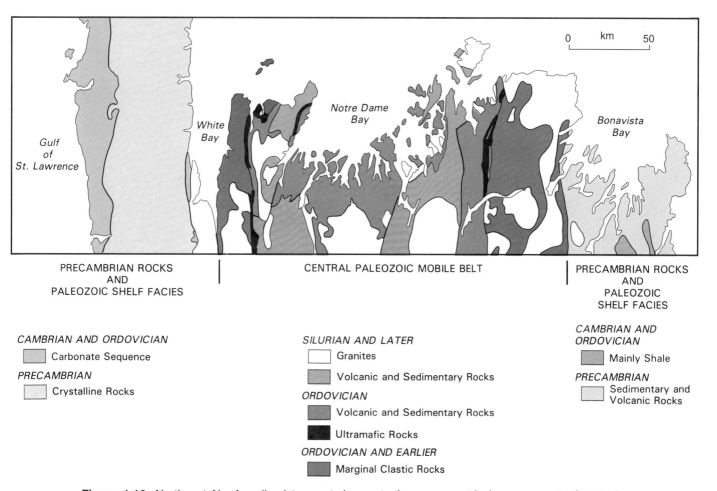

Figure 1.16. Northeast Newfoundland transect demonstrating a symmetrical arrangement of tectonic elements and a two-sided orogenic system (modified from Williams, 1964).

Since then there has been a plethora of conceptual plate models for the Appalachians. Their multitude and frequency make it increasingly difficult to retain a youthful enthusiasm for new attempts to incorporate the latest refinements as the database expands.

Now, with an appreciation of the array of complexities inherent in any Wilson cycle model, the emphasis has shifted to recognizing the major tectonostratigraphic divisions in the orogen, establishing structural and stratigraphic relationships, and interpreting the accretion history. More objective accounts based on geographic tectonstratigrahic zones (Williams et al., 1972; Williams, 1979) are forerunners of present suspect terrane analyses (Williams and Hatcher, 1982, 1983; Zen, 1983; Keppie, 1985, 1989). Their precept is that once the miogeocline is identified all outboard elements are viewed as paleogeographically suspect until proven otherwise (Coney et al., 1980). This is an objective approach to the analysis of any orogen. Still more cautiously, there is a move afoot to revert to Appalachian zonal analyses (Williams et al., 1988), with the descriptive term "zone" recommended over "terrane" in cases were the nature of boundaries is doubtful.

APPRAISAL OF EARLIER IDEAS

The problems of geosynclines were always inextricably interwoven with problems of orogenic belts – their siting and processes of mountain building. The idea of a geosyncline therefore brought attention to some of these important problems in tectonics. Problems of the controls of "geosynclinal" subsidence are still addressed today, although expressed differently in models of rifting and thermal subsidence of passive continental margins (Keen, 1979). Following the establishment of the geosynclinal concept by Hall and Dana, it was a natural progression for the next

generation of geologists to classify and subdivide geosynclines (Schuchert, 1923; Kay, 1951). The resulting fixist idea of continental accretion found support in North America because of the Paleozoic symmetry of the continent and hints of Precambrian symmetry within the Canadian Shield (Wilson, 1949). Of course the asymmetry of most other continents and the location of some Phanerozoic orogenic belts within continents, e.g. Urals, clearly spoke against the North American fixist accretionary model. Obviously there could be no reconciliation of these concepts without accepting some form of continental drift.

Plate models based upon the Wilson cycle of opening and closing oceans are by far the most viable and actualistic. They not only explain the siting of miogeosynclinal parts of orogens as continental margins and eugeosynclinal parts as accreted terranes but provide a mechanism for mountain building through subduction and collisional events.

However, the wave of subjective, poorly-constrained plate models that flourished for more than two decades has waned in the wake of an awareness of the array of complexities and paleogeographic uncertainties in palinspastic restorations of orogenic belts. This is displayed in the current popularity of the suspect terrane outlook and reversions to zonal subdivisions for meaningful "as is" analyses with emphasis on timing and mechanisms of accretion. A shift toward objective analyses, rather than hypothetical scenarios is refreshing, especially as expressed by students who were educated during and after the advent of plate tectonics – a time when it was more fashionable to produce a conceptual model then to dwell on rocks and relationships. And where would we be today without plate tectonics? Probably writing the very best papers on geosynclines or tectonostratigraphic zones, while still wrestling with the problems of fixed continents and permanent oceans.

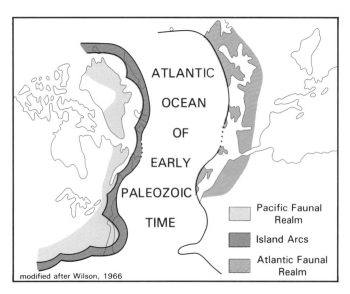

Figure 1.17. Proto-Atlantic ocean of early Paleozoic time (after Wilson, 1966).

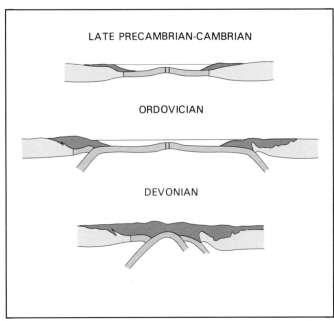

Figure 1.18. Schematic cross-sections illustrating the plate tectonic evolution of the Appalachian-Caledonian Orogen (modified from Dewey, 1969).

STANDARDS, SYMBOLS, AND NAMES

The geological time scale used in this volume is that adapted for the Decade of North American Geology (DNAG) Project (Palmer, 1983). Structural symbols are those in common usage by the Geological Survey of Canada. Most text figures use a Lambert projection; maps 1 to 6 use the DNAG base, a modified transverse Mercator projection.

ACKNOWLEDGMENTS

Thanks are extended to R.F. Blackwood, L.B. Chorlton, S.P. Colman-Sadd, L.R. Fyffe, J.P. Hibbard, E.R.W. Neale, and John Rodgers for reviews of earlier versions of this chapter. Thanks are also extended to John Rodgers for supplying copies of the original Hall and Dana papers, and to Charles Conway for producing figures and maps.

REFERENCES

Avramtchev, L. (comp.)
1985: Carte géologique du Québec; Ministère de l'Énergie et des Ressources du Québec, Carte n° 2000 du DV 84-02, scale 1:1 500 000.

Bird, J.M. and Dewey, J.F.
1970: Lithosphere plate-continental margin tectonics and the evolution of the Appalachian Orogen; Geological Society of America Bulletin, v. 81, p. 1031-1060.

Church, W.R. and Stevens, R.K.
1971: Early Paleozoic ophiolite complexes of the Newfoundland Appalachians as mantle-oceanic crust sequences; Journal of Geophysical Research, v. 76, no. 5, p. 1460-1466.

Colman-Sadd, S.P., Hayes, J.P., and Knight, I. (comp.)
1990: Geology of the Island of Newfoundland; Geological Survey Branch, Newfoundland Department of Mines and Energy, Map 90-01, scale 1:1 000 000.

Coney, P.J., Jones, D.L., and Monger, J.W.H.
1980: Cordilleran suspect terranes; Nature, v. 288, no. 5789, p. 329-333.

Dallmeyer, R.D. (ed.)
1989: Terranes in the Circum-Atlantic Paleozoic orogens; Geological Society of America, Special Paper 230, 277 p.

Dana, J.D.
1873: On some results of the earth's contraction from cooling including a discussion of the origin of mountains and the nature of the earth's interior; American Journal of Science, v. 5, p. 423-443; v. 6, p. 6-14; 104-115; 161-172.

Dewey, J.F.
1969: Evolution of the Appalachian/Caledonian Orogen; Nature, v. 222, no. 5189, p. 124-129.

Dewey, J.F. and Bird, J.M.
1971: Origin and emplacement of the ophiolite suite: Appalachian ophiolites in Newfoundland; Journal of Geophysical Research, v. 76, p. 3179-3206.

Dietz, R.S.
1963: Collapsing continental rises: An actualistic concept of geosynclines and mountain building; Journal of Geology, v. 71, no. 3, p. 314-333.

Dietz, R.S. and Holden, J.C.
1966: Miogeoclines in space and time; Journal of Geology, v. 74, p. 566-583.

Douglas, R.J.W. (ed.)
1970: Geology and Economic Minerals of Canada; Geological Survey of Canada, Economic Geology Report no. 1, 838 p.

Drake, C.L., Ewing, M., and Sutton, J.
1959: Continental margins and geosynclines: the east coast of North America north of Cape Hatteras; in Physics and Chemistry of the Earth, v. 5; New York, Pergamon Press, p. 110-198.

Engel, A.E.J.
1963: Geologic evolution of North America; Science, v. 140, no. 3563, p. 143-152.

Geikie, A.
1905: The Founders of Geology; reprinted in 1962, Dover Publications, New York, 486 p.

Geological Association of Canada
1971: A Newfoundland decade; Proceedings of the Geological Association of Canada, v. 24, no. 1, 118 p.

Grant, D.R.
1989: Quaternary geology of the Atlantic Appalachian region of Canada; in Chapter 5 of Quaternary Geology of Canada and Greenland, (ed.) R.J. Fulton; Geological Survey of Canada, Geology of Canada, no. 1 (also Geological Society of America, The Geology of North America, v. K-1).

Hall, J.
1883: Contributions to the geological history of the North American continent; American Association for the Advancement of Science Proceedings, 31st Annual Meeting, Montreal, 1882, p. 31-69; abstract in Canadian Naturalist and Geologist, v. 2, p. 284-286, 1857 (also Natural History of New York, Division 6, Paleontology, v. 3, p. 1-96, 1859 (after Rodgers, 1970).

Harland, W.B. and Gayer, R.A.
1972: The Arctic Caledonides and earlier oceans; Geological Magazine, v. 109, p. 289-314.

Hatcher, R.D., Jr., Thomas, W.A., and Viele, G.W. (ed.)
1989: The Appalachian-Ouachita Orogen in the United States; Boulder, Colorado, Geological Society of America, The Geology of North America, v. F-2.

Hatcher, R.D., Jr., Williams, H., and Zietz, I. (ed.)
1983: Contributions to the Tectonics and Geophysics of Mountain Chains; Geological Society of America, Memoir 158, 223 p.

Haworth, R.T. and Jacobi, R.D.
1983: Geophysical correlation between the geological zonation of Newfoundland and the British Isles; in Contributions to the Tectonics and Geophysics of Mountain Chains, (ed.) R.D. Hatcher, Jr., H. Williams, and I. Zietz; Geological Society of America, Memoir 158, p. 25-32.

Haworth, R.T., Daniels, D.L., Williams, H., and Zietz, I.
1980a: Bouguer gravity anomaly map of the Appalachian Orogen; Memorial University of Newfoundland, Map no. 3, scale 1:1 000 000 .
1980b: Bouguer gravity anomaly map of the Appalachian Orogen; Memorial University of Newfoundland, Map no. 3a, scale 1:2 000 000.

Haworth, R.T., Keen, C.E., and Williams, H.
1985: D-1 Northern Appalachians: (West Sheet) Grenville Province, Quebec, to Newfoundland; Geological Society of America, Centennial Continent/Ocean Transect #1.
in press: Transects of the ancient and modern continental margins of Eastern Canada; in North American Continental Margins, (ed.) R. Speed; Geological Society of America, Memoir.

Howley, J.P.
1925: Geological map of Newfoundland; Geological Survey of Newfoundland, scale 1:1 000 000 (approximate).

Kay, M.
1951: North American geosynclines; Geological Society of America, Memoir 48, 143 p.

Kay, M. (ed.)
1969: North Atlantic – geology and continental drift; American Association of Petroleum Geologists, Memoir 12, 1082 p.

Keen, C.E.
1979: Thermal history and subsidence of rifted continental margins - evidence from wells on the Nova Scotian and Labrador shelves; Canadian Journal of Earth Sciences, v. 16, p. 502-522.

Keen, C.E., Keen, M.J., Nichols, B., Ried, I., Stockmal, G.S., Colman-Sadd, S.P., O'Brien, S.J., Miller, H., Quinlan, G., Williams, H., and Wright, J.
1986: Deep seismic reflection profile across the northern Appalachians; Geology, v. 14, p. 141-145.

Keen, M.J. and Williams, G.L. (ed.)
1990: Geology of the Continental Margin of Eastern Canada; Geological Survey of Canada, Geology of Canada, no. 2 (also Geological Society of America, The Geology of North America, v. I-1).

Keppie, J.D.
1985: The Appalachian collage; in The Caledonide Orogen-Scandinavia and Related Areas, Part 2, (ed.) D.G. Gee and B.A. Sturt; John Wiley and Sons, p. 1217-1226.
1989: Northern Appalachian terranes and their accretionary history; in Terranes in the Circum-Atlantic Paleozoic Orogens, (ed.) R.D. Dallmeyer; Geological Society of America, Special Paper 230, p. 159-193.

Keppie, J.D. (comp.)
1979: Geological map of the Province of Nova Scotia; Nova Scotia Department of Mines and Energy, scale 1:500 000.

Keppie, J.D. (co-ord.)
1982: Structural map of the Appalachian Orogen in Canada; Memorial University of Newfoundland, Map no. 4, scale 1:2 000 000.

Keppie, J.D. and Dallmeyer, R.D.
1989:   Tectonic map of pre-Mesozoic terranes in Circum-Atlantic Phanerozoic orogens; International Geological Correlation Programme, Project #233, Terranes in Circum-Atlantic Paleozoic Orogens, scale 1:5 000 000.

Kerr, J.W. and Fergusson, A.J. (ed.)
1981:   Geology of the North Atlantic Borderlands; Canadian Society of Petroleum Geologists, Memoir 7, 743 p.

King, L.H.
1972:   Physiographic evolution of the Canadian Appalachian Province; in The Appalachian Structural Province, (co-ord.) H. Williams, M.J. Kennedy, and E.R.W. Neale; in Variations in Tectonic Styles in Canada, (ed.) R.A. Price and R.J.W. Douglas; Geological Association of Canada, Special Paper 11, p. 248-253.

King, P.B.
1959:   The Evolution of North America; Princeton University Press, New Jersey, 190 p.

Krynine, P.D.
1948:   The megascopic study and field classification of sedimentary rocks; Journal of Geology, v. 56, p. 130-165.

Logan, W.E.
1863:   Geology of Canada; Dawson, Montreal, 983 p.

Marillier, F., Keen, C.E., Stockmal, G.S., Quinlan, G., Williams, H., Colman-Sadd, S.P., and O'Brien, S.J.
1989:   Crustal structure and surface zonation of the Canadian Appalachians: implications of deep seismic reflection data; Canadian Journal of Earth Sciences, v. 26, p. 305-321.

McKenzie, D.P. and Parker, R.L.
1967:   The North Pacific: an example of tectonics on a sphere; Nature, v. 216, p. 1276-1280.

Murray, A. and Howley, J.P.
1881:   Report of the Geological Survey of Newfoundland for 1864-1880; Edward Stanford, London, 536 p.
1918:   Reports of the Geological Survey of Newfoundland for 1881-1909; Edward Stanford, London, 725 p.

Neale, E.R.W. and Williams, H. (ed.)
1967:   Geology of the Atlantic region; Lilly Memorial Volume; Geological Association of Canada, Special Paper 4, 292 p.

Nelson, K.D., McBride, J.H., Arnow, J.A., Oliver, J.E., Brown, L.D., and Kaufman, S.
1985:   New COCORP profiling in the southeastern United States. Part ll Brunswick and east coast magnetic anomalies, opening of the north-central Atlantic Ocean; Geology, v. 13, p. 718-721.

Occhietti, S.
1989:   Quaternary geology of the St. Lawrence Valley and adjacent Appalachian subregion; in Chapter 4 of Quaternary Geology of Canada and Greenland, (ed.) R.J. Fulton; Geological Survey of Canada, Geology of Canada, no. 1 (also Geological Society of America, The Geology of North America, v. K-1).

Palmer, A.R.
1983:   The Decade of North American Geology, 1983 Geologic Time Scale; Geology, v. 11, p. 503-504.

Piasecki, M.A.J., Williams, H., and Colman-Sadd, S.P.
1990:   Tectonic relationships along the Meelpaeg, Burgeo and Burlington Lithoprobe transects in Newfoundland; in Current Research (1990), Newfoundland Department of Mines and Energy, Geological Survey Branch, Report 90-1, p. 327-339.

Poole, W.H., Stanford, B.V., Williams, H., and Kelley, D.G.
1970:   Geology of southeastern Canada; in Geology and Economic Minerals of Canada, (ed.) R.J.W. Douglas; Geological Survey of Canada, Economic Geology Report no. 1, p. 227-304.

Potter, R.R., Hamilton, J.B., and Davies, J.L. (comp.)
1979:   Geological map of New Brunswick; New Brunswick Department of Natural Resources, Map no. N.R.-1, second edition, scale 1:500 000.

Rodgers, J.
1968:   The eastern edge of the North American continent during the Cambrian and Early Ordovician; in Studies of Appalachian Geology Northern and Maritime, (ed.) E-an Zen, W.S. White, J.B. Hadley, and J.B. Thompson, Jr.; Interscience Publishers, p. 141-149.
1970:   The Tectonics of the Appalachians; Wiley-Interscience, New York, 271 p.

Ruiz, J., Patchett, P.J., and Ortega-Gutierrez, F.
1988:   Proterozoic and Phanerozoic basement terranes of Mexico and Nd isotopic studies; Geological Society of America Bulletin, v. 100, p. 274-281.

St-Julien, P. and Béland, J. (ed.)
1982:   Major structural zones and faults of the Northern Appalachians; Geological Association of Canada, Special Paper 24, 280 p.

St-Julien, P., Slivitzky, A., and Feininger, T.
1983:   A deep structural profile across the Appalachians of southern Quebec; in Contributions to the tectonics and geophysics of mountain chains, (ed.) R.D. Hatcher, H. Williams, and I. Zietz; Geological Society of America, Memoir 158, p. 103-111.

Schenk, P.E.
1978:   Synthesis of the Canadian Appalachians; in IGCP Project 27, Caledonian-Appalachian Orogen of the North Atlantic region; Geological Survey of Canada, Paper 78-13, p. 111-136.

Schuchert, C.
1923:   Sites and natures of the North American geosynclines; Geological Society of America Bulletin, v. 34, p. 151-229.

Schuchert, C. and Dunbar, C.O.
1934:   Stratigraphy of western Newfoundland; Geological Society of America, Memoir 1, 123 p.

Sheridan, R.E. and Grow, J.A. (ed.)
1988:   The Atlantic Continental Margin U.S.; Geological Society of America, The Geology of North America, v. K-2.

Shih, K.G., Williams, H., and Macnab, R. (comp.)
1993a: Gravity anomalies and major structural features of southeastern Canada and the Atlantic continental margin; Geological Survey of Canada, Map 1777A, scale 1:3 000 000.
1993b: Magnetic anomalies and major structural features of southeastern Canada and the Atlantic continental margin; Geological Survey of Canada, Map 1778A, scale 1:3 000 000.

Stevens, R.K.
1970:   Cambro-Ordovician flysch sedimentation and tectonics in west Newfoundland and their possible bearing on a proto-Atlantic ocean; in Flysch Sedimentology in North America, (ed.) J. Lajoie; Geological Association of Canada, Special Paper 7, p. 165-177.
1974:   History of Canadian geology; Geoscience Canada, v. 1, no. 2, p. 40-44.

Tankard, A.J. and Balkwill, H.R. (ed.)
1989:   Extensional tectonics and stratigraphy of the North Atlantic margins; American Association of Petroleum Geologists, Memoir 46, 641 p.

Thomas, W.A.
1977:   Evolution of Appalachian-Ouachita salients and recesses from reentrants and promontories in the continental margin; American Journal of Science, v. 277, p. 1233-1278.

Tozer, E.T. and Schenk, P.E. (ed.)
1978:   Caledonian-Appalachian Orogen of the North Atlantic region; Geological Survey of Canada, Paper 78-13, 242 p.

van de Poll, H.W.
1983:   Geology of Prince Edward Island; P.E.I. Department of Energy and Forestry, Energy and Minerals Branch, Report 83-1, 66p. Accompanying geological map, scale 1:250 000 (approximate).
1989:   Lithostratigraphy of the Prince Edward Island redbeds; Atlantic Geology, v. 25, p. 23-25. Geological map on page 24, scale 1:1 777 778 (approximate).

Vogt, P.R. and Tucholke, B.E. (ed.)
1986:   The western North Atlantic region; Geological Society of America, The Geology of North America, v. M.

Williams, H.
1964:   The Appalachians in northeastern Newfoundland - a two-sided symmetrical system; American Journal of Science, v. 262, p. 1137-1158.
1971:   Mafic-ultramafic complexes in western Newfoundland Appalachians and the evidence for their transportation: A review and interim report; Geological Association of Canada Proceedings, v. 24, p. 9-25.
1976:   Tectonic stratigraphic subdivision of the Appalachian Orogen; Geological Society of America, Abstracts with Programs, v. 8, no. 2, p. 300.
1978a: Tectonic lithofacies map of the Appalachian Orogen; Memorial University of Newfoundland, Map no. 1, scale 1:1 000 000.
1978b: Tectonic lithofacies map of the Appalachian Orogen; Memorial University of Newfoundland, Map no. 1a, scale 1:2 000 000.
1979:   Appalachian Orogen in Canada; Canadian Journal of Earth Science, Tuzo Wilson volume, v. 16, p. 792-807.
1984   Miogeoclines and suspect terranes of the Caledonian-Appalachian Orogen: tectonic patterns in the North Atlantic region; Canadian Journal of Earth Sciences, v. 21, p. 887-901.

Williams, H. and Doolan, B.L.
1979: Evolution of Appalachian-Ouachita salients and recesses from reentrants and promontories in the continental margin. Discussion; American Journal of Science, v. 279, p. 92-95.

Williams, H. and Hatcher, R.D., Jr.
1982: Suspect terranes and accretionary history of the Appalachian Orogen; Geology, v. 10, p. 530-536.
1983: Appalachian Suspect terranes; in Contributions to the tectonics and geophysics of mountain chains, (ed.) R.D. Hatcher, H. Williams, and I. Zietz; Geological Society of America, Memoir 158, p. 33-53.

Williams, H. and Haworth, R.T.
1984a: Bouguer gravity anomaly map of Atlantic Canada; Memorial University of Newfoundland, Map no. 5, scale 1:2 000 000.
1984b: Magnetic anomaly map of Atlantic Canada; Memorial University of Newfoundland, Map no. 5, scale 1:2 000 000.

Williams, H. and St-Julien, P.
1982: The Baie Verte-Brompton Line: continent-ocean interface in the Northern Appalachians; in Major Structural Zones and Faults of the Northern Appalachians, (ed.) P. St-Julien and J. Béland; Geological Association of Canada, Special Paper 24, p. 177-207.

Williams, H., Colman-Sadd, S.P., and Swinden, H.S.
1988: Tectonic-stratigraphic subdivisions of central Newfoundland; in Current Research, Part B; Geological Survey of Canada, Paper 88-1B, p. 91-98.

Williams, H., Kennedy, M.J., and Neale, E.R.W.
1972: The Appalachian Structural Province; in Variations in Tectonic Styles in Canada, (ed.) R.A. Price and R.J.W. Douglas; Geological Association of Canada, Special Paper 11, p. 181-261.

Williams, H., Piasecki, M.A.J., and Colman-Sadd, S.P.
1989: Tectonic relationships along the proposed central Newfoundland Lithoprobe transect and regional correlations; in Current Research, Part B; Geological Survey of Canada, Paper 89-1B, p. 55-66.

Wilson, J.T.
1949: Some major structures in the Canadian Shield; Transactions of the Canadian Institute of Mining and Metallurgy, v. 52, p. 231-242.
1954: The development and structure of the crust; in The Earth as a Planet, Volume 2, (ed.) G.P. Kuiper; University of Chicago Press.
1966: Did the Atlantic close and then re-open?; Nature, v. 211, no. 5050, p. 676-681.

Wones, D.R. (ed.)
1980: Proceedings "The Caledonides in the USA"; Virginia Polytechnic Institute and State University, Department of Geological Sciences, Memoir 2, 329 p.

Yorath, C.J., Parker, E.R., and Glass, D.J. (ed.)
1975: Canada's Continental Margins and Offshore Petroleum Exploration; Canadian Society of Petroleum Geologists, Memoir 4, 898 p.

Zen, E-an
1983: Exotic terranes in the New England Appalachians - limits, candidates, and ages: a speculative essay; in Contributions to the tectonics and geophysics of mountain chains, (ed.) R.D. Hatcher, H. Williams, and I. Zietz; Geological Society of America, Memoir 158, p. 55-81.

Zietz, I., Haworth, R.T., Williams, H., and Daniels, D.L.
1980a: Magnetic anomaly map of the Appalachian Orogen; Memorial University of Newfoundland, Map no. 2, scale 1:1 000 000.
1980b: Magnetic anomaly map of the Appalachian Orogen; Memorial University of Newfoundland, Map no. 2a, scale 1:2 000 000.

Author's Address

H. Williams
Department of Earth Sciences
Memorial University of Newfoundland
St. John's, Newfoundland
A1B 3X5

Printed in Canada

Chapter 2

TEMPORAL AND SPATIAL DIVISIONS

Chapter 2

TEMPORAL AND SPATIAL DIVISIONS

Harold Williams

PREAMBLE AND PREVIOUS ANALYSES

A wealth of detailed data exists for the Canadian Appalachian region. Presentation of data for an orogen as complex as the Appalachian can be problematical. In this chapter the presentation of these data is based on temporal and spatial divisions of rock units. Previous analyses of the Canadian Appalachian region reflected theories in vogue at the time of their undertaking. Thus prior to the wide acceptance of plate tectonics and continental drift, analyses reflected geosynclinal theory. These varied in detail, from the early work of Schuchert (1923) to the sophisticated compound geosynclines that served as a theme for the fifth edition of the Geology and Economic Minerals of Canada (Poole et al., 1970). Subsequent to plate tectonics, conceptual plate models have abounded. Many were contrived on local relations and abandoned as new information on the geology was uncovered.

Analyses based on "as is" tectonostratigraphic spatial divisions are those that have met with most success. Thus divisions in the U.S. Appalachians (King, 1950, 1959) such as Appalachian Foreland, Valley and Ridge, Blue Ridge and Piedmont provinces, are still in wide usage. Similarly, the tripartite division of Newfoundland (Williams, 1964, see Fig. 1.16), Western Platform, Central Volcanic Belt and Avalon Platform (Kay and Colbert, 1965) remains little changed, though introduced well before the plate tectonic revolution. These are meaningful geographic divisions used in objective syntheses that describe rocks and structures, while attempting to separate what is known, from what is interpreted.

In any objective "as is" account, decisions must be made at the outset on the relative importance of temporal versus geographic divisions. Time-slice analyses work best for simple orogens with few tectonostratigraphic divisions, or where there are established linkages between rocks of one area and those of another. Geographic divisions are necessary for orogens that are complex and made up of terranes or zones that maintain a distinctiveness over long periods. A combination of temporal and spatial divisions is required in the case of the Canadian Appalachians where early Paleozoic geographic entities eventually lose their distinctiveness and share a later Paleozoic history.

A detailed tectonostratigraphic zonation for lower Paleozoic and older rocks of the Canadian Appalachians was introduced in 1972, based mainly on the northeast Newfoundland coastal section. The region was divided into nine zones, designated alphabetically A to I and later given local names in their type areas (Williams et al., 1972, 1974). While some of the zones were meaningful, such as zones H (Avalon) and I (Meguma), others were premature and extended without a proper understanding of rocks and relationships beyond the type areas. Consequently most of these zones were never widely accepted outside Newfoundland, and some such as zones B and C, were ephemeral even in their type areas. While not defined as suspect terranes, it was implicit that rocks of some zones travelled great distances from their places of origin to their present positions.

Subsequent trends have been toward fewer zones of broader definition allowing extrapolation across the Canadian Appalachians and extensions into the United States and elsewhere (Williams, 1976, 1978a, b, 1979). The Humber, Dunnage, Gander, Avalon, and Meguma zones are those of widest current usage. They are the theme for the Tectonic Lithofacies Map of the Appalachian Orogen (Williams, 1978a) and they are well known and widely used in Canada (e.g. St-Julien and Béland, 1982). Some zone names, e.g. Avalon, are adapted for correlatives far beyond the Canadian segment of the orogen (Williams and Max, 1980; Williams and Hatcher, 1982, 1983). Zonal divisions of early Paleozoic and older rocks for New Brunswick (Ruitenberg et al., 1977; Fyffe and Fricker, 1987) and Cape Breton Island (Barr and Raeside, 1986, 1989) correspond, for the most part, with divisions in Newfoundland.

There were few attempts for geographic divisions of middle Paleozoic rocks in the Canadian Appalachians, except for conceptual models for parts of the region (Berry and Boucot, 1970; McKerrow and Ziegler, 1971) and a former division in Newfoundland (Williams, 1967). Analyses of upper Paleozoic rocks and Mesozoic rocks generally focus on treatment of separate basins or graben.

Williams, H.
1995:   Temporal and spatial divisions; Chapter 2 in Geology of the Appalachian-Caledonian Orogen in Canada and Greenland, (ed.) H. Williams; Geological Survey of Canada, Geology of Canada, no. 6, p. 21-44 (also Geological Society of America, The Geology of North America, v. F-1).

TEMPORAL AND SPATIAL DIVISIONS

Clarity in classification demands clarity of thought and an insight into the important features of the orogen. Useful divisions should embody the current level of geological understanding and provide a framework for description and a model for future studies.

Rocks of the Canadian Appalachian region are divided into four broad temporal categories; early Paleozoic and older, middle Paleozoic, late Paleozoic, and Mesozoic. There are lithological distinctions and in most places unconformities between rocks of each temporal category (Table 2.1). Names of divisions within the temporal categories are taken from geographical localities.

The zonal division for lower Paleozoic and older rocks (Humber, Dunnage, Gander, Avalon, and Meguma), introduced more than 15 years ago (Williams, 1976), remains functional and still serves as a useful framework for new models and future studies. It is therefore retained as the first order division of the Canadian Appalachian region,

Table 2.1. Table of temporal and spatial subdivisions for rocks of the Canadian Appalachian Region.

MESOZOIC (Graben)						
Fundy			Chedabucto			
LATE PALEOZOIC (Basins)						
Ristigouche St. Andrews Plaster Rock Carlisle Marysville Central	Moncton Sackville Cumberland	Magdalen	Western Cape Breton Central Cape Breton Sydney Antigonish Stellarton Minas		Bay St. George Deer Lake	
MIDDLE PALEOZOIC (Belts/Divisions)						
Gaspé	Fredericton	Mascarene	Arisaig	Cape Breton	Annapolis	
Aroostook-Percé Chaleurs Bay Connecticut Valley-Gaspé		Nevepis Letete Ovenhead Campobello Oak Bay				
Clam Bank	Springdale	Cape Ray	Badger	Botwood	La Poile	Fortune
	White Bay Halls Bay Micmac Lake Cape St. John		Eastern New World Island Bay of Exploits New Bay Badger Bay			
EARLY PALEOZOIC AND OLDER (Zones/Subzones)						
(northwest)					(southeast)	
Humber	Dunnage		Gander	Avalon		Meguma
External Internal Blair River	Notre Dame Exploits Twillingate Indian Bay Dashwoods Belledune Elmtree Popelogan	Armstrong Brook Bathurst Hayesville Estrie-Beauce Mégantic Témiscouata Gaspésie	Gander Lake Mount Cormack Meelpaeg Aspy Miramichi St. Croix	Burgeo Bras d'Or Mira Antigonish Cobequid		

with important breakdowns into subzones in some places (Fig. 2.1 and Map 2). Rocks of these zones display the sharpest contrasts in lithology, stratigraphy and thickness across the orogen. The zones also exhibit structural, faunal, geophysical, plutonic and metallogenic contrasts, thus enhancing their distinctiveness. As well, the surface rocks of some zones coincide with lower crustal blocks. Because zones are based on the oldest rocks of the orogen, they are fundamental to understanding its protoliths and controls of subsequent development. Obviously, the zone boundaries are defined best where rocks of appropriate age are exposed over broad areas. This is the case in Newfoundland, but each zone can be traced by limited exposure across the entire Canadian Appalachian region (Fig. 2.1 and Map 2).

Middle Paleozoic rocks also exhibit lithological, stratigraphic and thickness variations across the orogen, and as in the case of early Paleozoic zones, they define middle Paleozoic belts (Fig. 2.2). Some of these divisions are new and few are as well known as zones. The belts are clearest among the extensive middle Paleozoic rocks exposed in Quebec, New Brunswick and Nova Scotia. These are Gaspé, Fredericton, Mascarene, Arisaig, Cape Breton, and Annapolis. The middle Paleozoic record in Newfoundland is fragmentary and there is no belt-for-belt equivalence with the mainland. Different names are used, which are Clam Bank, Springdale, Cape Ray, Badger, Botwood, La Poile, and Fortune. The broad offshore area of the Newfoundland Grand Banks has middle Paleozoic rocks unlike those of the onland orogen. These are discussed in Chapter 11 with offshore extensions of the Appalachian Orogen.

The contrasts among rocks of middle Paleozoic belts are less distinctive, compared to those of early Paleozoic zones. Lithological and stratigraphic contrasts are most important, but eastern mainland belts also have faunal, plutonic, and metallogenic distinctiveness. In areas affected by Ordovician deformation, middle Paleozoic belts cross early Paleozoic zones. Thus in the Gaspésie-New Brunswick area, the Gaspé Belt crosses the Humber, Dunnage, and Gander zones, indicating that these early Paleozoic zones were already in proximity during the middle Paleozoic. Similarly in Newfoundland, the Springdale Belt straddles the Humber-Dunnage zone boundary. In areas unaffected by Ordovician deformation, lower and middle Paleozoic rocks are conformable and middle Paleozoic belts are coincident with early Paleozoic zones. Thus the Annapolis Belt and Meguma Zone of Nova Scotia define the same area, and the Badger Belt of Newfoundland lies within the Dunnage Zone.

Upper Paleozoic rocks of the Canadian Appalachian region are mainly terrestrial cover sequences that unconformably overlie deformed middle Paleozoic and older rocks. The rocks are everywhere essentially the same, dominantly redbeds with local volcanic units toward the bases of the sections. Their thicknesses define a number of basins, chiefly located in New Brunswick and Nova Scotia (Fig. 2.3). Their distribution is unrelated to the geometry of belts or zones, but some coincide with ancestral structural boundaries.

Mesozoic rocks, mainly Triassic and Jurassic, are redbeds and basalts that unconformably overlie deformed upper Paleozoic and older rocks. These are restricted onland to the Fundy Graben in the Bay of Fundy area and the Chedabucto Graben to the east (Fig. 2.3). Coeval dykes are more widespread.

The temporal rock divisions are in most places separated by structural events, although these are not everywhere coeval and controls may be different from one area to another. Taconic Orogeny (Early and Middle Ordovician) affected the Humber Zone and western parts of the Dunnage Zone (Map 5). Ordovician deformation also affected parts of the Gander Zone and the eastern Dunnage Zone, at least in Newfoundland. It is absent in the Avalon and Meguma zones and parts of the northern Newfoundland Dunnage Zone. Middle Paleozoic orogeny affected the entire orogen, except for parts of the Newfoundland Avalon Zone (Map 6). It has been generally considered Acadian and of Devonian age but there is growing evidence for Silurian orogenesis in Newfoundland and in Cape Breton Island (Dunning et al., 1988, 1990; Addendum, note 1). Alleghanian Orogeny of Carboniferous-Permian age affected a narrow area of thick deposition (Map 6). It is partly coincident with the Avalon-Meguma zone boundary in Nova Scotia and the Humber-Dunnage zone boundary in Newfoundland.

In a general way, variations in structural style, plutonism, metamorphism, and other features correspond to lithic and stratigraphic contrasts exhibited among rocks of the four temporal divisions, thus adding to their importance. This is expected, as each temporal division represents a change in depositional environments and tectonic settings: entirely marine rocks of early Paleozoic zones, mixed marine and terrestrial rocks of middle Paleozoic belts, mainly terrestrial rocks of late Paleozoic basins, and entirely terrestrial rocks of Mesozoic graben.

Table 2.1 provides a list of subdivisions and the necessary systematics to describe stratigraphic and structural relationships. Zones are discrete entities. The Humber Zone represents that part of the orogen that is linked stratigraphically to the North American craton, or the Appalachian miogeocline. Others represent accreted parts of the orogen or suspect terranes. Belts may represent a number of tectonic elements. Some trend acutely across the miogeocline and already accreted terranes as cover sequences. Others are sited along zone boundaries, either above earlier structural junctions or possibly separating early Paleozoic zones/terranes. Still others form internal parts of zones/terranes, in cases where the older rocks are unaffected by early Paleozoic orogeny. Late Paleozoic basins represent a mainly terrestrial cover across the whole orogen with local anomalously thick sections controlled by subsidence and uplifted blocks. The Mesozoic Fundy and Chedabucto graben occur as a single rift coincident with a thick Carboniferous basin and the earlier Avalon-Meguma zone boundary.

NATURE OF BOUNDARIES

Rocks that define zones were in most cases widely separated at deposition. Since they are telescoped into a mountain belt now, their boundaries are tectonic. This applies to all except some Dunnage-Gander zone boundaries where mixed sedimentary and volcanic rocks assigned to the Dunnage Zone are stratigraphically above rocks of the Gander Zone (van Staal, 1987; O'Neill and Knight, 1988).

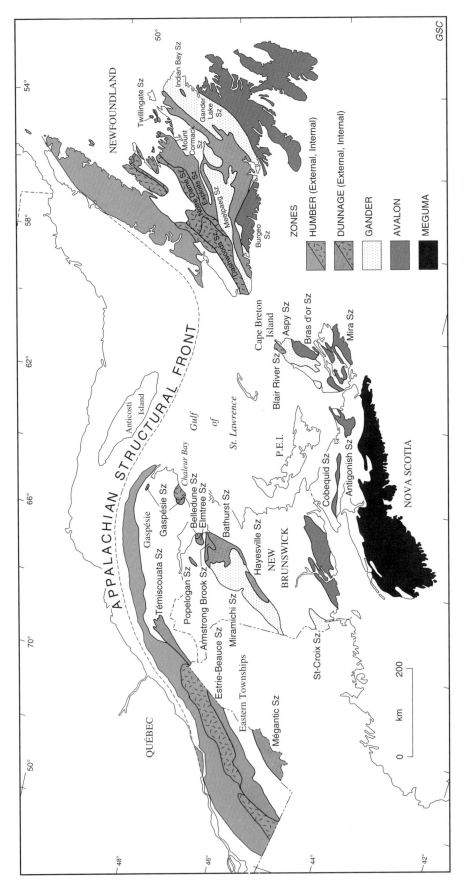

Figure 2.1. Outcrop areas of lower Paleozoic and older rocks that define zones and subzones of the Canadian Appalachian region.

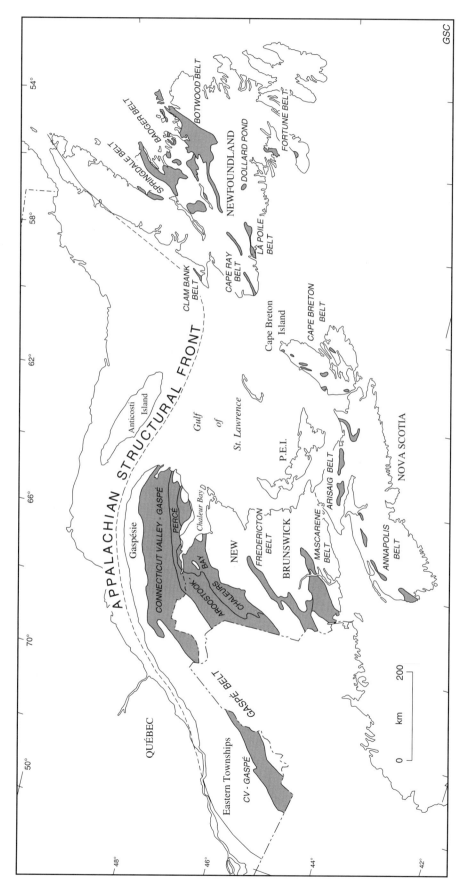

Figure 2.2. Outcrop areas of Middle Paleozoic rocks that define belts of the Canadian Appalachian region.

27

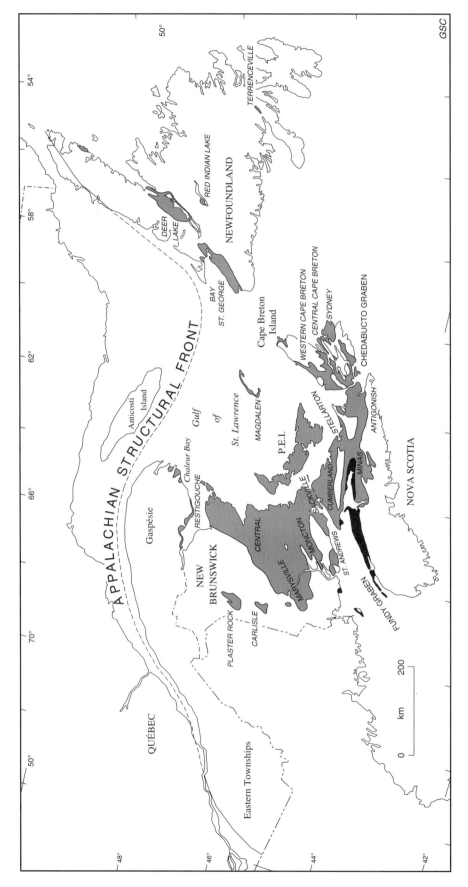

Figure 2.3. Outcrop areas of upper Paleozoic and Mesozoic rocks that define basins and graben, respectively, of the Canadian Appalachian region.

The Humber and Dunnage zones were first juxtaposed in the Ordovician (Williams and St-Julien, 1982). Dunnage-Gander zone boundaries are also interpreted as Ordovician tectonic junctions in most places but they are commonly overprinted by middle Paleozoic ductile deformation (Piasecki, 1988; Currie and Piasecki, 1989). Most Ordovician boundaries are marked by ophiolite complexes and mélanges, and the boundaries are concordant with respect to adjacent lithic units or structures. Gander-Avalon and Avalon-Meguma zone boundaries are later, of Silurian-Devonian age. They are marked by steep mylonite zones or brittle faults that are discordant with respect to adjacent lithic units and structures (Blackwood and Kennedy, 1975; Dallmeyer et al., 1981; Keppie, 1982).

The early Paleozoic zonation applies to the total area of the orogen. Expressed in another way, there is no evidence in the Canadian Appalachians of structural blocks or terranes that are made up entirely of middle Paleozoic or younger rocks.

Middle Paleozoic belts are in most places separated by exposed older rocks of underlying zones. Their present outlines and extent are largely the result of erosion.

Late Paleozoic basins developed upon deformed rocks of belts and zones. They are continuous across all zones and belts from northern Gaspésie to eastern Nova Scotia. Present exposures are erosional remnants of a once more continuous cover.

The Mesozoic Fundy Graben contains redbeds and mafic volcanic rocks that overlie deformed Carboniferous and older rocks. The graben coincides with the Avalon-Meguma zone boundary, the locus of deep Carboniferous basins, and projects offshore through the Chedabucto Graben toward the Orpheus Graben. Mesozoic dykes cross early Paleozoic zones, middle Paleozoic belts, and late Paleozoic basins indiscriminately.

EARLY PALEOZOIC AND OLDER ZONES

The zonal division of Humber, Dunnage, Gander, Avalon, and Meguma is used here because of its familiarity, wide usage, and practicality (Table 2.1, Fig. 2.1, and Map 2). Recent work has led to subdivision of the Dunnage and Gander zones in Newfoundland (Williams et al., 1988, 1989; Piasecki et al., 1990; Williams, Chapter 3; Williams et al., Chapter 3), New Brunswick (van Staal, 1987; van Staal and Fyffe, Chapter 3), Quebec (Tremblay et al., Chapter 3), and to refinements of divisions and boundaries in Cape Breton Island of Nova Scotia (Barr and Raeside, 1989).

Although the zones are defined, first and foremost, on rocks and stratigraphy, they fit well with present conceptual models of the orogen. Thus according to a Wilson cycle model, the Humber Zone records the development and destruction of an Atlantic-type passive margin, the ancient continental margin of eastern North America; parts of the Dunnage Zone represent vestiges of Iapetus with volcanic sequences and mélanges built upon oceanic crust; and the Gander, Avalon, and Meguma zones are a sampling of tectonic elements that lay outboard or on the opposing side of Iapetus.

The zonal division also fits suspect terrane analyses of the Appalachian Orogen (Williams and Hatcher, 1982, 1983; Zen, 1983; Keppie, 1985, 1989). These analyses emphasize the important distinction between the miogeocline, which was linked to the North American craton, and outboard terranes of unknown paleogeography that were accreted during the evolution of the orogen. Thus the Humber Zone represents the Appalachian miogeocline and the Dunnage, Gander, Avalon, and Meguma zones are all accreted terranes.

Further to the suspect terrane viewpoint, it should be emphasized that some zones are composite. The Humber Zone includes structural elements transported across it, such as volcanic groups and ophiolite complexes that are really Dunnage outliers. Similarly, the Dunnage and Avalon zones are composite or superterranes. Terrane division into impractical and unmappable sizes is avoided.

The following definitions and descriptions of the early Paleozoic zones, middle Paleozoic belts, late Paleozoic basins, and Mesozoic graben are mainly summarized from Chapters 3 to 6. More complete references are given there. Correlations across the Atlantic are taken from Williams (1978b, 1984) and Keppie and Dallmeyer (1989).

Humber Zone

The Humber Zone (Fig. 2.1 and Map 2) is recognized in western Newfoundland and Quebec. A small area of Grenville gneisses (Blair River Complex) in northern Cape Breton Island is either part of the Humber Zone (Barr et al., 1987; Barr and Raeside, 1989) or basement to the Avalon Zone (Keppie et al., 1990, Chapter 3).

The Humber Zone is separated into external and internal parts based on structural and metamorphic styles. In the external Humber Zone a crystalline basement is correlated with Precambrian rocks of the nearby Grenville Province of the Canadian Shield. The basement rocks are cut by mafic dykes, and toward the west the basement is overlain unconformably by an arkosic clastic unit with local mafic flows. The basal clastic/volcanic unit is overlain by Cambrian shales and quartzites and a thick Cambrian to Middle Ordovician carbonate sequence. This is capped by a Middle Ordovician shale-sandstone unit, in turn overlain by chaotic mélanges and transported sedimentary and igneous rocks of Taconic allochthons.

Deformation increases from west to east so that stratigraphic and structural relationships are less clear in the internal Humber Zone. There, rocks are mainly psammitic and pelitic schists with some marble units and chloritic schists. Basement rocks are recognized in some places but they are highly deformed and commonly indistinguishable from schistose cover. The structural style of the external Humber Zone is that of a foreland fold and thrust belt. In Quebec, and western Newfoundland, deformed and metamorphosed rocks of the internal Humber Zone are thrust above less deformed and less metamorphosed rocks of the external Humber Zone.

The western margin of the Humber Zone is defined by the limit of Appalachian deformation (Fig. 2.1, Maps 5 and 6). Its eastern margin is a steep structural belt (Williams and St-Julien, 1982), marked by ophiolite occurrences that separate polydeformed schists of the internal Humber Zone to the west from less deformed volcanic sequences of the

Dunnage Zone to the east (Fig. 2.1). The Humber-Dunnage boundary is well defined in northeast Newfoundland and in the Quebec Eastern Townships, where ophiolitic rocks at the boundary form the world's richest asbestos belt. The boundary is offset and hidden by younger cover rocks throughout much of Gaspésie and it is enigmatic and difficult to delineate among the heavily intruded and metamorphosed crystalline rocks and ophiolite complexes of southwest Newfoundland.

Taconic Orogeny is recognized as the first major event to affect the Humber Zone, wherever stratigraphic relationships or isotopic data allow its definition. Acadian Orogeny (Silurian-Devonian) also affected most of the zone, and locally in Newfoundland its eastern parts are affected by Alleghanian deformation (Maps 5 and 6).

The Humber Zone has a uniform width of approximately 100 km and its facies belts and tectonic elements follow the curvilinear course of the zone from western Newfoundland to Quebec. The Taconic structural front (Logan's Line) and the Acadian structural front are almost coincident along its western margin. The course of the Alleghanian structural front is erratic, cutting across older structures (Maps 5 and 6).

Taconic allochthons of the external Humber Zone are made up of a number of structural slices, many of which contain distinctive rock assemblages. These assemblages and structural slices are arranged in a consistent order of structural stacking. Lowest slices contain sedimentary rocks, mainly sandstones, shales, limestones, and limestone breccias. These rocks range in age from late Precambrian to Early and Middle Ordovician. In some places it can be shown that the lowest structural slices contain the stratigraphically youngest sedimentary rocks. Higher structural slices contain volcanic rocks and ophiolite complexes typical of the Dunnage Zone. These are dated isotopically as Late Cambrian to Early Ordovician (Dunning and Krogh, 1985).

Different structural slices exhibit different deformational styles, and most structures were imprinted prior to or during the assembly and transport of the allochthons. These vary from intense foliations, tectonic banding, and folded schistosities in ophiolitic rocks, to scaly cleavages, rootless folds, and overturned beds in sedimentary rocks. Lower slices have internally complex geometries of rock units and the slices are rarely morphologically distinct. Higher structural slices are of simpler internal makeup and some have marked morphological expression. Stratigraphic relationships, palinspastic restorations, and structural considerations all indicate that the allochthons were assembled from east to west and that the structurally highest slices travelled the farthest (Williams, 1975).

Rocks typical of the Canadian Humber Zone extend the full length of the Appalachian Orogen (see Fig. 1.8). Grenville inliers like those of western Newfoundland reappear in the U.S. Appalachians of Vermont and extend along the Blue Ridge Province to Georgia (Williams, 1978a). The cover is everywhere the same; clastic and volcanic rocks at the base overlain by a thick Cambrian-Ordovician carbonate sequence. A clastic unit everywhere overlies the carbonate rocks. Taconic allochthons like those of western Newfoundland and Quebec continue through the type area of New York to the Hamburg Klippe of Pennsylvania. Restoration of the North Atlantic indicates

that the Cambrian-Ordovician carbonate rocks of Scotland and east Greenland are continuations of the North American Humber Zone (see Fig. 1.4).

Dunnage Zone

The Dunnage Zone (Fig. 1.3, 2.1 and Maps 1 and 2) is distinguished by its lower Paleozoic dominantly mafic volcanic rocks, ophiolite suites, mélanges and associated greywackes, slates, cherts, and minor limestones. Some of these marine rocks rest on an ophiolitic substrate. Others have stratigraphic relationships with rocks of the Gander Zone (van Staal, 1987). Many more are of unknown basement relationship. A few rocks within the Dunnage Zone of Newfoundland and Quebec seem to be out of structural, metamorphic, and temporal context with respect to nearby volcanic rocks and ophiolite suites.

Recent work in central Newfoundland has led to a major two-fold division of the Dunnage Zone into a northwestern Notre Dame Subzone (containing the Twillingate Subzone) and a southeastern Exploits Subzone (Williams et al., 1988; Fig. 2.1). The Indian Bay Subzone is a small area of Exploits-type rocks within the northeast Gander Zone (Wonderly and Neuman, 1984). Another division, Dashwoods Subzone or Central Gneiss Terrane of van Berkel and Currie (1988) in southwestern Newfoundland, has rocks and structures characteristic of both the Dunnage and Humber zones and appears to be a tectonic mixture.

In New Brunswick, the Dunnage Zone is divided into six new subzones (van Staal and Fyffe, Chapter 3). The Elmtree, Belledune, and Popelogan subzones occur in two discrete Ordovician inliers that are surrounded by middle Paleozoic rocks in northern New Brunswick. The Armstrong Brook, Bathurst, and Hayesville subzones are recognized in central New Brunswick (Fig 2.1 and Map 2).

In Quebec, the Dunnage Zone is divided into four new subzones, which are from southwest to northeast, Estrie-Beauce, Megantic, Témiscouata, and Gaspésie. Rocks of the Estrie-Beauce, Témiscouata, and Gaspésie subzones are interpreted as laterally equivalent. The Megantic Subzone has different rocks and stratigraphy. Separation of rocks of the Témiscouata and Gaspésie subzones from rocks of the Gaspé Belt is problematic where stratigraphic sections are structurally conformable and possibly continuous (Bourque et al., Chapter 4).

Deformation across the Newfoundland Dunnage Zone is less intense than that in adjacent parts of the Humber and Gander zones. Sub-Silurian unconformities are everywhere present across the Notre Dame Subzone but rare in the Exploits Subzone, where Ordovician-Silurian sections are in places continuous. Similar relationships occur in Quebec where the main phases of deformation are interpreted as Acadian (Tremblay and St-Julien, 1990). In northern New Brunswick, rocks of the Popelogan, Belledune, and Elmtree Subzones are mildly deformed and overlain unconformably by Silurian strata. In central New Brunswick, Dunnage and Gander zone rocks are intensely deformed.

Ophiolitic complexes along the western margin of the Newfoundland Dunnage Zone are imbricated, and although steeply dipping to overturned, the sequences of units face eastward. At the Humber-Dunnage boundary, the ophiolite complexes are overlain by olistostromes and coarse conglomerates, in turn overlain by volcanic sequences. Immediately eastward, the ophiolites are succeeded by volcanic rocks without intervening olistostromes. Intensity of deformation increases westward toward the Humber-Dunnage boundary and, at the boundary, Dunnage ophiolite complexes are structurally and metamorphically gradational with ophiolitic mélanges and metaclastic rocks that are part of the eastern Humber Zone.

The Dunnage Zone is widest (150 km) and best defined in northeast Newfoundland. It is narrow and ill defined in the southwest portion of insular Newfoundland, and it is absent in Cape Breton Island. Apart from exposures in New Brunswick and Quebec, most of its mainland course is hidden by younger cover rocks.

Ophiolitic complexes and Paleozoic oceanic rocks can be traced southward into the New England Appalachians, and a large mafic-ultramafic complex at Baltimore, Maryland lies along a natural continuation of the Humber-Dunnage boundary (Williams, 1978a). Ophiolitic rocks occur farther south, but these occurrences are small and the oceanic rocks are commingled with high-grade metaclastic rocks. On the eastern side of the Atlantic, ophiolitic complexes and volcanic sequences like those of the Canadian Dunnage Zone occur in the Midland Valley of Scotland and there are numerous examples throughout the Scandinavian Caledonides (see Fig. 1.4).

Gander Zone

The Gander Zone (Fig. 2.1 and Map 2) is characterized by a thick sequence of pre-Middle Ordovician arenaceous rocks that are in most places polydeformed and metamorphosed. Relationships with adjacent migmatites and granitic gneisses of the zone have been interpreted as both unconformable and gradational. Most evidence points to a continental crystalline basement beneath the zone.

In Newfoundland, the Gander Zone is divided into the Gander Lake, Mount Cormack, and Meelpaeg subzones (Fig. 2.1). The nature and timing of Dunnage-Gander boundaries are controversial. The Exploits (Dunnage)-Gander Lake (Gander) boundary is tectonic where ophiolitic rocks of the Gander River Complex (O'Neill and Blackwood, 1989) abut sandstones of the Gander Lake Subzone, but the contact has been interpreted as stratigraphic where mafic-ultramafic rocks are absent (Currie et al., 1979, 1980; Pajari et al., 1979; Blackwood, 1982; Blackwood and Green, 1982). On the south coast of Newfoundland, the boundary has been interpreted as a thrust (Colman-Sadd, 1976), a conformable contact (Blackwood, 1985), and a major sinistral ductile shear zone (Piasecki, 1988). Exploits-Mount Cormack and Exploits-Meelpaeg boundaries are interpreted almost everywhere as tectonic (Colman-Sadd and Swinden, 1984; Williams et al., 1988, 1989).

The analysis of the Mount Cormack Subzone as a structural window in central Newfoundland (Colman-Sadd and Swinden, 1984) indicates that the Exploits Subzone is allochthonous, at least in part, above the Gander Zone. Quartzites and psammitic schists of the Mount Cormack Subzone are surrounded by upward- and/or outward-facing ophiolite suites. The ophiolitic rocks and their cover belong to the Exploits Subzone. Quartzites and psammites resemble those at Gander Lake, type area of the Gander Zone.

The Meelpaeg Subzone is also interpreted as a structural window, although ophiolite complexes are absent at most Meelpaeg-Exploits boundaries. Colman-Sadd and Swinden (1984) estimated a minimum displacement of 60 km on the Exploits-Mount Cormack boundary, and subsequent recognition of Gander equivalents at the northwestern Meelpaeg boundary, (Colman-Sadd, 1987, 1988), requires that this be increased to over 100 km. In most places, Meelpaeg-Exploits boundaries are major ductile shear zones that also coincide with sharp increases in metamorphic intensity toward the Meelpaeg Subzone (Colman-Sadd, 1984, 1985, 1987, 1988; Williams et al., 1988, 1989). A model that incorporates Dunnage-Gander zone relationships in Newfoundland has been proposed recently (Williams and Piasecki, 1990 ; Addendum, note 2).

In New Brunswick, the Gander Zone is divided into the Miramichi and St. Croix subzones. In the Miramichi Subzone, the clastic rocks are overlain by a thick Middle Ordovician volcanic sequence with contacts interpreted as both stratigraphic and structural (van Staal, 1987). The volcanic rocks are part of the Dunnage Zone (Bathurst and Hayesville subzones) and stratigraphic relationships with Gander rocks imply Dunnage-Gander proximity at deposition.

At its type area in northeast Newfoundland, rocks of the Gander Zone have a flat schistosity related to recumbent structures. Lineations indicate mainly orogen-parallel transcurrent movements. The latest studies support the interpretations of Kennedy and McGonigal (1972) and Kennedy (1975) for pre-Middle Ordovician deformation overprinted by middle Paleozoic events (Williams et al., 1991).

Local stratigraphic contacts between Lower to Middle Ordovician Dunnage and Gander zone rocks are not incompatible with the structural relationships (Williams and Piasecki, 1990). However at deeper levels, the ultimate Gander-Dunnage distinction presumably contrasts a granitic continental basement and arenaceous cover (Gander) with an ophiolitic oceanic basement and volcanic cover (Dunnage).

The Gander Zone is about 50 km wide in northeast Newfoundland but may be much wider in subsurface. Psammites and amphibolites of the Meelpaeg Subzone are possibly continuous with the Port aux Basques Gneiss at the southwest corner of Newfoundland (Williams et al., 1989; Colman-Sadd et al., 1990; Fig. 2.1 and Map 2).

The narrow Aspy Subzone of Cape Breton Island, which lies between Humber and Avalon correlatives, is treated with the Gander Zone mainly because of metamorphic and structural styles (Addendum, note 3). Relationships and correlations at the constriction in the orogen between southwest Newfoundland and Cape Breton Island are still problematic. The Gander Zone is much wider in New Brunswick, about 200 km.

The Grand Pitch Formation of Maine is a possible correlative of Gander Zone sandstones of Canada (Williams, 1978a). Gneiss dome complexes of New England may be extensions of metamorphic rocks of the Gander Zone in Canada. These can be traced southward from the United States-Canada border to Long Island Sound. The New England domes are covered by volcanic rocks and shales, similar to rocks of the Dunnage Zone. On the east side of the Atlantic, Lower Ordovician shales and sandstones above a crystalline basement in southeast Ireland, and correlatives northeastward in England may be Gander Zone equivalents (Williams, 1978b; Williams and Max, 1980).

Avalon Zone

Rocks of the Avalon Zone (Fig. 2.1 and Map 2) are mainly upper Precambrian sedimentary and volcanic rocks overlain by Cambrian to Lower Ordovician shales and sandstones. All of these rocks are relatively undeformed and unmetamorphosed compared to nearby parts of the Gander Zone. The oldest rocks of the Avalon Zone are marbles, quartzites, and gneisses in New Brunswick and Nova Scotia. These are interpreted to underlie the upper Precambrian sedimentary and volcanic rocks with structural unconformity, but contacts are mainly faults.

In some places the upper Precambrian sedimentary and volcanic rocks pass upward with structural conformity into Cambrian shales. In other places, unconformities and late Precambrian intrusions interrupt stratigraphic sections. These late Precambrian structural and intrusive events define the Avalonian Orogeny, which is unique to this zone (Map 5).

The Avalon Zone encompasses a variety of diverse geological elements among its upper Precambrian rocks, implying a composite makeup. If composite, the elements of the zone were assembled in the late Precambrian, as its Cambrian rocks have similar stratigraphy and distinctive Atlantic realm trilobite faunas throughout its length. In fact, the continuity of its Cambrian rocks and faunas gave rise to the notion of an "Avalon Platform" along the eastern margin of the orogen.

In Newfoundland, the boundary between the Avalon and Gander zones is marked by major faults (Fig. 2.1). The boundary in New Brunswick is also faulted. Silurian rocks are a cover sequence to a late Precambrian-Cambrian basement of the Burgeo Subzone (Dunning and O'Brien, 1989; O'Brien, 1989). The Bras d'Or Subzone of Cape Breton Island (Fig. 2.1 and Map 2) has plutonic and metamorphic affinities like those of the Burgeo subzone (Dunning et al., 1990), implying links between Cape Breton Island and southwest Newfoundland (see also Keppie and Dallmeyer, 1989). The Mira, Antigonish and Cobequid subzones of Nova Scotia contain elements similar to the Avalon Zone in its type area in Newfoundland.

The width of the Avalon Zone varies considerably from eastern Newfoundland to southeast New Brunswick. In Newfoundland, Avalonian rocks extend offshore across the Grand Banks to Flemish Cap (Fig. 1.1), making the Avalon Zone twice as wide as the rest of the orogen to the west. In New Brunswick, the zone is narrower, less than half the width of the rest of the orogen.

Rocks and structures like those of the Avalon Zone extend the full length of the Appalachian Orogen through the New England area to the Slate Belt of the southern U.S. Appalachians (Williams, 1978a). Similar rocks occur in a number of places on the opposite side of the Atlantic (see Fig. 1.4).

Meguma Zone

The Meguma Zone (Fig. 2.1 and Map 2) is restricted to mainland Nova Scotia but its geophysical expression suggests that it underlies a large part of the nearby Atlantic continental shelf (Map 4). Its boundary with the Avalon Zone is a major fault (Keppie, 1982), which is traceable offshore by linear geophysical anomalies (Maps 3 and 4).

The Meguma Zone has a conformable succession of greywackes and shales up to 13 km thick (Schenk, 1975, 1976, 1978, Chapter 3). Where fossiliferous, the rocks are of Cambrian and Early Ordovician age. The sandstones and shales are overlain conformably by a poorly dated sequence of sedimentary and volcanic rocks that includes a possible Upper Ordovician tillite at its base (Schenk, 1972). This sequence is in turn overlain by Devonian sedimentary rocks, mainly terrestrial.

Provenance studies (Schenk, 1970, 1971) of the lower Paleozoic rocks of the Meguma Zone indicate a low-lying metamorphic source to the southeast. The volume of clastic sedimentary rocks suggests a source of continental dimensions. Basement relationships are unknown.

The Meguma Zone was first deformed by Acadian Orogeny, and Alleghanian deformation is important along its boundary with the Avalon Zone.

Lower Paleozoic rocks of the Meguma Zone are unknown elsewhere in North America but equivalents occur in Morocco, northwest Africa (Fig. 1.4, Schenk, Chapter 3).

Other contrasting features of Canadian Appalachian zones

The early Paleozoic zonation of the orogen based on rocks and stratigraphy is also expressed in a variety of other contrasts including geophysics, structural style, metamorphism, plutonism, metallogeny, and faunas. These are mentioned briefly here and discussed more fully in ensuing thematic chapters.

Most features that enhance zonal distinctiveness are those related to late Precambrian-early Paleozoic development. However, some younger plutonic, structural, and metallogenic features are also typical of certain zones, implying inheritance and a prolonged influence of older rocks and deep structure on later orogenic development.

Geophysics

The form of the Canadian Appalachians and the shapes of its early Paleozoic zones are expressed in the Bouguer and magnetic anomaly fields of the region (Maps 3 and 4; Addendum, note 4). Geophysical expression also defines the extent and geometry of the orogen across offshore regions. Anomalies that reflect structural trends follow the sinuous course of the Humber Zone from western Newfoundland through the Gulf of St. Lawrence to southeast Quebec. The Humber-Dunnage boundary has a clear magnetic expression and it is coincident, or nearly so, with a gradient in the Bouguer anomaly field from negative values in the west to positive values in the east. The Dunnage Zone of northern Newfoundland has a strong positive Bouguer anomaly. Gravity values decrease eastward across the Gander Zone and they increase in places along the Avalon Zone. Offshore, the Avalon Zone has large "S"-shaped positive magnetic anomalies that are thought to express the form of late Precambrian volcanic belts (Haworth and Jacobi, 1983; Miller, Chapter 7).

The geophysical expression of the Humber, Dunnage, and Gander zones is less distinct in Quebec and New Brunswick, possibly because of a greater extent and thickness of younger cover rocks. The Meguma Zone has negative gravity values and its openly folded rocks are expressed in magnetic anomalies. The Avalon-Meguma zone boundary has the strongest geophysical expression of any zone boundary and its Bouguer and magnetic anomalies are traceable offshore to the continental shelf edge.

Seismic reflection data recorded along marine transects off northeast Newfoundland (Keen et al., 1986) and in the Gulf of St. Lawrence (Marillier et al., 1989) provide the first clear insight into the deep structure of the Canadian Appalachian region. The marine profiles are augmented by shorter onland traverses in Quebec (St-Julien et al., 1983) and the Lithoprobe East onland transects completed in 1989. This information allows comparisons between surface geology and the geometry of lower crustal blocks.

Three lower crustal blocks are distinguished (Fig. 1.7): the Grenville, Central, and Avalon blocks from west to east. The Grenville lower crustal block is wedge-shaped, and its subsurface edge follows the form of the Appalachian structural front. It abuts the Central lower crustal block at mid-crustal to mantle depths. The Avalon lower crustal block meets the Central lower crustal block at a steep junction that penetrates the entire crust.

The Grenville, Central, and Avalon lower crustal blocks correspond with the Humber, Gander, and Avalon tectonostratigraphic zones. The Dunnage Zone is allochthonous above the opposing Grenville and Central lower crustal blocks. The Gander Zone may be the surface expression of the Central lower crustal block, or it too may be allochthonous. A seismic distinction between the Avalon and Meguma zones, and/or corresponding lower crustal blocks remains uncertain. The spatial coherence, if not seismic continuity, between lower crustal blocks and surface zones implies genetic links and common controls during the early Paleozoic development of the Appalachian Orogen (Addendum, note 5).

In general, the deepest basins of Carboniferous and Mesozoic cover rocks crossed by seismic lines overlie thinner crust with a shallower Moho.

Structure

The earliest Appalachian structures of the Humber Zone are directed westward and are associated with the Ordovician emplacement of Taconic allochthons (Map 5). Intense polyphase deformation occurs along its eastern margin, and its rocks there are among the most deformed within the orogen. In contrast, the Dunnage Zone has mainly upright structures and some of its rocks are less deformed than nearby parts of the Humber Zone. Intensity of deformation increases eastward across the Gander Lake Subzone in Newfoundland and structures are commonly recumbent. Structural windows of Gander Zone rocks in the southeastern Dunnage Zone of Newfoundland (Fig. 2.1)

are also more metamorphosed and deformed than surrounding Dunnage Zone rocks. In central New Brunswick, Dunnage and Gander zone rocks are polydeformed. Structures of the Avalon and Meguma zones are upright and deformation is less intense.

The Humber, Dunnage, and Gander zones exhibit Ordovician deformation effects that are absent in the Avalon and Meguma zones (Map 5). All zones were affected by middle Paleozoic deformation, except for parts of the Newfoundland Avalon Zone (Map 6).

Metamorphism

Intensity of regional metamorphism coincides with intensity of Ordovician (Taconic) and middle Paleozoic (Acadian) deformation. Metamorphism related to Taconic deformation increases in intensity from west to east across the Humber Zone, from subgreenschist to upper greenschist and amphibolite facies. Dunnage Zone rocks are of greenschist or lower grade, except locally near its boundaries with the Humber and Gander zones. Intensity of regional metamorphism, mainly of Silurian age (Dunning et al., 1988, 1990), increases eastward across the Newfoundland Gander Lake Subzone, and rocks of the Mount Cormack and Meelpaeg subzones are higher grade than surrounding rocks of the Dunnage Zone. Metamorphism is low-grade throughout most of the Avalon Zone, and there is a sharp metamorphic contrast in most places across the Gander-Avalon zone boundary, especially in northeast Newfoundland. Most of the Meguma Zone has low-grade metamorphism, but metamorphism reaches amphibolite facies where plutons are abundant in some eastern areas.

Where zones are wide, such as in northeast Newfoundland and Quebec-New Brunswick, metamorphism occurs in relatively narrow belts of the eastern Humber and Gander zones. Where zones are narrow, as in southwest Newfoundland and Cape Breton Island, amphibolite facies metamorphism affects most rocks and zones.

Plutonism

There is a spatial relationship between the distribution of plutonic rocks and early Paleozoic zones in Newfoundland (Williams et al., 1989), although there are exceptions (Currie, Chapter 8). Precambrian plutons unrelated to Appalachian orogenesis are restricted to the Humber and Avalon zones. These are mainly 1000 Ma granites in Grenville inliers of the Humber Zone and 600 Ma calc-alkalic plutonic suites of the Avalon Zone. Late Precambrian alkali and peralkali plutons occur at the eastern margin of the Humber Zone and western margin of the Avalon Zone. Deformed early Paleozoic tonalites, trondjhemites, and quartz diorites occur throughout the Notre Dame Subzone of the Dunnage Zone and small bodies are allochthonous across the Humber Zone.

Middle Paleozoic plutons extend across the orogen from eastern parts of the Humber Zone to western parts of the Avalon Zone in Newfoundland and across the entire Avalon and Meguma zones in Nova Scotia. These make up about one third of all exposed rocks. Deformed middle Paleozoic biotite granites and foliated to massive garnetiferous muscovite leucogranites are typical of the Newfoundland Gander Zone and occur in all three (Gander Lake, Mount Cormack, and Meelpaeg) subzones. Some posttectonic middle Paleozoic plutons occur within zones whereas others cut zone boundaries or follow coeval volcanic belts that transgress zone boundaries. Composite batholiths with early peripheral mafic phases and later granitic phases are most common in the Exploits Subzone of the Newfoundland Dunnage Zone. Middle Paleozoic alkali plutons are most common in the Notre Dame Subzone of the Dunnage Zone, whereas coarse grained, porphyritic biotite granites cut other plutons and zone boundaries. The Meguma Zone has a middle Paleozoic plutonic history that is distinct from the adjacent Avalon Zone. A few small plutons elsewhere within the orogen have peculiarities related to local settings (Currie, Chapter 8).

Metallogeny

Mineral deposits that occur in late Precambrian and early Paleozoic rocks across the orogen are as distinct as the lithostratigraphic units on which the zones are defined (Swinden et al., Chapter 9). These vary from Mississippi Valley-type lead-zinc deposits in the Humber Zone; Cyprus-type copper-pyrite deposits, asbestos, and stratiform polymetallic base metal deposits in the Dunnage Zone; stratiform sulphides and vein deposits of the Gander Zone; local oddities such as late Precambrian pyrophyllite, Cambrian manganese, and Ordovician oolitic hematite in the Avalon Zone; and middle to late Paleozoic gold and tin deposits in the Meguma Zone.

Faunas

Some zones have distinct faunal characteristics. Early Paleozoic shelly faunas of North American affinity occur in the Humber Zone and Notre Dame Subzone of the Dunnage Zone. Ordovician brachiopod faunas of the Celtic realm occur in the Exploits Subzone of the Dunnage and Gander Zone. Ediacara-type late Precambrian faunas and Atlantic realm Cambrian trilobite faunas occur in the Avalon Zone (Nowlan and Neuman, Chapter 10).

MIDDLE PALEOZOIC BELTS

The middle Paleozoic temporal division includes rocks of Middle Ordovician (Caradoc) to Early and Middle Devonian age, although most rocks are of Silurian age. Spatial subdivisions are based mainly on lithofacies variations among Silurian-Devonian rocks. These rocks cover broad areas from Gaspésie to Nova Scotia, and more restricted areas across Newfoundland. Whereas the definitions of early Paleozoic zones of the Canadian Appalachians emanate from the extensive exposures in insular Newfoundland, the definitions of middle Paleozoic belts are most readily made where contrasting Silurian and Devonian rocks are widespread in the maritime provinces of the mainland Appalachians. In Newfoundland, Silurian conglomerates, terrestrial volcanic rocks, and redbeds dominate most belts.

The spatial divisions and descriptions of middle Paleozoic rocks are based on the following references: Williams, 1967; Boucot, 1968, 1969; Ayrton et al., 1969; Béland, 1969; Potter et al., 1969; Berry and Boucot, 1970; Rodgers, 1970; McKerrow and Ziegler, 1971; Ruitenberg et al., 1977; Schenk, 1978; Fyffe and Fricker, 1987; O'Brien, 1987, 1989;

and Bourque et al., Chapter 4. From west to east, mainland belts as outlined on Fig 2.2 and Map 2 are named Gaspé (Connecticut Valley-Gaspé, Aroostook-Percé, and Chaleurs Bay divisions), Fredericton, Mascarene, Arisaig, Cape Breton, and Annapolis. The Newfoundland belts are named Clam Bank, Springdale, Cape Ray, Badger, Botwood, La Poile, and Fortune.

The early Paleozoic zonation of the orogen is not expressed in middle Paleozoic belts, except possibly for the Badger Belt, which is peculiar to the Exploits Subzone of the Newfoundland Dunnage Zone; the Mascarene, Arisaig, and Fortune belts, which are contained within the Avalon Zone; and the Annapolis Belt, which is unique to the Meguma Zone. Thick marine greywackes of the Fredericton Belt may represent the deposits of a significant Silurian seaway within the New Brunswick Gander Zone, and those of the Badger Belt, a Silurian seaway within the Newfoundland Dunnage Zone. Middle Paleozoic rocks toward the west, e.g. those of the Gaspé and Springdale belts, cross early Paleozoic zone boundaries and are thus overlap assemblages of successor basins. Angular unconformities mark the bases of sections in most belts. Conformable contacts are present beneath parts of the Gaspé Belt and the Annapolis and Badger belts.

Middle Paleozoic faunas are more cosmopolitan compared to early Paleozoic faunas. However, there are contrasts between Silurian and Devonian faunas across the mainland belts, with those of the Gaspé Belt having North American affinities and those of the Fredericton and more eastern belts having European or "Old World Realm" affinities (Berry and Boucot, 1970; Boucot, 1989; Nowlan and Neuman, Chapter 10).

Middle Paleozoic belts of the mainland Appalachians

Gaspé Belt

The Gaspé Belt has three well-known divisions; Connecticut Valley-Gaspé, Aroostook-Percé, and Chaleurs Bay (Fig . 2.2, Map 2 and figures of Bourque et al., Chapter 4). The oldest rocks occur in the Aroostook-Percé Division and they are overlain by younger rocks of the Connecticut Valley-Gaspé Division toward the north, and the Chaleurs Bay Division toward the south.

The Connecticut Valley-Gaspé Division extends from the Quebec Eastern Townships to the tip of Gaspésie. Its rocks extend much farther to the southwest in the U.S. Appalachians along the Connecticut Valley Synclinorium (Williams, 1978a), a total distance of more than 1000 km. The Connecticut Valley-Gaspé Division is narrow and only a few kilometres wide at its eastern end but is more than 60 km wide to the southwest. Along its northern and northwestern portions, rocks are platformal in aspect, although much thicker (6 km). They are well-sorted Silurian quartz sands at the base followed by Silurian reefbearing limestone and topped by Silurian-Devonian calcareous siltstones. A small area of mainly Silurian rocks in the Eastern Townships to the southwest is included in this division. Toward the southeast, rocks of the Connecticut Valley-Gaspé division are mainly Devonian and they contain less limestone and much more shale compared to equivalent rocks to the north. Upper parts of the section are shaly sandstone and limy shale. Fossils are few. The rocks are judged to be exceptionally thick (6 to 10 km), and intermediate to acidic volcanic rocks are present locally.

Rocks of the northwestern portion of the Connecticut Valley-Gaspé Division have a structural style like that of the Valley and Ridge Province of the U.S. Appalachians. This region was previously referred to as the "Northern Belt of the Acadian Orogen" (Beland, 1969) and the "Gaspé Folded Zone" (Rodgers, 1970). Folds have wavelengths and amplitudes measured in kilometres, and thrust faults occur on the north flanks of some anticlines. Ordovician serpentinite and volcanic rocks are brought to the surface locally near the northern boundary of the division. Across a zone of steep faults, rocks of the southeastern Connecticut Valley-Gaspé Division are tightly folded with slaty cleavage.

Silurian rocks are unconformable on Ordovician rocks to the north, where the boundary is not faulted. The Silurian section is thick and complete in western Gaspésie and it thins eastward and disappears at the east end of the peninsula. There, resistant Lower Devonian limestones form the scenic cliffs. These are overlain by Lower Devonian rocks that change upward from grey or green marine shales and sandstones to red continental arkosic sandstones.

Rocks of the northern Connecticut Valley-Gaspé Division have many features, both lithological and faunal, in common with roughly coeval rocks of the Helderberg and Catskill groups of New York State (Rodgers, 1970), which lie almost entirely outside the Appalachian Orogen. Undeformed platy limestones and shales of Anticosti Island may be continuous with deformed Silurian rocks of the northern Gaspé Belt beneath the waters of the Gulf of St. Lawrence.

The evidence for two Paleozoic pre-Carboniferous orogenies, Taconic and Acadian, is especially clear in northern Gaspésie. Silurian and Devonian strata unconformably overlie more intensely deformed and locally metamorphosed Cambrian and Ordovician rocks of the Humber Zone, and folded Silurian and Devonian strata are overlain by subhorizontal Carboniferous redbeds. A change from marine to continental conditions preceded the onset of Acadian deformation.

The Aroostook-Percé Division of the Gaspé Belt is defined on distinctive thin-bedded limestones and limy shales of Middle Ordovician to Early Silurian age. The rocks can be traced from Percé to central New Brunswick and they continue through Maine as the "ribbon rocks" of the Aroostook-Matapédia Anticlinorium (Williams, 1978a). Correlatives occur in the central portion of the Connecticut Valley-Gaspé Division. The calcareous Ordovician-Silurian rocks of the Aroostook-Percé Division are separated from Devonian slates of the southern Connecticut Valley-Gaspé Division either by faults or intervening Silurian strata.

In central Gaspésie, deformed Silurian ribbon limestones are overlain by west-dipping silty limestones and limy shales, which extend into the Upper Silurian and include a kilometre of mafic volcanic rocks. Farther north, Devonian rocks of the Connecticut Valley-Gaspé Division are shale. In this area the sequence of rocks from Middle Ordovician to Lower Devonian appears continuous and conformable.

Toward the east, isolated exposures of Upper Cambrian (Murphy Creek Formation) and Middle Cambrian (Corner of the Beach Formation) limestones appear as inliers beneath the Middle Paleozoic rocks. The Cambrian limestones are typical of those along the west flank of the orogen and indicate that the Gaspé Belt is developed here on a Humber Zone basement.

In New Brunswick, rocks of the Aroostook-Percé Division overlie mafic tuffs and cherts, correlated with the Tetagouche Group of the New Brunswick Dunnage Zone (Ruitenberg et al., 1977). Southwestward along strike the tuffs and cherts are replaced by conglomerates and greywackes of the Caradoc Grog Brook Group, and the contact with the Middle Paleozoic rocks is conformable.

Rocks of the Aroostook-Percé Division of the Gaspé Belt are thick, of deep marine turbiditic aspect, and of unbroken continuity from Middle Ordovician to Devonian. Tight to isoclinal, nearly upright folds on all scales are also a characteristic feature. Some Middle Paleozoic rocks of the Aroostook-Percé Division seems to cross and link the Humber, Dunnage, and Gander zones. However, facies and structural styles indicate that the present width of the Aroostook-Percé Division is but a small fraction of its original width. An initially wide division is also demanded by the nature of its fine grained sedimentary rocks that are now in sight of volcanic rocks on either side. The analysis by Bourque et al. (Chapter 4) indicates that rocks of the Aroostook-Percé Division are offset dextrally by major faults.

The Chaleurs Bay Division of the Gaspé Belt (Fig 2.2) contains the highly fossiliferous Silurian Chaleurs Group (Bourque, 1975) consisting of quartz sandstone and siltstone, argillaceous and calcareous siltstone, and shallow water nodular reefy limestone. Upper parts of the group consist of amygdaloidal andesite and basalt, red arkosic sandstone, volcanic boulder conglomerate, and rhyolite of Late Silurian age. Folding is open, about northeast axes with wavelengths of a few kilometres. The Silurian section of the Chaleurs Bay Division contrasts with that of the nearby Aroostook-Percé Division of the Gaspé Belt, which contains deeper water, shalier, less fossiliferous rocks, with fewer volcanic units, and displays more penetrative deformation. The lithic and structural features of the Chaleurs Bay Division resemble, in part, those of the northern Connecticut Valley-Gaspé Division.

North of Chaleur Bay, the Silurian rocks overlie shales and sandstones. In northern New Brunswick, argillaceous and calcareous siltstones of the Chaleurs Group grade downward into limestones (Ruitenberg et al., 1977). At the east end of the Chaleurs Bay Division, two older groups are unconformable beneath the Silurian. The Mictaw Group of volcanic greywacke, siltstone, and shale has Middle to Late Ordovician graptolites and is thus partly contemporaneous with ribbon limestones of the Aroostook-Percé Division. the Mictaw Group, in turn, is unconformable on polydeformed and metamorphosed rocks of the interior Humber Zone (Williams and St-Julien, 1982).

On the north side of Chaleur Bay and in the nearby Restigouche Syncline (Williams, 1978a), Silurian rocks are overlain by Lower and lower Middle Devonian sandstones and conglomerates that are grey to buff below and red and coarser upward, portraying another marine to continental environmental change like that recorded in similar rocks of the Connecticut Valley-Gaspé Division of the Gaspé Belt. These beds are overlain unconformably by 100 m of Upper Devonian calcareous shale and sandstone containing the well-known Escuminac fish fauna. Flat-lying Carboniferous red sandstones (Bonaventure Formation) rest unconformably on all older units.

South of Chaleur Bay in northern New Brunswick, the Chaleurs Group unconformably overlies Ordovician rocks of the Elmtree and Belledune subzones. The Chaleurs Group is unconformably overlain by the Dalhousie Group consisting of calcareous sandstone, siltstone, and limestone with intercalated andesite, basalt, and tuff, and upper parts consisting of rhyolite and tuffaceous sandstone with local coal seams. These rocks, in turn, are unconformably overlain by continental greenish-grey conglomerates and sandstones of Campbellton Formation (Siegenian-Emsian). Along the northwestern margin of the belt in New Brunswick, a sequence of grey and red sandstones, siltstones, rhyolite, and basalt is correlated with the Dalhousie Group (Potter, 1965).

Fredericton Belt

The Fredericton Belt is defined on its preponderance of greywackes which lie between the Miramichi and St. Croix subzones of New Brunswick (Fig. 2.2, Map 2 and Fyffe, Chapter 4). These contain Silurian graptolites in central parts of the outcrop belt. The Silurian rocks are faulted against Cambrian-Ordovician rocks of the Miramichi and Bathurst subzones to the northwest and the St. Croix Subzone to the southeast (Williams, 1978a; Fyffe, Chapter 4).

Silurian rocks of the Fredericton Belt are assigned to the Kingsclear Group in the north and to the Flume Ridge Formation in the south separated by the Fredericton Fault. Distinctive conglomerates along the northwestern margin of the belt contain pebbles of basalt and slate derived from the adjacent Ordovician Tetagouche Group of the Bathurst Subzone. No such link has been established to the southeast with the Ordovician Cookson Group of the St. Croix Subzone.

Silurian turbidites of the Fredericton Belt cannot be demonstrated to be laterally or vertically continuous with shallow water Silurian strata of the Mascarene Belt (Fyffe, Chapter 4). Rhenish-Bohemian brachiopod assemblages occur in Silurian-Devonian rocks of the Mascarene Belt but are unknown in the graptolite-bearing rocks of the Fredericton Belt. Early Devonian Eastern Americas brachiopod assemblages are found in the Aroostook-Percé division of the Gaspé Belt to the northwest of the Fredericton Belt (Fyffe and Fricker, 1987). Basement relationships of the Fredericton Belt are unknown.

The Fredericton Fault has been designated a Silurian terrane boundary (Keppie, 1985). However, the similarities between Silurian rocks on either side of the fault indicate continuity, although significant transcurrent movement is possible (Ludman, 1987). Isotopic studies on granites may provide some constraints on basement relationships across the Fredericton Belt.

Rocks of the Fredericton Belt continue southwestward into the Central Maine Slate Belt (Williams, 1978a). The southern part of the belt has been affected by Acadian polyphase deformation, whereas only one phase of folding is recognized in most of its northern part.

Mascarene Belt

The Mascarene Belt of southern New Brunswick is characterized by vast quantities of Silurian to Lower Devonian volcanic rocks. These rocks can be traced southwestward into and along the coast of Maine where they exceed 3 km in thickness and are known as the Coastal Volcanic Belt. Mafic and local ultramafic intrusions are associated with the volcanic rocks, perhaps genetically, and discordant granitic plutons are common. The Mascarene Belt is developed mainly upon rocks of the Avalon Zone. At the Canada-United States border, rocks of the Mascarene Belt, as well as older Precambrian rocks nearby, are overlain unconformably by about 1000 m of posttectonic redbeds (Perry Formation) containing Late Devonian plants.

The margins of the Mascarene Belt are defined by faults. Internally, it is separated into five subbelts (Nerepis, Letete, Ovenhead, Campobello, and Oak Bay; McLeod and McCutcheon, Chapter 4). Rocks of the first four subbelts were deposited on the Avalon Zone but the Late Silurian Waweig Formation of the Oak Bay subbelt rests unconformably on the St. Croix Subzone of the Gander Zone.

General lithological correlations can be established within the Mascarene Belt with the aid of fossil control, but direct correlations of units are difficult. Rocks of the Ovenhead subbelt have been correlated directly with the Lower Devonian Eastport Formation in Maine but parts may be considerably older. Regardless of age, they form part of a distinct lithotectonic package that also incorporates the Silurian-Devonian sequences in southeastern Maine north of the Campobello subbelt. The Silurian Quoddy Formation of the Campobello subbelt also extends to the southwest into Maine. Although similar faunas are present throughout the Mascarene Belt (Boucot et al., 1966), lithological units in the Ovenhead and Campobello subbelts have no apparent direct correlatives in the Letete, Nerepis, and Oak Bay subbelts, and structural styles are different.

Chronological and, possibly, direct lithological correlations can be established between parts of the Letete and Nerepis subbelts.

Arisaig Belt

Upper Ordovician to Lower Devonian rocks in northern mainland Nova Scotia are assigned to the Arisaig Belt. These are the Arisaig Group in the east and the Wilson Brook, Portapique River, and Murphy Brook formations in the west (Murphy and Keppie, Chapter 4).

The Arisaig Group unconformably overlies the Upper Precambrian Georgeville Group and the Cambrian-Lower Ordovician McDonalds Brook and Iron Brook groups. The base of the Arisaig Group is not exposed in the type area along the coast. The Silurian coastal section is complete and consists of shallow marine to brackish water fossiliferous shales and siltstones. In general, the Arisaig Group consists of bimodal volcanic rocks with interlayered redbeds overlain with local unconformity by about 2000 m of fossiliferous siliciclastic rocks and minor carbonates. The Arisaig Group is unconformably overlain by Middle to Upper Devonian redbeds. The stratigraphic succession of the Arisaig Group indicates that a marine transgression occurred at the base of the Silurian.

The Wilson Brook Formation consists of fossiliferous siliciclastic rocks and minor carbonate rocks. The formation conformably overlies the Upper Precambrian Jeffers Group and is unconformably overlain by Middle to Upper Devonian redbeds of the Murphy Brook Formation.

The Wilson Brook Formation is broadly correlative with the Arisaig Group, although structural complications and the lack of continuous exposures make detailed correlations difficult. The age of the Wilson Brook Formation is late Llandovery to Pridoli. Redbeds of the Portapique River Formation containing Emsian to Eifelian fossils conformably overlie the Wilson Brook Formation.

The Arisaig Group is deformed into shallow southeast-plunging upright folds. The Middle Devonian McAras Brook Formation postdates these structures suggesting a mid-Devonian age of deformation. This deformation occurred at a shallow crustal level as indicated by the associated subgreenschist metamorphism. The counterclockwise orientation of the folds suggests that they were produced by dextral motions on nearby faults.

In southern parts of the Arisaig Belt, deformation is heterogeneous and is restricted to shear zones, with local evidence of thrusting and development of positive flower structures. At least some of this deformation affects Upper Devonian rocks and is therefore younger than that which affected Upper Ordovician-Lower Devonian rocks.

Toward the northwestern parts of the Arisaig Belt, folding of Silurian-Lower Devonian sedimentary rocks also predates deposition of Upper Devonian redbeds and volcanic rocks and is probably also of mid-Devonian age.

Cape Breton Belt

Some Silurian and Devonian rocks in Cape Breton Island are of high metamorphic grade, poorly dated, and described with other high grade poorly dated rocks of the Aspy Subzone (Barr et al., Chapter 3). However, others include sedimentary rocks, the Ingonish Island Rhyolite, the Fisset Brook Formation, and the McAdam Lake Formation. These are lower grade, better dated, and assigned to the Cape Breton Belt (see Barr and Jamieson, Chapter 4 and Fig. 4.30).

Annapolis Belt

The Annapolis Belt is defined by thick sequences, collectively called the Annapolis Supergroup, that range in age from Early Ordovician to Early Devonian (Schenk, Chapter 4). These sequences consist of fine grained, shallow marine, siliciclastic sedimentary rocks as well as volcaniclastic rocks. The Annapolis Supergroup overlies the Meguma Supergroup. The separation of a Meguma Zone and Annapolis Belt is artificial as: (1) both zone and belt are in the same geographic area; (2) some rocks of the Annapolis Supergroup are possibly Early Ordovician and therefore older than those normally found in middle Paleozoic belts; (3) Acadian (Devonian) Orogeny was the first major event to affect rocks of both the Meguma Zone and Annapolis Belt; and (4) the Meguma Supergroup and Annapolis Supergroup are viewed as integral parts of the same stratigraphic section and interpreted in the same

model of an African continental margin. Therefore, the Meguma Zone and Annapolis Belt are defined only to maintain the systematics of the time-slice subdivision used throughout this volume.

Discrete parts of Annapolis Belt occur in widely spaced, synclinal areas separated by broad expanses of postfolding granitoid plutons in the east, and the underlying Meguma Supergroup in the west.

The Annapolis Supergroup has a sharp contact with the underlying Meguma Supergroup. This surface is in general a presumed paraconformity but in places it is an angular unconformity and in other places a disconformity. The upper boundary is an angular unconformity beneath gently dipping Carboniferous and/or Triassic redbeds.

Middle Devonian Acadian Orogeny is the first major deformational, metamorphic, and intrusive event to affect rocks of the Annapolis Belt. Broad, shallowly plunging, upright folds describe an arc trending northerly in the west, to easterly in the eastern part of the belt. Associated slaty cleavage is subvertical, parallel to axial surfaces. The metamorphic facies is greenschist, with an increase to amphibolite facies in the west. Hornblende-hornfels facies is developed close to granitic batholiths (Taylor and Schiller, 1966). Large granitic plutons that cut rocks of the Annapolis Belt are Late Devonian and Early Carboniferous. The Meguma Zone and Annapolis Belt are well defined by metallogenic, plutonic, and tectonic characteristics.

The Annapolis Supergroup is composed of three groups, in ascending order, the White Rock, Kentville, and Torbrook. The White Rock and Kentville groups are divided into formal formations, but formations of the Torbrook Group are informal. All formations are subdivided into informal members (Schenk, Chapter 4).

Middle Paleozoic belts of the Newfoundland Appalachians
Clam Bank Belt

The Clam Bank Belt of western Newfoundland is defined on the single occurrence of Silurian-Devonian rocks of the Clam Bank Group (Fig. 2.2 and Map 2). The group consists of crossbedded red sandstones, red and grey shales, and grey limy siltstones and sandstones, about 500 m thick. Grey limy siltstones in its central part contain a Pridoli brachiopod fauna (Boucot, 1969).

The Clam Bank Group lies with structural conformity (most likely disconformity) above Middle Ordovician (Caradoc) limestones, limy sandstones, and shales of the Long Point Group. The two groups form a west-facing, gently west-dipping homocline, which is overturned to the west in southern parts of the outcrop belt. Penetrative cleavage is absent and the structural style resembles that of the Connecticut Valley-Gaspé Division of the Gaspé Belt. the Long Point Group overlies chaotic sedimentary rocks of the Ordovician Humber Arm Allochthon. The contact has been interpreted traditionally as an unconformity (Rodgers, 1965). However, the Clam Bank Belt may be the upper portion of a structural triangle zone (Stockmal and Waldron, 1990). Nearby Carboniferous rocks are flat lying.

Springdale Belt

Middle Paleozoic rocks of the Springdale Belt are dominated by silicic volcanic rocks in places overlain by terrestrial red conglomerates and sandstones (Sops Arm, Cape St. John, Micmac Lake, and Springdale groups). None of the rocks is dated by fossils except for parts of the Sops Arm Group which includes limy sandstones and shales (Natlins Cove Formation) containing Silurian brachiopod and conodont faunas. The volcanic rocks are spectacular red rhyolites and welded tuffs, purplish to red and green amygdaloidal lavas, and coarse volcanic breccias. Thicknesses are in the order of 3 km and overlying redbeds are more than 1000 m thick locally. Almost all of the rocks are of terrestrial deposition.

The rocks are folded and cleaved in most places with the intensity of deformation varying between moderate for most of the Sops Arm and Springdale groups to intense with local amphibolite facies metamorphism in the Cape St. John Group. The folded and deformed rocks are cut by large granite intrusions.

Basal relationships are everywhere unconformable. the Sops Arm Group lies unconformably upon the transported Coney Head Complex of the Humber Zone, the Cape St. John Group lies unconformably upon the Lower Ordovician Snooks Arm Group and Betts Cove ophiolite complex of the Dunnage Zone, the Micmac Lake Group lies unconformably upon the Ordovician Flat Water Group of the Dunnage Zone, and the Springdale Group lies unconformably on Ordovician volcanic rocks of Roberts Arm Group of the Dunnage Zone. The Springdale Belt therefore overlaps the Humber-Dunnage zone boundary.

Red sandstones like those of the Springdale Belt occur on the southeast side of Red Indian Lake in central Newfoundland.

Cape Ray Belt

Silurian volcanic rocks and associated redbeds in the northeast and Devonian sandstones, shales, and volcanic rocks of the the Windsor Point Group in the southwest define the Cape Ray Belt. Silurian redbeds in the northeast are unconformable above the Ordovician Annieopsquotch ophiolite complex. Farther southwest, the Windsor Point Group is unconformable on Ordovician tonalite.

The long, narrow Cape Ray Belt coincides with the Notre Dame-Exploits subzone boundary in the north and the confluence of this boundary and the Dashwoods-Meelpaeg subzone boundary in the south.

Badger Belt

The Badger Belt is distinguished by its preponderance of greywackes and overlying polymictic conglomerates (Sansom and Goldson formations), which exceed 3 km in thickness. The rocks are dated by shelly faunas of middle to late Llandovery age. The thick marine clastic sequences are conformable above Middle Ordovician (Caradoc) black graptolitic shales (Shoal Arm, Lawrence Harbour, and Dark Hole formations) of the northeast Exploits Subzone. Basal parts of the sections immediately above the Caradoc

shales are thought to include Upper Ordovician rocks, and Ashgill graptolites occur locally (S.H. Williams, 1991). Red conglomerates and sandstones occur toward the top of some sections.

Toward the west, coastal exposures of greywackes overlie Caradoc shales, in turn conformable upon a thick mafic volcanic sequence (Wild Bight Group). Toward the east, similar greywackes are conformable above a thin Caradoc shale unit that directly overlies the Ordovician Dunnage Mélange. Farther north, Silurian conglomerates, locally red, are unconformable above Ordovician volcanic rocks, implying a southeast polarity of the marine basin there.

Silurian mélanges, locally with fossiliferous shale matrices and containing huge Ordovician volcanic, limestone, and black argillite blocks are a distinctive lithology. These can be traced 80 km in separated coastal exposures across the northern part of the Badger Belt. Steeply dipping repeated sections in the northeast are interpreted as thrust imbricated.

Rocks of the Badger Belt are everywhere above rocks of the Exploits Subzone of the Newfoundland Dunnage Zone. Their presence in the Exploits Subzone and their absence in the adjacent Notre Dame Subzone strengthens the major two-fold division of the Newfoundland Dunnage Zone (Williams et al., 1988).

The lithic contrast between marine greywackes and conglomerates of the Badger Belt and terrestrial volcanic rocks and redbeds of the bordering Springdale and Botwood belts is pronounced, but the marine rocks of the Badger Belt are mainly older than the terrestrial rocks.

Botwood Belt

Rocks of the Botwood Belt are mainly terrestrial volcanic rocks overlain by fluviatile red and grey crossbedded sandstones. Polymictic conglomerates and greywackes like those of the Badger Belt occur with stratigraphic contact beneath its volcanic rocks at several localities in the northeast.

Widely separated fossil occurrences throughout the belt are of the same general Llandovery to Wenlock age, except for a single Ludlow graptolite reported from sandstones toward the southeast (Berry and Boucot, 1970).

Toward the southwest in central Newfoundland, a narrow continuous belt of conglomerates, Rogerson Lake Conglomerate, is included in the Botwood Belt. The conglomerates are more deformed than rocks of the northern parts of the belt, and they are locally unconformable above intrusive rocks.

The western boundary of the Botwood Belt is marked by faults. Greywackes and conglomerates along its eastern boundary appear conformable to Ordovician shales, but the lithological contrast and age difference imply a significant disconformity or faulted contact (Addendum, note 6).

An isolated occurrence of polymictic conglomerate at Dollard Pond resembles the Silurian conglomerates and greywackes of the Badger and Botwood belts. It forms a synclinal remnant surrounded by much higher grade metamorphic rocks. Contacts are interpreted as gradational.

La Poile Belt

The La Poile Belt of southwest Newfoundland contains felsic volcanic rocks, associated epiclastic rocks, and cross-bedded quartz sandstones, all of the La Poile Group (Cooper, 1954; Chorlton, 1978; O'Brien, 1982; O'Brien, 1988, 1989). Its thickness is estimated to be between 3.5 and 5 km. Some of the felsic volcanic units are dated isotopically as Silurian (Dunning et al., 1988). The isotopic ages support correlation with similar Silurian assemblages elsewhere in Newfoundland, and refute previous isotopic ages and correlations with Ordovician rocks of the Dunnage Zone (Chorlton and Dallmeyer, 1986).

Along its northern boundary, greenschist facies rocks of the La Poile Belt are thrust northwest above higher grade Ordovician rocks of the Dunnage Zone. The southern boundary of the La Poile Belt is a northwest directed thrust that brings upper Precambrian-Cambrian rocks of the Burgeo Subzone above the Silurian La Poile Group. Alternating volcanic and clastic sedimentary units of the La Poile Group outline regional folds that trend northeast with upright axial surfaces, and local overturning to the northwest.

Conglomerates at the base of the La Poile Group lie nonconformably on the Roti Granite dated at 563 ± 4 Ma at the northeast part of the belt (Dunning et al., 1988; Dunning and O'Brien, 1989). Rocks that are cut by the Roti Granite include amphibolite (Cinq Cerf gneiss) and less metamorphosed green siltstones, argillites, and volcanic rocks. These are part of the Burgeo Subzone and possibly of Avalonian affinity.

The regional geological setting and basal relationships of the La Poile Group contrast with those of other Newfoundland belts. The structural style of the Silurian rocks, thrust to the northwest above Ordovician rocks of the Dunnage Zone and themselves overthrust by southerly belts of mainly Precambrian rocks, is unique to southwestern Newfoundland.

Fortune Belt

The Middle Paleozoic units of the Fortune Belt overlie rocks of the Newfoundland Avalon Zone. The oldest, Cinq Isle Formation toward the north, consists of grey limestone with lenses and pockets of bright red sandstone, shale, and conglomerate. It is nonconformable upon intrusive rocks of the late Precambrian Simmons Brook Batholith and it is overlain by coarse plutonic boulder conglomerate of the Pools Cove Formation. Some granite boulders in conglomerates of the Pools Cove Formation resemble nearby Devonian granites, but the conglomerates are also cut by Devonian granite. The Cinq Isle and Pools Cove formations are therefore assigned a Silurian-Devonian age. A younger and less deformed conglomerate unit nearby, Great Bay de l'Eau Formation, contains Late Devonian plants. It too is cut by Devonian granite of the adjoining Belleoram stock. Volcanic rocks in a discrete small area toward the south are dated isotopically as Devonian (Krogh et al., 1988) and they are interpreted as cover to upper Precambrian rocks.

These Middle Paleozoic occurrences are significant, as some toward the north contain abundant metamorphic detritus from the adjacent Gander Zone. They therefore establish the earliest sedimentological linkage between the Gander and Avalon zones in Newfoundland.

LATE PALEOZOIC BASINS

Upper Paleozoic rocks of the Canadian Appalachian region are mainly of Carboniferous age but they include Upper Devonian beds at the bases of some sections and Permian strata at the top of some others. The rocks are mainly terrestrial sandstones and conglomerates. Volcanic units are present at the bases of some sections, and Upper Mississippian limestones and evaporites mark an important marine incursion. The rocks occupy a number of depocentres or basins (van de Poll et al., Chapter 5) and they occur as an unconformable cover on almost all of the rocks that define the early Paleozoic zones and middle Paleozoic belts (Fig. 2.3).

The rocks are everywhere the same and extend across the entire orogen, thus negating spatial divisions comparable to those for lower and middle Paleozoic rocks. All of the upper Paleozoic rocks are part of the Maritimes Basin (van de Poll et al., Chapter 5). It is transected obliquely by a narrow northeast-trending rift, the Maritimes Rift, which contains the earliest deposits and thickest sections, up to 12 km. It is also the locus of most intense deformation (Map 6). Thinner and less deformed upper Paleozoic strata border the Maritimes Rift in western New Brunswick, southern Nova Scotia, and western Newfoundland. Thickness variations allow the definition of numerous smaller basins, which are discussed and analyzed by van de Poll et al. (Chapter 5).

The oldest rocks are coarse fanglomerates close to boundary faults of the Maritime Rift. Deformation affects the thickest Lower Carboniferous rocks, with upright folds about northeast axes. Carboniferous rocks of Newfoundland are also deformed in narrow fault-bounded basins. Deformation is much more intense along the shoreline section of southern New Brunswick. There the Carboniferous rocks are involved in thrusts with polyphase deformation and development of subhorizontal penetrative cleavage.

In Nova Scotia, the entirely nonmarine Upper Carboniferous strata, locally including commercial coal seams, lie unconformably upon the deformed rocks of the Maritimes Rift and locally they overlap faults that bound the rift. Other evidence of unconformable overlap and local episodic deformation is common within both lower and upper parts of Carboniferous sections. The youngest rocks are exposed on Prince Edward Island and these are Upper Carboniferous and Lower Permian redbeds.

None of the upper Paleozoic rocks are intruded by granite, except for some rocks assigned to the Mispek Group of southern New Brunswick and the volcanic rocks that host tin mineralization nearby. Granites dated isotopically as Carboniferous, but surrounded by older rocks, occur in the Meguma Zone of Nova Scotia (Cormier et al., 1988).

Carboniferous rocks underlie the Gulf of St. Lawrence and wide areas of the continental margin around Newfoundland (Haworth et al., 1985). They are absent throughout the northern U.S. Appalachians, except for those of eastern Massachusetts and Rhode Island (Williams, 1978a). There the rocks are intensely deformed, cut by granites, and regionally metamorphosed to upper amphibolite facies. This structural style is more akin to that in eastern parts of the southern U.S. Appalachians.

Mississippian marine European faunas are present in Newfoundland, Nova Scotia, and coastal New Brunswick. North American faunas occur in Carboniferous rocks west of the Appalachian Orogen (Boucot, 1989).

MESOZOIC GRABEN

The Mesozoic record of the onland Canadian Appalachians is fragmentary (Fig. 2.3 and Map 1), although a full complement of Mesozoic and younger rocks is represented offshore (Keen and Williams, 1990). Triassic and Lower Jurassic redbeds and basalts occur in the Fundy Graben and mafic dyke swarms and small intrusions of Permian-Triassic to Cretaceous age occur in widely separated areas from southern Quebec to northeastern Newfoundland (Greenough, Chapter 6). In the Fundy Graben, the Triassic rocks are faulted and gently folded, but essentially undeformed; they unconformably overlie deformed Carboniferous and older rocks.

Triassic and Jurassic rocks are best exposed along the southeastern side of the Fundy Graben. They form the Fundy Group, which comprises two sedimentary units separated by basalt flows. Thicknesses reach about 1000 m. Fossils are nonmarine and features typical of continental deposition are common.

Mafic dykes of Triassic to Cretaceous age occur in southern Nova Scotia, Newfoundland, Prince Edward Island, and Anticosti Island. Small alkali Monteregian intrusions occur along a west-trending line for about 200 km in the Quebec Eastern Townships. These cut folded Paleozoic rocks, flat-lying strata of the St. Lawrence Platform, and crystalline rocks of the Canadian Shield. Monzonitic and granitic rocks form part of the easternmost intrusions and the Oka intrusion is a carbonatite that is mined for niobium-bearing pyrochlore.

About 20 m of Cretaceous sediments containing pollen and spores occur in central Nova Scotia. These sediments are a Coastal Plain outlier of the submerged continental shelf.

ACKNOWLEDGMENTS

Thanks are extended to E.R.W. Neale, Pierre-André Bourque, L.B. Chorlton, S.P. Colman-Sadd, and R.F. Blackwood for reviews of earlier drafts of this chapter.

REFERENCES

Ayrton, W.G., Berry, W.B.N., Boucot, A.J., Lajoie, J., Lespérance, P.J., Pavlides, L., and Skidmore, W.B.
1969: Lower Llandovery of the Northern Appalachians and adjacent regions; Geological Society of America Bulletin, v. 80, p. 459-484.
Barr, S.M. and Raeside, R.P.
1986: Pre-Carboniferous Tectonostratigraphic subdivisions of Cape Breton Island, Nova Scotia; Maritime Sediments and Atlantic Geology, v. 22, p. 252-263.
1989: Tectono-stratigraphic terranes in Cape Breton Island, Nova Scotia: Implications for the configuration of the northern Appalachian orogen; Geology, v. 17, p. 822-825.
Barr, S.M., Raeside, R.P., and van Breemen, O.
1987: Grenvillian basement in the northern Cape Breton Highlands, Nova Scotia; Canadian Journal of Earth Sciences, v. 24, p. 992-997.

Béland, J.
1969: The Geology of Gaspé; Canadian Institute of Mining Transactions, v. 67, p. 213-220.
Berry, W.B.N. and Boucot, A.J.
1970: Correlation of the North American Silurian rocks; Geological Society of America Special Paper 102, 289 p.
Blackwood, R.F.
1982: Geology of the Gander Lake (2D/15) and Gander River (2E/2) area: Newfoundland Department of Mines and Energy, Mineral Development Division, Report 82-4, 56 p.
1985: Geology of the Facheux Bay area (11P/9) Department of Mines and Energy, Mineral Development Division, Report 85-4, 56 p.
Blackwood, R.F. and Green, L.
1982: Geology of the Great Gull Lake (2D/6)-Dead Wolf Pond area (2D/10), Newfoundland; in Current Research, (ed.) C.F. O'Driscoll and R.V. Gibbons; Newfoundland Department of Mines and Energy, Mineral Development Division, Report 82-1, p. 51-64.
Blackwood, R.F. and Kennedy, M.J.
1975: The Dover Fault: western boundary of the Avalon Zone in northeastern Newfoundland; Canadian Journal of Earth Sciences, v. 12, p. 320-325.
Boucot, A.J.
1968: Silurian and Devonian of the Northern Appalachians; in Studies of Appalachian Geology: Northern and Maritime, (ed.) E-an Zen, W.S. White, J.B. Hadley, and J.B. Thompson, Jr.; Interscience, New York, p. 83-94.
1969: Silurian-Devonian of Northern Appalachians-Newfoundland; in North Atlantic-Geology and Continental Drift, (ed.) M. Kay; American Association of Petroleum Geologists Memoir 12, p. 477-483.
1989: Acadian Orogeny: Biogeographic constraints; Geological Society of America, Abstracts with Programs, v. 21, no. 2, p. 6.
Bourque, Pierre-André
1975: Lithostratigraphic Framework and Unified Nomenclature for Silurian and Basal Devonian Rocks in Eastern Gaspe Peninsula, Quebec; Canadian Journal of Earth Sciences, v. 12, p. 858-872.
Chorlton, L.
1978: The geology of the La Poile map area (110/9), Newfoundland; Newfoundland Department of Mines and Energy, Mineral Development Division, Report 78-5, 14 p.
Chorlton, L.B. and Dallmeyer, R.D.
1986: Geochronology of early to middle Paleozoic tectonic development in the southwest Newfoundland Gander Zone; Journal of Geology, v. 94, p. 67-69.
Colman-Sadd, S.P.
1976: Geology of the St. Albans map-area, Newfoundland (1M/13); Newfoundland Department of Mines and Energy, Mineral Development Division, Report 76-4, 19 p.
1984: Geology of the Cold Spring Pond map area (west part) 12A/1, Newfoundland; in Current Research, Newfoundland Department of Mines and Energy, Mineral Development Division, Report 84-1, p. 211-219.
1985: Geology of the west part of Great Burnt Lake (12A/8) area; in Current Research, Newfoundland Department of Mines and Energy, Mineral Development Division, Report 85-1, p. 105-113.
1987: Geology of part of the Snowshoe Pond (12A/7) map area; in Current Research, Newfoundland Department of Mines and Energy, Mineral Development Division, Report 87-1, p. 297-310.
1988: Geology of the Snowshoe Pond (12A/7) map area; in Current Research, Newfoundland Department of Mines, Mineral Development Division, Report 88-1, p. 127-134.
Colman-Sadd, S.P. and Swinden, H.S.
1984: A tectonic window in central Newfoundland? Geological evidence that the Appalachian Dunnage Zone is allochthonous; Canadian Journal of Earth Sciences, v. 21, p. 1349-1367.
Colman-Sadd, S.P., Hayes, J.P., and Knight, I.
1990: Geology of the Island of Newfoundland; Newfoundland Department of Mines and Energy, Map 90-01, 1:1,000,000 scale.
Cooper, J.R.
1954: La Poile-Cinq Cerf map-area, Newfoundland; Geological Survey of Canada, Memoir 276, 62 p.
Cormier, R.F., Keppie, J.D., and Odom, A.L.
1988: U-Pb and Rb-Sr geochronology of the Wedgeport granitoid pluton, south-western Nova Scotia; Canadian Journal of Earth Sciences, v. 25, p. 255-261.

Currie, K.L. and Piasecki, M.A.J.
1989: Kinematic model for southwestern Newfoundland based upon Silurian sinistral shearing; Geology, v. 17, p. 938-941.

Currie, K.L., Pajari, G.E., Jr., and Pickerill, R.K.
1979: Tectono-stratigraphic problems in the Carmanville area, northeastern Newfoundland; in Current Research, Part A, Geological Survey of Canada, Paper 79-1A, p. 71-76.
1980: Comments on the boundaries of the Davidsville Group, northeastern Newfoundland, in Current Research, Part A, Geological Survey of Canada, Paper 80-1A, p. 115-118.

Dallmeyer, R.D., Blackwood, R.F., and Odom, A.L.
1981: Age and origin of the Dover Fault: tectonic boundary between the Gander and Avalon Zones of the northeastern Newfoundland Appalachians; Canadian Journal of Earth Sciences, v. 18, p. 1431-1442.

Dunning, G.R. and Krogh, T.E.
1985: Geochronology of ophiolites of the Newfoundland Appalachians; Canadian Journal of Earth Sciences, v. 22, p. 1659-1670.

Dunning, G.R. and O'Brien, S.J.
1989: Late Proterozoic-early Paleozoic crust in the Hermitage flexure, Newfoundland Appalachians: U/Pb ages and tectonic significance; Geology, v. 17, p. 548-551.

Dunning, G.R., Barr, S.M., Raeside, R.P., and Jamieson, R.A.
1990: U-Pb zircon, titanite, and monazite ages in the Bras d'Or and Aspy terranes of Cape Breton Island, Nova Scotia: Implications for igneous and metamorphic history; Geological Society of America Bulletin, v. 102, p. 322-330.

Dunning, G.R., Krogh, T.E., O'Brien, S.J., Colman-Sadd, S.P., and O'Neill, P.P.
1988: Geochronologic framework for the Central Mobile Belt in southern Newfoundland and the importance of Silurian orogeny; Geological Association of Canada, Program with Abstracts, v. 13, p. 34.

Fyffe, L.R. and Fricker, A.
1987: Tectonostratigraphic terrane analysis of New Brunswick; Maritime Sediments and Atlantic Geology, v. 23, p. 113-122.

Haworth, R.T. and Jacobi, R.D.
1983: Geophysical correlation between the geological zonation of Newfoundland and the British Isles; in Contributions to the Tectonics and Geophysics of Mountain Chains, (ed.) R.D. Hatcher, Jr., Harold Williams, and Isidore Zietz; Geological Society of America Memoir 158, p. 25-32.

Haworth, R.T., Keen, C.E., and Williams, H.
1985: D-1 Northern Appalachians: (West Sheet) Grenville Province, Quebec, to Newfoundland; Geological Society of America, Centennial Continent/Ocean transect #1.

Kay, M. and Colbert, E.H.
1965: Stratigraphy and Life History; John Wiley and Sons, Inc. New York, 736 p.

Keen, M.J. and Williams, G.L. (ed.)
1990: Geology of the Continental Margin of Eastern Canada; Geological Survey of Canada, Geology of Canada, no. 2, 855 p. (also Geological Society of America, The Geology of North America, v. I-1).

Keen, C.E., Keen, M.J., Nichols, B., Ried, I., Stockmal, G.S., Colman-Sadd, S.P., O'Brien, S.J., Miller, H., Quinlan, G., Williams, H., and Wright, J.
1986: Deep seismic reflection profile across the northern Appalachians; Geology, v. 14, p. 141-145.

Kennedy, M.J.
1975: Repetitive orogeny in the northeastern Appalachians-New plate models based upon Newfoundland examples; Tectonophysics, v. 28, p. 39-87.

Kennedy, M.J. and McGonigal, M.
1972: The Gander Lake and Davidsville groups of northeastern Newfoundland: new data and geotectonic implications; Canadian Journal of Earth Sciences, v. 9, p. 452-459.

Keppie, J.D.
1982: The Minas Geofracture; in Major Structural Zones and Faults of the Northern Appalachians, (ed.) P. St-Julien and J. Béland: Geological Association of Canada Special Paper 24, p. 263-280.
1985: The Appalachian collage; in The Caledonide Orogen-Scandinavia and related areas, Part 2, (ed.) D.G. Gee and B.A. Sturt; John Wiley and Sons, p. 1217-1226.
1989: Northern Appalachian terranes and their accretionary history; in Terranes in the Circum-Atlantic Paleozoic Orogens, (ed.) R.D. Dallmeyer; Geological Society of America, Special Paper 230, p. 159-193.

Keppie, J.D. and Dallmeyer, R.D.
1989: Tectonic map of pre-Mesozoic terranes in Circum-Atlantic Phanerozoic orogens; International Geological Correlation Programme, Project #233, Terranes in Circum-Atlantic Paleozoic Orogens, 1: 5,000,000 scale.

Keppie, J.D., Dallmeyer, R.D., and Murphy, J.B.
1990: Tectonic implications of ^{40}Ar/^{39}Ar hornblende ages from late Proterozoic-Cambrian plutons of the Avalon Composite Terrane, Nova Scotia, Canada; Geological Society of America Bulletin, v. 102, p. 516-528.

King, P.B.
1950: Tectonic framework of southeastern United States; American Association of Petroleum Geologists, v. 34, p. 635-671.
1959: The Evolution of North America; Princeton University Press, New Jersey, 190 p.

Krough, T.E., Strong, D.F., O'Brien, S.J., and Papezik, V.S.
1988: Precise U-Pb zircon dates from the Avalon terrane in Newfoundland; Canadian Journal of Earth Sciences, v. 25, p. 442-453.

Ludman, Allan
1987: Pre-Silurian stratigraphy and tectonic significance of the St. Croix Belt, southeastern Maine; Canadian Journal of Earth Sciences, v. 24, p. 2459-2469.

Marillier, F., Keen, C.E., Stockmal, G.S., Quinlan, G., Williams, H., Colman-Sadd, S.P., and O'Brien, S.J.
1989: Crustal structure and surface zonation of the Canadian Appalachians: implications of deep seismic reflection data; Canadian Journal of Earth Sciences, v. 26, p. 305-321.

McKerrow, W.S. and Ziegler, A.M.
1971: The Lower Silurian paleogeography of New Brunswick and adjacent areas; Journal of Geology, v. 79, p. 635-646.

O'Brien, B.H.
1987: The lithostratigraphy and structure of the Grand Bruit-Cinq Cerf area (parts of NTS areas 11O/9 and 11O/16), southwestern Newfoundland; in Current Research, Newfoundland Department of Mines and Energy, Mineral Development Division, Report 88-1, p. 311-334.
1988: Relationships of phyllite, schist and gneiss in the La Poile Bay-Roti Bay area (parts of 11O/9 and 11O/16), southwestern Newfoundland; in Current Research, Newfoundland Department of Mines, Mineral Development Division, Report 88-1, p. 109-125.
1989: Summary of the geology between La Poile Bay and Couteau Bay (11O/9 and 11O/16), southwestern Newfoundland; in Current Research (1989) Newfoundland Department of Mines, Geological Survey of Newfoundland, Report 89-1, p. 105-119.

O'Brien, S.J.
1982: Peter Snout (east half), Newfoundland; Newfoundland Department of Mines and Energy, Mineral Development Division, Map 82-58.

O'Neill, P.P. and Blackwood, R.F.
1989: A proposal for revised stratigraphic nomenclature of the Gander and Davidsville groups and the Gander River Ultrabasic Belt of northeastern Newfoundland; in Current Research, Newfoundland Department of Mines, Mineral Development Division, Report 88-1, p. 165-176.

O'Neill, P. P. and Knight, I.
1988: Geology of the east half of the Wier's Pond (2E/1) map area and its regional significance; in Current Research (1988) Newfoundland Department of Mines, Mineral Development Division, Report 88-1, p. 165-176.

Pajari, G.E., Pickerill, R.K., and Currie, K.L.
1979: The nature, origin and significance of the Carmanville ophiolitic mélange, Northeastern Newfoundland; Canadian Journal of Earth Sciences, v. 16, p. 1439-1451.

Piasecki, M.A.J.
1988: A major ductile shear zone in the Bay d'Espoir area, Gander Terrane, southeastern Newfoundland; in Current Research, Newfoundland Department of Mines, Mineral Development Division, Report 88-1. p. 135-144.

Piasecki, M.A.J., Williams, H. and Colman-Sadd, S.P.
1990: Tectonic relationships along the Meelpaeg, Burgeo and Burlington Lithoprobe transects in Newfoundland; in Current Research (1990), Newfoundland Department of Mines, Geological Survey Branch, Report 90-1, p. 327-339.

Poole, W.H., Stanford, B.V., Williams, H., and Kelley, D.G.
1970: Geology of Southeastern Canada; in Geology and Economic Minerals of Canada, (ed.) R.J.W. Douglas; Geological Survey of Canada, Economic Geology Report No. 1, p. 227-304.

Potter, R.R.
1965: Upsalquitch Forks, New Brunswick; Geological Survey of Canada, Map 14-1964.

Potter, R.R., Ruitenberg, A.A., and Davies, J.L.
1969: Mineral exploration in New Brunswick in 1968; Canadian Mining Journal, v. 90, No. 4, p. 68-73.

41

Rodgers, J.
1965: Long Point and Clam Bank Formations, Western Newfoundland; Geological Association of Canada Proceedings, v. 16, p. 83-94.
1970: The Tectonics of the Appalachians; Wiley-Interscience, New York, 271 p.

Ruitenberg, A.A., Fyffe, L.R., McCutcheon, S.R., St. Peter, C.J., Irrinki, R.R., and Venugopal, D.V.
1977: Evolution of pre-Carboniferous tectonostratigraphic zones in the New Brunswick Appalachians; Geoscience Canada, v. 4, No. 4, p. 171-181.

St-Julien, P. and Béland, J. (ed.)
1982: Major structural zones and faults of the Northern Appalachians; Geological Association of Canada Special Paper 24, 280 p.

St-Julien, P., Slivitzky, A., and Feininger, T.
1983: A deep structural profile across the Appalachians of southern Quebec; in Contributions to the tectonics and geophysics of mountain chains, (ed.) R.D. Hatcher, Harold Williams and Isodore Zietz; Geological Society of America, Memoir 158, p. 103-111.

Schenk, P.E.
1970: Regional variation of the flysch-like Meguma Group (lower Paleozoic) of Nova Scotia compared to recent sedimentation off the Scotian Shelf; in Flysch Sedimentology in North America, (ed.) J. Lajoie; Geological Association of Canada Special Paper No. 7, p. 127-153.
1971: Southeastern Atlantic Canada, northwestern Africa and continental drift; Canadian Journal of Earth Sciences, v. 8, p. 1218-1251.
1972: Possible Late Ordovician glaciation of Nova Scotia; Canadian Journal of Earth Sciences, v. 9, p. 95-107.
1975: Paleozoic evolution of African Nova Scotia-polar and deep to equatorial and continental; International Congress of Sedimentologists, IXth, Nice, Theme 1, p. 181-186.
1976: A regional synthesis (of Nova Scotia geology); Maritime Sediments, v. 12, p. 17-24.
1978: Synthesis of the Canadian Appalachians; in IGCP Project 27, Caledonian-Appalachian Orogen of the North Atlantic region, (ed.) E.T. Tozer and P.E. Schenk; Geological Survey of Canada, Paper 78-13, p. 111-136.

Schuchert, C.
1923: Sites and natures of the North American geosynclines; Geological Society of America Bulletin, v. 34, p. 151-229.

Stockmal, G.S. and Waldron, J.W.F.
1990: Structure of the Appalachian deformation front in western Newfoundland: implications of multichannel seismic reflection data; Geology, v. 18, p. 765-768.

Taylor, F.C. and Schiller, E.A.
1966: Metamorphism of the Meguma Group of Nova Scotia; Canadian Journal of Earth Sciences, v. 3, p. 959-974.

Tremblay, A. and St-Julien, P.
1990: Structural style and evolution of a segment of the Dunnage Zone from the Quebec Appalachians and its tectonic implications; Geological Society of America Bulletin, v. 102, p. 1218-1229.

van Berkel, J.T. and Currie, K.L.
1988: Geology of the Puddle Pond (12A/5) and Little Grand Lake (12A/12) map areas, southwestern Newfoundland; in Current Research (1988) Newfoundland Department of Mines, Mineral Development Division, Report 88-1, p. 99-107.

van Staal, C.R.
1987: Tectonic setting of the Tetagouche Group in northern New Brunswick: implications for plate tectonic models of the northern Appalachians; Canadian Journal of Earth Sciences, v. 24, p. 1329-1351.

Williams, H.
1964: The Appalachians in Northeastern Newfoundland - a two-sided symmetrical system; American Journal of Science, v. 262, p. 1137-1158.
1967: Silurian rocks of Newfoundland; in Geology of the Atlantic Region, (ed.) E.R.W. Neale and H. Williams; Geological Association of Canada, Special Paper No. 4, p. 93-137.
1975: Structural succession, nomenclature, and interpretation of transported rocks in western Newfoundland; Canadian Journal of Earth Sciences, v. 12, p. 1874-1894.
1976: Tectonic stratigraphic subdivision of the Appalachian Orogen; Geological Society of America, Abstracts with Programs, v. 8, no. 2, p. 300.

Williams, H. (cont.)
1978a: Tectonic lithofacies map of the Appalachian Orogen; Memorial University of Newfoundland, Map no. 1, scale 1:1 000 000.
1978b: Geological development of the northern Appalachians: its bearing on the evolution of the British Isles; in Crustal evolution in northwestern Britain and adjacent regions, (ed.) D.R. Bowes and B.E. Leake; Seal House Press, Liverpool, U.K. Geological Journal, Special Issue No. 10, p. 1-22
1979: Appalachian Orogen in Canada; Canadian Journal of Earth Sciences, Tuzo Wilson volume, v. 16, p. 792-807.
1984: Miogeoclines and suspect terranes of the Caledonian-Appalachian Orogen: tectonic patterns in the North Atlantic region; Canadian Journal of Earth Sciences, v. 21, p. 887-901.

Williams, H. and Hatcher, R.D., Jr.
1982: Suspect terranes and accretionary history of the Appalachian Orogen; Geology, v. 10, p. 530-536.
1983: Appalachian Suspect terranes; in Contributions to the tectonics and geophysics of mountain chains, (ed.) R.D. Hatcher, H. Williams, and I. Zietz; Geological Society of America, Memoir 158, p. 33-53.

Williams, H. and Max, M.D.
1980: Zonal subdivision and regional correlation in the Appalachian-Caledonian Orogen; in "the Caledonides in the USA", (ed.) D.R. Wones; Virginia Polytechnic Institute and State University, Department of Geological Sciences, Memoir 2, p. 57-62.

Williams, H. and Piasecki, M.A.J.
1990: The Cold Spring Mélange and its significance in central Newfoundland; Canadian Journal of Earth Sciences, v. 27, p. 1126-1134.

Williams, H. and St-Julien, P.
1982: The Baie Verte-Brompton Line: Continent-Ocean interface in the Northern Appalachians; in Major Structural Zones and Faults of the Northern Appalachians, (ed.) P. St-Julien and J. Béland; Geological Association of Canada, Special Paper No. 24, p. 177-207.

Williams, H., Colman-Sadd, S.P., and Swinden, H.S.
1988: Tectonic-stratigraphic subdivisions of central Newfoundland; in Current Research, Part B; Geological Survey of Canada, Paper 88-1B, p. 91-98.

Williams, H., Dickson, W.L., Currie, K.L., Hayes, J.P., and Tuach, J.
1989: Preliminary report on a classification of Newfoundland granitic rocks and their relation to tectonostratigraphic zones and lower crustal blocks; in Current Research, Part B; Geological Survey of Canada, Paper 89-1B, p. 47-53.

Williams, H., Kennedy, M.J., and Neale, E.R.W.
1972: The Appalachian Structure Province; in Variations in Tectonic Styles in Canada, (eds.) R.A. Price and R.J.W. Douglas; Geological Association of Canada, Special Paper No. 11, p. 181-261.
1974: The northeastward termination of the Appalachian Orogen; in The Ocean Basins and Margins, volume 2, the North Atlantic, (ed.) A.E.M. Nairn and F.G. Stehli; Plenum Press, New York, p. 79-123.

Williams, H., Piasecki, M.A.J., and Colman-Sadd, S.P.
1989: Tectonic relationships along the proposed central Newfoundland Lithoprobe transect and regional correlations; in Current Research, Part B; Geological Survey of Canada Paper, 89-1B, p. 55-66.

Williams, H., Piasecki, M.A.J., and Johnston, J.
1991: The Carmanville Mélange and Dunnage-Gander relationships in northeast Newfoundland; in Current Research, Part D; Geological Survey of Canada, Paper 91-1D, p. 15-23.

Williams, S.H.
1991: Stratigraphy and graptolites of the Upper Ordovician Point Leamington Formation, Central Newfoundland; Canadian Journal of Earth Sciences, v. 28, p. 581-600.

Wonderley, P.F. and Neuman, R.B.
1984: The Indian Bay Formation: fossiliferous Early Ordovician volcanogenic rocks in the northern Gander Terrane, Newfoundland, and their regional significance; Canadian Journal of Earth Sciences, v. 21, p. 525-532.

Zen, E-an
1983: Exotic terranes in the New England Appalachians - limits, candidates, and ages: A speculative essay; in Contributions to the tectonics and geophysics of mountain chains, (ed.) R.D. Hatcher, Harold Williams and Isodore Zietz; Geological Society of America, Memoir 158, p. 55-81.

ADDENDUM

(Note 1) Further evidence for Silurian deformation in Newfoundland is contained in Dunning et al. (1990), Elliott et al. (1991), Lafrance and Williams (1992) and Willliams et al. (1993a). An overview of Acadian Orogeny and the extent and effects of middle Paleozoic metamorphism, plutonism, and deformation in Newfoundland is found in Williams (1993a).

(Note 2) Additional evidence for Ordovician Dunnage-Gander interaction and Ordovician deformation in Newfoundland is contained in Colman-Sadd et al. (1992) and Piasecki (1992).

(Note 3) The differences between the Aspy Subzone, consisting mainly of Ordovician-Silurian metavolcanic and metasedimentary rocks, and the Bras d'Or Subzone, consisting of Upper Precambrian metasedimentary rocks and abundant late Precambrian intrusions, are less important than first envisaged. An unconformity between a basal conglomerate of the Cheticamp Lake Gneiss of the Aspy Subzone and deformed diorite of the Bras d'Or Subzone indicates a depositional contact (Lin, 1993). Furthermore, (1) the Clyburn Brook formation, an unconformable cover on rocks of the Bras d'Or Subzone, is correlated with Ordovician-Silurian sedimentary rocks of the Aspy Subzone, (2) the late Precambrian Cheticamp Pluton is an integral part of the Aspy Subzone and not a Bras d'Or Subzone klippe, (3) plutonic clasts in Ordovician-Silurian sedimenary rocks of the Aspy Subzone were derived from the Bras d'Or Subzone or correlative plutons, such as the Cheticamp Pluton, and (4) the Eastern Highlands shear zone is not a subzone tectonic boundary, but possibly a shear zone related to the Late Silurian-Early Devonian closure of a small back-arc basin (Lin, 1993). This analysis implies that the main Iapetus suture crosses Cape Breton Island and lies between the Aspy Subzone and the Grenvillian Blair River Complex to the northwest. If the Aspy Subzone is a Gander Zone equivalent, and the Bras d'Or Subzone an Avalon Zone equivalent, Gander and Avalon zones in Nova Scotia locally share the same late Precambrian plutons and Ordovician Silurian cover rocks.

(Note 4) The Bouguer gravity anomaly Map 3 and magnetic anomaly Map 4 used in this volume are now superceded by digitized data sets published by the Geological Survey of Canada (Shih et al., 1993a, b; and Williams et al., 1993b).

(Note 5) Onland Lithoprobe East seismic results in central Newfoundland fail to recognize a Central lower crustal block (Quinlan et al., 1992). The Avalon lower crustal block may directly abut the Grenville lower crustal block at mid crustal depths (Quinlan et al., 1992). Both refraction and deep reflection data on the southern Grand Banks indicate that the Avalon-Meguma zone boundary is vertical and cuts the deep crust. Other reflection data from the Bay of Fundy and Gulf of Maine further imply that the Meguma Zone has a distinctive basement that defines the Sable lower crustal block (Keen et al., 1990).

(Note 6) Recognition of the Dog Bay Line as a major structural junction that separates different Silurian rock groups requires subdivision of the Botwood Belt into western and eastern parts (Williams, 1993b; Williams et al., 1993a). The name Botwood Belt is retained for the wide area of middle Paleozoic rocks northwest of the Dog Bay Line and a new name, the Indian Islands Belt, is introduced for a much narrower belt southeast of the Dog Bay Line. West of the line, marine greywackes and conglomerates are assigned to the Badger Group. The overlying terrestrial volcanics and sandstones are assigned to the Botwood Group. East of the Dog Bay Line, Silurian calciferous shales, sandstones, conglomerates, and redbeds are assigned to the Indian Islands Group.

The Dog Bay Line is marked by a wide zone of disrupted Ordovician rocks. The Badger and Botwood groups northwest of the line were deposited on Ordovician rocks already accreted to Laurentia. The Indian Islands Group southeast of the line was deposited on Ordovician rocks already amalgamated with the continental Gander Zone. Timing of major movement and a Silurian marine to terrestrial depositional change recorded on both sides of the line, agree within error with isotopic ages for the onset of middle Paleozoic plutonism, regional deformation, and metamorphism in central Newfoundland. The Dog Bay Line may mark the terminal Iapetus Suture.

Colman-Sadd, S.P., Dunning, G.R., and Dec, T.
1992: Dunnage-Gander relationships and Ordovician orogeny in central Newfoundland: a sediment provenance and U/Pb age study; American Journal of Science, v. 292, p. 317-355.

Dunning, G.R., O'Brien, S.J., Colman-Sadd, S.P., Blackwood, R.F., Dickson, W.L., O'Neill, P.P., and Krogh, T.E.
1990: Silurian orogeny in the Newfoundland Appalachians; Journal of Geology, v. 98, p. 895-913.

Elliott, C.G., Dunning, G.R., and Williams, P.F.
1991: New U-Pb zircon age constraints on the timing of deformation in north-central Newfoundland and implications for early Paleozoic Appalachian orogenesis; Geological Society of America Bulletin, v. 103, p. 125-135.

Keen, C.E., Kay, W.A., Keppie, D., Marillier, F., Pe-Piper, G, and Waldron, J.W.F.
1990: Deep seismic reflection data from the Bay of Fundy and Gulf of Maine; Lithoprobe East Transect, Report No. 13, Memorial University of Newfoundland, p. 111-116.

Lafrance, B. and Williams, P.F.
1992: Silurian deformation in eastern Notre Dame Bay, Newfoundland; Canadian Journal of Earth Sciences, v. 29, p. 1899-1914.

Lin, S.
1993: Relationship between the Aspy and Bras d'Or "terranes" in the northeastern Cape Breton Highlands, Nova Scotia; Canadian Journal of Earth Sciences, v. 30, p. 1773-1781.

Piasecki, M.A.J.
1992: Tectonics across the Gander-Dunnage boundary in northeast Newfoundland; in Current Research, Geological Survey of Canada, Paper 92-1E, p. 259-268.

Quinlan, G.M., Hall, J, Williams, H., Wright, J.A.,
Colman-Sadd, S.P., O'Brien, S.J., Stockmal, G.S., and Marillier, F.
1992: Lithoprobe onshore seismic reflection transects across the Newfoundland Appalachians; Canadian Journal of Earth Sciences, v. 29, p. 1865-1877.

Shih, K.G., Williams, H., and Macnab, R. (comp.)
1993a: Gravity anomalies and major structural features of southeastern Canada and the Atlantic continental margin; Geological Survey of Canada, Map 1777A, scale 1: 3 000 000.
1993b: Magnetic anomalies and major structural features of southeastern Canada and the Atlantic continental margin; Geological Survey of Canada, Map 1778A, scale 1:3 000 000.

Williams, H.
1993a: Acadian orogeny in Newfoundland; in The Acadian Orogeny: Recent Studies in New England, Maritime Canada, and the Autochthonous Foreland, (ed.) D.C. Roy and J.W. Skehan, Geological Society of America, Special Paper 275, p. 123-133.
1993b: Stratigraphy and structure of the Botwood Belt and definition of the Dog Bay Line in northeastern Newfoundland; in Current Research, Part D, Geological Survey of Canada, Paper 93-1D, p. 19-27.

Williams, H., Currie, K.L., and Piasecki, M.A.J.
1993a: The Dog Bay Line - a major Silurian tectonic boundary in northeast Newfoundland; Canadian Journal of Earth Sciences, v. 30, p. 2481-2494.

Williams, H., Macnab, R., and Shih, K.G.
1993b: Major structural features of southeastern Canada and the Atlantic continental margin portrayed in regional gravity and magnetic maps; Geological Survey of Canada, Paper 90-16.

Author's Address

H. Williams
Department of Earth Sciences
Memorial University of Newfoundland
St. John's, Newfoundland
A1B 3X5

Printed in Canada

Chapter 3

LOWER PALEOZOIC AND OLDER ROCKS

Chapter 3

LOWER PALEOZOIC AND OLDER ROCKS

The five-fold zonation of Humber, Dunnage, Gander, Avalon and Meguma is the first order geographic division of this temporal category. Within each zone the oldest rocks are treated first. Where continuity or correlation is established among distinctive rock packages throughout a zone, each rock package is treated separately, from north to south and province by province. In other cases, all of the lower Paleozoic and older rocks of a zone are treated by provinces, from north to south.

HUMBER ZONE

Introduction

Harold Williams

The Humber Zone, or Appalachian miogeocline, takes its name from Humber Arm of western Newfoundland. It has two divisions of contrasting structural style and metamorphic grade: a western or external division where deformation is moderate, regional metamorphism is low grade, and where stratigraphic sections are preserved or easily restored; and an eastern or internal division where deformation is intense, regional metamorphism is medium to high grade, and where stratigraphic and structural relationships are commonly in doubt (Fig. 3.1). Contacts between external and internal divisions of the Humber Zone are important structural junctions where not hidden by younger cover rocks.

Rocks of the external Humber Zone are subdivided into distinct stratigraphic and structural packages as follows: Grenville basement rocks, upper Precambrian to Lower Cambrian clastic sedimentary and volcanic rocks, Cambrian-Ordovician carbonate sequence, Middle Ordovician clastic rocks stratigraphically above the carbonate sequence and/or integral parts of Taconic allochthons, and allochthonous rocks structurally overlying the Middle Ordovician clastic rocks.

These stratigraphic and structural packages also occur in internal parts of the Humber Zone. Where correlations are reasonably clear, descriptions of the deformed and metamorphosed (internal) examples follow treatment of the less deformed (external) examples, from north to south and province by province. Where correlations are uncertain or unknown, the deformed rocks are treated under the general heading "Humber Zone internal".

Williams, H.
1995: Introduction (Humber Zone); in Chapter 3 of Geology of the Appalachian-Caledonian Orogen in Canada and Greenland, (ed.) H. Williams; Geological Survey of Canada, Geology of Canada, no. 6, p. 47-49 (also Geological Society of America, The Geology of North America, v. F-1).

Allochthonous rocks of the Humber Zone are of late Precambrian to Middle Ordovician age. These are mainly sedimentary rocks in lower structural slices that are coeval with the carbonate sequence and underlying sedimentary and volcanic rocks. They are viewed as eastern or offshore facies equivalents. Higher structural slices are a sampling of Dunnage Zone marine volcanic sequences and ophiolite suites. They are described here as they are integral parts of Ordovician allochthons emplaced upon the autochthonous sequence.

The oldest cover rocks above Humber Zone allochthons are Middle Ordovician (Caradoc) limestones, sandstones and shales of the Long Point Group in western Newfoundland. These are treated with the Clam Bank Belt of middle Paleozoic rocks as they are part of an unconformable overlap sequence. In Quebec, rocks of Caradoc age are integral parts of Ordovician allochthons and are therefore included in the Humber Zone. This situation arises because of the diachronous nature of events between Newfoundland and Quebec. Silurian and younger rocks, where present, are unconformable on rocks of the Humber Zone.

The external part of the Humber Zone forms the Great Northern Peninsula of western Newfoundland and its rocks and structures extend southward to Port au Port Peninsula. All tectonostratigraphic elements that define the zone are represented. On the opposite side of the Gulf of St. Lawrence, the external Humber Zone is defined almost entirely on its allochthonous rocks that overlie the Middle Ordovician clastic unit at the top of the underlying autochthon. These extend from Gaspésie along the south shore of the St. Lawrence River to the United States border.

Rocks of the internal Humber Zone are traceable in Newfoundland from White Bay to the southern end of Grand Lake. Relationships are debatable farther southwest in Newfoundland where the Dashwoods Subzone apparently includes both Humber and Dunnage zone rocks (Map 2). In this area, metamorphism, plutonism and structural styles make a Humber-Dunnage distinction impractical. Reworked Grenville gneisses form a small area at the northwest tip of Cape Breton Island, Nova Scotia, and metaclastic rocks of the internal Humber Zone reappear in the Maquereau Dome of Gaspésie, Quebec. Farther west, rocks of the internal Humber Zone are present in the Mount Logan Nappe of the Chic-Chocs Mountains (Hibbard et al., this chapter) and they form a wide continuous belt along the Notre-Dame Mountains and Sutton Mountains of Quebec.

The western limit of the Humber Zone is the Appalachian structural front, which separates deformed rocks of the orogen from undeformed rocks of the St. Lawrence Platform. It lies beneath the Gulf of St. Lawrence and St. Lawrence River for most of its length, except in parts of western Newfoundland and between Québec City and the

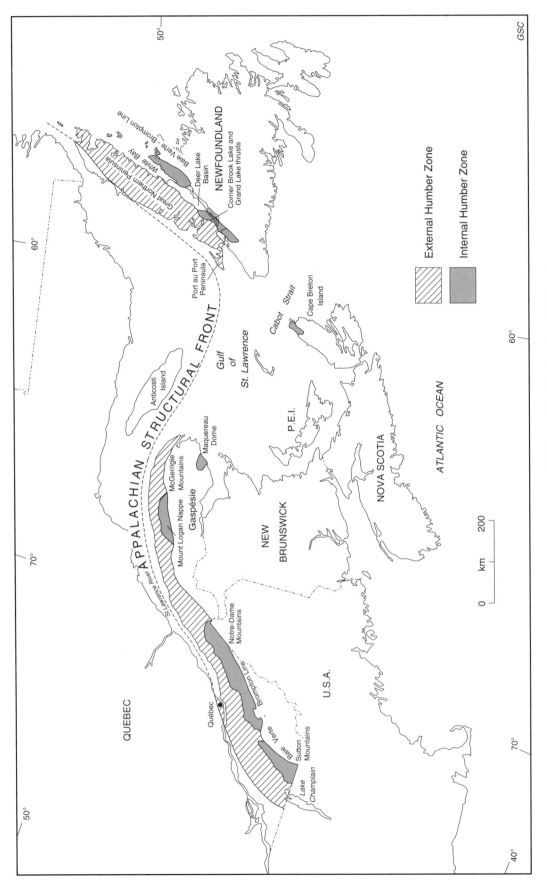

Figure 3.1. External and internal divisions of the Humber Zone in Canada.

United States border. In western Newfoundland the structural front is defined by thrusts or steep reverse faults, and it is marked by thrusts in southwestern Quebec.

The boundary between external and internal divisions of the Humber Zone in Newfoundland is the Corner Brook Lake and Grand Lake thrusts. To the north it is hidden by Carboniferous rocks of the Deer Lake Basin and to the south by the Gulf of St. Lawrence. In Quebec, it is defined by major thrust faults along the northern flanks of the Notre-Dame Mountains and Sutton Mountains. Farther east, it is hidden by middle Paleozoic rocks in Gaspésie.

The eastern boundary of the internal Humber Zone is the Baie Verte-Brompton Line, a steep structural zone of discontinuous ophiolite occurrences that defines the Humber-Dunnage zone boundary (Williams and St-Julien, 1982). It is well defined in northeast Newfoundland and in southwestern Quebec, but it is covered by younger rocks throughout Gaspésie.

The Humber Zone corresponds with the St. Lawrence Geosyncline of Schuchert (1923) and Schuchert and Dunbar (1934), who interpreted the geosyncline as an inland seaway between the North American Craton and the New Brunswick Geanticline. Long before that, it was viewed, much as it is today, as a North American marine "continental plateau" open to a "Palaeo-Atlantic" ocean (Logan, 1863; after Stevens, 1974). In Newfoundland, and still before the wide acceptance of plate tectonics, the Humber Zone was the western Newfoundland Shelf (Williams, 1964a) or the Western Platform (Kay and Colbert, 1965). Since then, all studies support the paradigm of an ancient continental margin that originated through rifting, evolved as a passive Atlantic-type margin, and was destroyed by ophiolite obduction and accretion of outboard terranes (Rodgers, 1968: Dewey, 1969; Bird and Dewey, 1970; Stevens, 1970; Williams, 1971a; Williams and Stevens, 1974; St-Julien and Hubert, 1975; and many others). In fact, stratigraphic and tectonic analyses of the late Precambrian and early Paleozoic development of the Humber Zone rival those of any ancient continental margin.

Rodgers (1968) suggested that the eastern limit of Grenville inliers in the Appalachian Orogen is roughly coincident with the edge of the early Paleozoic continental shelf, so that the declivity at the shelf edge lay above the basement edge. Palinspastic restoration implies continental crust well east of the shelf edge (Williams and Stevens, 1974). Seismic reflection studies off northeast Newfoundland (Keen et al., 1986) and in the Gulf of St. Lawrence (Marillier et al., 1989) indicate that Grenville basement extends in subsurface to central parts of the Newfoundland Dunnage Zone. Similarly in Quebec, onland seismic reflection profiles indicate a relatively undeformed basement and autochthonous cover that extends in subsurface to the Baie Verte-Brompton Line and beyond (St-Julien et al., 1983; Stewart et al., 1986).

The Humber Zone has negative Bouguer gravity anomalies compared to positive anomalies of the adjacent Dunnage Zone (Map 3). A gravity gradient expressed along the full length of the Appalachian miogeocline, from negative values on the continent side to positive values eastward, is one of the major geophysical features of the Appalachian Orogen (Haworth et al., 1980a, b). Whether the gradient represents the edge of a Paleozoic passive continental margin, a Paleozoic collisional zone, or Mesozoic extensional effects related to the modern Atlantic (Nelson et al., 1986), is unclear. The regional negative gravity field extends well south of the Humber Zone in Quebec, and a marked gradient in Newfoundland crosses acutely the exposed Long Range Grenville inlier (Map 3). There is no clear relationship between the gravity gradient and the Baie Verte-Brompton Line or between the gradient and the edge of the Grenville lower crustal block as defined by deep seismic reflection experiments.

Faunas of the Humber Zone are typically North American. Some show local provincialism controlled by shallow marine conditions at the ancient continental shelf or deeper marine environments at the slope and rise. Shallow marine shelly faunas of the Ordovician Toquima-Table Head Faunal Realm occur in the carbonate sequence of the Humber Zone and are found all along the ancient periphery of North America (Ross and Ingham, 1970). Diverse shelly, graptolite, and conodont faunas occur in deeper water carbonate breccias and shales. Examples in the Cow Head Group of Newfoundland provide ample opportunity for Cambrian-Ordovician biostratigraphic zonation and correlation with strata of the continental interior (e.g. Williams and Stevens, 1988). This is one of several reasons that make the Cow Head Group a leading contender for the Cambrian-Ordovician boundary stratotype (Barnes, 1988).

The Humber Zone was affected by Taconic (Middle Ordovician) and Acadian (Silurian-Devonian) deformation. Taconic deformation is indicated by stratigraphic and structural analyses, and confirmed by sub-Silurian unconformities in western Newfoundland and Quebec. Acadian deformation is demonstrated by unconformities between deformed Silurian-Devonian rocks and less deformed Carboniferous cover. It is also indicated by Devonian intrusions that cut deformed Silurian-Devonian rocks. Alleghanian (Permian-Carboniferous) deformation, so important in the southern U.S. Appalachians, is mild or absent in the Humber Zone of Canada. Intensities of both Taconic and Acadian deformations increase eastward across the Humber Zone.

There are few Paleozoic plutons in the exposed Humber Zone, except for its internal parts in Newfoundland. One exception is the Devonian McGerrigle Mountains pluton of Quebec (DeRomer, 1977), which cuts transported rocks near the Appalachian structural front.

Mineral deposits of the Humber Zone are those unique to Grenville basement (titanium, iron) or the carbonate cover (lead-zinc deposits of Mississippi Valley type). Others are integral parts of transported ophiolite suites (asbestos, chromite, nickel, copper-pyrite), or related to Devonian plutons (skarn deposits at the periphery of the McGerrigle Mountains pluton).

Humber Zone external

Grenville basement rocks

P. Erdmer and Harold Williams

Introduction

Crystalline basement rocks occur in both the external and internal divisions of the Humber Zone (Fig. 3.2). These are gneisses, schists, granitoids and metabasic rocks that are correlated with rocks of the Grenville Structural Province of the nearby Canadian Shield. Those of the external Humber Zone are for the most part unaffected by penetrative Paleozoic deformation, although they are faulted and in places retrograded. Those of the internal Humber Zone are deformed with their cover rocks and display Paleozoic fabrics. External occurrences form the cores of northeast-trending anticlines or they are brought to the surface by high angle faults or west-directed thrusts. Internal occurrences are thrust slices or structural culminations in high-grade metamorphic areas where their distinction from cover rocks is in places difficult. Most metamorphic and plutonic isotopic ages are about 1200 to 1000 Ma, except where the rocks are affected by intense Paleozoic deformation. Basement gneisses in the Long Range Inlier of Newfoundland crystallized at or before 1550 Ma (H. Baadsgaard, pers. comm., 1990).

Occurrences of basement rocks in the external Humber Zone are restricted to western Newfoundland. The Long Range Inlier is by far the largest with smaller inliers to its north and south at Belle Isle and Indian Head, respectively. Other small faulted examples occur nearby at Ten Mile Lake and Castors River.

Internal Humber Zone occurrences in Newfoundland are the East Pond Metamorphic Suite at Baie Verte Peninsula, the Cobble Cove Gneiss at Glover Island of Grand Lake, the Steel Mountain Inlier south of Grand Lake, and possibly the Cormacks Lake Complex of the Dashwoods Subzone.

Metamorphic and plutonic rocks of Cape Breton Island, Nova Scotia, with a metamorphic age of 1040 +10/-40 Ma and an igneous age of 1500-1100 Ma, are known as the Blair River Complex (Barr et al., 1987b; Barr and Raeside, 1989). They are separated from metamorphic rocks to the east by a wide zone of mylonite developed along the Wilkie Brook Fault.

Precambrian crystalline rocks occur in three widely-separated areas of the Quebec internal Humber Zone. From northeast to southwest these are located in the Port-Daniel area of Gaspésie, the Saint-Malachie area south of Québec City, and near Thetford Mines of the Eastern Townships. These occurrences are named the Port-Daniel Inlier, the Sainte-Marguerite Fault Slice and the Sainte-Hélène-de-Chester Fault Slice, respectively.

The occurrences are described separately, first for the external Humber Zone of Newfoundland (north to south), then for the internal Humber Zone of Newfoundland, Nova Scotia, and Quebec, in that order. This is followed by a summary overview and discussion.

Humber Zone external, Newfoundland

Belle Isle Inlier

The Belle Isle Inlier and its cover rocks form a small island of 70 km² in the Strait of Belle Isle (Fig. 3.3). Crystalline Precambrian rocks form the core of the island with an overlying discontinuous veneer of cover rocks dipping seaward along its periphery (Williams and Stevens, 1969; Bostock, 1983). Mafic columnar flows (Lighthouse Cove Formation) directly overlie pink gneisses on the west side of the island and an eastward-thickening unit of boulder conglomerate, coarse sandstone, and quartzite (Bateau Formation) occurs between basement and volcanic rocks on its eastern side. Arkosic sandstones (Bradore Formation) and fossiliferous Lower Cambrian shales (Forteau Formation) occur stratigraphically above the volcanic rocks.

Biotite-quartz-feldspar gneiss is the dominant rock type. Gneissic foliation is contorted and cut by granite pegmatites. Steep northeast-trending mafic dykes of the Long Range Swarm (Bostock, 1983) cut the gneisses and pegmatites. The melanocratic dykes form a conspicuous ramifying network among leucocratic gneisses in coastal exposures, and locally the dykes are as abundant as host gneisses. Relationships are especially clear at the southwest corner of the island where mafic dykes cut pink gneisses and blend upward into gently-dipping flows that overlie the gneisses. The dykes and volcanic rocks are coeval and this is confirmed by an absence of dykes in Lower Cambrian strata above the volcanic rocks. The dykes are affected by retrograde greenschist metamorphism (Bostock, 1983).

Faults of at least two generations cut the Precambrian basement rocks at Belle Isle. A younger set of northeast-trending faults cuts the gneisses and cover rocks, including fossiliferous Lower Cambrian strata. An older set on the east side of the island separates structural wedges of steeply-dipping beds of the Bateau Formation from the gneisses. These have no obvious effect on mafic dykes, indicating that the faults are older (Bostock, 1983).

Long Range Inlier

The Long Range Inlier forms the highlands of the Great Northern Peninsula of western Newfoundland with an area of 8500 km². It is the largest external basement massif of the Appalachian Orogen. Its rocks are known as the Long Range Complex (Baird, 1960; Williams, 1985a) or Basement Gneiss Complex (Bostock, 1983) and they extend across most of the width of the external Humber Zone. Upper Precambrian to Lower Cambrian unconformable cover rocks dip gently away from the basement rocks at the north and south ends of the inlier and an equivalent unconformity that dips eastward is preserved locally along its eastern side at White Bay (Fig. 3.4). Its western boundary is largely marked by thrusts or steep reverse faults (Williams et al., 1985a, b; Cawood and Williams, 1986, 1988; Grenier and Cawood, 1988). These juxtapose crystalline basement with the

Erdmer, P. and Williams, H.
1995: Grenville basement rocks (Humber Zone); in Chapter 3 of Geology of the Appalachian-Caledonian Orogen in Canada and Greenland, (ed.) H. Williams; Geological Survey of Canada, Geology of Canada, no. 6, p. 50-61 (also Geological Society of America, The Geology of North America, v. F-1).

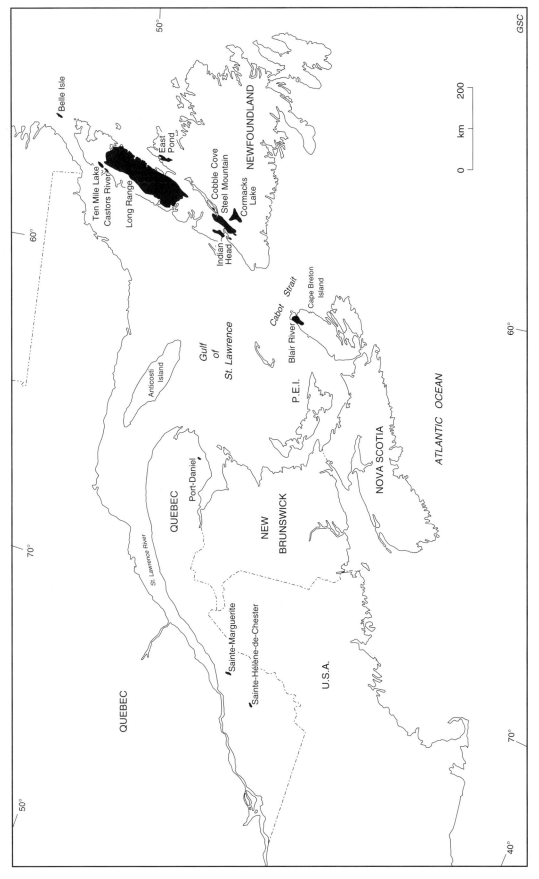

Figure 3.2. Distribution of Precambrian basement inliers in the Humber Zone.

Cambrian-Ordovician sedimentary cover, although basement-cover stratigraphic relationships are preserved locally within hanging-wall segments of the faulted boundary. At Western Brook Pond, St. Pauls Inlet, and south of Portland Creek Pond, the crystalline rocks are thrust above both the carbonate cover sequence and rocks of the Humber Arm Allochthon. At the southeast edge of the inlier, Cambrian-Ordovician cover rocks are detached from the crystalline basement (Owen, 1986) and Carboniferous rocks are unconformable on the lower Paleozoic and Precambrian basement rocks.

Parts of the Long Range Inlier were investigated by Foley (1937) and Fritts (1953) and mapped in reconnaissance fashion by Baird (1960), Neale and Nash (1963) and the British Newfoundland Exploration Company. Results of this work are summarized by Clifford and Baird (1962). The inlier is covered by systematic mapping at 1:100 000 and 1:125 000 scales (Bostock, 1983; Erdmer, 1986a, b) and locally at 1:50 000 scale (Owen, 1986).

Lithologies are nearly uniformly high-grade quartz-feldspar gneisses and granites. At the inlier's northern portion the rocks are biotite-quartz-feldspar gneiss and hornblende-biotite-quartz-feldspar gneiss with lesser amounts of quartz-rich gneiss, pelitic schist, amphibolite and calc-silicate gneiss (Bostock, 1983). Most of these are thought to be of sedimentary protolith. Small mafic plutons, now amphibolite, hypersthene amphibolite, metaperidotite and metadiorite cut the leucocratic gneisses. The mafic plutons are in turn cut by distinctive megacrystic foliated to massive granites of at least two suites. Pegmatites that occur in most outcrops probably relate to the granites. Northeast-trending diabase dykes of the Long Range Swarm (Bostock, 1983) cut all other rocks of the inlier. These are up to 50 m wide and they are coeval with basalt flows that locally overlie the gneisses at the north end of the inlier (Clifford, 1965; Williams and Stevens, 1969). Mafic dykes are thinner and they decrease in frequency toward the west.

In southern parts of the inlier, rocks and relationships are similar to those in the north. The commonest foliated rocks are pink to grey quartz-feldspar gneiss, quartz-rich gneiss, pelitic to psammitic schist, amphibolite and minor marble and calc-silicate rock. The bulk of the gneisses appear to have plutonic protoliths (Owen and Erdmer, 1986). A gabbro-anorthosite pluton, up to 15 km in diameter, occurs near the south end of the inlier. All these rocks are cut by large plutons of massive to foliated megacrystic granite to charnockite. Diabase dykes occur locally but they are less common than in the north.

Plutonism both accompanied and followed Precambrian regional metamorphism of the basement gneiss complex. The emplacement of mafic plutons in the northern part of the inlier accompanied regional metamorphism of amphibolite to granulite facies. Pelitic rocks contain microcline-sillimanite-cordierite (or garnet) assemblages, and remnants of hypersthene are preserved in several localities (Bostock, 1983). Some megacrystic granites cut the high grade metamorphic rocks and locally imprint lower amphibolite facies aureoles on the country rocks. Andalusite overprints sillimanite in pelitic aureole rocks and cummingtonite and anthophyllite occur sparingly. Other massive, mostly equigranular granites lack metamorphic aureoles.

Retrograde greenschist facies metamorphism affected eastern parts of the inlier with development of epidote and chlorite, recrystallization of biotite, replacement of sillimanite by muscovite, and sericitization of feldspars. It occurred in at least two phases. The first preceded the emplacement of the Long Range mafic dykes, indicating a Precambrian age, the second postdated the dykes and related volcanic rocks indicating a Paleozoic age (Bostock, 1983; Owen and Erdmer, 1986).

In central and southern portions of the Long Range Inlier, amphibolite and locally granulite facies mineral assemblages (two-pyroxene gneisses) occur in many places, with metamorphic grade increasing from amphibolite in the northeast to granulite in the southwest. The metamorphism predates most of the large granite plutons and possibly the gabbro-anorthosite suite. Metamorphic temperatures increase from 600-650°C to nearly 800°C southwestward across the amphibolite-granulite transition; metamorphic pressures vary between 5 and 8 kilobars throughout the inlier (Owen and Erdmer, 1990). Retrograde effects are greatest at the southeast tongue of the inlier and along its southeast margin. This retrogression is Paleozoic as it involves nearby Cambrian-Ordovician cover rocks.

Precambrian structures within the inlier are complex. In its northern part, the basement gneisses display three episodes of folding (Bostock, 1983). Large areas of shallowly-dipping foliation are separated by steep structural zones or by granitic plutons and faults. Most folds trend northeast with gently-plunging axes. These range from early isoclines to late open structures. A few folds of different orientation may be controlled by late granitic plutons. Gneissic banding, schistosity and mineral lineations are present in nearly every rock type, except in the granitic plutons.

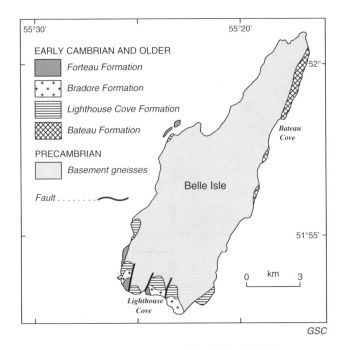

Figure 3.3. Geology of the Belle Isle Inlier.

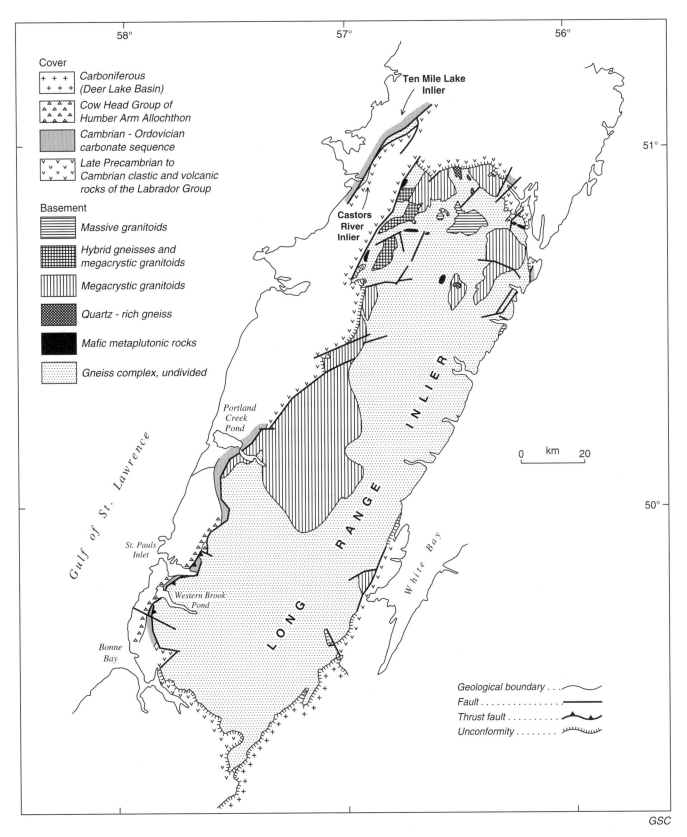

Figure 3.4. Geology of inliers of the Great Northern Peninsula: Long Range, Ten Mile Lake, and Castors River.

In the rest of the Long Range Inlier, multiple fabrics similar to those in the north are developed in the basement gneisses. Large amplitude (15 km) folds occur in the southwestern part of the inlier. A northwest-striking fabric predominates elsewhere and regional interference folds are common (Erdmer, 1986a, b). These fabrics are rotated to parallel the margins of large granite plutons. Southwest-trending shallowly-dipping mylonite zones mark ductile Precambrian thrusts (Erdmer, 1984). Local northeast-trending mylonite zones, breccia zones, and discrete brittle shear zones may also be Precambrian. A conjugate set of fractures present in the southern part of the inlier affects Cambrian-Ordovician cover rocks.

Isotopic ages are known for some of the granites and metamorphic rocks. A Rb-Sr isochron age of 1130 ± 90 Ma was determined for a megacrystic granite pluton (Pringle et al., 1971). Recent work (H. Baadsgaard, pers. comm., 1990) shows that basement gneisses crystallized at or about 1550 Ma. A voluminous suite of Grenvillian granites was intruded about 1056 ± 4 Ma; an age which matches zircon U-Pb ages from correlative basement rocks of the Cape Breton Highlands. Minerals from a few Grenvillian granites give concordant U-Pb ages of 970 ± 4 Ma, whereas sphene from granites and gneisses yields a Pb-Pb age of 978 ± 10 Ma, which is taken as the age of a separate metamorphism.

K-Ar ages of 960 ± 65 Ma, 840 ± 20 Ma, and 945 ± 65 Ma for biotite and hornblende in megacrystic and massive granite of the western part of the inlier (Pringle et al., 1971; Lowdon et al., 1963) date cooling and Precambrian uplift. The Long Range mafic dykes are dated by ^{40}Ar-^{39}Ar methods at 605 ± 10 Ma (Stukas and Reynolds, 1974a). A related mafic dyke about 300 km to the north at Sandwich Bay of Labrador yields a U-Pb zircon and baddeleyite age of 615 ± 2 Ma (Kamo et al., 1989).

In the eastern part of the inlier, K-Ar ages of 903 ± 37 Ma and 903 ± 38 Ma for hornblende in metagabbro, 843 ± 24 Ma for muscovite in pegmatite, and 434 ± 18 Ma and 512 ± 20 Ma for biotite in the gneiss complex (Wanless et al., 1973) record Precambrian cooling and Paleozoic resetting.

Paleozoic deformation has had limited effect on Precambrian rocks of the Long Range Inlier, except for its eastern parts. At its southeast exposure on the Bonne Bay Highway, a northeast-trending schistosity with chlorite and biotite growth is present in basement gneisses and it affects mafic dykes that cut the gneisses. A parallel schistosity occurs in isoclinally folded Paleozoic rocks that are detached from their basement in this locality (Nyman et al., 1984). Similarly at Sugarloaf Cove at the northeast end of the inlier, a basement-cover unconformity is repeated by westward thrusting and the basement rocks display a granulation fabric produced by cold reworking (Williams and Smyth, 1983).

Along Main River near White Bay, the Devils Room Granite (Smyth and Schillereff, 1982) occurs at the eastern margin of the inlier. It cuts all fabrics in adjacent gneisses and is dated by U-Pb zircon at 391 ± 3 Ma (H. Baadsgaard, pers. comm., 1990). The Gull Lake pluton that cuts Silurian rocks immediately to the east is dated at 372 ± 10 Ma. Gold mineralization in both the Precambrian basement rocks and nearby Silurian rocks may relate to this phase of middle Paleozoic plutonism.

Apatite fission track ages indicate that the present topography of the Long Range Inlier is the result of late Paleozoic to middle Mesozoic uplift (Hendriks et al., 1990). Depending upon the contemporary geothermal gradient, 3-5 km of overburden may have been stripped from the present surface since the late Paleozoic.

Ten Mile Lake and Castors River inliers

These two small inliers of less than 20 km^2 are thin wedges of Precambrian crystalline rocks like those of the Long Range Inlier (Fig. 3.4). The Ten Mile Lake Inlier consists of quartz-feldspar gneisses and minor mafic gneisses. The Castors River Inlier comprises fine- to coarse-grained granite that is locally foliated and contains inclusions of quartz-feldspar gneiss (Knight and Boyce, 1984).

The inliers occur in the coastal lowlands along a northeast-trending fault zone that lies about 10 km west of the Long Range morphological front. The inliers are faulted against Ordovician carbonates to the west and they are unconformably overlain by Cambrian strata along their eastern sides. Their positions therefore seem controlled by a steep reverse fault or frontal thrust near the western limit of Paleozoic deformation in this part of the orogen (Grenier, 1990).

Indian Head Inlier

The Indian Head Inlier consists of two separate massifs with a combined area of 80 km^2 in the vicinity of Stephenville (Fig. 3.5). Its rocks are assigned to the Indian Head Complex (Williams, 1985a). Early studies of the Precambrian rocks were initiated because of their magnetite deposits, discovered before 1920 and mined between 1941 and 1943 (Heyl and Ronan, 1954; Colman-Sadd, 1969). Riley (1962) and Williams (1985a) described the rocks in regional studies of the Stephenville area with subsequent structural studies by Cawood and Williams (1988) and Williams and Cawood (1989). The basement massifs form elongate northeast-trending domes, each plunging beneath an unconformable cover of mildly deformed Lower Cambrian strata at their northern ends (Fig. 3.5). The sides of the domes are faulted against deformed Cambrian-Ordovician rocks or undeformed Carboniferous rocks. A high-angle fault between basement rocks and Carboniferous strata along the west margin of the complex is downthrown on its western side. Faults that separate basement rocks and Paleozoic rocks of the Humber Arm Allochthon along the eastern side of the inlier are older and downthrown on their eastern sides (Williams, 1985a). An unconformity with Carboniferous rocks of the Bay St. George Basin (Knight, 1982; Hyde, Chapter 5) is exposed at the southeast edge of the southern massif.

Metamorphic rocks of the Indian Head Complex are almost all of plutonic protolith. Coarse grained white anorthosite and banded hypersthene gabbro are conspicuous in coastal headlands but less common inland. These are intruded by massive to foliated pink granites that are well-exposed in roadcuts near Stephenville. Disseminated to massive magnetite occurs as layers in banded gabbro and mafic gneisses. Distinctive quartz-rich gneisses locally resemble quartzites.

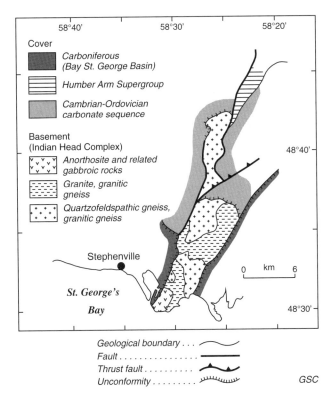

Figure 3.5. Geology and relationships of the Indian Head Inlier. Symbols as in Figure 3.4.

The northern massif consists mainly of grey to pink hornblende granodiorite, granite and quartz diorite. Intrusive relationships between mafic rock types of the Indian Head Complex were reported by Heyl and Ronan (1954) and Riley (1962). Cumulate textures in banded gabbro and anorthositic gabbro are preserved at Indian Head. Massive granite cuts banding in metagabbros and mafic gneisses, and foliated granite is cut by exceptionally coarse-grained pegmatite exposed in a roadcut at Gull Pond. A northeast-trending mafic dyke cuts banded gabbro and anorthosite at Indian Head, and Heyl and Ronan (1954) reported mafic dykes cutting all other rock types. The mafic dykes contain primary pyroxene and plagioclase, partially retrograded to greenschist mineral assemblages (Martineau, 1980). Dykes are absent in surrounding Paleozoic cover rocks.

Metamorphism in mafic gneisses reaches amphibolite and granulite facies. Biotite-hornblende-plagioclase gneisses are common and two-pyroxene plagioclase gneisses occur locally. Sapherine occurs at one locality (Richard Herd, pers. comm., 1988). The high-grade rocks are weakly retrograded to greenschist assemblages and epidote is common on joint surfaces. Greenschist retrogression is probably late Precambrian as nearby Cambrian fossiliferous shales and limestones are only mildly deformed and not metamorphosed.

Banding and gneissic foliation in rocks of the southern massif trend northwest, perpendicular to the length of the inlier and to Paleozoic structures (Williams, 1985a). Heyl and Ronan (1954) interpreted relationships at the type locality of Indian Head as an anorthosite core surrounded by gabbro.

Hornblende and biotite in samples from the Stephenville area yield undisturbed $^{40}Ar/^{39}Ar$ spectra with ages of 880 Ma and 825 Ma, respectively (Dallmeyer, 1978). These agree with K-Ar ages of 830 ± 42 Ma (Lowdon, 1961) for biotite in granitic gneiss, and 900 ± 45 Ma for biotite in a pegmatite dyke (Lowdon et al., 1963)

Plutonic rocks of the Indian Head Complex and later granites are correlated with similar rocks in the Long Range Inlier and with rocks in the Steel Mountain Inlier to the east. Mafic dykes are correlated with the Long Range Swarm. As in western parts of the Long Range Inlier, a lack of metamorphism and significant deformation in Paleozoic cover rocks, and ^{40}Ar-^{39}Ar plateau ages imply little thermal disturbance and retrogression during Paleozoic tectonism.

The Indian Head Inlier has a structural setting to the south of the Long Range Inlier like that of the Belle Isle Inlier to the north. Both form structural culminations equidistant from opposite ends of the Long Range Inlier and separated by depressions occupied by the Hare Bay Allochthon in the north and the Humber Arm Allochthon in the south (Fig. 3.2).

Humber Zone internal, Newfoundland

East Pond Metamorphic Suite

The East Pond Metamorphic Suite (Hibbard, 1983) consists of polydeformed schists and gneisses with an area of 250 km² on the Baie Verte Peninsula (Fig. 3.6). The suite is surrounded by metaclastic rocks of the upper Precambrian-lower Paleozoic Fleur de Lys Supergroup and lies just west of the Baie Verte-Brompton Line. Intense Paleozoic deformation affects both the East Pond Metamorphic Suite and surrounding Fleur de Lys Supergroup, and toward the south these rocks are cut by the Devonian Wild Cove Pond

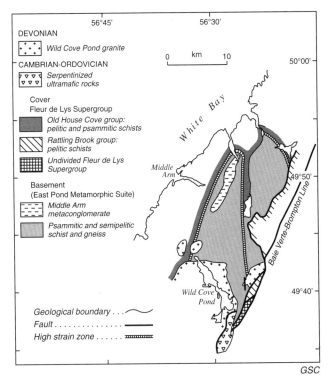

Figure 3.6. Geology and relationships of the East Pond Metamorphic Suite.

granite. At least parts of the East Pond Metamorphic Suite are interpreted as Grenville basement beneath the Fleur de Lys Supergroup (Neale and Nash, 1963; Church, 1969; de Wit, 1972, 1974, 1980; Kidd, 1974; Kennedy, 1975a, b; Williams and Hibbard, 1977; Kennedy, 1980; Hibbard, 1983; Piasecki, 1987; Hibbard et al., this chapter).

Four lithic assemblages are recognized: (1) migmatitic and banded gneiss, (2) metamorphosed conglomerate (the Middle Arm metaconglomerate, Hibbard, 1983), (3) psammitic and semipelitic schists, and (4) granitic gneiss. The psammitic and semipelitic schists make up more than 80 per cent of the area, with most of the remainder underlain by metaconglomerate. The gneisses and migmatites thus represent only a small portion of the metamorphic suite.

The banded gneisses are of granitic to granodioritic composition and are spatially associated with migmatites. Layers less than a centimetre wide are defined by biotite and muscovite concentrations. The Middle Arm metaconglomerate is locally well-bedded with rounded clasts of quartzite, psammite, pelite, schist, granite, gneiss, migmatite and amphibolite. In other places it is highly strained with elongate clasts. Fine grained psammitic and semipelitic schists are interlayered with quartzite, pelite, calc-silicate, and pebbly metaconglomerate. Two-mica granitic gneiss, surrounded by metaconglomerate, is exposed over an area of less than 100 m². Amphibolitic dykes, sills, and pods occur throughout the metamorphic suite.

Contacts between units of the East Pond Metamorphic Suite and surrounding rocks are largely gradational as a result of intense Paleozoic deformation and metamorphism. De Wit (1972, 1974, 1980) interpreted the migmatites, gneisses and schists as remobilized basement unconformably beneath the Middle Arm metaconglomerate. Hibbard (1983) considered only the small patches of gneiss as basement, and schists and metaconglomerates as a cover sequence. Some boulders in the metaconglomerate are clearly derived from a foliated gneissic source.

Complex gneissic banding and migmatitic banding with intrafolial fold interference patterns are probably Precambrian structures preserved in local areas of the East Pond Metamorphic Suite. Metamorphic textures also contrast with those of the surrounding Fleur de Lys Supergroup. In the latter, porphyroblasts are large and widely distributed, indicating a dominantly prograde metamorphic peak. In the former, the rocks are marked by complex microscopic intergrowths and chaotic recrystallization textures that suggest textural disequilibrium and retrogression from earlier higher grade assemblages. In other places, earlier structures are completely obliterated by Paleozoic effects.

Peak Paleozoic metamorphism in both the East Pond Metamorphic Suite and surrounding Fleur de Lys Supergroup was in the amphibolite facies with growth of a wide range of minerals including andalusite, kyanite, staurolite, oligoclase, garnet and pyroxene. Intensity of metamorphism decreases eastward to greenschist facies in the vicinity of the Baie Verte-Brompton Line. Eclogite assemblages are present in central areas of amphibolite pods in the East Pond Metamorphic Suite whereas only amphibolite assemblages are present in mafic pods and dykes in the surrounding Fleur De Lys Supergroup. Contrasts between eclogite and amphibolite assemblages are interpreted to reflect anhydrous conditions in basement rocks and hydrous conditions in the cover (de Wit and Strong, 1975).

Regional metamorphism in the Fleur de Lys Supergroup is of Taconic age and ^{40}Ar-^{39}Ar studies indicate that temperatures did not drop below 300°C until Middle Silurian time (Dallmeyer, 1977a; Dallmeyer and Hibbard, 1984). Intrusion of the Wild Cove Pond granite imprinted a narrow hornfels aureole on the East Pond Metamorphic Suite and Fleur de Lys Supergroup. The aureole contains potash feldspar, andalusite and sillimanite in pelitic rocks, biotite and hornblende in mafic rocks, and diopside in marble.

Banded gneisses and migmatites of the East Pond Metamorphic Suite are probably reworked Grenville basement. The Middle Arm metaconglomerate may be a correlative of the oldest cover rocks on the Belle Isle Inlier (Bateau Formation), and psammitic and pelitic schists possibly represent parts of a cover sequence (Hibbard, 1983). Amphibolite pods, dykes and sills may represent metamorphic equivalents of the Long Range Swarm.

Cobble Cove gneiss

A small anticlinal core of strongly foliated pink gneisses occurs on the west side of Glover Island, Grand Lake (Fig. 3.2). The rocks are referred to informally as the Cobble Cove gneiss and interpreted as basement to a cover sequence of metasedimentary rocks, capped by an ophiolite complex (Knapp, 1982). These rocks lie east of the Cabot Fault, which separates Glover Island from Paleozoic metaclastic rocks of the internal Humber Zone to the west.

The Cobble Cove gneiss is a medium grained homogeneous pink granitic orthogneiss with dark, thin, steeply dipping bands of biotite-microcline schist. Gneissic banding in the Cobble Cove gneiss is truncated at acute angles by the dark, biotite-rich bands. The latter display a faint schistosity subparallel to their margins. Dark bands are absent in the adjacent cover rocks. The contact between the Cobble Cove gneiss and metasedimentary cover rocks is interpreted as a faulted unconformity (Knapp, 1982).

The Cobble Cove gneiss is interpreted as Grenville basement with the biotite-rich bands representing mafic dykes that were affected by strong potash metasomatism during Paleozoic metamorphism (Knapp, 1982).

Steel Mountain Inlier

Crystalline Precambrian rocks, partly reworked by Paleozoic deformation, occupy an area of 1200 km² between the southwest end of Grand Lake and Fischells Brook (Fig. 3.7). The rocks are mainly quartz-feldspar gneisses and granites in the north and anorthosites in the south. In northern localities the rocks are referred to the Long Range Complex (Martineau, 1980), Tonalitic gneiss complex (Kennedy, 1981), and Disappointment Hill and Steel Mountain complexes (Currie, 1987a). In the south the rocks are known as the Steel Mountain anorthosite. All are part of the Steel Mountain Inlier. The Precambrian rocks are thrust above Paleozoic carbonate rocks at Grand Lake and are overlain unconformably by Carboniferous strata farther south. The contact of the Precambrian rocks with Paleozoic metamorphic and intrusive rocks to the east is marked by the Cabot Fault. Two smaller inliers to the southwest form anticlinal cores beneath Carboniferous rocks. These are the Mount Howley and Journois Brook inliers (Fig. 3.7).

The southwest portion of the inlier at Steel Mountain consists of coarse grained grey to pale purple anorthosite, gabbroic anorthosite, norite and minor pyroxenite. Pods of magnetite and ilmenite are abundant in places and are of economic interest. The northern part of the inlier consists of granitic, granodioritic, and amphibolitic gneisses with crosscutting bodies of granite and syenite (Knapp et al., 1979; Martineau, 1980; Currie, 1987a). The anorthositic and quartz-feldspar gneisses of the inlier are separated by steep faults. Small areas of marble and quartzite near Grand Lake are of uncertain age and unknown relationship to surrounding Precambrian rocks. Some are part of the Precambrian complex (Piasecki, 1991), others may be Paleozoic structural enclaves. The Mount Howley and Journois Brook satellite inliers consist of anorthosite similar to that at Steel Mountain.

Diabase dykes cut anorthosites and granitic gneisses. Massive to mildly foliated granites and syenites in the northern portion of the inlier are mainly fault-bounded and of unknown relationship to surrounding rocks. One example at Hare Hill, interpreted to cut surrounding metamorphic and plutonic rocks, is dated isotopically at 617 ± 8 Ma (van Berkel and Currie, 1988). This indicates that intense deformation and metamorphism in western parts of the inlier are Precambrian. A long narrow unit of metasedimentary rocks that extends from Grand Lake to Bottom Brook in the central to eastern part of the inlier is interpreted as cover to nearby gneisses (Currie, 1987a). Arkosic, reddish conglomerates with white quartz pebbles and cobbles in this unit resemble the basal Bradore Formation of the Strait of Belle Isle area.

The granitic gneisses and schists contain amphibolite- to granulite-facies mineral assemblages. These are affected by a pervasive retrograde metamorphism that affects the mafic dykes and increases in intensity from greenschist facies in the northwest to amphibolite facies in the southeast (Martineau, 1980). Paleozoic rocks north of the inlier exhibit a corresponding southeastward prograde metamorphism.

Along the Burgeo Road at Southwest Brook, anorthosite in the west is mildly foliated with internal fabrics trending northwest. The rocks are progressively more deformed eastward with strong northeast fabrics developed in banded mylonitic anorthosite and amphibolite (Williams, 1985a). Mafic dykes with northwest or variable orientations in the west have a northeastern alignment in the east. Folds with a strong northeast-trending axial planar fabric affect quartz-feldspar gneisses and produce a foliation in amphibolite facies mafic dykes.

Anorthosites of the Steel Mountain Inlier are correlated with similar rocks in the Indian Head Complex and both contain magnetite and titaniferous magnetite deposits. Relationships in the Long Range Inlier suggest that the high-grade granitic gneisses in the northeast portion of the Steel Mountain Inlier are older than nearby anorthosites. West of the Long Range Fault, salic granulite has an upper intercept U-Pb zircon age of 1498 ± 4 Ma and foliated gabbro associated with anorthosite has an upper intercept age of 1254 ± 14 Ma (Currie et al., 1989). Mafic dykes that cut the anorthosites and granitic gneisses are equated with the Long Range Swarm. A K-Ar isotopic age of 452 ± 20 Ma (Wanless et al., 1965) for metamorphic biotite of the Steel Mountain Complex is interpreted to date Paleozoic reworking. The Cabot Fault that bounds the Steel Mountain Inlier along its eastern side is a late rectilinear fault that cuts Carboniferous rocks.

Cormacks Lake complex

The Cormacks Lake complex (Herd and Dunning, 1979) is part of an igneous and metamorphic zone southeast of the Cabot Fault and about 40 km southwest of Grand Lake (Fig. 3.8). It occupies an area of 450 km² and consists of schists and gneisses with poorly known relationships to surrounding meta-igneous rocks.

Two lithic groups are distinguished: (a) biotite-quartz-feldspar orthogneiss and biotite-magnesium amphibole orthogneiss, and (b) pelitic and magnesium-rich schist, amphibolite, calc-silicate gneiss, and quartzite. All rock types appear to be structurally interleaved. The complex was affected by Paleozoic regional metamorphism with growth of biotite, cordierite, sillimanite, gedrite, hornblende, actinolite, garnet, and diopside (Herd and Dunning, 1979).

The Cormacks Lake complex displays at least three phases of folding with early isoclines overprinted by more open folds. The dominant trend of fold axes is northwest, like Grenville structural trends in the nearby Steel Mountain

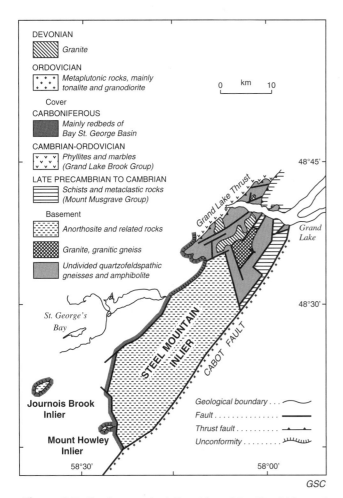

DEVONIAN
  Granite

ORDOVICIAN
  Metaplutonic rocks, mainly tonalite and granodiorite

Cover
CARBONIFEROUS
  Mainly redbeds of Bay St. George Basin

CAMBRIAN-ORDOVICIAN
  Phyllites and marbles (Grand Lake Brook Group)

LATE PRECAMBRIAN TO CAMBRIAN
  Schists and metaclastic rocks (Mount Musgrave Group)

Basement
  Anorthosite and related rocks
  Granite, granitic gneiss
  Undivided quartzofeldspathic gneisses and amphibolite

0   km   10

Grand Lake Thrust

48°45'

Grand Lake

48°30'

St. George's Bay

STEEL MOUNTAIN INLIER

CABOT FAULT

Journois Brook Inlier

Mount Howley Inlier

Geological boundary . . .
Fault . . . .
Thrust fault . . . .
Unconformity . . . .

58°30'   58°00'

GSC

Figure 3.7. Geology and relationships of the Steel Mountain Inlier and nearby Mount Howley and Journois Brook inliers.

and Indian Head inliers, and therefore suggesting a Precambrian age for the complex. It is also possible that all of the rocks are Paleozoic and represent a mixed metavolcanic-metasedimentary assemblage (van Berkel and Currie, 1988; see also the Dashwoods Subzone, Williams, this chapter). Small mafic-ultramafic complexes throughout the general area may represent ophiolite suites that are structurally comingled with all other rocks (Fox and van Berkel, 1988). The Cormacks Lake complex lies at a constriction in the Newfoundland Appalachians where structures are expectedly complicated and relationships are difficult to decipher.

Humber Zone internal, Nova Scotia

The northwestern Cape Breton Highlands of Nova Scotia are composed of a distinctive assemblage of basement rocks including felsic and mafic gneisses, monzodiorite, anorthosite, and syenite (Barr et al., 1987a; Barr and Raeside, 1989). These rocks, the Blair River Complex (Raeside et al., 1986), are separated by major fault zones from gneisses and schists on the south and east (Fig. 3.9). The Blair River Complex has an irregular pattern of concentric to northerly trending magnetic anomalies that contrasts with the relatively low magnetic relief and regular anomaly field of crystalline units to the east. Syenite from the complex has a metamorphic age of 1040 +10/-40 Ma and an igneous age of 1500-1100 Ma (Barr et al., 1987b).

The dominant lithology is quartzo-feldspathic gneiss with mild to strong foliation and local zones of intense shearing. Amphibolite, granitic gneiss, and minor calcareous rocks are associated with the quartzo-feldspathic gneisses. Small bodies of gabbroic to granitic rocks, syenite, and anorthosite are intimately mixed with the gneisses. In other places, anorthosite cuts the gneisses and is in turn cut by syenite. Two metamorphic events appear to have affected the Blair River Complex. An early high-grade

metamorphism, resulting in assemblages characteristic of upper amphibolite to granulite facies, is overprinted by a greenschist to lower amphibolite facies metamorphism. Foliation patterns in the Blair River Complex are extremely varied and do not conform to Appalachian trends.

The isotopic age, unusual association of lithologies, magnetic expression, and tectonic boundaries all suggest that the Blair River Complex is an older and distinct unit from adjacent metamorphic rocks. It is interpreted as Grenville basement and correlated with the Long Range and Indian Head complexes of western Newfoundland. The rocks are assigned to the internal domain of the Humber Zone because they are intruded by Paleozoic granites and reworked by Paleozoic deformation. Another interpretation has the Blair River Complex as basement to a composite Avalon Zone in Cape Breton Island (Murphy et al., 1989).

Humber Zone internal, Quebec
Port-Daniel Inlier

A wedge of orthogneiss and amphibolite 250 m wide occurs in a major fault zone bordering the Maquereau Dome of Gaspésie (De Broucker, 1984, Fig. 3.2). Along its southern and southeastern margins, the metamorphic wedge is juxtaposed with brecciated rocks of the Ordovician Mictaw Group and the Silurian Chaleurs Group. Contacts are steep faults of uncertain age and displacement. The crystalline rocks of the Port-Daniel Inlier are unconformably overlain by the upper Precambrian-lower Paleozoic Maquereau Group along their northwestern margin (De Broucker, 1984).

The main rock type is medium- to coarse-grained orthogneiss that intrudes finely banded amphibolites and contains amphibolite inclusions. The amphibolites were regionally metamorphosed before the intrusion of the orthogneiss, and the orthogneiss has a later cataclastic mortar texture. Foliation attitudes are variable within the Precambrian rocks and do not conform with those in surrounding rocks.

The rocks are more metamorphosed and deformed than the overlying Maquereau Group and zircons from the orthogneiss yield an isotopic age of 998 ± 26 Ma (De Broucker et al., 1986). The rocks are therefore interpreted as a window of Grenville basement.

Another example of crystalline rocks, which are out of structural and metamorphic context compared to surrounding rocks, occurs in ophiolitic mélange 3 km northwest of the Maquereau Dome. These rocks are tuffaceous metasandstones and amphibolites. Zircons are of several populations with ages of 1400 Ma, 1519 Ma, and 1812 Ma. The ages indicate derivation from a Precambrian source other than the nearby Grenville basement. The lithologies and zircon ages suggest correlation with the Chain Lakes Massif at the Quebec-Maine border.

Sainte-Marguerite Fault Slice

An assemblage of basement rocks called the Sainte-Marguerite Complex (Vallières et al., 1978) underlies an area of less than 2 km² in three faulted slivers near Saint-Malachie, 30 km southeast of Québec City. The slivers are up to 3 km long and 0.5 km wide. They are exposed within the Richardson Fault Zone, which forms the boundary between the internal

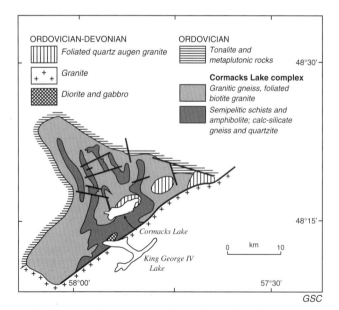

Figure 3.8. Geology and relationships of the Cormacks Lake complex.

and external divisions of the Humber Zone in this area (Fig. 3.10). The crystalline rocks structurally overlie Lower Ordovician clastic rocks of the Sainte-Hénédine Nappe in the west and are structurally overlain by Cambrian parautochthonous strata of the Armagh Group in the east. The crystalline rocks are interpreted as basement to a nearby fault-bounded sequence of spilitic volcanic rocks and interbedded sandstones of the Montagne du Saint-Anselme Formation (Vallières et al., 1978).

The largest sliver includes granoblastic gneissic biotite granite interlayered with thin bands of amphibolite and granitic pegmatite. The amphibolites are compositionally banded and fabrics in the amphibolite and gneissic granite are truncated by fine grained mafic dykes, less than a metre thick, that do not penetrate younger rocks. A smaller near-by sliver contains similar gneissic granite and pegmatite, and a third sliver contains massive metaquartzite with sutured grains and strained fabrics.

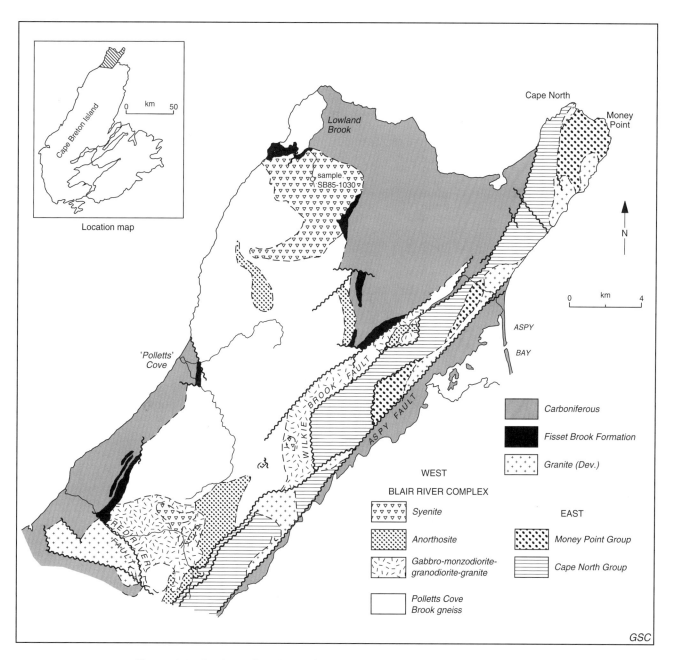

Figure 3.9. Geology of northwest Cape Breton Island, after Barr et al. (1987a).

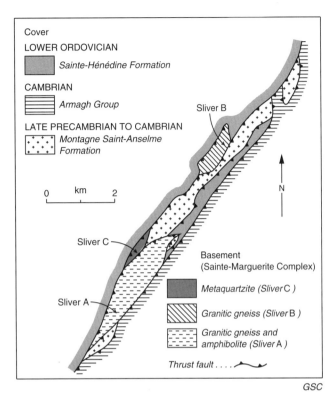

Figure 3.10. Geology and relationships of the Sainte-Marguerite fault slice.

Legend within figure:

Cover

LOWER ORDOVICIAN
Sainte-Hénédine Formation

CAMBRIAN
Armagh Group

LATE PRECAMBRIAN TO CAMBRIAN
Montagne Saint-Anselme Formation

Sliver B

0   km   2

N

Sliver C

Basement (Sainte-Marguerite Complex)

Metaquartzite (Sliver C)

Granitic gneiss (Sliver B)

Granitic gneiss and amphibolite (Sliver A)

Sliver A

Thrust fault . . . .

GSC

Hornblende compositions in amphibolites suggest equilibration under granulite facies conditions before retrogression (Vallières et al, 1978). Later retrograde minerals are epidote, chlorite and sericite.

Biotite from gneissic granite yields a Rb-Sr age of 954 ± 40 Ma, and hornblende from amphibolite yields ages of 900 Ma to 855 Ma (Vallières et al., 1978). These are minimum ages that date retrograde metamorphism and cooling. The Sainte-Marguerite Complex is therefore regarded as Grenville basement and correlated with occurrences elsewhere. Mafic dykes are correlated with late Precambrian equivalents in Newfoundland.

Sainte-Hélène-de-Chester Fault Slice

About 6 km east of Sainte-Hélène-de-Chester and 20 km west of Thetford Mines, a sliver of high-grade crystalline rocks 5 km long occurs within the Lower Paleozoic Bennett Schists of the Notre-Dame Mountains Anticlinorium. The rocks are exposed along a thrust fault a few kilometres northwest of the Baie Verte-Brompton Line (Fig. 3.2).

Banded amphibolite and garnetiferous amphibolite are the main rock types (Caron, 1982). Tectonic banding and metamorphic grade in the amphibolites are more intense and therefore out of context with those of surrounding Paleozoic rocks. A Grenville age for the amphibolites is also suggested by their similarities with rocks of the Sainte-Marguerite Complex.

Summary and discussion

All basement inliers and structural slices of Precambrian rocks in the Humber Zone, with the single exception of an exotic crystalline block in mélange at Port-Daniel, are similar to rocks of the Grenville Structural Province of the nearby Canadian Shield. The rocks are mainly of igneous protolith with minor occurrences of marble, quartzite and metaclastic sedimentary rocks. The age of the oldest rocks is unknown, other than that they were crystallized at or before 1550 Ma and metamorphosed as early as 1230 Ma. The emplacement of mafic intrusions accompanied or followed the earliest recognized phase of regional metamorphism and the deformed rocks are cut by granites dated at about 1000 Ma. Regional metamorphism is mainly of amphibolite facies and locally reaches granulite facies. This implies deep erosion of Precambrian rocks before deposition of late Precambrian-early Paleozoic cover sequences.

Mafic dykes cut the basement rocks in almost all inliers and they feed basalts that unconformably overlie the basement rocks in Newfoundland. This igneous activity is dated isotopically as latest Precambrian and stratigraphic relationships provide an upper age limit of Early Cambrian. Where best developed in the northeastern part of the Long Range Inlier, the dykes decrease in width and frequency from southeast to northwest. Relationships in the Belle Isle Inlier indicate that late Precambrian faulting affected coarse clastic cover rocks before the emplacement of the mafic dyke swarm. This implies rift-related deformation. Late Precambrian retrogression may relate to this extensional tectonism.

The present positions of the inliers are controlled by Paleozoic structures so that their dimensional orientations are parallel to Appalachian structural trends and facies belts. Internal fabrics and fold axes trend northwest in some examples, perpendicular to Appalachian structures. As northwesterly trends are common in the Grenville Structural Province of the Canadian Shield, this suggests minimal or no rotation during Paleozoic deformations.

Grenville inliers of the external Humber Zone are unaffected by Paleozoic deformation except for faulting and gentle folding. Retrograde metamorphism increases from west to east across the width of the Long Range Inlier, and cover rocks are intensely deformed and detached from their basement in the east. Some thrusts affect only the cover rocks. Others are deep shears that bring the basement to the surface. Major detachment zones must therefore lie deep within the basement.

Grenville rocks in the internal Humber Zone are deformed with their cover into tight isoclines and later open folds. Precambrian structures are for the most part obliterated, except in inliers like Steel Mountain and the Sainte-Marguerite Fault Slice that lie near the internal-external Humber Zone boundary. Where Paleozoic metamorphism reaches amphibolite and eclogite facies, the distinction between basement and cover is difficult. Thus the recognition of some Grenville rocks in the internal Humber Zone is relatively new, and doubts exist whether examples such as the Cormacks Lake complex are Grenville basement or high grade Paleozoic rocks.

The eastern limit of surface exposures of defined Precambrian basement is the Baie Verte-Brompton Line, but seismic reflection studies indicate that Grenville basement extends well east of this surface boundary (Keen et al., 1986; Marillier et al., 1989). Indirect geological reasoning and the presence of large granite batholiths support the geophysical data and indicate a crystalline basement beneath the entire Dunnage Zone. Plutonism and isotopic data in southwestern Newfoundland imply that Grenville basement extends in subsurface to the Cape Ray Fault Zone, and possibly beyond (Wilton, 1985). The Ordovician Coaker Porphyry, in the centre of the widest part of the Newfoundland Dunnage Zone is of continental affinity, although not necessarily Grenvillian (Lorenz, 1984a), and nearby Mesozoic plutons contain gneissic inclusions (Greenough et al., 1990). The absence of Dunnage Zone rocks in southwest Newfoundland and Cape Breton Island implies an exposed suture between the Humber Zone and metamorphic rocks to the east (Gander Zone and equivalents).

Upper Precambrian-Lower Cambrian clastic sedimentary and volcanic rocks

Harold Williams, P.S. Kumarapeli, and I. Knight

Introduction

These rocks overlie Grenville basement and they are overlain by Lower to Upper Cambrian carbonates, shales, and sandstones of the miogeoclinal carbonate sequence. They are mainly terrestrial red arkosic sandstones, conglomerates, columnar basalts, marine quartzites, greywackes, and pillow lavas. Abrupt changes in thickness and stratigraphic order are characteristic and these features contrast with uniform stratigraphy, constant thickness and broad lateral continuity of overlying formations. Extensive swarms of dark mafic dykes are coeval with mafic flows.

The rocks are unfossiliferous, except for trace fossils in some upper units. Isotopic ages of volcanic rocks, mafic dykes and related small anorogenic plutons range between 620 and 550 Ma.

Stratigraphic and intrusive relationships are well-preserved in the Strait of Belle Isle area of southeast Labrador and northwest Newfoundland (Fig. 3.11). Flat-lying sedimentary and volcanic rocks of southeast Labrador lie outside the Appalachian deformed zone. These are virtually continuous and correlate with mildly deformed nearby rocks of the orogen in northwest Newfoundland. Stratigraphic relationships are also preserved in deformed and metamorphosed equivalents in the internal Humber Zone at Hughes Lake, Newfoundland and in the Sutton Mountains Anticlinorium of Quebec.

Redbeds with associated mafic dykes, dated isotopically as late Precambrian (Murthy et al., 1988; Kamo et al., 1989), occur 300 km north of the Appalachian Structural Front at Lake Melville and Sandwich Bay of Labrador

Williams, H., Kumarapeli, P.S., and Knight, I.
1995: Upper Precambrian-Lower Cambrian clastic sedimentary and volcanic rocks (Humber Zone); *in* Chapter 3 of Geology of the Appalachian-Caledonian Orogen in Canada and Greenland, (ed.) H. Williams; Geological Survey of Canada, Geology of Canada, no. 6, p. 61-67 (*also* Geological Society of America, The Geology of North America, v. F-1).

(Gower et al., 1986). In Quebec, the Grenville dyke swarm of similar age (Fahrig and West, 1986) extends 700 km inland beyond the Appalachian Structural Front along the Ottawa Valley. Coeval anorogenic intrusions along the Saguenay Valley, St. Lawrence and Ottawa valleys lie outside the Appalachian Orogen (Kumarapeli, 1985; Fig. 3.11).

Correlative sedimentary and volcanic rocks with mafic dyke swarms are known throughout the miogeocline of the U. S. Appalachians, in the British Isles of Scotland, and in the Sarv Nappe of Sweden (Williams, 1984).

In the following treatment, the rocks are described first for the external and internal Humber Zone of Newfoundland, then for the internal Humber Zone of Quebec. Examples outside the Appalachian deformed zone are discussed later, first for Newfoundland, then Quebec. Intensely deformed and metamorphosed equivalents in eastern parts of the internal Humber Zone, such as at the Baie Verte Peninsula of Newfoundland and Notre-Dame Mountains of Quebec are treated elsewhere under the general heading, Humber Zone internal (Hibbard et al., this chapter). Correlatives in Taconic allochthons of both Newfoundland and Quebec are described with other rocks of the allochthons (Williams, this chapter).

Humber Zone external-Newfoundland

Three formations are recognized among the upper Precambrian-Lower Cambrian clastic sedimentary and volcanic rocks in the Strait of Belle Isle area of western Newfoundland and southeast Labrador (Schuchert and Dunbar, 1934; Williams and Stevens, 1969; Bostock, 1983). A basal unit of conglomerate and quartzite (Bateau Formation) is overlain by mafic volcanic rocks (Lighthouse Cove Formation), in turn overlain by red arkosic sandstones (Bradore Formation). The formations are transgressive northwestward with all units locally resting upon Grenville basement. The Bradore Formation is overlain by fossiliferous shale, siltstone and limestone of the Lower Cambrian Forteau Formation; the basal formation of the succeeding miogeoclinal carbonate sequence. Stratigraphic relations are displayed best at Belle Isle (Fig. 3.3; see also Fig. 3.14 for summary).

Mafic dykes of the Long Range Swarm (Bostock, 1983) are coeval with the Lighthouse Cove Formation (Williams and Stevens, 1969; Strong and Williams, 1972). They form prominent northeast-trending ridges which are the dominent physiographic feature over much of the northeast end of Long Range and Belle Isle inliers (Fig. 3.2).

The Bateau Formation (Williams and Stevens, 1969) lies unconformably above Grenville basement at the eastern side of the Belle Isle Inlier (Fig. 3.3). It consists of plutonic boulder conglomerate, quartzite, arkosic sandstone, siltstone, slate and minor volcanic rocks in the order of 200 m thick. Clasts in the conglomerates are derived from Grenville basement, except for intraformational quartzite clasts. Quartzites occur above the conglomerates in the northeast. At another locality toward the southwest, 15 m of sheared green volcanic rocks occur at the top of the section. A local occurrence of boulder conglomerate at Canada Bay and more than 400 m of white quartz sandstones at White Islands (Fig. 3.11) are correlated with the Bateau Formation (Williams and Smyth, 1983). The sandstones at White Islands are extremely well-sorted with frosted spherical grains suggesting wind-blown deposits.

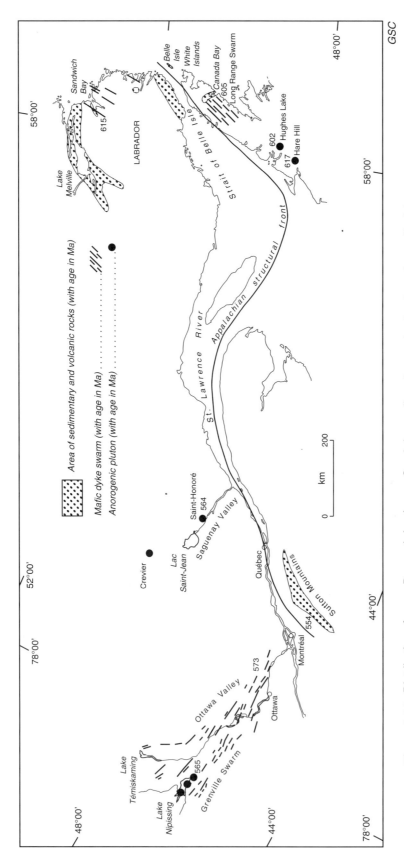

Figure 3.11. Distribution of upper Precambrian-Lower Cambrian sedimentary and volcanic rocks, mafic dyke swarms, and anorogenic plutons in eastern Canada.

Coarse conglomerate is also present beneath mafic volcanic rocks near Canada Bay. Diabase dykes of the Long Range Swarm cut the Bateau Formation.

The Lighthouse Cove Formation (Williams and Stevens, 1969) consists of tholeiitic basaltic flows and pyroclastic rocks that vary in thickness from a few metres in southeast Labrador to more than 300 m at the southeast end of Belle Isle. At its type area of Belle Isle, the volcanic rocks lie directly on Grenville basement and they are overlain by the Bradore Formation, in turn followed by fossiliferous Lower Cambrian strata of the Forteau Formation (Fig. 3.3). Black dykes in a ramifying network cut pink basement gneisses and feed Lighthouse Cove flows in the well exposed sea cliffs at Belle Isle. The dykes do not cut stratigraphically higher rocks, except possibly for a single thin dyke in the Bradore Formation at Canada Bay (Knight and Saltman, 1980).

Volcanic rocks of the Lighthouse Cove Formation are black to dark green and purple to reddish brown basalts and pyroclatic rocks. Dark green flows are most abundant and these vary from massive to columnar and locally pillowed in eastern exposures on Belle Isle. The rocks are fine to medium grained and locally amygdaloidal with calcite, quartz, chlorite, and epidote amygdules. Volcanic breccias with green and purple amygdaloidal basalt fragments occur at the southeast end of Belle Isle.

Volcanic rocks also occur a few kilometres west of Canada Bay (Bostock, 1983). Here, several discontinuous flows are present along the basement unconformity, at most 75 m thick and appearing to thin southeastward. Locally, arkosic sandstones and conglomerates underlie the flows. Arkosic sandstones up to 15 m thick similarly lie between basement and flows at Table Head in southeast Labrador (see Fig. 3.14).

The Bradore Formation (Schuchert and Dunbar, 1934) contains red, brown, green, and black subarkosic and arkosic sandstones, pebbly sandstones and conglomerates, and locally pink and gray shaly, glauconitic, micaceous sandstones, hematitic sandstones, and white to red quartzose sandstones and pebble conglomerates (Knight and Saltman, 1980; Bostock, 1983; Hiscott et al., 1984).

In its type area of southern Labrador, the formation is 70 to 105 m thick. A lower member (Blanc Sablon) of crossbedded, red-brown arkosic sandstone was derived from the west (Hiscott et al., 1984). Large trough and planar cross stratification is common. Similar red sandstones of the overlying Crow Head Member are extensively bioturbated principally by *Skolithus linearis*. The top of the formation is a thin unit of mature, white quartz-rich, coarse sandstone and rusty weathering pebbly sandstone and conglomerate of the L'Anse-au-Clair Member. It has polymodal trough and planar crossbeds and a variety of ichnofossils including *Skolithus, Conichnus, Dolopichnus,* and *?Monocraterion*.

In northern Newfoundland, the Bradore Formation is 90-150 m thick (Cumming, 1983). It consists mostly of crossbedded, red-brown arkosic sandstones and conglomerates with some thin green-grey sandstones and *Skolithus*-bearing quartzose sandstones near the top. In places, a basal unit of crossbedded, brown to black conglomerate, sandstone and red arkosic sandstone, 9 m thick, overlies the basement. Grey to pink, glauconitic and micaceous

sandstones, siltstones and shales, 50 m thick, displaying thin bedding, small-scale, cross lamination and burrows form a middle member below 30 m of red and locally white arkosic and quartzose sandstones. The lithologies of the middle member are extensive around the northern end of the Long Range Inlier (Bostock, 1983) where they lie locally upon basal pockets of coarse conglomerate.

In Canada Bay, about 60 m of sandstone lies between basement and the Forteau Formation. Trough crossbedded red arkoses and pebbly sandstones with a consistent east and northeast paleoflow form the lower 40 m of the formation. Planar bedded, well-sorted quartzose sandstones, wavy bedded, locally bioturbated, hematitic siltstones and sandstones and well-sorted conglomerate and grit with carbonate pebbles are intercalated with arkose in the upper part of the formation.

Thick red sandstones can be traced about 100 km southward. Farther south, redbeds are absent, probably as a result of structural omission or rapid southward thinning at deposition. At the south end of the Long Range Inlier, the formation consists of only 8-13 m of massive thin bedded, black and green-grey and minor red sandstones and shale resting at the unconformity on a zone of broken basement boulders (Nyman et al., 1984).

Immature red arkosic sandstones and conglomerates with easterly directed paleoflows are interpreted as braided fluvial deposits (Knight and Saltman, 1980). In Labrador, the streams passed into estuaries where the sands were intensively burrowed. The sands were finally reworked in a series of migrating tidal inlets of a barrier coastline. In northwest Newfoundland, glauconitic sandstones, quartzose sandstones and hematitic fine siliciclastics are shallow marine deposits (Knight and Saltman, 1980).

A regolith up to 2 m thick occurs between the Bradore Formation and basement in southeast Labrador and at the north end of the Long Range Inlier (Bostock, 1983; Hiscott et al., 1984). No regolith occurs beneath the Bateau or Lighthouse Cove formations, nor is it develped where the Bradore Formation rests upon Lighthouse Cove volcanic rocks.

The surface on which the Bradore Formation accumulated was irregular with large scale undulations and sharp local relief that led to significant thickness variations (Hiscott et al., 1984; Knight and Boyce, 1984). Some northeast-trending faults in the basement coincide with thickness changes in the Bradore Formation. Other contemporary faults at Belle Isle (Bostock, 1983) affected the Bateau Formation but not nearby mafic dykes. The late Precambrian faulting may also account for the patchy, thick deposits of the Lighthouse Cove volcanic rocks near Canada Bay (Bostock, 1983).

Mafic dykes of the Long Range Swarm vary from massive to polygonally jointed to slightly schistose. They vary in width from less than a metre to 100 m and some are traceable for several kilometres. They are steeply-dipping to vertical, trending northeast in the Long Range Inlier and slightly east of north at Belle Isle (Fig. 3.2). Their trend in the Long Range Inlier is parallel to the prominent gravity gradient across the inlier and they are most abundant at the change from negative to positive values. A more northerly trend of the gravity gradient at Belle Isle follows a change in the direction of the dykes. Possibly the locus of dyke injection and the gravity gradient have the same late Precambrian control in the deeper crust (Weaver, 1967; Bostock, 1983).

In the eastern part of the northern Long Range Inlier, dykes are altered and are green to dark green. Near the western margin of the swarm, they are less altered and grey. A similar change in dyke alteration is evident across Belle Isle with the most schistose, green dykes present along its eastern side.

Textures vary from equigranular to plagioclase porphyritic. The least altered dykes are composed of augite and andesine-labradorite with accessory magnetite, ilmenite and quartz. In altered varieties the principal minerals are amphibole, chlorite, epidote, and saussuritized plagioclase. Chemical and petrographic data show that they are typical tholeiites and there is no detectable difference between the Long Range dykes and Lighthouse Cove volcanic rocks of southeast Labrador, Belle Isle, and northeast Newfoundland (Strong and Williams, 1972).

The age of igneous activity is late Precambrian to Early Cambrian based on stratigraphic relationships. The most reliable isotopic age for the Long Range Swarm is 605 ± 10 Ma (Stukas and Reynolds, 1974a).

Humber Zone internal-Newfoundland

At Hughes Lake, a deformed and metamorphosed bimodal volcanic suite and associated granite are overlain by psammitic and pelitic schists (Fig. 3.11, 3.12). The rocks occur in a thrust slice that lies east of clastic rocks of the Old Mans Pond Allochthon (Williams et al., 1985a; Waldron and Milne, 1991). The volcanic and granitic rocks are referred to the informal Hughes Lake complex and their metasedimentay cover rocks are the Mount Musgrave Group. All are contained within the Hughes Lake structural slice. A mylonite zone up to 20 m wide separates the Hughes Lake slice and the Old Mans Pond Allochthon. Metamorphism and structural intensity are contrasted across this boundary between foliated granitic rocks and schistose metavolcanic rocks on the Hughes Lake side and argillites and phyllites on the Old Mans Pond side. The contact dips moderately northwestward, in accordance with northwesterly dips of bedding, cleavage, and schistosity throughout the general area. This inversion of structures is a later feature affecting all the rocks that were first assembled from east to west (Cawood and Williams, 1988). Subhorizontal Carboniferous conglomerates and sandstones unconformably overlie all older rocks west of Deer Lake.

Within the Hughes Lake structural slice there is a stratigraphic progression of units from northwest to southeast (Fig. 3.12). The oldest rocks are granite and volcanic rocks. A narrow continuous mafic flow, informally the Deer Pond volcanics, occurs at the stratigraphic top of the Hughes Lake complex. This is overlain by about a kilometre of arkosic metagreywacke (Little North Pond formation) and a thick sequence of psammitic to pelitic schists (South Brook formation).

Mildly foliated to schistose, pink granite (Round Pond granite) and pink fragmental volcanic rocks contain melanocratic bands of hornblende-biotite-chlorite schist interpreted as original mafic dykes. Amygdaloidal and

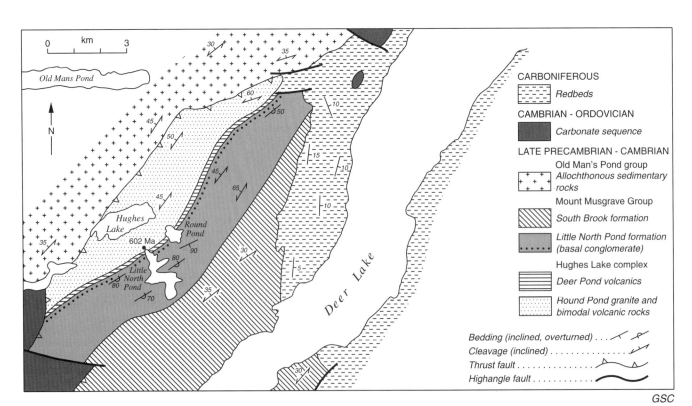

GSC

Figure 3.12. General geology of the Hughes Lake structural slice, after Williams et al. (1985a).

porphyritic dark green volcanic rocks, now biotite-chlorite schists, occur above the granite and pink volcanic rocks, followed by a pebble to cobble conglomerate at the base of the Mount Musgrave Group. The conglomerate and overlying arkosic metagreywackes (Little North Pond formation) form a thick continuous resistant unit that strikes northeast, dips steeply northwest, and everywhere faces southeast based on grading and ubiquitous crossbedding.

The Round Pond granite varies from medium to fine grained and massive to mildly foliated and schistose. Metadiabases and dark bands of biotite hornblende schist within schistose pink granite appear to increase in abundance from northwest to southeast. However, they are absent in the overlying arkosic metagreywackes of the Little North Pond formation. Textural and mineralogical variations across the Round Pond granite, from coarse equigranular and porphyritic varieties with less potash feldspar to medium- and fine-grained potassic varieties with common exsolution intergrowths, imply deeper to shallower crustal levels from northwest to southeast. Common occurrences of perthite and micropegmatite, combined with low modal anorthite and an excess of potassium over sodium, and iron over magnesium, indicate a hypersolvus high-level anorogenic pluton. The silicic volcanic rocks of the Hughes Lake complex have trace-element characteristics typical of rhyolitic end members of alkali igneous suites. Chemical analyses of mafic flows and dykes indicate they are tholeiites similar to the Lighthouse Cove Formation and dykes of the Long Range Swarm.

A relatively undeformed sample from the southeast portion of the Round Pond granite (Fig. 3.12), consisting of delicate perthite and micropegmatite intergrowths, yields a U-Pb zircon age of 602 ± 10 Ma (Williams et al., 1985a). The age of the granite, its hypersolvus textural features, its association with a bimodal volcanic suite, the alkaline affinities of associated silicic volcanic rocks, and the presence of mafic dykes and flows all suggest a consanguineous igneous suite that formed in a rift setting (Williams et al., 1985a).

Volcanic rocks and mafic dykes of the Hughes Lake complex are correlated with the Lighthouse Cove Formation and Long Range Swarm, respectively. Crossbedded arkosic metagreywackes of the Little North Pond formation are correlated with the Bradore Formation, although the Little North Pond is coarser and much thicker. Rocks of the Mount Musgrave Group are also correlated with the Fleur de Lys Supergroup of the internal Humber Zone at Baie Verte. These correlations indicate continuity of rocks from outside the Appalachian deformed zone, across the external Humber Zone, into the internal Humber Zone of the orogen.

The Hare Hill granite, which cuts Grenville basement rocks at the north end of the Steel Mountain Inlier, is dated at 617 ± 8 Ma (van Berkel and Currie, 1988). It has alkali affinities and is correlated with the Round Pond granite of the Hughes Lake complex (Fig. 3.11).

Humber Zone internal-Quebec

The Tibbit Hill volcanics in the Sutton Mountains Anticlinorium of Quebec (Fig. 3.11) are equated with occurrences in Newfoundland and nearby examples in Vermont (Pintson et al., 1985). The volcanic rocks are exposed in anticlinal culminations and they are contained within a thrust slice. The Tibbit Hill volcanic rocks are predominantly altered basalts that are polydeformed and metamorphosed. They are the lowest formation of the Oak Hill Group (Clark, 1934), and they are presumed to lie directly on Grenville basement, although basal relations are not exposed. They are overlain by the Pinnacle Formation in Quebec and correlatives occur within the Pinnacle Formation in Vermont. It is a lithic greywacke, with green sandstone, arkosic sandstone and conglomerate horizons. Rapid changes in thickness and facies are typical.

An early Cambrian age for the middle part of the Oak Hill Group sets an upper stratigraphic age limit for the volcanic rocks. Comenditic metafelsites that form a minor unit in the volcanic sequence yield a reliable U-Pb zircon age of 554 +4/-2 Ma (Kumarapeli et al., 1989).

Greenschist metamorphism is typical, but locally crossitic amphibole is abundant, indicating blueschist-greenschist transitional facies (Trzcienski, 1976).

Two main lithologies are recognized among the volcanic rocks: greenstone and epidosite (epidote-quartz-bearing greenstone). Greenstones constitute over half of the exposed metavolcanic rocks. They are fine grained, massive to schistose, locally amygdaloidal and porphyritic. They consist of albite, chlorite, epidote, crossitic and/or actinolitic amphibole, phengite, sphene, and Fe-Ti oxides (Pintson et al., 1985). Sparse cores of kaersutite within crossite and Fe-Ti oxides are the only primary minerals. Epidosite-bearing greenstones occur in isolated bodies or in large continuous zones. These consist of nodules of epidosite set in a groundmass like that of the greenstones. Contacts between the epidosites and host greenstones are sharp to gradational.

Abundances of immobile elements and enrichment patterns indicate transitional or alkalic basalts of "Within Plate" affinity. This is consistent with the presence of relict kaersutite, despite tholeiitic Nb/Y ratios. Anomalously low abundances of Nb and P suggest contamination by continental crust (Pintson et al., 1985). Rhyolitic rocks at Waterloo are comendites (Kumarapeli et al., 1989).

Gravity and magnetic anomalies associated with the exposed Tibbit Hill volcanic rocks suggest a much more voluminous volcanic mass in subsurface; up to 250 km long, 45 km wide, and 8 km thick (Kumarapeli et al., 1981). It is convex toward the northwest and lies at a structural salient in the Appalachian Orogen.

Related rocks outside the Appalachian deformed zone

Redbeds, mafic dykes, and carbonatite intrusions occur in widely separated areas of the Grenville Structural Province along the northwest flank of the Appalachian Orogen (Fig. 3.11). These are coeval, at least in part, with upper Precambrian to Lower Cambrian redbeds, mafic dykes and volcanic rocks of the Appalachian Humber Zone. Red sandstones and mafic dykes occur in the Lake Melville area and elsewhere in eastern Labrador. An impressive swarm of mafic dykes, the Grenville Swarm, extends inland presumably from the Sutton Mountains along the Ottawa Valley and beyond. Carbonatites and related intrusions occur mostly along the Ottawa and Saguenay valleys. These rocks are described here because of their similarity, synchroneity, geometry, and possible genetic links to rocks of the Appalachian Humber Zone.

The underlined Double Mer Formation (Stevenson, 1967a, b; Gower et al., 1986) is a sequence of red arkosic sandstones, conglomerates, siltstones and shales in eastern Labrador. The formation occurs in a series of faulted basins or graben at Lake Melville and Double Mer that extend inland for at least 300 km from the Labrador coast. Similar redbeds occur in an isolated faulted basin 100 km to the southeast at Sandwich Bay (Gower et al., 1986; Fig. 3.13).

Bedding attitudes in the Lake Melville and Double Mer graben suggest broad open folding with thicknesses of sediment possibly reaching 5 km. Conglomerates at the northern margin of the Double Mer Graben are interpreted as fanglomerates. Coupled with northerly paleocurrents elsewhere, the pattern suggests a half graben with a fault-bounded northern paleo-edge and a north-sloping floor (Gower et al., 1986). Relationships to the south, suggest that the Mealy Mountains formed a contemporary upland at the southern palaeo-edge of the Lake Melville Graben.

The Double Mer Fault at the north margin of the Double Mer Graben is a broad zone of brecciation and brittle fracture. Retrograded mylonitic blocks in the fault zone are thought to be much older and derived from the Grenville basement. The lobate regional pattern of the Double Mer and related faults, marking the Lake Melville and Double Mer graben, mimics the form of north-northwest thrusts in the Grenville Structural Province, suggesting ancestral controls.

The Double Mer Formation is traditionally correlated with the Bradore Formation of southeast Labrador and the Appalachian Orogen. Its depositional graben are parallel to the northeast direction of late Precambrian-Cambrian mafic dykes of southeast Labrador and the Appalachian Orogen (Fig. 3.11). One northeast-trending dyke, which crosses a fault at Sandwich Bay, yields a U-Pb zircon and baddeleyite age of 615 ± 2 Ma (Kamo et al., 1989).

The Grenville Swarm of mafic dykes extends 700 km inland from the Sutton Mountains of Quebec along the morphological depression of the Ottawa Graben (Kay, 1942; Kumarapeli, 1985; Fig. 3.11). The age of the swarm is given as 575 Ma (Fahrig and West, 1986). Its geometric relationship to the Sutton Mountains salient of the Appalachian Orogen suggests a genetic link with the Tibbit Hill volcanic rocks (Rankin, 1976; Kumarapeli, 1985). The Tibbit Hill volcanic rocks, however, are alkalic to transitional, and the dyke rocks are tholeiitic. The mafic dyke-volcanic episode is roughly coeval with the emplacement of alkali complexes, including several carbonatite complexes dated at 565 Ma (Gittins et al., 1967; Doig and Barton, 1968). Examples occur at Lake Nipissing, near the northwest extremity of the Grenville dyke swarm, and in the Saguenay Valley to the northeast (Fig. 3.11).

Coarse reddish and greyish sandstones, feldspathic sandstones, and conglomerates at the base of the Potsdam Group (Covey Hill Formation) in the vicinity of Montreal resemble the Bradore Formation of Newfoundland. Unidirectional trough-crossbedding indicates currents to the east and southeast. Thicknesses vary from a few metres to 600 m. The inferred age of the Covey Hill Formation is late

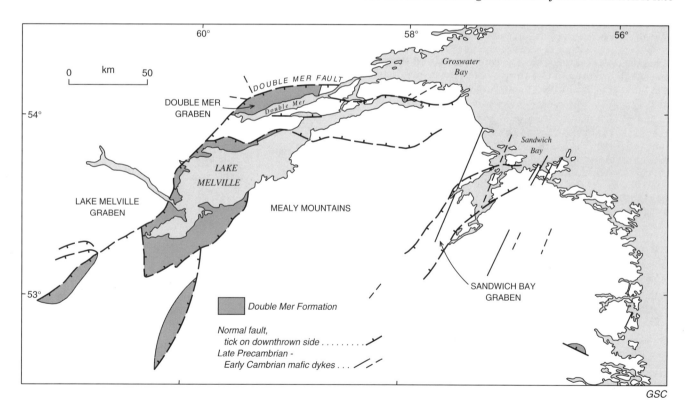

GSC

Figure 3.13. General setting and geometry of the Double Mer Formation and graben in eastern Labrador, after Gower et al. (1986).

Precambrian or early Cambrian. Similar clastic rocks lie on Grenville basement nearby in the Adirondacks Mountains of New York. They are overlain by well-sorted quartz sandstones (Châteauguay Formation) of Late Cambrian age (Globensky, 1987).

Interpretation and significance

Features of the upper Precambrian-Lower Cambrian sedimentary rocks, volcanic rocks, mafic dyke swarms and associated plutons are all consistent with a model of continental rifting and the initiation of the Appalachian cycle (Williams and Stevens, 1969, 1974; Strong and Williams, 1972; Pintson et al., 1985; Kumarapeli, 1985; Williams et al., 1985a; and many others). The rifting affected a broad area that extended inland several hundred kilometres from the present Appalachian Structural Front.

The timing and duration of rifting is defined by isotopic ages of igneous rocks and stratigraphic analysis of overlying fossiliferous strata. Rifting began at least 600 Ma ago and persisted to about 550 Ma ago.

In the Humber Zone, sedimentation was mainly terrestrial or mixed marine and terrestrial. Rapid lateral changes in facies and thickness of sedimentary rocks and rift affinities of coeval flows fed by dykes, imply deposition in rift basins. Faults, contemporaneous with sedimentation and locally predating mafic dyke intrusion, may have been listric normal faults like those so common beneath breakup unconformities at modern continental margins. Marine shales and limestones of the Forteau Formation rest on Bradore redbeds in western Newfoundland and southeast Labrador. An abrupt termination of coarse clastic deposition and absence of coarse detritus in the succeeding Forteau Formation signals a broad submergence and permanent change in the tectonic regime. This is interpreted to mark the transition from continental rifting to ocean spreading (Williams and Hiscott, 1987; Fig. 3.14). The marine transgression and onset of carbonate deposition is regarded as the result of cooling and thermal subsidence, rather than a eustatic rise of sealevel, as it is synchronous with the cessation of rifting.

Sedimentation and mafic dyke intrusion outside the orogen were localized in graben, which are evident today by reactivation of their faulted margins. Deposition was mainly terrestrial, and sedimentologic analyses indicate contemporary faulting at palaeo-basin edges. Widths of the Ottawa Graben and Lake Melville-Double Mer graben, in the order of 50 to 100 km, are typical of modern rift valleys (Kumarapeli and Saull, 1966).

The Ottawa Graben has the geometry of an aulacogen or failed arm of a triple rift system (Kumarapeli, 1985). Its setting at the Sutton Mountains salient is interpreted as an original deep reentrant in the rifted continental margin. The Long Range Swarm, and Lake Melville and Double Mer graben parallel the ancient continental margin, much as Triassic-Jurassic dyke swarms and graben parallel the modern Atlantic margin. The shapes of the Lake Melville and Double Mer graben may have been controlled by older tectonic boundaries. The Cartwright Fracture Zone at the modern Atlantic margin off Labrador may reflect the Lake Melville rift system. Other examples of ancestral controls and micmicry abound.

Cambrian-Ordovician carbonate sequence

I. Knight, N.P. James, and Harold Williams

Introduction

These rocks occur above the rift related rocks of the Humber Zone or they directly overlie Grenville basement. They also occur outside the Appalachian deformed zone in the St. Lawrence Platform. They are mainly limestones and dolomites with lesser siliciclastic rocks that range in age from Early Cambrian to Middle Ordovican. They record the development of a passive continental margin, and the early phases of Taconic Orogeny.

The rocks are preserved best in western Newfoundland throughout the external Humber Zone and nearby (Fig. 3.15). They are mainly hidden by Taconic allochthons along the Quebec segment of the Humber Zone. Small examples occur at Gaspésie and more deformed examples occur in the Sutton Mountains and Lake Champlain area to the west. The rocks occur in the St. Lawrence Lowlands west of Québec City, and they occur at Mingan Islands on the north shore of the St. Lawrence River.

In Newfoundland, the carbonate sequence is divided into the Labrador, Port au Port, St. George, and Table Head groups. In the St. Lawrence Lowlands of Quebec, the succession includes the Potsdam, Beekmantown, Chazy, Blackriver and Trenton groups, and the Romaine and Mingan formations on the north shore of the St. Lawrence River.

Where the rocks occur in thrust slices of the Sutton Mountains and Lake Champlain areas they include the Oak Hill Group, Milton Dolomite, Rock River Formation, and Phillipsburg Group. Cambrian carbonates of Gaspésie are the Corner-of-the-Beach and Murphy Creek formations. Present exposures of the carbonate sequence were part of a continental margin platform that was disposed about the St. Lawrence Promontory and the Newfoundland and Quebec reentrants (Fig. 3.15; Williams, 1978a). The landward margin of the platform is preserved only in Quebec where Upper Cambrian to Middle Ordovician rocks onlap Precambrian basement in the St. Lawrence Lowlands. There, a rolling and faulted paleotopography that was tilted southwards, is onlapped by progressively younger rock units northwards across the St. Lawrence Lowlands (St-Julien et al., 1983; Globensky, 1987). At Montmorency Falls near Québec City, the Middle Ordovician Trenton Group lies directly on Grenville basement. At Mingan Islands and in the subsurface of Anticosti Island, upper Lower Ordovician carbonates overlie Precambrian basement (Roliff, 1968; Shaw, 1980; Desrochers, 1985). The landward edge of Cambrian carbonates in the Quebec Reentrant therefore lay between the Gaspésie and Anticosti Island and southeast of the St Lawrence Lowlands. This implies that a relatively narrow Cambrian shelf was overstepped by Ordovician rocks. The offshore platform edge

Knight, I., James, N.P., and Williams, H.
1995:   Cambrian-Ordovician carbonate sequence (Humber Zone); in Chapter 3 of Geology of the Appalachian-Caledonian Orogen in Canada and Greenland, (ed.) H. Williams; Geological Survey of Canada, Geology of Canada, no. 6, p. 67-87 (also Geological Society of America, The Geology of North America, v. F-1).

was destroyed during Taconic Orogeny. However, coeval rocks of the continental slope are preserved in Taconic allochthons, and some deepwater carbonate breccias are present in the internal Humber Zone (Hibbard, 1983).

The proliferation of both macro- and micro-faunas throughout the carbonate sequence allows fairly accurate dating of the succession. Well-dated, transgressive-regressive mega-cycles are identified in both Cambrian and Lower Ordovician strata in Newfoundland (Chow and James, 1987; Knight and James, 1987; James et al., 1989) and there are sufficient geological data to fit the Quebec Ordovician rocks within this depositional framework.

Carbonate sedimentation occurred in two tectonic settings: (1) trailing passive margin sedimentation from the late Early Cambrian to the late Early Ordovician, and (2) Middle Ordovician carbonate sedimentation along the cratonic edge of the Taconic foreland basin. The Early Cambrian

passive margin history in western Newfoundland is preserved in the upper part of the Labrador Group (Forteau and Hawke Bay formations) and comprises mixed siliciclastic and carbonate marine strata laid down above non-marine to nearshore sandstones (Bradore Formation, early rift facies). Equivalent strata of the Oak Hill Group and Milton Dolomite occur only in thrust slices in Quebec (Globensky, 1978, 1981). Carbonate deposition was established throughout western Newfoundland during the Middle Cambrian and continued to the Early Ordovician. During this period of approximately 70 million years, strata belonging to the Port au Port and St. George groups formed the major part of the 1500 m sequence. In Quebec, carbonate sedimentation was not widespread until the Early Ordovician when the Beekmantown and Phillipsburg groups were deposited above sandstones of the Potsdam Group. Cambrian carbonate sedimentation is recorded in the southern part of the Quebec Reentrant by Middle and

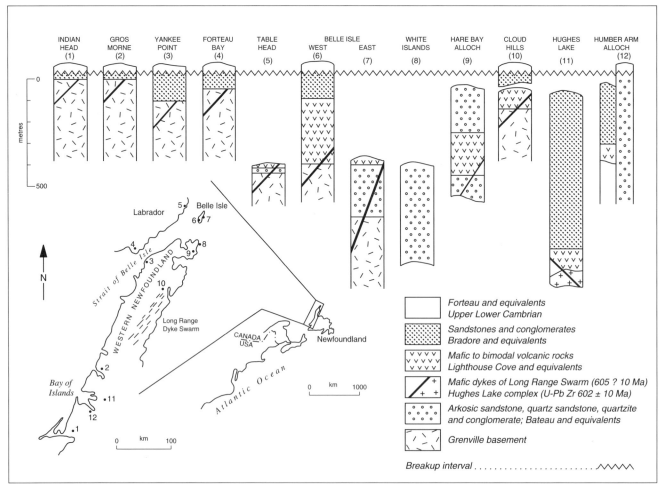

GSC

Figure 3.14. Stratigraphic relations of synrift rocks in western Newfoundland and southeast Labrador, and definition of the Iapetus rift-drift transition, after Williams and Hiscott (1987).

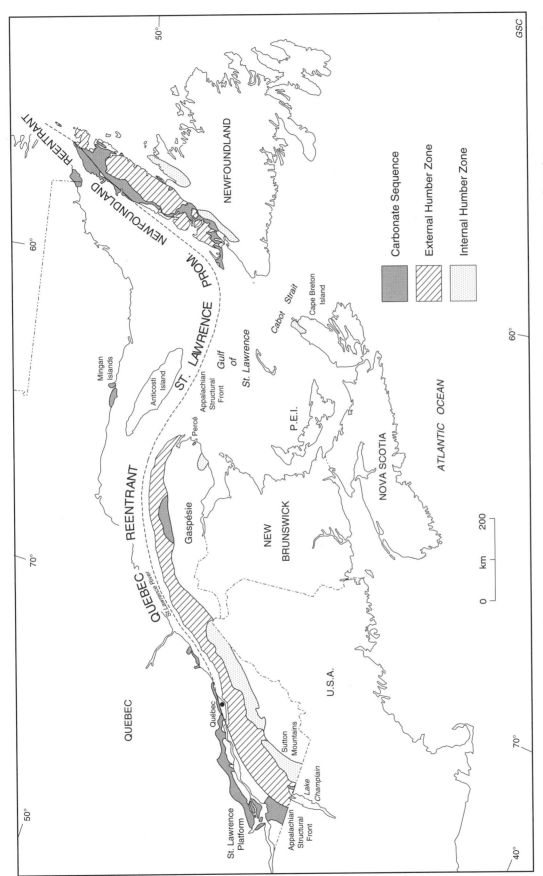

Figure 3.15. Distribution of the Cambrian-Ordovician carbonate sequence in the Canadian Appalachians and nearby St. Lawrence Platform.

Upper Cambrian carbonates near Percé, and shallow water, Cambrian carbonate clasts in allochthonous Ordovician deepwater sediments.

Foreland basin carbonate sedimentation spanned less than 30 million years. In Newfoundland, the Table Head Group (approximately 100 to 300 m thick) was deposited in a few million years. In Quebec, the dominantly carbonate rocks of the Chazy, Black River and Trenton groups (approximately 430 to 700 m thick) accumulated over a period of 20 million years.

The foreland basin carbonates are overlain by Middle Ordovician terrigenous flysch. In Newfoundland, the dominantly silty to shaly flysch with local sandstone-dominated sequences and some resedimented carbonates (Goose Tickle Group, Knight, 1986; Quinn, 1988a, b, c; Stenzel et al., 1990) was deposited during the Llanvirn. In Quebec, the shaly flysch sequences (Utica, Lorraine, and Macasty groups) range in age from late Middle Ordovician (Llandeilo) to early Late Ordovician (Caradoc). They are largely contemporaneous with carbonates of the Trenton Group, and they are thinner and younger to the north (Riva, 1969; Hofmann, 1972; Clark, 1977; Belt et al., 1979; Bussières et al., 1977; Riva et al., 1977; Globensky and Riva, 1982).

Stratigraphy and platform evolution

The stratigraphy of the carbonate sequence and the evolution of the continental margin platform is based on the best exposures of these rocks, and transported sequences of the Cow Head and Curling groups in the overlying Humber Arm Allochthon, in western Newfoundland. The following is summarized from the field guide of James et al. (1988) and companion papers (James et al., 1989; Knight et al., 1991). The distribution of the rock units is shown in Figure 3.16 and their stratigraphy is summarized in Figure 3.17. These rocks are correlated with sequences preserved in Quebec.

The rocks are described in four temporal divisions that correspond to four distinct phases in the evolution of the continental platform: (1) Early to Middle Cambrian (pre-platform shelf); (2) Middle to Late Cambrian (narrow, high energy platform), (3) Early to early Middle Ordovician (wide, low energy platform), and (4) Early Middle Ordovician to Late Ordovician (foundered platform).

Pre-platform shelf

The continental shelf was initiated on rifted and block faulted Grenville basement, intruded by late Precambrian mafic dykes. In Newfoundland, this irregular topography was veneered and buried by a sequence of siliciclastic and carbonate sediments of the Labrador Group (Fig. 3.17). The basal units of this division, the Bateau, Lighthouse Cove, and Bradore formations have been described already (Williams et al., this chapter). The upper part of the Bradore Formation is mineralogically more mature and consists of sands that filled laterally migrating tidal inlets along a barrier coastline, parts of which were removed by subsequent shoreface migration (James and Hiscott, 1982; Hiscott et al., 1984). Eastward, around the northern end of the Long Range Inlier, sandstones were also deposited in fluvial and tidal flat settings (Knight, 1987, 1991). Sediments

south of the Long Range Inlier are thinner green-grey sandstone and thin shale or red, white and grey quartz arenite.

Northeastward paleocurrents around the northern end of the Long Range Inlier and marked thickness variations probably reflect structural topographic control (Knight, 1987). Thickness variations suggest a faulted, northward-tilted upland that was onlapped, initiating a narrow open shelf. In some localities east of Labrador, the top of this sequence is capped by laminated and pisolitic hematite indicating brief subaerial exposure and lateritic weathering.

Everywhere north of Bonne Bay, the Bradore Formation and equivalents are overlain by limestone, shale, siltstone and minor sandstone (Forteau Formation) which is progressively more shale-dominated southward and eastward. There, the formation is dominated by dark shales and minor ribbon limestones, limestone concretions, intraclastic lime breccias, and oolite-oncolite lime grainstones. Poorly studied allochthonous deep water sediments, mainly laminated green and red shales (upper Summerside Formation, Curling Group), suggest tranquil deep water sedimentation under conditions of varying oxygenation. Together these facies indicate a narrow shelf with clean quartz sands at the strandline, a narrow belt of carbonate close to shore in the west, and a sea floor covered with silts and muds that sloped gently basinward to the east and south (Fig. 3.18).

Trilobites and archaeocyathans from different facies indicate deposition entirely within the *Bonnia-Olenellus* Zone of the Early Cambrian (Debrenne and James, 1981). The presence of *Wanneria* and *Salterella* suggests that most deposition was restricted to the middle part of this zone (Palmer and James, 1979; Fritz and Yochelson, 1988).

North of Bonne Bay, the sequence exhibits many attributes of a Grand Cycle (Aitken, 1966), dominated in the north by an upper carbonate half cycle and in the south and east by a lower shaly half cycle. The base of the Grand Cycle is everywhere marked by subtidal, transgressive carbonate (Devils Cove Member of the Forteau Formation; James and Debrenne, 1980). Shaly, nodular, red to white fossiliferous lime wackestone with solitary archeocyathans in western Newfoundland grades westward and northward into cross-bedded varieties. The subtidal, shale-dominated lower half cycle varies from burrowed calcareous siltstone with nodules and layers of *Salterella* and trilobite-rich limestone to dark shales punctuated by storm layers rich in shallow water carbonate clasts in northwest Newfoundland, to archaeocyathan bioherm complexes in southeast Labrador (James and Kobluk, 1978). These mudrocks are thicker eastward and southward and are progressively deeper water in aspect, locally containing synsedimentary slump folds.

The upper half cycle represents a narrow belt of high energy shoals and tidal flats, with much thinner strata eastward and southward (Fig. 3.18). The first deposits are skeletal and ooid calcarenites and archaeocyathan biostromes (Hughes, 1977) that colonized the shelf as far south as Hawke Bay (Knight and Boyce, 1984). The rest of the half cycle is typified by crossbedded ooid grainstones and oncolite rudstones with dolostone lenses, oolitic sandstones and sandstones which display rapid lateral variation and subordinate shale interbeds. A number of coarsening-upward shale to sandstone/grainstone sequences occur in the upper half cycle.

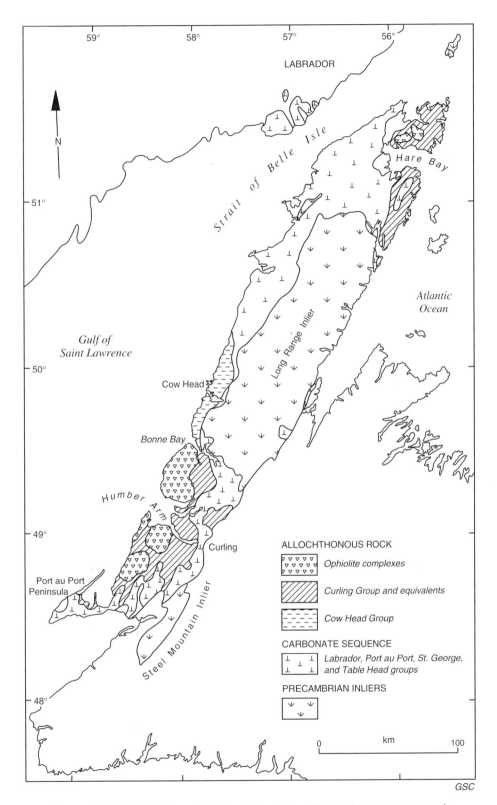

Figure 3.16. Distribution of the Cambrian-Ordovician carbonate sequence in western Newfoundland and the extent of the Cow Head Group and the Curling Group and equivalents in the Humber Arm and Hare Bay allochthons.

Terrigenous clastic rocks and minor carbonates (Hawke Bay and Penguin Cove formations) characterize the final phase of offlap. Contained trilobites range in age from the upper part of the *Bonnia-Olenellus* Zone of the Early Cambrian to the *Bathyuriscus-Elrathina* Zone of the early Middle Cambrian (Boyce, 1977; Knight, 1977; Knight and Boyce, 1987, 1991). Carbonates contain trilobites from all intervening zones (Knight and Boyce, 1987, 1991).

Basal deposits are subtidal to intertidal shallowing-upward shale to sandstone sequences on the inner shelf and subtidal fine grained thin sandstone, shale and lenticular skeletal limestone on the outer shelf.

Thick overlying sandstones on the inner shelf are mostly massive crossbedded quartz arenites deposited in tidal-dominated strandline and barrier island settings. Interbeds of glauconitic, phosphatic sandstone, burrowed

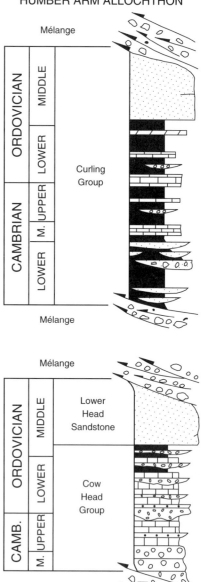

GSC

Figure 3.17. Major divisions of the carbonate sequence in western Newfoundland and comparitive stratigraphy with the Cow Head and Curling groups, modified from James and Stevens (1986).

mudstone and sandstone, and oolitic and pisolitic hematite represent bioturbated intertidal sand and mudflat deposits. On the outer part of the shelf, sediments are metre-scale shallowing-upward sequences of subtidal to intertidal shale and channel to sheet sands overlain by oolitic and oncolitic sand bodies containing stromatolite banks and

tidal flat dololaminites (Knight and Boyce, 1987). Some sequences are capped by fenestral carbonate, quartz and phosphate-rich dolostone and paleokarst.

Equivalent upper slope, parautochthonous strata in Canada Bay (Fig. 3.18) are grey, pyritiferous shale with ribbon quartz sandstone and minor limestone (Knight, 1987).

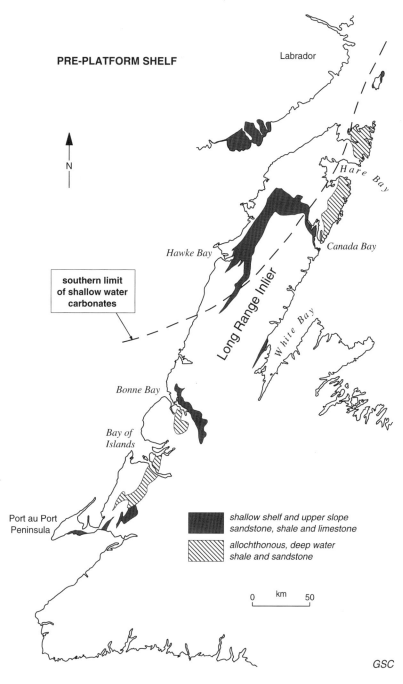

PRE-PLATFORM SHELF

Labrador

N

Hare Bay

Hawke Bay

Canada Bay

southern limit
of shallow water
carbonates

Long Range Inlier

White Bay

Bonne Bay

Bay of
Islands

Port au Port
Peninsula

shallow shelf and upper slope
sandstone, shale and limestone

allochthonous, deep water
shale and sandstone

0   km   50

GSC

Figure 3.18. Distribution of Lower Cambrian strata in western Newfoundland. The southern limit of shallow-water carbonates refers to the extent of ooid sands, archaeocyathan biostromes, and tidal flat facies, after James et al. (1989). Phase 1 of Figure 3.26.

Allochthonous lower slope sediments are a succession of conglomerate, sandstone, siltstone, and shale; mainly turbiditic with characteristics of a submarine fan. Conglomerates are polymict and contain clasts of all the older shallow shelf lithologies and basement, indicating profound erosion accompanying deposition.

In the St. Lawrence Lowlands, northwest of the Appalachian Structural Front, undated sandstones of the Covey Hill Formation, Potsdam Group possibly correlate with the lower part of the Labrador Group. The crossbedded, red, green and white arkosic sandstones, pebbly sandstone, conglomerate and thin siltstone have been correlated with the Ausable Formation in New York State (Clark et al., 1972; Fisher, 1977, 1982; Globensky, 1987). Fringing the craton in the inner part of the Quebec Reentrant, the Covey Hill Formation was probably laid down by fluvial processes similar to correlatives in Ontario, interpreted by Wolf and Dalrymple (1984) as sediments of an alluvial fan – braided fluvial system. If the Covey Hill Formation is a time equivalent of the Labrador Group, it suggests that fluvial sedimentation dominated the inner edge of the Quebec shelf.

In Quebec thrust slices, the Bonsecour Formation, Dunham Dolomite and Scottsmore Sandstone of the Oakhill Group are, in ascending order, Lower Cambrian phyllites, dolomites, and dolomitic quartz arenites similiar to the succession in the upper Labrador Group. They overlie the Tibbit Hill volcanics and Call Mill slates, and white to grey sandstones of the Pinnacle Formation, possibly equivalent of rift facies volcanic and sedimentary rocks of the lower Labrador Group. The rocks of the Oakhill Group are metamorphosed to chlorite grade and sedimentary structures are obliterated.

The Bonsecours Formation (Osberg, 1969; Globensky, 1978) consists of grey and green-grey, chlorite-mica, and quartz-mica phyllite that attains a thickness of 150 to 600 m (Osberg, 1965). Overlying the Bonsecours Formation, the Dunham Dolomite is the thin (10 m) northern extension of a thicker unit (450 m) that outcrops extensively in Vermont. It comprises thin-bedded, sandy dolostone with thin micaceous beds. The Scottsmore Sandstone is a dolomitic quartz arenite that is probably equivalent to the Monkton Quartzite of Vermont. Olenellid trilobites in the Dunham Dolomite in Vermont indicate an Early Cambrian age (Osberg, 1965).

The Milton Formation (Clark, 1934; Globensky, 1981), a unit of dolomite and sandstone up to 400 m thick, is the lowest rock unit within the Phillipsburg thrust slice. Containing the trilobites *Ptychoparia adamsi* Billings and *Olenoides marcoui* in adjacent American states, it is believed to be mostly Early Cambrian and equivalent of the Dunham Dolomite of Vermont. The formation consists of massive, well bedded, crystalline dolomite with abundant chert, sandy dolomite and dolomitic quartz sandstone. Locally it contains intraformational conglomerates.

The general succession of the Oakhill Group resembles the Labrador Group. The same onlap-offlap profile is dominated by siliciclastics, with the Dunham Dolomite, time equivalent of the Forteau Formation, probably the northern feather edge of a thicker carbonate unit to the south in Vermont. The Dunham Dolomite also correlates with the Milton Formation of the Phillipsburg Thrust Slice. This thick unit of intercalated dolostone and sandstone is interpreted by Gilmore (1971) as a beach-coastal dune complex.

Carbonate clasts containing a *Bonnia-Olenellus* fauna (Rasetti, 1944, 1946, 1948, 1955) that occur in Ordovician deepwater conglomerates on the north shore of Gaspésie indicate that the proto-shelf existed within the Quebec Reentrant.

Narrow, high energy platform

Onlap of carbonate sediments (Port au Port Group) across the platform was rapid during the late Middle Cambrian (*Ehmaniella* Zone) (just below the base of the Marjumian Stage, ie. Late Cambrian, *sensu* Ludvigsen and Westrop, 1985). Lack of sandy strandline facies points to a shoreline somewhere west of present insular Newfoundland, but not as far west as Quebec, indicating a shelf less than 200 km wide. Although dominated by carbonates, the sequence contains a high proportion of fine grained terrigenous clastic sediment. Middle Cambrian sediments were silty muds, muddy and burrowed carbonates, and locally oolitic grainstones of a subtidal shelf, and a belt of carbonate sand shoals in the east (Knight, 1977; Knight and Boyce, 1987). Three northeast-trending facies belts are recognized in the upper Cambrian (Fig. 3.19), all of which are dominated by peritidal sedimentation. In the **inner belt**, Upper Cambrian strata are mainly dolostones. The sediments were originally deposited in a complex mosaic of low energy muddy tidal flats, stromatolite-dominated tidal flats and intervening shallow subtidal environments (Knight, 1977, 1980; Knight and Saltman, 1980). The **mixed belt** contains sediments similar to those in the north intercalated with thick sequences of oolite including oolites arranged in metre-scale cycles with stromatolites and dololaminites (Knight and Boyce, 1991). The **outer belt** is narrow and contains dolomitized massive carbonates with relict ooid fabrics. These are believed to represent stacked ooid sand shoals. Traced eastward through an imbricate thrust stack, the shallow water Cambrian carbonates are replaced by ribbon limestones and shales, slump beds and carbonate conglomerates. In lower thrusts containing more proximal sequences, the ribbon limestones and sparce conglomerates are overlain by peritidal carbonates of the Petit Jardin and Berry Head formations. In higher thrusts containing more distal facies the succession is totally dominated by deeper water slope facies (Knight and Boyce, 1991; Knight, 1992; Boyce et al., 1992). These outer shelf facies preserved in the Reluctant Head Formation and the Weasel Group and time equivalent strata such as the Pinchgut Group suggest that the shelf was a prograding eastward deepening ramp to distally steepened ramp until the Franconian when the shelf widened to a true platform with a distally steepened ramp margin (Knight and Boyce, 1991; Boyce et al., 1992). Coeval rocks of the Shallow Bay Formation (Cow Head Group) comprise redeposited conglomerates, indicating that the adjacent margin was characterized by ooid and peloid sand shoals with numerous buildups of *Epiphyton* and *Girvanella* between shoals and on the upper slope (James, 1981; James and Stevens, 1986).

Cyclicity is a feature of Cambrian sedimentation in much of this area. Three Grand Cycles (Fig. 3.20), similar to those which typify Cambrian sedimentation in western North America, are recognized in the mixed belt from Bonne Bay south to Port au Port Peninsula (Chow and James, 1987), but they are not obvious in the St. Barbe region where two sequences may be present. Four cycles present in the parautochthon at Canada Bay lack fossils precluding their integration into the regional analysis.

Comparisons of platformal Grand Cycles with detailed stratigraphy of deep water lower slope sediments indicates a linked history of platform-slope growth (Fig. 3.21). Deep water sediments equivalent to Dresbachian Grand Cycles A and B of Chow and James (1987) are a sequence of limestone conglomerates with few calcarenite interbeds in the Cow Head region (Downes Point Member, Shallow Bay Formation) and conglomerate with interbedded calcarenite

turbidites in the Humber Arm region (Cooks Brook Formation). Such a sequence indicates that carbonate sedimentation at the adjacent platform margin generally exceeded the relative rate of sea level rise and resulted in more or less continuous margin progradation or offlap (James and Mountjoy, 1983). Blocks of early lithified sediment and calcified algal buildups were shed repeatedly into deep water, forming a wide debris apron of amalgamated conglomerates and breccias. Strata of Franconian to Trempealeauan age (equivalent to the upper Grand Cycle of Chow and James, 1987) are thicker and indicate vertical

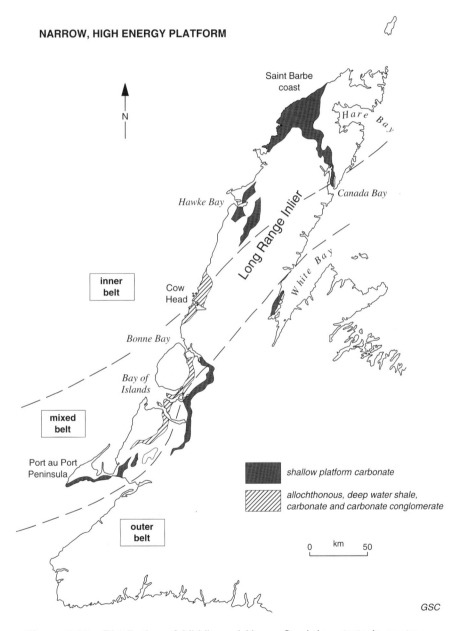

Figure 3.19. Distribution of Middle and Upper Cambrian strata in western Newfoundland and the distribution of three facies belts: (1) an inner mixed siliciclastic carbonate belt, (2) an intermediate belt in which inner facies and ooid sands are mixed, and (3) an outer belt of stacked ooid-sand shoals. Grand Cycles are best developed in the mixed belt, after James et al. (1989).

rather than lateral accretion. This is reflected in lower slope facies of the Cow Head region by a sequence of overlapping beds composed of quartzose carbonate sand turbidites (Broom Point Member, Shallow Bay Formation). Increasing relief between shelf and basin led to progressive narrowing of the deep water carbonate sand apron and onlap of anaerobic to dysaerobic basinal shales upslope (Martin Point Member, Green Point Formation). These sediments in the Humber Arm region (Cooks Brook Formation) are also quartzose calcarenite and shale.

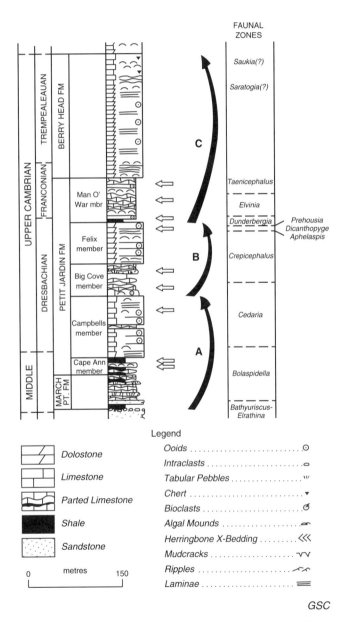

Figure 3.20. Middle and Upper Cambrian stratigraphy of the mixed carbonate-siliciclastic belt in western Newfoundland and its three Grand Cycles (A, B, and C). Open arrows indicate fossil control, after Chow and James (1987). Phase 2, events 4 and 5 of Figure 3.26.

In Quebec, the evidence for a carbonate shelf throughout the Middle and Upper Cambrian is restricted to clasts of shallow water limestones in the Lévis and Grosse Roche deepwater, Ordovician conglomerates, the Murphy Creek and Corner-of-the-Beach formations at Percé, Gaspésie, and the Rock River Formation of the Philipsburg thrust slice. Ribbon limestones and conglomerates of the late Middle to early Late Cambrian Corner-of-the-Beach and Murphy Creek formations of Percé, Gaspésie (Fritz et al., 1969; Kindle, 1942, 1948; North, 1971) match lithologically and faunally outer shelf, ramp facies of the Reluctant Head Formation and Weasel Group in Newfoundland. The sediments of the allochthonous Rock River Formation are poorly known. A thin, basal sandstone member overlain by carbonates interbedded with thin sandstone suggests an inner shelf setting close to basement or the inner siliciclastic belt. Gilmore (1971) interpreted the sediments as the deposits of high- to low-energy peritidal shoreline settings. The dolostone member probably reflects restricted tidal flat deposition like that common in Newfoundland.

The Cairnside Formation, Potsdam Group of the St Lawrence Lowlands probably correlates with the carbonate shelf in Newfoundland. These mature Upper Cambrian quartz arenites were derived by the reworking of aeolian sediments (Clark et al., 1972). They dominate the inner margin of the Quebec Reentrant. Sedimentary structures, ichnofauna and inarticulate brachiopods indicate a relatively high-energy shallow marine setting.

Other evidence that the carbonate shelf was more widespread in the Quebec Reentrant is shown by Middle to Upper Cambrian carbonate clasts in Ordovician conglomerates (Rasetti, 1944, 1945, 1946, 1963; Ludvigsen et al., 1989). Deep water Cambrian strata however, are predominantly siliciclastic suggesting the shelf was never very important and was either very narrow or intermittently developed.

Wide, low energy platform

A dramatic change in the style of sedimentation to widespread muddy carbonates coincides roughly with the Cambrian-Ordovician boundary and extensive inundation of the craton. The facies belts of the Cambrian are no longer discernable but the platform sequence has a distinct architecture recognized as two unconformity-bounded megacycles (Fig. 3.22). Each megacycle has a basal sequence of shallowing-upward peritidal cycles, a middle sequence of subtidal carbonate, and an upper sequence of shallowing-upward peritidal cycles (Knight and James, 1987). The sediments of the uppermost Berry Head Formation, Port au Port Group, the St. George Group and correlatives of the Romaine Formation, Beekmantown Group and lower Philipsburg Group, Quebec were part of a wide, low-energy rimmed platform that may have reached at least 500 km in width at its maximum development in the Arenig (James et al., 1989). The sediments were deposited throughout a general period of eustatic sea level rise that in Quebec overstepped the landward edge of Cambrian carbonates to onlap both Cambrian siliciclastics and Precambrian basement along the inner edge of the Quebec Reentrant. The Newfoundland carbonates accumulated along the eastern side of the St. Lawrence Promontory, close to its apex. The Romaine Formation lay at the far west of the southern side of the promontory adjacent to the Quebec Reentrant, the site of the Beekmantown and Philipsburg groups. Whereas the carbonate facies and importance of siliciclastics in the Beekmantown Group reflect its location in the inner part of the Quebec Reentrant, the facies of the allochthonous Philipsburg Group are quite like the Newfoundland succession, suggesting that it originated offshore.

The **lower megacycle** is mostly peritidal and is roughly equivalent to the Tremadoc Series as defined in Britain or the lower Canadian (Ibexian) Series as defined in North America. In Newfoundland, it comprises the Watts Bight Formation which is totally dolomitized over much of the Great Northern Peninsula and most of the Boat Harbour Formation (Fig. 3.22). Basal beds, which occupy the lower part of the Watts Bight Formation, are largely peritidal with burrowed, thrombolitic and stromatolitic carbonates capped by dololaminites. The middle subtidal portion consists of the middle and upper part of the Watts Bight Formation of burrowed carbonate and small to large thrombolitic mound complexes. *Renalcis*-thrombolite-coral mounds and reefs surrounded by skeletal sands mark the top of the middle portion of the lower megacycle as for example the Green Head Complex (Pratt and James, 1982) and spectacular mounds near Canada Bay (Knight, 1987). Thin dololaminites are more abundant in the top of the

middle part of the megacycle. The upper part of the megacycle is dominated by 100 m of repetitive, shallowing-upward, metre-scale cycles. The cycles comprise burrowed lime wackestone, wavy stratified lenticular and flaser bedded dolomitic lime wackestone and mudstone, and dololaminite or laminated lime mudstone. Thrombolite and stromatolite mounds are scattered throughout. The sequences represent sedimentation in a mosaic of tidal flat islands that grew upon the shallow shelf (Pratt and James, 1986). Overall the record is one of carbonate production keeping pace with relative sea level rise.

The top of the lower megacycle is marked by a disconformity of subaerial exposure, with solution surfaces, dolomitization, silicification, intraformational dolomite and/or chert breccias, and lag deposits of intraformational quartz sand and pebbles. Paleo-caves with collapse breccias occur beneath this horizon on Port au Port Peninsula. Extensive collapse breccias which occur at the base of the Boat Harbour Formation throughout the Great Northern Peninsula may be related to the disconformity. They are locally mineralized with sphalerite and galena. Trilobites typical of the underlying rocks are absent at this break (Boyce, 1979, 1983) and conodonts change dramatically (Stouge, 1982).

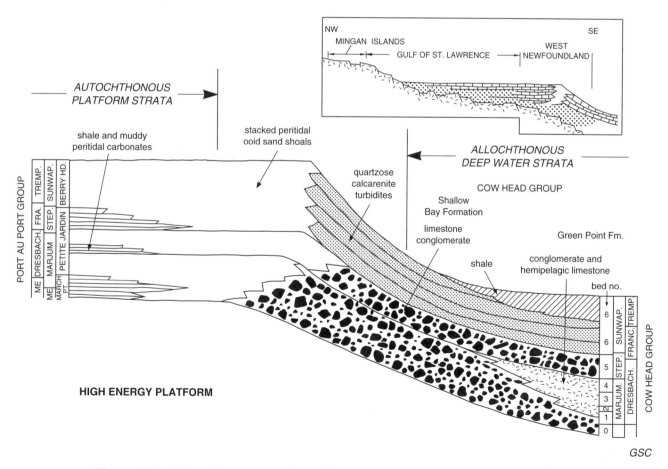

Figure 3.21. Interpretive cross-section of the restored high-energy, narrow, Middle-Upper Cambrian platform and coeval deep-water sediments in western Newfoundland (Port-au-Port Group and Cow Head Group), after James et al. (1989). Phase 2, events 4 and 5 of Figure 3.26.

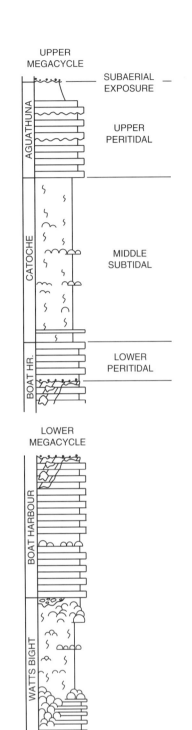

Figure 3.22. Two megacycles of the Lower Ordovician St. George Group, after Knight and James (1987). Phase 3, events 6 and 7 of Figure 3.26.

Coeval deep water lower slope facies are massive limestone conglomerates of the proximal Shallow Bay Formation (Stearing Island Member) and extensive parted to ribbon limestones of the more-distal Green Point Formation (Broom Point Member) in the Cow Head region (Fig. 3.23). The conglomerates are localized to a relatively narrow band in the distal facies, suggesting a toe of slope debris apron adjacent to a wide, gently dipping plain of carbonate silt and mud. There is abundant evidence of slope instability throughout in the form of intraformational truncation surfaces and slumped horizons, even in the distal facies (Coniglio, 1986). The Cooks Brook Formation at Humber Arm is also characterized by hemipelagic limestone deposition at this stage.

In Quebec, the megacycle is not present at Mingan Islands but rocks that resemble and correlate with these Newfoundland formations occur in the Strites Pond, Wallace Pond, Morgans Corner formations and most of the Hastings Creek Formation of the Philipsburg Group. In the St. Lawrence Lowlands, dolostone and sandstone of the Theresa Formation and the lower Beauharnois Formation, Beekmantown Group (Ogdenburg Member, Bernstein, 1991b) also belong to this megacycle. Sparse fauna and dolostones of peritidal aspect interbedded with laminated and cross-laminated, high energy quartz sands suggest that the Theresa Formation was deposited in shoreline to tidal flat settings. Rare conodonts (Sandi, 1978) hint that the formation was deposited at the same time as rocks of the regressive phase of the megacycle in Newfoundland. Conodonts from the dolostones and minor sandstones of the overlying lower Beauharnois Formation indicate that it probably correlates with the lower and middle members of the Boat Harbour Formation. Sedimentary structures, stromatolites and dolostones interbedded with quartz sands suggest that the lower Beauharnois was also deposited in very shallow shelf, shoreline and tidal flat settings influenced by the proximity to the inner silicilastic belt. At Ottawa and southeast Ontario, these sands are probably represented by non-marine to marine sandstones of the lower Nepean and March formations (Bond and Greggs, 1973, 1976; Greggs and Bond, 1971, 1972; Brand and Rust, 1977).

Farther south and more outboard in the Quebec Reentrant, rocks linked to the Tremadocian megacycle occur in the lower Philipsburg Group. The general details of the Philipsburg succession compare to those in Newfoundland, i.e. basal peritidal, middle subtidal and upper peritidal unit; although detailed sedimentological and stratigraphic aspects of the lower Philipsburg Group are still too vague to draw meaningful conclusions about the evolution of their depositional setting. In addition, the faunas listed by Globensky (1981a, b) generally give inconclusive ages that hinder correlation (W.D. Boyce, pers. comm., 1988). A basal cyclic peritidal sequence, the Strites Pond Formation, consists of intercalated limestone, shale, sandstone and dolostone that was probably deposited in shallow peritidal settings with subtidal thrombolite mound complexes. Crossbedded sand-rich beds accumulated in high-energy shoreline settings. Subaerial exposure with accompanying surficial karstification mark the top of several shoaling-upward sequences.

Subtidal to intertidal, bioturbated and fossiliferous limestones of the Wallace Creek Formation mark the middle subtidally dominated part of the megacycle and are succeeded by a thin interval of intertidal-supratidal dolostone of the Morgans Corner Formation (Gilmore, 1971). This dolostone is succeeded by a thick interval of repetitive small scale peritidal sequences of limestone, sandy limestone and dolostone of the Hastings Creek Formation. Local brecciation at the top of the Morgans Corner Formation and a basal sandy interval in the Hastings Creek Formation suggest a disconformity. Gilmore (1971) interpreted the dolomitic lower part of the Hastings Creek Formation to be dominantly supratidal and the upper part intertidal. The presence of an abundant, mixed skeletal fauna, the algae Nuia and microbial mounds in the limestones suggest a more open setting with tidal flats prograding over shallow, lagoonal shelf sediments. The lower Philipsburg Group probably correlates with the Shelburn, Cuttings and lower Bascom formations of Vermont (Globensky, 1981a, b).

The **upper megacycle** corresponds to roughly the Arenig Series of Britain and the upper Canadian (Ibexian) to lower Whiterock Series of North America, and is part of a world-wide Arenig transgression (Fortey, 1984). In Newfoundland, it comprises the upper part of the Boat Harbour Formation (Barbace Cove member), Catoche Formation, and Aguathuna Formation (Fig. 3.22). Onlap was extensive during this cycle, which represents the highest stand of sea level in the Early Ordovician (Barnes, 1984). Similar, coeval shallow water sediments are present over 300 km west of Newfoundland on the Mingan Islands, Quebec (Romaine Formation; Fig. 3.24) (Desrochers and James, 1988). Thus, the platform was extensive at this time and covered by an epeiric sea (sensu Shaw, 1964). Rocks of the uppermost part of the Hastings Creek Formation and the overlying Naylor's Ledge Formation are the only part of this megacycle preserved in the Philipsburg Group. Outside the Appalachian Structural Front, the upper Beauharnois Formation (Huntingdon Member) and the Carillon Formation of the St. Lawrence Lowlands (Globensky, 1987; L. Bernstein, 1991a, b, and pers. comm.), and the March and Oxford formations of southeast Ontario (Bond and Greggs, 1973, 1976; Greggs and Bond, 1971, 1972; Brand and Rust, 1977) onlapped the crystalline basement north and west of Montréal.

In Newfoundland, the lower peritidal sediments at the base of this cycle are distinctly coarser and more fossiliferous than most of the others in the St. George Group. The sediments of the Barbace Cove Member suggest that carbonate sediment production was greater than relative sea level rise, and that high energy conditions during this flooding stage extended well onto the shelf.

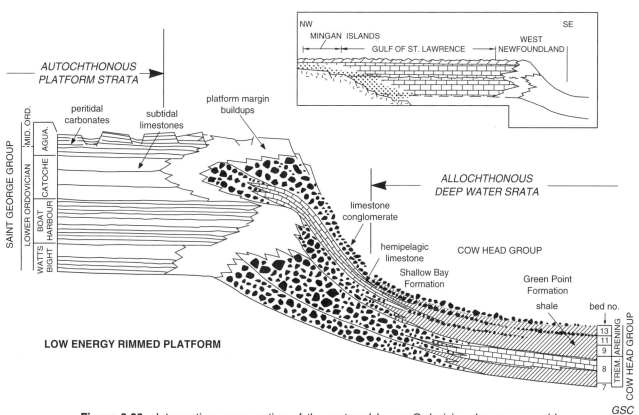

Figure 3.23. Interpretive cross-section of the restored Lower Ordovician, low-energy, wide-rimmed platform and coeval deep-water sediments in western Newfoundland (St. George Group and Cow Head Group), after James et al. (1989). Phase 3, events 6 and 7 of Figure 3.26.

The succeeding thick sequence of shallow subtidal sediments (Catoche Formation) indicates a period when relative sea level rise was equal to, or even greater than, the rate of carbonate sediment production. This subtidal facies extends to the Mingan Islands (Desrochers, 1985), indicating a vast shallow subtidal shelf of similar character (Fig. 3.24). Isolated mounds to large bank complexes of thrombolites, *Renalcis*, sponges, primitive corals, pelmatozoans, and other skeletal invertebrates studded the shelf at this time. Towards the margin, such mounds locally appear to form a major barrier complex as seen in the parautochthonous rocks of Hare Bay (Stevens and James, 1976), Pistolet Bay, and Canada Bay (Knight, 1986, 1987).

A gradual change in the balance between sea level rise and sediment production is reflected in the upper part of this subtidal sediment package, as fossils become less diverse and peloidal grainstones tend to dominate, reflecting less open marine conditions. In particular, a regionally developed complex of white, peloidal grainstone and fenestral limestone, designated the Costa Bay Member, marks the top of the Catoche Formation (Knight and James, 1987).

Flooding of the Precambrian basement in the Mingan Islands area, produced a classic transgressive onlap sequence. Mature sandstones of the Sauvage Member, Romaine Formation are deposits of actively-migrating, subtidal, nearshore sand bars that reworked an unconsolidated regolith, or they are fluvial sediments of the Ordovician sea that flooded Precambrian basement (Desrochers,1985). The member is equivalent of the Barbace Cove Member, Newfoundland. During marine highstand, open shelf, fossiliferous, subtidal carbonates and thrombolite mounds (Ste-Genevieve Member) were deposited mostly below wave base. Dolomitization of the member was early, as the burrowed dolomites are truncated by the post-Romaine unconformity (Desrochers, 1985). This dolomitization does not affect the member in the Anticosti Island sections.

The shelf in the inner part of the Quebec Reentrant remained very shallow throughout the Arenig. Dolostones and limestones of the upper Beauharnois Formation, are shallow subtidal to intertidal, muddy to grainy rocks arranged in no systematic way (Bernstein, 1991b) perhaps reflecting an inboard location on the shelf. The landward margin of the transgression consists of the sandstone dominated shoreline sequences of the March and Nepean formations. Their conodont faunas suggest correlation of their upper part with the Barbace Cove Member in Newfoundland.

Siltstones and silty dolostones of the top of the Hastings Creek Formation and the overlying subtidal burrowed and microbial mounded limestones of the Naylors Ledge Formation were laid down during the transgressive phase of this Arenig megacycle. The former is correlated with the Barbace Cove Member and with similiar siltstones of the Bascom Formation in Vermont, and the Ward Siltstone of the Fort Cassin Formation in New York (Fisher, 1977).

Coeval deep water sediments (Factory Cove Member, Shallow Bay Formation; Fig. 3.23) are a sequence of interbedded ribbon limestone, dark laminated shale, and conglomerate indicating platform margin progradation. Distal facies (St. Pauls Member, Green Point Formation) are mostly red bioturbated shales. Most proximal deep water sediments are fine grained ribbon limestones with a few limestone conglomerate horizons. Sediments equivalent to the middle of the subtidal (Catoche) facies are distinctive; silica replaces many ribbon limestone layers, phosphate pebble conglomerates are common, and graptolite zones are condensed. These characteristics suggest starved deep water carbonate sedimentation, likely because of the high position of sea level and backstepping of the platform margin. Distal facies are characteristically red, bioturbated shale with minor green and black horizons and thin limestone. This sequence indicates a stratified ocean in which proximal facies, deposited on the slope, accumulated

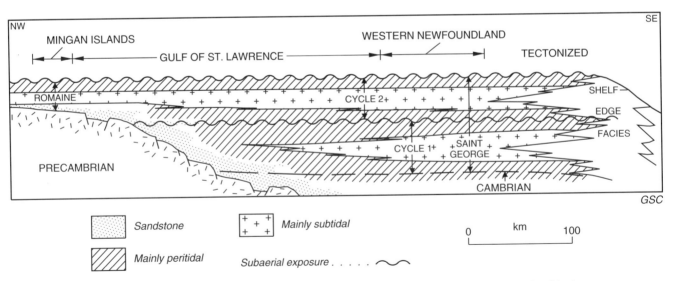

Figure 3.24. Generalized restoration of the two Lower Ordovician megacycles recognized in the St. George Group of Newfoundland, after Knight and James (1987).

in the lower part of an oxygen-minimum zone while distal facies on the lower slope were deposited under aerobic sea floor conditions (James and Stevens, 1986).

The microcrystalline peritidal dolostones, dolomitic shales, and minor limestones (Aguathuna Formation) which cap this cycle reflect the slowing of relative sea level rise and the onset of tectonism associated with the Taconic Orogeny. In Quebec, similiar peritidal sediments occur in the top of the Romaine Formation at Mingan Islands (Desrochers, 1985) and in the top of the Beekmantown Group in the St Lawrence Lowlands (Globensky, 1987; Bernstein, 1991a, b). The dolostones are younger westward into the Quebec Reentrant (Knight et al., 1991). In Newfoundland, the dolostones span the latest Arenig (Latest Canadian) to early Whiterock. Recent conodont studies by G.S. Nowlan show that the Whiterock Series is represented only in the uppermost strata of the Romaine Formation.

The predominance of dolostone, the importance of cryptalgal laminite and shale, the absence of fauna except locally in easternmost facies, and the presence of silicified evaporites suggest restricted and locally hypersaline environments. Intraformational breccias at several levels commonly contain quartz pebbles or sand grains, indicative of subaerial exposure. The erosional St. George Unconformity (Knight et al., 1991) marks the top of the megacycle. In some northern localities, this break is cryptic and recorded only by a change in the conodont fauna and by the local presence of thin chert pebble layers. Local erosion removed part or all of the peritidal dolomite sequence. The dramatic, local variations in thickness reflect a combination of faulting, uplift, and erosion (Knight et al., 1991). On the Mingan Islands, the Romaine Formation was tilted before erosion so that the unconformity is a planar karst surface which bevels several units (Desrochers and James, 1988). The unconformity is diachronous westward, younger in Quebec (Knight et al., 1991).

At Mingan Islands, the Grande-Île Member with its repetitive, metre-scale, shoaling-upward sequence displaying abundant evidence of subaerial exposure and displacive, diagenetic evaporite minerals in laminites suggests a sabhka-type tidal flat (Desrochers, 1985). The fine grained cyclic burrowed dolostones and dololaminites and shales of the Carillon and possibly Beldens formations display desiccation and microbial mats, and they have a very sparse fauna. These features suggest that very shallow lagoons, tidal flats, and pools characterised the shelf at its extreme inboard location in the Quebec Reentrant (Hofmann, 1963; Bernstein, 1991b). Although lithologically very similar to the Aguathuna Formation in Newfoundland, the Carillon and Beldens formations differ markedly in age. Conodonts (Sandi, 1978) and shelly fauna suggest that large parts of these two formations were deposited in the late Whiterock. This is at the same time that open shelf limestones of the Table Head Group were being deposited in the Taconic foreland basin in Newfoundland.

Coeval deep water facies in Newfoundland are a particularly sensitive record of the upper Lower Ordovician. Up to this time all proximal facies of the Cow Head Group were similar, but afterward they are dramatically different (James et al., 1987). In the Cow Head region the succession is one of hemipelagic ribbon limestones, minor calcarenites, shale, interbedded fine grained dolostones and two spectacular megaconglomerates (Beds 12 and 14 of Kindle and Whittington, 1958). Although the abundance of fine grained carbonates suggest that the margin was prograding, the megaconglomerates are traceable into the most distal deep water facies and they sample margin facies as old as Upper Cambrian, indicating widespread margin collapse (James and Stevens, 1984). In the Humber Arm Allochthon near Bonne Bay, deep water sediments change abruptly at this time to a sequence of alternating shale and dolostone with almost no carbonate, suggesting that the upslope margin was below the zone of rapid carbonate production. Drowning of the margin is postulated to be the result of faulting (James et al., 1987).

Shallow water strata in Newfoundland and Quebec suggest that the diachronous lithofacies of the regressive phase of the megacycle in Newfoundland and Quebec was directly influenced by Taconic reshaping of the passive margin. However, a synthesis of stratigraphy, facies and chronostratigraphy of shallow and deep water rocks led Knight et al. (1991) to conclude that this shallowing event and the restricted peritidal sedimentation resulted from the impending Taconic overthrust loading of the margin and migration of a peripheral bulge across the shelf margin (Jacobi, 1981; Quinlan and Beaumont, 1984). This effectively interrupted margin subsidence causing abrupt shallowing of the shelf (Costa Bay Member, Catoche Formation). Because the outer part of the passive margin was uplifted as the bulge impinged on the shelf edge, oceanic circulation was restricted on the inner shelf encouraging the widespread deposition of very shallow, restricted peritidal carbonates of the Aguathuna Formation and the Grande Île Member at Mingan Islands. This style of sedimentation also influenced the Quebec Reentrant as uplift and exposure of the shelf formed the St. George Unconformity in Newfoundland. The lack of peritidal facies above the Naylors Ledge Formation in the Philipsburg Group may reflect this uplift and erosion outboard in the Quebec Reentrant roughly sychronous with the unconformity in Newfoundland. The westward younging of the unconformity and bounding sedimentary packages into the Quebec Reentrant confirm its tectonic control rather than eustatic sea level fall (Sloss, 1963).

Foundered platform

Like the Taconic unconformities, such as the St. George Unconformity, Middle Ordovician carbonate and overlying flysch sediments are younger westward. The foreland shelf onlapped the foundered Early Ordovician platform, and then it foundered to be smothered by Taconic flysch. As this occurred in Newfoundland and outboard in southern Quebec, the foreland shelf continued to onlap the inner Quebec Reentrant where the shelf sequence is thicker and longer-lived. This suggests that sedimentation maintained equilibrium with subsidence perhaps in response to gradual slowing of accretionary processes and loading so that the foreland basin gradually deepened and narrowed within the Quebec Reentrant (Quinlan and Beaumont, 1984).

The foreland basin that evolved during the Middle Ordovician throughout the Canadian Appalachians was characterized by a carbonate shelf that lay along the foreland and a flysch basin adjacent to an uplifted Taconic hinterland. The foredeep developed above the ancient passive margin and with time onlapped Precambrian basement in Quebec, at least as far north as lac Saint-Jean. The chronostratigraphic relationships between autochthonous foreland shelf sediments and flysch indicate a west to northwestward, time transgressive relationship between each of these facies. The relationship between the flysch and underlying slope deposits preserved in the Taconic allochthons is also diachronous.

In spite of the markedly older and shorter-lived events in the Newfoundland part of the foreland basin compared to that in Quebec, there is a common pattern of shelf sedimentation in the basin in both areas. The shelf sediments are everywhere categorized as deepening-upward sequences. However, a significant interval of shallow, including peritidal, sedimentation preceded the deepening of the shelf in both areas. The shelf at any one time interval was probably never very wide, and likely formed a ramp in Quebec (cf. Parker, 1986; Mehrtens, 1989). However, local tectonism within the foredeep, particularly during the ultimate demise of the shelf, created different paleo-shelf morphologies from place to place. An underlying irregular, faulted paleo-karst basement topography added to active growth faults, and block faulting rendered thickness and distribution of shallow to deepwater facies of the Table Head Group less predictable. Newfoundland, which lay mid-shelf on the St. Lawrence Promontory, was dominated entirely by carbonates. Siliciclastics however formed an important element of the Quebec successions fringing Precambrian basement but also forming a basal unit above the Taconic unconformity of the area.

Deep water sedimentation – The youngest deep water carbonates of the allochthonous Cow Head Group including Bed 14 megaconglomerate are coeval with the peritidal carbonates at the top of the upper megacycle and the St. George Unconformity (Middle Ordovician Whiterock Series, *Orthidiella* Zone). Fine grained carbonate sediments and the lithofacies of carbonate clasts in the conglomerate indicate that a proto-shelf had developed along the foreland edge of the westward migrating foreland basin. The *Shumardia* limestone at the top of the Lévis Formation has Whiterock faunas and may represent similar deposits along the Quebec margin. The last deepwater carbonates in the allochthons are everywhere overlain by green lithic wackes and arenites (Lower Head and Eagle Island formations in Newfoundland; the Tourelle Formation and possibly Middle Ordovician wildflysch in Quebec; Hiscott, 1978; St-Julien, 1979). In Newfoundland, the basal beds are locally conglomeratic and sole marks indicate a north to northeast provenance. Ophiolite detritus suggests oceanic lithosphere in the source region, possibly also with Grenville basement slices (Stevens, 1970; Quinn, 1988a, b, c). The top of the Lower Head Formation is everywhere in thrust contact with an overlying allochthonous slice, indicating the sediments were soon caught up in the westward moving thrust complexes of an accretionary prism.

Taconic Unconformity – Taconic unconformities of the Canadian miogeocline, such as the St. George Unconformity in Newfoundland, are characterized by significant variations in the timing, duration and amount of erosion. Coupled with the bounding stratigraphy, the unconformity is clearly diachronous (Knight et al., 1991; Knight and Cawood, 1991), younging toward the craton, and the time represented by the hiatus decreases in the same direction. Tectonism shaped the unconformity in Newfoundland and to a lesser extent in the Mingan Islands. These characteristics are interpreted to be due to the westward passage

of a peripheral forebulge across the ancient continental margin during the development of the Taconic foreland (Knight et al., 1991). Although significant erosion probably occurred at the original site of the Philipsburg Group, the effect of peripheral forebulge uplift in the inner Quebec Reentrant is unclear. It is a younger event in the St. Lawrence Lowlands, where unlithified Beekmantown sediments beneath and minimal erosion at the unconformity in some sections suggests the time involved was small, with continuous sedimentation in some places (Bernstein, 1989, 1991a, b) and significant erosion in other places (Hofmann, 1963). This may indicate a faulted platform, as in Newfoundland, at a time when the forebulge amplitude affecting the inner Reentrant was less. In addition, sea level rise during this time would have counteracted the diminished effect of the forebulge on the Quebec shelf.

Middle to Upper Ordovician shelf sedimentation – In Newfoundland, autochthonous carbonates and shales of the Table Head Group are equivalent to the Lower Head Formation. They record breakup and foundering of the foreland shelf (Klappa et al., 1980; Stenzel et al., 1990). Faulting, uplift, and erosion of the Lower Ordovician platform (Fig. 3.25) during early Middle Ordovician time (Whiterock Series; *Orthidiella* Zone; Ross and James, 1987) created an irregular topography of horsts and graben. This landscape was flooded periodically with peritidal carbonates accumulating in the lows (basal Table Point Formation-Springs Inlet Member; Ross and James, 1987), while the highs remained above water. Periods of non-deposition and prolonged subaerial exposure are reflected now by several unconformities within this peritidal succession (Knight, 1986; Knight et al., 1991). Eventual complete inundation of this block faulted terrane (Fig. 3.25), led to widespread subtidal carbonate deposition (Whiterock Series, *Anomalorthis* Zone; Ross and Ingram, 1970; Stouge, 1984). The sediments are monotonous dark grey, lumpy to nodular, bioturbated wackestone to packstone with numerous lenses of fossiliferous grainstone, similar to the subtidal underlying carbonates of the Lower Ordovician. Small reefs built by lithistid sponges (Klappa and James, 1980) dotted an otherwise relatively uniform sea floor. The thickest sections, over 300 m, span less than one graptolite zone (Whittington and Kindle, 1963; Williams et al., 1987), indicating that rapid carbonate sedimentation never caught up to relative sea level rise. The rocks are characterized by repeated instability in the form of slumps, slides, deformed beds, and massive intraformational limestone conglomerates. The top of the sequence is everywhere abrupt, and marked either by lithistid sponge biostromes and oncolite rudstones or broken and eroded limestone cemented by dolomite, phosphate and chert (Knight, 1986; Stenzel et al.,1990).

Overlying sediments are variable in composition and reflect rapid subsidence within different settings (Stenzel and James, 1987, 1988; Stenzel et al., 1990). In some places, the subtidal carbonates are overlain by hemipelagic slope deposits of parted to ribbon limestones (Table Cove Formation) which are extremely fossiliferous at the base. These are interbedded with limestone plate conglomerates at various levels and are characteristically deformed by synsedimentary slumps and slides. In many localities these slope deposits are capped by black, graptolitic, laminated shales (Black Cove Formation) but in other places the black shales directly overlie the shallow subtidal carbonates

(Table Point Formation). In still other areas, the top of the subtidal carbonates are eroded and overlain by massive conglomerates (Cape Cormorant Formation) with clasts of both underlying subtidal lithologies and older Cambrian and Ordovician platform carbonates. These variations suggest that the platform foundered as separate blocks and at different rates.

The deep water euxinic shales and conglomerates were finally buried by northeasterly-derived synorogenic flysch (Mainland Sandstone, Goose Tickle Formation) as the foredeep migrated across the drowned and foundered platform

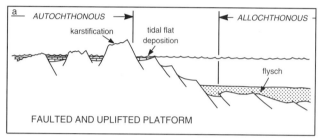

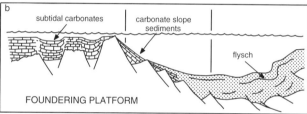

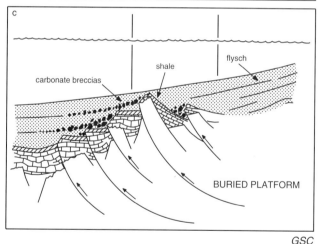

GSC

Figure 3.25. Three interpretive diagrams illustrating the evolution of the carbonate platform during the final foundering phase: **a)** development of a foreland basin to the east and uplift, faulting, and karstification in the west (Aguathuna Formation); **b)** thrusting and deformation of the foreland basin flysch in the east, foundering and rapid subsidence of the faulted platform in the west (Table Point and Table Cove formations of the Table Head Group); **c)** thrusting and transport of strata in the east, subsidence into deep water below level of carbonate deposition (shale of Black Cove Formation), thrusting and collapse of scarps to form conglomerate (Cape Cormorant Formation), burial by flysch (Mainland Sandstone and Goose Tickle Formation), after James et al. (1989). Phase 4, events 8 and 9 of Figure 3.26.

(Fig. 3.25). Exposure and erosion of the Middle Ordovician carbonates are reflected by massive conglomerate horizons, composed almost exclusively of Table Head limestone clasts, the Daniel's Harbour Member (Stenzel et al., 1990), in the basal units of the flysch.

In Quebec, sediments partly time equivalent of and younger than the Table Head Group, occur in the Lukes Hill, Solomons Corner and Corey formations. Faunas suggest a Whiterock to Chazyan age (Llanvirn to Llandeilo; Globensky, 1981; Boyce, pers. comm., 1988; Knight, unpublished data, 1988); althought the same problems that plague the lower Philipsburg also hinder unequivocal age determination and correlation. Subtidal and intertidal fine grained calcareous siliciclastics of the Lukes Hill Formation (Gilmore, 1971) indicate that energy conditions varied from high-energy, laminated and swaly bedded, fine grained siliciclastics of nearshore settings to extensively bioturbated, quiet water, subtidal muddy carbonate settings. Erosion surfaces possibly indicate periodic, subaerial exposure. Storm events ripped up limestone in the generally shallow setting and redeposited it as thin pebble beds.

Quiet water, lagoonal, nodular, bioturbated muddy limestones, together with sponge-microbial mounds typify the lower Solomons Corner Formation. As depositional rates matched sea level rise, peritidal carbonate sedimentation of repetitive limestone, dolomitic limestone and dolostone sequences and stromatolite mounds dominated much of the upper Solomons Corner Formation. Fossiliferous, bioturbated limestone and black shale of the overlying Corey Formation, which is rich in a molluscan fauna, show that the shelf then deepened. The lithofacies and the general sequence has the characteristics of a gently deepening ramp shelf.

The deepening trend may have continued with the deposition of the younger part of the Stanbridge Group which included *N. gracilis* zone shales and fine-grained limestones of the Basswood Formation deposited in a lower slope setting. Locally, the Mystic Conglomerate and similiar rudites contain resedimented clasts of shallow water origin that were transported down slope in debris flows. The rudite facies may compare to the Cape Cormorant Formation or Daniel's Harbour Member of Newfoundland.

The inner platform, as characterized by outcrops on the Mingan Islands and the St. Lawrence Lowlands, was exposed to karst erosion during most of this period. At Mingan Islands, the Taconic unconformity is covered by strandline sandstone and shale which grade upward into subtidal limestone (Mingan Formation; Chazyan Series), that are all younger than the west Newfoundland carbonates. Desrochers (1985) interpreted basal siliciclastics of the Corbeau Member as shallow water sand bars, tidal channels, sandy and muddy intertidal flats and silty carbonate supra-tidal flats. The sands were probably derived by reworking of eolian deposits that blanketed the unconformity during subaerial exposure. Fossiliferous limestones, cross bedded grainstones and metazoan biostromes of the Perroquet Member suggested to Desrochers (1985) a shallow open shelf with local carbonate sand shoals. Deposition of the succeeding Fantôme Member limestones occurred in semi-restricted lagoons, tidal flats traversed by tidal channels, and supratidal algal marshs (Desrochers, 1985). The Grande Pointe Member was deposited upon the intra-Mingan disconformity which formed during regional sea level fall (Desrochers and James, 1988). The latter formed a system of islands with rocky shorelines that controlled the distribution of the Grande Pointe lithofacies. Subtidal muddy limestone marks a restricted, low-energy shelf in the lee of the islands, grainstones mark tidal deltas and channels between the islands, and well-burrowed, skeletal-rich muddy limestones mark an outer, open marine subtidal shelf (Desrochers, 1985). The intra-Mingan karst surface also controlled the distribution of the metazoan reefs and associated fringe sands scattered through the member.

The sediments of the overlying Black River and Trenton groups intersected in several exploratory wells (Roliff, 1968) are too poorly known to interpret. The presence of graptolitic shale interbedded with the Trenton limestone indicates a deepening, open shelf environment that lasted from the Chazyan (Llandeilo) to the Trenton (early Caradoc, *Diplograptus multidens* zone). The unconformity of the Trenton Group on the Romaine Formation, the local disconformity at the top of the Trenton Group at the east end of Anticosti Island (Roliff, 1968), and the variable thickness of the Macasty Formation shale suggest that active faulting may have characterized this area that lay peripheral to the St. Lawrence Promontory.

Elsewhere in Quebec, shelf sedimentation began in the Chazyan and continued to the Trenton and locally the Eden (Llandeilo to late Caradoc, *Climacograptus spiniferous-- C. pygmaeus* zones). Sandstones fringe Precambrian basement from Québec City to Trois-Rivières; sandstones were most widespread in the Montréal area. Well sorted, well rounded, medium- to very coarse-grained, basal sandstones of the Chazy Group were reworked from weathered Precambrian basement and deposited as a complex of shallow marine estuarine or deltaic deposits that prograded eastwards into a deeper water embayment (Hofmann, 1963, 1979; Harland and Pickerill, 1982). Westward, the environment was dominated by high-energy, littoral nearshore sediments that passed into possible non-marine strata (Hofmann, 1972, 1979, 1982, 1989).

Although immature, the sandstone of the Cap-à-l'Aigle Formation, Black River Group and the underlying Cap-aux-Oies Formation (Rondot, 1972), northeast of Montmorency Falls (Harland and Pickerill, 1982), and the crossbedded quartz arenites of La Gabelle Formation, southeast to Trois-Rivières (Clark and Globensky, 1976a, b, c) are both high-energy marginal marine sediments. The Precambrian basement at Montmorency Falls was fringed by basal Trenton sands, the Saint Marc Formation. Although poorly sorted and of low maturity the sands are interpreted as nearshore clastics deposited during rapid burial of the Precambrian highs or islands. The presence of *Solenopora* gravels implies energetic conditions.

Shallow water Chazy Group carbonates that onlapped these fringing sands are characterized by mostly low energy facies with local high energy deposits. Fossiliferous, mixed, fine grained limestone and crossbedded grainstone facies of the Laval Formation occupy an embayment in the area of Montréal where high- to moderate-energy shelf conditions prevailed. A belt of coral-bryozoan bioherms proliferated in high-energy settings close to the southern edge of the embayment (Hofmann, 1979; Harland and Pickerill, 1982; Globensky, 1987). Carbonate muds with few organisms and an upward increase in dolomite indicate low-energy, restricted shelf conditions (Beaconsfield Member). The presence of quartz sand indicates a nearby source of clastics from Precambrian highs and islands in the northeast (Harland and Pickerill, 1982, 1984).

The carbonates of the Black River Group overlie the Chazy southwest of Trois-Rivières and reflect upward shoaling and progradation at the top of the Laval Formation. The Pamélia Formation was probably deposited in intertidal to supratidal carbonate flats with frequent exposure and influx of siliciclastic sand from nearby marginal basement sources (Hofmann, 1972; Harland and Pickerill, 1982). Low- to moderate-energy, shallow lagoonal sediments characterize the overlying Black River limestones and the basal carbonate units of the Trenton Group (Harland and Pickerill, 1982; Pickerill et al., 1984). These include the Leray and Lowville formations of the Black River Group and the Mile End, Ouareau and Fontaine formations from Montréal to Trois-Rivières, and the Sainte-Alban and Pont-Rouge formations northeast of Trois-Rivières to Montmorency Falls (Clark, 1952, 1972; Clark and Globensky, 1973, 1975; Harland and Pickerill, 1982). Other names given to basal beds include the Rockland beds near Montréal (Clark, 1972). Skeletal-rich beds, coral-stromatoporoid colonies and oolitic and skeletal grainstone typified higher-energy areas of the shelf. Storms reworked quartz sand and limestone intraclasts from nearshore sand and desiccated tidal flats. Abundance of shelly layers and of quartz sand in the northeast in the basal Trenton Group suggest a depositional setting influenced by more open marine conditions as well as proximity to a coastal clastic zone adjacent to Montmorency Falls.

The lagoonal sediments of the basal Trenton Group probably lay behind high-energy, intermittently migrating, subtidal carbonate sand shoals with local beaches that mark the Deschambault Formation (Harland and Pickerill, 1982). The facies of the Deschambault Formation are interpreted as a mid ramp shoal pivotal between near shore lagoons (basal Trenton carbonates) and deeper ramp sequences represented by the upper Trenton carbonates (Parker, 1986; Mehrtens, 1989). Northeast of Montmorency Falls, finer grained limestones with fewer fossils accumulated where lagoonal settings remained (Riva et al., 1977; Harland and Pickerill, 1982). Contemporaneous quartz sands accumulated along an adjacent rocky, high-energy shoreline and were derived by reworking of eolian deposits that once blanketed the unconformity during subaerial exposure.

Upper Trenton carbonates comprise low- to moderate-energy, fossiliferous, shallow subtidal, normal marine limestones of the overlying Montréal Formation and the Saint Casimir Member, Neuville Formation (Harland and Pickerill, 1982; Pickerill et al., 1984). Local carbonate sand shoals are common toward the shore in the northeast. The fine grained lithologies, ichnofauna and upward-changing restricted skeletal fauna of the Tétreauville Formation and the Grondines Member, Neuville Formation indicate the predominance of suspension deposition as the shelf deepened (Harland and Pickerill, 1982; Pickerill et al., 1984; Globensky, 1987). Sedimentation rates were low and infrequent storms reworked skeletal debris into lenticular coarse beds. Slump folds indicate that the shelf was frequently unstable. Significant thickness variations in these and the underlying basal Trenton formations suggest several depocentres with local growth faults (Mehrtens, 1989).

The transition to a deep water shale basin was marked by pelagic to gravity-generated sedimentation of a gentle slope, lower slope and basin margin, and deep basin settings represented by the Rivière-du-Moulin facies of the Neuville Formation, the Saint-Irénée Formation, and the Utica Group, respectively (Belt et al., 1979; Mehrtens, 1989). Pelagic sedimentation included thin-bedded fine grained limestone and shale deposited on the deeper distal portions of the ramp. Tectonism initiated frequent slumping. The siliciclastics of the Saint-Irénée Formation are interpreted as turbidites and basin muds that accumulated on submarine fans that built from the southeast. At the same time, limestone debris flows were generated on the steep unstable shelf slope to the northwest and carried limestone conglomerates into the basin (Belt et al., 1979). Taconic siliciclastic flysch of the Vauréal, Lotbinière, Beaupré and Nicolet formations smothered the shelf.

The final phases of the Taconic Orogeny resulted in the emplacement of the Humber Arm and Hare Bay allochthons in Newfoundland and similar allochthons in Quebec. Various lines of evidence, including conodont coloration indices (Nowlan and Barnes, 1987), indicate that the allochthons were never much more extensive than their present areas (Fig. 3.16). The first post-orogenic phase of sedimentation is represented by the Long Point Group. The lower part of this unit is Middle Ordovician (early Mohawk Series, Black River Stage; possibly latest Whiterock Series, Chazyan Stage; Bergström et al., 1974). Its age sets an upper limit to the time of Taconic Orogeny. Equivalent carbonates and clastics in the subsurface of Anticosti Island (Roliff, 1968), together with the carbonates of the Long Point Group to the east, represent a new phase of sedimentation in the Anticosti Basin over the relatively low relief of the foundered and deformed continental margin.

Summary

The history of the western Newfoundland carbonate platform can be summarized according to its four phases of development and nine events (Figure 3.26). Events in Quebec largely reflect the same pattern.

Phase 1 – Pre-platform shelf

Late Proterozoic?-Middle Cambrian – terrigenous clastic and minor carbonate sediments formed a narrow shelf on rifted plutonic basement, in the form of a pronounced onlap-offlap sequence, primary control was tectonic.

Event 1: Late Proterozoic-Early Cambrian – block faulting of Precambrian basement following continental rifting; sand deposition began in depressions and eventually covered underlying topography; shelf sedimentation occurred in braided fluvial to strandline settings; primary control was eustatic sea level rise against a background of minor tectonism and moderate thermal flux related to rifting.

Event 2: Late Early Cambrian – onlap-offlap in the form of a Grand Cycle; variable coarse and fine terrigenous clastic and fossiliferous carbonate deposition produced distinct facies belts; a subtidal ramp with no effective organic rim; most conspicuous carbonates were ooid sand shoals, archaeocyathan buildups, and muddy tidalites; deep water deposits were terrigenous muds and silts; primary control was eustatic.

Event 3: Latest Early Cambrian-middle Middle Cambrian – pronounced offlap; quartzose sands prograded to shelf edge; local carbonate deposition continued at shelf edge throughout; probable cutting of submarine canyons; deep water slope muds, silts and conglomeratic submarine fans; primary controls were tectonic and eustatic.

Phase 2 – High energy narrow platform

Middle-Late Cambrian – evolution to a carbonate platform less than 200 km wide and covered by terrigenous clastic and carbonate sediments with fewer biotic constituents than in the previous phase; distinct facies belts; shelf margin of ooid/peloid sand shoals; primary control was eustasy. This phase is largely absent in Quebec. There the shelf and deepwater sediments are predominantly siliciclastics and the presence of a perhaps poorly developed carbonate shelf is indicated by carbonate clasts in deepwater Ordovician rocks.

Event 4: Middle to middle Late Cambrian – marked cyclicity throughout; two Grand Cycles in transition zone between outer ooid sand shoals and inner terrigenous/carbonate belt; pronounced margin progradation; *Epiphyton-Girvanella* buildups on upper slope; extensive conglomerate debris flows in deep water; primary control was eustasy but generation of two Grand Cycles, rather than one, was probably the result of local tectonism.

Event 5: Middle Late to Latest Cambrian – similar to the preceding event but only one Grand Cycle developed; proximal quartzose carbonate turbidites and distal terrigenous muds characterized lower slope deposition; margin accretion was mostly vertical and deep water shales prograded upslope; primary control was eustatic. Sands blanket the inner Quebec shelf.

Phase 3 – Low energy, wide, rimmed platform

Upper Ordovician to early Middle Ordovician – evolution into a huge platform more than 500 km across, overall onlap; entirely carbonate deposition with a diverse fauna and buildups at shelf edge; primary control was eustatic representing climax of Sauk Sequence.

Event 6: early Early Ordovician to middle Early Ordovician – onlap-offlap cycle with karst unconformity at top; quiet muddy subtidal to peritidal deposition, local and marginal microbial-metazoan buildups; margin progradation; deep water carbonate toe-of-slope apron with minor debris sheets which accumulated under anaerobic conditions within, and aerobic conditions below an oxygen minimum zone, primary control was eustatic. Sands important in Quebec.

Event 7: Middle Early to early Middle Ordovician – onlap-offlap cycle with karst unconformity at top, same style as event 6 except that onlap was more rapid and extensive, marginal buildups were larger with more diverse metazoan fauna, and platform was blockfaulted during offlap; margin backstepping and then progradation; deep water sediments formed a narrow toe-of-slope debris apron fronted by a wide shale slope deposited in the lower part of a fluctuating oxygen minimum zone; deep water sediments were covered by synorogenic flysch; primary controls were initially eustatic but later tectonic.

Phase 4 – Foundered platform

(Early Middle Ordovician to Late Ordovician); collapse and burial of platform; carbonate deposition followed by flysch sedimentation; primary control was tectonic.

Event 8: Early Middle Ordovician to Late Ordovician – block faulting and uplift with passage of peripheral bulge across platform; karstification of highs, peritidal carbonate deposition in lows; subsequent general subsidence and rapid subtidal deposition on an irregular series of platforms

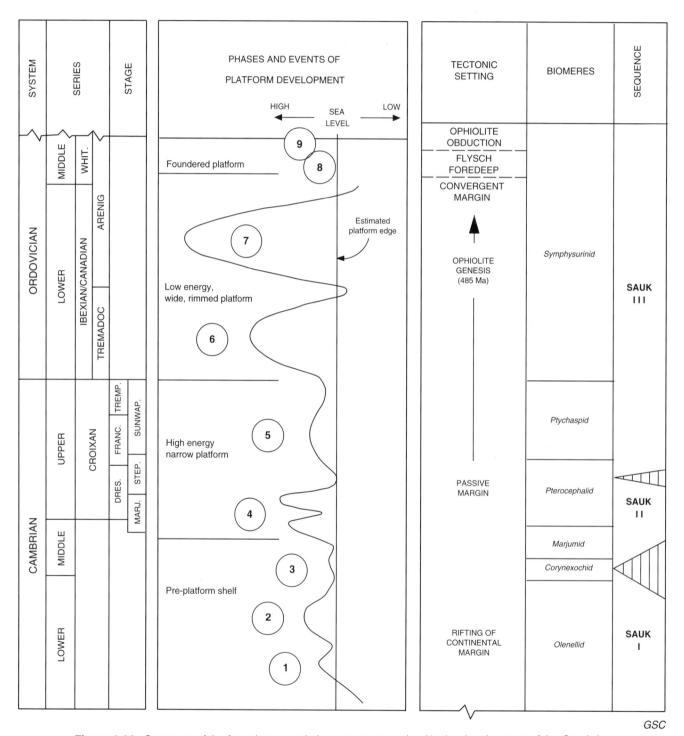

GSC

Figure 3.26. Summary of the four phases and nine events recognized in the development of the Cambrian-Ordovician carbonate platform, after James et al. (1988). See text for more details on events.

and depressions subject to continuing seismicity; eventual subsidence below zone of carbonate deposition; primary control was tectonic against a background of eustatic sea level rise.

Event 9: middle Middle to Late Ordovician – burial by flysch; contemporaneous erosion of carbonates from highs created by thrust faulting; primary control was tectonic.

Middle Ordovician clastic rocks of the Humber Zone and St. Lawrence Platform

R.N. Hiscott

Introduction

Subaerial erosion of rocks that lay mainly east of the Humber Zone produced abundant sand- and mud-sized detritus that forms distinctive flysch units in both autochthonous and allochthonous successions of chiefly Middle Ordovician age. The sandstones in these deposits are lithic arenites (Table 3.1), and contain recycled quartz grains, fresh and weathered feldspars, a variety of sedimentary rock fragments, mafic to felsic volcanic rock fragments, local serpentinite grains, and a heavy mineral suite containing green and green-brown hornblende, hypersthene and chromite (Enos, 1969a, b; Stevens, 1970; Hiscott, 1978; Beaulieu et al., 1980; Belt and Bussières, 1981). The detrital chromite is responsible for unusually high chromium concentrations of 300-1000 ppm (Hiscott, 1984). These sedimentary rocks represent a major reversal in the direction of sediment supply to the Humber Zone, from the Canadian Shield to outboard elements of the Dunnage Zone. The close genetic relationship between autochthonous and allochthonous flysch of the Humber Zone, and flysch just beyond the western limit of Appalachian deformation on the St. Lawrence Platform, justifies discussion of all examples of these rocks in this section.

Autochthonous Ordovician flysch units are found within the Humber Zone, or locally as overspill from the Humber Zone onto the St. Lawrence Platform, particularly between Montréal and Québec City (Fig. 3.27). The flysch lies stratigraphically above the Cambrian-Ordovician carbonate sequence, except in parautochthonous sections along the front of the major allochthons of Quebec (e.g., Cloridorme Formation, Fig. 3.27), where original stratigraphic relationships have been obscured by thrusting. Autochthonous and parautochthonous flysch ranges in thickness from about 250-4000 m (Table 3.2), and in age from Llanvirn in Newfoundland to Caradoc and Ashgill in Quebec (Fig. 3.28).

Sandy flysch successions are also found as the upper units in sedimentary allochthons, where they lie stratigraphically above Lower Ordovician shales and limestone breccias, interpreted as continental slope and rise deposits.

Hiscott, R.N.
1995: Middle Ordovician clastic rocks (Humber Zone and St. Lawrence Platform); in Chapter 3 of Geology of the Appalachian-Caledonian Orogen in Canada and Greenland, (ed.) H. Williams; Geological Survey of Canada, Geology of Canada, no. 6, p. 87-98 (also Geological Society of America, The Geology of North America, v. F-1).

These occurrences of flysch range in thickness from about 200-2500 m (Table 3.2), and in age from Arenig to Caradoc (Fig. 3.29).

There are equivalents of both the autochthonous and allochthonous flysch deposits to the south, in the U.S. Appalachians. For the Appalachian Orogen as a whole, allochthonous flysch is oldest (Arenig to Llanvirn) in Newfoundland and Quebec, and youngest in Pennsylvania, New York and Vermont (Llandeilo to Caradoc; Hiscott, 1984); autochthonous flysch is oldest in Newfoundland (Llanvirn to Caradoc), and Tennessee and Virginia (Llandeilo to Caradoc), and significantly younger (Caradoc to Ashgill) from Maryland through to Gaspésie of Quebec (Hiscott, 1984; Bradley, 1989). Where traverses can be made across the strike of the mountain belt, as in New York State (Rickard and Fisher, 1973), allochthonous flysch is always significantly older than structurally underlying autochthonous flysch, and the base of the autochthonous flysch is younger toward the craton.

Autochthonous and parautochthonous flysch

The Cambrian-Ordovician carbonate sequence of the Humber Zone and immediately adjacent St. Lawrence Platform is in most places overlain by (a) argillaceous limestones that may be characterized by wet-sediment sliding, and that may contain variable proportions of limestone turbidites and limestone conglomerates, (b) black, graptolitic shales, and (c) a succession of interbedded shales and turbidites, forming flysch units up to 4 km thick (Table 3.2, Fig. 3.30). Where present, the lowest black shale units, like the flysch, vary in age along the length of the orogen (Hiscott, 1984). The main exposures of autochthonous and

Table 1. Grain percentages for four Canadian flysch units and, for comparison, central and southern Appalachian flysch.

Component (%)	Unit designation (Fig. 3.27)					
	1	3	5	C	Martinsburg	Blount
Quartz	71	58	60	54	50	74
Feldspar	9	3	10	10	15	12
(K-spar)	(6)	?	(0)	(6)	(2)	(2)
SRF[2]	17	30	10	12	21	12
MRF	1	1	2	0	7	2
VRF	2	7	15[3]	4	7	0
Number of Samples	52	15	12	19	18	52

[1] Data sources by column, left to right: Beaulieu et al. (1980), Belt and Bussières (1981), Enos (1969a), Hiscott (1978), Mack (1985), Mack (1985).

[2] SRF, MRF, VRF = sedimentary, metamorphic, volcanic rock fragments, respectively.

[3] Cloridorme Formation contains 4% serpentine grains, here included with VRF.

parautochthonous flysch are indicated in Figure 3.27; ages are given in Figure 3.28. In western Newfoundland, autochthonous flysch is of Llanvirn to Caradoc age (Goose Tickle Group, including the Mainland Sandstone), whereas in Quebec, autochthonous and parautochthonous flysch is of Caradoc and Ashgill age (Cloridorme, Saint-Irénée, Lotbinière, Beaupré and Nicolet formations; Fig. 3.28). Paleoflow in the flysch is generally away from the St. Lawrence Promontory and toward the Quebec Reentrant (Thomas, 1977; Fig. 3.31).

The autochthonous and parautochthonous flysch units consist of a variety of hemipelagite, silty and sandy turbidite (Fig. 3.32), and sandy debrite facies (Tuke, 1968; Enos, 1969a, b; Beaulieu et al., 1980; Belt and Bussières, 1981; Pickering and Hiscott, 1985; Hiscott et al., 1986). Pioneer work on the geometry of turbidite beds in the parautochthonous Cloridorme Formation (Enos, 1969b) prompted later detailed studies of depositional mechanisms for: (a) classic turbidites (Walker, 1967; Parkash, 1970; Parkash and Middleton, 1970), and (b) unusual,

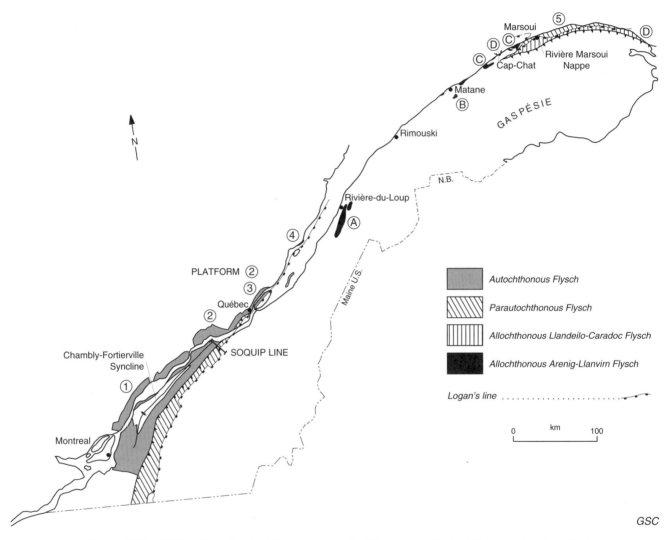

GSC

Figure 3.27. Distribution of autochthonous, parautochthonous and allochthonous flysch units in Quebec and western Newfoundland. Autochthonous flysch between Montréal and Québec City is on the St. Lawrence Platform, not in the Appalachian Orogen. The location of the Soquip seismic reflection profile shown in Figure 3.35 is indicated to the west of Québec City. This map is based on Biron (1972, 1974), Liard (1973), St-Julien and Hubert (1975), Vallières (1977), Williams (1978a), Williams et al. (1985), and Botsford (1987). Autochthonous and parautochthonous units are : 1 = Nicolet Formation; 2 = Lotbinière Formation; 3 = Beaupré Formation; 4 = Saint-Irénée Formation; 5 = Cloridorme Formation; 6 = Mainland Sandstone (Goose Tickle Group); 7 = American Tickle Formation (Goose Tickle Group). Allochthonous units are: A = Saint-Modeste Formation; B = Métis Formation; C = Tourelle Formation; D = Deslandes Formation; E = Eagle Island formation; F = Lower Head Formation.

thick, mud-rich beds (Skipper, 1971; Hand et al., 1972; Skipper and Middleton, 1975; Skipper and Bhattacharjee, 1978; Hiscott and Pickering, 1984; Pickering and Hiscott, 1985).

Newfoundland

In western Newfoundland (Fig. 3.27), autochthonous flysch overlies the Table Head Group, a predominantly shallow-water limestone unit (Fig. 3.30). The upper part of the Table Head Group consists variably of argillaceous limestones with wet-sediment slide folds (Table Cove Formation) and an erosively based succession of interbedded dark shales and sheets of limestone boulder conglomerate (Cape Cormorant Formation; Stenzel et al., 1990). In central and northern localities, including just beneath the Hare Bay Allochthon (Fig. 3.27), the Table Head Group is overlain by black shales of the Black Cove Formation, originally assigned to the Table Head Group (Klappa et al., 1980), but recently reassigned to Goose Tickle Group (Stenzel et al., 1990). The flysch in these areas is relatively shaly and is called the American Tickle Formation (unit 8; Tuke, 1968; Stenzel et al., 1990). On Port-au-Port Peninsula, the flysch of the Goose Tickle Group is called the Mainland Sandstone (Fig. 3.27, unit 6), and is about 1700 m thick (Klappa et al., 1980; see Addendum for Chapter 3).

Gulf of St. Lawrence

Shaly flysch occurs in the subsurface beneath Anticosti Island in the Gulf of St. Lawrence, an extension of the St. Lawrence Platform. The lowest silty unit of the Vauréal Formation (lower Ashgill English Head formation of Roliff, 1968) resembles the Nicolet Formation of the St. Lawrence Lowlands (INRS-Pétrole, 1976), and represents the distal edge of the flysch wedge. The Vauréal Formation overlies black bituminous shales of the upper Caradoc Macasty Formation, equivalent in age to the Cloridorme Formation of Gaspésie.

Quebec

Parautochthonous flysch of the Cloridorme Formation (Enos, 1969a) outcrops between Marsoui and the tip of Gaspésie (Fig. 3.27, unit 5). The Cloridorme was slightly redefined by Hiscott et al. (1986), because part of the

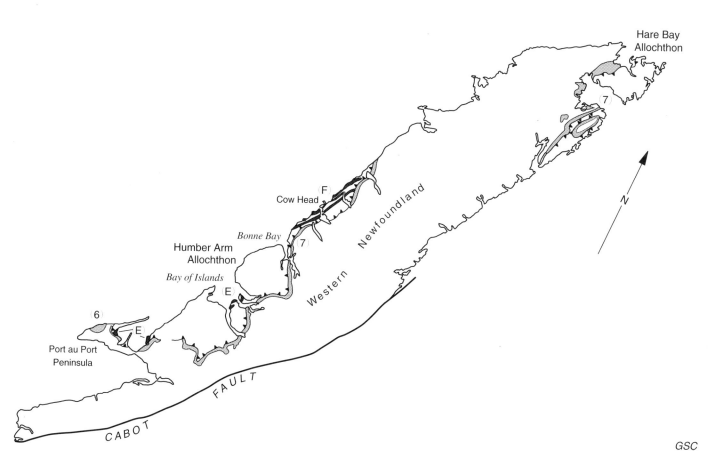

GSC

Figure 3.27. (cont.)

formation as defined by Enos (1969a) was allochthonous (Riva, 1968; Stevens, pers. comm., 1969), and is equivalent to the Deslandes Formation defined farther to the west by Biron (1974). This part of the Cloridorme Formation of Enos (1969a) is discussed below as part of the Deslandes Formation. The rest of the Cloridorme Formation is parautochthonous. Its base is covered by thrust sheets of the Marsoui River Nappe (St-Julien and Hubert, 1975), and its top is truncated by the modern erosion surface along the south shore of the St. Lawrence River. Informal members are recognized (Hiscott et al., 1986), based on proportions of sand, and presence and positions of laterally extensive megaturbidites (Fig. 3.33). The megaturbidites contain sole markings and cross laminations that indicate up to three reversals in flow direction during deposition of single graded beds (Pickering and Hiscott, 1985). These unusual beds are for the most part restricted to the Saint Hélier and Manche-d'Épée members, where they are associated with shales and fine grained, thin- to medium-bedded turbidites. Sandy parts of the Pointe-à-la-Frégate, Petite-Vallée and Marsoui members consist of 10-30 m-thick packets of medium- to coarse-grained, medium- to thick-bedded, massive to graded sandstones that alternate with 30-80 m-thick units of shale and siltstone turbidites (Hiscott et al., 1986). The sandstone packets thin, and in some cases, are replaced entirely by siltstone/shale facies, over lateral distances of 5-10 km (Enos, 1969a, b). Paleocurrents in the Cloridorme

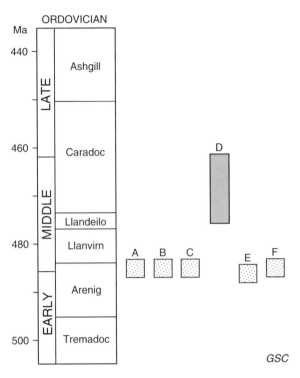

Figure 3.29. Ages of allochthonous flysch, adapted from Barnes et al. (1981) and Williams et al. (1985). For a key to unit letters, see Figure 3.27. Lithological symbols are explained in Figure 3.28.

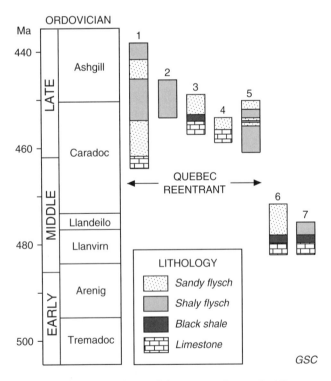

Figure 3.28. Ages of autochthonous and parautochthonous flysch, adapted from Barnes et al. (1981) and primary references. Faunal zones specified in primary references were related to European series using sheet 1 in Ross et al. (1982). For a key to unit numbers, see Figure 3.27. Units 3, 5 and 7 are locally overlain by thrust sheets.

Table 2. Maximum preserved thicknesses of flysch units (F), black shales (BS), and limestone conglomerates (CGL).

Locality and unit name (Fig. 3.27)	Thickness (m)	References
AUTOCHTHON AND PARAUTOCHTHON		
1. Nicolet Fm (F)	2175	Beaulieu et al. (1980)
2. Lotbinière Fm (F)	250	Belt et al. (1979)
3. Utica Shale (BS)	80	Belt et al. (1979)
Beaupré Fm (F)	240	Belt et al. (1979)
4. Saint-Irénée Fm (F)	320	Belt et al. (1979)
5. Cloridorme Fm (F)	4000	Hiscott et al. (1986)
6. Cape Cormorant Fm (BS + CGL)	255	Stenzel et al. (1990)
Mainland Sandstone (F)	1700	Klappa et al. (1980)
7. Black Cove Fm (BS)	3	Tuke (1968)
American Tickle Fm (F)	500	Tuke (1968); Stenzel et al. (1990)
ALLOCHTHONS		
A,B,C. Saint-Modeste Fm (F), Métis Fm (F), Tourelle Fm (F)	500	Biron (1972); Liard (1973); Vallières (1977); Hiscott (1980)
D. Deslandes Fm (F)	2300	Enos (1969a, α block)
E. Eagle Island fm (F)	200	Botsford (1987)
F. Lower Head Fm (F)	200	James and Stevens (1986)

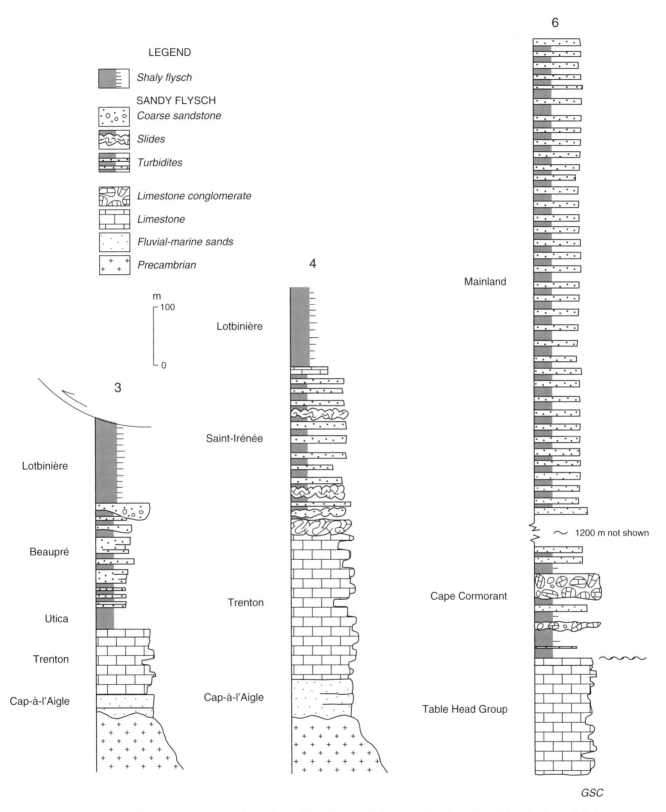

Figure 3.30. Simplified measured sections through autochthonous flysch at localities 3, 4, and 6 in Figure 3.27. Flysch units represented are the Beaupré, Saint-Irénée and Lotbinière formations, and Mainland Sandstone. Bed thicknesses are schematic. Sources are Belt et al. (1979), Klappa et al. (1980) and Stenzel et al. (1990). See also Addendum for Chapter 3.

Formation follow the axis of the foreland basin, predominantly to the west in the lower to middle part of the formation, and to the east at its top (Fig. 3.33).

In the vicinity of Québec City, the autochthonous flysch is divided into sandy and overlying shaly units: the sandy Beaupré (unit 3) and Saint-Irénée (unit 4) formations, and the overlying shaly Lotbinière Formation (unit 2; Fig. 3.30). Both the Beaupré and Saint-Irénée formations are lenticular bodies (Fig. 3.34), and pass laterally into shaly flysch, or marginal facies of the adjacent carbonate platform. Comparisons of stratigraphic sections of Belt

et al. (1979) and Beaulieu et al. (1980) indicate that the Utica Shale and Lotbinière Formation near Québec City pass westward into the lower part of the Nicolet Formation. The Nicolet Formation (Fig. 3.27, unit 1) extends along tectonic strike from the Canada-United States border to a distance of about 50 km east of Québec City (Beaulieu et al., 1980), and across strike to the northwest for about 50 km onto the St. Lawrence Platform, beyond which it is truncated by the modern erosion surface. In the core of the Chambly-Fortierville Syncline (Fig. 3.27), the Nicolet Formation is overlain conformably by redbeds of the Bécancour Formation (Clark, 1964).

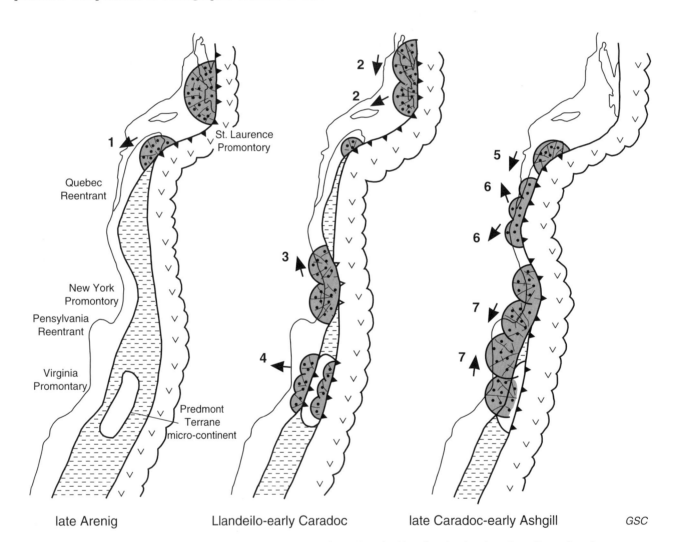

late Arenig   Llandeilo-early Caradoc   late Caradoc-early Ashgill   GSC

Figure 3.31. Summaries of paleocurrent data from Appalachian flysch showing the effect of early collision at promontories of the ancient continental margin of North America. The scissors-like collision of the continental margin and an offshore volcanic arc is inferred from the ages of flysch throughout the orogen (Hiscott, 1984; Bradley, 1989). Initial collision was in the north. Arrows summarize paleoflow obtained from the following sources: (1) Tourelle Formation, Hiscott (1978); (2) Goose Tickle Group, including Mainland Sandstone, R.K. Stevens (unpub. data); (3) Austin Glen Member of Normanskill Formation, Middleton (1965) and Fagan and Edwards (1969); (4) Tellico and Knobs formations of Tennessee and Virginia, Walker and Keller (1977) and Shanmugam and Walker (1980); (5) Cloridorme Formation, Enos (1969a); (6) Beaupré and Saint-Irénée formations, Belt and Bussières (1981); (7) Martinsburg Formation, McBride (1962). Paleoflow is generally away from promontories and toward reentrants. Piedmont Terrane microcontinent (Williams and Hatcher, 1983) was responsible for production of flysch at two different times in the southern Appalachians (Rodgers, 1971; Quinlan and Beaumont, 1984).

This is the only place in the Canadian Appalachians where terrestrial molasse sediments are preserved directly above Ordovician flysch. The Nicolet Formation has a composite thickness of 2175 m (Beaulieu et al., 1980), whereas the flysch strata near Québec City are thinner (Table 3.2). A deep seismic-reflection profile taken through the eastern end of the Chambly-Fortierville Syncline (Laroche, 1983) shows the relationship of the flysch of the Lotbinière and Nicolet formations to the Taconic thrust sheets to the south (Fig. 3.35). The flysch is in part buried by thrust sheets, and is itself somewhat telescoped by folding and thrusting.

Allochthonous flysch

Thrust slices that overlie the autochthon locally contain units of sandy flysch at the top of their stratigraphic sections. These deposits are older than flysch of the adjacent autochthon (Fig. 3.29). They overlie Lower Ordovician shales, limestone breccias, and older rocks that are interpreted as slope and rise deposits of the Cambrian-Ordovician passive margin (St-Julien and Hubert, 1975).

Newfoundland

In western Newfoundland, allochthonous flysch is found in lower structural slices of the Humber Arm Allochthon in the Port-au-Port area (Fig. 3.27, unit E, Eagle Island formation of Botsford, 1987), in the Bay of Islands area (unit E, Eagle Island formation) and north of Bonne Bay (unit F, Lower Head Formation of James and Stevens, 1986) where the basal beds of the flysch are interbedded with limestone conglomerates at the top of the Cow Head Group. The age of the flysch is Arenig to Llanvirn (Fig. 3.29). Thickness exceeds 200 m. Sole markings indicate derivation from the north to northeast (James et al., 1989). The flysch has a high sandstone:shale ratio, and resembles the allochthonous flysch of Quebec.

Quebec

In Quebec, the oldest allochthonous flysch units are the correlative Tourelle (Biron, 1972, 1974; Hiscott, 1980), Métis (Liard, 1973) and Saint-Modeste (Vallières, 1977) formations, units C, B and A, respectively, in Figure 3.27. All these units are of late Arenig to early Llanvirn age. Preserved thicknesses are no more than 500 m. The only formation studied in detail is the Tourelle. Its base is abrupt, with slides, slide scars and sandstone injections (Hiscott, 1979). The deformed underlying deposits are mapped as mélange (Biron, 1972, 1974; St-Julien and Hubert, 1975), but are best called broken formations because primary stratigraphy can be recognized, and truly exotic material is absent (Hiscott, 1977).

The Tourelle Formation generally consists of alternating packets, each many beds thick, of: (a) thinly bedded, fine grained turbidites, and (b) coarse grained, structureless or crudely stratified sandstones (Fig. 3.36). Fluid-escape structures are common. Four sandstone-filled channels, 7 m, 9 m, 10 m, and 13 m deep, are described by Hiscott (1980). The formation contains several slide horizons and some of the thicker sandstone beds contain enormous slide blocks, the largest 31 m long by 6 m thick. Some sandstone beds are in excess of 15 m thick.

In spite of its relatively proximal character, conglomerates are volumetrically minor in the Tourelle Formation, as they are in all other Canadian flysch units. Hydrodynamic calculations (Hiscott and Middleton, 1979) indicate

Figure 3.32. Interbedded shales, thin bedded sandstone turbidites, and two prominent sandstone packets (to the left) in the Petite-Vallée member of Cloridorme Formation at Petite-Vallée. Top is to the left. The section from the base of the bluff to the edge of the wavecut platform is 100 m thick.

that the turbidity currents and submarine debris flows that carried sands into the basin were capable of carrying larger gravel clasts, but these were apparently unavailable in the source area.

The Llandeilo to Caradoc Deslandes Formation (Biron, 1972, 1973) includes, in part, rocks in the "alpha" block of the Cloridorme Formation as defined by Enos (1969a). These rocks are now known to be separated from the rest of the Cloridorme by Logan's Line, the boundary between the autochthonous and allochthonous rocks in the Quebec Reentrant (Riva, 1968; St-Julien and Hubert, 1975). The Deslandes Formation forms the Marsoui River Nappe to the south of the Cloridorme Formation outcrops (Fig. 3.27). The Deslandes Formation has a stratigraphic thickness of about 2300 m (Table 3.2). Most exposures of the formation are inland (Fig. 3.27, unit D), where detailed facies analysis is difficult. Coastal outcrops can be examined, at Cap-Chat and near the tip of Gaspésie. At these localities, the Deslandes Formation consists mainly of thinly bedded turbidites and thick- to very thick-bedded, mud-rich megaturbidite beds like those described by Pickering and Hiscott (1985).

Structural and sedimentological setting

The flysch was deposited in an evolving foreland basin that extended the full length of the Appalachian Orogen from Newfoundland to Alabama. The superposition of these deep-sea deposits on top of the shallow marine Cambrian-Ordovician carbonate sequence indicates rapid subsidence to depths in the order of kilometres under the load of Taconic thrust sheets (Quinlan and Beaumont, 1984). Important elements of the foreland basin evolution also come from the U.S. Appalachians; and these are incorporated into the following model and overview.

A prominent Middle Ordovician unconformity in many parts of the Appalachian miogeocline is interpreted to represent erosion along the crest of a migrating peripheral bulge associated with loading of the continental margin as it approached an offshore trench system (Jacobi, 1981; Quinlan and Beaumont, 1984; Tankard, 1986; Knight and James, 1987; Lash, 1988; Bradley, 1989). In Newfoundland, this unconformity separates the St. George and Table Head groups (Knight and James, 1987; James et al., 1989; Knight et al., this Chapter). In contrast, an equivalent unconformity is only weakly developed or absent in the St. Lawrence Lowlands (Harland and Pickerill, 1982).

Erosion along the peripheral bulge was followed by downbending of the continental margin to form a foreland basin. Black shale deposits probably represent local sites of low siliciclastic sediment supply during the early phases of basin filling. True black shales, like the Utica Shale of southwestern Quebec, are rich in organic carbon and, by analogy with equivalent strata in New York State (Hay and Cisne, 1984), were probably deposited under anaerobic and dysaerobic conditions. There is no obvious relationship between times of early black shale deposition and times of Ordovician eustatic sea-level rise, as has been suggested for British black shales (Leggett, 1978). Nevertheless, unusually fine grained units within thick flysch formations like the Cloridorme Formation may have been deposited during times of high sea level (Hiscott et al., 1986), as proposed for turbidite systems in general (Mutti, 1985).

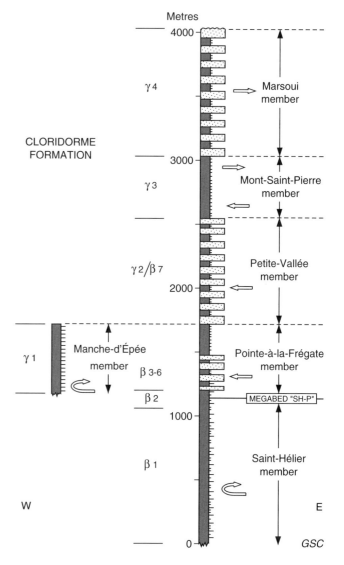

Figure 3.33. Simplified composite section through the Cloridorme Formation showing the informal members proposed by Hiscott et al. (1986), and correlation with the Greek-letter based members of Enos (1969a). The top of the Saint-Hélier member is the top of a distinctive thick megaturbidite described by Pickering and Hiscott (1985). Sandstone packets (stipple) are schematic. Black = shale; short ticks = thin bedded turbidites (Fig. 3.32); longer ticks = megaturbidites (Pickering and Hiscott, 1985). Large arrows crudely summarize paleoflow. In the Saint-Hélier and Manche-d'Épée members, megaturbidites show flow reversals because of reflections from basin margins, whereas the upper part of the Mont-Saint-Pierre member and the Marsoui member were deposited by a submarine fan that built into the foreland basin from the west (Beeden, 1983). Thick fine grained intervals in the Pointe-à-la-Frégate and Mont-Saint-Pierre members are interpreted to have formed during high stands of sea level.

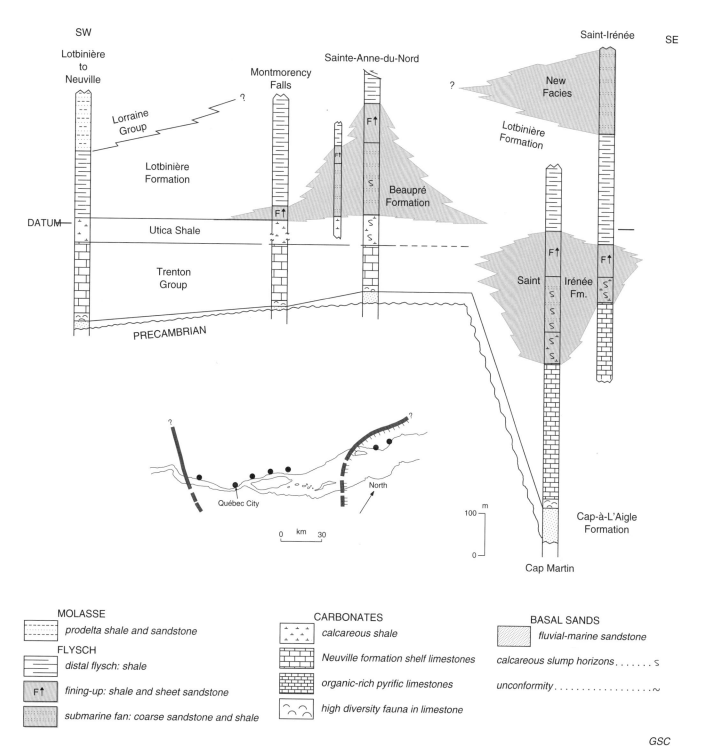

Figure 3.34. Correlation of upper Middle Ordovician strata near Québec City, with datum being the contact between the *Climacograptus spiniferus* and *Orthograptus ruedemanni* graptolite zones (simplified from Belt et al., 1979). The sandy flysch units, including an unnamed unit exposed only at the top of the most easterly section, are lenticular in shape.

Paleoflow data for flysch units throughout the Appalachians support a model for earliest collision at promontories of the ancient continental margin of North America. In the north, the oldest allochthonous flysch of southwestern Newfoundland and eastern Gaspésie was deposited close to the St. Lawrence Promontory, a site of early collision (Fig. 3.31). The ages of other allochthonous and autochthonous units can be explained by progressive collision of an island arc or arcs with the irregular continental margin (Fig. 3.31; Hiscott et al., 1983; Bradley, 1989). The young autochthonous flysch of Quebec was deposited in the Quebec Reentrant, an area that was protected from the early effects of the Taconic Orogeny.

Bradley (1989) interpreted the difference in age between allochthonous and autochthonous flysch in the same part of the orogen as a function of the distance that the allochthons were transported tectonically toward the continental margin during that time interval. He calculated distances of 270 km and 450 km, respectively, for movement of the allochthons of western Newfoundland and Gaspésie, during which time flysch was continually being supplied to the foreland basin.

The stratigraphic position of the allochthonous flysch of Newfoundland and Quebec indicates a relatively sudden change to turbidite deposition, directly on top of the undeformed deep-water fringe of a carbonate slope that extended seaward from the Cambrian-Ordovician carbonate platform. The Tourelle Formation has been interpreted as the deposit of several overlapping, small, sand-rich submarine fans (radius about 20 km; Hiscott, 1980). Most exposures show the characteristic high sandstone:shale ratios and erosional features of middle fan non-channelled and channelled lobes. The thicker beds probably represent submarine debris flow deposits (Hiscott and Middleton, 1979, 1980).

The scarcity of gravel is attributed to derivation from a source area of poorly lithified, fine grained siliciclastic sediments originally deposited on the older passive margin slope and rise (Hiscott, 1978).

Intense deformation and production of mélange below the allochthonous flysch units of Gaspésie may partly predate flysch deposition, although field relationships do not clarify timing of mélange formation (Biron, 1974). The abrupt basal contact is incompatible with fan progradation, and instead suggests deposition of the Tourelle, and by analogy the Métis and Saint-Modeste flysch, in small basins, probably on a slope formed by stacking of Taconic thrust sheets (Hiscott et al., 1986).

The Llandeilo to Caradoc Deslandes Formation, like the younger Cloridorme Formation (see below), is interpreted as a basin-floor succession (Hiscott et al., 1986). The thick megaturbidite beds represent major failures along the southern margin of the contracting foreland basin.

The parautochthonous Cloridorme Formation of Quebec is interpreted to include deposits of elongate turbidite systems (?fans) with isolated sandstone lobes (Hesse, 1982; Beeden, 1983; Hiscott et al., 1986), and deposits of confined basin floors (St-Hélier and Manche-d'Épée members, Fig. 3.33) where large turbidity currents produced by massive slope failures were reflected and deflected several times from basin slopes before final deposition (Pickering and Hiscott, 1985). In order to effectively contain these large flows in the deeper parts of the foreland basin, there must have been structural highs (saddles) along the basin axis. Crude calculations suggest spacing of about 100 km between highs (Hiscott et al., 1986). Similar segmentation of the foreland basin in Pennsylvania (Lash, 1988) has been attributed to the irregular shape of the eastern margin of ancient North America.

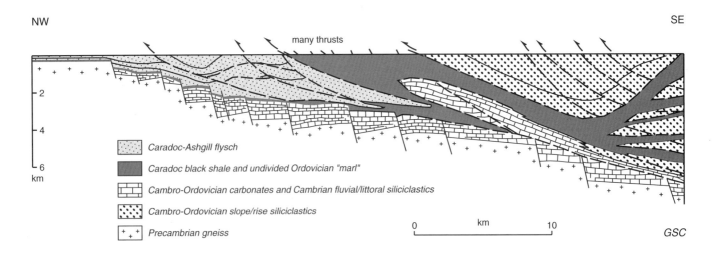

Figure 3.35. Simplified interpretation of part of a SOQUIP seismic reflection line fully described by Laroche (1983). The location of the line is shown in Figure 3.27. Foreland-basin flysch of the Lotbinière and Nicolet formations lies stratigraphically above Caradoc black shales, which lie above Cambrian-Ordovician carbonates of a passive margin sequence. The flysch was subsequently overthrust by nappes containing older rocks originally deposited farther southeast.

Figure 3.36. Alternating packets of sandstone beds (right and far left) and siltstone/shale beds (left foreground) in the Tourelle Formation at Cap Sainte-Anne. The 10-40 m-thick sandstone packets are characteristic of mid-fan deposits (Hiscott, 1980). Note the channelled base of the sandstone packet at right. Scale is 1 m long.

The transition from carbonate platform to deep foreland basin in the autochthon of western Newfoundland includes a complicated record of platform collapse, fault movements, and derivation of limestone boulder conglomerates from the crests of fault blocks (James et al., 1989; Stenzel et al., 1990). In particular, the Cape Cormorant boulder conglomerates contain a mixture of Cambrian and Ordovician carbonate clasts shed into the developing black shale and flysch environment from the tops of uplifted blocks. Elsewhere in western Newfoundland and also in the Québec City area, deepening and tilting of the foundering platform edge generated wet-sediment slides in argillaceous facies (Klappa et al., 1980; Belt and Bussières, 1981).

The autochthonous Beaupré and Saint-Irénée formations in the Québec City area have been interpreted as small submarine fans (Belt and Bussières, 1981). The older fan (Saint-Irénée Formation) is intercalated with major

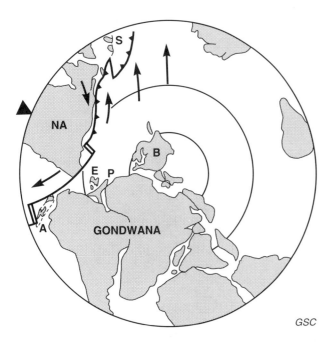

GSC

Figure 3.37. Llandeilo-early Caradoc paleocontinental base map (stereographic polar projection) from Scotese et al. (1979, their fig. 11). Southeastward-dipping subduction zone (barbed line) along eastern North America (NA) is consistent with Figure 3.31 and with current models for Taconic orogenesis; northwestward-dipping subduction zone beneath Ireland and Scotland (S) is from Leggett et al. (1979) and Leggett (1980). Vectors of relative plate motions (along small circles) and location of spreading ridges (double lines) provide one possible path to arrive at the position proposed by Scotese et al. (1979, their fig. 18) for North America relative to Gondwana and the Baltic Shield (B) in Middle Silurian (Wenlock) time. The relative motion is produced by rotating Gondwana (or North America) about a fixed point on the equator (solid triangle). E = England; A = Avalon of the northern Appalachians. P = Piedmont Terrane.

wet-sediment slides that came from the sinking margin of the carbonate platform to the northwest (Fig. 3.30). The younger fan (Beaupré Formation) contains three well-defined channel-fill sequences. In contrast, the Nicolet Formation between Montréal and Québec City appears to have been deposited on a basin floor controlled by differential subsidence, and characterized by active intrabasinal horsts (Beaulieu et al., 1980). Because of strong fault control, morphological submarine fans were apparently not formed. Steep active faults also controlled deposition of the nearby Trenton Group in New York State (Mehrtens and Parker, 1984).

Contrasts with Taconic flysch in the U.S. Appalachians

There are major compositional differences between the Ordovician flysch units in the Canadian and U.S. Appalachians. In particular, flysch of the U.S. Appalachians apparently contains little or no detrital chromite, and has chromium concentrations an order of magnitude less than those of Canadian units (Hiscott, 1984). Metamorphic rock fragments are locally more abundant (McBride, 1962), but volcanic rock fragments are less abundant, particularly in the southern U.S. Appalachians where they are essentially absent (Mack, 1985). Hiscott (1984) interpreted these differences to reflect relatively minor ophiolite obduction in the U.S. Appalachians compared to the Canadian Appalachians, perhaps as a result of more oblique plate convergence during Taconic orogenesis. Scotese et al. (1979) proposed, in a set of paleocontinental base maps, that North America rotated clockwise relative to Baltica and Gondwana from the Llandeilo/early Caradoc to the Middle Silurian (Wenlock). If one accepts the existence of a southeastward-dipping subduction zone along the eastern side of North America during much of this interval, and chooses a relative rotation pole for North America that permits the motion suggested by Scotese et al. (1979), inferred vectors of relative plate movement (Fig. 3.37) support the hypothesis of more oblique plate convergence in the U.S. Appalachians than in the Canadian Appalachians.

The second major difference between Canadian and some U.S. Ordovician flysch units is basin paleodepth. Flysch of the Martinsburg Formation of Tennessee and Virginia, and the Reedsville Formation of Pennsylvania, consists of monotonously interbedded sandstones, siltstones and shales like Canadian units, but here these facies were deposited at shelf depths under the influence of storms (Kreisa, 1981; Conrad, 1984). In the case of the Martinsburg Formation of Tennessee and Virginia, the anomalously shallow water deposition can be attributed to an early collisional event, in which a buoyant micro-continental block, the Piedmont Terrane (Williams and Hatcher, 1983), collided with the ancient continental margin of North America. This phase of deposition has been equated with the "Blountian phase" of the Taconic Orogeny (Rodgers, 1971). Subsequent deposition of the Martinsburg Formation during the main phase of the Taconic Orogeny was on top of, or just cratonward of, this accreted micro-continent, at isostatically-controlled water depths much less than those over the thinner attenuated continental crust present elsewhere along the continental margin.

Taconic allochthons

Harold Williams

Introduction

Transported rocks of Taconic allochthons, above the Middle Ordovician clastic unit of the autochthonous section, are a spectacular structural feature of the Humber Zone. Contorted marine shales, sandstones, and mélanges of the allochthons contrast sharply with the underlying mildly deformed carbonate sequence. Volcanic rocks and ophiolite suites seem particularly out-of-context with respect to Grenville basement and its cover rocks. Taconic allochthons make up a large proportion of the exposed parts of the external Humber Zone, especially in Quebec (Fig. 3.38).

Two well-known discrete allochthons occur in western Newfoundland: the Hare Bay Allochthon at the tip of the Great Northern Peninsula and the Humber Arm Allochthon between St. George's Bay and Portland Creek (Williams, 1975a). The Old Mans Pond Allochthon (Williams et al., 1982) is regarded as an eastern outlier of the Humber Arm Allochthon. The Southern White Bay Allochthon (Smyth and Schillereff, 1982) is another example on the east side of the Great Northern Peninsula (Fig. 3.38).

Transported rocks in Quebec form a series of structural slices, the Lower St. Lawrence River nappes (St-Julien and Hubert, 1975), that extend uninterrupted from Gaspésie to Québec City and the U.S. border in northern Vermont. The largest nappes from north to south are the Sainte Anne River, Marsoui River, Mount Logan, Lower St. Lawrence Valley, Chaudière, Sainte-Hénédine, Bacchus, Pointe-de-Lévy, Québec Promontory, Granby, and Stanbridge (Fig. 3.38).

The western edge of the transported rocks is Logan's Line, first defined at Québec City (Logan, 1863; Stevens, 1974) and traceable southward to Vermont (Williams, 1978a). Its northern course is covered by the St. Lawrence River, but it is exposed near the tip of Gaspésie. In Newfoundland, Logan's Line is discontinuous and defined by the leading edges of the Humber Arm and Hare Bay allochthons (Fig. 3.38).

Taconic allochthons of the external Humber Zone are described for Newfoundland and serve as an example of structural styles found all along the west flank of the Appalachian Orogen in Canada. Allochthons are absent or poorly defined in internal parts of the Humber Zone, except for occurrences of small disrupted ophiolitic rocks and mélanges. These are treated under the general heading "Humber Zone internal" (Hibbard et al., Chapter 3).

Newfoundland

Harold Williams

Taconic allochthons of western Newfoundland exhibit a variety of igneous and metamorphic rocks in upper structural slices that are rare or absent in most Appalachian examples. Especially well-known are the ophiolite suites and metamorphic soles of the Bay of Islands Complex in the Humber Arm Allochthon and the St. Anthony Complex of the Hare Bay Allochthon (Williams and Smyth, 1973; Williams, 1975a). A complete stratigraphy from late Precambrian to Early Ordovician has been deciphered among sedimentary rocks of the Humber Arm Allochthon and some facies, such as the Cambrian-Ordovician Cow Head breccias, are renowned for their sedimentological and paleontological characteristics (James and Stevens, 1986; Barnes, 1988). Gros Morne National Park, situated in the Humber Zone of western Newfoundland, was declared a World Heritage Site in 1987 based mainly on its rocks and geological relationships.

The Humber Arm and Hare Bay allochthons occupy structural depressions. The Humber Arm Allochthon is surrounded to the east by the Cambrian-Ordovician carbonate sequence and separated from the Hare Bay Allochthon by a major structural culmination that exposes Grenville basement of the Long Range Inlier. The Old Man's Pond Allochthon is separated from the Humber Arm Allochthon by a narrow culmination that exposes the carbonate sequence. The Southern White Bay Allochthon occurs on the east side of the broad anticlinal Long Range Inlier. Because these structural depressions and culminations are later than the emplacement of the allochthons, the present exposures are erosional remnants of once-more-continuous sequences that covered larger areas of western Newfoundland.

History of ideas

When Schuchert and Dunbar (1934) summarized the geology of western Newfoundland, they attempted to set up a single stratigraphic column that included transported rocks of the Humber Arm and Hare Bay allochthons at the top of the Cambrian-Ordovician carbonate sequence. There were difficulties, however, as some of their upper stratigraphic units contained fossils older than those of underlying units. At that time, plutonic rocks of the allochthons were viewed as intrusions (Ingerson, 1935; Buddington and Hess, 1937; Smith, 1958). Suggestions of significant lateral transport to explain the position and contrasts between the Cambrian-Ordovician carbonate sequence and coeval sedimentary and volcanic

Williams, H.
1995:   Taconic allochthons: Introduction (Humber Zone); in Chapter 3 of Geology of the Appalachian-Caledonian Orogen in Canada and Greenland, (ed.) H. Williams; Geological Survey of Canada, Geology of Canada, no. 6, p. 99 (also Geological Society of America, The Geology of North America, v. F-1).

Williams, H.
1995:   Taconic allochthons in Newfoundland (Humber Zone); in Chapter 3 of Geology of the Appalachian-Caledonian Orogen in Canada and Greenland, (ed.) H. Williams; Geological Survey of Canada, Geology of Canada, no. 6, p. 99-114 (also Geological Society of America, The Geology of North America, v. F-1).

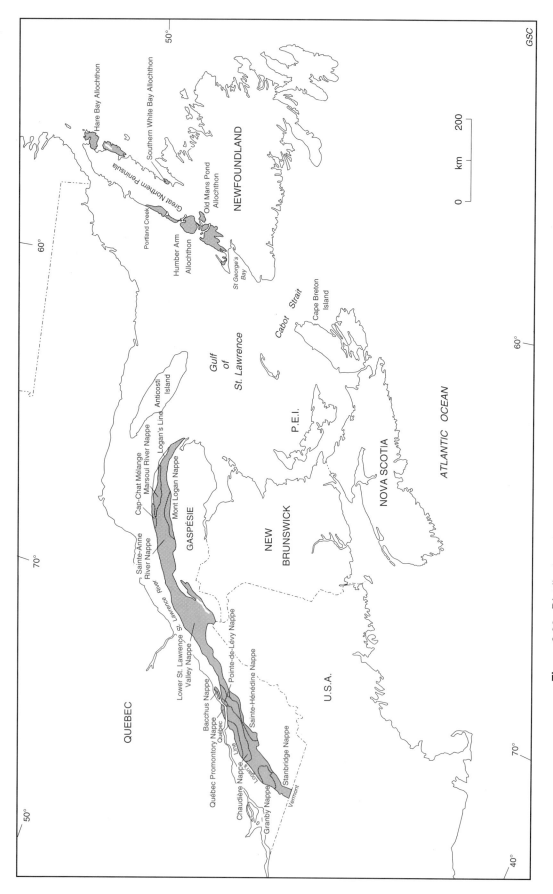

Figure 3.38. Distribution of Taconic allochthons in the Canadian Appalachian region.

rocks of the overlying sequences are credited to Johnson (1941) and Kay (1945). However, the implications of this idea and first meaningful analysis is that of Rodgers and Neale (1963), based mainly on parallel relationships in New York and the demonstration of allochthonous rocks there. They reasoned that the deep-water clastic rocks were deposited far to the east of the shallow water carbonate sequence and subsequently transported westward to over-lie the carbonate sequence. They further interpreted the plutonic rocks, such as those of the Bay of Islands Complex, as integral parts of the transported sequences; both because of their spatial affinity with the transported rocks, and like the sedimentary and volcanic rocks, they appear out-of-place within the famework of the carbonate sequence of western Newfoundland. According to the view of Rodgers and Neale (1963), the plutonic rocks of the ophiolite suites were intrusions, emplaced into the sedimentary and vol-canic rocks at their place of deposition and all transported together.

Lilly (1963) and Stevens (1965) worked out the strati-graphy of the Humber Arm Supergroup and Stevens (1970) implied that the ophiolite suites represented oceanic crust in separate structural slices (see also Cooper, 1936; 1937; Tuke, 1968). Steven's analysis of the Humber Arm Allochthon as a sampling of rocks from a continental mar-gin and adjacent ocean has been substantiated by all sub-sequent work. For more insight into this topic the reader is referred to Rodgers and Neale (1963), Stevens (1970), Williams (1971a), and Williams and Stevens (1974).

Previous work

The first comprehensive mapping and studies of parts of the Humber Arm and Hare Bay allochthons were by Cooper (1936, 1937); Betz (1939); Troelsen (1947); Walthier (1949); Smith (1958); Lilly (1963); Stevens (1965, 1968); Gillis (1966); Tuke (1968); and Smyth (1971). The southern por-tion of the Humber Arm Allochthon is included in the reconnaissance map of the Stephenville area (Riley, 1962) and its northern part in the reconnaissance map of the Sandy Lake area (Baird, 1960). Since the advent of plate tectonics, a multitude of studies focus on transported rocks of the Humber Arm and Hare Bay allochthons. These are listed in the bibliographies of Williams (1975a, 1985a) and Williams and Smyth (1983). All of the Humber Arm and Hare Bay allochthons have been remapped at 1:100 000 and 1:50 000 scales (Williams, 1973, 1985a, b; Williams et al., 1983, 1984, 1985b; Williams and Smyth, 1983; Williams and Cawood, 1986; Cawood and Williams, 1986; Cawood et al., 1987). A map of the entire Hare Bay Allochthon is available at 1:125 000 scale (Williams and Smyth, 1983) and a compilation map of the Humber Arm Allochthon at 1:250 000 scale (Williams and Cawood, 1989).

The Southern White Bay Allochthon was defined during reconnaissance studies in northwest Newfoundland (Williams, 1977a) and it was named and studied in more detail by Smyth (1981) and Smyth and Schillereff (1982). The Old Mans Pond Allochthon was defined by studies in the Pasadena area (Williams et al., 1982; 1983; Gillespie, 1983; Waldron and Milne, 1991).

Williams (1975a) summarized relationships within the Humber Arm and Hare Bay allochthons as they were known at that time. The present account is based largely on that summary, augmented by more current information.

The Humber Arm Allochthon is treated first as many of our present concepts and stratigraphic relationships were developed from this example.

Humber Arm Allochthon

The Humber Arm Allochthon is 200 km long and about 50 km across at its widest part (Fig. 3.39). Its structural thickness is in the order of a few kilometres. Because the intensity of post-emplacement deformation increases east-ward across the allochthon, relationships are clearest in the west. At its western leading edge, sedimentary rocks of the allochthon are separated by mélange from gently-dipping sandstones and shales at the top of the autochthonous sequence. Toward the east, its boundaries are steep and locally faulted with stratigraphic omission at the top of the autochthonous sequence. Relationships are confused by post-emplacement folds and thrusts of eastward polarity between Georges Lake and Corner Brook (Bosworth, 1985; Waldron, 1985; Williams and Cawood, 1986), and by west-erly thrusts that bring cover rocks and basement above the allochthon north of Bonne Bay (Williams et al., 1985b; Williams, 1985b; Williams et al., 1986; Cawood and Williams, 1988; Grenier and Cawood, 1988; Grenier, 1990). At Port au Port Peninsula, the Middle Ordovician (Caradoc) Long Point Group overlies sedimentary rocks of the Humber Arm Allochthon. The contact has been interpreted as an unconformity, following the lead of Rodgers (1965). Offshore seismic data suggest a fault, with the Long Point Group thrust eastward above the allochthon as the upper portion of a triangle zone (Stockmal and Waldron, 1990).

Sedimentary rocks of lower structural slices of the Humber Arm Allochthon are assigned to the Humber Arm Supergroup (Stevens, 1970). It is subdivided into the Curling Group (Stevens, 1970) in the vicinity of Humber Arm and southward, the informal Bonne Bay group (Quinn and Williams, 1983; Williams et al., 1984; Quinn, 1985) between Bay of Islands and Bonne Bay, and the Cow Head Group and Lower Head Formation (Kindle and Whittington, 1958; Williams et al., 1985b; James and Stevens, 1986) to the north of Bonne Bay. Other units are the informal Weasel and Pinchgut groups in separate slices at the base of the allochthon along its eastern margin (Williams et al., 1984; Williams and Cawood, 1986).

Four volcanic and plutonic units occur in higher slices of the allochthon. From structurally lowest to structurally highest, these are: (1) Skinner Cove Formation, Fox Island Group and related volcanic rocks (Troelsen, 1947; Williams, 1973; 1985a; Williams et al., 1984; Williams and Cawood, 1989), (2) Old Man Cove Formation (Williams, 1973), (3) Little Port Complex (Williams and Malpas, 1972; Williams, 1973) and the related Mount Barren Complex (Karson, 1977, 1979; Williams, 1985a), and (4) Bay of Islands Complex (Cooper, 1936; Smith, 1958; Williams, 1973; 1985a; Williams and Cawood, 1989). A complete section of stacked units is nowhere exposed and probably nowhere exists. In most places the higher slices directly overlie sedimentary rocks but locally they overlap and overlie one another. The stacking order is built up from local relationships.

Two conspicuous carbonate slivers occur beneath the Bay of Islands Complex and above mélange of lower structural levels. These are the Serpentine Lake and Fox Island River slices (Godfrey, 1982; Williams and Godfrey, 1980a, b).

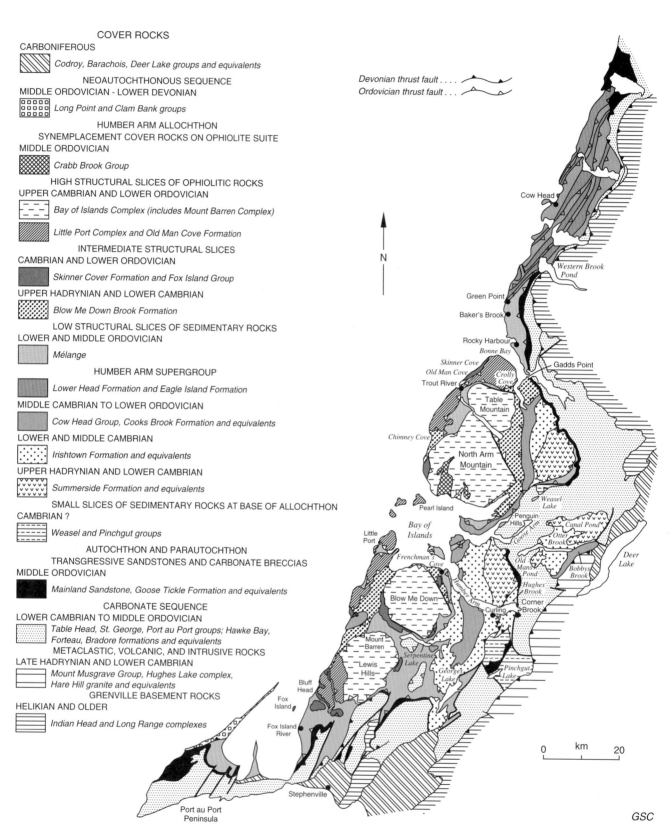

COVER ROCKS
CARBONIFEROUS
Codroy, Barachois, Deer Lake groups and equivalents
NEOAUTOCHTHONOUS SEQUENCE
MIDDLE ORDOVICIAN - LOWER DEVONIAN
Long Point and Clam Bank groups
HUMBER ARM ALLOCHTHON
SYNEMPLACEMENT COVER ROCKS ON OPHIOLITE SUITE
MIDDLE ORDOVICIAN
Crabb Brook Group
HIGH STRUCTURAL SLICES OF OPHIOLITIC ROCKS
UPPER CAMBRIAN AND LOWER ORDOVICIAN
Bay of Islands Complex (includes Mount Barren Complex)
Little Port Complex and Old Man Cove Formation
INTERMEDIATE STRUCTURAL SLICES
CAMBRIAN AND LOWER ORDOVICIAN
Skinner Cove Formation and Fox Island Group
UPPER HADRYNIAN AND LOWER CAMBRIAN
Blow Me Down Brook Formation
LOW STRUCTURAL SLICES OF SEDIMENTARY ROCKS
LOWER AND MIDDLE ORDOVICIAN
Mélange
HUMBER ARM SUPERGROUP
Lower Head Formation and Eagle Island Formation
MIDDLE CAMBRIAN TO LOWER ORDOVICIAN
Cow Head Group, Cooks Brook Formation and equivalents
LOWER AND MIDDLE CAMBRIAN
Irishtown Formation and equivalents
UPPER HADRYNIAN AND LOWER CAMBRIAN
Summerside Formation and equivalents
SMALL SLICES OF SEDIMENTARY ROCKS AT BASE OF ALLOCHTHON
CAMBRIAN ?
Weasel and Pinchgut groups
AUTOCHTHON AND PARAUTOCHTHON
TRANSGRESSIVE SANDSTONES AND CARBONATE BRECCIAS
MIDDLE ORDOVICIAN
Mainland Sandstone, Goose Tickle Formation and equivalents
CARBONATE SEQUENCE
LOWER CAMBRIAN TO MIDDLE ORDOVICIAN
Table Head, St. George, Port au Port groups; Hawke Bay, Forteau, Bradore formations and equivalents
METACLASTIC, VOLCANIC, AND INTRUSIVE ROCKS
LATE HADRYNIAN AND LOWER CAMBRIAN
Mount Musgrave Group, Hughes Lake complex, Hare Hill granite and equivalents
GRENVILLE BASEMENT ROCKS
HELIKIAN AND OLDER
Indian Head and Long Range complexes

Devonian thrust fault . . . .
Ordovician thrust fault . . .

Figure 3.39. General geology of the Humber Arm Allochthon, modified after Williams and Cawood (1989). For more detail the reader is referred to GSC Map 1678A – Geology of the Humber Arm Allochthon.

They are gently dipping, upright sections of thick bedded, white and grey limestones and dolomites of the St. George and Table Head groups. A smaller example occurs at Penguin Hills (Lilly, 1967; Williams et al., 1983). These are interpreted either as integral parts of the allochthon, entrained during Middle Ordovician assembly and transport, or features of later gravity collapse and late stage transport of ophiolites and parts of the carbonate sequence across the already-assembled allochthon (Cawood, 1989).

Humber Arm Supergroup. The Curling, Bonne Bay, and Cow Head groups are all partly correlative and their successions of stratigraphic units are for the most part relatively clear. The rocks occur in a number of lower structural slices, especially evident in the case of the Cow Head Group (Fig. 3.40).

The Curling Group (former Humber Arm Series, Schuchert and Dunbar, 1934; Humber Arm Group, Smith, 1958) comprises the Summerside, Irishtown, Cooks Brook, Middle Arm Point, and Eagle Island formations, from bottom to top. There is no single complete section and the overall succession is built up from partial sections exposed at or near Humber Arm. The Summerside Formation consists of monotonous quartz greywackes with purple slate interbeds. Its thickness is estimated at a kilometre or more with base unexposed. The Irishtown Formation consists of dark grey shale with prominent thick white quartzite units and local polymictic conglomerates containing plutonic clasts of Grenville basement and a variety of sedimentary rocks including fossiliferous Lower Cambrian limestones. In contrast, the middle formations of the Curling Group represent a condensed sequence of thin bedded shales and platy limestones with prominent limestone breccia units (Cooks Brook Formation) overlain by black and green shales and buff siltstones (Middle Arm Point Formation). Thicker polymictic clastic rocks reappear at the top of the group (Eagle Island formation) containing chromite and other ophiolite detritus (Stevens, 1970; Botsford, 1987).

The Summerside Formation is considered Early Cambrian or late Precambrian, by correlation with similar units along the west flank of the Humber Zone (Williams and Stevens, 1974) and similar rocks (Blow Me Down Brook Formation) in an overlying structural slice contain *Oldhamia* (Lindholm and Casey, 1989). Early Cambrian limestone boulders in Irishtown conglomerates indicate the formation is no older than Early Cambrian. Limestone at the base of the Cooks Brook Formation is Middle Cambrian and the Eagle Island and Middle Arm Point formations contain Early Ordovician graptolites (Stevens, 1970; Erdtmann and Botsford, 1986).

The Bonne Bay group was defined because of uncertain correlations between its formations and those of the Curling Group. The Mitchells and Barters formations correlate with the Summerside and Irishtown formations of the Curling Group. The McKenzies formation is a combined Cooks Brook-Middle Arm Point correlative that lacks limestone breccias so prominent at Humber Arm.

The Cow Head Group (Kindle and Whittington, 1958; Williams et al., 1985b; James and Stevens, 1986) is a thin (300 m) sequence of coarse limestone breccias, platy limestones, and dolomitic shales of Middle Cambrian to late Early Ordovician age. It is overlain by polymictic sandstones and conglomerates of the Lower Head Formation.

The limestone breccias and sandstones occur in east-dipping, east-facing stratigraphic sections that are repeated in northeast-trending belts. The coarsest limestone breccias occur in western belts with finer, thinner and fewer breccias in eastern belts (Fig. 3.40). The Cow Head Group correlates with the Cooks Brook and Middle Arm Point formations of the Curling Group and the McKenzies formation of the Bonne Bay group. Lower Head sandstones are correlatives of the Eagle Island formation at the top of the Curling Group.

The Cow Head Group is famous for its coarse limestone breccias and contained faunas. In a general way, fossiliferous limestone blocks are of the same age as enclosing shales (Kindle and Whittington, 1958; James and Stevens, 1986; Pohler, 1987). The stratigraphic sections contain shelly faunas, graptolites, and conodonts that allow comparisons among biostratigraphic subdivisions using different fossil assemblages. For this reason, the Cow Head Group is a potential world stratotype for the Cambrian-Ordovician boundary (Barnes, 1988).

The Weasel and Pinchgut groups that occur in separate slices at the base of the allochthon near Humber Arm are possible correlatives of the autochthonous carbonate succession and the allochthonous Cow Head Group. Grey to buff dolomitic shales and platy grey limestones resemble the Reluctant Head Formation of the autochthon. Carbonate breccias, some with a sandy lime matrix, resemble breccias of the Cow Head Group.

The Blow Me Down Brook Formation and correlative Sellars formation form an extensive and continuous structural slice of clastic sedimentary rocks above the Curling and Bonne Bay groups, respectively. The rocks are coarse quartz-feldspar sandstones derived from a crystalline source. Red pillow lavas and breccias occur at the stratigraphic base of the Blow Me Down Brook Formation at Woods Island. The sandstones at Blow Me Down Brook were first interpreted as part of the highest stratigraphic unit of the Curling Group (Stevens, 1970). The name Sellars formation was introduced because the rocks at Bonne Bay were interpreted as late Precambrian or Early Cambrian and, while lithologically similar and almost physically continuous with the Blow Me Down Brook Formation, they were correlated with the Summerside and Mitchells formations (Quinn, 1985). Recent discovery of *Oldhamia* in the Blow Me Down Brook Formation indicates an Early Cambrian age (Lindholm and Casey, 1989).

Mélanges are extensively developed among sedimentary rocks of the Humber Arm Allochthon (Fig. 3.39). Between Bonne Bay and Stephenville, western portions of the sedimentary allochthon are mainly mélange whereas eastern portions comprise intact stratigrahic sections. The largest area of mélange occurs between Fox Island River and Port aux Port Peninsula. Another large area extends from Green Point to Bonne Bay and continues as a narrow zone at the base of the allochthon from Bonne Bay to Bay of Islands. Some of the mélanges were given local names: Companion Mélange at Frenchmans Cove (Williams, 1973), Gadds Point mélange (Williams et al., 1984), Rocky Harbour mélange (Williams, 1985b), and Crolly Cove mélange (Williams, 1985b; Williams and Cawood, 1989).

Mélanges consist mainly of greywacke, quartzite, dolomitic shale, chert, and limestone blocks in a black, green, and red scaly shale matrix. Volcanic blocks occur in

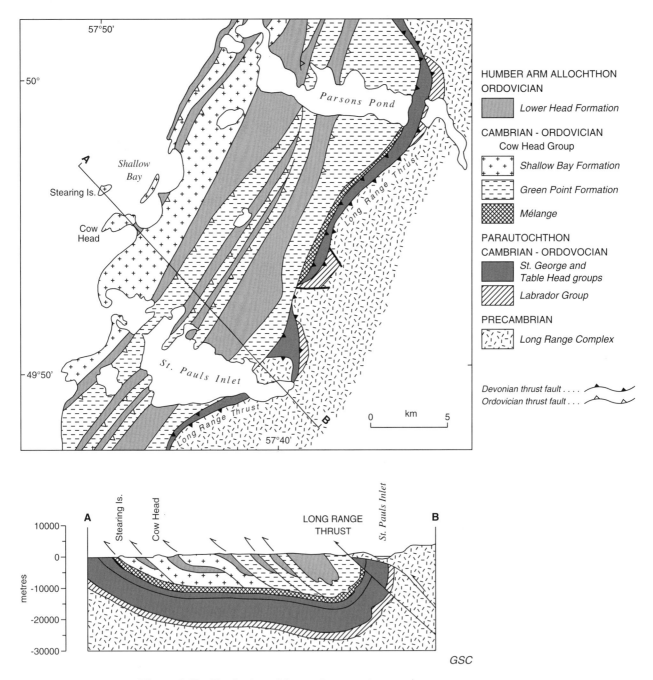

Figure 3.40. Distribution of thrust slices and facies belts of the Cow Head Group.

some structurally lower mélanges and they are common in structurally higher mélanges. A mélange along the east side of Lewis Hills contains a sampling of ophiolitic blocks including amphibolite and greenschist of the Bay of Islands metamorphic sole (Williams, 1985a). Mélange with serpentinite and gabbro blocks in a greasy green serpentinite matrix is exposed in a roadcut at the base of the Blow Me Down ophiolite slice at Humber Arm (Cawood et al., 1988). Serpentinite mélange also occurs below the Little Port Complex farther west. North of Bonne Bay, the Rocky Harbour mélange contains huge blocks of limestone breccia, and at Bakers Brook it contains large areas of shale, quartzite, and conglomerate with Lower Cambrian fossiliferous limestone clasts.

The Rocky Harbour mélange forms a continuous zone at the base of the allochthon from Western Brook Pond to Humber Arm. It is concordant and gradational with the underlying clastic unit at the top of the autochthon. Steep stratigraphic contacts between formations of the overlying Bonne Bay group and a steep structural contact between the McKenzies and Sellars formations all terminate abruptly and at high angles against the Rocky Harbour mélange.

On the scale of an outcrop, blocks in mélange range from a centimetre or less to a metre or more in diameter. Most are round or equidimensional. At Fox Island River, volcanic blocks range in diameter from a few metres to a few kilometres. Blocks beyond a kilometre across are oblate to discoidal in form and the largest are the flat ophiolite massifs of the Bay of Islands Complex. There is every gradation between the smallest equidimensional blocks and the largest structural slices.

Skinner Cove Formation, Fox Island group and related rocks. The Skinner Cove Formation (Williams, 1973), formerly the Skinner Cove Volcanics (Troelsen, 1947), is a remarkably unaltered and relatively undeformed sequence of layered volcanic rocks with minor sedimentary interbeds. Correlatives occur as far south as Bluff Head (Fig. 3.39). They overlie sedimentary rocks and mélanges of the Humber Arm Supergroup and they are overlain by higher igneous slices. In the type area at Skinner Cove, the formation constitutes a uniform southeast-dipping succession beneath the Old Man Cove Formation and Little Port Complex (Fig. 3.41). At localities farther south, the Little Port Complex and Bay of Islands Complex overlie the volcanic rocks. Local exposures of volcanic rocks along the west side of Table Mountain are interpreted as large blocks torn from the main Skinner Cove slice and embedded in mélange.

The rocks include black pillow basalts with white or pink limestones between pillow interstices, red volcanic breccias, latitic flows, distinctive agglomerates with black basalt fragments in a white limestone matrix, pink to grey shales and minor limestones. The Skinner Cove volcanic rocks have pronounced alkali chemical characteristics. Detailed descriptions and chemical data are given by Baker (1978) and descriptions of possible correlatives by Schillereff (1980), Godfrey (1982), and Quinn (1985). Shelly fossils and graptolites reported previously as dating the Skinner Cove Formation (Williams, 1975a) are from mélange that is not in stratigraphic continuity with other rocks of the formation.

Large discrete volcanic blocks, some a few kilometres across, occur along the east side of the Bay of Islands massifs from Bonne Bay to Fox Island River (Fig. 3.42). These are mainly green and red pillow lavas and pillow breccias with minor pink limestone and shale. Some resemble the Skinner Cove Formation and others resemble pillow lava of ophiolite suites. Occurrences at Bonne Bay are named the Crouchers formation (Williams et al., 1984) and those farther south, the Fox Island group (Williams and Cawood, 1989). A spatial association with Sellars sandstones between Bonne Bay and Bay of Islands suggests that some of the volcanic blocks may be dislodged from a stratigraphic section such as the Blow-Me-Down Brook Formation at Woods Island of Humber Arm.

Old Man Cove Formation. The Old Man Cove Formation (Williams, 1973), previously included in the Skinner Cove Volcanics (Troelsen, 1947), constitutes a single small slice that occurs between the Skinner Cove Formation and overlying Little Port Complex (Fig. 3.41). Its rocks are polydeformed greenschists with minor marble beds, interpreted as original tuffs and limestones. Undeformed mafic dykes are an integral part of the structural slice and they cut the deformed and metamorphosed rocks. The dykes are of Skinner Cove affinity, and therefore suggest a link between the Skinner Cove volcanic rocks and predeformed rocks of the Old Man Cove slice (W.S.F. Kidd, pers. comm., 1988). The age of the Old Man Cove Formation is unknown. It resembles lithologies in the metamorphic soles of the Bay of Islands and St. Anthony complexes (Williams and Smyth, 1973), and also greenschists of the Birchy Complex

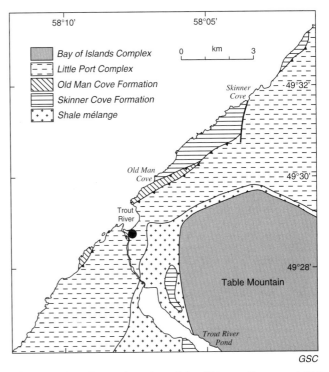

Figure 3.41. General setting of the Skinner Cove and Old Man Cove formations with respect to nearby groups and structural slices.

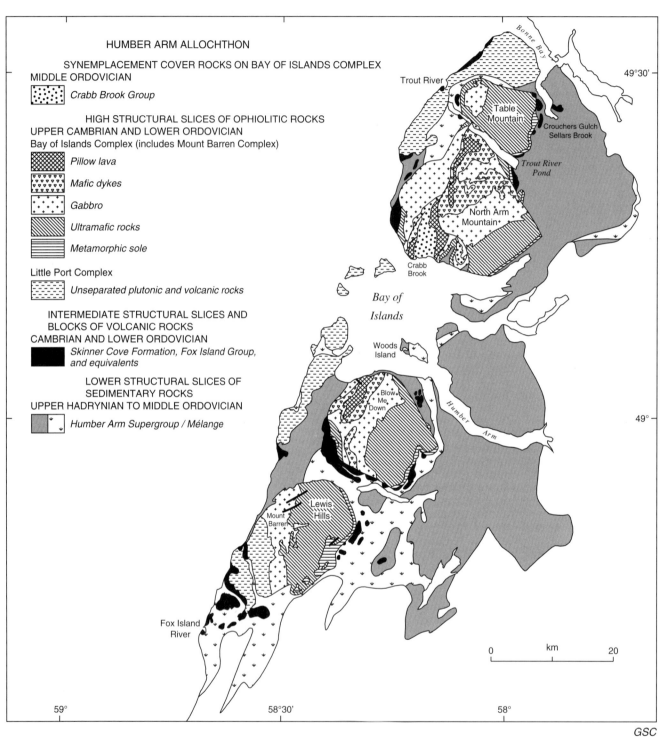

Figure 3.42. Distribution of discrete volcanic blocks and slices within the Humber Arm Allochthon and internal lithic units of the Bay of Islands Complex.

(Williams et al., 1977; Hibbard, 1983) at the Baie Verte Peninsula. Its structural position beneath the Little Port Complex suggests it may represent a partially preserved metamorphic sole related to the tranport of the Little Port Complex.

Little Port Complex. The Little Port Complex (Williams and Malpas, 1972; Williams, 1973) forms narrow northeast-trending coastal slices between Lewis Hills and Bonne Bay (Fig. 3.39). Its plutonic rocks were first regarded as part of the Bay of Islands Igneous Complex (Smith, 1958). Its rocks have also been called the Coastal complex (Karson and Dewey, 1978), an informal field term, now dropped. The Little Port Complex overlies the Humber Arm Supergroup in most places. Between Trout River and Bonne Bay it overlies the Skinner Cove and Old Man Cove formations (Fig. 3.41). The Bay of Islands Complex is separated from the Little Port Complex by a wide valley, except at Lewis Hills. There, deformed gabbros, peridotites and tonalite gneisses of the Mount Barren Complex (Karson, 1979; Williams, 1985a) occur between the Little Port Complex to the west and the Bay of Islands Complex to the east; all part of the Lewis Hills massif.

The oldest rocks of the Little Port Complex are foliated layered gabbros, amphibolites and minor peridotites. Quartz diorites or tonalites cut deformed amphibolites and produce intrusion breccia on Big Island of the Bay of Islands. The granitic rocks vary from massive to well-foliated and occur as northeast-trending bodies that parallel the form of the Little Port structural slices. On the south side of the Bay of Islands, the succession of Little Port slices trends northwest and the long dimension of granitic bodies and main foliations in gabbroic rocks all trend in the same northwest direction (Comeau, 1972).

Massive mafic dykes cut foliated gabbros, amphibolites and granitic rocks and the dykes are inseparable from mafic volcanic rocks that are also part of the complex. The dykes trend northeast in most places and form thin sheeted sets that separate deformed plutonic rocks from nearby, relatively undeformed volcanic rocks. The dykes are almost everywhere brecciated (Williams and Malpas, 1972).

Volcanic rocks of the Little Port Complex are mainly green and red mafic pillow lavas and pillow breccias. Porphyritic dacite and grey silicic flows occur within the northern slice between Chimney Cove and Bonne Bay. Volcanic boulder conglomerate and sandstone are included in the complex at Little Port. Locally, the volcanic rocks contain prehnite, pumpellyite and analcime (Zen, 1974).

The Little Port Complex has the components of an ophiolite suite. However, the proportions of its rock types, especially an abundance of tonalite, and intricate internal structures, contrast with the orderly succession and proportions of pristine units in the nearby Bay of Islands Complex.

Tonalites of the Little Port Complex at Trout River yield a U-Pb zircon ages of 508 ± 5 Ma (Mattinson, 1975; Williams, 1975b) and 505 +3/-2 Ma (Jenner et al., 1991).

Bay of Islands Complex. The Bay of Islands Complex occurs in four separate massifs (Fig. 3.42). The two northernmost massifs, Table Mountain and North Arm Mountain, are separated by a left lateral tear fault at Trout River Pond. A similar structural style and internal makeup

of the Blow-Me-Down massif suggests it was once part of the same northern slice. The Lewis Hills massif has an internal structure that contrasts with northern massifs. The Bay of Islands Complex overlies the Humber Arm Supergroup and locally overlies Skinner Cove correlatives. It is interpreted to form the highest structural slice at Trout River (Fig. 3.42) although its contact with the Little Port Complex is not exposed.

The Bay of Islands Complex (Cooper, 1936; Williams, 1973; Malpas, 1976), also termed the Bay of Islands Igneous Complex (Cooper, 1936; Smith, 1958), has a complete ophiolite suite of rock units from ultramafic rocks, through gabbros and sheeted dykes to mafic pillow lavas. As well, it includes a metamorphic sole of polydeformed amphibolites and greenschists welded to the stratigraphic base of its ultramafic unit (Williams and Smyth, 1973; Malpas, 1979, Jamieson, 1986). Complete sections are present at Blow-Me-Down and North Arm Mountain where the ophiolite suites are disposed in synclines with northeast-trending subhorizontal axes (Fig. 3.42). At North Arm Mountain, the ophiolite suite is unconformably overlain by Llandeilo breccias of the Crabb Brook Group, which is an integral part of the North Arm Mountain massif (Casey and Kidd, 1981).

The tectonic bases of the Bay of Islands massifs are subhorizontal, so that the ophiolite units are truncated structurally at depth in much the same manner as they are truncated erosionally at the surface (Fig. 3.42). Thus the approximately 10 km stratigraphic sections of the ophiolite suite occur in structural slices that are less than a kilometre thick (Williams, 1975a, 1985a). The morphological expressions of the massifs refute the idea of steeply dipping basal contacts at North Arm Mountain (Casey and Kidd, 1981) and steeply dipping contacts beneath the Little Port Complex at Trout River (Idleman, 1986).

Trondjemite of the Bay of Islands Complex is dated at 504 ± 10 Ma (Mattinson, 1976) and 505 Ma (Jacobsen and Wasserburg, 1979), which are identical within the limits of analytical error to a date of 508 ± 5 Ma for the Little Port Complex (Mattinson, 1975; Williams, 1975b). More precise U-Pb zircon ages for the Bay of Islands Complex are 485.7 +1.9/-1.2 Ma (Dunning and Krough, 1985) and 484 ± 5 Ma (Jenner et al., 1991). ^{40}Ar-^{39}Ar ages on hornblendes from the metamorphic sole of the complex are 469 ± 5 Ma (Dallmeyer and Williams, 1975) and 464 ± 9 Ma (Archibald and Farrar, 1976; recalculations after Dunning and Krough, 1985).

The Bay of Islands Complex has prehnite-pumpellyite metamorphism in its basaltic pillow lavas. Local areas of volcanic rocks with zeolite minerals (mainly laumontite) at the top of the pillow lava unit are interpreted as remnants of a zeolite facies zone that once existed above the prehnite-pumpellyite zone. Metamorphism increases to greenschist and low amphibolite facies at the sheeted dyke level and upper parts of the underlying gabbro. This metamorphism is a static hydration without penetrative fabrics.

Old Mans Pond Allochthon

The Old Mans Pond Allochthon lies to the east of the Humber Arm Allochthon and occupies a circular area of about 300 km^2 between Bay of Islands and Deer Lake (Fig. 3.39). The two allochthons are less than 10 km apart where they are separated by exposures of the carbonate sequence from Hughes Brook to Goose Arm. The Old Mans Pond Allochthon can be viewed therefore as an erosional outlier of the Humber Arm Allochthon.

The Old Mans Pond Allochthon is surrounded by rocks of the carbonate sequence, except along its southeastern margin where it is faulted against the Hughes Lake Complex. Structures are complex and a post emplacement penetrative cleavage that affects all rocks in the area dips moderately northwest. Contacts between the Old Mans Pond Allochthon and the surrounding carbonate sequence are poorly exposed Nowhere are the allochthonous rocks above the stratigraphic top of the carbonate sequence; rather different units within the allochthon are juxtaposed with carbonate rocks from different levels of the autochthonous section. Most contacts are marked by valleys, suggesting steep faults. Two isolated occurrences of altered gabbro and serpentinized ultramafic rocks, obviously tectonic blocks, occur at the northern periphery of the allochthon (Williams et al., 1983).

Rocks of the Old Mans Pond Allochthon are assigned to the informal Old Man's Pond group (Williams et al., 1982, 1983), which is divisible into three formations, namely Canal Pond, Otter Brook and Bobbys Brook formations. The Canal Pond formation occupies the northern portion of the allochthon and occurs in a small outlier at its northern periphery. It is an arenaceous unit consisting of grey to green and pink greywackes, grey to white quartzites, quartz pebble conglomerates, and green to purple slates. the Otter Brook formation is the most extensive and underlies the broad central portion of the allochthon. It consists primarily of dark grey slates and siltstones with prominent units of white thick bedded quartzites and quartz pebble conglomerates. The Bobbys Brook formation, along the southeast margin of the allochthon, consists of thin-bedded grey marbles and grey phyllites with local occurrences of oolitic limestone and limestone breccia, some of which contain button algae in their matrices.

The order of stratigraphic units of the Old Man's Pond group has not been worked out within the allochthon and the ages of the units are unknown. However, lithic similarities between rocks of the Old Man's Pond group and the Curling Group imply correlation. The Canal Pond formation resembles the Summerside Formation at the base of the Curling Group, the Otter Brook formation is virtually identical to the succeeding Irishtown Formation, and the Bobbys Brook resembles the overlying Cooks Brook Formation. These correlations imply an ascending stratigraphy in the Old Man's Pond group from northwest to southeast. Button algae and local oolitic beds in the Bobbys Brook formation further imply correlation with the Cambrian Penguin Cove and Reluctant Head formations of the carbonate sequence (Williams et al., 1982, 1983).

Hare Bay Allochthon

The Hare Bay Allochthon extends from Canada Bay to Cape Bald at the tip of the Great Northern Peninsula and has an area of about half that of the Humber Arm Allochthon. It contains six contrasting rock groups that make up separate structural slices (Fig. 3.43). As in the case of the Humber Arm Allochthon, no single vertical section exhibits all six rock groups, so that the order of structural stacking is built up from observations and relationships throughout the allochthon. Again, the lowest structural slices comprise

sedimentary rocks separated by mélange (Northwest Arm, Maiden Point, and Irish formations) overlain by a structurally higher mélange (Milan Arm Mélange) and volcanic rocks (Cape Onion Formation), all capped by an ophiolite complex (St. Anthony Complex). Intensity of post-emplacement deformation increases from west to east. North of Hare Bay the slices are subhorizontal, especially the White Hills ophiolite slice whose basal contact is everywhere at or near sea level. South of Hare Bay, the Maiden Point slice is domed or broken by faults to expose the autochthonous Goose Tickle Formation (Cooper, 1937) and Table Head Group at Whites Arm (Williams and Smyth, 1983).

Northwest Arm Formation. The Northwest Arm Formation (Cooper, 1937) is of limited areal extent and occurs only at the western margin of the allochthon. It has a chaotic internal structure, resulting from assembly and transport of the allochthon. Stratigraphic contacts are absent and the chaotic nature of the formation precludes thickness estimates. The formation consists of thin-bedded black and green shales, in most places graphitic and pyritic, with interbeds of grey and buff-weathering limy siltstone, grey to brown sandstone, white to grey limestone, limestone breccia, buff to dark grey and green chert, and siliceous argillite. Shale beds form a smeared and streaky matrix around more resistant boudins or blocks.

The contact between the Northwest Arm Formation and underlying autochthonous rocks of the Goose Tickle Formation is exposed on the west side of Triangle Point at Pistolet Bay and at several localities along the north shore of Northwest Arm in Hare Bay. In all localities, an abrupt but uneven contact separates well-bedded greywackes of the Goose Tickle Formation and chaotic dark shaly rocks with sandstone and limestone blocks of the Northwest Arm Formation.

Shales of the Northwest Arm Formation contain *Staurograptus dichotomus*, which indicates an Early Ordovician age (Tuke, 1968), older than the underlying Middle Ordovician Goose Tickle Formation. Correlatives in the Humber Arm Allochthon are the Middle Arm Point and McKenzies formations and parts of the Cow Head Group.

Maiden Point Formation. The Maiden Point Formation (Tuke, 1968) includes the former Maiden Point Sandstone (Cooper, 1937) and Canada Head Formation (Betz, 1939). Volcanic rocks are abundant toward the base of the formation at St. Lunaire and northward (Williams and Smyth, 1983). Monotonous greywackes, slates, and quartz-pebble conglomerates compose about 90 per cent of the formation. Graded beds up to a metre thick are common in most sections, and purple to red slate units are prominent locally.

Diorite and gabbro sills and plugs are associated with the sedimentary and volcanic rocks. The intrusive rocks are commonest at the structural top of the formation, where it is in contact with the overlying St. Anthony Complex. The thickness of the formation is estimated at 2 km in its type section along the south side of Hare Bay (Cooper, 1937; Betz, 1939; Tuke, 1968). The volcanic rocks at St. Lunaire are about 150 m thick. These are altered mafic agglomerates, tuffs, and basaltic lavas that are chemically similar to tholeiites of the Lighthouse Cove Formation (Smyth, 1973).

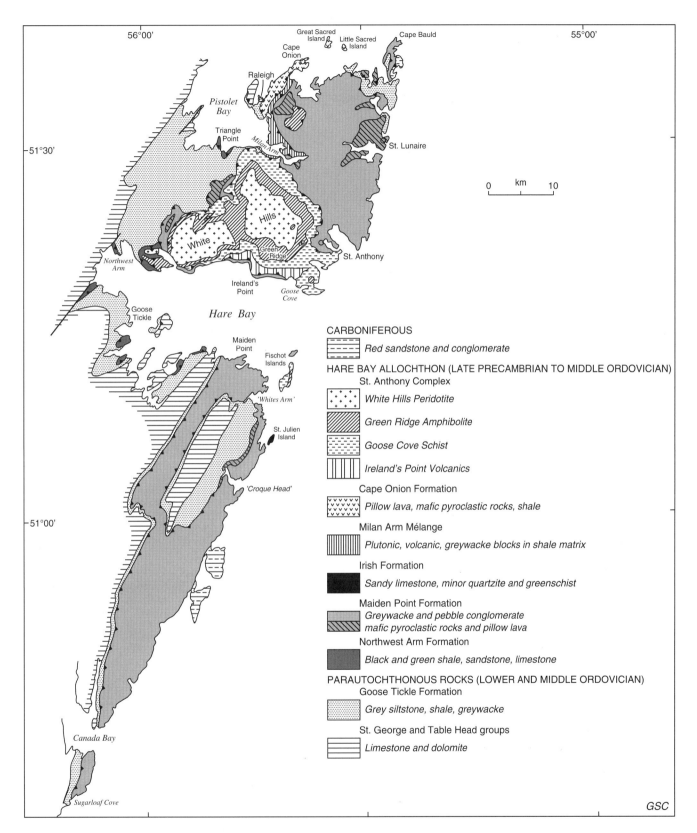

Figure 3.43. General geology of the Hare Bay Allochthon.

109

The Maiden Point Formation constitutes two separate slices. The larger, 130 km by 30 km, is the most extensive slice within the Hare Bay Allochthon. At Croque Head, it is overlain by a smaller slice (Croque Head slice of Smyth, 1973), and the two are separated by mélange. In most places the main slice overlies the autochthonous Goose Tickle Formation. Elsewhere, it overlies the Northwest Arm Formation. Its structural base is marked by mélange, except for local sharp contacts at Croque and at Canada Bay. Metamorphic rocks (Sugarloaf Schist Member of Goose Tickle Formation, Williams and Smyth, 1983) form a distinct unit at the base of the Maiden Point slice south of Canada Bay.

Early recumbent folds about subhorizontal axes face westerly and in the general direction of tectonic transport. Post-emplacement folds are upright with steep east-dipping axial surfaces (Williams and Smyth, 1983).

A preponderance of blue quartz in Maiden Point grey-wackes, and granite and metamorphic rock fragments in coarser beds, indicate a provenance from nearby Grenville basement. A late Precambrian to Early Cambrian age for the formation is suggested by correlation with the Summerside Formation of the Humber Arm Supergroup.

Irish Formation. The Irish Formation (Williams and Smyth, 1983) occurs in a small slice at St-Julien Island. It consists of brecciated quartzites overlain by siliceous lime-stones and slates, and a thicker upper unit of sandy lime-stones. Conglomerates in faulted contact to the east are interpreted as Silurian (Smyth and Schillereff, 1982), rather than an integral part of the Ordovician allochthon (Williams and Smyth, 1983).

The age of the Irish Formation is unknown and there are no direct correlatives in the Humber Arm Allochthon. The admixture of carbonate and quartz sand suggests derivation from Lower and Middle Cambrian impure limestones and quartz sandstones near the base of the autochthonous sequence.

Milan Arm Mélange. The Milan Arm Mélange (Williams and Smyth, 1983) is in most respects similar to the mélanges that separate all structural slices of the Hare Bay Allochthon. However, it is areally more extensive, it has a variety of exotic blocks, it occupies a distinct structural position, and some of its blocks are mappable at 1:25 000 scale. The mélange overlies the Maiden Point Formation toward the east and the autochthonous Goose Tickle Formation along its western margin at Pistolet Bay. South of Pistolet Bay, the mélange is overlain by the structurally and topographically higher St. Anthony Complex. The Cape Onion Formation occurs at the same structural level as the largest tabular igneous blocks within the Milan Arm Mélange, so that the Cape Onion slice could be viewed as an exotic at the same structural level.

The commonest blocks of the Milan Arm Mélange are sepentinized peridotite, mafic volcanic rocks, amphibolite, foliated gabbro, greywacke, diorite, and exceptionally coarse grained pyroxenite and hornblendite associated with tonalite and hornblende-biotite schist. Nephrite also occurs as small blocks at Milan Arm (R.K. Stevens, pers. comm., 1988). Some of these lithologies are unique to the Milan Arm Mélange.

Many of the amphibolite, gabbro, and diorite blocks along the north shore of Milan Arm are encased in a relatively thin, hard rind of light grey calc-silicate alteration products (rodingite). The tough and resistant alteration halos form wave-washed outcrop surfaces where the matrix shales are eroded. In other places, the rodingite alteration halos are surrounded by a thinner serpentinite coating, implying that the rodingite represents an alteration zone between mafic rock and serpentinite. The blocks were therefore once surrounded by serpentinite or serpentinite mélange, so that they are recycled where they are now embedded in shale. The Little Port Complex also includes amphibolites with rodingite reaction products where Little Port amphibolites are in contact with serpentinized peridotite.

Cape Onion Formation. The Cape Onion Formation (Williams and Smyth, 1983) consists of black to green basaltic pillow lavas with local agglomerate and tuff units and minor black shales. It constitutes a structural slice at Cape Onion Peninsula that overlies the Maiden Point Formation. A small erosional outlier at Raleigh overlies the Goose Tickle Formation. The best section is exposed along the east side of Pistolet Bay and impressive exposures of pillow lavas occur at Great Sacred Island and Little Sacred Island, northeast of Cape Onion. The moderately northeast-dipping section at Pistolet Bay is about a kilometre thick. Carbonate fills pillow intersticies and vesicules, and it forms the matrix of some tuffs.

Rocks of the Cape Onion Formation are relatively unde-formed compared to those of the underlying Maiden Point Formation and overlying St. Anthony Complex. Pillows have their original shapes and fossiliferous shales inter-layered with the volcanic rocks are essentially uncleaved. A lack of post-emplacement deformation is attributed to the westerly position of the formation. *Dictyonema flabelliforme* and other graptolites from black shales interlayered with mafic pillow lavas at Cape Onion date the formation as Tremadoc (Williams, 1971a).

Pillow lavas of the Cape Onion Formation are more alkalic than those of the Bay of Islands Complex. They compare more closely with volcanic rocks of the Little Port Complex or Skinner Cove Formation (Jamieson, 1977).

St. Anthony Complex. The St. Anthony Complex (Williams and Smyth, 1983) constitutes the structurally high-est slice of the Hare Bay Allochthon. The White Hills Peridotite at its top is typical of ultramafic rocks that occur at the stratigraphic bases of ophiolite suites. The underlying Green Ridge Amphibolite, Goose Cove Schist, and Ireland Point Volcanics form a metamorphic sole beneath the peridotite.

The St. Anthony Complex has a structural history that predates its emplacement. Structural contrasts with underlying rocks are everywhere pronounced, and locally, blocks of greenschist and volcanic rocks occur in black shale mélange beneath the St. Anthony slice. The St. Anthony Complex overlies the Maiden Point Formation of the allochthon in most places and it overlies other formations including the Goose Tickle Formation of the autochthon west of White Hills. A separate area of greenschists and amphibolites near Cape Onion may be an erosional outlier or a block in the Milan Arm Mélange. The St. Anthony

Complex is contained in a subhorizontal slice north of Hare Bay. Outcrop patterns of its formations are the topographic expression of subhorizontal units. Eastward at Fischot Islands, the slice dips moderately east and an offshore aeromagnetic anomaly suggests the presence of a steep ultramafic sheet.

The Ireland Point Volcanics (Cooper, 1937) consists of red and green volcanic breccias and pillow lavas. The rocks vary from undeformed at the structural base of the unit to schistose at its structural top.

The Goose Cove Schist (Cooper, 1937) occurs above the Ireland Point Volcanics. It is a polydeformed and metamorphosed sequence of green tuffs, agglomerates, and mafic pillow lavas, with thin units of greywacke, black pyritic slate, and marble. Primary features of the rocks are retained only in areas of relatively mild deformation. The thickness of the unit is estimated at 180 m. Metamorphic mineral assemblages include chlorite, albite, epidote, muscovite, tremolite, and actinolite. A first schistosity defined by platy minerals is affected by later recumbent folds. Axial surfaces are subhorizontal and roughly parallel to the first schistosity and the structural base of the White Hills Peridotite. Second phase folds in greywackes of the Goose Cove Schist face west and southwest at Fischot Islands (Smyth, 1973).

The Green Ridge Amphibolite (Williams and Smyth, 1983) overlies the Goose Cove Schist and occurs beneath the White Hills Peridotite. Green to black hornblende-plagioclase schists are the commonest rock type. These are derived mainly from mafic volcanic rocks and gabbros, although no primary features are preserved. Some of the rocks are heterogeneous, suggesting a coarse pyroclastic derivation, and some banded varieties probably represent layered tuffs. The hornblende-plagioclase schists commonly display increasing grain size across the lithic unit from structural base to top. Garnetiferous amphibolites and pyroxene-bearing amphibolites are common toward its top. The structural thickness of the unit is estimated at 120 m.

The White Hills Peridotite (Williams and Smyth, 1983) forms tabular erosional remnants of a single peridotite sheet at White Hills and westward. The rocks are less than 300 m in structural thickness. Harzburgites, dunites and minor pyroxenites make up the unit. Harzburgites are medium- to coarse-grained and everywhere display a foliation that is produced by flattened orthopyroxenes. Dunites are finer grained and equigranular, with less pronounced foliation. Irregular bands of dunite and pyroxenite in harzburgites are parallel to foliation, except where the bands are affected by recumbent folding. Other pyroxenite bands cross an earlier banding and tectonic fabric.

Finely banded, hard, recrystallized ultramafic rocks occur at the base of the White Hills Peridotite. Leucocratic bands of deep brown amphibole, colourless clinopyroxene, brown biotite, and garnet alternate with darker bands of serpentinite. The presence of attenuated relict orthopyroxenes with glide lamellae, relict chromite, and ceylonite indicates that these rocks are a metamorphosed basal part of the White Hills Peridotite. Similar rocks occur at the base of a small ultramafic outlier near St. Anthony, and they are associated with ultramafic rocks in the Milan Arm Mélange.

Rocks of the St. Anthony Complex are undated except for ^{40}Ar-^{39}Ar cooling ages of 490-485 Ma for amphiboles of its metamorphic sole (Dallmeyer, 1977b). These ages are somewhat older than determinations on equivalent rocks of the Bay of Islands Complex, suggesting that ophiolite emplacement progressed from north to south along the Newfoundland Humber Zone (see also Lux, 1986).

Southern White Bay Allochthon

The Southern White Bay Allochthon (Smyth and Schillereff, 1982) occurs on the west side of White Bay between Coney Head and Jacksons Arm, with narrow thrust slices extending southward along Doucers Valley (Fig. 3.44). Most allochthonous rocks are tonalites, gabbros and granites of the Coney Head Complex (Williams, 1977a). These are bordered to the west by a narrow steep zone of greenschists, Murrays Cove Schist (Lock, 1972); greywackes, Maiden Point Formation (Smyth and Schillereff, 1982); and ophiolitic mélange, Second Pond Mélange (Williams, 1977a). Slates and argillites of the Taylors Pond Formation (Heyl, 1937) in the same steep

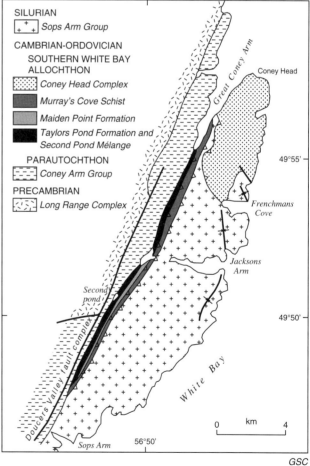

Figure 3.44. General geology of the Southern White Bay Allochthon.

zone may be part of the allochthon (Smyth and Schillereff, 1982) or represent the deformed top of the autochthon that borders the allochthon to the west (Williams, 1977a).

The Coney Head Complex consists of an intrusive suite of trondhjemites and tonalites, gabbros and quartz gabbros, biotite granites and muscovite leucogranites. Gabbros are intruded by tonalites producing spectacular intrusion breccias. The tonalite-gabbro phases are cut by pink biotite granites and coarse grained muscovite granites. Zircons from granitic rocks are dated at 474 ±2 Ma (Dunning, 1987a). Tuffs of the Silurian Sops Arm Group lie unconformably on the Coney Head Complex north of Frenchmans Cove.

The Murray's Cove Schist occurs to the west and therefore structurally below the Coney Head Complex. It is a polydeformed greenschist derived from mafic tuffs and breccias with thin calcareous tuffs, red chert, and pods of metagabbro. Williams (1977a) assigned the Murray's Cove Schist to the Coney Head Complex and interpreted the contact between schists and intrusive rocks as gradational. This suggested that the Murray's Cove Schist was a dynamothermal metamorphic sole to transported igneous rocks of the Coney Head Complex (Williams, 1977a). Smyth and Schillereff (1982) interpreted the contact between schists and intrusive rocks as a brittle fault with a thrust wedge of Silurian rocks present between the Murray's Cove Schist and igneous rocks of the Coney Head Complex. West of Jacksons Arm, rocks mapped as Murray's Cove Schist are gradational with Silurian volcanic rocks of the Sops Arm Group (Smyth and Schillereff, 1982). This raises the possibility that the Murray's Cove Schist is Silurian and structurally imbricated with older components of the Southern White Bay Allochthon.

Medium- to fine-grained green schistose greywackes form a narrow steep band and tectonic slivers to the west of the Murray's Cove Schist in the Doucers Valley fault complex. The greywackes contain blue quartz grains and interbeds of quartz pebble conglomerate. They are interpreted as part of the Southern White Bay Allochthon and assigned to the Maiden Point Formation as they are identical to rocks of the Hare Bay Allochthon to the north (Smyth and Schillereff, 1982).

The Second Pond Mélange occurs in the same steep structural zone at Doucers Valley and consists of green sandstone, quartzite, serpentinite and talc blocks in a black slate matrix. It is interpreted to occur at or near the western margin and base of the allochthon.

The occurrence of dated Ordovician, Silurian and Carboniferous rocks in the vicinity of the Southern White Bay Allochthon provides a rare insight into the effects of the various Paleozoic orogenies that affected eastern parts of the Humber Zone.

The presence of the Second Pond Mélange, combined with structural and metamorphic styles, lithic contrasts, and regional geological relationships at southern White Bay all indicate Ordovician emplacement of the Southern White Bay Allochthon above the Cambrian-Ordovician carbonate sequence (Coney Arm Group). West-facing recumbent folds with a penetrative muscovite-biotite schistosity in clastic rocks of the carbonate sequence, and a greenschist schistosity in allochtonous rocks, are related to Ordovician transport and emplacement (Lock, 1972; Williams, 1977a; Smyth and Schillereff, 1982). The Silurian Sops Arm Group, unconformably above the Southern White Bay

Allochthon, was affected by upright folding and it is cut by the syntectonic to post-tectonic Gull Lake granite to the south, dated at 372 ±3 Ma (H. Baadsgaard, pers. comm., 1990). Open upright folds in the eastern Sops Arm Group are tighter and westerly inclined at Doucers Valley, and Silurian rocks are imbricated with Ordovician rocks in the Doucers Valley fault complex. These are effects of middle Paleozoic Acadian Orogeny.

Greywackes and gabbros that occur farther south on the west side of White Bay at Hampden are interpreted as part of the Southern White Bay Allochthon (Smyth and Schillereff, 1982). These are juxtaposed with folded Carboniferous rocks of the Anguille Group along steep faults. Farther west, subhorizontal Carboniferous rocks of the Deer Lake Group overlap the Doucers Valley fault complex. Locally, Silurian rocks of the Sops Arm Group are thrust upon the Deer Lake Group.

Summary, interpretation and significance

Taconic allochthons of the Canadian Appalachians are composite. The lower structural slices contain upper Precambrian-Middle Ordovician sedimentary rocks. Higher structural slices are volcanic rocks and ophiolite suites. Mélanges, consisting of resistive sedimentary, volcanic and plutonic blocks in a scaly shale matrix, are developed between the slices.

Rocks of lower sedimentary slices of the Humber Arm Allochthon (Curling, Bonne Bay, Cow Head groups) are mainly eastern deeper water correlatives of the Humber Zone carbonate sequence, but relationships at deposition are nowhere preserved. All rocks fit the model of an evolving continental margin from its initiation through its passive development, to eventual destruction (Stevens, 1970; Williams and Stevens, 1974; James and Stevens, 1986). The oldest allochthonous clastic rocks (Summerside, Mitchells, Canal Pond, Blow-Me-Down Brook, Sellars, Maiden Point) are correlated with autochthonous rift-related clastics (Bateau, Bradore, Mount Musgrave). They were derived from a crystalline Grenville source before and during continental breakup and before the development of the carbonate sequence. Volcanic rocks, such as those at the base of the Blow-Me-Down Brook Formation and within the Maiden Point Formation, are possible Lighthouse Cove correlatives. Overlying shales, quartzites and polymictic conglomerates with Lower to Middle Cambrian carbonate clasts (Irishtown, Barters, Otter Brook) are coeval with breakup and the initiation of the carbonate sequence (Williams and Hiscott, 1987). The condensed limestone breccia, thin platy limestone, shale units (Cow Head, Cooks Brook, McKenzies, Bobbys Brook) are the continental slope/rise equivalents of the carbonate shelf sequence. At Cow Head, repeated belts are interpreted as thrust slices developed during transport and emplacement. The pattern of abundant coarse limestone breccias in western slices and finer, thinner and fewer breccias in eastern slices implies a proximal (west) to distal (east) facies arrangement that is still discernible across the thrust slices (Fig. 3.40). The Old Man's Pond group may provide a link between basal parts of the carbonate sequence (Reluctant Head Formation) and the condensed facies of the Cow Head and Curling groups. Overlying sandstones (Eagle Island, Lower Head) with chromite and other ophiolite detritus were derived from the east, signaling a major reversal in sedimentary provenance. The sandstones also transgress the carbonate sequence (Mainland, Goose Tickle) but they appear earliest in the allochthonous sequences (Hiscott, this chapter).

Volcanic rocks and ophiolite suites of higher structural slices are of unknown paleogeography with respect to other rocks of the Humber Zone.

The Skinner Cove volcanic rocks have been interpreted as oceanic islands (Strong, 1974), possibly seamounts (Baker, 1978; Meaux et al., 1988). Alternatively, they could represent rift-facies volcanic rocks like those at the base of the Blow-Me-Down Brook Formation.

The Bay of Islands Complex is interpreted as oceanic crust and mantle by all recent workers. Metamorphism in volcanic rocks and dykes is interpreted as a depth-controlled static seafloor hydration. A lack of metamorphism in deeper gabbros possibly reflects an absence of surficial fluids necessary to accomplish the hydration.

The metamorphic soles of the ophiolite suites, now subhorizontal or steeply dipping surfaces frozen into the ophiolite sequences, are interpreted as high temperature shear zones resulting from transport of hot mantle and oceanic crust (Williams and Smyth, 1973; Malpas, 1979; Jamieson, 1986). The best example occurs beneath the White Hills Peridotite of the Hare Bay Allochthon where the Ireland Point Volcanics, Goose Cove Schist and Green Ridge Amphibolite were accreted to the base of the peridotite during its displacement from an oceanic setting to its emplacement across a continental margin. Contacts between rock units within the accreted metamorphic sole were first interpreted as gradational, implying a profound metamorphic pressure/temperature gradient (Williams and Smyth, 1973). More recent work indicates that the Goose Cove Schist-Green Ridge Amphibolite contact is a ductile shear zone and that the Ireland Point Volcanics-Goose Cove Schist contact is a complex structural boundary (Jamieson and Talkington, 1980; Jamieson and Strong, 1981; Craw, 1983; Jamieson, 1986). The Green Ridge Amphibolite is retrograded toward its base; the Goose Cove Schist is prograded toward its top. The overall inverted metamorphic gradient is therefore fortuitous. The lack of an obvious break is explained by the high temperatures and ductile style of deformation that overprinted both higher and lower grade rocks in a shear zone where epidote-amphibolite mineral assemblages reflect the ambient conditions (temperatures of 500 to 600°C and pressures of 3 kilobars, Jamieson, 1986). The juxtaposition of oceanic and continental margin rocks along this shear zone suggests that it represents the interface between down-going and overriding plates in a subduction zone. The lithologies and pressure/temperature paths suggest assembly of the St. Anthony Complex at a depth of 10 km in a subduction zone where the geothermal gradient was abnormally high because the overriding plate was hot. Rocks of the upper plate were metamorphosed somewhere else and they were cooling when juxtaposed with the continental plate (Jamieson, 1986).

The Mount Barren Complex of Lewis Hills may provide an important genetic link between the Little Port and Bay of Islands complexes. One interpretation relates intense deformation in the Mount Barren and Little Port complexes to an oceanic fracture zone or transform fault that juxtaposed deformed rocks of the Little Port Complex with

pristine rocks of the Bay of Islands Complex (Karson and Dewey, 1978). Other studies indicate similar structural histories for parts of the Little Port and Bay of Islands complexes (Dunsworth et al., 1986; Dunsworth and Calon, 1988). An island arc setting for the Little Port Complex is suggested by its tonalite geochemistry (Malpas, 1979).

A recent integrated geochronological, isotopic, and geochemical study of the Little Port and Bay of Islands complexes shows: (1) a significant age difference between a U-Pb zircon age of 505 +3/-2 Ma for the Little Port Complex and a U-Pb zircon and baddaleyite age of 484 ± 5 Ma for the Bay of Islands Complex, (2) Little Port trondhjemites are characterized by initial ε values of -1 to +1 whereas those in the Bay of Islands Complex are +6.5, and (3) geochemical signatures in mafic and felsic volcanic rocks of the complexes are diverse and show a complete gradation between volcanic arc and non-volcanic arc patterns (Jenner et al., 1991). These data suggest that the complexes were not coeval parts of the same ophiolite suite connected by a mid-ocean ridge transform fault. An alternative interpretation relates the Little Port Complex to a volcanic arc and the Bay of Islands Complex to a supra-subduction zone ophiolite suite (Jenner et al., 1991).

Structural stacking orders within the allochthons indicate that the highest volcanic and ophiolitic slices are the farther traveled. The occurrence of ophiolitic detritus in allochthonous Lower Ordovician sandstones of lower slices and the local presence of volcanic and ophiolitic blocks in basal mélanges, indicate assembly from east to west and emplacement as already assembled allochthons.

Western parts of the Humber Arm Allochthon have extensive mélanges developed among the lower slices of sedimentary rocks. These are interpreted as the result of surficial processes with later structural overriding. The leading edges of the higher structural slices in both the Humber Arm and Hare Bay allochthons have a tendency to disrupt, repeat, and disintegrate.

Complex internal structures of the lower sedimentary slices were developed during transport and emplacement. As a generality, the lowest sedimentary slices contain the stratigraphically youngest rocks. The slices have little morphological expression and their outlines and external geometries are in places poorly known. Higher igneous slices have clearer morphological expression and sharper boundaries. Some of their internal structures predate transport. Others relate to transport and emplacement.

The earliest indication of assembly of the allochthons is the reversal in sedimentary provenance as recorded in the stratigraphy of lower structural slices. This and other features indicate diachroneity of assembly and emplacement along the length of the Canadian Appalachians with earlier events in Newfoundland compared to Quebec (Hiscott, this chapter). In Newfoundland, the easterly derived clastic rocks are as old as Arenig. The youngest rocks of the underlying autochthon are the Llanvirn Goose Tickle Formation and equivalents. The North Arm Mountain massif of the Bay of Islands Complex has an unconformable cover of Llandeilo age, interpreted as coeval with transport (Casey and Kidd, 1981), and the Caradoc Long Point Group was deposited after emplacement of the Humber Arm Allochthon.

Ordovician structures are confused in places where middle Paleozoic thrusts bring Grenville basement rocks and the Cambrian-Ordovician carbonate sequence above the Taconic allochthons. Polarity of structures is mainly westward. In a few places middle Paleozoic structures have eastward polarity. Final emplacement of uppermost slices of ophiolitic and/or volcanic rocks and carbonate slivers may relate to gravity collapse of an overthickened allochthon (Cawood, 1989).

Preservation of allochthons such as those in western Newfoundland suggests burial soon after emplacement. A lack of ophiolitic and other detritus in Silurian, Devonian and Carboniferous rocks suggests prolonged burial. Present exposure is the result of Mesozoic and Tertiary uplift.

Humber Zone internal

J.P. Hibbard, P. St-Julien and
W.E. Trzcienski, Jr.

Introduction

The internal Humber Zone is a well-defined area of intensely deformed and metamorphosed rocks (Fig. 3.45). These metamorphic rocks contrast sharply with less deformed and relatively unmetamorphosed rocks of flanking areas. They are mainly polyphase deformed, greenschist to amphibolite facies metaclastic rocks that are equivalent, in large part, to mildly deformed allochthonous and authochthonous clastic sequences of the external Humber Zone to the west. The internal Humber Zone is tectonically juxtaposed with volcanic-plutonic assemblages of the Dunnage Zone to the east along a steep, narrow boundary, the Baie Verte-Brompton Line (Williams and St-Julien, 1982). This tectonic juncture of Humber and Dunnage zones has strongly influenced the stratigraphic and tectonothermal evolution of the internal Humber Zone.

The internal Humber Zone extends discontinuously for the entire length of the Canadian Appalachians, from Grey Islands in northwest Newfoundland, southward to the Quebec-Vermont border (Fig. 3.45). It follows the sinuous course of the orogen, from the Newfoundland Reentrant, southwest into the St. Lawrence Promontory and Québec Reentrant (Williams, 1978a). In Newfoundland, the continuity of the zone is disrupted by Carboniferous rocks that unconformably overlie the deformed rocks; in Quebec, Silurian-Devonian rocks locally overlie the older deformed rocks. The maximum width of the internal Humber Zone is approximately 45 km at the Quebec-Vermont border. The belt is narrower in Newfoundland.

In most places, the Baie Verte-Brompton Line separates intensely deformed metaclastic rocks from mafic-ultramafic assemblages. In southwest Newfoundland, the line may coincide with the Cabot Fault. An extensive Ordovician plutonic terrane (Dashwoods Subzone, Map 2), immediately east of

Hibbard, J.P., St-Julien, P., and Trzcienski, W.E., Jr.
1995: Humber Zone internal; *in* Chapter 3 of Geology of the Appalachian-Caledonian Orogen in Canada and Greenland, (ed.) H. Williams; Geological Survey of Canada, Geology of Canada, no. 6, p. 114-139 (<u>also</u> Geological Society of America, The Geology of North America, v. F-1).

the Cabot Fault, envelops large tracts of paragneiss that may be equivalent to similar rocks in the internal Humber Zone to the northwest (A. Howse, P. Dean, pers. comm., 1983).Westward-directed thrusts bring metaclastic rocks of the internal Humber Zone onto lesser deformed parts of the external Humber Zone in Quebec and parts of Newfoundland. In the Grand Lake-Sandy Lake area of Newfoundland, the steeply-dipping Cabot Fault transects the internal Humber Zone; northward at southern White Bay, the Cabot Fault defines its western margin (Fig. 3.45).

The internal Humber Zone comprises two major pre-Middle Ordovician lithotectonic elements: (1) an infrastructure of gneisses and subordinate schists, and (2) a cover sequence of metaclastic rocks with intercalations of metavolcanic rocks and marble, and tectonic slivers of mafic-ultramafic rocks. Structural contrasts between the infrastructure and the cover mainly relate to early Paleozoic events.

Unravelling the stratigraphic and tectonothermal history of the internal Humber Zone has led to some of the liveliest and ongoing controversies in Appalachian geology (reviewed in St-Julien and Hubert, 1975; Hibbard, 1982, 1983). We therefore include a history of geological thought before presenting descriptions and contemporary interpretations.

History of ideas

The history of geological ideas on the internal Humber Zone is nearly as complex as its geology. Most regional syntheses have been based upon structural-stratigraphic relationships in the Baie Verte Peninsula of Newfoundland and thence extrapolated along strike for the length of the Canadian Appalachians. In retrospect, this starting point for regional syntheses in the late 1960s and early 1970s was unfortunate; for the Baie Verte Peninsula is in many respects atypical.

The major point of contention in geological interpretations of the Baie Verte Peninsula has been lithic and structural correlation between its western and eastern parts. Important in these correlations was a sequence of multideformed metaclastic rocks, the Ming's Bight Group, that outcrops in the midst of ophiolitic-volcanic assemblages, now assigned to the Dunnage Zone (Fig. 3.46).

Some workers correlated the Ming's Bight Group with the main outcrop area of the Fleur de Lys Supergroup and assigned the group as well as nearby, intensely-deformed Ordovician and Silurian rocks (now of Dunnage Zone and Springdale Belt, respectively) to an eastern division of the Fleur de Lys Supergroup. Because they considered all rocks as conformable, these workers proposed a common tectonic history for all of the Fleur de Lys Supergroup and nearby rocks of the Dunnage Zone and Springdale Belt (Church, 1969; Dewey and Bird, 1971; Kennedy et al., 1972; Kidd, 1977). On circumstantial evidence, all rocks were considered pre-Lower Ordovician and tectonized in the Late Cambrian or earliest Ordovician. The geographic separation of the Ming's Bight Group and tectonized metavolcanic rocks, on the one hand, from the main Fleur de Lys outcrop belt, on the other hand, was rationalized by interpreting intervening, lesser deformed ophiolitic-volcanic assemblages as vestiges of an Ordovician marginal ocean that formed by rifting the already deformed Ming's Bight Group and deformed

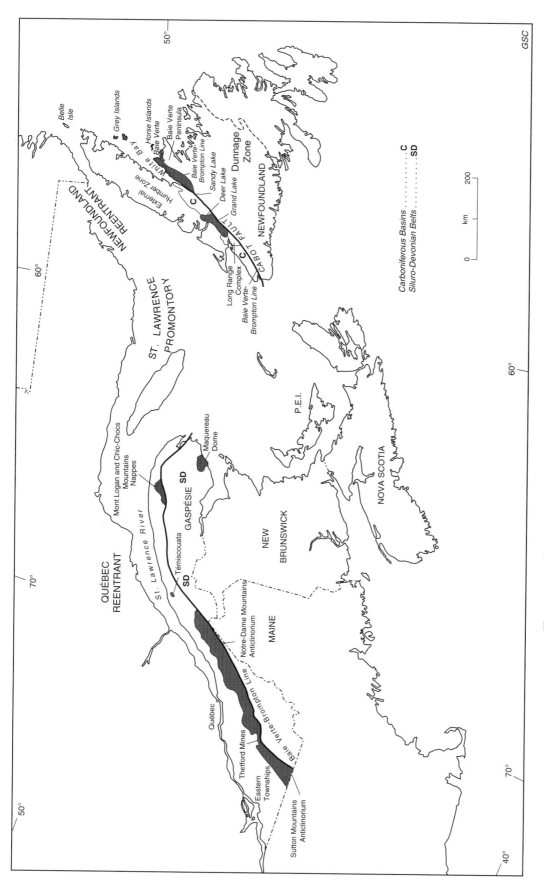

Figure 3.45. The internal Humber Zone in the Canadian Appalachians.

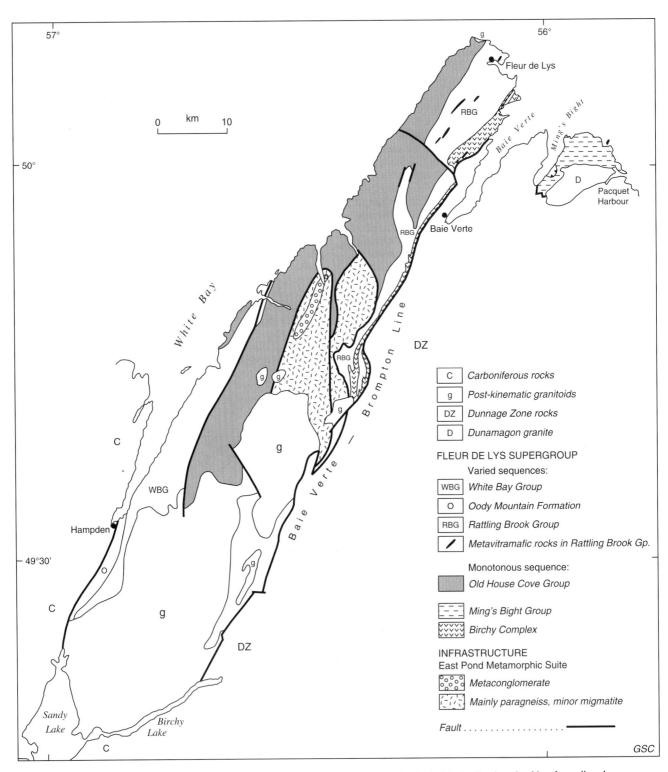

Figure 3.46. General geology of the internal Humber Zone on the Baie Verte Peninsula, Newfoundland. See Fig. 3.45 for location.

Dunnage Zone rocks from the main Fleur de Lys outcrop belt (Dewey and Bird, 1971; Kennedy et al., 1972; Kennedy, 1975b; Kidd, 1977).

This version of the regional tectonics belies evidence that some of the highly tectonized rocks are of Silurian age and they grade southward without a stratigraphic break, into relatively undeformed rocks of the Cape St. John Group that unconformably overlies the Lower Ordovician Snooks Arm Group (Neale, 1957; Neale et al., 1975; DeGrace et al., 1976; Williams et al., 1977). The Cape St. John Group above the unconformity is correlated with Silurian rocks nearby to the east, in Notre Dame Bay (Neale 1957; Neale and Kennedy, 1967; Williams, 1967; Degrace et al., 1976). Thus the idea of a Late Cambrian-Early Ordovician age of deformation for the Ming's Bight Group and some metavolcanic rocks of the eastern Baie Verte Peninsula conflicts with field evidence (DeGrace et al., 1976).

The notion of a Cambrian to earliest Ordovician age of deformation has also been challenged by regional structural constraints on the age of deformation of the Fleur de Lys Supergroup (Bursnall and de Wit, 1975; Williams et al., 1977; Williams, 1977b). These data led to another model; that the Fleur de Lys Supergroup is the equivalent of uppermost Precambrian-Lower Ordovician platformal strata of the external Humber Zone, deformed and metamorphosed by Lower to Middle Ordovician ophiolite obduction. Originally, ophiolite suites such as the Bay of Islands Complex in western Newfoundland were rooted to ophiolitic-volcanic suites of the Dunnage Zone, and the Baie Verte-Brompton Line thus marks a major disturbed boundary between a continental margin (Humber Zone) and adjacent ocean (Dunnage Zone). This model reasonably dispels the idea of a "mysterious" Cambrian-earliest Ordovician orogenic episode of unknown control which lacks any expression in the stratigraphic record of the external Humber Zone. However, the position of the Ming's Bight Group remained enigmatic.

The Ming's Bight Group is now interpreted as tectonically juxtaposed with the multideformed metavolcanic rocks of the Dunnage Zone, and continuous with the main Fleur de Lys outcrop belt along a submerged oroclinal bend, the Baie Verte Flexure (Hibbard, 1982, 1983; Dallmeyer and Hibbard, 1984). Thus, the Ming's Bight Group is viewed as being initially deformed with the Fleur de Lys Supergroup during Early to Middle Ordovician obduction. However, further deformation and metamorphism of the Ming's Bight Group and nearby metavolcanic rocks of the Dunnage Zone and Springdale Belt is viewed as Devonian in age, and localized along the east limb of the flexure. This later deformation masks the earlier tectonic juncture of the Fleur de Lys rocks (internal Humber Zone) and nearby metavolcanic rocks (Dunnage Zone).

Stratigraphy

Infrastructure

Structurally complicated gneisses and schists are exposed both in major anticlinoria and along major faults affecting the western margin of the internal Humber Zone. Intense tectonism has obliterated most primary features in these rocks; consequently interpretations concerning their origins have been controversial. Locally, as in Quebec, the rocks are known to be Grenville basement, remobilized during Paleozoic orogenesis (Vallières et al., 1978), in other places they are thought to represent intensely deformed, structurally lower parts of the cover sequence (Hibbard, 1983). Still larger areas of gneisses are of uncertain origin. Therefore, the tectonic term "infrastructure" is used here for this collection of intensely deformed, formerly deep-seated rocks, instead of the more confining stratigraphic term "basement". This distinction is especially critical in discussing Newfoundland examples below.

The infrastructure forms less than 5% of the internal Humber Zone. It is most extensive and varied in Newfoundland, whereas only small slivers of the infrastructure are recognized in Quebec. In both areas, the contact with the cover sequence is generally tectonic, although locally it has been interpreted as an unconformity.

Newfoundland

Rocks of the infrastructure outcrop on the Baie Verte Peninsula and in the Deer Lake and Grand Lake areas (Fig. 3.46, 3.47). Although similar lithic assemblages are present in each of these areas, the makeup of the infrastructure with respect to stratigraphic basement and cover is highly variable; in some places it contains only basement rocks, whereas in other areas it contains rocks equivalent to basement and cover sequences. This variability mainly reflects a heterogeneity of deformation and difficulty in distinguishing basement and cover. Rocks of the cover sequence immediately adjacent to the infrastructure have direct bearing on interpretations, and thus are included in the following description and discussion.

At the south end of the Baie Verte Peninsula (Fig. 3.46), immediately east of White Bay, the infrastructure is represented by large screens (kilometre scale) of granitic gneiss that are surrounded by amphibolite. The gneissosity in the granitic rocks is obviously truncated by both thin dykes and larger areas of amphibolite. To the east, the amphibolite is interlayered with metaconglomerate and gritty metaclastic schists that consistently young eastwards; this stratigraphic package has been assigned to the Oody Mountain Formation (Hibbard, 1983). The metaconglomerate contains granitic gneiss cobbles. In view of these relationships and because the amphibolite and metaclastics show the same deformational features, the granitic gneiss of the infrastructure has been interpreted as predeformed basement to a younger deformed cover sequence, the Oody Mountain Formation. The nature of the contact between the infrastructure and cover, here, is uncertain because Paleozoic deformation has obscured the primary nature of the Oody Mountain amphibolite. The contact could be intrusive or nonconformable.

The largest outcrop area of the infrastructure forms the cores of two major anticlinoria in the centre of the belt, on the Baie Verte Peninsula. In contrast to other occurrences, the infrastructure here is more varied; it consists of migmatite, banded gneiss, quartzo-feldspathic paragneiss, quartzite, epidosite, amphibolite, and metaconglomerate, all of which have been assigned to the East Pond Metamorphic Suite (Hibbard, 1983). Pods and concordant layers of amphibolite, most likely remnants of mafic dykes and sills, occur throughout the suite. Rocks of the suite are separated from the surrounding cover sequences by a steep tectonic zone characterized by coarse mica schists (de Wit, 1972, 1980). Within the suite, the migmatite and banded gneisses form small outcrop areas (less than 1 km^2) in the core of each anticlinoria. In the western outcrop area, the migmatite is surrounded by the metaconglomerate, which contains gneissic clasts; the metaconglomerate is, in turn, surrounded by the paragneisses. In the eastern area the banded gneiss is enveloped by paragneiss.

On structural evidence, the migmatite has been interpreted as representing an anatectic front within a basement sequence of paragneisses (de Wit, 1972, 1980); since the metaconglomerate contains gneissic clasts, it has been considered as representing the basal cover deposit on the basement (de Wit, 1972, 1974). This interpretation of the extent of basement has been challenged (Hibbard, 1979, 1983). The migmatites and banded gneisses display structures that apparently predate any tectonism in the remainder of the suite and the cover rocks surrounding the suite (de Wit, 1972, 1980; Hibbard, 1983). The migmatite layering is steeply dipping and trends northwestward, athwart the typical northeasterly trend of other rocks in the internal Humber Zone. This complex layering is overprinted by all the deformations evident in cover rocks surrounding the suite. Likewise, the banded gneisses display complex intrafolial fold interference patterns that are overprinted by the same deformations. The surrounding metaconglomerate and paragneiss units are void of any such complex early structures, and display only the regional events recorded in nearby cover rocks, albeit the deformation is generally much more intense in the suite. These observations suggest that the migmatites and banded gneisses may represent small windows of predeformed basement within an intensely deformed equivalent of the cover sequence (Hibbard, 1983).

Regional lithic correlation also supports this newer interpretation. The paragneisses, in the vicinity of the migmatite gneisses, are interlayered with the metaconglomerate and consist of amphibolite, epidosite, quartzite and psammitic gneiss. This stratigraphic package is seemingly an amalgamation of the lithic associations of the cover sequences adjacent to the infrastructure in other parts of Newfoundland. The metaconglomerate of the suite is identical to that found farther to the southwest on the peninsula, in the Oody Mountain Formation, which directly overlies infrastructure interpreted to be basement. The presence of amphibolite is a common feature of other infrastructure-cover junctures. It appears then, from both structural and stratigraphic evidence, that the infrastructure, represented by the East Pond Metamorphic Suite, comprises mainly intensely deformed equivalents of the cover sequences, and embraces a basement-cover relationship. The nature of this relationship has been tectonically-destroyed, but the character and position of the metaconglomerate suggests that it may have been a major unconformity.

Similar rock types and relationships are exposed in the area north of Deer Lake, on Glover Island, and at the southwest corner of Grand Lake (Fig. 3.47). However, in each of these areas, there is no evidence that deformation of the gneisses predated that of other nearby rocks. In the Deer Lake area, the infrastructure is composed of gneissic and foliated granitic rocks and metavolcanic rocks of the northeast-trending Hughes Lake Complex (Williams et al., 1985a). These are exposed beneath a westward-dipping, presumably overturned thrust fault (Williams et al., 1982). In the complex, granitoid gneiss surrounds outcrops of

foliated to massive granite, but the relationship between rock types is uncertain. Southeastwards, amphibolite and chlorite schists form conspicuous melanocratic units in the granite; these units grade southeastward into northwest-dipping mixed mafic schist-amphibolite and magnetite-quartz-feldspar schists and a sequence of southeast-younging coarse metaclastic schists.

This sequence was originally interpreted as representing Grenville basement gneisses and granite, with zones of intensely mobilized and retrograded gneiss (magnetite-quartz-feldspar schist), all cut by prekinematic mafic dykes (melanocratic units) and stratigraphically overlain by a metaclastic cover sequence (Williams et al., 1982). However, U-Pb zircon data and geochemical analyses indicate that the foliated granite is 602 ± 10 Ma and probably related to the mafic rocks (Williams et al., 1985a). In addition, field studies indicate that the magnetite-quartz-feldspar schists are intensely deformed metavolcanic rocks that are associated with the granite and the mafic units (Williams et al., 1985a). Thus, the Hughes Lake Complex is now interpreted as genetically related to the cover sequence.

On Glover Island and southwest of Grand Lake, the nature of infrastructural gneisses is also equivocal. On the island, granitic gneiss, exposed in the core of a major anticline, is concordant with structurally overlying amphibolite and metaclastic rocks (Knapp, 1980). The gneiss includes layers and bodies of amphibolite, not found in the cover sequence. At the southwest corner of Grand Lake, similar relationships are found, although the thick mafic unit found directly above the infrastructure elsewhere is absent; instead thin amphibolite bands are interlayered with the gneisses and these are structurally overlain by a sequence of epidosites and quartzites (Kennedy, 1981). The epidosites may represent highly deformed felsic volcanic rocks in this area (Hibbard, 1983). South of Grand Lake, similar gneisses merge with the regional Grenville Long Range Complex (Martineau, 1980) and they are interpreted as basement (Piasecki, 1991).

Age and correlation. The absolute age of rocks forming the infrastructure is uncertain; however, regional stratigraphic correlation provides reasonable indirect evidence for their age. The granitoid gneisses of the infrastructure, southeast of White Bay are lithically identical to nearby Grenville gneisses that form basement to miogeoclinal rocks on the west side of White Bay; likewise, gneisses at the southwest corner of Grand Lake merge with Grenville gneisses of the Long Range Complex. These lithic correlations suggest that the infrastructure in these areas represents Grenville basement.

The gneisses of uncertain origin at Glover Island, and in the core of the East Pond Metamorphic Suite provide little compelling evidence for regional correlation; however, the associated cover sequences and paragneisses of the suite consistently display a lithic association of metavolcanics, coarse metaclastics, and minor metaconglomerate, that is similar to lesser tectonized sequences of the miogeoline to the west (Fig. 3.48). Specifically this assemblage is remarkably reminiscent of the basal Labrador Group on Belle Isle and on the Great Northern Peninsula (Williams and Stevens, 1969; Bostock, 1983; Cumming, 1983). Here, a variable sequence of mafic rocks and coarse clastic rocks overlies Grenville basement gneisses. The

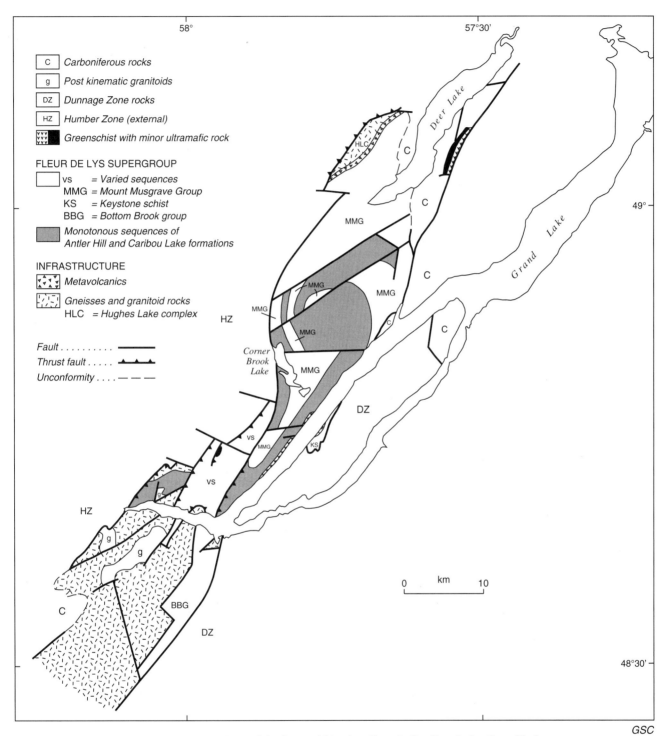

Figure 3.47. General geology of the internal Humber Zone in the Deer Lake-Grand Lake area, Newfoundland. See Figure 3.45 for location.

mafic volcanic rocks, locally termed Lighthouse Cove Formation, are generally tholeiitic massive flows that either directly overlie basement, or overlie a basal conglomerate. Mafic dykes that cut the basement are feeders to the basalts. The basal conglomerate, part of the Bateau Formation, is present only locally, and contains gneissic clasts. The overlying clastic rocks of the Bradore Formation are coarse quartz-feldspar grits and sandstones.

Most strikingly, metaconglomerate at White Bay and in the East Pond Metamorphic Suite is lithically identical to the Bateau Formation. Its association with amphibolite is similar to the Bateau-Lighthouse Cove association of the Labrador Group. Not surprisingly, amphibolite from the East Pond Metamorphic Suite, the Oody Mountain Formation, and the Deer Lake area are petrochemical correlatives of the tholeiitic Lighthouse Cove basalts (de Wit and Strong, 1975; Hibbard, 1983; Williams et al., 1985a). Epidosites and metafelsites associated with the infrastructure have no direct correlatives within the Labrador Group, but they appear to represent an easterly volcanic unit equivalent to the Lighthouse Cove Formation. Coarse metaclastic rocks and minor quartzites of the East Pond Metamorphic Suite, and similar rocks associated with the infrastructure in other areas, bear a resemblance, both stratigraphically and lithically to the Bradore Formation (Williams et al., 1982; Hibbard, 1983).

The basal Labrador Group immediately underlies the Lower Cambrian Forteau Formation and thus, has been considered to be latest Precambrian to earliest Cambrian. This suggests that paragneisses of the East Pond Metamorphic Suite and cover sequences associated with the infrastructure, elsewhere, are probably also latest Precambrian to earliest Cambrian. Where infrastructural gneisses show complex structures that predate Paleozoic tectonism, the correlations noted above suggest that these rocks are Grenville gneisses. Where infrastructural gneisses do not display these complex early structures, they may be intensely deformed rocks related to the cover sequence or represent intensely deformed Grenville gneisses in which any evidence of earlier tectonism has been totally eradicated.

Quebec

In the Quebec Appalachians, rocks of the infrastructure occur in the Port-Daniel region of southern Gaspésie, in the area of Saint-Malachie, 30 km southeast of Québec City, and in the Arthabaska region, 6 km east of Sainte-Hélène-de-Chester. In the Port-Daniel region (Fig. 3.49), DeBroucker (1984) mapped a wedge of orthogneiss and amphibolite 250 m wide occurring within a major fault zone bordering the Maquereau Group. The orthogneiss, a medium- to coarse-grained, grey and pink rock, is injected into the amphibolite with numerous amphibolite inclusions in the orthogneiss. The feldspar-rich bands are approximately 0.5 cm thick. A cataclastic mortar texture in the orthogneiss indicates that these rocks are highly deformed. We believe that the orthogneiss and amphibolite assemblage represents a window of Grenvillian basement underlying the Maquereau Group. U-Pb isotopic dating of the orthogneiss has yielded discordant ages of 998 ± 26 Ma on the lower intercept and 2464 ± 165 Ma on the upper intercept (De Broucker et al., 1986).

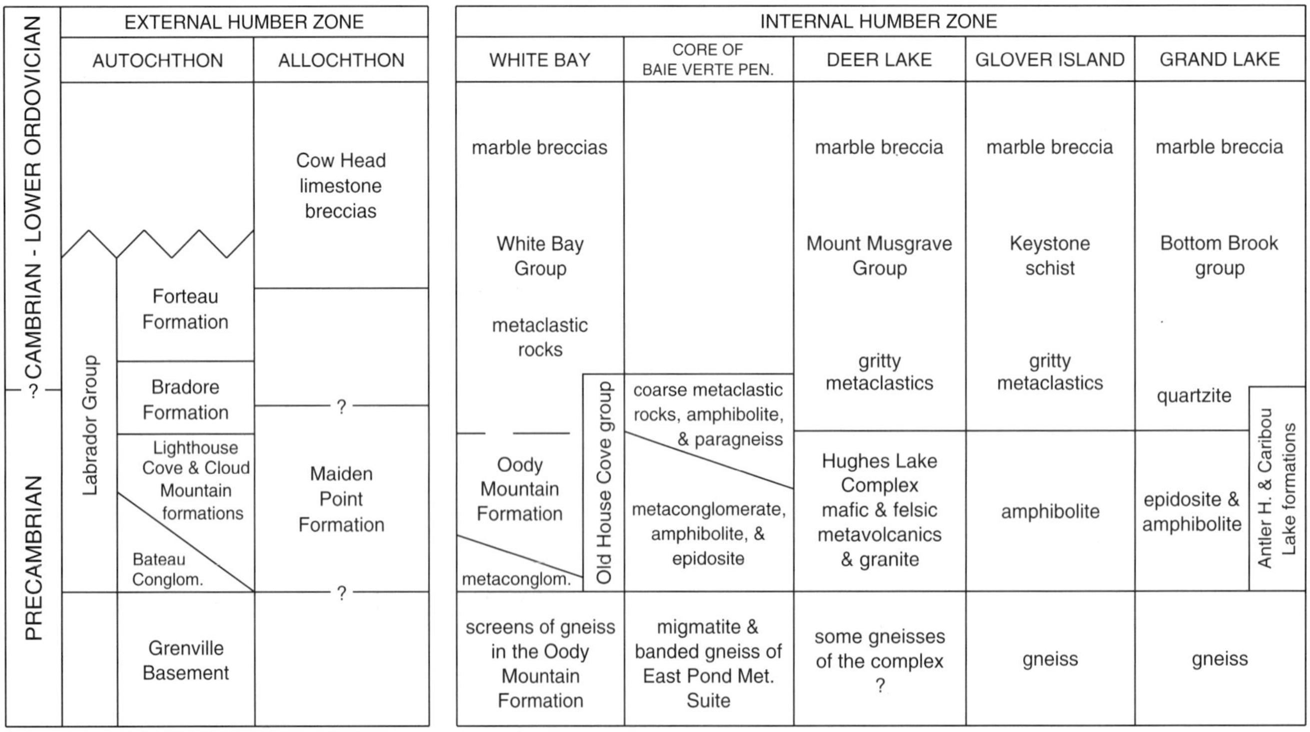

	EXTERNAL HUMBER ZONE		INTERNAL HUMBER ZONE				
	AUTOCHTHON	ALLOCHTHON	WHITE BAY	CORE OF BAIE VERTE PEN.	DEER LAKE	GLOVER ISLAND	GRAND LAKE
CAMBRIAN - LOWER ORDOVICIAN	Labrador Group: Forteau Formation; Bradore Formation; Lighthouse Cove & Cloud Mountain formations; Bateau Conglom.	Cow Head limestone breccias	marble breccias		marble breccia	marble breccia	marble breccia
			White Bay Group		Mount Musgrave Group	Keystone schist	Bottom Brook group
			metaclastic rocks		gritty metaclastics	gritty metaclastics	quartzite
		Maiden Point Formation	Oody Mountain Formation	Old House Cove group: coarse metaclastic rocks, amphibolite, & paragneiss	Hughes Lake Complex mafic & felsic metavolcanics & granite	amphibolite	Antler H. & Caribou Lake formations: epidosite & amphibolite
PRECAMBRIAN				metaconglomerate, amphibolite, & epidosite			
			metaconglom.				
	Grenville Basement		screens of gneiss in the Oody Mountain Formation	migmatite & banded gneiss of East Pond Met. Suite	some gneisses of the complex ?	gneiss	gneiss

Figure 3.48. Stratigraphic correlation chart for the internal Humber Zone in Newfoundland.

GSC

In the area of Saint-Malachie, 30 km southeast of Québec City (Fig. 3.49), Vallières et al. (1978) described three slivers of gneiss and amphibolite (termed the Sainte-Marguerite Complex) having a minimum age of 900 Ma; these are interpreted as representing thrust slices of Grenville type basement within the western margin of the internal domain of the internal Humber Zone. The complex consists of three northeasterly-trending slivers; one of interlayered gneissic granites with pegmatites and amphibolites, one of gneissic granites and pegmatites, and one of metaquartzite. These slices constitute the basement to a series of spilitized volcanic rocks and interbedded shallow-water sandstone, termed the Montagne Saint-Anselme Formation (Vaières et al., 1978).

The southwestern sliver is 3 km long and 600 m wide. It includes a 70 m thick amphibolite band composed of layers of amphibolite, 1 cm to 15 cm thick, alternating with layers of gneissic granites. These gneissic granites have a faint gneissosity and a granoblastic texture. Thin mafic dykes cut the gneissosity and foliation in both gneissic granites and amphibolites. The gneissic granites and the pegmatites in the northeastern sliver are similar to those in the southwestern sliver. A third small sliver is adjacent to, and northwest of, the southwestern sliver and consists of massive metaquarzites composed almost entirely of elongate quartz grains with a mean long axis dimension of 5.0 mm. Many of these quartz grains show undulose extinction and microscopic strain planes (Vallières et al., 1978).

In the Arthabaska region, 6 km east of Sainte-Hélène-de-Chester (Fig. 3.49), Caron (1982) described a 5 km long sliver of amphibolite associated with an important thrust fault in the centre of the Bennett Schists cover rocks in the Notre-Dame Mountains Anticlinorium. This amphibolite is banded and consists of alternating amphibolite-rich and quartz-feldspar-rich lamallae, from 0.5 to 1.0 cm thick. Garnets, 1.0 mm diameter, occur in some localities. Given the resemblance of these amphibolites to the amphibolite of the Sainte-Marguerite Complex, we consider them to be of Grenville age.

Cover sequences

The cover sequence of the internal Humber Zone is composed of mainly metaclastic schists with interlayered marble, marble breccia, amphibolite and greenschist. In contrast to the infrastructure of the zone, layering in most of the cover sequence is thought to represent bedding.

Newfoundland

All of the rocks of the cover sequence in Newfoundland have been assigned to the redefined Fleur de Lys Supergroup (Hibbard, 1983; Figs. 3.46, 3.47). It includes the following major lithostratigraphic units: the White Bay, Rattling Brook, Old House Cove and Ming's Bight groups and the Birchy Complex on the Baie Verte Peninsula; the insular Horse Islands and Grey Islands groups, the Mount Musgrave Group in the Deer Lake-Grand Lake area, and informal units such as the Bottom Brook group south of Grand Lake.

The total thickness of the supergroup is uncertain because of internal structural complexity. Maximum outcrop width is approximately 13 km, just south of Baie Verte, although the maximum width is probably much greater in the offshore area north of the Baie Verte Peninsula. The contact relationships between constituent units of the cover sequence are complex, and are discussed following a description of rock types.

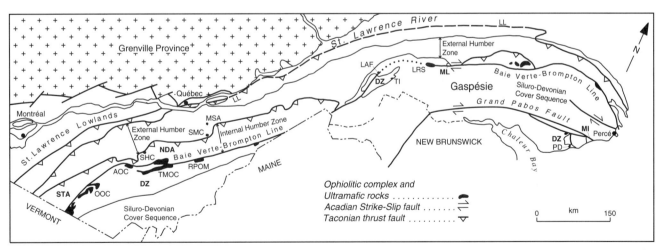

GSC

Figure 3.49. Simplified geological map of the internal Humber Zone in Quebec. OOC = Orford ophiolite complex; AOC = Asbestos ophiolite complex; SHC = Sainte-Hélène-de-Chester; PD = Port-Daniel; SM = Saint-Malachie; SMC = Sainte-Marguerite Complex; MSA = Montagne Sainte-Anselme; ML = Lac-Matapédia; TMOC = Thetford Mines ophiolite complex; RPOM = Rivière des Plante ophiolitic mélange; STA = Sutton Mountains Anticlinorium; NDA = Notre-Dame Mountains Anticlinorium; LL = Logan's Line; TI = Trinity Inlier; LAF = Lac-des-Aigles Fault; FTB = foreland thrust belt; LRS = La Rédemption serpentinite; MI = Maquereau Inlier; DZ = Dunnage Zone.

The units of the cover sequence can be divided into three divisions according to their lithic content: (1) the White Bay, Rattling Brook, and Mount Musgrave groups, as well as some informal sequences in the Grand Lake area that are composed of a variety of rock types, and hence are termed the varied sequences; (2) the Old House Cove Group and some informal sequences at Grand Lake that are less diverse in composition, and that are termed the monotonous sequences; and (3) ophiolitic rocks of the Birchy Complex. The insular Horse Islands and Grey Islands groups and the Ming's Bight Group display characteristics of both the varied and monotonous sequences.

On the Baie Verte Peninsula and in the Grand Lake area, the cover sequence is disposed in major anticlinoria. In each area, the monotonous sequences tend to form the core areas of these structures, with the varied sequences forming the flanks.

Components and stratigraphy of the cover sequence have been documented best on the Baie Verte Peninsula; the following descriptions will focus mainly on rocks of the peninsula and draw only lightly on information from other areas.

Varied sequences. These sequences, include the White Bay, Rattling Brook, and South Brook groups as well as the Mount Musgrave Group in the Grand Lake area and the informal Keystone schists on Glover Island (Knapp, 1980). They are composed of metaclastic schists, marble, greenschist, and amphibolite. The White Bay Group (Betz, 1948) is the best preserved and studied of these sequences, and serves as a model for the varied sequences.

On the west side of the Baie Verte Peninsula, the White Bay Group is disposed in a horizontal, north-northeast-trending synclinorium that has a steeply west-dipping axial surface. The metaclastic schists are mainly semipelitic and pelitic schists with abundant intercalations of sulfurous graphitic schists. Where psammitic rocks are present, they are most commonly thin bedded and fine grained. Marble and carbonate schist form volumetrically small portions of the group; however, they are the most spectacular and conspicuous rock types. Most striking are layers and large blocks of marble breccia within the pelitic schists (Fig. 3.50). These are generally characterized by numerous angular rip-up remnants of thin-bedded carbonates, as well as subordinate rounded clasts of carbonate, quartzite and vein quartz.

Amphibolite and greenschist locally form distinct formations up to 1 km thick within the metaclastic rocks. The basal unit of the group, the Oody Mountain Formation (Hibbard, 1983) is mainly massive amphibolite that may represent a thick sill or several massive flows. A second major mafic schist that occurs higher in the group is of extrusive origin. Petrochemically, these units resemble rift-type tholeiites (Hibbard, 1983).

Rocks of the Rattling Brook Group (Watson, 1947) resemble those of the White Bay Group, although the internal stratigraphy of the Rattling Brook Group is less consistent than that of the White Bay Group. This is probably a result of intense structural transposition because of the proximity of the unit to the Baie Verte-Brompton Line. One significant difference between the two units is that the Rattling Brook Group contains tectonic slivers of ultramafic rock (Fig. 3.46).

In the Deer Lake area, the cover sequence has a basal metavolcanic unit like the White Bay Group, but here, it is overlain by a thick sequence (2 km) of crossbedded metamorphosed arkose, greywacke, and pebble conglomerate that forms the base of the Mount Musgrave Group. This in turn is overlain by metaclastics and marble similar to the

Figure 3.50. Large pod of marble breccia (white) surrounded by metaclastic rocks (grey); White Bay Group, at the mouth of Bear Cove, White Bay. GSC 1994-766A

White Bay Group. Farther south, the Keystone schists and Bottom Brook group are lithically similar to the White Bay Group. The varied sequence west of Grand Lake locally contains tectonic slivers of ultramafic rock, similar to those in the Rattling Brook Group.

The White Bay Group contains the best preserved primary features in all of the Fleur de Lys Supergroup; consequently, it divulges the most information concerning the depositional environment. In particular, the spectacular calcareous metaconglomerates, large marble blocks, and calcareous schists interlayered with dominantly thin-layered metaclastics and sulfurous graphitic schist correspond with descriptions of carbonate slope facies that are marginal to a carbonate bank (McIlreath and James, 1979). The marble breccia layers most likely represent debris sheets emplaced on a slope environment; the thin flat nature of many of the carbonate clasts suggests that the deposits were slope-derived rather than platform-derived (McIlreath and James, 1979). The large isolated pods of marble breccia may be tectonized equivalents of either debris flow channels or submarine slump blocks. In light of these associations, the carbonate schists resemble hemipelagic deposits of peri-platformal ooze (McIlreath and James, 1979). On the basis of these observations, the varied sequences are interpreted as having been deposited on a continental slope, bordering a substantial carbonate bank.

Monotonous sequences. The Old House Cove Group and the informal Antler Hill and Caribou Lake formations (Kennedy, 1981) in the Grand Lake area outcrop extensively in the core of major anticlinoria. These units are monotonous sequences of alternating coarse feldspathic psammite and semipelite with local, minor interlayers of graphitic schist, pelite, pebbly metaconglomerate, garnetiferous semipelite, and quartzite. The Old House Cove Group contains numerous pods and sills of black-green amphibolite (Fig. 3.51), which locally cut bedding. Petrochemically, they are tholeiitic (de Wit and Strong, 1975; Hibbard, 1983). Mafic rocks also occur in the Grand Lake area, but they are not as abundant as in the Old House Cove Group. Primary features in these rocks have largely been obliterated by regional tectonism. The monotony of alternating layered psammite and semipelite is reminiscent of a turbidite sequence. This, considered with the medium- to thick-layering so prevalent in the unit, suggests that the group may represent a medial submarine fan facies. The clean, quartzo-feldspathic composition of these rocks suggests that they had a continental derivation.

Ophiolitic rocks. The Birchy Complex forms the easternmost flank of the Fleur de Lys Supergroup. It is a structural assemblage composed predominantly of greenschist, metagabbro, graphitic schist, mélange, and minor ultramafic rocks. The ultramafic rocks, metagabbro, and greenschists have collectively been interpreted as structurally dismembered fragments of an ophiolite suite (Bursnall, 1975); petrochemical data support this interpretation (Hibbard, 1983). Graphitic schist and mélange are interlayered with greenschist; Bursnall (1975) first recognized the mélanges and significantly noted that all of the occurrences are characterized by the inclusion of either serpentinized ultramafic rocks or bright green fuchsite-actinolitic schist of ultramafic derivation.

In direct contrast to other parts of the supergroup that appear to have a continental origin, the Birchy Complex is of oceanic affinity. Mélanges in the complex display all of the structures evident in surrounding rocks, hence it has been concluded that they mark the nascent stages of tectonism (Bursnall, 1975; Williams, 1977b; Williams et al., 1977).

Other sequences. The Horse Islands, Grey Islands, and Ming's Bight groups are all isolated groups that display characteristics of other sequences. They are all composed mainly of monotonous metaclastic rocks; however, they also contain minor variations not found in the monotonous sequences.

The Horse Islands Group contains greenschist and metafelsite infolded with its metaclastics; the metafelsites are locally fragmental. In the Grey Islands Group, one psammitic unit contains amphibolite pods similar to the monotonous sequences, but it also contains greenschist and mélange zones similar to those of the Birchy Complex. As well, graphitic schist and marble occur locally in the group. Likewise, the Ming's Bight Group contains amphibolite with mélange zones like those in the Birchy Complex.

Protoliths of these sequences are difficult to establish. They appear to represent turbidite sequences that have a volcanic component; the mélanges, similar to those in the Birchy Complex, may indicate that volcanism was ophiolite-related.

Contact relationships. The relationships between sequences appear to be complex (Fig. 3.52); contacts are exposed in many areas, although stratigraphic order is difficult to decipher because of the lack of primary structures.

The varied sequences overlie a variety of different substrates. The White Bay Group, disposed in a major synclinorium, overlies different units on opposing flanks of the structure. To the southwest, the group overlies and includes screens of infrastructural gneisses that have been interpreted as representing Grenville basement (Hibbard, 1983). To the west, the group conformably overlies the monotonous sequence, the Old House Cove Group (de Wit, 1972; Hibbard, 1983). Similar relationships are found at Deer Lake and Grand Lake. The Mount Musgrave Group has been interpreted as overlying the infrastructure at Deer Lake (Williams et al., 1982); the same relationship has been documented between the Keystone schists and the infrastructure on Glover Island (Knapp, 1980).

In contrast to these relationships, the Rattling Brook Group is conformably interlayered to the east with greenschists of the ophiolitic Birchy Complex; as well, the group contains tectonic slivers of ultramafic rock. These relationships suggest that the eastern part, if not all of the Rattling Brook overlies an ophiolitic basement. To the west, the Rattling Brook Group conformably interdigitates with the Old House Cove Group, although the stratigraphic order is uncertain. Similarly, west of Grand Lake, the eastern portions of the varied sequences contain tectonic slivers of ultramafic rock, suggesting that the sequence here may have originally had an ophiolitic substrate; to the west, the varied sequences overlie the monotonous sequence (Kennedy, 1981). This relationship, as well as the relationship of the varied White Bay

123

Figure 3.51. Typical psammite, semipelite, and amphibolite (dark pods) of the Old House Cove Group, Southern Arm, White Bay. GSC 1994-766C

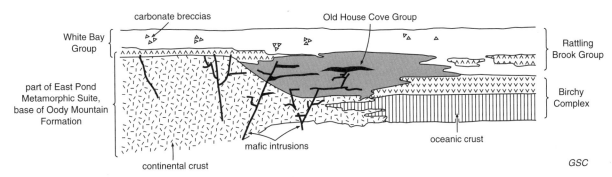

Figure 3.52. Schematic representation of contact relationships within the Fleur de Lys Supergroup, Newfoundland.

Group to the monotonous Old House Cove Group, suggests that the Rattling Brook Group likely overlies the Old House Cove Group.

The substrate to the monotonous sequences is uncertain. Both on the Baie Verte Peninsula and near Grand Lake these rocks are faulted against the infrastructure (de Wit, 1972, 1980; Kennedy, 1981). Relationships outlined above imply that the monotonous sequences, may be partially equivalent to the lowest portions of the varied sequences and underlie other portions. Since the varied sequences overlie continental basement to the west, and oceanic basement to the east, it would appear that the monotonous sequences may have originally overlain the transition of one substrate to the other.

Relationships of the Ming's Bight, Grey Islands, and Horse Islands groups are uncertain. However, both the Ming's Bight and Grey Islands groups contain mafic schists that are correlative with the Birchy Complex. This suggests that these more easterly groups may also overlie an ophiolitic substrate, analogous to the Rattling Brook Group.

In summary, the Fleur de Lys Supergroup overlaps two distinct substrates; one is of continental affinity and represented by portions of the infrastructure, whereas the other is the easterly ophiolitic Birchy Complex. The varied sequences of the supergroup overlap these in the eastern and western parts of the belt, whereas in the middle, they may interdigitate with and overlap the monotonous sequences.

Age and correlation. Direct evidence for the age of the Fleur de Lys Supergroup is lacking. A single brachiopod valve recovered from marble in the White Bay Group is the lone fossil occurrence known in the supergroup. It has an affinity to the Order Acrotretida and indicates that at least part of the group is of Paleozoic age (S. Stouge, pers. comm., 1979, reported in Hibbard, 1983).

Regional correlation provides indirect age constraints on parts of the supergroup. As described above with the infrastructure, lower portions of the supergroup associated with outcrops of the infrastructure are lithic correlatives of the lower portion of the Labrador Group (Fig. 3.48). This indicates that lower portions of the supergroup are probably late Precambrian to Early Cambrian.

Stratigraphically higher units within the varied sequences are mainly non-diagnostic; but, the carbonate breccias, prevalent in parts of the White Bay Group and in the Grand Lake area, are metamorphosed, lithic equivalents of carbonates of the Cow Head Group in western Newfoundland (de Wit, 1972; Bursnall and de Wit, 1975). The Cow Head Group ranges in age from Middle Cambrian to Middle Ordovician. This does not provide tight age constraints on parts of the Fleur de Lys Supergroup, but lithic correlation provides an important link to the lower Paleozoic miogeocline to the west.

The monotonous sequences laterally interdigitate with, and are overlain by stratigraphically higher parts of varied sequences. This suggests that the monotonous sequences are similar in age to the lower Labrador Group. Significantly, amphibolite sills and dykes in the Old House Cove Group are petrochemically identical to mafic rocks of the White Bay Group and basalts of the Lighthouse Cove Formation (Hibbard, 1983). The monotonous sequences are also lithically similar to allochthonous clastic rocks and mafic volcanic rocks of the Maiden Point Formation on the Great Northern Peninsula. The age of this formation is inferred to be late Precambrian to Cambrian (Smyth, 1973), similar to the inferred age for monotonous sequences.

The Birchy Complex, at the eastern margin of the supergroup, has correlatives both to the east and west. Ophiolitic mélanges of the complex have been correlated with unmetamorphosed mélanges related to allochthonous ophiolites on the miogeocline to the west (Williams, 1977b; Williams et al., 1977). Since these allochthons were emplaced by Middle Ordovician, the mélanges both under the allochthons, and their lithic correlatives in the Birchy Complex are probably older than Middle Ordovician. In addition, Bursnall (1975) demonstrated that the bulk of the Birchy Complex is correlative with ophiolitic rocks of the Dunnage Zone, to the east. The best age on the nearby Dunnage ophiolites is Early Ordovician (Dunning and Krough, 1983), suggesting a similar age for the Birchy Complex.

Based on regional correlation, the Fleur de Lys Supergroup appears to range in age from late Precambrian to Early Ordovician. Regional tectonic constraints outlined below put a probable upper age limit of Early Ordovician on the supergroup.

Quebec

In Quebec, cover sequences of the interior Humber Zone form a discontinuous belt from the international border to the southeastern tip of Gaspésie (Fig. 3.45, 3.49). In the Eastern Townships, the cover sequences include metasedimentary rocks colloquially termed the Sutton-Bennett schists and Rosaire Group; they form major anticlinoria, the Sutton Mountains to the south and the Notre-Dame Mountains to the north. In addition, distinctive shale-feldspathic sandstone cover sequences are found mainly along the eastern boundary of the anticlinoria, although locally, they also outcrop on their western sides. These sequences include the Mansonville and Brompton formations and the Armagh and Caldwell groups. In central Gaspésie, the cover is represented by the Shickshock Group, which constitutes the whole of the Mont Logan Nappe. At the southeastern corner of Gaspésie, cover rocks of the Maquereau Group are exposed in a dome, surrounded by Silurian-Devonian strata.

Eastern Townships. In the Eastern Townships, the Sutton Mountains Anticlinorium is formed mainly by the Sutton schists and Mansonville and Brompton formations, whereas the Notre-Dame Mountains Anticlinorium is composed of the Bennett Schists and Rosaire, Caldwell, and Armagh groups (Fig. 3.49).

The stratigraphy of the Sutton schists in the Sutton Mountains Anticlinorium (Fig. 3.49, 3.53) has been described by numerous workers (Clark, 1934, 1936; Osberg, 1965; Eakins, 1964; Charbonneau, 1980a, b; Marquis, 1989a; Colpron, 1990). Two assemblages are recognized in the Sutton schists: the Oak Hill Group, a sequence of volcanic and sedimentary rocks of Early Cambrian to Middle Ordovician (?) age, and the Sutton Metamorphic Suite, a sequence of highly deformed and metamorphosed metasedimentary and metavolcanic rocks interpreted as "distal" correlatives of the Oak Hill Group (Colpron, 1990; Marquis, 1989a). The Oak Hill Group is composed of, from bottom to top: (1) Tibbit Hill Greenstone; (2) Pinnacle Formation (including the Call Mill Slate, Pinnacle Greywacke and the White Brook Dolomite of Clark, 1934); (3) Bonsecours Formation (including the West Sutton Slate and the pelite of the Gilman Formation, described by Clark, 1934); (4) Gilman Quartzite; and (5) Sweetsburg Formation (consisting of the Dunham Dolomite, Oak Hill Slate, Scottsmore Quartzite and Sweetsburg Slate of Clark, 1934).

Trilobites and brachiopods found in the Scottsmore Quartzite, Dunham Dolomite and Gilman Quartzite indicate an Early Cambrian age (Clark, 1936; Osberg, 1965). The Sweetsburg Formation is considered Middle Cambrian because of the presence of a trilobite of the *Cedario* Zone (Shaw, 1958). Macrofossils found in the formation at Richmond, Quebec, have been attributed to the Late Cambrian to Early Ordovician (Mamet, pers. comm., 1987 in Marquis, 1989a). However, because the conformably overlying Melbourne Formation is of Middle Ordovician age, the Sweetsburg Formation is probably Early Ordovician.

The Oak Hill Group has been traced southward into Vermont (Doll et al., 1961) where it rests unconformably upon Precambrian, Grenville-like basement. This relationship is not exposed in Quebec. We regard the Oak Hill Group as one of the oldest sequences in the internal Humber Zone of Quebec, resting directly on Precambrian basement.

A recent structural interpretation of the Sutton Mountains Anticlinorium in the Sutton Valley has led to a reinterpretation of the local stratigraphy. Historically, rocks in the Sutton Valley have been interpreted as defining a major syncline, the Valcourt or Sutton/Richmond Syncline (Eakins, 1964; Osberg, 1965; Clark and Eakins, 1968). However, detailed investigation has shown that the sequences are disposed in four major thrust sheets. The name "Mansville Complex" has been introduced for rocks exposed within these sheets, east of the Brome Thrust and west of the Sutton Thrust (Colpron, 1990). The complex displays lithic and stratigraphic similarities with the Oak Hill Group, although it has experienced only low-grade metamorphism (Colpron, 1990).

The Notre-Dame Mountains Anticlinorium, southeast of Québec City (Fig. 3.49), is made up of the Rosaire Group, metamorphic schists, the Bennett Schists, and assemblages of feldspathic sandstones, the Caldwell and Armagh groups. The Bennett Schists are considered to be the equivalent of the Oak Hill, Rosaire and Caldwell groups (St-Julien et al., 1972, St-Julien and Hubert, 1975). Locally, as in the Mont Sainte-Marguerite area, there are units reminiscent of the lower part of the Oak Hill sequence (Charbonneau and St-Julien, 1981). In particular, these units include a basal volcanic member similar to Tibbit Hill lavas overlain successively by slates, greywackes, dolomite and red slates corresponding to the Call Mill Slate, Pinnacle Greywacke, White Brook Dolomite, and West Sutton slates.

The Rosaire Group (Béland, 1957) constitutes the bulk of the Notre-Dame Mountains Anticlinorium. The group (Béland, 1957; Benoit, 1958) comprises black, grey and green phyllitic shales and siltstones containing local beds of orthoquartzite, 25 cm to 3 m thick. Mudstones and red shales at the top of this stratigraphic unit are correlated with strata of the Lower Ordovician Rivière Ouelle Formation of the external Humber Zone (Hubert, 1965; Vallières, 1984). The Rosaire Group is correlated with the Kamouraska Formation of the external Humber Zone, which is Late Cambrian and/or Early Ordovician (Hubert, 1965; Vallières, 1984).

Shale-feldspathic sandstone assemblages with intercalations of basic volcanic rocks are represented all along the eastern flank of the Sutton Mountains and Notre-Dame Mountains anticlinoria; these include the Mansonville Formation (Clark, 1934), part of the Brompton Formation (Fortier, 1946) and all of the Caldwell Group (MacKay, 1919; Cooke, 1937, 1950; Béland, 1957). The Armagh Group (Béland, 1957) is lithically similar to these units but occurs on the northwest flank of the Notre-Dame Mountains Anticlinorium. Also, in the Bennett Schists, a unit composed of metafeldspathic sandstone with light green quartz-muscovite-ottrelite schist and some chlorite schist is very similar to the Caldwell and Armagh rocks, and considered to be correlative.

Shale-feldspathic sandstone units of the internal Humber Zone are unfossiliferous, but fossils are found in correlative rocks of the external Humber Zone. The Charny and Saint Roch formations, which are correlative to the Caldwell, Mansonville, Armagh, etc., are Early Cambrian as indicated by the presence of *Bonnia Austinvillia bicensis* and *Pagetides*, in the Charny (Rasetti, 1946), and *Bolboparia Canadensis, Calodiscus theokritoffi* and *Leptochilodiscus cuspunctulastus* in the Saint Roch (Hubert et al., 1969).

In the Saint-Malachie area (Vallières et al., 1978) slivers of Precambrian, Grenville-like gneisses (Sainte-Marguerite Complex), cited above, occur below a unit composed of a sequence of basic volcanic rocks interbedded with arkosic sandstone of the Montagne Saint-Anselme Formation. This formation occurs at the base of the Armagh Group. Lavas of the Montagne Saint-Anselme Formation

occur in several separate slivers within the Richardson fault zone and within klippe at Montagne Saint-Anselme (Vallières et al., 1978). At Montagne Saint-Anselme, the formation is 630 m thick, consisting, of 16 lava flows with interlayered arkosic sandstone (Vallières et al., 1978). Some flows are pillowed while others are scoriaceous (Vallières et al., 1978). The sandstone of the Montagne Saint-Anselme Formation contains sedimentary structures indicating a shallow-water depositional environment. The volcanic rocks are also interpreted as being deposited under shallow-water conditions.

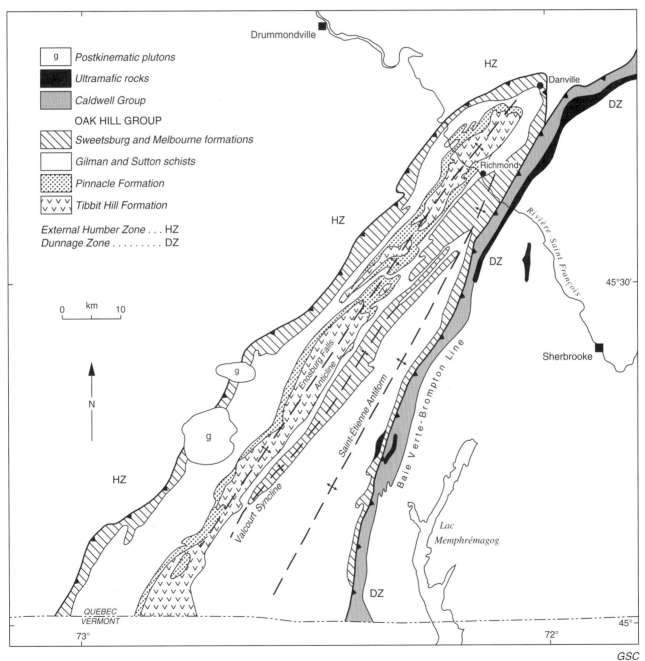

GSC

Figure 3.53. General geology of the internal Humber Zone on the Sutton Mountains Anticlinorium, Quebec, modified from Osberg (1965). See Figure 3.49 for location.

Although slivers of Precambrian gneisses have been brought up in a fault zone, their presence below the Armagh suggests that the shale-sandstone assemblage of the Armagh Group may have accumulated directly on Precambrian basement in that part of the internal Humber Zone.

Recent work suggests that the shale-orthoquartzite of the Rosaire Group rests directly, and apparently conformably, on the shale-feldspathic sandstone assemblage of the Caldwell Group (Vallières, 1984), which we consider to be the lateral equivalent of part of the Oak Hill Group, and which we believe rests on Precambrian Grenville gneisses (St-Julien et al., 1983).

Central Gaspésie. The Shickshock Group is named for the mountains in central Gaspésie, referred to in recent geographic lexicons as the Chic-Chocs. The group name was introduced by Crickmay (1932) for a stratigraphic unit that outcrops to the east of Lac Matapédia and is characterized by arkoses and basalts.

The Shickshock Group, except for the Lac Matapédia facies and several isolated outcrops in eastern Gaspésie, is confined to the Mont Logan Nappe (Fig. 3.49, 3.54). The lower contact of this nappe is the Lac Cascapédia Fault, structurally placing the Shickshock Group onto the external Humber Zone; the upper contact is the Marches du Géant Fault (Fig. 3.54) along which the ophiolitic Mont Albert Complex of Dunnage Zone affinity overrode the Shickshock Group. The thickness of the group is unknown, but an estimated minimum thickness is 1000 m. The basal metaconglomerate of the group is found locally, just above a ductile décollement that contains clasts of granite and granulite, probably derived from Grenville basement

(Beaudin, 1984); this observation suggests that the lower parts of the Shickshock Group were deposited on continental crust.

A number of geologists (Dresser and Dennis, 1944; Osborne and Archambault, 1948; McGerrigle, 1954; Mattinson, 1964; MacGregor, 1962, 1964; Girard, 1967; Ollerenshaw, 1967; Aumento, 1969; Carrara, 1972; and Beaudin, 1980, 1984) have mapped various areas of the Shickshock Group and have divided it into informal units. We follow Beaudin (1984) who has updated the earlier studies. Beaudin (1984) subdivided the group into four distinct assemblages: Lac Guelph, Bras au Saumon, Lac Matapédia, and Lac Cascapédia. The Lac Cascapédia assemblage is equivalent to the Orignal Formation of the external Humber Zone. The Lac Guelph assemblage is predominantly metasediments, the Bras au Saumon assemblage is predominantly metavolcanics, and the Lac Matapédia assemblage has alternating sediments and volcanics (Beaudin, 1980, 1984).

The Lac Guelph assemblage consists of meta-arkoses, conglomeratic meta-arkoses, chloritic schists, alternating siltstones and chloritic schists, and laminated metabasalts. The typical meta-arkose shows no sedimentary structures, but variably spaced metamorphic laminations. The conglomeratic meta-arkoses include rounded granite fragments and frayed pelitic fragments of principally red mudstone. Alternating layers of arkosic siltstone and reddish chloritic schist occur in places. Lepidoblastic to granolepidoblastic textures dominate the meta-arkose without having obliterated the primary clastic texture. With increasing metamorphism, the meta-arkoses pass into gneisses which are correlated with the arkoses of the Lac Matapédia assemblage.

The Bras au Saumon assemblage is composed of metabasalts which show diverse volcanic structures but which are compositionally homogeneous and of uniform metamorphic

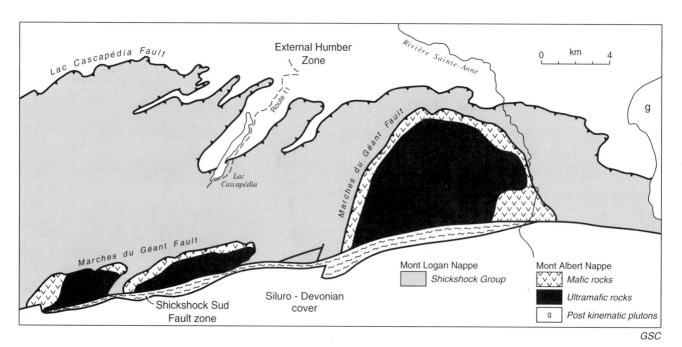

Figure 3.54. Geological map of the eastern portion of the Mont Logan Nappe (Shickshock Group) and the overlying Mont Albert Nappe, Quebec (Mont Paul, Mont Sud, and Mont Albert), from Beaudin (1984).

texture. In order of abundance these metavolcanics include amygdular and vesicular metabasalts, basic tuffs with hematite, and metabasalts with fragments of arkose and red mudstone.

The Lac Matapédia assemblage includes alternating basic volcanics, arkoses, and minor amounts of red slates. The basic volcanic rocks are greenish grey, generally massive and locally porphyritic with pillow structures. The arkose (quartz, pink and white feldspar) is coarse grained, and it is in places interstratified with red slate and minor conglomerate and tuff.

Metavolcanic rocks from the Shickshock Group have been analyzed for major and trace elements; many of the samples come from along the Rivière Cap-Chat, but numerous others are from within the Mont Logan Nappe where access is more difficult. Major and trace element patterns indicate that these rocks are similar to MORB-Type ophiolitic basalts (C. Wilson, pers. comm., 1990).

The Shickshock Group is considered latest Precambrian to Early Cambrian. It is equivalent, by virtue of its lithological similarities, stratigraphic position, and geotectonic setting, to the Tibbit Hill and Montagne Saint-Anselme formations, Saint-Flavien volcanics and Maquereau Group (St-Julien and Hubert, 1975; Vallières, 1984).

Southeastern Gaspésie. The Maquereau Group forms a portion of an inlier of Precambrian and lower Paleozoic rocks surrounded by Silurian rocks (Chaleurs Group; Fig. 3.55). The group outcrops in southeastern Gaspésie at Chaleur Bay, to the north of Port-Daniel. The Maquereau Group makes up the eastern portion of the inlier and is juxtaposed with the Mictaw Group of the Dunnage Zone. The group is underlain by Grenville basement, as described above.

The Maquereau Group is a volcanic-sedimentary sequence that has been folded and metamorphosed to greenshist facies (Ayrton, 1967; De Broucker, 1984). These rocks are unfossiliferous but are believed to be of Early Cambrian or

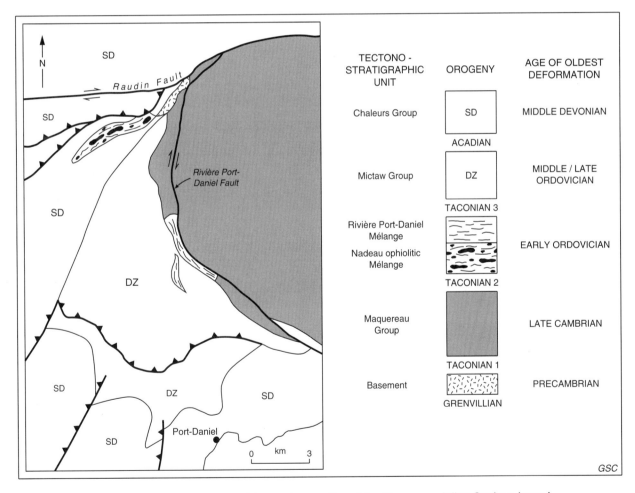

Figure 3.55. Geological map of the western portion of the Maquereau Inlier, Quebec, based on the oldest deformation in each of the tectonostratigraphic units, from De Broucker (1987). See Figure 3.49 for location.

129

late Precambrian age (De Broucker, 1986). They are bordered by the Raudin and Rivière Port-Daniel faults (Ayrton, 1967).

De Broucker (1984, 1987) recognized five lithological units in the Maquereau Group (Fig. 3.56). The most abundant lithology is a greenish arkosic wacke, spotted with small (millimetre-scale) pink feldspars, and associated with greenish and violet-red siltstones and mudstones. A second unit comprises dark green volcanic rocks (tholeiites) that vary in texture from aphanitic to diabasic. Violet amygdaloidal volcanics grade into the dark green theolites. These volcanic rocks are interstratified with the arkosic wackes of the first unit. A third unit is made up of a grey-green quartz-rich sandstone and a sericite-rich pink feldspar-bearing violet-grey sandstone. A fourth unit is composed of polygenetic conglomerates with fragments of granitic rocks, volcanic rocks and shale. These conglomerates have a greenish sandy-silt matrix which is similar to the arkosic wacke of the first unit. Finally, the fifth unit is formed of arkosic sandstone and brick-red arkose, both

locally conglomeratic, and of violet-red siltstone-mudstone. These rocks also seem to be interstratified with the first unit.

Tectonothermal history

Rocks of the internal Humber Zone are characterized by multiphase deformation and metamorphism. The tectonothermal history spans more than 100 Ma, with the most intense phases starting in the Early Ordovician and either episodically, or progressively continuing into the Devonian. This history has largely been controlled by the juxtaposition and interaction of the continental Humber Zone and the oceanic Dunnage Zone along the polygenetic Baie Verte-Brompton Line.

The internal Humber Zone is generally concordant with the structural trend of less deformed parts of the Humber Zone, although there are local departures from this structural grain. Structural dog-legs are found near Baie Verte, Newfoundland and Thetford Mines, Quebec, where structural trends abruptly flex to an orientation nearly perpendicular to that of the external Humber Zone (Fig. 3.45). These small flexures mimic the larger salients and recesses that affect the trend of the Baie Verte-Brompton Line (Hibbard, 1982, 1983).

Outcrop-scale structural and metamorphic features along the length of the internal Humber Zone are very similar; however, regional scale differences indicate that both deformation and metamorphic paths were not uniform.

Newfoundland

Numerous detailed structural and metamorphic analyses have been undertaken in the internal Humber Zone of Newfoundland (Church, 1969, Kennedy, 1971, 1975b; Kennedy et al., 1973; de Wit, 1972, 1974, 1980; Bursnall, 1975; Knapp, 1980; Kennedy, 1981; Hibbard, 1983; Jamieson and Vernon, 1987; Jamieson, 1990). Only on the Baie Verte Peninsula is the tectonic history well enough known to relate specific structures to discrete structural-metamorphic events; therefore, the following discussion centers on the peninsula.

The Baie Verte-Brompton Line arcs around a dog-leg in the internal Humber Zone on the Baie Verte Peninsula, the Baie Verte Flexure (Hibbard, 1982, 1983). At the flexure there is a change in orientation of the line and also a change in tectonic character.

Main outcrop area, Baie Verte Peninsula

Four major phases of deformation are recognized in the main outcrop area on the Baie Verte Peninsula. These include a basement deformation, D_B; an early deformation, D_E; a main phase deformation, D_M; and a late stage deformation, D_L. Metamorphism appears to have been essentially continuous from D_E to post D_L. D_B is confined to basement rocks in the infrastructure, whereas the other phases generally affect all rocks in the infrastructure and the cover.

Structures ascribed to D_B occur in limited areas in the core of the East Pond Metamorphic Suite and in basement rocks at the southwest corner of the peninsula. They are generally manifest as composite structures, including

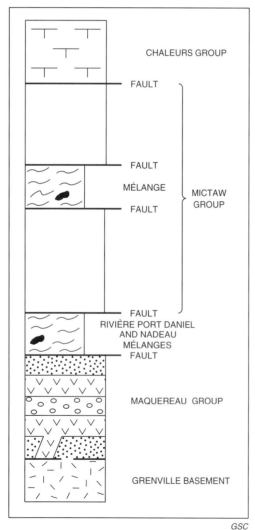

Figure 3.56. Schematic stratigraphic column of the Maquereau Inlier, Quebec, from De Broucker (1987).

either complex migmatitic or gneissic banding, that predate any other structures in the areas. Locally these structures trend northwesterly, askew to the north-northeasterly trend of other rocks in the belt. Any metamorphism associated with this deformation has subsequently been obliterated by later, Paleozoic thermal overprinting.

The early deformation, D_E, is characterized by ductile shear zones, that in many places contain associated ultramafic rocks. The ultramafic rocks were introduced into these zones before the locally preserved S_E fabric developed (Kennedy, 1971; Bursnall, 1975). One ductile shear, the Slaughter House Slide (Bursnall, 1975) may represent an extension of the ophiolitic mélange zones in the Birchy Complex. The incorporation of ultramafic rocks along the shear zones, as well as their association with ophiolitic mélanges, strongly suggests that the D_E deformation involved tectonic transport from the easterly oceanic Dunnage Zone onto the miogeocline (Bursnall, 1975). Most of the D_E ductile shear zones are distributed along the east side of the area near the Baie Verte-Brompton Line. As such, they probably represent the nascent stages of the line, and the early juxtaposition of the Humber and Dunnage zones.

The major structural grain was impressed upon the area during the D_M event. Minor D_M structures include tight to isoclinal folds (Fig. 3.57) with a penetrative axial planar schistosity, S_M. On a local scale, this event was very complex; in many places, D_M structures show evidence of progressive rotational strain and continued reactivation during the event (Kennedy, 1971; de Wit, 1972; Bursnall, 1975; Hibbard, 1983). Large scale folds, responsible for the outcrop pattern within the belt, are D_M structures.

Ultramafic rocks along the Baie Verte-Brompton Line appear to have been involved in the D_M event (de Wit, 1972, Kennedy, 1975b; Bursnall, 1975). In addition, F_M folds display a northwestward vergence, indicating that tectonic transport was east to west (Kennedy, 1971; Bursnall, 1975). Thus it appears that the D_M event may have been related to continued D_E transport of oceanic rocks, but with the interface between continental and oceanic rocks, the Baie Verte-Brompton Line, beginning to be localized into a narrower zone, defined by large ultramafic bodies.

Peak metamorphic conditions of middle pressure, upper greenschist to lower amphibolite grade were attained in most of the area (de Wit, 1972; Kennedy, 1971, 1975b; Bursnall, 1975; Hibbard, 1983); locally, in more central parts of the area, middle amphibolite grade conditions prevailed (de Wit, 1972; Kennedy, 1971, 1975b; Hibbard, 1983). Jamieson (1990) indicated that peak metamorphic conditions in garnet schists of the Fleur de Lys Supergroup were approximately 550°C and 6.5 kbar. Of particular note eclogite facies (omphacite-garnet ± quartz) has been reported from amphibolites of the East Pond Metamorphic Suite (Church, 1969; de Wit, 1972; de Wit and Strong, 1975). P-T determinations on the eclogites indicate early high-pressure conditions of 10-12 kbar and 450-500°C followed by re-equilibration at 7-9 kbar and 600-750°C (de Wit and Strong, 1975; Jamieson, 1990). These data corroborate structural evidence that the infrastructure was once deeper than the cover sequence. The timing of peak metamorphism has traditionally been considered to be post-D_M (de Wit, 1972; Kennedy, 1971, 1975b; Hibbard, 1983). However, detailed textural studies suggest that peak conditions were attained during D_M (Jamieson and Vernon, 1987).

Figure 3.57. An F_M isoclinal fold of interlayered magnetite-rich feldspathic psammite (light) and semipelite (dark); also note F_L minor folds to right of hammer; White Bay Group in Western Arm, White Bay. GSC 1994-766B

Textures related to peak metamorphism are distinctly different in the infrastructure and cover sequences on the peninsula. Infrastructural rocks are characterized by chaotic vermicular textures (de Wit, 1972, 1980) that are, in general, only evident microscopically, whereas the cover sequences display a widespread development of porphyroblasts.

The D_L event marked a major change in the tectonic style of the area, as D_L structures are generally less penetrative and more brittle in style than earlier structures. D_L structures are heterogeneous and include a collection of crenulation fabrics, folds, and kink bands; they are most intense along the Baie Verte-Brompton Line, where, locally, they nearly obliterate earlier structures.

In direct contrast to D_E and D_M tectonic polarity, D_L is marked by a west to east bulk translation; this is best displayed by eastward directed high angle thrusts along the Baie Verte-Brompton Line and in the immediately adjacent Dunnage Zone (Kidd, 1974; Bursnall, 1975; Hibbard, 1983). The Baie Verte-Brompton Line probably gained most of its present form during this event.

Ming's Bight-Pacquet Harbour area, Baie Verte Peninsula

Rocks of the internal Humber Zone as well as the Baie Verte-Brompton Line swing around the Baie Verte Flexure, to an east-west trend in this area (Hibbard, 1982, 1983). Three major phases of deformation are also evident in the Ming's Bight Group. D_E and D_M structures appear similar to those of the main outcrop area; D_E is marked by both ductile shear zones with associated ultramafic rocks and ophiolitic mélanges, and D_M is marked by isoclinal folds with a penetrative S_M fabric. D_L structures contrast with those to the west; here they are penetrative at outcrop scale and marked by intense southeastward-directed recumbent folds and a strong crenulation cleavage (Kennedy, 1971; 1975b; Hibbard, 1983). These structures also affect Dunnage Zone rocks immediately to the south.

The Baie-Verte-Brompton Line is also of different character than that of the line to the west; in the Ming's Bight-Pacquet Harbour area it is a broad zone of D_E ductile shears and ophiolitic mélanges, similar to the D_E zones in the main outcrop belt. However, the zone has been intruded by the Dunamagon Granite and subsequently subjected to D_M and D_L. These latter events did not concentrate the line into a narrow structural zone, as they did to the west. Instead, the D_E phase of the Baie Verte-Brompton Line was complexly folded, thus forming an ambiguous zone rather than a sharply defined lineament (Hibbard, 1982, 1983).

Age of tectonism, Baie Verte Peninsula

Some of the most decisive evidence for the timing and interpretation of the tectonothermal history comes from the D_E phase structures. In particular, the emplacement of ultramafic rocks along D_E shear zones and the presence of pre-kinematic ophiolitic mélanges indicate that ophiolites passed over the whole area including the Ming's Bight-Pacquet Harbour area (Bursnall, 1975; Williams, 1977b; Williams et al., 1977; Hibbard, 1983). These mélanges are lithically and structurally like those underlying the ophiolitic thrust sheets of the external Humber Zone, and they mark the early imbrication of the ophiolitic Birchy Complex at the

eastern margin of the miogeocline. Eastward of the Birchy Complex lies successively less deformed and less metamorphosed ophiolitic slices of the Dunnage Zone. Portions of the Birchy Complex are thus viewed as the structurally lowest remnants of an ophiolite in an imbricate stack that overrode the Fleur de Lys rocks and the miogeocline (Bursnall, 1975; Williams et al., 1977; Williams, 1977b; Hibbard, 1983). The Bay of Islands and Hare Bay ophiolitic allochthons of the external Humber Zone are interpreted to be westerly remnants of the overthrust oceanic crust (Bursnall and de Wit, 1975; Williams, 1977b; Williams et al., 1977). There, the timing of this event has been constrained to the Early Ordovician (Stevens, 1970). In the Ming's Bight-Pacquet Harbour area, D_E structures are intruded by the Dunamagon Granite which has been isotopically dated at 460 ± 12 Ma (Dallmeyer and Hibbard, 1984). Thus a similar Early Ordovician age is inferred for the event in the internal Humber Zone, although it is probable that tectonism started slightly earlier in this more eastern area (Fig. 3.58).

D_M structures on the peninsula have also been attributed to this event (Fig. 3.58), and have been related to the progressive deformation and burial of the Fleur de Lys Supergroup (de Wit, 1972; Bursnall, 1975). These structures have been regionally correlated with similar, westward directed structures in less metamorphosed thrust sheets, beneath the ophiolites of the external Humber Zone (Bursnall and de Wit, 1975).

The D_L reversal in structural polarity immediately follows peak metamorphism in the main outcrop area, strongly suggesting that this event relates to regional uplift of the main part of the Fleur de Lys Supergroup (de Wit, 1972; Bursnall, 1975; Hibbard, 1983; Fig. 3.58). The unusual

OROGENY	MAIN OUTCROP BELT	MING'S BIGHT - PACQUET HARBOUR AREA
"ACADIAN"	D_L - Reverse, w→e transport polarity - final exhumation of Fleur de Lys terrane (?)	D_L - Continued SE - thrusting and uplift of Fleur de Lys (?)
"TACONIC"	D_M - Continued e→w bulk transport and maximum burial of Fleur de Lys terrane	D_M - Probably encompasses 2 overlapping events — SE-directed thrusting from area to North, cause unknown, tightens "Taconic" structures — same as main outcrop belt
"TACONIC"	D_E - Initial westward obduction of ophiolites over Fleur de Lys terrane	D_E - Same as main outcrop belt
GRENVILLE	D_B - Grenville event ca. 1.1 Ga	

GSC

Figure 3.58. Tentative structural correlation chart for Baie Verte Peninsula, Newfoundland.

metamorphic textures of the infrastructure have been interpreted to indicate a sudden release of the D_M stress system by the initiation of D_L uplift (de Wit, 1972, 1980).

Incremental ^{40}Ar-^{39}Ar ages on the main outcrop belt on the peninsula indicate that regional metamorphism had subsided by the Middle Silurian (Dallmeyer, 1977a; Dallmeyer and Hibbard, 1984). These ages reflect cooling and uplift following peak metamorphism; thus, most of the metamorphic and structural character of the main outcrop area can broadly be ascribed to Taconic events.

In the Ming's Bight Group, D_L structures are not only different in form, but are of a different age, than those of the main outcrop area (Fig. 3.58). In the easterly area, they affect volcanic rocks that are considered to be Silurian to Devonian in age. ^{40}Ar-^{39}Ar ages indicate that regional metamorphism in this area did not subside until the Mid to Late Devonian (Hibbard, 1982, 1983; Dallmeyer and Hibbard, 1984). Considering stratigraphic and isotopic age constraints, late phase events in the Ming's Bight area are broadly Acadian in age.

The Acadian event on the peninsula involved south-southeastward-directed folds and thrust faulting in Dunnage Zone rocks (Kennedy, 1975b; Hibbard, 1982, 1983; P. Stella, pers. comm., 1983). Ophiolitic rocks that overlie and mask the Baie Verte-Brompton Line between Baie Verte and Ming's Bight have been involved in this thrusting (Hibbard, 1982, 1983); their position leads to the conclusion that these rocks originally were emplaced on top of the Fleur de Lys Supergroup during the Taconic obduction event, and later backthrust during the Acadian event. The extent of Acadian tectonism is uncertain, but the same event may be responsible for southeastward-directed thrusts farther to the east in the Dunnage Zone (Dean and Strong, 1977; Colman-Sadd, 1980; Nelson, 1981; Karlstrom et al., 1982).

Tectonothermal history elsewhere in Newfoundland

Metamorphic tectonites of the internal Humber Zone elsewhere in Newfoundland are similar in aspect to those of the Baie Verte Peninsula; generally they display three main phases of deformation and accompanying greenschist to lower amphibolite facies metamorphism. However, not enough evidence has been accumulated to constrain the age and mechanism of tectonothermal events. Locally, as on the Grey Islands, and west of Grand Lake, ophiolitic mélanges and ultramafic rocks localized along early faults (Kennedy et al., 1973; Williams, 1977b; Kennedy, 1981) strongly suggest that the Early Ordovician obduction event also affected these areas. The latter events in these areas are, at present, of uncertain origin.

There are regional contrasts between the metamorphic rocks in the Grand Lake-Deer Lake area and those on the Baie Verte Peninsula. Most strikingly, the Baie Verte rocks are disposed in upright folds with near horizontal axes, and are juxtaposed against rocks to the west along the steeply-dipping Cabot Fault. In contrast, rocks to the south mainly lie in moderately east-dipping thrust sheets. The simple, linear trends of the west-directed thrusts in this area suggests that the thrusts are late phase features; thus, the late phase transport events in this area may be diametrically opposed to those on the peninsula. There are no age

constraints on the late deformation in the Grand Lake area; however, minor west-directed thrusting affects nearby Carboniferous rocks, indicating long lived tectonism.

Quebec

The tectonothermal history of the internal Humber Zone in Quebec is variable along the trend of the zone. In particular, it appears that rocks in Gaspésie are, in general, at lower metamorphic grade and overall, are structurally simpler than internal Humber Zone rocks in the Eastern Townships. The origin of this heterogeneous tectonothermal history is not obvious. In the following section, we describe the structure and metamorphism from the three main geographic regions of the internal Humber Zone.

Eastern Townships

In general, three phases of deformation are recognized throughout the Sutton Mountains and Notre-Dame Mountains anticlinoria. However, there are structural contrasts both between and within each of these structures. The following section describes the structural and metamorphic schemes of stratigraphic units in the area, starting first in the Notre-Dame Mountains Anticlinorium and proceeding to the Sutton Mountains Anticlinorium.

The most recent structural studies in the Bennett Schists have been undertaken by Béland, 1957, 1962; Charbonneau, 1975; Charbonneau and St-Julien, 1981; St-Julien, 1987; and St-Julien and Hubert, 1975. Three thrust-sheets have been identified in the Bennett Schists and the rocks of all display at least three phases of post-emplacement folding. It is believed that ultramafic rocks (such as the Pennington body) mark the location of thrust faults that separate these thrust sheets. The outcrop pattern of the ultramafic rocks (mostly serpentinite) traces the complex geometry of folds within the Bennett Schists. The main phase deformation D_M involves tight to isoclinal recumbent folding with a penetrative axial planar schistosity. These folds trend northwest and verge southwest south of Québec City (St-Julien, 1987). An earlier phase of deformation D_E is also characterized by recumbent folds which possibly trend north or northeast. They are manifest by a fabric sub-parallel to bedding and mesoscopic folds. Bedding, S_E and S_M are sub-parallel except in the axial region of the F_M folds. On a broad scale all these planar elements are refolded to form a late upright anticlinorium, the Notre-Dame Mountains Anticlinorium, that plunges toward the northeast.

Southeast of the Bennett Schists in the Rosaire Group, the main schistosity, S_M, forms convergent fans trending northeast and dipping northwest in the southeast part of the Rosaire Group, and dipping southeast in the Bennett Schists toward the northwest. The early folds, F_E, are north-trending and west-verging. They have an axial plane schistosity, S_E, with a moderate dip to the east. These early folds are antiformal synclines and synformal anticlines (St-Julien, 1987). They are superposed on the overturned flank of a recumbent fold of unknown orientation.

To the southeast in the Caldwell Group, there is only one schistosity that trends northeast and dips steeply northwest. It is axial planar to the latest folds, which are variably plunging, either to the northeast or southwest. The variations in the plunge of these folds result from the superposition of this northeast-trending schistosity upon earlier, broad, open to tight, upright north-south folds without axial plane schistosity. However, these two sets of folds do not explain the regional overturning of the beds in the Caldwell Group. We believe that the north- and northeast-trending folds are superposed upon the overturned flank of a southeast-verging recumbent fold that covers an area of several hundred square kilometres. The hinge of that major recumbent fold is represented by the steeply-northwest-dipping Caldwell Group southeast of the Rosaire Group (St-Julien, 1987) and the overturned flank is observed in a window in the middle of the Rosaire Group.

The lower formations of the Oak Hill Group show three phases of deformation (Charbonneau, 1975; Charbonneau and St-Julien, 1981). These include an early northeast-trending phase of folding, an east-trending intermediate phase, and a late phase characterized by a crenulation cleavage, S_L, dipping approximately 30° northeast. The first two phases of folding produced domes and structural basins flattened and stretched along their plunge (30° to the northwest). They also produced interference patterns in the form of hooks and crescents along with complex multilobate forms (Charbonneau, 1975; Charbonneau and St-Julien, 1981).

No detail metamorphic studies have been done in the Bennett Schists. In general, the rocks have been metamorphosed to chlorite grade of the greenschist facies during D_E and D_M deformation. Ottrelite porphyroblasts are observed in some pelitic members and seem to be post D_M but pre D_L deformation. Quartz rods, stretching lineations, mineral lineations and most tight to isoclinal recumbent folds are assigned to the D_M deformation, whereas the crenulation lineations and open folds are assigned to D_L deformation.

Numerous detailed structural analyses have been undertaken in the Sutton Schists (DeRomer, 1960; Eakins, 1964; Osberg, 1965, 1967; Rickard, 1965; Clark and Eakins, 1968; Globensky, 1978; Charbonneau, 1975, 1980b; Lamothe, 1979, 1981a, b; Pintson et al., 1985; Marquis, 1989a; Colpron, 1990; and Slivitzky and St-Julien, 1987). The Oak Hill Group and the Sutton Metamorphic Suite have been deformed together. In general these rocks experienced three phases of deformation, including D_E, D_M, and D_L. D_E is characterized by a pervasive schistosity and west-verging early-ductile faults in the Sutton Metamorphic Suite and west-verging recumbent folds in the Oak Hill Group (Colpron, 1990).

D_M is marked by a major fold, the east-verging Enosburg Falls Anticline (Pinnacle Mountain Anticline of Clark and Eakins, 1968; Tibbit Hill Anticline of Richard, 1965; and Cooke, 1952). In the Mansville Complex, F_M folds are isoclinal and southeast-verging and plunge moderately to steeply along the S_M foliation. In the Sutton Metamorphic Suite, F_M folds are isoclinal, southeast verging and recumbent (Colpron, 1990). Most F_M hinges and mineral lineations are subparallel. This geometry of F_M folds suggests either sheath folds or reclined folds (Colpron, 1990).

D_L corresponds to the late arching that formed the Saint-Étienne Antiform (DeRomer, 1960; Eakins, 1964; Osberg, 1965; Marquis, 1989a; Colpron, 1990), a regional structure extending north some 75 km from the Quebec-Vermont border to Danville. F_L minor folds are almost

exclusively developed within the Sutton Schists and the Mansville Complex (Colpron, 1990). They are open folds and undulations, plunging gently to the north.

A recent structural interpretation of the Sutton Mountains Anticlinorium in the Sutton Valley has led to the recognition of four regional thrust faults. They are, from west to east: Oak Hill, Brome, Sutton, and Missisquoi faults. They are of different age, relative to peak metamorphism. The Sutton Thrust (Colpron, 1990) and Missisquoi Fault (Smith, 1982) are interpreted as synchronous with early metamorphism. The Sutton Thrust separates the Mansville Complex, to the west, from the Sutton Metamorphic Suite and is an east-directed thrust (Colpron, 1990). The Missisquoi Fault separates the Sutton Schists and Caldwell Group. It is interpreted as a steep east-directed thrust (Globensky, 1978; Slivitzky and St-Julien, 1987) or sinistral strike-slip fault (Marquis, 1989a). These early faults are usually marked by serpentinite slivers (Colpron, 1990).

According to Colpron (1990), most syn-metamorphic faults are located within the Mansville Complex. The Brome Thrust, marking the contact between the Oak Hill Group and the Mansville Complex, is one of the most significant syn-metamorphic faults (Colpron 1990). Kinematic indicators indicate west-over-east sense of shear (Colpron 1990). F_M axes in the Mansville Complex trend parallel to regional mineral stretching lineations, suggesting that the F_M folds are rotated toward the X axis of the strain ellipsoid and that the rotation is contemporaneous to shearing along the Brome Thrust (Colpron, 1990). The syn-metamorphic faults are usually marked by a well developed D_M mylonitic foliation, prominent mineral lineation, S-C fabrics, lithic truncations, and sericitized (phyllonite) and/or carbonatized zones (Colpron, 1990). The mylonitic foliation is parallel to the regional (S_M) cleavage (Colpron, 1990).

One possibly post-metamorphic thrust is the Oak Hill Thrust, marking the contact between the Oak Hill Group and Stanbridge Nappe of the external Humber Zone (Charbonneau, 1980b; Colpron, 1990). The westward transport of Oak Hill rocks over the Stanbridge Group is suggested by the presence of Sweetsburg klippen resting on Stanbridge strata in the vicinity of Meigs Corners (Charbonneau, 1975, 1980b).

In the Eastern Townships, metamorphism of the internal Humber Zone increases, in general, from north to south with evidence for polymetamorphism. In the Notre-Dame Mountains Anticlinorium, sub- to medium-greenschist facies assemblages are common. In the Thetford Mines area, Birkett (1981) has documented an early amphibolite facies metamorphism that has been overprinted by the regionally predominant greenschist facies metamorphism commonly seen in the metavolcanic rocks. Pumpellyite occurs in metavolcanics of the Montagne Saint-Anselme Formation, indicating a prehnite-pumpellyite facies locally overprinting the original igneous mineralogy and texture (Trzcienski and Birkett, 1982). On the north side of the Bécancour Dome, Birkett (1981) has described pumpellyite coexisting with actinolite; signifying a southwest transition from the prehnite-pumpellyite facies to the lower greenschist facies.

In the Sutton Mountains Anticlinorium, greenschist metamorphism is common and may reach lower amphibolite facies near the Quebec-Vermont border. Near the northern termination of the anticlinorium, Trzcienski (1976) reported sodic amphibole near Richmond, Quebec. This bluish amphibole was interpreted to be an indicator of a moderately high pressure metamorphism (about 6-7 kbar). Subsequent work on calc-silicate assemblages associated with ultramafic rocks in the region (Trzcienski, 1989) confirms this interpretation. Chloritoid occurs in a number of metapelites in the Oak Hill slice of the anticlinorium (Marquis, 1989a; Marquis et al., 1987) and is an indicator of greenschist or possibly higher pressure metamorphism. Chloritoid has also been reported at the southern end of the Notre-Dame Mountains Anticlinorium (Birkett, 1981). Throughout the Sutton and Bennett schists, white mica and poikiloblastic albite are common. Near Sutton Village (M. Colpron, pers. comm., 1990) and near the Quebec-Vermont border in the middle of the anticlinorium (M. Rickard, pers. comm., 1979) minor garnet has been observed in these schists. Also near the Quebec-Vermont border to the east of Abercorn, Poulin (1974) has observed garnet in the Lac Masten amphibolite and albite, quartz, aluminous amphibole and epidote. The general increase in metamorphism from north to south in the Sutton Mountains Anticlinorium agrees with the structural interpretation that shows the anticlinorium plunging northward and exposing more deeply-buried rocks southward.

Central Gaspésie

The Chic-Chocs area of the central Gaspésie comprises the Mont Albert and Mont Logan nappes (Fig. 3.49, 3.54, 3.59). The thrusting of the internal Humber Zone over the Orignal and Romieu formations of the external Humber Zone occurs along the Lac Cascapédia Fault. To the south, the Shickshock-Sud Fault separates the bedded Cambrian-Ordovician rocks of the interior Humber Zone from extensive Silurian-Devonian strata to the southeast.

The Mont Logan Nappe occupies the south-central part of the region, between the Rivière Matane and McGerrigle Mountains where it forms the Chic-Chocs Mountains. The nappe is bounded to the north by the Lac Cascapédia Fault and to the south by the Marches du Géant Fault and the Shickshock-Sud Fault.

The northern contact has been interpreted by some as an interdigitation of the Orignal Formation and the Trois-Pistoles Group with the Shickshock Group, accompanied by a gradual increase in metamorphism (Mattinson, 1964; Ollerenshaw, 1967; Beaudin, 1980; Biron, pers. comm., 1980). Nevertheless, for structural and metamorphic reasons we agree with McGerrigle (1954) and St-Julien and Hubert (1975) that the Shickshock Group constitutes a structural nappe. In effect, recent studies show that the

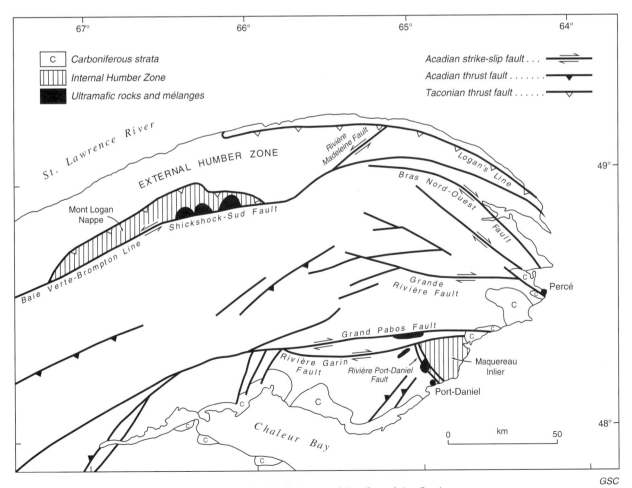

Figure 3.59. Geological map of the Gaspésie, Quebec.

GSC

135

boundary between the Shickshock Group and structurally underlying rocks to the north is a major metamorphic discontinuity. Moreover, the presence of a strongly penetrative stretching lineation in the Shickshock Group and the adjacent metapelites of the Orignal and Romieu formations, as well as the presence of S-C fabrics and shear bands in the metapelites, strongly suggest that the Shickshock Group is allochthonous upon the younger Orignal and Romieu formations. These structures are especially prominent in the outcrops along Route 11 leading to Lac Cascapédia south of Sainte-Anne-des-Monts. All Shickshock Group lithologies show a highly penetrative S_1 schistosity that is generally parallel to bedding (McGerrigle, 1954; Mattinson, 1964; Ollerenshaw, 1967; Beaudin, 1980, 1984). We believe that this schistosity, which was probably subhorizontal originally, is axial planar to the recumbent, northwest-verging folds. The S_1 schistosity is affected by a weak penetrative S_2 crenulation cleavage that is axial planar to relatively open macroscopic folds. These folds are upright to overturned with their axes plunging gently toward the northeast or southwest (Beaudin, 1980, 1984).

The highly penetrative, gently plunging, stretching lineation is a significant structural element in the Shickshock Group. It is also penetrative in the adjacent phyllites of the Orignal and Romieu formations. These lineations, along with the S-C fabrics and shear bands in the phyllites indicate that the Shickshock Group has been displaced horizontally in a westerly direction. These S-C fabrics and shear bands are superposed on the older S_1 schistosity. In effect, the simple shearing associated with the S-C fabrics has imprinted folds on the S_1 schistosity, the hinges of which are parallel to the stretching lineation and have characteristics of sheath folds.

The crenulation cleavage, S_2, has affected the schistosity S_1 and produced an intersection lineation L_2 parallel to the stretching lineation. In spite of the coaxiality of the lineations, we consider the advent of S_2 as late since this cleavage developed at a shallower structural level than the stretching lineation.

In summary, the Mont Logan Nappe was initially thrust toward the north or north-northwest. The recumbent folds and S_1 cleavage are considered to be contemporaneous with this movement. The stretching lineation and the S-C fabrics show that the nappe was then displaced westward. The last deformation involved northwest-southeast oriented shortening, producing macroscopic undulations and the crenulation cleavage S_2.

The Shickshock-Sud Fault, the southeastern border of the Mont Logan Nappe, has dextral strike-slip displacement of unknown magnitude (Fig. 3.54, 3.59). The fault trends about 060° in the Cuoq-Langis area (Ollerenshaw, 1967) and continues northeastward to the McGerrigle Mountains, where it is offset some 30 km toward the north by a sinistral strike-slip fault; from here, the fault continues eastward to Cap-des-Rosiers (Fig. 3.59). This new interpretation of a northerly offset and eastward continuation of the Shickshock-Sud Fault is based on the recently recognized sinistral strike-slip fault, the Rivière Madeleine Fault (Fig. 3.59). Toward the southwest, the Shickshock-Sud Fault continues beyond La Rédemption (south of Baie-des-Sables), but where the Rivière Montane trends east-west, branches of the fault continue on along the river for another 10 km on either side of Saint-Vianney.

The Shickshock-Sud Fault is easily recognized in the field by intense shearing and it is also evident from aeromagnetic and gravity anomalies. Along the fault an ophiolitic mélange is observed. It contains graphitic, graptolite-bearing slates, serpentinite, and dolomite produced by the carbonization of periodotites (Béland, 1960; Mattinson, 1964; Beaudin, 1984) in a chaotic shaly matrix. A marked example of this mélange and fault are found along the Rivière Cap-Chat Est.

According to some workers (Beaudin, 1984; Biron, pers. comm.) the Shickshock-Sud Fault is a normal fault, which formed the northern limit of the Connecticut Valley-Gaspé Synclinorium during Silurian-Devonian sedimentation. Nevertheless, Beaudin (1984) believed that this fault was reactivated during the Acadian deformation. Elsewhere, the work of Bourque et al. (1985) and that of De Broucker (1987) strongly suggest that the Baie Verte-Brompton Line (Williams and St-Julien, 1982) locally coincides in northern Gaspésie with the Shickshock-Sud Fault. This being the case, the Baie Verte-Brompton Line, which is a Taconic feature, would have had a local dextral, strike-slip movement during the Acadian Orogeny.

Metamorphism of the Shickshock Group is best displayed by the various assemblages in the metavolcanic rocks. In general, the assemblages record an increasing metamorphic gradient from north-northeast toward the southwest. Beneath the ophiolitic Mont Albert Nappe, the metavolcanics retain pillow structures in which chlorite, epidote, albite, and quartz coexist with relict clinopyroxene and an original igneous texture. Toward the Marches du Géant Fault separating the Mont Albert Nappe from the underlying Mont Logan Nappe, the metavolcanics show increasing deformation as displayed by a strong foliation and the loss of igneous features. The metamorphic assemblage is composed of chlorite, epidote, quartz, albite, opaque minerals, and minor actinolitic amphibole. West of the Mont Albert area, along the northern margin of the Chic-Chocs Mountains, the metavolcanics are well-foliated and lose most of their igneous features.

In the Cap-Chat River area immediately to the east of Figure 3.54, the metavolcanics at their northernmost extent are seen as layers, commonly defined by variable concentrations of epidote, quartz, chlorite, and albite. Southward across the Chic-Chocs area along the Cap-Chat River section, the metamorphic grade gradually increases; first, amphibole replaces chlorite and is followed by the increased tschermakitic component of the amphibole itself. Along the southern margin of the Chic-Chocs Mountains, chlorite is generally absent, except for retrograde grains associated with the Shickshock-Sud Fault, epidote is much less evident, and amphibole is commonly zoned with actinolitic cores and tschermakitic rims. The foliation that is well-defined by compositionally different beds in the north is completely absent in the south.

Because the metamorphism across the Chic-Chocs Mountains has occurred by a series of continuous reactions, definition of isograds has not been attempted; but the observed assemblages indicate lower greenschist facies to lower or middle amphibolite facies from north to south.

Southeast Gaspésie

The structure of the Maquereau Group has been studied by Ayrton (1967), Caron (1984), Ringele (1982) and De Broucker (1987). The intensity of deformation increases southward (toward Pointe au Maquereau, Fig. 3.60a). Two major phases of folding affect the Maquereau Group. The oldest phase produced a major south-verging anticlinorium (Fig. 3.60b) in which the parasitic folds (F_1) are asymmetric to overturned to the south with an axial plane cleavage dipping approximately 45° northwards (De Broucker, 1987; Caron, 1984). A second phase of deformation produced a crenulation cleavage and kink bands trending south-south-west (parallel to the coast) and a few southeast-verging folds (De Broucker, 1987).

Structural trends in the eastern part of the Maquereau Group are easterly (F_1 folds) whereas in the western part of the group the structural trend is south-southwest. This change in trend of the F_1 folds defines a late antiform plunging steeply to the south-southwest, which is possibly the result of shortening and dextral movement along the Rivière Port-Daniel Fault (Fig. 3.49) (De Broucker, 1987).

The positioning of the Maquereau Group to the southeast of the Baie Verte-Brompton Line appears to be a late structural feature. Post-Silurian wrench faulting involving

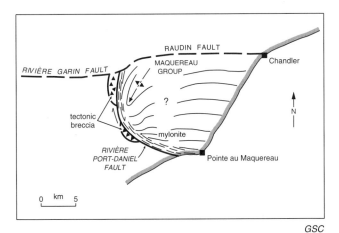

Figure 3.60a. Structural trend of F_1 folds defining, in plan view, a late antiform in the western Maquereau Group, Quebec, from De Broucker (1987). See Figure 3.59 for location.

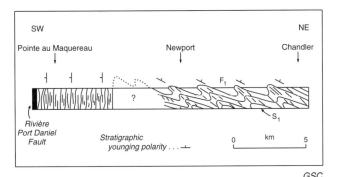

Figure 3.60b. Structural cross-section of the Maquereau Group along the coast of southern Gaspésie, from De Broucker (1987).

displacement of at least 250 km is postulated in Gaspésie (Bourque et al., 1985; Chapter 4, this volume). The main displacement occurred along the Grand Pabos Fault but significant movement also took place along associated faults such as the Grande Rivière Fault (Fig. 3.59). This faulting is responsible for the present position of the Maquereau Group to the southeast of the Baie Verte-Brompton Line. Prior to wrench faulting, it is believed that the Shickshock-Sud Fault in the central Gaspésie Peninsula was continuous with the Rivière Port-Daniel Fault in the Maquereau Dome (Fig. 3.59). This reconstruction allows a pre-Devonian alignment of Cambrian-Ordovician lithofacies of the Gaspésie in a single continuous belt.

The Maquereau Group has been metamorphosed to the lower greenschist facies. Sericite and chlorite are common in sandstones and shales. Locally, fine grained, hydrothermal biotite and small garnets (Blais, 1985) are found in the mylonitic sandstones of the group in the vicinity of the Rivière Port-Daniel Fault. This middle greenschist facies metamorphism is attributed to dynamic metamorphism related to movement along the Rivière Port-Daniel Fault.

Age of tectonism in Quebec

The timing of tectonism in the Quebec internal Humber Zone appears to vary along strike. Studies of the Maquereau-Mictaw Inlier of Gaspésie (De Broucker, 1986) put some constraints upon the paleogeographic evolution of the area.

The early deformations that affect the Maquereau Group are related to early phases of the Taconic Orogeny (Rodgers, 1967) known as "Taconic 1" (Fig. 3.55). These deformations of the Upper Cambrian and/or Lower Ordovician strata occurred prior to the tectonic accretion of the Rivière Port-Daniel mélange to the Maquereau Group (De Broucker, 1986). They correspond to the Gaspésienne Orogeny of Ayrton (1967) and probably to the Grampian Orogeny as described in the Caledonides (Mitchell, 1978; Williams, 1984). The earliest phase of "Taconic 1" is responsible for the south-verging F_1 fold observed in the eastern part of the Maquereau Group (De Broucker, 1986). These folds have an axial plane schistosity. A second phase of southeast-verging folds with a crenulation cleavage and kink-bands are also attributed to the "Taconic 1" (De Broucker, 1986).

A late event of "Taconic 1" is responsible for the mylonitization, cataclastic metamorphism, and brecciation related to the dextral stike-slip movement along the Rivière Port-Daniel Fault (De Broucker, 1986). Because the breccia contains only clasts of metamorphic Maquereau sandstone and Grenville basement, it is suggested that brecciation preceded tectonic accretion of the "mélange" with the Maquereau Group (De Broucker, 1986). This late event of the "Taconic 1" is also responsible for the major antiformal structure plunging southward in the western part of the Maquereau Group (De Broucker, 1986).

The "Taconic 2" of the Taconic Orogeny, which affected the Nadeau and the Rivière Port-Daniel ophiolitic mélanges predated the deposition of the Mictaw Group (pre-Llanvirn) and is assigned to the Early Ordovician (De Broucker, 1986). The development of mélanges in a subduction zone environment constituted the early part of "Taconic 2" whereas their accretion to the Maquereau Group along the Rivière Port-Daniel Fault is attributed to a late event of "Taconic 2" (De Broucker, 1986). Recurrent faulting along the Rivière Port-Daniel Fault is responsible for the local sedimentation of the tectonic breccia of the Maquereau Group at the top of the Rivière Port-Daniel Mélange (De Broucker, 1986).

A late deformation affecting the Mictaw Group (Middle Ordovician) and the Chaleurs Group on Gaspésie is related to the Middle Devonian Acadian Orogeny. This deformation produced a second phase of folding in the Mictaw Group and the first phase of folding in the Chaleurs Group. The folds are symmetric to asymmetric and locally overturned to the southeast. A highly penetrative axial plane cleavage is well developed in the Silurian rocks of the Chaleurs Group, except in the area southwest of the Mictaw Group where there is no cleavage.

In central Gaspésie, the F_1 and F_2 folds in the Shickshock Group are attributed to "Taconic 3", which is Middle Ordovician and postdates the obduction of the structurally overlying Mont Albert peridotite.

In the Eastern Townships, the age of tectonism is unknown, but since the Quebec ophiolites are subduction-related oceanic crust, subduction must have started before 478 Ma (age of the Thetford Mines complex; Laurent and Hébert, 1989).

The subduction zone was probably initiated during the Early Ordovician, if not in Late Cambrian as suggested by Stanley and Ratcliffe (1985) on the basis of the Early Ordovician age of the Moretown Formation. Taconic related, low-temperature high-pressure metamorphic amphiboles found in northern Vermont yield ^{40}Ar-^{39}Ar ages of 468 Ma and 490 Ma (Laird et al., 1984) suggesting that the Laurentian margin interacted with the subduction zone in Early Ordovician to early Middle Ordovician time (Tremblay, 1992b). In the Quebec Reentrant, stratigraphic and structural relationships within the Humber Zone indicate that the earliest deformation is post-Llanvirn (post-470 Ma; Osberg, 1965; Marquis, 1989b). Related metamorphic hornblendes yield a Middle Ordovician age (Laird, pers. comm. in Colpron, 1990). Thus, as suggested by Tremblay (1992b), available thermochronological data are in agreement with the inferred diachronism in the subduction of Laurentia in the Eastern Townships.

Significance and interpretation

The infrastructure is sparsely present along the length of the internal Humber Zone of the Canadian Appalachians (Fig. 3.45). Essentially, the infrastructure is a unit that recorded the deepest tectonism in the Humber Zone. Logically, rocks lowermost in the original stratigraphic succession have more potential for deeper burial during orogenesis than those higher in the succession. Thus, it is expected that the infrastructure incorporates elements of remobilized Grenville basement, as in all cases in Quebec, as well as intensely deformed equivalents of lower parts of the cover succession, as in Newfoundland. The presence of Grenville basement this far east in the Humber Zone is a strong indication that protoliths of the internal Humber Zone along the whole of the Canadian Appalachians, were deposited "within sight" of the less deformed miogeoclinal rocks to the west.

The cover sequence forms the bulk of the internal Humber Zone. It is a coherent depositional/magmatic package that extends for the entire length of the Canadian Appalachians. Most significantly, the sequence overlaps both continental and oceanic substrates in Newfoundland; in Quebec, it is interpreted to rest on continental basement. This observation, in conjunction with the recognition that the internal Humber Zone intervenes between platformal deposits of the external Humber Zone and the oceanic rocks of the Dunnage Zone is a decisive aid in reconstructing the pre-kinematic setting of the cover rocks.

The lower portions of the cover sequences throughout the internal zone are remarkably similar; in general they are characterized by metabasalts and coarse metasedimentary rocks that are locally shown to be of shallow water origin, and in some places in Newfoundland they include felsic metavolcanic rocks. In Newfoundland, these units include lithic correlatives of the lower Labrador Group of the external Humber Zone (such as the Oody Mountain Amphibolite), large portions of the infrastructure (including the granite at Deer Lake and part of the Keystone schist), lower portions of the monotonous sequences, and most amphibolites in the Fleur de Lys Supergroup. In Quebec, the Tibbit Hill, the Montagne Saint-Anselme, and the Shickshock metavolcanics, as well as the lower parts of the Shickshock and Maquereau groups are shallow water sequences that constitute the lower portion of the cover. All of the metabasalts in these units along the entire length of the zone have tholeiitic rift-related petrochemical signatures. In Newfoundland, the lower Labrador Group of the external Humber Zone is interpreted to represent the late Precambrian-Early Cambrian rift stage of the Laurentian continent (Williams and Hiscott, 1987; Williams et al., this chapter). Thus, on the basis of isotopic ages, petrochemical analyses, and regional correlation with rocks in the external Humber Zone, the oldest parts of the internal cover sequence are interpreted as recording a late Precambrian rift event; this regional event was most likely responsible for the formation of oceanic crust that floors the easternmost portions of the cover sequence in Newfoundland, as well as the western Dunnage Zone. The easterly oceanic crust in Newfoundland is represented by the Birchy Complex.

Younger strata of the cover sequence most reasonably represent the Early Paleozoic evolution of a slope and basin environment during the drift stage. These strata record the progressive transgression and building of a continental shelf system along a thermally subsiding passive margin. This stage is recorded by the varied sequences and the upper portions of the monotonous sequences in Newfoundland. Complex relationships between these units indicate that the slope and basin environments were most likely contemporaneous. In Quebec, the early Paleozoic drifting stage is represented by such units as the upper part of the Oak Hill, Armagh, Shickshock and Maquereau groups, the Rosaire and Caldwell groups, and parts of the Sutton Schists (St-Julien and Hubert, 1975; Beaudin, 1984; De Broucker, 1987). Marble breccia blocks within the varied sequences of Newfoundland indicate that the slope-basin geometry formed along the flank of a substantial carbonate bank; the extent of the varied sequences from Quebec through Newfoundland, indicate that the shelf-slope-basin system was of continental proportions. The apparent greater thickness of cover sequences in Quebec compared to Newfoundland examples may be related to location of the Quebec section at an indenture along the ancient continental margin, now represented by the Quebec Reentrant (Fig. 3.45).

The tectonothermal history of the internal Humber Zone records the demise of the established early Paleozoic North American continental margin. Initial deformation and metamorphism of the cover sequence above crystalline basement involved the tectonic juxtaposition of the internal Humber Zone and the oceanic Dunnage Zone along the Baie Verte-Brompton Line. This event involved the westward transport of ophiolite suites over the zone along pre- to syn-metamorphic shear zones and production of ophiolitic mélanges. The shear zones are commonly ornamented with dismembered fragments of the ophiolite suites. In Quebec, the transported ophiolites, represented by the Mont Albert Nappe, structurally overlie rocks of the internal Humber Zone. In Newfoundland, the structurally higher ophiolites are preserved to the west, in the Bay of Islands, and to the east, in the Birchy Complex and western Dunnage Zone. In the Baie Verte area, the distribution of early structural zones and later deformed zones indicates that the early form of the Baie Verte-Brompton Line, there, was arcuate in plan view; this dog-leg in the line has been interpreted as reflecting a morphological irregularity in the ancient continental margin (Hibbard, 1982, 1983).

The obduction of ophiolites is envisaged as the aborted subduction of the early Paleozoic North American continental margin beneath oceanic realms to the east (Stevens, 1974). In Newfoundland, the obduction event is manifested by D_E and D_M structures; regional correlation indicates that these events ranged from possibly Tremadoc to Caradoc. In Quebec, the timing of this event appears to have spanned the same time period.

The later stages of deformation, D_L in Newfoundland appear to be related to a release of the obduction stress regime and the initiation of easterly- directed structures along the Baie Verte-Brompton Line. This event is likely related to the uplift of the internal Humber Zone following obduction; ^{40}Ar-^{39}Ar cooling ages indicate that this event had probably ceased by the Middle Silurian in most areas of Newfoundland. In the Ming's Bight area, however, younger cooling ages indicate that either (1) cooling was less protracted than in the remainder of the zone or (2) there was a distinct Silurian-Devonian event in this area.

In Quebec, K-Ar cooling ages are Upper Ordovician in most areas; this cooling is likely related to the uplift of the internal Humber Zone. A crenulation cleavage associated with a late stage of deformation D_L in Quebec is related to the Acadian Orogeny.

Acknowledgments

The data from Newfoundland is published with the permission of the Director of the Geological Survey Branch of the Newfoundland Department of Mines and Energy. The first author thanks Harold Williams for originally introducing him to the internal Humber Zone in the 1970s, and for his editorial persistence in obtaining this manuscript. Brenda Batts at North Carolina State University is also thanked for her word-processing efforts.

DUNNAGE ZONE
Preamble
Harold Williams

The Dunnage Zone takes its name from Dunnage Island off northeast Newfoundland where the zone is 150 km wide with excellent exposures across islands and headlands. The zone is characterized by abundant volcanic assemblages, ophiolite suites and mélanges. Sedimentary rocks include slate, greywacke, epiclastic volcanic rocks, chert, and minor limestone all of marine deposition. Stratigraphic sequences are variable and formations are commonly discontinuous. Most rocks are of Late Cambrian to Middle Ordovician age; rocks younger than Middle Ordovician (Caradocian) are excluded.

The Dunnage Zone is traceable across most of Newfoundland, although it is narrow and disappears southwestward at Cape Ray. It is absent on the opposite side of Cabot Strait in Cape Breton Island but ophiolitic rocks, mélanges and volcanic sequences reappear in northern New Brunswick. Farther west, the Dunnage Zone is hidden by Silurian-Devonian cover rocks of the Gaspé Belt. On its western side, ophiolitic rocks, mélanges, and marine sedimentary and volcanic rocks occur throughout the Eastern Townships of Quebec (Fig. 3.61).

In the model of Schuchert (1923) and Schuchert and Dunbar (1934), the Dunnage (and Gander) zones fell into the general region of their New Brunswick Geanticline, a contrived land barrier that successfully separated the contrasting Cambrian trilobite faunas of the St. Lawrence Geosyncline (west) and Acadian Geosyncline (east). This notion of a landmass was replaced by the more realistic concept of an equally effective deep ocean basin, symmetrically disposed between sedimentary prisms built up at the margins of opposing platforms (Williams, 1964a). With the wide acceptance of plate tectonics, Dunnage Zone ophiolitic rocks were interpreted as oceanic crust and mantle (Stevens, 1970; Dewey and Bird, 1971), its mélanges as vestiges of subduction zones (Bird and Dewey, 1970; Kay, 1976) or major back-arc olistostromes (Hibbard and Williams, 1979), and its volcanic rocks as island arcs and oceanic islands (Bird and Dewey, 1970; Strong, 1973; Kean and Strong, 1975).

The Dunnage Zone is a suspect terrane, more precisely a composite suspect terrane (Williams and Hatcher, 1983; Keppie, 1985, 1989). Smaller suspect elements are recognized in Newfoundland, New Brunswick and Quebec. The latest tectonic analyses focus on terranes or subzone divisions and their important accretionary history. Controversies still exist over such basic questions as: (1) whether or not Dunnage ophiolitic rocks represent vestiges of a single ocean or a number of small ocean basins, (2) which of its volcanic sequences represent island arc volcanism and which are other oceanic volcanic products, (3) what are the

Williams, H.
1995: Preamble (Dunnage Zone); in Chapter 3 of Geology of the Appalachian-Caledonian Orogen in Canada and Greenland, (ed.) H. Williams; Geological Survey of Canada, Geology of Canada, no. 6, p. 139-142 (also Geological Society of America, The Geology of North America, v. F-1).

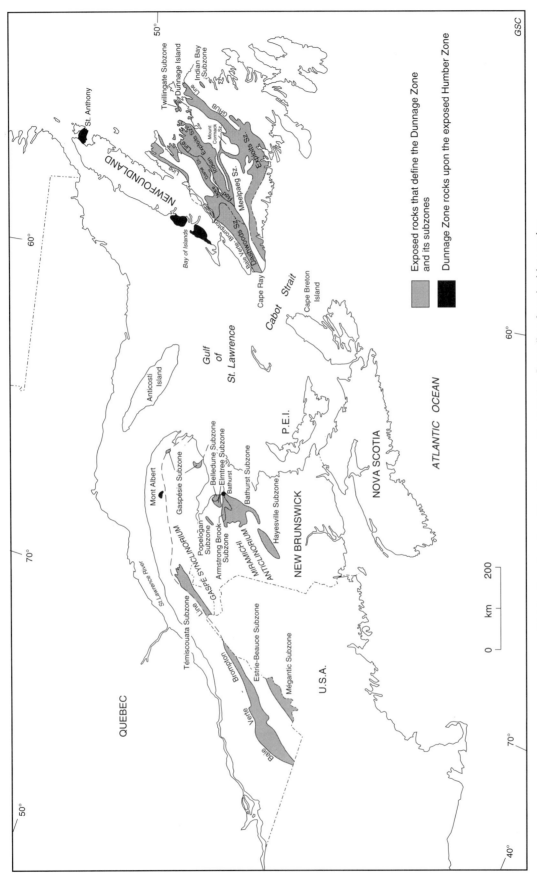

Figure 3.61. Distribution of rocks of the Dunnage Zone in the Canadian Appalachian region.

controls of obduction and polarity of subduction, and (4) what is the nature and significance of Dunnage Zone/subzone boundaries.

Aeromagnetic anomalies of the Dunnage Zone have short wavelengths, and high amplitudes that reflect its local volcanic belts. A positive Bouguer anomaly of 50 mGal in northeast Newfoundland is the broadest positive Bouguer anomaly in the interior part of the Appalachian Orogen (Haworth et al., 1980a, b). However, the anomaly is less than that predicted for mafic volcanic and ophiolitic rocks that extend to mantle depths (Karlstrom, 1983).

Deep seismic reflection transects off northeast Newfoundland (Keen et al., 1986) and in the Gulf of St. Lawrence (Marillier et al., 1989) imply that rocks of the Dunnage Zone are detached at mid crustal levels and overlie opposing Grenville and Central lower crustal blocks.

An ophiolitic basement to its volcanic-sedimentary sequences was part of the original Dunnage Zone definition (Williams, 1978a, b, 1979). However, basement relationships are commonly unknown, and in some places, volcanic-sedimentary sequences assigned to the Dunnage Zone are stratigraphically above Gander Zone clastic sedimentary rocks (van Staal, 1987; Williams and Piasecki, 1990).

The Humber-Dunnage boundary is interpreted everywhere as tectonic. It is the Baie Verte-Brompton Line (Williams and St-Julien, 1982), but highly allochthonous Dunnage Zone rocks occur well west of the line in Newfoundland and Quebec, and Humber Zone rocks occur well east of the line in subsurface.

The Gander-Dunnage boundary has always been problematic. In northeast Newfoundland, it is generally drawn at the eastern margin of the Gander River Ultrabasic Belt of Jenness (1958), the Gander River Complex of O'Neill and Blackwood (1989), or the Gander River Ultrabasic Belt (GRUB) Line of Blackwood (1978, 1982). There, Dunnage Zone sedimentary rocks are unconformable on plutonic rocks of the Gander River Complex and all are faulted against clastic rocks of the Gander Zone (Williams et al., 1991). However, in central Newfoundland there is no single boundary and Gander Zone rocks occur as inliers in the Dunnage Zone (Mount Cormack and Meelpaeg subzones of Williams et al., 1988) and Dunnage Zone rocks occur as an outlier in the northeast Newfoundland Gander Zone (Indian Bay Subzone of Williams et al. 1988). Recent studies in central and southern Newfoundland indicate structural boundaries and allochthonous Dunnage Zone rocks above Gander Zone rocks (Colman-Sadd and Swinden, 1984; Williams et al., 1988, 1989b; Piasecki, 1988; Piasecki et al., 1990; Williams and Piasecki, 1990). Boundaries that were interpreted as stratigraphic between Dunnage-Gander rocks north of Gander Lake (Pajari et al., 1979; Currie et al., 1979; Blackwood, 1982) and at Bay d'Espoir (Blackwood, 1985a) are reinterpreted as tectonic (Williams et al., 1991; Piasecki, 1988; respectively). Uncertainty exists in a few other places and the boundary of the Indian Bay Subzone, first viewed as possibly tectonic (Wonderley and Neuman, 1984), is reinterpreted as stratigraphic (O'Neill and Knight, 1988).

In New Brunswick, the Dunnage-Gander boundary was drawn at the Millstream Fault (Williams, 1978a) between the Bathurst and Elmtree subzones; but Dunnage Zone volcanic rocks overlie an ophiolitic substrate on both sides of the fault and farther south in the Bathurst area they overlie quartzites of the Gander Zone with a conglomerate unit marking the stratigraphic base (van Staal, 1987). Tectonic emplacement of northern belts above southern belts occurred in the Late Ordovician-Early Silurian and produced the largest known blueschist belt in the Appalachian Orogen (van Staal et al., 1990).

Where the same volcanic-sedimentary sequences overlie both an ophiolitic (Dunnage) and quartz clastic (Gander) substrate, they must span an ocean-continent transition, both in New Brunswick and Newfoundland. Although the volcanic-sedimentary sequences are assigned to the Dunnage Zone, they are really overstep sequences and therefore are inappropriate for making a true Dunnage-Gander distinction. This distinction is only possible where ophiolite suites are preserved and juxtaposed with sedimentary rocks of the Gander Zone.

Rocks of the Humber Zone are linked stratigraphically to crystalline Grenville basement and therefore to the Grenville lower crustal block. There are no established linkages between rocks of the Gander Zone and crystalline basement of the Central lower crustal block. Tectonic boundaries between the Dunnage and Gander zones allow stratigraphic relationships between surface rocks of the Gander Zone and the Central lower crustal block. Stratigraphic boundaries and a continuous cover that links an ophiolitic Dunnage substrate in one place with terrigenous Gander sedimentary rocks in another suggest that rocks of both the Dunnage and Gander zones are allochthonous above the Central lower crustal block.

Structure and metamorphism in rocks of the Dunnage Zone are generally less intense than in rocks of nearby parts of the Humber and Gander zones. In most places, notably along its western side in Newfoundland, New Brunswick and Quebec, Dunnage Zone rocks are affected by Taconic deformation and they are overlain unconformably by Silurian rocks. In central parts of the zone, such as northeast Newfoundland, Ordovician and Silurian rocks are conformable. Along the eastern margin of the Dunnage Zone in northeast Newfoundland, mafic-ultramafic complexes are overlain unconformably by Lower and Middle Ordovician rocks.

Mineral deposits of the Dunnage Zone are as distinctive as the rocks themselves and include asbestos, chromite, magnesite and nickel in ultramafic rocks of ophiolite suites, Cyprus-type copper-pyrite deposits in lower, ophiolitic volcanic rocks, and polymetallic deposits in upper volcanic units. Ultramafic rocks along the Baie Verte-Brompton Line are host to the world's largest asbestos deposits.

Granitic intrusions are common across the Dunnage Zone. Ordovician and Late Cambrian intrusions are tonalites, trondhjemites and plagiogranites related to its early constructional phases. Middle Paleozoic intrusions are also distinct in some places but they cross zone boundaries. A spatial coherence between types of plutons, tectonostratigraphic zones and lower crustal blocks in Newfoundland implies linkages between surface rocks and lower crustal blocks (Williams et al., 1989a). An allochthonous Dunnage Zone above opposing Grenville and Central lower crustal blocks rationalizes the presence and abundance of middle Paleozoic granites that cut the ophiolite-based sequences.

Ordovician brachiopod faunas of the Dunnage Zone are mainly of the Celtic realm and similar to those of the Gander Zone (Neuman, 1984; Nowlan and Neuman, Chapter 10, this volume). Along its western margin in Newfoundland, Ordovician conodont faunas of North American affinity are also known (Nowlan and Thurlow, 1984; Stouge, 1980; Boyce et al., 1988).

In the following treatment, rocks of the Dunnage Zone are described separately for Newfoundland, New Brunswick, and Quebec. This is because the rocks cannot be divided into convenient temporal or spatial packages recognized throughout the length of the Canadian Appalachians, as is the case in the Humber Zone. Interprovincial correlations are possible between some volcanic assemblages and ophiolite suites, but there is no assurance that Dunnage Zone rocks represent the same things in all places. Syntheses, appraisals of models and controls are given for each province.

Newfoundland

Harold Williams

Introduction

The Newfoundland Dunnage Zone, or central region of the Newfoundland Appalachians, was regarded as distinctive since the first regional syntheses. The area was the Central Mineral Belt of Snelgrove (1928) who recognized its abundant base metal occurrences hosted by mafic volcanic rocks. Combined with the Gander Zone and eastern parts of the Humber Zone, it was the Central Paleozoic Mobile Belt (Williams, 1964a) and the Central Volcanic Belt (Kay and Colbert, 1965) in tripartite divisions of the Newfoundland Appalachians. In analyses since the advent of plate tectonics, it included part of Zone C (Fleur de Lys) and zones D (Notre Dame), E (Exploits) and F (Botwood) of Williams et al. (1972, 1974). It was later defined as the Dunnage Zone (Williams, 1976, 1979) and extrapolated southward along the length of the Canadian and northern U.S. Appalachians (Williams, 1978a), and northeastward to the British Isles (Williams, 1978b; Kennedy, 1979). In suspect terrane analyses, it is the Dunnage Terrane (Williams and Hatcher, 1982, 1983).

Subdivisions

The Newfoundland Dunnage Zone is separated into two large geographic divisions or subzones, and several smaller ones, following the latest work of Williams et al. (1988, 1989b). The two large divisions are the Notre Dame Subzone in the west and the Exploits Subzone in the east, separated by the Red Indian Line, a major tectonic boundary traceable across Newfoundland (Fig. 3.62). The names Notre Dame and Exploits have been used before (Williams et al., 1974) but the present subdivisions and their definitions are new. Two small areas of volcanic and sedimentary rocks

are assigned to the Twillingate and Indian Bay subzones in northeast Newfoundland. A larger area of metamorphic rocks, tonalites and mafic/ultramafic complexes extends from Grand Lake to Port aux Basques. It was formerly known as the Tonalite Terrane (Whalen and Currie, 1983) or the Central Gneiss Terrane (van Berkel and Currie, 1988), and included in the Notre Dame Subzone (Williams et al., 1988). This division is renamed the Dashwoods Subzone.

The Notre Dame and Exploits subzones have numerous contrasts as follow: (a) Early to early Middle Ordovician faunas of the Notre Dame Subzone have North American affinities while those of the Exploits Subzone have Celtic affinities, (b) the Notre Dame Subzone was affected by the Taconic Orogeny whereas central parts of the Exploits Subzone had continuous deposition throughout the interval of the Taconic Orogeny, (c) the subzones have different plutonic histories, (d) the Exploits Subzone contains more sedimentary rocks in relation to volcanic rocks, and distinctive Ordovician mélanges, (e) the subzones have local contrasts in style and timing of structural development, (f) lead isotopic signatures in volcanogenic sulphide deposits of the Exploits Subzone contrast with those of the Notre Dame Subzone (Swinden, 1987), and (g) discrete areas of monotonous metaclastic rocks (Mount Cormack and Meelpaeg subzones) that are interpreted as structural inliers of the Gander Zone (Colman-Sadd and Swinden, 1984; Williams et al., 1988) are confined to the Exploits Subzone.

The Twillingate Subzone contains a large tonalite batholith dated at 510 Ma (Williams et al., 1976). Amphibolite facies metamorphism, intense deformation, and mylonitization along its southern margin are out of context with adjacent, younger, less metamorphosed and less deformed rocks of the Notre Dame Subzone.

The Indian Bay Subzone has volcanic rocks and fossiliferous sedimentary rocks typical of the Dunnage Zone, but lies within the northeastern Gander Zone.

The Dashwoods Subzone, although treated with the Dunnage Zone, has metaclastic rocks that are possible Humber Zone correlatives. These may represent Humber inliers in the western Dunnage Zone. Where Humber metaclastic rocks and Dunnage ophiolitic rocks are structurally commingled, metamorphosed, and intruded, a Humber-Dunnage distinction is impractical.

Subzone boundaries

The western boundary of the Dunnage Zone (Notre Dame Subzone) with the Humber Zone is the Baie Verte-Brompton Line (Williams and St-Julien, 1982). It is defined as the Ordovician tectonic contact between Humber Zone metaclastic rocks and Dunnage Zone ophiolitic rocks. It is everywhere overprinted and confused by later structures, later intrusions, and in places hidden by cover rocks. It is traceable as a steep structural zone from the Baie Verte Peninsula to the south end of Grand Lake. It is partly coincident with a faulted syncline of Silurian rocks at Baie Verte Peninsula and is hidden by Carboniferous basins farther south. At Deer Lake, it is coincident with a Carboniferous frontal thrust. At the south end of Grand Lake it may coincide with the Dashwoods-Notre Dame subzone boundary and it may be repeated structurally as irregular short boundaries between mafic-ultramafic complexes and metaclastic rocks in the Dashwoods Subzone. The Dashwoods Subzone is

Williams, H.
1995: Dunnage Zone-Newfoundland; in Chapter 3 of Geology of the Appalachian-Caledonian Orogen in Canada and Greenland, (ed.) H. Williams; Geological Survey of Canada, Geology of Canada, no. 6, p. 142-166 (also Geological Society of America, The Geology of North America, v. F-1).

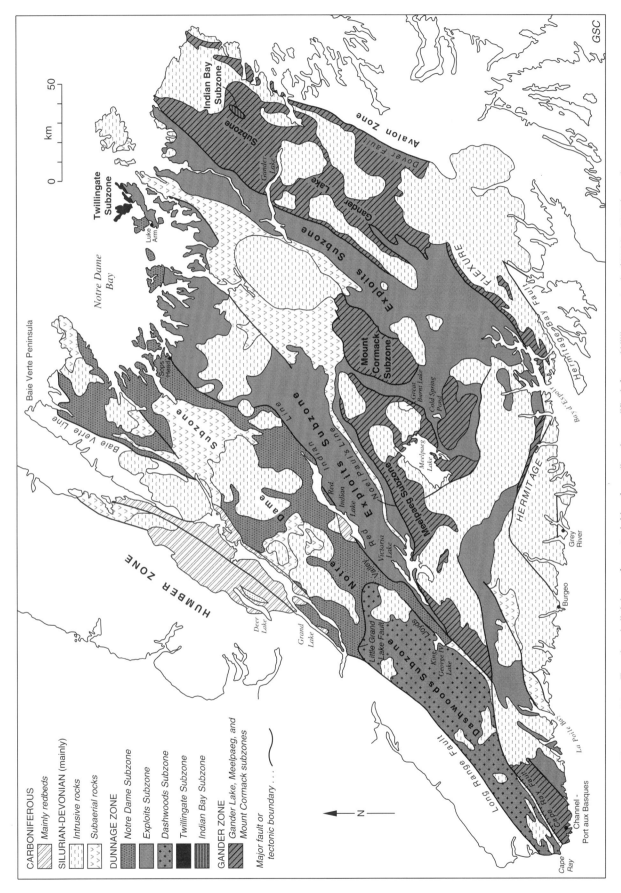

Figure 3.62. Zonal subdivision of central Newfoundland, modified from Williams et al. (1988, 1989) and Colman-Sadd et al. (1990).

143

bounded by the Long Range Fault to the west, the Cape Ray and Lloyds Valley faults to the east, and the Little Grand Lake Fault to the north.

The southern boundary of the Twillingate Subzone with the Notre Dame Subzone is marked by sheeted dykes which cut deformed tonalites and volcanic rocks. Toward the west the boundary has been interpreted as a fault (Williams and Payne, 1975; Williams et al., 1976) or a stratigraphic contact (Strong and Payne, 1973; Dean, 1978).

The eastern boundary (boundaries) of the Dunnage Zone (Exploits Subzone) with the Gander Zone is the east margin of the Gander River Complex (O'Neill and Knight, 1988) or the Davidsville-Gander group contact in northeastern Newfoundland, and the Bay d'Espoir shear zone at Bay d'Espoir in southern Newfoundland (Colman-Sadd, 1976; Piasecki, 1988).

Recognition of structural windows of psammitic rocks within the Exploits Subzone and correlation of the psammitic rocks with clastic rocks in the type area of the Gander Zone all imply structural Exploits-Gander boundaries. Major tectonic junctions are demanded where Dunnage ophiolite suites surround the Mount Cormack Subzone (Colman-Sadd and Swinden, 1984), and this model is supported by recent work in the Meelpaeg Subzone (Colman-Sadd, 1987; Williams et al., 1988, 1989b; Williams and Piasecki, 1990). Noel Pauls Line (Brown and Colman-Sadd, 1976; Williams et al., 1988, 1989b) is a major shear zone that marks the Exploits-Meelpaeg boundary in the northwest, and a shear zone traceable for 60 km and developed in granite at Great Burnt Lake marks the same boundary in the southeast. Ophiolitic mélange with black shaly matrix is developed locally at the Exploits-Meelpaeg boundary at Cold Spring Pond and its formation predates ductile shearing.

The Indian Bay Subzone-Gander Zone boundary has been interpreted as tectonic and stratigraphic (Wonderley and Neuman, 1984; O'Neill and Knight, 1988). The latest work suggests a conformable stratigraphic contact on its northwest side and a shear zone on its southeast side (O'Neill and Blackwood, 1989). The model of an allochthonous Exploits Subzone above the Gander Zone does not preclude conformable Dunnage-Gander boundaries, especially in eastern localities and beyond the effects of ophiolite transport. This may be the situation at Indian Bay Big Pond (Williams and Piasecki, 1990).

The Red Indian Line coincides with the Lukes Arm and Sops Head faults on the northern coast of Newfoundland. Inland, it coincides with faults along Red Indian Lake, Lloyds Valley, and the northwest side of Victoria Lake to Burgeo Road. These are relatively young brittle faults along a major junction of earlier significance (Williams et al., 1988). Farther south, the Red Indian Line coincides with Noel Pauls Line. This is a ductile mylonite zone that extends to Cape Ray on the south coast of Newfoundland. The Red Indian Line also coincides with an extremely narrow belt of Silurian rocks at Red Indian Lake and a narrow belt of Devonian rocks between King George IV Lake and Cape Ray. A local Carboniferous basin at Red Indian Lake and possibly another at Lloyds Valley (R.K. Herd, pers. comm., 1988) are located on the line.

Notre Dame Subzone

Rocks of the Notre Dame Subzone are mainly Lower Ordovician, thick, marine sequences dominated by mafic volcanic rocks. Sections are best exposed at Baie Verte Peninsula and Notre Dame Bay. From there, the rocks are traceable inland to Red Indian Lake and the south end of Grand Lake (Fig. 3.62). Plagiogranites of late Cambrian to middle Ordovician age are common throughout the subzone (Williams et al., 1989a). Terrestrial volcanic and sedimentary rocks of Silurian and Devonian age are unconformable on Ordovician rocks along its full length.

Ophiolite suites and mélanges

A prominent belt of ophiolite occurrences extends across Notre Dame Bay from Baie Verte to New World Island (Fig. 3.63). The Birchy, Advocate and Point Rousse complexes (Williams et al., 1977; Hibbard, 1983) all occur at or near the Humber-Dunnage boundary of the Baie Verte Peninsula. These vary in structural style from polydeformed and metamorphosed (Birchy) to imbricated and highly dismembered (Advocate) to almost complete suites (Point Rousse). Farther east, the Betts Cove Complex (Upadhyay et al., 1971) is a pristine example of an ophiolite suite with a conformable volcanic cover. Pillow lavas of the Lushs Bight Group are interpreted as ophiolitic (Smitheringale, 1972), sheeted dykes are exposed at Hall's Bay Head (Kean, 1983a), Pilleys Island (Strong, 1972), Sunday Cove Island (Kean, 1984), and as far east as Herring Neck (Williams et al., 1976), and gabbro of ophiolitic affinity occurs at Brighton Island (Hussey, 1974; Stukas and Reynolds, 1974b).

Inland, the Hungry Mountain Complex of mainly tonalite and gabbro (Thurlow, 1981; Whalen et al., 1987) is thrust eastward above the Buchans Group. Farther south, the disrupted Pynns Brook Complex (Williams et al., 1982) and the more complete Grand Lake Complex (Knapp, 1982) lie at or near the Baie Verte-Brompton Line. Eastward, the Annieopsquotch Complex (Dunning and Herd, 1980; Dunning, 1987b) is the largest intact example of an ophiolite suite in the Notre Dame Subzone. Other examples occur nearby at Star Lake (Dunning and Chorlton, 1985) and at King George IV Lake (Kean, 1983b).

The Birchy Complex (Williams, 1977b; Hibbard, 1983) consists mainly of green magnetite-chlorite schist containing boudins of preserved meta-gabbro and ophiolitic mélange, the Coachman's Mélange (Williams, 1977b). First defined as a stratigraphic unit, the Birchy Schist (Fuller, 1941), and regarded as part of the Fleur de Lys Supergroup by most workers (e.g. Kennedy, 1971), the Birchy Complex is clearly a structural assemblage. The history of ideas is involved and previous interpretations of structural and stratigraphic relationships across the Baie Verte-Brompton Line, locally the Baie Verte Lineament, are summarized in Figure 3.64 (after Hibbard, 1983).

The Coachman's Mélange has a black pelitic matrix surrounding conspicuous, deformed and recrystallized ultramafic blocks, now represented by bright green actinolite-fuchsite schist. Sedimentary blocks with ill-defined outlines are common and large blocks of serpentinized

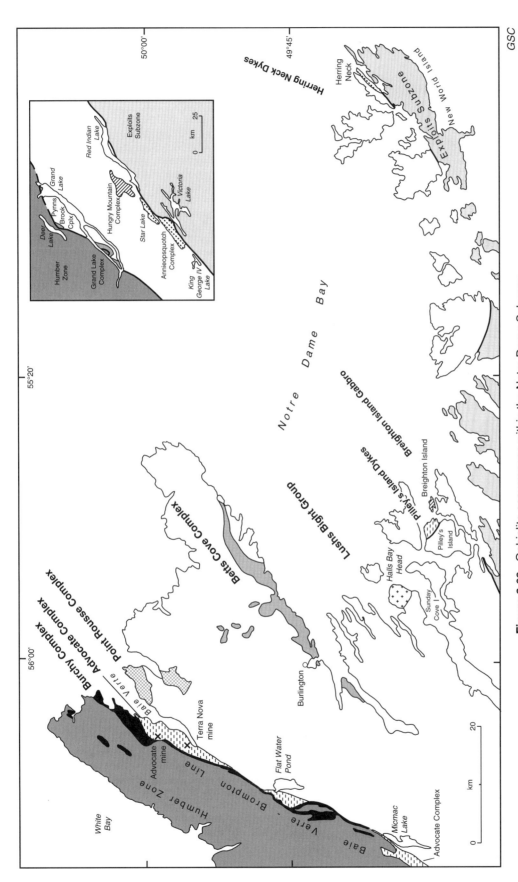

Figure 3.63. Ophiolite occurrences within the Notre Dame Subzone.

GSC

ultramafic rock, foliated gabbro, and marble are conspicuous in places. Actinolite-fuchsite schists exhibit folded schistosities identical to structures in the surrounding schistose matrix and nearby greenschists of the Birchy Complex. An ultramafic origin for these rocks is indicated by their mineralogy and because large nearby ultramafic blocks are recrystallized to similar metamorphic mineral assemblages at their margins. An internally brecciated ultramafic block at Coachman's Harbour is equidimensional and exceeds 50 m across. It is not known whether the Coachman's Mélange represents a single unit repeated by isoclines of 3 km amplitude or several folded units at different structural levels. Occurrences extend 50 km southward to Mic Mac Lake of the Baie Verte Peninsula and 100 km northward to northern Grey Island.

The Advocate Complex (Kennedy and Phillips, 1971; Bursnall, 1975; Williams et al., 1977; Hibbard, 1983) borders the Birchy Complex to the east and extends southward along the Baie Verte-Brompton Line. Ophiolitic rocks of the Advocate Complex are imbricated and in places intensely foliated. Separate slices contain one or more components of the ophiolite suite and they are separated in places by narrow bands of black slate or pelitic schist, incorporated tectonically along surfaces of major dislocation. An occurrence of garnetiferous amphibolite at the Advocate mine may represent a metamorphic sole related to earliest transport of the Advocate Complex.

South of Baie Verte, the Advocate Complex is bordered to the east by a steeply dipping east-facing sequence, the Flatwater Pond Group (Williams et al., 1977; Hibbard, 1983). The Kidney Pond Conglomerate (Kidd, 1974; Williams et al., 1977; Hibbard, 1983) at the base of the Flatwater Pond Group, and similar rocks above the Advocate Complex farther north (Bursnall, 1975; Hibbard, 1983), is a continuous unit about 50 m wide. Clasts range in size from tiny pebbles to huge blocks tens of metres in diameter, locally in a black shaly matrix. Most clasts are volcanic, but sedimentary rocks, diorite, gabbro, and altered ultramafic rocks are also common. An occurrence of the Kidney Pond Conglomerate at the former Terra Nova Mine at Baie Verte is typical mélange with blocks of massive sulphides, brecciated ultramafic rocks, and mafic volcanic rocks, all in a black slate matrix. A single 1 by 2 m block of fine grained muscovite-quartz-albite schist near Mic Mac Lake is interpreted as derived from the Fleur de Lys Supergroup to the west (Kidd, 1974; Williams et al., 1977). Boulders of granodiorite, which are also common in places, are equated with the Burlington Granodiorite (Hibbard, 1983) to the east.

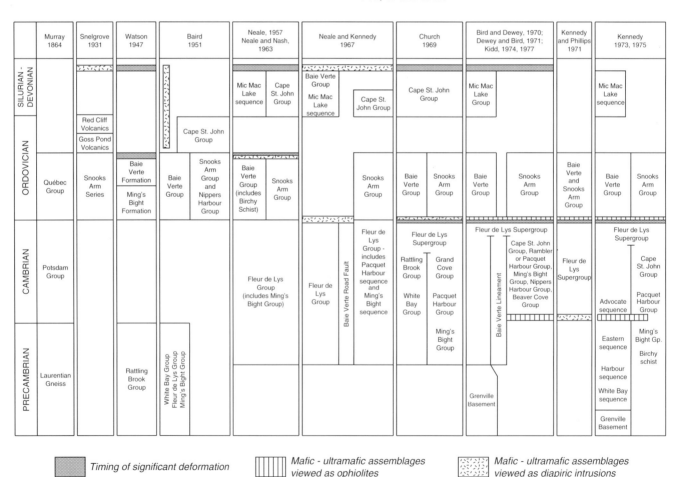

Figure 3.64. Summary of previous stratigraphic and structural interpretations at Baie Verte, after Hibbard (1983).

The contact between Flatwater Pond Group and Advocate Complex is faulted but it is thought to represent an original unconformity. Sedimentary evidence indicates that the Kidney Pond Conglomerate is younger than adjacent ophiolite complexes, younger than the Burlington Granodiorite, and younger than regional metamorphism of the Fleur de Lys metaclastic rocks. The Flatwater Pond Group is overlain unconformably by red rhyolites, ignimbrites and red sandstones of the Silurian Micmac Lake Group (Kidd, 1974). The Kidney Pond Conglomerate is correlated with conglomerates at the base of the Crabb Brook Group (Casey and Kidd, 1981) that lie unconformably on the Bay of Islands Complex (Hibbard, 1983). This correlation suggests a Llanvirn or Llandeilo age for the Flatwater Pond Group.

The Point Rousse Complex (Norman and Strong, 1975; Williams et al., 1977; Kidd et al., 1978; Hibbard, 1983) occurs east of Baie Verte in several structural blocks with tectonic boundaries marked by foliated serpentinite or foliated carbonate-talc-fuchsite alterations of ultramafic rocks. Three separate blocks at Point Rousse contain southeast-facing, overturned sections of gabbro, sheeted dykes and pillow lavas. Locally the pillow lavas are overlain by cherts, mafic volcanic rocks and volcaniclastic rocks. Conglomerates and olistostromes like those along the eastern margin of the Advocate Complex are absent.

Farther east, the Betts Cove Complex (Upadhyay et al., 1971) has a continuous stratigraphy from ultramafic rocks through gabbros, sheeted dykes and pillow lavas (Fig. 3.65). Its cover is the thick Snooks Arm Group (Hibbard, 1983), which comprises mainly volcanic rocks but also sedimentary rocks and local graptolitic shales of Arenig age. The combined Betts Cove-Snooks Arm section is steep and the rock units are successively younger to the southeast. Silurian rocks of the Cape St. John Group (Baird, 1951) are unconformable on the Snooks Arm Group and deeper levels of the Betts Cove ophiolite suite (Neale et al., 1975). Between Point Rousse and Betts Cove, porphyries related to the Cape St. John Group contain abundant mafic and ultramafic inclusions, implying an ophiolite substrate and subsurface continuity of ophiolitic rocks.

The Birchy-Advocate-Point Rousse-Betts Cove corridor is one of the widest ophiolite belts in the Appalachian Orogen. The stratigraphic order of ophiolitic units is toward the east or southeast. The overall structural arrangement indicates westward verging imbricate slices with eastward-facing stratigraphy. Intensity of deformation and metamorphism decreases progressively from west to east, or upward through the structural pile. An unconformable Silurian cover indicates that the earliest deformation is Ordovician. Later deformation affected the cover rocks and

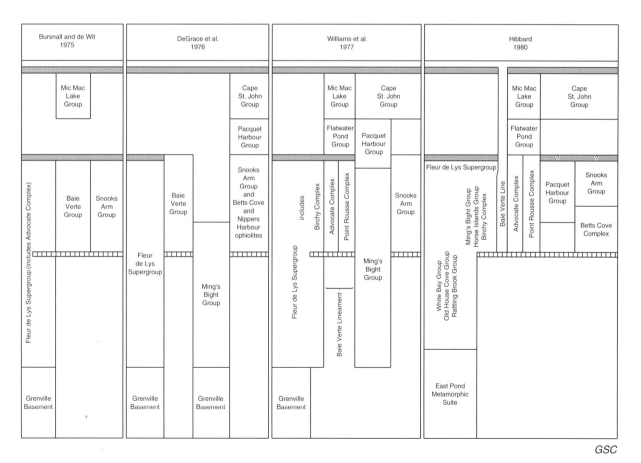

Figure 3.64 (cont.)

GSC

lcd to oversteepening and eastward thrusting (Hibbard, 1983). The enigma of Humber Zone metaclastic rocks (Ming's Bight Group) on the east side of the Baie Verte Lineament at Baie Verte (Williams et al., 1977) can be explained in several ways; the effect of a later "Z"-shaped fold that overprints Ordovician structures, structural doming of Ming's Bight metaclastic rocks through an ophiolitic Dunnage cover, or out-of-sequence imbrication (Hibbard, 1982; Williams and St-Julien, 1982).

The Pynn's Brook Complex east of Deer Lake resembles the Advocate Complex of the Baie Verte-Brompton Line in structural complexity. It overlies Humber Zone metaclastic rocks of the Mount Musgrave Group (Williams et al., 1982). The occurrence is significant in that Ordovician and older rocks are thrust westward above Carboniferous strata (Fig. 3.66) and define the Carboniferous structural front in this area.

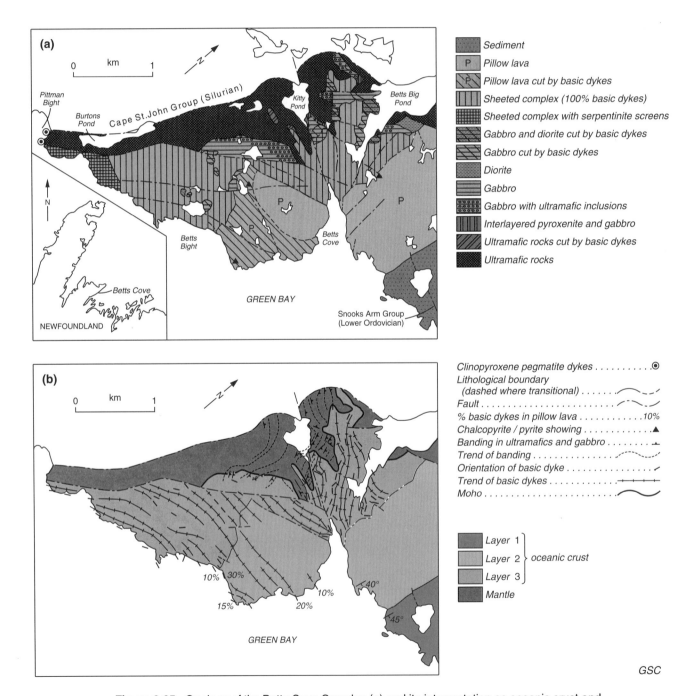

GSC

Figure 3.65. Geology of the Betts Cove Complex (**a**) and its interpretation as oceanic crust and mantle (**b**), after Upadhyay et al. (1971).

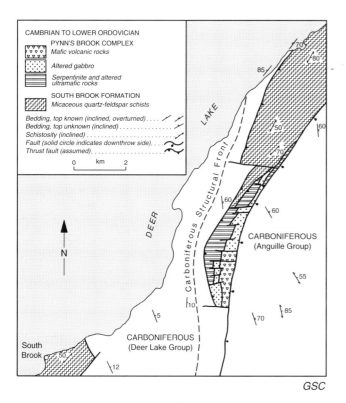

Figure 3.66. Pynn's Brook Complex and relationships to nearby groups, after Williams et al. (1982).

The Grand Lake Complex (Knapp, 1982) may also mark the Baie Verte-Brompton Line for its cover has coarse basal conglomerates that resemble olistostromes of the Flatwater Pond Group (Williams and St-Julien, 1982). A similar cover occurs above ophiolite complexes at the opposite end of the Baie Verte-Brompton Line in Quebec (Williams and St-Julien, 1978, 1982).

Recent U-Pb zircon geochronological studies indicate that the Notre Dame Subzone ophiolites are roughly coeval and about the same age as the Bay of Islands Complex (Dunning and Krogh, 1985). The Betts Cove Complex is dated at 488.6 +3.1/-1.8 Ma. and the Annieopsquotch Complex at 477.5 +2.6/-2.0 Ma and 481.4 +4.0/-1.9 Ma. Their age span is confined to the late Tremadoc and Arenig stages of the Early Ordovician.

Ophiolitic components, such as those of the Birchy Complex, are viewed as structurally commingled with the Fleur de Lys metaclastic rocks of the Humber Zone (Williams and Talkington, 1977), and some may have been oceanic basement to the eastern depositional edge of the Fleur de Lys Supergroup (Hibbard et al., this chapter). Occurrences of discrete ultramafic and gabbroic rocks farther west, such as those at Fleur de Lys village, are probably also ophiolitic and emplaced structurally. Similar discrete occurrences of mafic/ultramafic rocks throughout the metamorphic Dashwoods Subzone suggest a similar situation. Examples of small mafic/ultramafic plutons in metaclastic rocks are common all along the Appalachian miogeocline (Williams and Talkington, 1977).

Volcanic and sedimentary rocks

Volcanic rocks with minor sedimentary units are the predominant layered rocks throughout the Notre Dame Subzone. Two belts of contrasting lithology, stratigraphy and metallogenic characteristics are recognized: a belt of mainly mafic marine volcanic rocks along the northwestern part of the subzone, especially at its northern end; and a belt of bimodal volcanic rocks, partly terrestrial, along its southeastern margin (Fig. 3.67).

Rocks of the mafic volcanic belt are well exposed along the shoreline of Notre Dame Bay. From west to east these are the Baie Verte (Watson, 1947), Pacquet Harbour (Church, 1969), Snooks Arm (Baird, 1951), Western Arm (Marten, 1971), Lushs Bight (MacLean, 1947), Cutwell (Kean and Strong, 1975), Catchers Pond (Dewey and Bird, 1971), Moretons Harbour (Strong and Payne, 1973) and Herring Neck (Williams and Payne, 1975) groups. Steeply dipping, uninterrupted sections are about 3000 m thick at Snooks Arm and about twice that thickness at Moreton's Harbour. The Glover Group (Knapp, 1982) occurs 150 km southwest at Grand Lake, where it is thrust southward above metamorphic rocks of the Dashwoods Subzone (Whalen and Currie, 1983; van Berkel and Currie, 1988).

The stratigraphy, lithology and chemistry of the mafic volcanic sequences are well-known in places. The best preserved and best studied sections are those at Snooks Arm (Upadhyay, 1973; DeGrace et al., 1976), Western Arm (Marten, 1971), Long Island (Kean and Strong, 1975; Szybinski, 1988) and Moretons Harbour (Strong and Payne, 1973). Stratigraphic subdivisions, descriptions and proposed local correlations are given by Dean (1978) and Hibbard (1983) and summarized in Figure 3.68.

The overall stratigraphy of the lower Paleozoic rocks of the mafic volcanic belt is relatively simple and almost everywhere the same; thick mafic pillow lava units at the base alternating upward with felsic volcanic rocks, volcaniclastic and sedimentary units. Mafic dykes and sills, some up to 200 m thick, are common associates of the mafic volcanic rocks. Sedimentary rocks are mainly greywackes, slates, and cherts. Rare fossiliferous limestones occur as lenses or irregular masses within the volcanic rocks. The Catchers Pond Group contains a greater proportion of felsic volcanic rocks compared to nearby groups of the mafic volcanic belt, suggesting deposition in proximity to a volcanic centre (Dean, 1978).

The Cutwell Group at Long Island is a well exposed, thick volcanic succession imbricated by north directed thrusts (Szybinski, 1988). It has a lower unit of mafic to intermediate flows and pyroclastic rocks, a middle unit of dacitic-rhyodacitic rocks, and an upper unit of sedimentary rocks containing limestone breccias and Lower to Middle Ordovician black shales. Its volcanic stratigraphy and geochemical trends were interpreted as typical of an evolving island arc (Kean and Strong, 1975), with volcanism extending well into the Middle Ordovician (Caradoc). This interpretation is reassessed in view of structural complications (Szybinski, 1988) and an older age assignment for its conodonts (O'Brien and Szybinski, 1989) and graptolite faunas (S.H. Williams, 1989).

Black shales of the Cutwell Group, previously assigned to the Caradoc (Dean, 1978), belong to two separate units: a lower unit containing a late Arenig *I. v. maximus* Zone fauna, and an upper unit containing a rich Llanvirn fauna (S.H. Williams, 1989). These new graptolite ages for the Cutwell Group agree much better with ages established on conodonts (Williams, 1962; O'Brien and Szybinski, 1989) and cephalopods (Strong and Kean, 1972), and are closer to early Ordovician ages of brachiopods (Boucot, 1973) and trilobites (Dean, 1970) in the Catchers Pond Group. Black shales of the Snooks Arm and Glover groups contain early Arenig graptolites (probably *D. bifidus* Zone and *P. fruticosus* Zone, respectively), that are close, but not identical in age (S.H. Williams, 1989).

Several abandoned mines and numerous small mineral deposits throughout the mafic volcanic belt are Cyprus-type copper-pyrite deposits indigenous to ophiolitic rocks, or island arc deposits associated with discrete rhyolite domes (Fig. 3.69).

Rocks of the bimodal volcanic belt are traceable from Red Indian Lake to Robert's Arm and across the headlands and islands of Notre Dame Bay. The rocks are known as the Buchans Group in the south and the Roberts Arm Group in the north. Detailed descriptions are contained in Swanson et al. (1981) and Bostock (1988).

The Buchans Group (Thurlow, 1981) is a complex subaqueous to subaerial assemblage of volcanic, volcaniclastic and sedimentary rocks. Former thickness estimates of 5 to 9 km are probably excessive, as the rocks are now interpreted to occur in an east-directed imbricate stack (Calon and Green, 1987). Volcanic rocks range in composition from basalt to rhyolite with increasing proportions of felsic volcanic rocks toward the top of the section. The group shows calc-alkaline chemical trends but true andesites are minor. Pillowed lavas, and porphyritic and amygdaloidal lavas at the base of the sequence record quiet effusive conditions of deposition. Felsic pyroclastic rocks at its top exhibit features of explosive eruption. Clastic sedimentary rocks occur as discontinuous lenses and include conglomerates containing boulders of granite, massive sulphides, and rare limestone, all of local derivation. The Buchans Group is well known for its Kuroko-type baritic and polymetallic massive sulphide deposits.

The Roberts Arm Group to the northeast is regarded as coextensive with the Buchans Group (Dean, 1978; Bostock, 1988). Other correlatives are the Cottrells Cove, Chanceport, and Frozen Ocean groups. The Roberts Arm Group consists

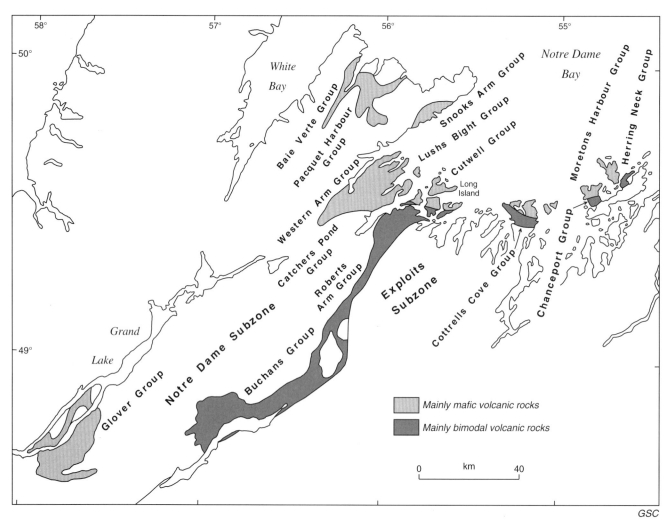

GSC

Figure 3.67. Distribution of mafic volcanic rocks and bimodal volcanic rocks in the Notre Dame Subzone.

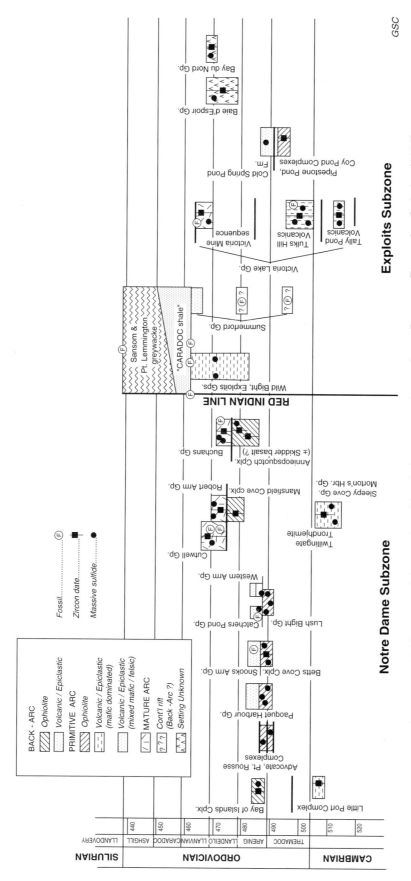

Figure 3.68. Table summarizing stratigraphic relations across the northeast Dunnage Zone, after Swinden (1990).

GSC

of a lower sequence of mafic pillow lavas containing lenses of chert and siliceous sedimentary rocks, and an upper sequence of mixed pillow lavas, mafic agglomerate, felsic flows, silicic pyroclastic rocks and interbedded volcanic greywacke, chert, and tuff. Diabase sills and dykes intrude the mafic sequence and fine grained granitic rocks are related spatially to areas of felsic volcanic rocks. The thickness of the group and eastern correlatives is about 5 km.

Polymetallic mineral deposits, like those of the Buchans Group but much smaller, are known in the Roberts Arm Group and correlatives (Dean, 1978; Bostock, 1988).

Both the Buchans and Roberts Arm groups were previously thought to overlie Upper Ordovician to Silurian sedimentary rocks of the adjacent Exploits Subzone (Dean, 1978, Thurlow, 1981). The rocks were therefore regarded as post-Caradoc, possibly related to volcanic rocks of the

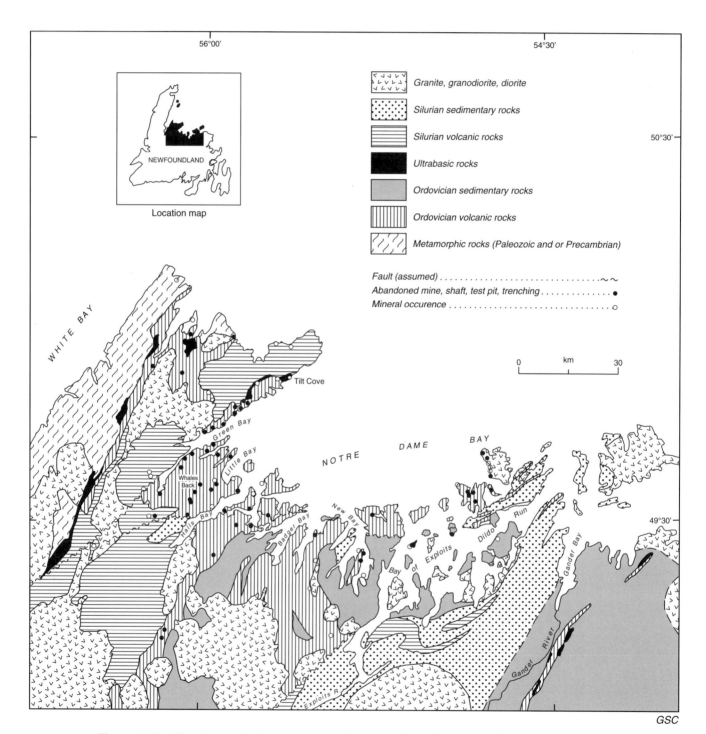

Figure 3.69. Mineral deposits throughout volcanic rocks of Notre Dame Bay, after Williams (1963).

Silurian Springdale and Botwood groups (Dean, 1978; Kean et al., 1981). Rb-Sr isotopic ages of 447 ± 7 Ma for the Roberts Arm Group (Bostock et al., 1979) and 447 ± 18 Ma for the Buchans Group (Bell and Blenkinsop, 1981) led to further debate (Nelson and Kidd, 1979; Dean and Kean, 1980).

New U-Pb zircon ages of 473 +3/-2 Ma for rhyolites of the Buchans Group and 473 ± 2 Ma for rhyolites of the Roberts Arm Group support correlation and indicate an Ordovician age (Dunning et al., 1987). Conodonts from carbonate clasts in intraformational breccia units of the Buchans Group are latest Arenig to early Llanvirn (Nowlan and Thurlow, 1984). Derivation of the clasts is considered local, so that they date the volcanism. Assignment of the conodont taxa to the Toquima-Table Head Faunal Realm (Ross and Ingham, 1970) implies a paleogeographic setting peripheral to the North American craton.

Evidence for Taconic deformation abounds throughout the Notre Dame Subzone. Unconformities, either dated or interpreted as sub-Silurian are recorded as follows: (a) Micmac Lake Group is nonconformable on the Burlington Granodiorite (Neale and Nash, 1963), dated at 463 ± 6 Ma (Hibbard, 1983 after Mattinson), (b) Micmac Lake Group is unconformable on Flatwater Pond Group (Kidd, 1974), (c) Cape St. John Group is unconformable on Snooks Arm Group (Neale, 1957; Neale et al., 1975), (d) Springdale Group is unconformable on Lushs Bight Group and Roberts Arm Group (Espenshade, 1937; Dean, 1978), (e) Silurian red sandstones and volcanic rocks are unconformable on Annieopsquotch Complex (Chandler, 1982; Chandler and Dunning, 1983), (f) sandstones, interpreted as Silurian, are nonconformable on tonalite that cuts ophiolitic rocks at Glover Island of Grand Lake (Knapp, 1982), and (g) redbeds discovered recently northeast of Red Indian Lake are possibly unconformable on the Buchans Group (Nowlan and Thurlow, 1984).

The contact of the Buchans Group with rocks of the Exploits Subzone is interpreted now as tectonic (Nowlan and Thurlow, 1984). Southeast-directed thrusts that affect the Buchans Group on the northwest side of Red Indian Lake terminate at a high angle fault that defines the Red Indian Line. In Notre Dame Bay, the Roberts Arm Group and correlatives are faulted against Silurian mélanges, greywackes and conglomerates of the Badger Belt (Nelson, 1981).

Exploits Subzone

Rocks of the Exploits Subzone are mainly Lower to Middle Ordovician with sedimentary rocks and mélanges more abundant than in the Notre Dame Subzone. Recent isotopic ages indicate volcanic rocks as old as Cambrian and granitic rocks as old as late Precambrian (Evans et al., 1990). In places, Caradoc shale and greywacke are conformable with underlying Ordovician rocks and in northeast Newfoundland they are continuous with Upper Ordovician and Lower Silurian marine greywacke and conglomerate of the Badger Belt. In other places, important sub-Middle Ordovician and sub-Silurian unconformities are present.

Volcanic rocks of the Exploits Subzone are mainly mafic and marine in the north and mainly felsic and possibly in part subaerial in the south. All are interlayered with or grade laterally into clastic sedimentary rocks, mainly turbidites, that contain abundant volcanic detritus. The thickest mafic volcanic suites occur in the northwestern part of the subzone and these grade eastward and southeastward into sections dominated by sedimentary rocks (Dean, 1978). Felsic and mafic volcanic suites in the south also grade eastward into sedimentary rocks (Colman-Sadd and Swinden, 1984; Blackwood, 1985a), although the most southeastern sedimentary belt at Bay d'Espoir contains volcanic rocks of the Isle Galet Formation (Colman-Sadd, 1976).

West of the Silurian Botwood Belt, the Exploits Subzone narrows toward the southwest and disappears south of Victoria Lake (Fig. 3.62). East of the Botwood Belt, its mainly sedimentary rocks extend from Gander Bay in the north to Bay d'Espoir in the south. From there they appear to continue westward around the Hermitage Flexure (O'Brien et al., 1986).

Ophiolite suites and mélanges

The Gander River Complex (O'Neill and Blackwood, 1989) at the eastern margin of the Exploits Subzone comprises relatively small bodies of mafic-ultramafic rocks that resemble incomplete ophiolite suites (Blackwood, 1978, 1982). Small, sparse, isolated ultramafic bodies occur well inland along strike toward Bay d'Espoir (Blackwood and Green, 1982). A small ultramafic exposure occurs in or near the North Bay granite near Bay d'Espoir and a much larger body of gabbro and lesser ultramafic rocks occurs within the Burgeo Granite farther west (O'Brien and Tomlin, 1984). Occurrences still farther west near La Poile (Chorlton, 1984) suggest discontinuous occurrences along the southeast margin of the Exploits Subzone (Fig. 3.70).

Much larger and more complete ophiolite complexes surround the Mount Cormack inlier. These are the Pipestone Pond, Coy Pond and Great Bend complexes (Colman-Sadd and Swinden, 1984; Zwicker and Strong, 1986). They consist mainly of ultramafic rocks and gabbros, with diabase and pillow lava in some examples. Their stratigraphic units face upward and outward, indicating a domal structure. Ordovician sedimentary and volcanic rocks overlie the ophiolite complexes (Colman-Sadd and Swinden, 1984). Nearby at Great Bend of the Northwest Gander River, ultramafic rocks at the Exploits-Mount Cormack subzone boundary are interpreted to occur in several blocks (Zwicker and Strong, 1986), and mélange with ophiolitic and sedimentary blocks in a serpentinite matrix occurs at the base of the Coy Pond Complex (Toby Rivers, pers. comm., 1989). Small ultramafic and gabbro bodies occur at the Exploits-Meelpaeg subzone boundary (Colman-Sadd, 1989) and ophiolitic mélange with a black phyllite matrix occurs at Cold Spring Pond (Williams and Piasecki, 1990).

The Pipestone Pond Complex is dated at 493.9 +2.5/-1.9 Ma and the Coy Pond Complex at a minimum of 489 Ma (Dunning and Krogh, 1985). These ages are slightly greater than those of ophiolite complexes in the Notre Dame Subzone and the Bay of Islands Complex of the Humber Zone.

Another incomplete ophiolite suite, the South Lake Complex (Dean, 1978; Lorenz and Fountain, 1982), occurs near Western Arm of Bay of Exploits.

The Dunnage Mélange (Kay and Eldredge, 1968; Hibbard and Williams, 1979) of the Exploits Subzone is one of the most extensive and best known mélanges of the Appalachian Orogen (Fig. 3.71). It is a strikingly heterogeneous deposit composed of blocks of mainly clastic sedimentary and mafic volcanic rocks enveloped in a dark scaly shale matrix. It is well exposed along the rugged coast and clusters of islands of Bay of Exploits, where it extends for 40 km along strike with a maximum outcrop width of 10 km. Its clasts vary in size from granules and cobbles to boulders and huge blocks up to a kilometre in diameter, thus producing a chaotic mosaic that contrasts sharply with nearby stratified volcanic and sedimentary rocks. Most blocks can be matched with formations of the Exploits and Summerford groups. A few are foreign. Middle Cambrian trilobites occur in a limestone lens in a large volcanic block at Dunnage Island (Kay and Eldredge, 1968) and a limestone block near Stanhope contains Arenig conodonts (Hibbard et al., 1977). Black shale at the Curtis Causeway contains Tremadoc graptolites.

The Dunnage Mélange overlies and interdigitates with the New Bay Formation of the Exploits Group in the southwest, and it has an apparent ghost stratigraphy comparable to that of the Exploits Group. Gabbro lumps are common where the mélange is adjacent to the profuse

gabbro sills at New Bay and the largest volcanic blocks in its northwest portion suggest continuity with the Lawrence Head volcanic rocks. Shale is more abundant in the mélange than in nearby intact stratigraphic sections. The mélange is overlain by Caradoc black shales to the northwest (Horne, 1969). These are succeeded by Sansom greywackes and Goldson conglomerates of the middle Paleozoic Badger Belt. The sequence of units above the mélange is viewed as the sedimentary fill of a marine trough, or an upward-shoaling and upward-coarsening sequence built upon a mélange basement.

A variety of small intrusions that are localized within the Dunnage Mélange are rare or absent in surrounding country rocks. These are mainly quartz-feldspar porphyries, of the Coaker Porphyry (Kay, 1972; Lorenz, 1984a, b); dated isotopically as Early to Middle Ordovician. The intrusions exhibit mud-magma relationships indicating contemporaneity with mélange formation (Williams and Hibbard, 1976; Lorenz, 1984b). Lobate, pillowed, corrugated, and pahoehoe-like igneous contact surfaces, complex interlayering of host

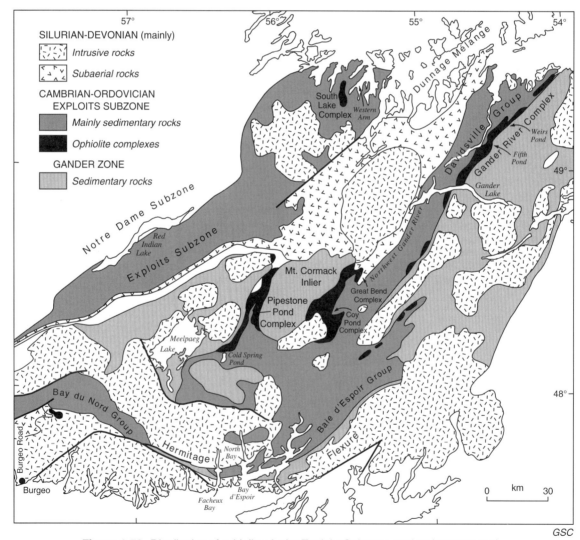

Figure 3.70. Distribution of ophiolites in the Exploits Subzone, and rock groups and relationships along its southeastern margin.

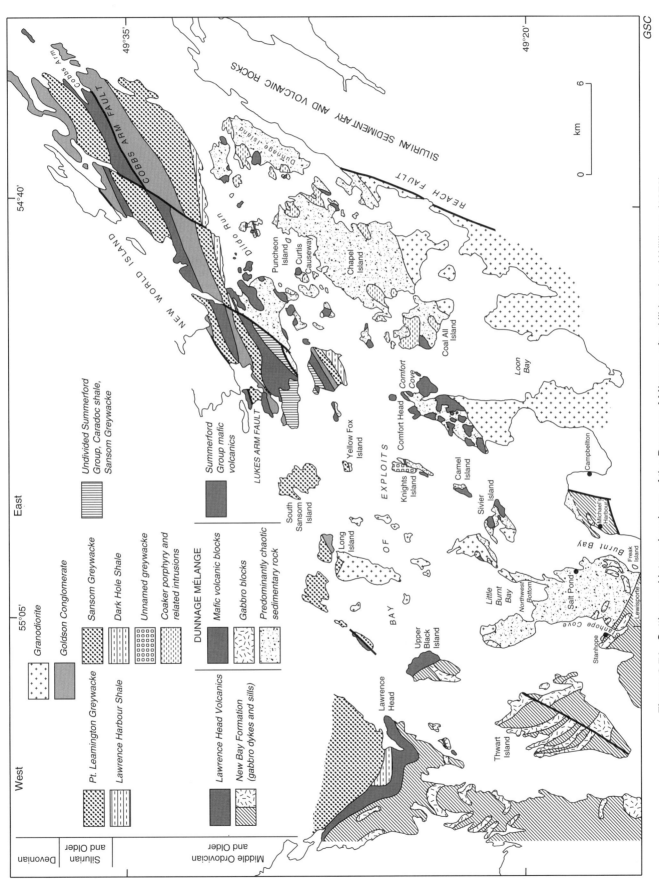

Figure 3.71. Setting and internal makeup of the Dunnage Mélange, after Hibbard and Williams (1979).

155

mudstone and dacite, and occurrences of peperite, all indicate that magma intruded unconsolidated mud, and the adjacent sediment was fluidized (Lorenz, 1984a, b). The intrusions and their setting imply a magmatic link between the surface mélange and deeper crustal levels. Mafic and ultramafic inclusions in xenolith phases of the Coaker Porphyry indicate that the porphyry penetrated an ophiolitic substrate enroute to the surface. Petrogenesis of the Coaker Porphyry requires a continental source still deeper in the crust (Lorenz, 1984a, b).

Major slumping is favoured as a controlling mechanism of Dunnage Mélange formation (Horne, 1969; Hibbard and Williams, 1979; Pajari et al., 1979). Brittle deformation and possible tectonic controls have also been emphasized (Karlstrom et al., 1982). Evidence for shallow emplacement of the Coaker Porphyry is inconsistent with the existence of a tectonic overburden during mélange formation, and supports an olistostromal origin for the Dunnage Mélange (Lorenz, 1984a, b).

The Carmanville Mélange (Pajari et al., 1979; Williams et al., 1991) occurs along the east margin of the Exploits Subzone (Fig. 3.72). It was first recognized in coastal exposures (Kennedy and McGonigal, 1972a) and later mapped and extended to include most of the area east of Gander Bay (Pickerill et al., 1978). It is more than 10 km wide in the north, but absent south of Gander Bay. The mélange consists of sedimentary, mafic volcanic, volcaniclastic, gabbro, trondhjemite, and limestone blocks in a matrix of pyritic black shale and siltstone. Contrasts in structural and metamorphic styles are the most obvious and problematic feature of the Carmanville Mélange. In places sedimentary rocks can be traced from coherently bedded sections, through deformed beds to disaggregated material and finally discrete olistoliths within a thixotropic matrix. In other places, the mélange consists of banded mafic schist, greenschist, attenuated pillow lava, psammitic schist, and a black pelitic to semipelitic matrix. Areas of intensely deformed and metamorphosed mélange are juxtaposed or surrounded by less deformed and metamorphosed mélange, with sharp contacts between the two. These contrasts in structural and metamorphic style, and the presence of discrete intensely deformed and metamorphosed blocks in a pebbly mudstone matrix, all indicate that an earlier generation of mélange was deeply buried, deformed, metamorphosed, and then exhumed or recycled where juxtaposed or embedded in pebbly mudstone. Any model for the Carmanville Mélange must therefore involve dynamic processes whereby surficial olistostromes are subjected to dynamothermal conditions and then returned to the surface. A spatial association of the Carmanville Mélange and volcanic rocks suggests that emplacement of the volcanic rocks as a major structural slice may have controlled mélange formation.

Mélanges occur all across the northeast Exploits Subzone from Carmanville to New World Island (Fig. 3.73). The Dog Bay Points Mélange has a similar makeup to the Dunnage

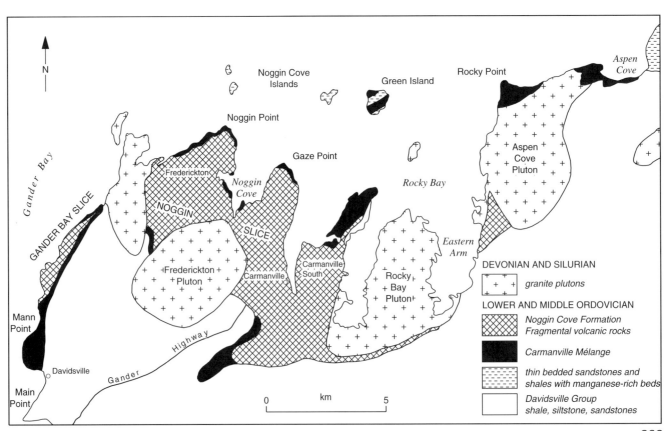

GSC

Figure 3.72. Distribution of the Carmanville Mélange, after Williams et al. (1991).

and Carmanville examples. All contain similar blocks in black shaly matrices, and all have associated dark grey shales with thin coticule layers and nodules. The intermediate position of the Dog Bay Point Mélange between the Dunnage and Carmanville mélanges (Fig. 3.73), and coticule occurrences in all examples, imply continuity of a coastal lithic belt for 50 km across the northeast Exploits Subzone (Williams, 1992).

Since their discovery, the various mélanges of northeast Newfoundland have been related to (1) subduction and filling of a marine trench (Bird and Dewey, 1970; Kay, 1976; McKerrow and Cocks, 1977), (2) transport of allochthons (Williams et al., 1991), (3) gravitational slumping (Horne, 1969; Kennedy and McGonigal, 1972a; Hibbard and Williams, 1979; Pajari et al., 1979), (4) tectonic disruption of already lithified units (Karlstrom et al., 1982), and combinations of these processes.

A model for the Dunnage Mélange as a major slump implied stratigraphic and sedimentological links with the Exploits Group to the west (Hibbard and Williams, 1979). A similar model for the Carmanville Mélange implied derivation from the east by collapse that followed obduction of the Gander River Complex (Pajari et al., 1979). However, other features are difficult to explain by simple surficial slumping, such as (1) structural and metamorphic variations in the Carmanville Mélange that imply tectonic

recycling of intensely deformed and metamorphosed mélange (Williams et al., 1991), and (2) the Dunnage Mélange is the locus for contemporary intrusions.

The much greater proportion of black shales and coticules associated with the mélanges, compared to their proportions in nearby groups, suggests a depositional setting that contrasted with that of surrounding groups. The coticules probably represent distal chemical precipitates associated with submarine volcanism within an extensive deep marine basin. A tectonically active basin is indicated by slumping of the shaly rocks and periodic influxes of greywacke, conglomerate and bouldery olistostromes. Tectonic processes may have been operative in convergent zones within the basin or at its periphery.

The Dunnage Mélange is overlain by Caradoc black shales that are also recognized in the adjacent Exploits Group, and Caradoc black shales are present in the Carmanville area. This implies cessation of tectonism across the marine basin and its opposing sides. Caradoc shales of the Exploits Group (Lawrence Harbour Formation) and Caradoc shales above the Dunnage Mélange (Dark Hole Formation) are overlain by a thick greywacke unit (Point Leamington and Sansom formations, respectively). Probable equivalents in the Carmanville area are the greywackes at Wings Point and along the east side of southern Gander Bay. This implies infilling and shallowing

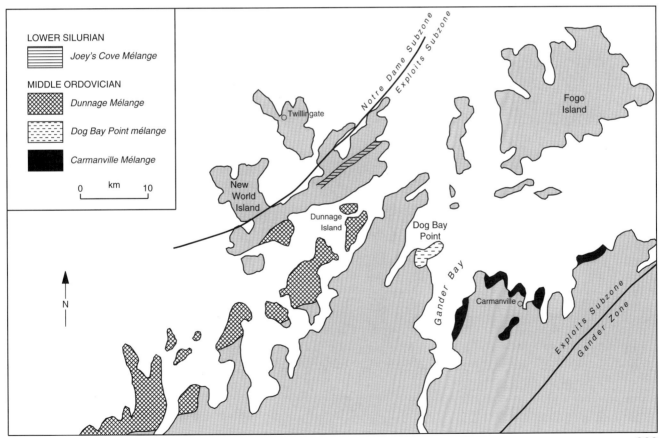

GSC

Figure 3.73. Distribution of mélanges across the northeast Exploits Subzone.

of the marine basin. The greywacke unit is absent in the centrally located Dog Bay area, where possible equivalents in the basal Indian Islands Group are finer and shalier. Silurian mélanges at New World Island, which are equated with thrusting and telescoping of Ordovician-Silurian sections represent destruction of the marine basin.

Volcanic and sedimentary rocks

The thickest and best known mafic volcanic assemblage of the Exploits Subzone is the Wild Bight Group (Dean, 1978; Swinden, 1987) of western Notre Dame Bay. A 10 km stratigraphic section is exposed in the northwest limb of the Seal Bay Anticline (Fig. 3.74). It consists of coarse mafic agglomerates and tuffs, pillow lava units up to 500 m thick, and well-bedded units of tuff, chert, tuffaceous sandstone and greywacke. Less common rock types are felsic flows and pyroclastic rocks. Abrupt facies and thickness changes are characteristic.

Gabbro and diabase sills occur among the volcanic and sedimentary rocks. One large sill, the Marks Lake Sill (Hayes, 1951), exhibits features of early soft-sediment intrusion and is a useful marker horizon near the top of the group. Tonalites of the South Lake Complex cut the volcanic rocks, but ophiolitic components of the complex are possible basement. The top of the group is defined as the base of the conformably overlying Caradoc shale and greywacke of the Shoal Arm Formation (Dean, 1978). The folded rocks are truncated inland by the middle Paleozoic Hodges Hill batholith.

Major and trace element whole rock analyses of volcanic and subvolcanic rocks, as well as clinopyroxene mineral chemistry, and isotopic studies reveal a complex association of island arc signatures for rocks in lower and middle parts of the group and transitional or non-arc varieties at the top (Swinden, 1987). The Wild Bight Group has important volcanogenic sulphide deposits of both massive and stockwork varieties.

The Exploits Group (Helwig, 1969) to the east has sedimentary and volcanic units overlain conformably by Caradoc shale and greywacke equivalent to those above the Wild Bight Group. Its oldest rocks, the Tea Arm Volcanics, are mafic pillow lavas, pillow breccias and mafic flows about 2 km thick. These are correlated with middle parts of the Wild Bight Group (Dean, 1978). The volcanic rocks are overlain by turbidites, tuffaceous sandstone, greywacke, conglomeratic turbidites, tuff and chert of the New Bay Formation. Its thickness is in excess of 2 km and it is viewed as a volcanic turbidite facies equivalent to upper parts of the Wild Bight Group. It is overlain by 800 m of mafic pillow lavas and flows (Lawrence Head Volcanics), in turn conformably overlain by Lower Caradoc cherts, black argillites and greywackes. Thick sills of diabase and gabbro are more numerous and larger in New Bay Formation than in any correlative section of the Exploits Subzone.

Toward the northeast at New World Island, correlatives of the Exploits Group (Summerford Group, Horne 1969, 1970) are mafic volcanic rocks overlain by a prominent 50 m unit of Llandeilo crystalline limestone (Cobbs Arm Formation, Bergström et al., 1974; Fåhreaus and Hunter, 1981)

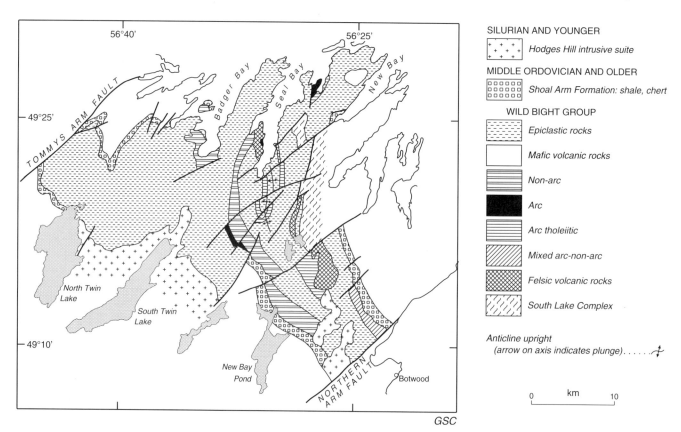

Figure 3.74. Structure and stratigraphy of the Wild Bight Group, after Swinden et al. (1990).

followed by Caradoc shale, and Lower Silurian marine greywacke and coarse conglomerate of the Badger Belt. Northward at Pikes Arm, Lower Silurian conglomerates lie unconformably on Summerford mafic volcanic rocks (Arnott, 1983a). Lower parts of the Summerford Group contain Late Arenig brachiopods (Neuman, 1976; Bergström et al., 1974) and one occurrence of limestone within mafic volcanic rocks at Summerford contains Tremadoc cephalopods (Kay, 1967; Boyce, 1987). These shelly faunas belong to the Celtic realm and indicate a paleogeographic setting distant from the North American craton (Neuman, 1984).

The Loon Bay Volcanics and overlying Luscomb Chert (Kay, 1975) at Loon Bay lie on the southeast or opposite side of the Dunnage Mélange. The stratigraphy is like that of the upper part of the Wild Bight Group with black shale and greywacke at the top of the section.

Inland, the Victoria Lake Group (Kean, 1977) is a correlative of the Wild Bight and Exploits groups (Fig. 3.75). It consists of mafic pillow lava, mafic and felsic pyroclastic rocks, chert, greywacke and shale. Laterally continuous linear belts of felsic pyroclastic rocks are a prominent feature. Volcanic rocks are more abundant toward the southwest and sedimentary rocks toward the northeast. Limestone lenses within the group contain conodonts of Llanvirn to Early Llandeilo age (Kean and Jayasinghe, 1980a, b). Black argillites at the top of the group along the Exploits River are profusely fossiliferous and of Caradoc age (Erdtmann,1976). These are overlain conformably by greywackes of the Badger Belt.

The latest studies indicate that the Victoria Lake Group is a composite and structurally complex collection of rocks of varying ages, geochemical groupings and tectonic environments (Evans et al., 1990). U-Pb zircon dating of the volcanic sequences within the Victoria Lake Group have identified three age-groupings of volcanic rocks; 513 ± 2 Ma for the Tally Pond volcanics, 498 +6/-4 Ma for the Tulks Hill volcanics, and 462 +4/-2 for the Victoria mine sequence (Fig. 3.75). Three intrusive bodies are also dated. The Roebucks quartz monzonite, intrusive into the Tulks Hill volcanic rocks, is dated at 495 ± 4 Ma and is probably coeval with the volcanism. Two large plutons, the Valentine Lake and Cripple Back Lake quartz monzonites, dated as 563 ± 2 Ma and 565 +4/-2 Ma, are late Precambrian, and interpreted to be structurally emplaced along the southern margin of the Victoria Lake Group. Geochemical studies indicate that the Victoria Lake Group is composed of a number of distinct chemical groupings, which appear to record the transition from island-arc to back-arc environments. Massive sulphide mineralization probably formed during the arc-rifting event, as is suggested by its spatial relationship to highly depleted mafic volcanic rocks.

Along its southeast margin, the Victoria Lake Group is bounded by the Rogerson Lake Conglomerate of the Botwood Belt. The conglomerate is polymictic and it is nonconformable on the Valentine Lake quartz monzonite (Kean, 1977; Kean and Jayasinghe,1980b; Evans et al., 1990). Southeast of the Rogerson Lake Conglomerate, a fault-bounded narrow band of mafic volcanic rocks, tuffs, and shales (Pine Falls Formation) and silicic volcanic rocks farther southeast (Carter Lake Volcanics and equivalents; Kean and Jayasinghe, 1980b) are also part of the Exploits Subzone. These rocks are faulted against metaclastic rocks

and foliated granite of the Meelpaeg Subzone (Colman-Sadd, 1987). Farther southwest, rocks along the southeast side of the Rogerson Lake Conglomerate at Victoria Lake are included in the Victoria Lake Group (Kean, 1977). Their contact with foliated granites of the Meelpaeg Subzone, Noel Pauls Line, is marked by intense mylonitization.

The eastern part of the Exploits Subzone has shales, sandstones, conglomerates and mafic volcanic rocks (Fig. 3.70). In the north, the rocks are the Davidsville Group (Kennedy and McGonigal, 1972a) and in the south, the Baie D'Espoir Group (Jewell, 1939; Colman-Sadd, 1976). Volcanic rocks in the Davidsville Group are coarse mafic pyroclastics in massive to thick graded units, pillow lavas, and cherts. Most of these are interpreted either as huge disrupted blocks in the Carmanville Mélange (Pajari et al., 1979) or erosional remnants of a structural slice above the Carmanville Mélange (Williams et al., 1991). Felsic and some mafic volcanic rocks also occur in the Baie D'Espoir Group. The Davidsville Group contains Caradoc graptolites in black shale units, Late Llanvirn to Early Llandeilo brachiopods, cephalopods and trilobites in limestones beneath the shales, and late Arenig trilobites and shelly fauna from nearby sandstones (Boyce et al., 1988). The Baie D'Espoir Group is considered Late Arenig to Caradoc on meagre fossil evidence (Colman-Sadd, 1976; S.H. Williams, pers. comm., 1990).

Upper Llanvirn to Lower Llandeilo limestone of the Davidsville Group lies directly on serpentinite at Weirs Pond (Stouge, 1979; O'Neill, 1987) and distinctive conglomerates overlie ultramafic rocks and tonalites of the Gander River Complex elsewhere (Kennedy, 1975b, 1976; Blackwood, 1982). The Weirs Pond limestone is bioclastic and contains quartz grains, volcaniclastic material and chromite, indicating a mixed provenance (O'Neill, 1987). Davidsville conglomerates are polymictic, matrix-supported with unsorted clasts ranging from pebbles to boulders. Clasts are mainly trondhjemite, gabbro, mafic volcanic and ultramafic rocks; corresponding to nearby lithologies of the Gander River Complex. They also contain quartz and jasper, and one example near Fifth Pond has almost all psammite and siltstone clasts, possibly derived from the Gander Zone.

South of Gander Lake, the Davidsville Group is dominated by shales and sandstones that continue to Bay d'Espoir on the south coast of the island. Contact relationships between the Baie D'Espoir Group and Coy Pond and Pipestone Pond ophiolite complexes to the north are interpreted as conformable (Colman-Sadd and Swinden, 1984) but an occurrence of ultramafic conglomerate with serpentinite sandstone interbeds at Northwest Gander River suggests an unconformity with the Coy Pond Complex, and local relations like those north of Gander Lake (Williams et al., 1988).

At Facheux Bay to the west, the Baie D'Espoir Group is exposed in a large overturned syncline-anticline pair (Blackwood, 1985a). The succession consists of pelitic sedimentary rocks with intercalated felsic volcanic rocks (Isle Galet Formation) that are conformable with psammitic and semipelitic schist (Riches Island Formation). The rocks are progressively metamorphosed southward. Their contact with the Little Passage Gneiss of the Gander Zone, variously interpreted as a thrust (Colman-Sadd, 1976) or a metamorphic gradation (Blackwood, 1985a), is regarded now as a mylonitic shear zone with left lateral displacement (Piasecki, 1988).

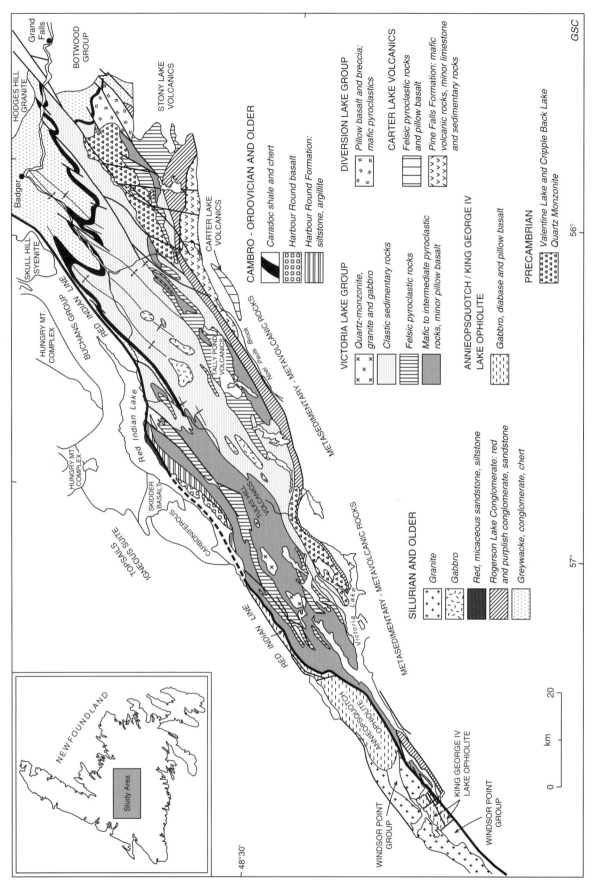

Figure 3.75. Distribution of rocks and relationships within the Victoria Lake Group, after Evans et al. (1990).

GSC

Farther west, correlatives of the Baie D'Espoir Group follow the Hermitage Flexure to merge with the Bay du Nord Group (Cooper, 1954). The Bay du Nord Group forms an eastward-tapering belt that consists of felsic meta-volcanic rocks intercalated with, and overlain by, pelitic and psammitic metasedimentary rocks. The volcanic rocks are most extensive toward the west, and in the Peter Snout area a lower unit of felsic crystal-lithic tuff has interbeds of psammite and boulder conglomerate containing gabbro clasts (O'Brien, 1983). Other Bay du Nord conglomerates on Burgeo Road contain mainly foliated granite and mylonite clasts. Upper parts of the volcanic sequence are rhyolite, and welded rhyolite tuff and felsite overlain by tuff and graphitic pelite. The commonest sedimentary rocks are grey sandstone and shale or their metamorphic equivalents. A U-Pb zircon age of 466 ± 3 Ma for felsic volcanic rocks (Dunning et al., 1990a) supports correlation of the Bay du Nord and Baie D'Espoir groups (O'Brien et al., 1986).

Small bodies of ultramafic rocks toward the west and a much larger gabbroic roof pendant in granite north of Burgeo (Fig. 3.70) are interpreted as possible ophiolitic basement to the Ordovician sedimentary and volcanic rocks (Chorlton, 1980; O'Brien and Tomlin, 1984; O'Brien et al., 1986). This is supported by gabbroic clasts in Bay du Nord Group conglomerates, but the presence of foliated granite and mylonite clasts suggests a nearby mixed source.

Rocks of the Bay du Nord-Baie D'Espoir groups contain base metals in mafic and felsic volcanic rocks, uranium in volcanic and sedimentary rocks, and they are hosts for epithermal-fumerolic gold mineralization. Granites in the same area are host to uranium, molybdenum, tin and tungsten mineralization (O'Brien et al., 1986).

Chronology of ophiolite emplacement in the Exploits Subzone and relationships to nearby groups are less clear than in the Notre Dame Subzone and Humber Zone. The Gander River Complex is overlain unconformably by limestone of Llandeilo age and nearby sandstones of Arenig age contain detrital chromite (Stouge, 1979; O'Neill, 1987; O'Neill and Blackwood, 1989). Similar unconformable relationships are suggested between ophiolitic rocks and Ordovician cover sequences at Coy Pond (Williams et al., 1988) and Cold Spring Pond (Williams and Piasecki, 1990). Ophiolitic mélanges like those at Cold Spring Pond imply ophiolite transport, and mélanges with black shale matrices are typically Ordovician elsewhere throughout the Dunnage Zone and Appalachian Orogen. The relationships indicate pre-Middle Ordovician ophiolite transport and emplacement, and that large tracts of Middle Ordovician rocks that define the Exploits Subzone are cover to already deformed and eroded ophiolite suites.

Twillingate Subzone

The Twillingate Subzone is a small area of mafic volcanic rocks cut by a tonalite or trondhjemite batholith, loosely referred to as the Twillingate Granite (Fig. 3.76). Amphibolite facies metamorphism, intense deformation, and mylonitization along the southern margin of the batholith are out of context with surrounding rocks of the Notre Dame Subzone. Mafic dykes coeval with nearby mafic volcanic rocks (Herring Neck Group) of the Notre Dame Subzone cut the mylonitic rocks. Tonalite is dated at 510 ± 16 Ma (U-Pb zircon) and mafic dykes that cut mylonitic tonalite have ^{39}Ar-^{40}Ar cooling ages up to 470 Ma (Williams et al., 1976). The 510 Ma age is significantly older than the Betts Cove and Annieopsquotch ophiolite complexes that are basement to volcanic rocks of the Notre Dame Subzone, and also older than ophiolitic basement complexes in the Exploits Subzone.

Volcanic rocks (Sleepy Cove Group) intruded by the Twillingate tonalite are pillow lavas and mafic dykes in the north, and a bimodal suite containing felsic pyroclastic rocks and flows toward the south. Their contact with the Moretons Harbour Group of the Notre Dame Subzone to the west has been interpreted both as stratigraphic (Strong and Payne, 1973; Dean, 1978) and a faulted zone of intrusive mafic dykes (Williams and Payne, 1975; Williams et al., 1976). One suggestion is that the Twillingate Subzone is a remnant island-arc, deformed and surrounded by intrusive products of a younger Notre Dame arc. According to this model, the subzone is viewed as suspect with respect to surrounding rocks of the Notre Dame Subzone. If its volcanic rocks are in stratigraphic continuity with the Moretons Harbour Group, it is merely a more deformed, metamorphosed basal part of the same volcanic sequence.

Indian Bay Subzone

The Indian Bay Subzone is defined on the occurrence of volcanic and fossiliferous sedimentary rocks of Dunnage aspect within the northeast Gander Zone at Indian Bay Big Pond (Fig. 3.62, 3.77). Their age and setting, well east of the Exploits Subzone, and lithic contrasts with surrounding monotonous sandstones of the Gander Zone (Gander Group) add to their significance. The rocks were named the Indian Bay Formation (Wonderley and Neuman, 1984), now changed to the Indian Bay Big Pond Formation (O'Neill and Blackwood, 1989). Brachiopods, trilobites and bryozoans in argillaceous tuffs indicate a Late Arenig (late Early Ordovician) age.

Three units are recognized in the type section at Indian Bay Big Pond (Wonderley and Neuman, 1984). Boundaries are unexposed so that stratigraphic order is doubtful. These are: (a) an eastern unit of tan argillaceous sandstone, purple siltstone and shale, grey quartz sandstone and shale, and fossiliferous argillites, (b) a central unit of green pillow lavas and massive flows, and (c) a western unit of grey to black silty shale with white sandstone beds and quartz pebble conglomerates. The rocks are mainly of low metamorphic grade and have the same structural and thermal history as surrounding rocks of the Gander Group (Wonderley and Neuman, 1984; O'Neill and Knight, 1988). A prominent northeast-trending southeast-dipping schistosity is axial planar to open and tight folds, overprinted by a second cleavage. Early pre-cleavage deformation is suggested by: (1) small scale isoclinal folds with limbs cut by the main cleavage, (2) conflicting facing directions on bedding-cleavage relationships, (3) downward facing folds, and (4) small-scale bedding thrust faults that are cut by the main cleavage (Wonderley and Neuman, 1984).

Brachiopods from the Indian Bay Big Pond Formation are characteristic of the Celtic biogeographic province. Similar brachiopod assemblages of Late Arenig to Early Llanvirn age occur in the Summerford Group of the Exploits Subzone at New World Island (Neuman, 1976) and in the basal Davidsville Group (McKerrow and Cocks, 1977; Boyce et al., 1988; O'Neill and Blackwood, 1989).

Relationships of the Indian Bay Big Pond Formation to surrounding sandstones of the Gander Zone are controversial (Fig. 3.77). As originally defined, the formation is about 1 km wide and bounded by faults (Wonderley and Neuman, 1984). Subsequent mapping shows an expanded outcrop width of 6 km or more and stratigraphic boundaries with the Gander Group (O'Neill and Knight, 1988).

The Indian Bay Big Pond Formation is correlated with basal beds of the Davidsville Group. Local stratigraphic relationships with rocks of the Gander Group (O'Neill and Knight, 1988; O'Neill and Blackwood, 1989) suggest it is part of an overlap assemblage that links diverse elements of the Exploits Subzone with an uninterrupted section of monotonous sandstones of the Gander Zone (Williams and Piasecki, 1990).

Dashwoods Subzone

The Dashwoods Subzone (Fig. 3.62, 3.78), formerly the Tonalite Terrane (Whalen and Currie, 1983) or Central Gneiss Terrane (van Berkel et al., 1986; van Berkel, 1987; van Berkel and Currie, 1988), takes its name from Dashwoods Pond in its central part. It consists of medium- to high-grade metasedimentary, plutonic and mafic-ultramafic rocks, all cut by tonalites and later granites. Its boundaries with the Humber Zone to the west, Meelpaeg Subzone to the east, and Notre Dame Subzone to the north are all faults. Locally, its contact with the Notre Dame Subzone is truncated by intrusions toward the northeast.

The Dashwoods Subzone has been previously assigned to either the Humber Zone (e.g. Williams, 1978a) or the Dunnage Zone (e.g. Williams et al., 1988). Its sedimentary rocks have been correlated with the Fleur de Lys Supergroup of the Humber Zone (van Berkel and Currie, 1988). Interpretation of its large mafic/ultramafic complexes as ophiolitic (Brown, 1976) and the presence of tonalites, dated locally at 456 ± 3 Ma (Dunning et al., 1989), and large mafic intrusions suggest links with the Notre Dame Subzone.

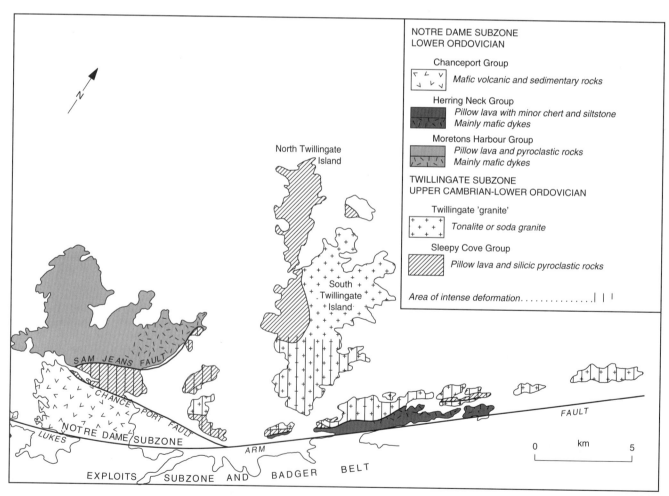

GSC

Figure 3.76. General geology of the Twillingate Subzone, after Williams and Payne (1975).

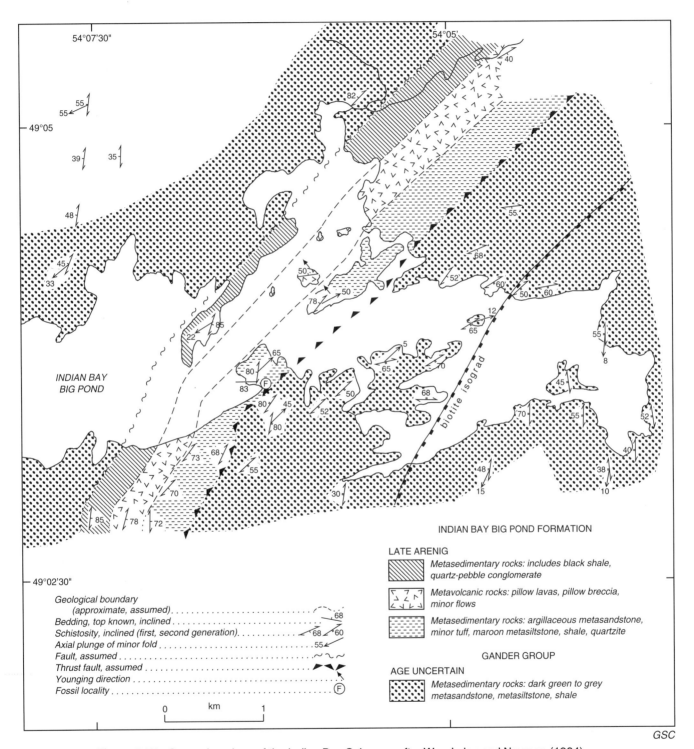

Figure 3.77. General geology of the Indian Bay Subzone, after Wonderley and Neuman (1984).

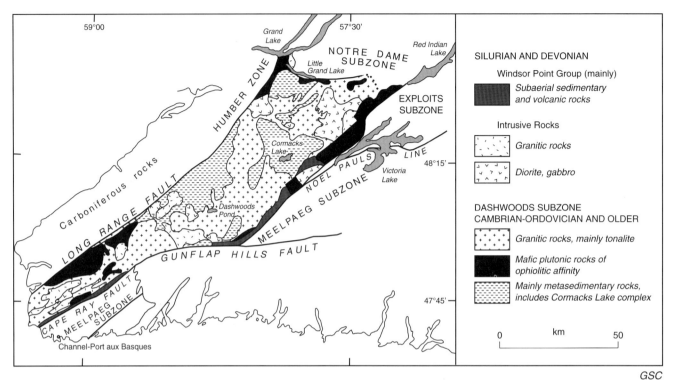

Figure 3.78. General geology of the Dashwoods Subzone, after Colman-Sadd et al. (1990).

GSC

The oldest rocks of the Dashwoods Subzone are metamorphic rocks of the Cormacks Lake complex (Herd and Dunning, 1979). The complex consists of quartz-feldspar gneisses, banded rocks with pyrite, gedrite, and garnet, and thick layers and lenses of amphibolite and coarse metagabbro. North of Cormacks Lake, the complex includes retrograded, two-pyroxene granulite. The variety of rocks suggests a basement-cover complex, possibly with exotic ophiolitic components. Quartzofeldspathic to semipelitic and calcareous gneisses to the west contain quartzite and calcareous beds, interpreted as a supracrustal cover sequence (van Berkel and Currie, 1988).

Schistosities and fold axes trend north and northeast, but at Cormacks Lake they trend northwest, like Precambrian trends in nearby Grenville inliers of the Humber Zone (Williams, 1985a). Latest deformation and metamorphism are interpreted as Ordovician, based on isotopic dates and crosscutting relationships. Along the southeast margin of the Dashwoods Subzone, Devonian rocks of the Windsor Point Group are nonconformable on foliated tonalites and granites (Chorlton, 1983; Williams et al., 1989b) and are themselves deformed.

The largest mafic/ultramafic complexes of the Dashwoods Subzone occur at its southern end. Their size and completeness rival the largest occurrences in the Dunnage Zone. Accordingly, they are interpreted as ophiolite suites (Brown 1976; Chorlton, 1983). Serpentinite lenses or layers of ultramafic rocks, gabbros and amphibolites bounded by metasedimentary or granitoid rocks are common throughout the Dashwoods Subzone. Some coincide with high strain zones and are thought to mark important thrust faults (van Berkel, 1987). Other small, equidimensional bodies have a "block in matrix" style typical of olistoliths in ophiolitic mélange (Fox and van Berkel, 1988).

The boundary of the Dashwoods Subzone and the Notre Dame Subzone is a north-dipping fault at Little Grand Lake with early ductile movement and later brittle movement. It is suggested that the Notre Dame Subzone is thrust southward above the Dashwoods Subzone (Whalen and Currie, 1983; van Berkel and Currie, 1988). Mylonites of two different ages along the Cape Ray fault zone are separated temporally by deposition of the Devonian Windsor Point Group. Both mylonites indicate westward thrusting of the Port aux Basques gneisses (Meelpaeg Subzone) above the Dashwoods Subzone (Williams et al., 1989b). The absence of Dunnage Zone rocks at this boundary and the juxtaposition of rocks normally found on opposite sides of the Dunnage Zone, suggest that the Cape Ray Fault is a major suture zone.

Summary and conclusions

The variety of marine rocks in the Dunnage Zone and structural relationships between its ophiolitic basement and adjacent metaclastic rocks of the Humber and Gander zones indicate that the Newfoundland Dunnage Zone represents the vestiges of an ancient ocean or several ocean basins now telescoped between opposing margins. The idea is not new (Williams, 1964a) but it is substantiated by structural and stratigraphic studies at its zone boundaries, and by deep seismic data that define an allochthonous Dunnage Zone between Grenville and Central lower crustal blocks.

Definition of the Baie Verte-Brompton Line and recognition of the Birchy Schist as a structural complex, complete with ophiolitic mélanges, implies major disruption and structural imbrication at the Humber-Dunnage boundary. This tectonism relates most reasonably to the transport of ophiolite suites such as the Bay of Islands Complex from their places of origin to their present positions within the Humber Zone (Williams, 1977b). This model (Fig. 3.79) relates ophiolite emplacement in western Newfoundland to events at the Humber-Dunnage zone boundary. It also furnishes a mechanism for polyphase deformation and metamorphism.

The Baie Verte-Brompton Line and the northern Exploits-Gander boundary on opposite sides of the Dunnage Zone bear several important similarities as follow: (a) both are marked by discontinuous mafic-ultramafic complexes; (b) there is a sub-Middle Ordovician unconformity developed above mafic-ultramafic rocks along the Exploits-Gander boundary and an inferred sub-Middle Ordovician unconformity above mafic-ultramafic rocks of the Baie Verte-Brompton Line, yet some ophiolite complexes in interior parts of the Dunnage Zone exhibit conformable cover relationships; (c) Ordovician conglomerates above the unconformities are of local derivation, unsorted, immature, and typical of rapidly evolving sources; (d) the conglomerates contain sedimentary clasts, deformed in some examples, that imply linkages with the Humber Zone to the west and the Gander Zone to the east; (e) mélanges occur locally at both boundaries; (f) in a general way, regional metamorphism increases in adjacent rocks outward and away from the Dunnage Zone; (g) the present steep to overturned boundaries are followed outwards from the Dunnage Zone by flatter structural belts of metaclastic rocks; and (h) polarity of first structural transport is outward from the Dunnage Zone with later transcurrent movements in both cases.

Unconformities on deeply eroded ophiolites, localized at zone boundaries, suggest uprooting of oceanic crust and mantle adjacent to bordering terranes. The relationships imply obduction. It is older than the Middle Ordovician age of unconformable cover rocks. Furthermore, the symmetry on opposite sides of the Newfoundland Dunnage Zone fits the model of compressed oceanic elements between colliding Grenville and Central lower crustal blocks. However, there are important differences between bordering terranes: no stratigrahic analysis exists for a Gander Zone continental margin comparable to that for the Humber Zone, emplacement of the Gander River Complex has little stratigraphic expression in rocks of the Gander Zone, ophiolite complexes of the Exploits Subzone lack dynamothermal soles of Bay of Islands type, and allochthons of Taconic style are absent in the east. Large parts of the Exploits Subzone are defined by Middle Ordovician rocks that are unconformable upon the Gander River Complex and other mafic-ultramafic complexes inland. Where coeval rocks of the Indian Bay Subzone are conformable above metaclastic rocks of the Gander Zone, a major change in depositional regime is implied. The change seems to coincide with earliest emplacement of Exploits ophiolitic suites above Gander metaclastic rocks.

Faunal and other distinctions between the Notre Dame and Exploits subzones suggest separation during their Early and early Middle Ordovician development. A sub-Silurian unconformity and absence of Caradoc shales and younger Ordovician rocks throughout the Notre Dame Subzone contrast with continuous deposition in north-central parts of the Exploits Subzone. The earliest subzone linkage is provided by volcanic clasts of the Roberts Arm Group, on the Notre Dame side, found in Upper Ordovician-Lower Silurian conglomerates and olistostrome on the Exploits side (Nelson, 1981; Nelson and Casey, 1979). The Red Indian Line is cut by intrusions correlated with the Devonian Hodges Hill batholith, and gabbro of Silurian age (Dunning et al., 1990a) cuts the Annieopsquotch Complex and the Victoria Lake Group on opposite sides of the line along the Burgeo Road.

Most features of the Ordovician volcanic rocks of the Notre Dame Subzone suggest they are the products of an ancient island arc (Strong, 1973, 1977; Strong and Payne, 1973; Kean and Strong, 1975; Szybinski, 1988). Other geochemical studies of the Snooks Arm volcanic rocks suggest marine volcanism other than that of island arcs

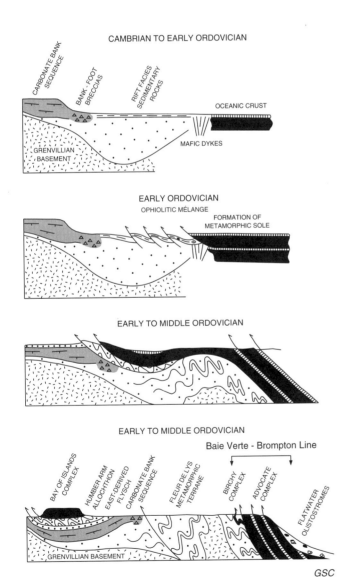

Figure 3.79. Model of ophiolite obduction at Humber miogeocline, after Williams and St-Julien (1982).

(Jenner and Fryer, 1980). It is noteworthy that nearby ophiolite suites are younger than volcanic rocks and tonalites of the Twillingate Subzone. Eastward subduction is favoured, as there is no evidence for a proximal island arc in the stratigraphic record of the Humber Zone.

Similar chemical trends suggesting island arc (Swinden, 1987) as well as other volcanic regimes (Reusch, 1983; Jacobi and Wasowski, 1985; Wasowski and Jacobi, 1986) are established for rocks of the Exploits Subzone. The geochemical and isotopic data for the Wild Bight Group, and comparisons with modern oceanic environments such as Fiji and the Mariana Trough, suggest a four-stage conceptual model (Swinden, 1987; Swinden et al., 1990). Stage one, the bottom of the group, records the last phases of volcanic activity in what may have been an originally more extensive island arc terrane. Stage two records the fragmentation of the arc, with attendant hydrous partial melting of both refractory mantle sources and basal arc crust. Stage three records early back-arc magmatism with alkalic and transitional alkalic/tholeiitic basalts. Stage four records more mature back-arc volcanism. The continued eruption of arc tholeiites at this stage suggests a narrow back-arc basin. Volcanogenic sulphide deposits in the Wild Bight Group include both massive sulphides and stockwork deposits formed during stage two, probably as a result of increased fracturing, heat flow, and hydrothermal circulation accompanying breakup of the arc.

Lead isotopic signatures of volcanogenic sulphide deposits in the Exploits Subzone are relatively radiogenic and contrast sharply with those in most of the Notre Dame Subzone, where relatively non-radiogenic lead is prevalent (Williams et al., 1988).

Volcanic rocks are dominant in western areas of the Exploits Subzone and they interdigitate eastward with mainly sedimentary sequences. Correlation of the Wild Bight and adjacent Exploits groups, established linkages between the Exploits Group and Dunnage Mélange, and possible equivalence of the Dunnage and Carmanville mélanges suggest gradation and continuity of rocks and structures across the northern Exploits Subzone. Extensive silicic volcanic rocks in the Bay du Nord Group and mainly shales in the Baie D'Espoir Group suggest a similar pattern across the Exploits Subzone in southern Newfoundland.

There is no evidence that the Dunnage Mélange was ever buried in a subduction zone, although it has been interpreted as a subduction phenomenon (Bird and Dewey, 1970; Williams and Hibbard, 1976; Kay, 1976). Most lines of reasoning suggest a back-arc setting, and mud-magma relations imply a surficial slump rather than brecciation beneath a tectonic cover (Lorenz, 1984a, b). Again, eastward subduction beneath the Exploits Subzone is favoured.

Lower Ordovician (Arenig-Llanvirn) black shale and limestone occurrences throughout both the Notre Dame and Exploits Subzones are sporadic, localized, widely separated, and all of somewhat different ages. The situation suggests evolving, multiple Early Ordovician tectonic elements without significant continuity, even within a particular tectonic subzone. The idea of a Dunnage collage is also supported by: (a) the transitional chemical affinities of most Dunnage ophiolite suites, implying that none represent the crust of a major ocean; (b) some volcanic rocks overlie

an ophiolitic substrate but others (Buchans, Roberts Arm, and Cutwell groups) contain zircons that show inheritance and therefore originated in older crust; and (c) late Precambrian ages for quartz monzonites (Valentine Lake and Cripple Back Lake) of the Exploits Subzone.

The notion of an exposed Iapetus suture within the Dunnage Zone is irrelevant to this analysis. An Iapetus suture is the deep collisional zone between Grenville and Central lower crustal blocks. Its surface definition is only possible where Dunnage rocks are absent, such as in southwest Newfoundland.

Acknowledgments

R.F. Blackwood, G.R. Dunning, S.P. Colman-Sadd, K.L. Currie, P. Erdmer, B.F. Kean, P.P. O'Neill, and P.F. Williams are thanked for reviews of earlier versions of this manuscript.

New Brunswick
C.R. van Staal and L.R. Fyffe
Introduction

The Dunnage Zone of New Brunswick is defined on its marine Lower and Middle Ordovician volcanic rocks. Locally, in the Elmtree River-Belledune River area of the Chaleur Uplands (Fig. 3.80), volcanic rocks overlie, or are part of, an ophiolitic substrate. Throughout most of the Miramichi Highlands, volcanic rocks overlie a thick terrigenous clastic sequence with at least local preservation of stratigraphic contacts. According to this definition of the New Brunswick Dunnage Zone, its rocks overlie both an ophiolitic and continental substrate and possibly link these contrasting basement elements from north to south. Previous placement of a Dunnage-Gander boundary at the Rocky Brook-Millstream Fault (Williams, 1978a, 1979; Fyffe and Fricker, 1987) was an attempt to separate ophiolite-based sequences to the north from clastic-based sequences to the south, rather than depict the distribution of Lower and Middle Ordovician volcanic rocks. However, detailed mapping (van Staal et al., 1990) and extensive geochemical studies (van Staal et al., 1991) have shown that identical volcanic and sedimentary suites, as well as local ophiolitic rocks, occur on opposite sides of the Rocky Brook-Millstream Fault. Therefore, the fault is not an important zone or terrane boundary.

Six subzones are recognized in the New Brunswick Dunnage Zone. The Elmtree, Belledune, and Popelogan subzones occur in two discrete Ordovician inliers that are surrounded by middle Paleozoic rocks of the Chaleur Uplands to the north of the Rocky Brook-Millstream Fault (Fig. 3.80). The inlier of Ordovician rocks in the Elmtree River-Belledune River area contains three major thrust sheets (Fig. 3.81). From bottom to top, rocks of successively higher sheets are the Elmtree Formation, Pointe Verte Formation, and the ophiolitic Devereaux Formation. The

van Staal, C.R. and Fyffe, L.R.
1995: Dunnage Zone-New Brunswick; in Chapter 3 of Geology of the Appalachian-Caledonian Orogen in Canada and Greenland, (ed.) H. Williams; Geological Survey of Canada, Geology of Canada, No. 6, p. 166-178 (also Geological Society of America, The Geology of North America, v. F-1).

Pointe Verte and Devereaux formations define the Belledune Subzone. The Elmtree Formation defines the Elmtree Subzone. Ordovician sedimentary and volcanic rocks of the Balmoral Group, exposed in a separate inlier to the west, define the Popelogan Subzone.

Detailed mapping at 1:10 000 scale (van Staal, unpublished data) south of the Rocky Brook-Millstream Fault led to three new subzones for the Middle Ordovician volcanic and sedimentary rocks in the Miramichi Highlands (Fig. 3.80). These are: the Armstrong Brook Subzone, which is thought to contain correlatives of the upper thrust sheets in the Belledune Subzone; the Bathurst Subzone, which has lithological similarities to the Elmtree Subzone; and the Hayesville Subzone of the southwestern Miramichi Highlands. The Armstrong Brook and Bathurst subzones

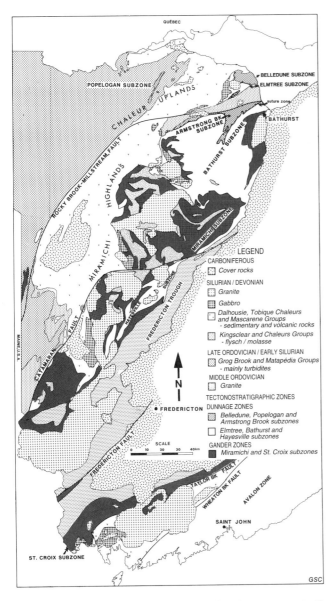

Figure 3.80. Distribution of the pre-Silurian tectonostratigraphic subzones in New Brunswick. Only the major faults are shown.

are separated by a major tectonic boundary that coincides with a 70 km long belt of blueschists that are traceable across most of the northern Miramichi Highlands. The important differences between the Armstrong Brook and Bathurst subzones, which justify their separation, are as follow: (1) north of the tectonic boundary, basalts are compositionally like those found at mid-ocean ridges (MOR) or are intermediate between mid-ocean ridges and island-arc tholeiites (van Staal et al., 1991); (2) no silicic volcanic rocks are interlayered with the basalts of the Armstrong Brook Subzone; (3) the sedimentary rocks overlying the basalts in the Armstrong Brook Subzone contain numerous limestone lenses and coarse, thick bedded quartz- and feldspar-rich lithic wackes as in the Pointe Verte Formation, which are rare or absent in the Bathurst Subzone; and (4) red phyllite, chert, jasper, and Zn-Pb-Cu-Ag type massive sulphides, typical of metalliferous sediments found in the Bathurst Subzone, are absent in the Armstrong Brook Subzone, which contains only one Cyprus type Cu-pyrite sulphide prospect, i.e., Middle River Copper (Davies, 1979). A syngenetic massive sulphide body in the Belledune Subzone, the Turgeon deposit (Davies et al., 1983), is also a Cyprus type Cu-pyrite body.

All of the Ordovician and older rocks of the Miramichi Highlands were assigned traditionally to the Tetagouche Group (Young, 1911; Alcock, 1941; Skinner and McAlary, 1952). Those of the present Armstrong Brook, Bathurst, and Hayesville subzones, dominated by volcanic and volcaniclastic sedimentary rocks, were referred to informally as the upper Tetagouche Group (Ruitenberg et al., 1977). The underlying non-volcanogenic quartzose clastic rocks were referred to as the lower Tetagouche Group. Since the entire Tetagouche Group has never been clearly defined, and since it contains distinct lithologies now assigned to the Dunnage and Gander zones, new names or redefinitions of existing names are required to distinguish the Dunnage volcanosedimentary rocks from underlying monotonous clastic rocks assigned to the Gander Zone. We retain the name Tetagouche Group for the volcanic and associated sedimentary rocks now assigned to the Dunnage Zone in the Bathurst and Hayesville subzones of the Miramichi Highlands. A new name, Miramichi Group (cf., Potter, 1969), designates the underlying clastic rocks assigned to the Gander Zone.

A few minor isolated bodies of Middle Ordovician rocks occur in west-central New Brunswick (Fyffe, 1982a). Small inliers of Middle Ordovician volcanic and sedimentary rocks also occur in the southern Gaspésie area of Quebec, e.g., the Arsenault Formation (Malo, 1988; Riva and Malo, 1988). A much larger occurrence forms the Munsungun-Winterville Anticlinorium in the state of Maine (Hall, 1970; Roy and Mencher, 1976; Osberg et al., 1985). These inliers suggest that Middle Ordovician volcanic and sedimentary rocks probably form a continuous basement to the Upper Ordovician-Devonian volcanic and sedimentary cover throughout this part of the northern Appalachians.

Belledune Subzone

Ordovician rocks of the Belledune Subzone (Fig. 3.81) were named the Fournier Group (Young, 1911). It was subdivided into: the Devereaux Formation, consisting of gabbro, mylonitic amphibolite, plagiogranite, and diabase, which locally defines a sheeted complex (Fig. 3.82-3.85); and the

167

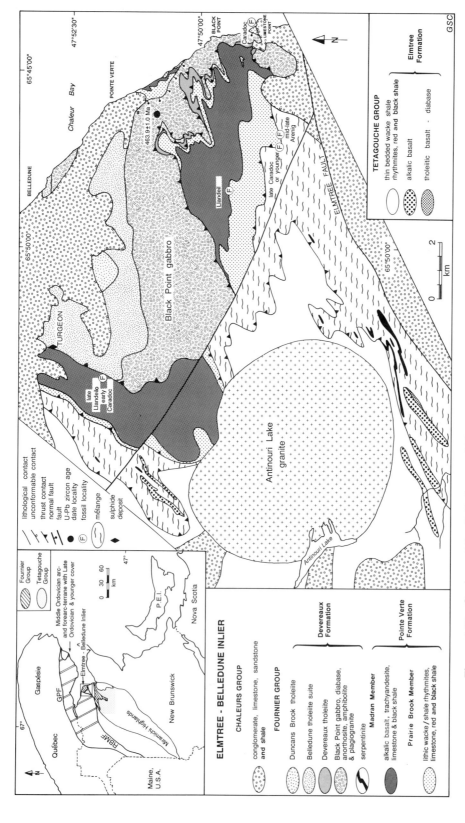

Figure 3.81. Geology of the Belledune and Elmtree subzones, modified from van Staal and Langton (1990).

Pointe Verte Formation, consisting of basalt, greywacke, shale, limestone, and chert (Pajari et al., 1977; Rast and Stringer, 1980). The combined Devereaux and Pointe Verte formations were interpreted as upper parts of an ophiolitic suite (Pajari et al., 1977). However, geochemical analyses (Winchester and van Staal, 1988) show that most basalts in the Pointe Verte Formation are alkalic, atypical of ophiolitic suites, in contrast to the tholeiitic basalts that overlie the gabbro. Furthermore, detailed mapping at 1:10 000 scale indicates that the alkalic basalts structurally underlie the Devereaux Formation; since the east-trending upright F_2 structures that fold the Ordovician rocks into a synform and antiform pair (Fig. 3.81, 3.86) have shallow to moderate eastward plunges (Langton and van Staal, 1989; Thomas et al., 1991). A major high strain zone, defined by low grade mylonites, phyllonites, and chaotically deformed sedimentary rocks (mélange) separates the contrasting alkalic and tholeiitic igneous suites. The Pointe Verte Formation, therefore, is redefined to

include only the alkalic basalts and associated sedimentary rocks, whereas the Devereaux Formation is expanded to include the tholeiitic basalts and sedimentary rocks that structurally and stratigraphically overlie its gabbroic rocks. The gabbroic rocks and tholeiitic basalts of the Devereaux Formation are interpreted as a fragment of back-arc oceanic crust (van Staal et al., 1991). Zircons from a coarse gabbro of the Devereaux Formation are dated at 463.9 ± 1 Ma (Sullivan et al., 1990), which falls in the Llandeilo of most time scales (Harland et al., 1982; Palmer, 1983; Snelling, 1985; Haq and van Eysinga, 1987). Recent calibration of the Ordovician time-scale (Tucker et al., 1990) suggests that this age is nearly coincident with the Llanvirn/Llandeilo boundary (Fig. 3.87). Sheared amphibolitized gabbro (Fig. 3.84) is locally cut by plagiogranite dykes (Fig. 3.82), which are themselves weakly deformed by the shear zones. One of these syntectonic plagiogranite dykes in the quarry near the Green Point railway station

Figure 3.82. Layered and foliated gabbroic rocks of the Devereaux Formation transected by a plagiogranite dyke (light grey in upper part of photo), Belledune Subzone. GSC 1994-767E

Figure 3.84. High-strain amphibolite zone within gabbroic rocks of the Devereaux Formation, Belledune Subzone. GSC 1994-767G

Figure 3.83. Massive diabase transecting layering within the gabbroic rocks of the Devereaux Formation, Belledune Subzone. GSC 1994-767F

Figure 3.85. Sheeted dykes within the Devereaux Formation, Belledune Subzone. GSC 1994-767H

yielded a U-Pb zircon age of 459.6 ± 1 Ma (Sullivan et al., 1990), whereas an age of 461 ± 3 Ma was obtained by Spray et al. (1990) for the same dyke. An ^{39}Ar/^{40}Ar hornblende cooling age of 459 ± 5 Ma for the amphibolite in the shear zone cut by the plagiogranite (C. Ravenhurst, pers. comm. 1989) supports the field interpretation that the intrusion is syntectonic with respect to the shearing, and indicates an approximate age difference of 4 Ma between the gabbro and the plagiogranite. This late Llandeilo/early Caradoc age defines a phase of intraoceanic deformation, since amphibolite facies metamorphic conditions and deformation are restricted to the sheared parts of the gabbro. The age difference between the crystallization of the gabbro and the plagiogranite suggest that the plagiogranite is an off-axis intrusion, either emplaced along a transform fault (van Staal et al., 1990) or in an embryonic volcanic arc developed on back-arc oceanic crust (Flagler and Spray, 1991). The Devereaux ophiolitic suite is thus significantly younger than the Lower Ordovician ophiolites of the Newfoundland Dunnage Zone (Dunning and Krogh, 1985).

The Pointe Verte Formation is subdivided into two members (Fig. 3.87; Langton and van Staal, 1989): a lower Prairie Brook Member, which mainly consists of thick bedded, quartz- and feldspar-rich lithic wacke (Fig. 3.86) with minor limestone, shale, felsic volcanic tuff, pebble conglomerate, and red siltstone; and an upper Madran Member, mainly consisting of alkalic basalts and minor interbedded rocks like those of the underlying Prairie Brook Member. Black shale occurs near the top of the Madran Member. All shales and siltstones are generally transformed into slates and phyllites. The conodont alteration index (CAI) in interbedded metalimestone ranges from 4 to 6 (Nowlan, 1986, 1988a), indicating metamorphic temperatures from 300 to 400°C. Thin limestone layers interbedded with the lithic wacke and slate near the structural base of the Prairie Brook Member contain middle to late Arenig conodonts (Nowlan, 1988a). Interpillow limestone near the base of the Madran Member has Llandeilo conodonts (*P. anserinus* zone; Nowlan, 1983a, 1986), whereas black shale near the top of the Madran Member has Llandeilo to early Caradoc graptolites of the *N. gracilis* zone (Riva <u>in</u> Fyffe, 1986). The Pointe Verte Formation thus

Figure 3.86. Upright, shallowly east-plunging F$_2$ folds in lithic wacke beds of the Prairie Brook Member, Pointe Verte Formation, Belledune Subzone. GSC 1994-767I

spans virtually the whole Middle Ordovician and the Madran Member is partly equivalent, or more likely younger, than the Devereaux ophiolitic suite according to the time scale calibration of Tucker et al. (1990). The presence of diabase dykes in the gabbro complex of the Devereaux Formation, with compositions similar to the alkalic basalts in the Pointe Verte Formation, also indicates that the Pointe Verte Formation is partly younger. The narrow zone of mylonite and phyllonite that separates the Pointe Verte and Devereaux formations, therefore, is interpreted as a thrust, with the Devereaux ophiolitic suite structurally above the slightly younger rocks of Pointe Verte Formation.

Isolated occurrences of typical Pointe Verte rocks associated with lenses of serpentinite occur in a 1-2 km thick black shale mélange that marks the tectonic contact between the Fournier and Tetagouche groups in the most southwestern corner of the Elmtree Inlier (Fig. 3.81). The presence of ultramafic rocks in the mélange suggests that the Fournier Group was once more extensive than is presently exposed and supports the earlier interpretations that the Fournier Group represents a dismembered ophiolite suite. Middle Ordovician rocks of the Belledune Subzone thus consist of two major thrust sheets emplaced upon the Elmtree Subzone during post-Caradoc thrusting. Initial emplacement was accompanied by the formation of a mélange.

Elmtree Subzone

The Elmtree Formation (Young, 1911) of the Elmtree Subzone structurally underlies the Pointe Verte Formation (Fig. 3.81). It consists of dark grey to black shale interlayered with thin bedded siltstones and wackes, alkalic and tholeiitic basalt and minor felsic volcanic rocks with interbeds of red and green phyllite and chert. The sedimentary and volcanic rocks of the Elmtree Formation are compositionally similar to rocks in the Tetagouche Group (Winchester and van Staal, unpublished results) of the Bathurst Subzone. They are regarded, therefore, as correlatives of the Bathurst Subzone and included within the Tetagouche Group. The age of the Elmtree Formation is not well constrained by fossils, but one *Orthograptus* sp., indicative of a Caradoc to Llandovery age (Dean, 1975), was found in relatively coherent black shale-siltstone rhythmites, typical of the Elmtree Formation along the Elmtree River, approximately 200 m structurally below the middle to late Arenig conodont locality at the base of the Pointe Verte Formation (Fig. 3.81). On the basis of the lithological similarities to other Caradoc black phyllites, these are also interpreted to be Caradoc, as Ashgill or Llandovery black phyllites are unknown in northern New Brunswick. The black shale-siltstone rhythmites are surrounded by mélange and probably represent a relatively low-strain pod in the mélange. The bimodal volcanic rocks in the structurally lower parts of the Elmtree Formation are tentatively interpreted to have a Llanvirn-early Caradoc age, because all isotopically and paleontologically dated Ordovician volcanic rocks in northern New Brunswick were deposited within this time range (see below), and because of compositional similarities to basalts present in the Tetagouche Group of the Bathurst Subzone (Winchester, pers. comm., 1988).

A 1-2 km thick mélange, characterized by chaotically deformed sedimentary rocks or rocks with strongly disrupted bedding, separates the Elmtree and overlying Pointe Verte formations. This mélange is also interpreted as a major thrust zone with older rocks above younger rocks (Fig. 3.81).

Popelogan Subzone

The Popelogan Subzone, exposed within the Popelogan Anticline (Ruitenberg et al., 1977), is a small, elongate inlier of Ordovician rocks that occurs in the Chaleur Uplands approximately 75 km northwest of Bathurst (Fig. 3.80). These rocks were considered to be equivalents of Middle Ordovician rocks within the Miramichi Highlands (Potter, 1964). They were later remapped by Irrinki (1976) and Philpott (1987, 1988) and assigned to the Balmoral Group. The lithological description and stratigraphic division summarized here are taken mainly from Philpott (1987, 1988). The Middle Ordovician Balmoral Group is subdivided into the Popelogan Formation, composed of black, graptolitic shale and chert, and the Goulette Brook Formation, made up of mafic and ultramafic volcanic rocks. The Popelogan Formation conformably overlies the Goulette Brook Formation and graptolites indicate a late Llandeillo-early Caradoc age (N. gracilis zone; Riva, in Philpott, 1987). The Popelogan Formation, therefore, is identical in age and lithology to the black slates that occur near the top of the Pointe Verte Formation. The Goulette Brook Formation is made up of an upper andesitic lapilli tuff member and a lower picritic hyaloclastite member. Volcanic rocks of the Goulette Brook Formation have a chemistry unlike that of any volcanic rocks of the Belledune Subzone or Miramichi Highlands. The Goulette Brook volcanic rocks are interpreted as part of a Middle Ordovician volcanic arc (van Staal et al., 1991). The Balmoral Group is disconformably overlain by Ashgill turbidites of the Grog Brook Group (St. Peter, 1978; Nowlan, 1983b, 1986), marking a short period of erosion and uplift in late Caradoc and/or early Ashgill time in this area.

Armstrong Brook Subzone

The Armstrong Brook Subzone occurs in the northernmost part of the Miramichi Highlands between a narrow belt of glaucophane- and epidote-bearing blueschists and the Rocky Brook-Millstream Fault (Fig. 3.80). The volcanic rocks of this subzone comprise the Sormany Formation and have been divided into three members: Armstrong Brook, Murray Brook, and Lincour basalts (van Staal et al., 1990, 1991) (Fig. 3.88). The Armstrong Brook basalts are primitive tholeiites with mid-ocean ridge characteristics, including flat REE patterns (van Staal et al., 1991). These basalts (Fig. 3.88) have a thin, discontinuous layer of mylonitic gabbro at their base along the contact with the underlying, very highly strained blueschists. The blueschists immediately beneath the Armstrong Brook basalts and gabbro form part of the Murray Brook alkalic basalt suite (see van Staal et al., 1991), which is chemically similar to the Llandeilo high-Cr alkalic basalts of the Pointe Verte Formation in the Belledune Subzone. The tectonostratigraphy of the Belledune Subzone is thus in part repeated in the Armstrong Brook Subzone (Fig. 3.87). Rocks in the Armstrong Brook Subzone therefore are also interpreted as

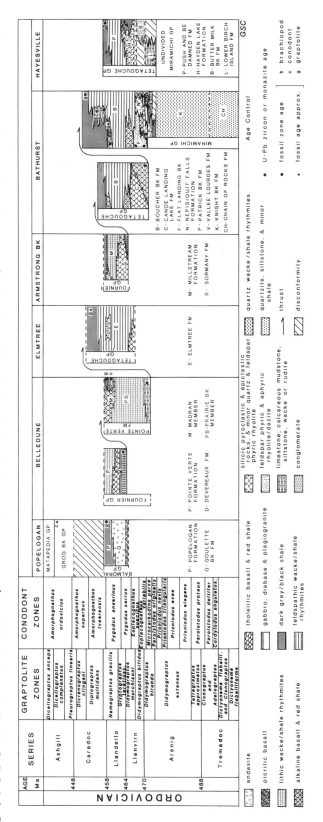

Figure 3.87. Tectonostratigraphy for the Dunnage Subzone of New Brunswick.

171

CHAPTER 3

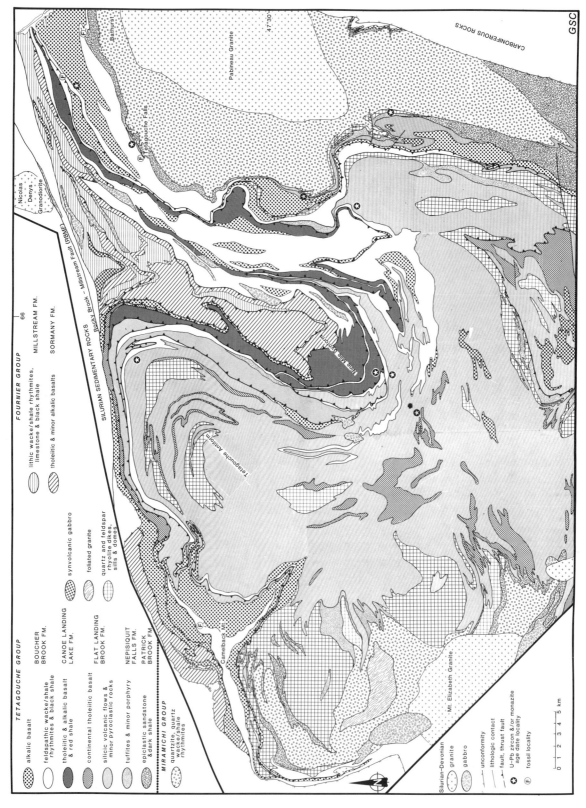

Figure 3.88. Geology of part of Armstrong Brook and Bathurst subzones in the northern Miramichi Highlands.

ophiolitic (van Staal et al., 1990, 1991). Structurally higher sheets of basalts, the Lincour basalts, are depleted in high-field-strength-elements (HFSE) and have compositions that are intermediate between mid-ocean ridge basalts and island-arc tholeiites (van Staal et al., 1991). Such basalts with transitional or arc-like compositions are also found at the highest structural levels in the Belledune Subzone and are typical of back-arc environments (Saunders and Tarney, 1984). The petrochemistry of the Lincour basalts, thus, supports the earlier interpretation of a back-arc setting proposed by van Staal (1987).

The Lincour basalts are overlain by, and in part interbedded with, quartz- and feldspar-rich lithic wackes and arkoses of the Millstream Formation (van Staal et al., 1988b). These are lithologically similar to the wackes of the Pointe Verte Formation in the Belledune Subzone. The Millstream Formation also includes a few layers of pebble conglomerate (with mainly igneous clasts), lenses of limestone, and black, shaly phyllite with interbedded chert. Although no biostratigraphically diagnostic fossils have been found in the Armstrong Brook Subzone, the black phyllites are interpreted as correlatives of the Caradoc, graptolite-bearing, black phyllites in the Belledune and Bathurst subzones (van Staal et al., 1988b). The limestone lenses, which are generally closely associated with the black phyllites, are interpreted to have a Llandeilo-early Caradoc age based on: (1) correlation with two small limestone lenses associated with black phyllite of the Boucher Brook Formation, near Camel Back Mountain in the Bathurst Subzone that yielded early Caradoc conodonts (Nowlan, 1981a); and (2) limestone lenses intercalated with basalts in the Pointe Verte Formation which yielded Llandeilo conodonts (Nowlan, 1983a). The Armstrong Brook basalts are interpreted to be Llandeilo on the basis of the Llandeilo, U-Pb zircon age of the ophiolitic Devereaux Formation, in the Belledune Subzone. If correct, the quartzofeldspathic wackes of the Millstream Formation must also be Llandeilo to Caradoc since they are interbedded with the basalts and, locally, with the black phyllites. This is the same age as similar rocks in the Madran and Prairie Brook members of the Pointe Verte Formation, but is contrary to earlier speculations suggesting an Ashgill age for the Millstream Formation (van Staal et al., 1988b).

In general, rock types and geochemistry of the basalts in the Armstrong Brook Subzone indicate that they corresponds more closely to rocks found in the Fournier Group than rocks found in the Tetagouche Group. We, therefore, assign the Sormany and Millstream formations to the Fournier Group. An obvious tectonic boundary, marked by the previously mentioned blueschist horizon, also supports exclusion of the rocks in the Armstrong Brook Subzone from the Tetagouche Group. Indeed, Young (1911) included volcanic rocks of the Armstrong Brook Subzone near Bathurst within his Fournier Group.

Bathurst Subzone

The Tetagouche Group of the Bathurst Subzone is dominated by a large complex of felsic metavolcanic rocks (Fig. 3.88) that have compositions mainly ranging from dacite to rhyolite (Whitehead and Goodfellow, 1978; van Staal, 1987; van Staal et al., 1991). The felsic volcanic rocks comprise a heterogeneous mixture of flows, shallow intrusions, pyroclastic rocks, and proximal epiclastic deposits (van Staal, 1987). Local sedimentary structures in the volcaniclastic sedimentary rocks indicate a dominantly subaqueous depositional environment. For mapping purposes, the rocks are generally divided into aphyric or feldspar-phyric rhyolite lavas and minor pyroclastic rocks of the Flat Landing Brook Formation, and mixed epiclastic/pyroclastic sedimentary rocks (tuffites) interbedded with tuffs and quartz and feldspar-phyric rhyolitic sills and/or flows of the Nepisiguit Falls Formation (Fig. 3.89; cf., Skinner, 1974; Saif, 1977; Davies, 1979). Numerous base metal Zn-Pb-Cu-Ag massive sulphide deposits are hosted by, or closely associated with, the felsic volcanic rocks, generally, the tuffites of the Nepisiguit Falls Formation (Smith and Skinner, 1958; McAllister, 1960; Davies, 1966; Harley, 1979). Thin, but generally laterally extensive bodies of iron-formation (Fig. 3.90), jasper, and a multicoloured (red, purple, green, and black) ferromanganiferous phyllite are commonly associated with the massive sulphides. Clastic sedimentary rocks other than epiclastic volcanic rocks are rare within the felsic volcanic succession.

The felsic metavolcanic rocks also contain two minor suites of tholeiitic basalt, either in the form of sills and dykes or as massive and pillowed flows, pyroclastic tuffs, breccias and agglomerates. These basalt suites are referred to as the Otter Brook and Forty Mile Brook tholeiites, and have compositions similar to continental tholeiites (van Staal et al., 1991). Basaltic agglomerates of the Otter Brook suite contain fragments of the surrounding felsic volcanic rocks into which compositionally similar dykes are intruded. The Otter Brook basalts were probably extruded contemporaneously with, or shortly after, deposition of the oldest felsic volcanic rocks. Pillows, common in all other basalt suites in the northern Miramichi Highlands, have never been observed among the Otter Brook basalts, possibly indicating a subaerial environment of deposition. If correct, at least some volcanic edifices must have been emergent at the time of their formation. Weathering of these subaerially exposed felsic volcanic edifices would also provide a simple explanation for the abundance of volcanic detritus in associated and coeval sedimentary rocks of the Patrick Brook and Boucher Brook formations (Fig. 3.87). Such an environment is not necessarily in conflict with the subaqueous depositional environment deduced for most of the felsic volcanic rocks (van Staal, 1987), since the fossils and sedimentary structures found in interbedded iron-formation and epiclastic, tuffaceous sedimentary rocks are consistent with a shallow water environment. These include stromatolite-like structures and ooids in carbonate facies iron-formation (McMillan, 1969; Saif, 1980; van Staal, 1987), and brachiopods and pelecypods in shaly sediments (Bolton, 1968; Fyffe, 1976; Gummer et al., 1978; Neuman, 1984). Pelecypods occurring in tuffaceous and/or epiclastic sedimentary rocks that host the Devils Elbow (Bolton, 1968) and Taylor Brook (Gummer et al., 1978) massive sulphide deposits are not biostratigraphically diagnostic (Pojeta, pers. comm., 1988). It is possible that these pelecypods represent unique faunas occurring only around hydrothermal vents. U-Pb zircon ages of 469 ± 2 Ma for a crystal tuff of the Nepisiguit Falls Formation and 466 ± 5 Ma for a spherulitic rhyolite at the top of the Flat Landing Brook Formation (Sullivan and van Staal, 1990) indicate a Llanvirn to early Llandeilo age for the felsic volcanism (Fig. 3.87). A similar age range, based on a few critical fossil localities and regional correlations, is deduced for felsic volcanic activity in central New Brunswick (Fyffe et al., 1983).

173

The Nepisiguit Falls Formation conformably overlies rocks of the Vallée Lourdes Formation (Fig. 3.87; van Staal et al., 1988b) below Tetagouche Falls and near the mouth of the Tetagouche River close to the town of Bathurst. The Vallée Lourdes Formation comprises crossbedded calcarenites (Fig. 3.91), calcareous siltstones, and conglomerates (Fig. 3.92, 3.93). This formation can be traced as lenses rather than as a continuous unit for more than 15 km southwestwards from the town of Bathurst. Both the Vallée Lourdes and Nepisiguit Falls formations are locally underlain and overlain by tuffaceous sandstones and dark shales of the Patrick Brook Formation (Fig. 3.94, 3.95) indicating that these formations interfinger lithostratigraphically (Fig. 3.87). The Vallée Lourdes Formation yielded middle Arenig to earliest Llanvirn conodonts and brachiopods near Tetagouche Falls (Fyffe, 1976; Nowlan, 1981a; Neuman, 1984) and at least in part represents shallow water deposits. These ages show that the Vallée Lourdes Formation is older than the conformably overlying formations as deduced from fossils and isotopic age determinations (van Staal et al., 1990).

The data set an upper limit to the age of the Miramichi Group, which must be Early Ordovician or older. Lithological correlatives of the Vallée Lourdes Formation (Fig. 3.95) in central New Brunswick (Lower Birch Island Formation of Hayesville Subzone) also contain brachiopods and conodonts suggesting a middle to late Arenig age (Poole, 1963; Neuman, 1968; Nowlan, 1981a), but the conodonts may extend into the earliest Llanvirn (Nowlan, 1981a). The brachiopods and conodonts belong to the Celtic and North Atlantic biogeographic provinces, respectively (Nowlan, 1981a), indicating a paleogeographic setting distant from the North American craton.

Understanding the relationships between the Dunnage Zone equivalents, i.e. Tetagouche Group, and the Gander Zone equivalents, i.e. the Miramichi Group, is critical to the tectonic interpretations of the northeastern Appalachians. As the lowest parts of the Tetagouche Group, i.e., Vallée Lourdes Formation, represent in part shallow water deposits that disconformably overlie the relatively deep water turbidites of the Miramichi Group (Ruitenberg et al., 1977),

Figure 3.89. Quartz- and feldspar-phyric intrusion or flow in the Nepisiguit Falls Formation at Devils Elbow, Tetagouche Group, Bathurst Subzone. GSC 1994-767A

Figure 3.91. Crossbedded, sandy limestone of the Vallée Lourdes Formation, Tetagouche Group, Bathurst Subzone. GSC 1994-767C

Figure 3.90. Upright, shallowly plunging F_2 fold in banded oxide-jasper iron-formation at the top of the Nepisiguit Falls Formation, Tetagouche Group, Bathurst Subzone. GSC 1994-767B

Figure 3.92. Thin bedded calcareous siltstone of the Vallée Lourdes Formation, Tetagouche Group, Bathurst Subzone. GSC 1994-767D

either the contact between these two groups is tectonic or some relative uplift has taken place in the Early Ordovician. A disconformity between the Vallée Lourdes Formation and Miramichi Group is marked by a thin layer of conglomerate east of Tetagouche Falls, 12 km west of Bathurst. The conglomerate contains well rounded quartzite and phyllite pebbles derived from the underlying Miramichi Group (Fig. 3.96), indicating uplift and local erosion of the Miramichi Group before deposition of the Tetagouche Group. Omission of the Vallée Lourdes Formation and the presence of a phyllonite along most Miramichi-Tetagouche contacts indicate that the original stratigraphic relationships between the groups has been tectonized subsequently. Quartzite and phyllite pebbles derived from the Miramichi Group are locally abundant in felsic agglomerates of the Flat Landing Brook Formation (Davies, 1972a, b), suggesting that the felsic volcanic magma came up through crust underlain by the Miramichi Group. Other

indications that both Miramichi and Tetagouche groups were underlain by a common continental crust are: (1) the presence of Middle Ordovician peraluminous granites (Fyffe et al., 1977; Whalen, 1987; Bevier, 1988) in areas underlain by the Miramichi Group; (2) the fact that the granites intruded locally consanguineous felsic volcanic rocks (Whalen, 1987) of the Tetagouche Group; and (3) the dominance of felsic volcanic over mafic volcanic rocks.

A conglomerate similar to the basal Vallée Lourdes Formation occurs locally at the base of the Tetagouche Group in central New Brunswick (see Buttermilk Brook Formation of Hayesville Subzone) and in Maine at the contact between the Grand Pitch and Shin Brook formations (Neuman, 1967). In Maine, this conglomerate marks the Penobscot unconformity of Neuman (1967, 1968) and it is the principal evidence for the Early Ordovician Penobscottian Orogeny in the central part of the northern Appalachians. There are several independent lines of evidence for the

Figure 3.93. Boulders of granitic gneiss in tuffaceous limestone of the Vallée Lourdes Formation, Bathurst Subzone. GSC 1994-767J

Figure 3.95. Northeast-trending F$_4$ crenulations refolding isoclinal F$_2$ folds within phyllite of the Patrick Brook Formation, Tetagouche Group, Bathurst Subzone. Note well developed S$_1$ differentiated layering in the tuffaceous rocks, which is folded by F$_2$. GSC 1994-767L

Figure 3.94. Angular felsic clasts in feldspathic wacke of the Patrick Brook Formation, Tetagouche Group at Tetagouche Falls, Bathurst Subzone. GSC 1994-767K

Figure 3.96. Quartz wacke pebbles in sandy limestone at the base of the Vallée Lourdes Formation, Tetagouche Group, Bathurst Subzone. GSC 1994-767M

existence of the Penobscottian Orogeny and its Ganderian equivalent in Newfoundland (Kennedy, 1976; Boone and Boudette, 1989; Williams and Piasecki, 1990; van Staal and Williams, 1991). However, the conglomerate at the base of the Tetagouche Group and Shin Brook Formation marks a phase of uplift that immediately preceded extrusion of back-arc volcanic rocks (van Staal et al., 1991). The unconformity may therefore also represent lithospheric doming associated with the onset of back-arc rifting during the middle to late Arenig (Fig. 3.97).

Although a part of the felsic volcanic rocks in the Bathurst Subzone are interpreted to be autochthonous or parautochthonous with respect to the underlying clastic sedimentary rocks of the Miramichi Group, most volcanic rocks are interpreted to be allochthonous (van Staal et al., 1990; Fig. 3.87, 3.88). The structurally lowest part of the succession consists of imbricated rocks of the Flat Landing Brook, Nepisiguit Falls, and Boucher Brook formations, which are structurally overlain by two major thrust sheets containing rocks of the Canoe Landing Lake Formation (Fig. 3.87, 3.88). The structurally lowest of these two thrust sheets consists mainly of pillowed, primitive alkalic basalts of the Canoe Landing Lake suite with minor intercalations of comendite and rhyolite, all interbedded with abundant red shale and chert. The overlying thrust sheet mainly consists of distinctive tholeiitic pillow basalts of the Nine Mile Brook suite interlayered with minor flows of alkalic basalts, red shale and chert. The alkalic basalts are compositionally identical to those in the Canoe Landing Lake suite (van Staal et al., 1991).

Narrow zones of phyllonites and mylonites, recording relatively high strains (van Staal, 1986, 1987), are present at the base of each thrust sheet and outline major truncations of units (Fig. 3.88). The volcanic formations of each thrust sheet are overlain by a remarkably similar package of dominantly sedimentary rocks that comprise thin-bedded feldspathic wacke, multicoloured shale, and three minor suites of alkalic basalt with trachyandesitic and comenditic differentiates (Fig. 3.88; van Staal et al., 1991). This largely sedimentary unit comprises the Boucher Brook Formation (Fig. 3.87). The black shale yielded Caradocian graptolites in three localities (van Staal et al., 1988b), the youngest of which belongs to the *D. clingani* or *C. spiniferus* zone (Riva and Malo, 1988). One trachyandesite yielded a U-Pb zircon age of 457 ± 1 Ma (R. Sullivan, pers. comm.), and confirms the Llandeilo-Caradoc age of the Boucher Brook Formation inferred from paleontological and other isotopic evidence (Sullivan and van Staal, 1990; Fig. 3.87). The presence of red shale and the similarities in age of zircons and monazites (van Staal and Sullivan, unpublished data), in each of the volcanic dominated formations indicates that the thrust sheets span the same short age range, the volcanic base being slightly older than the sedimentary top (Fig. 3.87). Where imbricated, this places older rocks over slightly younger rocks.

Detailed geochemistry of over 200 samples (van Staal et al., 1991) indicates that the volcanics of the Tetagouche Group represent a continental margin, rift sequence, produced in late Arenig-Llanvirn times during the opening of an ensialic back-arc basin, which subsequently developed into a marginal sea floored by oceanic crust (van Staal, 1987; Fyffe et al., 1990; van Staal et al., 1991). Zircon xenocrysts (L. Heaman, pers. comm., 1988) recovered from the youngest alkalic basalts (Fig. 3.98) of the Boucher Brook Formation lend further support to the interpretation that at least part of the Tetagouche Group was underlain by continental crust during deposition.

Hayesville Subzone

Middle Ordovician volcanic rocks of the Tetagouche Group occur throughout the central and southern parts of the Miramichi Highlands (Poole, 1963; Potter, 1969; Venugopal, 1978, 1979; Lutes, 1979; Irrinki, 1980; Crouse, 1981; Fyffe, 1982a, b). In general, Ordovician stratigraphy in the Hayesville Subzone (Fig. 3.80) is similar to that in the Bathurst Subzone of the northern Miramichi Highlands, except that mafic rocks are more abundant than felsic rocks. Geochemical studies of the mafic volcanic rocks in central New Brunswick (Fyffe et al., 1981) support this correlation. The stratigraphic sequence in the Hayesville area of central New Brunswick (Fig. 3.87) is best exposed along the Southwest Miramichi River on the southeastern flank of a northeasterly plunging anticline. Although outcrop is virtually continuous, the actual contact between the quartzose clastic rocks of the Miramichi Group and overlying carbonate-rich rocks of the Tetagouche Group is not exposed. Nonetheless, bedding attitudes suggest a conformable relationship. This relationship can be confirmed along strike to the southwest on Middle Hayden Brook where continuous stream-bed exposures reveal a gradational change from olive-green siltstone and shale of the Miramichi Group into fossiliferous calcareous beds of the Tetagouche Group.

The Tetagouche section exposed on the Southwest Miramichi River begins with calcareous siltstone of the Lower Birch Island Formation (Fig. 3.99). The siltstone is about 30 m thick and contains thin felsic tuff beds and numerous brachiopods dated as middle to late Arenig (Neuman, 1968, 1984). The Lower Birch Island Formation is overlain by 500 m of red, green, and black phyllite of the Hayden Lake Formation (Poole, 1963; Potter, 1969). A 1 m thick bed of red grit at the base of the Hayden Lake Formation contains clasts of underlying calcareous siltstone indicating a local post-Arenig erosional break (Irrinki, 1980). A period of extensive uplift is evident farther to the northwest on the limb of an adjacent synform where conglomerate of the Buttermilk Brook Formation directly overlies the Miramichi Group (Potter, 1969; Crouse, 1981) in place of the Lower Birch Island Formation. The conglomerate contains rounded cobbles of quartz wacke derived from the underlying Miramichi Group. Abundant mafic and lesser felsic volcanic rocks, are typically interbedded with the Hayden Lake Formation. The felsic volcanic rocks are mainly trachytes and typically associated with alkalic basalts. Extrusion of these volcanics apparently coincided with the onset of uplift recorded by the Buttermilk Brook Formation. Graptolites of the *G. terretiusculus* zone from black shale of the Hayden Lake Formation and of the *N. gracilis* zone from volcaniclastic wackes of the overlying Push and Be Dammed Formation (Fig. 3.100) and Belle Lake Formation of the Woodstock area, 100 km to the southwest, confirm the Llanvirn-Llandeilo age of volcanism throughout the Miramichi Highlands (Potter, 1969; Fyffe et al., 1983; Fyffe et al., 1988b).

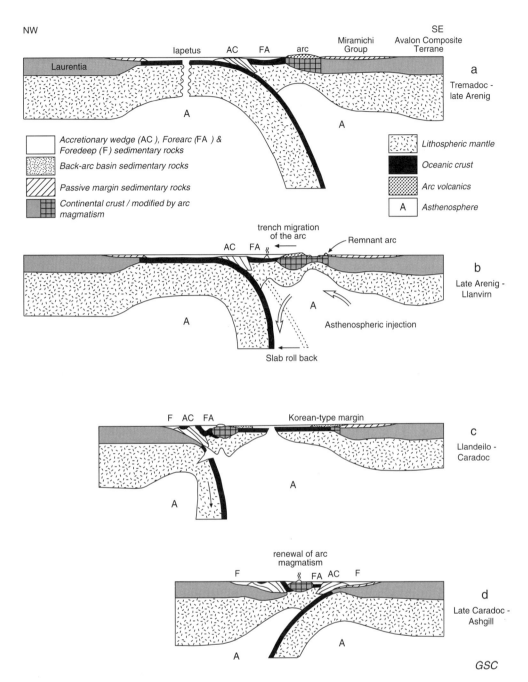

Figure 3.97. Schematic model of the tectonic evolution of the northern Appalachians in northern New Brunswick and adjacent Quebec. Adapted and expanded after van Staal et al. (1991). Although not shown, this model does not exclude earlier accretions of volcanic arcs (e.g. Taconic arc) to Laurentia, nor does it show the kinematic evolution of the Penobscottian or Ganderian orogeny, which was probably responsible for accretion of the Annidale Belt and GRUB in Newfoundland to the Gander Zone. Stages a, b, and c see van Staal et al. (1991). Stage d: closure of back-arc basin by northwest-directed subduction following a polarity reversal induced by the collision between the Popelogan arc and Laurentia. Late Ordovician-Early Silurian volcanism in adjacent Quebec (David and Gariépy, 1990) was probably initiated by this new Benioff zone. First arrival of the rocks of the Korean-type passive margin in the subduction zone probably took place in the Ashgill.

In the Woodstock area, black shales of the Bright Eye Brook Formation of the Miramichi Group (new name after brook flowing into Eel River) contain late Tremadoc to middle-late Arenig graptolites (Fyffe et al., 1983). The Bright Eye Brook Formation conformably underlies felsic and mafic volcanic rocks of the Oak Mountain Formation of the Tetagouche Group (formerly referred to as Pocomoonshine volcanics). These volcanics have calc-alkalic affinities and were interpreted as part of an ensialic arc (Dostal, 1989). They are intruded by the Benton granodiorite, which is lithologically and compositionally identical to the closely associated late Arenig Gibson granodiorite (U-Pb zircon age of 473 ± 1 Ma; Bevier, 1988). These two plutons contain porphyry Cu-type mineralization and they are possibly connected at depth (Fyffe, 1982a). The Oak Mountain Formation volcanics are therefore probably, at least in part, middle to late Arenig. They are interpreted as the remnants of a Lower Ordovician ensialic arc that was subsequently rifted in late Arenig-Llanvirn times (van Staal et al., 1991). The boundary between the conformable Miramichi and Tetagouche groups in this area is placed at the first occurrence of volcanic rocks. The Oak Mountain Formation is overlain by the Belle Lake Formation, which yielded graptolites of the *N. gracilis* zone (Fyffe et al., 1983). Conodonts of the *P. serrus-P. anserinus* zone (Llanvirn-Llandeilo boundary; Nowlan, 1981a; St. Peter, 1982) were extracted from the Waterville limestone member of the Belle Lake Formation (Venugopal, 1979). The conodont fauna of the Waterville limestone is identical to that of the Cobbs Arm limestone in the Exploits Subzone of Newfoundland (Nowlan, 1981a).

The contacts between the Tetagouche Group and the post-Middle Ordovician cover are generally faulted; however, an unconformity separates Middle Ordovician limestone from upper Ashgill conglomerate in the Becaguimec River area northeast of Woodstock (St. Peter, 1982) and Upper Silurian to Lower Devonian conglomerate overlies quartz sandstone of the Miramichi Group southwest of Woodstock (Venugopal, 1979; Lutes, 1979).

Structures of the New Brunswick Dunnage Zone are similar to those of the New Brunswick Gander Zone and they are described in a following section on the New Brunswick Gander Zone. Because rocks of the New Brunswick Dunnage and Gander zones are linked stratigraphically, a tectonic model that rationalizes all relationships is developed after the descriptions of the New Brunswick Gander Zone.

Figure 3.99. Tightly folded, fossiliferous, calcareous siltstone of the Lower Birch Island Formation, Tetagouche Group, Hayesville Subzone. GSC 1994-767-O

Figure 3.98. North-younging pillows in Beresford alkalic basalt of the Boucher Brook Formation, Tetagouche Group, Bathurst Subzone. GSC 1994-767N

Figure 3.100. Lithic wacke of the Push and Be Damned Formation containing rip-up clasts of dark grey siltstone, Hayesville Subzone. GSC 1994-767P

Quebec

*Alain Tremblay, Michel Malo, and
Pierre St-Julien*

Preamble

The Dunnage Zone of the Quebec Appalachians was originally part of the internal domain of St-Julien and Hubert (1975). Later, Williams (1979) and Williams and St-Julien (1982) recognized the Baie Verte-Brompton Line as a major tectonic boundary between continental and oceanic domains and assigned all Cambrian-Ordovician rocks lying east of the line to the Dunnage Zone. In Quebec, the Dunnage Zone is divided into four subzones which are from southwest to northeast: Estrie-Beauce, Mégantic, Témiscouata, and Gaspésie subzones (Fig. 3.101). Although post-Middle Ordovician rocks are generally excluded from the Dunnage Zone (Williams, Chapter 3, this volume), the Quebec segment locally includes Ordovician to Lower Silurian igneous and sedimentary rocks (Fig. 3.102).

In Quebec, rocks of the Dunnage Zone are covered, either disconformably or tectonically, by post-Taconic successor basins of the Gaspé Belt (Bourque et al., Chapter 4, this volume). The Dunnage Zone and its cover sequences are in tectonic contact with adjacent units. The best exposure of Dunnage rocks is in the Estrie-Beauce Subzone. Rocks of the Estrie-Beauce, Témiscouata, and Gaspésie subzones are interpreted as laterally equivalent. The Mégantic Subzone has different rocks and stratigraphy.

Estrie-Beauce Subzone

The Estrie-Beauce Subzone is bounded to the west by the Baie Verte-Brompton Line and to the east by the La Guadeloupe Fault (Fig. 3.103). The southwestern and northeastern limits are the Quebec-Vermont and the Quebec-Maine international borders, respectively (Fig. 3.103).

The Estrie-Beauce Subzone comprises three tectonostratigraphic units: (1) Saint-Daniel Mélange, including the southern Quebec ophiolite suites, (2) Ascot Complex, and (3) Magog Group (Fig. 3.102). Autochthonous post-Ordovician rocks of the Gaspé Belt lie unconformably upon rocks of the Estrie-Beauce Subzone (Fig. 3.103).

Saint-Daniel Mélange

The Saint-Daniel Mélange is a formal name introduced by Cousineau (1990) for an olistostromal mélange of the Quebec Appalachians previously known as the Saint-Daniel Complex (Lamothe, 1981a) or Saint-Daniel Formation (St-Julien, 1970; St-Julien and Hubert, 1975). In the Estrie-Beauce Subzone, the Saint-Daniel Mélange occurs along the Baie Verte-Brompton Line (Fig. 3.103). It was recognized on the eastern flank of the St-Victor Synclinorium

Tremblay, A., Malo, M., and St-Julien, P.
1995: Dunnage Zone-Quebec; in Chapter 3 of Geology of the Appalachian-Caledonian Orogen in Canada and Greenland, (ed.) H. Williams; Geological Survey of Canada, Geology of Canada, no. 6, p. 179-197 (also Geological Society of America, The Geology of North America, v. F-1).

(Fig. 3.103), in the Massawippi Lake area (de Römer, 1980; Tremblay, 1990), and Sherbrooke area (Tremblay and St-Julien, 1990; Tremblay, in press).

The Saint-Daniel Mélange was interpreted to stratigraphically overlie the ophiolite complexes of southern Quebec (Church, 1977; Williams and St-Julien, 1982; St-Julien, 1987). However, because of strong lateral discontinuities and apparently various geotectonic settings of the ophiolite complexes (Laurent and Hébert, 1989), they are here included as exotic tectonic slices within the Saint-Daniel Mélange. In the Estrie-Beauce Subzone, the Saint-Daniel Mélange forms a continuous unit, but with significant internal stratigraphic and structural complexities resulting from both olistostromal and tectonic processes (Cousineau, 1990; Tremblay and St-Julien, 1990).

Blocks and slices

Exotic blocks of limited lateral extent and tectonic or ambiguous relationships have been mapped within the Saint-Daniel Mélange. Dimensions of these blocks vary from a few tens of metres to several kilometres. They consist of ultramafic to granitic plutonic rocks, felsic to mafic volcanic rocks, high-grade metamorphic rocks, or continental- and oceanic-type sedimentary rocks (Fig. 3.103).

Ophiolitic rocks are dismembered and occur in three main bodies: Thetford-Mines, Asbestos, and Mount Orford complexes (Fig. 3.103). They are divided into two structural and petrological units: a lower Alpine-type peridotite unit, and an upper unit of ultramafic-mafic stratified cumulates overlain by gabbros, diabases, pillowed volcanic rocks and volcaniclastic rocks (Laurent, 1975). There are important lateral variations within each complex, as well as between complexes.

In the Thetford-Mines and Asbestos complexes, Alpine-type peridotite represents approximately 50 per cent of the sequence (Laurent, 1975). It is mainly composed of metamorphosed harzburgites crosscut by dunites and orthopyroxenite dykes (Laurent, 1975; Trottier, 1982). The rocks have deformation fabrics typical of mantle tectonites and subsolidus recrystallization textures. Peridotites are in faulted contact with surrounding greywackes of the Caldwell Group as well as with the cumulate portion of the ophiolite complexes. Serpentinized and/or carbonatized Alpine-type peridotites are also found as tectonic slivers along the Baie Verte-Brompton Line, within Saint-Daniel Mélange, or within the Sutton-Bennet schists (Humber Zone).

The upper unit of the ophiolite complexes consists of ultramafic-mafic cumulates grading to ophitic gabbros, sheeted sills or dyke swarms of diabases, shallow intrusive quartz-diorites and plagiogranites, overlain by pillowed volcanic rocks and sedimentary rocks. This sequence is typical of oceanic crust. U-Pb zircon ages on plagiogranites from Mount Ham date the Thetford-Mines complex at 478 Ma (Dunning, in Laurent and Hébert, 1989). ^{40}Ar-^{39}Ar dating of the amphibolite tectonic sole of the Thetford-Mines complex yields an age of 491 Ma (Clague et al., 1981). It was interpreted as the age of synemplacement metamorphism initiated during the obduction of oceanic crust. Considering the age of the plagiogranites (478 Ma), the amphibolite age may represent metamorphism on an upper mantle detachment fault initiated at a spreading centre.

179

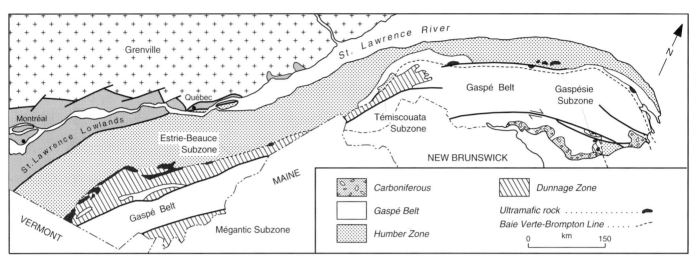

Figure 3.101. Tectonic divisions of the Quebec Appalachians and location of the Dunnage subzones.

GSC

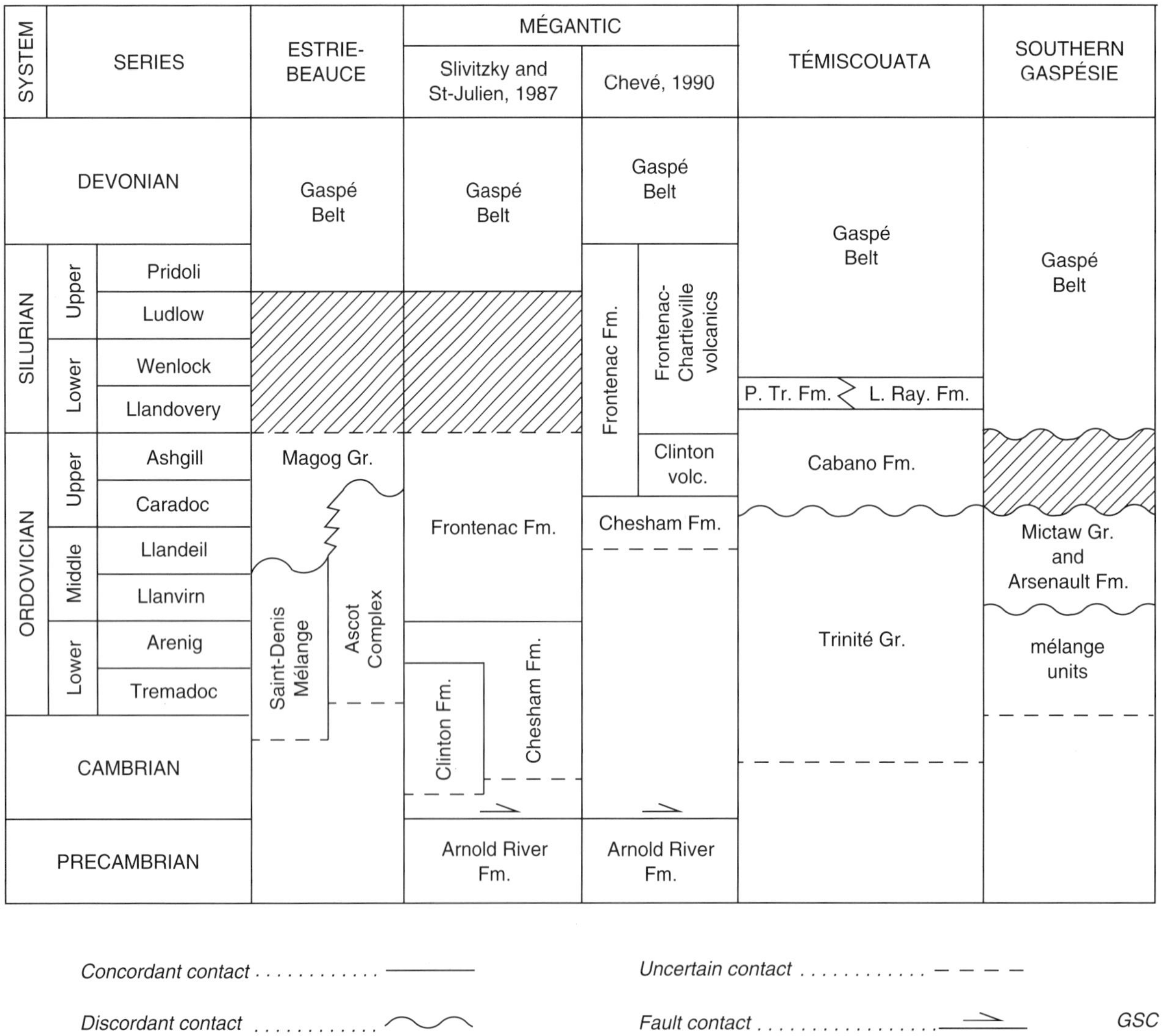

| SYSTEM | SERIES | | ESTRIE-BEAUCE | MÉGANTIC | | TÉMISCOUATA | SOUTHERN GASPÉSIE |
				Slivitzky and St-Julien, 1987	Chevé, 1990		
DEVONIAN			Gaspé Belt	Gaspé Belt	Gaspé Belt	Gaspé Belt	Gaspé Belt
SILURIAN	Upper	Pridoli					
		Ludlow			Frontenac Fm. / Frontenac-Chartieville volcanics		
	Lower	Wenlock				P. Tr. Fm. / L. Ray. Fm.	
		Llandovery					
ORDOVICIAN	Upper	Ashgill	Magog Gr.		Clinton volc.	Cabano Fm.	
		Caradoc		Frontenac Fm.	Chesham Fm.		Mictaw Gr. and Arsenault Fm.
	Middle	Llandeil	Saint-Denis Mélange / Ascot Complex				
		Llanvirn				Trinité Gr.	mélange units
	Lower	Arenig		Clinton Fm.	Chesham Fm.		
		Tremadoc					
CAMBRIAN							
PRECAMBRIAN				Arnold River Fm.	Arnold River Fm.		

Concordant contact ············ ——————   Uncertain contact ············ – – – –

Discordant contact ············ ∿∿∿   Fault contact ················ ⟶   GSC

Figure 3.102. Correlation chart of the Quebec Dunnage subzones. L. Ray. = Lac Raymond, P. Tr. = Pointe-aux-Trembles.

Recently, David et al. (1993) determined an U-Pb zircon age of 504 ± 3 Ma from a trondhjemite body of the Mount Orford Complex.

Volcanic and sedimentary rocks rest stratigraphically upon the Orford and Asbestos complexes and tectonically over the plutonic sequence of the Thetford-Mines complex. In the Thetford-Mines complex, it has been proposed that the volcanic cover forms a lower group of ocean floor basalts, overlain by an upper group of immature intraoceanic island arc volcanic products (Laurent and Hébert, 1977; Laurent et al., 1979). An island arc affinity has also been suggested for the lower group of volcanic rocks (Shaw and Wasserburg, 1984).

The upper contact of the southern Quebec ophiolite complexes with the Saint-Daniel Mélange is interpreted either as tectonic (Laurent and Hébert, 1977; Laurent et al., 1989) or stratigraphic (Williams and St-Julien, 1982; St-Julien, 1987). In the Thetford-Mines area, the contact is marked by a polygenic conglomerate and breccia unit (Coleraine-Burbank Hill breccia) interpreted as a debris flow close to an oceanic ridge (Hébert, 1981).

The **Rivière des Plante ophiolitic mélange** consists of a 30 km by 4 km tectonic slice occurring on both sides of the Chaudière River (Fig. 3.103). It is made up of composite slices of high grade metamorphic sandstone and gabbro, and ultramafic and granitic rocks within a schistose serpentinite matrix (Cousineau, 1991). The latter have a chaotic aspect characterized by metre-scale blocks of harzburgitic-dunitic composition set in a serpentinite or serpentinite-carbonate matrix. High grade metamorphic rocks (upper amphibolite facies) consist of granoblastic to brecciated metasandstones containing greenschist-retrograded subangular to subrounded centimetre-scale fragments of gneissic aspect and amphibolitic composition. Protoliths are interpreted as quartz-rich, feldspathic to lithic wackes (Cousineau, 1991). Ultramafic rock slivers have a harzburgitic to dunitic composition and a close spatial relationship with high grade metamorphic rocks. They are locally associated with black cherts. Ophiolitic mélanges similar to the Rivière des Plante are also reported from the Gaspésie Subzone (see below). Enclosed high grade metamorphic rock slices are correlated with some facies of the Chain Lakes Massif. The Rivière des Plante ophiolitic mélange was interpreted as the product of tectonism along a Taconic transform fault between an oceanic terrane and a Chain Lakes-type microcontinent (Cousineau, 1991) or as tectonic slices of continental rocks of the Humber Zone that were caught up during the obduction of the ophiolites (Pinet and Tremblay, in press).

Ware volcanics refers to a 15 km-long occurrence of felsic volcanic rocks west of the Chaudière River (Fig. 3.103; Cousineau, 1990). An absence of small volcanic fragments in the surrounding argillitic matrix of the Saint-Daniel Mélange suggests tectonic slices rather than large olistoliths. The Ware volcanics consist of three blocks separated by black shales. They are porphyritic fragmental felsic volcanic rocks, lapillistones, and hydrothermal breccia. Cousineau (1990) attributed a pre-Middle Ordovician age to these rocks which he interpreted as an accreted seamount.

The **Bolton lavas** (Ambrose, 1945) are large bodies of mafic volcanic rocks occurring along the west shore of Lake Memphrémagog (Fig. 3.103). These are brecciated, massive or pillowed metabasalts. They are relatively undeformed

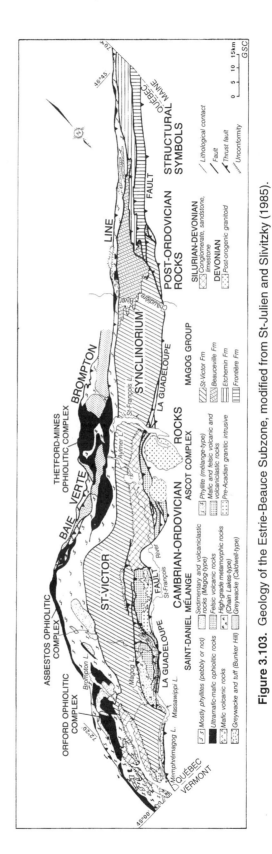

Figure 3.103. Geology of the Estrie-Beauce Subzone, modified from St-Julien and Slivitzky (1985).

although brecciation and mylonitization occur along some of the contacts with surrounding units (Lamothe, 1981b). Relationships of the Bolton lavas are controversial; contacts with the surrounding Saint-Daniel Mélange are interpreted either as tectonic (Lamothe, 1981b; Slivitzky and St-Julien, 1987) or stratigraphic (Doolan et al., 1982).

The **Bunker Hill sequence** (Blais et al., 1990) is an assemblage of felsic volcanic and sedimentary rocks occurring between Massawippi and Memphrémagog lakes (Fig. 3.103). It was historically included within the Ascot-Weedon volcanic belt (St-Julien and Lamarche, 1965; St-Julien and Hubert, 1975). More recent work shows that the Bunker Hill sequence differs from the Ascot Complex (de Römer, 1980; Tremblay, 1990; Blais et al., 1990). The Bunker Hill sequence is made up of greywackes underlain by felsic volcanic rocks (de Römer, 1980; Tremblay, 1990). Black phyllites and pebbly mudstone, included with the Saint-Daniel Mélange (de Römer, 1980), are in tectonic contact with the Bunker Hill sequence to the northwest. The southeastern contact is marked by the La Guadeloupe Fault (Fig. 3.103). More than 70 per cent of the sequence consists of greywackes; these are fine to coarse grained, quartz-rich, feldspathic to lithic wackes. Pyroclastic and volcaniclastic rocks lying at the base are well laminated and well bedded pyroclastic flow deposits interbedded with sedimentary debris flows. Metre-wide sills of gabbro occur within the pyroclastic deposits (de Römer, 1980; Tremblay, 1990). The continental character of the greywackes and their association with felsic volcanic rocks make the Bunker Hill sequence unique within the Estrie-Beauce Subzone. The greywackes bear strong similarities with those of the Caldwell Group (de Römer, 1980; Slivitzky and St-Julien, 1987; Tremblay, 1990). These rocks are unfossiliferous and their tectonic significance is unclear. According to Tremblay (1992a) and Tremblay and Pinet (in press), they belong to a structural inlier of the southern Quebec Humber Zone.

Caldwell-type and **Magog-type** blocks occur within the Saint-Daniel Mélange. Caldwell-type blocks are found southwest of the Thetford-Mines ophiolite complex, whereas Magog-type blocks occur along, and north of, the Mount Orford complex (Fig. 3.103). Caldwell-type greywackes occur as tectonic slices. In the Thetford-Mines area, a complex structural imbrication of Caldwell Group rocks within the Saint-Daniel pelitic matrix seems to result from out-of-sequence faulting (St-Julien, 1987; Tremblay, 1992a). West of the Chaudière River, large olistoliths of Caldwell sandstones occur within the Saint-Daniel Mélange (Cousineau, 1990). Magog-type blocks found near the Mount Orford complex consist mainly of feldspathic sandstone similar to the Etchemin Formation (see below) of the Magog Group. Regionally, the Magog Group rests unconformably on the Saint-Daniel Mélange.

Argillite matrix

The matrix of the Saint-Daniel Mélange is described as a graphitic pebbly mudstone (Slivitzky and St-Julien, 1987). It characterizes the Saint-Daniel Mélange in several localities of the Estrie-Beauce Subzone (Lamothe, 1981a; de Römer, 1980; St-Julien, 1987; Cousineau, 1990; Tremblay, 1990; Tremblay, 1992a, b). Centimetre-scale angular to subrounded fragments are dominated by shales, dolomitic siltstones, sandstones, quartzites, limestones and conglomerates. The surrounding matrix is a black-green laminated argillite. Metre-scale fragments of basaltic and felsic volcanic rocks are also present (St-Julien, 1987; Marquis, 1989b). Chaotic phyllites on the southeastern flank of the St-Victor Synclinorium (Fig. 3.103) are correlated with the Saint-Daniel Mélange (de Römer, 1980; Tremblay, 1990, 1992a, b). In some places, the Saint-Daniel matrix is a finely laminated argillite (St-Julien, 1987; Tremblay, 1992b) or consists of more or less continuous fine- to medium-grained turbiditic deposits (St-Julien, 1987; Cousineau, 1990).

Cousineau (1990) recognized seven lithofacies within the Saint-Daniel matrix with stratigraphic significance. Based on the nature of pebbles, they can be grouped into three informal stratigraphic units (Fig. 3.104): a pre-pebbly mudstone unit, a pebbly mudstone unit, and a post-pebbly mudstone unit. The pre-pebbly mudstone unit is made up of green and black argillite with a lesser amount of interbedded argillite and black sandstone or calcareous siltstone. Lithologies of this unit are mutually interbedded and more or less fragmental. They are characterized by slumping and synsedimentary deformation of partly consolidated deposits (Cousineau, 1990). The pebbly mudstone unit contains 20 to 40 per cent fragments within a black mudstone or a black to green laminated mudslate matrix. Fragments from centimetre to metre scale consist mainly of black and green argillite, black sandstone, and calcareous siltstone. Nodular pyrite is also characteristic and the unit appears to be similar to type III mélange of Cowan (1985). It is attributed to mud volcanism (Cousineau, 1990). The post-pebbly mudstone unit is made up of an assemblage of Magog-type argillite, blue-green sandstone and siliceous argillite, and a second assemblage of Caldwell-type argillite and greenish feldspathic sandstone.

The tectonostratigraphic framework of the Saint-Daniel Mélange suggests that it represents a relict accretionary prism formed in, or near a subduction zone (Tremblay and St-Julien, 1990; Cousineau and St-Julien, 1992). No fossils have been found in the Saint-Daniel Mélange and controversy still exists on its probable time range of formation. Based on the presence of Caldwell-type lithologies, St-Julien and Hubert (1975) proposed a Cambrian to Middle Ordovician age. Because the

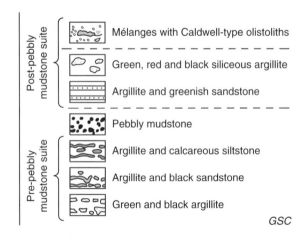

Figure 3.104. Stratigraphy of the Saint-Daniel Mélange (Early to Late Ordovician?), from Cousineau (1990).

Saint-Daniel Mélange seems to rest conformably upon the Thetford-Mines ophiolite complex (Williams and St-Julien, 1982), Cousineau (1990) inferred an Early to Late Ordovician age.

Ascot Complex

The name Ascot Complex was introduced by Tremblay and St-Julien (1990) to designate a volcanic and sedimentary unit previously known as the Ascot Formation (St-Julien and Lamarche, 1965). The Ascot Complex extends from the west shore of Massawippi Lake to Stoke Mountains (StJulien and Hubert, 1975; Fig. 3.103). Correlatives are found near St-François Lake (Weedon Formation of Duquette, 1961), and east of the Chaudière River (Cousineau, 1990). At each locality, the complex is in faulted contact with adjacent units (Labbé and St-Julien, 1989; Tremblay et al., 1989a; Cousineau, 1990).

Tremblay (1989, 1992b) divided the Ascot Complex into three distinctive lithotectonic domains separated by chaotic phyllites correlative with the Saint-Daniel Mélange; these are: Sherbrooke, Eustis, and Stoke domains.

The Sherbrooke domain is made up of an equal proportion of felsic and mafic volcanic rocks. Felsic rocks are pyroclastic breccia, crystal and aphanitic tuff, and foliated equivalents. Mafic rocks are vesicular to amygdaloidal, massive and pillowed basalts, chloritic schists, and a lesser amount of mafic tuff. The Eustis domain is characterized by quartz-plagioclase-sericite-chlorite schists originating from coarse grained to conglomeratic volcaniclastic rocks (Tremblay, 1992b) with lesser quartz-sericite schists, felsic pyroclastic rocks, and minor chlorite schists. Basalts are absent. In the Stoke domain, felsic rocks predominate over mafic volcanic rocks. Stoke felsites are homogeneous, porphyritic to fine grained pyroclastic rocks of rhyolitic composition. Mafic volcanic rocks are pillowed basalts and chloritic schists. The volcanic rocks of the Stoke domain are intruded by a granitic massif interpreted as a plutonic equivalent of the extrusive sequence (de Römer, 1985). Tremblay et al., 1994). K-Ar ages of the granite are 329 Ma (Wanless et al., 1968; Poole, 1980), and an Rb-Sr age is 620 Ma (Poole, 1980). The discordant ages are the result of Acadian metamorphic overprinting. Rocks of the Stoke domain are correlated with the plutonic and volcanic rocks at Stoke Mountains (de Römer, 1985), and rocks assigned to the Weedon Formation (Tremblay, 1989).

In the Sherbrooke area, the Ascot Complex includes argillitic rocks. These are graphitic phyllites containing clasts of shales, dolomitic siltstones, and black sandstones. They alternate with finely laminated grey-black phyllites locally interbedded with pyroclastic rocks. Tectonic slivers of carbonatized ultramafic rocks occur within the phyllites near major faults. Phyllites of the Ascot Complex are similar to lithologies of the Saint-Daniel Mélange.

In the St-François Lake area (Fig. 3.103), the Ascot Complex consists of ultramafic-mafic plutonic rocks, mafic and felsic volcanic rocks, and volcanogenic sedimentary rocks (Labbé, 1991). A 300 m-thick, fault-bounded, ophiolitic-like plutonic assemblage consists of ultramafic and mafic cumulates crosscut by quartz diorite and plagiogranite. This dismembered assemblage is unique within the Ascot Complex, it is interpreted as the plutonic basement of an immature volcanic arc (Labbé and Hébert, 1988). As in the Stoke domain, felsic rocks predominate over mafic varieties. Felsic rocks are porphyritic rhyolites and tuff, and quartz-sericite schists. Mafic rocks are pillowed basalts and schistose equivalents. The volcanic sequence is intruded by a foliated tonalitic-granodioritic body which resembles the Stoke Mountains intrusion. Volcanic conglomerates and sandstones, similar to those described by de Römer (1985), are included in the Ascot Complex in this area (Labbé, 1991).

In the Chaudière River area, Cousineau (1990) correlated a small slice of mafic and felsic volcanic rocks with the Ascot Complex. Mafic rocks overlie felsic varieties with a lesser amount of red mudslate at the base of the sequence. The felsic volcanic rocks are porphyritic and brecciated rhyolites, and rhyolitic tuff. The mafic rocks are vesicular to amygdaloidal, massive basalts and associated flow breccia.

The age of the Ascot Complex is poorly constrained. Based on the presence of felsic tuff in the Magog Group, St-Julien and Hubert (1975) suggested an Early to Middle Ordovician age. The complex and correlatives may represent a dismembered volcanic complex, or parts of a single or multiple volcanic belt (Tremblay and St-Julien, 1990). Recently, David et al. (1993) determined U-Pb zircon ages of 441+7/-12 Ma and 460 ± 3 Ma from the felsic extrusive sequence of the Sherbrooke and Stoke domains, respectively. However, the overall stratigraphic and paleotectonic framework of the Estrie-Beauce Subzone implies that the Ascot Complex is the remnant of a volcanic arc formed during Taconic Orogeny. As such, it cannot be older than the Saint-Daniel Mélange.

Magog Group

The Magog Group is a 10 km-thick flysch-dominated sequence unconformably overlying the Saint-Daniel Mélange within the St-Victor Synclinorium (Fig. 3.103). This unit was redefined by St-Julien (1970) in the St-François Lake area. Later, St-Julien (1987) divided the Magog Group into a lower Beauceville Formation and an upper Saint-Victor Formation. Recently, Cousineau and St-Julien (1994) recognized two more formations stratigraphically below the Beauceville. The Magog Group extends northeasterly up to the Témiscouata and Gaspésie subzones (see below). From base to top, it is made up of four formal units (Fig. 3.105): Frontière, Etchemin, Beauceville, and Saint-Victor formations.

The **Frontière Formation** is in fault contact with adjacent units (Fig. 3.103). It consists of unfossiliferous, interbedded green feldspathic sandstones and mudslates. Sandstones are graded lithic wackes (Cousineau, 1990). Most of the detrital components are felsic, intermediate, and mafic volcanic fragments and minor pelitic and granitic fragments. Petrographic analysis (Cousineau, 1990) indicates volcanic arc and recycled-orogen sources (Dickinson and Suzcek, 1979). The Frontière Formation is interpreted as the erosional products of a forearc volcanic apron associated with an ensialic volcanic arc. It is correlated with the Neckwick Formation of the Mictaw Group and the Arsenault Formation Inlier (see below) of the Gaspésie Subzone.

The **Etchemin Formation** includes interbedded, black and green mudslates overlain by green and siliceous mudrocks and a lesser amount of volcaniclastic rocks at the top of the sequence. Most contacts are faulted but locally gradational relations are preserved with the overlying Beauceville Formation. Volcaniclastic rocks occur in decimetre- to metre-thick beds, commonly graded. Coarser volcaniclastic rocks are porphyritic and with intermediate to felsic volcanic fragments. The age of the Etchemin Formation is unknown but it is constrained by the age of the Beauceville Formation (see below). The volcaniclastic rocks are interpreted as pyroclastic flow deposits, and the siliceous mudrocks as volcanic-related, silica-oversuturated, pelagic to hemipelagic sediments (Cousineau, 1990).

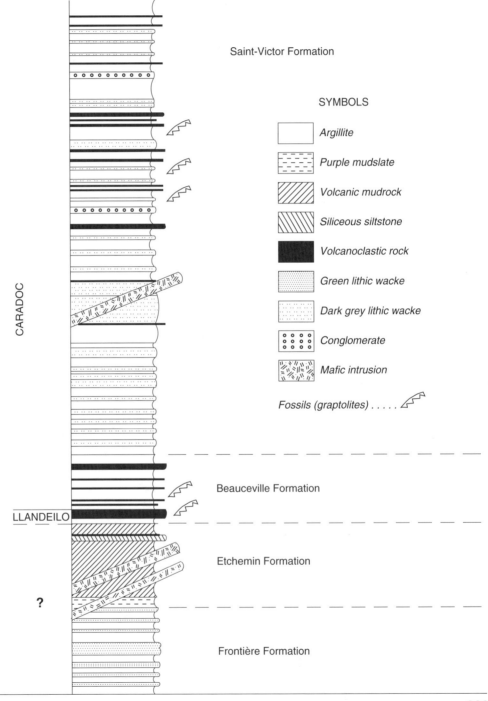

SYMBOLS

Argillite

Purple mudslate

Volcanic mudrock

Siliceous siltstone

Volcanoclastic rock

Green lithic wacke

Dark grey lithic wacke

Conglomerate

Mafic intrusion

Fossils (graptolites) . . . . .

Saint-Victor Formation

CARADOC

LLANDEILO

Beauceville Formation

Etchemin Formation

?

Frontière Formation

GSC

Figure 3.105. Stratigraphy of the Magog Group, from Cousineau (1990).

The **Beauceville Formation** as defined by St-Julien (1963, 1970) formed most of the St-Victor Synclinorium. It is now confined to the northeastern part of the Estrie-Beauce Subzone (Fig. 3.103; St-Julien, 1987; Cousineau, 1990) with a few occurrences in southern Quebec (St-Julien, 1963; Tremblay, 1990, 1992b). Based on several graptolite occurrences (St-Julien, 1963, 1970, 1987; Cousineau, 1990), the Beauceville and overlying Saint-Victor formations range from Late Llandeilo to Early Caradoc.

In the Chaudière River area, the Beauceville Formation is 800 m to 1 km thick and made up of interbedded graphitic clayslates, felsic volcaniclastic rocks, and siliceous claystones. Graphitic and siliceous claystones are homogeneous and in beds 2-10 cm thick with an overall thickness of several tens of metres. Thin, stratiform pyrite-rich interbeds are characteristic of the sequence (Cousineau, 1990). Massive volcaniclastic beds range from 10 cm to more than 10 m thick within the fine grained rocks. These commonly have a conglomeratic base (lapillistone) grading into a fine grained, crystal-rich medial tuff member, topped by tuffaceous mudrocks. Coarser beds are rich in aphanitic, felsic to intermediate volcanic fragments. Quartz and feldspar porphyroclasts and argillitic rip-up clasts are common.

Volcaniclastic rocks of the Beauceville Formation are interpreted as resedimented, submarine pyroclastic flows (Cousineau, 1990), with deposition of trailing ash clouds accounting for the fine grained parts of the volcaniclastic sequence.

The Middle Ordovician **Saint-Victor Formation** makes up more than 70 per cent of the Magog Group (Fig. 3.105). It is a turbidite sequence with rare interbeds of volcaniclastic rocks (St-Julien, 1987; Cousineau, 1990; Tremblay, 1992b). Its thickness is estimated between 7000 and 7500 m (Cousineau, 1990; St-Julien, 1987).

The turbidites consist of interbedded slates, feldspathic to lithic sandstones, and conglomerates. The slate-sandstone sequence is made up of 2 cm to 1 m argillite beds and fine- to medium-grained sandstones. The sandstones display parallel bedding, crossbedding, convolute laminations, and less common load casts and flute casts. The latter indicate west or northwest erosional sources (Cousineau, 1990). Thick conglomerate beds are 2-10 m with pebbles and boulders of siltstone, sandstone, chert, granite, and felsic volcanic rocks (St-Julien, 1970, 1987; Tremblay, 1992b). Some fragments reach 50 cm in diameter. Volcaniclastic rocks of the Saint-Victor Formation are similar to those of the Beauceville Formation, but less abundant. Beds are 2-10 m thick. Glass shards and dish structures occur locally.

The source of Saint-Victor Formation is believed to be siliciclastic rocks and fine grained felsic volcanic rocks (Cousineau, 1990). Detrital picotite and serpentinite, an ultramafic component in the source area (St-Julien, 1987), implies that the Magog Group rests unconformably on the Saint-Daniel Mélange. Petrographic analysis of the turbidites indicates a recycled-orogen affinity (Dickinson and Suzcek, 1979), which is attributed to the erosion of Taconic allochthons west of the Saint-Victor sedimentary basin. Fragments of felsic volcanic rocks and interbedded volcaniclastic rocks indicate a relatively proximal volcanic source. Cousineau (1990) concluded that these deposits originate from two sources: an uplifted accretionary prism (the Saint-Daniel Mélange) to the west, and a volcanic arc to the east.

Sill-like bodies of **intrusive rocks** are found in the northeastern part of the Magog Group (St-Julien, 1987; Cousineau, 1990). Three types of intrusions are distinguished (Cousineau, 1990); grey diorite, green diorite, and green-black gabbro. The grey diorite occurs within the Beauceville Formation. It is a fine grained, slightly carbonatized, diabasic rock. Green diorites and gabbros occur mostly as 20-30 m thick sills in the Etchemin and Beauceville formations. Similar, thinner sills, less than a metre, are also found in the Saint-Victor Formation.

Structure and metamorphism

The regional metamorphism of Cambrian-Ordovician rocks, and overlying post-Ordovician rocks of the Estrie-Beauce Subzone is attributed to the Middle Devonian Acadian Orogeny (Labbé and St-Julien, 1989; Tremblay and St-Julien, 1990; Cousineau and Tremblay, 1993). Thermochronologic data are scarce but indicate an Acadian-related metamorphic peak. Lowdon (1960) reported K-Ar biotite ages of 320 Ma and 379 Ma from granites of the Asbestos ophiolite complex and Weedon schist. Wanless et al. (1973) reported a metamorphic K-Ar biotite age of 358 Ma from the Orford ophiolite complex, and Leach et al. (1963) reported a K-Ar biotite age of 4 Ma from the Thetford-Mines ophiolite complex. The latter was attributed to a relict Taconic metamorphism preserved along the Baie Verte-Brompton Line (St-Julien and Hubert, 1975). Acadian ages are in agreement with those found in a sheared granite from the Ascot Complex (328 Ma; Wanless et al., 1968; Poole, 1980). Intensity of metamorphism varies from prehnite-pumpellyite facies in the northeast (Beauce area) to greenschist facies in the southwest (Estrie area).

Metamorphic ages and facies in the Estrie-Beauce Subzone contrast with those in the adjacent Humber Zone. West of the Baie Verte-Brompton Line, high-pressure, low-temperature minerals were found in southern Quebec (Trzciensky, 1976), and in northern Vermont (Laird and Albee, 1975). Humber Zone metamorphic data tend to indicate that major tectono-metamorphic events were Taconic-related (Rickard, 1965; St-Julien and Hubert, 1975; Laird et al., 1984). In the Dunnage correlatives of northern Vermont, the regional metamorphism is Acadian (Laird et al., 1984; Sutter et al., 1985) although some localities record an older metamorphism (Laird et al., 1984).

The intensity of faulting and folding also increases from northeast to southwest. The Estrie-Beauce Subzone is characterized by three major structures: Baie Verte-Brompton Line, St-Victor Synclinorium, and La Guadeloupe and related faults (Fig. 3.103). The Baie Verte-Brompton Line is defined by the narrow belt of discontinuous ophiolitic rocks and serpentinites at the contact between clastic rocks of the Caldwell Group, and argillitic chaotic rocks of the Saint-Daniel Mélange (Williams and St-Julien, 1982). The Caldwell-Saint-Daniel contact is interpreted as a northwest-directed thrust (St-Julien and Hubert, 1975; St-Julien, 1987; Cousineau, 1990). The occurrence of an Upper Silurian sedimentary basin (Cranbourne Formation) unconformably overlying the Baie Verte-Brompton Line (Fig. 3.103) is in agreement with its Taconic origin. Elsewhere, the Baie Verte-Brompton Line is a relatively sharp fault as in the Lac Memphrémagog area (Fig. 3.103), or an imbricated fault

185

zone probably resulting from out-of-sequence faulting, as in the Thetford-Mines area (St-Julien, 1987). Structures found along the Baie Verte-Brompton Line and in the Saint-Daniel Mélange are relatively homogeneous. The Saint-Daniel is characterized by a regional foliation (S_1) which is the same as that in the overlying Magog Group. Caldwell-type or Chain Lakes-type clasts with a pre-S_1 fabric indicate a pre-Acadian deformation (Taconic or older).

The Saint-Daniel Mélange-Magog Group contact is interpreted as a northwest-directed thrust (St-Julien, 1987; Cousineau, 1990). The Magog Group forms the base of the St-Victor Synclinorium (Labbé and St-Julien, 1989; Tremblay and St-Julien, 1990; Cousineau and Tremblay, 1993). It is characterized by open to tight folds, generally overturned to the northwest. Folds plunge gently or moderately to the southwest or northeast. Similar structures in overlying post-Ordovician rocks indicate that the deformation is Acadian. Taconic-related warping of Magog rocks was proposed previously (Robinson and Fyson, 1976; Cousineau, 1990) but the evidence is weak. Regional folds in the Magog Group are coeval with faults. In the Chaudière River area, Cousineau (1990) recognized two generations of faults. Older, Taconic faults are crosscut by Acadian thrusts related to the La Guadeloupe Fault. In southern Quebec, folds within the Magog Group are tighter and locally isoclinal (Tremblay, 1992b), especially near the contact with the Ascot Complex. A late, northwest-dipping crenulation cleavage is well-developed in this area.

The southeastern limit of the St-Victor Synclinorium is the La Guadeloupe Fault (Fig. 3.103; St-Julien et al., 1983). From the Quebec-Vermont border to Chaudière River, it was recognized as a northwest-directed high-angle reverse fault (Tremblay et al., 1989b; Labbé and St-Julien, 1989; Cousineau and Tremblay, 1993). The fault is correlative with the Rocky Mountain fault in western Maine (Roy, 1989), and with the Taconic Line in northern Vermont (Hatch, 1982). Cousineau and Tremblay (1993) recognized a dextral strike-slip component on the La Guadeloupe Fault northeast of Chaudière River. The La Guadeloupe and related faults have quartzofeldspathic and calc-silicate mylonites and shear zones in adjacent units on both sides (Tremblay et al., 1989b; Labbé and St-Julien, 1989; Cousineau, 1990).

The Ascot Complex is dominated by F_2 folds which are considered to be coeval with those of the St-Victor Synclinorium, as well as those of the Gaspé Belt. F_2 folds have a sheath-like geometry and are reoriented parallel to the stretching lineation along the La Guadeloupe Fault. D_2 structures are affected by southeast-verging, F_3 open folds. A northwest-dipping crenulation cleavage is axial-planar to F_3 folds. Intensity of D_3 deformation increases from northwest to southeast, and from northeast to southwest. These are late Acadian structures unrelated, and clearly post-dating the La Guadeloupe Fault and associated F_2 folds (Tremblay and St-Julien, 1990; Tremblay and Pinet, in press). Pre-D_2 folds and faults in the Ascot Complex were interpreted as accretionary-related pre-Acadian structures (Tremblay and St-Julien, 1990). Evidence for Taconic metamorphism and deformation in the Estrie-Beauce Subzone is rare. It has been stressed that the chaotic nature of the Saint-Daniel Mélange and correlative units must originate from Taconic-related accretionary processes (Cousineau, 1990; Tremblay and St-Julien, 1990; Tremblay, 1992a). North-south trending structural and

magnetic anomalies were attributed to Taconic sinistral strike-slip faulting (Gauthier et al., 1989; Tremblay and Malo, 1991). However, the structural analysis of the Estrie-Beauce Subzone and the adjacent Humber Zone in terms of Taconic versus Acadian structures is a current topic which will need more studies in the light of recent paleotectonic models and geochronological data from correlative or adjacent terranes of the Canadian and U.S. Appalachian regions (van Staal et al., 1990, Dunning et al., 1990a; Tucker and Robinson, 1990).

Mégantic Subzone

The Mégantic Subzone occurs along the southeastern flank of the Gaspé Belt (Fig. 3.101) in the easternmost part of the southern Quebec Appalachians. It is limited to the south and southeast by the Quebec-New Hampshire and Quebec-Maine boundaries. Most of its exposure lies in the New England Appalachians along the Boundary Mountains Anticlinorium (Williams, 1978a). In Quebec, its northwestern limit is the Victoria River Fault. The stratigraphy and structure of the Mégantic Subzone are controversial, especially with regard to the temporal and geometrical relations between the Clinton and Frontenac-Chartierville volcanic rocks (Fig. 3.102, see below). Originally, the Frontenac Formation was included in the Silurian-Devonian Saint-Francis Group (St-Julien and Hubert, 1975). Although its age is poorly constrained, the Frontenac Formation has been assigned to the oceanic domain of the Dunnage Zone (St-Julien and Slivitzky, 1985).

In the following descriptions, we follow the stratigraphy of Slivitzky and St-Julien (1987) and Boudette (1982). As shown in Figure 3.102, Chevé's (1990) interpretation is different. The Mégantic Subzone comprises three Ordovician-Silurian units: Clinton Formation, Chesham Mélange, and Frontenac Formation. These rocks are in fault contact with Precambrian rocks of the Chain Lakes Massif and post-Ordovician rocks of the Gaspé Belt. Locally, they are unconformably overlain by the Saint-Francis Group, and are crosscut by Middle Devonian felsic intrusions (Fig. 3.106).

Clinton Formation

The Clinton Formation lies in the Clinton River Syncline west of the Arnold River Formation (Marleau, 1959), a Quebec correlative to the Chain Lakes Massif (Chevé, 1978; Fig. 3.106). The Clinton River Syncline is interpreted as an anticline by Chevé (1990, see below). Hence, the western contact between the Clinton Formation and the Frontenac Formation is either interpreted as a fault (St-Julien and Slivitzky, 1985) or a gradational stratigraphic contact (Chevé, 1990).

The Clinton Formation consists of massive to foliated, interbedded mafic lavas and pyroclastic rocks overlain by an argillaceous epiclastic sequence topped by mafic, pillowed lavas. Basalts are plagioclase-bearing, porphyritic to glomeroporphyritic lavas. Associated pyroclastic rocks are mafic to intermediate cinder and lapilli tuff. Sedimentary rocks are metamorphic banded psammites. Fe-Mn-rich rocks occur on the top of the lava flows.

The Clinton Formation was originally included as a volcanic member within the Dixville Formation (Chevé, 1978). A Middle Ordovician age was based on a graptolite occurrence in adjacent Maine (Harwood and Berry, 1967). It was also correlated with the Ascot Complex (Slivitzky and St-Julien, 1987) and with the Jim Pond volcanics, the latter overlying the Boil Mountain ophiolite complex (Boudette, 1982) which is thrust over the Chain Lakes Massif. The latter correlation is favoured, following the interpretation of Boudette (1982). Thus, the base of the Clinton Formation should be a fault where the Boil Mountain ophiolite complex is absent. Correlation with the Jim Pond Formation also implies that the Clinton Formation is Cambrian to Early Ordovician (Aleinikoff and Moench, 1985). However, this contradicts the interpretation of Chevé (1990), who included Clinton volcanic rocks as an informal member within the Frontenac Formation.

Chesham Mélange

The Chesham Mélange is made up of graphitic schists, phyllites, wackes, conglomerates, and volcanic rocks. It lies in the core of the Clinton River Syncline and overlies the Clinton Formation (Bernard, 1987). It corresponds to the Megalloway Member (Chevé, 1978) of the Dixville Formation, and to the Chesham Formation of Chevé (1990).

Slivitzky and St-Julien (1987) described the Chesham as an ophiolitic mélange consisting of quartzite, volcanic rocks, serpentinite, gabbro blocks in a chaotic pebbly mudstone. According to Chevé (1978, 1990), the Chesham is made up of interbedded quartzitic metagreywacke, grey-green phyllite, red-brown argillite, conglomeratic black phyllite, pyroclastic rocks and serpentinite fault slivers. Chevé (1990) interpreted this unit as conformably underlying the Clinton volcanic rocks and its chaotic nature was attributed to tectonic disruption along the Clinton River fold hinge. Boudette (1982) correlated the Chesham with the Hurricane Mountain Mélange (Boone, 1985) that lies stratigraphically above the Jim Pond Formation in northwestern Maine.

The **Marble Mountain Complex** is a small body of quartz diorite and gabbro within the Chesham Mélange, at the contact with the underlying Clinton Formation (Fig. 3.106). It is spatially related to serpentinite slivers and all contacts are apparently faulted (Chevé, 1990). The Marble Mountain Complex is correlated with the East Inlet Complex, which occurs approximately 10 km southeast along the New Hampshire-Maine border. The East Inlet Complex has a U-Pb age of 430 Ma (Eisenberg, 1981; Lyons et al., 1986) and is correlated with the Attean Pluton (443 Ma; Boudette, 1982), which intrudes the Boil Mountain ophiolite complex.

The age of the Chesham Mélange is controversial. Marleau (1968) proposed an Early Devonian age and included it in the Frontenac Formation. Chevé (1978) assumed a Middle Ordovician age based on correlation with the Dixville Formation and regional relations. We follow Boudette (1982) and correlate the Chesham Mélange with the Hurricane Mountain Mélange, implying a Cambrian to Early Ordovician age.

Frontenac Formation

The Frontenac Formation is the main unit of the Mégantic Subzone and it occurs in the Frontenac Synclinorium (Chevé, 1990; Fig. 3.106). Its northwestern limit is the Victoria River Fault and it extends both northeasterly and southwesterly into Maine and New Hampshire.

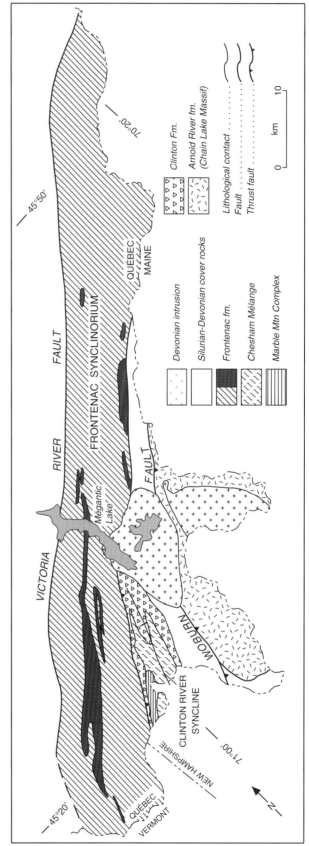

Figure 3.106. Geology of the Mégantic Subzone, modified from Chevé (1990).

The Frontenac Formation is made up of interbedded sedimentary and volcanic rocks. The volcanic rocks are the Frontenac-Chartierville volcanics of Chevé (1990). These are pillowed metabasalts structurally and texturally similar to volcanic rocks of the Clinton Formation. However, their transitional-alkalic composition is a distinctive feature. Sedimentary rocks are interbedded massive carbonaceous sandstones, greywackes and slates. Sills of gabbros, from a few metres to tens of metres thick, are found within the siliciclastic sequence.

Age assignments of the Frontenac Formation vary from Ordovician to Devonian. Its siliciclastic rocks are similar to the Compton Formation sandstones and slates belonging to the Silurian-Devonian cover sequence (Marleau, 1968; Bernard, 1987; Chevé, 1990). The Frontenac Formation has also been correlated with the Waits River Formation (Moench, 1984) that contains volcanic rocks in eastern Vermont (Hatch, 1988). Harron (1976) proposed an Early to Middle Ordovician age by correlation with the Ascot Complex. U-Pb dating of possible correlatives in New England (434-414 Ma; Lyons et al., 1986) indicates an Early to Late Silurian age. Chevé (1990) proposed that the Frontenac-Chartierville volcanics and adjacent sedimentary rocks are slightly younger than the Clinton River volcanics and as such are Late Ordovician to Early Silurian. Williams and St-Julien (1982) and Slivitzky and St-Julien (1987) attributed an Ordovician age to the Frontenac Formation based on its stratigraphic relations with the underlying Clinton Formation.

Structure and metamorphism

Regional metamorphism in the Mégantic Subzone is related to the Acadian Orogeny. It reaches upper greenschist facies (biotite zone). According to Boudette (1982), rocks of the Boil Mountain Complex, Jim Pond Formation and correlatives have been submitted to low grade (epidote-pumpellyite) pre-Taconic and Taconic metamorphism related to the tectonic juxtapositioning of the Boil Mountain ophiolite complex and Chain Lakes Massif. Lyons et al. (1982) recorded a major Acadian metamorphism and deformation in central New Hampshire and Maine that decreases westward.

Major structures characterizing the Mégantic Subzone are the Victoria River Fault, Frontenac Synclinorium, Clinton River Syncline, and Woburn Fault (Fig. 3.106). The Victoria River Fault marks the contact between the Frontenac Formation and the Gaspé Belt (Fig. 3.106). There are no clear kinematic indicators but deformation in adjacent units is strong. Folds verge northwest in the Frontenac Formation, whereas they verge southeast in the Gaspé Belt. This led Chevé (1990) to interpret the Victoria River Fault as northwest-directed and related to an antiformal culmination in the Frontenac Formation to the south. Based on seismic profiles and on the vergence of late folds, Bernard (1987) interpreted the Victoria River Fault as an out-of-sequence northwest-dipping high-angle reverse fault. The Woburn Fault is located west of the Chain Lakes Massif (Fig. 3.106). It is a 2 km-wide tectonic zone within the Arnold River Formation and adjacent units. It is interpreted as a reactivated sinistral strike-slip fault (Chevé, 1990) or as a northwest-directed thrust, like the Turner Pond fault in Maine (Osberg et al., 1985). There

is another thrust fault located between the Clinton River and Frontenac synclines which brings older rocks northwestward over younger rocks (Fig. 3.106).

The Clinton River Syncline is a southeast-overturned structure moderately plunging to the southwest. This interpretation is based on the regional map pattern, the southwest plunge of associated F_1 and F_2 folds, the inferred stratigraphic relation between the Clinton Formation with the Jim Pond volcanics and Hurricane Mountain Mélange, and the absence of associated antiformal-related reflectors in the seismic profile (Bernard, 1987). Chevé (1990) interpreted the structure as the northeastern continuation of the Second Lake Anticline (Harwood, 1969) in northern New Hampshire. The Frontenac Synclinorium to the northwest (Fig. 3.106) is more appropriately described as imbricate fault slices within the Frontenac Formation (Bernard, 1987).

Temiscouata Subzone

In the Estrie-Beauce Subzone, the Baie Verte-Brompton Line is characterized by a sharp gravity gradient and a strong magnetic signature (Williams and St-Julien, 1982). These geophysical anomalies can be traced northeastward across the Quebec-Maine border and into northern Gaspésie (Haworth et al., 1980a, b; Zietz et al., 1980a, b). There, the southern limit of the anomalies corresponds approximately to the southern limit of exposed Cambrian-Ordovician rocks, and to the Shickshock-Sud Fault in western Gaspésie (Fig. 3.101). Thus, the southern limit of the geophysical anomalies is believed to represent the trace of the Baie Verte-Brompton Line in northern Gaspésie (Malo et al., 1992) and the line must be present therefore in northern Maine and the Témiscouata region.

The Estrie-Beauce and Témiscouata subzones are separated by northern Maine where rocks typical of the Dunnage Zone were mapped by Osberg et al. (1985). The Saint-Daniel Mélange is also present in Maine where it is in fault contact with the Depot Mountain Formation, a Magog Group correlative (Roy, 1989).

In the Témiscouata area (Fig. 3.101), Cambrian-Ordovician rocks outcrop in the Trinité Inlier that is surrounded by Silurian-Devonian cover rocks of the Gaspé Belt (Fig. 3.107). Goutier (1989a) mapped the central part of the Trinité Inlier and interpreted its rocks as part of the Humber Zone. The Lac-des-Aigles Fault (Fig. 3.107), a northeasterly trending fault that cuts the inlier (Goutier, 1989a, b), is interpreted as the trace of the Baie Verte-Brompton Line (Malo et al., 1992). According to Goutier (1989a, b), this fault separates two distinct domains; rocks typical of the Québec Supergroup to the northwest, and mélange-type rocks to the southeast. To the northwest, Silurian-Devonian strata rest unconformably on rocks of the Québec Supergroup or are in thrust contact with older rocks (Goutier, 1989a). Southeast of the Lac-des-Aigles Fault, mélange-type rocks of the Trinité Group are unconformably overlain by the Middle Ordovician to Lower Silurian Cabano Formation (David and Gariépy, 1990). The Cabano Formation is conformably overlain by the Llandovery Pointe-aux-Trembles and Lac Raymond formations which are in turn conformably overlain by the Silurian-Devonian cover sequence of the Gaspé Belt. Mélange-type rocks of the Trinité Group are correlated with the Saint-Daniel Mélange of the Estrie-Beauce Subzone and the Cabano Formation is correlated with the Depot Mountain Formation in Maine (Roy, 1989) and with the Magog Group in the Eastern Townships of Quebec (Cousineau, 1990). The Pointe-aux-Trembles and Lac Raymond formations were interpreted as arc-related volcanic and sedimentary suites (David and Gariépy, 1990), much like the Ascot Complex and the Magog Group in the Eastern Townships. Based on correlations with similar rocks of the Estrie-Beauce Subzone, the package Cabano - Pointe-aux-Trembles - Lac Raymond is therefore included in the Témiscouata Subzone of the Dunnage Zone although a different view is expressed by Bourque et al. (Chapter 4).

Mélange units

Goutier (1989a) divided the Trinité Group into two units that contain mélanges with large blocks of lithic wacke. One unit is mainly brownish black to green claystone,

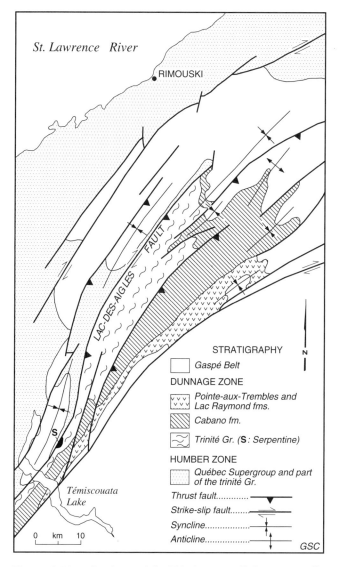

Figure 3.107. Geology of the Témiscouata Subzone, modified from David et al. (1985), Gouthier (1989b), and Bourque et al., 1993.

minor grey-blue slate, quartzo-dolomitic siltstone, and sandstone, with grey-blue calcareous mudstone, calcareous siltstone, sandstone, and massive calcisiltite at its base (Goutier, 1989a). The other unit is a brown laminated mudstone with quartzo-dolomitic siltstone and sandstone, calcirudite, calcisiltite, quartzitic wacke and black to green claystone. Mélanges consist of blocks of lithic wacke and various calcareous rocks of both units in a black and green shale matrix. The mélanges are bounded by faults and they may represent thrust-related broken formations. Large tectonic blocks of lithic wacke occur south of the Lac-des-Aigles Fault. Dark green to black serpentinite bodies also occur along the fault (Fig. 3.107; Lespérance and Greiner, 1969).

South of the Baie Verte-Brompton Line, the Trinité Group is lithologically similar to the Saint-Daniel Mélange. Claystones of the Trinité Group were deposited in a deep marine environment and they incorporate calcareous rocks derived from a Cambrian shelf (Goutier, 1989a). The lower calcareous unit of the Trinité Group contains a *Salterella* fauna of Early Cambrian age and overlying rocks contain Middle Cambrian trilobites (Goutier, 1989a). An upper age limit for the Trinité Group is given by Early Ordovician graptolites found in a block of lithic wacke.

Sedimentary rocks

The **Cabano Formation** (Lespérance and Greiner, 1969; Lajoie et al., 1968; David et al., 1985) consists of dark grey mudstone and siltstone interbedded with lithic wackes and conglomerates. Cherty horizons and black shales are present in the lower part of the formation. Sedimentary structures typical of distal turbidites are common (David et al., 1985). Clasts in fragmental rocks are sedimentary in origin, probably derived from a recycled orogen (Dickinson and Suczek, 1979). Paleocurrent analysis indicates a west to northwest source, but a southern source is not excluded (David et al., 1985). The thickness of the Cabano Formation is estimated at 1000-2500 m (David et al., 1985).

Cherts of the Cabano Formation rest unconformably on the Trinité Group. The upper Cabano units are conformably overlain by the Pointe-aux-Trembles and Lac Raymond formations. Black shales of the Cabano Formation contain graptolites of the *Nemagraptus gracilis* Zone of the Llandeilo-Caradoc boundary (Goutier, 1989a). Other graptolites and brachiopods indicate that its age ranges from Middle Ordovician (Llandeilo) to Early Silurian (Llandovery C3; David et al., 1985).

The **Pointe-aux-Trembles** and **Lac Raymond formations** (Lajoie et al., 1968; David et al., 1985) are laterally equivalent. Their ages are estimated as late Llandovery (C2-3 to C5; David and Gariépy, 1990). The Pointes-aux-Trembles Formation is a volcaniclastic unit consisting of fine- to coarse-grained tuffaceous rocks in its lower part, and is made up of agglomerate, lithic and crystal tuff, vitric tuff, and massive lava flows in its upper part (David and Gariépy, 1990). The Lac Raymond Formation is composed of mudstones and tuff interbedded with siltstone, sandstone and conglomerate. Gabbroic sills are found in this sequence. Primary structures in both formations indicate deposition by turbidity currents (David et al., 1985). Chemical compositions of volcanic material found in both formations are characteristic of orogenic andesites (David and Gariépy, 1990; see below). The thickness of the Pointe-aux-Trembles

189

Formation varies from 300-900 m, whereas the Lac Raymond Formation is estimated at a maximum thickness of 1800 m (Bourque et al., 1993).

Structure and metamorphism

Major structural features of the Témiscouata Subzone are east-northeast striking, dextral strike-slip faults and high-angle, west-northwest verging, reverse faults parallel to north-northeasterly-trending folds. These folds have an axial-planar cleavage parallel to folds of the Gaspé Belt, and are thus Acadian-related (Bourque et al., 1993; David and Gariepy, 1990). The Trinité Group records an earlier phase of deformation and the formation of mélanges (Goutier, 1989a). Since these events did not affect the Cabano and overlying formations, they are pre-Middle Ordovician and thus attributed to the Taconic Orogeny.

East-northeast striking faults, like the Lac-des-Aigles Fault, are Acadian dextral strike-slip faults which may be reactivated Taconic thrusts (Goutier, 1989a; Bourque et al., 1993). They control Acadian deformation which is related to a dextral transpressive regime (Lebel, 1985; Goutier, 1989a; Bourque et al., 1993) like that of Gaspésie (Malo and Béland, 1989).

Volcanic rocks of the Pointe-aux-Trembles and Lac Raymond formations were metamorphosed to prehnite-pumpellyite facies (David and Gariépy, 1990). There are no data on the metamorphic grade of the Trinité Group.

Gaspésie Subzone

Gaspésie forms the northernmost part of the Québec Reentrant (Williams, 1978a; Fig. 3.101). Most rocks are Silurian-Devonian cover rocks of the Gaspé Belt (Bourque et al., Chapter 4, this volume). Cambrian-Ordovician rocks are located mostly in the north, but those of the Maquereau-Mictaw and Murphy Creek inliers occur to the south (Fig. 3.108). The Baie Verte-Brompton Line, extrapolated beneath Silurian-Devonian cover rocks (Williams and St-Julien, 1982), is now better delineated using magnetic and gravity data as well as the distribution of the Cambrian-Ordovician rocks, particulary the mélanges and serpentinite bodies (Malo et al., 1992, Fig. 3.108). The Shickshock-Sud Fault is believed to be the Baie Verte-Brompton Line in the northwest (Fig. 3.108). To the east, the Baie Verte-Brompton Line follows the southern limit of the gravity high (Haworth et al., 1980a, b) and it is marked by magnetic anomalies (Zietz et al., 1980a, b). This trace corresponds broadly to the Bras Nord-Ouest Fault (Fig. 3.108). Southeastward, the Baie Verte-Brompton Line is interrupted by the Grand Rivière and Grand Pabos faults and reappears in the Maquereau-Mictaw inlier, as originally recognized (Williams and St-Julien, 1982). In the Percé area, south of the Grande Rivière Fault and north of the eastern extension of the Grand Pabos Fault, the Baie Verte-Brompton Line has been traced southwest of the Murphy Creek Inlier. This inlier contains the Cambrian Murphy Creek and Corner-of-the-Beach formations that are typical of the Humber Zone north of the Baie Verte-Brompton Line (Kirkwood, 1989).

Mélange units

Mélange units and serpentinite bodies mapped recently in Gaspésie mark the Baie Verte-Brompton Line as in the Estrie-Beauce Subzone. From northwest to southeast, the La Rédemption serpentinite, Ruisseau Isabelle mélange, Lady Step Complex, Rivière Port-Daniel mélange, Nadeau ophiolitic mélange, and McCrea mélange are found along the Baie Verte-Brompton Line (Fig. 3.108). The La Rédemption serpentinite is equated with other lensoid ultramafic bodies found along the Baie Verte-Brompton Line in the northeastern Estrie-Beauce Subzone (Cousineau, 1990). Ruisseau Isabelle, Rivière Port-Daniel, and McCrea mélanges are olistostromal mélanges (Rast and Horton, 1989) correlated with the Saint-Daniel Mélange of the Estrie-Beauce Subzone (Cousineau, 1990). The Nadeau ophiolitic mélange and Lady Step Complex are considered to be equivalent to the Rivière des Plante ophiolitic mélange (Cousineau, 1991; de Broucker, 1987).

The **La Rédemption serpentinite** is a fault-bounded block approximately 4 km by 1 km (Fig.3.108) that occurs along the western part of the Shickshock-Sud Fault. It is composed of serpentinized ultramafic rocks along with basic volcanic rocks. To the north, the La Rédemption serpentinite is in fault contact with the Lower Silurian Chaleurs Group, and to the south, with the Lower Devonian Cap Bon Ami Formation of the Upper Gaspé Limestones (Bourque et al., 1993).

The **Ruisseau Isabelle mélange** (Fig. 3.108) is located along the eastern part of the Shickshock-Sud Fault. It contains graphitic slate, dolomitized limestone, serpentinite and carbonatized serpentinite in a pebbly mudstone matrix (St-Julien et al., 1990). To the north, it is in fault contact with volcanic rocks of the Cambrian Shickshock Group, and to the south with shaly limestones of the Chaleurs Group.

The Mont Serpentine inlier is a small hill approximately 7 km by 1 km, 25 km northeast of the village of Gaspé (Fig. 3.108). Béland (1980) described Cambrian-Ordovician rocks of this inlier as a tectonic breccia similar to mélange units of Taconic allochthons. McGerrigle (1950) used the name Lady Step series for the volcanic rocks within the inlier. Berger and Ramsay (1990) changed the name to the **Lady Step Complex**. It comprises serpentinized ultramafic rocks, meta-arkoses and metagreywackes, metabasalts, amphibolitized gabbros, pyroxenites, granitic blocks and gneiss, chlorite schists, and other metamorphic schists (Berger and Ramsay, 1990; de Broucker, 1987; Béland, 1980). The matrix is strongly sheared and most fragments are brecciated. De Broucker (1987) interpreted the Lady Step breccia as ophiolitic mélange (Gansser, 1974) and correlated the mélange with the Nadeau ophiolitic mélange (see below). Some high grade metamorphic blocks are thought to be correlatives of the Chain Lakes Massif of Maine (Berger and Ramsay, 1990; de Broucker, 1987). To the north, the Bras Nord-Ouest Fault separates the Lady Step Complex from the Lower Devonian Gaspé Sandstones, whereas to the south, the mélange is unconformably overlain by Upper Silurian units of the Chaleurs Group, or it is in fault contact with the Upper Gaspé Limestones (Berger and Ramsay, 1990).

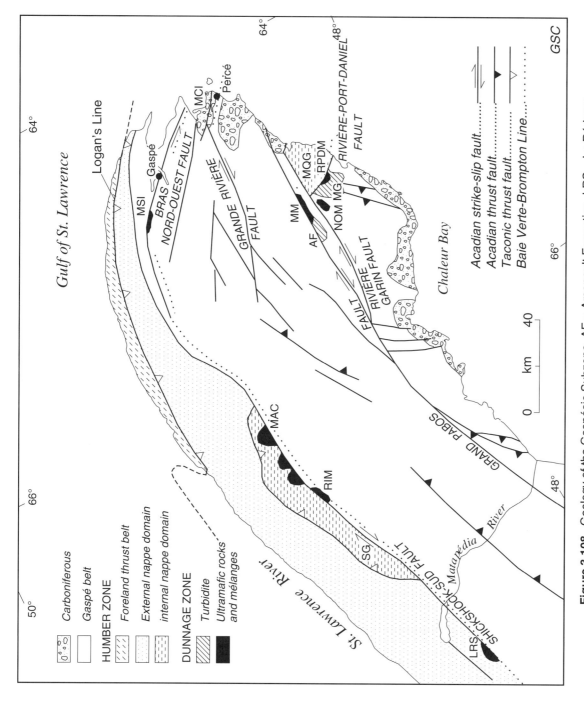

Figure 3.108. Geology of the Gaspésie Subzone. AF = Arsenault Formation, LRS = La Rédemption serpentinite, MAC = Mont Albert Complex, MCI = Murphy Creek Inlier, MM = McCrea mélange, MG = Mictaw Group, MQG = Maquereau Group, MSI = Mont Serpentine Inlier, NOM = Nadeau ophiolitic mélange, RIM = Ruisseau Isabelle mélange, RPDM = Rivière Port-Daniel mélange, SG = Shickshock Group.

The **Rivière Port-Daniel mélange** is an olistostromal mélange made up of igneous and sedimentary blocks in a shaly matrix (de Broucker, 1987). It corresponds to the Port-Daniel-River complex of Ayrton (1967). Blocks are composed of serpentinite, basic volcanic rocks, cherts, granitoids, metagreywackes, silty limestones, dolomites, red mudstones, and rare metasandstones (de Broucker, 1987; Williams and St-Julien, 1982). In the central and southern parts of the mélange, the matrix varies from green to black mudshale to black pebbly mudstone exhibiting phacoidal cleavage. In the northern part, the matrix is a polygenic conglomerate with radiolarites, cherts, and sandstones with various types of blocks (de Broucker, 1987). The mélange is intruded by a large diabase sill near its upper

contact with the Neckwick Formation. The Rivière Port-Daniel mélange is bordered to the east by the Port-Daniel Fault and farther east by the Cambrian Maquereau Group of the Humber Zone (Williams and St-Julien, 1982; de Broucker, 1987). To the west, it is unconformably overlain by the Llanvirn Neckwick Formation of the Mictaw Group, thus indicating a pre-Middle or Early Ordovician age for the Rivière Port-Daniel mélange (Fig. 3.109).

The **Nadeau ophiolitic mélange** (de Broucker, 1987) is the serpentinite mélange inlier of Williams and St-Julien (1982) and the Weir Townships serpentinite of Ayrton (1967). It is a large lensoid body 5 km by 0.6 km to the northwest of the Mictaw Group (Fig. 3.108). It contains huge blocks of serpentinized peridotites, amphibolites, tuffaceous

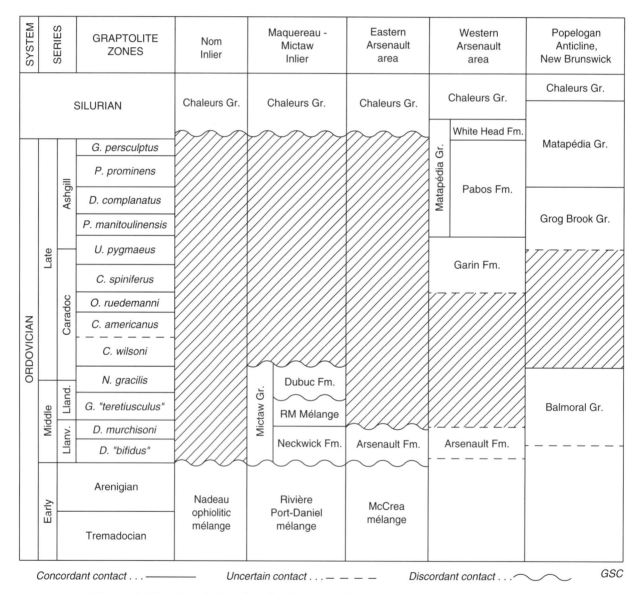

Figure 3.109. Correlation chart for Gaspésie Subzone units of southern Quebec and northern New Brunswick. NOM Inlier = Nadeau ophiolitic Mélange Inlier, RM Mélange = Rivière du Milieu Mélange, Lland = Llandeilo, Llanv = Llanvirn, modified from Malo and Bourque (1993), van Staal and Fyffe (this Chapter).

metagreywackes, quartzites, granites, and granodiorites in a matrix of schistose serpentinite (Williams and St-Julien, 1982; de Broucker, 1987). Tuffaceous metagreywackes and amphibolites were correlated with the Chain Lakes Massif (de Broucker, 1987). U-Pb zircon dating of metagreywackes indicates derivation from source rocks varying between 1812 to 1395 Ma (de Broucker et al., 1986) which is compatible with the age of the Chain Lakes Massif (1600 Ma; Naylor et al., 1973). The Nadeau ophiolitic mélange is correlated with the ophiolitic mélange of the Mont Serpentinite Inlier and with the Rivière des Plante ophiolitic mélange in the Estrie-Beauce Subzone (Cousineau, 1991; de Broucker, 1987). The Nadeau ophiolitic mélange is surrounded by Silurian units of the Chaleurs Group. The contact is either a fault or an unconformity. The mélange is believed to be pre-Middle Ordovician like the Rivière Port-Daniel mélange (de Broucker, 1987; Fig. 3.109).

The **McCrea mélange** (Malo and Moritz, 1991) is located along the Grand Pabos Fault, northwest of the Maquereau-Mictaw Inlier. It occurs on strike with the Rivière Port-Daniel mélange and Nadeau ophiolitic mélange (Fig. 3.108 and 3.110). It is 15 km by 1 km. The McCrea mélange is an olistostromal mélange containing blocks of granitic rocks, basic to intermediate volcanic rocks, red and green slates, sandstones, calcareous siltstones, dolomitized serpentinites and serpentinized peridotites (Malo and Moritz, 1991). One of the silty limestone blocks has Late Cambrian conodonts (Nowlan, 1988b). The matrix is a black pebbly mudstone. The mélange is intruded by a large gabbroic sill chemically similar to normal mid-ocean ridge basalts (N-MORB; Bédard, 1986). The McCrea mélange is limited to the north by the Grand Pabos Fault which marks its contact with the Upper Ordovician Garin Formation and the Upper Ordovician to Lower Silurian Matapédia Group. To the south, the mélange is bounded by an easterly trending fault which marks the contact between Lower Silurian beds of the Chaleurs Group and Middle Ordovician greywackes of the Arsenault Formation (Fig. 3.110). The contact between the mélange and the Llanvirn Arsenault Formation is interpreted as an unconformity (Fig. 3.110). The mélange is therefore considered to be Early Ordovician (Fig. 3.109). The McCrea mélange is believed to be correlative with the Rivière Port-Daniel mélange because both are lithologically similar and overlain by Llanvirn flysch.

Klippen of ophiolitic rocks within the Humber Zone are other displaced Dunnage Zone segments. The Mont Albert Complex is the largest example (Beaudin, 1980).

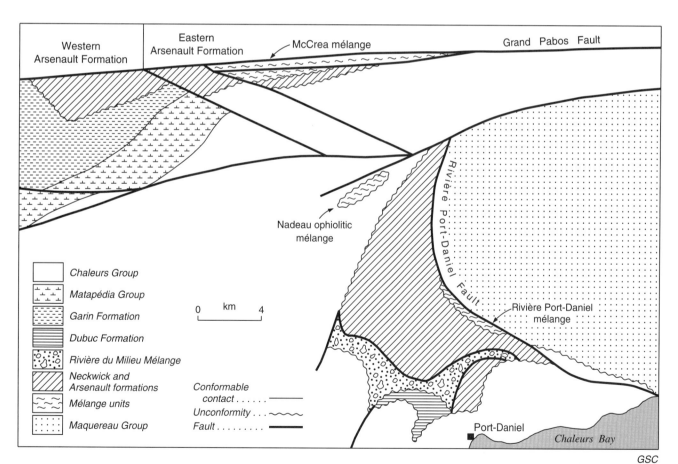

Figure 3.110. Geology of the Gaspésie Subzone in southern Gaspésie. Modified from Malo and Bourque (1993).

GSC

Mictaw Group and correlatives

The Mictaw Group (Badgley, 1956) is composed of three stratigraphic units: Neckwick Formation, a Llanvirn flysch sequence; Rivière du Milieu Mélange, an olistostromal mélange; and the Llandeilo to Caradoc turbidites of the Dubuc Formation (de Broucker, 1987). The Arsenault Formation is also a Llanvirn flysch sequence correlative with the Neckwick Formation (Malo, 1988; de Broucker, 1987).

The **Neckwick Formation** is composed of sandstones and shales with minor interbeds of felsic tuff and microconglomerate (de Broucker, 1987). Sandstones are greenish grey lithic wacke, medium- to coarse-grained, in beds of 10 cm-10 m thick. Most of the beds are massive, but graded bedding, loadcasts and flutecasts occur locally. Lithic fragments are mostly volcanic with minor sedimentary, plutonic, metamorphic and ophiolite-derived rock fragments. Shales are green to dark grey. Microconglomeratic beds are 50-100 cm thick and contain lithic fragments with quartz and feldspathic clasts. Tuffs are yellowish grey, fine grained, and 1-5 m bed thickness. These are acidic tuff with a high SiO_2 content (70 to 80%, de Broucker, 1987). Diabase sills, found in the lower part of the formation, have a tholeiitic affinity (Low-K tholeiite) like the diabase sill found in the underlying Rivière Port-Daniel mélange (see above). A petrographic provenance study (de Broucker, 1987) of sandstones and microconglomerates indicates that they were derived from a recycled orogen and an orogenic arc (Dickinson and Suczeck, 1979). Paleocurrent analysis suggests a southern source (de Broucker, 1987). The thickness of the Neckwick Formation ranges from 1500-2000 m (de Broucker, 1987). The Neckwick Formation rests unconformably above the Rivière Port-Daniel mélange and brecciated rocks of the Maquereau Group. Its upper contact with the Rivière du Milieu Mélange is faulted (Fig. 3.110). The Silurian Chaleurs Group unconformably overlies the Neckwick in the southeast (Fig. 3.110). The age of the Neckwick Formation is defined by graptolites (*Pseudoclimacograptus angulatus*) characteristic of the Llanvirn (de Broucker, 1987).

The **Rivière du Milieu Mélange** has a green to black mudshale matrix and blocks of calcareous to dolomitic siltstones and sandstones. Sandstone blocks are argillaceous lithic wackes similar to the greywackes of the Neckwick Formation. The contact between the Rivière du Milieu Mélange and Neckwick Formation is interpreted as a thrust fault. An unconformity is present between the mélange and the overlying Dubuc Formation (de Broucker, 1987). The age of the Rivière du Milieu Mélange is constrained by the fauna found in its sedimentary blocks, and by the age of the overlying Dubuc Formation. One block of Neckwick-type sandstone contains Llanvirn to Llandeilo graptolites, and some calcareous siltstone blocks yield Arenig conodonts (de Broucker, 1987). The overlying Dubuc Formation is Llandeilo to Caradoc, thus the age of the Rivière du Milieu Mélange is early Llandeilo.

The **Dubuc Formation** has two facies: a lower shale sequence and an upper, sandstone-rich turbidite sequence (de Broucker, 1987). Black graptolitic shales of the lower Dubuc Formation contain a few beds of sandstone and conglomerate as well as calcareous concretions aligned within bedding planes. Conglomeratic beds are polygenic

and contain rock fragments from the underlying Rivière du Milieu Mélange. The upper turbidite is a typical rhythmic sequence of sandstone and shale, with few conglomeratic beds. Sandstones are grey, 10-100 cm thick, fine- to coarse-grained, and are composed mainly of quartz with lesser lithic fragments and feldspars. Sedimentary and volcanic clasts are rare, whereas metamorphic clasts are abundant. Conglomerates are composed of sedimentary and metamorphic rock fragments, including greenish grey lithic wackes, calcareous to dolomitic siltstones, green and black mudshales, metasandstones, milky quartz, metamorphic schists, and quartzitic mylonites. Petrographic analysis (de Broucker, 1987) of sandstones and microconglomerates indicates that they were derived from a recycled orogen (Dickinson and Suczeck, 1979) of high grade metamorphic rocks. The minimum thickness of the Dubuc is estimated at 700 m (de Broucker, 1987). The Dubuc Formation is unconformably overlain by Lower Silurian rocks of the Chaleurs Group (Fig. 3.109, 3.110). Graptolites and conodonts found in the Dubuc Formation indicate a Llandeilo age (de Broucker, 1987).

The **Arsenault Formation** was originally described as the lower formation of the Honorat Group (Malo, 1988), and a correlative of the Neckwick Formation of the Mictaw Group (Malo, 1986). It is composed mainly of thick bedded, greenish grey, lithic wackes with black to olive-green claystones, minor grey to black fine grained tuff interbeds and containing gabbroic sills. Petrographically, the sandstones have the same composition as those of the Neckwick Formation. They are made up of volcanic rock fragments with minor sedimentary, plutonic, metamorphic, and ophiolite-derived rock fragments. A petrographic provenance study (Malo, 1986) indicates that they were derived from a recycled orogen and an orogenic arc (Dickinson and Suczeck, 1979), like the provenance of the Neckwick Formation. Moreover, tuff interbeds have a chemical composition similar to felsic tuff of the Neckwick Formation, and the gabbroic sills in the Arsenault Formation are also chemically similar to those occurring at the base of the Neckwick Formation (de Broucker, 1987). The minimum thickness of the Arsenault Formation is estimated at 600 m (Malo, 1988). A graptolite fauna from Arsenault greywackes belongs either to *Archiclimacograptus angulatus* (Bulman) or "*Climacograptus*" *confertus* (Lapworth), species characteristic of the lower Middle Ordovician of Scandinavia and Wales, respectively (Riva and Malo, 1988). Similar graptolites were found in the Neckwick Formation (de Broucker, 1987). In the eastern Arsenault area (Fig. 3.110), the formation is believed to overlie unconformably the McCrea mélange, but a faulted contact is not ruled out. At this locality, the Arsenault is in turn unconformably overlain by the Chaleurs Group. These relations are identical to those 15 km to the southeast where the Neckwick Formation rests unconformably on the Rivière Port-Daniel mélange and is in turn unconformably overlain by the Chaleurs Group (Fig. 3.110). The Rivière Port-Daniel mélange-Neckwick Formation sequence is thus considered to be correlative to the McCrea mélange-Arsenault Formation sequence.

The western part of the Aroostook-Percé Anticlinorium is the key area to understand the relationship between rocks of the Dunnage Zone and those of the Gaspé Belt

(Fig. 3.109). The Arsenault Formation is there in fault contact with the Lower Silurian Matapédia Group to the north, and overlain by the Caradoc to Asghill Garin Formation to the south. The Garin Formation is in turn conformably overlain by the Matapédia Group (Pabos and White Head formations) and Chaleurs Group (Fig. 3.110). The contact between the Arsenault and the Garin is structurally concordant but unexposed and there is a time hiatus between the two formations, the age of the Arsenault is Llanvirn and the oldest graptolite fauna recovered from the Garin is late Caradoc (Riva and Malo, 1988). The Llandeilo and much of the Caradoc is therefore missing (Fig. 3.109).

There are two possible explanations. First, the hiatus between the Arsenault and Garin formations in the western Arsenault area corresponds to the unconformity observed in the eastern Arsenault area (see above), in which case the Arsenault Formation should be removed from the Honorat Group and assigned to the Mictaw Group. The Garin Formation would then represent the basal lithostratigraphic unit of the Silurian-Devonian cover sequence of Gaspé Belt. Second, because the contact between the Garin and Arsenault formations is not exposed, the possibility of a conformable contact and of no time gap between the two cannot be ruled out. The graptolite faunas of the *C. spiniferus* Zone recovered in the Garin Formation, in the Chandler and New Richmond areas (Riva and Malo, 1988), come from its upper part. Therefore, the strata in the Garin under the *C. spiniferus* Zone could represent the interval between the *C. spiniferus* Zone and the Llanvirn graptolite zone of the Arsenault Formation. In this case, Dunnage Zone rocks and their cover would be a continuous sequence, as it is in the Témiscouata Subzone (see above). The Garin is lithologically similar and partly equivalent to the Cabano and Saint-Victor formations of the Estrie-Beauce and Témiscouata subzones, respectively. Distinguishing rocks of the Dunnage Zone and its cover sequence (Gaspé Belt) is therefore enigmatic in conformable stratigraphic sections.

In northern New Brunswick, Middle Ordovician volcanic and sedimentary rocks of the Dunnage Zone form the basement of the Gaspé Belt (van Staal and Fyffe, this chapter). In the Popelogan Subzone, a hiatus, similar to the one between the Arsenault and Garin formations, exists between the Middle Ordovician Balmoral Group and the Upper Ordovician Grog Brook Group (van Staal and Fyffe, this chapter). The Grog Brook Group, equivalent to the Garin Formation (Malo, 1988), is the basal unit of the Aroostook-Percé sequence, whereas the Balmoral Group is part of the Popelogan Subzone of the Dunnage Zone (Van Staal and Fyffe, this chapter). The Grog Brook is considered to be the basal unit of the cover sequence. The Grog Brook is conformably overlain by the Matapédia and Chaleurs groups (Bourque et al., Chapter 4, this volume) like the Garin Formation in Gaspésie. Therefore, according to regional correlations of equivalent rocks from northern New Brunswick and Gaspésie, we believe that a hiatus must exist between the Arsenault and Garin formations in the western Arsenault area, and we thus include the Garin Formation in the cover sequence.

Structure and metamorphism

The most significant structural feature of the Gaspésie Subzone is its boundary with the Humber Zone, the Baie Verte-Brompton Line. In Gaspésie, the line was reactivated by Middle Devonian Acadian strike-slip faults (Malo et al., 1992) and consequently shows a complex structural history. The history of the Shickshock-Sud Fault was interpreted as a succession of normal and reverse movements during the Taconic and Acadian orogenies (Beaudin, 1980). Berger (1985) and Lebel (1985) have shown that the fault had an Acadian dextral strike-slip component. Along the Shickshock-Sud Fault, at the Ruisseau Isabelle mélange (Fig. 3.108), there is a vertical stretching lineation in the volcanic rocks of the Cambrian Shickshock Group on its northern side, but no stretching lineations in Silurian-Devonian rocks on its southern side. However, vertical C-S fabrics indicate horizontal dextral strike-slip movement on both sides of the fault (St-Julien et al., 1990). The stretching lineation on its northern side (Humber Zone) is attributed to a Taconic thrust movement whereas superimposed C-S fabrics are associated with an Acadian strike-slip movement. Dextral C-S fabrics are also found in sheared ultramafic rocks of the La Rédemption serpentinite (Bourque et al., 1993). In northeastern Gaspésie, the Bras Nord-Ouest Fault (Fig. 3.108) is an Acadian dextral-slip fault (Béland, 1980), and is believed to have been active during sedimentation of the Gaspé Belt (Malo and Bourque, 1993) as well as during the Taconic Orogeny (Béland, 1980; Berger and Ramsay, 1990). In southeastern Gaspésie, the Rivière Port-Daniel Fault (Fig. 3.108) is thought to have four periods of movement from Cambrian to Late Devonian (de Broucker, 1987). Pre-Silurian movements along these faults are believed to be related to the accretion of the Dunnage Zone against the Humber Zone during the Taconic Orogeny.

The main deformational effects in the Mictaw Group are northeast-trending folds and associated penetrative cleavage that are also present in rocks of the Gaspé Belt, and therefore related to the Middle Devonian Acadian deformation. However, northwesterly trending pre-Acadian folds are also recognized in the Mictaw Group (de Broucker, 1987). The angular unconformity between the Silurian strata of the Chaleurs Group and the northwesterly folded Middle Ordovician Mictaw Group implies folding during Middle to Late Ordovician. De Broucker (1987) interpreted this deformation as a late Taconic event, probably contemporaneous with the final nappe emplacement in the Humber Zone. The formation of the Rivière du Milieu Mélange and the large northwest-trending fault at the base of the mélange are also related to this tectonic event (de Broucker, 1987).

In the McCrea mélange and Arsenault Formation, the penetrative cleavage trends east-northeast and is associated with the regional Middle Devonian Acadian deformation of the Gaspé Belt (Malo and Moritz, 1991). Fabrics found in the mélange along the Grand Pabos Fault are all indicative of Acadian dextral strike-slip (Malo, 1987; Kirkwood and Malo, 1991). In the McCrea-Arsenault area, Acadian deformation related to the Grand Pabos Fault has obliterated Taconic or earlier structures, except for the phacoidal cleavage in the mélange matrix that is attributed to Ordovician tectonism.

Metamorphic minerals in gabbroic intrusions found in the McCrea mélange indicate lower greenschist facies of regional metamorphism (Bédard, 1986).

Geochemistry of plutonic and volcanic rocks

During the last decade, a number of geochemical studies were conducted in the plutonic and volcanic sequences of the Quebec Dunnage Zone. Most of these concern the geochemical and petrological evolution of the ophiolite suites. The geochemistry of the Thetford-Mines and Orford complexes was documented by Rodrigue (1979), Harnois (1982), Trottier (1982), Oshin and Crocket (1986), Laurent and Hébert (1989), Hébert and Laurent (1989), Laurent et al. (1989), Harnois and Morency (1989), and Harnois et al. (1990). The ophiolitic plutonic and volcanic rocks are interpreted as cogenetic sequences related to the partial fusion of upper mantle sources, and associated fractional crystallization of mafic liquids (Harnois, 1982; Trottier, 1982; Hébert and Laurent, 1989; Laurent and Hébert, 1989). In the Estrie-Beauce Subzone, three distinct groups of extrusive and intrusive ophiolitic rocks were recognized (Laurent and Hébert, 1989). The first two groups are interlayered boninites and arc tholeiites of the Thetford-Mines and Asbestos complexes (Trottier, 1982; Laurent and Hébert, 1989; Laurent et al., 1989). A third class of alkaline basalts is restricted to the sheeted sills of the Orford complex. Laurent and Hébert (1989) proposed forearc rifting in a suprasubduction zone environment and associated immature arc volcanism to account for the petrology of the Thetford-Mines and Asbestos complexes. Their results indicate a within-plate affinity for the alkaline basalts of the Orford complex which they interpreted as the back-arc section of the same Ordovician island-arc system. In the Gaspésie Subzone, the Mont Albert Complex has not been as extensively studied as other Quebec ophiolite complexes. Geochemical studies focused on the metamorphic petrology of the complex. The Mont Albert Complex was first interpreted as a large intrusive plug (McGregor and Basu, 1979) and, more recently as an obducted slab of depleted, oceanic upper mantle (Beaudin, 1984; Trzciensky, 1988).

The Bolton lavas of the Estrie-Beauce Subzone are normal to transitional MORB (Laurent et al., 1989). Compared to the Thetford-Mines island-arc rocks, they are characterized by moderate to strong enrichments in high field strength elements, and by moderate to low concentrations of compatible elements. The Bolton lavas are also different from the Orford ophiolite diabases which are much more enriched in LREE and other incompatible elements. Laurent et al. (1989) interpreted the Bolton lavas as the product of plume-type spreading ridge segments in a back-arc setting. The Bunker Hill volcanic sequence is geochemically distinct from the Ascot Complex (Blais et al., 1990; Tremblay, 1992a). It is characterized by more LREE-enriched felsic volcanic rocks and mafic intrusions. Preliminary results suggest a within-plate affinity. The exact petrological and tectonic significance of these geochemical trends is unclear.

Based on its stratigraphic and structural position, the Ascot Complex was attributed to subduction-related volcanism (St-Julien and Hubert, 1975). Hynes (1980) considered the Ascot volcanic rocks as probably originating from arc volcanism but stressed that secondary remobilization precludes any significant interpretation based on geochemistry. This was reviewed by Tremblay et al. (1989b), who showed that the Ascot Complex is characterized by two distinctive geochemical groups of lavas. One group has coexisting boninitic and tholeiitic lavas and associated LREE-depleted felsic volcanic rocks (the Sherbrooke domain). The other group (the Stokes domain) is characterized by LREE-enriched felsic volcanic rocks and associated tholeiitic basalts. The former group is interpreted as the remnant of an intraoceanic volcanic arc, and the latter as arc volcanic rocks deposited upon a sialic basement (Tremblay et al., 1989a; 1994). Ascot Complex correlatives are geochemically similar to Stokes domain volcanic rocks (Labbé, 1991). Ophiolitic plutonic and volcanic rocks from the Weedon area have a geochemical signature similar to island arc tholeiites. They may represent the basement of an early stage of island arc formation, subsequently intruded and covered by more mature (sialic?) volcanic products (Labbé and Hébert, 1988). In the Témiscouata Subzone, arc-related volcanic rocks are found within the Pointe-aux-Trembles and Lac Raymond formations. David and Gariépy (1990) have emphasized the calc-alkaline geochemical affinity of these volcanic rocks along with their similarities to recent orogenic andesites. As these rocks are Early Silurian, subduction processes were active locally at least until this time.

In the Mégantic Subzone, the Clinton and Frontenac-Chartierville volcanic rocks were attributed to island-arc volcanism (Slivitzky and St-Julien, 1987). Recently, Chevé (1990) suggested that these two volcanic groups are related to the partial fusion of essentially the same mantle source. However, mantle heterogeneities or contamination have to be invoked to account for their distinctive Zr/Y ratios. The Clinton volcanics are more likely related to the Jim Pond volcanics which represent the extrusive part of the Boil Mountain ophiolite complex (Boudette, 1982). The Frontenac-Chartierville volcanics are tholeiitic to transitional basalts (Chevé, 1990) and interpreted as back-arc, or rift-related volcanic rocks.

In the Gaspésie Subzone, gabbros found in the Rivière Port-Daniel and McCrea mélanges, as well as those of the Neckwick and Arsenault formations, are all cogenetic, and bear some similarities with normal mid-ocean ridge basalts (N-MORB; Bédard, 1986). These MORB-like mafic rocks suggest a local tensional or rifting event between the Taconic and Acadian orogenies (Bédard, 1986). De Broucker (1987) interpreted the gabbroic sills as the product of tensional tectonism in a forearc setting during Middle Ordovician. Van Staal and Fyffe (this chapter) suggested a correlation of these gabbro sills with tholeiites found in the Fournier Group (Belledune Subzone in northern New Brunswick). These Belledune volcanics are interpreted to be related to back-arc rifting (van Staal and Fyffe, this chapter).

Tectonic interpretation

St-Julien and Hubert (1975) interpreted the Dunnage Zone of the Quebec Appalachians as the result of a northwest-dipping subduction zone with the Ascot Complex as a subduction-related volcanic arc, the Saint-Daniel Mélange as the distal equivalent of continental margin deposits, and the Magog Group as deposits of a backarc basin. Tectonic models based on Newfoundland (Williams, 1979) and the New England Appalachians (Osberg, 1978; Stanley and Ratcliffe, 1985) interpret Taconic Orogeny as the result of a continent-volcanic arc collision above an east-dipping subduction zone. These models fit the tectonostratigraphic framework of the Quebec Dunnage Zone. The Saint-Daniel, Trinité, and other mélanges of Gaspésie are viewed as remnants of a single subduction complex, and the Magog, Cabano-Pointe aux Trembles-Lac Raymond, and Mictaw-Arsenault units as overlying deposits of a forearc basin. The Ascot Complex and correlatives record the development of arc volcanism (Cousineau and St-Julien, 1986; Tremblay et al., 1989a; Tremblay and St-Julien, 1990; Cousineau, 1990). In the Mégantic Subzone, the Frontenac Formation records the development of subduction-related rift volcanism and sedimentation (Chevé, 1990), like the Middle Ordovician Tetagouche Group of New Brunswick (van Staal, 1987), although the age of the Frontenac Formation is still unknown. The tectonostratigraphic analysis of the Quebec Dunnage Zone suggests diachronous arc volcanism and forearc sedimentation from southwest to northeast (Estrie-Beauce Subzone to Gaspésie Subzone) with concomitant stratigraphic gaps between Dunnage rocks and overlying cover sequences (Tremblay et al., 1991).

However, many problems still exist. Geochronological data for the Quebec Dunnage Zone are sparse and reliable determinations are needed for the age of volcanic arc, backarc, and ophiolite sequences, as well as for mafic and intermediate intrusions in forearc basins. The nature of structural and lithological relationships on both sides of the Baie Verte-Brompton Line are also ambiguous; what is the extent of Acadian deformation west of the Baie Verte-Brompton Line? What is the extent and nature of Taconic deformation east of the Baie Verte-Brompton Line? Are any units laterally continuous between the Humber and Dunnage zones? A complete petrographic and geochemical comparison between inferred forearc sequences is necessary to improve tectonic models. Geochemical studies of the inferred backarc volcanic sequence are also necessary. These are only a few examples of future work required for a better understanding of the Dunnage Zone in the Quebec Appalachians.

Conclusions

In the Quebec Appalachians, the Dunnage Zone is subdivided into four subzones. Rocks of the Estrie-Beauce, Témiscouata, and Gaspésie subzone are correlative pre-Middle to Middle Ordovician mélanges and discontinuous volcanic arc sequences that are disconformably overlain by Middle Ordovician to Lower Silurian siliciclastic rocks recording the development of forearc basin sedimentation. Stratigraphic relations with overlying post-Taconic cover sequences vary along strike from disconformable in the southwest to conformable in the northeast. The Mégantic Subzone to the southeast, seems to represent Ordovician or Silurian rift magmatism and sedimentation.

The Estrie-Beauce, Témiscouata, and Gaspésie subzones show early to middle Ordovician interaction with the Humber Zone along the Baie Verte-Brompton Line. Although the main deformation of these rocks is Acadian, the mélange units were affected by Taconic deformation. These subzones have Silurian or earlier linkages with the Humber Zone and all lie above a Grenville lower crustal block which underlies Gaspésie (Marillier et al., 1989) and most of the Eastern Townships of Quebec (Stewart et al., 1986; Bernard, 1987). Sub-Silurian unconformities recognized in the Estrie-Beauce and Gaspésie subzones are also typical of the Newfoundland Dunnage Zone.

The Mégantic Subzone had early to middle Ordovician interaction with the Chain Lakes Massif (Boudette, 1982; Williams and Hatcher, 1983) which is believed to underlie the Frontenac Synclinorium (Bernard, 1987). This relationship suggests a correlation with the Exploits Subzone of Newfoundland which had early to middle Ordovician interaction with the Gander Zone (Williams et al., 1988, 1989; Williams and Piasecki, 1990).

In all Quebec subzones, the regional metamorphism and deformation is Acadian. The regional folding and faulting is dominated by dip-slip tectonics in southern Quebec and by strike-slip tectonics in Gaspésie. The structural style of the Témiscouata Subzone is characterized by reverse and strike-slip faults and associated folds, and is interpreted as a transition between dip-slip and strike-slip regimes. Intensity of metamorphism decreases from greenschist facies in southern Quebec to lower-greenschist facies in the Témiscouata and Gaspésie subzones.

The geochemical signature of the Dunnage Zone volcanic and intrusive rocks generally agrees with their inferred tectonic setting, although more work is needed in the case of mafic sequences. For the near future, it is important to focus on the need for geochronological data for the various oceanic rocks of the Dunnage Zone and to ascertain their interaction with adjacent rocks.

Acknowledgments

The authors thank H. Williams for his editorial review of this contribution. We also want to acknowledge P.A. Cousineau, S. Chevé, J. David, and G. de Broucker for discussions which were useful for the understanding of the Dunnage oceanic rocks. However, errors and misinterpretations are our own responsibility. Thanks are due to L. Dubé for drafting the original figures. NSERC subsidized part of this work in the form of operating grants to A. Tremblay (GP-105669) and M. Malo (GP-1908). This is Ministère de l'Énergie et des Ressources du Québec contribution no. 91-511020.

GANDER ZONE

Preamble

Harold Williams

The Gander Zone was first defined in northeast Newfoundland (Kennedy and McGonigal, 1972a; Williams et al., 1972, 1974) and takes its name from Gander Lake and the town of Gander (Fig. 3.111a). The type area comprises a thick monotonous sequence of polydeformed quartz greywacke, quartzite, siltstone and shale that grades eastward into psammitic schist, gneiss and migmatite. These sedimentary rocks contrast with the mixed volcanic-sedimentary assemblages of mainly lower grade rocks of the bordering Dunnage and Avalon zones. The age of the Gander Zone sedimentary rocks is poorly constrained. Only a few fossil occurrences are known and these are in lithologies more akin to those of the Dunnage Zone. If the fossiliferous rocks are in stratigraphic contact with the monotonous clastic sequence, they are presumably at the top of the succession. The fossils are Arenig to Llandeilo, and imply that Gander Zone clastics are mainly Early Ordovician and older.

In the original definition of the Gander Zone, its clastic rocks, or Gander Group, were interpreted as cover to a crystalline basement (Kennedy and McGonigal, 1972a; Blackwood, 1977). Furthermore, intrusion and polydeformation (Ganderian Orogeny, Kennedy, 1975b, 1976) were thought to affect the Gander Group before the deposition of adjacent Middle Ordovician rocks (Davidsville Group) of the Dunnage Zone (Kennedy and McGonigal, 1972a; Kennedy, 1975b, 1976). This interpretation has been challenged, because a basement-cover relationship has not been demonstrated and because metamorphism, plutonism and polydeformation that affect rocks of the Gander Zone also affect the Davidsville Group of the nearby Dunnage Zone (Currie and Pajari, 1977; Bell et al., 1977; Pajari and Currie, 1978; Pickerill et al., 1978; Currie et al., 1979; Blackwood, 1978, 1982). Still more recent studies in the type area of Gander Lake support the idea of Kennedy and McGonigal (1972a) of a structural contrast and time stratigraphic break between the Gander and Davidsville groups (Williams et al., 1991).

Clastic rocks of the Gander Zone were extrapolated from its type area southward to Bay d'Espoir (Colman-Sadd, 1976, 1980) and its crystalline rocks extended around the Hermitage Flexure to southwestern Newfoundland (Brown and Colman-Sadd, 1976; Colman-Sadd, 1980). The extent of the Gander Zone and relationships in central Newfoundland are updated (Colman-Sadd and Swinden, 1984; Williams et al., 1988), but the nature, or even the existence, of the Gander Zone in southwestern Newfoundland is debatable (O'Brien et al., 1986; Williams et al., 1989b). The Gander Zone was extended to Cape Breton Island (Williams et al., 1972; Neale and Kennedy, 1975) on tenuous evidence, and correlations are still possible between rocks of the Aspy Subzone (Barr and Raeside, 1986, 1989) and the Newfoundland Gander Zone, but boundaries have changed. In New Brunswick, the Gander Zone includes the clastic sequence of the lower Tetagouche Group (now Miramichi Group) in the Miramichi Anticlinorium (Rast et al., 1976a), and clastic rocks of the Cookson Inlier or St. Croix Subzone (Fyffe and Fricker, 1987) farther south, although the Cookson rocks have also been assigned to the Avalon Zone (Williams, 1978a; Ludman, 1986, 1987; Fyffe and Pickerill, 1986). Correlations are also suggested between rocks of the Gander Zone and the Grand Pitch Formation of the northern U.S. Appalachians (Neuman, 1967; Rast and Stringer, 1974; Williams, 1978a; Ludman, 1987) and with the Bray Series of Ireland and Skiddaw Group of Britain (Williams, 1978b; Kennedy, 1979).

The Gander-Avalon boundary is a sharp tectonic junction in Newfoundland – the Dover (Blackwood and Kennedy, 1975) and Hermitage Bay (Blackwood and O'Driscoll, 1976) faults. It is poorly defined in Cape Breton Island (Barr and Raeside, 1986, 1989) and hidden by Silurian greywackes of the Fredericton Belt in New Brunswick (Rast et al., 1976a). The Gander-Dunnage boundary has been discussed already under the Dunnage Zone.

Even before the wide acceptance of plate tectonics, Gander Zone clastic rocks were viewed as a prism of sediment built up parallel to an Avalon shoreline on the eastern side of a Paleozoic ocean (Williams, 1964a). The idea persisted (Kennedy, 1975b, 1976; Pajari et al., 1979; Colman-Sadd, 1980) chiefly because Gander clastic rocks are of continental affinity and Gander Zone gneiss and migmatite resemble crystalline basement of the Humber Zone. Accordingly, the Gander Zone is commonly termed the eastern margin of Iapetus (Kennedy, 1976; Williams, 1978a; Colman-Sadd, 1980). However, there are no confirmed basement relationships or stratigraphic analyses comparable to those for the Humber miogeocline. Thus, the Gander Zone has also been viewed as a suspect terrane (Williams and Hatcher, 1982, 1983) or composite suspect terrane (Keppie, 1985). Another recent model has the Gander and Dunnage zones evolving side by side with some Dunnage volcanic rocks deposited across Gander clastic rocks and continental basement (van Staal, 1987; van der Pluijm and van Staal, 1988). This model implies that the Dunnage and Gander zones lost their distinctiveness by late Early Ordovician (van Staal, 1987; van der Pluijm and van Staal, 1988; Williams and Piasecki, 1990).

The earliest linkage between Gander-Avalon rocks is Silurian or Devonian, based on the occurrence of Gander detritus in conglomerates of the Cinq Isle and Pools Cove formations in the southwestern Avalon Zone of Newfoundland (Williams, 1971b; Williams and Hatcher, 1983). Gander-Dunnage linkages are older; Middle Ordovician or earlier in Newfoundland and Late Ordovician in Quebec and New Brunswick, where the Aroostook-Percé ribbon limestones and shales of Late Ordovician and Silurian age cross the Humber, Dunnage and Gander boundaries.

The Gander Zone has mainly negative gravity and magnetic anomalies compared with mainly positive anomalies of the Dunnage Zone and mixed positive and negative

Williams, H.
1995: Preamble (Gander Zone); in Chapter 3 of Geology of the Appalachian-Caledonian Orogen in Canada and Greenland, (ed.) H. Williams; Geological Survey of Canada, Geology of Canada, no. 6, p. 198-199 (also Geological Society of America, The Geology of North America, v. F-1).

anomalies of the Avalon Zone (Haworth et al., 1980a, b; Zietz et al., 1980a, b). Deep seismic reflection transects off northeast Newfoundland (Keen et al., 1986) and in the Gulf of St. Lawrence (Marillier et al., 1989) suggest that the Gander Zone is underlain by a Central lower crustal block. The Central lower crustal block thins westward in Newfoundland and northward in New Brunswick and Quebec, and extends well beyond the Dunnage-Gander surface boundary. The eastern limit of the Central lower crustal block coincides with the Gander-Avalon boundary off northeast Newfoundland and in Cabot Strait. This boundary is steep and extends to mantle depths. Geomagnetic experiments suggest a major electrical-conductivity contrast across the Gander-Avalon boundary in northeast Newfoundland (Cochrane and Wright, 1977; Pal, 1982).

Almost one half the area of the Gander Zone consists of granitic intrusions and one half of the remainder comprises metamorphic rocks ranging from greenschist to upper amphibolite facies. Occurrences of foliated biotite granite, foliated to massive garnetiferous muscovite-biotite leucogranite and potassic megacrystic biotite granite are typical (Williams et al., 1989a, b). Since granites of this type and size are atypical of oceans or ophiolitic basement, it is obvious that the Central lower crustal block is continental. Results of regional gravity and magnetic surveys support this conclusion (Karlstrom, 1983).

Gravity data in northeast Newfoundland indicate the presence of local dense rocks, probably mafic-ultramafic complexes, at depths of a few kilometres or less (Miller, 1988). If the dense rocks are ophiolitic basement to Gander clastic rocks, as suggested by Miller (1988), then all are allochthonous with respect to the Central lower crustal block. If the dense rocks are intrusions, then there is no reason to divorce the clastic rocks of the Gander Zone from the underlying crustal block.

Rocks of the Gander Zone almost everywhere exhibit folded cleavages or schistosities and appear to be more deformed than rocks in the adjacent Dunnage and Avalon zones. Coupled with the presence of unconformities between Lower and/or lower Middle Ordovician conglomerates and mafic-ultramafic rocks of the Gander River Complex (O'Neill and Knight, 1988), relationships imply pre-Middle Ordovician deformation in the Gander Zone (Williams et al., 1991). The latest isotopic data suggest that the major deformation and regional metamorphism along the eastern margin of the Newfoundland Gander Zone is Silurian (Dunning et al., 1990a).

Mineral deposits of the Gander Zone are related mainly to plutonism and hydrothermal activity. These are vein deposits of tin, tungsten, uranium, molybdenum, fluorite and gold.

Faunas of the Gander Zone are all roughly contemporaneous, of Early to Middle Ordovician age, and of Celtic affinities (Neuman, 1984). Thus, there is no faunal distinction between the Dunnage (Exploits Subzone) and Gander zones. An occurrence of *Oldhamia* in the Grand Pitch Formation of Maine (Neuman, 1967) indicates Cambrian ages also.

Newfoundland

Harold Williams, S.P. Colman-Sadd and P.P. O'Neill

Introduction

Before the definition of the Gander Zone, the wide belt of sedimentary rocks on the north shore of Gander Lake were assigned to the Gander Lake Group (Jenness, 1963) and viewed as part of the Newfoundland Central Paleozoic Mobile Belt (Williams, 1964a) or Central Volcanic Belt (Kay and Colbert, 1965). The Gander Lake Group consisted of three units, all viewed as conformable: a lower unit of clastic rocks and metamorphic equivalents, a middle unit of mixed sedimentary and volcanic rocks that were "host" to the "plutons" of the Gander River Complex, and an upper unit of mainly shale containing Middle Ordovician (Caradoc) graptolites (Jenness, 1963). The lower unit of the Gander Lake Group (Jenness, 1963) is now the Gander Group (Kennedy and McGonigal, 1972a, b; Brueckner, 1972; McGonigal, 1973; Blackwood and Kennedy, 1975) of the Gander Zone, and the middle and upper units, the Davidsville Group (Kennedy and McGonigal, 1972a) of the Dunnage Zone. The Gander Group/Davidsville Group contact is tectonic, although relationships are debatable in some places (see discussion under Dunnage Zone).

The notion of a crystalline basement to a Gander Group cover dates back to the pioneer work of Alexander Murray (Murray and Howley, 1881). He viewed crystalline rocks of the eastern Gander Zone as "Laurentian Gneiss". Since then opinion has been divided; are the crystalline rocks of the Gander Zone pre-existing basement (Kennedy and McGonigal, 1972a, b; Blackwood, 1977) or are they prograde metamorphic and migmatitic equivalents of the Gander Group sedimentary sequence (Jenness, 1963; 1972; Williams, 1964b, 1968; Blackwood, 1978)?

Kennedy and McGonigal (1972a) also proposed a sequential structural model whereby the Gander Group was intruded and deformed before the deposition of the Davidsville Group. This was based partly upon a subdivision of granitic rocks of the Gander area into three categories: (1) foliated megacrystic biotite granites that were part of the "basement" to the Gander Group, (2) garnetiferous muscovite-biotite leucogranites that cut the Gander Group and were deformed with the Gander Group before Davidsville deposition, and (3) tonalites and granodiorites that cut the Davidsville Group. Subsequent isotopic studies indicate middle Paleozoic ages for the "basement" granites and garnetiferous leucogranites (Bell and Blenkinsop, 1975; Bell et al., 1977), and subsequent field work demonstrated that garnetiferous leucogranites also cut the Davidsville Group and Carmanville Mélange in northeast Newfoundland (Currie and Pajari, 1977; Pajari and Currie, 1978; Pickerill et al., 1978; Currie et al., 1979). Plutonic rocks are not part of the present Gander Zone

Williams, H., Colman-Sadd, S.P., and O'Neill, P.P.
1995:   Gander Zone-Newfoundland; in Chapter 3 of Geology of the Appalachian-Caledonian Orogen in Canada and Greenland, (ed.) H. Williams; Geological Survey of Canada, Geology of Canada, no. 6, p. 199-212 (also Geological Society of America, The Geology of North America, v. F-1).

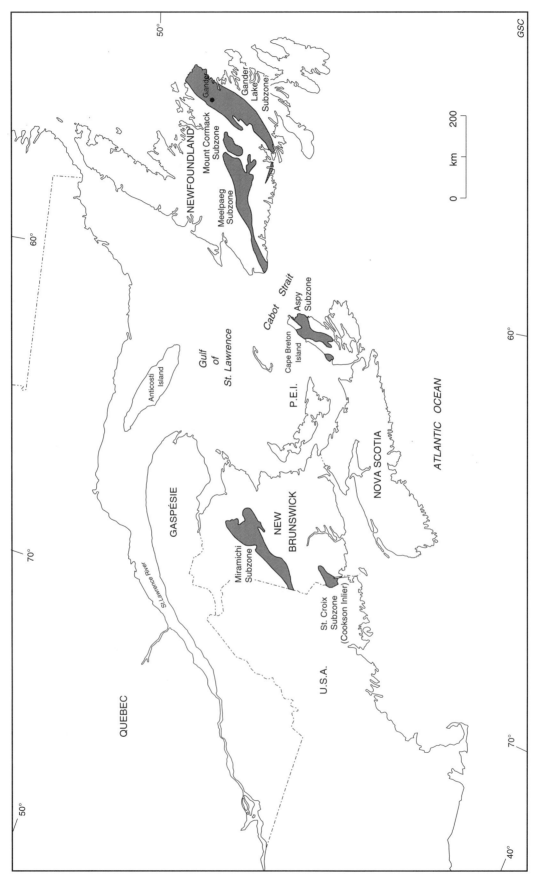

Figure 3.111a. Distribution of rocks of the Gander Zone in the Canadian Appalachians.

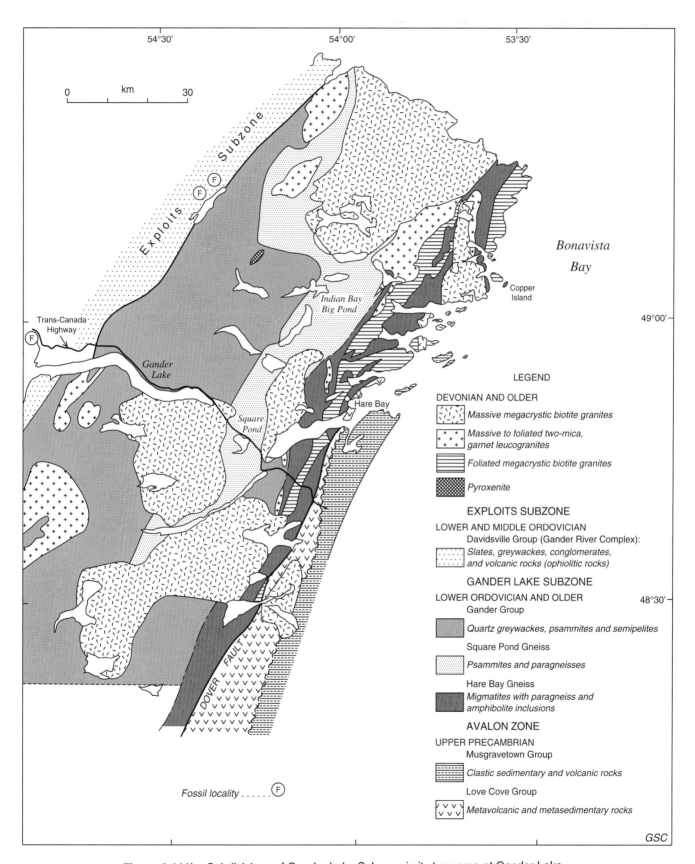

Figure 3.111b. Subdivisions of Gander Lake Subzone in its type area at Gander Lake.

definition, but certain plutons are concentrated in the Gander Zone (Williams et al., 1989a), and they have been used to extend the Gander Zone in the absence of other criteria (Brown and Colman-Sadd, 1976; Colman-Sadd, 1980; Williams et al., 1988).

Extending the Gander Zone from its type area into southern and central Newfoundland was always problematic. The wide belt of low grade Gander Group clastic rocks is missing on the south coast of the island at Bay d'Espoir, although some metamorphic rocks in the north have counterparts in the south. All of the Baie d'Espoir Group (Jewell, 1939; Colman-Sadd, 1976) is now assigned to the Dunnage Zone and its boundary with the Gander Zone metamorphic rocks, Little Passage Gneiss (Colman-Sadd, 1980), is a major ductile shear zone (Piasecki, 1988).

Recognition of the Mount Cormack terrane of central Newfoundland (Colman-Sadd and Swinden, 1984) as a structural inlier within the Dunnage Zone, and correlating its psammitic rocks with the Gander Group, provides the key for separating Dunnage-Gander rocks in central and southwestern Newfoundland. Another, larger inlier of Gander-type rocks is recognized in the Meelpaeg Lake area (Colman-Sadd, 1984, 1985, 1988), and it is traceable all the way to Cape Ray (Williams et al., 1988, 1989a; Colman-Sadd et al., 1990). Smaller structural windows, partly interrupted by intrusions, are anticipated south of Meelpaeg Lake. These relationships between Gander- and Dunnage-type rocks indicate a two-layer crust, with Dunnage Zone rocks above Gander Zone rocks throughout central Newfoundland (Colman-Sadd and Swinden, 1984; Williams et al., 1988).

The Newfoundland Gander Zone is divided spatially into three discrete subzones (Williams et al., 1988). The type area and its southward continuation is the Gander Lake Subzone. The inland areas to the west are the Mount Cormack Subzone and the Meelpaeg Subzone. Upper Precambrian and lower Paleozoic rocks between La Poile Bay and Grey River were included, provisionally, in the Gander Lake Subzone, chiefly because of metamorphic and plutonic styles, and because they occur along strike with the Gander Zone metamorphic and foliated plutonic rocks in the Hermitage Flexure. These rocks are now assigned to the Burgeo Subzone, that is more akin to the Avalon Zone of Newfoundland and the Bras d'Or Subzone of nearby Cape Breton Island.

In the following descriptions, rocks and relationships are treated separately for each subzone, followed by summaries of structure, metamorphism, plutonism, possible significance, and interpretation.

Gander Lake Subzone

Three lithic units are recognized in the Gander Lake Subzone in the vicinity of Gander Lake and eastward (Blackwood, 1977, 1978; Fig. 3.111b). From west to east, these are the low grade clastic rocks of the Gander Group, low- to medium-grade metamorphic rocks of the Square Pond Gneiss (also part of the Gander Group, O'Neill, 1987), and medium- to high-grade metamorphic rocks of the Hare Bay Gneiss.

The Gander Group clastic rocks are interbedded psammite and pelite with quartz-granule sandstone and quartzite. Sharp bedding surfaces separate psammite and pelite beds that vary from a few millimetres in pelitic units to 2 m in quartz-rich units. Clastic textures are preserved in low

grade rocks with sandstones containing 60-90 per cent quartz, commonly blue, and less than 10 per cent feldspar (O'Neill and Knight, 1988). The sandstones are immature to moderately mature with 10-30 per cent matrix of fine grained quartz and white mica. Sorting is moderate and grains are subangular. Detrital calc-silicate minerals are common in places as thin layers or isolated pods. Some sandstone beds show a subtle grading with ripple lamination toward their tops and scouring at their bases.

O'Neill and Knight (1988) included the Indian Bay Formation (Wonderley and Neuman, 1984) or Indian Bay Big Pond Formation (O'Neill and Blackwood, 1989) with the Gander Group and named the voluminous clastic rocks the informal Jonathan's Pond Formation (Fig. 3.112). Furthermore, the Indian Bay Big Pond Formation was expanded to include rocks gradational between the Jonathan's Pond Formation and the northwestern contact of the original Indian Bay Formation (compare Fig. 3.77 and Fig. 3.112). Black pelites are more abundant in the gradational rocks, forming up to 50 per cent of local exposures. These are interlayered with grey-green sandstones and shales typical of the Jonathan's Pond Formation (O'Neill and Knight, 1988). Six northeast-trending units are distinguished in the expanded Indian Bay Big Pond Formation (O'Neill and Knight, 1988). The southeastern contact of the Indian Bay Big Pond Formation is interpreted as a shear zone (O'Neill, 1987). In the absence of sedimentary tops and exposed stratigraphic contacts, relationships remain debatable.

The depositional setting of the Gander Group is uncertain. Sharp interbeds of quartz sandstone within pelite suggest distinct pulses of sand transport into a mud-dominated shallow marine or shelf environment. Thick, graded sandstone beds suggest turbidity currents and a deeper basinal setting (O'Neill and Knight, 1988). The Indian Bay Big Pond Formation has fossils that are coeval with those of the basal Davidsville Group of the Dunnage Zone. A major lithic change between the monotonous clastic sedimentary rocks of the Jonathan's Pond Formation and overlying mixed lithologies of the Indian Bay Big Pond Formation, including pillow lavas and conglomerates, coincides with the unconformity between the Davidsville Group and the Gander River Complex (O'Neill and Blackwood, 1989). The Indian Bay Big Pond Formation may have been deposited above the Gander Group after Dunnage-Gander juxtapositioning, and sited east of the leading edge of the overriding Dunnage Zone (Williams and Piasecki, 1990).

An occurrence of pyroxenite and gabbro in the vicinity of a positive gravity anomaly within the Gander Lake Subzone (O'Neill and Knight, 1988; Fig. 3.111b) suggests an ophiolitic substrate beneath part of the Gander Group (Miller, 1988). The lack of other possible ophiolitic components and poor exposure also allow tectonic or intrusive contacts.

The Square Pond Gneiss (Blackwood, 1977, 1978), or more eastern and higher grade equivalents of the Gander Group, occurs in a narrow belt 150 km long by 12 km wide between the lower grade Gander Group clastic rocks and the Hare Bay Gneiss. It consists of psammitic and semipelitic metasedimentary rocks with zones of schist and migmatite. Psammite with a "pinstripe" or "herring bone" banding is characteristic. The main fabric is a composite gneissosity/schistosity which transposes an earlier banding. Bedding and clastic textures are preserved locally.

202

Metamorphic facies vary from greenschist to upper amphibolite, generally from west to east across the outcrop belt.

The Hare Bay Gneiss (Blackwood, 1977, 1978) occurs in a linear zone about 140 km long by 10 km wide along the eastern margin of the Gander Lake Subzone. It consists of migmatitic and tonalitic gneiss containing xenoliths and rafts of paragneiss and amphibolite. Crudely banded biotite migmatites contain zones of tonalitic orthogneiss and inclusions of semipelite, psammite, and amphibolite. Green, banded amphibolite inclusions with discordant fabrics form augen surrounded by the gneissosity of the tonalitic host. A large area of banded metadiorite or amphibolite occurs at Copper Island on the extreme eastern margin of the subzone (Jenness, 1963; Williams, 1968). Complex interference fold patterns are common throughout the gneiss terrane. A variety of foliated granites intrude the migmatites.

Blackwood (1977) originally defined a tectonic break between the higher grade clastic rocks (Square Pond Gneiss) and lower grade clastic rocks of the Gander Group, and interpreted it as a basement-cover contact. A strong aeromagnetic expression for schists on its eastern side was thought to reflect a basement complex against a lower grade metaclastic cover sequence. This interpretation was changed to accomodate other relationships that suggest a conformable contact, coincident or nearly so with the biotite isograd of prograde metamorphism (Blackwood, 1978). All contacts are now regarded as gradational (Blackwood, 1978; O'Neill, 1987), reverting back to the views of earlier workers (Jenness, 1963; Williams, 1964b, 1968). A migmatite front, defined by the formation of granitic sweats, feldspathization, and lit-par-lit granitic veining, marks the boundary between higher grade parts of the eastern Gander Group (Square Pond Gneiss) and Hare Bay Gneiss. It is an abrupt contact in the north and more gradational southward.

Metamorphic rocks extend along the eastern margin of the Gander Lake Subzone from Bonavista Bay on the northeast coast of Newfoundland to Bay d'Espoir on the south coast (Jewell, 1939; Anderson and Williams, 1970; Blackwood, 1978; Colman-Sadd, 1980). On the south coast (Fig. 3.113), these are known as the Little Passage Gneiss (Colman-Sadd, 1980) and they are correlated with the Square Pond and Hare Bay gneisses to the north. The most common rock types are paragneiss, amphibolitic gneiss, and tonalitic gneiss. Paragneisses are composed of quartz, plagioclase, microcline, biotite, and muscovite. Garnet occurs as augen and sillimanite is present in semipelitic gneisses. Amphibolitic gneisses form xenoliths in the tonalitic gneisses and conformable layers in both the tonalitic gneisses and paragneisses. Tonalitic rocks intrude and truncate fabrics in amphibolitic gneisses.

Colman-Sadd (1980) proposed a structural history for the Little Passage Gneiss that predated deposition of Baie d'Espoir Group, following the model of Kennedy and

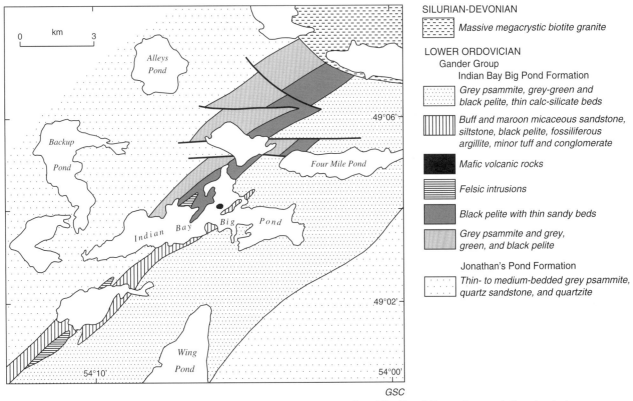

Figure 3.112. Relationships between the Indian Bay Big Pond Formation and Gander Lake Subzone according to O'Neill and Knight (1988).

203

CHAPTER 3

McGonigal (1972a). The Little Passage gneisses are reworked within 5 km of their boundary with the Baie d'Espoir Group (Dunnage Zone) and they are mylonitized within 50 m of the contact. The boundary is a major ductile shear with left lateral sense of displacement (Piasecki, 1988). Isotopic ages indicate Silurian tectonism (Dunning et al., 1990a).

Metamorphic rocks like those of the eastern Gander Lake Subzone, including foliated biotite granites, foliated to massive garnet muscovite leucogranites, tonalitic gneisses, migmatites, amphibolites, and paragneisses, extend from Bay d'Espoir to La Poile Bay. They are bordered to the north by the Baie d'Espoir and Bay du Nord groups of the Dunnage Zone and they are truncated by large circular plutons like the Francois and Chetwynd granites. Metasedimentary rocks, migmatites, and felsic volcanic rocks form an east-west belt at Grey River, the Grey River Enclave (Blackwood, 1985b). Farther west at Burgeo and Grand Bruit, amphibolitic complexes occur within the Burgeo and other granites. One example, the Cinq Cerf Complex (Chorlton, 1978) at Grand Bruit, is cut by Roti Granite dated by U-Pb zircon at 563 ± 4 Ma (Dunning et al., 1988; Dunning and O'Brien, 1989). The Silurian La Poile Group lies nonconformably on Roti Granite near Top Pond Brook (O'Brien, 1988).

Rocks and relationships west of Bay d'Espoir are atypical of both the Gander and Dunnage zones. Metasedimentary rocks of the Grey River Enclave have been correlated with Precambrian rocks of the Bras d'Or Subzone of the Avalon Zone of Cape Breton Island (Smyth, 1981; Barr and Raeside, 1989). The same rocks have been correlated with the Baie d'Espoir and Bay du Nord groups of the Dunnage Zone because of the association of felsic volcanic rocks and metasedimentary rocks (Blackwood, 1985b). Correlations with the Aspy Subzone (Barr and Raeside, 1989) of Cape Breton Island are also possible. More work is required in this complex part of the orogen where all zones converge into a narrow constriction with metamorphic and structural overprinting. A new subzone, the Burgeo Subzone, is proposed for the late Precambrian rocks in this area (Fig. 3-113).

The Dover Fault (Blackwood and Kennedy, 1975) marks the Gander Lake Subzone-Avalon Zone boundary in the north (Fig. 3.111b). It is a 300-500 m wide mylonite zone that separates high grade gneisses and mylonitic granites on the Gander Lake side from low grade Precambrian rocks on the Avalon side. It is an abrupt lithic, structural, and metamorphic break. Rocks on both sides are overprinted with a strong cataclastic fabric close to the fault. Seismic reflection studies offshore indicate that the

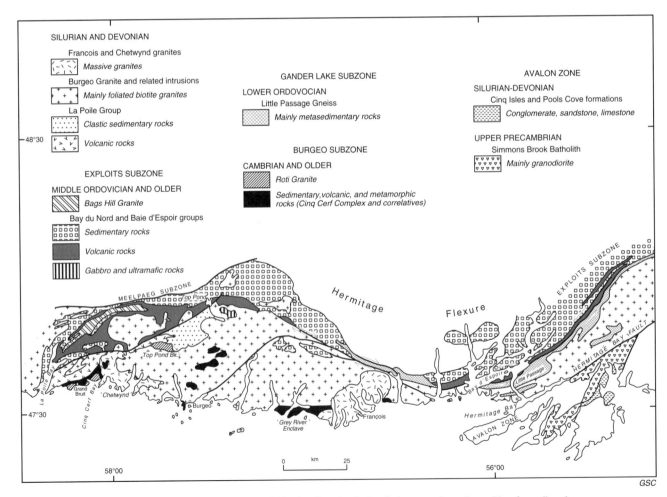

Figure 3.113. Relationships within the Gander Lake Subzone of southern Newfoundland.

Dover Fault is a steep junction that extends to mantle depths (Keen et al., 1986). The Hermitage Bay Fault (Blackwood and O'Driscoll, 1976; Kennedy et al., 1982) defines the Gander Lake Subzone-Avalon Zone boundary in the south (Fig. 3.113). It is a brittle fault with local brecciation. Conglomerates (Cinq Isle and Pools Cove formations) above Avalonian rocks east of the Hermitage Bay Fault contain a sampling of Gander Zone metamorphic rocks and granites (Williams, 1971b). The conglomerates are nonconformable on the late Precambrian Simmons Brook Batholith and they are cut by Devonian granite. They are Silurian or Devonian and furnish the oldest linkage between the Gander Lake Subzone and Avalon Zone.

Mount Cormack Subzone

The Mount Cormack Subzone (Williams et al., 1988), or Mount Cormack Terrane (Colman-Sadd and Swinden, 1984), is an oval area of clastic rocks and metamorphic equivalents about 30 km by 60 km (Fig. 3.114). It is surrounded, or almost so, by ophiolite complexes of the Dunnage Zone (Exploits Subzone). The three largest ophiolite occurrences are the Pipestone Pond, Coy Pond, and Great Bend complexes. They preserve enough stratigraphy to indicate their younging directions, which are everywhere outwards from the Mount Cormack Subzone. Accordingly, the Mount Cormack Subzone is interpreted as a window of Gander Zone rocks exposed through an allochthonous cover of ophiolitic and volcanic-sedimentary rocks of the Dunnage Zone (Colman-Sadd and Swinden, 1984). Most Exploits-Mount Cormack subzone contacts are modified by post emplacement faults. Lithological and structural considerations suggest that the psammitic rocks of the Mount Cormack Subzone (Spruce Brook Formation, Colman-Sadd and Swinden, 1984) are correlatives of the Gander Group in the type area of the Gander Lake Subzone. Fossiliferous limestone conglomerate, first interpreted as part of the Spruce Brook Formation (Colman-Sadd and Swinden, 1984), may be an unconformable or structural cover, and is reassigned to the Dunnage Zone (Dec and Colman-Sadd, 1990).

Where relatively unmetamorphosed, the Spruce Brook Formation is a monotonous sequence of medium- to thick-bedded, light grey sandstone and grey and black shales with laminae and thin interbeds of siltstone and fine grained sandstone. Typical sandstone beds have sharp bases, load structures, and they are graded. Upper parts of beds have faint lamination and cross lamination. They are interpreted as turbidites (Colman-Sadd and Swinden, 1984). Sandstones are composed mainly of quartz with accessory plagioclase, microcline, zircon, tourmaline, and opaque minerals. Sparse rock fragments are of sedimentary origin. Clasts within the sedimentary rocks of the Spruce Brook Formation and clasts within sedimentary rocks of the Dunnage Zone suggest separate tectonic and sedimentary environments, one without volcanic detritus, the other with abundant volcanic detritus, respectively.

Limestone conglomerate, first regarded as part of the Spruce Brook Formation (Colman-Sadd and Swinden, 1984) and now reassigned to the Dunnage Zone (Dec and Colman-Sadd, 1990), contains a brachiopod fauna similar to that found in the volcanic assemblages of the Exploits Subzone at New World Island and in the sedimentary rocks

at Indian Bay Big Pond. Boyce (1987) correlated trilobites of the conglomerate with those of the late Llanvirn-Llandeilo Cobbs Arm Limestone of the Dunnage Zone. The presence of quartzite and ophiolitic clasts in the limestone conglomerate requires input from rocks of both the Gander and Dunnage zones. The conglomerate therefore provides a sedimentary linkage between the Gander and Dunnage zones (Dec and Colman-Sadd, 1990).

Meelpaeg Subzone

The Meelpaeg Subzone consists of psammitic rocks and metamorphic or migmatitic equivalents. These are cut by large pretectonic granites that are confined to the subzone, and post-tectonic granites that in places cut the subzone boundaries (Fig. 3.115a). The Meelpaeg Subzone extends 280 km from Meelpaeg Lake toward the south coast of Newfoundland. Its maximum width is 40 km in the north, narrowing to about 10 km or less in the southwest. The subzone is extended to include the Port aux Basques Complex (Brown, 1977) between Cape Ray and Isle aux Morts and metasedimentary rocks farther east between Isle aux Morts and Harbour Le Cou (Williams et al., 1989b; Colman-Sadd et al., 1990). The largest area of low grade metaclastic rocks in the Meelpaeg Subzone occurs in its northwestern part. These are interlayered quartzite, psammite, semipelite, and pelite in beds from 5 cm to 3 m thick. The quartzite beds weather white and locally display grading and load casts at their sharp bases. Psammite weathers brownish and contains more mica and feldspar. Pelite is dark grey in finer and thinner beds, locally with thin interbeds of quartzite or psammite. These rocks are correlated with the Spruce Brook Formation of the Mount Cormack Subzone (Colman-Sadd, 1987), and therefore with the Gander Group of the type area.

In most other places, sedimentary rocks are of amphibolite facies and contain quartz, garnet, andalusite, and sillimanite. Amphibolite layers are common along the road to the Meelpaeg Reservoir and these locally truncate banding. Amphibolites are also abundant in the Port aux Basques Complex as parallel layers from 1 cm to 1 m, and locally up to several tens of metres, alternating with quartzitic psammitic layers. The majority of the amphibolites are probably original mafic dykes. Between Isle aux Morts and Rose Blanche, the rocks are quartz mica sillimanite schists without amphibolites but containing calc-silicate lenses.

Boundaries of the Meelpaeg Subzone are tectonic, where not truncated by intrusions. Its northwest boundary is Noel Pauls Line (Brown and Colman-Sadd, 1976; Colman-Sadd, 1987, 1988; Williams et al., 1988, 1989a). It separates volcanic and sedimentary rocks of the Exploits Subzone from metaclastic rocks, quartzite, mylonitic granite, and migmatite of the Meelpaeg Subzone. Farther southwest it is the Cape Ray Fault that separates the Port aux Basques gneisses from the Devonian Windsor Point Group and mylonitic granites of the Dashwoods Subzone. The eastern boundary of the Meelpaeg Subzone is a ductile shear zone, the Great Burnt Lake Fault (Colman-Sadd, 1985; Swinden, 1988), that separates mylonitic granites from lower grade rocks of the Exploits Subzone (Colman-Sadd and Swinden, 1988). Unlike the Mount Cormack Subzone boundaries, ophiolitic rocks are less evident in the surrounding Dunnage Zone.

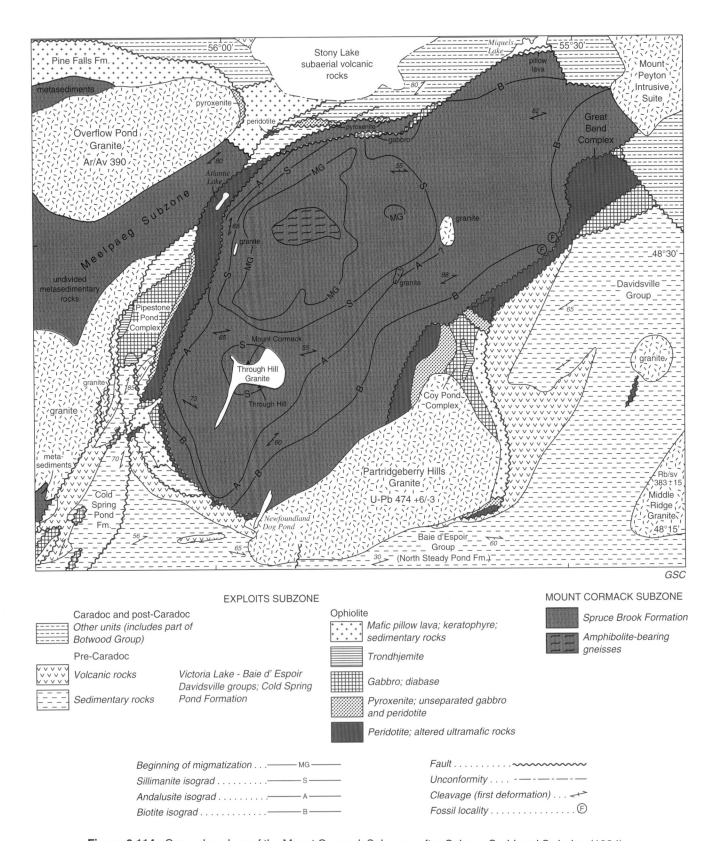

Figure 3.114. General geology of the Mount Cormack Subzone, after Colman-Sadd and Swinden (1984).

EXPLOITS SUBZONE

Caradoc and post-Caradoc
Other units (includes part of Botwood Group)

Pre-Caradoc

Volcanic rocks

Sedimentary rocks

Victoria Lake - Baie d' Espoir Davidsville groups; Cold Spring Pond Formation

Ophiolite

Mafic pillow lava; keratophyre; sedimentary rocks

Trondhjemite

Gabbro; diabase

Pyroxenite; unseparated gabbro and peridotite

Peridotite; altered ultramafic rocks

MOUNT CORMACK SUBZONE

Spruce Brook Formation

Amphibolite-bearing gneisses

Beginning of migmatization . . . ——— MG ———

Sillimanite isograd . . . . ——— S ———

Andalusite isograd . . . . ——— A ———

Biotite isograd . . . . ——— B ———

Fault . . . . ∿∿∿∿∿

Unconformity . . . . – ∙ – ∙ – ∙ –

Cleavage (first deformation) . . . ↰↲

Fossil locality . . . . Ⓕ

Ophiolitic mélange lies at or near the Exploits (Dunnage)-Meelpaeg (Gander) subzone boundary at Cold Spring Pond (fig. 3.115a; Williams and Piasecki, 1990). It comprises large ultramafic and smaller quartzitic sandstone-siltstone blocks in a matrix of black graphitic phyllite. The mélange is overprinted by a steep northwest-southeast cleavage or schistosity, and it is affected by ductile shearing at its boundries. Relationships here indicate that major ductile shear zones that define some Dunnage-Gander boundaries post-date ophiolite transport and first juxtapositioning of the Dunnage and Gander zones.

Structure

The effects of two main phases of deformation are apparent throughout the Gander Zone and its subdivisions. In the type area, a first penetrative foliation strikes north-northeast to northeast and is developed in all rock types. The foliation is steep at the Exploits-Gander Lake subzone boundary, shallow to subhorizontal across the wide central part of the Gander Lake Subzone, and steep along its eastern margin. Second phase folds are upright, open to tight, with associated crenulation cleavage. The north-northeast trend of the first foliation is intersected acutely by the northeast trends of the Davidsville-Gander group contact and trends in the Exploits Subzone, implying truncation and an Exploits- Gander Lake structural boundary (O'Neill, 1987).

At Bay d'Espoir, the main foliation in the Little Passage Gneiss projects westward to define the arcuate trend of the Hermitage Flexure. This foliation is composite. Tonalite cuts amphibolitic gneiss after the latter had acquired its foliation (Colman-Sadd, 1980) and both rock types were subsequently deformed; mylonitization reworked earlier fabrics.

Two phases of folding affect the Spruce Brook Formation of the Mount Cormack Subzone. The first deformation produced tight to isoclinal folds with a steep northeast-trending cleavage. Metamorphic minerals overgrow the first cleavage and isograds are unaffected by the early folds. Metamorphic rocks are refolded near the subzone margin along axes roughly parallel to boundary faults (Colman-Sadd and Swinden, 1984).

Deformation in the Meelpaeg Subzone increases in intensity toward its boundaries with the Exploits Subzone even though the metamorphic grade decreases in places. In the interior part of the subzone, most rocks are of sillimanite-grade, but the intensity of deformation is no more than that in lower grade rocks of the Exploits Subzone. The earliest fabric in metasedimentary rocks is commonly steep and subparallel to bedding, as associated folds are tight to isoclinal (Colman-Sadd, 1985). The earliest fabric is also developed in biotite granite that intrudes the sedimentary rocks. Second generation folds vary from open to isoclinal with local development of crenulation cleavage (Colman-Sadd, 1987).

There is no complete structural analysis of the Gander Zone that attempts to establish relationships among all subzones and polarity of structures and controls. Information is available for local areas and kinematic studies are in progress for others (Piasecki, 1988; Williams et al., 1989b).

First and second phase folds of the northeast Gander Lake Subzone were generally interpreted to face southeast, supporting a model of southeast thrusting of Dunnage Zone rocks above Gander Zone rocks (Kennedy, 1975b, 1976). Subsequent studies favour a model of sinistral transcurrent movements (Hanmer, 1981; Piasecki, 1988). In the type area, a northeast-southwest mineral lineation on a subhorizontal schistosity, and isoclinal folds with initial northwest-southeast axes, correspond to southwestward directed overthrusting (Hanmer, 1981). This agrees with a sinistral sense of shear where the gentle foliation turns down into a steep mylonite zone developed locally on the Exploits-Gander Lake contact north of Gander Lake (Piasecki, 1988). Furthermore, the steep foliation in the Square Pond Gneiss and Hare Bay Gneiss is also attributed to sinistral simple shear (Hanmer, 1981).

The same structural styles, metamorphic facies and intrusive rocks are found at Bay d'Espoir where the Baie d'Espoir Group is separated from the Little Passage Gneiss by a major zone of layer-parallel ductile shearing (Piasecki, 1988). Mylonites in the shear zone contain abundant kinematic indicators and a regional penetrative stretching lineation, which is subhorizontal and trends west-southwest. Where the inclination of foliation is gentle at the Exploits-Gander Lake contact, mylonitic fabrics indicate that the shear zone is a gently inclined strike-slip thrust with the Baie d'Espoir Group transported over the Little Passage Gneiss toward the west-southwest. Where the foliation is steeper toward the Avalon boundary, the layer-parallel shear zone is a steep zone of sinistral strike-slip (transcurrent) movement.

Mylonitic fabrics and kinematic indicators are well displayed along Noel Pauls Line from its northeast end all the way to the Cape Ray Fault. At Victoria Lake and northeastward, a strong mylonitic fabric with a down-dip lineation affects rocks of both the Exploits Subzone and Meelpaeg Subzone (Colman-Sadd, 1987, 1988; Williams et al., 1988, 1989a). Foliations dip steeply to moderately southeastward, probably as the result of overturned regional second phase folds that post date the mylonite. Kinematic indicators related to the mylonitic fabric indicate an oblique or up-dip southeast over northwest sense of displacement, that is northwest thrusting of Meelpaeg over Exploits. However, if the overturning effects of second folds are removed, northwest thrusting of Exploits over Meelpaeg is indicated.

Similar polarities are evident at the Cape Ray Fault. There, the Port aux Basques Complex is affected by northwest thrusting before deposition of the Devonian Windsor Point Group. Later thrusting with the same northwest polarity affected the Devonian Windsor Point Group (Williams et al., 1989b).

A long narrow body of megacrystic biotite granite along the east edge of the Meelpaeg Subzone, north and south from Great Burnt Lake, has a steep planar fabric and a subhorizontal lineation (Colman-Sadd, 1984, 1985; Piasecki et al., 1990). This indicates dextral, north-south transcurrent movement, in contrast to overthrusting at Noel Pauls Line. The equivalent Meelpaeg-Exploits boundary on the Burgeo Road is marked by steeply inclined mylonites with subhorizontal lineation indicating dextral trancurrent displacement (Piasecki et al., 1990).

207

lists of metamorphism

Polarities of structures for the Gander Lake and Meelpaeg subzones are summarized in Figure 3.115b. Information for the Mount Cormack Subzone is unavailable. The relationships suggest that penetrative deformation throughout the Gander Lake Subzone was controlled by sinistral shear in a ductile region up to 60 km wide. Where gentle foliations turn upward toward the east, the steepening coincides with an increase in intensity of regional metamorphism and the presence of foliated to massive granites. Sinistral shear is discernible as far west as the west side of Bay d'Espoir along the course of the Hermitage Flexure. Farther west at Facheau Bay, Dragon Bay, Hare Bay and westward, transcurrent movements are dextral (Piasecki et al., 1990). This regional pattern of transcurrent movements contrasts with northwest and west polarity of thrusting along the full length of Noel Pauls Line.

Metamorphism

In the Gander Zone as a whole, and especially in the Meelpaeg and Gander Lake subzones, amphibolite facies metamorphism is regional, and large areas of migmatite are unrelated spatially to granitoid plutons (Colman-Sadd, 1980, 1984, 1985, 1988; Colman-Sadd and Swinden, 1984). A wide range of metamorphic temperatures, in most places independent of plutons, is characteristic of Gander Zone metamorphism and contrasts with the common uniformity of low grade conditions that prevailed in the Exploits Subzone. The common progression is from andalusite- to sillimanite-muscovite-bearing assemblages, with partial melting occurring at the sillimanite isograd (Colman-Sadd and Swinden, 1984). Kyanite occurs in the Cape Ray area (Brown, 1977) and the northern part of Gander Lake Subzone (Kennedy, 1976; O'Neill and Lux, 1989).

In the type area at Gander Lake, metamorphism increases in a general sense from west to east. In the west, the assemblage muscovite-chlorite-calcite-epidote-plagioclase in mafic rocks is indicative of low greenschist facies. Farther east, biotite coexists with muscovite and chlorite. At higher grades, garnet, staurolite, andalusite, sillimanite, and kyanite are present. At the northeast extremity of the Gander Lake Subzone, isograds cross the Davidsville-Gander group contact with garnet, staurolite, and andalusite developed in the Davidsville Group (Currie and Pajari, 1977). In this area of unusually high grade Dunnage rocks, garnetiferous muscovite leucogranites cut the Davidsville Group and Carmanville Mélange.

At the Exploits-Gander Lake boundary in Bay d'Espoir there is a sharp jump in metamorphism from garnet grade on the Exploits side to muscovite-sillimanite on the Gander Lake side (Colman-Sadd, 1980). Less than 40 km to the northeast, the contrast diminishes and greenschist facies rocks occur on both sides of the contact (Dickson, 1988). However, in all parts of the Gander Lake Subzone there is a metamorphic progression southeastward that culminates in migmatitic gneisses intruded by syntectonic granitoid plutons (Blackwood, 1978; Colman-Sadd, 1980; Hanmer, 1981; Piasecki, 1988; Dickson, 1988).

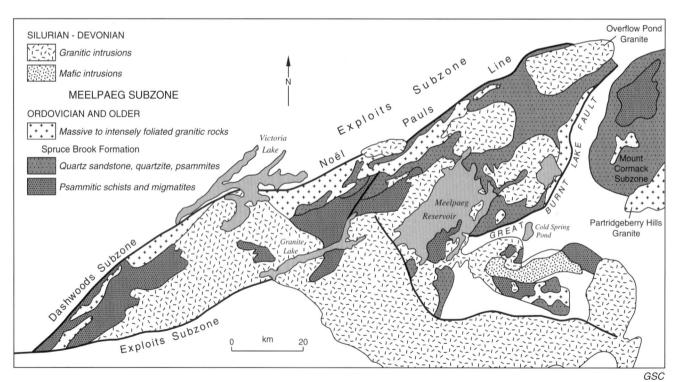

Figure 3.115a. General geology of the Meelpaeg Subzone.

A similar pattern holds farther west where metamorphism increases from north to south across the Grey River Enclave (Blackwood, 1985b). Amphibolites at Grey River have possible counterparts in mafic complexes within the Burgeo Granite and Cinq Cerf Complex farther west. Banded amphibolites near the Gander Lake-Avalon boundary on the northeast coast of Newfoundland, such as those of Copper Island (Fig. 3.111b), may represent additional correlatives; implying widespread occurrences of amphibolite complexes in high grade rocks at the periphery of the Gander Lake Subzone.

The juxtaposition of migmatites of the Gander Lake Subzone and greenschist or lower grade Precambrian rocks of the Avalon Zone is one of the sharpest metamorphic breaks at any zone boundary, and emphasizes the importance of the Dover and Hermitage Bay faults and their possible precursors.

A single episode of metamorphism affected the Spruce Brook Formation of the Mount Cormack Subzone. It caused the systematic progression of concentric isograds that are roughly conformable with the oval outline of the subzone (Fig. 3.114). From the periphery inward, they indicate a rapid increase in grade from greenschist to migmatitic upper amphibolite facies. In central areas of migmatization there is a further progression from the production of a granitic partial melt to the conversion of the entire rock to a homogeneous granodiorite (Colman-Sadd and Swinden, 1984).

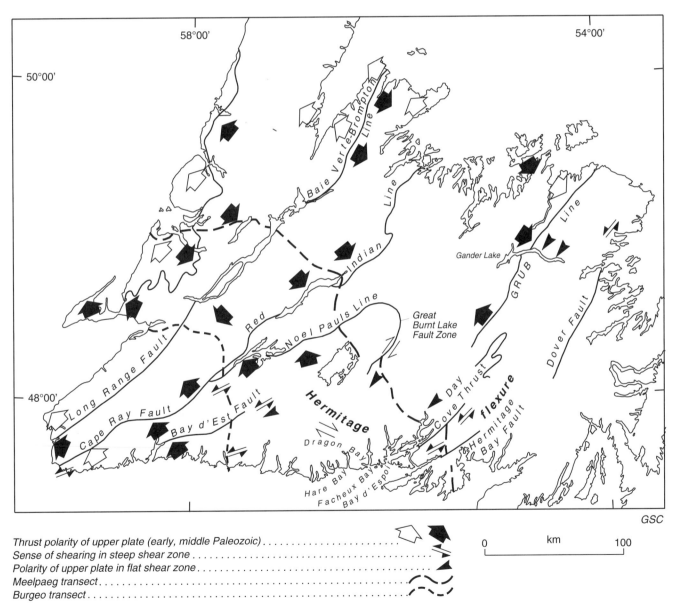

Thrust polarity of upper plate (early, middle Paleozoic)	
Sense of shearing in steep shear zone	
Polarity of upper plate in flat shear zone	
Meelpaeg transect	
Burgeo transect	

Figure 3.115b. Polarity of structures for the Gander Lake and Meelpaeg subzones, after Williams et al. (1989b).

Plutonism

In all the non-migmatized parts of the Spruce Brook Formation, bedding is continuous and sedimentary structures are visible in sandstone beds. Thiner bedding in shales and siltstones is destroyed at the andalusite isograd by disruptive porphyroblasts up to 3 cm long. It is further obliterated upgrade by the growth of large mica flakes, straurolite porphyroblasts and aggregates of fibrolitic sillimanite. With limited melting, sandstone beds form dismembered rafts. With more extensive melting, thinly banded gneisses retain no relict structures, although there are quartzitic xenoliths that may be remnants of sandstone beds. Locally, the gneisses are homogenized into a massive biotite-muscovite granodiorite that appears to have crystallized from a melt and grades into and cuts the banded gneisses. The progressive metamorphism is everywhere gradational, there is no marked compositional change, and there is no obvious structural discontinuity. The entire sequence is therefore considered to have the same protolith, although it is recognized that the migmatites are so reconstituted that they could easily contain an obscured boundary of major significance. The relationships seem to indicate in situ melting of the country rock rather than heating by an extraneous magma (Colman-Sadd and Swinden, 1984).

The Spruce Brook Formation cooled from peak metamorphic conditions at relatively low pressure estimated at 3.3 kbar for rocks at the andalusite-sillimanite transition and 4 kbar for those showing incipient melting (ca. 650°C, Colman-Sadd and Swinden, 1984).

Metamorphic isograds are poorly defined throughout the Meelpaeg Subzone, partly because its metasedimentary rocks are interrupted by large areas of migmatite and granite. These mixed metasedimentary, migmatitic and granitic rocks are especially common along the Burgeo Road at Peter Strides Pond and southeastward. Sillimanite is common all along the south coast from Cape Ray to Harbour Le Cou and kyanite is also reported (Brown, 1977). A sharp jump in metamorphic grade coincides with Noel Pauls Line along Victoria Lake. There, Exploits Subzone rocks of the Victoria Lake Group change from laminated sediments with abundant mafic intrusions of greenschist metamorphic facies to rusty garnetiferous schists and amphibolites that abut mylonitic granites of the Meelpaeg Subzone (Williams et al., 1988). Farther northeast at Noel Pauls Brook, rocks on both sides of Noel Pauls Line are in greenschist facies (Colman-Sadd, 1987). Southeastward, metamorphic grade rises rapidly where muscovite-sillimanite schists are cut by biotite granites and granodiorites. On the east side of the Meelpaeg Subzone at Great Burnt Lake, chlorite-grade sedimentary and volcanic rocks of the Exploits Subzone contrast with garnetiferous rocks of the Meelpaeg Subzone (Colman-Sadd, 1985; Swinden, 1988). At Cold Spring Pond, some 15 km farther south, the contrast is sharper with migmatites of the Meelpaeg Subzone separated from greenschist facies sedimentary and volcanic rocks of the Exploits Subzone by a narrow band of mylonitic granite (Colman-Sadd, 1984). Migmatitic sillimanite grade metasedimentary rocks at White Hill (Colman-Sadd, 1984) may be a Meelpaeg diapir surrounded by Exploits greenschist facies sedimentary rocks (Williams et al., 1989b).

Plutonism

Granitic rocks are abundant throughout the Gander Zone, although to a lesser extent in the Mount Cormack Subzone. They are divided into three broad categories based mainly upon composition, texture, and relative age (Kennedy and McGonigal, 1972a; Colman-Sadd, 1980; Williams et al., 1989a).

The first group consists of deformed coarse grained biotite granite with large feldspar phenocrysts (Burgeo type of Williams et al., 1989a). The granites form linear bodies within the Hare Bay and Little Passage gneisses along the eastern margin of the Gander Lake Subzone (Blackwood, 1977; Colman-Sadd, 1980). The granites cut, or are gradational with migmatites and have a strong northeast-trending fabric. In southern Newfoundland, the rocks continue westward into the Burgeo Granite and their tectonic fabrics are parallel to the outline of the Hermitage Flexure (O'Brien et al., 1988). Foliated biotite granites are also associated with migmatites of the Meelpaeg Subzone. A large foliated biotite granite, with gentle folding of the first fabric, occurs southeast of Noel Pauls Brook (Colman-Sadd, 1987; Williams et al., 1989b), and a remarkably linear body of megacrystic granite, some 60 km long, occurs at Great Burnt Lake along the east margin of the Meelpaeg Subzone (Colman-Sadd, 1984, 1985; Swinden, 1988). These deformed biotite granites are confined to the Gander Zone.

A second group of granites consists of medium- to coarse-grained hornblende-biotite granodiorite through muscovite-biotite granite to muscovite-biotite leucogranite, commonly garnetiferous with associated pegmatites (Middle Ridge type of Williams et al., 1989a). They vary from massive to well foliated and from small to large, linear to equidimensional bodies. They occur mainly along the eastern side of the Gander Lake Subzone and as smaller bodies in the Meelpaeg and Mount Cormack subzones. There are all gradations between massive examples that cut deformed metaclastic rocks to foliated examples that are deformed with Gander Zone metaclastic rocks and mylonitic migmatites. Examples along the western side of the Gander Lake Subzone cut rocks of the Exploits Subzone at Aspen Cove and Middle Ridge (Currie and Pajari, 1977; Blackwood and Green, 1982). The Through Hill Granite (Colman-Sadd and Swinden, 1984) is the largest example in the Mount Cormack Subzone and it cuts first deformation structures in the sillimanite grade Spruce Brook Formation. Sheets of muscovite-garnet-tourmaline leucogranite and related pegmatites occur within the Little Passage Gneiss (Colman-Sadd, 1980). These intrusions are deformed into mylonites and protomylonites with a penetrative mineral lineation parallel to that in host mylonitic metasedimentary rocks (Piasecki, 1988). A variety of features of the leucogranites suggest an origin through anatexis of supracrustal rocks; such as numerous xenoliths of partially assimilated country rock, a peraluminous chemistry, and high strontium isotope initial ratios (Colman-Sadd and Swinden, 1984; Williams et al., 1989a). This underscores their localization in high grade areas of the Gander Zone and nearby anomalously high grade areas of the Exploits Subzone.

The third group of plutons are undeformed massive coarse grained megacrystic granites in large equidimensional bodies (Ackley type of Williams et al., 1989a). These intrude deformed metasedimentary rocks and all other plutons. They also intrude the Exploits-Gander Lake subzone boundary near Bay d'Espoir, the Exploits-Meelpaeg subzone boundary south of Meelpaeg Lake, and the Gander Lake Subzone-Avalon Zone boundary in central Newfoundland.

Most of the granites are of Silurian-Devonian age (for review see Williams et al., 1989a). Lineations and foliations in deformed examples are everywhere parallel to those of host or nearby migmatites and mylonites, suggesting emplacement during shearing (Hanmer, 1981; Piasecki, 1988). Undeformed examples indicate that the intrusive activity outlasted ductile shearing.

The ages of plutons that cut the Dunnage-Gander boundary provide an upper limit to the time of structural superposing. Thus the Ragged Harbour (380 ± 20 Ma, Strong and Dickson, 1978), Middle Ridge (383 ± 15 Ma, Bell and Blenkinsop, 1977) and Terra Nova (352 ± 10 Ma, Bell et al., 1977) plutons stitch the Gander Lake and Exploits subzones; the North Bay (396 Ma, O'Brien et al., 1986) and Overflow Pond (390 Ma, Dallmeyer et al., 1983a) plutons stitch the Meelpaeg and Exploits subzones; and the Partridgeberry Hills (431 ± 5 Ma, Elias and Strong, 1982) stitches the Mount Cormack and Exploits subzones. A U-Pb zircon age of 474 +6/-3 Ma for the Partridgeberry Hills Granite indicates Early Ordovician juxtapositioning of the Mount Cormack and Exploits subzones (Colman-Sadd et al., 1992).

Summary and conclusions

The Newfoundland Gander Zone is defined on its clastic sedimentary rocks and metamorphosed equivalents. These rocks are, in general, more obviously polydeformed and metamorphosed than rocks of the Dunnage Zone, and their intrusive history is different. Only in a few places are structural styles of the Dunnage and Gander zones similar. Examples of regional metamorphic isograds crossing Dunnage-Gander boundaries are rare, and only locally do certain categories of granitic plutons occur outside the Gander Zone. This focus of structure, metamorphism, and plutonism within the Gander Zone suggests related controls and peculiarities of development that did not extend to adjacent zones.

Accumulating evidence supports the model of an allochthonous Dunnage Zone (Exploits Subzone) above the Gander Zone with the Mount Cormack and Meelpaeg subzones forming structural windows (Colman-Sadd and Russell, 1982; Colman-Sadd and Swinden, 1982, 1984; Williams et al., 1988). According to this model, the sedimentological differences between rocks of the Dunnage and Gander zones are explained by deposition in separate settings, although large distances between the depositional settings are not demanded.

The suggestion of an ophiolite substrate beneath the Gander Group (Miller, 1988) in the Gander Lake Subzone implies an allochthonous Gander Group and equivalents across the Central lower crustal block. If the single pyroxenite-gabbro occurrence in the central portion of the Gander Lake Subzone is intrusive, and if positive gravity anomalies here and elsewhere are the result of mafic intrusions, there is no reason to divorce the Gander Group and the Central lower crustal block. Regardless of the above inferences, the preponderance of middle Paleozoic granitic plutons indicates a Central lower crustal block of continental affinity.

The timing of Dunnage-Gander juxtapositioning is an important consideration in tectonic models. Colman-Sadd (1980) and Blackwood (1982) suggested a Silurian Dunnage-Gander linkage because Ordovician and Silurian rocks of the Exploits Subzone are conformable and the first deformation of the Ordovician rocks appears to be the same as the first deformation of the Silurian Botwood Group (Karlstrom et al., 1982). Colman-Sadd and Swinden (1984) reiterated this reasoning for emplacement of the Exploits Subzone on the Mount Cormack Subzone. However, other workers (Kennedy, 1976; Currie et al., 1979) have placed greater emphasis on the deformation of mafic-ultramafic rocks of the Gander River Complex during the Ordovician, as indicated by deformed gabbroic clasts in conglomerates at the base of the Davidsville Group and by angular unconformities between Davidsville conglomerates and serpentinite at Gander Lake and between Llandvirn limestone and serpentinite at Wiers Pond (Blackwood, 1978; Stouge, 1979). Ordovician Dunnage-Gander linkages are also indicated by Llanvirn conglomerates that contain mixed quartzite and ophiolitic clasts at the eastern periphery of the Mount Cormack Subzone (Dec and Colman-Sadd, 1990) and by the occurrence of ophiolitic mélange with a black shale matrix at the Exploits-Meelpaeg contact on the west side of Cold Spring Pond (Williams and Piasecki, 1990). The presence of a monomict ultramafic pebble conglomerate above the Coy Pond Complex also invites comparisons with Ordovician relationships between the Davidsville Group and Gander River Complex north of Gander Lake (Williams et al., 1988). Parts of the Gander Group and correlatives may have been deformed by ophiolite emplacement and Dunnage-Gander juxtapositioning. This is implicid in the early models of Kennedy (1975, 1976) and it is supported by structural studies at Gander Lake and northward (Williams et al., 1991). The latest U-Pb geochronological studies support Early Ordovician superpositioning of the Exploits Subzone above the Gander Zone with concomitant Ordovician metamorphism and plutonism (Colman-Sadd et al., 1992).

The expectable direction of initial transport of Dunnage Zone rocks above Gander Zone rocks is from west to east, that is, from the central oceanic tract outward across its marginal terranes. Ophiolitic mélange such as that at Cold Spring Pond may be an expression of this early transport. The lack of a uniform stratigraphy within Dunnage Zone rocks above the Gander Zone and the absence of Dunnage ophiolite suites at some boundaries are probably the result of structural omission or later structural rearrangement. Major transcurrent movements along the Gander Lake Subzone are later, coeval with plutonism and metamorphism dated as Silurian or Devonian (Bell et al., 1977; Dallmeyer et al., 1981a; Dunning et al., 1988, 1990). Northwest thrusting along the southwestward extension of Noel Pauls Line at Cape Ray is of two generations (Williams et al., 1989b): the first predating Devonian volcanic rocks of the Windsor Point Group and the second affecting the Devonian Windsor Point Group.

Structural styles, metamorphism and plutonism are all consistent with the model of a major Dunnage allochthon above the Gander Zone. A two layer crust explains the presence of widespread granitic plutons throughout the ophiolitic Dunnage Zone (Colman-Sadd and Swinden, 1984). A structural cover of Dunnage rocks above Gander rocks may have enhanced metamorphism, migmatization and plutonism by depressing the crust to a level where these processes were effective. Metamorphic conditions for rocks that cooled near the base of the Dunnage Zone are consistent with a depth of 13 km (Colman-Sadd and Swinden, 1984).

A uniform structural cover of Dunnage rocks on Gander rocks would not in itself explain variations in structural, metamorphic and plutonic styles within the Gander Zone. However, broad zones of sinistral ductile shear, while consistent with Dunnage-Gander relationships, could also localize plutonism and metamorphism. Thus from west to east across the Gander Lake Subzone, an increase in metamorphic grade, a change in structural style from flat to steep, and the presence of diapiric granitic plutons in the east are all consistent with a broad zone of sinistral shear that localized plutonism and metamorphism (Hanmer, 1981; Piasecki, 1988). Where the effects of the shearing are confined to the Gander Lake Subzone, its structural, metamorphic and plutonic styles contrast with those of the nearby Dunnage, which was not so affected. Where shearing affected rocks of both the Dunnage and Gander zones, structural, metamorphic and plutonic styles have no regard for zone boundary.

Acknowledgments

The authors thank R.F. Blackwood and B.F. Kean for reviews of an earlier draft of this contribution.

Nova Scotia

S.M. Barr, R.P. Raeside, and
R.A. Jamieson

Introduction

A description of rocks and structures of the Aspy Subzone of Cape Breton Island, or Aspy Terrane (Barr and Raeside, 1989), is included with the Gander Zone mainly because: (1) the Aspy Subzone lies between rocks of Humber affinity to the northwest and Avalon affinity to the southeast, (2) structural, metamorphic and plutonic styles of the Aspy Subzone are more akin to those of the Gander Zone than to any other zone, and (3) rocks of the Aspy Subzone may correlate with those of the Burgeo and Gander Lake subzones and La Poile Belt of southwest Newfoundland (Dunning et al., 1990b). However, no monotonous psammitic sequence like that of the Newfoundland and New Brunswick Gander zones is present and there are no indigenous layered

Barr, S.M., Raeside, R.P., and Jamieson, R.A.
1995: Gander Zone-Cape Breton Island, Nova Scotia; in Chapter 3 of Geology of the Appalachian-Caledonian Orogen in Canada and Greenland, (ed.) H. Williams; Geological Survey of Canada, Geology of Canada, no. 6, p. 212-216 (also Geological Society of America, The Geology of North America, v. F-1).

rocks of definite Ordovician or older age. In other analyses of Cape Breton Island, the Aspy Subzone has been interpreted as part of, or developed upon, the Avalon Zone (Keppie, 1989; Keppie et al., 1989, 1991a).

Aspy Subzone

The term Aspy Subzone is synonomous with the Highlands Zone or Aspy Terrane, terms introduced to refer to that part of the Cape Breton Highlands south of the Red River and Wilkie Brook fault zones and north and west of the Eastern Highlands shear zone (Barr and Raeside, 1986, 1989; Barr et al., 1987a; Dunning et al., 1990b; Raeside and Barr, 1992; Fig. 3.116). It is characterized by a variety of low- to high-grade metamorphic rocks, intruded by extensive suites of mainly Silurian-Devonian granitic rocks. The distinctiveness of the Aspy Subzone and the presence of its major bounding faults were recognized by systematic regional mapping (Barr et al., 1985, 1987a; Raeside et al., 1984; 1986; Jamieson et al., 1987, 1989, 1990).

The southwest extension of the Aspy Subzone beneath Carboniferous cover rocks to the Gillanders Mountain area and Mabou Highlands is inferred on lithological similarity of rock units (Fig. 3.116). Southwest of the Mabou Highlands, the Aspy Subzone may be offset to the northwest by a postulated northwest-trending fault (the Canso Fault). There are no correlatives in mainland Nova Scotia.

To the northeast, the Aspy Subzone may extend offshore to the Burgeo or Gander Lake subzones of southwestern Newfoundland. This correlation is based in part on geological similarities between these areas, which had also been noted by some earlier workers (e.g. Neale and Kennedy, 1975), but also on the basis of geophysical data (Loncarevic et al., 1989). The Aspy Subzone is characterized by subdued magnetic relief and negative anomalies, like the Gander Lake Subzone of Newfoundland.

The Eastern Highlands shear zone and associated faults, which form the southeastern boundary of the Aspy Subzone against the Bras d'Or Subzone, are not clearly imaged on a deep seismic reflection profile offshore in the Cabot Strait (Marillier et al., 1989; Loncarevic et al., 1989), but the data show some evidence that the boundary dips to the south, and that the Bras d'Or Subzone may be thrust against the Aspy Subzone. This is consistent with the presence of southerly dipping thrusts within the Eastern Highlands shear zone. The Chéticamp Pluton and associated diorites and gneisses in the western part of the Aspy Subzone are anomalously old and lithologically unlike other rocks of the subzone (Jamieson et al., 1987, 1989; Barr et al., 1986). They may be part of the Bras d'Or Subzone, tectonically emplaced above rocks of the Aspy Subzone.

Stratigraphy

The stratified rocks of the Aspy Subzone are greenschist to amphibolite facies phyllites and schists derived from volcanic and sedimentary protoliths and upper amphibolite facies gneisses of similar compositions. They are assigned local names (Fig. 3.116) because of uncertainties in correlation between areas separated by plutons or cover rocks (Macdonald and Smith, 1980; Jamieson et al., 1987, 1989, 1990; Barr et al., 1987b; Barr and Jamieson, 1991; Raeside and Barr, 1992). In a general way, mafic volcanic rocks are overlain by felsic pyroclastic rocks which interfinger with clastic sedimentary rocks. Felsic volcanic rocks are most abundant near Sarach Brook in the southeastern Aspy Subzone; the relative proportion of sedimentary rocks increases to the north and west.

Complex deformation and metamorphism, combined with poor exposure, make it difficult to determine whether tectonic or stratigraphic breaks exist within the metavolcanic-metasedimentary sequences, and the relations among phyllites, schists and gneisses are controversial. Overlapping metamorphic grades, similar structural and metamorphic histories, and the lack of any clear unconformity suggest that the phyllites/schists and the gneisses are equivalents (Craw, 1984; Jamieson et al., 1987; Plint and

Jamieson, 1989), although an unconformity has been suggested (Currie, 1982, 1987a, b). U-Pb zircon ages of felsic volcanic rocks indicate that they are early Silurian (Dunning et al., 1990b; Keppie et al., 1991). A similar age was obtained from tonalitic orthogneiss within the high grade gneissic complex (Jamieson et al., 1986). Hence the protolith ages of metamorphic rocks of the Aspy Subzone are inferred to be mainly Late Ordovician to Early Silurian (Barr and Jamieson, 1991). The mafic volcanic units have petrological characteristics like those of volcanic-arc tholeiites, and are interpreted to have formed in a subduction zone (Jamieson et al. 1989, 1990; Barr and Jamieson, 1991).

Unmetamorphosed Devonian-Carboniferous strata unconformably overlie the older rocks.

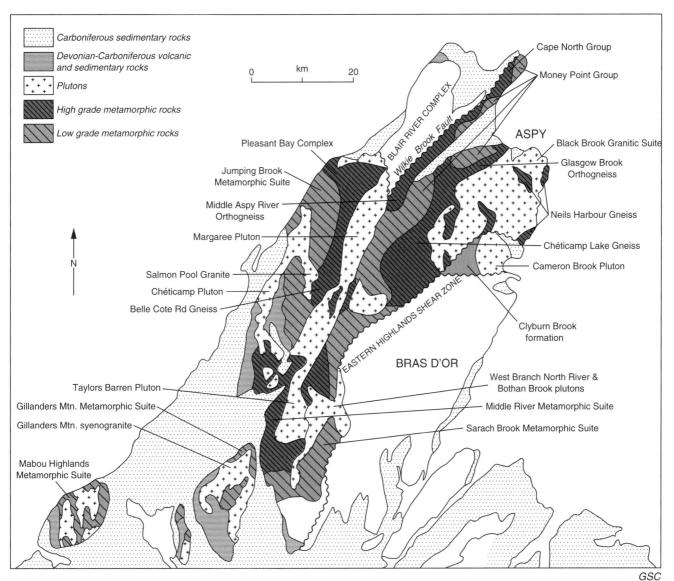

Figure 3.116. Simplified geological map of northern Cape Breton Island showing the locations of units referred to in the text within the Aspy and adjacent subzones.

Jumping Brook Metamorphic Suite and correlative units

The Jumping Brook Metamorphic Suite was defined by Jamieson et al. (1987) to include low-, medium-, and high-grade metasedimentary, metavolcanic, and metaplutonic rocks in the west-central Highlands, including those assigned to the Jumping Brook Complex (Currie, 1983, 1987a, b). The suite is subdivided into five lithologic units: (1) mafic to intermediate metavolcanic rocks (Faribault Brook metavolcanics); (2) metaconglomerate, quartz sericite schist, and minor quartzite (Barren Brook schist); (3) pelitic, semi-pelitic, and psammitic schist (Dauphinee Brook schist); (4) pelitic and semi-pelitic schist with minor marble and psammite (Corney Brook schist); and (5) medium- to coarse-grained metabasite, at least in part with gabbroic to dioritic protoliths (George Brook amphibolite). The first three units all have greenschist facies mineralogy. The Corney Brook schist is interpreted to be a higher grade equivalent of these three units.

The term Money Point Group was introduced for relatively low grade metamorphic rocks on Cape North Peninsula (Macdonald and Smith, 1980), and was extended to lithologically similar rocks farther west and south where the metamorphic grade is higher (Raeside et al., 1986; Barr et al., 1987a, b; Raeside and Barr, 1992). Major lithologies are hornblende-mica schist and amphibolite; at low grades on the Cape North Peninsula, Macdonald and Smith (1980) described these rocks as metabasalt, metatuff, and meta-rhyolite. The meta-rhyolite has a U-Pb zircon date of 427.5 ± 4 Ma, interpreted as the age of extrusion (Keppie et al., 1992). Metasedimentary lithologies form about 40 per cent of the Money Point Group and include pelites, semi-pelites, calc-silicate rocks, and rare marbles. Detailed stratigraphic relations have not been determined in the Money Point Group, but lithologies are similar to those in the Jumping Brook Metamorphic Suite and the units are therefore correlated.

The name Cape North Group was introduced by Macdonald and Smith (1980) for high grade metamorphic rocks on Cape North Peninsula. It was extended to the west and south by Raeside et al. (1986) and Barr et al. (1987a, b). On Cape North Peninsula, the Cape North and Money Point groups display differences in lithology and metamorphic grade. To the west and south, metamorphic contrasts are less apparent but subtle lithological differences persist, and hence the groups are considered to be separate. However, the Cape North Group is also considered a correlative of the Jumping Brook Metamorphic Suite. The major lithologies in the Cape North Group are semipelitic, pelitic, and calc-silicate gneisses, with minor amphibolite and marble. Pelitic lithologies are porphyroblastic mica schists with abundant kyanite, garnet, and locally sillimanite.

The Sarach Brook Metamorphic Suite corresponds to the Northern (Crowdis Mountain) volcanic unit of Jamieson and Doucet (1983). It consists of variably deformed greenschist-facies volcanic and volcaniclastic rocks bounded by shear zones and intruded by varied granitoid units. The western part of the suite consists of metasedimentary and interlayered metabasic rocks. The eastern part of the suite is dominantly felsic crystal-lithic tuff, epiclastic volcanigenic sedimentary rocks, and minor flow-banded rhyolite. A rhyolitic crystal tuff sample from near the structural top of the suite has a U-Pb zircon crystallization age of 433 +7/-4 Ma (Dunning et al., 1990b). These felsic rocks are interpreted to correlate with felsic tuffs and rhyolitic dykes and sills in the Jumping Brook Metamorphic Suite to the northwest. A rhyolite dyke in the Chéticamp Pluton adjacent to the Jumping Brook Metamorphic Suite has a similar zircon age (Currie et al., 1982).

The Middle River Metamorphic Suite corresponds to part of the Middle River unit of Jamieson and Doucet (1983), which was inferred to have a gradational relationship with the Sarach Brook Metamorphic Suite. In its southern part, the suite consists of low- to medium-grade metasedimentary rocks (quartz-biotite-muscovite phyllites and schists, quartz-biotite-oligoclase-muscovite gneisses, and rare marble and metaconglomerate) with metabasite horizons.

The Gillanders Mountain Metamorphic Suite was mapped by French (1985), who divided the metamorphic rocks into three units: (1) massive to foliated amphibolite, garnetiferous semipelitic schist, and minor graphitic schist; (2) semipelitic schist, quartzofeldspathic gneiss, impure quartzite, amphibolite, and mafic to felsic metavolcanic rocks; and (3) psammitic to pelitic schist, metabasite, and minor quartzofeldspathic augen gneiss. The overall lithological character suggests correlation with metamorphic rocks of the Middle River Metamorphic Suite.

The Mabou Highlands Metamorphic Suite is divided into five units (Barr and Macdonald, 1989): quartz schist and phyllite, lithic metatuff, mafic schist and amphibolite, banded schists (including metarhyolite), and an injection gneiss complex. Lithologically, the first four units are very similar to the Middle River and Sarach Brook metamorphic suites, with felsic volcanic units overlying more mafic volcanic units toward the top of the succession. The injection gneiss complex is an assemblage of quartzofeldspathic and amphibolitic gneisses with minor marble injected by a variety of dioritic to granitic sheets and dykes. It appears to have developed around large dioritic plutons and may be essentially the higher grade equivalent of the first four units (Barr and Macdonald, 1989).

Pleasant Bay Complex

The Pleasant Bay Complex was defined by Currie (1983, 1987a, b), although boundaries were modified by subsequent mapping (Jamieson et al., 1987; Raeside et al., 1986). It includes tonalitic to granodioritic orthogneiss, amphibolite, pelitic gneiss, minor quartzite and marble, and a variety of foliated granitic rocks and pegmatites. In its southern part, the Pleasant Bay Complex is dominated by orthogneiss, termed the Belle Côte Road Gneiss (Jamieson et al., 1986). This intrusive unit was deformed and metamorphosed with other high grade gneisses of the complex. U-Pb data from igneous and metamorphic zircons suggest that the age of the orthogneiss is 433 +20/-10 Ma (Jamieson et al., 1986).

Chéticamp Lake Gneiss

The Chéticamp Lake Gneiss occurs in the central part of the Cape Breton Highlands, near the boundary with the Bras d'Or Subzone to the southeast (Fig. 3.116). It was first recognized by Raeside et al. (1984), who informally referred to it as the central Cape Breton Highlands gneiss. It was subsequently described and named by Raeside and Barr

(1986), Barr et al. (1987b), and Raeside and Barr (1992). The most abundant lithologies are semipelitic gneiss, biotite schist, and massive quartzofeldspathic orthogneiss, with rare amphibolite and hornblende leucogneiss. Dykes and other small intrusions of granite and pegmatite related to the Black Brook Granitic Suite are abundant. The orthogneiss (which makes up about 20 per cent of the unit) is of monzogranitic to granodioritic composition, and may have formed during high-grade metamorphism and migmatization of the Chéticamp Lake unit. A U-Pb zircon age of 396 ± 2 Ma is interpreted to be the age of intrusion of the orthogneiss, and the age of accompanying metamorphism and migmatization (Dunning et al., 1990b).

Neils Harbour Gneiss

Heterogeneous gneissic rocks (the Coastal gneisses of Wiebe, 1972) occur along the northeastern edge of the Aspy Subzone, mainly as roof pendants and xenoliths in the Black Brook Granitic Suite (Yaowanoiyothin, 1988). They are mainly K-feldspar-megacrystic granodioritic orthogneisses with minor screens of pelitic schist. U-Pb zircon dating indicates an early Devonian age, like the adjacent Cameron Brook Granodiorite (Dunning et al., 1990b).

Metamorphism and deformation

Metamorphic rocks in the Aspy Subzone display mineral assemblages consistent with greenschist and amphibolite facies Barrovian metamorphism. A sequence of zones involving staurolite, kyanite, and sillimanite have been described (Phinney, 1963; Macdonald and Smith, 1980; Plint and Jamieson, 1989; Jamieson et al., 1989, 1990; Raeside and Barr, 1992). Lower grade assemblages include garnet, biotite, chloritoid, and chlorite. Kyanite-K-feldspar assemblages occur locally and indicate pressures of over 8.5 kbar (Plint and Jamieson, 1989). However, many of the rocks have been overprinted by retrograde metamorphism, including widespread extensive development of coarse grained muscovite which has mid-Devonian cooling ages (Reynolds et al., 1989).

The metamorphic units throughout the Aspy Subzone record at least two phases of pervasive deformation, as well as later ductile shearing and brittle shearing (Jamieson et al., 1987, 1989; Raeside and Barr, 1992).

Early deformation and metamorphism of the Jumping Brook Metamorphic Suite and inferred correlative units throughout the Aspy Subzone probably accompanied syntectonic emplacement of the Belle Cote Road orthogneiss in the early Silurian. This suggests that metamorphism is associated with intrusion of the orthogneiss (Jamieson et al., 1987). However, a Devonian age for orthogneiss in the Chéticamp Lake Gneiss, combined with foliated to unfoliated character of Devonian plutons (see below) in the Aspy Subzone indicates that deformation, metamorphism, and plutonism continued into the Devonian.

The Eastern Highlands shear zone is a zone of retrograded mylonite and strongly sheared chlorite schist up to 800 m wide. Towards the northeast and southwest, the shear zone splays into a series of low angle fault zones, intruded by Devonian plutons, and from there extends offshore to the northeast and under Carboniferous cover sequences to the southwest. Movement on the Eastern

Highlands shear zone was probably protracted (Lin and Williams, 1990), but U-Pb dates and ^{40}Ar-^{39}Ar cooling ages suggest that the Aspy and Bras d'Or subzones were amalgamating during the Silurian and were juxtaposed by the late Devonian, when rapid cooling and exhumation occurred, perhaps related to accretion with the Grenville block to the north (Reynolds et al., 1989; Dunning et al., 1990b).

Plutonism

A diverse suite of mainly Silurian and Devonian plutons, which range from strongly foliated (gneissic) to unfoliated, is a distinctive feature of the Aspy Subzone (Barr, 1990).

Gneissic and strongly foliated plutons (pre- to syntectonic)

Granitic to tonalitic orthogneisses are major components of the Pleasant Bay Complex and Chéticamp Lake Gneiss. In other areas they form separate mappable units and have been given separate names (Middle Aspy River Granitic Orthogneiss, Glasgow Brook Granodioritic Orthogneiss, Taylors Barren Pluton). The Belle Cote Road orthogneiss within the Pleasant Bay Complex has a U-Pb zircon age of 433 ± 20 Ma (Jamieson et al., 1986). In contrast, orthogneiss in the Chéticamp Lake Gneiss has a Devonian igneous age (396 ± 2 Ma, U-Pb zircon), as has orthogneiss within the Neils Harbour Gneiss (403 ± 3 Ma) in the northeastern Aspy Subzone (Dunning et al., 1990b). The strongly foliated Taylors Barren Pluton has a Rb-Sr isochron age of 419 ± 17 Ma (Gaudette et al., 1985). These plutons were intruded during Silurian-Devonian metamorphism and deformation of the Aspy Subzone.

Variably foliated plutons (syntectonic)

Extensive Devonian plutonism in the Cape Breton Highlands has long been known (Wiebe, 1975), based on pioneering Rb-Sr work of Cormier (1972), and the Devonian age has now been confirmed by U-Pb and ^{40}Ar-^{39}Ar dating (Dunning et al., 1990b; Reynolds et al., 1989). The Cameron Brook Granodiorite, located between splays of the Eastern Highlands shear zone, is early Devonian (402 ± 3 Ma), similar in age to orthogneisses within the Chéticamp Lake Gneiss and Neils Harbour Gneiss. However, the Cameron Brook Granodiorite is unaffected by the pervasive deformation present in the orthogneisses of the same age, perhaps because of its protected location between two branches of the Eastern Highlands shear zone.

Major plutonism continued to the middle Devonian, forming the extensive Black Brook Granitic Suite (Yaowanoiyothin, 1988). The major locus for plutonism was the Eastern Highlands shear zone but dykes and small plutons of Black Brook type are widespread throughout the northern part of the Aspy Subzone. For example, the Pleasant Bay Granite in the Pleasant Bay Complex is lithologically identical to parts of the Black Brook Granitic Suite and has a similar Rb-Sr age (Cormier, 1972). Some parts of the Black Brook Granitic Suite are foliated, and other parts are not, suggesting that deformation may have waned or become more localized during the plutonism. Widespread cooling ages of 370 to 390 Ma on hornblende, muscovite, and biotite in metamorphic and plutonic rocks of the Aspy Subzone suggest rapid cooling accompanying exhumation from more than 25 km to less than 10 km depth (Reynolds et al., 1989; Plint and Jamieson, 1989).

Unfoliated plutons (post-tectonic)

Mid- to late Devonian plutons are widespread in the Aspy Subzone. These are non-foliated, have A-type characteristics, and are of two main types: (1) coarse grained, commonly megacrystic hornblende-biotite and biotite granites, such as the Margaree, West Branch North River, and Bothan Brook plutons (O'Beirne-Ryan et al., 1986; O'Beirne-Ryan and Jamieson, 1986); and (2) fine grained syenogranites. Numerous small bodies of type (1) occur near the northern margin of the Aspy Subzone (Barr et al., 1986). The Salmon Pool pluton has a U-Pb zircon age of 365 +10/-5 Ma (Jamieson et al., 1986). The Gillanders Mountain syenogranite has been shown to be a subvolcanic intrusion, co-magmatic with rhyolites of the Devonian-Carboniferous Fisset Brook Formation (French, 1985).

These post-tectonic plutons probably formed in response to post-collisional extension. Prominent late brittle faults in the Aspy Subzone cut the Margaree and West Branch North River plutons, causing lateral offsets of up to a few kilometres (Raeside and Barr, 1992).

Summary

The Aspy Subzone is characterized by upper Ordovician to lower Silurian metasedimentary and meta-igneous rocks and abundant Silurian and Devonian plutonic rocks. Protoliths of the metamorphic units appear to have formed in or above a subduction zone. Metamorphism and protracted deformation occurred during Silurian to Devonian juxtapositioning of the Aspy Subzone and Blair River Complex (Subzone) to the north and the Bras d'Or Subzone to the south.

New Brunswick

Cees van Staal and L.R. Fyffe

Introduction

The New Brunswick Gander Zone is characterized by a thick, complexly deformed sequence of Cambrian-Ordovician quartz sandstone and pelite that is lithologically similar to the Gander Group (Kennedy, 1976) of Newfoundland. These quartzose rocks constitute the Miramichi and Cookson groups in the Miramichi and St. Croix subzones, respectively. The stratigraphy of the Miramichi Highlands or Miramichi Terrane (Fyffe and Fricker, 1987) has previously been compared solely to that of the Gander Zone in Newfoundland (Rast et al., 1976a), but the integral association of voluminous Ordovician volcanic rocks in northeastern New Brunswick gives the region characteristics of both the

Gander and Dunnage zones (Fyffe, 1977; van der Pluijm and van Staal, 1988). In order to correspond with established tectonostratigraphic nomenclature from Newfoundland, the Cambrian-Ordovician clastic sequence of the Miramichi Highlands is included in the Gander Zone as the Miramichi Subzone, while the younger volcanic sequence is included in the Dunnage Zone (see Fig. 3.80). High grade metamorphic complexes in the central Miramichi Highlands have been interpreted as crystalline basement to the clastic sequence by analogy with the Gander Zone of Newfoundland (Rast et al., 1976a), although no definite basement is known in Newfoundland. Recent U-Pb geochronology has shown that at least the intrusive components of these complexes in New Brunswick are Ordovician (Fyffe et al., 1988a).

Extension of the New Brunswick Gander Zone to include the St. Croix Subzone differs from that of Rast et al. (1976a), who included only rocks of the Miramichi Highlands. The stratigraphy within the St. Croix Subzone compares with that of the Newfoundland Gander Zone in that both contain an abundance of clastic sedimentary rocks with relatively minor volcanic rocks, intruded by garnet-muscovite-bearing granites. Inclusion of the St. Croix Subzone extends the Gander Zone of New Brunswick southward to the Avalon Zone boundary and, thus, corresponds with its limits in Newfoundland. This boundary in New Brunswick is mainly hidden by Silurian to Lower Devonian volcanic and sedimentary rocks of the Mascarene Belt. The St. Croix Subzone is separated from the Miramichi Subzone by Silurian turbidites of the Fredericton Belt (Fig. 3.80). A wedge of complexly deformed pelite and psammite of Gander aspect (Fig. 3.80) has recently been recognized along the Fredericton Fault, which transects the central portion of the Fredericton Belt (Fyffe, 1988).

Miramichi Subzone

The Miramichi Group (previously informally referred to as the lower Tetagouche Group) in the northern part of the Miramichi Highlands is represented by the Knights Brook Formation (new name after brook flowing into Nepisiguit River, east of the Nepisiguit Falls) and Chain of Rocks Formation (new name after rapids on Nepisiguit River). The Knights Brook Formation consists of dark grey to black, locally graphitic shales, rhythmically interbedded with medium- to thick-bedded, light grey to olive-green, graded quartz sandstones (Fig. 3.117). The Chain of Rocks Formation is compositionally and texturally similar to the Knights Brook Formation, but does not contain the dark grey to black shales (Fig. 3.118). The Knights Brook Formation structurally overlies the Chain of Rocks Formation. The boundary is placed where the black shale beds disappear and are replaced by siltier greenish grey phyllites (cf., Helmstaedt, 1971; van Staal et al., 1988b). Although no fossils have been found in either the Knights Brook or Chain of Rocks formations, Tremadoc to middle/late Arenig graptolites were discovered in the Bright Eye Brook Formation, a lithological correlative of the Knights Brook Formation in the Woodstock area of west-central New Brunswick (Fyffe et al., 1983).

Black slates of the Bright Eye Brook Formation conformably overlie olive-green quartz wackes and light green shales that are interbedded locally with maroon and green siltstones and shales. Grading locally varies from inverse at the base of a bed to normal in the rest of the bed. Flute

van Staal, C. and Fyffe, L.R.
1995: Gander Zone-New Brunswick; in Chapter 3 of Geology of the Appalachian-Caledonian Orogen in Canada and Greenland, (ed.) H. Williams; Geological Survey of Canada, Geology of Canada, no. 6, p. 216-223 (also Geological Society of America, The Geology of North America, v. F-1).

casts, current ripples, and flame structures are present locally (Fig. 3.119). Similar turbidites along strike in the Danforth area of adjacent Maine are known as the Baskahegan Lake Formation (Ludman, 1987).

The greenschist-facies quartz sandstones and shales in the central Miramichi Highlands (correlatives of Knights Brook and Chain of Rocks formations) pass westward through a steep metamorphic gradient into amphibolite facies rocks of the Trousers Lake Complex (Skinner, 1975; Fyffe and Pronk, 1985; Fyffe et al., 1988a). This complex has been traced south-southeastward from the Trousers Lake area to the Catamaran Fault, where it is dextrally offset by a few kilometres, and thence continues southward to the western part of the Hayesville area. The Trousers Lake Complex is characterized by thin banded, fine grained amphibolites which contain dark green hornblende-rich bands alternating with light grey plagioclase-rich bands. The amphibolites are interlayered with cordierite-andalusite-bearing pelites and psammites, and with granitic gneiss containing microcline augen. Locally, partial melting of the psammitic and pelitic beds has progressed to the extent that the more competent amphibolite layers occur as isolated rafts within the mobilized, sillimanite-bearing host (St. Peter, 1981). Although originally interpreted as Precambrian basement to the Miramichi Group (Rast et al., 1976a), U-Pb ages of 451 +15/-1 Ma and 434 +29/-6 Ma have recently confirmed previously published Ordovician Rb-Sr ages for the granitic gneisses of the complex (Fyffe et al., 1988a).

St. Croix Subzone

The St. Croix Subzone of southern New Brunswick underlies a northeast-southwest-trending belt that extends from Canaan River, 50 km west of Moncton, southwestward to St. Stephen on the border with Maine. Its extension into Maine, although interrupted by numerous Devonian plutons, can be traced over 100 km to the Penobscot Bay area (Ludman, 1987). Ruitenberg (1967) separated rocks in the St. Croix Highlands into Silurian and Ordovician successions, and included the Ordovician rocks in the Cookson Formation, named after exposures on Cookson Island at the head of Oak Bay. Ludman (1987), mapping on the Maine side of the border, elevated the formation to group status and established new formations within it. These have been extended into the St. Stephen area of New Brunswick and their stratigraphic order revised by Fyffe and Riva (1990).

The Calais Formation, the lowermost unit of the Cookson Group, comprises black carbonaceous slate interstratified with minor thin bedded wacke. A pillowed basalt member, about 100 m thick, occurs near the top of the formation (Fig. 3.120). The flows have been thermally metamorphosed to a plagioclase-actinolite-biotite-sphene-iron oxide assemblage during emplacement of nearby Devonian gabbroic plutons. The basalt is moderately evolved with trace element abundances similar to some passive continental margin tholeiites (Fyffe et al., 1988b). Black slates in the contact aureoles contain abundant cordierite and andalusite porphyroblasts. A thick sandstone-rich sequence, conformably overlying the Calais Formation, is divided into

Figure 3.117. Interference pattern between upright F$_2$ fold and northeast-trending open F$_4$ fold within phyllites of the Knights Brook Formation, Miramichi Subzone. GSC 1994-767Q

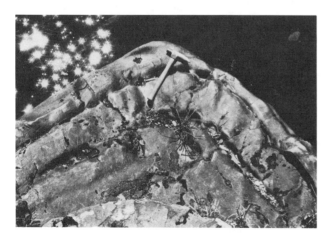

Figure 3.118. F$_2$ fold in medium-bedded quartz wacke of the Chain of Rocks Formation, Tetagouche Group, on Middle River, Miramichi Subzone. GSC 1994-767R

Figure 3.119. Flute casts in quartz wackes of the Miramichi Subzone. GSC 1994-767S

three formations. The Woodland Formation contains rhythmically interstratified, thin- to medium-bedded (2-20 cm), quartzofeldspathic wacke and slightly to highly carbonaceous slate (Fig. 3.121). The wacke is graded and about as equally abundant as the slate. The Digdeguash Formation comprises interbedded dark grey slate, greenish grey quartz wacke, lithic wacke, and granule conglomerate containing abundant volcanic clasts (Fig. 3.122). Although assigned to the Silurian by Ruitenberg (1967), it is considered by Fyffe and Riva (1990) to be a facies of the Woodland Formation. The overlying Kendall Mountain Formation is mainly thin- to thick-bedded (10 cm-2 m), quartzose arenite with interbeds of siltstone, black slate conglomerates, and mafic volcanic rocks (Fig. 3.123). The Kendall Mountain Formation is estimated to be between 500 to 700 m thick and the Woodland Formation, about 800 m thick (Ludman, 1987).

The Basswood Ridge-Pendar Brook Fault separates the St. Croix Subzone from Silurian wackes of the Fredericton Belt to the northwest. An unconformity between the Calais

Formation and Silurian conglomerate of the Oak Bay Formation is exposed on Cookson Island in the southeast. The graptolite *Clonograptus herrmanni* found immediately beneath the unconformity provides an Early Ordovician (Tremadoc) age for the Calais Formation (Cumming, 1967). Graptolites from black slate of the Kendall Mountain Formation belong to the *Climacograptus wilsoni* Zone of the Caradoc (Fyffe and Riva, 1990). The Cookson Group has been lithologically correlated with the Miramichi Group (Rast and Stringer, 1974; Fyffe et al., 1983). It has also been correlated with parts of the Saint John Group of the Avalon Zone (Ludman, 1987), although this has been disputed

Figure 3.122. Southward-verging F$_3$ folds in interbedded wacke and slate of the Digdeguash Formation, Cookson Group, St. Croix Subzone. GSC 1994-767V

Figure 3.120. Pillow basalt in the Calais Formation of the Cookson Group, St. Croix Subzone. GSC 1994-767T

Figure 3.121. Isoclinal F$_1$ folds in interbedded wacke and slate of the Woodland Formation, Cookson Group, St. Croix Subzone. GSC 1994-767U

Figure 3.123. Tight F$_1$ folds in thin bedded quartz arenite and slate of the Kendall Mountain Formation, Cookson Group, St. Croix Subzone. GSC 1994-767W

(Fyffe and Pickerill, 1986). Equivalents in coastal Maine are the Megunticook and Penobscot formations (Ruitenberg and Ludman, 1978; Ludman, 1987).

The St. Croix Subzone to the northeast of the Saint John River contains a thick, steeply dipping volcanic succession (Annidale belt) of Dunnage affinity that, on the basis of rare facing evidence, appears to young to the southeast (McLeod et al., 1990). In the lower part of the volcanic pile, pillow lava and massive mafic flows are overlain by felsic tuffs, wacke, and carbonaceous slate. Sheared mafic tuffs, extensively altered to calcite and chrome-mica, are interbedded with wacke and slate in the upper part of the pile. Serpentinized ultramafic rocks occupy shear zones within mafic tuffs. Zircons from a massive felsite, which apparently represents a dome intruded into the base of the volcanic pile, yielded a Tremadoc age of 493 ± 2 Ma (McLeod et al., 1991).

Structure and metamorphism (Gander and Dunnage zones)

Structural studies and regional mapping indicate that the principal elements of the tectonometamorphic history of the Gander and Dunnage zones of New Brunswick are remarkably similar, despite their complex, polyphase character (Fyffe, 1982b; van der Pluijm and van Staal, 1988). The Middle Ordovician and older rocks record evidence of several generations of folds and cleavages (e.g., Helmstaedt, 1971, 1973; Irrinki, 1979; Venugopal, 1979; Stringer and Burke, 1985; van Staal and Williams, 1984; van Staal, 1987) and a complex ductile-brittle faulting history (van Staal and Williams, 1988). The deformation is markedly heterogeneous on all scales (van Staal, 1985, 1986) and, as a result, the amount of strain recorded in the rocks as well as the spectrum of structures present varies from place to place. There is no correspondence between structural styles and stratigraphic divisions. A notion that older rocks exhibit greater strain and number of fold generations has led to tenuous tectonic interpretations. For instance, McBride (1976a, b), Pajari et al. (1977), and Rast and Stringer (1980) claimed the existence of a cryptic, angular unconformity within the Tetagouche Group on the basis of a lesser degree of deformation in the upper Caradoc (*D. clingani* zone) black phyllites at its very top. This unconformity was cited as evidence for the existence of the classical Middle Ordovician Taconic Orogeny. Subsequent work (van Staal et al., 1988b) failed to substantiate this observation and instead demonstrated that the earliest penetrative structures affecting the Miramichi and Tetagouche groups are also present in the upper Caradoc black phyllites.

The polyphase nature of the deformation has been studied most intensively in the northern part of New Brunswick, mainly because of relatively good outcrop, the presence of three dimensional exposure in mines and open pits (e.g., van Staal and Williams, 1984, 1986) and its immediate application to ongoing exploration. At least five generations of structures (F_1-F_5) have been recognized by overprinting relationships. They have been grouped into four phases of deformation (D_1-D_4), formed in response to one continous orogenic event that is constrained between the Late Ordovician and the Middle Devonian on paleontological data, ^{40}Ar-^{39}Ar age dates (van Staal et al., 1990; Ravenhurst and van Staal, unpublished data), and U-Pb zircon ages (see below). The earliest structures (D_1) are

interpreted to have formed by a progressive, thrusting in an accretionary wedge setting (Fig. 3.97d; van Staal 1987; van Staal et al., 1990). These structures comprise recumbent to steeply inclined, intrafolial, non-cylindrical folds (e.g. sheath folds) and a bedding-parallel, transposition foliation. Narrow zones of high strain mark the sites of ductile thrusts (van Staal and Williams, 1984; van Staal, 1987). The geometry of the thrust "cut-offs" and volcanic horses (Fig. 3.88) suggests that the thrusting was in a southerly or easterly direction. Such a sense of movement is consistent with the generally steeply plunging L_1 stretching lineation and vergence of the known F_1 folds.

The D_1 deformation is commonly accompanied by pumpellyite-actinolite- and/or high- pressure greenschist-facies metamorphism (M_1) in the Armstrong Brook Subzone (van Staal et al., 1990). A relatively narrow but extensive zone of M_1 blueschist-facies metamorphism (Skinner, 1974; Trzcienski et al., 1984) marks the tectonic contact between the Armstrong Brook and Bathurst subzones (van Staal et al., 1990). A few S_1 muscovites analyzed in the Bathurst Subzone are phengites as opposed to the more typical muscovites found in the S_2 foliation (van Staal, 1985). These variations in muscovite composition may reflect decompression between D_1 and D_2, and suggest that D_1 was associated with a relatively high pressure regional metamorphism (M_1), although blueschist-facies conditions were not reached in the Bathurst Subzone. M_1 metamorphic conditions in the Bathurst Subzone varied from prehnite-pumpellyite facies to greenschist facies. The increase in metamorphic grade (e.g., biotite isograd) towards the southwest (Helmstaedt, 1973; Skinner, 1974) appears to be mainly an M_2 effect, superimposed on M_1 (see below). The D_1 deformation and associated metamorphism (M_1) started in the Late Ordovician and probably lasted into the Early Silurian (van Staal et al., 1990; see below). Exhumation of parts of the accretionary wedge (Fig. 3.97) started at least in the Early Silurian (see also van Staal et al., 1990), since the Fournier Group is unconformably overlain by Upper Llandovery conglomerates and red beds in the Belledune Subzone (Noble, 1976) and by Upper Llandovery to Ludlow conglomerates, containing metagabbro pebbles (Fig. 3.124) of the Middle Ordovician Upsalquitch gabbro (Helmstaedt, 1971) in the Armstrong Brook Subzone.

Figure 3.124. Foliated gabbroic boulder within upper Silurian conglomerate of the Chaleurs Group. GSC 1994-767X

The D_2 deformation was penetrative and resulted in close to isoclinal, upright folds (F_2), which generally refolded all the earlier D_1 structures into steeply dipping attitudes (steep belts). The upright F_2 folds are refolded by open to tight, recumbent F_3 folds (referred to as F_5 by van Staal and Williams, 1984). Where the D_3 strain is high, the D_2 steep belts are deformed into flat belts (van Staal, 1987; de Roo et al., 1990; de Roo and van Staal, 1991). A well developed flat-lying S_3 transposition foliation is developed in the flat belts (Fig. 3.125), which are subsequently refolded into dome and basin structures by F_4 and F_5 folds (de Roo and van Staal, 1991). Although F_4 and F_5 folds represent two different generations of structures, they are believed to be kinematically related and therefore interpreted to have formed as a result of one distinct phase of deformation (D_4). The style of F_4 and F_5 folds is identical, comprising open to tight crenulations, conjugate and single kinkbands, and chevron and box folds. D_4 structures range from small, to large, kilometre-scale folds, which become tighter toward the Rocky Brook-Millstream Fault. The Tetagouche Antiform and Nine Mile Brook Synform (Skinner, 1974; van Staal, 1986, 1987) are major late D_4 structures with wavelengths and amplitudes in the order of several kilometres.

The deformation history of Silurian rocks in the narrow belt that separates the Armstrong Brook Subzone from the Belledune Subzone (Fig. 3.) is also polyphase and comprises at least four generations of structures (F_1-F_4). The F_1 structures have only been observed in Llandovery rocks of the Limestone Point Formation (Stringer, 1975; Noble, 1976). They comprise asymmetrical, isoclinal folds that are not accompanied by an axial planar cleavage. The F_1 folds are refolded by upright, centimetre to kilometre-scale F_2 folds that locally have a weakly developed axial planar cleavage (S_2). The F_2 folds are consistently transected by a strong domainal cleavage (S_3) in an anticlockwise fashion. This cleavage, which is the main tectonic fabric in the Silurian rocks (Stringer, 1975), is associated in a few places with asymmetrical, Z-shaped folds (F_3) that overprint the F_2 folds. Toward the major splays of the Rocky Brook-Millstream Fault system, the cleavage is rotated into parallelism with the trend of the F_2 axial planes, suggesting that F_3 and S_3 are a result of dextral transpression, with the strain partitioned into dextral shear in the narrow fault zones of the Rocky Brook-Millstream Fault system and a more or less coaxial shortening between the fault zones. Shallowly plunging stretching lineations in rocks close to the major splays of the Rocky Brook-Millstream Fault system (e.g., elongated pebbles and porphyroclasts), asymmetrical boudinage, and tension gashes in various stages of rotation further indicate that the Rocky Brook-Millstream Fault system accommodated a significant amount of dextral transcurrent shear. The S_3 cleavage is locally overprinted by kinkbands and brittle shears (F_4), which may represent the late brittle expression of a continuous progressive, transpression-related deformation. Another major ductile-brittle, dextral fault zone like the Rocky Brook-Millstream Fault is the Catamaran Fault in central New Brunswick (Anderson, 1972). Together these two fault zones divide the Miramichi Highlands into crude rhombohedra (van Staal and Williams, 1988). F_3-F_4 structures in the Silurian rocks are correlated with the large northeast-trending D_4 structures in the Ordovician rocks south of the Rocky Brook-Millstream Fault. D_4 deformation in

Figure 3.125. Recumbent F_2 folds transposed into a shallow dipping orientation by F_3 within the Nepisiguit Falls Formation at Devils Elbow, Bathurst Subzone. GSC 1994-767Y

the Ordovician rocks is thus also related to the dextral transpression. The F_3 structures affect Pridoli turbidites of the La Plante member of Petit Rocher Formation (Noble, 1985) and are cut by the Early-Middle Devonian Nicolas Denys stock (Davies et al., 1969; U-Pb zircon age of 381 ± 4 Ma; M.L. Bevier, pers comm., 1991). Porphyries associated with this pluton cut some of the subsidiary faults of the Rocky Brook-Millstream Fault system. A significant part of the dextral, ductile transpression thus took place in the Early Devonian. D_2 and D_3 in the Ordovician rocks, therefore, took place before the Early Devonian. Dextral motion may have continued after the Middle Devonian as a rhombohedral-shaped body of Carboniferous coarse clastic rocks is associated with a right-hand step in the main break of the Rocky Brook-Millstream Fault northwest of Bathurst (Fig. 3-2b.1), possibly indicating a small pull-apart basin.

F_1 and F_2 folds in the Silurian rocks are correlated on style and orientation with D_1 and D_2 structures in the Ordovician rocks, respectively. If correct, the F_1 in the Llandovery rocks should represent folds that formed in the upper levels of the accretionary wedge.

^{40}Ar-^{39}Ar dating of low greenschist-facies, muscovite porphyroblasts related to peak metamorphism (M_2) in the Bathurst Subzone yield cooling ages between 410 and 400 Ma (Ravenhurst and van Staal, unpublished results). These cooling ages suggest that M_2 only slightly predates the earliest known evidence for dextral transpression associated with the Rocky Brook-Millstream Fault, using the absolute time scale of Palmer (1983). The M_2 metamorphism is marked by: (1) the introduction of biotite at the expense of chlorite and K-feldspar in the felsic volcanic and epiclastic rocks; (2) hornblende at the expense of actinolite in mafic rocks; and (3) grunerite, garnet, and rare chloritoid in the iron-formation or Fe/Mn-rich metalliferous sediments (van Staal, 1985). These porphyroblasts grew across the S_1 cleavage and are locally deformed by F_2. Peak M_2 metamorphic conditions were thus reached after the main D_1 deformation but before the end of D_2 deformation. Recrystallization of biotite and clino-amphiboles accompanied F_4 but not F_5 folding. Widespread alteration of the porphyroblasts to chlorite, Fe-hydroxides and epidote took place before F_5 since these alteration products were deformed during this folding event, indicating that metamorphism was certainly retrograde and waning after D_4 deformation.

The cooling ages of muscovite porhyroblasts and other age constraints suggest that M_2 metamorphism is linked to the dominantly Upper Silurian granitic and associated gabbroic plutons of the bimodal Acadian magmatic suite (Bevier and Whalen, 1990) that intrudes the Miramichi Highlands and Chaleurs Uplands. This suggests a genetical relationship between plutonism and metamorphism. The relationship is further supported by the relative age of porphyroblast growth in the contact aureole of the Upper Silurian Mount Elizabeth biotite granite (U-Pb monazite age of 418 ± 1 Ma., Bevier, 1988). Porphyroblasts of cordierite overgrew relatively open F_2 crenulations but are wrapped-around by the S_2 crenulation cleavage. These microstructures indicate that D_2 and M_2 metamorphism were synchronous with the intrusion of the bulk of the Acadian magmatic suite in the Late Silurian. A similar age relationship was inferred for porphyroblast growth in the contact aureole of the younger, Early Devonian Pabineau granite (van Staal, 1987; U-Pb zircon age of 394 ± 1 Ma., Bevier, 1988). However, this age may not be relevant since the Pabineau granite appears to have intruded its own contact aureole. This relation is indicated by a very thin aureole or absence of an aureole along the southern contact of the granite. The U-Pb zircon age of Bevier (1988) may therefore represent a late intrusive phase that postdates the formation of the contact aureole.

Greenschist- and subgreenschist-facies mineral assemblages are also present in the Hayesville Subzone (Irrinki, 1979; Venugopal, 1979), but here the grade increases westwards into cordierite- and sillimanite-bearing pelites and migmatites and hornblende- and plagioclase-bearing amphibolites (St. Peter, 1981; Fyffe, 1982b; Fyffe et al., 1988a). The earliest growth of sillimanite preceded or accompanied tight folding (F_2) but continued during later open folding (F_3?). This high-temperature-low-pressure type of metamorphism probably correlates with M_2 metamorphism in the Bathurst Subzone. K-Ar age determinations on micas from these high grade rocks yield Devonian cooling ages (Wanless et al., 1972, 1973). Although some Early Devonian plutons postdate, at least in part, the M_2 metamorphism, there appears to be an overall spatial and temporal relationship between the abundance of Late Silurian plutonism and high grade metamorphism throughout central New Brunswick. However, there is no straightforward correlation between individual exposed plutons and metamorphism. Lux et al. (1986) explained the high temperature-low pressure metamorphism in the adjacent New England States as a result of regional scale contact effects produced by large granitoid plutons. These supposedly spread into sill-like bodies at intermediate crustal levels and were generated by melting of the lower continental crust during thermal relaxation following the Acadian collision. However, this model fails to explain the significant volume of mantle derived layered mafic-ultramafic intrusions, gabbroic rocks and the large volume of coeval basalts (e.g., McCutcheon and Bevier, 1990) that form part of the Acadian magmatic suite (Fyffe et al., 1981).

Tectonic interpretations

Plate tectonic models were first developed for the Appalachians by Bird and Dewey (1970). They proposed that the volcanic rocks of the Tetagouche Group represented the remnants of an island arc, developed above a northwest-dipping subduction zone, which implies that an ocean basin was present between the Avalon Zone of southern New Brunswick and the Miramichi Highlands. However, this is not consistent with the Celtic and North Atlantic faunas present in the lower parts of the Tetagouche Group (Nowlan, 1981a; Neuman, 1984). Rast and Stringer (1974) and Rast et al. (1976a) emphasized the lithological similarities of the Cookson and Miramichi groups, a correlation which was subsequently strengthened by Fyffe et al. (1983) and justifies their being combined into the Gander Zone (see above). Furthermore, the Lower Ordovician black phyllites, which generally occur at the bottom of the Cookson and top of the Miramichi groups, may onlap onto the Avalon Zone (Rast and Stringer, 1974; Ludman, 1987), which, if correct, would suggest that the Gander and Avalon zones were juxtaposed or in proximity by Early Ordovician times. Such an interpretation is consistent with the high southern paleolatitudes for the Tetagouche Group (van der Pluijm and van der Voo, 1991) and the Early Ordovician Celtic faunas (Neuman, 1984). This lends itself to the possibility that the Avalon Zone or Avalon Composite Terrane (Keppie, 1989) forms the continental basement for most of the Gander Zone (Dunning and O'Brien, 1989). Pb and Nd isotopic studies of granitoids in the Gander (Bevier, 1987) and Avalon zones are inconclusive, although they exclude juvenile Avalonian crust typical of the eastern part of the Avalon Zone as a potential basement. This interpretation is supported by Pan-African or Trans-Amazonian detrital zircons in the Miramichi Group (David et al., 1991). Combined, these data argue against the existence of an ocean basin between the Miramichi Highlands and the Avalon Zone.

An alternative model to that of Bird and Dewey (1970) was proposed by Pajari et al. (1977) and Rast and Stringer (1980) who suggested that the Tetagouche Group was an Andean-type magmatic arc built on Gander Zone rocks, i.e., Miramichi Group above a southeastward-dipping subduction zone. In this model, the Miramichi Group represents the slope/rise portion of a passive margin developed on the southeastern side of Iapetus (Ruitenberg et al., 1977; Williams, 1979). Such an interpretation is compatible with the sparsely preserved sedimentary structures, and overall turbiditic character of the Miramichi Group. Results of recent petrochemical studies have shown that only the volcanic rocks of the Lower Ordovician Oak Mountain Formation and correlatives in the Miramichi Highlands of Maine have arc-like compositions (Dostal, 1989; Winchester and van Staal, unpublished data). The Lower Ordovician andesites and felsites in the Annidale belt in the St. Croix Subzone may also belong to this arc, although the age relationships between the various volcanic suites are poorly understood. The associated serpentinites may be correlatives of the Lower Ordovician ophiolitic rocks that occur in the Exploits Subzone of Newfoundland and probably have been emplaced on Gander Zone rocks during the Penobscottian-Ganderian orogeny (Williams and Piasecki, 1990; van Staal and Williams, 1991). The Middle Ordovician

✗ Back arc basin

Reversal R

volcanic rocks of the Tetagouche Group in the central and northern Miramichi Highlands on the other hand have compositions typical of a rift suite (van Staal et al., 1991), whereas coeval or younger oceanic crust (Sullivan et al., 1990) preserved in the Armstrong Brook and Belledune subzones formed in a suprasubduction zone setting. Middle Ordovician arc rocks occur in the Popelogan Subzone. The Middle Ordovician rocks of the Dunnage Zone in New Brunswick are therefore interpreted as the remnants of a west facing magmatic arc (Popelogan arc) with a short-lived (≈15 Ma, Fig. 3.87) back-arc basin (Fig. 3.97c). This back-arc basin possibly developed into a small ocean basin, i.e., Iapetus II, van der Pluijm and van Staal (1988). Equivalents of the Popelogan arc in Newfoundland are probably represented by the Middle Ordovician arc volcanics in the Victoria Lake and Wild Bight groups (Dunning et al., 1991; Swinden et al., 1990) of the Exploits Subzone.

Closure of the back-arc basin by means of northwest-wards directed subduction (Fig. 3.97d) was probably initiated in the late Caradoc or early Ashgill, following a polarity reversal induced by the collision of the Popelogan arc with the Humber Zone or Laurentian continental margin (van Staal et al., 1990, 1991). Independent evidence for a Caradoc collision between the Popelogan arc and Humber Zone is scarce, but may be represented by a Caradoc hiatus between the Ashgill Grog Brook turbidites and the underlying Popelogan Formation (see section on Popelogan Subzone). A similar hiatus is present between the Middle Ordovician Arsenault Formation and Garin Formation, a correlative of the Grog Brook Group in southern Gaspésie of Quebec, and in northern Maine (Riva and Malo, 1988), suggesting that it has regional significance. Resubmergence of this part of the Dunnage Zone in the Late Ordovician is equated with the onset of turbidite deposition in a successor basin situated in the fore-arc area above the newly established northwest-dipping subduction zone. The accretion of the Popelogan arc is probably unrelated to a slightly earlier collision between the Taconic arc (such as the Notre Dame Subzone of Newfoundland) and Laurentia (Taconic Orogeny sensu stricto) according to recent paleomagnetic data (van der Pluijm and van der Voo, 1991).

The timing of the collision between the eastern margin of the back-arc basin and Laurentia, which by this stage included the Taconic and Popelogan arcs, is not well constrained. This is mainly because the original relationships between the Lower Silurian and Ordovician rocks are generally obscured by faulting. This also applies to the eastern boundary of the Miramichi Highlands with the Fredericton Belt (Fig. 3.80), where the remnants of a fore-deep are potentially preserved (van Staal et al., 1990). The following lines of evidence, however, suggest that collision and amalgamation of the Avalon, Gander and Dunnage zones took place before the Late Silurian, probably close to the Ordovician-Silurian boundary: (1) the Acadian magmatic suite of bimodal plutonic and extrusive rocks cross all boundaries between the Dunnage, Gander and Avalon zones, including a late Ordovician-Early Silurian blueschist suture (van Staal et al., 1990) that links the Dunnage and Gander zones by at least the Late Silurian; (2) the regional M_2, high-temperature-low-pressure metamorphism is Late Silurian in age, which should represent a minimum age for collision; and Late Ordovician-Early Silurian $^{40}Ar/^{39}Ar$ ages of phengite and crossite, developed in Tetagouche Group basalts caught up in the blueschist suture zone (van Staal et al., 1990),

suggest that parts of the eastern margin or transitional crust entered the subduction zone during the Ashgill; and (3) the oldest known rocks in the Fredericton Belt that link the Miramichi and St. Croix subzones and their correlatives in Maine (Vassalboro or Sangerville Formation) are early Llandovery (*M. cyphus* zone, see Fyffe, Chapter 4, this volume) sandstones and shales of the Kingsclear Group. The Fredericton Belt is bounded on both sides by rocks of the Gander Zone and its Silurian rocks contain detritus eroded form the nearby Tetagouche Group (Fyffe and Riva, 1990) and Nd-isotope studies on granites (Whalen et al., 1989) suggest that the crustal basement to the Fredericton Belt is similar to that of the St. Croix and Miramichi subzones. This early Llandovery age sets a minimum age for the start of collision (post-Taconic Orogeny), regardless of whether these turbidites represent foredeep deposits or an unconformable, post-collisional cover sequence.

Deformation continued after the collision started (see above). The multiply-deformed Lower Silurian rocks of the Chaleurs Group in northern New Brunswick are in turn unconformably overlain by Upper Silurian (Pridoli) or lowest Devonian sandstones and conglomerates (e.g., New Mills Formation; Fyffe and Noble, 1985), which appear to have undergone a less complex deformation. These conglomerates contain large boulders of the underlying Lower Silurian rocks, some foliated (McCutcheon and Bevier, 1990), indicating rapid exhumation and tectonic activity before the Late Silurian. In western New Brunswick, however, the Silurian sequence appears to be continous into the Lower Devonian. The central part of the northern Appalachians thus seems to record an intimate interplay between deformation, sedimentation and magmatic activity (see also Dunning et al., 1990a), including mantle derived melts. We suggest that this Silurian orogenic phase records continuous collision in the Early Silurian (probably having an oblique component) followed by delamination of the lithospheric mantle (Bird, 1978), causing upwelling and partial melting of the asthenosphere. Delamination of the lithospheric mantle should result in rapid uplift, possibly followed by gravitational collapse of the collision zone (England and Houseman, 1988). Gravititational collapse provides a mechanism for the formation of the flat belts during D_3. Subsequent cooling of the lithosphere would cause subsidence that could explain the formation of the Late Silurian-Devonian successor basins. D_4 and D_5 associated with orogen-wide transcurrent movements and transpression, probably record continuous oblique convergence into the Early Devonian.

Summary of Gander-Dunnage zone relationships

Regional analyses show that the Ordovician rocks in the Hayesville, Bathurst, and Armstrong Brook subzones of the Miramichi Highlands, along with the Belledune, Elmtree, and Popelogan subzones of the Chaleur Uplands have similar stratigraphy. Volcanism started in the middle-late Arenig, peaked in the Llanvirn and was finished by the earliest Caradoc. Coeval magmatism formed oceanic lithosphere in the Belledune and Armstrong Brook subzones, coeval with deposition of metalliferous sedimentary rocks. Deposition of red and black shales, volcaniclastic sediments and limestone was in part coeval with volcanic activity mainly in Llandeilo to early Caradoc times. The area was covered by an extensive, but diachronous blanket of black shale and chert, in Llandeilo-Caradoc time (*N. gracilis-D. clingani* zones), a period during which most of the central part of the New Brunswick Appalachians was starved of coarse clastic sediments. A similar graptolite fauna is present in the Point Leamington greywacke of the Exploits Group in Newfoundland (Riva and Malo, 1988). This black shale represents an excellent marker horizon, analogous to the black shale unit in the Dunnage Zone of Newfoundland (e.g., Dean, 1978; van der Pluijm et al., 1987). Lower to Middle Ordovician rocks in the Bathurst and Hayesville subzones are similar to rocks of the Exploits and Indian Bay subzones in Newfoundland, and all contain the same faunas. The correlation with the Exploits Subzone is further strengthened by a U-Pb zircon age of 468 ± 2 Ma for a felsic volcanic of the Bay du Nord Group (Dunning et al., 1990a) and similarities between lead isotopes from massive sulphides of the Tetagouche Group with those of the Bay du Nord and Baie d'Espoir groups (Swinden and Thorpe, 1984). This indicates that recognizable rock packages representing similar settings can be traced considerable lengths of the orogen as earlier emphasized by Williams (1979). Tectonic models should therefore be consistent with all critical data preserved in various parts of the orogen.

In contrast to the mainly tectonic Dunnage-Gander contacts in Newfoundland, the weight of geological evidence in New Brunswick favours the interpretation that most of the Tetagouche Group of the Dunnage Zone represents a conformable to disconformable stratigraphic cover to the Miramichi Group of the Gander Zone. Although contacts are commonly tectonized, field relationships do not support the idea that the numerous bodies of Middle Ordovician volcanic and sedimentary rocks represent the remnants of a once enormous klippe of Dunnage Zone rocks above Gander Zone rocks.

Most geologists who have worked in New Brunswick consider the rocks formerly referred to as the upper and lower Tetagouche Group as a single tectonostratigraphic unit. This practice is discontinued since the Tetagouche Group of the Miramichi Highlands includes unrelated rocks. We restrict the name Tetagouche Group to the Middle Ordovician volcanic and associated sedimentary rocks and consider these to be part of the Dunnage Zone. Underlying clastic rocks, which can be correlated with the Gander Group of Newfoundland, are named the Miramichi Group and are included in the Miramichi Subzone of the Gander Zone. Rocks of the Elmtree Subzone are very similar to those found in the Bathurst Subzone, but differ from those found in the Armstrong Brook and Belledune subzones. The tectonic boundary between the Pointe-Verte and Elmtree formations, therefore, is interpreted as a continuation of the Armstrong Brook-Bathurst subzone boundary, which has been dextrally offset along the Rocky Brook-Millstream Fault. This tectonic boundary is interpreted as the suture of "Iapetus II" (van Staal et al., 1990). Further support is given by Nd and Pb isotopic ratios of granites on opposite sides of the suture, which indicate differences in source areas (Bevier, 1987; Whalen et al., 1989, pers. comm., 1990).

The Middle Ordovician rocks of the New Brunswick Dunnage Zone were deposited on an ophiolitic substrate and on continental crust of the Gander Zone. A similar situation may exist in Newfoundland (van Staal and Williams, 1991).

AVALON ZONE
Preamble
Harold Williams

The Avalon Zone, named after the Avalon Peninsula in Newfoundland (Williams et al., 1974), is defined by its well preserved upper Precambrian sedimentary and volcanic rocks and overlying Cambrian-Ordovician shales and sandstones. Confidence in identifying its disjunct segments is expressed by wide usage of the name "Avalon" or "Avalonian" for correlatives along the entire length of the Appalachian Orogen and for some occurrences in the Caledonides on the opposite side of the Atlantic. The term "Avalon Platform" (Kay and Colbert, 1965), while synonomous with the Newfoundland Avalon Zone, is misleading as only its Cambrian rocks are platformal. The Avalon Terrane (Williams and Hatcher, 1982, 1983) or Avalon Composite Terrane (Keppie, 1985) are other names for the Avalon Zone. There is a tendency to regard metamorphic or plutonic rocks dated isotopically at about 700-550 Ma as "Avalonian". This is not part of the Avalon Zone definition.

The Avalon Zone has little in common with other zones of the Appalachian Orogen: its upper Precambrian rocks evolved during the Grenville-Appalachian time gap, although they are in part coeval with the early stages of the Appalachian cycle; its full compliment of Cambrian strata, mainly shales, contain Acado-Baltic trilobite faunas distinctive from those of the Humber miogeocline; the widest expanse of the Avalon Zone across the Grand Banks of Newfoundland is virtually unaffected by Paleozoic deformation (Williams, 1993), and the zone has unique mineral deposits such as pyrophyllite in Upper Precambrian volcanic rocks, manganese in Cambrian shales, and oolitic hematite in Ordovician sandstones. It is therefore a suspect terrane. If composite, diverse elements were amalgamated in the late Precambrian, as Cambrian strata are everywhere similar and contain the same faunas. Accretion to the North American margin was much later, in middle Paleozoic time.

The Avalon Zone extends well west of the Avalon Peninsula in Newfoundland to its boundary with the Gander Zone (Maps 1 and 2, and Fig. 3.126). Offshore it underlies the Grand Banks of Newfoundland and Flemish Cap. Its full width here is 750 km, making it the broadest zone of the Canadian Appalachians and more than twice the combined width of all other Newfoundland zones. It is much narrower in Nova Scotia where its rocks are exposed in Cape Breton Island and in the inliers of the Antigonish Highlands and Cobequid Mountains. Farther southwest, Avalonian

Williams, H.
1995: Preamble (Avalon Zone); in Chapter 3 of Geology of the Appalachian-Caledonian Orogen in Canada and Greenland, (ed.) H. Williams; Geological Survey of Canada, Geology of Canada, no. 6, p. 223-226 (also Geological Society of America, The Geology of North America, v. F-1).

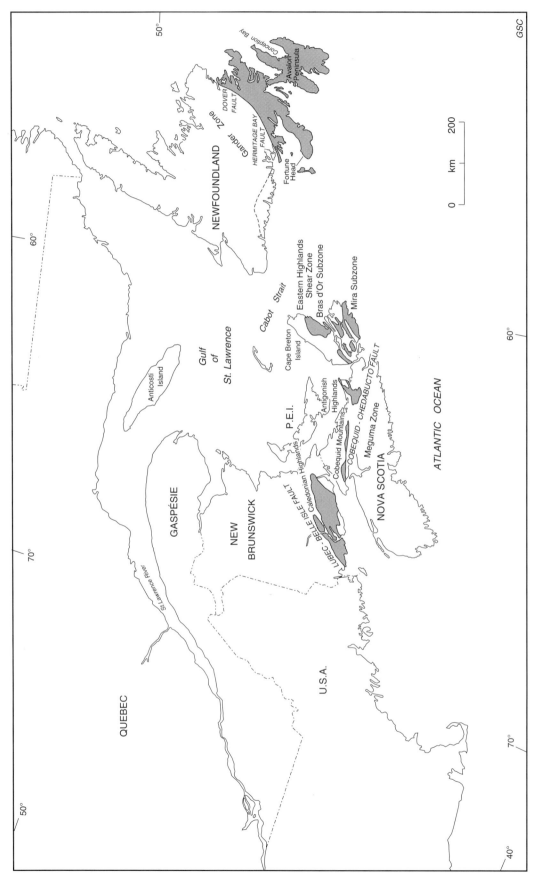

Figure 3.126. Distribution of Avalon Zone rocks in the Canadian Appalachians.

rocks are exposed in the Caledonian Highlands of New Brunswick and they continue to the United States border. In the northern U.S. Appalachians, correlatives occur in coastal Maine, eastern Massachusetts, and Rhode Island. The broad Slate Belt of the central and southern U.S. Appalachians, or Carolina Terrane (Secor et al., 1983), has similar rocks and structures. On the eastern side of the Atlantic, examples are known in Wales, Brittany, Iberian Peninsula, and northwest Africa (Rast et al., 1976b; Williams, 1984).

Boundaries of the Avalon Zone are major faults. In Newfoundland, its western boundary is the Dover and Hermitage Bay faults. In New Brunswick, its western boundary is the Lubec-Belleisle Fault. Ambiguity exists regarding the full extent of the Avalon Zone in Cape Breton Island, Nova Scotia. One interpretation places the northwestern Avalon boundary at the Eastern Highlands shear zone or farther southeast between the Bras d'Or and Mira subzones (Barr and Raeside, 1989); another suggests that all of Cape Breton Island, including Grenvillian basement at its northwest extremity (Barr et al., 1987b), belongs to the Avalon Zone (Keppie et al., this chapter). The choice between these interpretations is significant as the former implies an Iapetus suture that crosses Cape Breton Island whereas the latter implies a suture offhore to the north. The southern boundary of the Avalon Zone with the Meguma Zone is the Cobequid-Chedabucto Fault of Nova Scotia. All these boundaries are steep ductile shears or brittle faults that disrupt the continuity of lithic belts within the zone and account for its extreme variability in width from Newfoundland to New Brunswick. The Avalon Zone is coextensive with the Avalon lower crustal block and its western faulted boundary in Newfoundland extends to mantle depths (Keen et al., 1986; Marillier et al., 1989). Since there is no reason to divorce the surface rocks from the deep crust, the zone is viewed as a microplate embedded in the mantle.

The Avalon Zone has gravity (Map 3) and magnetic (Map 4) expressions that contrast with those of adjacent zones (Haworth and Lefort, 1979). Positive gravity anomalies are characteristic and contrast with negative anomalies of the adjacent Gander and Meguma zones. Precambrian volcanic rocks exposed in domes or faulted horsts are marked by positive magnetic anomalies that are also recognized offshore, particularly across the Grand Banks (Fig. 3.127). The western limits of the Avalon Zone are straight boundaries in New Brunswick and Newfoundland, with major dextral offsets in the Gulf of St. Lawrence and Cabot Strait (Fig. 3.127). The western boundary projects northeastward across the Newfoundland continental shelf to the western end of the Charlie-Gibbs Fracture Zone, truncating magnetic patterns along its course. Magnetic patterns are also truncated by the Chedabucto Fault, which is traceable offshore as the Collector magnetic anomaly (Map 4).

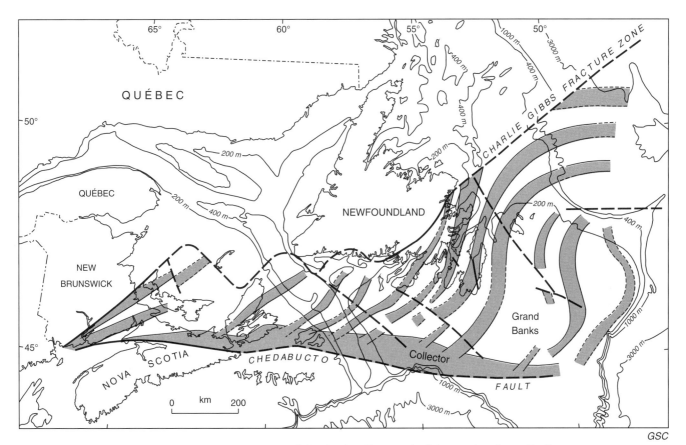

Figure 3.127. Magnetic anomalies of the Avalon Zone and offshore extensions. Faults are shown as solid or dashed heavy lines and magnetic zones correlated with structural highs are in red, after Haworth and Lefort, 1979.

Precambrian-Cambrian relationships are variable, though well known, throughout most of the Avalon Zone. Preserved sections of upper Precambrian rocks in places pass upwards into Cambrian strata without significant disconformity. This is exemplified by the section at Fortune Head in Newfoundland that is a contender for the Precambrian-Cambrian boundary stratotype (Narbonne et al., 1987). In other places, upper Precambrian volcanic and sedimentary rocks are cut by granites and all are overlain unconformably by gently dipping Cambrian strata.

The oldest rocks of the Avalon Zone in Canada are quartzites, marbles, and metamorphic rocks of Nova Scotia and New Brunswick. Relationships to younger Precambrian rocks are not well known, although unconformities are implied. If Grenvillian rocks of Cape Breton Island are basement to other rocks of the Avalon Zone, the Avalon Zone developed upon a basement like that of the Humber Zone on the opposite side of the orogen.

The Avalon Zone was affected by late Precambrian Avalonian Orogeny (Lilly, 1966; Rodgers, 1967; Hughes, 1970), expressed by granite plutonism and generally mild deformation. Middle Paleozoic orogeny was more intense, especially where the zone is narrow in Nova Scotia and New Brunswick. In Newfoundland, intensity of Paleozoic deformation - decreases eastward and locally at Conception Bay, Cambrian-Ordovician rocks are subhorizontal. Paleozoic rocks with mild deformation also extend eastward across the Grand Banks (King et al., 1986; Durling et al., 1987). A large Middle Paleozoic granite batholith cuts the Gander-Avalon boundary in Newfoundland and smaller plutons are common in nearby parts of the Newfoundland Avalon Zone but absent in the Avalon Peninsula and offshore (Williams, 1993).

In the early geosynclinal models of Schuchert (1923) and Schuchert and Dunbar (1934), The Avalon Zone was the Acadian Geosyncline, defined by its marine Cambrian-Ordovician rocks and faunas that contrast with those of the St. Lawrence Geosyncline on the North American side of an intervening landmass, the New Brunswick Geanticline. The Avalon "Platform" was viewed as a marine shelf linked to the Gander Zone on the eastern side of an ancient ocean (Williams, 1964a; Kay and Colbert, 1965). Since the advent of plate tectonics, models for the late Precambrian development of the Avalon Zone are ambivalent, favouring either convergence and island arc development unrelated to the Appalachian cycle (Rast et al., 1976b) or rifting and initiation of the Appalachian cycle (Strong, 1979). The zone is an excellent example of an Appalachian-Caledonian suspect terrane (Williams and Hatcher, 1982, 1983). A complete understanding must consider all of its presently dispersed and disjunct parts, especially those related to contemporary shields or platforms.

In the treatment that follows, upper Precambrian rocks of the Avalon Zone are described separately for Newfoundland, Nova Scotia, and New Brunswick. An overview of its Cambrian-Ordovician rocks is missing from this account.

Newfoundland

Harold Williams, S.J. O'Brien, A.F. King, and M.M. Anderson

Introduction

The Avalon Peninsula of Newfoundland is the type area of the Avalon Zone because of its excellent exposures, preservation of relatively undeformed complete stratigraphic sections, well known Precambrian-Cambrian relationships, profusely fossiliferous Cambrian and Lower Ordovician rocks, and a long history of investigation that provides a strong database. In places, distinctive upper Precambrian sedimentary and volcanic rocks are stratigraphically continuous with Cambrian and Lower Ordovician strata, with thicknesses in excess of 10 km. In other places, well-established unconformities occur within the upper Precambrian section and beneath the Cambrian cover. Other significant features are upper Precambrian plutons that are defined both stratigraphically and isotopically, upper Precambrian tillites, and a fauna of soft-bodied fossils of Ediacaran type. West of the Avalon Peninsula, the rocks are similar but more deformed. Continuity offshore is indicated by magnetic patterns and outcrops at Virgin Rocks, Eastern Shoal (Lilly, 1966), and Flemish Cap (Pelletier, 1971; King et al., 1985).

Avalon-Gander boundary

The Avalon-Gander boundary is an abrupt tectonic junction between Upper Precambrian low grade volcanic rocks and related granites of the Avalon Zone and amphibolite facies gneisses and granites of the Gander Zone. It is the Dover Fault in northeast Newfoundland and the Hermitage Bay Fault on the south coast of the island (Blackwood and Kennedy, 1975; Blackwood and O'Driscoll, 1976; Kennedy et al., 1982 ; Fig. 3.126, 3.128).

A regional steep fabric in rocks of the Avalon Zone increases in intensity toward the Dover Fault, culminating in a zone of mylonite 500 m wide. Subhorizontal lineations indicate mostly strike-slip movements, the latest of which is dextral (Caron, 1986; Caron and Williams, 1988). Brecciation affected the mylonite. Uplift on the Gander side of the fault is indicated by contrasts in metamorphic grade and crustal level.

The Hermitage Bay Fault is a 50 to 100 m wide cataclastic zone. Local occurrences of brecciated mylonite within the fault zone indicate earlier ductile shearing.

Inland extensions of the Dover and Hermitage Bay faults are truncated by the Ackley Granite, dated at 355 ± 10 Ma and 370 Ma (Dallmeyer et al., 1983b; Tuach, 1987; Kontak et al., 1988). The earliest sedimentological linkage between the Avalon and Gander zones is provided by a sampling of Gander lithologies in conglomerates and sandstones of the Cinq Isles and Pools Cove formations of

Williams, H., O'Brien, S.J., King, A.F., and Anderson, M.M.
1995:   Avalon Zone-Newfoundland; in Chapter 3 of Geology of the Appalachian-Caledonian Orogen in Canada and Greenland, (ed.) H. Williams; Geological Survey of Canada, Geology of Canada, no. 6, p. 226-237 (also Geological Society of America, The Geology of North America, v. F-1).

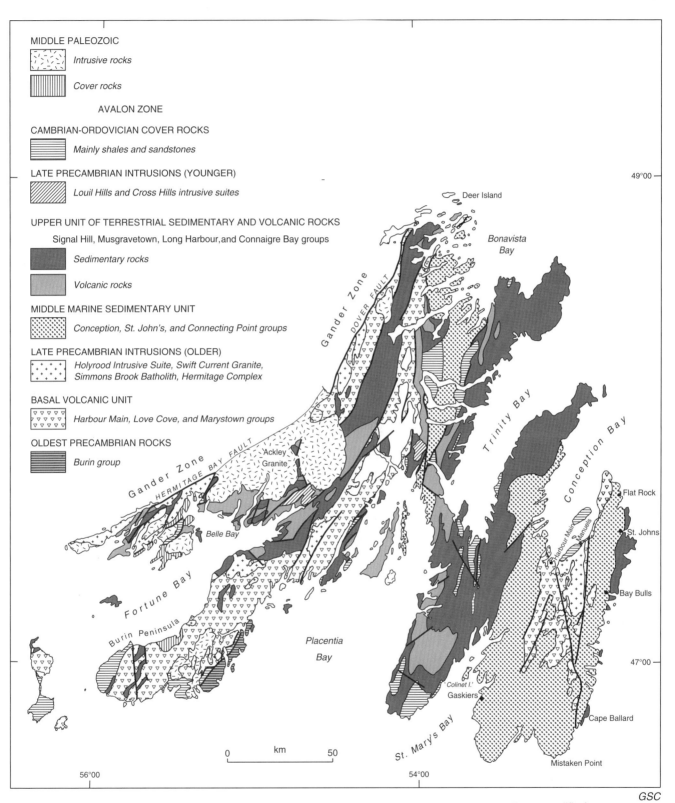

Figure 3.128. Distribution of upper Precambrian rocks of the Newfoundland Avalon Zone, modified from Colman-Sadd et al. (1990).

the southwestern Avalon Zone (Williams, 1971b; Fig. 3.113). These formations unconformably overlie upper Precambrian rocks and they are cut by Ackley type granite. Their age is probably Silurian or early Devonian. The present Avalon-Gander boundary may have been controlled by a late Precambrian crustal weakness (O'Brien et al., 1990). Deep seismic data indicate a vertical boundary that penetrates the entire crust (Marillier, et al., 1989).

Stratigraphy and general geology

The oldest rocks of the Newfoundland Avalon Zone are those of the Burin Group, a mixed assemblage of sedimentary and volcanic rocks with comagmatic gabbro (Strong et al., 1978). They underlie relatively small fault-bounded areas adjacent to the east coast of the Burin Peninsula (Fig. 3.128, 3.129).

Upper Precambrian rocks underlie most of the type area of the Avalon Peninsula and similar rocks occur west of the Avalon Peninsula (Fig. 3.128). They are separable into three lithological units that have at least local stratigraphic significance: a basal volcanic unit dominated by ignimbrites and volcanic breccias; a middle marine sedimentary unit of grey-green sandstones, siltstones, and siliceous argillites overlain by shales and sandstones; and an upper unit of terrestrial sedimentary and volcanic rocks dominated by red and grey sandstones and conglomerates. In the type area of the Avalon Peninsula, rocks of the three units are: (1) Harbour Main Group, (2) Conception and St. John's groups, and (3) Signal Hill and Musgravetown groups. The basal volcanic unit forms the core of the Avalon Peninsula, flanked to the east and west by northeast-trending belts of younger Precambrian rocks (Fig. 3.128).

West of the Avalon Peninsula, similar units occur in fault-bounded belts. Their stratigraphic order is everywhere the same, although correlations with the type area are in some cases uncertain. Western correlatives of the basal volcanic unit are the Love Cove (Jenness, 1963) and Marystown (Strong et al., 1978) groups; a western equivalent of the middle marine sedimentary unit is the Connecting Point Group (Hayes, 1948; Jenness, 1963); and western equivalents of the upper unit of terrestrial sedimentary and volcanic rocks are the Musgravetown (Hayes, 1948), Long Harbour (Williams, 1971b) and Connaigre Bay (Widmer, 1950; O'Driscoll and Strong, 1978) groups. Extensive volcanic rocks form lower parts of the Musgravetown Group (Bull Arm Formation, McCartney, 1967) and Long Harbour Group (Belle Bay Formation, White, 1939; Fig. 3.128).

In the following treatment, the oldest Precambrian rocks of the Burin Group are described first. Younger Precambrian groups are described for the type area of the Avalon Peninsula, then for the western areas. Intrusive rocks and structural history are later topics, followed by a concluding section on significance, interpretation and models.

Oldest Precambrian rocks: Burin Group

The Burin Group is about 5 km thick and restricted to a 50 by 5-10 km fault-bounded belt (Fig. 3.129). Its structural style suggests deformation before it was juxtaposed with upper Precambrian and Cambrian strata (Strong et al., 1978; O'Brien and Taylor, 1983). A gabbro sill, comagmatic with mafic volcanic and sedimentary rocks is dated isotopically

at 763 +2.2/-1.8 Ma (Krogh et al., 1988). This age negates correlation with other Precambrian rocks that have younger isotopic ages.

The Burin Group consists of pillow lavas, volcanic breccias, hyaloclastites, and fine- to coarse-grained volcanogenic sedimentary rocks. A spectacular stromatolite-bearing carbonate olistostrome occurs near its base. Locally, serpentinite slivers mark fault zones that cut the group. Basalts from the base of the succession are of alkalic affinity. Volcanic rocks in the rest of the group and its mafic intrusions are oceanic tholeiites (Strong et al., 1978; Strong and Dostal, 1980). Chemical characteristics and petrogenetic trends are similar to those of magmatic suites from modern ocean basins.

Basal volcanic unit

The Harbour Main Group (Rose, 1952; McCartney, 1967), with its type area in the Harbour Main-Avondale region of the central Avalon Peninsula, consists of at least 1800 m of felsic to mafic, subaerial to submarine volcanic rocks. Original textures of ignimbrites, welded tuffs, and flow-banded rhyolites are remarkably well preserved. The rocks occupy at least three blocks separated by faults. The eastern block contains submarine basaltic flows and pyroclastic rocks with intrusive rhyolitic domes. The central block has bimodal flows and fragmental silicic volcanic rocks and also includes local siliceous sedimentary rocks like those of the overlying

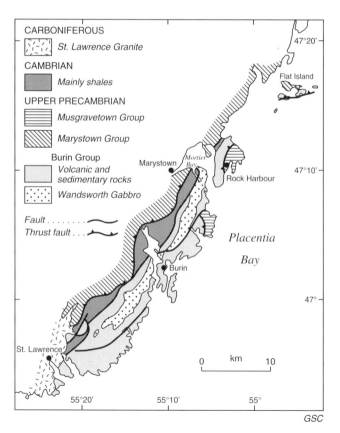

Figure 3.129. General geology and structural setting of the Burin Group, Burin Peninsula.

marine sedimentary unit. The western block has subaerial ash flows and associated red sandstones and shales succeeded by basalt.

The base of the Harbour Main Group is not exposed. Its contact with the overlying Conception Group adjacent to Conception Bay is an angular unconformity (McCartney, 1967); 50 km to the south, it is a disconformity (Williams and King, 1979). The presence of sedimentary rocks indistinguishable from those of the Conception Group within the Harbour Main Group, and of Harbour Main-type volcanic rocks (especially air-fall and resedimented tuffs) within the Conception Group, show that volcanism and marine sedimentation were contemporaneous, at least in part (Hughes and Brueckner, 1971). South of Conception Bay, felsic and mafic flows and pyroclastic rocks are cut by the Holyrood Intrusive Suite. Lower Cambrian strata, essentially undeformed in the central region of the Avalon Peninsula, overlie rocks of the Harbour Main and Conception groups and the Holyrood Intrusive Suite with pronounced unconformity.

The Love Cove Group (Jenness, 1963) and Marystown Group are correlated with the Harbour Main Group (Fig. 3.128). They form a major anticlinorium cored by felsic and mafic volcanic rocks that extends from the southern end of the Burin Peninsula to the northeast coast of Newfoundland. Both groups are lithologically diverse suites of flows and volcaniclastic rocks. Lower parts are felsic and intermediate to mafic rocks, middle parts are locally derived clastic and epiclastic continental deposits, and upper parts are basalts and rhyolites. In Bonavista Bay, the Love Cove Group is gradational and conformable with marine sedimentary rocks of the overlying Connecting Point Group (O'Brien and Knight, 1988). At the southern end of the Burin Peninsula, the Marystown Group is overlain unconformably at several places by redbeds of the upper unit of terrestrial sedimentary and volcanic rocks.

Samples of the Harbour Main Group yield the following U-Pb zircon ages from the three blocks recognized on the Avalon Peninsula: 589.5 ± 3 Ma for the eastern block, 631 ± 2 Ma for the central block, and 606 +3.7/-2.9 Ma and 622 +2.3/-2.0 Ma for the western block (Krogh et al., 1988). Isotopic ages for the Love Cove and Marystown groups are 590 ± 30 Ma (Dallmeyer et al., 1981b) and 608 +20/-7 Ma (Krogh et al., 1988), respectively.

Mafic volcanic rocks of the Harbour Main Group are high alumina, low titanium basalts of transitional to mildly alkaline chemical affinity (Nixon and Papezik, 1979). Lower parts of the Love Cove and Marystown groups are of calc-alkaline and tholeiitic affinity (Hussey, 1979; O'Brien and Taylor, 1983). Upper parts of the Marystown and Love Cove groups are bimodal and alkaline.

Middle marine sedimentary unit

This unit comprises the Conception Group (Rose, 1952) and overlying St. John's Group (Williams and King, 1979) of the Avalon Peninsula, and the Connecting Point Group (Jenness, 1963) westward from the isthmus of Avalon (Fig. 3.128).

The Conception Group is dominated by green to grey siliceous fine grained sedimentary rocks. It also includes conglomerate, mixtite of glacial origin, tuff, agglomerate, minor pillow lava, and mafic dykes. Its thickness in the southern Avalon Peninsula is about 4.5 km. An isotopic age of 565 ± 3 Ma (Benus, 1988) for a tuff near the top of the group, coupled with basal interfingering with the Harbour Main Group, imply a depositional span of about 50 Ma. The group is well exposed in the southern part of the Avalon Peninsula east of St. Mary's Bay where it has been divided into five formations (Williams and King, 1979).

The lower part of the Conception Group consists of three lithic units. In ascending order, these are siliceous turbidites (Mall Bay Formation, exposed thickness about 800 m), subaqueous mixtites and associated sedimentary rocks (Gaskiers Formation, 250-300 m), and turbidites (Drook Formation, 1500 m).

The Gaskiers Formation is the most distinct and easily recognizable unit of the Conception Group. North and south of Gaskiers, St. Mary's Bay, it consists of a sequence of unsorted and internally unlayered beds (mixtites) separated by thinner stratified units, most of which contain dropstones (Fig. 3.130A, B). Clasts are usually less than 10 cm in diameter but a few are much larger, including outsize granite boulders up to 2 m in diameter. Faceted and striated clasts occur locally. Clasts and dropstones in the mixtites are of silicic volcanic rocks, vesicular to massive basalt, a variety of coarse grained granites, porphyritic granite, granophyre, diorite, siltstone, quartz, volcanic-pebble conglomerate, quartzite, and foliated granite. Some of the clasts, such as those of quartzite and foliated granite, as well as detrital garnets (Gravenor, 1980), are exotic to the Avalon Zone of Newfoundland. On Colinet Islands, some mixtites contain contemporary volcanic detritus, including air-fall bombs and blocks, and red agglomerate is interlayered with red mixtite and sandstone at the top of the formation (Williams and King, 1979).

The Green siliceous volcaniclastic sedimentary rocks of Drook Formation are characteristic of the Conception Group throughout the Avalon Peninsula. Thin interbeds of tuff and cryptocrystalline chert of volcanogenic origin indicate active volcanism during sedimentation.

A deep water intrabasinal site for Gaskiers deposition means that glacial debris in the mixtites and the stratified units must have been derived, by redeposition and resedimentation, from the margin of the basin. Thus, glacial debris from the seaward edge of a grounded ice-shelf became unstable and flowed downslope as subaqueous debris flows (Gravenor, 1980; Anderson and King, 1981; Anderson, 1987). The inferred margin of the basin lay at an unknown distance to the west of the Avalon Peninsula (Williams and King, 1979) and a glaciated region may have extended beyond the present Avalon-Gander boundary. The Gaskiers Formation is widely distributed on the Avalon Peninsula and it occurs at Virgin Rocks, 200 km southeast of St. John's (Anderson and King, 1981; Anderson, 1987). The glacial episode recorded by the rocks of the Gaskiers Formation was, therefore, of regional extent.

East of St. Mary's Bay, the three lithic units of the lower part of the Conception Group are interpreted to represent: (1) progradation of a submarine fan from a volcanic source to the east or northeast, (2) glaciogenic debris flows, and (3) renewed submarine fan progradation, first from the northeast and later from the southwest as a new volcanic source was established (Gardiner and Hiscott, 1988).

The upper part of the Conception Group in the southern Avalon Peninsula consists of turbiditic sandstones and shales. Coarse grained, thick bedded greywackes (Briscal Formation, 1500 m) overlie the Drook Formation and, in turn, are succeeded by greenish grey and reddish purple tuffaceous siltstones, shales, and sandstones (Mistaken Point Formation, 400 m). Precambrian fossils are known from the top of the Drook Formation to the top of the Mistaken Point Formation (Anderson, 1978; Anderson and Conway Morris, 1982). These are casts and impressions of soft-bodied multicellular animals or metazoans. They were discovered first in the upper part of the Mistaken Point Formation at Mistaken Point, where some of the fossil-bearing beds are well exposed and covered with a profusion of different forms. The fossils occur on the upper surfaces of beds that are generally overlain by a thin layer of tuff (one example dated at 565 ± 3 Ma). The fossils have a variety of distinctive shapes, such as discoidal, bushlike, pectinate, and spindle-shaped (Misra, 1969; Williams and King, 1979; Fig. 3.131). The Mistaken Point Formation defines the top of the Conception Group, and it has the combined advantages of being a lithostratigraphic, biostratigraphic, and chronostratigraphic unit.

The Connecting Point Group (Jenness, 1963) is a 4-5 km sequence of clastic rocks that overlies volcanic rocks of the Love Cove Group west of the Avalon Peninsula. Whether or not the Conception and Connection Point groups were deposited in separate basins or a single sedimentary basin is presently unclear. The Connecting Point Group is divided into six mappable units in Bonavista Bay that form two distinct packages of turbidites separated by a major olistostrome (O'Brien, 1987; Knight and O'Brien, 1988).

Sphaeromorph acritarchs and filaments (empty sheaths of cyanophytes) occur at a number of stratigraphic levels within the Connecting Point Group. A sample from the upper part of the group on the east side of Placentia Bay also contains acanthomorph acritarchs (Hofmann et al., 1979). The Connecting Point Group is overlain by the Musgravetown Group, with angular unconformity in Bonavista Bay and conformity in eastern localities.

The St. John's Group (Rose, 1952; Williams and King, 1979) consists of marine shales and interbedded sandstones that conformably overlie the Conception Group of the Avalon Peninsula. The lowest part of the group, the Trepassey Formation, is a sequence of turbiditic sandstones similar to those of the underlying Mistaken Point Formation. Graded beds are thinner and finer toward the

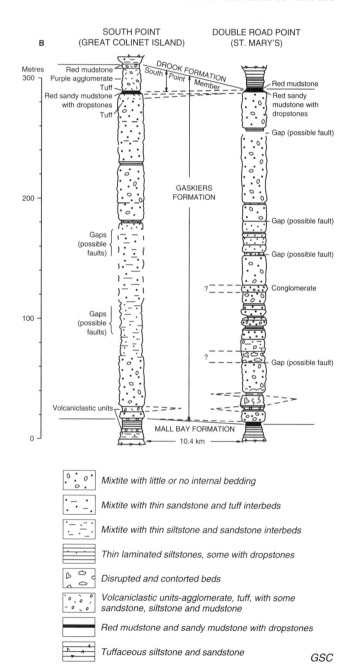

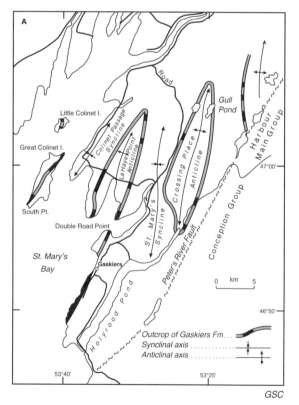

Figure 3.130. A) Distribution of the Gaskiers Formation. **B)** Thickness and lithic variation within the Gaskiers Formation at South Point and Double Road Point, after Wiliams and King (1979).

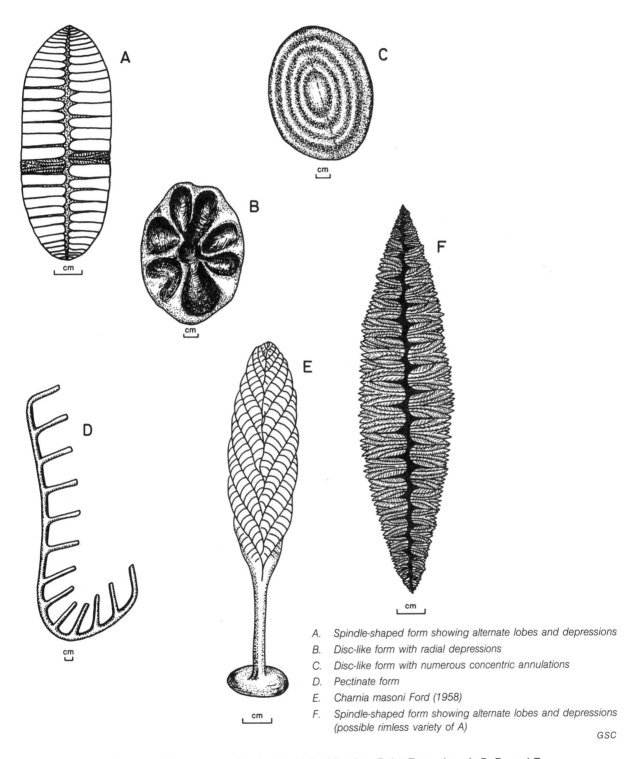

A. *Spindle-shaped form showing alternate lobes and depressions*
B. *Disc-like form with radial depressions*
C. *Disc-like form with numerous concentric annulations*
D. *Pectinate form*
E. *Charnia masoni Ford (1958)*
F. *Spindle-shaped form showing alternate lobes and depressions (possible rimless variety of A)*

GSC

Figure 3.131. Precambrian fossils of the Mistaken Point Formation. A, B, D, and F after Anderson (in King, 1982); E after Ford (1958).

top of the formation, and they are succeeded transitionally by shales and fine grained sandstones of the Fermeuse Formation. Slump folds are a conspicuous feature of the Fermeuse Formation. It is widely distributed on the Avalon Peninsula and required a depositional basin in excess of 10 000 km^2.

The upper part of the St. John's Group (Renews Head Formation) is an upward shoaling sequence of thin, lenticular-bedded, rusty-brown- weathering sandstones with dark shale laminae. Overlying, thin irregular beds of sandstone exhibit ripple-drift cross-lamination, interference ripples, and dessication cracks. Cross-stratified sandstones that infill channnels at several horizons near its top are harbingers of the overlying Gibbet Hill Formation of the Signal Hill Group. That part of the St. John's Group above the Trepassey Formation is prograded and thickens from about 400 m in the north to over 1700 m in the south (Fig. 3.132). Local tuff beds in the Trepassey, Fermeuse, and Renews Head formations record intermittent, waning volcanic activity.

Fossils of soft-bodied metazoans are present in the Trepassey Formation and the lowest part of the Fermeuse Formation. A variety of problematical markings, including those named *Aspidella terranovica* (Billings, 1872), occur near the top of the Fermeuse Formation. Some of these are concentric markings that resemble medusoid impressions, but they are probably of inorganic origin (Hsu, 1972). A variety of Vendian acritarchs is present throughout the St. John's Group of which the following are the commonest species: *Bavlinella faveolata*, *Protosphaeridium densum*, *Trachysphaeridium laminaritum*, and *Trematosphaeridium* sp. (Anderson et al., 1982a, b).

Upper unit of terrestrial sedimentary and volcanic rocks

A succession of mainly terrestrial sedimentary rocks conformably overlies the St. John's Group of the Avalon Peninsula. These are the Signal Hill Group on the east side of the Avalon Peninsula and the Musgravetown Group on the west side of the peninsula and farther west. The Long Harbour and Connaigre Bay groups are correlatives in the southwest part of the Avalon Zone.

The Signal Hill Group (Rose, 1952; Williams and King, 1979) has a thickness of 5 km and extends from Flat Rock in the north to Cape Ballard in the south (Fig. 3.128, 3.132, 3.133). It is divided into four formations in its type area; in ascending order, Gibbett Hill, Quidi Vidi, Cuckold, and Blackhead (King et al., 1988).

The Gibbett Hill Formation is mainly grey sandstone in beds that prograde southward and represent a transition from shallow marine to subaerial. Large- and small-scale tabular and trough crossbeds indicate paleoflow to the south and southwest. Numerous small cyclic sequences (sandstone-siltstone-shale) occur throughout the formation with grey-black shale units decreasing in thickness and frequency upwards. Mudcracks, major channels, and red sheet sands at the top of the formation are evidence for intermittent fluvial erosion and deposition.

The Quidi Vidi Formation consists of an alternation of thick bedded red arkosic sandstones and mudstones. The sandstones exhibit primary current lineation and unidirectional crossbedding. Trough crossbeds show that deposition was from currents that flowed south to southwestward. There is an abundance of red mudflakes in the sandstones and extensive scours in thick underlying mudstones.

The Cuckold Formation consists of red conglomerates and sandstones. Large scale trough and planar-tabular crossbeds indicate that unidirectional currents flowed to the south and southwest. The overall sequence coarsens and thickens upwards toward the central part of the formation. The upper part of the formation has finer and thinner beds. Clasts in the Cuckold Formation are mainly rhyolite. Local clasts of granite and quartz sericite schist are noteworthy as they are exotic to the Newfoundland Avalon Zone.

The Blackhead Formation has sheet-like, variegated sandstones with large trough-crossbeds, abundant mudstones, and mudflake breccias. Large load casts, deformed mudflake horizons, and oversteepened trough crossbeds are common.

Cuckold Formation conglomerates pass southward into a red sandy facies near Bay Bulls that resembles both the Quidi Vidi and Blackhead formations (Fig. 3.133). The formation is thinner to the south and absent at Cape Ballard. Another large clastic wedge north of St. John's, the Flat Rock Cove Formation, also disappears to the south. It overlies the Cuckold Formation, and rests unconformably on the Conception Group farther north. Relationships indicate a tectonically active, northern basin margin with maximum uplift during deposition of the upper units of the Signal Hill Group.

The Musgravetown Group (Hayes, 1948; Jenness, 1963; McCartney, 1967) is widely distributed in the Avalon Zone. In the western Avalon Peninsula, east of Trinity Bay and south to the head of St. Mary's Bay, marine shales at the base of the group are overlain by red sandstones and conglomerates. This sequence represents a major shoaling- and coarsening-upward cycle comparable in its development with that of the St. John's and Signal Hill groups. In the southwest part of the Avalon Peninsula, on the isthmus of Avalon, and in the western part of the Avalon Zone, the Musgravetown Group includes an extensive bimodal volcanic unit at or near its base, the Bull Arm Formation. This formation is a lithologically variable succession over 2 km thick of subaerial basaltic flows and pyroclastic rocks, rhyolitic ash flow tuffs, and related breccias. Some of the rhyolites are peralkaline. Felsic correlatives in Bonavista Bay have compositions ranging from alkali rhyolite to pantellerite and comendite. A peralkaline chemistry appears to be a signature of volcanic successions that post-date the Connecting Point Group (O'Brien et al., 1990). The Bull Arm Formation is overlain by a succession of marine to terrestrial, coarsening-upward clastic sedimentary rocks (McCartney, 1967; O'Brien and Knight, 1988). At Bonavista Bay, a coarse conglomeratic unit at the base of the Musgravetown Group (Cannings Cove Formation) overlies the Connecting Point Group with angular unconformity. In most areas, the Musgravetown Group is overlain disconformably by Lower Cambrian strata but at Deer Island in northwestern Bonavista Bay, the youngest rocks of the group are overlain conformably by a succession of marine sedimentary rocks 900 m thick (Deer End Formation) that is probably of Cambrian age (Younce, 1970; O'Brien and Knight, 1988).

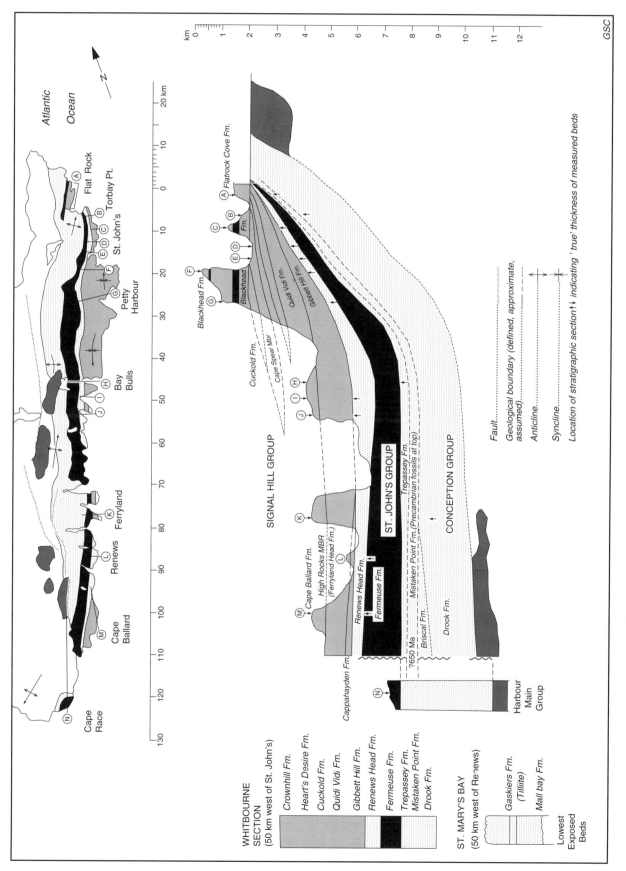

Figure 3.132. Stratigraphic profile of the upper Precambrian units of the eastern Avalon Peninsula, after King (1990).

233

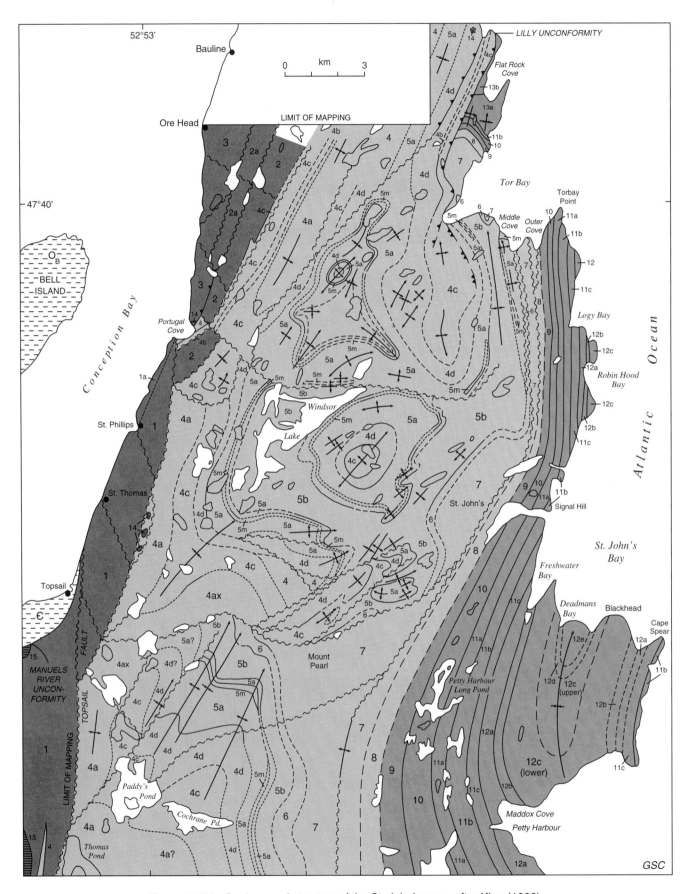

Figure 3.133. Geology and structure of the St. John's area, after King (1990).

The <u>Long Harbour</u> and <u>Connaigre Bay groups</u> are situated north of Fortune Bay in the most southwestern part of the Avalon Zone. Both groups have basal volcanic units; Belle Bay Formation of Long Harbour Group and Tickle Point Formation of Connaigre Bay Group. These are mainly silicic volcanic rocks, flow-banded to massive and spherulitic rhyolites, lithic and crystal tuffs, and agglomerates. Subordinate mafic volcanic rocks are massive and amygdaloidal basalt and mafic pyroclastic rocks. Minor volcanogenic sedimentary rocks are also present (Williams, 1971b). Subaerial volcanic rocks of the Belle Bay Formation constitute a bimodal calc-alkaline suite (O'Driscoll and Strong, 1978).

The Anderson's Cove Formation (White, 1939) conformably overlies the Belle Bay Formation north and northeast of Fortune Bay. The Anderson's Cove Formation consists of 1500 m of predominantly marine thinly laminated siltstones, sandstones, and conglomerates. These rocks are

ORDOVICIAN
       BELL ISLAND AND WABANA GROUPS
O_B *Sandstone, shale ironstone*

CAMBRIAN
       ADEYTON AND HARCOURT GROUPS
€ *Shale*

       UPPER PRECAMBRIAN (may include minor Lower Paleozoic rocks)
15 *HOLYROOD INTRUSIVE SUITE: granite (U-Pb zircon age* $620^{+2.2}_{-1.7}$ *Ma)*

14 *DOGBERRY HILL GABBRO: fine- to coarse-grained massive gabbro (age of intrusion uncertain)*

       SIGNAL HILL GROUP
12, 13 *FLAT ROCK COVE FORMATION(13): red and grey sandstone, siltstone, mudstone, conglomerate and breccia.*
13a, Knobby Hill Member; 13b, Piccos Brook Member
BLACKHEAD FORMATION (12): sandstone, mudstone;
12a, Petty Harbour Member; 12b, Maddox Cove Member; 12c, Spriggs Point Member; 12d, Deadman's Brook Member;
12e, Cliff Point Member

11 *CUCKOLD FORMATION: red conglomerate and sandstone.*
11a, Cabot Tower Member; 11b, Cape Spear Member; 11c, Skerries Bight Member

10 *QUIDI VIDI FORMATION: red and green sandstone, siltstone and mudstone*

9 *GIBBETT HILL FORMATION: thick bedded, greenish grey sandstone, siltstone and tuff*

       ST. JOHN'S GROUP
8 *RENEWS HEAD FORMATION: thin, lenticular bedded, dark-grey sandstone and minor shale*

7 *FERMEUSE FORMATION: grey to black shale containing thin lenses of buff-weathering sandstone and siltstone*

6 *TREPASSEY FORMATION: medium- to thin-bedded, graded, grey sandstone and shale; minor tuffaceous rocks*

       CONCEPTION GROUP
5 *MISTAKEN POINT FORMATION: medium bedded, grey to pink sandstone and green to purple and red shale,*
minor thin tuff horizons. 5a, Middle Cove Member. Includes 5m, a marker unit of red, waterlain tuff;
5b, Hibbs Cove Member: sandstone; contains Precambrian fossils

4 *DROOK FORMATION: green siliceous siltstone and sandstone, silicified tuff, grey, coarse grained turbiditic sandstone.*
4ax, Octagon Pond Member; 4a, Broad Cove River Member; 4b, Bauline Line Member; 4c, Torbay Member; 4d, Mannings Hill Member

       HARBOUR MAIN GROUP
3 *PORTUGAL COVE FORMATION: green siliceous sandstone and chert; minor red sandstone, siltstone, argillite and chert*

2 *PRINCES LOOKOUT FORMATION: massive and pillow basalt. 2a, Blast Hole Ponds Unit*

1 *ST. PHILLIPS FORMATION: green metatuff, agglomerate, mafic lava; siliceous volcaniclastic sedimentary rocks*
1a, Beachy Cove Member

Geological boundary
  (defined, approximate, assumed) . . . . .

Fault (defined, approximate, assumed) . .

Anticline (plunging) . . . .

Syncline . . . .

Legend for Figure 3.133.

typically grey to green but local basal units of boulder to pebble conglomerates and sandstone are red to purple (Williams, 1971b). There is considerable variation in the number, kinds, and proportions of lithotypes present in different places, and some sections contain mafic tuffs, diabase sills and basalts (O'Brien and Nunn, 1980; O'Brien et al., 1984).

The Mooring Cove Formation of the Long Harbour Group is a mixed assemblage of silicic and mafic flows and pyroclastic rocks that are interlayered with one another and with coarse- to fine-grained clastic sedimentary rocks that form a varying proportion of the group in different places (Williams, 1971b). The volcanic rocks display peralkaline affinities (O'Brien et al., 1983, 1984). The Mooring Cove Formation is conformable upon the Anderson's Cove Formation northeast of Fortune Bay. Farther west, north of Belle Bay, the Mooring Cove Formation overlies the Belle Bay Formation, suggesting local unconformity (Williams, 1971b).

The Rencontre Formation overlies the Mooring Cove Formation transgressively. It consists of pebble to bounder conglomerates, sandstones, and shales deposited in fluvial, marginal marine, and open marine settings. The final phase of deposition was marked by transgression and the spread of peritidal sedimentary rocks beyond the previous confines of the basin, notably to the south across the southern end of the Burin Peninsula (Smith and Hiscott, 1984). There, the upper part of the formation rests unconformbably on volcanic rocks of the Marystown Group. The Rencontre Formation is overlain conformably at several places by a sequence of marine sedimentary rocks (Chapel Island Formation) that is, in large part, of earliest Cambrian age (Narbonne et al., 1987).

Upper Precambrian intrusions

Intrusive rocks are common throughout the Newfoundland Avalon Zone. One suite, dated isotopically from 620 to 580 Ma, consists of hornblende and biotite granite, granodiorite, and diorite with calc-alkaline affinities. Examples are spatially associated with the basal volcanic unit; Harbour Main Group of the Avalon Peninsula and Love Cove and Marystown groups farther west. Another suite of smaller plutons, confined to the western part of the Avalon Zone, consists of gabbro, alkali granite, riebeckite peralkaline granite, and related hybrid rocks. Preliminary Rb-Sr ages for this suite are younger, from 560 to 540 Ma (John Tuach, pers. comm., 1988).

The Holyrood Intrusive Suite (Rose, 1952; King, 1988) intrudes the Harbour Main Group in the centre of the Avalon Peninsula. Its isotopic age of 620 +2.1/-1.8 Ma (Krogh et al., 1988) demonstrates a coeval relationship with the Harbour Main Group. West of the Avalon Peninsula, elongate intrusions cut the Marystown and Love Cove groups in a linear belt extending 120 km northeastward from the southern end of the Burin Peninsula to Bonavista Bay. The largest of these, the Swift Current Granite, is well-exposed west of the isthmus of Avalon. It has an age of 580 ± 20 Ma (Dallmeyer et al., 1981b). Examples of calc-alkaline complexes farther west are the Simmons Brook Batholith (Williams, 1971b) and the Hermitage Complex (O'Driscoll, 1977).

The younger suite of plutons includes the Louil Hills and Cross Hills intrusive suites (O'Brien et al., 1984; O'Brien, 1987). The Louil Hills Intrusive Suite of northwest Bonavista Bay intrudes the Love Cove and Connecting Point groups, and possibly, the lowest part of the Musgravetown Group (O'Brien, 1987). A nearly circular massive granite forms the core of this suite. Its pristine condition suggests that the Louil Hills suite is younger than the deformation that produced penetrative schistosity and cleavage in rocks of the Love Cove and Connecting Point groups. The Cross Hills Intrusive Suite, northeast of Fortune Bay, intrudes the Belle Bay and Anderson's Cove formations (O'Brien et al., 1984). Both the Louil Hills and Cross Hills suites are peralkaline complexes, probably related to the volcanic rocks of the Bull Arm and Mooring Cove formations.

Structural history

Evidence for late Precambrian Avalonian Orogeny in its type area of the Avalon Peninsula is provided by granite intrusion, block faulting, gentle folding, and low grade regional metamorphism. High strain zones and penetrative fabrics indicate local intense deformation (O'Brien, 1987; King, 1986). In some places, unconformities are common both within the upper Precambrian successions and below the Cambrian cover. In other places, contacts are conformable.

The stratigraphically lowest unconformity, restricted to a relatively small area south of Conception Bay, is between volcanic rocks of the Harbour Main Group and overlying marine clastic rocks of the Conception Group (McCartney, 1967). Conglomerate marking the base of the Conception Group truncates underlying steeply dipping beds of the Harbour Main Group. Other Precambrian unconformities occur between the marine clastic sedimentary rocks and overlying uppermost Precambrian terrestrial sandstones and conglomerates. These occur between the Conception Group and overlying Signal Hill Group of the eastern Avalon Peninsula (Anderson et al., 1975) and between the Connecting Point Group and overlying Musgravetown Group in the western Avalon Zone (Hayes, 1948; Jenness, 1963). Red conglomerates at the stratigraphic top of the Signal Hill Group lie on lower stratigraphic levels of the Conception Group. The stratigraphic gap is less profound between basal conglomerates of the Musgravetown Group and unspecified levels of the Connecting Point Group. Angular discordance is clear at both unconformities.

The best known unconformity of the Avalon Zone is that between the Cambrian cover and underlying Precambrian rocks at Manuels River. There, the Harbour Main Group and Holyrood Intrusive Suite are overlain by a thick bed of conglomerate of late Early Cambrian age. Nearby, Cambrian strata are unconformable on the Conception Group. These unconformities record a major depositional hiatus that corresponds with a long period of erosion that preceded the Cambrian transgression. The detail of the database for parts of the Avalon Zone in its type area is illustrated in Figure 3.133.

Paleozoic deformation affected almost the whole of the onland Avalon Zone, except for the northern part of the central Avalon Peninsula where Cambrian-Ordovician rocks are subhorizontal. This lack of Paleozoic deformation is attributed to a buttressing effect of the underlying Holyrood Intrusive Suite (Neale et al., 1961). Deformation was most intense in the western Avalon Zone where it resulted in thrusts, transcurrent faults, and upright to east-vergent folds, which overprint Precambrian structures (O'Brien, 1987). Isotopic ages between 400 and 400 Ma on whole rock phyllites, and 356 and 352 Ma for hornblende in crosscutting plutons (Dallmeyer et al., 1983b) suggest late Devonian deformation. The main movement on the Dover Fault is also regarded as Devonian (Dallmeyer et al., 1981a). Southeast of the Hermitage Bay Fault, undeformed Upper Devonian conglomerates (Great Bay de l'Eau Formation) overlie steeply dipping upper Precambrian and Cambrian strata (Williams, 1971b). Nearby, the Upper Devonian rocks are cut by the Belleoram granite.

On the Avalon Peninsula, Paleozoic deformation gave rise to open, periclinal folds and steep faults. In the Flat Rock area, east-directed thrusts place Conception Group rocks upon upper units of the Signal Hill Group (King et al., 1988). Offshore, Ordovician and Silurian rocks are mildly deformed, and they are overlain by Devonian red beds with mild unconformity (Durling et al., 1987). Mid-Paleozoic granites are common in the western Avalon Zone. They are absent on the Avalon Peninsula and unknown in the offshore Avalon Zone.

Significance, interpretation, and models

The Burin Group is approximately 130 Ma older than other Precambrian groups of the Avalon Zone, and it constitutes a separate fault-bounded terrane. Its igneous rocks display oceanic chemical affinities, implying the existence of ancient oceanic crust. Correspondence in age, composition, and regional setting with Pan-African ophiolitic rocks dated at 760 Ma (LeBlanc, 1981) suggest a Pan-African-Avalonian link (Strong, 1979; O'Brien, et al., 1983). The presence of stromatolitic limestones in the Burin Group, both as beds and clasts in olistostrome, suggests a shallow water environment for at least part of its deposition (Landing et al., 1988).

Upper Precambrian volcanic rocks of the Harbour Main Group and equivalents formed over a broader area. These products of subaerial and submarine eruptions vary widely in composition. They are intruded by voluminous calc-alkaline, diorite-granodiorite-granite magmas without significant metamorphism or structural complications. This volcanic-plutonic activity was at least in part contemporaneous with sedimentation in deep marine basins, and this phase of development may have lasted for at least 50 Ma.

The extremely thick basin-fills include complex successions of turbidites with tillites and olistostromes (Conception and Connecting Point groups) overlain by deltaic deposits (St. John's Group) and alluvial facies (Signal Hill and Musgravetown groups). The mild effects of Paleozoic and late Precambrian deformation on the wide Newfoundland Avalon Zone suggest that the present configuration of alternating Upper Precambrian volcanic and sedimentary belts is a relict of late Precambrian evolution. The pattern favours several discrete basins separated by volcanic ridges.

The upper Precambrian terrestrial clastic rocks were deposited as fan-deltas and alluvial fans within fault-bounded basins (McCartney, 1967; Smith and Hiscott, 1984).

Paleocurrent analyses indicate that some of these basins on the Avalon Peninsula were filled from their sides toward their centres. Associated volcanic rocks are terrestrial bimodal suites, and comagmatic plutons are strongly alkaline or peralkaline. The chemistry of the igneous rocks and styles of sedimentation are completely different compared to the older volcanic and sedimentary rocks.

The composition of the Harbour Main Group varies so widely that it does not provide a clear indication of its volcanic source or the nature of its basement; whether ensimatic, transitional, or ensialic. Exotic clasts in upper Precambrian marine sequences and overlying alluvial sequences suggest a nearby continental source. Ophiolitic detritus is unknown. Older quartzites, marbles, and foliated granites in the Nova Scotia and New Brunswick Avalon Zone suggest a continental substrate for the Avalon Zone in Newfoundland. Short synformal reflectors at mid crustal levels may mark the base of Newfoundland Avalonian sequences on a deeper crust (Keen et al., 1986). Bouguer gravity anomalies are much too low for a preserved ophiolitic substrate extending to mantle depths, just as the gravity anomalies of the Dunnage Zone are too low for such a model (Karlstrom, 1983).

A modern analogue for late Precambrian volcanic ridges flanked and separated by marine sedimentary basins may be the Marianas region of the Pacific Ocean (Hussong and Uyeda, 1981). This possibility receives support from a comparison between the succession of marine sedimentary rocks of late Precambrian age in the northwest Avalon Zone and recent counterparts in basins adjacent to existing island arcs (Knight and O'Brien, 1988). A heat source for Avalonian volcanism and plutonism is unknown.

Comparisons between the late Precambrian evolution of the Newfoundland Avalon Zone and Pan-African terranes of Gondwana (Strong, 1979; O'Brien et al., 1983, 1990), particularly the Afro-Arabian region and the Hijaz Arc (Stern et al., 1984), suggest similar tectonic controls. These comparisons imply that Avalonian Orogeny represents an accretionary event.

The chemical evolution of the Newfoundland Avalon Zone, after the main phase of Avalonian Orogeny, is analogous to that seen in some modern continental rifts. For example, in the southwestern United States there was a shift to alkalic compositions synchronous with onset of regional transtension related to the collision of the East Pacific Rise with North America (Atwater, 1970; Lipman et al., 1972; Wilson, 1988). According to this model, the upper Precambrian alluvial rocks of the Avalon Zone may have been deposited in pull-apart basins or rifts, like those of the U.S. Basin and Range Province. If the age of the alluvial rocks is 565 Ma or younger (Benus, 1988), then their deposition was penecontemporaneous with the rift-drift transition as defined in the stratigraphy of the Humber miogeocline (Williams and Hiscott, 1987), and a major episode of worldwide rifting (Bond et al., 1984). The choice here is between interpreting the alluvial rocks as molasse related to collisional tectonics, or the fill of rift basins formed in advance of imminent Cambrian subsidence and marine transgression.

Acknowledgments

G.R. Dunning and J.B. Murphy are thanked for reviews of this manuscript.

237

Nova Scotia

J.D. Keppie, J.B. Murphy, R.D. Nance, and J. Dostal

Introduction

The late Precambrian-early Paleozoic evolution of the Avalon Zone is generally held to have evolved separately from other zones in the Appalachian Orogen so that the Avalon Zone is a suspect terrane. Contrasts in the Precambrian stratigraphy in Nova Scotia led to subdivision of the Avalon Zone into several late Proterozoic terranes (Keppie, 1985, 1989). The main terranes were: (a) Cape Breton terrane: a cratonic magmatic arc; (b) Antigonish Highlands and Chéticamp terranes: inter-arc basins floored by oceanic lithosphere; and (c) Cobequid Highlands terrane: a cratonic magmatic arc. New data, however, have allowed lithostratigraphic correlations to be made across northern Nova Scotia which suggest to us the presence of just one terrane consisting of alternating basins and horsts, with the basins floored by thinned continental lithosphere (rather than oceanic lithosphere; Keppie et al., 1990). We interpret this composite terrane as a magmatic arc-intra-arc rift complex that lacks oceanic components.

In contrast, Barr and Raeside (1986) subdivided Cape Breton Island into four tectonostratigraphic divisions on the basis of apparent contrasts in the pre-Carboniferous histories across three major shear zones. The divisions were designated from southeast to northwest; Southeastern, Bras d'Or, Highlands and Northwestern Highlands. Some of these divisions were renamed (Barr and Raeside, 1989), so that their order from southeast to northwest is Mira, Bras d'Or, Aspy, and Blair River. The divisions are defined as follows: (1) Mira: late Precambrian volcanism and plutonism followed by Cambrian-Ordovician rift basin sedimentation and minor volcanism; (2) Bras d'Or: gneissic basement overlain by sedimentary rocks (George River Group) intruded by mainly Upper Precambrian and Ordovician(?) granitoid rocks; (3) Aspy: gneissic core flanked by low grade sedimentary and volcanic rocks of probable Precambrian age intruded by diverse Precambrian-Carboniferous plutons; and (4) Blair River Complex: gneissic basement intruded by varied plutonic rocks including anorthosite and Grenvillian syenite. These divisions are also referred to as subzones (Williams, Chapter 2 of this volume).

In extrapolating between Cape Breton Island and Newfoundland, the Blair River Subzone is correlated with the Humber Zone (the Cambrian-Ordovician cratonic margin of North America), and the Mira is placed within the Avalon Zone (Barr et al., 1990a; Table 3.3). The Aspy is variously correlated with all or part of the Gander Zone, although Reynolds et al. (1989) included part of the Aspy surrounding the ca. 550 Ma old Chéticamp pluton in the Bras d'Or (Table 3.3). Reynolds et al. (1989) suggested that the Chéticamp pluton could be an erosional klippe or a

Keppie, J.D., Murphy, J.B., Nance, R.D., and Dostal, J.
1995: Avalon Zone-Nova Scotia; in Chapter 3 of Geology of the Appalachian-Caledonian Orogen in Canada and Greenland, (ed.) H. Williams; Geological Survey of Canada, Geology of Canada, no. 6, p. 238-249 (also Geological Society of America, The Geology of North America, v. F-1).

window, although Currie (1987b) described it as intruding the Pleasant Bay Complex, a major component of the Aspy. Certainly, igneous activity at ca. 575-490 Ma is not uncommon in the Avalon Zone (Keppie et al., 1990). The Bras d'Or has been placed in either the Avalon or Gander zones (Table 3.3). Part of southern Newfoundland previously assigned to the Gander Zone (Burgeo Terrane of Keppie and Dallmeyer, 1989) has recently been correlated with the Avalon Zone (Dunning and O'Brien, 1989).

How much of Cape Breton Island belongs within the Avalon Zone depends on the definition of the Avalon Zone. Keppie et al. (1991) defined the Avalon Zone or Avalon Composite Terrane on the basis of overstep sequences that contain a Cambrian-Ordovician Acado-Baltic fauna (Fig. 3.134) and Silurian-Lockhovian, Rhenish-Bohemian fauna, that may be traced, albeit discontinuously, across the entire width of Cape Breton Island (Keppie, 1990). Outcrops of the Cambrian-Ordovician overstep sequence in the Bras d'Or Subzone are considered by Barr and Raeside (1990) and Raeside and Barr (1990) to be an overthrust remnant of the Mira Subzone. However, Hutchinson (1952) and Weeks (1954) described the contact between these rocks and the underlying George River Group (exposed on Long Island and in Gregwa Brook) as an angular unconformity and a disconformity, respectively. Hence, the Bras d'Or Subzone is part of the Avalon Zone. The presence of an angular unconformity was confirmed by Helmstaedt and Tella (1973) who recorded deformed and metamorphosed pebbles of the George River Group in the overlying Cambrian sedimentary rocks. Furthermore, in western Cape Breton Island, Lower Carboniferous rocks contain pebbles with Early Paleozoic, Acado-Baltic fauna that on the basis of current directions and associated pebble lithologies were derived from the adjacent Mabou Highlands well within the Aspy Subzone (Phinney, 1956; Keppie, 1990).

Critical to the interpretation is the distribution of a sequence of metasedimentary rocks including quartzite, marble and psammitic-pelitic schists that have commonly been assigned to the mid-upper Proterozoic George River Group (Fig. 3.134; Keppie, 1989 and references therein). This sequence can be followed discontinuously and with minor facies changes (Hill, 1990) from the Boisdale Hills (in the Bras d'Or Subzone) through Cape North (in the Aspy Subzone) to Meat Cove (in the Blair River Subzone). At Meat Cove, the metasediments are intruded by the Lowland Cove syenite whose crystallization age of 1172 +175/-73 Ma (Barr et al., 1987a) provides a minimum age for deposition of the George River Group, based on the lithostratigraphic correlation. Such a correlation is supported by similar sulphur and lead isotopes in syngenetic Zn deposits at Meat Cove and at Lime Hill in the George River Group of central Cape Breton Island which suggest that mineralization occurred at ca. 1.0 Ga (Sangster et al., 1990). The George River Group and the metasedimentary rocks correlated with it suggest that the Bras d'Or, Aspy and Blair River subzones were part of a single terrane at the time of the group's deposition. A further linkage between Upper Proterozoic, Avalonian sequences and the ca. 1.0 Ga basement like that of the Blair River Complex may be indicated by the presence of ca. 1.0-1.2 Ga detrital zircons in turbidites of the Georgeville Group in the Antigonish Highlands (Keppie and Krogh, 1990). These lithostratigraphic correlations would clearly require the whole of Cape Breton

Island to have been amalgamated at least by early Cambrian times. That the Blair River Subzon is not contiguous with the outer edge of cratonic North American basement is indicated on seismic reflection profile 86-4 between Newfoundland and Cape Breton Island which shows the top of the Grenville basement to be at a depth of >10 seconds two-way travel time in Cabot Strait just north of the subzone (Marillier et al., 1989).

Finally, published geochronological data show that Upper Proterozoic-Cambrian and Paleozoic plutons are present across most of Cape Breton Island (Fig. 3.135, 3.136). Although the basement beneath the Mira Subzone is not exposed, the geochemistry of the Upper Proterozoic volcanic rocks indicates the presence of continental crust (Keppie et al., 1979; Dostal et al., 1990). These volcanic rocks are probably extrusive equivalents of the plutons. On the basis of these data we conclude that the rocks of Cape Breton Island exhibit Precambrian linkages, and rather than representing distinct terranes in an entire cross-section of the Appalachian Orogen as proposed by Barr and Raeside (1986, 1989), expose an oblique section of the Avalon Zone with progressively deeper levels of the late Proterozoic magmatic arc complex outcropping northwestwards.

Stratigraphy and lithological description
Mid/Upper Proterozoic or older rocks

Given the arguments listed above, the oldest rocks in the Avalon Zone in Nova Scotia are likely to be represented by ortho- and para-gneisses and amphibolite sheets in the Blair River Subzone and equivalents in the Aspy Subzone (e.g. Pleasant Bay Complex), and the Economy River Gneiss and Mount Thom Complex in the Cobequid Highlands (Fig. 3.134-3.139).

The Pleasant Bay Complex underlies large areas in the western Cape Breton Highlands and is made up of biotite gneiss, amphibolitic gneiss, gneissic granodiorite, marble, quartzite and schist cut by lit-par-lit gneiss (Currie, 1987b). A syenite intrusion into gneisses of the Polletts Cove Brook Group (part of the Pleasant Bay Complex) in northern Cape Breton Island has yielded an U-Pb chord with a lower intercept age of 918 +78/-179 Ma and an upper intercept age of 1172 +135/-73 Ma; interpreted as the time of metamorphism and crystallization, respectively (Barr et al., 1987a; Keppie et al., 1991).

Table 3.3. Correlations between Cape Breton Island and Newfoundland made by proponents who subdivided Cape Breton Island into four terranes or subzones.

HUMBER ZONE	DUNNAGE ZONE		GANDER ZONE	AVALON ZONE	REFERENCE
	Notre Dame Subzone	Exploits Subzone			Williams and Hatcher, 1983 / Williams et al., 1988
Blair R.			Aspy	B + M	Barr and Raeside, 1986
Blair R.	← Aspy		→ + B	M	Loncarevic et al., 1989
Blair R.	← Aspy		→	B + M	Marillier et al.,1989
Blair R.	← Aspy		→ + B	M	Barr and Raeside, 1989
Blair R.		← Aspy	→ +B+M		Reynolds et al.,1989
Blair R.	← Aspy		→ + B	M	Dunning et al.,1990
					Barr et al., 1990[a]

Terranes and subzones in Cape Breton Island

GSC

Abbreviations: Blair R.= Blair River Subzone or Northwestern Highlands terrane; Aspy = Aspy Subzone or Highlands terrane; B = Bras d'Or Subzone; M = Mira Subzone or Southeastern terrane.

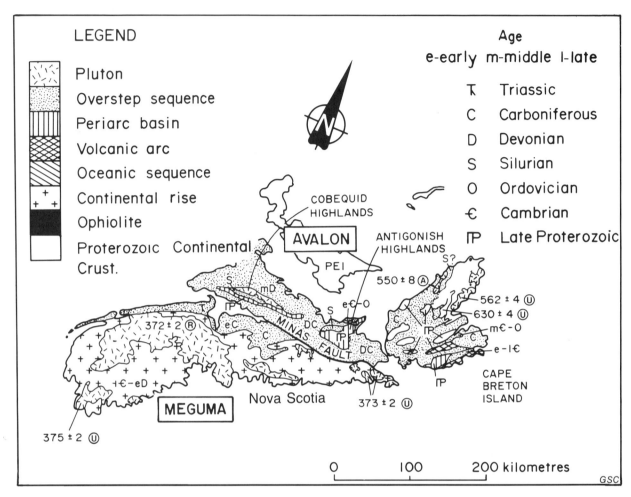

Figure 3.134. Tectonic map of the Avalon and Meguma Zones in Nova Scotia, after Keppie et al. (1991). Age determinations: U – U-Pb zircon, R – Rb-Sr whole rock, A – Ar-Ar hornblende.

The Mount Thom Complex occurs in the eastern end of the Cobequid Highlands as two small enclaves surrounded by intrusive rocks (Donohoe and Wallace, 1982). It consists of biotite-garnet-feldspar gneiss, hornblende-feldspar gneiss and granite gneiss. These gneisses have yielded a poorly defined Rb-Sr whole rock isochron of 934 ± 82 Ma and a protolith model age of 1150 Ma using an initial ratio of 0.711 (Gaudette et al., 1984; Doig et al., 1991). Farther west in the Cobequid Highlands, the Economy River gneiss is composed of biotite-garnet psammitic paragneisses, quartz-plagioclase layered amphibolite dykes and hornblende orthogneisses. The Economy River gneiss is dated at 734 ± 3 Ma (concordant U-Pb zircon age, Doig et al., 1989).

Mid/Upper Proterozoic metasedimentary rocks

Strongly deformed metasedimentary rocks consisting of quartzite, carbonate, greywacke, slate and minor metavolcanic rocks are represented in Cape Breton Island by the McMillan Flowage Formation, Meat Cove metasedimentary rocks, and George River and Cape North groups, and by the Gamble Brook Formation in the Cobequid Highlands (Fig. 3.134, 3.136, 3.139). At Meat Cove of the Blair River Subzone,

Figure 3.135. Summary of geochronological data previously reported for plutons in the Avalon Zone of Nova Scotia, adapted from Keppie et al. (1990). Locations of ages are shown in Figures 3.136-3.138. All ages are based upon the decay constants and isotopic abundance ratios listed in Steiger and Jager (1977). SOURCE: 1 = Fairbairn et al. (1960); 2, 3 = Wanless et al. (1967, 1968); 4, 5, 6, 7 = Cormier (1972, 1979, 1980, pers. comm.); 8 = Poole (1980); 9 = Stevens et al. (1982); 10, 11, 12 = Gaudette et al. (1983, 1984, 1985); 13 = Donohoe et al. (1986); 14 = Keppie and Halliday (1986); 15 = O'Beirne and Jamieson (1986); 16 = O'Beirne et al. (1986); 17 = Jamieson et al. (1986); 18 = French and Barr (1984); 20, 21, 22, 23, 24 = Barr et al. (1984, 1986, 1987a, b, 1988); 25 = Keppie and Dallmeyer (1989); 26 = Doig et al. (1989); 27 = Keppie et al. (1990); 28 = Barr et al. (1990); 29 = Sangster et al. (1990); 30 = Dunning et al. (1990b).

	PLUTON	AGE (Ma)	$\frac{87}{86}$ Sr$_o$	#	MSWD	SOURCE
(NW)	LOWLAND COVE	U–Pb z chord(918$^{+78}_{-179}$Ma:metamorphism)(1172$^{+175}_{-73}$:intrusive age)		5	—	23
	CAPE NORTH	M·B?⊞? W–P?⊞?	.705	2	—	6
	WILKIE SUGARLOAF	W–K–P	.7053	9	—	6
	GULCH BROOK	W	.7045	7	23.3	14
	WHITE POINT	Z | W	—	—	—	18,30
	PLEASANT BAY	W–B	.709	2	—	4
	FISHING COVE	M–B ⟷		2		3
	MARGAREE	W	.7067	6	—	16
	BELLE COTE	Z (1218$^{+335}_{-229}$:upper int.)		4	1.47	17
	SALMON POOL	Z		3	1.58	17
	CHETICAMP	W Z	.7048	7	6.4	17,21
	TAYLORS BARREN	W	.712	—		12
	W. BRANCH NORTH R.	⊞	.7039	8	2.78	15
	N BRANCH BADDECK R	Z (1814$^{+238}_{-217}$:upper int.)		3	0.31	17
	GILLANDERS MTN.	W⊞ W	.708,.707	6,4	——	18
	CAMERON BROOK	W	.7036	3	—	16,30
	CAPE SMOKEY	W	.7044	11	—	6,30
	INGONISH RIVER	W	—	—	—	22,30
	GISBOURNE	H⟷ T | Z	—	—		22,30
	KATHY ROAD	H–B ↔ |T | Z	—	—		22,30
	NEW GLEN	Z W	—	—		22,30
	INDIAN BROOK	Z				30
	ST. ANNS MTN.	Z⊞ W W	.7055	7	—	6 |28
	KELLYS MTN.	H ▶ W⊞ W	–,.699	–,6	——	1,4,12,25
	CREIGNISH	H W–M	.709	2	—	4,27
	NORTH MTN.	W–K–M W–B	.704,.716	2,3	——	4,27,29
	BOISDALE	W–B B⟶ W W	.704,.701,.720	7,2,2	———	3,4,27
	SHUNACADIE	B?|?|? K?+?Z ⊞ W	.7055	4	0.12	1,8,28
	COXHEATH	H⟷ W ⟷ |⟷ H	.706	2	—	3,4,9,27
	HUNTINGTON MTN.	W⊞ H	.7011	6	—	6,27
	LOCH LOMOND	⊞ W ▀H	.7070	14	—	4,20,27
	CAPELIN COVE	B?+? ⊞ W–B Z	.704	5	—	1,4 |28
	GILLIS MTN.	W–K–P–B⊞	.7032	14	—	5
(SE)	DEEP COVE	W–B⊞	.710	2	—	4
	PETIT DE GRAT	W⊞	.7024	6	—	6
ANTIGONISH HIGHLANDS	GEORGEVILLE	⊞ W ⟷M	.765	6	—	7
	GREENDALE	▀H				27
	JAMES RIVER	K?+?				1
	OHIO RIVER	⊞ W–K–P	.7033	12	—	5
	BARNEYS RIVER	W⊞ ⟷	.7054	4	—	3,5
	BLACK BROOK	▀H				27
	GUNN LAKE	W⊞	.7167	7	—	5
	EDEN LAKE	⟷ H				3
COBEQUID HIGHLANDS	McCALLUM SETTLEMENT	⊞ W	.7065	8	—	10,11
	DEBERT RIVER	⊞ W	.7059	6	81.6	13,26
	FROG LAKE	Z▀H				27
	PLEASANT HILLS	⊞ W–K–P	.7082	23	—	6
	BYERS/HART	⊞ W	.7046	7	5.8	13
	JEFFERS	B⟷ ⟷ H				2,5,27
	CAPE CHIGNECTO	W⊞ B⟶⟷	.7064	5	—	5
	GREAT VILLAGE R.	(granite) Z (orthogneiss)Z Z				26

GSC

MSWD = Mean Square Weighted deviates

▪ U–Pb age; ⊞ Rb–Sr age; ⟷ $^{40}Ar–^{39}Ar$ age; ▬ $^{40}Ar–^{39}Ar$ age (this paper); ↔K–Ar age
Z–zircon, T–titanite, W–whole rock, K– Kfeldspar, P–plagioclase, M–muscovite, B–biotite, H–hornblende

these metasediments are intruded by the Lowland Cove syenite whose intrusive age of 1172 +175/-73 Ma (Barr et al., 1987a) provides a minimum age for their deposition. In the Aspy Subzone, the upper age limit of the McMillan Flowage Formation is constrained only by the oldest of several Upper Proterozoic-Cambrian plutons which yielded a 564 ± 2 Ma U-Pb zircon age (Dunning et al., 1990b). Sulphur and lead isotopes in syngenetic Zn deposits at Meat Cove and Lime Hill (in the George River Group of central Cape Breton Island) suggest a ca. 1.0 Ga time of mineralization (Sangster et al., 1990). The similar geochemical characteristics of all of the marbles in these metasedimentary units in Cape Breton Island (Hill, 1990) supports their lithostratigraphic correlation. In the Cobequid Highlands, a minimum age for lithostratigraphically similar rocks is provided by the 734 ± 2 Ma Economy River orthogneiss which is intrusive into the Gamble Brook Formation (Doig et al., 1991). These units are also lithostratigraphically correlative with the Green Head Group in southern New Brunswick which contains stromatolites assigned a Middle Riphean age (Hofmann, 1974). This would correspond to a 1350-1050 Ma depositional age (Harland et al., 1989). In most places the contact between the gneisses and the metasediments is tectonic, and nowhere can original basement-cover relationships be unequivocally demonstrated.

Upper Proterozic volcanic-sedimentary rocks

The remaining Upper Proterozoic rocks of the Avalon Zone are characterized by volcanic and clastic metasedimentary rocks (Fig. 3.134-3.139). These uppermost Proterozoic rocks are exposed in the Mira Subzone (Fourchu Group and equivalents, Fig. 3.136), in the Antigonish Highlands (Georgeville Group, Fig. 3.137), in the northern Cobequid Highlands (Jeffers Group or Warwick Mountain Formation, Fig. 3.138) and in the southern Cobequid Highlands (Folly River Formation).

In southern Cape Breton Island, the Fourchu Group and its equivalents occur in four fault-bounded blocks (Fig. 3.136) where they are primarily composed of subaerial-shallow marine pyroclastic rocks (crystal tuff, ignimbrite, breccia, lithic tuff and ash), flows and small intrusive bodies. The base of the Fourchu Group is not exposed. However, the geochemistry of the Fourchu Group indicates the presence of thinned continental crust beneath the group during the late Proterozoic. The Stirling block is distinctive because of the presence of a massive sulphide deposit, turbiditic sedimentary rocks and chemical sediments (cherts and carbonates). The Upper Proterozoic volcanic rocks of the Fourchu Group are predominantly bimodal, although intermediate types are abundant and mafic rocks are generally higher in SiO_2 in the Coxheath and East Bay Hills fault blocks. The geochemistry of the mafic rocks from all four fault blocks is characteristic of lavas from orogenic zones (Dostal et al., 1990). The basaltic rocks of the Coastal block are tholeiitic, whereas those of the East Bay Hills and Coxheath blocks are calc-alkaline, and those of the Stirling block are intermediate between these two types. The basalts of the Coastal block have the lowest abundances of incompatible elements including high-field-strength elements and large-ion-lithophile elements and resemble island arc tholeiites. The abundances of these incompatible elements progressively increase from

the tholeiitic basalts of the Coastal block through intermediate basalts of the Stirling block to the calc-alkaline basalts of the East Bay and Coxheath blocks. The Stirling metabasalts have abundances of highly and moderately incompatible trace elements which overlap those of calc-alkaline basalts and tholeiites from evolved island arcs. The abundances of these elements in the East Bay Hills and Coxheath mafic rocks are notably higher and are similar to calc-alkaline basalts emplaced on continental margins. Assuming that these four volcanic suites are related, the progressive compositional changes in Fourchu mafic rocks across southeastern Cape Breton Island are similar to trends observed in modern island arcs (Gill, 1981) and would be consistent with a northwest-dipping subduction zone and a trench located southeast of the Coastal block (Dostal et al., 1990).

The volcanic rocks of the Fourchu Group have yielded concordant U-Pb zircon ages of (1) 574 ± 1 Ma in the Coastal block; (2) 676 ± 1 Ma in the Stirling block; and (3) ca. 660-600 Ma in the East Bay Hills block (Barr et al., 1990a). The Coxheath volcanic rocks were inferred to be synchronous with cross-cutting plutonic rocks that have yielded ^{40}Ar-^{39}Ar hornblende plateau ages between 635 and 600 Ma (Keppie et al., 1990). Plutons of 635-600 Ma ages are also present in the East Bay Hills and Stirling blocks (Keppie et al., 1990; Barr et al., 1990b). In each fault block, however, these U-Pb ages represent one datum in a thick pile of volcanic rocks such that volcanism overall appears to have lasted at least 100 Ma (Keppie et al., 1979; Keppie et al., 1990; Dostal et al., 1990).

Farther north in Cape Breton Island, possible volcanic equivalents of younger parts of the Fourchu Group are only preserved as ouliers, such as the Prices Point formation in the southeastern Cape Breton Highlands (Raeside and Barr, 1990). However, widespread calc-alkaline plutonic bodies that are presumeably roots to the volcanic edifice yield ages in the range 565-555 Ma (Barr et al., 1988; Dunning et al., 1990b).

Uppermost Proterozoic rocks in the Antigonish Highlands (Georgeville Group) consists of a bimodal sequence of basalt, basaltic andesite, and rhyolite overlain by a thick succession of turbidites with arc-derived clasts, and minor basalts (Murphy and Keppie, 1987). The turbidites are predominant in the central highlands flanked on the north and south by mainly volcanic assemblages. The age of these turbidites is constrained between the youngest detrital zircon grains that yielded U-Pb ages ranging between 625 and 605 Ma (Keppie and Krogh, 1990), and the 625-600 Ma ^{40}Ar-^{39}Ar hornblende plateau ages from plutons that posttectonically intrude the Georgeville Group (Keppie et al., 1990). Geochemically, the basalts are within-plate, continental rift tholeiites with some alkalic tendencies, the basaltic andesites are calc-alkalic, and the rhyolites have volcanic arc affinities (Murphy et al., 1990). The Georgeville Group is interpreted to represent an intra-arc rift. The paleogeography of this rift was influenced by contemporaneous motion on the bounding northeast-southwest faults (Murphy and Keppie, 1987). The continental tholeiitic lavas and abundant feeder dykes and sills coincide with a strong north-south vertical gradient magnetic anomaly and positive gravity anomaly, which may express the trend of the extensional zone. If this rift zone has preserved its original trend relative to the bounding faults, then a sinistral transtensional origin may be deduced for the intra-arc basin. However, subsequent (Paleozoic) dextral movements (Murphy et al., 1989) may have

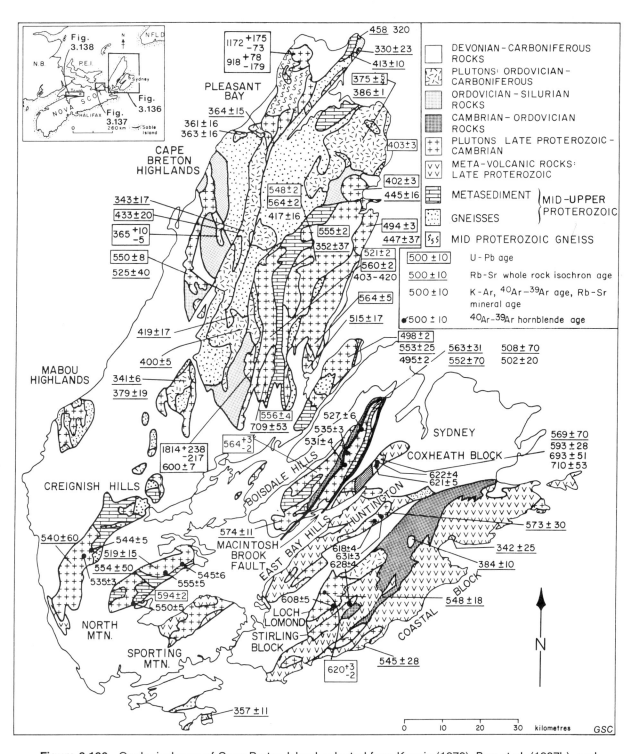

Figure 3.136. Geological map of Cape Breton Island, adapted from Keppie (1979), Barr et al. (1987b), and Keppie et al. (1990), showing published geochronological data.

rotated the rift in a clockwise sense from an original east-west orientation, in which case the original sense of shear would have been dextral.

In the northern and western Cobequid Highlands, the upper Proterozoic Jeffers Group records a similar stratigraphy and volcanic geochemistry to that of the Georgeville Group (Pe-Piper and Piper, 1987, 1989). In the eastern Cobequid Highlands, the upper Proterozoic Folly River Formation is similar lithologically to the upper part of the Georgeville Group (Murphy et al., 1989). Deposition of the formation probably coincides with the 630-605 Ma transtensional deformation in the underlying Gamble Brook

Formation (Nance and Murphy, 1989). The geochemistry is consistent with continental rifting within a volcanic arc (Pe-Piper and Murphy, 1989). Thus the transtensional deformation may herald the development of basinal conditions within the arc.

Lower Paleozoic sequences

In the Mira and Bras d'Or Subzones of southern Cape Breton Island, the Proterozoic rocks are overlain by Cambrian-Ordovician platformal rocks containing an Acado-Baltic (Avalonian) fauna. A major component of these rocks

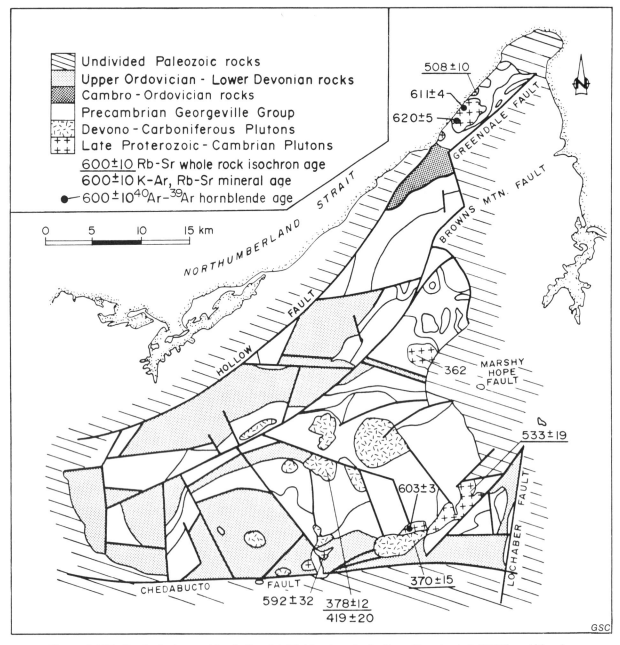

Figure 3.137. Geological map of the Antigonish Highlands, adapted from Keppie et al. (1990) and Murphy et al. (1990) showing published geochronological data.

in the Boisdale Hills of the Bras d'Or Subzone is Middle Cambrian, bimodal, within-plate, continental rift lavas (Murphy et al., 1985). Farther north in the Aspy and Blair River subzones of Cape Breton Island, Cambrian-Ordovician rocks are absent. However, pebbles in the Lower Carboniferous Horton Group derived from the neighbouring Mabou Highlands contain Cambrian-Ordovician and Silurian-Devonian fauna typical of the Avalon Zone (Norman, 1935; Keppie, 1990), suggesting the former presence of the overstep sequence across the Aspy Subzone. Precambrian rocks in the Aspy Subzone are overlain by Ordovician-Silurian metasedimentary and metavolcanic rocks (Jumping Brook Complex, Sarach Brook Metamorphic Suite, Money Point Group). The age of these units is given by (1) an U-Pb zircon date of 430 ± 3 Ma from a rhyolite dyke that feeds the Jumping Brook Complex (Currie et al., 1982); (2) a 433 +7/-4 Ma U-Pb zircon age from rhyolite in the Sarach Brook Metamorphic Suite (Dunning et al., 1990b); (3) an U-Pb zircon date of 427 ± 2 Ma from a rhyolite in the Money Point Group (Keppie et al., 1991); and (4) a 412 ± 15 Ma Rb-Sr whole rock isochron derived from rhyolitic flows on Ingonish Island (Keppie et al., 1986).

In the Antigonish Highlands, Cambrian-Ordovician platformal units with an Acado-Baltic (Avalonian) fauna (Landing et al., 1980) rest unconformably upon deformed Upper Proterozoic rocks. Interbedded with the Lower Cambrian sedimentary rocks are bimodal, within-plate, continental rift volcanic rocks (Murphy et al., 1985). The Cambrian-Ordovician rocks are, in turn, unconformably overlain by a Silurian-Devonian succession that is also typical of the Avalonian overstep sequence. In the Cobequid Highlands, only the Silurian-Devonian overstep sequence is preserved.

Upper Precambrian-Cambrian intrusions

Plutonic rocks are common throughout the Avalon Zone of Nova Scotia and they fall into two age groups: ca. 635-600 Ma and ca. 565-490 Ma (Keppie et al., 1990; Dunning et al., 1990b; Barr et al., 1990; Fig. 3.134-3.139). The Fourchu Group and its correlatives in Cape Breton Island are intruded by

hornblende-biotite, I-type, epizonal, calc-alkaline diorites, tonalites and granites (Barr et al., 1982) which give ages in the range 635-600 Ma (^{40}Ar-^{39}Ar hornblende plateau ages: Keppie et al., 1990; and U-Pb zircon age: Barr et al., 1990a). Rb-Sr whole rock isochrons previously gave a much longer age range of about 615-515 Ma and appear to be related to post-intrusion thermal resetting (Keppie and Smith, 1978; Keppie and Dallmeyer, 1991). The calc-alkaline geochemistry of the plutons suggests that they are genetically related to subduction that produced the Fourchu volcanic arc. Contact metamorphic minerals indicate a generally low pressure hornfels facies series. The North Branch Baddeck River leucotonalite (600 +6/-7 Ma: U-Pb lower intercept age: Jamieson et al., 1986) intrudes gneisses and schists in the Cape Breton Highlands.

The geochemistry of the 635-600 Ma plutons in the Antigonish and Cobequid Highlands appears to be quite variable. Thus, the Jeffers Brook pluton in the Cobequid Highlands has both calc-alkaline and alkaline affinities (Pe-Piper and Piper, 1989), whereas the Greendale Complex in the Antigonish Highlands may have a tholeiitic parental magma (Murphy et al., 1991). This matches the mixed affinity of the upper Proterozoic volcanic rocks in these areas. Unlike southern Cape Breton Island, the plutons in the Antigonish and Cobequid Highlands cut polyphase structures developed in the volcanic-sedimentary host rocks. However, age constraints indicate that sedimentation, deformation and intrusion closely followed each other: detrital zircons in the Georgeville Group are as young as 613 ± 5 Ma (Keppie and Krogh, 1990) whereas the oldest hornblende plateau age on the Greendale Complex is 620 ± 5 Ma (Keppie et al., 1990). Thus, both the late Proterozoic volcanism and the 635-600 Ma plutonism have been attributed to the same subduction event in which the transpressional rifting of the magmatic arc led to deposition, deformation and plutonism in a short time interval (Keppie et al., 1990).

The ca. 565-490 Ma plutonism occurs in Cape Breton Island north of the Macintosh Brook Fault (Keppie et al., 1990; Dunning et al., 1990b; Barr et al., 1990a). While most of these plutons appear to be concentrated in central Cape

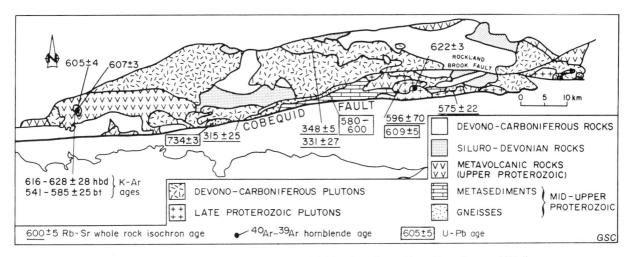

Figure 3.138. Geological map of the Cobequid Highlands, adapted from Donohoe and Wallace (1982) and Keppie et al. (1990) showing published geochronological data: hbd = hornblende; bt = biotite.

245

Breton Island and eastern Cape Breton Highlands, the Chéticamp pluton (550 ± 8 Ma: U-Pb zircon age: Jamieson et al., 1986) occurs in the western Cape Breton Highlands. Reynolds et al. (1989) and Dunning et al. (1990b) suggested that the Chéticamp pluton might be an overthrust klippe, although the steeply dipping attitude of its bounding faults, and the presence of xenoliths within it that have been correlated with the Pleasant Bay Complex (Currie, 1987b), do not support this suggestion.

In the Antigonish Highlands, the post-collisional, Ohio granodiorite (533 ± 19 Ma; Rb-Sr whole rock isochron age; Cormier, 1979) and several undated stocks associated with the Cambrian volcanic rocks are presumed to belong to this younger age group. Keppie et al. (1990) proposed that the ca. 570-490 Ma plutons might be the subvolcanic equivalents of the Eocambrian-Cambrian, tholeiitic-alkalic, continental rift volcanic rocks (Murphy et al., 1985; Keppie and Dallmeyer, 1991). On the other hand, Barr and Setter (1986), Justino (1985), and Raeside et al. (1988) suggested that these plutons formed in a magmatic arc. Raeside and Barr (1990) and Barr (1990) have subsequently related them to post-orogenic uplift. This apparent contradiction may be explained by one or both of the following: (1) melting of a crustal source contaminated by previous (ca. 630-600 Ma) subduction; or (2) oxidation and hydration of the plutonic rocks in the magma chamber which would superimpose calc-alkaline characteristics upon the original chemistry.

Structural history

The gneisses of the Pleasant Bay Complex display polyphase deformation and several episodes of metamorphism: the early metamorphism is largely obscured by a high pressure and high temperature upper amphibolite to granulite facies metamorphism overprinted by late retrograde metamorphism (Currie, 1987c). As stated above, the time of the high grade metamorphism is inferred to be 918 +78/-179 Ma, based upon the U-Pb data of Barr et al. (1987a; Keppie et al., 1991).

Gneisses of the Mount Thom Complex (Fig. 3.138) exhibit polyphase deformation consisting of an early foliation overprinted by a second fabric axial planar to isoclinal folds and associated with amphibolite facies metamorphism. The second fabric, in turn, is deformed by a crenulation cleavage (Donohoe, 1976; Donohoe and Cullen, 1983). The 934 ± 82 Rb-Sr whole-rock errorchron age (Gaudette et al., 1984) may date the high grade metamorphic event (Fig. 3.138) although isotopic mixing and/or resetting by the last event cannot be ruled out and could account for the large error.

While more geochronological data are essential, existing data in the Pleasant Bay and Mount Thom complexes tentatively suggest the existence of a metamorphic event near the Middle-Late Proterozoic boundary.

The George River-Gamble Brook metasedimentary rocks generally display polyphase deformation and a metamorphism that ranges from greenschist facies in southern Cape Breton Island and the Cobequid Highlands to amphibolite facies in northern Cape Breton Island (Keppie, 1979). The contacts between these metasedimentary rocks and the upper Proterozoic sequences are generally tectonic which has led to a divergence of opinion on the ages of the structures: pre-late Proterozoic, post-late Proterozoic/pre-Eocambrian

or Paleozoic (Helmstaedt and Tella, 1973; Keppie, 1982; Nance and Murphy, 1990). In most places the data are equivocal. However, in the central Cobequid Highlands, the Great Village River Gneiss is composed of polydeformed hornblende-bearing orthogneisses, biotite-garnet psammitic paragneisses and quartz-plagioclase layered amphibolite dykes. Syntectonic granite gneisses and amphibolites intrusive into the Great Village Gneiss have yielded 600-580 Ma U-Pb ages (Doig et al., 1991), and 640-625 Ma Rb-Sr isochron ages (Gaudette et al., 1984). Intense mylonitic fabrics are present in the Gamble Brook Formation: C-S fabrics, gently plunging east-southeast stretching lineations defined by dimensionally oriented hornblende, quartz and quartz-feldspar augen, and fold asymmetry in a moderate to steep southeast-dipping mylonitic fabric that indicates an oblique sinistral sense of shear (Nance and Murphy, 1990). These fabrics are truncated by a mafic dyke that is petrographically and chemically similar to the mafic lavas in the Folly River Formation and is inferred to be a feeder dyke (Nance and Murphy, 1990). Xenoliths of mylonitic Gamble Brook quartzite have been recorded in the post-tectonic (about 609 Ma old) Debert River pluton (Doig et al., 1991). This suggests that the contact between the Gamble Brook and Folly River formations is an unconformity representing deformation. Thus, the deformational events in the Cobequid Highlands are constrained between 615 and 580 Ma. In the Boisdale Hills of the Bras d'Or Subzone, Helmstaedt and Tella (1973) recorded polydeformed and metamorphosed pebbles of the George River Group in the overlying Cambrian rocks indicating a Late Proterozoic age for some of the deformation.

The Fourchu Group is deformed by northeast-trending folds associated with a steeply dipping cleavage. Fourchu clasts containing a predepositional fabric have been found in the overlying Cambrian rocks implying that some deformation occurred prior to the Cambrian.

Polyphase deformation of the Georgeville Group involving thrusts and recumbent east-vergent folds deformed by upright, north-south folds is attributed to dextral transpressional closure of the Georgeville basin (Murphy et al., 1989). This deformation was accompanied by greenschist-subgreenschist facies metamorphism. In the northern Antigonish Highlands, the earliest folds are cut by the gabbroic Greendale Complex dated at 611±5 Ma (^{40}Ar-^{39}Ar plateau ages on hornblende: Keppie et al., 1990). Layering within this complex has been interpreted in terms of syn-intrusion dextral shear on northeast-trending bounding faults (Murphy and Hynes, 1990), and suggests a continuation of the stress regime that produced the folds in the Georgeville Group.

Upper Proterozoic rocks of the Jeffers Group and Folly River Formation were deformed by thrusts and isoclinal folds with kinematics indicating dextral transpression (Nance and Murphy 1990). This deformation took place prior to the intrusion of the Debert River pluton at about 609 Ma (U-Pb zircon age; Doig et al., 1991) and 596 ± 70 Ma (Rb- Sr whole-rocks isochron; Donohoe et al. 1986).

Interpretation

The tectonic setting of the Avalonian basement gneisses in Nova Scotia is poorly understood. They apparently record a mid to late Proterozoic (Grenvillian?) tectonothermal event and/or a series of events that occurred prior to the

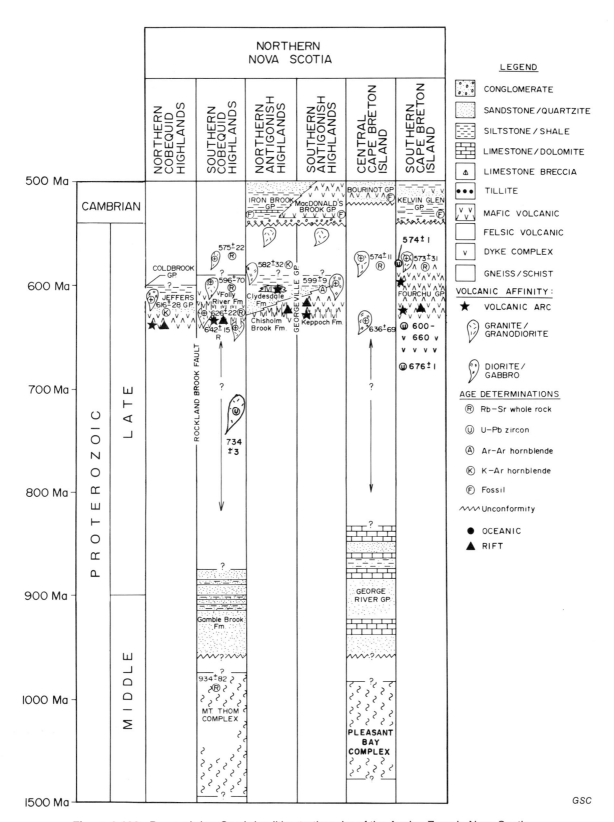

Figure 3.139. Precambrian-Cambrian lithostratigraphy of the Avalon Zone in Nova Scotia.

development of the characteristic Upper Proterozoic volcanic-sedimentary sequences. The middle Proterozoic (mid Riphean) metasedimentary rocks form an extensive succession that probably records the development of a passive margin. They are traditionally interpreted as a "cover" sequence deposited unconformably on the higher grade "basement". However, this cannot be demonstrated unequivocally because the present contacts, where exposed, are sheared.

The stratigraphic and petrological data for the Upper Proterozoic rocks indicate that they were formed in a rifted ensialic magmatic arc environment in which alternating horsts and graben were developed (Fig. 3.134, 3.139). Available data indicate that the basins were floored by continental basement. Deformation is associated with the opening and closing of these intra-arc rifts and reflects transcurrent motions on their bounding faults. Thus the age of deformation has only local significance and does not yield direct information as to the timing of the Avalonian Orogeny because in this type of setting, basin opening and closure may be diachronous. The opening and closing stages were presumably related to oblique convergence. Geochronological data suggest that the magmatic arc was active from about 680-570 Ma, and that deposition, deformation and plutonism were penecontemporaneous.

The geochemistry of the volcanic rocks of southern Cape Breton Island and southern New Brunswick (Dostal et al., 1990; Dostal and McCutcheon, 1990) indicates that they were erupted in a volcanic arc environment. In both regions, the mafic rocks display a remarkably similar compositional zonation which resembles the across-arc variations observed in modern volcanic arcs. The progressive compositional changes that include a transition from island arc tholeiites along the southeastern coasts to calc-alkaline rocks inland to the northwest, suggest a northwest-dipping subduction zone with a trench located to the southeast. In the Antigonish and Cobequid Highlands, continental tholeiites and calc-alkaline volcanic rocks are interbedded and overlain by turbidites that were probably deposited in a back-arc or intra-arc environment. The superimposed positive gravity and magnetic anomalies that may be traced discontinuously from the Grand Banks through Nova Scotia and along the northern side of the Gulf of Maine to near Boston, have been interpreted to represent Upper Proterozoic ophiolitic remnants preserved by incomplete suturing along a Late Proterozoic plate boundary. In this scenario, northeast-trending segments could be inferred to mark subduction zones, and the east-west portions might correspond to transform faults. Thus, the present geographical distribution of Upper Proterozoic volcanic rocks

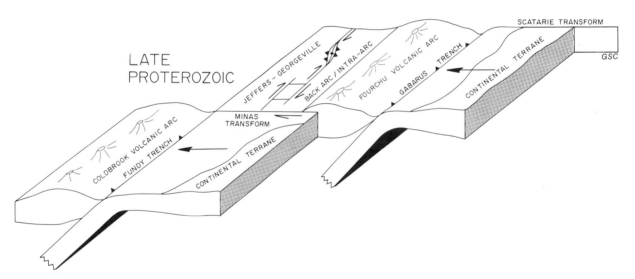

Figure 3.140. Three-dimensional, Late Proterozoic tectonic model for the volcanic rocks of the Avalon Zone in Maritime Canada.

Figure 3.141. Schematic representation of the latest Proterozoic-Early Cambrian tectonic setting in the Avalon Zone of Nova Scotia.

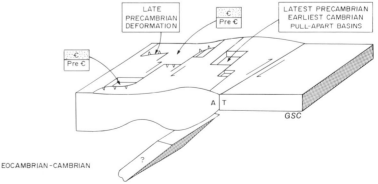

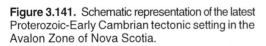

could be explained by a tectonic model involving a single subduction zone offset by two Late Proterozoic transform faults located approximately along the Minas Fault (Fig. 3.134 also called Cobequid-Cledabucto Fault) and Scatarie Ridge (Fig. 3.140). This would be consistent with dextral strike-slip origin and closure of the late Proterozoic basins. The Late Proterozoic rifted magmatic arc evolved to produce latest Proterozoic-Cambrian tholeiitic-alkaline volcanic rocks. The Cambrian rocks appear to have been deposited in pull-apart basins suggesting a continuation of the transpressional strike-slip regime operative in the Late Proterozoic (Fig. 3.141).

Paleomagnetic data indicate that the Avalon Zone had affinities with Gondwana and Armorica during Late Proterozoic-Cambrian times rather than with North America (Johnson and Van der Voo, 1985, 1986). However, the Avalon Zone had converged with North America by Silurian-Devonian times (Miller and Kent, 1988). This is consistent with: (1) the contrasting Cambrian faunal provinces; Acado-Baltic/Avalonian in the Avalon and Laurentian in North America, and (2) with the Rhenish-Bohemian fauna that was present only in the Avalon during the Late Silurian-Lochkovian but spread across onto Laurentia during the Pragian and Emsian (Boucot, 1975). In most circum-Atlantic regions the Late Proterozoic orogens are built upon Archaean and Lower Proterozoic basement, but in South America Late Proterozoic orogens are built upon a Mid-Upper Proterozoic (Grenville) basement (Keppie and Dallmeyer, 1989). Along the southwestern side of the Amazonian craton (ca. 2.5- >1.9 Ga) the Jurena-Rio Negro orogenic belt (ca. 1.75-1.5 Ga) lies inboard of the Rondonian orogenic belt (ca. 1.45-1.25 Ga), the Sunsas orogenic belt (ca. 1.1-0.9 Ga) and a Brasiliano orogenic belt (ca. 0.6-0.5 Ga; Teixeira et al., 1989). This region is inferred to be a source for the detrital zircons in the Georgeville Group of the the Antigonish Highlands (Keppie and Krogh, 1990) and may therefore represent the provenance of the Avalon Zone in Nova Scotia and New Brunswick. The

synchroneity of Late Proterozoic-Cambrian subduction beneath the Avalon Zone and the opening of Iapetus implies that the Avalon Zone was not located within the Iapetus Ocean. A possible Late Proterozoic-Cambrian reconstruction is shown in Figure 3.142. Preservation of the Avalon magmatic arc sequences in the Canadian Appalachian region contrasts with their general elimination in continent-continent collisional orogens where they are usually eroded because they lie on the upper plate. This suggests that termination of subduction was not the result of continent-continent collision, but instead may reflect global plate reorganization associated with the breakup of a late Proterozoic supercontinent.

Acknowledgments

This contribution is published with permission of the Minister of the Nova Scotia Department of Mines and Energy.

New Brunswick

S.R. McCutcheon and M.J. McLeod

Introduction

The Avalon Zone of New Brunswick, of which the Caledonia Highlands constitute a major part, has mainly upper Precambrian to Lower Cambrian volcanic and intrusive rocks. Most of the volcanic rocks have been assigned to the Coldbrook Group. Historically the Coldbrook has been inferred to overlie carbonates and clastic rocks of the Green Head Group, including the spatially associated Brookville Gneiss, and is overlain by the Cambrian-Ordovician Saint John Group. Isotopic dates indicate that the intrusive rocks are divisible into roughly 600 to 550 Ma and 547 to 510 Ma age groups.

Extent and boundaries

The lateral extent of typical Avalon rocks beneath younger Paleozoic cover and the Gulf of St. Lawrence is interpreted from regional gravity (Haworth et al., 1980a, b) and magnetic (Zietz et al., 1980a, b) maps (Fig. 3.143). Positive gravity and magnetic anomalies, typical of the Avalon Zone, underlie a triangular area approximately bounded by the northeast-trending Falls Brook-Taylor Brook Fault, the east-west trending Cobequid-Chedabucto Fault and the northwesterly trending Canso Fault that extends from southwestern Cape Breton Island into the Gulf of St. Lawrence (McCutcheon and Robinson, 1987).

The Avalon-Meguma zone boundary is the east-west trending Cobequid-Chedabucto Fault that traverses the Bay of Fundy and terminates south of Campobello Island (southwestern New Brunswick) against the north-northwest-trending Oak Bay Fault (McCutcheon and Robinson, 1987). However, the Avalon- Gander zone boundary is less clear because its surface and subsurface expressions are different.

Figure 3.142. Late Proterozoic palinspastic reconstruction of the circum-Atlantic region showing the locations of Pan-African (which would include the Avalon Zone) and Grenvillian orogens.

McCutcheon, S.R. and McLeod, M.J.
1995: Avalon Zone-New Brunswick; in Chapter 3 of Geology of the Appalachian-Caledonian Orogen in Canada and Greenland, (ed.) H. Williams; Geological Survey of Canada, Geology of Canada, no. 6, p. 249-261 (also Geological Society of America, The Geology of North America, v. F-1).

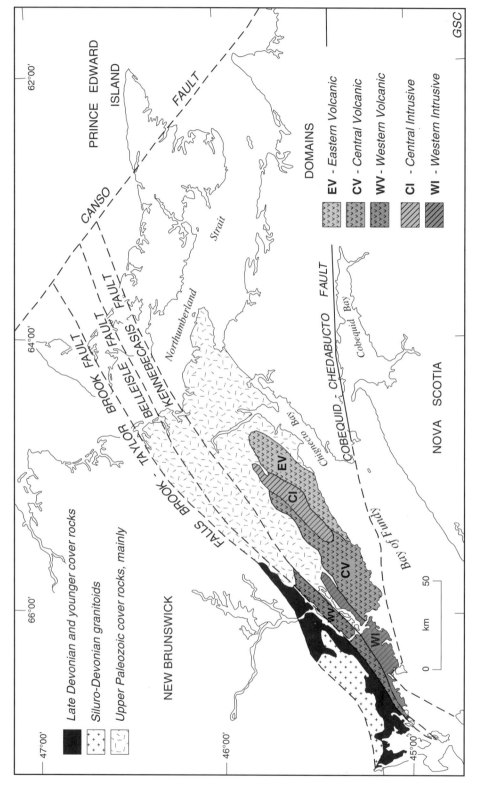

Figure 3.143. Location map showing the extent of the Avalon Zone in southern New Brunswick, and the major faults that extend into Prince Edward Island and Nova Scotia.

At the surface, the Avalon-Gander zone boundary has generally been considered to be the Belleisle Fault. This fault, which is predominantly a pre-Carboniferous (Acadian) structure (Leger and Williams, 1986), marks the northern limit of typical Coldbrook volcanic rocks. the Kingston Dyke Swarm (McCutcheon and Ruitenberg, 1987) is subparallel to, and not far south of, the Belleisle Fault. Both the fault and dyke swarm can be traced across southern New Brunswick by their topographic and magnetic expressions, respectively, to Campobello Island. At Campobello Island, a linear high magnetic anomaly that characterizes the dyke swarm appears to be offset sinistrally by the Oak Bay Fault and associated faults, to a point south of Grand Manan Island, suggesting that the Belleisle Fault is also offset (McCutcheon and Robinson, 1987). Alternatively, McLeod and Rast (1988) suggested that the abrupt termination of the linear magnetic anomaly reflects vertical (west-side-down) movement on the Oak Bay Fault and that the dyke swarm continues to the southwest beneath thick sedimentary cover of the Quoddy Formation.

In the subsurface, the northern boundary of the Avalon Zone is uncertain. There is little doubt that there are Lower Cambrian rocks north of the Belleisle Fault, but typical Coldbrook volcanic rocks are absent (McCutcheon, 1981). Rast and Currie (1976) suggested that there are late Precambrian granitoids north of the Belleisle Fault. Currie (1988a) demonstrated that they are about 555 Ma. Gravity signatures (Thomas and Willis, 1989) typical of the Avalon Zone extend beneath Paleozoic cover at shallow depths (a few kilometres) as far north as the Falls Brook-Taylor Brook Fault (Fig. 3.143). This fault is likely a continuation of the Turtle Head Fault in Maine, which may mark the northern limit of probable Avalon rocks as suggested by seismic data (Stewart et al., 1986) and regional correlations (Fyffe et al., 1988b). The two faults are sealed by the Silurian-Devonian Saint George Batholith. Others have drawn the subsurface Avalon-Gander zone boundary at the Fredericton Fault (Williams, 1979; Rast et al., 1976a).

Evidence that the Avalon Zone continues north of the Falls Brook-Taylor Brook Fault is inconclusive. Lead isotopic and gravity data (Bevier, 1987; Thomas and Willis, 1989) indicate northward continuation, but Nd-Sm isotopic studies (Whalen et al., 1989) do not. Models based on correlations between clastic rocks of the St. Croix Subzone (Gander) and Saint John Group (Avalon; e.g. Rast and Stringer, 1974; Ludman, 1987) are disputed by Fyffe and Pickerill (1986). For now, the subsurface Avalon-Gander zone boundary is most reasonably drawn at the Falls Brook-Taylor Brook Fault based on geophysical expression.

Stratigraphy

For purposes of stratigraphic description, rocks of the New Brunswick Avalon Zone are grouped under five headings; (1) Green Head Group and Brookville Gneiss, (2) Coldbrook Group, (3) Ratcliffe Brook Formation and other redbeds, (4) Saint John Group, and (5) rocks between the Belleisle and Falls Brook-Taylor Brook faults.

Green Head Group and Brookville Gneiss

Metacarbonates and clastic sedimentary rocks of the Green Head Group and Brookville Gneiss are best exposed in and around the city of Saint John but similar rocks outcrop intermittently to the southwest as far as Lepreau and to the northeast along the north side of the Caledonia Highlands (Fig. 3.144). Traditionally, they have been considered to be basement to the Coldbrook Group. However, Barr and White (1989) suggested that they constitute a separate terrane, unrelated to the Coldbrook Group, an interpretation disputed by Dallmeyer et al. (1990).

The Green Head Group (Leavitt, 1963), formerly Green Head Formation (Hayes and Howell, 1937), consists of a lower mainly carbonate unit (Ashburn Formation) and an upper mainly clastic unit (Martinon Formation). Wardle (1978) subdivided the group into four units. From apparent base upward they are: (1) a predominantly orthoquartzite-grey psammite sequence with minor limestone, dolomite and biotite schist; (2) a limestone-dolomite sequence with minor orthoquartzite and grey pelite; (3) a heterogeneous assemblage of banded calcareous pelite, orthoquartzite, limestone, dolomite, conglomerate and stromatolitic limestone; and (4) a homogeneous sequence of massive grey psammite, pelite, greywacke and quartzite. More recently, Currie (1984, 1986) has excluded the Martinon Formation, the upper clastic unit of Wardle (1978), from the Green Head Group because; (1) it unconformably overlies other rock units with the contact marked by a distinctive basal marble conglomerate (sedimentary breccia), (2) it is of turbiditic origin and therefore sedimentologically different from Ashburn rocks, and (3) it contains numerous basalt sills suggesting that the Martinon is more akin to the Coldbrook Group rather than to the Green Head Group. Currie (1987a, 1991) further suggested that the Martinon Formation is equivalent to the lower part of the Coldbrook Group.

Stromatolites in the Green Head Group suggest a Neohelikian (1000-1400 Ma) age (Hofmann, 1974). The group is therefore much older than the Coldbrook Group, although both groups may have been deformed and intruded at the same time.

The Brookville Gneiss (Wardle, 1978) consists of biotite-quartz-feldspar and biotite-hornblende-quartz-feldspar paragneiss with abundant intercalations of marble and calc-silicate, and "swirled" quartz-diorite orthogneiss. Currie et al. (1981) redefined the Brookville Gneiss and considered all the gneissic rocks to be intrusive, following the lead of Hayes and Howell (1937) who assigned the gneisses to the Golden Grove Intrusive Complex.

Different ages have been suggested for the Brookville Gneiss. Currie et al. (1981) considered the gneiss to be remobilized Aphebian (greater than 1700 Ma) basement, whereas Wardle (1978) believed the major part of the Brookville Gneiss was metamorphosed Green Head with a diapiric core of remobilized basement, i.e. his "swirled" quartz-diorite orthogneiss. The first U-Pb zircon ages (Olszewski and Gaudette, 1982) indicated that both the Brookville Gneiss and Green Head Group are older than 800 Ma but younger than 1200 Ma, thus substantiating Wardle's interpretation. An age of 1641 Ma for inherited zircons is thought to be the age of the source rocks (Olzewski and Gaudette, 1982); either basement rocks or a deeper part of the sedimentary pile. However, the latest zircon ages show that the Brookville orthogneiss is much younger than previously considered. Orthogneiss and paragneiss have yielded igneous and sedimentary protolith ages of about 605 Ma and a maximum age of 641 Ma, respectively (Bevier et al., 1990; Dallmeyer et al., 1990).

251

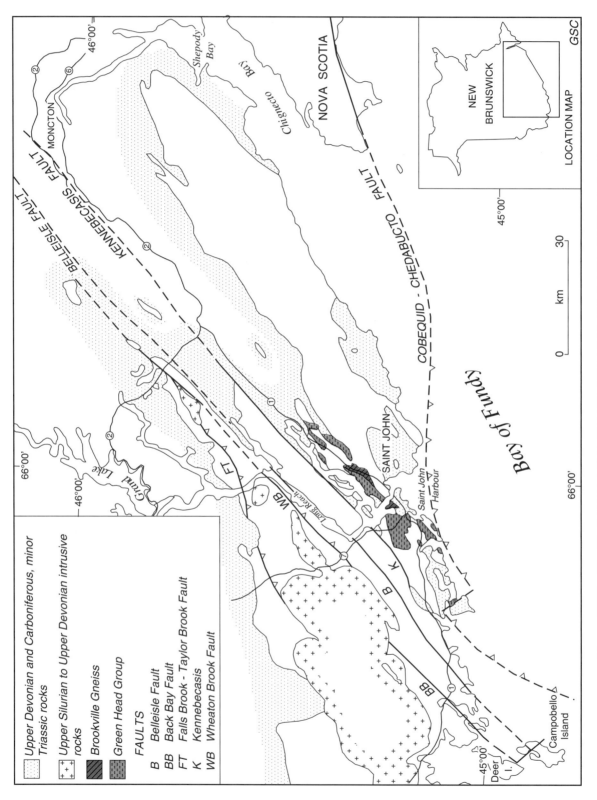

Figure 3.144. Simplified geology of southern New Brunswick showing the distribution of rocks belonging to the Precambrian Green Head Group and/or Brookville Gneiss.

These dates suggest that the contact between the Green Head Group and the Brookville Gneiss may be tectonic (see discussion by Bevier et al., 1990).

For more than a century, the contact between the Green Head Group and Coldbrook Group was considered to be an unconformity because of the more deformed and metamorphosed nature of the Green Head Group (Matthew, 1863; Bailey, 1865; Alcock, 1938; Rast et al., 1976b; Wardle, 1978). The exposed contact at Rockwood Park in Saint John is between carbonate rocks (Ashburn Formation) and massive rhyolite of the Coldbrook Group, and is quite clearly tectonic. If this is a major terrane boundary, as suggested by Barr and White (1989), then the contrast in deformational effects between the two units is readily explicable without requiring an unconformity. However, if the Martinon Formation is laterally equivalent to the lower part of the Coldbrook Group, as suggested by Currie (1987, 1991), an unconformity is expected. Along Highway 7 just north of Saint John, weathered clasts of Ashburn carbonate rocks occur in basal conglomerate of the Martinon Formation.

Coldbrook Group

The Coldbrook Group (Matthew, 1863; Alcock, 1938) was originally defined on rocks just east of Saint John, but it has been expanded to include most of the volcanic and sedimentary rocks in the Caledonia Highlands, the Kingston Peninsula, and the southwestern extension of the Kingston Peninsula (Ruitenberg et al., 1979). Also, much of the Mispeck group (name now abandoned) in the vicinity of the Fundy coast is assigned to the Coldbrook Group (McCutcheon, 1984; Currie, 1987d). The Coldbrook constitutes at least two groups (e.g. Barr and White, 1989; Barr et al., 1990b) and subdivision is anticipated, but structural complexities and poor exposure make the stratigraphy difficult to decipher and rock units hard to trace.

Three lithologically distinct assemblages are recognized, herein designated Assemblage I, II, and III. Each occurs in more than one of the three geographic domains, delineated by Giles and Ruitenberg (1977), i.e. the Eastern, Central and Western volcanic domains (Fig. 3.143), but only Assemblages I and II have enough areal extent to delineate in Figure 3.145.

Assemblage I is a subaqueous sequence, at least 5000 m thick, that is dominated by fine grained mafic and felsic tuffaceous rocks, massive and pillowed mafic flows, greenish grey and purple (rarely siliceous) siltstone, and greenish grey sandstone (McLeod, 1986, 1987; Barr and White, 1988). This sequence locally includes minor polymictic conglomerate, arkosic sandstone, limestone and chert. Disseminated pyrite is present in many of the fine grained rocks and some beds are highly pyritiferous. Rocks of Assemblage I underlie most of the Eastern domain where U-Pb ages range from 600-625 Ma (Barr and White, 1988; Bevier and Barr, 1990).

Lithologically similar rocks, provisionally equated with Assemblage I, underlie the northwestern side of the Western domain. There, a rhyolite gives an ^{40}Ar-^{39}Ar age of about 650 Ma (Stukas, 1978, sample VCB9).

Assemblage II is a subaerial sequence, several kilometres thick, composed of felsic and lesser mafic volcanic rocks with minor sedimentary rocks (Giles and Ruitenberg, 1977; McLeod, 1986, 1987; Barr and White, 1988). The type-Coldbrook

rocks just east of Saint John are part of this assemblage that underlies most of the Central domain and the northwestern edge of the Eastern domain. West of Saint John, along the Bay of Fundy coast, a U-Pb date of ca. 555 Ma (Zaineldeen et al., 1991) was obtained from rhyolites. Other rhyolites have yielded U-Pb zircon ages of about 550 Ma (Barr and White, 1988; Bevier and Barr, 1990) and a ^{40}Ar-^{39}Ar age of 557 Ma (Stukas, 1978; sample VCB20). Within the Central domain, there are 550 Ma syenogranitic and gabbroic plutons that cut and are co-magmatic with Assemblage II volcanic rocks (Barr and White, 1988). Subaerial volcanic rocks also underlie the southeastern side of the Western domain. However, these rocks appear to be much older with ^{40}Ar-^{39}Ar ages of 615 Ma and 660 Ma (Stukas, 1978; samples VCB 4 and VCB 6). There are three ways to interpret these isotopic ages: (1) there are two different age groups of subaerial volcanic rocks in Assemblage II, (2) the age range of Assemblage II volcanic rocks is much longer than suspected, or (3) the ^{40}Ar-^{39}Ar ages are erroneous. Until more U-Pb work is done, these rocks are all tentatively included in Assemblage II. They are separated from Assemblage I to the northwest by the Kingston Dyke Swarm (McCutcheon and Ruitenberg, 1987), from which a felsic intrusion has yielded a U-Pb date of 435 Ma (Doig et al., 1990). Older (Precambrian?) mafic dykes, which commonly are more abundant than felsic intrusions in the swarm, have been recognized (McLeod, 1979, Rast and Dickson, 1982).

Assemblages I and II are bimodal; the mafic rocks are basalts and basaltic andesites, whereas the felsic rocks are rhyolites (Ruitenberg et al., 1979; McLeod, 1986, 1987; Barr and White 1988; Bevier and Barr, 1990; Dostal and McCutcheon, 1990). The Eastern domain contains arc-related calc-alkalic volcanic rocks (McLeod, 1986; Barr and White, 1988; Bevier and Barr, 1990; Dostal and McCutcheon, 1990). The Central domain contains volcanic rocks interpreted to have formed in a rift environment on an older volcanic arc (Barr and White, 1988) or on continental crust (McLeod, 1986). The northern side of the Western domain contains calc-alkalic volcanic rocks (Dostal and McCutcheon, 1990).

Assemblage III comprises greenish grey and purple, thinly bedded, highly siliceous siltstone and fine grained sandstone of turbiditic origin. These turbidites are spatially associated with Assemblage II rocks, i.e. confined to the Central and Eastern domains, but they do not reflect the terrestrial depositional environment of Assemblage II. They are not areally extensive and consequently are not shown in Figure 3.145. These rocks presumably occur in small fault-bounded blocks, although in one locality they appear to underlie Assemblage II rocks.

Ratcliffe Brook Formation and other redbeds

The Ratcliffe Brook Formation is predominantly a redbed sequence, the "Etcheminian" of Matthew (1889), consisting of maroon, purplish red and greenish grey siltstone with minor limestone, sandstone, and conglomerate (Fig. 3.146). Volcanic rocks are absent or sparse and there is commonly distinctive quartzite-pebble conglomerate at, or near, the base of the sequence. This unit exhibits abrupt changes in thickness and lithology (Matthew, 1889, 1890; Tanoli and Pickerill, 1990). It lacks body fossils but contains trace fossils and tests of trilobites (Patel, 1976; Hofmann and Patel, 1988; Tanoli and Pickerill, 1988) comparable to those

253

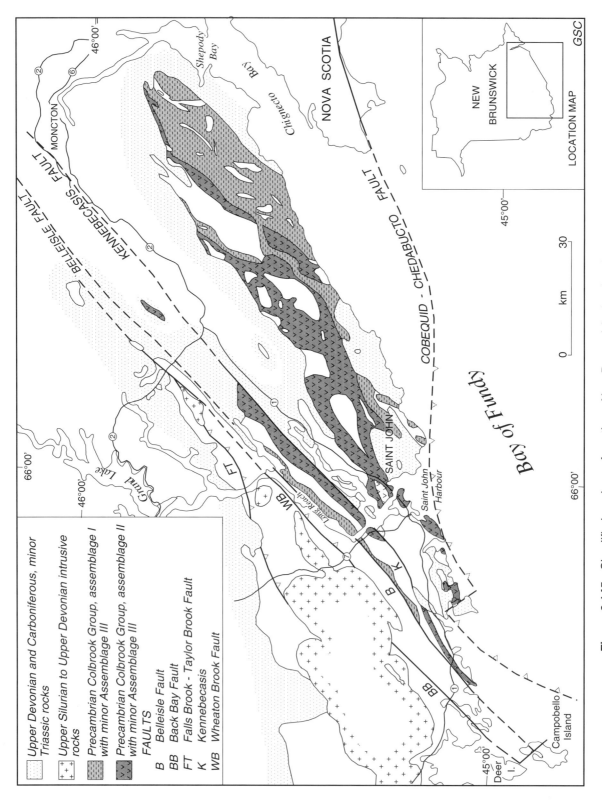

Figure 3.145. Simplified geology of southern New Brunswick showing the distribution of rocks belonging to the Precambrian Coldbrook Group.

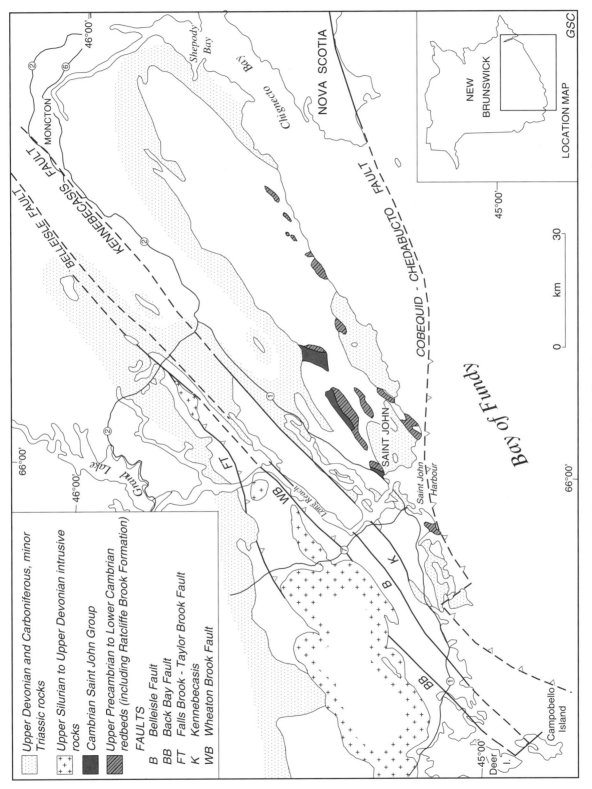

Figure 3.146. Simplified geology of southern New Brunswick showing the distribution of upper Precambrian to Lower Cambrian redbeds and the Cambrian Saint John Group.

in earliest Cambrian (Tommotian) rocks in Newfoundland. The lithology and stratigraphy of these rocks are distinct from younger trilobite-bearing Cambrian strata of the Saint John Group, and contrary to the opinions of Pickerill and Fyffe (1991), they deserve separate status as proposed by Matthew (1863) and Currie (1989).

Correlative rocks include the "Early Cambrian" strata of McLeod and McCutcheon (1981), the "Eocambrian red-beds" of Tanoli et al. (1985) and the "Cape Spencer redbeds" of Watters (1987). These rocks unconformably overlie pillow basalt in the Eastern domain (McLeod and McCutcheon, 1981), 625 Ma granite east of Saint John (Watters, 1987), and possibly rest unconformably on terrestrial volcanic rocks in the Central domain (see discussion by Tanoli and Pickerill, 1990). Tanoli et al. (1985) suggested that there are two redbed units, one late Precambrian and the other Cambrian in age. Most significantly, all the redbeds postdate the main Coldbrook volcanism and predate trilobite-bearing Cambrian rocks.

Saint John Group

As currently defined (Tanoli and Pickerill, 1988), the Saint John Group (Fig. 3.146) includes the Lower Cambrian Ratcliffe Brook, Glenn Falls and Hanford Brook formations; the Middle Cambrian Forest Hills Formation; the Upper Cambrian King Square and Silver Falls formations; and the Lower Ordovician Reversing Falls Formation. Following Matthew (1863) and Currie (1989), we recommend that the Ratcliffe Brook Formation be excluded from the group because (1) Glen Falls and younger rocks are remarkably similar throughout the New Brunswick Avalon Zone, (2) The Ratcliffe Brook Formation exhibits great diversity in lithology and thickness from one locality to another, and (3) there appears to be a faunal gap between the Ratcliffe Brook and Hanford Brook formations. These observations suggest a non-depositional or erosional disconformity between the Glen Falls and Ratcliffe Brook formations, as originally suggested by Matthew (1895).

The thicknesses and predominant lithologies of the remainder of the group are as follows, in ascending stratigraphic order (after Tanoli and Pickerill, 1988). The Glen Falls Formation (up to 43 m) consists of fine- to coarse-grained, greenish white quartzite overlain by coarse grained black sandstone. These rocks are unfossiliferous. The Hanford Brook Formation (up to 39 m) is composed of olive grey to dark grey, fine grained sandstone and minor shale. This unit contains the late Early Cambrian *Protolenus* fauna of Matthew (1895). The Forest Hills Formation (up to 20m) is predominantly a shale unit with the "Black Limestone" of Hayes and Howell (1937) at the base and minor fine grained sandstone and fossiliferous, lime-rich lenses toward the top. This unit contains the Middle Cambrian *Paradoxides* fauna. The King Square Formation (up to 380 m) is dominated by grey sandstone that is fine grained and thinly bedded at the base and top of the unit, but medium- to thick-bedded in the middle with a decreased proportion of grey to black shale and siltstone interbeds. This unit contains inarticulate brachiopods, trace fossils and in places medusoids. The Silver Falls Formation (up to 80 m) consists of dark grey to black, locally calcareous, shale with minor calcareous nodules and a few thin interbeds of fine grained sandstone. The calcareous parts of this unit contain a Late Cambrian trilobite (olenid)

fauna and a conodont fauna. The Reversing Falls Formation (about 20 m) comprises black carbonaceous and pyritiferous shales that commonly contain graptolites.

Rocks between the Belleisle and Falls Brook-Taylor Brook faults

All of the Cambrian-Ordovician and older rocks in this area are somewhat enigmatic. Two distinct stratigraphic sequences occur between the Falls Brook-Taylor Brook and Belleisle faults (Fig. 3.147). They are separated by the Wheaton Brook Fault and its probable extension, the Back Bay Fault. To the north, is the Silurian to Lower Devonian Mascarene Group. To the south, is a lithologically diverse assemblage of rocks that range in age from Early Cambrian and/or late Precambrian to Devonian. Four different units herein designated A, B, C and D are recognized. Granitic rocks are abundant in the southern area.

Unit A consists of grey polymictic conglomerate, dark grey, locally very quartz-rich sandstone, and grey to black siltstone. McCutcheon and Ruitenberg (1987) thought these rocks were Silurian. Currie (1987d), following MacKenzie (1964), equated them with the Precambrian Martinon Formation near Saint John. However, some of the conglomerates contain clasts of greenish grey, thinly bedded, siliceous siltstone, which were probably derived from the Coldbrook Group (McCutcheon and Ruitenberg, 1987), making a Martinon correlation unlikely. Subsequently, Currie (1988a) assigned these rocks to the "Eocambrian", because they are intruded by 555 Ma granitoids.

Unit B is dominated by redbeds but also includes subvolcanic felsic porphyries and mafic and felsic volcanic rocks. The redbeds comprise fine- to coarse-grained sandstone, granule conglomerate, polymictic pebble-cobble conglomerate, and mudstone. The sandstones generally contain abundant volcanic detritus rich in Fe-Ti oxides whereas the conglomerates have abundant quartz-eye porphyry clasts. The volcanic rocks are mostly mafic and consist of flows, breccias, and minor tuffs. McCutcheon (1981) and McCutcheon and Ruitenberg (1987) assigned these rocks to the Early Cambrian, i.e. equivalent to the Ratcliffe Brook Formation but lithogically different. Currie (1988b) assigned the rocks to the "Eocambrian". The relative ages of units A and B are uncertain but Unit A is considered to be older.

Unit C includes upper Lower Cambrian and younger sedimentary rocks that have been assigned traditionally to the Saint John Group, and some rift-related mafic volcanic rocks (Greenough et al., 1985). The sedimentary rocks, which are locally fossiliferous, are greenish grey to grey, fine- to coarse-grained sandstone and grey to black siltstone. In places they contain calcareous lenses (concretions). This unit is coeval with parts of the type-Saint John Group but differs in having a significant volcanic component.

Unit D encompasses all the post-Cambrian rock units. It includes several outliers and/or fault-bounded slivers of fossiliferous Silurian rocks, first mapped by Helmstaedt (1968) and Hay (1968), and the Devonian Blacks Harbour Conglomerate (Helmstaedt, 1968). Notably, Silurian rocks also occur south of the Belleisle Fault in a few localities, i.e. at Long Reach (MacKenzie, 1964; McCutcheon, 1981), on Campobello Island (McLeod, 1979; McLeod and Rast, 1988), and possibly at Beaver Harbour (Currie, 1988a, b).

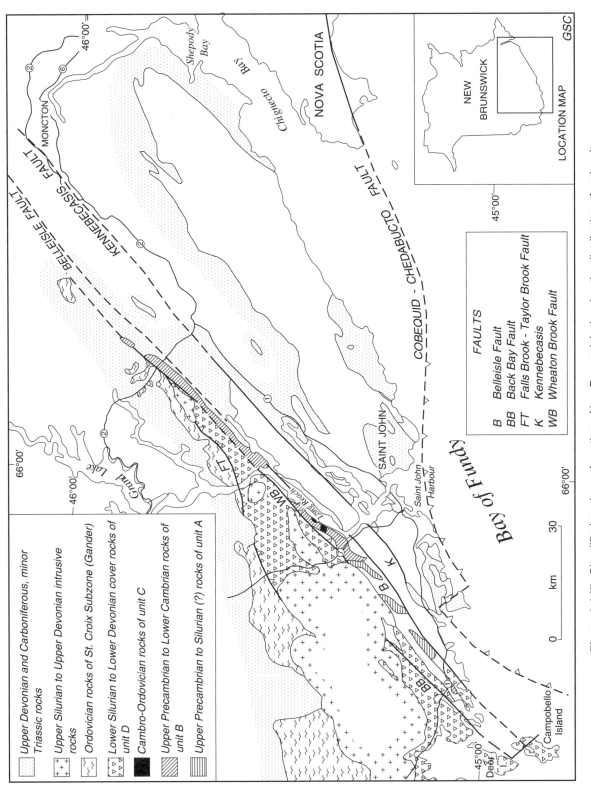

Figure 3.147. Simplified geology of southern New Brunswick showing the distribution of rock units between the Belleisle and Falls Brook-Taylor Brook faults. The outcrop area of the adjacent St. Croix Subzone (Gander) is also shown.

257

Intrusions

Intrusive rocks are concentrated in, but not restricted to, two areas in the Caledonia Highlands (Fig. 3.148). One separates the Central and Eastern volcanic domains and the second separates the Central and Western volcanic domains (Fig. 3.143). These were called the Central and Western intrusive belts by Ruitenberg et al. (1979). Intrusions also occur north of the Belleisle Fault (Currie, 1988a), and the Kingston Dyke Swarm runs the full length of the Western volcanic domain (McCutcheon and Ruitenberg, 1987). These intrusive rocks are broadly divisible into two age groups: late Precambrian about 600-550 Ma, and Cambrian about 547-510 Ma (Bevier and Barr, 1990; Currie, 1988b).

The Central Intrusive Belt of Ruitenberg et al. (1979) is largely made up of the Point Wolfe River and Bonnell Brook plutons. The Point Wolfe River pluton belongs to the 600-550 Ma group that ranges in composition from diorite to granite. The Bonnell Brook pluton belongs to the circa 547-510 Ma group that is mainly syenogranitic (Barr and White, 1988; Bevier and Barr, 1990). The 547-510 Ma group is considered to be comagmatic with at least some of the terrestrial volcanic rocks of Assemblage II of the Coldbrook Group. The 600-550 Ma group is approximately coeval with Assemblage I of the Coldbrook Group (Barr and White, 1988; Bevier and Barr, 1990). Notably, the 547-510 Ma group does not intrude the Ratcliffe Brook Formation and equivalent redbeds.

The Western Intrusive Belt of Ruitenberg et al. (1979) includes the Golden Grove Suite (Hayes and Howell, 1937), Milkish Head, Musquash, and Lepreau plutons (names introduced by Belyea, 1944), and the Red Head pluton. Cutting relationships and contrasting deformation effects show that there are at least two age groups (Belyea, 1944; Ruitenberg et al., 1979; Currie, 1987d). U-Pb zircon ages of 550 ± 11 and 538 ± 1 Ma have been reported, respectively, from a granitoid in the Musquash Harbour area (Grant et al., 1987), and from the Rockwood Park granodiorite that constitutes part of the Golden Grove Suite (Barr et al., 1990b).

The Golden Grove Suite is divisible lithologically and chronologically (oldest to youngest) into (1) early gabbros, (2) quartz diorites and granodiorites, (3) adamellites, and (4) late gabbros (Wardle, 1978). The early gabbros predate the first deformation (D1) and are mainly amphibolites. Wardle (1978) considered all the quartz diorites coeval, but there are probably two populations (Ruitenberg et al., 1979). Wardle also noted that his Spruce Lake and Rockwood Park plutons (the older populaton) are less pristine than the rest, and that the Spruce Lake pluton is intimately associated with the older gabbros. The younger quartz diorites postdate D1 but predate D2. The adamellites intrude the quartz diorites and Brookville Gneiss but predate D2. Also, the late gabbros predate D2.

The Kingston Dyke Swarm intrudes a major mylonite zone that separates Assemblage I and Assemblage II volcanic rocks of the Coldbrook Group in the Western volcanic domain. Variably deformed mafic and felsic dykes are intercalated with mylonites (Rast and Dickson, 1982; McCutcheon and Ruitenberg, 1987). U-Pb isotopic dates indicate that some of the dykes have a Silurian (435.5 ± 1.5 Ma) age (Doig et al., 1990).

There are at least two groups of intrusions between the Belleisle and Wheaton Brook faults (McCutcheon and Ruitenberg, 1987). One group is lithologically similar to the 600-550 Ma diorites and granites of the Caledonia Highlands. Rocks of this group intrude Unit A but do not intrude Unit B. The other group consists of hypabyssal, quartz-feldspar porphyries that are coeval with and constitute part of Unit B, suggesting relationships that are unlike those of the Central Intrusive Belt where the Ratcliff Brook Formation is unaffected by 547-510 Ma intrusions.

Paleozoic intrusions are numerous in the area between the Belleisle and Falls Brook-Taylor Brook faults but are absent south of the Belleisle Fault. Most of them are north of the Wheaton Brook Fault but at least one, the Late Devonian Mount Douglas Granite (McLeod et al., 1988), crosses the Wheaton Brook Fault. Isotopic dating will probably reveal other Paleozoic intrusions south of this fault.

Deformation history

Three deformation episodes are recognized in the Green Head Group (Wardle, 1978; Currie et al., 1981; Nance, 1982). The earliest episode is related to the emplacement of Brookville orthogneiss diapirs, an event that Wardle (1978) thought predated 800 Ma. However, a new U-Pb zircon age (Bevier et al., 1990) indicates that this orthogneiss was emplaced at about 600 Ma and the first deformation, therefore, is much younger than previously considered. Furthermore, detrital micas in the Ratcliffe Brook Formation, which were probably derived from Brookville gneisses rather than mylonite zones as suggested by Dallmeyer and Nance (1990), yield ^{40}Ar-^{39}Ar ages of 620-610 Ma. A second deformation episode may have occurred in the latest Precambrian or earliest Cambrian, coincident with amphibolite facies metamorphism, as indicated by a U-Pb age on titanite of 564 ± 6 Ma (Bevier et al., 1990) and by ^{40}Ar-^{39}Ar amphibole ages of 542 ± 4 and 538 ± 2 Ma (Dallmeyer et al., 1990). However, the fact that amphibolite grade metamorphism only occurs in Green Head rocks that are spatially associated with the Brookville Gneiss (Wardle, 1978) strongly suggests that this metamorphism and orthogneiss emplacement were coeval. If so, there is either an inheritance problem with the zircon age or the metamorphic ages have been reset. Other evidence of late Precambrian deformation is mostly restricted to mylonite zones (Rast et al., 1976b; Rast, 1979; Currie, 1987d).

Unconformities and intrusions, rather than contrasting structural styles, indicate Avalonian orogenesis. Lowermost Lower Cambrian redbeds (Ratcliffe Brook and similar rocks) unconformably overlie oxidized pillow basalts in the Eastern volcanic domain (McLeod and McCutcheon, 1981) and 600-625 Ma granitoid rocks southeast of Saint John (Watters, 1987). Also, they may unconformably overlie Coldbrook terrestrial volcanic rocks in the Central domain. Furthermore, there are numerous intrusions that cut the Coldbrook Group but do not intrude the Ratcliffe Brook Formation and equivalent redbeds.

Most of the deformation in Coldbrook rocks appears to be the result of middle Paleozoic (Acadian) tectonism, but the effects are different in the three domains. In the Eastern domain, a post-Cambrian but pre-Carboniferous, gently dipping, composite cataclastic fabric is generally subparallel to bedding but in places is axial planar to rare, open-to-tight shear folds of variable size and plunge

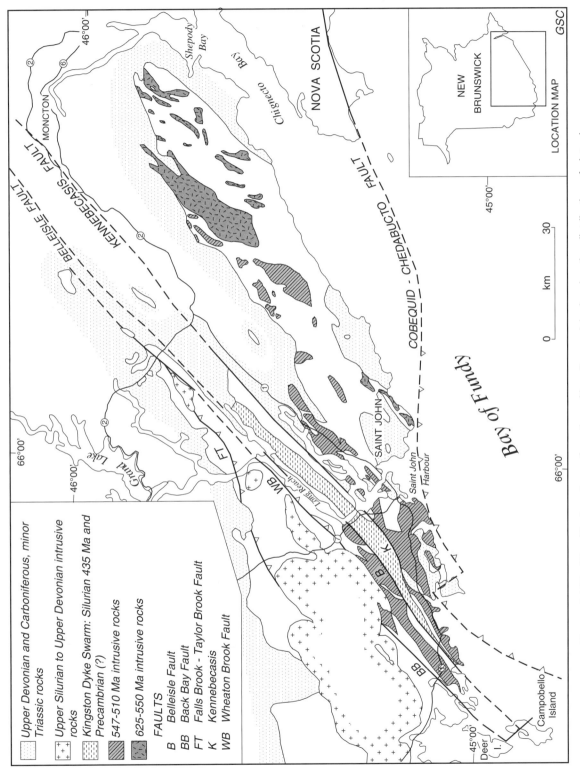

Figure 3.148. Simplified geology of southern New Brunswick showing the distribution of plutonic rock units in the Avalon Zone.

(Ruitenberg et al., 1973, 1979; McLeod, 1987). In the Central domain, cleavage folding is mostly absent, probably besause of the buttressing effects of large intrusions (Ruitenberg et al., 1979). Along the Fundy coast near the Cobequid-Chedabucto Fault, rocks of both domains are overprinted by Alleghanian fabrics (Nance, 1987) that are dated at 318 ± 1 Ma by ^{40}Ar-^{39}Ar on muscovites (Dallmeyer and Nance, 1990). In the Western domain, steeply dipping cataclastic cleavage parallels the Kingston mylonite zone and is axial planar to tight, northeast-trending, steeply plunging folds (McCutcheon and Ruitenberg, 1987). Nearby, Silurian rocks are less deformed, suggesting that at least some of the deformation in the Coldbrook Group is pre-Acadian. This is also suggested by undeformed 565 Ma granitoid rocks (Currie, 1988a) that intrude the Kingston mylonite zone.

Deformation in the Saint John Group is variable. However, in the city of Saint John these rocks exhibit phases of deformation that are equated with all (Nance, 1982) or only the latter (Currie et al., 1981; Wardle, 1978) deformational events in the Green Head Group.

Conclusions and interpretations

It is not clear whether the Green Head Group is older than, or time equivalent to, Coldbrook rocks. The carbonates and clastics in the lower part of the group (Ashburn Formation) were deposited prior to about 600 Ma but exactly how long before is not well constrained. The stromatolites in these rocks have a long time range, although they are most common in the Neohelikian (Hofmann, 1974). Notably, the 760 Ma-old Burin Group in Newfoundland (Krogh et al., 1988) contains large stromatolite-bearing carbonate blocks (Strong et al., 1978), indicating that carbonates were deposited prior to 760 Ma. In contrast, Martinon clastics could be considerably younger, even coeval with parts of the Coldbrook Group as suggested by Currie (1987d). If so, the Martinon Formation should be excluded from the Green Head Group (Currie, 1991).

The Brookville Gneiss is not basement to the Green Head Group as suggested by Currie et al. (1981), and it may not be basement to the Coldbrook Group (Bevier and Barr, 1990). It consists of a diapiric core of orthogneiss surrounded by a mantle of paragneiss, largely after Green Head rocks, like Wardle (1978) proposed. Amphibolite grade metamorphism in the Green Head Group is spatially related to the Brookville Gneiss, strongly suggesting that high grade metamorphism is coeval with orthogneiss emplacement. Whether emplacement and metamorphism occurred about 665-550 Ma or around 625-600 Ma is not resolved.

The Coldbrook rocks encompass more than one stratigraphic group but the divisions have never been formalized (if in fact they have all been recognized). They are thus designated as assemblages I to III. The oldest known rocks are those of Assemblage I, a subaqueous volcanic and sedimentary sequence that is intruded by 625 to 600 Ma plutons and underlies most of the Eastern, and part of the Central domain of the Caledonia Highlands. Another subaqueous sequence, which is tentatively correlated with, and may be about the same 650 Ma age as Assemblage I, underlies the northwestern side of the Western domain.

Assemblage I rocks are predominantly calc-alkaline and reflect an ensialic, convergent (volcanic arc) setting (Barr and White, 1988; Dostal and McCutcheon, 1990).

Younger terrestrial volcanic rocks, including the type-Coldbrook Group, comprise Assemblage II. They underlie the Central domain and parts of the Eastern domain. Their age is about 550 Ma and they are co-magmatic with syenogranitic and gabbroic plutons. Assemblage II rocks are predominantly tholeiitic and reflect a within-plate (intra-arc?) rift environment (McLeod, 1986; Barr and White, 1988; Dostal and McCutcheon, 1990). Terrestrial volcanic rocks also underlie the southeastern side of the Western domain, and although they are provisionally correlated with Assemblage II, they may be coeval with Assemblage I. Locally, there are greenish grey and purple siliceous, turbidites (Assemblage III) that are spatially associated with terrestrial volcanic rocks, but they reflect an entirely different depositional setting. Lithologically, these rocks are similar to the Conception Group in Newfoundland (Drook Formation of Williams and King, 1979).

The two sides of the Western domain are separated by a major mylonite zone that is intruded by, and is in part synchronous with, mafic and felsic dykes of the Kingston Dyke Swarm. The significance of the mylonites and dykes is debatable. Rast and Dickson (1982) and Rast (1979) interpreted the mylonite zone as a major shear zone near which continental separation began and in which the bulk of the dyke swarm was emplaced during the late Precambrian. Leger and Williams (1986) argued that the mylonite zone is Acadian (Silurian-Devonian) and that the dyke swarm is related to accretion of the Avalon Zone to North America (Doig et al., 1990). Recent U-Pb dates verify that Silurian dykes occur in the swarm, but Precambrian dates are also expected. The history of the zone is undoubtedly complex, as would be expected for a major structural boundary.

Redbeds (Ratcliffe Brook Formation and similar rocks), at least locally, unconformably overlie the volcanic and plutonic pile. Even the younger plutons (about 550 Ma) are not known to intrude these redbeds, which is an enigmatic feature considering that the Early-Middle Cambrian boundary is 545 Ma on the time scale of Haq and van Eysinga (1987). South of the Belleisle Fault these redbeds contain a few interbedded, rift-related volcanic rocks, but to the north interbedded volcanic rocks are common and the redbeds contain abundant volcanic detritus. These redbeds are reminiscent of the Musgravetown Group (McCartney, 1967) and equivalents in Newfoundland.

The Avalon Zone north of the Belleisle Fault differs from that to the south in at least four important respects: (1) typical Coldbrook rocks are absent (Currie, 1986; McCutcheon, 1981; McCutcheon and Ruitenberg, 1987), (2) the oldest granitoids, based on field relationships and one U-Pb date, are about 555 Ma (Currie, 1988a), (3) the Cambrian sequences contain a significant volcanic component (Greenough et al., 1985; McCutcheon and Ruitenberg, 1987), and (4) Silurian rocks that structurally/unconformably overlie this terrane, occur in abundance (Helmstaedt, 1968; Hay, 1968; Currie 1988a, b) whereas, south of the Belleisle Fault Silurian strata are sparse (Currie, 1986; McLeod and Rast, 1988).

Perhaps the rocks north of the Belleisle Fault were deposited upon crust that was attenuated during the opening of Iapetus. This could explain the apparent paucity of typical Coldbrook strata (they are deeply buried), the abundance of Lower Cambrian volcanic rocks (more volcanism closer to the axis of rifting), and the presence of abundant Silurian cover (the disrupted continental margin experienced greater subsidence). Such an explanation is compatible with the proposed "breakup unconformity" between the Glen Falls and Ratcliffe Brook formations of the Saint John Group (McCutcheon, 1987). The contrasts in late Precambrian-Cambrian stratigraphy north and south of the fault were enhanced by Devonian high-angle reverse and/or thrust movements, which probably juxtaposed rocks that were widely separated at the time of deposition (McCutcheon, 1981; McLeod and Rast, 1988).

Acknowledgments

We thank A. Leblanc for drafting the original figures.

MEGUMA ZONE

P.E. Schenk

Introduction

Definition

The Meguma Zone is defined by a thick siliciclastic sequence, the Meguma Supergroup, that ranges in age from Late Cambrian or older to Early Ordovician. It is the most eastern zone of the Canadian Appalachian region and it is a leading example of an Appalachian suspect terrane. The Meguma Supergroup is overlain by another thick siliciclastic succession, the Annapolis Supergroup, that has volcanic units at its base and fossiliferous Silurian and Lower Devonian rocks at its top. The Annapolis Supergroup defines the Annapolis Belt, and although excluded from the Meguma Zone in this chapter, it is an integral part of the Meguma terrane, since there are no established linkages until Carboniferous time. The separation of a Meguma Zone and Annapolis Belt is artificial as: (1) both zone and belt define the same geographic area; (2) some rocks of the Annapolis Supergroup are possibly Early Ordovician and therefore older than those normally found in middle Paleozoic belts; (3) the Acadian (Devonian) Orogeny was the first major event to affect rocks of both the Meguma Zone and Annapolis Belt; and (4) the Meguma Supergroup and the Annapolis Supergroup are viewed as integral parts of the same stratigraphic section and interpreted in the same model of an evolving continental margin. Therefore, the Meguma Zone and the Annapolis Belt are defined only to maintain the systematics of the time-slice subdivision used throughout this volume.

Schenk, P.E.
1995:   Meguma Zone; in Chapter 3 of Geology of the Appalachian-Caledonian Orogen in Canada and Greenland, (ed.) H. Williams; Geological Survey of Canada, Geology of Canada, no. 6, p. 261-277 (also Geological Society of America, The Geology of North America, v. F-1).

Extent and boundaries

The Meguma Zone occupies all of the southern mainland of Nova Scotia (Fig. 3.149). It extends beneath the Scotian Shelf to the south, and underlies parts of the continental shelf both southeast of Cape Breton Island and also south of Newfoundland to the east, and the Bay of Fundy to the northwest (Lefort and Haworth, 1981). Its total area is approximately 200 000 km².

The boundary between the Meguma Zone and the Avalon Zone is the Glooscap Fault (also The Cobequid, Cobequid-Chedabucto, and Minas faults). It has had a long and complex history of activity (Rodgers, 1970). Most recent workers suggest that it is a major dextral transcurrent fault (Webb, 1969; Eisbacher, 1969; Donohoe and Wallace, 1978; Keppie, 1982). Significant dip-slip is also possible (Donohoe and Wallace, 1978). The Glooscap Fault trends eastward as the reactivated Southeast Newfoundland Transform and thence along the Gibraltar Fracture Zone (King and MacLean, 1970, 1976). On geophysical grounds, Lefort and Haworth (1981) considered that it trends farther north, along the Collector Anomaly to cross the Tail of the Grand Banks, and thence across southern Iberia as the Guadalquivir Fault.

The southern boundary of the Meguma Zone may be the Maine-Rhard fracture zone (Lefort and Haworth, 1981). This is based on geophysics, including gravity, magnetics, and both seismic reflection and refraction. The zone begins in the Boston area and runs eastwards through the Gulf of Maine, south of the Meguma and along the western half of the Scotian Shelf to the Rhard Fault and associated Rabat Fault of northwestern Morocco (Fig. 3.149).

The base of the Meguma Supergroup is not exposed. Geochemical evidence suggests that its metasedimentary rocks and plutons rest on a thrust surface above Avalon basement (Eberz et al., 1991). The upper contact of the supergroup is an assumed paraconformity, but locally it is a disconformity or an angular unconformity (Schenk, 1991). The uppermost strata of the Meguma Supergroup have sedimentological evidence of upward shoaling toward the contact with the overlying Annapolis Supergroup (Schenk, 1991). The base of Annapolis Supergroup is marked by subaerial felsic volcaniclastic rocks.

History of investigation

The Meguma Zone has had a long history of mineral exploration because of accessibility and excellent coastal exposures. Some of the first geologists engaged in systematic mapping in Canada surveyed these rocks (e.g. Jackson and Alger, 1828; Gesner, 1836). At the turn of this century, most of the Meguma was mapped at a scale of one mile to the inch with the view of delineating favourable structures for gold deposits.

General features

Characteristics of the Meguma Zone such as structural style, metamorphism, plutonism, mineral deposits, and geophysical expression are Middle Paleozoic effects. Consequently they are treated in more detail in the discussion of the Annapolis Belt.

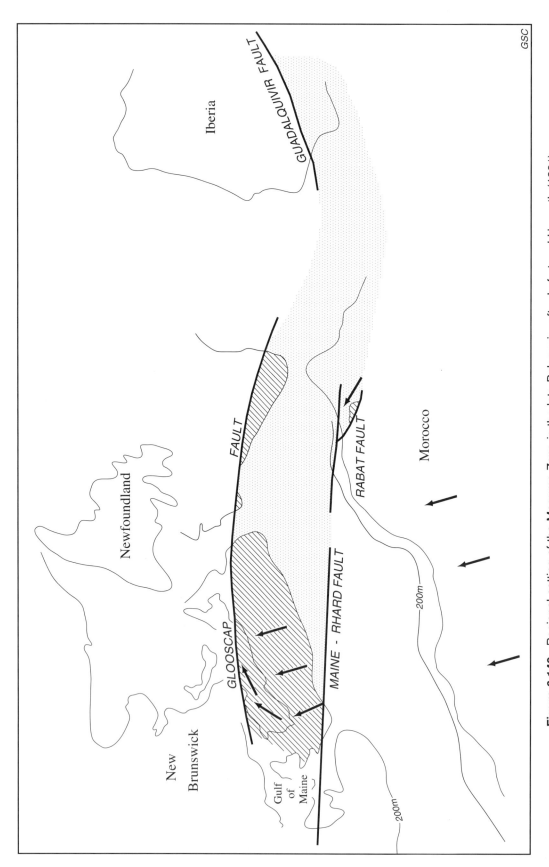

Figure 3.149. Regional setting of the Meguma Zone in the late Paleozoic, after Lefort and Haworth (1981). Cross-hatching indicates area of the Meguma Zone and possible equivalents. The stippled area shows the distinctive, fault-bound crustal strip of the Meguma Zone extending into southern Spain and northern Morocco. Arrows indicate Cambrian-Ordovician dispersal directions of the Meguma facies and southern Moroccon shelf-type siliciclastics, after Schenk (1981).

The Middle Devonian Acadian Orogeny is the first major deformational, metamorphic, and intrusive event to affect the Meguma Zone (and Annapolis Belt). Broad, low plunging, upright folds describe an arc trending northerly in the southwest to easterly in the eastern part of the zone (Fig. 3.150). Associated slaty cleavage is sub-vertical, parallel to axial surfaces. The metamorphic facies is greenschist, with an increase to amphibolite facies in the southwest. Hornblende-hornfels facies are developed close to granitic batholiths (Taylor and Schiller, 1966). Large granitic plutons in the Meguma Zone are of Late Devonian and Early Carboniferous age, and cut folded, Early Devonian strata (Clarke et al., 1980).

Both magnetic and gravity expressions show sharp differences between the Meguma Zone and Avalon Zone to the north. The Meguma Zone is characterized by low values in magnetic and gravity anomalies whereas the Avalon Zone has high values in both geophysical expressions (Lefort, 1983). High values result from a concentration of basic rock bodies in the Avalon Zone toward the Glooscap Fault (Lefort, 1983, 1989). The Meguma metasedimentary rocks are characterized by near zero gravity anomalies and negative magnetic anomalies. The Halifax Group is distinguished by higher magnetic anomalies, because of an abundance of sulphides (especially pyrrhotite). Granodioritic plutons are identified by low gravity anomalies. This distinctive combination of geophysical properties allows recognition of the Meguma Zone beneath the Scotian Shelf (McGrath et al., 1973).

Offshore seismic refraction studies indicate that the folded strata of the Meguma Zone are recognizable to a depth of 8 km (Keen and Haworth, 1986). The remainder of the continental crust consists of younger intrusions. The crust is approximately 30 km thick (Barrett et al., 1964).

Basement to the Meguma metasedimentary rocks may be represented by xenoliths in the younger intrusions. Lower crustal granulite xenoliths in a calc-alkaline lamprophyre dyke at Tangier have a mean K/Rb ratio (321) comparable to values of Archean crust (300). The protolith of the Meguma lower continental crust was probably a mixture of shale, calcareous shale, gabbro, and mafic cumulate (Eberz et al., 1991). Sm/Nd ratios indicate that the age of the sedimentary protolith is constrained between 1100 and 500 Ma (Eberz et al., 1991). Similar analyses of the Meguma metasedimentary rocks (Clarke and Halliday, 1985) indicate that their source age is 600 to 1200 Ma older than the source of the deep-seated xenoliths. That is, the lowermost portion of the crust beneath the Meguma Zone has a model source age significantly younger than the source age of the upper crustal sequences. Indeed, the model source age for the lower crust is very close to the depositional age of the overlying Meguma Supergroup. A tectonic break is required between underlying autochthonous younger (Tangier crust) material and overlying allochthonous older material (Meguma Supergroup; Eberz et al., 1991). The following points suggest that Meguma Zone is an allochthon resting on Avalon Zone basement (Eberz et al., 1991): (1) the similarity between the T_{CHUR} model ages of the Tangier xenoliths (ca. 500 Ma) and the isotopic ages from the Avalon Zone (800-600 Ma; Krogh et al., 1988); (2) the ca. 1 Ga TDM

model ages of many Tangier xenoliths and those of Avalon rocks from Massachusetts (Hon et al., 1984); and (3) the similarity in their tectonic settings (both volcanic arcs).

Stratigraphy
Name

The name Meguma is derived from the root of the native term for the Micmac Indian tribe. The Meguma Supergroup was also known as Quartz Rock and part of the Transitional Clay-Slate Formation (Jackson and Alger, 1828), the Atlantic Coast (metamorphic) Series (Dawson, 1868), the Gold-bearing series (e.g. Faribault, 1912), the Acadian group or division (Ami, 1900), the Meguma series (Woodman, 1904), and the Meguma Group (Stevenson, 1959). Recent mapping and subdivision require that the Meguma Group be elevated to the rank of supergroup (first mention). The Meguma Group was traditionally subdivided into the Goldenville and Halifax formations. These are now elevated to group status and are subdivided into formations (Fig. 3.151). Contact relations between all formations within the supergroup are gradational and conformable.

The Goldenville Group has no known base. The Halifax Group has no known top over almost its entire exposure. The maximum measured thickness of the Goldenville Group is 6.7 km near Liverpool (Faribault, 1914). The Halifax Group is 11.8 km thick at the stratotype in the Halifax area (Milligan, pers. comm., 1990). The overall minimal thickness of each group is about 7 km because thicknesses of 4-7 km occur above and below the Goldenville/Halifax contact and are uniformly distributed across southern Nova Scotia (Schenk, 1970). Local variations are expected. Thicknesses of individual formations are given in Figure 3.151. These vary greatly, as best shown by the Cunard Formation which changes from 500 to 8000 m over a distance of 4 km, with a corresponding thickening of the overlying Feltzen Formation (O'Brien, 1986). The Rockville Notch Formation thickens from less than a metre in the west to 35 m in the east.

The Meguma Supergroup is overlain by the Annapolis Supergroup (new name). The contact is a paraconformity except at two localities (Schenk, 1991). At Cape St. Mary (Fig. 3.150), the uppermost few metres of the Meguma Supergroup (Rockville Notch Formation) are deformed into isoclinal, recumbent folds, and these structures appear discordant with respect to overlying beds of the Annapolis Supergroup. The discordance has been interpreted as a local angular unconformity (Taylor, 1965), a disconformity (Rodgers, 1970), and a conformable contact (Lane, 1981). The contact has also been interpreted as a thrust fault, (1) either older than the overlying Annapolis Supergroup and the result of glacial tectonics (Schenk, 1972); or (2) younger than the Annapolis Supergroup and the result of Acadian tectonics (Keppie, 1982). Near Fales River (Fig. 3.150) the contact is a disconformity with eroded clasts of Rockville Notch Formation within basal strata of the overlying Annapolis Supergroup (Smitheringale, 1973). In general, uppermost strata of the Meguma Supergroup are marine whereas basal strata of the overlying Annapolis

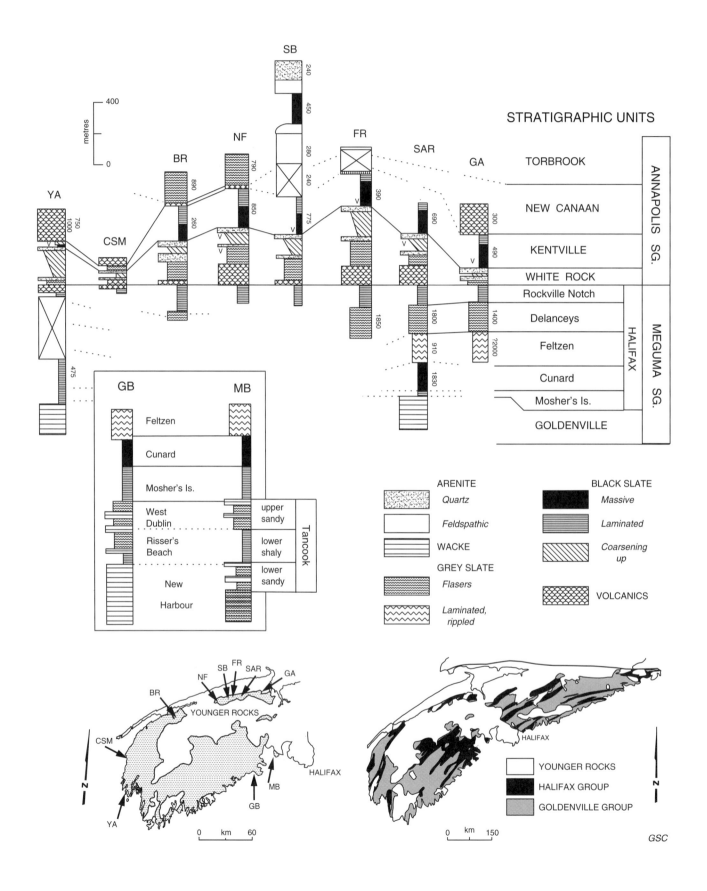

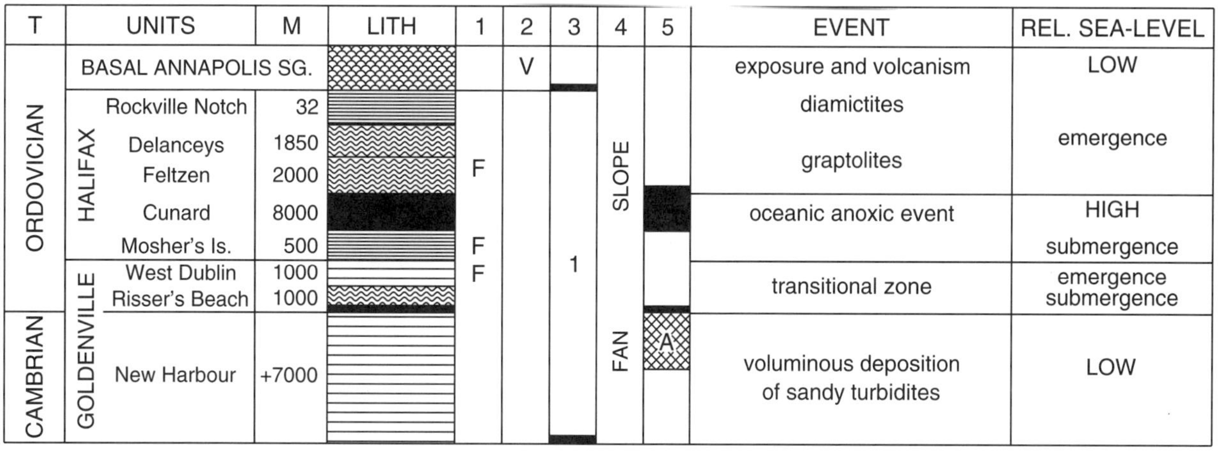

T	UNITS		M	LITH	1	2	3	4	5	EVENT	REL. SEA-LEVEL
ORDOVICIAN	HALIFAX	BASAL ANNAPOLIS SG.				V				exposure and volcanism	LOW
		Rockville Notch	32					SLOPE		diamictites	emergence
		Delanceys	1850		F						
		Feltzen	2000							graptolites	
		Cunard	8000							oceanic anoxic event	HIGH
		Mosher's Is.	500		F		1				submergence
	GOLDENVILLE	West Dublin	1000		F					transitional zone	emergence submergence
		Risser's Beach	1000								
CAMBRIAN		New Harbour	+7000					FAN	A	voluminous deposition of sandy turbidites	LOW

GSC

Figure 3.151. Summary of stratigraphy, event stratigraphy, and relative sea-level changes in the Meguma Supergroup. Compare with Figure 4.32 for the Annapolis Belt (Schenk, this volume). T refers to geological time and M to maximum measured thicknesses in metres. Thicknesses of stratigraphic units are scaled except where noted by numbers. Blocks with single crosses indicate measured covered intervals. The lithological column gives predominant lithologies represented by patterns used in Figure 3.150 except for the sandstones, where arenites are shown by grid. Column 1 indicates fossil-bearing intervals; column 2 shows episodes of volcanism; column 3 identifies unconformities by heavy horizontal lines (sequences are numbered); column 4 summarizes general depositional environments; and column 5 displays (1) times of global anoxic events (black) and (2) the Andean glacial episode (A). Major events and relative sea-level changes in the Meguma Zone are listed in the next two columns (after Schenk, 1991).

Supergroup are subaerial volcaniclastic rocks. Because fossils below the boundary are Tremadoc and those above are Caradoc or younger, a paraconformity is implied.

Goldenville Group

The Goldenville Group has been divided into formations only in the southwestern part of the Meguma Zone (O'Brien, 1985; modified by Schenk and Adams, 1986; and Waldron, 1987). In the Green Bay area, the stratigraphic order is New Harbour, Risser's Beach, and West Dublin formations (Fig. 3.150, 3.151). In the Mahone Bay area, the

Figure 3.150. Geology of Meguma Zone with localities and stratigraphic columns. GA = Gaspereau area; SAR = South Annapolis River; FR = Fales River; SB = Spinney Brook; NF = Nictaux Falls; BR = Bear River; CSM = Cape St. Mary; YA = Yarmouth area; GB = Green Bay; and MB = Mahone Bay. Groups are indicated by upper case letters, formations by upper and lower case letters, and informal subdivisions by lower case letters. Thicknesses of stratigraphic units in the Green Bay and Mahone Bay areas are given in Figure 3.151 and discussed in the text (Schenk, 1991).

New Harbour is overlain by the Tancook Island Formation. Rock units similar to these formations occur in the eastern part of the zone.

New Harbour Formation

The New Harbour Formation consists of thick beds of fine grained, quartzose to feldspathic metawacke and minor volcanic debris, interbedded with thin sandy strata of green to grey slate. Interformational clasts (mudstone-siltstone rip-ups) are components of many beds. The sands are very mature. Where metamorphism is slight, grain size is fine to very fine and grains are spherical. Larger grains of polycrystalline quartz and chert are locally abundant to produce a "millet-seed" texture. Fresh (diagenetic?) albite locally comprises up to 30 per cent of the grains, and detrital biotite and large muscovite flakes are also locally abundant. Volcanic clasts and metasedimentary and gneissic lithologies are generally rare. Rounded to angular pebbles of phosphate are locally abundant.

Because of the uniformity in grain size, graded bedding in sandstone strata is rare so that the upward succession of sedimentary structures corresponds usually to truncated Bouma sequences (Dzulynski and Walton, 1965). Sole marks are abundant. Slump structures, slide blocks, and slurried intervals are common. Channels that cut up to 3 m into pre-existing layers are evident at many outcrops. Larger channels occur that reach depths of at least 40 m and probable widths of 2-3 km (Waldron and Jensen, 1985). Vertical and lateral variations in the regional distribution of textures, structures, and compositions are remarkably persistent (see section on dispersal).

Strata tend to be discontinuous, and either segregated into a number of thinner beds or amalgamated into thicker strata (Harris and Schenk, 1975). Thick (50-150 m) packages of sandstones are lenticular or interdigitate with more slaty intervals over distances of several kilometres. More slaty successions contain conspicuous thinner (5-20 m), tabular packets of sandstone that are also discontinuous (Waldron and Jensen, 1985). Cyclic sedimentation is evident in both thickening-upward, large-scaled megarhythms from 200-400 m in thickness as well as smaller-scale megarhythms approximately 25 m thick. The latter are of both thickening-upward and thinning-upward varieties (Harris and Schenk, 1975).

Risser's Beach Formation

The Risser's Beach Formation consists of a basal, massive, black slate that is coarser upward by the addition of siltstone laminae followed stratigraphically by thinly stratified, very fine grained sandstones. Above these, very thickly stratified amalgamated sandstones pinch out to the west and are overlain by thinly stratified siltstones. Several thinning-upward successions have thicknesses of 2 m (Schenk, 1988). The Risser's Beach Formation may (O'Brien, 1985) or may not (Waldron, 1987) pinch out toward Mahone Bay to the east.

West Dublin and Tancook formations

The West Dublin Formation is predominantly thickly bedded, fine grained sandstone. Individual lithozones pinch out northeasterly in the down-current direction (O'Brien, 1985; Schenk and Adams, 1986; Waldron, 1987). These rocks are called the Tancook Formation (O'Brien, 1985) in the Mahone Bay area and they may include equivalents of the Risser's Beach Formation (Waldron, 1987).

The Tancook Formation is mainly sandstone. Waldron (1987) noted a general thickening, coarsening-upwards trend in the lower sandy part of the section. The upper sandy part consists of packages of amalgamated sandstones between more normal, Bouma turbidites. The upper contact is gradational and intercalated with black slate of the Halifax Group.

Halifax Group

The Halifax Group in the western half of the Meguma Zone has been divided into five conformable formations, several of which can be traced into the eastern half of the zone. From base to top these are the Mosher's Island, Cunard, Feltzen, Delanceys, and Rockville Notch formations (Fig. 3.151).

Mosher's Island Formation

The Mosher's Island Formation is a very distinctive, relatively thin, metalliferous, laminated, grey slate (O'Brien, 1985, 1986). It extends across eastern Nova Scotia at the base of the Halifax Group and is a useful marker horizon (Zentilli et al., 1986). Siltstone laminae are interlaminated with fine grained, uniform, and parallel-laminated dark grey slates. Both normal and inverse grading are common,

and some silt layers show sharp contacts at both upper and lower surfaces. A few of the coarser laminated units have small scale current-ripple cross-lamination, including both starved and fading ripples (Waldron, 1987). Manganese-rich carbonate laminae and concretions are common, in which metamorphism has generated spessartine garnets (coticules). Beside being enriched in manganese (2-12 weight per cent), the formation also contains relatively large amounts of lead, copper, zinc, and barium (Zentilli et al., 1986).

Cunard Formation

This formation is a very thick, pyritiferous, graphitic, fissile slate laced with thin, fine grained siltstones and very fine grained sandstones (O'Brien, 1985, 1986; Schenk, 1970). The mineralogy of the siltstones and sandstones is similar to those of the Goldenville Group. The high content of carbon in relatively silt-free layers so reduced structural competence that deformation is usually intense.

Sedimentary structures are dominated by laminated and finely cross laminated slates and siltstones that are almost everywhere normally graded. Beds may be grossly lenticular and contain cross-sets with amplitudes of 5-10 cm (Waldron, 1987). The vertical succession of sedimentary structures in individual sandy layers corresponds to base cut-out Bouma sequences. Sole marks are rare. The formation continues the uniform regional trends in sedimentary structures, textures, and composition of the underlying Goldenville Group (Schenk, 1970).

Vertical successions thin upward in some places and thicken upward in other places (Stow et al., 1984).

Feltzen Formation

The Feltzen Formation is much more silty than the underlying Cunard Formation. It consists of an intercalation of two lithologies: (1) banded grey slate with even laminae of light grey silty slates, and (2) black graptolitic slates intercalated with thin siltstones.

Sedimentary structures are abundant, and dominated by parallel lamination and cross-stratification. Vertical successions correspond to silty turbidites of Piper (1978). Regional and stratigraphic trends in sedimentary structures, textures, and composition are less well-defined than those in the Cunard Formation (Schenk, 1970).

Delanceys Formation

This formation appears to be confined to the northwestern boundary area of the Meguma Zone. It is a grey, laminated, silty slate with abundant fine-scaled lenses of quartz siltstone. Very thin beds and medium-scaled lenses of quartz metawacke are in places intercalated with the slates. The lower and upper boundaries are gradational and conformable with stratigraphically adjacent formations. Its thickness is about 1400 m. The formation corresponds in general with the lower two lithozones of Smitheringale (1973).

Regional trends in sedimentary structures, textures, and composition are the least well-defined of any in the Meguma Supergroup (Schenk, 1970).

Rockville Notch Formation

The Rockville Notch Formation is a dark grey to black, pelitic laminite with some siltstone microlenses. The upper boundary is conformable and sharp with the overlying Annapolis Supergroup except at Cape St. Mary and near Fales River (see above).

The laminite is unusual in consisting of alternating layers of pure, micaceous siltstone and coarser, quartzose siltstone (Lane, 1976, 1981). Lens-shaped pods of pebbly to bouldery mudstones (diamictite) up to 32 m thick occur at five localities over a distance of 240 km. They are difficult to distinguish, patchy in distribution, and may be intercalated with quartz arenites. In several places, pebble dents are visible. The mud-supported clasts range from 5 mm to 20 cm in diameter. Their lithologies include extraformational, finely laminated, and micro-crosslaminated (90 per cent of clasts) massive metaquartzite, metasiltstone, chert, felsite, and dark grey argillite. Clast shape is angular to sub-rounded. Some clasts are polished, commonly faceted and striated, and rarely grooved (Schenk, 1972; Lane, 1981). Where the clasts of the diamictite are absent or not recognized, the laminated slate is distinctive and contains much less silt than the underlying Delanceys Formation of the Halifax Group.

Paleontology and age

Goldenville Group

Trace fossils are common throughout the Goldenville Group (Pickerill and Keppie, 1981). The apparently biogenic structure *Astropolithon hindii* indicated a Phanerozoic rather than Cryptozoic age (Walcott, 1891; Woodman, 1908). Subsequently, it was used as a Cambrian index fossil (e.g. Crimes et al., 1977) and reinterpreted as a sand volcano (Pickerill and Keppie, 1981). *Nereites* ichnofauna occur in lower parts of the Goldenville Group (Schenk, 1982). *Paleodictyon (Glenodictyum) cf. imperfectum* Seilacher in the upper portion of the group suggests an Ordovician age (Pickerill and Keppie, 1981). The *Nereites* ichnofacies *Taphrohelminthopsis* is abundant in the Risser's Beach Formation (Schenk, 1988).

The New Harbour Formation is probably Late Cambrian (Merioneth). Maximum age of deposition is indicated both by radiometric and faunal evidence. Detrital zircons dated at 600 Ma provide a maximum isotopic age (Krogh and Keppie, 1986). Sedimentary and/or early diagenetic flakes of muscovite are dated at 476 ±19 and 496 ± 20 Ma (Late Cambrian to Early Ordovician; Poole, 1971). The uppermost part of the Tancook Formation contains a lens of poorly preserved trilobite and echinoderm fragments that are early Middle Cambrian (Pratt and Waldron, 1991). The lens may be a large, early diagenetic concretion that has preserved the fossils from metamorphism and deformation. Because the fossils have been transported, they provide a maximum biostratigraphic age (Schenk, 1991). If the trilobites date the time of deposition, then the underlying New Harbour Formation could conceivably be Early Cambrian. Minimum age of deposition is provided by conformably overlying, thick, Tremadoc slates of the Halifax Group (Crosby, 1962). These slates are intercalated with the Tancook Formation.

The Risser's Beach and West Dublin formations are the lateral equivalents of the Tancook Formation. For the reason stated above, the former two formations are conservatively Tremadoc or possibly Merioneth (Late Cambrian). The contacts between both the underlying West Dublin and Tancook formations with the overlying Mosher's Island Formation are conformable and intercalated. The Mosher's Island Formation is the most widespread unit in the Meguma Supergroup, and elsewhere contains Tremadoc acritarchs. Black slates at this stratigraphic level at Tangier contain poorly preserved Early Ordovician graptolites (Poole, 1971). As noted above, a rise in sea level, perhaps glacioeustatic, marks the beginning of Tremadoc time (Erdtmann and Miller, 1981). This relative rise in sea level may be signaled in the Meguma Supergroup by an upward change from sandy Goldenville to shaly Halifax sedimentation (Schenk, 1991).

Halifax Group

Bioturbation is almost entirely absent in the Mosher's Island and Cunard formations but is intense in the overlying formations. The Feltzen Formation commonly displays the vertical, U-shaped form *Arenicolites variabalis* Fursich. The Delanceys Formation is characterized by the *Zoophycus* ichnofauna. The Rockville Notch Formation is typified by minute horizontal burrows and *Tomaculum* (H. Hofmann, pers. comm., 1978) that has been found only in Ordovician strata (Hantzschel, 1975).

The Mosher's Island and Cunard formations contain Tremadoc acritarchs (W. A. M. Jenkins, pers. comm., 1977). An abundance of manganiferous concretions may indicate an age range from Middle Cambrian to mid-Caradoc, most likely Early Ordovician (Kennan and Kennedy, 1983).

The three upper formations of the Halifax Group are also Tremadoc. Black slates in the Feltzen Formation contain the Tremadoc graptolites *Dictyonema flabelliforme* (Eichwald) and *Anisograptus sp.* (Smitheringale, 1973; Cumming, 1985). Tremadoc acritarchs have been recovered from several areas in the Mosher's Island lithology and Rockville Notch Formation (W.A.M. Jenkins, pers. comm., 1977). Pebbly and bouldery mudstones of the Rockville Notch Formation have been interpreted as marine tillites (Schenk, 1972; Schenk and Lane, 1981), possibly related to continental glaciation of Gondwana (Erdtmann and Miller, 1981; Schenk, 1991).

Provenance

Dispersal

The dispersal pattern in the lower portion of the Meguma Supergroup is remarkably constant both regionally and stratigraphically (Schenk, 1970). In general, the sedimentary transport direction is northward in the southwest area and then gradually changes through 90 degrees in the central area to eastward in the far eastern extremity of the exposure area (Fig. 3.152). Major variations from this trend are confined to the present Atlantic shoreline, where transport directions are consistently toward the northwest. This general pattern as well as the major variations appear to be almost independent of other sedimentological scalar properties. The paleocurrents are parallel, rather than oblique, to such properties as grain size, layer thickness, per cent lithic clasts, etc. The northwestwards increase in clay-sized matrix and the shale/sand ratio suggest quieter perhaps deeper water, compared with the general decrease of other properties in this direction. Conversely the southeastward increase in grain size and quartz content and decrease in total feldspar suggest winnowing and better sorting toward the southeast. Size-grading and scour are more abundant in the southern (upstream) and Atlantic coast areas, and in the stratigraphically lower parts of the supergroup. Sediment was funnelled downslope from a sourceland that lay to the present southeast, and the sediment was perhaps redistributed by northeastwards flowing, geostrophic undercurrents parallel to ancient bathymetric contours (Fig. 3.152).

Stratigraphically, this overall uniformity breaks down in the upper formations of the Halifax Group (Schenk, 1970). Directions of vectoral sedimentary structures show increasing variance toward the uppermost units of the supergroup. This is accompanied by more diverse patterns in scalar sedimentary properties such as texture and composition. In turn, changes in these dispersal variables coincide with increases in silt content, cross-stratification, and shallow-water ichnofauna. Shoaling is suggested (Schenk, 1991).

Source

The volume of the Meguma Supergroup suggests the possible size of the source area. The source must have been at least the size of the Meguma Zone because the supergroup consists entirely of siliceous detritus. Conservative palinspastic calculations indicate that the volume of the Meguma Supergroup in Nova Scotia is at least equal to a block with an area of Ontario and a height of 5 km (Schenk, 1983). Undoubtedly the source continent was much larger than these estimates because: (1) detritus of the Meguma Supergroup is very fine grained; (2) more than a single dispersal system would be likely on the source land; and (3) the Meguma Zone is a tectonic and erosional remnant. If the age of the Meguma Supergroup is correct, this immense amount of sediment was rapidly deposited and perhaps rapidly produced. Continental glaciation is the most efficient agent in producing large amounts of detritus. Late Precambrian glaciation may have created the Goldenville sands (outwash) and Halifax silts (loess).

Petrography of the Meguma Supergroup provides insight into the nature of the source area. It was rich in quartz, biotite, and plagioclase but relatively poor in total feldspar, especially potassic varieties. The source area was probably vast, deeply eroded, cratonic, quartz-rich, granodioritic, and presumably of Precambrian age. It consisted of gneiss, granite, and metasedimentary rocks as well as volcanic rocks. Preliminary study of major, minor, and trace elements of the Meguma Supergroup substantiates this interpretation (Liew, 1979). Local volcanic detritus in the supergroup may have been eroded from a late Precambrian volcanic source, or have issued from contemporary volcanos. Bulk chemical analyses of the supergroup suggest a strong volcanic component (M. Zentilli, pers. comm., 1991). Detrital blue quartz, conceivably from granulite in the source area, is abundant in parts of the Goldenville, and may be useful as a tracer. Detrital graphite in the Cunard Formation suggests that the adjacent shelf was anoxic (Waldron, 1987).

The remarkable fineness of grain sizes in the Meguma Supergroup indicates extensive winnowing and sorting. Efficient separation of sand and silt is achieved best by the wind. The source-continent could also show an extensive distributary system, and for the Halifax Group a wide shelf (Fig. 3.152).

Resedimented phosphate and carbonate in Goldenville sandstones provide clues to the source continent. Phosphatic mud was precipitated on the upper slope and eroded by headward migrating canyons. Transportation by turbidity currents produced round to angular pebbles of the phosphate in the sandstones. Carbonate concretions are abundant in thick sandstones. They may result from dissolution of skeletal debris transported from shallow water to deeper water, beyond the carbonate compensation depth. Calcareous fossils in the centre of the sand sheets may be preserved by diagenetic precipitation of calcite in the form of concretions. These hard concretions would protect the fossils from subsequent deformation and metamorphism

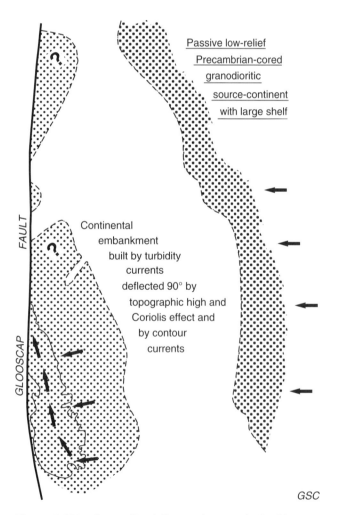

Figure 3.152. Generalized dispersal pattern in the Meguma Zone and nature of sourceland.

(Schenk, 1982). This is a possible explanation for the fossils found in the Tancook Formation (Waldron and Graves, 1987; Pratt and Waldron, 1991).

The uppermost part of the Meguma Supergroup along the northwest rim of the zone contains a diamictite which may be useful to discriminate possible source areas. The unit may be a glaciomarine tillite or drift-ice deposit (Schenk, 1972; Lane, 1981) related to Tremadoc glaciation of the South American portion of Gondwana (Erdtmann and Miller, 1981; Schenk, 1991). Although a near-polar and Gondwana origin of the Meguma Supergroup is implied, clasts may have been rafted great distances; the Greenland ice sheet today launches icebergs which travel 2000 km southward. A relative drop in sea level, suggested by shoaling in the upper part of the Halifax Group, may reflect significant global glacioeustatic sea-level changes (Schenk, 1991).

Sm-Nd isotopic data from both the Goldenville and Halifax groups have a mean crustal residence age of 1773 ± 95 Ma. This is interpreted as the time when the material, now represented as clastic debris in the Meguma Supergroup, was segregated from a depleted mantle. If the mantle was chondritic, then a 1358 ± 104 Ma age is indicated. A single source is assumed (Clarke and Halliday, 1985).

In summary, the source area for the Meguma Supergroup is interpreted as a low-lying, deeply eroded, stable, granodioritic, Precambrian continent marked by an extensive and long-term distributary system and in the Ordovician, a wide shelf (Fig. 3.152). The margin was essentially passive but volcanic rocks suggest periodic extensional tectonics. The mineralogical composition of sands indicates a very stable tectonic source which was deeply eroded. Sedimentary textures imply very efficient sorting and probable recycling. Thus, the source-land must have been continental in size, and the location of the depositional basin must have been at the end of a continental distributary system, perhaps similar to the present Mississippi delta and fan (Bouma et al., 1985). Conceivably, phosphate was precipitated on the upper slope. This source area was located somewhere to the present southeast of Nova Scotia (Schenk, 1970). Events recorded in the Meguma Supergroup should have at least some counterparts on the departed source continent. Identification of these events, most resulting from eustatic changes in sea level, offer a real hope of identifying the source of the Meguma Supergroup (Schenk, 1991).

Environment of deposition

Goldenville Group

New Harbour Formation

Each of the thick to very thick, fine grained, quartz wacke sheets of the New Harbour Formation was delivered during discrete instantaneous events by high-concentration turbidity currents (Schenk, 1970). Although most of the strata conform to the turbidite model, other kinds of sediment gravity flows are evident. Many of these thick layers in the New Harbour Formation display abundant dewatering features including dish structures, fluid-escape pipes and sheets, and sand volcanos. They are fluidized-flow deposits. Because of the excellent size sorting, it is in most places difficult to disregard grain flow processes which, if present,

might indicate local transport on slopes of 18 degrees or more (Middleton, 1970). Debris flows are relatively common as thick, pebbly sandstones in the New Harbour lithology (Schenk and Adams, 1986; Waldron and Jensen, 1985). Synsedimentary creeping and slumping are common.

A deep-water environment is indicated by: (1) deposition of thousands of metres of sandy and shaly turbidites apparently rapidly, with very little upward change in sedimentary properties or mineralogy; (2) persistence of a uniform dispersal pattern lasting at least 35 Ma over a large area; (3) absence of in situ invertebrate fauna except for pelagic forms; (4) trace fauna dominantly of the Nereites ichnofacies; (5) clasts of phosphate, eroded upflow presumably from the upper slope or outer shelf; and (6) carbonate concretions within thick New Harbour sandstones possibly formed during early diagenesis beneath the carbonate compensation depth (Schenk, 1970).

Risser's Beach, West Dublin, and Tancook formations

The Risser's Beach and West Dublin formations, and correlative Tancook Formation, constitute a submarine, channel-levee complex. They appear to be restricted to the southern (upstream) portion of the Meguma Zone. In total they are stratigraphically and lithologically intermediate between underlying deep-sea fan complexes and overlying, prodeltaic siltstones (Schenk and Adams, 1986; Schenk, 1988). These formations correspond to Type III/Stage III turbidite systems of Mutti (1985). The siltstones and black slates are interpreted as levee and overbank deposits. Thick channel-fill sandstones occur within the Risser's Beach Formation and dominate the West Dublin Formation. All lithosomes thin laterally in the down-current direction (Schenk and Adams, 1986).

Halifax Group

The basal two formations of the Halifax Group were deposited in relatively deep water, as indicated by similar sedimentological trends to the underlying Goldenville Group. These intercalated units represent the lower part of a continental-rise prism or prodeltaic wedge that prograded northwesterly over the Goldenville fan deposit.

Mosher's Island Formation

The Mosher's Island Formation was deposited by dilute silty turbidity currents. The relatively high content of metals could indicate relatively slow sedimentation. The longest depositional interval would prevail if the transported trilobites immediately below the formation were dating the true and not maximum age of the uppermost Tancook Formation. Because the overlying Cunard Formation is Tremadoc, deposition of the Mosher's Island Formation would span mid-Middle through Late Cambrian, about 25 Ma. Therefore its 500 m stratigraphic thickness would be deposited at the average rate of 2 mm per 100 years (or accumulation of 2 cm of unconsolidated mud per 100 years). This is not a slow rate relative to modern continental slopes. The high content of metals in the unit may be due to oxidation of organic matter during early diagenesis (Graves and Zentilli, 1988).

Cunard Formation

The Cunard Formation records an intense and prolonged stagnation. Stagnancy was maintained despite frequent influxes of thin bedded turbidites of very fine grained sandstones and silt. These also contain graphite to indicate that more shallow waters were also rich in organic matter (Waldron, 1987). Perhaps the beginning of anoxia is indicated by pebbles of black, phosphatic argillite in the uppermost sandstone beds of the underlying Goldenville Group. Phosphate mineralization may peak at the beginning (or end) of an oceanic anoxic event (Jenkyns, 1986). Anoxia spread into deeper water, and led to deposition of the Cunard Formation.

Feltzen and Delanceys formations

The differences in regional trends in the upper three formations are probably because of deposition in shallower waters than those of underlying ones. These upper formations are considered to be shelf deposits.

In the Feltzen Formation, alternating dark and light-coloured slates indicate that the long period of stagnancy during deposition of the Cunard Formation did not end abruptly. The distribution of graptolites suggests mass killing, perhaps because of upwelling of anoxic waters into a more shallow, oxygenated environment (Berry and Wilde, 1978). *Dictyonema* was the first of the planktonic graptolites, and may have evolved suddenly in polar, cold waters to exploit blooms of oceanic phytoplankton. Such blooms would result from the expansion of nutrient-rich, anoxic water masses during the Early Tremadoc submergence (Erdtmann and Miller, 1981).

The overlying Delanceys Formation was deposited mainly by silty turbidites in relatively shallow, oxygenated waters.

Rockville Notch Formation

The Rockville Notch Formation was interpreted as a marine tillite, possibly of Ashgill age (Schenk, 1972; Lane, 1981). However, W.A.M. Jenkins recovered Tremadoc acritarchs from the formation (J.D. Keppie, pers. comm., 1982). Similar diamictites occur in lower Tremadoc strata of Argentina and elsewhere (Erdtmann and Miller, 1981). The diamictite and also the uppermost parts of the Halifax Group are similar to Quaternary continental slope sequences in areas where there is a large fluvial or glacial discharge of fine sediment, but coarse sediment is trapped on the continental shelf (Hill, 1984).

Model for deposition of the Meguma Supergroup

Introduction

The following model for deposition of the Meguma Supergroup is based on stratigraphic and sedimentological observations. The model is one of overall shoaling upward from submarine fans, through a channel-levee complex, to a prograding wedge, and finally shelf and nearshore lithologies (Fig. 3.151). The supergroup consists entirely of siliciclastic debris that had to be eroded from, and so controlled by an adjacent continent. The general continuity and magnitude of the formations suggest that the rate of erosion on this source-continent controlled the vertical succession. This rate was in turn dependent on the elevation of base-level and so, relative sea level. Consequently the Meguma Supergroup (together with the Annapolis Supergroup, see following chapter) can be set in a framework of sequence stratigraphy (Fig. 3.151, 3.153). The identification of sequence boundaries and system tracts permits inter-regional correlations and location of the probable source-continent of the Meguma Zone.

Cambrian

In general the Goldenville Group is interpreted as an ancient abyssal-plain fan (Schenk, 1970). Thicker sandstone packages in the Goldenville Group are submarine-fan channel complexes flanked by raised levees; currents on the levees diverged from those in the channel axis. Thinner packages record both less persistent, shallow channels that probably lacked levees, as well as possibly also in small depositional lobes. Slate-dominated intervals (e.g. as in the Risser's Beach Formation) represent inter-channel/overbank deposits (Waldron and Jensen, 1985; Schenk, 1991). At least local slopes in the southwestern, relatively up-current area is indicated by large-scaled slumps, like those in the Risser's Beach Formation (O'Brien, pers. comm., 1985; Schenk and Adams, 1986).

The dispersal pattern seems very similar to recent deposition in the western North Atlantic (Horn et al., 1971; Elmore et al., 1979). There, channelled turbidites issuing from canyons in the continental slope, hook 90 degrees to the right because of deflection by the Bermuda topographic high. Thus the general dispersal trend is parallel to the adjacent slope of North America, the source continent. The intervening continental rise is swept by contour currents of the Western Boundary Undercurrent that transports fine grained sediment toward the south. Both currents are parallel to each other as well as the adjacent continental slope, which is incised by many distributary canyons. Apparently the continental rise is prograding over the abyssal plain. In a similar manner (Fig. 3.152), dispersal patterns in the Goldenville Group are seen as trending more or less parallel to its ancient continental margin (Schenk, 1982). Vectoral and scalar data indicate that the dispersal pattern was from the present southeast (Schenk, 1970).

Deposition of large amounts of relatively coarse grained sediment on deep-sea fans, like those of the Goldenville Group, can correlate directly with lowstands of sea level. At such times, base-level is relatively low, continental shelves are narrow, total organic productivity is low, and relatively coarse grained detritus reaches canyons feeding fan-systems (Shanmugam and Moiola, 1982). Conceivably continental glaciation or tectonism could lower relative sea level. Both agents can generate large volumes of siliciclastic debris that eventually would reach a continental margin. During a glacioeustatic lowstand, cold currents circulate away from ice-covered polar areas to ventilate deep waters by bottom currents and winnow turbidites by vigourous geostrophic currents (Berry and Wilde, 1978). A drop in relative sea level immediately before Tremadoc time may reflect glaciation of Gondwana (Erdtmann and Miller, 1981), although the glaciogene sediments are dated as only probably Cambrian (Cecioni, 1981; Rocha-Campos, 1981). However during the Cambrian Period, continents may not have been located close enough to the poles to support continental ice sheets (Caputo and Crowell, 1985). Possibly very early Phanerozoic glaciations may have occurred at low paleolatitudes, similar to latest Proterozoic ones (Chumakov and Elston, 1989).

Pliocene and Pleistocene sedimentation off northwest Africa may be an analogue (Stein and Littke, 1990). Glacial stages result in increased areas of high surface-water organic matter productivity. The amount of terrigenous organic matter is decreased, and the fluvial input of clay materials and nutrients ceases because of the arid climate. Trade wind intensity and the upwelling of coastal currents are increased. Phosphate, which precipitated on the upper slope by upwelling currents, could be eroded and carried down to the fan by turbidity currents.

On the other hand, the Cambrian influx of sediment onto the Goldenville fan could be an idiosyncrasy of the source, triggered by intraplate stresses (Cloetingh, 1988). Note that the Pan-African II Orogeny in northwestern Africa may be late Early Cambrian (Lecorche and Dallmeyer, 1987). The newly created margin of northwest Gondwana probably subsided because of thermal relaxation as the Iapetus Ocean widened. The combined effect was probably to tilt the West African Craton northwestwards, cause erosion of latest Precambrian glaciogenetic sediments in Mali, and flood northwestern Africa with fine grained detritus (see next section). Thus, the large quantity of Goldenville sandstone may have resulted from glaciation as well as local tectonism, both of which lowered relative sea-level to activate deposition of relatively coarse-grained sediment on marginal deep-sea fans. The return of thick sandstones in the West Dublin and uppermost Tancook formations is probably the consequence of a short-lived drop in relative sea-level. These sandstones record the last impulses of thick, deep-water sandstones into the Meguma Zone. The overlying, shaly Mosher's Island Formation (basal Halifax Group) is laterally persistent across the zone.

In terms of sequence stratigraphy (Posamentier and Vail, 1988), the New Harbour and underlying unnamed formations of the Goldenville Group conform to lowstand fan depositional systems of a major lowstand systems-tract (Fig. 3.153). Although the extent of exposures makes identification difficult, the strata are probably in the form of fan lobes and channels. They are assumed to rest on a regional unconformity (i.e. a type 1 sequence boundary). This unconformity should be apparent on the adjacent craton, especially on the adjacent shelf. Cratonic sandstones should overlie this surface and reflect renewed erosion because of the lowstand. These sandstones would record a large northwestwards directed distributary system. Presumably the unconformity is at least sub-Late Cambrian, but the hiatus probably extends into the late Precambrian (Sinian) because of the high degree of textural maturity of the Goldenville sands.

The Risser's Beach and West Dublin formations constitute the upper part of the lowstand systems-tract, namely the mass flow/channel/overbank complex (Fig. 3.153). This complex appears only in the southwestern, relatively up-current, proximal area of exposure. The sharp contact between the New Harbour and Risser's Beach formations is the "top of the basal fan surface" (Fig. 3.153). In the eastern area, the transition from the Goldenville to Halifax groups is much more abrupt as would be expected in this generally more down-current region (Fig. 3.152).

Early Ordovician

The depositional setting of the Halifax Group is interpreted to be the mid- or upper-fan area of a muddy deep-sea fan, passing upwards into a prograding continental slope and shelf (Schenk, 1981). The lower part of the group resembles modern deep-sea channel-levee complexes; the upper part may have accumulated principally from turbidity currents on a rapidly prograding continental slope. The uppermost formations were probably deposited on a muddy outer shelf or upper slope (Schenk, 1991).

The decrease in grain size from the Goldenville to Halifax groups probably results from submergence of the source-continent. The effect was sudden, both regionally as well as locally within the transition zone, where black slate of the Risser's Beach Formation abruptly overlies very thick, amalgamated sandstone beds of the New Harbour Formation. The process could be autocyclic or allocyclic. In the former case, the contact could be due to either an abrupt switching in the distributary system, or a rapid drowning of the source-region. However, a package of similar thickness has not been observed in the thousands of metres of underlying Goldenville sandstones; moreover, thick bedded sandstones do not occur in the Halifax Group. In the latter case, it could be a response to rapid drowning of the source-region. That is, a relatively rapid rise in sea level achieved a higher elevation of base-level, a decrease in source-land area, a widening of the continental shelf, and a sharp decrease in the volume of sands reaching the Goldenville fan. Considering the magnitude of the units involved, this process appears to be most probable. The return of thick sandstones in the West Dublin Formation is probably the consequence of a short-lived drop in relative sea level. If so, such a minor change might record the pre-Tremadoc glacioeustatic change, or even identify the source-continent (see below). Deposition of the Mosher's Island Formation returns the regimen to fine grained sedimentation.

The Pliocene and Pleistocene section off northwest Africa may again be an analogue (Stein and Littke, 1990). Following a glacial episode the fluvial input of clay minerals and nutrients is increased because of the humid climate. Supply of terrigenous organic matter is increased, and adds to marine organic matter, to increase the total organic carbon content. The Cunard Formation has up to 5 per cent carbon (Graves and Zentilli, 1988). Trade wind intensity and coastal upwelling are reduced.

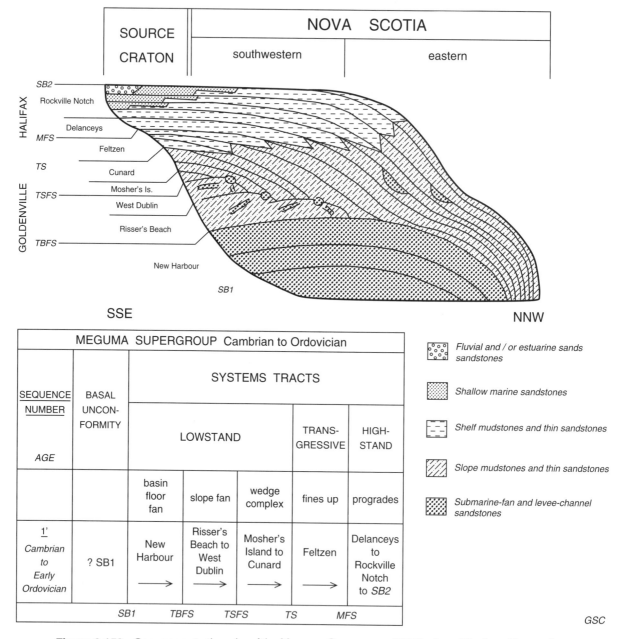

Figure 3.153. Sequence stratigraphy of the Meguma Supergroup. TBFS – top of the basal fan surface; TSFS – top of the slope fan surface; SB-1, 2 – type 1 or 2 sequence boundary; TS – transgressive surface; MFS – maximum flooding surface. The Meguma Supergroup is the first sequence of the Meguma terrane and is a type 1 sequence (indicated by the superscript). Arrows from left to right show stratigraphically upward changes. See Figure 4.34.

The Cunard Formation is similar to other world-wide laminated black shales of late Tremadoc through Arenig age that record an oceanic anoxic event related to a high-stand of sea level (Berry and Wilde, 1978; Jenkyns, 1986). Such events occur during a relatively warm, non-polar episode when an absence of ice sheets raises world sea level, decreases continental erosion, widens continental shelves, and diminishes both atmospheric and oceanic currents. Poorly-defined climatic zones reduce the frequency of storms so that coarse sediments are not carried off-shore. Water masses are poorly mixed so that density stratification is common. Consequently, cold, oxygenated bottom currents are not generated in polar areas to ventilate the ocean depths. As a result, the remains of rich planktonic fauna that flourish in warm surface waters are not oxidized but accumulate to form black muds. The anoxia prohibits infauna so that delicate sedimentary structures are preserved. The black shales accumulate mainly on continental slopes, but extend both into abyssal and epicontinental settings. The general conditions during the Tremadoc may have been somewhat similar to the better known, Middle Cretaceous oceanic anoxic event. Global climate was warm; eustatic sea levels were generally high; spreading rates were fast; pelagic sediments in the world's oceans were accumulating rapidly; and oceanic surface and bottom waters were relatively warm (Schlanger and Jenkyns, 1976; Jenkyns, 1980; Arthur et al., 1984).

The Feltzen, Delanceys, and Rockville Notch formations show a progressive, upward increase in silt and fine sand, such as silty turbidites in the Feltzen Formation, flaser structures in the Delanceys, and laminations in the Rockville Notch Formation. Bioturbation also increases upward, and changes in character from horizontal traces to oblique and vertical burrows. As noted previously, regional trends of sedimentary structures, textures, and composition are significantly different from those of the lower part of the Meguma Supergroup (Schenk, 1970). These vertical changes together suggest slow emergence to the unconformity at the top of the Meguma Supergroup.

In terms of sequence stratigraphy, the Mosher's Island Formation may be a condensed interval along the Goldenville/Halifax contact, as is indicated by the time lines in Figure 3.153. This surface in the eastern (up-current) area is the "top of the slope fan surface" (Fig. 3.153); to the east it is the "top of the basal fan surface". Prodeltaic black muds and turbidites of the Cunard Formation constitute the lower part of a prograding wedge (uppermost part of the lowstand systems tract). Intercalation of black and grey slate (Feltzen Formation) may record the transgressive systems tract. The prograding Delanceys and Rockville Notch formations developed immediately before evidence of subaerial exposure, erosional truncation, and a significant hiatus. This regional unconformity between the Meguma and Annapolis supergroups is a type 2 sequence boundary.

North Atlantic connection for the Meguma Supergroup

A suitable source continent for the Meguma Supergroup must satisfy the following considerations:

(1) It must be a continental land mass other than Laurasia. This is indicated by stratigraphy, sedimentology, dispersal, provenance, paleontology, igneous petrology,

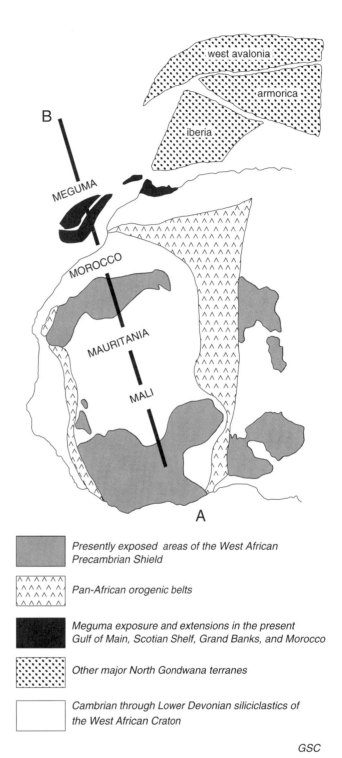

Presently exposed areas of the West African Precambrian Shield

Pan-African orogenic belts

Meguma exposure and extensions in the present Gulf of Main, Scotian Shelf, Grand Banks, and Morocco

Other major North Gondwana terranes

Cambrian through Lower Devonian siliciclastics of the West African Craton

GSC

Figure 3.154. Northwestern African portion of Gondwana during early Paleozoic time. Presently exposed areas of the West African Precambrian Shield shown in (red); Pan-African orogenic belts in (grey); Meguma exposure and extensions in the present Gulf of Maine, Scotian Shelf, Grand Banks, and Morocco shown in (black); other major North Gondwana terranes shown in (light grey); Cambrian through Lower Devonian siliciclastics of the West African Craton shown in (white). Cross-section A to B is the dispersal direction of Cambrian, Upper Ordovician, and Lower Devonian fluvial to shallow marine, cratonic sandstones, from Mali toward Morocco.

deformation, and geophysics. During the Paleozoic Era, the only other large continental area was Gondwana (Fig. 3.154). Also, the Meguma and Annapolis supergroups are interpreted in terms of sea-level changes that are in part the result of glaciation. Scarcity of fossils in the Meguma Supergroup may be a response to low temperatures (Schenk, 1982). During the Phanerozoic, Laurasia straddled the paleoequator while Gondwana was polar.

(2) The immediate source-area must have had some paleogeographic connection with the present Meguma Zone of Atlantic Canada. Transfer to Atlantic Canada involved southeast to northwest compression. The last collisional event in the Appalachian Orogen was the accretion of Gondwana. The Meguma Zone was the last terrane to dock against the North American margin (Williams and Hatcher, 1983). To the present southeast of the Meguma Zone lies Morocco. This part of Gondwana was located against Nova Scotia for approximately the middle third of Phanerozoic time. The source continent had to depart, leaving behind a chip of its ancient continental margin stuck to North America. Separation of Morocco from Nova Scotia occurred in the Early Jurassic.

(3) The craton should show evidence of stability during the depositional interval of the Meguma and Annapolis supergroups. The Pan-African Peneplain of the West African Craton is a remarkably stable surface (Fig. 3.154). It was produced during regional planation following the Sturtian Precambrian I Orogeny (about 650 Ma). Subsequently, it controlled dispersal patterns and provenance of northwestern Africa from latest Precambrian to Middle Devonian time. Today it still extends as an almost flat surface over an area of 2×10^6 km^2.

(4) The source area should show evidence of a powerful erosive agent to generate large amounts of detritus in perhaps a relatively short time. Continental glaciation is the ultimate of such agents. During the latest Precambrian (630 Ma) ice sheets moved down the Pan-African Peneplain

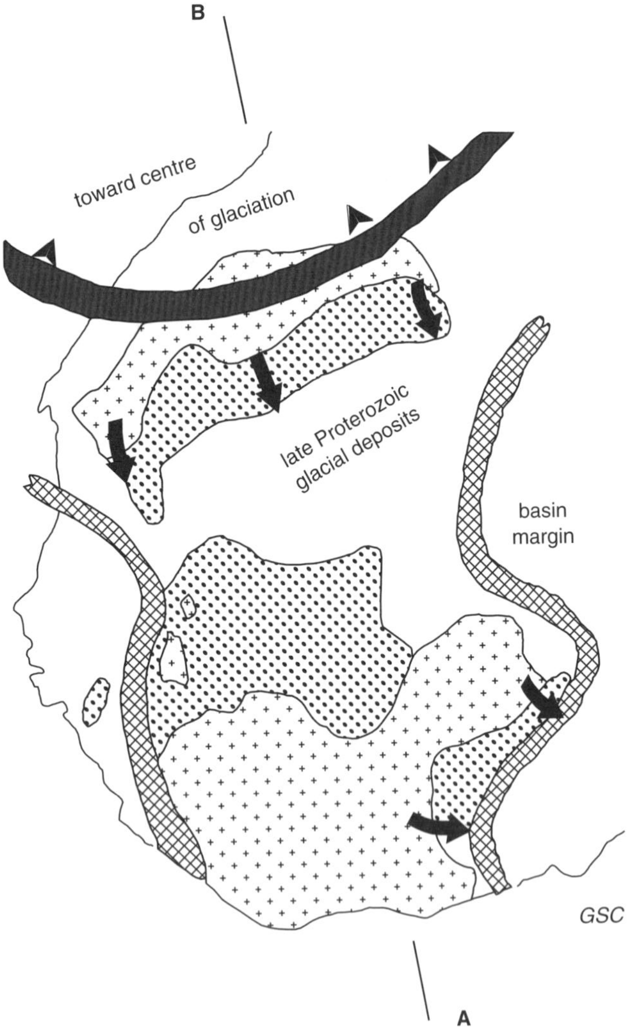

Figure 3.155. Late Precambrian (700 Ma) glaciation of northern Gondwana was directed south-southeast from Morocco (and the future site of the Meguma Supergroup) onto the West African Craton. In Mali, earlier sandstones (Sotuba Group – Fig. 3.156) were covered with thick glaciogenetic, eolian sand blankets that were in part reworked southeastward to form shallow marine and turbidite successions (Bakoye Group) (Deynoux, 1985).

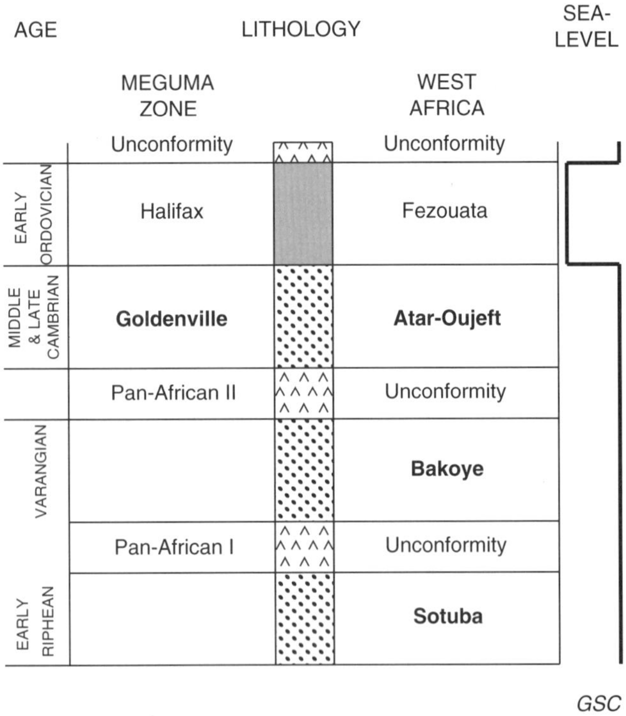

Figure 3.156. The stratigraphic section of the Meguma Supergroup is similar to that of the West African Craton in lithologies and timing of major hiatuses. Both sections consist of Cambrian thick, fine grained sandstone underlying Ordovician thick, silty, black shale. The sandstones result from erosion of Pan-African orogenic belts and upper Proterozoic sandstones (Sotuba and Bakoye groups). The Paleozoic sandstones record sea level lows mainly due to continental glaciation of West Africa; the shale marks a sea level high following this glaciation. Volcanism (inverted V's) occurred during major hiatuses.

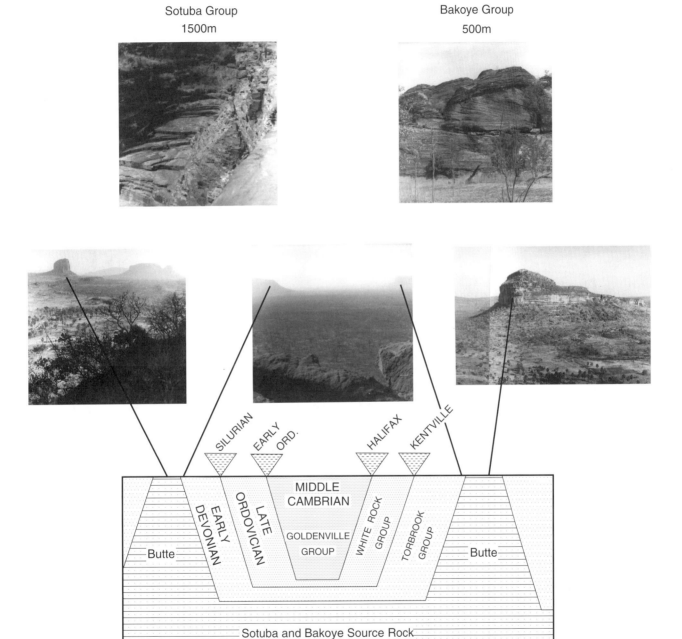

Figure 3.157. Buttes in western Mali are erosional remnants of the Neoproterozoic Bakoye and Sotuba groups (Deynoux et al., 1991). Fine grained sands, eroded from these extensive, once-continuous sheets, were carried northward across vast braidplains to the continental margin of North Gondwana (Fig. 3.158). Three episodes of erosion (Middle and Late Cambrian, Late Ordovician, Early Devonian) were divided by two submergences (Middle Ordovician, Silurian).

from a centre in Morocco across Mauritania to Mali in the southeast (Fig. 3.155) (Deynoux, 1985; Clauer and Deynoux, 1987).

(5) A reservoir of detritus on the craton could be repeatedly tapped to supply sands of the Meguma Supergroup. The thickest and distal deposit of this glaciation is the 500 m thick Bakoye Group of Mali (Figs. 3.156, 3.157). It is still undeformed, unmetamorphosed, and exposed over an area of thousands of square kilometres in western Mali and

eastern Mauritania (Deynoux, 1980). Stratigraphically below this unit and beneath the Pan-African Peneplain is another reservoir of detritus with even greater distribution – the Sotuba Group (Fig. 3.156, 3.157). It consists of approximately 2 km of unmetamorphosed and little deformed sandstones deposited between approximately 1000 and 670 Ma. Both sequences of sandstones are preserved as widely scattered buttes, the erosional remnants of more extensive sheets (Fig. 3.157).

(6) An efficient sorting agent would be required to concentrate the detritus into fine sand and silt. The best such agent is eolian transport. Almost all of the Bakoye Group consists of fine grained feldspathic quartz arenites in the form of eolian dunes (Fig. 3.157) in places recycled to shallow marine environments. Dispersal was toward the southeast, down the Pan-African Peneplain (Fig. 3.158). The underlying sandstones are also fine grained lithic and feldspathic arenites, well sorted in fluvial to shallow marine environments (Deynoux, 1980).

(7) An extensive and long-termed distributary system would be necessary to carry the detritus across the craton to its margin. Following glaciation, the Pan-African Peneplain reversed it's regional, slight tilt from southeast to northwest, perhaps as a consequence of the Pan-African II Orogeny, perhaps as a result of thermal subsidence following birth of the Iapetus Ocean (Fig. 3.158). This slope was maintained until the Middle Devonian. Postglacial submergence capped the glacial deposits with thin but widespread carbonate and shale containing distinctive chert layers. These were eroded as well as the Bakoye and underlying groups of fine grained sandstone. The sands were carried northward across Mauritania by braided streams to the shallow sea in Morocco, and presumably on to the margin of Gondwana. The capacity and duration of this cratonic distributary system must have built an immense deep-water fan along the northwest margin of Gondwanan Africa. The distance from southern Mali to northern Morocco is approximately the length of the Mississippi River and delta, that continue as the deep-water Mississippi fan. The cratonic distributary system for the Meguma Supergroup would be a braid plain because of the absence of land plants. Northwestwards prevailing winds must have carried vast amounts of silt as desert loess toward the continental margin.

(8) The source-continent should show evidence of a deepwater fan that began during the Cambrian Period, was similar to the Meguma fan, and possibly is segmented. Extensive carbonate deposition immediately following latest Precambrian glaciation was thickest in southern Morocco where the regional, northwestwards tilt of the Pan-African Peneplain permitted more sedimentary accommodation. From Late Vendian to late Early Cambrian a wedge thickening westward to more than 12 km of mainly carbonate rocks was deposited in the western Anti-Atlas (Michard, 1976, Fig. 21). Shallow water sands from the south first reached the area in latest Early Cambrian time. Thereafter fine-grained siliciclastics issuing from the extensive distributary system crossed the shelf at least until the Early Devonian (Michard, 1976, p. 58). Presumably this event started deposition of the Gondwanan fan. The Bou Regreg Formation in northern Morocco is a remnant of this fan (Pique et al., 1990). It includes a thick succession of lower Paleozoic turbidites similar to the Meguma Supergroup in depositional environment, dispersal pattern, age, provenance, diagenesis, and deformation (Schenk, 1982). Unfortunately it is

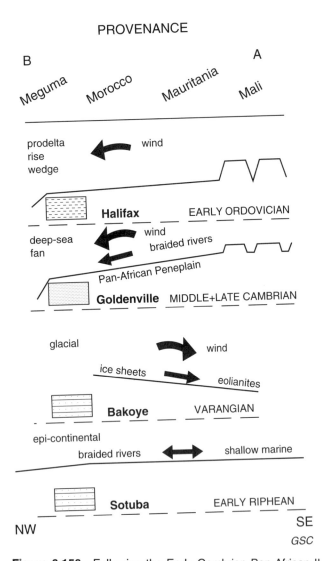

PROVENANCE

Figure 3.158. Following the Early Cambrian Pan-African II orogeny, the West African Craton reversed its regional slope to down to the north and north-northwest, probably as a result of thermal subsidence following birth of the Iapetus Ocean. Consequently these cratonic sandstones were eroded by north and north-northwest moving braided rivers and prevailing winds. Early Ordovician submergence of the West African Craton reduced sand influx onto the margin. See also Figure 4.38 of the Annapolis Belt (in Chapter 4).

also a suspect terrane (the Sehoul block) that was at least 50 km to the north of its present location during Late Cambrian. Strata of northwestern Africa are affected by both late Paleozoic and Cenozoic orogenies. Paleozoic rocks along the western coast of northwest Africa are overlapped by sediments of the Atlantic Coastal Plain.

(9) Detrital sediments in the Meguma Supergroup should show some similarity to source rocks and coeval sediments remaining on the craton. Lithoclasts of chert are conspicuous in Cambrian sandstones of Mauritania, Morocco, and the Meguma Supergroup. They decrease in abundance and grain size from south to north. The chert is identical to the chert beds above glaciogenetic strata that are widely distributed across western Mali and Mauritania. Volcanism was widespread during the Early Cambrian in the Atlas and Anti-Atlas areas of Morocco. During Middle Cambrian time several trachyandesitic volcanos were active along the Coastal Meseta. Volcanic detritus occurs within the Bou Regreg Formation of the Sehoul Block and is suspected in the Meguma Supergroup. Bulk chemical analyses of the Meguma Supergroup suggest a strong volcanic component (M. Zentilli, pers. comm., 1991). The Meguma Supergroup has a Nd model age of 1.74 ± 0.1 Ga. This is possibly a mixed age representing inputs from the West African Craton (>2.0 Ga) and the Pan-African orogenic belts (ca. 0.6 Ga) in West Africa (Clarke and Halliday, 1985). Ages and varieties of zircons from the Goldenville Group have been explained in a similar manner (Krogh and Keppie, 1986).

(10) The stratigraphy of the source continent's margin should exhibit an extensive lacuna that is represented by the Goldenville Group (Schenk, 1983). Upper Cambrian and Tremadoc strata are absent in Morocco. This would be expected if the Goldenville Group is a lowstand fan depositional system. Morocco was deeply eroded in places at this time but the sands continued to be supplied from the south. Morocco was essentially a bypass as fine grained quartz sands crossed the shelf and spilled directly down to the Goldenville deep-sea fan. Subsequent rise in sea-level in the Tremadoc reduced the supply of sands so that the Ordovician in both West Africa and the Meguma Zone is shaly.

Summary and significance

The Meguma Zone is the last of the Appalachian's five, exposed, major terranes to accrete against eastern North America. The derivation of the last to arrive should be the easiest to solve, presumably by trans-Atlantic correlation. It is the largest terrane in the Canadian Appalachians, comprising a folded area of 200 000 km². If its strata were unfolded, the restored width would exceed the presently exposed breadth of the Canadian Appalachians. The composite thickness of the stratigraphic succession exceeds 21 km; however, nowhere do all of the units occur at one locality, nor are their thicknesses constant. The apparent volume of the Meguma Supergroup is equal to a block with the area of Ontario (or Mali) and a height of 4 km.

Strata of the Meguma Zone consist entirely of metamorphosed and deformed, fine grained siliciclastic sedimentary rocks. The succession is divided into the Goldenville and Halifax groups of the Meguma Supergroup. The Goldenville Group was deposited from high-concentration turbidity flows as a number of fan lobes and channels on the mid-fan area of submarine fan systems. Vectoral and scalar sedimentary structures indicate that the source-area lay to the present southeast. The Halifax Group is interpreted as a large, prograding wedge of slate and siltstone. The basal formations were deposited in relatively deep water, as indicated by similar sedimentological trends to the underlying Goldenville Group. The succession of the upper, silty Halifax Group indicates slow emergence through the upper part of a prograding wedge. The upper boundary of the Meguma Supergroup is locally an angular unconformity or disconformity. The overlying Nictaux Volcanics of the White Rock Group (Annapolis Supergroup) are in part subaerial and probably indicate a significant hiatus, although fossil control is unavailable.

In terms of sequence stratigraphy, the Meguma Supergroup is a type 1 sequence, equivalent to a supercycle. The Goldenville and basal part of the Halifax groups comprise a low-stand systems tract of basin-floor fan (New Harbour Formation) overlain by a slope fan (Risser's and New Dublin formations) and prograding wedge (Mosher's Island and Cunard formations). The upper part of the Halifax Group consists of transgressive systems (Feltzen Formation) and highstand systems tracts (Delanceys and Rockville formations). Although the Meguma Supergroup was deposited on a continental margin, its cycle corresponds in time with the Sauk cratonic sequence (Sloss, 1963).

Vectoral and scalar sedimentary structures indicate that the source-area lay to the present southeast. For a third of Phanerozoic time (at least the Carboniferous to Middle Jurassic) northwest Africa was joined to the Meguma Zone. Stratigraphic sections, provenance, sequence stratigraphy, paleontology, igneous petrology, and geophysics suggest that the Meguma Zone is a part of the continental margin of Gondwana, stranded against North America following Jurassic rifting. Sand and silt grains presently comprising the supergroup were winnowed and stored in thick and extensive Malian sand sheets to be periodically tapped by northward moving streams, ice, and wind. Remnants of the reservoir sheets remain in Mali as tall buttes; remnants of the northward-moving detritus remain as mesas and buttes of fluvial (Mauritania) and shallow marine (southern Morocco) facies. The most far-travelled sand and silt reached the margin of North Gondwana where they formed thick, deep-water fan-complexes. Rifting of this margin created a plethora of microcontinents or microplates, including present parts of North Africa, the Middle East, central and southwestern Europe, eastern and western Avalonia, Florida, and the Meguma Zone. Thus the Meguma Zone is not a provincial curiosity but is related to now far-distant areas as well as being a significant element in the Appalachian Orogen.

Acknowledgments

Operating grants over the last 25 years from the Natural Sciences and Engineering Research Council of Canada have funded my study of sedimentary strata in the Meguma Zone. T. E. Lane made significant contributions to knowledge of the White Rock Formation. Comments by R. Boyd and J.W.F. Waldron improved early drafts of the manuscript. I especially appreciate the review by H. Williams.

REFERENCES

Aitken, J.D.

1966: Middle Cambrian to Middle Ordovician cyclic sedimentation, southern Rocky Mountains of Alberta; Canadian Petroleum Geology Bulletin, v. 14, p. 405-441.

Alcock, F.J.

1938: Geology of Saint John region, New Brunswick; Geological Survey of Canada, Memoir 216, 65 p.

1941: Jacquet River and Tetagouche River map-areas; Geological Survey of Canada, Memoir 227, 46 p.

Aleinikoff, J.N. and Moench, R.H.

1985: Metavolcanic stratigraphy in Northern New England- U-Pb zircon geochronology (abstract); Geological Society of America, Abstracts with Programs, v. 17, no. 1, p. 1.

Ambrose, J.W.

1945: Stanstead, Comtés de Stanstead et de Brome, Québec; Commission géologique du Canada; Étude 43-12.

Ami, H.M.

1900: Synopsis of the geology of Canada (being a summary of the principal terms employed in Canadian geological nomenclature); in Annual Report 1900; Royal Society of Canada, Proceedings and Transactions, 203 p.

Anderson, F.D.

1972: The Catamaran Fault, north-central New Brunswick; Canadian Journal of Earth Sciences, v. 9, p. 1278-1286.

Anderson, F.D. and Williams, H.

1970: Gander Lake (West half), Newfoundland; Geological Survey of Canada, Map 1195A, final map with descriptive notes.

Anderson, M.M.

1978: Ediacaran Faunas; in McGraw-Hill Yearbook of Science and Technology, (ed.) D.N. Lapedes; McGraw-Hill Book Company, p. 146-149.

1987: Late Precambrian glaciomarine sequence, Conception Group, Double Road Point, St. Mary's Bay, southeast Newfoundland; in Northeastern Section, (ed) D.C. Roy; Geological Society of America Centennial Field Guide, v. 5, p. 473-476.

Anderson, M.M. and Conway-Morris, S.

1982: A review, with descriptions of four unusual forms of the soft bodied fauna of the Conception and St. John's Groups, Avalon Peninsula, Newfoundland; Proceedings I, Third North American Paleontological Convention.

Anderson, M.M. and King, A.F.

1981: Precambrian tillites of the Conception Group on the Avalon Peninsula, southeastern Newfoundland; in Earth's Pre-Pliestocene Glacial Record, (ed.) M.J. Hambrey and W.B. Harland; International Geological Correlation Program Project 38, Pre-Pliestocene tillites, p. 760-766.

Anderson, M.M., Brueckner, W.D., King, A.F., and Maher, J.B.

1975: The Late Proterozoic "H.D. Lilly Unconformity" at Red Head, northeastern Avalon Peninsula, Newfoundland; American Journal of Science, v. 275, p. 1012-1027.

Anderson, M.M., Choubert, G., Faure-Muret, A., and Timofeiev, B.V.

1982a: Les couches à *Bavlinella* de part et d'autre de l'Atlantique; Société géologique de France (Bulletin), v. XXIV, no. 2, p. 389-392.

1982b: Découverte du Vendien à Newfoundland; Société géologique de France, 9ième réunion annuelle des sciences de la terre, Paris, 1982, p. 10.

Archibald, D.A. and Farrar, E.

1976: K-Ar ages of amphibolites from the Bay of Islands ophiolite and the Little Port Complex, western Newfoundland, and their geological implications; Canadian Journal of Earth Sciences, v. 13, p. 520-529.

Arnott, R.J.

1983a: Sedimentology of Upper Ordovician-Silurian sequences on New World Island, Newfoundland: separate fault-controlled basins?; Canadian Journal of Earth Sciences, v. 20, p. 345-354.

Arthur, M.A., Dean, W.E., and Stow, D.A.V.

1984: Models for the deposition of Mesozoic-Cenozoic fine-grained organic-rich sediment in the deep sea; in Fine-grained Sediments: deep water processes and facies, (ed.) D.A.V. Stow and D.J.W. Piper; Geological Society of London, Special Publication 15, p. 527-560.

Atwater, T.

1970: Implication of plate tectonics for the Cenozoic evolution of western North America; Geological Society of America Bulletin, v. 81, p. 3513-3536.

Aumento, F.

1969: Serpentine mineralogy of ultramafic intrusions in Canada and on the Mid- Atlantic Ridge; Geological Survey of Canada, Paper 69-53, p. 1-45.

Ayrton, W.G.

1967: Région de Chandler-Port-Daniel, Québec; Ministère des Richesses naturelles du Québec; GR 120, 97 p., 1 carte.

Badgley, P.C.

1956: New-Carlisle map-area, Québec; Quebec Department of Mines, GR 70, 36 p.

Bailey, L.W.

1865: Observations on the geology of southern New Brunswick; Queens Printer, Fredericton, 158 p.

Baird, D.M.

1951: The geology of Burlington Peninsula, Newfoundland; Geological Survey of Canada, Paper 51-21, 70 p.

1960: Sandy Lake (west half), Newfoundland; Geological Survey of Canada, Map 47-1959.

Baker, D.F.

1978: Geology and geochemistry of an alkali volcanic suite (Skinner Cove Formation) in the Humber Arm Allochthon, Newfoundland; M.Sc. thesis, Memorial University of Newfoundland, St. John's, Newfoundland, 314 p.

Barnes, C.R.

1984: Early Ordovician eustatic events in Canada; in Aspects of the Ordovician System, (ed.) D.L. Bruton; Paleontological Contributions No. 295, University of Oslo, Universitetsforlaget, p. 51-63.

1988: The proposed Cambrian-Ordovician global boundary stratotype and point (GSSP) in Western Newfoundland, Canada; Geological Magazine, v. 125, p. 381-414.

Barnes, C.R., Norford, B.S., and Skevington, D.

1981: The Ordovician System in Canada; International Union of Geological Sciences, Publication no. 8, 27 p., 2 charts.

Barr, S.M.

1990: Granitoid rocks and terrane characterization: an example from the northern Appalachian Orogen; Geological Journal, v. 25, p. 295-304.

Barr, S.M. and Jamieson, R.A.

1991: Tectonic setting and regional correlation of Ordovician-Silurian rocks of the Aspy terrane, Cape Breton Island, Nova Scotia; Canadian Journal of Earth Sciences, v. 28, p. 1769-1779.

Barr, S.M. and Macdonald, A.S.

1989: Geology of the Mabou Highlands; Nova Scotia Department of Mines and Energy, Paper 89-2, 65 p.

Barr, S.M. and Raeside, R.P.

1986: Pre-Carboniferous tectonostratigraphic subdivisions of Cape Breton Island, Nova Scotia; Maritime Sediments and Atlantic Geology, v. 22, p. 252-263.

1989: Tectono-stratigraphic terranes in Cape Breton Island, Nova Scotia: implications for the configuration of the northern Appalachian Orogen; Geology, v. 17, p. 822-825.

1990: Geology and tectonic development of the Bras d'Or suspect terrane, Cape Breton Island, Nova Scotia; Canadian Journal of Earth Sciences, v. 27, p. 1371-1381.

Barr, S.M. and White, C.E.

1988: Petrochemistry of contrasting Late Precambrian volcanic and plutonic associations, Caledonia Highlands, southern New Brunswick; Maritime Sediments and Atlantic Geology, v. 24, p. 353-372.

1989: Re-interpretation of Precambrian stratigraphy, Kings and Saint John counties; in Project summaries for 1989, Fourteenth Annual Review of Activities, (ed.) S.A. Abbott; New Brunswick Department of Natural Resources, Minerals and Energy Division, Information Circular 89-2, p. 182-189.

Barr, S.M., Dunning, G.R., Raeside, R.P., and Jamieson, R.A.

1990a: Contrasting U-Pb ages from plutons in the Bras d'Or and Mira terranes of Cape Breton Island, Nova Scotia; Canadian Journal of Earth Sciences, v. 27, p. 1200-1208.

Barr, S.M., Macdonald, A.S., and Blenkinsop, J.

1986: The Cheticamp pluton: an early Cambrian granodioritic intrusion in the western Cape Breton Highlands, Nova Scotia; Canadian Journal of Earth Sciences, v. 23, p. 1686-1699.

Barr, S.M., O'Reilly, G.A., and O'Beirne, A.M.

1982: Geology and geochemistry of selected granitoid plutons of Cape Breton Island; Nova Scotia Department of Mines and Energy, Paper 82-1, 176 p.

Barr, S.M., Raeside, R.P., and Jamieson, R.A.
1987a: Geological map of the igneous and metamorphic rocks of northern Cape Breton Island, Geological Survey of Canada, Open File 1594, six sheets, scale 1:50 000.

Barr, S.M., Raeside, R.P., and Macdonald, A.S.
1985: Geology of the southeastern Cape Breton Highlands, Nova Scotia; in Current Research, Part B; Geological Survey of Canada, Paper 85-1B, p. 103-109.

Barr, S.M., Raeside, R.P., and van Breemen, O.
1987b: Grenvillian basement in the northern Cape Breton Highlands, Nova Scotia; Canadian Journal of Earth Sciences, v. 24, p. 992-997.

Barr, S.M., Raeside, R.P., Dunning, G.R., and Jamieson, R.A.
1988: New U-Pb ages from the Cape Breton Highlands and correlations with southern Newfoundland: a Lithoprobe East contribution (abstract); Geological Association of Canada, Program with Abstracts, v. 13, p. A5.

Barr, S.M., White, C.E., and Bevier, M.L.
1990b: Contrasting petrochemistry and age of plutonic rocks in the Saint John area and Caledonia Highlands, Kings and Saint John counties, southern New Brunswick; in Fifteenth Annual Review of Activities, (ed.) S.A. Abbott; New Brunswick Department of Natural Resources, Minerals and Energy Division, Information Circular 90-2, p. 178-186.

Barrett, D.L., Berry, M., Blanchard, J.E., Keen, M.J., and McAllister, R.E.
1964: Seismic studies on the eastern seaboard of Canada: the Atlantic coast of Canada; Canadian Journal of Earth Science, v. 1, p. 10-22.

Beaudin, J.
1980: Région du Mont-Albert et du Lac Matapédia; Ministère de l'Énergie et des Ressources, Québec; DPV-705, 83 p.
1984: Analyse structurale du Groupe de Shickshock et de la péridotite alpine du Mont Albert, Gaspésie; thèse de doctorat, Université Laval, Québec (Québec), 241 p.

Beaulieu, J., Lajoie, J., and Hubert, C.
1980: Provenance et modèle de dépôt de la Formation de la rivière Nicolet: Flysch taconique du Domaine autochtone et du Domaine externe des Appalaches du Québec; Journal canadien des sciences de la Terre, v. 17, p. 855-865.

Bédard, J.H.
1986: Pre-Acadian magmatic suites of the southeastern Gaspé Peninsula; Geological Society of America Bulletin, v. 97, p. 1177-1191.

Beeden, D.R.
1983: Sedimentology of some turbidites and related rocks from the Cloridorme Group, Ordovician, Quebec; M.Sc. thesis, McMaster University, Hamilton, Ontario.

Béland, J.
1957: St-Magloire and Rosaire-St-Pamphile areas, southern Quebec; Quebec Department of Natural Resources, GR 76, 49 p.
1960: Preliminary report on Rimouski-Matapédia areas, Québec; Department of Natural Resources, PR 430, 20 p.
1962: Région de Ste-Perpétue, Comtés de Kamouraska et de l'Islet, Ministère des Ressources Naturelles, GR 98, 23 p.
1980: Faille du Bassin Nord-ouest et faille du Troisième Lac dans la partie est de la Gaspésie; Ministère de l'Énergie et des Ressources, Québec, DP-740, 20 p.

Bell, K. and Blenkinsop, J.
1975: Geochronology of eastern Newfoundland; Nature, v. 254, p. 410-411.
1977: Geochronological evidence of Hercynian activity in Newfoundland; Nature, v. 265, p. 616-618.
1981: A geochronological study of the Buchans area, Newfoundland; in The Buchans Orebodies: Fifty Years of Geology and Mining, (ed.) E.A. Swanson, D.F. Strong, and J.G. Thurlow; Geological Association of Canada, Special Paper 22, p. 91-111.

Bell, K., Blenkinsop, J., and Strong, D.F.
1977: The geochronology of some granite bodies from eastern Newfoundland and its bearing on Appalachian evolution; Canadian Journal of Earth Sciences, v. 14, p. 456-476.

Belt, E.S. and Bussières, L.
1981: Upper Middle Ordovician submarine fans and associated facies northeast of Québec City; Canadian Journal of Earth Sciences, v. 18, p. 981-994.

Belt, E.S., Riva, J., and Bussières, L.
1979: Revision and correlation of late Middle Ordovician stratigraphy northeast of Québec City; Canadian Journal of Earth Sciences, v. 16, p. 1467-1483.

Belyea, H.R.
1944: Plutonic rocks of the Musquash area, New Brunswick; Acadian Naturalist, v. 1, p. 87-102.

Benoit, F.
1958: Geology of the St. Sylvestre and St. Joseph West Half Area, Quebec; Ph.D. thesis, Laval University, Quebec, Quebec, 116 p.

Benus, A.P.
1988: Sedimentological context of a deep-water Ediacaran fauna (Mistaken Point Formation), Avalon Zone, eastern Newfoundland (abstract); in Bulletin No. 463 of the New York State Museum and Geological Survey, (ed.) Ed Landing, G.M. Narbonne, and P. Myrow, p. 8-9; Abstracts of papers presented at the Precambrian-Cambrian boundary working group meeting, Memorial University of Newfoundland, 1987.

Berger, J.
1985: Analyse structurale de la faille de Schickshock-Sud en Gaspésie occidentale, Québec; thèse de maîtrise, Université de Montréal, Montréal (Québec).

Berger, J. and Ramsay, E.
1990: Géologie de la région du Mont de la Serpentine, Gaspésie, Québec (abstract); in Program with abstracts of the Québec-Maine-New Brunswick Appalachian Workshop, (ed.) M. Malo, D. Lavoie, and D. Kirkwood; Geological Survey of Canada, Open File 2235, p. 35.

Bergström, S.M., Riva, J., and Kay, M.
1974: Significance of conodonts, graptolites, and shelly faunas from the Ordovician of western and north-central Newfoundland; Canadian Journal of Earth Sciences, v. 11, p. 1625-1660.

Bernard, D.
1987: Étude structurale et stratigraphique du synclinorium de Connecticut Valley-Gaspé dans le sud-est du Québec; thèse de maîtrise, Université Laval, Québec (Québec).

Bernstein, L.
1989: Stratigraphy and sedimentology of the Lower Ordovician Beekmantown Group, St. Lawrence Lowlands, Quebec and Ontario; Geological Association of Canada, Program with abstracts, v. 14, p. A28.
1991a: Depositional history of the Lower Ordovician Beekmantown Group, St. Lawrence Lowlands, Quebec and Ontario: sea level, tectonics, and the Sauk-Tippecanoe boundary; Geological Association of Canada, Program with abstracts, v. 16, p. A11.
1991b: The Lower Ordovician Beekmantown Group, Quebec and Ontario; Ph.D. thesis, Université de Montréal, Montréal, Québec, 330 p.

Berry, W.B.N. and Wilde, P.
1978: Progressive ventilation of the oceans: an explanation for the distribution of the Lower Paleozoic black shales; American Journal of Science, v. 278, p. 257-275.

Betz, F. Jr.
1939: Geology and mineral deposits of the Canada Bay area, northern Newfoundland; Newfoundland Geological Survey Bulletin, v. 16, 53 p.
1948: Geology and mineral deposits of southern White Bay; Newfoundland Geological Survey Bulletin, v. 24, 26 p.

Bevier, M.L.
1987: Pb isotopic ratios of Paleozoic granitoids from the Miramichi terrane, New Brunswick, and implications for the nature and age of the basement rocks; in Radiogenic Age and Isotopic Studies, Report 1; Geological Survey of Canada, Paper 87-2, p. 43-50.
1988: U-Pb geochronologic studies of igneous rocks in New Brunswick; in Thirteenth Annual Review of Activities, Project Résumés, (ed.) S.A. Abbott; New Brunswick Department of Natural Resources and Energy, Minerals and Energy Division, Information Circular 88-2, p. 134-140.

Bevier, M.L. and Barr, S.M.
1990: U-Pb age constraints on the stratigraphy and tectonic history of the Avalon Terrane, New Brunswick, Canada; Journal of Geology, v. 98, p. 53-63.

Bevier, M.L. and Whalen, J.B.
1990: Tectonic significance of Silurian magmatism in the Canadian Appalachians; Geology, v. 18, p. 411-414.

Bevier, M.L., White, C.E., and Barr, S.M.
1990: Late Precambrian U-Pb ages for the Brookville Gneiss, southern New Brunswick; Journal of Geology, v. 98, p. 955-965.

Billings, E.
1872: On some fossils from the primordial rocks of Newfoundland; Canadian Naturalist, v. 6, no. 4, p. 465-479.

Bird, J.M. and Dewey, J.F.
1970: Lithosphere plate-continental margin tectonics and the evolution of the Appalachian Orogen; Geological Society of America Bulletin, v. 81, p. 1031-1060.

Bird, P.
1978: Initiation of intracontinental subduction in the Himalaya; Journal of Geophysical Research, v. 83, p. 4975-4987.

279

Birkett, T.C.
1981: Metamorphism of a Cambro-Ordovician sequence in southeastern Quebec; Ph.D. thesis, University of Montreal, Montreal, Quebec, 268 p.

Biron, S.
1972: Géologie de la région de Ste-Anne-des-Monts; Ministère des Richesses naturelles , Québec, DP 243, 8 p.
1973: Géologie de la région de Marsoui; Ministère des Richesses naturelles, Québec, DP 243, 10 p.
1974: Géologie de la région des Méchins; Ministère des Richesses naturelles, Québec, DP 299, 15 p.

Blackwood, R.F.
1977: Geology of the Hare Bay area, northwestern Bonavista Bay; in Report of Activities; Department of Mines and Energy, Mineral Development Division, Report 77-1, p. 7-14
1978: Northeastern Gander Zone, Newfoundland; in Report of Activities for 1977, (ed.) R.V. Gibbons; Newfoundland Department of Mines and Energy, Mineral Development Division, Report 78-1, p. 72-79.
1982: Geology of the Gander Lake (2D/15) and Gander River (2E/2) area; Newfoundland Department of Mines and Energy, Mineral Development Division, Report 82-4, 56 p.
1985a: Geology of the Facheux Bay area (11P/9); Department of Mines and Energy, Mineral Development Division, Report 85-4, 56 p.
1985b: Geology of the Grey River Area, southwest coast of Newfoundland; in Current Research, (ed.) K. Brewer, D. Walsh, and R.V. Gibbons; Newfoundland Department of Mines and Energy, Mineral Development Division, Report 85-1, p. 153-164.

Blackwood, R.F. and Green, L.
1982: Geology of the Great Gull Lake (2D/6)-Dead Wolf Pond area (2D/10), Newfoundland; in Current Research, (ed.) C.F. O'Driscoll and R.V. Gibbons; Newfoundland Department of Mines and Energy, Mineral Development Division, Report 82-1, p. 51-64.

Blackwood, R.F. and Kennedy, M.J.
1975: The Dover Fault: western boundary of the Avalon Zone in northeastern Newfoundland; Canadian Journal of Earth Sciences, v. 12, p. 320-325.

Blackwood, R.F. and O'Driscoll, C.F.
1976: The Gander Avalon Zone boundary in southeastern Newfoundland; Canadian Journal of Earth Sciences, v. 13, p. 1155-1159.

Blais, D., Jebrac, M., and Tremblay, A.
1990: Petrography and geochemistry of the Bunker Hill volcanic sequence, Québec Appalachians (abstract); Geological Association of Canada, Program with Abstracts, v. 15, p. A11.

Blais, P.
1985: Étude de textures cataclastiques associées à la Faille de la Rivière Port-Daniel; Projet colloque, Université Laval, Québec (Québec) 40 p.

Bolton, T.E.
1968: Report on fossils from northern New Brunswick; Geological Survey of Canada, Internal Fossil Report 05-1-1968-TEB.

Bond, I.J. and Greggs, R.G.
1973: Revision of the March Formation (Tremadocian) in southeastern Ontario. Canadian Journal of Earth Sciences, v. 10, p. 1140-1155.
1976: Revision of the Oxford Formation (Arenig) of southeastern Ontario and New York State. Canadian Journal of Earth Sciences, v.13, p. 19-26.

Bond, G.C., Nickeson, P.A., and Kominz, M.A.
1984: Breakup of a supercontinent between 625 Ma and 555 Ma: new evidence and implications for continental histories; Earth and Planetary Science Letters, v. 70, p. 325-345.

Boone, G.M.
1985: Hurricane Mountain Formation mélange: history of a Cambro-Ordovician accretion of the Boundary Mountains terrane within the northern Appalachians orthotectonic zone (abstract); Geological Society of America, Abstracts with Programs, v. 21, p. 528.

Boone, G.M. and Boudette, E.L.
1989: Accretion of the Boundary Mountains terrane within the northern Appalachian orthotectonic zone; Geological Society of America, Special Paper 228, p. 17-42.

Bostock, H.H.
1983: Precambrian rocks of the Strait of Belle Isle area; in Geology of the Strait of Belle Isle area, northwestern insular Newfoundland, southern Labrador and adjacent Quebec; Geological Survey of Canada, Memoir 400, Part 1, p. 1-73.
1988: Geology and petrochemistry of the Roberts Arm Group, Notre Dame Bay, Newfoundland; Geological Survey of Canada Bulletin 369, 84 p.

Bostock, H.H., Currie, K.L., and Wanless, R.K.
1979: The age of the Roberts Arm Group, north-central Newfoundland; Canadian Journal of Earth Sciences, v. 16, p. 599-606.

Bosworth, W.
1985: East-directed imbrication and oblique-slip faulting in the Humber Arm Allochthon of western Newfoundland: structural and tectonic significance; Canadian Journal of Earth Sciences, v. 22, p. 1351-1360.

Botsford, J.W.
1987: Depositional history of Middle Cambrian to Lower Ordovician deep water sediments, Bay of Islands, western Newfoundland; Ph.D. thesis, Memorial University of Newfoundland, St. John's, Newfoundland, 534 p.

Boucot, A.J.
1973: The Lower Ordovician brachiopod Syntrophia sp., cf. S. arethusa (Billings 1962) from South Catcher Pond, northeastern Newfoundland; Canadian Journal of Earth Sciences, v. 10, p. 427-430.
1975: Evolution and extinction rate controls; in Developments in Paleontology and Stratigraphy l, Elsevier.

Boudette, E.L.
1982: Ophiolite assemblage of early Paleozoic age in central western Maine; in Major Structural Zones and Faults of the Northern Appalachians, (ed.) P. St-Julien and J. Béland; Geological Association of Canada, Special Paper 24, p. 209-230.

Bouma, A.H., Normark, W.R., and Barnes, N.E.
1985: Section VI: Mississippi Fan, DSDP Leg 96 seismic surveys and drilling results; in Submarine fans and related turbidite systems, (ed.) A.H. Bouma, W.R. Normark, and N.E. Barnes; Springer-Verlag, p. 247-340.

Bourque, P.A., Laurent, R., and St-Julien, P.
1985: Acadian wrench faulting in southern Gaspe Peninsula, Quebec Appalachians (abstract); Geological Association of Canada, Annual Meeting, Fredericton, Program with Abstracts, v. 10, p. A6.

Bourque, P.-A., Gosselin, C., Kirkwood, D., Malo, M., and St-Julien, P.
1993: Le Silurien du segment appalachien Gaspésie-Matapédia-Témiscouata, Québec: stratigraphie, géologie structurale et paléogéographie; Ministère de l'Énergie et des Ressources, Québec; MB 93-25.

Boyce, W.D.
1977: New Cambrian trilobites from western Newfoundland; B.Sc. thesis, Memorial University of Newfoundland, St. John's, Newfoundland, 66 p.
1979: Further developments in western Newfoundland Cambro-Ordovician biostratigraphy; in Report of Activities for 1978, (ed.) R.V. Gibbons; Mineral Development Division, Department of Mines and Energy, Government of Newfoundland and Labrador, Report 79-1, p. 7-10.
1983: Early Ordovician trilobite faunas of the Boat Harbour and Catoche Formations (St. George Group) in the Boat Harbour-Cape Norman area, Great Northern Peninsula, western Newfoundland; M.Sc. thesis, Memorial University of Newfoundland, St. John's, Newfoundland, 272 p.
1987: Cambrian-Ordovician trilobite biostratigraphy in central Newfoundland; in Current Research (1987); Newfoundland Department of Mines and Energy, Mineral Development Division, Report 87-1, p. 335-341.
1989: Early Ordovician trilobite faunas of the Boat Harbour and Catoche Formations (St. George Group) in the Boat Harbour-Cape Norman area, Great Northern Peninsula, western Newfoundland; Newfoundland Department of Mines and Energy Report 89-2, 169 p.

Boyce, W.D., Ash, J.S., O'Neill, P., and Knight, I.
1988: Ordovician biostratigraphic studies in the Central Mobile Belt and their implications for Newfoundland tectonics; in Current Research (1988); Newfoundland Department of Mines and Energy, Mineral Development Division, Report 88-1, p. 177-182.

Boyce, W.D., Knight, I., and Ash, J.S.
1992: The Weasel Group, Goose Arm area, western Newfoundland: Lithostratigraphy, biostratigraphy, correlation and implications; in Current Research; Newfoundland Department of Mines and Energy Report 92-1, p. 69-83.

Bradley, D.C.
1989: Taconic plate kinematics as revealed by foredeep stratigraphy, Appalachian Orogen; Tectonics, v. 8, p. 1037-1049.

Brand, U. and Rust, B.R.
1977: The age and upper boundary of the Nepean Formation in its type section near Ottawa; Canadian Journal of Earth Sciences, v. 14, p. 2002-2006.

Brown, P.A.

1976: Ophiolites in southwestern Newfoundland; Nature, v. 264, p. 712-715.

1977: Geology of the Port aux Basques map-area (110/10), Newfoundland; Newfoundland Department of Mines and Energy, Mineral Development Division, Report 77-2, 11 p.

Brown, P.A. and Colman-Sadd, S.P.

1976: Hermitage Flexure: figment or fact?; Geology, v. 4, p. 561-564.

Brueckner, W.D.

1972: The Gander Lake and Davidsville groups of northeastern Newfoundland: new data and geotectonic implications (discussion); Canadian Journal of Earth Sciences, v. 9, p. 1778-1779.

Buddington, A.F. and Hess, H.H.

1937: Layered peridotite laccoliths in the Trout River area, Newfoundland: comment; American Journal of Science, v. 33, p. 380-388.

Bursnall, J.T.

1975: Stratigraphy, structure and metamorphism west of Baie Verte, Burlington Peninsula, Newfoundland; Ph.D. thesis, Cambridge University, Cambridge, England, 337 p.

Bursnall, J.T. and de Wit, M.J.

1975: Timing and development of the orthotectonic zone in the Appalachian Orogen of northwest Newfoundland; Canadian Journal of Earth Sciences, v. 12, p. 1712-1722.

Bussières, L., Mehrtens, C.J., Belt, E.S., and Riva, J.

1977: Late Middle Ordovician shelf, slope and flysch facies between Baie-St-Paul and La Malbaie; in 69th Annual Meeting New England Intercollegiate Geological Congress, Laval University, Quebec, excursion B9, p. 1-26.

Calon, T.J. and Green, F.K.

1987: Preliminary results of a detailed structural analysis of the Buchans Mine area; in Buchans Geology, Newfoundland, (ed.) R.V. Kirkham; Geological Survey of Canada, Paper 86-24, Report 17, p. 273-288.

Caputo, M.V. and Crowell, J.C.

1985: Migration of glacial centers across Gondwana during Paleozoic Era; Geological Society of America Bulletin, v. 96, p. 1020-1036.

Caron, A.

1982: Région d'Arthabaska (SE), Québec; Ministère des Richesses naturelles, Québec, DP 83-07.

1984: Géologie de la région de Chandler, Gaspésie; Ministère de l'Énergie et des Ressources, Québec, ET 83-04, 25 p.

1986: Microstructural study of the Dover Fault, northeastern Newfoundland (abstract); Geological Association of Canada, Program with Abstracts, v. 11, 52 p.

Caron, A. and Williams, P.F.

1988: The multistage development of the Dover Fault in northeastern Newfoundland: the late stages (abstract); Geological Association of Canada, Program with Abstracts, v. 13, p. A-17.

Carrara, A.

1972: Structural geology of lower Paleozoic rocks, Mont Albert area, Gaspé Peninsula; Ph.D. thesis, University of Ottawa, Ottawa, Ontario, 206 p.

Casey, J.F. and Kidd, W.S.F.

1981: A parallochthonous group of sedimentary rocks unconformably overlying the Bay of Islands ophiolite complex, North Arm Mountain, Newfoundland; Canadian Journal of Earth Sciences, v. 18, p. 1035-1050.

Cawood, P.A.

1989: Acadian remobilization of a Taconian ophiolite, Hare Bay allochthon, northwestern Newfoundland; Geology, v. 17, p. 257-260.

Cawood, P.A. and Williams, H.

1986: Northern extremity of the Humber Arm Allochthon in the Portland Creek area, western Newfoundland, and relationships to nearby groups; in Current Research, Part A; Geological Survey of Canada, Paper 86-1A, p. 675-682.

1988: Acadian basement thrusting, crustal delamination, and structural styles in and around the Humber Arm Allochthon, western Newfoundland; Geology, v. 16, p. 370-373.

Cawood, P.A., Williams, H., and Grenier, R.

1987: Geology of Portland Creek area (12I/4), western Newfoundland; Geological Survey of Canada, Open File 1435, scale 1:50 000.

Cawood, P.A., Williams, H., O'Brien, S.J., and O'Neill, P.P.

1988: Geologic cross-section of the Appalachian Orogen; Geological Association of Canada, Field Trip Guidebook, 160 p.

Cecioni, G.

1981: Paleozoic varve-like sediments in the Patagonia Archipelago, Chile; in Earth's Pre-Pleistocene Glacial Record, (ed.) M.J. Hambrey and W.B. Harland; Cambridge University Press, Cambridge, U.K., 912 p.

Chandler, F.W.

1982: Sedimentology of two Middle Paleozoic terrestrial sequences, King George IV Lake area, Newfoundland, and some comments on regional paleoclimate; in Current Research, Part A; Geological Survey of Canada, Paper 82-1A, p. 213-219.

Chandler, F.W. and Dunning, G.R.

1983: Fourfold significance of an Early Silurian U-Pb zircon age from rhyolite in redbeds, southwest Newfoundland; in Current Research, Part B; Geological Survey of Canada, Paper 83-1B, p. 419-421.

Charbonneau, J-M.

1975: Analyse structurale des tectonites métamorphiques du Groupe d'Oak Hill dans la région de St-Sylvestre, Appalaches du Quebec; thèse de maîtrise, Université Laval, Québec (Québec), 65 p.

1980a: Le front Appalachien au nord immediat de la frontière Québec-Vermont; thèse de doctorat, Université Laval, Québec (Québec), 193 p.

1980b: Region de Sutton (W); Ministère de l'Énergie et des Ressources, Québec, DPV-681, 89 p.

Charbonneau, J-M. and St-Julien, P.

1981: Analyse structurale et relations déformation-métamorphisme, Groupe d'Oak Hill, region du mont Sainte-Marguerite, Appalaches du Québec; Journal canadien des sciences de la Terre, v. 18, no. 6, p. 1051-1064.

Chevé, S.

1978: Région du sud-est des Cantons de l'Est; Ministère des Richesses naturelles, Québec; DP-613, 80 p.

1990: Cadre géotectonique de la région de Lac Mégantic, Québec: implications métallogéniques; thèse de doctorat, Université de Montréal, Montréal (Québec).

Chorlton, L.B.

1978: The geology of the La Poile map area (110/9), Newfoundland; Newfoundland Department of Mines and Energy, Mineral Development Division, Report 78-5, 14 p.

1980: The geology of Peter Snout (11P/13 west half), Newfoundland; Newfoundland Department of Mines and Energy, Mineral Development Division, Map 80-201 with marginal notes.

1983: Geology of the Grandy Lake area (110/15), Newfoundland; Newfoundland Department of Mines and Energy, Mineral Development Division, Report 83-7, Part I, p. 1-116.

1984: Geological development of the southern Long Range Mountains, Newfoundland: a regional synthesis; Ph.D. thesis, Memorial University of Newfoundland, St. John's, Newfoundland, 579 p.

Chow, N.

1986: Sedimentology and diagenesis of Middle and Upper Cambrian platform carbonates and siliciclastics, Port au Port Peninsula, western Newfoundland; Ph.D. thesis, Memorial University of Newfoundland, St. John's, Newfoundland, 458 p.

Chow, N. and James N.P.

1987: Facies specific calcite and bimineralic ooids from Middle and Upper Cambrian platformal carbonates, western Newfoundland, Canada; Journal of Sedimentary Petrology, v. 57, p. 907-921.

Chumakov, N.M. and Elston, D.P.

1989: The paradox of late Proterozoic glaciations at low latitudes; Episodes, v. 12, p. 115-119.

Church, W.R.

1969: Metamorphic rocks of Burlington Peninsula and adjoining areas of Newfoundland and their bearing on continental drift in North Atlantic; in North Atlantic Geology and Continental Drift, (ed.) M. Kay; American Association of Petroleum Geologists, Memoir 12, p. 212-233.

1977: The ophiolites of southern Québec: oceanic crust of Betts Cove type; Canadian Journal of Earth Sciences, v. 14, p. 1668-1673.

Clague, D., Rubin, J., and Brackett, R.

1981: The age and origin of the garnet amphibolite underlying the Thetford-Mines ophiolite, Québec; Canadian Journal of Earth Sciences, v. 18, p. 1257-1261.

Clark, T.H.

1934: Structure and stratigraphy of southern Quebec; Geological Society of America Bulletin, v. 45, p. 1-20.

1936: A Lower Cambrian series from southern Quebec; Royal Canadian Institute, Transactions, v. 21, part I, p. 135-151.

1952: Montréal area; Quebec Department of Mines, GR 46, 159 p.

1972: Région de Montréal; Montréal area; Ministère des Richesses Naturelles, Québec, RG 152, 244 p.

1964: Région de Yamaska-Aston; Ministère des Ressources naturelles, Québec, RG 102, 192 p.

1977: Région de Granby (W); Ministère des Richesses naturelles, Québec, RG 177, 109 p.

Clark, T.H. and Eakins, P.R.
1968:   The stratigraphy and structure of the Sutton area of Southern Quebec; in Tectonics of the Appalachians, Northern and Maritime, (ed.) E-an Zen, W.S. White, J.B. Hadley, and J.B. Thompson, Jr.; Wiley Interscience, New York, p. 163-173.

Clark T.H. and Globensky, Y.
1973:   Région de Portneuf et parties de St-Raymond et de Lyster; Ministère des Richesses naturelles, Québec, RG 148, 101 p.

1975:   Région de Grondines; Ministère des Richesses naturelles, Québec, RG 154, 159 p.

1976a: Région de Sorel; Ministère des Richesses naturelles, Québec, RG 155, 151 p.

1976b: Région de Laurentides; Ministère des Richesses naturelles, Québec, RG 157, 112 p.

1976c: Région de Trois-Rivières; Ministère des Richesses naturelles, Québec, RG 164, 87 p.

1976d: Région de Bécancour; Ministère des Richesses naturelles, Québec, RG 165, 66 p.

1977:   Région de Verchères; Ministère des Richesses naturelles, Québec, RG 190, 64 p.

Clark, T.H., Globensky, Y., Riva, J., and Hofmann, H.J.
1972:   Stratigraphy and structure of the St.Lawrence Lowland of Quebec; 24th International Geological Congress, Montreal, Canada, Fieldtrip Guidebook C-52, 82 p.

Clarke, D.B. and Halliday, A.N.
1985:   Sm/Nd isotopic investigation of the age and origin of the Meguma Zone metasedimentary rocks; Canadian Journal of Earth Sciences, v. 22, p. 102-107.

Clarke, D.B., Barr, S.M., and Donahoe, H.V.
1980:   Granitoid and other plutonic rocks of Nova Scotia; Canadian Journal of Earth Science, v. 14, p. 2858-2864.

Clauer, N. and Deynoux, M.
1987:   New information on the probable isotopic age of the Late Proterozoic glaciation in West Africa; Precambrian Research, v. 37, p. 89-94.

Clifford, P.M.
1965:   Palaeozoic flood basalts in northern Newfoundland and Labrador; Canadian Journal of Earth Sciences, v. 2, p. 183-187.

Clifford, P.M. and Baird, D.M.
1962:   Great Northern Peninsula of Newfoundland, Grenville inlier; Canadian Mining and Metallurgical Bulletin, v. 65, p. 95-102.

Cloetingh, S.
1988:   Intraplate stresses: a tectonic cause for third-order cycles in apparent sea level; in Sea-level Change: an Integrated Approach, (ed.) C.K. Wilgus, B.S. Hastings, C.G. Kendall, H.W. St. C. Posamentier, C.A. Ross, and J.C. Van Wagoner; Society of Economic Paleontologists and Mineralogists, Special Publication 42, p. 19-29.

Cochrane, N.A. and Wright, J.A.
1977:   Geomagnetic sounding near the northern termination of the Appalachian System; Canadian Journal of Earth Sciences, v. 14, p. 2858-2864.

Colman-Sadd, S.P.
1969:   Geology of the iron deposits near Stephenville, Newfoundland; M.Sc. thesis, Memorial University of Newfoundland, St. John's, Newfoundland, 97 p.

1976:   Geology of the St. Albans map area, Newfoundland (1M/13); Newfoundland Department of Mines and Energy, Mineral Development Division, Report 76-4, 19 p.

1980:   Geology of south central Newfoundland and evolution of the eastern margin of Iapetus; American Journal of Science, v. 280, p. 991-1017.

1984:   Geology of the Cold Spring Pond map area (west part) 12A/1, Newfoundland; in Current Research; Newfoundland Department of Mines and Energy, Mineral Development Division, Report 84-1, p. 211-219.

1985:   Geology of the west part of Great Burnt Lake (12A/8) area; in Current Research; Newfoundland Department of Mines and Energy, Mineral Development Division, Report 85-1, p. 105-113.

1987:   Geology of part of the Snowshoe Pond (12A/7) map area; in Current Research; Newfoundland Department of Mines and Energy, Mineral Development Division, Report 87-1, p. 297-310.

1988:   Geology of the Snowshoe Pond (12A/7) map area; in Current Research; Newfoundland Department of Mines, Mineral Development Division, Report 88-1, p. 127-134.

1989:   Miguels Lake area (2D/12): An update of the geology; in Current Research; Newfoundland Department of Mines, Geological Survey of Newfoundland, Report 89-1, p. 47-53.

Colman-Sadd, S.P. and Russell, H.A.J.
1982:   Geology of the Miguels Lake map area (2D/12), Newfoundland; in Current Research, (ed.) C.F. O'Driscoll and R.V. Gibbons; Newfoundland Department of Mines and Energy, Mineral Development Division, Report 82-1, p. 30-50.

Colman-Sadd, S.P. and Swinden, H.S.
1982:   Geology and mineral potential of south-central Newfoundland; Newfoundland Department of Mines and Energy, Mineral Development Division, Report 82-8, 102 p.

1984:   A tectonic window in central Newfoundland? Geological evidence that the Appalachian Dunnage Zone is allochthonous; Canadian Journal of Earth Sciences, v. 21, p. 1349-1367.

1988:   Cold Spring Pond, Newfoundland; Newfoundland Department of Mines and Energy, Geological Survey Branch, Map 89-107.

Colman-Sadd, S.P., Dunning, G.R., and Dec, T.
1992:   Dunnage-Gander relationships and Ordovician orogeny in central Newfoundland: a sediment provenance and U/Pb age study; American Journal of Science, v. 292, p. 317-355.

Colman-Sadd, S.P., Hayes, J.P., and Knight, I.
1990:   Geology of the Island of Newfoundland; Newfoundland Department of Mines and Energy, Geological Survey Branch, Map 90-01, scale 1:1 000 000.

Colpron, M.
1990:   Rift and collisional tectonics of the Eastern Townships Humber zone, Brome lake area, Quebec; M.Sc. thesis, University of Vermont, Burlington, U.S.A., 278 p.

Comeau, R.L.
1972:   Transported slices of the coastal complex, Bay of Islands, western Newfoundland; M.Sc. thesis, Memorial University of Newfoundland, St. John's, Newfoundland, 105 p.

Coniglio, M.
1986:   Synsedimentary submarine slope failure and tectonic deformation in deep-water carbonates, Cow Head Group, western Newfoundland; Canadian Journal of Earth Sciences, v. 23, p. 476-490.

Conrad, J.A.
1984:   Shelf sedimentation above storm wave base in Upper Ordovician Reedsville Formation in central Pennsylvania (abstract); American Association of Petroleum Geologists Bulletin, v. 68, p. 1917.

Cooke, H.C.,
1937:   Thetford, Disraëli, and Eastern Half of Warwick map area, Quebec; Geological Survey of Canada, Memoir 211, 160 p.

1950:   Geology of a southwestern part of the Eastern Townships of Quebec; Geological Survey of Canada, Memoir 257, 142 p.

1952:   Geology of parts of Richmond and Drummondville map-area, Eastern Townships, Quebec; Ministère des Richesses naturelles, Québec, DP 467, 45 p.

Cooper, J.R.
1936:   Geology of the southern half of the Bay of Islands Igneous Complex; Newfoundland Department of Natural Resources, Geological Section, Bulletin, no. 4, 62 p.

1937:   Geology and mineral deposits of the Hare Bay area; Newfoundland Department of Natural Resources, Geological Section, Bulletin, No. 9, 36 p.

1954:   La Poile-Cinq Cerf map area, Newfoundland; Geological Survey of Canada, Memoir 276, 62 p.

Cormier, R.F.
1972:   Radiometric ages of granitic rocks, Cape Breton Island, Nova Scotia; Canadian Journal of Earth Sciences, v. 9, p. 1074-1086.

1979:   Rubidium/strontium ages of Nova Scotian granitoid plutons; Nova Scotia Department of Mines and Energy, Report 79-1, p. 143-147.

1980:   New rubidium/strontium ages in Nova Scotia; Nova Scotia Department of Mines and Energy Report 80-1, p. 223-234.

Cousineau, P.A.
1990:   Le Groupe de Caldwell et le domaine océanique entre St-Joseph-de-Beauce et Sainte-Sabine; Ministère de l'Énergie et des Ressources, Québec, MM 87-02, 165 p.

1991:   The Rivière des Plantes ophiolitic mélange: tectonic setting and mélange formation in the Québec Appalachians; Journal of Geology, v. 99, p. 81-96.

Cousineau, P.A. and St-Julien, P.
1986:   The St-Daniel Formation: a mélange along the Taconic suture in southern Québec, Canada (abstract); Geological Society of America, Northeastern section annual meeting, Abstracts with Programs, v. 9, p. A12.

1992:   The Saint-Daniel melange: evolution of an accretionary complex in the Dunnage terrane of the Quebec Appalachians; Tectonics, v. 11, p. 898-909.

Cousineau, P.A. and St-Julien, P. (cont.)

1994: Stratigraphie et paléogéographie d'un bassin d'avant-arc ordovicien, Estrie-Beauce, Appalaches du Québec; revue canadienne des sciences de la Terre, v. 31, p. 435-446.

Cousineau, P.A. and Tremblay, A.

1993: Acadian deformations in the southwestern Québec Appalachians; in The Acadian Orogeny: Recent studies in New England, Maritime Canada, and the Autochthonous Foreland, (ed.) D.C. Roy and J.W. Skehan; Geological Society of America, Special Paper 275, p. 85-99.

Cowan, D.S.

1985: Structural styles in Mesozoic and Cenozoic mélanges in the western Cordillera of North America; Geological Society of America Bulletin, v. 96, p. 451-461.

Craw, D.

1983: Structural evolution of a section along the base of the St. Anthony Complex, northwest Newfoundland; Canadian Journal of Earth Sciences, v. 20, p. 1713-1724.

1984: Tectonic stacking of metamorphic zones in the Cheticamp River area, Cape Breton Highlands, Nova Scotia; Canadian Journal of Earth Sciences, v. 21, p. 1229-1244.

Crickmay, G.W.

1932: Evidence of Taconic Orogeny in the Matapedia-Valley, Québec; American Journal of Science, v. 24, p. 368-386.

Crosby, D.G.

1962: Wolfville map-area, Nova Scotia; Geological Survey of Canada, Memoir 325, 67 p.

Crouse, G.W.

1981: Geology of parts of Burnthill, Clearwater and McKiel brooks, map-areas K-14, K-15, K-16 (Parts of 21J/15 and 21J/10); New Brunswick Department of Natural Resources, Mineral Resources Branch, Map Report 81-5, 46 p.

Cumming, L.M.

1967: Geology of the Passamaquoddy Bay region, Charlotte County, New Brunswick; Geological Survey of Canada, Paper 65-29, 36 p.

1983: Lower Paleozoic autochthonous strata of the Strait of Belle Isle area; in Geology of the Strait of Belle Isle area, northwestern insular Newfoundland, southern Labrador and adjacent Quebec; Geological Survey of Canada, Memoir 400, Part 2, p. 75-108.

1985: A Halifax Slate graptolite locality, Nova Scotia; in Current Research, Part A; Geological Survey of Canada, Paper 85-1A, p. 215-221.

Currie, K.L.

1982: Paleozoic supracrustal rocks near Cheticamp, Nova Scotia; Maritime Sediments and Atlantic Geology, v. 18, p. 94-103.

1983: Repeated basement reactivation in the northeast Appalachians; Geological Journal, v. 18, p. 223-239.

1984: A reconsideration of geological relations near Saint John, New Brunswick; in Current Research, Part A; Geological Survey of Canada, Paper 84-1A, p. 193-201.

1986: The boundaries of the Avalon tectonostratigraphic zone, Musquash Harbour-Loch Alva region, southern New Brunswick; in Current Research, Part A; Geological Survey of Canada, Paper 86-1A, p. 33-341.

1987a: A preliminary account of the geology of Harrys River map area, southern Long Range of Newfoundland; in Current Research, Part A; Geological Survey of Canada, Paper 87-1A, p. 653-662.

1987b: Relations between metamorphism and magmatism near Chéticamp, Cape Breton Island, Nova Scotia; Geological Survey of Canada, Paper 85-23.

1987c: Contrasting metamorphic terranes near Cheticamp, Cape Breton Highlands, Nova Scotia; Canadian Journal of Earth Sciences, v. 24, p. 2422-2435.

1987d: Late Precambrian igneous activity and its tectonic implications, Musquash-Loch Alva region, southern New Brunswick; in Current Research, Part A; Geological Survey of Canada, Paper 87-1A, p. 663-671.

1988a: The western end of the Avalon Zone in southern New Brunswick; Maritime Sediments and Atlantic Geology, v. 24, p. 339-352.

1988b: Saint George map area: the end of the Avalon Zone, southern New Brunswick; in Current Research, Part B; Geological Survey of Canada, Paper 88-1B, p. 9-16.

1991: A note on the stratigraphy of the Martinon Formation, Saint John, New Brunswick; in Current Research, Part D; Geological Survey of Canada, Paper 91-1D, p. 9-13.

Currie, K.L. and Pajari, G.E.

1977: Igneous and metamorphic rocks between Rocky Bay and Ragged Harbour, northeastern Newfoundland; in Report of Activities, Part A; Geological Survey of Canada, Paper 77-1A, p. 341-346.

Currie, K.L., Loveridge, W.D., and Sullivan, R.W.

1982: A U-Pb age on zircon from dykes feeding basal rhyolitic flows of the Jumping Brook Complex, northwestern Cape Breton Island, Nova Scotia; in Current Research, Part C; Geological Survey of Canada, Paper 82-1C, p. 125-128.

Currie, K.L., Nance, R.D., Pajari, G.E., and Pickerill, R.K.

1981: Some aspects of the pre-Carboniferous geology of Saint John, New Brunswick; in Current Research, Part A; Geological Survey of Canada, Paper 81-1A, p. 23-30.

Currie, K.L., Pajari, G.E., and Pickerill, R.K.

1979: Tectonostratigraphic problems in the Carmanville area, northeastern Newfoundland; in Current Research, Part A; Geological Survey of Canada, Paper 79-1A, p. 71-76.

Currie, K.L., van Breemen, O., Hunt, P.A., and van Berkel, J.T.

1992: The age of high-grade gneisses south of Grand Lake, Newfoundland; Atlantic Geology, v. 28, no. 2, p. 153-161.

Dallmeyer, R.D.

1977a: ^{40}Ar/^{39}Ar age spectra of minerals from the Fleur de Lys terrane in northwest Newfoundland: their bearing on chronology of metamorphism within the Appalachian orthotectonic zone; Journal of Geology, v. 85, p. 89-103.

1977b: Diachronous ophiolite obduction in western Newfoundland: evidence from ^{40}Ar/^{39}Ar ages of the Hare Bay metamorphic aureole; American Journal of Science, v. 277, p. 61-72.

1978: ^{40}Ar/^{39}Ar incremental-release ages of hornblende and biotite from Grenville basement rocks within the Indian Head Range Complex, southwest Newfoundland: their bearing on late Proterozoic-early Paleozoic thermal history; Canadian Journal of Earth Sciences, v. 15, p. 1374-1379.

Dallmeyer, R.D. and Hibbard, J.P.

1984: Geochronology of the Baie Verte Peninsula, Newfoundland: implications for the tectonic evolution of the Humber and Dunnage zones of the Appalachian Orogen; Journal of Geology, v. 92, p. 489-512.

Dallmeyer, R.D. and Nance, R.D.

1990: ^{40}Ar/^{39}Ar ages of detrital muscovite within early Paleozoic overstep sequences, Avalon composite terrane, southern New Brunswick: implications for extent of late Paleozoic tectonothermal overprint; Canadian Journal of Earth Sciences, v. 27, p. 1209-1214.

Dallmeyer, R.D. and Williams, H.

1975: ^{40}Ar/^{39}Ar ages from the Bay of Islands metamorphic aureole: their bearing on the timing of Ordovician ophiolite obduction; Canadian Journal of Earth Sciences, v. 12, p. 1685-1690.

Dallmeyer, R.D., Blackwood, R.F., and Odom, A.L.

1981a: Age and origin of the Dover Fault: tectonic boundary between the Gander and Avalon zones of the northeastern Newfoundland Appalachians; Canadian Journal of Earth Sciences, v. 18, p. 1431-1442.

Dallmeyer, R.D., Doig, R., Nance, R.D., and Murphy, J.B.

1990: ^{40}Ar/^{39}Ar and U-Pb mineral ages from the Brookville Gneiss: Implications for terrane analysis and evolution of Avalonian "basement" in southern New Brunswick; Atlantic Geology, v. 26, p. 247-257.

Dallmeyer, R.D., Hussey, E.M., O'Brien, S.J., and O'Driscoll, C.F.

1983b: Chronology of tectonothermal activity in the western Avalon Zone of the Newfoundland Appalachians; Canadian Journal of Earth Sciences, v. 20, p. 355-363.

Dallmeyer, R.D., Kean, B.F., Odom, A.L., and Jayasinghe, N.R.

1983a: Age and contact metamorphic effects of the Overflow Pond Granite: an undeformed pluton in the Dunnage Zone of the Newfoundland Appalachians; Canadian Journal of Earth Sciences, v. 20, p. 1639-1645.

Dallmeyer, R.D., Odom, A.L., O'Driscoll, C.F., and Hussey, E.M.

1981b: Geochronology of the Swift Current Granite and host volcanic rocks of the Love Cove Group, southwestern Avalon Zone, Newfoundland: evidence for a late Proterozoic volcanic-subvolcanic association; Canadian Journal of Earth Sciences, v. 18, p. 699-707.

David, J. and Gariépy, C.

1990: Early Silurian orogenic andesites from the central Quebec Appalachians; Canadian Journal of Earth Sciences, v. 27, p. 632-643.

David, J., Chabot, N., Marcotte, C., Lajoie, J., and Lespérance, P.J.

1985: Stratigraphy and sedimentology of the Cabano, Pointe aux Trembles, and Lac Raymond Formations, Témiscouata and Rimouski Counties, Québec; in Current Research, Part B; Geological Survey of Canada, Paper 85-1B, p 481-497.

David, J., Gariépy, C., and Machado, N.
1991:   U-Pb geochronology of detrital zircons from the Tetagouche Group, New Brunswick (abstract); Geological Association of Canada, Program with Abstracts, v. 16, p. A 28.

David, J., Marquis, R., and Tremblay, A.
1993:   U-Pb geochronology of the Dunnage Zone in the southwestern Quebec Appalachians (abstract); Geological Society of America Annual Meeting, Boston, Abstracts with programs, p. A-485.

Davies, J.L.
1966:   Geology of Bathurst-Newcastle area, New Brunswick; in Guidebook: Geology of parts of Atlantic Provinces, (ed.) W.H. Poole; Annual Meeting, Geological Association of Canada, Halifax, p. 33-43.
1972a: The geology and geochemistry of the Austin Brook area, Gloucester County, New Brunswick, with special emphasis on the Austin Brook iron formation; Ph.D. thesis, Carleton University, Ottawa, Ontario, 254 p.
1972b: The Bathurst-Newcastle area; 24th International Geological Congress, Field Guide A-58, p. 50-58.
1979:   Geological map of northern New Brunswick; New Brunswick Department of Natural Resources, Mineral Resources Branch, Map NR-3.

Davies, J.L., Fyffe, L.R., and McAllister, A.L.
1983:   Geology and massive sulphides of the Bathurst area, New Brunswick; in Field Trip Guidebook to Stratabound Sulphide Deposits, Bathurst Area, New Brunswick, Canada and West-Central New England, U.S.A., (ed.) D.E. Sangster; IGCP-CCSS Symposium, Ottawa, Canada, Miscellaneous Report 36, p. 1-30.

Davies, J.L., Tupper, W.M., Bachinski, D., Boyle, R.W., and Martin, R.
1969:   Geology and mineral deposits of the Nigadoo River-Millstream River area, Gloucester County, New Brunswick; Geological Survey of Canada, Paper 67-49, 70 p.

Dawson, J.W.
1868:   Acadian geology (second edition); MacMillan and Company, London, 694 p.

De Broucker, G.
1983:   Géologie des groupes de Mictaw et de Maquereau, région de Port-Daniel (Gaspésie); Ministère des Richesses naturelles, Québec, MB 84-01, 56 p.
1984:   Groupes de Mictaw et de Maquereau, région de Port-Daniel; Ministère de l'Énergie et des Ressources, Québec, DP 84-09.
1986:   Évolution tectonostratigraphique de la boutonnière de Maquereau-Mictaw (Cambro-Ordovicien) Gaspésie, Québec; thèse de doctorat, Université Laval, Québec (Québec), 322 p.
1987:   Stratigraphie, pétrographie et structure de la boutonnière de Maquereau-Mictaw (Région de Port-Daniel, Gaspésie); Ministère de l'Énergie et des Ressources, Québec, MM 86-03, 160 p.

De Broucker, G., St-Julien, P., Krogh, T.E., and Machado, N.
1986:   Geochronology and tectonic implications of basement rocks in the Cambro-Ordovician Macquereau and Mictaw Inlier, Gaspé Peninsula, Quebec (abstract); Geological Society of America, Abstracts with Programs, v. 22, p. 13.

De Römer, H.S.
1960:   Geology of the Eastman-Orford Lake area, Eastern Townships, Province of Quebec; Ph.D. thesis, McGill University, Montreal, Quebec, 397 p.
1977:   Région des monts McGerrigle; Ministère des Richesses naturelles, Québec, 174.
1980:   Région de Baie Fitch-Lac Massawippi; Ministère de l'Énergie et des Ressources, Québec; RG 196, 59 p.
1985:   Géologie des monts Stoke; Ministère de l'Énergie et des Ressources, Québec; MM 85-03, 57 p.

De Roo, J.A. and van Staal, C.R.
1991:   The structure of the Half Mile Lake region, Bathurst Camp, New Brunswick; in Current Research, Part D; Geological Survey of Canada, Paper 91-1D, p. 179-186.

De Roo, J.A., Moreton, C., Williams, P.F., and van Staal, C.R.
1990:   The structure of the Heath Steele Mines region, Bathurst Camp, New Brunswick; Atlantic Geology, v. 26, p. 27-41.

de Wit, M.J.
1972:   The geology around Bear Cove, eastern White Bay, Newfoundland; Ph.D. thesis, Cambridge University, Cambridge, England, 232 p.
1974:   On the origin and deformation of the Fleur de Lys metaconglomerate, Appalachian fold belt, northwest Newfoundland; Canadian Journal of Earth Sciences, v. 11, p. 1168-1180.

de Wit, M.J. (cont.)
1980:   Structural and metamorphic relationships of pre-Fleur de Lys and Fleur de Lys rocks of the Baie Verte Peninsula, Newfoundland; Canadian Journal of Earth Sciences, v. 17, p. 1559-1575.

de Wit, M.J. and Strong, D.F.
1975:   Eclogite-bearing amphibolites from the Appalachian mobile belt, northwest Newfoundland: dry versus wet metamorphism; Journal of Geology, v. 83, p. 609-627.

Dean, P.L.
1978:   The volcanic stratigraphy and metallogeny of Notre Dame Bay Newfoundland; Memorial University of Newfoundland, Geology Report 7, 205 p.

Dean, P.L. and Kean, B.F.
1980:   The age of the Roberts Arm Group, north-central Newfoundland: discussion; Canadian Journal of Earth Sciences, v. 17, p. 800-804.

Dean, P.L. and Strong, D.F.
1977:   Folded thrust faults in Notre Dame Bay, central Newfoundland; American Journal of Science, v. 277, p. 97-108.

Dean, W.T.
1970:   Lower Ordovician trilobites from the vicinity of South Catchers Pond, northeastern Newfoundland; Geological Survey of Canada, Paper 70-44, 15 p.
1975:   Report on sample from GSC locality 91889; Geological Survey of Canada, Internal Fossil Report 0-S-2-1975-WTD-RBR.

Debrenne, F. and James, N.P.
1981:   Reef associated archaeocyathans of the Forteau Formation, southern Labrador; Palaeontology, v. 24, p. 343-378.

Dec, T. and Colman-Sadd, S.P.
1990:   Timing of ophiolite emplacement onto the Gander Zone: evidence from provenance studies in the Mount Cormack Subzone; in Current Research (1990); Newfoundland Department of Mines and Energy, Geological Survey Branch, Report 90-1, p. 289-303.

DeGrace, J.R., Kean, B.F., Hsu, E., and Green, T.
1976:   Geology of the Nippers Harbour map area (2 E/13), Newfoundland; Newfoundland Department of Mines and Energy, Mineral Development Division, Report 76-3, 73 p.

Desrochers, A.
1985:   The Lower and Middle Ordovician platform carbonates of the Mingan Islands, Quebec: stratigraphy, paleokarst and limestone diagenesis; Ph.D. thesis, Memorial University of Newfoundland, St. John's, Newfoundland, 342 p.

Desrochers, A. and James, N.P.
1988:   Early Paleozoic surface and subsurface paleokarst: Middle Ordovician carbonates: Mingan Islands, Quebec: in Paleokarst, (ed.) N.P. James and P.W. Choquette; Springer-Verlag, New York.

Dewey, J.F.
1969:   The evolution of the Caledonian/Appalachian Orogen; Nature, v. 222, p. 124-128.

Dewey, J.F. and Bird, J.M.
1971:   Origin and emplacement of the ophiolite suite: Appalachian ophiolites in Newfoundland; Journal of Geophysical Research, v. 76, p. 3179-3206.

Deynoux, M.
1980:   Les formations glaciaires du Précambrien terminal et de la fin de l'Ordovicien en Afrique de l'Ouest: Deux exemples de glaciation d'inlandsis sur une plate-forme stable; Travaux des laboratoires des sciences de la terre, Saint-Jérome, Marseille, B 17, 554 p.
1985:   Terrestrial or waterlain glacial diamictites? Three case studies from the late Precambrian and Late Ordovician glacial drifts in West Africa; Palaeogeography, Palaeoclimatology, and Palaeoecology, v. 51, p. 97-141.

Dickinson, W.R. and Suczek, C.A.
1979:   Plate tectonics and sandstone compositions; American Association of Petroleum Geologists Bulletin, v. 63, p. 2164-2182.

Dickson, W.L.
1988:   Geology and mineralization in the Hungry Grove Pond (1M/14) map area, southeastern Newfoundland; in Current Research; Newfoundland Department of Mines, Mineral Development Division, Report 88-1, p. 145-164.

Doig, R. and Barton, J.M., Jr.
1968:   Ages of carbonatites and other alkaline rocks in Quebec; Canadian Journal of Earth Sciences, v. 5, p. 1401-1407.

Doig, R., Murphy, J.B., and Nance, R.D.
1989:   Preliminary results of U-Pb geochronology, Cobequid Highlands, Avalon Terrane, Nova Scotia (abstract); Geological Association of Canada, Program with Abstracts, v. 14, p. A126.
1991:   U-Pb geochronology of late Proterozoic rocks of the eastern Cobequid Highlands, Avalon Composite Terrane, Nova Scotia; Canadian Journal of Earth Sciences, v. 28, p. 504-511.

Doig, R., Nance, R.D., Murphy, J.B., and Cassedy, P.
1990: Evidence for Silurian sinistral accretion of Avalon Composite Terrane in Canada; Geological Society of London, Journal, v. 147, p. 927-930.

Doll, C.G., Cady, W.M., Thompson, J.B., Jr., and Billings, M.P.
1961: Centennial geologic map of Vermont; Vermont Geological Survey; map scale 1:250 000.

Donohoe, H.V.
1976: The Cobequid Mountains Project; Nova Scotia Department of Mines, Report 76-2, p. 113-124.

Donohoe, H.V. and Cullen, M.
1983: Deformation, age and regional correlation of the Mt. Thom and Bass River complexes, Cobequid Highlands, Nova Scotia (abstract); Geological Society of America, Abstracts with Programs, v. 15, no. 3.

Donohoe, H.V. and Wallace, P.I.
1978: Anatomy of the Cobequid Fault and its relations to the Glooscap and Cabot fault systems (abstract); Geological Association of Canada/Geological Society of America, Abstracts with Programs, v. 3, p. 391.
1982: Geological map of the Cobequid Highlands, Colchester, Cumberland and Pictou Counties, Nova Scotia; Nova Scotia Department of Mines and Energy, Halifax, Nova Scotia, Maps 82-6, 82-7, 82-8 and 82-9.

Donohoe, H.V., Halliday, A.N., and Keppie, J.D.
1986: Two Rb-Sr whole rock isochrons from plutons in the Cobequid Highlands, Nova Scotia, Canada; Maritime Sediments and Atlantic Geology, v. 22, p. 148-155.

Doolan, B.L., Gale, M.H., Gale, P.N., and Hoar, R.S.
1982: Geology of the Québec re-entrant: possible constraints from early rifts and the Vermont-Québec serpentine belt; in Major Structural Zones and Faults of the Northern Appalachians, (ed.) P. St-Julien and J. Béland; Geological Association of Canada, Special Paper 24, p. 87-116.

Dostal, J.
1989: Geochemistry of Ordovician volcanic rocks of the Tetagouche Group of southwestern New Brunswick; Atlantic Geology, v. 25, p. 199-209.

Dostal, J. and McCutcheon, S.R.
1990: Geochemistry of late Proterozoic basaltic rocks from southeastern New Brunswick, Canada; Precambrian Research, v. 47, p. 83-98.

Dostal, J., Keppie, J.D., and Murphy, J.B.
1990: Geochemistry of late Proterozoic basaltic rocks from southeastern Cape Breton Island, Nova Scotia; Canadian Journal of Earth Sciences, v. 23, p. 619-631.

Dresser, J.A. and Denis, T.
1944: Gaspé Peninsula: Geology of Quebec; Quebec Department of Mines, GR 20, 302 p.

Dunning, G.R.
1987a: U-Pb geochronology of the Coney Head Complex, Newfoundland; Canadian Journal of Earth Sciences, v. 24, p. 1072-1075.
1987b: Geology of the Annieopsquotch Complex, southwest Newfoundland; Canadian Journal of Earth Sciences, v. 24, p. 1162-1174.

Dunning, G.R. and Chorlton, L.B.
1985: The Annieopsquotch ophiolite belt of southwest Newfoundland: geology and tectonic significance; Geological Society of America Bulletin, v. 96, p. 1466-1476.

Dunning, G.R. and Herd, R.K.
1980: The Annieopsquotch ophiolite complex, southwest Newfoundland and its regional relationships; in Current Research, Part A; Geological Survey of Canada, Paper 80-1A, p. 227-234.

Dunning, G.R. and Krogh, T.E.
1983: Tightly-clustered precise U/Pb Zircon ages of ophiolites from the Newfoundland Appalachians (abstract); in Abstracts with Programs, Northeastern Section; Geological Society of America, v. 15, no. 3, p. 136.
1985: Geochronology of ophiolites of the Newfoundland Appalachians; Canadian Journal of Earth Sciences, v. 22, p. 1659-1670.

Dunning, G.R. and O'Brien, S.J.
1989: Late Proterozoic-early Paleozoic crust in the Hermitage Flexure, Newfoundland Appalachians: U-Pb ages and tectonic significance; Geology, v. 17, p. 548-551.

Dunning, G.R., Barr, S.M., Raeside, R.P., and Jamieson, R.A.
1990b: U-Pb zircon, titanite, and monazite ages in the Bras d'Or and Aspy terranes of Cape Breton Island, Nova Scotia: implications for igneous and metamorphic history; Geological Society of America Bulletin, v. 102, p. 322-330.

Dunning, G.R., Kean, B.F., Thurlow, J.G., and Swinden, H.S.,
1987: Geochronology of the Buchans, Roberts Arm, and Victoria Lake groups and Mansfield Cove complex, Newfoundland; Canadian Journal of Earth Sciences, v. 24, p. 1175-1184.

Dunning, G.R., Krogh, T.E., O'Brien, S.J., Colman-Sadd, S.P., and O'Neill, P.P.
1988: Geochronologic framework for the Central Mobile Belt in southern Newfoundland and the importance of Silurian orogeny (abstract); Geological Association of Canada, Program with Abstracts; v. 13, p. 34.

Dunning, G.R., O'Brien, S.J., Colman-Sadd, S.P., Blackwood, R.F., Dickson, W.L., O'Neill, P.P., and Krogh, T.E.
1990a: Silurian orogeny in the Newfoundland Appalachians; Journal of Geology, v. 98, p. 895-913.

Dunning, G.R., Swinden, H.S., Kean, B.F., Evans, D.T.W., and Jenner, G.A.
1991: A Cambrian island arc in Iapetus; geochronology and geochemistry of the Lake Ambrose volcanic belt, Newfoundland Appalachians; Geological Magazine, v. 128, p. 1-17.

Dunning, G.R., Wilton, D.H.C., and Herd, R.K.
1989: Geology, geochemistry and geochronology of a Taconic calc-alkaline batholith, southwest Newfoundland; Transactions of the Royal Society of Edinburgh; Earth Sciences, v. 80, p. 159-168.

Dunsworth, S. and Calon, T.
1988: Field relations in the Lewis Hills Massif, Bay of Islands ophiolite: a history of synkinematic, multiple intrusive oceanic crustal accretion (abstract); Geological Association of Canada, Program with Abstracts, v. 13, p. A34.

Dunsworth, S., Calon, T., and Malpas, J.
1986: Structural and magmatic controls on the internal geometry of the plutonic complex and its chromite occurrences in the Bay of Islands ophiolite, Newfoundland; in Current Research, Part A; Geological Survey of Canada, Paper 86-1B, p. 471-482.

Duquette, G.
1961: Geology of the Weedon Lake area and its vicinity, Wolfe and Compton counties, Québec; Ph.D. thesis, Université Laval, Québec, Quebec.

Durling, P.W., Bell, J.S., and Fader, G.B.J.
1987: The geological structure and distribution of Paleozoic rocks on the Avalon Platform, offshore Newfoundland; Canadian Journal of Earth Sciences, v. 24, p. 1412-1420.

Dzulynski, S. and Walton, E.K.
1965: Sedimentary features of flysch and greywackes; Elsevier Publishing Company, Amsterdam, 274 p.

Eakins, P.R.
1964: Région de Sutton, Québec; Commission géologique du Canada, Étude 63-34, 3 p.

Eberz, G.W., Clarke, D.B., Chatterjee, A.K., and Giles, P.S.
1991: Chemical and isotopic composition of the lower crust beneath the Meguma lithotectonic zone, Nova Scotia: evidence from granulite facies xenoliths; Contributions to Mineralogy and Petrology, v. 109, p. 69-88.

Eisbacher, G.H.
1969: Displacement and stress field along part of the Cobequid Fault, Nova Scotia; Canadian Journal of Earth Science, v. 6, p. 1095-1104.

Eisenberg, R.A.
1981: Chronostratigraphy of metavolcanic and associated intrusive rocks of the Boundary Mountains Anticlinorium (abstract); Geological Society of America, Abstracts with Programs, v. 13, p. 131.

Elias, P. and Strong, D.F.
1982: Paleozoic granitoid plutonism of southern Newfoundland: contrasts in timing, tectonic setting and level of emplacement; Transactions of the Royal Society of Edinburgh; Earth Sciences, v. 73, p. 43-57.

Elmore, R.D., Pilkey, O.H., Cleary, W.J., Curran H.A.
1979: Black shell turbidite, Hatteras Abyssal plain, western Atlantic Ocean; Geological Society of America Bulletin, v. 90, p. 1165-1176.

England, P.C. and Houseman, G.A.
1988: The mechanics of the Tibetan Plateau; Royal Society of London, Philosophical Transactions, ser. A, v. 326, p. 301-320.

Enos, P.
1969a: Cloridorme Formation, Middle Ordovician flysch, northern Gaspé Peninsula, Quebec; Geological Society of America, Special Paper 117, 66 p.
1969b: Anatomy of a flysch; Journal of Sedimentary Petrology, v. 39, p. 680-723.

Erdmer, P.
1984: Summary of field work in the northern Long Range Mountains, western Newfoundland; in Current Research, Part A; Geological Survey of Canada, Paper 84-1A, p. 521-530.

Erdmer, P. (cont.)

1986a: Geology of the Long Range Inlier in the Sandy Lake map area, western Newfoundland; in Current Research, Part B; Geological Survey of Canada, Paper 86-1B, p. 19-29.

1986b: Geology of the Long Range Inlier in Sandy Lake map area (12H); Geological Survey of Canada, Open File 1310.

Erdtmann, B.D.

1976: Die Graptolithenfauna der Exploits Gruppe von Zentral-Neufundland; Mitteilungen aus dem Geologisch-Palaeontologischen Institut der Universitat Hamburg, Heft 45, S. 65-140, Hamburg.

Erdtmann, B.D. and Botsford, J.W.

1986: A new early Tremadoc (La1) graptolite faunule from western Newfoundland: its Australian affinity and biofacies relations; Canadian Journal of Earth Sciences, v. 23, p. 766-773.

Erdtmann, B.D. and Miller, J.F.

1981: Eustatic control of lithofacies and biofacies changes near the base of the Tremadocian; United States Geological Survey, Open-File Report 81-743, p. 78-81.

Espenshade, G.H.

1937: Geology and mineral deposits of the Pilleys Island area; Newfoundland Department of Natural Resources, Geology Section, Bulletin, v. 9, 36 p.

Evans, D.T.W., Kean, B.F., and Dunning, G.R.

1990: Geological studies, Victoria Lake Group, central Newfoundland; in Current Research (1990); Newfoundland Department of Mines and Energy, Geological Survey Branch, Report 90-1, p. 131-144.

Fagan, J.J. and Edwards, M.

1969: Paleocurrents in Normanskill Formation, south of Hudson and Catskill, New York; Geological Society of America Bulletin, v. 80, p. 121-124.

Fåhraeus, L.E. and Hunter, D.R.

1981: Paleoecology of selected conodontophorid species from the Cobbs Arm Formation (Middle Ordovician), New World Island, north-central Newfoundland; Canadian Journal of Earth Sciences, v. 18, p. 1653-1665.

Fahrig, W.F. and West, T.D.

1986: Diabase dyke swarms of the Canadian Shield; Geological Survey of Canada, Map 1627A.

Fairbairn, H.W., Bottino, W.H., Pinson, W.H., and Hurley, P.M.

1966: Whole-rock age and initial $^{87}Sr/^{86}Sr$ of volcanic rocks underlying fossiliferous Lower Cambrian in the Atalntic Provinces of Canada; Canadian Journal of Earth Sciences, v. 3, p. 509-521.

Faribault, E.R.

1912: Gold-bearing series of the basins of Medway River, Nova Scotia; Geological Survey of Canada, Summary Report 1911, Sessional Paper 26, p. 334-340.

1914: Greenfield and Liverpool Town map-areas, Nova Scotia; Geological Survey of Canada, Summary Report 1912, p. 372-382.

Fisher, D. M.

1977: Correlation of Hadrynian, Cambrian and Ordovician rocks of New York State; New York State Museum map and chart series, folio 25, 75 p.

1982: Cambrian and Ordovician straitigraphy and paleontology of the Champlain Valley; 3rd North American Paleontological Convention, Excursion Guide A, p. A1-A27.

Flagler, P.A. and Spray, J.G.

1991: Generation of plagiogranite by amphibolite anatexis in oceanic shear zones; Geology, v. 19, p. 1-96.

Foley, F.C.

1937: Geology and mineral deposits of Hawke Bay-Great Harbour Deep area, northern Newfoundland; Newfoundland Department of Natural Resources, Geological Section, Bulletin, v. 10, 22 p.

Ford, T.D.

1958: Precambrian fossils from Charnwood Forest; Yorkshire Geological Society Proceedings, v. 31, p. 211-217.

Fortey, R.A.

1979: Early Ordovician trilobites from the Catoche Formation (St. George Group), western Newfoundland; in Contributions to Canadian Paleontology, (ed.) P.J. Griffin; Geological Survey of Canada, Bulletin 321, p. 61-114.

1984: Global early Ordovician transgression and regressions and their biological implications; in Aspects of the Ordovician System: Paleontological Contributions from the University of Oslo, (ed.) D.L. Bruton; Universitetsforlaget, No. 295, p. 37-50.

Fortier, Y.O.

1946: Geology of the Orford map area in the Eastern Township, Quebec; Ph.D. thesis, Stanford University, U.S.A., 248 p.

Fox, D. and van Berkel, J.T.

1988: Mafic-ultramafic occurrences in metasedimentary rocks of southwestern Newfoundland; in Current Research, Part B; Geological Survey of Canada, Paper 88-1B, p. 41-48.

French, V.A.

1985: Geology of the Gillanders Mountain Intrusive Complex and satellite plutons, Lake Ainslie area, Cape Breton Island, Nova Scotia; M.Sc. thesis, Acadia University, Wolfville, Nova Scotia, 236 p.

French, V.A. and Barr, S.M.

1984: Age and petrology of the Gillanders Mountain intrusive complex, Lake Ainslie area, Cape Breton Island, Nova Scotia; Geological Association of Canada-Mineralogical Association of Canada, Program with Abstracts, v. 9, p. 98.

Fritts, C.E.

1953: Geological reconnaissance across the Great Northern Peninsula of Newfoundland; Newfoundland Geological Survey, Report no. 4, 27 p.

Fritz, W.H. and Yochelson, E.L.

1988: The status of Salterella as a Lower Cambrian index fossil; Canadian Journal of Earth Sciences, v. 25, p. 403-416.

Fritz, W.H., Kindle, C.H., and Lespérence, P.J.

1970: Trilobites and stratigraphy of the Middle Cmbrian Corner-of-the-Beach Formation, eastern Gaspé Peninsula, Quebec; in Contributions to Canadian Paleontology; Geological Survey of Canada, Bulletin 187, p. 43-58.

Fuller, J.O.

1941: Geology and mineral deposits of the Fleur de Lys area; Newfoundland Geological Survey, Bulletin, v. 15, 41 p.

Fyffe, L.R.

l976: Correlation of geology in the southwestern and northern parts of the Miramichi zone; New Brunswick Department of Natural Resources, 139th Annual Report, p. 137-141.

1977: Comparison of some tectono-stratigraphic zones in the Appalachians of Newfoundland and New Brunswick: discussion; Canadian Journal of Earth Sciences, v. 14, p. 1468-1469.

1982a: Geology of Woodstock (sheet 21J); New Brunswick Department of Natural Resources, Map NR-4.

1982b: Taconian and Acadian structural trends in central and northern New Brunswick; in Major Structural Zones and Faults of the Northern Appalachians, (ed.) P. St-Julien and J. Béland; Geological Association of Canada, Special Paper 24, p. 117-130.

1986: A recent graptolite discovery from the Fournier Group of northern New Brunswick; in Eleventh Annual Review of Activities, Project Resumés, (ed.) S.A. Abbott; New Brunswick Department of Natural Resources and Energy, Minerals and Energy Division, Information Circular 86-2, p. 43-45.

1988: Bedrock geology of the McAdam-St. Croix River area; in Thirteenth Annual Review of Activities, Project Résumés, (ed.) S.A. Abbott; New Brunswick Department of Natural Resources and Energy, Minerals and Energy Division, Information Circular 88-2, p. 70-75.

Fyffe, L.R. and Fricker, A.

1987: Tectonostratigraphic terrane analysis of New Brunswick; Maritime Sediments and Atlantic Geology, v. 23, p. 113-123.

Fyffe, L.R. and Noble, J.P.B.

1985: Stratigraphy and structure of the Ordovician, Silurian and Devonian of the northern New Brunswick; Geological Association of Canada, Annual Meeting, Fredericton, New Brunswick, Guidebook to Excursion 4, 56 p.

Fyffe, L.R. and Pickerill, R.

1986: Timing of terrane accretion in eastern and east-central Maine: comment; Geology, v. 14, p. 1051-1052.

Fyffe, L.R. and Pronk, A.G.

1985: Bedrock and surficial geology rock and till geochemistry in the Trousers Lake area, Victoria County, New Brunswick; New Brunswick Department of Natural Resources, Mineral Resources Division, Report of Investigation 20, 74 p.

Fyffe, L.R. and Riva, J.

1990: Revised stratigraphy of the Cookson Group based on new paleontological, sedimentological, and structural evidence; Atlantic Geology, v. 26, p. 271-276.

Fyffe, L.R., Barr, S.M., and Bevier, M.L.

1988a: Origin and U-Pb geochronology of amphibolite-facies metamorphic rocks, Miramichi Highlands, New Brunswick; Canadian Journal of Earth Sciences, v. 25, p. 1674-1686.

Fyffe, L.R., Forbes, W.H., and Riva, J.

1983: Graptolites from the Benton area of west-central New Brunswick and their regional significance; Maritime Sediments and Atlantic Geology, v. 19, p. 117-125.

Fyffe, L.R., Irrinki, R.R., and Cormier, R.F.
1977: A radiometric age of deformed granitic rocks in north-central New Brunswick; Canadian Journal of Earth Sciences, v. 14, p. 1687-1689.

Fyffe, L.R., Pajari, G.E., and Cherry, M.E.
1981: The Acadian plutonic rocks of New Brunswick; Maritime Sediments and Atlantic Geology, v. 17, p. 23-36.

Fyffe, L.R., Stewart, D.B., and Ludman, A.
1988b: Tectonic significance of black pelites and basalts in the St. Croix Terrane, coastal Maine and New Brunswick; Maritime Sediments and Atlantic Geology, v. 24, p. 281-288.

Fyffe, L.R., van Staal, C.R., and Winchester, J.A.
1990: Late Precambrian-early Paleozoic volcanic regimes and associated massive sulphide deposits in the northeastern mainland Appalachians; Canadian Institute of Mining and Metallurgy Bulletin, v. 83, no. 938, p. 70-78.

Gansser, A.
1974: The ophiolitic mélange, a world-wide problem on Tethyan examples; Eclogae Geologicae Helvetiae, v. 67-3, p. 479-507.

Gardiner, S. and Hiscott, R.N.
1988: Deep-water facies and depositional setting of the lower Conception Group (Hadrynian), southern Avalon Peninsula, Newfoundland; Canadian Journal of Earth Sciences, v. 25, p. 1579-1594.

Gaudette, H.E., Olszewski, W.J., and Donohoe, H.V.
1983: Age and origin of basement rocks, Cobequid Highlands, Nova Scotia; Geological Society of America, Abstracts with Programs, v. 15, p. 136.
1984: Rb-Sr isochrons of Precambrian age from plutonic rocks in the Cobequid Highlands, Nova Scotia; Nova Scotia Department of Mines and Energy, Report 84-1, p. 285-292.

Gaudette, H.E., Olszewski, W.J., and Jamieson, R.A.
1985: Rb-Sr ages of some basement rocks, Cape Breton Highlands, Nova Scotia (abstract); Geological Association of Canada, Program with Abstracts, v. 10, p. A20.

Gauthier, M., Auclair, M., Bardoux, M., Blain, M., Boisvert, D., Brassard, B., Chartrand F., Darimont, A., Dupuis, L., Durocher, M., Gariépy, C., Godue, R., Jebrak, M., et Trottier, J.
1989: Synthèse gîtologique de l'Estrie et de la Beauce; Ministère de l'Énergie et des Ressources, Québec, MB 89-20, 631 p.

Gesner, A.
1836: Remarks on the geology and mineralogy of Nova Scotia; Gossip and Code, Halifax, Nova Scotia.

Giles, P.S. and Ruitenberg, A.A.
1977: Stratigraphy, paleogeography and tectonic setting of the Coldbrood Group in the Caledonia Highlands of southern New Brunswick; Canadian Journal of Earth Sciences, v. 14, p. 1263-1275.

Gill, J. B.
1981: Orogenic Andesites and Plate Tectonics; Springer-Verlag, Berlin.

Gillespie, R.T.
1983: Stratigraphic and structural relationships among rock groups at Old Mans Pond, western Newfoundland; M.Sc. thesis, Memorial University of Newfoundland, St. John's, Newfoundland, 198 p.

Gillis, J.W.
1966: Great Northern Peninsula, Newfoundland; in Report of Activities, May to October, 1965; Geological Survey of Canada, Paper 66-1, p. 179-181.

Gilmore, R.G.
1971: Stratigraphy of the Phillipsburg, Rosenberg thrust sheets, southern Quebec; M.Sc. thesis, McGill University, Montreal, Quebec.

Girard, P.
1967: Mount Richardson area, Gaspe Nord Co., Quebec; Quebec Department of Natural Resources, PR 563, 11 p.

Gittins, J., Macintyre, R.M., and York, D.
1967: The ages of carbonatite complexes in eastern Canada; Canadian Journal of Earth Sciences, v. 4, p. 651-655.

Globensky, Y.
1978: Région de Drummondville; Ministère des Richesses naturelles, Québec, RG-192, 107 p.
1981a: Région de Lacolle Saint-Jean(S); Ministère des Richesses naturelles, Québec, RG-197, 197 p.
1981b: Région de Huntingdon; Ministère des Richesses naturelles, Québec, RG-198, 53 p.
1982: Région de Lachute; Ministère des Richesses naturelles, Québec, RG-200, 67 p.
1987: Géologie des Basses-Terres du Saint-Laurent; Ministère de l'Énergie et des Ressources, Québec, MM 85-02, 63 p.

Globensky, Y. and Riva, J.
1982: Ordovician Stratigraphy and paleontology of the St. Lawrence Lowlands and the frontal Appalachians near Quebec City; 3rd North American Paleontological Convention, Montreal, Excursion Guide C, p. C1-C47.

Godfrey, S.C.
1982: Rock groups, structural slices and deformation in the Humber Arm Allochthon at Serpentine Lake, western Newfoundland; M.Sc. thesis, Memorial University of Newfoundland, St. John's, Newfoundland, 182 p.

Goutier, J.
1989a: Géologie de la région de Biencourt-Lac-des-Aigles (Témiscouata); Ministère de l'Énergie et des Ressources, Québec, ET 88-02, 25 p.
1989b: Géologie de la région de Biencourt-Lac-des-Aigles, Témiscouata, Québec; thèse de maîtrise, Université de Montréal, Montréal (Québec).

Gower, C.F., Erdmer, P., and Wardle, R.J.
1986: The Double Mer Formation and the Lake Melville rift system, eastern Labrador; Canadian Journal of Earth Sciences, v. 23, p. 359-368.

Grant, R.H., McCutcheon, S.R., and Nance, R.D.
1987: A reassessment of Carboniferous stratigraphy and structure in southern New Brunswick, Canada (abstract); 11th International Congress on Carboniferous stratigraphy and geology, Section I, Programs with Abstracts, p. 12.

Gravenor, C.P.
1980: Heavy minerals and sedimentological studies on the glaciogenic Late Precambrian Gaskiers Formation of Newfoundland; Canadian Journal of Earth Sciences, v. 17, p. 1331-1341.

Graves, M.C. and Zentilli, M.
1988: The Goldenville-Halifax transition of the Meguma Group of Nova Scotia: the lithochemistry of metal-rich coticules; in Current Research, Part B; Geological Survey of Canada, Paper 88-1B, p. 251-261.

Greggs, R.G. and Bond, I.J.
1971: Conodonts from the March and Oxford Formations in the Brockville area, Ontario; Canadian Journal of Earth Sciences, v. 8, p. 1455-1471.
1972: A principle reference section for the Nepean Formation of probable Tremadocian age near Ottawa, Ontario; Canadian Journal of Earth Sciences, v. 9, p. 933-941.

Greenough, J.D., Fryer, B.J., and Owen, J.V.
1990: Evidence for sialic basement to the Dunnage Zone, Notre Dame Bay, Newfoundland; in Lithoprobe, (ed.) J. Hall; Memorial University of Newfoundland, Lithoprobe East Transect Report 13, p. 85-87.

Greenough, J.D., McCutcheon, S.R., and Papezik, V.S.
1985: Petrology and geochemistry of Cambrian volcanic rocks from the Avalon Zone in New Brunswick; Canadian Journal of Earth Sciences, v. 22, p. 881-892.

Grenier, R.
1990: The Appalachian fold and thrust belt, northwestern Newfoundland; M.Sc. thesis, Memorial University of Newfoundland, St. John's, Newfoundland, 214 p.

Grenier, R. and Cawood, P.A.
1988: Variations in structural style along the Long Range Front, western Newfoundland; in Current Research, Part B; Geological Survey of Canada, Paper 88-1B, p. 127-133.

Gummer, P.K., Von Guttenberg, R., and Ferguson, L.
1978: A recent fossil discovery in the Tetagouche volcanic complex of New Brunswick; Paper presented at the New Brunswick Annual Meeting of the Canadian Institute of Mining and Metallurgy, Bathurst, New Brunswick; New Brunswick Department of Natural Resources and Energy, Minerals and Energy Division, Author File.

Hall, B.A.
1970: Stratigraphy of the southern end of the Munsungun Anticlinorium, Maine; Maine Geological Survey, Bulletin, v. 22, 63 p.

Hand, B.M., Middleton, G.V., and Skipper, K.
1972: Antidune cross-stratification in a turbidite sequence, Cloridorme Formation, Gaspé, Quebec; Sedimentology, v. 18, p. 135-138.

Hanmer, S.
1981: Gander Zone, Newfoundland: an Acadian ductile shear zone; Canadian Journal of Earth Sciences, v. 128, p. 120-135.

Hantzschel, W.
1975: Miscellanea Supplement; in Treatise on Invertebrate Paleontology, Part W, (ed.) C. Teichert; University of Kansas and Geological Society of America, p. W9.

Haq, B.U. and Van Eysinga, W.B.
1987: Geological Time Table, 4th edition; Elsevier, Amsterdam, Holland.

Harland, T.L. and Pickerill, R.K.

1982: A review of Middle Ordovician sedimentation in the St. Lawrence Lowland, eastern Canada; Geological Journal, v. 17, p. 135-156.

1984: Ordovician rocky shoreline deposits – the basal Trenton Group around Quebec City; Geological Journal, v. 19, p. 271-298.

Harland, W.B., Armstrong, R.L., Cox, A.V., Craig, L.E., Smith, A.G., and Smith, D.G.

1989: Geologic time scale 1989; Cambridge University Press, 256 p.

Harland, W.B., Cox, A.V., Llewellyn, P.G., Pickton, C.A.G., Smith, A.G., and Walters, R.

1982: A geologic time scale; Cambridge University Press, Cambridge, United Kingdom.

Harley, D.N.

1979: A mineralized Ordovician resurgent caldera complex in the Bathurst-Newcastle mining district, New Brunswick, Canada; Economic Geology, v. 74, p. 786-796.

Harnois, L.

1982: Géochimie des terres rares dans le complexe ophiolitique du Mont Orford (Province de Québec, Canada); thèse de maîtrise, Université du Québec à Montréal, Montréal (Québec).

Harnois, L. and Morency, M.

1989: Geochemistry of Mount Orford ophiolite complex, Northern Appalachians, Canada; Chemical Geology, v. 77, p. 133-147.

Harnois, L., Trottier, J., and Morency, M.

1990: Rare earth element geochemistry of Thetford Mines ophiolite complex, Northern Appalachians, Canada; Contribution to Mineralogy and Petrology, v. 105, p. 433-445.

Harris, I.M. and Schenk, P.E.

1975: The Meguma Group; in Ancient sediments of Nova Scotia, (ed.) I.M. Harris; Eastern Section Society of Economic Paleontologists and Mineralogists, Guidebook, p. 17-38.

Harron, G.A.

1976: Métallogénèse des gîtes de sulfures des Cantons de l'Est; Ministère des Richesses naturelles, Québec, ES27, 42 p.

Harwood, D.S.

1969: The Second Lake Anticline, a major structure of the northwest limit of the Boundary Mountains Anticlinorium, northern New Hampshire, west-central Maine, and adjacent Québec; United States Geological Survey, Professional Paper 650-D, p. D106-D115.

Harwood, D.S. and Berry, W.B.N.

1967: Fossiliferous lower Paleozoic rocks in the Cupsuptic quadrangle, west-central Maine; United States Geological Survey, Professional Paper 575-D, p. D16-D23.

Hatch, N.L., Jr.

1982: The Taconian Line in western New England and its implication to Paleozoic tectonic history; in Major Structural Zones and Faults of the Northern Appalachians, (ed.) P. St-Julien and J. Béland; Geological Association of Canada, Special Paper 24, p. 67-85.

1988: Some revisions to the stratigraphy and structure of the Connecticut Valley trough, eastern Vermont; American Journal of Science, v. 288, p. 1041-1059.

Haworth, R.T. and Lefort, J.P.

1979: Geophysical evidence for the extent of the Avalon Zone in Atlantic Canada; Canadian Journal of Earth Sciences, v. 16, p. 552-567.

Haworth, R.T., Daniels, D.L., Williams, H., and Zietz, I.

1980a: Bouguer gravity anomaly map of the Appalachian Orogen; Memorial University of Newfoundland, Map No. 3, scale 1:1 000 000.

1980b: Bouguer gravity anomaly map of the Appalachian Orogen; Memorial University of Newfoundland, Map No. 3a, scale 1:2 000 000.

Hay, B.J. and Cisne, J.L.

1984: Deposition in anoxic Taconic foreland basin, late Middle Ordovician, New York (abstract); American Association of Petroleum Geologists Bulletin, v. 68, p. 1919-1920.

Hay, P.W.

1968: Geology of the St. George-Seven Mile Lake area, southwestern New Brunswick; New Brunswick Department of Natural Resources, Mineral Resources Division, Map Series 68-1, 7 p.

Hayes, A.O.

1948: Geology of the area between Bonavista and Trinity bays, eastern Newfoundland; Geological Survey of Newfoundland, Bulletin, v. 32, Part 1, p. 1-37.

Hayes, A.O. and Howell, B.F.

1937: Geology of Saint John, New Brunswick; Geological Society of America, Special Paper no. 6, 146 p.

Hayes, J.J.

1951: Marks Lake, Newfoundland; Geological Survey of Canada, Paper 51-20, map and marginal notes.

Hébert, R.

1981: Étude pétrologique des roches ophiolitiques: anciens éboulis de talus de fonds océaniques?; Journal canadien des sciences de la Terre, v. 18, p. 619-623.

Hébert, R. and Laurent, R.

1989: Mineral chemistry of ultramafic and mafic plutonic rocks of the Appalachians ophiolites, Québec, Canada; Chemical Geology, v. 77, p. 265-285.

Helmstaedt, H.

1968: Structural and metamorphic analysis in Beaver Harbour region, Charlotte County, New Brunswick; Ph.D. thesis, University of New Brunswick, Fredericton, New Brunswick, 196 p.

1971: Structural geology of Portage Lakes area, Bathurst-Newcastle District, New Brunswick; Geological Survey of Canada, Paper 70-28, 52 p.

1973a: Structural geology of the Bathurst-Newcastle district; in Geology of New Brunswick, Field Guide to Excursions, (ed.) N. Rast; New England Intercollegiate Geological Conference, p. 34-46.

1973b: Discussions: deformational history of the Caribou strata-bound deposits, Bathurst, New Brunswick, Canada; Economic Geology, v. 68, p. 571-572.

Helmstaedt, H. and Tella, S.

1973: Pre-Carboniferous structural history of southeast Cape Breton Island, Nova Scotia; Maritime Sediments, v. 9, p. 88-99.

Helwig, J.A.

1969: Redefinition of Exploits Group, Lower Paleozoic, northeast Newfoundland; in North Atlantic-Geology and Continental Drift, (ed.) M. Kay; American Association of Petroleum Geologists, Memoir 12, p. 408-413.

Hendriks, M., Jamieson, R., Zentilli, M., and Beaumont, C.

1990: Apatite fission track analyses of the Great Northern Peninsula, western Newfoundland: Data and preliminary interpretation; in Lithoprobe, (ed.) J. Hall; Lithoprobe East Transect Report 13, Memorial University of Newfoundland, p. 95-96.

Herd, R.K. and Dunning, G.R.,

1979: Geology of Puddle Pond map area, southwestern Newfoundland; in Current Research, Part A; Geological Survey of Canada, Paper 79-1A, p. 305-310.

Hesse, R.

1982: Cloridorme Formation; in Field Trip Guidebook, Excursion 7B: Paleozoic continental margin sedimentation in the Quebec Appalachians, (ed.) R. Esse, G.V. Middleton, and B.R. Rust; International Association of Sedimentologists, 11th International Congress on Sedimentology, McMaster University, Hamilton, Canada, p. 126-138.

Heyl, A.V. and Ronan, J.J.

1954: The iron deposits of Indian Head area; Geological Survey of Canada, Bulletin 27, p. 42-65.

Heyl, G.R.

1937: The Geology of the Sops Arm area, White Bay, Newfoundland; Newfoundland Department of Natural Resources, Geological Section, Bulletin, v. 8, 42 p.

Hibbard, J.P.

1979: Geology of the Baie Verte map area, west (12H/16W), Newfoundland; in Report of Activities for 1978, (ed.) R.V. Gibbons; Newfoundland Departmemnt of Mines and Energy, Mineral Development Division, Report 79-1, p. 58-63.

1982: Significance of the Baie Verte Flexure, Newfoundland; Geological Society of America Bulletin, v. 93, p. 790-797.

1983: Geology of the Baie Verte Peninsula, Newfoundland; Newfoundland Department of Mines and Energy, Mineral Development Division, Memoir 2, 279 p.

Hibbard, J.P. and Williams, H.

1979: The regional setting of the Dunnage Melange in the Newfoundland Appalachians; American Journal of Science, v. 279, p. 993-1021.

Hibbard, J.P., Stouge, S., and Skevington, D.

1977: Fossils from the Dunnage Melange, north central Newfoundland; Canadian Journal of Earth Sciences, v. 14, p. 1176-1178.

Hill, J.R.

1990: A geological and geochemical study of metacarbonate rocks and contained mineral deposits, Cape Breton Island, Nova Scotia; Geological Survey of Canada, Paper 90-8, p. 3-30.

Hill, P.R.

1984: Facies and sequence analysis of Nova Scotia slope muds: turbidites versus "hemipelagic" deposition; in Fine-grained Sediments: Deep Water Processes and Facies, (ed.) D.A.V. Stow and D.J.W. Piper; The Geological Society, Blackwell Scientific Publications, Oxford, p. 311-318.

Hiscott, R.N.

1977: Sedimentology and regional implications of deep-water sandstones of the Tourelle Formation, Ordovician, Quebec; Ph.D. thesis, McMaster University, Hamilton, Ontario.

1978: Provenance of Ordovician deep-water sandstones, Tourelle Formation, Quebec, and implications for initiation of the Taconic Orogeny; Canadian Journal of Earth Sciences, v. 15, p. 1579-1597.

1979: Clastic sills and dikes associated with deep-water sandstones, Tourelle Formation, Ordovician, Quebec; Journal of Sedimentary Petrology, v. 49, p. 1-10.

1980: Depositional framework of sandy mid-fan complexes of Tourelle Formation, Ordovician, Quebec; American Association of Petroleum Geologists Bulletin, v. 64, p. 1052-1077.

1984: Ophiolitic source rocks for Taconic-age flysch; Trace-element evidence; Geological Society of America Bulletin, v. 95, p. 1261-1267.

Hiscott, R.N. and Middleton, G.V.

1979: Depositional mechanics of thick-bedded sandstones at the base of a submarine slope, Tourelle Formation (Lower Ordovician), Quebec; in Geology of Continental Slopes, (ed.) L.J. Doyle and O.H. Pilkey; Society of Economic Paleontologists and Mineralogists, Special Publication 27, p. 307-326.

1980: Fabric of coarse deep-water sandstones, Tourelle Formation, Quebec; Journal of Sedimentary Petrology, v. 50, p. 703-722.

Hiscott, R.N. and Pickering, K.T.

1984: Reflected turbidity currents on an Ordovician basin floor, Canadian Appalachians; Nature, v. 311, p. 143-145.

Hiscott, R.N., James, N.P., and Pemberton, S.G.

1984: Sedimentology and ichnology of the Lower Cambrian Bradore Formation, coastal Labrador; Fluvial to shallow-marine transgressive sequence; Canadian Petroleum Geology Bulletin, v. 32, p. 11-26.

Hiscott, R.N., Pickering, K.T., and Beeden, D.R.

1986: Progressive filling of a confined Middle Ordovician foreland basin associated with the Taconic Orogeny, Quebec; in Foreland Basins, (ed.) P. Homewood and P.A. Allen; International Association of Sedimentologists, Special Publication 8, p. 309-325.

Hiscott, R.N., Quinlan, G.M., and Stevens, R.K.

1983: Analogous tectonic evolution of the Ordovician foredeeps, southern and central Appalachians: comment; Geology, v. 11, p. 732.

Hofmann, H.J.

1963: Ordovician Chazy Group in southern Quebec; Bulletin of the American Association of Petroleum Geologists, v. 47, p.270-301.

1972: Stratigraphy of the Montréal area; 24th International Geological Congress, Montréal, Guidebook for excursion B-03, 32 p.

1974: The stromatolite Archaezoon Acadiense from the Proterozoic Green Head Group of Saint John, New Brunswick; Canadian Journal of Earth Sciences, v. 11, p. 1098-1115.

1979: Chazy (Middle Ordovician) trace fossils in the Ottawa-St. Lawrence Lowlands; Geological Survey of Canada, Bulletin 321, p. 27-59.

1982: Middle and Late Ordovician fossiliferous rocks of the Montréal area; 3rd North American Paleontological Convention, Montréal, Guide for excursion D, p. D1-D41.

1989: Stratigraphy of the Montréal area; Geological Association of Canada-Mineralogical Association of Canada, Fieldtrip Guidebook A4, 36 p.

Hofmann, H.J. and Patel, I.M.

1988: Trace fossils from the Lower Cambrian Ratcliffe Brook Formation, Saint John area, New Brunswick (abstract); Geological Association of Canada, Program with Abstracts, v. 13, p. A57.

Hofmann, H.J., Hill, J., and King, A.F.

1979: Late Precambrian microfossils, southeast Newfoundland; in Current Research, Part B; Geological Survey of Canada, Report 79-1B, p. 83-88.

Hon, R., Hill, M., Hepburn, J.C., Smith, C.

1984: Composition and age of source materials for the late Proterozoic magmas in the Avalon Terrane of eastern North America: evidence from the Boston Platform (abstract); Geological Society of America, Abstracts with Programs, v. 16, p. 543.

Horn, D.R., Ewing, M., Horn, B.M., and Delach, M.N.

1971: Turbidites of the Hatteras and Sohm abyssal plains, Western North Atlantic; Marine Geology, v. 11, p. 287-323.

Hornc, G.S.

1969: Early Ordovician chaotic deposits in the Central Volcanic Belt of northeast Newfoundland; Geological Society of America Bulletin, v. 80, p. 2451-2464.

1970: Complex volcanic-sedimentary patterns in the Magog Belt of northeastern Newfoundland; Geological Society of America Bulletin, v. 81, p. 1767-1788.

Hsu, E.Y.C.

1972: The stratigraphy and sedimentology of the late Precambrian St. John's and Gibbett Hill formations and the upper part of the Conception Group in the Torbay map area, Avalon Peninsula, Newfoundland; M.Sc. thesis, Memorial University of Newfoundland, St. John's, Newfoundland, 116 p.

Hubert, C.

1965: Stratigraphy of the Quebec Complex in the I'Islet-Kamouraska area, Quebec; Ph.D. thesis, McGill University, Montreal, Quebec.

Hubert, C., St-Julien, P., Lajoie, J., and Léonard, M.A.

1969: Flysch sedimentology in the Appalachians; Geological Association of Canada, Guidebook 1, Montreal, 38 p.

Hughes, C.J.

1970: The late Precambrian Avalonian Orogeny in Avalon, southeast Newfoundland; American Journal of Science, v. 269, p. 183-190.

Hughes, C.J. and Brueckner, W. D.

1971: Late Precambrian rocks of eastern Avalon Peninsula, Newfoundland: a volcanic island complex; Canadian Journal of Earth Sciences, v. 8, p. 899-915.

Hughes, S.

1977: Facies anatomy of a Lower Cambrian archaeocyathid biostrome complex, southern Labrador; M.Sc. thesis, Memorial University of Newfoundland, St. John's, Newfoundland, 276 p.

Hussey, E.M.

1974: Geological and petrochemical data on the Brighton Complex, Notre Dame Bay, Newfoundland; B.Sc. thesis, Memorial University of Newfoundland, St. John's, Newfoundland, 66 p.

1979: Geology of the Clode Sound area, Newfoundland; M.Sc. thesis, Memorial University of Newfoundland, St. John's, Newfoundland, 312 p.

Hussong, D.M. and Uyeda, S.

1981: Tectonics in the Mariana arc: results of recent studies including DSDP Leg 60; Oceanology Acta, p. 203-211.

Hutchinson, R.D.

1952: The stratigraphy and trilobite faunas of the Cambrian sedimentary rocks of Cape Breton Island, Nova Scotia; Geological Survey of Canada, Memoir 263, 124 p.

Hynes, A.

1980: Carbonatization and mobility of Ti, Y, and Zr in Ascot Formation metabasalts, southeast Québec; Contribution to Mineralogy and Petrology, v. 75, p. 1208-1224.

Idleman, B.D.

1986: Structural evolution of the coastal complex, Western Newfoundland (abstract); Geological Society of America, Abstracts with Programs, v. 18, no. 1, 24 p.

Ingerson, E.

1935: Layered peridotitic laccoliths in the Trout River area, Newfoundland; American Journal of Science, v. 29, p. 422-440.

INRS-Pétrole

1976: Stratigraphie, potentiel roche-mère, diagenèse minérale-organique du forage New Associated Consolidated Paper (N.A.C.P.), Anticosti no. 1; Ministère des Ressources naturelles du Québec, DP-360, 78 p.

Irrinki, R.R.

1976: Geology of Charlo (21 O/16) and parts of Campbellton (21 O/15), Tetagouche Lake (21 O/9) and Upsalquitch Forks (21 O/10); New Brunswick Department of Natural Resources, Mineral Resources Branch, Map Plate 76-8.

1979: Geology of North and South Little Sevogle Rivers-North Branch, Little Southwest Miramichi River-McKendrick and Catamaran lakes region, map areas 0-12, N-12, and N-13 (Part of 21 J/16); New Brunswick Department of Natural Resources, Mineral Resources Branch, Map Report 79-1, 36 p.

1980: Geology of Kennedy Lakes-Little Dungarvon and South Renous rivers region (map areas M-13, M-14, M-15 and Part of M-16); New Brunswick Department of Natural Resources, Mineral Resources Branch, Map Report 80-2, 39 p.

Jackson, C.T. and Alger, F.

1828: Mineralogy and geology of a part of Nova Scotia; American Journal of Science, v. XIV, p. 305-330.

Jacobi, R.D.
1981: Peripheral bulge-A causal mechanism for the Lower-Middle Ordovician unconformity along the western margin of the northern Appalachians; Earth and Planetary Science Letters, v. 56, p. 245-251.

Jacobi, R.D. and Wasowski, J.J.
1985: Geochemistry and plate-tectonic significance of the volcanic rocks of the Summerford Group, north-central Newfoundland; Geology, v. 13, p. 126-130.

Jacobsen, S.B. and Wasserburg, G.J.
1979: Nd and Sr isotopic study of the Bay of Islands ophiolite complex and the evolution of the source of mid ocean ridge basalts; Journal of Geophysical Research, v. 84, p. 7429-7445.

James, N.P.
1981: Megablocks of calcified algae in the Cow Head Breccia, western Newfoundland: vestiges of a Lower Paleozoic continental margin; Geological Society of America Bulletin, v. 92, p. 799-811.

James, N.P. and Debrenne, F.
1980: Regular archaeocyaths from the Forteau Formation, west Newfoundland; Canadian Journal of Earth Sciences, v. 17, p. 1609-1615.

James, N.P. and Hiscott, R.N.
1982: Lower Cambrian bioherms and sandstones, southern Labrador; 11th International Congress on Sedimentology, Hamilton, Ontario, Field Excursion Guidebook 1A.

James, N.P. and Kobluk, D.R.
1978: Lower Cambrian patch reefs and associated sediments, southern Labrador, Canada; Sedimentology, v. 25, p. 1-35.

James, N.P. and Mountjoy, E.W.
1983: The shelf slope break in fossil carbonate platforms; in The Shelf Slope Boundary, a critical interface on continental margins, (ed.) D.J. Stanley and G.T. Moore; Society of Economic Paleontologists and Mineralogists, Special Publication No. 36, p. 189-207.

James, N.P. and Stevens, R.K.
1986: Stratigraphy and correlation of the Cambro-Ordovician Cow Head Group, western Newfoundland; Geological Survey of Canada, Bulletin 366, 143 p.

James, N.P., Botsford, J., and Williams, S.H.
1987: Allochthonous slope sequence at Lobster Cove Head: evidence for a complex Middle Ordovician platform margin in western Newfoundland; Canadian Journal of Earth Sciences, v. 24, p. 1199-1211.

James, N.P., Knight, I., Stevens, R.K., and Barnes, C.R.
1988: Sedimentology and paleontology of an Early Paleozoic continental Margin, western Newfoundland; in Field Trip Guidebook; Geological Association of Canada, St. John's, Newfoundland, 121 p.

James, N.P., Stevens, R.K., Barnes, C.R., and Knight, I.
1989: Evolution of a Lower Paleozoic continental-margin carbonate platform, northern Canadian Appalachians; in Controls on Carbonate Platform and Basin Development, (ed.) P.D. Crevello, J.L. Wilson, J.F. Sarg, and J.F. Read; Society of Economic Paleontologists and Mineralogists, Special Publication 44, p. 123-146.

Jamieson, R.A.
1977: A suite of alkali basalts and gabbros associated with the Hare Bay Allochthon of western Newfoundland; Canadian Journal of Earth Sciences, v. 14, p. 346-356.
1986: P-T paths from high temperature shear zones beneath ophiolites; Journal of Metamorphic Geology, v. 4, p. 3-22.
1990: Metamorphism of an early Palaeozoic continental margin, western Baie Verte Peninsula, Newfoundland; Journal of Metamorphic Geology, v. 8, p. 269-288.

Jamieson, R.A. and Doucet, P.
1983: The Middle River-Crowdis Mountain area, southern Cape Breton Highlands; in Current Research, Part A; Geological Survey of Canada, Paper 83-1A, p. 269-276.

Jamieson, R.A. and Strong, D.F.
1981: A metasomatic mylonite zone within the ophiolite aureole, St. Anthony Complex, north-western Newfoundland; American Journal of Science, v. 281, p. 264-281.

Jamieson, R.A. and Talkington
1980: A jacupirangite-syenite assemblage beneath the White Hills Peridotite, northwestern Newfoundland; American Journal of Science, v. 280, p. 459- 477.

Jamieson, R.A. and Vernon, R.
1987: Timing of porphyroblast growth in the Fleur de Lys Supergroup, Newfoundland; Journal of Metamorphic Geology, v. 5, p. 273-288.

Jamieson, R.A., Tallman, P., Marcotte, J.A., Plint, H.E., and Connors, K.A.
1987: Geology of the west-central Cape Breton Highlands, Nova Scotia; Geological Survey of Canada, Paper 87-13, 11 p.

Jamieson, R.A., Tallman, P.D., Plint, H.E., and Connors, K.A.
1989: Geological setting of pre-Carboniferous mineral deposits in the western Cape Breton Highlands, Nova Scotia; Geological Survey of Canada, Open File 2008.
1990: Geological setting of pre-Carboniferous mineral deposits in the western Cape Breton Highlands; in Mineral Deposit Studies in Nova Scotia, Volume 1, (ed.) A. Sangster; Geological Survey of Canada, Paper 90-8, p. 77-99.

Jamieson, R.A., van Breemen, O., Sullivan, R.W., and Currie, K.L.
1986: The age of igneous and metamorphic events in the western Cape Breton Highlands, Nova Scotia; Canadian Journal of Earth Sciences, p. 1891-1901.

Jenkyns, H.C.
1980: Cretaceous anoxic events: from continents to oceans; Journal of the Geological Society of London, v. 137, p. 171-188.
1986: Pelagic environments; in Sedimentary Environments and Facies (Chapter 11), (ed.) H.C. Reading; Blackwell Scientific, p. 343-398.

Jenner, G.A. and Fryer, B.J.
1980: Geochemistry of the upper Snooks Arm basalts, Burlington Peninsula, Newfoundland: evidence against formation in an island arc; Canadian Journal of Earth Sciences, v. 17, p. 888-900.

Jenner, G.A., Dunning, G.R., Malpas, J., Brown, M., and Brace, T.
1991: Bay of Islands and Little Port complexes, revisited: age, geochemical and isotopic evidence confirm suprasubduction-zone origin; Canadian Journal of Earth Sciences, v. 28, p. 1635-1652.

Jenness, S.E.
1958: Geology of the lower Gander River ultrabasic belt, Newfoundland; Geological Survey of Newfoundland, Report 11, 58 p.
1963: Terra Nova and Bonavista map areas, Newfoundland (2D E1/2 and 2C); Geological Survey of Canada, Memoir 327, 184 p.
1972: The Gander Lake and Davidsville groups of northeastern Newfoundland: new data and geotectonic implications: discussion; Canadian Journal of Earth Sciences, v. 9, p. 1779-1781.

Jewell, W.B.
1939: Geology and mineral deposits of the Baie d'Espoir area; Newfoundland Geological Survey, Bulletin 17, 29 p.

Johnson, H.
1941: Paleozoic lowlands of western Newfoundland; Transactions of the New York Academy of Science, series 2, v. 3, p. 141-145.

Johnson, R.J.E. and Van der Voo, R.
1985: Middle Cambrian paleomagnetism of the Avalon terrane in Cape Breton Island, Nova Scotia; Tectonics, v. 4, p. 629-651.
1986: Paleomagnetism of the late Precambrian Fourchu Group, Cape Breton Island, Nova Scotia; Canadian Journal of Earth Sciences, v. 23, p. 1673-1685.

Justino, M.F.
1985: Geology and petrogenesis of the plutonic rocks of North Mountain, Cape Breton Island; in Ninth annual open house and review of activities; Program and summaries; Nova Scotia Department of Mines and Energy, ,p. 111.

Kamo, S.L., Gower, C.F., and Krogh, T.E.
1989: Birthdate for the Iapetus Ocean? A precise U-Pb zircon and baddeleyite age for the Long Range dikes, southeast Labrador; Geology, v. 17, p. 602-605.

Karlstrom, K.E.
1983: Reinterpretation of Newfoundland gravity data and arguments for an allochthonous Dunnage Zone; Geology, v. 11, p. 263-266.

Karlstrom, K.E., van der Pluijm, B.A., and Williams, P.F.
1982: Structural interpretation of the eastern Notre Dame Bay area, Newfoundland: regional post-Middle Silurian thrusting and asymmetrical folding; Canadian Journal of Earth Sciences, v. 19, p. 2325-2341.
1983: Sedimentology of Upper Ordovician-Silurian sequences on New World Island, Newfoundland: separate fault-controlled basins?: discussion; Canadian Journal of Earth Sciences, v. 20, p. 1757-1758.

Karson, J.A.
1977: Geology of the Northern Lewis Hills, western Newfoundland; Ph.D. dissertation, State University of New York, Albany, New York, U.S.A., 474 p.
1979: Geologic map and descriptive notes of Lewis Hills Massif, western Newfoundland; Geological Survey of Canada, Open File 628.

Karson, J.A. and Dewey, J.F.
1978: Coastal complex, western Newfoundland: an Early Ordovician oceanicfracture zone; Geological Society of America Bulletin, v. 89, p. 1037-1049.

Kay, M.

1942: Ottawa-Bonnechère Graben and Lake Ontario Homocline; Geological Society of America Bulletin, v. 53, p. 585-646.

1945: Paleogeographic and palinspastic maps; American Association of Petroleum Geologists Bulletin, v. 29, p. 426-450.

1967: Stratigraphy and structure of northeastern Newfoundland-bearing on drift in North Atlantic; American Association of Petroleum Geologists Bulletin, v. 51, p. 579-600.

1972: Dunnage Melange and lower Paleozoic deformation in northeast Newfoundland; 24th International Geological Congress, Montréal, 1972, Section 3, p. 122-133.

1975: Campbellton sequence, manganiferous beds adjoining the Dunnage Melange, northeastern Newfoundland; Geological Society of America Bulletin, v. 86, p. 105-108.

1976: Dunnage Melange and subduction of the Protacadic Ocean, northeast Newfoundland; Geological Society of America, Special Paper 175, 49 p.

Kay, M. and Colbert, E.H.

1965: Stratigraphy and Life History; John Wiley and Sons, Inc., New York, 736 p.

Kay, M. and Eldredge, N.

1968: Cambrian trilobites in central Newfoundland volcanic belt; Geological Magazine, v. 105, No. 4, p. 372-377.

Kean, B.F.

1977: Geology of the Victoria Lake map area, southwestern Newfoundland; Newfoundland Department of Mines and Energy, Mineral Development Division, Report 77-4.

1983a: Geology and mineral deposits of the Lushs Bight Group in the Little Bay Head-Halls Bay area; in Current Research; Newfoundland Department of Mines and Energy, Mineral Development Division, Paper 83-1, p. 157-174.

1983b: Geology of the King George IV Lake map area (12A/4); Newfoundland Department of Mines and Energy, Mineral Development Division, Report 83-4, 67 p.

1984: Geology and mineral deposits of the Lushs Bight Group, Notre Dame Bay, Newfoundland; in Current Research; Newfoundland Department of Mines and Energy, Mineral Development Division, Report 84-1, p. 141-156.

Kean, B.F. and Jayasinghe, N.R.

1980a: Badger map area (12A/16) Newfoundland; in Current Research, (ed.) C.F. O'Driscoll and R.V. Gibbons; Newfoundland Department of Mines and Energy, Mineral Development Division, Report 80-1, p. 37-43.

1980b: Geology of the Lake Ambrose (12A/10) Noel Paul's Brook (12A/9) map area, central Newfoundland; Newfoundland Department of Mines and Energy, Mineral Development Division, Report 80-2, 29 p.

Kean, B.F. and Strong, D.F.

1975: Geochemical evolution of an Ordovician island arc of the central Newfoundland Appalachians; American Journal of Science, v. 275, p. 97-118.

Kean, B.F., Dean, P.L., and Strong, D.F.

1981: Regional geology of the Central Volcanic Belt of Newfoundland; in The Buchans Orebodies: Fifty years of Geology and Mining, (ed.) E.A. Swanson, D.F. Strong, and J.G. Thurlow; Geological Association of Canada, Special Paper 22, p. 65-78.

Keen, C.E. and Haworth, R.

1986: Rifted margin offshore northeast Newfoundland; Geological Society of America, Centennial Continent/Ocean Transect #1, D-1 Northern Appalachians.

Keen, C.E., Keen, M.J., Nichols, B., Ried, I., Stockmal, G.S., Colman-Sadd, S.P., O'Brien, S.J., Miller, H., Quinlan, G., Williams, H., and Wright, J.

1986: Deep seismic reflection profile across the northern Appalachians; Geology, v. 14, p. 141-145.

Kennan, P.S. and Kennedy, M.J.

1983: Coticules: a key to correlation along the Appalachian-Caledonian Orogen?; in Regional Trends in the Geology of the Appalachian-Caledonian-Hercynian- Mauritanide Orogen, (ed.) P.E. Schenk; D. Reidel Publishing Co., p. 355-361.

Kennedy, D.P.

1980: Geology of the Corner Brook Lake area, western Newfoundland; in Current Research, Part A; Geological Survey of Canada, Paper 80-1A, p. 235-240.

1981: Geology of the Corner Brook Lake area, western Newfoundland; M.Sc. thesis, Memorial University of Newfoundland, St. John's, Newfoundland, 370 p.

Kennedy, M.J.

1971: Structure and stratigraphy of the Fleur de Lys Supergroup in the Fleur de Lys area, Burlington Peninsula, Newfoundland; Geological Association of Canada, Proceedings: a Newfoundland Decade, v. 24, p. 59-71.

1973: Pre-Ordovician polyphase structure in the Burlington Peninsula of the Newfoundland Appalachians; Nature, v. 241, p. 114-116.

1975a: The Fleur de Lys Supergroup: stratigraphic comparison of Moine and Dalradian equivalents in Newfoundland with the British Caledonides; Journal of the Geological Society of London, v. 131, p. 305-310.

1975b: Repetitive orogeny in the northeastern Appalachians- New plate models based upon Newfoundland examples; Tectonophysics, v. 28, p. 39-87.

1976: Southeastern margin of the northeastern Appalachians: late Precambrian orogeny on a continental margin; Geological Society of America Bulletin, v. 87, p. 1317-1325.

1979: The continuation of the Canadian Appalachians into the Caledonides of Ireland and Britain; in The Caledonides of the British Isles-Reviewed, (ed.) A.L. Harris, C.H. Holland, and B.E. Leake; Geological Society of London, p. 33-64.

Kennedy, M.J. and McGonigal, M.H.

1972a: The Gander Lake and Davidsville groups of northeastern Newfoundland: new data and geotectonic implications; Canadian Journal of Earth Sciences, v. 9, p. 452-459.

1972b: The Gander Lake and Davidsville groups of northeastern Newfoundland: new data and geotectonic implications: Reply; Canadian Journal of Earth Sciences, v. 9, p. 1781-1783.

Kennedy, M.J. and Philips, W.E.

1971: Ultramafic rocks of Burlington Peninsula, Newfoundland; Geological Association of Canada, Proceedings, v. 24, p. 35-46.

Kennedy, M.J., Blackwood, R.F., Colman-Sadd, S.P., O'Driscoll, C.F., and Dickson, W.L.

1982: The Dover-Hermitage Bay Fault: boundary between the Gander and Avalon zones, eastern Newfoundland; in Major Structural Zones and Faults of the Northern Appalachians, (ed.) P. St-Julien and J. Béland; Geological Association of Canada, Special Paper 24, p. 231-247.

Kennedy, M.J., Phillips, W.E., and Neale, E.R.W.

1972: Similarities in the early structural development of the northwestern margin of the Newfoundland Appalachians and Irish Caledonides; 24th International Geological Congress, Montréal Report, Section 3, p. 516-531.

Kennedy, M.J., Williams, H., and Smyth, W.R.

1973: Geology of the Grey Islands, Newfoundland: northernmost extension of the Fleur de Lys Supergroup; Geological Association of Canada, Proceedings, v. 25, p. 79-90.

Keppie, J.D.

1979: Geological, structural and metamorphic maps of the Province of Nova Scotia; Nova Scotia Department of Mines and Energy, Map 79-1.

1982: The Minas Geofracture; in Major Structural Zones and Faults of the Northern Appalachians, (ed.) P. St-Julien and J. Béland; Geological Association of Canada, Special Paper 24, p. 263-280.

1985: The Appalachian collage; in The Caledonide Orogen-Scandinavia and related areas, Part 2, (ed.) D.G. Gee and B.A. Sturt; John Wiley and Sons Ltd, p. 1217-1226.

1989: Northern Appalachian terranes and their accretionary history; in Terranes in the Circum-Atlantic Paleozoic Orogens, (ed.) R.D. Dallmeyer; Geological Society of America, Special Paper 230, p. 159-192.

1990: Tectono-stratigraphic terranes in Cape Breton Island, Nova Scotia: implications of the configuration of the northern Appalachian Orogen (comment); Geology, v. 18, p. 669-671.

Keppie, J.D. and Dallmeyer, R.D.

1989: Tectonic map of pre-Mesozoic terranes in Circum-Atlantic Phanerozoic orogens; International Geological Correlation Programme, Project 233, Terranes in Circum-Atlantic Paleozoic Orogens, scale 1:5 000 000.

1991: Contrasting U-Pb ages from plutons in the Bras d'Or and Mira terranes of Cape Breton Island, Nova Scotia: discussion; Canadian Journal of Earth Sciences, v. 28, p. 1493-1494.

Keppie, J.D. and Halliday, A.N.

1986: Rb-Sr isotopic data from three suites of igneous rocks, Cape Breton Island, Nova Scotia; Maritime Sediments and Atlantic Geology, v. 22, p. 162-171.

Keppie, J.D. and Krogh, T.E.
1990: Detrital zircon ages from late Precambrian conglomerate, Avalon Composite Terrane, Antigonish Highlands, Nova Scotia (abstract); Geological Society of America, Abstracts with Programs, v. 22, p. 27-28.

Keppie, J.D. and Smith, P.K.
1978: Compilation of isotopic age data of Nova Scotia; Nova Scotia Department of Mines, Report 78-4.

Keppie, J.D., Dallmeyer, R.D., and Krogh, T.E.
1991: U-Pb and ^{40}Ar/^{39}Ar mineral ages from the Cape North area, Cape Breton Island: implication for docking of the Avalon Composite Terrane (abstract); Geological Society of America, Abstracts with Programs; v. 23, no. 1, p. 52.
1992: U-Pb and ^{40}Ar/^{39}Ar mineral ages from the Cape North area, Cape Breton Island: implication for accretion of the Avalon Composite Terrane; Canadian Journal of Earth Sciences, v. 29, p. 295.

Keppie, J.D., Dallmeyer, R.D., and Murphy, J.B.
1990: Tectonic implications of ^{40}Ar/^{39}Ar hornblende ages from late Proterozoic-Cambrian plutons of the Avalon Composite Terrane, Nova Scotia; Geological Society of America Bulletin, v. 102, p. 516-528.

Keppie, J.D., Dostal, J., and Murphy J.B.
1979: Petrology of the late Precambrian Fourchu Group in the Louisburg area, Cape Breton Island; Nova Scotia Department of Mines, Paper 79-1.

Keppie, J.D., Murphy, J.B., Nance, R.D., and Dostal, J.
1989: Terranes in Nova Scotia: their characteristics and accretionary histories; Nova Scotia Department of Mines and Energy, Report 89-3, p. 117-122.

Keppie, J.D., Nance, R.D., Murphy, J.B., and Dostal, J.
1991a: Northern Appalachians: Avalon and Meguma Terranes; in The West African orogens and circum-Atlantic correlatives, (ed.) R.D. Dallmeyer and J.P. Lécorché; Springer-Verlag, Berlin, p. 315-333.

Kidd, W.S.F.
1974: The evolution of the Baie Verte Lineament, Burlington Peninsula, Newfoundland; Ph.D. thesis, Cambridge University, 294 p.
1977: The Baie Verte Lineament: ophiolite complex floor and mafic volcanic fill of a small Ordovician marginal basin; in Island Arcs, Deep Sea Trenches and Back Arc Basins, (ed.) M. Talwani and W.C. Pitman; Maurice Ewing Series, v. 1, p. 407-418.

Kidd, W.S.F., Dewey, J.F., and Bird, J.M.
1978: The Mings Bight ophiolite complex, Newfoundland: Appalachian oceanic crust and mantle; Canadian Journal of Earth Sciences, v. 15, p. 781-804.

Kindle, C.H.
1942: A Lower (?) Cambrian fauna from eastern Gaspé, Quebec; American Journal of Science, v. 240, p.633-641.
1948: Crepicephalid trilobites from Murphy Creek, Quebec, and Cow Head, Newfoundland; American Journal of Science, v. 246, p. 441-451.

Kindle, C.H. and Whittington, H.B.
1958: Stratigraphy of the Cow Head region, western Newfoundland; Geological Society of America Bulletin, v. 69, p. 315-342.

King, A.F.
1986: Geology of the St. John's area, Newfoundland; in Current Research; Newfoundland Department of Mines and Energy, Mineral Development Division, Report 86-1, p. 209-218.
1988: Geology of the Avalon Peninsula, Newfoundland (parts of 1K, 1L, 1M, 1N and 2C); Newfoundland Department of Mines, Mineral Development Division, Map 88-01, scale 1:250 000.
1990: Geology of the St. John's area; Newfoundland Department of Mines and Energy, Geological Survey Branch, Report 90-2, 88 p. and Map at 1: 20 000.

King, A.F., Anderson, M.M., and Benus, A.P.
1988: Late Precambrian sedimentation and related orogenesis of the Avalon Peninsula, eastern Avalon Zone; Geological Association of Canada, Field Trip Guidebook, Trip A4, 84 p.

King, A.F., Brueckner, W.D., Anderson, M.M., and Fletcher, T.
1974: Field Trip Manual B-6; Geological Association of Canada-Mineralogical Association of Canada, St. John's, Newfoundland, 59 p.

King, L.H. and MacLean, B.
1970: Seismic reflection study, Orpheus gravity anomaly; American Association of Petroleum Geologists Bulletin, v. 54, p. 2007-2031.
1976: Geology of the Scotian Shelf; Geological Survey of Canada, Paper 74-31, 31 p.

King, L.H., Fader, G.B.J., Jenkins, W.A.M., and King, E.L.
1986: Occurrence and regional geological setting of Paleozoic rocks on the Grand Banks of Newfoundland; Canadian Journal of Earth Sciences, v. 23, p. 504-526.

King, L.H., Fader, G.B., Poole, W.H., and Wanless, R.K.,
1985: Geological setting and age of the Flemish Cap granodiorite, east of the Grand Banks of Newfoundland; Canadian Journal of Earth Sciences, v. 22, p. 1286-1298.

Kirkwood, D.
1989: Géologie structurale de la région de Percé; Ministère de l'Énergie et des Ressources, Québec, ET 87-17, 42 p.

Kirkwood, D. and Malo, M.
1991: Geometry and strain patterns of the Grand Pabos Fault, Gaspé Peninsula, Québec Appalachians (abstract); Geological Society of America, Abstracts with Programs; v. 23, no. 1, p. 53.

Klappa, C.F. and James, N.P.
1980: Lithistid sponge bioherms of Middle Ordovician age, Table Head Group, west Newfoundland; Canadian Petroleum Geology Bulletin, v. 28, p. 425-451.

Klappa, C.F., Opalinski, P.R., and James, N.P.
1980: Middle Ordovician Table Head Group of western Newfoundland: a revised stratigraphy; Canadian Journal of Earth Sciences, v. 17, p. 1007-1019.

Knapp, D.A.
1980: The stratigraphy, structure, and metamorphism of central Glover Island, western Newfoundland; in Current Research, Part B; Geological Survey of Canada, Paper 80-1B, p. 89-96.
1982: Ophiolite emplacement along the Baie Verte-Brompton Line at Glover Island, western Newfoundland; Ph.D. thesis, Memorial University of Newfoundland, St. John's, Newfoundland, 338 p.

Knapp, D., Kennedy, D.P., and Martineau, Y.
1979: Stratigraphy, structure, and regional correlation of rocks at Grand Lake, western Newfoundland; in Current Research, Part A; Geological Survey of Canada, Paper 79-1A, p. 317-325.

Knight, I.
1977: The Cambro-Orodovician platformal rocks of the Northern Peninsula, Newfoundland; Newfoundland Department of Mines and Energy, Mineral Development Division, Report 77-6, 27 p.
1980: Cambro-Ordovician carbonate stratigraphy of western Newfoundland: sedimentation, diagenesis and zinc-lead mineralization; Newfoundland Department of Mines and Energy, Mineral Development Division, Open File 1154, 43 p.
1982: Geology of the Carboniferous Bay St. George Sub Basin; Newfoundland Department of Mines and Energy, Mineral Development Division, Map 82-1.
1986: Ordovician sedimentary strata of the Pistolet Bay and Hare Bay area, Great Northern Peninsula, Newfoundland; in Current Research; Newfoundland Department of Mines and Energy, Mineral Development Division, Report 86-1, p. 147-160.
1987: Geology of the Roddickton (12I/16) map area; in Current Research; Newfoundland Department of Mines and Energy, Mineral Development Division, Report 87-1, p. 343-357.
1991: Geology of the Cambro-Ordovician rocks in the Port Saunders (NTS 12I/11), Castors River (NTS 12I/15), St. John Island (NTS 12I/14) and Torrent River (NTS 12I/10) map areas; Newfoundland Department of Mines and Energy, Report 91-4, 138 p.
1992: Geology of marmorized, lower Paleozoic, platformal carbonate rocks, "Pye's Ridge", Deer Lake; in Current Research; Newfoundland Department of Mines and Energy, Report 92-1, p. 141-158.

Knight, I. and Boyce, W.D.
1984: Geological mapping of the Port Saunders (12I/11), St. John Island (12I/14), and parts of the Torrent River (12I/10) and Bellburns (12I/16) map sheets, northwestern Newfoundland; in Current Research; Newfoundland Department of Mines and Energy, Mineral Development Division, Report 84-1, p. 114-124.
1987: Lower to Middle Cambrian terrigenous-carbonate rocks of Chimney Arm, Canada Bay: lithostratigraphy, preliminary biostratigraphy and regional significance; in Current Research, (ed.) R.F. Blackwood, D.G. Walsh, and R.V. Gibbons; Newfoundland Department of Mines and Energy, Mineral Development Division, p. 359-365.
1991: Deformed lower Paleozoic platform carbonates, Goose Arm-Old Man Pond; in Current Research; Newfoundland Department of Mines and Energy, Report 91-1, p. 141-154.

Knight I. and Cawood, P.A.
1991: Paleozoic geology of western Newfoundland: an exploration of a deformed Cambro-Ordovician passive margin and foreland basin, and Carboniferous successor basin; Centre for Earth Resources Research, Short Course and Field Guide, 403 p.

Knight, I. and James, N.P.
1987: The stratigraphy of the Lower Ordovician St. George Group, western Newfoundland: the interaction between eustasy and tectonics; Canadian Journal of Earth Sciences, v. 24, p. 1927-1951.

Knight, I. and O'Brien, S.J.
1988: Stratigraphy and sedimentological studies of the Connecting Point Group, portions of the Eastport (2C/12) and St. Brendans (2C/13) map areas, Bonavista Bay, Newfoundland; in Current Research, (ed.) R.S. Hyde, D. Walsh, and R.F. Blackwood; Newfoundland Department of Mines and Energy, Mineral Development Division, Report 88-1, p. 207-228.

Knight, I. and Saltman, P.
1980: Platformal rocks and geology of the Roddickton map area, Great Northern Peninsula; in Current Research, (ed.) C.F. O'Driscoll and R.V. Gibbons; Newfoundland Department of Mines and Energy, Mineral Development Division, Report 80-1, p. 10-28.

Knight, I., James, N.P., and Lane, T.E.
1991: The Ordovician St. George Unconformity, northern Appalachians: the relationship of plate convergence at the St. Lawrence Promontory to the Sauk/Tippecanoe sequence bounfary; Geological Society of America Bulletin, v. 103, p. 1200-1225.

Kontak, D.J., Tuach, J., Strong, D.F., Archibald, D.A., and Farrar, E.
1988: Plutonic and hydrothermal events in the Ackley Granite, southeast Newfoundland, as indicated by total-fusion $^{40}Ar/^{39}Ar$ geochonology; Canadian Journal of Earth Sciences, v. 25, p. 1151-1160.

Kreisa, R.D.
1981: Storm-generated sedimentary structures in subtidal marine facies with examples from the Middle and Upper Ordovician of southwestern Virginia; Journal of Sedimentary Petrology, v. 51, p. 823-848.

Krogh T.E. and Keppie, J.D.
1986: Detrital zircon ages indicating a North African provenance for the Goldenville Formation of Nova Scotia (abstract); Geological Association of Canada, Program with Abstracts, v. 11, p. 9.

Krogh, T.E., Strong, D.F., O'Brien, S.J., and Papezik, V.S.
1988: Precise U-Pb zircon dates from the Avalon terrane in Newfoundland; Canadian Journal of Earth Sciences, v. 25, p. 442-453.

Kumarapeli, P.S.
1985: Vestiges of Iapetan rifting in the craton west of the northern Appalachians; Geoscience Canada, v. 12, no. 2, p. 54-59.

Kumarapeli, P.S. and Saull, V.A.
1966: The St. Lawrence Valley System: A North American equivalent of the East African rift valley system; Canadian Journal of Earth Sciences, v. 3, p. 639-658.

Kumarapeli, P.S., Goodacre, A.K., and Thomas, M.D.
1981: Gravity and magnetic anomalies of the Sutton Mountain region, Quebec and Vermont: expression of rift volcanics related to the opening of Iapetus; Canadian Journal of Earth Sciences, v. 18, p. 680-692.

Kumarapeli, P.S., Dunning, G.R., Pintson, H., and Shaver, J.
1989: Geochemistry and U-Pb zircon age of comenditic metafelsites of the Tibbit Hill Formation, Quebec Appalachians; Canadian Journal of Earth Sciences, v. 26, p. 1374-1383.

Labbé, J-Y.
1991: Géologie de la région de Weedon, Estrie; Ministère de l'Énergie et des Ressources, Québec, ET 88-05, 44 p.

Labbé, J-Y. and Hébert, R.
1988: The Weedon Igneous Complex in Ascot-Weedon Formation of Québec Appalachians: the plutonic foundation of an Ordovician island-arc (abstract); Geological Association of Canada, Program with Abstracts, v. 13, p. A70.

Labbé, J-Y. and St-Julien, P.
1989: Évidence de failles de chevauchement acadiennes dans la région de Weedon, Québec; Journal canadien des sciences de la Terre, v. 26, p. 2268-2277.

Laird, J. and Albee, A.L.
1975: Polymetamorphism and the first occurrence of glaucophane and omphacite in northern Vermont (abstract); Geological Society of America, Abstracts with Programs; v. 7, p. 1159.

Laird, J., Lanphere, M.A., and Albee, A.L.
1984: Distribution of Ordovician and Devonian metamorphism in mafic and peltic schists from northern Vermont; American Journal of Science, v. 284, p. 376-413.

Lajoie, J., Lespérance, P.J., and Béland, J.
1968: Silurian stratigraphy and paleogeography of Matapédia-Témiscouata region, Québec; American Association of Petroleum Geologists Bulletin, v. 52, p. 615-640.

Lamothe, D.
1979: Région de Bolton Centre; Ministère de l'Énergie et des Ressources, Québec, DPV-687, 14 p.
1981a: Région de Mansonville; Ministère de l'Énergie et des Ressources, Québec, DP-833, 12 p.
1981b: Région du Mont Sugar Loaf; Ministère de l'Énergie et des Ressources, Québec, DPV 839, 12 p.

Landing, E., Nowlan, G.S., and Fletcher, T.P.
1980: A microfauna associated with early Cambrian trilobites of the Callavia Zone, northern Antigonish Highlands, Nova Scotia; Canadian Journal of Earth Sciences, v. 17, p. 400-418.

Landing, E., Narbonne, G.M., Myrow, P., Benus, A.P., and Anderson, M.M.
1988: Faunas and depositional environments of the Upper Precambrian through Lower Cambrian, southeastern Newfoundland; in Trace Fossils, Small Shelly Fossils, and the Precambrian-Cambrian Boundary, (ed.) E. Landing, G.M. Narbonne, and P. Myrow; New York State Museum Bulletin No. 463.

Lane, T.E.
1976: Stratigraphy of the White Rock Formation; Maritime Sediments, v. 12, p. 87-106.
1981: The stratigraphy and sedimentology of the White Rock Formation (Silurian), Nova Scotia, Canada; M.Sc. thesis, Dalhousie University, Halifax, Nova Scotia, 270 p.

Langton, J.P. and Van Staal, C.R.
1989: Geology of the Ordovidian Elmtree terrane (abstract); Geological Association of Canada, Program with Abstracts, v. 14, p. A11.

Laroche, P.J.
1983: Appalachians of southern Quebec, seen through Seismic Line no. 2001; in Seismic Expression of Structural Styles, v. 3, (ed.) A.W. Bally; American Association of Petroleum Geologists, Studies in Geology, v. 15, p. 3, 2, 1-7 to 3, 2, 1-24.

Lash, G.G.
1988: Along-strike variations in foreland basin evolution: possible evidence for continental collision along an irregular margin; Basin Research, v. 1, p. 71-83.

Laurent, R.
1975: Occurrences and origin of the ophiolites of southern Québec, Northern Appalachians; Canadian Journal of Earth Sciences, v. 12, p. 443-45.

Laurent, R. and Hébert, R.
1989: The volcanic and intrusive rocks of the Quebec Appalachians ophiolites and their island-arc setting; Chemical Geology, v. 77, p. 265-286.

Laurent, R. and Hébert, Y.
1977: Features of submarine volcanism in ophiolites from the Québec Appalachians; Geological Association of Canada, Special Paper 16, p. 91-109.

Laurent, R., Hébert, R., and Dostal, J.
1989: Geochemistry of Ordovician island-arc and ocean-floor rock assemblages from Québec ophiolites, Canadian Appalachians; in Abstracts of the 28th International Geological Congress, v. 2, p. 2-263.

Laurent, R., Hébert, R., and Hébert, Y.
1979: Tectonic setting and petrological features of the Québec Appalachians ophiolites; in Ophiolites of the Canadian Appalachians and Soviet Urals, (ed.) J. Malpas and R.W. Talkington; Memorial University of Newfoundland, Geology Department, Report 8, p. 53-77.

Leach, G.B., Lowdon, J.A., Stockwell, C.H., and Wanless, R.K.
1963: Age determinations and geological studies, including isotopic ages; Report 4, Geological Survey of Canada, Paper 63-17.

Leavitt, E.M.
1963: The geology of the Precambrian Green Head Group in the Saint John, New Brunswick area; M.Sc. thesis, University of New Brunswick, Fredericton, New Brunswick.

Lebel, D.
1985: Analyse structurale de la déformation acadienne, principalement la faille de Shickshock-Sud dans la région de Rimouski-Matapédia; thèse de maîtrise, Université de Montréal, Montréal (Québec).

LeBlanc, M.
1981:   The late Proterozoic ophiolites at Bou Azzer, Morocco: evidence for Pan-African plate tectonics; in Precambrian Plate Tectonics, (ed.) A. Kroner; Elsevier, Amsterdam, p. 435-441.

Lecorche, J.P. and Dallmeyer, R.D.
1987:   An introduction to the West African orogens: in Geotraverse excursion across the Central Mauritanide Orogen, (ed.) R.D. Dallmeyer, J.P. Lecorche, and O. Dia; International Geological Correlation Programme, Project 233: Terranes in the Circum-Atlantic Paleozoic Orogens, p. 113-135.

Lefort, J.P.
1983:   A new geophysical criterion to correlate the Acadian and Hercynian orogenies of western Europe and eastern North America; in Contributions to the Tectonics and Geophysics of Mountain Chains, (ed.) R.D. Hatcher, H. Williams, and I. Zietz; Geological Society of America, Memoir 158, p. 3-18.
1989:   The submerged part of the Ligerian (Eo-Hercynian): Acadian mobile belt; in Basement Correlation across the North Atlantic, (ed.) J.P. Lefort; Springer-Verlag, Berlin, p. 52-69.

Lefort, J.P. and Haworth, R.
1981:   Geophysical correlations between basement features in North Africa and eastern New England: and their control over North Atlantic structural evolution; Société géologique et minéralogique de Bretagne, v. 13 no. 2, p. 103-116.

Leger, A. and Williams, P.F.
1986:   Transcurrent faulting history of southern New Brunswick; in Current Research, Part B; Geological Survey of Canada, Paper 86-1B, p. 111-120.

Leggett, J.K.
1978:   Eustacy and pelagic regimes in the Iapetus Ocean during the Ordovician and Silurian; Earth and Planetary Science Letters, v. 41, p. 163-169.
1980:   The sedimentological evolution of a Lower Paleozoic accretionary fore-arc in the Southern Uplands of Scotland; Sedimentology, v. 27, p. 401-417.

Leggett, J.K., McKerrow, W.S., and Eales, M.H.
1979:   The Southern Uplands of Scotland: a lower Palaeozoic accretionary prism; Geological Society of London, Journal, v. 136, p. 755-770.

Lespérance, P.J. and Greiner, H.G.
1969:   Région de Squatec-Cabano, comtés de Rimouski, Rivière-du-Loup et Témiscouata; Ministère des Richesses naturelles, Québec, RG 128, 122 p.

Liard, P.
1973:   Legend for map sheets of Mont-Joli, Matane, Sayabec, Ste-Blandine E; Ministère des Richesses naturelles, Québec, DP-290.

Liew, M.Y.C.
1979:   Geochemical studies of the Goldenville Formation at Taylor Head, Nova Scotia; M.Sc. thesis, Dalhousie University, Halifax, Nova Scotia, 204 p.

Lilly, H.D.
1963:   Geology of Hughes Brook-Goose Arm area, west Newfoundland; Memorial University of Newfoundland, Geology Report no. 2, 123 p.
1966:   Late Precambrian and Appalachian tectonics in the light of submarine exploration on the Great Bank of Newfoundland and in the Gulf of St. Lawrence. Preliminary views; American Journal of Science, v. 264, p. 569-574.
1967:   Some notes on stratigraphy and structural style in central west Newfoundland; in Geology of the Atlantic Region, (ed.) E.R.W. Neale and H. Williams; Geological Association of Canada, Special Paper 4, p. 201-211.

Lin, S. and Williams, P. F.
1990:   The structural evolution of the Eastern Highlands Shear Zone in Cape Breton Island, Nova Scotia; in Programme and Summaries: Fourteenth Annual Open House and Review of Activities; Nova Scotia Department of Mines and Energy, Report 90-3, 42.

Lindholm, R.M. and Casey, J.F.
1989:   Regional significance of the Blow Me Down Brook Formation, western Newfoundland: New fossil evidence for an Early Cambrian age; Geological Society of America Bulletin, v. 101, p. 1-13.

Lipman, P.W., Prostka, H.J., and Christiansen, R.L.
1972:   Cenozoic volcanism and plate tectonic evolution of the western United States; Royal Society of London, Philosophical Transactions A, v. 271, p. 217-284.

Lock, B.E.
1972:   Lower Paleozoic history of a critical area: eastern margin of the St. Lawrence Platform in White Bay, Newfoundland, Canada; 24th International Geological Congress, Montreal 1972, section 6, p. 310-324.

Logan, W.E.
1863:   Geology of Canada; Dawson, Montreal, 983 p.

Loncarevic, B.D., Barr, S.M., Raeside, R.P., Keen, C.E., and Marillier, F.
1989:   Northeastern extension and crustal expression of terranes from Cape Breton Island, Nova Scotia, based on geophysical data; Canadian Journal of Earth Sciences, v. 26, p. 2255-2267.

Lorenz, B.E.
1984a: A study of the igneous intrusive rocks of the Dunnage Melange, Newfoundland; Ph.D. thesis, Memorial University of Newfoundland, St. John's, Newfoundland, 220 p.
1984b: Mud-magma interactions in the Dunnage Melange, Newfoundland; in Marginal Basin Geology, (ed.) E.P. Kokelaar and M.F. Howells; Geological Society, Special Publication, no. 16, p. 271-277.

Lorenz, B.E. and Fountain, J.C.
1982:   The South Lake Igneous Complex, Newfoundland: a marginal basin-island arc association; Canadian Journal of Earth Sciences, v. 19, p. 490-503.

Lowdon, J.A.
1960:   Age determinations by the Geological Survey of Canada; in Isotopic ages; Geological Survey of Canada, Report 1, Paper 60-17.
1961:   Age determinations by the Geological Survey of Canada; Geological Survey of Canada, Paper 61-17.

Lowdon, J.A., Stockwell, C.H., Tipper, H.W., and Wanless, R.K.
1963:   Age determinations and geological studies; Geological Survey of Canada, Paper 62-17.

Ludman, A.
1986:   Timing of terrane accretion in eastern and east-central Maine; Geology, v. 14, p. 411-414.
1987:   Pre-Silurian stratigraphy and tectonic significance of the St. Croix Belt, outheastern Maine; Canadian Journal of Earth Sciences, v. 24, p. 2459-2469.

Ludvigsen, R. and Westrop, S.R.
1985:   Three new Upper Cambrian stages for North America; Geology, v. 13, p. 139-143.

Ludvigsen, R., Westrop, S.R., and Kindle, C.H.
1989:   Sunwaptan (Upper Cambrian) trilobites of the Cow Head Group, western Newfoundland, Canada; Palaeontographica Canadiana, No. 6, 175 p.

Lutes, G.
1979:   Geology of Fosterville-North and Eel lakes, map area G-23 and Cantebury- Skiff Lake map area H-23; New Brunswick Department of Natural Resources, Mineral Resources Branch, Map Report 79-3, 22 p.

Lux, D.R.
1986: $^{40}Ar/^{39}Ar$ ages for minerals from the amphibolite dynamothermal aureole, Mont Albert, Gaspé, Quebec; Canadian Journal of Earth Sciences, v. 23, p. 21-26.

Lux, D.R., De Yoreo, J.J., Guidotti, C.V., and Decker, E.R.
1986:   Role of plutonism in low-pressure metamorphic belt formation; Nature, v. 323, no. 6091, p. 794-797.

Lyons, J.B., Aleinikoff, J.N., and Zartman, R.E.
1986:   Uranium-thorium-lead ages of the Highlandcroft Plutonic Suite, northern New England; American Journal of Science, v. 286, p. 489-509.

Lyons, J.B., Boudette, E.L., and Aleinikoff, J.N.
1982:   The Avalonian and Gander zones in central eastern New England; in Major Structural Zones and Faults of the Northern Appalachians, (ed.) P. St-Julien and J. Béland; Geological Association of Canada, Special Paper 24, p. 43-66.

Macdonald, A.S. and Smith, P.K.
1980:   Geology of the Cape North area, northern Cape Breton Island, Nova Scotia; Nova Scotia Department of Mines and Energy, Paper 80-1, 60 p.

MacGregor, I.D.
1962:   Geology, petrology and geochemistry of the Mount Albert and associated ultramafic bodies of central Gaspe, Quebec; M.Sc. thesis, Queens University, Kingston, Ontario, 177 p.
1964:   A study of the contact metamorphic aureole surrounding the Mont Albert ultramafic intrusions; Ph.D. thesis, Princeton University, 195 p.

Mack, G.H.
1985:   Provenance of the Middle Ordovician Blount clastic wedge, Georgia and Tennessee; Geology, v. 13, p. 299-302.

Mackay, B.R.
1919:   Beauceville map-area, Quebec; Geological Survey of Canada, Memoir 127, 105 p.

MacKenzie, G.S.
1964: Geology, St. John, New Brunswick; Geological Survey of Canada, Map 1113A.

MacLean, H.J.
1947: Geology and mineral deposits of the Little Bay area; Newfoundland Geological Survey Bulletin No. 22, 45 p.

Malo, M.
1986: Stratigraphie et structure de l'anticlinorium d'Aroostook-Percé en Gaspésie, Québec; thèse de doctorat, Université de Montréal, Montréal (Québec).

1987: Structural evidence for Acadian wrench faulting in the southeastern Gaspé Peninsula, Quebec (abstract); Geological Association of Canada, Program with abstracts, v. 12, p. 70.

1988: Stratigraphy of the Aroostook-Percé Anticlinorium in the Gaspé Peninsula, Quebec; Canadian Journal of Earth Sciences, v. 25, p. 893-908.

Malo, M. and Béland, J.
1989: Acadian strike-slip tectonics in the Gaspé region, Québec Appalachians; Canadian Journal of Earth Sciences, v. 26, p. 1764-1777.

Malo, M. and Bourque, P.A.
1993: Timing of the deformation events from Late Ordovician to Mid-Devonian in the Gaspé Peninsula; in The Acadian Orogeny: Recent studies in New England Maritime Canada, and the Autochthonous Foreland, (ed.) D.C. Roy and J.W. Skehan; Geological Society of America, Special Paper 275, p. 101-122.

Malo, M. and Moritz, R.
1991: Géologie et métallogénie de la faille du Grand Pabos, région de Raudin-Weir; Ministère de l'Énergie et des Ressources, Québec, MB 91-03, 47 p.

Malo, M., Moritz, R., Chagnon, A., and Roy, F.,
1991: Géologie et métallogénie du segment oriental de la faille du Grand Pabos, Gaspésie; Québec, Ministère de l'Energie et des Ressources, MB 93-55.

Malo, M., Kirkwood, D., de Broucker, G., and St-Julien, P.
1992: A reevaluation of the position of the Baie Verte-Brompton Line in the Québec Appalachians – the influence of Middle Devonian Strike-slip faulting in Gaspé Peninsula; Canadian Journal of Earth Sciences, v. 29, p. 1265-1273.

Malpas, J.G.
1976: The geology and petrogenesis of the Bay of Islands ophiolite suite, west Newfoundland; Ph.D. thesis, Memorial University of Newfoundland, St. John's, Newfoundland, 435 p.

1979: The dynamothermal aureole of the Bay of Islands ophiolite suite; Canadian Journal of Earth Sciences, v. 16, p. 2086-2101.

Marillier, F., Keen, C.E., Stockmal, G.S., Quinlan, G., Williams, H., Colman- Sadd, S.P., and O'Brien, S.J.
1989: Crustal structure and surface zonation of the Canadian Appalachians: implications of deep seismic reflection data; Canadian Journal of Earth Sciences, v. 26, p. 305-321.

Marleau, R.A.
1959: Age relations in the Lake Megantic Range, southern Québec; Geological Association of Canada, Proceedings, v. 11, p. 129-139.

1968: Régions de Woburn-Mégantic-est, comtés de Frontenac et de Beauce; Ministère des Richesses naturelles, Québec, RG131, 55 p.

Marquis, R.
1989a: L'Anticlinorium des Monts Sutton, Richmond, Québec; thèse de doctorat, Université de Montréal, Montréal (Québec)

1989b: Géologie de la région de Windsor, Estrie; Ministère de l'Énergie et des Ressources, Québec, MB 89-12, 22 p.

Marquis, R., Béland, J., and Trzcienski, W.E., Jr.
1987: The Oak Hill Group, Richmond, Quebec: Termination of the Green Mountains-Sutton Mountains Anticlinorium; in Centennial Field Guide, Northeast Section, (ed.) D.Roy; Geological Society of America, p. 363-368.

Marten, B.E.
1971: Stratigraphy of volcanic rocks in the Western Arm area of the Central Newfoundland Appalachians: a Newfoundland Decade; Geological Association of Canada, Proceedings, v. 24, no. 1, p. 73-84.

Martineau, Y.
1980: The relationships among rock groups between the Grand Lake Thrust and Cabot Fault, west Newfoundland; M.Sc. thesis, Memorial University of Newfoundland, St. John's, Newfoundland, 150 p.

Matthew, G.F.
1863: Observations on the geology of St. John County, New Brunswick; Canadian Naturalist and Geologist, v. l, no. 8, p. 241-260.

1889: On the classification of the Cambrian rocks in Acadia; Canadian Record of Science, v. 3, p. 71-81.

Matthew, G.F. (cont.)
1890: On Cambrian organisms in Acadia; Transactions of the Royal Society of Canada, v. 7, Section 4, p. 135-143.

1895: The Protolenus fauna; New York Academy of Sciences, Transactions, v. 14, p. 101-153.

Mattinson, C.R.
1964: Mount Logan area, Matane and Gaspé Nord Co., Quebec; Quebec Department of Natural Resources, GR 118, 97 p.

Mattinson, J.M.
1975: Early Paleozoic ophiolite complexes of Newfoundland: isotopic ages of zircons; Geology, v. 3, p. 181-183.

1976: Ages of zircons from the Bay of Islands ophiolite complex, western Newfoundlnand; Geology, v. 4, p. 393-394.

McAllister, A.L.
1960: Massive sulphide deposits in New Brunswick; Canadian Institute of Mining and Metallurgy Bulletin, v. 53, no. 573, p. 88-98.

McBride, D.E.
1976a: The structure and stratigraphy of the B-zone, Heath Steele Mines, Newcastle, New Brunswick; Ph.D. thesis, University of New Brunswick, Fredericton, New Brunswick, 227 p.

1976b: Tectonic setting of the Tetagouche Group, host to the New Brunswick polymetallic massive sulphide deposits; in Metallogeny and Plate Tectonics, (ed.) D.F. Strong; Geological Association of Canada, Special Paper 14, p. 473-485.

McBride, E.F.
1962: Flysch and associated beds of the Martinsburg Formation (Ordovician), central Appalachians; Journal of Sedimentary Petrology, v. 32, p. 39-91.

McCartney, W.D.
1967: Whitbourne map-area, Newfoundland; Geological Survey of Canada, Memoir 341, 135 p.

McCutcheon, S.R.
1981: Revised stratigraphy of the Long Reach area, southern New Brunswick: evidence for major northwestward-directed Acadian thrusting; Canadian Journal of Earth Sciences, v. 18, p. 646-656.

1984: Geology of the gold-bearing rocks in the Lorneville-Lepreau area; in Ninth Annual Review of Activities, (ed.) B.M.W. Carroll; New Brunswick Department of Natrual Resources, Mineral Resources Division, p. 2-6.

1987: Cambrian stratigraphy in the Hanford Brook area, southern New Brunswick, Canada; in Centennial Field Guide, v. 5, (ed.) D.C. Roy; Northeastern Section of the Geological Society of America, p. 399-402.

McCutcheon, S.R. and Bevier, M.L.
1990: Implications of field relations and U-Pb geochronology for the age of gold mineralization and timing of Acadian deformation in northern New Brunswick; Atlantic Geology, v. 26, p. 237-246.

McCutcheon, S.R. and Robinson, P.T.
1987: Geological constraints on the genesis of the Maritimes Basin, Atlantic Canada; in Sedimentary Basins and Basin-Forming Mechanisms, (ed.) C. Beaumont and A.J. Tankard; Canadian Society of Petroleum Geologists, Memoir 12, p. 287-297.

McCutcheon, S.R. and Ruitenberg, A.A.
1987: Geology and mineral deposits of the Annidale-Nerepis area; New Brunswick Department of Natural Resources, Geological Surveys Branch, Memoir 2, 141 p.

McGerrigle, H.W.
1950: The geology of eastern Gaspé; Québec Department of Mines, GR 35, 168 p.

1954: Les régions de Tourelle et de Courcellette, Péninsule de Gaspé; Ministère des Ressources naturelles, Québec, RG 62, 63 p.

McGonigal, M.H.
1973: The Gander and Davidsville Groups: major tectonostratigraphic units in the Gander Lake area, Newfoundland; M.Sc. thesis, Memorial University of Newfoundland, St. John's, Newfoundland, 121 p.

McGrath, P.H., Hood, P.J., and Cameron, G.W.
1973: Magnetic surveys of the Gulf of St. Lawrence and the Scotian Shelf: in Earth Science Symposium on offshore Eastern Canada, (ed.) P.J. Hood; Geological Survey of Canada, Paper 71-23, p. 339-358.

McGregor, I.D. and Basu, A.R.
1979: Petrogenesis of the Mount Albert ultramafic massif, Québec (summary); Geological Society of America, Part I, v. 90, p. 898-900.

McIlreath, I.A. and James, N.P.
1979: Carbonate slopes; in Facies Models, (ed.) R.G. Walker; Geoscience Canada, Reprint Series 1, p. 133-143.

McKerrow, W.S. and Cocks, L.R.M.
1977: The location of the Iapetus Ocean suture in Newfoundland; Canadian Journal of Earth Sciences, v. 14, p. 488-495.

McLeod, M.J.

1979:   The geology of Campobello Island, southwestern New Brunswick; M.Sc. thesis, University of New Brunswick, Fredericton, Canada, 181 p.

1986:   Contrasting geology across the Cradle Brook thrust zone: subaerial vs marine Precambrian environments, Caledonia Highlands, New Brunswick; Maritime Sediments and Atlantic Geology, v. 22, p. 296-307.

1987:   Geology, geochemistry and mineral deposits of the Big Salmon River-Goose River area, New Brunswick; New Brunswick Department of Natural Resources, Geological Surveys Branch, Report of Investigation 21, 47 p.

McLeod, M.J. and McCutcheon, S.R.

1981:   A newly recognized sequence of possible early Cambrian age in southern New Brunswick: evidence for major southward-directed thrusting; Canadian Journal of Earth Sciences, v. 18, p. 1012-1017.

McLeod, M.J. and Rast, N.

1988:   Correlations and fault systematics in the Passamoquoddy Bay area, southwestern New Brunswick; Maritime Sediments and Atlantic Geology, v. 24, p. 289-300.

McLeod, M.J., Johnson, S.C., and Ruitenberg, A.A.

1990:   Compilation and correlation of southern New Brunswick geology, Charlotte, Queens Kings and Sunbury counties; in Fifteenth Annual Review of Activities, Project Resumes 1990, (ed.) S.A. Abbott; New Brunswick Department of Natural Resources and Energy, Minerals and Energy Division, Information Circular 90-2, p. 137-147.

1991:   Geological compilation of the Hampstead map area (21G/9), Sussex map area (21H/12), and Codys map area (21H/13), southern New Brunswick; New Brunswick Department of Natural Resources and Energy, Mineral Resources Division, Plates 90-152, 90-153, 90-154.

McLeod, M.J., Taylor, R.P., and Lux D.R.

1988:   Geology, $^{40}Ar/^{39}Ar$ geochemistry and Sn-W-Mo-bearing sheeted veins of the Mount Douglas Granite, southwestern New Brunswick; Canadian Mining and Metallurgical Bulletin, v. 81, no. 81, p. 70-77.

McMillan, R.H.

1969:   A comparison of the geological environments of base metal sulfide deposits of the B Zone and North Boundary Zone at Heath Steele Mines, New Brunswick; M.Sc. thesis, University of Western Ontario, London, Ontario, 192 p.

Meaux, D.P., Casey, J.F., and Hart, S.R.

1988:   Origin of subophiolitic volcanic slices associated with the Bay of Islands ophiolite complex of western Newfoundland (abstract); Geological Society of America, Abstracts with Programs, v. 20, no. 7, A215 p.

Mehrtens, C.J.

1989:   Comparison of foreland basin sequences: the Trenton Group in southern Quebec and central New York; in The Trenton Group (Upper Ordovician Series) of Eastern North America. Depostion, Diagenesis and Petroleum; American Association of Petroleum Geologists, Studies in Geology, no. 29, p. 139-158.

Mehrtens, C. and Parker, R.

1984:   Comparison of foreland basin sequences: Trenton Group in southern Quebec and central New York (abstract); American Association of Petroleum Geologists Bulletin, v. 68, p. 1924.

Michard, A.

1976:   Élements de géologie marocaine; Notes et Mémoires du Service géologique du Maroc 252, 408 p.

Middleton, G.V.

1965:   Paleocurrents in Normanskill graywackes north of Albany, New York; Geological Society of America, Bulletin, v. 76, p. 841-844.

1970:   Experimental studies related to problems of flysch sedimentation; in Flysch Sedimentology in North America, (ed.) J. Lajoie; Geological Association of Canada, Special Paper 7, p. 253-272.

Miller, H.G.,

1988:   Geophysical interpretation of the geology of the northeast Gander Terrane, Newfoundland; Canadian Journal of Earth Sciences, v. 25, p. 1161-1174.

Miller, J.D. and Kent, D.V.

1988:   Paleomagnetism of the Siluro-Devonian Andreas redbeds: evidence for an Early Devonian supercontinent; Geology, v. 16, p. 195-198.

Misra, S.B.

1969:   Late Precambrian (?) fossils from southeastern Newfoundland; Geological Society of America Bulletin, v. 82, p. 979-988.

Mitchell, A.H.G.

1978:   The Grampian Orogeny in Scotland: arc-continent collision and polarity reversal; Journal of Geology, v. 86, p. 643-646.

Moench, R.H.

1984:   Geological map of the Sherbrooke-Lewiston area, Maine, New Hampshire, and Vermont; United States Geological Survey, Open-File Report 84-0650.

Murphy, J.B. and Hynes, A.J.

1990:   Tectonic control on the origin and orientation of igneous layering: an example from the Greendale Complex, Antigonish Highlands, Nova Scotia, Canada; Geology, v. 18, p. 403-406.

Murphy J.B. and Keppie J.D.

1987:   The stratigraphy and depositional environment of the late Precambrian Georgeville Group, Antigonish Highlands, Nova Scotia; Maritime Sediments and Atlantic Geology, v. 23, p. 49-61.

Murphy, J.B., Cameron K., Dostal, J., Keppie, J.D., and Hynes, A.J.

1985:   Cambrian volcanism in Nova Scotia, Canada; Canadian Journal of Earth Sciences, v. 22, p. 599-606.

Murphy, J.B., Keppie, J.D., and Hynes, A.J.

1991:   Geology of the Antigonish Highlands; Geological Survey of Canada, Paper 89-10, 115 p.

Murphy, J.B., Keppie, J.D., Dostal, J., and Hynes, A.J.

1990:   The geochemistry and petrology of the late Precambrian Georgeville Group: a volcanic arc-rift succession in the Avalon terrane of Nova Scotia; Geological Society of London, Special Publication No. 51, p. 383-393.

Murphy, J.B., Keppie, J.D., Nance, R.D., and Dostal, J.

1989:   Reassessment of terranes in the Avalon composite terrane of Atlantic Canada (abstract); Geological Society of America, Abstracts with Programs, v. 21, no. 2, 54 p.

Murray, A.

1881:   Report for 1864; in Report of the Geological Survey of Newfoundland from 1864 to 1880; Geological Survey of Newfoundland, Publication, 536 p.

Murray, A. and Howley, J.P.

1881:   Report of the Geological Survey of Newfoundland for 1864-1880; Edward Stanford, London, 536 p.

Murthy, G., Gower, C.F., Tubrett, M., Patzold, R., and Kamo, S.

1988:   Two late Precambrian-Early Cambrian paleomagnetic results from eastern Labrador (abstract); Geological Association of Canada, Program with Abstracts, v. 13, A89 p.

Mutti, E.

1985:   Turbidite systems and their relations to depositional sequences; in Provenance of Arenites, (ed.) G.G. Zuffa; NATO Advanced Scientific Institute, D. Reidel Publishing Company, Boston, p. 65-93.

Nance, R.D.

1982:   Structural reconnaissance of the Green Head Group, St. John, New Brunswick; in Current Research, Part A; Geological Survey of Canada, Paper 82-1A, p. 37-43.

1987:   Model for the Precambrian evolution of the Avalon Terrane in southern New Brunswick, Canada; Geology, v. 15, p. 753-756.

Nance, R.D. and Murphy, J.B.

1990:   Kinematic history of the Bass River Complex, Nova Scotia: Cadomian tectonostratigraphic relations in the Avalon Terrane of the Canadian Appalachians; in The Cadomian Orogeny, (ed.) R.S. D'Lemos, R.A. Strachan, and C.G. Topley; Geological Society of London, Special Publication No. 51, p. 395-406.

Narbonne, G.M., Myrow, P., Landing, E., and Anderson, M.M.

1987:   A candidate stratotype for the Precambrian-Cambrian boundary, Fortune Head, Burin Peninsula, southeast Newfoundland; Canadian Journal of Earth Sciences, v. 24, p. 1277-1293.

Naylor, R.S., Boone, G.M., Boudette, E.L., Ashende, D.D., and Robertson, P.

1973:   Precambrian rocks in the Bronson Hill and Boundary Mountain anticlinoria; EOS, v. 54, p. 495.

Neale, E.R.W.

1957:   Ambiguous intrusive relationships of the Betts Cove-Tilt Cove Serpentinite Belt, Newfoundland; Geological Association of Canada, Proceedings, v. 9, p. 95-107.

Neale, E.R.W. and Kennedy, M.J.

1967:   Relationships of the Fleur de Lys Group to younger groups of the Burlington Peninsula, Newfoundland; in Geology of the Atlantic Region, (ed.) E.R.W. Neale and H. Williams; Geological Association of Canada, Special Paper 4, p. 139-169.

1975:   Basement and cover rocks at Cape North, Cape Breton Island; Maritime Sediments, v. 11, p. 1-5.

Neale, E.R.W. and Nash, W.A.

1963:   Sandy Lake (east half) map-area, Newfoundland; Geological Survey of Canada, Paper 62-28, 40 p.

Neale, E.R.W., Kean, B.F., and Upadhyay, H.D.
1975: Post-ophiolite unconformity, Tilt Cove-Betts Cove area, Newfoundland; Canadian Journal of Earth Science, v. 12, p. 880-886.

Neale, E.R.W., Beland, J., Potter, R.R., and Poole, W.H.
1961: A preliminary tectonic map of the Canadian Appalachian region based on age of folding; Canadian Mining and Metallurgical Bulletin, Transactions, v. LXIV, p. 405-412.

Nelson, K.D.
1981: Mélange development in the Boones Point Complex, north-central Newfoundland; Canadian Journal of Earth Sciences, v. 18, p. 433-442.

Nelson, K.D. and Casey, J.D.
1979: Ophiolitic detritus in the Upper Ordovician flysch of Notre Dame Bay and its bearing on the tectonic evolution of western Newfoundland; Geology, v. 7, p. 27-31.

Nelson, K.D. and Kidd, W.S.F.
1979: The age of the Roberts Arm Group, north-central Newfoundland: discussion; Canadian Journal of Earth Sciences, v. 16, p. 2068-2070.

Nelson, K.D., McBride, J.H., and Arnow, J.A.
1986: Deep reflection character, gravity gradient, and crustal thickness variation in the Appalachian Orogen: Relation to Mesozoic extension and igneous activity (abstract); Geological Society of America, Abstracts with Programs, v. 18, no. 6, p. 704-705.

Neuman, R.B.
1967: Bedrock geology of the Shin Pond and Staceyville quadrangles, Penobscot County, Maine; United States Geological Survey, Professional Paper 524-1, 47 p.
1968: Paleogeographic implications of Ordovician shelly fossils in the Magog Belt of the Northern Appalachian Region; in Studies of Appalachian Geology: Northern and Maritime, (ed.) E-an Zen, W.S. White, J.B. Hadley, and J.B. Thompson, Jr.; Interscience Publishers, p. 35-48.
1976: Early Ordovician (Late Arenig) brachiopods from Virgin Arm, New World Island, Newfoundland; Geological Survey of Canada, Bulletin 261, p. 10-61.
1984: Geology and paleobiology of islands in the Ordovician Iapetus Ocean: Review and implications; Geological Society of America Bulletin, v. 94, p. 1188-1201.

Nixon, G.T. and Papezik, V.S.
1979: Late Precambrian ash-flow tuffs and associated rocks of the Harbour Main Group near Colliers, eastern Newfoundland: chemistry and magmatic affinities; Canadian Journal of Earth Sciences, v. 16, p. 167-181.

Noble, J.P.A.
1976: Silurian stratigraphy and paleogeography, Pointe Verte area, New Brunswick, Canada; Canadian Journal of Earth Sciences, v. 13, p. 537-546.
1985: Occurrence and significance of Late Silurian reefs in New Brunswick, Canada; Canadian Journal of Earth Sciences, v. 22, p. 1518-1529.

Norman, G.W.H.
1935: Lake Ainslie map-area, Nova Scotia; Geological Survey of Canada, Memoir 177, 103 p.

Norman, R.E. and Strong, D.F.
1975: The geology and geochemistry of ophiolitic rocks exposed at Mings Bight, Newfoundland; Canadian Journal of Earth Sciences, v. 12, p. 777-797.

North, F.K.
1971: The Cambrian of Canada and Alaska; in Cambrian of the New World, (ed.) C.H. Holland; Wiley-Interscience, London, New York, Sydney, Toronto, p. 219-324.

Nowlan, G.S.
1981a: Some Ordovician conodont faunules from the Miramichi Anticlinorium, New Brunswick; Geological Survey of Canada, Bulletin 345, 35 p.
1981b: Stratigraphy and conodont faunas of the Lower and Middle Ordovician Romaine and Mingan formations, Mingan Islands, Quebec; Maritime Sediments and Atlantic Geology, v. 12, p. 67.
1983a: Report on three samples from limestone in pillow basalts from the Pointe Verte Formation (northern New Brunswick); Geological Survey of Canada, Internal Fossil Report 002-GSN-1983, 2 p.
1983b: Early Silurian conodonts of eastern Canada; Fossils and Strata, v. 15, p. 95-110.
1986: Report on fifteen samples from Ordovician and Silurian strata of northern New Brunswick; Geological Survey of Canada, Internal Fossil Report 005-GSN-1986.
1988a: Report on twelve samples from Lower Paleozoic strata of northern New Brunswick; Geological Survey of Canada, Internal Fossil Report 011-GSN-1988.

Nowlan, G.S. (cont.)
1988b: Report on two samples from the Lower Paleozoic of Québec and New Brunswick; Geological Survey of Canada, Internal Fossil Report 012-GSN-1988.

Nowlan, G.S. and Barnes, C.R.
1987: Application of conodont colour alteration indices to regional and economic geology; in Conodonts: Investigative techniques and applications, (ed.) R.L. Austin; British Micropaleontological Society, Ellis Horwood Ltd., p.188-202.

Nowlan, G.S. and Thurlow, J.G.
1984: Middle Ordovician conodonts from the Buchans Group, central Newfoundland, and their significance for regional stratigraphy of the Central Volcanic Belt; Canadian Journal of Earth Sciences, v. 21, p. 284-296.

Nyman, M., Quinn, L., Reusch, D.N., and Williams, H.
1984: Geology of Lomond map area, Newfoundland; in Current Research, Part A; Geological Survey of Canada, Paper 84-1A, p. 157-164.

O'Beirne-Ryan, A.M. and Jamieson, R.A.
1986: Geology of the West Branch North River and the Bothan Brook plutons of the south-central Cape Breton Highlands, Nova Scotia; in Current Research, Part B; Geological Survey of Canada, Paper 86-1B, p. 191-200.

O'Beirne-Ryan, A.M., Barr, S.M., and Jamieson, R.A.
1986: Contrasting petrology and age of two megacrystic granitoid plutons, Cape Breton Island, Nova Scotia; in Current Research, Part B; Geological Survey of Canada, Paper 86-1B, p. 179-190.

O'Brien, B.H.
1985: Preliminary report on the geology of the LaHave River area, Nova Scotia; in Current Research, Part A; Geological Survey of Canada, Paper 85-1A, p. 789-794.
1986: Preliminary report on the geology of the Mahone Bay area, Nova Scotia; in Current Research, Part A; Geological Survey of Canada, Paper 86-1A, p. 439-444.
1988: Relationships of phyllite, schist and gneiss in the La Poile Bay-Roti Bay area (parts of 110/9 and 110/16), southwestern Newfoundland; in Current Research; Newfoundland Department of Mines, Mineral Development Division, Report 88-1, p. 109-125.

O'Brien, F.H.C. and Szybinski, Z.A.
1989: Conodont faunas from the Catchers Pond and Cutwell groups, central Newfoundland; in Current Research; Geological Survey of Newfoundland, Report 89-1, p. 121-125.

O'Brien, S.J.,
1983: Geology of the eastern half of the Peter Snout map area (11P/13E), Newfoundland; in Current Research; Newfoundland Department of Mines and Energy, Mineral Development Division, Report 83-1, p. 57-67.
1987: Geology of the Eastport (west half) map area, Bonavista Bay, Newfoundland; in Current Research, (ed.) R.F. Blackwood and R.V. Gibbons; Newfoundland Department of Mines and Energy, Mineral Development Divison, Report 87-1, p. 257-270.

O'Brien, S.J. and Knight, I.
1988: The Avalonian geology of southwest Bonavista Bay: Portions of the St. Brendan's (2C/13) and Eastport (2c/12) map areas; in Current Research (1988); Newfoundland Department of Mines, Mineral Development Divison, Report 88-1, p. 193-205.

O'Brien, S.J. and Nunn, G.A.G.
1980: Terrenceville (1M/10) and Gisborne Lake (1M/15) map areas, Newfoundland; in Current Research; Newfoundland Department of Mines and Energy, Mineral Development Division, Report 80-1, p. 120-133.

O'Brien, S.J. and Taylor, S.W.
1983: Geology of the Baine Harbour (1M/7) and Point Enragee (1M/6) map areas, southeast Newfoundland; Newfoundland Department of Mines and Energy, Mineral Development Division, Report 83-5, 70 p.

O'Brien, S.J. and Tomlin, S.
1984: Geology of the White Bear River map area (11P/14), southern Newfoundland; in Current Research, (ed.) M.J. Murray, J.G. Whelan, and R.V. Gibbons; Newfoundland Department of Mines and Energy, Mineral Development Division, Report 84-1, p. 220-231.

O'Brien, S.J., Dickson, W.L., and Blackwood, R.F.
1986: Geology of the central portion of the Hermitage Flexure area, Newfoundland; in Current Research, (ed.) R.F. Blackwood, D.G. Walsh, and R.V. Gibbons; Newfoundland Department of Mines and Energy, Mineral Development Division, Report 86-1, p. 189-208.

O'Brien, S.J., Nunn, G.A.G., Dickson, W.L., and Tuach, J.

1984: Geology of the Terrenceville (1M/10) and Gisborne Lake (1M/15) map areas, southeast Newfoundland; Newfoundland Department of Mines and Energy, Mineral Development Divison, Report 84-4, 54 p.

O'Brien, S.J., O'Neill, P.P., King A.F., and Blackwood, R.F.

1988: Eastern margin of the Newfoundland Appalachians: A cross-section of the Avalon and Gander zones; Geological Association of Canada, Field Trip Guidebook, Trip B4, 126 p.

O'Brien, S.J., Strong, D.F., and King, A.F.

1990: The Avalon Zone type area: southeastern Newfoundland Appalachians; in Avalon and Cadomian Geology of the North Atlantic, (ed.) R.A. Strachan and G.K. Taylor; Blackie, Glasgow, p. 166-194.

O'Brien, S.J., Wardle, R.J., and King, A.F.

1983: The Avalon Zone: A Pan-African terrane in the Appalachian Orogen of Canada; Geological Journal, v. 18, p. 195-222.

O'Driscoll, C.F.

1977: Geology, petrology and geochemistry of the Hermitage Peninsula, southern Newfoundland; M.Sc. thesis, Memorial University of Newfoundland, St. John's, Newfoundland, 144 p.

O'Driscoll, C.F. and Strong, D.F.

1978: Geology and geochemistry of late Precambrian volcanic and intrusive rocks of southwestern Avalon Zone in Newfoundland; Precambrian Research, v. 8, p. 19-48.

Ollerenshaw, N.C.

1967: Région de Cuoq-Langis (comtés de Matapédia et de Matane); Ministère des Ressources naturelles du Québec, RG 121, 230 p.

Olszewski, W.J. and Gaudette, H.E.

1982: Age of the Brookville Gneiss and associated rocks, southeastern New Brunswick; Canadian Journal of Earth Sciences, v. 1, no. 19, p. 2158-2166.

O'Neill, P.P.

1987: Geology of the west half of the Weir's Pond (2E/1) map area; in Current Research (1987); Newfoundland Department of Mines and Energy, Mineral Development Division, Report 87-1, p. 271-281.

O'Neill, P.P. and Blackwood, F.

1989: A proposal for revised stratigraphic nomenclature of the Gander and Davidsville groups and the Gander River Ultrabasic Belt of northeastern Newfoundland; in Current Research; Newfoundland Department of Mines, Mineral Development Division, Report 89-1, p. 165-176.

O'Neill, P.P. and Knight, I.

1988: Geology of the east half of the Wier's Pond (2E/1) map area and its regional significance; in Current Research (1988); Newfoundland Department of Mines, Mineral Development Division, Report 88-1, p. 165-176.

O'Neill, P.P. and Lux, D.

1989: Tectonothermal history and $^{40}Ar/^{39}Ar$ geochronology of northeastern Gander Zone, Weir's Pond area (2E/1); in Current Research; Newfoundland Department of Mines, Geological Survey of Newfoundland, Report 89-1, p. 131-139.

Osberg, P.H.

1965: Structural geology of the Knowlton-Richmond area, Quebec; Geological Society of America Bulletin, v. 76, p. 223-250.

1967: Lower Paleozoic stratigraphy and structural geology, Green Mountain-Sutton Mountain Anticlinorium, Vermont and southern Quebec; American Association of Petroleum Geologists, Memoir 12, p. 687-700.

1969: Lower Paleozoic stratigraphy and structural geology, Green Mountain-Sutton Mountain Anticlinorium, Vermont and southern Quebec; American Association of Petroleum Geologists, Memoir 12, p. 687-700.

1978: Synthesis of the geology of the Northern Applachians, United States; Geological Survey of Canada, Paper 78-13, p. 137-147.

Osberg, P.H., Hussey, A.M., and Boone, G.M. (ed.)

1985: Bedrock geological map of Maine; Maine Geological Survey, Department of Conservation.

Osborne, F.F. and Archambeault, M.

1948: Chromiferous chlorite from Mont Albert, Quebec; Royal Society of Canada Transactions, v. XLII, Section IV, p. 61-67.

Oshin, I.O. and Crockett, J.H.

1986: The geochemistry and petrogenesis of ophiolitic volcanic rocks from Lac de l'Est, Thetford Mines Complex, Québec, Canada; Canadian Journal of Earth Sciences, v. 23, p. 202-213.

Owen, J.V.

1986: Geology of the Silver Mountain area, western Newfoundland; in Current Research, Part A; Geological Survey of Canada, Paper 86-1A, p. 515-522.

Owen, J.V. and Erdmer, P.

1986: Precambrian and Paleozoic metamorphism in the Long Range Inlier, western Newfoundland; in Current Research, Part B; Geological Survey of Canada, Paper 86-1B, p. 29-38.

1990: Middle Proterozoic geology of the Long Range Inlier, Newfoundland: regional significance and tectonic implications; in Mid-Proterozoic Laurentia-Baltica, (ed.) C.F. Gower, T. Rivers, and B. Ryan; Geological Association of Canada, Special Paper 38, p. 215-231.

Pajari, G.E. and Currie, K.L.

1978: The Gander Lake and Davidsville Groups of northeastern Newfoundland: a re-examination; Canadian Journal of Earth Sciences, v. 15, p. 708-714.

Pajari, G.E., Pickerill, R.K., and Currie, K.L.

1979: The nature, origin and significance of the Carmanville ophiolitic mélange, northeastern Newfoundland; Canadian Journal of Earth Sciences, v. 16, p. 1439-1451.

Pajari, G.E., Rast, N., and Stringer, P.

1977: Paleozoic volcanicity along the Bathurst-Dalhousie geotraverse, New Brunswick, and its relations to structure; in Volcanic Regimes in Canada, (ed.) W.R.A. Baragar, L.C. Coleman, and J.M. Hall; Geological Association of Canada, Special Paper 16, p. 111-124.

Pal, B.K.

1982: Geomagnetic induction studies in eastern Newfoundland; M.Sc. thesis, Memorial University of Newfoundland, St. John's, Newfoundland.

Palmer, A.R.

1983: The Decade of North American Geology, 1983, Geologic time scale; Geology, v. 11, p. 503-504.

Palmer, A.R. and James, N.P.

1979: The Hawke Bay disturbance: a circum-Iapetus event of Lower Cambrian age; in The Caledonides in the USA, (ed.) D.R. Wones; Virginia Polytechnic Institute and State University, Memoir 2, p. 15-18.

Parkash, B.

1970: Downcurrent changes in sedimentary structures in Ordovician turbidite greywackes; Journal of Sedimentary Petrology, v. 40, p. 572-590.

Parkash, B. and Middleton, G.V.

1970: Downcurrent textural changes in Ordovician turbidite greywackes; Sedimentology, v. 14, p. 259-293.

Parker, R.L.

1986: Paleoenvironments of a Middle Ordovician carbonate ramp: Deschambault limestone, of southeastern Quebec; Geological Society of America, Abstracts with Programs, v. 18, no. 1, p. 60.

Patel, I.M.

1976: Lower Cambrian of southern New Brunswick and its correlations with successions in northeastern Appalachians and parts of Europe (abstract); 11th Annual Meeting, Northeastern/Southeastern Sections; Geological Society of America, Abstracts with Programs, p. 243.

Pelletier, B.

1971: A granodiorite drill core from the Flemish Cap, eastern Canadian continental shelf; Canadian Journal of Earth Sciences, v. 8, p. 1499-1503.

Pe-Piper, G. and Murphy, J.B.

1989: Geochemistry and tectonic environment of the late Precambrian Folly River Formation, Cobequid Highlands, Avalon Terrane, Nova Scotia: a continental rift within a volcanic-arc environment; Atlantic Geology, v. 25, p. 143- 152.

Pe-Piper, G. and Piper, D.J.W.

1987: The Pre-Carboniferous rocks of the western Cobequid Hills, Avalon Zone, Nova Scotia; Maritime Sediments and Atlantic Geology, v. 23, p. 41-48.

1989: The Late Hadrynian Jeffers Group, Cobequid Highlands, Avalon Zone of Nova Scotia: a back-arc volcanic complex; Geological Society of America Bulletin, v. 101, p. 364-376.

Philpott, G.R.

1987: Precious-metal and geological investigation of the Charlo River area; in Twelfth Annual Review of Activities, Project Résumés, (ed.) S.A. Abbott; New Brunswick Department of Natural Resources and Energy, Minerals and Energy Division, Information Circular 87-2, p. 13-16.

1988: Precious-metal and geological investigation of the Charlo River area, New Brunswick; in Thirteenth Annual Review of Activities, Project Résumés, (ed.) S.A. Abbott; New Brunswick Department of Natural Resources and Energy, Minerals and Energy Division, Information Circular 88-2, p. 20-31.

Phinney, W.C.
1963: Phase equilibria in the metamorphic rocks of St. Paul Island and Cape North, Nova Scotia; Journal of Petrology, v. 4, p. 90-130.

Piasecki, M.A.J.
1987: Possible basement-cover relationships in the Fleur de Lys terrane, western Newfoundland; in Current Research, Part A; Geological Survey of Canada, Paper 87-1A, p. 391-397.
1988: A major ductile shear zone in the Bay d'Espoir area, Gander Terrane, southeastern Newfoundland; in Current Research; Newfoundland Department of Mines, Mineral Development Division, Report 88-1. p. 135-144.
1991: Geology of the southwest arm of Grand Lake, western Newfoundland; in Current Research, Part D; Geological Survey of Canada, Paper 91-1D, p. 1-8.

Piasecki, M.A.J., Williams, H., and Colman-Sadd, S.P.
1990: Tectonic relationships along the Meelpaeg, Burgeo and Burlington Lithoprobe transects in Newfoundland; in Current Research; Newfoundland Department of Mines, Mineral Development Division, Report 90-1, p. 327- 339.

Pickerill, R.K. and Fyffe, L.R.
1991: Revised late Precambrian stratigraphy near Saint John, New Brunswick: Discussion; in Current Research, Part D; Geological Survey of Canada, Paper 91-1D, p. 187-188.

Pickerill, R.K. and Keppie, J.D.
1981: Observations on the ichnology of the Meguma Group (?Cambro-Ordovician) of Nova Scotia; Maritime Sediments and Atlantic Geology, v. 17, p. 130-138.

Pickerill, R.K., Fillion, D., and Harland, T.L.
1984: Middle Ordovician trace fossils in carbonates of the Trenton Group between Montreal and Quebec City, St. Lawrence Lowland, eastern Canada; in Trace Fossils and Paleoenvironments: Marine Carbonate, Marginal Marine Terrigenous and Continental Terrigenous Settings, (ed.) M.O. Miller, A.A. Eckdale, and M.D. Picard; Journal of Paleontology, v. 58, No. 2, p. 416-439.

Pickerill, R.K., Pajari, G.E., and Currie, K.L.
1978: Carmanville map area, Newfoundland; the northeastern end of the Appalachians; in Current Research, Part A; Geological Survey of Canada, Paper 78-1A, p. 209-216.

Pickering, K.T. and Hiscott, R.N.
1985: Contained (reflected) turbidity currents from the Middle Ordovician Cloridorme Formation, Quebec, Canada: an alternative to the antidune hypothesis; Sedimentology, v. 32, p. 373-394.

Pinet, N. and Tremblay, A.
in press: Tectonic evolution of the Québec Maine Appalachians: from oceanic spreading to obduction and collision in the northern Appalachians; American Journal of Science.

Pintson, H., Kumarapeli, P.S., and Morency, M.
1985: Tectonic significance of the Tibbit Hill Volcanics: Geochemical evidence from Richmond area, Quebec; in Current Research, Part A; Geological Survey of Canada, Paper 85-1A, p. 123-130.

Piper, D.J.W.
1978: Turbidite muds and silts on deep-sea fans and abyssal plains; in Sedimentation in Submarine Canyons, Fans, and Trenches, (ed.) D.J. Stanley and G. Kelling; Dowden, Hutchinson, and Ross, Stroudsburg, Pennsylvania, p. 163-176.

Pique, A., O'Brien, S.J., King, A.F., Schenk, P.E., Skehan, J.W., and Hon, R.
1990: La marge nord-occidentale du Paléo-Gondwana (Maroc occidental et zones orientales des Appalaches); rifting au Précambrien terminal et au Paléozoïque inférieur, et compression hercynienne et alléghanienne au Paléozoïque supérieur; Compte Rendu, Académie des Sciences, Paris, v. 310, p. 411-416.

Plint, H.E. and Jamieson, R.A.
1989: Microstructure, metamorphism, and tectonics of the western Cape Breton Highlands, Nova Scotia; Journal of Metamorphic Geology, v. 7, p. 407-424.

Pohler,
1987: Conodont biofacies and carbonate lithofacies of Lower Ordovician megaconglomerates, Cow Head Group, western Newfoundland; Ph.D. thesis, Memorial University of Newfoundland, St. John's, Newfoundland, 545 p.

Poole, W.H.
1963: Geology, Hayesville, New Brunswick; Geological Survey of Canada, Map 6-1963.
1971: Graptolites, copper and potassium-argon in Goldenville Formation, Nova Scotia; Geological Survey of Canada, Paper 71-1A, p. 9-11.
1980: Rb-Sr age study of the Moulton Hill granite, Sherbrooke area, Québec; in Current Research, Part C; Geological Survey of Canada, Paper 80-1C, p. 185-189.

Posamentier, H.W. and Vail, P.R.
1988: Sequence stratigraphy; sequences and systems tract development; in sequence; stratigraphy, sedimentology: surface and subsurface; (ed.) D.P. James and D. A. Leckie, Canadian Society of Petroleum Geologists, v. 15, p. 571-572.

Potter, R.R.
1964: Upsalquitch Forks map area; in Summary of Activities, Field, 1963; Geological Survey of Canada, Paper 64-1, p. 64-65.
1969: The geology of Burnt Hill area and ore controls of the Burnt Hill tungsten deposit; Ph.D. thesis, Carleton University, Ottawa, Ontario, 136 p.

Potter, R.R., Ruitenberg, A.A., and Davies, J.L.
1969: Mineral exploration in New Brunswick in 1968; Canadian Mining Journal, v. 90, no. 4, p. 68-73.

Poulin, R.
1974: La pétrologie du complexe méta-igné du Lac Masten, Québec; thèse de maîtrise, Université de Montréal, Montréal (Québec), 67 p.

Pratt, B.R. and James, N.P.
1982: Cryptalgal-metazoan bioherms of Early Ordovician age in the St. George Group, western Newfoundland; Sedimentology, v. 29, p. 543-569.
1986: The tidal flat island model for peritidal shallowing-upward sequences; St. George Group, western Newfoundland; Sedimentology, v. 33, p. 313-345.

Pratt, B.R. and Waldron, J.W.F.
1991: A Middle Cambrian trilobite faunule from the Meguma Group of Nova Scotia; Canadian Journal of Earth Science, v. 28, p. 1843-1853.

Pringle, I.R., Miller, J.A., and Warrell, D.M.
1971: Radiometric age determinations from the Long Range Mountains, Newfoundland; Canadian Journal of Earth Sciences, v. 8, p. 1325-1330.

Quinlan, G. and Beaumont, C.
1984: Appalachian overthrusting, lithospheric flexure and the development of Paleozoic stratigraphy in the eastern interior region, United States; Canadian Journal of Earth Sciences, v. 21, p. 973-996.

Quinn, L.A.
1985: The Humber Arm Allochthon at South Arm, Bonne Bay, with extensions in the Lomond area, west Newfoundland; M.Sc. thesis, Memorial University of Newfoundland, St. John's, Newfoundland, 188 p.
1988a: Distribution and significance of Ordovician flysch units in western Newfoundland; in Current Research, Part B; Geological Survey of Canada, Paper 88-1B, p. 119-126.
1988b: Easterly derivation of Ordovician flysch in western Newfoundland; Geological Association of Canada-Mineralogical Association of Canada, Joint Annual Meeting, St. John's, Newfoundland, Program with Abstracts, v. 13, p. A 101.
1988c: Significance of Ordovician flysch in western Newfoundland; 5th International Symposium on the Ordovician System, St. John's, Newfoundland, Program with Abstracts, p. 76.

Quinn, L.A. and Williams, H.
1983: Humber Arm Allochthon at South Arm, Bonne Bay, west Newfoundland; in Current Research, Part A; Geological Survey of Canada, Paper 83-1A, p. 179-182.

Raeside, R.P. and Barr, S.M.
1986: Stratigraphy and structure of the southeastern Cape Breton Highlands, Nova Scotia; Maritime Sediments and Atlantic Geology, v. 22, p. 264-277.
1990: Geology and tectonic development of the Bras d'Or suspect terrane, Cape Breton Island, Nova Scotia; Canadian Journal of Earth Sciences, v. 27, p. 1371-1381.
1992: Geology of the northern and eastern Cape Breton Highlands, Cape Breton Island, Nova Scotia; Geological Survey of Canada, Paper 89-14, 39 p.

Raeside, R.P., Barr, S.M., and Jong, W.,
1984: Geology of the Ingonish River-Wreck Cove area, Cape Breton Island, Nova Scotia; Nova Scotia Department of Mines and Energy, Report of Activities, Report 84-1, p. 249-258.

Raeside, R.P., Barr, S.M., White, C.E., and Dennis, F.A.R.
1986: Geology of the northernmost Cape Breton Highlands, Nova Scotia; in Current Research, Part A; Geological Survey of Canada, Paper 86-1A, p. 291-296.

Rankin, D.W.
1976: Appalachian salients and recesses: late Precambrian continental breakup and the opening of the Iapetus Ocean; Journal of Geophyical Research, v. 81, p. 5605-5619.

Rasetti, F.
1944:   Upper Cambrian trilobites from the Levi conglomerate; Journal of Paleontology, v. 18, p. 229-258.
1945:   New Upper Cambrian trilobites from the Levi conglomerate; Journal of Paleontology, v. 19, p. 462-478.
1946:   Cambrian and Early Ordovician stratigraphy of the Lower St. Lawrence Valley; Geological Association of America Bulletin, v. 57, p. 687-706.
1948a: Lower Cambrian trilobites from the conglomerates of Quebec (exclusive of the Ptychopariidea); Journal of Paleontology, v. 22, p. 1-29.
1948b: Middle Cambrian trilobites from the conglomerates of Quebec (exclusive of the Ptychopariidea); Journal of Paleontology, v. 22, p. 315-339.
1955:   Lower Cambrian Ptychopariid trilobites from the conglomerates of Quebec; Smithsonian Miscellaneous Collections, v. 128, no. 7, 35p.
1963:   Middle Cambrian Ptychoparioid trilobites from the conglomerates of Quebec; Journal of Paleontology, v. 37, p. 575-594.

Rast, N.
1979:   Precambrian meta-diabases of southern New Brunswick: the opening of the Iapetus Ocean?; Tectonophysics, v. 59, p. 127-137.

Rast, N. and Currie, K. L.
1976:   On the position of the Variscan Front in southeastern New Brunswick and its relation to Precambrian basement; Canadian Journal of Earth Sciences, v. 13, p. 194-196.

Rast, N. and Dickson, W.L.
1982:   The Pocologan mylonite zone; in Major Structural Zones and Faults of the Northern Appalachians, (ed.) P. St-Julien and J. Béland; Geological Association of Canada, Special Paper 24, p. 249-261.

Rast, N. and Horton, J.W.
1989:   Mélanges and olistostromes in the Appalachians of the United States and mainland Canada: An assessment; in Mélanges and Olistostromes of the U.S. Appalachians, (ed.) J.W. Horton and N. Rast; Geological Society of America, Special Paper 228, p. 1-15.

Rast, N. and Stringer, P.
1974:   Recent advances and the interpretation of geological structure of New Brunswick; Geoscience Canada, v. 1, no. 4, p. 15-25.
1980:   A geotraverse across a deformed Ordovician ophiolite and its Silurian cover, northern New Brunswick, Canada; Tectonophysics, v. 69, p. 221-245.

Rast, N., Kennedy, M.J., and Blackwood, R.F.
1976a: Comparison of some tectonostratigraphic zones in the Appalachians of Newfoundland and New Brunswick; Canadian Journal of Earth Sciences, v. 13, p. 868-875.

Rast, N., O'Brien, B.H., and Wardle, R.J.
1976b: Relationships between Precambrian and Lower Paleozoic rocks of the "Avalon Platform" in New Brunswick, the northeast Appalachians and the British Isles; Tectonophysics, v. 30, p. 315-338.

Reusch, D.N.
1983:   The New World Island Complex and its relationships to nearby formations, north central Newfoundland; M.Sc. thesis, Memorial University of Newfoundland, St. John's, Newfoundland, 248 p.

Reynolds, P.H., Jamieson, R.A., Barr, S.M., and Raeside, R.P.
1989:   A $^{40}Ar/^{39}Ar$ dating study in the Cape Breton Highlands, Nova Scotia: thermal histories and tectonic implications; Canadian Journal of Earth Sciences, v. 26, p. 2081-2091.

Rickard, M.J.
1965:   Taconic Orogeny in the western Appalachians: experimental application of microtextural studies to isotopic dating; Geological Society of America Bulletin, v. 76, p. 523-536.

Rickard, L.V. and Fisher, D.W.
1973:   Middle Ordovician Normanskill Formation, eastern New York: Age, stratigraphic and structural position; American Journal of Science, v. 273, p. 580-590.

Riley, G.C.
1962:   Stephenville map area, Newfoundland; Geological Survey of Canada, Memoir 323, 72 p.

Ringele, H.
1982:   La déformation taconienne en Gaspésie du Sud: le Groupe de Maquereau et son encaissant; thèse de maîtrise, Université de Montréal, Montréal (Québec), 68 p.

Riva, J.
1968:   Graptolite faunas from the Middle Ordovician of the Gaspé north shore; Canadian Naturalist, v. 95, p. 1379-1400.
1969:   Middle and Upper Ordovician graptolites of the St.Lawrence Lowlands of Quebec and Anticosti Island; in North Atlantic Geology and Continental Drift, (ed.) G.M. Kay; American Association of Petroleum Geologists, Memoir 12, p. 513-556.

Riva, J. and Malo, M.
1988:   Age and correlation of the Honorat Group, southern Gaspé Peninsula; Canadian Journal of Earth Sciences, v. 25, p. 618-1628.

Riva, J., Belt, E.S., and Mehrtens, C.J.
1977:   The Trenton, Utica and flysch successions of the platform near Québec City, Canada; 69th Annual Meeting of the the New England Intercollegiate Geological Conference, Laval University, Québec (Quebec), Guidebook for excursion A8, p. 1-37.

Robinson, D.S. and Fyson, W.K.
1976:   Fold structures, southern Stokes Mountains area, Eastern Townships, Canada-Taconic or Acadian?; Canadian Journal of Earth Sciences, v. 13, p. 66-74.

Rocha-Campos, A.C.
1981:   The Cambrian(?) Limbo Group of Bolivia; in Earth's pre-Pleistocene glacial record, (ed.) M.J. Hambrey and W.B. Harland; Cambridge University Press, Cambridge, United Kingdom, p. 910-911.

Rodgers, J.
1965:   Long Point and Clam Bank formations, western Newfoundland; Geological Association of Canada, Proceedings, v. 16, p. 83-94.
1967:   Chronology of tectonic movements in the Appalachian region of eastern North America; American Journal of Science, v. 265, p. 408-427.
1968:   The eastern edge of the North American continent during the Cambrian and Early Ordovician; in Studies of Appalachian Geology: Northern and Maritime, (ed.) E-an Zen, W.S. White, J.B. Hadley, and J.B. Thompson; Interscience Publishers, p. 141-149.
1970:   The Tectonics of the Appalachians; John Wiley and Sons, New York, 271 p.
1971:   The Taconic Orogeny; Geological Society of America Bulletin, v. 82, p. 1141-1178.

Rodgers, J. and Neale, E.R.W.
1963:   Possible "Taconic" klippen in western Newfoundland; American Journal of Science, v. 261, p. 713-730.

Rodrigue, G.
1979:   Étude pétrologique des roches ophiolitiques du Mont Orford; thèse de maîtrise, Université Laval, Québec (Québec).

Roliff, W.A.,
1968:   Oil and gas exploration: Anticosti Island, Quebec; Geological Association of Canada, Proceedings, v. 19, p. 31-36.

Rondot, J.
1972:   La transgression ordovicienne dans le comté de Charlevoix, Québec; Journal canadien des sciences de la Terre, v. 9, p. 1187-1203.

Rose, E.R.
1952:   Torbay map area, Newfoundland; Geological Survey of Canada, Memoir 265, 64 p.

Ross, R.J., Jr. and Ingham, J.K.
1970:   Distribution of the Toquima-Table Head (Middle Ordovician Whiterock) faunal realm in the Northern Hemisphere; Geological Society of America Bulletin, v. 81, p. 393-408.

Ross, R.J. and James, N.P.
1987:   Brachiopod biostratigraphy of the Middle Ordovician Cow Head and Table Head groups, western Newfoundland; Canadian Journal of Earth Sciences, v. 24, p. 70-95.

Ross, R.J., Jr., et al.
1982:   The Ordovician System in the United States; International Union of Geological Sciences, Publication, no. 12, 73 p., 3 charts.

Roy, D.C.
1989:   The Depot Mountain Formation: Transition from syn-to post-Taconian Basin along the Baie Verte-Brompton Line in northwestern Maine; in Studies in Maine Geology; Maine Geological Survey, v. 2, p. 85-99.

Roy, D.C. and Mencher, E.
1976:   Ordovician and Silurian stratigraphy of northeastern Aroostook County, Maine; in Contributions to the Stratigraphy of New England, (ed.) L.R. Page; Geological Society of America, Memoir 148, p. 25-52.

Ruitenberg, A.A.
1967:   Stratigraphy, structure and metallization Piskahegan-Rolling Dam area, northern Appalachians, New Brunswick, Canada; Leidse Geologische Mededelingen, v. 40, p. 79-120.

Ruitenberg, A.A. and Ludman, A.
1978:   Stratigraphy and tectonic setting of early Paleozoic sedimentary rocks of the Wirral-Big Lake area, southwestern New Brunswick and southeastern Maine; Canadian Journal of Earth Sciences, v. 15, p. 22-32.

Ruitenberg, A.A., Fyffe, L.R., McCutcheon, S.R., St. Peter, C.J., Irrinki, R.R., and Venugopal, D.V.
1977: Evolution of pre-Carboniferous tectonostratigraphic zones in the New Brunswick Appalachians; Geoscience Canada, v. 4, no. 4, p. 171-181.

Ruitenberg, A.A., Giles, P.S., Venugopal, D.V., Buttimer, S.M., McCutcheon, S.R., and Chandra, J.
1979: Geology and mineral deposits, Caledonia area; New Brunswick Department of Natural Resources, Mineral Resources Branch, Memoir 1, 213 p.

Ruitenberg, A.A., Venugopal, D.V., and Giles, P.S.
1973: "Fundy Cataclastic Zone", New Brunswick: evidence for post-Acadian penetrative deformation; Geological Society of America Bulletin, v. 84, p. 3029-3044.

Saif, S.I.
1977: Identification, correlation and origin of the Key Anacon-Brunswick mines ore horizon, Bathurst, New Brunswick; Ph.D. thesis, University of New Brunswick, Fredericton, New Brunswick, 292 p.

1980: Petrographic and geochemical investigation of iron formation and other iron-rich rocks in Bathurst district, New Brunswick; in Current Research, Part A; Geological Survey of Canada, Paper 80-1A, p. 309-317.

St-Julien, P.
1963: Géologie de la région d'Orford; thèse de doctorat, Université Laval, Québec (Québec).

1970: Région d'Orford-Sherbrooke; Ministère des Richesses naturelles, Québec, Carte 1619, échelle 1/50 000.

1987: Géologie des régions de Saint-Victor et de Thetford-Mines (moitié est); Ministère de l'Énergie et des Ressources, Québec, MM 86-01, 66 p.

1979: Structure and stratigraphy of platform and Appalachian sequences near Québec City; Geological Association of Canada - Mineralogical Association of Canada, Field Trip Guide A-9, 31 p.

St-Julien, P. and Hubert, C.
1975: Evolution of the Taconian Orogen in the Quebec Appalachians; in Tectonics and Mountain Ranges; American Journal of Science, v. 275-A, p. 337-362.

St-Julien, P. and Lamarche, R-Y.
1965: Géologie de la région de Sherbrooke; Ministère des Richesses naturelles, Québec, RP 530, 36 p.

St-Julien, P. and Slivitzky, A.
1985: Compilation géologique de la région de l'Estrie-Beauce; Ministère de l'Énergie et des Ressources, Québec, Carte 2030, échelle 1/250 000.

St-Julien, P., Slivitsky, A., and Feininger, T.
1983: A deep structural profile across the Appalachians of southern Quebec; in Contributions to the tectonics and geophysics of mountain chains, (ed.) R.D. Hatcher, H. Williams, and I. Zietz; Geological Society of America, Memoir 158, p. 103-111.

St-Julien, P., Hubert, C., Skidmore, B., and Béland, J.
1972: Appalachian structure and stratigraphy, Quebec; 24th International Geological Congress, Montréal, 1972, Guidebook A56-C56, 99 p.

St-Julien, P., Trzcienski, W.E., and Wilson, C.
1990: A structural, petrological and geochemical traverse of the Schickshock Terrane, Gaspésie; in Guidebook for field trips in La Gaspésie, Quebec, (ed.) W.E. Trzcienski; New England Intercollegiate Geological Conference, 82nd annual meeting, Gîte du Mont Albert, Gaspésie, Quebec, p. 248-285.

St. Peter, C.J.
1978: Geology of parts of Restigouche, Victoria and Madawask counties, northwestern New Brunswick; New Brunswick Department of Natural Resources, Mineral Resources Branch, Report of Investigation 17, 69 p.

1981: Geology of North Branch Southwest Miramichi River (map areas J-14, J-15, J-16); New Brunswick Department of Natural Resources, Mineral Resources Branch, Map Report 80-1, 61 p.

1982: Geology of Juniper-Knowlesville-Carlisle area, New Brunswick; New Brunswick Department of Natural Resources, Map Report 82-l, 82 p.

Sandi, E.M.
1978: Conodont biostratigraphy of the Lower and Middle Ordovician of the Montreal area; M.Sc. thesis, University of Ottawa, Ottawa, Ontario, 41 p.

Sangster, A.L., Hunt, P.A., and Mortensen, J.K.
1990: U-Pb geochronology of the Lime Hill gneissic complex, Cape Breton Island; Atlantic Geology, v. 26, p. 229-236.

Saunders, A.D. and Tarney, J.
1984: Geochemical characteristics of basaltic volcanism within back-arc basins; in Marginal Basin Geology, (ed.) B.P. Kokelaar and M.F. Howells; Geological Society of London, Special Publication 16, p. 59-76.

Schenk, P.E.
1970: Regional variation of the flysch-like Meguma Group (lower Paleozoic) of Nova Scotia compared to recent sedimentation off the Scotian Shelf; in Flysch Sedimentology in North America, (ed.) J. Lajoie; Geological Association of Canada, Special Paper 7, p. 127-153.

1972: Possible Late Ordovician glaciation of Nova Scotia; Canadian Journal of Earth Sciences, v. 9, p. 95-107.

1981: The Meguma Zone of Nova Scotia – a remnant of Western Europe, South America, or Africa?; in Geology of North Atlantic Borderlands, (ed.) J.M. Kerr, A.J. Ferguson, and L.C. Machan; Canadian Society of Petroleum Geologists Memoir, v. 7, p. 119-148.

1982: Stratigraphy and sedimentology of the Meguma Zone and part of the Avalon Zone; in Field guide for Avalon and Meguma zones, (comp.) A.F. King; IGCP Project 27, NATO Advance Study Institute, Atlantic Canada, Report 9, Department of Earth Sciences, Memorial University of Newfoundland, p. 189-224.

1983: The Meguma Terrane of Nova Scotia, Canada: an aid in trans-Atlantic correlation: in Regional Trends in the Geology of the Appalachian- Caledonian-Hercynian-Mauritanide Orogen, (ed.) P.E. Schenk; D. Reidel Publishing Company, p. 121-130.

1988: Happenings on the West African Craton and events in the Meguma Zone: cause and effect (abstract); Geological Association of Canada, Program with Abstracts, v. 13, p. A109.

1991: Events and sea-level changes on Gondwana's margin: the Meguma Zone (Cambrian to Devonian) of Nova Scotia, Canada; Geological Society of America Bulletin, v. 103, p. 512-521.

Schenk, P.E. and Adams, P.J.
1986: Sedimentology of the Risser's Beach Member of the Meguma Group (lower Paleozoic) of Nova Scotia: a highly efficient deep-sea fan system (abstract); Geological Society of America, Northeastern Section, Abstracts with Programs, v. 18, p. 64.

Schenk, P.E. and Lane, T.E.
1981: Early Paleozoic tillite of Nova Scotia, Canada; in Earth's Pre-Pleistocene Glacial Record, (ed.) M.J. Hambrey and W.B. Harland; Cambridge University Press, p. 707-710.

Schillereff, H.S.
1980: Relationships among rock groups within and beneath the Humber Arm Allochthon at Fox Island River, western Newfoundland; M.Sc. thesis, Memorial University of Newfoundland, St. John's, Newfoundland, 166 p.

Schlanger, S.O. and Jenkyns, H.C.
1976: Cretaceous oceanic anoxic events: causes and consequences; Geologie en Mijnbow, v. 55, p. 179-184.

Schuchert, C.
1923: Sites and natures of the North American geosynclines; Geological Society of America Bulletin, v. 34, p. 151-229.

Schuchert, C. and Dunbar, C.O.
1934: Stratigraphy of western Newfoundland; Geological Society of America, Memoir 1, 123 p.

Scotese, C.R., Bambach, R.K., Barton, C., van der Voo, R., and Ziegler, A.M.
1979: Paleozoic base maps; Journal of Geology, v. 87, p. 217-277.

Secor, D.T., Jr., Samson, S.L., Snoke, A.W., and Palmer, A.R.
1983: Confirmation of the Carolina Slate Belt as an exotic terrane; American Association for the Advancement of Science, v. 221, p. 649-650.

Shanmugam, G. and Moiola, R.J.
1982: Eustatic control of turbidites and winnowed turbidites; Geology, v. 10, p. 231-235.

Shanmugam, G. and Walker, K.R.
1980: Sedimentation, subsidence, and evolution of a foredeep basin in the Middle Ordovician, southern Appalachians; American Journal of Science, v. 280, p. 479-496.

Shaw, A.B.,
1958: Stratigraphy and structure of the St-Albans area, northwestern Vermont; Geological Society of America Bulletin, v. 69, p. 519-568.

1964: Time in Stratigraphy; McGraw-Hill, New York, 365 p.

Shaw, F.C.
1980: Shallow water lithofacies and trilobite biofacies of the Mingan Formation (Ordovician), eastern Quebec; naturaliste canadien, v. 107, p. 227-242.

Shaw, H.F. and Wasserburg, G.J.
1984: Isotopic constraints on the origin of Appalachian mafic complexes; American Journal of Science, v. 284, p. 319-349.

Skinner, R.
1974: Geology of Tetagouche Lakes, Bathurst and Nepisiguit Falls map areas, New Brunswick; Geological Survey of Canada, Memoir 371, 133 p.
1975: Geology of Tuadook Lake map area (21 J/15), New Brunswick; Geological Survey of Canada, Paper 74-33, 9 p.

Skinner, R. and McAlary, J.D.
1952: Preliminary map, Nepisiquit Falls, Gloucester and Northumberland counties, New Brunswick; Geological Survey of Canada, Paper 53-23.

Skipper, K.
1971: Antidune cross-stratification in a turbidite sequence, Cloridorme Formation, Gaspé, Quebec; Sedimentology, v. 17, 51-68.

Skipper, K. and Bhattacharjee, S.B.
1978: Backset bedding in turbidites: a further example from the Cloridorme Formation (Middle Ordovician), Gaspé, Quebec; Journal of Sedimentary Petrology, v. 48, p. 193-202.

Skipper, K. and Middleton, G.V.
1975: The sedimentary structures and depositional mechanics of certain Ordovician turbidites, Cloridorme Formation, Gaspé Peninsula, Quebec; Canadian Journal of Earth Sciences, v. 12, p. 1934-1952.

Slivitzky, A. and St-Julien, P.
1987: Compilation géologique de la région de l'Estrie-Beauce; Ministère de l'Énergie et des Ressources, Québec; MM 85-04, 40 p.

Sloss, L.L.
1963: Sequences in the cratonic interior of North America; Geological Society of America Bulletin, v. 74, p. 93-114.

Smith, C.H.
1958: Bay of Islands Igneous Complex, western Newfoundland; Geological Survey of Canada, Memoir 290, 132 p.

Smith, C.H. and Skinner, R.
1958: Geology of the Bathurst-Newcastle mineral district, New Brunswick; Canadian Institute of Mining and Metallurgy Bulletin, v. 51, no. 551, p. 150-155.

Smith, S.A. and Hiscott, R.N.
1984: Latest Precambrian to Early Cambrian basin evolution, Fortune Bay, Newfoundland: fault-bounded basin to platform; Canadian Journal of Earth Sciences, v. 21, p. 1379-1392.

Smitheringale, W.G.
1972: Low potash Lushs Bight tholeiites: ancient oceanic crust in Newfoundland; Canadian Journal of Earth Sciences, v. 9, p. 574-588.
1973: Geology of parts of Digby, Bridgetown, and Gaspereau map-areas, Nova Scotia; Geological Survey of Canada, Memoir 375, 78 p.

Smyth, W.R.
1971: Stratigraphy and structure of part of the Hare Bay Allochthon, Newfoundland; Geological Association of Canada, Proceedings, v. 24, no. 1, p. 47-57.
1973: The stratigraphy and structure of the southern part of the Hare Bay Allochthon, northwest Newfoundland; Ph.D. dissertation, Memorial University of Newfoundland, St. John's, Newfoundland, 172 p.
1981: The Grey River orthoquartzites and related rocks, southern Newfoundland: a slice of Avalon Zone Precambrian basement tectonically positioned along the southern margin of the Central Mobile Belt? (abstract); Geological Association of Canada; in Program with Abstracts, v. 6, p. 52.

Smyth, W.R. and Schillereff, H.S.
1982: The pre-Carboniferous geology of southwest White Bay; in Current Research, (ed.) C.F. O'Driscol and R.V. Gibbons; Newfoundland Department of Mines and Energy, Mineral Development Division, Report 82-1, p. 78-98.

Snelgrove, A.K.
1928: The geology of the Central Mineral Belt of Newfoundland; Canadian Mining and Metallurgical Bulletin, no. 197, p. 1057-1127.
1931: Geology and ore deposits of Betts Cove-Tilt Cove area, Notre Dame Bay, Newfoundland; Canadian Mining and Metallurgical Bulletin, v. 24, No. 4, 43 p.

Snelling, N.J.
1985: An interim time scale; in The Chronology of the Geological Record, (ed.) N.J. Snelling; British Geological Survey, Memoir 10, p. 261-265.

Spray, J.G., Flagler, P.A., and Dunning, G.R.
1990: Crystallization and emplacement chronology of the Fournier oceanic fragment, Canadian Appalachians; Nature, v. 344, p. 232-235.

Stanley, R.S. and Ratcliff, N.M.
1985: Tectonic synthesis of the Taconian Orogeny in western New England; Geological Society of America Bulletin, v. 96, p. 1227-1250.

Stein, R. and Littke, R.
1990: Organic-carbon-rich sediments and paleoenvironment: results from Baffin Bay (ODP- Leg 105) and the upwelling area off northwest Africa (ODP-Leg 108); in Deposition of Organic Facies, (ed.) A.Y. Huc; American Association of Petroleum Geologists, Studies in Geology, No. 30, Tulsa, U.S.A., p. 41-56.

Stenzel, S.R. and James, N.P.
1987: Death and destruction of an early Paleozoic carbonate platform, western Newfoundland (abstract); Abstracts with Programs; Society of Economic Paleontologists and Mineralogists, Annual Meeting, Austin, Texas, p. 80.
1988: Foundering and burial of an early Paleozoic carbonate platform, western Newfoundland; Joint Annual Meeting Geological Association of Canada - Mineralogical Association of Canada, Program with Abstract, v. 13, p. A117.

Stenzel, S.R., Knight, I., and James, N.P.
1990: Carbonate platform to foreland basin: revised stratigraphy of the Table Head Group (Middle Ordovician), western Newfoundland; Canadian Journal of Earth Sciences, v. 27, p. 14-26.

Stern, R.J., Gottfried, D., and Hedge, C.E.
1984: Late Precambrian rifting and crustal evolution in the Northeastern Desert of Egypt; Geology, v. 12, p. 168-172.

Stevens, R.D., Delabio, R.N., and Lachance, G.R.
1982: Age determinations and geological studies; K-Ar isotopic ages, Report 15; Geological Survey of Canada Paper 81-2, 56 p.

Stevens, R.K.
1965: Geology of the Humber Arm, West Newfoundland; M.Sc. thesis, Memorial University of Newfoundland, St. John's, Newfoundland.
1968: Taconic klippen of Western Newfoundland; in Report of Activities, Part A; Geological Survey of Canada, Paper 68-1A, p. 8-10.
1970: Cambro-Ordovician flysch sedimentation and tectonics in west Newfoundland and their possible bearing on a Proto-Atlantic Ocean; in Flysch Sedimentology in North America, (ed.) J. Lajoie; Geological Association of Canada, Special Paper 7, p. 165-178.
1974: History of Canadian geology; Geoscience Canada, v. 1, no. 2, p. 40-44.

Stevens, R.K. and James, N.P.
1976: Large sponge-like mounds from the Lower Ordovician of western Newfoundland (abstract); in Abstracts with Programs; Geological Society of America, v. 8, p. 1122.

Stevenson, I.M.
1959: Shubenacadie and Kennetcook map-areas, Colchester, Hants and Halifax counties, Nova Scotia; Geological Survey of Canada, Memoir 302, 88 p.
1967a: Goose Bay map area, Labrador (13F); Geological Survey of Canada, Paper 67-33, 12 p.
1967b: Minipi Lake; Geological Survey of Canada, Map 6-1967.

Stewart, D.B., Unger, J.D., Phillips, J.D., Goldsmith, R., Poole, W.H., Spencer, C.P., Green, A.G., Loiselle, M.C., and St-Julien, P.
1986: The Quebec-western Maine seismic reflection profile: setting and first year results; in Reflection Seismology: the Continental Crust, (ed.) M. Barazangi and L. Brown; American Geophysical Union, Geodynamics Series, v. 14, p. 189-199.

Stockmal, G.S. and Waldron, J.W.F.
1990: Structure of the Appalachian deformation front in western Newfoundland: implications of multichannel seismic reflection data; Geology, v. 18, p. 765-768.

Stouge, S.
1979: Conodonts from Davidsville Group of Botwood Zone, Newfoundland; in Current Research, (ed.) R.V. Gibbons; Newfoundland Department of Mines and Energy, Mineral Development Division, Report 79-1, p. 43-44.
1980: Lower and Middle Ordovician conodonts from central Newfoundland and their correlatives in western Newfoundland; in Current Research; Newfoundland Department of Mines and Energy, Mineral Development Division, Report 80-1, p. 134-142.
1982: Preliminary conodont biostratigraphy and correlation of Lower to Middle Ordovician carbonates of the St. George Group, Great Northern Peninsula, Newfoundland; Newfoundland Department of Mines and Energy, Mineral Division, Report 80-3, 59 p.
1984: Conodonts of the Middle Ordovician Table Head Formation, western Newfoundland; Fossils and Strata, No. 16, 145 p.

Stow, D.A.V., Alam, M., and Piper, D.J.W.
1984:   Sedimentology of the Halifax Formation, Nova Scotia: Lower Palaeozoic fine-grained turbidites; in Fine-grained Sediments: Deep Water Processes and Facies, (ed.) D.A.V. Stow and D.J.W. Piper; Geological Society of London, Special Publication 15, p. 127-144.

Stringer, P.
1975:   Acadian slaty cleavage noncoplanar with fold axial surfaces in the Northern Appalachians; Canadian Journal of Earth Sciences, v. 12, p. 949-961.

Stringer, P. and Burke, K.B.S.
1985:   Structure in the southwest New Brunswick, Excursion 9; Geological Association of Canada, University of New Brunswick, Fredericton, New Brunswick, 34 p.

Strong, D.F.
1972:   Sheeted diabases of central Newfoundland: new evidence for Ordovician sea floor spreading; Nature, v. 235, p. 102-104.

1973:   Lushs Bight and Roberts Arm groups of central Newfoundland: possible juxtaposed oceanic and island arc volcanic suites; Geological Society of America Bulletin, v. 84, p. 3917-3928.

1974:   An 'off-axis' alkali volcanic suite associated with the Bay of Islands ophiolites, Newfoundland; Earth and Planetary Science Letters, v. 21, p. 301-309.

1977:   Volcanic regimes of the Newfoundland Appalachians; in Volcanic Regimes of Canada, (ed.) W.R.A. Baragar; Geological Association of Canada, Special Paper 16, p. 61-90.

1979:   Proterozoic tectonics of Northwestern Gondwanaland: new evidence from eastern Newfoundland; Tectonophysics, v. 54, p. 81-101.

Strong, D.F. and Dickson, W.L.
1978:   Geochemistry of Paleozoic granitoid plutons from contrasting tectonic zones of northeast Newfoundland; Canadian Journal of Earth Sciences, v. 15, p. 145-156.

Strong, D.F. and Dostal, J.
1980:   Dynamic partial melting of Proterozoic upper mantle: evidence from rare earth elements in oceanic crust of eastern Newfoundland; Contributions to Mineralogy and Petrology, v. 72, p. 165-173.

Strong, D.F. and Kean, B.F.
1972:   New fossil localities in the Lush's Bight terrane of central Newfoundland; Canadian Journal of Earth Sciences, v. 9, p. 1572-1576.

Strong, D.F. and Payne, J.G.
1973:   Early Paleozoic volcanism and metamorphism of the Moretons Harbour-Twillingate area, Newfoundland; Canadian Journal of Earth Sciences, v. 10, p. 1363-1379.

Strong, D.F. and Williams, H.
1972:   Early Paleozoic flood basalts of northwest Newfoundland: their petrology and tectonic significance; Geological Association of Canada, Proceedings, v. 25, no. 2, p. 43-54.

Strong, D.F., O'Brien, S.J., Taylor, S.W., Strong, P.G., and Wilton, D.H.C.
1978:   Marystown (1M/3) and St. Lawrence (1L/14) map areas, Newfoundland; Newfoundland Department of Mines and Energy, Mineral Development Division, Report 77-8, 81 p.

Stukas, V.
1978:   Plagioclase release patterns: a high resolution $^{40}Ar/^{39}Ar$ study; Ph.D. thesis, Dalhousie University, Halifax, Nova Scotia.

Stukas, V. and Reynolds, P.H.
1974a: $^{40}Ar/^{39}Ar$ dating of the Long Range dikes, Newfoundland; Earth and Planetary Science Letters, v. 22, p. 256-266.

1974b: $^{40}Ar/^{39}Ar$ dating of the Brighton Gabbro Complex, Lushs Bight terrane, Newfoundland; Canadian Journal of Earth Sciences, v. 11, p. 1485-1488.

Sullivan, R.W. and van Staal, C.R.
1990:   Age of a metarhyolite from the Tetagouche Group, Bathurst, New Brunswick, from U-Pb isochron analyses of zircons enriched in common Pb; in Radiogenic Age and Isotopic Studies, Report 3; Geological Survey of Canada, Paper 89-2, p. 109-117.

Sullivan, R.W., van Staal, C.R., and Langton, J.P.
1990:   U-Pb zircon ages of plagiogranite and gabbro from the ophiolitic Devereaux Formation, Fournier Group, northeastern New Brunswick; in Radiogenic Age and Isotopic Studies, Report 3; Geological Survey of Canada, Paper 89-2, p. 119-122.

Sutter, J.F., Ratcliffe, N.M., and Musaka, S.B.
1985: $^{40}Ar/^{39}Ar$ and K-Ar data bearing on the metamorphic and tectonic history of western New England; Geological Society of America Bulletin, v. 96, p. 123-136.

Swanson, E.A., Strong, D.F., and Thurlow, J.G. (ed.)
1981:   The Buchans Orebodies: fifty years of geology and mining; Geological Association of Canada, Special Paper 22, 305 p.

Swinden, H.S.
1987:   Ordovician volcanism and mineralization in the Wild Bight Group, central Newfoundland: A geological, petrological, geochemical and isotopic study; Ph.D. thesis, Memorial University of Newfoundland, St. John's, Newfoundland, 452 p.

1988:   Geology and economic potential of the Pipestone Pond area (12A/1 NE; 12A/8 E), central Newfoundland; Newfoundland Department of Mines, Geological Survey Branch, Report 88-2, 88 p.

Swinden, H.S. and Thorpe, R.I.
1984:   Variations in style of volcanism and massive sulfide deposition in Early-Middle Ordovician island arc sequences of the New Brunswick Central Mobile Belt; Economic Geology, v. 79, p. 1596-1619.

Swinden, H.S., Jenner, G.A., Fryer, B.J., Hertogen, J., and Roddick, J.C.
1990:   Petrogenesis and paleotectonic history of the Wild Bight Group, an Ordovician rifted island arc in central Newfoundland; Contributions to Mineralogy and Petrology, v. 105, p. 219-241.

Szybinski, Z.A.
1988:   New interpretation of the structural and stratigraphic setting of the Cutwell Group, Notre Dame Bay, Newfoundland; in Current Research, Part B; Geological Survey of Canada, Paper 88-1B, p. 263-270.

Tankard, A.J.
1986:   On the depositional response to thrusting and lithospheric flexure: examples from the Appalachian and Rocky Mountain basins; in Foreland Basins, (ed.) P. Homewood and P A. Allen; International Association of Sedimentologists, Special Publication 8, p. 369-392.

Tanoli, S.K. and Pickerill, R.K.
1988:   Lithostratigraphy of the Lower Cambrian-Lower Ordovician Saint John Group, southern New Brunswick; Canadian Journal of Earth Sciences, v. 25, p. 669-690.

1990:   Lithofacies and basinal development of the type "Etchcheminian Series" (Lower Cambrian Ratcliffe Brook Formation), Saint John area, southern New Brunswick; Atlantic Geology, v. 26, p. 57-78.

Tanoli, S.K., Pickerill, R.K., and Currie, K.L.
1985:   Distinction of Eocambrian and Lower Cambrian redbeds, Saint John area, southern New Brunswick; in Current Research, Part A; Geological Survey of Canada, Paper 85-1A, p. 699-702.

Taylor, F.C.
1965:   Silurian stratigraphy and Ordovician-Silurian relationships in southwestern Nova Scotia; Geological Survey of Canada, Paper 64-13, 24 p.

Taylor, F.C. and Schiller, E.A.
1966:   Metamorphism of the Meguma Group of Nova Scotia; Canadian Journal of Earth Sciences, v. 3, p. 959-974.

Teixeira, W., Tassinari, C.C.G., Cordani, U.G., and Kawashita, K.
1989:   A review of the geochronology of the Amazonian craton: tectonic implications; Precambrian Research, v. 42, p. 213-227.

Thomas, M.D. and Willis, C.
1989:   Gravity models of the Saint George Batholith, New Brunswick Appalachians; Canadian Journal of Earth Sciences, v. 26, p. 561-576.

Thomas, M.D., Tanczyk, W.I., Cioppa, M., and O'Dowd, D.V.
1991:   Ground magnetic and rock magnetism studies near the Appalachian Dunnage-Gander terrane boundary, northern New Brunswick; in Current Research, Part D; Geological Survey of Canada, Paper 91-1D, p. 169-178.

Thomas, W.A.
1977:   Evolution of Appalachian-Ouachita salients and recesses from reentrants and promontories in the continental margin; American Journal of Science, v. 277, p. 1233-1278.

Thurlow, J.G.
1981:   The Buchans Group: its stratigraphic and structural setting; in The Buchans Orebodies: Fifty Years of Geology and Mining, (ed.) E.A. Swanson, D.F. Strong, and J.G. Thurlow; Geological Association of Canada, Special Paper 22, p. 79-89.

Tremblay, A.
1989:   Géologie structurale et géochimie des roches volcaniques et sédimentaires du Complexe d'Ascot, Sherbrooke, Québec, Canada; thèse de doctorat, Université Laval, Québec (Québec).

1990:   Géologie de la région d'Ayers Cliff (partie Est); Ministère de l'Énergie et des Ressources, Québec, MB 90-30, 95 p.

Tremblay, A. (cont.)

1992a: Tectonic and accretionary history of Taconian oceanic rocks of the Quebec Appalachians; American Journal of Science, v. 292, p. 229-252.

1992b: Synthèse géologique de la région de Sherbrooke (Estrie); Ministère de l'Énergie et des Ressources, Québec, ET 90-02, 71 p.

Tremblay, A. and Bergeron, M.

1990: The Ascot Complex granitoid as a source for granite-bearing conglomerate of the Magog Group, Québec Appalachians: comparative petrography and geochemistry (abstract); Geological Association of Canada, Program with Abstracts; v. 15, p. A132.

Tremblay, A. and Pinet, N.

in press: Distribution and characteristics of Taconian and Acadian deformation, southern Québec Appalachians; Geological Society of America Bulletin.

Tremblay, A. and St-Julien, P.

1990: Structural style and evolution of a segment of the Dunnage Zone from the Quebec Appalachians and its tectonic implications; Geological Society of America Bulletin, v. 102, p. 1218-1229.

Tremblay, A., Hebert, R., and Bergeron, M.

1989a: Le Complexe d'Ascot des Applaches du sud du Québec: pétrologie et géochimie; Journal canadien des sciences de la Terre, v. 26, p. 2407-2420.

Tremblay, A., Laflèche, M.R., McNutt, R.H., and Bergeron, M.

1994: Petrogenesis of Cambro-Ordovician subduction-related granitic magmas of the Québec Appalachians; Chemical Geology, v. 113, p. 205-220.

Tremblay, A., St-Julien, P., and Labbé, J-Y.

1989b: Mise à l'évidence et cinématique de la faille de La Guadeloupe, Appalaches du sud du Québec; Journal canadien des sciences de la Terre, v. 26, p. 1932-1943.

Tremblay, A., Malo, M., Kirkwood, D., and Cousineau, P.A.

1991: Tectonic and structural evolution of Cambro-Ordovician oceanic rocks and post-Ordovician cover rocks of the Québec Appalachians (abstract); Geological Society of America, Northeastern-Southeastern sections meeting, Abstracts with Programs; v. 23, p. A140.

Troelsen, J.

1947: Stratigraphy and structure of the Bonne Bay-Trout River area; Ph.D. dissertation, Yale University, New Haven, Connecticut, U.S.A.

Trottier, J.

1982: Géochimie et pétrologie du complex ophiolitique de Thetford Mines, Québec; thèse de maîtrise, Université du Québec à Montréal, Montréal (Québec).

Trzcienski, W.E., Jr.

1976: Crossitic amphibole and its possible tectonic significance in the Richmond area, southeastern Québec; Canadian Journal of Earth Sciences, v. 13, p. 711-714.

1988: Retrograde eclogite from Mount Albert, Gaspé, Québec; Canadian Journal of Earth Sciences, v. 25, p. 30-37.

1989: Studies of the Sterrett Mine area, Eastern Townships, Quebec; Geological Survey of Canada, Open File 2015, 40 p.

Trzcienski, W.E. and Birkett, T.C.

1982: Pumpellyite compositional variations along the western marign of the Quebec Appalachians; Canadian Mineralogist, v. 20, p. 203-209.

Trzcienski, W.E., Carmichael, D.M., and Helmstaedt, H.

1984: Zoned sodic amphibole: petrologic indicator of changing pressure and temperature during tectonism in the Bathurst area, New Brunswick, Canada; Contributions to Mineralogy and Petrology, v. 85, p. 311-320.

Tuach, J.

1987: The Ackley high-silica magmatic-metallogenic system and associated post-tectonic granites, southeast Newfoundland; Ph.D. thesis, Memorial University, St. John's, Newfoundland.

Tucker, R.D. and Robinson, P.

1990: Age and setting of the Bronson Hill magmatic arc: a re-evaluation based on U-Pb zircon ages in southern New England; Geological Society of America Bulletin, v. 102, p. 1404-1419.

Tucker, R.D., Krogh, T.E., Ross, R.J., and Williams, S.H.

1990: Time-scale calibration by high precision U-Pb zircon dating of interstratified volcanic ashes in the Ordovician and Lower Silurian stratotypes of Britain; Earth and Planetary Science Letters, v. 100, p. 51-58.

Tuke, M.F.

1968: Autochthonous and allochthonous rocks in the Pistolet Bay area in northernmost Newfoundland; Canadian Journal of Earth Sciences, v. 5, p. 501-513.

Upadhyay, H.D.

1973: The Betts Cove ophiolite and related rocks of the Snooks Arm Group, Newfoundland; Ph.D. thesis, Memorial University of Newfoundland, St. John's, Newfoundland, 224 p.

Upadhyay, H.D., Dewey, J.F., and Neale, E.R.W.

1971: The Betts Cove ophiolite complex, Newfoundland: Appalachian oceanic crust and mantle; Geological Association of Canada, Proceedings, v. 24, no. 1, p. 27-34.

Vallières, A.

1977: Géologie de la région de Cacouna à Saint-André-de-Kamouraska, Comté de Rivière-du-Loup et de Kamouraska; Ministère des Richesses naturelles, Québec, DPV-513, 31 p.

1984: Stratigraphie et structure de l'orogénie taconique de la région de Rivière-du-Loup, Québec; thèse de doctorat, Université Laval, Québec (Québec).

Vallières, A., Hubert, C., and Brooks, C.

1978: A slice of basement in the western margin of the Appalachian Orogen, Saint-Malachie, Quebec; Canadian Journal of Earth Sciences, v. 15, p. 1242-1249.

van Berkel, J.T.

1987: Geology of the Dashwoods Pond, St. Fintans and Main Gut map areas, southwest Newfoundland; in Current Research, Part A; Geological Survey of Canada, Paper 87-1A, p. 399-408.

van Berkel, J.T. and Currie, K.L.

1988: Geology of the Puddle Pond (12A/5) and Little Grand Lake (12/a/12) map areas, southwestern Newfoundland; in Current Research (1988); Newfoundland Department of Mines, Mineral Development Division, Report 88-1, p. 99-107.

van Berkel, J.T., Johnston, H.P., and Currie, K.L.

1986: A preliminary report on the geology of the southern Long Range, southwest Newfoundland; in Current Research, Part B; Geological Survey of Canada, Paper 86-1B, p. 157-170.

van der Pluijm, B.A. and van der Voo, R.

1991: Paleogeography and accretion of terranes in the northern Appalachians: paleomagnetic and tectonostratigraphic evidence (abstract); Geological Society of America, Abstracts with Programs; v. 23, no. 1, p. 143.

van der Pluijm, B.A. and van Staal, C.R.

1988: Characteristics and evolution of the Central Mobile Belt, Canadian Appalachians; Journal of Geology, v. 96, p. 535-547.

van der Pluijm, B.A., Karlstrom, K.E., and Williams, P.F.

1987: Fossil evidence for fault derived stratigraphic repetition in the northeastern Newfoundland Appalachians; Canadian Journal of Earth Sciences, v. 24, p. 2337-2350.

van Staal, C.R.

1985: Structure and metamorphism of the Brunswick Mines area; Bathurst, New Brunswick, Canada; Ph.D. thesis, University of New Brunswick, Fredericton, New Brunswick, 484 p.

1986: Preliminary results of structural investigations in the Bathurst Camp of northern New Brunswick; in Current Research, Part A; Geological Survey of Canada, Paper 86-1A, p. 193-204.

1987: Tectonic setting of the Tetagouche Group in northern New Brunswick: implications for plate tectonic models of the northern Appalachians; Canadian Journal of Earth Sciences, v. 24, p. 1329-1351.

van Staal, C.R. and Williams, H.

1991: Dunnage Zone-Gander Zone relationships in the Canadian Appalachians (abstract); Geological Society of America, combined Northeastern and Southeastern sections, Abstracts with Programs; v. 23, no. 1, p. 143.

van Staal, C.R. and Williams, P.F.

1984: Structure, origin and concentration of the Brunswick no. 6 and no. 12 ore bodies; Economic Geology, v. 79, p. 1669-1692.

1986: Structural interpretation of the Brunswick ore bodies; in Geology in the Real World-The Kingsley Dunham Volume, (ed.) R.W. Nesbitt and J. Nichol; The Institute of Mining and Metallurgy, London, United Kingdom, p. 451-462.

1988: Collision along an irregular margin: a regional plate tectonic interpretation of the Canadian Appalachians; Canadian Journal of Earth Sciences, v. 25, p. 1912-1916.

van Staal, C.R., Langton, J.P., and Sullivan, R.W.

1988a: A U-Pb zircon age for the ophiolitic Deveraux Formation, Elmtree Terrane, northeastern New Brunswick; in Radiogenic Age and Isotopic Studies: Report 2; Geological Survey of Canada, Paper 88-2, p. 37-40.

van Staal, C.R., Winchester, J.A., and Bedard, J.H.

1991: Geochemical variations in Middle Ordovician volcanic rocks of the northern Miramichi Highlands and their tectonic significance; Canadian Journal of Earth Sciences, v. 28, p. 1031-1049.

van Staal, C.R., Winchester, J.A., and Cullen, R.
1988b: Evidence for D₁-related thrusting and folding in the Bathurst-Millstream River area, New Brunswick; in Current Research, Part B; Geological Survey of Canada, Paper 88-1B, p. 135-148.

van Staal, C.R., Ravenhurst, C.E., Winchester, J.A.,
Roddick, J.C., and Langton, J.P.
1990: Post-Taconic blueschist suture in the northern Appalachians of northern New Brunswick, Canada; Geology, v. 18, p. 1073-1077.

Venugopal, D.V.
1978: Geology of Benton-Kirkland, Upper Eel River Bend, map area G-22 (21 G/13); New Brunswick Department of Natural Resources, Mineral Resources Branch, Map Report 78-3, 16 p.
1979: Geology of Debec Junction-Gibson Millstream-Temperance Vale-Meductic region, map areas G-21, H-21, I-21, and H-22 (Parts of 21 J/3, 21 J/4, 21 G/13, 21 G/14); New Brunswick Department of Natural Resources, Mineral Resources Branch, Map Report 79-5, 36 p.

Walcott, C.D.
1891: Correlation papers-Cambrian; United States Geological Survey Bulletin, v. 81, 447 p.

Waldron, J.W.F.
1985: Structural history of continental margin sediments beneath the Bay of Islands ophiolite, Newfoundland; Canadian Journal of Earth Sciences, v. 22, p. 1618-1632.
1987: Sedimentology of the Goldenville-Halifax transition in the Tancook Island area, South Shore, Nova Scotia; Geological Survey of Canada, Open File 1535, 49 p.

Waldron, J.W.F. and Graves, M.
1987: Preliminary report on sedimentology of sandstones, slates, and bioclastic material in the Meguma Group, Mahone Bay, Nova Scotia; in Current Research, Part A; Geological Survey of Canada, Paper 87-1A, p. 409-414.

Waldron, J.W.F. and Jensen, L.R.
1985: Sedimentology of the Goldenville Formation, eastern shore, Nova Scotia; Geological Survey of Canada, Paper 85-15, 31 p.

Waldron, J.W.F. and Milne, J.V.
1991: Tectonic history of the central Humber Zone, western Newfoundland Applachians: post-Taconian deformation in the Old Man's Pond area; Canadian Journal of Earth Sciences, v. 28. p. 398-410.

Walker, K.R. and Keller, F.B.
1977: Stop 17: Tellico Formation-submarine fan, proximal to distal turbidite environments; in The Ecostratigraphy of the Middle Ordovician of the Southern Appalachians (Kentucky, Tennessee, and Virginia), U.S.A. (a field excursion), (ed.) S.C. Ruppel and K.R. Walker; University of Tennessee, Knoxville, Studies in Geology 77-1, p. 134-140.

Walker, R.G.
1967: Turbidite sedimentary structures and their relationship to proximal and distal depositional environments; Journal of Sedimentary Petrology, v. 37, p. 25-43.

Walthier, T.N.
1949: Geology and mineral deposits of the area between Corner Brook and Stephenville, western Newfoundland; Geological Survey of Newfoundland, Bulletin, v. 35, Part 1, p. 1-62.

Wanless, R.K., Stevens, R.D., Lachance, G.R., and Delabio, R.N.
1972: Age determinations and geological studies: K-Ar isotopic ages, Report 10; Geological Survey of Canada, Paper 71-2.
1973: Age determinations and geological studies: K-Ar isotopic ages, Report 11; Geological Survey of Canada, Paper 73-2.

Wanless, R.K., Stevens, R.D., Lachance, G.R., and Edmond, C.M.
1968: Age determinations and geological studies, K-Ar isotopic ages, Report 8; Geological Survey of Canada, Paper 67-2A.

Wanless, R.K., Stevens, R.K., Lachance, G.R., and Rimsaite, J.Y.H.
1965: Age determinations and geological studies, K-Ar isotopic ages, Report 5; Geological Survey of Canada, Paper 64-17, Part 1.

Wardle, R.J.
1978: The stratigraphy and tectonics of the Greenhead (sic) Group: its relationship to Hadrynian and Paleozoic rocks, southern New Brunswick; Ph.D. thesis, University of New Brunswick, Fredericton, New Brunswick.

Wasowski, J.J. and Jacobi, R.D.
1986: The tectonics and depositional history of the Ordovician and Silurian rocks of Notre Dame Bay, Newfoundland: discussion; Canadian Journal of Earth Sciences, v. 23, p. 583-585.

Watson, K. de P.
1947: Geology and mineral deposits of the Baie Verte-Ming's Bight area, Newfoundland; Geological Survey of Newfoundland, Bulletin, v. 21, 48 p.

Watters, S.E.
1987: Gold-bearing rocks-Bay of Fundy coastal zone; in Twelfth annual review of activities, (ed.) S.A. Abbott; New Brunswick Department of Natural Resources and Energy, Mineral Resources Division, Information Circular 87-2, p. 41-44.

Weaver, D.F.
1967: A geological interpretation of the Bouguer anomaly field of Newfoundland; Publication of the Dominion Observatory, Ottawa, v. XXXV, no. 5, p. 223-251.

Webb, G.W.
1969: Paleozoic wrench faults in Canadian Appalachians: in North Atlantic- geology and continental drift, (ed.) M. Kay; American Association of Petroleum Geologists, Memoir 12, p. 754-786.

Weeks, L.J.
1954: Southeast Cape Breton Island, Nova Scotia; Geological Survey of Canada, Memoir 277, 112 p.

Whalen, J.B.
1987: Geology of a northern portion of the Central Plutonic Belt, New Brunswick; in Current Research, Part A; Geological Survey of Canada, Paper 87-1A, p. 209-217.

Whalen, J.B. and Currie, K.L.
1983: The Topsails igneous terrane of western Newfoundland; in Current Research, Part A; Geological Survey of Canada, Paper 83-1A, p. 15-23.

Whalen, J.B., Hegner, E., and Jenner, G.A.
1989: Nature of Canadian Appalachians basement terranes as inferred from a Nd isotopic transect (abstract); in Abstracts with Programs, Geological Society of America, Annual Meeting, St. Louis Missouri; v. 21, no. 6, p. A-201.

Whalen, J.B., Currie, K.L., and van Breeman, O.
1987: Episodic Ordovician-Silurian plutonism in the Topsails igneous terrane, western Newfoundland; Transactions of the Royal Society of Edinburgh, Earth Sciences, v. 78, p. 17-28.

White, D.E.
1939: Geology and molybdenite deposits of the Rencontre East area, Fortune Bay, Newfoundland; Ph.D. thesis, Princeton University, Princeton, U.S.A., 119 p.

Whitehead, R.E.S. and Goodfellow, W.D.
1978: Geochemistry of volcanic rocks from the Tetagouche Group, Bathurst, New Brunswick, Canada; Canadian Journal of Earth Sciences, v. 15, p. 207-219

Whittington, H.B. and Kindle, C.H.
1963: Middle Ordovician Table Head Formation, western Newfoundland; Geological Society of America Bulletin, v. 74, p. 745-758. .

Widmer, K.
1950: The geology of the Hermitage Bay area, Newfoundland; Ph.D. thesis, Princeton Univeristy, Princeton, U.S.A., 439 p.

Wiebe, R.A.
1972: Igneous and tectonic events in northeastern Cape Breton Island, Nova Scotia; Canadian Journal of Earth Sciences, v. 9, p. 1262-1277.
1975: Origin and emplacement of Acadian granitic rocks, northern Cape Breton Island; Canadian Journal of Earth Sciences, v. 12, p. 252-262.

Williams, H.
1962: Botwood (west-half) map-area, Newfoundland; Geological Survey of Canada, Paper 62-9, 16 p.
1963: Relationship between base metal mineralization and volcanic rocks in northeastern Newfoundland; Canadian Mining Journal, v. 84, no. 8, p. 39-42.
1964a: The Appalachians in northeastern Newfoundland: a two-sided symmetrical system; American Journal of Science, v. 262, p. 1137-1158.
1964b: Botwood, Newfoundland; Geological Survey of Canada, Map 60-1963, scale 1:253 440.
1967: Silurian Rocks of Newfoundland; in Geology of the Atlantic Region, (ed.) E.R.W. Neale and H. Williams; Geological Association of Canada, Special Paper 4, p. 93-137.
1968: Wesleyville, Newfoundland; Geological Survey of Canada, Map 1227A, scale 1:253 440 (with descriptive notes).
1971a: Mafic-ultramafic complexes in western Newfoundland Appalachians and the evidence for their transportation: a review and interim report; Geological Association of Canada, Proceedings, v. 24, p. 9-25.
1971b: Geology of Belleoram map-area, Newfoundland (1M/11); Geological Survey of Canada, Paper 70-65, 39 p.

Williams, H. (cont.)

1973: Bay of Islands map-area (12G), Newfoundland; Geological Survey of Canada, Paper 72-34, 7 p. (includes Map 1355A, scale 1:125 000).

1975a: Structural succession, nomenclature, and interpretation of transported rocks in western Newfoundland; Canadian Journal of Earth Sciences, v. 12, p. 1874-1894.

1975b: Early Paleozoic ophiolite complexes of Newfoundland: isotopic age of zircons: comment; Geology, v. 3, no. 8, p. 479.

1976: Tectonic stratigraphic subdivision of the Appalachian Orogen (abstract); in Abstracts with Programs; Geological Society of America, v. 8, no. 2, p. 300.

1977a: The Coney Head Complex: another Taconic Allochthon in west Newfoundland; American Journal of Science, v. 277, p. 1279-1295.

1977b: Ophiolitic melange and its significance in the Fleur de Lys Supergroup, northern Appalachians; Canadian Journal of Earth Sciences, v. 13, p. 987-1003.

1978a: Tectonic-lithofacies map of the Appalachian Orogen; Memorial University of Newfoundland, St. John's, Newfoundland, Map No. 1, scale 1:1 000 000, Map No. 1a, scale 1: 2 000 000.

1978b: Geological development of the northern Appalachians: its bearing on the evolution of the British Isles; in Crustal evolution in northwestern Britain and adjacent regions, (ed.) D.R. Bowes and B.E. Leake; Seal House Press, Liverpool, U.K. Geological Journal, Special Issue, no. 10, p. 1-22.

1979: Appalachian Orogen in Canada; Canadian Journal of Earth Sciences, Tuzo Wilson volume, v. 16, p. 792-807.

1984: Miogeoclines and suspect terranes of the Caledonian-Appalachian Orogen: tectonic patterns in the North Atlantic region; Canadian Journal of Earth Sciences, v. 21, p. 887-901.

1985a: Geology, Stephenville map area, Newfoundland; Geological Survey of Canada, Map 1579A, scale 1:100 000 (with descriptive notes and bibliography).

1985b: Geology of Gros Morne Area, 12H/12 (west half) western Newfoundland; Geological Survey of Canada, Open File 1134, scale 1:50 000.

1992: Mélanges and coticule occurrences in the northeast Exploits Subzone, Newfoundland; in Current Research, Part D; Geological Survey of Canada, Paper 92-1D, p. 121-127.

1993: Acadian Orogeny in Newfoundland; in The Acadian Orogeny: recent studies in New England, Maritime Canada, and the Autochthonous Foreland, (ed.) D.C. Roy and J.W. Skehan; Geological Society of America, Special Paper, 275, p. 123-133.

Williams, H. and Cawood, P.A.

1986: Relationships along the eastern margin of the Humber Arm Allochthon between Georges Lake and Corner Brook, western Newfoundland; in Current Research, Part A; Geological Survey of Canada, Paper 86-1A, p. 759-765.

1989: Geology, Humber Arm Allochthon, western Newfoundland; Geological Survey of Canada, Map 1678A, scale 1:250 000.

Williams, H. and Godfrey, S.C.

1980a: Carbonate slivers within the Humber Arm Allochthon, Newfoundland (abstract); Geological Society of America, Abstracts with Programs; v. 12, no. 2, p. 89.

1980b: Geology of Stephenville map area, Newfoundland; in Current Research, Part A; Geological Survey of Canada, Paper 80-1A, p. 217-221.

Williams, H. and Hatcher, R.D.

1982: Suspect terranes and accretionary history of the Appalachian Orogen; Geology, v. 10, p. 530-536.

1983: Appalachian suspect terranes; in Contributions to the Tectonics and Geophysics of Mountain Chains, (ed.) R.D. Hatcher, H. Williams, and I. Zietz; Geological Society of America, Memoir 158, p. 33-53.

Williams, H. and Hibbard, J.P.

1976: The Dunnage Melange, Newfoundland; in Report of Activities, Part A; Geological Survey of Canada, Paper 76-1A, p. 183-185.

1977: Geology of the Baie Verte Peninsula and western White Bay; Mineral Development Division, Newfoundland Department of Mines and Energy, Map 77-29.

Williams, H. and Hiscott, R.N.

1987: Definition of the Iapetus rift-drift transition in western Newfoundland; Geology, v. 15, p. 1044-1047.

Williams, H. and King. A.F.

1979: Trepassey map area, Newfoundland; Geological Survey of Canada, Memoir 389, 24 p.

Williams, H. and Malpas, J.G.

1972: Sheeted dikes and brecciated dike rocks within transported igneous complexes, Bay of Islands, western Newfoundland; Canadian Journal of Earth Sciences, v. 9, p. 1216-1229.

Williams, H. and Payne, J.G.

1975: The Twillingate Granite and nearby volcanic groups: an island arc complex in northeast Newfoundland; Canadian Journal of Earth Sciences, v. 12, p. 982-995.

Williams, H. and Piasecki, M.A.J.

1990: The Cold Spring Melange and a possible model for Dunnage-Gander zone interaction in central Newfoundland; Canadian Journal of Earth Sciences, v. 27, p. 1126-1134.

Williams, H. and St-Julien, P.

1978: The Baie Verte-Brompton Line in Newfoundland and regional correlations in the Canadian Appalachians; in Current Research, Part A; Geological Survey of Canada, Paper 78-1A, p. 225-229.

1982: The Baie Verte-Brompton Line: Early-Paleozoic Continent-Ocean interface in the Canadian Appalachians; in Major Structural Zones and Faults of the Northern Appalachians, (ed.) P. St-Julien and J. Béland; Geological Association of Canada, Special Paper 24, p. 177-207.

Williams, H. and Smyth, W.R.

1973: Metamorphic aureoles beneath ophiolite suites and Alpine peridotites: Tectonic implications with west Newfoundland examples; American Journal of Science, v. 273, p. 594-621.

1983: Geology of the Hare Bay Allochthon; in Geology of the Strait of Belle Isle area, northwestern insular Newfoundland, southern Labrador, and adjacent Quebec; Geological Survey of Canada, Memoir 400, p. 109-141.

Williams, H. and Stevens, R.K.

1969: Geology of Belle Isle: northern extremity of the deformed Appalachian miogeosynclinal belt; Canadian Journal of Earth Sciences, v. 6, p. 1145-1157.

1974: The ancient continental margin of eastern North American; in The Geology of Continental Margins, (ed.) C.A. Burke and C.L Drake; Springer-Verlag Publisher, p. 781-796.

Williams, H. and Talkington, R.W.

1977: Distribution and tectonic setting of ophiolites and ophiolitic melanges in the Appalachian Orogen; in North American Ophiolites, (ed.) R.G.Coleman and W.P. Irwin; Oregon Department of Geology and Mineral Industries, Bulletin 95, p. 1-11.

Williams, H., Cawood, P.A., James, N.P., and Botsford, J.W.

1986: Geology of St. Pauls Inlet (12H/13), western Newfoundland; Geological Survey of Canada, Open File 1238, scale 1:50 000.

Williams, H., Colman-Sadd, S.P., and Swinden, H.S.

1988: Tectonic-stratigraphic subdivisions of central Newfoundland; in Current Research, Part B; Geological Survey of Canada, Paper 88-1B, p. 91-98.

Williams, H., Dallmeyer, R.D., and Wanless, R.K.

1976: Geochronology of the Twillingate Granite and Herring Neck Group, Notre Dame Bay, Newfoundland; Canadian Journal of Earth Sciences, v. 13, p. 1591-1601.

Williams, H., Dickson, W.L., Currie, K.L., Hayes, J.P., and Tuach, J.

1989a: Preliminary report on a classification of Newfoundland granitic rocks and their relation to tectonostratigraphic zones and lower crustal blocks; in Current Research, Part B; Geological Survey of Canada, Paper 89-1B, p. 47-53.

Williams, H., Gillespie, R.T., and Knapp, D.A.

1982: Geology of Pasadena map area, Newfoundland; in Current Research, Part A; Geological Survey of Canada, Paper 82-1A, p. 281-288.

1983: Geology of Pasadena map area, 12H/4, western Newfoundland; Geological Survey of Canada, Open File 928, scale 1:50 000.

Williams, H., Gillespie, R.T., and van Breemen, O.

1985a: A late Precambrian rift-related igneous suite in western Newfoundland; Canadian Journal of Earth Sciences, v. 22, p. 1727-1735.

Williams, H., Hibbard, J.P., and Bursnall, J.J.

1977: Geologic setting of asbestos-bearing ultramafic rocks along the Baie Verte Lineament, Newfoundland; in Report of Activities, Part A; Geological Survey of Canada, Paper 77-1A, p. 351-360.

Williams, H., James, N.P., and Stevens, R.K.

1985b: Humber Arm Allochthon and nearby groups between Bonne Bay and Portland Creek, western Newfoundland; in Current Research, Part A; Geological Survey of Canada, Paper 85-1A, p. 399-406.

Williams, H., Kennedy, M.J., and Neale, E.R.W.

1972: The Appalachian Structural Province; in Variations in Tectonic Styles in Canada, (ed.) R.A. Price and R.J.W. Douglas; Geological Association of Canada, Special Paper 11, p. 181-261.

1974: The northeastward termination of the Appalachian Orogen; in The Ocean Basins and Margins, v. 2: The North Atlantic, (ed.) A.E.M. Nairn and F.G. Stehli; Plenum Press, New York, p. 79-123.

Williams, H., Piasecki, M.A.J., and Colman-Sadd, S.P.

1989b: Tectonic relationships along the proposed central Newfoundland Lithoprobe Transect and regional correlations; in Current Research, Part B; Geological Survey of Canada, Paper 89-1B, p. 55-66.

Williams, H., Piasecki, M.A.J., and Johnston, D.

1991: The Carmanville Melange and Dunnage-Gander relationships in northeast Newfoundland; in Current Research, Part D; Geological Survey of Canada, Paper 91-1D, p. 15-23.

Williams, H., Quinn, L.A., Nyman, M., and Reusch, D.N.

1984: Geology of Lomond map area, 12H/5, western Newfoundland; Geological Survey of Canada, Open File 1012, scale 1:50 000.

Williams, S.H.

1989: New Graptolite discoveries from the Ordovician of central Newfoundland; in Current Research (1989); Newfoundland Department of Mines, Geological Survey of Newfoundland, Report 89-1, p. 149-157.

Williams, S.H. and Stevens, R.K.

1988: Early Ordovician (Arenig) graptolites from the Cow Head Group, western Newfoundland; Palaeontographica Canadiana, no. 5, 167 p.

Williams, S.H., Boyce, W.D., and James, N.P.

1987: Graptolites from the Middle Ordovician St. George and Table Head Groups, western Newfoundland, and implications for the correlation of trilobite, brachiopod and conodont zones; Canadian Journal of Earth Sciences; v. 24, p. 456-470.

Wilson, J.T.

1988: Convection tectonics: some possible effects upon the earth's surface of flow from the deep mantle; Canadian Journal of Earth Sciences, v. 25, p. 1199-1208.

Wilton, D.H.C.

1985: Tectonic evolution of southwestern Newfoundland as indicated by granitoid petrogenesis; Canadian Journal of Earth Sciences, v. 22, p. 1080-1092.

Winchester, J.A. and Van Staal, C.R.

1988: Middle Ordovician volcanicity in northern New Brunswick: geochemistry and probable tectonic setting (abstract); Geological Association of Canada, Program with Abstracts; v. 13, p. A135.

Wolf, R.R., and Dalrymple, R.W.

1984: Sedimentology of the Cambro-Ordovician sandstones of eastern Ontario; Ontario Geological Survey, Miscellaneous Paper 121, p. 240-252.

Wonderley, P.F. and Neuman, R.B.

1984: The Indian Bay Formation: fossiliferous Early Ordovician volcanogenic rocks in the northern Gander Terrane, Newfoundland, and their regional significance; Canadian Journal of Earth Sciences, v. 21, p. 525-532.

Woodman, J.E.

1904: The sediments of the Meguma Series of Nova Scotia; American Geologist, v. 34, p. 13-34.

1908: Probable age of the Meguma (Gold-bearing) series of Nova Scotia; Geological Society of America Bulletin, v. 19, p. 99-112.

Yaowanoiyothin, W.

1988: Petrology of the Black Brook Granitic Suite and associated gneisses, northeastern Cape Breton Island, Nova Scotia; M.Sc. thesis, Acadia University, Wolfville, Nova Scotia.

Younce, G.B.

1970: Structural geology and stratigraphy of the Bonavista Bay region, Newfoundland; Ph.D. thesis, Cornell University, Ithaca, U.S.A., 188 p.

Young, G.A.

1911: Bathurst district, New Brunswick; Geological Survey of Canada, Memoir 18E, 105 p.

Zen, E-an

1974: Prehnite- and pumpellyite-bearing mineral assemblages, west side of the Appalachian metamorphic belt, Pennsylvania to Newfoundland; Journal of Petrology, v. 15, Part 2, p. 197-242.

Zentilli, M., Graves, M.C., Mulja, T., and MacInnis, I.

1986: Geochemical characterization of the Goldenville-Halifax transition of the Meguma Group of Nova Scotia: preliminary report; in Current Research, Part A; Geological Survey of Canada, Paper 86-1A, p. 423-428.

Zietz, I., Haworth, R.T., Williams, H., and Daniels, D.L.

1980a: Magnetic Anomaly Map of the Appalachian Orogen; Memorial University of Newfoundland, Map, no. 2, scale 1:1 000 000.

1980b: Magnetic Anomaly Map of the Appalachian Orogen; Memorial University of Newfoundland, Map, no. 2a, scale 1:2 000 000.

Zwicker, E.J. and Strong, D.F.

1986: The Great Bend ophiolite, eastern Newfoundland; field investigations; in Current Research, Part A; Geological Survey of Canada, Paper 86-1A, p. 393-397.

ADDENDA

Middle Ordovician clastic rocks of Humber Zone and St. Lawrence Platform (R.N. Hiscott)

Quinn (1992) measured bed-by-bed sections through most western Newfoundland flysch units and point-counted the major components of flysch sandstones (Table 3.A1). She questioned previous estimates of the thickness of the Mainland Sandstone (e.g. Table 3.2), which has a poorly defined, perhaps faulted upper contact (Waldron and Stockmal, 1991), and is cut internally by several bedding-parallel thrust faults. Her conservative thickness estimate for the Mainland Sandstone was about 650 m. Because of structural complexity, Quinn (1992) also believed it impossible to determine the pre-deformation thickness of the American Tickle Formation. Quinn (1992) measured paleoflow to the south in the American Tickle Formation, and to the southwest in the Mainland Sandstone (except to the northwest at Cape Cormorant).

Quinn, L.

1992: Foreland and trench slope basin sandstones of the Goose Tickle Group and Lower Head Formation, western Newfoundland; Ph.D. thesis, Memorial University of Newfoundland, St. John's, Newfoundland, 574 p.

Waldron, J.W.F. and Stockmal, G.S.

1991: Mid-Paleozoic thrusting at the Appalachian deformation front: Port au Port Peninsula, western Newfoundland; Canadian Journal of Earth Sciences, v. 28, p. 1992-2002.

Table 3.A1. Grain percentages for three western Newfoundland flysch units (Quinn, 1992).

Component (%)	Unit designation (Fig. 3.27)		
	6	7	F
Quartz	56	57	60
Feldspar	18	20	25
SRF[1]	20	17	10
MRF	1	1	1
VRF	5	5	4
Number of Samples	16	16	18

[1]SRF, MRF, VRF = sedimentary, metamorphic, volcanic rock fragments, respectively

Quinn, L.
1992:   Foreland and trench slope basin sandstones of the Goose Tickle Group and Lower Head Formation, western Newfoundland; Ph.D. thesis, Memorial University of Newfoundland, St. John's, Newfoundland, 574 p.

Waldron, J.W.F. and Stockmal, G.S.
1991:   Mid-Paleozoic thrusting at the Appalachian deformation front: Port au Port Peninsula, western Newfoundland; Canadian Journal of Earth Sciences, v. 28, p. 1992-2002.

Humber Zone internal, *(J.P. Hibbard, P. St-Julien, and W.E. Trzcienski, Jr.)*

Since the initial writing of this manuscript in 1985 and subsequent revision in 1991, new data have been collected concerning the nature and timing of deformation of the internal Humber Zone in Newfoundland. In the Grand Lake-Corner Brook Lake area, Cawood et al. (1994) obtained a U-Pb zircon age of 434+3/-2 Ma on a synkinematic pegmatite, a U-Pb monazite age of 430 ± 2 Ma from a schist, and a U-Pb rutile age of 437 ± 6 Ma from the same schist. They also reported new $^{40}Ar/^{39}Ar$ cooling ages ranging from 430-410 Ma on amphibole and muscovite. Collectively, these ages are interpreted as indicating a Silurian age for regional deformation and metamorphism of the internal Humber Zone in the Grand Lake-Corner Brook Lake area (Cawood et al., 1994). These data are supported by a previous 420 ± 20 Ma K-Ar age on a late synkinematic granitoid pegmatite in the area (Kennedy, 1981).

On the Baie Verte Peninsula, the Wild Cove Pond Igneous Suite (Hibbard, 1983) is a dominantly post-kinematic composite batholith that intrudes the Fleur de Lys Supergroup. A foliated muscovite-garnet granitoid from the suite has yielded a U-Pb zircon age of 427 ± 2 Ma (Cawood et al., 1994). This age, considered in conjunction with previous Silurian $^{40}Ar/^{39}Ar$ cooling ages for the regionally metamorphosed Fleur de Lys Supergroup (Dallmeyer, 1977; Dallmeyer and Hibbard, 1984) has led to the interpretation that deformation and peak metamorphism on the Baie Verte Peninsula is also Silurian (Cawood et al., 1994). However, the nature of the foliation in the dated muscovite granitoid is uncertain. The muscovite granitoid phase is isolated from the Fleur de Lys Supergroup by other phases of the batholith. The other phases demonstrably intrude the supergroup post-kinematically with respect to the regional penetrative deformation. These observations suggest that any Silurian event was not accompanied by a regional penetrative deformation in the Fleur de Lys Supergroup in the area.

In addition, the compelling regional tectonic constraints provided by the ophiolitic mélanges in the Fleur de Lys Supergroup as described in the main text above, indicate that initial intense deformation of the internal Humber Zone relates to Taconic obduction. Thus, any Silurian tectonism has been imprinted on a previously deformed terrane.

In summary, the case for a penetrative Silurian tectonothermal event in the internal Humber Zone of Newfoundland is stronger in the Grand Lake-Corner Brook Lake area than on the Baie Verte Peninsula. It is possible that differences in tectonic styles of the two areas may be related to different intensities of post-Ordovician tectonism in these areas (Hibbard, 1983a).

The nature and controls of this Silurian tectonism in the internal Humber zone have yet to be convincingly revealed. Some workers relate it to collisional tectonics between Laurentia and Avalon (Cawood et al., 1994) whereas others have related it to regional extension following Taconic thrust loading (Jamieson et al., 1993).

Cawood, P.A., Dunning, G.R., Lux, D. , and van Gool, J.A.M.
1994:   Timing of peak metamorphism and deformation along the Appalachian margin of Laurentia in Newfoundland: Silurian, not Ordovician; Geology, v. 22, 399-402.

Hibbard, J.P.
1983a: Notes on the metamorphic rocks in the Corner Brook Lake area and regional correlation of the Fleur de Lys belt, western Newfoundland; in Current Research, (ed.) M.J. Murray, P.D. Saunders, W.D. Boyce, and R.V. Gibbons; Newfoundland Department of Mines and Energy Report 83-1, 41-50.

Jamieson, R.A., Anderson, S., and McDonald, L.
1993:   Slip on the Scrape - an extensional allochthon east of the Baie Verte Line, Newfoundland; Geological Society of America, Abstracts with Programs, v. 25, no. 2, p. 26.

Dunnage Zone-Quebec *(A. Tremblay, M. Malo, and P. St-Julien)*

Since the completion of this contribution, more field work and detailed studies were completed on stratigraphic and structural relationships between the Gaspé Belt and rocks included in the Mégantic Subzone (Lebel and Tremblay, 1993; Lafrance and Tremblay, 1993; Tremblay et al., 1993; Tremblay and Pinet, in press) and on the geochemistry of volcanic rocks from the Frontenac and Clinton formations (Desjardins et al., 1994).

Based on stratigraphic relations, Label and Tremblay (1993) and Lafrance and Tremblay (1993) suggested that the Frontenac Formation most probably overlies or represents a lateral equivalent of uppermost units belonging to the Gaspé Belt of southern Québec. The Victoria River Fault, presented in Chapter 3 as a major fault between the Mégantic Subzone and cover rocks of the Gaspé Belt (see Fig. 3.106), is relocated in the Frontenac Formation (Lafrance and Tremblay, 1993; Tremblay and Pinet, in press) and is a northwest-directed reverse fault (Lafrance and Tremblay, 1993). The geochemistry of volcanic rocks from the Frontenac and Clinton formations indicates that they are distinct volcanic sequences which most probably originated from different mantle sources (Desjardins et al., 1994). The Frontenac volcanics are within-plate basalts similar in composition to Silurian-Devonian volcanic rocks of Gaspésie and northern New Brunswick (Desjardins et al., 1994). Finally, a preliminary analysis of microfauna (*Chitinozoan*) from the Frontenac Formation suggests a Late Silurian-Early Devonian age (Tremblay et al., 1993). We therefore suggest removal of the Frontenac Formation from the Dunnage Zone and to include it in the cover sequence of the Gaspé Belt. The Mégantic Subzone will then be made up of two rock units, the Clinton Formation and the Chesham Mélange.

Desjardins, D., Tremblay, A., and Paradis, S.
1994:   Petrological and geochemical characteristics of volcanic rocks of the Frontenac and Clinton formations, southern Québec Appalachians [abs.]; Geological Society of America, Northeastern Section. Annual Meeting, Abstracts with programs, v. 26, p. 14.

Lafrance, B. and Tremblay, A.
1993:   Nature and significance of the Rivière Victoria fault, southern Quebec Appalachians [abs.]; Geological Society of America, Northeastern Section Annual Meeting, Abstracts with programs, v. 25, p. 31.

Lebel, D. and Tremblay, A.
1993:   Géologie de la région de Lac Mégantic; Québec, Ministère de l'Energie et des Ressources, DV 93-02.

Tremblay, A., Lebel, D., and van Grootel, G.
1993:   Stratigraphy and structure of the Connecticut Valley-Gaspé trough in southern Quebec [abs.]; Geological Society of America, Northeastern Section Annual Meeting, Abstracts with programs, v. 25, p. 85.

Gander Zone, Nova Scotia, *(S.M. Barr)*

Aspy Subzone

The model of Barr and Raeside (1989) for terranes in Cape Breton Island, in particular the Aspy and Bras d'Or terranes, remains controversial. As a result of a detailed field and structural study of the Eastern Highlands shear zone and adjacent areas in the eastern Cape Breton Highlands, Lin (1992, 1993) has proposed that the relationship between the Aspy and Bras d'Or subzones was originally depositional. He interpreted a metaconglomerate in the Chéticamp Lake Gneiss of the Aspy Subzone as a basal conglomerate, derived in part from Bras d'Or subzone rocks, and hence concluded that the Eastern Highlands shear zone is not a terrane boundary. Lin (1992, 1993) provided a detailed analysis of the shear zone, showing that it is characterized by an amphibolite-facies deformation zone over 5 km in width, with late-stage greenschist-facies deformation localized in a central zone about 1 km wide. Shear-sense indicators suggest east-over-west dip-slip movement for Late Silurian to Early Devonian deformation on the shear zone, and oblique-slip with a dextral horizontal component and west-side-up dip slip component for Late Devonian to Early Carboniferous movement. Lin (1992, 1993) also defined the Clyburn Brook Formation, a Silurian low-grade metavolcanic and metasedimentary unit south of the main Eastern Highlands shear zone (Fig. 3.116). Barr and Raeside (1989) included this unit in their Bras d'Or terrane, but it may be similar to low-grade metamorphic units in the Aspy Subzone.

Other new work in the Aspy Subzone includes mapping by Mengel et al. (1991), Horne (1993), and Lynch and co-workers (e.g. Lynch and Tremblay, 1992a, b; Lynch et al., 1993). The latter workers have interpreted the Silurian rocks of the Aspy Subzone as an overlap assemblage deposited on accreted terranes. According to their model, basement and cover were complexly imbricated during an Acadian regime of reverse oblique shear, and then exhumed by Late Carboniferous to Devonian extension that formed a major low-angle mylonitic fault complex in the southwestern Cape Breton Highlands.

Keppie et al. (1991a, 1992) continued to maintain that all of Cape Breton Island is part of an "Avalon Composite Terrane".

Horne, R.J.

1993:   Bedrock geology of the south-central Cape Breton Highlands, Nova Scotia; Nova Scotia Department of Natural Resources, Mines and Energy Branch, Halifax, N.S., scale 1:25 000.

Lin, S.

1992:   The stratigraphy and structural geology of the southeastern Cape Breton Highlands National Park and its implications for the tectonic evolution of Cape Breton Island, Nova Scotia, with emphasis on lineations in shear zones; Ph.D. thesis, University of New Brunswick, Fredericton, New Brunswick.

1993:   Relationship between the Aspy and Bras d'Or "terranes" in the northeastern Cape Breton Highlands, Nova Scotia; Canadian Journal of Earth Sciences, v. 30, p. 1773-1781.

Lynch, J.V.G. and Tremblay, C.

1992a: Geology of the Cheticamp River map (11K/10), central Cape Breton Highlands, Nova Scotia; Geological Survey of Canada, Open File 2448.

1992b: Imbricate thrusting, reverse-oblique shear, and ductile extensional shear in the Acadian Orogen, central Cape Breton Highlands, Nova Scotia; in Current Research, Part D; Geological Survey of Canada, Paper 92-1D, p. 91-100.

Lynch, G., Tremblay, C., and Rose, H.

1993:   Compressional deformation and extensional denudation of Early Silurian volcanic overlap assemblages in western Cape Breton Island, Nova Scotia; in Current Research, Part D; Geological Survey of Canada, Paper 93-1D, p. 103-110.

Mengel, F., Godue, R., Sangster, A., Dubé, B., and Lynch, G.

1991:   A progress report on the structural control of gold mineralizations in the Cape Breton Highlands; in Current Research, Part D; Geological Survey of Canada, Paper 91-1D, p. 117-127.

Avalon Zone – New Brunswick *(S.R. McCutcheon and M.J. McLeod)*

Recent detailed geological mapping and numerous U/Pb and ^{40}Ar/^{39}Ar dates necessitate changes in interpretations as follows; (Barr et al., 1992).

1. The Green Head Group and Brockville Gneiss constitute a distinct Avalonian terrane (Barr and White, 1991). This has been verified by the presence of a unique suite of late Precambrian-Cambrian intrusions (White et al., 1990; Dallmeyer and Nance, 1992).

2. Most rocks of the Western volcanic domain, formerly assigned to the Precambrian Coldbrook Group are Silurian (McLeod, unpublished information). Assemblage I is now referred to as the Broad River Group and is restricted to the Eastern volcanic domain. The name Coldbrook Group is retained for the Central volcanic domain only. Assemblage II contains a younger "Eocambrian" redbed-volcanic sequence equated in part with the Ratcliffe Brook Formation (e.g. Barr et al., 1992).

3. Rocks between the Belleisle and Falls Brook-Taylor Brook faults constitute another distinct Avalonian division in southern New Brunswick (Johnson, 1994; McLeod et al., 1994). Unit A is Late Cambrian in age based on its conformable relationship with U-Pb dated Upper Cambrian rocks (Johnson, 1994). This unit contains abundant volcanic rocks and likely rests unconformably on units B and C (Johnson, 1994). Unit D encompasses an extensive unit of Upper Ordovician rocks in the southwestern portion of the belt in addition to Silurian and Devonian rocks. The age of these rocks, which is unique to the belt, is based on detailed mapping and fossil information (McLeod et al., 1994; Nowlan et al., 1994).

4. Most intrusive rocks in the Western intrusive domain are late Precambrian to Cambrian based on extensive isotopic dating (Dallmeyer and Nance, 1992; White et al., 1990).

Barr, S.M. and White, C.E.

1991:   Late Precambrian-Early Cambrian geology, Saint John-St. Martins area, southern New Brunswick (parts of 21 H/5, 21 H/12 and 21 G/8); Geological Survey of Canada, Open File 2353, 1 map.

Barr, S.M., Currie, K.L., Nance, R.D., and White, C.E.

1992:   New Brunswick; in Wolfville '92, Field Trip Guidebook, (ed.) J.B. Murphy, S.M. Barr, K.L. Currie, J.D. Keppie, R.D. Nance, and C.E. White; Geological Association of Canada, Mineralogical Association of Canada, Acadia University, Wolfville, Nova Scotia, Excursion C-12, p. 70-87.

Dallmeyer, R.D. and Nance, R.D.

1992:   Tectonic implications of ^{40}Ar/^{39}Ar mineral ages from Late Precambrian-Cambrian plutons, Avalon composite terrane, southern New Brunswick, Canada; Canadian Journal of Earth Sciences, v. 29, p. 2445-2462.

Johnson, S.C.

1994:   Geological investigation of the Pocologan River area, Charlotte County, New Brunswick: bedrock geology and regional implications; in Current Research 1993, (ed.) S.A. Abbott Merlini; New Brunswick Department of Natural Resources and Energy, Mineral Resources, Miscellaneous Report 12, p. 71-81.

McLeod, M.J., Nowlan, G.S., and McCracken, A.D.

1994:   Tectonic and economic significance of an Ordovician conodont discovery in southwestern New Brunswick; in Current Research 1993; (ed.) S.A. Abbott Merlini; New Brunswick Department of Natural Resources and Energy, Mineral Resources, Miscellaneous Report 12, p. 121-128.

Nowland, G.S., McCracken, A.D., and McLeod, M.J.

1994:   Tectonic and paleogeographic significance of Ordovician conodonts from the Letang limestone, Avalon Terrane, southwestern New Brunswick (abstract); Geological Association of Canada, Mineralogical Association of Canada, Program with Abstracts, p. A83.

White, C.E., Barr, S.M., Bevier, M.L., and Deveau, K.A.

1990:   Field relations, petrography, and age of plutonic units in the Saint John area of southern New Brunswick; Atlantic Geology, v. 26; p. 259-270.

Meguma Zone *(P.E. Schenk)*

Stratotypes

Delanceys Formation of the Halifax Group

South Annapolis River at intersection with Ridge Road near Factorydale Pond (NTS 21-A/15; UTM 20 T LE595827) 1 km west of Delanceys and 6.25 km southeast of Aylesford.

Rockville Notch Formation of the Halifax Group

Rockville Notch on Fales River southeast of bridge (NTS 21A/15; UTM 20 T LE508790)

South American Connection?

Paleomagnetic reconstruction suggests that the Meguma terrane could be displaced Gondwana margin of southwestern South America. Laurentia could have moved clockwise around Gondwana during the early and middle Paleozoic from a position near Antarctica (Dalziel, 1991). This hypothesis raises the possibility of an Ordovician (Taconic) collision between the eastern margin of Laurentia (including the Canadian Appalachians) and the southwestern margin of South America (Argentina) (Dalla Salda et al., 1992). However, the sedimentology and stratigraphy of the Meguma terrane provide direct evidence of the juxtaposition of northwestern Africa and southeastern Laurentia prior to the late Paleozoic. The Meguma terrane is unlike others in the Appalachians in showing no structural evidence of an Ordovician collisional event (J.R. Henderson, pers. comm., 1993).

Dalla Salda, L.H., Dalziel, I.W.D., Cingolani, C.A., and Varela, R.
1992: Did the Taconic Appalachians continue into South America?; Geology, v. 20, p. 1059-1062.

Dalziel, I.W.D.
1991: Pacific margin of Laurentia as a conjugate rift pair: evidence and implications for an Eocambrian supercontinent; Geology, v. 19, p. 598-601.

Authors' Addresses

Harold Williams
Department of Earth Sciences
Memorial University of Newfoundland
St. John's, Newfoundland
A1B 3X5

Philippe Erdmer
Department of Geology
University of Alberta
Edmonton, Alberta
T6G 2E3

P.S. Kumarapeli
Geology Department, Concordia University
7141, rue Sherbrooke Ouest
Montréal, (Québec)
H4B 1R6

Ian Knight
Geological Survey Branch
Newfoundland Department of Mines and Energy
P.O. 8700
St. John's, Newfoundland
A1B 4J6

N.P. James
Department of Geological Sciences
Queen's University
Kingston, Ontario
K7L 3N6

R.N. Hiscott
Department of Earth Sciences
Memorial University of Newfoundland
St. John's, Newfoundland
A1B 3X5

J.P. Hibbard
Department of Marine, Earth and Atmospheric Sciences
North Carolina State University
Raleigh, North Carolina
27695-8208
U.S.A.

Pierre St-Julien
Département de géologie,
Université Laval
Sainte-Foy (Québec)
G1K 7P4

W.E. Trzcienski
Département de géologie
Université de Montréal
Montréal (Québec)
H3C 3J7

C.R. van Staal
Geological Survey of Canada
601 Booth St.
Ottawa, Ontario
K1A 0E8

L.R. Fyffe
Minerals and Energy Division
New Brunswick Department of Natural Resources
P. O. Box 6000
Fredericton, New Brunswick
E3B 5H1

Alain Tremblay
Quebec Geoscience Centre
INRS - Géoressources
C.P. 7 500
Sainte-Foy (Québec)
G1V 4C7

Michel Malo
Quebec Geoscience Centre
INRS - Géoressources
C.P. 7 500
Sainte-Foy (Québec)
G1V 4C7

S.P. Colman-Sadd
Geological Survey Branch
Newfoundland Department of Mines and Energy
P.O. 8700
St. John's, Newfoundland
A1B 4J6

P.P. O'Neill
Geological Survey Branch
Newfoundland Department of Mines and Energy
P.O. 8700
St. John's, Newfoundland
A1B 4J6

S.M. Barr
Department of Geology
Acadia University
Wolfville, Nova Scotia
B0P 1X0

R.P. Raeside
Department of Geology
Acadia University
Wolfville, Nova Scotia
B0P 1X0

R.A. Jamieson
Department of Geology
Dalhousie University
Halifax, Nova Scotia
B3H 3J5

S.J. O'Brien
Geological Survey Branch
Newfoundland Department of Mines and Energy
P.O. 8700
St. John's, Newfoundland
A1B 4J6

A.F. King
Department of Earth Sciences
Memorial University of Newfoundland
St. John's, Newfoundland
A1B 3X5

M.M. Anderson
Department of Earth Sciences
Memorial University of Newfoundland
St. John's, Newfoundland
A1B 3X5

J.D. Keppie
Nova Scotia Department of Mines and Energy
Halifax, Nova Scotia
B3J 1X1

J.B. Murphy
Department of Geology
St. Francis Xavier University
Antigonish, Nova Scotia
B2G 1C0

R.D. Nance
Department of Geological Sciences
Porter Hall, Ohio University
Athens, Ohio
45701
U.S.A.

J. Dostal
Department of Geology
St. Mary's University
Halifax, Nova Scotia
B3H 3C3

S.R. McCutcheon
Department of Natural Resources and Energy
P.O. Box 50
Bathurst, New Brunswick
E2A 3Z1

M.J. McLeod
Department of Natural Resources and Energy
P.O. Box 50
Bathurst, New Brunswick
E2A 3Z1

P.E. Schenk
Department of Geology
Dalhousie University
Halifax, Nova Scotia
B3H 3J5

Printed in Canada

Chapter 4

MIDDLE PALEOZOIC ROCKS

Chapter 4

MIDDLE PALEOZOIC ROCKS

INTRODUCTION

Harold Williams

Middle Paleozoic belts are recognized across the Canadian Appalachian region from Quebec to Nova Scotia, and across the island of Newfoundland. Middle Paleozoic rocks are most extensive along the Quebec-Nova Scotia transect and from northwest to southeast they define the Gaspé, Fredericton, Mascarene, Arisaig, Cape Breton, and Annapolis belts. Middle Paleozoic rocks are less abundant in Newfoundland. From west to east they define the Clam Bank, Springdale, Cape Ray, Badger, Botwood, La Poile, and Fortune belts (Map 2).

Rocks of the middle Paleozoic belts do not express the early Paleozoic zonation of the orogen, except in a few cases where middle Paleozoic rocks occur within the confines of a particular zone.

The middle Paleozoic belts are less continuous than early Paleozoic zones and there have been few attempts to correlate middle Paleozoic rocks and interpret their significance for the entire Canadian Appalachian region. This is because (a) there are few complete sections with fossiliferous rocks so that the record is fragmentary, (b) occurrences in some places are small and of unknown age, thus precluding broad correlations and linkages across the orogen, (c) there are no middle Paleozoic ophiolite suites, few mélanges, and other rocks that can be related to plate boundaries, (d) contrasts among middle Paleozoic faunas are less pronounced than those among early Paleozoic faunas, and (e) many middle Paleozoic rocks are terrestrial redbeds and associated volcanic rocks that are post accretionary cover sequences, some of which overstep the early Paleozoic elements of the orogen.

Williams, H.
1995:   Introduction: Middle Paleozoic rocks; in Chapter 4 of Geology of the Appalachian-Caledonian Orogen in Canada and Greenland, (ed.) H. Williams; Geological Survey of Canada, Geology of Canada, no. 6, p. 313-315 (also Geological Society of America, The Geology of North America, v. F-1).

Stratigraphic sections of most middle Paleozoic belts show an upward change from marine to terrestrial rocks with all rocks deformed together and cut by plutons. Middle Paleozoic regional metamorphism is important locally, but in general, its highest grades are developed in Ordovician or older rocks. The middle Paleozoic plutons are small, sparse, or absent in Quebec but large and more abundant in New Brunswick, Nova Scotia, and Newfoundland. The relationships indicate a change from mainly marine to terrestrial conditions immediately preceding middle Paleozoic orogenesis. Some middle Paleozoic rocks, such as those of the Grand Banks of Newfoundland, lie outside the limits of Paleozoic metamorphism, plutonism, and deformation.

The most complete middle Paleozoic sections are found in the Gaspé Belt of Quebec. A wealth of new stratigraphic and structural data allow palinspastic restoration and indicate a northwest shelf region linked to Anticosti Island in the Gulf of St. Lawrence with deeper water facies southeastward (Bourque et al., this chapter). The oldest rocks of the Gaspé Belt (Honorat and equivalents) are marine clastics that predate Ordovician deformation in the nearby Humber Zone. This presents an interesting scenario of local marine basins unaffected by Ordovician deformation. A parallel situation occurs in the Newfoundland Badger Belt where marine greywackes lie conformably upon Caradoc shales of the Exploits Subzone and continue upward into Silurian conglomerates and eventually terrestrial volcanic rocks and red sandstones (Williams et al., this chapter). In most other places there are important angular unconformities between rocks of pre-Caradoc zones and Silurian-Devonian belts.

The stratigraphic record and degree of preservation of middle Paleozoic rocks in the Canadian Appalachians allows an analysis of accretionary history that is not possible in the United States Appalachians where middle Paleozic rocks are absent, or where present, they are much more deformed and metamorphosed.

In the following descriptions, all of the mainland belts are described first, from northwest to southeast, because their rocks are most extensive and some stratigraphic sections are relatively complete. Descriptions of all the Newfoundland belts follow, treated from west to east.

QUEBEC, NEW BRUNSWICK, AND NOVA SCOTIA

Gaspé Belt

Pierre-André Bourque, Daniel Brisebois, and Michel Malo

Middle Paleozoic rocks of the Gaspé Belt in Quebec and adjacent New Brunswick range in age from Late Ordovician (Caradoc) to Late Devonian (Frasnian). They belong to three major structural divisions, from north to south, the Connecticut Valley-Gaspé Synclinorium, the Aroostook-Percé Anticlinorium, and the Chaleurs Bay Synclinorium (Fig. 4.1). The Aroostook-Percé Anticlinorium (Béland et al., 1979) corresponds with the Aroostook-Matapédia Anticlinorium (Pavlides et al., 1964; Williams, 1978), the Matapédia-Aroostook Zone (Fyffe, 1982a), and partly with the Matapédia Belt (Ferguson and Fyffe, 1985) and Matapédia Basin (Fyffe and Noble, 1985). The Chaleurs Bay Synclinorium corresponds with the Tobique-Chaleur Zone (Fyffe, 1982a) and partly with the Matapédia Belt (Ferguson and Fyffe, 1985) and Matapédia Basin (Fyffe and Noble, 1985).

The Middle Paleozoic rocks are separable into four broad temporal and lithological packages (Fig. 4.2a, b): (1) Upper Ordovician-lowermost Silurian (Caradoc to Llandovery) deep water fine grained siliciclastic and carbonate facies composed of the Honorat, Grog Brook and Matapédia groups that occur chiefly but not exclusively in the Aroostook-Percé Anticlinorium, and the Cabano Group in the Connecticut Valley-Gaspé Synclinorium of the Témiscouata region; (2) Silurian-lowermost Devonian (Llandovery to Lochkovian) shallow to deep shelf facies, principally made up of the Chaleurs Group in Gaspésie and nearby equivalents, that occur in both the Connecticut Valley-Gaspé and the Chaleurs Bay synclinoria; (3) Lower Devonian (Pragian-Emsian) mixed siliciclastic and carbonate fine grained deep shelf facies, including several groups (Upper Gaspé Limestones, Fortin, Témiscouata, upper part of Sainte Francis) that occur mainly in the Connecticut Valley-Gaspé Synclinorium; and (4) upper Lower to Upper Devonian (Emsian to Frasnian) nearshore to terrestrial coarse grained facies composed of the Gaspé Sandstones Group and its equivalents, occurring mainly in the Connecticut Valley-Gaspé Synclinorium, but also in the Chaleurs Bay Synclinorium.

Rocks of all four packages occur in the Connecticut Valley-Gaspé Synclinorium, whereas only Upper Ordovician-lowermost Silurian rocks occur in the Aroostook-Percé Anticlinorium (Fig. 4.2 and 4.3). Rocks of the Connecticut Valley-Gaspé and Chaleurs Bay synclinoria are either in stratigraphic continuity with those of the Aroostook-Percé Anticlinorium, or contacts are faults. Along the northern margin of the Connecticut Valley-Gaspé Synclinorium, the middle Paleozoic sequence is unconformable above

pre-Taconic rocks of the Humber Zone, although contacts are poorly exposed. In the Chaleurs Bay Synclinorium, the Silurian sequence is unconformable above the pre-Taconic Maquereau and Mictaw groups of the Humber and Dunnage Zones and rocks of the Elmtree Inlier in the Dunnage Zone. Rocks of the three structural units were deformed by Acadian Orogeny. In southern Gaspésie and northern New Brunswick, middle Paleozoic sequences are overlain unconformably by Carboniferous rocks.

The middle Paleozoic rocks were interpreted to link the Humber, Dunnage and Gander zones (Williams, 1979), but pre-Acadian palinspastic restoration was not considered.

Upper Ordovician-lowermost Silurian rocks lie unconformably or disconformably above older units of the Humber and Dunnage zones. Some of the Upper Ordovician rocks of the Aroostook-Percé Anticlinorium (base of Honorat Group), while unaffected by Taconic Orogeny, are correlatives of Taconic-deformed rocks of the Humber Zone of northern Gaspésie, namely, the upper part of the Cloridorme Formation (Fig. 4.4; see further under Aroostook-Percé Anticlinorium).

In the descriptions that follow, the Aroostook-Percé Anticlinorium is treated first as it contains the oldest rocks. The Chaleurs Bay Synclinorium is described next because it is the type area of several of the Silurian-lowermost Devonian units that are referred to when finally describing the Connecticut Valley-Gaspé Synclinorium.

Aroostook-Percé Anticlinorium

Definition and limits

The Aroostook-Percé Anticlinorium extends from Percé to Matapédia, and across northwestern New Brunswick to Aroostook County, Maine (Fig. 4.2a). At the northern limit of the anticlinorium its Upper Ordovician-lowermost Silurian sequences are conformably overlain by Silurian-Devonian rocks of the Connecticut Valley-Gaspé Synclinorium. Locally, as in the western part of the anticlinorium, the Ristigouche Fault juxtaposes rocks of the Matapédia Group with Devonian units of the Connecticut Valley-Gaspé Synclinorium. Similarly in the northeastern part of the anticlinorium, the Grande Rivière fault system separates stratigraphic units of the Connecticut Valley-Gaspé Synclinorium from those of the Aroostook-Percé Anticlinorium (Fig. 4.2a). The southern limit of the Aroostook-Percé Anticlinorium is also a conformable contact, locally faulted. In northern New Brunswick, the Rocky Brook-Millstream Fault separates Ordovician rocks of the Aroostook-Percé Anticlinorium from Devonian units of the Chaleurs Bay Synclinorium (Fig. 4.2a). In the southernmost part of the Aroostook-Percé Anticlinorium, the Catamaran Fault separates its rocks from those of the Miramichi Anticlinorium. In Quebec, a major strike-slip fault, the Grand Pabos Fault, crosses the Aroostook-Percé Anticlinorium (Skidmore and McGerrigle, 1967) and merges with the Ristigouche Fault.

The contact between Upper Ordovician-lowermost Silurian rocks of the Aroostook-Percé Anticlinorium and the underlying Cambrian-Ordovician rocks is interpreted as an unconformity or a disconformity where it is exposed. In the northern part of the anticlinorium, in the Percé area (column 1 of Figures 4.2 and 4.3), the Upper Ordovician-lowermost Silurian sequence rests unconformably on rocks

Bourque, P.-A., Brisebois, D., and Malo, M.
1995: Gaspé Belt; *in* Chapter 4 of Geology of the Appalachian-Caledonian Orogen in Canada and Greenland, (ed.) H. Williams; Geological Survey of Canada, Geology of Canada, no. 6, p. 316-351 (*also* Geological Society of America, The Geology of North America, v. F-1).

correlative with the Humber Zone, the Murphy Creek Formation. In the middle part of the anticlinorium, in the New Richmond-Honorat and Patapédia River areas (columns 20 and 24 of Fig. 4.3), the Upper Ordovician-lowermost Silurian sequence disconformably overlies rocks of the Dunnage Zone (Tremblay et al., Chapter 3; van Staal and Fyffe, Chapter 3). To the south, in the Carleton County area (column 27 of Fig. 4.3), the Upper Ordovician-lowermost Silurian sequence rests unconformably or disconformably on rocks of the Miramichi Subzone of the Gander Zone (van Staal and Fyffe, Chapter 3).

Structural geology

In Quebec, rocks of the Aroostook-Percé Anticlinorium record two phases of folding (Théberge, 1979; Vennat, 1979; Malo, 1986). The first folds (F_1) are large, low-amplitude, open, upright structures trending northwest and lacking penetrative cleavage. The second folds (F_2) are tighter and have an associated slaty cleavage that trends northeast. The F_2 folds are doubly plunging as a result of superposition of F_2 on F_1 folds and production of large domes and basins. This is well illustrated in the Mount Alexandre-Pellegrin area (Fig. 4.5) where the oldest rocks of the

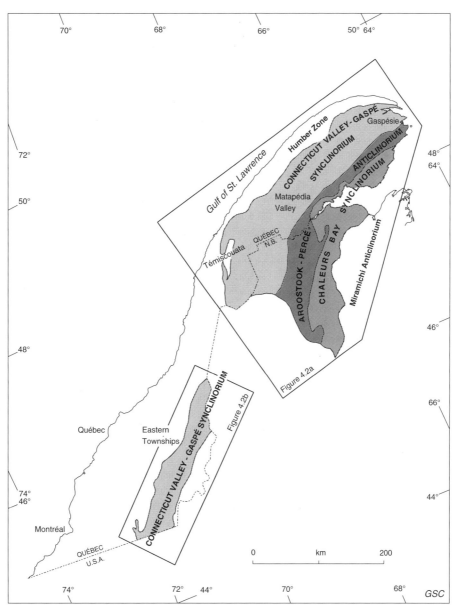

Figure 4.1. Structural divisions of the Gaspé Belt in Quebec and adjacent New-Brunswick.

317

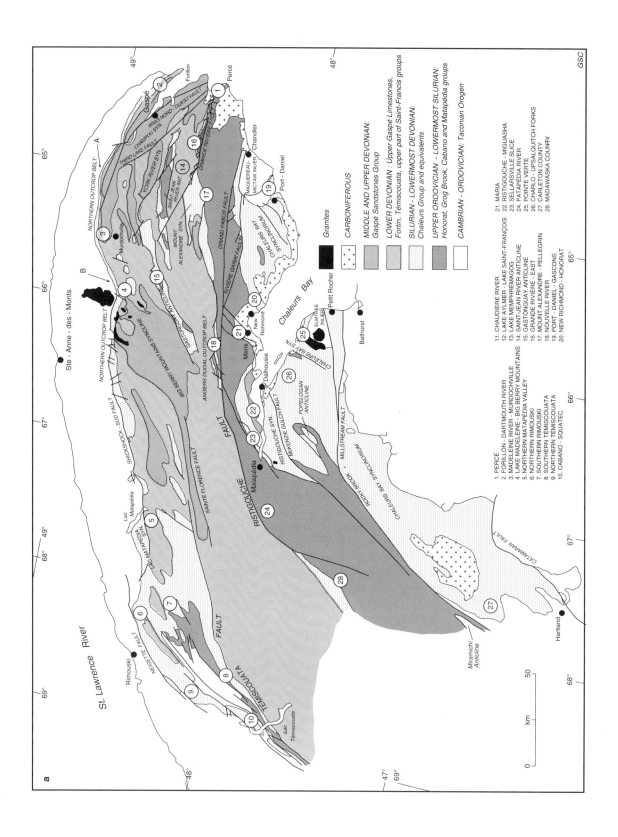

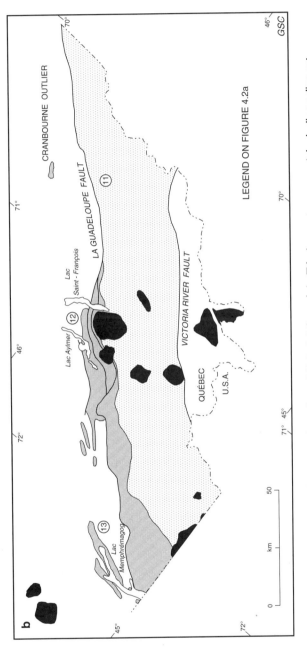

Figure 4.2. Geological map of the Gaspé Belt: **(a)** Gaspésie-Témiscouata segment, including adjacent New Brunswick; **(b)** Quebec Eastern Townships. Locations of a and b are shown in Figure 4.1.

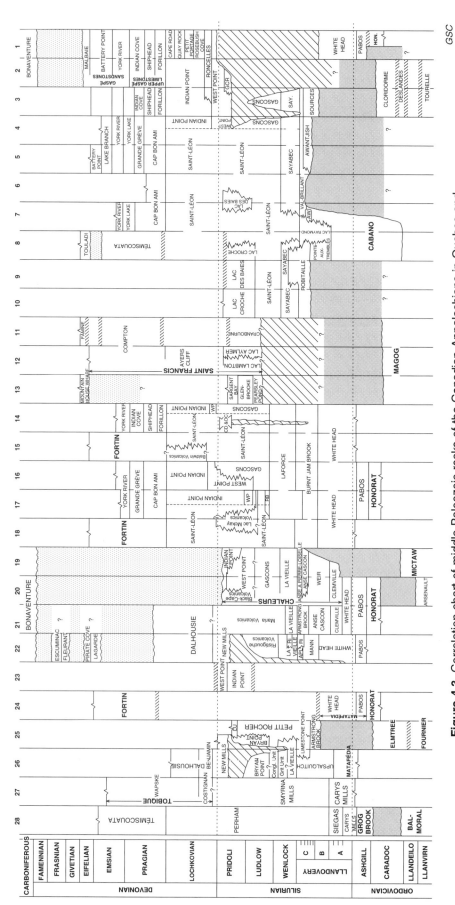

Figure 4.3. Correlation chart of middle Paleozoic rocks of the Canadian Appalachians in Quebec and adjacent New-Brunswick. See text for source of data. Location of sections shown in Figure 4.2.

GSC

Honorat Group outcrop in a large dome in the core of the anticlinorium. The F_2 folds are open and upright in the core of the anticlinorium, but tight, upright and inclined to the northwest or to the southeast near the major faults. Slaty S_2 cleavage trends 030° to 060° in the core of the Aroostook-Percé Anticlinorium and 060° to 080° adjacent to major faults (Grand Pabos, Grande Rivière, Rivière Garin faults). Second deformation is the most intense in the Aroostook-Percé Anticlinorium, and corresponds to the main deformation in the Connecticut Valley-Gaspé and Chaleurs Bay synclinoria.

A S_3 crenulation cleavage trends west-northwest and crosses S_2 with no associated major folds. Some other late fabrics are well-developed adjacent to major faults. The fault zones consist of highly foliated and sheared rocks where a vertical northeasterly trending penetrative cleavage subparallel to the regional S_2 cleavage, is deformed by discrete ductile micro-shear zones or secondary S_3 cleavages. These are east-trending dextral sets subparallel to the main fault zones, west-northwest-trending dextral sets, and conjugate sets of shear bands with opposite sense of shear, trending west-northwest (dextral) and north-northeast (sinistral). All of these secondary cleavages are subvertical and their orientations are indicative of an overall dextral sense of shear in the fault zones (Malo, 1987). These micro-shear zones and the shear bands are typical of ductile deformation (Platt, 1984; White et al., 1980; Weijermars and Rondeel, 1984). There are also conjugate mesoscopic faults reflecting a late brittle-ductile episode of deformation.

In New Brunswick, the major Acadian folds trend generally north-northeast to northeast, and are related to the same D_2 deformation as in Quebec. These are open to tight, upright and gently plunging to the northeast and southwest. A vertical slaty cleavage is well-developed and, in turn, locally deformed by a crenulation cleavage or kink bands (St. Peter, 1977). Rast (1980) has recognized earlier folds that he interpreted as large recumbent structures without cleavage. Hamilton-Smith (1970) also suggested that the major northeasterly trending Acadian folds are superposed on earlier folds.

The Grand Pabos Fault is the most important in southern Gaspésie (Fig. 4.2a). Other major faults are the Rivière Garin and Grande Rivière. These east-west faults have dextral strike-slip displacement as indicated by offset of some lithostratigraphic units (Malo and Béland, 1989). Associated subsidiary faults are either west-northwest-trending dextral faults, or northwest- to northeast-trending sinistral faults. These are interpreted as synthetic and antithetic riedels on the major east-west faults. Near major faults, the regional northeast cleavage is rotated eastward and late cleavages are better developed. This rotation, together with the geometry of the secondary cleavages in the fault zones, indicate a dextral movement. Indeed, structural evidence at all scales confirms the dextral strike-slip movement indicated by offsets along major east-trending faults. A model of deformation associated with wrench tectonics (Wilcox et al., 1973) applies to the deformation history of the Aroostook-Percé Anticlinorium (Malo and Béland, 1989; Malo, 1986).

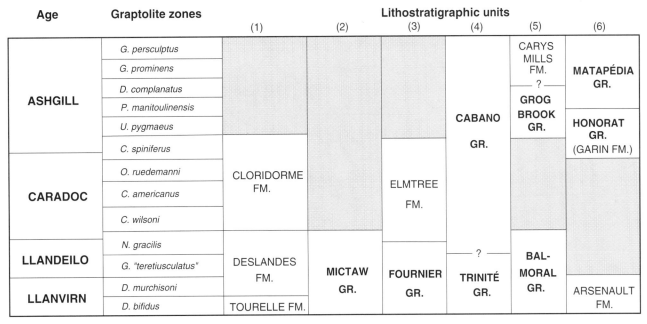

Age	Graptolite zones	Lithostratigraphic units					
		(1)	(2)	(3)	(4)	(5)	(6)
ASHGILL	G. persculptus					CARYS MILLS FM.	MATAPÉDIA GR.
	G. prominens				CABANO GR.	?	
	D. complanatus					GROG BROOK GR.	
	P. manitoulinensis						HONORAT GR. (GARIN FM.)
	U. pygmaeus						
	C. spiniferus	CLORIDORME FM.		ELMTREE FM.			
CARADOC	O. ruedemanni						
	C. americanus						
	C. wilsoni						
LLANDEILO	N. gracilis	DESLANDES FM.	MICTAW GR.	FOURNIER GR.	?	BAL-MORAL GR.	
	G. "teretiusculatus"				TRINITÉ GR.		ARSENAULT FM.
LLANVIRN	D. murchisoni						
	D. bifidus	TOURELLE FM.					

GSC

Figure 4.4. Age of lower Middle Ordovician and Upper Ordovician units of the Gaspé Belt and Humber and Dunnage zones with respect to graptolite zones. Data from Riva and Malo (1988), Fyffe and Fricker, (1987), DeBroucker, (1986), David et al. (1985), St. Peter (1977) and Malo and Bourque (1993). (1) northern Gaspésie, (2) Maquereau-Mictaw Inlier, southeastern Gaspésie, (3) Elmtree Inlier, northeastern New Brunswick, (4) Témiscouata region, (5) Madawaska County, northwestern New Brunswick, (6) southern Gaspésie.

Ristigouche, McKenzie Gulch, Rocky Brook-Millstream, and Catamaran are other major faults in the New Brunswick part of the Aroostook-Percé Anticlinorium (Fig. 4.2a). The east-west segments of the Rocky Brook-Millstream, Catamaran, and McKenzie Gulch faults also have dextral strike-slip displacement (Fyffe, 1982a), whereas their northeast segments may be thrust faults (St. Peter, 1978a, b, 1982; Greiner, 1973). The Ristigouche Fault is described as a thrust fault (St. Peter, 1977; Lachance, 1979), but vertical C-S fabrics in the fault zone indicate a dextral strike-slip component like that of the Grand Pabos Fault in eastern Gaspésie.

In summary, an episode of Acadian folding is recognized throughout the Aroostook-Percé Anticlinorium. Associated cleavage (S_2) trends northeast and is superposed on earlier folds (F_1). This main northeast cleavage (S_2) is locally overprinted by later cleavages (S_3). The northeast folds and cleavage are oblique to major east-west faults. Both geometry and structural analyses indicate large-scale dextral displacement along east-west strike-slip faults.

Stratigraphy

The Aroostook-Percé Anticlinorium is composed chiefly of Upper Ordovician to Lower Silurian sedimentary rocks (Fig. 4.2a). In Quebec, rocks of the anticlinorium are divided into two groups, the Honorat and overlying Matapédia. The Honorat Group comprises only the Garin Formation (described under New Richmond-Honorat area), whereas the Matapédia Group is composed of a lower Pabos Formation and upper White Head Formation. In northwestern New Brunswick the anticlinorium comprises the Grog Brook and overlying Matapédia groups. In southwestern New Brunswick and in Maine, all of its rocks are assigned to the Carys Mills Formation.

The stratigraphy of the Aroostook-Percé Anticlinorium is summarized for six areas (Fig. 4.2a, 4.3). Detailed stratigraphic studies in the eastern part of the Aroostook-Percé Anticlinorium between Percé and New Richmond areas (Malo, 1979, 1986, 1988a, b; Skidmore and Lespérance, 1981; Lespérance et al., 1987), establish a lithostratigraphy that applies to the entire Quebec part of the anticlinorium as well as to northwestern New Brunswick. This is enhanced by the mapping of Lachance (1974, 1975, 1977, 1979), Gosselin and Simard (1983), Simard (1986), Malo (1988b, 1989), Gosselin (1988), and Kirkwood (1989).

Percé area (column 1, Fig. 4.3).

The earliest work in the Percé area is that of Clarke (1908), Schuchert and Cooper (1930), and Cooper and Kindle (1936) on the White Head Formation. These studies were

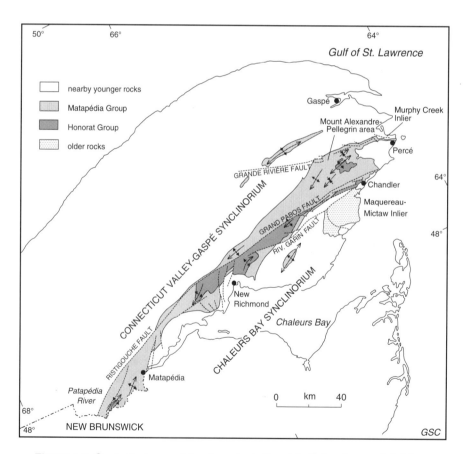

Figure 4.5. Geological map of the Aroostook-Percé Anticlinorium and distribution of Upper Ordovician-lowermost Silurian rocks in Gaspésie. Rocks of the Matapédia Group also occur in the adjacent Connecticut Valley-Gaspé and Chaleurs Bay synclinoria.

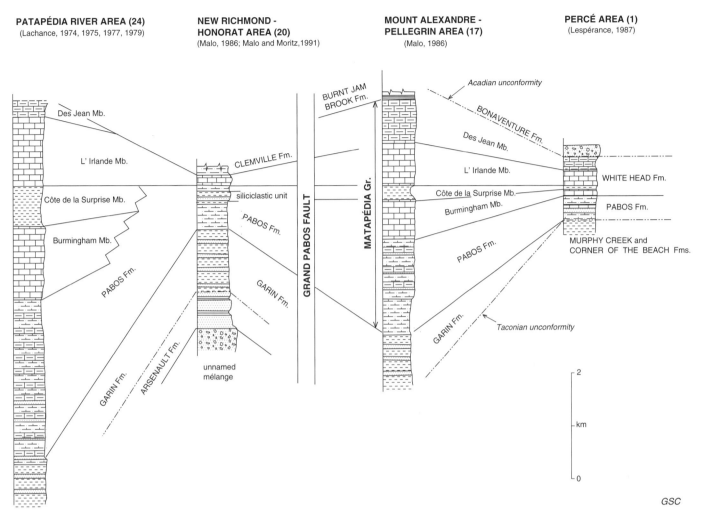

Figure 4.6. Lithostratigraphy of Honorat and Matapédia groups in the Aroostook-Percé Anticlinorium, and correlation across the Grand Pabos Fault. Locations of sections shown in Figure 4.2a.

mostly paleontological, but Clarke (1908) also described the stratigraphy of the type section. Subsequently, the White Head was divided into nine informal lithostratigraphic units (Skidmore and Lespérance, 1981). Malo (1988b) later assigned the three lower siliciclastic-dominated units to the Pabos Formation (Fig. 4.6), thus redefining the base of the White Head Formation and extending the Pabos Formation, formerly known only in the Mount Alexandre-Pellegrin area (column 17 of Fig. 4.2a and 4.3), into the Percé area. This practice was followed by Lespérance et al. (1987) and Malo (1988a).

In the Percé area, the oldest unit of the anticlinorium, the Honorat Group, is absent on the southwestern side of the Murphy Creek Inlier (Fig. 4.5), and the Matapédia Group overlies unconformably the Cambrian Murphy Creek Formation (Fig. 4.6), which correlates with the carbonate sequence of the external Humber Zone (Kirkwood, 1989). The lower beds of the Pabos Formation of the Matapédia Group consist of a quartz pebble conglomerate or very coarse sandstone, overlain by mudshales with thin-bedded bioclastic calcarenite, sandy limestone and

calcilutite. The mudstone-limestone assemblage is followed by thin beds of sandy mudshales and calcareous feldspathic wackes. The upper part of the Pabos Formation consists of mudshales and calcilutites. The base of the White Head Formation is defined by a predominance of calcilutites.

The White Head Formation has four members (Lespérance et al., 1987). The lower Burmingham Member consists of 130 m of thin-bedded grey calcilutites with thinner interbeds of calcareous mudshales and a few thin beds of calcarenites. The overlying Côte de la Surprise Member is composed of 50 m of dark green calcareous mudstones. A 450 m recurrent calcilutite member, the L'Irlande Member, overlies the Côte de la Surprise Member and consists chiefly of grey to brown calcilutites in regularly bedded 7-15 cm layers with thinner calcareous mudshale interbeds and a few rare thin-bedded bioclastic calcarenites. At its stratotype, this member contains a 32 m middle unit of clayshales. The upper Des Jean Member, is made up of 250 m of thin-bedded silty and argillaceous limestone, calcareous shale and lenticular calcarenite. The

323

basal part of the member consists of limestone conglomerate, calcarenite, calcilutite and sandy limestone in thin to very thick beds. Overall thickness of the White Head Formation is 880 m.

Faunas of this sequence have been studied by Lespérance (1968, 1974), Lespérance and Sheehan (1976), Sheehan and Lespérance (1981), Lespérance and Tripp (1985), Nowlan (1981), and reviewed in Lespérance et al. (1987). The Pabos Formation and the Burmingham Member of the White Head Formation are Ashgill. The Côte de la Surprise Member has a Hirnantian fauna (latest Ordovician) and the upper L'Irlande and Des Jean members are not younger than Llandoverian C5 (Nowlan, 1983).

Mount Alexandre-Pellegrin area
(column 17, Fig. 4.3).

Kindle (1936 and in Alcock, 1935) was the first to describe the Pabos Formation and designated a type area in the southern part of the Grande Rivière and Petit Pabos rivers in the Mount Alexandre-Pellegrin region. The Pabos Formation of Kindle (1936) also included the Ordovician rocks north of the Chaleurs Bay Synclinorium, from the Grande Rivière to the Petite Cascapédia River. Malo (1988a) redefined the lower and upper limits of the Pabos Formation in this area, recognizing the White Head Formation in the upper part of the sequence, and the Garin Formation of the Honorat Group in the lower part of the sequence. He selected an accessory stratotype for the Pabos Formation in the northern sector of the Grande Rivière Ouest River where contacts with the underlying Honorat Group and the overlying White Head Formation are exposed.

The Garin Formation is a terrigenous unit of fine grained sandstones to conglomerates. In the Mount Alexandre-Pellegrin area, only its upper 800 m is exposed. It consists of greenish grey non calcareous mudstone interbedded with thin- to thick-bedded, fine to medium grained, lithic wacke and thin-bedded fine grained calcareous quartz wacke. The sandstones exhibit numerous sedimentary structures (graded bedding, parallel and cross-laminations, load and flute casts). The base of the overlying Pabos Formation consists of calcareous mudstone with very thin interbeds of calcareous siltstone. This assemblage is followed by a sequence of calcareous rocks; calcareous mudstone, argillaceous limestone, calcareous siltstone, silty limestone, calcareous sandstone, calcareous conglomerate, sandy calcarenite, and calcilutite. Siltstone and sandstone are thin- to medium-bedded and exhibit parallel and cross-laminations as well as flute casts. Calcareous conglomerates are thick bedded and contain clasts composed of Pabos and Honorat lithologies together with exotic clasts of sericite and chlorite schists, milky quartz and foliated sandstone. Thickness of the Pabos is estimated at 1400 m. The four members of the White Head Formation of the Percé area are also recognized above the Pabos Formation in the Mount Alexandre-Pellegrin area where they are nearly 2000 m thick. Their lithological composition does not differ significantly from that of the Percé area.

Dating of the sequence in this area is based mainly on graptolites, brachiopods, and conodonts. The Garin Formation contains two graptolite assemblages (Riva and Malo, 1988). The oldest assemblage belongs to the *C. spiniferus*

Zone of Caradoc age, whereas the youngest assemblage ranges from the *Climacograptus pygmaeus* to *Paraclimacograptus manitoulinensis* zones of Caradoc to Ashgill age. The Pabos Formation has Ashgill brachiopods (P.J. Lespérance in Malo, 1979). The Burmingham Member of the White Head Formation is Ashgill based on a conodont fauna (G.S. Nowlan in Malo, 1988b). The Côte de la Surprise Member is unfossiliferous, but considered Hirnantian as in the Percé area. The upper member of the White Head Formation, the Des Jean Member, contains graptolites of Llandovery A4 to B3 (Riva and Malo, 1988), and conodonts of the *Distimodus kentuckyensis* Zone (G.S. Nowlan in Malo, 1988b). The member is therefore not younger than Llandovery C3-C4. The overlying Burnt Jam Brook Formation has graptolites of Llandovery C1-C2 age (Bourque and Lachambre, 1980), thus setting an upper age limit for the White Head Formation.

New Richmond-Honorat area
(column 20, Fig. 4.3).

Skidmore (1965a) assigned the Ordovician rocks of Honorat and Reboul counties to two groups, the Matapédia Group on the north side of the Grand Pabos Fault and the Honorat Group on its south side (Fig. 4.5). Subsequently, Malo (1988a) divided the Honorat Group into two formations, a lower Arsenault Formation and an upper Garin Formation, but Malo and Bourque (1993) later excluded the Arsenault Formation from the Honorat Group. The Arsenault Formation, of Llanvirn age, is composed mainly of thick bedded, greenish grey lithic wacke with black to olive-green claystone, minor grey to black fine grained tuff interbeds and some gabbroic sills (Malo, 1988a). Recent mapping south of the Grand Pabos Fault (Malo and Moritz, 1991) shows that the Arsenault Formation rests unconformably on a mélange unit. The Llanvirn Arsenault Formation is in turn overlain unconformably by the Silurian Chaleurs Group in its eastern part, whereas in its western part, it is disconformably overlain by the Caradoc to Ashgill Garin Formation, which is in turn, conformably overlain by the complete Ashgill to Llandovery Matapédia Group and Silurian Chaleurs Group (Fig. 4.3, 4.4). Field relations show, on the one hand, that the contact between the Arsenault and Garin formations is structurally concordant, but on the other hand, that there is a hiatus between the two formations, since the age of the Arsenault is Llanvirn whereas the oldest graptolite fauna recovered from the Garin is late Caradoc (*Climacograptus spiniferus* Zone, Riva and Malo, 1988). The Llandeilo and much of the Caradoc is therefore missing (column 20, Fig. 4.3). The mélange and Arsenault Formation assemblage is included in the Gaspésie Subzone of the Québec Dunnage Zone (Tremblay et al., Chapter 3) and the late Caradoc to Ashgill Garin Formation is considered as the basal unit of the Gaspé Belt that overlies rocks of the Dunnage Zone (Malo and Bourque, 1993). The Garin Formation is widespread throughout the Aroostook-Percé Anticlinorium in Gaspésie. Its thickness is estimated at 1200 m and it is made up of various rocks, chiefly terrigenous; black claystone, non calcareous grey mudstone, greenish grey siltstone, calcareous siltstone, calcareous quartz wacke, lithic wacke, conglomerate, and silty dolomitic limestone. The fine grained rocks with which sandstones and conglomerates are interbedded constitute the major part of the unit. The lithic wacke and conglomerate occur in the lower part of the unit, whereas the calcareous silt-stone, calcareous wacke and dolomitic limestone are in the upper part.

The Pabos Formation above the Garin Formation consists of three lithological units. The lower unit is 300 m of calcareous laminated mudstone interbedded with very thin siltstone. The middle unit is composed mainly of greenish grey mudstone and siltstone and is similar to the fine grained terrigenous rocks of the upper Honorat Group. Sandstone and conglomerate also occur locally in this assemblage that reaches a thickness of 300 m. The upper unit consists of 100 m of calcareous mudstone with a few beds of calcilutite near the top of the unit. The White Head Formation is 150 m in the New Richmond-Honorat area compared to 880 m in the Percé area and 2000 m in the Mount Alexandre-Pellegrin area. It is made up of medium bedded calcilutite with shale interbeds.

Graptolites from the Garin Formation are Caradoc *Climacograptus spiniferus* Zone (Riva and Malo, 1988). An Hirnantian brachiopod fauna (Ashgill) occurs in the middle unit of the Pabos Formation (P.J. Lespérance in Malo, 1986). The White Head Formation is poorly dated in the New Richmond-Honorat area, but its age is bracketed by the underlying Ashgill Pabos Formation and the overlying Llandovery A3-A4 Clemville Formation (Bourque and Lachambre, 1980). In the Carleton area, west of the New Richmond-Honorat area, the occurrence of the trilobite *Acernaspis* sp. in the calcilutite beds of the White Head Formation indicates a probable Silurian age.

Patapédia area (column 24, Fig. 4.3).

This is the type-area of the Matapédia Series (Crickmay, 1932), now Matapédia Group (Alcock, 1935). Rocks here are correlated with the White Head Formation of the Percé area (Alcock, 1935). Recent mapping of the area has been done by Lachance (1974, 1975, 1977, 1979).

Three main lithostratigraphic units of the Aroostook-Percé Anticlinorium between the Percé and New Richmond-Honorat areas, namely the Garin, Pabos and White Head formations (Malo, 1988a), are recognized in the Patapédia area. The Garin Formation of the Honorat Group is composed of dark grey non calcareous mudstone interbedded with thin- to medium-bedded siltstone and medium- to thick-bedded wacke (Lachance, 1974, 1975). Its thickness is estimated at 800 m. The overlying Pabos Formation consists of 3000 m of calcareous terrigenous facies. The lower part of the formation consists of fine- to coarse-grained lithic sandstone with interbeds of calcareous and non calcareous mudstone, and local thick beds of quartz pebble conglomerate with a calcareous matrix. The upper part of the formation is made up of massive calcareous mudstone and argillaceous limestone with thin interbeds of calcareous siltstone. The overlying White Head Formation consists of 4000 m of carbonate facies with all four members of the Percé area represented (Malo, 1986).

The Honorat Group is undated in this area. The Pabos Formation is Ashgill, based on the presence of *Dicellograptus complanatus* (Riva and Malo, 1988). Nowlan (1981) reported an Asgill conodont fauna from the Burmingham Member of the White Head Formation, and *Glyptograptus persculptus* was found in the Côte de la Surprise Member (Riva and Malo, 1988). The ostracode *Bolbinoessia* sp. and

the trilobite *Acernaspis* sp. (P.J. Lespérance, pers. comm. to Malo, 1985) in the upper part of White Head Formation indicate a Silurian age.

Madawaska County and northern New Brunswick area (column 28, Fig. 4.3).

Hamilton-Smith (1970) extended the ribbon limestone of the Carys Mills Formation of Maine (Pavlides, 1968) to the Siegas area of Madawaska County in northwestern New Brunswick. This formation overlies an unnamed terrigenous unit. Farther north, in northwestern New Brunswick, St. Peter (1977) recognized two groups, the Grog Brook and the Matapédia. The older Grog Brook Group is correlated with the Honorat Group of Gaspésie and the unnamed unit of the Siegas area. The overlying Matapédia Group is correlated with the Carys Mills Formation of the Siegas area. The lower contact between the Upper Ordovician-lowermost Silurian sequence and the Cambrian-Ordovician rocks is exposed in the Popelogan Anticline (Ruitenberg et al., 1977). A hiatus similar to the one between the Arsenault and Garin formations exists between the Middle Ordovician Balmoral Group and the Upper Ordovician Grog Brook Group (Fig. 4.4; van Staal and Fyffe, Chapter 3). The Grog Brook Group, equivalent to the Garin Formation (Malo, 1988a), is the basal unit of the Aroostook-Percé Anticlinorium in northern New Brunswick, whereas the Balmoral Group represents the Popelogan Subzone of the New Brunswick Dunnage Zone (van Staal and Fyffe, Chapter 3).

The Grog Brook Group is made up of various terrigenous rocks; slate, siltstone, sandstone and conglomerate (St. Peter, 1977). Its thickness is estimated at 7600 m. Recent investigations in the Ristigouche River area suggest that the Grog Brook Group includes also the lower Pabos Formation equivalents (Malo, 1988a).

The Matapédia Group is composed of various carbonate rocks (argillaceous limestone, calcareous shale, sandy limestone and limestone conglomerate) from 1250-2750 m thick. Some rocks of this carbonate sequence are similar to the Pabos and White Head lithologies of the Gaspé sequence.

The lower, unnamed unit of the Siegas area is composed mainly of dark grey, non calcareous, laminated mudstone interbedded with fine grained calcareous sandstones about 600 m thick. The unit is correlated with the Grog Brook Group (St. Peter, 1977) and the Garin Formation of the Honorat Group of the Gaspé sequence (Malo, 1988a; Fig. 4.7). The Carys Mills Formation has three members: lower and upper limestone members made up mainly of thin- to medium-bedded argillaceous calcilutite with shale interbeds, and an intervening member of slate with minor quartz sandstone. The thickness of the formation is about 400 m.

Lithostratigraphic relationships in Madawaska County and comparisons with Gaspésie are summarized in Figure 4.7.

Carleton County area (column 27, Fig. 4.3).

Upper Ordovician-lowermost Silurian rocks of this area are assigned to the Carys Mills Formation (Hamilton-Smith, 1972; St. Peter, 1982) and consist of thinly interbedded dark grey calcareous slates and light to dark grey fine grained limestones. Rocks beneath the Carys Mills Formation in an anticline just west of the Rocky Brook-Millstream Fault (Fig. 4.2a), across the Maine-New Brunswick border, are included in the New Brunswick Dunnage Zone (van Staal and Fyffe, Chapter 3).

Summary of stratigraphy

The Aroostook-Percé Anticlinorium in Quebec and adjacent New Brunswick has a lower siliciclastic assemblage (Grog Brook and Honorat groups) and an upper carbonate assemblage (Matapédia Group and Carys Mills Formation). A transitional assemblage (Pabos Formation) is identified in Quebec. The lower siliciclastic assemblage is Late Ordovician, the carbonate assemblage is mainly Early Silurian.

The stratigraphy of the Aroostook-Percé Anticlinorium is different on opposite sides of the Grand Pabos Fault (compare sections 24 and 20 with section 17 of Fig. 4.6). North of the fault, the sequence is dominated by calcareous rocks of the Matapédia Group (Pabos Formation and all four members of the White Head Formation), with very few representatives of the Honorat Group (Fig. 4.5). South of the fault, terrigenous facies predominate. In the New Richmond-Honorat area for example, only the L'Irlande calcilutite member of White Head Formation is present (Fig. 4.6). Therefore, the carbonate sedimentation which started during Late Ordovician time, north of the fault, first occurred in Early Silurian time south of the fault. Moreover, thicknesses of the carbonate facies that reach 2000 m north of the fault, are only 150 m south of the fault in the New Richmond-Honorat area. Farther west, in the Patapédia area, the stratigraphic succession south of the Ristigouche Fault (southern area) is similar to that of the Mount Alexandre-Pellegrin area north of the Grand Pabos

Fault. Juxtaposing the two areas of similar carbonate stratigraphy demands a lateral translation of at least 85 km along the Grand Pabos-Ristigouche faults.

Provenance and depositional environments

Sandstones of the Honorat and Grog Brook groups and the Pabos Formation are typical turbidites (St. Peter, 1977; Ducharme, 1979; Malo, 1988a). Features of turbidites also characterize the silty limestones and calcarenites of the White Head Formation in Gaspésie, as well as rocks of the Matapédia Group and Carys Mills Formation in New Brunswick (Hamilton-Smith, 1970, 1972; St. Peter, 1977, 1982). Therefore, the Upper Ordovician-lowermost Silurian rocks of the Aroostook-Percé Anticlinorium occupied a relatively deep marine basin. This interpretation is also supported by a general lack of shelly faunas, except for the Percé area, the presence of graptolites in the Honorat Group, and deep marine trace fossils (Pickerill, 1980) in the Grog Brook and Matapédia groups of northern New Brunswick.

A progressive change from deeper terrigenous deposition (Honorat and Grog Brook) to shallower carbonate deposition (Matapédia and Carys Mills) reflects basin infilling. Composition of the Garin Formation in the New Richmond-Honorat area, suggests a source containing volcanic, metamorphic, sedimentary, plutonic and ultrabasic rocks (Malo, 1986). Paleocurrent analysis of the Honorat sandstones in the Patapédia area indicates currents from the southeast (Ducharme, 1979). Paleocurrent measurements from the Grog Brook Group in western New Brunswick indicate currents from the northeast (St. Peter, 1977). The Fournier Group of the Elmtree Inlier is a possible source for mafic volcanic fragments of the Grog Brook conglomerates (Rast and Stringer, 1980).

Carbonate detritus of the Matapédia Group was probably derived from the contemporary Anticosti Platform to the northeast. The conodont fauna of the Percé sequence is very similar to that of the coeval Anticosti sequence (Nowlan, 1981).

In the Patapédia and Mount Alexandre-Pellegrin areas, paleocurrents indicate a northern source for the Matapédia Group (Ducharme, 1979; Malo, 1988b). This implies former proximity to the Anticosti Platform and supports large scale strike-slip movement along the Grand Pabos-Ristigouche fault system (Bourque et al., 1985).

Chaleurs Bay Synclinorium

Definition and limits

The Chaleurs Bay Synclinorium is located between the Aroostook-Percé Anticlinorium to the northwest and the Miramichi Anticlinorium to the southeast. The boundary between the Aroostook-Percé Anticlinorium and Chaleurs Bay Synclinorium, where stratigraphic, is the conformable contact between the Upper Ordovician-lowermost Silurian (Matapédia Group) and the Silurian-lowermost Devonian (Chaleurs Group). The boundary is the Rocky Brook-Millstream Fault in much of New Brunswick. The Chaleurs Bay Synclinorium-Miramichi Anticlinorium boundary is faulted in most places, but locally it is recognized as an unconformity (Helmstaedt, 1971; Fyffe and Noble, 1985).

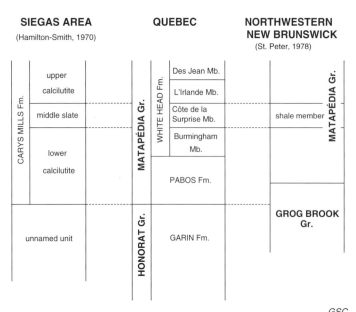

Figure 4.7. Correlations between the Quebec and New Brunswick parts of the Aroostook-Percé Anticlinorium.

Three of the four stratigraphic packages previously described occur in the Chaleurs Bay Synclinorium (Fig. 4.2a); the Upper Ordovician-lowermost Silurian assemblage, the Silurian-lowermost Devonian assemblage, and the Middle and Upper Devonian assemblage. The Silurian-lowermost Devonian assemblage (Chaleurs Group and equivalents) is predominant. The Upper Ordovician-lowermost Silurian asssemblage occurs in two subsidiary anticlines within the Chaleurs Bay Synclinorium; the Duval Anticline north of Chaleurs Bay, and the Nash Creek Anticline in New Brunswick. The Middle and Upper Devonian assemblage occurs chiefly in the subsidiary Ristigouche Syncline.

Two inliers of older rocks are known in the Chaleurs Bay Synclinorium: the Cambrian (?) and Ordovician rocks of the Maquereau-Mictaw Inlier of southern Gaspésie, and the Ordovician rocks of the Elmtree Inlier of northern New Brunswick. The middle Paleozoic rocks are either unconformable or faulted against older rocks of the Maquereau-Mictaw and Elmtree inliers. Devonian granites cut the middle Paleozoic rocks in New Brunswick, and Carboniferous rocks are unconformable on rocks of the Chaleurs Bay Synclinorium in Chaleurs Bay, and inland to the south.

Middle Paleozoic rocks of the Chaleurs Bay Synclinorium are mainly siliciclastites with a few intercalated limestone units. Locally, felsic and mafic volcanic rocks are abundant. Although stratigraphic names are different on opposite sides of Chaleur Bay, fossil occurrences and similar lithologies permit correlation.

Structural geology

Structures of the Chaleurs Bay Synclinorium trend northeast. The synclinorium is dissected by at least two major dextral strike-slip faults, the McKenzie Gulch and the Rocky Brook-Millstream faults (Fig. 4.2a). The main axis of the synclinorium extends from the Maquereau-Mictaw Inlier where it plunges gently to the southwest, to Hartland in New Brunswick where it plunges northeast. Subsidiary folds occur on both sides of the main axis, such as the Ristigouche Syncline.

The axial trace of the Chaleurs Bay Synclinorium is offset by about 25 km in a dextral sense on opposite sides of Chaleur Bay. Displacement may have occurred along the extension of the McKenzie Gulch Fault. Farther south, the axial trace of the synclinorium is displaced approximately 35 km dextrally by the Rocky Brook-Millstream Fault. These displacements are probably associated with the Grand Pabos-Ristigouche strike-slip fault system of the Aroostook-Percé Anticlinorium. Reverse faults occur between the Grand Pabos and the McKenzie Gulch faults, principally in the Maria area and at the western extremity of the Ristigouche Syncline. Numerous minor easterly striking faults occur in northern New Brunswick (Greiner, 1970, 1973).

Folds are open to moderately tight with subhorizontal or gently plunging axes. Box folds in the Port-Daniel area (Bourque and Lachambre, 1980, Map 1929), may reflect interference between regional folding and accomodation of the Maquereau-Mictaw Inlier.

Stratigraphy

Rocks of the Chaleurs Bay Synclinorium are described in four areas: Quebec, New Brunswick, Ristigouche Syncline, and Maria faulted zone.

Quebec (columns 19 and 20, Fig. 4.3).

The stratigraphic nomenclature of the Silurian-lowermost Devonian rocks in the Gaspésie-Témiscouata Valley segment of the Quebec Appalachians is chiefly derived from the Chaleurs Bay Synclinorium (columns 19 and 20 of Fig. 4.3) and the northern part of the Matapédia Valley (column 5 of Fig. 4.3) in the Connecticut Valley-Gaspé Synclinorium.

After the pioneering work of Sir William Logan (1863) along the Gaspésie coast, Schuchert and Dart (1926) proposed a formal stratigraphy for the Silurian sequence of the Port-Daniel area in the Chaleurs Bay Synclinorium. Their nomenclature was subsequently modified (Alcock, 1935; Northrop, 1939; Badgley, 1956; Burk, 1964; Ayrton, 1967; Bourque, 1975; Bourque and Lachambre, 1980). Mapping in this area was done by Badgley (1956), Skidmore (1965b), Ayrton (1967) and Bourque and Lachambre (1980).

The stratigraphic nomenclature of Bourque (1975) and Bourque and Lachambre (1980) is followed here. In the Port-Daniel area (Fig. 4.2a, and 4.3), the sequence above the Mictaw Group (Middle Ordovician) and below the Bonaventure Formation (Carboniferous) is assigned to the Chaleurs Group (Burk, 1964; Bourque, 1975; Bourque and Lachambre, 1980), previously the Chaleur Group, Chaleur Formation (Ami, 1900), Chaleur Series (Schuchert and Dart, 1926; Northrop, 1939), Chaleur Bay Series (Alcock, 1935; Badgley, 1956), or Chaleurs Bay Group (Skidmore, 1958). In the New Richmond and Duval anticline areas (Fig. 4.2 , 4.6), the Chaleurs Group is conformable above the White Head Formation of the Matapédia Group. The oldest middle Paleozoic rocks occur in the Duval Anticline, northwest of the main axis of the synclinorium, and consist of two units; a lower unit of mustone and siltstone with 3-5 m interbeds of sandstone and conglomerate, and an upper unit of calcilutite in 4-5 cm regular layers interbedded with thin argillaceous limestone. The lower unit is correlated with the terrigenous unit of the Pabos Formation in the New Richmond-Honorat area. Brachiopods from the sandstones are representative of the uppermost Ordovician Hirnantian fauna (P.J. Lespérance, pers. comm., to Malo, 1985). The calcilutite unit is correlated with the Silurian part of the White Head Formation in the New Richmond-Honorat area (column 20, Fig. 4.3).

A detailed stratigraphy of the Chaleurs Group, references, location and log-description of the type-section, and lithological variations throughout the synclinorium are given by Bourque (1975) and Bourque and Lachambre (1980).

The Clemville Formation is a fine grained siliciclastic unit with minor calcilutites. The lower part of the formation is a thick bedded, light-coloured quartz sandstone, with greenish mudstone and siltstone interbeds. The upper part is composed of thick bedded greenish mudstone and siltstone with intercalations of parallel and cross-laminated sandstones and minor thin bedded calcilutites. The formation is locally very fossiliferous with offshore brachiopod

327

and coral assemblages. The Clemville Formation is unconformable upon the Mictaw Group with only the upper sixth of the Clemville Formation present in the type section). In the western part of the synclinorium, the entire Clemville Formation is present. Thicknesses range from 105 m in the east to 600 m in the west.

The overlying Weir and Anse Cascon formations are related units. They consist of medium to coarse grained sandstones and conglomerates interbedded with fine grained siliciclastites, commonly parallel- and cross-stratified. The coarse grained siliciclastites of the Weir Formation are pinkish grey and green feldspathic, arkosic sandstones and conglomerates with distinctive pink feldspars. Siliciclastites of the Anse Cascon Formation are buff quartz sandstones and conglomerates. In the Weir Formation, the fine grained siliciclastites are not significantly different from those of the underlying Clemville Formation. The proportion of the fine- to coarse-grained siliciclastites in the Weir Formation is 2:1, compared to 1:1 in the Anse Cascon Formation. Both formations contain nearshore brachiopods, and the Anse Cascon is thoroughly bioturbated. Combined thicknesses range from 73 m in the east to 1410 m in the west.

The Anse à Pierre-Loiselle Formation is transitional between the underlying Anse Cascon sandstones and the overlying La Vieille limestones. It is a muddy siliciclastic unit with a few beds of Anse Cascon-type sandstones at the base, and nodular limestones like the La Vieille in its upper part. It contains a middle marker unit of cross-stratified conglomerate and coarse sandstone. The formation is thoroughly bioturbated and contains abundant offshore brachiopods, corals, stromatoporoids and algae in its upper part.

La Vieille Formation is the lowest limestone horizon in the Silurian-lowermost Devonian sequence and is a useful marker. It has three members (Bourque and Lachambre, 1980): a lower, very fossiliferous nodular limestone; a middle oncolite-stromatolite-algal-fenestral limestone with coral-bryozoan-stromatoporoid-algal bioherms, and peloidal and intraclast calcarenites; and an upper, unfossiliferous nodular limestone. Its thickness ranges from 200-450 m.

The overlying Gascons Formation is thick, monotonous, commonly structureless, fine grained sandstones, siltstones and mudstones, with minor nodular mudstones and calcareous shales. In several localities, it has an abundant trace fossil *Zoophycos*, indicative of intermediate water depths (Seilacher, 1964, 1967). It also contains a few brachiopods, corals and graptolites. Thickness ranges from 600 m in the east to 1200 m in the west.

The top of the Gascons is defined by the first occurrence of limestones of the West Point Formation. The West Point is nearly 900 m thick and composed of three superposed complexes: lower reef, middle bank, and upper reef complex (Bourque et al., 1986). The overlying *Zoophycos*-rich fine grained siliciclastic Indian Point Formation does not differ significantly from the Gascons. Detailed correlations within the Port-Daniel and New Richmond areas are given by Bourque and Lachambre (1980) and Bourque et al. (1986).

Volcanic rocks occur in the upper part of the Chaleurs Group at Black Cape near New Richmond. They are porphyritic and amygdular mafic lava flows, volcanic breccia and conglomerate, with minor intercalations of red sandstones and limestone conglomerates (Laurent and Bélanger, 1984). A few corals and stromatoporoids are trapped in the lava flows. Minimum thickness is 725 m.

Thus, the stratigraphy of the Quebec part of the Chaleurs Bay Synclinorium is viewed as three broad assemblages: a lower siliciclastic assemblage, ranging from fine- to coarse-grained, made up of the Clemville, Weir, Anse Cascon and Anse à Pierre-Loiselle; a distinctive middle limestone assemblage, the La Vieille; and an upper fine grained siliciclastic *Zoophycos* assemblage, the Gascons and Indian Point, with intervening reef limestones, the West Point, and mafic lava flows, the Black Cape Member. The overall thickness of the sequence ranges from 2588 m in the eastern part of the synclinorium, up to 5119 m in the western part.

These lithostratigraphic units contain sufficient brachiopods (Boucot and Bourque, 1981) along with fewer graptolites and conodonts to adequately bracket their ages. The only poorly dated rocks are in the Ludlow-Pridoli interval. No Devonian rocks are known in the Québec part of the Chaleurs Bay Synclinorium; the youngest fossils are Late Pridoli conodonts in the Indian Point Formation a few metres above the West Point Formation (C.B. Rexroad in Bourque and Lachambre, 1980).

New Brunswick
(columns 25, 26 and 27, Fig. 4.3).

The New Brunswick part of the Chaleurs Bay Synclinorium constitutes its southwestern extension from Quebec. It is wedged between the Aroostook-Percé Anticlinorium to the northwest and the Miramichi Anticlinorium to the southeast. Three regions are described: Pointe Verte, Charlo-Upsalquitch Forks and Carleton County.

The **Pointe Verte area** was mapped by Greiner (1960) and Fyffe (1974), and its stratigraphy summarized by Noble (1976, 1985) and Fyffe and Noble (1985). The lowest unit is the Armstrong Brook Formation which lies unconformably on the Ordovician Fournier Group. It is coarse grained red and green sandstone and boulder conglomerate with minor shale. Sandstones are feldspathic, lithic and commonly have a clay matrix. Clasts are dominantly volcanic. Large scale dune cross bedding, parallel-laminations and ripple bedding are common, together with chanelling, irregular lensing of beds and fining-upward sequences. Nearshore brachiopods occur locally (Noble, 1976). The Armstrong Brook has a maximum thickness of 365 m.

The overlying Limestone Point Formation has interbedded siliciclastites and limestones. Siliciclastites are fine- to coarse-grained green calcareous quartz sandstones that are parallel-laminated, rarely crossbedded, and bioturbated. Limestones consist of irregular lenses of calcarenite. The formation is very fossiliferous, containing a nearshore to shallow shelf brachiopods fauna, corals and crinoids. Thicknesses range from 183-253 m. The succeeding La Vieille Formation has three units (Noble, 1976; Fyffe and Noble, 1985) that correspond to its three members in Quebec. The only notable differences are the occurrence of red sandstones in the lower unit in New Brunswick and the absence of the Quebec coral-bryozoan-stromatoporoid-algal bioherms in the middle unit. Maximum thickness of the La Vieille is 287 m.

The Petit Rocher Formation, above the La Vieille in the southern part of the Pointe Verte area, is partly equivalent to both the Limestone Point and the La Vieille in the northern part of the area. The Petit Rocher Formation has red, green and dark shales and siltstones, with minor turbiditic sandstones (Noble, 1976; Fyffe and Noble, 1985) and sparse marine fauna debris. Its thickness is 428 m in the south, including the portion laterally equivalent to the Limestone Point and the La Vieille, and 137 m in the north. Beds referred to the Laplante unit (Noble, 1985) and the Laplante Formation (Fyffe and Noble, 1985) occur in the upper part of the Petit Rocher Formation. The Laplante has transported reefal debris in siltstones and dark shales. Descriptions suggest that the debris was derived from the reef margin limestone of the upper reef complex of the West Point Formation in the Port-Daniel area of Quebec (Noble, 1985). Matrix siliciclastites contain Pridoli conodonts (Nowlan, 1982), compatible with the age of the West Point reefs.

In the northern part of the Pointe Verte area, volcanic rocks of the Bryan Point Formation overlie the Petit Rocher. These are porphyritic and amygdaloidal basalts and minor agglomerates, up to 183 m thick. Above the volcanic rocks, the New Mills Formation consists of boulder conglomerate and associated red sandstone with minor volcanic rocks. Fossiliferous pebbles in the conglomerate are of lower La Vieille age, indicating local uplift (Fyffe and Noble, 1985). Pebbles of underlying basaltic flows are also common. Minimum thickness of the New Mills is 130 m (Noble, 1976).

The Silurian sequence of the northern part of the Pointe Verte area is similar to that of the Quebec part of the Chaleurs Bay Synclinorium. It has three broad lithofacies: lower siliciclastics that range from fine- to coarse-grained (Armstrong Brook and Limestone Point); middle distinctive limestone (La Vieille); and upper fine grained siliciclastics (Petit Rocher) with overlying lava flows (Bryan Point), capped by Pridoli to lowermost Devonian conglomerate. The sequence in the southern part of the Pointe Verte area is different. Except for the Armstrong Brook conglomerates at its base, it consists of rather uniform basinal fine grained turbidites of the Petit Rocher. The Pointe Verte sequence is relatively thin (1355 m), compared to the Quebec part of the Chaleurs Bay Synclinorium (2588-5119 m).

Ages are constrained from the base of the succession to the La Vieille Formation by brachiopods (Noble, 1976). However, ages of the upper units are poorly defined. The New Mills is dated approximately as Pridoli to earliest Devonian (Fyffe and Noble, 1985).

Stratigraphy of the **Charlo-Upsalquitch Forks area** was studied mainly by Alcock (1935), Greiner (1970) and Lee and Noble (1977). The oldest rocks occur in a small anticline, the Nash Creek Anticline, northwest of the Chaleurs Bay Synclinorium axis (Fig. 4.2a). They consist of well bedded limestone and mudstone. Calcilutites are dominant with minor calcarenites (Lee and Noble, 1977). These rocks are probably equivalents of the Matapédia Group. They are poorly dated, but seem to range from Middle Ordovician to Early Llandovery (Lee and Noble, 1977).

Rocks above the Matapédia Group belong to the Limestone Point Formation in the northeasternmost part of the area, and to the Upsalquitch Formation in central and western parts. The limestone Point Formation is generally finer grained and contains fewer limestones than that in the Pointe Verte area. The siltstone unit of Lee and Noble (1977) is included in the Upsalquitch Formation of St. Peter (1977; fig. 2 of Fyffe and Noble, 1985). The Upsalquitch Formation is a coarsening-upward sequence divided into three parts (Lee and Noble, 1977): a lower part consisting of alternating 10-50 cm sandstone turbidites and much thicker non-turbiditic siltstones; a middle part composed of bioturbated and burrowed siltstones, shales and sandstones with minor allodapic limestones; and an upper part consisting of coarser grained siltstones and sandstones with common large scale parallel and cross-laminations and dune crossbedding, and fossiliferous arenaceous calcarenites. Intensity of burrowing increases upwards. Fossils are rare or absent in the lower part. The middle Upsalquitch contains a few basinal brachiopods and the upper Upsalquitch is rich in shallow shelf to nearshore brachiopods. The formation has a maximum thickness of 1600 m.

Only the lower unit of the overlying La Vieille Formation is present. It has a thickness of 275 m and is composed of nodular fossiliferous limestones and siltstones with abundant corals, stromatoporoids and brachiopods. It is like the lower member of the La Vieille in Quebec. A unit of graded, parallel- and cross-bedded, coarse grained sandstones and maroon to red grits, the Grit Unit (Lee and Noble, 1977) overlies the incomplete La Vieille Formation. It contains marine fossils at the base, but the top is unfossiliferous. Thickness is 136 m. Above the Grit Unit there is a 200 m unfossiliferous Conglomerate Unit (Lee and Noble, 1977) with basic volcanic boulders in a poorly sorted silty matrix, and a few mafic lava flows. The overlying Bryan Point Formation is more widespread in the Charlo-Upsalquitch Forks area than in the Pointe Verte area. It consists of greenish grey to maroon basalts, in places vesicular and mottled, and feldspar porphyries, tuffs, agglomerates, volcanic breccias, welded crystal tuffs, and andesites (Greiner, 1970). Conglomerates with volcanic pebbles and boulders occur near the base of the formation, as well as red sandstones (Greiner, 1970; Lee and Noble, 1977). Thickness of the Bryan Point Formation is about 300 m.

Lee and Noble (1977, table 1) indicate that the age of the lower unit of the La Vieille Formation in this area is older than that in the Pointe Verte area. However, the lower La Vieille of both areas is assigned to the *Costistricklandia gaspensis* (=*lirata*) Zone (Noble, 1976; Lee and Noble, 1977). This age corresponds to that of the lower La Vieille in Gaspésie (Bourque and Lachambre, 1980; Boucot and Bourque, 1981). The age of the Grit Unit, Conglomerate Unit, and Bryan Point Formation is no more precise than Early Wenlock or younger (Lee and Noble, 1977).

As in the Pointe Verte area, the unfossiliferous New Mills redbeds and volcanic-boulder conglomerate contain reworked boulders of the underlying formations (Greiner, 1970). Its thickness ranges from nearly zero to almost

120 m. A unit of felsic extrusive rocks above the New Mills, the Benjamin Formation (Greiner, 1970), consists of flow-banded rhyolites, quartz-feldspar crystal tuffs and agglomerates (Fyffe and Noble, 1985). Thickness is estimated at "several hundred feet" (Greiner, 1970). The Benjamin felsic volcanic rocks are overlain by mafic extrusive and sedimentary rocks of the Dalhousie Formation that are the stratigraphically highest beds of the Chaleurs Bay Synclinorium in northern New Brunswick.

Stratigraphy of the **Carleton County area** at the southwestern extremity of the Chaleurs Bay Synclinorium (column 27, Fig. 4.3) is summarized from St. Peter (1978a, 1982) and Hamilton-Smith (1972). The lowest unit, the Smyrna Mills Formation, conformably overlies the Carys Mills Formation of the Aroostook-Percé Anticlinorium. It is divided into two informal members (Hamilton-Smith, 1972): a lower member of thick bedded dark grey to black non-calcareous and non- laminated slates with minor red and green slates, and iron- and manganese-rich siliceous siltstones; and an upper member of dark grey laminated slates, fine grained quartz sandstones with Bouma sequences, and dark micaceous shales. Thickness of the formation ranges from 600-2000 m. It is overlain by the Tobique Group divided into two formations; the volcanic Costigan Formation, and the sedimentary and volcanic Wapske Formation. The Costigan consists of pink sub-aerial felsic volcanic rocks and subordinate sedimentary rocks, and minor mafic volcanic rocks. It may be as thick as 3000 m. The Wapske Formation is composed of dark grey laminated slates, thin- to very thick-bedded fine grained siliciclastites, and minor coarse grained sandstones and conglomerates with isolated lenses of mafic volcanic rocks. Its thickness possibly reaches 4000 m. Local pebbles of the Smyrna Mills Formation in the Wapske conglomerates indicate local uplifts. An earliest Devonian age is attributed to the Costigan Formation, whereas spores indicate a late Emsian age for the upper part of the Wapske Formation.

The rather uniform Silurian Smyrna Mills sedimentary facies in the southwestern extremity of the Chaleurs Bay Synclinorium is time equivalent to a number of much more diversified sedimentary facies along the Chaleurs Bay coast to the northeast. The only similar and possibly correlative facies is the Petit Rocher Formation of the southern Pointe Verte area. The Wapske Formation of the Tobique Group is in part equivalent to the Dalhousie Formation. Its volcanic rocks are similar to the basalt, basaltic tuff and palagonite tuff of the Dalhousie Formation at its type section (St. Peter, 1978a). The underlying Costigan felsic volcanic rocks resemble the Benjamin felsic rocks underlying the Dalhousie Formation.

In summary, the Chaleurs Bay Synclinorium in northern New Brunswick, exclusive of the Ristigouche Syncline, contains four broad assemblages: a lower fine- to coarse-grained siliciclastic assemblage, the Upsalquitch, Armstrong Brook and Limestone Point formations; overlain by the distinctive limestone assemblage of the La Vieille Formation; followed by the fine grained assemblage of the Petit Rocher Formation with its associated Grit and Conglomerate units, and volcanic rocks (Bryan Point Formation); and an upper assemblage of a basal conglomerate (New Mills Formation) overlain by the thick volcanic and fine grained sedimentary rocks of the Dalhousie Formation. An erosional surface separates the third and fourth assemblages.

Ristigouche Syncline (columns 22 and 23, Fig. 4.3).

The Ristigouche Syncline is a subsidiary structure of the Chaleurs Bay Synclinorium on opposite sides of the Ristigouche River in Quebec and New Brunswick (Fig. 4.2a). In Quebec, the syncline has two parts (Bourque and Lachambre, 1980): the Ristigouche Syncline proper (referred to as the Ristigouche-Nouvelle Autochthon) whose northern limit coincides with the Aroostook-Percé Anticlinorium-Chaleurs Bay Synclinorium boundary; and a small structural slice, the Sellarsville Slice, separating the Ristigouche Syncline from the Aroostook-Percé Anticlinorium (Fig. 4.8).

Rocks of the Ristigouche Syncline in Quebec conformably overlie the Matapédia Group of the Aroostook-Percé Anticlinorium. The overlying sequence contains an unconformity (Bourque and Lachambre, 1980).

The succession below the unconformity belongs to the Chaleurs Group. Its base is the Mann Formation – grey, structureless, fine grained siliciclastites with intercalated thinly bedded parallel and convolute-laminated bioturbated fine grained sandstones, and minor nodular calculutites and calcarenites. It is locally rich in shallow shelf brachiopods. Thickness ranges from 230-680 m. The overlying Anse à Pierre-Loiselle Formation is similar to its occurrence in the Chaleurs Bay Synclinorium to the east and constitutes a transition between the Mann and La Vieille formations, with a relatively uniform thickness of 130 m. Although the overlying La Vieille Formation is not characterized by a threefold division as in the Chaleurs Bay Synclinorium, lithologies typical of the formation are recognized. Typical nodular limestones are common, with locally abundant large smooth-shell pentamerid brachiopods. Recent work in the Ristigouche Syncline (Lavoie and Bourque, 1986) has shown that reefoid limestones assigned to the West Point Formation (Bourque and Lachambre, 1980, columnar sections 30-33 and 39-40 of their fig. 6 and Map 1958) below the unconformity, actually belong to La Vieille Formation and as such are similar to the bioherm lithologies of the middle member of the Port-Daniel area. Maximum thickness of the La Vieille is 140 m. Mafic volcanic rocks of the informal Ristigouche member occur below, within, and above the La Vieille Formation. They are basic lava flows, tuffs and volcanic breccias. The rocks beneath the unconformity were folded before Late Silurian erosion (Fig. 4.8). The folded beds are not younger than early Ludlow.

The succession above the unconformity begins with 30-300 m of basal conglomerate of the New Mills Formation (Bourque and Gosselin, 1986). Lithology of the conglomerate is laterally variable. In the western part of the syncline, it consists mainly of pebbles and boulders of dark mafic volcanic rocks in a volcanic sandy matrix, while in the central part, pebbles and boulders (up to 1 m) are dark grey calculutite, contained in red fine grained sandstone. The limestone conglomerate is in very thick beds interlayered with red sandstones. Minor pink felsic volcanic rocks occur in the western part of the syncline. The conglomerate unit disappears eastward.

The overlying Dalhousie Formation is composed of sedimentary and volcanic rocks. In the western part of the syncline, 1500 m of thin- to medium-bedded dark grey parallel-laminated argillaceous fine grained sandstones interbedded with mudstones underlie 5000 m of mafic and minor felsic volcanic rocks. In the eastern part of the syncline, sedimentary units are more massive, muddier, more calcareous, and they alternate with volcanic rocks. The volcanic rocks are vesicular and massive or brecciated lava flows and pyroclastic rocks varying in composition from basaltic to rhyolitic (Bélanger, 1982; Laurent and Bélanger, 1984). Andesites predominate. Textures of lavas are in most places aphyric or microporphyric. The pyroclastic rocks occur as thin layers of well-bedded pumice or scoria and ash deposits, and as thick units of unsorted block and ash deposits. Consanguineous intrusive bodies are minor. Fossils of the Dalhousie Formation in Quebec and New Brunswick are the same age (Boucot and Johnson, 1967; A.J. Boucot in Bourque and Lachambre, 1980).

In northern New Brunswick, the Dalhousie Formation is a mafic volcanic unit of basalt and andesite, tuff and breccia, including some felsic volcanic rocks and palagonite tuff, with interlayered sedimentary rocks composed of grey calcareous and arenaceous shale and local fossiliferous limestone. Orange to pink felsic extrusive rocks, correlative with the Benjamin Formation to the east, occur in places below the Dalhousie Formation (Greiner, 1973). Stratigraphy of the Chaleurs Group below the Dalhousie sedimentary and volcanic rocks in New Brunswick is not well enough known to permit correlation with the Quebec side of the syncline nor to demonstrate the sub-New Mills erosional surface. However, Greiner (1973) indicated a disconformity at the base of the Dalhousie Formation, and below the disconformity at the top of the Chaleurs Group, he noted an occurrence of the brachiopod *Costistricklandia*

typical of the lower La Vieille Formation. Therefore, the stratigraphy of the New Brunswick side of the Ristigouche Syncline is in some respects the same as that of the Quebec side.

Emsian to Frasnian terrestrial coarse grained rocks occur above the Dalhousie Formation, in the core of the Ristigouche Syncline (Fig. 4.8). These rocks are chiefly in Quebec, but a few outcrops are also known in New Brunswick. Inland exposure is poor. The sequence (column 22, Fig. 4.3) is divided into four units: the Lagarde, Pirate Cove, Fleurant, and Escuminac formations (Alcock, 1935; Dinely and Williams, 1968).

The Lagarde Formation lies disconformably on the Dalhousie Formation (Alcock, 1935; Béland, 1958; Dineley and Williams, 1968). The contact is exposed only near Atholville in New Brunswick where the basal beds are coarse grained breccias. Elsewhere, the Lagarde Formation is composed of thick beds of well rounded pebble and cobble conglomerates interlayered with grey or greenish grey, coarse- to fine-grained crossbedded sandstones and a few mudstones. Clasts of Matapédia Group limestone occur in conglomerates in the eastern part of the area. The sequence coarsens upward and contains very few fossils. The thickness of the unit is estimated to be less than 1500 m (Dineley and Williams, 1968). The overlying conformable Pirate Cove Formation (Alcock, 1935; Dineley and Williams, 1968; Zaitlin and Rust, 1983) consists of red and grey sandy siltstones and mudstones with channelized lenses of sandstone. Although minor, the most marked lithology is limestone conglomerate with clasts derived

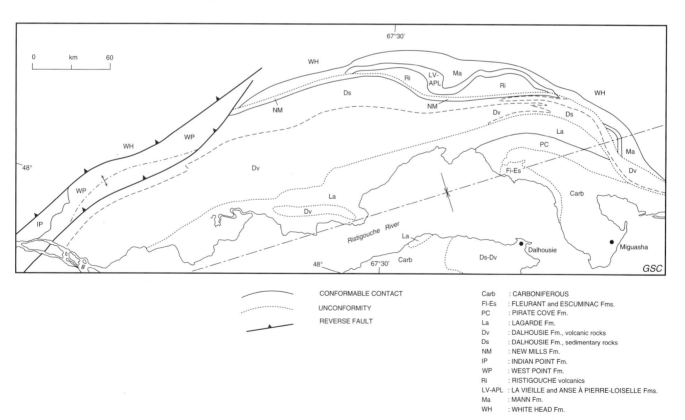

Figure 4.8. Geological map of the northern flank of the Ristigouche Syncline (after Bourque and Lachambre, 1980; Dineley and Williams, 1968).

mainly from the Matapédia Group to the north. Clasts of the underlying Devonian units are also present. Except for the occurrence of palynomorphs, the unit is unfossiliferous. Its thickness is estimated at 500 m (Dineley and Williams; 1968).

The Fleurant Formation (Alcock, 1935; Dineley and Williams, 1968; Zaitlin and Rust, 1983) is of Frasnian age and lies disconformably upon the Pirate Cove Formation. It is a grey, pebble and cobble conglomerate. Clasts are dominantly limestone, with lesser volcanic, sandstone, and minor plutonic clasts (Rust, 1982). The overlying Escuminac Formation is conformable upon the Fleurant. It consists of greenish grey thin- to thick-bedded sandstone, siltstone and varve-like mudstone. Sandstones and siltstones exhibit abundant sole marks, as well as parallel-laminations and current ripples. Fossil fish and plants are abundant and have been the subject of several paleontological studies.

At the western extremity of the Ristigouche Syncline, the Sellarsville Slice (Bourque and Lachambre, 1980) is limited by two reverse faults, trending northeast, and dipping to the northwest. The eastern fault connects with the Rafting Ground Fault in New Brunswick. The slice contains sedimentary rocks of Ludlow to Lochkovian age, consisting of approximately 4000 m of mudcracked red and green fine grained siliciclastites, minor limestone conglomerates with marine fossils, and an upper unit of small stromatoporoid bioherms and biostromes. The redbeds, as well as the bioherms and biostromes, correspond in age and lithology to the upper part of the West Point Formation of the Port Daniel-Black Cape area. The overlying 500 m of massively bedded, greenish grey, calcareous fine grained siliciclastites, locally rich in brachiopods, belong to the Indian Point Formation. The age interval of the thick West Point-Indian Point sequence of the Sellarsville Slice spans the unconformity and the New Mills conglomerates of the nearby Ristigouche Syncline and the northern New Brunswick sequence. This indicates accelerated sedimentation in local basins.

Maria faulted zone (column 21, Fig. 4.3).

Rocks of the Chaleurs Group occur in a fault-bounded small area north of the village of Maria in southern Gaspésie. The stratigraphic section (Gosselin, 1988) begins with fine grained siliciclastites typical of the Clemville Formation conformably overlying the Matapédia Group and reaching a thickness of nearly 200 m. The Clemville is succeeded by 500 m of buff, strongly bioturbated sandstones of the Anse Cascon Formation. A distinctive conglomerate unit above the Anse Cascon is divided into four lithofacies; a lower buff conglomerate and coarse grained sandstone: a red argillaceous conglomerate and coarse grained sandstone with large-scale cross-laminations: a buff sandstone locally rich in Eocoelid brachiopods: and an upper volcaniclastic lithofacies. The conglomerate unit is correlative with the Armstrong Brook Formation of the Pointe Verte area of northern New Brunswick. It reaches thicknesses of 500 m. Fifteen metres of mudcracked microbial laminites and stromatolitic limestone of the La Vieille Formation overlie the Armstrong Brook. The La Vieille is overlain by nearly 1200 m of mafic volcanic rocks similar to those of the Ristigouche member of the Ristigouche Syncline. A minimum thickness of 900 m of Dalhousie-type lithologies and fauna occur above the volcanic rocks. Poor exposure does

not permit clear demonstration of relationships between the Dalhousie beds and the underlying volcanic rocks, so that presence or absence of the sub-New Mills erosional surface, known in northern New Brunswick and in the Ristigouche Syncline, is uncertain.

Depositional environments

Figure 4.9 summarizes the Silurian and Devonian lithofacies for representative sections of the Chaleurs Bay Synclinorium. Depositional environments belong to two broad domains; a shelf and terrestrial domain (successions A, B and C), and a basinal domain (succession D). In the shelf domain, the entire succession is composed of three successive shallowing upward phases, all three ending with nearshore and terrestrial facies, separated by transgressions.

Following deposition of the lime turbidites of the White Head Formation (Matapédia Group), shallowing phase I (Llandovery-Early Wenlock) was dominantly siliciclastic, but ended with limestones. In the Quebec part of the Chaleurs Bay Synclinorium (Port Daniel-Black Cape area), the lithological and faunal succession of the Clemville-Weir-Anse Cascon ranges from offshore muds with sandy storm layers, to lower shoreline sands (Bourque, 1981). After a minor transgression, represented by nodular muds and *Costistricklandia* benthic community of the Anse à Pierre-Loiselle and the lower La Vieille, this phase ended with peritidal limestones of the La Vieille rimmed by a narrow belt of metazoan knob reefs (Bourque et al., 1986). In northern New Brunswick, the Upsalquitch-Limestone Point succession also coarsens and shallows upward, ranging from basinal muds to cleanly washed shallow-water sands (Noble, 1976; Lee and Noble, 1977). The occurrence of nodular mud and *Costistricklandia* community at the base of the La Vieille represents the same minor transgression recorded in Quebec. The end of shallowing phase I in northern New Brunswick is recorded by peritidal limestones of the La Vieille in the northern Pointe Verte area or by subaerial Grit and Conglomerate units in the Charlo-Upsalquitch Forks area. In the Ristigouche Syncline, the Mann Formation has not been analyzed in detail, but is interpreted as shallowing upward between the basinal White Head limestones and the peritidal La Vieille facies. The source of the siliciclastic facies was located likely at two uplifts, the Maquerau-Mictaw and Elmtree inliers. During that time, dark siliciclastic muds (Smyrna Mills) were deposited in the southern basinal area. Therefore, shallowing upward phase I led to the establishment of a carbonate-dominated platform, onto which (Grit Unit), shallow shelf (La Vieille), and basinal facies (Smyrna Mills) are traceable (Bourque et al., 1986).

Subsequent to shallowing phase I, Ludlow-Pridoli time corresponds to a period of uplift and erosion in the Ristigouche Syncline and Charlo-Upsalquitch Forks areas. Thick terrestrial redbeds (New Mills and associated facies) and subaerial volcanic rocks (Black Cape, Ristigouche, Bryan Point) are associated with the uplifted areas. During that time, in the Port Daniel-Black Cape area, a second shallowing upward phase (II), followed a transgression (T1) and led to a siliciclastic-dominated platform (Gascons) rimmed at the end of the phase, by a 100 km reef tract (West Point). Several facies of the reefs record tectonic instability and in some cases, developed in response to instability (Bourque, 1979; Bourque et al., 1986). Allochthonous reef

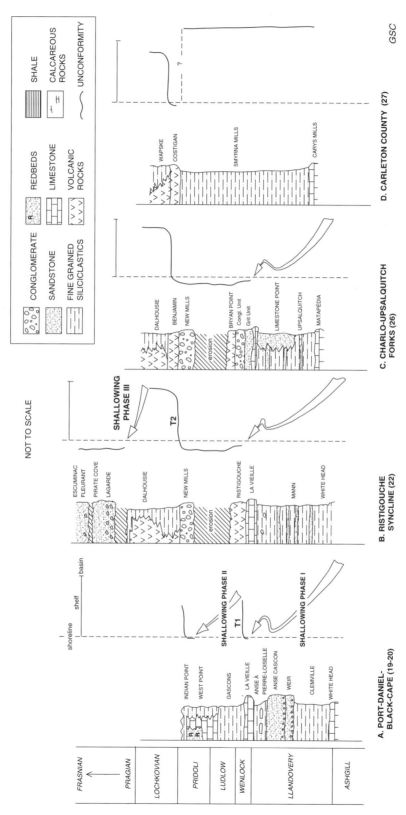

Figure 4.9. Progression of depositional environments during Silurian and Devonian time, with respect to distance from shoreline, for four rock successions in the Chaleurs Bay Synclinorium. T1 and T2 are transgressive episodes.

blocks are recorded in basinal sediments of the Petit Rocher area (Noble, 1985; column 25, Fig. 4.3). Farther south, the dark muds and silts of the Smyrna Mills represent basinal sedimentation.

In the northern area of the Chaleurs Bay Synclinorium, a second transgression (T2), dominated by volcanism, followed shallowing phase II during Lochkovian time. Mafic and locally felsic lava flows are commonly associated with marine calcareous siliciclastic muds containing shelf fauna (Dalhousie). Felsic volcanism (Benjamin, Costigan) characterized the beginning of the transgression in New Brunswick. In the southern area of the Chaleurs Bay Synclinorium, the occurrence of sub-aerial felsic volcanic rocks of the Costigan (St.Peter, 1982) above the basinal graptolitic muds of the Smyrna Mills implies significant shallowing.

Shallowing upward phase III is poorly recorded, except that a terrestrial facies overlies the deep shelf facies of the Dalhousie Formation. Sands and gravels of the Lagarde, Pirate Cove, Fleurant and Escuminac of the Emsian to Frasnian sequence represent various aspects of terrestrial sedimentation. The conglomerates of the Lagarde Formation are interpreted as alluvial deposits (Rust, 1982), and the Pirate Cove represents an alluvial fan to flood plain (Dineley and Williams, 1968). The Fleurant was deposited on a gravelly flood plain and paleocurrent measurements indicate a source toward the northwest (Zaitlin and Rust, 1983). Fossil fish of the Escuminac Formation are commonly considered fresh water, and the unit interpreted as lacustrine (Dinely and Williams, 1968). Elements of the fauna have also been interpreted as brackish or marine (Schultze and Arsenault, 1985). The Escuminac sandstones and siltstones may be lake turbidites (Hesse and Sawh, 1982). Paleocurrent measurements indicate a source area to the north-northeast. Material composing the sequence is of local derivation and several of the underlying Silurian and Devonian lithologies are recognized in the clasts of conglomerates and sandstones.

Connecticut Valley-Gaspé Synclinorium
Definition and limits

The Connecticut Valley-Gaspé Synclinorium (Williams, 1978) extends from the eastern coast of Gaspésie to the Quebec-United States border (Fig. 4.1, 4.2). It lies between Taconic allochthons to the north and northwest, and a number of anticlinoria to the south and southeast; from west to east, the Boundary Mountains and Pennington Mountains anticlinoria in the United States, and the Aroostook-Percé Anticlinorium in New Brunswick and Quebec. The northern contact of the Connecticut Valley-Gaspé Synclinorium with older rocks is an unconformity or a fault; the Shickshock-Sud Fault in central and western Gaspésie, the Neigette Fault in the Matapédia Lake area, and the Guadeloupe Fault in southern Quebec. A few outliers of middle Paleozoic rocks lie unconformably on Cambrian-Ordovician rocks north of its northern limb. The southern limit of the synclinorium, with rocks of the Aroostook-Percé Anticlinorium, is either a conformable contact or a fault.

The Connecticut Valley-Gaspé Synclinorium includes fine- to very coarse-grained siliciclastic rocks, various types of limestones, felsic to mafic volcanic rocks, and intrusive rocks. All four time-stratigraphic packages are present (Fig. 4.2). Devonian rocks predominate.

Structural geology

All that area of the Connecticut Valley-Gaspé Synclinorium comprising the eastern and central Gaspésie, and north of the Sainte-Florence and Guadeloupe faults is characterized by several broad and flat synclines of Devonian rocks trending northeasterly, and relatively narrow and steep-flanked anticlines of Silurian-lowermost Devonian rocks (Fig. 4.2a). Several of the anticlines are asymmetrical, with their northern limbs slightly steeper than their southern limbs. Most of the folds are gently plunging toward the east-northeast and west-southwest. Northeast of Lake Témiscouata, a complex anticlinorial structure exposes Cambrian-Ordovician rocks of the Trinité Group in its core. Several folds of lesser amplitude are superimposed on that major structure. South of the Chic-Chocs Mountains, Lower to Middle(?) Devonian rocks form a large open syncline, the Big Berry Mountains Syncline. In central and eastern Gaspésie, four structures are prominent: the Gastonguay and Saint-Jean River anticlines; the Champou Syncline; a large basin of Gaspé Sandstones; and the Mount Alexandre Syncline, steeply plunging to the west and wedged between the Grande Rivière Fault and the Aroostook-Percé Anticlinorium.

The area of the Connecticut Valley-Gaspé Synclinorium south of the Témiscouata and Sainte-Florence faults is a broad synclinorial structure whose northwestern limit is the Témiscouata and Sainte-Florence faults, and whose southeastern margin is in part truncated by the Ristigouche Fault. It is made up of a series of northeast-trending, gently plunging, open upright folds, with a well-developed slaty cleavage (Kirkwood and St-Julien, 1987). It is composed of the Fortin Group in Gaspésie and the Matapédia Valley, the Témiscouata Group in the Témiscouata region, and the Compton Formation in the Eastern Townships of Quebec. Together with the Gile Mountain and Littleton formations in Vermont and New Hampshire, and the Seboomook Formation in Maine, the Fortin-Témiscouata-Compton is an Early Devonian "slate belt" 160 km wide and 1300 km long (Boucot, 1970; Hall et al., 1976).

The Connecticut Valley-Gaspé Synclinorium is dissected in places by right-lateral strike-slip faults. High angle reverse faults are commonly associated with major anticlines. In eastern Gaspésie, there are three major strike-slip faults: the Grande Rivière, Bras Nord-Ouest, and Third Lake. The Grande Rivière Fault separates the Aroostook-Percé Anticlinorium and Connecticut Valley-Gaspé Synclinorium from Percé to the eastern end of the Mount Alexandre Syncline. From there, the Grande Rivière Fault splits: one branch follows the northern flank of the Mount Alexandre Syncline to the west (Morin and Simard, 1987), and the other extends to the northwest and crosses the Saint-Jean River Anticline. Right-lateral displacement along the fault has been evaluated at about 30 km (Bourque et al., 1985; Kirkwood and Béland, 1986). The Bras Nord-Ouest and Third Lake faults are also right-lateral strike-slip, but with less displacement — 7 km for the Bras

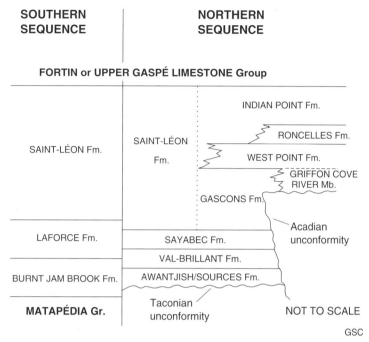

Figure 4.10. Summary of lithostratigraphic relationships of Silurian-lowermost Devonian sequences in the Gaspésie region of the Connecticut Valley-Gaspé Synclinorium.

Nord-Ouest and 8 km for the Third Lake (Brisebois, 1981; Kirkwood, 1986). In central and western Gaspésie, the Shickshock-Sud Fault is a dextral strike-slip fault, but its displacement is unknown (Berger, 1985). In southern Quebec (see Fig. 4.2b), the La Guadeloupe Fault is a thrust (St-Julien et al., 1983; Tremblay et al., 1989a). It can be traced toward the northeast and merges into the Témiscouata Fault in the Témiscouata region. The southern limit of the Connecticut Valley-Gaspé Synclinorium in southern Quebec has been interpreted as a thrust contact (Victoria River Fault).

Stratigraphy

Stratigraphy of the Connecticut Valley-Gaspé Synclinorium is described in four areas: Gaspésie; Matapédia Valley-Témiscouata; Madawaska County-Assémetquagan River; and the Eastern Townships.

Gaspésie (columns 2, 3, 4, 14, 16, 17, 18, Fig. 4.3).

Regional aspects of the stratigraphy of the Connecticut Valley-Gaspé Synclinorium in this area are from McGerrigle (1950), Burk (1964), and Bourque (1973, 1975, 1977). More local studies include work by Russell (1946), McGerrigle (1959), Cumming (1959), Skidmore (1965), Lespérance and Bourque (1970), Bourque and Lespérance (1977), Lespérance (1980a, b), Brisebois (1981), Rouillard (1984b, 1986), and Lachambre (1987).

Descriptions of Late Ordovician-Silurian stratigraphy are keyed to two sequences: a northern shallow water sequence, including chiefly the Northern Outcrop Belt, but also the eastern tip of the Saint-Jean River Anticline and the Percé area to the southeast; and a southern deeper water sequence, comprising the Saint-Jean River Anticline, Gastonguay Anticline, Mount Alexandre Syncline and Angers-Dugal Outcrop Belt (Fig. 4.2a). Lithostratigraphic relationships of the two sequences are summarized in Figure 4.10.

Stratigraphy of the **northern sequence** (columns 1 to 4, part of 14, Fig. 4.3, 4.9) along the Northern Outcrop Belt between Lake Matapédia and point B of Figure 4.2a, is identical to that of the northern Matapédia Valley (see below), with the exception that the biohermal and biostromal limestones of the upper assemblage of the Sayabec Formation are absent (Bourque et al., 1986; Lachambre, 1987). To the east, between point B and point A, the lower part of the succession is very similar, with the Sayabec and Val-Brillant formations recognizable, but claystones of the Awantjish Formation replaced by calcilutites of the Sources Formation. The Sources is a thin bedded brownish grey calcilutite with shaly interbeds, locally very fossiliferous (Lespérance and Bourque, 1970). Above the Sayabec, the thick sequence of fine grained siliciclastites is similar to those of the Saint-Léon Formation, and contains reefoid limestone bodies named the Lefrançois Member (Lespérance and Bourque, 1970), but later assigned to the West Point Formation (Bourque, 1975). Farther east, between point A and the Forillon Peninsula, only the upper part of the sequence is known and it rests unconformably above Cambrian-Ordovician rocks.

Between points A and B, the West Point Formation consists of carbonate buildups reaching thicknesses up to 325 m (Lespérance and Bourque, 1970; Bourque, 1972, 1977; Bourque et al., 1986; Lachambre, 1987). Between point A and the Forillon Peninsula, it consists of well bedded, very fossiliferous calcarenite and calcirudite units, cladoporid biostromes and small local stromatoporoid bioherms, ranging in thickness from 4-40 m. The carbonate buildups are pinnacle reefs composed of various limestone lithologies: well bedded calcarenites and calcirudites, quartzose crinoid calcarenite, massive crinoid and stromatoporoid calcirudites, stromatoporoid boundstones, and limestone debris units embedded in fine grained siliciclastites. The reef bodies are totally surrounded by siliciclastic facies.

The Griffon Cove River Member at the base of the West Point is a terrigenous- dominated unit ranging from fine grained siliciclastites to conglomerate, with local limestones (Bourque et al., 1986). It lies unconformably on severely folded pre-Taconic rocks and varies in thickness from 10-70 m. It is composed of two main facies: a pebble and cobble petromictic conglomerate facies with clasts of quartz, chert, mafic and felsic extrusive rocks, and various sedimentary rocks; and a local, mudcracked green and red fine grained siliciclastic facies with intercalations of microbial laminites and a few stromatoporoid rubbles. The siliciclastic facies is similar to that of the intertidal-supratidal redbed facies of the upper reef complex of the West Point Formation of the Chaleurs Bay Synclinorium.

The Roncelles Formation, overlying the West Point, is a uniform unit of thick bedded, very calcareous, grey siliciclastic mudstone, with local fossiliferous argillaceous limestone and nodular mudstone. It ranges in thickness from 40-300 m. The overlying Indian Point Formation is a fine grained siliciclastite which, in the eastern part of the Northern Outcrop Belt, is divided into four members (Russell, 1946). These are: the Rosebush Cove, a massive and thick bedded green mudstone with local red mudstones; the Petit Portage, a turbiditic unit of interbedded green mudstones and fine grained sandstones; the Quay Rock, a thin- to medium-bedded, commonly slumped, muddy fine grained limestone; and the Cape Road, a turbiditic unit of mudstones and fine grained sandstones. Total thickness ranges from 275 m in the east to about 650 m in the west. West of the Petite Fourche River, the undivided Indian Point Formation consists of thick bedded greenish grey mudstones, siltstones and fine grained sandstones, with local, thin bedded, parallel-laminated, fine grained sandstones and nodular mudstones. Lithologies similar to those of the Rosebush Cove and Petit Portage-Cape Road members occur in the upper part of the formation. Maximum thickness of the Indian Point is 1200 m.

Stratigraphy of the **southern sequence** (columns 14, 16 to 18, Fig. 4.3) is more uniform than that of the northern sequence. Four formations are recognized. The oldest rocks belong to the White Head Formation of the Matapédia Group and occur in the core of the Saint-Jean River and Gastonguay anticlines. The White Head Formation of the Saint-Jean River Anticline is divided into two members (Bourque, 1977): a lower dark green mudstone, and an upper grey to brownish grey thin bedded calcilutite with shaly interbeds containing local clayshale units. The lower member correlates with the Côte de la Surprise Member of the White Head Formation of the Aroostook-Percé Anticlinorium,

whereas the upper member corresponds to both the L'Irlande and the Des Jean limestone members. In the Gastonguay Anticline, only the Des Jean Member (upper member) of the White Head Formation is present, consisting of thinly bedded silty and argillaceous limestones, and calcareous shales.

The Burnt Jam Brook Formation above the White Head is a very thick bedded dark green to greenish grey claystone with minor mudstone and siltstone. It contains a middle member of thin- to medium-bedded, parallel- and cross-laminated siltstones and fine grained sandstones that thickens toward the southwest. These siltstones and fine grained sandstones are similar to lithologies of the Mann Formation of the Ristigouche Syncline in the Chaleurs Bay Synclinorium. Thickness of the Burnt Jam Brook varies from 20 m to nearly 700 m, locally including intrusive sills.

The overlying Laforce Formation is composed of medium- to thick-bedded, parallel- and cross-laminated, lithoclastic calcarenite and calcirudite containing up to 50 per cent of well rounded coarse quartz grains, and minor limestone conglomerate. The composition of lithoclasts is micritic, silty, peloidal, skeletal or oolitic limestones, indicate derivation from a carbonate platform. A graptolite fauna is locally abundant. Thickness of the Laforce varies from 260 m in the Saint-Jean River Anticline, 65 m in the Mount Alexandre Syncline, to only a few metres in the southwestern part of the Gastonguay Anticline. Farther west, it is absent in the Angers-Dugal Outcrop Belt.

The Saint-Léon Formation above the Laforce is composed of various fine grained siliciclastites. Most commonly, it is thick bedded, structureless greenish grey mudstone, siltstone and fine grained sandstone, but locally it has other facies. In the Saint-Jean River Anticline, the lower part of the Saint-Léon consists of thin-bedded and parallel-laminated graptolite-bearing, fine grained sandstones, with a distinctive conglomerate higher in the formation, the Owl Capes Member. The Owl Capes is a very thick bedded, coarse grained petromictic conglomerate composed of quartz particles and fragments of various limestones, corals, stromatoporoids, feldspathic wacke and volcanic rock with an abundant argillaceous matrix. A distinctive unit of parallel- and cross-laminated siltstones and fine grained sandstones was mapped in the Angers-Dugal Outcrop Belt, (Bourque and Gosselin, 1986). Green and red mudstones like those of the Rosebush Cove Member of the Indian Point Formation of the northern sequence, together with turbiditic facies like those of the Petit Portage-Cape Road members commonly occur in the upper Saint-Léon. Thickness of the Saint-Léon Formation in the southern sequence ranges from 700-1750 m.

The Saint-Léon Formation locally contains volcanic rocks; the Lac McKay Member in the Mount Alexandre Syncline and the Cedar Barn Member in the Saint-Jean River Anticline. The volcanic rocks are green and purple porphyritic and amygdaloidal mafic lava flows, volcanic breccias, and lapilli tuffs. Mudstone, siltstone, and volcanic conglomerate are associated. The volcanic rocks reach thicknesses up to 3000 m (Skidmore, 1965a, b).

Ages assigned to the units of the northern and southern sequences are based on brachiopods, and to a lesser extent, on graptolites (Lenz, 1972; Bourque, 1977; Bourque and Lachambre, 1980; Bourque and Gosselin, 1986). Correlations between the two sequences are easy. The green claystone of the Burnt Jam Brook Formation is identical to that of the Awantjish Formation, and the greenish fine-grained siliciclastites of the Saint-Léon are much the same in both sequences. Volcanic rocks are absent in the northern sequence. The Val-Brillant quartz arenite of the northern sequence has no equivalent in the southern sequence.

Nomenclature of the rocks above the Indian Point or Saint-Léon formations has been inherited from the Devonian sequence of the Forillon Peninsula. The rocks in that area are divided into two groups; the Upper Gaspé Limestones and the Gaspé Sandstones. The term **Gaspé Limestones** was introduced by Logan (1863) to designate limestones and siliciclastites which rest unconformably on transported Cambrian-Ordovician rocks and occur below the Gaspé Sandstones at Forillon. Subsequently, the group was divided into three formations; the Saint Alban, Cap Bon Ami, and Grande Grève formations (Ami, 1900; Clarke, 1908; McGerrigle, 1950). The Saint Alban and lower part of the Cap Bon Ami were later assigned to the West Point, Roncelles and Indian Point formations of the Chaleurs Group described above (Bourque, 1975, 1977), thus abandoning the term Saint Alban. The remaining part of the Cap Bon Ami and the Grande Grève Formation were divided into three formations; the Forillon, Shiphead and Indian Cove formations and grouped under the Upper Gaspé Limestones "Sub-Group" (Lespérance, 1980a, after mapping by Russell, 1946), abandoning the widespread but somewhat ambiguous Cap Bon Ami and Grande Grève terms. The term **Gaspé Sandstones** was coined by Logan (1863) to designate the Devonian terrigenous rock sequence above his Gaspé Limestones. Subsequent work formalized the Gaspé Sandstones Group and divided it into four formations; the York Lake (Jones, 1935), York River (H.S. Williams, 1910; McGerrigle, 1950), Battery Point (McGerrigle, 1950), and Malbaie (Alcock, 1935). The York Lake and Malbaie formations are however absent on the Forillon Peninsula.

In northeastern Gaspésie (columns 1 to 3, Fig. 4.3), east of the Big Berry Mountains Syncline, the Forillon, Shiphead and Indian Cove formations are traceable along the northern edge of the Connecticut Valley-Gaspé Synclinorium, from Forillon Peninsula to Murdochville (Lespérance, 1980a, b; Rouillard, 1984a). To the west and south, the three formations still occur, but with major facies changes (Rouillard, 1984b, 1986). The Forillon Formation conformably overlies the Indian Point Formation. It is a monot-onous sequence of more or less shaly, dolomitic and siliceous calcilutite or limy mudstone. In the eastern part of the area, the sequence is more calcareous and more siliceous. The Shiphead Formation is distinguished from the underlying Forillon Formation and the overlying Indian Cove Formation, by its higher siliciclastic content. In the northern half of the area, the formation consists of thin- to very thick-bedded siliceous and dolomitic limestone and mudstone, with minor calcarenite, sandstone and bentonite beds. In the southern half of the area, the Shiphead is composed of the same limestone and mudstone, but also contains units of coarse grained sandstone and mudstone with slump structures, a lithology very similar to that of the Fortin Group of the Madawaska County-Assémetquagan River area. Near Percé, the base of the sequence is a 50 m unit of thick bedded, cross laminated, quartz granule conglomerate and coarse grained quartz arenite that grade upward to silty limestone. Thickness of the formation ranges from 100-160 m along the Northern Outcrop Belt and reaches 690 m in the Saint-Jean River Anticline. The overlying Indian Cove Formation is a homogeneous unit of thin- to medium-bedded cherty to siliceous or silty calcilutite. A few calcarenites and calcirudites are scattered throughout the sequence, as well as sandstones and siltstones in its upper part. Its thickness ranges from 100 to 250 m along the Northern Outcrop Belt and increases to 660 m in the Saint-Jean River Anticline.

In the Big Berry Mountains Syncline and adjacent areas (column 4, Fig. 4.3), the division into Forillon, Shiphead, and Indian Cove formations is vague in the northernmost part of the area (Rouillard, 1984b). The Indian Cove is recognizable, but the Forillon and Shiphead formations show significant changes. To date, no attempt has been made to trace the formations farther south. Therefore, the division into the Cap Bon Ami and Grande Grève formations (McGerrigle, 1950) in that area is followed here. Further work will probably show that like most other areas of interior Gaspésie, the Cap Bon Ami corresponds to both the Forillon and the Shiphead formations, and the Grande Grève to the Indian Cove Formation. Cap Bon Ami Formation consists of a monotonous sequence of shaly dolomitic and siliceous calcilutites to calcareous and dolomitic shales with a conspicuous horizon of quartz sandstones and calcarenites in its upper part. Mafic lava flows and pyroclastic rocks are abundant. Thickness of the unit ranges from 500-1000 m. The Grande Grève Formation is similar to the Indian Cove Formation of northeastern Gaspésie. It is a homogeneous sequence of thin- and evenly-bedded shaly to silty dolomitic and siliceous limestones. Quartz arenites occur in the uppermost part of the formation. Like the Cap Bon Ami Formation, mafic volcanic rocks are an important component of the formation in the northeastern part of the Big Berry Mountains Syncline. Thickness of the Grande Grève ranges from 500 m in the east to 1000 m in the west.

The lowest unit of the Gaspé Sandstones in northeastern Gaspésie, east of the Big Berry Mountains Syncline, is the York Lake Formation. It consists of alternating siliceous calcilutites with minor quartz arenites and wackes, like those of the underlying Indian Cove Formation, and greenish grey medium grained feldspathic wackes similar to those of the overlying York River Formation, thus constituting a transition between the Indian Cove and the York River. The alternation of sandstone and limestone is more commonly measured in metres or in tens of metres. Maximum thickness of the formation occurs at its type-section on the hillside west of Lake York near Murdochville where it reaches 500 m. The sequence thins progressively to zero eastward and southward. The overlying York River Formation is composed of a lower mudstone-siltstone- sandstone assemblage with a few calcarenites, and an upper sandstone assemblage with minor mudstones. The sandstones are medium- to thick-bedded, greenish grey, medium- to coarse-grained, feldspathic and lithic wackes with large scale cross-bedding. Thickness of the formation ranges from 400 m in the Forillon Peninsula to about 1000 m in the York River Syncline west of Gaspé. The Battery Point Formation above, can be divided into four informal assemblages. The lower assemblage consists of superposed fining upward sequences of conglomeratic sandstone, medium- to coarse-grained sandstone, and minor siltstone and mudstone (Cant and Walker, 1976). It is overlain by another assemblage of fining upward sequences into which red mudstone and siltstone are abundant (Rust, 1981; Walker and Cant, 1979). The third assemblage is similar to the first, but with coarser facies and less clearly defined fining upward sequences. The upper assemblage is a redbed unit of sandstone, siltstone and mudstone which outcrops mostly in the southeastern part of the area. The top of the formation is rarely exposed so that known thicknesses are few. Northwest of Percé, the formation is about 2500 m thick, whereas to the west, east of Murdochville, the unit is only 500 m thick. The Malbaie Formation is a conglomerate unit conformably overlying the Battery Point Formation (Rust, 1976, 1981). The conglomerate is thick bedded and composed of pebbles and cobbles of limestone, siliciclastite and volcanic fragments derived from the older Matapédia and Chaleurs groups. It is interbedded with medium- to coarse-grained red sandstone. The Malbaie is unconformably overlain by Carboniferous rocks. Maximum thickness of the formation is 1300 m.

In the Big Berry Mountains Syncline area, Gaspé Sandstones have roughly the same stratigraphic sequence as in northeastern Gaspésie, except for the occurrence of the Lake Branch Formation between the the York River and Battery Point formations, and for the absence of the Malbaie Formation on top of the sequence. The York Lake Formation is the same as in northeastern Gaspésie, except that the limestones are muddier and the sandstones are finer grained and more argillaceous, and the upper part of the formation contains mafic lava flows and pyroclastic rocks (Sikander, 1975). Its thickness ranges from 300 m in the north to 600 m in the east, and thins to zero in the south. The overlying York River Formation consists of crossbedded, fine- to medium-grained, grey, feldspathic sandstone interbedded with grey mudstone and siltstone (Sikander, 1975). The grey rocks contain intercalations of red or brown siltstone, mudstone and sandstone with mudcracks and ripple marks. Both red and grey facies are fossiliferous (brachiopods and bivalves). The York River Formation contains thick bimodal volcanic sequences (Doyon and Valiquette, 1986, 1987; Doyon, 1988). A complete sequence consists of basal basaltic flows and mafic pyroclastic rocks, followed by rhyolitic flows and felsic pyroclastic rocks, and capped by rhyolitic flows. In the northeastern part of the Big Berry Mountains Syncline, mafic volcanic rocks occur on top of the unit. Thickness of the York River is about 600 m in the north and up to 3000 m in the south (Carbonneau, 1959). The Lake Branch Formation above the York River is limited to the north limb of the Big Berry Mountains Syncline and is poorly exposed (Carbonneau, 1959). It is a 1500 m thick unit of red to brownish red shale with siltstone and fine grained sandstone. Mudcracks and ripple-marks are sparse, fossils are absent. The overlying Battery Point Formation constitutes the core of the Big Berry Mountains Syncline. Sikander (1975) subdivided the formation into two informal units: a 2500 m lower "typical Battery Point" unit of crossbedded, fine- to medium-grained, light grey feldspathic sandstones with minor grey or reddish shales, siltstones and granule conglomerates; and a 500 m upper redbed unit of fine- to medium-grained silty sandstones with minor shales and common ripple-marks, rain-prints and mudcracks. Except for palynomorphs, fossils are absent in both units.

The age of the Upper Gaspé Limestones and Gaspé Sandstones groups ranges from Pragian to possibly Eifelian. The only reliable age obtained from the Upper Gaspé Limestones is from the Forillon Formation which contains brachiopods and trilobites of Pragian age (*Rennsaelaria* Zone) in the Forillon and the Percé areas; fossils are rare and not diagnostic in the two other formations. Fossils, mainly brachiopods, from limestones and sandstones of the York Lake Formation (Gaspé Sandstones Group) and from the lower assemblage of the York River have an Emsian age. The upper assemblage of the York River contains abundant but undiagnostic plant debris. Except for a few brachiopods and cephalopods, the lower part of the Battery Point is unfossiliferous, but fish and eurypterids are known from the second assemblage of the formation. The Malbaie Formation is unfossiliferous.

Matapédia Valley-Témiscouata area (columns 5 to 10, Fig. 4.3).

This area has been studied by Lespérance and Greiner (1969), Lajoie et al. (1968), David et al. (1985), and Bourque et al. (1993). Rocks of the Cabano Group (Lajoie et al., 1968; David et al., 1985; Bourque and Gosselin, 1988) are the oldest. It consists of two siliciclastic facies: fine grained dark grey lithic wacke, and coarse conglomerate. Both wacke and conglomerate are interbedded with siltstones and mudstones. Primary sedimentary structures typical of distal turbidites are common in the wackes (David et al., 1985). The Cabano ranges in age (Fig. 4.4), from late Middle Ordovician (Llandeilo) to Early Silurian (Llandovery C3) (Bourque and Gosselin, 1988), which is the age span of the Honorat, Grog Brook and Matapédia sequences of the Aroostook-Percé Anticlinorium. However, the Cabano only resembles the lower siliciclastic Honorat and Grog Brook groups and lacks limestones like those of the Matapédia Group. Some of the Cabano conglomerates may be equivalent to conglomerates of the Siegas Formation of the Madawaska County area to the south. Basal relationships of the Cabano Group are poorly known. At one locality, Llandeilo chert beds of the Cabano are apparently unconformable above the Cambrian-Ordovician Trinité Group.

The Cabano Group is overlain by two, laterally equivalent, terrigenous formations; the Pointe-aux-Trembles and Lac Raymond formations. The Pointe-aux-Trembles Formation is a volcaniclastic unit consisting of fine- to coarse-grained tuffaceous sedimentary rocks in the lower part of the formation, and agglomerate, lithic and crystal tuff, vitric tuffs, and massive lava flows in the upper part (David and Gariépy, 1990). The Lac Raymond is composed of mudstones and tuffs interbedded with siltstone, sandstone and rare conglomerate. Gabbroic sills occur in the Lac Raymond Formation. Primary structures in both formations indicate deposition by turbidity currents (David et al., 1985). Chemical compositions of volcanic materials of both formations are characteristic or orogenic andesites (David and Gariepy, 1990; see below). Thickness of the Pointe-aux-Trembles Formation varies from 300-900 m, whereas thickness of the Lac Raymond Formation is estimated at a maximum of 1800 m (Bourque et al., 1993).

The rock package consisting of the Cabano-Pointe-aux-Trembles-Lac Raymond is here considered as the basal unit of the Gaspé Belt in the Connecticut Valley-Gaspé Synclinorium.

This is based on age and lithological equivalence of the Cabano with the Honorat and Grog Brook, and also on age equivalence of the upper part of the Lac Raymond Formation with the Robitaille and Val-Brillant formations (Fig. 4.3).

A different view is expressed by Tremblay et al. (Chapter 3) based on the following arguments. The Cabano is correlated with the Depot Mountain Formation in Maine (Roy, 1989) and the Magog Group in the Eastern Townships of Quebec (Cousineau, 1990), whereas the Pointe-aux-Trembles and Lac Raymond formations are arc-volcanic suites like volcanic rocks of the Ascot Complex in the Eastern Townships (Tremblay et al., 1989b). The Magog Group and the Ascot Complex are part of the Estrie-Beauce Subzone of the Quebec Dunnage Zone (Tremblay et al., Chapter 3). The Cabano Group and, Pointe-aux-Trembles and Lac Raymond formations were consequently assigned to the Dunnage Zone in the Eastern Townships and included in the Temiscouata Subzone farther northeast in the Quebec Dunnage Zone (Tremblay et al., Chapter 3). In the absence of detailed internal stratigraphy of the Cabano, the problem remains unresolved.

In the region northeast of Lake Témiscouata (columns 8-10, Fig. 4.3), the Robitaille Formation is either conformable above the Pointe-aux-Trembles Formation or unconformable on severely deformed pre-Taconic rocks of the Humber Zone to the north. The Robitaille Formation is a redbed unit of mudcracked, red and green, siltstones and fine grained sandstones with minor white quartzose sandstones, quartz and volcanic pebble conglomerates, and limestones. It ranges in thickness from 250- 800 m. To the northeast (columns 5 and 6, Fig. 4.3), the lateral equivalents of the Robitaille and Lac Raymond formations are the Val-Brillant and Awantjish formations. Except for a limited area where the Val-Brillant is conformable above the Cabano, both rest unconformably on the pre-Taconic Québec Group. The Awantjish Formation (Béland, 1960) occurs either below or is laterally equivalent to the Val-Brillant. It is a greenish grey shale, with a few sandy beds at the top and locally abundant offshore brachiopods. Thickness ranges from 27-750 m. The Val-Brillant Formation is a medium- to thick-bedded, parallel- and cross-laminated, whitish quartzose sandstone, with minor quartz conglomerate (Crickmay, 1932; Lajoie, 1968). It contains nearshore to offshore brachiopods, and its thickness ranges from 27 m in the Lake Matapédia area to 150 m toward the southwest.

The Val-Brillant and Robitaille formations are overlain conformably by the Sayabec Formation, a widespread limestone unit in the Lake Matapédia-Témiscouata area. In the Lake Matapédia Syncline, the Sayabec is made up of two limestone assemblages: a lower assemblage of peritidal limestones (microbial laminites, and stromatolite-oncolite-fenestral limestones), sandy limestones, dolomites, and nodular limestones (about 40 m); and an upper assemblage of the same peritidal limestones, with biohermal and biostromal coral-stromatoporoid-algal limestones (about 340 m; Héroux, 1975; Lavoie, 1988). These two Sayabec assemblages have many similarities with the peritidal laminite-oncolite-stromatolite-fenestral limestone facies and the biohermal-biostromal limestone facies of the coeval La Vieille Formation of the Chaleurs Bay Synclinorium (Bourque et al., 1986; Lavoie and Bourque, 1986; Lavoie, 1988). In the Témiscouata region, the Sayabec is a

near shore (intertidal to supratidal) facies, chiefly composed of mudcracked microbial laminites, nodular limestones, and a few bryozoan biostromes, locally interlayered with red mudstones and sandstones similar to those of the Robitaille Formation. Thickness is in the range of 100-500 m.

The stratigraphic framework of the Sayabec Formation and underlying units presented in Figure 4.3 for the Matapédia Valley-Témiscouata area (columns 5 to 10) differs from that of Lajoie et al. (1968). It is based on a revision of the litho- and biostratigraphy of the Silurian rocks of the area (Bourque et al., 1993). The Sayabec is Llandovery C6 to Late Wenlock. The underlying Val-Brillant is Llandovery C4-C5, while the Awantjish has brachiopods of Llandovery C3-C4. The base of the Robitaille is no older than Llandovery C4, and except for that part of the Robitaille laterally equivalent to the Sayabec (columns 8 and 9, Fig. 4.3), its top is older than Llandovery C6. The Pointe-aux-Trembles Formation is poorly dated. Brachiopods in its lower part are not older than Llandovery C3 (Lajoie et al., 1968). The basal part of the Lac Raymond Formation is dated as Llandovery C1-C2.

The Sayabec Formation is overlain everywhere by the Saint-Léon Formation (Crickmay, 1932). It is composed of thick bedded greenish grey, slightly calcareous siltstones interbedded with parallel and crossbedded fine grained sandstones and minor dark grey shales. In the region south of Rimouski (columns 6 and 7), the Saint-Léon Formation contains a conglomerate member, the Lac des Baies Member (Lajoie et al., 1968; Lajoie, 1971), composed of polymictic conglomerates, quartz and feldspathic wackes, quartz arenites and minor green calcareous siltstones. Adjacent to the Neigette Fault, south of Rimouski, a 100 m limestone conglomerate unit, the Neigette breccia, is composed of large (up to several metres) clasts of reef limestone identical to that of the West Point reefs of the Chaleurs Bay Synclinorium (Dansereau and Bourque, 1988). In the northern Témiscouata region (column 9, Fig. 4.3), the Lac des Baies Member constitutes the upper part of the Saint-Léon Formation. To the south (columns 8 and 10), the Saint-Léon is overlain by a limestone formation, the Lac Croche Formation (Lespérance and Greiner, 1969), composed of mudcracked microbial laminites, coral and stromatoporoid boundstones, stromatoporoid rubbles, nodular limestones, and fine grained sandstones (Bourque et al., 1993). Several facies of the Lac Croche Formation are similar to those of the upper reef complex of the West Point Formation of the Chaleurs Bay Synclinorium.

The Saint-Léon Formation contains local offshore brachiopods and in places, abundant graptolites (Lenz, 1972; Bourque et al.,1993). The graptolites indicate an age span of early Ludlow to Lochkovian (*Monograptus scanicus-M. nilssoni* to *M. uniformis* Zone). Brachiopods in the Lac Croche Formation are latest Silurian (Pridoli; Lajoie et al., 1968). A Pridoli age for the Lac des Baies Member is constrained by graptolites in the Saint-Léon Formation.

In the Matapédia Valley, east of the Matapédia River, rocks correlated with the Upper Gaspé Limestones of Gaspésie have been mapped above the Saint-Léon Formation, as the Cap Bon Ami and Grande Grève formations (Stearn, 1965; Ollerenshaw, 1967). West of the river, the two formations are less clearly defined and have not been separated (Béland, 1960; Lajoie, 1971; Stearn, 1965). However, recent exploratory work (Rouillard, 1986) suggests that the three divisions of the Upper Gaspé Limestones of eastern Gaspésie are present. The situation is therefore the same as in the Big Berry Mountains area, and for the same reasons, the Upper Gaspé Limestones are described under the Cap Bon Ami and the Grande Grève divisions (columns 5-7, Fig. 4.3).

The Cap Bon Ami consists of thin- to thick-bedded dark grey shaly limestones with interbeds of calcareous mudstones and siltstones, and andesitic tuffs in the upper part of the unit. Thickness is estimated between 600-1000 m. The Grande Grève Formation has been mapped east of the Matapédia River (Ollerenshaw, 1967; Stearn, 1965; Huff, 1970), but Béland (1960) and Lajoie (1971) could not distinguish it from the Cap Bon Ami west of the river. It is composed of thin-bedded, commonly parallel-laminated, calcareous siltstone and siliceous limestone. The limestone becomes more silty and sandy toward the top of the unit. It is 1100 m thick east of the Matapédia River. It disappears toward the west.

The overlying Gaspé Sandstones have the same units as in Gaspésie. In the Matapédia Valley, the complete sequence of York Lake, York River, Lake Branch, and Battery Point formations is recognized. To the west, the two upper units are apparently missing. The York Lake Formation consists of thin-to thick-bedded, greenish grey feldspathic arenites and wackes, siltstones and mudstones, reaching a maximum thickness of 1400 m. The York River Formation is similar to its lower part in eastern Gaspésie, but is thicker, about 4000 m (Béland, 1960). Some 1300 m of red, brown and green feldspathic sandstones and siltstones of the Lake Branch Formation are exposed in a small syncline (Béland, 1960; Stearn, 1965). The Battery Point Formation is known only from a few outcrops above the Lake Branch Formation (Huff, 1970). It is composed of the same coarse feldspathic sandstones as in its type-region.

In summary, the stratigraphy of the Matapédia Valley-Témiscouata area of the Connecticut Valley-Gaspé Synclinorium consists of six assemblages: (1) fine grained turbidites (Cabano) with volcanic conglomerates (Pointe-aux- Trembles); (2) nearshore to terrestrial mudstones and sandstones (Val-Brillant and Robitaille); (3) a widespread distinctive unit of platformal limestones (Sayabec); (4) a thick, fine grained siliciclastic sequence (Saint-Léon) with local conglomerate bodies (Lac des Baies; Neigette breccia) and reef limestones (Lac Croche); (5) fine grained deep water carbonates (Cap Bon Ami and Grande Grève); and (6) nearshore to terrestrial sandstones and conglomerates (York Lake, York River, Lake Branch and Battery Point). This is the most complete sequence of middle Paleozoic rocks in the Quebec Appalachians. Relationships with the underlying Dunnage Zone are, at present, poorly understood.

Madawaska County-Assémetquagan River area (column 28, Fig. 4.3).

This area comprises middle Paleozoic rocks that occur between the Témiscouata and Sainte-Florence faults to the north and the Aroostook-Percé Anticlinorium and Ristigouche Fault to the south, from Madawaska County in New Brunswick to Assémetquagan River east of the Matapédia River in Quebec.

In Madawaska County, stratigraphy was studied by Hamilton-Smith (1970) and St. Peter (1977). The base of the sequence consists of the Grog Brook Group and Carys Mills Formation of the Matapédia Group (Aroostook-Percé Anticlinorium). The Siegas Formation that conformably overlies the Carys Mills Formation is a 100-245 m turbidite sequence of thin- to thick-bedded, fine- to coarse-grained, calcareous, quartz and lithic sandstones with Bouma sequences and various sole marks. It also contains limestone conglomerates with coarse- to medium-grained sandstone matrices, and minor calcilutites and dark grey chert beds. Conformably overlying the Siegas, is the fine grained siliciclastic Perham Formation, divided in nearby northeastern Maine into lower and upper members (Boucot et al., 1964). The lower member consists of about 200 m of grey calcareous laminated slate containing a distinctive 10 m horizon of red and green slates, and iron and manganese-rich siliceous siltstones. The upper member is composed of shales, siltstones and slates with minor limestones, together with graded, parallel- and cross-laminated fine grained sandstones reaching thicknesses of 600 m.

The age of the Siegas Formation has been established as Early Llandovery (Ayrton et al., 1969; Hamilton-Smith, 1970). The age of the lower member of the Perham Formation is poorly constrained, being no more precise than post-Early Llandovery and pre-Ludlow, whereas the upper member of the Perham ranges from Late Wenlock through at least Early Ludlow (Pavlides, 1968) and probably to Pragian (St. Peter and Boucot, 1981).

The Madawaska County-Assémetquagan River area is chiefly characterized by the occurrence of a thick distinctive Devonian dark mudstone and fine- to medium-grained sandstone sequence, called the Témiscouata Formation in Madawaska County and near Lake Témiscouata (McGerrigle, 1943; Lespérance and Greiner, 1969; St. Peter, 1977), and the Fortin Group in the Matapédia River-Assémetquagan River area (McGerrigle, 1946). The Témiscouata Formation conformably overlies the Perham Formation in Madawaska County, but the sequence below the Fortin Group in the Matapédia River-Assémetquagan River area is unknown. However, the presence of the Saint-Léon Formation below the Fortin Group east of the Assémetquagan River (column 18) suggests rocks sharing attributes of both the Saint-Léon and Perham formations.

In its type-area in the vicinity of Lake Témiscouata, the Témiscouata Formation is divided into three assemblages (Lespérance and Greiner, 1969); a lower assemblage of quartz arenite and a few volcanic rocks, a middle sandstone assemblage, and an upper assemblage of mudslate to muddy limestone. The two lowest assemblages occur in the type area only, so that the Témiscouata elsewhere is mainly composed of rocks of the upper assemblage. The mudslates and the limestones are thin bedded with ubiquitous slumps. In New Brunswick, the Témiscouata contains thick, pebble conglomerate horizons. The Fortin Group of the Matapédia River-Assémetquagan River area is lithologically identical to the upper Témiscouata Formation. The Fortin Group has a minimum thickness of 2650 m in the Matapédia River Valley (Kirkwood and St-Julien, 1987). The Témiscouata Formation is about 3600 m thick (Lespérance and Greiner, 1969).

The age of the Témiscouata Formation and its Fortin equivalent ranges from Lockhovian to Emsian (St. Peter and Boucot, 1981), an age span corresponding to that of the Upper Gaspé Limestones and Gaspé Sandstones groups. The lateral equivalence of the Témiscouata and Fortin with the Upper Gaspé Limestones and Gaspé Sandstones is also supported by lithological evidence. Indeed, the facies change occurs in the area east of Assémetquagan River, between the Saint-Jean River Anticline and Mount Alexander Syncline (Fig. 4.2a). South of the eastern part of the Saint-Jean River Anticline, the Fortin Group conformably overlies silty to cherty limestones of the Upper Gaspé Lime- stones Group. It is composed of dark calcareous mudstones and minor limestones, pebble conglomerates and fine grained sandstones, that are gradually replaced upward by brownish, coarse grained lithic wackes with minor quartz arenites, pebble conglomerates and calcareous mudstones. York River-type green sandstones are interbedded in the upper part of the group. To the west, the underlying Upper Gaspé Limestones disappear gradually and are replaced by thick bedded, dark grey to brownish grey calcareous mudstones and fine grained sandstones typical of the Fortin Group (McGerrigle, 1950; Skidmore, 1965b), conformably overlying the Chaleurs Group. The upper coarse grained facies also disappears westward, being replaced by finer grained facies.

East of Lake Témiscouata, a narrow belt of Middle Devonian rocks outcrops north of the Temiscouata Fault, the Touladi Formation (Boucot et al., 1969; Lespérance and Greiner, 1969). The Touladi Formation rests unconformably on Silurian or older rocks. It consists predominantly of thin- to thick-bedded, dark grey shaly to sandy fossiliferous limestones, but it also contains sandstones and conglomerates.

Eastern Townships (columns 11 to 13, Fig. 4.3).

Middle Paleozoic rocks of the Eastern Townships of southern Quebec occur in the Connecticut Valley-Gaspé Synclinorium, and in a few outliers north of the synclinorium, among which, are the Lake Memphrémagog and Cranbourne outliers. The more complete sequence is in the Lake Aylmer and Lake Saint-François area.

In the Lake Aylmer-Lake Saint-François area (column 12), the Silurian rocks outcrop in a belt that is 80 km long. Stratigraphy of the belt was studied by Lavoie (1985). The belt is transected along its length by the La Guadeloupe thrust fault. North of the fault, the Silurian Lac Aylmer Formation unconformably overlies rocks of the Dunnage Zone. South of the fault, the Lac Lambton Formation, a lateral equivalent of the Lac Aylmer Formation, is wedged in a small structural slice and is for its most part thrust above Cambrian-Ordovician rocks or the Lac Aylmer Formation. Locally within the slice, the Lac Lambton is unconformable above pre-Taconicrocks.

The Lac Aylmer Formation is divided into three members: a lower member of thick bedded chanellized petromictic conglomerates composed of medium- to well-rounded porphyry, granite, granophyre, schist, quartz and sedimentary rock fragments up to half a metre in diameter, alternating with lithic sandstones; a middle sandstone and

siltstone member; and an upper silty calcarenite, calcilutite and calcareous siltstone member with ubiquitous slumps. At the western extremity of the outcrop belt, near the village of Marbleton, the upper member contains 200 m of thick bedded, very pure, massive, reefoid limestone, rich in stromatoporoids. Thickness of the formation is nearly 1200 m.

The Lac Lambton Formation is composed chiefly of conglomerate, dolostone and fine grained siliciclastites. Eight informal members have been recognized in the formation (Lavoie, 1985) which from base to top are as follow: (1) a petromictic conglomerate and quartz arenite unit; (2) a silty dolostone unit laterally grading to fossiliferous limestone; (3) a shale unit with dolostone and limestone beds; (4) a second conglomerate and sandstone unit; (5) a shale unit with silty calcilutite beds; (6) a fossiliferous silty dolostone unit; (7) a siltstone unit with shale; and (8) a shale unit with minor calcilutite. Thickness of the Lac Lambton Formation is about 1300 m.

The age of both the Lac Aylmer and Lac Lambton formations is poorly constrained. Although fossils occur throughout the Lac Lambton formation, only unit 6 is dated reliably. It is Pridoli, based on brachiopods (Boucot and Drapeau, 1968). The Lac Aylmer Formation is almost devoid of fossils, but the limestone unit of its upper member has brachiopods also dated as Pridoli (Boucot and Drapeau, 1968). Correlating the Lac Lambton and Lac Aylmer formations (column 12) is based on these ages and similar stratigraphic sequences (Lavoie, 1985). The lower and upper age limits of the two formations are uncertain.

The overlying Ayers Cliff Formation (Doll, 1951) consists of thin bedded sandy limestones to calcareous sandstones, with minor argillaceous or silty limestones. Thickness of the unit is unknown. Doll (1951) and Cooke (1950) assigned the Ayers Cliff Formation to the Ordovician, based on its occurrence above the Lac Lambton Formation, at that time considered to be Ordovician. Since the age of the Lac Lambton is now known to be Pridoli, the Ayers Cliff can be no older.

The oldest Silurian rocks southwest of the Lake Aylmer-Lake Saint-François area belong to the Ayers Cliff Formation. To the northeast, Silurian rocks are absent. The rock assemblage overlying the Ayers Cliff Formation constitutes most of the Connecticut Valley-Gaspé Synclinorium of the Eastern Townships. Its stratigraphy is complex and in places ambiguous. Recently however, Slivitzky and St-Julien (1987) proposed a more unified nomenclature. All the Silurian and Devonian rocks south of the La Guadeloupe Fault are included in the Saint-Francis Group (Cooke, 1937), which is divided into three formations: the Lac Lambton, Ayers Cliff, and Compton formations. The Compton Formation, in the Chaudière River area, includes all of the Saint-Francis Group, the Saint- Juste Group of Béland (1957) and Gorman (1954, 1955), and part of the Frontiére Group of Béland (1957). Thus defined, the Compton Formation is an uniform sequence of dark grey mudstones and slates interbedded with thin- to thick-bedded, fine- to medium-grained sandstones, with minor quartzites and limestones. Its thickness is unknown. The Compton is a lateral equivalent of the Témiscouata Formation and the Fortin Group of the Témiscouata and Gaspésie regions, respectively.

341

The Famine Formation above the Compton is a fault-bounded unit composed of a basal conglomerate overlain by quartzose sandstones, and fossiliferous Eifelian sandstones (Boucot and Drapeau, 1968). It is an equivalent of the Touladi Formation of the Témiscouata Valley area.

Silurian rocks of the Lake Memphrémagog Outlier are divided into three formations (Boucot and Drapeau, 1968). The lower, Peasley Pond Formation consists of polymictic pebble conglomerates and quartz sandstones unconformably above metavolcanic rocks of the pre-Taconic sequence. The formation reaches a maximum thickness of 60 m. The overlying Glenbrooke Formation has a thickness of about 550 m. Its lower third is commonly shale with little or no carbonate, whereas its upper two-thirds is a calcareous siltstone. A 90 m volcanic member of greenish grey and pinkish white amygdaloidal lava with minor agglomerate and tuff occurs near the top of the formation. The overlying Sargent Bay Formation is a limestone unit composed of bluish grey massive calcilutites, shaly in places. Thickness of the formation is estimated at 150 m.

The Peasly Pond Formation is unfossiliferous, whereas the Glenbrooke Formation has yielded fossils no more precisely dated than "Middle or Upper Silurian" (H.B. Whittington, in Boucot and Drapeau, 1968). The Sargent Bay Formation is fossiliferous, but its age is not more precise than Late Silurian. The lithological succession of the Lake Memphrémagog outlier ressembles that of the Lake Aylmer-Lake Saint-François area. Both successions begin with basal polymictic conglomerate and both contain a rather pure limestone unit within or above a fine grained siliciclastic unit in its upper part. The age assigned to the Lake Memphrémagog succession is based on lithic correlation.

The Sargent Bay Formation is overlain by the Mountain House Wharf Formation, a fossiliferous Eifelian limestone (Boucot and Drapeau, 1968). The Mountain House Wharf Formation is equivalent to the Famine and Touladi formations.

East of the Chaudière River, Silurian rocks occur in the small Cranbourne Outlier, and are referred to as the Cranbourne Formation, a monotonous sequence of fine grained siliciclastites, in places nodular, and lying unconformably above Dunnage Zone rocks. The upper part of the formation is limited by a fault. The Cranbourne contains Pridoli brachiopods (Boucot and Drapeau, 1968).

The Silurian sequence of the Eastern Townships is not easily correlated with that of the Gaspésie, Matapédia Valley, and Témiscouata regions. Since only Pridoli ages are known and since the succession begins with basal conglomerate, unconformably above rocks of the Dunnage Zone, it is possible that the Eastern Township sequence is equivalent to that above the Ludlow-Pridoli unconformity in Gaspésie. The stromatoporoid reef limestone of the upper member of the Lac Aylmer Formation has been correlated with the West Point reefs of the Chaleurs Bay Synclinorium (Hughson and Stearn, 1988). Lithologies of the Ayers Cliff Formation are in general similar to the Saint-Léon Formation. The Compton Formation obviously belongs to the same slate belt as the Témiscouata Formation and the Fortin Group.

Middle Devonian granodiorite plutons cut the Silurian-Devonian sequence of the Connecticut Valley-Gaspé Synclinorium in the Eastern Townships.

Depositional environments

A progression of depositional environments for representative Silurian and Devonian successions of the Gaspé-Témiscouata segment of the Connecticut Valley-Gaspé Synclinorium is shown in Figure 4.11. As in the Chaleurs Bay Synclinorium (Fig. 4.9), depositional environments can be grouped into two broad domains: a terrestrial and shelf (successions A, C and D), and a basin (succession B). The same three shallowing upward phases observed here are like those recognized in the Chaleurs Bay Synclinorium.

Shallowing upward phase I is detectable in the Lake Témiscouata region and Madeleine River area (successions A and C). In the Lake Témiscouata region, the deep water turbidite facies of the Cabano Group progressively changes upward to shallow water sediments and then to subaerial lava flows and volcaniclastic gravels of the Pointe-aux-Trembles (David et al., 1985), and finally, to terrestrial redbeds of the Robitaille Formation. The overlying Sayabec Formation has alternating redbeds and peritidal limestones. In the Madeleine River area, facies evolved progressively from calcareous clays and lime muds of the Awantjish-Sources formations containing offshore brachiopods (Stricklandid community; Bourque, 1977), to nearshore clean quartz sands of the Val-Brillant Formation (Lajoie, 1968). The overlying Sayabec Formation consists of peritidal and shallow subtidal limestones with associated metazoan knob reefs (Héroux, 1975; Bourque et al., 1986; Lavoie, 1988). The basinal equivalent of this phase corresponds to clays and silts of the Burnt Jam Brook Formation with chanel sands and gravels of the Laforce Formation (succession B).

The first transgression (T1) occurred by the end of Wenlock time and brought deep shelf siliciclastic muds of the Saint-Léon or Gascons formations. Shallowing upward phase II (Ludlow-Pridoli) corresponds to a period of uplift that likely resulted in erosion of the entire Lower Silurian sequence in northeastern Gaspésie. As in the Chaleurs Bay Synclinorium, terrestrial redbeds and conglomerates are associated with uplifted areas (Griffon Cove River beds of the West Point Formation). In the Madeleine River and Lake Témiscouata regions (successions C and A), the shallowing upward phase ended with reef construction (West Point and Lac Croche formations). Basinward, the equivalent of shallowing upward phase II is a thick monotonous sequence of deep shelf fine grained siliciclastic sedimentary rocks (Saint-Léon Formation). Unlike the Chaleurs Bay Synclinorium, no volcanic rocks are associated with terrestrial and shelf successions of phase II. Mafic volcanism, however, occurred in the deep shelf environment (Lac McKay Member).

The second transgression (T2; Lochkovian-Pragian) following uplift, erosion and local reef construction of shallowing phase II, resulted in a deepening upward sequence of deep shelf sediments, firstly siliciclastic muds of the Saint-Léon and Indian Point formations and secondly, carbonate and siliciclastic muds of the Upper Gaspé Limestones Group. This deepening upward is exemplified by the sequence of northeastern Gaspésie (succession D). From reef shelf limestones of the West Point Formation, facies evolved to relatively deep-water lime muds of the Upper Gaspé Limestones Group, with thick bedded, structureless, argillaceous and silty lime muds of the Roncelles Formation and fine grained siliciclastic sedimentary rocks

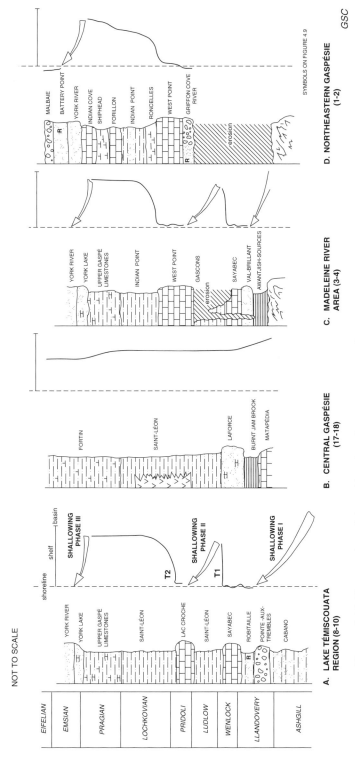

Figure 4.11. Progression of depositional environments during Silurian and Devonian time, with respect to distance from shoreline, for four rock successions of the Gaspé-Témiscouata segment of the Connecticut Valley-Gaspé Synclinorium. T1 and T2 are transgressive episodes. Compare with Figure 4.2a. Location of sections shown in Figure 4.9.

of the Indian Point Formation as an intermediate facies. Indeed, the stromatoporoid and coral bioherms and biostromes of the West Point shelf reef facies were progressively buried under fine grained siliciclastic sediments (Roncelles and Indian Point). This resulted from drowning of the platform concurrently with a westward progradation of delta deposits (Bourque et al., 1986). The Upper Gaspé Limestones of eastern Gaspésie record a period of uniform sedimentation on a deep stable platform (Lespérance, 1980a), a conclusion based on the predominance of laminated fine grained siliceous limestones, the ubiquity of sponge spicules, and the rarity of shelly fauna. The greater abundance of fossils on the Forillon Peninsula compared to other areas, is indicative of shallower environments. These conclusions seem to apply generally throughout northeastern Gaspésie, at least for the Forillon and Indian Cove formations, which are relatively homogeneous. To the west, in the Madeleine River area, the occurrence of fine grained siliciclastites together with slumps at the base of the Forillon Formation possibly indicate a slightly deeper environment. The Shiphead Formation probably reflects a more complex sedimentary history. The occurrence of bentonites and other volcanic rocks is probably related to volcanism in an area south of Monts McGerrigle.

The southern equivalent of trangression episode T2 is the rather uniform and monotonous sequence of fine grained siliciclastic-dominated sediments of the Saint-Léon and Fortin-Témiscouata units. A number of facies, such as structureless thick bedded muds, well bedded parallel-and cross-laminated silts and fine grained sands, and thinly laminated graptolitic muds and silts, occur in the Saint-Léon Formation (Bourque and Lachambre, 1980; Bourque and Gosselin, 1986). These are compatible with a deep shelf environment. The sedimentology of the Fortin Group of central Gaspésie (succession B) and its western equivalent, the Témiscouata Formation, are poorly known. The sandstones of the Fortin-Témiscouata of the Madawaska County-Assémetquagan River area are interpreted as channel-levee/interchannel turbidites, and the siltstones and slates, as spillover turbidites (Dalton, 1987). Paleocurrents are toward the northwest, indicating a source to the southeast. In northern Maine, the slates and associated turbiditic sandstones of the Seboomook Formation, a lateral equivalent of the Témiscouata-Fortin, are interpreted as a northwestward spreading deltaic wedge (Hall and Stanley, 1973; Hall et al,. 1976). It is probable that the Fortin-Témiscouata sequence represents the prodelta marine slope deposits of a northwesterly prograding delta, like the Seboomook.

Shallowing upward phase III (Emsian) is much better developed in the Connecticut Valley-Gaspé Synclinorium than in the Chaleurs Bay Synclinorium. It is represented by the Gaspé Sandstones Group (successions A, C and D, Fig. 4.11). It corresponds to a rapid shoaling event, from shallow marine to terrestrial facies. The change from deep water facies of the Upper Gaspé Limestones to shallow water sands of the Gaspé Sandstones is gradational (York Lake) in the northern part of the central Connecticut Valley-Gaspé Synclinorium and Lake Témiscouata region (successions C and A). It is abrupt in northeastern Gaspésie (succession B). Both the York Lake and the York River contain marine fauna. The York Lake is transitional between limestones of the Indian Cove Formation and

sandstones of the York River Formation. The York River sandstones coarsen upward, reflecting shallowing of the basin and proximity to the source area. Deltaic or estuarine environments were proposed for the York River (Mason, 1971; Sikander, 1975). The first assemblage of the overlying Battery Point Formation was interpreted as distal, braided stream deposits (Cant and Walker, 1976). However, marine cephalopods and brachiopods indicate that the braided stream environment was at least periodically subjected to marine influx. The second assemblage of the Battery Point conforms to the classical fining upward sequence of a meandering river (Rust, 1981). This assemblage also contains units of dolomite and dolomitic siltstone, in places deeply cracked by dessication, which can be interpreted as playa or ephemeral lake deposits. The presence of fresh-water fossil fish support this model (Pageau, 1968). The third assemblage is very similar to the first but coarser grained, and is also interpreted as braided stream deposits, but possibly nearer the source area and on a steeper slope. The upper, fourth assemblage of the Battery Point may be laterally equivalent to the lower part of the Malbaie Formation, which in turn is interpreted as a proximal braided plain deposit (Rust, 1981).

In the Eastern Townships of southern Quebec, only the Ludlow(?)-Pridoli Lac Aylmer and Lac Lambton formations of the Lake Aylmer-Lake Saint-François area were analysed sedimentologically (Lavoie, 1985). Depositional environments of all other units are poorly understood. Conglomerate units of the Lac Aylmer and Lac Lambton formations, are interpreted as nearshore or even terrestrial, with the Lac Aylmer representing more proximal facies. The two formations show the same progression of two deepening upward cycles separated by a short regressive episode. Because of the poorly constrained ages of these units, correlation with the sedimentary phases of the Gaspé-Témiscouata segment of the Connecticut Valley-Gaspé Synclinorium is speculative, but in a general way, it may correspond to the beginning of transgression T2 (Fig. 4.11). Depositional environments of the overlying Ayers Cliff and Compton formations resemble those of the Saint-Léon and Fortin-Témiscouata formations, respectively.

Summary and conclusions

Middle Paleozoic rocks of the Quebec and adjacent New Brunswick Appalachians have been described according to four broad assemblages: an Upper Ordovician-lowermost Silurian assemblage of deep water fine grained siliciclastic and carbonate facies comprising the Honorat, Grog Brook, Cabano, and Matapédia groups; a Silurian-lowermost Devonian shallow to deep shelf assemblage, principally made up of the Chaleurs Group and nearby equivalents; a Lower Devonian deep shelf, mixed fine grained siliciclastic and carbonate assemblage including several groups (Upper Gaspé Limestones, Fortin, Témiscouata, upper part of Saint-Francis); and a Middle and Upper Devonian nearshore to terrestrial coarse grained assemblage comprising the Gaspé Sandstones Group. These middle Paleozoic rocks occur in three structural divisions of the Gaspé Belt, from north to south: the Connecticut Valley-Gaspé Synclinorium, the Aroostook-Percé Anticlinorium, and the Chaleurs Bay Synclinorium.

Sequence analysis

Three distinctive unconformities occur in the middle Paleozoic sequence of Quebec and adjacent New Brunswick (Fig. 4.3). The oldest unconformity corresponds to the Taconic Orogeny. The next unconformity is within the Chaleurs Group. It is dated as late Ludlow-early Pridoli and corresponds to the Salinic Disturbance recognized in Maine (Boucot, 1962). It is either an angular unconformity (Ristigouche Syncline), or an erosional disconformity that in places cuts deep into the underlying Silurian and Ordovician rocks (Northern Outcrop Belt). The third unconformity is angular and separates Middle or Upper Devonian rocks from Carboniferous rocks. It corresponds to the Acadian Orogeny.

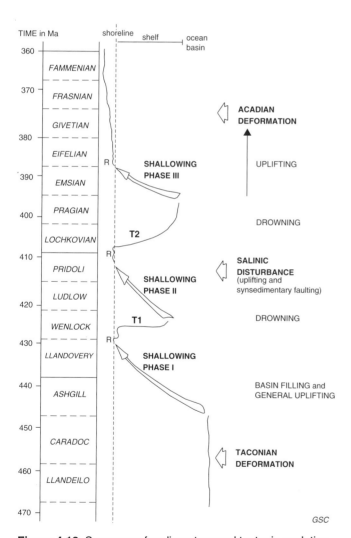

Figure 4.12. Summary of sedimentary and tectonic evolution of the upper Middle Ordovician to Upper Devonian sequence of the Gaspé Belt. T1 and T2 are transgressive episodes. R, denotes redbeds.

Three shallowing upward phases, separated by two transgressions, are recorded in the middle Paleozoic rock sequence (Fig. 4.9, 4.11, 4.12). The shallowing phases are more or less linked to the three tectonic pulses recorded by the unconformities. Figure 4.12 depicts the main tectonic events with respect to the sedimentary evolution of the Gaspé Belt. Shallowing phase I corresponds to filling of the basin following Taconic Orogeny. Shallowing phase II culminated with the Salinic Disturbance. Shallowing phase III culminated with Acadian Orogeny.

Paleogeography and tectonics

Silurian-Early Devonian stratigraphy, structural geology and paleogeography of the Gaspésie-Témiscouata segment of the Canadian Appalachians was recently summarized, based on over 200 measured sections (Bourque et al., 1993). In order to construct reliable paleogeographical maps, effects of Acadian (Middle Devonian) folding and faulting were erased and each Silurian-Devonian section restored to its position at time of deposition. The resulting palinspastic map (Fig. 4.13) provides a more realistic portrayal of depositional settings and tectonic elements.

The amount of shortening because of folding was obtained by detailed structural analysis along a number of structural sections across the Gaspésie-Témiscouata segment of the Quebec Appalachians (Kirkwood and St-Julien, 1987; Kirkwood et al., 1988a, b). It is estimated to be as great as 55 per cent in places. Acadian Orogeny also produced strike-slip faults with large right-lateral displacements (Bourque et al., 1985; Malo, 1986; Malo et al., 1988). Cumulative lateral displacements along the Grand Pabos-Ristigouche, Grande Rivière, and Rivière Garin faults (Fig. 4.2a) is at least 150 km. Smaller displacements occurred along the Third Lake and Bras Nord-Ouest faults (8 km and 7 km, respectively).

Restoration shows that the three structural units, Connecticut Valley-Gaspé Synclinorium, Aroostook-Percé Anticlinorium, and Chaleurs Bay Synclinorium are spatially related integral parts of the Gaspé Belt (Fig. 4.13).

Paleogeographical evolution of the Gaspé-Témiscouata segment of the Gaspé Belt during Late Ordovician-Early Devonian time is illustrated in Figure 4.14. These maps are based on the sections measured and described by Bourque (1977), Bourque and Lachambre (1980), and Bourque et al. (1993). In northwestern Gaspésie, along the Northern Outcrop Belt, no sections were directly measured, but detailed mapping by Lachambre (1987) provides reliable facies control. The paleogeographical map of Emsian time (Fig. 4.14f) is based chiefly on known regional distribution of facies. The New Brunswick part of the Gaspé Belt is not included because palinspastic reconstruction is uncertain. For instance, based on offset of the main axis of the Chaleurs Bay Synclinorium, right-lateral displacement of at least 35 km and 25 km probably occurred along the Rocky Brook-Millstream and McKenzie Gulch faults, respectively.

The Gaspé Belt evolved south and southwest of the Quebec Reentrant and St. Lawrence Promontory, respectively (Fig. 4.13b). During Llanvirn to early Ashgill time, flysch sedimentation (Tourelle, Deslandes, Cloridorme, Mictaw, Arsenault, Fournier, Trinité) took place in a marginal basin/arc complex (Dunnage Zone) south of the North American craton. In early Ashgill time, units of the northern part of the basin (Tourelle, Deslandes, Cloridorme, and Mictaw) were already deformed and uplifted by Taconic Orogeny to form the Taconic allochthons (Fig. 4.14a). The trace of the Baie Verte-Brompton Line (Williams and St-Julien, 1982), which separates the Humber and Dunnage zones, parallels the edge of the Quebec Reentrant-St. Lawrence Promontory. South of the Taconic allochthons, deposition of fine grained silici-clastic turbidites with local coarser sands and gravels (Honorat, Cabano and Grog Brook groups) was taking place in a deep water basin.

At the beginning of the Silurian (Llandovery B), the Taconic allochthons were emergent south of the Quebec Reentrant and in the Maquereau-Mictaw Inlier south of the St. Lawrence Promontory (Fig. 4.14b). Occurrences of near-shore sands (Weir and Anse Cascon formations) reflect

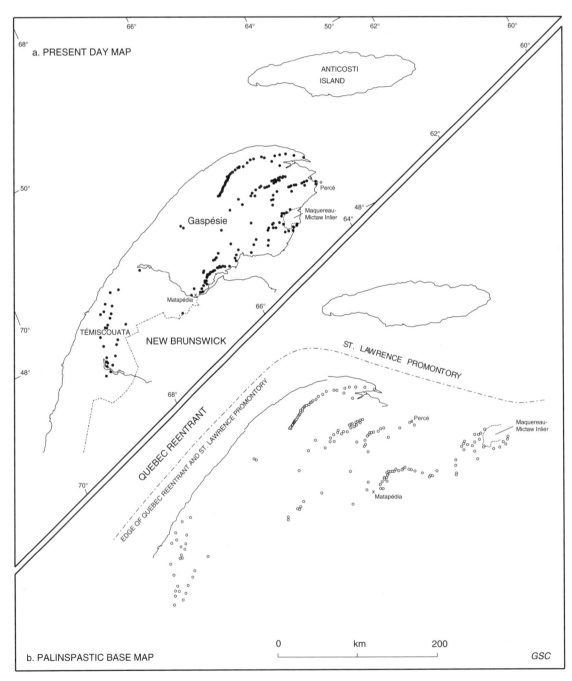

Figure 4.13. Palinspastic reconstruction of the Gaspé-Témiscouata segment of the Gaspé Belt: **(a)** present distribution of control points and stratigraphic sections; **(b)** palinspastic distribution of same control points. From Bourque et al. (1993).

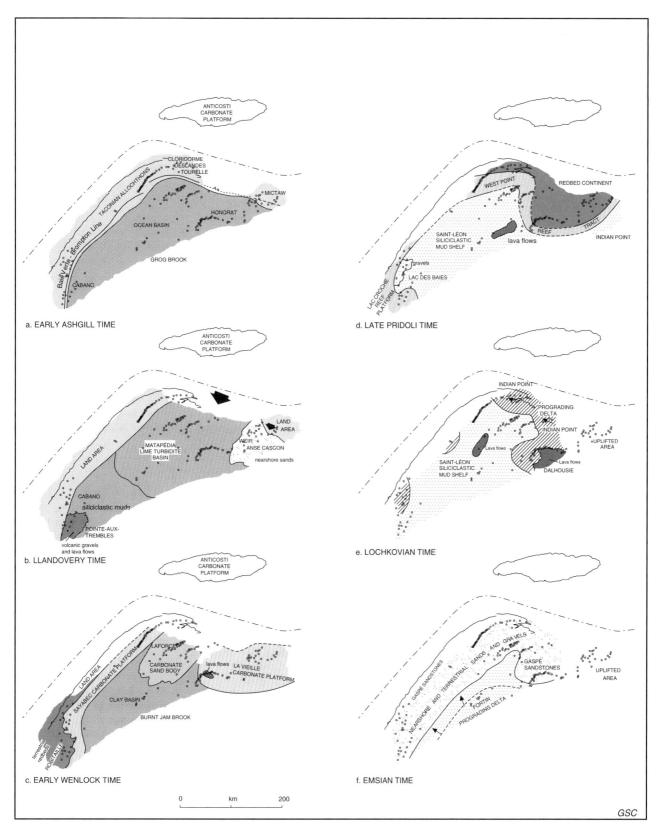

Figure 4.14. Paleogeography of the Gaspé-Témiscouata segment of the Gaspé Belt, from post-Taconic Orogeny through Acadian Orogeny. Facies were plotted on palinspastic base map of Figure 4.13b. Arrows indicate direction of sedimentary transport.

erosion in that area. Between these two land areas, lime turbidites of the Matapédia Group and Carys Mills Formation, derived from the Anticosti platform, filled the eastern half of the basin. Northeast to southwest lateral facies zonation in the Matapédia Group, from proximal to distal, together with paleocurrent analysis, support this interpretation. In the western half of the basin, volcanic conglomerates with a few associated lava flows (Pointe-aux-Trembles Formation) indicate erosion of an active volcanic area (David and Gariépy, 1986).

During early Wenlock time, and corresponding to the end of the first shallowing phase, limestone deposition dominated shallow water areas of the Gaspé Belt; the Sayabec platform in the west and northwest, and the La Vieille platform in the southeast (Fig. 4.14c). It is not clear whether these two very similar platforms were connected, but the occurrence of a carbonate sand and gravel body (Laforce Formation) in the clay basin (Burnt Jam Brook Formation) that succeeded the lime turbidites of the Matapédia Group suggests the platforms were separated. The Sayabec platform bordered a terrestrial redbed area, whereas the land facies north of the La Vieille platform is unknown.

Late Pridoli time (Fig. 4.14d) corresponds to the peak of the second shallowing phase. A reef tract, the West Point Formation in the east and the Lac Croche Formation in the west, made up of barrier reefs, a cluster of individual reefs, or isolated patch and pinnacle reefs (Bourque et al., 1986), can be followed over 700 km in the Gaspésie-Témiscouata region. Reef limestone of the same age (upper member of the Lac Aylmer Formation) occurs in the Eastern Townships, 300 km farther west, giving the reef tract a possible total extension of about 1000 km in the Gaspé Belt. The West Point reef tract developed at the margin of a wide siliciclastic redbed landmass (Plage Woodmans Member of the West Point Formation). It seems likely that during that time, the Anticosti Platform was totally emerged and probably covered by redbeds. Redbeds of the same age, Clam Bank Group of the Clam Bank Belt, occur in western Newfoundland on the southeast side of the St. Lawrence Promontory. The reef tract developed on and at the margin of a nearshore siliciclastic bank (part of Saint-Léon and Gascons formations), and was controlled by tectonics (Salinic Disturbance) and the rate of siliciclastic sedimentation (Bourque, 1987). South of the siliciclastic bank and reef tract, fine grained siliciclastic facies (part of the Saint-Léon and Indian Point formations) were deposited in an outer shelf environment.

In Early Devonian time (Lochkovian), drowning of the shelf and westward progradation of a delta system (Rosebush Cove, Petit Portage, Quay Rock, and Cape Road members of the Indian Point Formation) formed in response to uplifting of the southeastern part of the basin (Fig. 4.14e). Depth indicators suggest that no Devonian rocks were deposited in the Quebec part of the Chaleurs Bay Synclinorium (Nowlan and Barnes, 1987; Lavoie, 1988; Savard and Bourque, 1989; Y. Héroux, pers. comm. to P. A. Bourque). Submarine volcanism (Dalhousie Formation, Baldwin Member) occurred in the south and centre of the shelf. In late Early Devonian time (Emsian), at the peak of the third shallowing upward phase, nearshore and then terrestrial sands and gravels (Gaspé Sandstones Group)

developed in response to the onset of Acadian Orogeny (Fig. 4.14f). The advance of a prograding delta (Fortin and Témiscouata groups) progressively filled the basin.

In summary, the middle Paleozoic sedimentary sequence of the Gaspé-Témiscouata segment of the Gaspé Belt evolved upon an unstable shelf at the margin of the Quebec Reentrant-St. Lawrence Promontory. Indeed, the pre-Acadian palinspastic restoration of the basin permits definition of its geometry at the time of deposition and indicates that the basin developed not only at the margin of the Quebec Reentrant, as the present facies distribution suggests, but also at the margin of the St. Lawrence Promontory and Anticosti Platform. This important conclusion explains the presence and nature of the distinctive facies deposited during Silurian time.

Acadian deformation

Folding and faulting that affected the Upper Ordovician to Middle Devonian rocks of the Gaspé Belt did not affect the unconformably overlying Carboniferous rocks, therefore indicating that Acadian deformation occurred during or after Middle Devonian and before Carboniferous time.

The principal structural features of the Aroostook-Percé Anticlinorium can be explained by strike-slip tectonics (Malo and Béland, 1989). The major northeasterly structural trend in the Chaleurs Bay and Connecticut Valley-Gaspé synclinoria is similar to the structural trend of the Aroostook-Percé Anticlinorium and both are related to the D_2 deformation, which was mainly controlled by major strike-slip faults.

The Shickshock-Sud, Grande Rivière, Grand Pabos, and Rivière Garin faults are Acadian dextral strike-slip faults (Berger, 1985; Malo and Béland, 1989). They are conspicuous as major straight longitudinal lineaments on LANDSAT images. In outcrop, the faults are high strain zones with C-S fabrics indicating dextral strike-slip movement. Major Acadian folds trend northeast, usually oblique to the major easterly trending faults or parallel to some northeast-trending faults (Fig. 4.15). The obliquity to the easterly trending faults suggests that the folds originated by dextral transcurrent movement along the faults. A regional northeast cleavage sub-parallel to fold axes is well-developed throughout the Aroostook-Percé Anticlinorium, the lowest structural level, and in the shaly Silurian and Devonian rocks. This northeast regional cleavage is rotated in a clockwise sense near the major dextral strike-slip faults (Bernard and St-Julien, 1986; Kirkwood and St-Julien, 1987; Malo and Béland, 1989), compatible with a dextral sense of shear. Secondary faults are riedel-type, compatible with major dextral easterly trending strike-slip faults (such as the low- and high-angle faults in the area along the Grand Pabos Fault north of the Maquereau-Mictaw Inlier, Fig. 4.15).

Most of these Acadian structural elements were generated in a tectonic regime of major dextral, easterly transcurrent movements which occurred after the emplacement of the north verging Taconic allochthons. The transcurrent regime also explains the contrasting structural styles often noted between the Taconic and Acadian deformed zones in the Gaspésie segment of the Quebec Appalachains (Malo and Béland, 1989).

Tectonic setting

The Appalachian Orogen resulted from accretion of outboard terranes to the North American craton during Paleozoic time (Williams and Hatcher, 1983; Keppie, 1985). The three main accretionary events coincide with the Taconic, Acadian, and Alleghanian orogenies. The Taconic Orogeny is explained by the collision of an island arc against the North American margin in Middle Ordovician time (Williams, 1979; Williams and Hatcher, 1983). In the Canadian part of the orogen, the Baie Verte-Brompton Line with its mélanges and ophiolite complexes represents the boundary between the deformed North American miogeocline (Humber Zone) and the outboard volcanic terrane (Dunnage Zone).

Some workers (Dewey, 1969; McKerrow and Cocks, 1977; Osberg, 1978; Hatch, 1982) relate Acadian Orogeny to the final head-on closure of the Iapetus Ocean and the docking of the Avalon terrane (continent-continent collision). This model is not supported by the stratigraphic facts as there are no oceanic features such as ophiolites, mélanges, and island arc sequences in the Silurian-Devonian record that prove the persistence of the Iapetus Ocean tract in Silurian-Devonian time (Williams, 1979). Williams and Hatcher (1983) explained the Devonian accretion of more outboard terranes by major transcurrent motion. This view is supported by steep mylonitic zones and brittle faults at Gander-Avalon and Avalon-Meguma boundaries and the recognition of major transcurrent movements (Blackwood and Kennedy, 1975; Hanmer, 1981; Keppie, 1982; Mawer and White, 1987).

Our structural analysis supports a wrench tectonics model for the Acadian deformation of Silurian-Devonian rocks of the Gaspé Belt. Coeval volcanism was generated in an intraplate tectonic environment, unrelated to subduction (Laurent and Bélanger, 1984; Bédard, 1986). There are no Acadian mélanges and/or ophiolitic rocks, nor is there any evidence of a Devonian suture.

The continental convergence in the Canadian Appalachians was along an irregular margin of which the Quebec Reentrant-St. Lawrence Promontory represents one of its major irregularities. The sedimentation of the Gaspé Belt took place in a successor basin located in the gap formed by the Quebec Reentrant west of the St. Lawrence Promontory. This Late Ordovician to Middle Devonian basin was mostly over the Dunnage Zone between the already destroyed Taconic margin (Humber Zone, Fig. 4.14a), and the Miramichi Subzone of the Gander Zone. The middle Paleozoic rocks of the Gaspé Belt overlap the Humber, Dunnage, and Gander (Miramichi) zones, thus proving that these zones (terranes) were already accreted by the end of Middle Ordovician time (Williams and Hatcher, 1983; Fyffe and Fricker, 1987).

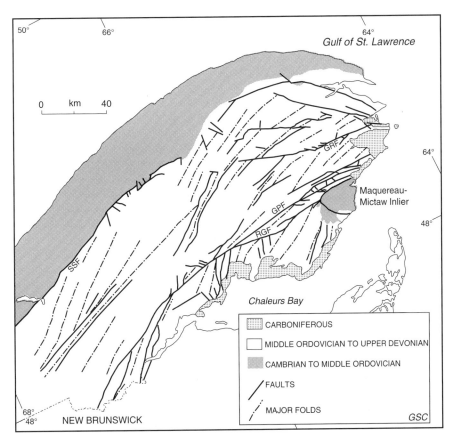

Figure 4.15. Major Acadian folds (F₂) and faults in Gaspésie. GPF: Grand Pabos Fault, GRF: Grande Rivière Fault, RGF: Rivière Garin Fault, SSF: Shickshock-Sud Fault.

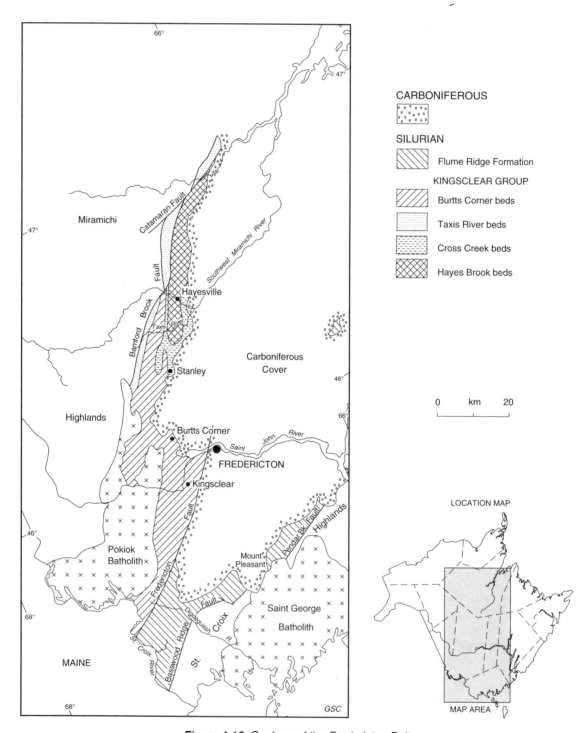

Figure 4.16. Geology of the Fredericton Belt.

The deformation associated with accretion on the North American margin is the Taconic Orogeny. The main deformation of the Gaspé Belt is Acadian with sedimentation taking place in a tectonically active environment (Fyffe and Noble, 1985) which was also the site of intraplate volcanism (Laurent and Bélanger, 1984; Bédard, 1986).

Thomas (1977) suggested that the promontories of the Appalachian Orogen were sites of active transform faults during the opening of the Iapetus Ocean. The convergence of more outboard terranes on this irregular margin created a major dextral shear along the St. Lawrence Promontory, with renewed movement along this northwest-trending ancient transform (Stockmal et al., 1987). Lithospheric delamination and tectonic wedging during Acadian Orogeny is implied at the St. Lawrence Promontory (Stockmal et al., 1987). In the Quebec Reentrant, the southwestern wall of the St. Lawrence Promontory moved northwest, creating a stress field in the basement with a northwesterly directed principal stress.

Acadian deformation is summarized in six stages: (1) during D₂ deformation, northwesterly directed forces induced dextral easterly trending strike-slip faults in the basement and northeast-trending folds in the sedimentary cover of the Gaspé Belt; (2) shortening was achieved with the development of regional cleavage and some northeast-trending high-angle reverse faults; (3) ductile deformation with creation of late cleavage and rotation of the regional cleavage over high strain zones destined to be the sites of future faults; (4) beginning of brittle deformation with secondary riedel-type faults and conjugate mesoscopic faults; (5) master easterly-trending faults appeared at the surface followed by displacements along these fracture zones; and (6) late stage faulting, represented by northwest-trending dextral strike-slip faults in the eastern part of the peninsula and northeast-trending oblique-slip and high-angle reverse faults in the western part of Gaspésie.

The Acadian Orogeny, which created strike-slip faults in the southeastern area of Gaspésie, is expressed by northwest thrusting along the major La Guadeloupe Fault in the Eastern Townships (St-Julien et al., 1983; Tremblay et al., 1989a).

Carboniferous rocks unconformably overlying the middle Paleozoic rocks of the Gaspé Belt are affected by minor faults, interpreted as dip-slip faults related to the Alleghanian Orogeny. They may have controlled sedimentation. In addition, some Acadian strike-slip faults were reactivated during Alleghanian Orogeny.

Acknowledgments

We are much indebted to A.J. Boucot (Oregon State University), P.J. Lespérance (University of Montréal), G.S. Nowlan (Geological Survey of Canada) and J. Riva (Laval University) for brachiopod, trilobite, conodont, and graptolite identifications, respectively, and for their important contributions to the biostratigraphy and chronostratigraphy of this area. We benefited greatly from discussions, comments and suggestions of several workers in the Québec Appalachians, among which are: our Laval University colleagues D. Kirkwood, R. Laurent, and P. St-Julien; W.B. Skidmore formerly with the Ministère de l'Énergie et des Ressources du Québec; and J. Béland and J. David of the University of Montréal. Critical review and editing by H. Williams (Memorial University, St. John's) were invaluable. Financial support at various stages of the study was provided by the Ministère de l'Énergie et des Ressources du Québec (contribution no. 91-5110-19), as well as a Natural Sciences and Engineering Research Council of Canada Operating Grant (A9142) and Quebec Department of Education FCAR Grant (EQ-1005) to P.-A. Bourque. This paper is published with the kind permission of the Ministère de l'Énergie et des Ressources du Québec.

Fredericton Belt
L.R. Fyffe

The Fredericton Belt comprises a thick sequence of Silurian turbidites lying between the Miramichi and St. Croix highlands of New Brunswick (Poole et al., 1970; McKerrow and Ziegler, 1971; Rast et al., 1976). As no systematic study of the stratigraphy or structure of these rocks has yet been undertaken, only a brief description is given here. The Silurian rocks are in contact with Cambrian-Ordovician rocks (Miramichi and Tetagouche groups) of the Miramichi Highlands to the northwest along the Bamford Brook Fault, a part of which is intruded by the Pokiok Batholith. To the southeast, their contact with the Ordovician Cookson Group of the St. Croix Highlands is marked by the Basswood Ridge-Pendar Brook Fault and the Saint George Batholith (McCutcheon, 1981; Fyffe and Riva, 1990). The Fredericton Fault, a continuation of the Norumbega Fault of Maine (Stewart and Wones, 1974; Unger et al., 1987), is within the Fredericton Belt (Fig. 4.16).

Stratigraphy

Silurian rocks of the Fredericton Belt have been included in the Flume Ridge Formation to the south of the Fredericton Fault (Ruitenberg, 1967; Ludman, 1978) and the Kingsclear Group to the north of the Fredericton Fault (Freeze, 1936; Smith, 1966). It is recommended that all of the rocks be included in the Kingsclear Group. The Flume Ridge Formation typically contains greyish green, thin bedded, calcareous, flaggy wacke, siltstone, and green to grey, generally pyritiferous slate (Fig. 4.17). The slate and siltstone generally grade vertically into the wacke rather than occurring as discrete beds. Where weathered, iron carbonate imparts a distinctive limonitic, brown colouration to the sequence, which characteristically contains large muscovite flakes on bedding surfaces (Ruitenberg, 1967; Ruitenberg and Ludman, 1978). Although calcareous siltstones predominate at the type-section on the Digdeguash River (Fig. 4.18), medium- to thick-bedded, slightly calcareous to noncalcareous lithic wackes form a significant proportion of the sequence farther to the northwest (Ludman, 1978, 1981). The Flume Ridge Formation is apparently unfossiliferous; the reported presence of poorly preserved plant fragments of *Lepidodendron* (Bailey and Matthew, 1872), has not been verified by subsequent surveys.

Fyffe, L.R.
1995: Fredericton Belt; *in* Chapter 4 of Geology of the Appalachian-Caledonian Orogen in Canada and Greenland, (ed.) H. Williams; Geological Survey of Canada, Geology of Canada, no. 6, p. 351-354 (also Geological Society of America, The Geology of North America, v. F-1).

The Kingsclear Group in the Hayesville area of central New Brunswick (Fig. 4.16) has been divided into four units with a combined thickness of at least 3 km (Poole, 1957, 1963; Poole et al., 1970). These units are referred to the informal Hayes Brook (after exposures just north of Hayesville), Cross Creek (after exposures northeast of Stanley), Burtts Corner, and Taxis River beds. The Cross Creek beds form a relatively thin sequence of red and green siltstone and slate that serves as a marker horizon separating the underlying Hayes Brook quartz wacke beds from the overlying Burtts Corner lithic wacke beds. The Taxis River beds, which occur only along the Bamford Brook Fault on the northwestern margin of the Fredericton Belt (Anderson and Poole, 1959), contain distinctive lenses of conglomerate with pebbles of basalt and slate derived from the adjacent Ordovician Tetagouche Group of the Miramichi Highlands. The Taxis River beds grade upward into the Burtts Corner beds and may represent a local debris-flow facies deposited along a fault scarp.

Turbidites of the Burtts Corner beds, as exposed along the Trans-Canada Highway west of Fredericton, are medium- to thick-bedded, noncalcareous, and exhibit internal grading and flute casts indicating a paleocurrent direction from the northeast along the axis of the Fredericton Belt (Fig. 4.19). The wackes are composed of quartz (19-48%), volcanic (2-36%), metamorphic (3.5-18%), and sedimentary (0.5-15%) rock fragments, feldspar (1.5-7.5%), and matrix (14.46%) (Wake, 1984). Graptolites recovered from interbedded slates along the Trans-Canada Highway and in the Hayesville area indicate that the Burtts Corner beds range from the *Cyrtograptus linnarssoni* Zone of the Wenlock to the *Monograptus nilsonni* Zone of the early Ludlow (Cumming, 1966; Anderson, 1970; Berry and Boucot, 1970; Gordon, 1973). Graptolites from slates interbedded with the Hayes Brook beds belong to the *Monograptus cyphus* Zone of the early Llandovery (Lenz, written communication to W.H. Poole concerning GSC Location 99360, 1982).

Figure 4.17. Thin bedded, calcareous, siltstones of the Flume Ridge Formation at the Flume on Digdeguash River, southwestern New Brunswick. GSC 1994-768A

Figure 4.19. Flute casts (15 cm in width) in the Burtts Corner beds exposed on the Trans-Canada Highway west of Fredericton, New Brunswick. GSC 1994-768B

Figure 4.18. Medium bedded, slightly calcareous wackes and slates of the Flume Ridge Formation on Canoose River, southwestern New Brunswick. GSC 1994-768C

Figure 4.20. First generation folds in turbidites of the Burtts Corner beds on the Trans-Canada Highway west of Fredericton, New Brunswick. GSC 1994-768D

Figure 4-21. Overturned first generation fold in Hayes Brook beds east of Hayesville, central New Brunswick. GSC 1994-768E

Thick beds of lithic wacke within the Burtts Corner beds along the St. Croix River just to the north of the Fredericton Fault are interstratified with calcareous siltstones identical to those within the Flume Ridge Formation but in less abundance. It is for this reason that the Flume Ridge Formation be included in the Kingsclear Group and that it is more likely Wenlock to Ludlow in age than Early Devonian as was suggested by Ruitenberg (1967). A similar facies relationship has been recognized between Silurian wackes of the Sangerville (formerly Vassalboro) Formation and slates of the Waterville Formation in adjacent Maine (Osberg, 1968, 1988).

Structure and metamorphism

First generation structures in the Fredericton Belt are mesoscopic tight, upright folds with well developed axial planar slaty cleavage (Fig. 4.20) that varies in trend from north-northeasterly to northeasterly. Bedding-cleavage intersections indicate shallow plunges to the northeast and southwest. Close to tight, south-verging, mesoscopic second folds associated with a shallow-dipping crenulation cleavage are common in the Flume Ridge Formation along the southern margin of the Fredericton Belt (Ruitenberg, 1967), whereas first and second generation structures tend to verge to the north along its northern margin (Fig. 4.21). A later conjugate set of vertical kink folds is present in most exposures; sinistral kink bands trend easterly and dextral bands trend southeasterly, indicative of shortening approximately parallel to bedding. Subhorizontal kink bands are common in the Flume Ridge Formation. Lower greenschist facies regional metamorphism accompanied the main deformation. Higher grade metamorphic rocks are found only in the contact aureoles of major intrusions (Ruitenberg, 1967; Ruitenberg and Ludman, 1978).

Brittle deformation along the Fredericton Fault is indicated by narrow zones of quartz veining and brecciation. En echelon wedges of Ordovician phyllite and Carboniferous conglomerate along the fault trace, and counterclockwise orientation of slaty cleavage in Silurian rocks relative to the trend of the fault zone, are consistent with dextral-transcurrent motion (Fyffe, 1988). Post-Devonian displacement is not extensive since Middle Devonian granite is offset only by a few hundred metres along the Norumbega Fault in Maine (Wones, 1980). The Basswood Ridge-Pendar Brook Fault, which separates the Flume Ridge Formation from the Ordovician Cookson Group to the south, has numerous quartz veins and it is intruded by granite porphyry dykes of probable Late Devonian age (Fyffe and Riva, 1990). Southward-verging, second folds in both the Silurian and Ordovician rocks suggest that this fault is a southward-directed thrust similar to those identified on seismic profiles in Maine (Unger et al., 1987).

Tectonic significance

Silurian turbidites of the Fredericton Belt cannot be demonstrated to be laterally or vertically continuous with shallow-water Silurian strata of the Mascarene Belt, since the two sequences are separated by the Cookson Group and Basswood Ridge-Pendar Brook Fault (Fig. 4.16). Previous interpretations assumed that the Flume Ridge and Waweig formations were facies equivalents deposited in the same basin (Ruitenberg, 1967; Poole et al., 1970; Poole, 1976). Rhenish-Bohemian brachiopod assemblages occur in Silurian-Devonian rocks of the Mascarene Belt (Boucot et al., 1966; Pickerill, 1976), but are unknown in the graptolite-bearing rocks of Fredericton Belt. Early Devonian Eastern Americas brachiopod assemblages are found in the Aroostook-Perce Division of Gaspé Belt on the northwestern margin of the Fredericton Belt (Fyffe and Fricker, 1987). Basalt pebbles within the Taxis River beds provide a sedimentary link with the Ordovician Tetagouche Group of the Miramichi Highlands but no such link has been established to the southeast with the Ordovician Cookson Group of the St. Croix Highlands. Silurian rocks of the Fredericton and Mascarene belts, therefore, may have been deposited in separate basins. Although the Mascarene Belt developed at least in part upon an Avalonian basement, basement relationships of the Fredericton Belt are unknown.

The Fredericton Belt has been interpreted as a remnant of a middle Paleozoic ocean that closed by subduction during Acadian orogenesis (McKerrow and Ziegler, 1971; Rast and Stringer, 1974; Bradley, 1983; Berry and Osberg, 1989). This model is supported by the presence of distinct faunal provinces on opposite sides of the Fredericton Belt (as noted above), although the absence of contemporary arc-volcanic rocks and other features do not support the model. Van Staal et al. (1990) have suggested that the Fredericton Belt is a foredeep formed by tectonic loading in front of southward-advancing nappes. In this model, the Wenlock to Ludlow Burtts Corner lithic wacke beds would represent a clastic wedge that prograded over the more mature Llandovery Hayes Brook quartz wacke beds as the tectonic land mass of the Miramichi Highlands emerged to the northwest. The finer grained, calcareous Flume Ridge Formation would represent a distal facies of the Burtts Corner beds. The foredeep could have developed either in response to crustal delamination (Stockmal et al., 1987), or westward subduction (van Staal, 1987), and was apparently the locus of transcurrent movement throughout the mid- to late Paleozoic.

The Fredericton Fault has been designated as an Acadian terrane boundary with distinctive Silurian rocks on opposite sides (Keppie, 1985). However, the similarity of the Flume Ridge Formation, south of the fault, to the Burtts Corner beds, north of the fault, indicates continuity, although significant transcurrent movement is possible (Ludman, 1987). Isotopic studies on granites may provide some constraints on basement relationships across the Fredericton Belt (Whalen et al., 1989).

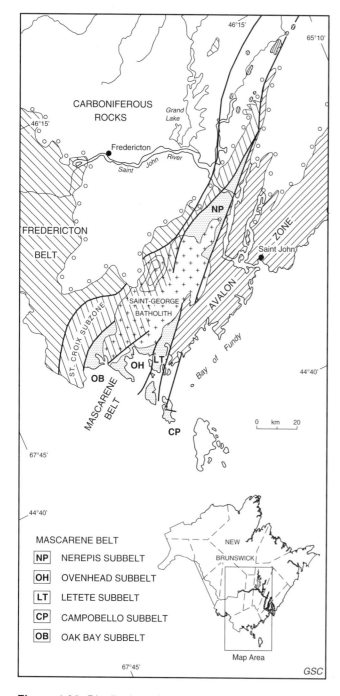

Figure 4.22. Distribution of the Mascarene Belt in southern New Brunswick.

MASCARENE BELT

NP NEREPIS SUBBELT
OH OVENHEAD SUBBELT
LT LETETE SUBBELT
CP CAMPOBELLO SUBBELT
OB OAK BAY SUBBELT

Acknowledgments

Thanks are extented to H. Williams for critically reading this manuscript.

Mascarene Belt

M.J. McLeod and S.R. McCutcheon

The Mascarene Belt of southern New Brunswick consists of a relatively thin cover sequence deposited mainly upon the Avalon Zone. It is a lithologically and structurally diverse sequence of Silurian to Lower Devonian volcanic and sedimentary rocks intruded by locally voluminous, Silurian-Devonian mafic plugs and dykes, granitoid complexes and felsic dykes. It is unconformably overlain by Upper Devonian to Carboniferous subaerial sedimentary and minor volcanic rocks that formed in response to subsidence after the Acadian Orogeny (McCutcheon and Robinson, 1987).

The Mascarene Belt is about 15 km by 150 km and is a continuation of the Coastal Volcanic Belt in Maine (Gates, 1969). It extends northeast from Campobello and Deer Islands to the Annidale area where it terminates along northeast-trending faults (Fig. 4.22). The northwestern margin of unequivocal rocks of the Mascarene Belt is generally defined by a series of northeast-trending faults, including the Sawyer Brook Fault to the southwest and several unnamed faults to the northeast (McCutcheon and Ruitenberg, 1987; Fyffe, 1991; McLeod et al., 1991a) and by the southern contact of the Saint George Batholith (McLeod et al., 1989). The southeastern margin is roughly delineated by the trace of the Wheaton Brook Fault (McCutcheon and Ruitenberg, 1987) and Back Bay Fault (Donohoe, 1978). Inliers of Avalon Zone rocks and outliers of Mascarene Belt rocks with faulted or unknown contact relations occur north and south of this trace, respectively (e.g. Helmstaedt, 1968; McCutcheon and Ruitenberg, 1987). The inliers and outliers may be also bounded by unconformities (Hay, 1968; McLeod, 1979; McCutcheon, 1981; Currie, 1988a; McLeod and Rast, 1988).

The Mascarene Belt is divisible into five subbelts herein termed the Nerepis, Letete, Ovenhead, Campobello and Oak Bay (Fig. 4.22). Rocks of the first four subbelts were deposited on the Avalon Zone but rocks of the Oak Bay Subbelt rest unconformably on the St. Croix Subzone of the Gander Zone (Ruitenberg, 1967). Rocks in the Oak Bay Subbelt provide the first link between the Avalon Zone and the St. Croix Subzone (Fyffe and Fricker, 1987). Although most subbelt boundaries are faulted, general lithological correlations have been established (Cumming, 1967a, b; Ruitenberg, 1968; Donohoe, 1978; McLeod, 1979; McCutcheon and Boucot, 1984; McLeod and Rast, 1988). Correlating structural histories, however, is rarely possible (Donohoe, 1978; Ruitenberg and McCutcheon, 1982; McLeod and Rast, 1988).

McLeod, M.J. and McCutcheon, S.R.
1995: Mascarene Belt; in Chapter 4 of Geology of the Appalachian-Caledonian Orogen in Canada and Greenland, (ed.) H. Williams; Geological Survey of Canada, Geology of Canada, no. 6, p. 354-360 (also Geological Society of America, The Geology of North America, v. F-1).

Stratigraphy and structure

Nerepis Subbelt

The Nerepis Subbelt makes up the northeastern part of the Mascarene Belt east of the Saint George Batholith and is bounded by a series of northeast-trending faults (McLeod et al., 1991a) and the Wheaton Brook Fault (McCutcheon and Ruitenberg, 1987) to the northwest and southeast, respectively (Fig. 4.23). Most of the following descriptions are based on the detailed work of McCutcheon and Ruitenberg (1987).

The oldest fossiliferous rocks, herein referred to as the Henderson Brook beds, are typically light to dark grey sandstone and siltstone with minor granule to pebble conglomerate that are dated precisely as Llandovery C5 (McCutcheon and Boucot, 1984). This unit is restricted in distribution to the northeastern part of the subbelt and is characterized by sandstone that commonly occurs in rusty-weathering, medium to thin beds and locally very thick beds. Similar lithologies and polymictic pebble to cobble conglomerate also occur in a small area north of Evandale (McLeod et al., 1991a).

The Long Reach Formation is Llandovery C6 to Wenlock (Boucot et al., 1966) and is, in part, equivalent in age to the nearby Henderson Brook beds. It is dominated by variable amounts of green to purple mafic flows, breccias, hyalotuffs and tuffs that are locally interbedded with limestone, felsic volcanic rocks, and thin bedded, dark grey sandstone and siltstone.

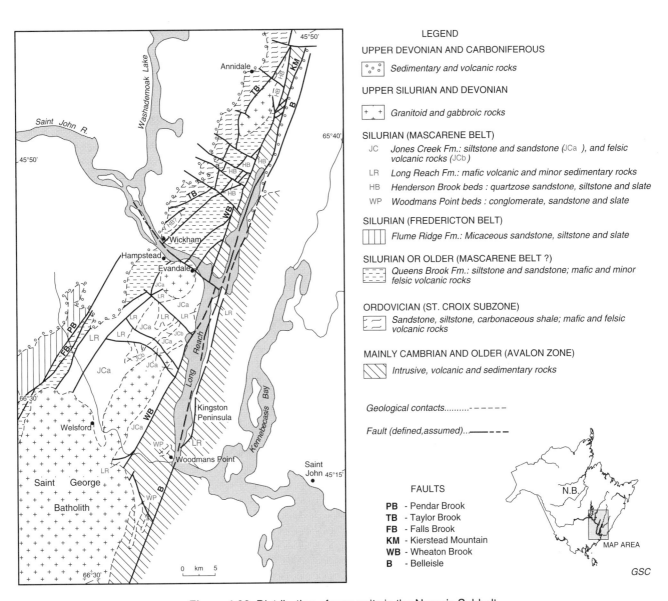

LEGEND

UPPER DEVONIAN AND CARBONIFEROUS

[° ° °] *Sedimentary and volcanic rocks*

UPPER SILURIAN AND DEVONIAN

[+ +] *Granitoid and gabbroic rocks*

SILURIAN (MASCARENE BELT)

JC   *Jones Creek Fm.: siltstone and sandstone (*JCa*), and felsic volcanic rocks (*JCb*)*

LR   *Long Reach Fm.: mafic volcanic and minor sedimentary rocks*

HB   *Henderson Brook beds : quartzose sandstone, siltstone and slate*

WP   *Woodmans Point beds : conglomerate, sandstone and slate*

SILURIAN (FREDERICTON BELT)

[|||] *Flume Ridge Fm.: Micaceous sandstone, siltstone and slate*

SILURIAN OR OLDER (MASCARENE BELT ?)

[----] *Queens Brook Fm.: siltstone and sandstone; mafic and minor felsic volcanic rocks*

ORDOVICIAN (ST. CROIX SUBZONE)

[≂] *Sandstone, siltstone, carbonaceous shale; mafic and felsic volcanic rocks*

MAINLY CAMBRIAN AND OLDER (AVALON ZONE)

[////] *Intrusive, volcanic and sedimentary rocks*

Geological contacts..........- - - - -

Fault (defined,assumed)... —— - - - -

FAULTS

PB - Pendar Brook
TB - Taylor Brook
FB - Falls Brook
KM - Kierstead Mountain
WB - Wheaton Brook
B - Belleisle

N.B.

MAP AREA

GSC

Figure 4.23. Distribution of map units in the Nerepis Subbelt.

355

The upper part of the succession in the Nerepis Subbelt is represented by the Jones Creek Formation that contains two units of light to dark grey, thin bedded sandstone and siltstone separated by red to grey flow banded and porphyritic rhyolite. The Jones Creek Formation contains abundant Pridoli fossils near its top (Berry and Boucot, 1970; McCutcheon and Ruitenberg, 1987). The conformable sequence constituted by the Long Reach (as old as Llandovery C6) and the Jones Creek formations (as young as Pridoli) indicates that parts of one or both formations must be of Wenlock and Ludlow age.

Outliers of grey polymictic conglomerate, overlain by dark grey siltstone and sandstone, possibly equivalent to the Long Reach Formation, occur immediately south of the

Wheaton Brook Fault. These rocks, which are referred to as the Woodmans Point beds, are inferred to lie unconformably on Avalon Zone rocks (McCutcheon, 1981). Farther to the southeast, the Long Reach Formation outcrops in a fault sliver along the Belleisle Fault.

The effects of deformation in the Nerepis Subbelt are highly variable. In general, its northwestern part has shallow plunging, close to tight folds with slaty cleavage. The cleavage is best developed in tuffaceous rocks of the Long Reach Formation and in finer grained rocks of the Jones Creek Formation. In most of the subbelt, however, folds are generally open and an associated coplanar slaty cleavage is developed intermittently in favourable lithologies.

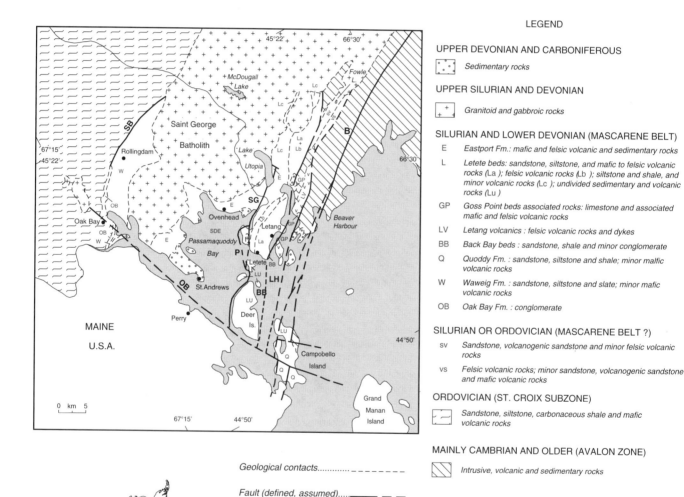

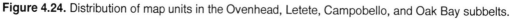

Figure 4.24. Distribution of map units in the Ovenhead, Letete, Campobello, and Oak Bay subbelts.

The Queens Brook Formation comprises two main units of undetermined age that separate unequivocal Mascarene Belt rocks from the St. Croix Subzone in the Nerepis area. The first unit contains light to dark green to grey and purple, fine grained sandstone, siltstone and minor amounts of volcanogenic sedimentary rocks and ortho-quartzite (Queens Brook Formation of McCutcheon and Ruitenberg, 1987). The second unit (McLeod et al., 1991a) consists of thin bedded, dark grey and locally, black or purple siltstone, sandstone and shale intercalated with thick beds of mafic tuffaceous rocks and minor felsic tuffaceous rocks.

The formation is typified by a well developed, steeply dipping and, locally, refolded cleavage that is associated with close to isoclinal and shallowly plunging folds. These folds are upright or slightly overturned to the south and transposed bedding is common along fold limbs.

The affiliation of the Queens Brook Formation is problematic. It has lithological similarities to both Cambrian rocks of the Avalon Zone to the south and Ordovician rocks of the St. Croix Subzone to the north; and it is structurally similar to the Ordovician rocks. Furthermore, it has been suggested (McCutcheon and Ruitenberg, 1987) that the Queens Brook Formation is gradational with the overlying Long Reach Formation, and that it may be, in part, Ordovician in age. Until reliable dates are obtained, its affiliation is equivocal.

Letete Subbelt

The Letete Subbelt makes up the central part of the Mascarene Belt southwest of the Saint George Batholith in the Passamaquoddy Bay area (Fig. 4.24). On the mainland, it is bounded by the St. George and Back Bay faults to the northwest and southeast, respectively. On Campobello Island, its southeastern boundary is defined by the Belleisle-Beaver Harbour fault system. Most of the following descriptions are based on the detailed mapping of Donohoe (1978) who worked predominantly on the mainland. He also gave informal names to some units and determined the structural history of the area.

The Back Bay beds, which are Llandovery C4-C5 (Boucot et al., 1966), constitute the base of the section in this area and are composed of light to dark grey, thick- to thin-bedded sandstone, shale and lesser amounts of calcareous siltstone, volcanogenic sedimentary rocks and polymictic pebble conglomerate. Contacts with younger units are either faulted or unknown.

The Goss Point beds and associated, unnamed volcanic units apparently overlie the Back Bay beds based on the presence of a Late Llandovery to Ludlow fauna (Donohoe, 1978). The Goss Point beds are characterized by limestone, calcareous siltstone and shale. These are locally intercalated with thin- to medium-bedded, and locally thick-bedded sequences of sandstone, siltstone and quartz pebble conglomerate. Unnamed units of mafic tuffaceous rocks and flows conformably underlie and overlie the Goss Point beds. The Goss Point beds and associated mafic volcanic rocks are in fault contact with most other units in the area. One such unit composed predominantly of tan volcanogenic sedimentary rocks is inferred to lie stratigraphically above or to be partially equivalent with the Goss Point beds based mainly on lithological correlations. Another unit containing felsic tuffaceous rocks and dykes, known as the Letang

rhyolite, is partly of Silurian age. Dykes, similar to those in the Letang rhyolite, intrude the Goss Point beds at a few localities.

The top of the Silurian sequence in the Letete Subbelt is represented by the Letete beds which are fault bounded and contain three main units. These are, from bottom to top, (1) grey, medium- to thick-bedded sandstone, grey to orange felsic tuff, and mafic tuff and agglomerate, (2) mainly orange to grey felsic tuff and agglomerate, and (3) grey, thin- to thick-bedded siltstone and sandstone, and minor volcanic rocks. Fossils in the lower unit suggest a Pridoli age (Donohoe, 1978).

Rocks, lithologically similar to the Letete beds extend southwest from the mainland to Deer Island (Cumming, 1967a, b; Ruitenberg, 1968) and to Campobello Island (McLeod, 1979). Sedimentary and volcanic rocks of probable Silurian age similar to those in the Letete Subbelt also extend to the northeast along the southern margin of the Saint George Batholith (Hay, 1968; Currie, 1988a, b; McLeod et al., 1990a, b). Poorly preserved fossils and complex structural relationships preclude establishment of detailed stratigraphy in this area. One occurrence of fossiliferous sedimentary rocks comparable to rocks in the Jones Creek Formation (Nerepis Subbelt) has also been identified in this area (McCutcheon et al., 1976; Currie, 1988b).

South of the main Nerepis Subbelt, a narrow, faulted block of questionable Silurian rocks occurs in the Belleisle-Beaver Harbour fault zone west of Beaver Harbour. These rocks have traditionally been considered to be mainly Late Devonian to Early Carboniferous (Cumming, 1967a, b; Helmstaedt, 1968; Greenough et al., 1985), but recent work (Currie, 1988b) suggests a Silurian age.

Structurally, the Letete Subbelt on the mainland is characterized by polyphase deformation. First phase folds, that are accompanied by a well developed cleavage, are isoclinal to tight, upright, and plunge steeply northeast. Second phase folds, that are locally accompanied by a strong crenulation cleavage, are tight to close, upright and plunge moderately to steeply northeast or southwest. Cleavages related to first and second phase folds are deformed by two generations of kink bands. The structures in equivalent rocks on Campobello Island are also characterized by polyphase deformation but first phase folds plunge shallowly towards the northeast and second phase folds or crenulation cleavages associated with the second deformation are rarely developed. Two generations of kink bands, which deform the well-developed cleavage associated with first phase folds, are similar to those on the mainland (McLeod, 1979).

Ovenhead Subbelt

The Ovenhead Subbelt is immediately south and west of the Saint George Batholith (Fig. 4.24). It is intruded to the north by the batholith and is bounded to the southeast by the St. George Fault or by intrusive rhyolite (Donohoe, 1978). Most lithological descriptions and the stratigraphy of the northern part of this subbelt are based on the detailed work of van Wagoner et al. (1988). That part of the subbelt to the north of the Perry Fault is divided into eastern or lower and western or upper sequences within which there are three (or four; van Wagoner et al., 1989) cycles of bimodal volcanism (mafic followed by felsic activity) with associated sedimentary rocks. Descriptions of the rocks and structures in the southern part of the subbelt, south of the Perry Fault, are taken from Donohoe (1978).

Basaltic flows and rhyolitic flows and domes are present throughout the succession north of the Perry Fault. Basalts occur in thick to very thick sheets of massive, variably vesiculated flows with pahoehoe toes and as breccias. Thick bedded, fine- to coarse-grained units of mafic scoria also occur in the upper part of the succession. Pink to grey rhyolite flows and domes are massive, porphyritic or flow banded and occur in very thick units with variable lateral extents.

Pyroclastic rocks in the lower part of the succession north of the Perry Fault are felsic, mostly green to grey and vary from massive to bedded. Massive types are weakly to strongly welded and vary from lapilli tuff to tuff. Bedded types are air fall deposits composed of thin- to thick-bedded tuff, lapilli tuff and accretionary lapilli tuff. Pyroclastic rocks in the upper part of the succession are felsic, vary from brown and red to green and grey, and occur as a variety of types. These types include accretionary lapilli tuff, crystal and lithic tuff, well-bedded tuffs and welded tuffs. Features of the various types indicate deposition by air fall and pyroclastic flow mechanisms.

Volcaniclastic rocks are common in the lower part of the succession but rare in its upper part north of the Perry Fault. In the lower part, these rocks are either predominantly composed of mafic or felsic debris, occur in poorly bedded, thin to thick beds and are commonly intercalated with green and grey siltstone. In the upper part, volcaniclastics are composed of mafic debris that occurs in thin to medium beds intercalated with red siltstone.

Clastic sedimentary rocks were deposited throughout the succession and are commonly intercalated with all the above volcanic rock types. Sedimentary rock types include thin- to thick-bedded red sandstone, mudstone and conglomerate, grey to green mudstone, and minor buff to green sandstone. The redbeds, particularly the coarse clastics, mostly occur in the upper part of the succession.

South of the Perry Fault, the Ovenhead Subbelt is composed of three members. The lower member contains thin- to thick-bedded red tuffaceous to feldspathic and quartzose sandstone, and orange felsic, crystal to accretionary lapilli tuff. Minor amounts of dark green agglomerate also occurs in this member. The middle member contains thin- to thick-bedded, green to grey, micaceous sandstone and siltstone, and minor conglomerate. The upper member contains red siltstone interbedded with thin to thick beds of green sandstone, red conglomerate, buff felsic tuff and tuffaceous sandstone.

Based on lithological and paleontological similarities, rocks of the Ovenhead Subbelt are correlated with the Lower Devonian Eastport Formation in Maine (Pickerill and Pajari, 1976; Pickerill et al., 1978). A recent U-Pb zircon age (430 ± 3 Ma; Bevier, 1988) from the Saint George Batholith indicates that part of the succession may be considerably older as suggested by other workers (e.g. Ruitenberg, 1968). Work is in progress to resolve this problem.

Structures in the Ovenhead Subbelt differ across the Perry Fault. Most rocks north of the fault are gently folded with a single, locally developed, northeast-trending cleavage that appears in part to be non coplanar to the folds. No later folds have been found. South of the fault, the major structure is defined by open to close, inclined, shallow northeast- and southwest-plunging, second generation folds that deform a pre-existing moderate- to well-developed cleavage. The second folds have no cleavage. Kink bands are also developed south of the fault in favourable lithologies.

Campobello Subbelt

The Campobello Subbelt is the southernmost of the Mascarene Belt southwest of the Saint George Batholith (Fig. 4.24). It is bounded to the northwest by a complex fault zone that includes the Belleisle Fault in New Brunswick and the Lubec Fault in Maine (Gates, 1977; McLeod and Rast, 1988). It is bounded by faults and an inferred unconformity to the northeast, and continues into Maine to the southwest. The following descriptions are mainly taken from McLeod and Rast (1988).

Rocks of the Campobello Subbelt are assigned to the Quoddy Formation. It represents the oldest fossiliferous Silurian strata in the Mascarene Belt. Fauna collected in Maine indicate a Llandovery C3 or younger age (Boucot et al., 1966). In New Brunswick, this formation consists of predominantly black to grey, siliceous and commonly pyritiferous sandstone, siltstone and shale that occur in medium- to thick-bedded turbidite sequences. Mafic volcanic rocks, that include hyaloclastite, hyalotuff, flows, agglomerate, pillow lavas and tuffaceous rocks, are intercalated with the turbidites in the lower part of the succession. Variable, but usually minor, amounts of mafic volcanic debris are commonly present in the turbiditic sequences throughout the succession.

Structurally, the Campobello Subbelt is characterized by a single phase of deformation that produced broad, open folds. The folds plunge shallowly towards the southwest and are accompanied by a weak- to well-developed, axial plane cleavage that diverges towards fold limbs.

Oak Bay Subbelt

The Oak Bay Subbelt is the only division of the Mascarene Belt north of the Saint George Batholith and contains the Oak Bay and Waweig formations (Ruitenberg, 1967; Fig. 4.24). It is bounded to the south by the Saint George Batholith and to the north by faults and/or an unconformity against Ordovician rocks of the St. Croix Subzone. The following lithological and structural descriptions are taken mainly from Ruitenberg (1967).

At most localities, the base of the sequence is defined by the base of the Oak Bay Formation that consists of a thick-bedded succession of grey polymictic, pebble- to cobble-conglomerate intercalated with coarse grained sandstone. A limestone pebble from the conglomerate has Early Silurian fossils (Cumming, 1967a, b). The lower contact of the Oak Bay Formation has been variously interpreted as a fault or unconformity (see discussion by Ruitenberg, 1967) and the conglomerate itself likely represents a product of debris flows in a tectonically active environment (Gates, 1989).

The Waweig Formation contains abundant Late Silurian fossils (Pickerill, 1976) and conformably overlies the Oak Bay Formation. It consists of thin- to thick-bedded, dark green to grey and slightly calcareous sandstone, siltstone, slate and minor tuffaceous rocks.

The Waweig Formation is typified by a single, steeply dipping and northeast-trending penetrative cleavage that is locally deformed by open, steeply northeast- or southwest-plunging folds. The Oak Bay Formation, however, contains no penetrative fabrics but is affected by the later, open folding.

Geochemistry

The most comprehensive geochemical studies in the Mascarene Belt have been conducted in the Ovenhead Subbelt (van Wagoner et al., 1989) and in correlative rocks in Maine (Gates and Moench, 1981). In New Brunswick, the Lower Devonian volcanic rocks are distinctly bimodal and range in composition from rhyolite to basalt. The rhyolitic rocks are calc-alkaline but abundances of Y, Nb and Zr suggest alkalic to peralkalic affinities. The basaltic rocks are tholeiitic with some variations in trace element abundances progressing both vertically, and in some cases laterally in the volcanic sequence (van Wagoner et al., 1989). In Maine, the entire Silurian-Devonian sequence of volcanic rocks is strongly bimodal and ranges in composition from basalts and basaltic andesites to silicic dacites and rhyolites. The Silurian mafic volcanic rocks are high-alumina tholeiites whereas the Devonian mafic volcanic rocks are normal tholeiites but are richer in Fe and Ti, and poorer in Mg and Ni than their Silurian counterparts. The felsic volcanic rocks are calc-alkaline throughout with elevated SiO_2 and K_2O contents and higher FeO/MgO ratios (Gates and Moench, 1981). These studies indicate that the Silurian-Devonian volcanism occurred in a within-plate tectonic environment, possibly in an extensional setting.

Studies in other parts of the Mascarene Belt support these conclusions. In the Campobello Subbelt, analyses of Silurian diabase dykes that feed volcanic rocks near the base of the Silurian sequence are also tholeiitic (McLeod, 1979). In the Nerepis Subbelt, preliminary studies indicate that the Silurian volcanic rocks are subalkaline rhyolites and tholeiitic basalts which are indistinguishable from those in the Letete Subbelt (McCutcheon and Ruitenberg, 1987). Recent work has shown that some rhyolitic rocks in the Nerepis Subbelt have peralkaline affinities (Payette and Martin, 1988). These authors suggest that the rhyolites formed in a tensional rifting environment.

Correlations

General lithological correlations can be established within the Mascarene Belt with the aid of fossil control but direct correlations of units are difficult (Ruitenberg, 1968; Donohoe, 1978; McCutcheon and Ruitenberg, 1987). Rocks of the Ovenhead Subbelt have been correlated directly with the Lower Devonian Eastport Formation in Maine (Pickerill and Pajari, 1976) but parts may be considerably older as discussed above. Regardless of age, they form part of a distinct litho-tectonic package (referred to as the Ovenhead Subzone by McLeod and Rast, 1988) that also incorporates the Silurian-Devonian sequences in southeastern Maine north of the Campobello Subbelt. The Quoddy Formation

of the Campobello Subbelt also extends to the southwest into Maine (Cumming, 1967a, b; McLeod and Rast, 1988). Although similar faunas are present throughout the Mascarene Belt (Boucot et al., 1966), lithological units in the Ovenhead and Campobello subbelts have no apparent, direct, correlatives in the Letete, Nerepis and Oak Bay subbelts and structural styles are different (Donohoe, 1978; McLeod and Rast, 1988).

Chronological and, possibly, direct lithological correlations can be established between parts of the Letete and Nerepis subbelts (McCutcheon and Boucot, 1984). The Back Bay beds (Letete Subbelt) are similar to the Henderson Brook beds in the Nerepis Subbelt. The Goss Point beds, which are underlain and overlain by mafic volcanic rocks (Letete Subbelt), may be equivalent to all or part of the Long Reach Formation (Nerepis Subbelt). Equivalents of younger strata (the Jones Creek Formation) in the Nerepis Subbelt have not been recognized in the Letete Subbelt.

Correlation of units in the Mascarene Belt with possible Silurian rocks to the north, (e.g. the Digdeguash and Flume Ridge formations of the Fredericton Belt) have been proposed by several authors (e.g. Ruitenberg, 1968; Donohoe, 1978). These attempts may be unwarranted because it has since been established that the northern units were mostly deposited in a different setting (Fyffe and Fricker, 1987; Fyffe, 1989; McLeod et al., 1989). Faunal evidence from the Waweig Formation (Pickerill, 1976) indicates that the St. Croix Subzone and Avalon Zone were linked in the Late Silurian (uppermost Ludlow-Pridoli; Fyffe, 1989). In this respect, the Waweig Formation likely represents part of the Mascarene Belt that overstepped the St. Croix Subzone in Late Silurian time as suggested by Fyffe (1989).

Synopsis

The present distribution of Silurian-Devonian subbelts in southern New Brunswick is controlled by a series of northeast- and northwest-trending faults. The complexity of faulting, which involves activation and reactivation of faults in both directions, has been demonstrated by McLeod and Rast (1988) in the Passamaquoddy Bay area. Early movements on the major northeast-trending faults produced mylonites that exhibit evidence of dextral (Leger and Williams, 1986) and, locally, sinistral (Currie, 1988a, b) motion. Since mylonites of undisputed Middle Paleozoic age have not been recognized in southern New Brunswick, it is assumed that this style of deformation occurred during pre-Silurian time (Currie, 1988a, b). Brittle faults that were active during Silurian-Devonian time provided avenues of ascent for the magmas which fed the voluminous volcanic accumulations (Gates, 1984; Currie, 1988a, b) and plutonic rocks (McLeod, 1990) of the Mascarene Belt.

Lithological differences among subbelts are easily explained as different facies within a large basin(s) as suggested by Gates (1984) and McLeod and Rast (1988), and as modelled by Currie (1988a, b). Features common to all subbelts are similarity of fauna (Boucot et al., 1966) and depositional environments. The Silurian-Devonian sequences reflect filling of the basin(s) in which fairly deep-water desposition, in the Early Silurian, gave way to shallow water and eventually subaerial deposition in the Early Devonian (Gates, 1977; Pickerill et al., 1978; McCutcheon and Ruitenberg, 1987). Although evidence is meager, diversity in source areas is indicated by the volcanic rocks in the Ovenhead Subbelt that were derived from the north and northeast (van Wagoner et al., 1988) and by the turbidites in the Campobello Subbelt that were probably derived from the southwest (McLeod, 1979). The source area for the Oak Bay Formation must have lain to the south of the Oak Bay Belt (Fyffe and Fricker, 1987; Gates, 1989).

Following deposition of the Silurian-Devonian successions, the Mascarene Belt was subjected to compressional events that culminated in activation or reactivation of several northeast-trending faults and juxtapositioning of the Avalon Zone and St. Croix Subzone more or less in their present position. Shortening was primarily accomplished by northwest directed and mainly high angle reverse faulting, thrusting and folding as proposed by numerous authors (e.g. McCutcheon, 1981; Ruitenberg and McCutcheon, 1982; McLeod and Rast, 1988). Although penetrative deformation may be attributed to fault movements locally (Gates, 1977) or possibly regionally (Currie, 1988a, b), fundamental structural differences among subbelts are likely a function of their pre-fault positions. If deposition occurred in a large basin or separate interconnected basins, such factors as basin geometry, nearby major fault zones or proximity to basement could control the structural style of a given subbelt.

Minor spacial readjustments to the basic configuration of the Mascarene Belt produced by the deformational events described above occurred during Devonian-Carboniferous time. These readjustments resulted from activation of northwest-trending faults that exhibit both sinistral and dextral movements, and reactivation of the northeast-trending faults that exhibit complex and varied dip-slip, strike-slip and thrust movements.

Arisaig Belt

J. Brendan Murphy and J. Duncan Keppie

Upper Ordovician to Lower Devonian rocks in northern mainland Nova Scotia are assigned to the Arisaig Belt developed upon the Avalon Zone. In the Antigonish Highlands (Fig. 4.25), they are predominantly low grade continental, bimodal, within-plate, rift volcanic rocks (Keppie et al., 1978; Murphy, 1987a, b) overlain by shallow marine fossiliferous siliciclastic rocks (Boucot et al., 1974; Pickerill and Hurst, 1983). In the Cobequid Highlands (Fig. 4.26), volcanic rocks are scarce and the sequence is dominated by sedimentary rocks similar to those of the Antigonish Highlands.

The relationship between the Arisaig Belt and Laurentia is controversial. Chandler et al. (1987) suggested that the ca. 430 Ma (Llandovery) volcanism constituted an overstep sequence across the Canadian Appalachian Orogen however these Silurian rocks are generally preserved only in discrete outliers. Paleomagnetic data indicate that wide paleolatitude separation between the Avalon Zone and

Murphy, J.B. and Keppie, J.D.
1995:   Arisaig Belt; in Chapter 4 of Geology of the Appalachian-Caledonian Orogen in Canada and Greenland, (ed.) H. Williams; Geological Survey of Canada, Geology of Canada, no. 6, p. 360-365 (also Geological Society of America, The Geology of North America, v. F-1).

Laurentia diminished during the Ordovician and disappeared by Late Ordovician or Silurian (e.g. Briden et al., 1988). On the other hand, Silurian-Gedinnian, Rhenish-Bohemian fauna are restricted to the Avalon and Meguma zones and only spread westwards onto the Laurentian margin during the Siegenian and Emsian (Boucot, 1975; Keppie et al., 1991a).

Stratigraphy

Upper Ordovician-Lower Devonian rocks occur in the Antigonish Highlands (Arisaig Group; Murphy, 1987a, b), and in the Cobequid Highlands (Wilson Brook and the overlying Murphy Brook formations; Donohoe and Wallace, 1982, 1985).

In the Antigonish Highlands (Fig. 4.25), the Arisaig Group unconformably overlies the Upper Precambrian Georgeville Group and the Cambrian-Lower Ordovician McDonalds Brook and Iron Brook groups. The base of the Arisaig Group is not exposed in the type area along the coast. In general, the Arisaig Group consists of bimodal volcanic rocks and interlayered redbeds overlain with local unconformity by about 2000 m of siliciclastic rocks and minor carbonates. The sedimentary rocks contain Rhenish-Bohemian fauna (Boucot, 1975). The Arisaig Group is unconformably overlain by Middle to Upper Devonian redbeds (Keppie et al., 1978).

In the Cobequid Highlands (Fig. 4.26), Silurian to Lower Devonian rocks consist of grey-green fossiliferous siliciclastics and minor carbonates (Wilson Brook Formation) conformably overlain by red siltstones and sandstones (Portapique River Formation). These formations unconformably overlie the Upper Precambrian Jeffers Group and are faulted against Middle to Upper Devonian redbeds of the Murphy Brook Formation (Donohoe and Wallace, 1982, 1985).

The oldest rocks of the Arisaig Group are the Dunn Point Formation (in the type section at Arisaig) and Bears Brook Formation (in the Antigonish Highlands). At Arisaig, the Dunn Point Formation consists of at least 220 m of amygdaloidal, generally aphyric, mafic flows and flow-banded ignimbritic rhyolites. The tops of the mafic flows are highly amygdaloidal and show evidence of deep lateritic weathering and the cutting of deep erosional channels. These features indicate subaerial deposition of the volcanic rocks. The Dunn Point Formation is disconformably overlain by the Beechill Cove Formation (60-75 m thick). Pickerill and Hurst (1983) interpreted the lower Beechill Cove conglomerates and red shales as transgressive beach lag and shoreface deposits, respectively. These are succeeded by mudstones and siltstones which are thought to represent deposition in subtidal and shelf environments and alternating periods of low-energy and storm activity. The Beechill Cove Formation is conformably overlain by the Llandovery Ross Brook Formation (280-300 m thick) which consists of silty mudstones, shales and siltstones deposited on a shallow marine storm-influenced, subtidal, inner-middle muddy shelf (Hurst and Pickerill, 1986). Upper Silurian rocks (French River and Moydart formations) consist of 360-400 m of predominantly shallow marine, fossiliferous siltstones and mudstones. The French River Formation also contains a fossiliferous oolitic ironstone (Boucot et al., 1974). The Moydart Formation contains an upper red member representing a temporary return to subaerial conditions. Lower Devonian rocks consist of 600 m of blue-grey fossiliferous shallow marine siltstones (Stonehouse Formation) overlain by at least 500 m of red clastic sedimentary rocks that contain calcareous nodules (Knoydart Formation), thought to have been deposited in a coastal floodplain.

The succession in the Antigonish Highlands is similar to that exposed on the Arisaig coast except for the basal units. The Bears Brook Formation varies from 200-400 m in thickness and consists of bimodal volcanic rocks interlayered with red, fluviatile, arkoses and conglomerates. The Bears Brook Formation is conformably overlain by 80 m of fluviatile, red arkoses and conglomerates followed by 20 m of marine, blue-green micaceous, fossiliferous siltstones of the Beechill Cove Formation.

With the exception of the volcanic rocks at the base of the succession, the age of the Arisaig Group is based on paleontological evidence (Maehl, 1961; Boucot et al., 1974; Benson, 1974; Smith, 1979). Thus, the Beechill Cove and

Ross Brook formations are assigned to the Early Silurian, the French River, Doctors Brook, McAdam and Moydart formations to the Late Silurian, and the Stonehouse and Knoydart formations to the Early Devonian. The age of the volcanic rocks at the base of the succession is less precisely known. Rb-Sr studies on felsic volcanic rocks in the Arisaig area and Antigonish Highlands indicate ages of 434 ± 15 Ma (Fullager and Bottino 1968, recalculated in Keppie et al., 1978) and 425 ± 37 Ma (R.F. Cormier, pers. comm., 1986). These ages and their stratigraphic position indicate that the volcanic rocks are latest Ordovician or earliest Silurian, and that they are broadly correlative.

The stratigraphic succession at Arisaig and in the Antigonish Highlands indicates that a marine transgression occurred at the base of the Silurian. This has been associated with a worldwide eustatic marine transgression (Cant, 1980). The synchroneity of intra-continental rifting and the eustatic rise in sea level may be responsible for the almost complete sequence of upper Ordovician to lower

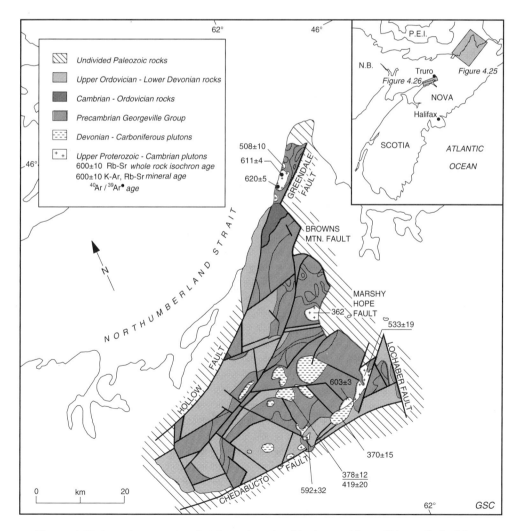

Figure 4.25. Location, structural style and grade of metamorphism of Upper Ordovician-Lower Devonian rocks in mainland Nova Scotia. References to the age data may be found in Keppie et al. (1990). Antigonish Highlands (adapted from Murphy et al., in press)

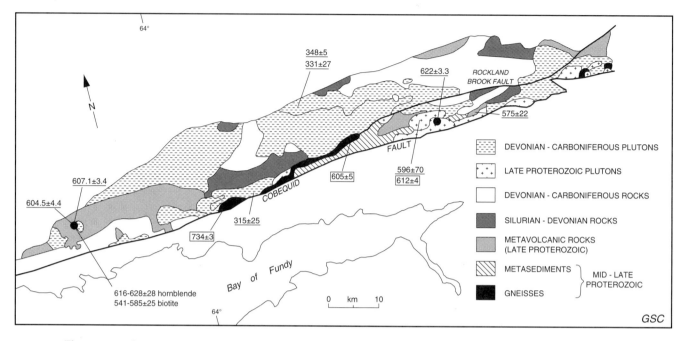

Figure 4.26. Geology of Cobequid Highlands, (adapted from Donohoe and Wallace, 1982, Murphy et al., 1988), Age data may be found in Keppie et al., (1990).

Devonian rocks. The lack of large-scale coarsening- or fining-upward sequences suggests that sedimentation kept pace with changes in sea level in the early Silurian. The increasing abundance of redbeds at the top of the Arisaig Group suggests a return to more subaerial deposition.

The Wilson Brook Formation and the overlying Portapique River Formation (Donohoe and Wallace, 1985) are broadly correlated with the Arisaig Group although structural complications and lack of continuous exposures in the Cobequid Highlands make detailed correlations difficult. The age of the Wilson Brook Formation is indicated by fossils which range from Late Llandovery to Pridoli (Donohoe and Wallace, 1982). The Portapique River Formation is not fossiliferous, but it is apparently conformable above the Wilson Brook Formation and is therefore probably Early Devonian. If so, it may correlate with the lithologically similar Knoydart Formation of the Antigonish Highlands. The unconformably overlying Murphy Brook Formation is also dominated by redbeds and contains Emsian to Eifelian fossils.

Geochemistry

Volcanic rocks predominantly occur towards the base of the Arisaig Group. The geochemistry of the Dunn Point Formation (Keppie et al., 1978) and the stratigraphically equivalent Bears Brook Formation and overlying Beechill Cove Formation (Murphy, 1987a, b) is very similar. The rocks are bimodal with a SiO_2 gap between approximately 58-66 per cent (Fig. 4.27a). Mafic rocks are within plate (Fig. 4.27b), tholeiitic (Fig. 4.27c) although they display some alkalic tendencies on discrimination plots such as Zr/TiO_2 vs. Nb/Y (Fig. 4.27d). The concentration of FeO_{total} in some samples reaches more than 15 per cent, TiO_2 more than 3.5 per cent, and P_2O_5 more than 1.0%. These major

element variations, together with the high FeO_t/MgO are typical of an Fe-enriched continental tholeiite which underwent fractionation of the gabbroic assemblage: clinopyroxene+plagioclase+olivine (Cox, 1980). An intra-continental tectonic environment is consistent with the subaerial to shallow marine depositional environment for the group that was deposited on continental crust.

The felsic volcanic rocks of the Bears Brook Formation and the lower felsic member of the Dunn Point Formation are rhyodacitic in composition (Fig. 4.27a). On the other hand, the felsic volcanic rocks of the Beechill Cove Formation and the upper felsic member of the Dunn Point Formation are high-silica rhyolites. In addition to higher SiO_2, they contain, on average, higher K_2O, Nb, and La/Yb; and lower Na_2O, Al_2O_3, CaO, P_2O_5, TiO_2, and Sr. The geochemistry of the rhyodacites is attributed to crustal anatexis. The genetic relationship of these rocks to the rhyolites is uncertain. Trace element geochemistry is consistent with a volcanic arc setting (Fig. 4.27e). However, this signature may be inherited from an earlier (late Precambrian) subduction event (Murphy, 1988).

Deformation and metamorphism

The Arisaig Group in the type area and in the northern Antigonish Highlands is deformed into shallow southeast-plunging upright folds. The Middle Devonian McAras Brook Formation postdates this structure suggesting a mid-Devonian age of deformation. This deformation occurred at a shallow crustal level as indicated by the associated sub-greenschist metamorphism. The counterclockwise orientation of the folds relative to the Hollow Fault suggests that they were produced by dextral motions on the Hollow Fault (Keppie, 1982; Murphy et al., 1989).

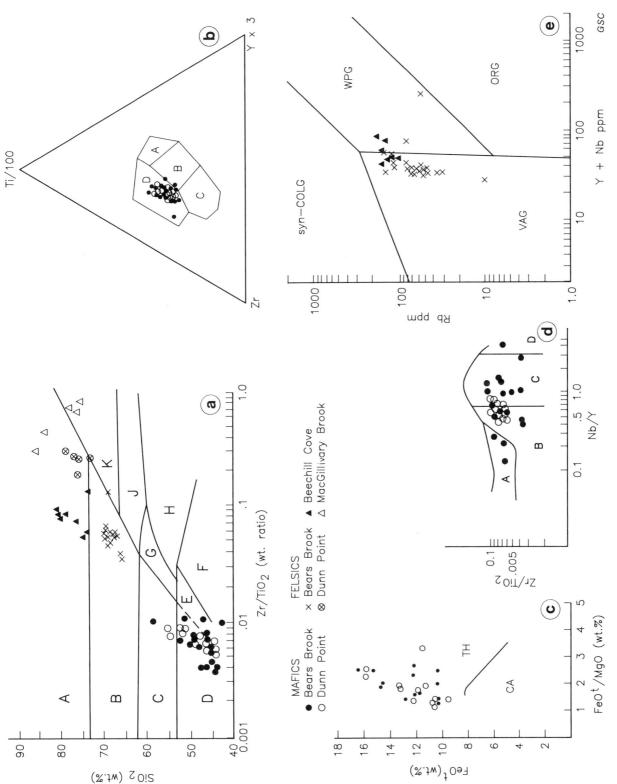

Figure 4.27. Tectonic setting and magnetic affinity of Lower Silurian volcanic rocks of the Antigonish Highlands. **(a)** after Winchester and Floyd (1977): A, rhyolite; B, rhyodacite; C, andesite; D, subalkalic basalt; E, alkalic basalt; F, basanite;G, trachyandesite; H, phonolite; J, trachyte; K, pantellerite. **(b)** after Pearce and Cann (1973): A, Low K tholeiite; B, Low K tholeiite, ocean floor basalt or calc-alkaline basalt; C, calc-alkaline basalt; D, within-plate basalt. **(c)** after Miyashiro (1974). CA, calc-alkaline; TH, theolite. **(d)** after Winchester and Floyd (1977): A, andesite, basalt; C, alkalic basalt; D, nephelinite, basalt. **(e)** after Pearce et al. (1984). VAG, volcanic arc granite; ORG, ocean ridge granite; WPG, within plate granite; syn-COLG, syn-collisional granite.

363

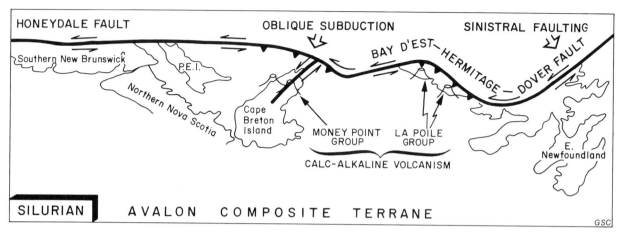

Figure 4.28. Tectonic interpretation of the Upper Ordovician – lower Devonian rocks associated with sinistral accretion of the Avalon Composite terrane with Laurentia during the Silurian-Early Devonian (after Keppie et al., 1991c).

In the southern Antigonish Highlands deformation is heterogenous and is restricted to shear zones, with local evidence of thrusting and development of positive flower structures (Murphy et al., 1989). At least some of this deformation affects Upper Devonian rocks and is therefore younger than that which affected Upper Ordovician-Lower Devonian rocks in the northern Antigonish Highlands.

In the northern Cobequid Highlands (i.e. north of the Rockland Brook Fault, Fig. 4.26), folding of Silurian-lower Devonian sedimentary rocks also pre-dates deposition of Upper Devonian redbeds and volcanic rocks and is probably also of mid-Devonian age. However, this deformation is overprinted by deformation adjacent to, and south of the Rockland Brook and Kirkhill faults, which penetratively deform Lower Carboniferous plutons (Miller et al., 1989; Pe-Piper and Turner, 1988). Isoclinal, en echelon folds with dextral sense and associated thrusts deform Upper Viséan to Lower Namurian rocks adjacent to the Cobequid Fault (Donohoe and Wallace, 1982; Murphy et al., 1988). In contrast, Upper Carboniferous rocks adjacent to the fault are only mildly deformed. Thus, localized penetrative deformation associated with dextral fault motions continued in the Cobequid Highlands until the Late Carboniferous.

Tectonic interpretation

In mainland Nova Scotia, the geochemistry of the volcanic rocks indicates deposition of the upper Ordovician-lower Devonian rocks in an intra-continental rift-related environment. Further constraints on the tectonic environment may be derived from a comparison with similar aged rocks in other parts of the Avalon Zone.

In Newfoundland, deposition of Upper Ordovician to Lower Devonian rocks is synchonous with the accretion of the Avalon Zone with the Gander Zone (Dunning et al., 1990a). The Silurian-Early Devonian deformation in the Avalon Zone may be related to sinistral accretion approximately parallel to the northeast trend of the orogen (Fig. 4.28, after Keppie et al., 1991c). Upper Ordovician to

Lower Devonian rocks in mainland Nova Scotia and Cape Breton Island have been correlated with the La Poile Group in the La Poile Belt of southern Newfoundland (Barr and Jamieson, 1991; Keppie et al., 1991c) which lay at a promontory in the western boundary of the Newfoundland Avalon Zone (Keppie and Dallmeyer, 1989a, b). Structures in the Burgeo Subzone indicate that it was obducted over the Gander or Dunnage zones (Fig. 4.28). This is consistent with the paleomagnetic data which suggests that a sinistral megashear existed between the Gander and Avalon zones in the Silurian and Early Devonian (Briden et al., 1988; Kent and van der Voo, 1990). These paleomagnetic data suggest that at least 1500 km of relative sinistral displacement took place during Silurian times. Kent and van der Voo (1990) have suggested that this coincided with collision between Laurentia and Gondwana (South America).

The Avalonian status of Upper Ordovician to Lower Devonian rocks in Cape Breton Island is controversial (c.f. Barr and Jamieson, this chapter; Keppie et al., 1991c), but are considered herein to comprise part of the Avalon Zone for reasons outlined by Keppie (1990). The general term Western Highlands volcano-sedimentary complex (Barr et al., 1985b; Jamieson et al., 1986) has been used to describe all these sequences. The complex is characterized by low to high-grade metamorphic rocks intruded by Silurian to Devonian granitoid rocks (Barr et al., 1985c). Although the detailed stratigraphy is unknown due to structural complexities, in general it is similar to that of the Arisaig Group. The geochemistry of the volcanic rocks suggests that they are subduction-related (MacDonald and Smith, 1980; Connors, 1976; Jamieson et al., 1990; Barr and Jamieson, 1991). Although Barr and Jamieson (1991) inferred that these Silurian arc rocks were extruded on North America with subduction beneath the Laurentian continent, Keppie et al. (1991c) argued that they formed in the Avalon Zone above a local subduction zone located around the Cabot Promontory (Williams, 1979) (Fig. 4.28). The contemporaneous intra-continental rift rocks in the Antigonish Highlands may be related to splay rifts from the sinistral shear zone.

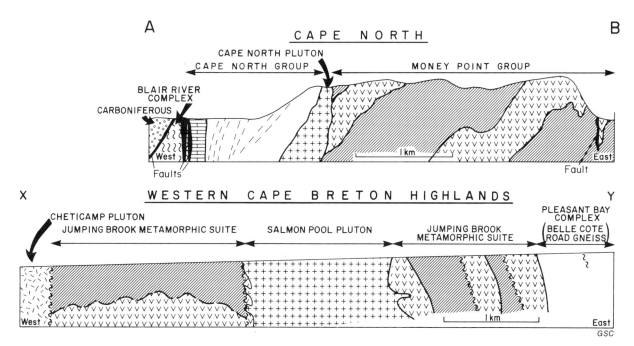

Figure 4.29. Schematic (west-east) structural cross-section (AB) of Cape Breton Highlands showing a positive flower structure (after Keppie et al., 1991c). The cross-section (XY) of the western Cape Breton Highlands is modified after Craw (1984).

The late Ordovician deformation associated with steeply dipping shear zones in the northern Antigonish Highlands may be early manifestations of this accretionary event (Keppie and Murphy, 1988). Structural cross-sections across the Cape Breton Highlands (Fig. 4.29, after Keppie et al., 1991c) reveal a north-trending vertically-dipping high metamorphic grade, central zone bounded by outward-verging asymmetric to recumbent, low metamorphic grade, marginal zones and is interpreted as a positive flower structure associated with this accretionary event (Keppie et al., 1991c). Similar P-T-t paths were calculated for the Cheticamp (Plint and Jamieson, 1989) and Cape North areas (Keppie et al., 1991c) which suggest rapid subsidence of Upper Ordovician to Lower Silurian rocks followed by rapid uplift and cooling through muscovite blocking temperatures (ca. 350°C) by 380 Ma (Reynolds et al., 1989; Keppie et al., 1991c). Although tectonic interpretations on the significance of this path differ (see Plint and Jamieson, 1989), Keppie et al. (1991c) pointed out that the path is consistent with a transpressional accretionary regime in which homogeneous thickening was followed by rapid uplift, erosion and basin inversion. The Middle Devonian dextral deformation of the Arisaig Group may be related to the dextral motion of the Meguma terrane and Gondwana relative to the Avalon Zone along the Minas Geofracture which forms the southern boundary of the Antigonish Highlands.

Acknowledgments

Supported by Natural Sciences and Engineering Research Council of Canada, and Canada-Nova Scotia Mineral Development Agreements. We wish to acknowledge the contributions of our colleagues, especially Sandra Barr, Ken Currie, Howard Donohoe, Becky Jamieson, Rob Raeside, Ron Pickerill, and Peter Wallace to the understanding of Upper Ordovician to Lower Devonian rocks of Nova Scotia. This contribution is published with the permission of the minister of the Nova Scotia Department of Mines and Energy.

Cape Breton Belt

S.M. Barr and R.A. Jamieson

Most Silurian and Devonian rocks in Cape Breton Island are within the Aspy Subzone and were described with the Gander Zone-Cape Breton Island. However, some other Silurian and Devonian units are present (Cape Breton Belt), and are described in this section. They include a sedimentary unit near Mabou Highlands, the Ingonish Island Rhyolite, the Fisset Brook Formation, and the McAdam Lake Formation (Fig. 4.30).

Barr, S.M. and Jamieson, R.A.
1995: Cape Breton Belt; *in* Chapter 4 of Geology of the Appalachian-Caledonian Orogen in Canada and Greenland, (ed.) H. Williams; Geological Survey of Canada, Geology of Canada, no. 6, p. 365-366 (also Geological Society of America, The Geology of North America, v.F-1).

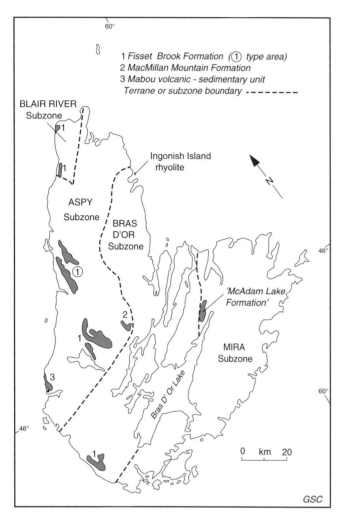

Figure 4.30. Distribution of unmetamorphosed sedimentary and volcanic rocks of Middle Paleozoic age in Cape Breton Island.

Mabou Sedimentary unit

Several small isolated outcrops of quartz arenite, siltstone, and shale occur along the western edge of the Mabou Highlands. They have been tentatively assigned to the Silurian on the basis of a local occurrence of rhynchonellid brachiopods and other fauna (Phinney, 1956; Barr and Macdonald, 1989). They are unlike other Silurian units in the Aspy Subzone because they are fossiliferous and unmetamorphosed. Correlation with units in the Antigonish Highlands has been suggested (Phinney, 1956; Keppie et al., 1989).

Ingonish Island rhyolite

Banded rhyolites unconformably overlain by limestone of the Carboniferous Windsor Group are exposed on Ingonish Island (Fig. 4.30). The rhyolites are considered Silurian-Devonian, based on a Rb-Sr isochron age of 412 ± 15 Ma

(Keppie and Halliday, 1986), but stratigraphic relationships are unknown. The volcanic rocks include dark brown to black flows with sparse phenocrysts of alkali feldspar and quartz interbedded with tuffaceous units containing sparse crystal and lithic fragments in a flow-banded matrix (Raeside and Barr, 1992).

Fisset Brook Formation

The Devonian-Carboniferous Fisset Brook Formation consists of unmetamorphosed volcanic and sedimentary rocks that occur in restricted areas on the periphery of the highland areas (Fig. 4.30). The formation consists of a bimodal basalt-rhyolite sequence associated with redbeds. It is overlain directly by conglomerates and sandstones of the Lower Mississippian Horton Group. It was defined by Mackasey (1963) and Kelley and Mackasey (1965), and individual areas have been described by Smith and Macdonald (1981), Blanchard et al. (1984), French (1985), and Raeside and Barr (1992). In some areas, probable correlatives have been assigned local names: e.g. MacMillan Mountain Formation on the southern margin of Cape Breton Highlands (More, 1982; Jamieson and Doucet, 1983; Raeside and Barr, 1992). Possible correlatives in the Mabou Highlands (Mabou volcanic-sedimentary unit) have been described by Barr and Macdonald (1989). Volcanic rocks in the St. Anns area, originally included in the formation, have been shown to be Precambrian (Dunning et al., 1990b; Raeside and Barr, 1990, 1992).

In the type area near Chéticamp, the Fisset Brook Formation contains Tournasian spores near its base and volcanic components have been dated isotopically at 376 ± 12 Ma (Cormier and Kelly, 1964; recalculated by Keppie and Smith, 1978). The MacMillan Mountain Formation has a similar age (384 ± 10 Ma; Jamieson and Doucet, 1983), but Mississippian isotopic ages have been reported from the Lake Ainslie area (French, 1985; R.F. Huard, 1984).

The chemical characteristics of the volcanic rocks in the Fisset Brook Formation indicate a within-plate setting (e.g. Blanchard et al., 1984). The unit is generally considered to have formed in post-Acadian pull-apart basins, and correlative units are widespread in Atlantic Canada (Bradley, 1982; Dostal et al., 1983; Fyffe and Barr, 1986; Barr et al., 1985b).

Post-tectonic middle to upper Devonian plutons, with typical "A-type" features, are inferred to be plutonic equivalents of rhyolites in the Fisset Brook Formation (French, 1985). Hence the plutonic rocks in Cape Breton Island record a change from Silurian-Devonian compression to local Devonian-Carboniferous extension (Barr, 1990).

McAdam Lake Formation

The McAdam Lake Formation is located in the Bras d'Or Lake area (Barr and Raeside, 1989). It consists of interbedded greywacke, shale, and pebble conglomerate and contains early to middle Devonian plant fossils (Bell and Goranson, 1938). Bell and Goranson (1938) reported a local tuff unit but it may be an intrusive porphyritic andesite (Thicke, 1987).

Annapolis Belt

P.E. Schenk

Definition

The Annapolis Belt is defined by thick sequences, collectively called the Annapolis Supergroup, that range in age from Early Ordovician to Early Devonian. These sequences consist of fine grained shallow marine siliciclastic sedimentary rocks as well as volcaniclastic rocks. The Annapolis Supergroup and underlying Meguma Supergroup comprise the Meguma terrane. This terrane has also been called the Meguma Zone, but for the purposes of this publication, the Meguma Zone includes only the Meguma Supergroup. Middle Paleozoic rocks that overlie the Meguma Supergroup are assigned to the Annapolis Belt. As stated previously, the separation of a Meguma Zone and Annapolis Belt is artificial as: (1) both zone and belt are the same geographic area; (2) some rocks of the Annapolis Supergroup are possibly Early Ordovician and therefore older than those normally found in middle Paleozoic belts; (3) Acadian (Devonian) Orogeny was the first major event to affect rocks of both the Meguma Zone and Annapolis Belt; and (4) the Meguma Supergroup and Annapolis Supergroup are viewed as integral parts of the same stratigraphic section and interpreted in the same model of an evolving continental margin. Therefore, the Meguma Zone and Annapolis Belt are defined only to maintain the systematics of the time-slice subdivision used in this volume.

Extent and boundaries

The Annapolis Belt is 215 km in length, from the Cornwallis Valley in the east, through the Annapolis Valley and along the Acadian shore of the Bay of Fundy to Yarmouth in the west (Fig. 4.31). Maximum width is 23 km. It extends offshore beneath the Bay of Fundy to the northwest, and underlies the Gulf of Maine to the west. Because it is now an erosional remnant, the Annapolis Belt must have once been more widespread, possibly extending across the entire Meguma Zone to the east and southeast. Discrete parts of the belt occur now in widely spaced, synclinal areas. These are separated by broad expanses of post-folding granitoid plutons in the east, and the underlying Meguma Supergroup in the west (Fig. 4.31). For orientation, the areas are named from northeast to southwest: the Gaspereau, Nictaux Falls, Bear River, Cape St. Marys, and Yarmouth (Fig. 4.31). The Nictaux, Bear River, and Yarmouth areas are depressions along one projected, synclinal hinge. The northern boundary of the belt is presumably the Glooscap Fault. Characteristics of this fault are summarized elsewhere.

The Annapolis Supergroup has a sharp contact with the underlying Meguma Supergroup. This surface is in general a presumed paraconformity but in places it is an angular unconformity and in other places a disconformity. The upper boundary is an angular unconformity under gently dipping Carboniferous and/or Triassic redbeds.

Schenk, P.E.
1995:   Annapolis Belt; in Chapter 4 of Geology of the Appalachian-Caledonian Orogen in Canada and Greenland, (ed.) H. Williams; Geological Survey of Canada, Geology of Canada, no. 6, p. 367-383 (also Geological Society of America, The Geology of North America, v. F-1).

General features

The Middle Devonian Acadian orogeny is the first major deformational, metamorphic, and intrusive event to affect rocks of the Annapolis Belt. Broad, low-plunging, upright folds describe an arc trending northerly in the west to easterly in the eastern part of the belt. Associated slaty cleavage is sub-vertical, parallel to axial surfaces. The metamorphic facies is greenschist, with an increase to amphibolite facies in the west. Hornblende-hornfels facies are developed close to granitic batholiths (Taylor and Schiller, 1966). Large granitic plutons that cut rocks of the Annapolis Belt are Late Devonian and Early Carboniferous (Clarke et al., 1980).

The Meguma Zone and Annapolis Belt are well defined by metallogenic, plutonic, and tectonic characteristics (Schenk, 1981). They form a distinct metallogenic domain in the Canadian Appalachians until Devonian time (Zentilli, 1977). This domain is distinguished by concentrations of gold, arsenic, tungsten, antimony, and tin whereas the adjacent Avalon Zone has base metal sulphides with deposits of molybdenum and bismuth. After the Devonian, complex structural-magmatic and hydrothermal effects on the intervening Glooscap Fault overprinted these two domains. All granitic plutons of the Meguma Zone and Annapolis Belt are highly alumina-saturated and thus differ greatly from those in the Avalon Zone (Clarke et al., 1980). Intense deformation in the vicinity of the Glooscap Fault points to significant post-intrusive slip to juxtapose two very different plutonic regions (Clarke et al., 1980).

The Annapolis Belt may be allochthonous and rest on Avalon basement (Eberz et al., 1991). Geochemical evidence is summarized in the description of the Meguma Zone (Schenk, Chapter 3). Seismic reflection data show that the Avalon lower crustal block underlies the Avalon Zone (Marillier et al., 1989) and may also underlie the Meguma Zone-Annapolis Belt (Stockmal et al., 1990). Conceivably, the Meguma Zone-Annapolis Belt was preserved by structural transport. This event may have depressed the lithosphere leading to melting of the crust. Resulting extensive granitoid plutonism began circa 375 Ma, approximately 40 to 25 Ma following the age of first penetrative deformation (Reynolds and Muecke, 1978; Dallmeyer and Keppie, 1987). This deformation may provide an upper limit on the time of initial docking of the Meguma and Avalon zones. The 40 to 25 Ma time interval between initial docking and the onset of granitoid plutonism is consistent with the slow rise of isotherms which typically follows tectonic thickening of continental crust (Keppie and Dallmeyer, 1987).

The Annapolis Belt has had a long history of exploration. Some of the first geologists engaged in systematic mapping in Canada surveyed these rocks (e.g. Dawson, 1855; Honeyman, 1879; Bailey, 1898). Much of the early attention focused on iron deposits at Torbrook Mines. More recently, granite plutons have been mapped in detail for economic deposits of tin and uranium.

Stratigraphy

Name

The informal term "Supra-Meguma strata" was used to designate all stratigraphic units of the Annapolis Belt (Schenk, 1981). These strata form a single, stratigraphic

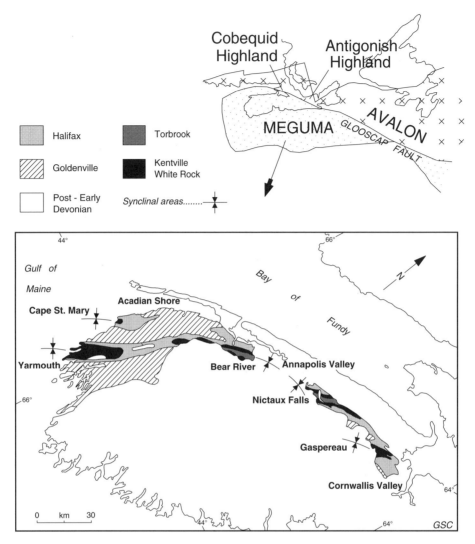

Figure 4.31. Northwestern rim of the Meguma Zone showing synclinal areas with an insert of the Meguma and Avalon terranes. The latter locates the Meguma terrane south of its faulted border with the Avalon terrane.

entity, restricted in exposure to a relatively small area, separated from the underlying Meguma Supergroup by a significant hiatus, and distinguished by shallow marine siliciclastic as well as volcaniclastic and minor carbonate rocks. The "supra-Meguma strata" are here called the Annapolis Supergroup (see descriptions of the Meguma Supergroup). Furthermore, recent mapping and subdivision of the original formations require that they be elevated to the rank of groups (Fig. 4.32). Contact relations between groups within the Annapolis Supergroup are conformable.

The Annapolis Supergroup is named for its type area, the Annapolis Valley of northwestern Nova Scotia (Fig. 4.31). Other stratigraphic names are taken from villages and towns in the Annapolis Valley (see Addenda).

The Annapolis Supergroup is composed of three groups, in ascending order the White Rock, Kentville, and Torbrook (Fig. 4.32). The lower two groups are divided into formal formations, but the Torbrook formations are still informal. All formations are subdivided into informal members.

White Rock Group

The White Rock Group is relatively uniform in thickness (Lane, 1981): approximately 100 m (Gaspereau area), 480 m (Nictaux Falls area), 300 m (both Bear River and Yarmouth areas), and 100 m (Cape St. Marys area). It consists of three formations, in ascending order, the Nictaux Volcanics, Fales River Formation, and Deep Hollow Formation.

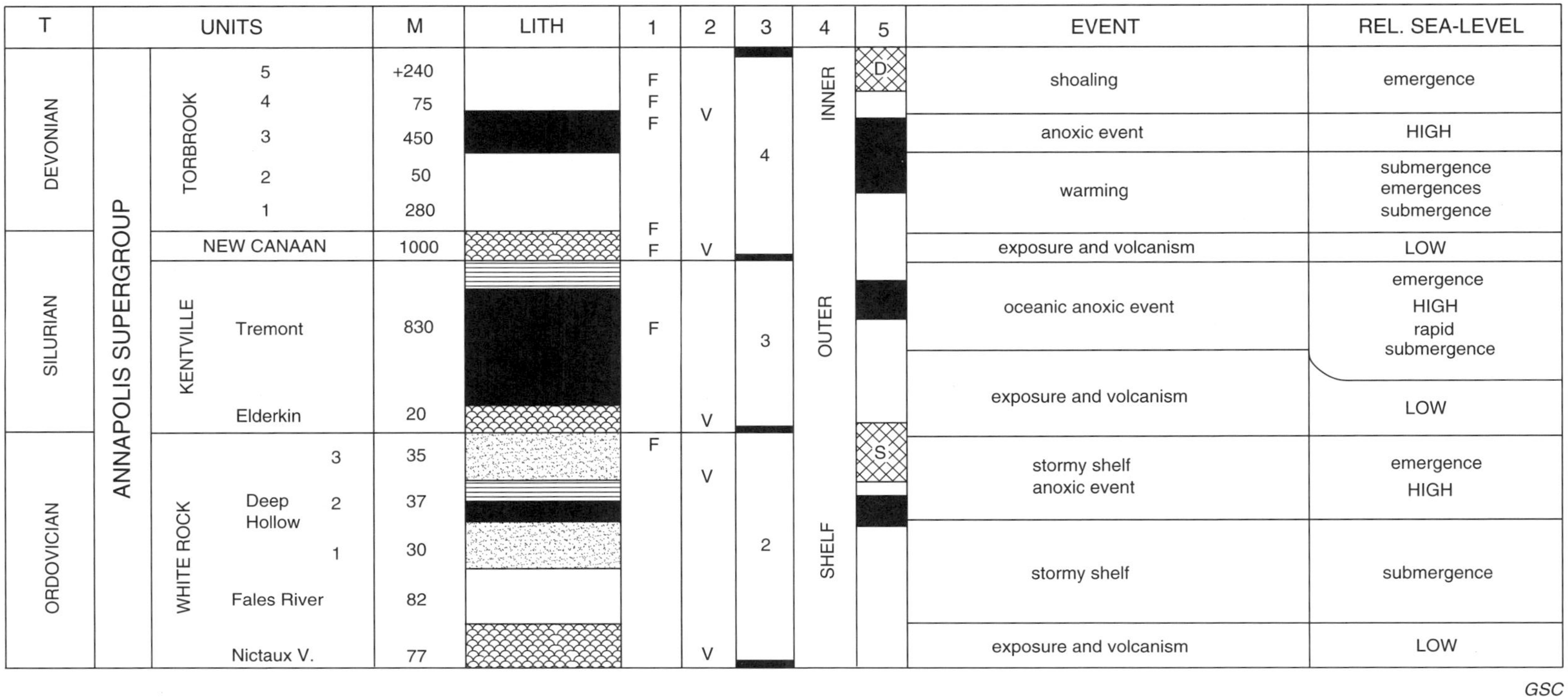

GSC

T		UNITS		M	LITH	1	2	3	4	5	EVENT	REL. SEA-LEVEL
DEVONIAN	TORBROOK	5		+240		F			INNER	⊠ D	shoaling	emergence
		4		75	■	F	>	4		■	anoxic event	HIGH
		3		450		F					warming	submergence / emergences / submergence
		2		50			>					
		1		280		F				■	exposure and volcanism	LOW
	NEW CANAAN			1000	▨	F						emergence / HIGH / rapid submergence
SILURIAN	KENTVILLE	Tremont		830	■	F	>	3	OUTER	■	oceanic anoxic event	LOW
		Elderkin		20	▨						exposure and volcanism	
ORDOVICIAN	WHITE ROCK	Deep Hollow	3	35	⠿	F	>	2	SHELF	⊠ S	stormy shelf / anoxic event	emergence / HIGH
			2	37			>			■		
			1	30	⠿						stormy shelf	submergence
		Fales River		82			>					
		Nictaux V.		77	▨		>				exposure and volcanism	LOW
	ANNAPOLIS SUPERGROUP											

Figure 4.32. Summary of stratigraphy, event stratigraphy, and relative sea-level changes in the Annapolis Supergroup. Compare with Figure 3.151 for the Meguma Supergroup (Schenk, chapter 3). T refers to geological time and M to maximum measured thicknesses in metres. The lithological column gives predominant lithologies represented by patterns used in Figure 4.33. Column 1 indicates fossil-bearing intervals (sequences are numbered); column 2 shows episodes of volcanism; column 3 identifies unconformities by heavy horizontal lines (sequences are numbered); column 4 summarizes general, depositional environments; and column 5 displays (1) times of global anoxic events (black) and (2) major Paleozoic glacial episodes (S for Saharan, and D for Devonian). Major events and relative sea-level changes in the Annapolis Belt are listed in the next two columns (after Schenk, 1991).

369

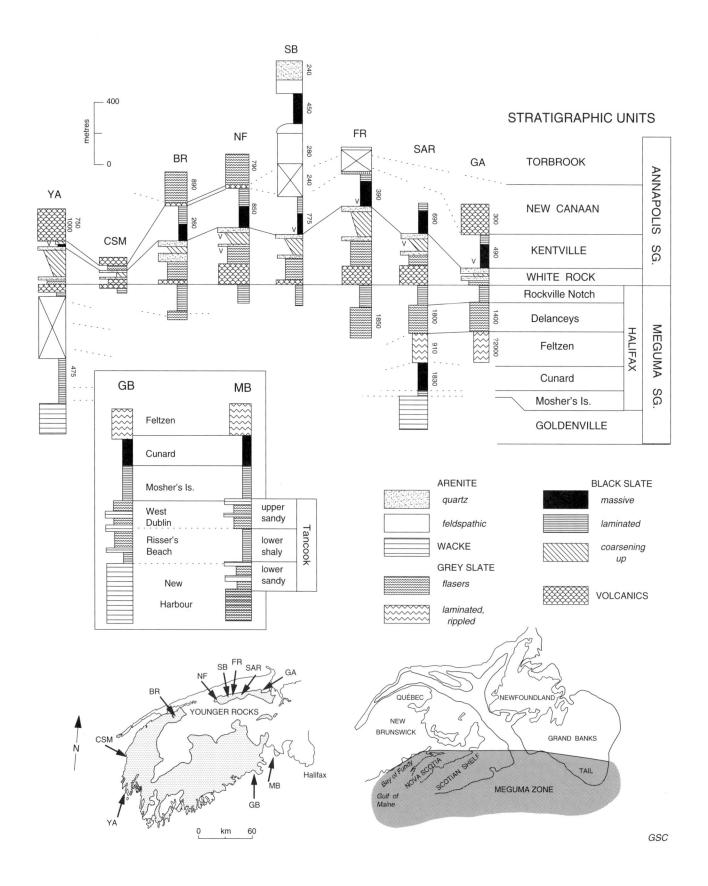

Nictaux Volcanics

In the Nictaux Falls area this formation is a bimodal couplet, 77 m thick of felsic ashflow and overlying basaltic flows or tuff (Fig. 4.33). The felsic rocks form massive outcrops of devitrified, vesiculated, and accretionary lapilli tuffs and ashflow tuffs. They grade laterally into bedded tuffs. Ash flows show flow banding and eutaxitic structures. The felsite rocks are recognized from the central part of the Annapolis Belt to Yarmouth, a distance of 180 km. The overlying basaltic rocks consist of pillow basalts with weathered tops and mafic tuffs. They occur only in the Cape St. Marys and western part of the Nictaux Falls areas. The Nictaux Formation, as well as the overlying Fales River Formation, are absent in the Gaspereau and eastern part of the Nictaux Falls areas. A local occurrence of sheared carbonate grainstone is interstratified with felsite in the Bear River area (Lane, 1981).

The lower boundary of the Nictaux Volcanics with the Meguma Supergroup is sharp and structurally conformable. In general this contact is a presumed paraconformity (see description of the Meguma Zone). Two exceptions are at Cape St. Marys and near Fales River (Fig. 4.33). At Cape St. Marys, the contact is an angular unconformity; near Fales River, it is a disconformity. The upper contact of the Nictaux Volcanics is sharp and conformable at the upper surface of basalt or thick felsite, whichever lithohorizon is stratigraphically highest.

Fales River Formation

The Fales River Formation varies considerably in both lithology and thickness, but is a recognizable stratigraphic entity between bounding, widely correlative formations. It is 82 m thick in the Nictaux Falls area. The formation consists of unbioturbated beds of quartz arenite with sharp upper and lower contacts, alternating with thoroughly bioturbated shales and siltstones. Some of the arenites occur as amalgamated beds. Most of them have flat to gently-inclined sets of parallel-laminae. Trough and planar cross-sets of ripple, mega-ripple, and sand-wave bedding are also present along with wave-formed straight-crested ripples and cross-laminae. Thin- to thick-bedded, quartz arenite strata are parallel laminated to low-angled (less than 10 degrees) cross-stratified, and show hummocky cross-stratification. Black slate is conspicuously absent; instead, slates are green or light grey (Lane, 1981).

Figure 4.33. Location maps and stratigraphic correlations of the Annapolis Belt. The accompanying map shows distribution of metasedimentary and metavolcanic rocks in the western half of the Meguma Zone, and localities of stratigraphic sections (GA, Gaspereau area; SAR, South Annapolis River; FR, Fales River; SB, Spinney Brook; NF, Nictaux Falls; BR, Bear River; CSM, Cape St. Mary; YA, Yarmouth area; GB, Green Bay; and MB, Mahone Bay). Groups are indicated by upper case letters, formations by upper and lower case letters, and informal subdivisions by lower case letters. Thicknesses of stratigraphic units in the Green Bay and Mahone Bay sections are given in Figure 3.151. (Schenk, 1991).

Facies relations are most complex in the western part of the Nictaux Falls area where strata of the Fales River Formation locally occur beneath the Nictaux Volcanics. If the volcanic rocks are considered as a chronozone, the environment responsible for the Fales River Formation began earlier here than elsewhere. The formation includes mafic volcanic rock in the Cape St. Mary area.

The upper boundary of the formation is sharp and conformable with the overlying Deep Hollow Formation.

Deep Hollow Formation

Thickness of the Deep Hollow Formation is 110 m in the Gaspereau area but regionally it ranges from 28 to 160 m. The formation consists of three distinctive members: two ridge-forming sandstones (members 1 and 3) separated by a thick dark slate (member 2).

The sandstones are quartz arenites forming lenticular packages of amalgamated strata generally 15 to 40 m thick, and 40 to several hundred metres in length. The packages are massive or faintly laminated with gentle dips (less than 10 degrees). Grain size coarsens upward and cross-lamination changes upward from current-formed to wave-rippled and wave-formed. In the Cape St. Marys area, the sandstones are unusual in commonly displaying cross-set beds, both tabular and wedge-shapes. Thin successions of siltstones and intercalated, thin bedded, fine grained arenites lie immediately on top of the sheet-arenites. The sandstones are fine grained and supermature in both texture and composition. Grains are spherical and well rounded. Member 3 sandstone changes laterally at a few localities in the Gaspereau and Bear River areas to thinner quartz wackes and quartz siltstone layers, but otherwise is physically continuous across the entire Annapolis Belt. In contrast, member 1 sandstone is more variable, changing laterally to concentrations of thin quartz arenite as in the Bear River and Yarmouth areas or even disappearing locally. Where it is absent the formation can be identified by the distinctive middle member of the formation.

Member 2 is a dark grey, silty slate that coarsens upward with even laminae of siltstone. Lithologies are similar to the Tremont Formation. The member is 168 m thick in the Gaspereau area. A felsitic tuff layer 15 m thick occurs at the top of member 2 in the Nictaux Falls area. Member 2 slate accompanies member 3 sandstone across the Annapolis Belt.

The lower contact of the formation is sharp and conformable. It is defined at the base of either the member 1 sandstone or member 2 slate, whichever is stratigraphically lower. The upper contact is gradational and conformable at the top surface of member 3 sandstone. The top of the formation is defined by the top of member 3 across the Annapolis Belt.

Kentville Group

The Kentville Group is 670 m thick along Fales River and 1170 m thick to the southwest (Smitheringale, 1960). It contains the Elderkin and overlying Tremont formations.

Elderkin Formation

This formation consists of two members, a silty slate (15 m) overlain by a thick felsite (1 to 32 m; Lane, 1981). The slate member consists of two main lithologies: (1) dark grey, slightly silty nonlaminated slate; and (2) laminated slate composed of thin, alternating laminae of medium grey, impure arenite and dark grey slate. The felsite member is distinctive in the Gaspereau area and reappears in the Nictaux Falls area. There, it is a laminated felsic tuff 8 m thick and located 10 to 20 m above the White Rock Group and immediately beneath the *Monograptus nilsonni* biozone of the Tremont Formation. Farther west in the Yarmouth area the Elderkin Formation is indistinguishable from the basal part of the New Canaan Formation because of pinch out of the intervening Tremont Formation.

The upper contact is sharp and conformable at the upper lithohorizon of the tuff member.

Tremont Formation

This formation also consists of two members, a dark grey slate and an overlying greenish grey slate. The lower member is a mixture of massive and laminated, slightly silty slate. The latter is composed of thin alternating laminae of impure quartz arenite and dark grey slate. Rare lenses of black limestone contain a fairly abundant shelly fauna (Smitheringale, 1973). Phosphatic nodules are present. The upper member is greenish grey, non-silty and sparsely laminated slate.

The formation thins westward across the Annapolis Belt from 600 m at the type locality to 257 m in the Bear River and 25 m in the Cape St. Marys areas before pinch out in the Yarmouth area (Lane, 1981). The upper contact is sharp and conformable at the base of the lowest volcanic lithozone of the New Canaan Group, or where the New Canaan is absent (Bear River and Nictaux Falls areas), the contact is transitional and gradational over 33 m with the Torbrook Group.

New Canaan Group

The New Canaan Group is best developed in the Gaspereau (300 m thick) and Yarmouth (1000-2000 m thick) areas. It pinches out eastward to a thin erosional remnant in the Cape St. Mary area and is absent in the Bear River and Nictaux Falls areas. The group is not yet divided into formations.

In the Gaspereau area the group is poorly exposed and consists of diamictites with volcanic and fossiliferous sedimentary clasts in a dark grey, slaty matrix (Crosby, 1962). In the Yarmouth area it is much better exposed but metamorphic and deformational effects are more intense. Two successions of greenstone lapilli tuffs are intercalated with marine siliciclastic facies, which are in turn overlain by lava, ash flows, and volcaniclastic conglomerates. The tuffs occur as very thin to thin, even, continuous strata that extend over a minimum distance of 100 m. Lapilli in these strata are graded and are of the same composition as the tuffs. The primary, magmatic flows are a bimodal suite of alkaline to peralkaline metabasalts and metatrachytes, and less voluminous olivine tholeiite-rhyolite series. Some flows have flow-top breccias that may result from subaerial

weathering (Sarkar, 1978). Interstratified diamictites contain boulders of meta-volcanic rocks and meta-arenites. Some reverse grading is evident. Metabasites alternate with diamictites to form seven successive lithozones (Lane, 1981).

The lower boundary in the Yarmouth area is conformable and gradational with the underlying quartz arenites of the White Rock Group. The upper boundary is an angular unconformity with Triassic strata. Elsewhere this contact is conformable and gradational.

Torbrook Group

The Torbrook Group is more than 1500 m thick in the Nictaux Falls area. It is divided into informal formations following lithozones along Spinney Brook (Fig. 4.33). This section has been described by Hickox (1958), Smitheringale (1973), and Jensen (1975). Gabbroic sills intrude the Torbrook Group and underlying strata.

The group consists of interstratified silt-rich to clay-rich mudstone, coarse siltstone, and very fine grained sandstone with subsidiary arenaceous coquinite as well as oolitic and skeletal iron-formation. Single sandstone strata are lenticular and range in thickness from 3-40 cm with composite beds reaching 2 m. The sandstones are mainly lithic to quartz arenites. Primary sedimentary structures are abundant, especially parallel and low-angle cross lamination. Under surfaces are mainly flat to gently undulating but some have distinct scour-depressions. Coquina layers consisting predominantly of unbroken, disarticulated brachiopod shells commonly occur at the base of sandy beds. Small mudstone clasts are interspersed in places with the shell debris. The sandy material immediately above each coquina layer tends to have either (1) parallel-lamination locally associated with scoured bases, or (2) low-amplitude ripple-lamination. In general the sandstones are not graded. Bioturbation is common on the upper surfaces of many beds, but where absent, the tops are usually undulatory and sharp. Mudstones are extensively burrowed. Structures indicating subaerial exposure are absent (Jensen, 1975).

Formations 1-5 of the Torbrook Group

The two lowest formations are sandstones. Formation 1 consists of three coarsening upward sandstone members totalling 280 m in thickness. The contact with the underlying Kentville Group is gradational. In contrast formation 2 fines upward in grain size over a thickness of 50 m. Both formations show parallel stratification and very low-angled cross-stratification. Formation 3 consists of two members, a basal black slate with fetid black limestones and an overlying member of grey mudstones and tuffs. The contact between members is gradational. Thin siltstones contain phosphatic pebbles. Laterally discontinuous, oolitic, fossiliferous ironstones reach 3 m in thickness (Smitheringale, 1973). Formations 4 and 5 are sandstones. Formation 4 consists of 75 m of sandy mudstones, muddy sandstones, and purple quartz arenites. Formation 5 consists of 240 m of thick-bedded cross-stratified sandstone. It is overlain by a covered interval of 440 m, which is in turn overlain with angular unconformity by shallowly dipping Triassic redbeds.

Paleontology and age

White Rock Group

The White Rock Group contains few fossils so that its age has been inferred from stratigraphic position and apparent facies relations with other fossiliferous units. Its position between the underlying Halifax Group (Tremadoc) and overlying Kentville Group (Ludlow) suggests an age ranging from late Early Ordovician to Middle Silurian. Inferred facies relations with the Kentville Group suggested a Late Silurian (Ludlow) age (Smitheringale, 1960, 1973; Taylor, 1965). Lithic correlation with the Glencoe Brook Formation of the nearby Avalon Zone suggested an Early Silurian age (Boucot et al., 1974). However, the White Rock Group can be mapped along the Annapolis Belt stratigraphically beneath the Kentville Group (Lane, 1981). Event stratigraphy suggests that the White Rock Group is Late Ordovician with the lowermost units perhaps extending down into the Arenig (Schenk, 1991).

The Nictaux Volcanics is unfossiliferous except for undatable ostracodes in grainstone of the Bear River area. Determination of its age was attempted by establishing the location of its paleomagnetic pole (Spariosu et al., 1984). Unfortunately, the pole position proved to be post-folding, i.e. Devonian. A hiatus of unknown length separates the formation from underlying Tremadoc strata. By stratigraphic position, the Nictaux volcanics could be as old as Middle or Late Arenig, which would agree with conclusions based on event stratigraphy (Schenk, 1991).

Bioturbation is common in the Fales River Formation. *Chondrites* is prevalent in the silty slates. *Scolithus* dominates in the sandy layers and may reflect opportunistic conditions that followed storms. *Spirophyton* occurs in siltstones in the Bear River area. Only one example of a *Cruziana* trace has been found. Fragmented gastropods occur in thin beds of quartz arenites in the Cape St. Mary area. By stratigraphic position and event stratigraphy, the Fales River Formation could be as old as Late Arenig and as young as Llanvirn or Llandeilo (Schenk, 1991).

Member 1 sandstone of the Deep Hollow Formation in the Gaspereau area contains a 20 to 30 cm thick biozone *of Tigillites* (*Scolithus* ichnofacies). Member 3 sandstone at the same locality has thin lags of fragmented shelly fauna; brachiopods, crinoids, gastropods, and pelecypods in order of abundance. Unfortunately, these are unidentifiable because of fragmentation and both tectonic strain and shear. The only dated fossil is a costellate rhynchonellid brachiopod that ranges in age from Caradoc to Pennsylvanian. (Boucot, pers. comm., in Lane, 1981).

Kentville Group

The Elderkin Formation of this group overlies strata interpreted as Late Ordovician, and underlies fossiliferous Upper Silurian slates; thus a latest Ashgill or Early Silurian age is probable. The Elderkin Formation immediately underlies the Early Ludlow graptolite zone of *Monograptus nilsonni*.

The Tremont Formation has a *Chondrites* ichnofauna. The formation also contains fairly abundant fossils of Early Ludlow age: graptolites (*Monograptus nilssoni* and *colonus*); crinoids (*Scyphocrinites*), molluscs (*Modiolopsis, Actinopteria,*

Nuculites, Hormotoma, Plectonotus, Orthoceras, and *Kionoceras*), and brachiopods (*Camarotechia* and *Delthyris*). In the Bear River area it contains fragments of vertebrates including acanthodians (spiny sharks) and scales of thelodonts, thus confirming a Late Silurian age (Bouyx et al., 1985). Where the overlying New Canaan Group is absent, the Tremont is gradational into the fossiliferous Lower Devonian Torbrook Group. Thus, the age of the Tremont Formation extends from Ludlow through Pridoli.

New Canaan Group

The New Canaan Group contains crinoid columnals, pentamerid brachiopods (?*Conchidium*), and tabulate corals in the Gaspereau area. They indicate a probable Middle Silurian age (Wilson in Crosby, 1962), although Boucot et al. (1974) placed the group in the latest Silurian (Pridoli). Paleomagnetic poles are post-folding (J.M. Hall, pers. comm., 1990). Ludlow graptolites occur beneath the formation.

Torbrook Group

The Torbrook Group is unusual for its rich shelly fauna. Brachiopods are of the Rhenish-Bohemian faunal province (Boucot, 1960). They date the group as Gedinnian to Siegenian/Early Emsian (Hickox, 1958). The uppermost 30 per cent of the group is covered by overburden and the strata continue beneath overlying younger rock; thus the minimum age of the Torbrook Group is unknown.

Provenance

Dispersal

Sedimentary structures indicating paleocurrent direction in the White Rock Group are restricted to cross-stratification and are so rare that conclusions on dispersal directions are statistically invalid. In general, distributions are polymodal with some possible preferred movement toward the west and either north or south (Lane, 1981). Decreases in thickness of sandstone members imply a source either to the east (Crosby, 1962) or northwest (Lane, 1981). Smitheringale (1973) saw no evidence of provenance in either the coarse- or fine-grained lithologies of the White Rock. Volcanism would result in complications during sedimentation so that paleoenvironments should vary from place to place (Taylor, 1965).

Sedimentary structures indicating dispersal are quite rare in the Kentville Group. Faint cross-stratification in siltstone laminae show variable directions of dispersal.

Directions of dispersal are unknown in the New Canaan Group. Exposures are best in the Yarmouth area but the degree of strain and metamorphism makes any analysis suspect. Metabasites there and in the Cape St. Marys area are geochemically related to a gabbroic plug in the Cape St. Marys area (Kendall, 1981).

Paleocurrent directions are quite variable in the Torbrook Group. Long, low-amplitude, ripple cross-lamination indicates currents trending northwest and also southeast. Most of the large bottom-scours are directed toward either the north or the south. Thus the dispersal pattern is similar to that of the White Rock Group.

Source

As is true for the Meguma Supergroup, the volume of the Annapolis Supergroup indicates that the source area must have been continental in size. Total thickness of the supergroup exceeds 3000 m. Although its present exposure is confined to a belt measuring approximately 5×10^3 km^2, this is a folded, strained, erosional remnant. Projection of the synclinal axes of the Gaspereau and Cape St. Marys areas would increase this area to 8×10^3 km^2. Unfolding of the strata by the conservative circular-arc method increases this number to more than 10×10^3 km^2. Conceivably the volume of the supergroup would exceed by several times the apparent 30×10^3 km^3. Even so, this volume corresponds to the area of Ontario stripped to a depth of 25 m (or more than 700 m if the supergroup extended over the area of the Meguma Supergroup).

Siliciclastic rocks in the supergroup are remarkably fine grained, indicating extensive winnowing and sorting on the source continent. Whereas sandstones in the White Rock Group are mature quartz arenites with spherical grains, those in the Torbrook Group are more lithic with subangular grains. Mineralogy of sandstones in the Annapolis Supergroup is similar to that of the Meguma Supergroup. Both groups were probably derived from the same source area (Schenk, 1981, 1991).

Environment of deposition

White Rock Group

Subaerial deposition of the Nictaux Volcanics is indicated by ash flows with flow banding and eutaxitic structures, and mafic flows with weathered tops. However a mainly subaqueous environment is implied by pillow basalts, vesiculated and accretionary lapilli tuffs, and basaltic pyroclastic rocks without interstratified flows. The subaerial features probably represent periods of temporary emergence during deposition of the formation (Lane, 1981). Volcaniclastic rocks are more abundant toward the west. Presumably alkali-rich vents were located in the far western part of the belt, although the basaltic lavas must have had a more local source. The amount of volcanism at this time was minor, relative to Late Silurian volcanism that also may have had a western source.

Arenites of the Fales River Formation were rapidly deposited on a substrate of otherwise slowly accumulating silt and mud. Presumably they accumulated during periods of high energy (storms) when waves and currents were capable of winnowing and transporting such sediments. This is supported by the distribution and variety of ichnofauna. Slates are light-coloured, indicating oxidation (Lane, 1981). In situ coated grains of carbonates (pisolites and algal oncolites) associated with volcanic tuff beds at the base of the formation probably formed in waters less than 20 m deep. The formation was deposited between storm wave base and fair-weather wave base on a storm-swept, muddy inner shelf (Fig. 4.32).

Sandstones of members 1 and 3 of the Deep Hollow Formation show all the characteristics of storm deposits (Hunter and Clifton, 1982; Aigner, 1982). There is no evidence of subaerial exposure or shoreface development typical of a shoreline. The large sets of low-angled cross-stratification are interpreted as hummocky cross-stratification. The members were deposited above storm wave base on a continental shelf that was swept by intense and frequent storms generated during a time of well-defined climatic zones, perhaps in high-latitudes (Lane, 1981; Schenk, 1991). A very sparse shelly fauna as well as abundant traces of the *Cruziana* and *Zoophycus* ichnofauna indicate a shallow, sublittoral environment. The paleocurrent directions are polymodal, characteristic of shallow marine deposits. By stratigraphic position, both sandstone members were deposited offshore with respect to the Fales River Formation. Member 2 slate of the Deep Hollow Formation was deposited in deeper water below storm-weather wave base (Fig. 4.32). Phosphate nodules in black shales suggest water depths of less than 200 m (the maximum depth of present-day phosphate precipitation).

Kentville Group

Tuffs of the Elderkin Formation indicate a limited amount of volcanism perhaps related to activity to the southwest.

In general, black shales of the Tremont Formation were deposited in stagnant, anoxic waters during the Late Silurian oceanic anoxic event. Stratigraphic position between shelf sands of the White Rock and Torbrook groups suggests that the Tremont Formation probably also represents a shelf environment, but deeper than storm-weather wavebase (Schenk, 1991).

New Canaan Group

Volcanic rocks of this group are primarily subaqueous in origin; however, some subaerial deposition is indicated. This conclusion may be supported by the great thickness of the volcanic rocks, their lateral extent, and presence of large boulders.

Volcaniclastic conglomerates lie above flow units and are the products of erosion of the local volcanic-sedimentary rocks. In general they are diamictites that have characteristics of resedimented grain flow, debris flow, and slump deposits, based on some reverse size-grading. The high metamorphic grade and intense strain makes sedimentologic interpretation uncertain (Lane, 1981). Fossils within clasts in the Gaspereau area indicate that parental strata were deposited in marine, fairly shallow, warm waters (Crosby, 1962).

Geochemical and sedimentary analyses suggest that the tectonic environment of the volcanism was probably extensional. In the Yarmouth area, Sarkar (1978) postulated a "failed" continental rift and/or back-arc basin, whereas in the Cape St. Marys area, Kendall (1981) indicated a within-plate setting. Erosion of underlying quartz arenite lithologies indicates that contemporaneous active faults probably existed. Eruptions may have been centred along these faults (Lane, 1981). A gabbroic plug in the Cape St. Mary area is geochemically related to metabasites in the Yarmouth and Cape St. Mary areas (Kendall, 1981). High magnetic but low gravity anomalies offshore of the Yarmouth area appear to be unrelated to New Canaan volcanic rocks (Pe-Piper and Loncarevic, 1989).

Torbrook Group

The probable environment for formations 1 and 2 of the Torbrook Group was a storm-dominated shelf with initial progradation followed by submergence (Schenk, 1991). Waters were initially oxygenated, variable as to energy, and shallow (Smitheringale, 1973).

Formation 3 is similar to other black lithozones in the Meguma Supergroup, with differences perhaps due to the more inshore position and warmer conditions of the Torbrook environment. Waters were stagnant, oxygen-starved, and protected from currents (Smitheringale, 1973). The anoxic event could record a highstand, but ironstone commonly marks the late stages of regional regression (Hallam and Bradshaw, 1979). Oolitic ironstones form in shallow seas at no great distance from extensive, low-lying, well-vegetated land. The climate was warm and humid so that iron is preconcentrated by extensive leaching. During late stages of regional regressions, terrigenous input to the depositional area diminishes in many places as the sea shallows (Smitheringale, 1973).

The coarsening-upward profile of the upper three formations of the group form an offlap succession. The abundance of hummocky stratification suggests frequent storms, as is true of correlatives in the adjacent Avalon Zone of northern Nova Scotia (Hurst and Pickerill, 1986). There, the Lower Devonian section culminates in thick continental redbeds immediately before the Acadian Orogeny (Boucot et al., 1974). The Torbrook Group shows no evidence of subaerial exposure, but abundant plant fossils in the upper formations may indicate proximity to land (Hickox, 1958; Smitheringale, 1973).

Model for deposition

The model for deposition of the Annapolis Supergroup is one of relatively shallow-water sedimentation on the Gondwanan shelf, interrupted by two instances of subaerial exposure. The supergroup, in common with the underlying Meguma Supergroup, consists entirely of siliciclastic debris that had to be eroded from, and so controlled by an adjacent continent. The general continuity and magnitude of the formations suggest that the rate of erosion on this source-continent controlled the vertical succession. This rate was in turn dependent on the elevation of base-level and so, relative sea-level. Consequently, the Annapolis Supergroup (together with the Meguma Supergroup) can be set in a framework of sequence stratigraphy. Each sequence first fines upward and then coarsens to subaerial volcaniclastic rocks that mark a bounding unconformity (Fig. 4.32). The Meguma Supergroup comprises the first sequence of the Meguma terrane; the Annapolis Supergroup is composed of sequences two (the White Rock Group), three (the Kentville Group), and four (the Nictaux Volcanics and Torbrook Group; Schenk, 1991).

Ordovician

The Nictaux Volcanics of the White Rock Group signal an important event in the Meguma terrane (Meguma Zone and Annapolis Belt). They are: (1) the first definite volcanic event in the terrane; (2) a reliable chronostratigraphic marker from the eastern Nictaux to the western Yarmouth areas

(Figs. 4.31, 4.33), a distance of 180 km (Smitheringale, 1973; Lane, 1981); and (3) the strongest evidence of subaerial exposure (Smitheringale, 1973). The contact between the Nictaux Volcanics and the underlying Meguma Supergroup is an unconformity whose hiatus may well span the Early Ordovician. For conservative reasons, Figure 4.32 shows it to be minimal. It coincides with the postulated world-wide unconformity separating the Sauk and Tippecanoe cratonic sequences (Sloss, 1963). In terms of sequence stratigraphy the unconformity is a type 2 sequence boundary. The Nictaux Volcanics represent the late stage of subaerial exposure of the highstand systems tract (Fig. 4.34).

The Fales River and Deep Hollow formations of the White Rock Group record a single transgressive-regressive cycle with maximum submergence recorded by the black slate of member 3 (Schenk, 1991). This slate is very similar both to member 3 of the Torbrook Formation and the Tremont Formation. Maximum sea-level probably occurred near its base, and a hiatus is suspected. Sandstones of the White Rock Group show no evidence of subaerial exposure or shoreface development. Their sedimentary structures and successions indicated deposition on a stormy, shallow shelf (Lane, 1981). In terms of sequence stratigraphy, The Fales River Formation represents the shelf-margin systems tract and its sharp upper contact would be the transgressive surface (Fig. 4.34). The overlying member 1 sandstone represents the transgressive-systems tract and its upper surface is both the downlap surface and surface of maximum flooding (Schenk, 1991). Black shales at the base of member 2 correspond to a condensed section. The member coarsens upward, culminating in the widespread member 3 sandstone, which together form the highstand systems tract. The Deep Hollow Formation is overlain by subaerial volcaniclastic rocks of the Elderkin Formation that conforms to a type 2 sequence boundary. The highstand systems tract probably was a response to a global drop in sea level because of Saharan glaciation (Schenk, 1972).

Alternatively, deposition of the White Rock Group could involve only a fraction of Ordovician time and not follow the scenario of sequence stratigraphy. Time control on the group is not rigorous; the basal unconformity could encapsulate all of Early Ordovician time and deposition of the group would be confined to the Late Ordovician. Member 2 slate is identical in lithology to the Tremont Formation and so should record a similar environment. The Tremont is interpreted as a result of a drop in global sea level, a consequence of melting of the Saharan ice sheet. Saharan glaciation was multiple in nature, global in effect, and late Caradoc and younger in age (Beuf et al., 1971). Member 3 sandstone contains Caradoc or younger fauna. Member 2 slate could record an interglacial highstand between shallow water sandstones of members 1 and 3 that indicate glacioeustatic lows.

Silurian

Volcaniclastic rocks of the Elderkin Formation indicate at least nearby subaerial exposure. They overlie a regressive succession probably controlled by Saharan glacioeustatic lowering of sea level. The volcaniclastic rocks underlie widespread, Ludlow black slates suggestive of post-glacial submergence. The formation probably marks the sea-level minimum due to Saharan glaciation. As in the case of

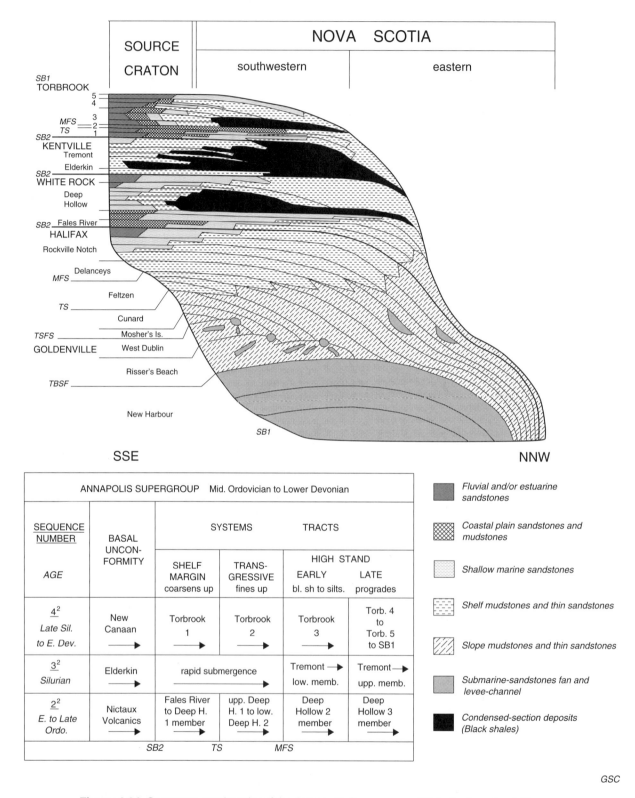

Figure 4.34. Sequence stratigraphy of the Annapolis Supergroup. SB 1, 2 – type 1 or 2 Sequence Boundary; TS – Transgressive surface; TBFS – Top of the basal fan surface; TSFS – Top of the slope fan surface; MFS – Maximum flooding surface. Three sequences (2, 3, and 4 of the Meguma terrane) are type 2 sequences (indicated by superscripts). Arrows from bottom left to top right show stratigraphically upward changes. See Figure 3.153.

somewhat similar, ashflow volcaniclastic rocks in both the New Canaan Group and member 1 of the Deep Hollow Formation of the White Rock Group, a paraconformity is presumed. The hiatus could span the Llandovery and Wenlock; for conservative reasons Figure 4.32 shows it to be minimal.

The sharp lower contact of the Tremont Formation signifies rapid and widespread submergence. This is typical of many Silurian black slates worldwide, and records a rapid rise in sea level following melting of the Saharan ice sheet (Beuf et al., 1971; McKerrow, 1979). The lower sections of these slates are commonly very thin and encompass much of the Early Silurian, and also mark maximum transgression. This may also be true in the Annapolis Belt, where a hiatus is expected either at the base of, or within, the Elderkin Formation. Slow emergence is indicated by the upward increase in silt and lightening of colour. Elsewhere, relative sea level declined through the Ludlow and perhaps Pridoli (McKerrow, 1979).

The base of sequence 3 occurs either at the base of, or within the Elderkin Formation (Fig. 4.32). It is a type 2 sequence boundary. The assumed rapid rise of eustatic sea level precluded deposition of the shelf-margin systems tract in the Annapolis Belt so that the lower contact of the Tremont Formation would be the transgressive surface (Fig. 4.34). The lower member of the Tremont Formation is a condensed section that includes the surface of maximum flooding. Middle and upper members coarsen upward in response to shoaling, and are aggradational and progradational parasequences respectively of the highstand systems tract. Sequence 3 is terminated by a type 2 sequence boundary at or within the New Canaan Group. This boundary corresponds to the unconformity separating the Tippecanoe and Kaskasia cratonic sequences (Sloss, 1963).

A significant volcanic episode occurred in the Meguma terrane during the Late Silurian. This may also be the upper time limit on initial docking of the Meguma and Avalon terranes (Keppie and Dallmeyer, 1987). Most of the volcaniclastic rocks are water-lain but their thickness and lateral extent, as well as the presence of large boulders, suggest at least nearby subaerial exposure. The diamictites may be debris flows and slumps, based on some reverse size-grading (Lane, 1981). This sea-level low would coincide with the boundary unconformity between the cratonic Tippecanoe and Kaskasia sequences (Sloss, 1963).

Early Devonian

The Torbrook Group marks the first appearance of abundant shelly fossils in the Annapolis Belt of the Meguma terrane, and the last return of large amounts of fine grained sandstones before the Acadian Orogeny. The dark shales and fetid limestones in member 3 mark an anoxic event that may have regional significance.

Each of the three members of formation 1 are progradational parasequences of the shelf-margin systems tract (Fig. 4.34). The upper surface of formation 1 is a transgressive surface overlain by the fining upward, formation 2 sandstone. This reverse parasequence must represent the transgressive systems tract recording a rise in eustatic sea level. Black shale and fetid limestone of formation 3 indicate an anoxic interval, as in underlying groups. In a similar manner, formation 3 represents a condensed section either at the top of the transgressive tract or within

the distal part of the highstand systems tract. Conceivably, it would contain the surface of maximum flooding. Formations 4 and 5 record aggradational and then widespread progradational parasequences, respectively, of the highstand systems tract. Sequence four is terminated by the angular unconformity of the Acadian orogeny.

The offlap succession recorded by formations 3, 4, and 5 reflects a regional lowering of sea level, conceivably because of tectonic uplift before collision between the Meguma terrane and Laurasia (Fig. 4.32). The ironstones probably record a low stand of sea level, in common with a few other Peri-Gondwanan blocks including north-central, northwest, and central-west Africa (van Houten and Hou, 1990). In general, the lowest sea-level during the Paleozoic may have occurred in Late Gedinnian time (Hallam, 1984). Renewed Gondwanan glaciation is a consideration (Caputo and Crowell, 1985).

North Atlantic connection

Arguments for a West African source for the Meguma Supergroup are applicable also to the Annapolis Supergroup.

The Upper Cambrian through Lower Devonian stratigraphic column of West Africa is remarkably similar to a cratonic equivalent of the Meguma terrane in provenance, dispersal pattern, lithology, and age of stratigraphic units (Fig. 4.35). Moreover the regional unconformities on the West African craton are of sufficient scope and timing to rationalize the large quantities of sands now in the Meguma terrane. These sands originated by repeated tapping of the late Proterozoic reservoir of fine-grained detritus in western Mali and eastern Mauritania (see Fig. 3.157).

The African succession consists almost entirely of siliciclastic sediments arranged in general as three major intervals of thick sandstones alternating with two, equally thick, dominantly shaly or silty units (Beuf et al., 1971; Deynoux, 1980, 1983; Legrand, 1985). The three sandstone intervals are Late Cambrian to earliest Ordovician, Late Ordovician, and Early Devonian. They correspond to the Goldenville, White Rock, and Torbrook groups, respectively. The two shales are latest Cambrian through Early Ordovician, and Silurian. These correspond to the Halifax and Kentville groups, respectively. The Late Ordovician sandstones are glaciogenetic. The other two sandstones are fluvial, grading geographically northwestwards and stratigraphically upward into epicontinental marine sandstones. The older shale interval is the product of marine transgression from the north-northwest and north onto the West African craton. The black to blue Silurian shale is the result of post-glacial submergence (Beuf et al., 1971).

The Meguma terrane was not alone on the northern periphery of Gondwana (Cocks and Fortey, 1982; Soper and Woodcock, 1990; Robardet et al., 1990). Instead, a plethora of microcontinents or microplates rimmed North Gondwana (Fig. 4.36), including parts of northern Africa, the Middle East, southwestern Europe, central Europe, and Florida (Paris and Robardet, 1990), including both Eastern and Western Avalonia (Soper and Woodcock, 1990). The Meguma terrane may show some similarities to these areas because from Cambrian through Early Devonian time North Gondwana received detritus from the huge Saharan platform. Two areas are examples, eastern Avalonia and the Arabian Peninsula.

Sequence stratigraphy of the Paleozoic Welsh basin is remarkably similar to that of the Meguma and Annapolis supergroups (Woodcock, 1990). Four major unconformities divide the Cambrian through Lower Devonian succession into three megasequences, each defined by a supergroup. The three supergroups were deposited from Cambrian through Tremadoc, from Arenig through earliest Ashgill, and from Middle Ashgill through Early Devonian time, with some division at the Silurian-Devonian boundary. These correspond to the Meguma Supergroup, the White Rock Group, and the Kentville/Torbrook Groups. Lithologies are

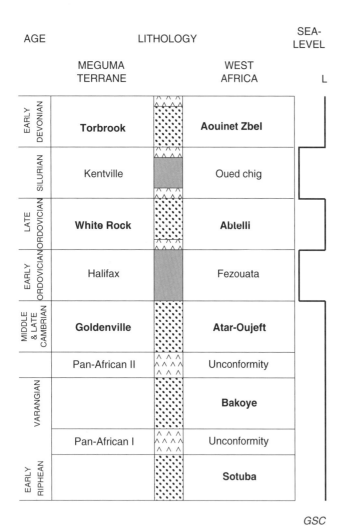

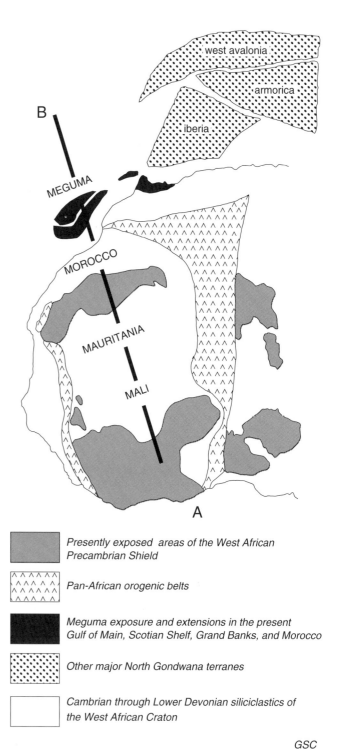

Figure 4.35. The stratigraphic section of the Meguma and Annapolis supergroups is similar to that of the West African Craton in lithologies and timing of major hiatuses. Both sections consist of three, thick, fine grained sandstones (Late Cambrian, Late Ordovician, Early Devonian) alternating with two, thick, silty, black shales (Early Ordovician, Silurian). The sandstones result from erosion of Pan-African orogenic belts and upper Proterozoic sandstones (Sotuba and Bakoye groups). The Paleozoic sandstones record sea level lows mainly due to continental glaciations of West Africa; the shales mark sea level highs between these glaciations. Volcanism is associated with major hiati in both areas.

Figure 4.36. Northwestern African portion of North Gondwana during Early Paleozoic time. Cross-section A to B is the dispersal direction of Cambrian, Upper Ordovician, and Lower Devonian fluvial to shallow marine, cratonic sandstones, from Mali toward Morocco and on to the continental margin of the Meguma terrane.

similar, including important zones of volcanic rocks. Of note are volcanic rocks at the base of supergroup 2 (equivalent of the Nictaux Volcanics?) and ash layers at its top (Elderkin Formation?). There is no evidence of Late Silurian (New Canaan Group) volcanism. Dewey (1969) first suggested a relationship between the Welsh basin and the Meguma terrane. This similarity is enhanced by reconstructions that indicate separation of Eastern Avalonia from Gondwana in the Early to Middle Ordovician, initial collision with Laurentia in the Early Silurian, and docking from Middle Silurian to Early Devonian (Soper and Woodcock, 1990).

The White Rock succession is similar to that of the Arabian Peninsula, another terrane on the periphery of Gondwana (Vaslet, 1990). The Cambrian through Lower Devonian succession consists of more than 1 km of almost entirely fine grained siliciclastic sediments. Cambrian (?) through Arenig sediments consist of coarse- to fine-grained sediments that fine up to medium- to fine-grained prodeltaic sandstones with siltstones (equivalent to the Goldenville Group?). Above a possible unconformity are transgressive, mainly argillaceous rocks of middle to late Llanvirn age (Fales River Formation?). Overlying fine grained sandstones may be equivalent to member 1 sandstone of the Deep Hollow Formation; both units have distinctive Tigillites ichnofauna. Maximum submergence is marked by possibly late Caradoc, offshore dark claystone that coarsens upward to alternating fine grained sandstone, siltstone, and claystone in both the Annapolis Belt (member 2 slate) and Arabian Peninsula. Disconformably overlying Ashgill glaciogenetic sediments consist mainly of widespread low-angled cross-stratified fine- to medium-grained sandstones that overflow paleovalleys approximately 100 m deep (member 3 sandstone?). After a hiatus, Lower Silurian offshore graptolitic shales coarsen upwards with the addition of siltstone and fine grained sandstone (Tremont Formation?). A disconformity separates this Silurian section from overlying fining- then coarsening-upward siliciclastic rocks of Early Devonian age (Torbrook Group?). The section is terminated by a disconformity beneath Carboniferous to Lower Permian strata. The Annapolis Belt is now distant from the Arabian Peninsula; however the South Portuguese Zone (eastern continuation of the Meguma terrane – Lefort and Haworth, 1981; Lefort, 1989) may have lain offshore Turkey on the southern margin of the rapidly closing Rheic Ocean during the Early Devonian (Robardet et al., 1990).

A few other features that indicate links between the Meguma terrane and northwest Africa are discussed below.

Member 3 sandstone of the Deep Hollow Formation is remarkably similar to the uppermost Ordovician (Ashgill) strata of northwestern Africa and southwestern Europe (Schenk, 1972). There, this stage records extensive continental glaciation that is well established by sedimentology, stratigraphy, paleomagnetism, and paleontology (Fig. 4.37). The African and South American portions of Gondwana were extensively glaciated from Caradoc to Llandovery times, respectively (Caputo and Crowell, 1985). A near-polar or at least high-latitude setting of the Meguma terrane during the Early Ordovician might explain the interpreted storm-swept nature of the overlying White Rock lithosomes. The ichnogenus Tomaculum

Groom (H. Hofmann, 1980 pers. comm.) in this matrix has been previously found only in Spain, France, Britain, Germany, and Czechoslovakia (Hantzschel, 1975).

The Armorican Massif as well as Morocco shows a lacuna that spans most of the Llandovery Epoch. The Wenlock strata consist of a condensed section, only a few metres thick, of widespread graptolitic black shales that are typical of the whole North Gondwana marine platform (Robardet et al., 1990). In Morocco, upper Wenlock through Ludlow strata consist of dark, laminated shales distinguished by black, fossiliferous calcite concretions (Destombes et al., 1985). These are similar to the Kentville Group. In the extreme north of Morocco the shales change to fossiliferous black limestones. Basalts are intercalated with these upper Wenlock through Ludlow strata at several localities in the Coastal Meseta (Destombes et al., 1985), and may be equivalent to the New Canaan Group of the Annapolis Supergroup. The Lower Ordovician Armorican arenite facies is widespread in Western Europe and northwestern Africa and could be correlative with the Fales River Formation and lower member of the Deep Hollow Formation. Upper Ordovician sheet arenites are widespread over northwestern Africa and Western Europe as outwash deposits of Saharan glaciation.

The Silurian shales may derive from the reworking and recycling of the finest fraction of the glacial deposits during marine sedimentation. These shales constitute almost the entire condensed succession of the Silurian in which all epochs are represented in a section that is less than 100 m thick. The Late Ordovician glaciation sorted the sediment in a gigantic manner. The finest particles were transported to the margins of the continent by melt-waters as the ice-cap shrank, and were then reworked during the eustatic Silurian marine transgression (Deynoux, 1980). Eolian transportation must also have been very important over this vast, unvegetated outwash plain. The large quantity of angular silt and very fine grained arenite in the Kentville and Torbrook groups of the Meguma terrane is probably derived form glacial loess, reworked during such a submergence (Fig. 4.38). Similar dark, graptolitic shales and mudstones of the Kentville Group developed on the continental shelf-like areas of North Africa, southern Europe and the Middle East (Berry and Boucot, 1973). In Portugal, Spain, the Pyrenees, and Normandy, the Early Ashgill was a time of widespread basic and acidic volcanism, both submarine and subaerial (Tamain, 1978), possibly equivalent to the Elderkin Formation.

The central Sahara doleritic sills or flows are intercalated with uppermost Silurian and/or lowermost Devonian strata (Beuf et al., 1971), reminiscent of the New Canaan Group.

The abundant fauna of Rhenish affinity in the Torbrook Group offers the best paleontological connection with other areas. It is part of the Rhenish-Bohemian subprovince that includes middle and southern Europe and North Africa (Boucot et al., 1968). Dictyonema flabelliforme is an index fossil of the Atlantic province, which includes the area of Scandinavia to North Africa (Skevington, 1978). Monograptus nilssoni and M. colonus are also characteristic of the Ludlow in western Europe and North Africa (Bouyx et al., 1985). The crinoid Scyphocrinites is particularly abundant in a zone that extends from middle Europe to North Africa (Termier and Termier, 1964). Further evidence of a European faunal association is the presence of

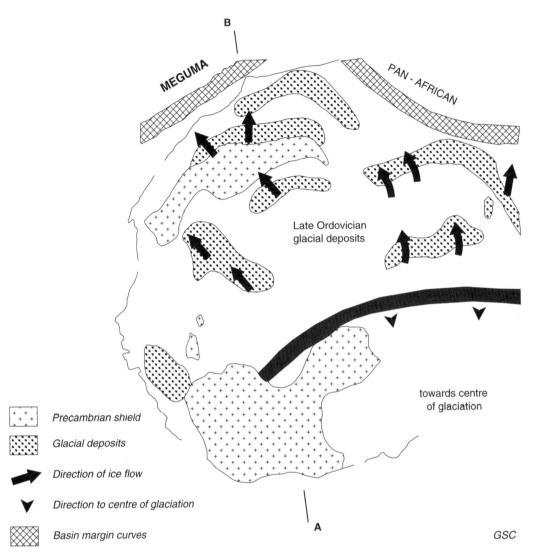

Figure 4.37. Late Ordovician (440 Ma) glaciation of northern Gondwana was directed north-northwestward toward Morocco (and the ancient site of the Annapolis Supergroup) from the West African Craton (Deynoux, 1985). Terranes of North Gondwana are characterized by outwash sandstones and ice-rafted erratics.

several species of the trilobite *Acastella*, not previously reported in North America (A. Ormiston, pers. comm; Jensen, 1975). Thus the fauna of the Meguma terrane has a distinct Eurafrican stamp (Bouyx et al., 1985).

Structure

At least three generations of folds and faults affect sedimentary strata of the Meguma terrane (Fyson, 1966). The main folds are low plunging and upright, and describe an arc that trends northerly in the southwest to easterly in the east. The axes of some folds are traceable for at least 55 km. Fold axes have local variations in plunge which develop smaller domes and basins that control gold occurrences (O'Brien, 1986). The wave-lengths of the folds are very

regular at approximately 15 km. Fold plunges are generally shallow, but steep plunges are locally recorded along the axial traces. The axial surfaces of the major folds are near vertical so that only locally are limbs overturned. These surfaces parallel pervasive, steep axial-plane foliation. A second generation of cross-folds are common in the Halifax Group. Axial planes are steep and trend north to northeast; plunges are mainly steep. Granites are younger than these folds. A third generation of kink-folds and kink-bands trends northwesterly and plunges steeply to the southeast. They appear to be younger than the granites (Fyson, 1966). They parallel a principal joint set and so may be of similar origin (Smitheringale, 1973).

Several types of foliation are developed. Sub-vertical slaty cleavage that is parallel with axial surfaces of folds is most common. In some localities crenulation cleavage is

PROVENANCE

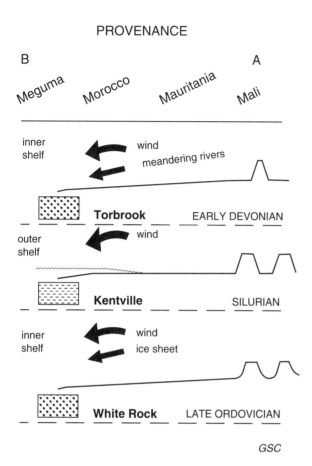

GSC

Figure 4.38. The northerly slope of the West African Craton maintained its stability from Cambrian to Early Devonian time. Incremental erosion in the source area of Malian Proterozoic fine grained sandstones left scattered buttes and mesas. Sands were carried by Late Ordovician braided streams and Early Devonian meandering rivers to the margin of North Gondwana. Saharan deglaciation caused extensive flooding of the craton from the north during Silurian time. Northwestwards eolian transport of silts and fine grained sands was extensive, especially during the Late Ordovician due to glacier winds. This may account for the large quantity of very fined grained sediments in the Annapolis Supergroup. See also Figures. 3.157 and 3.158 in chapter 3.

developed. Spaced cleavage in sandstones is better developed in the eastern rather than southwestern part of the terrane. Sub-vertical fracture cleavage is associated with steeply plunging kink folds that deform the slaty cleavage (O'Brien, 1986).

Certain map units in theFeltzen thicken and thin across particular folds. Significant structural thickening or thinning is unlikely because this effect is normally seen only in one of several members affected by the folding. Some anticlines may be sited on areas that were isostatically unstable during deposition and later became sites of Carboniferous basins (O'Brien, 1986).

Whereas some granitoid plutons were implaced by passive magmatic stoping, metamorphic aureoles around others are structurally complex. Folds are highly discordant to the regional pattern and have axial traces that parallel granite contacts.

Faults are relatively common in the Meguma terrane and generally trend northwesterly. They are both normal and reverse; however strike-slip movements have also occurred (Fyson, 1966). Large sinistral faults strike northwest. These faults are similar in strike and displacement to kinked zones and may be genetically related (Fyson, 1966). The sinistral faults and possibly the kinks were formed both before and after the emplacement of granites. Dextral faults trend northeast (Fyson, 1966). Underground workings at various gold properties show numerous small-scale, northwest-trending faults that do not continue for great distances along strike, and are of shallow depth (Stevenson, 1959). The proposed Tobiatic Fault Zone of Giles (1985) is a major sinistral fault that would extend from Yarmouth to Canso, effectively dividing the Meguma terrane into two equal parts. Giles suggested that reconstruction of 110 km of post-Viséan sinistral slip would merge the two largest Carboniferous basins in the province, unite batholiths of the Meguma terrane into a near-circular body, combine gravity anomalies, and rationalize metal occurrences. This fault may extend over 700 km southwest into the Gulf of Maine and continue as the Georges Basin Fault System of Ballard and Uchupi (1975).

The basal contact of the White Rock Group is an important horizon for determining whether Ordovician orogeny affected the Meguma Supergroup. It is everywhere conformable and usually gradational except at Cape St. Marys. The uppermost few metres of the Halifax Group at this locality only contain isoclinal, recumbent folds. The contact has been interpreted as a local angular unconformity (Taylor, 1965), a disconformity (Rodgers, 1970), and a conformable contact (Lane, 1981). Schenk (1972) suggested that the local folds may be due to drag folding before deposition of the overlying White Rock Group (i.e. ice-thrust features, or involutions). Lane (1981) concluded that they were due to the interface between structurally incompetent and competent strata. Keppie (1982) considered them to be younger than the White Rock and the result of local, low-angle thrusting. Taylor (1965) concluded that on structural grounds, no Ordovician effects are evident. However he suggested that the increase in thickness and proportion of volcanic rocks southwestward in the White Rock Group indicated a disturbance

Possibly the maximum horizontal compression across the Meguma terrane was originally oriented northwest to north, perpendicular to the first generation of folds to result in vertical extension and regional metamorphism (Fyson, 1966). The stress direction then changed to east-west, forming the second fold generation and continued both before and after granites were intruded. This led to the third generation folds and sinistral faults striking northwestwards, and to dextral faults striking northeast to eastward.

Metamorphism

Both the Meguma and Annapolis supergroups exhibit regional and contact metamorphism. Greenschist facies occurs in the central and eastern parts of the terrane, with amphibolite facies in its southernmost part. All of the significant mineral deposits, including gold, are restricted to rocks of greenschist facies, with the exception of some beryl and molybdenum (Taylor and Schiller, 1966). Regional metamorphic patterns show increasing grades marked by chlorite, biotite, almandine garnet, andalusite-cordierite-staurolite, and sillimanite zones. Regional isograds outline a thermal dome in the southwest. The isograds show no apparent geometric relationship to batholiths or associated plutons, and commonly strike perpendicular to contacts. Regional metamorphism therefore predates the granitoid intrusions and low grade contact aureoles, up to 3 km wide, clearly overprint the regional isograds. Contact metamorphism is in the hornblende-hornfels facies, with temperatures of 550-700°C and water pressures between 1000 and 3000 bars (Taylor and Schiller, 1966). In higher grades, evidence of overprinting is less clear and probably indicates a reduced country-intrusion temperature contrast at deeper crustal levels (Clarke and Muecke, 1980). The age of regional metamorphism has been set at between 412 and 400 Ma (Reynolds and Muecke, 1978); the granitoid plutons begin to intrude at 408 Ma and reach a climax after folding at 370 to 360 Ma (Cormier and Smith, 1973). The metamorphic assemblages of the Meguma and Annapolis Supergroups indicate conditions of low-medium pressure and high temperatures, akin to those of Pyrenean-type metamorphic belts. Peak conditions are estimated to have reached P_{total} = 3.5-4.5 kbar and 650°C (Muecke, 1973; Clarke and Muecke, 1980). Effects of metasomatism and retrogressive metamorphism are slight (Taylor and Schiller, 1966). Taylor and Schiller (1966) summarized the metamorphic-deformational history as: (1) deformation producing the major folds; (2) contemporaneous low grade and locally higher grade metamorphism; (3) after folding had essentially ceased, the major period of metamorphism took place; and (4) later deformation producing cross-folds.

Granulite xenoliths occur in a mafic dyke at Tangiers. Xenolith textural and compositional data reveal a three stage P-evolution. Temperatures were 1030-1166°C for stage one and 710-820°C for stage two. Pressure estimates range from 14.2-14.8 kbar for stage one, 5.6-8.7 kbar for stage two. Peak metamorphic conditions for stage one evolved through moderate decompression (3 kbar) with cooling to approximately 380°C. Stage two records rapid decompression and heating of the xenoliths following incorporation by the ascending dyke magma. Stage three represents a later (low pressure-temperature) hydration event affecting both xenoliths and dyke matrix (Chatterjee and Giles, 1988).

Plutonism

Both the Meguma and Annapolis supergroups have been intruded by the South Mountain Batholith and its associated satellite plutons. These intrusions have been the object of intensive geological, geochemical, and geophysical investigations over the past decade. Much of this initial work provided the framework for extensive mineral exploration which culminated in the discovery and development of North America's largest tin deposit as well as the delineation of numerous prospects with appreciable tin, tungsten, and uranium mineralization. The South Mountain Batholith outcrops over 10 000 km² in the Meguma Zone and is the largest peraluminous granitoid complex in the Appalachian Orogen. It has a three-dimensional mushroom shape, 5 km thick on the margins and as much as 25 km thick at its centre (Garland, 1953). It is a typical massive, post-tectonic, undeformed, epizonal granitoid characterized by a discordant relationship with the country rocks, a pronounced thermal metamorphic aureole, and miarolitic cavities. A variety of techniques tightly constrain the age of the intrusions to an interval between 372 and 360 Ma (Late Devonian; Clarke et al., 1980; Clarke and Halliday, 1980; Reynolds et al., 1981). An exception to this general rule is a number of so-called southern satellite plutons with ages of ca. 316 Ma (middle Carboniferous). Petrologically the batholith consists largely of biotite granodiorite, two-mica monzogranite, and leucomonzogranite showing a variety of textural types. It contains a suite of minerals characteristic of peraluminous granitoids (andalusite, garnet, cordierite, tourmaline, topaz, etc.). Major and trace elements show that several similar comagmatic suites exist, as well as smaller volumes of rock subsequently affected by strong metasomatic alteration (para-intrusive suites), which are in places associated with economic mineral deposits. Evidence from both radiogenic and stable isotope geochemistry points to a primitive crustal, but not a Meguma flyschoid metasedimentary, source for the magmas. The possibility of a mantle component in the lithodeme cannot yet be excluded. That is, the batholith was derived from crustal rocks more primitive than the Meguma Supergroup, with or without some component of mantle input, either energy and material, or energy alone. Much of the compositional diversity of the batholith is attributable to three processes: fractional crystallization, reaction with the country rocks, and late-stage interaction with fluids (Clarke and Muecke, 1985).

The Liscomb Complex (190 km northeast of Halifax) consists of high grade gneisses, peraluminous granitoid intrusions, and two quartz-gabbro to quartz-diorite breccia pipes. The gneisses occur as a domal uplift, cut by the igneous intrusions. Sr-Nd isotopic study indicates: (1) intermediate gneisses have the highest epsilon Nd, calcic gneisses have lower values, and a third isotopically lowest group is similar to the Meguma Supergroup; (2) Liscomb granitoids in part isotopically overlap the intermediate gneisses; and (3) the mafic intrusive rocks represent magmas derived from slightly depleted to slightly enriched mantle sources. Other peraluminous granitoids of the Meguma terrane may have been derived from intermediate gneisses similar to those of the Liscomb Complex, modified by assimilation of Meguma Supergroup metawackes and metapelites (Clarke and Chatterjee, 1988).

Summary and significance

Strata of the Annapolis Belt record three, major, transgressive-regressive cycles (Schenk, 1991). Stratigraphically they are equivalent to the White Rock, Kentville/New Canaan, and Torbrook groups. The cycles correspond to supercycles of Vail et al. (1977).

Each cycle begins with a basal sandstone, followed by a black shale, and ends with a siltstone and/or sandstone. An exception is the third cycle (Silurian) in which basal sandstones are absent. Each cycle is terminated by igneous activity, usually in the form of subaerial volcaniclastics but also extrusive or intrusive sheets. These units indicate subaerial exposure and probable significant hiati, although fossil control is unavailable. They are interpreted as unconformities, specifically paraconformities, so that the cycles are sequences. Although the Annapolis Belt was deposited on a continental margin, its three sequences are similar in age to cratonic sequences; namely, the Tippecanoe, and Kaskasia (Sloss, 1963), with two modifications: (1) subdivision of the Tippecanoe by the Saharan, glacioeustatic sea-level low; and (2) interruption in the lower Kaskasia by the Acadian unconformity.

Four, relatively coarse grained units were deposited during times of low sea-level (Fig. 4.32). The sandstones in the White Rock and Torbrook formations were deposited in storm-swept, shallow water of the upper part of the Gondwanan continental embankment. Sedimentological characteristics and stratigraphic position of the White Rock sandstones suggest an outer shelf origin, whereas those of the Torbrook indicate inner shelf or estuarine settings (Jensen, 1975). The Deep Hollow Formation is lithologically and stratigraphically correlative with Saharan glacial deposits of Gondwana. The absence of a basal sandstone in the overlying Silurian sequence may reflect rapid submergence caused by melting of the ice sheet, as has been interpreted elsewhere (e.g. Beuf et al., 1971; McKerrow, 1979).

Four thick units of black shale are important units in the Annapolis Belt. The Tremont Formation represents an oceanic anoxic event that would result from poor circulation of deep water in the absence of cold, well-oxygenated bottom currents downwelling from ice-covered polar areas (Jenkyns, 1986). Shales of the remaining two anoxic episodes may have regional importance. They are relatively thin, and occur within predominantly sandy formations; the thick dark slate (member 2) in the Deep Hollow Formation and member 3 of the Torbrook Formation (Siegenian; Schenk, 1991).

Volcanism was also important despite the previous interpretation of the Meguma terrane as a passive margin (Schenk, 1981). Widespread sheets of both volcaniclastic rocks and lavas line the sub-White Rock, sub-Kentville, and sub-Torbrook unconformities. Because all of these are mainly ashflows, at least nearby subaerial conditions are indicated. Their stratigraphic positions between underlying regressive and overlying transgressive successions suggest paraconformities (Schenk, 1991).

Acknowledgments

Operating grants over the last 25 years from the Natural Sciences and Engineering Research Council of Canada have funded my study of sedimentary strata in the Meguma Zone. T. E. Lane made significant contributions to knowledge of the White Rock Group. Comments by R. Boyd and J.W.F. Waldron improved early drafts of the manuscript. I especially appreciate the review by H. Williams.

NEWFOUNDLAND
Introduction
Harold Williams

Rocks included in this temporal category in Newfoundland range in age from Middle Ordovician (Caradoc) to Middle Devonian, although the majority are of Silurian age. Rocks of the Clam Bank Belt were viewed traditionally as unconformable upon the Humber Arm Allochthon, linking the allochthon with platformal rocks farther west. However, another interpretion (Stockmal and Waldron, 1990) favours a faulted contact, with rocks of the Clam Bank Belt thrust eastward as the upper level of a structural triangle zone. The Springdale Belt links the deformed Humber miogeocline with deformed ophiolitic rocks of the Notre Dame Subzone of the Dunnage Zone. Rocks of the Cape Ray Belt are unconformable upon the plutonic and metamorphic rocks of the Dashwoods Subzone of the Dunnage Zone and farther north they are unconformable upon ophiolitic rocks of the Notre Dame Subzone of the the Dunnage Zone. Rocks of the Badger and Botwood belts contain lower units of turbidites that are conformable upon Caradoc shales of the Exploits Subzone of the Dunnage Zone. Rocks of the La Poile Belt are unconformable upon the Roti Granite (560 ± 4 Ma; Dunning et al. 1988; O'Brien, 1989) that may be an Avalon correlative. The Fortune Belt has conglomerates that contain detritus from the Gander Lake Subzone of the the Gander Zone and these are unconformable upon upper Precambrian and other rocks of the Avalon Zone.

Most rocks of the Clam Bank Belt are dated by shelly faunas, and brachiopods and other forms are relatively abundant in rocks of the Badger Belt. Fossils are sparse in rocks of the other belts and correlations are based on isotopic ages or tenuous lithic comparisons. Silurian graptolites have been discovered recently in rocks of the Badger Belt (S.H. Williams and O'Brien, 1991).

The stratigraphic sections of all belts include red sandstones and conglomerates toward their tops, and terrestrial volcanic rocks are abundant in the Springdale, Botwood, Cape Ray and La Poile belts. All of these rocks are dated or inferred to be of Silurian or Devonian age. Lower Silurian or Ordovician rocks, where present, display sharp contrasts from one belt to another, e.g. limestones of the Clam Bank Belt compared to coarse turbidites of the Badger Belt. Net stratigraphic thicknesses in places exceed several kilometres.

Much of the available information on the Newfoundland middle Paleozoic rocks is found in government reports and on accompanying geological maps. University theses studies have been carried out in all the belts, and the geology and conceptual models for rocks in some belts are topics of papers in current geological periodicals.

Williams, H.
1995:   Introduction: Middle Paleozoic rocks-Newfoundland; in Chapter 4 of Geology of the Appalachian-Caledonian Orogen in Canada and Greenland, (ed.) H. Williams; Geological Survey of Canada, Geology of Canada, no. 6, p. 383-384 (also Geological Society of America, The Geology of North America, v. F-1).

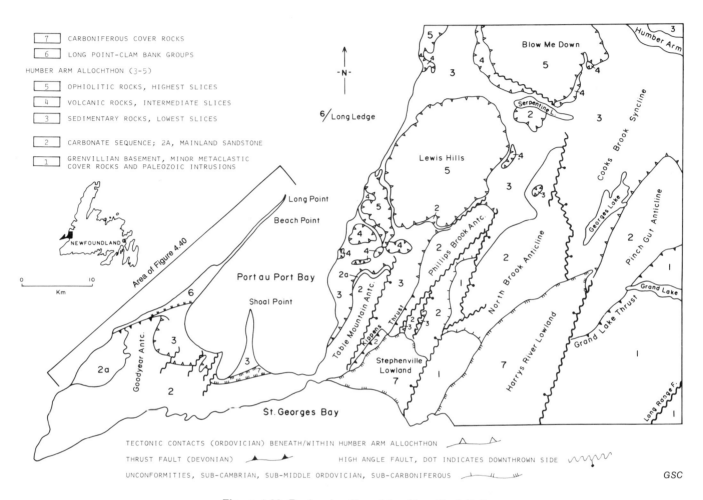

Figure 4.39. Regional setting of the Clam Bank Belt.

References prior to 1967, which deal with middle Paleozoic rocks in Newfoundland, are given in Williams (1967). Only the more recent references are included with the analyses here. The belts are described from west to east.

Clam Bank Belt

Harold Williams, L.E. Fåhraeus, and R.K. Stevens

Definition and setting

The Clam Bank Belt of western Newfoundland is defined on a small outcrop area of Middle Ordovician to Devonian rocks along the western shore of the Port au Port Peninsula

Williams, H., Fåhraeus, L.E., and Stevens, R.K.
1995: Clam Bank Belt; *in* Chapter 4 of Geology of the Appalachian-Caledonian Orogen in Canada and Greenland, (ed.) H. Williams; Geological Survey of Canada, Geology of Canada, no. 6, p. 384-388 (*also* Geological Society of America, The Geology of North America, v. F-1).

(Fig. 4.39 and 4.40). Middle Paleozoic rocks there form a homoclinal sequence that is separated into the Long Point and the Clam Bank groups. The Long Point Group contains the youngest Ordovician fauna in western Newfoundland and the Clam Bank Group contains the only Silurian-Devonian fauna in this region. Although diminutive, rocks of the belt provide an important record of deposition and deformation between the Middle Ordovician emplacement of the Humber Arm Allochthon and deposition of Carboniferous cover rocks. Because Port au Port Peninsula projects 20 km seaward of the west Newfoundland shoreline, the Clam Bank Belt provides a record of rocks and relationships otherwise hidden in the Gulf of St. Lawrence.

The Long Point Group is dominated by platy fossiliferous limestone, grey limy sandstone, and shale. The Clam Bank Group consists mainly of redbeds. No complete section is exposed and thicknesses are estimated from partial shoreline sections and map patterns. Thickness estimates for the Long Point Group are 500 m (Schuchert and Dunbar, 1934), 600 m (Rodgers, 1965; Fåhraeus, 1973), and 400 m (Bergstrom et al., 1974). The thickness of the Clam Bank Group is approximately 450 m with top unexposed (Rodgers, 1965). The Long Point Group extends 32 km from Three Rock Point (Fig. 4.40) to the tip of Long Point, and it

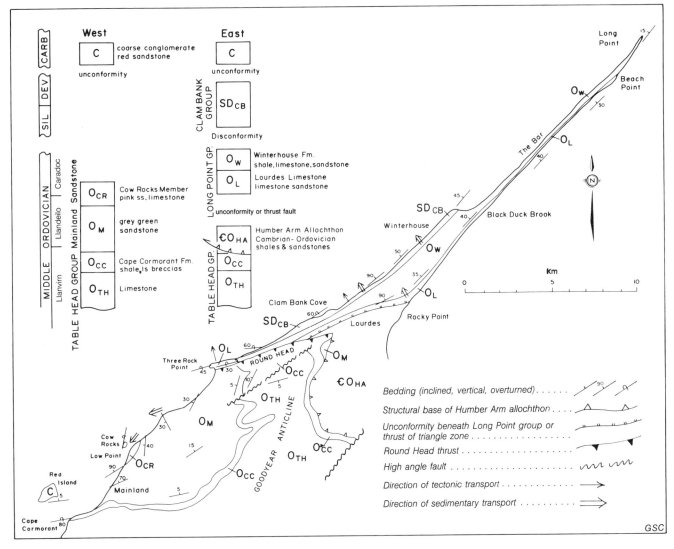

Figure 4.40. General geology of the Clam Bank Belt.

outcrops 12 km farther northeastward along strike as a series of narrow islands and shoals, Long Ledge (Fig. 4.39). Lilly (1966) suggested that the Clam Bank Group extends northeastward from its type area for an additional 200 km where correlative subhorizontal redbeds occur less than 15 km from the west Newfoundland shoreline.

The Long Point Group dips moderately northwestward above Lower Ordovician largely chaotic rocks of the Humber Arm Allochthon at Rocky Point (Fig. 4.40). Toward the southwest, the sequence is overturned and interpreted as overthrust by lower Middle Ordovician rocks of the Table Head Group. The Clam Bank Group overlies the Long Point, but the contact is poorly exposed. Toward the northeast at Winterhouse (Fig. 4.40), the Clam Bank Group dips moderately northwest. Toward the southwest at Clam Bank Cove, its beds are overturned.

The Long Point Group has been viewed as an unconformable cover upon the Humber Arm Allochthon; thus providing an upper limit to the time of emplacement of the Allochthon (Rodgers and Neale, 1963; Rodgers, 1965; Stevens, 1965, 1970). New offshore seismic reflection profiles suggest that the Long Point-Clam Bank groups form the upper structural level of a triangle zone, like similar structures that mark the foreland limit of other orogens (Stockmal and Waldron, 1990). This suggests that the Long Point-Clam Bank groups are thrust eastward above the Humber Arm Allochthon and that the underlying Cambrian-Ordovician carbonate sequence and Humber Arm allochthon are all transported with respect to underlying rocks. The presence of slickensides at the basal surface of the Long Point Group at Rocky Point and the absence of a basal conglomerate support the structural interpretation (Stockmal and Waldron, 1990).

West of the Goodyear Anticline and south of Three Rock Point (Fig. 4.40), the Llanvirn Table Head Group is overlain conformably by about 500 m of greenish sandstone, the Mainland Sandstone. Dips are gentle toward the west and they are steeper at the coast where bedding is locally overturned. Crossbedded pink sandstones with limestone beds, Cow Rocks Member, occur at the top of the section at Cow Rocks. Conodonts from Cow Rocks limestones are the same general Caradoc age as the lower Long Point Group. This is interpreted as a key relationship that limits the Ordovician leading edge of the Humber Arm Allochthon in the local stratigraphy (Stevens, 1970).

Conglomerates and sandstones at Red Island (Fig. 4.40) are subhorizontal and unaffected by the latest deformation of the Silurian-Devonian Clam Bank Group. They are interpreted as Carboniferous and the youngest rocks in the region. The Red Island conglomerates contain pebbles, cobbles, and boulders of mainly silicic volcanic rocks. Similar boulders, exotic to the Port au Port region, are strewn along the western beaches of the peninsula and imply the presence of Carboniferous conglomerates immediately offshore. The provenance of the volcanic boulders is unknown but they are most likely intraformational and derived from Carboniferous volcanic units like those present in stratigraphic section to the southwest of the area, e.g. Îles de la Madeleine.

A distinctive magnetic anomaly, the Odd Twins magnetic anomaly (Ruffman and Woodside, 1970), extends 65 km northward from the tip of Long Point and it is coincident, or nearly so, with the northeastward projection of the Long Point-Clam Bank section, and the eastern depositional edge of offshore Carboniferous rocks. The anomaly has been interpreted as two parallel mafic dykes dipping gently westward and located near the base of the Long Point Group (Ruffman and Woodside, 1970). Because the anomaly is parallel to depositional trends, it could also represent inclined mafic volcanic units, either within the Clam Bank Group or at the base of the overlying Carboniferous section.

Stratigraphy and lithology

The Long Point Group (Riley, 1962; Bergström et al., 1974; Schillereff and Williams, 1979; Williams, 1985), formerly the Long Point Series (Schuchert and Dunbar, 1934) or Long Point Formation (Rodgers, 1965; Fåhraeus, 1973), has been the subject of a number of stratigraphic and paleontological studies. It consists of two formations; the Lourdes Limestone and overlying Winterhouse Formation (Bergström et al., 1974). The Lourdes formation consists of nodular to platy limestones with coralline (Labyrinthites) reef mounds, and thin bedded platy limestones with thin shale and sandstone interbeds. Its base is marked by a calcareous quartz sandstone with interlayered green shale and siltstone. The stratigraphic section, about 80-90 m, has been measured along the east shore escarpment of the Long Point peninsula (Rodgers, 1965; Fåhraeus, 1973; Bergström et al., 1974) and faunas and age are reported by Fåhraeus (1973), Bergström et al. (1974), Copeland and Bolton (1977), and Dean (1979). The Winterhouse Formation consists of shale, limy sandstone, and fossiliferous sandstone; and it includes coarse, but thin boulder conglomerate beds on the northwest shoreline of the Long Point peninsula.

The basal strata of the Lourdes Limestone overlie red and green clayey shales like those of the nearby chaotic Humber Arm Allochthon. This relationship has been seen on the coastline at Rocky Point (Rodgers, 1965; Brueckner, 1966; Stockmal and Waldron, 1990), but periodic landslides hide the critical contact. Farther southwest, the Lourdes Limestone is steep to overturned and it is interpreted as overthrust by the Cape Cormorant Formation of the Table Head Group at Round Head. Previous interpretations of relations between the Long Point and nearby groups have been summarized by Rodgers (1965) and Stockmal and Waldron (1990).

The contact between the Lourdes and Winterhouse Formations is described as gradational (Bergström et al., 1974) but it has not been outlined on geological maps. As defined by Bergström et al., (1974), the Lourdes Limestone includes the complete coastal cliff section east of Black Duck Brook. For purposes of mapping and practicality, Williams (1985) selected a planar surface of erosional disconformity in the Black Duck Brook section, between nodular reefy limestone below and thin bedded limestone, siltstone and shale above, as a sharp, easily identified boundary between lithic units of the Long Point Group. However, this boundary as shown in Figure 4.40 is somewhat below the Lourdes-Winterhouse contact as described by Bergström et al. (1974). The top of the Winterhouse Formation contains grey sandstones in a stream section at Clam Bank Cove that pass upward into maroon to reddish sandstones. The relationship suggests a gradation into the overlying Clam Bank Group but a continuous section with adjacent Clam Bank redbeds is not exposed.

The Clam Bank Group (Riley, 1962; Williams, 1985), earlier the Clam Bank Series (Schuchert and Dunbar, 1934) or the Clam Bank Formation (Rodgers, 1965; Fåhraeus, 1973; Bergström et al., 1974), consists of red crossbedded sandstone, red to grey and green sandstone, siltstone and shale, and fossiliferous limy siltstone and shale at the south end of Clam Bank Cove. North of Clam Bank Cove the beds are uniformly coarser and redder, compared to those to the south. Trough crossbeds indicate currents that flowed from southeast to northwest and Rodgers (1965) has reported serpentinite fragments in the coarse sandstones.

The Mainland Sandstone is Llanvirn at its base and it is older than the Long Point Group, except for its uppermost beds at Cow Rocks (James et al., 1980). The sandstone is conformable above the Cape Cormorant Formation of the Table Head Group and most of the section consists of grey to green friable sandstone. Chromite detritus has been reported by Stevens (1970) and sole markings indicate currents that flowed from east to west. The uppermost unit of the Mainland Sandstone is crossbedded quartz sandstone, Cow Rocks Member, with interlayers of limestone at its top. The Mainland section represents a transition from the top of the Table Head Group through a turbiditic sequence coeval with the emplacement of the Humber Arm Allochthon, into a quartz sandstone-limestone unit equivalent to basal beds of the Long Point Group.

Age

The Long Point Group contains a diverse fauna including brachiopods, corals, bryozoans, gastropods, cephalopods, ostracods, trilobites, graptolites, and conodonts. The age of the formation has been reviewed by Bergström et al. (1974). The Lourdes Limestone is assigned to the Porterfieldian-Bolarian that is equivalent to portions of the *Diplograptus multidens* and *Nemagraptus gracilis* zones. Lower parts of the Winterhouse Formation are of the same Middle Ordovician age (Fåhraeus, 1973; Bergström et al., 1974), whereas graptolites from its upper parts indicate a Cincinnatian or Late Ordovician age (Bergström et al., 1974).

South of Clam Bank Cove, the Clam Bank Group contains a brachiopod fauna of latest Silurian-earliest Devonian (Pridoli) age (Rodgers, 1965; Corkin, 1965; Berry and Boucot, 1970). North of Clam Bank Cove the red sandstones contain fresh water shells and O'Brien (1975) has reported an Ordovician fossil. Pending further work and confirmation, two sections of different ages are possible within the Clam Bank Group.

The Mainland Sandstone contains Llanvirn graptolites at its base and a younger Llandeilo graptolite assemblage toward its top. Conodonts from the Cow Rocks member are of Midcontinent zone 6 to 7 age, equivalent to late Llandeilo or early Caradoc. This establishes the Cow Rocks member as partly coeval with basal beds of the Long Point Group.

Structure

North of Lourdes, the Long Point-Clam Bank section dips uniformly northwest, beds are upright and dips are progressively more gentle toward the northeast (Fig. 4.40). At Lourdes, bedding is steep to vertical, the strike of the section swings more westerly, and farther southwestward the section is overturned toward the northwest. Rodgers (1965) has suggested that the overturned part of the section has been pushed forward, i.e. northwestward, by the competent Cape Cormorant breccias of Round Head in response to the rise of the Goodyear Anticline. This interpretation is supported by the local relief of Round Head that lies southeast of the overturned section. Thrusting also explains the narrow width of the Long Point outcrop belt west of Round Head, and it is supported by slickensides and outcrop-scale faults in the Long Point Group that indicate southeast over northwest structural polarity (Fig. 4.40). Furthermore, the degree of overturning diminishes northwestward and away from Round Head, and in local cliff exposures it diminishes downward, thus implying only local overturning controlled by the Round Head thrust.

Where the Mainland section is overturned at Cow Rocks, the zone of overturning is roughly along trend of the Round Head Thrust (Fig. 4.40). These structures are interpreted as Devonian (Acadian) since they affect the Clam Bank rocks but have no apparent effect on nearby Carboniferous rocks at Red Island. In the model of a triangle zone (Stockmal and Waldron, 1990), the Round Head Thrust is interpreted as a late feature that overprints structures of the triangle zone.

Depositional environments

Sandstones at the base of the Long Point Group are interpreted as crossbedded beach sands with foresets dipping northwest and facing the axis of a depositional basin (Fåhraeus, 1973). Overlying micritic limestones, shales and reefy limestones suggest tidal flats, intertidal carbonate banks and local fringing reefs. The length and continuity of Labyrinthites reefs along the full length of the Lourdes Limestone suggest that the present structural strike is parallel to depositional trends. Erosional transgression above the reefy limestones, increasing shales in the Winterhouse Formation, and the appearance of trilobites and local graptolites imply more open marine conditions. An increase in sandstone at the top of the Winterhouse Formation and a transition into coarser reddish sands implies uplift and a major change to a continental alluvial environment (Fåhraeus, 1973).

Red crossbedded sandstones of the Clam Bank Group indicate a continuation of continental fluviatile deposition, and trough crossbeds and sedimentary makeup imply derivation of the clastic material from the east.

The Mainland Sandstone was deposited in a marine environment in response to the emplacement of the Humber Arm Allochthon. Its turbiditic sandstones pass upwards into crossbedded fluviatile sandstones at Low Point (Fig. 4.40) and limestones at Cow Rocks. These lithologies are more akin to the Long Point Group and typical of shallow water deposition. Crossbedding in sandstones of the Cow Rocks Member indicate currents that flowed toward the southwest.

Conodonts from the top of the Lourdes Limestone are representative of the Atlantic Faunal Province (Sweet et al., 1959; Fåhraeus, 1973). Conodonts from the top of the Winterhouse Formation are typical of the North American Midcontinental Faunal Province, and mixed conodont faunas occur between these stratigraphic levels. Different faunas in the same relatively thin stratigraphic section warn against implications of major geographic separations based on conodont faunal realms alone.

Tectonic significance

Rocks of the Clam Bank Belt provide continuity between the St. Lawrence Platform and deformed rocks of the Appalachian Orogen. The original eastward extent of rocks of the Clam Bank Belt is unknown but they may have covered much of the external Humber Zone before Acadian deformation.

During deposition of the intertidal to marine Long Point Group, the nearby presently elevated Humber Arm Allochthon must have been at or below sealevel. A change to continental conditions indicated by red sandstones at the top of the Long Point Group and throughout the Clam Bank Group, coupled with provenance and sedimentary makeup of the Clam Bank sandstones, indicate rising land to the east.

Correlation of the Cow Rocks Member of the Mainland Sandstone and basal parts of the Lourdes Limestone provides a key linkage between a section without transported rocks and a nearby section that is unconformable or thrust above the Humber Arm Allochthon. The continuity of the Long Point Group and northeastward submarine extensions of the Clam Bank Group, approximate the Ordovician leading edge of the Humber Arm Allochthon, or Logan's Line.

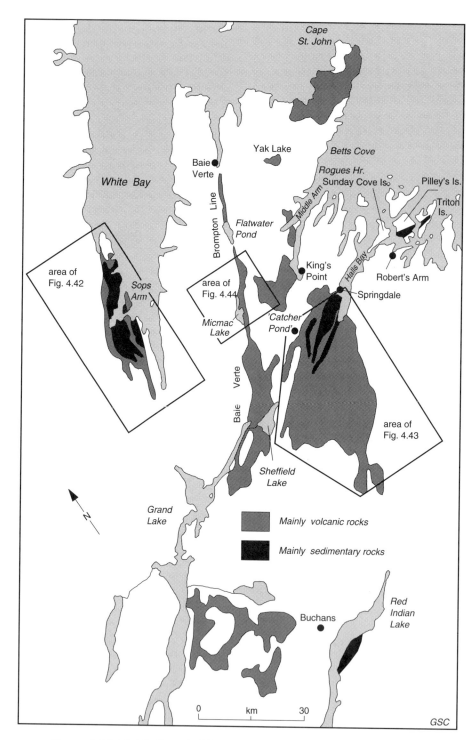

Figure 4.41. Distribution of middle Paleozoic rocks of the Springdale Belt.

Springdale Belt

Introduction

Harold Williams

The Springdale Belt is defined on its extensive terrestrial volcanic rocks and associated terrestrial to shallow marine clastic sedimentary rocks. The rocks are unfossiliferous except for occurrences of Middle Silurian fossils at Sops Arm of White Bay. Unconformities between rocks of the Springdale Belt and underlying Ordovician or older rocks are recorded wherever basal relationships are preserved. Granites of Silurian and Devonian age cut rocks of the Springdale Belt and all are locally overlain by undeformed Carboniferous rocks. Lithic similarities among rocks throughout the belt, isotopic ages, and similar regional settings, suggest that most rocks are of Silurian age.

Rocks of the Springdale Belt extend from White Bay to Cape St. John and southward to Grand Lake and Red Indian Lake (Fig. 4.41). Those at White Bay occur upon rocks of the Humber Zone. Those along the eastern part of the Baie Verte Peninsula and southward occur upon rocks of the Notre Dame Subzone of the Dunnage Zone. Thus, the Humber-Dunnage zonal distinction is not evident among the middle Paleozoic rocks, and although exposures are discontinuous, the correlations imply a linkage across the early Paleozoic Humber-Dunnage boundary, the Baie Verte-Brompton Line.

Rocks of the Springdale Belt occur in four separate areas: the White Bay area, Mic Mac Lake-Grand Lake area, Cape St. John area, and the Halls Bay area (Fig. 4.41). Rocks of the White Bay area are referred to the Sops Arm Group (Lock, 1969a, b, 1972), those of the Mic Mac-Grand Lake area are known as the Mic Mac Lake Group (Neale and Kennedy, 1967; Kidd, 1974) in the north and the Sheffield Lake Group (Coyle et al., 1986) in the south, those of the Cape St. John area are known as the Cape St. John Group (Baird, 1951), and those of the Halls Bay area are referred to the Springdale Group (MacLean, 1947). Correlations among these rock groups at a formational level are not currently possible as controls of deposition are local. Each area is treated separately in the following descriptions, followed by a summary and discussion.

White Bay area

Harold Williams and W.R. Smyth

Earliest studies of Silurian rocks of the White Bay area (Fig. 4.42) were conducted by Murray and Howley (1881), Heyl (1937), Betz (1948), and Neale and Nash (1963). Since then most of the rocks were restudied and remapped as a doctorate thesis by Lock (1969a, b, 1972), and local investigations bearing on regional relationships were made by Williams (1977). More recent mapping and overall interpretations were made by Smyth (1981) and Smyth and Schillereff (1982) and their work forms the basis for most of the following account.

The Sops Arm Group consists of mixed volcanic and clastic sedimentary rocks in the order of 3 km thick. The group is broadly subdivided into four units of at least formational status that partly coincide with earlier subdivisions (Williams, 1977; Smyth and Schillereff, 1982). From bottom to top, a lower volcanic formation is succeeded by conglomerates and sandstones (Jacksons Arm and Frenchmans Cove formations), siltstones and shales (Simms Ridge Formation), and an uppermost unit of sandstones and silicic volcanic rocks (Natlins Cove Formation).

The lower volcanic unit consists mainly of felsic ashy tuffs, rhyolite flows, minor mafic flows, and interflow conglomerates and sandstones. Exposures at western Sops Arm include altered rhyolites, vesicular mafic flows, flow-banded rhyolites, welded tuffs, volcanic breccias and minor conglomerate. The lower volcanic unit extends the full length of the Silurian outcrop belt and it is cut and partly terminated by the Gull Lake intrusive suite toward the south. At its north end, a buff felsic tuffaceous unit at the base of the section lies nonconformably upon the Ordovician Coney Head Complex (Williams, 1977), dated by U-Pb zircon at 474 ± 2 Ma (Dunning, 1987). Elsewhere its boundary with older rocks is marked by the Doucers Valley fault complex. Close to the fault, the formation is sheared and its rhyolites are altered to quartz-sericite schists (Smyth and Schillereff, 1982). A dolomitic member along the western margin of the lower volcanic unit is host to lead mineralization (Smyth and Schillereff, 1982).

Conglomerates and sandstones of the Jacksons Arm and Frenchmans Cove formations are confined to the area north of Sops Arm where they overlie the lower volcanic unit. The Jacksons Arm conglomerates at the base of the unit are 600 m thick in the north. They are massive, unsorted, polymictic, cobble to boulder conglomerates that resemble the thinner interflow conglomerates of the lower volcanic unit. Intercalated mafic volcanic flows less than 4 m thick and local sandstone beds occur in the area of Frenchmans Cove. Clasts are dominantly felsic volcanic lithologies, with less common argillite, sandstone, shale, chert, granite, gneiss, and gabbro. The overlying Frenchmans Cove Formation consists of interbanded pebble to boulder polymictic conglomerate and medium to coarse sandstone. The contact with the Jacksons Arm Formation is gradational with a decrease in clast size, increased sorting, and well-defined bedding in the overlying clastic rocks. Difficulty in assigning isolated inland exposures to lower and upper conglomerate units suggest that the Jacksons Arm and Frenchmans Cove conglomerates should be regarded as members of a single formation. Conglomerates like those at Jacksons Arm occur well over 100 km northward at St. Julien Island. There they are faulted against quartzites and sandy limestones (Irish Formation) of the Hare Bay

Williams, H.
1995: Introduction: Springdale Belt; in Chapter 4 of Geology of the Appalachian-Caledonian Orogen in Canada and Greenland, (ed.) H. Williams; Geological Survey of Canada, Geology of Canada, no. 6, p. 389 (also Geological Society of America, The Geology of North America, v. F-1).

Williams, H. and Smyth, W.R.
1995: Springdale Belt-White Bay area; in Chapter 4 of Geology of the Appalachian-Caledonian Orogen in Canada and Greenland, (ed.) H. Williams; Geological Survey of Canada, Geology of Canada, no. 6, p. 389-393 (also Geological Society of America, The Geology of North America, v. F-1).

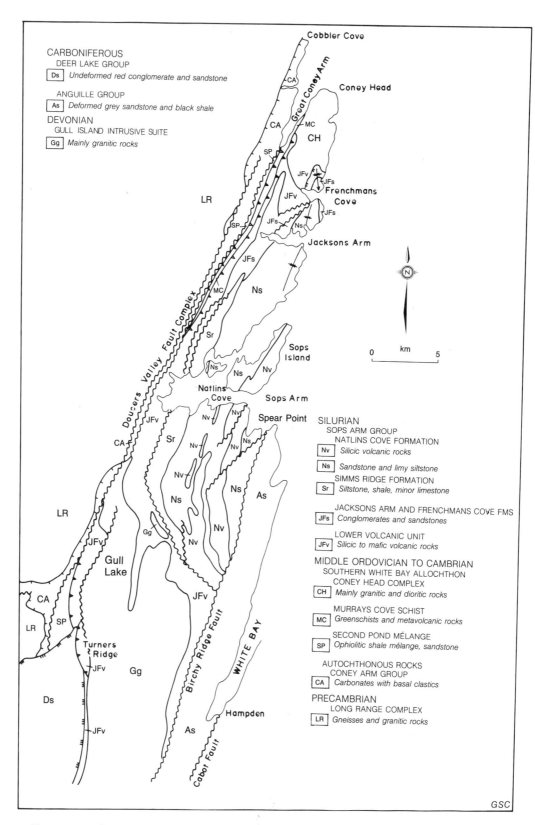

Figure 4.42. Geological setting and general geology of the Sops Arm Group in the White Bay area, after Smyth and Schillereff (1982).

Allochthon (Williams and Smyth, 1983). An absence of mappable conglomerates south of Sops Arm, and a thinner conglomerate unit between Frenchmans Cove and Sops Arm, imply significant facies change. A pronounced topographic lineament at the stratigraphic position of the conglomerate unit south of Sops Arm also suggests structural omission (Smyth and Schillereff, 1982).

The Simms Ridge Formation consists of brown-weathering slate, siltstone, minor sandstone, and fossiliferous limestone. The unit is best displayed south of Sops Arm where it is faulted against the lower volcanic unit to the west. North of Sops Arm, the unit is faulted toward the west and farther north it appears to pinch out stratigraphically between the Jacksons Arm and Natlins Cove formations. Fossiliferous limestones and marbles in beds up to 3 m thick contain a shallow marine fauna of gastropods, crinoids, corals and scarce brachiopods. Quartz veins in the Simms Ridge Formation contain base metals and minor gold (Snelgrove, 1935; Heyl, 1937).

The uppermost unit of the Sops Arm Group, the Natlins Cove Formation, is a mixed sequence of shallow marine sandstones and subaerial felsic and minor mafic volcanic rocks. Prominent volcanic units occur to the south of Sops Arm and at Sops Island. Top determinations indicate that the sedimentary and volcanic rocks are interbedded and all form an east-facing conformable sequence. Limestone beds throughout the formation contain poorly preserved macrofossils and a limestone bed at the top of the section near Spear Point contains conodonts (Smyth and Schillereff, 1982).

Sedimentary rocks of the Natlins Cove Formation are fine grained, grey sandstone with lesser limy siltstone, shale, and limestone beds. Toward the north, the sandstones are transitional with coarser clastic rocks of the underlying Frenchmans Cove Formation. Toward the south, the sandstones are gradational above spotted slates of the Simms Ridge Formation. A band of well-bedded, white quartz sandstone at Jacksons Arm displays large-scale, trough crossbeds that indicate currents from the north.

Pink to dark grey, flow-banded rhyolites and welded lithic and crystal-lithic ash-flow tuffs are common in the Natlins Cove volcanic units. Individual flows, separated by air fall tuffs or bedded epiclastic rocks can be distinguished in some places. Mafic volcanic rocks occur in the central and eastern volcanic belts and consist of green, chloritized, vesicular and amygdaloidal flows with minor interflow mafic tuff and local dykes. Large parts of the volcanic belt on the east shore of Sops Island are volcanic breccias of proximal character.

The presence of fossiliferous Ordovician, Silurian and Carboniferous rocks in the White Bay area affords a unique opportunity to unravel the effects of Paleozoic deformations along the eastern margin of the Newfoundland Humber Zone and subsequent development of the Springdale Belt and Deer Lake Basin.

The Middle Ordovician Taconic Orogeny affected the area before deposition of the Silurian Sops Arm Group. It produced west-facing recumbent folds with a penetrative muscovite-biotite schistosity in clastic rocks of the Cambrian-Ordovician Coney Arm Group and a strong penetrative schistosity in allochthonous greenschists (Murrays Cove Formation) of the Southern White Bay

Allochthon. Taconic Orogeny is defined by the nonconformity between the Silurian Sops Arm Group and the Lower Ordovician Coney Head Complex (Fig. 4.42).

In its type area at Sops Arm, the Sops Arm Group forms a steep, east-facing succession with its base faulted against the Coney Arm Group to the west and its top faulted against Mississippian rocks of the Anguille Group to the east (Fig. 4.42). Between Frenchmans Cove and Jacksons Arm, the Sops Arm Group is disposed in a south-plunging, upright syncline with an associated steep axial plane cleavage. At the nonconformity with the Coney Head Complex, the underlying granitic rocks are unaffected by the steep cleavage, except for a narrow zone immediately beneath the nonconformity. Open, upright folds in the Sops Arm Group are tight and inclined near the Doucers Valley fault complex, and the Sops Arm Group and Ordovician rocks are imbricated in the fault complex.

Granitoid dykes and sills that cut the Simms Ridge and Natlins Cove formations are affected by the steep cleavage in adjacent rocks. These intrusions are probably feeders to Silurian volcanic rocks of the Natlins Cove Formation. Some phases of the Gull Lake intrusive suite vary from massive to foliated and are interpreted as pre- or syn-tectonic with respect to structures in the Sops Arm Group. Other phases, especially megacrystic biotite granites, cut cleaved Sops Arm sedimentary rocks. A U-Pb zircon age of 373 ± 10 Ma for the Gull Lake intrusive suite (Baadsgaard, pers. comm., 1990) indicates a Devonian age for deformation of the Sops Arm Group. The nearby Devils Room granite may be consanguineous but its age of 391 ± 3 Ma (Baadsgaard, pers. comm., 1990) is somewhat older. Possibly these separate occurrences are parts of a single pluton now displaced by dextral movement on the Doucers Valley fault complex (Lock, 1969a, b). Steep lineations in Silurian conglomerates near the fault complex suggest thrusting.

The effects of Alleghanian deformation on the Sops Arm Group are less clear. This event produced tight upright folds in the Lower Mississippian Anguille Group to the east, but beds of the Viséan Deer Lake Group to the west are gently dipping and poorly indurated, suggesting the Deer Lake Group is younger than this phase of Carboniferous deformation. However, at Turners Ridge (Fig. 4.42), the Sops Arm Group is locally thrust westward above the Deer Lake Group (Smyth and Schillereff, 1982).

The Doucers Valley fault complex does not affect the Viséan Deer Lake Group. Farther southwest, the Birchy Ridge and Cabot faults affect the Carboniferous Anguille Group and they are viewed as major strike-slip faults (Hyde, 1979).

Fossils collected from the Natlins Cove Formation include *Halysites, Heolites, Isorthis* cf. *I. arcuaria, Hyattidina, Howellella* and *cystoids* (Berry and Boucot, 1970). *I. arcuaria* suggests an age span of late Wenlock to Pridoli. Other forms, though wide ranging, are Silurian. Conodonts reported by Smyth and Schillereff (1982) from limestones of the Natlins Cove Formation are also of Silurian age. Lithic similarities among volcanic rocks of the Sops Arm Group, and an overall contrast with Cambrian-Ordovician rocks, indicate that the entire Sops Arm Group is of Silurian age.

The preponderance of terrestrial volcanic rocks throughout the Sops Arm Group, their association with coarse conglomerates and fluviatile crossbedded sandstones,

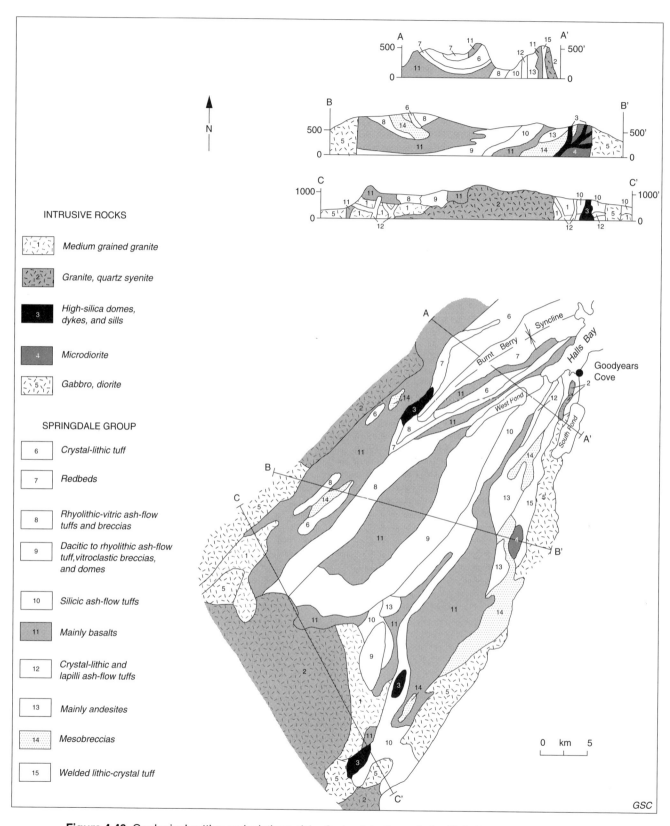

Figure 4.43. Geological setting and relations of the Springdale Group in the Halls Bay area, after Coyle and Strong (1987).

marked thickness variations in conglomerate and sandstone units, and thin limestone horizons with shallow marine faunas suggest deposition in a subaerial evolving basin with local marine incursions. Several lines of evidence indicate paleocurrents from north to south with appreciable relief in the source area. These include: (a) the coarseness of the conglomerate units in the north, (b) profound thinning of the conglomerates from north to south, (c) northward disappearance of the Simms Ridge sandstones that may reflect shoaling of a basin toward the source area of the clastic rocks, and (d) trough crossbeds in the Natlins Cove Formation that indicate currents from the north. The bimodal aspect of the Silurian volcanic rocks suggests extension and the major steep northeast-trending transcurrent faults that are so prominent now, may be sited along ancestral transcurrent faults that controlled mid Paleozoic tectonism.

Halls Bay area

Harold Williams, M. Coyle, and D.F. Strong

Middle Paleozoic rocks of the Halls Bay area (Fig. 4.41) are assigned to the Springdale Group. Earliest studies were made by Espenshade (1937), MacLean (1947), Kalliokoski (1953, 1955), Neale and Nash (1963) and Williams (1967, 1972). Redbeds of the Springdale Group were studied by Wessel (1975) and volcanic rocks by Coyle et al. (1985), Coyle and Strong (1987), and Coyle (1990). Local occurrences of red sandstones along the southeast side of Red Indian Lake (Fig. 4.43) are regarded as correlatives of the Springdale Group. Rocks previously assigned to the Silurian at Catchers Pond (Neale and Nash, 1963; Williams, 1967) are reassigned to the Lower Ordovician (Dean, 1970; Bergström et al., 1972; Boucot, 1973; O'Brien and Szybinski, 1989). Similarly, the Roberts Arm and Buchans groups are now shown to be Ordovician rather than Silurian (Williams, 1967; Dean, 1978; Bostock et al., 1979; Nelson and Kidd, 1979; Nowlan and Thurlow, 1984; Dunning et al., 1987). Redbeds at Kings Point, previously correlated with the Springdale Group (Neale and Nash, 1963; Williams, 1967) are now assigned to the Carboniferous (Marten, 1971; Dean, 1978).

The Springdale Group is dominantly volcanic with less than 20 per cent exposed sedimentary rocks. It can be broadly subdivided into a volcanic sequence at the base, about 3 km thick (Neale and Nash, 1963), overlain by red sandstones and conglomerates with a maximum exposed thickness of 1700 m (MacLean, 1947). Kalliokoski (1953) recognized 12 mappable units, and 10 units are distinguished by Coyle and Strong (1987), most of which are of potential formational status (Fig. 4.43).

The volcanic rocks are mainly silicic tuffs and breccias with chiefly volcanic fragments and rare fragments of granite and red sandstone. Coarse, unsorted vent facies are

prominent in places, such as at Goodyears Cove of Halls Bay. Associated silicic flows are pale red to grey, commonly exhibiting flow layering and local spherulitic structure. Also included are quartz-feldspar porphyries that are partly of intrusive origin. Some of these high level intrusions contain inclusions of chromite and serpentinite (Dean, 1978).

Mafic volcanic rocks alternate with silicic units. The mafic flows are chiefly purplish, green, maroon, reddish brown, and black. Calcite and quartz amygdales are common. Pillows are absent.

Bimodal volcanic roof pendants in the Topsails igneous complex west of Buchans (Fig. 4.41) are possible correlatives of the Springdale Group (Taylor et al., 1980).

Sedimentary rocks at the top of the Springdale Group in the northern part of the Halls Bay area are red sandstones and pebble conglomerates. The red sandstones are commonly micaceous and the conglomerates contain clasts of fresh volcanic rocks and quartz-feldspar porphyry of local derivation. They also contain clasts of altered mafic volcanic rocks, chert, and granite. Depositional features of sandstone beds include crossbedding, ripple marks, scour-and-fill structures, mud cracks, rain prints, and mud pellets.

At Pilleys Island (Fig. 4.41), conglomerates and red sandstones of the Springdale Group directly overlie marine mafic volcanic rocks of the Roberts Arm Group. As the Roberts Arm is Ordovician, and whereas the lower volcanic units of the Springdale Group are absent at the contact, the relationship implies significant unconformity. Unconformities between the Springdale Group and the Ordovician Lushs Bight and Catchers Pond groups are also reported (Coyle and Strong, 1987).

In the widest part of the Halls Bay area (Fig. 4.43), stratigraphic units of the Springdale Group are displayed in a broad, open, gently-northeast-plunging syncline, the Burnt Berry Syncline (Dean, 1978). This structure is truncated to the southwest by the Topsails intrusive suite. In the northeast, the northeast-trending sandstones at Springdale can be traced seaward into east- and southeast-trending beds at Pilleys Island. The western boundary of the Springdale Group is faulted and its northern boundary is marked by the Lobster Cove Fault that is traceable across Sunday Cove, Pilleys and Triton Islands (Fig. 4.41). The southeast boundary of the group is marked by intrusion in the north and faults toward the south. Open folds and a general lack of penetrative cleavage in the Springdale Group contrast with tighter structures and more intense deformation in Lower Ordovician rocks to the north of the Lobster Cove Fault. There is also a metamorphic contrast across the fault from sub-greenschist facies in Springdale Group rocks to the south and greenschist facies in Ordovician rocks to the north.

None of the rocks of the Springdale Group are dated by fossils. The rocks were assigned to the Silurian where mapped by Espenshade (1937) and MacLean (1947), but later assigned to the Devonian (Twenhofel, 1947; Kalliokoski, 1953, 1955). Neale and Nash (1963) and Williams (1967) reassigned the rocks to the Silurian because of their apparent correlation with dated Silurian rocks of the Botwood Belt to the east. The Springdale Group is cut by intrusions dated isotopically as Devonian (Wessel, 1975; Taylor et al., 1980) and Springdale clasts occur in Carboniferous

Williams, H., Coyle, M., and Strong, D.F.
1995:   Springdale Belt-Halls Bay area; in Chapter 4 of Geology of the Appalachian-Caledonian Orogen in Canada and Greenland, (ed.) H. Williams; Geological Survey of Canada, Geology of Canada, no. 6, p. 393-394 (also Geological Society of America, The Geology of North America, v. F-1).

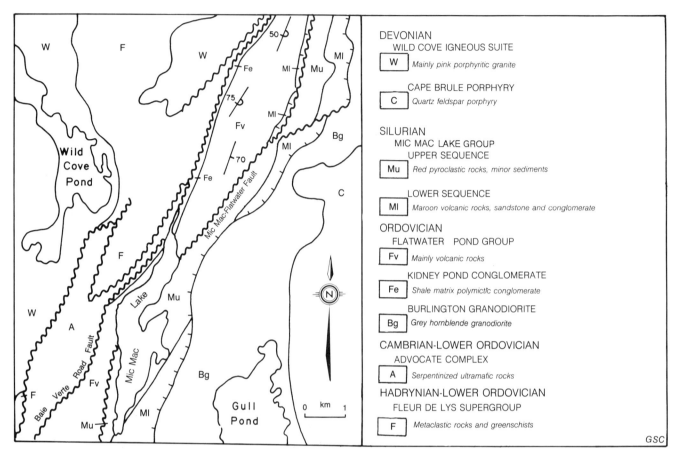

Figure 4.44. Geological setting and relations of the Micmac Lake Group in its type area, after Kidd (1974).

conglomerates. A U-Pb zircon age of 429 +6/-5 Ma on volcanic rocks near the top of the group supports a Silurian age assignment (Chandler et al., 1987).

Red micaceous sandstones on the southeast shore of Red Indian Lake were assigned to the Carboniferous in map compilations by Baird (1954) and Weeks (1955). The rocks are now considered as Silurian because they are more deformed and more highly indurated than nearby subhorizontal Carboniferous beds, and pebbles of the red micaceous sandstones occur in nearby Carboniferous conglomerates.

A lack of pillowed units in the thick volcanic section of the Springdale Group provides a sharp contrast with profuse pillows in marine Ordovician units nearby. An association of the volcanic rocks with redbeds indicates terrestrial conditions of accumulation. This is further supported by mud cracks and rain prints in the sedimentary rocks and a bimodal aspect for the volcanic rocks. Red sandstones and conglomerates that cap the volcanic sequence may represent deltaic fans shed from the eroding volcanic topography (Dean, 1978). Local serpentinite fragments in Springdale Group volcanic rocks indicate an underlying substrate of ophiolitic rocks of the Dunnage Zone.

Recent studies of the Springdale Group relate its volcanic stratigraphy and distribution of units to the evolution of an epicontinental caldera (Coyle and Strong, 1987; Coyle, 1990). Large areas of laharic breccia and pyroclastic breccia along the margins of the Halls Bay volcanic belt have lobate forms suggesting debris flows off steep scarps, and faulted boundaries of the volcanic belt are equated with a topographic wall or collapsed caldera margin (Fig. 4.43).

Micmac Lake area

Harold Williams, W.S.F. Kidd, and J.P. Hibbard

Subaerial volcanic and sedimentary rocks that occupy a narrow northeast-trending belt in the central part of the Baie Verte Peninsula are considered Silurian or Devonian

Williams, H., Kidd, W.S.F., and Hibbard, J.P.
1995:   Springdale Belt-Micmac Lake area; *in* Chapter 4 of Geology of the Appalachian-Caledonian Orogen in Canada and Greenland, (ed.) H. Williams; Geological Survey of Canada, Geology of Canada, no. 6, p. 394-395 (*also* Geological Society of America, The Geology of North America, v. F-1).

in age. The rocks are unfossiliferous and their age, first based on regional relationships and lithic resemblance to nearby groups (Neale and Nash, 1963; Neale and Kennedy, 1967; Williams, 1967) is supported by Rb-Sr isotopic dating of volcanic rocks at 404 ± 24 Ma (quoted in Neale and Kennedy, 1967) and 375 ± 15 Ma (Pringle, 1978). The rocks are assigned to the Mic Mac Lake Group (Neale and Kennedy, 1967; Kidd, 1974), now Micmac Lake Group (Hibbard, 1983) and they extend northeastward from their type area to slightly beyond Flat Water Pond and southwestward to Birchy Lake (Fig. 4.41). Similar volcanic rocks on the east side of Grand Lake are possible extensions of this belt truncated by intervening Devonian intrusions or hidden by Carboniferous cover rocks.

Neale and Nash (1963) were the first to recognize probable Silurian rocks at Micmac Lake. A much more detailed and comprehensive study was conducted as part of a doctorate dissertation by Kidd (1974) and a summary and overview of rocks and relationships is given by Hibbard (1983).

The Micmac Lake Group is disposed in a steep syncline with its axial zone truncated by the major longitudinal Micmac-Flat Water Fault (Fig. 4.44). Most of the rocks lie on the east side of the fault where a clear nonconformable relationship is well exposed between the Micmac Lake Group and underlying Burlington Granodiorite (Neale and Kennedy, 1967). There, the sequence is west-dipping with dips steepening toward the faulted synclinal axis. Smaller areas west of the Micmac-Flat Water Fault lie unconformably on the Flatwater Pond Group with beds younger toward the east. The overall thickness of the Micmac Lake Group is about 2 km.

The erosional nonconformity between the Micmac Lake Group and the Burlington Granodiorite is irregular with pockets of coarse conglomerate above the granodiorite in some places and red rhyolite, with basal intrusive relationships, overlying the granodiorite in other places. A sharp contact between the Flatwater Pond Group and overlying Micmac Lake Group appears conformable in outcrop but regionally it is a low angle unconformity (Kidd, 1974). An occurrence of fine grained brittle rhyolite south of Micmac Lake on the Baie Verte highway contains numerous closely spaced green serpentinite inclusions, implying an ophiolitic substrate.

The Micmac Lake Group is divided into lower and upper sequences, separated by an erosional unconformity in the type area (Kidd, 1974). The lower sequence has an abundance of fine grained felsic volcanic rocks with prominent conglomerate units toward its base and a relative paucity of mafic volcanic rocks. It has a maximum thickness of 750 m east of the Micmac-Flat Water Fault whereas correlatives west of the fault are no more than 300 m thick. Felsic volcanic rocks are dominated by bright reddish flow-banded non-porphyritic rhyolite with associated eutaxitic ignimbrite. Welded glass-shard textures are preserved in a few ignimbrites. Associated conglomerates have cobbles and boulders of local derivation; Burlington Granodiorite, massive to flow-banded maroon rhyolite, red quartz-feldspar porphyry, purple mafic lava, and eutaxitic ignimbrite. Mafic lavas that occur higher in the lower sequence west of the Micmac-Flat Water Fault are green to purple, massive to amygdaloidal, and in places porphyritic with large plagioclase phenocrysts.

The upper sequence has mainly mafic lavas and conglomerates at its base with associated porphyritic ignimbrites and minor trachytic flows. Total thickness in the type locality is estimated at 1600 m (Kidd, 1974). Its top is mainly porphyritic ignimbrite with cooling units recognizable in places. Associated red to grey sandstones and siltstones display planar beds with graded bedding and erosional scouring, and less commonly crossbedding, ripple marks, and mud cracks. East of Micmac Lake and near Flat Water Pond, the upper sequence lies directly on the Burlington Granodiorite, thus emphasizing the local importance of the erosional surface that separates the upper and lower sequences elsewhere. Sparse mafic dykes that cut polydeformed schists of the Fleur de Lys Supergroup near Micmac Lake are either related to the Wild Cove Pond granite/diorite plutonic complex (Kidd, 1974) or to volcanic rocks of the Micmac Lake Group.

The Micmac Lake Group has a metamorphic and structural history like that of the underlying Flatwater Pond Group (Neale and Kennedy, 1967; Kidd, 1974). Regional metamorphism is lower greenschist facies. A single steep slaty cleavage is apparent throughout the outcrop area, although it is not visible in resistive rhyolite flows and ignimbrites. Less resistive lithologies, such as unwelded tuffs, exhibit a subhorizontal crenulation cleavage that affects the early steep cleavage. Carboniferous rocks nearby are unaffected by these structures.

An abundance of ignimbrites, widespread hematitic oxidation of mafic flows, major erosional contacts within stratigraphic sections, and terrestrial alluvial fan features of associated sedimentary rocks all point to a subaerial environment of deposition with substantial topographic relief.

Cape St. John and nearby areas

Harold Williams and J.R. DeGrace

Volcanic and sedimentary rocks that are similar to occurrences elsewhere in the Springdale Belt outcrop in the vicinity of Cape St. John (Fig. 4.41). Other smaller occurrences of mainly felsic volcanic rocks, with associated high level intrusions in some examples, occur to the southwest at Yak Lake, Rogues Harbour, Middle Arm, and Kings Point (Fig. 4.41).

At the eastern end of the Baie Verte Peninsula the rocks are known as the Cape St. John Group (Baird, 1951). They were first assigned to the Ordovician and interpreted as older than nearby ultramafic "intrusions" (Snelgrove, 1931; Douglas et al., 1940; Baird, 1951). A major unconformity between the Ordovician Snooks Arm Group and overlying Cape St. John Group was recognized by Neale (1957), and ambiguous intrusive relationships between the Cape St. John Group and Ordovician serpentinites was indicated by the occurrence of serpentinite clasts in Cape St. John conglomerates (Neale, 1957).

Williams, H. and DeGrace, J.R.
1995: Springdale Belt-Cape St. John and nearby areas; in Chapter 4 of Geology of the Appalachian-Caledonian Orogen in Canada and Greenland, (ed.) H. Williams; Geological Survey of Canada, Geology of Canada, no. 6, p. 395-396 (also Geological Society of America, The Geology of North America, v. F-1).

The intensity of deformation and metamorphism increases from south to north across the eastern end of the Baie Verte Peninsula from low greenschist to amphibolite facies. Some of the deformed, higher grade rocks of the Cape St. John Group in northern exposures were assigned to the Grand Cove Group and correlated with the much older Fleur de Lys Supergroup (Church, 1969). This definition of an "Eastern Fleur de Lys", spawned problems that occupied the talents of geologists working in this area for the next ten years.

The Cape St. John-Snooks Arm unconformity was revisited and revitalized (Neale et al., 1975) and original views were supported by subsequent regional mapping (DeGrace et al., 1976; Williams et al., 1977; Hibbard, 1983). For a more complete summary of previous work and history of ideas the reader is referred to Williams et al. (1977) and Hibbard (1983).

The Cape St. John Group occupies a major upright syncline, vertically shortened, with a sinuous east-trending axis resulting from later flexuring (Hibbard, 1982). Its thickness is estimated at 3500 m (DeGrace et al., 1976). Conglomerates at the base of the group contain clasts of mafic volcanic rocks, argillite and chert of the Snooks Arm Group, ultramafic rocks, a variety of felsic volcanic rocks, and quartz-feldspar porphyry. Associated sandstones are locally crossbedded with mud cracks and they are interpreted as fluviatile (DeGrace et al., 1976).

Volcanic rocks of the group vary from mafic to silicic. The mafic varieties are mainly dark green to purple, fine grained and equigranular, with vesicles and amygdales common in most examples. Pillows that are so well developed in nearby Ordovician rocks of the Snooks Arm Group are absent in the Cape St. John volcanic rocks. Andesitic flows and pyroclastic rocks are common in northern exposures. The flows are dark to light green, massive, vesicular, and locally porphyritic. Flow-banded, dark, fine grained to aphanitic dacites and rhyolites are common associates. Fragmental acid volcanic rocks include pink lapilli tuff, ash-flow tuff, welded crystal tuff, ignimbrite, and agglomerate. These alternate with flow banded rhyolites, trachytes, and quartz-feldspar porphyries. Pink quartz-feldspar crystal lithic tuffs are abundant in southern exposures. Some are spherulitic and lithic varieties contain rhyolite and ultramafic inclusions. Northeast of Betts Cove, tuffs can be traced into near-vent coarse explosive facies (DeGrace et al., 1976).

An outlier of felsic volcanic rocks at Yak Lake (Fig. 4.41), now correlated with the Cape St. John Group (Hibbard, 1983), has mainly quartz-feldspar crystal tuffs that resemble nearby porphyries. Another small outlier at Rogues Harbour contains conglomerates that are unconformable upon Ordovician ophiolitic rocks of the Betts Cove Complex (Schroeter, 1971). Intrusive breccias near Rogues Harbour cut ophiolitic rocks and they are interpreted as related to Cape St. John volcanic activity. The breccias occur as dykes and irregular masses containing subangular to rounded fragments and blocks of ophiolitic rocks, granodiorite, and acid volcanic rocks. Nearby conglomerates may be reworked equivalents of the intrusive breccias (DeGrace et al., 1976). Volcanic rocks at Middle Arm and Kings Point (Fig. 4.41) are possible Cape St. John correlatives.

Isotopic ages for volcanic rocks of the Cape St. John Group span an interval from early Ordovician to early Carboniferous (see Hibbard, 1983, for details). The group is most reasonably interpreted as Silurian or Early Devonian according to regional relationships and setting.

Large areas of quartz-feldspar porphyry (not shown in Fig. 4.41) occur in a belt between Micmac Lake and Cape St. John of the Baie Verte Peninsula (Cape Brule Porphyry and equivalents, Baird, 1951). The porphyries contain rafts and inclusions of felsic volcanic rocks and in some places it is difficult to distinguish fine grained phases of the porphyry from porphyritic rhyolite flows and welded crystal tuffs. Small mafic and ultramafic inclusions are common and much larger, outcrop-size blocks of serpentinite and gabbro are also known. Arcuate porphyry dykes, like those of ring complexes, cut the Burlington Granodiorite near Middle Arm.

The spatial relationships and commingling of Cape St. John silicic volcanic rocks and Cape Brule porphyries imply a genetic link. Neale (unpublished data, 1962) suggested that the intrusive porphyries mark volcanic centres which supplied the material for the flanking Cape St. John and related groups. He also suggested that the volcanic centres terminated as cauldron subsidence features with the volcanic piles collapsing into their intrusive lower levels. Later studies near Kings Point support this viewpoint (Mercer et al., 1985; Coyle and Strong, 1987; Coyle, 1990). There, a complex of porphyries and volcanic rocks provides a section through a high level pluton and overlying volcanic carapace. The volcanic rocks are dominantly welded silicic ash-flow tuffs, and the plutonic rocks are mainly granite and syenite which are all gradational through variations in proportions of quartz and alkali feldspar phenocrysts. The volcanic rocks are peralkaline with prominent riebeckite oikocrysts, and their presence is interpreted to reflect metasomatism and hydrothermal alteration resulting from intrusive activity. Occurrences of composite dykes and intermixed mafic and silicic volcanic rocks are typical of the high level granite suites.

Cauldron subsidence is thought to have been a common phenomenon during mid Paleozoic magmatism and volcanism throughout the Springdale Belt of western Newfoundland (Mercer et al., 1985; Coyle and Strong, 1987).

Summary and discussion
Harold Williams

Rocks of the Springdale Belt are everywhere dominated by volcanic units, except for the White Bay area. The volcanic rocks are mainly rhyolites and pyroclastic equivalents. Associated mafic rocks are green to purplish vesicular and amygdaloidal flows of entirely different aspect compared to predominantly pillow lavas in underlying Ordovician sections. Associated sedimentary rocks are polymictic

Williams, H.
1995: Springdale Belt: summary and discussion; in Chapter 4 of Geology of the Appalachian-Caledonian Orogen in Canada and Greenland, (ed.) H. Williams; Geological Survey of Canada, Geology of Canada, no. 6, p. 396-397 (also Geological Society of America, The Geology of North America, v. F-1).

conglomerates and fluviatile sandstones, commonly red-beds, with shallow water depositional features. Individual units are discontinuous, even in local areas, and evidence of erosion is either present or implied in most stratigraphic sections. Overall thicknesses contrast from one area to another and local variations in both thickness and facies are in places abrupt. These features are indicative of terrestrial deposition. They also explain the difficulties in regional correlations throughout the belt.

Fossils are rare and the few known occurrences are all in marine sedimentary rocks of the White Bay area. However, the evidence for Silurian to Early Devonian ages in other areas is strong. Unconformities with underlying Ordovician rocks occur throughout the entire belt. Rocks assigned to the middle Paleozoic are everywhere deformed and the deformed rocks are truncated by intrusions that do not affect nearby Carboniferous strata. Isotopic ages support the stratigraphic data and fall within the Silurian-Early Devonian time frame. Granites that cut the stratigraphic sections are Silurian and Devonian.

The latest studies indicate that depositional patterns within the Springdale Belt were controlled by volcanic centres that terminated as cauldon subsidence features (Mercer et al., 1985; Coyle and Strong, 1987; Coyle, 1990). The most extensive areas of volcanic rocks and the thickest volcanic sections occur side by side with consanguineous high level plutons. Thus the Cape St. John volcanic rocks border the Cape Brule porphyries, the Kings Point Complex (Mercer et al., 1985) contains both volcanic and plutonic components, and the Springdale volcanic rocks, southerly extensions of the Micmac Lake volcanic rocks, and volcanic rocks at Grand Lake are all peripheral to distinctive intrusions of the Topsails igneous suite. Intrusive breccias that cut ophiolitic rocks at Rogues Harbour and diatreme intrusions that cement conglomerates above the Burlington Granodiorite are consistent with this model. Arcuate ring dykes, composite dykes, and intermixed mafic and silicic globules are other features of high level intrusive-extrusive suites; and cauldron subsidence is a viable model for both the stratigraphic order and morphological expression of some units in the Halls Bay area.

Peralkaline volcanic rocks occur at Sheffield Lake (Fig. 4.41) and peralkaline intrusions and related volcanic rocks are common in the Topsails igneous complex (Whalen and Currie, 1983) and in the Kings Point Complex (Mercer et al., 1985). Peralkaline affinities possibly relate to metasomatism and hydrothermal activity at intrusive contacts (Mercer et al., 1985). A lack of peralkaline volcanic rocks in the Springdale Group could mean that its volcanic rocks represent higher structural levels of a volcanic-plutonic pile.

The distribution of the volcanic-plutonic complexes in the Springdale Belt indicates four or more nested volcanic centres, all partly interlocking and aligned northeasterly between Grand Lake and Cape St. John (Fig. 4.45). Ophiolitic rocks of the the Dunnage Zone occur beneath the Cape St. John Group and their subsurface presence is indicated beneath the Micmac Lake and Halls Bay areas by serpentinite inclusions in volcanic and intrusive rocks.

Volcanic and sedimentary rocks at White Bay are akin to those farther east in the Springdale Belt. However, their basement rocks are part of the Humber Zone. Rhyolite and porphyry dykes, interpreted as feeders to the Silurian volcanic rocks, cut the Natlins Cove Formation, and plutonic rocks of the Gull Island intrusive suite may be genetically related to Silurian volcanism.

Devonian and Carboniferous thrusting, coupled with steep dips and deformation in the Carboniferous rocks at White Bay, indicate shortening and imply that Silurian rocks west of White Bay were once more distant from other Silurian occurrences of the Springdale Belt than at present. Regardless of paleogeography, the important Humber-Dunnage zone boundary has no expression in the mid Paleozoic rocks. This, and the fact that the mid Paleozoic rocks are everywhere unconformable on Ordovician rocks affected by the Taconic Orogeny, indicates that the mid Paleozoic rocks are a cover on the already juxtaposed Humber-Dunnage zones.

Middle Paleozoic volcanic rocks and alkali plutonic suites dominate the Springdale Belt. Where similar middle Paleozoic intrusions are absent in other parts of Newfoundland, so too are the thick mid Paleozoic volcanic accumulations. Faulting has been suggested to explain the local sedimentary and volcanic patterns (Williams, 1967; Kidd, 1974; Smyth and Schillereff, 1982; Strong and Coyle, 1989); and major faults may have controlled mid Paleozoic plutonism in the Appalachian Orogen (Strong, 1980). The numerous, long, linear, steep, northeast-trending faults that transect the Springdale Belt may follow middle Paleozoic zones of weakness that controlled volcanism and sedimentation at or near the Humber-Dunnage boundary.

Subduction has also been suggested as a controlling mechanism for mid Paleozoic plutonism and volcanism (Whalen, 1989; Bevier and Whalen, 1990). However, the lithologies and chemistry of the volcanic rocks seem to refute a subduction model (Strong, 1980; Coyle and Strong, 1987). It is also difficult to define an open oceanic tract in the stratigraphic analysis of the orogen during the Silurian period (Williams, 1979).

Cape Ray Belt

L.B. Chorlton, Harold Williams, and D.H.C. Wilton

Definition

The Cape Ray Belt is defined by discontinuous occurrences of Silurian and Devonian terrestrial sedimentary and bimodal volcanic rocks that are traceable from Cape Ray 150 km northeastward to Victoria Lake (Fig. 4.46). Toward the southwest, the rocks are steeply dipping and localized along a kilometre wide zone of faulting and mylonitization that defines the Cape Ray Fault. Toward the northeast, the belt is wider, dips are in places gentle, and deformation is less intense. An unconformity with underlying Ordovician rocks is exposed in the vicinity of King George IV Lake and near the Burgeo Road along the northwestern margin and northern boundary of the belt, respectively. An unconformity

Chorlton, L.B., Williams, H., and Wilton, D.H.C.
1995: Cape Ray Belt; in Chapter 4 of Geology of the Appalachian-Caledonian Orogen in Canada and Greenland, (ed.) H. Williams; Geological Survey of Canada, Geology of Canada, no. 6, p. 397-403 (also Geological Society of America, The Geology of North America, v. F-1).

along its northwestern margin is locally present in the southwest. Where contacts are unexposed, unconformable relationships are inferred on sedimentological evidence. The southeastern margin of the belt is everywhere faulted.

Previous work and nomenclature

The earliest geological studies that included rocks of the Cape Ray Belt were regional mapping projects by the Geological Survey of Canada. Plant-bearing Devonian rocks were discovered by Cooper (1954) in the central part of the belt during routine mapping of the La Poile Bay area. Other rocks of the belt at King George IV Lake were first noted by Riley (1957) who suggested a middle Paleozoic age

and correlation with the fossiliferous rocks to the southwest. DeGrace (1974) reinvestigated these same exposures and suggested a Carboniferous age for the rocks because of their mild deformation compared to Devonian rocks outlined by Cooper (1954). Gillis (1972) was the first to recognize the Cape Ray Fault and its continuity along the southwestern half of the belt. His reconnaissance map depicts Devonian rocks outlined by Cooper (1954) but mid Paleozoic rocks farther southwest were unseparated.

Rocks of the southwestern coastal portion of the belt were outlined during a doctorate thesis study by Brown (1975) and later remapped during followup studies for the Newfoundland Department of Mines (Brown, 1977). Rocks of the central portion of the belt, including those mapped

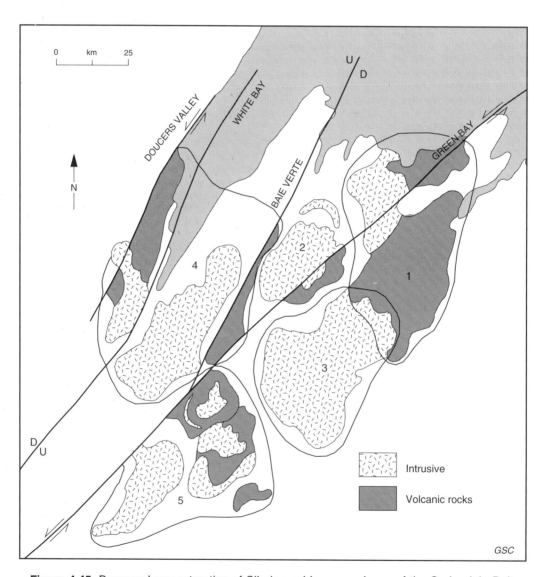

Figure 4.45. Proposed reconstruction of Silurian caldera complexes of the Springdale Belt. 1– includes Springdale and Cape St. John groups and Cape Brule Porphyry; 2 – includes King's Point Complex and Sheffield Lake Group; 3 – mainly plutonic northern lobe of Topsails intrusive suite; 4 – includes Wild Cove Pond, Gull Lake, and Devil's Room plutons and Sops Arm and Micmac Lake groups; 5 – includes a number of volcano-plutonic circular structures that may represent smaller nested complexes, after Coyle and Strong (1987).

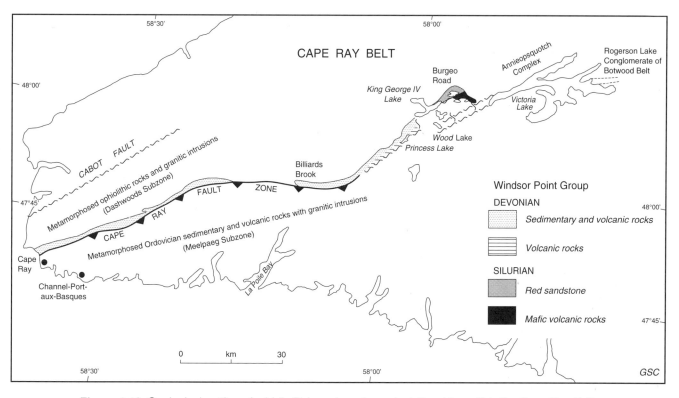

Figure 4.46. Geological setting of middle Paleozoic rocks and relationships within the Cape Ray Belt.

by Cooper (1954), formed part of a doctorate thesis study by Chorlton (1984) who also remapped the area for the Newfoundland Department of Mines (Chorlton, 1980, 1983). Rocks of the northern extremity of the belt were studied locally by Dunning and Herd (1980) and Chandler (1982). Since then they were remapped by Kean (1983) so that all rocks of the belt are covered now by 1:50 000 scale maps of the Newfoundland Department of Mines. More recent studies near Cape Ray were made by Wilton (1983, 1984) and Dubé et al. (1991) in connection with gold mineralization discovered along the Cape Ray Fault, and by Piasecki (1989) as part of a study of ductile shearing in Newfoundland. Remapping of the Cape Ray area at 1:50 000 scale is now in progress (van Staal et al., 1992).

Cooper (1954) assigned the Devonian rocks of the Cape Ray Belt to the Bay du Nord Group, which also included much higher grade and more complexly deformed rocks to the southeast. The fossiliferous Devonian rocks were later informally named the Billiards Brook formation (Chorlton, 1980) and separated from the unrelated parts of the original Bay du Nord Group. Near Cape Ray, the rocks are known as the Windsor Point Group (Brown 1975, 1977) and it has been expanded to include the Billiards Brook formation (Chorlton, 1983). Rocks to the northeast are also assigned to the Windsor Point Group (Kean, 1983), although they include a Silurian redbed-volcanic sequence that is not present toward the southwest (Chandler and Dunning, 1983).

Red and purple conglomerates of the Botwood Belt, the Rogerson Lake Conglomerate (Kean, 1983), occur within a few kilometres of the Cape Ray Belt at Victoria Lake and Burgeo Road, but on the opposite side of the Red Indian

Line. Deformation of the middle Paleozoic rocks southeast of the Red Indian Line is much more intense and relations between the middle Paleozoic rocks of the Cape Ray and Botwood belts are unknown.

Stratigraphy

The stratigraphy of the Windsor Point Group is unknown. Sedimentary rocks predominate in some places and volcanic rocks and related sills predominate in others. An ignimbrite unit occurs at the base of the group near Cape Ray and it yields a Rb-Sr isotopic age of 377 ± 21 Ma (Wilton, 1983). In the southwest portion of the belt, felsic volcanic rocks are more common along its northwest side and tuffaceous mafic volcanic rocks along its southeast side. In other places sedimentary and volcanic rocks are interlayered, although in some intensely deformed areas of the Cape Ray Fault, alternating lithic units may represent structural interleaving. The contact between foliated tonalite and grey ignimbrite at Cape Ray is interpreted as an unconformity (Wilton, 1983; Williams et al. 1989a).

Where deformation is less intense toward the north, Chandler (1982) recognized two separate sequences; an older sequence of redbeds and mixed volcanic rocks and a younger plant-bearing sequence of grey sandstone, conglomerate and mudstone that is devoid of volcanic rocks. The redbed sequence is well exposed in road cuts along the Burgeo Road and similar rocks extend southwest to Princess Lake. This sequence lies unconformably above the Ordovician Annieopsquotch Complex, its volcanic rocks yield a Silurian U-Pb zircon isotopic age of 431 ± 5 Ma

399

(Chandler and Dunning, 1983), and there is a strong resemblance between rocks of this sequence and those of the Silurian Springdale and Botwood groups to the north. The grey sequence occurs mainly to the south of King George IV Lake and is presumably continuous with the Billiards Brook formation and Windsor Point Group farther southwest. Chandler (1982) suggested a faulted contact between the red and grey sequences. Kean (1983) suggested a gradation between the two west of Princess Lake. There, conglomerates of the grey sequence lie directly upon granite to the west, implying westward stratigraphic onlap.

Lithology

Toward the northeast at Burgeo Road, sedimentary rocks of the red sequence are mainly micaceous sandstones with argillite and pebble conglomerate interbeds. Crossbedding is common and ripple marks, mudcracks, channel scours, and mud-flake beds are present. Pumice clasts up to 3 cm long were noted in red sandstone at one locality (Dunning and Herd, 1980). Volcanic rocks occur to the southeast and below the redbeds, although redbeds and volcanic rocks are interlayered in some places. Volcanic rocks are flow-banded to massive rhyolite, rhyolite breccia, and maroon to purplish and green mafic flows with vesicles and calcite amygdules. Where deformed toward the southwest, volcanic rocks are altered to chlorite-actinolite schists. A coarse volcanic breccia unit 10 m thick is interpreted as a lahar where it occurs at the redbed-volcanic contact in a quarry on Burgeo Road (Dunning and Herd, 1980).

Conglomerates at the base of the redbed sequence dip gently southwest and directly overlie the Annieopsquotch Complex north of the Burgeo Road. The contact is sharp with pebble to boulder red conglomerate lying on dykes of the ophiolite suite. Clasts represent a sampling of the ophiolite suite and also include felsic volcanic rocks, granite, and non-ophiolitic gabbro and diorite boulders. Imbrication in conglomerate clasts indicate sedimentary transport toward the northeast (Chandler, 1982).

About 200 m of redbeds and volcanic rocks occur east of Princess Lake. This section has intermediate to mafic volcanic rocks, cobble conglomerate with felsic volcanic clasts, red fissile siltstone and argillite, and crossbedded sandstones.

The grey sequence of the Windsor Point Group in the vicinity of King George IV Lake and Princess Lake consists of conglomerate, grey sandstone, arkose, and plant-bearing black siltstone and shale. Clasts in conglomerate beds include porphyritic volcanic rocks, black to purple and red flow banded rhyolite, green and red chert, red sandstone and siltstone, black argillite and siltstone, paragneiss, granite, gabbro, diabase, quartz, and possible ultramafic rocks. Associated grey sandstone exhibits trough crossbeds. West of Princess Lake, conglomerates lie unconformably above granitic rocks of the Lloyds River Intrusive Suite (Kean, 1983). A 10 m unit of grey-green, limy, fine grained massive rock at the contact in one locality is a possible regolith (Chandler, 1982). Nearby, 75 m of pebble to cobble conglomerate with rounded clasts of granite and felsic to mafic volcanic rocks overlie pink porphyritic granite with a sharp contact. Grey arkosic grit with interlayers of black mudstone occur above the conglomerate.

Along the Cape Ray fault zone, the Windsor Point Group consists of mixed volcanic and sedimentary rocks. The volcanic rocks are strongly bimodal and occurrences of subvolcanic leucogranophyre and mafic sills or dykes are probably coeval intrusions. Felsic volcanic and volcaniclastic rocks include rhyolite porphyry, ashflow tuff, aphyric rhyolite, rhyolite tuff, rhyolite breccia, quartz crystal tuff, bedded siliceous tuff, and volcanic conglomerate. Mafic volcanic rocks include vesicular mafic flows, tuffs, and related dykes and sills. Dark red siltstone and polymictic conglomerate are interlayered with mafic flows in the Billiards Brook area. Chlorite schists along the southeast side of the Cape Ray Fault contain abundant carbonate and are interpreted as highly deformed mafic tuffs.

Hypabyssal intrusions include mafic dykes and sills, a few of which represent uralitized pyroxenites and hornblendites. Felsic intrusions include rhyolite porphyry dykes and sills, and two small leucogranophyric bodies, the Windowglass Hill and Nitty Gritty Brook granites (Chorlton, 1983). The Windowglass Hill granite cuts nearby volcanic rocks of the Windsor Point Group whereas the Nitty Gritty Brook granite contributed detritus to Windsor Point sedimentary rocks.

Flaggy, locally calcareous sandstone interbedded with thick units of alternating clast-supported polymictic conglomerate, lithic grey sandstone, and graphitic mudstone form a distinctive association in the Billiards Brook area. The mudstones contain fragmented plant fossils. Conglomerates have variable clast populations that include mafic and felsic volcanic rocks, jasper, red siltstone, quartz, limestone, trondhjemite, and quartz porphyry. Farther southwest toward the coast, the sediments contain a much greater volcanogenic component and mudstone is less common.

The presence of redbeds somewhere below or within the Windsor Point Group is suggested by red siltstone fragments in many of its polymictic conglomerates, and in sedimentary rocks overlying crossbedded grey sandstone of the fossiliferous sequence in Billiards Brook.

The Windsor Point Group is highly deformed within the Cape Ray fault zone and three phases of deformation are recognized locally (Chorlton, 1983; Wilton, 1983). In places the foliations in the Windsor Point Group are statistically indistinguishable from the fabrics in older rocks on either side of the Cape Ray fault zone (Wilton, 1983). Thus, intensely deformed Windsor Point mafic volcanic rocks and associated chlorite schists may be difficult to separate from mylonitized amphibolite of the Bay du Nord Group to the southeast or metamorphosed mafic igneous rocks of the tonalite terrane to the northwest. Along the northwest boundary of the belt, foliated tonalite with porphyroclasts of quartz, forming augen in the cataclastic foliation, may be confused with much younger quartz porphyry.

Age

Devonian plant fossils were known in sedimentary rocks of the Windsor Point Group at Billiards Brook since the earliest geological studies in this area (Dorf and Cooper, 1943; Cooper, 1954). More recent collections at the same locality (Chorlton, 1983) indicate a latest Early Devonian to earliest Middle Devonian age. New fossil discoveries in the grey beds of the Princess Lake area are of similar age (McGregor, 1981; Chandler, 1982; Kean, 1983). A Silurian isotopic age for volcanic rocks in association with redbeds to the north (Chandler and Dunning, 1983) indicates the presence of both Silurian and Devonian rocks within the Cape Ray Belt as defined here.

The redbed-volcanic sequence in the northeast lies unconformably on the Ordovician Annieopsquotch Complex that is dated isotopically at approximately 480 Ma (Dunning and Krogh, 1985). Volcanic rocks of the redbed sequence are dated isotopically as Silurian (431 ± 5 Ma) indicating a gap of about 50 Ma between ophiolite generation and its eventual emplacement and erosion. The Windsor Point Group overlies tonalites and mylonites to the northwest of the Cape Ray Fault (Brown, 1977; Chorlton, 1983; Wilton, 1983; Williams et al., 1989a; Piasecki et al., 1990) and it overlies the Lloyds River Intrusive Suite (Chandler, 1982; Kean, 1983). The latter is probably coeval with mafic intrusions at Boogie Lake that are dated at 435 +5/-2 Ma (Dunning et al., 1988, 1990a). These mafic intrusions are interpreted as the first intrusions related to faulting (extension) following ophiolite emplacement (Dunning et al., 1989).

The Windowglass Hill granite, which is viewed as a subvolcanic equivalent of the Windsor Point Group, yields a Rb-Sr isotopic age of 368 ± 12 Ma and the Strawberry and Isle aux Morts Brook granites, which post-date the three main deformations of the Windsor Point Group, yield Rb-Sr isotopic ages of 362 ± 27 Ma and 352 ± 6 Ma, respectively (Wilton, 1983). Carboniferous rocks in southwestern Newfoundland are unaffected by granite intrusions.

Metamorphic rock fragments such as mylonitic tonalite in Devonian conglomerates indicate that deposition followed an important phase of intrusion and metamorphism that affected adjacent rocks.

The red Silurian and grey Devonian sequences in the northeast have a similar structural and metamorphic style. Whether their contact is tectonic or stratigraphic, there is no reason to suspect a significant structural break between them.

Sedimentology and depositional environment

Sedimentary features of rocks of the Cape Ray Belt, the bimodal character and lithological features of its volcanic rocks, and structural-stratigraphic relationships all indicate a terrestrial depositional environment.

The red sequence in the northeast is alluvial, with possible fan and braided stream facies (Chandler, 1982). The absence of plant fossils is compatible either with a Silurian age or a dry climate. The Devonian grey sedimentary rocks of the Windsor Point Group may be of meandering fluviatile origin with grey grits as channel sands and black mudstones as overbank deposits (Chandler, 1982). A terrestrial environment is supported by abundant vascular plant tissue, trilete spores in the black mudstone, and an absence of marine palynomorphs (Chandler, 1982). The sedimentary attributes of the sequence, combined with paleobotonical data, their grey to black colour, and the presence of a local regolith suggest a humid climate and deposition on a vegetated terrane.

Preliminary geochemical studies on volcanic rocks of the Windsor Point Group indicate a within-plate affinity (Chorlton, 1983, 1984), which is also supported by their bimodality and field relationships. These features further support a terrestrial depositional environment.

Correlation

Redbeds and volcanic rocks of the northeastern Cape Ray Belt are identical to those of the Springdale and Botwood belts. A Silurian isotopic age of 431 ± 5 Ma for the volcanic rocks of the northern Cape Ray Belt matches ages of 429 +6/-5 Ma for the volcanic rocks of the Springdale Belt (Chandler et al., 1987) and 423 ± 3 Ma for the Stoney Lake volcanics of the Botwood Belt (Dunning et al., 1988, 1990a). Sedimentological features of the redbeds of the Cape Ray Belt and those of the Springdale and Botwood belts are the same, even to the point of pumice and volcanic rock fragments occurring in sandstones of all localities.

Grey plant-bearing Devonian sandstones of the Windsor Point Group have no correlatives in the Springdale and Botwood belts.

Significance and interpretation

The Cape Ray fault zone has an important role in tectonic models for southwestern Newfoundland. In early plate tectonic models, it was viewed as a major suture between continental rocks of the Humber Zone to the northwest and the Gander Zone to the southeast; a suture where the oceanic Dunnage Zone is completely obliterated (Brown, 1973; Kennedy, 1976; Williams, 1978).

Subsequent workers viewed metamorphosed tonalites that cut ophiolitic rocks northwest of the Cape Ray fault zone, and Ordovician sedimentary and volcanic rocks southeast of the fault zone, as highly deformed and metamorphosed Dunnage equivalents (Chorlton, 1983; Wilton, 1983).

Isotopic and petrogenetic studies of the Devonian Strawberry and Isle au Morts Brook granites on opposite sides of the fault zone suggest derivation from the same underlying continental crust, most likely Grenville basement (Wilton, 1985). Thus, surface rocks may be allochthonous with respect to deeper crust.

The latest studies suggest that the Dashwoods Subzone to the northwest of the fault zone includes rocks typical of both the Humber and the Dunnage zones, all structurally commingled, intruded, and metamorphosed together (Piasecki et al., 1990). Southeast of the fault zone, the Port aux Basques Gneiss is correlated with rocks of the Meelpaeg Subzone (Colman-Sadd et al., 1990), implying Gander affinities. Relationships of surface rocks to lower crustal blocks remain enigmatic but deep seismic reflection studies indicate a collisional boundary in the lower crust between Grenville and Central lower crustal blocks at or near the Cape Ray fault zone (Marillier et al., 1989). Regional analysis of Newfoundland granitic rocks supports the existence of Grenville and Central lower crustal blocks (Williams et al., 1989b).

Unconformable relationships everywhere beneath rocks of the Cape Ray Belt are typical of relationships found northwest of the Red Indian Line farther north. Southeast of the Red Indian Line, Ordovician-Silurian sections are

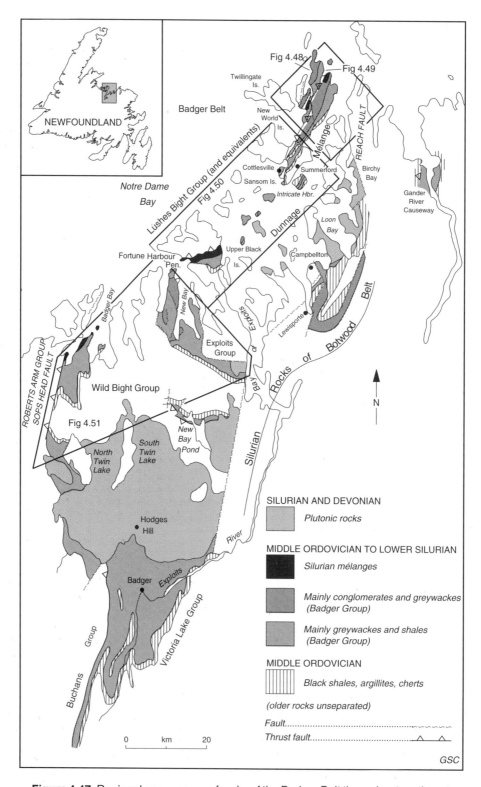

Figure 4.47. Regional occurrences of rocks of the Badger Belt throughout northeast Newfoundland.

conformable in most places. The distribution of terrestrial sandstone-volcanic facies in the Springdale and Botwood belts suggests that these belts converge southwestward into the Cape Ray Belt. This implies a paleogeography for middle Paleozoic rocks completely unrelated to Ordovician tectonic elements.

The narrow outcrop belt of the Windsor Point Group coincides, at least in part, with an ancestral zone of intense deformation and mylonitization. Chorlton (1983) suggested an intracratonic wrench basin as a precursor to Devonian deposition. After deposition, the Windsor Point Group was deformed and mylonitized (Wilton, 1983). The later deformation is much more intense in the southern Cape Ray Belt than elsewhere in the Springdale and Botwood belts.

Badger Belt

Harold Williams, Bruno Lafrance, P.L. Dean, P.F. Williams, K.T. Pickering, and B.A. van der Pluijm

Introduction

The Badger Belt comprises Upper Ordovician-Silurian greywackes, sandstones, conglomerates, and shales with local Silurian mélanges in the upper parts of stratigraphic sections. The rocks range from deep to shallow marine and they contrast with the dominantly terrestrial rocks of other Newfoundland belts. The marine clastic units overlie Caradoc black shales with gradational and conformable contacts, except for a local unconformity between Silurian conglomerates and Ordovician volcanic rocks at New World Island (Arnott, 1983b; Reusch, 1987). Rocks like those of the Badger Belt are represented only in middle Paleozoic sections developed upon the Exploits Subzone. Silurian shelly faunas are common, especially at New World Island; Middle to Late Ordovician graptolites occur in New Bay and elsewhere (S.H. Williams, 1989, 1991); and Silurian graptolites are now known at Upper Black Island of Bay of Exploits (S.H. Williams and O'Brien, 1991). The consistant stratigraphic order of Caradoc black shales overlain by silty to sandy turbidites, then conglomerates and mélanges allows correlation throughout the belt (Dean, 1978; Kean et al., 1981). However, conglomerate-filled channels occur within the lower turbidite unit and sandstone and shale occur locally in the overlying conglomerate unit (e.g. Pickering, 1987a, b).

Rocks of the Badger Belt are located in separated areas throughout Notre Dame Bay and inland to the southwest. The main outcrop areas are at New World Island, in the Bay of Exploits, New Bay, and Badger Bay (Fig. 4.47). Small outliers of similar rocks occur at New Bay Pond and Campbellton. Turbidites at Badger Bay are almost continuous southward with a much larger area of similar rocks above Caradoc shales along the Exploits River.

Williams, H., Lafrance, B., Dean, P.L., Williams, P.F., Pickering, K.T., and van der Pluijm, B.A.
1995: Badger Belt; in Chapter 4 of Geology of the Appalachian-Caledonian Orogen in Canada and Greenland, (ed.) H. Williams; Geological Survey of Canada, Geology of Canada, no. 6, p. 403-413 (also Geological Society of America, The Geology of North America, v. F-1).

The Badger Belt is bounded by steep faults along its western and northern margins. Structural investigations indicate that the faults are transcurrent with a complex deformation history. Some may have been initial thrust faults that were later modified, but evidence of early thrusting, if it ever existed, is obscured by the overprinting trancurrent movement. The southeast margin of the belt is marked by high-angle, possibly transcurrent faults. Bedding-parallel and transverse faults have modified the initial configuration of the Badger Belt, and their effect on the distribution of the rock units must be removed before elaborating on tectonic and depositional models.

Greywackes and conglomerates like those of the Badger Belt also occur in the Botwood Belt where they underlie volcanic-redbed sequences (Fig. 4.47). Examples in the Botwood Belt at Lewisporte and immediately east of the Reach Fault may be equivalent to those of New World Island. The intervening Dunnage Mélange is largely bounded by faults and is now an anticlinal area of older rocks between the Badger and Botwood belts.

Nomenclature is complicated by a long history of changing ideas in this well-studied area of northeast Newfoundland. A new stratigraphic group name is required to denote and distinguish the Ordovician-Silurian marine greywacke-conglomerate facies of the Badger and Botwood belts from the Silurian terrestrial volcanic-sandstone facies of the Botwood Belt. In the ensuing treatment of the Badger and Botwood belts, the Ordovician-Silurian marine greywacke-conglomerate facies is referred to informally as the Badger group. The Silurian terrestrial volcanic-sandstone facies, overlying the Badger group in the Botwood Belt, is the Botwood Group (Williams, 1962).

In the following account, lithology, stratigraphy, and structure of rocks in the largest and most informative areas are described first. Other aspects, such as correlation, depositional environments, and interpretations are treated for all rocks of the entire belt.

Eastern New World Island area

Previous work and ideas

The stratigraphy and age of the rocks in the eastern New World Island area are the best known of any in the Badger Belt. The rocks therefore furnish a middle Paleozoic history for the interior part of the Newfoundland Appalachians that is unrecorded or poorly known in other parts of the orogen. Since the earliest reconnaissance studies (Heyl, 1936; Twenhofel and Shrock, 1937; Baird, 1953; Patrick, 1956; Williams, 1957, 1963; Kay and Williams, 1963) there have been a number of stratigraphic and structural studies (Harris, 1966; Kay, 1967, 1976; Horne and Helwig, 1969; Helwig and Sarpi, 1969; Helwig, 1969; Horne, 1969, 1970, 1976; Bergström et al., 1974; Dean and Strong, 1977; McKerrow and Cocks, 1977, 1978, 1981; Dean, 1978; Fåhraeus and Hunter, 1981; Karlstrom et al., 1982; Arnott, 1983a, b; Arnott et al., 1985; Karlstrom, 1985; van der Pluijm, 1986, van der Pluijm et al., 1987; Pickering, 1987a, b; Elliott and Williams, 1988). As well, the middle Paleozoic rocks of this area are also the topic of many theses studies (Horne, 1968; Hunter, 1978; Jacobi, 1980; Watson, 1981; Arnott, 1983b; Reusch, 1983; van der Pluijm, 1984; Antonuk, 1986; Elliott, 1988). A complex deformation history and repetition of lithological units in the area has led to several conceptual models.

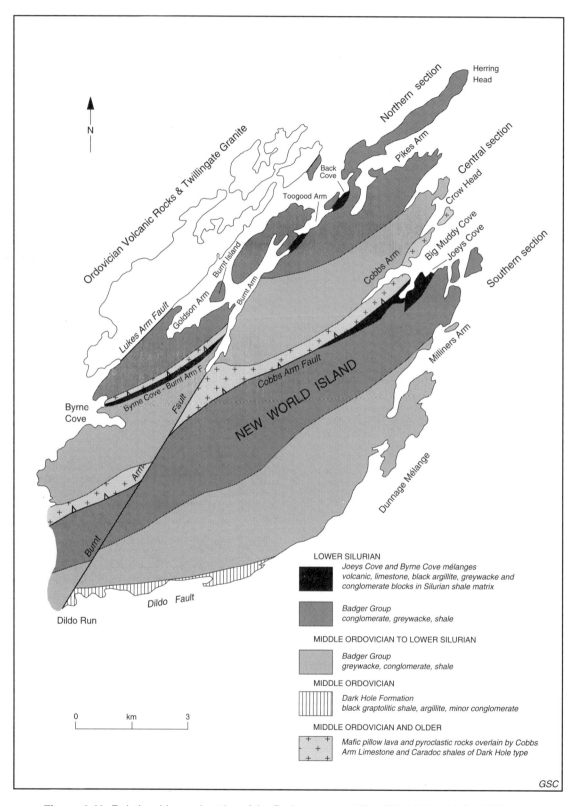

Figure 4.48. Relationships and setting of the Badger group at New World Island (after Williams, 1963; Kay, 1976; Reusch, 1983).

Middle Paleozoic rocks of eastern New World Island trend east-northeast and they are steep to overturned with stratigraphic younging mainly toward the northwest (Fig. 4.48). Ordovician and Silurian rocks occur in three stratigraphic sections: a southern section between the Dildo Fault and Cobbs Arm Fault, a central section between the Cobbs Arm Fault and Byrne Cove-Burnt Arm faults, and a northern section between the Byrne Cove-Burnt Arm faults and the Lukes Arm Fault. All of the rocks were first thought to be part of a continuous stratigraphic section (Twenhofel and Shrock, 1937) with progressively younger rocks represented from south to north. Routine mapping (Williams, 1963) and the discovery of Silurian fossils to the south of the Ordovician rocks at Cobbs Arm suggested correlation with similar Silurian units to the north of Cobbs Arm and structural repetition (Kay and Williams, 1963; Kay, 1976; Dean and Strong, 1977; Dean, 1978). The original idea of a continuous section was revived by Jacobi and Schweickert (1976) and McKerrow and Cocks (1978) when Ordovician rocks, like those at Cobbs Arm, were interpreted as huge blocks in Silurian olistostromes. The latest studies support structural repetition, but there are differing viewpoints concerning paleogeographic setting. Thus the nomenclature and correlation of middle Paleozoic rocks at New World Island, and for that matter throughout the Badger Belt, are dependent on whether or not lithologically similar units are viewed as; (a) fault repetitions of a single stratigraphic section (Kay and Williams, 1963; Kay, 1976; Jacobi and Schweichert, 1976; Dean and Strong, 1977; Dean, 1978; Karlstrom et al., 1982, 1983; Reusch, 1983; van der Pluijm, 1986; van der Pluijm et al., 1987; P.F. Williams et al., 1988; Elliott et al., 1989); (b) cyclic parts of a continuous Silurian section that contains olistostromes with huge Ordovician blocks (McKerrow and Cocks, 1978, 1981), or (c) the deposits of discrete small basins separated by contemporary tectonic ridges (Arnott, 1983a; Arnott et al., 1985). Many lithological contacts are either unexposed or faulted, and depositional contacts are rarely observed. Fossil occurrences at New World Island are helpful, but refinement of the stratigraphy is hampered by structural complexities.

Nomenclature, distribution, and thickness

According to the model of structural repetition, the greywacke units were correlated and assigned to the Sansom Formation (Heyl, 1936) and conglomerate units to the Goldson Formation (Twenhofel and Shrock, 1937); all part of the present Badger group. A new terminology was introduced with the revival of a continuous section model (McKerrow and Cocks, 1978) but it has had limited usage (Arnott, 1983a; Reusch, 1983). The recent studies that support fault repetition propose a simple stratigraphy and retain the names Sansom for the mainly greywacke unit and Goldson for the overlying mainly conglomerate unit (Dean, 1978; Kean et al., 1981; Nelson, 1981; Reusch, 1983; Karlstrom et al., 1983; van der Pluijm, 1986; van der Pluijm et al., 1987; Elliott et al., 1989). Two Early Silurian mélanges have been mapped separately at New World Island (Fig. 4.48): the Joeys Cove and Byrne Cove mélanges (Arnott, 1983a; Reusch, 1983; van der Pluijm, 1986). They are located at the boundaries of the faulted stratigraphic sequences.

Rocks assigned to the Sansom and Goldson formations occur in the three sections exposed along the eastern shore of New World Island. In the southern section, both the Sansom greywackes and Goldson conglomerates were reassigned to the Milliners Arm Formation (McKerrow and Cocks, 1978; Watson, 1981; Arnott, 1983a). In the central section, the Sansom greywackes were reassigned to the Big Muddy Cove Formation (Arnott, 1983a) and the Crow Head Formation (Reusch, 1983), but the name Goldson was retained for the conglomerates. The northern section is the type area of the Goldson Formation (Twenhofel and Shrock, 1937). Thin shale units interlayered with the conglomerates, that were originally assigned to the Pikes Arm Formation (Twenhofel and Shrock, 1937), are now considered to be members of the Goldson Formation (Williams, 1967; Dean, 1978).

The thickest exposed sequence of the Badger group occurs in the southern section; more than 2000 m of Sansom greywackes and 1500 m of overlying Goldson conglomerates. Structural complications at Milliners Arm may exaggerate thicknesses.

Stratigraphic relationships

In the southern section northward from the Dildo Fault to the Cobbs Arm Fault, Caradoc black shale (Dark Hole Formation of Kay, 1976) is overlain conformably by Sansom greywackes with Silurian fossils at the top of the section, in turn overlain conformably and gradationally by Goldson conglomerates (Watson, 1981). The top of the Goldson Formation is faulted, either against Ordovician rocks or against greywackes and chaotic Silurian rocks of the Joeys Cove Mélange (Arnott, 1983a). In the central section, Ordovician rocks at Cobbs Arm are followed northward by Sansom greywackes. Contacts are poorly exposed but the common occurrence of steep, north-younging Ordovician volcanic rocks, limestone and black shale, then Sansom greywacke, in that stratigraphic order, suggests a conformable sequence. Goldson conglomerates at the top of the central section are also conformable and gradational above the Sansom greywackes. In the northern section, chaotic Ordovician rocks are interpreted to occupy an anticlinal core at Toogood Arm and they are overlain directly by Goldson conglomerate (Arnott, 1983a). Nearby, a significant unconformity is present between Goldson conglomerates and Ordovician volcanic rocks at Back Cove (Arnott, 1983b; Reusch, 1987).

Along the road from Summerford to Cottlesville in southwest New World Island, a sequence of Ordovician volcanic rocks is bounded to the north and to the south by north-younging Sansom greywackes (Elliott et al., 1989). The Ordovician volcanic sequence is internally faulted and repeated with, from north to south, Llanvirn volcanic rocks structurally above Caradoc black shale, and upper Arenig-Llanvirn volcanic rocks above Upper Llanvirn volcanic rocks.

The deposition of Sansom greywackes and Goldson conglomerates was diachronous over a considerable time interval on New World Island with transgression toward the northeast (Elliot et al., 1989). On the southern side of Intricate Harbour in southwest New World Island, the age

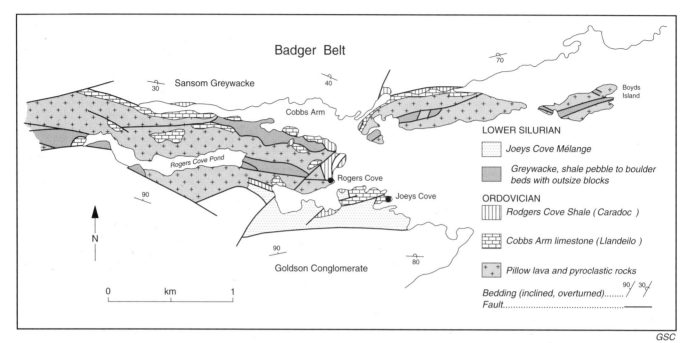

Figure 4.49. Complexities of the Cobbs Arm fault zone and location of the Joeys Cove Mélange, after Reusch (1983).

of the Goldson conglomerate is interpreted as Late Caradoc (Horne and Johnson, 1970; Elliott et al., 1989). Sansom greywackes below the conglomerate are therefore Late Caradoc or older (Elliott et al., 1989). Eastward in the southern section near Milliners Arm, the Sansom Formation contains Ashgill and Llandovery faunas, and in the central and northern sections, the Goldson Formation contains Llandovery faunas which may be as young as late Late Llandovery.

Lithology

The Sansom Formation consists of thick- to medium-bedded greywackes with slate interbeds and pebble conglomerate units. Graded bedding, sole marks, and small scale crossbedding are common. Other sedimentological features include scours, load casts, convolute bedding, slide and slump deposits, shale rip-up breccias, and channels. The greywackes consist of poorly sorted, angular plagioclase and quartz grains, chert, volcanic and intrusive rock fragments, and shale chips, all texturally immature. Coarser varieties contain well-rounded pebbles of granite, granodiorite, chert, and volcanic rocks.

Goldson conglomerates are grey to buff, brown, or red, and composed of sub-angular to well-rounded pebbles, cobbles, and boulders of intrusive and volcanic rocks, chert, sandstone, and plutonic rocks. Polymictic grey varieties are most common but an example at Pikes Arm is red and contains mainly diorite boulders. A nearby conglomerate has mainly metamorphic and granite fragments (Helwig and Sarpi, 1969). Angular shale clasts and large limestone boulders are prominent in some examples and a serpentinite

clast was noted at Pikes Arm. The conglomerates are generally graded, with scouring and channeling common at the base of some units.

Red to brownish and pink varieties of Goldson conglomerate are common in the northern belt of New World Island. Most examples are coarse, poorly bedded, and polymictic. At Burnt Island of Goldson Arm, about 100 m of red sandstone is included in the Goldson Formation. Stratigraphic relationships are unclear, but the sandstones resemble terrestrial redbeds of the Springdale and Botwood belts.

The Joeys Cove Mélange forms a narrow fault-bounded belt along the northern border of the Goldson Formation at Cobbs Arm (Fig. 4.49) and separates the southern and central sections. Outcrop size blocks of mafic volcanic rocks, limestone, and black shale can be matched with Ordovician lithologies that border the mélange to the north. A south-younging section of turbidites on the south shore of Joeys Cove resembles Sansom greywackes, but they are more disrupted and locally contain sandstone boulders and pebble to cobble mudstone beds. In Joeys Cove, Silurian turbidites adjacent to the mélange contain large limestone clasts similar to the mélange olistoliths, indicating a transition from turbidite to mélange. Blocks in the mélange at Rogers Cove are surrounded by green to grey siltstone and shale that contain Llandovery brachiopods and corals (McKerrow and Cocks, 1978). These fossils are slightly younger than those from the nearby Goldson and Sansom formations (Arnott, 1983a). The size and angularity of blocks indicate debris flows. Shallow marine fossils of the mélange matrix are typical of a marine shelf.

The Byrne Cove Mélange, which separates the central and northern sections, consists of slumped sandstones and shales with boulders and blocks of Ordovician volcanic and intrusive rocks, and large lenses of conglomerate.

Structure

The structure of New World Island is of special interest as it bears on stratigraphic analysis. Four major stages have been recognized in the deformation history (P.F. Williams et al., 1988); three were recognized previously (Karlstrom et al., 1982; Arnott, 1983a; Reusch, 1983; van der Pluijm, 1986). The first phase of deformation is equated with the Late Llandovery formation of the Joeys Cove Mélange and the repetition of lithic belts. It is attributed either to normal faulting (Arnott, 1983a, b) or to thrust faulting (Kay and Williams, 1963), but recent studies favour the latter (Karlstrom et al., 1982; Reusch, 1983; van der Pluijm, 1984, 1986; Elliott, 1985, 1988; Arnott et al., 1985; P.F. Williams et al., 1988).

The second phase of deformation produced a ubiquitous regional foliation, which on New World Island dips consistently to the southeast. Although regional F_2 folds are absent on New World Island, D_2 was probably largely responsible for the present steep attitudes of bedding and lithological contacts. D_2 has been interpreted as the climactic Acadian deformation by various workers (Horne, 1968; Nelson, 1981; Arnott, 1983a; Eastler, 1971; Karlstrom et al., 1982). The regional foliation is overprinted by Late Silurian intrusions, and D_1 and D_2 are therefore Silurian events (Elliott et al., 1991). P.F. Williams et al (1988) reported a third deformation that is associated with dextral east-northeast trending ductile faults which overprint the regional foliation. Multiple generations of folds are generated in D_3 movement zones as a result of progressive deformation. The early ductile phase of the Chanceport/Lukes Arm faults, Dildo Fault, and Indian Islands Fault are examples of D_3 movement zones. The fourth stage of deformation comprises brittle faults that overprint earlier structures. A late sinistral brittle movement is observed in the Chanceport/Lukes Arm fault zone.

Although there is good evidence for thrust faulting on New World Island, the direction of tectonic transport is controversial. Van der Pluijm (1984), following the original idea of Kay and Williams (1963), proposed thrusting towards the northwest, based on the vergence of mesostructures and on the present geometry and location of the three stratigraphic sections on eastern New World Island. Reusch (1983) proposed thrusting toward the southeast, based on sedimentological and structural analyses of the three sections. He pointed out that conglomerate clasts in the northern section matched lithologies of the Notre Dame Subzone to the northwest, and therefore initially lay northwest of the central and southern sections. Elliott (1988) proposed southwest-directed thrusting based on the orientation and geometry of outcrop scale and map scale duplexes. Southeast tectonic transport is favoured elsewhere in central Newfoundland (Dean and Strong, 1977; Nowlan and Thurlow, 1984; Colman-Sadd and Swinden, 1984).

The importance of bedding-parallel faults in the New World Island area cannot be overemphasized. Repetitions occur both at the scale of individual beds and at the regional scale, so that stratigraphic sequences cannot be assumed to be unmodified. Mesoscopic thrust faults are abundant but they are only recognizable at ramps since their flat segments are parallel to undeformed beds. Numerous late faults also affect the area. These are both dextral and sinistral transcurrent faults trending either parallel or oblique to the regional trends of lithic units. On the outcrop scale, bedding-parallel transcurrent faults, similar to thrust faults, are recognized at ramps (van der Pluijm et al., 1987; Elliott et al., 1989).

Caution is also recommended in paleocurrent analysis. Since bedding is steeply dipping, paleocurrent determinations are dependent on the axis of rotation chosen to restore bedding to the horizontal. Several phases of faulting and folding must be removed in these analyses.

Bay of Exploits area

Greywackes and conglomerates of the Badger Group can be traced from western New World Island across the islands of Exploits Bay to the Fortune Harbour Peninsula (Fig. 4.50). The first mapping of this area was done by Heyl (1936) and Patrick (1956) with reconnaissance studies by Williams (1964; 1972). More detailed work was done by Horne (1970) in western New World Island, Helwig (1967, 1969) at Fortune Harbour Peninsula, and Antonuk (1986) on Farmers Island. Recent compilations and syntheses are those of Dean (1978) and Kean et al. (1981). Descriptions of new fossil localities are given by McKerrow and Cocks (1981) and Karlstrom (1985).

South Sansom Island is the type locality of the Sansom Formation (Heyl, 1936). Conformable contacts between its greywackes and overlying conglomerates are exposed on its northeastern extremity and on its southern side. Northward from its southern side, the rock units are: Sansom greywackes (Caradoc or Ashgill age, Karlstrom, 1985), conglomerates (containing a Late Ordovician rugose coral clast, R.B. Neuman, pers. comm., 1989), pillow lava, limestone (probable Llandeilo age, McKerrow and Cocks, 1981), volcanic breccia, calcareous tuff, and Badger group lithologies farther north. The Badger group to the south of the Ordovician rocks are north-younging. A fault contact is therefore demanded.

McKerrow and Cocks (1981) identified three mélange units at South Sansom Island that locally contain Ordovician blocks. They interpreted the mélange units as olistostromes, and since the South Sansom Island rocks appear continuous with those of western New World Island, all occurrences of Ordovician rocks at western New World Island were also interpreted as blocks in olistostrome (McKerrow and Cocks, 1981). The mélanges are characterized by extreme boudinage of sandstone beds and blocks within a cleaved argillite matrix (Karlstrom, 1985). On northern South Sansom Island, a penetrative cleavage in mélange flows around a 30 m thick horizon of conglomerate, indicating that the conglomerate was consolidated prior to the deformation. This and other structural evidence suggest that some mélanges are tectonic (Lafrance, 1989).

At nearby Farmers Island, north-facing greywackes overlain by conglomerates have been correlated traditionally with the Sansom and Goldson formations, and a Llandeilo (McKerrow and Cocks, 1981) volcanic-limestone unit farther north was interpreted as structurally above the Goldson conglomerate (Williams, 1964; Horne, 1970; Kean et al., 1981). McKerrow and Cocks (1981) viewed the Ordovician volcanic unit as stratigraphically above

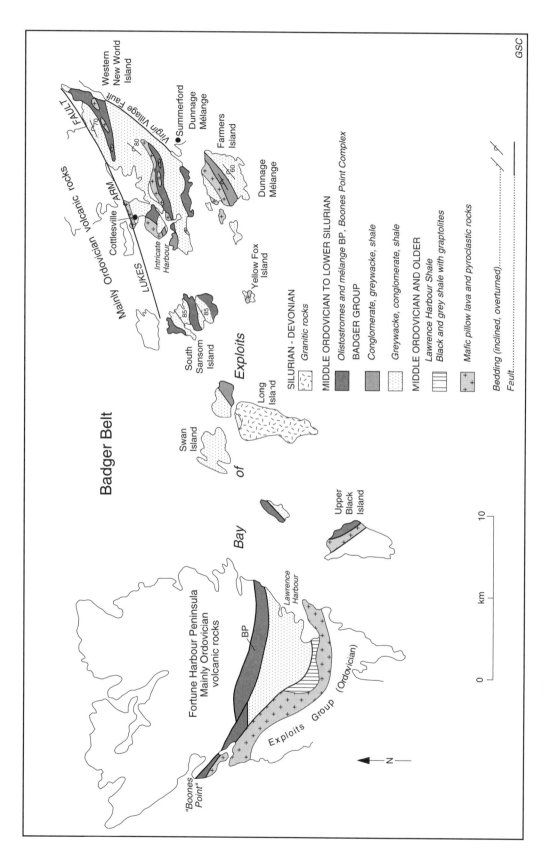

Figure 4.50. Relationships and setting of the Badger group in the Bay of Exploits area, after McKerrow and Cocks (1981) and Kean et al. (1981).

conglomerate, thus demanding a Llandeilo or older age for the conglomerates and greywackes to the south. Accordingly, McKerrow and Cocks (1981) assigned all of the rocks at Farmers Island to the Farmers Island Formation, which they interpreted as below their Sansom Formation. Greywackes and volcanic rocks at Yellow Fox Island (Fig. 4.50) were correlated with the Farmers Island Formation and also interpreted as a conformable Ordovician section (McKerrow and Cocks, 1981). As there is a narrow exposure gap between the north-facing conglomerates and Ordovician volcanic rocks on Farmers Island, a thrust imbrication and structural mixing of lithological units seem more plausible (Antonuk, 1986). Thrusting at Yellow Fox Island is also favoured where east-northeast-trending narrow fault zones are defined by thin phyllonites and calcite mylonites beneath volcanic sequences (van der Pluijm et al., 1987).

Sansom greywackes, immediately north of the Dunnage Mélange on Farmers Island, are fine grained and thinly bedded in the south with grey silt and dark grey shale interbeds. The beds are thicker and coarser northward and isoclinal folds are well developed. Conglomerates above the greywackes occur in beds up to 3 m thick and they contain pebbles and cobbles of basalt, granophyre and chert. Many of the conglomerate beds are graded.

At Fortune Harbour Peninsula, thin- to medium-bedded greywackes assigned to the Point Leamington Formation (Helwig, 1969; Bergström et al., 1974; S.H.Williams, 1991) are correlatives of the Sansom Formation. The greywackes overlie Caradoc shales (Lawrence Harbour Shale of Helwig, 1969) and their northern boundary is marked by a mélange, the Boones Point Complex (Helwig, 1969; Nelson, 1981). It consists of large blocks of pillow lava, dacite, and limestone in a cleaved argillaceous matrix. The blocks are typical of nearby Ordovician rocks. Large lenses of coarse conglomerate, resembling Goldson conglomerate, are also present. Corals from limestone blocks are of late Ordovician age (Horne and Helwig, 1969; Arnott et al., 1985). The Boones Point Complex is separated by a major fault from Ordovician volcanic rocks farther north.

The Boones Point Complex can be correlated across the Bay of Exploits, with the mélanges on Swan Island, South Sansom Island, and western New World Island (Lafrance, 1989). On eastern Fortune Harbour Peninsula, the Boones Point Complex has a strong penetrative foliation that envelops and flows around bed fragments and blocks in mélange. Foliation in the mélange also transposes the main regional cleavage, indicating that the mélange and shear zone formed late in the deformation history of the area. This interpretation relates the regional foliation and the formation of the Boones Point Complex to progressive deformation in a dextral movement zone (Lafrance, 1989). On South Sansom Island, Late Silurian intrusions overprint structures associated with a similar dextral movement zone, and since this movement zone overprints Llandovery conglomerate, its structures are Silurian.

Limestone blocks associated with mafic volcanic rocks of Upper Black Island contain a shelly fauna that was interpreted as Silurian (Twenhofel and Shrock, 1937; Williams, 1967; Boucot and Smith, 1978). This fauna is now reassigned a Middle to Late Ordovician age (Arnott et al., 1985). Silurian graptolites also occur at Upper Black Island and this is the first well-documented discovery of Silurian gratolites in Newfoundland (S.H. Williams and O'Brien, 1991).

New Bay area

The Point Leamington Formation (Helwig, 1967) of the New Bay area (Fig. 4.51) is another correlative of Badger greywakes. It is conformable above Caradoc argillites and it is overlain by Badger conglomerates. Maximum thickness is 3 km but the greywackes are locally less than 1 km. The beds are thinner and the rocks are finer than those of the type area, and silty argillites alternate with greywacke beds in rhythmic couplets. Pebbly to conglomeratic greywackes, pebbly mudstones and black argillites are less common lithologies. Slide and slump folds, graded-bedding, ripple lamination, convolute bedding, load casts, flutes, and grooves are common sedimentary features. Sediment transport was toward the south and southeast (Dean, 1978), with some local variation (Pickering, 1987b). A conglomerate that was previously mapped as Goldson Formation near Point Leamington (Williams, 1964) is an example of coarse conglomerate that occurs in the lower part of the succession (Dean, 1978).

Late Caradoc and Ashgill graptolites and Ashgill corals occur in the Point Leamington formation(Helwig, 1967; S.H. Williams, 1989, 1991). Graptolite faunas at the base of the formation in the west and south of New Bay are slightly older than those on the Fortune Harbour Peninsula (Bergström et al., 1974; Dean, 1978). However, one of the localities referred to by Bergström et al. (1974) lies well within black shale and far from the contact with the overlying greywacke, and at another locality the contact is faulted (S.H. Williams, 1989).

Conglomerates overlie and interdigitate with Point Leamington greywackes in the West Arm and Osmonton Arm areas of New Bay. Maximum thickness is in the order of 600 m. The conglomerates are grey, poorly bedded, with greywacke and shale interbeds near their base. Angular sedimentary clasts, rounded plutonic boulders, and coraline limestone boulders are common. A block of Silurian coraline limestone, 3 m by 6 m, occurs in conglomerate at Besom Cove.

Badger Bay area

Rocks correlated with the Sansom Formation and assigned to the Badger group in the Badger Bay area are the previous Gull Island Formation and parts of the Julies Harbour and Burtons Head groups (Espenshade, 1937). Greywackes of the Gull Island Formation overlie black Caradoc argillites of the Shoal Arm Formation at Gull Island and on the northwest shore of Shoal Arm (Fig. 4.51). The section has units that are progressively younger northward and it is approximately 2 km thick. The base of the Gull Island Formation is defined by the appearance of greywacke beds above the Shoal Arm argillites. The lower 100 m consist of interbedded greywacke and argillite, with thicker bedded and coarser grained greywackes occurring higher in the section. Fine grained arenaceous greywackes and argillite in the upper part of the section contain poorly preserved Late Ordovician graptolites (Williams, 1972; Dean, 1978).

Badger conglomerates are absent at Badger Bay. Instead, greywackes are overlain by mélange of the Sops Head Complex (Dean, 1978). The northern contact between the Sops Head Complex and the pre-Caradoc Roberts Arm Group is a prominent fault with clear morphological

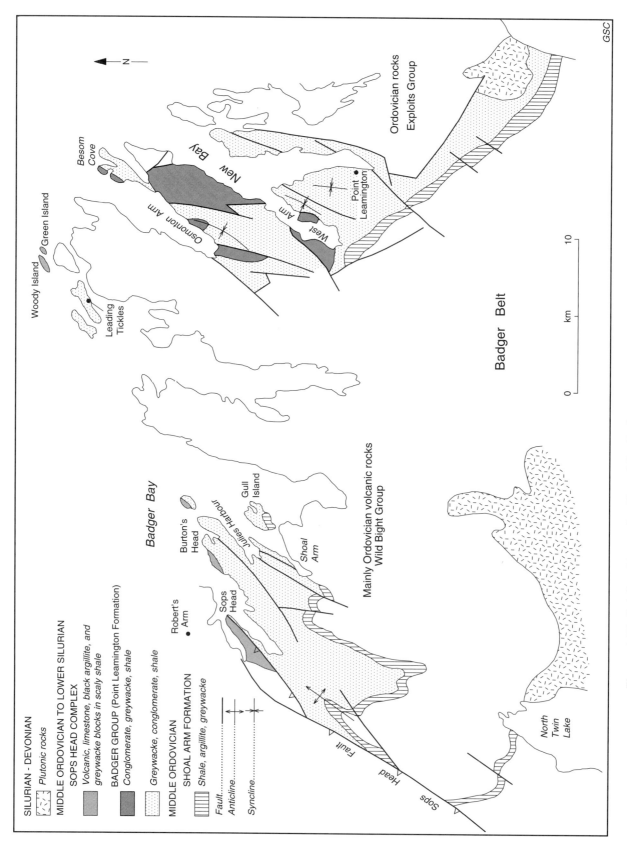

Figure 4.51. Relationships and setting of the Badger group in the New Bay and Badger Bay areas, after Kean et al. (1981).

expression. At the contact, mélange lithologies and mafic and silicic volcanic rocks are intimately sheared and commingled in a zone several tens of metres wide (Nelson, 1981). The southern contact between the Sops Head Complex and the north-younging greywackes is gradational with the chaotic rocks passing into regularly bedded greywackes. Locally, the underlying section has conglomerate and debris flow horizons that are identical to those within the Sops Head Complex, though less deformed. Similar chaotic rocks occur between Badger Bay and Fortune Harbour Peninsula on Green Island and Woody Island to the north of Leading Tickles (Fig. 4.51).

The Sops Head Complex consists of pebble and cobble size clasts in a dark grey-silty mudstone matrix. Clasts include volcanic rocks resembling the Roberts Arm Group, greywackes of Badger type, plutonic cobble conglomerates, silicic plutonic rocks, diabase, green volcaniclastic sandstone, chert, limestone, and black shale. Conodonts from limestone blocks are Late Llanvirn to Llandeilo (Nelson, 1981).

Green Island and Woody Island consist of blocks or slices up to 100 m wide of interbedded greywacke and slate, mafic volcanic rocks, and green silicic porphyry. Thin horizons with smaller blocks and clasts within a sheared sedimentary matrix separate the larger blocks.

An early deformation which affects both matrix and resistive blocks in the Sops Head Complex has an associated cleavage that is restricted to the mélange and follows its outcrop trend. A second steep cleavage trends northeast and affects both the mélange and strata outcropping on either side (Nelson, 1981).

The Sops Head Complex is correlated with the Boones Point Complex (Nelson, 1981), but controversy still remains regarding their interpretation either as tectonized olistostromes (Nelson, 1981) or tectonic mélanges (Lafrance, 1989).

Other areas of the Badger Belt

A large area of greywacke occurs above Caradoc argillites in the vicinity of Badger (Kean and Jayasinghi, 1980; Fig. 4.47). Pebble to cobble conglomerates also occur in the area but none has been mapped separately. Ordovician volcanic rocks occur to the south and the rocks are truncated by the Hodges Hill batholith to the north. Ordovician rocks of the Buchans Group to the northwest are faulted against the greywackes (Nowlan and Thurlow, 1984).

On the northeast shore of New Bay Pond (Fig. 4.47), greywackes overlie Caradoc argillites. The section faces southwest and has a maximum exposed thickness of about 1 km. The thickness appears to increase southward where the section is truncated by a granodioritic intrusion. The rocks are thick bedded lithic greywackes with volcanic and chert clasts. Volcanic rocks of the Frozen Ocean Group to the west were interpreted to overlie the greywackes (Dean, 1978), but the contact is more likely a northwest-trending fault across which the Frozen Ocean Group youngs to the northeast and the greywackes young toward the southwest (Swinden, 1988). A thrust fault (Kusky, 1985; Kusky and Kidd, 1985) or transcurrent fault are possible (see Elliott, 1988 for further discussion).

An overturned syncline at Campbellton (Fig. 4.47) contains a kilometre of Badger-type greywackes conformably above Caradoc shales. To the north, the greywackes are metamorphosed by the Loon Bay granite and metamorphic biotite and garnet are common near the intrusive contact.

Interpretation and discussion

Stratigraphic relationships and correlations among mid Paleozoic rocks of the Badger Belt are summarized in Figure 4.52. All of the mainly greywacke or mainly conglomerate successions of the Badger group overlie Caradoc cherts and/or graptolitic black shales, except for the northern section of eastern New World Island where Goldson conglomerates are unconformable on Ordovician volcanic rocks. All of the sections exhibit the same pattern of large scale coarsening upward which is intrepreted as a shoaling and infilling of a sedimentary basin. The greywackes are marine turbidites. Those in the southern section of eastern New World Island are interpreted as part of a southeast and southwest prograding submarine fan system (Watson, 1981), fed from the north and northwest with lower mid-fan greywacke facies and upper inner-fan conglomerate facies. Conglomerates accumulated under unstable conditions probably on a narrow shelf, with near-shore deposits involved in later downslope transport and accumulation in submarine depressions (Helwig and Sarpi, 1969). Where diachroneity is apparent, the Badger group is generally older in the west and southwest indicating tectonic/depositional controls that migrated east and northeast; though this pattern is not uniform for the full length of the Badger Belt. In some cases, thicknesses increase from north to south, or northwest to southeast, indicating a basin margin toward the north and a source area in that direction. Most paleocurrents are toward the south and southeast. Coarse, poorly bedded, red conglomerates in the northern section of New World Island appear to be the most proximal and sedimentological evidence indicates derivation from the north (Arnott, 1983a; Reusch, 1983, 1987). Red sandstones of nearby Burnt Island indicate a change from marine to terrestrial conditions. These relations, coupled with sub-Silurian unconformities throughout the adjacent Springdale Belt, indicate an emergent landmass to the west and northwest.

The Silurian Joeys Cove Mélange is interpreted as an olistostrome. It is in stratigraphic contact with adjacent Silurian sedimentary rocks to the south and it contains clasts of Ordovician rocks to the north (Reusch, 1983; P.F. Williams et al., 1988). Since it separates the southern and central sections of eastern New World Island, its most plausible control is thrust faulting that brought the central section above the southern section. The Sops Head Complex is also interpreted as an olistostrome and proximal facies of the Badger group; formed in response to southeast-directed thrusting of the Roberts Arm Group at the northwest edge of a contemporary sedimentary basin (Nelson, 1981).

The significance of some other melanges is debatable. The Byrne Cove Mélange was interpreted as a slump horizon laterally equivalent to Badger greywackes (Arnott, 1983a, b), but two of us (P.F. Williams and Bruno Lafrance) think it is a tectonic mélange related to late transcurrent faulting, since it is unfossiliferous and is situated where the major Lukes Arm and Chanceport faults converge and

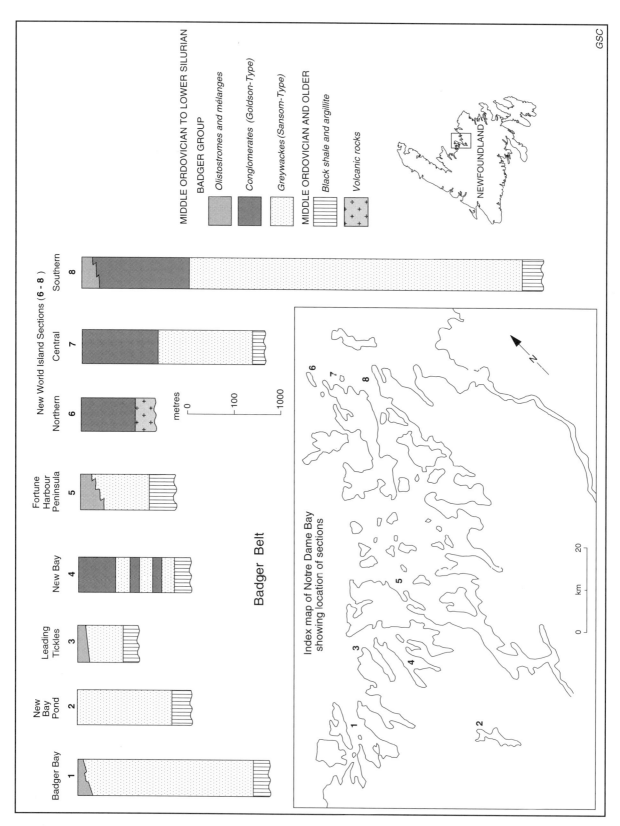

Figure 4.52. Summary of stratigraphic sections throughout the Badger Belt, modified after Arnott et al. (1985).

MIDDLE ORDOVICIAN TO LOWER SILURIAN

BADGER GROUP

Olistostromes and mélanges

Conglomerates (Goldson-Type)

Greywackes (Sansom-Type)

MIDDLE ORDOVICIAN AND OLDER

Black shale and argillite

Volcanic rocks

NEWFOUNDLAND

GSC

New World Island Sections (6 - 8)

Southern
8

Central
7

Northern
6

Fortune Harbour Peninsula
5

New Bay
4

Leading Tickles
3

New Bay Pond
2

Badger Bay
1

Badger Belt

metres

0

100

1000

Index map of Notre Dame Bay showing location of sections

km

0   20

coalesce. Other mélanges, such as the Boones Point Complex and those of South Sansom Island, are also considered as tectonic because of overprinting structural relationships and because they coincide with zones of ductile shearing that obliterate any evidence of a sedimentary origin (Lafrance, 1989).

McKerrow and Cocks (1977, 1978) suggested a major oceanic suture and middle Paleozoic trench at the site of the Reach Fault, but the evidence is weak and short-comings in this model have been pointed out already (Nelson and Kidd, 1979; Karlstrom et al., 1982; Elliott and Williams, 1986; Wasowski and Jacobi, 1986; Boyce et al., 1988).

Watson (1981) and Arnott et al. (1985) suggested that the Badger Belt was the site of several small basins, all bounded by faults, and of similar scale to Tertiary basins of California and North Island of New Zealand. By analogy with these modern examples they reason that the Badger Belt was a wide zone of oblique slip movement that was virtually open to a contemporary ocean toward the southeast.

Structural styles of the Badger Belt with southeast thrusting, or northwestward underplating of outboard elements from the southeast, coupled with mélange formation, resemble structural styles of accretionary prisms either in forearc or foreland basin settings (Reusch, 1983; van der Pluijm, 1986). The latest interpretation of volcanic rocks of the Ordovician Summerford Group as oceanic islands (Reusch, 1983; Jacobi and Wasowski, 1985) rather than an island arc, fits this model. The setting and structural history of the Badger Belt has some similarities to the accretionary prism of the Southern Uplands of Scotland (McKerrow et al., 1977). Both areas have thrust sheets of post-Caradoc flysch underlain by Caradoc and older volcanic rocks, which are predominantly northwest-facing and are folded asymmetrically (Reusch, 1983). However, a chronological progression of successively younger faults with younger flysch packages to the southeast cannot be demonstrated across the Badger Belt. Indeed, the younger (Ashgill/Llandovery) Goldson conglomerates occur north of the older (post Caradoc/Ashgill) greywacke successions; the converse of what would be predicted in a simple accretionary prism model. Furthermore, late ductile and brittle faults, which trend both parallel and oblique to the local stratigraphy, have completely modified the original configuration of the area.

Greywackes and conglomerates like those of the Badger group occur along the northwest and east sides of the adjacent Botwood Belt. Undated greywackes of Badger type also occur farther east above the Davidsville Group at the Gander River Causeway. No equivalent middle Paleozoic rocks are known above rocks of the Newfoundland Gander or Avalon zones. Whereas in the Badger Belt, the greywackes and conglomerates almost everywhere overlie Caradoc shales, the shales overlie various older assemblages such as the Victoria Lake Group, Wild Bight Group, Exploits Group, Summerford Group, Dunnage Mélange, and deformed mafic-ultramafic rocks of the Gander River Complex, from west to east. The obvious concern here is whether or not the Caradoc marine basin was open to the southeast or bordered by an emergent and already coupled eastern Dunnage and Gander zones. An unconformable Ordovician cover on ophiolitic rocks of the Gander River Complex supports the idea of a limiting eastern margin to the marine basin that localized Badger deposition. Interpretation of timing and style of Dunnage-Gander boundaries in

central Newfoundland suggest lower Ordovician Dunnage-Gander linkages (Williams and Piasecki, 1990; Piasecki et al., 1990; Dec and Colman-Sadd, 1990) and support the notion of a bordering positive element limiting the Badger basin to the east. Even if the California Borderland analogy is valid, there is no reason to suspect a Pacific-scale ocean basin to the southeast.

The model of southeast polarity of Silurian thrusting contrasts with northwest polarity of Ordovician thrusting in the western Dunnage Zone and Humber Zone. The Silurian structural style was possibly controlled by an encroaching Gander terrane that underthrust rocks of the Exploits Subzone and led to southeast thrusting as the heretofore undeformed Paleozoic rocks of the Exploits Subzone were compressed against the stabilized Taconic deformed parts of the Notre Dame Subzone (Reusch, 1983; Pickering, 1987a, b; Pickering et al., 1988). Geochronological evidence of an important Silurian metamorphic and plutonic event along the eastern Gander Zone also supports this model (Dunning et al., 1988, 1990a).

Acknowledgments

We thank Peter Stringer for a review of this manuscript.

Botwood Belt

Harold Williams, P.L. Dean, K.T. Pickering

Definition

The Botwood Belt is the largest single area of uninterrupted middle Paleozoic rocks in Newfoundland, extending 300 km between Victoria Lake in the southwest and Fogo Island in the northeast (Fig. 4.53). The belt is 55 km wide at Grand Falls and narrows to about 20 km in the northeast and to less than a kilometre in the southwest. All of its rocks are Silurian, except for possible Late Ordovician greywackes and conglomerates in the vicinity of Lewisporte and northeastward. The belt is defined mainly on its subaerial volcanic rocks and red to grey sandstones of the Botwood Group. It also includes marine greywackes and conglomerates of the Badger group along its northwest margin at Lewisporte, Port Albert Peninsula, and Change Islands. Similar rocks occur on its southeast side at Horwood Bay, Indian Islands, and Rocky Pond. Silurian marine shales also occur in association with greywackes and conglomerates at Indian Islands and southward.

Steep faults mark the northwest boundary of the Botwood Belt with Ordovician rocks in the north but unconformable relationships between Silurian conglomerates and Ordovician rocks of the Victoria Lake Group (Kean, 1977) and with the late Precambrian-Cambrian Valentine Lake quartz monzonite are reported toward the south near Victoria Lake (Fig. 4.53; Kean and Jayasinghe, 1980; Evans et al., 1990). Its southeast boundary is faulted in most places but a conformable stratigraphic contact with Ordovician

Williams, H., Dean, P.L., and Pickering, K.T.
1995: Botwood Belt; in Chapter 4 of Geology of the Appalachian-Caledonian Orogen in Canada and Greenland, (ed.) H. Williams; Geological Survey of Canada, Geology of Canada, no. 6, p. 413-420 (also Geological Society of America, The Geology of North America, v. F-1).

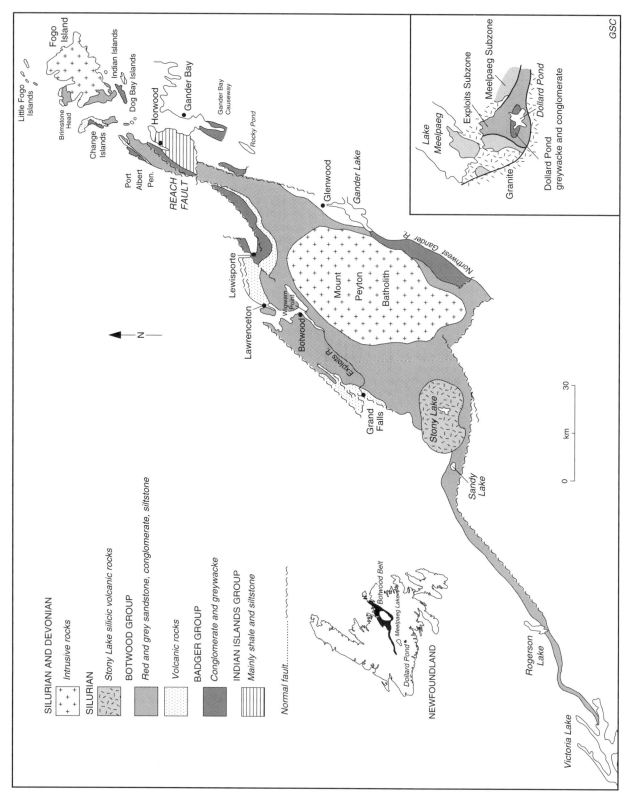

Figure 4.53. Distribution of rocks and relationships in the Botwood Belt, and at Dollard Pond of central Newfoundland.

shales was suggested on the west side of Gander Bay (Wu, 1979; Karlstrom et al., 1982) and Ordovician and Silurian rocks are structurally concordant near Glenwood and southward (Williams, 1967; Anderson and Williams, 1970). These boundaries are also interpreted as faults (Blackwood, 1982; Williams, 1992a). An isolated occurrence of conglomerate at Dollard Pond was assigned to the Silurian because of its lithological similarity to Silurian conglomerates at New World Island (Williams, 1967, 1970).

Previous work

Most early studies were carried out in coastal areas (Twenhofel and Shrock, 1937; Twenhofel, 1947; Patrick, 1956; Baird, 1958) and inland areas to the south of the Exploits River (Hriskevitch, 1950; Peters, 1953; Mullins, 1961). Many of the Silurian rocks were assigned to the Ordovician and Devonian during this phase of study. The entire belt was mapped subsequently in reconnaissance fashion (Williams, 1964, 1970; Anderson and Williams, 1970) and an attempt was made at overall stratigraphic correlation (Williams, 1967). This reconnaissance work demonstrated the continuity of units and emphasized regional similarities, rather than differences, along the length of the Botwood Belt. Since then, more detailed studies have been conducted in coastal areas (Kay, 1969; Eastler, 1969, 1971; McCann and Kennedy, 1974; Wu, 1979; Currie et al., 1980; Karlstrom et al., 1982; Williams, 1992a). Some inland parts of the belt have been remapped also (Kean and Jayasinghe, 1980; Blackwood, 1980, 1981; Colman-Sadd, 1981; Colman-Sadd and Russell, 1982).

Nomenclature and review of stratigraphic relationships

Where first studied by Twenhofel and Shrock (1937), red sandstones of the Botwood Belt at the Exploits estuary were assigned to the Botwood Formation and correlated with the Springdale Formation (Twenhofel, 1947). Four separate groups were recognized in the northeast: the Farewell, Indian Islands, Springdale, and Fogo groups (Patrick, 1956; Baird, 1958). Hriskevitch (1950) assigned the rocks to the Breakheart, Mortons, Fortune, and Springdale formations; Peters (1953) used the names Roberts Arm, Springdale, and Rattling Brook formations; and Mullins (1961) referred conglomerates farther southwest to the Lake Ambrose formation.

During these early studies, volcanic rocks were traditionally assigned to the Ordovician whereas overlying sandstones were interpreted as Silurian or Devonian. The sharp contrast between subaerial volcanic rocks of Silurian age and marine Ordovician volcanic rocks was noted during reconnaissance mapping (Williams, 1962, 1964). Conformable contacts and local interlayering of subaerial volcanic rocks and red sandstones near Botwood indicated coeval deposition of related formations and therefore they were assigned to the Botwood Group (Williams, 1962). The Botwood Group was expanded to include the correlative Farewell and Fogo groups at Port Albert Peninsula, Change Islands, and Fogo Island (Williams, 1964, 1967), so that the Botwood Group included all rocks of the present Botwood Belt in the northeast with the exception of the Indian Islands Group. Conglomerates and greywackes at Port Albert Peninsula and Change Islands were interpreted as a

lower unit of the group and correlated with the Goldson Formation of New World Island (Williams, 1967). Subaerial volcanic rocks were interpreted to overlie the conglomerates and greywackes and named the Lawrenceton Formation. Overlying sandstones were named the Wigwam Formation (Williams, 1972; Dean, 1978). More detailed work at Change Islands (Eastler, 1969, 1971) supported these divisions and stratigraphic relations. A new name, Change Islands Formation, was introduced for the greywacke-conglomerate unit (Eastler, 1969), and successively younger volcanic and sandstone units were called the North End and South End formations (Baird, 1958; Eastler, 1969). The Botwood Group was also extended southwestward toward Victoria Lake where its mainly sedimentary rocks are now known as the Rogerson Lake Conglomerate (Kean and Jayasinghe, 1980). A narrow belt of volcanic rocks to the northwest was also assigned to the Botwood Group (Williams, 1970) but these rocks have been reassigned to the Ordovician Victoria Lake Group (Kean, 1977; Kean and Jayasinghe, 1980). Volcanic rocks in a large oval area at Stony Lake (Fig. 4.53), interpreted as younger than the Botwood Group (Anderson and Williams, 1970), yield a U-Pb zircon age of 423 ± 3 Ma (Dunning et al., 1988, 1990a).

The Silurian Indian Islands Group (Patrick, 1956; Baird, 1958) was excluded from the Botwood Group because its rocks are shalier and more calcareous. Both Williams (1964) and Eastler (1969) proposed a fault boundary between rocks of the two groups. It extended from the north side of Horwood Bay through Middle Dog Bay Island and between Indian Islands and Fogo Island. Eastler (1971) referred to the fault as the Indian Islands Thrust and believed the Indian Islands Group was thrust northwestward above the Botwood Group. Williams (1972) subdivided the Indian Islands Group and pointed out that its highest unit of deformed conglomerate and greywacke resembles the Goldson Formation. He also raised the possibility that the Indian Islands conglomerates at Horwood Bay were equivalent to conglomerates and greywackes at Port Albert, on the opposite side of a northeast-trending syncline. The large proportion of shales in the Indian Islands Group suggested it was a distal facies of the Goldson and Sansom formations. At Horwood Bay, conglomerates, greywackes, and siltstones, formerly part of the Indian Islands Group (Patrick, 1956; Williams, 1964), have been named the Stoneville Formation (McCann and Kennedy, 1974). On the southeast side of the Port Albert Peninsula, Silurian shales and siltstones appear to underlie the conglomerates and greywackes but a mélange at Dog Bay Point implies structural complexity (Karlstrom et al., 1982; Williams, 1992a). A suggestion that the Dunnage Mélange toward the northwest was exposed in an anticline (McCann and Kennedy, 1974; Karlstrom et al., 1982), which corresponds with the syncline at Port Albert Peninsula, implies stratigraphic continuity of at least some rocks on opposite sides of the Reach Fault and enhances correlations between the greywackes and conglomerates of the Badger Belt and similar rocks of the Botwood Belt. Another view is that the Dunnage Mélange is a distinct fault-bounded Lower Ordovician terrane, and continuity of rocks and structures on its opposite sides cannot be presumed (Elliott, 1988; P.F. Williams et al., 1988). A proposed fault boundary (Williams, 1964, 1972) or unconformity (Currie et al., 1980) between rocks of the Indian Islands Group and the Ordovician Davidsville Group on the west side of Gander Bay has been reinterpreted as a conformable

SILURIAN - DEVONIAN

Granite diorite

Silicic to mafic small intrusions

SILURIAN

BOTWOOD GROUP

Volcanic rocks within or above sandstone

Red and grey sandstone

BADGER GROUP

Conglomerate

Greywacke

INDIAN ISLANDS GROUP

Siltstone and shale

Fault (assumed)

Bedding (inclined, overturned)

Syncline (upright)

Anticline (upright, overturned)

Figure 4.54. Relationships at the northeast extremity of the Botwood Belt.

contact (Wu, 1979; Karlstrom et al., 1982). New regional mapping in the Gander Bay area suggests that the Dunnage Mélange is essentially continuous in subsurface with mélanges at Dog Bay Point and Carmanville, and that all contacts with rocks of the Botwood Belt are faults (Williams, 1992a; see addendum).

Classification and nomenclature of rocks of the Botwood Belt present a problem in the middle Paleozoic stratigraphy of central Newfoundland. The Botwood Belt contains rocks typical of both the marine greywacke-conglomerate assemblage of the Badger Belt and the subaerial volcanic-sandstone assemblage of the Springdale and other Newfoundland belts. The assignment of both assemblages to a single group in the Botwood Belt, e.g. Farewell Group (Patrick, 1956; Baird, 1958), Fogo Group (Baird, 1958), Botwood Group (Williams, 1963, 1964, 1967), has served only to obscure their basic differences, both depositional and chronological. The distinct assemblages require separate stratigraphic group names. Accordingly, the informal Badger group is extended to include the marine greywacke-conglomerate assemblage of the Botwood Belt, and the name Botwood Group is retained in its original usage of Williams (1962) for the subaerial volcanic-sandstone assemblage.

The contact between the marine greywacke-conglomerate assemblage (Badger group) and the subaerial volcanic-sandstone assemblage (Botwood Group) was interpreted as conformable (Patrick, 1956; Baird, 1958; Williams, 1963, 1964; Eastler, 1969; McCann and Kennedy, 1974) with the marine rocks below the subaerial rocks. Unconformities are predicted where the change is abrupt. Recent structural studies at Change Islands indicate discordance of fold axes across the boundary between marine greywackes (Badger group) and subaerial volcanic rocks (Botwood Group), and the boundary is interpreted as an unconformity (van der Pluijm et al., 1989). At the northeast end of the Botwood Belt at Port Albert Peninsula, Karlstrom et al. (1982) proposed that the subaerial volcanic rocks and sandstones of the Botwood group were thrust above marine conglomerates and greywackes of the Badger group.

Lithology and correlation

The distribution of rocks and relationships within the Botwood Belt are summarized in Figures 4.53, 4.54, and 4.55. Conglomerates and greywackes of the Badger group occur mainly along the northwest and southeast sides of the belt. Sandstones and volcanic rocks of the Botwood Group occupy the axial region of the belt. Sandstones occur above the volcanic rocks in most places but sandstones and volcanic rocks are interlayered at Brimstone Head of Fogo Island and at Little Fogo Islands. At Fogo Island the Botwood Group is intruded by a large composite batholith. Another similar intrusion cuts the Botwood Group in the vicinity of Mount Peyton (Fig. 4.53).

Badger group

Greywackes of the Badger group are predominant at Change Islands. They are thin- to medium-bedded with siltstone and conglomerate beds. They exhibit abundant sole marks, convolute bedding, and grading. Sole marks indicate currents toward the northwest. Thickness is estimated at 1000 m (Eastler, 1969) but structure is complex.

Conglomerates and greywackes of the Badger group on the west side of the Port Albert Peninsula are deformed with elongate pebbles and boulders. Clasts include chert, sheared to phyllitic siltstone, jasper, quartz, granite, a variety of intermediate to acidic volcanic rocks, and a few highly deformed and elongate limestone boulders rich in coral fragments. Conglomerate units are up to a hundred metres thick and separated by interbedded sandstones or sequences of sandstone and siltstone.

Conglomerates, greywackes, and siltstones at Horwood Bay are about 600 m thick. Coarse matrix-supported beds with dispersed clasts of variable size occur in the middle part of the sequence. Deformation is intense, so that most clasts are flattened, but some volcanic and granite clasts are equidimensional and outsize with diameters in excess of bed thicknesses. Rhyolite-dacite and quartz-feldspar porphyry clasts are most common, followed by mafic volcanic clasts, granodiorite, quartz diorite, jasper, greywacke, quartz, and quartzite. Outsize and poorly sorted clasts supported in thin, roughly alternating siliceous and silty laminae are uncommon in turbidite sequences like those of the Badger group. One interpretation is that the unsorted beds were deposited in a glaciomarine environment, largely by a process of ice rafting with associated turbidity current activity (McCann and Kennedy, 1974). An absence of glaciomarine features elsewhere in the Badger group, coupled with occurrences of warm-water corals and brachiopods, are contrary to the glacial model. The unsorted beds are possibly debris flows like those present in the Badger Belt.

Conglomerates and greywackes at Horwood Bay can be traced northeastward to Dog Bay Islands, Indian Islands, and Cann Island (Fig. 4.54). The beds are steeply dipping to slightly overturned and they are younger mainly toward the northwest; supporting the interpretation of a major syncline whose axis follows Port Albert Peninsula and extends to the southern tip of Change Islands and beyond to Fogo Island (Karlstrom et al., 1982). Shalier and calcerous parts of the Indian Islands sequence contain poorly preserved Silurian fossils, and the rocks may be an eastern and shalier equivalent of Sansom greywackes. Intensity of deformation at Horwood Bay and northeastward hinders correlation.

Badger conglomerates, about 1 km thick, occur at Lewisporte. Contacts are faulted, but Badger greywackes and Caradoc argillites occur nearby to the north. The conglomerates are coarse with plutonic boulders and fossiliferous limestone boulders. Relationships with terrestrial volcanic rocks of the Botwood Group to the southeast are unknown, but the outcrop pattern suggests a conformable contact with the conglomerates below the volcanic rocks. The occurrence of typical Badger lithologies in this area is important as they imply previous continuity between rocks of the Badger and Botwood belts. Greywackes of Badger type at the Gander Bay Causeway (Fig. 4.53) appear to overlie dark grey shales of the Davidsville Group (Williams, 1992a).

Botwood Group

Subaerial volcanic rocks of the Botwood Group are most abundant on the northwest side of the belt. They are mainly flows and pyroclastic rocks, in roughly equal proportions, that are purplish green to purple, red, and green. Calcite amygdules are common and some flows are porphyritic

417

with feldspar phenocrysts up to 2 cm long. Purple to dark red varieties, best seen at Change Islands, are distinctive with both calcite amygdules and feldspar phenocrysts. Coarse unsorted breccias with round lava bombs exceeding a metre in diameter are well displayed at Little Fogo Islands. Red sandstone interbeds occur among the volcanic rocks at Change Islands and coarse volcanic breccia overlies sandstones at Little Fogo Islands. Most of the volcanic rocks are mafic, but grey silicic varieties, of a type not represented elsewhere throughout the Botwood Belt, are interlayered with crossbedded grey sandstones at Brimstone Head of Fogo Island. Fossiliferous Silurian limestone blocks occur in volcanic rocks where they are in contact with greywackes on the western shore of Change Islands (Eastler, 1969).

Sandstones of the Botwood Group overlie the volcanic rocks with local conglomerates present at their bases. The conglomerates contain clasts of the underlying volcanic rocks. The sandstones are typically red or grey but also pink, buff, and brown. Pebble conglomerates and micaceous siltstones and shales show corresponding colour variations. The rocks consist mainly of well-sorted quartz grains and detrital muscovite, commonly with hematitic matrix that causes the red colouration. Crossbedding, ripple lamination, current and oscillatory ripple marks are common. Shrinkage cracks, raindrop prints, and dendritic erosional channels occur in some places.

In its type area opposite Botwood (Fig. 4.55), volcanic rocks (Lawrenceton Formation) and sandstones (Wigwam Formation) occur in stratigraphic conformity and comprise

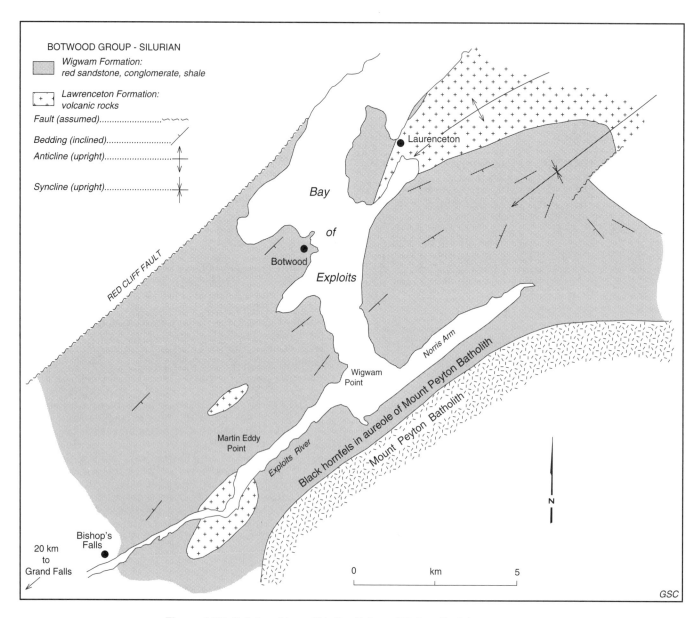

Figure 4.55. Relationships within the Botwood Belt at Exploits estuary.

an anticline-syncline pair. Each formation is in excess of 1000 m thick. The conformable stratigraphic contact between the Lawrenceton and Wigwam formations is well-exposed at Lawrenceton and south of Grand Falls on the Exploits River. Farther southwest (Fig. 4.53), the Rogerson Lake Conglomerate consists of reddish to purplish and grey pebble conglomerate with associated sandstone, siltstone and shale. Conglomerate is common in the southwestern part of the unit with sandstone common in the northeast. Clasts in conglomerates are mostly red siltstone, sandstone, shale, volcanic rocks, quartz porphyry, quartz, limestone, and granitic rocks. The Rogerson Lake Conglomerate has faulted boundaries in most places but it is nonconformable on quartz monzonite dated at 563 ± 2 Ma (Evans et al., 1990) at Valentine Lake, and an unconformity is inferred with the Victoria Lake Group (Kean and Jayasinghe, 1980).

Structural studies have shown that the earliest period of regional deformation to affect rocks of the Botwood Belt was thrusting (Karlstrom et al., 1982), followed by major folding that deformed the thrusts. The structural styles of the Botwood Belt are similar to those of the Badger Belt, and the four phases of deformation defined on New World Island are recognized on Port Albert Peninsula (P.F. Williams et al., 1988).

Volcanic rocks at Stony Lake (Fig. 4.53) appear unaffected by steep folding of the Botwood Group and they are thought to lie unconformably above the folded Botwood sandstones (Anderson and Williams, 1970). The Stony Lake volcanic rocks are distinctive grey to pink flow-banded rhyolites, rhyolitic tuff and agglomerate, porphyritic (feldspar) grey and purple rhyolite, crystal tuff, and flow breccia. They contrast with nearby volcanic rocks of the Ordovician Victoria Lake Group and volcanic rocks of the Botwood Group. Clasts of Botwood sedimentary rocks are reported to occur within the volcanic rocks (Anderson and Williams, 1970), and this coupled with a distinctive lithology, pristine appearance, and structural relationships all suggest that the Stony Lake volcanic rocks are younger than the Botwood Group. A U-Pb zircon age of 423 ± 3 Ma indicates that the Stony Lake volcanic rocks are Silurian. If they are unconformable on Botwood sandstones, their age defines a Silurian deformational event that affected the Botwood Group (Dunning et al., 1988, 1990a).

Rocks 20 km south of Meelpaeg Lake at Dollard Pond are a possible outlier of rocks of the Botwood Belt. They are conglomerates, greywackes, siltstones, and argillites (Williams, 1967). They appear to form a preserved synclinal remnant within the surrounding regional metamorphic terrane. The conglomerates contain pebbles of chert, argillite, grey to buff limestone, and granite. Some varieties have a shaly or limy shale matrix and the most abundant clasts are dark grey to black graphitic argillite of a variety that is typically Ordovician in this part of Newfoundland. The occurrence is interesting because of its extreme southeasterly position in central Newfoundland.

Age

Upper Ordovician to Lower Silurian limestone boulders occur in conglomerates at Lewisporte, and Llandovery limestone boulders occur at the top of the greywackes at Change Islands (Eastler, 1969). Llandovery fossils also occur in shales of the Indian Islands Group (Berry and Boucot, 1970).

A local occurrence of grey conglomerate with fossiliferous matrix and boulders at Martin Eddy Point of Exploits River (Fig. 4.55), and fossiliferous grey shales nearby to the south, are included in the Botwood Group (Williams, 1962). These rocks are of Late Llandovery to Early Wenlock age. A brachiopod fauna from grey Botwood sandstones near Glenwood (Fig. 4.53) on the east side of the belt is of Wenlock age and a single monograptid from the same general locality is of Early Ludlow age (Berry and Boucot, 1970). This paleontological evidence is consistent with a younger Botwood Group stratigraphically above the Badger group (see addendum).

Conglomerates and greywackes of the Badger group in the Botwood Belt are of the same general age as those in the New World Island area of the Badger Belt. Rocks of the Botwood Group, although slightly younger, are also of Early to Middle Silurian age. The youngest Silurian fossils from the Badger Belt occur in rocks associated with red sandstones at Burnt Island in the New World Island area (Fig. 4.48). The Burnt Island sandstones resemble those of the Wigwam Formation of the Botwood Group, suggesting correlation of the youngest rocks of the Badger Belt with subaerial sandstones of the Botwood Belt.

The Botwood Group is a reasonable lithological correlative of the Springdale Group and other rocks of the Springdale Belt.

Significance and interpretation

Correlation and implied continuity of rocks of the Badger group across the Badger and Botwood belts is significant to recent models for the Silurian development of the northeast Newfoundland Appalachians. Whether the Badger group represents marine deposits of separate fault-bounded basins (Watson, 1981; Arnott, 1983a; Arnott et al., 1985) or the deposits of a single basin (Karlstrom et al., 1982; Reusch, 1983; van der Pluijm, 1986), similar conditions must have prevailed across the Badger and Botwood belts. A detailed sedimentological study at Milliners Arm of New World Island (Watson, 1981) suggests a model of small, proximal submarine fans. If contemporaneous Silurian rocks to the southeast at Port Albert Peninsula and farther eastward at Indian Islands were coextensive, the southeasterly increase in shale and greywacke and accompanying decrease in conglomerate imply that the Port Albert and Indian Islands rocks represent distal, outer-fan or basinal deposits. Furthermore, their occurrence to the southeast fits the model of a fan fed from the north and northwest.

Stratigraphic relationships between greywackes of Badger type and the Ordovician Davidsville Group to the east of the Botwood Belt indicate that the Davidsville Group was part of the conformable substrate beneath the Silurian deposits of eastern Notre Dame Bay. An unconformity between the Davidsville Group and ophiolitic rocks of the Gander River Complex (O'Neill and Blackwood, 1989) at the the Dunnage Zone-Gander Zone boundary, and an absence of Silurian rocks above the Gander Zone suggest an eastern limit to the depositional regime of the Badger group.

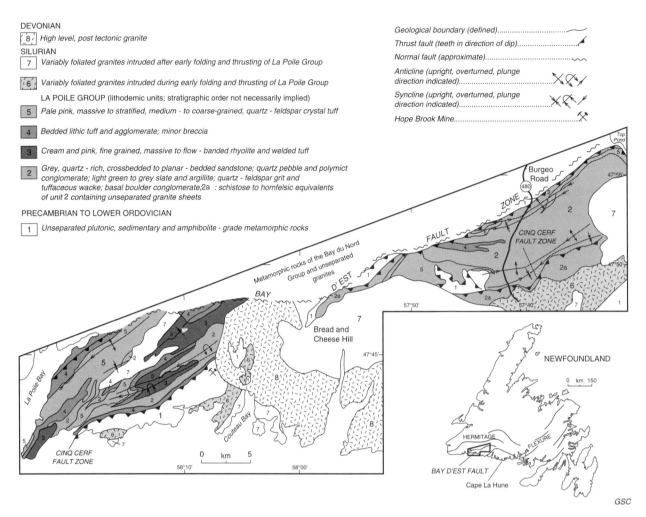

DEVONIAN

8 — *High level, post tectonic granite*

SILURIAN

7 — *Variably foliated granites intruded after early folding and thrusting of La Poile Group*

6 — *Variably foliated granites intruded during early folding and thrusting of La Poile Group*

LA POILE GROUP (lithodemic units; stratigraphic order not necessarily implied)

5 — *Pale pink, massive to stratified, medium - to coarse-grained, quartz - feldspar crystal tuff*

4 — *Bedded lithic tuff and agglomerate; minor breccia*

3 — *Cream and pink, fine grained, massive to flow - banded rhyolite and welded tuff*

2 — *Grey, quartz - rich, crossbedded to planar - bedded sandstone; quartz pebble and polymict conglomerate; light green to grey slate and argillite; quartz - feldspar grit and tuffaceous wacke; basal boulder conglomerate;2a : schistose to hornfelsic equivalents of unit 2 containing unseparated granite sheets*

PRECAMBRIAN TO LOWER ORDOVICIAN

1 — *Unseparated plutonic, sedimentary and amphibolite - grade metamorphic rocks*

Geological boundary (defined)..............................

Thrust fault (teeth in direction of dip)...........................

Normal fault (approximate).............................

Anticline (upright, overturned, plunge direction indicated)..............................

Syncline (upright, overturned, plunge direction indicated)..............................

Hope Brook Mine.......................................

Figure 4.56. Regional setting and general geology of the La Poile Belt at the southern portion of the Hermitage Flexure, after O'Brien and O'Brien (1990).

Acknowledgments

We thank B.A. van der Pluijm, Bruno Lafrance, and P.E. Williams for reviews of this manuscript.

La Poile Belt

B.H. O'Brien, S.J. O'Brien, and Harold Williams

The La Poile Belt forms an integral part of the Hermitage Flexure in southwestern Newfoundland (Fig. 4.56). Two separate outcrop areas consist of subaerial felsic volcanic and associated epiclastic rocks interdigitated with crossbedded quartz sandstones of nonmarine shallow water

O'Brien, B.H., O'Brien, S.J., and Williams, H.
1995: La Poile Belt; in Chapter 4 of Geology of the Appalachian-Caledonian Orogen in Canada and Greenland, (ed.) H. Williams; Geological Survey of Canada, Geology of Canada, no. 6, p. 420-423 (also Geological Society of America, The Geology of North America, v. F-1).

origin. In the type area between La Poile Bay and Couteau Bay, some of the felsic volcanic units are dated isotopically as Early-Late Silurian (Dunning et al., 1988; O'Brien et al., 1991). This supports correlation with similar Silurian assemblages elsewhere in Newfoundland, and argues against previous correlations with marine Ordovician rocks of the the Dunnage Zone (e.g., Chorlton and Dallmeyer, 1986). Rocks in both outcrop areas are assigned to the La Poile Group (Cooper, 1954; O'Brien, 1982). Thickness in the type area is estimated at between 3.5 and 5 km (Cooper, 1954; Chorlton, 1978; O'Brien, 1988, 1989).

Following the early work of Cooper (1954), the southwestern outcrop area (type area) of the La Poile Group was remapped by Chorlton (1978), who slightly expanded the unit to include metabasic rocks and associated granitoids near the coast. The northeastern outcrop area (near Burgeo Road) was first outlined by Riley (1959) and remapped by Chorlton (1980) and O'Brien (1982). Since the development of the Hope Brook Gold Mine, the type area has been mapped in detail with emphasis on structure and stratigraphy (B.H. O'Brien, 1987, 1988, 1989; O'Brien and O'Brien, 1990).

The Silurian La Poile Group contrasts with the Ordovician Bay du Nord Group along its northern boundary in sedimentary facies, metallogenic features, metamorphic grade, and structural style. Cooper (1954) suggested that, prior to faulting of the Silurian strata, the La Poile Group unconformably overlay the Bay du Nord Group. Chorlton (1978, 1980) suggested that the La Poile and Bay du Nord groups are time stratigraphic equivalents, based mainly on Rb-Sr isotopic ages of volcanic rocks. However, precise U-Pb zircon ages of 429 ± 2 Ma – 422 ± 2 Ma for the La Poile Group are now taken to confirm a Silurian age (Dunning et al., 1988; O'Brien et al., 1991).

The northern boundary of the La Poile Belt is the Bay d'Est Fault, a southeast-dipping thrust that places greenschist facies rocks of La Poile Group above higher grade rocks that include the Bay du Nord Group. Relationships of the La Poile Group to rocks along its southern boundary have always been problematic (see O'Brien, 1988 for review; Dunning and O'Brien, 1989; O'Brien and O'Brien, 1990). Amphibolites (Cinq Cerf gneiss) that are multiply deformed and cut by the 563 ± 4 Ma Roti Granite (Dunning and O'Brien, 1989) occur in a narrow zone along the shoreline and islands of the type area. Other less deformed and metamorphosed sedimentary and volcanic rocks, while also cut by the Roti Granite, were erroneously included with the La Poile Group (Chorlton, 1978; McKenzie, 1986; O'Brien, 1987, 1988). Recent work shows that the southern boundary of the La Poile Group is a major northwest-directed thrust, the Cinq Cerf Fault Zone (O'Brien, 1987, 1989). Its recognition rationalizes otherwise conflicting relations among rock groups. The extension of the Cinq Cerf Fault Zone is now recognized, in places, along the southern margin of the northeastern outcrop area of the La Poile Belt. There, La Poile Group conglomerate, sandstone, and tuff nonconformably overlie foliated granodiorite correlated with the Roti Granite (O'Brien et al., 1989; O'Brien and O'Brien, 1990). The Hope Brook gold mine lies on the southeast side of the Cinq Cerf Fault Zone.

The La Poile Group is divided into twenty four lithostratigraphic units in the type area between La Poile Bay and Couteau Bay (O'Brien, 1988, 1989). Although considered as members, many are of formational rank (Fig. 4.56). Felsic fragmental volcanic rocks are predominant, especially in upper parts of the group. Conglomerate, sandstone and shale are most common toward its base, but thin shale and tuffaceous sandstone units are represented throughout most of the group. A very thick sequence of crossbedded sandstones makes up most of the La Poile Group in the northeastern outcrop area. The sandstone beds interdigitate southwestward with volcanic horizons in lower parts of the succession.

The main volcanic rock types of the La Poile Group are air fall tuffs, lava flows, ash-flow tuffs, lahars, and related finer grained, epiclastic rocks. In the type area, the ash flows are variably welded and quartz-phyric, and are rhyolitic and rhyodacitic in composition. They are associated with a northeastward-fining sequence of lahar deposits and northeastward thinning successions of breccia and agglomerate (Chorlton, 1979; O'Brien, 1987, 1988), implying a volcanic centre in the southwestern outcrop area of the La Poile Group. Neither welded ash flows nor rhyolite flows are developed in the northeastern outcrop area, where subordinate volcanic rocks are mainly of airfall and epiclastic origin (O'Brien, 1983).

Lower parts of the La Poile clastic succession, especially east of the Burgeo Road, consist of fine- and medium-grained, immature quartz arenite, feldspathic lithic arenite, pebbly sandstone, and granule conglomerate, all rich in blue quartz (O'Brien, 1983). The sandstones are massive to thin-bedded, internally cross-stratified, and locally planar bedded. Interlayered, granule to boulder conglomerate beds are 1 to 5 m thick and poorly sorted. Conspicuous angular clasts of vein quartz and rare fragments of highly altered volcanic rocks are up to 15 cm in diameter. A thick-bedded, polymictic conglomerate unit occurs at or near the stratigraphic base of the group. It contains boulders and cobbles of local and exotic volcanic, sedimentary, plutonic, and metamorphic rocks.

The Ordovician and Silurian rocks of the Hermitage Flexure are affected by Middle Paleozoic orogenesis (Dunning et al., 1988, 1990a), although the Silurian La Poile Group is less deformed and less metamorphosed than juxtaposed Ordovician rocks. Three episodes of regional deformation accompanied greenschist facies dynamothermal metamorphism of the La Poile Group. Deformation is most intense at bounding shear zones, whereas primary features are preserved in steeply dipping internal areas. Alternating volcanic and clastic sedimentary units of the La Poile Group outline regional folds that trend northeast and have upright axial surfaces. These folds are locally overturned to the northwest where the La Poile Group is faulted against older rocks. Since the major folds display a consistent sense of asymmetry and have relatively short middle limbs, the La Poile Group can be considered, in regional terms, as a northwest-dipping, northwest-younging succession.

Biotite, quartz-feldspar porphyry (the Hawks Nest Pond Porphyry, unit 5 of Fig. 4.56) forms several northeast-trending bodies in the type area of the La Poile Group. Some sheet intrusions of porphyry are concordant and emplaced along the limbs of regional folds. Others are discordant and truncate the hinge zones of these same fold structures. In places, the porphyry is folded with the La Poile Group and both have the same foliation. The Hawks Nest Pond porphyry is viewed, therefore, as a syntectonic intrusion like most of the Burgeo Intrusive Suite and La Poile batholith (O'Brien et al., 1986, 1991; O'Brien, 1988). All are cut by the post-tectonic (Devonian) Chetwynd Granite.

The structural styles and basal relationships of the La Poile Group contrast with those in other Newfoundland belts. Northwest thrusting of Silurian rocks above the Ordovician Bay du Nord Group, northwest thrusting of southerly belts of mainly upper Precambrian rocks above the La Poile Group, and an unconformity between Silurian and upper Precambrian-Cambrian rocks are unique to southwestern Newfoundland.

The alternation of volcanic and sedimentary units of the La Poile Group contrasts with the stratigraphic succession of the Badger and Botwood belts. Whereas the stratigraphic order of units in the latter belts define upward shoaling, and a change from marine to terrestrial conditions, the La Poile Group has subaerial volcanic rocks throughout that alternate with fluvial-alluvial and nonmarine sandstone-shale units. Crossbedded, quartz-rich sandstones of the lower La Poile succession, in particular those exposed east of the Burgeo Road, are similar to sandstones (Wigwam Formation) of the Botwood Group.

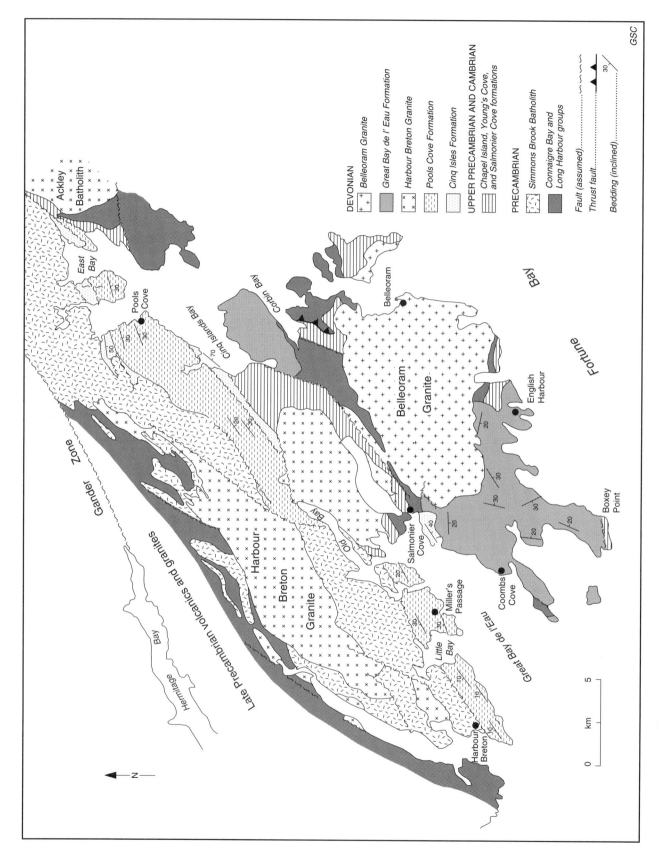

Figure 4.57. Distribution of rocks and relationships in the Fortune Belt on the north shore of Fortune Bay.

The Bay du Nord Group is correlated with the Baie d'Espoir Group and these Ordovician units appear to be colinear around the northern portion of the Hermitage Flexure (O'Brien, et al., 1986). As both groups are assigned to the Exploits Subzone (Williams et al., 1988), the La Poile Belt separates the the Dunnage Zone from upper Precambrian rocks of uncertain zonal affinity to the south (Burgeo Subzone). Gander Group correlatives are absent in this area, although syntectonic plutons and Paleozoic structures typical of the Gander Zone extend around the Hermitage Flexure from Bay d'Espoir to La Poile Bay.

Sedimentological evidence always required a metamorphic and granitic source terrane for the La Poile Group. Some of the basement rocks south of the Cinq Cerf Fault Zone, such as the Cinq Cerf gneiss, were deformed and metamorphosed prior to intrusion of the 563 ± 4 Ma Roti Granite (O'Brien, 1987; Dunning and O'Brien, 1989). Others are preserved grey green siltstones, argillites and volcanic rocks (for example, the Whittle Hill sandstone of O'Brien, 1989; O'Brien and O'Brien, 1990). Their late Precambrian age, preservation, and relation to Silurian rocks invite correlation with the Avalon Zone, particularly the Avalon Zone of southern New Brunswick (McCutcheon, 1981). This implies that a major zone boundary lies immediately northwest of the La Poile Group and is omitted elsewhere by intrusions. The La Poile Group, therefore, may have linked the Avalon Zone to western or northern zones in the Early-Late Silurian before subsequent northwest-directed thrusting, syntectonic intrusion and regional metamorphism (cf. Hutton and Murphy, 1987).

Other areas of middle Paleozoic rocks in southern Newfoundland

A small area of gently dipping, upright, pale green, grey and reddish brown granule to pebble conglomerate occurs at the entrance to La Hune Bay (Fig. 4.56, inset). The conglomerate is interbedded with sandstone and siltstone which grade into white weathering carbonate-rich layers. The contact between the sedimentary rocks and the Devonian Francois granite has been interpreted as intrusive (Williams, 1971a) and nonconformable (Blackwood, 1985). These rocks are post-tectonic with respect to deformation in nearby rocks and they are probably Devonian or later in age (O'Brien et al., 1986).

Fortune Belt

Harold Williams and S.J. O'Brien

Definition

Rocks of the Fortune Belt occur as small areas along the north shore of Fortune Bay (Fig. 4.57) and a larger area on its southeast side at Grand Beach (Fig. 4.58). On the north shore of Fortune Bay the rocks lie unconformably upon

Williams, H., and O'Brien, S.J.
1995: Fortune Belt; in Chapter 4 of Geology of the Appalachian-Caledonian Orogen in Canada and Greenland, (ed.) H. Williams; Geological Survey of Canada, Geology of Canada, no. 6 p. 423-425 (also Geological Society of America, The Geology of North America, v. F-1).

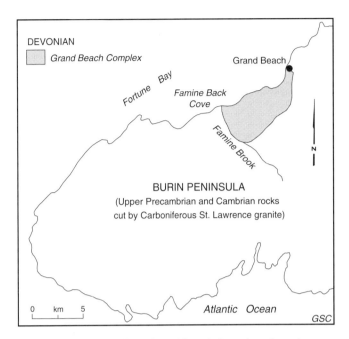

Figure 4.58. Extent of the Grand Beach Complex of southeast Fortune Bay.

upper Precambrian and Cambrian rocks. These middle Paleozoic rocks are mainly terrestrial conglomerates and sandstones, with minor mafic volcanic rocks. Fossils are scarce, except for Late Devonian plants in the youngest conglomerate unit. Volcanic rocks at Grand Beach are dated isotopically as Devonian (Krogh et al., 1988). The Fortune Belt is significant as its rocks furnish an onland middle Paleozoic record of the Newfoundland Avalon Zone. As well, its clastic rocks provide the earliest sedimentary linkage between the Gander and Avalon zones.

Previous work and nomenclature

North of Fortune Bay, some middle Paleozoic rocks of the Fortune Belt are described in local dissertations (Taylor, 1940; Widmer, 1950; Calcutt, 1971; Smith, 1976) and regional dissertations and work concerned with molybdenite deposits (White, 1939; Smith and White, 1954; Widmer, 1950). The rocks were mapped in reconnaissance fashion by Anderson (1965) with more detailed regional mapping by Williams (1971b), Greene (1974) and Colman-Sadd et al. (1979).

On the southeast side of Fortune Bay, the middle Paleozoic rocks were first included with the upper Precambrian rocks of the Avalon Zone (e.g. Anderson, 1965). They were remapped in more detail by O'Brien et al. (1977) and Strong et al. (1978) and assigned to the Carboniferous mainly because of preliminary isotopic ages and chemical and structural contrasts with the upper Precambrian rocks.

The oldest rocks of the Fortune Belt north of Fortune Bay are the Cinq Isles Formation (Williams, 1971b), formerly Cinq Isles Series (White 1939, 1940; Widmer, 1950) or Cinq Isles Group (Smith and White, 1954; Anderson, 1965). It is overlain by conglomerates of the Pools Cove

423

Formation (White, 1939, 1940; Widmer, 1950; Williams, 1971b). The youngest rocks north of Fortune Bay are the Great Bay de l'Eau Formation (Widmer, 1950). A review of previous work and nomenclature in this area is given by Williams (1971b). On the southeast side of Fortune Bay, felsic volcanic and volcaniclastic rocks with associated porphyry are the Grand Beach Complex (O'Brien et al., 1977).

Stratigraphy and relationships

The Cinq Isles Formation (Fig. 4.57) forms a northeast-trending, steeply to moderately dipping succession of red sandstone, limy shale, limestone and conglomerate in excess of 300 m thick. The formation lies nonconformably on the Simmons Brook Batholith (Williams, 1971b) that has been assigned to the late Precambrian. The Cinq Isles Formation is overlain by more than 1000 m of conglomerate and coarse arkosic sandstone of the Pools Cove Formation. In places, the Pools Cove Formation transgresses the Cinq Isles Formation to lie directly on the Simmons Brook Batholith. The Great Bay de l'Eau Formation occurs to the south and east of Cinq Isles and Pools Cove exposures. It consists mainly of poorly bedded coarse conglomerates in excess of 300 m thick. Beds are gently dipping to sub-horizontal and they overlie steeply dipping upper Precambrian and Cambrian strata with significant unconformity. All of the middle Paleozoic rocks north of Fortune Bay are cut by granites and a profusion of dykes that are interpreted as mainly late Devonian.

Layered rocks of the Grand Beach Complex are sub-horizontal and there is less than 200 m of exposed section. A local conformable contact with underlying basalts, assigned to the late Precambrian, suggests the basalts may also be part of the mid Paleozoic assemblage.

Lithology

The Cinq Isles Formation consists of red micaceous siltstone and shale, grey and reddish crossbedded sandstone, grey micritic limestone, and quartz pebble conglomerate. Abrupt lithological changes are common both vertically and laterally so that subdivision of the formation is impractical. White and pink micritic limestone and mixed redbeds are characteristic of the formation. Limestone beds reach 6 m in thickness but are usually less than a metre. Most are featureless, but in a few places, the limestone has structures similar to algal growths. Limestone also occurs as nodules and lenses in red siltstone beds, as a matrix in mixed sand and pebble beds, and as irregular interlayers in sandstone and siltstone sections.

Conglomerates of the formation are red and grey, commonly with matrix-supported, well-rounded white quartz pebbles in arkosic red sandy matrix. Clasts of granite, volcanic rock, and red sandstone are also common.

A nonconformity between the Cinq Isles Formation and Simmons Brook Batholith is exposed near Pools Cove. At the nonconformity, about 60 cm of unsorted poorly indurated arkose grades upward over 30 cm into harder arkosic sandstone, siltstone, and silty micrite. The upper contact with the Pools Cove Formation is drawn where the fine grained clastic rocks of the Cinq Isles Formation are overlain by coarse, poorly sorted conglomerate.

The Pools Cove Formation has a basal unit of coarse red conglomerate (150 m), a middle unit of coarse orange arkosic sandstone and conglomerate (120 m), and an upper unit of boulder conglomerate with arkosic sandstone interbeds (700 m). The basal red conglomerate contains mainly granite and diorite boulders derived from the Simmons Brook Batholith but also has clasts of coarse grained porphyritic biotite granite, medium grained pink alaskitic granite, foliated granite, granite gneiss, amphibolitic gneiss, muscovite-quartz-feldspar schist, quartzite, red sandstone, and a variety of volcanic rocks. Arkosic sandstones of the middle unit directly overlie the Simmons Brook Batholith where the red conglomerate unit thins southwestward from its type locality at Pools Cove. Farther southwest, the red basal unit reappears on the west side of Great Bay de l'Eau.

Arkoses of the middle unit contain up to 90 per cent orange potash feldspar fragments. The rocks are crossbedded with isolated outsize well rounded granite boulders. Minor occurrences of reddish to pink mixed conglomerate and nodular limestone resemble rocks of the Cinq Isles Formation.

The upper unit consists mainly of coarse conglomerate with well rounded and poorly sorted boulders up to 2 m in diameter. Most are an assortment of granites and metamorphic rocks of the nearby Gander Zone. Other plutonic boulders are; Simmons Brook granite and diorite, mafic to silicic volcanic rocks and red to purple sandstones of the Precambrian Long Harbour Group (Williams, 1971b), micaceous siltstone and quartzite of the Cambrian Youngs Cove Group, and sandstones and limestones of the Cinq Isles Formation. Boulders are progressively smaller from northeast to southwest, suggesting a northeast source. Clasts of cleaved micaceous Cambrian sandstone indicate a Paleozoic deformation before conglomerate deposition.

Where the Pools Cove Formation overlies the Simmons Brook Batholith, conglomerates are separated from underlying granite by local pockets of red shale and siltstone typical of the Cinq Isles Formation. The presence of Cinq Isles fragments in Pools Cove conglomerate indicates that the Cinq Isles Formation was removed by erosion.

The Great Bay de l'Eau Formation consists of poorly bedded purple to red, grey and buff cobble and boulder conglomerate. Clasts in the conglomerate are mostly derived from immediately underlying rocks. Its matrix is a coarse sandstone. White Cambrian quartzite and Cinq Isles limestone cobbles are conspicuous in places. Other clasts are volcanic rocks, biotite granite, quartz, and foliated muscovite granite. The conglomerates resemble the Pools Cove Formation where granite clasts are abundant and where matrices are arkosic. An occurrence of mafic igneous rocks at Boxey Point may represent a vesicular flow at the top of the formation (Smith, 1976).

The Grand Beach Complex has abrupt facies variations as suggested by stratigraphic differences where its base is exposed at Famine Brook and Famine Back Cove (Fig. 4.58). At Famine Brook, 5 m of flat-lying epiclastic rocks overlie vesicular basalts. Red sandstones and interbedded red sedimentary breccia with minor conglomerate occur at the base of the section. The breccias contain fragments of vesicular basalt, red clastic sedimentary rocks and minor rhyolite tuffs. The basal unit is overlain by about 25 m of immature volcaniclastic conglomerates and poorly sorted volcaniclastic breccias and associated agglomerates with large felsic and mafic blocks. The coarse fragmental rocks may represent mudflow or landslide deposits of laharic origin. In the northern parts of the complex, similar breccias are intercalated with welded tuffs.

Near Famine Back Cove, the base of the Grand Beach Complex is represented by a few metres of agglomerate and volcaniclastic conglomerate that overlies massive and vesicular basalts. The agglomerates contain blocks and bombs of vesicular basalt and lesser rhyolitic tuffaceous rocks. The agglomerates are overlain by a metre of porphyritic rhyolite tuffs and welded ash-flow tuffs which grade upwards into a coarse mafic breccia. This is followed by a thicker section of felsic ash-flow tuffs, devitrified rhyolites, massive crystal tuffs, crystal lithic tuffs, and welded and unwelded lithic tuffs. Coarse, porphyritic phases of rhyolites locally show intrusive relationships with the ash-flows. However, most porphyries contain isolated flattened lithic fragments, suggesting an extrusive origin.

Age

The youngest rocks that lie unconformably beneath the Cinq Isles and Pools Cove formations are Upper Cambrian shales of the Salmonier Cove Formation. The Pools Cove conglomerates contain a sampling of granite and metamorphic rock clasts from the nearby Gander Zone. The age of metamorphism in the Gander Zone is considered Silurian in the south (Dunning et al., 1988,1990a) and Devonian in the northeast (Bell et al., 1977; Blackwood, 1981; O'Neill and Lux, 1989). The Pools Cove conglomerates also contain boulders of megacrystic porphyritic biotite granite of Ackley type (Williams et al., 1989b). The nearby Ackley Granite yields Ar^{40}-Ar^{39} cooling ages of 410 to 355 Ma (Dallmeyer et al., 1983; Tuach, 1987; Kontack et al., 1988). At Old Bay, the Pools Cove Formation is cut and thermally metamorphosed by medium- to coarse-grained biotite granite, the Harbour Breton granite, that resembles the nearby Ackley Granite and is dated at 360 ± 20 Ma (Bell, in Green and O'Driscoll, 1976). A report of poorly preserved plant fossils in the Pools Cove Formation (Widmer, 1950) has not been confirmed by subsequent studies. Collectively the relationships indicate a Devonian age for the Cinq Isle and Pools Cove formations.

The Great Bay de l'Eau Formation has black bituminous shale and dark grey siltstone interbeds near Coombs Cove. Plant remains identified as *Protolepidodendron* sp. and *Eospermatopteris* sp. are typical of the Late Devonian (Widmer, 1950). The Great Bay de l'Eau Formation is gently dipping to subhorizontal in most places. In contrast, the Cinq Isles and Pools Cove formations occur in moderate to steep homoclinal sections with local repetition by thrusts or steep reverse faults. These structural contrasts suggest that the Great Bay de l'Eau Formation is the younger. Reports of granite clasts in the Great Bay de l'Eau Formation that resemble Harbour Breton granite also imply an age gap with the Pools Cove Formation. The Great Bay de l'Eau Formation, in turn, is cut by the Belleoram granite, with K-Ar ages of 400 and 342 Ma (Wanless et al., 1965, 1967).

Silicic volcanic rocks of the Grand Beach Complex yield a U-Pb zircon age of 394 +6/-4 Ma (Krogh et al., 1988).

Interpretation

Crossbedded sandstones and conglomerates of the Cinq Isles Formation are typical fluviatile deposits. Its limestones may be intertidal (Calcutt, 1971) or freshwater deposits of a dry climate (Chandler, 1982). The latter interpretation is supported by the irregular admixtures of red siltstone, sandstone and limestone and a lack of marine polynomorphs in the limestone. The Pools Cove and Great Bay de l'Eau formations are typical coarse alluvial deposits. Abrupt vertical and lateral facies changes, poorly developed bedding, coarseness, and erosional disconformity beneath the Pools Cove suggest unstable depositional conditions, possibly within local intermontane basins.

The sedimentology and intrusive relationships among middle Paleozoic rocks of the Fortune Belt indicate deposition between periods of granite intrusion and contemporary deformation. This is supported by the synchronous ages of silicic volcanic rocks of the Grand Beach Complex with Acadian plutonism. The clastic rocks north of Fortune Bay provide a firm sedimentological linkage between the Gander and Avalon zones. Conceivably, the synorogenic deposits of the Fortune Belt reflect transcurrent faulting and uplift that was associated with the accretion of the Avalon Zone against the Gander Zone.

Offshore-Grand Banks

Seismic-stratigraphy, drill cores, and palynological analyses indicate that 8000 m of Cambrian to Devonian rocks cover some 50 000 km^2 east of the Avalon Peninsula on the Grand Banks of Newfoundland (King et al., 1986; Durling et al., 1987). These rocks are mainly marine and they contrast with mainly terrestrial onshore sections. Formline structural mapping revealed a 4000 m thick Ordovician-Silurian section that is gently folded about north-northwest axes and is unconformably overlain by a synclinal outlier of Devonian strata 700 m thick (Durling et al., 1987).

Upper Ordovician and Silurian rocks are grey laminated fissile siltstones with burrows and fragments of brachiopods and bryozoa, light grey calcareous siltstones which are laminated and bioturbated with well-sorted beds of fossil fragments, and non-fossiliferous, bioturbated siltstones. The rock types and fauna indicate deposition in a variety of shallow marine environments. Devonian rocks are redbeds of continental affinity.

Mild deformation in the Paleozoic rocks offshore supports the onland observation of decreasing intensity of Paleozoic deformation from west to east across the Newfoundland Avalon Zone (Williams, 1993). Furthermore, there is no middle Paleozoic plutonism or regional metamorphism recorded in the offshore sections. The upper Ordovician-Silurian marine siltstones contrast with the coarser marine and/or thick continental sequences that make up most of the equivalent onland sections. The presence of an offshore Devonian basin of terrestrial rocks records a change to subaerial conditions as noted in the onland successions, although this change is much later in the case of the offshore.

REFERENCES

Aigner, T.
1982: Calcareous tempestites: storm-dominated stratification Upper Muschelkalk limestones (Middle Trias, southwest Germany); in Cyclic and Event Stratigraphy, (ed.) G. Einsele and A. Seilacher; Springer-Verlag, p. 180-198.

Alcock, F.J.
1935: Geology of Chaleurs Bay region; Geological Survey of Canada, Memoir 183, 146 p.

Ami, H.M.
1900: Synopsis of the geology of Canada: a summary of the principal terms employed in the Canadian geological nomenclature; Royal Society of Canada, Transactions, ser. 2, v. 4, p. 187-225.

Anderson, F.D.
1965: Belleoram, Newfoundland; Geological Survey of Canada, Map 8-1965 (with marginal notes).
1970: Geology of McKendrick map area, New Brunswick; Geological Survey of Canada, Paper 69-12, 16 p.

Anderson, F.D. and Poole, W.H.
1959: Geology of Woodstock-Fredericton, York, Carleton, Sunbury, and Northumberland counties, New Brunswick; Geological Survey of Canada, Map 37-1959 (with marginal notes).

Anderson, F.D. and Williams, H.
1970: Gander Lake (West half), Newfoundland; Geological Survey of Canada, Map 1195A (final map with descriptive notes).

Antonuk, C.
1986: Geology, structure and microstructure of Farmer's Island and adjacent islands, north-central Newfoundland, with a discussion of the Dunnage Melange; M.Sc. thesis, University of New Brunswick, Fredericton, New Brunswick, 195 p.

Arnott, R.J.
1983a: Sedimentology of Upper Ordovician-Silurian sequences on New World Island, Newfoundland: separate fault-controlled basins?; Canadian Journal of Earth Sciences, v. 20, p. 345-354.
1983b: Sedimentology, structure and stratigraphy of northeast New World Island, Newfoundland; Ph.D. thesis, University of Oxford, Oxford, England.

Arnott, R.J., McKerrow, W.S., and Cocks, L.R.M.
1985: The tectonics and depositional history of the Ordovician and Silurian rocks of Notre Dame Bay, Newfoundland; Canadian Journal of Earth Sciences, v. 22, p. 607-618.

Ayrton, W.G.
1967: Chandler-Port Daniel area, Bonaventure and Gaspe South Counties; Quebec Department of Natural Resources, Geological Report 120, 91 p.

Ayrton, W.G., Berry, W.B.N., Boucot, A.J., Lajoie, J., Lespérance, P.J., Pavlides, L., and Skidmore, W.B.
1969: Lower Llandovery of the northern Appalachians and adjacent regions; Geological Society of America Bulletin, v. 80, p. 459-484.

Badgley, P.C.
1956: New-Carlisle map-area; Quebec Department of Mines, Geological Report 70, 36 p.

Bailey, L.W.
1898: Report on the geology of southwest Nova Scotia, embracing the counties of Queen's, Shelburne, Yarmouth, Digby, and part of Annapolis; in Annual Report for 1896; Geological Survey of Canada, v. 9, part M, 154 p.

Bailey, L.W. and Matthew, G.F.
1872: Geology of southern New Brunswick; in Report of Progress for 1870-71, Part 2; Geological Survey of Canada, Preliminary Report, p. 13-240.

Baird, D.M.
1951: The geology of Burlington Peninsula, Newfoundland; Geological Survey of Canada, Paper 51-21, 70 p.
1953: Reconnaissance geology of part of the New World Island-Twillingate area; Geological Survey, Newfoundland, Report 1, 20 p.
1954: Geological map of Newfoundland; Newfoundland Department of Mines and Resources, scale 1:760 320.
1958: Fogo Island map-area, Newfoundland; Geological Survey of Canada, Memoir 301, 62 p.

Ballard, R.D. and Uchupi, E.
1975: Triassic rift structure in the Gulf of Maine; American Association of Petroleum Geologists Bulletin, v. 59, p. 1041-1072.

Barr, S.M.
1990: Granitoid rocks and terrane characterization: an example from the northern Appalachian orogen; Geological Journal, v. 25, p. 295-304.

Barr, S.M. and Jamieson, R.A.
1991: Tectonic setting and regional correlation of Ordovician-Silurian rocks of the Aspy Terrane, Cape Breton Island, Nova Scotia; Canadian Journal of Earth Sciences, v. 28, p. 1769-1779.

Barr, S.M. and Macdonald, A.S.
1989: Geology of the Mabou Highlands; Nova Scotia Department of Mines and Energy, Paper 89-2, 65 p.

Barr, S.M. and Raeside, R.P.
1989: Tectono-stratigraphic terranes in Cape Breton Island, Nova Scotia: implications for the configuration of the northern Appalachian orogen; Geology, v. 17, p. 822-825.

Barr, S.M., Brisebois, D., and Macdonald, A.S.
1985a: Carboniferous volcanic rocks of the Magdalen Islands, Gulf of St. Lawrence; Canadian Journal of Earth Sciences, v. 22, p. 1679-1688.

Barr, S.M., Jamieson, R.A., and Raeside, R.P.
1985b: Igneous and metamorphic geology of the Cape Breton Highlands; Geological association of Canada/Mineralogical Association of Canada, Annual Meeting, Fredericton, New Brunswick, Field Guide 10.

Barr, S.M., Raeside, R.P., and Macdonald, A.S.
1985c: Geology of the southeastern Cape Breton Highlands, Nova Scotia; in Current Research, Part B; Geological Survey of Canada, Paper 85-1B, p. 103-109.

Bédard, J.H.
1986: Pre-Acadian magmatic suites of the southeastern Gaspé Peninsula; Geological Society of America Bulletin, v. 97, p. 1177-1191.

Béland, J.
1957: St-Magloire and Rosaire-St-Pamphile areas, southern Quebec; Quebec Department of Natural Resources, Geological Report 76, 49 p.
1958: Oak Bay map area; Québec Department of Mines, Preliminary Report 375, 12 p.
1960: Preliminary report on Rimouski-Matapédia areas, Québec; Quebec Department of Natural Resources, Preliminary Report 430, 20 p.

Béland, J., Hubert, C., Ducharme, D., Thebérge, R. et Vennat, G.
1979: Un segment de l'anticlinorium d'Aroostook-Percé en Gaspésie, Québec (abstract); Geological Association of Canada, Annual Meeting, Québec City, Program with Abstracts; v. 4, p. 39.

Bélanger, J.
1982: Roches volcaniques dévoniennes de la bande de Ristigouche, Québec; Ministère de l'Énergie et des Ressources, DP-939, 13 p.

Bell, K., Blenkinsop, J., and Strong, D.F.
1977: The geochronology of some granite bodies from eastern Newfoundland and its bearing on Appalachian evolution; Canadian Journal of Earth Sciences, v. 14, p. 456-476.

Bell, W.A. and Goranson, E.A.
1938: Sydney Sheet, West Half, Cape Breton and Victoria Countries, Nova Scotia; Geological Survey of Canada, Map 360A.

Benson, D.G.
1974: Geology of the Antigonish Highlands, Nova Scotia; Geological Survey of Canada, Memoir 343, 58 p.

Berger, J.
1985: Analyse structurale de la faille de Schickshock-Sud en Gaspésie occidentale, Québec; thèse de maîtrise, Université de Montréal, Montréal (Québec).

Bergström, S.M., Epstein, A.G., and Epstein, J.B.
1972: Early Ordovician North Atlantic Province conodonts in eastern Pennsylvania; United States Geological Survey, Professional Paper 800-D, p. D37-D44.

Bergström, S.M., Riva, J., and Kay, M.
1974: Significance of conodonts, graptolites, and shelly faunas from the Ordovician of western and north-central Newfoundland; Canadian Journal of Earth Sciences, v. 11, p. 1625-1660.

Bernard, D. et St-Julien, P.
1986: Analyse structurale du Siluro-Dévonien du centre de la Gaspésie et du Carbonifère du Sud et de l'Est de la Gaspésie, Québec; Ministère de l'Énergie et des Ressources (Québec), MB 86-36, 33 p.

Berry, H.N. and Osberg, P.H.
1989: Stratigraphy of eastern Maine and western New Brunswick; in Structure and Stratigraphy, (ed.) R.D. Tucker and R.G. Marvinney; Maine Geological Survey, Department of Conservation, Augusta, Maine, Studies in Maine Geology, v. 2, p. 1-32.

Berry, W.B.N. and Boucot, A.J.
1970: Correlation of the North American Silurian rocks; Geological Society of America, Special Paper 102, 289 p (For New Brunswick, see p. 53-71, 203, and 232).
1973: Glacio-eustatic control of Late Ordovician-Early Silurian platform sedimentation and faunal changes; Geological Society of America Bulletin, v. 84, p. 275-284.

Betz, F.
1948: Geology and mineral deposits of southern White Bay; Newfoundland Geological Survey Bulletin, v. 24, 26 p.

Beuf, S., Biju-Duval, B., Charpal, O. de, Rognon, P., Gariel, O., et Bennacef, A.
1971: Les grès du Paléozoïque inférieur au Sahara: sédimentation et discontinuités évolution structurale d'un craton; Publications de l'Institut français du Pétrole, no. 18, Éditions Technip, Paris, 464 p.

Bevier, M.L.
1988: U-Pb geochronologic studies of igneous rocks in New Brunswick; in Thirteenth Annual Review of Activities, Project Résumés, (ed.) S.A. Abbott; New Brunswick Department of Natural Resources and Energy, Minerals and Energy Division, Information Circular 88-2, p. 134-140.

Bevier, M.L. and Whalen, J.B.
1990: Tectonic significance of Silurian magmatism in the Canadian Appalachians; Geology, v. 18, p. 411-414.

Blackwood, R.F.
1980: Geology of the west Gander area (12D/15), Newfoundland; in Current Research, (ed.) C.F. O'Driscoll and R.V. Gibbons; Newfoundland Department of Mines and Energy, Mineral Development Division, Report 80-1, p. 53-61.
1981: Geology of the west Gander Rivers area, Newfoundland; in Current Research, (ed.) C.F. O'Driscoll and R.V. Gibbons; Newfoundland Department of Mines and Energy, Mineral Development Division, Report 81-1, p. 50-56.
1982: Geology of the Gander Lake (2D/15) and Gander River (2E/2) area; Newfoundland Department of Mines and Energy, Mineral Development Division, Report 82-4, 56 p.
1985: Geology of the Grey River area, southwest coast of Newfoundland; in Current Research, (ed.) K. Brewer, D. Walsh, and R.V. Gibbons; Newfoundland Department of Mines and Energy, Mineral Development Division, Report 85-1, p. 153-164.

Blackwood, R.F. and Kennedy, M.J.
1975: The Dover Fault: western boundary of the Avalon Zone in northeastern Newfoundland; Canadian Journal of Earth Sciences, v. 12, p. 320-325.

Blanchard, M.C., Jamieson, R.A., and More, E.B.,
1984: Late Devonian-Early Carboniferous volcanism in western Cape Breton Island, Nova Scotia; Canadian Journal of Earth Sciences, v. 21, p. 762-774.

Bostock, H.H., Currie, K.L., and Wanless, R.K.
1979: The age of the Robert Arm Group, north-central Newfoundland; Canadian Journal of Earth Sciences, v. 16, p. 599-606.

Boucot, A.J.
1960: Implication of Rhenish Lower Devonian brachiopods from Nova Scotia; 21th International Geological Congress, part 12, p. 129-137.
1962: Appalachian Siluro-Devonian; in Some aspects of the Variscan fold belt, (ed.) K. Coe; Manchester University Press, Manchester, England, p. 155-163.
1970: Devonian slate problems in the northern Appalachians; Maine Geological Survey Bulletin, v. 23, p. 42-48.
1973: The Lower Ordovician brachiopod Syntrophia sp., cf. S. arethusa (Billings 1962) from south Catcher Pond, northeastern Newfoundland; Canadian Journal of Earth Sciences, v. 10, p. 427-430.
1975: Evolution and extinction rate controls; in Developments in Paleontology and Stratigraphy l, Elsevier.
1989: Acadian Orogeny: Biogeographic constraints (abstract); Geological Society of America, Abstracts with Programs, v. 21, no. 2, p. 6.

Boucot, A.J. and Bourque, P.-A.
1981: Brachiopod biostratigraphy of the Llandoverian rocks of the Gaspé Peninsula; in Field Meeting Anticosti-Gaspé, Québec; Subcommission on Silurian Stratigraphy, Ordovician-Silurian Boundary Working Group, (ed.) P.J. Lespérance; v. 2: Stratigraphy and Paleontology.

Boucot, A.J. and Drapeau, G.
1968: Siluro-Devonian rocks of Lake Memphrémagog and their correlatives in the eastern townships; Québec Department of Natural Resources, Special Paper 1, 44 p.

Boucot, A.J. and Johnson, J.G.
1967: Paleogeography and correlation of Appalachian Province Lower Devonian sedimentary rocks; Tulsa Geological Society, Digest 35, p. 35-87.

Boucot, A.J. and Smith, R.E.
1978: First Upper Silurian fossils from northern Newfoundland; Journal of Paleontology, v. 52, p. 267-274.

Boucot, A.J., Johnson, J.G., and Talent, J.A.
1968: Lower and Middle Devonian faunal provinces based on brachiopods; in International Symposium on the Devonian System; Alberta Association of Petroleum Geologists, v. 11, p. 1239-1254.
1969: Early Devonian brachiopod zoogeography; Geological Society of America, Special Paper 119, 111 p.

Boucot, A.J., Johnson, J.G., Harper, C., and Walmsley, V.G.
1966: Silurian brachiopods and gastropods of southern New Brunswick; Geological Survey of Canada, Bulletin 140, 45 p.

Boucot, A.J., Field, M.T., Fletcher, R., Forbes, W.H., Naylor, R.S, and Pavlides, L.
1964: Reconnaissance bedrock geology of the Presque Isle quadrangle, Maine; Maine Geological Survey, Quadrangle Mapping, ser. 2.

Boucot, A.J., Dewey, J.F., Dineley, D.L., Fletcher, R., Fyson, W.K., Griffin, J.G., Hickox, C.F., McKerrow, W.S., and Ziegler, A.M.
1974: Geology of the Arisaig area, Antigonish County, Nova Scotia; Geological Society of America, Special Paper 139, p. 315-321.

Bourque, P.-A.
1972: Le complexe carbonaté de Lefrançois (Membre de la Formation de Saint-Léon) dans sa région-type, Québec; Ministère des Richesses naturelles, Québec, GM-27886, 46 p.
1973: Stratigraphie du Silurien et du Dévonien basal du Nord-Est de la Gaspésie, avec une illustration de la faune à Brachiopodes; thèse de doctorat, Université de Montréal, Montréal (Québec)
1975: Lithostratigraphic framework and unified nomenclature for Silurian and basal Devonian rocks in eastern Gaspé Peninsula, Quebec; Canadian Journal of Earth Sciences, v. 12, p. 858-872.
1977: Silurian and basal Devonian of northeastern Gaspé Peninsula, Québec; Department of Natural Resources, ES-29, 23 p.
1979: Facies of the Silurian West Point reef complex, Baie des Chaleurs, Gaspésie, Québec; Geological Association of Canada, Annual Meeting, Québec City, Guidebook to Field Trip B-2, 29 p.
1981: Introduction to Gaspé Peninsula and Baie des Chaleurs area; in Subcommission on Silurian Stratigraphy, (ed.) P.J. Lespérance; Ordovician-Silurian Boundary Working Group, Field Meeting Anticosti-Gaspé, Québec, Guidebook, v. 1, p. 27-29 and 42-53.
1987: The West Point Formation of Gaspé Basin, Québec Appalachians: Late Silurian bank and reef complexes controlled by regime of siliciclastic sedimentation on a major paleo-delta system (abstract); Canadian Society of Petroleum Geologists, Reef Research Symposium, Banff, Alberta, p. 19.

Bourque, P.-A. et Gosselin, C.
1986: Stratigraphie du Silurien et du Dévonien basal du bassin de Gaspésie, Québec; Ministère de l'Énergie et des Ressources, Québec, MB 86-34, 49 p.
1988: Stratigraphie du Siluro-Dévonien de la bande du lac Auclair et de la région de Cabano-Squatec, sud-ouest du Témiscouata, Québec; Ministère de l'Énergie et des Ressources, Québec, MB 88-03, 21 p.

Bourque, P.-A. and Lachambre, G.
1980: Stratigraphie du Silurien et du Dévonien basal du Sud de la Gaspésie, Québec; Ministère de l'Énergie et des Ressources Québec, ES-30, 123 p.

Bourque, P.-A. and Lespérance, P.J.
1977: The Silurian-Devonian boundary in northeastern Gaspé Peninsula, Québec; in the Silurian-Devonian Boundary, (ed.) A. Martinsson; International Union of Geological Sciences, ser. A, no. 5, p. 245-255.

Bourque, P.-A., Laurent, R., and St-Julien, P.
1985: Acadian wrench faulting in southern Gaspé Peninsula, Quebec Appalachians (abstract); Geological Association of Canada, Annual Meeting, Fredericton, Program with Abstracts, v. 10, p. A6.

Bourque, P.-A., Gosselin, C., Kirkwood, D., Malo, M., et St-Julien, P.
1993: Le Silurien du segment appalachien Gaspésie-Matapédia-Témiscouata: stratigraphie, géologie structurale et paléogéographie, Québec; Ministère de l'Énergie et des Ressources, Québec, MB 93-25, 115 p., et 23 cartes.

Bourque, P.-A., Amyot, G., Desrochers, A., Gignac, H. Gosselin, C., Lachambre, G., and Laliberté, J.-Y.
1986: Silurian and Lower Devonian reef and carbonate complexes of the Gaspé Basin, Québec (summary); Canadian Petroleum Geology Bulletin, v. 34, p. 452-489.

Bouyx, E., Boucarut, M., Clin, M., et Zeilinga de Boer, J.
1985: Le Paléozoïque anté-carbonifère de la zone de Meguma, en Nouvelle-Écosse (provinces maritimes canadiennes) comparaison avec l'Europe occidentale et implications paléogéographiques; Extrait des annales de la Société géologique du Nord, v. 104, p. 147-166.

Boyce, W.D., Ash, J.S., O'Neill, P., and Knight, I.
1988: Ordovician biostratigraphic studies in the Central Mobile Belt and their implications for Newfoundland tectonics; in Current Research (1988); Newfoundland Department of Mines, Mineral Development Division, Report 88-1, p. 177-182.

427

Bradley, D.C.
1982: Subsidence in late Paleozoic basins in the northern Appalachians; Tectonics, v. 1, p. 107-123.
1983: Tectonics of the Acadian Orogeny in New England and adjacent Canada; Journal of Geology, v. 91, p. 381-400.

Briden, J.C., Kent, D.B., Lapoint, P.L., Livermore, R.A., Roy, J.L., Seguin, M.K., Smith, A.G., van der Voo, R., and Watts, D.R.
1988: Paleomagnetic constraints on the evolution of the Caledonide-Appalachian orogen; in The Caledonide-Appalachian Orogen, (ed.) A.L. Harris and D.C. Fettes; Geological Society, London, Special Publication 38, p. 35-48.

Brisebois, D.
1981: Géologie de la région de Gaspé, Québec; Ministère de l'Énergie et des Ressources, Québec, DPV-824, 19 p.

Brown, P.A.
1973: Possible cryptic suture in south-west Newfoundland; Natural Physical Science, v. 245, no. 140, p. 9-10.
1975: Basement-cover relations in southwest Newfoundland; Ph.D. thesis, Memorial University of Newfoundland, 221 p.
1977: Geology of the Port aux Basques map-area (110/10), Newfoundland; Newfoundland Department of Mines and Energy, Mineral Development Division, Report 77-2, 11 p.

Brueckner, W.D.
1966: Stratigraphy and Structure of west-central Newfoundland; in Geology of part of Atlantic Provinces, (ed.) W.H. Poole; Geological Association of Canada/Mineralogical Association of Canada, Guidebook, p. 137-151.

Burk, C.F.
1964: Silurian stratigraphy of Gaspé Peninsula, Québec; American Association of Petroleum Geologists, Bulletin, v. 48, p. 437-464.

Calcutt, M.
1971: The stratigraphy and sedimentology of the Cinq Isles Formation, Fortune Bay, Newfoundland; M.Sc. thesis, Memorial University of Newfoundland, 104 p.

Cant, D.J.
1980: Storm-deposited shallow marine sediments of the Arisaig Group (Silurian-Devonian) of Nova Scotia; Canadian Journal of Earth Sciences, v. 5, p. 311-317.

Cant, D.J. and Walker, R.G.
1976: Development of a braided-fluvial facies model for the Devonian Battery Point Sandstone, Québec; Canadian Journal of Earth Sciences, v. 13, p. 102-119.

Caputo, M.V. and Crowell, L.C.
1985: Migration of glacial centers across Gondwana during Paleozoic era; Geological Society of America Bulletin, v. 96, p. 1020-1036.

Carbonneau, C.
1959: Région de Richard-Gravier, péninsule de Gaspé, Québec; Ministère des Mines, Québec, RG 90, 75 p.

Chandler, F.W.
1982: Sedimentology of two Middle Paleozoic terrestrial sequences, King George IV Lake area, Newfoundland, and some comments on regional paleoclimate; in Current Research, Part A; Geological Survey of Canada, Paper 82-1A, p. 213-219.

Chandler, F.W. and Dunning, G.R.
1983: Fourfold significance of an Early Silurian U-Pb zircon age from rhyolite in redbeds, southwest Newfoundland; in Current Research, Part B; Geological Survey of Canada, Paper 83-1B, p. 419-421.

Chandler, F.W., Sullivan, R.W., and Currie, K.L.
1987: The age of the Springdale Group, western Newfoundland, and correlative rocks: evidence for a Llandovery over assemblage in the Canadian Appalachians; Royal Society of Edinburgh, Transactions, v. 78, p. 41-49.

Chatterjee, A.K. and Giles, P.S.
1988: Meguma Zone basement II: P-evolution of granulite xenoliths (abstract); Geological Association of Canada, Annual Meeting, Program with Abstracts, v. 13, p. A-19.

Chorlton, L.B.
1978: The geology of the La Poile map area (110/9), Newfoundland; Newfoundland Department of Mines and Energy, Mineral Development Division, Report 78-5, 14 p.
1979: La Poile River map area (11O/16), Newfoundland; in Report of Activities 1978; Newfoundland Department of Mines and Energy, Mineral Development Division, Report 79-1, p. 45-53.
1980: The geology of the La Poile River area (110/16), Newfoundland; Newfoundland Department of Mines and Energy, Mineral Development Division, Report 80-3, 86 p.
1983: The geology of the Grandy Lake area (110/15), Newfoundland; Newfoundland Department of Mines and Energy, Mineral Development Division, Report 83-7, Part I, p. 1-116.

Chorlton, L.B. (cont.)
1984: Geological Development of the southern Long Range Mountains, Newfoundland: a regional synthesis; Ph.D. thesis, Memorial University of Newfoundland, St. John's, Newfoundland, 579 p.

Chorlton, L.B. and Dallmeyer, R.D.
1986: Geochronology of Early to Middle Paleozoic tectonic development in the southwest Newfoundland Gander Zone; Journal of Geology, v. 94, p. 67-69.

Church, W.R.
1969: Metamorphic rocks of Burlington Peninsula and adjoining areas of Newfoundland and their bearing on continental drift in North Atlantic; in North Atlantic Geology and Continental Drift, (ed.) M. Kay; American Association of Petroleum Geologists, Memoir 12, p. 212-233.

Clarke, D.B. and Chatterjee, A.K.
1988: Meguma Zone basement I: Sr-Nb isotopic study of the Liscombe Complex and the origin of peraluminous granites in the Meguma Zone (abstract); Geological Association of Canada, Annual Meeting, Program with Abstracts, v. 13, p. A-21.

Clarke, D.B. and Halliday, A.N.
1980: Strontium isotope geology of the South Mountain batholith, Nova Scotia; Canadian Journal of Earth Science, v. 22, p. 102-107.

Clarke, D.B. and Muecke, G.K.
1980: Igneous and metamorphic geology of southern Nova Scotia; Geological Association of Canada, Field Trip 21, Guidebook, 101 p.
1985: Review of the petrochemistry and origin of the South Mountain batholith and associated plutons, Nova Scotia, Canada; Geochemica et Cosmochimica Acta, v. 44, p. 1045-1058.

Clarke, D.B., Barr, S.M., and Donahoe, H.V.
1980: Granitoid and other plutonic rocks of Nova Scotia; Canadian Journal of Earth Science, v. 14, p. 2858-2864.

Clarke, J.M.
1908: Early Devonic history of New York and eastern north America; New York State Museum of Natural History, Memoir 9, 366 p.

Cocks, L.R.M. and Fortey, R.A.
1982: Faunal evidence for oceanic separation in the Palaeozoic of Britain; Geological Society of London, Quarterly, v. 139, p. 465-478.

Colman-Sadd, S.P.
1981: Geology of the Burnt Hill map area (2D/5), Newfoundland; in Current Research, (ed.) C.F. O'Driscoll and R.V. Gibbons; Newfoundland Department of Mines and Energy, Mineral Development Division, Report 81-1, p. 40-49.

Colman-Sadd, S.P. and Russell, H.A.J.
1982: Geology of the Miguels Lake map area (2D/12), Newfoundland; in Current Research, (ed.) C.F. O'Driscoll and R.V. Gibbons; Newfoundland Department of Mines and Energy, Mineral Development Division, Report 82-1, p. 30-50.

Colman-Sadd, S.P. and Swinden, H.S.
1984: A tectonic window in central Newfoundland? Geological evidence that the Appalachian Dunnage Zone is allochthonous; Canadian Journal of Earth Sciences, v. 21, p. 1349-1367.

Colman-Sadd, S.P., Greene, B.A., and O'Driscoll, C.F.
1979: Graultois, Newfoundland; Newfoundland Department of Mines and Energy, Mineral Development Division, Map 79-104.

Colman-Sadd, S.P., Hayes, J.P., and Knight, I.
1990: Geology of the Island of Newfoundland; Newfoundland Department of Mines and Energy, Geological Survey Branch, Map 90-01, scale 1:1 000 000.

Connors, K.A.
1976: Relationship between sulphide minerals, metamorphism and deformation in the Faribault Brook area of the Cape Breton Highlands, Nova Scotia; B.Sc. thesis, Dalhousie University, Halifax, Nova Scotia, 105 p.

Cooke, H.C.
1937: Thetford, Disraëli, and eastern half of Warwick map area, Quebec; Geological Survey of Canada, Memoir 211, 160 p.
1950: Geology of a southwestern part of the Eastern Townships of Quebec; Geological Survey of Canada, Memoir 257, 142 p.

Cooper, G.A and Kindle, C.H.
1936: New brachiopods and trilobites from the Upper Ordovician of Percé, Québec; Journal of Paleontology, v. 10, p. 348-372.

Cooper, J.R.
1954: La Poile-Cinq Cerf map-area, Newfoundland; Geological Survey of Canada, Memoir 276, 62 p.

Copeland, M. and Bolton, T.
1977: Additional paleontological observations on the age of the Lourdes Formation (Ordovician), Port au Port Peninsula, Western Newfoundland; in Report of Activities, Part B; Geological Survey of Canada, Paper 77-1B, p. 1-13.

Corkin, H.
1965:   The Petroleum Geology of the Port au Port Peninsula Newfoundland; Golden Eagle Oil and Gas Limited, confidential report, 185 p.

Cormier, R.F. and Kelley D.G.
1964:   Absolute age of the Fisset Brook Formation and the Devonian-Mississipian boundary, Cape Breton Island, Nova Scotia; Canadian Journal of Earth Sciences, v. 1, p. 159-166.

Cormier, R.F. and Smith, T.E.
1973:   Radiometric ages of granitic rocks, southwestern Nova Scotia; Canadian Journal of Earth Science, v. 10, p. 1201-1210.

Cousineau, P.A.
1990:   Le Groupe de Caldwell et le domaine océanique entre Saint Joseph-de-Beauce et Sainte-Sabine; Ministère de l'Énergie et des Ressources, Québec, MM 87-02, 165 p.

Cox, K.G.
1980:   A model for flood basalt volcanism; Journal of Petrology, v. 21, p. 629-650.

Coyle, M.
1990:   Geology, geochemistry and geochronology of the Springdale Group: an Early Silurian caldera in Central Newfoundland; Ph.D. thesis, Memorial University, St. John's, Newfoundland, 310 p.

Coyle, M. and Strong, D.F.
1987:   Geology of the Springdale Group: a newly recognized Silurian epicontinental type caldera in Newfoundland; Canadian Journal of Earth Sciences, v. 24, p. 1135-1148.

Coyle, M., Strong, D.F., and Dingwell, D.B.
1986:   Geology of the Sheffield Lake area; in Current Research, Part A; Geological Survey of Canada, Paper 86-1A, p. 453-459.

Coyle, M., Strong, D.F., Gibbons, D., and Lambert, E.
1985:   Geology of the Springdale Group, central Newfoundland; in Current Research, Part A; Geological Survey of Canada, Paper 85-1A, p. 157-163.

Crickmay, G.W.
1932:   Evidence of Taconic Orogeny in the Matapédia-Valley, Québec; American Journal of Science, v. 24, p. 368-386.

Crosby, D.G.
1962:   Wolfville map area, Nova Scotia; Geological Survey of Canada, Memoir 325.

Cumming L.M.
1959:   Silurian and Lower Devonian formations in eastern part of Gaspé Peninsula, Québec; Geological Survey of Canada, Memoir 304, 45 p.
1966:   Report on graptolites from Mactaquac, New Brunswick, collected by D.A. Gordon, University of New Brunswick; Geological Survey of Canada, internal paleontological report, 4 p.
1967a: Geology of the Passamaquoddy Bay region, Charlotte County, New Brunswick; Geological Survey of Canada, Paper 65-29, 36 p.
1967b: Geology of the Passamaquoddy Bay region, Charlotte County, New Brunswick; Maritime Sediments and Atlantic Geology, v. 24, p. 339-352.

Currie, K.L.
1988a: The western end of the Avalon Zone in southern New Brunswick; Maritime Sediments and Atlantic Geology, v. 24, p. 339-352.
1988b: Saint George map area: the end of the Avalon Zone, southern New Brunswick; in Current Research, Part B; Geological Survey of Canada, Paper 88-1B, p. 9-16.

Currie, K.L., Pajari, G.E., and Pickerill, R.K.
1980:   Comments on the boundaries of the Davidsville Group, north-eastern Newfoundland; in Current Research, Part A; Geological Survey of Canada, Paper 80-1A, p. 115-118.

Dallmeyer, R.D. and Keppie, J.D.
1987:   Polyphase late Paleozoic tectonothermal evolution of the southwest Meguma Terrane, Nova Scotia: evidence from $^{40}Ar/^{39}Ar$ mineral ages; Canadian Journal of Earth Sciences, v. 24, p. 1242-1254.

Dallmeyer, R.D., Hussey, E.M., O'Brien, S.J., and O'Driscoll, C.F.
1983:   Chronology of tectonothermal activity in the western Avalon Zone of the Newfoundland Appalachians; Canadian Journal of Earth Sciences, v. 20, p. 355-363.

Dalton, E.
1987:   Sedimentary facies and diagenesis of the Lower Devonian Temiscouata and Fortin Formations, Northern Appalachians, Québec and New Brunswick; M.Sc. thesis, McGill University, Montreal, Quebec, 228 p.

Dansereau, P. and Bourque, P.-A.
1988:   A peculiar limestone breccia related to Late Silurian synsedimentary faulting in the Gaspé Basin, Québec Appalachians (abstract); Geological Association of Canada, Annual Meeting, St. John's, Newfoundland, Program with Abstracts, v. 13, p. A-28.

David, J. and Gariépy, C.
1986:   Geochemistry of the Lower Pointe aux Trembles and Lac Raymond Formations, central Québec Appalachians (a preliminary report); in Current Research, Part B; Geological Survey of Canada, Paper 86-1B, p. 131-140.
1990:   Early Silurian orogenic andesites from the central Quebec Appalachians; Canadian Journal of Earth Sciences, v. 27, p. 632-643.

David, J., Chabot, N., Marcotte, C., Lajoie, J., and Lespérance, P.J.
1985:   Stratigraphy and sedimentology of the Cabano, Pointe aux Trembles, and Lac Raymond Formations, Témiscouata and Rimouski Counties, Québec; in Current Research, Part B; Geological Survey of Canada, Paper 85-1B, p 481-497.

Dawson, W.D.
1855:   Acadian geology: an account of the geological structure and mineral resources of Nova Scotia, and portions of the neighbouring provinces of British America; Oliver and Boyd, Edinburgh, London and Pictou, 388 p.

De Broucker, G.
1986:   Evolution tectonostratigraphique de la boutonnière de Maquereau-Mictaw (Cambro-Ordovicien) Gaspésie, Québec; thèse de doctorat, Université Laval, Québec (Quebec), 322 p.

Dean, P.L.
1978:   The volcanic stratigraphy and metallogeny of Notre Dame Bay, Newfoundland; Memorial University of Newfoundland, Geology Report 7, 205 p.

Dean, P.L. and Strong, D.F.
1977:   Folded thrust faults in Notre Dame Bay, central Newfoundland; American Journal of Science, v. 277, p. 97-108.

Dean, W.T.
1970:   Lower Ordovician trilobites from the vicinity of South Catcher Pond, Northeastern Newfoundland; Geological Survey of Canada, Paper 70-44, 15 p.
1979:   Trilobites from the Long Point Group (Ordovician), Port au Port Peninsula, Southwestern Newfoundland; Geological Survey of Canada, Bulletin 290, 53 p.

Dec, T. and Colman-Sadd, S.P.
1990:   Timing of ophiolite emplacement onto the Gander Zone: evidence from provenance studies in the Mount Cormack Subzone; in Current Research (1990); Newfoundland Department of Mines and Energy, Geological Survey Branch, Report 90-1, p.289-303.

DeGrace, J.R.
1974:   Notes on the geology of the King George IV Lake area, southwest central Newfoundland; in Report of Activities 1973; Newfoundland Department of Mines and Energy, Mineral Development Division, p. 43-49.

DeGrace, J.R., Kean, B.F., Hsu, E., and Green, T.
1976:   Geology of the Nippers Harbour map-area (2 E/13), Newfoundland; Newfoundland Department of Mines and Energy, Mineral Development Division, Report 76-3, 73 p.

Destombes, J., Hollard, H, and Willefert, S.
1985:   Lower Paleozoic rocks of Morocco; in Lower Palaeozoic of Northwestern and West-Central Africa, (ed.) C.B. Holland; John Wiley and Sons, London, p. 91-336.

Dewey, J.F.
1969:   The evolution of the Caledonian/Appalachian Orogen; Nature, v. 222, p. 124-128.

Deynoux, M.
1980:   Les formations glaciaires du Précambrien terminal et de la fin de l'Ordovicien en Afrique de l'Ouest: deux exemples de glaciation d'inlandsis sur une plate-forme stable; Travaux des laboratoires des sciences de la Terre, Saint-Jérome, Marseille, B17, 554 p.
1983:   Late Precambrian and Upper Ordovician glaciations in the Taoudeni Basin, West Africa: an introduction to the field excursion; in Abstracts and Introduction to the Field Excursion of Symposium Till Mauretania, (ed.) M. Deynoux; Universite de Poitiers, p. 43-86.

Dineley, D.L. and Williams B.P.J.
1968:   The Devonian continental rocks of the lower Restigouche River, Québec; Canadian Journal of Earth Sciences, v. 5, p. 945-953.

Doll, C.G.
1951:   Geology of the Memphrémagog Quadrangle and the southeastern portion of the Irasburg Quadrangle, Vermont; Vermont Development Commission, Bulletin, no. 3, p. 113.

Donohoe, H.V.
1978:   Analyses of structures in the St. George area, Charlotte County, New Brunswick; Ph.D. thesis, University of New Brunswick, Fredericton, New Brunswick, 227 p.

Donohoe, H.V. and Wallace, P.I.
1982: Geological map of the Cobequid Highlands, Colchester, Cumberland and Pictou counties, Nova Scotia; Nova Scotia Department of Mines and Energy, Maps 82-6, 82-7, 82-8 and 82-9, Scale 1:50 000.

1985: Repeated orogeny, faulting, and stratigraphy in the Cobequid Highlands, Avalon Terrain of northern Nova Scotia; Geological Association of Canada/ Mineralogical Association of Canada, Field Trip Guide, Excursion 3, Fredericton, New Brunswick, 77 p.

Dorf, E. and Cooper, J.R.
1943: Early Devonian plants from Newfoundland; Journal of Palaeontology, v. 17, no. 3, p. 264-270.

Dostal, J., Keppie, J.D., and Dupuy, C.
1983: Petrology and geochemistry of Devono-Carboniferous volcanic rocks in Nova Scotia; Maritime Sediments and Atlantic Geology, v. 19, p. 59-71.

Douglas, G.V., Williams, D., and Rove, O.N.
1940: Copper deposits of Newfoundland; Geological Survey of Newfoundland, Bulletin, v. 20, 176 p.

Doyon, M.
1988: Synthèse géologique des roches volcaniques du Centre Nord de la Gaspésie; thèse de maîtrise, École polytechnique, Montréal (Québec) 244 p.

Doyon, M. and Valiquette, G.
1986: Synthèse géologique des volcanites du Centre Nord de la Gaspésie, Québec; Ministère de l'Énergie et des Ressources, Québec, MB 86-48.

1987: Devonian and Tertiary volcanism in the Lemieux dome area, central Gaspé, Québec (abstract); Geological Association of Canada, Annual Meeting, Saskatoon, Saskatchewan, Program with Abstracts, v. 12, p. 38.

Dubé, B., Lauzière, K., and Tremblay, A.
1991: Observations on the structural control and tectonic setting of gold mineralization in the Cape Ray fault zone, southwestern Newfoundland; in Current Research, Part D; Geological Survey of Canada, Paper 91-1D, p. 135-145.

Ducharme, D.
1979: Pétrographie du flysch de l'Ordovicien supérieur et du Silurien inférieur-Anticlinorium d'Aroostook-Percé, Gaspésie, Québec; Université de Montréal, Montréal (Quebec).

Dunning, G.R.
1987: U-Pb geochronology of the Coney Head Complex, Newfoundland; Canadian Journal of Earth Sciences, v. 24, p. 1072-1075.

Dunning, G.R. and Herd, R.K.
1980: The Annieopsquotch Ophiolite Complex, southwest Newfoundland and its regional relationships; in Current Research, Part A; Geological Survey of Canada, Paper 80-1A, p. 227-234.

Dunning, G.R. and Krogh, T.E.
1985: Geochronology of ophiolites of the Newfoundland Appalachians; Canadian Journal of Earth Sciences, v. 22, p. 1659-1670.

Dunning, G.R. and O'Brien, S.J.
1989: Late Proterozoic-Early Paleozoic crust in the Hermitage Flexure, Newfoundland Appalachians: U-Pb ages and tectonic significance; Geology, v. 17, p. 548-551.

Dunning, G.R., Wilton, D.H.C., and Herd, R.K.,
1989: Geology, geochemistry and geochronology of a Taconic calc-alkaline batholith, southwest Newfoundland; Royal Society of Edinburgh, Transactions, Earth Sciences, v. 80, p. 159-168.

Dunning, G.R., Barr, S.M., Raeside, R.P., and Jamieson, R.A.
1990b: U-Pb zircon, titanite, and monazite ages in the Bras d'Or and Aspy terranes of Cape Breton Island, Nova Scotia: implications for igneous and metamorphic history; Geological Society of America, Bulletin, v. 102, p. 322-330.

Dunning, G.R., Kean, B.F., Thurlow, J.G., and Swinden, H.S.,
1987: Geochronology of the Buchans, Roberts Arm, and Victoria Lake groups and Mansfield Cove Complex, Newfoundland; Canadian Journal of Earth Sciences, v. 24, p. 1175-1184.

Dunning, G.R., Krogh, T.E., O'Brien, S.J., Colman-Sadd, S.P., and O'Neill, P.
1988: Geochronologic framework for the Central Mobile Belt in southern Newfoundland and the importance of Silurian orogeny (abstract); Program with Abstracts; Geological Association of Canada, v. 13, p. 34.

Dunning, G.R., O'Brien, S.J., Colman-Sadd, S.P., Blackwood, R.F., Dickson, W.L., O'Neill, P.P., and Krogh, T.E.
1990a: Silurian orogeny in the Newfoundland Appalachians; Journal of Geology, v. 98, p. 895-913.

Durling, P.W., Bell, J.S., and Fader, G.B.J.,
1987: The geological structure and distribution of Paleozoic rocks on the Avalon Platform, offshore Newfoundland; Canadian Journal of Earth Sciences, v. 24, p. 1412-1420.

Eastler, T.E.
1969: Silurian geology of Change Islands and eastern Notre Dame Bay, Newfoundland; in North Atlantic: Geology and Continental Drift, (ed.) M. Kay; American Association of Petroleum Geologists, Memoir 12, p. 425-432.

1971: Geology of Silurian rocks, Change Islands and easternmos Notre Dame Bay, Newfoundland; Ph.D. thesis, Columbia University, New York, 143 p.

Eberz, G.W., Clarke, D.B., Chatterjee, A.K., and Giles, P.S.
1991: Chemical and isotopic composition of the lower crust beneath the Meguma Lithotectonic Zone, Nova Scotia: evidence from granulite facies xenoliths; Contributions to Mineralogy and Petrology, v. 109, p. 69-88.

Elliott, C.G.
1985: Stratigraphy, structure and timing of deformation of southwestern New World Island, Newfoundland; in Current Research, Part B; Geological Survey of Canada, Paper 85-1B, p. 43-49.

1988: The depositional, intrusive and deformational history of southwest New World Island and its bearing on orogenesis in Central Newfoundland; Ph.D. thesis, University of New Brunswick, Fredericton, New Brunswick, 270 p.

Elliott, C.G. and Williams, P.F.
1986: The tectonics and depositional history of the Ordovician an Silurian rocks of Notre Dame Bay, Newfoundland (Discussion); Canadian Journal of Earth Sciences, v. 23, p. 586-588.

1988: Sediment slump structures: a review of diagnostic criteria, and application to an example from Newfoundland; Journal of Structural Geology, v. 10, p. 171-182.

Elliott, C.G., Barnes, C.R., and Williams, P.F.
1989: Southwest New World Island stratigraphy: new fossil data, new implications for the history of the Central Mobile Belt, Newfoundland; Canadian Journal of Earth Sciences, v. 26, p. 2062-2074.

Elliott, C.G., Dunning, G.R., and Williams, P.F.
1991: New U-Pb zircon age constraints on the timing of deformation in north-central Newfoundland and implications for early Paleozoic Appalachian orogenesis; Geological Society of America Bulletin, v. 103, p. 125-135.

Espenshade, G.H.
1937: Geology and mineral deposits of the Pilleys Island area; Newfoundland Department of Natural Resources, Geology Section, Bulletin, v. 9, 36 p.

Evans, D.T.W., Kean, B.F., and Dunning, G.R.
1990: Geological studies, Victoria Lake Group, central Newfoundland; in Current Research (1990); Newfoundland Department of Mines and Energy, Geological Survey Branch, Report 90-1, p. 131-144.

Fåhraeus, L.E.
1973: Depositional environments and conodont-based correlation of the Long Point Formation (Middle Ordovician), Western Newfoundland; Canadian Journal of Earth Sciences, v. 10, p. 1822-1833.

Fåhraeus, L.E. and Hunter, D.R.
1981: Paleoecology of selected conodontophorid species from the Cobb Arm Formation (Middle Ordovician), New World Island, north-central Newfoundland; Canadian Journal of Earth Sciences, v. 18, p. 1653-1665.

Ferguson, L. and Fyffe, L.R.
1985: Geological Highway Map of New Brunswick and Prince Edward Island; Atlantic Geoscience Society, Special Publication, no. 2.

Freeze, A.C.
1936: Geology of the Fredericton sheet, New Brunswick; M.Sc. thesis, University of New Brunswick, Fredericton, New Brunswick, 62 p.

French, V.A.
1985: Geology of the Gillanders Mountain Intrusive Complex and satellite plutons, Lake Ainslie area, Cape Breton Island, Nova Scotia; M.Sc. thesis, Acadia University, Wolfville, Nova Scotia, 236 p.

Fullager, P.D. and Bottino, M.L.
1968: Radiometric age of the volcanics at Arisaig, Nova Scotia and the Ordovician-Silurian boundary; Canadian Journal of Earth Sciences, p. 5, p. 311-317.

Fyffe, L.R.
1974: P-3 Belledune-Pointe Verte-Guitar, New Brunswick; Department of Natural Resources, Mineral Branch, Preliminary Geological Map.

1982a: Taconian and Acadian structural trends in central and northern New Brunswick; in Major Structural Zones and Faults of the Northern Appalachians, (ed.) P. St-Julien and J. Béland; Geological Association of Canada, Special Paper 24, p. 117-130.

1982b: Geology of Woodstock (sheet 21J); New Brunswick Department of Natural Resources, Map NR-4.

Fyffe, L.R. (cont.)
1986: A recent graptolite discovery from the Fournier Group of northern New Brunswick; in Eleventh Annual Review of Activities, Project Resumés, (ed.) S.A. Abbott; New Brunswick Department of Natural Resources and Energy, Minerals and Energy Division, Information Circular 86-2, p. 43-45.
1988: Bedrock geology of the McAdam-St. Croix River area; in Thirteenth Annual Review of Activities, Project Résumés, (ed.) S.A. Abbott; New Brunswick Department of Natural Resources and Energy, Minerals and Energy Division, Information Circular 88-2, p. 70-75.
1989: Bedrock geology of the Moores Mills area, Charlotte County, New Brunswick; in Fourteenth Annual Review of Activities, Project Summaries for 1989, (ed.) S.A. Abbott; New Brunswick Department of Natural Resources and Energy, Minerals and Energy Division, Information Circular 89-9, p. 28-39.
1991: Tectonic significance of the St. Croix terrane, southwestern New Brunswick (abstract); Atlantic Geology, v. 27, p. 152.

Fyffe, L.R. and Barr, S.M.
1986: Petochemistry and tectonic significance of Carboniferous volcanic rocks in New Brunswick; Canadian Journal of Earth Sciences, v. 23, p. 1243-1256.

Fyffe, L.R. and Fricker, A.
1987: Tectonostratigraphic terrane analysis of New Brunswick; Maritime Sediments and Atlantic Geology, v. 23, p. 113-123.

Fyffe, L.R. and Noble, J.P.3.
1985: Stratigraphy and structure of the Ordovician, Silurian and Devonian of the northern New Brunswick; Geological Association of Canada, Annual Meeting, Fredericton, New Brunswick, Excursion 4, Guidebook, 56 p.

Fyffe, L.R. and Riva, J.
1990: Revised stratigraphy of the Cookson Group based on new paleontological, sedimentological, and structural evidence; Atlantic Geology, v. 26, p. 271-276.

Fyson, W.K.
1966: Structures in the Lower Paleozoic Meguma Group, Nova Scotia; Geological Society of America Bulletin, v. 77, p. 931-944.

Garland, G.D.
1953: Gravity measurements in the Maritime provinces; Publication of the Dominion Observatory, v. 16, no. 7, p. 185-275.

Gates, O.
1969: Lower Silurian-Lower Devonian volcanic rocks of New England coast and southern New Brunswick; in North Atlantic-Geology and Continental Drift, (ed.) M. Kay; American Association of Petroleum Geologists, Memoir 12, p. 484-503.
1977: Geological map and cross sections of the Eastport 15 Quadrangle, Washington County, Maine; Maine Geological Survey, Preliminary Report to Map GM-3, Eastport, Maine, 19 p.
1984: The geology of the Passamaquoddy Bay area, Maine and New Brunswick; Maine Geological Survey, Open File 84-10, 22 p.
1989: Silurian roundstone conglomerates of coastal Maine and adjacent New Brunswick: in Structure and Stratigraphy, (ed.) R.D. Tucker and R.G. Marvinney; Maine Geological Survey, Studies in Maine Geology, v. 2, p. 127-144.

Gates, O. and Moench, R.H.
1981: Bimodal Silurian and Lower Devonian volcanic assemblages in the Machias-Eastport area, Maine; United States Geological Survey, Professional Paper 1184, 32 p.

Giles, P.S.
1985: A major Post-Visean sinistral shear zone: new perspectives on Devonian and Carboniferous rocks of southern Nova Scotia; in Guide to the Granites and Mineral Deposits of Southwestern Nova Scotia, (ed.) A.K. Chatterjee and D.B. Clarke; Department of Mines and Energy, Paper 85-3.

Gillis, J.W.
1972: Geology of the Port aux Basque map area, Newfoundland; Geological Survey of Canada, Paper 71-42, 6 p.

Gordon, A.J.
1973: Silurian rocks of the Fredericton area; in Geology of New Brunswick (Field Guide to Excursions, New England Intercollegiate Geological Conference), (ed.) N. Rast; University of New Brunswick, Fredericton, New Brunswick, p. 125-131.

Gorman, W.A.
1954: Rapport préliminaire sur la région de Sainte-Justine, districts électoraux de Montmagny, Bellechasse et Dorchester; Ministère des Mines, Québec, RP 297.
1955: Rapport préliminaire sur la région de Saint Georges-Saint Zacharie, districts électoraux de Beauce et Dorchester; Ministère des Mines, Québec, RP 314.

Gosselin, C.
1988: Géologie de la région de Maria, Gaspésie; Ministère de l'Énergie et des Ressources, Québec, ET-87-01, 28 p.

Gosselin, C. et Simard, M.
1983: Évaluation du potentiel minéral de la région de Nouvelle; Ministère de l'Énergie et des Ressources, Québec, DP-83-13, 24 p.

Greene, B.A.
1974: Harbour Breton, Newfoundland; Newfoundland Department of Mines and Energy, Mineral Development Division, Map (with marginal notes).

Greene, B.A. and O'Driscoll, C.F.
1976: Gaultois map-area; in Current Research; Newfoundland Department of Mines and Energy, Mineral Development Division, Report 76-1, p. 56-63.

Greenough, J.D., McCutcheon, S.R., and Papezik, V.S.
1985: Petrology and geochemistry of Cambrian volcanic rocks from the Avalon Zone in New Brunswick; Canadian Journal of Earth Sciences, v. 22, p. 881-892.

Greiner, H.R.
1960: Pointe Verte-Gloucester and Restigouche Counties, New Brunswick; New Brunswick Mines Branch, Preliminary Map 60-2.
1970: Geology of the Charlo area 21-0/16, Restigouche County, New Brunswick; New Brunswick Department Natural Resources, Map Series 70-2, 18 p.
1973: Pointe Verte to Tide Head, Chaleur Bay area, New Brunswick; in Geology of New Brunswick, (ed.) N. Rast; New England Intercollegiate Geological Conference, Field guide to excursions, p. 58-70.

Hall, B.A. and Stanley, D.J.
1973: Levee-bounded submarine base-of-slope channels in the Lower Devonian Seboomook Formation, Northern Maine; Geological Society of America, Bulletin, v. 84, p. 2101-2110.

Hall, B.A., Pollock, S.G., and Dolan, K.M.
1976: Lower Devonian Seboomook Formation and Matagamon sandstone, northern Maine: a flysch basin-margin delta complex; in Contributions to the Stratigraphy of New England, (ed.) L.R. Page; Geological Society of America, Memoir 148, p. 57-63.

Hallam, A.
1984: Pre-Quaternary sea-level changes; Earth and Planetary Sciences, Annual Review, v. 12, p. 205-243.

Hallam, A. and Bradshaw, M.J.
1979: Bituminous shales and oolitic ironstones as indicators of transgressions and regressions; Geological Society of London, Journal, v. 136, p. 157-164.

Hamilton-Smith, T.
1970: Stratigraphy and structure of Ordovician and Silurian of the Siegas area, New Brunswick; New Brunswick Department of Natural Resources, Report Investigation 12, 55 p.
1972: Stratigraphy and structure of Silurian rocks, the McKenzie Corner area, New Brunswick; New Brunswick Department of Natural Resources, Report Investigation 15, 26 p.

Hanmer, S.
1981: Gander Zone, Newfoundland: an Acadian ductile shear zone; Canadian Journal of Earth Sciences, v. 128, p. 120-135.

Hantzschel, W.
1975: Miscellanea supplement; in Treatise on Invertebrate Paleontology, Part W, (ed.) C. Teichert; University of Kansas/Geological Society of America, p. W-9.

Harris, I.M.
1966: Geology of the Cobbs Arm area, New World Island, Newfoundland; Newfoundland Department of Mines, Agriculture and Resources, Bulletin, v. 37, 38 p.

Hatch, N.L.
1982: The Taconian Line in western New England and its implication to Paleozoic tectonic history; in Major structural zones and faults of the Northern Appalachians, (ed.) P. St-Julien and J. Béland; Geological Association of Canada, Special Paper 24, p. 67-85.

Hay, P.W.
1968: Geology of the St. George-Seven Mile Lake area, southwestern New Brunswick; New Brunswick Department of Natural Resources, Mineral Resources Division, Map Series 68-1, 7 p.

Helmstaedt, H.
1968: Structural and metamorphic analysis in Beaver Harbour region, Charlotte County, New Brunswick; Ph.D. Thesis, University of New Brunswick, Fredericton, New Brunswick, 196 p.
1971: Structural geology of Portage Lakes area, Bathurst-Newcastle District, New Brunswick; Geological Survey of Canada, Paper 70-28, 52 p.

Helwig, J.A.

1967: Stratigraphy and structural history of the New Bay area, north central Newfoundland; Ph.D. thesis, Columbia Univeristy, U.S.A., 211 p.

1969: Redefinition of Exploits Group, Lower Paleozoic, Northeast Newfoundland; in North Atlantic: Geology and Continental Drift, (ed.) M. Kay; American Association of Petroleum Geologists, Memoir 12, p. 408-413.

Helwig, J.A. and Sarpi, E.

1969: Plutonic-pebble conglomerates, New World Island, Newfoundland, and the history of eugeosynclines; in North Atlantic Geology and Continental Drift, (ed.) M. Kay; American Association of Petroleum Geologists, Memoir 12, p. 443-466.

Héroux, Y.

1975: Stratigraphie de la Formation de Sayabec (Silurien) dans la Vallée de la Matapédia (Québec); thèse de doctorat, Université de Montréal, Montréal (Québec).

Hesse, R and Sawh, H.

1982: Escuminac Formation; in Excursion Guidebook 7B: Paleozoic continental margin sedimentation in the Québec Appalachians, (ed.) R. Hesse, G.V. Middleton, and B.R. Rust; International Association of Sedimentologists, Eleventh International Congress, Hamilton, p. 72-80.

Heyl, G.R.

1936: Geology and mineral deposits of the Bay of Exploits area; Newfoundland Department of Natural Resources, Geological Section, Bulletin, v. 3, 66 p.

1937: The Geology of the Sops Arm area, White Bay, Newfoundland; Newfoundland Department of Natural Resources, Geological Section, Bulletin, v. 8, 42 p.

Hibbard, J.P.

1982: Significance of the Baie Verte Flexure, Newfoundland; Geological Society of America Bulletin, v. 93, p. 790-797.

1983: Geology of the Baie Verte Peninsula, Newfoundland; Newfoundland Department of Mines and Energy, Mineral Development Division, Memoir 2, 279 p.

Hickox, C.F.

1958: Geology of the central Annapolis Valley, Nova Scotia; Ph.D. thesis, Yale University, U.S.A.

Honeyman, D.

1879: Nova Scotia geology-Kings County; Nova Scotia Institute of Science, Proceedings and Transactions, v. 5, p. 21-31.

Horne, G.S.

1968: Stratigraphy and structural geology of southwestern New World Island area, Newfoundland; Ph.D. thesis, Columbia University, U.S.A.

1969: Early Ordovician chaotic depositis in the Central Volcanic Belt of Northeast Newfoundland; Geological Society of America Bulletin, v. 80, p. 2451-2464.

1970: Complex volcanic-sedimentary patterns in the Magog Belt of northeastern Newfoundland; Geological Society of America Bulletin, v. 81, p. 1767-1788.

1976: Geology and Lower Ordovician fossiliferous strata between Virgin Arm and Squid Cove, New World Island, Newfoundland; Geological Survey of Canada, Bulletin, v. 261, p. 1-9.

Horne, G.S. and Helwig, J.

1969: Ordovician stratigraphy of Notre Dame Bay; in North Atlantic Geology and Continental Drift, (ed.) M. Kay; American Association of Petroleum Geologists, Memoir 12, p. 388-407.

Horne, G.S. and Johnson, J.H.

1970: Ordovician algae from boulders in Silurian deposits of New World Island, Newfoundland; Journal of Paleontology, v. 44, p. 1055-1059.

Hriskevitch, M.E.

1950: Little Rattling Brook, Newfoundland; Geological Survey of Canada, Paper 50-17.

Huard, A.A.,

1984: The Carboniferous volcanic rocks at Lake Ainslie, Nova Scotia: implications for tectonic regime and barite mineralization; B.Sc. thesis, St. Francis Xavier University, Antigonish, Nova Scotia.

Huff, G.E.

1970: SAREP, Geological report on Central Gaspé, Surface party, Québec; Department of Natural Resources, Québec, Open File Report GM-27078.

Hughson, R.C. and Stearn, C.W.

1988: Upper Silurian reefal facies of the Memphrémagog-Marbleton area, Eastern Townships, Québec Appalachians; (ed.) H.H. Geldsetzer, N.P. James, and G. Tebbut; in Canadian Society of Petroleum Geologists, Memoir 13, p. 306-315.

Hunter, D.R.

1978: Conodonts of the Cobbs Arm Formation (Middle Ordovician), north-central Newfoundland; M.Sc. thesis, Memorial University of Newfoundland, St. John's, Newfoundland, 180 p.

Hunter, R.E. and Clifton, H.E.

1982: Cyclic deposits and hummocky-cross-stratification of probable storm origin in Upper Cretaceous rocks of the Cape Sebastian area, southwestern Oregon; Journal of Sedimentary Petrology, v. 52, p. 127-143.

Hurst, J.M. and Pickerill, R.K.

1986: The relationship between sedimentary facies and faunal associations in the Llandovery siliciclastic Ross Brook Formation, Arisaig, Antigonish County, Nova Scotia; Canadian Journal of Earth Science, v. 23, p. 705-726.

Hutton, D.H.W. and Murphy, F.C.

1987: The Silurian of the Southern Uplands and Ireland as a successor basin to the end-Ordovician closure of Iapetus; Geological Society of London, v. 144, p. 765-772.

Hyde, R.S.

1979: Geology of Carboniferous strata in portions of the Deer Lake Basin, western Newfoundland; Newfoundland Department of Mines and Energy, Mineral Development Division, Report 79-6, 43 p.

Jacobi, R.D.

1980: Geology of part of the terrane north of the Lukes Arm Fault, north-central Newfoundland (Part 1), Modern submarine slides and their geological implications (Part II); Ph.D. thesis, Columbia University, New York, 422 p.

Jacobi, R.D. and Schweickert, R.A.

1976: Implication of new data on stratigraphic and structural relations of Ordovician rocks on New World Island, north-central Newfoundland (abstract); Geological Society of America, Abstracts with Programs, v. 8, p. 206.

Jacobi, R.D. and Wasowski, J.J.

1985: Geochemistry and plate-tectonic significance of the volcanic rocks of the Summerford Group, north-central Newfoundland; Geology, v. 13, p. 126-130.

James, N.P., Klappa, C.F., Skevington, D., and Stevens, R.K.

1980: Cambro-Ordovician of west Newfoundland: Sediments and Faunas; Geological Association of Canada, Field Trip 13, Guidebook, 88 p.

Jamieson, R.A. and Doucet, P.,

1983: The Middle River-Crowdis Mountain area, southern Cape Breton Highlands; in Current Research, Part A; Geological Survey of Canada, Paper 83-1A, p. 269-275.

Jamieson, R.A., Tallman, P.D., Plint, H.E., and Connors, K.A.

1990: Geological setting of pre-Carboniferous mineral deposits in the western Cape Breton Highlands; in Mineral Deposit Studies in Nova Scotia, Volume 1, (ed.) A. Sangster; Geological Survey of Canada, Paper 90-8, p. 77-99.

Jamieson, R.A., van Breemen, O., Sullivan, R.W., and Currie, K.L.

1986: The age of igneous and metamorphic events in the western Cape Breton Highlands, Nova Scotia; Canadian Journal of Earth Sciences, v. 3, p. 1891-1901.

Jenkyns, H.C.

1986: Pelagic environments; in Sedimentary Environments and Facies (Chapter 11), (ed.) H.C. Reading; Blackwell Scientific, p. 343-398.

Jensen, L.R.

1975: The Torbrook Formation; in Ancient Sediments of Nova Scotia, (ed.) I.M. Harris; Society of Economic Paleontologists and Mineralogists, Eastern Section, v. 123, p. 63-74.

Jones, I.W.

1935: Upper York River map area, Gaspé Peninsula, Quebec; Department of Mines, Annual Reports, part D, 1934-1936, p. 3-28.

Kalliokoski, J.

1953: Springdale, Newfoundland; Geological Survey of Canada, Paper 53-5, 4 p.

1955: Gull Pond, Newfoundland; Geological Survey of Canada, Paper 54-4, geological map (with descriptive notes).

Karlstrom, K.E.

1985: Structural reconnaissance of South Samson Island, northeastern Newfoundland; in Current Research, Part B; Geological Survey of Canada, Paper 85-1B, p. 95-101.

Karlstrom, K.E., van der Pluijm, B.A., and Williams, P.F.

1982: Structural interpretation of the eastern Notre Dame Bay area, Newfoundland: regional Post-Middle Silurian thrusting and asymmetrical folding; Canadian Journal of Earth Sciences, v. 19, p. 2325-2341.

1983: Sedimentology of Upper Ordovician-Silurian sequences on New World Island, Newfoundland: separate fault-controlled basins? (discussion); Canadian Journal of Earth Sciences, v. 20, p. 1757-1758.

Kay, G.M.
1967: Stratigraphy and structure of northeastern Newfoundland: bearing on drift in north Atlantic; American Association of Petroleum Geologists, Bulletin, v. 51, p. 579-600.
1969: Silurian of northeast Newfoundland coast; in North Atlantic-Geology and Continental Drift, (ed.) M. Kay; American Association of Petroleum Geologists, Memoir 12, p. 414-424.
1976: Dunnage Melange and subduction of the Protoacadic Ocean, northeast Newfoundland; Geological Society of America, Special Paper 175, 49 p.

Kay, G.M. and Williams, H.
1963: Ordovician-Silurian relations on New World Island, Notre Dame Bay, northeast Newfoundland (Abstract); Geological Society of America, Bulletin, v. 74, 807 p.

Kean, B.F.
1977: Geology of the Victoria Lake, map area, southwestern Newfoundland; Newfoundland Department of Mines and Energy, Mineral Development Division, Report 77-4.
1983: Geology of the King George IV Lake map area (12A/4); Newfoundland Department of Mines and Energy, Mineral Development Division, Report 83-4, 67 p.

Kean, B.F. and Jayasinghe, N.R.
1980: Badger map area (12A/16), Newfoundland; in Current Research, (ed.) C.F. O'Driscoll and R.V. Gibbons; Newfoundland Department of Mines and Energy, Mineral Development Division, Report 80-1, p. 37-43.

Kean, B.F., Dean, P.L., and Strong, D.F.
1981: Regional geology of the central volcanic belt of Newfoundland; in The Buchans Orebodies: Fifty years of Geology and Mining, (ed.) E.A. Swanson, D.F. Strong, and J.G. Thurlow; Geological Association of Canada, Special Paper 22, p. 65-78.

Kelley, D.G. and Mackasey, W.O.
1964: Basal mississippian volcanic rocks in Cape Breton Island, Nova Scotia; Geological Survey of Canada, Paper 64-34, 10 p.

Kendall, A.J.D.
1981: The White Rock Formation metavolcanics at Cape St. Mary, Nova Scotia: petrography, geochemistry, and geologic affinities; B.Sc. thesis, Dalhousie University, Halifax, Nova Scotia, 115 p.

Kennedy, M.J.
1976: Southeastern margin of the northeastern Appalachians: Late Precambrian orogeny on a continental margin; Geological Society of America Bulletin, v. 87, p. 1317-1325.

Kent, D.V. and van der Voo, R.
1990: Paleozoic paleogeography from paleomagnetism of the Atlantic-bordering continents; in Paleogeography and Biogeography (ed.), W.S. McKerrow and C.R. Scotese; Geological Society of London, Memoir 12, p. 49-56.

Keppie, J.D.
1982: The Minas Geofracture; in Major Structural Zones and Faults of the Northern Appalachians, (ed.) P. St-Julien and J. Béland; Geological Association of Canada, Special Paper 24, p. 263-280.
1982: Tectonic map of the Province of Nova Scotia; Nova Scotia Department of Mines and Energy, Halifax, Nova Scotia.
1985: The Appalachian collage; in The Caledonide Orogen-Scandinavia and Related Areas, Part 2, (ed.) D.G. Gee and B.A. Sturt; John Wiley and Sons Ltd, p. 1217-1226.
1990: Tectono-stratigraphic terranes in Cape Breton Island, Nova Scotia: Implications of the configuration of the northern Appalachian orogen. Comment; Geology, v. 18, p. 669-671.

Keppie, J.D. and Dallmeyer, R.D.
1987: Dating transcurrent terrane accretion: an example from the Meguma and Avalon composite terranes in the northern Appalachians; Tectonophysics, v. 6, p. 831-847
1989a: ^{40}Ar/^{39}Ar mineral ages from Kellys Mountain, Cape Breton Island, Nova Scotia: implications for the tectonothermal evolution of the Avalon Composite Terrane; Canadian Journal of Earth Sciences, v. 26, p. 1509-1516.
1989b: Tectonic map of Pre-Mesozoic Terranes in Circum-Atlantic Phanerozoic orogens; in Terranes in Circum-Atlantic Paleozoic Orogens; International Geological Correlation Programme, Project 233, scale 1:5 000 000.

Keppie, J.D. and Halliday, A.N.
1986: Rb-Sr isotopic data from three suites of igneous rocks, Cape Breton Island, Nova Scotia; Maritime Sediments and Atlantic Geology, v. 22, p. 162-171.

Keppie, J.D. and Murphy, J.B.
1988: Anatomy of a telescoped pull-apart basin: The stratigraphy and structure of the Antigonish Highlands; Maritime Sediments and Atlantic Geology, v. 24, p. 123-138.

Keppie, J.D. and Smith, P.K.
1978: Compilation of isotopic age data of Nova Scotia; Nova Scotia Department of Mines, Report 78-4.

Keppie, J.D., Dallmeyer, R.D., and Krogh, T.E.
1991a: U-Pb and ^{40}Ar/^{39}Ar mineral ages from the Cape North area, Cape Breton Island: implication for docking of the Avalon Composite Terrane (abstract); Geological Society of America, Abstracts with Programs, v. 23, no. 1, p. 52.

Keppie, J.D., Dallmeyer, R.D., and Murphy, J.B.
1990: Implications of 40Ar/39Ar hornblende ages from late Proterozoic-Cambrian plutons in the Avalon composite terrane of Nova Scotia; Atlantic Geology, v. 26, p. 175-176.
1991b: Tectonic implications of ^{40}Ar/^{39}Ar hornblende ages from late Proterozoic- Cambrian plutons of the Avalon Composite Terrane, Nova Scotia (Reply); Geological Society of America Bulletin, v. 103, p. 1380-1383.

Keppie, J.D., Dostal J., and Zentilli, M.
1978: Petrology of the early Silurian Dunn Point and McGillivray Brook Formations; Nova Scotia Department of Mines and Energy, Paper 78-5, 20 p.

Keppie, J.D., Murphy, J.B., Nance, R.D., and Dostal, J.
1989: Terranes in Nova Scotia: their characteristics and accretionary histories; Nova Scotia Department of Mines and Energy, Report 89-3, p. 117-122.

Keppie, J.D., Nance, R.D., Murphy, J.B., and Dostal, J.
1991c: Northern Appalachians: Avalon and Meguma Terranes; in The West African Orogens and Circum-Atlantic Correlatives, (ed.) R.D.Dallmeyer and J.P. Lécorche; Springer-Verlag, Heidelberg, p. 315-333.

Kidd, W.S.F.
1974: The evolution of the Baie Verte Lineament, Burlington Peninsula, Newfoundland; Ph.D. thesis, Cambridge University, Cambridge, England, 294 p.

Kindle, C.H.
1936: A geologic map of southeastern Gaspé; The Eastern Geologist, no. 1, 8 p.

King, L.H., Fader, G.B.J., Jenkins, W.A.M., and King, E.L.
1986: Occurrence and regional geological setting of Paleozoic rocks on the Grand Banks of Newfoundland; Canadian Journal of Earth Sciences, v. 23, p. 504-526.

Kirkwood, D.
1986: Géologie structurale de la région de Percé, Gaspésie, Québec; thèse de maîtrise, Université de Montréal, Montréal (Quebec).
1989: Géologie structurale de la région de Percé; Ministère de l'Énergie et des Ressources, Québec, ET 87-17, 42 p.

Kirkwood, D. and Béland, J.
1986: Structural geology of the Percé area, Gaspésie (abstract); Geological Association of Canada, Annual Meeting, Ottawa, Program with Abstracts, v. 11, p. 89-90.

Kirkwood, D. et St-Julien, P.
1987: Analyse structurale du Siluro-Dévonien dans la Vallée de la Matapédia, Québec; Ministère de l'Énergie et des Ressources, Québec, MB 87-33, 17 p.

Kirkwood, D., Malo, M., and St-Julien, P.
1988a: Strain patterns, strain integration and shortening in Middle Paleozoic rocks of western Gaspé Peninsula, northern Appalachians (abstract); Geological Society of America, Northeastern Section, Twenty-Third Annual Meeting, Portland, Maine, Abstracts with Programs, v. 20, no. 1, p. 30.
1988b: Palinspastic reconstruction of the Gaspé-Témiscouata Basin during Silurian time, northern Appalachians (abstract); Geological Association of Canada, Annual Meeting, St. John's, Newfoundland, Program with Abstracts, v. 13, p. A-66.

Kontak, D.J., Tuach, J., Strong, D.F., Archibald, D.A., and Farrar, E.
1988: Plutonic and hydrothermal events in the Ackley Granite, southeast Newfoundland, as indicated by total-fusion ^{40}Ar/^{39}Ar geochonology; Canadian Journal of Earth Sciences, v. 25, p. 1151-1160.

Krogh, T.E., Strong, D.F., O'Brien, S.J., and Papezik, V.S.
1988: Precise U-Pb zircon dates from the Avalon terrane in Newfoundland; Canadian Journal of Earth Sciences, v. 25, p. 442-453.

Kusky, T.M.
1985: Geology of the Frozen Ocean Lake-New Bay Pond area, north-central Newfoundland; M.Sc. thesis, State University of New York at Albany, 214 p.

Kusky, T.M. and Kidd, W.S.F.
1985: Middle Ordovician conodonts from the Buchans Group, central Newfoundland and their significance for regional stratigraphy of the central volcanic belt (discussion); Canadian Journal of Earth Sciences, v. 22, p. 484-485.

Lachambre, G.
1987:   Le Silurien et le Dévonien basal du Nord de la Gaspésie; Ministère de l'Énergie et des Ressources, Québec, ET 84-06, 83 p.

Lachance, S.
1974:   Région de l'Ascension-de-Patapédia, comté de Bonaventure; Ministère des Richesses naturelles, Québec, DP-273, 19 p.
1975:   Région de Saint-François-d'Assises, comté de Bonaventure; Ministère des Richesses naturelles, Québec, DP-328, 16 p.
1977:   Région de Saint-Alexis-de-Matapédia, comté de Bonaventure; Ministère des Richesses naturelles, Québec, DPV-458, 23 p.
1979:   Géologie de la région de Saint-André-de-Ristigouche, comté de Bonaventure; Ministère des Raturelles, Québec, DPV-667, 19 p.

Lafrance, B.
1989:   Structural evolution of a transpression zone in north central Newfoundland; Journal of Structural Geology, v. 11, p. 705-716.

Lajoie, J.
1968:   Dispersal and petrology of the Silurian Val-Brillant and Robitaille sandstones, Appalachians, Québec; Journal of Sedimentary Petrology, v. 38, p. 643-647.
1971:   Région des Lacs Prime et des Baies (comté de Rimouski); Ministère des Richesses naturelles, Québec, RG 139, 85 p.

Lajoie, J., Lespérance, P.J.,and Béland, J.
1968:   Silurian stratigraphy and paleogeography of Matapédia-Témiscouata region, Québec; American Association of Petroleum Geologists Bulletin, v. 52, p. 615-640.

Lane, T.E.
1981:   The stratigraphy and sedimentology of the White Rock Formation (Silurian), Nova Scotia, Canada; M.Sc. thesis, Dalhousie University, Halifax, Nova Scotia, 270 p.

Laurent, R. and Bélanger, J.
1984:   Geochemistry of Silurian-Devonian alkaline basalt suites from the Gaspé Peninsula, Québec Appalachians; Maritime Sediments and Atlantic Geology, v. 20, p. 67-78.

Lavoie, D.
1985:   Stratigraphie, géologie structurale, sédimentologie et paléomilieux de la bande silurienne supérieure des lacs Aylmer et Saint-François; thèse de maîtrise, Université Laval, Québec (Québec).
1988:   Stratigraphie, sédimentologie et diagenèse du Wenlockien (Silurien) du Bassin de Gaspésie-Matapédia; thèse de doctorat, Université Laval, Québec (Québec).

Lavoie, D. et Bourque, P.-A.
1986:   Stratigraphie, pétrographie et faciès du niveau Sayabec-Laforce-La Vieille, Gaspésie et vallée de la Matapédia; Ministère de l'Énergie et des Ressources, Québec, MB 86-35, 31 p.

Lee, H.J. and Noble, J.P.A.
1977:   Silurian stratigraphy and depositional environments: Charlo-Upsalquitch Forks area, New Brunswick; Canadian Journal of Earth Sciences, v. 14, p. 2533-2542.

Lefort, J.P.
1989:   The submerged part of the Ligerian (Eo-Hercynian)-Acadian mobile belt; in Basement correlation across the North Atlantic; Springer-Verlag, Berlin, p. 52-69.

Lefort, J.P. and Haworth, R.
1981:   Geophysical correlation between basement features in northwest Africa and North America, and their control over structural evolution; Société géologique et minéralogique de Bretagne, v. 13, no. 2, p. 103-116.

Leger, A. and Williams, P.F.
1986:   Transcurrent faulting history of southern New Brunswick; in Current Research, Part B; Geological Survey of Canada, Paper 86-1B, p. 111-120.

Legrand, P.
1985:   Lower Palaeozoic rocks of Algeria; in Lower Palaeozoic of North-Western and West-Central Africa, (ed.) C.H. Holland; John Wiley and Sons, Chichester, England, v. 512, p. 5-90.

Lenz, A.C.
1972:   Graptolites from the Laforce and the St. Léon formations, northern Gaspé, Québec; Canadian Journal of Earth Sciences, v. 9, p. 1148-1162.

Lespérance, P.J.
1968:   Ordovician and Silurian Trilobites faunas of the White Head Formation, Percé region, Québec; Journal of Paleontology, v. 143, no. 3, p. 811-826.
1974:   The Hirnantian fauna of the Percé area (Québec) and the Ordovician-Silurian boundary; American Journal of Science, v. 274, p. 10-30.
1980a: Calcaires Supérieurs de Gaspé: Les aires-types et le prolongement ouest; Ministère de l'Énergie et des Ressources, Québec, DPV-595, 92 p.

Lespérance, P.J. (cont.)
1980b: Les Calcaires Supérieurs de Gaspé (Dévonien Inférieur) dans le nord-est de la Gaspésie; Ministère de l'Énergie et des Ressources, Québec, DPV-751, 35 p.

Lespérance, P.J. and Bourque, P.-A.
1970:   Silurian and basal Devonian stratigraphy of northeastern Gaspé Peninsula, Québec; American Association of Petroleum Geologists Bulletin, v. 54, p. 1868-1886.

Lespérance, P.J. et Greiner, H.G.
1969:   Région de Squatec-Cabano, comtés de Rimouski, Rivière-du-Loup et Témiscouata; Ministère des Richesses naturelles, Québec, RG 128, 122 p.

Lespérance, P.J. and Sheehan, P.M.
1976:   Brachiopods from the Hirnantian stage (Ordovician-Silurian) at Percé, Quebec; Paleontology, v. 19, p. 719-731.

Lespérance, P.J. and Tripp, R.P.
1985:   Encrinurids (Trilobita) from the Matapédia Group (Ordovician), Percé, Québec; Canadian Journal of Earth Sciences, v. 22, p. 205-213.

Lespérance, P.J., Malo, M., Sheehan, P.M., and Skidmore, W.B.
1987:   A stratigraphical and faunal revision of the Ordovician-Silurian strata of the Percé area, Québec; Canadian Journal of Earth Sciences, v. 24, p. 117-134.

Lilly, H.D.
1966:   Late Precambrian and Appalachian tectonics in light of submarine exploration of the Great Bank of Newfoundland in the Gulf of St. Lawrence; American Journal of Science, v. 264, p. 569-574.

Lock, B.E.
1969a: The Lower Paleozoic geology of western White Bay, Newfoundland (2 volumes); Ph.D. thesis, Cambridge University, Cambridge, England, 343 p.
1969b: Silurian rocks of west White Bay area, Newfoundland; in North Atlantic Geology and Continental Drift, (ed.) M. Kay; American Association of Petroleum Geologists, Memoir 12, p. 433-442.
1972:   Lower Paleozoic history of a critical area: eastern margin of the St. Lawrence Platform in White Bay, Newfoundland, Canada; 24th international Geological Congress, section 6, Montreal, p. 310-324.

Logan, W.E.
1863:   Report on the geology of Canada; Dawson Brothers, Montreal, 983 p.

Ludman, A.
1978:   Stratigraphy, structure and progressive metamorphism of Lower Paleozoic rocks in the Calais area; in Guidebook for Field Trips in Southeastern Maine and Southwestern New Brunswick, (ed.) A. Ludman; New England Intercollegiate Geological Conference, 70th Annual Meeting, Queen's College Press, Flushing, New York, p. 78-101.
1981:   Significance of transcurrent faulting in eastern Maine and location of the suture between Avalonia and North America; American Journal of Science, v. 281, p. 463-483.
1987:   Pre-Silurian stratigraphy and tectonic significance of the St. Croix Belt, southeastern Maine; Canadian Journal of Earth Sciences, v. 24, p. 2459-2469.

Macdonald, A.S. and Smith, P.K.
1980:   Geology of the Cape North area, northern Cape Breton Island, Nova Scotia; Nova Scotia Department of Mines and Energy, Paper 80-1, 60 p.

Mackasey, W.O.
1963:   Petrography and stratigraphy of a Lower Mississippian, pre-Horton volcanic succession in north-west Cape Breton Island; M.Sc. thesis, Carleton University, Ottawa, Ontario.

MacLean, H.J.
1947:   Geology and mineral deposits of the Little Bay area; Newfoundland Geological Survey Bulletin No. 22, 45 p.

Maehl, R.D.
1961:   The older Palaeozoic of Pictou County, Nova Scotia; Nova Scotia Department of Mines, Memoir 4.

Malo, M.
1979:   L'axe Aroostook-Matapédia au nord de Chandler, Gaspésie, Québec; thèse de maîtrise, Université Laval, Québec (Québec).
1986:   Stratigraphie et structure de l'anticlinorium d'Aroostook-Percé en Gaspésie, Québec; thèse de doctorat, Université de Montréal, Montréal (Québec).
1987:   Structural evidence for Acadian wrench faulting in the south-eastern Gaspé Peninsula, Quebec (abstract); Geological Association of Canada, Program with Abstracts; v. 12, p. 70.
1988a: Stratigraphie et structure de l'anticlinorium d'Aroostook-Percé au nord de la faille du Grand Pabos, régions de Chandler et de Grande Rivière, Gaspésie; Ministère de l'Énergie et des Ressources, Québec, ET 87-06, 42 p.

Malo, M. (cont.)

1988b: Stratigraphy of the Aroostook-Percé Anticlinorium in the Gaspé Peninsula, Quebec; Canadian Journal of Earth Sciences, v. 25, p. 893-908.

1989: Stratigraphie et structure de l'anticlinorium d'Aroostook-Percé de la région d'Honorat (Gaspésie); Ministère de l'Énergie et des Ressources, Québec, ET 88-01, 48 p.

Malo, M. and Béland, J.

1989: Acadian strike-slip tectonics in the Gaspé Region, Québec Appalachians; Canadian Journal of Earth Sciences, v. 26, p. 1764-1777.

Malo, M. and Bourque, P.-A.

1993: Timing of the deformation events from Late Ordovician to Mid-Devonian in the Gaspé Peninsula; in The Acadian Orogeny: Recent studies in New England, Maritime Canada, and the Autochthonous Foreland, (ed.) D.C. Roy and J.W. Skehan; Geological Society of America, Special Paper 275, p. 101-122.

Malo, M. et Moritz, R.

1991: Géologie et métallogénie de la faille du Grand Pabos, région de Raudin-Weir; Ministère de l'Énergie et des Ressources, Québec; MB 91-03, 47 p.

Malo, M., St-Julien, P., Bourque, P.-A., and Kirkwood, D.

1988: Tectonic evolution of the Acadian Orogeny in the Gaspé Basin, northern Appalachians (abstract); Geological Association of Canada, Annual Meeting, St. John's, Newfoundland, Program with Abstracts; v. 13, p. A78.

Marillier, F., Keen, C.E., Stockmal, G.S., Quinlan, G., Williams, H., Colman- Sadd, S.P., and O'Brien, S.J.

1989: Crustal structure and surface zonation of the Canadian Appalachians: implications of deep seismic reflection data; Canadian Journal of Earth Sciences, v. 26, p. 305-321.

Marten, B.E.

1971: Stratigraphy of volcanic rocks in the Western Arm area of the central Newfoundland Appalachians: a Newfoundland decade; Geological Association of Canada, Proceedings, v. 24, no. 1, p. 73-84.

Mason, G.D.

1971: A stratigraphical and paleoenvironmental study of the Upper Gaspé limestone and Lower Gaspé sandstone groups (Lower Devonian) of eastern Gaspé peninsula, Québec; Ph.D. thesis, Carleton University, Ottawa, Ontario.

Mawer, C.K. and White, J.C.

1987: Sense of displacement on the Cobequid-Chedabucto fault system, Nova Scotia, Canada; Canadian Journal of Earth Sciences v. 24, p. 217-223.

McCann, A.M. and Kennedy, M.J.

1974: A probable glacio-marine deposit of Late Ordovician-Early Silurian age from the north central Newfoundland Appalachian belt; Geological Magazine, v. 111, p. 549-564.

McCutcheon, S.R.

1981: Revised stratigraphy of the Long Reach area, southern New Brunswick: evidence for major northwestward-directed Acadian thrusting; Canadian Journal of Earth Sciences, v. 18, p. 646-656.

McCutcheon, S.R. and Boucot, A.J.

1984: A new Lower Silurian fossil locality in the northeastern Mascarene-Nerepis Belt, southern New Brunswick; Maritime Sediments and Atlantic Geology, v. 20, p. 121-126.

McCutcheon, S.R. and Robinson, P.T.

1987: Geological constraints on the genesis of the Maritimes Basin, Atlantic Canada; in Sedimentary basins and basin-forming mechanisms, (ed.) C. Beaumont and A.J. Tankard; Canadian Society of Petroleum Geologists, Memoir 12, p. 287-297.

McCutcheon, S.R. and Ruitenberg, A.A.

1987: Geology and mineral deposits of the Annidale-Nerepis area; New Brunswick Department of Natural Resources, Geological Surveys Branch, Memoir 2, 141 p.

McCutcheon, S.R., Pearce, G.A., and Bagnell, B.

1976: Preliminary geological map of M-29, scale 1"=1/4 mile; New Brunswick Department of Natural Resources, Minerals Resources Division, Plate 76-56.

McGerrigle, H.W.

1943: Reconnaissance geological survey of south central and western Gaspé peninsula; Québec Department of Mines, DP-479, 106 p.

1946: A revision of the Gaspé Devonian; Transaction of the Royal Society of Canada, Ser. 3, section IV, v. XL, p. 41-54.

1950: The geology of eastern Gaspé; Québec Department of Mines, Geological Report 35, 168 p.

1959: Madeleine River area; Québec Department of Mines, Geological Report 77, 50 p.

McGregor, D.C.

1981: Geological Survey of Canada, Internal Palentogical Report F1-8-1981-DCM.

McKenzie, C.B.

1986: Geology and mineralization of the Chetwynd deposit, southwestern Newfoundland, Canada; in Proceedings of Gold '86, (ed.) A.J. MacDonald; An International Suymposium on the Geology of Gold, Toronto, 1986, p. 137-148.

McKerrow, W.S.

1979: Ordovician and Silurian changes in sea level; Journal of the Geological Society of London, v. 136, p. 137-145.

McKerrow, W.S. and Cocks, L.R.M.

1977: The location of the Iapetus Ocean suture in Newfoundland; Canadian Journal of Earth Sciences, v. 14, p. 488-495.

1978: A lower Paleozoic trench-fill sequence, New World Island, Newfoundland; Geological Society of America, Bulletin, v. 89, p. 1121-1132.

1981: Stratigraphy of eastern Bay of Exploits, Newfoundland; Canadian Journal of Earth Sciences, v. 18, p. 751-764.

McKerrow, W.S. and Ziegler, A.M.

1971: The Lower Silurian paleogeography of New Brunswick and adjacent areas; Journal of Geology, v. 79, p. 635-646.

McKerrow, W.S., Leggett, J.K., and Eales, M.H.

1977: Imbricate thrust model of the Southern Uplands of Scotland; Nature, London, v. 267, p. 237-239.

McLeod, M.J

1979: The geology of Campobello Island, southwestern New Brunswick; M.Sc. thesis, University of New Brunswick, Fredericton, New Brunswick, 181 p.

1990: Geology, geochemistry, and related mineral deposits of the Saint George Batholith, Charlotte, Queens and Kings counties, New Brunswick; New Brunswick Department of Natural Resources and Energy, Mineral Resources, Mineral Resource Report 5, 169 p.

McLeod, M.J. and Rast, N.

1988: Correlations and fault systematics in the Passamaquoddy Bay area, southwestern New Brunswick; Maritime Sediments and Atlantic Geology, v. 24, p. 289-300.

McLeod, M.J., Johnson, S.C., and Ruitenberg, A.A.

1989: Compilation and correlation of southern New Brunswick geology, Charlotte, Queens and Kings counties; in 14th Annual Review of Activities, Project Resumes, 1989, (ed.) S.A. Abbott; New Brunswick Department of Natural Resources and Energy, Mineral Resources Division, Information Circular 89-9, p. 131-141.

1990a: Compilation and correlation of southern New Brunswick geology, Charlotte, Queens Kings and Sunbury counties; in 15th Annual Review of Activities, Project Resumes 1990, (ed.) S.A. Abbott; New Brunswick Department of Natural Resources and Energy, Minerals and Energy Division, Information Circular 90-2, p. 137-147.

1990b: Compilation and correlation of southern New Brunswick; in Project Summaries for 1989, 14th Annual Review of Activities, (ed.) S.A. Abbott; New Brunswick Department of Natural Resources and Energy, Minerals and Energy Division, Information Circular 89-2, p. 151-159.

1991a: Geological compilation of the Hampstead map area (21G/9), Sussex map area (21H/12), and Codys map area (21H/13), southern New Brunswick; New Brunswick Department of Natural Resources and Energy, Mineral Resources, Plates 90-152, 90-153, 90-154.

McLeod, M.J., Ruitenberg, A.A., and Krogh, T.E.

1991b: A previously unrecognized lower Ordovician sequence in southern New Brunswick: evidence for multistage development of Iapetus? (abstract); Atlantic Geoscience Society, Program with Abstracts, Atlantic Geology, v. 27, p. 157.

Mercer, B., Strong, D.F., Wilton, H.D.C., and Gibbons, D.

1985: The Kings Point volcano-plutonic complex, Western Newfoundland; in Current Research, Part A; Geological Survey of Canada, Paper 85-1A, p. 737-741.

Miller, B.V., Nance, R.D., and Murphy, J.B.

1989: Preliminary kinematic analysis of the Rockland Brook Fault, Cobequid Highlands, Nova Scotia; in Current Research, Part B; Geological Survey of Canada, Paper 89-1B, p. 7-14.

Miyashiro, A

1974: Volcanic rock series in island arcs and active continental margins, American Journal of Science, v. 274, p. 321-355.

More, E.B.

1982: A Lower Carboniferous sedimentary-volcanic succession, north Baddeck River, Nova Scotia; B.Sc. thesis (Hons.), Dalhousie University, Halifax, Nova Scotia, 72 p.

Morin, R. et Simard, M.

1987: Géologie des régions de Sirois et Raudin, Gaspésie; Ministère de l'Énergie et des Ressources, Québec, ET-86-06, 69 p.

Muecke, G.K.

1973: Meguma Group metamorphism-a progress report; Geological Association of Canada, Newfoundland Section.

Mullins, J.
1961: Geology of the Noel Paul's Brook area, central Newfoundland; M.Sc. thesis, Memorial University of Newfoundland, St. John's, Newfoundland.

Murphy, J.B.
1987a: The stratigraphy and depositional environment of Upper Ordovician to Lower Devonian rocks in the Antigonish Highlands, Nova Scotia; Maritime Sediments and Atlantic Geology, v. 23, p. 63-75.
1987b: Petrology of Upper Ordovician-Lower Silurian rocks in the Antigonish Highlands, Nova Scotia; Canadian Journal of Earth Sciences, v. 24, p. 752-759.
1988: Late Precambrian to Late Devonian mafic magmatism in the Antigonish Highlands, Nova Scotia: multistage melting of a hydrated mantle; Canadian Journal of Earth Sciences, v. 25, p. 473-485.

Murphy, J.B., Keppie, J.D., Nance, R.D., and Dostal, J.
1989: Reassessment of terranes in the Avalon composite terrane of Atlantic Canada (abstract); Geological Society of America, Abstracts with Programs, v. 21, no. 2, 54 p.

Murphy, J.B., Pe-Piper, G., Nance, R.D., and Turner, D.
1988: A preliminary report on the geology of the eastern Cobequid Highlands, Nova Scotia; in Current Research, Part B; Geological Survey of Canada, Paper 88-1B, p. 99-107.

Murray, A. and Howley, J.P.
1881: Report of the Geological Survey of Newfoundland for 1864-1880; Edward Stanford, London, 536 p.

Neale, E.R.W.
1957: Ambiguous intrusive relationships of the Betts Cove-Tilt Cove serpentinite belt, Newfoundland; Geological Association of Canada, Proceedings, v. 9, p. 95-107.

Neale, E.R.W. and Kennedy, M.J.
1967: Relationships of the Fleur de Lys Group to younger groups of the Burlington Peninsula, Newfoundland; in Geology of the Atlantic Region, (ed.) E.R.W. Neale and H. Williams; Geological Association of Canada, Special Paper, no. 4, p. 139-169.

Neale, E.R.W. and Nash, W.A.
1963: Sandy Lake (east half) map-area, Newfoundland; Geological Survey of Canada, Paper 62-28, 40 p.

Neale, E.R.W., Kean, B.F., and Upadhyay, H.D.
1975: Post-ophiolite unconformity, Tilt Cove-Betts Cove area, Newfoundland; Canadian Journal of Earth Science, v. 12, p. 880-886.

Nelson, K.D.
1981: Melange development in the Boones Point Complex, north-central Newfoundland; Canadian Journal of Earth Sciences, v. 18, p. 433-442.

Nelson, K.D. and Kidd, W.S.F.
1979: A lower Paleozoic trench-fill sequence, New World Island, Newfoundland. Discussion; Geological Society of America Bulletin, v. 90, p. 985-986.

Noble, J.P.A.
1976: Silurian stratigraphy and paleogeography, Pointe Verte area, New Brunswick, Canada; Canadian Journal of Earth Sciences, v. 13, p. 537-546.
1985: Occurrence and significance of Late Silurian reefs in New Brunswick, Canada; Canadian Journal of Earth Sciences, v. 22, p. 1518-1529.

Northrop, S.A.
1939: Paleontology and stratigraphy of the Silurian rocks of the Port-Daniel-Black Cape region, Gaspé; Geological Society of America, Special Paper 21, 302 p.

Nowlan, G.S.
1981: Late Ordovician-Early Silurian conodont biostratigraphy of the Gaspé Peninsula (a preliminary report); in Stratigraphy and Paleontology, (ed.) P.J. Lespérance; Ordovician-Silurian Boundary Working Group, Field Meeting Anticosti-Gaspé, Québec, Subcommission on Silurian Stratigraphy, v. 2, p. 257-291.
1982: Report on 19 samples collected for conodonts analysis from a single section in northern New Brunswick, NTS 21P/13; Geological Survey of Canada, Internal Paleontological Report 09-GSN-1982.
1983: Early Silurian conodonts of eastern Canada; Fossils and Strata, v. 15, p. 95-110.

Nowlan, G.S. and Barnes, C.R.
1987: Thermal maturation of Paleozoic strata in eastern Canada from conodont colour alteration index (CAI) data with implication for burial history, tectonic evolution, hotspot tracks and mineral and hydrocarbon exploration; Geological Survey of Canada, Bulletin 367, 47 p.

Nowlan, G.S. and Thurlow, J.G.
1984: Middle Ordovician conodonts from the Buchans Group, central Newfoundland, and their significance for regional stratigraphy of the central volcanic belt; Canadian Journal of Earth Sciences, v. 21, p. 284-296.

O'Brien, B.H.
1986: Preliminary report on the geology of the Mahone Bay area, Nova Scotia; in Current Research, Part A; Geological Survey of Canada, Paper 86-1A, p. 439-444.
1987: The lithostratigraphy and structure of the Grand Bruit-Cinq Cerf area (parts of NTS areas 11O/9 and 11O/16), southwestern Newfoundland; in Current Research; Newfoundland Department of Mines and Energy, Mineral Development Division, Report 88-1, p. 311-334.
1988: Relationships of phyllite, schist and gneiss in the La Poile Bay-Roti Bay area (parts of 11O/9 and 11O/16), southwestern Newfoundland; in Current Research; Newfoundland Department of Mines, Mineral Development Division, Report 88-1, p. 109-125.
1989: Summary of the geology between La Poile Bay and Couteau Bay (11O/9 and 11O/16), southwestern Newfoundland; in Current Research; Newfoundland Department of Mines, Geological Survey of Newfoundland, Report 89-1, p. 105-119.

O'Brien, B.H. and O'Brien, S.J.
1990: Re-investigation and re-interpretation of the Silurian La Poile Group of southwestern Newfoundland; in Current Research (1990), Newfoundland Department of Mines and Energy, Geological Survey Branch, Report 90-1, p. 305-316.

O'Brien, B.H., O'Brien, S.J., and Dunning, G.R.
1991: Silurian cover, late Precambrian-early Ordovician basement, and the chronology of Silurian orogenesis in the Hermitage Flexure (Newfoundland Appalachians); American Journal of Science, v. 291, p. 760-799.

O'Brien, F.H.C.
1975: The stratigraphy and paleontology of the Clam Bank Formation and the upper part of the Long Point Formation of the Port au Port Peninsula on the west coast of Newfoundland; M.Sc. thesis, Memorial University of Newfoundland, St. John's, Newfoundland, 156 p.

O'Brien, F.H.C. and Szybinski, Z.A.
1989: Conodont faunas from the Catchers Pond and Cutwell groups, central Newfoundland; in Current Research; Geological Survey of Newfoundland, Report 89-1, p. 121-125.

O'Brien, S.J.,
1982: Peter Snout (east half), Newfoundland; Newfoundland Department of Mines and Energy, Mineral Development Division, Map 82-58.
1983: Geology of the eastern half of the Peter Snout map area (11P/13E), Newfoundland; in Current Research; Newfoundland Department of Mines and Energy, Mineral Development Divison, Report 83-1, p. 57-67.
1987: Geology of the Eastport (west half) map area, Bonavista Bay, Newfoundland; in Current Research, (ed.) R.F. Blackwood and R.V. Gibbons; Newfoundland Department of Mines and Energy, Mineral Development Divison, Report 87-1, p. 257-270.

O'Brien, S.J., Dickson, W.L., and Blackwood, R.F.
1986: Geology of the central portion of the Hermitage Flexure area, Newfoundland; in Current Research, (ed.) R.F. Blackwood, D.G. Walsh, and R.V. Gibbons; Newfoundland Department of Mines and Energy, Mineral Development Division, Report 86-1, p.189-208.

O'Brien, S.J., Strong, D.F., and King, A.F.
1977: The geology of the Grand Bank (1M/4) and Lamaline (1L/13) map areas, Burin Peninsula, Newfoundland; Newfoundland Department of Mines and Energy, Mineral Development Division, Report 77-7, 16 p.
in press: The Avalon Zone in Newfoundland; in Pan African Terranes of the North Atlantic Borderlands.

Ollerenshaw, N.C.
1967: Région de Cuoq-Langis (comtés de Matapédia et de Matane); Ministère des Richesses naturelles, Québec, RG 121, 230 p.

O'Neill, P.P. and Blackwood, F.
1989: A proposal for revised stratigraphic nomenclature of the Gander and Davidsville groups and the Gander River Ultrabasic Belt of northeastern Newfoundland; in Current Research; Newfoundland Department of Mines, Mineral Development Division, Report 88-1, p. 165-176.

O'Neill, P.P. and Lux, D.
1989: Tectonothermal history and $^{40}Ar/^{39}Ar$ geochronology of northeastern Gander Zone, Weir's Pond area (2E/1); in Current Research; Newfoundland Department of Mines, Geological Survey of Newfoundland, Report 89-1, p. 131-139.

Osberg, P.H.
1968: Stratigraphy, structural geology, and metamorphism of the Waterville-Vassalboro area, Maine; Maine Geological Survey, Bulletin, v. 20, 64 p.
1978: Synthesis of the geology of the northwestern Appalachians, United States; in Calaedonian-Appalachian Orogen of the North Atlantic Region, IGCP Project 27: Caledonide Orogen; Geological Survey of Canada, Paper 78-13, p. 137-147.
1988: Geologic relations within the Shale-Wacke Sequence in south-central Maine; in Structure and Stratigraphy, (ed.) R.D. Taylor and R.G. Marvinney; Maine Geological Survey, Department of Conservation, Augusta, Maine, Studies in Maine Geology, v. 1, p. 51-73.

Pageau, Y.
1968: Nouvelle faune ichthyologique du Dévonien Moyen dans les Grès de Gaspé, Québec; Naturaliste Canadien, v. 95, p. 1459-1497.

Paris, F. and Robardet, M.
1990: Early Paleozoic palaeobiogeography of the Variscan regions; Tectonophysics, v. 177, p. 193-213.

Patrick, T.O.H.
1956: Comfort Cove, Newfoundland; Geological Survey of Canada, Paper 55-31.

Pavlides, L.
1968: Stratigraphic and facies relationship of the Carys Mills Formation northeast Maine adjoining New Brunswick; United States Geological Survey, Bulletin, v. 1264, 44 p.

Pavlides, L., Mencher, E., Naylor, R.S., and Boucot, A.J.
1964: Outline of the stratigraphic and tectonic features of northeastern Maine; in Geological Survey Research 1964; United States Geological Survey, Professional Paper 501-C, p. C28-C38.

Payette, C. and Martin, R.F.
1988: The Welsford anorogenic igneous complex, southern New Brunswick: rift-related Acadian magmatism; Geological Survey of Canada, Open File 1727, 346 p.

Pearce, J.A. and Cann, J.R.
1973: Tectonnic setting of basic volcanic rocks determined using trace element analysis; Earth and Planetary Science Letters, v. 19, p. 290-300.

Pearce, J.A., Harris, N.W.B., and Tindle, A.G.
1984: Trace element determination diagrams for the tectonic interpretation of granitic rocks; Journal of Petrology, v. 25, p. 956-983.

Pe-Piper, G. and Loncarevic, B.D.
1989: Offshore continuation of Meguma Terrane, southwestern Nova Scotia; Canadian Journal of Earth Sciences, v. 26, p. 176-191.

Pe-Piper, G. and Turner, D.S.
1988: History of movement on the Kirkhill fault, Minas geofracture: evidence from the Hanna Farm pluton; Maritime Sediments and Atlantic Geology, v. 24, p. 171-183.

Peters, H.R.
1953: Geology of the Stony Lake area, central Newfoundalnd; M.Sc. thesis, Dalhousie University, Halifax, Nova Scotia, 39 p.

Phinney, W.E.
1956: Structural relationships around the southern extension of the Mabou Highlands, Inverness Coutty, Cape Breton Island, Nova Scotia; M.Sc. thesis, Massachusetts Institute of Technology, U.S.A. 75 p.

Piasecki, M.A.J.
1989: A new look at the Port aux Basques region; Geological Survey of Canada, unpublished report, 22 p. Contract 23233-8-001/01-SZ.

Piasecki, M.A.J., Williams, H., and Colman-Sadd, S.P.
1990: Tectonic relationships along the Meelpaeg, Burgeo and Burlington Lithoprobe transects in Newfoundland; in Current Research; Newfoundland Department of Mines, Mineral Development Division, Report 90-1.

Pickerill, R.K.
1976: Significance of a new fossil locality containing a Salopina community in the Waweig Formation (Silurian-Uppermost Ludlow/Pridoli) of southwest New Brunswick; Canadian Journal of Earth Sciences, v. 13, p. 1329-1331.
1980: Phanerozoic flysch trace fossil diversity-observations based on an Ordovician flysch ichnofauna from the Aroostook-Matapédia Carbonate Belt of northern New Brunswick; Canadian Journal of Earth Sciences, v. 17, p. 1259-1270.

Pickerill, R.K. and Hurst, J.M.
1983: Sedimentary facies, depositional environments, and faunal associations of the lower Llandovery (Silurian) Beechill Cove Formation, Arisaig, Nova Scotia; Canadian Journal of Earth Sciences, v. 20, p. 1761-1779.

Pickerill, R.K. and Pajari, G.E.
1976: The Eastport Formation (Lower Devonian) in the northern Passamaquoddy Bay area, southwest New Brunswick; Canadian Journal of Earth Sciences, v. 13, p. 266-270.

Pickerill, R.K., Pajari, G.E., and Dickson, W.L.
1978: Geology of the Lower Devonian rocks of Passamaquoddy Bay, southwestern New Brunswick; in Guidebook for Field Trips in Southeastern Maine and Southwestern New Brunswick, (ed.) A. Ludman; 70th Annual Meeting, New England Intercollegiate Geological Conference, Trip A-3, p. 38-56.

Pickering, K.
1987a: Deep-marine foreland basin and forearc sedimentation: a comparitive study from the Lower Paleozoic northern Appalachians, Quebec and Newfoundland; in Marine Clastic Sedimentology, (ed.) J.K. Leggett and G.G. Zuffa; Graham and Trotman, p. 190-211.
1987b: Wet-sediment deformation in the Upper Ordovician Point Leamington Formation: an active thrust-imbricate system during sedimentation, Notre Dame Bay, north-central Newfoundland; in Deformation of Sediments and Sedimentary Rocks, (ed.) M.E. Jones and R.M.F. Preston; Geological Society, Special Publication 29, p. 213-239.

Pickering, K.T., Bassett, M.G., and Siveter, D.J.
1988: Late Ordovician-early Silurian destruction of the Iapetus Ocean: Newfoundland, British Isles and Scandinavia- a discussion; Transactions of the Royal Society of Edinburgh: Earth Sciences, v. 79, p. 361-382.

Platt, J.P.
1984: Secondary cleavages in ductile shear zones; Journal of Structural Geology, v. 6, p. 439-442.

Plint, H.E. and Jamieson, R.A.
1989: Microstructure, metamorphism, amd tectonics of the western Cape Breton Highlands, Nova Scotia; Journal of Metamorphic Geology, v. 7, p. 407-424.

Poole, W.H.
1957: Geology of Burtts Corner (west half); Geological Survey of Canada, Map 7-1957 (with marginal notes).
1963: Geology, Hayesville, New Brunswick; Geological Survey of Canada, Map 6-1963.
1976: Plate tectonic evolution of the Canadian Appalachian region; in Report of Activities, Part B; Geological Survey of Canada, Paper 76-1B, p. 113-126.

Poole, W.H., Stanford, B.V., Williams, H., and Kelley, D.G.
1970: Geology of Southeastern Canada; in Geology and Economic Minerals of Canada, (ed.) R.J.W. Douglas; Geological Survey of Canada, Economic Geology Report 1, p. 227-304.

Pringle, I.R.
1978: Rb-Sr ages of silicic igneous rocks and deformation, Burlington Peninsula, Newfoundland; Canadaian Journal of Earth Sciences, v. 15, p. 293-300.

Raeside, R.P. and Barr, S.M.
1990: Geology and tectonic development of the Bras d'Or suspect terrane, Cape Breton Island, Nova Scotia; Canadian Journal of Earth Sciences, v. 27, p. 1371-1381.
1992: Geology of the northern and eastern Cape Breton Highlands, Cape Breton Island, Nova Scotia; Geological Survey of Canada, Paper 89-14 , 39 p.

Rast, N.
1980: The geology and deformation history of the southern part of the Matapédia zone and its relationship to the Miramichi zone and Canterbury basin; in The geology of Northeastern Maine and neighboring New Brunswick, (ed.) D.C. Roy and R.S. Naylor; New-England Intercollegiate Geological Conference, 72th Annual Meeting, Presque Isle, Maine, Guidebook, p. 191-201.

Rast, N. and Stringer, P.
1974: Recent advances and the interpretation of geological structure of New Brunswick; Geoscience Canada, v. 1, no. 4, p. 15-25.
1980: A geotraverse across a deformed Ordovician ophiolite and its Silurian cover, northern New Brunswick, Canada; Tectonophysics, v. 69, p. 221-245.

Rast, N., Kennedy, M.J., and Blackwood, R.F.
1976: Comparison of some tectonostratigraphic zones in the Appalachians of Newfoundland and New Brunswick; Canadian Journal of Earth Sciences, v. 13, p. 868-875.

Reusch, D.N.
1983: The New World Island Complex and its relationships to nearby formations, north central Newfoundland; M.Sc. thesis, Memorial University of Newfoundland, St. John's, Newfoundland, 248 p.
1987: Silurian stratigraphy and melanges, New World Island, north central Newfoundland; in Centennial Field Guide, v. 5, (ed.) D.C. Roy; Geological Society of America, p. 463-466.

Reynolds, P.H. and Muecke, G.K.
1978: Age studies on slates: applicability of the $^{40}Ar/^{39}Ar$ stepwise outgassing method: Earth and Planetary Science Letters, v. 40, p. 111-118.

Reynolds, P.H., Zentilli, M., and Muecke, G.K.
1981: K-Ar and $^{40}Ar/^{39}Ar$ geochronology of granitoid rocks from southern Nova Scotia: its bearing on the geological evolution of the Meguma Zone of the Appalachians; Canadian Journal of Earth Science, v. 18, p. 386-394.

Reynolds, P.H., Jamieson, R.A., Barr, S.M., and Raeside, R.P.
1989: A $^{40}Ar/^{39}Ar$ dating study in the Cape Breton Highlands, Nova Scotia: thermal histories and tectonic implications; Canadian Journal of Earth Sciences, v. 26, p. 2081-2091.

Riley, G.C.
1957: Red Indian Lake (west half), Newfoundland; Geological Survey of Canada, Map 8-1957.
1959: Geology, Burgeo-Ramea (11P, west half), Newfoundland; Geological Survey of Canada, Map 22-1959.
1962: Stephenville map-area, Newfoundland; Geological Survey of Canada, Memoir 323, 72 p.

Riva, J. and Malo, M.
1988: Age and correlation of the Honorat Group, southern Gaspé Peninsula; Canadian Journal of Earth Sciences, v. 25, p. 1618-1628.

Robardet, M., Paris, F., and Racheboeuff, P.R.
1990: Palaeogeographic evolution of southwestern Europe during Early Palaeozoic times; in Palaeozoic Palaeogeography and Biogeography, (ed.) W.S. McKerrow and C.R. Scotese; Geological Society of London, Memoir 12, p. 411-420.

Rodgers, J.
1965: Long Point and Clam Bank Formations, Western Newfoundland; Geological Association of Canada, Proceedings, v. 16, p. 83-94.
1970: The tectonics of the Appalachians; Wiley-Interscience, New York, 271 p.

Rodgers, J. and Neale, E.R.W.
1963: Possible "Taconic" klippen in western Newfoundland; American Journal of Science, v. 261, p. 713-730.

Rouillard, M.
1984a: Les Calcaires Supérieurs de Gaspé dans les cantons de Lesseps, Lemieux et Richard; Ministère de l'Énergie et des Ressources, Québec, DP 84-14.
1984b: Stratigraphie des Calcaires Supérieurs de Gaspé entre Murdochville et le ruisseau Lesseps; Ministère de l'Énergie et des Ressources, Québec, DP 84-12.
1986: Les Calcaires Supérieurs de Gaspé (Dévonien inférieur), Gaspésie; Ministère de l'Énergie et des Ressources, Québec, MB 86-15, 94 p.

Roy, D.C.
1989: The Depot Mountain Formation: transition from syn-to post-Taconian basin along the Baie Verte-Brompton Line in northwestern Maine; Studies in Maine Geology; Maine Geological Survey, v. 2, p. 85-99.

Ruffman, A. and Woodside, J.
1970: The Odd-Twins magnetic anomaly and its possible relationships to the Humber Arm Klippe of western Newfoundland, Canada; Canadian Journal of Earth Sciences, v. 7, p. 326-337.

Ruitenberg, A.A.
1967: Stratigraphy, structure and metallization Piskahegan-Rolling Dam area, northern Appalachians, New Brunswick, Canada; Leidse Geologische Mededelingen, v. 40, p. 79-120.
1968: Geology and mineral deposits, Passamaquoddy Bay area; New Brunswick Department of Natural Resources, Mineral Resources Branch, Report of Investigation 7, 47 p.

Ruitenberg, A.A. and Ludman, A.
1978: Stratigraphy and tectonic setting of Early Paleozoic sedimentary rocks of the Wirral-Big Lake area, southwestern New Brunswick and southeastern Maine; Canadian Journal of Earth Sciences, v. 15, p. 22-32.

Ruitenberg, A.A. and McCutcheon, S.R.
1982: Acadian and Hercynian structural evolution of southern New Brunswick; in Major Structural Zones and Faults of the Northern Appalachians, (ed.) P. St-Julien and J. Béland; Geological Association of Canada, Special Paper 24, IGCP Project 27, p. 131-148.

Ruitenberg, A.A., Fyffe, L.R., and McCutcheon, S.R.
1977: Evolution of pre-Carboniferous tectonostratigraphic zones in the New Brunswick Appalachians; Geoscience Canada, v. 4, p. 171-181.

Ruitenberg, A.A., Fyffe, L.R., McCutcheon, S.R., St. Peter, C.J., Irrinki, R.R., and Venugopal, D.V.
1977: Evolution of Pre-Carboniferous tectonostratigraphic zones in the New Brunswick Appalachians; Geoscience Canada, v. 4, no. 4, p. 171-181.

Russell, L.S.
1946: Stratigraphy of the Gaspé Limestone Series, Forillon Peninsula, Cap-des-Rosiers Township, County of Gaspé South; Québec Department of Natural Resources, DPV-347, 96 p.

Rust, B.R.
1976: Stratigraphic relationships of the Malbaie Formation (Devonian), Gaspé, Québec; Canadian Journal of Earth Sciences, v. 13, p. 1556-1559.
1981: Alluvial deposits and tectonic style: Devonian and Carboniferous successions in eastern Gaspé; in Sedimentology and Tectonics in Alluvial Basins, (ed.) A.D. Miall; Geological Association of Canada, Special Paper 23, p. 49-76.
1982: Continental Devonian and Carboniferous sedimentation of eastern Gaspé Peninsula; in Paleozoic Continental Margin Sedimentation in the Québec Appalachians, (ed.) R.E. Hesse, G.V. Middleton, and B.R. Rust; International Association of Sedimentologists, 11th International Congress, Hamilton, Ontario, Field Excursion 7B, Guidebook, p. 107-125.

Sarkar, P.K.
1978: Petrology and geochemistry of the White Rock metavolcanic suite, Yarmouth, Nova Scotia; Ph.D. thesis, Dalhousie University, Halifax, Nova Scotia, 350 p.

Savard, M. and Bourque, P.-A.
1989: Diagenetic evolution of a Late Silurian reef platform, Gaspé Basin, Quebec, based on cathodoluminescence petrography; Canadian Journal of Earth Sciences, v. 26, p. 791-806.

Schenk, P.E.
1972: Possible Late Ordovician glaciation of Nova Scotia; Canadian Journal of Earth Sciences, v. 9, p. 95-107.
1981: The Meguma Zone of Nova Scotia - a remnant of western Europe, South America, or Africa?; in Geology of North Atlantic Borderlands, (ed.) J.M. Kerr, A.J. Ferguson, and L.C. Machan; Canadian Society of Petroleum Geologists, Memoir, v. 7, p. 119-148.
1991: Events and sea-level changes on Gondwana's margin: the Meguma Zone (Cambrian to Devonian) of Nova Scotia, Canada; Geological Society of America Bulletin, v. 103, p. 512-521.

Schillereff, H.S. and Williams, H.
1979: Geology of the Stephenville map-area, Newfoundland; in Current Research, Part A; Geological Survey of Canada, Paper 79-1A, p. 327-332.

Schroeter, T.G.
1971: Geology of the Nippers Harbour area, Newfoundland; M.Sc. thesis, University of Western Ontario, London, Ontario, 88 p.

Schuchert, C. and Cooper, G.A.
1930: Upper Ordovician and Lower Devonian stratigraphy and paleontology of Percé, Québec; American Journal of Science, v. 20, p. 161-176.

Schuchert, C. and Dart, J.D.
1926: Stratigraphy of the Port-Daniel-Gascons area of southeastern Québec; Geological Survey of Canada, Bulletin 44, Geological Series 46, p. 35-38 and 116-121.

Schuchert, C. and Dunbar, C.O.
1934: Stratigraphy of western Newfoundland; Geological Society of America, Memoir 1, 123 p.

Schultze, H.P. and Arsenault, M.
1985: The Panderichthyid fish Elpistostege: a close relative of tetrapods?; Palaeontology, v. 28, p. 293-309.

Seilacher, A.
1964: Biogenic sedimentary structures; in Approaches to Paleoecology, (ed.) J. Imbrie and N.D. Newells; John Wiley and Sons Incorporated, New York, p. 296-316.
1967: Bathymetry of trace fossils; Marine Geology, v. 5, p. 413-428.

Sheehan, P.M. and Lespérance, P.J.
1981: Brachiopods from the White Head Formation (Late Ordovician-Early Silurian) of the Percé region, Québec, Canada; in Stratigraphy and Paleontology, (ed.) P.J. Lespérance; Ordovician-Silurian Boundary Working Group, Field Meeting Anticosti-Gaspé, Québec, Subcommission on Silurian Stratigraphy, v. 2, p. 247-256.

Sikander, A.H.
1975: Geology for hydrocarbon potential of the Berry Mountain Syncline, central Gaspé (Matane, Matapédia, Gaspé W and Bonaventure counties); Québec Department of Natural Resources, DP-376, 119 p.

Simard, M.
1986: Géologie et évaluation du potentiel minéral de la région de Carleton; Ministère de l'Énergie et des Ressources, Québec, ET 84-11, 27 p.

Skevington, D.
1978: Latitudinal surface water temperature gradients and Ordovician fauna provinces; 25th International Geological Congress, Sydney, Alcheringa 2, p. 21-26.

Skidmore, W.B.
1958:   Honorat-West area; Québec Department of Mines, Preliminary Report 366, 7 p.
1965a: Honorat-Reboul area; Québec Department of Natural Resources, Geological Report 107, 30 p.
1965b: Gastonguay-Mourier area; Québec Department of Natural Resources, Geological Report 105, 74 p.

Skidmore, W.B. and Lespérance, P.J.
1981:   Percé area: The White Head Formation; in Stratigraphy and Paleontology, (ed.) P.J. Lesperance; Ordovician-Silurian Boundary Working Group, Field Meeting Anticosti-Gaspé, Québec, Subcommission on Silurian Stratigraphy, v. 2, p. 31-40.

Skidmore, W.B. and McGerrigle, H.W.
1967:   Geologic map, Gaspé Peninsula, Quebec; Quebec Department of Natural Resources, Map 1642.

Slivitzky, A. and St-Julien, P.
1987:   Compilation géologique de la région de l'Estrie-Beauce; Ministère de l'Énergie et des Ressources, Québec; MM 85-04, 40 p. (1 250 000).

Sloss, L.L.
1963:   Sequences in the cratonic interior of north America; Geological Society of America Bulletin, v. 74, p. 93-114.

Smith, B.L. and White, D.E.
1954:   Geology of the Rencontre East area; Geological Survey of Canada, Open File, 46 p.

Smith, J.C.
1966:   Geology of southwestern New Brunswick; in Geology of Parts of the Atlantic Provinces; Geological Association of Canada/Mineralogical Association of Canada, Guidebook, p. 1-18.

Smith, M.M.
1975:   Geology of an area around Mose Ambross-English Harbour West, south coast, Newfoundland; B.Sc. thesis, Memorial University of Newfoundland, St. John's, Newfoundland, 38 p.

Smith, P.K.
1979:   A note on the geology of Silurian fossil occurrences south and west of Kenzieville, Antigonish Highlands; Nova Scotia Department of Mines, Report 79-1, p. 89-94.

Smith, P.K. and Macdonald, A.S.
1981:   The Fisset Brook Formation at Lowland Cove, Inverness County, Nova Scotia; Nova Scotia Department of Mines and Energy, Paper 81-1, 18 p.

Smitheringale, W.G.
1960:   Geology of the Nictaux-Torbrook map-area, Annapolis and Kings counties, Nova Scotia; Geological Survey of Canada, Paper 60-13, 32 p.
1973:   Geology of parts of Digby, Bridgetown, and Gaspereau map-areas, Nova Scotia; Geological Survey of Canada, Memoir 375 , 78 p.

Smyth, W.R.
1981:   Lower Paleozoic geology of southwestern White Bay; in Current Research, (ed.) C.F. O'Driscol and R.V. Gibbons; Newfoundland Department of Mines and Energy, Mineral Development Division, Report 81-1, p. 70-79.

Smyth, W.R. and Schillereff, H.S.
1982:   The pre-Carboniferous geology of southwest White Bay; in Current Research, (ed.) C.F. O'Driscol and R.V. Gibbons; Newfoundland Department of Mines and Energy, Mineral Development Division, Report 82-1, p. 78-98.

Snelgrove, A.K.
1931:   Geology and ore deposit of Betts Cove-Tilt Cove area, Notre Dame Bay; Canadian Mining and Metallurgical Bulletin, v. 228, p. 447-519.
1935:   Geology of the gold deposits of Newfoundland; Newfoundland Department of Natural Resources, Geology Section, Bulletin, v. 2, 46 p.

Soper, N.J. and Woodcock, N.H.
1990:   Silurian collision and sediment dispersal patterns in southern Britain; Geological Magazine, p. 527-542.

Spariosu, D.J., Kent, D.V., and Keppie, J.D.
1984:   Late Paleozoic motions of the Meguma terrane: new paleomagnetic evidence; American Geophysical Union, Geodynamics Series, v. 12, p. 82-98.

St-Julien, P., Slivitsky, A., and Feininger, T.
1983:   A deep structural profile across the Appalachians of southern Quebec; in Contributions to the Tectonics and Geophysics of Mountain Chains, (ed.) R.D. Hatcher, H. Williams, and I. Zietz; Geological Society of America, Memoir 158, p. 103-111.

St. Peter, C.J.
1977:   Geology of parts of Restigouche, Victoria and Madawaska counties, Northwestern New Brunswick; New Brunswick Department of Natural Resources, Report Investigation 17, 69 p.
1978a: Geology of parts of Restigouche, Victoria and Madawask counties, Northwestern New Brunswick; New Brunswick Department of Natural Resources, Mineral Resources Branch, Investigation Report 17, 69 p.
1978b: Geology of Head of Wapske River map area J-13 (21 J/14); New Brunswick Department of Natural Resources, Mineral Resources Branch, Map Report 78-1, 24 p.
1982:   Geology of Juniper-Knowlesville-Carlisle area, map areas I-16, I-17, I-18 (Parts of 21 J/11 and 21 J/06); New Brunswick Department of Natural Resources, Mineral Resources Branch, Map Report 82-1, 82 p.

St. Peter, C.J. and Boucot, A.J.
1981:   Age and regional significance of brachiopods from the Témiscouata Formation of Madawaska County, New Brunswick; Maritime Sediments and Atlantic Geology, v. 17. p. 88-95.

Stearn, C.W.
1965:   Région de Causapscal (comtés de Matapédia et de Matane); Ministère des Ressources Naturelle, Québec, RG 117, 52 p.

Stevens, R.K.
1965:   Geology of the Humber Arm, west Newfoundland; M.Sc. thesis, Memorial University of Newfoundland, St. John's, Newfoundland.
1970:   Cambro-Ordovician flysch sedimentation and tectonics in west Newfoundland and their possible bearing on a Proto-Atlantic Ocean; in Flysch sedimentology in North America, (ed.) J. Lajoie; Geological Association of Canada, Special Paper 7, p. 165-178.

Stevenson, I.M.
1959:   Shubenacadie and Kennetcook map-areas, Colchester, Hants and Halifax counties, Nova Scotia; Geological Survey of Canada, Memoir 302, 88 p.

Stewart, D.B. and Wones, W.R.
1974:   Bedrock geology of northern Penobscot Bay area; in Geology of East-Central and North-Central Maine, (ed.) P.H. Osberg; New England Intercollegiate Geological Conference, Guidebook, p. 223-239.

Stockmal, G.S. and Waldron, J.W.F.
1990:   Structure of the Appalachian deformation front in western Newfoundland: implications of multichannel seismic reflection data; Geology, v. 18, p. 765-768.

Stockmal, G.S., Colman-Sadd, S.P., Keen, C.E., O'Brien, S.J., and Quinlan, G.
1987:   Collision along an irregular margin: a regional plate tectonic interpretation of the Canadian Appalachians; Canadian Journal of Earth Sciences, v. 24, p. 1098-1107.

Stockmal, G.S., Colman-Sadd, S.P., Keen, C.E., Marillier, F., O'Brien, S.J., and Quinlan, G.M.
1990:   Deep seismic structure and plate tectonic evolution of the Canadian Appalachians; Tectonics, v. 9, p. 45-62.

Strong, D.F.
1980:   Granitoid rocks and associated mineral deposits of eastern Canada and western Europe; in The Continental Crust and its Mineral Deposits, (ed.) D.W. Strangway; Geological Association of Canada, Special Paper 20, p. 741-769.

Strong, D.F. and Coyle, M.
1989:   Evolution of Taconic to Variscan fault styles and associated Acadian magmatism in west Newfoundland, with implications for driving mechanisms (abstract); Geological Society of America, Abstracts with Programs; v. 21, no. 2, p. 69-70.

Strong, D.F., O'Brien, S.J., Taylor, S.W., Strong, P.G., and Wilton, D.H.C.
1978:   Marystown (1M/3) and St. Lawrence (1L/14) map areas, Newfoundland; Newfoundland Department of Mines and Energy, Mineral Development Division, Report 77-8, 81 p.

Sweet, W.C., Turco, C.A., Warner, E., and Wilkie, L.C.
1959:   The American Upper Ordovician Standard I: Eden conodonts from the Cincinnati region of Ohio and Kentucky; Journal of Paleontology, v. 33, p. 1029-1068.

Swinden, H.S.
1988:   Re-examination of the Frozen Ocean Group: juxtaposed middle Ordovician-and Silurian volcanic sequences in central Newfoundland; in Current Research, Part B; Geological Survey of Canada, Paper 88-1B, p. 221-225.

Tamain, A.L.G.
1978:   L'évolution calédono-varisque des Hespérides; in Caledonian-Appalachian Orogen of the North Atlantic Region, Commission géologique du Canada, Étude 78-13, p. 183-212.

Taylor, F.C.
1965: Silurian stratigraphy and Ordovician-Silurian relationships in south-western Nova Scotia; Geological Survey of Canada, Paper 64-13, 24 p.

Taylor, F.C. and Schiller, E.A.
1966: Metamorphism of the Meguma Group of Nova Scotia; Canadian Journal of Earth Sciences, v. 3, p. 959-974.

Taylor, R.P., Strong, D.F., and Kean, B.F.
1980: The Topsail's Igneous Complex: Silurian-Devonian per alkaline magmatism in western Newfoundland; Canadian Journal of Earth Sciences, v. 17, p. 425-439.

Taylor, T.H.
1940: The geology and lithology of the Smiths Hole conglomerate, Fortune Bay, Newfoundalnd; B.Sc. thesis, Princeton University, U.S.A.

Termier, H. et Termier, G.
1964: Les temps fossilifères; in Paléozoïque inférieur (1), (ed.) H. Termier et G. Termier; Masson, Paris, 689 p.

Théberge, R.
1979: Étude tectono-stratigraphique des roches ordoviciennes et siluriennes de l'anticlinorium d'Aroostook-Percé à Matapédia, Appalaches du Québec; thèse de maîtrise, Université de Montréal, Montréal (Québec).

Thicke, M.J.
1987: The geology of Late Hadrynian metavolcanic and granitoid rocks of the Coxheath Hills-northeastern East Bay Hills areas, Cape Breton Island, Nova Scotia; M.Sc. thesis, Acadia University, Wolfville, Nova Scotia, 300 p.

Thomas, W.A.
1977: Evolution of Appalachian-Ouachita salients and recesses from reentrants and promontories in the continental margin; American Journal of Science, v. 277, p. 1233-1278.

Tremblay, A., Hebert, R., and Bergeron, M.
1989b: Le Complexe d'Ascot des Appalaches du sud du Québec: pétrologie et géochimie; Journal canadien des sciences de la Terre, v. 26, p. 2407-2420.

Tremblay, A., St-Julien, P., et Labbé, J.-Y.
1989a: Mise à l'évidence et cinématique de la faille de La Guadeloupe, Appalaches du sud du Québec; Journal canadien des sciences de la Terre, v. 26, p. 1932-1943.

Tuach, J.
1987: Mineralized environments, metallogenesis, and the Doucers Valley fault complex, western White Bay: a philosophy for gold exploration in Newfoundland; in Current Research, (ed.) R.F. Blackwood, D.G. Walsh, and R.V. Gibbons; Newfoundland Department of Mines, Mineral Development Division, Report 87-1, p. 113-128.

Twenhofel, W.H.
1947: The Silurian of eastern Newfoundland with some data relating to Physiography and Wisconsin Glaciation of Newfoundland; American Journal of Science, v. 245, p. 65-122.

Twenhofel, W.H. and Shrock, R.R.
1937: Silurian Strata of Notre Dame Bay and Exploits Valley, Newfoundland; Geological Association of America Bulletin, v. 48, p. 1743-1772.

Unger, J.D., Stewart, D.B., and Phillips, J.D.
1987: Interpretation of migrated seismic reflection profiles across the northern Appalachians in Maine; Geophysical Journal of the Royal Astronomical Society, v. 89, p. 171-176.

Vail, P.R., Mitchum, R.M., Todd, R.G., Widmier, J.M.,
Thompson, S.,III, Sangree, J.B., Bubb, J.N., and Hatfield, W.G.
1977: Seismic stratigraphy and global changes of sea level; in Seismic Stratigraphy - Applications to Hydrocarbon Exploration, (ed.) C.E. Payton; American Association of Petroleum Geologist, Memoir 26, p. 49-212.

van der Pluijm, B.A.
1984: Geology and microstructures of eastern New World Island, Newfoundland and implications for the northern Appalachians; Ph.D. thesis, University of New Brunswick, Frederiction, New Brunswick, 223 p.

1986: Geology of eastern New World Island, Newfoundland: an accretionary terrane in the northeastern Appalachians; Geological Society of America Bulletin, v. 97, p. 932-945.

van der Pluijm, B.A., Karlstrom, K.E., and Williams, P.F.
1987: Fossil evidence for fault derived stratigraphic repetition in the northeastern Newfoundland Appalachians; Canadian Journal of Earth Sciences, v. 24, p. 2337-2350.

Van der Pluijm, B.A., van der Voo, R., and Johnson, R.J.E.
1989: Middle Ordovician to Lower Devonian evolution of the northern Appalachians: The Acadian phase? (abstract); Geological Society of America, Abstracts with Programs, v. 21, no. 2, p. 72.

van Houten, F.B. and Hou, Hong-Fel
1990: Stratigraphic and palaeogeographic distribution of Palaeozoic oolitic ironstones; in Palaeozoic Palaeogeography and Biogeography, (ed.) W.S. McKerrow and C.R. Scotese; The Geological Society of London, Memoir, v. 12, p. 87-96.

van Staal, C.R.
1987: Tectonic setting of the Tetagouche Group in northern New Brunswick: implications for plate tectonic models of the northern Appalachians; Canadian Journal of Earth Sciences, v. 24, p. 1329-1351.

van Staal, C.R., Winchester, J.A., Brown, M., and Burgess, J.L.
1992: A reconnaissance geotraverse through southwestern Newfoundland; in Current Research, Part D; Geological Survey of Canada, Paper 92-1D, p. 133-1453.

van Staal, C.R., Ravenhurst, C.E., Winchester, J.A., Roddick, J.C.,
and Langton, J.P.
1990: Post-Taconic blueschist suture in the northern Appalachians of northern New Brunswick, Canada; Geology, v. 18, p. 1073-1077.

van Wagoner, N.A., McNeil, W., and Fay, V.
1988: Early Devonian bimodal volcanic rocks of southwestern New Brunswick: petrography, stratigraphy, and depositional setting; Maritime Sediments and Atlantic Geology, v. 24, p. 301-319.

van Wagoner, N.A., Dadd, K.A., Baldwin, D.K., and Mcneil, K.
1989: Cyclic, bimodal volcanic zone, Charlotte County, New Brunswick; in 14th Annual Review of Activities, Project Resumes, 1989, (ed.) S.A. Abbott; New Brunswick Department of Natural Resources and Energy, Mineral Resources Division, Information Circular 89-9, p. 215-216.

Vaslet, D.
1990: Upper Ordovician glacial deposits in Saudi Arabia; Episodes, v. 13, p. 147-161.

Vennat, G.
1979: Structure et stratigraphie de l'anticlinorium d'Aroostook-Percé dans la région de Saint-Omer Carleton, Gaspésie, Appalaches du Québec; thèse de maîtrise, Université de Montréal, Montréal (Québec)

Wake, C.
1984: Sedimentology and provenance of the Kingsclear Group turbidites of the Fredericton Trough, New Brunswick; B.Sc. thesis, University of Ottawa, Ottawa, Ontario, 59 p.

Walker, R.G. and Cant, D.J.
1979: Facies model 3: Sandy fluvial system; in Facies Models, (ed.) R.G. Walker; Geoscience Canada, Reprint Series 1, p. 23-31.

Wanless, R.K., Stevens, R.D., Lachance, G.R., and Edmond, C.M.
1967: Age determinations and geological studies: K-Ar isotopic ages (Report 7); Geological Survey of Canada, Paper 66-17.

1968: Age determinations and geological studies: K/Ar isotopic ages (Report 8); Geological Survey of Canada, Paper 67-2 Part A.

Wanless, R.K., Stevens, R.K., Lachance, G.R., and Rimsaite, J.Y.H.
1965: Age determinations and geological studies: K-Ar isotopic ages (Report 5); Geological Survey of Canada, Paper 64-17, Part 1.

Wasowski, J.J. and Jacobi, R.D.
1986: The tectonics and depositional history of the Ordovician and Silurian rocks of Notre Dame Bay, Newfoundland (discussion); Canadian Journal of Earth Sciences, v. 23, p. 583-585.

Watson, M.P.
1981: Submarine fan deposits of the Upper Ordovician-Lower Silurian Milliners Arm Formation, New World Island, Newfoundland; Ph.D. thesis, Oxford University, England.

Weeks, L.J.
1955: Geological map of the Island of Newfoundland; Geological Survey of Canada, Map 1047A, scale 1: 760 320.

Weijermars, R. and Rondeel, H.E.
1984: Sheer band foliation as an indication of sense of shear: field observations in central Spain; Geology, v. 12, p. 603-606.

Wessel, J.M.
1975: Sedimentary petrology of the Springdale and Botwood Formations, central mobile belt, Newfoundland, Canada; Ph.D. thesis, University of Massachusetts, U.S.A. 216 p.

Whalen, J.B.
1989: The Topsails igneous suite, western Newfoundland: an Early Silurian subduction-related magmatic suite?; Canadian Journal of Earth Sciences, v. 26, p. 2421-2434.

Whalen, J.B. and Currie, K.L.
1983: The Topsails igneous terrane of western Newfoundland; in Current Research, Part A; Geological Survey of Canada, Paper 83-1A, p. 15-23.

Whalen, J.B., Hegner, E., and Jenner, G.A.
1989:	Nature of Canadian Appalachians basement terranes as inferred from a Nd isotopic transect (abstract); Geological Society of America, Annual Meeting, St. Louis Missouri, Program With Abstracts; v. 21, no. 6, p. A-201.

White, D.E.
1939:	Geology and molybdenite deposits of the Rencontre East area, Fortune Bay, Newfoundland; Ph.D. thesis, Princeton University, U.S.A. 119 p.
1940:	The molybdenite deposits of the Rencontre East area, Newfoundland; Economic Geology, v. 35, p. 967-995.

White, S.H., Burrows, S.E., Carreras, J., Shaw, M.D., and Humphreys, F.J.
1980:	On mylonites in ductile shear zones; Journal of Structural Geology, v. 2, p. 175-187.

Widmer, K.
1950:	The geology of the Hermitage Bay area, Newfoundland; Ph.D. thesis, Princeton University, U.S.A. 439 p.

Wilcox, R.E., Harding, T.P., and Seeley, D.R.
1973:	Basic wrench tectonics; American Association of Petroleum Geologists Bulletin, v. 57, p. 74-96.

Williams, H.
1957:	The geology and limestone deposits of the Cobbs Arm area, Newfoundland; Newfoundland Department of Mines and Resources, Mines Branch, Open File Report.
1962:	Botwood (west-half) map-area, Newfoundland; Geological Survey of Canada, Paper 62-9, 16 p.
1963:	Twillingate map-area, Newfoundland; Geological Survey of Canada, Paper 63-36, 30 p.
1964:	Botwood, Newfoundland; Geological Survey of Canada, Map 60-1963.
1967:	Silurian Rocks of Newfoundland; in Geology of the Atlantic Region, (ed.) E.R.W. Neale and Harold Williams; Geological Association of Canada, Special Paper 4, p. 93-137.
1970:	Red Indian Lake (east half), Newfoundland; Geological Survey of Canada, Map 1196A (with description notes).
1971a:	Burgeo (east half), Newfoundland; Geological Survey of Canada, Map 1280A.
1971b:	Geology of Belleoram map-area, Newfoundland (1M/11); Geological Survey of Canada, Paper 70-65, 39 p.
1972:	Stratigraphy of Botwood map-area, northeastern Newfoundland; Geological Survey of Canada, Open File 113, 117 p.
1977:	The Coney Head Complex: another Taconic allochthon in west Newfoundland; American Journal of Science, v. 277, p. 1279-1295.
1978:	Tectonic-lithofacies map of the Appalachian orogen (1:1 000 000); Department of Earth Sciences, Memorial University of Newfoundland, St. John's, Newfoundland, Map 1.
1979:	Appalachian Orogen in Canada; Canadian Journal of Earth Science, Tuzo Wilson Volume, v. 16, p. 792-807.
1985:	Geology, Stephenville map-area, Newfoundland; Geological Survey of Canada, Map 1579A, scale 1:1 000 000 (with descriptive notes and bibliography).
1992a:	Melanges and coticule occurrences in the northeast Exploits Subzone, Newfoundland; in Current Research, Part D; Geological Survey of Canada, Paper 92-1D, p. 121-127.
1993:	Acadian Orogeny in the Newfoundland Appalachians; in The Acadian Orogen: Recent tsudies in New England Maritime Canada and the Autochthonous Foreland, (ed.) D.C. Roy and S. Skehan; Geological Society of America, Special Paper 275, p. 123-133.

Williams, H. and Hatcher, R.D.
1982:	Suspect terranes and accretionary history of the Appalachian Orogen; Geology, v. 10, p. 530-536.
1983:	Appalachian suspect terranes, in Contributions to the tectonics and geophysics of mountain chains, (ed.) R.D. Hatcher and others; Geological Society of America, Memoir 158, p. 33-53.

Williams, H. and Piasecki, M.A.J.
1990:	The Cold Spring Melange and a possible model for Dunnage-Gander zone interaction in central Newfoundland; Canadian Journal of Earth Sciences, v. 27, p. 1126-1134.

Williams, H. and Smyth, W.R.
1983:	Geology of the Hare Bay Allochthon; in Geology of the Strait of Belle Isle Area, Northwestern Insular Newfoundland, Southern Labrador, and adjacent Quebec; Geological Survey of Canada, Memoir 400, p. 109-141.

Williams, H. and St-Julien, P.
1982:	The Baie Verte-Brompton Line: Early-Paleozoic continent-ocean interface in the Canadian Appalachians; in Major Structural Zones and Faults of the Northern Appalachians, (ed.) P. St-Julien and J. Béland; Geological Association of Canada, Special Paper 24, p. 177-207.

Williams, H., Colman-Sadd, S.P., and Swinden, H.S.
1988:	Tectonic-stratigraphic subdivisions of central Newfoundland; in Current Research, Part B; Geological Survey of Canada, Paper 88-1B, p. 91-98.

Williams, H., Hibbard, J.P., and Burnsall, J.J.
1977:	Geologic setting of abestos-bearing ultramafic rocks along the Baie Verte Lineament, Newfoundland; in Report of Activities, Part A; Geological Survey of Canada, Paper 77-1A, p. 351-360.

Williams, H., Piasecki, M.A.J., and Colman-Sadd, S.P.
1989a:	Tectonic relationships along the proposed central Newefoundland lithoprobe transect and regional correlations; in Current Research, Part B; Geological Survey of Canada, Paper 89-1B, p. 55-66.

Williams, H., Dickson, W.L., Currie, K.L., Hayes, J.P., and Tuach, J.
1989b:	Preliminary report on a classification of Newfoundland granitic rocks and their relation to tectonostratigraphic zones and lower crustal blocks; in Current Research, Part B; Geological Survey of Canada, Paper 89-1B, p. 47-53.

Williams, H.S.
1910:	Age of the Gaspé Sandstones; Geological Society of America Bulletin, v. 20, p. 688-698.

Williams, P.F., Elliott, C.G., and Lafrance, B.
1988:	Structural geology and melanges of eastern Notre Dame Bay, Newfoundland; Geological Association of Canada, Trip B2, Guidebook, 60 p.

Williams, S.H.,
1989:	New Graptolite discoveries from the Ordovician of central Newfoundland; in Current Research; Newfoundland Department of Mines, Geological Survey of Newfoundland, Report 89-1, p. 149-157.
1991:	Stratigraphy and graptolites of the Upper Ordovician Point Leamington Formation, central Newfoundland; Canadian Journal of Earth Sciences, v. 28, p. 581-900.

Williams, S.H. and O'Brien, B.H.
1991:	Silurian (Llandovery) graptolites from the Bay of Exploits, north-central Newfoundland, and their geological significance; Canadian Journal of Earth Sciences, v. 28, p. 1534-1540.

Wilton, D.H.C.
1983:	The geology and structural history of the Cape Ray fault zone in southwestern Newfoundland; Canadian Journal of Earth Sciences, v. 20, p. 1119-1133.
1984:	Metallogenic, tectonic and geochemical evolution of the Cape Ray Fault Zone with emphasis on electron mineralization; Ph.D. thesis, Memorial University of Newfoundland, St. John's, Newfoundland, 618 p.
1985:	Tectonic evolution of southwestern Newfoundland as indicated granitoid petrogenesis; Canadian Journal of Earth Sciences, v. 22, p. 1080-1092.

Winchester, J.A. and Floyd. P.A.
1977:	Geochemical discrimination of different magma series and their differentiation products using immobile elements; Chemical Geology, v. 20, p. 325-343.

Wones, D.R. (ed.)
1980:	The Caledonides in the USA; Virginia Polytechnic Institute and State University, Department of Geological Sciences, Memoir 2, 329 p.

Woodcock, N.H.
1990:	Sequence stratigraphy of the Palaeozoic Welsh basin; Geological Society of London, v. 147, p. 537-547.

Wu, T.W.
1979:	Structural, stratigraphic and geochemical studies of the Horwood Peninsula-Gander Bay area, northeast Newfoundland; M.Sc. thesis, Brock University, St. Catherines, Ontario, 185 p.

Zaitlin, B.A. and Rust, B.R.
1983:	A spectrum of alluvial deposits in the Lower Carboniferous Bonaventure Formation of western Chaleurs Bay area, Gaspé and New Brunswick, Canada; Canadian Journal of Earth Sciences, v. 20, p. 1098-1110.

Zentilli, M.
1977:	Evolution of metallogenic domains in Nova Scotia; Canadian Institute of Mining and Metallurgy, Bulletin, v. 70, p. 69.

ADDENDA

Gaspé Belt
(P.-A. Bourque, D. Brisebois and M. Malo)

This manuscript was submitted in 1986 and since that time only minor revisions were made despite much new work in the Quebec Appalachians and an expanding database. Conclusions presented here are in general agreement with the latest data, with the exception of the palinspastic reconstruction. Work by Kirkwood (1993) showed that shortening across the Gaspé Belt reached up to 75 per cent, instead of 55 per cent. We now have a better palinspastic map, although our conclusions on paleogeography are not significantly changed. A set of updated paleogeographical maps is available from P.-A. Bourque.

Kirkwood, D.
1993:   Étude qualitative et quantitative de la déformation acadienne du bassin siluro-dévonien de la péninsule gaspésienne, Appalache du Nord; thèse de doctorat, Université Laval, Québec (Québec), 178 p.

Mascarene Belt
(M.J. McLeod and S.R. McCutcheon)

Recent investigations in the Mascarene Belt necessitate the following changes:

1) *Nerepis Subbelt:* the outliers of grey polymictic conglomerate south of the Wheaton Brook Fault assigned to the Woodmans Point beds, are more likely part of the Cambrian Saint John Group;

2) *Letete Subbelt:* the Goss Point beds and Letang volcanics as defined herein are mostly Late Ordovician (Nowland et al., 1994) and do not form part of the Mascarene Belt. Only small faulted slivers of sandstone and shale with fossils of Silurian age in this area remain in the Mascarene Belt. These include the Back Bay beds and rocks currently referred to as the Frye Island beds.

The stratigraphy in much of the area along the southern margin of the Saint George Batholith has been established and most of the rocks are late Precambrian to Cambrian. Fault bounded blocks of fossiliferous Silurian Mascarene rocks, however, are present. These are referred to as the Fowle Lake beds and are Llandovery C3 to Ludlow (Johnson, 1994).

3) Since the Goss Point beds and Letang volcanics are mostly Ordovician, the correlation of these beds with the Long Reach Formation is untenable.

4) Geological mapping south of the Saint George Batholith by Johnson (1994) has demonstrated that undisputed Paleozoic rocks are involved in episodes of mylonitization.

Johnson, S.C.
1994:   Geological investigation of the Pocologan River area, Charlotte County, New Brunswick: bedrock geology and regional implications; in Current Research 1993, (ed.) S.A. Abbott Merlini; New Brunswick Department of Natural Resources and Energy, Mineral Resources, Miscellaneous Report 12, p. 71-81.

Nowland, G.S., McCracken, A.D., and McLeod, M.J.
1994:   Tectonic and paleogeographic significance of Ordovician conodonts from the Letang limestone, Avalon Terrane, southwestern New Brunswick (abstract); Geological Association of Canada/ Mineralogical Association of Canada, Program with Abstracts, v. 19, p. A83.

Annapolis Belt
(P.E. Schenk)

New Canaan Group

Seven cores drilled in 1992 to depths ranging from 90 to 520 m provide a complete stratigraphic record of the poorly exposed New Canaan and underlying Kentville groups (Smith, 1992). Cores of the New Canaan Group consist predominantly of volcaniclastic sediments and minor flows, sulphide-rich, marine, fossiliferous shale and three limestone units. The later range in thickness from 3 to 22 m and act as reliable markers amidst common facies variations in the volcanic rocks. Smith (1992) interpreted the paleoenvironment as a volcanic island with reef-like facies.

Stratotypes

Nictaux Volcanics of the White Rock Group
Nictaux Canal (NTS 21-A/14; UTM 20 T LE393734) 5.6 km southeast of Middleton and 1.6 km south of Nictaux Falls, outcrop on west side on north end of canal at penstock entrance.

Fales River Formation of the White Rock Group
Rockville Notch on Fales River 84 m south (downstream) of bridge (NTS 21-A/15; UTM 20 T LE508790), section of approximately 3.5 km southeast of New Minas and 6 km southwest of Wolfville. Section approximately LE888900 (base) to LE887892 (top).

Elderkin Formation of the Kentville Group
East branch of Elderkin Brook (NTS 21-H/1; UTM 20 T LE846889) 2.5 km southwest of New Minas and 4.4 km west of White Rock.

Tremont Formation of the Kentville Group
Rockville Notch on Fales River with base 366 m and top projected at 700 m south (downstream) of bridge (NTS 21-A/15; UTM 20 T LE508790) 3.3 km east of Tremont.

South American Connection?

Paleomagnetic reconstruction suggests that the Meguma terrane could be a displaced Gondwana margin not of northwestern Africa but southwestern South America. Laurentia could have moved clockwise around Gondwana during the early and middle Paleozoic from a position near Antarctica (Dalziel, 1991). This hypothesis raises the possibility of an Ordovician (Taconic) collision between the eastern margin of Laurentia (including the Canadian Appalachians) and the southwestern margin of South America (Argentina) (Dalla Salda et al., 1992). However, the sedimentology and stratigraphy of the Meguma terrane provide direct evidence of the juxtaposition of northwestern Africa and southeastern Laurentia prior to the late Paleozoic. The terrane is dissimilar to others in the Appalachians in showing no structural evidence of an Ordovician collisional event (J.R. Henderson, pers. comm., 1993).

Dalla Salda, L.H., Dalziel, I.W.D., Cingolani, C.A., and Varela, R.
1992: Did the Taconic Appalachians continue into South America?; Geology, v. 20, p. 1059-1062.

Dalziel, I.W.D.
1991: Pacific margin of Laurentia as a conjugate rift pair: evidence and implications for an Eocambrian supercontinent; Geology, v. 19, p. 598-601.

Smith, P.K.
1992: Results of diamond-drilling program in the Upper Silurian New Canaan Formation: mineral potential of Mills limestone and sulphide-rich black shale units; in Program and Summaries, Sixteenth Annual Open House and Review of Activities, (ed.) D.R. McDonald and K.A. Mills; Nova Scotia Department of Natural Resources, Mines and Energy Branches, Report 92-04, p. 29.

Botwood Belt
(H. Williams)

Recognition of the Dog Bay Line as a major structural junction that separates different Silurian rock groups requires subdivision of the Botwood Belt into western and eastern parts (Williams, 1993; Williams et al., 1993). The name Botwood Belt is retained for the wide area of middle Proterozoic rocks northwest of the Dog Bay Line and a new name, the Indian Islands Belt, is introduced for a much narrower belt southeast of the Dog Bay Line. East of the line, Silurian calciferous shales, sandstones, conlgomerates and redbeds are assigned to the Indian Islands Group (Fig. 3.A1; see Williams et al., 1993, for a more detailed description of the stratigraphy).

The Dog Bay Line is marked by a wide zone of disrupted Ordovician rocks. The Badger and Botwood groups northwest of the line, were deposited on Ordovician rocks already accreted to Laurentia. The Indian Islands Group southeast of the line, was deposited on Ordovician rocks already amalgamated with the continental Gander Zone. Timing of major movement and a Silurian marine to terrestrial depositional change recorded on both sides of the line, agree within error with isotopic ages for the onset of middle Paleozoic plutonism, regional deformation, and metamorphism in central Newfoundland. The Dog Bay Line may mark the terminal Iapetus Suture.

Williams, H.
1993: Stratigraphy and structure of the Botwood Belt and definition of the Dog Bay Line in northeastern Newfoundland; in Current Research, Part D; Geological Survey of Canada, Paper 93-1D, p. 19-27.

Williams, H., Currie, K.L., and Piasecki, M.A.J.
1993: The Dog Bay Line – a major Silurian tectonic boundary in northeast Newfoundland; Canadian Journal of Earth Sciences, v. 30, p. 2481-2494.

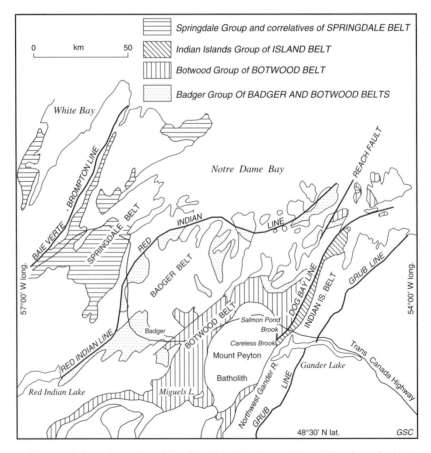

Figure 3.A1. Location of the Dog Bay Line in northeast Newfoundland with respect to tectonostratigraphic belts defined on middle Paleozoic rocks (after Williams et al., 1993).

Authors' Addresses

S.M. Barr
Department of Geology
Acadia University
Wolfville, Nova Scotia
B0P 1X0

Pierre-André Bourque
Département de géologie
Université Laval
Sainte-Foy (Québec)
G1K 7P4

Daniel Brisebois
Ressources Naturelles
Service Géologique de Québec
5700, 4e Avenue
Charlesbourg, Québec
G1H 6R1

L.B. Chorlton
1138 Deer Park Road
Nepean, Ontario
K2E 6H2

P.L. Dean
Newfoundland Department of Mines and Energy
P.O. 8700
St. John's, Newfoundland
A1B 4J6

J.R. DeGrace
Director Research Development
University of Prince Edward Island
550 University Avenue
Charlottetown, P.E.I.
C1A 4P3

L.E. Fåhraeus (deceased)
Department of Earth Sciences
Memorial University of Newfoundland
St. John's, Newfoundland
A1B 3X5

L.R. Fyffe
Minerals and Energy Division
New Brunswick Department of Natural Resources
P. O. Box 6000
Fredericton, New Brunswick
E3B 5H1

R.A. Jamieson
Department of Geology
Dalhousie University
Halifax, Nova Scotia
B3H 3J5

J.D. Keppie
Nova Scotia Department of Mines and Energy
Halifax, Nova Scotia
B3J 1X1

B.D. Lafrance
Department of Geology
University of New Brunswick
Box 4400
Fredericton, New Brunswick
E3B 5A3

Michel Malo
Quebec Geoscience Centre
INRS - Géoressources
C.P. 7 500
Sainte-Foy (Québec)
G1V 4C7

S.R. McCutcheon
Department of Natural Resources and Energy
P.O. Box 50
Bathurst, New Brunswick
E2A 3Z1

M.J. McLeod
Department of Natural Resources and Energy
P.O. Box 50
Bathurst, New Brunswick
E2A 3Z1

J.B. Murphy
Department of Geology
St. Francis Xavier University
Antigonish, Nova Scotia
B2G 1C0

B.H. O'Brien
Geological Survey Branch
Newfoundland Department of Mines and Energy
P.O. 8700
St. John's, Newfoundland
A1B 4J6

S.J. O'Brien
Geological Survey Branch
Newfoundland Department of Mines and Energy
P.O. 8700
St. John's, Newfoundland
A1B 4J6

K.T. Pickering
Department of Geology
Leicester University
Leicester
LE1 7RH
U.K.

P.E. Schenk
Department of Geology
Dalhousie University
Halifax, Nova Scotia
B3H 3J5

R.K. Stevens
Department of Earth Sciences
Memorial University of Newfoundland
St. John's, Newfoundland
A1B 3X5

B.A. van der Pluijm
Department of Geological Sciences
University of Michigan
Ann Arbor, Michigan
48109
U.S.A.

H. Williams
Department of Earth Sciences
Memorial University of Newfoundland
St. John's, Newfoundland
A1B 3X5

P.F. Williams
Department of Geology
University of New Brunswick
Box 4400
Fredericton, New Brunswick
E3B 5A3

D.H.C. Wilton
Department of Earth Sciences
Memorial University of Newfoundland
St. John's, Newfoundland
A1B 3X5

Printed in Canada

Chapter 5

UPPER PALEOZOIC ROCKS

Chapter 5

UPPER PALEOZOIC ROCKS

H.W. van de Poll, M.R. Gibling, and R.S. Hyde

INTRODUCTION

H.W. van de Poll, M.R. Gibling, and R.S. Hyde

Preamble and regional setting

The Silurian to mid Devonian Acadian Orogeny resulted in an extensive region of uplands across Atlantic Canada. Isolated occurrences of Lower to Middle Devonian redbed conglomerates, arkosic sandstones and mudstones, locally with intercalated felsic and/or mafic volcanic flows, are the earliest post-orogenic deposits. These became gradually thicker and more widespread so that by Middle Carboniferous time, large tracts of the orogen were buried by sedimentary rocks including coarse to fine red and grey terrigenous clastics, oil shales, coal, carbonates, and evaporites. Variation in sediment accumulation rates, syndepositional uplift, and extensive reworking of earlier deposits have greatly complicated the upper Paleozoic lithostratigraphy. This situation has been complicated further by widespread subsequent faulting, tilting, folding, and erosion.

Tilted and/or downfaulted regions in which upper Paleozoic rocks accumulated and/or have been preserved are referred to as "basins" (Fig. 5.1). The name "Maritimes Carboniferous Basin" (Roliff, 1962) or "Maritimes Basin" (Williams, 1974) has in practice become a collective term for the total area in eastern Canada that is presently underlain by upper Paleozoic rocks contained in structural basins, including the offshore. In this sense the "Maritimes Basin" is essentially a basin complex comprising structural remnants of a formerly more extensive area of upper Paleozoic rocks. The designation "Maritimes Basin" is useful however in that it serves to distinguish the area of Upper Paleozoic rocks in Atlantic Canada from the Appalachian, Illinois, and other basins in the United States.

The upper Paleozoic rocks contained in the Maritimes Basin complex constitute an approximately east-west-trending region with local thicknesses in excess of 12 000 m (Sanford and Grant, 1990). A relatively narrow, northeast-trending, fault-bounded and fault-dominated central region underlies the Bay of Fundy and Gulf of St. Lawrence and adjacent inland areas. It has a thick sequence of Upper Devonian to Lower Permian rocks that are extensively deformed. It extends from southwestern New Brunswick to beyond northwestern Newfoundland (Belt, 1968a) and is flanked by platformal areas in western New Brunswick, southern Nova Scotia, and western Newfoundland with relatively thin, slightly tilted to mildly deformed upper Paleozoic strata (Neale et al., 1961; Belt, 1965, 1969; Poole, 1967; van de Poll, 1970; Hacquebard, 1972; Bradley, 1982; McCutcheon and Robinson, 1987). The central region has been variously referred to as the "Fundy Basin" (Bell, 1958), the "Late Paleozoic Folded Zone" (Neale et al., 1961) the "Fundy Geosyncline: (Poole, 1967), the "Fundy Basin Rift" (Belt, 1968a) the "Maritime Rift" (Belt, 1969) and the "Fundy Basin" and "Fundy Rift" (Webb, 1969; Hacquebard, 1972).

For descriptions and correlation of the upper Paleozoic rocks in eastern Canada, and in particular in New Brunswick, it is useful to differentiate between the relatively thin and undeformed rocks of the stable platformal areas and those of the "mobile" central region, where thick deposits were periodically extensively deformed and reworked. For this reason the name "Maritimes Rift" is used as a major element of the Maritimes Basin, modified slightly after Belt (1969). The Maritimes Rift is generally believed to have had a complex history of east-west wrench movement and related northeasterly transtension and transpression throughout the late Paleozoic. It is characterized by a "horst and graben" topography of subsiding basins separated by intervening source areas in the form of "arches", "uplifts" or "massifs". Local activity on normal, thrust, and strike-slip faults resulted in a complex temporal and spatial pattern of erosion and deposition. Most of the important syn- and post-depositional folding and fault movements were confined to the Maritimes Rift, whereas adjacent platformal areas were largely unaffected (Neale et al., 1961; Poole, 1967). The Maritimes Rift in northern Nova Scotia and southeastern New Brunswick lies close to the Avalon-Meguma zone boundary and may in part reflect continued activity on structural features associated with accretion during the Acadian Orogeny.

Initial deposition of upper Paleozoic rocks was largely confined to the Maritimes Rift, a few other areas such as the Ristigouche, St. Andrews and Marysville basins of New Brunswick, and to local fault-bounded depressions elsewhere in the Atlantic region (Durling and Marillier, 1990; Sanford and Grant, 1990). By Middle Carboniferous time sedimentary rocks covered platformal areas adjacent to the Maritimes Rift.

van de Poll, H.W., Gibling, M.R., and Hyde, R.S.
1995: Introduction: Upper Paleozoic rocks; *in* Chapter 5 of Geology of the Appalachian-Caledonian Orogen in Canada and Greenland, (ed.) H. Williams; Geological Survey of Canada, Geology of Canada, no. 6, p. 449-455 (*also* Geological Society of America, The Geology of North America, v. F-1).

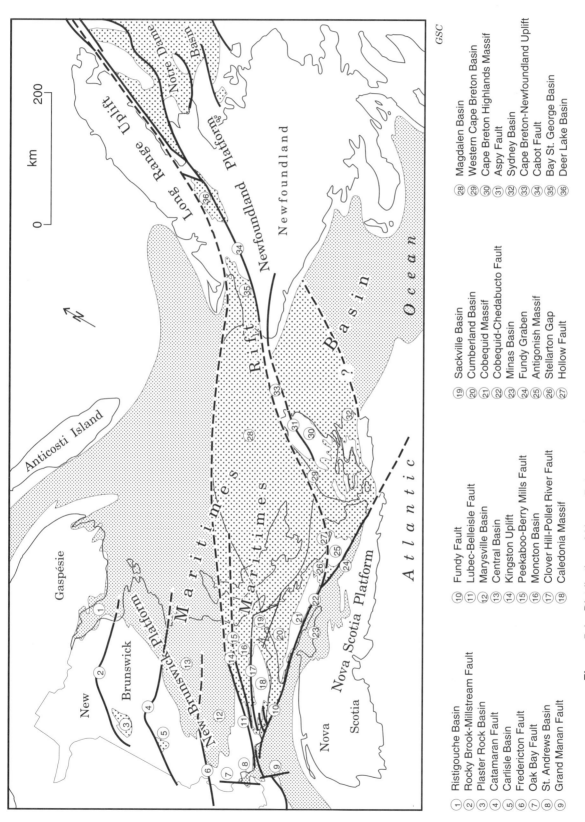

Figure 5.1. Distribution of Upper Paleozoic rocks in eastern Canada and the location of the major structural elements and faults (after Williams, 1978; Sanford and Grant, 1990; Howie, 1988).

1. Ristigouche Basin
2. Rocky Brook-Millstream Fault
3. Plaster Rock Basin
4. Catamaran Fault
5. Carlisle Basin
6. Fredericton Fault
7. Oak Bay Fault
8. St. Andrews Basin
9. Grand Manan Fault
10. Fundy Fault
11. Lubec-Belleisle Fault
12. Marysville Basin
13. Central Basin
14. Kingston Uplift
15. Peekaboo-Berry Mills Fault
16. Moncton Basin
17. Clover Hill-Pollet River Fault
18. Caledonia Massif
19. Sackville Basin
20. Cumberland Basin
21. Cobequid Massif
22. Cobequid-Chedabucto Fault
23. Minas Basin
24. Fundy Graben
25. Antigonish Massif
26. Stellarton Gap
27. Hollow Fault
28. Magdalen Basin
29. Western Cape Breton Basin
30. Cape Breton Highlands Massif
31. Aspy Fault
32. Sydney Basin
33. Cape Breton-Newfoundland Uplift
34. Cabot Fault
35. Bay St. George Basin
36. Deer Lake Basin

GSC

(Ryan et al., 1991b; Ryan and Zentilli, 1993) concluded that up to 4000 m of Upper Paleozoic sedimentary cover has been eroded since Early Permian time. This great thickness of missing strata implies that, prior to erosion, most if not all of the Atlantic region and much of the contiguous United States was covered by upper Paleozoic strata (van de Poll, 1973a; Friedman and Sanders, 1982; Johnsson, 1986; Gibling et al., 1992).

Facies evidence suggests a shift in the paleoclimate from warm subhumid or humid to semi-arid or arid during the Early Carboniferous (Schenk, 1969; Varma, 1969; van de Poll, 1978) and a subsequent return to subhumid or humid tropical conditions during the Middle and Late Carboniferous (Hacquebard, 1972; van de Poll, 1973a, 1983; D'Orsay and van de Poll, 1985b, c). Coincident with the paleoclimatic return to more humid conditions was also a tectonic change that contributed to the change in style of sedimentation and burial. The localized relatively thick redbed accumulations of the Early Carboniferous basins were replaced by more widespread, grey and red fluvio-clastic deposits locally with coal measures across basinal and adjacent platformal areas. Sediment maturity increased as proximal source areas were eroded, buried and replaced by more distal source terranes, except in locally subsiding Late Carboniferous coal basins (e.g. Cumberland and Sydney basins of Nova Scotia) where subhumid, tropical alluvial-fan and braid-plain deposits from nearby source areas replaced the arid piedmont deposits of Early Carboniferous times.

General stratigraphy

The upper Paleozoic rocks of eastern Canada occupy a series of northeast- and east-trending basins separated in part by uplifted basement blocks. Periodic differential movement between adjacent basement blocks resulted in rapid facies changes, local and regional unconformities, diastems, and a complex lithostratigraphy that for most units cannot be readily extended beyond the limits of individual basins. Volcanic rocks of bimodal aspect occur locally, most commonly at the bases of stratigraphic sections. Granitic intrusions of Early Carboniferous (?) age occur locally at Mount Pleasant (Sinclair et al., 1988; McCutcheon, 1990), in the Saint John area of southern New Brunswick (Alcock, 1938; Rast et al., 1984), and in the Cobequid Highlands of northwestern Nova Scotia (Donohoe et al., 1986, Pe-Piper et al., 1989a, b; Piper et al., 1993).

The original stratigraphic subdivision of the Maritimes Basin Carboniferous sequence into six major units, based on stratigraphic work by W.A. Bell over a period of 50 years (1912-1962) is to a large extent still in wide usage. The modifications that have been made reflect an improved knowledge of lithofacies distribution in combination with improved palynological information. This has allowed a better stratigraphic control for sequences where ages were previously unknown or poorly constrained. The group names currently in use and their biostratigraphic zonation based on European stages is given in Table 5.1.

Bell's stratigraphic work was mainly based on paleobotanical evidence although he generally adhered to recognizable stratigraphic breaks to delineate his units. Subsequent work showed, however, that the stratigraphic breaks are in most cases local, are concentrated along basin margins, and recur at different levels throughout the stratigraphic record. Where fossils are absent, subsequent workers were forced to rely increasingly on lithological correlations. As a result, Bell's original units are defined on a mixture of lithostratigraphic (mainly facies), and biostratigraphic (mainly palynological) criteria resulting in much broader age limits than were originally envisaged by Bell (Kelley, 1967).

The Horton Group is essentially of Late Devonian-Early Carboniferous (Late Famennian-Tournaisian) age and consists of red and grey-green polymictic conglomerates, arkosic sandstones, mudstones, oil shales, and minor non-marine evaporites. In most areas the Horton Group has been subdivided into three formations, although they have only local significance. The type area of the Horton Group is the western part of the Minas Basin (Windsor area).

Several fault bounded basins filled with Horton strata have been tentatively identified from seismic data at depth in the western and central part of the Gulf of St. Lawrence (Durling and Marillier, 1990). Horton Group strata overlie pre-Acadian metamorphic basement with angular unconformity and in places the group is overlain conformably by basal limestone of the Windsor Group. Ostracodes and fish fossils are locally common in thin-laminated shales. Fossilized plant remains are also locally common in the Horton Group, but coal is rare. Mafic volcanic rocks are interbedded with basal Horton Group conglomerates in several areas (Howie and Barss, 1975; Blanchard et al., 1984). Onland sections up to 3291 m thick (Kelley, 1967) occur in southeastern New Brunswick, mainland Nova Scotia, Cape Breton Island, and Newfoundland. A thickness of 4000 m has been inferred from seismic data in the Magdalen Basin (Bell and Howie, 1990). The Horton equivalent in Newfoundland is the Anguille Group.

The Windsor Group is also of Early Carboniferous (Viséan-early Namurian) age (Utting, 1987b). It consists of marine limestones, evaporites, and intercalated redbeds. The group represents the only known marine incursion into the Maritimes Basin prior to the Late Carboniferous, at a time when the paleoclimate was arid. In the type area

Table 5.1. Lithostratigraphic subdivision by groups of upper Devonian-Carboniferous-Permian strata of the Maritimes Basin.

Main group names Maritimes Basin	Eurasian stages (general)	Epochs and Periods
Prince Edward Island Group (upper part Northumberland Strait Supergroup)	Asselian-Artinskian	Early Permian
Pictou Group (lower part Northumberland Strait Supergroup)	Early Westphalian C-Stephanian	Middle-Late Carboniferous
Cumberland Group	late Namurian-Westphalian C-D	Middle Carboniferous
Canso-Riversdale Group (Mabou Group in Nova Scotia)	late Namurian-Early Westphalian A	Middle Carboniferous
Windsor Group	Viséan-Early Namurian	Early Carboniferous
Horton Group	Famennian-Tournaisian	Late Devonian-Early Carboniferous

(western Minas Basin of Nova Scotia), Windsor strata overlie Horton strata conformably and grade upward into redbeds of the Canso Group (renamed Mabou Group; Belt, 1964, 1965; Ryan et al., 1991a). Windsor redbeds are commonly coarse and conglomeratic where they border former highlands, and they are finer grained towards the interior regions of the basin.

The Windsor Group was subdivided into two zones and five subzones (Bell, 1929). The cyclical carbonate-evaporite-redbed lithofacies in the Windsor section may represent marine transgression-regression cycles (Schenk, 1967a) or seasonal and climatic fluctuations in the degree of aridity (Schenk, 1969).

The Windsor Group is widespread at the surface and in subsurface onland areas adjacent to the Gulf of St. Lawrence (e.g., southern New Brunswick, mainland Nova Scotia, Cape Breton Island, Prince Edward Island, the Îles de la Madeleine, and Newfoundland). The Windsor Group is generally poorly exposed and is best known from drillholes. Several salt structures have been drilled in New Brunswick, Nova Scotia, Prince Edward Island, and Îles de la Madeleine. Although intersections of up to 3600 m of highly contorted Windsor strata have been reported, the true thickness of the Windsor section may not exceed 600-800 m in most places (Howie, 1988). The most extensive and thickest Windsor sequence occurs in the Magdalen Basin where wells and seismic profiles have shown the presence of many swells, ridges, and diapiric sturctures involving Windsor strata 5000 m or more thick over an area of some 60 000 km^2 (Bell and Howie, 1990).

Windsor limestone and associated clastics locally contain important base metal concentrations, especially along the basal Horton-Windsor contact in Nova Scotia (Boyle, 1972; Binney and Kirkham, 1974,1975; MacEachern and Hannon, 1974; Kirkham, 1978; van de Poll, 1978; Smith and Collins, 1984; MacLeod, 1984). Best known are the Gays River and the now depleted Walton deposits of western Nova Scotia. Local concentrations of barite (Boyle, 1972) and celestite (Crowell, 1971) are also present. Manganese appears to be the most common metal associated with Windsor carbonates in New Brunswick.

Potash and road salt are locally mined from Windsor strata in southern New Brunswick. Salt is also mined in northwest Nova Scotia. Gypsum has been quarried for decades in both provinces. Gypsum also occurs in Newfoundland where comparable Windsor marine strata are referred to as the Codroy Group.

The Canso-Riversdale Group (Mabou Group in Nova Scotia) is of Early-Middle Carboniferous (mainly Namurian) age. Its type localities are near Riversdale Station and Canso Strait areas of central Nova Scotia (Bell, 1944). Following the final marine withdrawal, a variety of red and grey terrestrial strata were deposited. Coarse conglomerates and arkosic sandstones were laid down adjacent to faulted basin-margins becoming gradually replaced by fine grained grey and red sandy beds and mudstones in the axial regions of the basin. This was also a time of gradual paleoclimatic return from arid to more subhumid conditions as evidenced by the proliferation of coalified plant remains and the first appearance of thin coal seams near the top of the Canso-Riversdale sequence. As a result, the Canso-Riversdale Group displays a considerable variety of facies that range from evaporitic to lacustrine. Riversdale

strata were thought to overlie Canso strata (Bell, 1944) but recent palynological work indicates that in the type areas of mainland Nova Scotia and Cape Breton Island the two are, at least in part, time equivalent (Howie and Barss, 1975). They appear to represent proximal (Riversdale) and distal (Canso) facies of the same sequence (Bell and Howie, 1990). Based on recent lithostratigraphic work of the Cumberland Basin of Nova Scotia, Ryan et al. (1991a) suggest that the names Canso-Riversdale be dropped and that the upper and mainly coarse strata be assigned in part to the overlying and younger Cumberland Group, whereas the finer grained lower beds be assigned to the Mabou Group, a division initially proposed by Belt (1964, 1965). It is not yet certain whether these changes are valid outside the Cumberland Basin. For the western (New Brunswick) part of the basin the Canso-Riversdale division, as suggested by Howie (1979) to include all strata overlying the Windsor Group and lying below the Pictou Group (or Cumberland Group where present), will be retained here until a more formal stratigraphic review of the Chegnecto Bay area of New Brunswick, currently in progress by H.W. van de Poll is completed. However, we adopt the suggested revisions of Mabou and Cumberland groups in Nova Scotia.

Low-grade concentrations of sedimentary copper-silver are locally present in Canso-Riversdale strata of southeastern New Brunswick (van de Poll, 1978; van de Poll and Sutherland, 1976).

The Canso-Riversdale Group, as used initially, is widely present throughout the Maritimes Basin and varies in thickness from a few hundred metres in platformal areas (Hacquebard, 1972; Bell and Howie, 1990) to perhaps as much as 3500 m or more in the Canso Strait-George Bay area between mainland Nova Scotia and Cape Breton Island (Hacquebard, 1972).

The Cumberland Group as originally defined by Bell (1944) is restricted to the Cumberland Basin of Nova Scotia and just offshore along the Fundy coast of New Brunswick. The type section includes the classical coastal cliffs of the Joggins area, northwestern Nova Scotia. Here the Cumberland Group consists of approximately 2700 m of red and grey fluvial conglomerates, sandstones, coal measures, and mineable coal. The sequence is gradational with the Canso-Riversdale (Mabou Group) below, and the Pictou Group above, and locally oversteps older Carboniferous strata onto basement (Bell and Howie, 1990). A Middle Carboniferous (Westphalian B) age was initially assigned to the Cumberland Group (Bell, 1944; Hacquebard and Barss, 1958; Hacquebard, 1960). Subsequent work has shown that its age extends from middle Westphalian A through Westphalian B into early Westphalian C (Kelley, 1967; Howie and Barss, 1975), and even early Westphalian D (Ryan et al., 1990a). Strata of known Westphalian B age are not known to occur anywhere else in the Maritimes Basin outside the Cumberland Basin, and their absence suggests a paraconformity of considerable lateral extent. Cumberland equivalent strata in the Sydney Basin of Cape Breton Island are assigned to the Morien Group of Westphalian C to Stephanian age (Bell, 1938; Macquebald, 1972). Equivalent strata in Newfoundland are referred to as the Barachois Group (Bell, 1948; Hacquebard, 1972). Coal has been extensively mined from Cumberland Group strata of which the Joggins and Springhill areas were the major producers.

The Northumberland Strait Supergroup contains the Pictou Group and overlying Prince Edward Island Group (van de Poll, 1989). The age of the supergroup extends from Middle Carboniferous to late Early Permian (early Westphalian C to Artinskian). Its type area is contained in the coastal exposures and cliffs of Northumberland Strait.

In New Brunswick, the Pictou Group has been subdivided into four formations (Gussow, 1953) or four fining-upward cyclic sequences (van de Poll, 1970, 1973a). The Prince Edward Island Group is subdivided also on the basis of fining upward megacycles into five formations and two members (van de Poll, 1989). In western Nova Scotia, the Pictou Group has been subdivided into three formations (Ryan et al., 1990a).

Because of post-depositional northeasterly tilting and erosion of the strata, the mid Carboniferous-Lower Permian sequence is younger and thicker to the northeast. The thickness increases from a few tens of metres along the margin in southwest New Brunswick to more than 3500 m beneath eastern Prince Edward Island (van de Poll, 1983) and 5000 m in the Magdalen Basin (Bell and Howie, 1990).

The Pictou Group has its type section along River John of northwestern Nova Scotia (Bell, 1944). Pictou strata in Nova Scotia occur in two major sequences, a mainly grey, coal-bearing facies and a predominantely red, essentially non coal-bearing facies (Bell, 1944). The red Pictou facies occurs predominantly in the Cumberland and Sackville basins whereas the grey facies occurs everywhere else in Nova Scotia (see Hacquebard, Fig. 7, Columns 3, 7, 8, 12, 14, 17 and 18). Both the grey and red facies of the Pictou Group comprise fine- to medium-grained sediments that were deposited in lacustrine, paludal, and flood plain environments (Hacquebard, 1972). The cause of the colour difference is not known.

Pictou strata of the New Brunswick Platform and Moncton Basin differ from those in Nova Scotia in that they consist of alternating grey and red facies. The grey facies consists in ascending order of fining upward grey-buff fluvial conglomerates, coarse- to fine grained sandstones, locally with coal measures, minor lacustrine limestone, and thin coals. The grey facies is capped in cyclic order by laterally extensive fine-grained red facies sequences comprising maroon fine- to very fine-grained sandstones, mudstones and locally minor lacustrine limestones (Gussow, 1953; van de Poll, 1970, 1973a; Ball et al., 1981). Coal is mined from Pictou strata, notably from the Sydney-Glace Bay area of Cape Breton, the Stellarton area of mainland Nova Scotia, and the Minto-Chipman area of New Brunswick. Sub economic copper-silver and uranium occurrences are locally present, of which those of the Tatamagouche area in Nova Scotia are best known.

Prince Edward Island Group strata are gradational above the Pictou Group. Their age ranges from late Stephanian to Artinskian. Coastal exposures in cliffs on Prince Edward Island are the type area. The Prince Edward Island rocks are in all respects similar to the Pictou Group of New Brunswick except that they are red, rather than grey-and-red, and they lack coal. Plant-fossil imprints and fossil tree trunks, locally in growth position, are present, predominantly in the lower part of the redbeds (van de Poll, 1983; van de Poll and Forbes, 1984). The transition from grey coarse to fine facies (New Brunswick)

to redbeds (Prince Edward Island) is contained within a diachronous sequence in which the coarse facies may be either grey or red. The transitional sequence below ranges in thickness from approximately 100 m in western Prince Edward Island to nearly 2300 m in the east. This increase in thickness is accompanied by a west to east expansion of the time limits from an interval that is confined to the Late Carboniferous in the west to an interval that spans the upper part of the Middle Carboniferous to the Early Permian in the east (van de Poll, 1983). The diachronous grey-red colour change complicates the lithostratigraphy where colour is used as a criterion in differentiating units. Lower Pictou equivalents in western Nova Scotia, for example, are now assigned to the Cumberland Group because of their predominantely grey colour (Ryan et al., 1990a, b), whereas the overlying redbeds are retained in the Pictou Group. In New Brunswick, however, time equivalent strata are grey and red and have been assigned to the Pictou Group (Gussow, 1953), not because of lithology or colour, but because of age.

Structural geology

Four principal periods of deformation affected upper Paleozoic strata of the Canadian Appalachians. The first affected mainly Horton age strata of the Moncton Basin (Gussow, 1953; Ruitenberg and McCutcheon, 1982; Nickerson 1991, 1992; St. Peter, 1992, 1993), the Mabou Basin (Kelly, 1967), the Bay St. George and Deer Lake basins (Belt, 1969; Hyde, 1984a), and the Gulf of St. Lawrence region (Durling and Marillier, 1990, 1993). Deformation comprising folding, faulting and uplift, resulted in widespread erosion prior to deposition of the Windsor Group. In the Moncton and Deer Lake basins fold axes are oblique to regional faults, suggesting a dextral strike-slip setting (Webb, 1963; Hyde, 1979). Seismic evidence from the Moncton Basin suggests that pre-Windsor deformation is either thrust-related along northwest-dipping listric faults (Gussow, 1953; Nickerson, 1991, 1992), or transpressive (St. Peter, 1987, 1992; Foley, 1989). Webb (1963) interpreted the Windsor fault movement in the Moncton Basin as dextral strike-slip.

The second period of deformation is regional (St. Peter, 1987) and commenced during Riversdale time. Its upper time limit is not accurately known. It involved Windsor strata, with flowage of salt into pillows, ridges, and diapirs. It is possible that the effects of the second deformation have in many areas been overprinted by prolonged flowage of salt. In the Moncton Basin, an angular unconformity within the Riversdale Group (Gussow, 1953) provides an upper time limit. Elsewhere, salt tectonics have affected all post-Windsor strata (Howie, 1988), including those of Early Permian age in eastern Prince Edward Island (Frankel; 1966; van de Poll, 1983). The ambiguous upper time limit for the second deformation has led some workers to conclude that it lasted throughout the mid Carboniferous (e.g., Ruitenberg and McCutcheon, 1982). It is not known how much of the observed deformation reflects mid Carboniferous tectonism related to dextral strike-slip faulting, how much is tectonically induced by salt flow, and how much reflects a subsequent (post Early Permian) period of deformation.

453

The second period of deformation was followed by local erosion and planation of uplifted strata. It was in turn followed by regional subsidence and sedimentary onlap onto adjacent platformal areas. During this time, the Maritimes Basin reached its maximum extent and probably covered most if not all of the northeastern Appalachians, except for a narrow north-south-trending divide centred on central Maine and New Hampshire (van de Poll, 1973; Gibling et al., 1992).

The third period of deformation involved regional uplift and tilting of basement blocks along reactivated basement faults during Middle to Late (?) Permian time (van de Poll, 1970). It marks the onset of the destructional stage in the evolution of the Maritimes Basin that eventually led to the present structural outline and horst-and-graben morphology. Structures of the third deformation are gentle, open

synclinal folds representing fault-bounded graben or half graben that formed over earlier-formed basinal areas (e.g., Marysville and the Moncton basins of southwestern New Brunswick, and the Magdalen Basin). It also caused widespread tilting of strata, and the development of broad anticlinal warps across buried basement highs (e.g., Egmont Bay in western Prince Edward Island, van de Poll, 1983). Timing is believed to be post Early Permian-pre Late Triassic (van de Poll, 1983; Hacquebard and Cameron, 1989) because gentle folding affected Early Permian redbeds of Prince Edward Island and because of the existence of an angular unconformity between mid Carboniferous and Triassic strata in the Bay of Fundy region.

The fourth and final period of deformation in the Maritimes Basin involved subsidence, tilting, and normal faulting that also affected Triassic rocks.

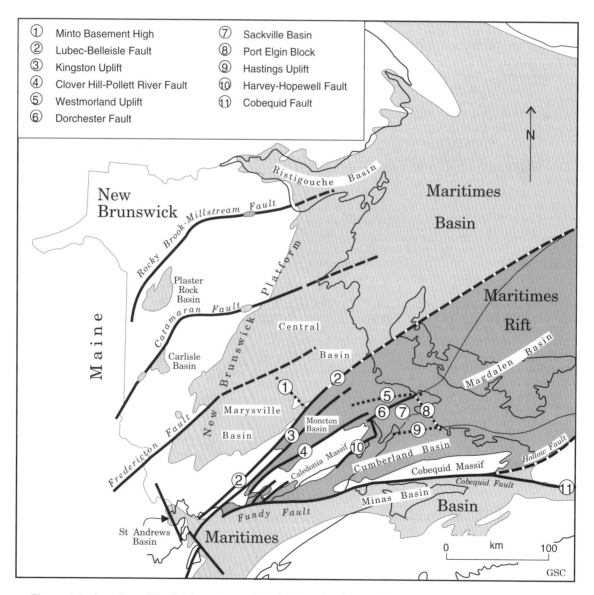

Figure 5.2. Location of the Ristigouche and St. Andrews basins and the western (New Brunswick) part of the Maritimes Basin and Maritimes Rift as well as the location of its major structural elements and faults.

There are two main interpretations concerning the structural controls of the Maritimes Basin. One favours a stress regime that was largely extensional (continental rifting) punctuated by periods of transverse compression; the other favours a dextral wrench tectonic setting. Evidence cited in support of transverse rifting and compression includes the apparent "basin and range", block-faulted topography of the basement floor (Howie and Barss, 1974, 1975; Howie, 1984), the presence of thick alluvial fanglomerates adjacent to upland source areas (Gussow, 1953; Shroder, 1963; Popper, 1965; Belt, 1968b), the presence of northeast-trending normal and high angle reverse faults in southern New Brunswick (Gussow, 1953; van de Poll, 1970, 1972; Rast, 1983, 1984; Ruitenberg and McCutcheon, 1982; Plint and van de Poll, 1982; Rast et al., 1984) and locally in Nova Scotia (Boehner 1991), and the lack of evidence for large-scale lateral offset of Windsor Group strata on opposite sides of the Belleisle Fault (McCutcheon and Robinson, 1987).

A dextral strike-slip model was presented by Bradley (1982), who envisaged a longitudinal pull-apart structure. This model is based on dextral strike-slip motion on the major faults that transect upper Paleozoic strata (Bradley, 1982; St. Peter, 1987): including the Belleisle, Kennebecasis, Cloverdale, Cobequid, and Harvey-Hopewell faults in southern New Brunswick (Webb 1963, 1969); the east-west-trending Cobequid-Chedabucto Fault (Eisbacher, 1969; Webb, 1969; Keppie, 1982); the Hollow Fault of Nova Scotia (Yeo and Ruixiang, 1987); the Big Pond Fault System of Cape Breton Island (Bradley and Bradley, 1986); and the major faults bounding Newfoundland Carboniferous basins (Popper, 1970; Knight, 1983; Hyde, 1984a; Hyde et al., 1988).

The age of faulting is in most cases unknown, except where unconformable overstep is preserved (e.g. Durling and Marillier, 1993; St. Peter, 1993) or provenance relations are known. Most faults had important post-Carboniferous movement, which complicates structural analysis. It remains unclear whether the Maritimes Basin evolved partially or entirely in a wrench-tectonic setting or whether it could have commenced initially as a wrench basin that changed subsequently to a rift basin or vice versa.

Most workers would probably agree with the following observations and interpretations: (1) compressional and tensional forces during the late Paleozoic were directed to the west or northwest and to the east or southeast; (2) northeast- and east-trending reactivated basement faults (Acadian Orogeny) played an important role. They imparted a component of anisotropy to the fractured basement and syn- and/or post-Carboniferous movement may or may not accurately reflect the principal stress regime of that time; (3) not all parts of the Maritimes Basin reacted similarly, nor at the same time, to the dominant regional stress; and (4) dextral strike-slip along the east-trending Cobequid-Chedabucto Fault led to thrusting in southern New Brunswick (e.g., Rast, 1979; Brown, 1986a) and the Moncton Basin (e.g., Gussow, 1953; Webb, 1963) and/or to transpression in these areas (e.g., Nance and Warner, 1986; Martel, 1987; St. Peter, 1987; Foley, 1989).

In the sections that follow, upper Paleozoic rocks of the Canadian Appalachians are discussed first for New Brunswick, Prince Edward Island, and the Îles de la Madeleine (H.W. van de Poll), then for Nova Scotia (M.R. Gibling), and Newfoundland (R.S. Hyde). This division reflects the affiliation

and local expertise of the authors. It also reflects differences in tectonic histories and sedimentation in the different provinces.

Much of northern and central New Brunswick formed a broad, steadily subsiding platform with relatively thin and little deformed upper Paleozoic strata. Coeval strata extend beneath the Gulf of St. Lawrence and occur in the subsurface of Prince Edward Island, where a broadly similar stratigraphic succession is recognized. Thick and deformed strata are present along the Bay of Fundy coast in southern New Brunswick where they are associated with major faults that were active during the late Paleozoic.

During the late Paleozoic, Nova Scotia was transected by a tectonically active zone (Maritimes Rift) where numerous interconnected depocentres were generated with sedimentation related to motion on major boundary faults. As in the contiguous structural region of southern New Brunswick, the strata are highly deformed locally.

The Carboniferous rocks of Newfoundland occupy two major basins, the Bay St. George and Deer Lake basins, where strike-slip faults exerted major controls on sedimentation. These faults are a continuation of fault systems characterizing the Maritimes Rift in southern New Brunswick and Nova Scotia. Several small outliers of upper Paleozoic rocks are also present.

A major part of the upper Paleozoic succession of the Canadian Appalachians lies beneath the Gulf of St. Lawrence and present Atlantic continental shelf (Scotian Shelf and Grand Banks). Strata in these offshore areas are incompletely known, and are referred to only briefly in this review. Present knowledge of these strata is summarized by Sanford and Grant (1990) and Bell and Howie (1990).

Economic deposits (including metalliferous ores, coal, evaporites and hydrocarbons) are mentioned in their stratigraphic context in each provincial treatment. Other descriptions are found in Chapter 9 by H. S. Swinden and others.

An overview of the upper Paleozoic rocks within the regional framework of the Canadian Appalachians is found in Chapter 11 by H. Williams.

UPPER PALEOZOIC ROCKS, NEW BRUNSWICK, PRINCE EDWARD ISLAND, AND ÎLES DE LA MADELEINE

Introduction

H.W. van de Poll

Upper Paleozoic rocks of New Brunswick and Prince Edward Island are almost continuous across the southeastern half of the province from Chaleur Bay to the Bay of Fundy and the adjacent Gulf of St. Lawrence area. They define a series of basins (Fig. 5.2). Some of these were discrete depocentres at their initiation in the Devonian and

van de Poll, H.W.
1995: Upper Paleozoic rocks: New Brunswick, Prince Edward Island and Îles de la Madeleine; *in* Chapter 5 of Geology of the Appalachian-Caledonian Orogen in Canada and Greenland, (ed.) H. Williams; Geological Survey of Canada, Geology of Canada, no. 6, p. 455-492 (also Geological Society of America, The Geology of North America, v. F-1).

Early Carboniferous sharing little subsequent development, others shared a later history with adjacent basins during the Middle Carboniferous or commenced at that time. The rocks of northern and central New Brunswick are relatively thin and less deformed than those of southeastern New Brunswick. They are assigned to the Plaster Rock, Carlisle, Marysville, and Central basins. This area has been referred to as the New Brunswick Platform (Poole, 1967), although the thicknesses and kinds of upper Paleozoic rocks involved are not typically "platformal" facies. In southeastern New Brunswick, the rocks are thicker and more deformed, and assigned to the Moncton, Sackville, Cumberland, and Magdalen basins. This area is referred to as the Maritimes Rift, comprising a central region of thick deposition and relatively intense deformation within the Maritimes Basin. The upper Paleozoic rocks at Chaleur Bay in the north and Passamaquoddy Bay in the south are isolated occurrences assigned to the Ristigouche and St. Andrews basins, respectively.

Isolated basins
Ristigouche Basin

The name Ristigouche Basin is informally applied here to designate the basinal setting of a Lower to Middle Devonian and Lower Carboniferous (?) molasse suite in the Chaleur Bay region of northern New Brunswick and eastern Gaspésie (Fig. 5.3). Exposures of post Acadian terrestrial conglomerates, sandstones and mudstones along both coasts of Chaleur Bay suggest that the basin was a relatively narrow east- to southeast-trending structure. Its eastern limit is not accurately known. Possibly, the several fault bounded east-southeast-trending basins with rocks of approximately the same age and extending along strike in the Gulf of St. Lawrence towards the Îles de la Madeleine (see Durling and Marillier, 1990, Fig. 3b) belong to the same structure.

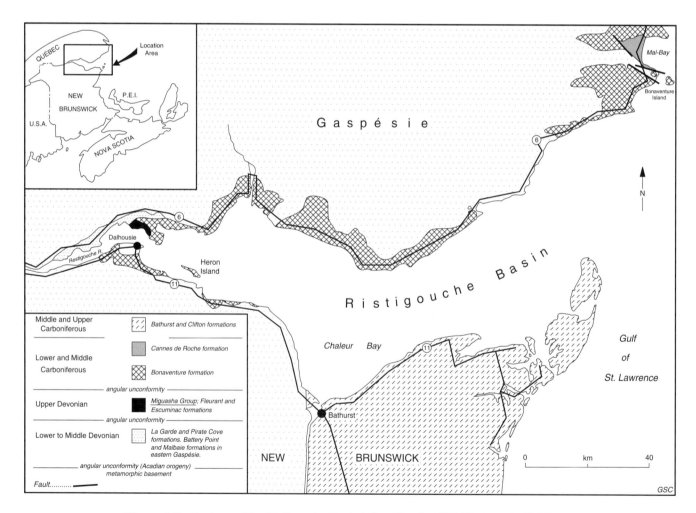

Figure 5.3. Geology of the Ristigouche Basin (after Alcock, 1935; Rust et al., 1984).

Table 5.2. Lithostratigraphic subdivision of post Acadian rocks of the Ristigouche Basin (after Dinely and Williams, 1968a).

System	Group	Formation	Lithology
upper Lower? Carboniferous		Bonaventure Fm. (>180m thick) (includes the Cannes de Roche in eastern Gaspésie) (>75 m thick)	red conglomerate sandstone, mudstone, minor grey strata at top of Cannes de Roche Fm.
		---------------------------angular unconformity---------------------------	
Upper Devonian	Miguasha Group	Escuminac Formation (40 m thick?)	fossiliferous, grey-green sandstone and shale
		Fleurant Formation (18 m thick)	mainly grey conglomerate and sandstone
		---------------------------angular unconformity---------------------------	
Lower to Middle Devonian		Pirate Cove Fm. (500 m thick) Malbaie Fm. in eastern Gaspésie (estimated 1300 m thick)	red conglomerate sandstone, shale, fossil plant remains
		La Garde Fm. Battery Point Fm. in eastern Gaspésie (estimated up to 1500 m thick)	grey and buff sandstone, shale and conglomerate, plant, vertebrate and invertebrate remains
		---------------------------basal angular unconformity (Acadian Orogeny)---------------------------	

The Ristigouche Basin developed in a platformal area. Sedimentological evidence suggests that it commenced as a separate structure in Devonian time and became part of the Maritimes Basin in the Early Carboniferous.

Post Acadian strata of the Ristigouche Basin have been studied by Alcock (1935), Dinely and Williams (1968a, b), Rust (1981, 1984a, b), Legun and Rust (1982), Zaitlin and Rust (1983) and others. This summary is largely based on their reports and references quoted therein.

The Ristigouche Basin contains mainly Middle Devonian to Middle Carboniferous terrestrial red and minor grey beds that include at least one grey lacustrine interval (Escuminac Formation). The strata are folded, faulted and include two angular unconformities. The sequence has been subdivided as shown in Table 5.2: (after Dinely and Williams, 1968a).

The oldest continental, post Acadian orogenic strata are assigned to the La Garde and Pirate Cove formations (Battery Point and Malbaie formations in eastern Gaspésie). These overlie metamorphic basement with angular unconformity and are in turn unconformably overlain by Upper Devonian strata of the Miguasha Group.

La Garde Formation (Battery Point Formation of eastern Gaspésie) includes beds that were previously assigned to the Middle Devonian Gaspé Sandstone (Dineley and Williams, 1968a; see also Alcock, 1935 for a detailed description). The latter is extensively exposed in the interior of Gaspésie. La Garde Formation consists of a basal breccia (Athollville beds) grading upwards into buff, grey and yellowish conglomerate, sandstone, and shale. Conglomerate clasts consist of locally derived volcanics, granodiorite, jasper, vein quartz, and quartzite. Limestone clasts are absent, a feature that distinguishes La Garde Formation from the overlying Pirate Cove Formation. This distinction also separates Battery Point from Malbaie conglomerates in eastern Gaspésie (Rust, 1981). La Garde Formation is interpreted to be mainly fluvial-deltaic in origin with minor shallow marine incursions (Alcock, 1938) and is well known for its Devonian plant remains, vertebrate fragments (ostracoderm and placoderm), and small invertebrates (Dawson, 1882). Fossils suggest an Early to Middle Devonian age (Dineley and Williams, 1968a).

It becomes coarser upwards and rests conformably on Devonian marine-shelf strata (Rust, 1981).

The Battery Point Formation defines the base of the post orogenic sequence in eastern Gaspésie. The sequence consists mainly of sandstone representing distal braided and meandering streams flowing from the east-southeast (Cant and Walker, 1976). The Battery Point Formation grades upward into the coarser, more conglomeratic Malbaie Formation (McGerrigle, 1950).

The Pirate Cove Formation (Malbaie Formation of eastern Gaspésie) includes the top part of the Gaspé Sandstone of Alcock (1935). The unit comprises red conglomerates, coarse to fine sandstones, and mudstones.

The conglomerates are poorly sorted and are made up of angular to subangular clasts of limestone up to 30 cm in diameter. Sparse plant remains are locally present; miospores indicate a possible Early to Middle Devonian age (D.C. McGregor, op. cit.; Dineley and Williams, 1968a). Facies relationships suggest that the Pirate Cove represents an alluvial fan complex derived from the northwest (Dineley and Williams, 1968a).

The Malbaie Formation overlies the Battery Point Formation in eastern Gaspésie with an abrupt conformable contact. Like the Pirate Cove conglomerates, Malbaie conglomerates display an abundance of Lower Paleozoic limestone clasts. The Malbaie Formation is estimated to be 1300 m thick (Rust, 1976). It is interpreted as a proximal alluvial-fan, fed from two sources: from the south for its conglomerates, and west-southwest for its sandstones (Rust, 1981, 1984a). The Malbaie Formation is overlain by the mid-Carboniferous Cannes de Roche Formation with angular unconformity.

The Miguasha Group is of Late Devonian age (Alcock, 1935; Dineley and Williams, 1968a) and is made up of the Fleurant and Escuminac formations. It overlies the Pirate Cove Formation with angular unconformity and is in turn unconformably overlain by basal conglomerates of the Lower to Middle Carboniferous Bonaventure Formation. The Miguasha Group is approximately equivalent or slightly older than the Horton Group of southern New Brunswick and Nova Scotia, and the Anguille Group of Newfoundland. Miguasha strata are apparently absent in eastern Gaspésie.

The Fleurant Formation is fluvial in origin and consists almost entirely of grey conglomerates containing subrounded to rounded clasts of Lower Devonian and older volcanic and metasedimentary rocks. Boulders up to 1 m in diameter occur at the base of the sequence. Paleocurrent directions based on crossbedding suggest a Fleurant source-area to the east and northeast (Dineley and Williams, 1968a).

The Escuminac Formation consists of grey-green, thin bedded sandstones, sandy shale and mudstone. Its lithology is quite distinct from other continental strata of the Ristigouche Basin. The Escuminac is noted for its well preserved Late Devonian fish fossils and excellent examples of Late Devonian ferns. Several of the ferns are said to be identical to the ones found in the Perry Formation of northeastern Maine and in strata of the Catskill Group of New York (Alcock, 1935). Facies relationships and fossil content suggest a lacustrine setting with periodic deposition from density underflows. Paleocurrent directions are unimodal from the east (Dineley and Williams, 1968b).

The Bonaventure Formation (Cannes de Roche Formation in eastern Gaspésie) is extensive in coastal exposures mainly on the north side of Chaleur Bay. It extends from the head of the Bay around Dalhousie to Bonaventure Island in eastern Gaspésie. The Bonaventure Formation is mainly flat-lying and overlies all older strata with angular unconformity. It is characterized by a predominance of coarse clastics and a prevailing deep red colour. Conglomerate units with boulders up to 50 cm in diameter are not uncommon near its base. Local derivation of the clasts is indicated by their heterogeneous lithologies which include volcanic rocks, limestone, quartzite, and intrusive rocks.

The western exposures of the Bonaventure Formation have been studied by Zaitlin and Rust (1983) who interpret the facies relationships to reflect tectonic rejuvenation and alluvial fan progradation under semi-arid conditions. Paleocurrents suggest axial drainage to the east; an almost complete reversal of the northwesterly transverse drainage of the underlying Devonian continental strata of the Ristigouche Basin.

The age of the Bonaventure is not accurately known. Its redbeds are unfossiliferous and there are no overlying strata. The formation differs lithologically from the mid Carboniferous Bathurst and Clifton formations exposed east of Bathurst on the opposite shore of Chaleur Bay. An early-Mid Carboniferous age has been assigned by Alcock (1935).

The Cannes de Roche Formation is confined to the southwestern and western shore of MalBaie and inland along the Beatty and Portage rivers in eastern Gaspésie. Exposures of the Cannes de Roche are separated from Bonaventure strata by a narrow, upthrust ridge of Ordovician limestone (Alcock, 1935). The lower part of the Cannes de Roche Formation consists of a 10-12 m thick sequence of red conglomerates containing angular clasts of siliceous limestone derived from the Lower Devonian Grant Grève Formation (Rust, 1981). This is overlain by 30-35 m of red and green fine sandstone and mudstone which in turn is overlain by a 30-35 m thick upper sequence of grey-buff conglomerate, sandstone, and mudstone; the latter contains plant remains. The Cannes de Roche Formation overlies the Middle Devonian Malbaie Formation with angular unconformity (Rust, 1976) and is generally thought to be equivalent to the Bonaventure Formation (Alcock, 1935). Although the red and grey strata of the Cannes de Roche Formation are structurally conformable, there are important differences in lithology, morphometrics, and fossil content. In addition, Rust (1981, 1984b) reported an important change in paleocurrent directions and facies contrasts between the red and grey strata. The redbeds are interpreted as alluvial-fan deposits with a principal derivation from a nearby source area to the northeast accumulating in a northwest-southeast paleovalley. In contrast, the grey sequence is interpreted as having been deposited by a perennial trunk river system originating from the northwest and flowing southeast through this valley (Rust 1984b).

Rust (1984b) briefly reviewed two possible causes of the facies and colour change; a change in groundwater level, or a paleoclimatic change to more humid conditions. Such an apparently sudden change in facies, provenance, colour, sediment maturity, and degree of preservation of plant remains also occurs elsewhere in the Maritimes Basin, notably in Canso-Riversdale strata of southern New Brunswick (e.g., van de Poll, 1966, 1970, 1973c; McCabe and Schenk, 1982; Belt, 1965; Ryan et al., 1990a). There, it coincides approximately with the Lower to Middle Carboniferous transition. Based on plant-fossil evidence Bell (op. cit., Alcock, 1935) concluded that the grey beds of the Cannes de Roche Formation were probably mid Carboniferous in age, which is similar to the Canso-Riversdale.

Correlation of the Bonaventure and Cannes de Roche formations is tenuous. It is based on lithological similarity between relatively fine grained Bonaventure strata at Heron Island near the head of Chaleur Bay and red Cannes de Roche strata at Malbaie, some 160 km distant (Fig. 5.3).

Because the red to grey transition within the Cannes de Roche Formation is probably more significant than the apparent similarity with the Heron Island beds (which do not include grey strata), the Bonaventure Formation may be older (e.g., Early to Mid Carboniferous). In terms of Maritimes Basin stratigraphy, the Bonaventure Formation could be a Windsor-Canso correlative, whereas the Cannes de Roche might be a Canso-Riversdale correlative. The interpreted arid paleoclimatic setting for the Bonaventure Formation (Zaitlin and Rust, 1983) supports a Windsor-Canso correlation.

Paleocurrent directions for the Ristigouche Basin, though highly variable, were predominantly to the north, northwest, and west during Devonian time (Cant and Walker, 1976; Rust, 1981, 1984a; Dinely and Williams, 1968a, b) and predominantly east and southeast during Carboniferous time (van de Poll, 1973a; Rust, 1981; Legun and Rust, 1982; Zaitlin and Rust, 1983). This raises an important question concerning the Late Devonian paleogeography of the region and the relationship of the Ristigouche Basin to the Maritimes Basin. It seems that the angular unconformity at the base of the Bonaventure Formation and its prevailing easterly transport directions may reflect not just local uplift and basinal subsidence but more significantly, a complete reversal in paleoslope from northwest to southeast. This reversal implies two stages of development for the Ristigouche Basin: a Devonian initial stage as a separate structure, followed by an Early Carboniferous stage as part of adjacent basins.

St. Andrews Basin

The St. Andrews Basin (Schluger, 1973) contains Devonian strata and, like the Ristigouche Basin, may have developed as a separate structure. It is situated in the Passamaquoddy Bay region of southwestern New Brunswick and adjacent Maine (Fig. 5.4) and contains a single stratigraphic unit named the Perry Formation (Smith and White, 1905). The Perry Formation is known from drillhole intersections to underlie a large part of Passamaquoddy Bay (Cumming, 1967). The redbeds have most recently been studied by McIlwaine (1962, 1967), Cumming (1967), Schluger (1973, 1976), Gates (1984, 1989) and Stringer et al. (1991). This summary is largely based on their reports and references quoted therein.

The Perry Formation is Late Devonian and consists of red conglomerates, coarse- to fine-grained arkosic sandstones, mudstones, and minor nonmarine limestone. Three layers of basalt are present in the middle of the sequence. In addition, pre-syn(?) and post-Perry dykes are distinguished of which the east-northeasterly trending Ministers Island Dyke is best known (Stringer and Burke, 1985, Dunn and Stringer, 1990, Stringer et al., 1991). The belt of upper Devonian (?), dark grey and red plant-bearing conglomeratic sedimentary rocks of uncertain stratigraphic position in the Beaver Harbour area are included in the formation (Cumming, 1967, Stringer et al., 1991).

Coastal sections along St. Croix River in Maine and the opposite shore on St. Andrews Peninsula are the only exposures that offer continuous sections. Dips are variable from subhorizontal to subvertical with overall dip directions towards the centre of Passamaquoddy Bay. Noncylindrical folds are locally extensively developed in the upper part of the Perry Formation (Stringer et al., 1991).

Schluger (in part after Smith and White, 1905) recognized five members in the Perry Formation of the St. Andrews area with a total thickness of 757 m (revised to 1076 m by Stringer et al., 1991), and six members in the Blacks Harbour area with a combined thickness of 1170 m.

Earlier workers thought the entire Perry Formation to have been deposited in one basin. Schluger (1973) on the other hand concluded that the formation was deposited in two basins, the St. Andrews Basin to the northwest and the Blacks Harbour Basin to the southeast, each with different source areas and separated by a northeast-trending highland divide with its axis extending from Deer Island to Mascarene Peninsula (Fig. 5.4). However, the notion of a highland divide is not supported by the overall facies distributions that change laterally and upwards from coarse and conglomeratic alluvial-fan deposits along the present basin margin, through arkosic sandstone and sandy mudstone (channel-overbank facies) to mainly mudstone (lacustrine facies) in the central basin area (Gates, 1989; Stringer et al., 1991). The present separation of the St. Andrews Basin and so-called Blacks Harbour Basin could equally well be explained by later uplift. The centripedal paleocurrent directions (McIlwaine, 1962, 1967; Schluger, 1973) lend additional support to the interpretation of a single basin centred on Passamaquoddy Bay (Fig. 5.4).

In most areas the present margin of the basin is marked by a basal angular unconformity or nonconformity. The Lubec-Belleisle Fault marks the present southeastern limit of the St. Andrews Basin.

The Oak Bay Fault is a major north-northwest-trending feature transecting the basin along the St. Croix River. The mapped offset of the unconformable basal Perry contact and Perry basalt layers across the Oak Bay Fault on opposite shores of St. Croix River is small (Cumming, 1967; Gates, 1989; Stringer et al., 1991). In addition there is little or no detectable offset of the Early Jurassic Ministers Island Dyke across the Oak Bay Fault. These observations suggest that the fault movement was mainly pre-Perry and had essentially ceased by Early Jurassic time. Williams and Hy (1990) on the other hand interpreted the Oak Bay Fault as a transfer fault associated with the opening of the Atlantic Ocean and that displacement on the fault was approximately coeval with the intrusion of late Triassic-Early Jurassic dykes (Williams and Hy, 1990; Stringer, 1991).

The Late Devonian age assigned to the Perry Formation (Smith and White, 1905) is based on poorly preserved comminuted plant remains occurring in sandstone and mudstone of adjacent Maine. Although the preservation of the plant fossils is poor, the Late Devonian age is said to have been verified by F.M. Hueber, Curator, Division of Paleobotany, National Museum, Washington, D.C. (Smith and White, 1905; Schluger, 1973).

Basins of the New Brunswick Platform

The name New Brunswick Platform (Poole, 1967) refers to all Upper Devonian to Permian-Carboniferous deposits lying to the west and northwest of the Maritimes Rift (Fig. 5.2). In contrast to the relatively thick and rather extensively deformed strata of the rift zone, New Brunswick platformal rocks are mainly mid Carboniferous in age, relatively thin and undeformed except for rocks, in part Devonian, in the Marysville Basin. The pre-Carboniferous

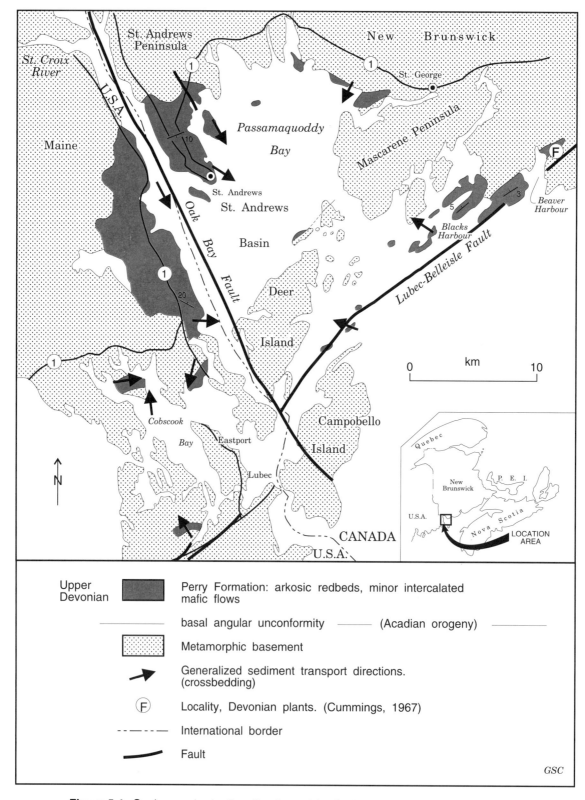

Figure 5.4. Geology and paleoflow directions of the St. Andrews Basin (after Cummings, 1967; McIlwaine, 1967; Schluger, 1973).

basement underlying the platform is interpreted from seismic evidence to be a gently undulating peneplain exhibiting a few hundred metres of relief except perhaps beneath the Newcastle Miramichi Bay region where a graben structure some 1700 m deep and 3-5 km wide may be present (Wade et al., 1977; Kingston and Steeves, 1984). The New Brunswick Platform was initially thought to extend northeast into the Strait of Belle Isle region of Newfoundland (Poole, 1967) but recent seismic evidence suggests that the offshore extension of the platform beneath the Gulf of St. Lawrence is probably more complex than previously envisaged. Durling and Marillier (1990) indicated the presence of several east-southeast-trending fault-bounded molasse basins of Horton age between Chaleur Bay and the Îles de la Madeleine suggesting that early rifting in that part of the Maritimes Basin may have been important. As used here, the New Brunswick Platform is essentially confined to New Brunswick and the adjacent Northumberland Strait.

Upper Paleozoic strata of the New Brunswick Platform are contained in the Marysville and Central basins as well as several outliers preserved in local structural depressions referred to as the Plaster Rock and Carlisle basins, and smaller unnamed outliers preserved along faults.

Plaster Rock Basin

The Plaster Rock Basin is a north-northeast-trending erosional remnant or outlier of Lower Carboniferous rocks centred on the town of Plaster Rock in western New Brunswick (Fig. 5.5). The basin covers approximately 825 km² and occupies part of a structural depression in the axial region of the northeast-trending Chaleurs Bay Synclinorium. The latter consists of folded Silurian-Devonian metasedimentary and metavolcanic rocks lying between the Rocky Brook- Millstream Fault to the west and the Miramichi Highlands (Anticlinorium) to the east (St. Peter, 1979).

The rocks of the Plaster Rock Basin are only mildly deformed. Dips in excess of 20° are uncommon except near faults; more commonly, dips are 0-10° towards the centre of the basin.

Parts of the Plaster Rock Basin have most recently been remapped by Hamilton (1965) and St. Peter (1979). The following account is mainly a summary of their reports and references quoted therein.

The rocks of Plaster Rock Basin have been informally subdivided as shown in Table 5.3).

The Gladwyn Basalt overlies pre-Acadian basement (lower Devonian Wapski Formation) with angular unconformity. The basalt is local and where absent, basal conglomerates of the Arthurette Redbeds occur at the base of the section. The basalt is brown-weathering, porphyritic, and amygdaloidal. It is extensively altered throughout and deeply weathered in the upper 2 m. A red lithic sandstone lies locally at the base of the basalt.

The Arthurette Redbeds overlie the basalt and consist mainly of red and grey conglomerates, sandstones and minor mudstones. They probably represent debris flow and braided stream deposits (St. Peter., 1979).

The Plaster Rock Formation overlies the Arthurette Redbeds conformably and consists of nodular limestone and/or red mudstone at the base, overlain by nodular gypsum and red mudstone. The transition from limestone-dominated to gypsum-dominated facies is gradational. The unit appears unfossiliferous nor have microfossils been found in the formation (Globensky, 1962). The evaporites are calcrete and gypcrete probably formed by diagenetic replacement from intrastratal evaporation within a basin of internal drainage during a period of excessive evaporation (St. Peter., 1979).

The Unnamed red shale unit consists of red mudstone capped by red conglomerate. It overlies the Plaster Rock Formation conformably.

Table 5.3. Lithostratigraphic subdivision of Carboniferous rocks of the Plaster Rock Basin (after St. Peter, 1979).

System	Formation	Lithology
Middle Carboniferous (?)	unnamed unit	Mainly red shale, red conglomerate at the top (>240 m thick)
Lower Carboniferous	Plaster Rock Fm.	Red shale, nodular limestone, calcareous and gypsiferous red shale (150-300 m thick)
	Arthurette Redbeds	Consists of a grey conglomerate-sandstone-shale member and a red conglomerate-sandstone-shale member (total 550 m thick)
	Gladwyn Basalt	Tholeiitic basalt (30 m thick), minor lithic arenite at base
---angular unconformity (Acadian Orogeny)---		
Lower Devonian	Wapske Formation	Metasedimentary and volcanic rocks

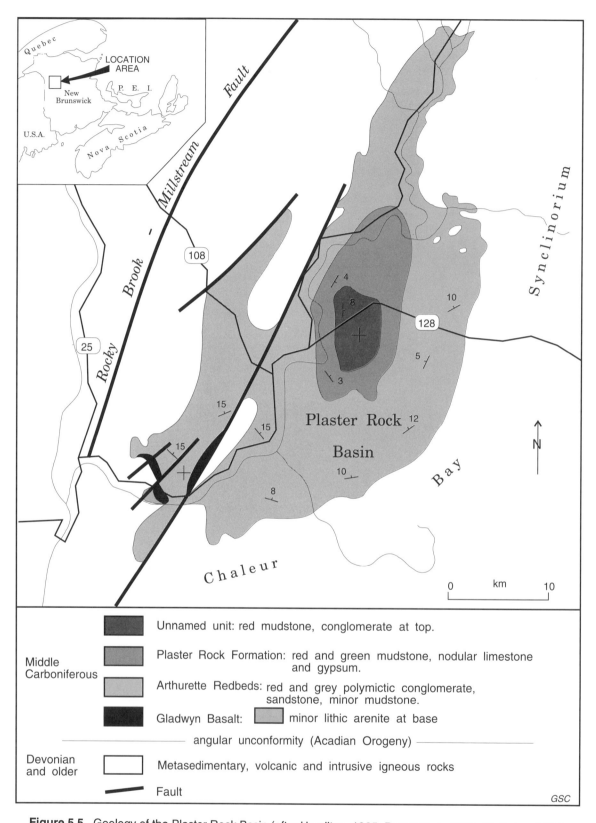

Figure 5.5. Geology of the Plaster Rock Basin (after Hamilton, 1965; Potter et al., 1979; St. Peter, 1979).

The age of the sequence within the Plaster Rock Basin is not accurately known. Previous workers (e.g., Rose, 1936; Sund, 1958; Webb and Kingston, 1975) thought the evaporites to be marine and lithogenetic correlatives of the Mississippian Windsor Group of southern New Brunswick. This raises the question, as to where and how the "Windsor sea" might have crossed the Central Basin and the adjacent Miramichi Highlands without leaving a trace of its presence. Sund (1958) thought the Plaster Rock evaporites originated in an arm of the "Windsor sea" through northern New Brunswick. This could have been linked via the Ristigouche Basin (Fig. 5.2) except that the approximately equivalent Bonaventure redbeds have no known associated marine evaporites. Similarly, the apparent absence of marine strata in the Marysville Basin of southern New Brunswick suggests that the Plaster Rock Basin was not connected to the sea at any time. Based on these constraints, Plaster Rock evaporites are probably nonmarine (as suggested by St. Peter, 1979) and represent a playa setting in a region of internal drainage. They represent an arid paleoclimatic interval that may have been synchronous with the arid Windsor paleoclimatic interval of Early Carboniferous time.

The original size and shape of the Plaster Rock Basin are unknown. The presence of a few small outliers of probable Arthurette redbeds preserved along faults to the south of the present basin-limits indicates that the Plaster Rock Basin was at one time larger and extended southward. It is quite likely that the Plaster Rock and Carlisle basins, the latter some 50 km to the south (Fig. 5.2), were connected and that both were part of the area of platformal strata that probably covered all of the region during the Carboniferous (e.g., van de Poll, 1973a; Gibling et al., in press).

St. Peter (1979) suggested that the Plaster Rock Basin was initiated as an isolated fault-generated intermontane graben. However, judging from the generally uniform thicknesses of its formations, the gentle basinward dips, the circular facies distributions, and the prevalence of fine grained strata, it is also possible that the basin commenced as a small sag-basin or platform depression adjacent to an elevated source terrane (Miramichi Highlands).

The Plaster Rock Basin is transected by several northeast-trending reverse and normal faults (St. Peter, 1979).

Carlisle Basin

The Carlisle Basin is a north-trending structural remnant or outlier (Fig. 5.6). It covers approximately 230 km² and lies some 15 km east of the town of Hartland in western New Brunswick. Most of the recent geological work on the Carlisle Basin has been performed by St. Peter (1982a) and Venugopal (1982). This brief review is essentially a summary of their reports and the references quoted therein. The basin consists of nearly flat-lying to very gently, west-tilted Lower (?) and Middle Carboniferous strata that partially cover a northwest-trending down-faulted and folded segment of Silurian and Devonian rocks; separating the southwestern from the northeastern

extensions of the Miramichi Anticlinorium (Fig. 5.6). The latter consist of Cambrian and Ordovician metasedimentary strata intruded by Devonian and older igneous rocks.

The western boundary of the Carlisle Basin is delineated in part by the northeast-trending Howard Brook (high-angle reverse) Fault. The fault is a subsidiary of the extensive northeast-trending Catamaran-Woodstock Fault system that separates the Miramichi Anticlinorium from the Chaleurs Bay Synclinorium to the northwest.

Rocks of the Carlisle Basin have been subdivided as shown in Table 5.4.

An Unnamed mafic flow unit of tholeiitic composition lies at the base of the Carboniferous sequence (Fig. 5.6). It has a similar composition to other flows of the New Brunswick Platform including the Gladwyn Basalt of the Plaster Rock Basin and the South Oromocto Basalt near Hoyt of the Central Basin (Fyffe and Barr, 1986).

The mafic volcanic unit is only locally present in the southeastern part of the Carlisle Basin. Where absent, the more extensive redbeds of the Carlisle Formation directly overlie Devonian and older rocks with angular unconformity.

The Carlisle Formation consists mainly of red conglomerates and interbedded red lithic sandstones. Minor red mudstone with intercalations of greenish white nodular limestone (calcrete) occur locally near the top of the formation. The conglomerates are polymictic and matrix supported. They contain subangular to subrounded clasts of locally derived felsic and mafic volcanic rocks, quartzite, vein quartz, slate and schist.

The Carlisle Formation and underlying volcanic unit may correlate with the Arthurette Redbeds and Gladwyn Basalt, respectively, of the Plaster Rock Basin some 50 km to the north.

Facies and compositional characteristics of the redbeds suggest a prograding alluvial-fan setting with a nearby source terrane. Sparse crossbedding suggests that sedimentary transport was to the northwest (St. Peter, 1982a), subparallel to the local northwest structural trend of the underlying pre-Acadian basement rocks.

A small down-faulted outlier of Carlisle redbeds is preserved along the Woodstock Fault some 15 km to the southwest of the Carlisle Basin (Anderson, 1968). In contrast to other Carboniferous outliers in western New Brunswick, which tend to overlie Silurian and Devonian rocks of the Chaleurs Bay Synclinorium, these redbeds directly overlie Cambrian-Ordovician strata of the Miramichi Anticlinorium.

The Carlisle Formation is apparently unfossiliferous. Its Early Carboniferous age is inferred from lithological comparisons with similar strata in the Plaster Rock and Central basins, its unconformable base, and Middle Carboniferous cover of the Mountain View Formation.

The Mountain View Formation consists of grey-buff conglomerates and sandstones which overlie Carlisle redbeds with an abrupt conformable contact.

The conglomerates contain subangular to subrounded clasts of quartzite, felsic volcanic rocks, and minor chert. Paleocurrent directions (crossbedding) are predominantly

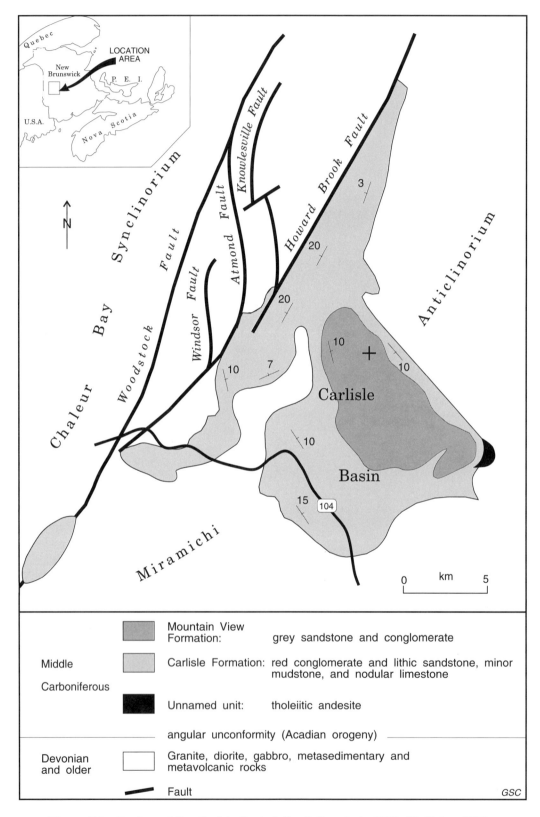

Figure 5.6. Geology of the Carlisle Basin (after Potter et al., 1979; St. Peter, 1982a; Venugopal, 1982).

Table 5.4. Lithostratigraphic subdivision of Carboniferous rocks of the Carlisle Basin (after St. Peter 1982a; Venugopal, 1982).

Middle Carboniferous	Mountain View Formation	Grey conglomerate and sandstone (>100 m thick)
Lower (?) to Middle Carboniferous	Carlisle Formation	Red conglomerate and sandstone, minor mudstone and nodular limestone (150-200 m thick)
	Unnamed mafic volcanic unit (only locally present)	Tholeiitic andesite (20-30 m thick)
-------------------------- Angular Unconformity (Acadian Orogeny) --		
Lower Devonian and older metasedimentary, volcanic and intrusive rocks		

to the northeast and subparallel to the regional northeast trend of the Appalachian Orogen (St. Peter, 1982a). This contrasts with the northwest paleotransport directions of the underlying Carlisle redbeds, suggesting that the sharp contact between the two formations may represent a substantial change in paleotopography and/or in paleoslope and possibly a disconformity of some duration. The few fossil plant-remains that have been found in the Mountain View Formation suggest a general Middle Carboniferous age (M. Copeland, pers. comm., in St. Peter, 1982a).

Marysville and Central basins

The Marysville and Central basins form a contiguous northeast-trending and diverging structure that underlies most of central and eastern New Brunswick and adjacent parts of the Gulf of St. Lawrence (Fig. 5.7). They contain gently folded and tilted strata ranging in age from Late Devonian to Late Carboniferous.

The basins are remnants of a mid Carboniferous to Permian platformal cover that probably extended across most, if not all, of southwestern and western New Brunswick and the adjacent New England States (van de Poll, 1973a; Gibling et al., 1992). A number of small Carboniferous outliers in southern and western New Brunswick (e.g., occurrences along the Fredericton Fault, the small outlier at Prince William, the Plaster Rock and Carlisle basins, as well as the small outliers along the Catamaran and Rocky Brook-Millstream faults) give some indication of the extent of post Carboniferous erosion in this region (Fig. 5.7).

The Marysville and Central basins lie across the axial region of the northeast-trending Fredericton Belt to the southeast of the Miramichi Anticlinorium. The basins are bounded to the southeast by the Lubec-Belleisle Fault that separates them from the Moncton Basin of the Maritimes Rift. To the north and northeast, the Central Basin is bounded by the Rocky Brook-Millstream Fault separating it from the Ristigouche Basin (Fig. 5.7).

Rocks of the Marysville and Central basins were considered previously as a single platformal sequence (e.g., van de Poll, 1970, 1973a). More recently, the name Marysville Basin is used for the southwestern part of the cover rocks because of substantial differences between the two basinal areas (e.g., McCutcheon, 1981; Fyffe and Barr, 1986). The Marysville Basin contains the only Upper Devonian rhyolites and associated redbeds of the New Brunswick Platform and is considerably older than the Central Basin where the oldest platformal strata are late Early to early Middle Carboniferous (top of Windsor Group redbeds). The Marysville Basin may have been initiated as a platformal depression or local sag-basin, perhaps as a result of pressure release and subsidence over a magmatic source-terrane following volcanic extrusion. The Central Basin by contrast lacks rhyolites and tuffs. Deep drilling indicates that it has many areas of Middle Carboniferous rocks that lie directly on pre Acadian metamorphic basement, with all of the Upper Devonian and Lower Carboniferous strata absent. The rocks of the Marysville and Central basins have been studied most recently by MacKenzie (1964), van de Poll (1967, 1970, 1973a), Hacquebard (1972); Gemmel (1975), McCutcheon (1981, 1990), Ball et al. (1981), Le Gallais (1983), Fyffe and Barr (1986), McGregor and McCutcheon (1988), and Hacquebard and Cameron (1989). The present review is essentially based on their publications and references cited therein.

Marysville Basin

The Marysville Basin is a northeast-trending and diverging structure with its apex centred on Oromocto Lake in southwestern New Brunswick. An angular unconformity forms the southeastern margin of the basin. The Fredericton Fault forms its northwest margin, and together with the Minto Basement High separates the Marysville Basin from the Central Basin (Fig. 5.8).

The basin-fill includes Upper Devonian rhyolites, welded tuffs and associated conglomeratic redbeds, mafic flows, minor Lower Carboniferous marine limestone and associated redbeds and middle-Upper Carboniferous grey conglomerates, sandstones, grey and red mudstones, and coal. The change from Lower Carboniferous redbeds to Middle Carboniferous grey and red strata, locally with coal, not only reflects the previously discussed paleoclimatic shift from arid (Windsor marine evaporites) to sub humid (coal measures) but also represents a major change in the evolution of the New Brunswick Platform. The rather sudden appearance of mature Middle Carboniferous quartz-quartzite

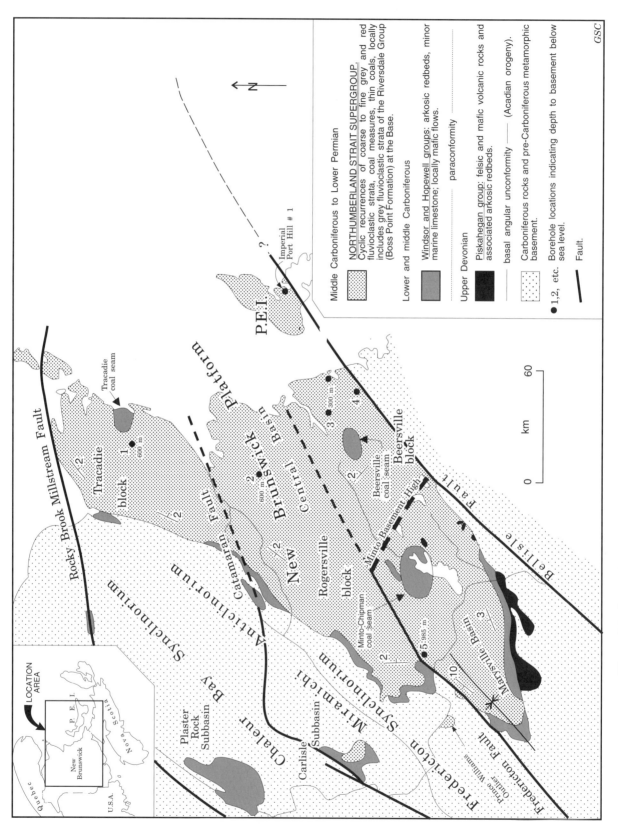

Figure 5.7. General geology of the Marysville and Central basins (modified after Potter et al., 1979).

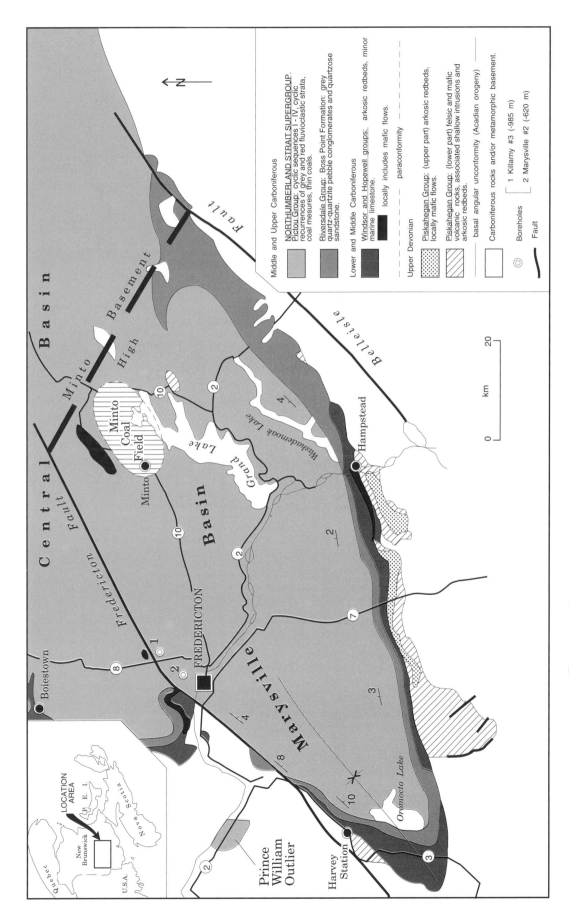

Figure 5.8. Geology of the Marysville Basin (modified after Potter et al., 1979).

467

Table 5.5. Correlation chart of Devonian-Carboniferous strata of the Marysville Basin.

Stratigraphic Position	Age	Group Names New Brunswick Platform	Lithological Description	Minto area (Muller, 1951)	Fredericton area (Freeze 1956; van de Poll, 1973a)	Harvey Station area (Pajari, 1973; van de Poll, 1973a)	Mount Pleasant area (van de Poll, 1967, 1973a; McGregor and McCutcheon, 1988)	Hampstead area (MacKenzie, 1964; van de Poll, 1973a; McCutcheon, 1981)
Upper Carboniferous	Stephanian	Pictou Group (Petitcodiac Group)	grey conglomerate, sandstone, mudstone, minor coal, red mudstone, megacyclic sequence IV		Cyclic Sequence IV			
	Westphalian D	Pictou Group (Petitcodiac Group)	grey conglomerate, sandstone, mudstone, red mudstone, megacylic sequence I, II and III		Cyclic Sequence III			
Middle Carboniferous	Westphalian C	Pictou Group (Petitcodiac Group)	grey conglomerate, sandstone, grey and red mudstone, coal measures, (Minto coal) meagacyclic sequence I, II	Sunbury Creek Formation (grey sandstone and conglomerate, minor coal, sandstone). Hurley Creek Formation (red sandstone and mudstone). Minto Formation (grey sandstone, conglomerate, Minto coal seam at top)	Cyclic Sequences I, II / or / Division B and C of Petitcodiac Group	Megacyclic Sequence I, II / or / Division B and C of Petitcodiac Group	Megacyclic Sequence I, II / or / Division B and C of Petitcodiac Group	Megacyclic Sequence I, II / or / Division B and C of Petitcodiac Group
	Late Westphalian A, B, lower C	Cumberland Group						
	Late Namurian-Westphalian A	Riversdale Group (Boss Point Fm. in New Brunswick)	grey quartz:quartzite pebble conglomerate, sandstone, minor grey and red mudstone	Division A Petitcodiac Group? (Boss Point Fm.)	Division A of Petitcodiac Group (Boss Point Fm)	Division A of Petitcodiac Group? (Boss Point Fm.)	Division A of Petitcodiac Group? (Boss Point Fm.)	Division A of Petitcodiac Group (Boss Point Fm.)
	Late Viséan-Early Namurian	(Hopewell Group)	red polymictic conglomerate, arkosic sandstone, locally mafic flows	unnamed redbeds. Hardwood Ridge mafic flows. Rhyolite-trachite plug at Cumberland Bay. Newcastle Creek Formation (redbeds)	unnamed redbeds. Royal Road mafic flows. McKinley Formation (redbeds)	unnamed redbeds	Shin Formation (redbeds)	unnamed redbeds. Queenstown mafic flows. Hopewell Group (redbeds)
Lower Carboniferous	Viséan	Windsor Group	Minor marine limestone, associated redbeds locally evaporites(?)					Windsor Group Parleeville Formation (thin unit of marine limestone)
	Tournaisian	Horton Group						
Upper Devonian	Famennian	Piskahegan Group	Rhyolite flows and tuffs, ash-flow tuffs and related granophyric intrusions. Redbeds			Harvey Formation, Harvey Mtn. Member (rhyolite flows and tuffs). York Mills Member (redbeds). Unnamed red siltstone unit as base	Kleef Formation (redbeds). Piskahegan Group (rhyolite flows and tuffs, interbedded redbeds (includes South Oromocto and Hoyt mafic flows at base. Endo and Exo Caldeira units)	Hampstead volcanics (rhyolite flows, tuffs and redbeds). unnamed red conglomerate-arkosic sandstone unit at base

Angular unconformity pre-Acadian basement rocks

pebble conglomerates of the Boss Point Formation overlying immature fanglomerates of the Hopewell Group, represents a rapid shift from local to more distal source areas in response to widespread platformal expansion by overstep across older strata in local Early Carboniferous platform depressions (e.g., Marysville Basin) directly onto basement rocks (e.g., parts of the Central Basin and Prince William Outlier, Fig. 5.7, 5.8).

The oldest rocks of the Marysville Basin (Upper Devonian, Piskahegan Group) lie along its southern margin; the youngest strata (Upper Carboniferous Pictou Group) lie along its northwestern faulted margin (Fredericton Fault). There, in Marysville #2 and Killarney #3 drillholes (Fig. 5.8) a total Carboniferous section from Late Viséan to Stephanian is represented with a thickness of about 1030 m.

The age and thickness distributions of the strata within the Marysville Basin, becoming younger and thicker northwestward from the southeastern margin towards the Fredericton Fault , suggest that the basin consists of a north-northwest tilted half-graben dipping toward the Fredericton Fault. As well, a small southwest- trending and deepening graben structure may lie beneath the Fredericton area along the Fredericton Fault (van de Poll, 1973b). The large north-northeast-trending and plunging asymmetrical syncline that forms the southwestern apex (i.e., fold hinge) around Oromocto Lake reflects the combined effects of this post Carboniferous north-northwest tilting and southeast directed dip reversal by drag adjacent to the Fredericton Fault (Fig. 5.7, 5.8).

Rocks of the Marysville Basin and their stratigraphic positions and names are summarized in five sections from around the basin (Table 5.5).

Piskahegan Group

The oldest rocks are rhyolite flows, tuffs, ash-flow tuffs, breccias and related shallow intrusive rocks, redbeds, and locally mafic flows of the Piskahegan Group (van de Poll, 1967; Gemmel, 1975; McCutcheon, 1983, 1990; McGregor and McCutcheon, 1988). They are well exposed along the unconformable southern and southwestern margins of the basin (Fig. 5.8) and are generally thought to be continuous in the subsurface below Carboniferous strata of the apex area between Mount Pleasant and Harvey Station where they have been informally named the Harvey Formation (Pajari, 1973).

Felsic volcanic rocks also occur on the southwestern flank of the Minto Basement High just east of Grand Lake. They were previously thought to be either older than the adjacent grey sandstones of the Pictou Group (van de Poll, 1970; Ball, 1981) or to be pre-Pictou (Muller, 1951; Fyffe and Barr, 1986). A recent preliminary date of 335 Ma on the Cumberland volcanics obtained by D. Boyle of the Geological Survey of Canada (Fyffe, pers. comm., 1991) indicates a pre-Pictou Viséan-Namurian age. They are trachytes and peralkaline rhyolites and differ in composition from the Piskahegan felsic rocks which are mainly peralkaline (Fyffe and Barr, 1986). The Piskahegan Group has been subdivided into formations related to intra- and exo-caldera evolution (McGregor and McCutcheon, 1988; McCutcheon, 1990).

Molybdenum, tin, tungsten, bitsmuth, and base metals occur in association with Late Devonian - Early Carboniferous granitic rocks intruded into rhyolitic volcanic rocks of the Piskahegan Group along the southwestern margin of the Mount Pleasant Caldera (Sinclair et al., 1988; McCutcheon, 1990). The latter is a partially fault-bounded depression underlain by Piskahegan volcanic rocks and associated shallow intrusions, flanked along its northern margin by Lower Carboniferous redbeds (Fig. 5.8).

The Piskahegan Group thins away from the Mount Pleasant Caldera and is estimated to be at least 450 m thick (van de Poll, 1967). The age of the Piskahegan Group and its lithostratigraphic equivalents at Harvey Station and Hampstead was initially interpreted as late Tournaisian-early Viséan. This age was based on stratigraphic proximity to overlying Windsor marine fossiliferous limestone on-strike at Hampstead, and the presence of similar rhyolitic volcanic rocks of late Tournaisian age (Moncton Formation) south of Moncton (Gussow, 1953). Recent palynological work (McGregor and McCutcheon, 1988) contradicts this interpretation. Spores obtained from the top of one of the exo-caldera redbed units (Carrow Formation), near the top of the Piskahegan Group, indicate a Famennian (Late Devonian) rather than a Late Tournaisian (Early Carboniferous) age. This creates a relatively long time-span between Piskahegan Group and Windsor Group strata suggesting that a paraconformity or disconformity separates the two units. Redbeds of the Shin Formation (Hopewell Group) above the Piskahegan Group contain relatively few rhyolite clasts suggesting a change in source area(s) for the latter.

The Piskahegan Group overlies pre-Acadian basement rocks with angular unconformity. It is overlain disconformably or paraconformably by marine limestone of the Windsor Group, or in its absence along strike, by redbeds assigned to the (post Windsor) Hopewell Group (van de Poll, 1967).

Windsor Group

The Parleeville Formation of the Windsor Group (McCutcheon, 1981) is restricted to the Hampstead area along the southern margin of the Marysville Basin (MacKenzie, 1964; McCutcheon, 1981). Its thickness is not given but probably does not exceed 50 m. The Parleeville consists mainly of light coloured coquina limestone made up of fragments of a single brachiopod and minor flaggy argillaceous strata (MacKenzie, 1964). Over its strike length the limestone rests either directly on Hampstead rhyolite, or rests with angular unconformity on basement rocks, or overlies Piskahegan equivalent red conglomerates with paraconformity.

The limited extent of the Windsor limestone may represent: (1) the western limit of the Windsor marine transgression, (2) reflect post-Windsor uplift and erosion, or (3) the Windsor limestone may be covered within the Marysville Basin by overstep. In the latter situation Windsor redbeds and possibly even sabkha-type marine evaporites could be present in the subsurface between Hampstead and Fredericton (McCutcheon, 1981). Red and grey clastics but no marine carbonates or evaporites of Late Viséan (Windsor) age occur at a depth of approximately 850 m below sea level in the Killarney #3 well just north of Fredericton (Fyffe and Barr, 1986).

Hopewell Group

Redbeds occupying the stratigraphic position of the Hopewell Group, locally with intercalated mafic flows of Namurian age, overlie and/or overstep Windsor limestone in the Marysville Basin, and also extend into the Central Basin.

The Boss Point Formation and Pictou Group are grey-buff conglomerates and sandstones of Late Namurian-Early Westphalian age that overlie Hopewell equivalent redbeds disconformably. Strata of the Hopewell Group and Boss Point Formation will be discussed more fully with rocks of the Central Basin.

Central Basin

The Central Basin, as defined here, includes all cover rocks of the New Brunswick Platform that lie to the northwest and north of the Fredericton Fault and east of the Minto Basement High (Fig. 5.7).

The Central Basin is made up of several northeast-trending fault-bounded domains, comprising basement blocks and overlying Carboniferous strata that have been tilted slightly towards either the east-southeast or northeast during post Carboniferous times. From north to south these are the Tracadie, Rogersville and Beersville blocks (informal names introduced here for convenience, see Fig. 5.7).

The Tracadie block is bounded by the Rocky Brook-Millstream and Catamaran faults. The western margin of the Carboniferous cover rocks is an angular unconformity. The Tracadie block was tilted slightly to the southeast prior to erosion and its partial exhumation. This is indicated by its orientation, by mid Carboniferous rocks that become gradually younger to the east (van de Poll, 1970; Ball et al., 1981; Hacquebard and Avery, 1984), and by the presence of several Carboniferous outliers over the block to the west of the present margin. The gentle east to southeast tilt of approximately 2° can be observed in the almost continuous Carboniferous coastal exposures east of Bathurst. Pre-Carboniferous basement was intersected in the St. Isodore #2 well west of Tracadie at a depth of approximately 600 m below sea level (Ball et al., 1981).

The Rogersville block is bounded by the Catamaran and Fredericton faults. The Rogersville block is tilted to the east-southeast at approximately 2°. This is indicated by the orientation of the basal Carboniferous unconformity along its western margin, the age progression of mid Carboniferous strata that is successively younger to the east-southeast, and the presence of Carboniferous outliers over the western (basement) extension of the block (e.g., along the Catamaran and Fredericton faults, at Prince William, and Carlisle Basin). The Rogersville #1 well near Rogersville (Fig. 5.7) intersected metamorphic basement at a depth of approximately 650 m below sea level (Howie and Cumming, 1963).

The Beersville block is a northeast-trending structure extending from east of Minto into the Northumberland Strait area. It is bounded by the Fredericton and Belleisle faults. The western limit of the Beersville block is known from the presence of several basement inliers, informally referred to here collectively as the Minto Basement High, occurring at or near the surface and also known from coal-drilling in the area (Ball et al., 1981). The uplift history

of the Minto Basement High is not clear. Windsor and Hopewell strata are absent; overlying the high are lower Westphalian C beds (Minto and Hurley Creek formations or Cyclic Sequence I) which appear overstepped by upper Westphalian C strata (Sunbury Creek Formation, Cyclic Sequence II) directly onto the basement high. Furthermore, early Westphalian C transport directions are deflected to the north and around the high suggesting that the Minto Basement High was a positive structure at that time (van de Poll, 1970). On the other hand, the age distribution showing that Pictou strata become younger to the northeast suggests that the Minto Basement High represents the upturned margin of a gently northeast tilted basement block (informally referred to here as the Beersville block). Two drillholes to the west of Buctouche (Fig. 5.7) intersected basement at approximately 250 and 300 m below sea level, indicating that the Beersville block is tilted or slopes to the northeast at about 1° or less.

Post depositional break-up of the New Brunswick Platform by slight differential tilting of basement blocks along rotational faults seemingly best explains the present geometry, lithostratigraphy and age distribution of the Permian-Carboniferous rocks including the distribution of Carboniferous outliers. There is no evidence for strike slip faulting during this interval in the evolution of the New Brunswick Platform.

Hopewell Group (Marysville and Central basins combined).

Red polymictic conglomerates, arkosic sandstones, and red mudstones of the Hopewell Group overlie Windsor limestone or older upper Paleozoic rocks with apparant conformity (MacKenzie, 1964), and overlie basement rocks with angular unconformity. The name Hopewell Group has been extended from the Moncton Basin (McKenzie, 1964) but has not been subdivided. Different stratigraphic names have been applied to these redbeds in different parts of the Marysville and Central basins (see Table 5.5). Mafic flows, present at or near the top of the group occur in surface exposures at Hardwood Ridge (near Minto), Queenstown (near Hampstead), Currie Mountain, Royal Road and Manzer Brook (near Fredericton), and at Boiestown. Mafic flows of the Hopewell Group also occur at depth in the Killarney #3 well and drillhole 62-1. The former probably represents the down-faulted extension of the nearby Manzer Brook basalt, whereas the latter is the down-faulted extension of the Hardwood Ridge basalt.

Fyffe and Barr (1986) classified the mafic flows of the Hopewell Group as alkalic basalts and trachy andesites, in contrast to the older Piskahegan mafic flows which are tholeiitic. They attributed the change in composition to changes in tectonic evolution and interpreted the change to be related to the percentage of partial melting in the mantle. This, according to them, suggests greater crustal extension in the Late Devonian followed by less crustal extension in the early-Middle Carboniferous (Fyffe and Barr 1986).

Hopewell redbeds, although thin (estimated here to be up to 200 m in thickness), are relatively common on the New Brunswick Platform and give the first indication of regional subsidence and overstep of Carboniferous strata that occurred throughout the Middle Carboniferous to

Early Permian (Bradley, 1982). Their distribution, composition, and general appearance is that of alluvial fan deposits in areas of low topographic relief.

The Hopewell Group is described more fully with rocks of the Moncton and Cumberland basins.

Riversdale Group.

The Boss Point Formation (also referred to as Division A of the Petitcodiac Group by earlier workers, MacKenzie, 1964) consists of grey-buff quartz-quartzite pebble conglomerates and quartzose sandstones. It overlies the Hopewell Group redbeds with a sharp and well defined contact and is paraconformably or disconformably overlain by grey and red strata of the Pictou Group. The abrupt transition from immature polymictic and arkosic redbeds with subangular clasts to grey mature quartz-quartzite pebble conglomerates with subrounded to rounded clasts is an important junction in the tectonostratigraphic and thermal (?) development of the Central Basin. Bradley (1982) incorporated the change into a transition interval between an initial phase of rapid fault-controlled subsidence and the onset of a subsequent phase of gradual, more widespread thermal subsidence and expansion of the basin by sedimentary overstep onto adjacent platformal areas. In terms of lithostratigraphic development, the junction represents the relatively sudden change from proximal alluvial-fan to distal alluvial-plain, from braid-plain to meander-plain, and from playa lake to coal-swamp sedimentation (McCabe and Schenk, 1982). Paleoclimatically it represents the rapid change from semi-arid to subhumid conditions, and once established, subhumid to perhumid conditions probably prevailed throughout the Middle Carboniferous-Early Permian (van de Poll 1978, 1983; D'Orsay and van de Poll, 1985b, c).

The Boss Point Formation appears thin (estimated here to be up to 100 m thick) and intermittently exposed on the New Brunswick Platform. Its rather characteristic quartz-quartzite pebble conglomerates have been recognized locally along the southern margin of the Marysville Basin as well as near Fredericton (van de Poll, 1970) and also near Bathurst (Ball et al., 1981). This common view has been challenged by Browne (1990) however, who concluded from paleosol evidence, lithological composition, and bedforms that the Boss Point represents a sandy braided river system with a not too distant source and deposited under semi-arid conditions.

The Boss Point Formation is of Late Namurian (early Middle Carboniferous) age and lies at the base of younger Middle-Upper Carboniferous strata (Pictou Group) that form the most widespread platformal unit of this part of the Maritimes Basin (Gussow, 1953). Because of the similarity in colour and lithology, some earlier workers combined the Boss Point and Pictou strata into a single unit named the Petitcodiac Group (, MacKenzie, 1964). It was subsequently realized from palynological evidence that the Petitcodiac Group includes a substantial depositional hiatus that lasted through most of Westphalian A and B (Gussow, 1953). Whilst Boss Point strata (Petitcodiac Division A) throughout central and southern New Brunswick are late Namurian, perhaps early Westphalian A in age, the next youngest Petitcodiac (Division B) strata are early Westphalian C. There is no paleontological evidence for the presence of Westphalian B strata anywhere in New Brunswick except at the southern tip of Cape Maringouin which falls within the Cumberland Basin (see also Ball et al., 1981, plate 80-87; Plint and van de Poll, 1982). Recent work has indicated the presence of Westphalian B polynomorphs in the redbeds of Salisbury Formation in the Moncton Basin (St. Peter, 1993). This Suggests that the Westphalian B strata do occur outside the Cumberland Basin and that Westphalian B hiatus may not be as extensive as first thought. Consequently, the name Petitcodiac Group is now dropped and many workers attempt to make a distinction between the Boss Point and Pictou strata on maturity and morphometrics (e.g. van de Poll, 1970; Ball et al., 1981).

Northumberland Strait Supergroup

The name Northumberland Strait Supergroup was introduced by van de Poll (1989) to formally incorporate the Prince Edward Island redbeds (Late Carboniferous to Permian age) into the regional lithostratigraphy of the New Brunswick Platform. The supergroup consists of two groups, the Pictou Group below and the Prince Edward Island Group above. Further revisions are probably needed to make the Permian-Pennsylvanian lithostratigraphy of New Brunswick and Prince Edward Island compatible with that of Nova Scotia, particularly in light of recent proposed changes for the Cumberland Basin (e.g., Ryan et al., 1991a).

Pictou Group

The Pictou Group of the New Brunswick Platform consists of fining-upward recurrences of grey-buff pebble and cobble conglomerates at the base (not necessarily everywhere present), overlain by lithic sandstones of similar colour, which in turn fine upwards into grey and purplish red fine grained sandstones and mudstones, locally with a single coal seam up to 50 cm thick. The conglomerates are mainly submature with approximately 50% subrounded quartz and quartzite clasts. This feature distinguishes the Pictou from the older Boss Point conglomerates which contain a greater percentage of quartz and quartzite clasts (up to 90%, van de Poll, 1970) that are, on the whole, also better rounded than those of the Pictou. The Pictou conglomerates vary in thickness from a few metres to massive, laterally continuous units, tens of metres thick. Drilling results suggest that the total Pictou section overlying the New Brunswick Platform is up to 1000 m thick.

Local subdivision of the Pictou Group is essentially based on the recognition of these alternating recurrences of relatively coarse grained grey strata and finer grained red sequences (see Table 5.6). The repetitive nature of these units in New Brunswick was initially recognized by Muller (1951) and Gussow (1953) who assigned formation names to mainly grey and red units based largely on colour and grain size. This is turn led to subsequent subdivision of all of the Pictou Group in southern New Brunswick, as well as the redbeds of Prince Edward Island, into a total of 8 megacyclic sequences (van de Poll, 1970, 1973a, 1989). In each sequence, the relatively coarse grained grey and overlying finer grained red lithofacies were not given separate formational status, as done by previous workers, but are combined into a single mappable unit of formational status with palynological age given by coal seams where present.

Table 5.6. Age and stratigraphic subdivision of Middle and Upper Carboniferous strata, Central Basin and adjacent Prince Edward Island.

System	Age	Group names	Western part, Maritimes Basin. General (after van de Poll 1973a, 1989)	Bathurst - Shippigan area (after Ball et al., 1981; Hacquebard and Avery, 1984; fig. 3.2)	Minto - Chipman Area (after Muller, 1956)
Lower Permian	Sakmarian / Asselian	Prince Edward Island Group	Orby Head Formation	Concealed interval Northumberland Strait	overlying strata eroded
Upper Carboniferous	Stephanian		Hillsborough River Formation Malpeque and Wood Islands members	Clifton Formation Member C	
			Kildare Capes Formation		
			Egmont Bay Formation	disconformity?	
			Miminigash Formation (upper red lithofacies only)		
		Northumberland Strait Supergroup	Concealed interval, Northumberland Strait		Sunbury Creek Formation Member C
Middle Carboniferous	Westphalian C and D		Cyclic Sequence IV (basal grey lithofacies only)	Clifton Formation Member B	Sunbury Creek Formation Member A and B (Moffat Brook coal seam)
		Pictou Group	Cyclic Sequence III (Beersville coal seam)		Hurley Creek Formation
			Cyclic Sequence II (Moffat Brook and Lakestream coal seams)		Minto Formation (Minto coal seam)
			Cyclic Sequence I (Minto coal seam)		
	Late Westphalian A and Westphalian B	Cumberland Group	Not present	Not present	Not Present
	Late Namurian - Early Westphalian A	Riversdale Group	Boss Point Formation	Clifton Formation, Member A (Boss Point Formation)	Not present ?
Lower Carboniferous	Late Viséan - Middle Namurian	Hopewell Group	Shin and McKinley formations	Bathurst Formation	Newcastle Creek Formation

older Devonian-Carboniferous rocks and/or pre-Acadian metamorphic basement

The redbeds of Prince Edward Island display identical fining-upward megacyclic sequences, except that the coarse facies are red, not grey, and coal is absent. Subdivision of the New Brunswick Pictou Group into cyclic sequences (in the order of 150-250 m thick) is difficult where inland exposure is poor and the finer grained, mainly red strata are preferentially eroded. Another problem is that the large cyclic sequences are difficult to recognize in drill cuttings (Ball et al., 1981), although they are more easily identified in diamond drill core (van de Poll, 1970; Le Gallais, 1983). The usefulness of megacyclic sequences as mappable lithostratigraphic units is best demonstrated in Prince Edward Island, where more or less continuous coastal sections in combination with compositional differences of the basal conglomeratic units provide improved lithostratigraphic control (van de Poll, 1989). The various subdivisions of the Northumberland Strait Supergroup in use today are shown in Table 5.6.

The interpretive significance of the alternating relatively coarse grey and finer grained red facies in cyclic order is not clear. They were first thought to be related to paleoclimatic variations, ranging from somewhat more humid (coarser grey lithofacies) to less humid (finer red lithofacies, van de Poll, 1970, 1973a; Legun and Rust 1982) with or without periodic uplift in the source area(s). However, the red sequences in New Brunswick are now known to widely contain pisolitic ferricretes (hematite nodules, van de Poll, 1983), ranging from a few millimetres (so called buckshot-ore, Mohr and van Baren, 1954) to several centimetres in diameter. Modern analogs of pisolitic ferricretes form in lateritic soils of tropical regions. They are a common feature in warm to hot subhumid or perhumid areas where periods of considerable rainfall alternate with periods of lower rainfall (Mohr and van Baren, 1954; Frankel and Bayliss, 1966; Nahon et al., 1977). It would appear therefore that the paleoclimate during the depositional intervals of the red fine sandstone-mudstone facies was monsoonal and still quite humid (see also D'Orsay and van de Poll, 1985b, c). Alternatively, if the grain size and colour changes are related to climate, the grey strata might reflect somewhat lower evaporation rates (i.e., lowering of temperatures) or more uniform humidity levels rather than seasonal changes in humidity. Another possibility is that the relatively coarse grained grey lithofacies are not climatically controlled but reflect the development of multi-story, multi-lateral channel facies of braid plains and/or meander belts traversing a large alluvial plain of mainly lateritic soils. If this is true, the utility of the magacyclic sequences as mappable units on a regional scale would be severely curtailed. Finally, the observed cyclicity could perhaps be a combination of all these factors. Whatever their origin, the megacyclic sequences have proven to be useful map units on a local and perhaps even a regional scale (e.g., Prince Edward Island). Their lateral continuity on an even larger scale across platformal and basinal elements of the Maritimes Basin is highly unlikely however. For instance, they are absent in (proximal?) time-equivalent strata of the Cumberland Basin, suggesting that the megacyclic development is either restricted to platformal areas (van de Poll and Ryan, 1985) and/or require perhaps a certain minimum distance from their source area(s) to allow development.

The Middle and Late Carboniferous age distribution of the Pictou Group from palynological data, summarized by Ball et al. (1981) is shown in Figure 5.9. The data show the strata to become generally younger in the directions of tilt, i.e., to the north-northwest in the Marysville Basin and generally easterly elsewhere (see also Hacquebard and Avery, 1984, fig 3.2). There is presently insufficient palynological information to conclude whether the Boss Point and/or early Pictou strata are present or not over the northeasterly extension of the Kingston Uplift. The apparently continuous time lines on either side of the Kingston Uplift as shown in Figure 5.9 displays no evidence for post Late-Carboniferous strike-slip movement. It could be however that the Kingston Uplift itself has either shifted laterally to the northeast relative to time equivalent strata on either side in the Central and Moncton basins, or that it became an uplifted and northeast tilted block during post Carboniferous time (probably post Early Permian time, van de Poll, 1983). The latter interpretation is favoured on borehole and structural evidence (van de Poll, 1983), magnetic data (Bhattacharyya and Raychandhuri, 1967), and on structural trends and basement subdivisions (St. Peter and Fyffe, 1990).

The presence of upper Carboniferous strata at the surface in the Fredericton area just south of the Fredericton Fault and adjacent to Middle Carboniferous strata immediately northwest of the fault, indicates that rocks of perhaps even younger age extended into southwestern New Brunswick. The possible magnitude of post depositional tilting and erosion of Carboniferous strata between central New Brunswick and Prince Edward Island is indicated by the vertical distribution of spores in the Imperial PortHill No. 1 well in western Prince Edward Island. Drilling results record the same stratigraphic order at depth as is present at the surface between New Brunswick and Prince Edward Island (Barss et al., 1963). Restoration for tilt of the strata suggests that a minimum of 800 m of Carboniferous-Permian strata has been eroded from the Minto-Chipman area since Late Carboniferous time (van de Poll, 1970, fig. 16; see also Hacquebard and Avery, 1984, fig. 3.3).

Paleocurrents over the New Brunswick Platform indicate predominantly northeast to east sediment transport. The major source area(s) is/are inferred to have lain to the south and southwest (van de Poll, 1973a, 1983; Gibling et al., 1992).

Relatively thin coal seams of limited areal extent characterize Pictou strata of the New Brunswick Platform (Ball et al., 1981). Of these, only the Minto-Chipman seam of the Grand Lake area has been mined for over a century. The seam is approximately 50 cm thick and yields coal of poor quality (high ash and sulphur contents). It is presently used only for generating electricity.

Basins of the Maritimes Rift

The western part of the "Maritimes Rift" consists of the Moncton, Sackville, southwestern part of the Cumberland, and Magdalen basins, separated in part by uplifted basement blocks or massifs, including the Kingston, Westmorland and Hastings uplifts, and the Caledonia and Cobequid massifs (Fig. 5.10). This region is fragmented into a number of positive and negative generally northeast-tilted and fault-bounded domains comprising basement

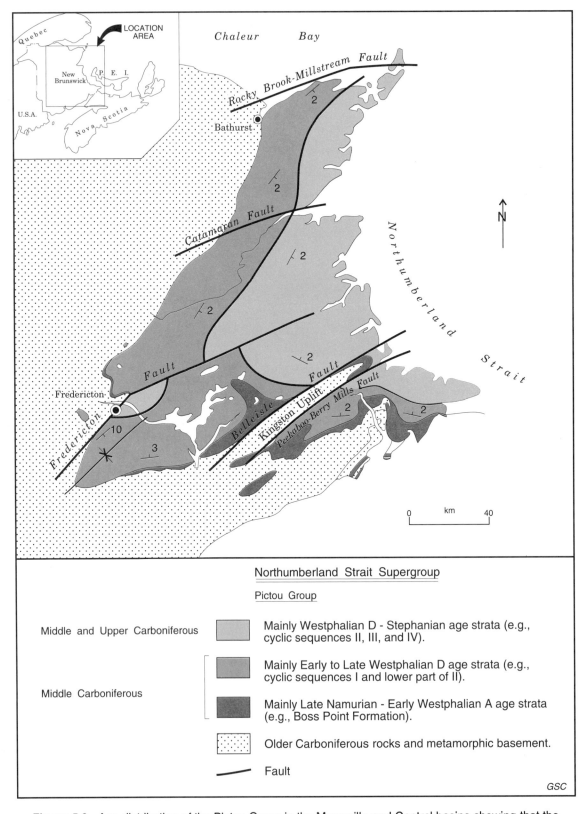

Figure 5.9. Age distribution of the Pictou Group in the Marysville and Central basins showing that the strata become younger in the direction of tilt, mainly to the south-southeast and east-northeast, except in the Marysville Basin where the strata become younger to the north-northwest (after van de Poll, 1970, 1973a; Ball et al., 1981; Hacquebard and Avery, 1984).

blocks overlain in part by Permian-Carboniferous strata. In the central part of the rift, basement fragmentation and tectonic instability took place over a longer period (commencing in the Late Devonian and lasting at least into the mid Permian) and was considerably more severe than on the adjacent New Brunswick Platform. Differential uplift of basement massifs and subsidence of basins at different times in different places led to a complex and highly variable lithostratigraphy that can usually be resolved within individual basins but cannot be readily extended from one basin to another, or from basinal to platformal areas (Kelley, 1967). To demonstrate this intrabasinal variability and avoid duplication in lithological descriptions, the stratigraphy of the Moncton, Sackville and Cumberland basins (New Brunswick part) will be discussed together. The stratigraphy of the Magdalen Basin beneath the central Gulf of St. Lawrence is still incompletely known and is discussed separately.

The Moncton Basin is a northeast-trending and diverging structure with its present apex positioned near the city of Saint John in southern New Brunswick. Its northeastern limit has not been adequately defined and may in fact merge with the Magdalen Basin.

The Moncton Basin is bounded to the northwest by the New Brunswick Platform from which it is separated by the Belleisle Fault, and to the southeast by the Caledonian Massif. The latter is an uplifted domain of metamorphic basement that separates the Moncton Basin from the Cumberland Basin to the east (Fig. 5.10).

Northeast of the Caledonian Massif, the northeast-trending Dorchester Fault separates the Moncton Basin from the Sackville Basin to the southeast.

Structures and lithofacies of Carboniferous strata of the Moncton Basin show that most of the tectonic activity took place during the Early and Middle Carboniferous and that the Late Carboniferous was mainly a period of relative quiescence. Differential uplift and associated northeast-tilting and partial exhumation of basement blocks during post Early Permian times led to the present horst-and-graben topography (alternating areas of basement rocks and Upper Devonian and Permian Carboniferous cover strata). The following elements make up the Moncton Basin:

1) The Kingston Uplift is a narrow northeast-trending domain (Fig. 5.10) that extends from southern New Brunswick into Prince Edward Island (van de Poll, 1983; St Peter and Fyffe, 1990). It is bounded in part by the northeasterly trending Belleisle and Peekaboo-Berry Mills faults to the northwest and southeast, respectively (Gussow, 1953). The uplift consists of metamorphic basement, overlain with angular unconformity by folded and faulted Lower Carboniferous rocks (Horton, Windsor, and Hopewell Groups). These are in turn unconformably overlain by gently northeast-dipping (<5°) Middle Carboniferous to Permian strata. The easterly strike of the basal unconformity and the gradual younging of the sequence to the northeast suggest that the Kingston Uplift was elevated and slightly rotated (southwest side up, northeast side down) during post Early Permian time prior to erosion.

2) The Kennebecasis block (informal name), lies to the southeast of the Kingston Uplift. It is bounded on the northwest by the Peekaboo-Berry Mills Fault and to the

southeast by the Clover Hill-Pollett River Fault (Fig. 5.10). The Kennebecasis block is downfaulted with respect to the Kingston Uplift and Caledonia Massif, and consists of metamorphic basement, overlain with angular unconformity by steeply-dipping, folded, and faulted Lower Carboniferous strata. These are in turn unconformably overlain by mildly deformed Middle Carboniferous beds that become gradually younger to the north-northeast toward Prince Edward Island. This younging trend suggests that this block, like the adjacent Kingston Uplift, was rotated a few degrees to the north and northeast during post Early Permian time. The strata are downfaulted against the Peekaboo-Berry Mills Fault and now form a northerly tilted half-graben against the Kingston Uplift (see also Howells and Roulston, 1991).

3) The Westmorland Uplift is an east-northeast-trending domain of basement rocks lying at shallow depth beneath Middle Carboniferous cover strata (Fig. 5.10). Basement is exposed in the Gaytons Quarry near Calhoun, is displayed in seismic data and has also been intersected along strike in several wells and drillholes (Gussow, 1953). Near vertical Lower Carboniferous strata can be seen to lie along the faulted southern margin of the Westmorland Uplift. The uplift is itself unconformably overlain by basal Boss Point conglomerates of Middle Carboniferous age; the timing of uplift therefore was probably late Early Carboniferous.

4) The Hillsborough Basin lies between the Westmorland Uplift and Caledonia Massif (Fig. 5.10). It is a relatively small basin that contains folded and faulted Lower Carboniferous red and grey terrestrial strata and oil-shales of the Horton Group. Lacustrine beds of the Albert Formation contain oil and natural gas in the small and mainly depleted Stoney Creek oil and gas field, and a small mass of rock salt including glauberite. Horton strata are deformed in response to uplift of the Westmorland block (St. Peter, 1992).

5) The Petitcodiac Basin lies between the Westmorland Uplift and the New Brunswick Platform (Foley, 1989). Little is known about this basinal structure (Fig. 5.10).

The Caledonia Massif is an uplifted domain of metamorphic basement that forms the southern boundary of the Moncton Basin (Fig. 5.10). It is locally unconformably overlain by, and faulted against Carboniferous strata. Coarse Lower Carboniferous fanglomerates along its margins at different stratigraphic levels suggest that this basement block has been repeatedly upfaulted and rejuvenated (Carter and Pickerill, 1985a). Provenance studies show that by early Middle Carboniferous time the Caledonian Massif ceased to be a source area and was probably buried beneath a relatively thin cover of Late Carboniferous rocks (van de Poll, 1966).

The Sackville Basin (Martel, 1987) is a relatively small fault-bounded structure on the northeast sub-surface extension of the Caledonia Massif (Fig. 5.10). It is bounded on the west side by the Memramcook River Fault, on the northwest by the Westmorland uplift, on the east by the Port Elgin Block, separating the Sackville from the Magdalen Basin and on the south side by the Hastings Uplift (Martel, 1987). The latter separates the Sackville Basin from the adjacent Cumberland Basin to the south.

475

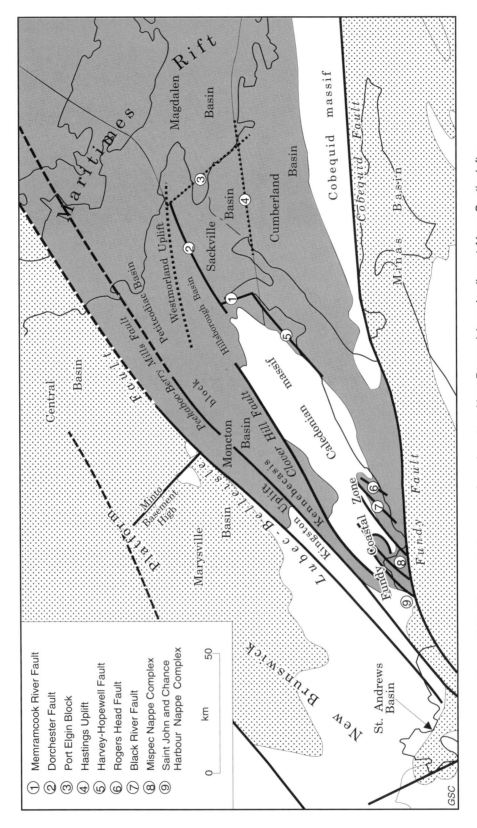

Figure 5.10. Structural elements of southeastern New Brunswick and adjacent Nova Scotia (after Martel, 1987; Foley, 1989).

Table 5.7. Age, stratigraphic position, and lithostratigraphic subdivision of the Horton Group, Moncton Basin (after Carter and Pickerill, 1985a).

	Windsor Group	Subzones A & B	Marine limestone, evaporites and redbeds		
Lower Carboniferous	Horton Group	Moncton Formation	Hillsborough Member (redbeds) Boyd Creek Tuff (felsic) Weldon Member (redbeds)		
		Albert Formation	Round Hill Member	Gautreau Member non marine evaporites	
				Hiram Brook Member calcareous shale sandstone	
				Frederick Brook Member oil shale, calcareous siltstone	
				Dawson Settlement Member grey sandstone, shale conglomerate	
		Memramcook Formation		Arkosic redbeds	
Upper Devonian	————— angular unconformity————————(Acadian Orogeny)——————————				
	pre-Acadian metamorphic basement				

The thick section of mainly Westphalian B strata characteristic of the Cumberland Basin appear to be absent in the adjacent Sackville Basin.

The Cumberland Basin is a relatively large east-trending structure extending from southeastern New Brunswick into Nova Scotia (Fig. 5.10). It is bounded in part on the west by the Caledonia Massif, on the north by the Hastings Uplift, and on the south by the Cobequid Massif. Sedimen-tological and stratigraphic aspects of the Cumberland Basin are discussed later in this chapter by M.R. Gibling.

The Magdalen Basin is modified somewhat from the definition of Bradley (1982) and Howie (1988) and defined here as a northeast-trending structure that extends from Baie Verte (Port Elgin block of New Brunswick and Nova Scotia) to southwestern Newfoundland, and from the western margin of the Maritimes Rift to just west and offshore of Cape Breton Island (Fig. 5.11). The Magdalen Basin is the largest and probably the deepest late Paleozoic basin within the Canadian Appalachian region (Bradley, 1982).

Moncton, Sackville, and Cumberland basins

The southwestern part of the Maritimes Rift contains a relatively thick section of uppermost Devonian and Lower to Upper Carboniferous rocks. The oldest strata are exposed in the southwestern part of the Moncton Basin as well as along its margins, becoming gradually younger towards the northeast (Fig. 5.12). Deep drilling on Prince Edward Island demonstrates that a thick section comprising all the Carboniferous-Early Permian groups

except for Westphalian B strata lies on strike in the subsurface extension of the Moncton Basin (van de Poll, 1983).

Horton Group

The Horton Group consists of three formations. In ascending order these are the Memramcook, Albert, and Moncton formations (Table 5.7).

The Memramcook Formation contains the oldest post Acadian strata of the Moncton Basin. This brief review is largely based on the publications and references of Gussow (1953), Popper (1965), McLeod (1980), and Carter and Pickerill (1985a, b). The Memramcook Formation is thought to vary between 135 and 2350 m in thickness (Gussow, 1953). It consists of red basal conglomerates, overlain by arkosic sandstones and mudstones. The conglomerates comprise subangular to subrounded clasts of mainly granite, rhyolite, and mafic volcanic rocks. They are coarse alluvial fan deposits originating from nearby source areas such as the Caledonia Massif. Its late Devonian age is based on its conformable relationship with the overlying Albert Formation which is of Tournaisian age (Early Carboniferous; Hacquebard, 1972) and the presence of Late Devonian miospores in the type section of Stony Creek (Carr, 1968).

The Albert Formation conformably overlies the Memramcook Formation and consists of several terrestrial lithofacies including alluvial fan, fluvial delta, playa lake, evaporite carbonate-mudflat, and perennial lacustrine (Carter and Pickerill, 1985a). Stratigraphic work was done by a substantial number of people including Gussow, 1953; Greiner, 1962, 1974; Howie, 1968; Pickerill and Carter,

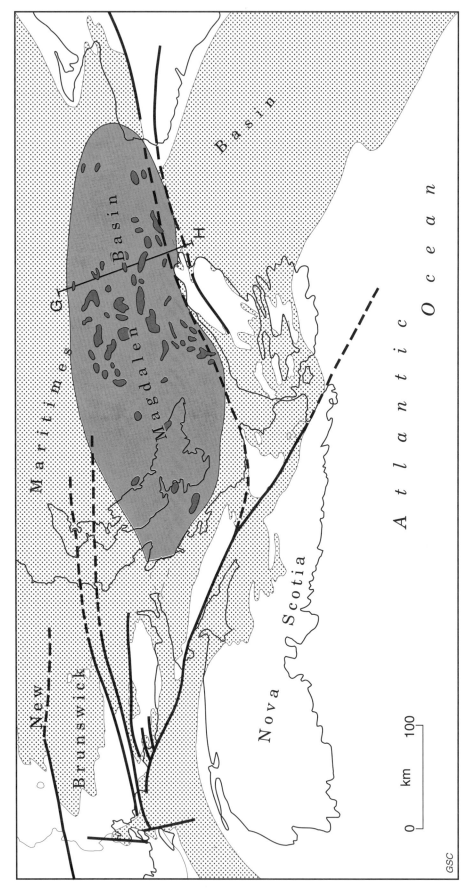

Figure 5.11. Location of the Magdalen Basin (modified after Howie, 1984; Bell and Howie, 1990). Dark areas within the Magdalen Basin indicate salt domes. See Figure 5.15 for details on cross-section G-H.

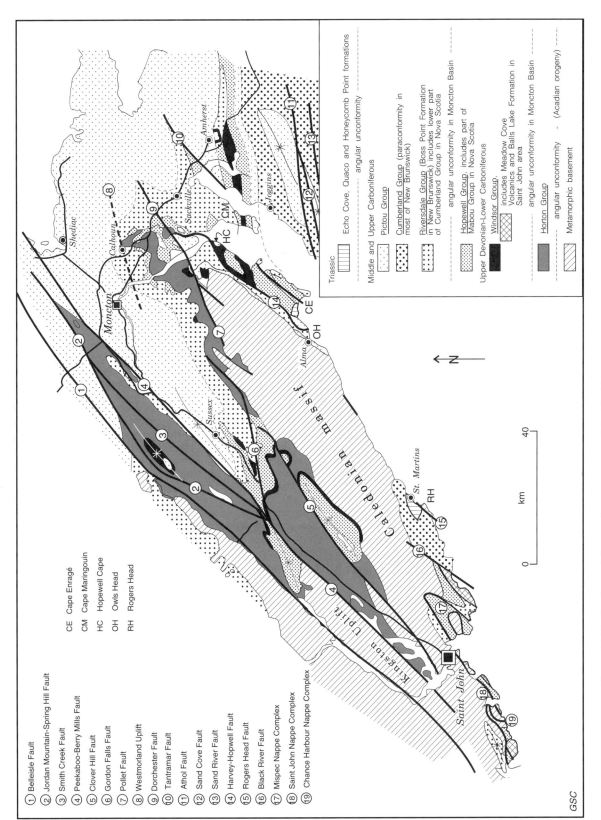

Figure 5.12. Geology of the southwestern part of the Maritimes Rift (Moncton Basin, Sackville Basin and western part of the Cumberland Basin including the Fundy coastal zone of the Saint John area (after Potter et al., 1979; Rast et al., 1984; Ryan et al., 1990a).

Legend:

1. Belleisle Fault
2. Jordan Mountain-Spring Hill Fault
3. Smith Creek Fault
4. Peekaboo-Berry Mills Fault
5. Clover Hill Fault
6. Gordon Falls Fault
7. Pollet Fault
8. Westmorland Uplift
9. Dorchester Fault
10. Tantramar Fault
11. Athol Fault
12. Sand Cove Fault
13. Sand River Fault
14. Harvey-Hopewell Fault
15. Rogers Head Fault
16. Black River Fault
17. Mispec Nappe Complex
18. Saint John Nappe Complex
19. Chance Harbour Nappe Complex

CE Cape Enragé
CM Cape Maringouin
HC Hopewell Cape
OH Owls Head
RH Rogers Head

Triassic
- Echo Cove, Quaco and Honeycomb Point formations
--- angular unconformity ---

Middle and Upper Carboniferous
- Pictou Group
- Cumberland Group (paraconformity in most of New Brunswick)
- Riversdale Group (Boss Point Formation in New Brunswick) includes lower part of Cumberland Group in Nova Scotia
--- angular unconformity in Moncton Basin ---
- Hopewell Group includes part of Mabou Group in Nova Scotia

Upper Devonian-Lower Carboniferous
- Windsor Group,
- includes Meadow Cove Volcanics and Balls Lake Formation in Saint John area
--- angular unconformity in Moncton Basin ---
- Horton Group
--- angular unconformity ---
- Metamorphic basement --- (Acadian orogeny) ---

GSC

479

1980; MacAuley and Ball, 1982; St. Peter, 1982a or b, 1986; MacAuley et al., 1984; Carter and Pickerill, 1985a, 1985b; Smith and Gibling, 1987; Utting, 1987a. The Albert Formation is best known for its lacustrine sequence consisting of oilshale and containing the only known oil and natural gas field (now depleted) in the Maritimes Basin. It has been subdivided into five members (Greiner, 1962; Carter and Pickerill, 1985a) of which the Round Hill Member represents the basin margin lithofacies, the other four are interior basin lithofacies (Table 5.7). The total sequence is thought to vary between 170-1650 m in thickness (Gussow, 1953).

The Dawson Settlement Member conformably overlies Memramcook redbeds. It consists mainly of coarse to fine feldspathic sandstones and intercalated dark pyritic shale.

The Frederick Brook Member overlies the Dawson Settlement Member conformably. It comprises brown kerogeneous shale (paper shale) with minor interbeds of sandstone. Palaeoniscid fish fossils are locally present in the Frederick Brook strata.

The Frederick Brook Member grades upward into the Hiram Brook Member which is the upper member of the Albert Formation over most of the Moncton Basin. The exception lies to the south and southeast of the Stoney Creek Oilfield where the Hiram Brook Member grades laterally into non marine evaporites of the Gautreau Member. The Hiram Brook Member consists of dark kerogenous and calcareous siltstones, shale and sandstones.

The Gautreau Member comprises rock salt, glauberite, anhydrite, and shale that accumulated under increasingly brinal conditions in a local and restricted area in the Hillsborough Basin. The unit represents the remaining salt-flat lithofacies of a drying-up "Lake Albert". The glauberite of the Gautreau Member is interpreted as nonmarine because 1) its association with typical shallow clastic lacustrine kerogenous lithofacies, 2) no marine strata are known anywhere within the Albert Formation, 3) because the glauberite is overlain by terrestrial redbeds (Carter and Pickerill, 1985a), and 4) all the palynomorphs found in the Albert are entirely non marine (Utting, 1987a).

The Round Hill Member (Carter and Pickerill, 1985a) is a coarse grained conglomeratic unit restricted mainly to the southern margin of the Moncton Basin. The Round Hill Member is interpreted as the coarse near-source alluvial-fan facies-equivalent of the basin centre deltaic-lacustrine facies of the Albert Formation (McLeod, 1980).

Natural gas and oil in the Albert Formation have been known for the past century (Henderson, 1940) and the presence of extensive oil shales is of commercial interest. The richest sections of oilshale yield an average of 93.5 litres of oil per tonne (MacAuley and Ball, 1982). The best oil shale deposit contains an estimated reserve of 270 000 000 barrels in situ to a depth of 600 m (Macauley and Ball, 1982; Wilson and Ball, 1983). However, a more realistic figure based on recoverable reserves to a mineable depth of 100 m is 5 800 000 barrels (St. Peter, 1988).

The Albert Formation is of Tournaisian (Early Carboniferous) age (Varma, 1969; Hacquebard, 1972; Utting, 1987a).

The Moncton Formation (Carter and Pickerill, 1985a) consists mainly of redbed conglomerates, arkosic sandstones, and minor red mudstones. It comprises two members, the Weldon Member below and the Hillsborough Member above with a combined thickness of up to 2200 m (Gussow, 1953). A local unconformity separates the two members. This brief review is largely based on the reports of Gussow (1953), Shroder (1963), McLeod (1980), and Carter and Pickerill (1985a).

The Weldon Member consists mainly of thin bedded, red to reddish brown mudstones and siltstones, locally with intercalations of greenish non marine limestone. The unit is commonly conglomeratic where the Weldon directly overlies pre Carboniferous basement.

The Hillsborough Member consists of red to greyish red conglomerates, coarse to fine arkosic sandstones, and minor mudstone. Gussow (1953) mapped a lithic felsic tuff at the base of the Hillsborough Member named Boyd Creek Tuff by McLeod (1980). The conglomerates consist mainly of subangular clasts of granite, gneiss, diorite, chert, and volcanic rocks. The Moncton redbeds are unfossiliferous. A late Tournaisian or Early Viséan (Early Carboniferous) age is based on the conformable relationship with the Albert Formation below and the Windsor marine limestone above, which are Tournaisian and Late Viséan, respectively (Hacquebard, 1972). In the absence of either one or both of these units the Moncton redbeds cannot readily be distinguished from the Memramcook redbeds below and/or the Hopewell redbeds above. The Moncton Formation contains alluvial fan, braided and meandering stream facies (Shroder, 1963; Carter and Pickerill, 1985a).

Windsor Group

The Windsor Group (Fig. 5.12) comprises fossiliferous marine limestones, evaporites, and transitional terrestrial-marine redbeds. The marine strata represent the only marine incursion into this part of the Maritimes Basin. This marine transgression coincided with an arid paleoclimatic interval and advanced from, and withdrew to, the east. It advanced across the Cumberland Basin into southern New Brunswick and the Moncton Basin and marginally onto the adjacent New Brunswick Platform. The marine Windsor beds in southern New Brunswick probably represent the western limit of the "Windsor Sea". The thickness of the Windsor Group is difficult to estimate because of salt flowage; Gussow (1953) estimated an average thickness of 350 m.

Table 5.8. Age, stratigraphic position, and lithostratigraphic subdivision of the Windsor Group, Moncton Basin (after McCutcheon, 1981).

Lower Carboniferous	Hopewell Group		Hopewell Conglomerate and Maringouin Formation. Mainly redbeds.
	Windsor Group	Subzones A and B	"Evaporite dominated facies sequence" Upperton, Cassidy Lake, Clover Hill formations and overlying red mudstones.)
			"Carbonate dominated facies sequence" (Macumber, Gays River, Parleeville formations, Swamp Hill Beds, Demoiselle Creek Beds.)
	Horton Group		Moncton Formation, red conglomerates, coarse to fine arkosic sandstone, mudstone.

The Windsor Group was initially subdivided on paleontological evidence into two zones and five subzones; named A, B, C, D, and E (Bell, 1929). Each subzone included a basal limestone, overlain by red mudstone, evaporites, and capped by redbeds. Only the lower part of the Windsor Group (subzones A and B) is represented in New Brunswick. Thus while subzones C, D, and E were deposited in eastern Nova Scotia and Cape Breton, coarse to fine alluvial-fan facies of the Hopewell Group were deposited in New Brunswick and western Nova Scotia. This change reflects a period of renewed uplift in the source areas of southern New Brunswick and the development of prograding terrestrial and subaqueous fans as well as distal fluviolacustrine facies in Nova Scotia which replaced the retreating "Windsor Sea".

The Windsor Group strata of New Brunswick have been previously described by Gussow (1953), Hamilton (1961, 1965), Hamilton and Barnett (1970), Worth (1972), Webb (1976), Anderle et al. (1979),; Utting (1987b); Howie (1988); and others. McCutcheon (1981) published an extensive account of the stratigraphy and paleogeography of the Windsor Group of southern New Brunswick and the following is based mainly on his report.

The Windsor Group is best exposed in the Moncton Basin. It is also known to occur (mainly in the subsurface) in the Sackville and Cumberland basins, and in a few small outliers near Saint John, where the Windsor also includes thick sections of volcanic rocks (e.g. Meadow Cove Volcanics and Balls Lake Formation, Rast et al., 1984) and also in thin exposures along the southern margin of the Marysville Basin. According to McCutcheon (1981), the Windsor Group of the Moncton Basin consists of two principal lithofacies units: a shoreface to offshore fossiliferous, carbonate-dominated unit, and an overlying offshore evaporite-dominated unit locally with redbeds. In New Brunswick, redbeds that overlie the evaporites are assigned to the Hopewell Group.

McCutcheon (1981) subdivided the Windsor Group into six formal formations of regional significance: three in the carbonate dominated lithofacies, and three in the evaporite dominated lithofacies. There are three additional informal units but these have only local significance (Table 5.8).

The Macumber, Gays River, and Parleeville formations make up the carbonate dominated lithofacies. They are in part superposed and in part laterally equivalent and range from relatively deep water anoxic to shallow water subtidal to intertidal basin-margin lithofacies.

The Macumber Formation is correlated with the Macumber Formation of Nova Scotia (McCutcheon, 1981). It is characterized by parallel laminated drab coloured wackestone-packstone containing few fossils. It conformably overlies redbeds of the underlying Moncton Group and is interpreted to represent the relatively deep water lithofacies of the carbonate dominated facies sequence. The Macumber grades laterally into, and is overstepped by the Gays River Formation. Macumber lithologies have a fetid odour on freshly broken surfaces.

The Gays River Formation is correlated with the Gays River Formation of Nova Scotia (McCutcheon, 1981). It consists of drab to buff to black algal boundstone locally with stromatolites. The unit occurs in thick wavy bedded to massive lenses locally overlying basal bafflestones and

thin bedded wackes and a packstone containing smooth-shelled brachiopods. The Gays River Formation locally overlies basement rocks with angular unconformity and is interpreted as a series of basement-fringing algal carbonate build-ups in a shallow sub-littoral to low intertidal depositional environment.

The Parleeville Formation is lithologically similar to the Gays River Formation and contains the same range of fossils but is generally thinner and contains a much greater siliciclastic component. It is interpreted as a basin margin facies adjacent to an active basement source area, whereas the Gays River strata represent basin-margin facies adjacent to passive basement.

The Upperton, Cassidy Lake, and Clover Hill formations make up the evaporite dominated lithofacies. They are in part laterally equivalent and in part superposed.

The Upperton Formation consists of stratiform replacive and displacive nodular to mozaic gypsum-anhydrite in either greenish grey mudstone, and/or Macumber type host rock. The gypsum-anhydrite is probably laterally equivalent to the Gays River carbonates.

The Cassidy Lake and Clover Hill formations (Anderle et al., 1979) comprise homogeneous halite, sylvinite, mixed halite and sylvinite, and minor argillaceous halite and sylvinite locally with intercalations of claystone and anhydrite. The salt units are probably lateral equivalents of the Upperton anhydrite but tend to be confined to the axes of subsiding troughs or depositional synclines within the Moncton Basin.

In addition to the aforementioned formations, McCutcheon (1981) recognized three local units. These include: 1) the Samp Hill Beds (siliciclastic wackestone) near Havelock, which could be part of the Gays River Formation; 2) Local redbeds associated with marine evaporites; and 3) the Demoiselle Creek Beds consisting of two local shallow marine, oolitic siliciclastic grainstone beds containing stromatolites and oncolites interbedded with coarse basal fanglomerates of the Hopewell Group at Hopewell Cape. The Demoiselle Creek Beds are assigned by McCutcheon (1981) to the lower Windsor Group (Subzone B) on the basis of paleontological evidence.

The overall interpretation for the Windsor strata of the Moncton Basin is that of increasing brinal conditions and density stratification leading to evaporite deposition in relatively deep-water axial regions of the basin. Subsequent uplift in the Moncton area (renewed Westmorland uplift) caused a reduction in the supply of seawater to the point that potash salts were deposited. This interpretation differs considerably from the coastal playa-lake model for the Windsor evaporites of Nova Scotia (e.g., Schenk, 1969). According to McCutcheon (1981) there is no evidence that coastal playa facies were part of the Windsor sequence in the Moncton Basin except at the top of the group where the evaporites are overlain by fine grained redbeds of the Hopewell Group.

The Windsor Group in New Brunswick is of Viséan age (Hacquebard, 1972; Utting, 1987b) and overlies the Moncton Group redbeds conformably, and metamorphic basement with angular unconformity. The group is overlain by Hopewell redbeds. The contact is in part conformable, with local angular unconformity, and in part karsted (McCutcheon, 1981).

Post Windsor Group lithostratigraphy

Uplift and the re-establishment of red coarse-to-fine alluvial-fan (bajada-playa) facies was associated with (and perhaps caused) the early withdrawal of the "Windsor Sea" from the Moncton Basin. The post Windsor redbeds are named the Hopewell Group in New Brunswick and are at least in part equivalent to Windsor Subzones C, D, and perhaps E in Nova Scotia. The Windsor marine regression, in combination with renewed uplift and local subsidence at different intervals during post Windsor times, has led to a complicated lithostratigraphy in southern New Brunswick and adjacent Nova Scotia that has not as yet been satisfactorialy resolved on a regional scale.

The problem is essentially one of local uplift and periodic stabilization, subsidence, platform development, subsequent platform expansion and renewed local subsidence and uplift that took place at various times in different parts of the region. This resulted in the deposition of (1) a wholly terrestrial marine-regression redbed sequence (Hopewell Group) in New Brunswick but redbeds interbedded with Windsor marine strata in Nova Scotia and Cape Breton; (2) Hopewell Group conglomerates in Moncton, Sackville and western margin of Cumberland basins that are in the remainder of the Cumberland Basin; (3) grey and redbeds with plant fragments of the Shepody Formation in New Brunswick that thin to the east and are mainly absent in the adjacent Cumberland Basin of Nova Scotia (e.g. Ryan et al., 1990b) and are apparently also absent in the Moncton Basin (Gussow, 1953); (4) thick Claremont conglomerates in Nova Scotia that thin to the west (called the Enragé Formation in southeastern New Brunswick) and are apparently absent in the Moncton and Sackville basins (Gussow, 1953); (5) a thick section of upper Westphalian A and Westphalian B strata (Cumberland Group in adjacent Nova Scotia) that is essentially absent in the Moncton Basin and the New Brunswick platform; and (6) a relatively thick section of Westphalian C, D, and Stephanian redbeds without coal in Nova Scotia named the Pictou Group that in New Brunswick is also called Pictou Group but is made up of grey as well as red fluvioclastic strata that locally contains coal.

Conventional lithostratigraphy, demanding lateral continuity of mappable units, is not designed to deal with such rapidly changing tectosomes, source areas, variable drainage patterns, and merging and overlapping lithofacies at this scale. In the following account no attempt is made to reconcile the lithostratigraphy of the Cumberland Basin of Nova Scotia with that of the Moncton Basin, but rather recognizes that the two basins had differing tectono-stratigraphic histories and lithofacies development.

Hopewell Group

The Hopewell Group covers large areas of the Moncton and Sackville basins, and also occurs along the western side of the Cumberland Basin (Fig. 5.12). It consists of purplish to greyish red pebble and cobble conglomerates, and coarse to fine arkosic sandstones with intercalations of mudstone. The conglomerate-arkosic sandstone lithofacies represent bajada depositional environments whereas accumulations of red, fine-grained arkosic sandstones, mudstones, minor lacustrine limestone, caliche and calcrete horizons represent laterally equivalent playa and playa-lake lithofacies (van de Poll and Sutherland, 1976). The following account

is essentially a summary of the reports and references of Flaherty and Norman (1941), Gussow (1953), Webb (1963), Kindle (1965), Leger (1969), van de Poll (1972, 1973c), van de Poll and Sutherland (1976), and McCabe and Schenk (1982).

In the Moncton Basin, the Hopewell Group is not formally subdivided although its two principal lithofacies have been informally referred to as "Hopewell Coarse" and "Hopewell Fine" (e.g., McCutcheon, 1981). In the Chignecto Bay area, the "Hopewell Coarse" or "Undifferentiated Hopewell" (Leger, 1969) has been informally referred to by some as "Hopewell conglomerate" (with the tentative status of formation; e.g., van de Poll and Sutherland, 1976), whereas the "Hopewell Fine" has previously been named the "Maringouin Formation" (Gussow, 1953). The latter forms the basal unit of the original Hopewell Group and in the type area of Cape Maringouin of southeastern New Brunswick is conformably overlain by the Shepody and Enrage formations as shown in Table 5.9.

The Hopewell Group incorporates several diverse lithologies that are contained in the Maringouin, Shepody, and Enragé formations.

From a New Brunswick lithostratigraphic viewpoint, there is a problem as to where to place the Hopewell conglomerate. Some workers (e.g., Gussow, 1953; McCabe and Schenk, 1982) equate it with the conglomeratic Claremont Formation of the Cumberland Basin which overlies the Maringouin Formation, whereas van de Poll and Sutherland (1976) viewed the Hopewell conglomerate as the course facies- equivalent of the Maringouin. This problem is currently being resolved (van de Poll, in prep.) and although changes are forthcoming, it is proposed to adhere to established stratigraphy.

Undivided Hopewell Group strata

The undivided Hopewell is estimated to range up to 1650 m in thickness. It is widely present in the Moncton Basin where the Hopewell is largely made up of coarse polymictic conglomerates with clast lithologies that can be related to nearby source areas on the New Brunswick Platform, Caledonia Massif, and Westmorland Uplift. They are

Table 5.9. Lithostratigraphic subdivision of the Hopewell Group in southeastern New Brunswick (after Gussow, 1953).

	Riversdale Group	Boss Point Formation	grey-buff quartz-quartzite conglomerate and/or sandstone at base, grey and/or red mudstone.
Lower to Middle Carboniferous	Hopewell Group	Enragé Formation	mainly reddish grey polymictic conglomerates, arkosic sandstones, red mudstone.
		──────────angular unconformity──────────	
		Shepody Formation	locally red and grey quartz-miscellaneous pebble conglomerates or red sandstones at base, overlain by grey sandstones containing coalified plant remains, and red and grey mudstones.
		Maringouin Formation	fine grained playa-type redbeds.

essentially a basin-margin bajada facies that grade later-ally and upwards into finer grained playa facies. The con-glomerates locally contain clasts of Windsor limestone (Gussow, 1953) indicating that local post-Windsor uplift caused erosion and reworking of older Carboniferous rocks. A thick accumulation of very coarse Hopewell conglomerate also occurs at Hopewell Cape and in the Chignecto Bay area along the southeastern flank of the Caledonia Massif (west-ern limit of the Cumberland Basin). Their presence sug-gests substantial normal fault movement (southeast side down) along the Harvey-Hopewell Fault (St. Peter, 1987). On the footwall (northwest) side of the fault, Hopewell redbeds thin dramatically over the Caledonia Massif and are overstepped onto basement by grey strata of the Boss Point Formation. Evidently, the Caledonia Massif was gradually being buried by the Hopewell and younger strata during Middle Carboniferous time. Provenance studies on Hopewell and younger conglomerates clearly demonstrate that from Shepody time onward the Caledonia Massif ceased to be a source terrane of any consequence through-out the Middle and Late Carboniferous (van de Poll, 1966) and perhaps even early Permian. Its present-day elevated position with respect to adjacent Carboniferous strata rep-resents a post-Early Permian uplift that had no bearing on the Middle and Upper Carboniferous lithostratigraphy of the region.

Maringouin Formation

In southeastern New Brunswick, not the coarse conglom-erates, but the fine grained Maringouin Formation forms the base of the Hopewell Group. The Maringouin consists of a coarsening upward sequence of red mudstones and interbedded red, fine grained sandstones. Climbing ripples are ubiquitous throughout the sequence and small chan-nels are also common. The Maringouin is thought to be structurally conformable with the Windsor beds below and is in turn conformably overlain by the Shepody Formation (Gussow, 1953). The Maringouin is up to 750 m thick (Gussow, 1953) and its typical fine grained redbed lithologies extend into adjacent Nova Scotia as the Middle-borough Formation. The latter is recognized as late Viséan facies equivalents of the upper part of the Windsor Group (Bell, 1944). On this basis the redbeds were previously assigned to the top of the Windsor Group (Bell, 1944) but recently they have been reassigned to a new and separate unit, the Mabou Group (Ryan et al., 1990b, 1991a). The latter is reintroduced after Belt (1964, 1965) to include all red lutites and minor grey fluvioclastic strata (Shepody Formation) between Windsor marine strata (below) and Claremont conglomerates (above).

The Maringouin Formation is unfossiliferous. Its late Viséan age is based on the early Viséan age of the Windsor Group below and the Early Namurian age of the Shepody Formation above (Hacquebard, 1972).

Shepody Formation

The Shepody Formation comprises grey and red sandstones with interbedded fine grained sandstone, and red and grey mudstones up to 500 m thick (Gussow, 1953). Plant remains are locally present in the grey sandstones. Inter-calations of fine-grained redbeds near the base of the Shepody Formation, which show lithogenetic affinity with the Maringouin redbeds below, gradually change upwards into mudstone intervals that are darker purplish red to dark grey with increasing organic content.

The Shepody Formation is a transitional sequence in which fluvial channel-overbank facies apparently spread over, and gradually replaced, pre-existing playa-flat envi-ronments of the Maringouin Formation. The Shepody Formation is regarded as a basin-axis deposit of limited lateral extent. It is mainly confined to the western part of the Cumberland Basin. Transport directions are essen-tially northeast to east. The Shepody Formation extends into the Cumberland Basin of Nova Scotia where it is gradually overstepped by Claremont conglomerates (Ryan et al., 1990a, b).

Enragé Formation

The Enragé Formation consists mainly of red shale, mot-tled pinkish and whitish red quartzose sandstone and conglomerate. The unit is friable and commonly weathers into low cliffs and bays. The Enragé overlies the Shepody with an angular unconformity in the Dorchester area (Gussow, 1953) but can be seen to be conformable with the Shepody Formation at Cape Maringouin and Cape Enragé. In New Brunswick, the Enragé is estimated to be up to 350 m thick (Gussow, 1953), although at Cape Maringouin it is approximately 70 m thick (van de Poll, unpub. data, 1991).

The Enragé extends into adjacent Nova Scotia as the Claremont Formation (Gussow, 1953) where it gradually increases from 50 m to over 1500 m in thickness (Ryan et al., 1990a, b, Ryan and Boehner, 1990). Here the Claremont consists mainly of reddish alluvial fanglomerates and arkosic sandstones deposited along the Cobequid Highlands. It reflects post Windsor uplift of the Cobequid Massif of Nova Scotia, shedding its fanglomerate facies northward into the Cumberland Basin.

Enragé strata are not known to be present west of the Fundy coastal region of New Brunswick (Gussow, 1953) although the Middle Carboniferous angular unconformity has been recognized in seismic data in the Moncton Basin (St. Peter, pers. comm., 1992).

Poorly fossiliferous redbed conglomerates, sandstones and mudstones of the Tynemouth Creek Formation just southwest of St. Martin's, New Brunswick (Fig. 5.12), though possibly slightly younger than the Enragé (Bell, 1927; Ryan, pers. comm., 1991), could reflect the same uplift-event in the Cobequid Mountains as the one that is recorded by the Claremont-Enragé conglomerates. Much like the Enragé beds, the Tynemouth Creek strata repre-sent alluvial-fan progradation from the south and south-east (Plint and van de Poll, 1982).

Both the Tynemouth Creek and Shepody-Enragé strata are essentially regarded as Cumberland Basin facies. The climate was evidently warm and probably seasonally subhumid (Plint and van de Poll, 1982). Much of the red colouration in the Tynemouth Creek may not have been primary but includes the effects of extensive reworking of pre-existing (presumably Carboniferous) redbeds, as evidenced by the common presence of red sandstone pebbles in the Tynemouth Creek conglomerates (van de Poll, 1970).

Table 5.10. Stratigraphic position of the Riversdale Group (Boss Point Formation), southeastern New Brunswick (after Gussow, 1953).

	Pictou Group	Megacyclic sequence 1 or Salisbury Formation
Middle Carboniferous		Depositional hiatus
	Riversdale Group	Boss Point Formation (grey conglomerates, sandstones, grey and red mudstones, coalified plant remains)
	Hopewell Group	Enragé Formation (red conglomerates, sandstones and minor mudstones)
		------------------unconformity----------------------

Riversdale Group

The name Riversdale Group is considered by some Nova Scotia workers to be outdated and may in time be abandoned in New Brunswick as well. It was initially introduced by Bell (1944) to combine the Claremont and Boss Point formations of western Nova Scotia into a single unit, but it is now abandoned there in favour of assigning the Shepody Formation to the top of the Mabou Group and the Claremont Formation and Boss Point Formation to the base of the overlying Cumberland Group (Ryan et al., 1990a,b, 1991a). The name Riversdale Group is, retained here for historical reasons and to accommodate the Boss Point at separate group status. Middle Carboniferous stratigraphic revisions need to be made for New Brunswick and are presently under consideration by the writer, but until they are finalized the currently accepted lithostratigraphy will be retained to avoid confusion. As mentioned previously, considerable tectonostratigraphic and paleoclimatic importance is placed on the gradual reappearance of mainly grey fluvioclastic strata overlying the Windsor-Hopewell redbeds (see also McCabe and Schenk, 1982). This heralded an important change in the tectonic development of New Brunswick accompanied by widespread platform subsidence, overstep, and the exchange of proximal source areas for more distal ones.

As used here, the Riversdale Group lies between the Hopewell Group (below) and the late Westphalian A, Westphalian B, paraconformity (above). The group is essentially of Namurian, early Westphalian A age (Hacquebard, 1972) and presently consists of only the Boss Point Formation as shown in Table 5.10.

Boss Point Formation

The Boss Point Formation consists of sub-mature to mature quartz-quartzite pebble conglomerates and sub-lithic to quartzose sandstones with intercalations of dark purplish red and grey mudstone, minor dark grey to black lacustrine limestone, but little or no coal. The conglomerates differ considerably from the typical Hopewell conglomerates below by their greater compositional material, (quartz content), better sorting, and pebble roundness. The Boss Point is estimated to be up to 1000 m thick in southeastern New Brunswick (Gussow, 1953), which is considerably thicker than on the New Brunswick Platform. The formation extends eastward into the Cumberland Basin under the same name. Recent stratigraphic revisions in western

Nova Scotia have led to the inclusion of the Boss Point Formation with the basal part of the Cumberland Group (Ryan et al., 1990a, b, 1991a). The problem of extending the Cumberland Group into New Brunswick this way is that Westphalian B age rocks are apparently absent (i.e., is represented by a "non-sequence" or paraconformity) and that because of colour and presence of coal, all of the Pictou grey and red strata of New Brunswick would also be included as part of the Cumberland Group. Until these stratigraphic problems and differences have been brought to a satisfactory conclusion, it is best to retain the established lithostratigraphy for New Brunswick and leave the changes for the future.

The Boss Point Formation is extensive in the Moncton Basin and the western part of the Cumberland Basin. Its characteristic well-rounded quartz-quartzite pebble conglomerates are confined to New Brunswick where they help differentiate the Boss Point Formation from younger Pictou Group conglomerates. The Boss Point Formation is not easily differentiated, however, where diagnostic conglomerates are scarce. Gussow (1953) mapped all the grey beds in the Moncton Basin below the red Salisbury Formation of the overlying Pictou Group as the Boss Point Formation thereby effectively placing the Boss Point-Pictou paraconformity (some 4000 m of the Cumberland Basin strata containing late Westphalian A and Westphalian B palynomorphs are missing in the Moncton Basin) at the base of the Salisbury.

The grey Boss Point conglomerates of southern New Brunswick change laterally into grey Boss Point sandstones and grey and red mudstones of the Cumberland Basin, where conglomerates are either rare or absent. Fossil plant-remains are common in the Boss Point Formation. Palynology suggests a late Namurian to early Westphalian A age (Hacquebard, 1972). The Boss Point sediment-transport directions are northeast to east (Lawson, 1962; van de Poll, 1970; Ryan et al., 1990a).

Riversdale strata along the Fundy coastal region of New Brunswick (Cumberland Basin) are extensively deformed and show a variety of soft-sediment structures including mud intrusions, brecciforms, convolute bedding, ball and pillow structures, and structures resembling flute molds formed by rheoplasis (van de Poll and Patel, 1981, 1988, 1989).

The basal grey fluvioclastic strata of the Shepody Formation and/or Boss Point Formation above Hopewell redbeds locally contain concentrations of copper (chalcocite, malachite) around comminuted plant remains (e.g., Dorchester, Horton, and Midway Copper occurrences of southeastern New Brunswick; Brown, 1975; Wilson and Ball, 1983; van de Poll, 1978). Mineralization found to date has been variable and too local in extent to warrant development.

Northumberland Strait Supergroup

The Pictou Group in the Moncton Basin overlies the Boss Point Formation paraconformably. It is estimated here to be up to 1500 m thick and consists of recurrences of grey-buff quartz miscellaneous pebble conglomerates, grey-buff lithic sandstones, red and minor grey-black, fine sandstones and mudstones, and minor coal. The Pictou is lithologically very similar to the Boss Point below and as a consequence Boss Point-Pictou strata have in the past been combined into a single unit, the Petitcodiac Group (e.g., Norman, 1941; Alcock and Evans, 1945). Subsequently, it was recognized that there is probably a hiatus between the Boss Point and overlying Pictou strata as no Westphalian B age fossils have been found in New Brunswick. Because of this hiatus, Gussow (1953) adopted the name Pictou Group from Nova Scotia for the beds in New Brunswick that are of Westphalian C age and younger and subdivided the unit into four formations based largely

on grain size and colour. Subsequently, van de Poll (1970, 1973a) applied the cyclic-sequence concept to the Pictou of the Moncton Basin and eventually placed the Pictou Group into the lower part of the Northumberland Strait Supergroup (van de Poll, 1989) as shown in Table 5.11. Each cyclic sequence is in the order of 150-200 m thick. They are fining-upward from a grey-buff conglomeratic and/or sandstone lower part to a finer grained redbed sequence at the top, locally containing dark grey coal

Table 5.11. A provisional Permian-Carboniferous correlation table between the Moncton and Cumberland basins. The heavy black horizontal and vertical line indicates the time transgressive nature of the "grey to red transition zone" across the two basins.

System	Age	Group		Moncton basin and adjacent Prince Edward Island	Cumberland basin and adjacent Prince Edward Island		Colour
Lower Permian	Artinskian	Northumberland Strait Supergroup	Prince Edward Island Group	Orby Head Formation	Orby Head Formation		"All redbeds"
	Sakmarian			Hillsborough River Formation	Hillsborough River Formation		
				Malpeque Member	Wood Islands Member		
				Kildare Capes Formation	Kildare Capes Formation		
					concealed interval (Northumberland Strait)		
				Egmont Bay Formation			
	Asselian			Miminigash Formation	Cape John Formation	Pictou Group	
Upper Carboniferous	Stephanian			Concealed interval (Northumberland Strait)			
					Tatamagouche Formation		
			Pictou Group	Tormentine Formation (?)			
Middle Carboniferous	Westphalian D			Cyclic sequences I, II, III and lower part of IV or Salisbury, Scoudouc and Richibucto formations	Balfron Formation		"Grey to red transition zone"
	Westphalian C				Malagash Formation	Cumberland Group	
	Westphalian B	Cumberland Group		Not present in Moncton Subbasin	Ragged Reef Formation		"Grey and red beds" Locally coal measures and coal.
					Springhill Mines Formation		
					Joggins Formation		
	Westphalian A				Polly Brook Formation		
		Riversdale Group		Boss Point Formation	Boss Point Formation		
	Namurian			------------Unconformity------------	Claremont Formation	Mabou Group	
Lower Carboniferous					Shepody Formation		
	Late Viséan	Hopewell Group		Hopewell "coarse" and Hopewell "fine"	Middleborough Formation		"All redbeds"
				Older Carboniferous rocks and/or pre-Acadian metamorphic basement			

485

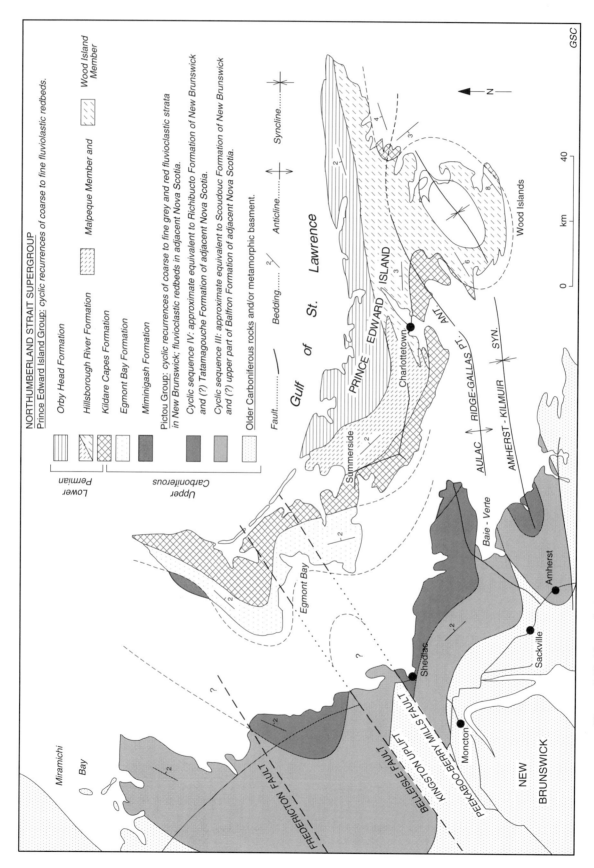

Figure 5.13. Upper Carboniferous-Lower Permian geology of Prince Edward Island and adjacent parts of New Brunswick and Nova Scotia (after van de Poll, 1970, 1973a, 1983; 1989; Ryan et al., 1990a).

ures and minor coal. The recognition that grey and red alternations are characteristic of the Pictou sequence of the Moncton Basin also formed the basis for subdividing the sequence into formations as mapped by Gussow (1953). However, poor inland exposures make regional subdivision of the Pictou Group by either scheme a tenuous exercise.

The Tormentine Formation (Gussow, 1953) is the first appearance of medium- grained sandstones in New Brunswick that are red rather than grey. They are on this basis assigned to the Prince Edward Island Group and possibly represent the equivalent to the basal part of the Miminigash Formation of Prince Edward Island (Fig. 5.13). The relationship of this unit with adjacent Pictou strata in Nova Scotia has not yet been fully established. The "grey-red" to "all red" transition between the Pictou and Prince Edward Island strata is known from drillhole information beneath Prince Edward Island to be time transgressive (van de Poll, 1983). The time-transgressive nature of the transition is also well displayed between Pictou strata of the New Brunswick Platform and the Cumberland Basin. On the New Brunswick Platform, the transition takes place within strata of approximately Late Carboniferous (Stephanian) age, whereas in the Cumberland Basin the transition occurs at the boundary between the Cumberland and Pictou groups and involves Westphalian C-D strata. All younger Pictou strata of the Cumberland Basin are red-brown and coal is absent (Ryan et al., 1990a, b). The Moncton Basin, between New Brunswick Platform and Cumberland Basin, is in a transitional position and contains Pictou strata that are more reddish and brownish-buff than on the platform, but also contain a few thin Late Carboniferous coal seams.

Prince Edward Island redbeds and also those of the Pictou Group of Nova Scotia contain fossilized plant remains in the form of imprints (partially replaced by hematite) and fossil tree-stumps (replaced by silica and/or barite). Reptilian fossil trackways are locally present in the redbeds of Prince Edward Island (Mossman and Place, 1989). The redbeds contain rather mature conglomerates, display meandering channel patterns, show well established subfacies differentiation, and locally contain lacustrine facies including limestone (van de Poll, 1983). In short, they are in every respect similar to the underlying grey and red strata of the Pictou Group, except for the evidence of strong and pervasive oxidation that gradually increases upward, and a concomitant decline in fossilized plant remains. This colour change from grey sandstone to redbeds and an apparent simultaneous reduction in plant remains is puzzling. The similarity of lithofacies, degree of facies differentiation, channel-form and pattern (van de Poll, 1983), and quartz-grain surface textures (van de Poll 1983, D'Orsay and van de Poll, 1985b) indicate little or no change from the warm and humid paleoclimate established during the Late Carboniferous. Van de Poll and Forbes (1984) noted an apparent decline in the size of the Early Permian plants which they interpreted to reflect climatic stress. The apparent stress was evidently not from aridity, however. It may be that the colour change is not paleoclimatically related at the time of deposition but is post depositional, reflecting downward oxidation below the Permian-Triassic unconformity (Ryan, pers. comm., 1989). This would mean that the apparent gradual decline in fossilized plant remains does not reflect a paleoclimatic change but increased upward oxidation of the redbeds. The

cause for the eastward increase in the depth of reddening is not known but if secondary it could reflect, at least in part, variations in the amount of organic material originally retained in the strata and/or differences in depth of penetration of oxygenated surface waters.

The Prince Edward Island Group is known from drill-hole intersections up to 1000 m thick (van de Poll, 1983). It is subdivided into five formations and two members on the basis of their mainly fining-upward megacycles (Foley, 1984; Place and van de Poll, 1988; van de Poll, 1990) and the characteristic compositions of the basal conglomerates of the cyclic sequences (van de Poll, 1983, 1989) as follows (see also Fig. 5.13): Orby Head Formation, pebble and cobble conglomerates with low rhyolite content (<14%), subrounded to rounded clasts (Rho .45-.60); Hillsborough River Formation, pebble conglomerates with high rhyolite content (25-40%) subrounded clasts Malpeque Member, and pebble conglomerates with relatively high miscellaneous-low rhyolite content (<8%), subangular clasts Wood Islands Member; Kildare Capes Formation, pebble conglomerates with high rhyolite content (25-40%) subrounded clasts ; Egmont Bay Formation few small lenses of basal conglomerate with high rhyolite content; and Miminigash Formation, fine grained mudstones, top of unit only exposed in cliffs around Miminigash, western Prince Edward Island.

The strata consist of fluvioclastic redbeds deposited mainly in a low lying, broad, alluvial plain. The exception to this is the Wood Islands Member which is interpreted to represent an advancing (subhumid to humid ?) alluvial fan from the south or southeast (van de Poll, 1983; Foley, 1984).

The origin and source of the relatively high rhyolite content in some of the Prince Edward Island conglomerates are enigmatic. Paleocurrents for the high rhyolite units suggest derivation from the southwest. However, rhyolite clasts are not particularly common in the older Pictou strata of central and southern New Brunswick (van de Poll, 1970). The stratigraphic restriction in combination with southwest derivation of the high rhyolite conglomerates tend to favour, a nearby intrabasinal source somewhere in central or eastern New Brunswick. We do not know of a late Pennsylvanian intrabasinal uplift of rhyolitic rocks, nor did Jamer (1980), find any indication of late Pennsylvanian volcanicity in the 222 lithological and natural gamma drillhole logs covering all of the area in New Brunswick underlain by Pictou strata. Resolving the source of the rhyolite clasts could have economic implications. Local anomalous concentrations of radioactive minerals in the Prince Edward Island redbeds might be related to the high-rhyolite conglomerates, possibly by leaching (van de Poll, 1983).

A Stephanian to late Early Permian age has been assigned to the redbeds based on its flora (Holden, 1913; Darrah, 1936, palynological data (Barss et al., 1963; Hacquebard, 1972), vertebrate bone fragments (Langston, 1963) including the vertebrate fossil remains *Bathygnathus borealis* Leidy (Case, 1905, Huene 1905) and pelycosaur fossil trackways (Mossman and Place, 1909).

The redbeds of Prince Edward Island are gently folded, mainly on northeast to east-northeast-trending axes (Fig. 5.13). Deformation is believed to be related in part to slight movement on the Kingston Uplift and in part to the migration of Windsor salt at depth. Maximum dips do not exceed 12° (van de Poll, 1983). The strata display a wide variety of soft-sediment deformation structures reflecting the effects of strong physical diagenesis. Dewatering, fluidization, liquefaction, and intrastratal collapse phenomena (van de Poll and Plint, 1983) as well as a wide range of sediment intrusion structures occur at a frequency and scale throughout the redbeds that suggests induced compaction from periodic syn-depositional seismic activity (van de Poll and Ryan, 1985). The apparent syn-depositional instability may have been related to the migration of the Windsor salt, or it may reflect a recurrence of regional tectonism that includes uplift in central Nova Scotia, or it may be a combination of the two (van de Poll 1983, 1990).

Transport directions for the Northumberland Strait Supergroup are generally northeast, north-northeast, and north (van de Poll, 1983).

Structure in the Moncton and Sackville basins

Structural deformation of the Moncton and Sackville basins is discussed separately from that of the Cumberland Basin (Fundy Bay coastal region) because of the differing tectonic histories, styles and responses between them.

Carboniferous-Permian strata of the Moncton and the Sackville basins have undergone four periods of deformation. Deformation consisted of folding and faulting that almost everywhere involved renewed basement uplift, erosion, and associated recurrences of coarse alluvial-fan facies in the stratigraphic record.

The first period of deformation is shown by Horton strata (Memramcook, Albert, and lower part of Moncton formations) which, in the Moncton Basin, are folded on east-northeast and north-northwest-trending axes. Folding is associated with small faults in the area and may reflect lateral movement of the Caledonia Massif (Ruitenberg and McCutcheon, 1982). This was also the time of initial uplift of the Westmorland Basement High, separating the Hillsborough Basin from the Petitcodiac Basin to the north (Foley, 1989; St. Peter, 1992). This initial period of deformation is interpreted by Gussow (1953) and Nickerson (1991, 1992) as one of southeast-directed thrusting against the Caledonia Massif. Webb (1963, 1969) suggested the Early Carboniferous deformation to have been related to dextral wrench-faulting along northeast-trending master faults. St. Peter (1987), Martel (1987), and Foley (1989) have recognized positive flower structures in Horton strata of the Moncton Basin on the basis of which they reinterpreted the movement as transpressional. This Early Carboniferous deformation event did not affect the Windsor Group and is approximately of Late Tournaisian-Early Viséan age. It represents the first major compressive event after the Acadian Orogeny and according to Ruitenberg and McCutcheon (1982) may be time equivalent to early Hercynian deformation in Western Europe (see also Haq and van Eysinga, 1987).

The second period of deformation in southern New Brunswick is post Windsor and affected late Namurian and older strata (Gussow, 1953; Ruitenberg and McCutcheon, 1982). Deformation produced northeast-trending open folds and much of the major fault movement in the area is thought to have taken place at this time (Webb, 1963, 1969). Widespread erosion removed large segments of Windsor and lower Hopewell strata except in low or synclinal areas (Gussow, 1953). Webb (1963) thought this to have been the latest significant movement on the northeasterly

trending master faults transecting southern New Brunswick north of the Caledonia Massif. He noted the parallelism of major faults and fold axes and agreed with Gussow (1953) that during this interval thrust faulting was probably the dominant mode of deformation in the Moncton Basin.

Sedimentary rocks of Middle Carboniferous-Early Permian age (upper Riversdale and younger) that overlie Lower Carboniferous and older deformed rocks with angular unconformity are in turn deformed by the migration and diapiric intrusion of Windsor salt (Howie, 1988). The timing of the salt migration is not certain and could have been over a prolonged period and at different times in different places. Ryan (1985) noted an east deflection in the northerly drainage pattern in Westphalian strata on the south side of the salt-cored Malagash Anticline in the Cumberland Basin of Nova Scotia. He interpreted this to indicate that the salt started to migrate during the Westphalian. On the other hand, salt pillows at depth beneath Prince Edward Island have domed-up Lower Permian strata without showing evidence of attenuation of strata on their flanks (van de Poll, 1983). This suggests that initiation of salt migration beneath Prince Edward Island either commenced later than in Nova Scotia, (probably not before post Early Permian time) or that salt migration lasted throughout the Late Carboniferous and Early Permian.

The fourth period of deformation affected both rift and platformal strata of New Brunswick and is reflected by widespread uplift, slight, mainly northeast, tilting of basement blocks together with Permian-Carboniferous cover rocks, and erosion leading to the present "horst-and-graben" topography and structural outline of the Maritimes Basin. The tilting also affected Lower Permian strata of Prince Edward Island. This suggests that it may have been a Late Permian event (Alleghanian Orogeny?) and it is possible that the migration of Windsor salt below Prince Edward Island was initiated by this movement.

Structure in the Cumberland Basin

Carboniferous strata along the Chignecto Bay and Fundy coast of New Brunswick are assigned here to the Cumberland Basin because they lie to the north and northwest of the western extension of the Cobequid-Chedabucto Fault (named Fundy Fault by Keen et al., 1991) and are separated from the Moncton Basin by the Caledonia massif (Figs. 5.10 and 5.12).

The Carboniferous coastal exposures can be subdivided on structural criteria into two main sequences separated by a northeast-trending fault centred on Black River.

To the east of the Black River Fault lies a belt of open- to steeply- folded and faulted strata belonging to the Windsor, Hopewell and Riversdale groups. These are locally overlain with angular unconformity, or overthrust by, Triassic redbeds. This belt extends from Black River to Hopewell Cape and probably includes beneath Chignecto Bay and Bay of Fundy, a thick offshore section of Cumberland Group rocks (including coal bearing strata) unconformably overlain by Triassic redbeds. Deformation shows the effects of northwest directed thrusting (e.g., the Harvey-Hopewell Fault between Alma and Hopewell Cape, and the Rogers Head and Quaco Head faults in the St. Martin's area). Thrust-related deformation in this belt is more intense towards the southwest. The

Harvey-Hopewell Fault has high angle reverse movement and is only spectacularly exposed where thick and competent Hopewell conglomerates are involved, such as at Owl Head, near Alma. Stratigraphic separation is minimal, however, with red Hopewell conglomerate on the hanging wall and overturned red Entagé and grey Boss Point conglomeratic sandstone-mudstone on the footwall. The northeast extension of the Harvey-Hopewell Fault beyond Hopewell Cape in uncertain. It does not intersect the opposite coastal section of the Petitcodiac River and probably terminates into the north-trending Memramcook River Fault (Martel, 1987). The age of the post Boss Point movement along the Harvey-Hopewell Fault is not known. By contrast, the age of northwest-directed thrusting at Rogers Head, St. Martin's area is more closely constrained. At Rogers Head, Hopewell-Boss Point strata (Plint and van de Poll, 1984) are overthrust by Precambrian basement (volcanic) rocks, spreading alluvial-fan facies of the Tynemouth Creek Formation (Claremont-Polybrook equivalent in Nova Scotia ?) to the west of the thrust-front during late Namurian-Early Westphalian (e.g., Cumberland Group) time. Subsequent renewed northwest-directed thrusting involved Tynemouth Creek strata and the whole imbricate thrust sequence is overlain with angular unconformity by the Triassic redbeds of the Wolfville Formation (Plint and van de Poll, 1984).

The timing of deformation corresponds to about the mid-Hercynian Orogeny of Europe (Ruitenberg and McCutcheon, 1982).

Carboniferous strata to the southwest of the Black River Fault reach upper greenschist facies in the Saint John area (Rast and Skehan, 1991) and have been extensively investigated by Hayes and Howell (1937), Alcock (1938, 1959), Ruitenberg and McCutcheon (1982), Rast and Grant (1973, 1977), Rast et al.(1978, 1984), Caudill and Nance (1986), Nance and Warner (1986), Nance (1987a) and references therein. The sequence is intensely deformed into a series of recumbent folds (nappe complexes) each resting on a major overthrust and includes strata thought to range from Viséan (Windsor equivalent) to Westphalian C (Pictou equivalent, Rast et al., 1984). Redbeds with intercalated volcanic rocks characterize the Lower Carboniferous sequence (e.g., Dipper Harbour Beds and West Beach Formation, Balls Lake Formation, Meadow Cove Volcanics, Belding Cove Beds, and Chance Harbour Beds). These are overlain by grey strata containing black shale, minor coal and plant fossils of the mid-Carboniferous Lancaster Formation. The age of the Lancaster Formation is not accurately known. It was erroneously regarded as Riverdale (Westphalian A) because of its general lithological similarity with the Riverdale Group (Bell, 1927). According to Bell (1944) the Lancaster, together with "Fern Ledges" strata at Duck Cove near Saint John, N.B. (Stopes, 1914), is of the same (Cumberland, Westphalian B) age as the Joggins section of Nova Scotia. More recently however, Lancaster rocks in the Lepreau area have locally yielded an abundant flora indicating a Westphalian C age (W.B. Forbes, pers. comm., 1978; Bell, 1944; Rast et al., 1984).

The total Carboniferous sequence has been informally subdivided into three low-angle nappe complexes, the Mispec Complex just east of Saint John, the Saint John Complex just west of Saint John, and the Chance Harbour Complex to the southwest of the latter (Rast et al., 1984).

Carboniferous intrusions of both felsic and mafic compositions occur in association with the three nappe complexes. They are deformed and must have been emplaced prior to thrusting (Rast et al., 1984).

The Mid Carboniferous fault movement in the Bay of Fundy coastal zone (Saint John area) of New Brunswick suggests northwest-southeast compression (Rast et al., 1978, 1984; Skehan and Rast, 1984; Plint and van de Poll, 1984) or transpression (Nance, 1987a). The movement was probably contemporaneous with the Westphalian convergent wrench tectonics resulting in dextral strike-slip movement on the Cobequid-Chedabucto fault zone (Nance, 1987a). The latter (also called the Minas Geofracture, Keppie, 1982, and the Glooscap Fault, Schenk, Chapter 3) is one of the most important faults transecting the Maritimes Basin. It is a deep crustal structure that marks the plate boundary between the Avalon and Meguma zones (Keppie, 1982; Mosher and Rast, 1984). The dominant motion along the fault during mid Carboniferous-Permian time was dextral (Eisbacher, 1969; Keppie, 1982) with the Meguma block of Nova Scotia moving westward against the Avalon block of southern New Brunswick (Brown, 1986a; Keen et al., 1991). Seismic evidence from the Bay of Fundy shows the presence of a large thrust ramp dipping gently to the southeast from the New Brunswick side of the Bay of Fundy to beneath the Meguma block of Nova Scotia. Brown (1986a) believed the Meguma block to represent a large thrust sheet overriding the ramp of which the latter is the compressional component to the dextral Cobequid-Chedabucto fault zone. Stacked, northwest-directed thrust sheets seen in seismic sections beneath the ramp near the New Brunswick coast are interpreted to represent thrust slices that formed ahead of the westward moving Meguma block (Brown, 1986a). These are, according to Brown (1986a), probably the subsurface extensions of the onshore nappe-complexes mapped by Rast (e.g., Rast et al., 1984) in the coastal region of southwestern New Brunswick. The collision between the Meguma and Avalon blocks and the resulting uplift is thought to be represented onshore by the syntectonic sedimentary deposits of the Mispec, Tynemouth Creek and Lancaster formations and also accounts for the development of the Mid Carboniferous and younger folds and thrusts in the Moncton Basin of southern New Brunswick (Nance, 1987a).

Magdalen Basin

The Magdalen Basin is a northeast-trending structure centred on the Îles de la Madeleine that, as defined here, extends from the Baie Verte area of New Brunswick and Nova Scotia, to the southwest coast of Newfoundland (Fig. 5.11).

The basin forms the central part of the Maritimes Rift and lies offshore from the Cumberland, Sackville and Moncton basins. Its present configuration is that of a slightly east-tilted and deepening half graben down-faulted along its eastern margin against the Hollow Fault, or a parallel subsidary (Durling and Marillier, 1990). A basement ridge extending from the northern tip of Cape Breton Island to southwestern Newfoundland separates the Magdalen Basin from the Sydney Basin to the east (Bell and Howie, 1990).

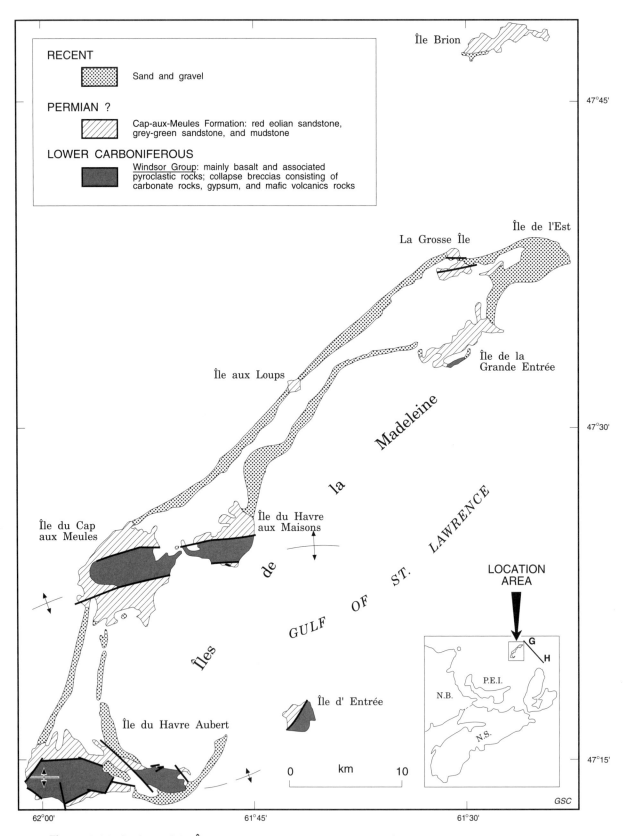

Figure 5.14. Geology of the Îles de la Madeleine, Gulf of St. Lawrence, (modified after Bell and Howie, 1990; Brisebois, 1981).

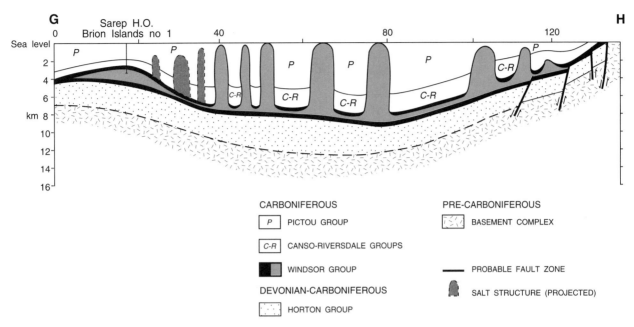

Figure 5.15. Section GH showing salt structures in the Magdalen Basin (adapted from Bell and Howie, 1990). Location shown in Figure 5.11.

GSC

The Magdalen Basin is thought to have originated in response to crustal stretching during Late Devonian-Early Carboniferous time (Bradley, 1982; Durling and Marillier, 1990). It contains a thick sequence of Windsor marine strata, in particular evaporites overlain by younger fluvio-clastic beds of Middle Carboniferous to Early Permian age (Fig. 5.14). Horton strata (Late Devonian-Early Carbonif-erous) are inferred to lie beneath the Windsor (Durling and Marillier, 1990). The following review of the geology of the Magdalen Basin and that of the Îles de la Madeleine is based on a recent compilation by Howie (1988) and refer-ences therein, and an account by Barr et al. (1985) on the Carboniferous basaltic volcanics of the Îles de la Madeleine.

The Îles de la Madeleine lie in the central part of the Gulf of St. Lawrence and the Magdalen Basin. The islands consist of a number of relatively small rock exposures that are joined together by sand spits and barrier beaches (Fig. 5.14).

The Havre-aux-Maisons Formation comprises the old-est exposed rocks and consists of Windsor Group mud-stones, siltstones, collapse breccias, carbonates, gypsum, and volcanic rocks.

Plint and von Bitter (1986) concluded that the conodont fauna of Windsor strata from the Islands supports Brisebois (1981) interpretation that carbonate-evaporite deposition was confined to restricted, possibly hypersaline-littoral, lagoonal and transitional lagoonal-continental supratidal paleoenvironments.

The Cap-au-Diable Formation overlies the Havre-aux-Maisons Formation. It consists of basalts and pyroclastic rocks which have associated gabbroic dykes and stocks believed to be of Viséan age. The Windsor sedimentary and basaltic volcanic and intrusive rocks have been brought up from considerable depth to the surface by salt diapirism (Barr et at., 1985).

The basaltic rocks vary in texture and mineralogy. Most are highly altered, fine- to medium-grained, display inters-ertal to intergranular textures, and consist mainly of plagioclase, clinopyroxene, and a cryptocrystalline matrix. The high alteration, which is thought to be the interaction of brinal seawater during extrusion, masks their original chemical character. Although ambiguous, the data suggest a continental "within plate" tectonic setting (Barr et al., 1985).

The lower Carboniferous (Viséan) section is unconfor-mable or disconformable beneath Permian (?) strata of the Cap-aux-Meules Formation.

The Cap-aux-Meules Formation consists of grey-green sandstones, mudstones and calcareous conglomerates at the base, L'Étang-du-Nord Member, overlain by red eolean sandstones of the L'Étang des Cap Member (Brisebois, 1981). The Cap-aux-Meules is apparently unfossiliferous and of uncertain age.

Salt diapirs are numerous in the central area of the Magdalen Basin. A number have been drilled yielding sections of near pure halite (Howie, 1988). Drilling results show that the diapirs are flanked by Windsor sedimentary and volcanic rocks, overlain by Canso, Riversdale and Pictou strata, (Fig. 5.15).

Summary

Post orogenic upper Paleozoic rocks of New Brunswick and Prince Edward Island range in age from late Early Devonian to late Early Permian. They consist mainly of coarse- to fine-grained continental red and grey sedimentary rocks that include fluvial and fluvio-lacustrine strata, coal meas-ures, marine limestone, evaporites, felsic to mafic vol-canics, and minor intrusive rocks. They were deposited following the Late Silurian-Early Devonian Acadian Orogeny and are presently contained within several structural basins and platform outliers. Basins that underwent a similar structural evolution and display a certain degree of stratigraphic unity make up a basin complex collectively referred to as the Maritimes Basin. There are two late Paleozoic structural basins in New Brunswick that do not fit these criteria and lie outside the Maritimes Basin. These are the Ristigouche Basin and the St. Andrews Basin of northern and southwestern New Brunswick, respectively.

A central rift zone (Maritimes Rift) and adjacent platform (New Brunswick Platform) are recognized in New Brunswick as major elements of the Maritimes Basin. The distinction between rift and platformal areas is mainly based on thickness and structural style. Carboniferous platformal rocks are mildly deformed and form a relatively thin sequence of mainly coarse- to fine grained clastics, felsic and mafic volcanic flows and tuffs, and associated redbeds that are preserved in synclinorial and/or down faulted areas. Rift strata by contrast contain thick sequences of bahada-playa facies, fine grained lacustrine beds, fluvioclastic strata, coal measures and a marine sequence including evaporites that have been extensively deformed.

The origin and evolution of the Maritimes Basin continues to be a topic of considerable debate. The current interpretations may be summarized as follows:

1) It formed as a result of differential subsidence reflecting continental rifting in an east-west strike-slip regime centred on the Cobequid-Chedabucto fault system of Nova Scotia (Belt, 1968a).

2) It reflects a horst-and-graben tectonic regime over a highly fractured basement following the Acadian Orogeny (Howie and Barss, 1975; Howie, 1984, 1988).

3) It reflects extensional subsidence in a dominantly right-lateral strike-slip regime followed by widespread thermal subsidence (Bradley, 1982).

4) It formed as a failed rift along the margin of a late Paleozoic ocean (Fyffe and Barr, 1986).

5) It formed as a natural consequence of Acadian continental collision leading to localized rapid initial subsidence and subsequent less rapid, more widespread subsidence (McCutcheon and Robinson, 1987).

The actual timing of fault movement(s) that have affected upper Paleozoic strata in New Brunswick is not accurately known, except for situations where local geological detail and seismic data are available. The majority of these faults probably had important post-Carboniferous movement and since they intersect and affected Carboniferous strata, this raises an important question as to the actual timing of the onset of fault related deformation. Except for dextral movement on the Cobequid-Chedabucto Fault during early Middle Carboniferous time, that led to reasonably well dated thrusting and thrust-front related deposition in southern New Brunswick, it cannot be said for certain that the Maritimes Basin evolved in, or partially in, a wrench-tectonic setting. The available evidence is contradictory and still rather inconclusive. Evidence for substantial Early Carboniferous strike-slip motion on the northeast-trending faults in southeastern New Brunswick

as suggested by Webb (1963) remains controversial (e.g. Brown and Helmstaedt, 1969). Most recently a consensus seems to be emerging that the seismic evidence from the Moncton Basin suggests that Early Carboniferous fault movement was dextral transpressional rather than strictly dextral (Nickerson, 1992; St. Peter, 1992). It also appears generally accepted that subsequent Carboniferous movement on the master faults of southeastern New Brunswick was essentially thrusting (Gussow, 1953; Webb, 1963). Widespread uplift and low angle tilting of the fragmented basement and erosion of overlying Permian-Carboniferous strata during post middle Permian time (van de Poll, 1970, 1983) characterized the destructional phase in the evolution of the western part of the Maritime Basin and led to the present "horst and graben" topography.

Stratigraphic evidence indicates that the extent of post Early Permian uplift, tilting and erosion must have been considerable. Restoration for tilt in combination with vitrinite reflectance and palynological evidence suggests that an additional minimum thickness of approximately 800 m of Middle Carboniferous to Lower Permian strata once covered the Marysville and Central basins (van de Poll, 1970; Hacquebard and Avery, 1984). Even greater thicknesses have apparently been removed by erosion from other parts of the basin. Depth of burial interpretations based on coal rank in western Cape Breton suggests that some 2300 m of additional strata were present in the Mabou Mines area during the Permian (Hacquebard, 1986). Fission track analyses indicate that up to 3000 m of sedimentary cover may have been eroded from the Maritimes Basin since Permian time (Ryan et al., 1991). Projection of these thicknesses of eroded strata across the regional Carboniferous paleotopography suggests that prior to erosion most, if not all, of the Atlantic region and probably also adjacent New England States must have been covered by continental rocks during the late Paleozoic (van de Poll, 1973a; Ryan et al., 1991; Gibling et al., 1992).

Upper Paleozoic strata have played a significant role in the economic development of the Atlantic region. Oil, natural gas, oil shale, coal, salt, gypsum, potash, lime, building stone and, on a small scale, also base metals (mainly copper and lead) have been in production intermittently for over a century and some of the commodities may still hold promise for new discoveries. Tin, tungsten, and molybdenum occur in Upper Devonian-Lower Carboniferous felsic volcanic and intrusive rocks at Mount Pleasant, southern New Brunswick.

Acknowledgments

Sincere thanks are extended to S. Townsend, D. Basque, P. Racette and D. Tabor for typing the manuscript, to P. Racette for preparing the tables and to A. Gomez for drafting the original diagrams.

I am particularly indebted to C. St. Peter, whose careful reading and editing of the manuscript, and innumerable helpful comments, based on his extensive knowledge of the Carboniferous and its literature, have greatly improved the original version of this chapter. H. Williams made further improvements.

UPPER PALEOZOIC ROCKS, NOVA SCOTIA

Martin R. Gibling

Regional setting

Upper Paleozoic strata of Nova Scotia occur in numerous isolated or partially connected areas (Fig. 5.16). The depocentres are here termed "basins", following traditional usage in the province, although, in a broader context, all the rocks are part of the Maritimes Basin (Williams, 1974). Presently disjunct, but formerly contiguous, outcrops of upper Paleozoic strata outcrop also in New Brunswick, Prince Edward Island, Newfoundland, and Quebec. In addition, much of the Gulf of St. Lawrence, the Laurentian Channel, the southern Grand Banks and the northeastern Newfoundland Shelf are underlain by upper Paleozoic strata that show close correspondence with those onshore in Nova Scotia (Bell and Howie, 1990; Sanford and Grant, 1990). Late Paleozoic paleogeographic compilations that show the Maritimes Basin in a regional and global context have been presented by Wilson (1981), Haszeldine (1984), and Scotese and McKerrow (1990).

Component depocentres of Nova Scotia

The disjunct distribution of upper Paleozoic rocks in Nova Scotia reflects in part the vagaries of post-Paleozoic erosion. Some basins are believed to be structural remnants of formerly distinct depocentres within the regional Maritimes Basin. This inference is based principally on significant local lithostratigraphic and facies variation. In addition, parts of the succession have been removed locally at unconformities and disconformities, especially adjacent to late Paleozoic uplands. Despite this local variation, the major rock groups described in this review can be identified in most of the component depocentres. The depocentres are inferred to have been partially connected, intermontane basins separated by uplands of moderate elevation.

Structural activity that has contributed to the disjunct stratal distribution includes syndepositional motion on transcurrent faults and associated thrust zones, halokinesis (both syn- and post-depositional), and extensional tectonism associated with Mesozoic seafloor spreading in the North Atlantic. Most basins were sites of active subsidence only periodically during the late Paleozoic.

Seven basins are recognized, excluding depocentres associated with the Minas Geofracture, which are grouped with the Minas and Cumberland basins (Fig. 5.16). Table 5.12 lists important references and maps dealing with the fundamental litho- and biostratigraphy of each basin.

1) The Cumberland Basin occupies most of northern Nova Scotia north of the Cobequid Highlands, including the area between the Cobequid and Antigonish highlands (the Stellarton Gap, not including the Stellarton Basin)

Martin R. Gibling
1995: Upper Paleozoic rocks, Nova Scotia; *in* Chapter 5 of Geology of the Appalachian-Caledonian Orogen in Canada and Greenland, (ed.) H. Williams; Geological Survey of Canada, Geology of Canada, no. 6, p. 493-523 (*also* Geological Society of America, The Geology of North America, v. F-1).

and north of the Hollow Fault. Upper Paleozoic strata rest upon rocks of the Avalon Zone (Williams, 1978), as do those of all other basins except the Minas Basin. The basinal strata underlie parts of the Bay of Fundy and are contiguous with strata in southern New Brunswick and Prince Edward Island. Complex outcrop belts located along the Cobequid-Chedabucto fault zone include small, structurally disjunct entities that are here considered part of the Cumberland and Minas basins but may originally have been substantially isolated.

Table 5.12. Main component depocentres of the Maritimes Basin in Nova Scotia, with list of major stratigraphic studies. The references listed are mainly recent studies and maps which list important earlier work. Regional summaries covering many depocentres include Bell (1958), Kelley (1967), Poole (1967), Belt (1968a, b), Hacquebard (1972), Howie and Barss (1975), and McCabe and Schenk (1982). A selected bibliography covering stratigraphy and economic geology was published by Boehner (1990). A 1:500 000 scale map of the province, showing outcrop belts, is available (Keppie, 1979).

Basin	References
Cumberland	Ryan et al., 1987, 1990b, 1991a; Browne, 1990; Calder, 1991
Minas (Windsor-Shubenacadie-Musquodoboit-Mahone Bay)	Bell, 1929, 1960; Moore, 1967; Giles and Boehner, 1982; Ferguson, 1983; Boehner, 1984, 1986; Moore and Ferguson, 1986; Martel, 1990
Stellarton Basin (& Stellarton Gap)	Bell, 1940; Fralick and Schenk, 1981; Giles, 1982; Yeo, 1985; Yeo and Ruixiang, 1987; Naylor et al., 1989
Antigonish	Belt, 1964, 1965; Boehner and Giles, 1982; McCabe and Schenk, 1982; Prime, 1987
Western Cape Breton (marginal to Gulf of St. Lawrence)	Belt, 1964, 1965; McCabe and Schenk, 1982; Hacquebard et al., 1989; Hamblin, 1989; Hamblin and Rust, 1989
Central Cape Breton (including Loch Lomond & Glengarry)	Weeks, 1954; Prime and Boehner, 1984; Boehner and Prime, 1985
Sydney	Hacquebard, 1983; Boehner and Giles, 1986; Gibling et al., 1987; Rust et al., 1987
Depocentres associated with the Minas Geofracture (grouped with Minas, Antigonish, and Cumberland basins)	Belt, 1964, 1965; Donohoe and Wallace, 1982, 1985; McCabe and Schenk, 1982

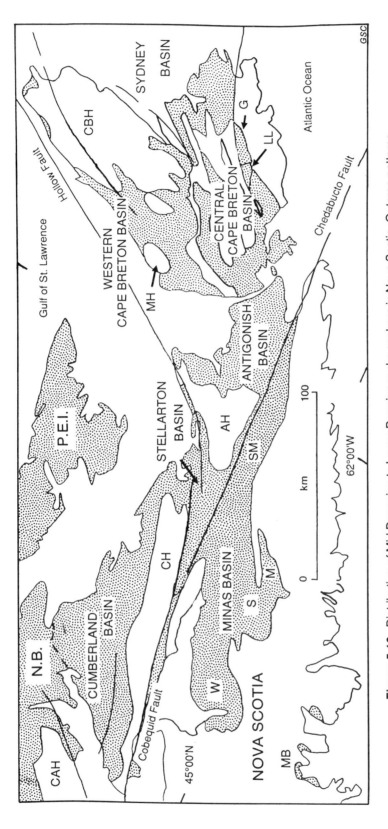

Figure 5.16. Distribution of Mid Devonian to Lower Permian rocks onshore in Nova Scotia. Outcrop patterns simplified from Keppie (1979). AH = Antigonish Highlands; CAH = Caledonia Highlands; CBH = Cape Breton Highlands; CH = Cobequid Highlands; G = Glengarry Basin; LL = Loch Lomond Basin; M = Musquodoboit Basin; MB = Mahone Bay outcrops; MH = Mabou Highlands; N.B. = New Brunswick; P.E.I. = Prince Edward Island; S = Shubenacadie Basin; SM = St. Mary's Basin. Note that component depocentres (basins) of the Maritimes Basin were formerly interconnected, and that present basinal areas reflect in part structural and erosional factors.

2) The Minas Basin lies to the south of the Cobequid Fault where the upper Paleozoic rocks rest upon rocks of the Meguma Zone. Structurally distinct areas known locally as the Windsor, Shubenacadie and Musquodoboit basins probably were contiguous parts of a single depocentre during the late Paleozoic. An easterly continuation of the outcrop belt forms the St. Mary's basin which covers a narrow belt south of the Cobequid Fault and reaches Chedabucto Bay; little exposure is available in this basin. Small outliers of upper Paleozoic rocks near Mahone Bay on the southern Nova Scotia coast are considered part of the Minas Basin.

3) The Stellarton Basin is a small pull-apart structure, about 20 km by 6 km in extent, that lies adjacent to the Cobequid and Hollow faults. Strata of the Stellarton Formation (Cumberland Group) cover the area, and no information is available concerning underlying strata. Coeval Cumberland Basin strata adjacent to this structure show markedly divergent facies.

4) The Antigonish Basin lies east of the Antigonish Highlands and north of the Chedabucto Fault. It is here arbitrarily separated from the basins of Cape Breton Island along the Strait of Canso. Large areas of little studied strata, generally assigned a Devonian-Carboniferous age, border Chedabucto Bay and the Strait of Canso.

5) The Western Cape Breton Basin includes strata adjacent to the Cape Breton and Mabou highlands and several other inliers of pre-upper Paleozoic rocks. The strata pass offshore beneath the Gulf of St. Lawrence where they are cut by the Hollow Fault, and probably pass southwestward into those of the Antigonish Basin. Coal-bearing strata in western Cape Breton were included in the Gulf of St. Lawrence Basin (Hacquebard, 1986), which was considered to include upper Paleozoic rocks bordering the gulf in New Brunswick and elsewhere in Nova Scotia.

6) The Central Cape Breton Basin includes strata in central and southern Cape Breton Island. Much of the area lies beneath Bras d'Or Lake and is poorly known; the area includes numerous small inliers of pre-upper Paleozoic rocks. The presently disjunct Loch Lomond and Glengarry basins are located in this area.

7) The Sydney Basin occupies eastern Cape Breton Island, and extends offshore almost to Newfoundland and about 300 km eastward to the western approaches of Placentia Bay (Bell and Howie, 1990). The present-day outcrop belt is separate from those of central and western Cape Breton. Upper Paleozoic strata that underlie large areas of the Grand Banks may formerly have been contiguous with rocks of the Sydney Basin.

Recent work by many geoscientists has led to an improved understanding of correlation between upper Paleozoic strata in the component depocentres of the Maritimes Basin. Consequently, in this review, it is considered appropriate to describe the strata in terms of six tectonostratigraphic packages (Fig. 5.17), broadly similar to those discussed by Hacquebard (1972), Gibling et al. (1987) and Ryan et al. (1987). No complete basin-by-basin description is provided, although each package is described

for most basins. Emphasis is placed on complete and/or well-described successions where important recent work (stratigraphic, sedimentological and paleontological) has been carried out. Biostratigraphic zonation is based on stages defined in western Europe.

Fountain Lake Group and equivalent strata

Intercalated volcanic and sedimentary rocks overlie Acadian and older basement rocks and underlie the Horton Group conformably or unconformably at numerous locations across Nova Scotia. Outcrops are discontinuous and the strata are deformed locally, and no comprehensive regional study has been made. The strata are generally considered to be continental in origin, although the Upper Devonian (partially coeval) Escuminac Formation of the Gaspésie, Québec may be marine in part (Vézina, 1991).

The Cobequid Highlands

The Fountain Lake Group is widely distributed in this area (Donohoe and Wallace, 1982, 1985), and comprises bimodal volcanics (rhyolite and basalt) with terrigenous units. Its thickness was estimated as up to 2.9 km (P. Wallace, in Williams et al., 1985). The strata are largely undeformed, although they are cut locally by fractures associated with the Cobequid Fault and deformed along discrete thrust zones near Cape Chignecto (Waldron et al., 1989). The Fountain Lake Group is overlain unconformably by the Falls Formation, probably a Horton Group equivalent (Donohoe and Wallace, 1985).

Evidence for the age of the Fountain Lake Group is largely indirect. Several plutons, particularly the Pleasant Hills and Cape Chignecto, can be linked directly to the group through subvolcanic lithology; Rb-Sr dates on the plutons have yielded early Carboniferous (338-356 Ma) ages (Pe-Piper et al., 1989a), although the pervasive hydrothermal activity that has affected some areas adds uncertainty to these dates (G. Pe-Piper, pers. comm., 1991). A preliminary zircon date of 360 Ma, close to the Devonian-Carboniferous boundary, was obtained from the Pleasant Hills pluton (Doig et al., 1991). Rb-Sr dating of the Fountain Lake volcanics in the eastern Cobequid Highlands suggests a Late Devonian to Early Carboniferous age (Donohoe et al., 1986), whereas palynomorphs suggest Mid Devonian and Early Carboniferous ages (S. Barss, in Donohoe and Wallace, 1985). Volcanic rocks associated with plutons in the western Cobequid Highlands yield Early Carboniferous dates (Pe-Piper et al., 1989a).

Western Cape Breton Basin

In western Cape Breton Island and near St. Paul Island offshore northern Cape Breton, the Fisset Brook Formation, at least 400 m thick, unconformably overlies Avalonian rocks, lies unconformably beneath the Horton Group and contains bimodal volcanic rocks (basalt and rhyolite). The formation is dated as Late Devonian to Early Carboniferous (Tournaisian) on the basis of isotopic, palynological and megafloral dating (Blanchard et al., 1984), although

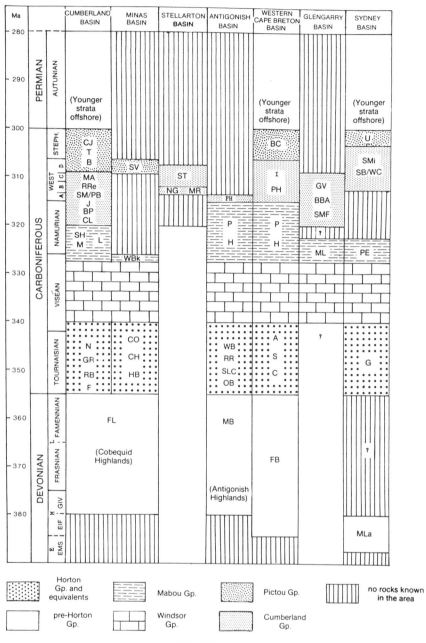

Figure 5.17. Summary stratigraphic chart for component depocentres of the Maritimes Basin in Nova Scotia. See Figure 5.16 for locations of basins. Formations along the Cobequid-Chedabucto fault (Cobequid and Antigonish highlands area) are included with the Cumberland, Minas and Antigonish basins. Age and formational data derived from references cited in the text, as well as regional surveys by Barss and Hacquebard (1967) and Barss et al. (1979). Isotopic ages based on Cowie and Bassett (1989) for Devonian to Viséan rocks, and on Hess and Lippolt (1986) for younger strata. Definition of the Mabou, Cumberland and Pictou groups in accord with criteria suggested by Ryan et al. (1991a). Formations of the Windsor Group are listed by Giles (1981).

FORMATIONS

A	Ainslie	FL	Fountain Lake Gp.
B	Balfron	G	Grantmire
BBA	Big Barren	GR	Greville River
BC	Broad Cove	GV	Glengarry Valley
BP	Boss Point	H	Hastings
C	Craignish	HB	Horton Bluff
CH	Cheverie	I	Inverness
CJ	Cape John	J	Joggins
CL	Claremont	L	Londonderry
CO	Coldstream	M	Middleborough
F	Falls	MA	Malagash
FB	Fisset Brook	MB	McAras Brook
		RR	Rights River

ML	MacKeigan Lake	S	Strathlorne
MLa	McAdam Lake	SB	South Bar
MR	Middle River	SH	Shepody
N	Nuttby	SLC	South Lk. Crk.
NG	New Glasgow	SM	Springhill Mines
OB	Ogden Brook	SMF	Silver Mine
P	Pomquet	SMi	Sydney Mines
PB	Polly Brook	ST	Stellarton
PE	Point Edward	SV	Scotch Village
PH	Port Hood	T	Tatamagouche
RB	Rapid Brook	U	Unnamed
RRe	Ragged Reef	WBk	Watering Brook
WB	Wilkie Brook	WC	Waddens Cove

GSC

496

these authors noted considerable discrepancy between the different dating techniques. The volcanic rocks have yielded whole-rock Rb-Sr dates of 376 ± 12 and 370 ± 20 Ma (mid to late Devonian), although recent redating has yielded Early Carboniferous (Viséan) ages. The strata contain the Late Devonian plant fossil *Archaeopteris*, as well as Tournaisian spores. A pluton near Cheticamp, considered subvolcanic to the Fisset Brook Formation, yielded a zircon date of 365 +10/-5 Ma, a Late Devonian age (Jamieson et al., 1986).

Sydney Basin

The McAdam Lake Formation in the Coxheath Hills comprises 340 m of grey sandstone, conglomerate and shale with felsic tuffs, coaly shale and oil shale. The strata have been dated as Emsian to Eifelian (D.C. McGregor, in Williams et al., 1985).

Antigonish Basin

The McAras Brook Formation, known near Antigonish and offshore, consists of 150-340 m of red conglomerate with minor sandstone and shale, and includes basic lava flows overlain by caliche (Keppie et al., 1978; Fralick and Schenk, 1981; Pe-Piper and Jansa, 1986). The formation was assigned a mid-Late Devonian age by Fralick and Schenk (1981); probably equivalent volcanic rocks drilled in the Fina F-25 well located between Cape Breton and Prince Edward Island were dated at 362 ± 15 Ma (Pe-Piper and Jansa, 1986).

Horton Group

Alluvial and lacustrine rocks of the Horton Group rest conformably to unconformably on Fountain Lake and equivalent rocks, or with nonconformity on Acadian and older basement rocks. Palynological dating of Horton strata at several localities indicates a Tournaisian age (Utting et al., 1989; Fig. 5.18), with possible Late Devonian palynomorphs obtained in one sample at the base of the Horton Bluff Formation (J. Utting, pers. comm., 1991). In the Minas Basin, Horton Group strata nonconformably overlie granitic rocks of the South Mountain Batholith, dated as Late Devonian (Clarke and Halliday, 1980).

An apparent hiatus that locally includes latest Tournaisian and early Viséan has been widely identified between the Horton and Windsor groups; however, in view of the widespread redbeds lacking palynomorphs at the top of the Horton Group, such a hiatus may be absent (Utting et al., 1989). Horton Group equivalents in New Brunswick, including oil shales of the Albert Formation, are dated as Late Devonian to Early Carboniferous (Pickerill and Carter, 1980; Utting, 1987a).

A tripartite division of the Horton Group into a lacustrine unit (Horton Bluff, South Lake Creek and Strathlorne formations in individual basins) bounded by alluvial units seems to apply throughout the outcrop belt. The age of the lacustrine unit appears to be Tournaisian2 to Tournaisian3 throughout the area (Fig. 5.18). The lacustrine unit probably reflects a regional phase of rapid subsidence during

which the local depocentres were underfilled and organic-rich shales widely deposited. McMahon et al. (1986), Hamblin (1989), and Smith and Naylor (1990) have discussed hydrocarbon seeps and the petrology and hydrocarbon potential of the shales at numerous locations. Horton Group strata in general (Bell, 1958), and those that occupy small, fault-bounded basins beneath the Gulf of St. Lawrence in particular (Sanford and Grant, 1990), are considered prospective reservoir rocks for hydrocarbons.

Minas Basin

The type area of the Horton Group is located in the Minas Basin (Fig. 5.16, 5.19) where Bell (1929, 1960) recognized the predominantly lacustrine Horton Bluff Formation overlain by the alluvial Cheverie Formation. Martel (1990) and Martel and Gibling (1991) divided the Horton Bluff Formation into four (presently informal) members (Fig. 5.20), at least 560 m thick in the Avon River Valley and 1015 m thick in the Soquip Noel #1 well near Kennetcook. The basal Harding Brook member, a braided-fluvial sandstone unit, rests unconformably on rocks of the Meguma Zone. The overlying Curry Brook member is a mudstone-sandstone succession deposited in a lacustrine delta. The Blue Beach member, exposed in superb cliffs near Blue Beach, consists of coarsening- and shallowing-upward cycles with hummocky cross-stratification (HCS) (Fig. 5.21). Such cycles represent repeated progradation of nearshore, wave-dominated lacustrine sediments, capped by paleosols, across offshore sediments. The Hurd Creek member contains wave-reworked, sandy deltaic deposits in addition to lacustrine shoreline deposits.

The Cheverie Formation, about 200 m thick, rests disconformably on the Horton Bluff and consists of sandstone and mudstone, attributed to coarse grained meandering rivers (Conrod, 1987). In the Shubenacadie and Musquodoboit areas, 50-90 m of interbedded conglomerate, sandstone and, more rarely, shale that lies disconformably beneath the Windsor Group is recognized as the Coldstream Formation (Giles and Boehner, 1982; Utting et al., 1989).

A biota that includes fish, amphibians, ostracodes and in situ vascular plants testifies to nonmarine conditions during Horton Group deposition.

Both the Horton Bluff Formation in total and shallowing-upward cycles within the Blue Beach and Hurd Creek members thicken northward (Fig. 5.20) toward the present-day position of the Cobequid Fault. Martel (1990) suggested that the Minas Basin was a half-graben during the Early Carboniferous. The Cobequid Highlands region contains faulted outcrops of Lower Carboniferous Nuttby Formation, an alluvial and lacustrine unit that is apparently coeval with the Horton Group of the Minas Basin (Donohoe and Wallace, 1985); its presence in this area raises the possibility that Horton strata formerly extended over a considerable area north of the present Cobequid Fault.

Western Cape Breton Basin

The Horton Group in western Cape Breton Island is estimated to be up to 3 km thick (Hamblin and Rust, 1989). The group consists of red and grey conglomerate, sandstone and mudstone, and shows a tripartite division into alluvial and lacustrine deposits of the Craignish, Strathlorne and Ainslie formations (Fig. 5.22). Dark grey mudstones of the Strathlorne Formation are about 300 m in cumulative thickness.

Hamblin and Rust (1989) inferred from facies distribution and paleoflow data that the Horton Group was deposited in two fault-bounded, asymmetric depocentres in western and northern Cape Breton, the Ainslie and Cabot basins, respectively. These half-graben had opposed asymmetry, were approximately 100 by 50 km in size, and were separated by a narrow zone of elevated basement rocks. The basal strata of the Craignish Formation (unit C3: Fig. 5.22) appear to be common to both depocentres and may reflect a broad, pre-rift sag prior to the formation of the separate half-graben (Fig. 5.23).

Antigonish Basin

In the Antigonish Basin, four formations are recognized within the Horton Group (Fig. 5.17) (Prime, 1987). The basal Ogden Brook Formation is a coarse grained alluvial fan deposit about 2000 m thick. The overlying South Lake Creek Formation (Late Devonian to ?Early Carboniferous), at least 500-600 m thick, consists of alluvial and lacustrine strata, including uninterrupted black shale sequences about 150 m thick near Big Marsh. The Rights River Formation is a coarse grained alluvial fan deposit at least 1200 m thick. Finally, the Wilkie Brook Formation consists of conglomerate, sandstone and shale with carbonate biostromes and algal bioherms (Boehner and MacBeath, 1989); the formation is 194 m thick, rests with angular discordance on the underlying South Lake Creek strata, and

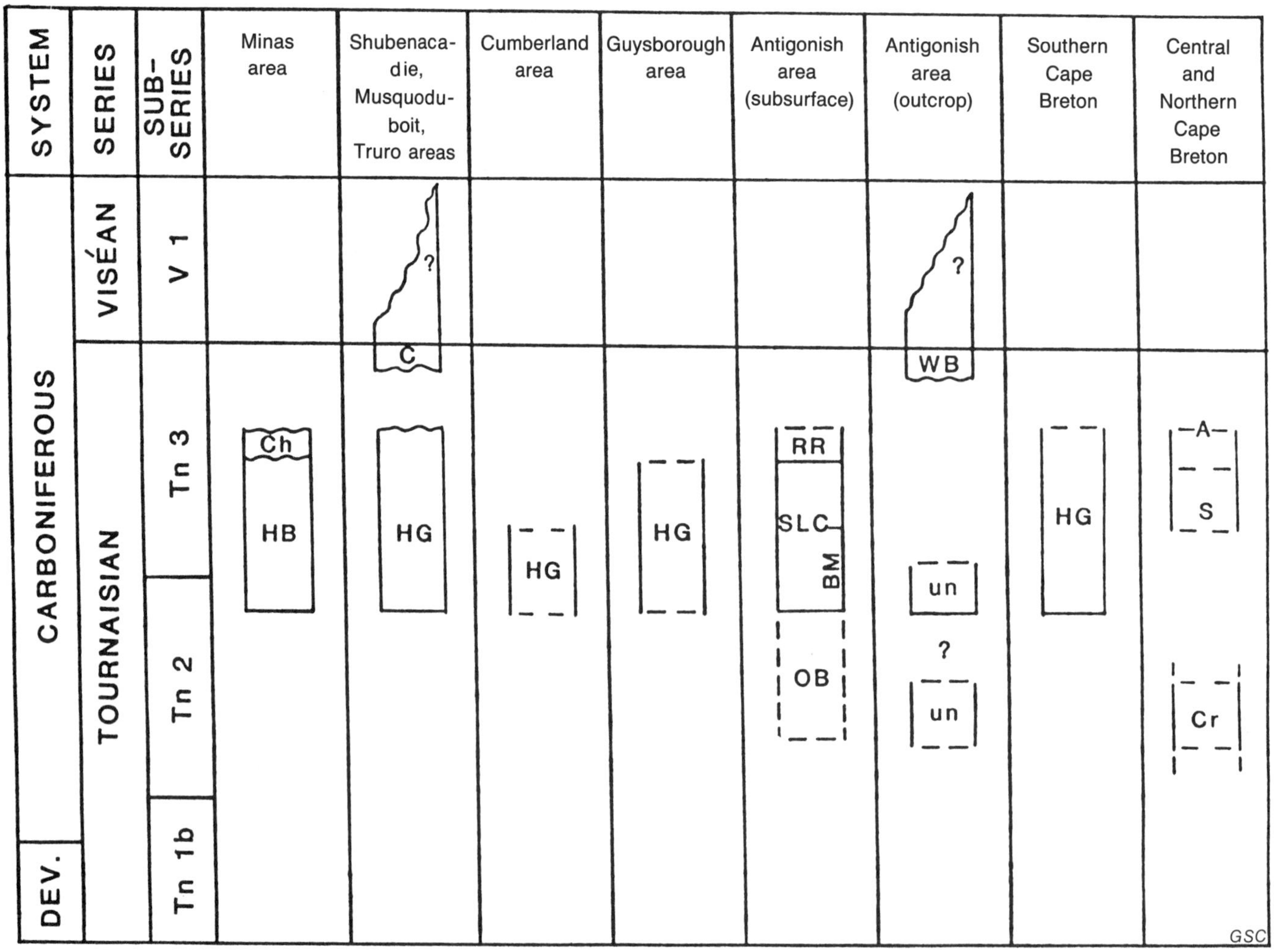

Figure 5.18. Comparison of age determinations for the Horton Group in Nova Scotia. Palynostratigraphic correlation is with the spore zones of Western Europe, and probable ages are in terms of British and European stages. A = Ainslie; BM = Big Marsh; C = Coldstream Formation; Ch = Cheverie Formation; Cr = Craignish Formation; HB = Horton Bluff Formation; HG = Horton Group (undifferentiated); OB = Ogden Brook Formation; RR = Rights River Formation; S = Strathlorne Formation; SLC = South Lake Creek Formation; un = unnamed formation; WB = Wilkie Brook Formation. Data from Utting et al. (1989, Fig. 5.9) and J. Utting (pers. comm., 1991).

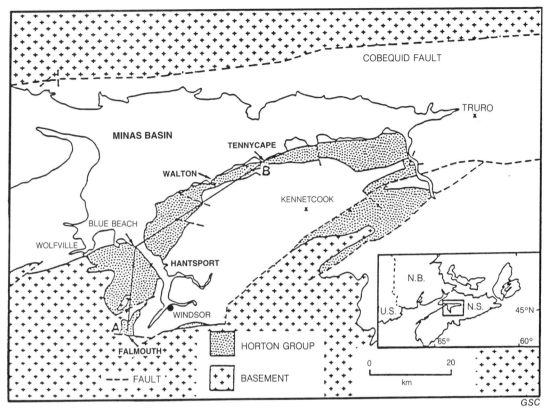

Figure 5.19. Distribution of Horton Group strata in the Windsor area of the Minas Basin. Basement is composed of undifferentiated Meguma Supergroup and granitic rocks of the South Mountain Batholith south of the Cobequid Fault (Meguma Zone), and granitic, metamorphic and sedimentary rocks of the Avalon Zone to the north. Blank areas onshore include younger Carboniferous and Mesozoic rocks. Stratigraphic sections along A-B shown in Figure 5.20. Modified from Martel (1990, fig. 1.3).

represents an interfingering of alluvial fan and lacustrine strata. The total thickness of the Horton Group at any location in the Antigonish Basin is uncertain.

Sydney and Central Cape Breton basins

Adjacent to the Coxheath Hills, up to 1000 m of red conglomerate, sandstone and shale with minor pedogenic limestone underlie evaporites and carbonates of the Windsor Group. These strata, termed the Grantmire Formation (Weeks, 1954), are poorly dated, but may be Tournaisian to early Viséan. They were mapped as Horton Group equivalents (Boehner and Giles, 1986).

Windsor Group

The Windsor Group, originally defined by Bell (1929, 1958), is the only marine unit recognized in the upper Paleozoic succession, apart from minor brackish intervals in the upper Westphalian Cumberland Group of the Sydney Basin (see below). A detailed regional account of the Windsor Group has been given by Boehner (1986) and Howie (1988). The strata are assigned a Middle to Late Viséan age (Utting, 1978; Giles, 1981). Exposures are generally scattered because of the soluble nature of its gypsum, anhydrite and halite; gypsum karst is developed locally.

The Windsor Group is here described from its type area in the Minas Basin, specifically in the Windsor area (Fig. 5.24-5.26), although the strata have been described more fully from the Shubenacadie area where the structure is less complicated (Boehner, 1984, 1986). Thickness is about 920 m in the type area, and drilling in the Shubenacadie area has confirmed about 800 m of strata. The group is thicker to the northeast and appears to be thickest in central Cape Breton Island (R. Moore, in Williams et al., 1985).

Numerous formational names have been applied to Windsor strata in mainland Nova Scotia. A simplified system of five major cycles of marine transgression and regression was introduced by Giles (1981) to facilitate correlation (Fig. 5.25, 5.26). A dominant facies trend apparent in all major (and constituent minor) cycles is an overall upward decrease in the proportion of evaporites. The evaporite component increases in volume and inferred salinity of basinal waters toward the probable centre of individual depocentres. The proportion of carbonate varies from 10-5% in Cycle 1 to 25-50% in cycles 2-5 (R.C.Boehner, pers. comm., 1991). Minas Basin cycles are in part separated by disconformities (Fig. 5.25).

The lower boundaries of major cycles show reasonable correlation with biostratigraphic boundaries, and correlation has been suggested with Dinantian stages recognized in Britain and Belgium (Fig. 5.27). The cycles may reflect glacioeustasy; Veevers and Powell (1987) have presented evidence for Viséan glaciation.

The following description is based on the Minas Basin succession, with brief reference to the Windsor Group elsewhere in the province.

Stratigraphic Succession

Major Cycle 1 is the thickest (300-500 m) and comprises a basal carbonate and an overlying evaporite suite that changes progressively upward from anhydrite to halite and locally potash salts; terrigenous rocks lie adjacent to paleo-basin margins locally. The basal Macumber Formation (equivalent to the Ship Cove Formation of the Codroy Group in Newfoundland) is a dark grey to black laminite,

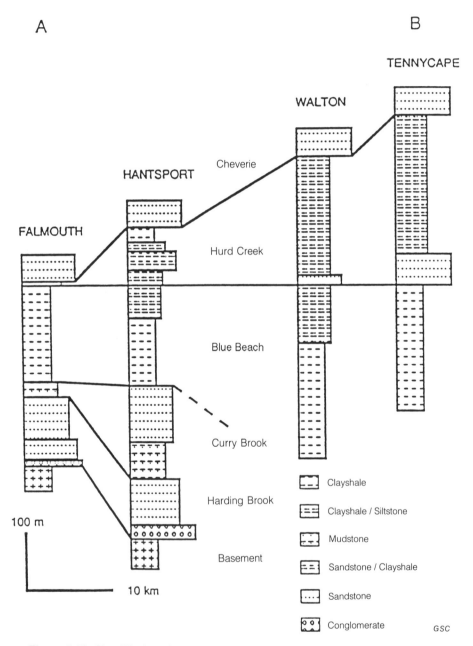

Figure 5.20. Simplified southwest to northeast stratigraphic sections of the northern part of the Minas Basin (Horton Bluff Formation). Both Hantsport and Falmouth sections are composites of several measured sections and drillholes. The Hantsport composite section includes the Blue Beach section. Note thickening of the Hurd Creek Member and probably the Blue Beach Member to the northeast. Line of locations A-B shown in Figure 5.19. Members are presently informal. After Martel (1990, fig. 1.4).

generally <2 m and locally up to 50 m thick (Schenk, 1967a, 1969). The Macumber can be traced over an area of 250 by 100 km, and is present at most localities (Schenk, 1967a; Geldsetzer, 1977, 1978; Schenk et al., 1990). The rock is composed of alternate laminae of peloidal lime wackestone to grainstone and bituminous shale. The shale typically contains 4-6% of organic carbon in the form of alginite, liptinite and liptodetrinite (McMahon, 1988). Associated fossils include crustaceans, brachiopods and conodonts. The laminite is considered a relatively deep-marine deposit (Geldsetzer, 1978; Schenk et al., 1990) based on

association with turbidites, debris-flow deposits, and *Nereites ichnofauna*. Depositional conditions were predominantly anoxic.

The laminite is associated with, and passes laterally into, lime mud-rich buildups that reach 50 m designated the Gays River Formation. In Nova Scotia, ten areas each contain up to 28 large mounds (Boehner, 1987). A prominent example is the mound at the type area of the Gays River Formation, located where the Windsor Group overlaps pre-Carboniferous basement in the Musquodoboit area. This mound hosts an economic Pb-Zn deposit at Gays

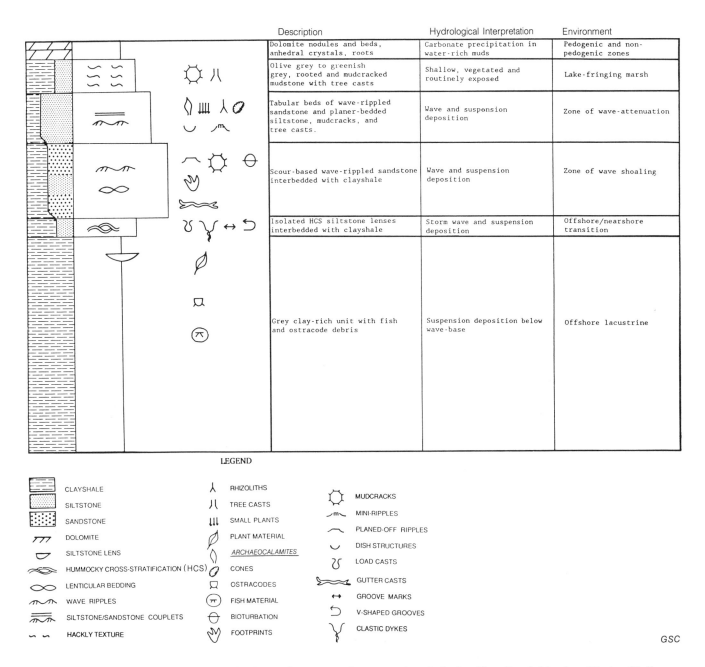

Figure 5.21. Summary section of a lacustrine coarsening-upward cycle in the Blue Beach Member (Horton Bluff Formation). Cycle thickness is typically 1-12 m at Blue Beach. Modified from Martel (1990) and Martel and Gibling (1991).

FORMATION	FACIES ASSEMBLAGES	POSITION
AINSLIE	A1 red/grey pebbly coarse sandstone - conglomerate (distal alluvial fan/proximal braidplain)	↑ basin margin
	A2 red fine coarse sandstone and siltstone (low sinuosity fluvial)	
	A3 grey/green fine sandstone and siltstone (high sinuosity fluvial)	↓ basin centre
STRATHLORNE	S1 dark grey mudstone (open lacustrine)	↑ basin centre
	S2 grey/green very fine - fine sandstone (deltaic nearshore/shoreline)	
	S3 grey/green medium sandstone - boulder conglomerate (fan delta)	
	S4 red siltstone - fine sandstone (coastal mudflat)	↓ basin margin
CRAIGNISH	C3 grey/green coarse sandstone - granulestone (distal gravelly braid plain)	↑ basin centre
	C2 brick red silstone - fine sandstone (mudflat/playa)	
	C1 red/orange coarse sandstone - conglomerate (distal-medial gravel braidplain)	↓ basin margin

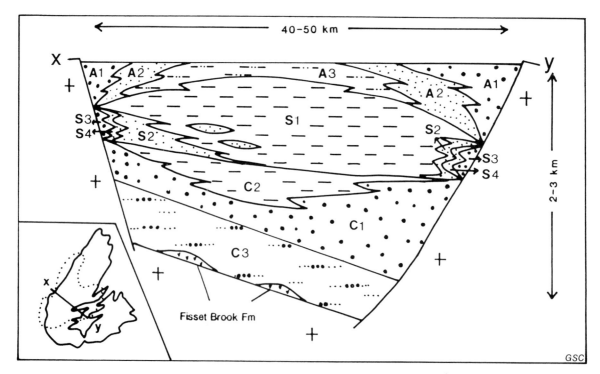

Figure 5.22. Formations and facies distribution of the Horton Group in western Cape Breton Island. Modified from Hamblin (1989, tables 4a, b).

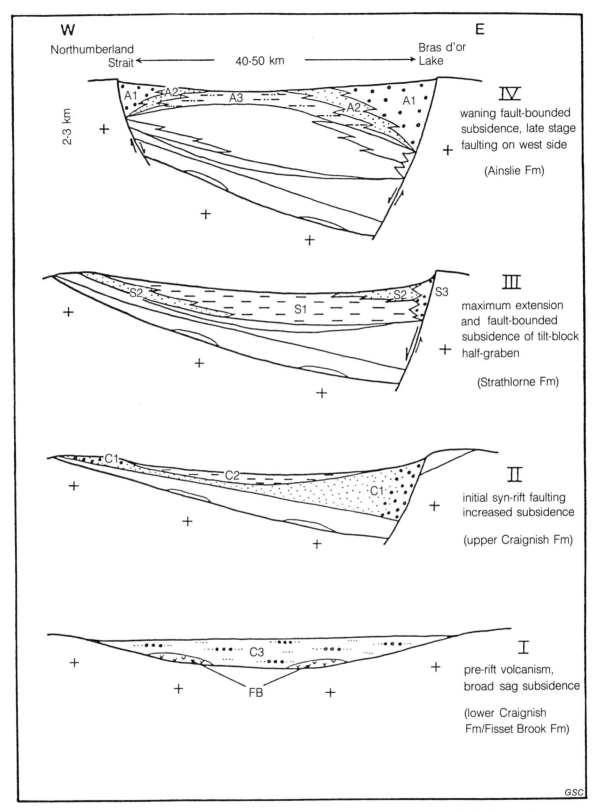

Figure 5.23. Four phase tectonic evolution of Horton depositional systems, Ainslie basin. From Hamblin (1989, fig. 166). See Figure 5.22.

503

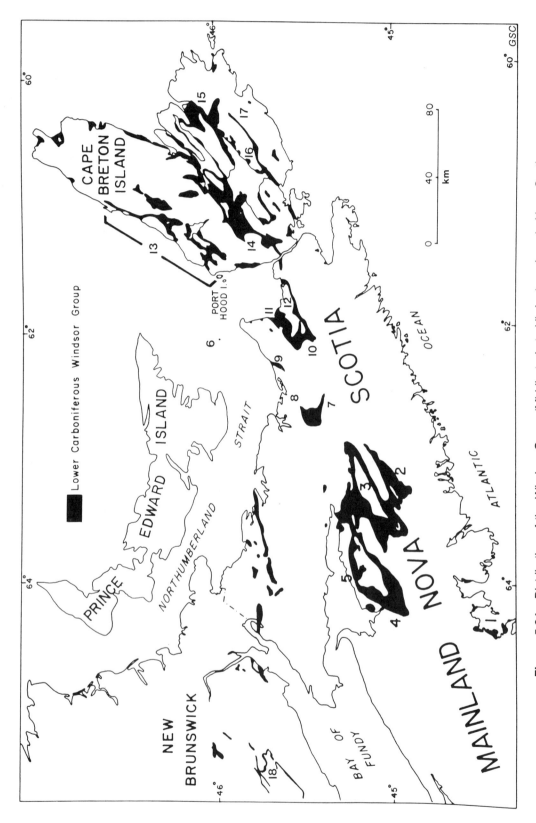

Figure 5.24. Distribution of the Windsor Group (Middle to Late Viséan) onshore in Nova Scotia. Numbers refer to section locations shown in Figure 5.26. After Boehner (1986, fig. 1-7).

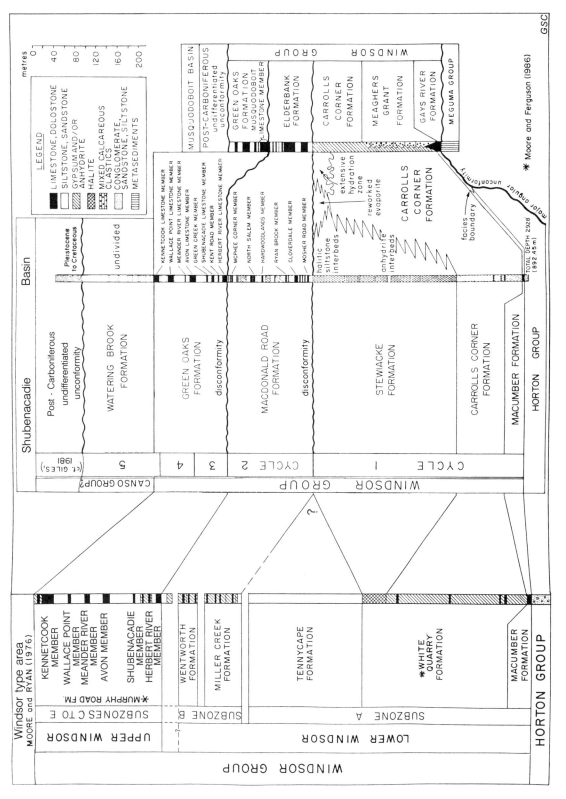

Figure 5.25. Stratigraphic summary for the Windsor Group in the Windsor, Shubenacadie and Musquodoboit regions, Minas Basin, Nova Scotia. After Boehner (1986, fig. 2-1).

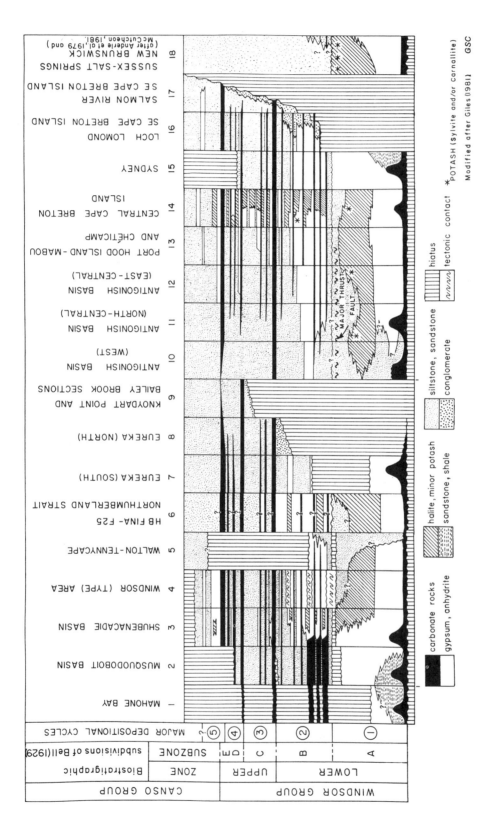

Figure 5.26. Lithostratigraphy of the Windsor Group in mainland Nova Scotia and Cape Breton Island, to show relationship of rock units to major cycles. The locations of sections are shown in Figure 5.24. The vertical scales are distorted for ease of illustration. After Boehner (1986, fig. 1-6; modified from Giles, 1981, fig. 2).

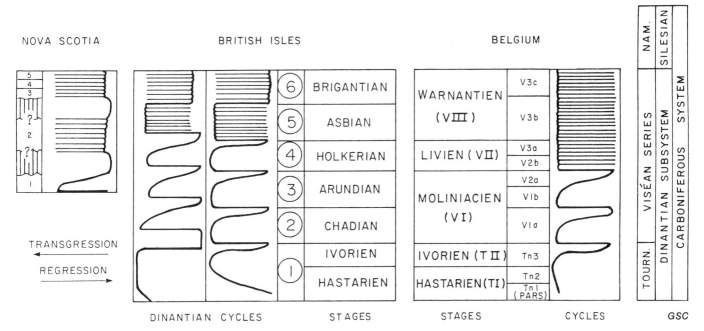

Figure 5.27. Tentative correlation of major cycles in the Windsor Group of Nova Scotia with transgressive-regressive cycles in the Dinantian of Britain and Belgium. Vertically ruled areas indicate possible breaks in the stratigraphic record produced by subaerial erosion during regressive maxima. Small-scale cyclicity in Nova Scotia is shown as largely regressive with small transgressive pulses. This contrasts with Belgium, where small-scale cyclicity is defined by regressive pulses in a regionally transgressive situation. Note that small scale sedimentary rhythms appear earlier in Belgium (and in Nova Scotia?) than in the British Isles. After Giles (1981, fig. 4).

River (Giles et al., 1979; Schenk and Hatt, 1984; Akande and Zentilli, 1984; Boehner et al., 1989b; Arne et al., 1990). Substantial relief (60 m+) has been demonstrated in the Gays River mound complex (Schenk and Hatt, 1984), and the Macumber Formation lies basinward and downdip from the complex. Subtidal, below wave-base, to supratidal conditions were inferred for the mound by Boehner et al. (1989b).

The Carrolls Corner Formation (basal anhydrite) overlies the Macumber in the Shubenacadie area (Fig. 5.25). The formation consists of 160 to 300 m of massive anhydrite with thin partings of dolostone and calcareous siltstone. It is overlain by the Stewiacke Formation, 300 m of banded halite with anhydrite and halitic siltstone interbeds. Similar, correlative anhydrite and halite units occur throughout Atlantic Canada.

Major Cycles 2, 3, 4 and 5 are more variable across Nova Scotia and differ from Major Cycle 1 in containing numerous minor units of marine carbonate, with varying proportions of evaporites and redbeds. These higher major cycles comprise up to twelve internal minor cycles 5-30 m thick. Individual minor cycles show, from base to top, shaly limestone-dolostone, local bioherms overlain by nodular mosaic to massive anhydrite, bedded halite, local sylvite-carnallite and redbeds.

In the Shubenacadie area, the Macdonald Road (Major Cycle 2) and Green Oaks (Major Cycles 3-5) formations consist of about 300 m of intercalated fossiliferous marine carbonates, evaporites (anhydrite, halite) and fine-grained, red and green terrigenous rocks. Distinctive fossiliferous marine carbonate members, a few to 20 m thick, in the

Green Oaks Formation can be traced through much of mainland Nova Scotia and Cape Breton Island (Moore and Austin, 1984). Bioherms are present at several levels in the younger formations (Boehner, 1987; Boehner et al., 1989a). Marine carbonates are absent in the overlying Watering Brook Formation, which is assigned to the Mabou (Canso) Group. Two disconformities are recognized within the succession. Facies variations are present in the Musquodoboit area, where the strata are thinner and evaporites are less prominent, and in the Windsor area (Fig. 5.25).

Potash-bearing units are developed at different stratigraphic levels in many parts of Nova Scotia. A full economic assessment of the Windsor evaporites is provided by Boehner (1986). Continental redbeds predominated in the upper Windsor Group in southern New Brunswick while evaporitic-marine deposition prevailed in southern Nova Scotia.

Conglomerates, interbedded with Windsor Group marine carbonates of Viséan age, were mapped near Big Pond by Bradley and Bradley (1986) and near Loch Lomond by Boehner and Prime (1985).

Breccias composed of Windsor rocks in a carbonate-rich matrix are present at several levels in the group. The Pembroke Breccia (Weeks, 1948) lies between the Macumber Formation and overlying evaporites at many localities, and is up to 30 m thick. It has been interpreted as the product of post-Windsor, intrastratal solution collapse (Clifton, 1967; Geldsetzer, 1978), karstification (Schenk, 1984), and hydrofracturing by overpressured hydrothermal brines, probably in the Mid-Late Carboniferous (Ravenhurst and Zentilli, 1987). The Windsor Group in the Sydney Basin is capped locally by up to 180 m of breccia, attributed to karstic processes operative during a period of Late Carboniferous erosion and karstification (Boehner, 1985, 1986).

Salt diapirs are widespread onshore in Nova Scotia and offshore in the southern Gulf of St. Lawrence (Boehner, 1986; Howie, 1988). Paleoflow data from Cumberland Group strata (Westphalian A-B) in the Cumberland Basin suggest intra-Carboniferous diapirism (Bell, 1944; Ryan et al., 1987).

Mabou Group

The mid-Carboniferous strata of Atlantic Canada have proven difficult to define and correlate. The continental strata are transitional with the underlying marine Windsor Group, and appear to be associated with some important

tectonic events, discussed later. The strata generally have been assigned to the Canso and Riversdale groups (Bell, 1944). However, Ryan et al. (1991a) recently reviewed nomenclature for the Upper Carboniferous and Permian strata of the region and, following the precedent of Belt (1964, 1965), assigned the strata to the Mabou Group and the revised Cumberland Group. The terms Canso and Riversdale are abandoned. Numerous component formations have been recognized (Belt, 1964, 1965; McCabe and Schenk, 1982). The Mabou Group spans the latest Viséan, Namurian and early Westphalian A (Fig. 5.17).

Thickness of the Mabou Group is about 2100 m, and it was probably overestimated at about 6100 m in the Antigonish Basin (Belt, 1964) because of faults (Prime, 1987). The group is everywhere conformable or disconformable with the underlying Windsor Group, but locally onlaps older rocks. Its upper contact is conformable or unconformable, and the strata have been removed locally by mid-late Carboniferous erosion. The type section is located at Southwest Mabou River, Cape Breton Island. Several well-described localities are discussed below.

Antigonish, Western Cape Breton, and Cumberland basins

Belt (1964, 1965, 1968a) described the Mabou Group over a large area of Nova Scotia as a nonmarine unit consisting of silicilastic and carbonate mudstone with up to 50% of fine- to medium-grained sandstone. Two regional lithofacies assemblages were recognized: (1) a red assemblage of red and grey-green calcareous mudstone, with subordinate sandstone; this assemblage was interpreted as fluvial interchannel deposits with thin channel bodies, and (2) a grey assemblage of dark- to medium-grey, laminated shale, with subordinate mudstone and carbonate-cemented sandstone, and with some reddish mudstone-rich intervals. This assemblage was interpreted as lacustrine with red fluvio-lacustrine intercalations, and the lakes may have covered at least 28 000 km^2. Comparison was drawn between the calcilutite ("cementstone") beds in the Mabou Group and those in roughly coeval strata of the Midland Valley of Scotland and northern Ireland (Belt et al., 1967).

Belt noted a strong textural contrast between sandstones of his Mabou Group and those of the overlying "Coarse Fluvial Facies", now considered part of the Cumberland Group and/or Pictou Group.

Brown (1986b) measured a 375 m section of the Mabou Group at Broad Cove, western Cape Breton, and recognized three facies assemblages (Fig. 5.28). A carbonate facies assemblage, attributed to ephemeral (playa) lakes, consists of calcareous shale, dolostone, stromatolitic and oolitic limestone in 8-30 m packages (average 17 m). An intercalated siliciclastic facies assemblage consists of red, mainly massive mudstone with sandstone units up to 10 m thick. Packages are 6-60 m thick (average 20 m), and deposition took place on lake-marginal flats subject to sheet floods and channelization. Carbonate and siliciclastic layers are locally thinly-interbedded (mixed facies assemblage).

Thick alluvial units (the Lismore and Pomquet formations) were described by Fralick and Schenk (1981) and Prime (1987), respectively, in the eastern Cumberland and Antigonish basins. Both formations contain red floodplain deposits, and the rivers were interpreted as meandering to ephemeral.

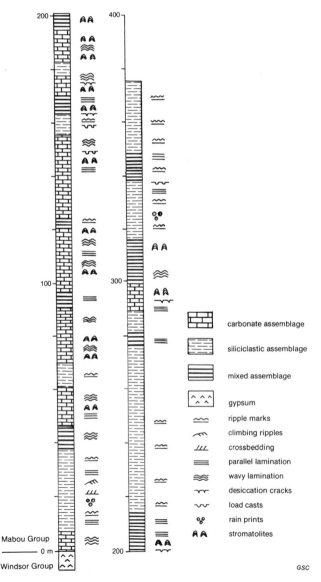

Figure 5.28. Simplified stratigraphic section of the Mabou (Canso) Group at Broad Cove, Cape Breton Island. After Brown (1986b, fig. 14).

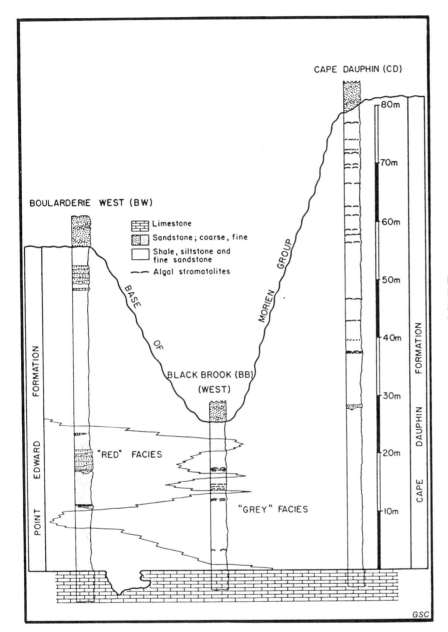

Figure 5.29. Sections of the Point Edward Formation measured on Boularderie Island, Sydney Basin. Modified from Giles (1983, fig. 5).

The Maringouin Formation (equivalent to the Middleborough Formation of Cumberland Basin: Fig. 5.17) was described by McCabe and Schenk (1982) from Cape Maringouin just west of the Nova Scotia-New Brunswick border. The formation is up to 810 m thick (Gussow, 1953), and belongs to the red assemblage of Belt (1968a). At Cape Maringouin, it consists of alternate units of red mudstone and very fine grained sandstone with horizontal lamination and ripple-drift cross-lamination, interpreted as sheet-flood deposits. Sandstones higher in the formation, which show channelization, are probably the deposits of ephemeral meandering rivers. The red colouration, deep desiccation cracks and gypsum casts suggest an arid paleoclimate (McCabe and Schenk, 1982). Sand-sized mud aggregates are common in the rippled sandstone (Rust and Nanson, 1989). They are probably pedogenic, and resemble mud aggregates transported as bedload in modern, seasonally flowing rivers of the Lake Eyre Basin, Australia.

Minas Basin

The lacustrine West Bay Formation near Parrsboro (McCabe and Schenk, 1982), located adjacent to the Cobequid Fault, consists of four major facies. Wavy-bedded grey sandstone and mudstone contains symmetrical ripples and some deep (up to 50 cm) desiccation cracks, and probably represents "deep" lakes subject to occasional exposure. Red and green mudstone is massive to moderately laminated, shows abundant desiccation cracks, and is interpreted as fringing mudflat deposits. A few beds of cross-laminated sandstone, up to 1 m thick, represent influxes derived from sheet floods or ephemeral streams. Granule- and pebble-bearing sandstones less than 1 m thick are probably transgressive lags; clast types include caliche, ooids and pisolites, sparry carbonates with replaced gypsum and halite, ostracodes and bivalves, and sandstone and mudstone blocks. Amphibian tracks are common.

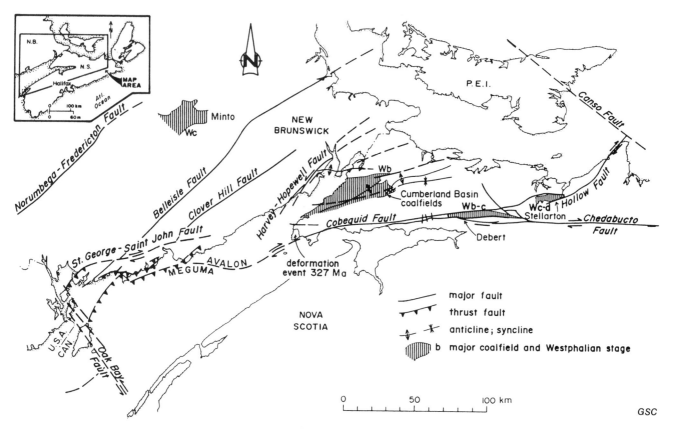

Figure 5.30. Geological setting of the Cumberland Basin of Nova Scotia and eastern New Brunswick. After Calder et al. (1991, fig. 1).

Sydney Basin

In the northwestern Sydney Basin, the lacustrine Point Edward Formation consists of 213 m of red, green and grey shale with minor sandstone and limestone (Giles, 1983). Stromatolitic limestone forms a minor but distinctive part of the succession. Grey and red facies show lateral equivalence (Fig. 5.29).

Cumberland and Pictou groups

The upper Paleozoic succession in Nova Scotia is capped by a thick alluvial sequence represented by the Cumberland and Pictou groups. Braided-, meandering- and anastomosed-fluvial deposits are all represented, with local alluvial fan deposits which are especially prominent in the Cumberland Basin. Lacustrine strata form relatively thin units in most depocentres but predominate in the Stellarton Basin. The strata range from late Namurian to Stephanian onshore in Nova Scotia (Fig. 5.17), and range into the Early Permian on Prince Edward Island. Coal deposits in the Cumberland Group form a major economic resource of Nova Scotia. Regional summaries of coal occurrences can be found in Calder (1985a) and Smith (1989).

Namurian to Stephanian strata in Nova Scotia have been assigned local group and formation names that have tended to obscure regional correlation. However, recent systematic mapping of most basinal areas by the Nova Scotia Department of Mines and Energy, coupled with

palynological analysis, has led to a fundamental revision of the stratigraphic nomenclature (Ryan et al., 1991a). The revision is of particular relevance to the Cumberland Basin but can be applied to other parts of the province (Ryan et al., 1991a), and is followed here.

Two groups are recognized, both with type areas in the Cumberland Basin. The Cumberland Group comprises coal-bearing, predominantly grey, strata, and includes strata formerly assigned to the Riversdale Group. The Stellarton Formation and Morien Group of the Stellarton Basin and Sydney Basin, respectively, also are provisionally reassigned to the Cumberland Group. The Pictou Group comprises predominantly red, non coal-bearing strata that overlie the Cumberland Group. Because this revision is recent and has not been applied formally to all areas of the province, the two groups are considered together in this review.

Cumberland Basin

The tectonic setting of the Cumberland Basin and the distribution of Cumberland Group coalfields are shown in Figure 5.30. During deposition of the Cumberland Group, alluvial fans and braidplains periodically fringed the bounding Cobequid Highlands and Caledonia Highlands (Fig. 5.15), with axial braided and meandering river systems (Bell, 1938; Calder, 1985b, 1991; Salas, 1986). Syn-Westphalian diapirism of Windsor Group evaporites is inferred from paleoflow analyses (Bell, 1944; Ryan et al., 1987).

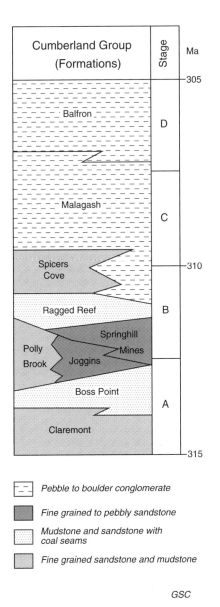

Figure 5.31. Schematic representation of formational relationships in the Cumberland Group and basal Pictou Group (Balfron Formation), Cumberland Basin. Modified from Calder (1991).

The Cumberland Group consists of red and grey conglomerate, sandstone and mudstone, with coal seams and thin bituminous limestone and shale. The group reaches thicknesses of about 5 km based on stratigraphic measurement (Ryan et al., 1991a) and seismic analysis (Calder and Bromley, in press). Seven predominantly alluvial formations have been defined (Ryan et al., 1990b, 1991a; Calder, 1991) (Fig. 5.31, 5.32). The Claremont and Boss Point formations were formerly assigned to the Riversdale Group.

The Claremont Formation is a conglomerate unit, typically about 100 m thick, deposited in alluvial fans adjacent to the Cobequid Highlands. Elsewhere in the basin, equivalents include the Millsville Formation in north-central Nova Scotia and the Enragé Formation in southern New Brunswick.

The Boss Point Formation, about 750 m thick in the type section at Boss Point (Fig. 5.32), is restricted to the Moncton and Cumberland basins. The formation consists predominantly of trough crossbedded sandstone, deposited in braided rivers, and intercalated fine grained sandstone, ostracode-bearing limestone and coaly shale deposited in poorly drained floodplains and shallow lakes (Browne, 1990). Paleosols that include nodular calcrete probably represent ancient aridisols and vertisols, and suggest a semi-arid paleoclimate.

Polly Brook Formation is an alluvial-fan conglomerate up to 600 m thick, principally stream-flow deposits, that fringes the northern margin of the Cobequid Highlands near Springhill (Calder, 1991). The formation passes basinward into the Joggins and Springhill Mines formations (Fig. 5.33), which are exposed in magnificent coastal cliffs near Joggins where some of the earliest studies of coalbearing strata were conducted (Logan, 1845; Dawson, 1878). The Joggins and Springhill Mines formations are about 2100 m thick near Joggins, and consist of grey and red mudstone, grey sandstone, coals up to 4.3 m thick near Springhill, and thin, bivalve-rich black shales and limestones which are a diagnostic feature of the Joggins Formation. Some of the world's oldest reptiles have been found in the Joggins Formation where they commonly occur in trunks of the lycopsid tree *Sigillaria* (Carroll, 1967; Carroll et al., 1972). Duff and Walton (1973) ascribed part of the Joggins Formation to a delta plain with channel and overbank facies. The bivalve assemblages may indicate marine connections based on comparison with European assemblages. Geochemical analyses of some bivalve shells suggest freshwater conditions and indicate the presence of aragonite, possibly the oldest freshwater occurrence of aragonite yet identified (U. Brand, pers. comm., 1990). Reworked Viséan palynomorphs are present in the Springhill Mines Formation near Springhill (Dolby, 1991a), indicating erosion of Lower Carboniferous strata during this period.

Coal seams in the Springhill Mines Formation were deposited in a piedmont setting between the toes of alluvial fans fringing the Cobequid Highlands (Polly Brook Formation) and an axial meandering-fluvial system marked by sedimentary features indicative of highly variable discharge. This setting reflects groundwater recharge from the fans (Calder et al., 1991; Calder, 1993). The seams progressively onlap the Polly Brook Formation (Fig. 5.33). Maceral and palynological analyses of the No. 3 Seam at Springhill suggest predominantly rheotrophic (groundwater-fed) conditions during the life of the paleomire, which may have experienced moderate raising and subsequent deflation during its final stages (Calder et al., 1991; Calder, 1993) (Fig. 5.34). Coal/clastic intercalations (cyclothems) at Springhill represent about 40 ka on average (Calder, 1991).

Carbonaceous limestone and shale, in units up to 2 m thick, are widespread in the Joggins Formation and to a lesser degree in the Springhill Mines Formation (Smith and Naylor, 1990), where they represent lakes with aerobic to anaerobic bottom waters (Gibling and Kalkreuth, 1991). Samples from a unit overlying the Forty Brine Seam at Joggins contain up to 43% TOC (total organic carbon), mainly vitrinite and sporinite (Gibling and Kalkreuth, 1991). Detailed geochemical and microscopic analysis of Joggins coal and coaly shale by Mukhopadhyay et al. (1991) indicate their excellent potential as hydrocarbon source rocks.

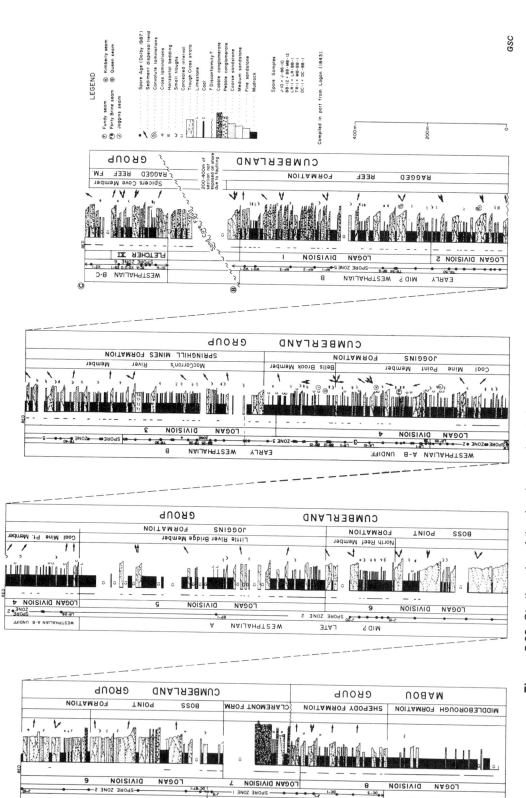

Figure 5.32. Stratigraphy of the Joggins section (adapted from Logan, 1845): type section of the Cumberland Group, Boss Point, Joggins and Ragged Reef formations, and reference section of the Claremont and Springhill Mines formations. After Ryan et al. (1991a, fig. 5).

Correlation of coals in the Joggins and Springhill coalfields, about 30 km apart, has received considerable attention (Bell, 1938; Shaw, 1951; Copeland, 1959; Hacquebard and Donaldson, 1964). Results of recent seismic (Calder and Bromley, in press) and palynological (Dolby, 1991a) analyses indicate that the Fundy to Joggins seam interval at Joggins is slightly older than the main economic seams at Springhill; the latter correlate approximately with the MacCarron's River Member of the Springhill Mines Formation at Joggins (Calder, 1991).

In the Joggins coastal section, Rust et al. (1984) ascribed the MacCarron's River Member to anastomosed rivers, based on the presence of laterally discontinuous, multistorey channel deposits lacking well-developed lateral accretion sets.

The Ragged Reef Formation, about 1000 m thick, forms extensive cliffs southwest of Joggins and underlies much of the Athol Syncline. Alluvial fan, transverse and trunk braidplain deposits were identified by Salas (1986) and Deal (1990). Alluvial fan deposits border the Cobequid Highlands and pass basinward into alluvial plain deposits; the fanglomerates locally are difficult to distinguish from the Polly Brook Formation. Coarse grained sediments also bordered the Caledonia Highlands.

The Malagash Formation, youngest of the Cumberland Group and transitional to the overlying Pictou Group redbeds, is an alluvial succession of grey and red sandstone and mudstone, with thin limestones and minor coals, about 250-450 m thick.

In the Stellarton Gap region of the eastern Cumberland Basin, located between the Cobequid Highlands and Antigonish Highlands (Fig. 5.16), the fluvial Middle River Formation, about 1150 m of grey and red sandstone, mudstone and minor conglomerate, may be equivalent to the upper Boss Point Formation (Yeo and Ruixiang, 1987). The New Glasgow Conglomerate, a coarsening-upward sequence more than 700 m thick of red and grey conglomerate with minor sandstone and siltstone, probably overlies the Middle River Formation and was deposited in alluvial fans and braided rivers. Both formations were assigned to the Cumberland Group by Bell (1940).

The Pictou Group differs from the underlying Cumberland Group in the predominant red colour of the alluvial sandstones and mudstones and the scarcity of conglomerate and coal (Ryan et al., 1991a). Thin grey limestones are present locally. The group is 1650 m thick at West Branch River John, the type section (Fig. 5.35), where the Balfron, Tatamagouche and Cape John formations are alluvial deposits that show a broad upward fining. The group is thicker northward, about 3000 m on Prince Edward Island. It is widespread both onshore and offshore and underlies much of the Gulf of St. Lawrence (Sanford and Grant, 1990) and Sydney Basin. It rests conformably on the Cumberland Group or unconformably on older Carboniferous rocks and basement.

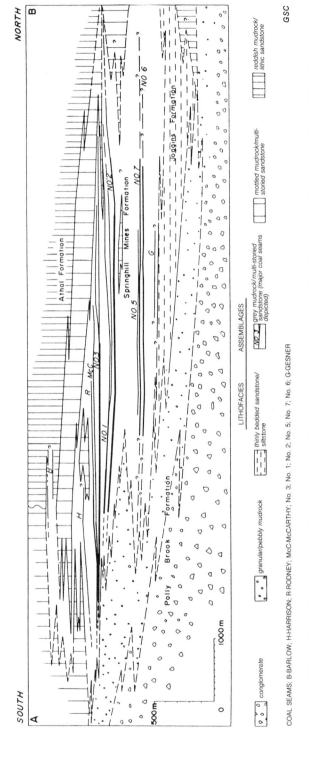

Figure 5.33. Cross-section of the Cumberland Group in the southern part of the Cumberland Basin near Springhill. After Calder (1991).

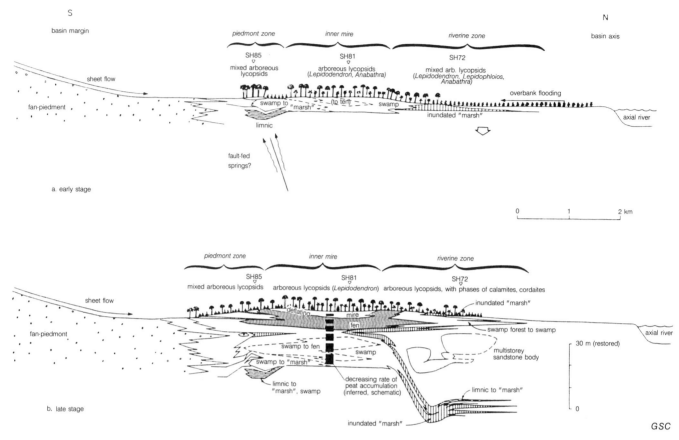

Figure 5.34. Evolutionary development of paleomire types for the ancestral peat-forming ecosystem of the No. 3 Seam, Springhill Coalfield. **a)** Early, rheotrophic stage; **b)** Late, meso-ombrotrophic stage. After Calder (1993, fig. 17).

Stellarton Basin

The Stellarton Basin, although only 6 km by 20 km in extent (Fig. 5.36), contains a lacustrine-dominated succession that is unique onshore in Nova Scotia. The depocentre formed as a pull-apart by dextral transcurrent motion on the Cobequid and Hollow faults (Fralick and Schenk, 1981; Yeo and Ruixiang, 1987). Westphalian B-D (Dolby, 1986, 1987) strata of the Stellarton Formation (Cumberland Group) are about 2600 m thick and are divided into six members (Bell, 1940; Yeo and Ruixiang, 1987; Naylor et al., 1989; Fig. 5.36, 5.37).

The Skinner Brook and Plymouth members are red siliciclastic units that consist of alluvial fan, fan apron and lacustrine deposits (Yeo and Ruixiang, 1987), locally overprinted by well-developed paleosols. These redbed members pass basinward, over a 1-3 km distance, into grey sandstones and mudrocks of the coal- and oil shale-bearing Westville and Albion members. The sandstones and mudrocks of these latter formations are commonly bioturbated by roots, and are attributed to lacustrine deltas that prograded into shallow lakes. The associated coals are thickest (maximum 13.4 m) near the basin centre and pass

laterally toward the basin margin into carbonaceous shales. The coals formed in basin-centre mires which coexisted with poorly drained swamps, where peat did not accumulate, near basin margins. The 13.4 m Foord Seam was inferred to have originated as a eutrophic, flow-fed mire (Calder, 1979).

The coal-bearing Thorburn and Coal Brook members are composed predominantly of grey, coarsening-upward mudrock to sandstone sequences, 5-40 m thick, bounded by oil shales. The coarsening-upward sequences are finer toward the centre of the basin. Bioturbation by roots is less common than in Albion and Westville strata, suggesting that lacustrine deltas formed in relatively deep lakes. The oil shales (maximum 35 m thick) are freshwater lacustrine deposits and contain ostracodes, bivalves and fish remains. Thorburn and Coal Brook coals are thicker toward the basin margins and pass into oil shales basinward. The coals probably represent mires that developed on abandoned lacustrine deltas along the margin of a basin-centre lake.

Naylor et al. (1989) suggested that basin-fill patterns within the Stellarton Basin were largely controlled by basin subsidence and variation in sediment supply. During deposition of the Skinner Brook, Westville, Albion and Plymouth

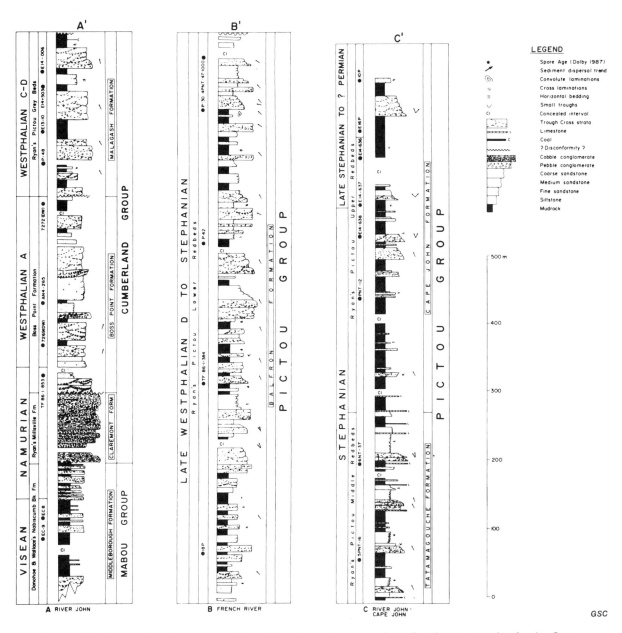

Figure 5.35. Composite stratigraphy of the Pictou Group type area, including the type section for the Cape John Formation, River John – Tatamagouche area. After Ryan et al. (1991a, fig. 12).

members, subsidence rates were relatively low. Shallow lakes were infilled by lacustrine deltas on which poorly drained swamps became established; the swamps restricted sediment supply to the basin so that peats accumulated over much of the area. In contrast, subsidence rates during deposition of the Thorburn and Coal Brook members were relatively high. The basin floor remained well below the water table, and lacustrine deltas that advanced into the relatively deep, perennial lakes were unable to fill the available accommodation space. Coals are thus restricted to basin-margin areas.

Stellarton coals and oil shales have been discussed in detail by numerous authors (Hacquebard and Donaldson, 1969; Calder, 1979; Macauley et al., 1985; Kalkreuth and Macauley, 1987; Paul et al., 1989; Kalkreuth et al., 1990). A good summary is provided by Smith and Naylor (1990). Thirty-five coal seams have been identified, as well as 60 oil shales up to 35.3 m thick (average 5.1 m). In the Coal Brook Member oil shales form 45% of the section in the central basin area. The oil shales are silicate-rich, and yield up to 250 l/t on pyrolysis. Shale types include cannel and boghead (stellarite), the latter rich in highly fluorescing telalginite (Kalkreuth et al., 1990).

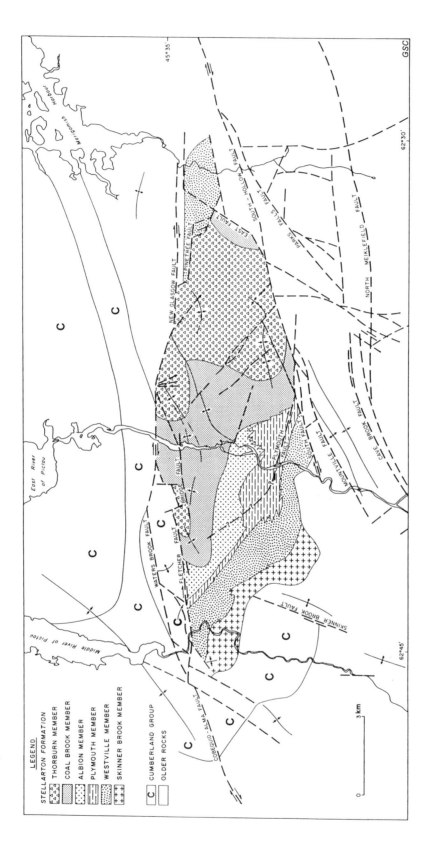

Figure 5.36. Simplified geological map of the Stellarton Basin to show distribution of members of the Stellarton Formation. Modified from Naylor et al. (1989) and Yeo and Ruixiang (1987).

Minas Basin

South of the Cobequid Fault and in the Minas Basin, Cumberland and Pictou groups (Parrsboro and Scotch Village formations, respectively) outcrop locally (McCabe and Schenk, 1982; Boehner, 1986). They are alluvial formations, with some probable lacustrine intervals in the Parrsboro Formation (R. Naylor, pers. comm., 1991). SEM analysis of quartz grains in the Parrsboro Formation shows grain surface features consistent with glacial transport, possibly reflecting the presence of mid-Carboniferous, high-altitude glaciers in the vicinity (D'Orsay and van de Poll, 1985a). In the Debert Coalfield (Westphalian B-C: Dolby, 1991b) west of Truro, coals occur in association with alluvial-fan deposits (Calder, 1985c), a relationship analogous to that at Springhill in the Cumberland Basin.

Antigonish, Western and Central Cape Breton basins

Cumberland (Riversdale) strata are widespread and are locally deformed adjacent to the Hollow Fault which runs roughly parallel to the coast a short distance offshore under the Gulf of St. Lawrence. The Port Hood Formation on western Cape Breton Island (Westphalian A to mid B: Dolby, 1989a, 1991b) consists of an estimated 2000 m of grey and red alluvial and lacustrine strata, with coals up to 2 m thick (Hacquebard et al., 1989). Gersib and McCabe (1981) interpreted strata in the upper, coal-bearing part of the formation as meandering-fluvial with prominent floodplain lake deposits. Black shales with bivalves and ostracodes are present locally. The strata show considerable similarity to the Boss Point and Joggins formations in the Cumberland Basin.

At Broad Cove, north of Inverness, the basal 70 m of the Cumberland Group (unconformable on the Mabou Group) consists of braided-fluvial, conglomerate-sandstone bodies, 2.3 m thick on average, with massive red mudstone forming 25% of the succession (Waterfield, 1986). Small, faulted Port Hood outcrops at St. Rose and Chimney Corner, north of Inverness, consist of grey sandstone, grey/red mudstone and black shales, with coals up to 2.5 m thick (Hacquebard et al., 1989). About 200 m of Port Hood strata are present in the Antigonish Basin (Prime, 1987), and the formation is widely distributed near Richmond in the Central Cape Breton Basin where organic-rich shales are present (Smith and Naylor, 1990).

In the Glengarry outcrop belt of Central Cape Breton, a sandstone-dominated succession, with minor thin coals, of Namurian to Westphalian C and D age (Fig. 5.17) is attributed to braidplains and meandering rivers (Prime and Boehner, 1984). In the Sydney Basin, only a few tens of kilometres distant (Fig. 5.16), an unconformity spans the late Namurian to Westphalian B. These nearby outcrop belts with differing tectonostratigraphic history may have been juxtaposed along transcurrent faults.

The Inverness Formation (Westphalian C to D) outcrops in isolated and faulted areas, principally at Inverness and Mabou in western Cape Breton (Hacquebard et al., 1989). It is about 950 m thick and contains coals up to 4 m thick. Onshore at Mabou, the formation includes packages of grey

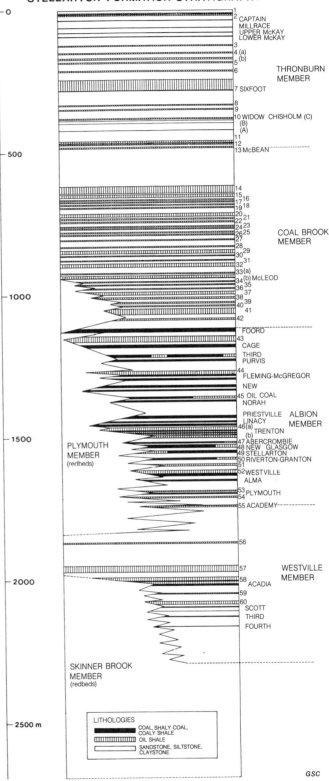

Figure 5.37. Stratigraphy of the Stellarton Formation. After Naylor et al. (1989, fig. 3).

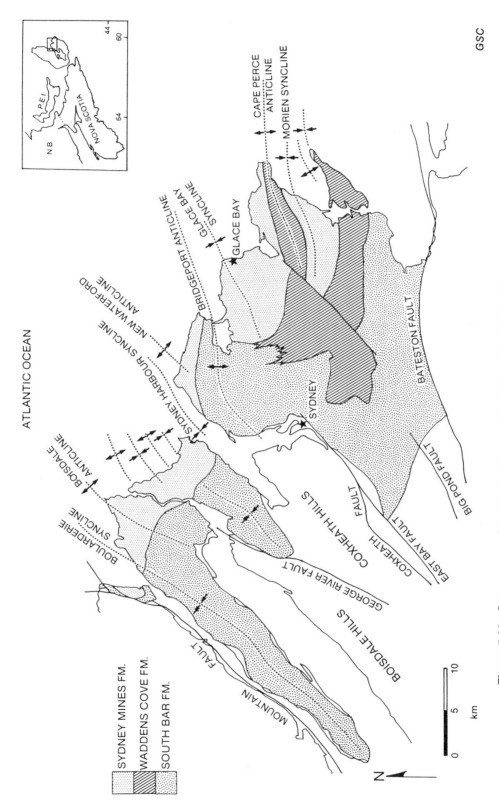

Figure 5.38. Outcrop area of the Morien Group in the Sydney Basin. The Waddens Cove Formation in the southeast part of the basin is equivalent to parts of the South Bar and Sydney Mines formations to the northwest. From Gibling and Rust (1990a, fig. 1).

mudstone and sandstone with coal, carbonaceous lime-stone and shale (Smith and Naylor, 1990; Gibling and Kalkreuth, 1991) and banded siderite, at least 125 m thick, ascribed to shallow lakes and poorly drained floodplains (Deal, 1986; Dickie, 1986). Intercalated, multistorey sand-stones at least 93 m thick, are ascribed to braided rivers. The alternation of strata that are typical of lakes and braidplains in this and numerous other formations in the Late Carboniferous of Nova Scotia is an unusual phenome-non (a British example was discussed by Haszeldine and Anderton, 1980).

The Broad Cove Formation, at least 180 m of Stephanian red siliciclastic strata (Norman, 1935), is located north of Inverness. It forms the only onshore out-crop of Pictou Group strata in Nova Scotia outside the Cumberland Basin.

Sydney Basin

The Sydney Basin lies between the Cobequid-Chedabucto Fault and the Cape Breton Highlands. In the basin, the Morien (Cumberland) Group of Westphalian B to Stephanian age (Zodrow and Cleal, 1985; Dolby, 1988, 1989b) rests unconformably on the Mabou and Windsor groups. The name "Morien Group" (Hayes and Bell, 1923) has been applied exclusively to strata in the Sydney Basin. A major hiatus spans the late Namurian to Westphalian B (Fig. 5.17).

Morien strata (Fig. 5.38) form a broadly upward-fining succession about 2 km thick (Rust et al., 1987). The basal South Bar Formation (up to 1300 m) is a braidplain deposit dominated by trough crossbedded sandstone with local antidune bedforms (Rust and Gibling, 1990a, b). Coals are rare but coal fragments, abundant at many levels, testify to the former abundance of peat on the braidplain.

The overlying Sydney Mines Formation (about 500 m) was deposited from meandering rivers that traversed well- to poorly-drained floodplains upon which extensive peats formed periodically (Hacquebard and Donaldson, 1969; Masson and Rust, 1990). The channel deposits show well-developed lateral accretion sets and exhumed ridge-and-swale topography attributed to scroll-bar accretion (Gibling and Rust, 1987, 1993). Tough carbonaceous shales and limestones up to 1 m thick, with sporinite as the predomi-nant maceral (Smith and Naylor, 1990; Gibling and Kalkreuth, 1991) contain a rich freshwater biota domi-nated by ostracodes and bivalves but including domal stro-matolites and xenacanthid shark remains (Masson and Rust, 1983, 1984). The presence of sparse agglutinated ("marsh") foraminifera (Thibaudeau and Medioli, 1986) in associated shales implies a periodic marine connection, although marine embayments were probably no more than brackish.

Coals in the Sydney Mines Formation are up to 4.3 m thick (Hacquebard, 1983) and represent the largest coal reserve in eastern Canada. The high sulphur content (generally 2-8%) of the coals, commonly inferred elsewhere to reflect marine influence, could reflect in part derivation of sulphur from the thick, karstified evaporites of the Windsor Group which underlie the basin and may have formed part of the hinterland (Bell, 1928; Gibling et al., 1989). The major hiatus that underlies the Morien Group and the

abundance of reworked Viséan palynomorphs in Morien strata (Dolby, 1988) support this hypothesis. The coal-bearing strata at Sydney can be correlated with those that underlie the Gulf of St. Lawrence (Hacquebard, 1986) (Fig. 5.39), and the combined area of the coal-bearing strata may exceed 80 000 km^2.

The Sydney Mines Formation shows alternate coal- and red mudstone-bearing lithosomes (Fig. 5.40) that can be traced along depositional strike and downdip for tens of kilometres. These define cyclothems 20-70 m thick (Gibling and Bird, 1994). Cyclothem duration is estimated at 200-300 ka, based on palynological dating, and the cyclothems probably reflect glacioeustatic fluctuations in response to the coeval Gondwanan Glaciation (see Heckel, 1986). Simi-lar, alluvial- and lacustrine-dominated cyclothems were described from the Appalachian Basin (Beerbower, 1961). Several Sydney cyclothems contain paleovalleys that reflect deep incision during sea-level lowering and are filled with stacked channel deposits.

In the southeastern part of the Sydney Basin, the Waddens Cove Formation (about 840 m thick) interfingers with the South Bar and Sydney Mines formations. The Waddens Cove Formation is predominantly red, with few economic coals and with channel deposits of low width:thickness ratio. Deposition took place from sinuous channels in incised valleys, the channels being confined in part by tough silica-cemented paleosols which developed in associated floodplain and channel-top sandstones (Gibling and Rust, 1990a, 1992).

Red strata with few, thin coals predominate in younger sections offshore (Hacquebard, 1983; Boehner and Giles, 1986) and are provisionally ascribed to the Pictou Group. Their age is unknown.

Tectonic history

The following review considers the late Paleozoic tectonic history of Nova Scotia, with reference to events in other parts of the Atlantic region. A good review of the general tectonic setting was provided by Rast (1983).

Basement rocks

Deep-crustal seismic studies of several transects across the Canadian Appalachians show that Nova Scotia and con-tiguous parts of Atlantic Canada are underlain by three major lower crustal blocks, termed the Grenville, Central and Avalon blocks (Keen et al., 1986; Marillier et al., 1989; Marillier and Verhoef, 1989; Stockmal et al., 1990). The blocks are identified by their deep seismic reflection char-acter especially at or near Moho depths. Surface zones composed of Precambrian and lower Paleozoic rocks (Williams, 1978; Barr and Raeside, 1986; Loncarevic et al., 1989; Dunning et al., 1990) correspond only locally with lower crustal boundaries, and probably reflect in part dis-tinctive upper crustal terranes now juxtaposed as a conse-quence of successive, mainly early to mid Paleozoic, collisional events (Keppie, 1989). A variety of facts testifies to the widespread and intense nature of Acadian deforma-tion in the region (Chorlton and Dallmeyer, 1986; Plint and Jamieson, 1989; Dunning et al., 1990).

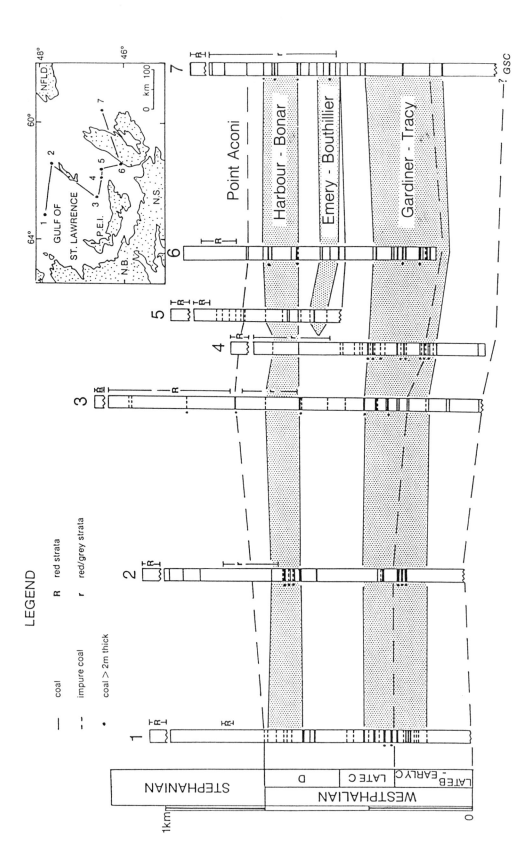

Figure 5.39. Correlation of Westphalian and Stephanian coal-bearing strata in the Gulf of St. Lawrence and Sydney Basin. Inset shows the location of the drillholes. Possible correlations of seams and seam intervals with those in the Sydney Basin (location 7) is shown. After Gibling and Bird (1994, fig. 17), modified from Hacquebard (1986, fig. 6).

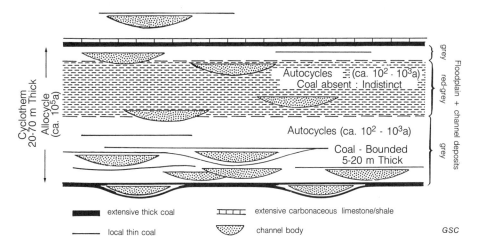

Figure 5.40. Schematic representation of allocycles (cyclothems) and autocycles in the Sydney Mines Formation. The cyclothems are bounded by extensive coals, with carbonaceous shale and limestone. Autocycles are bounded by thin coals or paleosols, and reflect emplacement of channels and associated levees and splays, or drowning under lakes or bays. After Gibling and Bird (1994, fig. 16).

Overlap of younger sedimentary and volcanic formations across terrane boundaries, the commonality of intrusions in adjacent terranes, and dating of deformational events indicate the time of accretion. Middle to Upper Devonian sedimentary and volcanic rocks overlie most terranes and, in the Antigonish area, the Lower Devonian Knoydart Formation shows well developed cleavage which is absent from the Mid to Upper Devonian McAras Brook Formation (Fralick and Schenk, 1981). The Meguma and Avalon terranes are inferred to have been juxtaposed by the Early Devonian (Schenk, 1981; Keppie, 1989). The Aspy and Bras d'Or terranes of Cape Breton Island have contrasting pre-Carboniferous histories (Dunning et al., 1990), and the Bras d'Or-Mira (Avalon) terrane boundary is covered by strata at least as old as Early Carboniferous. Stratigraphic and paleoflow data from Gaspésie indicate active Acadian collision in the Early to Mid Devonian (Lawrence and Williams, 1987; Rust et al., 1989). These, and many other, observations indicate that Acadian terrane accretion was substantially complete by the mid Devonian, to be followed by renewed subsidence and accumulation of strata in Carboniferous successor basins.

Middle Devonian to Early Viséan

Subsidence in Nova Scotia may reflect re-extension of earlier thrusts, such as those identified in deep-crustal profiles, during one or more episodes of strike-slip motion (Quinlan et al., 1988). Marillier et al. (1989) suggested that south-dipping reflectors at mid-crustal depths represent a detachment associated with the formation of the Sydney Basin. The early (mid Devonian to mid Carboniferous) history of the region has been interpreted as one of extensional rifting (McCutcheon and Robinson, 1987). Texture, chemistry and stratigraphic relations of Mid Devonian to Lower Carboniferous volcanics, including those of the Fisset Brook Formation and Fountain Lake Group of Nova Scotia and Viséan volcanics on the Îles de la Madeleine,

suggest eruption in a terrestrial setting related to intraplate continental rifting (Dostal et al., 1983; Blanchard et al., 1984; Barr et al., 1985; Murphy et al., 1988). A change in basaltic composition from tholeiitic to alkalic between the Tournaisian and Namurian in New Brunswick suggests a decreased rate of extension (Fyffe and Barr, 1986).

Tournaisian to lower Viséan strata of the Horton Group in Nova Scotia and New Brunswick were deposited in a series of rift basins, probably half-graben, as indicated by the presence of basin-margin facies such as alluvial fans and patterns of stratal thickening and paleoflow (Pickerill and Carter, 1980; Hamblin and Rust, 1989; Martel and Gibling, 1991). Analysis of seismic sections shows that fault-bounded structures filled with probable Horton strata are abundant beneath much of the southern Gulf of St. Lawrence (Sanford and Grant, 1990). Similar structures were identified on deep-crustal seismic lines through the Sydney Basin and Gulf of St. Lawrence in the vicinity of the Îles de la Madeleine (Marillier et al., 1989). The Deer Lake Basin in Newfoundland shows evidence of Tournaisian to Viséan dextral strike-slip motion associated with the Cabot Fault Zone (Hyde, 1984a, b; Hyde et al., 1988). The widespread and apparently coeval occurrence of organic-rich lacustrine strata (Albert, Horton Bluff, South Lake Creek, Strathlorne formations) in the Horton Group implies a measure of interconnection between the depocentres and suggests a period of accelerated regional subsidence (Hamblin and Rust, 1989).

Middle to Late Viséan

Strong topographic relief (much of it inherited) persisted throughout deposition of the mid to lower Viséan Windsor Group, as indicated by facies relations in the Macumber and Gays River formations (Schenk and Hatt, 1984) and the presence of basin-margin facies. Correlative Lower Carboniferous marine strata were deposited widely in the tropical seas of western Europe and Appalachian to mid-continental U.S.A. However, the unusually thick and extensive evaporites of Nova Scotia and offshore probably reflect an embayed and semi-enclosed basin, remote from Viséan oceans; minor evaporites are known also from Dinantian embayments in Belgium (e.g. Swennen et al., 1990) and Ireland. The onset of Windsor sedimentation is believed to have been sudden (Schenk, 1967a; Geldsetzer, 1977; Kirkham, 1978), with the Viséan sea inundating a broad, complex terrain largely below sea level. Windsor cycles identified by Giles (1981) may reflect glacioeustatic events.

Late Viséan to Namurian

The conformable change from marine (Windsor Group) to evaporitic lacustrine (Mabou Group) conditions indicates a progressive loss of marine connection and a return to prevailing nonmarine conditions. This may correlate with the onset of a major phase of the Gondwanan Glaciation and concomitant sea-level lowering which took place at about the Viséan-Namurian boundary (Veevers and Powell, 1987). Aridity is indicated for both groups (McCabe and Schenk, 1982) so that a regional climatic change is improbable.

An important tectonic phase, accompanied by widespread strike-slip activity (Webb, 1963, 1969), local uplift and concomitant basin development, affected much of Nova Scotia and adjacent areas in the mid Carboniferous. The effects of this activity include:

1) A change from lacustrine and fine grained alluvial strata of the Mabou Group to the "coarse clastic facies" (Belt, 1964) of the predominantly alluvial Cumberland (Riversdale) and Pictou groups. In most areas, this change took place during the late Namurian through Westphalian A (Fig. 5.17). A remarkably thick Westphalian A-B section, including fanglomerates, was laid down in the Athol Syncline of the Cumberland Basin.

2) In southern New Brunswick, structural and stratigraphic data indicate Westphalian A-C basin inversion and thrusting at a major, right-lateral compressive bend in the Cobequid Fault system (Plint and van de Poll, 1984; Nance, 1987a). Synsedimentary fault scarps generated by earthquake activity were identified by Plint (1985) in the Westphalian A-B Tynemouth Creek Formation.

3) The Early Carboniferous Cape Chignecto pluton at the western end of the Cobequid Highlands shows mylonitization and thrust relationships indicative of Namurian to possibly early Westphalian deformation; the thrusting may reflect a major component of convergent movement between the Meguma and Avalon terranes (Waldron et al., 1989). Eisbacher (1969) inferred Westphalian B to Stephanian dextral motion on the western Cobequid Fault. Intensive deformation, including overturning of Lower Carboniferous strata, was documented in the Minas Basin and areas adjacent to the Cobequid Fault (Fyson, 1964a, b, 1967). The deformational episode has not been dated, but some events were inferred to have taken place prior to mid Carboniferous sedimentation.

4) Browne (1991) inferred strike-slip motion on the Harvey-Hopewell fault zone coeval with deposition of the Boss Point Formation of Westphalian A age. This

motion, in conjunction with motion along the Cobequid Fault, probably accounts for Late Carboniferous subsidence of the Cumberland Basin.

5) Structural and sedimentary analyses indicate that the Stellarton Basin formed as a pull-apart structure in association with the Cobequid-Hollow fault system (Fralick and Schenk, 1981; Yeo and Ruixiang, 1987). Associated sedimentation was dated as Westphalian B-D (Naylor et al., 1989).

6) Unconformities, commonly angular, between the Mabou and Cumberland/Pictou groups are developed at numerous locations, with Namurian and Westphalian A-C strata missing locally (Fig. 5.17). Hiatuses are especially prominent in the Sydney and Minas basins (Boehner, 1984; Gibling et al., 1987) and at locations adjacent to the Cobequid Fault. Reworked Devonian to Namurian palynomorphs are common in many Westphalian formations (e.g. Dolby, 1988), indicating widespread uplift of Lower Carboniferous rocks.

7) A widespread thermal event within the Meguma Zone has been dated at 300-320 Ma, approximately mid Namurian to Stephanian (Reynolds et al., 1987). The evidence includes recrystallization of plutons, alteration in the vicinity of veins and shear zones, and mineralization. A Rb-Sr isochron for a hydrothermal clay suite in Horton strata below the Pembroke Breccia gave an age of 300 ± 6 Ma, possibly indicative of the timing of a major hydrofracturing and mineralization event (Ravenhurst and Zentilli, 1987; Arne et al., 1990).

Carboniferous rocks of Atlantic Canada formed part of a regional transcurrent zone through eastern North America and southern Europe during intervals of the late Paleozoic. This structural zone was associated with oblique collision between the North American continent and the African region of the Gondwanan continent (Arthaud and Matte, 1977; Badham, 1982). Tankard (1986) correlated the onset of extensive fluvial sedimentation (Lee Formation and equivalents) in the Appalachian Basin of the United States with the initiation of Alleghanian thrusting. The Lee Formation is probably Namurian in age (Patchen et al., 1985). It is probable that Alleghanian thrusting in the central to southern Appalachians was accommodated in part by dextral strike-slip motion in the Canadian Appalachians where the plate boundary showed a marked curvature, thus accounting for the mid-Carboniferous change in tectonic style. Models for the region that involve rifting succeeded by a broad thermal sag (Bradley, 1982) should be considered partial representations of a complex basinal history.

Westphalian to Early Permian

Westphalian to Early Permian sedimentation was predominantly alluvial. The 3 km thick succession of Stephanian to Lower Permian redbeds that underlies the Gulf of St. Lawrence is commonly overlooked in analyses of late Paleozoic sedimentary history. Strike-slip faulting probably continued to be associated with basinal subsidence throughout this period; a K-Ar whole-rock age of 280 ± 34 Ma was obtained from phyllitic rocks within the Cobequid fault zone (Wanless et al., 1967; Eisbacher, 1969). However, Upper Carboniferous sediments encroached on most adjacent basement areas, for example in the

Cumberland Basin, and Glengarry Basin, and on the New Brunswick Platform (Ball et al., 1981; Boehner and Prime, 1985; Calder, 1991), suggesting a progressive diminution of tectonic activity through the Late Carboniferous. In the Sydney Basin, faults that cut Lower Carboniferous strata cannot be traced into Upper Carboniferous strata (Boehner and Giles, 1986), and paleoflow data suggest progressively reduced intrabasinal relief during Morien Group deposition (Gibling et al., 1992); however, minor differential subsidence has been documented across fault-bounded basement blocks beneath the Morien Group, using thickness and facies variation (Gibling and Rust, 1990b). Sparse agglutinated foraminifera and the presence of well developed cyclothems in Westphalian D to Stephanian strata in the Sydney Basin indicate the presence of marine waters for the first time since the late Viséan.

Late Carboniferous and Permian thermal and structural events, as well as local intrusions, are being recognized widely in eastern New England, Gulf of Maine, and southwestern Nova Scotia (Skehan et al., 1986; Cormier et al., 1988; Pe-Piper and Loncarevic, 1989; Dallmeyer et al., 1990). Rocks of the Narragansett Basin of southern New England contain Westphalian A to Stephanian C coarse grained alluvium, including thick conglomerates, which are intensely deformed (Skehan et al., 1986). Peralkalic volcanism took place at Minto, New Brunswick during the late Westphalian (Fyffe and Barr, 1986), and conglomerates are present locally in Lower Permian strata of Prince Edward Island (van de Poll and Forbes, 1984). It is probable that tectonic activity continued into the Permian in the Atlantic region of Canada, possibly reflecting the final stages of North American-Gondwanan collision.

Paleoflow data for strata of late Westphalian A to Early Permian age in Nova Scotia and New Brunswick indicate predominantly northeasterly paleoflow throughout this period (Fig. 5.41) (Gibling et al., 1992). Regional drainage is inferred to have originated in the fold-and-thrust belt of the central Appalachians and parts of the northern Appalachians (Fig. 5.42), traversing older Acadian mountains to reach the Gulf of St. Lawrence region and probably draining eastward into the Rheic Ocean of western Europe.

The development of redbeds (Pictou Group) across most of Nova Scotia during the Westphalian D to Stephanian (Fig. 5.17, 5.39) probably reflects a regionally lowered water table. This in turn probably reflects regional climatic factors associated with the final assembly of Pangea and a northward drift, as inferred for Midcontinental North America by Schutter and Heckel (1985). A similar trend of upward reddening is observed in Upper Carboniferous to Permian successions in western Europe (Besly, 1988).

The youngest strata yet documented in the Atlantic region are late Early Permian (late Autunian) redbeds of Prince Edward Island, dated using vertebrate footprints by Mossman and Place (1989). The presence of coals of bituminous rank in many Westphalian strata implies substantial post-Westphalian burial, possibly to a depth of several kilometres (Hacquebard and Donaldson, 1970; Hacquebard and Cameron, 1989). The former presence of such additional strata further implies that most upland areas were buried during the Permian, at which time the Atlantic area formed an extensive alluvial plain.

Later history

Subsidence in the Atlantic Region probably terminated by the early Triassic. Fission-track analyses of apatites in a variety of rocks in Nova Scotia indicate that the rocks cooled below ca. 100°C, probably due to uplift, in the Permian to early Mesozoic (Arne et al., 1990; Grist, 1990; Ravenhurst et al., 1990; R. Ryan, pers. comm., 1991). Early Mesozoic fission track ages were obtained also by Johnsson (1986) in New York State, suggesting that uplift during this period was widespread along the eastern margin of North America. Uplift probably was associated with the onset of Atlantic seafloor spreading and the breakup of Pangea. Minor structural deformation, as well as a few intrusions, was associated with Mesozoic activity, which included further movement along the Cobequid-Chedabucto fault system, rifting, dyke intrusion and volcanic eruption locally. Mesozoic and Cenozoic strata are widespread on the Scotian Shelf and locally onshore in Nova Scotia.

Acknowledgments

Bob Boehner, John Calder, Rob Naylor, Georgia Pe-Piper and Walter van de Poll provided thoughtful comments on earlier versions of this review. I am much in their debt, although errors and omissions remain my own. I thank also Hank Williams for his valuable editorial comments. Mai Nguyen and Roger Edgecombe drafted some of the figures. Financial assistance from the Natural Sciences and Engineering Research Council is gratefully acknowledged. This paper is dedicated to the memory of Brian Rust, colleague and friend, who first introduced me to the upper Paleozoic rocks of Atlantic Canada.

UPPER PALEOZOIC ROCKS, NEWFOUNDLAND

Richard S. Hyde

General stratigraphy

All of the upper Paleozoic rocks of Newfoundland have been traditionally regarded as Carboniferous but some may be Late Devonian or Early Permian. The rocks are almost entirely sedimentary and are mostly contained within two basins, the Bay St. George and Deer Lake basins. Outliers of Carboniferous sedimentary rocks occur at Conche, Red Indian Lake, Spanish Room, and Terrenceville (Fig. 5.43). Outliers of suspected Carboniferous age occur at Kings Point, Sheffield Lake, near Birchy Lake, and along South Brook (Fig. 5.43). It is likely that the smaller outcrop areas are erosional remnants of larger depositional areas, but this has not been firmly established.

Volcanic rocks of known Carboniferous age occur locally within the Deer Lake Basin (Hyde, 1984a). Reports of isotopic Carboniferous ages for volcanic and plutonic rocks are in

Hyde, R.S.
1995: Upper Paleozoic rocks, Newfoundland; in Chapter 5 of Geology of the Appalachian-Caledonian Orogen in Canada and Greenland, (ed.) H. Williams; Geological Survey of Canada, Geology of Canada, no. 6, p. 523-552 (also Geological Society of America, The Geology of North America, v. F-1).

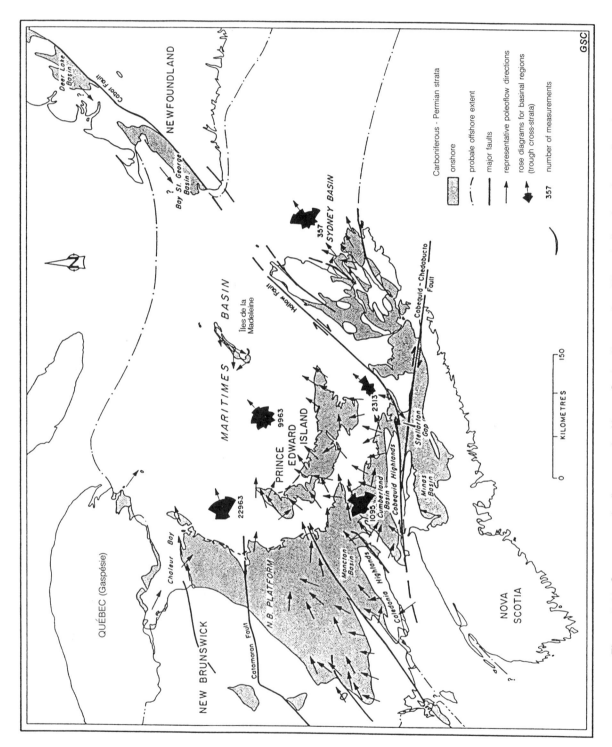

Figure 5.41. Summary paleoflow map for Upper Carboniferous (late Westphalian A) to Lower Permian strata of the Maritimes Basin. Solid arrows show representative paleoflow directions. Rose diagrams show all measurements of trough cross-strata within local areas of the basin total number of measurements in excess of 36 000). Note the predominantly northeasterly paleoflow, subparallel to the line of major faults. From Gibling et al. (1992).

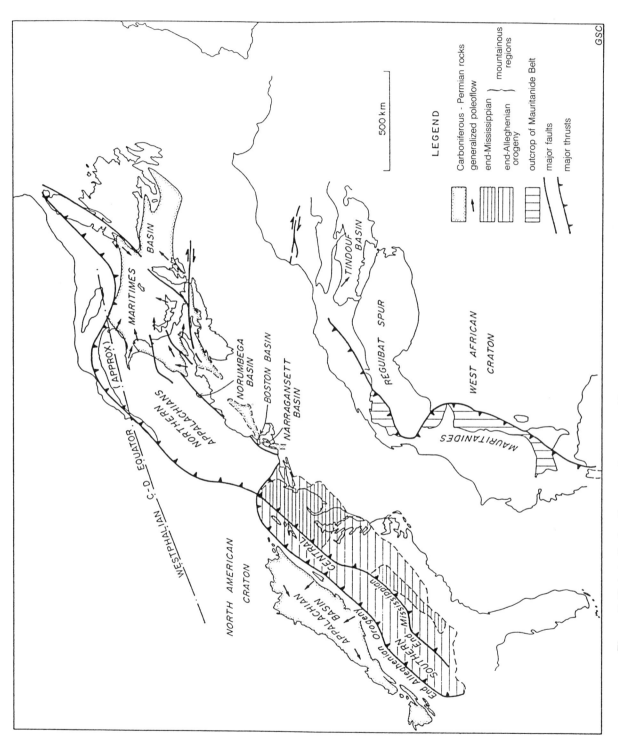

Figure 5.42. Late Carboniferous paleogeographic reconstruction for eastern North America and adjacent areas. Continental disposition from Scotese and McKerrow (1990). From Gibling et al. (1992).

525

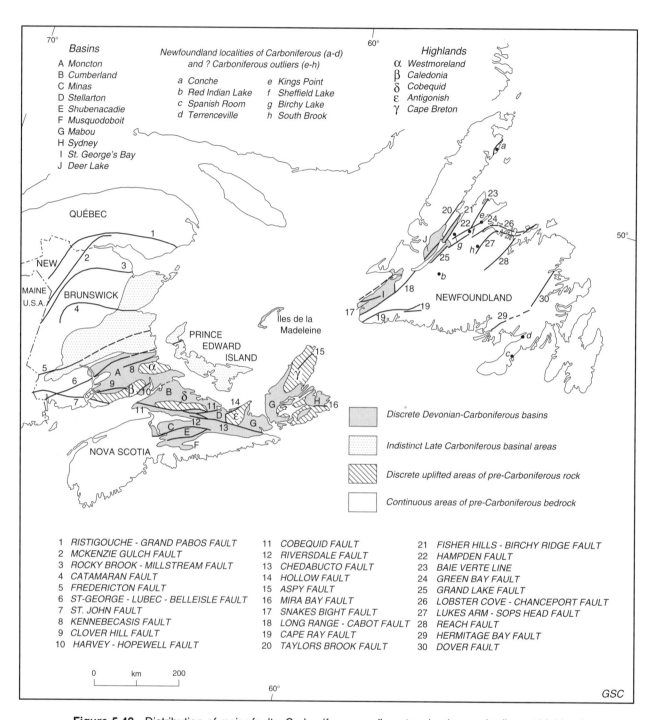

Figure 5.43. Distribution of major faults, Carboniferous sedimentary basins, and adjacent highland areas in eastern Canada and known and suspected Carboniferous outliers in Newfoundland.

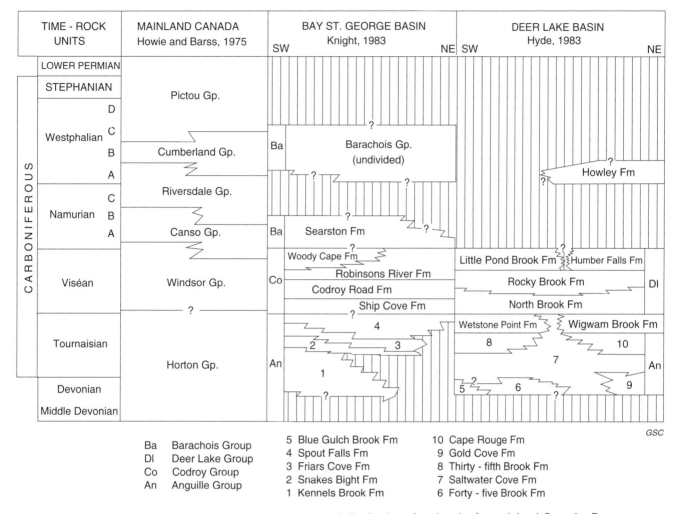

TIME - ROCK UNITS			MAINLAND CANADA Howie and Barss, 1975	BAY ST. GEORGE BASIN Knight, 1983		DEER LAKE BASIN Hyde, 1983

Figure 5.44. Stratigraphic nomenclature, ages, and distribution of rock units for mainland Canada, Bay St. George Basin, and Deer Lake Basin.

some cases doubtful because they tend to fall near the Devonian-Carboniferous boundary (360 Ma adopted by Palmer, 1983; Haq and van Eysinga, 1987). Some of the isotopic dates could also be cooling ages.

Offshore areas underlain by rocks of known and inferred Carboniferous age almost completely surround the island of Newfoundland (Haworth et al., 1985; Bell and Howie, 1990). In some cases drilling and grab sampling have permitted fossil dating (e.g., Jansa and Mamet, 1984). However, most samples interpreted to be Carboniferous are unfossiliferous. A full account of offshore Carboniferous rocks is given by Bell and Howie (1990).

All of the Carboniferous rocks in Newfoundland are interpreted to rest unconformably on rocks variably affected by Acadian deformation. Carboniferous basins in Newfoundland are considered to have developed as a consequence of wrench movements along major fault systems (Wilson, 1962; Belt, 1968b; Webb, 1969; Bradley, 1982; Hyde et al., 1988; Fig. 5.43). In many cases the ages of Newfoundland Carboniferous strata cannot be confidently bracketed for any given succession. Dating is problematical

for older (Tournaisian) and younger (Namurian-Stephanian) Carboniferous rocks, because these rocks are mainly, if not entirely, nonmarine and, thus, sparsely fossiliferous. Dating of these rocks is best achieved from spores, but the spores are commonly destroyed by oxidation or burial carbonization.

Bay St. George Basin

The Bay St. George Basin developed concurrently with the Deer Lake Basin, with basin-fill sedimentary rocks ranging in age from Late Devonian (?) to Late Carboniferous. Three major groups of strata are known from the Bay St. George Basin (Fig. 5.44, 5.45): the Anguille, Codroy, and Barachois groups. There are substantial thicknesses of fluvial and lacustrine strata, but in addition, and in contrast to the Deer Lake Basin, there are many units of marine origin.

The main branch of the Cabot Fault Zone is the Long Range Fault, which forms most of the eastern border of the Bay St. George Basin. Another major fault with the same orientation is evidently present a few kilometres from land,

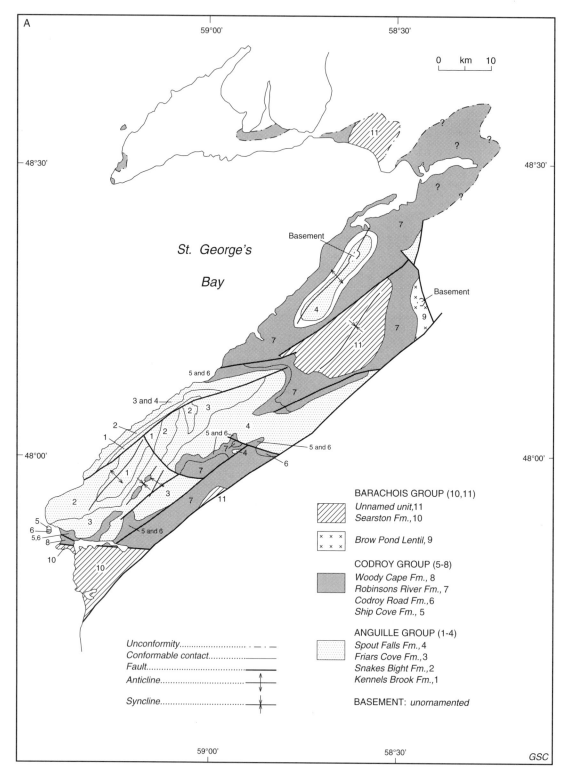

Figure 5.45. Geology map (A) and localities (B) of the Bay St. George Basin. Geology is simplified from Knight (1983). Codroy Group strata on the Port au Port Peninsula are undivided.

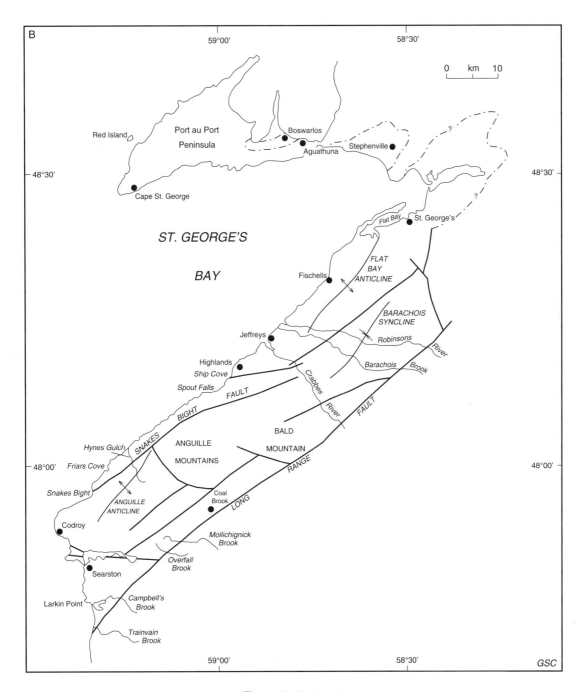

Figure 5.45. (cont.)

GSC

Table 5.13. Chronological listing of work done in establishing the geology of the Bay St. George Basin.

Reference	Contribution
Murray (1881a,b,c)	General observations on the geology near St. George's Bay; author erects a crude stratigraphy and confirms the presence of coal seams.
Howley (1913, 1917b)	Description and summary of coal seams and coal-bearing strata in St. George's Bay area.
Hayes and Johnson (1938)	Stratigraphic foundations; first formal description of Carboniferous units in St. George's Bay area; structural and economic geology; geological maps and sections.
Bryan (1938)	Detailed study of coal seams and adjacent strata; pessimistic about chances for development of coal resource.
Bell (1948)	Detailed measured sections with listing and description of fauna and flora; biostratigraphic correlations with sequences in mainland Canada.
Hayes (1949)	Summary of coal and coal-bearing strata with maps, sections, and laboratory analyses of coal.
Riley (1962)	Geological mapping (1:250 000 scale) includes the northern part of Bay St. George Basin.
Coté (1964)	Stratigraphic analysis with subdivision of Anguille Group; stratigraphic comparisons with the Lower Carboniferous of Cape Breton Island, Nova Scotia.
Baird and Coté (1964)	Publication based on Coté (1964) Ph.D. thesis; interpretation of Anguille Group as nonmarine.
Utting (1966)	Stratigraphy and paleontology (marine invertebrates, palynology) of the Codroy and Barachois groups.
Mamet (1968)	Study of foraminifera in Codroy Group; dating and paleobiogeography.
Belt (1968a)	Description and discussion of Carboniferous nonmarine facies in eastern Canada; examples from Bay St. George Basin.
Belt (1968b)	Tectonic setting of Bay St. George Basin; recognition of graben structure in relation to wrench faulting.
Belt (1969)	Stratigraphy of Bay St. George Basin and discussion of major faults.
Dix (1982)	Sedimentology and paleontology of Codroy Group on Port au Port Peninsula; paleokarstic rocks in a shallow marine setting.
von Bitter and Plint-Geberl (1982)	Conodont zonation of Codroy Group; some revision of stratigraphy.
Knight (1983)	Comprehensive study of geology of Bay St. George Basin; geological map (1:125 000), cross-sections, and detailed measured sections; places basin in a strike-slip tectonic regime.
Dewey (1983, 1985)	Faunal assemblages and paleoecology of ostracodes and peracarids in Codroy Group and their place in global paleobiogeography.
Murthy (1985)	Paleomagnetic study of redbeds in Bay St. George Basin; author finds no evidence for large-scale displacements along Cabot Fault Zone.
Peavy (1985)	Study of gravity and magnetics in the northern part of the basin; depth to basement estimates in the area underlain by Barachois Group.
Solomon (1986)	Physical sedimentology, petrography, clay mineralogy, and organic geochemistry of Barachois Group.
Utting (1987b)	Palynology and thermal maturation of spores in Codroy Group.
Dix and James (1987)	Description and origin of bryozoan and microbial carbonate build-ups in Codroy Group of the Port au Port Peninsula.
Kilfoil (1988)	Study of gravity and magnetics in the southern part of the basin and underneath St. George's Bay.
Dix and James (1989)	Stratigraphy and depositional environments of the Codroy Group on Port au Port Peninsula.
Miller et al. (1990)	Gravity, magnetics and seismic reflection studies of Bay St. George Bay Basin (onshore and offshore).
Hyde et al. (1991)	Composition, rank, age and depositional environments of Barachois Group coals.

under the water of Bay St. George (Knight, 1983; Kilfoil, 1988; Miller et al., 1990). It also appears that this fault is buried by later deposition of the Codroy Group. During Codroy time, the basin was open to the west, joining with the Magdalen Basin (Bradley, 1982) and forming an embayment to an extensive shallow sea (Williams, 1974; Dewey, 1985). The southern part of the basin may be bounded by a north-south trending arch of pre-Carboniferous basement rock (Howie and Barss, 1975; Knight, 1983) that lies off the southwest coast of Newfoundland. The northern margin of the Carboniferous outcrop area is marked by faults and an unconformity (Knight, 1983).

Table 5.13 lists works devoted to the geology of Bay St. George Basin. The most comprehensive work to date is the memoir by Knight (1983).

Anguille Group

Subdivision of the Anguille Group in the Bay St. George Basin is taken from Knight's (1983) mapping and is shown in Figure 5.44. The type area for the Anguille Group is the Anguille Mountains (Fig. 5.45A, B), an upland plateau in the southwestern part of the basin where the rocks were studied and named by Hayes and Johnson (1938). All four of the formations of Knight (1983) occur in the Anguille Mountains, where the Anguille Group is at least 5000 m thick. Elsewhere in the basin, however, only the Spout Falls Formation is present. The Anguille Group is only 1-2 m thick along the northern margin of the basin. In the Bay St. George Basin, the Anguille Group can be considered to range in age from Late Devonian(?) to Viséan(?), based on plant fragments and fossil fish.

The Kennels Brook Formation (Knight, 1983): constitutes the basal unit of the Anguille Group in the type area, and was defined as a replacement for the Cape John Formation of Baird and Coté (1964). Well-indurated, green-grey sandstones, pebbly sandstones, conglomerates and red sandstones with red and green slates (carbonate nodules) represent the older strata of the formation. These lithologies are exposed in the centre of the Anguille Anticline (Fig. 5.46), and were encountered in the subsurface in the Union-Brinex exploration well drilled on the anticline (Knight, 1983). Brown silty mudstones, very fine sandstones, and thin, impersistent limestone beds occur in the upper part of the unit.

The Kennels Brook Formation is at least 3200 m thick; the total depth of the Union-Brinex well (Fig. 5.46), which terminated in the Kennels Brook Formation, was 2305 m. The base of the formation is not exposed. There is a sharp, but conformable contact with black shales of the overlying Snakes Bight Formation, although in places the contact is faulted. The age of the formation can only be broadly bracketed between Late Devonian(?) and Viséan(?). Belt (1968b, 1969) and Knight (1983) suggested a Late Devonian age based, in part, on Cote's (1964) finding of miospores of this age from a single sample thought to have been collected from the Kennels Brook Formation. The formation is poorly exposed, so that no type section was designated by Knight (1983).

Coté (1964), Belt (1968a, 1969), and Knight (1983) are all in agreement that the Kennels Brook Formation is fluvial in origin, based on coarse grained lithologies, scouring, fining-upward sequences, and inferred nodular calcretes. Fine grained strata near the top of the unit were interpreted by Knight (1983) as lacustrine. There are no paleocurrent data available, but Knight (1983) noted that the arkosic composition and presence of abundant felsic plutonic/volcanic clasts suggest derivation of the sediment from pre-Carboniferous rocks of these types from central Newfoundland.

The Snakes Bight Formation (Hayes and Johnson, 1938; Baird and Coté, 1964): (replacing Snakes Bight Shale of Hayes and Johnson, 1938) was proposed by Baird and Coté (1964) for a sequence of black mudstones and shales interbedded with grey sandstones. More rarely occurring lithologies are dolomitic and calcareous mudstones and shales, argillaceous and arenaceous carbonates, and quartz-dolostone intraclastic conglomerates.

In the upper two-thirds of the formation individual units of shale and sandstone are thicker so that the scale of interbedding is larger. Coarsening-upward sequences are present in the upper half of the formation. Many of the thinner sandstone beds and intercalated shales contain partial Bouma sequences; some of the sandstones in the upper part of the formation are crossbedded.

Knight (1983) measured 785 m of Snakes Bight strata at the type section at Hynes Gulch in Snakes Bight (Fig. 5.45), but the formation appears to be around 1000 m thick southeast of the Snakes Bight Fault (Fig. 5.46). The upper part of the formation has been dated by Boyce (1986) as Late Tournaisian based on paleoniscoid fish and vascular plant remains.

Baird and Coté (1964) and Belt (1968a, 1969) suggested a lacustrine origin. Knight (1983) agreed and expanded the interpretation considerably, explaining the deposition in terms of a deep, perennial, bottom-anoxic lake filled with muds, sandy turbidites, and inorganically precipitated calcite. A systematic search for fossils (Boyce, 1986) in the upper part of the formation failed to uncover marine invertebrates, which supports the lacustrine interpretation. Orientation measurements on flute casts by Knight (1983, his fig. 7) showed a northwestward-directed, fan-shaped distribution, with a marked absence of paleoflow directions toward the southeast. This suggests variably oriented turbidity flows moving down a general northwestward-dipping, sub-lacustrine paleoslope.

The Friars Cove Formation (Knight, 1983): is defined as the sequence of grey sandstones, conglomerates, and shales with minor redbeds and carbonates that conformably overlies the Snakes Bight Formation. Previously these strata were called the Seacliffs Formation (Baird and Coté, 1964). The type section was designated as Friars Gulch at Friars Cove, St. George's Bay (Fig. 5.45), where the strata are about 500 m thick. The formation is approximately 1300 m in the northeastern part of the Anguille Mountains. The age of the Friars Cove Formation is unknown, as it has so far proved to be paleontologically barren except for undatable ichnofossils and plant scraps.

The base of the formation is marked by a widespread polymictic conglomerate-sandstone unit, several to tens of metres thick, containing rounded clasts up to 130 cm size. Grey sandstones, which can be either crossbedded, parallel-laminated, rippled, or massive, with lesser amounts of grey to black shale, comprise the bulk of the formation. Red arkosic sandstones are confined to the upper part of the unit, whereas a few arenaceous and stromatolitic dolostones occur sporadically throughout the formation.

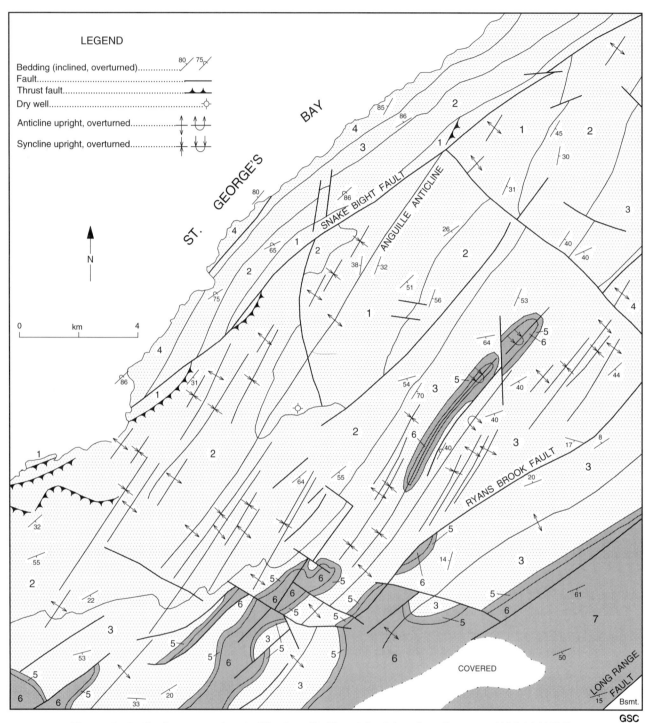

Figure 5.46. Geology map of part of the Anguille Mountains taken from the map of Knight (1983). Stratigraphic units are numbered as in Figure 5.45. Orientation of fold axes between the Snakes Bight and Ryans Brook faults suggests dextral movements.

Knight (1983) interpreted the Friars Cove Formation to be the result of fluvial-deltaic deposition in a nonmarine basin containing the Snakes Bight paleolake. In particular, sandstones and conglomerates represent ephemeral, braided stream deposits and deltaic and fan-deltaic sediments which built out into the paleolake. Shales and carbonates represent offshore and interdistributary bay deposition.

Orientation measurements on ripple marks, parting lineations, and crossbeds reveal a complicated pattern of paleocurrents, although in general terms paleoflow into the basin appears to be from both the north and south. This conclusion is supported by the assemblage of lithoclasts in the conglomeratic rocks. Carbonates, quartzites, and chert were probably derived from the Cambrian-Ordovician platformal sequence that lies north of the Friars Cove Formation. In contrast, clasts of equigranular and porphyritic granitic rocks and felsic volcanic rocks were probably derived from the south and east.

The Spout Falls Formation (Knight, 1983): was defined in reference to predominantly red sandstones near the top of the Anguille succession. In addition, the Spout Falls Formation contains red siltstones, conglomerates and grey sandstones. Knight (1983) included the Fischells Conglomerate Member (Baird, 1951) of the northern part of the Bay St. George Basin in the Spout Falls, and he also placed the lower part of a thick sequence of arkosic, reddish sandstone and pebble conglomerate in the Spout Falls Formation. These arkosic rocks, informally known as the Brow Pond lentil (Knight, 1983), are exposed in the northeastern part of the Bay St. George Basin. The lower part of the Brow Pond lentil is tilted and gently folded, and appears to be overlain unconformably by additional arkosic rocks.

The type section for the Spout Falls Formation was designated by Knight (1983) at Spout Falls, a waterfall situated where Spout Gulch forms a hanging valley along the cliffed shoreline of St. George's Bay (Fig. 5.45). Northwest of the Snakes Bight Fault near the type section, the formation is about 800 m thick, but it is at least 2250 m thick on the other side of the Snakes Bight Fault and in the eastern part of Anguille Mountains. The Fischells Conglomerate Member is normally 100-200 m thick, but it is only several metres thick on the Port au Port Peninsula at the northern extremity of the basin.

In the Anguille Mountains and on Bald Mountain, the Spout Falls Formation overlies the Friars Cove Formation conformably and gradationally. The Fischells Conglomerate Member, however, rests unconformably or is faulted against pre-Carboniferous crystalline basement. Similarly, the basal arkoses of the Brow Pond lentil lie unconformably, or are faulted, against basement anorthosites of Grenville age. Except for the basal arkoses of the Brow Pond lentil, which are succeeded by additional arkosic sandstones, the Spout Falls Formation is conformably and sharply overlain by the Ship Cove Formation, the basal unit of the Lower Carboniferous Codroy Group.

There are no known fossils which would allow an age to be established for the Spout Falls Formation. The unit cannot be younger than Viséan, the age of the overlying Ship Cove Formation.

Lithologically, the Spout Falls Formation is predominantly red to grey sandstones that contain scoured contacts, intraclasts, parallel lamination, and crossbeds.

Conglomerates with clasts up to cobble size occur sporadically, and are more abundant in the northeast, particularly along the Long Range Fault on the eastern side of Bald Mountain (Fig. 5.45). Mudstones and shales are rare in the formation. In contrast, the Fischells Conglomerate Member is mostly grey and red, polymictic, pebble-cobble conglomerates with only a small proportion of flaggy, grey sandstone.

All rocks in the Spout Falls Formation are interpreted to represent basin-marginal alluvial fans and sandy, braided channels in a basin-centre alluvial plain (Knight, 1983). This alluvial fan/alluvial plain complex developed above the earlier lake deposits. Only a few paleocurrent measurements are available; these suggest dominantly southerly paleoflow. Clasts in conglomerates are predominated by a two-fold assemblage: quartzites and limestones, probably derived from the north and west, and granites, granitic gneisses, and felsic volcanic rocks, probably derived from the east.

Codroy Group

The Codroy Group represents the middle stratigraphic unit in the three-fold division of strata in the Bay St. George Basin (Fig. 5.44). Knight (1983) defined the group as all the strata above the Anguille Group and marked by a basal limestone (Ship Cove Formation) to the first continuous deposits of fining-upward, coarse grained, grey, arkosic sandstones and red siltstones (Barachois Group). The Codroy Group consists of 4000-6000 m of nonmarine and marine siliciclastic, carbonate, and evaporitic rocks which conformably overlie the Anguille Group. Locally, the Codroy Group also overlies Precambrian and lower Paleozoic rocks in the northern part of the Bay St. George Basin. In turn, the Codroy Group is overlain (disconformably?) by the Barachois Group, although faults separate these two groups in several places. Macrofossils and microfossils in the Codroy Group both suggest a Middle to Late Viséan age, implying a correlation with the Windsor Group of Maritime Canada (Mamet, 1968; von Bitter and Plint-Geberl, 1982; Knight, 1983; Utting, 1987b). Knight (1983) has provided the most recent subdivision of the group (Fig. 5.44), and by far the best description and interpretation of lithofacies. Dix (1982) and Dix and James (1987, 1989) have provided further stratigraphic and sedimentological details for the Port au Port Peninsula, and Dewey (1983, 1985) provided information on paleoecology of Codroy faunas with emphasis on ostracodes.

The Ship Cove Formation (Knight, 1983): is the basal unit of the Codroy Group, and is marked by the presence of bituminous and carbonaceous, millimetre-scale, laminated (cryptalgal) limestones. In addition to these laminated wackestones, there are thin beds (centimetre scale) of pelletal and peloidal packstones occurring in the basal part of the formation. The upper half of the Ship Cove Formation is dominated by argillaceous laminite with moldic porosity that originated by dissolution of displacive gypsum. More rarely occurring carbonate facies are intraformational breccias, stromatolitic growths, and oolitic-oncolitic-peloidal grainstones and packstones. Siliciclastic lithologies form only a very small part of the formation and consist of fine- to coarse-grained arkosic sandstones, fine grained, argillaceous sandstones, and red, grey, and green mudstones.

Like its counterpart in Maritime Canada (Macumber Formation), the Ship Cove is a very distinctive and easily mapped unit in the Bay St. George Basin. This has helped to outline regional structures. The thickness varies from about 13-25 m, but in places it may be less than 10 m where massive gypsum layers have displacively overgrown original lithofacies. An abundant, but restricted, conodont fauna in the basal few centimetres has been assigned to the *Diplognathodus* zone by von Bitter and Plint-Geberl (1982). This correlates with the macrofaunal A Subzone of the Windsor Group, which has an Early Viséan age (Giles, 1981). Knight (1983) has interpreted the depositional environment of this unit in terms of tidal flat conditions in the southern and northernmost parts of the basin together with shallow, subtidal deposition in the central and northern parts of the basin. After the initial transgression, sea level stabilized to allow for seaward tidal flat progradation and increasing seawater salinity (Dewey, 1983).

The Codroy Road Formation (Knight, 1983): Formation was defined for a lithologically diverse and much faulted section of the Codroy Group exposed along the coast southeast of Codroy (Fig. 5.45). It is difficult to estimate a thickness for the formation because of faulting, but Knight (1983) suggested up to 300 m at the type section.

The Codroy Road Formation consists mainly of fine sandstones to mudstones which can be of various colours (red, grey, black) and evaporitic, dolomitic, or calcareous. There are also evaporites consisting of surface gypsum, subsurface anhydrite, and minor halite, all disposed in two main evaporitic layers (up to 150 m thick) that appear to thin toward the south. Carbonate rocks occur less commonly, and are black to buff, argillaceous, microsparic dolostone and biohermal limestones. These carbonates are richly fossiliferous with brachiopods, pelycypods, gastropods, goniatites, crinoids, bryozoans, and algae as macrofossils and conodonts and forams as microfossils.

With such a variety of facies it can be expected that the sedimentary environments of the Codroy Road Formation consisted of a complex mosaic of related but distinct subenvironments. Knight (1983) suggested a fine grained alluvial plain of redbeds, and hypersaline pools for evaporitic siliciclastics and laminated evaporites. These were probably landward of carbonate lagoons of hypersaline (massive, nearly unfossiliferous dolostones) and normal marine salinities (fossiliferous strata) with associated bioherms (see also Dewey, 1983).

The Codroy Road Formation conformably and gradationally overlies the Ship Cove Formation, and is conformably overlain by the Jeffreys Village Member of the Robinsons River Formation. Microfossils and shelly fauna indicate a Viséan age (von Bitter and Plint-Geberl, 1982; Knight, 1983; Utting, 1987b). The first authors correlated some of the carbonate layers in the upper part of the formation with carbonates in the overlying Jeffrey's Village Member; however, Knight (1983) did not accept this correlation. With respect to the Windsor Group of mainland eastern Canada, the Codroy Road Formation probably correlates with Windsor Subzones A and B of Bell (1948): see Knight (1983) and Utting (1987b). Rocks mapped by Knight (1983) as Codroy Road Formation appear to fall into the following conodont zones of von Bitter and Plint-Geberl (1982): *Taphrognathus, Cavusgnathus,* and *Gnathodus.* Utting

(1987b) assigned palynomorph assemblages to his NS concurrent range zone, which correlates with Windsor Subzones A and/or B.

The Robinsons River Formation (Knight, 1983): is again a lithologically diverse stratigraphic unit which has been subdivided into the following members: Jeffrey's Village, Highlands, Mollichignick, and Overfall Brook. In the eastern part of the Bay St. George Basin, the Robinsons River Formation is undivided. Total thickness for the formation, is about 5000 m. The following account is a summary of the work of Knight (1983).

The Jeffrey's Village Member is restricted to the western part of the basin between Highlands and St. George's (Fig. 5.45). It consists of grey marine and brown nonmarine mudstones with minor evaporites (including potash) in a lower unit. There is an upper unit of dominantly red mudstones, sandstones, and pebbly sandstones with intercalated fossiliferous carbonates and shales and minor evaporites. This upper, mainly red fluviatile unit coarsens upward, and is overlain gradationally near Highlands by the Highlands Member.

The Highlands Member contains red pebble conglomerates, sandstones, and siltstones, which are all organized into recurring, fining-upward sequences. Calcrete nodules and calcrete beds are common. There are also a few beds of fossiliferous green sandstones and grey limestones.

The Mollichignick Member occurs in the southern part of the basin and close to the Long Range Fault (Fig. 5.45), where it represents a diverse lithological assemblage of: 1) red to grey siltstones, sandstones, and pebbly sandstones of fluvial and alluvial-fan origin, 2) grey mudstones, carbonates, and micaceous sandstones of lacustrine origin, and 3) carbonates, shales, and sandstones of shallow marine origin. These rocks are overlain by fluviatile (humid, alluvial fan setting), red, pink, and brown, thick bedded, coarse to very coarse sandstone and pebble conglomerate of the Overfall Brook Member.

The Overfall Brook Member also contains thin layers of grey micaceous and carbonaceous claystones and siltstones of probable lacustrine origin near the base of the member.

Conodonts, forams, macrofaunal elements, and spores all indicate a Viséan age (Knight, 1983; Utting, 1987b). More particularly, the fossil evidence suggests an age ranging from Windsor Subzone B to at least Subzone D. Von Bitter and Plint-Geberl (1982) have assigned rocks of the Jeffrey's Village Member to the same conodont zones as in the Codroy Road Formation. Utting (1987b) assigned spores to his AT zone which is equivalent to Subzones C and D of the Windsor Group.

The Woody Cape Formation (Knight, 1983): is structurally complicated succession of green to grey mudstones, shales, sandstones, and minor-carbonates, exposed along the coast between Codroy and Searston. An estimated thickness of 700 m is thought to be present in the type section.

Mudstones, shales, and associated thin layers of very fine grained sandstones constitute about 60% of the formation and are abundantly fossiliferous (brachiopods, pelycypods, gastropods, orthocone cephalopods, ostracodes, brachiopods). Coarsening-upward sequences especially near the base of the unit were reported by Knight (1983). Thick bedded sandstones have scoured bases and occur near the tops of coarsening-upward sequences. Carbonate rocks consist of fossiliferous limestone concretions, unfossiliferous black, nodular dolostones and limestones, stromatolitic carbonates, lime mudstones, and grainstones.

Knight (1983) interpreted the Woody Cape Formation as the deposits of a delta that established itself after a rise in sea level flooded the basin. Gradual progradation led to the establishment of a subaerial deltaic plain containing distributaries, lakes, marshes, and coastal brackish water ponds.

Fossil evidence from both macrofossil and microfossil faunas and spores indicates a correlation with Windsor Sub-zones D and E (late Viséan) (Knight, 1983). Von Bitter and Plint-Geberl (1982) assigned a conodont assemblage to the *Gnathodus* zone. Utting (1987b) assigned spores to his AT zone.

The Brow Pond lentil (Knight, 1983): is a thick succession (1000-2000 m according to Peavy, 1985) of brown, arkosic, pebbly sandstones the lower portion of which he included in the Spout Falls Formation (Anguille Group). In the northeastern part of Bay St. George Basin Knight (1983) also recognized an arkosic sequence above an erosional surface within the Brow Pond lentil. He included the upper pebbly arkoses with the Robinsons River Formation, in contrast to Fong (1976) who mapped all of the arkosic succession as part of the overlying Barachois Group. The stratigraphic position of the Brow Pond lentil is still unclear.

Codroy Group on Port au Port Peninsula

Carbonate rocks, redbeds, and plant-bearing, green to grey sandstones on the Port au Port Peninsula (Fig. 5.45) have been assigned to the Codroy Group (Bell, 1948; von Bitter and Plint-Geberl, 1982; Knight, 1983; Dix and James, 1987, 1989). Based on conodont assemblages in the carbonate rocks (von Bitter and Plint-Geberl, 1982) and miospore assemblages (Dix, 1982), these rocks are correlated with the Ship Cove Formation.

Dix and James (1989) named the carbonate rocks and green, plant-bearing sandstones, the Big Cove Formation. Redbeds, which they interpreted to be coeval with the carbonates, were named the Lower Cove Formation. The Ship Cove and Codroy Road formations also occur in local outcrops.

The carbonate rocks were categorized into two broad lithofacies by Dix and James (1987): 1) bryozoan and microbial build-ups that were constructed within drowned, linear, paleokarstic valleys adjacent to a rocky coastline, and 2) well-bedded carbonate and evaporitic rocks deposited in broad, shallow basins with restricted water circulation.

Barachois Group

The Barachois Group (Baird and Coté, 1964) is exposed in the southern and eastern parts of the Bay St. George Basin (Fig. 5.45), and is subdivided into the Searston Formation (Knight, 1983) and an unnamed succession that has only been informally subdivided into a lower unit without coal seams and an upper unit with coal seams (Knight, 1983; Solomon, 1986). All or part of the unnamed sequence is younger (Westphalian A; Hayes and Johnson, 1938; Bell, 1948; Hacquebard et al., 1960; Utting, 1966; Knight, 1983; Solomon, 1986; Hyde et al., 1991) than the Searston

Formation (Namurian A; Hayes and Johnson, 1938; Bell, 1948; Hacquebard et al., 1960; Utting, 1966, 1987b; Knight, 1983). A coal-bearing sequence near Stephenville was dated as Westphalian C (Riley, 1962; Hyde et al., 1991); these strata were assigned to the Barachois Group by Knight (1983). Other outcrops of inferred Barachois strata occur at Trainvain Brook, Campbell's Brook, and Coal Brook (Fig. 5.45). The last two exposures contain coal seams. Undated conglomerates at Trainvain Brook nonconformably overlie pre-Carboniferous crystalline rocks. These rocks are provisionally correlated with the Barachois Group on lithological criteria, and have been included in the Searston Formation by Knight (1983).

The Searston Formation (Knight, 1983): is well exposed along the coast between Searston and Larkin Point. The Namurian A age has been established on the basis of megafloral remains and spores (Knight, 1983; Utting, 1987b); no marine fossils have been discovered from the Searston Formation. Knight (1983) estimated a thickness of about 2500 m. The unit is in fault contact with older Carboniferous strata. Lithologies are mainly grey, cross-bedded feldspathic sandstones and pebbly sandstones, and red to brown and grey to green mudstones. These are arranged in repetitive fining-upward sequences, which also contain basal erosion and lateral accretion surfaces, abundant transported and broken woody plant material, and calcrete nodules. Based on these lithologies, structures and sequences, Knight (1983) inferred meandering stream deposition. Coarser conglomerates and muddy sandstones were interpreted as proximal alluvial-fan deposits. Knight (1983) also inferred drainage toward the west and northwest, and from a terrane consisting of igneous and metamorphic rocks.

Barachois Group undivided

These strata are areally distinct from the Searston Formation, and outcrop in the St. George's Coalfield, forming the gentle, doubly plunging Barachois syncline (Fig. 5.45). There is a fault boundary to the west and north, but the contact to the south and east appears conformable on the Robinsons River Formation (Knight, 1983). A thickness for this part of Barachois Group has not been firmly established; Hayes and Johnson (1938) suggested approximately 1500-1600 m and Peavy (1985) estimated 1200 m from gravity modelling.

Solomon (1986) investigated the sedimentology of the unnamed unit of the Barachois Group. Sequences consist of grey, feldspathic sandstones and pebble conglomerates that alternate with interbedded reddish brown to grey mudstone and sandstone. There are also carbonaceous shales, thin bituminous coals, and cannel shales and limestones containing nonmarine ostracodes and gastropods (Solomon, 1986). These sequences and lithofacies define two informal units (Knight, 1983; Solomon, 1986): a lower "coarse" unit consisting of about 50% conglomerate and sandstone, and an overlying "fine" unit with about 20% conglomerate and sandstone. Mudstone is more abundant in the "fine" unit than the "coarse" unit, and the former also contains thin coal seams. Overall, the succession appears to be a fining-upward megasequence at least 1000 m thick.

Table 5.14. Listing of major research (in chronological order) in establishing the geology of the Deer Lake Basin.

Reference	Contribution
Murray (1881a,b,c,d)	Reconnaissance mapping; established presence of Coal Measures.
Howley (1913, 1917a)	General stratigraphy; summary of coal exploration.
Hatch (1919)	Oil shales; first to recognize lacustrine strata.
Landell-Mills (1922)	Mapping, further discussion of oil shale and fossil fish-bearing lacustrine unit.
Heyl (1937)	Mapping; account of the stratigraphy and structure in the northern part of the basin.
Grossman (1946)	Mapping; outcrop descriptions near Deer Lake.
Betz (1948)	Mapping; further discussion of stratigraphy and structure in the White Bay area.
Hayes (1949)	Coal assessment; coal prospects are poor.
Werner (1956)	Mapping; establishes Deer Lake Group and older Carboniferous strata in the basin.
Riley (1957)	Reconnaissance mapping in the southern part of the basin.
Baird (1960)	Reconnaissance mapping in the northern part of the basin.
Neale and Nash (1963)	Reconnaissance mapping in the northeastern part of the basin.
Belt (1968a)	Description of fluvial, alluvial fan, and lacustrine facies.
Belt (1968b)	Tectonics; tectonic setting of Deer Lake Basin; recognition of graben structure in relation to strike-slip faults.
Belt (1969)	Stratigraphy and structure of Deer Lake Basin.
Popper (1970)	Mapping; best available description of Anguille Group in the basin; good discussion of tectonic setting.
Hyde (1983)	Mapping; first published map covering the entire basin.
Hyde (1984a)	Stratigraphy, facies; description of stratigraphic units; tectonics.
Miller and Wright (1984)	Interpretation of gravity, aeromagnetic, and seismic refraction data; estimated depths to basement.
Irving and Strong (1984)	Paleomagnetism; demonstration that lateral displacements along the Cabot Fault Zone are not large.
Hyde (1985b)	Uranium mineralization; classification of uranium showings in the Deer Lake Group.
Gall and Hiscott (1986)	Diagenesis; diagenesis and its relationship to uranium mineralization in sandstones of the Deer Lake Group.
Hyde et al. (1988)	Tectonics, thermal maturation; integration of tectonic evolution and thermal maturation.
Hyde (1989)	Discussion of the North Brook Formation (Deer Lake Group) and why it appears to contain evidence for deposition in a strike-slip tectonic regime at one locality and extensional tectonics elsewhere.
Kalkreuth and Macauley (1989)	Petrographic and geochemical study of the lacustrine oil shales of the Rocky Brook Formation (Deer Lake Group).

Knight (1983) and Solomon (1986) inferred fluviatile conditions marked by channels that cross a highly vegetated floodplain/swamp complex (Hyde et al., 1991). The floodplain evidently contained small and scattered ponds to account for the cannel shales and carbonates.

According to Solomon (1986), the lower part of the succession appears to show paleoflow towards a general southerly direction, but drainage in the upper, finer unit was more variable and directed toward the north. Source rocks include plutonics spanning the compositional spectrum, felsic volcanics, and low grade metamorphic rocks.

Bay St. George Basin as a strike-slip basin

The Bay St. George Basin is interpreted as having formed in response to dextral strike-slip faulting (Knight, 1983) along the Long Range Fault (Fig. 5.43), but there is no direct evidence for strike-slip movements from rock bodies that have been offset. Instead, the evidence is indirect, and includes aspects of structure, stratigraphy, and petrology.

There is obliquity between fold orientations and some of the northeasterly trending fault traces in a manner implying dextral fault movements (Wilcox et al., 1973; Odonne and Vialon, 1983). For example, folds between the Snakes Bight and Ryans Brook faults are en echelon and oblique to the fault traces by about 30°. These particular folds involve both the Anguille and Codroy groups, but other folds in the Barachois Group are also oblique to northeasterly trending faults. This suggested to Knight (1983) that wrenching persisted until at least Late Carboniferous time.

Thrust faults are present in the Anguille Mountains, and are especially prominent in the shaly Snakes Bight Formation. These faults, which have westerly directed movements, are consistent with larger-scale strike-slip faulting, as in the model illustrated by Lowell (1972).

Another line of evidence for dextral wrenching along the Long Range Fault is the observation by Knight (1983) that the maximum thickness of progressively younger stratigraphic units is shifted northeastwards (i.e., along the length of the basin) through time. This means the depocentre moved northeasterly in response to basin elongation in that direction.

Deer Lake Basin

Some of the more important works on the Deer Lake Basin are listed in Table 5.14 with a statement on the nature of the study and any notable advances.

The Deer Lake Basin is a composite basin in that it consists of depositional and structural elements of varying styles that developed at discrete time periods spanning the interval between Late Devonian(?) and Late Carboniferous. Evidence is presented later that the geological history of major faults trending along the length of the basin (Fig. 5.47) included strike-slip movements during the Carboniferous. These major faults comprise the Cabot Fault Zone (Hyde et al., 1988), along which the Deer Lake and Bay St. George basins developed (Fig. 5.43). Localized zones of crustal extension along segments of the Cabot Fault Zone created graben that served as "seeds" for later expansion of the basins. All of the Deer Lake Basin fill is

regarded as nonmarine. Stratigraphic units currently recognized in the basin are shown in Fig. 5.44 along with correlations with Carboniferous units in the Bay St. George Basin and in mainland Canada.

Anguille Group

Rocks assigned to the Anguille Group (Belt, 1969) in the Deer Lake Basin range in age from Devonian(?) to Tournaisian (Fig. 5.44). The presence of spores (Hamilton et al., in press), paleoniscoid fish remains, and a *Lepidodendropsis* flora suggest a Tournaisian age. Attempts at recovering spores have generally not met with success, probably because of the relatively high levels of thermal maturation (Hyde et al., 1988). The Anguille Group outcrops in upland areas that form elongated, but disjunct and end-on structural blocks (Birchy Ridge and Fisher Hills blocks of Fig. 5.47). These structural blocks form a central "spine" or backbone, flanked by topographically lower areas underlain by younger Carboniferous strata. An aggregate thickness of the Anguille Group is probably in the order of 3000 m. The Anguille Group is only observed to be in fault contact with pre-Carboniferous rocks in the Deer Lake Basin.

The Gold Cove Formation (Hyde, 1983, 1985a): forms the apparent base of the Anguille Group in the Birchy Ridge block. The type section is along the beach at Gold Cove, White Bay (Fig. 5.47), where a westward-facing succession of grey and red, fluvial, calcrete-bearing, arkoses and arkosic conglomerates is found. Clasts in the conglomerates are marked by the occurrence of schists and phyllites that were probably derived from the Fleur-de-Lys Supergroup exposed to the east. The Gold Cove Formation reaches a maximum thickness of about 500 m, but is faulted against pre-Carboniferous rocks and overlying Anguille units. It is exposed intermittently along the western shoreline of White Bay (Fig. 5.47; Hyde, 1983). The Gold Cove Formation is part of Hyde's (1984a) informal unit 1; other grey sandstones showing well-developed graded beds and a fracture cleavage included in unit 1 (Hyde, 1984a) are now regarded as pre-Carboniferous.

The Saltwater Cove Formation (Hyde, 1983, 1985a): is the most extensive formation within the Anguille Group, and is exposed in both the Birchy Ridge and Fisher Hills blocks. It corresponds to what Popper (1970) called the grey sandstone suite in the Fisher Hills block (Glide Mountains in Popper, 1970) and to Hyde's (1984a) unit 3. The type section is designated as the southeastern shoreline of Saltwater Cove, White Bay (Fig. 5.47, 5.48), where a westward-facing, steeply dipping succession of dark grey mudstones and grey micaceous sandstones is well exposed (Fig. 5.49). These alternating lithologies typify the entire formation. Mudstones and fine grained sandstones contain abundant plant debris in the form of stems and twigs, whereas coarser sandstones tend to contain drifted tree branches and logs, especially of the genus *Lepidodendropsis*. Together with a well-defined portion of a paleoniscoid fish (collected from float in the Fisher Hills block), these fossils suggest a Tournaisian age. Pebble to cobble conglomerates are also present and contain prominent volcanic and feldspar clasts, although a thin section study of some of the clasts shows a wide diversity of rock fragments. Impure limestones are also present, particularly in the Grand Lake

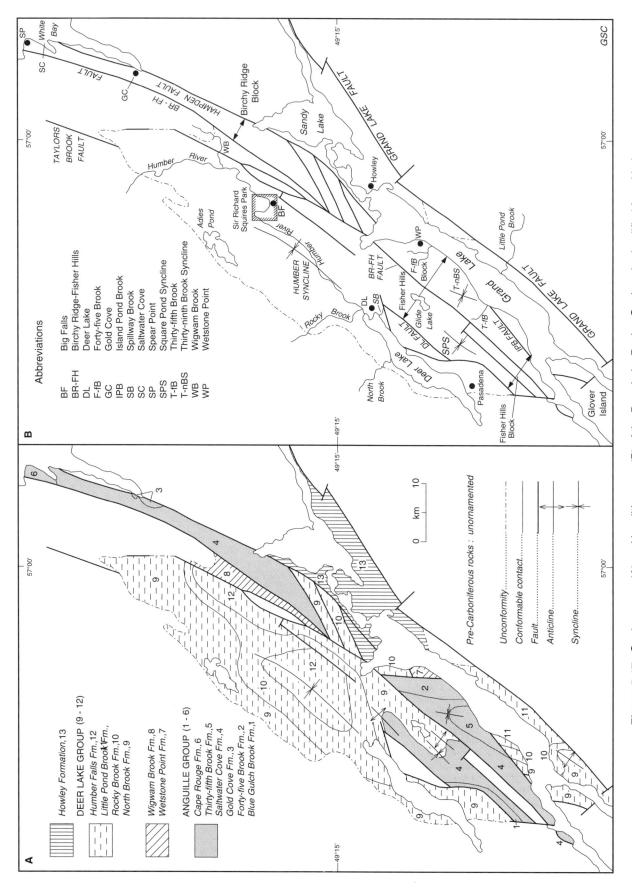

Figure 5.47. Geology map (A) and localities map (B) of the Deer Lake Basin. Geology is simplified from Hyde (1983). Unpatterned areas in A represent pre-Carboniferous basement rocks.

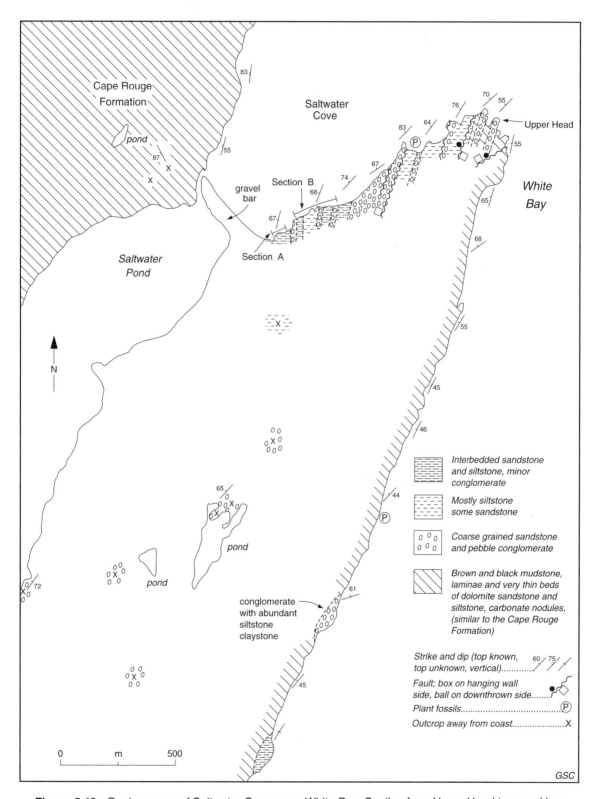

Figure 5.48. Geology map of Saltwater Cove area, White Bay. Section from Upper Head to gravel bar is the stratotype for the Saltwater Cove Formation.

Figure 5.49. Steeply dipping Saltwater Cove strata at the type section consisting of shaly units (recessive) and protruding sandstone-dominated layers. Tops toward the right. GSC 1993-216

region. These carbonates are thin (less than 1 m thick) and vary from fine grained micrites/microsparites to pebble breccias.

At the southern end of White Bay, the Hampton Bay Member was shown on Hyde's (1983) map and described in Hyde (1984a) as part of unit 3 (Facies B). This member consists of graded sandstones, which together with alternating shales, define partial Bouma sequences. Synsedimentary deformation is characteristic.

The Saltwater Cove Formation is in fault contact with the Gold Cove Formation, but appears to conformably overlie the Blue Gulch Brook Formation in the Fisher Hills block. It appears to intertongue with the Forty-five Brook and Thirty-fifth Brook formations in Fisher Hills and the Cape Rouge Formation in Birchy Ridge. A thickness of up to 2700 m occurs on the western limb of the Square Pond Syncline (Fig. 5.47).

A lacustrine paleoenvironment is inferred (Belt, 1968a, 1969; Popper, 1970; Hyde, 1984a) based on the abundant terrestrial plant debris, the paleoniscoid fish, and an apparent absence of marine invertebrates. This last evidence is compelling because even carbonate rocks in the Saltwater Cove Formation do not contain visible remains of marine fossils. Much of the lacustrine facies are organized into deltaic units, which prograded outward. Coarsening-upward deltaic sequences are prominently displayed on the western shoreline of Grand Lake. Fining-upward sequences in the Saltwater Cove Formation are interpreted as abandoned distributary channels on the subaerial part of deltas. Sediments of the Hampden Bay Member were probably deposited in deeper water in the lacustrine basin by turbidity currents.

The Cape Rouge Formation (Baird, 1966): has its type locality on the Cape Rouge Peninsula, White Bay, near Conche (Fig. 5.43). It was extended southwesterly to the northern part of Birchy Ridge by Hyde (1984a) in reference to grey to brown, dolomitic mudstones and sandstones. At Spear Point (Fig. 5.47) and westward to the faulted contact with Silurian strata, the section is overturned but faces westward, as shown by numerous graded sandstone layers. These sandstones are organized into partial Bouma sequences together with overlying siltstones. Dolomitic nodules and visible pyrite are common in the Cape Rouge Formation. A thickness of at least 500 m is estimated for this unit. A siltstone-dominated succession, which appears to overlie the Saltwater Cove Formation near Upper Head, White Bay (Fig. 5.48), is included in the Cape Rouge Formation. No fossils are known in Birchy Ridge block. At the type locality, fossil spores and plants indicate a Tournaisian age (Baird, 1966; Hamblin et al., in press).

From the numerous partial Bouma sequences, a turbidite origin is inferred for this formation. It is also probable that turbidity currents were generated in a lacustrine setting, given the apparent absence of marine fossils. Collectively, the Gold Cove, Saltwater Cove, and Cape Rouge formations form a generalized fining-upward megasequence, which developed from an initial alluvial plain to a deepening lake system.

The Blue Gulch Brook Formation (Hyde, 1983, 1985a): was first noted, informally, by Popper (1970). Hyde (1984a) referred to these rocks informally as unit 2. The type section is at Blue Gulch Brook (Fig. 5.47) on the western side of the Birchy Ridge block. The formation is approximately 400 m thick, and consists of a distinctive marble- and quartz-pebble to cobble conglomerate. These conglomerates are interbedded with feldspathic grey to brown sericitic sandstones and siltstones. Grey impure carbonates are rare.

The Blue Gulch Brook Formation is undated, and is in fault contact with schists and phyllites of the Fleur-de-Lys Supergroup. It appears to be conformably overlain by the Saltwater Cove Formation, so that it might be a time equivalent of the Gold Cove Formation. An absence of schistose Fleur-de-Lys clasts suggests it was deposited on top of, rather than being derived from, the Fleur-de-Lys Supergroup. The prominent marble and quartz clast assemblage was probably derived from Cambrian-Ordovician carbonates to the west. An alluvial fan paleo-environment was suggested by Popper (1970) and Hyde (1984a).

Forty-five Brook Formation (Hyde, 1983, 1985a): In his study of Anguille Group in the Grand Lake-Fisher Hills area, Popper (1970) designated a group of carbonate-bearing rocks as "the black siltstone-orange dolomite suite". These rocks were mapped by Hyde (1983) and termed the Forty-five Brook Formation. The type section is here designated as the stretch of Forty-five Brook from its mouth to a point along the west branch about 1.3 km upstream of the fork (Fig. 5.47). The carbonate units referred to above are laminated or unlaminated, thin beds of light brown and orange dolostone that are interbedded with micaceous sandstones and shales very similar to those in the Saltwater Cove Formation. Some of the sandstones contain graded beds, sole marks, and bioturbation (Planolites?) structures. A lacustrine origin has been inferred (Popper, 1970; Hyde, 1984a).

The Forty-five Brook Formation probably intertongues with the overlying Saltwater Cove Formation. As such, the Forty-five Brook Formation is probably no younger than Tournaisian, and it may be as old as Devonian. Its lower contact with older stratigraphic units is not exposed, and it may be laterally equivalent to the Blue Gulch Brook and Gold Cove formations. The Forty-five Brook Formation is at least 500 m thick, and may be up to 900 m thick.

The Thirty-fifth Brook Formation (Hyde, 1983, 1985a): was termed the red arkose suite by Popper (1970). It consists mainly of a very arkosic grey to pink coarse grained sandstone and pebble to cobble, arkosic conglomerate. The unit was named by Hyde (1983) after a well-exposed south-easterly dipping section along Thirty-fifth Brook (Fig. 5.47). These rocks are interbedded with sandstones and shales similar to those in the Saltwater Cove Formation, with which Thirty-fifth Brook Formation is thought to intertongue and overlie. The Thirty-fifth Brook Formation forms the core of the southerly plunging Thirty-ninth Brook Syncline near Grand Lake (Fig. 5.47), so that some of the original thickness has been removed by erosion. Nonetheless, the formation is at least 500 m and may be up to 1000 m thick as there is no or little repetition of steeply dipping strata on the western limb of the syncline.

The underlying Saltwater Cove Formation is probably mostly Tournaisian in age, and rocks now known as the Wetstone Point Formation, which is thought to overlie the Thirty-fifth Brook Formation, have been dated as Tournaisian (Barss, 1981a). This suggests that the Thirty-fifth Brook Formation is also Tournaisian.

From its conglomeratic nature and burial of upright tree trunks, Popper (1970) inferred an alluvial fan origin. This seems likely, but from the apparent intertonguing with the Saltwater Cove Formation and the close strati-graphic association of conglomerate and lacustrine shale, it also seems likely that the alluvial fans were prograding out into a lake in the manner of fan deltas.

Wetstone Point Formation

This formation is named for exposures of conglomerate and sandstone at Wetstone Point on the western shoreline of Grand lake (Hyde, 1983, 1985a; Fig. 5.47). The rocks here, and in the nearby vicinity, have been mapped as the North Brook Formation (e.g., Baird, 1960; Popper, 1970) and as the Howley Formation (Hyde and Ware, 1981). These rocks, however, are now known to contain a well-preserved spore assemblage of Late Tournaisian age. Barss (1981a) reported that the spore assemblage is similar to the Vallatisporites vallatus — Pustulatisporites pretiosus Zone of the Horton Group (zone g in Hacquebard, 1972).

Grey, pebble to cobble conglomerates and grey, green, and red sandstones are the most abundant rock types, but there are also mudrocks and more rarely occurring lime-stone beds. A few measurements on crossbed orientations (Hyde and Ware, 1981) indicate paleoflow directions towards the northeast. A fluvial interpretation was suggested by Hyde and Ware (1981).

The Wetstone Point Formation is in fault contact with other Carboniferous units (Forty-five Brook Formation, Rocky Brook Formation, see Hyde, 1983). Only a small part of the original vertical and lateral extent of the formation is preserved, so that probably no more than a few hundred metres is now exposed. The geographic position of the Wetstone Point Formation is such that it flanks the Anguille Group to the east within the Fisher Hills struc-tural block (Fig. 5.47).

Wigwam Brook Formation

The Wigwam Brook Formation (Hyde, 1983, 1985a) is in fault contact with older rocks of the Anguille Group to the east and younger rocks of the Deer Lake Group to the west. It nonconformably overlies felsic to mafic plutonic rocks of the Silurian-Devonian Gull Lake intrusive suite (Smyth and Schillereff, 1982; Hyde, 1983). It is assigned a Tournaisian age on the basis of a single sample, which contained a typical Tournaisian spore assemblage (Barss, 1980a). These rocks consist of grey, brown, and red con-glomerates, sandstones, and subordinate siltstones, with a few relatively thin (up to 60 cm thick) carbonate units. The abundant conglomerate and coarse sandstone lithologies are similar to rocks in the North Brook Formation of the Deer Lake Group and Anguille Group, so that the Wigwam Brook Formation has been variously mapped as these latter units (e.g., Baird, 1960; Hyde and Ware, 1980). However, finer grained, plant-bearing tan sandstones along the east-trending reach of Wigwam Brook at the type section are a distinctive facies from those found in other Carboniferous units in the Deer Lake Basin.

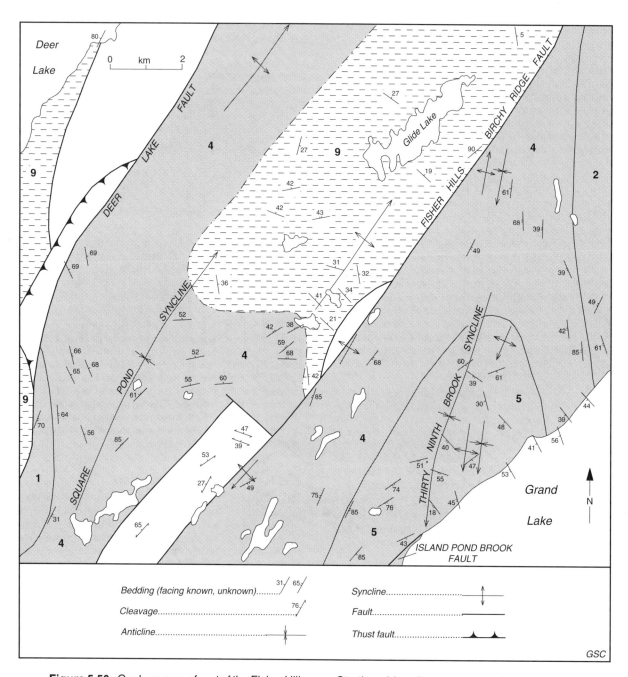

Figure 5.50. Geology map of part of the Fisher Hills area. Stratigraphic units are numbered and patterned as in Figure 5.47. An angular unconformity is inferred between the Saltwater Cove Formation (4) and North Brook Formation (9) because of the substantial difference in strike directions between the Fisher Hills-Birchy Ridge Fault and Square Pond Syncline.

Sandstone clasts in a conglomerate of the Wigwam Brook Formation are petrographically similar to Anguille Group rocks in the southern end of the Birchy Ridge block. In particular, both the sandstone clasts and Anguille Group sandstones contain fairly abundant grains of garnet, epidote, and epidote-quartz intergrowths. This is taken as evidence that the sandstone clasts were derived from an uplifted Birchy Ridge structural block, which exposed part of Saltwater Cove Formation.

The original thickness of the Wigwam Brook Formation cannot be determined because of faulting, but it must be at least 300 m. It is viewed as being predominantly fluvial in origin; the carbonate layers probably originated as deposits of small lakes and ponds on an alluvial plain.

Deer Lake Group

The North Brook Formation (Werner, 1956; Belt, 1969): is the basal unit of the Deer Lake Group and the name was originally assigned by Werner (1956) for gently dipping conglomerates and sandstones along North Brook (Fig. 5.47). It unconformably overlies pre-Carboniferous basement rocks on the western and northern margins of the Deer Lake Basin (Hyde, 1983, 1989). In the Glide Lake region of Fisher Hills (Fig. 5.47), the North Brook Formation is interpreted to unconformably overlie the Saltwater Cove Formation (Fig. 5.50), based on nearly orthogonal strike directions seen in scattered outcrops (actual unconformity is unexposed). The Rocky Brook Formation conformably overlies the North Brook Formation. This contact was redefined by Hyde and Ware (1981), placing the base of the Rocky Brook Formation at the first appearance of grey or green mudstone above a tens-of-metres-thick succession of red and brown beds.

The North Brook Formation has not been dated paleontologically. The thickness of the unit varies from its feather edge in the westernmost outcrop area (interpreted as the original basin margin) to close to 2000 m on Glover Island and in the Glide Lake areas (Fig. 5.47).

Rock types in North Brook Formation are mainly red, brown and grey pebble to boulder conglomerates and sandstones. Redbeds are particularly rich in calcite (cements, nodules, veins, replaced fossil wood) and are mainly exposed in the Grand Lake and western basin margin areas. Thin beds of pink to grey micrite limestones, limestone breccias, and stromatolitic carbonates are also present, especially in the western part of the Deer Lake Basin. A single outcrop of amygdaloidal basalt with pillow-like forms is exposed on Lanes Brook east of Deer Lake ; this lava flow is underlain and overlain by sandstone beds.

All of the North Brook rocks are interpreted to be the product of alluvial fan and alluvial plain systems (Belt, 1968a, 1969; Hyde, 1984a). Carbonate layers are thought to represent deposition as inorganic precipitates in small ponds and lakes and, in some cases, as pedogenic calcretes. Carbonate breccias are also regarded as debris flow deposits, but certain breccias and finer limestones west of Cormack (mapped as North Brook Formation, Hyde, 1983) have attributes of paleokarstic origins. Some of the fine grained carbonates were deposited geopetally in metre-sized caverns in Cambrian-Ordovician marbles. The amygdaloidal basalt is also interpreted as being deposited in a nonmarine setting.

It is clear from paleocurrent measurements and clast lithologies (Belt, 1969; Hyde, 1984a) that the Deer Lake Basin was filled from nearly all sides. Of special significance is the observation that conglomerates in its southern part near Pasadena (Fig. 5.47) contain abundant schist, phyllite, and quartzite clasts of the adjacent Fleur-de-Lys equivalents to the southwest. This implies that the region to the southwest was a positive area during North Brook deposition, with the further implication that the Deer Lake and Bay St. George basins were separate at this time.

The Rocky Brook Formation (Werner, 1956; Belt, 1969; Hyde and Ware, 1981; Hyde, 1983, 1985a): conformably overlies the North Brook Formation. However, it is also interpreted to be partly equivalent to the North Brook Formation, because red to brown, calcareous siltstones and very fine sandstones that are included in the Rocky Brook Formation are very similar to North Brook lithologies, and could actually represent tongues of North Brook Formation. In the Grand Lake area, the Rocky Brook Formation is overlain gradationally by the Little Pond Brook Formation, but it is sharply overlain by the Humber Falls Formation in the Humber River area.

In addition to the (1) red-brown beds in the lower part of the formation, there are (2) grey to green, dolomitic claystones and (3) aphanitic calcareous dolostones. These three lithologies are organized into symmetrical cycles as: $(1) \rightarrow (2) \rightarrow (3) \rightarrow (2) \rightarrow (1)$. The cycles, together with thicker accumulations of interbedded (2) and (3), help to define the Spillway Member (Hyde, 1983, 1985a), with its type section along the upper part of Spillway Brook at Deer Lake (Fig. 5.47).

The upper part of the Rocky Brook Formation lacks redbeds and has been called the Squires Park Member (Hyde, 1983, 1985a); the type section is designated as the bluffs along the Humber River in Sir Richard Squires Memorial Park (Fig. 5.47). In relation to the Spillway Member, the Squires Park Member contains (2) and (3), but: (a) more stromatolitic dolostones, (b) more allochemical dolostones, (c) more oil shales, (d) more disarticulated paleoniscoid fish debris (scales, bones, fins, headplates), (e) more carbonate and sulphide nodules, and (f) no evidence for desiccation in the form of mudcracks, fenestrae, and tepee structures.

The Rocky Brook Formation is a widespread and distinctive formation in the Deer Lake Basin. It varies in thickness, but appears to have a maximum thickness in the Grand Lake area of about 1000 m. Spore assemblages (Barss, 1980a, b, 1981a, b) indicate a Viséan age. This formation was studied early in the century (Hatch, 1919; Landell-Mills, 1922) as a potential oil shale resource (Hyde, 1984c; Kalkreuth and Macauley, 1989). The name Rocky Brook Formation first appeared on a map by Hayes (1949), but it was not discussed in a wider stratigraphic context until several years later (Werner, 1956).

The lithologies, the presence of paleoniscoid fish and terrestrial plant fossils, and an absence of a marine fauna led Belt (1968a) to propose a dominantly lacustrine, but also alluvial, origin. This was Hyde's (1984a) interpretation as well. An additional line of evidence for a lacustrine origin is the presence of analcime, a typical lake mineral, in almost all Rocky Brook lithologies (Gall and Hyde, 1989).

The Humber Falls Formation (Belt, 1969; Hyde, 1983, 1985a): in the western part of Deer Lake Basin overlies the Rocky Brook Formation with abrupt contact. This is possibly a disconformity because the contact appears to gently undulate at Sir Richard Squires Park, where it is discontinuously exposed for a few hundred metres. The Humber Falls Formation is mostly Viséan, judging from spore assemblages reported by Barss (1980a), so there is no significant time gap between it and the Rocky Brook Formation.

The Humber Falls Formation consists of red, grey, and pink, arkosic sandstone and siltstone interbedded with grey, pebble to cobble conglomerates. The latter rock type is especially common near the base of the formation at Sir Richard Squires Park, where volcanic clasts are abundant. Drilling, as part of an uranium-exploration program by Westfield Minerals from 1978-1981 revealed that there is considerably more siltstone and claystone than is evident from surface exposures. In fact, siltstone and claystone comprise about 50% of the formation. Drilling also showed that the lithologies are organized into fining-upward sequences. This, together with abundant trough and planar cross-stratification, suggests a fluvial origin for the bulk of the formation. Conglomerates and lesser amounts of sandstone near the base of the formation probably developed on alluvial fans. Gall and Hiscott (1986) discussed the diagenetic sequence in the Humber Falls Formation and related this to the development of uraninite cements (see also Hyde, 1985b).

Crossbeds and clast lithologies point toward a southwestward-dipping paleoslope (Hyde, 1984a). A basinal tilt towards the southwest at the onset of Humber Falls deposition perhaps led to draining down the regional paleoslope, followed by the development of an alluvial plain.

The Humber Falls Formation is disposed in the regional, doubly-plunging Humber Syncline (Fig. 5.47) in the western part of the Deer Lake Basin, where it forms the core of this major fold. It also reappears to the northeast as the core of a synclinal structure that has its eastern limb truncated by the Wigwam Fault (Hyde, 1983). The upper part of the Humber Falls has been lost to erosion, so that at present the maximum thickness is about 250 m along the axis of the Humber Syncline.

The name Humber Falls stems from Hayes' (1949) map, and was named for what was then informally known as Humber Falls on the Humber River. Werner (1956) subsequently named the unit the Big Falls Formation after the new(?) and formal name Big Falls, the same location as the informal Humber Falls. Yet the name Humber Falls has persisted (Baird, 1960; Belt, 1969; Hyde, 1983, 1984a, 1985a).

The Little Pond Brook Formation (Hyde, 1983, 1985a): is a succession of sandstones, mudstones, and conglomerates gradationally above the Rocky Brook Formation (Belt, 1969). These rocks were mapped by Riley (1957), Belt (1969), and Hyde and Ware (1981), all of whom correlated them with the Howley Formation (Westphalian A) in the northeastern part of the Deer Lake Basin (Fig. 5.44). However, Barss (1981a) recovered a Viséan-Namurian? spore assemblage, which led Hyde (1983) to map these rocks as the Little Pond Brook Formation. It now appears that these rocks correlate with the Humber Falls Formation in the Humber Valley area.

The type section is designated as the exposures along Little Pond Brook on the eastern side of Grand Lake (Fig. 5.47), from the faulted eastern boundary with pre-Carboniferous volcanic rocks to the mouth of the brook. Good exposures also occur along the western and eastern shores of Grand Lake. At one locality along the eastern shore, the Little Pond Brook Formation seems to overlie pre-Carboniferous volcanic rocks (Hyde, 1983), apparently overstepping the Rocky Brook Formation. The Little Pond Brook Formation is at least 750 m thick.

A wide variety of lithofacies characterizes the formation. Near the eastern boundary fault at the type section, there is a boulder conglomerate facies containing abundant felsic, porphyritic clasts. Elsewhere, conglomerates in the formation are finer and probably make up less than 10% of the unit. Grey to red sandstones are interbedded with red to brown mudstones in either poorly or well-defined fining-upward sequences. Some of these rocks have very poorly sorted grains, and are markedly ferruginous, suggesting that they are ancient soil zones. Other sandstones are grey and plant-bearing and are interbedded with carbonaceous mudstones. A distinctive sandstone facies along the type section is very micaceous, green, parallel-laminated, and

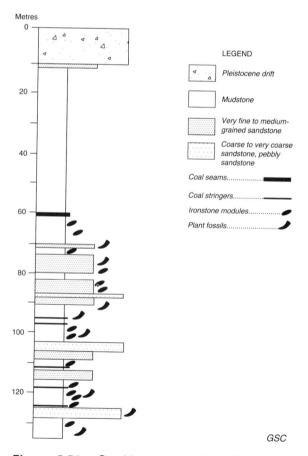

GSC

Figure 5.51. Graphic representation of stratigraphically logged, single drillhole from 1908 and 1909 near Howley in the Howley Formation. Drillhole data from Howley (1918a, Hole #5, 1908, Hole #6, 1909).

ripple cross-laminated. Some sandstones contain black, vitreous bitumen as streaks that outline stratification or as irregular patchy zones.

Measurements on the orientation of current-produced structures (crossbeds, parting lineation) in fluvial facies, in combination with abundant volcanic and plutonic clasts, point toward a westerly dipping paleoslope. The presence of the boulder conglomerate close to the eastern bounding fault suggests that the fault, in fact, marks or is very close to the original eastern basin margin.

The Howley Formation

The Howley Formation (Hacquebard et al., 1960; Belt, 1969; Hyde, 1983, 1985a) has poorly exposed coal-bearing sandstones and mudstones near Howley (Fig. 5.47) dated as Westphalian A (Hacquebard et al., 1960; Barss, 1981a), thus making these rocks the youngest known strata in the Deer Lake Basin. No specific type section has been designated. The Howley Formation consists of plant-bearing sandstones that are interbedded with grey to black carbonaceous mudstone (Fig. 5.51) and lesser amounts of pebble to cobble conglomerate. Bituminous coals are very poorly exposed (Howse and Fleischmann, 1982).

The Howley Formation is mostly in fault contact with pre-Carboniferous crystalline rocks, but an unconformity appears to be present at one locality on Grand Lake (Hyde, 1983). Because of the poor exposure, a thickness for the unit has been difficult to establish or even estimate. Hyde (1984a) suggested a potential homoclinal thickness of 3100 m, however a gravity survey (Miller and Wright, 1984) of the Deer Lake Basin indicated a vertical depth to basement of around 1500 m. Although it is not well exposed, the Howley Formation is interpreted to represent a predominantly fluvial unit, in which high-sinuosity channels traversed a westerly to southerly dipping, vegetated alluvial plain.

Deer Lake Basin as a strike-slip basin

Popper (1970), Hyde (1979, 1984a) and Hyde et al. (1988) presented evidence that the Deer Lake Basin was genetically associated with dextral wrenching along the Cabot Fault Zone. This supported earlier suggestions (Wilson, 1962; Belt, 1968b; Webb, 1969) that the Cabot Fault Zone is fundamentally a zone of wrenching, although it is only recently that there has been evidence for strike-slip offset of basement rocks.

Earlier statements that mafic bodies in southwestern Newfoundland are offset (Belt, 1969; Webb, 1969) along the Cabot Fault Zone appear to be incorrect (Knight, 1983, p. 277) in that rocks of different age and composition were compared. North of the Deer Lake Basin, however, there are two similar granitoid bodies, the Devils Room granite and Moose Lake granite, on opposite sides of the Taylors Brook Fault (synonymous with Doucers Valley Fault) that have been dated at about 398 Ma (Erdmer, 1986; Tuach, 1987; Coyle and Strong, 1987). These granitic bodies are offset by about 15 km and placed in a position that implies dextral offset (Lock, 1969; Coyle and Strong, 1987). Two separate groups of volcanic rocks also appear to have been dextrally offset along the Green Bay Fault, a northeasterly-trending splay of the Cabot Fault Zone. These offset rocks are: 1) the Lower Ordovician Snooks Arm and Western Arm groups that have been offset from each other by about

25 km (Bradley, 1982: Hibbard, 1983), and 2) Silurian pyroclastic rocks of the Springdale Group and Kings Point Complex, which appear to have been offset by about 50 km (Coyle and Strong , 1987).

The continuity and apparent steep dips of the faults comprising the fault zone, and their tendency to splay and enclose elongated blocks of basement rock (Fig. 5.47, 5.50) are taken as identifying characteristics of strike-slip faults (De Sitter, 1964; Freund, 1974; Crowell, 1974; van de Fliert et al., 1980). The elongated structural blocks containing the Anguille Group (Birchy Ridge, Fisher Hills blocks) were shown by Popper (1970) to be analogous to what have been termed piercement structures (Kingma, 1958) along the Alpine Fault in New Zealand. The piercement structures are downward-tapering horsts bounded by reverse faults, and are arranged in an end-on fashion and separated by topographically low areas. These features can be seen in the Deer Lake Basin (Popper, 1970; Hyde, 1983, 1984a; Hyde et al., 1988; Fig. 5.47 this paper). Similar horsts in wrench zones have been termed welts (Lowell, 1972), structural domains (Aydin and Page, 1984), and positive flower structures (Harding, 1985).

Within the elongated structural blocks and within the Anguille Group there are generally tight, flexural-slip folds with north-northeasterly trending axes that are oblique to northeasterly trending fault traces (Fig. 5.52). This arrangement of structures is well known from wrench zones (Wilcox et al., 1973; Odonne and Vialon, 1983); in this particular situation the geometrical arrangement of faults and folds indicates dextral movements along the faults. The unconformity in the Fisher Hills block between the Anguille Group and North Brook Formation, if correctly inferred, would mean that these folds formed prior to deposition of North Brook sediments. Also in the Fisher Hills block, the folds are bounded by the Fisher Hills-Birchy Ridge and Island Pond Brook faults. The north-northeastly-trending folds in Birchy Ridge are bounded by the Fisher Hills-Birchy Ridge and Hampden faults (Fig. 5.47). Outside of these faults, folds in all stratigraphic units are subparallel to the fault traces, and were probably produced by dip-slip movements on the bounding faults (Hyde, 1984a; Hyde et al., 1988). Transpressive movements (combined compression and translation) along the wrench zone resulted in reverse movements on the faults bounding the now uplifted structural blocks. Synclinal folds in the Deer Lake Group adjacent to these faults contain overturned beds, with facing directions in the overturned layers away from the faults.

At the northern end of Birchy Ridge, northeasterly trending fractures show clear evidence of dextral displacement (Hyde et al., 1988). Similar fractures were mapped in pre-Carboniferous rocks a few tens of kilometres north of Birchy Ridge (Lock, 1969; Smyth and Schillereff, 1982). All of these fractures are interpreted as Riedel shears that developed concurrently with lateral movements along the major through-going faults (Wilcox et al., 1973). In the case at hand, the inferred Riedel shears are in the Anguille Group, and indicate dextral displacement on the Fisher Hills-Birchy Ridge Fault.

Another feature in the Deer Lake Basin that is characteristic of strike-slip basins is the pronounced arrangement of strata in a shingled fashion (Reading, 1980; Crowell, 1982). In this arrangement, beds consistently dip in a direction parallel to the long dimension of the basin, and

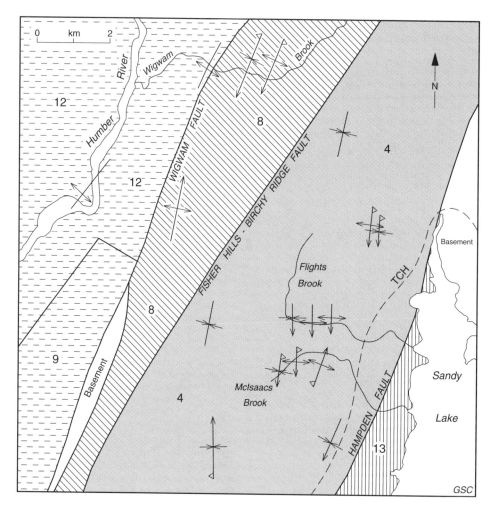

Figure 5.52. Geology map of part of Birchy Ridge showing north-south trend of fold axes between the Fisher Hills-Birchy Ridge and Hampden faults. Stratigraphic units are numbered and patterned as in Figure 5.47. Arrowheads on fold-axis symbols indicate plunge direction; triangles at end of fold-axis symbols indicate dip direction of axial surface. TCH = Trans Canada Highway.

the depocentre shifts along the length of the basin, keeping step with lateral fault movements. As a consequence, homoclinal thicknesses (as measured by the sine of the dip angle times length of outcrop) far exceeds the vertical basinal depth. The North Brook Formation presents a similar situation in the Glide Lake area in Fisher Hills (Fig. 5.50), where a northeastward-dipping section of conglomerate and sandstone (Fig. 5.53) has an approximate homoclinal thickness of about 5-7 km (Hyde, 1989). A basin this deep, however, is not indicated by the gravity map of Miller and Wright (1984), suggesting that the North Brook strata are shingled in the manner outlined by Reading (1980) and Crowell (1982).

Active strike slip along the Fisher Hills-Birchy Ridge Fault (Fig. 5.47) in the Glide Lake area is not accompanied by similar movements along the western basin margin. This margin is marked by unconformities between basement and Carboniferous cover (Hyde, 1983). Clast types

within North Brook fanglomerates along the western border match adjacent basement lithologies, implying no lateral movement of Carboniferous strata (Hyde, 1989).

There is no evidence for strike-slip fault movements after deposition of the North Brook Formation, so it seems that most lateral movements ceased during the Viséan. The inclusion of Anguille clasts in the Wigwam Brook Formation implies at least some uplift, resulting from transpression during the Tournaisian.

Geology of Carboniferous outliers
Conche area

At and near the town of Conche on the eastern side of the Great Northern Peninsula (Fig. 5.43), there are two small peninsulas which project eastward from an otherwise nearly straight coastline. The relatively straight coastline can be attributed to an unnamed, but obvious major strand of the Cabot Fault Zone.

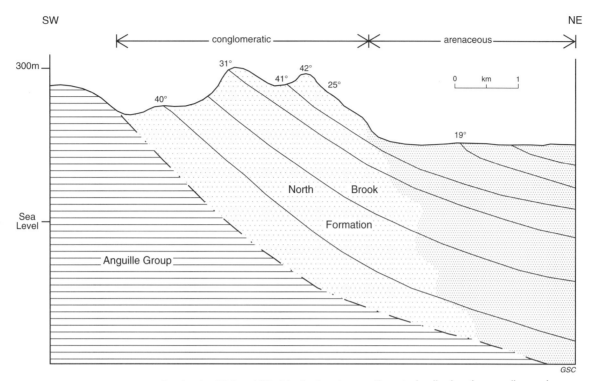

Figure 5.53. Cross-section in the Fisher Hills block showing northeasterly dipping (some dip angles given) and shingled North Brook rocks unconformably overlying Anguille strata (Saltwater Cove Formation). The section here is subparallel to the strike of Anguille Group strata so layering is shown as gently dipping. Near-source facies in the North Brook Formation are conglomeratic; a sandier distal facies occurs to the northeast. Topography is not exaggerated.

Eocambrian metasedimentary rocks of the Maiden Point Formation are exposed west of this fault and strata belonging to the Anguille Group are exposed east of the fault (Fig. 5.54). There is a group of islands and rocks (Groais Island), 13-24 km east of the peninsulas, where an angular unconformity is exposed (Fig. 5.54) (Baird, 1966; Kennedy et al., 1973). This unconformity involves metasedimentary rocks of the Fleur-de-lys Supergroup (Grey Island Schist of Baird, 1966) overlain by reddish sandstones and conglomerates of the Anguille Group.

Baird (1966) subdivided the Carboniferous strata into the Crouse Harbour and Cape Rouge formations. According to Baird (1966) both of these units comprise about 1524 m of section with the latter unit comprising about 1250 m. A Tournaisian age is assigned to the Cape Rouge Formation on the basis of fossil plants and spores (Baird, 1966; Hamblin, et al., in press), and from this, together with lithological criteria, it is included within the Anguille Group. The Crouse Harbour Formation is undated, but appears to intertongue with the Cape Rouge. Together these formations define a distinctive, seismically transparent unit that can be traced southwestwards under White Bay to onland exposures of the Anguille Group in the Birchy Ridge block of the Deer Lake Basin (Haworth et al., 1976).

The Crouse Harbour Formation consists of angular to rounded cobbles and boulders set in a matrix that varies both in colour (red on Groais Island, greenish brown on the peninsula) and texture (variably sandy, muddy, and pebbly).

There are also intercalated coarse grained to very coarse grained sandstones and thin layers of limestone (calcrete?). Baird (1966) noted, that some of the conglomerates are interbedded with shaly rocks that are similar to fine grained facies of the Cape Rouge Formation. This suggests intertonguing relationships between the two formations. Clasts in the Crouse Harbour conglomerates are clearly of local derivation; on Groais Island clasts are schists and phyllites of the Fleur-de-lys Supergroup, whereas on the peninsulas, clasts appear to be derived from the Maiden Point Formation. This distribution of clast rock types implies derivation of debris from both sides of what must have been a very narrow depositional trough.

The Cape Rouge Formation consists of silicified, fine- to medium-grained sandstones interbedded with grey to reddish-brown mudstones. Sandstones are marked by the presence of graded beds, ripple marks, cross lamination, and, more rarely, large scale crossbeds and down-cutting erosional features. The mudstones are commonly mudcracked and contain flattened "nodules" of tan-weathering, aphanitic dolostone that can be confused with primary bedding. These "nodules" also mimic mudcracks (pseudomudcracks) on bedding surfaces where irregularities have led to a mosaic of displaced mudstone and secondary dolostone.

The presence of abundant graded layers suggests repetitive turbidity current activity, but this must have taken place in shallow water because of the associated

547

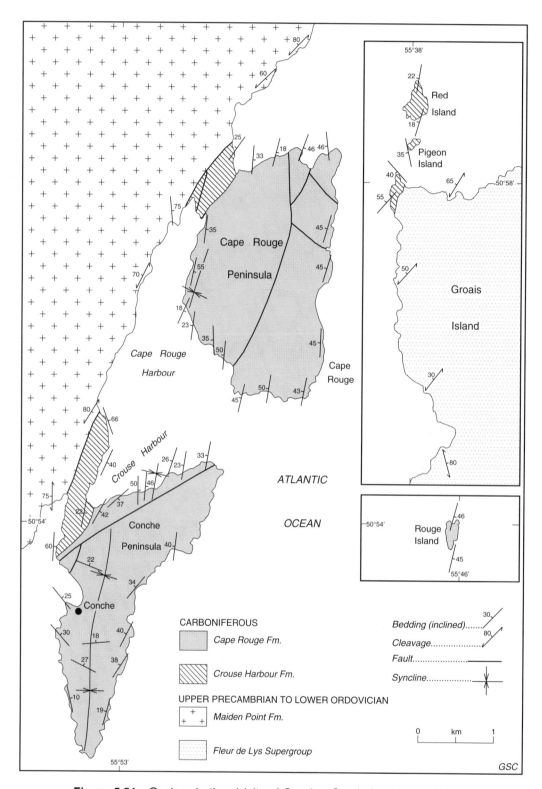

Figure 5.54. Geology in the vicinity of Conche, Groais Island, and Rouge Island, northern Newfoundland, simplified from Baird (1966).

desiccation cracks. In some instances, these cracks occur at the top of Bouma sequences. Abundant evidence for desiccation is in marked contrast to the Cape Rouge Formation in southern White Bay, where graded beds were evidently deposited in deeper water.

In the Conche region, it is unclear whether or not deposition occurred in a fully marine setting, brackish estuary, or a lacustrine environment. A finding of the "coffe bean" trace fossil *Isopodichnus* suggests nonmarine deposition (Trewin, 1976). Intertonguing of Crouse Harbour and Cape Rouge formations probably represents a rapid facies change from a standing water body environment (Cape Rouge) into a fan delta environment (Crouse Harbour).

There is little available evidence for strike-slip faulting in the Conche region, although given the relatively small area of Carboniferous exposure this should not be surprising. The fault separating pre-Carboniferous rocks from the Anguille Group (Fig. 5.54) has not been investigated to establish its movement history. There is a smaller northeasterly trending fault on the southern peninsula that apparently has dextrally offset a north-trending synclinal axis by about 900 m (Fig. 5.54). This is reminiscent of the northeasterly trending fractures in the Birchy Ridge area discussed previously in this chapter that were interpreted as Riedel shears (Hyde et al., 1988). If this northeasterly trending fault at Conche is also a Riedel shear to

the larger north-northeasterly fault that bounds the Carboniferous section on the west, it implies a similar dextral motion on the latter fault: this is the same sense that was established for the Deer Lake Basin.

Red Indian Lake area

The name Shanadithit Formation (also Shanadithit Brook Formation; Kean, 1978; Kean and Jayasinghe, 1980) is given to a succession of mainly lithic sandstones (very fine- to very coarse-grained, grey, green, brown, red) and mudstones (grey, green, and brown) that unconformably overlie pre-Carboniferous volcanic and plutonic rocks at Red Indian Lake in central Newfoundland (Fig. 5.43, 5.55). In drillcore studied by Deering (1987), conglomerate is clearly subordinate to the above lithologies, although brown and red conglomerates are common in surface exposures along and at the mouth of Shanadithit Brook (Kean, 1977, 1978). A few other exposures occur on the east side of Red Indian Lake (Fig. 5.55). Thin layers of carbonaceous shales with vitrain seams (<1 cm thick) are present in the drillcore studied by Deering (1987). This core also contains thin limestones, which grade vertically into clusters of limestone nodules, suggesting that these carbonates are calcretes. Similar limestone layers occur in outcrop (Kean, 1978).

A total thickness is difficult to establish because of relatively poor exposure and the nearly horizontal attitude, but drilling by ASARCO penetrated 216 m (Deering, 1987). Detrital sediments of this formation appear to have been derived primarily from volcanic sources based on sandstone composition and clast types (Kean, 1978; Deering, 1987). There is also a component of plutonic source rocks, but personal observations showed very few (about 1%) plutonic clasts. Gravel-clast imbrication indicates a strong northerly component of paleoflow.

Deering (1987) interpreted the successions as probably representing meandering stream deposition. This is suggested by the thick mudstones, which commonly cap fining-upward sequences.

The age of the Shanadithit Formation, based on spores, is Tournaisian (Barss, 1974). It, therefore, correlates roughly with the Anguille Group along the Cabot Fault Zone and Horton Group in the Maritime Provinces. However, the Shanadithit Formation has not been affected by the transpressive deformation seen in the Anguille Group, which suggests that the depositional basin at Red Indian Lake developed in a different tectonic regime from that represented by the Cabot Fault Zone.

Terrenceville area

Gently dipping sedimentary rocks at Terrenceville, Fortune Bay, (Fig. 5.43, 5.56), underlying an area of only about 2 km^2, were named the Terrenceville Formation (Bradley, 1962). These rocks were dated as Tournaisian by Barss (1974) based on a spore assemblage, although Bradley (1962) assigned a Late Devonian age from fossil plants. The Terrenceville Formation is at least 300 m thick (Bradley, 1962) and is faulted against Precambrian volcanic and sedimentary rocks (O'Brien et al., 1984).

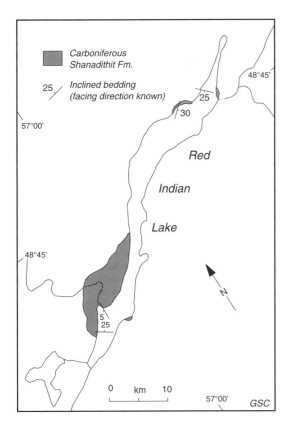

Figure 5.55. Distribution of the Shanadithit Formation in the Red Indian Lake area, modified from Deering (1987).

Bradley (1962) and O'Brien et al. (1984) have provided the best descriptions of the formation. It consists of reddish sandstones (with irregular erosional contacts, trough cross-beds, ripple marks, and desiccation cracks) that are interbedded with red mudstones and rare grey/green mudstones. There are also brown, maroon, and grey, poorly sorted pebble to boulder conglomerates. These conglomerates are interbedded with the sandstones in a lenticular fashion.

Bradley (1962) interpreted the Terrenceville Formation as a fluvial deposit. It is likely that the original depositional basin encompassed a larger area than the present exposures. Clasts in the conglomerates are primarily felsic volcanics, and were apparently derived from the nearby Precambrian volcanic rocks.

Spanish Room area

At Spanish Room, Newfoundland (Fig. 5.57), there is a sandy tombolo that connects an island to the land surrounding Mortier Bay. On the island there is a succession of gently southward (20-30°) and vertically dipping strata that are clearly different in facies and structure from the surrounding Precambrian volcanic and sedimentary rocks. Both the base and the top of the succession are not exposed; the top is inferred by Strong et al. (1978) to be in fault contact with Precambrian volcanic rocks. These authors reported a spore assemblage of Tournaisian age from a vertically dipping black shale. The succession is about 300 m thick.

Laracy and Hiscott (1982) divided the section into two units: 1) vertically dipping limestones, black shale, and sandstone, and 2) gently dipping red conglomerates, siltstones, sandstones, and thin limestones. They interpreted the gently dipping strata (age unknown but suggested to be Viséan based on lithological similarity to known Viséan strata) to unconformably overlie the vertically dipping beds of Tournaisian age. The vertical beds were informally called the Cashel Cove beds (Laracy and Hiscott, 1982) and the overlying redbeds were designated the Spanish Room Formation (originally named by Strong et al., 1978 for the entire succession including the Cashel Cove beds).

The Cashel Cove beds consist of about 15 m of bioturbated, micritic limestones, black, pyritiferous shale, and a 1-m thick bed of structureless, medium- to coarse-grained sandstone. Laracy and Hiscott (1982) interpreted this unit as lacustrine, and noted, in particular, an absence of marine fossils.

The re-defined and diminished Spanish Room Formation consists of about 120 m of poorly consolidated, pebble to cobble conglomerate overlain by about 170 m of red siltstone, sandstone, and minor conglomerate and limestone. The contact between the lower conglomeratic unit and the upper silty unit is covered by colluvium. Conglomerates in the lower part contain small-scale fining-upward units. Clasts are set in a poorly sorted, sandy matrix, are in places imbricated, and are dominantly volcanic. Red siltstone is the dominant lithology in the overlying beds. These are characterized by the presence of carbonate nodules, which in two parts of the section have coalesced to form layers. There are also thin beds of sandstone (pebble conglomerate) disseminated throughout the silty section.

Laracy and Hiscott (1982) interpreted the lower conglomeratic unit as an alluvial fan deposit and the upper silty section as representing deposition within a floodplain-dominated, meandering river system. These authors also suggested that Carboniferous deposition at Spanish Room occurred within a fault-bounded, strike-slip basin.

Economic geology

The following remarks on the mineral and fossil fuel resources, both economic and uneconomic, are taken from Knight (1983, 1984), Howse (1984) and Hyde (1984b).

Metallic minerals

In the Bay St. George Basin base metal mineralization occurs in black shales of the Snakes Bight Formation as clots and disseminations in small faults, tension joints, and fractures. Minerals observed are, chalcopyrite, sphalerite, galena, and pyrite. Base metal are also found near the contact between the Anguille and Codroy groups; in these cases, various combinations of chalcopyrite, chalcocite,

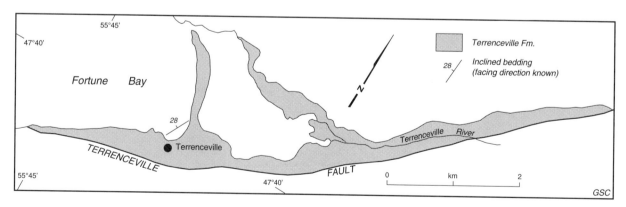

Figure 5.56. Distribution of the Terrenceville Formation in the Terrenceville area modified from O'Brien et al. (1984).

galena, sphalerite, and pyrite are associated with barite, hematite, fluorite and malachite in centimetre-wide calcite veins as well as coatings on joint surfaces. Stratiform zinc-lead mineralization also occurs near the Anguille-Codroy contact, where black Anguille sandstone contains up to 0.50 % zinc and 0.37% lead.

Base metal mineralization also occurs in Codroy Group carbonates (Big Cove Formation) on the Port au Port Peninsula at the north end of the basin. Here galena, sphalerite, and marcasite occur as vug fillings in carbonate build-ups and breccias and as diagenetic replacements.

Uranium occurs with minor copper mineralization in the Robinsons River Formation (Codroy Group). The highest uranium content discovered to date (323 g/t) occurs in grey, carbonaceous sandstones.

In the Deer Lake Basin, uranium appears to represent the most promising metallic mineralization. Sandstone boulders derived from the Humber Falls Formation contain uraninite cements and uranium contents up to 104 kg/t. Hematite and uranophane are also present as secondary minerals. There are, however, only a few examples of in situ uranium mineralization; this is probably because of periods of oxidation from the time of deposition to the present. Uneconomic uranium also occurs as stratiform bodies in mudstones and carbonate rocks of the Rocky Brook Formation and in paleokarstic carbonate breccias in the North Brook Formation.

There are only scattered occurrences of base metal minerals in the Deer Lake Basin. There is minor copper mineralization in a limestone bed of the Wigwam Brook Formation. Up to 1.4% copper also occurs in centimetre-scale, solid hydrocarbon nodules in the Rocky Brook Formation.

Precious metals have only been reported in a few localities in the Deer Lake Basin. Silver (up to 1250 g/t) occurs as acanthite in some of the uraniferous till boulders of the Humber Falls Formation. Gold has been reported (Howley, 1918b) in quartz veins in what could be the Anguille Group in southernmost White Bay.

Industrial minerals

Deposits of industrial minerals in Carboniferous basins (excluding sand and gravel) are known only from the Bay St. George Basin, and consist entirely of various salts in the Codroy Group. These formed mainly as evaporites, but there are also replacement and vein deposits.

The most important deposits are gently dipping layers of gypsum and anhydrite, which are chiefly found in the Codroy Road Formation. A deposit of some 38 million tonnes (Knight, 1984) is being mined in the northern part of the basin. There are two other gypsum-anhydrite deposits of 10 million tonnes in the Codroy Road Formation, and many smaller deposits. In the larger deposits an irregular contact separates gypsum from the underlying anhydrite. Impurities consist of dolostone and dolomitic shale; the gypsum-anhydrite bodies are also associated stratigraphically with redbeds and conglomerates.

There is also an association of salt and potash that occurs mainly in the Jeffreys Village Member of the Robinsons River Formation, although there is some uncertainty about the true stratigraphic position. There are three non-productive deposits, of which the thickest (Fischells Brook deposit) is at least 390 m thick. In the Fischells Brook deposit there is a well-defined potassic (sylvite) layer near the top of a halite section, in which K_2O contents approach 20 percent. In the other two major deposits, potassic zones consisting of sylvite, carnallite and polyhalite are interlayered with and disseminated within halite, and occur throughout the thickness of the salt section.

An association of the sulphate minerals barite and celestite occurs in the Bay St. George Basin. The largest deposits occur on Port au Port Peninsula in carbonates of the Codroy Group; for example, the Ronan deposit contains at least 150 000 tonnes of mixed barite and subordinate celestite. A smaller deposit, predominantly bluish celestite, contains about 17 000 tonnes and is also located on Port au Port Peninsula. These two sulphate minerals, and barite alone, also occur together with sulphides (primarily galena) as small lenses, veins and disseminations on Port au Port Peninsula. Barite (sulphides) also occurs in veins (up to 40-cm wide) in the Ship Cove Formation elsewhere in the Bay St. George Basin.

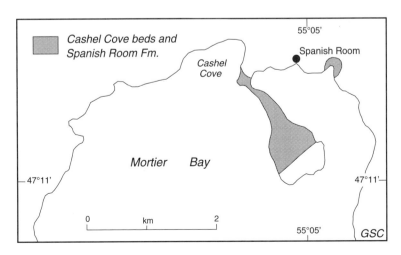

Figure 5.57. Distribution of the combined Carboniferous Spanish Room Formation and Cashel Cove beds, in Mortier Bay and surrounding area, modified from Strong et al. (1978).

551

Fossil fuels

Bituminous coal seams are known from both Carboniferous basins in Newfoundland, but there has been only limited production of coal dating back to the turn of the twentieth century. Mining has been hampered by the lenticular nature of the coals, steeply dipping seams, and faulting. It appears that depositional conditions on alluvial floodplains were not favourable for the formation of thick and widespread peat layers (Solomon and Hyde, 1985).

In the Bay St. George Basin, all of the coals occur in the Barachois Group in three separate areas. Most of the coal (Westphalian A age) occurs in an area (known as the St. George's coalfield) underlain by a doubly plunging, periclinal syncline in the Barachois Group. Proximate analyses (Hayes, 1949; Bonnell, 1984; Hyde et al.,1991) show ash contents as low as 3% and sulphur contents in the range 0.2-14.0%.

Coal seams in the Deer Lake Basin are also Westphalian A in age; these are present in the Howley Formation. Again, the coals are lenticular and structurally complicated, so that mining attempts have not been successful. Analyses show that most of these coals have ash contents as low as 4% and sulphur contents from 0.5 to 1.5 % (Hayes, 1949).

Oil shales are known from the Rocky Brook Formation and have been the subject of several recent studies (Macauley, 1981, 1987; Hyde, 1984c; Kalkreuth and Macauley, 1989). Although some grades are quite high (up to 155 litres/tonne), individual oil shale layers are relatively thin (<1 m thick), and are separated by kerogen-poor mudstones and carbonates. Oil shales or cannel shales are also known from the Barachois Group in the Bay St. George Basin (Solomon and Hyde, 1985; Solomon, 1986), but these have not been investigated in any detail. Unlike the Deer Lake oil shales, which are related to an extensive paleolake, the organic-rich shales in the Barachois Group are associated with coal, and were probably deposited in small open-water bogs in a poorly drained floodplain.

There has been no oil or gas production from either of the two Carboniferous basins and few attempts at exploration. All of the significant exploration occurred in 1955-56. Four wells were drilled in the Deer Lake Basin, two of which struck gas in the Rocky Brook and North Brook formations (gas was also struck in drilling for oil shales in the period 1919-20). No additional drilling was done in spite of recommendations to the contrary. Fleming (1970) has summarized these exploration attempts and shown lithological logs and the location of well sites.

There has been limited exploration for oil and gas in the Bay St. George Basin. During drilling for evaporites, gas was encountered in an anhydrite zone in the lower Codroy Group. In another drillhole, recovered core seeped oil in a conglomeratic unit (probably the Fischells Conglomerate Member of Anguille Group). These oil and gas showings are reported and discussed further in Fleming (1970). Subsequent to his report, an exploration well by the joint venture Union-Brinex in an anticlinal structure in the Kennels Brook Formation (Anguille Group) was dry.

Summary

In Newfoundland most Carboniferous strata are situated in the Bay St. George and Deer Lake basins along the Cabot Fault Zone, a zone of active dextral wrench faulting during the Carboniferous. The Bay St. George and Deer Lake basins were probably discrete, and developed at localized zones of tension along the Cabot Fault Zone. The basins became wider and, in the case of the Bay St. George Basin, elongate through time. These growth phases have led to overstepping stratigraphic relations in which younger strata extend beyond the boundaries of older basin fill. It is quite conceivable that initial deposition along the Cabot Fault Zone began during the Devonian. The youngest known rocks in the Deer Lake Basin are dated as Westphalian A; in the Bay St. George Basin the youngest known rocks are Westphalian C.

Depositional environments within the Deer Lake Basin are entirely nonmarine, consisting of basin-margin alluvial fans and fan-deltas, together with basin-interior alluvial plains and lake systems. Similar environments occur in the Bay St. George Basin, but, in addition, marine transgressions during the Viséan led to deposition of shallow-marine carbonate and evaporitic sediments. The only known volcanic activity is represented by a thin basaltic flow in the North Brook Formation.

Folds are attributed in some cases to transpressive movements along components of the Cabot Fault Zone. In the Deer Lake Basin, such folds are restricted to Tournaisian strata (Anguille Group) within elongate structural blocks. In the Bay St. George Basin, transpressive folding affected both Tournaisian and Viséan strata (Anguille Group and Codroy Group) and, perhaps, the Barachois Group of Westphalian age. Later folds in these strike-slip basins can be related to dominantly vertical movements on the major northeasterly trending faults.

Acknowledgments

I want to thank all of the people who have assisted me in my work on the Carboniferous basins over the many years. Michael Ware and Brian Alexander contributed some valuable data through their independent mapping as senior field assistants. Ian Knight is thanked for his time in discussing many geological problems of mutual interest. This paper benefitted from the critical reading of Hank Williams and Walter van de Poll.

REFERENCES

Akande, S.O. and Zentilli, M.
1984: Geologic, fluid inclusion and stable isotope studies at Gays River lead-zinc deposit, Nova Scotia, Canada; Economic Geology, v. 79, p. 1187-1211.

Alcock, F.J.
1935: Geology of Chaleur Bay region; Geological Survey of Canada, Memoir 183, 146 p.
1938: Geology of Saint John region, New Brunswick; Geological Survey of Canada, Memoir 216, 65 p.
1959: Geology of Musquash, New Brunswick; Geological Survey of Canada, Map 1084A (with marginal notes).

Alcock, F.J. and Evans, C.S.
1945: Waterford sheet; Geological Survey of Canada, Map 829A (with marginal notes).

Anderle, J.P., Crosby, K.S., and Waugh, D.C.E.
1979: Potash at Salt Springs, New Brunswick; Economic Geology, v. 74, p. 389-396.

Anderson, F.D.
1968: Woodstock, Millville and Coldstream map areas, Carleton and York counties, New Brunswick; Geological Survey of Canada, Memoir 353, 69 p.

Arne, D.C., Duddy, I.R., and Sangster, D.F.
1990: Thermochronologic constraints on ore formation at the Gays River Pb-Zn deposit, Nova Scotia, Canada, from apatite fission-track analysis; Canadian Journal of Earth Sciences, v. 27, p. 1013-1022.

Arthaud, F. and Matte, P.
1977: Late Paleozoic strike-slip faulting in southern Europe and northern Africa: result of a right-lateral shear zone between the Appalachians and the Urals; Geological Society of America Bulletin, v. 88, p. 1305-1320.

Aydin, A. and Page, B.M.
1984: Diverse Pliocene-Quaternary tectonics in a transform environment, San Francisco Bay region, California; Geological Survey of America Bulletin, v. 95, p. 1303-1317.

Badham, J.P.N.
1982: Strike-slip orogens: an explanation for the Hercynides; Geological Society of London, Journal, v. 139, p. 493-504.

Baird, D.M.
1951: Gypsum deposits of southwestern Newfoundland; Canadian Institute of Mining and Metallurgy, Mining Society of Nova Scotia, Transactions, v. 54, p. 85-94.
1960: Sandy Lake (west half), Newfoundland; Geological Survey of Canada, Map 47-1959, scale 1:250 000.
1966: Carboniferous rocks of the Conche-Groais Island area, Newfoundland; Canadian Journal of Earth Sciences, v. 3, p. 247-257.

Baird, D.M. and Coté, P.R.
1964: Lower Carboniferous sedimentary rocks in southwestern Newfoundland and their relations to similar strata in western Cape Breton Island; Canadian Institute of Mining and Metallurgy Bulletin, v. 57, p. 509-520.

Ball, F.D., Sullivan, R.M., and Peach, A.R.
1981: Carboniferous drilling project; New Brunswick Department of Natural Resources, Report of Investigation 18, 109 p.

Barr, S.M. and Raeside, R.P.
1986: Pre-Carboniferous tectonostratigraphic subdivisions of Cape Breton Island, Nova Scotia; Atlantic Geology, v. 22, p. 252-263.

Barr, S.M., Brisebois, D., and MacDonald, A.S.
1985: Carboniferous volcanic rocks of the Magdalen Islands, Gulf of St. Lawrence; Canadian Journal of Earth Sciences, v. 22, p. 1679-1688.

Barss, M.S.
1974: Palynological zonation of the Carboniferous and Permian rocks of the Atlantic Provinces; in Report of Activities, Part B; Geological Survey of Canada, Paper 74-1B, p. 135-136.
1980a: Palynological analyses of samples submitted for age determinations; Geological Survey of Canada, Internal Paleontological Report No. EPGS-PAL, 3-80 MSB, 2 p.
1980b: Palynological analyses of samples from Sandy Lake and Red Indian Lake areas, Newfoundland; Geological Survey of Canada, Internal Paleontological Report No. EPGS-PAL, 44-80 MSB, 3 p.
1981a: Palynological analyses of samples from Deer Lake and Rainy Lake map areas, Newfoundland; Geological Survey of Canada, Internal Paleontological Report No. EPGS-PAL, 4-81 MSB, 3 p.
1981b: Palynological analyses of samples from borehole in the Rocky Brook Formation, Sandy Lake area, Newfoundland; Geological Survey of Canada, Internal Paleontological Report No. EPGS-PAL, 18-81 MSB, 1 p.

Barss, M.S. and Hacquebard, P.A.
1967: Age and the stratigraphy of the Pictou Group in the Maritime Provinces as revealed by fossil spores; in Geology of the Atlantic Region, (ed.) E.R.W. Neale and H. Williams; Geological Association of Canada, Special Paper 4, p. 267-282.

Barss, M.S., Bujak, J.P., and Williams, G.L.
1979: Palynological zonation and correlation of sixty-seven wells, eastern Canada; Geological Survey of Canada, Paper 78-24, 103 p.

Barss, M.S., Hacquebard, P.A., and Howie, R.D.
1963: Palynology and stratigraphy of some Upper Pennsylvanian and Permian rocks of the Maritime Provinces; Geological Survey of Canada, Paper 63-3, 13 p.

Beerbower, J.R.
1961: Origin of cyclothems of the Dunkard Group (Upper Pennsylvanian-Lower Permian) in Pennsylvania, West Virginia and Ohio; Geological Society of America Bulletin, v. 72, p. 1029-1050.

Bell, J.S. and Howie, R.D.
1990: Paleozoic geology; in Geology of the Continental Margin of Eastern Canada, (ed.) M.J. Keen and G.L. Williams; Geological Survey of Canada, Geology of Canada, no. 2, p. 141-165 (also Geological Society of America, the Geology of North America, v. I-1).

Bell, W.A
1927: Outline of Carboniferous stratigraphy and geologic history of the Maritime Provinces of Canada; Royal Society of Canada Transactions, Section IV, p. 75-108.
1928: Discussion on paper of F.W. Gray; Canadian Institute of Mining and Metallurgy, Transactions, v. 30, p. 1144-1150.
1929: Horton-Windsor district, Nova Scotia; Geological Survey of Canada, Memoir 155, 268 p.
1938: Fossil flora of Sydney coalfield, Nova Scotia; Geological Survey of Canada, Memoir 215, 334 p.
1940: The Pictou coalfield, Nova Scotia; Geological Survey of Canada, Memoir 225, 161 p.
1944: Carboniferous rocks and fossil floras of northern Nova Scotia; Geological Survey of Canada, Memoir 238, 276 p.
1948: Early Carboniferous strata of St. George's Bay area, Newfoundland; Geological Survey of Canada, Bulletin 10, 45 p.
1958: Possibilities for occurrence of petroleum reservoirs in Nova Scotia; Nova Scotia Department of Mines, Halifax, Nova Scotia, 177 p.
1960: Mississippian Horton Group of type Windsor-Horton district, Nova Scotia; Geological Survey of Canada, Memoir 314, 58 p.

Belt, E.S.
1964: Revision of Nova Scotia Middle Carboniferous units; American Journal of Science, v. 262, p. 653-673.
1965: Stratigraphy and paleogeography of Mabou Group and related middle Carboniferous facies, Nova Scotia, Canada; Geological Society of America Bulletin, v. 76, p. 777-802.
1968a: Post Acadian rifts and related facies, eastern Canada; in Studies of Appalachian Geology, Northern and Maritime, (ed.) E.A. Zen, W.S. White, J.B. Hadley, and J.B. Thompson; Interscience Publishers, New York, Chapter 7, p. 95-113.
1968b: Carboniferous continental sedimentation, Atlantic Provinces, Canada; in Late Paleozoic and Mesozoic Continental Sedimentation, Northeastern North America, (ed.) G. de Vries Klein; Geological Society of America, Special Paper 106, p. 127-176.
1969: Newfoundland Carboniferous stratigraphy and its relation to the Maritimes and Ireland; in North Atlantic: Geology and Continental Drift, (ed.) M. Kay; American Association of Petroleum Geologists, Memoir 12, p. 734-753.

Belt, E.S., Freshney, E.C., and Read, W.A.
1967: Sedimentology of Carboniferous cementstone facies, British Isles and eastern Canada; Journal of Geology, v. 75, p. 711-721.

Besly, B.M.
1988: Paleogeographic implications of late Westphalian to early Permian red-beds, Central England; in Sedimentation in a synorogenic basin complex, (ed.) B.M. Besly and G. Kelling; Blackie, London, p. 200-222.

Betz, F., Jr.
1948: Geology and mineral deposits of southern White Bay; Geological Survey of Newfoundland Bulletin, v. 24, 26 p.

Bhattacharyya, G.K. and Raychaudhuri, B.
1967: Aeromagnetic and geological interpretations of a section of the Appalachian Belt in Canada; Canadian Journal of Earth Sciences, v. 4, p. 1015-1037.

Binney, W.P. and Kirkham, R.V.
1974: A study of copper mineralization in Mississippian rocks of Nova Scotia; in Report of Activities, Part A; Geological Survey of Canada, Paper 74-1, Part A, p. 129-130.
1975: A study of copper mineralization in Mississippian rocks of the Atlantic Provinces; in Report of Activities, Part A; Geological Survey of Canada, Paper 75-1, Part A, p. 245-248.

Blanchard, M.C., Jamieson, R.A., and More, E.B.
1984: Late Devonian-Early Carboniferous volcanism in western Cape Breton Island, Nova Scotia; Canadian Journal of Earth Sciences, v. 21, p. 762-774.

Boehner, R.C.
1984: Stratigraphy and depositional history of marine evaporites in the Windsor Group, Shubenacadie and Musquodoboit structural basins, Nova Scotia, Canada; in Atlantic Coast Basins, (ed.) H.M.J. Geldsetzer; 9th International Congress of Carboniferous Stratigraphy and Geology, Compte Rendu, v. 3, p. 163-178.

1985: Carboniferous basin studies: salt, potash, celestite and barite-new exploration potential in the Sydney Basin, Cape Breton Island; Nova Scotia Department of Mines and Energy, Report 85-1, p. 153-164.

1986: Salt and potash resources in Nova Scotia; Nova Scotia Department of Mines and Energy, Bulletin, v. 5, 346 p.

1987: Carbonate buildups in the Early Carboniferous Windsor and Horton Groups, Nova Scotia (a compilation); Nova Scotia Department of Mines and Energy, Open File Report 87-010, 99 p.

1990: Selected bibliography of the geology (including economic geology) of Carboniferous basins in Nova Scotia; Nova Scotia Department of Mines and Energy, Open File Report 90-023, 33 p.

1991: Seismic interpretation potential overthrust geology and mineral deposits in the Kennetcook Basin, Nova Scotia (Abstract); in Atlantic Geology, v. 27, p. 147.

Boehner, R.C. and Giles, P.S.
1982: Geological map of the Antigonish Basin, Nova Scotia; Nova Scotia Department of Mines and Energy, Map 82-2, scale 1:50 000.

1986: Geological map of the Sydney Basin; Nova Scotia Department of Mines and Energy, Map 86-1.

Boehner, R.C. and MacBeath, B.E.
1989: A lacustrine limestone mound (algal reef) in the Wilkie Brook Formation (Horton Group), Cape George area, northeastern mainland Nova Scotia; in Reefs of Canada and Adjacent Areas, (ed.) H.H.J. Geldsetzer, N.P. James, and G.E. Tebbutt; Canadian Society of Petroleum Geologists, Memoir 13, p. 626-630.

Boehner, R.C. and Prime, G.
1985: Geological map of the Loch Lomond Basin and Glengarry Half Graben; Nova Scotia Department of Mines and Energy, Map 85-2.

Boehner, R.C., Horne, R.J., and Ryan, R.J.
1989a: Carbonate bioherms in the Kennetcook and Herbert River limestone members, upper part of the Windsor Group, central mainland Nova Scotia; in Reefs of Canada and Adjacent Areas, (ed.) H.H.J. Geldsetzer, N.P. James, and G.E. Tebbutt; Canadian Society of Petroleum Geologists, Memoir 13, p. 622-625.

Boehner, R.C., Giles, P.S., Murray, D.A., and Ryan, R.J.
1989b: Carbonate buildups of the Gays River Formation, Lower Carboniferous Windsor Group, Nova Scotia; in Reefs of Canada and Adjacent Areas, (ed.) H.H.J. Geldsetzer, N.P. James, and G.E. Tebbutt; Canadian Society of Petroleum Geologists, Memoir 13, p. 609-621.

Bonnell, G.W.
1984: Coal analysis results to assess the fossil fuel potential of Carboniferous basins in western Newfoundland; Coal Research Laboratories Division Report ERP/CRL 84-57 (CF), Canada Centre for Mineral and Energy Technology (Energy, Mines and Resources, Canada), 4 p.

Boyce, W.D.
1986: Biostratigraphic and paleoenvironmental significance of palaeoniscid fish and vascular plant remains from the Snakes Bight Formation (Anguille Group), Codroy area, southwestern Newfoundland; Newfoundland Department of Mines and Energy, Mineral Development Division, Report 86-1, p. 169-171.

Boyle, R.W.
1972: The geology, geochemistry and origin of the barite, manganese and lead-zinc-copper-silver deposits of the Walton Cheverie area, Nova Scotia; Geological Survey of Canada, Bulletin 166, 181 p.

Bradley, D.A.
1962: Gisborne Lake and Terrenceville map-areas, Newfoundland; Geological Survey of Canada, Memoir 321, 56 p.

Bradley, D.C.
1982: Subsidence in Late Paleozoic basins in the northern Appalachians; Tectonics, v. 1, p. 107-123.

Bradley, D.C. and Bradley, L.M.
1986: Tectonic significance of the Carboniferous Big Pond Basin, Cape Breton Island, Nova Scotia; Canadian Journal of Earth Sciences, v. 23, p. 2000-2011.

Brisebois, D.
1981: Lithostratigraphie des strates Permo-Carboniferes, De L'Archipel Des Iles de la Madeleine; Ministère des Ressources naturelles du Québec, DVP 796, 48 p.

Brown, A.C.
1975: A study of stratiform copper deposits in Carboniferous strata of New Brunswick and Nova Scotia; in Report of Activities, Part A; Geological Survey of Canada, Paper 75-1, Part A, p. 598-599.

Brown, D.E.
1986a: The Bay of Fundy: thin skinned tectonics and resultant Early Mesozoic sedimentation; Presented at Atlantic Geoscience Society, Canadian Society of Petroleum Geologists and the Inter Union Commission on the Lithosphere Symposium: Basins of Eastern Canada and Worldwide Analogues, 44 p.

Brown, R.L. and Helmstaedt, M.
1969: Paleozoic wrench Faults in Canadian Appalachians: Discussion; in North Atlantic Geology and Continental Drift. (ed.) M. Kay, American Association of Petroleum Geologists, Memoir 12, p. 787.

Brown, Y.
1986b: Sedimentology of the Carboniferous Canso Group at Broad Cove, Nova Scotia; B.Sc. Honours thesis, Dalhousie University, Halifax, Nova Scotia, 63 p.

Browne, G.H.
1991: The sedimentology of the Boss Point Formation (Pennsylvanian), eastern New Brunswick and northern Nova Scotia; Ph.D. thesis, University of Western Ontario, London, Ontario.

Bryan, A.M.
1938: St. George's coalfield, Newfoundland; Geological Survey of Newfoundland, Information Circular 5, 23 p.

Calder, J.H.
1979: Effects of subsidence and depositional environment on the formation of lithotypes in a hypautochthonous coal of the Pictou coalfield; Nova Scotia Department of Mines and Energy, Report 79-6, 23 p.

1985a: Coal in Nova Scotia; Nova Scotia Department of Mines and Energy, Special Publication, 79 p.

1985b: Depositional environment of the Westphalian B Cumberland Basin coals of Springhill, Nova Scotia; Nova Scotia Department of Mines and Energy, Report 85-1, p. 11.

1985c: Preliminary observations on the sedimentology of selected sites within the Debert-Kemptown Coal Basin; Nova Scotia Department of Mines and Energy, Information Series No. 9, Program and Summaries, 9th Annual Open House and Review of Activities, p. 3-8.

1991: Controls on Westphalian Peat Accumulation: the Springhill Coalfield, Nova Scotia; Ph.D. thesis, Dalhousie University, Halifax, Nova Scotia.

1993: The evolution of a groundwater-influenced (Westphalian B) peat-forming ecosystem in a piedmont setting: the No. 3 seam, Springhill coalfield, Cumberland Basin, Nova Scotia; in Modern and Ancient Environments of Coal Formation; Geological Society of America, Special Paper 286.

Calder, J.H. and Bromley, D.S.,
in press: Seismic stratigraphy of the Athol Syncline, Cumberland Basin, Nova Scotia: implications for coal and coal-bed methane exploration; Geological Survey of Canada, Bulletin

Calder, J.H., Gibling, M.R., and Mukhopadhyay, P.K.
1991: Peat formation in a Westphalian B piedmont setting, Cumberland basin, Nova Scotia: implications for the maceral-based interpretation of rheotrophic and raised paleomires; Geological Society of France, Bulletin, v. 162, p. 283-298.

Cant, D.J. and Walker, R.G.
1976: Development of a braided-fluvial facies model for the Devonian Battery Point Sandstone, Québec; Canadian Journal of Earth Sciences, v. 13, p. 102-119.

Carr, P.A.
1968: Stratigraphy and spore assemblages, Moncton Map area, New Brunswick; Geological Survey of Canada, Paper 67-29, 47 p.

Carroll, R.L.
1967: Labyrinthodonts from the Joggins Formation; Journal of Paleontology, v. 41, p. 111-142.

Carroll, R.L., Belt, E.S., Dineley, D.L., Baird, D., and McGregor, D.C.
1972: Vertebrate paleontology of eastern Canada; 24th International Geological Congress, Montreal, Guidebook to Excursion A59, p. 64-80.

Carter, D.C. and Pickerill, R.K.
1985a: Lithostratigraphy of the Late Devonian-Early Carboniferous Horton Group of the Moncton Subbasin, southern New Brunswick; Maritime Sediments and Atlantic Geology, v. 21, p. 11-24.

1985b: Algal swamp, marginal and shallow evaporitic lacustrine lithofacies from the late Devonian-early Carboniferous Albert Formation, southeastern New Brunswick, Canada; Maritime Sediments and Atlantic Geology, v. 21, p. 69-86.

Case, E.E.
1905: Bathygnathus borealis Leidy, and the Permian of Prince Edward Island; Science, v. 22, p. 52-53.

Caudill, M.R. and Nance, R.D.
1986:   Variscan tectonostratigraphy of the Mispec Group, southern New Brunswick: stratigraphy and depositional setting; in Current Research, Part A; Geological Survey of Canada, Paper 86-1A, p. 343-350.

Chorlton, L.B. and Dallmeyer, R.D.
1986:   Geochronology of Early to Middle Paleozoic tectonic development in the southwest Newfoundland Gander Zone; Journal of Geology, v. 94, p. 67-89.

Clarke, D.B. and Halliday, A.N.
1980:   Strontium isotope geology of the South Mountain batholith, Nova Scotia; Geochimica et Cosmochimica Acta, v. 44, p. 1045-1058.

Clifton, H.E.
1967:   Solution-collapse and cavity filling in the Windsor Group, Nova Scotia, Canada; Geological Society of America Bulletin, v. 78, p. 819-832.

Conrod, D.
1987:   Fluvial sedimentation in the Lower Member of the Cheverie Formation at Blue Beach, Nova Scotia; B.Sc. Honours thesis, Dalhousie University, Halifax, Nova Scotia, 91 p.

Copeland, M.J.
1959:   Coalfields, west half Cumberland County, Nova Scotia; Geological Survey of Canada, Memoir 298, 89 p.

Cormier, R.F., Keppie, J.D., and Odom, A.L.
1988:   U-Pb and Rb-Sr geochronology of the Wedgeport granitoid pluton, southwestern Nova Scotia; Canadian Journal of Earth Sciences, v. 25, p. 255-261.

Coté, P.F.
1964:   Lower Carboniferous sedimentary rocks of the Horton Group in parts of Cape Breton Island, and their relation to similar strata of the Anguille Group in southwestern Newfoundland; Ph.D. thesis, University of Ottawa, Ottawa, Ontario, 279 p.

Cowie, J.W. and Bassett, M.G.
1989:   Global stratigraphic chart; Episodes, v. 12 (supplement).

Coyle, M. and Strong, D.F.
1987:   Geology of the Springdale Group: a newly recognized Silurian epicontinental-type caldera in Newfoundland; Canadian Journal of Earth Sciences, v. 24, p. 1135-1148.

Crowell, G.D.
1971:   The Kaiser celestite operation at Loch Lomond; Canadian Institute of Mining and Metallurgy Transactions, v. 74, p. 224-228.

Crowell, J.C.
1974:   Origin of Late Cenozoic basins in southern California; in Tectonics and Sedimentation, (ed.) W.R. Dickinson; Society of Economic Paleontologists and Mineralogists, Special Publication 22, p. 190-204.
1982:   The tectonics of Ridge Basin, southern California; in Geologic History of Ridge Basin, Southern California, (ed.) J.C. Crowell and M.H. Link; Pacific Section, Society of Economic Paleontologists and Mineralogists, p. 25-42.

Cumming, L.M.
1967:   Geology of the Passamaquoddy Bay region, Charlotte County, New Brunswick, 21B, 21G (parts of); Geological Survey of Canada, Paper 65-29, 36 p.

Dallmeyer, R.D., Hermes, O.D., and Gil-Ibarguchi, J.I.
1990:   40Ar/39Ar mineral ages from the Scituate Granite, Rhode Island: implications for Late Palaeozoic tectonothermal activity in New England; Journal of Metamorphic Geology, v. 8, p. 145-157.

Darrah, W.C.
1936:   Permian elements in the fossil flora of the Appalachian Province; II. Walchia, Harvard University, Botany Museum Leaflets, v. 4, p. 9-19.

Dawson, J.W.
1878:   Acadian Geology; MacMillan and Company, 3rd Edition, 694 p.
1882:   The fossil plants of the Erian (Devonian) and Upper Silurian formations of Canada; Part 2: Erian and Upper Silurian plants from the Restigouche River and Baie de Chaleur; Part 3: New ferns from the Middle Erian of Saint John, New Brunswick; Geological Survey of Canada, Separate Report 429, p. 96-113, 114-118.

De Sitter, L.V.
1964:   Structural Geology, 2nd edition; McGraw-Hill, New York, 587 p.

Deal, A.J.
1986:   Carboniferous fluvial strata of the Inverness Formation at Finlay Point, Southwest Cape Breton Island; B.Sc. Honours thesis, Dalhousie University, Halifax, Nova Scotia, 73 p.
1990:   The stratigraphy and depositional environments of the Ragged Reef Formation in the Athol Syncline, Cumberland Basin, Nova Scotia, Canada; M.Sc. thesis, Acadia University, Wolfville, Nova Scotia, 253 p.

Deering, P.R.
1987:   Paleodepositional environment of the Shanadithit Formation, central Newfoundland; B.Sc. thesis, Memorial University, St. John's, Newfoundland, 75 p.

Dewey, C.P.
1983:   The taxonomy and palaeoecology of Lower Carboniferous ostracodes and peracarids (Crustacea) from southwestern Newfoundland and central Nova Scotia; Ph.D. thesis, Memorial University, St. John's, Newfoundland, 383 p.
1985:   The palaeobiogeographic significance of Lower Carboniferous crustaceans (ostracodes and peracarids) from western Newfoundland and central Nova Scotia, Canada; Palaeogeography, Palaeoclimatology, Palaeoecology, v. 49, p. 175-188.

Dickie, J.R.
1986:   Upper Carboniferous Fluvial Sedimentation in the Gulf of St. Lawrence Coal Basin, Mabou Mines, Nova Scotia; B.Sc. Honours thesis, Dalhousie University, Halifax, Nova Scotia, 118 p.

Dineley, D.L. and Williams, B.P.J.
1968a: Sedimentation and paleoecology of the Devonian Escuminac Formation and related strata, Escuminac Bay, Quebec; in Late Paleozoic and Mesozoic Continental Sedimentation, (ed.) G. de Vries Klein; Geological Society of America, Special Paper 106, p. 241-264.
1968b: The Devonian continental rocks of the Lower Restigouche River, Quebec; Canadian Journal of Earth Sciences, v. 5(1), p. 945-953.

Dix, G.R.
1982:   The Codroy Group (Upper Mississippian) on the Port au Port Peninsula, western Newfoundland: stratigraphy, palaeontology, sedimentology and diagenesis; M.Sc. thesis, Memorial University, St. John's, Newfoundland, 219 p.

Dix, G.R. and James, N.P.
1987:   Late Mississippian bryozoan/microbial build-ups on a drowned karst terrain: Port au Port Peninsula, western Newfoundland; Sedimentology, v. 34, p. 779-793.
1989:   Stratigraphy and depositional environments of the Upper Mississippian Codroy Group: Port au Port Peninsula, western Newfoundland; Canadian Journal of Earth Sciences, v. 26, p. 1089-110.

Doig, R., Murphy, J.B., Nance, R.D., and Stokes, T.
1991:   Review of the geochronology of the Cobequid Highlands, Avalon composite terrane, Nova Scotia; in Current Research, Part D; Geological Survey of Canada, Paper 91-1D, p. 71-78.

Dolby, G.
1986:   Palynological analysis of samples from the AP-83-0372 corehole, Stellarton Basin, Nova Scotia; Robertson Research Canada Limited, Exploration Report 2321, 6 p.
1987:   Palynological analysis of samples from the Stellarton Basin, Nova Scotia; Internal Paleontological report, Geological Survey of Canada, 48 p.
1988:   The palynology of the Morien Group, Sydney Basin, Cape Breton Island, Nova Scotia; unpublished report, Nova Scotia Department of Mines and Energy, 21 p.
1989a: Palynological analysis of samples from the Port Hood and Chimney Corner coal fields, Cape Breton Island, and other miscellaneous samples; unpublished report, Nova Scotia Department of Mines and Energy, 11 p.
1989b: The palynology of the Morien Group, Sydney Basin, Cape Breton Island, Nova Scotia; unpublished report, Nova Scotia Department of Mines and Energy, 23 p.
1991a: The palynology of the western Cumberland Basin, Nova Scotia; Nova Scotia Department of Mines and Energy, Open File Report 91-006, 39 p.
1991b: Palynology of samples from the Debert and Port Hood coalfields, Nova Scotia; unpublished report, Nova Scotia Department of Mines and Energy, 8 p.

Donohoe, H.V., Jr. and Wallace, P.I.
1982:   Geology maps of the Cobequid Highlands, Nova Scotia; Nova Scotia Department of Mines and Energy, Maps 82-6 to 82-9, scale 1:50 000.
1985:   Repeated orogeny, faulting, and stratigraphy in the Cobequid Highlands, Avalon Terrain of northern Nova Scotia; Geological Association of Canada/Mineralogical Association of Canada, Fieldguide for Excursion 3, University of New Brunswick, 77 p.

Donohoe, H.V., Jr., Halliday, A., and Keppie, J.D.
1986:   Two Rb-Sr whole-rock isochrons from plutons in the Cobequid Highlands, Nova Scotia; Atlantic Geology, v. 22, p. 148-155.

D'Orsay, A.M. and van de Poll, H.W.
1985a: Quartz-grain surface textures: evidence for middle Carboniferous glacial sediment input to the Parrsboro Formation of Nova Scotia; Geology, v. 13, p. 285-287.
1985b: Quartz grain surface textures: evidence for a tropical climate during the Middle Pennsylvanian of eastern Canada; Canadian Journal of Earth Sciences, v. 22, p. 786-790.
1985c: Quartz grain surface textures and their significance in paleo-climatic interpretation of Permo-Carboniferous strata of eastern Canada (abstract); Geological Association of Canada/ Mineralogical Association of Canada, Annual Meeting, Program with Abstracts, v. 10, p. A14.

Dostal, J., Keppie, J.D., and Dupuy, C.
1983:   Petrology and geochemistry of Devono-Carboniferous volcanic rocks in Nova Scotia; Atlantic Geology, v. 19, p. 59-71.

Duff, P.McL.D. and Walton, E.K.
1973:   Carboniferous sediments at Joggins, Nova Scotia; 7th International Congress on Carboniferous Stratigraphy and Geology, Compte Rendu, v. 2, p. 365-379.

Dunn, J.T. and Stringer, P.
1990:   Petrology and petrogenesis of the Ministers Island dike southwest New Brunswick, Canada; Contributions to Mineralogy and Petrology, v. 105, p. 55-65.

Dunning, G.R., Barr, S.M., Raeside, R.P., and Jamieson, R.A.
1990:   U-Pb zircon, titanite, and monazite ages in the Bras d'Or and Aspy terranes of Cape Breton Island, Nova Scotia: implications for igneous and metamorphic history; Geological Society of America Bulletin, v. 102, p. 322-330.

Durling, P.W. and Marillier, F.J.Y.
1990:   Structural trends and basement rock sub-divisions in the western Gulf of St. Lawrence, Northern Appalachians; Atlantic Geology, v. 26, p. 79-95.
1993:   Tectonic setting of Middle Devonion to Lower Carboniferous rocks in the Magdalen Basin; Atlantic Geology, v. 29, p. 199-217.

Eisbacher, G.H.
1969:   Displacement and stress field along part of the Cobequid Fault, Nova Scotia; Canadian Journal of Earth Sciences, v. 6, p. 1095-1104.

Erdmer, P.
1986:   Geology of the Long Range Inlier in the Sandy Lake map area, western Newfoundland; in Current Research, Part B; Geological Survey of Canada, Paper 86-1B, p. 19-29.

Ferguson, S.A.
1983:   Geological map of the Hantsport area, Nova Scotia; Nova Scotia Department of Mines and Energy, Map 83-1, scale 1:25 000.

Flaherty, G.F. and Norman, G.W.H.
1941:   Albert map sheet; Geological Survey of Canada, Map 648A (with marginal notes).

Fleming, J.M.
1970:   Petroleum exploration in Newfoundland and Labrador; Newfoundland Department of Mines, Agriculture and Resources, Mineral Resources Report 3, 118 p.

Foley, P.L.
1984:   Depositional Setting of the Permo-Carboniferous redbeds around Hillsborough Bay, Prince Edward Island, Canada; M.Sc. thesis, University of New Brunswick, Fredericton, New Brunswick, 130 p.

Foley, S.L.
1989:   Geology of the Stoney Creek Oil and Gas Field, and its implications regarding the Tectonic Evolution of the Eastern Moncton Subbasin, New Brunswick; New Brunswick Department of Natural Resources and Energy, Minerals and Energy Division, Geoscience Report 89-1, 77 p.

Fong, C.C.K.
1976:   Geological mapping, northern half of the St. George's Bay Carboniferous basin; Newfoundland Department of Mines and Energy, Mineral Development Division, Report 76-1, p. 2-9.

Fralick, P.W. and Schenk, P.E.
1981:   Molasse deposition and basin evolution in a wrench tectonic setting: the Late Paleozoic, eastern Cumberland Basin, Maritime Canada; in Sedimentation and Tectonics in Alluvial Basins, (ed.) A.D. Miall; Geological Association of Canada, Special Paper 23, p. 77-97.

Frankel, J.J. and Bayliss, P.
1966:   Ferruginized surface deposits from Natal and Zululand, South Africa; Journal of Sedimentary Petrology, v. 36, p. 193-201.

Frankel, L.
1966:   Geology of southeastern Prince Edward Island; Geological Survey of Canada, Bulletin 145, 70 p.

Freeze, A.C.
1936:   Geology of the Fredericton Sheet, New Brunswick; M.Sc. thesis, University of New Brunswick, Fredericton, New Brunswick.

Freund, R.
1974:   Kinematics of transform and transcurrent faults; Tectonophysics, v. 21, p. 93-134.

Fyffe, L.R. and Barr, S.M.
1986:   Petrochemistry and tectonic significance of Carboniferous volcanic rocks in New Brunswick; Canadian Journal of Earth Sciences, v. 23, p. 1243-1256.

Fyson, W.K.
1964a: Folds in the Carboniferous rocks near Walton, Nova Scotia; American Journal of Science, v. 262, p. 513-522.
1964b: Repeated trends of folds and cross-folds in Palaeozoic rocks, Parrsboro, Nova Scotia; Canadian Journal of Earth Sciences, v. 1, p. 167-183.
1967:   Gravity sliding and cross folding in Carboniferous rocks, Nova Scotia; American Journal of Science, v. 265, p. 1-11.

Gall, Q. and Hiscott, R.N.
1986:   Diagenesis of locally uraniferous sandstones of the Deer Lake Group, and sandstones of the Howley Formation, Carboniferous Deer Lake Subbasin, western Newfoundland; Canadian Petroleum Geology Bulletin, v. 34, p. 17-29.

Gall, Q. and Hyde, R.S.
1989:   Analcime in lake and lake-margin sediments of the Carboniferous Rocky Brook Formation, western Newfoundland, Canada; Sedimentology, v. 36, p. 875-887.

Gao, R.
1987:   Deformation characteristics of the eastern Cobequid and Hollow Fault Zones and Stellarton Basin, Nova Scotia; M.Sc. thesis, University of New Brunswick, Fredericton, New Brunswick, 225 p.

Gates, O.
1984:   The geology of the Passamaquaddy Bay area, Maine and New Brunswick; Maine Geological Survey, Open-File no. 84-10, 22 p.
1989:   The geology and geophysics of the Passamaquaddy Bay area, Maine and New Brunswick and their bearing on local subsidence; in Neotectonics of Maine Studies in seismicity crustal warping and sealevel change, (ed.) W.A. Anderson and M.W. Barns; Maine Geological Survey, Bulletin, v. 40, p. 11-24.

Geldsetzer, H.H.J.
1977:   The Windsor Group of Cape Breton Island, Nova Scotia, in Report of Activities, Part A; Geological Survey of Canada, Paper 77-1A, p. 425-428.
1978:   The Windsor Group in Atlantic Canada-an update; in Current Research, Part C; Geological Survey of Canada, Paper 78-1C, p. 43-48.

Gemmell, D.E.
1975:   Carboniferous volcanic and sedimentary rocks of the Mount Pleasant caldera and Hoyt appendage, New Brunswick; M.Sc. thesis, University of New Brunswick, Fredericton, New Brunswick, 110 p.

Gersib, G.A., and McCabe, P.J.
1981:   Continental coal-bearing sediments of the Port Hood Formation (Carboniferous), Cape Linzee, Nova Scotia, Canada; in Recent and Ancient Nonmarine Depositional Environments: Models for Exploration, (ed.) F.G. Ethridge, and R.M. Flores; Society of Economic Paleontologists and Mineralogists, Special Publication 31, p. 95-108.

Gibling, M.R. and Bird, D.J.
1994:   Late Carboniferous cyclothems and alluvial paleovalleys in the Sydney Basin, Nova Scotia; Geological Society of America Bulletin, v. 106, p. 105-117.

Gibling, M.R. and Kalkreuth, W.D.
1991:   Petrology of selected carbonaceous limestones and shales in Late Carboniferous coal basins of Atlantic Canada; International Journal of Coal Geology, v. 17, p. 239-271.

Gibling, M.R. and Rust, B.R.

1987: Evolution of a mud-rich meander belt in the Carboniferous Morien Group, Nova Scotia, Canada; Canadian Society of Petroleum Geologists Bulletin, v. 35, p. 24-33.

1990a: Ribbon sandstones in the Pennsylvanian Waddens Cove Formation, Sydney Basin, Atlantic Canada: the influence of siliceous duricrusts on channel-body geometry; Sedimentology, v. 37, p. 45-65.

1990b: Tectonic influence on alluvial sedimentation in the coal-bearing Sydney Basin, Nova Scotia (abstract); in Abstracts, 13th International Sedimentological Congress, Nottingham, U.K., p. 188.

1992: Silica-cemented paleosols (ganisters) in the Pennsylvanian Waddens Cove Formation, Nova Scotia, Canada; in Diagenesis, (ed.) K.H. Wolf and G.V. Chilingarian; Elsevier Publishing Company, Amsterdam, v. 3, p. 621-655.

1993: Alluvial ridge-and-swale topography: a case study from the Morien Group of Atlantic Canada; in Alluvial Sedimentation, (ed.) M. Marzo, and C. Puigdefabregas; International Association of Sedimentologists, Special Publication 17, p. 133-150..

Gibling, M.R., Boehner, R.C., and Rust, B.R.

1987: The Sydney Basin of Atlantic Canada: an Upper Paleozoic strike-slip basin in a collisional setting; in Sedimentary Basins and Basin-forming Mechanisms, (ed.) C. Beaumont and A.J. Tankard; Canadian Society of Petroleum Geologists, Memoir 12, p. 269-285.

Gibling, M.R., Zentilli, M., and McCready, R.G.L.

1989: Sulphur in Pennsylvanian coals of Atlantic Canada: geologic and isotopic evidence for a bedrock evaporite source; International Journal of Coal Geology, v. 11, p. 81-104.

Gibling, M.R., Calder, J.H., Ryan, R., van de Poll, H.W., and Yeo, G.M.

1992: Late Carboniferous and Early Permian drainage patterns in Atlantic Canada; Canadian Journal of Earth Sciences, v. 29, p. 338-352.

Giles, P.S.

1981: Major transgressive-regressive cycles in Middle to Late Viséan rocks of Nova Scotia; Nova Scotia Department of Mines and Energy, Paper 81-2, 27 p.

1982: Geological map of the Eureka area, Nova Scotia; Nova Scotia Department of Mines and Energy, Map 82-3, scale 1:50 000.

1983: Sydney Basin Project; Nova Scotia Department of Mines and Energy, Report 83-1, p. 57-70.

Giles, P.S. and Boehner, R.C.

1982: Subdivision and regional correlation of strata of the Upper Windsor Group, Cape Breton Island and central Nova Scotia; Nova Scotia Department of Mines and Energy, Report 82-1, p. 69-78.

Giles, P.S., Boehner, R.C., and Ryan, R.J.

1979: Carbonate banks of the Gays River Formation in central Nova Scotia; Nova Scotia Department of Mines and Energy, Paper 79-7, 57 p.

Globensky, Y.R.

1962: Upper Mississippian conodonts from the Windsor Group of the Maritime Provinces; M.Sc. thesis, University of New Brunswick, Fredericton, New Brunswick, 145 p.

Greiner, H.R.

1962: Facies and sedimentary environments of Albert shale, New Brunswick; American Association of Petroleum Geologists Bulletin, v. 46, no. 2, p. 219-234.

1974: The Albert Formation of New Brunswick: a Paleozoic lacustrine model; Geologische Rundschau, v. 63, p. 1102-1113.

Grist, A.M.

1990: Provenance and thermal history of detrital sandstones of the Scotian Basin, offshore Nova Scotia, using the apatite fission track and $^{40}Ar/^{39}Ar$ methods; M.Sc. thesis, Dalhousie University, Halifax, Nova Scotia, 196 p.

Grossman, I.

1946: Preliminary report on the geology of the Deer Lake area; Geological Survey of Newfoundland, unpublished report, 6 p.

Gussow, W.C.

1953: Carboniferous stratigraphy and structural geology of New Brunswick, Canada; American Association of Petroleum Geologists Bulletin, v. 37, p. 1713-1816.

Hacquebard, P.A.

1960: Contribution to stratigraphical colloquium: a summary of Carboniferous stratigraphy and palaeontology of the Maritime Provinces; 4th International Congress on Carboniferous Stratigraphy and Geology; Compte Rendu, v. 1, p. 233-235.

1972: The Carboniferous of Eastern Canada; 7th International Congress on Carboniferous Stratigraphy and Geology, Compte Rendu, v. 1, p. 69-90.

Hacquebard, P.A. (cont.)

1983: Geological development and economic evaluation of the Sydney Coal Basin, Nova Scotia; in Current Research, Part A; Geological Survey of Canada, Paper 83-1A, p. 71-81.

1986: The Gulf of St. Lawrence Carboniferous Basin; the largest coal field of Eastern Canada; Canadian Institute of Mining, Bulletin, v. 79, p. 67-78.

Hacquebard, P.A. and Avery, M.P.

1984: Geological and geothermal effects on coal rank variations in the Carboniferous Basin of New Brunswick; in Current Research, Part A; Geological Survey of Canada, Paper 84-1A, p. 17-28.

Hacquebard, P.A. and Barss, M.S.

1958: Progress Report on the spore study of the coal deposits in the Minto, Chipman and Beersville areas of New Brunswick; Geological Survey of Canada, report, 35 p.

Hacquebard, P.A. and Cameron, A.R.

1989: Distribution and coalification patterns in Canadian bituminous and anthracite coals; International Journal of Coal Geology, v. 13, p. 207-260.

Hacquebard, P.A. and Donaldson, J.R.

1964: Stratigraphy and palynology of the Upper Carboniferous coal measures in the Cumberland Basin of Nova Scotia, Canada; 5th International Congress on Carboniferous Stratigraphy and Geology, Compte Rendu, v, 3, p. 1157-1169.

1969: Carboniferous coal deposition associated with flood-plain and limnic environments in Nova Scotia; in Environments of Coal Deposition, (ed.) E.C. Dapples and M.E. Hopkins; Geological Society of America, Special Paper 114, p. 143-191.

1970: Coal metamorphism and hydrocarbon potential in the Upper Paleozoic of the Atlantic Provinces, Canada; Canadian Journal of Earth Sciences, v. 7, p. 1139-1163.

Hacquebard, P.A., Barss, M.S., and Donaldson, J.R.

1960: Distribution and stratigraphic significance of small spore genera in Upper Carboniferous of the Maritime Provinces of Canada; 14th International Carboniferous Stratigraphy and Geology Congress, Heerlen, The Netherlands, 1958, Compte Rendu, v. 1, p. 237-245.

Hacquebard, P.A., Gillis, K.S., and Bromley, D.S.

1989: Re-evaluation of the coal resources of western Cape Breton Island; Nova Scotia Department of Mines and energy, Paper 89-3, 47 p.

Hamblin, A.P.

1989: Sedimentology, tectonic control and resource potential of the Upper Devonian-Lower Carboniferous Horton Group, Cape Breton Island, Nova Scotia; Ph.D. thesis, University of Ottawa, Ottawa, Ontario, 300 p.

Hamblin, A.P. and Rust, B.R.

1989: Tectono-sedimentary analysis of alternate-polarity half-graben basin-fill successions: Late Devonian-Early Carboniferous Horton Group, Cape Breton Island, Nova Scotia; Basin Research, v. 2, p. 239-255.

Hamilton, J.B.

1961: Salt in New Brunswick; New Brunswick Department of Lands and Mines, Mines Branch, Mineral Resource Report No. 1, 73 p.

1965: Limestone in New Brunswick (Plaster Rock); New Brunswick Department of Lands and Mines, Mines Branch, Mineral Resource Report 2, p. 17-23.

Hamilton, J.B. and Barnett, D.E.

1970: Gypsum in New Brunswick; New Brunswick Department of Natural Resources, Mineral Resources Branch, Report of Investigation 10, 62 p.

Haq, B.U. and van Eysinga, F.W.B.

1987: Geological Time Table (4th edition,); Elsevier, Science Publishing Company, Amsterdam, The Netherlands Publishers BV.

Harding, T.P.

1985: Seismic characteristics and identification of negative flower structures, positive flower structures, and positive structural inversion; American Association of Petroleum Geologists Bulletin, v. 69, p. 582-600.

Haszeldine, R.S.

1984: Carboniferous North Atlantic palaeogeography: stratigraphic evidence for rifting, not megashear or subduction; Geological Magazine, v. 121, p. 443-463.

Haszeldine, R.S. and Anderton, R.

1980: A braidplain facies model for the Westphalian B Coal Measures of north-east England; Nature, v. 284, p. 51-53.

Haworth, R.T., Keen, C.E., and Williams, H.

1985: D-1 Northern Appalachians: (West Sheet) Grenville Province, Quebec, to Newfoundland; Geological Society of America, Centennial Continent/Ocean transect #1.

Haworth, R.T. and Lefort, J.P.
1979:   Geophysical evidence for the extent of the Avalon Zone in Atlantic Canada; Canadian Journal of Earth Sciences, v. 16, p. 552-567.

Haworth, R.T., Poole, W.H., Grant, A.C., and Sandford, B.V.
1976:   Marine geoscience survey northeast of Newfoundland; in Report of Activities, Part A; Geological Survey of Canada, Paper 76-1A, p. 7-15.

Hayes, A.O.
1949:   Coal possibilities of Newfoundland; Geological Survey of Newfoundland, Information Circular 6, 30 p.

Hayes, A.O. and Bell, W.A.
1923:   The southern part of the Sydney coal field, Nova Scotia; Geological Survey of Canada, Memoir 133, 108 p.

Hayes, A.O. and Howell, B.F.
1937:   Geology of Saint John, New Brunswick; Geological Society of America, Special Paper 5, 146 p.

Hayes, A.O. and Johnson, H.
1938:   Geology of the St. George's Bay Carboniferous area; Geological Survey of Newfoundland, Bulletin, v. 12, 62 p.

Heckel, P.H.
1986:   Sea-level curve for Pennsylvanian eustatic marine transgressive-regressive depositional cycles along midcontinent outcrop belt, North America; Geology, v. 14, p. 330-334.

Henderson, J.A.L.
1940:   The development of oil and gas in New Brunswick; Canadian Institute of Mining and Metallurgy, Transactions 43, p. 159-178.

Hess, J.C. and Lippolt, H.J.
1986: $^{40}Ar/^{39}Ar$ ages of tonstein and tuff sanidines: new calibration points for the improvement of the Upper Carboniferous time scale; Chemical Geology, v. 59, p. 143-154.

Heyl, G.R.
1937:   The geology of the Sops Arm area, White Bay, Newfoundland; Geological Survey of Newfoundland, Bulletin, v. 8, 42 p.

Hibbard, J.
1983:   Geology of the Baie Verte Peninsula, Newfoundland; Newfoundland Department of Mines and Energy, Mineral Development Division, Memoir 2, 279 p.

Holden, R.
1913:   Some fossil plants from eastern Canada; Ann. Bot. v. 27, p. 243-255.

Howells, K. and Roulston, B.V.
1991:   Seismic Reflection Interpretation for Evaporites in the Moncton Subbasin, New Brunswick (Abstract); Atlantic Geology, v. 27, p. 154.

Howie, R.D.
1968:   Stoney Creek gas and oil field, New Brunswick; American Association of Petroleum Geologists, Memoir 9, v. 2, p. 1819-1832.
1984:   Carboniferous evaporites in Atlantic Canada; in Atlantic Coast Basins, (ed.) H.H.J. Geldsetzer; 9th International Congress on Carboniferous Stratigraphy and Geology, Compte Rendu, v. 3, p. 131-142.
1988:   Upper Paleozoic evaporites of southeastern Canada; Geological Survey of Canada, Bulletin 380, 120 p.

Howie, R.D. and Barss, M.S.
1974:   Upper Paleozoic rocks of the Atlantic Provinces, Gulf of St. Lawrence and adjacent Continental Shelf; in Offshore Geology of Eastern Canada; Geological Survey of Canada, Paper 74-30, v. 2, p. 35-50.
1975:   Paleogeography and sedimentation in the Upper Paleozoic, eastern Canada; in Canada's Continental Margins and Offshore Petroleum Exploration, (ed.) C.J. Yorath, E.R. Parker, and D.J. Glass; Canadian Society of Petroleum Geologists, Memoir 4, p. 45-57.

Howie, R.D. and Cumming, L.M.
1963:   Basement features of the Canadian Appalachians; Geological Survey of Canada, Bulletin 89, 18 p.

Howley, J.P.
1913:   The coal deposits of Newfoundland; in The Coal Resources of the World, v. II, p. 692-701.
1917a: Report on the Humber Valley and central Carboniferous area of the Island (for the years 1891 and 1892, Geological Survey of Newfoundland); Robinson and Company Ltd., Press, St. John's, Newfoundland, 51 p.
1917b: Report on coal deposits in the Codroy River Valley (for the year 1897, Geological Survey of Newfoundland); Robinson and Company Ltd. Press, St. John's, Newfoundland, 26 p.
1918a: Report on coal boring operations near Spruce Brook, 1909; in Reports of the Geological Survey of Newfoundland from 1881 to 1909, (ed.) A.P. Murray and J.P. Howley; Robinson and Company Ltd., St. John's.

Howley, J.P. (cont.)
1918b: Report for 1902-geological exploration in the District of White Bay; in Reports of the Geological Survey of Newfoundland from 1881 to 1909, (ed.) A.P. Murray and J.P. Howley; Robinson and Company Ltd., St. John's, p. 484-501.

Howse, A.
1984:   Mineral deposits in Carboniferous rocks, Port au Port area; in Mineral Deposits of Newfoundland-a 1984 Perspective, (comp.) H.S. Swinden, (ed.) M.J. Murray; Newfoundland Department of Mines and Energy, Mineral Development Division, Report, 84-3, p. 14-18.

Howse, A. and Fleischmann, J.
1982:   Coal assessment in the Deer Lake Carboniferous Basin; Newfoundland Department of Mines and Energy, Mineral Development Division, Report 82-1, p. 208-213.

Hyde, R.S.
1979:   Geology of Carboniferous strata in portions of the Deer Lake Basin, western Newfoundland; Newfoundland Department of Mines and Energy, Open File Report No. 1012, 43 p.
1983:   Geology of the Carboniferous Deer Lake Basin; Newfoundland Department of Mines and Energy, Mineral Development Division, Map 82-7, scale 1:100 000.
1984a: Geologic history of the Carboniferous Deer Lake Basin, west-central Newfoundland, Canada; in Atlantic Coast Basins, (ed.) H.H.J. Geldsetzer; 9th International Congress on Carboniferous Stratigraphy and Geology, Compte Rendu, v. 3, p. 85-104.
1984b: Geology and mineralization of the Carboniferous Deer Lake Basin, western Newfoundland; in Mineral Deposits of Newfoundland-a 1984 Perspective, (comp.) H.S. Swinden, (ed.) M.J. Murray; Newfoundland Department of Mines and Energy, Mineral Development Division, Report, 84-3, p. 19-26.
1984c: Oil shales near Deer Lake, Newfoundland; Geological Survey of Canada, Open File 1114, 10 p.
1985a: Entries for stratigraphic units in the Deer Lake Basin; in Lexicon of Canadian Stratigraphy, v. VI: Atlantic Region, (ed.) G.L. Williams, L.R. Fyffe, R.J. Wardle, S.P. Colman-Sadd, R.C. Boehner, and J.A. Watt; Canadian Society of Petroleum Geologists, Calgary, Alberta, 572 p.
1985b: Geological setting and aspects of uranium mineralization in the Carboniferous Deer Lake Basin, western Newfoundland; in Geology of Uranium Deposits, (ed.) T.I.I. Sibbald and W. Petruk; Canadian Institute of Mining and Metallurgy, Special Volume 32, p. 186-191.
1989:   The North Brook Formation: a temporal bridge spanning contrasting tectonic regimes in the Deer Lake Basin, western Newfoundland; Atlantic Geology, v. 25, p. 15-22.

Hyde, R.S. and Ware, M.J.
1980:   Geology of Carboniferous strata in the Cormack (12H/6) and Silver Mountain (12H/11) map areas; Newfoundland Department of Mines and Energy, Mineral Development Division, Report 80-1, p. 29-36.
1981:   Geology of Carboniferous strata in the Deer Lake (12H/3) and Rainy Lake (12A/14) map areas, Newfoundland; Newfoundland Department of Mines and Energy, Mineral Development Division, Report 81-1, p. 17-31.

Hyde, R.S., Kalkreuth, W.D., and Utting, J.
1991:   Coal petrology, palynology and depositional environments of Upper Carboniferous Barachois Group (Westphalian A and C) coals of the southwestern Newfoundland; Canadian Journal of Earth Sciences, v. 28, p. 1905-1924.

Hyde, R.S., Miller, H.G., Hiscott, R.N., and Wright, J.A.
1988:   Basin architecture and thermal maturation in the strike-slip Deer Lake Basin, Carboniferous of Newfoundland; Basin Research, v. 1, p. 85-105.

Irving, E. and Strong, D.F.
1984:   Paleomagnetism of the Early Carboniferous Deer Lake Group, western Newfoundland: no evidence for mid-Carboniferous displacement of "Acadia"; Earth and Planetary Science Letters, v. 69, p. 379-390.

Jamer, D.B.
1980:   A preliminary investigation of tuff horizon presence in the Pennsylvanian rocks of New Brunswick; B.Sc. thesis, University of New Brunswick, Fredericton, New Brunswick, 34 p.

Jamieson, R.A., Van Breemen, O., Sullivan, R.W., and Currie, K.L.
1986:   The age of igneous and metamorphic events in the western Cape Breton Highlands, Nova Scotia; Canadian Journal of Earth Sciences, v. 23, p. 1891-1901.

Jansa, L.F. and Mamet, B.L.
1984:   Offshore Viséan of eastern Canada: paleogeographic and plate tectonic implications; in Atlantic Coast Basins, (ed.) H.H.J. Geldsetzer; 9th International Congress on Carboniferous Stratigraphy and Geology, Compte Rendu v. 3, p. 205-214.

Johnsson, M.J.
1986: Distribution of maximum burial temperatures across northern Appalachian Basin and implications for Carboniferous sedimentation patterns; Geology, v. 14, p. 384-387.

Kalkreuth, W.D. and Macauley, G.
1987: Organic petrology and geochemical (Rock-Eval) studies on oil shales and coals from the Pictou and Antigonish areas, Nova Scotia, Canada; Canadian Petroleum Geology Bulletin, v. 35, p. 263-295.
1989: Organic petrology and Rock-Eval studies on oil shales from the Lower Carboniferous Rocky Brook Formation, western Newfoundland; Canadian Petroleum Geology Bulletin, v. 37. p. 31-42.

Kalkreuth, W.D., Naylor, R., Pratt, K., and Smith, W.D.
1990: Fluorescence properties of alginite-rich oil shales from the Stellarton Basin, Canada; Fuel, v. 69, p. 139-144.

Kean, B.F.
1977: Geology of the Lake Ambrose sheet, west half (12A/10W); Newfoundland Department of Mines and Energy, Mineral Development Division, Report 77-1, p. 21-25.
1978: Geology of the Star Lake east half sheet (12A/11E), Newfoundland; Newfoundland Department of Mines and Energy, Mineral Development Division, Report 78-1, p. 129-134.

Kean, B.F. and Jayasinghe, N.R.
1980: Geology of the Lake Ambrose (12A/10)-Noel Paul's Brook (12A/9) map-areas, central Newfoundland; Newfoundland Department of Mines and Energy, Mineral Development Division, Report 80-2, 29 p.

Keen, C.E., MacLean, B.C., and Kay, W.A.
1991: A deep seismic reflection profile across the Nova Scotia continental margin, offshore eastern Canada; Canadian Journal of Earth Sciences, v. 28, p. 1112-1120.

Keen, C.E., Keen, M.J., Nichols, B., Reid, I., Stockmal, G.S., Colman-Sadd, S.P., O'Brien, S.J., Miller, H., Quinlan, G., Williams, H., and Wright, J.
1986: Deep seismic reflection profile across the northern Appalachians; Geology, v. 14, p. 141-145.

Kelley, D.G.
1967: Some aspects of Carboniferous stratigraphy and depositional history in the Atlantic Provinces; in Geology of the Atlantic Region, (ed.) E.R.W. Neale and H. Williams; Geological Association of Canada, Special Paper 4, p. 213-228.

Kennedy, M.J., Williams, H., and Smyth, W.R.
1973: Geology of the Grey Islands, Newfoundland: northernmost extension of the Fleur de Lys Supergroup; Geological Association of Canada, Proceedings, v. 25, p. 79-91.

Keppie, J.D.
1979: Geological map of Nova Scotia; Nova Scotia Department of Mines and energy, scale 1:5 000 000.
1982: The Minas Geofracture; in Major Structural Zones and Faults of the Northern Appalachians, (ed.) P. St-Julien and J. Beland; Geological Association of Canada, Special Paper 24, p. 263-280.
1989: Northern Appalachians terranes and their accretionary history; Geological Society of America, Special Paper 230, p. 159-192.

Keppie, J.D., Giles, P.S., and Boehner, R.C.
1978: Some Middle Devonian to Lower Carboniferous rocks of Cape George, Nova Scotia; Nova Scotia Department of Mines and Energy, Paper 78-4, 37 p.

Kilfoil, G.
1988: An integrated gravity, magnetic, and seismic interpretation of the Carboniferous Bay St. George Subbasin, western Newfoundland; M.Sc. thesis, Memorial University, St. John's, Newfoundland, 168 p.

Kindle, E.D.
1962: Geology of Pointe Wolfe, Albert, Kings and Saint John counties, New Brunswick; Geological Survey of Canada, Map 1109A.

Kingma, J.T.
1958: Possible origin of piercement structures, local unconformities, and secondary basins in the Eastern Geosyncline, New Zealand; New Zealand Journal of Geology and Geophysics, v. 1, p. 269-274.

Kingston, P.W.E. and Steeves, B.A.
1984: Shallow seismic data from the eastern part of the New Brunswick Platform: Stratigraphic and structural implications; in Atlantic Coast Basins, (ed.) H.H.J. Geldsetzer, 9th International Congress on Carboniferous Stratigraphy and Geology; Compte Rendu, v. 3, p. 36-46.

Kirkham, R.V.
1978: Base metal and uranium distribution along the Windsor-Horton contact, central Cape Breton Island, Nova Scotia; in Current Research, Part B; Geological Survey of Canada, Paper 78-1B, p. 121-135.

Knight, I.
1983: Geology of the Carboniferous Bay-St. George Subbasin, western Newfoundland; Newfoundland Department of Mines and Energy, Mineral Development Division, Memoir 1, 358 p.
1984: Mineral deposits of the Carboniferous Bay-St. George subbasin; in Mineral Deposits of Newfoundland-a 1984 Perspective, (comp.) H.S. Swinden, (ed.) M.J. Murray; Newfoundland Department of Mines and Energy, Mineral Development Division, Report 84-3, p. 1-13.

Landell-Mills, T.
1922: The Carboniferous rocks of the Deer Lake district of Newfoundland; paper presented to Geological Society of London, 12 p.

Langston, W., Jr.
1963: Fossil vertebrates and the late Palaeozoic redbeds of Prince Edward Island; National Museum of Canada, Bulletin, v. 187.

Laracy, P.J. and Hiscott, R.N.
1982: Carboniferous redbeds of alluvial origin, Spanish Room Formation, Avalon Zone, southeastern Newfoundland; Bulletin of Canadian Petroleum Geology, v. 30, p. 264-273.

Lawrence, D.A. and Williams, B.J.P.
1987: Evolution of drainage systems in response to Acadian deformation: the Devonian Battery Point Formation, eastern Canada; in Recent Developments in Fluvial Sedimentology, (ed.) F.G. Ethridge, R.M. Flores, and M.D. Harvey; Society of Economic Paleontologists and Mineralogists, Special Publication 39, p. 287-300.

Lawson, D.E.
1962: Sedimentology of the Boss Point Formation in southeastern New Brunswick; M.Sc. thesis, University of New Brunswick, Fredericton, New Brunswick, 141 p.

Le Gallais, C.J.
1983: Stratigraphy, sedimentation and basin evolution of the Pictou Group (Pennsylvanian), Oromocto Subbasin, New Brunswick; unpublished M.Sc. thesis, McGill University, Montreal, Quebec.

Leger, A.R.
1969: Stratigraphy of the Hopewell Formation of Nonmarine Redbeds, Southeastern New Brunswick; The Compass of Sigma Gamma Epsilon, v. 46, The University of Oklahoma, p. 99-108.

Legun, A.S. and Rust, B.R.
1982: The Upper Carboniferous Clifton Formation of northern New Brunswick: coal-bearing deposits of a semi-arid alluvial plain; Canadian Journal of Earth Sciences, v. 19, p. 1775-1785.

Lock, B.E.
1969: Paleozoic wrench faults in Canadian Appalachians: discussion; in North Atlantic-Geology and Continental Drift, (ed.) M. Kay; American Association of Petroleum Geologists, Memoir 12, p. 789-790.

Logan, W.E.
1845: A section of the Nova Scotia coal measures as developed at Joggins on the Bay of Fundy, in descending order, from the neighbourhood of the west Ragged Reef to Minudie, reduced to vertical thickness; Geological Survey of Canada, Report of Progress 1843, Appendix, p. 92-153.

Loncarevic, B.D., Barr, S.M., Raeside, R.P., Keen, C.E., and Marillier, F.
1989: Northeastern extension and crustal expression of terranes from Cape Breton Island, Nova Scotia, based on geophysical data; Canadian Journal of Earth Sciences, v. 26, p. 2255-2267.

Lowell, J.D.
1972: Spitsbergen Tertiary orogenic belt and the Spitsbergen fracture zone; Geological Society of America Bulletin, v. 83, p. 3091-3102.

MacAuley, G.
1981: Geology of the oil shale deposits of Canada; Geological Survey of Canada, Open File 754, 156 p.
1987: Geochemical investigation of Carboniferous oil shales along Rocky Brook, western Newfoundland; Geological Survey of Canada, Open File 1438, 12 p.

MacAuley, G. and Ball, F.D.
1982: Oil shales of the Albert Formation, New Brunswick; New Brunswick Department of Natural Resources, Mineral Resources Division, Open File Report 82-12, 173 p.

MacAuley, G., Ball, F.D., and Powell, T.G.
1984: A review of the Carboniferous Albert Formation Oil Shales, New Brunswick; Canadian Petroleum Geology Bulletin, v. 32, p. 27-37.

MacAuley, G., Snowdon, L.R., and Ball, F.D.
1985: Geochemistry and geological factors governing exploitation of selected Canadian oil shale deposits; Geological Survey of Canada, Paper 85-13, 65 p.

MacEachern, S.B and Hannon, P.
1974: The Gays River discovery of a Mississippi Valley type lead-zinc deposit in Nova Scotia; Canadian Mining and Metallurgy Bulletin, v. 67, p. 61-66.

MacKenzie, G.S.
1964:   Geology of Hampstead, New Brunswick; Geological Survey of Canada, Map 1114A (with marginal notes).

MacLeod, J.L.
1984:   Diagenesis and its effects on base metal mineralization within a Mississippian Carbonate Complex, Gays River, Nova Scotia, Canada; in Atlantic Coast Basins, (ed.), H.H.J. Geldsetzer; 9th International Congress on Carboniferous Stratigraphy and Geology, Compte Rendu, v. 3, p. 193-204.

Mamet, B.L.
1968:   Sur une microfaune du Viséan Superior de Terre Neuve; Naturaliste Canadien, v. 95, p. 1357-1372.

Marillier, F. and Verhoef, J.
1989:   Crustal thickness under the Gulf of St. Lawrence, northern Appalachians, from gravity and deep seismic data; Canadian Journal of Earth Sciences, v. 26, p. 1517-1532.

Marillier, F., Keen, C.E., Stockmal, G.S., Quinlan, G., Williams, H., Colman-Sadd, S.P., and O'Brien, S.J.
1989:   Crustal structure and surface zonation of the Canadian Appalachians: implications of deep seismic reflection data; Canadian Journal of Earth Sciences, v. 26, p. 305-321.

Martel, A.M.
1990:   Stratigraphy, fluviolacustrine sedimentology and cyclicity of the Late Devonian/Early Carboniferous Horton Bluff Formation, Nova Scotia, Canada; Ph.D. thesis, Dalhousie University, Halifax, Nova Scotia, 297 p.

Martel, A.M. and Gibling, M.R.
1991:   Wave-dominated lacustrine facies and tectonically controlled cyclicity in the Lower Carboniferous Horton Bluff Formation, Nova Scotia, Canada; International Association of Sedimentologists, Special Publication 13, p. 223-243.

Martel, A.T.
1987:   Seismic Stratigraphy and Hydrocarbon Potential of the Strike-Slip Sackville Sub-Basin, New Brunswick; in Sedimentary Basins and Basin-Forming Mechanisms, (ed.) C. Beaumont and A.J. Tankard; Canadian Society of Petroleum Geologists, Memoir 12, p. 319-334.

Masson, A.G. and Rust, B.R.
1983:   Lacustrine stromatolites and algal laminites in a Pennsylvanian coal-bearing succession near Sydney, Nova Scotia, Canada; Canadian Journal of Earth Sciences, v. 20, p. 1111-1118.
1984:   Freshwater shark teeth as paleoenvironmental indicators in the Upper Pennsylvanian Morien Group of the Sydney Basin, Nova Scotia; Canadian Journal of Earth Sciences, v. 21, p. 1151-1155.
1990:   Alluvial plain sedimentation in the Pennsylvanian Sydney Mines Formation, eastern Sydney Basin, Nova Scotia; Canadian Society of Petroleum Geologists Bulletin, v. 38, p. 89-105.

McCabe, P.J. and Schenk, P.E.
1982:   From Sabkha to coal swamp: the Carboniferous sediments of Nova Scotia and southern New Brunswick; International Association of Sedimentologists, 11th International Congress, on Sedimentology Field Excursion Guidebook, Excursion 4A, 169 p.

McCutcheon, S.R.
1981:   Stratigraphy and paleogeography of the Windsor Group in southern New Brunswick; New Brunswick Department of Natural Resources, Mineral Resources Division, Geological Surveys Branch, Open File Report 81-31, 210 p.
1983:   Geology of the Mount Pleasant Caldera; in Eighth Annual Review of Activities, Project Resumes; New Brunswick Department of Natural Resources, Mineral Resources Division, Information Circular 83-3, p. 6-11.
1990:   The late Devonian Mount Pleasant Caldera Complex, stratigraphy, mineralogy, geochemistry and geologic setting of a S-W deposit in southwestern New Brunswick; Ph.D. thesis, Dalhousie University Halifax, Nova Scotia, 609 p.

McCutcheon, S.R. and Robinson, P.T.
1987:   Geological constraints on the genesis of the Maritimes Basin, Atlantic Canada; in Sedimentary Basins and Basin-forming Mechanisms, (ed.) C. Beaumont and A.J. Tankard; Canadian Society of Petroleum Geologists, Memoir 12, p. 287-297.

McGerrigle, H.W.
1950:   The geology of eastern Gaspé; Quebec Department of Mines Geological Report 35, 168 p.

McGregor, D.C. and McCutcheon, S.R.
1988:   Implications for spore evidence for Late Devonian age of the Piskahegan Group, southwestern New Brunswick; Canadian Journal of Earth Sciences, v. 25, p. 1349-1364.

McIlwaine, W.H.
1962:   Age and origin of the Perry Formation (Devonian), Charlotte County, New Brunswick; M.Sc. thesis, University of New Brunswick, Fredericton, New Brunswick, 86 p.
1967:   Age and origin of the Perry Formation, Charlotte County, New Brunswick, Canada; Maritime Sediments, v. 3, p. 56-60.

McLeod, M.J.
1980:   Geology and mineral deposits of the Hillsborough area, map area V-22 and V-23; New Brunswick Department of Natural Resources, Mineral Resources Branch, Map Report 79-6, 35 p.

McMahon, P.G.
1988:   Petroleum source rock study, onshore Nova Scotia: a progress summary; Nova Scotia Department of Mines and Energy, Report 88-3, Part A, p. 3-7.

McMahon, P.G., Short, G., and Walker, D.
1986:   Petroleum wells and drillholes with petroleum significance; Nova Scotia Department of Mines and Energy, Information Series, no. 10, 194 p.

Miller, H.G. and Wright, J.A.
1984:   Gravity and magnetic interpretation of the Deer Lake Basin, Newfoundland; Canadian Journal of Earth Sciences, v. 21, p. 10-18.

Miller, H.G., Kilfoil, G.J., and Peavy, S.T.
1990:   An integrated geophysical interpretation of the Carboniferous Bay St. George Subbasin; Canadian Petroleum Geology Bulletin, v. 38, p. 320-331.

Mohr, E.J.C. and van Baren, F.A.
1954:   Tropical soils; Interscience Publishers Inc., New York, New York, 498 p.

Moore, R.G.
1967:   Lithostratigraphic units in the upper part of the Windsor Group, Minas Sub-basin, Nova Scotia; in Collected Papers on Geology of the Atlantic region, (ed.) E.R.W. Neale and H. Williams; Geological Association of Canada, Special Paper 4, p. 245-266.

Moore, R.G. and Austin, I.A.
1984:   The Herbert River and Musquodoboit Limestone Members, keys to the reconstruction of the Fundy depositional trough in the Upper Windsor, Late Mississippian times in the Atlantic area of Canada; in Atlantic Coast Basins, (ed.) H.H.J. Geldsetzer; 9th International Congress of Carboniferous Statigraphy and Geology, Compte Rendu, v. 3, p. 179-192.

Moore, R.G. and Ferguson, S.A.
1986:   Geological map of the Windsor area, Nova Scotia; Nova Scotia Department of Mines and Energy, Map 86-2, scale 1:25 000.

Moore, R.G. and Ryan, R.J.
1976:   Guide to the invertebrate fauna of the Windsor Group in Atlantic Canada; Nova Scotia Department of Mines, Paper 76-5, 57 p.

Mosher, S. and Rast, N.
1984:   The deformation and metamorphism of Carboniferous rocks in Maritime Canada and New England; in Variscan Tectonics of the North Atlantic Region, (ed.) D.H.W. Hutton and D.J. Sanderson; Blackwell Scientific Publications, Boston, p. 233-243.

Mossman, D.J. and Place, C.H.
1989:   Early Permian fossil vertebrate footprints and their stratigraphic setting in megacyclic sequence II red beds, Prim Point, Prince Edward Island; Canadian Journal of Earth Sciences, v. 26, p. 591-605.

Mukhopadhyay, P.K., Hatcher, P.G., and Calder, J.H.
1991:   Hydrocarbon generation from deltaic and intermontane fluvio-deltaic coal and coaly shale from the Tertiary of Texas and Carboniferous of Nova Scotia; Organic Geochemistry, v. 17, p. 765-783.

Muller, J.E.
1951:   Geology and coal deposits of Minto and Chipman map areas, New Brunswick; Geological Survey of Canada, Memoir 260, 40 p.

Murphy, J.B., Pe-Piper, G., Nance, R.D., and Turner, D.
1988:   A preliminary report on geology of the eastern Cobequid Highlands, Nova Scotia; in Current Research, Part B; Geological Survey of Canada, Paper 88-1B, p. 99-107.

Murray, A.
1881a: Report for 1865: being a narrative of exploration from the eastern to the western shores of the island, by the valley of the Indian Brook, Hall's Bay, and hence by the Grand Pond and tributaries to St. George's Bay; in Geological Survey of Newfoundland, Reports 1864-1880, p. 51-70; Edward Stanford, London, 536 p.
1881b: Report for 1866: surveys of the Codroy and Humber Rivers, notes on the Carboniferous rocks of St. George's Bay and central country; in Geological Survey of Newfoundland, Reports 1864-1880, p. 73-101; Edward Stanford, London, 536 p.

Murray, A. (cont.)
1881c: Report for 1873: description of the country surrounding St. George's Bay (divided into three areas), distribution of the Carboniferous formations in the same region; in Geological Survey of Newfoundland, Reports 1864-1880, p. 298-349; Edward Stanford, London, 536 p.
1881d: Report for 1879: boring operations near the Grand Pond, survey of the west branch of the Humber River; in Geological Survey of Newfoundland, Reports 1864-1880, p. 512-531; Edward Stanford, London, 536 p.

Murthy, G.S.
1985: Paleomagnetism of certain constituents of the Bay-St. George subbasin, western Newfoundland; Physics of the Earth and Planetary Interiors, v. 39, p. 89-107.

Nahon, D., Janot Karpoff, A.M., Paquet, H., and Tardy, Y.
1977: Mineralogy, petrography and structure of iron crusts (ferricretes) developed on sandstones in the western part of Senegal; Geoderma, v. 19, p. 263-277.

Nance, R.D.
1987a: Dextral transpression and Late Carboniferous sedimentation in the Fundy Coastal Zone of southern New Brunswick; in Sedimentary Basins and Basin-forming Mechanisms, (ed.) C. Beaumont and A.J. Tankard; Canadian Society of Petroleum Geologists, Memoir 12, p. 363-377.
1987b: Model for the Precambrian evolution of the Avalon terrane in southern New Brunswick; Geology, v. 15, p. 127-137.

Nance, R.D. and Warner, J.B.
1986: Variscan Tectonostratigraphy of the Mispec Group, southern New Brunswick: Structural geometry and deformational history; in Current Research, Part A; Geological Survey of Canada, Paper 86-1A, p. 357-358.

Naylor, R.D., Kalkreuth, W., Smith, W.D., and Yeo, G.M.
1989: Stratigraphy, sedimentology and depositional environments of the coal-bearing Stellarton Formation, Nova Scotia; Geological Survey of Canada, Paper 89-8, p. 2-13.

Neale, E.R.W. and Nash, W.A.
1963: Sandy Lake (east half), Newfoundland; Geological Survey of Canada, Paper 62-28, 40 p.

Neale, E.R.W., Beland, J., Potter, R.R., and Poole, W.H.
1961: A preliminary Tectonic Map of the Canadian Appalachian Region based on age of folding; Canadian Institute of Mining and Metallurgy, Bulletin, v. 54, p. 687-694.

Nickerson, W.A.
1991: Seismic clues to Carboniferous Basin development in the Moncton Subbasin, New Brunswick (abstract); Atlantic Geology, v. 27, p. 159.
1992: Horton basin inversion event in the Moncton Subbasin, New Brunswick (abstract); in Atlantic Geoscience Society, 1992 Colloquium Symposium Fredericton, New Brunswick.

Norman, G.W.H.
1935: Lake Ainslie Map-area, Nova Scotia; Geological Survey of Canada, Memoir 177, 103p.
1941: Geology of Hillsborough, Albert and Westmorland Counties, New Brunswick; Geological Survey of Canada, Map 647 A with marginal notes, scale 1 inch = 10 miles.

O'Brien, S.J., Nunn, G.A.G., Dickson, W.L., and Tuach, J.
1984: Geology of the Terrenceville (1M/10) and Gisborne Lake (1M/15) map areas southeast Newfoundland; Newfoundland Department of Mines and Energy, Mineral Development Division, Report 84-4, 54 p.

Odonne, F. and Vialon, P.
1983: Analogue models of folds above a wrench fault; Tectonophysics, v. 99, p. 31-46.

Pajari, G.E.
1973: The Harvey volcanic area; in New England Intercollegiate Geological Conference, Field Guide to Excursions, (ed.) N. Rast; Department of Geology, University of New Brunswick, Fredericton, New Brunswick, p. 119-123.

Palmer, A.R.
1983: The Decade of North American Geology, 1983 Geologic Time Scale; Geology, v. 11, p. 503-504.

Patchen, D.G., Avary, K.L., and Erwin, R.B.
1985: Correlation of stratigraphic units in North America: Southern Appalachian Region Correlation Chart, Northern Appalachian Region Correlation Chart; American Association of Petroleum Geologists, Tulsa, Oklahoma.

Paul, J., Kalkreuth, W., Naylor, R., and Smith, W.D.
1989: Petrology, Rock-Eval and facies analyses of the McLeod coal seam and associated beds, Pictou Coalfield, Nova Scotia, Canada; Atlantic Geology, v. 25, p. 81-92.

Pe-Piper, G. and Jansa, L.F.
1986: Triassic olivine-normative diabase from Northumberland Strait, eastern Canada: implications for continental rifting; Canadian Journal of Earth Sciences, v. 23, p. 1013-1021.

Pe-Piper, G. and Loncarevic, B.D.
1989: Offshore continuation of Meguma Terrane, southwestern Nova Scotia; Canadian Journal of Earth Sciences, v. 26, p. 176-191.

Pe-Piper, G., Cormier, R.F., and Piper, D.J.W.
1989a: The age and significance of Carboniferous plutons of the western Cobequid Highlands, Nova Scotia; Canadian Journal of Earth Sciences, v. 26, p. 1297-1307.

Pe-Piper, G., Murphy, J.B., and Turner, D.S.
1989b: Petrology geochemistry and tectonic setting of some Carboniferous plutons of the eastern Cobequid Hills; Atlantic Geology, v. 25, p. 37-49.

Peavy, S.T.
1985: A gravity and magnetic interpretation of the Bay St. George Carboniferous Subbasin in western Newfoundland; M.Sc. thesis, Memorial University, St. John's, Newfoundland, 207 p.

Pickerill, R.K. and Carter, D.
1980: Sedimentary facies and depositional history of the Albert Formation; New Brunswick Department of Natural Resources, Open File Report 80-3, 132 p.

Piper, D.J.W., Pe-Piper, G. and Loncarevic, B.D.
1993: Devonian-Carboniferous igneous intrusions and their deformation, Cobequid Highlands, Nova Scotia; Atlantic Geology, v. 29, p. 219-232

Place, C.H. and van de Poll, H.W.
1988: Facies and stratigraphic subdivision of the Prince Edward Island redbeds in southeastern Prince Edward Island; in Geological Association of Canada, Program with Abstracts, v. 13, p. A99.

Plint, A.G.
1985: Possible earthquake-induced soft-sediment faulting and remobilization in Pennsylvanian alluvial strata, southern New Brunswick, Canada; Canadian Journal of Earth Sciences, v. 22, p. 907-912.

Plint, A.G. and van de Poll, H.W.
1982: Alluvial fan and piedmont sedimentation in the Tynemouth Creek Formation (Lower Pennsylvanian) of southern New Brunswick; Maritime Sediments and Atlantic Geology, v. 18, p. 104-128.
1984: Structural and sedimentary history of the Quaco Head area, southern New Brunswick; Canadian Journal of Earth Sciences, v. 21, p. 753-761.

Plint, H.A. and von Bitter, P.H.
1986: Windsor Group (Lower Carboniferous) conodont biostratigraphy and palaeoecology, Magdalen Islands, Quebec; Canadian Journal of Earth Sciences, v. 23, p. 439-453.

Plint, H.E. and Jamieson, R.A.
1989: Microstructure, metamorphism, and tectonics of the western Cape Breton Highlands, Nova Scotia; Journal of Metamorphic Geology, v. 7, p. 407-424.

Poole, W.H.
1967: Tectonic evolution of Appalachian region of Canada; in Geology of the Atlantic Region, (ed.) E.R.W. Neale and H. Williams; Geological Association of Canada, Special Paper 4, p. 9-51.

Popper, G.H.P.
1965: Stratigraphy and tectonic history of the Memramcook terrestrial redbeds of New Brunswick, Canada; M.Sc. thesis, University of Massachusetts, Amherst, Massachusetts.
1970: Paleobasin analysis and structure of the Anguille Group, west-central Newfoundland; Ph.D. thesis, Lehigh University, Bethlehem, Pennsylvania, 215 p.

Potter, R.R., Hamilton, J.B., and Davies, J.L.
1979: Geological Map, New Brunswick; New Brunswick Department of Natural Resources, Map Number NR1, second edition, scale 1:500 000.

Prime, G.A.
1987: The Antigonish Basin of Maritime Canada: a sedimentary tectonic history of a Late Paleozoic fault-wedge basins; M.Sc. thesis, Dalhousie University, Halifax, Nova Scotia, 223 p.

Prime, G.A. and Boehner, R.C.
1984: Preliminary report on the geology of the Glengarry half graben and vicinity, Cape Breton Island, Nova Scotia; Nova Scotia Department of Mines and Energy, Report 84-1, p. 61-69.

Quinlan, G.M. and Lithoprobe East Transect Members
1988: Structure and evolution of the Magdalen Basin: constraints from Lithoprobe seismic data (abstract); Geological Association of Canada, Program with Abstracts, v. 13, p. A101.

561

Rast, N.

1979:   Precambrian meta-diabases of southern New Brunswick: the opening of the Iapetus Ocean?; Tectonophysics, v. 59, p. 127-137.

1983:   The Northern Appalachian traverses in the Maritimes of Canada; in Profiles of Orogenic Belts, (ed.) N. Rast and F.M. Delany; American Geophysical Union, Geodynamic Series, v. 10, p. 243-274.

1984:   The Alleghenian orogeny in eastern North America; in Variscan Tectonics of the North Atlantic Region, (ed.) D.H.W. Hutton and D.J. Sanderson; Geological Society of London, Special Publication 14, p. 197-218.

Rast, N. and Grant, R.H.

1973:   The Variscan front in southern New Brunswick; in Geology of New Brunswick, (ed.) N. Rast; New England Intercollegiate Geological Conference, 65th Annual Meeting, Field Guide to Excursions, p. 4-11.

1977:   Variscan Appalachian and Alleghanian deformation in the northern Appalachians; in Colloquium Volume: Contributions to the International Colloquium on the Variscan System, Rennes, France; 1974 Paris Editions du CNRS No. 243, p. 583-586.

Rast, N. and Skehan, J.W.

1991:   Tectonic relationships of Carboniferous and Precambrian rocks in southwestern New Brunswick; in Geology of the Coastal Geotectonic Block and Neighboring Terranes, eastern Maine and southern New Brunswick, (ed.) A. Ludman, New England Intercollegiate Geological Conference; 83rd Annual Meeting Guidebook, p. 209-221.

Rast, N., Grant, R.H., Parker, J.S.D. and Teng, H.C.

1978:   The Carboniferous deformed rocks west of Saint John, New Brunswick; in Guidebook for field trips in southeastern Maine and southwestern New Brunswick, (ed.) A. Ludman, New England Intercollegiate Geological Conference, 70th Annual Meeting, p. 162-173.

1984:   The Carboniferous succession in southern New Brunswick and its state of deformation; in Atlantic Coast Basins, (ed.) H.H.J. Geldsetzer; 9th International Congress on Carboniferous Stratigraphy and Geology, Compte Rendu, v. 3, p. 13-22.

Ravenhurst, C. and Zentilli, M.

1987:   A model for the evolution of hot (>200°C) overpressured brines under an evaporite seal: the Fundy/Magdalen Carboniferous Basin of Atlantic Canada and its associated Pb-Zn-Ba deposits; in Sedimentary Basins and Basin-forming Mechanisms, (ed.) C. Beaumont and A.J. Tankard; Canadian Society of Petroleum Geologists, Memoir 12, p. 335-349.

Ravenhurst, C., Donelick, R., Zentilli, M., Reynolds, P., and Beaumont, C.

1990:   A fission track pilot study of the thermal effects of rifting on the onshore Nova Scotian margin, Canada; Nuclear Tracks and Radiation Measurements, v. 17, p. 373-378.

Reading, H.G.

1980:   Characteristics and recognition of strike-slip fault systems; in Sedimentation in Oblique-slip Mobile Zones, (ed.) P.F. Ballance and H.G. Reading; International Association of Sedimentologists, Special Publication 4, p. 7-26.

Reynolds, P.H., Elias, P., Muecke, G.K., and Grist, A.M.

1987:   Thermal history of the southwestern Meguma Zone, Nova Scotia, from an $^{40}Ar/^{39}Ar$ and fission track dating study of intrusive rocks; Canadian Journal of Earth Sciences, v. 24, p. 1952-1965.

Riley, G.C.

1957:   Red Indian Lake (west half), Newfoundland; Geological Survey of Canada, Map 8-1957, scale 1:250 000.

1962:   Stephenville map-area, Newfoundland; Geological Survey of Canada, Memoir 323, 72 p.

Roliff, W.A.

1962:   The Maritimes Carboniferous Basin of eastern Canada; Geological Association of Canada, Proceedings, v. 14, p. 21-41.

Rose, B.

1936:   Preliminary Report, Plaster Rock area, New Brunswick; Geological Survey of Canada, Paper 36-19, 10 p.

Ruitenberg, A.A. and McCutcheon, S.R.

1982:   Acadian and Hercynian Structural Evolution of Southern New Brunswick; in Major Structural Zones and Faults of the Northern Appalachians, (ed.) P. St.-Julien and J. Beland; Geological Association of Canada, Special Paper 24, p. 131-148.

Rust, B.R.

1976:   Stratigraphic relationships of the Malbaie Formation (Devonian), Gaspé, Quebec; Canadian Journal of Earth Sciences, v. 13, p. 1556-1559.

Rust, B.R. (cont.)

1981:   Alluvial deposits and tectonic style: Devonian and Carboniferous successions in eastern Gaspé; in Sedimentation and Tectonics in Alluvial Basins, (ed.) A.D. Miall; Geological Association of Canada, Special Paper No. 23, p. 49-76.

1984a: Proximal braidplain deposits in the Middle Devonian Malbaie Formation of eastern Gaspé, Quebec, Canada; Sedimentology, v. 31, p. 645-695.

1984b: The Cannes de Roche Formation: Carboniferous alluvial deposits in eastern Gaspé; in Atlantic Coast Basins, (ed.) H.H.J. Geldsetzer; Canada 9th International Congress on Carboniferous Stratigraphy and Geology, Compte Rendu, v. 3, p. 13-22.

1985:   The Upper Carboniferous strata of the Tatamagouche Syncline, Cumberland Basin, Nova Scotia; in Current Research, Part B; Geological Survey of Canada, Paper 85-1B, p. 481-490.

Rust, B.R. and Gibling, M.R.

1990a: Braidplain evolution in the Pennsylvanian South Bar Formation, Sydney Basin, Nova Scotia, Canada; Journal of Sedimentary Petrology, v. 60, p. 59-72.

1990b: Three-dimensional antidunes as HCS mimics in a fluvial sandstone: the Pennsylvanian South Bar Formation near Sydney, Nova Scotia; Journal of Sedimentary Petrology, v. 60, p. 540-548.

Rust, B.R. and Nanson, G.C.

1989:   Bedload transport of mud as pedogenic aggregates in modern and ancient rivers; Sedimentology, v. 36, p. 291-306.

Rust, B.R., Gibling, M.R., and Legun, A.S.

1984:   Coal deposition in an ananastomosing-fluvial system: the Pennsylvanian Cumberland Group, south of Joggins, Nova Scotia, Canada; in Sedimentology of Coal and Coal-bearing Sequences, (ed.) R.A. Rahmani and R.M. Flores; International Association of Sedimentologists, Special Publication 7, p. 105-120.

Rust, B.R., Lawrence, D.A., and Zaitlin, B.A.

1989:   The sedimentology and tectonic significance of Devonian and Carboniferous terrestrial successions in Gaspe, Quebec; Atlantic Geology, v. 25, p. 1-13.

Rust, B.R., Gibling, M.R., Best, M.A., Dilles, S.J., and Masson, A.G.

1987:   A sedimentological overview of the coal-bearing Morien Group (Pennsylvanian), Sydney Basin, Nova Scotia, Canada; Canadian Journal of Earth Sciences, v. 24, p. 1869-1885.

Ryan, R.J.

1985:   The Upper Carboniferous strata of the Tatamagouche Syncline, Cumberland Basin, Nova Scotia; in Current Research, Part B; Geological Survey of Canada, Paper 85-1B, p. 481-490.

Ryan, R.J. and Boehner, R.C.

1990:   Nova Scotia Department of Mines and Energy, Cumberland Basin Geology Map, Map 90-14, Tatamagouche and Malagash.

Ryan, R.J. and Zentill, M.,

1993:   Allocyclsc and thermo chronological constraints on the evolution of the Maritimes Basin of eastern Canada; Atlantic Geology, v. 29, p. 187-197.

Ryan, R.J., Boehner, R.C., and Calder, J.H.

1991a: Lithostratigraphic revisions of the upper Carboniferous to lower Permian strata in the Cumberland Basin, Nova Scotia and the regional implications for the Maritimes Basin; Bulletin of Canadian Petroleum Geology, v. 39, p. 289-314.

Ryan, R.J., Boehner, R.C., and Deal, A.

1990a: Cumberland Basin Geology Map and Apple River and Cape Chignecto Map; Nova Scotia Department of Mines and Energy, Oxford and Pugwash, Maps 90-13, 90-11.

Ryan, R.J., Grist, A., and Zentilli, M.

1991b: The thermal evolution of the Maritimes Basin: evidence from apatite fission track analysis and organic maturation data (abstract); Atlantic Geology, v. 27, p. 162.

Ryan, R.J., Boehner, R.C., Deal, A. and Calder, J.H.

1990b: Cumberland Basin Geology Map, Amherst, Springhill and Parsborough; Nova Scotia Department of Mines and Energy, Map 90-12.

Ryan, R.J., Calder, J.H., Donohoe, H.V., Jr., and Naylor, R.

1987:   Late Paleozoic sedimentation and basin development adjacent to the Cobequid Highlands Massif, eastern Canada; in Sedimentary Basins and Basin-forming Mechanisms, (ed.) C. Beaumont and A.J. Tankard; Canadian Society of Petroleum Geologists, Memoir 12, p. 311-317.

St. Peter, C.
1979:   Geology of the Wapske-Odell River-Arthurette region, New Brunswick, map areas I-13, I-14, H-14 (Parts of 21 J/11, 21 J/12, 21 J/13, 21 J/14); New Brunswick Department of Natural Resources, Geological Surveys Branch, Map Report 79-2, 32 p.
1982a: Geology of Juniper-Knowlesville-Carlisle area, Map areas I-16, I-17, I-18, parts of 21J/11 and 21J/06; New Brunswick Department of Natural Resources, Geological Surveys Branch, Map Report 82-1, 82 p.
1982b: Geology of the Albert Formation, New Brunswick, Canada; 1982 Eastern Oil Shale Symposium; Kentucky Department of Energy and University of Kentucky, Lexington, Kentucky, p. 39-47.
1986:   Oil shale investigations; in Eleventh Annual Review of Activities, Project Résumés, (ed.) S.A. Abbott; New Brunswick Department of Natural Resources and Energy, Mineral Resources Division, Information Circular, 86-2, p. 94-96.
1987:   Geotectonic evolution of the late Paleozoic Maritimes Basin; New Brunswick Department of Natural Resources and Energy, Minerals and Energy Division, Open File Report 87-14, 142 p.
1988:   Horton Group Facies in the Moncton Subbasin; in Thirteenth Annual Review of Activities, Project Résumés, (ed.) S.A. Abbott; New Brunswick Department of Natural Resources and Energy, Minerals and Energy Division, Information Circular 88-2, p. 125-128.
1992:   Lithologic facies, seismic facies, and strike-slip setting of the Lower Carboniferous alluvial/fluvial/lacustrine Albert Formation of New Brunswick; M.Sc. theses, University of New Brunswick, Fredericton, New Brunswick, 214 p.
1993:   Maritimes Basin evolution: key geologic and seismic evidence from the Moncton Subbasin of New Brunswick, Atlantic Geology, v. 29, p. 233-270.

St. Peter, C. and Fyffe, L.R.
1990:   Structural trends and basement rock subdivisions in the western Gulf of St. Lawrence: Discussion; Atlantic Geology, v. 26, p. 277.

Salas, C.J.
1986:   Braided fluvial architecture within a rapidly subsiding basin: the Pennsylvanian Cumberland Group, southwest of Sand River, Nova Scotia; M.Sc. thesis, University of Ottawa, Ottawa, Ontario, 300 p.

Sanford, B.V. and Grant, A.C.
1990:   Bedrock geological mapping and basin studies in the Gulf of St. Lawrence; in Current Research, Part B; Geological Survey of Canada, Paper 90-1B, p. 33-42.

Schenk, P.E.
1967a: The Macumber Formation of the Maritime Provinces, Canada — a Mississippian analogue to Recent standline carbonates of the Persian Gulf; Journal of Sedimentary Petrology, v. 37, p. 365-376.
1967b: The significance of algal stromatolites to palaeoenvironmental and chronostratigraphic interpretations of the Windsorian stage (Mississippian), Maritime Provinces; in Geology of the Atlantic Region, (ed.) E.R.W. Neale and H. Williams; Geological Association of Canada, Special Paper No. 4, p. 229-243.
1969:   Carbonate-sulphate-redbed facies and cyclic sedimentation of the Windsorian stage (Middle Carboniferous), Maritime Provinces; Canadian Journal of Earth Sciences, v. 6, p. 1037-1066.
1981:   The Meguma Zone of Nova Scotia-a remnant of western Europe, South America or Africa?; in Geology of the Atlantic Borderlands, (ed.) J.W. Kerr and A.J. Fergusson; Canadian Society of Petroleum Geologists, Memoir 7, p. 119-148.
1984:   Carbonate-sulfate relations in the Windsor Group, central Nova Scotia, Canada; in Atlantic Coast Basins, (ed.) H.H.J. Geldsetzer; 9th International Congress on Stratigraphy and Geology of the Carboniferous, Compte Rendu, v. 3, p. 143-162.

Schenk, P.E. and Hatt, B.L.
1984:   Depositional environment of the Gays River Reef, Nova Scotia, Canada; 9th International Congress on Carboniferous Stratigraphy and Geology, Compte Rendu, v. 3. p. 117-130.

Schenk, P.E., Matsumoto, R.R., and von Bitter, P.H.
1990:   Deep-water, bacterial precipitated peloidal laminite containing chemosynthetic, microbial mounds, Carboniferous, Atlantic Canada; in Geological Association of Canada, Minerlogical Association of Canada, Program with Abstracts, v. 15, p. A117-118.

Schluger, P.R.
1973:   Stratigraphy and sedimentary environments of the Devonian Perry Formation, New Brunswick, Canada and Maine, United States of America; Geological Society of America Bulletin, v. 84, p. 2533-2548.
1976:   Petrology and origin of the redbeds of the Perry Formation, New Brunswick, Canada and Maine, United States of America; Journal of Sedimentary Petrology, v. 46, p. 22-37.

Schutter, S.R. and Heckel, P.H.
1985:   Missourian (Early Late Pennsylvanian) climate in midcontinent North America; International Journal of Coal Geology, v. 5, p. 111-140.

Scotese, C.R. and McKerrow, W.S.
1990:   Revised world maps and introduction; in Palaeozoic palaeogeography and biogeography, (ed.) W.S. McKerrow and C.R. Scotese; Geological Society of London, Memoir 12, p. 1-21.

Shaw, W.S.
1951:   Preliminary map, Springhill, Cumberland and Colchester Counties, Nova Scotia; Geological Survey of Canada, Paper 51-11.

Shroder, J.F.
1963:   Stratigraphic and tectonic history of the Moncton Group of non-marine redbeds of New Brunswick, Canada; M.Sc. thesis, University of Massachusetts, Amherst, Massachusetts, 81 p.

Sinclair, W.D., Kooiman, G.J.A., and Martin, D.A.
1988:   Geological setting of granites and related tin deposits in the North Zone, Mount Pleasant, New Brunswick; in Current Research, Part B; Geological Survey of Canada, Paper 88-1B, p. 201-208.

Skehan, J.W., Rast, N.
1984:   Correlation of Carboniferous Tectonostratigraphic zones in Europe and North America; in Atlantic Coast Basins, (ed.) H.H.J. Geldsetzer; 9th International Congress on Carboniferous Stratigraphy and Geology, Compte Rendu 3, p. 5-12.

Skehan, J.W., Rast, N., and Mosher, S.
1986:   Paleoenvironmental and tectonic controls of sedimentation in coal-forming basins of southeastern New England; in Paleoenvironmental and Tectonic Controls in Coal-forming Basins of the United States, (ed.) P.C. Lyons and C.L. Rice; Geological Society of America, Special Paper 210, p. 9-30.

Smith, G.G.
1989:   Coal resources of Canada; Geological Survey of Canada, Paper 89-4, 146 p.

Smith, G.O. and White, D.
1905:   Geology of the Perry Basin; United States Geological Survey, Professional Paper 35, 102 p.

Smith, L. and Collins, J.A.
1984:   Unconformities, sedimentary copper mineralization and thrust faulting in the Horton and Windsor groups, Cape Breton Island and Central Nova Scotia; in Atlantic Coast Basins, (ed.) H.H.J. Geldsetzer; Ninth International Congress on Carboniferous Stratigraphy and Geology, Compte Rendu, v. 3, p. 105-116.

Smith, W.D. and Gibling, M.R.
1987:   Oil shale composition related to depositional setting: a case study from Albert Formation, New Brunswick, Canada; Canadian Petroleum Geology Bulletin, v. 35, p. 469-487.

Smith, W.D. and Naylor, R.D.
1990:   Oil shale resources of Nova Scotia; Nova Scotia Department of Mines and Energy, Economic Geology Series 90-3, 73 p.

Smyth, W.R. and Schillereff, H.S.
1982:   The pre-Carboniferous geology of southwest White Bay; Newfoundland Department of Mines and Energy, Mineral Development Division, Report 82-1, p. 78-98.

Solomon, S.M.
1986:   Sedimentology and fossil-fuel potential of the Upper Carboniferous Barachois Group, western Newfoundland; M.Sc. thesis, Memorial University, St. John's, Newfoundland, 256 p.

Solomon, S.M. and Hyde, R.S.
1985:   Stratigraphy and sedimentology of some coal seams in the Carboniferous Bay- St. George Basin, southwestern Newfoundland; Newfoundland Department of Mines and Energy, Mineral Development Division, Report 85-1, p. 168-176.

Stockmal, G.S., Colman-Sadd, S.P., Keen, C.E., Marillier, F., O'Brien, S.J., and Quinlan, G.M.
1990:   Deep seismic structure and plate tectonic evolution of the Canadian Appalachians; Tectonics, v. 9, p. 45-62.

Stopes, M.C.
1914:   The "fern ledges" Carboniferous flora of St. John, New Brunswick; Geological Survey of Canada, Memoir 41, 167 p.

Stringer, P. and Burke, K.B.S.
1985:   Structure in southwest New Brunswick; Geological Association of Canada, Mineralogical Association of Canada, Field Guide for Excursion, University of New Brunswick, 34 p.

Stringer, P., Burke, K.B.S. and Dunn, T.
1991: Stratigraphy, structure and associated igneous rocks of the upper Devonian Perry Formation in the St. Andrews area, southwest New Brunswick and adjacent coastal Maine; in Geology of the Coastal Lithotectonic Block and Neighboring Terranes, Eastern Maine and Southern New Brunswick, (ed.) A. Ludman; New England Intercollegiate Geological Conference, 83rd Annual Meeting Guidebook, p. 222-264.

Strong, D.F., O'Brien, S.J., Taylor, S.W., Strong, P.G., and Wilton, D.H.
1978: Geology of the Marystown (1M/3) and St. Lawrence (1L/14) map areas, Newfoundland; Department of Mines and Energy, Mineral Development Division, Report 77-8, 81 p.

Sund, J.O.
1958: Origin of New Brunswick Gypsum Deposits; M.Sc. thesis, University of New Brunswick, Fredericton, New Brunswick, 122 p.

Swennen, R., Viaene, W., and Cornelissen, C.
1990: Petrography and geochemistry of the Belle Roche breccia (lower Visean, Belgium): evidence for brecciation by evaporite dissolution; Sedimentology, v. 37, p. 859-878.

Tankard, A.J.
1986: Depositional response to foreland deformation in the Carboniferous of eastern Kentucky; American Association of Petroleum Geologists Bulletin, v. 70, p. 853-868.

Thibaudeau, S.A. and Medioli, F.S.
1986: Carboniferous thecamoebians and marsh foraminifera: new stratigraphic tools for ancient paralic deposits (abstract); Abstract with Programs; Geological Society of America, Annual Meeting, San Antonio, Texas, p. 771.

Trewin, N.H.
1976: Isopodichnus in a trace fossil assemblage from the Old Red Sandstone; Lethaia, v. 9, p. 29-37.

Tuach, J.
1987: Mineralized environments, metallogenesis, and the Doucers Valley fault complex, western White Bay: a philosophy for gold exploration in Newfoundland; Newfoundland Department of Mines and Energy, Mineral Development Division, Report 87-1, p. 129-144.

Utting, J.
1966: Geology of the Codroy Valley, southwestern Newfoundland, including results of a preliminary palynological investigation; M.Sc. thesis, Memorial University, St. John's, Newfoundland, 92 p.
1978: Palynological investigation of the Windsor Group (Mississippian) of Port Hood Island and other localities on Cape Breton Island, Nova Scotia; in Current Research, Part A; Geological Survey of Canada, Paper 78-1A, p. 205-207.
1987a: Palynostratigraphic investigation of the Albert Formation (Lower Carboniferous) of New Brunswick, Canada; Palynology, v. 11, p. 75-98.
1987b: Palynology of the Lower Carboniferous Windsor Group and Windsor-Canso boundary beds of Nova Scotia, and their equivalents in Quebec, New Brunswick and Newfoundland; Geological Survey of Canada, Bulletin 374, 93 p.

Utting, J., Keppie, J.D., and Giles, P.S.
1989: Palynology and stratigraphy of the Lower Carboniferous Horton Group, Nova Scotia; in Contributions to Canadian Paleontology; Geological Survey of Canada, Bulletin 396, p. 117-143.

van de Fliert, J.R., Graven, H., Hermes, J.J., and de Smet, M.E.M.
1980: On stratigraphic anomalies associated with major transcurrent faulting; Ecologae Geologicae Helvetiae, v. 73, p. 223-237.

van de Poll, H.W.
1966: Sedimentation and paleocurrents during Pennsylvanian time in the Moncton Basin, New Brunswick; New Brunswick Department of Natural Resources, Mines Division, Report of Investigation 1, 33 p.
1967: Carboniferous volcanic and sedimentary rocks of the Mount Pleasant area, New Brunswick; New Brunswick Department of Natural Resources, Mineral Resources Branch, Report of Investigation 3, 52 p.
1970: Stratigraphic and sedimentological aspects of Pennsylvanian strata in southern New Brunswick; Ph.D. thesis, University of Wales, 140 p.
1972: Stratigraphy and economic geology of Carboniferous Basins in the Maritime Provinces; XXIV International Geological Congress; Excursion A-60 Guidebook, 96 p.
1973a: Stratigraphy, sediment dispersal and facies analysis of the Pennsylvanian Pictou Group in New Brunswick; Maritime Sediments, v. 9, p. 72-77.
1973b: Feasibility report concerning the hydrocarbon and coal potential of the Fredericton area; New Brunswick Department of Natural Resources, Mineral Resources Branch Report, 35 p.

van de Poll, H.W. (cont.)
1973c: Carboniferous stratigraphy and sedimentology of the Chignecto Bay Area, southern New Brunswick; in 1973 Field Guide to Excursions, (ed.) N. Rast; New England Intercollegiate Geological Conferenc (N.E.I.G.C.), p. 25-33.
1978: Paleoclimatic control and stratigraphic limits of syn sedimentary mineral occurrences in Mississippian-Early Pennsylvanian strata of eastern Canada; Economic Geology, v. 73, p. 1069-1081.
1983: The Upper Carboniferous Clifton Formation of Northern New Brunswick: coal-bearing deposits of a semi-arid alluvial plain (Discussion); Canadian Journal of Earth Sciences, v. 20, p. 1212-1215.
1989: Lithostratigraphy of the Prince Edward Island redbeds; Atlantic Geology, v. 25, p. 23-35.

van de Poll, H.W. and Forbes, W.H.
1984: On the lithostratigraphy, sedimentology, structure and paleobotany of the Stephanian-Permian redbeds of Prince Edward Island; in Atlantic Coast Basins, (ed.) H.H.J. Geldsetzer; 9th International Congress on Carboniferous Stratigraphy and Geology, Compte Rendu, v. 3, p. 47-60.

van de Poll, H.W. and Patel, I.M.
1981: Flute casts and related structures on moulded silt injection surfaces in continental sandstone of the Boss Point Formation, southeastern New Brunswick, Canada; Maritime Sediments and Atlantic Geology, v. 17, p. 1-22.
1989: Slump blocks, intraformational conglomerates and associated erosional structures in Pennsylvanian fluvial strata of eastern Canada (Discussion); Sedimentology, v. 36, p. 137-150.
1990: An ornamented mudstone cavity, Boss Point Formation, Sackville, New Brunswick, Canada: evidence for mud intrusion and rheoplasis; Sedimentology, v. 37, p. 931-942.

van de Poll, H.W. and Plint, A.G.
1983: Secondary sedimentary structures associated with fluidization zones in Permo-Carboniferous redbeds of Prince Edward Island, Canada; Maritimes Sediments and Atlantic Geology, v. 19, p. 49-58.

van de Poll, H.W. and Ryan, R.J.
1985: Lithostratigraphic, physical diagenetic and economic aspects of the Pennsylvanian-Permian transition sequence of Prince Edward Island and Nova Scotia; Geological Association of Canada/ Mineralogical Association of Canada, Field Guide for Excursion 14, University of New Brunswick, 109 p.

van de Poll, H.W. and Sutherland, J.K.
1976: Cupriferous reduction spheres in Upper Mississippian redbeds of the Hopewell Group at Dorchester Cape, New Brunswick; Canadian Journal of Earth Sciences, v. 13, p. 781-789.

Varma, C.P.
1969: Lower Carboniferous miospores from the Albert oil shales (Horton Group) of New Brunswick, Canada; Micropaleontology, v. 15, p. 301-324.

Veevers, J.J. and Powell, C. McA.
1987: Late Paleozoic glacial episodes in Gondwanaland reflected in transgressive-regressive depositonal sequences in Euramerica; Geological Society of America Bulletin, v. 98, p. 475-487.

Venugopal, D.V.
1982: Geology of upper parts Begaguimec, Keswick and Nashwaak Rivers, Cloverdale-Millville map areas I-19, J-19, J-20 (Parts of 21J/3 and 21J/6); New Brunswick Department of Natural Resources, Geological Surveys Branch, Map Report 82-2, 35 p.

Vézina, D.
1991: Nouvelles observations sur l'environment sédimentaire de la Formation d'Escuminac (Dévonien supérieur, Frasnien), Québec, Canada; Revue canadienne des sciences de la Terre, v. 28, p. 225-230.

von Bitter, P.H. and Plint-Geberl, H.A.
1982: Conodont biostratigraphy of the Codroy Group (Lower Carboniferous), southwestern Newfoundland, Canada; Canadian Journal of Earth Sciences, v. 19, p. 193-221.

Wade, J.A., Grant, A.C., Sanford, B.V., and Barss, M.S.
1977: Basement structure, eastern Canada and adjacent areas; Geological Survey of Canada, Map 1400 A, scale 1:2 000 000.

Waldron, J.W.F., Piper, D.J.W., and Pe-Piper, G.
1989: Deformation of the Cape Chignecto Pluton, Cobequid Highlands, Nova Scotia: thrusting at the Meguma-Avalon boundary; Atlantic Geology, v. 25, p. 51-62.

Wanless, R.K., Stevens, R.D., Lachance, G.R., and Edmonds, C.M.
1967: Age determinations and geological studies, K-Ar isotopic ages (Report 7); Geological Survey of Canada, Paper 66-17, p. 104.

Waterfield, J.J.
1986: Sedimentology and stratigraphy of Canso and Riversdale fluvial strata at Broad Cove, Cape Breton, Nova Scotia; B.Sc. Honours thesis, Dalhousie University, Halifax, Nova Scotia, 99 p.

Webb, G.W.
1963: Occurrence and exploration significance of strike slip faults in southern New Brunswick, Canada; American Association of Petroleum Geologists Bulletin, v. 47, p. 1904-1927.
1969: Paleozoic wrench faults in Canadian Appalachians; in North Atlantic Geology and Continental Drift, (ed.) M. Kay; American Association of Petroleum Geologists Memoir 12, p. 754-786.

Webb, T.C.
1976: Limestone resources of the Havelock area, Kings and Westmorland counties, New Brunswick; New Brunswick Department of Natural Resources, Mineral Resources Division, Topical Report 76-12, 119 p.

Webb, T.C. and Kingston, P.W.
1975: Limestone and gypsum resources of Plaster Rock, Victoria County; New Brunswick Department of Natural Resources, Mineral Resources Branch, Topical Report 74-14, 13 p.

Weeks, L.J.
1948: Londonderry and Bass River map-areas, Colchester and Hants counties, Nova Scotia; Geological Survey of Canada, Memoir 245, 86 p.
1954: Southeast Cape Breton Island, Nova Scotia; Geological Survey of Canada, Memoir 277, 112 p.

Werner, H.J.
1956: The geology of Humber Valley, Newfoundland; Ph.D. thesis, Syracuse University, Syracuse, New York, 111 p.

Wilcox, R.E., Harding, T.P., and Seely, D.R.
1973: Basic wrench tectonics; American Association of Petroleum Geologists Bulletin, v. 57, p. 74-96.

Williams, E.P.
1974: Geology and petroleum possibilities in and around Gulf of St. Lawrence; American Association of Petroleum Geologists Bulletin, v. 58, p. 1137-1155.

Williams, G.L., Fyffe, L.R., Wardle, R.J., Colman-Sadd, S.P., and Boehner, R.C.
1985: Lexicon of Canadian Stratigraphy, Volume VI Atlantic Region; Canadian Society of Petroleum Geologists, Calgary, 572 p.

Williams, H.
1978: Tectonic lithofacies map of the Appalachian Orogen; Memorial University of Newfoundland, St. John's, Newfoundland, Map no. 1a, scale 1:2 000 000.

Williams, P.F. and Hy, C.
1990: Origin and deformational and metamorphic history of gold bearing quartz veins on the eastern shore of Nova Scotia; in Mineral Deposit Studies in Nova Scotia; (ed.) Al Sangster; Geological Survey of Canada, Paper 90-8, v. 1, p. 169-194.

Wilson, J.T.
1962: Cabot Fault, an Appalachian equivalent of the San Andreas and Great Glen Faults and some implications for continental displacement; Nature, v. 195, p. 135-138.

Wilson, L.M.
1981: Circum-North Atlantic Tectono-stratigraphic reconstruction; in Geology of the North Atlantic Borderlands, (ed.) J.W. Kerr and A.J. Fergusson; Canadian Society of Petroleum Geologists, Memoir 7, p. 167-184.

Wilson, R.A. and Ball, F.D.
1983: Carboniferous compilation (Second Edition), v. IV: Uranium and Base Metals; New Brunswick Department of Natural Resources, Mineral Resources Division, Topical Report 75-22 (revised 1983), 140 p.

Worth, J.K.
1972: Final report on drilling, Sussex area, New Brunswick (1971); New Brunswick Department of Natural Resources, Mineral Resources Branch, Topical Report 72-3, 108 p.

Yeo, G.M.
1985: Upper Carboniferous sedimentation in northern Nova Scotia and the origin of Stellarton Basin; in Current Research, Part B; Geological Survey of Canada, Paper 85-1B, p. 511-518.

Yeo, G.M. and Ruixiang, G.
1987: Stellarton Graben: an Upper Carboniferous pull-apart basin in northern Nova Scotia; in Sedimentary Basins and Basin-forming Mechanisms, (ed.) C. Beaumont and A.J. Tankard; Canadian Society of Petroleum Geologists, Memoir 12, p. 299-309.

Zaitlin, B.A. and Rust, B.R.
1983: A spectrum of alluvial deposits in the Lower Carboniferous Bonaventure Formation of western Chaleur Bay area, Gaspé and New Brunswick, Canada; Canadian Journal of Earth Sciences, v. 20, p. 1098-1110.

Zodrow, E.L. and Cleal, C.J.
1985: Phyto-and chronostratigraphical correlations between the late Pennsylvanian Morien Group (Sydney, Nova Scotia) and the Silesian Pennant Measures (South Wales); Canadian Journal of Earth Sciences, v. 22, p. 1465-1473.

ADDENDUM

Upper Paleozoic Rocks, Newfoundland

Based on an apatite fission-track study of granitic and gneissic clasts from conglomerates in the Deer Lake Basin, Hendricks et al. (1993) suggested minimum burial temperatures of about 120°C. Based on vitrinite reflectance measurements, clay mineral assemblages and little crystallinity measurements Hyde et al. (1988) suggested burial temperatures of about 100°C for the Rocky Brook Formation and about 200° C for the Anguille Group.

Hamblin et al. (in press) confirmed a Tournaisian age for strata exposed as an outlier in the Conche area. These authors also established facies and facies associations and provided additional evidence for nonmarine deposition. They also located, sampled, and geochemically analyzed oil seeps in the Crouse Harbour Formation.

Hendricks, H., Jamison, R.A., Willett, S.D., and Zentilli, M.
1993: Burial and exhumation of the Long Range Inlier and its surroundings, western Newfoundland; results of an apatite fission - track study; Canadian Journal of Earth Sciences, v. 30, p.1594-1006.

Hamblin, A.P., Fowler, M.G., Utting, J.P., Hawkins, D., and Riediger, C.L.
in press: Sedimentology, palynology and source rock potential of Lower Carboniferous (Tournaisian) rocks, Conche area, Great Northern Peninsula, Newfoundland; Bulletin of Canadian Petroleum Geology.

Authors' Addresses

M.R. Gibling
Department of Geology
Dalhousie University
Halifax, Nova Scotia
B3H 3J5

R.S. Hyde
3115 Allan Rd.
North Vancouver
British Columbia
V7J 3C3

H.W. van de Poll
Department of Geology
University of New Brunswick
Box 4400
Fredericton, New Brunswick
E3B 5A3

Printed in Canada

Chapter 6

MESOZOIC ROCKS

Chapter 6

MESOZOIC ROCKS

John D. Greenough

INTRODUCTION

Mesozoic rocks of Atlantic Canada (New Brunswick, Nova Scotia, Prince Edward Island, and Newfoundland) are mainly red continental clastic rocks, tholeiitic basalts, and mafic dykes of Triassic and Early Jurassic age. They occur throughout the Bay of Fundy area and locally at Chedabucto Bay (Fig. 6.1), and form sequences up to 3500 m thick in faulted graben; the Fundy Graben and Chedabucto Graben (formerly the Cape Split Trough of Poole et al., 1970). Mafic dykes are more extensive and occur beyond the boundaries of the graben; (Fig. 6.2). Cretaceous rocks include small alkaline intrusions at Notre Dame Bay, Newfoundland and Ford's Bight, Labrador as well as local unconsolidated clay and sand deposits in central Nova Scotia (Fig. 6.1, 6.2) and minor breccias at Ford's Bight.

The Mesozoic rocks are related to the early stages of rifting and drifting that led to the opening of the Atlantic Ocean, and they overlie a wide variety of Paleozoic rocks related to the Appalachian or Iapetus cycle of orogenesis. The Fundy Graben is typical of Newark-type graben represented along the length of the Appalachian Orogen as well as offshore (Williams, 1978a). It is bounded by faults and sited, in part, above a deep Carboniferous basin (Minas Basin), in turn parallel to the major Avalon-Meguma zone boundary.

Over 150 years have passed since publication of the first report (Jackson and Alger, 1829) on Mesozoic rocks of Atlantic Canada. Many scientists have worked on these rocks over the years but the comprehensive observations of Powers (1916) serve as a useful reference to this day. Current thinking as to the events that formed the Mesozoic rocks may be summarized as follows. During the Permian, parts of Pangea that included the present southeastern United States and eastern Canada, as well as western North America, were uplifted causing widespread erosion (Manspeizer et al., 1978). Associated with this stage of domal uplift were small amounts of predominantly alkaline volcanism (McHone and Butler, 1984) represented in Atlantic Canada by the Malpeque Bay sill (Greenough et al., 1988, Fig. 6.2) and diabase in the exploratory oil well Northumberland Strait F-25 (Pe-Piper and Jansa, 1986). A

period of lithospheric extension and collapse ensued during the Middle to Late Triassic and formed many graben, one of the most northerly being the Fundy Graben. These graben or half-graben structures resulted from reactivation of major Paleozoic faults (Swanson, 1986) such as the Cobequid Fault of Nova Scotia. They were filled with continental sediments (Klein, 1962; Hubert and Hyde, 1982; Nadon and Middleton, 1985) but marine and restricted marine conditions prevailed in graben closest to the incipient rift axis by Early Jurassic time (Jansa and Wade, 1974). Large volumes of tholeiitic magma formed extensive dyke swarms (Caraquet, Minister Island, Shelburne, Anticosti Island, and Avalon dykes in Atlantic Canada) and flood basalt sequences (North Mountain Basalt) during the Early Jurassic (Dostal and Dupuy, 1984; Greenough and Papezik, 1986; Papezik et al., 1988; Bédard, 1992). The volcanism and onland basin subsidence ceased abruptly by the Middle Jurassic when North Africa separated from eastern North America (Olsen and Galton, 1977; McHone and Butler, 1984), with the rifted margin parallel to Paleozoic structural trends of the Appalachians (Williams, 1984). Newfoundland and Labrador remained attached to Europe and Greenland until some 60 to 70 Ma later when further rifting led to their separation. This later cycle of extension produced small volumes of alkaline volcanic rocks (Strong and Harris, 1974; Jansa and Pe-Piper, 1985), affected a narrower band of continental crust, and produced a rift axis that cut across Paleozoic structural trends (Williams, 1978b, 1984).

Only the generalities of the rifting event are known and the underlying tectonic processes remain poorly understood. Perusal of review articles on the geology of Atlantic Canada shows that the onshore Mesozoic rocks were largely ignored. However, recent realization that they may hold valuable oil reserves has spawned a flurry of research activity. Academics have also responded, noting that a complete understanding of rifting events recorded in ancient rocks requires a knowledge of margins that have not gone through a complete Wilson (1966) cycle (Haworth and Keen, 1979). The heightened research activity has already provided exciting new discoveries in such diverse fields of study as the origin of magmas (Dostal and Dupuy, 1984) and the evolution of life (Olsen et al., 1987).

This chapter reviews research on the Mesozoic rocks of Atlantic Canada, emphasizing the onshore rocks. Offshore equivalents are described in a companion volume (Keen and Williams, 1990). Igneous and sedimentary rocks together with associated structures are briefly compared with Mesozoic correlatives elsewhere in eastern North America and placed in a plate tectonic framework. Mention

Greenough, J.D.
1995: Mesozoic rocks; Chapter 6 in Geology of the Appalachian-Caledonian Orogen in Canada and Greenland, (ed.) H. Williams; Geological Survey of Canada, Geology of Canada, no. 6, p. 567-600 (also Geological Society of America, The Geology of North America, v. F-1).

of some current research projects serves to outline the direction of new investigations, and areas in need of study are pointed out.

STRATIGRAPHY AND AGE

Sedimentary rocks

Distribution and nomenclature

Mesozoic sedimentary rocks occur in Nova Scotia, New Brunswick, and Labrador but have not been identified in either Prince Edward Island or on the island of Newfoundland. Most exposures occur in scattered and highly faulted sections around the shores of the Bay of Fundy, with isolated outcrops at Chedabucto Bay (Fig. 6.1a, b, f), and the rocks are Triassic to Early Jurassic in age. The Fundy Graben exposures mark the periphery of a major synclinal structure, with an axis in the centre of the Bay of Fundy. Geophysical surveys indicate that Mesozoic strata underlie much of the Bay of Fundy and Gulf of Maine (Swift and Lyall, 1968; Ballard and Uchupi, 1975). Similarly, seismic information suggests that the Triassic rocks of the Chedabucto Graben, Nova Scotia, can be traced offshore into the Orpheus Graben (Jansa and Wade, 1974). Small isolated outcrops of Cretaceous deposits occur outside the Fundy and Chedabucto graben in central Nova Scotia, in southwestern Cape Breton Island (Fig. 6.1b, g), and in Labrador.

In this and subsequent sections on sedimentary rocks, the Triassic-Lower Jurassic rocks representing Klein's (1962) Fundy Group are discussed first, followed by Cretaceous deposits. Because the Triassic-Lower Jurassic rocks are much more voluminous and have received more attention, they are described separately for Nova Scotia and New Brunswick.

Table 6.1 summarizes Mesozoic formation names in Atlantic Canada. The first account of Mesozoic rocks in Nova Scotia was given by Jackson and Alger (1829). Lithostratigraphic names presently applied to sedimentary rocks of the Fundy Graben in Nova Scotia are essentially those proposed by Powers (1916), except that Klein (1962) discarded the term Annapolis Formation and raised its Wolfville and Blomidon members to formation status. The following sequence unconformably overlies lower and upper Paleozoic rocks (from bottom to top): Wolfville Formation, Blomidon Formation, North Mountain Basalt, and Scots Bay Formation. Keppie (1979) used the name "McCoy Brook Formation" for redbeds lithologically distinct from the Scots Bay Formation but overlying North Mountain Basalt on the north side of Minas Basin. Triassic rocks at Chedabucto Bay were assigned to the Chedabucto Formation (Klein, 1962).

In New Brunswick, Mesozoic sedimentary rocks occur on Grand Manan Island (Fig. 6.1d) and at Point Lepreau, Saint John Harbour, St. Martins, Martin Head, and Rocher Bay (Fig. 6.1c), the most important location, and most studied being St. Martins. All rocks occur close to the northwest faulted margin of the Fundy Graben.

In a publication providing the first measured section of the Triassic at St. Martins, Powers (1916) divided the sedimentary rocks into three informal units: lower redbeds, the

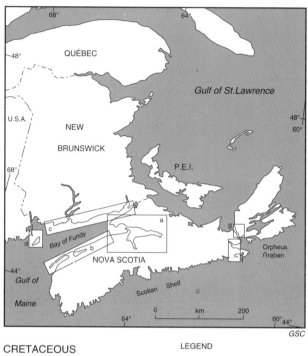

Figure 6.1. Location maps and stratigraphic sections for the Lower Mesozoic sedimentary rocks of Atlantic Canada. Information compiled from Sarjeant and Stringer (1978), Keppie (1979), Potter et al. (1979), Olsen (1981), Donohoe and Wallace (1982), Hubert and Mertz (1984), and Nadon and Middleton (1985).

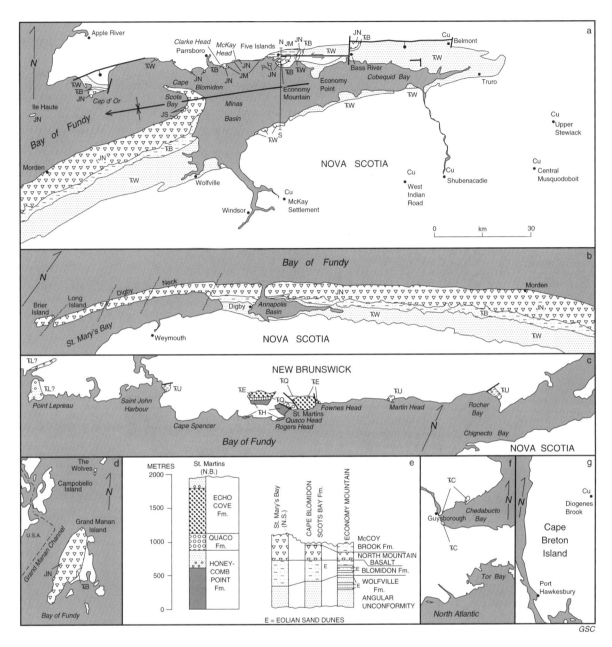

Figure 6.1. (cont.)

Quaco conglomerate, and upper redbeds. Various workers redefined and debated the extent of these units (e.g. Hayes and Howell, 1937; Alcock, 1938; Magnusson, 1955; Klein, 1962), with the latest comprehensive work by Nadon and Middleton (1985) suggesting division of the local New Brunswick section into three formal units, which are from bottom to top the Honeycomb Point, Quaco, and Echo Cove formations (Fig. 6.1c, e).

Two separate clastic rock sections in the Point Lepreau area (Fig. 6.1c) show lithological similarities with the Triassic rocks 150 km to the east at St. Martins and were regarded by Klein (1962) as time-equivalent. Belyea (1939) assigned these rocks to the Point Lepreau Formation, subsequently named the Lepreau Formation (Wright and Clements, 1943). The more northerly rocks are in part

Carboniferous (Stringer and Burke, 1985) and the Triassic age ascribed to southerly rocks at Point Lepreau has not yet been confirmed.

Cretaceous deposits of central Nova Scotia and in southwestern Cape Breton Island (Fig. 6.1a, g) have not been given formal names (Dickie, 1987). King and McMillan (1975) assigned the name Ford's Bight Breccia to three small isolated outcrops of breccia at Ford's Bight, Labrador (Fig. 6.2).

Triassic-Lower Jurassic of Nova Scotia

The Wolfville Formation (early Middle Triassic to Late Triassic) lies unconformably upon mainly Pennsylvanian sedimentary rocks on the north side of Minas Basin,

571

Mississippian sedimentary rocks on the south side of Minas Basin, and pre-Mississippian metamorphic rocks and Devonian granites along North Mountain. Exposed thicknesses range from 60 to 750 m but the unit is apparently thicker in the centre of the Bay of Fundy (Fig. 6.3d). The rocks comprise a red sequence of coarse breccias, conglomerates, mudstones, and poorly sorted and well-sorted sandstones (e.g. Powers, 1916; Klein, 1962; Hubert and Mertz, 1980, 1984).

The Wolfville Formation grades laterally and vertically into the dominantly Upper Triassic Blomidon Formation, which has exposed thicknesses from 7 to 370 m (Fig. 6.3c).

Rock types in the latter include planar, crossbedded, and crosslaminated sandstones together with horizontal and crosslaminated siltstones, mudstones, and claystones.

The Lower Jurassic North Mountain Basalt, locally up to 400 m thick, overlies the Blomidon Formation. Stratigraphy and distribution of the basalts are discussed with the igneous rocks.

At Scots Bay (Fig. 6.1a), several isolated small outcrops consisting of calcareous sandstones, calcareous siltstones, limestones, silicified nodules of limestone (chert and jasper), and silicified tree trunks make up the Lower Jurassic

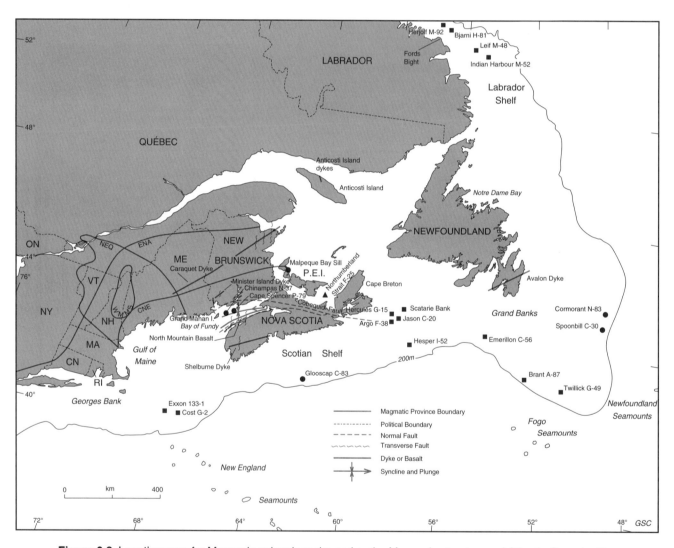

Figure 6.2. Location map for Mesozoic volcanic rocks and major Mesozoic structures of Atlantic Canada. The exploratory oil wells shown contain volcanic rocks belonging to the Permian-Triassic (triangles), Early Jurassic (circles), and Jurassic-Cretaceous (squares) cycles of magmatic activity. The rocks at Ford's Bight and Notre Dame Bay belong to the Jurassic-Cretaceous cycle. The Caraquet, Minister Island, Shelburne, and Avalon dykes as well as North Mountain Basalt formed during the Early Jurassic cycle. The Malpeque Bay Sill formed during the Permian-Triassic cycle. Boundaries for the magmatic provinces of McHone and Butler (1984) are labelled as follows: NEQ, New England-Quebec province; WMMS, White Mountains magmatic series; ENA, Eastern North American dolerite province (boundary location modified by the author); and CNE, Coastal New England province (extended into Canada, Greenough et al., 1988). Data sources for the location of faults, folds, and igneous rocks are given in the text and in Figure 6.1.

Table 6.1. Mesozoic stratigraphy for Atlantic Provinces graben and offshore Atlantic Canada.

Age			Atlantic Provinces			Offshore Atlantic Canada	
			Fundy Graben		Chedabucto Graben[4]		
			Nova Scotia[2]	New Brunswick[3]		Scotian Shelf[5]	Grand Banks[5]
Cretaceous	U	Maastrichtian				Banquereau Fm.	Wyandot Fm.
		Campanian				Wyandot Fm.	
		Santonian				Dawson Canyon Fm.	Dawson Canyon Fm.
		Coniacian					
		Turonian					
		Cenomanian				Logan Canyon Fm.	Logan Canyon Fm.
	L	Albian					
		Aptian	Cretaceous deposits[1]				
		Barremian				Mississauga Fm. / Verrill Canyon	Mississauga Fm. / Verrill Canyon
		Hauterivian					
		Valanginian					
		Berriasian					
Jurassic	U	Tithonian				Mic Mac Fm. / Aben-aki Fm. — Fm.	Mic Mac Fm. / Canyon Fm.
		Kimmeridgian					
		Oxfordian					
	M	Callovian				Mohawk Fm.	
		Bathonian					
		Bajocian					
		Aalenian					
	L	Toarcian				Iroquois Fm.	
		Pliensbachian					Iroquois Fm.
		Sinemurian	McCoy Brook Fm.			Argo Fm.	Argo Fm.
		Hettangian	Scots Bay Fm. / North Mountain Basalt		?		
Triassic	U	Noria	Blomidon Fm.		Chedabucto Fm.	Eurydice Fm. — ? —	Eurydice Fm.
		Carnian	Wolfville Fm. — ?[6]	Echo Cove Fm. — ? — Quaco Fm. — ? —	— ? —		
	M	Ladinian					
		Anisian		Honeycomb Point Fm. — ? —			
		[7]					— ? —
	L	Scythian					

[1.] Cretaceous deposits occur outside the Fundy Graben. Age data from Stevenson and McGregor (1963) and Dickie (1987).
[2.] Age data (except Cretaceous deposits) mainly from Olsen et al. (1987) but see details in section on Age Data.
[3.] Ages and formation names from Nadon and Middleton (1984).
[4.] Age of Chedabucto Formation, Baird in Klein (1962).
[5.] Formation names and ages as summarized in Jansa and Wade (1974) with minor changes in Jansa (1986).
[6.] Question marks indicate age relations not well known.
[7.] Dashed lines indicate unconformity.

Scots Bay Formation. It is 2 to 7 m thick, with uppermost beds nowhere exposed, and overlies the North Mountain Basalt (Powers, 1916; Klein, 1962; Colwell and Cameron, 1985; Cameron, 1986). Lower Jurassic redbeds of the McCoy Brook Formation (Fig. 6.1a) occur above North Mountain Basalt on the north side of Minas Basin (Hubert and Mertz, 1981, 1984; Olsen, 1981) and have exposed

thicknesses not exceeding 200 m (Fig. 6.3a). They are in part time-equivalent to the Scots Bay Formation but resemble the Blomidon Formation.

Stevenson (1960) first reported the coarse breccias of the Triassic Chedabucto Formation at Chedabucto Bay (Fig. 6.1f). The sequence of breccias exceeds 61 m and the rocks resemble those of the Wolfville Formation on the north shore of Minas Basin (Klein, 1962).

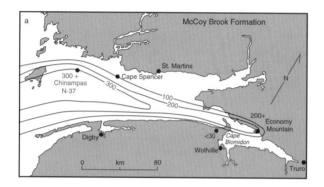

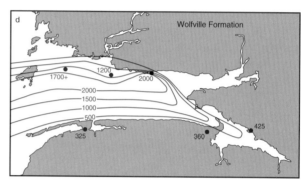

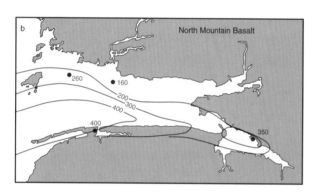

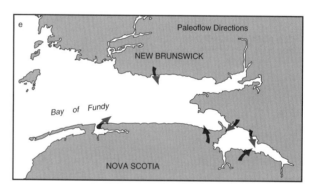

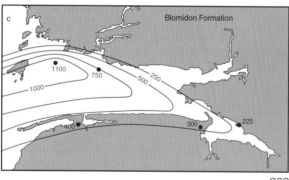

GSC

Figure 6.3. Isopach maps (a-d) and paleoflow directions (e) for Mesozoic rocks of the Fundy Graben. Isopach maps are: **(a)** McCoy Brook Formation and correlative Scots Bay Formation (contour interval, 100 m); **(b)** North Mountain Basalt (contour interval, 100 m); **(c)** Blomidon Formation (contour interval, 250 m); and **(d)** Wolfville Formation including the Honeycomb Point, Quaco, and Echo Cove formations at St. Martins (contour interval, 500 m). Approximate thicknesses are shown for each formation at localities indicated. **Figure 6.3e** summarizes and interprets paleoflow direction data presented in Klein (1963a), Hubert and Hyde (1982), and Nadon and Middleton (1984). Data sources for the isopach maps are as follows: Cape Spencer P-79, Chevron Canada Resources Ltd. (1983); Chinampas N-37, Keown and Bint (1975); Digby, Cape Blomidon, and Economy Mountain, Hubert and Mertz (1980), Olsen (1981), Olsen et al. (1987); St. Martins, Nadon and Middleton (1984). Additional information for North Mountain Basalt from Greenough and Papezik (1987), Papezik et al. (1988), and Greenough et al. (1989a).

Triassic-Lower Jurassic of New Brunswick

Mesozoic rocks at St. Martins occur in several isolated and geologically complicated fault blocks (Fig. 6.1c). The lowermost unit, the Honeycomb Point Formation (~990 m), unconformably overlies Carboniferous sedimentary rocks and consists of red conglomerate breccias and parallel-laminated and trough-crossbedded red sandstones that are finer toward the east. The overlying, conglomeratic Quaco Formation (190-300 m) is lithologically distinct with well-rounded quartzite clasts (Klein, 1962). Above these rocks are red and green breccias of the Echo Cove Formation (850-1300 m) containing numerous clasts of metasedimentary and metaigneous rocks resembling clasts in the Honeycomb Point Formation.

Southerly outcrops of the Lepreau Formation, at Point Lepreau (Fig. 6.1c) consist of about 3000 m of coarse, red arkosic conglomerates, sandstones, and minor shale.

Cretaceous

Lower Cretaceous, unconsolidated, red, grey, and white kaolinitic clays as well as white, unconsolidated, almost pure (95% SiO_2) silica sands occur at Shubenacadie, Upper Stewiac, Central Musquodoboit Valley, West Indian Road, McKay Settlement (all in central Nova Scotia, Fig. 6.1a), Belmont (north of Truro, Fig. 6.1a), and Diogenes Brook (southwestern Cape Breton Island, Fig. 6.1g) (Stevenson and McGregor, 1963; Stea and Fowler, 1981; Dickie, 1987). The clays are economically important sources of pottery clay (Lin, 1971; Dickie, 1987) and the purity and whiteness of the silica sands give them various industrial uses (e.g. manufacturing silicon metal, filtration, production of glass, outdoor decoration, and sand blasting). The sedimentary rocks occur in depressions and sinkholes that have escaped erosion. Coal compaction studies indicate that over 700 m of Cretaceous sedimentary rocks may have covered Nova Scotia (Hacquebard, 1984). Lithologically similar but undated silica sands also occur 25 km north of St. Martins in southern New Brunswick (Lockhart, 1984; Dickie, 1987).

The total area of the three breccia outcrops at Ford's Bight, Labrador is 7 m². The breccias consist of angular to subangular blocks of lamprophyre, felsic intrusive rocks, and metamorphosed mafic igneous rocks, with rounded clasts of amphibolite, arkose, and quartzite. The matrix is fossiliferous and has a lamprophyric-carbonatite component restricted to microfractures. A vertical lamprophyre dyke cuts the breccias. Vitrinite reflectance on carbon from a breccia sample indicates heating to at least 170°C. Pebbles in the breccia largely resemble quartzite, amphibolite, and conglomerate from the underlying Precambrian Allik Group.

Igneous rocks
Distribution and nomenclature

Igneous rocks related to Mesozoic tectonism occur in all four Atlantic Provinces (Fig. 6.2) and fall into three distinct age categories: Late Permian-Early Triassic, Late Triassic-Early Jurassic, and Middle Jurassic-Early Cretaceous. Upper Permian-Lower Triassic rocks are represented by the Malpeque Bay Sill in Prince Edward Island, and mafic rocks in Northumberland Strait F-25 exploratory oil well

(Fig. 6.2). Upper Triassic-Lower Jurassic rocks include the North Mountain Basalt and Shelburne Dyke of Nova Scotia, the Minister Island and Caraquet dykes in New Brunswick, two Anticosti Island dykes and the Avalon Dyke in Newfoundland. Middle Jurassic-Lower Cretaceous rocks include lamprophyric dykes in Labrador and the alkaline Budgell Harbour pluton and surrounding lamprophyre dykes of Notre Dame Bay, Newfoundland. Igneous rocks from each category are described in the following three sections.

Upper Permian-Lower Triassic

The Malpeque Bay Sill, first described by Milligan (1949), is less than 5 m thick and outcrops for about 60 m along the shore of George Island on Malpeque Bay. There, it intrudes red sandstones probably of Permian age (Larochelle, 1967; Prest, 1972). The sill is very fine grained with microphenocrysts of olivine (<Fo_{91}) set in a groundmass of augite, high-Ti phlogopite (8.5 wt.% TiO_2), and finer grained devitrified glass and felsic minerals (Greenough et al., 1988). Spinel lherzolite xenoliths entrained in the sill indicate, along with sill mineral compositions and the presence of phlogopite, that the sill has "lamprophyric" affinities.

A Permian-Triassic diabase from the Northumberland Strait F-25 well occurs within an Upper Devonian section of red and grey clastic rocks (Pe-Piper and Jansa, 1986). It is 130 m thick and is composite with at least three separate phases, each of equal thickness. Lateral extent of the diabase is not known. It has a holocrystalline, ophitic texture with primary plagioclase (An_{44}-An_{72}), clinopyroxene, and opaques. Chemical criteria indicate that the diabase is an olivine tholeiite.

Upper Triassic-Lower Jurassic

Igneous activity reached a volumetric maximum during Early Jurassic time yielding the North Mountain Basalt of Nova Scotia and New Brunswick and mafic dykes that occur in all Atlantic Provinces except Prince Edward Island. The quartz-normative tholeiitic North Mountain Basalt has a total thickness of about 400 m near Digby, 200 m or more in the Wolfville area, and possibly 350 m in the Parrsboro-Five Islands area (Fig. 6.1a, b, 6.3b; Crosby, 1962; J.A. Colwell, pers. comm., 1987; Papezik et al., 1988; Greenough et al., 1989a). Stratigraphy of the flows can be summarized as follows. The lower (about 190 m at Digby) and upper (about 160 m at Digby) units consist of massive, coarse grained, columnar-jointed basalt, each comprising a single flow and traceable along North Mountain to within 30 km of Truro. Numerous thin amygdaloidal flows make up the middle unit, approximately 50 m thick at Digby, and a 75 m thick unit overlying the upper unit in the Parrsboro area. On land there is little evidence for any appreciable sedimentation between flows (≤1 m; Crosby, 1962), but in the exploratory oil well Mobil Chinampas N-37, a 25 m thick sedimentary section is a distinct unit not recognized on land, and represents local, rapid sedimentation between eruption episodes (Greenough and Papezik, 1987). Klein (1962) suggested that basalt at McKay Head represents a separate basalt unit lower in the stratigraphy, but Olsen (1981, 1983) presents paleontological evidence from accompanying sedimentary rocks that supports time equivalence with North Mountain Basalt.

North Mountain Basalt constitutes a major portion of
Grand Manan Island (Fig. 6.1d) and has also been recognized in the exploratory oil well Chinampas N-37 (Fig. 6.2),
implying that the flows underlie most of the Bay of Fundy
(Greenough and Papezik, 1987). The basalts originally
covered an area of at least 9400 km². Given an average
thickness of 250 m, 2350 km³ of basalt must have erupted
(Greenough and Papezik, 1987), most of it in volcanic
episodes that produced two of the largest lava flows on
earth.

Mafic dykes of Late Triassic-Early Jurassic age include
the Caraquet (Burke et al., 1973; Seguin et al., 1981;
Greenough and Papezik, 1986) and Minister Island
(Stringer and Burke, 1985) dykes of New Brunswick, the
Shelburne Dyke (Papezik and Barr, 1981) of Nova Scotia,
two Anticosti Island dykes in Quebec (Bédard, 1992), and
the Avalon Dyke in Newfoundland (Hodych and Hayatsu,
1980; Papezik and Hodych, 1980; Fig. 6.2). A strong magnetic signature characterizes these dykes and allows
delineation: Caraquet, 475 km (Hamilton and Gupta, 1972);
Minister Island, greater than 10 km (Stringer and Burke,
1985); Shelburne, 200 km (Papezik and Barr, 1981, 1982);
and Avalon, 180 km (Papezik et al., 1975). The Anticosti
Island dykes have not been mapped in detail. Ruffman and
Woodside (1970) reported a similar magnetic lineament,
the Odd-twins anomaly, 65 km long in the Gulf of
St. Lawrence off western Newfoundland that could
represent another dyke. Maximum widths for the Atlantic
Provinces dykes range from 180 m for the Shelburne Dyke
to only 6 and 17 m for the two Anticosti Island dykes. The
dykes dip vertically and trend 060°, except for the local 140°
trend of the Anticosti Island dykes (Fig. 6.4a). They are
roughly parallel to the trend of North Mountain and the
inferred feeder dyke to the North Mountain Basalt flows
(Greenough et al., 1989a). The dykes cut rocks of diverse
age and origin. Shelburne Dyke intrudes the Meguma Zone
(Williams, 1979), Minister Island and Avalon dykes cut the
Avalon Zone, Caraquet Dyke cuts the Gander Zone, and
Anticosti Island dykes intrude Ordovician limestones of the
St. Lawrence Platform.

Petrography of the Avalon (Papezik and Hodych, 1980),
Shelburne (Papezik and Barr, 1981; Smith and Huang,
1982), Caraquet (Greenough and Papezik, 1986), and
Anticosti Island (Bédard, 1992) dykes as well as upper and
lower units of the North Mountain Basalt (Stevens, 1980;
Wark and Clarke, 1980; Dostal and Dupuy, 1984; J.A. Colwell,
pers. comm., 1987; Greenough and Papezik, 1987; Papezik
et al., 1988; Dostal and Greenough, 1992; Greenough and
Dostal, 1992a, b; Pe-Piper et al., 1992) can be summarized
as follows. All rocks show diabasic, fine- to coarse-grained,
subophitic to ophitic textures. They typically display
glomeroporphyritic aggregates of augite phenocrysts (or
microphenocrysts) and contain augite in the matrix. Zoned
microphenocrysts or phenocrysts of plagioclase (An₆₀-An₈₀)
occur in all of the rocks with the most calcic compositions
in the Caraquet Dyke. Pigeonite occurs in all of the units,
but olivine phenocrysts have only been reported from the
Avalon, Caraquet, and Anticosti Island dykes. Orthopyroxene phenocrysts only appear in the Shelburne and
Minister Island dykes and the North Mountain Basalt. The
percentage of orthopyroxene phenocrysts in the North
Mountain Basalt decreases toward the northeast. Tachylite

glass is commonly preserved in the basalt. The middle and
overlying units as well as upper portions of the lower and
upper units of the North Mountain Basalt tend to be vesicular to highly vesicular. The thick lower and upper units
display well-developed layering in their upper 40 m. Mafic
pegmatite layers averaging 25 cm thick and containing
2 cm thick rhyolite bands are separated by 130 cm thick
layers of fine grained basalt. The pegmatites grade into
25 cm thick vesicular layers toward the top of the flows.

Zeolite facies to greenschist facies metamorphism
affected both the dykes and North Mountain Basalt. The
middle unit of North Mountain Basalt (Colwell, 1980) has

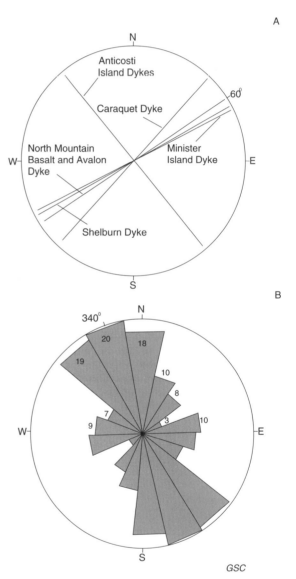

Figure 6.4. Orientations of Mesozoic rocks: **(a)** Orientations
of the five Early Jurassic dykes and North Mountain Basalt;
(b) Orientations of 104 lamprophyre dykes in Notre Dame Bay,
Newfoundland in Strong and Harris (1974). The number of
dykes occurring within each 20 degree interval is shown.

the zeolites stilbite, heulandite, analcite, apophyllite, laumontite, chabazite, natrolite, and less commonly gmelinite, mesolite, mordenite, scolecite, and thomsonite (Walker and Parsons, 1922; Aumento, 1966). The zeolites record maximum temperatures of between 200 and 260° C (Aumento, 1966) and apparently formed from solutions passing along fractures and through highly vesicular portions of the basalt. The massive dykes and lower portions of the thick upper and lower units of North Mountain Basalt are relatively unaffected by the metamorphism (Greenough et al., 1989a, b).

Middle Jurassic-Lower Cretaceous

Middle Jurassic-Cretaceous volcanic activity is represented by lamprophyric rocks at Ford's Bight, Labrador and Notre Dame Bay, north-central Newfoundland (Fig. 6.2). Approximately 15 km to the north-northwest of Ford's Bight at Cape Makkovik, undated lamprophyric dykes studied geochemically by Foley (1984) and Malpas et al. (1986) are probably Jurassic to Early Cretaceous. The rocks at Ford's Bight have not been examined petrographically.

Rocks in Notre Dame Bay (Strong and Harris, 1974) have been examined petrographically in detail. Intrusions at Notre Dame Bay include the Budgell Harbour pluton, lamprophyre dykes, and a diatreme pipe (Helwig et al., 1974; Strong and Harris, 1974). The Budgell Harbour pluton covers an area of about 3 km² and intrudes Ordovician sedimentary and volcanic rocks of the Dunnage Zone. The more important rock types are analcite, hornblende, biotite gabbros and hornblende pyroxenites, all showing cumulus textures. Other minerals include plagioclase (An$_{30}$-An$_{60}$), zoned titanaugite, olivine, biotite, opaques, and minor sphene and apatite.

The over 100 lamprophyre dykes that radiate for distances of 5 km from the Budgell Harbour pluton have widths up to 1 m and mainly trend north-northwest southsoutheast (Fig. 6.4b). Olivine, titanaugite, and biotite occur as phenocrysts concentrated in the centres of the dykes and as matrix minerals together with Fe-Ti oxides, plagioclase, biotite, analcite, apatite, and calcite. Leucocratic ocelli are common and show an enrichment in calcite and analcite. Zeolite facies to greenschist facies minerals such as chlorite, iddingsite, albite, and calcite are common in the pluton and lamprophyres; some may be primary as the rocks are alkaline.

Age information from fauna and flora

The fossil discoveries of Olsen et al. (1987) focused world attention on the Mesozoic sedimentary rocks of Atlantic Canada. However, fossil discoveries have not been common or without controversy. The following is a brief outline of the more important information from fossils.

A Mesozoic age was first established for rocks of the Fundy Graben of Nova Scotia with the discovery of fossil fish remains in the Scots Bay Formation (Haycock, 1900, as discussed in Powers, 1916). The work of Baird (1972, 1976) confirmed a Triassic age for Wolfville and Blomidon sedimentary rocks with the discovery of pelecypods, amphibian and reptilian remains, and footprints as well as fish scales and bone fragments. More recent research revealed a variety of Late Carnian to Early Norian trace fossils in the upper Wolfville Formation with part of the assemblage possibly containing the oldest dinosaur footprints in Canada (Olsen and Baird, 1986; Cameron and Jones, 1989). A Late Triassic age is also indicated for the beds at Chedabucto Bay by reptilian remains found by Baird (in Klein, 1962). The oldest rocks discovered thus far from the Fundy Graben are Anisian, or early Middle Triassic (Olsen et al., 1987) and occur at Economy in the Wolfville Formation (P.E. Olsen, pers. comm., 1986). Cameron and Jones (1987, 1988) reported semionotid(?) fish, ostracod, conchostracan, and various plant fossils including ferns, cycadophytes (?), and lycopsids from the Blomidon Formation. The lycopsids indicate that a dry climate existed during deposition of these Upper Triassic sedimentary rocks.

Olsen and Galton (1977) reported tetrapod footprints that established an Early Jurassic (Hettangian) or younger age for the North Mountain Basalt and overlying, uppermost sedimentary rocks in the section (McCoy Brook Formation). This is substantiated by pollen and spore assemblages dominated by Corollina meyeriana (B. Cornet quoted in Olsen et al., 1987) as well as vertebrate bones (Olsen, 1981; Olsen et al., 1982). This work also indicates that the uppermost 20 m or so of the Blomidon Formation may be Hettangian (Olsen et al., 1987). Microgastropods from the Scots Bay Formation on the other side of the Bay of Fundy are consistent with an Early Jurassic age (Cameron, 1986). Other fossils from the Scots Bay Formation include ostracodes, charophytic algae, algal stromatolites, and conifer twigs and tree trunks (Cameron et al., 1985; Cameron, 1986; Cameron and Jones, 1988).

The McCoy Brook Formation along the northern shores of Minas Basin is well dated by an Early Jurassic (Hettangian) tetrapod fauna which includes protosuchian and sphenosuchian crocodylomorphs, prosauropod, theropod, and ornithischian dinosaurs, as well as tritheledont and mammal-like reptiles (Olsen et al., 1987). Although important to resolving local time-stratigraphic problems, the much publicized fossil finds (see Hiscott, 1986; Thurston, 1986) also may help explain worldwide faunal extinctions at the Triassic-Jurassic boundary (Olsen et al., 1987). The absence of "typical Triassic" forms suggests a major extinction event, presumably around the Triassic-Jurassic boundary and just prior to deposition of the McCoy Brook Formation. Olsen et al. (1987) suggested that the extinction event may be related to a meteorite impact at Manicouagan, Quebec which has isotopic ages compatible with the Triassic-Jurassic boundary.

Of the areas in southern New Brunswick mapped as Mesozoic (Fig. 6.1c, d), only the rocks at St. Martins and Martin Head have fossils that clearly constrain their age. Dawson (as reported in Bailey and Matthew, 1865) identified silicified plant remains at St. Martins that indicated a Carnian (Late Triassic) age. Similar plant remains were reported by Holden (1913) from sandstones and shales at Martin Head. Recently J. Bujak (quoted in Nadon and Middleton, 1985) recovered pollen from the Echo Cove Formation that indicate a Carnian or older age based on correlations with American assemblages reported by Cornet (1977).

Klein (1962) proposed that sedimentary deposits at Point Lepreau are equivalent in age to those at St. Martins on the basis of similarities in conglomerate clast lithologies. Fossil footprints preserved in red sandstone at the more northerly outcrop of the Lepreau Formation (Fig. 6.1c) were interpreted to belong to the ichnogenus Isocampe and to resemble prints of the same ichnogenus from Upper Triassic to Lower Jurassic rocks in the Connecticut Valley (Sarjeant and Stringer, 1978; Stringer and Lajtai, 1979). More recently a palynological study of outcrops 750 m southwest of and apparently along strike with those containing the footprints indicates an Early Carboniferous (Late Visean) age (M.S. Barss quoted in Stringer and Burke, 1985). It thus appears that the northerly rocks, at least, of the Lepreau Formation are Carboniferous (P. Stringer, pers. comm., 1988).

Table 6.1 summarizes time-stratigraphic correlations of Mesozoic formations in Atlantic Canada. Within the Fundy Graben, rocks of the Honeycomb Point, Quaco, and Echo Cove formations (New Brunswick) correlate with upper beds (Upper Triassic) of the Wolfville Formation and equivalent beds of the Blomidon Formation (Nova Scotia). Similarly the Upper Triassic Chedabucto Formation of the Chedabucto Graben can be correlated with these units.

Stevenson and McGregor (1963) used spores in a lignite bed to establish an Early Cretaceous age for the unconsolidated clays at Shubenacadie. Dickie (1987) reported Late Aptian to Early Albian ages for spores recovered from lignite, sand and clay beds at Musquodoboit (central Nova Scotia) and Diogenes Brook (southwestern Cape Breton Island). Lower beds at Diogenes Brook gave Valanginian to Barremian ages (Dickie, 1987).

Nannofossils recovered from the Ford's Bight Breccia indicate initial deposition of breccia components during the Early or Middle Jurassic. Other nannofossils from the same unit, however, are much younger and suggest that the sediment was remobilized and redeposited during the Early Cretaceous (King and McMillan, 1975).

Isotopic dates

Table 6.2 summarizes isotopic age data for Mesozoic igneous rocks of Atlantic Canada, including the offshore. Most dates are K-Ar determinations, which are easily reset by low-grade metamorphism. In general, isochron ages, as opposed to whole rock ages, and more recently determined ages reflecting improvements in analytical techniques, are the most reliable. The 247 Ma age (Late Permian) for the Malpeque Bay sill is supported by paleomagnetic data (Larochelle, 1967). Mafic rocks in the Northumberland Strait are Late Permian to Early Triassic with ages between 239 ± 10 and 214 ± 9 Ma.

The Shelburne, Avalon, Minister Island, Caraquet, and Anticosti Island dykes, as well as North Mountain Basalt, define a relatively small spread of ages generally between 180 and 200 Ma, making them Bajocian to Sinemurian (Early Jurassic to early Middle Jurassic) according to recent time scales (e.g. Harland et al., 1982; Kent and Gradstein, 1986). Olsen et al. (1987) argued on paleontological grounds that the North Mountain Basalt (dated by K-Ar at 191 ± 2 Ma) is Hettangian, and therefore 205 to 214 Ma (Harland et al., 1982) or 204 to 208 Ma (Kent and Gradstein, 1986). Recent studies (some utilizing U-Pb systematics) indicate that most of the Late Triassic-Early Jurassic igneous rocks have not acted as closed (Ar) systems, and that nearly all of the igneous activity occurred at 201 ± 2 Ma (Sutter, 1988; Dunning and Hodych, 1990). A U-Pb zircon date of 202 ± 1 Ma for the North Mountain

Basalt supports these conclusions (Hodych and Dunning, 1992). In a paleomagnetic study of the dykes (excluding the Minister Island Dyke) and North Mountain Basalt, Hodych and Hayatsu (1988) concluded that virtual paleopole positions (66-69°N, 72-104°E) and the normal polarity of all rocks is consistent with a Hettangian age.

The wide range of ages recorded for lamprophyres of Notre Dame Bay may reflect analytical problems or resetting of the K-Ar systematics in some samples. The oldest age for these rocks (144 ± 12 Ma) is similar to ages for the Budgell Harbour pluton (135 ± 8 Ma) and lamprophyric rocks at Ford's Bight, Labrador (145 ± 6 Ma). These straddle the Jurassic-Cretaceous boundary (Harland et al., 1982; Kent and Gradstein, 1986). Virtual paleomagnetic poles and normal remnance also suggest a Late Jurassic (Deutsch and Rao, 1977) or Cretaceous (Lapointe, 1979) age for the lamprophyre dykes of Notre Dame Bay. The impressive swarm of alkaline lamprophyric dykes at Cape Makkovik, 15 km north-northwest of Ford's Bight, (Foley, 1984; Malpas et al., 1986) is probably Cretaceous but needs to be dated using modern methods.

SEDIMENTOLOGY AND PALEOENVIRONMENT

The following section summarizes information on the conditions of sedimentation, paleoflow and wind directions, and paleoenvironment for the Mesozoic rocks of Atlantic Canada. During the Late Permian to Middle Triassic, when the Malpeque Bay Sill and Northumberland Strait diabase were intruded, the area experienced uplift and erosion and sedimentary rocks were not deposited. This discussion therefore begins with the Upper Triassic-Lower Jurassic rocks of the Fundy and Chedabucto graben.

Upper Triassic-Lower Jurassic

Clastic rocks of the Wolfville Formation have been interpreted as deposits of a proximal alluvial fan (Klein, 1962) together with braided river and eolian (barchan) sand dune settings (Hubert and Mertz, 1980, 1984). The eolian sands occur on the north shore of Minas Basin and indicate paleowind directions between southwest and northwest (Hubert and Mertz, 1984). Paleoflow directions (summarized in Fig. 6.3e) show that alluvial fan detritus was derived from highlands bordering the graben and generally moved south on the north side of the basin and north on the south side (Klein, 1963a; Hubert and Hyde, 1982). Waters from these fans entered braided rivers that flowed to the southwest, west of Cape Blomidon and to the east on the east of Cape Blomidon because of a topographic high just east of Wolfville. Coarse breccias of the Chedabucto Formation resemble alluvial fan deposits of the Wolfville Formation on the north shore of Minas Basin (Klein, 1962), and probably represent proximal fan deposits.

Facies relationships and sedimentary structures indicate that crosslaminated and crossbedded clastic rocks of the Blomidon Formation represent the distal sheet-flood deposits of alluvial fans as well as sand flats, playa mudflats, and lacustrine deposits (Hubert and Hyde, 1982; Mertz and Hubert, 1990). Cameron and Jones (1987) identified fining-upward cycles and lateral accretion surfaces at several localities along North Mountain, and argued that meandering river deposits are also common. Periodic movement on the Cobequid Fault, which bounds the northern edge of the Fundy Graben, produced more than 100 depositional cycles of sand-flat sandstones followed by playa mudstones and/or lacustrine claystones (Mertz and Hubert, 1990). Flow directions and thus sources for the sedimentary detritus appear similar to those for the Wolfville Formation, except that additional data from the Digby area indicate flow to the northeast. The lithologically similar McCoy Brook Formation on the north side of the Fundy Graben represents a continuation of Blomidon Formation depositional conditions after extrusion of the North Mountain Basalt. However, eolian sandstones are more common in the McCoy Brook Formation and give paleowind directions to the southwest (Hubert and Mertz, 1981, 1984; Olsen, 1981).

Calcareous clastic rocks, limestones, and silicified organic material composing the Scots Bay Formation represent carbonate sedimentation in a nearshore lacustrine environment. Silicification probably resulted from hot spring activity associated with cooling of the underlying North Mountain Basalt flows (Cameron et al., 1985; Colwell and Cameron, 1985).

Nadon and Middleton (1985) interpreted coarse clastic rocks of the Honeycomb Point Formation at St. Martins as alluvial fan deposits with western outcrops representing proximal fan, and finer grained eastern sheet-flood deposits the mid to distal fan. Eolian sands in the same formation indicate a paleowind direction from the northeast. Well-rounded clasts of the overlying conglomeratic Quaco Formation (190-300 m) were deposited after prolonged transportation by a large braided river system flowing north along the Fundy Graben axis (Klein, 1962, 1963a, b). The distal provenance of the quartzite clasts has not been determined but similar rocks occur in the Carboniferous and Precambrian of southern New Brunswick. The overlying red and green breccias of the Echo Cove Formation (850-300 m) represent renewed alluvial fan deposition with the detritus source to the west. A preponderance of metasedimentary and meta-igneous rocks as clasts in both the Honeycomb Point and Echo Cove formations indicates a local, northwesterly source from the Precambrian Coldbrook Group (Nadon and Middleton, 1985).

Nadon (1981) interpreted the coarse clastic rocks in southerly outcrops of the Lepreau Formation as (from bottom to top) an alluvial fan/playa lake sequence, braided river deposits, and debris flows of alluvial fan origin.

To summarize lithological similarities, the coarse grained beds of the Honeycomb Point and Echo Cove formations correlate most closely with the Wolfville and Chedabucto formations and lowermost and uppermost southerly beds of the Lepreau Formation. All of these rocks largely represent alluvial fan deposits. The Quaco Formation has no lithological equivalents but has similarities with the Blomidon Formation and middle portions of the Lepreau Formation, as all three groups of rocks are composed, at least in part, of braided river deposits.

Age and lithological correlations allow reconstruction of paleogeographical and environmental conditions within the Fundy Graben during Middle to Late Triassic time. Paleoflow direction data show that water drained off the highlands bordering both sides of the graben and out onto alluvial fans (Fig. 6.3e). Sediments of the Blomidon and Quaco formations suggest that water in braided rivers flowed along the graben axis toward a topographic low west of Cape Blomidon and away from a topographic high just east of Cape Blomidon. Apparently a large portion of the graben had no external drainage system, but a deep permanent lake has not been identified suggesting that evaporation rates were high and/or precipitation rates low.

The inferred paleoenvironmental conditions for the Mesozoic redbeds of the Fundy and Chedabucto graben vary from semiarid (e.g. Powers, 1916; Crosby, 1962), as indicated by the paucity of weathered grains, scarcity of carbon, and the oxidized state of iron in the sediments, to hot and humid with high precipitation and an annual dry season (e.g. Magnusson, 1955; Klein, 1962, 1963a), also based on colour of the beds. Hubert and Mertz (1980, 1984) pointed out that red colouration is not a definitive environmental indicator. However, their recognition of eolian sandstones, caliche paleosols, alluvial fan conglomerates, and playa gypsiferous mudstones confirms an arid to semiarid environment with a few hundred millimetres of seasonal precipitation. Lycopsid megaspores discovered in the Blomidon Formation support a desert-like climate (Cameron and Jones, 1988). The southwest to northwest paleowind directions inferred from eolian sands of the Fundy Graben reflect the direction and effect of subtropical trade winds at 25°N paleolatitude (Hubert and Mertz, 1980, 1984) that predictably would lead to an arid environment.

The identification of organic-rich lacustrine sediments could have important hydrocarbon implications (Brown, 1986). Lack of a deep anoxic lake that could produce and preserve the organic material is in keeping with the inferred arid to semiarid climatic conditions. Nevertheless there may have been long-term variations in the average annual rainfall in the Fundy Graben. Nadon and Middleton (1985) suggested that propagation of the Quaco Formation conglomerates over the Honeycomb Point Formation fan deposits could reflect an increase in aridity resulting in a decreased supply of detritus to the fan. Similarly, facies changes, the abundance of plant material, and green colouration in one member of the Echo Cove Formation imply a period of increased rainfall. The Blomidon Formation shows an upsection increase in the ratio of lacustrine claystone to playa sandy mudstone and decrease in evaporite mineralization, indicating that climatic conditions became progressively wetter during latest Triassic to earliest Jurassic time (Mertz and Hubert, 1990).

Cretaceous

Primary sedimentary structures in the Musquodoboit, Shubenacadie, and Diogenes Brook deposits show that they are deltaic (Stea and Fowler, 1981). Heavy mineral studies suggest a diverse provenance for the central Nova Scotia deposits which includes local Carboniferous rocks of the Windsor Group and lower Paleozoic metasedimentary rocks of the Meguma Group, as well as distal volcanic rocks from the Cobequid Highlands (Dickie, 1987). The purity of the clay and silica sand may be the result of intense weathering during the Early Tertiary in a humid subtropical climate with high rainfall (Dickie, 1987).

King and McMillan (1975) suggested a complicated origin for the Ford's Bight Breccia. Nannofossils indicate initial deposition of clasts derived from the adjacent Precambrian Allik Group during the Jurassic, with reworking

Table 6.2. Summary of ages and rock types for Mesozoic volcanism in the Atlantic Canada region and adjacent areas.

Location	Age and Method		Rock Type	Reference
Permian-Triassic volcanism				
Malpeque Bay sill (Prince Edward Island)	247	K/Ar wr	Lamprophyre	1
	211 ± 8	K/Ar wr		2
Fina F-25 (Northumberland Strait)	214 ± 9	K/Ar wr	Olivine tholeiite	4
	239 ± 10			
Coastal New England	240-210	(review)	Olivine dolerites, alkali granite, syenites	3
Early Jurassic volcanism				
Shelburne Dyke (Nova Scotia)	≈201	K/Ar isochron	Qtz. norm. tholeiite	5
	193 ± 2	K/Ar isochron		21
Avalon Dyke (Newfoundland)	201.1 + 2.6	K/Ar isochron	Qtz. norm. tholeiite	5
	≈191	K/Ar isochron		
	189 ± 3	K/Ar isochron		21
North Mountain Basalt (Nova Scotia)	≈196(mean)	K/Ar wr	Qtz. norm. tholeiite	6
	190.9 ± 2.4	K/Ar isochron		7
	≈192(mean)	K/Ar wr		7
	202 ± 1	U-Pb zircon		23
Palisades Sill (New Jersey)	196-186 range	K/Ar biotite		8
	192 ± 9	Ar/Ar	Qtz. norm. tholeiite	9
	193 ± 9			
Minister Island Dyke (New Brunswick)	189 ± 8	K/Ar wr	Qtz. norm. tholeiite	10
Caraquet Dyke (New Brunswick)	180 ± 31	K/Ar wr	Qtz. norm. tholeiite	11
	167 ± 28			
	191 ± 2	K/Ar isochron		21
Anticosti Island dykes (Quebec)	178 ± 8	K/Ar wr	Qtz. norm. tholeiite	22
Eastern North American dolerites	205-165	(review)	Qtz. and ol. norm. tholeiites	3
White Mountains magma series	200-155	(review)	Alkali granite, syenite, and gabbro	3

Table 6.2. (cont.)

Location	Age and Method		Rock Type	Reference
Jurassic-Cretaceous volcanism				
Exxon Lydonia Canyon 133-1 (Georges Bank)	134 ± 4 137 ± 7	K/Ar wr	Trachybasalt (alkaline intrusive)	12
Budgell Harbour pluton (Newfoundland)	135 ± 8	K/Ar biotite	Alkaline gabbro	13
Notre Dame Bay dykes (Newfoundland)	144 ± 12 129 ± 7 115 ± 20	K/Ar biotite	Lamprophyres	14,15
Ford's Bight (Labrador)	145 ± 6	K/Ar wr	Lamprophyres	16
Brant sill (Grand Banks)	135 ± 6	K/Ar wr	Alkali basalt	17
Bjarni H-81 (Labrador Shelf)	139 ± 7 122 ± 6	K/Ar wr	Basalt flow Basalt flow	16
Leif M-48 (Labrador Shelf)	131 ± 6 104 ± 5	K/Ar wr	Basalt flow	16
Herjolf M-92 (Labrador Shelf)	122 ± 2	K/Ar wr	Basalt flow	16
Scaterie Bank (Scotian Shelf)	127 ± 15	K/Ar wr	Diabase	18
Hesper I-52 (Scotian Shelf)	125.9 ± 3.2	K/Ar wr	Alkali basalt	19
Hercules J-15 (Scotian Shelf)	119 ± 5 to 102.9 ± 2.6		Alkali basalt	19
Twillick G-49 (Grand Banks)	117 ± 5	K/Ar wr	Diabase	17
Argo F-38 (Scotian Shelf)	Aptian palynology (116-108)		Alkali basalt	19
Jason C-20 (Scotian Shelf)	Aptian palynology (116-108)		Alkali basalt	19
Emerillon C-56 (Grand Banks)	96.4 ± 3.8	K/Ar wr	Trachyandesite	20
Indian Harbour M-52 (Labrador Shelf)	90 ± 4	K/Ar wr	Intermediate to mafic tuff	16
New England-Quebec province	190-130	(review)	Lamprophyres, alkali gabbros, syenites	3

Notes: qtz. norm. = quartz normative; ol. norm. = olivine normative; review = range of ages in cited review; wr = whole rock

References:
1. Greenough et al., 1988; 2. Snelling in Poole et al., 1970; 3. McHone and Butler, 1984; 4. Pe-Piper and Jansa, 1986; 5. Hodych and Hayatsu, 1980; 6. Carmichael and Palmer, 1968; 7. Hayatsu, 1979; 8. Ericson and Kulp, 1961; 9. Dallmeyer, 1975; 10. Stringer and Burke, 1985; 11. Wanless et al., 1972; 12. Exxon in Jansa and Pe-Piper, 1986; 13. Helwig et al., 1974; 14. Wanless et al., 1965; 15. Wanless et al., 1967; 16. Umpleby, 1979; 17. Amoco Canada Petroleum Company Ltd., 1974; 18. Wanless et al., 1979; 19. Jansa and Pe-Piper, 1985; 20. Jansa and Pe-Piper, 1986; 21. Hodych and Hyatsu, 1988; 22. Wanless and Stevens, 1971; 23. Hodych and Dunning, 1992.

Table 6.3. Average geochemical compositions of Mesozoic volcanic rocks from the Atlantic Provinces region.

	Permian-Triassic						Early Jurassic							Jurassic-Cretaceous					
	1	2	3	4	5	6	7	8	9	10	11	12	13	14	15	16	17	18	19
SiO_2	43.78	47.58	51.93	53.49	51.69	52.23	53.72	53.79	53.72	52.96	51.51	52.03	50.97	43.88	39.06	47.75	52.11	47.46	47.16
TiO_2	2.86	1.61	0.87	1.06	0.99	0.93	0.98	1.15	1.08	1.27	0.76	1.12	0.93	4.78	4.58	4.27	3.16	3.38	4.14
Al_2O_3	9.74	16.56	15.28	14.94	14.58	14.58	13.11	15.07	13.77	15.07	14.91	14.20	15.29	14.31	12.18	14.78	16.09	14.85	14.30
Fe_2O_3	1.95	1.77	1.65	1.58	1.56	1.72	1.48	1.56	1.54	1.61	1.83	1.82	1.47	1.99	2.42	2.04	1.76	2.29	1.57
FeO	10.56	9.57	8.95	8.51	8.43	9.29	8.00	8.41	8.33	8.72	9.90	9.82	7.95	10.72	13.10	11.03	9.51	12.37	8.50
MnO	0.18	0.78	0.19	0.17	0.19	0.20	0.18	0.18	0.16	0.17	0.20	0.19	0.16	0.20	0.35	0.22	0.14	0.23	0.02
MgO	14.91	9.25	7.76	6.86	8.45	8.02	9.20	6.65	8.08	7.08	7.42	7.40	8.89	8.16	8.67	4.87	4.25	5.58	8.59
CaO	10.61	8.79	10.54	10.28	11.65	10.76	10.53	9.65	10.36	9.93	10.77	10.64	12.20	11.21	14.23	9.43	6.26	8.58	8.10
Na_2O	3.62	3.70	2.19	2.21	1.91	1.74	2.02	2.48	2.06	2.35	2.22	2.12	2.08	2.43	2.16	3.66	4.05	3.28	4.58
K_2O	0.87	0.27	0.50	0.75	0.40	0.44	0.70	0.91	0.81	0.68	0.48	0.66	0.06	1.69	1.75	1.12	2.14	1.35	1.66
P_2O_5	0.92	0.12	0.14	0.15	0.12	0.10	0.08	0.15	0.09	0.13	–	–	–	0.56	1.52	0.82	0.52	0.63	1.37
Mg'	0.71	0.62	0.60	0.58	0.63	0.59	0.66	0.57	0.62	0.58	0.59	0.56	0.66	0.56	0.53	0.43	0.43	0.43	0.63
Rb	10	10	15	29	13	15	20	22	24	16	15	21	–	41	38	17	45	21	17
Sr	1003	275	134	206	164	160	158	180	169	202	127	186	108	819	739	622	558	711	1278
Ba	646	353	134	207	128	159	191	214	216	163	–	–	18	1216	1434	384	481	987	670
Zr	227	84	66	104	73	69	87	106	96	107	60	92	54	259	344	258	272	222	625
Nb	98	6	6	6	6	5	8	11	9	12	–	–	18	–	–	45	42	30	80
Y	26	24	27	24	21	24	26	30	26	27	–	–	15	–	–	36	35	31	50
Ga	16	–	14	21	18	–	13	–	15	–	–	–	15	–	–	26	23	27	26
V	268	256	256	254	253	289	255	258	244	268	–	–	257	–	–	322	268	331	160
Cr	348	227	118	206	199	187	353	71	225	178	218	277	494	149	105	16	20	31	232
Ni	398	156	50	78	100	58	99	47	64	60	48	81	175	75	72	26	14	28	182
Cu	68	189	65	113	122	64	82	210	100	129	68	111	181	36	24	38	18	38	64
Zn	110	776	65	78	79	81	71	91	63	108	86	84	129	88	138	218	121	132	123

Notes: Mg' = Mg/(Mg + 0.9 Fe) atomic; Major-element oxides in wt.%, recalculated to 100% volatile-free, with Fe_2O_3/FeO adjusted to 0.185; Trace elements in ppm; Number of analyses given in brackets.

1. Malpeque Bay sill (Greenough et al., 1988) (average of 4 analyses). 2. Northumberland Strait F-25 (Pe-Piper and Jansa, 1986), upper diabase (4). 3. Caraquet Dyke (Greenough and Papezik, 1986) (6). 4. Shelburne Dyke (Papezik and Barr, 1981) (8). 5. Avalon Dyke (Papezik and Hodych, 1980) (6). 6. Anticosti Island dykes (Bédard, 1992) (5). 7. Average of upper unit, North Mountain Basalt, Digby area (Papezik et al., 1988) (7 analyses excluding pegmatite P82-13). 8. Average upper unit, North Mountain Basalt, Wolfville area (J.A. Colwell, pers. comm., 1987) (2). 9. Average lower unit, North Mountain Basalt, Digby area (Papezik et al., 1988) (6 analyses excluding pegmatite P82-25). 10. Average lower unit, North Mountain Basalt, Wolfville area (J.A. Colwell, pers. comm., 1987) (8 analyses, major elements; 4 to 8 analyses, trace elements). 11. Low-Ti quartz normative dolerite (Weigand and Ragland, 1970) (37 analyses, major elements; 17 analyses, trace elements). 12. High-Ti quartz normative dolerite (Weigand and Ragland, 1970) (20 analyses, major elements; 17 analyses, trace elements). 13. Jurassic seafloor basalt, Cape Hatteras (Bryan et al., 1977, Table 3C). 14. Budgells Harbour pluton (Strong and Harris, 1974) (5). 15. Twillingate lamprophyres (Strong and Harris, 1974) (6 analyses excluding felsic sample DFS 37). 16. Brant P-87 well, Scotian Shelf (Jansa and Pe-Piper, 1986), lower unit basalt flows (9 analyses, major elements; 2 analyses, trace elements). 17. Twillick G-49 well, Grand Banks, diabase sill(?) (Jansa and Pe-Piper, 1986) (2). 18. Hercules J-15 well, Orpheus Graben on Scotian Shelf, basaltic sill or dyke (Jansa and Pe-Piper, 1986) (4 analyses, major elements; 2 analyses, trace elements). 19. Exxon 133-1 well on Georges Bank, diabase dykes, lower unit (Jansa and Pe-Piper, 1986) (4 analyses, major elements; 2 analyses, trace elements).

and younger nannofossils suggesting later mass movement and proximal deposition during the Early Cretaceous. The reworking is attributed to a diatreme eruption within the Jurassic and Lower Cretaceous marine sediments. The high temperatures indicated by vitrinite reflectance were produced by subsequent intrusion of lamprophyric-carbonatite dykelets and are possibly related to intrusion of an accompanying lamprophyre dyke. The rocks indicate that shallow seas extended west of the present Labrador coast during the Early Jurassic forming the initial deposits. These were reworked and intruded by alkaline igneous rocks during the Early Cretaceous (King and McMillan, 1975).

GEOCHEMISTRY AND PETROGENESIS OF IGNEOUS ROCKS

Average chemical compositions (major and selected trace elements) of the Atlantic Provinces igneous rocks appear in Table 6.3. For rocks with requisite geochemical data, Figure 6.5 shows the abundance ratios of various incompatible elements (element/Yb) normalized to element/Yb ratios in chondrites. These plots allow comparison of the Atlantic Provinces volcanic rocks (Fig. 6.5a, b, c), and selected reference samples, in terms of partial melting and crustal assimilation processes. The use of element/Yb ratios helps eliminate the overprinting effects of low-pressure crystal fractionation processes (Dostal and Dupuy, 1984).

Upper Permian-Lower Triassic

High-Ti phlogopite in the Malpeque Bay Sill indicates an affinity with alkaline igneous rocks such as lamprophyres (Greenough et al., 1988). The sill is chemically similar to nephelinites (Fig. 6.5a, compare Hawaiian nephelinite) with its high Nb (~100 ppm), Zr (~225 ppm) and light rare earth element (REE) concentrations (e.g. La ~50 ppm; Greenough et al., 1988) together with high Nb/Y (3.7) and Nb/Zr (0.5) ratios (Table 6.3). High Mg' values (0.71 where Mg'=Mg/Mg+0.9Fe atomic), Ni/MgO ratios near 31, and the presence of spinel lherzolite nodules and Fo-rich olivine phenocrysts indicate an unevolved magma. A Sr and Nd isotopic study indicates that the extreme high field strength element (HFSE) enrichment typical of alkaline magmas but unusual depletion of some large ion lithophile elements (LILE, e.g. Rb), reflects mantle metasomatism of a long-term large ion lithophile element and high field strength element depleted source region just prior to magma generation (Greenough and Fryer, 1991).

In contrast with the Malpeque Bay Sill, rocks from the Northumberland Strait F-25 well (Pe-Piper and Jansa, 1986) show extremely low, tholeiitic REE (La ~5 ppm), Nb (~5 ppm), and Zr (~85 ppm) concentrations together with high field strength element ratios resembling some mid-ocean ridge basalts (MORB) (Table 6.3, Fig. 6.5a). High Al_2O_3 (16.6 wt.%) indicates derivation through complete melting of plagioclase in a plagioclase peridotite, and intermediate Ni (~150 ppm) and Mg' (0.62) values suggest a significant amount of magma evolution.

Upper Triassic-Lower Jurassic

The North Mountain Basalt flows (Stevens, 1980; Wark and Clarke, 1980; Dostal and Dupuy, 1984; J.A. Colwell, pers. comm., 1987; Greenough and Papezik, 1987; Papezik et al., 1988; Dostal and Greenough, 1992; Pe-Piper et al., 1992) are high-Ti quartz normative tholeiites according to the Weigand and Ragland (1970) classification scheme. The upper and lower units have remarkably constant strontium isotopic compositions but the upper unit has a higher initial $^{87}Sr/^{86}Sr$ isotopic ratio than the lower unit (Jones and Mossman, 1988; Greenough et al. 1989a). A northeastward decrease in orthopyroxene phenocrysts has been attributed to lateral intrusion of magma along a dyke system (with phenocrysts removed by flowage differentiation), that fed a 300 km long fissure cutting the basin (Papezik et al., 1988). Studies of layering in the upper and lower units indicate that crystal fractionation cannot entirely account for variations in the concentration of elements such as Fe and Ti, and that this may reflect movement of volatile complexes during the early stages of flow cooling (Greenough and Dostal, 1992b). Thin rhyolite bands occurring within mafic pegmatite layers in the upper 40 m of the thick flows have been attributed to silicate liquid immiscibility (Greenough and Dostal, 1992a).

Like North Mountain Basalt, the Avalon (Papezik and Hodych, 1980), Shelburne (Papezik and Barr, 1981; Smith and Huang, 1982), Caraquet (Greenough and Papezik, 1986), and Anticosti Island (Bédard, 1992) dykes are classified as high-Ti quartz normative tholeiites, but the Caraquet Dyke is actually transitional between high-Ti and low-Ti types (Table 6.3). Subtle geochemical differences distinguish each of the dykes and basalt but they are amazingly similar (Fig. 6.5b). The high field strength element concentrations show similarity to MORB (Bryan et al., 1977; Holm, 1982; Dostal and Dupuy, 1984) but the shape of geochemical patterns resembles average continental crust with elevated Ba, Rb, Th, and K and a negative Nb anomaly (Fig. 6.5a, b; Dostal and Dupuy, 1984). Modelling suggests that the patterns reflect assimilation of continental crust by crystallizing N-MORB (normal MORB) magmas prior to final emplacement (Dostal and Dupuy, 1984; Greenough and Papezik, 1986). Compared to the lower unit of North Mountain Basalt, the higher initial $^{87}Sr/^{86}Sr$ isotopic ratio of the upper unit indicates that the latter assimilated more continental crust (Jones and Mossman, 1988). Similarly, dyke interiors have higher initial ratios than chilled margins, supporting the idea that the high large ion lithophile elements of these rocks are largely the result of crustal contamination and not derivation from an enriched subcontinental mantle (Greenough et al., 1989b). Nevertheless, a geochemical study of the Minister Island Dyke incorporating Sr and Nd isotopic data shows that mantle metasomatism may have played a role in genesis of the dykes (Dunn and Stringer, 1990). Pe-Piper et al. (1992) argued that the trace element, and Pb, Nd and Sr isotopic composition of the rocks can be explained by melting mantle material with a crustal component. Fabric, magnetic anisotropy, and contact-rock studies of the dykes demonstrate that magma, as in the North Mountain Basalt feeder dyke, was probably injected toward the northeast (Greenough and Hodych, 1990).

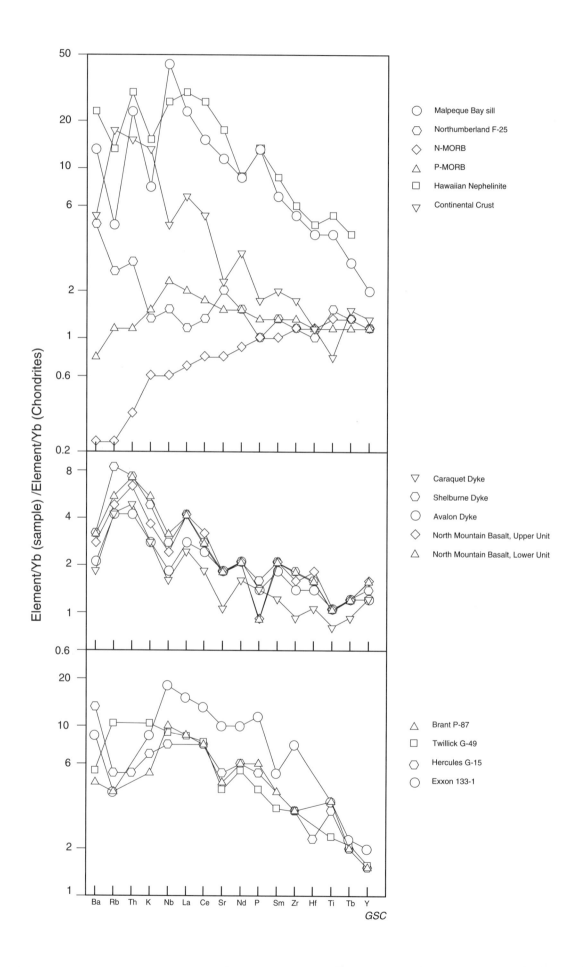

GSC

Middle Jurassic-Lower Cretaceous

The intermediate to low Mg', Ni, and Cr values of lamprophyric rocks from the Notre Dame Bay area indicate evolved compositions as a result of high-pressure eclogite fractionation followed by low-pressure precipitation of olivine, titan-augite, plagioclase, biotite, and/or kaersutite (Strong and Harris, 1974). This is seen by comparing unevolved Malpeque Bay Sill, analysis 1, with Notre Dame Bay, analyses 14 and 15, Table 6.3) Many of the Notre Dame Bay examples show leucocratic globules in a dark host matrix that strongly support coexistence of two silicate liquids. High concentrations of TiO_2, P_2O_5, Rb, Sr, Ba, and Zr attest to the alkaline nature of these rocks and may reflect extreme source-region metasomatism.

Figure 6.5. Trace element patterns for Mesozoic volcanic rocks and selected reference rocks, as discussed in the text: **(a)** Permian-Triassic volcanic rocks (Malpeque Bay and Northumberland F-25) together with reference analyses of average N-MORB and P-MORB (Sun et al., 1979), a Hawaiian nephelinite (Greenough, 1979), and continental crust (Taylor, 1964); **(b)** Lower Jurassic volcanic rocks; **(c)** Jurassic-Cretaceous volcanic rocks. Data sources for the Atlantic Provinces rocks appear in Table 6.3. Normalizing values are from Thompson et al. (1982).

STRUCTURE AND AGE OF DEFORMATION

The Fundy Graben is a major asymmetrical syncline (steeper north limb) plunging gently to the southwest with closure near Scots Bay (Fig. 6.1). The graben is bounded to the north by faults of the extensive Cobequid system (Fig. 6.2) and cut transversely by north-south faults that offset Lower Jurassic strata and all other faults (Donohoe and Wallace, 1980). Folding and faulting relationships are in part illustrated in a north-south cross-ection through the graben at the Minas marine basin (Fig. 6.6). The accurate dating of folding and faulting is of vital importance to unravelling the Mesozoic tectonic evolution of the area. The absence of strata younger than Early Jurassic makes chronological analysis difficult.

Faulting

Various phases of faulting affected the Bay of Fundy region. An early phase is represented by vertical movement along the Cobequid Fault, a major zone boundary that can be traced along the entire north shore of the Bay of Fundy to Chedabucto Bay and out onto the continental shelf along the Orpheus Graben (Keppie, 1982). The fault shows a long history of activity dating back to the Middle Devonian. Keppie (1982) proposed hundreds of kilometres of dextral movement along the fault during the late Paleozoic, based on a correlation of displaced Proterozoic volcanic strata. Although not exposed, a similar fault is thought to border the southern margin of the basin (Fig. 6.2, 6.6) as indicated by magnetic anomalies and morphology of the Annapolis

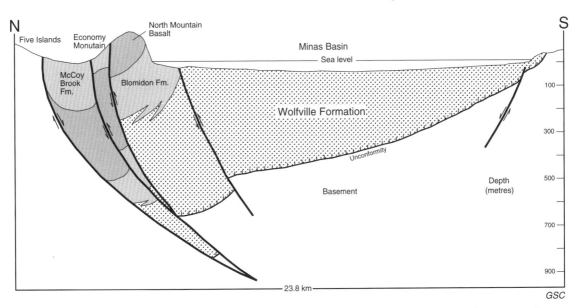

Figure 6.6. Hypothetical cross-section showing structural and lithological relationships across the Minas Basin. Location of the section is given by the line marked N-S in Figure 6.1a. All faults shown may bear a significant (predominant) component of strike-slip movement and their subsurface extent and orientation are hypothetical. The greatest vertical movement occurred on the most northerly fault which acted as a normal fault. Other faults on the north side of Minas Basin appear to show reverse movement. The diagram also illustrates stratigraphic relationships and interfingering of Wolfville and Blomidon sedimentary rocks.

Valley (Keppie, 1977; Zietz et al., 1980). Movement along these normal faults during the Triassic created a major graben similar in shape to the present-day Bay of Fundy (Olsen and Schlische, 1990). Under the Gulf of Maine, at the south end of the structure, a central horst associated with three smaller graben complicates the structure (Ballard and Uchupi, 1975). Total vertical movement on the east-west faults was probably at least 3500 m as given by the thickness of Mesozoic strata in the Chinampas exploratory oil well in the Bay of Fundy (Keown and Bint, 1975). Greater movement may have occurred at the southern end of the graben where seismic studies indicate 9000 m of Mesozoic strata (Brown, 1986).

The detailed stratigraphy of alluvial fan deposits adjacent to the faults reflects, at least in part, the timing and nature of fault movement. For example, Hubert and Hyde (1982) related megacycles in fining-upward sequences of alluvial sand-flat deposits in the Digby area to periodic tectonic movement along the southern border fault. Movement on the northern fault is represented by metre-scale cycles in claystone and mudstone deposition in Blomidon beds near Wolfville and Economy Mountain (Mertz and Hubert, 1990). Similarly, Nadon and Middleton (1985) assigned facies changes in Triassic sedimentary rocks at St. Martins to movement on the northern fault and tilting of the graben floor. These east-west faults are of more than academic interest as various hydrothermal sulphide, magnetite, and native copper showings occur where the faults cut Mesozoic igneous rocks (Crosby et al., 1990).

Apparent displacement of North Mountain Basalt along the Economy Mountain fault (Fig. 6.1a) indicates 3 km of dextral movement on east-west faults (Donohoe and Wallace, 1980). Olsen and Schlische (1990) argued that movement on the fault is dominantly sinistral with the apparent offset the result of contemporaneous vertical movement. Keppie (1982) proposed that sinistral movements along the Cobequid Fault were in the order of 75 km during the Mesozoic. In addition to the normal and strike-slip movement, detailed mapping indicates reverse faulting (Cameron, 1956; Liew, 1976). Swift et al. (1967) pointed out the importance of reverse faulting and proposed that it occurred concurrently with folding of the Mesozoic strata. All of the Mesozoic rocks have been faulted as a result of reactivation of the Cobequid Fault demonstrating that fault movement continued into the Early Jurassic or later. A seismic reflection study of the offshore extension of the Cobequid Fault along the Orpheus Graben indicates that movement continued into the Early Cretaceous (King and MacLean, 1970).

In addition to the east-west faults, the Atlantic Provinces area is cut by north-south to northwest-southeast wrench or transverse faults. These faults are represented by horizontal (sinistral) displacements on the vertical Caraquet (2.5 km, Burke et al., 1973), Minister Island (2.5 km, Stringer and Burke, 1985), and Shelburne dykes (3 km, Papezik and Barr, 1981), as well as less than 2 km of apparent dextral movement on North Mountain Basalt (Fig. 6.1b, 6.2). Some of the displacements on the dykes may be apparent and the product of diverging trajectories for "tongues" of magma (Greenough and Hodych, 1990). The apparent dextral movement on faults cutting the gently dipping North Mountain Basalt near Digby could result

from an upward component of movement on the west side of the faults (Fig. 6.1b). In the Five Islands area of Minas Basin the relative movement of strata between islands indicates both vertical and (later?) horizontal movement along north-south faults (Powers, 1916; Liew, 1976). Brittle transcurrent faults with similar orientations influence topography and control drainage on the Eastern Shore of Nova Scotia (Williams and Hy, 1990). They overprint all earlier structures and exploit older, Devonian joint systems. As with other Mesozoic deformational features the time of formation and last movement on the transcurrent faults is poorly constrained but some activity must have postdated formation of the Lower Jurassic rocks that are cut by faults.

Folding

Gentle to open folds with wavelengths of a few kilometres are superimposed on the larger synclinal structure of the Fundy Graben (e.g. Powers, 1916; Swift and Lyall, 1968). Ballard and Uchupi (1975) recognized the same folds in the Gulf of Maine and also identified open folds with axes at right angles to the border faults. The regional significance of the latter folds is not known. Origin of the east-west folds has been related to concomitant normal faulting along the basin through a variety of mechanisms (see review in Stringer and Lajtai, 1979). Cleavage in northerly outcrops of the Lepreau Formation led Stringer and Lajtai (1979) to suggest that a compressional event caused the folding. Although the Mesozoic age for the northerly outcrops is uncertain, and cleavage has not been identified in the southerly outcrops (Stringer and Wardle, 1973), Stringer and Lajtai (1979) pointed out that features of compressional deformation occur in Triassic redbeds elsewhere, such as Pennsylvania. Likewise, orthogonal joint sets in upper Paleozoic sandstones from New Brunswick and Prince Edward Island appear related to a compressional post-Triassic stress regime (Lajtai and Stringer, 1981).

Stress-field orientation

McHone (1978) argued on theoretical grounds that the orientation of mafic dykes is at right angles to the maximum extensional stress direction in the upper crust at the time of intrusion. The northeast-southwest orientation of the Caraquet, Minister Island, Shelburne, and Avalon dykes (Fig. 6.4a) indicates a northwest-southeast direction of maximum extension during the Early Jurassic. The orientation of Early Cretaceous lamprophyre dykes in the Notre Dame Bay area is north-northwest south-southeast (Fig. 6.4b) locally indicating an east-northeast west-southwest maximum extension direction at the time of emplacement.

In summary, the following Mesozoic structural features are of regional importance: 1) east-west-trending faults with a normal component of movement; 2) dextral and/or sinistral movement on the east-west faults; 3) reverse movement on the east-west faults; 4) folds with east-west axes that plunge to the west; 5) sinistral transverse faults that have a vertical component of movement; 6) northeast-southwest trend of Early Jurassic dykes; and 7) the north-northwest south-southeast trend of Early Cretaceous dykes.

CORRELATION

Mesozoic intracontinental graben bearing igneous and sedimentary rocks similar to those in Atlantic Canada can be traced from Georgia to Nova Scotia (e.g. Sanders, 1963; King, 1971; Robbins, 1981) and also occur on the continental shelves of eastern North America (Fig. 6.7), Northern Africa, and Western Europe. These rocks comprise the Newark Supergroup (Froelich and Olsen, 1985) of the Appalachian region. In this section, formations within the Fundy and Chedabucto graben, and igneous rocks not everywhere restricted to the graben, are compared with equivalents in offshore Atlantic Canada and southward. Finally, aspects of these rocks such as thickness, age, environment of deposition, and character of volcanism are compared with other rocks of the Newark Supergroup.

Correlation with offshore Atlantic Canada

Sedimentary rocks

Formations in offshore Atlantic Canada that are approximately time-equivalent to those on land include the Eurydice, and overlying Argo and Iroquois formations on the Scotian Shelf and Grand Banks (Table 6.1) (Jansa and Wade, 1974). Rocks of equivalent age were not deposited on the Labrador Shelf (Umpleby, 1979). Deposition took place in discrete graben or basins: the Grand Banks, Scotian, and Orpheus in Figure 6.7. The Eurydice Formation consists of reddish, anhydrite-bearing siltstones and shales, in the order of 2000 m thick, that unconformably overlie basement rocks of unknown age. The sedimentary rocks range in age from Middle Triassic to earliest Jurassic, reflect alluvial to lacustrine depositional conditions, and formed under arid desert conditions similar to those inferred for the onshore deposits (Jansa and Wade, 1974). Ocean waters entered the graben during the latest Triassic producing the dominantly Hettangian/Sinemurian shallow water evaporite (halite) deposits of the Argo Formation (1000-2000 m thick). These conditions shifted westward with time across the Scotian Shelf area (Jansa and Wade, 1974; Manspeizer et al., 1978). Following a brief period of continental deposition, a marine transgression started in the Sinemurian and deposited up to 610 m of anhydrite and dolomite composing the Iroquois Formation in a sabkha and shallow marine environment.

During late Early to early Middle Jurassic, a regional uplift yielded continental clastics of the Mohawk Formation (250 m) on the Scotian Shelf and a shallow water carbonate facies of the Verrill Canyon Formation on the Grand Banks (Jansa and Wade, 1974). The rocks at Ford's Bight indicate that a marine incursion affected the Labrador area during Early to Middle Jurassic (King and McMillan, 1975). On the Scotian Shelf and Grand Banks, a late Middle to early Late Jurassic marine transgression produced shales, sandstones, and limestones of the Mic Mac Formation, carbonates of the Abenaki Formation, and deeper water shales of the Verrill Canyon Formation, with a total thickness exceeding 1800 m. Tectonic uplift caused a marine regression during the latest Jurassic that lasted for most of the Early Cretaceous. Nearly 2.5 km of terrigenous clastic sediments of the Mississauga and Logan Canyon formations were deposited in deltaic sequences during this interval (Jansa and Wade, 1974). These rocks are time-equivalent to the clay and sand deposits in Nova Scotia (Dickie, 1987).

Igneous rocks

Early Jurassic basalt and diabase dykes were encountered in the Glooscap C-63 (Scotian Shelf), Spoonbill C-30, and Cormorant N-83 (Grand Banks) exploratory oil wells (Jansa and Pe-Piper, 1986) in the offshore of Eastern Canada (Fig. 6.2). These are equivalents to the North Mountain Basalt and Caraquet, Minister Island, Shelburne, Anticosti Island, and Avalon dykes and are high-Ti quartz normative tholeiites, chemically similar to most other Early Jurassic igneous rocks in the northern Appalachians (Pe-Piper et al., 1992).

Middle Jurassic to Cretaceous mafic to intermediate rocks were intersected in exploratory oil wells (Fig. 6.2) on the Labrador Shelf (Herjolf M-92, Bjarni H-81, Indian Harbour M-52, and Leif M-48), Grand Banks (Twillick G-49, Brant P-87, and Emerillon C-56), Scotian Shelf (Hercules G-15, Jason C-20, Argo F-38, and Hesper I-52), and Georges Bank (Cost G-2 and Exxon 133-1) (Table 6.2). These are related to the lamprophyric rocks at Ford's Bight (Labrador) and Notre Dame Bay (Newfoundland). Only those on the Labrador Shelf have not been examined petrographically and geochemically.

Jansa and Pe-Piper (1986, 1988) recognized two mafic units in the Brant P-87 well on the Grand Banks, each composed of four or five porphyritic, 1 to 42 m thick, vesicular lava flows. A 15 m thick diabase sill or flow was intersected in the Twillick G-49 well whereas monzodioritic dykes with a total thickness of 21 m were encountered in the Emerillon C-56 well.

Two igneous units are recognized in the Scotian Shelf wells (Jansa and Pe-Piper, 1985). The lower unit consists of thick (less than ten to tens of metres) holocrystalline mafic rocks (flows, sills, or dykes). In the upper unit, tens of metres of pyroclastic and volcaniclastic rocks are interbedded with sedimentary rocks and thin basalt and trachyte lava flows. Dips up to 36° indicate that eruptions produced volcanic cones.

Three igneous units were recognized by Jansa and Pe-Piper (1986, 1988) in the Exxon 133-1 well on Georges Bank. The lower unit (78 m) probably represents a diabase dyke and a middle unit contains limestones intercalated with two volcanic tuff beds. The upper unit is composed of pyroclastic rocks interbedded with thin lava flows. The sequence is interpreted as a strombolian-type volcanic cone with seismic data indicating dimensions of 3.6 km by at least 1 km. Similar volcanic features occur in older rocks on the Scotian Shelf.

Intermediate to low Mg', Ni, and Cr values of all offshore middle Jurassic to Cretaceous rocks indicate evolved compositions (Jansa and Pe-Piper, 1985, 1986; Pe-Piper and Jansa, 1987; compare unevolved Malpeque Bay Sill, analysis 1, with analyses 16 through 19, Table 6.3). Their alkaline nature is shown by high concentrations of TiO_2, P_2O_5, Rb, Sr, Ba and Zr with the highest values (most extreme alkalinity) in the Exxon 133-1 volcanic rocks (analysis 19, Table 6.3). Their alkaline character is also illustrated by elevated geochemical

Figure 6.7. Early Mesozoic graben (Newark system) of Eastern North America. Numbers beside the graben refer to East Newfoundland and Grand Banks (1), Orpheus and Chedabucto (2), Scotian (3), Fundy (4), Georges Bank (5), Deerfield (6), Hartford (7), Pomperaug (8), Baltimore Canyon (9), Newark (10), Gettysburg (11), Culpeper (12), Barboursville (13), Taylorsville (14), Richmond (15), Scottsville (16), Farmville (17), graben north of Scottsburg (18), Scottsburg (19), Dan River (20), Durham (21), Sanford (22), Davie County (23), Wadesboro (24), and the Blake Plateau (25). Graben not assigned numbers have not been named in the literature. The red line on the west side of the map marks the western boundary for nearly all dykes of the Eastern North American magmatic province (Papezik and Greenough, 1985; McHone et al., 1987). It closely corresponds with the eastern boundary of the Appalachian miogeocline shown as a dotted and dashed black line. The dashed black line is the western boundary of the Coastal New England magmatic province of McHone and Butler (1984) extended to the Atlantic Provinces (Greenough et al., 1988). Sources for basin distribution include Sheridan (1976), de Boer and Snider (1979), and Froelich and Olsen (1985).

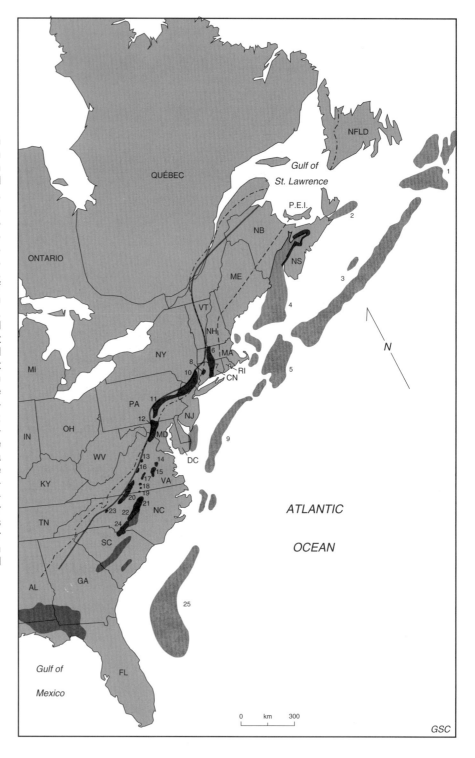

patterns between Nb and Y with the extreme Exxon 133-1 pattern (Fig. 6.5c) somewhat resembling the highly alkaline Malpeque Bay Sill (Fig. 6.5a).

The similarities of the Scotian Shelf and Grand Banks geochemical patterns (Brant P-87, Twillick G-49, and Hercules G-15) indicate derivation from comparable sources through similar processes. To account for the contrasting alkalinity of the Scotian Shelf and Georges Bank volcanic rocks, Jansa and Pe-Piper (1986) suggested that the Scotian Shelf volcanic rocks were derived by small percentages of melting of normal or nonenriched mantle whereas the Georges Bank rocks were derived from enriched sources resembling those associated with mantle plumes.

Correlation south of Atlantic Canada

Sedimentary rocks

Newark graben of eastern North America (Fig. 6.7) show many similarities, and some significant differences, with the Fundy Graben in terms of sedimentology, structure, chronology, and volcanology. Total thicknesses for Triassic through Lower Jurassic sedimentary rocks in onland Newark graben range from 9000 m for the Gettysburg Graben to 1670 m for the Richmond-Taylorsville Graben (reviewed in Manspeizer and Cousminer, 1988) with an average for the major graben of 5200 m. These are comparable to those of the Fundy Graben (3500 m). Sedimentary rocks in the graben include conglomerates, arkosic and lithic arenites, and grey, black, and red siltstones and shale, with interbedded basaltic volcanic rocks (Manspeizer and Cousminer, 1988). Most of the sedimentary rocks represent lacustrine deposits with grey to black siltstones the dominant lithology. Some of the graben had deep anoxic lakes that produced organic-rich sediments that could be important hydrocarbon sources (Manspeizer and Cousminer, 1988). Coarser lithologies representing alluvial fan deposits flank, and lie at the bottom of, the graben (e.g. LeTourneau, 1985; Turner-Peterson and Smoot, 1985). Most graben lacked external drainage (Manspeizer and Cousminer, 1988).

Differences in the style of sedimentation between northern and southern graben may reflect environmental controls. The presence of eolian sandstones, caliche paleosols, alluvial fan conglomerates, and playa gypsiferous deposits suggests a more arid climate for the Fundy Graben than for the southerly graben (e.g. Deep River Graben of North Carolina), where coal and grey mudstones imply intermittent humid conditions (Wheeler and Textoris, 1978; Hubert and Mertz, 1980; Tanner and Hubert, 1992).

During the latest Triassic to Early Jurassic, continental sedimentation was supplanted by marine deposition producing thick evaporite sequences (mainly halite) in the offshore graben of Atlantic Canada as well as in Morocco (Manspeizer et al., 1978). Onland graben lack these Lower Jurassic marine evaporite deposits. Flood basalts, chemically and chronologically very similar to the North Mountain Basalt, were erupted during earliest Jurassic in most graben north of Virginia (Froelich and Olsen, 1985; Olsen et al., 1987).

Determination of the age of the Newark graben has long been a problem because most sedimentary rocks are nonfossiliferous. Estimates range from late Permian-Triassic to Middle Triassic-Early Jurassic (Froelich and Olsen, 1985). Various types of fossils have proven useful for biostratigraphic correlation, the most important being palynomorphs, fishes, and footprints and remains of reptilian and amphibian forms (Olsen et al., 1982). However, actual correlation and dating of these nonmarine fossil assemblages with the marine biostratigraphic standard of Europe and North America is critically dependent on palynomorphs. The palynomorphs are the only fossils that occur in both the Newark strata and marine type-sections of Europe. Palynomorph biostratigraphy has largely been carried out by Cornet and co-workers (e.g. Cornet et al., 1973; Cornet and Traverse, 1975; Cornet, 1977; Traverse, 1987). Unfortunately the palynomorphs are not ideal biostratigraphic markers, especially in Jurassic rocks because of the large time range for most species (Manspeizer and Cousminer, 1988).

The oldest beds in the Newark graben, from Morocco to the Grand Banks, are Middle Triassic, with the sedimentary rocks unconformable above rocks of diverse ages (Manspeizer et al., 1978). However, sedimentation generally started in onland graben during the Carnian (Late Triassic), although the Fundy Graben is unique in this regard having Anisian (Middle Triassic) basal sedimentary rocks. The youngest beds preserved in the onland graben (including the Fundy Graben) are Early Jurassic (Froelich and Olsen, 1985). The absence of younger strata is probably the result of tectonic uplift and nondeposition (Jansa and Wade, 1974), and possible erosion of younger Jurassic rocks. In offshore areas such as the Scotian Basin, sedimentation continued into the Jurassic and Cretaceous (Jansa and Wade, 1974).

Igneous rocks

In terms of age, the Malpeque Bay Sill and Northumberland Strait F-25 diabase correlate with a series of plutonic complexes in coastal New England as well as the Seabrook dykes of New Hampshire (McHone and Butler, 1984; Bellini et al., 1982; Table 6.3). They also correlate with these rocks in that they occur east of the boundary for the Eastern North American dolerite province (discussed below) suggesting a continuation of the Coastal New England province of McHone and Butler (1984) to the north (CNE in Fig. 6.2). The scanty geochemical information on the New England rocks indicates alkalic to mildly alkalic compositions for plutonic complexes (McHone and Butler, 1984; McHone et al., 1987), implying a geochemical association with the alkaline Malpeque Bay Sill. Similarly, the Seabrook dykes have mildly alkaline characteristics (McHone, 1992; Pe-Piper et al., 1992).

The North Mountain Basalt and equivalent dykes of Atlantic Canada are the most northerly occurrences of igneous rocks of the Eastern North American dolerite province (King, 1971; Papezik and Hodych, 1980). This province extends all along the Appalachian Orogen and is the most extensive magmatic province related to opening of the Atlantic Ocean. Ages for the Eastern North American dolerites range from 205 to 165 Ma with a peak at 190 Ma (McHone and Butler, 1984). The province's western boundary (Fig. 6.7) is almost coincident with the eastern boundary of the Appalachian miogeocline. Dykes constituting a major portion of the province tend to decrease in frequency to the north so that only five dykes

589

have been identified in Atlantic Canada. Time-equivalent and chemically similar rocks also occur in Morocco (Bertrand et al., 1982). The White Mountains magma series in New Hampshire (200-155 Ma) overlaps with the Eastern North American dolerite province in space (WMMS and ENA, respectively, in Fig. 6.2) and time but consists of syenitic, monzonitic, and granitic plutonic rocks (McHone, 1978, 1981).

Atlantic Provinces rocks belonging to the Eastern North American dolerite province are high-Ti quartz normative tholeiites. This is also the most common magma type observed in the U.S.A. (Philpotts and Martello, 1986). Other American magma types include low-Ti and high-Fe quartz normative tholeiites, olivine normative tholeiites in the southernmost Appalachians (Ragland et al., 1968; Weigand and Ragland, 1970), and volumetrically small alkali basalt (McHone and Trygstad, 1982; Hermes et al., 1984). Weigand and Ragland (1970) suggested that the high-Fe types represent low-pressure fractionation derivatives of the high-Ti types. Slight differences in the composition of primitive olivine normative magmas (Ragland et al., 1983) may yield both the high-Ti and low-Ti magma types through crystal fractionation and crustal assimilation processes discussed by various authors (e.g. Weigand and Ragland, 1970; Ragland et al., 1983; Dostal and Dupuy, 1984; Greenough and Papezik, 1986). Greenough and Hodych (1990) suggested that the radiating geometry of Mesozoic dykes, the southerly occurrence of "primitive" rocks, and information on injection directions support derivation of magma from a hot spot near Florida with progressive magma evolution resulting from north to northeastward movement through the crust. Still other authors contend that the quartz normative types represent primitive magmas derived from an Fe-rich subcontinental mantle (Philpotts and Martello, 1986). The alkali basalts cannot be related to any of the tholeiites by low-pressure fractionation processes (Yoder and Tilley, 1962).

Middle Jurassic to Cretaceous alkaline igneous rocks analogous to the lamprophyres at Notre Dame Bay, Newfoundland, Ford's Bight, Labrador, and volcanic cones in the offshore of Atlantic Canada, compose the New England-Quebec magmatic province (130-90 Ma) of McHone and Butler (1984). This province includes alkaline intrusions of the Monteregian Hills in Quebec (Currie, 1975). Also of similar age are local occurrences of alkaline igneous rocks off southwestern England (Harrison et al., 1977) and in the northern North Sea (Gibb and Kanaris-Sotiriou, 1976; Fall et al., 1982). Apparently the Middle Jurassic to Cretaceous igneous activity was largely restricted to more northerly areas around the Atlantic Ocean and produced small, localized eruptions of predominantly alkaline to strongly alkaline igneous rocks.

Structural comparisons

Newark graben in the eastern United States show many structural characteristics observed in the Fundy Graben. The graben margins are bounded by normal faults with a significant component of dextral strike-slip movement (Aydin and Nur, 1982; Ratcliffe and Burton, 1985). The Mesozoic faults formed through reactivation of major Paleozoic faults that developed during the Taconic, Acadian, and Alleghanian orogenies (Robbins, 1981; Swanson, 1986; Olsen and Schlische, 1990). The graben are

generally sited east of the Appalachian miogeocline and trend northeast-southwest parallel to major Appalachian structural trends and terrane boundaries. Fault movement apparently accompanied sedimentation and filling of the graben though some movement certainly continued after sedimentation, as some faults are still active (Robbins, 1981). Unlike classical graben, these examples are asymmetrical half-graben with sediments adjacent to reactivated faults on the west and draped over earlier deformed rocks to the east (Manspeizer and Cousminer, 1988). As in Atlantic Canada, mineralization of economic interest tends to occur where faults cut Early Jurassic mafic rocks (Robinson and Sears, 1988; Crosby et al., 1990).

Faults at right angles to the graben axes have both horizontal and vertical components (Sanders, 1963; Van Houten, 1969) and may have formed contemporaneously with sedimentation (Cloos and Pettijohn, 1973). From south to north, on-land graben in the southern United States step westward with left-lateral offsets. Some of these offsets are collinear to Atlantic ocean-floor transform faults (Manspeizer and Cousminer, 1988). North of Virginia the graben tend to step eastward (Fig. 6.7).

Gentle folds, characteristic of the graben, are generally assumed to have formed contemporaneously with faulting and subsidence (e.g. Ratcliffe and Burton, 1985). Locally a tight cleavage, compressed mudcracks, and other deformational features indicate horizontal compression during fold development (Stringer and Lajtai, 1979; Burton and Ratcliffe, 1985; Harrison, 1985; discussion in Manspeizer and Cousminer, 1988).

Orientations of Lower Jurassic dykes that occur all along the Appalachian Orogen (the Eastern North American dolerite province) provide information on the stress regime responsible for graben formation (McHone, 1978). The northeast-southwest orientations of the Lower Jurassic volcanic rocks of Atlantic Canada (Fig. 6.4a) resemble those of the New England states (McHone, 1978) and indicate a northwest-southeast direction of maximum extension. To the south, the orientation of the stress field changed as shown by the north-south trends of volcanic rocks in the Carolinas through Pennsylvania, and dominantly northwest-southeast orientations in the southern Appalachians (McHone, 1978), which cut across older structural trends and the graben axes. In a study of the Appalachian dykes as well as dykes in northwestern Africa and northern South America, May (1971) suggested a causative, single radial stress system related to a triple junction centred on the present south-central Atlantic Ocean.

TECTONIC SYNTHESIS

The rocks of Atlantic Canada play a critical role in reconstructing the Atlantic Ocean, as the North Atlantic opened like a "zipper" with the northern portions last (Pitman and Talwani, 1972) such that some stages in the rifting event are only represented in Atlantic Canada. The stages or cycles in the latest Paleozoic-Mesozoic dismemberment of Pangea appear to have lasted between 50 and 70 Ma. Although the precise controls for these cycles remain unknown, a review of phenomena preceding major rifting led Sheridan (1984) to propose that pulses in mantle convection produce 60 Ma cycles in volcanic, sedimentological, and structural events that are recognizable in the stratigraphic record of rifted areas.

Each stage in the Mesozoic ocean-opening process tends to be characterized by its own style of sedimentation (or lack of it) and deformation, as well as geochemistry of volcanic rocks, geographic distribution of volcanism, and volcanic style. The isotopic age data for Mesozoic igneous rocks on the eastern edge of the North American continent reveal a periodicity in magmatic activity (McHone, 1978; McHone and Butler, 1984; Jansa and Pe-Piper, 1985; Pe-Piper and Jansa, 1986; Table 6.3) that helps define the rifting stages. Four cycles of activity are recognized. From oldest to youngest they are: Late Permian-Middle Triassic (240-210 Ma); Middle Triassic-Early Jurassic (200-165 Ma); Middle Jurassic-Early Cretaceous (145-90 Ma); and Late Eocene to Miocene (50-20 Ma). Products of the last cycle have only been identified (in Canada) in the offshore (northern Labrador Sea and Davis Strait; Clarke, 1977; Umpleby, 1979) and are not discussed here.

Late Permian-Middle Triassic

During the Late Permian, areas later affected by graben formation and crustal thinning were uplifted and eroded. The uplift is commonly attributed to a period of crustal doming perhaps as a result of upwelling asthenosphere beneath eastern North America and Northern Africa (McHone and Butler, 1984). Igneous rocks of the Central New England magmatic province (represented by the Malpeque Bay Sill and Northumberland Strait diabase in Atlantic Canada, Fig. 6.2) are commonly related to this earliest stage in the rifting process (e.g. Manspeizer et al., 1978; Pe-Piper and Jansa, 1986). The alkaline tendencies of some of these rocks are in keeping with models for continental rifting (e.g. Le Pichon and Sibuet, 1981). During the earliest stages of rifting, small amounts of lithospheric stretching allow only diminutive amounts (and percentages) of mantle melting and thus small batches of alkaline magmas.

Late Triassic-Early Jurassic
Tectonic controls on sedimentation

A second stage in the Atlantic-opening process began in Middle to Late Triassic time as the Newark graben began to form. As a result of tectonic uplift and erosion during the Late Permian to Early Triassic, the continental sedimentary rocks in graben from Morocco to the Grand Banks rest unconformably on older rocks of diverse age (Manspeizer et al., 1978; Swanson, 1986). As yet it is not established whether graben formation migrated eastward to where the Atlantic eventually opened (Nadon and Middleton, 1984), whether extension simply affected a broad zone (several hundred kilometres) with all graben forming concurrently (Manspeizer et al., 1978), or whether the location of basin formation migrated westward away from the primary axis of extension. The latter hypothesis is supported by the observation that offshore graben have older (Middle Triassic) basal sedimentary rocks than onland graben (Late Triassic), though the Fundy Graben is an exception to this pattern.

Sedimentation styles in the graben reflect both tectonic and environmental controls. Coarse clastic rocks representing alluvial fan deposits formed adjacent to active dip-slip border faults with fan detritus derived locally from adjacent highlands. Periodic movement on the faults may be reflected in local stratigraphic sections (e.g. Hubert and Hyde, 1982; Nadon and Middleton, 1985). Perhaps as a result of cross faults (continental extensions of ocean transform faults?) delimiting graben ends, external drainage systems did not develop, a condition that helped control the types and character of sediments in the centres of graben. The closed system drainage led to the formation of deep permanent lakes in some more southerly graben (e.g. Newark), and playa lakes in more northerly graben (e.g. Fundy). Differences in the style of sedimentation between northerly and southerly graben are thought to reflect, at least in part, the more arid environment prevalent in the northerly graben (Wheeler and Textoris, 1978; Hubert and Mertz, 1980; Tanner and Hubert, 1992).

In offshore graben, such as the Scotian Graben, continental sedimentation conditions were supplanted by restricted marine conditions during latest Triassic to Early Jurassic time. The resulting salt deposits reflect an incursion of ocean water probably from the Tethys sea to the north (Jansa, 1986) into deeply subsided graben close to the axis of eventual rifting (Manspeizer et al., 1978). Onland graben, including the Fundy Graben, remained isolated from these ocean waters.

Sedimentation continued in the onland graben into the Early Jurassic. Cessation of sedimentation probably denotes an end to fault movement and extensional tectonic conditions as a result of separation of Northern Africa from North America at the end of the second stage in the rifting process. In marked contrast, subsidence and sedimentation continued in offshore graben, such as the Scotian Graben, during most of the Jurassic in response to thermal cooling of lithosphere formed during rifting and separation of plates (Keen, 1979; Royden and Keen, 1980).

Structural models for graben formation

Given the few restrictions on relative timing and relationships between graben structural features, it is not surprising that models explaining their origin are diverse and poorly constrained. Some authors propose that the Mesozoic graben developed as simple downwarps at right angles to extension directions with postdepositional faulting (e.g. Faill, 1973). Other workers recognize that in most cases normal faulting accompanied graben formation (Klein, 1963; Sanders, 1963) and that the faults formed through reactivation of major Paleozoic faults (Robbins, 1981; Swanson, 1986; Olsen and Schlische, 1990). Still others argue for the importance of shear couples, representing onland extensions of oceanic transform faults, in developing the Mesozoic graben as pull-aparts (Le Pichon and Fox, 1971; Ballard and Uchupi, 1975; Sheridan, 1976; Manspeizer and Cousminer, 1988). Most recently, the shape of detachment faults underlying graben, together with transfer faults (the continental analogues of, but not necessarily extensions of, oceanic transform faults), have been called on to explain graben geometries (Gibbs, 1984; Lister et al., 1986a, 1986b; Etheridge et al., 1988). In the following discussion the Early Mesozoic stress field is analyzed using the orientation and distribution of diabase dykes of the Eastern North American dolerite province (McHone and Butler, 1984), and then structural models for the graben are examined in light of the inferred stress field.

Mesozoic graben may have formed in response to an extensional tectonic regime established by one or more "triple junctions" or "mantle plumes" approximately situated where the present Atlantic Ocean opened (de Boer and Snider, 1979; Nadon and Middleton, 1984). May (1971) related his radial system of dykes to thermal doming around a hot spot centred on the present south-central Atlantic Ocean. In the southern Appalachians, all dykes are at high angles to the orientation of the Mesozoic rift graben and Appalachian structural trends, reflecting local control of the stress regime by the hot spot at the time of dyke emplacement (McHone, 1978). In the northern Appalachians the dykes parallel the incipient Atlantic Ocean rift axis and define the regional orientation of maximum extension. The more northerly dykes are parallel to the overall orientation of the Mesozoic graben though the most important controls on graben orientations are Appalachian structural trends. It would appear then that there are problems with a model that proposes graben opening simply through extension (King, 1971).

Although Mesozoic graben follow Appalachian structural trends, whereas Early Jurassic dykes cut those trends, the distribution of both (with few exceptions) is limited to areas east of the Appalachian miogeocline. In the case of the dykes, this may reflect a basement boundary that acted as a Mesozoic strain limit (Papezik and Greenough, 1985; McHone et al., 1987). Alternatively, thinner, uncooled lithosphere below the newly accreted Paleozoic terranes of eastern North America possibly allowed magmas to reach upper crustal levels more easily, or determined the area where crustal stretching and thinning were greatest (during rifting), thus providing both less distance for the magma to travel in reaching the upper crust and routes (faults) along which magma could move. Dyke density also shows a dramatic decrease in the Canadian Appalachians but the dykes that do exist are large compared with most dykes in eastern North America (de Boer et al., 1988). Although more dykes may be found in the future (McHone et al., 1987), de Boer et al. (1988) suggested that the decrease in dyke density indicates lower amounts of extension in the Canadian Appalachians. The decrease in density more likely reflects a greater distance from the triple junction (Greenough and Papezik, 1986; Greenough and Hodych, 1990) inferred by May (1971).

Correlation between location of the western extent of the Mesozoic graben and the eastern boundary of the Appalachian miogeocline may result from absorption of extensional stresses by, and along, unannealed zones of weakness formed during the accretion of Paleozoic terranes. Alternatively a thinner lithosphere for the terranes may have permitted a concentration of extension effects. The first hypothesis is supported by positioning of graben along pre-existing fault systems reactivated as normal faults (Swanson, 1986).

A feature of the graben that is difficult to explain by pre-existing fault control is the offsets of graben along the orogen. The offsets may be caused by shearing that accompanied extension as a result of the differential rotation of North America away from Africa about a rotational pole situated in the Sahara (Swanson, 1982). In a model explaining inferred fault patterns in the Gulf of Maine, Ballard and Uchupi (1975) proposed that graben boundary faults (comparable to the Cobequid Fault along the Fundy Graben) and inferred dextral transverse faults represent

coupled shears prior to detachment (former) and after separation (latter) of North America from Africa. This shear couple resulted in west-northwest east-southeast extension.

Sheridan (1976) related inferred boundary faults for offshore graben, oceanic fracture systems, and onland fracture systems along the east coast of North America to three stress regimes (Atlantic Ocean, White Mountains, and Labrador Sea) that overlapped in space but acted independently in time. His Atlantic Ocean stress regime was active during the separation of the African plate from the North American plate. East-west extension produced north-south boundary faults (bordering most of the Mesozoic graben) with complementary northwest-southeast dextral shearing, represented by ocean floor fracture zones, the result of relative rotation of the plates during separation. A problem of at least local concern with the Ballard and Uchupi (1975) and Sheridan (1976) models is that the observed transverse faulting in the Fundy Graben and offsets on and within many southerly graben suggest sinistral (as opposed to dextral) shear. Nevertheless a similar model proposed by Manspeizer and Cousminer (1988) apparently relates the orientation of shear couples with observed sinistral displacement of the southerly Mesozoic graben but also suggests that movements on boundary faults and cross faults could be either sinistral or dextral depending on graben orientation relative to shear couples of the strain ellipse.

In detachment models for graben formation, low-angle faults, in some cases extending to the upper mantle, and pure shear in the underlying basement rocks, absorb the effects of lithospheric extension (e.g. Lister et al., 1986a; Etheridge et al., 1988). In the simplest case, graben form at right angles to the direction of maximum extension (Lister et al., 1986b). The specific structure of individual graben is dependent on the shape of underlying master detachment faults and the amount of pure shear in basement rocks below the detachment (Etheridge et al., 1988). Offsets between graben, and faults cutting graben, are explained by motion along transfer faults which may in some cases represent onland extensions of oceanic transform faults (Gibbs, 1984; Lister et al., 1986b). Relative displacements between graben can be either sinistral or dextral because of the interfingering of, and movement between, lithospheric segments separated by the transfer faults. Williams and Hy (1990) interpreted northwest-southeast-trending faults on the southeastern shore of Nova Scotia, as well as faults offsetting dykes in the Atlantic Provinces area, as transfer faults formed during opening of the Atlantic Ocean. Brown (1986) described seismic evidence that the Cobequid Fault dips eastward below the Bay of Fundy, apparently representing a major detachment surface formed through reactivation of a Paleozoic basement thrust fault. Ratcliffe and Burton (1985) and Olsen and Schlische (1990) showed that graben geometry can be accounted for by activation, through extension alone, of variably shaped faults dipping below the graben. Their data and conclusions appear consistent with general aspects of the detachment model.

Interpretation of the formation of fold structures in the graben rests on an assessment of the relative importance and effect of (for example) transcurrent movement along border faults, border fault geometry, and fold development. Swift et al. (1967) suggested that asymmetry of the Fundy syncline could be related to strike-slip movement on the Cobequid Fault. Ballard and Uchupi (1975) proposed that folding accompanied Late Triassic-Early Jurassic normal faulting (with a sinistral component of movement) of basement rocks underlying the Gulf of Maine. In shear models for graben formation, fold structures have been related to the buttressing effect of horst blocks formed during the evolution of right-slip pull-apart graben bounded by dextral strike-slip faults (Ratcliffe, 1980; Aydin and Nur, 1982). In these models folding accompanies sedimentation and formation of the pull-apart graben.

Ratcliffe and Burton (1985) proposed that the geometry of fold structures in the Newark graben can be related to variations in the shape and dip of reactivated basement faults that border and dip beneath the graben. Their model involves simple extension and suggests that although significant amounts of strike-slip movement occurred on boundary faults this movement played a small role in development of fold geometry. The model has advantages over other models in that it is simple and testable without a knowledge of Mesozoic stress regimes which are difficult to ascertain. It is also most consistent with the broader scale detachment models for graben formation.

It is possible that the formation of folds, and related features such as cleavage, joints, and distorted mud cracks (Stringer and Lajtai, 1979; Lajtai and Stringer, 1981; Burton and Ratcliffe, 1985; Harrison, 1985) occurred after graben formation (Lajtai and Stringer, 1981). De Boer et al. (1988) suggested that emplacement of a compressional tectonic regime accompanied the change from rifting to drifting as North America separated from Africa during the Early to Middle Jurassic. This is supported by reverse compressional faulting observed in the southeastern United States (Rader et al., 1986). Robbins (1981) reported that the fault systems defining the Newark graben in the eastern United States are presently active as high-angle reverse faults resulting from east-west compression. An east-west compressional stress field is also indicated by in situ strain measurements in New England sandstones (Engelder and Sbar, 1976) and numerical stress field solutions from earthquake data (Sykes and Sbar, 1973). It would seem reasonable that a similarly oriented stress field has remained in effect since the separation of North Africa and Europe from North America. If so, there is a long period of time (Jurassic-Cretaceous to present) during which folding may have occurred in Newark graben including the Fundy Graben.

Relationship of volcanism to tectonism

Magmatic activity associated with the second cycle of ocean opening produced igneous rocks of the Eastern North American dolerite province and the White Mountains magma series (McHone and Butler, 1984). The latter rocks have no lithological correlatives in Atlantic Canada. Hypotheses for origin of the White Mountains magma series include activation of crisscrossing fractures (Chapman, 1968), hotspot activity (e.g. Morgan, 1983; de Boer et al., 1988), activation of a deep-basement structure (McHone and Butler, 1984), and onland transform faulting (Ballard and Uchupi, 1975) during separation of North America from Africa.

The Eastern North American dolerites form the most extensive magmatic province associated with rifting and opening of the Atlantic Ocean. Volcanic rocks in continental rifts commonly show an increase in alkalinity away from the rift axis (Neumann and Ramberg, 1978) but in general such a relationship has not been recognized for the Eastern North American dolerites. Hermes et al. (1984) suggested that alkaline rocks east of tholeiites in the Connecticut Graben represent flank volcanic activity with rifting centred on the graben itself. This model does not apply to the Atlantic Provinces, as dykes outside the Fundy Graben are no more alkaline than North Mountain Basalt within the Fundy Graben (Greenough and Papezik, 1986).

Models for continental rifting suggest that the dominantly tholeiitic composition and high volume of eastern North American magmas are the result of rapid and extensive lithospheric stretching (e.g. Le Pichon and Sibuet, 1981). Thinned lithosphere is replaced by hot mantle material which undergoes large percentages of melting as a result of decompression. White and McKenzie (1989) proposed that major dyke and flood basalt provinces form when the stretching location coincides with a mantle hot spot such as that proposed by May (1971) for the Eastern North American dolerite province. High percentages of melting ensue from pressure decreases in thermally anomalous upwelling mantle and this produces large volumes of tholeiitic magma (McKenzie and Bickle, 1988). The geochemical similarities of these rocks to MORB tholeiites (Wark and Clarke, 1980) support the hypothesis for large percentages of melting and substantial thinning of the crust. Such hot spot, rift-related igneous outbursts have occurred at least six times in the last 200 Ma (White and McKenzie, 1989). They are associated with mass extinction events such as recorded in Newark graben sedimentary rocks at the Triassic-Jurassic boundary (White, 1989), thus providing an alternative to the meteorite impact hypothesis of Olsen et al. (1987).

Second cycle volcanism contrasts markedly with first cycle (Late Permian-Triassic) igneous activity. The first cycle produced small volumes of alkaline to tholeiitic magmas indicating lesser amounts of crustal thinning. It is important to note that there is no evidence for large-scale crustal extension below the onland graben during the second cycle. High gravity and magnetic anomalies related to mafic intrusions along the axes of rifts do not occur below the onland Mesozoic graben as is best illustrated by the absence of gravity or magnetic highs associated with the large Fundy Graben (see maps by Haworth et al., 1980; Zietz et al., 1980). Further, cooling of thermally perturbed lithosphere below thinned crust results in prolonged subsidence following rifting such as is observed for the Scotian Shelf (Keen, 1979; Royden and Keen, 1980). Subsidence in the onland Mesozoic graben stopped about the time that North America separated from Africa and apparently never resumed to any great extent, suggesting little thermal perturbation and associated crustal thinning.

A major impediment to constructing a detailed chronological account of the relationships between faulting, graben formation, magmatism, and rifting is that the time of drifting is poorly constrained. The oldest recorded age for ocean floor basalt off the southeast coast of North America (DSDP Sites 100 and 105) is about 150 Ma (Bryan et al., 1977), which is much younger than the time of drifting. It has been suggested that the peak period for Eastern North American province magmatism, 190 Ma, marks the separation of North America from Africa (e.g. McHone and Butler, 1984). This crudely corresponds with age estimates for the time of drifting from magnetic lineations and inferred seafloor spreading rates ranging from 170 to 190 Ma (see review in Klitgord and Schouten, 1986). However, most isotopic dates for the Eastern North American tholeiitic basalts (190 Ma) do not correspond with 204 to 214 Ma ages for Hettangian rocks (Hodych and Hayatsu, 1988), inferred from paleontological data (e.g. North Mountain Basalt, Olsen et al., 1987; Harland et al., 1982; Kent and Gradstein, 1986). The 15 to 20 Ma age discrepancy indicates one or more of the following: that most of the Ar-based age dates have been reset thus giving low values, that the paleontological assignment of these rocks to the Hettangian is incorrect as a result of imprecise age calibration of the palynological floral zones, or that widely accepted time scales for the Mesozoic are in error. Recent studies, some utilizing U-Pb systematics, support the first suggestion and indicate that nearly all of the igneous activity occurred at 201 ± 2 Ma (Sutter, 1988; Dunning and Hodych, 1990; Hodych and Dunning, 1992).

Middle Jurassic-Early Cretaceous

Tectonic controls on sedimentation

From Middle Jurassic through Early Cretaceous, tectonic activity in the Atlantic Provinces reflected adjustments to Early Jurassic opening of the southern North Atlantic. This was a prolonged period of extension that separated Iberia from eastern Newfoundland, started opening of the Labrador Sea, and saw crustal adjustment to these rifting events through tectonic uplift. The uplift resulting from separation of Africa from North America led to erosion and nondeposition in most onland graben, but in the offshore it yielded late Early to early Middle Jurassic continental clastics of the Mohawk Formation on the Scotian Shelf (Jansa and Wade, 1974). Marine transgression during late Middle to early Late Jurassic, produced shales, sandstones, and limestones of the Mic Mac Formation and carbonates of the Abenaki Formation (Jansa and Wade, 1974). These and the Ford's Bight Breccias in Labrador may represent the initial effects of extension that separated Iberia from Newfoundland during Early Cretaceous (Valanginian) time (Srivastava and Tapscott, 1986).

Tectonic uplift in response to decoupling of Iberia and Newfoundland caused a marine regression that lasted for most of the Early Cretaceous. Deltaic sedimentary rocks of the Mississauga and Logan Canyon formations are products of this regression (Jansa and Wade, 1974). Dickie (1987) showed that Lower Cretaceous clays and silica sands of Nova Scotia, which are time and lithological correlatives of the Mississauga Formation, provide evidence for a vast area affected by deltaic sedimentation.

Little is known of structural activity (faulting and folding) that almost certainly occurred during this period as a result of rifting east and northeast of Newfoundland. Dyke orientations suggest that the direction of maximum extension varied dramatically from the New England area, where it was approximately north-south (McHone, 1978), to the Scotian Shelf (Jansa and Pe-Piper, 1988) and Notre Dame Bay (Fig. 6.4b) areas where it was dominantly east-northeast-west-southwest. This apparently reflects control of the stress regime in the New England area by oceanic fracture zones (McHone, 1978) with more northerly areas affected by the north-northwest-south-southeast axis of incipient rifting (Jansa and Wade, 1974) east and northeast of Newfoundland.

Relationship of volcanism to tectonism

Rifting in the northern North Atlantic occurred sequentially (Sullivan, 1983); Iberia separated from the Grand Banks, Britain rotated away from northeastern Newfoundland, and Greenland rotated away from Labrador through activation of a series of fracture zones. Rifting processes that led to separation of plates occurred throughout most of the Early Cretaceous and into the Late Cretaceous (e.g. Jansa and Wade, 1974; Srivastava and Tapscott, 1986). Volcanism associated with this rifting activity encompassed a similar time span (135 to 90 Ma, Table 6.2). The igneous activity was for the most part restricted to northern North Atlantic areas (north of New England in North America and north of Iberia in Europe) reflecting the areas affected by extension. Along with their timing, the alkaline character of the Jurassic-Cretaceous igneous rocks has been used to relate them to rifting in the northern North Atlantic (e.g. Strong and Harris, 1974; Lapointe, 1979; Umpleby, 1979; Jansa and Pe-Piper, 1986; Pe-Piper and Jansa, 1987), but there is considerable difference of opinion on the specific causes of formation. Some of the volcanic rocks appear to be on strike with oceanic shear zones represented by the New England, Fogo, and Newfoundland seamounts (Fig. 6.2). As a result, the volcanism has been related to activation (or reactivation) of continental extensions of these "leaky" transform faults during rifting (e.g. Le Pichon and Fox, 1971; Foland and Faul, 1977; McHone, 1981; Jansa and Pe-Piper, 1985; McHone et al., 1987). The hypothesis does not explain why volcanism is restricted in many cases to certain sections of the fault extensions. Kumarapeli (1978) related igneous activity of the New England-Quebec magmatic province to reactivation of the St. Lawrence rift system, but this model does not help explain time-correlative volcanic rocks such as on the Scotian Shelf.

Other authors noted that some of the volcanic rocks show a linear distribution (e.g. New England-Quebec province; Crough, 1981; Morgan, 1983) or realized that plate reconstructions place various volcanic rocks (e.g. Notre Dame Bay lamprophyres) above mantle plumes active today, such as the Azores (de Boer et al., 1988). These workers suggested that the volcanism resulted from hotspot activity. The hot-spot model predicts that volcanic rocks should show a regular age progression because of movement of the lithosphere over thermal anomalies, but in at least some cases (e.g. New England-Quebec province and extensions on Georges Bank to the New England Seamounts) the available age data do not fit the model or there is evidence for several episodes of volcanic activity (McHone, 1981; Jansa and Pe-Piper, 1988).

Concluding remarks

The Middle Jurassic-Early Cretaceous cycle of rifting had a remarkably different effect on the continental lithosphere (and on resulting volcanism) than the Late Triassic-Early

Jurassic cycle, in that it affected a narrower band of lithosphere with graben only forming off eastern and north-eastern Newfoundland. The resulting incipient rift axis cut across Paleozoic structural trends, as opposed to paralleling them as in the previous cycle (Williams, 1978a, 1984). Similarly Jurassic-Cretaceous volcanic rocks even occur on pre-Paleozoic continental crust (e.g. Ford's Bight, Labrador). Furthermore, the localized style of volcanic eruptions, explosive activity forming volcanic cones, and alkaline character of the rocks contrast sharply with the tholeiitic dykes and lava flows of Late Triassic-Early Jurassic time, which stretch for hundreds of kilometres. Both the Late Triassic-Early Jurassic and Middle Jurassic-Early Cretaceous rifting cycles culminated in the separation of plates but the reason(s) for these important differences in the effects of rifting particularly with regard to the character of volcanism await explanation.

ACKNOWLEDGMENTS

I wish to thank P. Stringer for thoughtful comments and criticism on various aspects of the paper. H. Williams provided helpful recommendations on content and organization, and the thoughtful comments of E.R.W. Neale led to numerous improvements in the manuscript. D. Hattie contributed many important references as well as encouragement. References and reprints were provided by B. Cameron and G. Dickie. Stimulating discussions with J. Colwell, J. Dostal, J. Hodych, J.V. Owen, G. Pe-Piper, G. Stevens, and P. Williams also added to the paper. B. Webb helped with the preparation of tables and diagrams. GSC personnel provided technical editing and prepared final versions of the tables and figures. The author's research was made possible through an operating grant from the Natural Sciences and Engineering Research Council of Canada.

REFERENCES

Alcock, F.J.
1938: Geology of Saint John region, New Brunswick; Geological Survey of Canada, Memoir 216, 65 p.

Amoco Canada Petroleum Company Ltd.
1974: Well history report, Amoco-Imperial-Skelly Twillick G-49; released by Canada Oil and Gas Land Administration in 1976.

Aumento, F.
1966: Zeolite minerals, Nova Scotia; in Geology of Parts of the Atlantic Provinces; Geological Association of Canada, Guidebook, p. 71-77.

Aydin, A. and Nur, A.
1982: Evolution of pull-apart basins and their scale independence; Tectonics, v. 1, p. 91-105.

Bailey, L.W. and Matthew, G.F.
1865: Observations on the geology of southern New Brunswick, made principally during the summer of 1864 by Professor L.W. Bailey and Messrs. G.F. Matthew and C.F. Hart; Queen's Printer, Fredericton, New Brunswick, 185 p.

Baird, D.M.
1972: Burntcoat, Upper Triassic; in Vertebrate Paleontology of Eastern Canada, (ed.) R.L. Carroll, E.S. Belt, D.L. Dineley, D.M. Baird, and D.C. McGregor; 24th International Geological Congress, Montreal, Excursion A59, p. 22-30.
1976: Dinosaur footprints; Ichnology Newsletter, v. 9, p. 5.

Ballard, R.D. and Uchupi, E.
1975: Triassic rift structure in the Gulf of Maine; American Association of Petroleum Geologists Bulletin, v. 59, p. 1041-1072.

Bedard, J.H.
1992: Jurassic quartz-normative tholeiite dikes from Anticosti Island, Quebec; in Eastern North American Mesozoic Magmatism, (ed.) J.H. Puffer and P. Ragland; Geological Society of America, Special Paper 268, p. 161-167.

Bellini, F.X., Corkum, D.H., and Stewart, A.J.
1982: Geology of foundation excavations at Seabrook Station, Seabrook, New Hampshire; in Geotechnology in Massachusetts Proceedings of a Conference in March 1980, University of Massachusetts, Amherst, Massachusetts, p. 109-117.

Belyea, H.R.
1939: The geology of the Musquash area, New Brunswick; Ph.D. thesis, Northwestern University, Evanston, Illinois, 109 p.

Bertrand, H., Dostal, J., and Dupuy, C.
1982: Geochemistry of Early Mesozoic tholeiites from Morocco; Earth and Planetary Science Letters, v. 58, p. 225-239.

Brown, D.W.
1986: The Bay of Fundy: thin-skinned tectonics and resultant Early Mesozoic sedimentation (abstract); in Basins of Eastern Canada and Worldwide Analogues Program with Abstracts, Symposium, August 13-15, 1986; Atlantic Geoscience Society, Halifax, Canada, p. 28.

Bryan, W.B., Frey, F.A., and Thompson, G.
1977: Oldest Atlantic sea-floor: Mesozoic basalt from western North Atlantic margin and Eastern North America; Contributions to Mineralogy and Petrology, v. 64, p. 223-242.

Burke, K.B.S., Hamilton, J.B., and Gupta, V.K.
1973: The Caraquet dike: its tectonic significance; Canadian Journal of Earth Sciences, v. 10, p. 1760-1768.

Burton, W.C. and Ratcliffe, N.M.
1985: Compressional structures associated with right-oblique normal faulting of Triassic-Jurassic strata of the Newark Basin near Flemington, New Jersey (abstract); Geological Society of America, Abstracts with Programs, v. 17, no. 1, p. 9.

Cameron, B.
1986: Jurassic fossils from the Scots Bay Formation; in Tenth Annual Open House and Review of Activities, Program and Summaries; Nova Scotia Department of Mines and Energy, Information Series no. 12, p. 167-169.

Cameron, B. and Jones, J.R.
1987: Discovery of fossils and meandering stream deposits in the Late Triassic Blomidon Formation of Nova Scotia; in Mines and Minerals Branch Report of Activites 1987, Part A, (ed.) J.L. Bates and D.R. MacDonald; Nova Scotia Department of Mines and Energy, Report 87-5, p. 179-181.
1988: Plant fossils from the Early Mesozoic Fundy Group of the Annapolis Valley region of Nova Scotia; in Mines and Minerals Branch Report of Activites 1988, Part A, (ed.) D.R. MacDonald and Y. Brown; Nova Scotia Department of Mines and Energy, Report 88-3, p. 173-177.
1989: Late Triassic footprints and other trace fossils from the Wolfville Formation, eastern Annapolis Valley, Nova Scotia; in Mines and Minerals Branch Report of Activities 1989, Part A, (ed.) D.R. MacDonald and K.A. Mills; Nova Scotia Department of Mines and Energy, Report 89-3, p. 153-157.

Cameron, B., Rogers, D., Grantham, R., and Jones, J.R.
1985: Silicification and silicified microfossils and stromatolites from the Scots Bay Formation, Fundy Basin, Nova Scotia: a progress report; Maritime Sediments and Atlantic Geology, v. 21, p. 41-42.

Cameron, H.C.
1956: Tectonics of the Maritime area; Royal Society of Canada Transactions, v. 50, series 3, section 4, p. 45-51.

Carmichael, C.M. and Palmer, H.C.
1968: Paleomagnetism of the Late Triassic North Mountain basalt of Nova Scotia; Journal of Geophysical Research, v. 13, p. 2811-2833.

Chapman, C.A.
1968: A comparison of the Maine coastal plutons and the magmatic central complexes of New Hampshire; in Studies of Appalachian Geology: Northern and Maritime, (ed.) E-An Zen, W.S. White, J.B. Hadley, and J.B. Thompson, Jr.; Interscience Publishers, p. 385-396.

Chevron Canada Resources Ltd.
1983: Well history report, Chevron Cape Spencer P-79; unpublished company report released to the public by Canada Oil and Gas Land Administration in 1985, 38 p.

Clarke, D.B.
1977: The Tertiary volcanic province of Baffin Bay; in Volcanic Regimes in Canada, (ed.) W.R.A. Baragar, L.C. Coleman, and J.M. Hall; Geological Association of Canada, Special Paper 16, p. 445-460.

Cloos, E. and Pettijohn, F.J.
1973: Southern border of the Triassic basin west of York, Pennsylvania; fault or overlap?; Geological Society of America Bulletin, v. 84, p. 523-536.

Colwell, J.A.
1980: Zeolites in the North Mountain Basalt, Nova Scotia; Geological Association of Canada-Mineralogical Association of Canada, Field Trip Guide Book, Trip 18, 16 p.

Colwell, J.A. and Cameron, B.
1985: North Mountain Basalt. Notes for the 1985 Atlantic Universities Geologic Conference; Acadia University, Wolfville, Nova Scotia, 6 p.

Cornet, B.
1977: The palynostratigraphy and age of the Newark Supergroup; Ph.D. thesis, Pennsylvania State University, University Park, Pennsylvania, 505 p.

Cornet, B. and Traverse, A.
1975: Palynological contributions to the chronology and stratigraphy of the Hartford Basin in Connecticut and Massachusetts; Geoscience and Man, v. II, p. 1-33.

Cornet, B., Traverse, A., and McDonald, N.G.
1973: Fossil spores, pollen, and fishes from Connecticut indicate Early Jurassic age for part of the Newark Group; Science, v. 182, p. 1243-1247.

Crosby, D.G.
1962: Wolfville map-area, Nova Scotia; Geological Survey of Canada, Memoir 325, 67 p.

Crosby, R.M., Greenough, J.D., Hattie, D., and Venugopal, D.V.
1990: Post-Triassic mineralization in central New Brunswick: implications of the McBean Brook Zn-Pb-Ag occurrence; Atlantic Geology, v. 26, p. 1-9.

Crough, S.T.
1981: Mesozoic hotspot epeirogeny in Eastern North America; Geology, v. 9, p. 2-6.

Currie, K.L.
1975: The alkaline rocks of Canada; Geological Survey of Canada, Bulletin 239, 228 p.

Dallmeyer, R.D.
1975: The Palisades sill: a Jurassic intrusion? Evidence from $^{40}Ar/^{39}Ar$ incremental release ages; Geology, v. 3, p. 243-245.

de Boer, J. and Snider, F.G.
1979: Magnetic and chemical variations of Mesozoic diabase dikes from Eastern North America: evidence for a hot-spot in the Carolinas?; Geological Society of America Bulletin, v. 90, p. 185-198.

de Boer, J., McHone, J.G., Puffer, J.H., Ragland, P.C., and Whittington, D.
1988: Mesozoic and Cenozoic magmatism; in The Atlantic Continental Margin, (ed.) R.E. Sheridan and J.A. Grow; The Geology of North America, Geological Society of America, v. I-2, p. 217-241.

Deutsch, E.R. and Rao, K.V.
1977: Paleomagnetism of Mesozoic lamprophyres from central Newfoundland (abstract); EOS, Transactions, American Geophysical Union, v. 58, p. 745.

Dickie, G.B.
1987: Cretaceous deposits of Nova Scotia; Nova Scotia Department of Mines and Energy, Paper 86-1, 54 p.

Donohoe, H.V. and Wallace, P.I.
1980: Structure and stratigraphy of the Cobequid Highlands, Nova Scotia; Geological Association of Canada-Mineralogical Association of Canada, Field Trip Guidebook, Trip 19, 64 p.
1982: Geological map of the Cobequid Highlands, Colchester, Cumberland and Pictou counties, Nova Scotia; Nova Scotia Department of Mines and Energy, Halifax, Nova Scotia, Maps 82-6, 82-7, 82-8, and 82-9.

Dostal, J. and Dupuy, C.
1984: Geochemistry of the North Mountain basalt (Nova Scotia, Canada); Chemical Geology, v. 45, p. 245-261.

Dostal, J. and Greenough, J.D.
1992: Geochemistry and petrogenesis of the early Mesozoic North Mountain basalts of Nova Scotia, Canada; in Eastern North American Mesozoic Magmatism, (ed.) J.H. Puffer and P. Ragland; Geological Society of America, Special Paper 268, p. 149-159.

Dunn, T. and Stringer, P.
1990: Petrology and petrogenesis of the Ministers Island dike, southwest New Brunswick, Canada; Contributions to Mineralogy and Petrology, v. 105, p. 55-65.

Dunning, G. and Hodych, J.P.
1990: U/Pb zircon and baddeleyite ages for the Palisades and Gettysburg sills of the northeastern United States: implications for the age of the Triassic/Jurassic boundary; Geology, v. 18, p. 795-798.

Engelder, J.T. and Sbar, M.L.
1976: Evidence for uniform strain orientation in the Potsdam Sandstone, northern New York, from in situ measurements; Journal of Geophysical Research, v. 81, p. 3013-3017.

Ericson, G.P. and Kulp, K.L.
1961: Potassium-argon measurements on the Palisades sill, New Jersey; Geological Society of America Bulletin, v. 72, p. 649-652.

Etheridge, M.A., Symonds, P.A., and Lister, G.S.
1988: Application of detachment extensional models to the interpretation of conjugate margin pairs (abstract); Geological Association of Canada, Mineralogical Association of Canada and Canadian Society of Petroleum Geologists, Joint Annual Meeting, Program With Abstracts, v. 13, p. A38.

Faill, R.T.
1973: Tectonic development of the Triassic Newark-Gettysburg Basin in Pennsylvania; Geological Society of America Bulletin, v. 84, p. 725-740.

Fall, H.G., Gibb, F.G.F., and Kanaris-Sotiriou, R.
1982: Jurassic volcanic rocks of the northern North Sea; Journal of the Geological Society of London, v. 139, p. 277-292.

Foland, K.A. and Faul, H.
1977: Ages of the White Mountains intrusives - New Hampshire, Vermont, and Maine, U.S.A.; American Journal of Science, v. 277, p. 888-904.

Foley, S.F.
1984: Liquid immiscibility and melt segregation in alkaline lamprophyres from Labrador; Lithos, v. 17, p. 127-137.

Froelich, A.J. and Olsen, P.E.
1985: Newark Supergroup, a revision of the Newark Group in Eastern North America; in Proceedings of the Second United States Geological Survey Workshop on the Early Mesozoic Basins of the Eastern United States, (ed.) G.R. Robinson and A.J. Froelich; United States Geological Survey, Circular 946, p. 1-3.

Gibb, F.G.F. and Kanaris-Sotiriou, R.
1976: Jurassic igneous rocks of the Forties Field; Nature, v. 260, p. 23-25.

Gibbs, A.D.
1984: Structural evolution of extensional basin margins; Journal of the Geological Society of London, v. 141, p. 609-620.

Greenough, J.D.
1979: The geochemistry of Hawaiian lavas; M.Sc. thesis, Carleton University, Ottawa, Ontario, 104 p.

Greenough, J.D. and Dostal, J.
1992a: Layered rhyolite bands in a thick North Mountain Basalt flow: the products of silicate liquid immiscibility?; Mineralogical Magazine, v. 56, p. 309-318.
1992b: Cooling history and differentiation of a thick North Mountain Basalt flow (Nova Scotia, Canada); Bulletin of Volcanology, v. 55, p. 63-73.

Greenough, J.D. and Fryer, B.J.
1991: Nd and Sr isotopic composition of the lamprophyric Malpeque Bay sill, Prince Edward Island, Canada; Canadian Mineralogist, v. 29, p. 311-317.

Greenough, J.D. and Hodych, J.P.
1990: Evidence for lateral magma injection in the Early Mesozoic dykes of Eastern North America; in Mafic Dykes and Emplacement Mechanisms, Proceedings of the Second International Dyke Conference, (ed.) A.J. Parker, P.C. Rickwood, and D.H. Tucker, Balkema, Rotterdam, p. 35-46.

Greenough, J.D. and Papezik, V.S.
1986: Petrology and geochemistry of the Early Mesozoic Caraquet dyke, New Brunswick, Canada; Canadian Journal of Earth Sciences, v. 23., p. 193-201.
1987: Note on the petrology of North Mountain basalt from the wildcat oil well Mobil Gulf Chinampas N-37, Bay of Fundy, Canada; Canadian Journal of Earth Sciences, v. 24, p. 1255-1260.

Greenough, J.D., Hayatsu, A., and Papezik, V.S.
1988: Mineralogy, petrology and geochemistry of the alkaline Malpeque Bay sill, Prince Edward Island, Canada; Canadian Mineralogist, v. 26, p. 97-108.

Greenough, J.D., Jones, L.M., and Mossman, D.J.
1989a: Petrochemical and stratigraphic aspects of North Mountain basalt from the north shore of the Bay of Fundy, Nova Scotia, Canada; Canadian Journal of Earth Sciences, v. 26, p. 2710-2717.
1989b: The Sr isotopic composition of Early Jurassic mafic rocks of Atlantic Canada: implications for assimilation and injection mechanisms affecting mafic dykes; Chemical Geology (Isotope Geoscience Section), v. 80, p. 17-26.

Hacquebard, P.A.
1984: Composition, rank and depth of two Nova Scotia lignite deposits; Current Research, Part A; Geological Survey of Canada, Paper 84-1A, p. 11-15.

Hamilton, J.B. and Gupta, V.K.
1972: The Caraquet Dike; Mineral Development Branch, New Brunswick Department of Natural Resources, Topical Report 72-1, 20 p.

Harland, W.B., Cox, A.V., Llewellyn, P.G., Pickton, C.A.G., Smith, A.G., and Walters, R.
1982:   A Geologic Time Scale; Cambridge University Press, Cambridge, England, 131 p.

Harrison, D.H.
1985:   An analysis of the tectonic strain in Triassic rocks around the Jacksonwald syncline, Jacksonwald, Pennsylvania; Senior thesis, Bryn Mawr College, Bryn Mawr, Pennsylvania, 47 p.

Harrison, R.K., Snelling, N.J., Merriman, R.J., Morgan, G.E., and Goode, A.A.J.
1977:   The Wolf Rock, Cornwall; Geological Magazine, v. 114, p. 249-264.

Haworth, R.T. and Keen, C.E.
1979:   The Canadian Atlantic Margin: a passive continental margin encompassing an active past; Tectonophysics, v. 59, p. 83-126.

Haworth, R.T., Daniels, D.L., Williams, H., and Zietz, I.
1980:   Bouguer gravity anomaly map of the Appalachian Orogen; Memorial University of Newfoundland Map no. 3a, Memorial University of Newfoundland, St. John's, Newfoundland.

Hayatsu, A.
1979:   K-Ar isochron age of the North Mountain basalt, Nova Scotia; Canadian Journal of Earth Sciences, v. 16, p.973-975.

Haycock, E.
1900:   Records of post-Triassic changes in Kings County, Nova Scotia; Transactions of the Nova Scotia Institute of Science, v. 10, p. 287-302.

Hayes, A.O. and Howell, B.F.
1937:   Geology of Saint John, New Brunswick; Geological Society of America, Special Paper 6, 146 p.

Helwig, J., Aronson, J., and Day, D.S.
1974:   A Late Jurassic mafic pluton in Newfoundland; Canadian Journal of Earth Sciences, v. 11, p. 1314-1319.

Hermes, O.D., Rao, J.M., Dickenson, M.P., and Pierce, T.A.
1984:   A transitional alkalic dolerite dike suite of Mesozoic age in southeastern New England; Contributions to Mineralogy and Petrology, v. 86, p. 386-397.

Hiscott, E.J.
1986:   Nova Scotia's stupendous fossil find; Atlantic Advocate, June, p. 42-44.

Hodych, J.P. and Dunning, G.R.
1992:   Did the Manicouagan impact trigger end-of-Triassic mass extinction?; Geology, v. 20, p. 51-54.

Hodych, J.P. and Hayatsu, A.
1980:   K-Ar isochron age and paleomagnetism of diabase along the trans-Avalon aeromagnetic lineament: evidence of Late Triassic rifting in Newfoundland; Canadian Journal of Earth Sciences, v. 17, p. 491-499.
1988:   Paleomagnetism and K-Ar isochron dates of Early Jurassic basaltic flows and dikes of Atlantic Canada; Canadian Journal of Earth Sciences, v. 25, p. 1972-1989.

Holden, R.
1913:   Fossil plants from Eastern Canada; Annals of Biology, v. 27, p. 243-255.

Holm, P.E.
1982:   Non-recognition of continental tholeiites using the Ti-Y-Zr diagram; Contributions to Mineralogy and Petrology, v. 79, p. 308-310.

Hubert J.F. and Hyde, M.G.
1982:   Sheet-flow deposits of graded beds and mudstones on an alluvial sandflat-playa system: Upper Triassic Blomidon redbeds, St. Mary's Bay, Nova Scotia; Sedimentology, v. 29, p. 457-474.

Hubert, J.F. and Mertz, K.A.
1980:   Eolian dune field of Late Triassic age, Fundy Basin, Nova Scotia; Geology, v. 8, p. 516-519.
1981:   Reply on "Eolian dune field of Late Triassic age, Fundy Basin, Nova Scotia"; Geology, v. 9, p. 559.
1984:   Eolian sandstones in Upper Triassic-Lower Jurassic red beds of the Fundy Basin, Nova Scotia; Journal of Sedimentary Petrology, v. 54, p. 798-810.

Jackson, C.T. and Alger, F.
1829:   A description of the mineralogy and geology of a part of Nova Scotia; American Journal of Science, v. 15, p. 132-160.

Jansa, L.F.
1986:   Paleoceanography and evolution of the North Atlantic Ocean basin during the Jurassic; in The Western North Atlantic Region, (ed.) P.R. Vogt and B.E. Tucholke; Geological Society of America, The Geology of North America, v. M, p. 603-616.

Jansa, L.F. and Pe-Piper, G.
1985:   Early Cretaceous volcanism on the northeastern American margin and implications for plate tectonics; Geological Society of America Bulletin, v. 96, p. 83-91.
1986:   Geology and geochemistry of Middle Jurassic and Early Cretaceous igneous rocks on the Eastern North American continental shelf; Geological Survey of Canada, Open File 1351, 72 p.
1988:   Middle Jurassic to Early Cretaceous igneous rocks along Eastern North American continental margin; American Association of Petroleum Geologists Bulletin, v. 72, p. 347-366.

Jansa, L.F. and Wade, J.A.
1974:   Geology of the continental margin off Nova Scotia and Newfoundland; in Offshore Geology of Eastern Canada, Volume 2 Regional Geology, (ed.) W.J.M. Van der Linden and J.A. Wade; Geological Survey of Canada, Paper 74-30, v. 2, p. 51-105.

Jones, L.M. and Mossman, D.J.
1988:   The isotopic composition of strontium and the source of the Early Jurassic North Mountain basalt, Nova Scotia; Canadian Journal of Earth Sciences, v. 25, p. 942-944.

Keen, C.E.
1979:   Thermal history and subsidence of rifted continental margins - evidence from wells on the Nova Scotian and Labrador shelves; Canadian Journal of Earth Sciences, v. 16, p. 505-522.

Keen, M.J. and Williams, G.L. (ed.)
1990:   Geology of the Continental Margin of Eastern Canada; Geological Survey of Canada, Geology of Canada, no. 2, 855 p. (also Geological Society of America, The Geology of North America, v. I-2).

Kent, D.V. and Gradstein, F.M.
1986:   A Jurassic to recent chronology; in The Western North Atlantic Region, (ed.) P.R. Vogt and B.E. Tucholke; Geological Society of America, The Geology of North America, v. M, p. 45-50.

Keown, M.E. and Bint, B.W.
1975:   Well history report, Mobile Gulf Chinampas N-37; unpublished company report released to the public by Canada Oil and Gas Land Administration in 1977, 57 p.

Keppie, J.D.
1977:   Tectonics of southern Nova Scotia; Nova Scotia Department of Mines, Paper 77-1, 34 p.
1979:   Geologic Map of the Province of Nova Scotia; Nova Scotia Department of Mines and Energy.
1982:   The Minas geofracture; in Major Structural Zones and Faults of the Northern Appalachians, (ed.) P. St-Julien and J. Beland; Geological Association of Canada, Special Paper 24, p. 263-280.

King, A.F. and McMillan, N.J.
1975:   A mid-Mesozoic breccia from the coast of Labrador; Canadian Journal of Earth Sciences, v. 12, p. 44-51.

King, L.H. and MacLean, B.
1970:   Continuous seismic reflection study of Orpheus gravity anomaly; American Association of Petroleum Geologists Bulletin, v. 54, p. 1127-1146.

King, P.B.
1971:   Systematic pattern of Triassic dikes in the Appalachian region. Second Report; United States Geological Survey, Professional Paper 750-D, p. 84-88.

Klein, G.D.
1962:   Triassic sedimentation, Maritime Provinces, Canada; Geological Society of America Bulletin, v. 73, p. 1127-1146.
1963a: Regional implications of Triassic paleocurrents, Maritime Provinces, Canada; Journal of Geology, v. 71, p. 801-808.
1963b: Boulder surface markings in Quaco Formation (Upper Triassic), St. Martins, New Brunswick, Canada; Journal of Sedimentary Petrology, v. 33, p. 49-52.

Kumarapeli, P.S.
1978:   The St. Lawrence paleo-rift system: a comparative study; in Tectonics and Geophysics of Continental Rifts, (ed.) I.B. Ramberg and E.R. Neumann; D. Reidel Publishing Company, New York, p. 367-384.

Lajtai, E.Z. and Stringer, P.
1981:   Joints, tensile strength and preferred fracture orientation in sandstones, New Brunswick and Prince Edward Island, Canada; Maritime Sediments and Atlantic Geology, v. 17, p. 70-87.

Lapointe, P.L.
1979:   Paleomagnetism of the Notre Dame Bay lamprophyre dikes, Newfoundland, and the opening of the North Atlantic Ocean; Canadian Journal of Earth Sciences, v. 16, p. 1823-1831.

Larochelle, A.
1967:   Palaeomagnetic directions of a basic sill in Prince Edward Island; Geological Survey of Canada, Paper 67-39, p. 1-6.

Le Pichon, X. and Fox, P.J.
1971: Marginal offsets, fracture zones, and the early opening of the North Atlantic; Journal of Geophysical Research, v. 76, p. 6294-6308.

Le Pichon, X. and Sibuet, J-C.
1981: Passive margins: a model of formation; Journal of Geophysical Research, v. 86, p. 3708-3720.

LeTourneau, P.M.
1985: Alluvial fan development in the Lower Jurassic Portland Formation, Central Connecticut - implications for tectonics and climate; in Proceedings of the Second United States Geological Survey Workshop on the Early Mesozoic Basins of the Eastern United States, (ed.) G.R. Robinson and A.J. Froelich; United States Geological Survey, Circular 946, p. 17-26.

Liew, M.Y.-C.
1976: Structure, geochemistry and stratigraphy of Triassic rocks, north shore of Minas Basin, Nova Scotia; M.Sc. thesis, Acadia University, Wolfville, Nova Scotia.

Lin, C.L.
1971: Cretaceous deposits in the Musquodoboit River Valley, Nova Scotia. Canadian Journal of Earth Sciences, v. 8, p. 1152-1154.

Lister, G.S., Etheridge, M.A., and Symonds, P.A.
1986a: Detachment faulting and the evolution of passive continental margins; Geology, v. 14, p. 246-250.
1986b: Reply on "Detachment faulting and the evolution of passive continental margins"; Geology, v. 14, p. 891-892.

Lockhart, A.W.
1984: Sussex silica and aggregate; New Brunswick Department of Natural Resources and Energy, Assessment File No. 473023, 17 p.

Magnusson, D.H.
1955: The Triassic sedimentary rocks at St. Martins, New Brunswick; M.Sc. thesis, University of New Brunswick, Fredericton, New Brunswick, 97 p.

Malpas, J., Foley, S.F., and King, A.F.
1986: Alkaline mafic and ultramafic lamprophyres from the Aillik Bay area, Labrador; Canadian Journal of Earth Sciences, v. 23, p. 1902-1918.

Manspeizer, W. and Cousminer, H.L.
1988: Late Triassic-Early Jurassic synrift basins of the U.S. Atlantic margin; in The Atlantic Continental Margin, (ed.) R.E. Sheridan and J.A. Grow; Geological Society of America, The Geology of North America, v. I-2, p. 197-216.

Manspeizer, W., Puffer, J.H., and Cousminer, H.L.
1978: Separation of Morocco and Eastern North America: a Triassic-Liassic stratigraphic record; Geological Society of America Bulletin, v. 89, p. 901-920.

May, P.R.
1971: Pattern of Triassic-Jurassic diabase dikes around the North Atlantic in context of predrift position of the continents; Geological Society of America Bulletin, v. 82, p. 1285-1291.

McHone, J.G.
1978: Distribution, orientations and ages of mafic dikes in central New England; Geological Society of America Bulletin, v. 89, p. 1645-1655.
1981: Comment on "Mesozoic hotspot epeirogeny in eastern North America"; Geology, v. 9, p. 341-343.
1992: Mafic dike suites within Mesozoic igneous provinces of New England and Atlantic Canada; in Eastern North American Mesozoic Magmatism, (ed.) J.H. Puffer and P. Ragland; Geological Society of America Special Paper 268, p. 1-11.

McHone, J.G. and Butler, J.R.
1984: Mesozoic igneous provinces of New England and the opening of the North Atlantic Ocean; Geological Society of America Bulletin, v. 95, p. 757-765.

McHone, J.G. and Trygstad, J.C.
1982: Mesozoic mafic dikes of Southern Maine; Maine Geology Bulletin, no. 2, p. 16-32.

McHone, J.G., Ross, M.E., and Greenough, J.D.
1987: Mesozoic dyke swarms of Eastern North America; in Mafic Dyke Swarms, (ed.) H.C. Halls and W.F. Fahrig; Geological Association of Canada, Special Paper 34, p. 279-288.

McKenzie, D. and Bickle, M.J.
1988: The volume and composition of melt generated by extension of the lithosphere; Journal of Petrology, v. 29, p. 625-679.

Mertz, K.A. and Hubert, J.A.
1990: Cycles of sand-flat sandstone and playa-lacustrine mudstone in the Triassic-Jurassic Blomidon redbeds, Fundy rift basin, Nova Scotia: implications for tectonic and climatic controls; Canadian Journal of Earth Sciences, v. 27, p. 442-451.

Milligan, G.C.
1949: Geological survey of Prince Edward Island; Prince Edward Island Department of Industry and Natural Resources, p. 1-83.

Morgan, W.J.
1983: Hotspot tracks and the early rifting of the Atlantic; Tectonophysics, v. 94, p. 123-139.

Nadon, G.C.
1981: The stratigraphy and sedimentology of the Triassic at St. Martins and Lepreau, New Brunswick; M.Sc. thesis, McMaster University, Hamilton, Ontario, 279 p.

Nadon, G.C. and Middleton, G.V.
1984: Tectonic control of Triassic sedimentation in southern New Brunswick: local and regional implications; Geology, v. 12, p. 619-622.
1985: The stratigraphy and sedimentology of the Fundy Group (Triassic) of the St. Martins area, New Brunswick; Canadian Journal of Earth Sciences, v. 22, p. 1183-1203.

Neumann, E.R. and Ramberg, I.B.
1978: Paleorifts - concluding remarks; in Tectonics and Geophysics of Continental Rifts, (ed.) I.B. Ramberg and E.R. Neumann; D. Reidel Publishing Company, Dordrecht, Holland, p. 409-424.

Olsen, P.E.
1981: Comment on "Eolian dune field of Late Triassic age, Fundy Basin, Nova Scotia"; Geology, v. 9, p. 557-559.
1983: Relationship between biostratigraphic subdivisions and igneous activity in the Newark Supergroup; Geological Society of America, Southeastern Section, 32nd Annual Meeting, Abstracts With Programs, v. 15, p. 93.

Olsen, P.E. and Baird, D.M.
1986: The ichnogenus Atreipus and its significance for Triassic biostratigraphy; in The Beginning of the Age of Dinosaurs, (ed.) D. Padian; Cambridge University Press, p. 61-87.

Olsen, P.E. and Galton, P.M.
1977: Triassic-Jurassic tetrapod extinctions: are they real?; Science, v. 197, p. 983-986.

Olsen, P.E., McCune, A.R., and Thomson, K.S.
1982: Correlation of the Early Mesozoic Newark Supergroup by vertebrates, principally fishes; American Journal of Science, v. 282, p. 1-44.

Olsen, P.E. and Schlische, R.W.
1990: Transtensional arm of the early Mesozoic Fundy rift basin: penecontemporaneous faulting and sedimentation; Geology, v. 18, p. 695-698.

Olsen, P.E., Shubin, N.H., and Anders, M.H.
1987: New Early Jurassic tetrapod assemblages constrain Triassic-Jurassic tetrapod extinction event; Science, v. 237, p. 1025-1029.

Papezik, V.S. and Barr, S.M.
1981: The Shelburne dike, an Early Mesozoic diabase dike in Nova Scotia: mineralogy, chemistry and regional significance; Canadian Journal of Earth Sciences, v. 18, p. 1346-1355.
1982: The Shelburne dike, an Early Mesozoic diabase dyke in Nova Scotia: mineralogy, chemistry, and regional significance: Reply; Canadian Journal of Earth Sciences, v. 19, p. 1709.

Papezik, V.S. and Greenough, J.D.
1985: Early Mesozoic dikes of Atlantic Canada; in International Conference on Mafic Dike Swarms (Extended) Abstracts; University of Toronto, Erindale Campus, June 4-7, 1985, p. 119-123.

Papezik V.S. and Hodych, J.P.
1980: Early Mesozoic diabase dikes of the Avalon Peninsula, Newfoundland: petrochemistry, mineralogy, and origin; Canadian Journal of Earth Sciences, v. 17, p. 1417-1430.

Papezik, V.S., Greenough, J.D., Colwell, J., and Mallinson, T.
1988: North Mountain basalt from Digby, Nova Scotia: models for a fissure eruption from stratigraphy and petrochemistry; Canadian Journal of Earth Sciences, v. 25, p. 74-83.

Papezik, V.S., Hodych, J.P., and Goodacre, A.K.
1975: The Avalon magnetic lineament - a possible continuation of the Triassic dike system of New Brunswick and Nova Scotia; Canadian Journal of Earth Sciences, v. 12, p. 332-335.

Pe-Piper, G. and Jansa, L.F.
1986: Triassic olivine-normative diabase from Northumberland Strait, Eastern Canada: implications for continental rifting; Canadian Journal of Earth Sciences, v. 23, p. 1013-1021.
1987: Geochemistry of late Middle Jurassic-Early Cretaceous igneous rocks on the Eastern North American margin; Geological Society of America Bulletin, v. 99, p. 803-813.

Pe-Piper, G., Jansa, L.F., and Lambert, R.St.J.
1992: Early Mesozoic magmatism on the eastern Canadian margin: Petrogenetic and tectonic significance; in Eastern North American Mesozoic Magmatism, (ed.) J.H. Puffer and P. Ragland; Geological Society of America, Special Paper 268, p. 13-36.

Philpotts, A.R. and Martello, A.
1986: Diabase feeder dikes for the Mesozoic basalt in Southern New England; American Journal of Science, v. 286, p. 105-126.

Pitman, W.C. and Talwani, M.
1972: Sea floor spreading in the North Atlantic; Geological Society of America Bulletin, v. 83, p. 619-646.

Poole, W.H., Sanford, B.V., Williams, H., and Kelley, D.G.
1970: Geology of south-eastern Canada; in Geology and Economic Minerals of Canada, (ed.) R.J.W. Douglas, p. 298.

Potter, R.R., Hamilton, J.B., and Davies, J.L.
1979: Geological Map of New Brunswick, Second edition; New Brunswick Department of Natural Resources, Map Number N.R.-1.

Powers S.
1916: The Acadian Triassic; Journal of Geology, v. 24, p. 1-26, 105-122, 254-268.

Prest, V.K.
1972: Geology of Malpeque-Summerside area, Prince Edward Island; Geological Survey of Canada, Paper 71-45, 21 p.

Rader, E.K., Newell, W.L., and Mixon, R.B.
1986: Mesozoic and Cenozoic compressional faulting along the Coastal Plain margin, Fredericksburg, Virginia; in Geological Society of America, Centennial Field Guide - Southeastern Section, (ed.) T.L. Neathery; p. 309-314.

Ragland, P.C., Hatcher, R.D., Jr., and Whittington, D.
1983: Juxtaposed Mesozoic diabase dike sets from the Carolinas: a preliminary assessment; Geology, v. 11, p. 394-404.

Ragland, P.C., Rogers, J.J.W., and Justus, P.S.
1968: Origin and differentiation of Triassic dolerite magmas, North Carolina, USA; Contributions to Mineralogy and Petrology, v. 20, p. 57-80.

Ratcliffe, N.M.
1980: Brittle faults (Ramapo fault) and phyllonitic ductile shear zones in basement rocks of the Ramapo seismic zone, New York and New Jersey, and their relationship to current seismicity; in Field Studies of New Jersey Geology and Guide to Field Trips, (ed.) W. Manspeizer; 52nd Annual Meeting of the New York State Geological Association, Rutgers University, Newark, New Jersey, p. 278-311.

Ratcliffe, N.M. and Burton, W.C.
1985: Fault reactivation models for origin of the Newark Basin and studies related to eastern U.S. seismicity; in Proceedings of the Second United States Geological Survey Workshop on the Early Mesozoic Basins of the Eastern United States, (ed.) G.R. Robinson and A.J. Froelich; United States Geological Survey, Circular 946, p. 36-45.

Robbins, E.I.
1981: A preliminary account of the Newark rift system; in Papers Presented to the Conference on the Processes of Planetary Rifting, (ed.) K. Hrametz; Lunar and Planetary Institute, Houston, p. 107-109.

Robinson, G.R. and Sears, C.M.
1988: Inventory of metal mines and occurrences associated with the Early Mesozoic basins of the eastern United States - summary tables; in Studies of the Early Mesozoic Basins of the Eastern United States, (ed.) A.J. Froelich and G.P. Robinson; United States Geological Survey Bulletin, v. 1776, p. 265-303.

Royden L. and Keen, C.E.
1980: Rifting process and thermal evolution of the continental margin of Eastern Canada determined from subsidence curves; Earth and Planetary Science Letters, v. 51, p. 343-361.

Ruffman, A. and Woodside, J.
1970: The Odd-twins magnetic anomaly and its possible relationship to the Humber Arm Klippe of Western Newfoundland, Canada; Canadian Journal of Earth Sciences, v. 7, p. 326-337.

Sanders, J.F.
1963: Late Triassic tectonic history of northeastern United States; American Journal of Science, v. 261, p. 501-524.

Sarjeant, W.A.S. and Stringer, P.
1978: Triassic reptile tracks in the Lepreau Formation, Southern New Brunswick, Canada; Canadian Journal of Earth Sciences, v. 15, p. 594-602.

Seguin, M.K., Rao, K.V., Venugopal, D.V., and Gahe, E.
1981: Paleomagnetism of parts of the Late Triassic diabase dike system associated with the trans-New Brunswick aeromagnetic lineament; Canadian Journal of Earth Sciences, v. 18, p. 1776-1787.

Sheridan, R.E.
1976: Sedimentary basins of the Atlantic Margin of North America; Tectonophysics, v. 36, p. 113-132.
1984: Phenomena of pulsation tectonics related to the breakup of the Eastern North American continental margin; Tectonophysics, v. 94, p. 169-185.

Smith, T.E. and Huang, C.H.
1982: The Shelburne dike, an Early Mesozoic diabase dike in Nova Scotia: mineralogy, chemistry, and regional significance: discussion; Canadian Journal of Earth Sciences, v. 19, p. 1707-1708.

Srivastava, S.P. and Tapscott, C.R.
1986: Plate kinematics of the North Atlantic; in The Western North Atlantic Region, (ed.) P.R. Vogt and B.E. Tucholke; Geological Society of America, The Geology of North America, v. M, p. 379-404.

Stea, R. and Fowler, J.H.
1981: Petrology of Lower Cretaceous sands of Brazil Lake, Hants County, Nova Scotia; Mineral Resources Division, Report of Activities, 1980, Nova Scotia Department of Mines and Energy, Report 81-1, p. 47-64.

Stevens, G.R.
1980: Mesozoic volcanism and structure - northern Bay of Fundy region, Nova Scotia; Geological Association of Canada-Mineralogical Association of Canada, Field Trip Guidebook, Trip 8, 41 p.

Stevenson, I.M.
1960: New occurrences of Triassic sedimentary rocks in Chedabucto Bay area, Nova Scotia; Geological Society of America Bulletin, v. 71, p. 1807-1808.

Stevenson, I.M. and McGregor, D.C.
1963: Cretaceous sediments in central Nova Scotia; Geological Society of America Bulletin, v. 74, p. 355-356.

Stringer, P. and Burke, K.B.S.
1985: Structure in southwest New Brunswick; Geological Association of Canada-Mineralogical Association of Canada, Field Excursions, v. 3, Excursion 9, p. 1-34.

Stringer, P. and Lajtai, E.Z.
1979: Cleavage in Triassic rocks of southern New Brunswick, Canada; Canadian Journal of Earth Sciences, v. 16, p. 2165-2180.

Stringer, P. and Wardle, R.J.
1973: Post-Carboniferous and post-Triassic structures in southern New Brunswick; in Geology of New Brunswick, (ed.) N. Rast; New England Intercollegiate Geological Conference, Field Guide to Excursions, p. 88-95.

Strong, D.F. and Harris, A.
1974: The petrology of Mesozoic alkaline intrusives of central Newfoundland; Canadian Journal of Earth Sciences, v. 11, p. 1208-1219.

Sullivan, K.D.
1983: The Newfoundland Basin: ocean-continent boundary and Mesozoic seafloor spreading history; Earth and Planetary Science Letters, v. 62, p. 321-339.

Sun, S.S., Nesbitt, R.W., and Sharaskin, A.Y.
1979: Geochemical characteristics of mid-ocean ridge basalt; Earth and Planetary Science Letters, v. 44, p. 119-138.

Sutter, J.F.
1988: Innovative approaches to the dating of igneous events in the Early Mesozoic basins of the Eastern United States; in Studies of the Early Mesozoic Basins of the Eastern United States, (ed.) A.J. Froelich and G.R. Robinson; United States Geological Survey Bulletin, v. 1776, p. 194-200.

Swanson, M.T.
1982: Preliminary model for an early transform history in central Atlantic rifting; Geology, v. 10, p. 317-320.
1986: Preexisting fault control for Mesozoic basin formation in Eastern North America; Geology, v. 14, p. 419-422.

Swift, D.J.P. and Lyall, A.K.
1968: Reconnaissance of bedrock geology by sub-bottom profiler, Bay of Fundy; Geological Society of America Bulletin, v. 79, p. 639-646.

Swift, D.J.P., Jagodits, F.L., Ministre, B.L., and Paterson, N.R.
1967: Structure of the Minas Passage, Bay of Fundy: a preliminary report; Maritime Sediments, v. 3, p. 112-118.

Sykes, L.R. and Sbar, M.L.
1973: Intraplate earthquakes, lithospheric stresses and the driving mechanism of plate tectonics; Nature, v. 245, p. 298-302.

Tanner, L.H. and Hubert, J.F.
1992: Depositional facies, palaeogeography and palaeoclimatology of the Lower Jurassic McCoy Brook Formation, Fundy rift basin, Nova Scotia; Palaeogeography, Palaeoclimatology, Palaeoecology, v. 96, p. 261-280.

Taylor, S.R.
1964: Abundance of chemical elements in the continental crust: a new table; Geochimica et Cosmochimica Acta, v. 28, p. 1273-1285.

Thompson, R.N., Dickin, A.P., Gibson, I.L., and Morrison, M.A.
1982: Elemental fingerprints of isotopic contamination of Hebridean Palaeocene mantle-derived magmas by Archean sial; Contributions to Mineralogy and Petrology, v. 79, p. 159-168.

Thurston, H.
1986: The fossils; Atlantic Insight, July, p. 20-23.

Traverse, A.
1987: Pollen and spores date origin of rift basins from Texas to Nova Scotia as early Late Triassic; Science, v. 236, p. 1469-1472.

Turner-Peterson, C.E. and Smoot, J.P.
1985: New thoughts on facies relationships in the Triassic Stockton and Lockatong formations, Pennsylvania and New Jersey; in Proceedings of the Second United States Geological Survey Workshop on the Early Mesozoic Basins of the Eastern United States, (ed.) G.R. Robinson and A.J. Froelich; United States Geological Survey, Circular 946, p. 10-17.

Umpleby, D.C.
1979: Geology of the Labrador Shelf; Geological Survey of Canada, Paper 79-13, 34 p.

Van Houten, F.B.
1969: Late Triassic Newark Group, north-central New Jersey and adjacent Pennsylvania and New York; in Geology of Selected Areas in New Jersey and Eastern Pennsylvania, (ed.) S. Subinsky; Rutgers University Press, New Brunswick, New Jersey, p. 314-347.

Walker, T.L. and Parsons, A.L.
1922: The zeolites of Nova Scotia; University of Toronto, Series 14, Contributions to Canadian Mineralogy, p. 13-73.

Wanless, R.K. and Stevens, R.D.
1971: Note on the age of diabase dykes, Anticosti Island, Quebec; The Geological Association of Canada, Proceedings, v. 23, p. 77-78.

Wanless, R.K., Stevens, R.D., Lachance, G.R., and Delabio, R.N.
1972: Age determinations and geological studies, K-Ar isotopic ages, Report 10; Geological Survey of Canada, Paper 71-2, p. 72-73.
1979: Age determinations and geological studies, K-Ar isotopic ages, Report 14; Geological Survey of Canada, Paper 79-2, 67 p.

Wanless, R.K., Stevens, R.D., Lachance, G.R., and Edmond, C.M.
1967: Age determinations and geological studies, K-Ar ages, Report 7; Geological Survey of Canada, Paper 66-17.

Wanless, R.K., Stevens, R.D., Lachance, G.R., and Rimsaite, R.Y.H.
1965: Age determinations and geological studies, part I - isotopic ages. Report 5; Geological Survey of Canada, Paper 64-17.

Wark, J.M. and Clarke, D.B.
1980: Geochemical discriminators and the palaeotectonic environment of the North Mountain basalt, Nova Scotia; Canadian Journal of Earth Sciences, v. 17, p. 1740-1745.

Weigand, P.W. and Ragland, P.C.
1970: Geochemistry of Mesozoic dolerite dikes from Eastern North America; Contributions to Mineralogy and Petrology, v. 29, p. 195-214.

Wheeler, W.H. and Textoris, D.A.
1978: Triassic limestone and chert of playa origin in North Carolina; Journal of Sedimentary Petrology, v. 48, p. 765-776.

White, R.S.
1989: Igneous outbursts and mass extinctions; EOS, v. 70, p. 1480-1491.

White, R. and McKenzie, D.
1989: Magmatism at rift zones: the generation of volcanic continental margins and flood basalts; Journal of Geophysical Research, v. 94, no. B6, p. 7685-7729.

Williams, H.
1978a: Tectonic-Lithofacies Map of the Appalachian Orogen; Memorial University of Newfoundland, Map No. 1.
1978b: Geological development of the northern Appalachians: its bearing on the development of the British Isles; in Crustal Evolution in Northwestern Britain and Adjacent Areas, (ed.) D.R. Bowes and B.E. Leake; Geological Journal, Special Issue Number 10, p. 1-22 plus maps.
1979: Appalachian Orogen in Canada; Canadian Journal of Earth Sciences, v. 16, p. 792-807.
1984: Miogeoclines and suspect terranes of the Caledonian-Appalachian Orogen: tectonic patterns in the North Atlantic region; Canadian Journal of Earth Sciences, v. 21, p. 887-901.

Williams, P.F. and Hy, C.
1990: Origin and deformational and metamorphic history of gold-bearing quartz veins on the Eastern Shore of Nova Scotia; in Mineral Deposit Studies in Nova Scotia, Volume 1, (ed.) A.L. Sangster; Geological Survey of Canada, Paper 90-8, p. 169-194.

Wilson, J.T.
1966: Did the Atlantic close and then re-open?; Nature, v. 211, p. 676-681.

Wright, W.J. and Clements, C.S.
1943: Coal deposits of Lepreau-Musquash district, New Brunswick; Acadian Naturalist, v. 1, p. 5-27.

Yoder, H.S. and Tilley, C.E.
1962: Origin of basaltic magmas: an experimental study of natural and synthetic systems; Journal of Petrology, v. 3, p. 342-532.

Zietz, I., Haworth, R.T., Williams, H., and Daniels, D.L.
1980: Magnetic anomaly map of the Appalachian Orogen; Memorial University of Newfoundland Map No. 2a, Memorial University of Newfoundland, St. John's, Newfoundland.

Author's Address

J.D. Greenough
Geology Department,
Okanagan University College,
3333 College Way,
Kelowna,
British Columbia
V1V 1V7

Printed in Canada

Chapter 7

GEOPHYSICAL CHARACTERISTICS

Chapter 7

GEOPHYSICAL CHARACTERISTICS

H.G. Miller

INTRODUCTION

Preamble

Geophysical studies of the Canadian Appalachian region have contributed primarily to an understanding of the structural features at depth through the application of gravity, magnetic, and seismic techniques supplemented to a lesser extent by geomagnetic depth sounding and magnetotelluric and heat flow studies. In addition, gravity and magnetic maps allow comparisons between geological and geophysical features of the region. These comparisons may be used to extend the geological features to regions of poor or no exposure, for example to the northern offshore and Gulf of St. Lawrence. Paleomagnetic results have generated discussion regarding the relative horizontal displacements of large parts of the orogen. Offshore geophysical surveys provided the first evidence of the existence of the thick Tertiary and Mesozoic sedimentary sequence on the continental shelf and slope which are the target of extensive petroleum exploration and activity.

The geophysical data have been collected over the last 30 years but the most exciting interpretations are arising from the integration of the older data sets with the results from the new, deep, seismic reflection data. New aeromagnetic data are also adding immensely to our understanding of the nature of the crust in water-covered areas, particularly the critical area between Newfoundland and Nova Scotia where geophysics is the only method of tracing the geological features across the Cabot Strait. The interpretation of each data set as it was collected and published was undertaken in the context of contemporary geological models. Many such models are now obsolete. The present compilation documents the data and reviews interpretations in the context of the present geological models, i.e. the Wilson cycle of orogenic development (Wilson, 1966), definition of continental margins and suspect terranes, and the accretionary history of the orogen. The objective of the present work is to provide the first comprehensive compilation of the geophysical data for the Canadian Appalachian

region. The historical sequence of data acquisition is summarized, the salient features of the data are reviewed, and an attempt is made to synthesize the geophysical data into a coherent description of the crust.

History

The earliest geophysical studies in the Canadian Appalachian region for scientific as opposed to exploration purposes were gravity surveys conducted along railways (Garland, 1953), airborne magnetic surveys conducted in the 1950s (Hood, 1983), and seismic refraction studies on the Grand Banks in the late 1940s and early 1950s (Press and Beckmann, 1954). In the 1960s, systematic regional surveys were conducted on land (Weaver, 1967), at sea (Goodacre, 1964; Ewing et al., 1966; Sheridan and Drake, 1968; Goodacre et al., 1969; Haworth and MacIntyre, 1975), and by air. Paleomagnetic studies at selected sites began in the 1950s (Nairn et al., 1959). Geomagnetic, magnetotelluric, and heat flow data have been collected since 1970. Basin modelling techniques using geological and geophysical data are new (Quinlan and Beaumont, 1984). Most important is the recent acquisition of deep seismic reflection data from the Quebec-Maine portion of the system by the Geological Survey of Canada and the United States Geological Survey and the collection of similar data offshore, northeast of Newfoundland and in the Gulf of St. Lawrence, by the Atlantic Geoscience Centre of the Geological Survey of Canada as part of the Frontier Geoscience Program. Onland seismic reflection transects have recently been conducted in Newfoundland as part of the Lithoprobe East Project. Some of these data sets are now processed and various interpretations have been published (Keen et al., 1986; Marillier et al., 1989; Stockmal et al., 1990; Quinlan et al., 1992). Others are still in processing stages.

The earliest seismic refraction surveys, conducted from 1948 to 1961 on the continental shelf and slope, revealed the presence of thick Mesozoic and Cenozoic sediments. Subsequent publications (e.g. Press and Beckman, 1954; Sherwin, 1973 after Sheridan and Drake, 1968) elucidated the nature of specific basins and the crust in general and demonstrated the similarity of some of the Canadian basins to those of the Gulf of Mexico, thereby stimulating exploration activity.

At the same time, aeromagnetic and shipborne magnetic data were being collected across Labrador, the Labrador Sea, and Greenland. These data revealed the presence of a thick wedge of sediments on the Labrador and Baffin Island shelves (McMillan, 1973).

Miller, H.G.
1995: Geophysical characteristics; Chapter 7 in Geology of the Appalachian-Caledonian Orogen in Canada and Greenland, (ed.) H. Williams; Geological Survey of Canada, Geology of Canada, no. 6, p. 603-627 (also Geological Society of America, The Geology of North America, v. F-1).

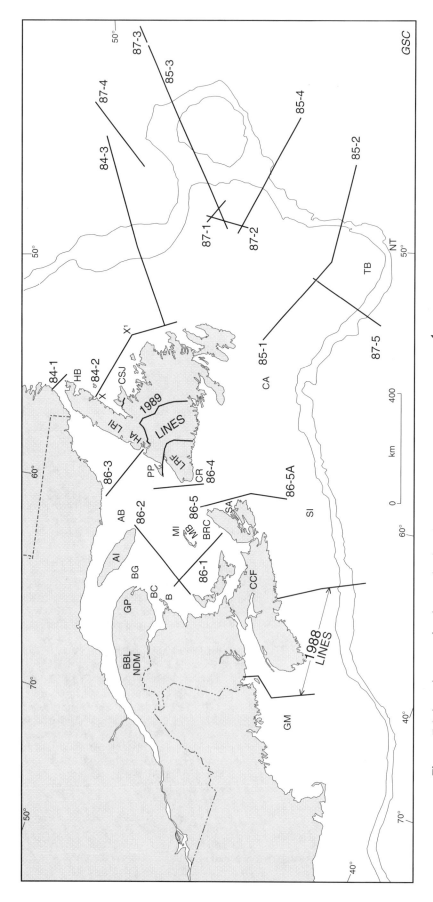

Figure 7.1. Location map for deep seismic reflection lines to December 1988. XX¹ shows location of portion of Line 84-2 modelled in Figure 7.7. AB – Anticosti Basin, AI – Anticosti Island, BBL – Baie Verte Brompton Line, B – Bathurst, BC – Chaleur Bay, BRC – Blair River Complex, BG – Gaspé Bay, CA – Collector Anomaly, CCF – Cobequid-Chedabucto Fault, CR – Cape Ray, CSJ – Cape St. John, GM – Gulf of Maine, GP – Gaspésie, HA – Humber Arm, HB – Hare Bay, LRF – Long Range Fault, LRI – Long Range Inlier, MB – Magdalen Basin, MI – îles de la Madeleine Islands, ND – MNotre Dame Mountains, NT – Southeast Newfoundland Transform, PP – Port au Port Peninsula, SA – St. Ann's Bay, SI – Sable Island, TB – Tail of the Bank.

Aeromagnetic surveys by the Geological Survey of Canada commenced over Newfoundland in the mid-1950s and were concentrated over areas of known mineral occurrences and high mineral potential. Similar surveys were conducted in Nova Scotia and New Brunswick. Only recently have regional interpretations of these data been undertaken.

Database

The geophysical databases vary considerably in detail and spatial extent. Gravity and magnetic data are available for the entire Appalachian Orogen and Atlantic region at various scales ranging from 1:1 000 000 and 1:5 000 000 (Zeitz et al., 1980; Haworth et al., 1980; Williams and Haworth, 1984a, b; Maps 3 and 4 of this volume) to 1:50 000 and larger scale for local surveys of zone boundaries and other features.

Seismic refraction data were collected along selected profiles in the Gulf of St. Lawrence, and east of Nova Scotia and Newfoundland for the determination of crustal dimensions. The initial interpretation of these data was conducted without the benefit of modern ray tracing programs or geological models, which are now used in reinterpretations. The Lithoprobe East Project includes refraction and magnetotelluric experiments to augment recent onland reflection transects conducted in Newfoundland.

Major parts of the offshore have been surveyed using seismic reflection techniques. These data are useful in analyzing the Mesozoic history of the offshore parts of the region, the subject of a separate volume of this series (Keen and Williams, 1990), as well as the Cenozoic. The quality of the data has improved dramatically from the earliest surveys. Much of the best industry data are now available, as their five-year confidentiality period has expired. The major contributions of seismic reflection to crustal studies have come from the detailed deep seismic data (Fig. 7.1) acquired in 1984 northeast of Newfoundland, in 1985 on the Grand Banks, in 1986 in the Gulf of St. Lawrence, and in 1989 in Newfoundland. Earthquake location data are available for the area (Fig. 7.2).

Geomagnetic and magnetotelluric data have been collected at fewer than 40 sites (Fig. 7.3). Most of the sites are part of two transects across the Appalachian Orogen. One transect crosses all tectonostratigraphic zonal subdivisions from north of the St. Lawrence River to Sable Island. The other transect crosses Newfoundland. Additional stations are on both sides of the St. Lawrence River and on Prince Edward Island.

Heat flow measurements were made at sites in New Brunswick, Nova Scotia, Newfoundland, and offshore (Fig. 7.4). These studies utilized existing drillholes and were made as opportunities arose. The older studies (Hyndman et al., 1979; Wright et al., 1980) were land based. Newer data have been obtained from offshore petroleum exploration (Drury et al., 1987).

Data from numerous paleomagnetic sites have been used to decipher horizontal displacements (Fig. 7.5). In no other field of geophysics is there such an exhaustive array of publications addressing problems ranging from the width of the Paleozoic Iapetus Ocean to the magnitude of transcurrent Carboniferous movements. This review concentrates on the evolution of paleomagnetic ideas and assesses the likelihood of meaningful results. For this reason a historical approach is adopted.

GRAVITY AND MAGNETIC EXPRESSIONS

The potential field methods, gravity and magnetics, are considered together in assessing their contribution to our comprehension of the Canadian Appalachian region. Both reflect physical properties in a cumulative or integrated fashion, i.e. the net gravity or magnetic field is the sum of the contributions from each of the rock units in a particular area. Magnetic anomalies are better indicators of the near surface boundaries and of the attitude of units. The gravity data provide valuable information on the deeper structure with vertical dimensions up to crustal scale and large horizontal dimensions because density varies less over large dimensions than does magnetic susceptibility.

Gravity signatures

Gravity mapping on land at 13 km station spacing in Newfoundland and 6 km elsewhere encompasses the whole region. A significant part of the offshore area has been mapped by shipborne gravimetry using a line spacing of 5 or 10 km. Locally the station network is denser in areas of specific interest. All the land and marine data available to 1984 have been compiled by Williams and Haworth (1984b; Map 3). Since then several underwater surveys have filled in gaps in the nearshore coverage (Miller, 1983; Miller et al., 1985; Miller, 1987). These new data do not change the regional picture. A new gravity map of the combined onshore and offshore areas (Geological Survey of Canada, 1988a) has also been produced which has rationalized the discrepancies that existed in the earlier data sets.

The correlation of specific gravity and magnetic features with tectonostratigraphic zones or terranes is complicated by the integrated or cumulative nature of these fields. Miller (1977) attempted to divide Newfoundland into a number of distinct gravity zones and to deduce the crustal structure from these signatures.

The gravity field of the Appalachian Orogen is significantly higher than the adjoining Grenville Structural Province where typical Bouguer anomalies are between - 60 mGal and 0 mGal. A strong eastward gravity gradient from predominantly negative to predominantly positive Bouguer anomalies is found throughout the Canadian Appalachians (Map 3). In general, the gradient is located in the eastern part of the Humber Zone, especially northeast of Newfoundland and throughout the island. A similar feature is also recognized across the Gulf of St. Lawrence and through Quebec (Haworth, 1978).

The Humber Zone gravity field is typified by negative Bouguer anomalies (Map 3). Locally this negative Bouguer anomaly field is disturbed and attains positive values over obducted ophiolitic rocks in southern Quebec and over a major Precambrian dioritic body on the eastern margin of the zone in Newfoundland (Miller and Wright, 1984). Over similar obducted ophiolitic rocks in Gaspésie and western Newfoundland the field remains negative but is perturbed towards positive values (Weaver, 1967; Miller, 1977; Seguin, 1982, 1983). As the station spacing for the Newfoundland surveys is large, it is not possible to determine if the field may locally attain positive values. Although the gravity gradient across the Humber-Dunnage

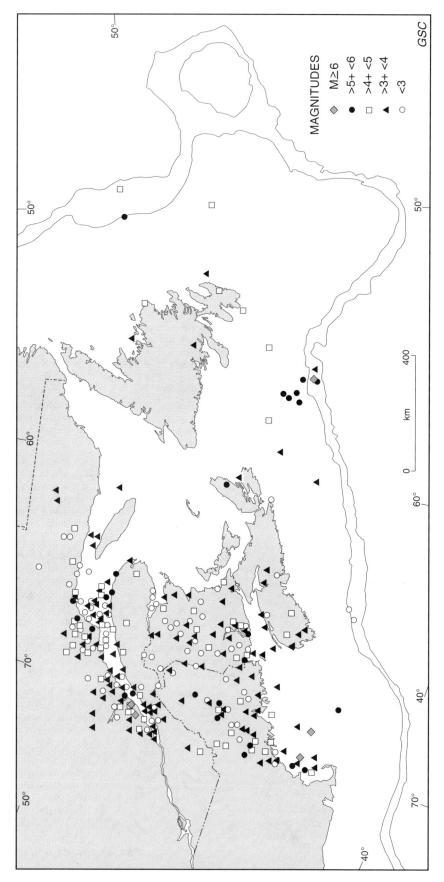

Figure 7.2. Seismicity of the Canadian Appalachian region based on epicentres for earthquakes prior to 1979 and for 1982.

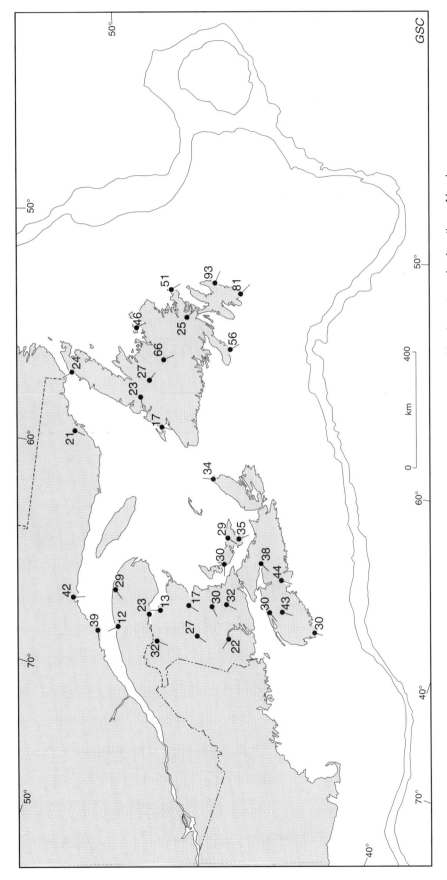

Figure 7.3. Geomagnetic and magnetotelluric sites with 2000 second in phase transfer functions. Numbers indicate the magnitude of the transfer function in per cent, arrows point towards the inferred conductor.

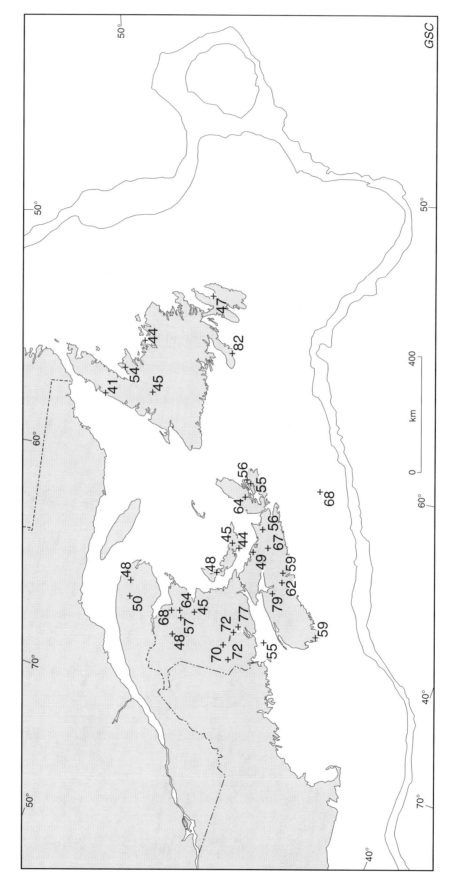

Figure 7.4. Heat flow measurement sites with heat flow in metres per square metre per second.

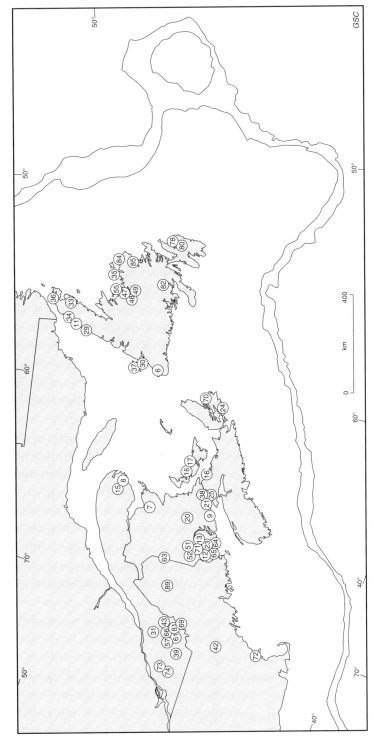

Figure 7.5. Paleomagnetic sampling sites of the Canadian Appalachian region from the Canadian Catalogue of Paleomagnetic Directions and Paleopoles (Irving et al., 1990).

2222

22222222222222222222222222222222222

2222222222222222

2222222222222222

22222222222222222

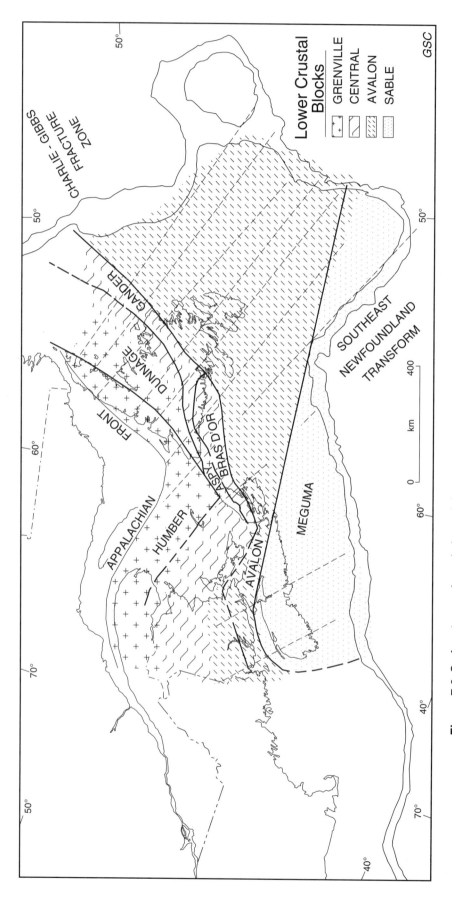

Figure 7.6. Surface terrane boundaries as inferred from various geological sources particularly Williams and Hatcher (1983), Williams et al. (1988) and Loncarevic et al. (1989) with modifications discussed in the text after Miller (1990), indicated by solid lines. Major geophysically inferred lower crustal block boundaries after Marillier et al. (1989) and Miller (1990) with modifications based on interpretation in the present paper. Lower crustal blocks denoted by various shading patterns. Subparallel dashed lines represent the inferred north-west-trending lineations discussed in the text.

zone boundary is perturbed by local features (Haworth, 1978; Miller and Wright, 1984), its general form, especially in Newfoundland, exhibits the classic paired anomaly signature attributed to a transition from an older to a younger crust with the positive values over the younger region (Thomas, 1983).

The positive gravity gradient associated with the Humber-Dunnage transition is most pronounced in and northeast of Newfoundland where it can be traced northward parallel to the coast of the island and southern Labrador. At approximately 52.5°N this pattern merges with a southeast gravity trend from Labrador and swings northeast to east across the continental shelf in the vicinity of the Cartwright Arch at 53.5°N (Haworth and Jacobi, 1983).

The region of strong positive Bouguer anomalies east of the gravity gradient in general coincides with the Dunnage Zone in Newfoundland. The positive Bouguer anomalies have magnitudes as large as +50 mGal along the coast of Notre Dame Bay and immediately offshore. This region of high values is coincident with the Notre Dame Subzone (Williams et al., 1988). There is a linear belt of anomalies in excess of +30 mGal from Cape St. John southwest to Cape Ray. The eastern edge of this belt generally follows the 0 mGal contour northeastward. At the northeast coast of Newfoundland it approximately coincides with the position of the Gander River Ultramafic Belt (GRUB Line) which marks the Dunnage-Gander zone boundary. In the interior of the island it generally defines the boundary between the Notre Dame and Exploits subzones. The boundaries of the Dunnage Zone converge in southwest Newfoundland, and the zone is absent farther southwest in Cape Breton Island.

The presence of the Dunnage Zone beneath the Gulf of St. Lawrence is difficult to document using gravity data (Map 3), especially since the zone is very narrow in this region. Also the positive anomaly expected from oceanic rocks is obscured by the negative gravity effects of the upward of 9 km of Carboniferous cover rocks (Hobson and Overton, 1973). Marillier and Verhoef (1989) have corrected for the effects of the water and the sediments and demonstrated that the corrected anomaly exhibits the general positive nature associated with the Dunnage Zone in Newfoundland. Haworth and Jacobi (1983) delineated the boundary between the Dunnage and Humber zones as passing in an arcuate path which emerges from southern Newfoundland and passes north of the Îles de la Madeleine and then into Gaspé Bay and on through Quebec. There, the positive anomalies tend to be on the Humber Zone side and they are associated with obducted ophiolite suites. This is at variance with the recently documented presence of Grenville age rocks in northwestern Cape Breton Island (Barr and Raeside, 1986). The gravity signature of the western edge of the Dunnage Zone must therefore be the gravity gradient that follows the trace of the Long Range Fault in southwestern Newfoundland and continues with little offset across the Cabot Strait into Cape Breton Island from which it must emerge on the western side and curve northwest of Prince Edward Island (Map 3).

The absence of a large amplitude, long wavelength Bouguer gravity anomaly in New Brunswick is expected since the Dunnage Zone rocks there are overlain by middle Paleozoic cover rocks (Williams, 1984). The middle Paleozoic sequences have a lower density than the oceanic rocks of the Dunnage Zone, reducing the amplitude of the gravity anomaly to the point where it is negative in New Brunswick. In general the Bouguer anomalies within the Dunnage Zone are higher than those in the adjacent Humber and Gander zones. They correlate reasonably well with the features of the New Brunswick Elmtree and Miramichi terranes (Fyffe and Fricker, 1987; Miller, 1990).

The correlation of a distinct gravity signature with the Dunnage Zone may be made with certainty from the Atlantic continental margin off Labrador through Newfoundland where the zone narrows appreciably (Map 3 and Fig. 7.6). Beneath the Gulf of St. Lawrence and into New Brunswick the correlation is more speculative as is its disposition in New Brunswick when gravity data alone are considered. The boundaries depicted in Figure 7.6 are consistent with the geological boundaries chosen by Williams (1984), the recent mapping in Cape Breton Island (Barr and Raeside, 1986), and the nature of the patterns in the Gulf of St. Lawrence. In Quebec, the boundary was drawn to place the exposed ophiolites there within the Dunnage Zone and to follow the Brompton part of the Baie Verte-Brompton Line. This places the Notre Dame Mountains within the Dunnage Zone, consistent with the interpretations of Kumarapeli et al. (1981) and Seguin (1982, 1983). It also suggests that sediments in Gaspésie are responsible for the reduction of the gravity field there.

The characteristic positive gravity gradient of the Humber-Dunnage transition is a distinctive feature that enables the western margin of the Dunnage Zone to be readily traced throughout the Canadian Appalachian region. The eastern boundary of the Dunnage Zone with the Gander Zone in Newfoundland is typified by a much gentler negative gravity gradient (Map 3; Jacobi and Kristoffersen, 1976; Miller, 1977; Haworth and Miller, 1982; Haworth and Jacobi, 1983) which is easily recognized throughout Newfoundland and northeastward to the continental edge.

The Gander Zone in Newfoundland is typified by significantly negative, long wavelength, Bouguer gravity anomalies. Locally in the northern part of the zone in Newfoundland there are positive gravity anomalies that may be indicative of an ophiolitic substrate (Miller, 1977; Miller and Weir, 1982; Miller, 1988). Similar positive anomalies are also found in the Gander Zone offshore to the northeast, although the anomalies there are negative relative to the high positive anomalies of the Dunnage and Avalon zones on either side. This generally negative Bouguer gravity field can be traced in an arcuate pattern southward through Newfoundland. The location of the Gander Zone and its western boundary beneath the Gulf of St. Lawrence is less certain since the negative gravity characteristics of the zone are blended with the negative Bouguer gravity effects of Carboniferous sediments in the Gulf of St. Lawrence.

New constraints on the likely extent of the Gander Zone in Cape Breton Island are provided by recent geological interpretations (Barr and Raeside, 1986). Although Haworth and Jacobi (1983) attempted to trace the Gander Zone beneath the Gulf of St. Lawrence, their boundary is questionable in light of new information that suggests the presence of Gander-type rocks in central Cape Breton Island (Barr and Raeside, 1986; Barr et al., 1988).

In New Brunswick the Gander Zone gravity anomalies are negative except for a boomerang shaped area beginning near Bathurst and following the coast to terminate near

Chatham. The anomalies are negative relative to the Avalon Zone to the south and east but are more positive than those in the Humber Zone in the vicinity of the Gaspé Belt. There is no obvious correlation between the gravity anomalies and the New Brunswick terranes (Fyffe and Fricker, 1987).

The eastern boundary of the Gander Zone with the Avalon Zone is delineated in Newfoundland and offshore to the north and south by a transition from the predominantly negative gravity anomalies of the Gander Zone to the alternating positive and negative anomalies of the Avalon Zone. Tracing the boundary between southern Newfoundland and Cape Breton Island has become easier with the recognition of probable Gander equivalent rocks in central Cape Breton Island and Avalon correlatives to the south. This suggests that the gravity gradient from St. Ann's Bay in Cape Breton Island to the mapped Gander-Avalon zone boundary on the south coast of Newfoundland is the appropriate boundary (Map 3 and Fig. 7.6). The location of the boundary between the Gander and Avalon zones in New Brunswick and in the Gulf of St. Lawrence is not well defined. Haworth and Jacobi (1983) considered that in New Brunswick it coincides with the Fredericton-Norembega Fault and continues north through Prince Edward Island. However, Williams and Haworth (1984b) did not feel that there was sufficient evidence to depict a Gander-Avalon zone boundary on their gravity map. The location of the boundary in Prince Edward Island is speculative as the gravity high on the western part of the island has been ascribed to Carboniferous volcanic rocks (Jones and Garland, 1986). The preferred location on the basis of mapping in Cape Breton Island, in conjunction with the magnetically defined boundary discussed below, is shown in Figure 7.6.

The Avalon Zone has a unique geophysical signature (Map 3) comprising arcuate belts of alternating positive and negative gravity and magnetic anomalies that coincide with late Precambrian volcanic belts (Haworth and Lefort, 1979; Miller, 1983b; Miller et al., 1985). This signature has been used to trace the extent of the Avalon Zone beneath the Grand Banks and northeast of Newfoundland (Haworth, 1977; Haworth and Lefort, 1979; Haworth and Jabobi, 1983). The major difference in orientation of the arcuate trends northeast of Newfoundland enabled Haworth (1977) to trace the Gander-Avalon zone boundary (Dover Fault) across the continental margin toward the Charlie-Gibbs Fracture Zone (Haworth, 1977). South of Newfoundland these arcuate anomaly patterns are terminated by the Collector Anomaly which Haworth and Lefort (1979) considered the expression of the Avalon-Meguma zone boundary.

The Avalon-Meguma zone boundary is marked by a transition from positive to negative gravity anomalies coincident with major faults on land and at sea north of the Gulf of Maine. In the Gulf of Maine, there is a transition from more positive to less positive anomalies from west to east. The Meguma Zone is typified by negative gravity anomalies in Nova Scotia. This has led Pe-Piper and Loncarevic (1988) to reinterpret the location of the Avalon-Meguma zone boundary southwest of Nova Scotia and to propose that it trends more to the southeast (Fig. 7.6).

East of Nova Scotia and on to the Grand Banks there is a zone of less positive anomalies. Haworth and Lefort (1979) drew the boundary between the Avalon and Meguma zones as following these trends. Haworth and Lefort (1979) termed the east-west anomaly separating the two zones,

the Collector Anomaly. Recent deep reflection data, refraction data, and additional magnetic data have corroborated the placement of this boundary in the Tail of the Banks area (C.E. Keen, pers. comm., 1987).

Magnetic signatures

Aeromagnetic surveys at a scale of 1:63 360 are available for all onland areas of the Canadian Appalachian region. The same data have been used to produce 1:250 000 and 1:1 000 000 compilations; these data are now available in digital format. The results of marine surveys are compiled at 1:250 000 with more detailed maps available for parts of the Scotian Shelf. The data are presented in the compilation maps of the Appalachian Orogen (Zeitz et al., 1980; Williams and Haworth, 1984b, Map 4; Geological Survey of Canada, 1988b). In recent years the Geological Survey of Canada has commissioned several high resolution aeromagnetic surveys northeast and south of Newfoundland with detailed surveys made over specific areas on land. These data are also incorporated into the publicly available data sets.

The magnetic signatures are much more complicated than the gravity expressions in the various zones. The complexity arises from the very nature of the magnetic anomalies with their greater dependence on depth to source and the rapid changes in the magnetic susceptibility that may occur within the same rock type. As a result, the marine anomalies tend to be smoother than those for land areas as the magnetic sensor is farther from the source. These effects make it difficult to correlate anomalies from land areas with those from adjacent marine areas without data processing.

The magnetic signatures are particularly valuable in determining the major tectonostratigraphic divisions of the orogen. False colour magnetic relief maps show a vivid difference between the characters of the Humber and Dunnage zones and the Gander and Avalon zones in Newfoundland. Thus, the boundaries between these zones are easily determined, as well as the structural fabric within particular zones. Several major plutons in Newfoundland are recognized by their elliptical signatures or a peripheral high surrounding a magnetic low over the pluton itself; this contrasts with more typical magnetic highs over the plutons in New Brunswick. The diagnostic arcuate patterns of the Avalon Zone are unmistakable. The general trends of the magnetic anomalies also outline the major promontories and re-entrants in the orogen (Map 4).

In the Humber Zone the magnetic anomalies tend to be smoothly varying and of low amplitude indicating deep magnetic sources beneath the Cambrian-Ordovician carbonate sequence. Pronounced high amplitude, short wavelength magnetic anomalies occur particularly over exposed and shallow mafic dykes in the Long Range Inlier of Newfoundland (Map 4 and Fig. 7.6) and over the Grenville (Blair River Complex) of Cape Breton Island (Barr and Raeside, 1986). Indeed one of Barr and Raeside's (1986) criteria for identifying Grenville rocks was the nature of the magnetic pattern. Moderate amplitude magnetic anomalies occur over major obducted ophiolitic bodies in Quebec and Newfoundland. Seguin (1982, 1983) applied spectral analyses to the magnetic data for Quebec to determine the depth of Grenville basement beneath allochthonous cover south of the St. Lawrence River. He demonstrated that the basement depth increases to more than 6 km just south of the river. The 0 nT contour (Williams and Haworth, 1984a; Map 4) coincides with Seguin's (1982) 0 km depth contour. South of this the positive magnetic anomalies occur in a belt containing the obducted sequences visible at the surface and those interpreted to exist in the upper crust (Kumarapeli et al., 1981; Seguin, 1982, 1983). The boundary between the Humber and Dunnage zones in Quebec and New Brunswick is expressed by the arcuate pattern of high magnetic anomalies that can be traced from the Brompton area to the Gulf of St. Lawrence in the vicinity of Chaleur Bay (Map 4, Fig. 7.1). The boundary beneath the sediments of the Anticosti Basin is difficult to follow magnetically because of the depth to source.

The Dunnage Zone magnetic data exhibit several horizontal wavelengths and variable intensities. In general the magnetic anomalies are positive. The highest amplitude magnetic anomalies, in excess of 6000 nT, occur over known Jurassic plutons in central Newfoundland (Miller, 1976). Anomalies over mafic and ultramafic plutons emplaced along both sides of the zone are less intense, typically having amplitudes of a few hundred nanoteslas and wavelengths of a few kilometres to tens of kilometres. These anomaly patterns can be used to constrain the geometry of the various features and to trace them offshore northeast of Newfoundland (Haworth and Miller, 1982; Haworth and Jacobi, 1983; Miller and Morris, 1988; Miller, 1990). The fundamental differences in the magnetic character between the western and eastern parts of the Dunnage Zone in Newfoundland support its subdivision into Notre Dame and Exploits subzones (Williams et al., 1988).

The Dunnage Zone in New Brunswick is typified by similar magnetic anomalies (Bhattacharyya and Raychaudhuri, 1967). The most distinctive pattern is one of positive anomalies which begins at Chaleur Bay and continues inland to the Maine border. Extrapolation of boundaries depicted on Figure 7.6 is a compromise among gravity, magnetic, and geological information.

The Gander Zone in Newfoundland is a zone of moderate amplitude, long wavelength, negative magnetic anomalies which can be traced throughout the island and offshore to the vicinity of 52°N, 52°W. The overall negative anomalies are interrupted by the narrow belt of positive anomalies associated with the Gander River Ultrabasic Belt (GRUB) along the eastern margin of the Dunnage Zone (Williams, 1984). Subsequent detailed geophysical studies in the geologically classic Gander Zone of northeast Newfoundland reveal two subtle positive anomaly belts east of the GRUB Line. These too have been ascribed to the presence of mafic and ultramafic material in local thrusts (Miller, 1988).

In New Brunswick, the Gander Zone is also typified by negative magnetic anomalies with minor positive areas in the northeastern part of the zone. Barr et al. (1988) have reprocessed the Cape Breton Island data to remove the higher frequency effects and render the results similar in wavelength to that over the Cabot Strait between Newfoundland and Cape Breton Island. Combined with the new zoning for Cape Breton Island, it is possible to follow the Gander-Avalon zone boundary across the Cabot Strait. This location of the boundary also agrees with the new deep seismic data (Marillier et al., 1989) and is consistent with gravity data.

The arcuate belts of alternating positive and negative magnetic anomalies of the Avalon Zone are 20 to 100 km wide with amplitudes of 200 to 400 nT (Haworth and Lefort, 1979). This signature may be traced northeast of Newfoundland to the continental shelf in the vicinity of the Charlie-Gibbs Fracture Zone (Haworth, 1977). South of Newfoundland these arcuate anomaly patterns are terminated by the Collector Anomaly (Haworth and Lefort, 1979) which expresses the Avalon-Meguma zone boundary (Map 4). Onshore in Newfoundland the positive anomalies correlate with mafic Precambrian volcanic rocks (Haworth and Lefort, 1979; Miller, 1983b; Miller et al., 1985). More recent mapping south of Newfoundland has led to the recognition of distinctive Avalon magnetic signatures which Loncarevic et al. (1989) have traced southward into a major northwest-striking fault in approximately the same position as the Gulf Fault of Keppie (1985). Several other northwest-trending features are interpreted as Carboniferous or younger faults (Loncarevic et al., 1989). Similar features have been recognized in both the gravity and magnetic data in the southern Avalon Zone of Newfoundland (Miller and Tuach, 1988) and across the Avalon-Gander zone boundary in Bonavista Bay (Miller, 1988).

Similar alternating bands of magnetic anomalies are found in Nova Scotia and New Brunswick (Haworth and Jacobi, 1983; Map 4). The Gander-Avalon zone boundary in New Brunswick coincides with the magnetic signature adjacent to the Fredericton-Norembega Fault and crosses the Carboniferous New Brunswick basin. An extension of the Collector Anomaly denoting the Avalon-Meguma zone boundary coincides with the trace of the Cobequid-Chedabucto Fault (Fig. 7.1). These broad positive anomalies mark the southeastern extent of the Avalon Zone on land and are interpreted as the southern boundary of the zone offshore (Haworth and Lefort, 1979).

The Avalon-Meguma zone boundary has already been noted as a transition to negative magnetic anomalies marked by a major positive magnetic anomaly coincident with major faults on land and at sea. The Meguma Zone has negative gravity anomalies and distinctive northeast-trending linear magnetic anomaly belts (McGrath, 1970; Cameron and Hood, 1975) of moderate amplitude. These correlate with the general structural trends of the folded Meguma Supergroup sedimentary rocks. Cameron and Hood (1975) and Pe-Piper and Loncarevic (1988) used the characteristic linear patterns to extend the Meguma Zone offshore, especially near the Nova Scotian coast. The zone is not found in Newfoundland but is interpreted to underlie the southeastern parts of the Grand Banks (Haworth and Lefort, 1979; Haworth and Jacobi, 1983; Fig. 7.6).

In addition to magnetic expressions of structural trends in the major Appalachian zones, there are other features that appear on large scale magnetic maps and on individual track records offshore. Haworth et al. (1976) noted the presence of linear, short wavelength, low amplitude anomalies northeast of Newfoundland which they ascribed to the presence of Mesozoic dykes, based on the presence of similar dykes on land. Hodych and Hayatsu (1980) and Papezik et al. (1975) also noted a magnetic signature associated with the Avalon Dyke. The magnetic signature of the dyke extends offshore but drilling attempts to prove its existence were unsuccessful (Haworth, 1977). Loncarevic et al. (1989) reported the presence of a similar offshore feature in the Avalon Zone south of Newfoundland, which they have called the St. Pierre Dyke. Papezik et al. (1975) commented on the presence of similar magnetic signatures over dykes in Nova Scotia and New Brunswick. All of these features along with the strongly magnetic Mesozoic plutons of central Newfoundland provide evidence of tectonic activity related to the Mesozoic opening of the present Atlantic.

Also evident on magnetic maps, especially magnetic shadowgrams, and to a lesser extent on the gravity map of Newfoundland, are regional northwest-trending lineations, which appear to truncate major gravity and magnetic anomaly patterns (Maps 3 and 4). These patterns were reported by Loncarevic et al. (1989) south of Newfoundland, by Miller (1988) and Miller and Tuach (1989) in the Gander-Avalon boundary area, by Miller (1990) in the Avalon Zone, and by Miller et al. (1990) in the Humber Zone.

SEISMIC DATA

Introduction

Seismic studies in the Canadian Appalachian region provide the least ambiguous information on deep crustal structure (Fig. 7.1). These studies consist of both conventional exploration and deep crustal investigations as well as refraction studies, seismicity studies, and analyses using teleseismic data from distant earthquakes reflected at the surface.

The earliest reported seismic studies are the Lamont refraction projects of the late 1940s and early 1950s on the Grand Banks and Scotian Shelf (Press and Beckman, 1954) and in the Gulf of St. Lawrence (Willmore and Scheidigger, 1956). Refraction studies to investigate Moho depths were initiated by a combined Bedford Institute-Dalhousie University group in the mid-1960s (Barrett et al., 1964; Dainty et al., 1966; Ewing et al., 1966; Fenwick et al., 1968). This work was carried out in the Gulf of St. Lawrence, on the Scotian Shelf, and north and south of Newfoundland. Subsequent refraction work was conducted by Sheridan and Drake (1968) and Hobson and Overton (1973).

Thousands of line kilometres of conventional exploration reflection data for petroleum exploration have been collected since the mid-1960s on various parts of the Labrador, Northeast Newfoundland, and Nova Scotian shelves, as well as on the Grand Banks and in the Gulf of St. Lawrence. The older data are of poor quality and shallow penetration, unsuitable for deep crustal interpretations as multiples are not suppressed and recording times are short. Cutt and Laving (1977) have interpreted some of the industry data north of the Grand Banks in terms of the general tectonics of the area. Durling and Marillier (1990) have conducted a similar investigation in the southern Gulf of St. Lawrence and identified several distinct basement regions.

The most informative deep seismic reflection data were collected under the Frontier Geoscience Program on the continental shelf northeast of Newfoundland in 1984, on the Grand Banks in 1985, and in the Gulf of St. Lawrence in 1986. Two-way-travel times exceed 15 seconds, thus portraying the entire crust. The results from these surveys provide significant insights into deep crustal structure (Keen et al., 1986; Marillier et al., 1988; Tankard et al., 1988). Similar data have been collected on a line in Quebec

(Seguin, 1982, 1983), and this line has now been extended and deeper data collected (Green et al., 1986). The latest studies are Lithoprobe East transects completed across Newfoundland in 1989.

Earthquake occurrence data for the Canadian Appalachian region date back to the earliest days of settlement. Only recently has the network of recording stations enabled events to be well located. Even now the low density and irregular geographic distribution of recording sites severely limit the minimum magnitude event that may be detected, and only large events are recognizable from major parts of the area (Basham et al., 1979). The geographic distribution of earthquake epicentres (Fig. 7.2) is controlled in part by the limitations of the detection network, although similarity between the general patterns of old earthquake epicentres and more recent data suggest the patterns of earthquake locations are the same.

Teleseismic methods have been used to infer general crustal structure (Stewart, 1976, 1978, 1984; Stewart and Keen, 1978).

Seismic refraction studies

The earliest refraction studies on the Grand Banks and Scotian Shelf (Press and Beckmann, 1954) and in the Gulf of St. Lawrence (Willmore and Scheidigger, 1956) estimated sediment thickness but were unable to determine crustal thickness since the lines were short. The Press and Beckman (1954) study demonstrated significant change in upper crustal structure across the Laurentian Channel.

Studies to determine Moho depths by the Bedford Institute-Dalhousie University group in the mid-1960s provided the first evidence for an intermediate crustal layer having velocities in the 7.0 to 7.5 km/s range. This layer was detected on the reversed refraction line across the New Brunswick-Nova Scotia- Prince Edward Island section of the system in the Gulf of St. Lawrence and on the reversed line from St. Anthony to Cape Freels off north-central Newfoundland. Along these lines and on unreversed lines across southern Newfoundland and off the northeast coast of Newfoundland, the inferred Moho depths vary from 30 to 35 km beneath the Humber Zone (Ewing et al., 1966; Berry, 1973) to 45 km beneath the Dunnage and Gander zones (Ewing et al., 1966), and 33 to 36 km beneath the Meguma Zone on the Scotian Shelf (Barrett et al., 1964). No reliable Moho depth was obtained for the Avalon Zone.

Subsequent refraction surveys (Sheridan and Drake, 1968; Hobson and Overton, 1973) elucidated the upper crustal structure in the region. The area underlain by the intermediate velocity layer first detected by Dainty et al. (1966) and Ewing et al. (1966) was delineated in more detail by Sheridan and Drake (1968). Fenwick et al. (1968) corroborated this interpretation that the intermediate layer extended across the continental shelf to the continental margin northeast of Newfoundland. Sheridan and Drake (1968) also demonstrated that this intermediate velocity crustal layer could be traced northward from central Newfoundland to a region off southern Labrador. Subsequent re-analysis of the St. Anthony-Cape Freels refraction data using ray tracing methods has provided more detail on the disposition of the intermediate velocity layer (Mitchelmore, 1985) in the Dunnage Zone.

Sheridan and Drake (1968) confirmed the presence of the intermediate velocity layer in other localities off Newfoundland. The thickness and boundaries of the Magdalen Carboniferous Basin in the Gulf of St. Lawrence were determined in gross form (Sheridan and Drake, 1968) and later detail was provided (Hobson and Overton, 1973). These studies agreed on the distribution of "basement" velocities greater than 6.0 km/s. The older Dalhousie refraction data which were approximately along line 86-1 (Fig. 7.1) from New Brunswick to Nova Scotia have also been subjected to ray tracing modelling. This technique affirms the presence of the intermediate velocity material in the area where the transect crosses the Dunnage and Gander zones (Campbell, 1984).

Refraction data collected from lines across Newfoundland as part of the Lithoprobe East study suggest variations in crustal thickness from north to south across Newfoundland. (Hughes et al., in press)

Seismic reflection studies

The most important information on the crustal configuration of the Canadian Appalachian region has come from the recently collected deep reflection data obtained as part of the Frontier Geoscience Program and Lithoprobe East Project. These data from Newfoundland and offshore (Fig. 7.1) have dramatically changed our interpretation of the subsurface of the orogen.

The initial survey northeast of Newfoundland (Keen et al., 1986) demonstrated a Grenville lower crustal block, interpreted as the ancient North American continent, extending in subsurface beneath an allochthonous Dunnage Zone. At approximately 100 km along the profile the Grenville lower crustal block abuts the Central lower crustal block (Marillier et al., 1989; Stockmal et al., 1990). This central block is overlain by rocks of the eastern Dunnage Zone and the Gander Zone. The boundary between the Gander and Avalon zones, the Dover Fault, is seen to be a major, steep feature having distinctly different seismic character on either side. The Moho is inferred to be greater than 45 km beneath the asymmetrical Dunnage Zone, corroborating the interpretation provided from the reinterpreted refraction data (Mitchelmore, 1985). The subsequent surveys, especially those in the Gulf of St. Lawrence, confirm the general interpretation of the basic deep crustal blocks and provide information on the extent of deep features possibly related to the Grenville Front (Marillier et al., 1989; Stockmal et al., 1990). The Grand Banks profiles mainly elucidate the extensional nature of the Mesozoic graben on their eastern parts. The sections also demonstrate that the whole of the Grand Banks are underlain by rocks typical of the Avalon Zone. The Grand Banks area has numerous extensional features (Tankard et al., 1988) with evidence that some of these were active even in the Precambrian and have been reactivated during succeeding periods with some reversal of polarity (Enachescu, 1987). The Lithoprobe East data collected on Newfoundland in 1989 have led to a dramatic rethinking of the block models proposed earlier (Quinlan et al., 1992).

Similar deep reflection seismic data were acquired across the Quebec Appalachians (Seguin, 1982, 1983; Ando et al., 1983; St. Julien et al., 1983). These data are interpreted as delineating a transitional region on the eastern side of the Humber Zone continental margin miogeocline with its characteristic shelf carbonates overlain by several westward transported nappes. The region beneath the Notre Dame Mountains is considered to represent stacked slices of ophiolitic crust at shallow depth, i.e. the upper 15 km of the crust. The seismic data only extend to 6.0 seconds two-way-travel time so the southward extent of the base of the Dunnage Zone is not resolved. Seguin (1982) and Ando et al. (1983) noted the similarity between the Quebec interpretation and that for the southern United States Appalachians. The maximum width of Iapetus in Quebec was 1500 km according to Seguin's (1982) interpretation of the degree of telescoping discernible from the gravity and seismic data. Additional deep reflection seismic data to 15 seconds were collected in this area in 1984; these data show that the extent of the miogeocline can be traced in the subsurface well to the south of the region delineated on the shallower data. There is evidence that this buried surface of the miogeocline served as a décollement zone at 4 to 6 seconds depth (12-15 km) along which ophiolitic material from Iapetus was thrust to the north (Green et al., 1986). Both these and the northeast Newfoundland marine data demonstrate the same general pattern of obduction above an east-dipping continental shelf of the Paleozoic American continent.

Industry reflection data from north of the Grand Banks to the Labrador Shelf have been analyzed by Cutt and Laving (1977) who noted the presence of basement highs in the Gander Zone and off southern Labrador. Cutt and Laving (1977) also demonstrated the presence of Carboniferous and younger sediments in the general synclinorium of the offshore Dunnage Zone, and pointed out the existence of salt diapirs and salt withdrawal features east of Hare Bay (Fig. 7.1) in what they termed the Belle Isle Basin. These features were also noted by Haworth et al. (1976) on data collected by single channel seismic methods. Subsequent drilling results from the Verrazano and Hare Bay wells confirmed the presence offshore of upper Paleozoic redbeds.

The industry data for the Grand Banks have been analyzed by Tankard and Welsink (1987) and Enachescu (1987). Although these authors have differing views about the relative amounts of extension and the absolute timing of specific events relevant to the formation of Mesozoic graben on the eastern Grand Banks, they agreed in their recognition of the role of Precambrian structural features of the Avalon Zone in controlling the subsequent development of the area. They also recognized several major northwest-trending transfer faults in the southern parts of the Grand Banks. These they considered to be associated with the Mesozoic separation of North America and Europe. The trends are identified on gravity and magnetic data as well as on the seismic sections.

The presence of Permian-Carboniferous rocks in the Gulf of St. Lawrence was inferred from Hobson and Overton's (1973) refraction studies. The internal structure and disposition of these rocks were mapped by Shearer (1973) using single channel seismic reflection methods. Haworth and Sanford (1976) collected additional single channel reflection data and provided "ground truthing" by recovering several cores that were used to correlate with units onshore. Considerable industry seismic data have been collected in various areas of the Gulf of St. Lawrence near

Anticosti Island, Prince Edward Island, and Port aux Port. The limited data near Port aux Port are of poor quality with many multiple reflections obscuring the nature of deeper reflectors, but the general basement structure is discernible (Miller et al., 1990). Reflection data from three industry lines off the northwest coast of Newfoundland provide information on the extent of the Humber Arm Allochthon offshore and suggest that the Long Range Inlier may be allochthonous (Stockmal and Waldron, 1990).

Seismicity

The Canadian Appalachian region is the oldest populated region of Canada. The area, especially New Brunswick and Quebec, has been the site of numerous historic earthquakes dating back to the seventeenth century (Smith 1962, 1966). Only since the late 1920s has there been modern instrumental monitoring of earthquake activity (Rast et al., 1979; Basham et al., 1979) and up to now the uneven distribution of the few instruments limits the detectability and positioning of events to those larger than magnitude 3.5 with magnitude 3 events resolvable in particular areas.

Earthquake epicentres for events occurring up to 1979 and those for 1982 are presented in Figure 7.2. It is evident that earthquakes occur in restricted areas. The mouth of the Laurentian Channel is the site of the 1929 Grand Banks earthquake and of several subsequent events. This area is on the southern flank of the Collector Anomaly and Cobequid-Chedabucto Fault considered to be the Avalon-Meguma zone boundary. The activity in this region is considered to be Recent, possibly associated with movements related to the Southeast Newfoundland Transform off the southern edge of the Grand Banks.

A second region of earthquakes is along the New Brunswick shore of the Bay of Fundy where there is also a spatial relationship to the Avalon-Meguma zone boundary. This may indicate that there is still stress release along that boundary but no fault plane solutions determining the sense of movement are available.

The third major concentration is in a northwest-trending belt in New Brunswick (Rast et al., 1979). The 1982 Miramichi earthquake sequence (Basham and Adams, 1984) occurred northwest of this area. Earthquakes in the St. Lawrence Valley occur outside the Appalachian region near La Malbaie (the Charlevoix Zone; Basham et al., 1979) and northwest of Gaspésie. The Charlevoix Zone is associated with documented modern movements. The area northwest of Gaspésie is parallel to the general Appalachian trends and the earthquakes there may indicate stress release is still occurring. Localization of the events is unexplained (Basham et al., 1979).

Atlantic Canada earthquakes have been ascribed to medium to deep crustal sources (Stevens et al., 1973; Rast et al., 1979). J. Adams (pers. comm., 1985) pointed out that the earlier determinations of focal depths are questionable, since recordings were made at nearby stations and broad assumptions were used in the calculations of epicentre parameters. The general conclusion is that the area north of Gaspésie and the Avalon-Meguma zone boundary are the only two major areas that are seismically active at present.

Teleseismic methods

Teleseismic methods have been used by Stewart (1976, 1978, 1984) and Stewart and Keen (1978) to examine Canadian Appalachian crustal structure. Data from Central America earthquakes recorded at European observatories and at a series of temporary stations in Newfoundland, supplementing the permanent observatories at St. John's and Corner Brook, were analyzed to determine the difference in arrival time between P and PP waves. The PP arrivals at European observatories were those that had midpoint surface reflections in Atlantic Canada. These analyses indicate that arrivals from reflections beneath the Dunnage and Gander zones are delayed relative to those having ray paths through the Humber, Avalon, and Meguma zones. This suggests that the crust beneath the Newfoundland Dunnage and Gander zones is thicker, the extra thickness of lower velocity crustal material contributing to the delay relative to travel times in the mantle of the adjoining zones. The crust in the Dunnage Zone could be up to 10 km thicker than in the Humber and Avalon zones.

GEOMAGNETIC, MAGNETOTELLURIC, AND HEAT FLOW STUDIES

Geomagnetic studies

Geomagnetic induction anomalies are caused by inducted electric currents along boundaries across which there are significant electrical conductivity contrasts. In Atlantic Canada, an irregular shoreline inhibits the isolation and interpretation of the subsurface conductivity structure using electric currents induced in the ocean at the land-sea interface. Analogue and numerical techniques have been used to predict the amplitude and direction of this effect at various frequencies for the intricate shoreline of Atlantic Canada (Dosso et al., 1982; Hebert et al., 1982) and to separate it from the total response thereby isolating the effects of other conductors in the crust and upper mantle.

Major geomagnetic studies have been reported for various regions of the Canadian Appalachian region (Fig. 7.3). Hyndman and Cochrane (1971) and Srivastava et al. (1973) collected and analyzed data from the mainland part of the Appalachian region in Quebec, New Brunswick, and Nova Scotia. They initially interpreted the presence of a conductive crustal layer between Dartmouth and Sable Island and explained the geomagnetic variations observed in New Brunswick as effects of currents induced in the Gulf of St. Lawrence.

Cochrane and Hyndman (1974) re-examined some of these data and determined that a conductive layer exists in the lower crust extending from south of the Appalachian Structural Front to central New Brunswick. A second conductive zone (Hyndman and Cochrane, 1971) was detected beneath the Scotian Shelf.

Similar geomagnetic depth sounding studies in Newfoundland (Cochrane and Wright, 1977; Wright and Cochrane, 1980) indicate the presence of a conductivity anomaly beneath the Gander Zone. This anomaly was investigated in more detail using a higher frequency range and conductive material hypothesized west of the Dover Fault at fairly shallow depth (Pal, 1982). The

Newfoundland studies did not detect any significant change in conductivity structure across the Humber Dunnage zone boundary.

Another geomagnetic investigation along the St. Lawrence River concluded that a major conductivity contrast exists between the craton and the Appalachian Orogen (Bailey et al., 1974). This interpretation locates the conductive crust in the Appalachian Orogen but does not provide any crustal models for the geometry or depth of the feature.

Data for the 2000 second in-phase transfer functions are presented in Figure 7.3. The induction arrows point toward boundaries along which currents flow, and the amplitude of the transfer function is indicative of the strength of the induced current and the conductivity contrast across the boundary. The sparse data distribution makes it difficult to contour the data over such a large area and introduces considerable bias in correlating from Newfoundland to the Maritime Provinces.

The major feature is the steep gradient in the amplitude near the boundary between the Gander and Dunnage zones and also between the Gander and Avalon zones in northeast Newfoundland. This localized high gradient is defined on data collected from a single data set (Wright and Cochrane, 1980) for which the uncertainty limits are as large as the amplitude of the transfer function (J.A. Wright, pers. comm., 1985). Wright and Cochrane (1980) suggested an east-dipping subduction zone beneath the Gander Zone on the basis of this feature of questionable amplitude. Reconsideration of this feature in light of later work (Pal, 1982) suggests that the gradient exists across the whole Gander Zone and that the important contrast is between the Gander-Dunnage zones and the Avalon Zone (J.A. Wright, pers. comm., 1985).

A less pronounced gradient is found in the mainland part of the Appalachian region, near the western Avalon zone boundary (Fig. 7.3, 7.6). A gradient of similar magnitude is also located near the Humber-Dunnage zone boundary in northern New Brunswick. Between these two gradients the transfer function amplitudes are smaller than elsewhere, indicating the presence of a more conductive crust (Cochrane and Hyndman, 1974).

There is no distinct geomagnetic signature associated with the Humber-Dunnage zone boundary in Newfoundland. Whether this is because of the complicated induced current geometry or whether it implies that there is no conductivity contrast between the Grenville province and the Dunnage Zone in Newfoundland has never been resolved.

Magnetotelluric studies

Magnetotelluric studies provide estimates of the vertical resistivity structure beneath a station from the monitoring of time variations in the electric and magnetic fields at the station. Two orthogonal horizontal directions for the magnetic and electric field are measured and used to determine the resistivity in two orthogonal directions and to examine anisotropy in the resistivity structure.

Magnetotelluric measurements are reported by Cochrane and Hyndman (1974), Kurtz and Garland (1976), and Jones and Garland (1986). Kurtz and Ostrowski (1984) conducted a magnetotelluric study in the vicinity of the Miramichi earthquake swarm of 1982 (see Basham and Adams, 1984).

Cochrane and Hyndman (1974) incorporated the magnetotellurics into their geomagnetic interpretation discussed above. Kurtz and Garland (1976) inferred from a few stations in the Appalachian part of their study area that the upper crust in the Dunnage-Gander zones is more resistive and the lower crust more conductive than the crust in zones on either side.

Cochrane and Hyndman (1974) discussed the possible origin of the enhanced conductivity of the lower crustal material they detected in New Brunswick and suggested that modified ancient oceanic crust would have the appropriate conductivity. In Newfoundland the highly conductive material is interpreted to lie beneath the Gander Zone and to be of oceanic origin (Wright and Cochrane, 1980).

Jones and Garland (1986) conducted a high frequency magnetotelluric survey across Prince Edward Island and found significant differences in the electrical resistivity structure beneath the eastern and western parts of the island. In particular they found relatively higher resistivity rocks at shallower depth beneath the western half of the island and ascribed the increased resistivity to Carboniferous volcanic rocks, which Garland (1953) deemed necessary to explain the gravity anomaly in the area. Jones and Garland (1986) confirmed this by demonstrating that the thickness of volcanic rocks determined from the magnetotellurics could explain the gravity anomaly.

A regional magnetotelluric study as part of the Lithoprobe East investigation of the crust of Newfoundland was conducted in the summer of 1991. The results have not yet been published.

The geomagnetic and magnetotelluric studies indicate that significant changes in crustal electrical properties occur across the Humber-Dunnage zone boundary in New Brunswick and across the Gander-Avalon zone boundary throughout the Canadian Appalachian region. These studies also demonstrate the difference in the crustal structure and composition of the Dunnage Zone and possibly the Gander Zone.

Heat flow studies

Reliable heat flow estimates require the accurate determination of the temperature gradient and the thermal conductivity. The temperature gradient is measured on land in deep, near vertical drillholes in which the temperature regime has returned to equilibrium after the drilling disturbance. The required thermal conductivity measurements are obtained from laboratory measurements on samples from the drill core. Few sites meeting these two criteria are available in Atlantic Canada. Hyndman et al. (1979) reported heat flow values for 12 sites in the Maritime Provinces and Wright et al. (1980) reported data from five sites in Newfoundland. Jessop et al. (1984) reported these data and added additional data derived from offshore petroleum exploration well log information. Several additional sites have been occupied in the Maritime Provinces (Drury et al., 1987) as part of a geothermal resource evaluation program.

The measurement of heat flow at sea is accomplished by inserting a thin temperature probe into the unconsolidated sediment of the seafloor. Modern instruments also measure the in situ thermal conductivity. Lewis and Hyndman (1976) reported data from 16 sites on the Scotian Shelf and

Grand Banks. Wright et al. (1984) measured the heat flow at four sites in the bays of northeast Newfoundland and considered the measurements at only two of these sites as reliable. Estimates of heat flow can also be obtained from an integration of temperature and lithology logs from petroleum exploration wells (Jessop et al., 1984). The sites of heat flow measurements are shown in Figure 7.4.

The raw data collected during a heat flow measurement are corrected for glacial effects, thermal effects of the bedrock topography, and water column effects. These corrections are uncertain in some situations leading to significant uncertainty in the final heat flow estimates.

Although the data are sparse and the uncertainty in both the measurements and the corrections is significant, nevertheless, there is a discernible general trend of increasing heat flow progressing from the ancient continental margin in the west to the modern margin in the east (Fig. 7.4). The hint of a zone of low heat flow centred over the Dunnage and Avalon zones is suggested by decreased heat flow values in central Newfoundland, Prince Edward Island, and New Brunswick. The offshore data on the shelf are considered by Jessop et al. (1984) to belong to a different heat flow regime and are not considered further here.

It must be recognized that the paucity of data with the attendant uncertainties in the heat flow values permits considerable freedom in interpretation. If the data are analyzed according to the zone in which the site is located, there is no statistical difference between the Dunnage and Avalon data when the highly anomalous St. Lawrence granite is excluded. There is significant difference between the data from each of the Dunnage and Avalon zones and those for the Humber Zone which are markedly lower. Heat flow values for the Meguma Zone are significantly higher than those for the Avalon Zone.

The major conclusion of the studies in the Maritime Provinces and Newfoundland is that the observed heat flow is related to local heat production in the crust and that the Canadian Appalachian region belongs to a heat flow province similar to that of the eastern United States (Hyndman et al., 1979; Wright et al., 1980). Differences in measurements between zones appear to reflect the local conditions rather than significant crustal differences.

PALEOMAGNETIC AND ROCK MAGNETIC STUDIES

Paleomagnetism is the only geophysical method that can, under ideal circumstances, provide information on major horizontal movements. Other geophysical methods provide constraints on the vertical and horizontal extent of various geological features in the subsurface. The objective of this section is to summarize the major problems that have been addressed by the paleomagnetic method. This is done historically to show how interpretations have changed as the database expanded and particularly as more magnetization components could be isolated.

Paleomagnetic studies are conducted on the premise that each orogenic, diagenetic, or thermal event affecting a rock mass has left a unique magnetic signature in the magnetic minerals of the rock which can be isolated by the appropriate demagnetization techniques. The subsequent isolation and identification of these separate remanence components is the objective of all paleomagnetic studies.

Since the net remanence recorded in the rock is the vector sum of all the individual remanences acquired by the rock since its formation, each of the remanence components will be diagnostic of the magnetic field conditions that prevailed at the time of acquisition of that component. Only when the time of acquisition can be firmly established can the appropriate component be used to provide information regarding the paleolatitude of the rock unit at the time of acquisition of the remanence. This fundamental point is often ignored and the age of a remanence component is often erroneously assumed to be the age of formation of the rock. Roy et al. (1983) have discussed the problems this causes in interpreting the paleomagnetic data from the Canadian Appalachian region. Irving and Strong (1984a, b) illustrated how neglect of this basic principle has led to erroneous conclusions about the nature and magnitude of motions in the Appalachian Orogen. Hodych et al. (1984) have demonstrated that the time of acquisition of a remanence component may be significantly younger than the age of the rock. Finally, Van der Voo (1988) has provided a concise summary of the often forgotten fundamental assumptions in paleomagnetic studies and their interpretation.

Since the Paleozoic rocks of the Canadian Appalachian region have experienced structural events in Ordovician, Devonian, and Permian-Carboniferous times it is not surprising that the apparent polar wander curves are only considered accurate back to the latest Carboniferous (Irving, 1981; Van der Voo, 1988). For older rocks, Late Carboniferous-Early Permian paleomagnetic overprints must be removed (Roy and Morris, 1983; Irving and Strong, 1984a, b). Murthy (1983a, b) has suggested that Devonian overprints are also possible. Thus the older paleopole positions and the polar wander curves based on them are less reliable.

Some paleomagnetic studies (Nairn et al., 1959; Black, 1964) in the Canadian Appalachian region predate the major acceptance of plate tectonic concepts. One such early study led to the suggestion that western Newfoundland had rotated by 30° counterclockwise in pre-Carboniferous time about a pivot situated in the Strait of Belle Isle (Black, 1964). With the general acceptance of plate tectonics, paleomagnetic studies focused on the question of the width of the Iapetus Ocean. Deutsch (1969) pointed out that only if the two sides of that ocean changed paleolatitude could the paleomagnetic method yield results that could be interpreted in terms of the width of the intervening ocean.

Recently the nature and magnitude of post-Devonian (Morris, 1976) and Carboniferous movements (Kent and Opdyke, 1978, 1979, 1980; Roy et al., 1983) have been the subject of intense debate. There have also been studies on the youngest rocks in the area, the Mesozoic dykes (Lapointe, 1979; Hodych and Hayatsu, 1980; Prasad, 1981).

The paleomagnetic contributions to the study of these problems, which have generated an impressive array of paleomagnetic literature, are summarized below. For a detailed analysis of a particular problem and specific details regarding sampling and demagnetization procedures the reader is referred to the original papers.

Rotation of Newfoundland

The hypothesis of post-Devonian but pre-Carboniferous rotation of Newfoundland by approximately 30° counterclockwise about a pivot situated in the Strait of Belle Isle is credited to Black (1964), although the question was posed by Wegener (1929) on geographic grounds (Deutsch, 1969) and addressed by Nairn et al. (1959). Black's (1964) hypothesis was based upon the difference in pole positions between: (1) the Lower Cambrian Ratcliffe Brook Formation of New Brunswick compared with the Lower Cambrian Bradore Formation of western Newfoundland; (2) the sedimentary rocks of the Upper Devonian Perry Formation and volcanic rocks of New Brunswick compared with the Lower Devonian Clam Bank Group of western Newfoundland; and (3) several Permian-Carboniferous redbed localities of New Brunswick and Prince Edward Island compared with the Carboniferous Codroy Group of western Newfoundland.

Since the publication of the 1964 paper, the question of the nature and magnitude of such a rotation has been addressed either directly or indirectly by many authors (Robertson et al., 1968; Murthy and Rao, 1976; Rao and Deutsch, 1976; Deutsch and Rao, 1977; Buchan and Hodych, 1982; Rao et al., 1986). Their papers have criticized specific parts of the original work or have provided new data bearing on the topic. The arguments for or against the rotation are summarized below.

Robertson et al. (1968) questioned Black's (1964) conclusion by demonstrating that Black's (1964) Cambrian poles based on the Ratcliffe Brook Formation from New Brunswick and the Bradore Formation of Newfoundland are not statistically significantly different as the error ovals intersect. More importantly, Robertson et al. (1968) demonstrated that Black's (1964) work did not detect a later remanence which they isolated, thus the uncertainty in his Devonian poles for the Perry Formation of New Brunswick is 15°, not the 9° that he quoted. Robertson et al. (1968) also questioned the age of the remanence in the Perry Formation and pointed out that it may have been acquired as late as the Carboniferous, a conclusion that is consistent with more recent evidence on the nature of remanence acquisition by redbeds (Hodych et al., 1984).

Subsequent studies have either suggested that the rotation could be interpreted as a measure of the width of Iapetus (Rao and Deutsch, 1976; Deutsch and Rao, 1977) or could result from a much smaller, unresolvable rotation (Irving and Strong, 1984b; Murthy, 1985). These authors suggested that a 5° to 10° rotation cannot be discounted, although they do not specify whether this is a rotation of the whole of western Newfoundland or only of the local area sampled.

The paleomagnetic studies have not unequivocally demonstrated that Newfoundland rotated about a pivot in the Strait of Belle Isle. Sheridan and Drake (1968) commented that their seismic data provided evidence against such a rotation. Gravity and magnetic data collected in the St. George's Bay area of western Newfoundland (Miller et al., 1990) suggest that local block rotation may have occurred. Such rotations would explain the apparent differences in paleopoles of rocks from different locations on the periphery of the Magdalen Basin in Newfoundland and New Brunswick and would also explain the rotations that Murthy (1985) and Murthy and Rao (1976) implied. Deutsch and Prasad (1987) and Deutsch and Storetvedt (1988) discounted the possibility of tectonic rotation and demonstrated that the polar wander curves for Norway, Ireland, and Scotland contain similar apparent rotations as the apparent 30° rotation of Newfoundland.

The width of the early Paleozoic ocean (Iapetus)

Central to all models explaining the presence of ophiolite suites in the Appalachian-Caledonian Orogen is the idea of an early Paleozoic ocean, Iapetus. Its closure resulted in obduction of ophiolitic suites across adjacent continental margins (Williams, 1984). In understanding the history of Iapetus, the paleomagnetic data have a unique contribution to make in determining its initial width.

The type of paleomagnetic evidence needed to resolve the width of Iapetus was discussed by Deutsch (1969). Since then numerous pole positions have been determined from Cambrian-Ordovician rocks on the western side of the Appalachian Orogen (Seguin, 1978, 1980; Dankers and Lapointe, 1979), from sites in the Dunnage Zone (Rao and Deutsch, 1976; Lapointe, 1979), and from the Avalon Zone (Roy et al., 1979; Rao et al., 1981; Roy and Anderson, 1981; Buchan and Hodych, 1982; Seguin et al., 1982). Additional authors have contributed data which have been used in other discussions of this problem.

There is no agreement on the significance of the differences between pole positions determined from rocks of similar age in the same regions. Dankers and Lapointe (1979) noted the discrepancy between Ordovician pole positions and suggested that an unspecified amount of plate motion may have occurred between the craton and parts of the orogen. Roy et al. (1983) noted that Ordovician pole positions from Newfoundland are intermingled with Ordovician pole positions from the craton, a fact that Buchan and Hodych (1982) interpreted as indicating that the Avalon Zone of Newfoundland had approximately the same paleolatitude as the craton during the Ordovician. There is the possibility that pole positions from Ordovician rocks may represent a more recent remagnetization (Buchan and Hodych, 1982; Roy et al., 1983). Further complicating the interpretation is the possibility that some of the Dunnage Zone results were determined on isolated terranes, or islands, in Iapetus (Van der Voo, 1988).

Estimates of the width of Iapetus range from 4000 km (Morel and Irving, 1978) to approximately 1500 km (Seguin, 1982). Rao et al. (1986) reported an intermediate value of 3600 with an uncertainty in the estimate quoted at 2000 km. Rao et al. (1986) also gave an angular estimate of 20° which is equivalent to 2500 km linear distance. The Seguin (1982) estimate is closest to a minimum width of 1500 km based on geological considerations (Williams, 1984). The final conclusion is that the width is not determined with any accuracy because of remagnetizations and problems in isolating precise Ordovician pole positions. The width may also have been different at different parts of the orogen, just as the width of modern oceans varies from place to place.

Transcurrent movements since the Devonian

The idea of major transcurrent movement in the northern Appalachians since the Devonian and the hypothesis that eastern parts of the orogen were displaced northward relative to the craton were proposed by Kent and Opdyke (1978, 1979, 1980) and Van der Voo and Scotese (1981). Roy and Anderson (1981), Seguin et al. (1982), and Roy et al. (1983) also addressed various aspects of the theory. Roy et al.

(1983) noted that one of the critical pole positions was derived from two partially resolved remanences and that uncertainty in the pole position may be 15° or more. Rao et al. (1983) also pointed out that the true ages of the magnetizations used to deduce the separation are in doubt and that although the sampling locations are west of the major early Paleozoic deformation areas they could have been affected by Hercynian events. The major problems in ascertaining this movement are the classic conundrums of paleomagnetism, namely the age of the various remanence components and the many movements that can be inferred if the ages are incorrect. As a result of these uncertainties there is evidence for differences between Silurian-Devonian poles and Permian-Carboniferous poles which may be indicative of polar wander, block, or plate motions during the interval. The major difficulty in resolving the question is the lack of data from the interior of North America which can be unequivocally identified as Silurian-Devonian. A further complication is the latest evidence from Devonian and Carboniferous rocks of western Newfoundland which indicates a possible southward movement during that interval (Murthy, 1985).

Irving and Strong (1984a) provided Carboniferous pole positions from the Avalon and Humber zones of Newfoundland that contradict the notion of 1500 km of northward displacement of "Acadia" during the Early Carboniferous, as proposed by Kent and Opdyke (1978, 1979, 1980). They did not eliminate the possibility of mid-Devonian motion as suggested by Morris (1976). Murthy (1985), using Carboniferous rocks from the Bay St. George Basin of western Newfoundland, also concluded that no Carboniferous motion occurred and pointed out that paleomagnetic evidence suggests a general northward movement of all of North America during the Carboniferous.

The overall evidence indicates that there has not been significant transcurrent movement during the Carboniferous and that major movements since the Devonian are questionable. This conclusion is now generally accepted (Van der Voo, 1988).

Post-Carboniferous results

Jurassic to Cretaceous dykes of central Newfoundland (Lapointe, 1979; Prasad, 1981) and Permian-Triassic rocks of the Avalon Peninsula (Hodych and Hayatsu, 1980) are interpreted as intrusions related to the opening of the present Atlantic Ocean. Their age and paleomagnetic expression provide evidence for the difference in opening ages from Nova Scotia to Newfoundland. These data indicate that the Nova Scotia margin began to rift in Triassic-Jurassic while the Newfoundland margin began rifting in Late Jurassic-Early Cretaceous, in agreement with geological evidence from offshore wells.

Paleomagnetic summary

Paleomagnetic studies have led to several tantalizing hypotheses independent of other geological and geophysical information. In the final analysis, the paleomagnetic studies fail to provide unambiguous evidence on the width of the early Paleozoic Iapetus Ocean. This is hardly surprising as the width of Iapetus was probably not constant and the ocean probably had irregular margins like the present Atlantic. Alternatively, it may have been of limited extent

similar to modern back arc seas such as the Sea of Japan. Major sinistral transcurrent motions in the Carboniferous have now been discounted. The nature and timing of motion in the Devonian have not been resolved and it is probably this motion that led to the joining of the Gander and Avalon zones (Van der Voo, 1988). Paleomagnetic data have provided some evidence on the timing of the opening of the present Atlantic at various locations along the margin.

INTERPRETATION

Qualitative

The general qualitative interpretation of the data, especially the gravity and magnetic data, has been discussed in the presentation of each data set. The zone boundaries inferred from geophysical signatures do not always coincide with those drawn from the geology. This is not surprising as the geophysical boundaries are representative of integrated effects whereas the geological evidence represents the surface expressions. The deep seismic reflection programs have resulted in fundamental changes in our understanding of the deep crustal structure of the region.

Marillier et al. (1989) have identified three major lower crustal blocks on the northeast Newfoundland and Gulf of St. Lawrence seismic lines, and Keen et al. (1990) have identified another lower crustal block beneath the Meguma Zone on the Grand Banks. These blocks, the Grenville, Central, Avalon and Sable, have distinctive seismic signatures and are separated by clearly identifiable boundaries. The configuration and interpretation of these blocks has been revised by Quinlan et al. (1992) based on the 1989 Lithoprobe East lines on Newfoundland.

The inferred boundaries (Marillier et al., 1989) between these blocks are presented in Figure 7.6 along with their extrapolation throughout the region based on gravity and magnetic signatures. The Grenville-Central block boundary is interpreted as a deep collisional zone or suture, along which the Iapetus Ocean has been completely obliterated. On the northeast Newfoundland line (Fig. 7.7) it correlates with the positive eastward gravity gradient at 100 km from point X. This also corresponds with the seaward extension of the Red Indian Line (Williams et al., 1988). South of Newfoundland the inferred position of the suture (Marillier et al., 1989) coincides with the east side of the positive gravity anomaly belt. The extrapolated position for the suture through the Gulf of St. Lawrence is parallel to a set of gravity and magnetic anomalies which in turn are parallel to the Appalachian Structural Front (Marillier et al., 1989). These trends are more consistent than earlier extrapolations deduced by Haworth and Jacobi (1983) from the potential field data alone. The position of this suture is also consistent off northeast Newfoundland with the location of the greatest crustal thickness inferred from the reinterpreted refraction data (Mitchelmore, 1985) and subsequently corroborated by the deep reflection data.

The Central-Avalon block boundary as delineated in and northeast of Newfoundland coincides with the Dover-Hermitage Bay Fault. The recognition of this boundary northeast of Cape Breton Island (Marillier et al., 1989) allows a definitive correlation with a distinctive gravity anomaly pattern, which may be traced northward to the area in Newfoundland where the boundary is identified geologically. Recognition of the Central-Avalon block boundary northeast of Cape Breton Island, new geological mapping (Barr et al., 1987, 1988), and interpretation of aeromagnetic data (Barr et al., 1988) all indicate that the Central-Avalon block boundary may be drawn as shown in Figure 7.6. This eliminates the need for locating the boundary north of Cape Breton Island, which previous interpretations attempted (Haworth and Jacobi, 1983). It also suggests that the Gander-Avalon zone boundary may follow the anomaly pattern in the Northumberland Strait south of Prince Edward Island, and then extend southward along the Fredericton-Norembega Fault. The source of the gravity high over western Prince Edward Island is probably Carboniferous volcanic rocks (Jones and Garland, 1986).

Marillier et al. (1989) were unable to determine the nature of the boundary between the Avalon lower crustal block and basement to the Meguma Zone. C.E. Keen and I. Reid (pers. comm., 1990), on the basis of both refraction and deep reflection data on the southern Grand Banks, considered that the Avalon-Meguma zone boundary is vertical and cuts through the crust. They agreed with the interpretation of Haworth and Lefort (1979) that the Collector Anomaly defines a major zone boundary. Onland, the boundary is the Cobequid-Chedabucto Fault (Keppie, 1985). Stockmal et al. (1987, 1990) proposed a model for structural evolution following the sequential obduction of an island arc complex over an east-dipping subducting Grenville plate. Their scenario explains the parallelism of the major boundaries deduced here and the pre-existing Grenville margin. In some respects their interpretation follows a model based on potential field data alone (Haworth, 1974b, 1978). Their latest interpretation (Stockmal et al., 1990) also explains the siting of Carboniferous basins on reactivated faults and accommodates the known strike-slip motions. Central to this model is the existence of a major zone of strike-slip motion that joins the southeast trend of the Appalachian Structural Front in the Gulf of St. Lawrence to the general trend of the Newfoundland Fracture Zone (Fig. 7.6). It is in this area that Keppie (1985) has also postulated the presence of a major fault between Newfoundland and Nova Scotia. Recent high resolution aeromagnetic data from that area show the presence of such a feature and demonstrate that there may be numerous similar features present (Loncarevic et al., 1989). Such features are recognized also throughout the Grand Banks (Haworth and Lefort, 1979; Enachescu, 1987; Tankard and Welsink, 1987). In the extreme offshore area, these are considered to be transfer faults related to the Mesozoic opening of the Atlantic. Miller (1988, 1990) reported similar trends nearshore and onland. They are interpreted as two sets of lineaments that cut and apparently offset the Dover Fault in northeast Newfoundland. Miller and Tuach (1988a) also recognized northwest-trending features in the Ackley Granite Suite that stitches the Gander and Avalon zones in southern Newfoundland. Using gravity, magnetic, topographic, and geological maps it is possible to identify a series of such features that occur almost regularly across Newfoundland (Fig. 7.6). Similar features can be recognized in Nova Scotia, New Brunswick, and in the adjacent offshore regions (Fig. 7.6). At present there is no analysis of movement or timing. Their association with bathymetric and topographic features imply that they are late features unrelated to Appalachian structural trends. Their parallelism to the Newfoundland Fracture Zone on the southern nose of the Grand Banks is noteworthy. The features are also perpendicular to the general trends of the known

Mesozoic dykes (Papezik et al., 1975). On the Grand Banks, transfer faults related to Cretaceous and younger basin formation have similar orientation; this suggests that these features may be younger than the Late Triassic dykes (Hodych and Hayatsu, 1980).

Quantitative

Deep reflection data, although offering a tantalizing cross-section of the crust in any particular region, are unable to resolve the nature of the upper few kilometres of crust, especially on offshore profiles. The combined interpretation of the gravity, magnetic, and seismic data in conjunction with the surface geology assist in near surface analyses. The refraction data, when reinterpreted using ray tracing packages, can provide the absolute depth information which cannot be extracted from the reflection data because of the poor resolution of velocities. Magnetotelluric results can provide constraints on the crustal thickness to supplement the refraction data.

An example of the use of the gravity, magnetic, and seismic data is presented in Figure 7.7 for the northeast Newfoundland reflection line 84-2 (Miller, 1990). The deep reflection (seismic) picks define the geometry of the lower crustal blocks; the gravity and in particular the magnetic anomalies provide information on the location of the edges of ultramafic floored thrust sheets, which in turn can be correlated with known geological features on land. The interpreted model from the ray tracing analysis of the refraction data (Mitchelmore, 1985), when projected onto the same profile, corroborates the geometry and provides some depth control. The model presented is consistent with all available information and the gravity and magnetic anomalies calculated from it match the observed anomalies. This enables us to associate particular features of the gravity and magnetic fields with specific geological features. The magnetic anomalies at 20 and 40 km from point X correlate with the gravity gradient marking the leading edge of the allochthonous Dunnage Zone overlying the Grenville lower crustal block. This correlation strengthens the interpretation of the Humber-Dunnage boundary as coincident with this gravity signature as discussed earlier. Similarly at the other end of the profile, the feature at 190 km from point X is associated with the Gander River Ultrabasic Belt and in this area it does not have a magnetic expression, in contrast to the signature on land (Miller, 1988). The total thickness of the Dunnage Zone in this model is approximately 24 km which is in agreement with the upper 6.0 and 6.5 km.s^{-1} layer of the reinterpreted refraction data (Mitchelmore, 1985). This implies that the 7.5 km.s^{-1} material correlates with the Grenville and Central lower crustal blocks of Marillier et al. (1989), which have the same geometry defined by the "intermediate" layer of the reinterpreted refraction data. The general shape of the Dunnage Zone is a compromise between the two models deduced from gravity alone (Weaver, 1967; Miller, 1977). The model is in general agreement with the geomagnetic depth sounding and heat flow data which serve to corroborate its validity.

A similar model has not been constructed for the equivalent seismic line northeast of Prince Edward Island although that interpretation is in progress (F. Marillier, pers. comm., 1990). In order to proceed with such a model the free air gravity anomalies must be converted to Bouguer anomalies. Preliminary results indicate that

LITHOPROBE LINE 84-2

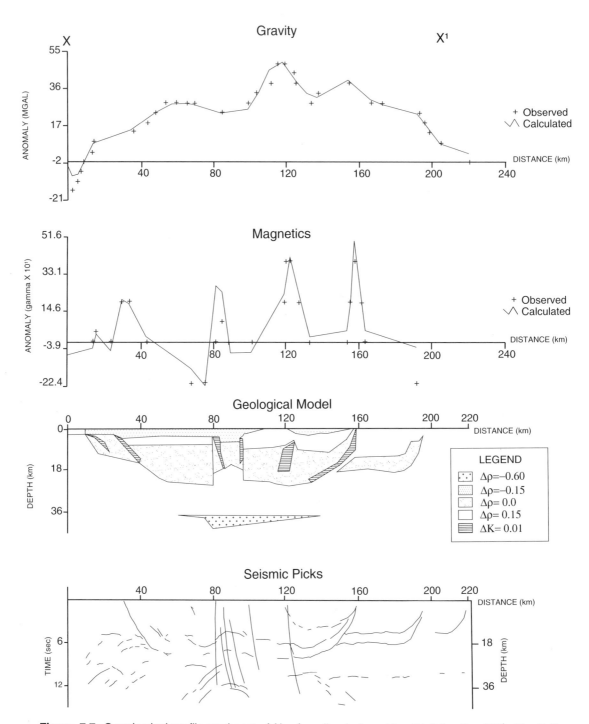

Figure 7.7. Geophysical profile northeast of Newfoundland along Line 84-2 (portion XX[1], Fig. 7.1) modified from Miller (1990). Observed gravity and magnetic anomalies, denoted by +, taken from Maps 3 and 4. Calculated anomalies, denoted by solid lines, computed from the geological model using the physical properties depicted in the legend. ρ = density contrast in g•cm[-3] and K = magnetic susceptibility in MKS units. Seismic picks from line 84-2 in the bottom panel.

initial estimates of crustal thickness from inverted Bouguer anomalies are consistent with the thinned crust seen on the deep reflection line (Marillier et al., 1989) and crustal thicknesses as indicated by the ray tracing of the older refraction data (Campbell, 1984). The electromagnetic and heat flow data are not incompatible with these models.

SUMMARY AND FUTURE STUDIES

The major conclusion to be drawn from the geophysical studies is that there are significant differences in geophysical signatures reflecting the differing crustal structures in the five major zones in the Canadian Appalachian region. The geophysical data demonstrate the fundamental difference between the Gander and Dunnage zones, and the Humber and Avalon zones. These differences in geophysical signatures are apparent in the compilations for the transects northeast of Newfoundland (Fig. 7.7). The Dunnage and Gander zones are typified by material having higher than normal density causing positive gravity anomalies in Newfoundland where there are no overlying upper Paleozoic sedimentary rocks, and negative anomalies in New Brunswick where there are such cover rocks. Enhanced electrical conductivity in the Dunnage and Gander zones produces the geomagnetic induction anomalies at their boundaries and suppressed-transfer-functions in the centre of the area. Both zones have magnetic anomalies of varying wavelength and intensity associated with sheets of mafic and ultramafic rocks. The Humber and Avalon zones are typified by thinner crust (35 km), compared with the thicker (about 45 km) crust beneath the Dunnage Zone and parts of the Gander Zone, deduced from refraction and teleseismic studies. A combined interpretation of deep reflection data with the gravity and magnetic data off northeast Newfoundland indicates that the Dunnage Zone may contain imbricate structures down to 25 km. The Gander Zone may also contain sheets of obducted ophiolitic material underlain by a possible continental rise prism and continental basement, the Central lower crustal block. The Moho beneath the Avalon Zone is deeper than that beneath the Gander Zone at the projected Gander-Avalon boundary (Keen et al., 1986).

The Avalon Zone has a series of alternating gravity and magnetic highs that are indicative of major Precambrian block faulting and rifting. The basins formed by these rifts were subsequently filled. In the Humber Zone, Carboniferous sedimentary basins formed primarily by transcurrent motions. The deep seismic data show that the Moho depths beneath the Humber and Avalon zones are similar at around 10 seconds two-way-travel time (33-35 km). There is a series of apparently east-dipping reflectors at the Moho in the Avalon part of the deep reflection section that have not been explained. The nature of the Avalon Zone, especially beneath the Grand Banks, and its ancestral role as a floor to the Mesozoic petroliferous basins is another major subject for continuing research (Tankard and Welsink, 1987; Enachescu, 1987). The Meguma Zone has a similar crustal thickness to the Avalon Zone and is typified by a gravity low and a series of linear magnetic anomalies. The Meguma Zone also has a heat flow unlike that of other zones because of the preponderance of radiogenic plutons

and the high heat production in its thick sedimentary sequence. On the basis of the refraction and reflection data, its total crustal thickness appears to be similar to that of the Avalon Zone (I. Reid, pers. comm., 1988).

The paleomagnetic data suggest that whatever motion caused the formation of the Magdalen Basin in the Gulf of St. Lawrence must have occurred before the Carboniferous. The formation of the Magdalen Basin is a subject of continuing study. Its existence and the orientation of the Acadian Structural Front in New Brunswick were explained by a collisional plate model (Haworth, 1974a, b; Stockmal et al., 1990). The relationship to the plate tectonic regime after the closing of Iapetus is the focus for future study in and around the Gulf of St. Lawrence.

The overwhelming need for a series of good, high quality, deep crustal seismic lines similar to those of COCORP in the United States and BIRPS in the United Kingdom has been partially met by the Frontier Geoscience Program and the Lithoprobe East Project. New refraction experiments and magnetotelluric studies have been conducted as follow-up to the Lithoprobe East transects in Newfoundland. These data will not answer all the remaining questions about the Canadian Appalachian region but they will provide a more continuous picture of the subsurface which will stimulate the next generation of questions.

ACKNOWLEDGMENTS

The preparation of this chapter was made possible with the co-operation of many people. The contributors listed below are thanked for providing reprints of their articles on various geophysical studies in and adjacent to the Canadian Appalachian region, and for communications regarding specific aspects of the geophysics of the region: E.R. Deutsch, S.P. Srivastava, A.K. Goodacre, J.P. Hodych, K. Howell, M.D. Thomas, G.S. Murthy, K.B.S. Burke, A. Jessop, J.A. Wright, M.K. Seguin, R. Bailey, G. Quinlan, J.L. Roy, R.D. Hyndman, N.A. Cochrane, P. Lapointe, P.H. McGrath, I.C.F. Stewart, P.J. Hood, R. Kurtz, G.D. Garland, M. Drury, E. Tanczyk, A.G. Jones, R.T. Haworth, R.D.Jacobi, J. Adams, and A. Green.

The References list is by no means complete. The author's objective has been to provide a review of the major geophysical contributions to our understanding of the Canadian Appalachian region. Individuals wishing more details are referred to the papers listed in the References and through them to the breadth of literature on specific topics.

The diagrams were compiled from several sources. J. Adams of the Seismology Division, Earth Physics Branch arranged for the seismicity plots; the late J. Roy of the Geomagnetism Division, Earth Physics arranged for the paleomagnetic sampling site map.

The author wishes to express special thanks to E.W.R. Neale for his initial review of the paper and H. Williams for his meticulous editing of the final version. Their comments helped clarify the presentation. The final draft of the chapter was written and final reviews completed in 1991. A few newer references were added at the proof stage but the text was not altered at that stage.

REFERENCES

Ando, C.J., Cook, F.A., Oliver, J.E., Brown, L.D., and Kaufman, S.
1983: Crustal geometry of the Appalachian Orogen from seismic reflection studies; Geological Society of America, Memoir 158, p. 83-102.

Bailey, R.C., Edwards, R., Garland, G.D., Kurtz, R., and Pitcher, D.
1974: Electrical conductivity studies over a tectonically active area in Eastern Canada; Journal of Geomagnetism and Geoelectricity, v. 26, p. 125-146.

Barr, S.M. and Raeside, R.P.
1986: Precarboniferous tectonostratigraphic subdivisions of Cape Breton Island, Nova Scotia; Maritime Sediments and Atlantic Geology, v. 22, p. 252-263.

Barr, S.M., Raeside, R.P., Keen, C.E., Loncarevic, B.D., and Marillier, F.
1988: Crustal expression of tectonostratigraphic zones in Cape Breton Island, Nova Scotia, using deep reflection seismic and aeromagnetic data; Program with abstracts, Geological Association of Canada, v.13, p. A5.

Barr, S.M., Raeside, R.P., and van Breemen, O.
1987: Grenvillian basement in the northern Cape Breton Highlands, Nova Scotia; Canadian Journal of Earth Sciences, v. 24, p. 992-997.

Barrett, D.L., Berry, M., Blanchard, M.E., Keen, M.J., and McAllister, R.E.
1964: Seismic studies on the eastern seaboard of Canada: the Atlantic coast of Nova Scotia; Canadian Journal of Earth Sciences, v. 1, p. 10-22.

Basham, P.W. and Adams, J.
1984: The Miramichi, New Brunswick earthquakes - Near surface thrust faulting in the Appalachian terrain; Geoscience Canada, v. 11, p. 115-121.

Basham, P.W., Weichert, H., and Berry, M.J.
1979: Regional assessment of seismic risk in eastern Canada; Bulletin of the Seismological Society of America, v. 69, p. 1567-1602.

Berry, M.
1973: Structure of the crust and upper mantle in Canada; Tectonophysics, v. 20, p. 183-201.

Bhattacharyya, B.K. and Raychaudhuri, R.
1967: Aeromagnetic and geological interpretation of a section of the Appalachian belt in Canada; Canadian Journal of Earth Sciences, v. 4, p. 1016-1037.

Black, R.F.
1964: Paleomagnetic support of the theory of rotation of the western part of the island of Newfoundland; Nature, v. 202, p. 945-948.

Buchan, K.L. and Hodych, J.P.
1982: Paleomagnetic re-examination of the Lower Ordovician Wabana and Bell Island groups of the Avalon Peninsula, Newfoundland; Canadian Journal of Earth Sciences, v. 19, p. 1055-1069.

Cameron, G.W. and Hood, P.J.
1975: Residual aeromagnetic anomalies associated with the Meguma Group of Nova Scotia and their relationship to gold mineralization; in Report of Activities, Part C, Geological Survey of Canada, Paper 75-1C, p. 197-211.

Campbell, D.G.
1984: Re-interpretation of refraction data from the Magdalen basin; B.Sc. thesis, Memorial University of Newfoundland, St. John's, Newfoundland, 61 p.

Cochrane, N.A. and Hyndman, R.D.
1974: Magnetotelluric and magnetovariational studies in Atlantic Canada; Geophysical Journal, Royal Astronomical Society, v. 39, p. 385-406.

Cochrane, N.A. and Wright, J.A.
1977: Geomagnetic sounding near the northern termination of the Appalachian system; Canadian Journal of Earth Sciences, v. 14, p. 2858-2864.

Cutt, B.J. and Laving, J.G.
1977: Tectonic elements and geologic history of the South Labrador and Newfoundland continental shelf, eastern Canada; Bulletin, Canadian Society of Petroleum Geologists, v. 25, p. 1037-1058.

Dainty, A.M., Keen, C.E., Keen, M.J., and Blanchard, J.E.
1966: Review of geophysical evidence on the crust and upper mantle structure on the eastern seaboard of Canada; in Geophysical Monograph No. 10, American Geophysical Union, p. 349-369.

Dankers, P. and Lapointe, P.
1979: Paleomagnetism of Lower Cambrian volcanics and a cross-cutting Cambro-Ordovician diabase dyke from Buckingham (Quebec); Canadian Journal of Earth Sciences, v. 18, p. 1174-1186.

Deutsch, E.R.
1969: Paleomagnetism and North Atlantic paleogeography; in North Atlantic - Geology and continental drift, (ed.) M. Kay; American Association of Petroleum Geologists, Memoir 2, p. 931-954.
1980: Magnetism of the mid-Ordovician volcanics, SE Ireland and the question of a wide proto Atlantic Ocean; Journal of Geomagnetism and Geoelectricity, v. 32, p. 77-98.
1984: Mid-Ordovician paleomagnetism and the proto-Atlantic ocean in Ireland; in Permo-Triassic Continental Configurations and Pre-Permian Plate Tectonics; Geodynamics Series No. 12, American Geophysical Union, p. 116-119.

Deutsch, E.R. and Prasad, J.N.
1987: Paleomagnetic discordance within the Ordovician St. George Group, western Newfoundland, indicating a post mid-Ordovician 40° westward swing of apparent polar wander; Abstracts, v. 2, International Association of Geomagnetism and Aeronomy), International Union of Geodesy and Geophysics 19th General Assembly, Vancouver, p. 502.

Deutsch, E.R. and Rao, K.V.
1977: New paleomagnetic evidence fails to support the rotation of western Newfoundland; Nature, v. 266, p. 314-318.

Deutsch, E.R. and Storetvedt, K.M.
1988: Magnetism of igneous rocks from the Tourmakeady and Glensaul inliers, W. Ireland: mode of emplacement and aspects of the Ordovician field pattern; Geophysics Journal, v. 92, p. 223-234.

Dosso, H.W., Neinaber, W., and Wright, J.A.
1982: Analogue model study of electromagnetic induction in the Newfoundland region; Physics of the Earth and Planetary Interiors, v. 32, p. 65-84.

Drury, M.J., Jessop, A.M., and Lewis, T.J.
1987: The thermal nature of the Canadian Appalachian crust; Tectonophysics, v. 133, p. 1-14.

Durling, P.W. and Marillier, F.J.Y.
1990: Structural trends and basement subdivision in the western Gulf of St. Lawrence, northern Appalachians; Atlantic Geology, v. 26, p. 79-95.

Enachescu, M.E.
1987: Extended basement beneath the intracratonic rifted basins of the Grand Banks of Newfoundland; presented at the 7th International Conference on Basement Tectonics, Kingston, Canada.

Ewing, G.N., Dainty, A.M., Blanchard, J.E., and Keen, M.J.
1966: Seismic study of the eastern seaboard of Canada: the Appalachian system I; Canadian Journal of Earth Sciences, v. 3, p. 89-109.

Fenwick, D.K.B., Keen, M.J., and Lambert, A.
1968: Geophysical studies of the continental margin northeast of Newfoundland; Canadian Journal of Earth Sciences, v. 5, p. 483-500.

Fyffe, L.R. and Fricker, A.
1987: Tectonostratigraphic terrane analysis of New Brunswick; Maritime Sediments and Atlantic Geology, v.23, p. 113-122.

Garland, G.D.
1953: Gravity measurements in the Maritime Provinces; Publications of Dominion Observatory, v. 16, p. 185-275.

Geological Survey of Canada
1988a: Gravity anomaly map of the Continental Margin of Eastern Canada; Geological Survey of Canada, Map 1708A, scale 1:5 000 000.
1988b: Magnetic anomaly map of the Continental Margin of Eastern Canada; Geological Survey of Canada, Map 1709A, scale 1:5 000 000.

Goodacre, A.K.
1964: Preliminary results of underwater gravity surveys in the Gulf of St. Lawrence; Gravity Map Series No. 46, Energy, Mines and Resources, Ottawa.

Goodacre, A.K., Brule, B.G., and Cooper, R.V.
1969: Results of underwater gravity surveys in the Gulf of St. Lawrence; Gravity Map Series No. 86, Earth Physics Branch, Ottawa.

Green, A.G., Berry, M.J., Spencer, C.P., Kanesewich, E.R., Chiu, S., Clowes, R.M., Yorath, C.J., Stewart, D.B., Unger, J.D., and Poole, W.H.
1985: Recent seismic reflection studies in Canada; in Reflection Seismology: A global perspective, (ed.) M. Barazanai and L. Brown, American Geophysical Union, Geodynamic Series, v. 13, p. 85-97.

Haworth, R.T.

1974a: Paleozoic continental collision in the northern Appalachians in light of gravity and magnetic data in the Gulf of St. Lawrence; in Offshore geology of Eastern Canada, Volume 2 - Regional Geology; Geological Survey of Canada, Paper 74-30, p. 1-10.

1974b: The development of Atlantic Canada as a result of continental collision - evidence from gravity and magnetic data; Canadian Society of Petroleum Geologists, Memoir 4, p. 59-77.

1977: The continental crust northeast of Newfoundland and its ancestral relationship to the Charlie Fracture Zone; Nature, v. 266, p. 59-77.

1978: Interpretation of geophysical data in the northern Gulf of St. Lawrence and its relevance to lower Paleozoic geology; Geological Society of America, v. 89, p. 1091-1110.

Haworth, R.T. and Jacobi, R.D.

1983: Geophysical correlation between the geological zonation of Newfoundland and the British Isles; Geological Society of America, Memoir 158, p. 25-32.

Haworth, R.T. and Lefort, J.P.

1979: Geophysical evidence for the extent of the Avalon Zone in Atlantic Canada; Canadian Journal of Earth Sciences, v. l6, p. 552-567.

Haworth, R.T. and MacIntyre, J.B.

1975: Gravity and magnetic fields of Atlantic Offshore Canada; Geological Survey of Canada, Paper 75-9.

Haworth, R.T. and Miller, H.G.

1982: The structure of Paleozoic oceanic rocks beneath Notre Dame Bay, Newfoundland; in Major structural zones and faults of the Northern Appalachians; (ed.) P. St-Julien and J. Béland; Geological Association of Canada, Special Paper 24, p. 149-173.

Haworth, R.T. and Sanford, B.V.

1976: Paleozoic geology of the northeast Gulf of St. Lawrence; in Report of Activities, Part A; Geological Survey of Canada, Paper 76-1A, p. 1-6.

Haworth, R.T., Daniels, D.L., Williams, H., and Zietz, I.

1980: Bouguer gravity map of the Appalachian Orogen; Memorial University of Newfoundland, Map no. 3.

Haworth, R.T., Grant, A.C., and Folinsbee, R.A.

1976: Geology of the continental shelf off southeastern Labrador; in Report of Activities, Part C; Geological Survey of Canada, Paper 76-1C, p. 61-70.

Haworth, R.T., Poole, W.H., Grant, A.C., and Sanford, B.V.

1976: Marine geoscience survey northeast of Newfoundland; in Report of Activites, Part A; Geological Survey of Canada, Paper 76-1A, p. 7-15.

Hebert, D., Dosso, H.W., Neinaber, W., and Wright, J.A.

1982: Analogue model study of electromagnetic induction in the Newfoundland region; Physics of the Earth and Planetary Interiors, v. 32, p. 65-84.

Hobson, G.D. and Overton, A.

1973: Sedimentary refraction seismic surveys, Gulf of St. Lawrence; in Earth Science Symposium on Offshore Eastern Canada, (ed.) P.J. Hood; Geological Survey of Canada, Paper 71-23, p. 325-336.

Hodych, J.P. and Hayatsu, A.

1980: K-Ar isochron age and paleomagnetism of diabase along the trans-Avalon aeromagnetic lineament - evidence of Late Triassic rifting in Newfoundland; Canadian Journal of Earth Sciences, v. 17, p. 491-499.

Hodych, J.P., Patzold, R.R., and Buchan, K.L.

1984: Paleomagnetic dating of the transformation of oolitic goethite to hematite in iron ore; Canadian Journal of Earth Sciences, v. 21, p. 127-130.

Hughes, S.J., Hall, J., and Leutgert

in press: The seismic velocity structure of the Newfoundland Appalachian orogen: results from land component of the Lithoprobe East 1991 experiment; Journal of Geophysical Research.

Hyndman, R.D. and Cochrane, N.A.

1971: Electrical conductivity structure by geomagnetic induction at the continental margin of Atlantic Canada; Geophysical Journal, Royal Astronomical Society, v. 25, p. 425-446.

Hyndman, R.D., Jessop, A.M., Judge, A.S., and Rankin, D.S.

1979: Heat flow in the Maritime Provinces of Canada; Canadian Journal of Earth Sciences, v. 16, p. 1154-1165.

Irving, E.

1981: Phanerozoic continental drift; Physics of the Earth and Planetary Interiors, v. 24, p. 197-204.

Irving, E. and Strong, D.F.

1984a: Paleomagnetism of the Early Carboniferous Deer Lake Group, western Newfoundland: no evidence for mid-Carboniferous displacement of Acadia; Earth and Planetary Science Letters, v. 69, p. 379-390.

1984b: Evidence against large-scale Carboniferous strike-slip faulting in the Appalachian-Caledonian Orogen; Nature, v. 310, p. 762-764.

Irving, E., Wheadon, P., and Horel, G.

1990: Catalogue of paleomagnetic directions and paleopoles (sixth issue): Results from Canada and the Laurentian Shield (1953-1989); Geological Survey of Canada, Open File 2247, 205 p.

Jacobi, R. and Kristoffersen, Y.

1976: Geophysical and geological trends on the continental shelf off northeastern Newfoundland; Canadian Journal of Earth Sciences, v. 13, p. 1039-1051.

Jessop, A.M., Lewis, T.J., Judge, A.S., Taylor, A.E., and Drury, M.J.

1984: Terrestrial heat flow in Canada; Tectonophysics, v. 103, p. 239-261.

Jones, A.G. and Garland, G.D.

1986: Preliminary interpretation of upper crustal structure beneath Prince Edward Island; Annales Geophysicae, v. 4, p. 157-164.

Keen, C.E., Kay, W.A., Keppie, D., Marillier, F., Pe-Piper, G., and Waldron, J.W.F.

1990: Deep seismic reflection data from the Bay of Fundy and Gulf of Maine; in Lithoprobe East, Report 13, (ed.) Jeremy Hall; Memorial University of Newfoundland, p. 111-116.

Keen, C.E., Keen, M.J., Nichols, B., Reid, I., Stockmal, G.S., Colman-Sadd, S.P., O'Brien, S.J., Miller, H., Quinlan, G., Williams, H., and Wright, J.

1986: Deep seismic reflection profile across the northern Appalachians; Geology, v. 15, p. 141-145.

Keen, M.J. and Williams, G.L. (ed.)

1990: Geology of the Continental Margin of Eastern Canada; Geological Survey of Canada, Geology of Canada, no. 2, 855 p. (also Geological Society of America, The Geology of North America, v. I-2).

Kent, D.V. and Opdyke, N.D.

1978: Paleomagnetism of the Devonian Catskill red beds: evidence for motion of the coastal New England-Canadian Maritime region relative to cratonic North America; Journal of Geophysical Research, v. 83, p. 4441-4450.

1979: The Early Carboniferous paleomagnetic field for North America and its bearing on the tectonics of the northern Appalachians; Earth and Planetary Science Letters, v. 44, p. 365-372.

1980: Paleomagnetism of Siluro-Devonian rocks from southeastern Maine; Canadian Journal of Earth Sciences, v. 17, p. 1653-1665.

Keppie, J.D.

1985: The Appalachian collage; in The Caledonide Orogen, Scandinavia and Related Areas, (ed.) D.G. Gee and B. Sturt; J. Wylie and Sons, p. 1217-1226.

Kumarapeli, P.S., Goodacre, A.K., and Thomas, M.D.

1981: Gravity and magnetic anomalies of the Sutton Mountains region, Quebec and Vermont: expressions of rift volcanics related to the opening of Iapetus; Canadian Journal of Earth Sciences, v. 18, p. 680-692.

Kurtz, R.D. and Garland, G.D.

1976: Magnetotelluric measurements in eastern Canada; Geophysical Journal, Royal Astronomical Society, v. 45, p. 321-347.

Kurtz, R.D. and Ostrowski, J.A.

1984: Preliminary results of the magnetotelluric survey in the Miramichi, N.B. earthquake zone; Earth Physics Branch Open File 84-18, 9 p.

Lapointe, P.L

1979: Paleomagnetism of Notre Dame Bay lamphrophyre dykes, Newfoundland and the opening of the North Atlantic Ocean; Canadian Journal of Earth Sciences, v. 16, p. 1823-1831.

Lewis, J.T. and Hyndman, R.D.

1976: Oceanic heat flow measurements over the continental margins of eastern Canada; Canadian Journal of Earth Sciences, v. 13, p. 1031-1038.

Loncarevic, B.D., Barr, S.M., Raeside, R.P., Keen, C.E., and Marillier, F.

1989: Northeastern extension and crustal expression of terranes from Cape Breton Island, Nova Scotia, based on geophysical data; Canadian Journal of Earth Sciences, v. 26, p. 2255-2267.

McMillan, N.J.

1973: Shelves of Labrador Sea and Baffin Bay, Canada; in The future petroleum provinces of Canada - their geology and potential, (ed.) R.G. McCrossan; Canadian Society of Petroleum Geologists, Memoir 1, p. 473-517.

Marillier, F. and Verhoef, J.

1989: Crustal thickness under the Gulf of St. Lawrence, northern Appalachians, from gravity and seismic data; Canadian Journal of Earth Sciences, v. 26, p. 1517-1532.

Marillier, F., Keen, C.E., Stockmal, G.S., Quinlan, G., Williams, H., Colman-Sadd, S.P., and O'Brien, S.J.

1989: Crustal structure and surface zonation of the Canadian Appalachians: implications of deep seismic reflection data; Canadian Journal of Earth Sciences, v. 26, p. 305-321.

McGrath, P.H.
1970: Aeromagnetic interpretation Appalachia, New Brunswick and Nova Scotia (11D, E, F, K; 20O, P; 21A, G, H, J); Geological Survey of Canada, Paper 70-1A, p. 79-82.

Miller, H.G.
1976: A magnetic model for the Budgell Harbour stock, Newfoundland; Canadian Journal of Earth Sciences, v. 13, p. 231-236.
1977: Gravity zoning in Newfoundland; Tectonophysics, v. 38, p. 317-326.
1988: Geophysical interpretation of the geology of the northeast Gander Terrane, Newfoundland; Canadian Journal of Earth Sciences, v. 25, p. 1161-1174.
1990: A synthesis of the geophysical characteristics of terranes in eastern Canada; Tectonophysics, v. 177, p. 171-191.

Miller, H.G. and Morris, C.N.
1988: Modelling of gravity and magnetic data along the Lithoprobe lines 84-1 and 84-2 northeast of Newfoundland; Program with Abstracts, Geological Association of Canada, v. 13, p. A85.

Miller, H.G. and Tuach, J.
1989: Gravity and magnetic signatures of the Ackley Granite Suite, southeastern Newfoundland: implications for magma emplacement; Canadian Journal of Earth Sciences, v. 26, p. 2697-2709.

Miller, H.G. and Weir, H.C.
1982: The northwest portion of the Gander Zone - a geophysical interpretation; Canadian Journal of Earth Sciences, v. 19, p. 1371-1381.

Miller, H.G. and Wright, J.A.
1984: Gravity and magnetic interpretation of the Deer Lake Basin, Newfoundland; Canadian Journal of Earth Sciences, v. 21, p. 10-18.

Miller, H.G., Goodacre, A.K., Cooper, R.V., and Halliday, D.
1985: Offshore extensions of the Avalon Zone of Newfoundland; Canadian Journal of Earth Sciences, v. 22, p. 1163-1170.

Miller, H.G., Kilfoil, G.J., and Peavy, S.T.
1990: An integrated geophysical interpretation of the Carboniferous Bay St. George Subbasin, western Newfoundland; Bulletin of the Canadian Society of Petroleum Geologists, v. 38, p. 320-331.

Mitchelmore, L.
1985: Reinterpretation of the refraction data from Notre Dame Bay, Newfoundland; B.Sc. Thesis, Memorial University of Newfoundland, 50 p.

Morel, P. and Irving, E.
1978: Tentative paleocontinental maps for the early Phanerozoic and Proterozoic; Journal of Geology, v. 86, p. 535-561.

Morris, W.A.
1976: Transcurrent motion determined paleomagnetically in the northern Appalachians and Caledonides and the Acadian Orogeny; Canadian Journal of Earth Sciences, v. 13, p. 1236-1243.

Murthy, G.S.
1983a: Paleomagnetism of the Deadman's Bay diabase dikes from northeastern central Newfoundland; Canadian Journal of Earth Sciences, v. 20, p. 195-205.
1983b: Paleomagnetism of diabase dikes from the Bonavista Bay area of northeastern central Newfoundland; Canadian Journal of Earth Sciences, v. 20, p. 206-216.
1985: Paleomagnetism of ceratin constituents of the Bay St. George sub-basin, western Newfoundland; Physics of the Earth and Planetary Interiors, v. 39, p. 89-107.

Murthy, G.S. and Rao, K.V.
1976: Paleomagnetism of Steal Mountain and Indian Head anorthosites, western Newfoundland; Canadian Journal of Earth Sciences, v. 13, p. 75-84.

Nairn, A.E.M., Frost, D.V., and Light, B.G.
1959: Paleomagnetism of certain rocks from Newfoundland; Nature, v. 183, p. 596-597.

Pal, B.K.
1982: Geomagnetic induction studies in eastern Newfoundland; M.Sc. thesis, Memorial University of Newfoundland, St. John's, Newfoundland, 93 p.

Papezik, V.S., Hodych, J.P., and Goodacre, A.K.
1975: The Avalon magnetic lineament: a possible continuation of the Triassic dike system of New Brunswick and Nova Scotia; Canadian Journal of Earth Sciences, v. 12, p. 332-335.

Pe-Piper, G. and Loncarevic, B.D.
1989: Offshore continuation of the Meguma Terrane, southwest Nova Scotia; Canadian Journal of Earth Sciences, v. 26, p. 176-191.

Prasad, J.N.
1981: Paleomagnetism of Mesozoic lamprophyre dykes in north-central Newfoundland; M.Sc. thesis, Memorial University of Newfoundland, St. John's, Newfoundland, 119 p.

Press, F. and Beckman, W.C.
1954: Geophysical investigations in the emerged and submerged Atlantic coastal plain, pt. 8, Grand Banks and adjacent shelves; Geological Society of America Bulletin, v. 65, p. 299-314.

Quinlan, G. and Beaumont, C.
1984: Appalachian thrusting, lithospheric flexure and Paleozoic stratigraphy of the eastern interior of North America; Canadian Journal of Earth Sciences, v. 21, p. 973-996.

Quinlan, G.M., Hall, J., Williams, H., Wright, J.A., Colman-Sadd, S.P., O'Brie7n, S.J., Stockmal, G.S., and Marillier, F.
1992: Lithoprobe onshore reflection transects across the Newfoundland Appalachians; Canadian Journal of Earth Sciences, v. 29, p. 1865-1877.

Rao, K.V. and Deutsch, E.R.
1976: Paleomagnetism of Lower Cambrian Bradore sandstones and the rotation of Newfoundland; Tectonophysics, v. 29, p. 337-357.

Rao, K.V., Seguin, M.K., and Deutsch, E.R.
1981: Paleomagnetism of Siluro-Devonian and Cambrian granitic rocks from the Avalon Zone in Cape Breton Island, Nova Scotia; Canadian Journal of Earth Sciences, v. 18, p. 1187-1210.
1986: Paleomagnetism of Early Cambrian redbeds on Cape Breton Island, Nova Scotia, and the question of a displaced Avalon Zone; Canadian Journal of Earth Sciences, v. 23, p. 1233-1242.

Rast, N., Rast, D., and Burke, K.B.S.
1979: The earthquakes of Atlantic Canada and their relationship to structure; Geoscience Canada, v. 6, p. 173-180.

Robertson, W.A., Roy, J.L., and Park, J.S.
1968: Magnetization of the Perry Formation of New Brunswick and the rotation of Newfoundland; Canadian Journal of Earth Sciences, v. 5, p. 1175-1181.

Roy, J.L. and Anderson, P.
1981: An investigation of the remanence characteristics of three sedimentary units of the Silurian Mascarene Group of New Brunswick, Canada; Journal of Geophysical Research, v. 86, p. 6351-6368.

Roy, J.L. and Morris, W.A.
1983: Paleomagnetic results from the Carboniferous of North America: the concept of Carboniferous geomagnetic field marker horizons; Earth and Planetary Science Letters, v. 65, p. 167-177.

Roy, J.L., Anderson, P., and Lapointe, P.
1979: Paleomagnetic results from three rock units of New Brunswick and their bearing on lower Paleozoic tectonics of North America; Canadian Journal of Earth Sciences, v. 16, p. 1210-1227.

St-Julien, P., Slivitsky, A., and Feininger, T.
1983: A deep structural profile across the Appalachians of southern Quebec; Geological Society of America, Memoir 158, p. 103-112.

Seguin, M.K.
1978: Paleomagnetism of Cambrian volcanics in the Quebec Appalachians; Geomagnetism and Aeronomy, v. 18, p. 218-224.
1980: Paleomagnetism of the Upper Silurian-Lower Devonian units in the Gaspé-New Brunswick region; Bulgarian Geophysical Journal, v. 6, p. 86-100.
1982: Geophysics of the Quebec Appalachians; Tectonophysics, v. 81, p. 1-50.
1983: Tectonic style of the Appalachian allochthonous zone of southern Quebec; Seismic and gravimetric evidence; Tectonophysics, v. 96, p. 1-18.

Seguin, M.K., Rao, K.V., and Pineault, R.
1982: Paleomagnetic study of Devonian rocks from Ste. Cecile-St. Sebastian region, Quebec Appalachians; Journal of Geophysical Research, v. 87, p. 7853-7864.

Shearer, J.M.
1973: Bedrock and surficial geology of the northern Gulf of St. Lawrence as interpreted from continuous seismic reflection profiles; in Earth Science Symposium on Offshore Eastern Canada, (ed.) P.J. Hood; Geological Survey of Canada, Paper 71-23, p. 285-303.

Sheridan, R.E. and Drake, C.L.
1968: Seaward extension of the Canadian Appalachians; Canadian Journal of Earth Sciences, v. 5, p. 337-373.

Sherwin, D.F.
1973: Scotian Shelf and Grand Banks; in The future petroleum provinces of Canada, (ed.) R.G. McCrossan; Canadian Society of Petroleum Geologists, Memoir 1, p. 519-525.

Smith, W.E.T.
1962: Earthquakes of eastern Canada and adjacent areas: 1534-1927; Publications of Dominion Observatory, v. 21, p. 271-301.
1966: Earthquakes of eastern Canada and adjacent areas: 1928-1959; Publications of Dominion Observatory, v. 32, p. 87-121.

Srivastava, S.P., Hyndman, R.D., and Cochrane, N.A.
1973: Magnetic and telluric measurements in Atlantic Canada; in Earth Science Symposium on Offshore Eastern Canada, (ed.) P.J. Hood; Geological Survey of Canada, Paper 71-23, p. 359-370.

Stevens, A.E., Milne, W.G., Westmiller, R.I., and LeBlanc, G.
1973: Canadian earthquakes 1967; Seismology Series, Earth Physics Branch, no. 65, 65 p.

Stewart, I.C.F.
1976: Travel time residuals of PP waves reflected under Atlantic Canada; Bulletin, Seismological Society of America, v. 66, p. 1203-1219.
1978: Teleseismic reflections and the Newfoundland lithosphere; Canadian Journal of Earth Sciences, v. 15, p. 175-180.
1984: P-wave travel-time residuals in the Newfoundland Appalachians; Canadian Journal of Earth Sciences, v. 21, p. 1278-1285.

Stewart, I.C.F. and Keen, C.E.
1978: Anomalous upper mantle structure beneath the Cretaceous Fogo seamounts indicated by P wave reflections; Nature, v. 274, p. 788-791.

Stockmal, G.S., Colman-Sadd, S.P, Keen, C.E., Marillier, F., O'Brien, S.J., and Quinlan, G.M.
1990: Deep seismic structure and plate tectonic evolution of the Canadian Appalachians; Tectonics, v. 9, p. 45-62.

Stockmal, G.S., Colman-Sadd, S.P., Keen, C.E., O'Brien, S.J., and Quinlan, G.
1987: Collision along an irregular margin: a regional plate tectonic interpretation of the Canadian Appalachians; Canadian Journal of Earth Sciences, v. 24, p. 1098-1107.

Tankard, A.J. and Welsink, H.J.
1987: Extensional tectonics and stratigraphy of Hibernia oil field, Grand Banks, Newfoundland; American Association of Petroleum Geologists Bulletin, v. 71, p. 1210-1232.

Tankard, A.J., Welsink, H.J., Keen, C.E., and Stockmal, G.S.
1988: Styles of Mesozoic extension of Canada's eastern seaboard and the effect of inherited Paleozoic fabrics; Program with Abstracts, Geological Association of Canada, v. 13, p. A123.

Thomas, M.D.
1983: Tectonic significance of paired gravity anomalies in the southern and central Appalachians; Geological Society of America, Memoir 158, p. 113-124.

Van der Voo, R. and Scotese, C.R.
1981: Paleomagnetic evidence for a large (2000 km) sinistral offset along the Great Glen Fault during Carboniferous time; Geology, v. 9, p. 583-589.

Weaver, D.F.
1967: A geological interpretation of the Bouguer anomaly field of Newfoundland; Publication of Dominion Observatory, v. 35, p. 223-250.

Williams, H.
1984: Miogeoclines and suspect terranes of the Caledonian-Appalachian Orogen: tectonic patterns in the North Atlantic region; Canadian Journal of Earth Sciences, v. 21, p. 887-901.

Williams, H. and Hatcher, R.D.
1983: Appalachian suspect terranes; in Contributions to the tectonics and geophysics of mountain chains, (ed.), R.D. Hatcher, H. Williams, and I. Zietz; Geological Society of America, Memoir 158, p. 33-64.

Williams, H. and Haworth, R.T.
1984a: Magnetic anomaly map of Atlantic Canada; Memorial University of Newfoundland, Map no. 5.
1984b: Bouguer gravity anomaly map of Atlantic Canada; Memorial University of Newfoundland, Map no. 6.

Williams, H., Colman-Sadd, S.P., and Swinden, H.S.
1988: Tectonic-stratigraphic subdivisions of central Newfoundland; in Current Research, Part B; Geological Survey of Canada, Paper 88-1B, p. 91-98.

Willmore, P.L. and Scheidigger, A.E.
1956: Seismic observations in the Gulf of St. Lawrence; Royal Society of Canada Proceedings and Transactions, v. 50, p. 21-38.

Wilson, J.T.
1966: Did the Atlantic close and then reopen?; Nature, v. 211, p. 676-681.

Wright, J.A. and Cochrane, N.A.
1980: Geomagnetic sounding of an ancient plate margin in the Canadian Appalachians; Journal of Geomagnetism and Geoelectricity, v. 32, p. 133-140.

Wright, J.A., Jessop, A.M., Judge, A.S., and Lewis, T.J.
1980: Geothermal measurements in Newfoundland; Canadian Journal of Earth Sciences, v. 17, p. 1370-1376.

Wright, J.A., Keen, C.E., and Keen, M.J.
1984: Marine heat flow along the northeast coast of Newfoundland; in Current Research, Geological Survey of Canada, Paper 1984-1B, p. 93-100.

Zeitz, I., Haworth, R.T., Williams, H., and Daniels, D.L.
1980: Magnetic anomaly map of the Appalachian Orogen; Memorial University of Newfoundland, Map no. 2.

Author's Address

H.G. Miller
Department of Earth Sciences
Memorial University of Newfoundland
St. John's, Newfoundland
A1B 3X5

Printed in Canada

627

Chapter 8

PLUTONIC ROCKS

Chapter 8

PLUTONIC ROCKS

K.L. Currie

INTRODUCTION

Plutonic rocks make up about one quarter of the exposed Canadian Appalachians, occuring in all tectonostratigraphic zones. Plutons range in age from Middle Proterozoic to Cretaceous, but ages of plutons in particular zones are restricted to small parts of this range. Compositions range from ultramafic to high-silica granites. Overall there is a noticeable deficiency of mafic compositions (dioritic and more mafic) compared to other well-studied orogenic belts such as the Cordillera. Late Precambrian, early Paleozoic, and Mesozoic magmatism produced mafic dyke swarms of major dimensions, and significant amounts of mafic plutonic rocks. Most other periods of magmatism produced few dykes, and mafic plutonic rocks are either subordinate to salic phases or virtually absent.

Despite their large area, plutonic rocks received relatively little attention prior to 1965 because of the stratigraphic and structural focus of much of the early geological work in this region. Interest in the plutonic rocks increased markedly with the realization that they hold important clues to the tectonic history of the region, and may contain significant mineral deposits. The pioneering compilation of Neale and Pajari (1972) served as a prototype for provincial compilations (Strong, 1980; Clarke et al., 1980; Fyffe et al., 1981) and stimulated studies of individual plutons. During the past decade, mainly under the sponsorship of provincial and federal governments, a great deal of work has been undertaken on plutonic rocks of the Canadian Appalachian region. The general outlines and petrography of most plutons are now known, and bulk rock chemical analyses are relatively abundant. Precise age determinations, and investigations of isotope systems of Pb, Nd, Sr, and O are increasing in number. Integrated studies including mineral and isotopic chemistry, trace element and structural data remain sparse, although several plutons have been studied in detail in Newfoundland, Nova Scotia and New Brunswick.

The accompanying compilation figures (Fig. 8.1, 8.2, 8.4, 8.6, 8.8, 8.9, 8.10) are based mainly upon the following sources for location, distribution, and nomenclature: Newfoundland – Hayes et al. (1987); northern Cape Breton – Barr et al. (1987); Antigonish Highlands – Murphy et al. (1991); southern Nova Scotia – MacDonald et al. (1987) and Hill (1991); central and northern New Brunswick – Fyffe (unpublished data); and southern New Brunswick – Currie (1988) and Barr and White (1988). These figures are supplemented with a large amount of published and unpublished material listed in the references. The primary data search was completed in September 1988, but references have been added in less systematic fashion up to February 1991. In this overview I have classified a vast amount of disparate (and in some cases conflicting) data into a manageable number of orogen-wide categories, and then related these categories to the tectonic evolution of the orogen. As many names assigned to igneous rocks in this region have been used in both formal and informal senses or have been invalidated in later studies, formal terminology is avoided in this chapter. Therefore, an individual body is here termed a "pluton", an assemblage of consanguinous plutons a "batholith", and an asssemblage of igneous rocks is called an "igneous suite" where the ages or relations of component bodies are complex or uncertain. Individual rocks have been named according to the internationally accepted system of nomenclature of Streckeisen (1976), as outlined in Table 8.1. Most current workers in the region use this system, and older descriptions have been converted as accurately as possible. Age data have been converted to the constants recommended by Steiger and Jager (1977) and adopted to the DNAG geological time scale of Palmer (1983). Many of the pre-1980 isotopic dates from this region are of poor quality. Even more recent data should be used with caution. Rb-Sr ages are stratigraphically both too old and too young in particular cases (McCutcheon et al., 1981). Ar-Ar and K-Ar dates have been reset in many cases by metamorphism or hydrothermal activity. U-Pb dates appear to be the most reliable, but even these may be slightly too old in some cases (Roddick and Bevier, 1988). Ophiolitic suites are not treated in this compilation, but granitoids possibly related to such suites are included.

No division of the plutons into genetic, chemical or temporal groups is completely satisfactory, since marginal and disputable cases always occur. Both genetic I-A-S-M division (for example Pitcher, 1983) and division by chemistry (for example Pearce et al., 1984) are reasonable if mafic phases are present, but break down for evolved granites, leading to confusion about the origin of large and economically significant granites. A field-based and empirical classification is outlined in Table 8.2. This system agrees in general with that proposed by Hayes et al. (1987) for Newfoundland, although there are differences in detail, and can also be related to the genetic classification of Pitcher (1983) as noted at various points in the following text. The raw data on which this classification is based appear in Table 8.6 at the end of this chapter.

Currie, K.L.
1995: Plutonic Rocks; Chapter 8 in Geology of the Appalachian-Caledonian Orogen in Canada and Greenland, (ed.) H. Williams; Geological Survey of Canada, Geology of Canada, no. 6, p. 629-680 (also Geological Society of America, The Geology of North America, v. F-1).

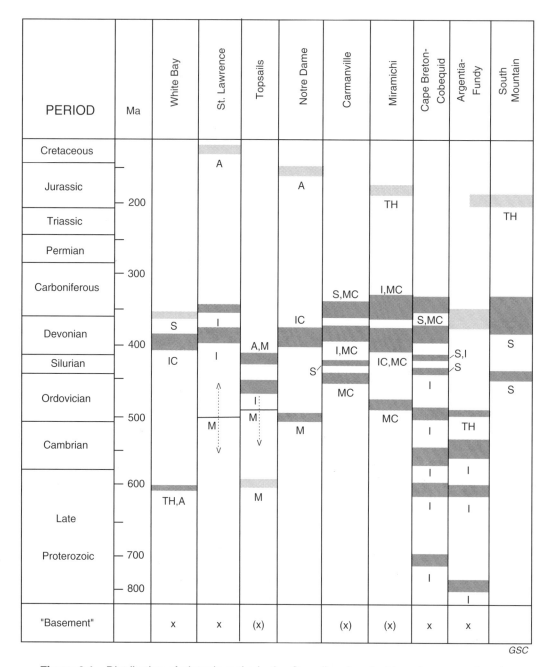

Figure 8.1. Distribution of plutonic rocks in the Canadian Appalachians by age and region. Regions with abundant plutons are listed across the top of the figure. Cape Breton-Cobequid includes all of Cape Breton Island and the Antigonish and Cobequid highlands. Argentia-Fundy includes the Avalon Peninsula of Newfoundland and the Fundy coast of New Brunswick. South Mountain comprises southern Nova Scotia. Red bars indicate major emplacement of plutons; pink bars indicate minor emplacement. Duration of activity indicated by depth of bar; age uncertainty indicated by dotted lines with arrows. Type of magmatic activity indicated as follows: I = I-type plutons in the terminology of Pitcher (1983); IC=composite acid-mafic plutons (I-type Caledonian of Pitcher (1983)); S = S-type (peraluminous) plutons; A = A-type plutons (cf. Whalen et al.,1987); M = M-type plutons (typical of oceanic crust, compare Pitcher (1983); TH = tholeiitic dyke suite; MC = evolved biotite granites with prominent or dominant megacrystic texture. Basement identified in region is indicated by x; if evidence for basement, by (x).

DISTRIBUTION IN SPACE AND TIME

Table 8.2 proposes a division of the plutonic rocks into 10 categories, based on age and petrography, some of which can be subdivided. Further work probably will require additional subdivision, or new divisions. As with all descriptive classifications of natural phenomena, a few examples do not fit well anywhere in the framework. Different groups of plutons in Table 8.2 occur in restricted parts of the Canadian Appalachian region, as summarized below.

Group 1 (middle Proterozoic and older) plutons comprise plutons of Grenville inliers in western Newfoundland and northern Cape Breton Island. Rock types include amphibole and pyroxene-bearing granite and syenite, some of alkaline affinities (Owen et al., 1987; Barr et al., 1987), large anorthosite complexes (van Berkel, 1987; Barr et al., 1988), mafic complexes (Owen, 1986), and plutons of charnockitic affinity (Owen et al., 1987; Currie, 1987a). The granitic rocks give U-Pb zircon emplacement dates of 1020 to 1040 Ma. One anorthosite gave a dubious date of about 1275 Ma, but correlation with other massif anorthosites suggests ages up to 1450 Ma (Emslie, 1978). Charnockitic complexes have ages of about 1500 Ma (Baadsgaard et al., pers. comm., 1989; Currie et al., 1992).

Grenvillian plutons have not been documented in Newfoundland east of the Long Range Fault. Stockmal et al. (1987) drew the southern limit of Grenvillian rocks at the northwestern corner of Cape Breton Island, but ages greater than 900 Ma have also been reported from the Cobequid Highlands of Nova Scotia (Gaudette et al., 1983). Zircons with Grenvillian inheritance have been recovered from granites of the Miramichi region of central New Brunswick (Roddick and Bevier, 1988) and the Saint John region of southern New Brunswick (Currie et al., 1981). Grenvillian plutons occur in thrust slices of the Humber Zone south of the St. Lawrence River (Vallieres et al., 1978).

Group 2 plutons include gabbro-diorite-granodiorite-granite complexes of late Proterozoic age such as the Holyrood intrusive suite of eastern Newfoundland, and are characteristic of the Avalon Zone. Typically such complexes exhibit a distinct elongation parallel to tectonic grain, and intrude roughly coeval volcanic and sedimentary rocks. The most voluminous phases tend to be dioritic to granodioritic, although granites are commonly present and locally dominant. Cognate mafic enclaves, schliers and mixed phases are characteristic of Avalonian plutonic complexes. These complexes are a typical I-type suite in the terminology of Pitcher (1983). Radiometric ages fall in the range 610 ± 20 Ma (Dallmeyer et al., 1981; Krogh et al., 1988).

Table 8.1. Nomenclature of plutonic rocks (after Streckeisen, 1976).

Rocks are named according to measured or estimated volume percent of minerals present (modes). Plagioclase refers to plagioclase more calcic than An_{10}. Colour index (CI) refers to percentage of strongly coloured minerals (including opaques).

1. Quartz > 20 per cent of salic minerals, CI<90
   (a) Plagioclase 0-10% of total feldspar - alkali granite
   (b) Plagioclase 10-35% of total feldspar - syenogranite
   (c) Plagioclase 35-65% of total feldspar - monzogranite
   (d) Plagioclase 65-90% of total feldpsar - granodiorite
   (e) Plagioclase 90-100% of total feldspar - tonalite
           (trondhjemite=tonalite with CI<10)

2. Quartz 0-20 per cent of salic minerals, CI<90
   (Prefix "quartz" added if quartz >5%)
   (a) Plagioclase 0-10% of total feldspar - alkali syenite
   (b) Plagioclase 10-35% of total feldspar - syenite
   (c) Plagioclase 35-65% of total feldspar - monzonite
   (d) i Plagioclase 65-90% and An <50 - monzodiorite
       ii Plagioclase 65-90% and An >50 - monzogabbro
   (e) i Plagioclase 90-100% and An<50 - diorite
       ii Plagioclase 90-100% and An>50 - gabbro
           (anorthosite=gabbro or diorite with CI<10)

3. Feldspathoid-bearing (including nepheline) rocks use names as under 2. If the amount of feldspathoid is 0-10% of salic minerals, use the prefix "foid-bearing", where foid stands for the name of the dominant feldspathoid. If the amount of feldspathoid is >10% of salic minerals, use the prefix "foid".

4. Rocks with colour index >90 are here termed "ultramafic". See Streckeisen (1976) for breakdown.

Table 8.2. Classification of plutonic rocks.

Group	Age (Ma)	Form	Structure	Rock types	Localization	Remarks	Example
1a	>1500	plutons	gneissic	charnockite to enderbite	Grenville inliers	Found as intruded hosts	Disappointment Hill
1b	1300 ±	massifs	brecciated	anorthosite, gabbro	Grenville inliers	large masses, massive but severely brecciated	Steel Mtn.
1c	?	plutons	massive	gabbro, diorite	Grenville inliers	massive but metamorphosed cut by Group 1d	Taylor Brook
1d	1030 ± 30	plutons	massive to foliated	granite, granodiorite, syenite (hn, bi, px)	Grenville inliers	locally megacrystic, may be recrystallized	Lake Michel
2a	615 ± 25	elongate plutons	massive to foliated	diorite, granodiorite granophyre (hb, bi)	Avalon Zone	chlorite-epidote altered	Holyrood
2b	780 ± 30	dykes, plutons	foliated, brecciated	diorite, gabbro	Avalon Zone	high-temperature metamorphism in Cape Breton	Wandsworth
3a	550 ± 25	dykes, plutons	massive	bimodal granite, granodiorite, gabbro	Avalon Zone Exploits Subzone	local high U+Th, may be mildly metamorphosed	Musquash, Valentine Lake
3b	605 ± 15	plutons, dykes	massive to lineated	bimodal alkali granite gabbro	Humber Zone	peralkaline phases	Hare Hill
4	495 ± 20	plutons	foliated	diorite, tonalite, granodiorite, trondhjemite	Dunnage Zone	found in allochthons	Twillingate
5a	460 ± 20	plutons	massive to fractured	granodiorite, monzonite, granite, minor mafics	Dunnage, Gander Zone	rare outside Notre Dame Subzone	Burlington
5b	440 ± 25	composite plutons	massive to foliated	granodiorite, granite, migmatite, minor mafics	Gander Zone	<440 Ma in Newfoundland, >440 Ma in New Brunswick, megacrystic	Burgeo
6	425 ± 15	plutons volcanics	massive	bimodal alkali granite, rhyolite, basalt	Dunnage Zone	peralkali granite, many mixed phases	Topsails
7a	405 ± 25	plutons	massive to foliated	peraluminous granite, minor tonalite	Gander Zone	associated with migmatite	North Bay
7b	370 ± 10	plutons	massive	peraluminous granite, granodiorite, tonalite	Meguma Zone	older mafic phases foliated	South Mtn.
8	405 ± 20	plutons	massive	granite, granodiorite cutting gabbro, diorite	Dunnage, Gander, Humber, Avalon zones	rarely cut by Group 7a	Mt. Peyton
9a	365 ± 15?	plutons	megacrystic	granite, granodiorite	Gander Zone	emplaced into high-grade metamorphic rocks	Deadmans Bay
9b	360 ± 15?	plutons	massive	granite, alkali granite	all zones	mainly high silica granites	Ackley
10a	200 ± 15	dykes	massive	tholeiitic basalt	all zones except Humber zones	rare, large dykes	Shelburne
10b	145 ± 25	plutons	massive	alkalkine gabbro, syenite, nepheline syenite	Dunnage, Humber zones	includes alkaline central complexes	Megantic

However a subset (Group 2b) of generally more deformed mafic rocks, gives ages greater than 700 Ma. Some of these ages are difficult to interpret (Olszewski and Gaudette, 1982; Gaudette et al., 1985) but a date of 762 Ma on the Wandsworth gabbro (Krogh et al., 1988) appears to date emplacement.

Latest Precambrian to Cambrian igneous complexes (Group 3a) also occur throughout the Avalon Zone. This suite tends to be bimodal, locally with spectacular comingled or alternating phases, as in southern New Brunswick, and may be quite alkaline (Louil Hills, A17). Historically such plutons have been lumped with Group 2, but recent work makes it clear that they are distinct in chemistry, petrography and age. The best dates fall in the range 550 ± 20 Ma, or Cambrian on the DNAG scale. Such plutons, however, are nowhere known to cut Cambrian strata and locally are unconformably overlain by Cambrian strata (Currie, 1988). Ages like those of Group 3a plutons have recently been reported from allochthons of Chain Lakes type in northwestern Maine and adjacent Quebec (Boone and Heisler, 1988).

Small alkaline to peralkaline granitoid plutons (Group 3b) of known or probable late Precambrian age occur in the eastern Humber Zone of Newfoundland. Round Pond (602 Ma, Williams et al., 1985), a medium grained leucogranite in which ilmenite replaced original amphibole, and Hare Hill (608 Ma, Currie et al., 1992), a peralkaline leucogranite with phases similar to Round Pond, form parts of westerly transported allochthons on the Humber Zone. These plutons are distinctly older than alkaline Group 3a plutons of the Avalon Zone (Louil Hills, A17, Cross Hills, A8, Fig. 8.6). Fault slices of younger granites (565 Ma) have recently been identified in the Exploits Subzone of the Dunnage Zone (Evans et al., 1990).

Deformed tonalite to trondhjemite complexes of Cambrian to early Ordovician age, commonly associated with amphibolitic mafic rocks (Group 4) typify the Dunnage Zone. These complexes are poor in potash feldspar, even in very silicic varieties. The Twillingate pluton (N4, Fig. 8.2; U-Pb zircon age of 510 Ma, Williams et al., 1976) forms a typical example, and many smaller but similar bodies occur around Notre Dame Bay. Similar rocks occur in westerly transported allochthons along the length of the Humber Zone (Moulton Hill, L8), and rarely in easterly transported allochthons of the Exploits Subzone. These plutons form an M-type suite in the terminology of Pitcher (1983).

A sub-group of similar age but very different chemistry comprises leucogranites of A-type affinity on opposite sides of Cabot Strait. The full extent and significance of this sub-group recently defined by Dunning et al. (1990), are not yet known.

Large plutons of mid-Ordovician to Silurian age are now known to be common in central Newfoundland (Dunning et al., 1988), northwestern Cape Breton Island (Jamieson et al., 1986), and the Miramichi region of New Brunswick (Bevier, 1988). These plutons are difficult to classify because of petrographic diversity, varying deformation, and gradations between types. A three-fold division into groups 5 to 7, with subdivisions of groups 5 and 7, is here adopted. Group 5 plutons comprise those which do not show obvious S- or A-type affinities. Group 5a, which is confined to Newfoundland, includes mid-Ordovician (460 Ma) tonalitic to granodioritic plutons with associated mafic

phases (for example, Costigan Lake, N16, Hinds Lake, T10, and Southwest Brook, S1). All examples lie northwest of the Victoria River fault. These plutons form an I-type suite. Group 5b includes large deformed megacrystic plutons such as Burgeo, which may exhibit minor mafic phases, contain innumerable enclaves and screens suggesting a complex history with crustal involvement, and exhibit mylonitic deformation. Late muscovite-garnet aplite phases are characteristic. Ages of such plutons range from greater than 460 Ma (Dallmeyer et al., 1981; Bevier, 1988) to 420 Ma (O'Brien et al., 1986; Dunning et al., 1988), with

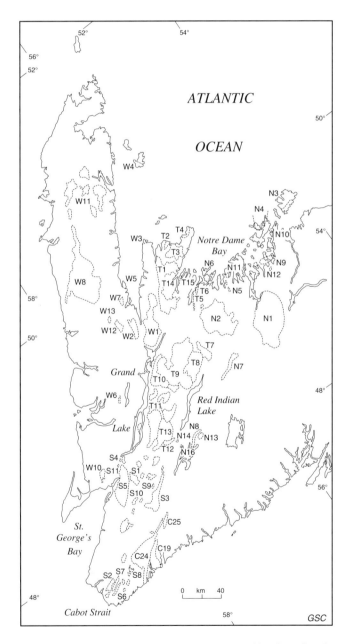

Figure 8.2. Index of plutons in northwestern Newfoundland. Letter-number identifiers are keyed to Table 8.6. W = White Bay-Grand Lake region; T = Topsails region; S = southwest Newfoundland region; N = Notre Dame Bay region.

most examples in Newfoundland and Cape Breton Island giving Silurian ages, whereas those in New Brunswick give mainly Ordovician ages.

Group 6 plutons form a well-defined, localized belt of alkali to peralkaline granite of Silurian age lying in the Notre Dame Subzone of the Dunnage in Newfoundland, and just north of the Avalon Zone in southern New Brunswick. The plutons consist predominantly of leucocratic, mildly peralkaline one-feldspar granites with minor amounts of basaltic rocks, commonly in the form of dykes, and a plethora of mixed, comingled and assimilated phases. The Topsails igneous suite (Whalen and Currie, 1990) is a typical example. Group 6 plutons are little deformed, and give ages in the range of 415 to 435 Ma. Group 6 plutons clearly cut Group 5a plutons. Chandler et al. (1987) showed that Group 6 plutons form an essentially continuous belt running the length of the Canadian Appalachians. These plutons are an A-type suite in the terminology of Pitcher (1983) and Whalen et al. (1987).

Group 7 plutons comprise peraluminous granites. They may be divided into two subdivisions on the basis of age and style of occurrence. Gander Zone peraluminous rocks in Newfoundland and northwestern Cape Breton Island, for example Middle Ridge (C11) and the southeast part of the North Bay Complex, are associated with high grade metamorphic rocks, and may exhibit significant deformation (Group 7a), although posttectonic phases occur in the North Bay igneous suite (C16, 17). Tourmaline and beryl (in muscovite-garnet pegmatites) characterize late phases of this suite. The peraluminous plutons of central New Brunswick are only locally deformed, and were emplaced into greenschist or lower grade hosts. Group 7a plutons give ages varying from 425 (Silurian) to 395 Ma (early Devonian). Group 7b plutons comprise clearly posttectonic peraluminous rocks, for example the South Mountain batholith (E1) of Nova Scotia, which generally crosscut relatively low-grade rocks. Group 7b plutons give ages of 360 to 380 Ma. Only Group 7b plutons of the Meguma Zone consistently contain cordierite or an aluminosilicate, and therefore qualify as S-type in the terminology of Pitcher (1983), but some Group 7a plutons are equally peraluminous and exhibit local evidence of anatexis of sedimentary protoliths (Currie and Pajari, 1981).

Large composite plutons with early mafic and late granitic phases (Group 8) cut Ordovician and Silurian strata of the northeast Exploits Subzone in Newfoundland, and also occur in the Miramichi region of New Brunswick. The Mount Peyton igneous suite (N1) is the best known and largest example. The mafic portions typically have tholeiitic chemistry, and may include ultramafic phases. The granitic portion cuts the mafic portion, although minor local transitions may be present. Detailed geochemistry on Mount Peyton suggests that the salic portion is crustally derived whereas the mafic portion is mantle-derived (Strong and Dupuy, 1982). These complexes give ages in the range 425 Ma to 390 Ma or Silurian to early Devonian. They commonly cut relatively low-grade strata, producing hornfels aureoles, and are essentially massive. At Copper Island near Wesleyville a Group 7 granite cuts the mafic phase of a Group 8 pluton (Williams, 1968). Salic phases of Group 8 plutons locally exhibit alkaline affinities (Mount Elizabeth, M12, New Brunswick), or even peralkaline undersaturated dykes (Mont McGerrigle, L15, Gaspésie). Salic and mafic phases of such plutons may show geographic separation with mafic phases partially rimming

salic ones. Pitcher (1983) erected a special category for such mafic-felsic composite plutons, which he called I-type, Caledonian.

Middle Devonian to Early Carboniferous igneous rocks of the Canadian Appalachians consist almost entirely of post-tectonic, massive evolved granites with mafic phases absent or very subordinate. These plutons may be divided into two types. Group 9a consists of very large, equant to lobate masses of biotite granite with large rounded crystals of microcline perthite up to 10 cm across (megacrysts). Such plutons are essentially confined to the Gander Zone of Newfoundland and Cape Breton Island, where they comprise a major element of the geology. The Deadmans Bay pluton (C4) is taken as a typical example. Such plutons commonly occur in high-grade metamorphic rocks and appear to be relatively deep-seated. The highly poikilitic megacrysts overgrow both internal (dyke) and external boundaries, but also locally show flow alignment. They result from a complex combination of igneous and metasomatic processes (Pickerill et al., 1978). These plutons typically contain minor hornblende-bearing phases, and late, crosscutting garnet-muscovite aplites. They have proved difficult to date because of inheritance and isotopic disequilibrium (compare Currie et al., 1982a) but ages of emplacement seem to fall in the range 350 to 380 Ma. Megacrystic texture with all the above peculiarities is not confined to Devonian plutons but is also common in deformed Ordovician and Silurian plutons of Group 5b, and is known locally in late Precambrian plutons of Group 2 and peraluminous plutons of Group 7. It characterizes a particular set of igneous and metasomatic conditions which remain poorly understood.

Circular or lobate masses of high-level, evolved granite (Group 9b) occur across the Appalachians, in many cases associated with volcanic equivalents. High silica granite (SiO_2>74%) forms the predominant phase. Megacrystic texture is common, but patchy, or confined to a particular phase. These bodies cut surrounding rocks and appear posttectonic, although in many cases influenced by major faults (Tuach et al., 1986; Currie, 1988). The Ackley Batholith (C23) of Newfoundland is a typical example. Many of these bodies were once thought to be mid-Carboniferous or younger, but more refined age measurements suggest ages in the range 340 to 380 Ma, or Devonian to early Carboniferous. Despite the limited range in composition, some plutons show a well-defined sequence of intrusive phases. These bodies have proved difficult to fit into standard granite classifications. Some have been referred to as A-type (Burnthill, M7; Whalen, 1986), and one (Saint Lawrence, A4) is clearly peralkaline. However most appear to be so highly evolved that reliable classification is difficult (Whalen et al., 1987). Classification is of some significance, since many of these bodies contain economically important concentrations of lithophile elements.

Post-Paleozoic magmatic rocks (Group 10) are of negligible volume but considerable tectonic significance. They fall into two groups. Large, but sparse tholeiitic dykes and flows of Triassic age (Group 10a) occur in northern New Brunswick, along the Bay of Fundy, and on the southeastern tips of Nova Scotia and Newfoundland. Jurassic to Cretaceous alkaline mafic rocks (Group 10b) occur in north-central Newfoundland and in southern Quebec. Newfoundland examples include strongly undersaturated mafic dykes and mildly alkaline gabbro plugs. The central complexes and dykes of Quebec form part of the Monteregian (Brome, L19, Mount Shefford, L18) and White Mountains (Mont Megantic, L20) alkaline igneous provinces.

In the following section brief descriptions of plutonic rocks by geographic regions have been arranged in three northwest to southeast transects, one across Newfoundland, another across Cape Breton Island and Nova Scotia, and the third across Gaspésie and New Brunswick. Within regions, descriptions proceed from older to younger plutonic rocks.

DESCRIPTIONS OF PLUTONIC ROCKS
Humber Zone of Newfoundland

The Humber Zone of Newfoundland contains a significant plutonic record (Fig. 8.2). The oldest plutonic rocks are middle Proterozoic or older crystalline rocks of the Long Range Inlier (Bostock, 1983; Owen et al., 1987) and probably equivalent rocks of the Indian Head Range (Riley, 1962) and southern Long Range Mountains (van Berkel, 1987; van Berkel and Currie, 1988). Charnockitic rocks, described by Owen et al. (1987) and Currie (1987) have U-Pb zircon ages of about 1500 Ma in the northern and southern Long Range (Baadsgaard et al., pers. comm., 1989; Currie et al., 1992). These rocks are intruded by anorthosites, of which the largest and most spectacular example is the Steel Mountain (S5, Fig. 8.2) massif (van Berkel, 1987), more than 50 km long, which consists mainly of massive but fractured andesine anorthosite (colour index <10) locally containing uralitized pyroxene crystals a metre in length. The massif contains minor amounts of layered anorthosite and gabbroic anorthosite, and bodies of foliated norite up to a kilometre in length occur around its margins. Similar rocks occur in the Indian Head Range (Riley, 1962), but are absent in the northern Long Range Inlier. A mylonitic norite from the faulted eastern margin of the Steel Mountain massif yielded a strongly discordant U-Pb zircon age suggesting emplacement, or extensive resetting, at 377 Ma with significant remanence of 1275 Ma (O. van Breemen, pers. comm., 1989). The remanence is interpreted to approximately date emplacement of the anorthosite. Massive gabbro bodies occur in the northern Long Range Inlier (Bostock, 1983; Owen, 1986). The Taylor Brook igneous suite (Owen, 1986) consists of olivine-two pyroxene gabbro, locally layered. The rock is quite massive but extensively amphibolitized. Bostock (1983) correlated such rocks with the anorthosite massifs, but there seems to be little evidence for this correlation. The mafic rocks are cut by and included in granitic rocks. Large granitic plutons emplaced about 1000 ± 50 Ma are common in the northern Long Range Inlier but have not been identified in the southern Long Range or Indian Head Inlier, although they may exist (van Berkel and Currie, 1988). Plutons of the northern Long Range Inlier consist of pale pink to grey, biotite-hornblende monzogranites of colour index 2 to 20 (Bostock, 1983; Owen et al., 1987). Clinopyroxene occurs locally in trace amounts. Many of the plutons display megacrystic texture, although more equigranular phases occur locally. Some data on the chemistry of these plutons are shown in Table 8.3. The largest pluton, Lake Michel (W8, Fig. 8.4) gave a U-Pb zircon age of 1042+22/-11 Ma (Erdmer, 1986). Some monazite and sphene ages give concordant results of about 970 Ma, representing high-grade metamorphism

(Baadsgaard, pers. comm., 1989). Somewhat younger metamorphic ^{40}Ar-^{39}Ar ages (~880 Ma) were obtained from the Indian Head Range (W10; Dallmeyer, 1978).

Middle Proterozoic and older rocks of the Long Range Inlier are cut by a swarm of north-northeast-trending basaltic dykes of late Precambrian age which locally form up to 5% of its volume. In the northern part of the inlier these Long Range dykes are large (some >10 m in width), but the dykes decrease in width and increase in metamorphic grade to the south and east (Owen and Machin, 1987), reaching amphibolite grade near White Bay. The dykes exhibit a transitional chemistry between continental within-plate basalts and oceanic tholeiites (Table 8.3). A related dyke at Sandwich Bay, Labrador has a precise baddeleyite-zircon age of 615 ± 2 Ma (Kamo et al., 1989), compared with an ^{40}Ar/^{39}Ar release spectra age of 605 ± 10 Ma obtained by Stukas and Reynolds (1974b).

Faintly foliated ilmenite leucogranite (Round Pond pluton, W6) associated with salic and mafic flows and sedimentary rocks of the Humber Zone occurs at the base of the Hughes Lake structural slice west of Deer Lake (Williams et al., 1985). This rock gave a U-Pb zircon age of 602 ± 10 Ma. Very similar Group 3b leucogranites, ranging from massive to strongly lineated, and strongly peralkaline to subalkaline, occur south of Grand Lake (Hare Hill pluton, S4, Fig. 8.2, and related rocks of Currie, 1987a). The pluton at Hare Hill gave a U-Pb age on zircon of 608 ± 2 Ma (Currie et al., 1992). Replacement of poikilitic mafic minerals by magnetite and ilmenite (as seen in the Round Pond pluton) is common. The full extent of this alkaline to peralkaline granite province is not yet known but it may be extensive. Like Round Pond, Hare Hill and related plutons occur in westerly-transported allochthons. The host rocks are latest Precambrian-early Paleozoic metasedimentary rocks of the Humber Zone.

The Coney Head igneous suite (W5, Fig. 8.4) on the west side of White Bay is another allochthonous pluton (Williams, 1977) which consists of massive pink to grey tonalite (quartz, oligoclase, altered biotite, accessory sphene and monazite) and monzonite (oligoclase, orthoclase, chlorite, accessory quartz) with large enclaves, probably cogenetic, of gabbro. Dunning (1987a) obtained a U-Pb age on zircon of 474 ± 2 Ma (Arenig), for these rocks, suggesting they belong to Group 4. This material is intruded by Silurian microgranite (432 ± 2 Ma according to Dunning, 1987a), possibly correlative to the nearby Silurian Sops Arm Group of sedimentary and felsic volcanic rocks.

The Wild Cove Pond igneous suite (W1; Hibbard, 1983) contains foliated hornblende-biotite gabbro, diorite and minor granodiorite intruded by younger granite, locally megacrystic. Small stocks and dykes of biotite-muscovite granite cut all other phases. Contacts of the pluton with country rocks tend to be irregular, sheeted or agmatitic although contact aureoles of andalusite-cordierite type occur at two localities. The pluton intrudes greenschist grade metasedimentary rocks of the Upper Proterozoic to lower Paleozoic Fleur de Lys Supergroup and a small area of higher grade, presumably older, amphibolite-rich gneiss. Both sedimentary and igneous inclusions occur in the pluton. The Wild Cove Pond igneous suite gave a U-Pb zircon age of 425 Ma (Cawood and Dunning, 1993). Correlation of the Wild Cove Pond igneous suite is

Table 8.3. Representative bulk-rock analyses of Middle and Late Proterozoic plutonic suites.

	1	2	3	4	5	6	7	8	9	10	11	12	13	14	15
SiO$_2$	61.0	67.4	54.4	66.5	59.01	69.15	74.44	66.5	68.5	73.3	49.7	75.3	61.23	58.2	50.7
TiO$_2$	1.34	1.22	0.14	0.59	1.10	0.24	0.24	0.49	0.43	0.18	2.42	0.22	0.62	0.68	1.60
Al$_2$O$_3$	16.6	14.3	26.3	15.9	14.88	15.07	12.73	15.3	15.6	14.4	13.2	12.0	16.72	17.2	15.6
Fe$_2$O$_3$	3.1	2.8	0.9	1.4	8.18*	2.10*	1.76*	3.9*	1.5	0.3	11.84	1.92*	5.74*	2.8	2.5
FeO	1.7	3.5	0.4	2.4	-	-	-	-	1.8	0.8	3.04	-	-	4.6	8.1
MnO	0.07	0.09	0.03	0.07	0.17	0.05	0.04	0.08	0.15	0.11	0.23	0.03	0.10	0.16	0.19
MgO	2.0	0.8	0.80	2.0	3.50	1.05	0.44	1.6	1.1	0.3	8.51	0.20	3.11	3.5	6.3
CaO	3.2	2.2	9.29	3.96	6.92	1.01	1.29	2.7	3.5	1.2	10.12	0.26	5.25	5.78	8.62
Na$_2$O	3.3	2.9	6.1	3.6	3.70	4.41	3.88	3.8	3.3	3.8	2.34	3.33	3.95	3.5	3.2
K$_2$O	5.9	2.0	0.73	1.77	2.57	5.09	4.26	2.3	2.9	3.2	0.35	5.16	1.86	1.72	1.09
H$_2$O	0.6	0.6	0.8	2.0	-	-	-	-	-	-	-	-	-	2.0	2.2
CO$_2$	<0.1	<0.1	0.4	0.1	-	-	-	-	-	-	-	-	-	0.1	0.1
P$_2$O$_5$	0.60	0.42	0.06	0.14	-	-	0.07	-	0.11	0.08	0.26	0.01	0.14	0.22	0.25
Trace elements (ppm)															
Rb	-	-	0	64	173	130	201	70	87	90	13	146	64	55	38
Ba	2500	800	160	466	785	-	632	-	811	420	92	666	538	332	157
Sr	1100	220	110	240	170	255	85	160	445	317	188	15	377	290	240
Zr	1200	1000	10	220	-	273	-	-	202	87	152	313	120	155	170
Nb	-	-	-	14	-	-	-	-	10	12	13	94	9	9	8
La	310	300	-	21	-	-	-	-	-	-	-	-	-	17	13
Ce	730	700	-	-	-	-	-	-	-	-	-	-	-	-	-
Yb	<4	52	2	2	-	-	-	-	-	-	-	-	-	3	3

- = not analyzed, * total Fe as Fe$_2$O$_3$

1. Monzonite, Hooping Harbour pluton, western Newfoundland (Bostock, 1983, Table 5, sample 336)
2. Monzogranite, Lake Michel pluton, western Newfoundland (Bostock, 1983, Table 5, sample 466)
3. Anorthosite, Red River complex, northwestern Cape Breton Island (Currie, 1987b, analysis 4-3)
4. Dioritic gneiss, Green Head Island, southern New Brunswick (Currie, unpublished data)
5. Diorite, Holyrood pluton, southeastern Newfoundland (Strong and Minatides, 1975)
6. Granite, Holyrood pluton, southeastern Newfoundland (Strong and Minatides, 1975)
7. Average leucogranite, Huntingdon Mountain pluton, southeast Cape Breton Island (Barr et al., 1982)
8. Average granodiorite, Boisdale Hills pluton, southeast Cape Breton Island (Barr and Setter, 1981)
9. Granodiorite, Cheticamp pluton, northwestern Cape Breton Island (Barr et al., 1986, analysis 20)
10. Biotite-muscovite granodiorite, Cheticamp pluton (Barr et al., 1986, analysis 82)
11. Long Range dyke, western Newfoundland (Owen and Machin, 1987, analysis 283)
12. Leucogranite, Round Pond pluton, western Newfoundland (Williams et al., 1985, analysis 2)
13. Tonalite, Point Wolfe River pluton, southern New Brunswick (Barr and White, 1988, analysis 1004A)
14. Tonalite, Renforth pluton, southern New Brunswick (Currie, unpublished data)
15. Mafic dyke, Kingston complex, southern New Brunswick (Currie, unpublished data)

uncertain. Hayes et al. (1987) correlated it with the Silurian alkali to peralkaline Topsails igneous suite (T9, Fig. 8.2). However, Topsails plutons typically contain only one feldspar, and have sparse mafic or intermediate components. The age and association of diverse phases in the Wild Cove Pond more closely resembles plutons like Mount Peyton (N1; Group 8).

The Gull Pond pluton (W2, Fig. 8.4); apparently resembles the Wild Cove Pond pluton. Saunders and Smyth (1990) reported gabbro, trondhjemite and granite, locally megacrystic, forming discrete phases. One phase of the granite contains riebeckite. The Devils Room pluton (W7) of megacrystic biotite and muscovite granite intrudes gneisses of the Long Range Inlier. Baadsgaard (pers. comm., 1989) reported U-Pb ages of about 390 Ma on zircon and sphene for both the Gull Pond and Devils Room plutons.

The Partridge Point pluton (W3) forms a small elongate mass of garnetiferous muscovite leucogranite which cuts the Fleur de Lys Supergroup. The pluton gave a K-Ar age on biotite of 368 Ma. The Bell Island pluton (W4) in northern White Bay consists of grey to pink, massive, equigranular biotite-muscovite granite (Williams and Smyth, 1983) with a marginal zone of intrusion breccia containing disoriented rafts of Fleur de Lys metasedimentary rocks and local development of skarn around marble inclusions. A K-Ar age on muscovite gave 368 Ma. Both plutons belong to Group 7a (S-type affinities).

Notre Dame Subzone of Newfoundland

The Notre Dame Subzone of Newfoundland, between Grand Lake and Red Indian Lake, contains a distinctive plutonic assemblage (Whalen and Currie, 1983) comprising plutons of groups 4, 5a, and 6, but dominated by voluminous alkaline to peralkaline granites of Group 6 (Fig. 8.2). A major fault with post-Silurian movement (Lobster Cove Fault) divides the region into northern and southern segments.

No Precambrian basement has been clearly identified in this region, although various enclaves of gneiss are possibly of Precambrian age, and Precambrian inheritance is ubiquitous in zircon ages (Whalen et al., 1985; Dunning et al., 1987). The oldest plutons probably formed by recycling of oceanic crust. The Twillingate pluton (N4, Fig. 8.2, Table 8.4), which gave a U-Pb age on zircon of 510 Ma (Williams et al., 1976), exceeds 20 km by 10 km in size. The pluton consists mainly of trondhjemite (about equal amounts of oligoclase and quartz with less than 10% of amphibole+biotite and accessory epidote, apatite, magnetite, carbonate, and sphene). Granodiorite occurs locally as well as lenticles and transposed dykes of amphibolite (Williams and Payne, 1975). Strong and Payne (1973) suggested that the Twillingate pluton originated by anatectic recycling of metamorphosed oceanic island arc material. The host rocks of the Twillingate pluton are mafic and salic volcanic rocks of the Sleepy Cove Group which were deformed and metamorphosed with the pluton. The Mansfield Cove trondhjemite (T5, 479 Ma; Dunning et al., 1987) is structurally associated with a lower Paleozoic oceanic volcano-sedimentary

complex (Dunning et al., 1987). The Hungry Mountain igneous suite (T8) contains mafic to ultramafic rafts, possibly of ophiolitic origin as well as large enclaves of metasedimentary gneisses (possibly fragments of basement) in a ubiquitous tonalitic to granodioritic matrix.

Plutons similar in age and lithology to Twillingate, occur in transported slices at Little Port (Table 8.4) on the west coast of the island (Mattinson, 1975).

Granitoid complexes giving U-Pb ages on zircon of about 470 to 460 Ma intrude Group 4 plutons and associated rocks in the Notre Dame Subzone. The Hinds Lake pluton (T1, T10) forms a large irregular mass of coarse grained hornblende-biotite granodiorite and granite which is essentially massive, although fractured and altered to chlorite and epidote. The Dunamagon pluton (T2) consists of coarse biotite granite which exhibits a moderate foliation in its northwestern part. All three plutons gave U-Pb ages on zircon of 460 Ma (Dallmeyer and Hibbard, 1984; Whalen et al., 1985). The Sunday Cove pluton (T6, 464 Ma by Rb-Sr, Bostock et al., 1979), a high level granophyric intrusion contains large nebulous enclaves of granite gneiss thought to represent older crust from which the pluton was derived.

The Rainy Lake igneous suite (T11; Whalen and Currie, 1983) forms an M-type suite, slightly older (438 Ma, U-Pb on zircon, Whalen et al., 1985) than peralkaline granites of the Topsails intrusive suite (T9) and apparently cogenetic to mafic plutons in the Dashwoods Subzone (Dunning, 1987; van Berkel and Currie, 1988; Currie and van Berkel, 1989). These plutons do not fit well into the classification scheme used in this chapter, but appear to be Group 6 plutons, and may also be related to Silurian volcanics of the Springdale Group described by Coyle and Strong (1987).

The most characteristic feature of this region is extensive high-level peralkaline Silurian magmatism, commonly associated with volcanic equivalents. These extensive A-type suites have been described in some detail by Whalen and Currie (1983, 1990). The cores of the Topsails igneous suite contains aegirine, arfvedsonite and aenigmatite, but marginal phases and satellite plutons contain sub-alkaline amphibole and amphibole-biotite granites (Table 8.3, column 4). Screens and roof pendants of cogenetic bimodal volcanic rocks are common. A characteristic feature of the plutons is spectacular evidence of magma mixing (Fig. 8.3, see also Whalen and Currie, 1984b). Extensive dating (Taylor et al., 1980; Bell and Blenkinsop, 1981; Whalen et al., 1985) has shown that igneous activity occurred in a narrow time interval between 420 and 435 Ma. The Kings Point igneous suite (T14; Kontak and Strong, 1986), north of the Lobster Cove Fault, closely resembles the Topsails intrusive suite. The most northerly plutons of this group occur at Cape Brule and La Scie (de Grace et al., 1976). The Cape Brule pluton (T3) consists of porphyry charged with mafic to ultramafic fragments, whereas the La Scie pluton (T4) consists of peralkaline granite, syenite, and gabbro. Both plutons are strongly deformed in their northern parts. Both have returned diverse and anomalous isotopic ages ranging from 320 to 475 Ma, but regional correlation suggests both are of Silurian age, correlative to the Topsails igneous suite.

Table 8.4. Representative bulk rock analyses of Cambrian to Silurian plutonic suites.

	1	2	3	4	5	6	7	8	9	10	11	12	13	14	15
SiO_2	72.7	70.3	72.8	62.72	77.10	72.80	73.5	62.59	74.90	48.21	74.61	74.25	70.88	71.27	70.95
TiO_2	0.24	0.53	0.23	0.77	0.12	0.21	0.22	0.93	0.28	2.68	0.16	0.05	0.36	0.33	0.19
Al_2O_3	12.8	14.3	12.8	15.65	12.20	13.95	14.3	16.49	13.20	14.99	10.21	14.75	15.10	14.67	12.71
Fe_2O_3	2.75*	3.9*	3.00*	1.25	0.17	2.02*	0.5	1.27	0.39	3.94	5.27*	0.00	0.14	2.26*	1.46
FeO	-	-	-	3.39	1.64	-	0.7	3.67	1.20	8.48	-	0.64	1.90	-	-
MnO	0.06	0.08	0.09	0.09	0.02	0.03	0.04	0.11	0.03	0.24	0.09	0.08	0.04	-	-
MgO	0.51	1.8	0.86	3.24	0.07	0.69	0.40	2.01	0.18	5.76	0.00	0.14	0.80	0.84	0.16
CaO	1.89	3.9	2.35	5.89	0.35	2.00	1.10	3.69	0.46	8.41	0.08	0.64	1.82	1.49	0.57
Na_2O	4.80	3.1	4.44	4.17	4.26	3.77	3.3	3.10	4.00	3.36	5.04	4.07	4.17	3.53	3.40
K_2O	0.57	1.9	1.32	2.57	3.98	3.76	4.21	4.24	5.20	1.25	4.09	4.43	4.19	3.97	4.92
H_2O	1.27	-	1.45	-	-	-	0.9	1.34	0.50	2.12	-	0.22	0.66	-	-
CO_2	-	0.16	-	-	-	-	0.0	-	0.10	0.19	-	-	-	-	-
P_2O_5	0.02	0.12	0.03	0.24	0.02	0.07	0.28	0.19	0.02	0.60	0.01	0.87	0.07	-	-
Trace elements (ppm)															
Rb	18	59	28	80	110	104	209	197	199	41	357	140	414	369	260
Ba	-	836	-	821	549	618	229	1018	405	410	4	53	143	114	40
Sr	87	386	63	542	45	222	125	356	62	344	0	109	967	279	80
Zr	102	151	80	173	267	-	0	233	462	230	612	27	194	131	198
Nb	-	8	-	-	34	-	-	-	36	10	53	-	-	-	-
Ce	-	49	-	-	105	-	43	-	114	51	-	-	-	-	-
La	-	27	-	-	40	-	26	-	61	27	83	-	-	-	-
Yb	-	-	-	-	-	-	2	-	6	3.4	-	-	-	-	-

- not analyzed, * total Fe as Fe_2O_3

1. Average trondhjemite, Twillingate Pluton, Notre Dame Bay, Newfoundland (Williams and Payne, 1975)
2. Average tonalite, Cape Ray Complex, southwest Newfoundland (Wilton, 1985, Table 2)
3. Trondhjemite, Little Port Complex, western Newfoundland (Williams and Payne, 1975)
4. Average granodiorite, Burlington Pluton, western Newfoundland (de Grace et al., 1976)
5. Baggs Hill granite, southwestern Newfoundland (Chorlton, 1980, analysis 088)
6. Average biotite granite, Cape Smoky Pluton, northwestern Cape Breton Island (Barr et al., 1982)
7. Biotite-muscovite granite, Ragged Harbour Pluton, Carmanville region, Newfoundland (Currie, unpublished data)
8. Average megacrystic granodiorite, Gaultois Pluton, southern Newfoundland (Elias and Strong, 1982)
9. Average Topsails hybrid granite, western Newfoundland, (Whalen and Currie, 1990)
10. Average diabase dike in Topsails Granite, western Newfoundland (Whalen and Currie, 1990)
11. Alkali granite, Welsford Complex, southern New Brunswick (Payette and Martin, 1987, sample AG)
12. Average muscovite-biotite granite, Through Hill Pluton, central Newfoundland (Elias and Strong, 1982)
13. Average muscovite-biotite granite, North Bay Pluton, southern Newfoundland (Elias and Strong, 1982)
14. Average muscovite-biotite granite, Miramichi region of New Brunswick (Fyffe et al., 1981, type F)
15. Average equigranular biotite granite, Miramichi region of New Brunswick (Fyffe et al., 1981, type H)

Dashwoods Subzone of southwestern Newfoundland

The Dashwoods Subzone, essentially the southern Long Range Mountains, appears to have experienced a plutonic history distinct from that of other regions. Essentially all plutons belong to Group 5. No Precambrian crystalline basement has been definitely identified, although offshore geophysical evidence suggests that Grenvillian crust may extend southeast to the Victoria River Fault (Stockmal et al., 1987; Marillier et al., 1988a). U-Pb zircon dating has failed to detect Precambrian igneous crystallization ages, although Precambrian inheritance is common (Dunning, 1987b; Currie et al., 1992).

The southern Long Range Mountains contain numerous fragments of ultramafic rocks thought to have formed parts of ophiolite suites (Chorlton and Knight, 1983; Dunning and Chorlton, 1985; Dunning, 1987b), and now forming the oldest rocks in this region. Most of these fragments are small slivers caught up in major ductile shear zones cutting quartzofeldspathic gneisses (Fox and van Berkel, 1988). An extensive suite of variably foliated tonalitic rocks (Table 8.4), consisting of medium- to coarse-grained saussuritized plagioclase, quartz, biotite, and locally amphibole, and

Figure 8.3. Magma comingling, Topsails pluton of western Newfoundland. Note intimate mix of basaltic and rhyolitic phases with lobate margins and chilled (darker) edges on the basic phase. GSC 203034-O

abundant amphibolite inclusions, extends more than 150 km from Cape Ray to Little Grand Lake (van Berkel, 1987), and makes up a significant part of the southern Long Range. These plutons intrude the ophiolite suites (Dunning and Chorlton, 1985) and give both U-Pb zircon and K-Ar ages of between 475 and 455 Ma (Dunning, 1987b; Stevens et al., 1982), close to the age of granulite facies metamorphism in this region (Currie et al., 1992). The tonalitic suite grades to and is cut by more potassic leucocratic rocks varying from biotite leucogranite to porphyritic or megacrystic biotite granite. The Red Rocks granite, a megacrystic pluton with local cataclastic to mylonitic fabric, gave a Rb-Sr "errorchron" age of 495 ± 57 Ma (Wilton, 1983) which is almost certainly too old. Accumulated geological and geochronological data suggest that essentially all of the granitoid rocks in this region were emplaced in a relatively short interval between 475 and 455 Ma ago. This suite differs from the Dunamagon-Hinds Lake suite in the Notre Dame Subzone because it exhibits pervasive deformation, and was emplaced under high-grade metamorphic conditions, but also belongs to Group 5. It may represent infrastructure to the higher level rocks of the Notre Dame Subzone.

Currie and van Berkel (1989) sampled and analyzed a group of posttectonic, locally layered gabbro-mafic dyke complexes which cut the tonalite plutons, and give either within-plate or volcanic arc chemical signatures. One of these plutons (Main Gut, S10, Fig. 8.2) gave a Silurian U-Pb zircon age of 431 ± 2 Ma (Dunning et al., 1988). Dunning (1987b) found a similar layered gabbro-diorite complex which cut the Annieopsquotch ophiolite suite which gave a U-Pb zircon age of 435 +5/-2 Ma, and Currie and van Berkel (1989) have drawn attention to similarities with the Rainy Lake igneous suite (438 Ma, T11) of the Notre Dame Subzone.

Devonian plutons occur in southwestern Newfoundland along major shear zones. The Windowglass Hill leucogranite (S6) of albitized alkali feldspar and quartz with spectacular graphic textures, lies in the centre of the Cape Ray Fault Zone (Wilton, 1983), and appears to be a subvolcanic equivalent to ignimbrites. The Strawberry (S7) and Isle aux Morts Brook (S7) plutons are coarse grained to porphyritic monzogranites to syenogranites, with red perthitic microcline, quartz, and up to 5% biotite. All three plutons gave Rb-Sr isochron ages between 352 and 369 Ma and initial Sr ratios between 0.7065 and 0.710 (Wilton, 1983). Reset Devonian ages associated with large ductile shear zones occur at least as far northwest as Steel Mountain (O. van Breemen pers. comm., 1989), and the suite of Silurian-Devonian plutons in northwestern Newfoundland (W3, Wild Cove Pond, W1, Gull Pond, W2, Devils Room, W7, Partridge Point, (W3), Bell Island, W4) also lies along the major Long Range (Cabot) fault system.

Exploits Subzone of Newfoundland

No autochthonous Precambrian rocks are known in the Exploits Subzone of Newfoundland (Fig. 8.4), but late Precambrian (565 Ma) U-Pb zircon dates have recently been obtained from fault slices of quartz monzonite in the western part of the subzone (Evans et al., 1990). These bodies (Valentine Lake and Crippleback Lake; N8, N7, Fig. 8.2) may be fragments of the infrastructure to the Exploits Subzone.

The Notre Dame Bay region exhibits a suite of small lower Ordovician mafic plutons (Group 4a). The South Lake igneous suite (N5, Fig. 8.2) forms a fault-bounded fragment of layered gabbro intruded by hornblende quartz diorite and tonalite. Lorenz and Fountain (1982) interpreted the pluton to result from remelting of amphibolite by mantle-derived mafic magma. The Colchester and Brighton plutons (T15, N6, Fig. 8.2) are small masses of massive but altered gabbro and subordinate tonalite. All three plutons were emplaced into oceanic crust. The Brighton pluton gave an Ar-Ar age of 495 Ma (Stukas and Reynolds, 1974a).

One or more small plutons of Group 4 occur in the southwest part of the Exploits Subzone associated with volcanic rocks of the Victoria Lake Group (Roebuck Lake quartz monzonite, 495 Ma., Evans et al., 1990, N17, Table 8.6).

The Coaker Porphyry (N10, Fig. 8.2) was emplaced into the Dunnage Mélange, contains numerous ultramafic nodules, and exhibits spectacular textures suggestive of emplacement into wet mud (Lorenz, 1982). This mid-Ordovician (468 Ma, Elliot et al., 1991) dacitic porphyry is probably best classified as a Group 5 pluton.

A suite of large, Late Silurian to Early Devonian, composite plutons (Group 8) typified by Mount Peyton (N1) forms one of the characteristic features of the Exploits Subzone. The Mount Peyton igneous suite (Strong, 1979; Strong and Dupuy, 1982) forms an elliptical mass about 60 km by 30 km with a narrow thermal aureole (<200 m) surrounding an incomplete outer ring of tholeiitic gabbro (plagioclase, poikilitic clinopyroxene and orthopyroxene, ilmenite, minor late biotite and amphibole). The central granitic core (Table 8.5) ranges from hornblende biotite granodiorite to granite and granophyre. Transitions from mafic to felsic phases are abrupt and the amount of intermediate rocks is negligible. Petrographic and geochemical evidence strongly suggest an origin for the Mount Peyton igneous suite by partial melting of deep crustal material due to upwelling of mafic magma from the mantle (Strong and Dupuy, 1982). The granodiorite portion of Mount Peyton gave a Rb-Sr isochron age of 390 Ma (Bell et al., 1977), whereas the gabbroic rim gave a significantly older Ar-Ar age of 420 Ma (Reynolds et al., 1981). The latter age seems more likely to represent the emplacement age, as the Hodges Hill igneous suite (N2; Hayes, 1950), a few kilometres to the northwest, which resembles Mount Peyton in petrography, gave a similar K-Ar age (415 Ma). Much of Fogo Island forms a composite complex of similar age. The

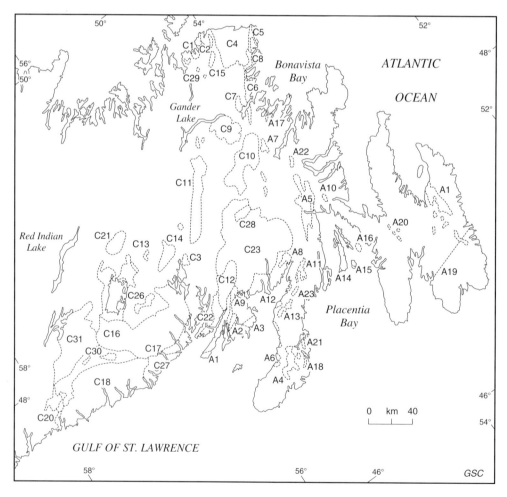

Figure 8.4. Index of plutons in central and eastern Newfoundland. Letter-number identifiers are keyed to Table 8.6. C = Carmanville-Burgeo region; A = Avalon Zone.

Table 8.5. Representative bulk rock analyses of Devonian and younger plutonic suites.

	1	2	3	4	5	6	7	8	9	10	11	12	13	14	15
SiO_2	53.90	68.50	55.1	71.7	68.11	74.55	64.82	64.59	68.40	74.57	75.12	73.65	61.21	52.5	44.93
TiO_2	1.74	0.42	1.54	0.19	0.59	0.09	0.75	0.73	0.56	0.26	0.21	0.25	0.67	0.93	3.67
Al_2O_3	15.50	15.50	17.00	14.4	15.52	14.32	17.34	17.14	14.53	12.78	11.74	14.00	18.85	14.9	18.10
Fe_2O_3	1.27	0.45	1.96	0.65	4.18	1.34	0.68	3.94	0.95	0.47	1.70	0.81	1.79	2.48	3.34
FeO	7.94	1.20	4.59	0.97	-	-	3.19	0.92	2.09	0.91	-	1.00	1.40	7.37	8.39
MnO	0.26	0.04	0.13	0.05	0.09	0.05	0.06	0.10	0.08	0.04	0.03	0.05	0.15	0.17	0.17
MgO	4.60	0.88	4.42	0.43	1.50	0.58	1.88	1.29	1.31	0.29	0.52	0.33	0.49	7.11	5.74
CaO	7.82	2.35	8.52	1.35	1.79	0.36	3.89	2.29	2.10	0.77	0.39	1.30	1.29	10.37	9.11
Na_2O	2.99	3.69	3.45	3.37	3.49	3.46	3.77	4.17	3.17	3.34	3.60	3.99	6.83	2.13	3.30
K_2O	1.45	4.55	1.98	4.47	3.80	4.42	2.33	4.39	4.90	5.21	4.98	4.30	5.49	0.65	2.07
H_2O	-	-	-	-	-	-	1.12	0.33	-	-	-	0.22	0.60	-	-
CO_2	-	-	-	-	-	-	-	-	-	-	-	0.16	0.13	-	-
P_2O_5	0.17	0.00	0.30	0.06	0.19	0.25	0.22	0.02	0.16	0.04	-	0.11	0.16	0.13	0.56
Trace elements (ppm)															
Rb	55	182	55	169	164	349	85	-	232	255	249	130	152	25	50
Ba	267	544	392	1038	588	59	445	620	504	293	216	-	-	194	1197
Sr	196	120	513	230	171	18	345	200	167	93	-	158	300	197	1318
Zr	164	354	139	126	177	46	-	-	185	153	450	177	845	97	299
Nb	15	20	28	18	12	11	-	-	23	31	-	46	202	10	-
Ce	28	57	39	210	-	-	62	-	-	-	-	82	155	28	-
La	12	24	23	25	34	5	30	-	-	-	-	47	92	22	-
Yb	1.3	2.1	-	-	-	-	1.8	-	-	-	-	2.6	3.8	-	-

- no data, * total Fe as Fe_2O_3

1. Mafic phase, Mount Peyton Complex, central Newfoundland (Strong and Dupuy, 1982, analysis CD 96)
2. Salic phase, Mount Peyton Complex, central Newfoundland (Strong and Dupuy, 1982, analysis CD 124)
3. Mafic phase, Wild Cove Pond Pluton, western Newfoundland (Hibbard, 1983, analysis 59)
4. Salic phase, Wild Cove Pond Pluton, western Newfoundland (Hibbard, 1983, analysis 71)
5. Average biotite granodiorite, South Mountain Batholith, Nova Scotia (MacDonald et al., 1987, unit gda)
6. Average monzogranite, South Mountain Batholith, Nova Scotia (MacDonald et al., 1987, unit mgTa)
7. Biotite tonalite, Barrington Passage Pluton, Nova Scotia (data from de Albuquerque, 1977)
8. Biotite monzogranite, Belleoram Pluton, southern Newfoundland (data from Ermanovics et al., 1967)
9. Biotite-hornblende granodiorite, Ackley Batholith, northwest part, southern Newfoundland (Tuach et al., 1986, unit 9)
10. Biotite granite, Ackley Batholith, eastern lobe, southern Newfoundland (Tuach et al., 1986, unit 13)
11. Peralkaline granite, Saint Lawrence Pluton, southern Newfoundland (data from Teng and Strong, 1976)
12. Granite, Mont McGerrigle Pluton, Gaspe (Whalen and Gariepy, 1986, analysis 83)
13. Nepheline syenite, McGerrigle Pluton, Gaspe (Whalen and Gariepy, 1986, analysis 16)
14. Diabase, Shelburne dyke, southern Nova Scotia (Papezik and Barr, 1981, analysis 112)
15. Gabbro, Budgells Harbour, north central Newfoundland (Strong and Harris, 1974, analysis AH651)

Table 8.6. Plutonic rocks.

This compilation gives a compact reference to the occurrence and character of plutonic rocks of the Canadian Appalachians. Blanks signify that no data are known to compiler. Data on bulk rock or mineral chemistry are not included. Where available, they can be found in the quoted references. The structure of the compilation is explained in the following remarks, numbered to correspond to the headings of the compilation.

(1) IDENTIFICATION NUMBERS are shown on index maps and keyed to geographical regions as follows: W = White Bay region; T = Topsails region; S = southwestern Newfoundland; N = Notre Dame Bay region; C = Carmanville-Burgeo region; A = Avalon Zone; B = northwestern Cape Breton Island; D = southeastern Cape Breton, Antigonish and Cobequid highlands; E = southern Nova Scotia; L = Gaspésie and Eastern Townships; M = central and northern New Brunswick; F = southern New Brunswick.

(2) NAME of the pluton was taken from the provincial glossary where possible. Nearby similar bodies have been described together, the relevant names given in the title or in a footnote.

(3) AGE is given in millions of years, with determinative method abbreviated: K = K/Ar (biotite, muscovite, hornblende), Z = U/Pb on zircon, U = U/Pb on sphene or monazite, R = Rb-Sr isochron, A = $^{40}Ar/^{39}Ar$.

(4) LITHOLOGY is given by mafic mineralogy and rock type (a period separates mineral from rock names): b = biotite, g = garnet, h = hornblende, m = muscovite, p = pyroxene, r = alkaline amphibole, o = olivine, n = nepheline. um = ultramafic rocks, ga = gabbro, no = norite, do = diorite, to = tonalite, tr = trondjhemite, gd = granodiorite, mg = monzogranite, gr = granite (includes syenogranite), lg = leucogranite, sy = syenite, md = monzodiorite, an = anorthosite.

(5) OPAQUE MINERALS are abbreviated: il = ilmenite mt = magnetite s = sulphides

(6) ACCESSORY MINERALS are abbreviated: a = andalusite s = sphene b = beryl t = topaz f = fluorite, g = garnet

(7) INCLUSIONS: s = sedimentary, i = igneous, m = metamorphic gneisses

(8) CONTACTS: ho = hornfelsed or cross-cutting, ca = cataclastic, mi = migmatitic, sh = sheeted, ag = agmatitic, me = metasomatized

(9) MINERALIZATION (by element)

(10) INITIAL Sr RATIO

(11) PLUTON STRUCTURE (m = massive, gn = gneissic, ca = cataclastic, mc = megacrystic, sh = sheeted, l = layered, c = composite, f = foliated)

(12) Pluton Group as outlined in Table 8.2

The second line gives the latitude and longitude of the centre or most typical exposure of each pluton, and appropriate references.

1	2	3	4	5	6	7	8	9	10	11	12
White Bay region (Fig. 8.4)											
W1	Wild Cove Pond (49°22'/54°45')	425Z (Hibbard, 1983; Cawood and Dunning, 1993)	bh.do,gd	--	--	i	ag	--	--	c	8
W2	Gull Pond (49°33'/56°58')	398Z (Saunders and Smyth, 1990; Baadsgaard, pers. comm., 1989)	bh.gr,do	--	ri	i	ag	--	--	c	8
W3	Partridge Point (50°09'/56°58')	368K (Hibbard, 1983)	bm.gr	--	--	i	sh	--	--	c	7b
W4	Bell Island (50°44'/55°34')	368K (Williams and Smyth, 1983; Wanless et al., 1973)	bm.gr	--	--	s	ag	--	--	m	7b
W5	Coney Head (49°46'/56°45')	484Z[1] (Williams, 1977; Dunning, 1987a)	do,to,tr	--	--	--	ca	--	--	gn	4
W6	Round Pond (49°04'/57°43')	602Z (Williams et al., 1985)	mt.lg	mt	--	--	ca	--	--	gn	3b
W7	Devils Room (49°42'/56°56')	398Z (Erdmer, 1986)	b.gd	--	f	i	ca	Au	--	mc	8
W8	Lake Michel (50°10'/57°00')	1042Z (Owen et al., 1987; Erdmer, 1986)	bh.gr,gd	--	--	m	ag	--	0.707	mc	1d
W10	Indian Head (48°36'/58°25')	880A (Riley, 1962; Dallmeyer, 1978)	bh.to,gd	--	--	m	--	--	--	m	1d
W11	Hooping Hbr.[2] (50°40'/56°12')	-- (Bostock, 1983)	bh.gr,gd	--	--	m	sh	--	--	mc	1d
W12	Taylor Brook (49°35'/57°10')	-- (Owen, 1986	op.ga	mt	--	--	ho	--	--	1	1c
W13	Potato Hill (49°38'/57°14')	-- (Owen, 1986)	p.do	--	--	--	me	--	--	gn	1a

[1] A younger microgranite with age 432Z intrudes the complex
[2] Includes Hooping Hbr. Satellite, Forche Hbr., Cloud River, Horse Chops, Leg Pond, Pigeon Cove, Little Cat Arm, Killdevil Hill, Lomond River, French-Childe and similar plutons of the northern Longe Range (Bostock, 1983; Owen, 1986; Erdmer, 1986)

1	2	3	4	5	6	7	8	9	10	11	12
Topsails region (Fig. 8.2)											
T1	Burlington (49°37'/56°00')	455Z (Hibbard, 1983; Dallmeyer and Hibbard, 1984)	bh.gd,gr	mt	s	i	ho	--	--	m,f	5a
T2	Dunamagon (49°58'/55°57')	459Z (Hibbard, 1983; Dallmeyer and Hibbard, 1984)	b.gr	--	--	i	ca	--	--	ca	5a
T3	Cape Brule (49°53'/55°55')	475Z (de Grace et al., 1976; Pringle, 1978)	bh.gr	mt	s	i	ho	--	--	p	6
T4	La Scie[1] (49°/55°02')	435Z (de Grace et al., 1976; Mattinson, 1977)	r.gr,sy,ga	--	f	--	ca	--	--	ca	6
T5	Mansfield Cove (49°27'/56°03')	479Z (Bostock et al., 1979; Dunning et al., 1987)	bh.to,tr	mt	--	i	ca	--	--	m	4
T6	Sunday Cove-Loon Pond (49°30'/55°55')	464R (Bostock et al., 1979)	bh.gr	--	--	i	ho	--	0.706	P	5a
T7	Skull Hill (49°00'/56°30')	415Z (Kean and Jayasinghe, 1982)	h.sy	mt	s	i	ho	--	--	m	6
T8	Hungry Mountain (49°00'/56°30')	469Z (Thurlow, 1981; Whalen et al., 1987)	do,ga,gd	--	s	i	ho	--	0.709	c	5a

Table 8.6. (cont.)

Topsails region (Fig. 8.2) (cont.)

1	2	3	4	5	6	7	8	9	10	11	12
T9	Topsails (49°09'/56°37') (Whalen and Currie, 1983; Whalen et al., 1987)	429Z	r.gr,bh.gr	mt	s	i	ho	Mo	0.704	m	6
T10	Hinds Lake (49°04'/57°05') (Whalen and Currie, 1983; Whalen et al., 1987)	465Z	bh.gd	--	--	--	ag	--	--	gn	5a
T11	Rainy Lake (48°52'/57°20') (Whalen and Currie, 1983; Whalen et al., 1987)	433Z	h.ga,do,gd	mt	s	i	ag	--	0.704	m	6
T12	Star Lake (48°39'/57°17') (Dunning et al., 1982)	465Z	do,to	--	--	i	ca	--	--	m	4
T13	Shandithit Brook (49°15'/57°21') (Whalen and Currie, 1983)	--	bh.gd	--	--	i	ag	--	--	gn	5a
T14	Kings Point (49°45'/56°10') (Kontak and Strong, 1986)	--	r.sy,lg	--	--	--	ho	--	--	m,po	6
T15	Colchester[2] (49°40'/56°02') (Kean, 1984)	--	do,gd,gr	--	--	i	ag	--	--	gn	4

[1] Includes Reddits Cove gabbro and Seal Island Bight syenite
[2] Includes Coopers Cove, Dolland Arm and Wellmans Cove plutons

Southwestern Newfoundland (Fig. 8.2)

1	2	3	4	5	6	7	8	9	10	11	12
S1	Southwest Brook[1] (48°33'/57°58') (Herd and Dunning, 1979; Dunning, 1987b)	456Z	ga,do	mt	s	--	ca	--	--	1	5a
S2	Cape Ray[2] (47°43'/59°13') (Wilton, 1983; Dube et al., 1993)	472Z	to,b.gr	mt	--	i	ca	--	0.710	ca	5a
S3	Lloyds River (48°11'/58°00') (Kean, 1983)	--	do,bh.gd	--	--	--	--	--	--	gn	5a
S4	Hare Hill (48°36'/58°05') (Currie, 1987a; Currie et al., 1992)	608Z	r.gr,h.gr	il	s	i	mi	--	--	m,f	3b
S5	Steel Mtn. (48°20'/58°15') (van Berkel, 1987)	1274Z	an,no	mt	--	--	ca	Ti	--	ca	1b
S6	Windowglass Hill (47°43'/59°05') (Wilton, 1985; Dube et al., 1993)	424Z	b.lg	--	--	--	ca	Au	0.710	ca	9b
S7	Strawberry (47°44'/59°07') (Wilton, 1985; Dube et al., 1993) (includes Isle aux Morts Brook)	384Z	b.gr	s	--	m	sh	--	0.707	m	9b
S8	Port aux Basques (47°40'/59°09') (Wilton, 1985; Dunning et al., 1988)	412Z	b.gr,bm.lg	mt	s	--	mi	--	--	gn	7a
S9	Cormacks Lake (48°20'/57°55') (van Berkel and Currie, 1988; Currie et al., 1992)	460Z	p.lg	--	g	--	me	--	--	gn	5a
S10	Main Gut[3] (48°10'/58°07') (Currie and van Berkel, 1989; Dunning et al., 1988)	431Z	ga.	mt	--	m	ho	--	--	m	6
S11	Disappointment Hill (48°10'/58°06') (Currie, 1987a; Currie et al., 1992)	1498Z	p.lg	--	--	m	me	--	--	gn	1a

[1] Southwest Brook is taken to be representative of the tonalite to granodiorite to leucogranite complexes in this region which are little known (van Berkel, 1987).
[2] Includes megacrystic biotite granite (see Wilton, 1985).
[3] Other complexes of this type occur (Currie and van Berkel, 1989)

1	2	3	4	5	6	7	8	9	10	11	12
Notre Dame Bay region (Fig. 8.2)											
N1	Mount Peyton (48°56'/55°15') (Strong, 1980; Strong and Dupuy, 1982)	390R	ga,bh.gr	mt	--	i	ho	--	0.709	c	8
N2	Hodges Hill-Twin Lakes (49°08'/55°46') (Hayes, 1950; Wanless et al., 1965)	415R	do,bh.gd	mt	--	i	ho	--	0.705	c	8
N3	Fogo Island (49°43'/54°12') (Williams, 1964; Cawthorn, 1978)	380K	um,ga,b.gr	mt	--	i	ho	--	--	c	8
N4	Twillingate (49°38'/54°43') (Williams et al., 1976; Strong and Payne, 1973)	510Z	tr,gd,do	mt	g	i	ca	--	--	ca	4
N5	South Lake (49°15'/55°28') (Lorenz and Fountain, 1982)	--	bh.do,tr	--	--	i	ca	--	--	m	4
N6	Brighton (49°33'/55°42') (Stukas and Reynolds, 1974a; Kean, 1984)	495A	ga,bh.to	mt	--	i	sh	--	--	ca	4
N7	Crippleback Lake (48°47'/56°01') (Kean and Jayasinghe, 1980; Evans et al., 1990)	565Z	ga,gd	--	--	--	ca	--	--	c	8
N8	Valentine Lake (48°26'/57°05') (Kean and Jayasinghe, 1980; Evans et al., 1990)	563Z	h.do,gd	--	--	--	ca	--	--	c	8
N9	Loon Bay-Long Island (49°20'/54°52') (Williams, 1964; Elliot et al., 1991)	408Z	bh.gd	mt	s	i	ca	--	--	m	8
N10	Coaker (49°28'/54°45') (Lorenz, 1982; Elliot et al., 1991)	467Z	gp,do	--	--	i	ho	--	--	p	5a
N11	Budgell Harbour (49°28'/55°25') (Strong and Harris, 1974)	155K	b.ga,um	--	--	--	ho	--	--	m	10b
N12	Dildo Pond (49°14'/54°46') (Strong and Harris, 1974)	--	b.ga,um	--	--	--	ho	--	--	m	10b
N13	Rodeross Lake (48°24'/56°59') (Strong and Harris, 1974)	--	b.ga,um	--	--	--	ho	--	--	m	10b
N14	Boogie Lake (48°17'/57°35') (Dunning, 1987b)	435Z	do	--	--	--	ho	--	--	m	6
N16	Costigan Lake-Tower Hill (48°30'/57°08') (Kean, 1977)	--	do,to,gd	m,s	--	m	ca	--	--	gn	5a
N17	Roebuck Lake (48°26'/57°14') (Evans et al., 1990)	495Z	ng	--	--	i	ho	--	--	n	4k
Carmanville Burgeo region (Fig. 8.4)											
C1	Frederickton-Rocky Bay (49°23'/54°15') (Currie et al., 1979; Wanless et al., 1973)	398K	hb.to,gd	mt	s	i	ho	--	--	m	8
C2	Aspen Cove-White Point (49°27'/54°10') (Currie et al., 1979)	--	bm.mg,lg	il	g	s	ho	--	--	m	7a
C3	Matthews Pond (48°11'/56°31') (Colman-Sadd, 1979)	--	bm.gd	--	g	s	--	--	--	m	7a
C4	Deadmans Bay (49°16'/53°51') (Jayasinghe and Berger, 1976; Currie et al., 1982a)	355Z	b.gr	mt	f	i	me	Mo	0.702	mc	9a
C5	Cape Freels[1] (49°14'/53°31') (Jayasinghe and Berger, 1976; Kerr et al., 1993)	417Z	b.gr	mt	--	i	ca	--	0.708	mc	5b
C6	Lockers Bay (48°56'/53°48') (Dallmeyer et al., 1981)	418Z	b.gr	mt	--	--	ca	--	0.715	mc	5b
C7	Middle Brook (48°53'/54°11') (Strong et al., 1974; Bell et al., 1979)	432R	b.gr	mt	--	i	me	--	0.708	mc	5b
C8	Newport (49°03'/53°40') (Kerr et al., 1993)	366R	b.gr	--	f	i	me	--	0.706	mc	9a
C9	Gander Lake (48°42'/54°26') (Strong et al., 1974)	357A	b.gr	--	f	--	ho	--	--	m	9b
C10	Maccles Pond (48°32'/54°27') (Bell et al., 1977)	354R	b.gr	mt	--	s	mi	--	0.704	mc	9a
C11	Middle Ridge (48°31'/55°00') (Strong et al., 1974; Kerr et al., 1993)	410U	m.gr	--	g	--	--	Be	0.709	m	7a
C12	Northwest Brook-Indian Point (47°42'/55°18') (Dickson, 1983)	--	bm.gf	--	g	s	ca	--	--	gn	7a

Table 8.6. (cont.)

1	2	3	4	5	6	7	8	9	10	11	12
	Carmanville Burgeo region (Fig. 8.4) (cont.)										
C13	Through Hill (48°24'/55°56') (Colman-Sadd, 1985; Colman-Sadd et al., 1992)	464Z	bm.gr	--	g	s	sh	--	0.721	m	7a
C14	Partridgeberry Hills (48°18'/55°45') (Colman-Sadd, 1985; Colman Sadd et al., 1992)	474Z	bm.gr	s	a	s	ho	Pb	0.715	m	7a
C15	Ragged Harbour (49°27'/59°01') (Currie and Pajari, 1981)	401K	b.gr,m.lg	--	g	s	sh	--	--	mc	7a
C16	NW North Bay (48°00'/56°00') (Dickson, 1982; Dunning et al., 1988)	396Z	b.gd,m.lg	il	g	s	mi	W	--	m	9a
C17	SE North Bay (47°50'/56°00') (Dickson, 1982; Kerr et al., 1993)	413Z	bm.gr	--	g	--	ho	--	0.706	mc	7a
C18	Burgeo (47°45'/57°15') (O'Brien et al., 1986; Dunning et al., 1988)	428Z	bh.gd, b.gr,to	--	--	i	ca	--	--	mc	5b
C19	La Poile (47°47'/58°30') (Chorlton, 1978, 1980)	432Z	bh.gr	mt	s	i	sh	--	--	mc	5b
C20	Chetwynd-Iron Bound (47°49'/57°58') (O'Brien et al., 1986, 1991)	390Z	b.gr,bh.gr	mt	g	i	ho	Au	--	m	9b
C21	Overflow Pond (48°33'/56°09') (Dallmeyer et al., 1983b)	390Z	bm.gr	--	a	s	ho	--	--	m	7a
C22	Gaultois[2] (47°44'/55°39') (Dunning, pers. comm., O'Brien et al., 1986)	421Z	b.gr	--	--	i	ca	--	0.710	mc	5b
C23	Ackley[3] (48°00'/58°00') (Dickson, 1983; Whalen, 1983)	355R	b.gr,lg	mt	--	i	ho	--	0.706	mc	9
C24	Rose Blanche (47°45'/59°26') (Chorlton, 1980)	--	to,gd	--	g	i	sh	Be	--	gn	5b
C25	Baggs Hill (48°57'/57°58') (Chorlton, 1978, 1980)	--	gr	--	--	--	ho	--	--	po	5
C26	Steel Pond (48°57'/56°15') (Colman-Sadd, 1984)	--	ga,do,gd	mt	--	--	ho	--	--	m	10a
C27	Francois (47°36'/56°41') (O'Brien et al., 1986; Kerr et al., 1993)	378Z	b.gr	mt	--	i	ho	--	--	m	9b
C28	Eastern Maelpaeg (48°10'/54°48') (Dickson, 1983)	--	bm.gr	--	g	s	ho	--	--	gn	7a
C29	Island Pond-Ocean Pond (49°19'/54°18'; 49°08'/54°12') (Currie et al., 1979)	--	bm.gr	--	--	s	sh	--	--	m	7a
C30	Cochrane Pond[4] (47°57'/57°08') (O'Brien et al., 1986)	--	mb.gr	--	t	m	sh	--	--	m	7a
C31	Buck Lake (48°12'/57°25') (Kean, 1984)	418Z	b.gr,mb.gr	--	--	m	sh	Be	--	mc	7a

[1] Includes Business Cove and Wareham
[2] Includes Piccaire and McCallum granites, age probably about 440 Ma
[3] Includes Hungry Grove, Rencontre Lake, Mt.Sylvester, Kepenkeck and Tolt
[4] Cochrane Pond is used as a representative of numerous moderately peraluminous, locally foliated plutons in the central and western Carmanville-Burgeo region including Dead Wolf Brook, Great Burnt Lake, Peter Snout, Terra Nova River, Terra Nova River west and Third Berry Hill Pond.

1	2	3	4	5	6	7	8	9	10	11	12
	Avalon Peninsula (Fig. 8.6)										
A1	Holyrood (47°19'/53°00') (Strong and Minatides, 1975; Krogh et al., 1988)	620Z	hb.gd,do,ga	mt	s	i	ho	py	0.704	m	2a
A2	Pass Island (47°29'/56°06') (Strong et al., 1974; Furey, 1985)	--	bh.gr	--	--	--	ho	Mo	--	m	9b
A3	Belleoram (47°31'/55°30') (Furey, 1985; Wanless et al., 1965, 1966)	400K	bhp.gr,gd	mt	f	s	ho	--	--	m	9b
A3a	Harbour Breton (47°05'/55°37') (Furey, 1985; Bell and Blenkinsop, 1975)	343R	bh.gr	--	--	--	ho	--	0.708	m	9b
A4	Saint Lawrence (47°08'/55°19') (Teng and Strong, 1976; Kerr et al., 1993)	374Z	r.gr	mt	f	i	ag	F	0.702	P	9b
A5	Swift Current (47°58'/54°07') (Dallmeyer et al., 1981)	590Z	b.gr,gd	mt	s	i	me	--	0.705	m	2a
A6	Grand Beach (47°06'/55°31') (O'Brien et al., 1977; Krogh et al., 1988)	393Z	r.po	--	f	i	ag	--	--	P	9b
A7	Terra Nova (48°30'/54°10') (Jenness, 1963; Bell et al., 1977)	352R	b.gr	--	f	i	ho	--	0.705	m	9b
A8	Cross Hills (47°13'/54°43') (O'Brien et al., 1984)	547Z	bh.gr,r.gr,do	--	--	--	ho	--	--	m	3a

1	2	3	4	5	6	7	8	9	10	11	12
Avalon Peninsula (Fig. 8.6) (cont.)											
A9	Simmons Brook-Furbys Cove (47°15'/55°21') (Colman-Sadd, 1976; Kerr et al., 1993)	620Z	bh.gd.gr	--	--	--	ca	--	0.706	ca	2a
A10	Powderhorn (47°55'/54°00') (Jenness, 1963; Strong et al., 1974)	--	do,b.gr	--	--	--	ho	--	--	m	3a
A11	Cape Rodger Mtn. (47°34'/54°07') (Bradley, 1962; O'Brien et al., 1984)	--	bh.gr	--	--	--	ho	--	--	gn	2a
A12	Deepwater Point (47°25'/55°05') (O'Brien et al., 1984)	--	bh.gr	--	--	--	ho	--	--	gn	2a
A13	Knee (47°28'/55°08') (Bradley, 1962; O'Brien et al., 1977)	--	lg	--	f	--	ag	--	--	m	3a
A14	Ragged Islands (47°44'/54°14') (Strong et al., 1974; Leech et al., 1964)	382K	bh.gr,do,sy	--	--	--	--	--	--	m	8
A15	Red Island (47°23'/54°00') (Hogg, 1954)	--	b.gr,do	--	--	--	ho	--	--	m	8
A16	Iona Islands (47°34'/53°58') (McCartney, 1967; Kerr et al., 1992)	364R	b.gr,do	--	--	--	ho	--	--	m	8
A17	Louil Hills (48°39'/53°55') (Strong et al., 1974; O'Brien et al., 1988)	570Z	r.gr	--	f	--	ho	--	--	m	3a
A18	Wandsworth[1] (47°06'/55°07') (Strong et al., 1978; Krogh et al., 1988)	762Z	ga	il	s	i	ho	--	--	m	2b
A19	Avalon (46°58'/53°25') (Papezik and Hodych, 1980)	201K	ga	--	--	--	ho	--	--	d	10a
A20	Spreadeagle (47°23'/53°40') (McCartney, 1967)	--	ga	--	--	--	ho	--	--	d	3a
A21	Anchor Drogue (47°05'/55°15') (Strong et al., 1978)	--	gd,q.do	--	--	--	ag	--	--	m	2a
A22	Georges Pond (48°17'/54°02') (Taylor et al., 1979)	--	bh.gr,gd	--	--	--	ho	--	--	m	3a
A23	Berry Hills (54°52'/47°31') (O'Brien et al., 1984)	--	b.gr	--	--	--	ho	--	--	m	9b

[1] Correlated to the Laughlin Hill gabbro and Grole diorite to the west, and Whalesback gabbro south of the Holyrood pluton by Hayes et al. (1987).

Northwestern Cape Breton Island (Fig. 8.6)

1	2	3	4	5	6	7	8	9	10	11	12
B1	Creignish Hills (45°48'/61°23') (Campbell, 1980)	446R	md,b.gr	--	--	i	ho	--	--	m	5
B2	Cheticamp (46°32'/60°55') (Barr et al., 1986; Jamieson et al., 1986)	550Z	bh.gd,bm.gr	mt	s	i	ca	--	0.707	ca	3a
B3	Whycocamagh (46°00'/61°06') (Barr et al., 1982)	402K	b.gr,h.mg	--	--	s	ho	Cu	--	P	8
B4	Mabou (46°09'/61°22') (Barr and MacDonald, 1983)	--	do,to,tr	--	--	s	sh	--	--	gn	3a
B5	Gillanders Mtn. (61°00'/46°16') (French and Barr, 1984)	379R	do,b.mg,sy	--	f	i	ho	Cu	0.708	m	8
B6	Belle Cote Road (46°40'/60°49') (Jamieson et al., 1986; Barr et al., 1982)	433Z	bm.gr	--	--	s	sh	--	--	gn	7a
B7	Salmon Pool (46°41'/60°53') (Jamieson et al., 1986; Currie, 1987b)	365Z	b.gr	--	f	i	ho	--	--	m	9b
B8	Margaree (46°11'/60°46') (O'Bierne-Ryan et al., 1986; Currie, 1987b)	343R	b.gr	mt	f	i	ho	--	0.706	mc	9a
B9	Pleasant Bay (46°49'/60°50') (Currie, 1987b)	364R	lg	--	f	i	sh	--	0.709	gn	7a
B10	Lowland Cove (47°00'/60°36') (Neale, 1963; Barr et al., 1987)	1040Z	bp.sy	--	ne	--	ho	Cu	--	m	1d
B11	Wilkie Sugarloaf (47°00'/60°24') (MacDonald and Smith, 1980; Cormier, 1979)	330R	b.lg	--	--	--	ho	--	0.705	m	9b
B12	Glasgow Brook (46°50'/60°30') (Wiebe, 1972)	--	bh.do,to	mt	--	i	ca	--	--	gn	5b
B13	White Point-Black Brook- Warren Brook (47°51'/60°22') (Barr et al., 1982, 1988)	373Z	bm.mg,gd	mt	g	i	ag	--	0.705	gn	7

Table 8.6. (cont.)

1	2	3	4	5	6	7	8	9	10	11	12
Northwestern Cape Breton Island (Fig. 8.6) (cont.)											
B14	Cameron Brook (46°41'/60°27')	403Z	b.gr	--	--	--	ho	--	--	mc	9a
	(O'Bierne-Ryan et al., 1986; Barr et al., 1988)										
B15	Gulch Brook (46°51'/60°30')	413R	bm.lg,gd	--	--	--	sh	--	0.705	gn	7a
	(MacDonald and Smith, 1980; Keppie and Halliday, 1986)										
B16	Cape Smoky (46°36'/60°24')	494Z	b.mg	--	--	--	--	--	0.704	m	5a
	(Barr et al., 1982; Barr et al., 1988)										
B17	Ingonish River[1] (46°34'/60°26')	555Z	do,to,gd	--	s	i	sh	--	--	gn	3a
	(Barr et al., 1982, 1988)										
B18	St. Ann's Mt. (46°17'/60°35')	515R	bh.gd,to,do	--	--	--	ca	Cu	0.706	m	3a
	(Barr et al., 1982)										
B19	North Branch Baddeck River (46°16'/60°44')	614Z	to,do	mt	--	i	ca	--	--	gn	2a
	(Jamieson et al., 1986)										
B20	North River (46°21'/60°52')	401R	bh.mg,gd	il	s	--	ho	--	0.704	m	8
	(Jamieson and Craw, 1983)										
B21	Red River (46°57'/60°41')	966Z	an	--	--	--	ca	--	--	c	1b
	(MacDonald and Smith, 1980; Miller et al., 1993)										
B22	Kathy Road (46°18'/60°41')	560Z	do,to	--	s	--	ca	--	--	gn	3a
	(Barr et al., 1987, 1988)										
B23	Gisborne Flowage (46°27'/60°32')	564Z	bh.do	mt	s	--	ca	--	--	gn	3a
	(Barr et al., 1987, 1988)										
B24	Indian Brook[2] (46°18'/69°31')	564U	bh.gd,gr	mt	s	m	ho	Cu	--	gn	3a
	(Barr et al., 1987, 1988)										
B25	Cross Mountain (46°18'/60°45')	--	bm.gr	il	--	m	ho	--	--	gn	7
	(Barr et al., 1987)										
B26	Taylors Barren (46°30'/61°08')	419R	bm.gr	--	--	m	sh	--	0.705	gn	7
	(Jamieson et al., 1986; Gaudette et al., 1985)										

[1] Includes Wreck Cove
[2] Includes Birch Plain, Kerrs Bk., Murray Bk., and similar rocks

Southeastern Cape Breton and northern Nova Scotia (Fig. 8.6)

1	2	3	4	5	6	7	8	9	10	11	12
D1	Jeffers Brook (45°29'/64°08')	620K	do,gd	--	--	--	ca	--	0.706	gn	2a
	(Donohoe and Wallace, 1979; Pe-Piper and Piper, 1987)										
D2	Eden Lake[1] (45°25'/62°17')	582K	do,ga	--	--	--	ca	--	--	gn	2a
	(Wanless et al., 1965; Murphy et al., 1988)										
D3	Ohio[2] (45°41'/62°02')	533R	bh.gd,lg	--	--	s	ho	--	0.703	m	3a
	(Murphy et al., 1988; Cormier, 1979)										
D4	Gunn Lake (45°24'/62°11')	370R	bh.gr,hp.gr	mt	--	--	ho	--	0.717	m	9b
	(Benson, 1974; Murphy et al., 1988; Cormier, 1979)										
D5	Salmon River (45°22'/63°04')	504R	bh.gr	--	--	--	--	--	0.704	gn	3a
	(Donohoe and Wallace, 1979; Cormier, 1979)										
D6	Cape Chignecto (45°22'/64°54')	339R	m.lg	--	--	--	--	--	0.706	m	9b
	(Donohoe and Wallace, 1979; Cormier, 1979)										
D7	Pleasant Hills (45°27'/63°50')	315Z	bh.gr	--	--	--	--	U	0.708	m	9b
	(Donohoe and Wallace, 1979; Doig et al., 1991)										
D8	Hart-Byers Lakes (45°34'/63°31')	348R	br.gr	--	--	--	--	--	0.705	m	9b
	(Donohoe and Wallace, 1979; Donohoe et al., 1986)										
D9	Barneys River (45°31'/62°17')	378K	b.gr	--	--	--	--	--	0.705	m	9b
	(Benson, 1974)										
D10	Coxheath Hills (46°05'/60°22')	569R	mg,do	mt	--	s	ag	Cu	0.706	sh	3a
	(Kirkham and Soregaroli, 1975; Cormier, 1972)										
D11	Boisdale Hills (46°46'/60°28')	563R	do,bh.gd,gr	--	--	i	ho	--	0.704	m	3a
	(Barr and Setter, 1981; Cormier, 1972)										
D12	Kellys Mountain I (46°46'/60°29')	498Z	lg	--	--	i	ho	--	0.699	m	3a
	(Barr et al., 1982; 1990)										
D13	Chisholm Brook (45°43'/60°34')	620	b.gr,gd,do	--	--	i	ho	--	0.708	m	3a
	(Barr et al., 1984; 1990)										
D14	Loch Lomond (45°45'/60°32')	368R	b.gr	--	--	--	ag	--	0.704	P	9b
	(Barr et al., 1984)										

1	2	3	4	5	6	7	8	9	10	11	12
	Southeastern Cape Breton and northern Nova Scotia (Fig. 8.6) (cont.)										
D15	Capelin Cove[3] (45°40'/60°26') 545R b.mg (O'Reilly, 1977; Cormier, 1972)	545R	b.mg	--	--	i	ag	--	0.704	m	3a
D16	North Mountain (45°48'/61°07') (Kelly, 1967; Cormier, 1972)	533R	b.gr,do	mt	--	--	--	--	0.704	m	3a
D17	Gillis Mtn.-Deep Cove (45°53'/60°10') (Barr and O'Beirne, 1981; Cormier, 1972)	394R	md,gr	mt	--	i	ag	Cu	0.703	m	8
D18	Huntingdon Mtn. (45°57'/60°23') (Barr et al., 1982; Cormier, 1979)	540R	b.gr,lg	--	s	i	ho	--	0.701	m	3a
D19	Sporting Mountain (45°42'/60°59') (Sexton and Cotie, 1985)	--	bh.gr,gd,do	mt	s	i	ca	Cu	--	ca	3a
D20	Petit de Grat (45°30'/60°56') (Barr et al., 1982; Cormier, 1979)	357R	lg	--	--	--	ca	--	0.702	ca	9b
D21	Shunacadie (46°01'/60°35') (Barr and Setter, 1981; Barr et al., 1990)	564Z	b.gd,do	--	s	i	ca	--	0.706	m	3a
D22	Irish Cove (45°49'/60°36') (O'Reilly, 1977)	--	gr,gd	--	--	--	--	--	--	m	3a
D23	Greendale (46°01'/61°37') (Murphy et al., 1988)	653A	h.do,gr	mt	o	s	ho	--	--	m	2a
D24	Debert River (45°30'/62°12') (Donohoe et al., 1986; Doig et al., 1991)	612Z	bh.gd,mg	--	s	m	ho	--	0.706	m	2a
D25	McCallum Settlement (45°29'/62°09') (Gaudette et al., 1983; Murphy et al., 1988)	575R	bh.gd	--	--	--	ho	--	0.705	m	2a
D26	Folly Lake (45°35'/62°16') (Murphy et al., 1988)	--	h.ga,bh.gd	--	s	--	ag	--	--	c	8
D27	Moose River-North River (45°29'/62°41') (Pe-Piper and Piper, 1989)	342R	do,b.gr	--	--	--	ho	--	--	c	8
D28	Kellys Mountain II (46°46'/60°29') (Gaudette et al., 1985)	636R	do	--	--	--	ho	--	0.702	gn	2a
D29	Mount Thom (45°34'/61°54') (Gaudette et al., 1983; Murphy et al., 1988)	934R	gr	--	--	--	ca	--	0.705	gn	1d

[1] + Black Brook
[2] + James River and Indian Lake
[3] + Economy River
[4] + Lower Saint Esprit

South Mountain belt (Fig. 8.8)

1	2	3	4	5	6	7	8	9	10	11	12
E1	South Mountain (44°41'/65°00') (MacDonald et al., 1987; Clarke and Halliday, 1980)	372R	b.gd	il	g	s	ho	--	0.708	m	7b
E2	Halifax (44°35'/63°45') (Clarke and Halliday, 1980; MacDonald and Horne, 1987)	364R	bm.mg,lg	il	g	s	ho	U	0.704	l	7b
E3	New Ross-Vaugn (44°41'/64°18') (Clarke and Halliday, 1980; O'Reilly et al., 1982)	395R	bm.mg,lg	il	g	s	ho	U	0.703	m	7b
E4	Lake George (44°54'/64°39') (McKenzie and Clarke, 1975; Taylor, 1969)	--	bm.mg,lg	il	g	s	ho	--	--	m	7b
E5	Springfield Lake (44°43'/64°49') (McKenzie and Clarke, 1975; Taylor, 1969)	--	bm.mg,lg	il	g	s	ho	U	--	m	7b
E6	West Dalhousie (44°43'/65°12') (McKenzie and Clarke, 1975; Taylor, 1969)	--	bm.mg,lg	il	g	s	--	U	--	m	7b
E7	Tombstone (44°52'/65°04') (Wray, 1972)	--	b.gd	il	--	s	ho	--	--	m	7b
E8	Brenton (43°56'/66°02') (Taylor, 1967; Krogh, 1984)	451Z	bm.gr	il	g	s	ca	Sn	--	ca	7a
E9	Wedgeport (43°45'/66°01') (Chatterjee and Keppie, 1981; Cormier et al., 1988)	323Z	b.gr	--	--	s	ca	W	0.708	ca	7b
E10	Barrington Passage (43°09'/65°42') (de Albuquerqu;e, 1977; Rogers, 1985)	348K	b.to	mt	s	i	mi	--	--	gn	7b
E11	Shelburne (43°47'/65°25') (de Albuquerque, 1977; Rogers, 1985)	--	bm.gr	il	--	s	sh	--	--	gn	7a

Table 8.6. (cont.)

1	2	3	4	5	6	7	8	9	10	11	12
	South Mountain belt (Fig. 8.8) (cont.)										
E12	Port Mouton (43°52'/64°53') (Cormier and Smith, 1973; Hope et al., 1988)	340R	b.gr	--	b	s	ho	Be	0.702	m	7b
E13	Liscombe (49°44'/62°31') (Giles and Chatterjee, 1987a; Fairbairn et al., 1960)	389K	b.gd	--	--	s	ho	--	--	m	8
E14	Musquedoboit (44°49'/63°00') (MacDonald and Clarke, 1985; Reynolds et al., 1981b)	377K	bm.mg	--	g	s	ho	--	--	m	7b
E15	Queensport[1] (45°16'/61°21') (Ham, 1984; Hill, 1991; Dalmeyer and Keppie, 1984)	362K	bm.mg,b.gd	--	--	s	--	U	--	gn	7b
E16	Forest Hills[2] (45°14'/61°37') (Hill, 1991)	--	bm.mg,b.gd	--	--	s	ca	--	--	gn	7b
E17	Sherbrooke (45°09'/61°43') (Smith, 1981; Fairbairn et al., 1960)	347K	bm.gr	--	--	s	ho	--	--	gn	7b
E18	Larrys River[2] (45°13'/61°39') (O'Reilly, 1983; Fairbairn et al., 1960)	369K	bm.mg,b.gd	--	--	s	ho	--	--	gn	7b
E19	Ellison Lake (44°36'/65°45') (Allen and Barr, 1983)	346K	bm.mg	--	--	s	ho	Cu	--	m	7b
E20	Chedabucto (45°12'/61°05') (Hill, 1991)	371U	bm.mg,b.gd,to	--	--	--	ho	--	0.707	gn	7b
E21	Shelburne dyke (43°51'/65°16') (Papezik and Barr, 1981)	201K	ga	mt	--	--	ho	--	--	d	10
E22	Davis Lake (44°01'/65°51') (Richardson et al., 1988; Kontak, 1987)	317R	bm.mg	il	g	s	me	Sn	0.733	m	7b

[1] + Halfway Cove
[2] Bull Ridge and Country Harbour
[3] + Sangster Lake

1	2	3	4	5	6	7	8	9	10	11	12
	Gaspésie-Eastern Townships region (Fig. 8.9)										
L1	Stanstead (Stanhope) (45°00'/71°45') (Erdmer, 1982)	346K	b.gd	--	--	--	ho	--	--	m	9b
L2	Graniteville (45°01'/72°11') (Erdmer, 1982)	--	b.gr,bm.gr	--	--	--	ho	--	--	m	9b
L3	Hereford Mountain (45°04'/71°35') (Erdmer, 1982)	--	b.gr	--	--	s	ho	--	--	m	9b
L4	Scotstown (45°34'/71°19') (Danis, 1984; Simonetti and Doig, 1990)	384Z	bm.gr	--	--	--	ho	--	--	m	9b
L5	Aylmer (45°44'/71°19') (Bourne, 1986a; Simonetti and Doig, 1990)	375Z	b.gr	--	--	--	ho	Cu	--	m	9b
L6	Winslow (45°45'/71°12') (Danis, 1985; Simonetti and Doig, 1990)	377Z	b.gd,do	--	--	--	ho	Cu	--	m	9b
L7	Sainte Cecile (45°43'/70°56') (Bourne, 1986b; Simonetti and Doig, 1990)	374Z	b.gr	--	--	--	ca	--	--	gn	9b
L8	Moulton Hill (45°26'/71°49') (St. Julien, 1970; Poole, 1980b)	505R	tr	--	--	--	ca	--	0.710	gn	4
L9	Salmon River (45°26'/71°49') (St. Julien, 1970)	--	tr	--	--	--	ca	--	--	ca	4
L10	Thetford Mines (46°07'/71°20') (Poole, 1980c; Laurent et al., 1984)	466R	bm.gd,tr	--	--	--	ca	--	0.717	gn	5
L11	Spider Lake (45°25'/70°46') (Bourne, 1984; Simonetti and Doig, 1990)	383Z	b.gr	--	--	i	ho	--	--	m	9b
L12	Murdochville (48°59'/65°30') (Whalen, 1993)	375K	b.gd	s	--	i	ho	Cu	--	p	9b
L13	Port Daniel (48°16'/64°59') (Wanless et al., 1966)	442K	to,do,tr	--	--	i	ca	--	--	gn	4
L14	Mont Breche (48°49'/66°04') (Robert, 1966a)	--	hb.gr,ga	s	--	i	ho	Cu	--	c	8
L15	Mont McGerrigle (48°57'/65°59') (Whalen, 1993; Whalen and Gariepy, 1986)	391Z	bh.gr,ne.sy	--	--	i	ho	Cu	--	c	8
L16	Mont Hogs Back (48°53'/66°02') (de Romer, 1977; Robert, 1966b)	342K	bh.gr	--	--	i	ho	Cu	--	p	9b
L17	Mont Vallieres (48°52'/66°00') (de Romer, 1977; Whalen, 1993)	371K	bh.gr,gd	--	--	i	ho	Cu	--	p	9b
L18	Mount Shefford (45°21'/72°35') (Eby, 1984)	120, 129R	b.ga,ne.sy,gr	--	--	i	ho	--	0.704	m	10
L19	Brome (45°16'/72°40') (Eby, 1984)	118, 136R	b.ga,sy,gr	--	--	i	ho	--	0.703	m	10
L20	Mont Megantic (45°28'/71°10') (Eby, 1984)	128, 132R	b.ga,gr	mt	--	i	ho	--	0.704	m	10

Central and northern New Brunswick (Fig. 8.12)

1	2	3	4	5	6	7	8	9	10	11	12
M1	Skiff Lake (45°45'/67°40')	409Z	bm.gr	il	--	s	sh	--	0.708	gn	7a
	(Lutes, 1987; Bevier, 1988)										
M2	Hawkshaw (45°50'/67°16')	411U	b.gr	--	--	--	ho	--	0.705	mc	7a
	(Lutes, 1987; Bevier, 1988)										
M3	Hartsfield (46°02'/67°18')	415U	h.to	mt	s	i	--	--	--	gn	7a
	(Venugopal, 1979)										
M4	Gibson-Benton (46°05'/67°33';46°02'/67°36')	479Z	bh.gr,gd	--	--	i	ag	Cu	--	m	8
	(Venugopal, 1978; Whalen, 1993)										
M5	Nashwaak (46°28'/67°03')	422R	bm.gr,b.gr	--	--	--	--	W	--	m	7a
	(Lutes, 1987; Whalen, 1993)										
M6	Becaguimec Lk. (46°22'/67°12')	--	ga,bh.gd	mt	--	i	--	--	--	m	8
	(Venugopal, 1982)										
M7	Burnthill-Trout Brook-Dungarvon (46°37'/66°53')	380A	b.gr,bm.gr	--	t	s	me	W	0.705	m	9b
	(Taylor et al., 1987)										
M8	Sweat Hill (46°18'/67°05')	454Z	b.gd	--	--	--	ca	--	--	gn	5b
	(Crouse, 1978; Bevier, 1988)										
M9	North Pole Stream (47°01'/66°31')	417U	ga,b.gr	--	--	i	ho	--	--	mc	8
	(Whalen, 1987; Bevier, 1988)										
M10	North Dungarvon (46°52'/66°41')	382K	bh.gd,bm.gr	--	--	s	ho	--	--	m	7a
	(Irrinki, 1978; Skinner, 1975)										
M11	Redstone Mountain[1] (46°49'/67°57')	409R	bh.gr,ga	--	--	--	ho	--	0.706	p	8
	(Crouse, 1978; Fyffe and Cormier, 1979)										
M12	Mount Elizabeth[2] (47°20'/66°35')	414Z	bh.gr,r.gr,ga	--	--	--	ho	--	--	c	8
	(Whalen, 1993; Bevier, 1988)										
M13	Charlo (47°50'/66°25')	370R	ga,bh.gr	mt	s	i	ho	Cu	0.706	p	8
	(Greiner, 1970; Stewart, 1979)										
M14	Bois Gagnon (47°32'/68°10')	399K	b.gr	--	--	--	--	--	--	m	8
	(Abbott and Barnett, 1969)										
M15	Antinouri Lake (47°49'/65°56')	372Z	b.gr	--	--	--	ho	--	--	m	8
	(Whalen, 1993; Walker et al., 1991)										
M16	Nicholas Denys (47°42'/65°53')	381Z	b.gd	--	--	--	ho	Pb	--	m	8
	(Whalen, 1993; Walker et al., 1991)										
M17	Pabineau Falls (47°42'/65°44')	392Z	b.gr	--	--	--	ho	--	--	m	8
	(Skinner, 1975; Roddick and Bevier, 1988)										
M18	Fox Ridge (46°52'/66°45')	452Z	b.gr	--	--	--	ca	--	0.703	gn	5
	(Crouse, 1978, 1979; Poole, 1980d; Bevier, 1988)										
M19	South Renous (46°50'/66°30')	441Z	b.gr	--	--	--	ho	--	--	gn	5
	(Irrinki, 1980; Bevier, 1988)										
M20	Meridian Brook (47°19'/67°45')	464Z	b.gr	--	--	--	ca	--	--	gn	5
	(Crouse, 1978, 1979; Bevier, 1988)										
M21	Mullen Stream Lk. (47°01'/66°18')	458Z	b.gr	--	--	--	ca	--	0.703	gn	5
	(Crouse, 1981b; Poole, 1980a; Bevier, 1988)										
M22	Serpentine River (47°14'/67°46')	455Z	b.gr	--	--	--	ca	--	--	gn	5
	(Crouse, 1978, 1979; Bevier, 1988)										
M23	Trousers Lake (46°48'/66°46')	434Z	b.gr	--	--	--	ca	--	--	gn	5
	(Crouse, 1978; Bevier, 1988)										
M24	Juniper Barren (45°33'/67°11')	421K	b.gr,m.gr	--	g	i	ca	--	--	m	7
	(St. Peter, 1981; Whalen, 1993)										
M25	Lost Lake (46°38'/67°00')	424R	bm.gr,b.gd	--	--	--	ho	--	0.703	gn	7
	(Whalen, 1993; Poole, 1980d)										
M26	Caraquet (47°48'/64°57')	180K	ga	mt	--	--	ho	--	--	d	11
	(Burke et al., 1973; Wanless et al., 1972)										

[1] + Clearwater Brook
[2] + Portage Brook
NOTE: The Pokiok batholith comprises Skiff Lake, Hawkshaw and Hartfield plutons

Table 8.6. (cont.)

1	2	3	4	5	6	7	8	9	10	11	12
Southern New Brunswick (Fig. 8.10)											
F1	Stewarton (45°41'/66°09')	-- (Bennett, 1965)	do,ga	mt	--	i	ho	Cu	--	l	8
F2	Evandale (45°44'/65°47')	364K (Ruitenberg, 1970)	bh.gr,gd	--	--	--	ho	Cu	--	m	9b
F3	Brookville (45°13'/65°17')	605Z (Bevier et al., 1990)	bh.do	mt	--	m	sh	--	0.702	gn	2b
F4	Magaguadavic (45°18'/66°50')	396Z (Ruitenberg and Fyffe, 1982; Bevier, 1988)	b.gr	mt	--	i	ho	--	0.707	m	8
F5	Bocabec[1] (45°13'/67°05')	394R (Ruitenberg, 1967; Pajari, 1977)	do,ga,to	mt	--	i	ho	Cu	0.707	m	8
F6	Saint Stephen (45°12'/67°21')	401K (Dunham, 1959; Wanless et al., 1973)	ga,no,um	mt	--	i	ho	Ni	--	l	8
F7	Tower Hill-Pleasant Ridge (45°17'/67°14';45°22'/66°00')	361A (McLeod, 1990)	bm.gr	il	--	--	ho	U	0.737	m	7b
F8	Sorrell Hill-Beech Hill (45°18'/67°03';45°23'/66°53')	337R (Butt, 1976; McLeod, 1990)	b.gr	--	f	i	ho	--	0.715	m	9b
F9	Welsford[1] (45°27'/66°15')	422Z (Payette and Martin, 1987; Bevier, 1988)	r.gr,sy	mt	--	i	ho	Zr	0.709	p	6
F10	Mount Douglas[1] (45°21'/66°26')	367Z (McLeod, 1990; Bevier, 1988)	b.gr	il	--	s	ho	W	0.708	m	9b
F11	Utopia[1] (45°14'/66°19')	430Z (McLeod, 1990)	b.gr	mt	--	--	ho	--	0.705	m	8
F12	Mount Pleasant (45°24'/66°32')	360K (Ruitenberg and Fyffe, 1982; McCutcheon, 1986)	b.gr	mt	--	--	ho	Sn	--	p	9b
F13	Bonnell Brook (45°30'/65°24')	550Z (Barr et al., 1987; Bevier, 1988)	do,h.gd	--	--	i	ca	--	--	ca	3a
F14	Point Wolfe River (45°38'/65°14')	625Z (Barr et al., 1987; Bevier, 1988)	do,h.gd	--	--	i	ca	Cu	--	ca	2a
F15	Musquash[2] (45°10'/66°17')	550Z (Ruitenberg et al., 1979; Currie, 1987a,b)	hb.gr	--	--	--	ca	--	0.706	ca	3a
F16	Lepreau (45°11'/66°24')	546R (Ruitenberg et al., 1979; Poole, 1980a)	th.do	mt	s	i	ca	--	0.705	ca	3a
F17	Forty Five River (45°41'/64°59')	598R (Barr et al., 1987)	h.do,h.gd	mt	s	i	ca	--	0.703	ca	2a
F18	Caledonia Road (45°50'/64°42')	-- (Ruitenberg et al., 1979)	ga,do,gr	--	--	i	ho	--	--	m	3a
F19	Mechanic Settlement (45°44'/65°15')	557Z (Ruitenberg et al., 1979; McLeod et al., 1994)	ga	--	--	--	--	Pt	--	ca	3a
F20	Fairville-Renforth (45°17'/66°05')	547A (Currie, 1984; McLeod et al., 1994)	bh.gr,to	mt	s	i	ho	--	0.706	m	2a
F21	Red Head (45°06'/66°08')	518A (Ruitenberg et al., 1979; McLeod et al., 1994)	h.gd,to	mt	s	i	ca	--	--	ca	2a
F22	Cape Spenser (45°12'/65°54')	625Z (Watters, 1987; Currie, 1984)	gd,lg	--	--	i	ca	Au	--	ca	2a
F23	Duck Lake (45°21'/65°59')	-- (Ruitenberg et al., 1979; Currie, 1984)	ga,um,do	mt	s	i	me	--	--	m	3a
F24	Upham Mtn. (45°28'/65°40')	554Z (Ruitenberg et al., 1979; McLeod et al., 1994)	hp.sy,gr	mt	--	--	ca	--	--	m	2a
F25	Chance Harbour (45°09'/66°18')	546R (Poole, 1980a)	gr,lg,to	mt	--	--	ca	--	0.705	ca	3a
F26	Kingston (45°07'/66')	435Z (Eby and Currie, 1993; Doig et al., 1990)	ga,mg	--	s	i	ho	--	--	d	3a

[1] Saint George batholith comprises the Mount Douglas, Utopia, Magaguadavic, Welsford and Bocabec plutons.
[2] + Ragged Falls and Hardwood Hill.

Tilting Harbour portion of the complex (Cawthorn, 1978) forms the mafic part, which differs from Mount Peyton in exhibiting cumulates of orthopyroxene, clinopyroxene and amphibole. The granitic portion of the complex covers much of the rest of Fogo Island. The Loon Bay pluton (N9) of biotite-hornblende granodiorite may represent the salic portion of another composite pluton whose mafic portion is unexposed. Probably correlative minor tonalitic to leucogranitic minor plutons (Frederickton-Rocky Bay, Aspen Cove (C1, C2, Fig. 8.4; Currie et al., 1979)) occur east of Notre Dame Bay as far as Copper Island near Wesleyville (Williams, 1968), and as far south as Top Pond (O'Brien, 1983). Most of these bodies exhibit distinctive large biotite flakes, poikilitic with tiny quartz and plagioclase grains.

The Notre Dame Bay region contains dykes and small plugs of strongly alkaline rocks of Jurassic (155 Ma, Group 10b) age (Strong and Harris, 1974), including alkaline biotite gabbro and feldspar-free lamprophyres. The Rodeross Lake pluton (N13, Fig. 8.2) has been assigned to this group but its age is unknown.

Gander Zone of Newfoundland

Granitoid plutons and high grade metamorphic rocks are abundant in the Gander Zone (Fig. 8.4). Silurian and younger plutons in this region also penetrate nearby portions of the Exploits Subzone which structurally overlie the Gander Zone. In the extreme southwestern portion distinction between the Gander and Avalon zones is uncertain. Precambrian rocks have been unequivocally identified only in this region ("old" Roti granite, 563 ± 4 Ma, Dunning et al., 1988). In age and composition this pluton is identical to Group 3a plutons of the Avalon Zone. Volcano-sedimentary rocks hosting the "old" Roti granite must also be of Precambrian age, but their extent is unknown. Parts of the psammitic-amphibolitic Hare Bay, Little Passage and Port aux Basques gneisses may be of similar age (Currie et al., 1982a; Currie, 1983), but these gneisses probably consist principally of metamorphosed volcano-sedimentary strata of Ordovician and older age.

Tonalitic plutons of Group 4 are unknown in this region, but leucogranites of Group 4 are present ("young" Roti granite, Dunning et al., 1988). Small subvolcanic plutons of plagioclase-porphyritic biotite granodiorite occur in the southwest part of the region (Baggs Hill granite) which are thought to be approximately contemporary to surrounding Ordovician volcanic rocks, and are assigned to Group 5a.

Foliated to mylonitic megacrystic plutons of Group 5b commonly give U-Pb zircon ages in the range of 420-440 Ma (Dunning et al., 1988), although the Lockers Bay pluton (C6) gave a U-Pb zircon age of 460 Ma (Dallmeyer et al., 1981). All these plutons contain microcline megacrysts which form 10 to 70% of the volume. Apart from megacrysts, the plutons consist of bluish quartz, oligoclase to andesine, and sparse matrix microcline. Biotite is the most common mafic mineral, but amphibole-bearing phases are locally present. Matrix microcline is sufficiently sparse that many of the rocks would be tonalitic in the absence of the megacrysts, and some examples have the appearance of potassium metasomatized rocks. Extensive migmatites and sheeted complexes fringe most megacrystic plutons, commonly accompanied by aureoles of metasomatic megacrysts. Sheets and dykes of aplitic muscovite-garnet

leucogranite intrude the megacrystic phases. The Burgeo batholith (C18) the largest and most complex of these bodies, forms a composite intrusion with phases ranging from hornblende gabbro, quartz diorite and biotite tonalite to muscovite-biotite granite and alaskite. Extensive screens of migmatite and reworked host rocks occur within the batholith (O'Brien et al., 1986).

Major peraluminous granite emplacement (Group 7a) appears to be somewhat younger. The North Bay igneous suite (C16, 17) forms an extensive composite batholith including phases of porphyritic biotite±muscovite granodiorite, muscovite dominant granite, and muscovite-garnet granite. The southern margin shows an extensive sheeted contact zone (O'Brien et al., 1986), whereas the northwestern contacts are cataclastic to mylonitic. An age of 396 Ma (U-Pb on zircon) for the granite phase of the North Bay suite (Dickson et al., 1985) may date emplacement of a late mobile phase, but initiation of the anatectic melting was somewhat older as shown by metamorphic dates on the Port aux Basques granite (412 Ma, Dunning et al., 1988) and Little Passage gneiss (423 Ma, Dunning et al., 1988). These anatectic melts show a characteristic progression of phases throughout the southwest part of the region, commencing with biotite-dominant granite with small, euhedral megacrysts, cut by muscovite-garnet granite, in turn cut by a spectacular late muscovite-garnet-tourmaline-beryl granite. In the northern part of the Gander Zone, along the boundary with the Exploits Subzone, detailed field and geochemical study of the process of anatexis showed that with increasing grade of metamorphism, segregations in the rock become progressively more granitic (Fig. 8.5) and the rock ultimately passed to migmatite and foliated granite (Currie and Pajari, 1981). The mafic mineralogy of such complexes shows a clear relation to that of the source. Hornblende-bearing plutons appear in hornblende-bearing hosts (metavolcanics) and muscovite-bearing plutons appear in metasedimentary hosts.

Posttectonic plutons of two different types occur in the Gander Zone. Megacrystic plutons exhibit all the igneous-metasomatic phenomena associated with Group 5b, but they are completely massive (Group 9a). The Deadmans Bay pluton (C4, Fig. 8.4) may be taken as a typical example. A U-Pb zircon age gave an upper intercept age of 1110 Ma and a lower intercept age of 360 Ma (Currie et al., 1982a). Poor Rb-Sr data suggested an age in the range of 600 Ma, whereas K-Ar ages ranged from 440 to 345 Ma. Presumably the complex interplay of metasomatic and magmatic factors explains the anomalous age results. The emplacement age of this and similar plutons (Newport, C8, Maccles Pond, C10) is thought to be about 350 to 360 Ma. The Lockers Bay pluton with a zircon age of 460 Ma gives late Devonian Ar-Ar and Rb-Sr ages with a high initial Sr ratio (0.7145) (summary in Dallmeyer et al., 1981) suggesting resetting of an Ordovician pluton by metamorphism or metasomatism.

The other major type of post-tectonic pluton may be typified by the Ackley batholith (C23, Fig. 8.4; Tuach et al., 1986; Dickson, 1983) a large, lobate batholith which seals the Dover Fault, the boundary between the Avalon and Gander zones. The northwestern part comprises biotite±hornblende granites which are less fractionated than the biotite granite, commonly megacrystic, to the southeast, although all units are relatively highly fractionated. The Dover Fault appears to exert a significant control on distribution of compositions

(Tuach et al., 1986) even though it is sealed by the batholith. Numerous miarolitic cavities occur in the southern part of the batholith along with significant Mo-Sn-W mineralization (Whalen, 1983). The batholith was emplaced in a number of stages at 410, ~375 and 355 Ma (Kontak et al., 1988; Dallmeyer et al., 1983a). A number of other, smaller high level plutons characterized by miarolitic cavities, extensive granophyric to porphyritic margins, local greisens, and narrow metamorphic aureoles indicating emplacement at high structural level occur along the south coast of Newfoundland (Belleoram (A3), Harbour Breton (A3a), Pass Island (A3, A2; Fig. 8.6), Francois (C27, Fig. 8.4; see Dickson et al., 1988). All have been prospected for lithophile elements.

Avalon Zone of Newfoundland

Much of the Avalon Zone of Newfoundland (Fig. 8.4) is underlain by late Precambrian volcano-sedimentary rocks. No crystalline basement is known, although geophysical data suggests a sialic basement (Stockmal et al., 1987). The oldest known plutons (Group 2b) occur in the southwestern corner of the region on the Burin Peninsula. The layered

Figure 8.5. Anatectic melt segregations, Ragged Harbour pluton, Newfoundland. These quartz-feldspar-biotite segregations, aligned on tectonic foliation, mark the culmination of a complex metasomatic process (Currie and Pajari, 1981). GSC 203034-P

Wandsworth gabbro (dated at 762 Ma by Krogh et al., 1988) occurs within submarine volcanic rocks and may form part of an ophiolitic suite (Strong and Dostal, 1980). Hayes et al. (1987) considered a number of similar gabbro bodies correlative to the Wandsworth gabbro.

The Avalon Zone exhibits a suite of late Precambrian I-type plutons (Group 2a) of which the Holyrood igneous suite (A1, Fig. 8.6), high-level biotite-hornblende granite and granodiorite with local dioritic and gabbroic phases, may be taken as a typical example. The suite intrudes roughly coeval Upper Proterozoic bimodal volcanic rocks (Strong and Minatides, 1975). Its debris is found in Upper Proterozoic to lower Paleozoic sedimentary rocks (McCartney et al., 1966). The suite is not foliated, but locally exhibits brecciation accompanied by intense alteration which developed a commercially exploited deposit of pyrophyllite in the host rocks. Krogh et al. (1988) obtained a U-Pb zircon age of 620 Ma for the Holyrood igneous suite. Similar plutons in the Burin Peninsula give slightly younger ages (590 Ma, Dallmeyer et al., 1981) and exhibit significant deformation, probably of Devonian age.

In the western part of the Avalon Zone stratified Precambrian rocks are intruded by small plutons of gabbro, alkali granite, peralkaline granite, and related hybrid rocks (O'Brien et al., 1984). These plutons are spatially associated with, and in some cases compositionally resemble, peralkaline volcanic rocks which yield U-Pb zircon and Rb-Sr whole rock and mineral ages of 540 to 570 Ma (O'Brien et al., 1988). The Cross Hills igneous suite (A8, Fig. 8.6; O'Brien et al., 1984; Saunders and Tuach, 1989) consists of gabbro, diorite, biotite granite and peralkaline granite, with high level granophyric, miarolitic granite phases net veining mafic phases. The Powder Horn igneous suite (A10) is described as dioritic (McCartney, 1967), and the Louil Hills pluton (A17) is a peralkaline leucogranite (Strong et al., 1974).

The western part of the Avalon Zone contains numerous dykes and minor plutons of gabbro (Spread Eagle gabbro) thought to have served as feeders to Cambrian volcanic rocks (McCartney, 1967). Similar Cambrian basaltic volcanism occurs in Avalonian terranes in southeastern Cape Breton, the Antigonish Highlands (Murphy et al., 1985), and southern New Brunswick (Currie, 1988).

Two Devonian(?) plutons, Red Island and Iona Island (A15, A16), occur in Placentia Bay. The Iona Island igneous suite (McCartney, 1967) comprises olivine gabbro, augite-biotite monzodiorite and massive syenogranite. Salic members intrude more mafic ones with sharp contacts, and formation of igneous breccias. The similar Red Island igneous suite intrudes Cambrian strata with spectacular development of skarn and breccia (Hogg, 1954). The age of these complexes is uncertain. They resemble early Devonian bimodal complexes of Group 8 more than late Devonian evolved granites of the Ackley type (Group 9b).

A large, northeast-trending tholeiitic dyke of Triassic age crosses the southeastern corner of the the Avalon Peninsula and has been traced for some kilometres offshore by geophysical means (Papezik and Hodych, 1980). This dyke is probably correlative to, and may be continuous with, the Shelburne Dyke of southern Nova Scotia (Papezik and Barr, 1981; Greenough, Chapter 6, this volume).

Humber, Gander, and/or Avalon zones of Cape Breton Island

Distinction of tectonostratigraphic zones in northwestern Cape Breton Island is difficult and uncertain. Elements of the Humber, Gander and Avalon zones are probably all present. Middle Proterozoic rocks, including severely deformed anorthosite plutons (Barr et al., 1987), correlative to parts of the Humber Zone of Newfoundland underlie a small region on the northwest tip of Cape Breton Island. Barr et al. (1987) considered the Wilkie Brook Fault separated

this region from the rest of the highlands, but Currie (1986) presented evidence suggesting a marble-quartzite sequence farther south shared the granulite facies metamorphism of these rocks, and unconformably underlay Silurian supracrustal strata. High-grade granitoid gneisses in the region between the Wilkie Brook Fault and Eastern Highlands Shear Zone give Silurian U-Pb zircon dates and Devonian metamorphic dates, whereas little deformed plutons commonly give early Devonian dates (Jamieson et al., 1986; Dunning et al., 1990). This region (Fig. 8.6) may correlate with the Gander Zone of Newfoundland, but a late Proterozoic,

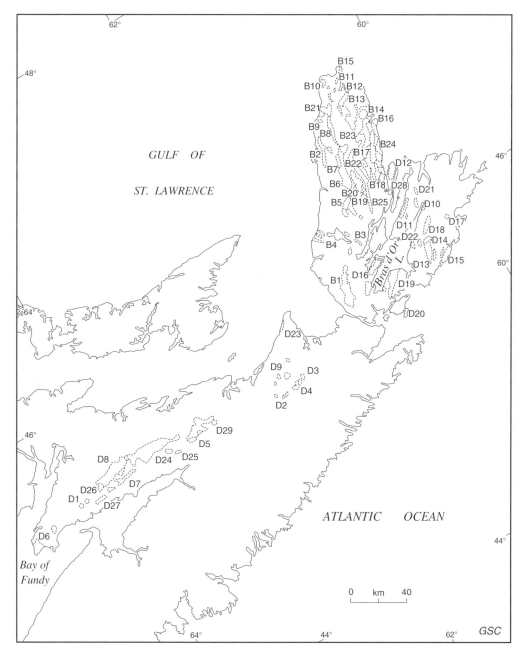

Figure 8.6. Index of plutons of Cape Breton Island and northern Nova Scotia. Letter-number identifiers are keyed to Table 8.6. B = northwestern Cape Breton Island; D = Southeastern Cape Breton Island and northern Nova Scotia.

cataclastically deformed pluton is also present (Cheticamp pluton B2, Fig. 8.6). Southeast of the Eastern Highlands Shear Zone plutonism is entirely of late Precambrian age (555-560 Ma), and probably lies within the Avalon Zone, although the character of the host rocks is unusual.

The Lowland Cove syenite (B10, Fig. 8.6), a pink to red rock consisting mainly of perthitic alkali feldspar with clinopyroxene and amphibole, gave a U-Pb zircon metamorphic age of 1040+10/-40 Ma, and an emplacement age of 1100 to 1500 Ma (Barr et al., 1987). A high content of Zr and local relicts of nepheline suggest alkaline affinities for this pluton. The rocks range from massive to strongly foliated but are pervasively recrystallized. The southeastern limit of Grenvillian ages is poorly known but may lie between the Wilkie Brook and Aspy faults. The Cheticamp pluton (550 Ma, Jamieson et al., 1986) may intrude a Grenville host, although this is very uncertain. The Cheticamp pluton forms an elongate mass which contains amphibole and biotite in the northern part (Currie, 1986) but biotite and muscovite in the southern part (Barr et al., 1986). Wall rocks around the pluton were metamorphosed to 650°, 9 kilobars at the time of emplacement, leading to minor anatexis (Currie, 1987b). Possibly admixture of an anatectic component explains the mixed character.

East of the Aspy Fault the volumetrically predominant granitoid gneisses are of Silurian age, but Upper Precambrian rocks are also present: North Branch Baddeck River diorite (B19) 614 Ma (Jamieson et al., 1986), New Glen megacrystic granite 708 Ma (Barr et al., 1988). The Belle Cote Road (B6, 433 Ma, Jamieson et al., 1986) and similar bodies between the Aspy Fault and the Eastern Highlands Shear Zone consist of foliated biotite-muscovite granite and were originally thought to form part of a basement complex. This belt between the Aspy Fault and the Eastern Highlands Shear Zone contains no volcanic rocks or high-level plutons. It may reflect an uplifted block of infrastructure, as suggested by the occurrence of volcanic rocks, and diorite plutons of Silurian age to the west (Currie et al., 1982b; Barr et al., 1988). However A.L. Sangster (pers. comm., 1990) has pointed out that Pb isotope data from this block suggest a juvenile Silurian source.

East of the Eastern Highlands Shear Zone voluminous elongate tonalite to granodiorite plutons gave U-Pb ages on zircon of 555 to 565 Ma, (Barr et al., 1988; Dunning et al., 1990). Older rocks appear to be present at the southern end of this block. Dioritic rocks from Kelly's Mountain (D28) gave a Rb-Sr age of 636 ± 69 (Gaudette et al., 1985), and intrude a host with a poorly defined age of metamorphism of about 750 Ma.

The Cape Smoky pluton (B16) of massive biotite leucogranite in the Bras d'Or Subzone has recently given a U-Pb zircon age of 496 ± 5 Ma, almost identical to the metamorphic age of 493 Ma for amphibolite from Kelly's Mountain (Dunning et al., 1990). This age is similar to that of the younger part of the Roti granite in Newfoundland (Dunning and O'Brien, 1989). The Creignish Hills igneous suite (B1) in the southwest highlands of Cape Breton Island consists mainly of massive biotite granite (Barr et al., 1982). White et al. (1990) interpreted this complex to be composite with a younger portion emplaced about 500 Ma into an older granitic rock probably emplaced about 550 Ma.

Silurian-Devonian plutons are not found east of the Eastern Highlands Shear Zone, although the Cameron Brook pluton (B14; 403 Ma), a posttectonic megacrystic granite (O'Beirne-Ryan et al., 1986) lies between two splays of the shear zone, and the White Point-Black Brook-Warren Brook-Clyburn Brook igneous suite (B13, sheeted to migmatitic muscovite-biotite granite, U-Pb zircon age of 375 ± 5 Ma (Dunning et al., 1990)) straddles the zone. These ages are almost the same as that of gneisses a few kilometres to the north and west (Neills Harbour gneiss 403 ± 3 Ma, Cheticamp Lakes gneiss 396 ± 2 Ma, and similar to metamorphic ages to the west (380-395 Ma, Jamieson et al., 1986; Barr et al., 1988). A slightly younger suite of plutons (Salmon Pool, Margaree, Gillanders Mountain B7, B8, B5, Fig. 8.6) west of the Eastern Highlands Shear Zone have high K contents, are commonly megacrystic (Fig. 8.7), and commonly possess bimodal volcanic correlatives of Late Devonian to Early Carboniferous age. Some plutons of this group give Rb-Sr dates as young as 330 Ma, but stratigraphy and zircon dating suggest that an age range of 350 to 365 Ma is more reasonable, since stratigraphic observations show that debris from the Margaree pluton appears in the Tournaisian Horton Group.

Extensive mapping and dating make it clear that the part of the Cape Breton Highlands southeast of the Eastern Highlands Shear Zone is little affected by Silurian or Devonian thermal activity. However the continuity of plutonism at about 560 Ma suggests both sides originally formed parts of an Avalonian block which has been dissected by a (post-Devonian) thrust or transcurrent faults.

Avalon Zone of southeastern Cape Breton Island and northern Nova Scotia

Crystalline basement in the Cobequid Highlands gave a Grenvillian Rb-Sr age of doubtful quality (934 ± 82 Ma, Gaudette et al., 1983). Crystalline basement is not recognized in the Antigonish Highlands or southeastern Cape Breton Island, but a Middle Proterozoic or younger marble-quartzite sequence occurs in southeastern Cape Breton Island (George River Group) and in highly deformed remnants in the Cobequid Highlands (Gamble Brook Formation, Murphy et al., 1988). Upper Precambrian volcanic-sedimentary sequences occur in all three regions.

The oldest dated plutons are of late Precambrian age. The oldest plutons in the Cobequid and Antigonish highlands are hornblende-rich bodies of appinitic affinities, varying from hornblendite to diorite, but commonly with some late, cross-cutting leucogranite phases (Greendale, D23, Eden Lake, D2, Murphy et al., 1988; Jeffers Brook, D1, and correlative plutons). This type of pluton, which has not been recognized in Cape Breton Island, gives ages greater than 600 Ma (Group 3a). In southeastern Cape Breton Island, biotite±hornblende-bearing suites exhibit a continuous range from diorite to monzogranite or leucogranite and were emplaced at relatively high levels (Sporting Mountain, Coxheath Hills, Chisholm Brook, D19, D10, D13, and correlative plutons, Barr et al., 1984) and may grade to volcanic members. Some of these plutons are associated with copper mineralization, and one (Capelin Cove, D15) is strongly deformed. Rb-Sr ages range from 530 to 575 Ma (Group 3a), but all of the plutons appear to be Precambrian, since in some cases they are unconformably overlain by Cambrian strata. Similar plutons are present in the Antigonish (Ohio, D3) and Cobequid highlands, but they are poorly exposed, and strongly deformed in the Cobequid Highlands.

Cambrian magmatism is known in southeastern Cape Breton and the Antigonish Highlands from the presence of mafic volcanics, and Murphy et al. (1985) have described minor sills and stocks of this age in the Antigonish Highlands. Minor intensely altered syenitic rocks intrude the mafic volcanics, and are inferred to be differentiation products (Murphy et al., 1991).

Silurian magmatism is not recognized in southeastern Cape Breton, but is abundant in the Antigonish and Cobequid highlands (see Chandler et al., 1987 for review). However few, if any, plutons give reliable ages in this range. The Indian Lake pluton in Antigonish Highlands gave an age of 432 ± 18 Ma by K-Ar on biotite (Wanless et al., 1966) but further work suggests that it is probably late Precambrian (Murphy et al., 1991). Some undeformed biotite granites in southeastern Cape Breton Island may be of this age, but in the absence of radiometric dates they cannot be distinguished from Devonian granites.

Devonian to Carboniferous plutons in Cape Breton Island tend to mimic the character of older granitoid rocks so that the presence of a Devonian porphyry in the Loch Lomond igneous suite was not realized prior to dating (Barr et al., 1984). Two groups of Devonian plutons may be present. I-type plutons with considerable composition range and associated Cu mineralization (Gillis Mountain, Deep Cove D17, Fig. 8.6) appear to be early Devonian (395 Ma) compared to late Devonian or Carboniferous ages for the evolved granite plutons such as Gunn Lake (D4) in Antigonish Highlands and Hart Lake-Byers Lake (D8) and others in the Cobequid Highlands (Pe-Piper et al., 1989). The latter bodies consist of syenogranite commonly containing

sphene and amphibole. Many of these plutons lie on or very close to major faults. In the Cobequid Highlands some of these bodies have given ages as low as 315 Ma. Comparison with other similar bodies suggests that these ages have been disturbed by faulting and fluid circulation, and that emplacement ages are more likely to be in the range 345 to 370 Ma, with the possible exception of Pleasant Hills for which Doig et al. (1991) have reported an imprecise U-Pb zircon age of 315 Ma.

Meguma Zone of southern Nova Scotia

No crystalline basement is known in southern Nova Scotia. Krogh and Keppie (1988) reported detrital Precambrian zircon populations with ages of 580 to 600 Ma and 1945 to 2700 Ma from the Cambrian Ordovician sediments of the Meguma Supergroup which underlies most of this region (Fig. 8.8). The former age could indicate derivation of some of the sedimentary pile from Avalonian terranes. Giles and Chatterjee (1987a,b) have recently reported gneiss thought to be remobilized basement in the Liscomb igneous suite, and inclusions of granulites brought up in a mafic dyke at Tangier. Clarke and Chatterjee (1988) reported that Sm-Nd isotope systematics suggest that the Liscomb gneisses could be similar to sources of granitoid plutons in this region. These observations suggest that Precambrian crust underlies this region and that some information can be gained about it from surface sampling.

The oldest known pluton is strongly foliated to mylonitic two-mica granite of Middle Ordovician age (Brenton 460 Ma, Krogh and Keppie, 1984) at the southwest corner of the region. Bimodal volcanism of similar age occurs in the nearby White Rock Formation. The extent of this Ordovician magmatism is not well understood either in space or time.

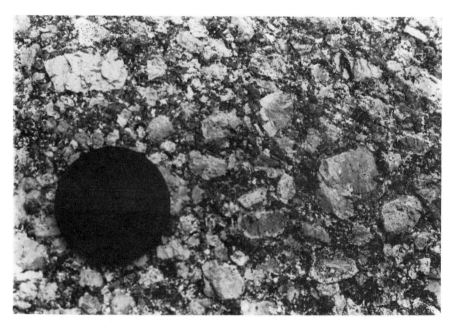

Figure 8.7. Megacrystic granite, Margaree pluton, Cape Breton Island. Oligoclase (pale) rims microcline (rapakivi texture), but some megacrysts are roughly euhedral whereas others are amoeboid and strongly poikilitic. This pluton has a volcanic equivalent (Fisset Brook Formation). GSC 188590

Roughly one third of the Meguma Zone is underlain by peraluminous Devonian plutonic rocks ranging from biotite tonalite through leucocratic granodiorite to monzogranite and granite. Age subdivisions, if any, of these rocks are not well established. Dallmeyer and Keppie (1987) dated major deformation and metamorphism at 385 to 415 Ma. Some dioritic to tonalitic plutons are syn-orogenic. More salic plutons are post-orogenic. Intermediate to mafic plutons are comparitively rare in southern Nova Scotia. The Barrington Passage pluton (E10, Fig. 8.8), on the south tip of Nova Scotia, is a moderately to strongly foliated tonalite (de Albuquerque, 1977) formed of zoned plagioclase (An$_{42-26}$), biotite, quartz and accessory allanite and apatite. The pluton lies within sillimanite-grade metamorphic rocks. Reynolds et al. (1987) reported an age of 385 Ma for the Barrington Passage pluton which would be consistent with syntectonic emplacement. Similar deformed tonalite and diorite occur in small amount on the eastern tip of Nova Scotia (Hill, 1991). U-Pb zircon ages for these occurrences are 378 to 381 Ma (Hill, 1991). Mafic rocks of uncertain age have also been reported from the Liscombe igneous suite (E13; Giles and Chatterjee, 1987a) and the Port Mouton igneous suite (E12; Hope et al., 1988).

The South Mountain batholith (E1, Fig. 8.8; McKenzie and Clarke, 1975), the largest igneous mass in the Appalachian Orogen, comprises numerous plutons, some of which have now been studied in considerable detail (Corey, 1987; MacDonald and Horne, 1987). The older parts of the batholith consist of leucocratic biotite granodiorite (roughly 33% each of quartz and andesine (An$_{30-35}$) and 14% each of biotite and potash feldspar). Muscovite and garnet appear sporadically in accessory amount, apparently produced by some combination

of assimilation and alteration. Numerous irregular to ovoid bodies of two-mica monzogranite intrude or fringe the granodiorite, and form about 25% of the exposed area of the batholith. These plutons consist of roughly 33% each of quartz and potash feldspar, 25% oligoclase to andesine (An$_{20-32}$), 5 to 10% biotite, and lesser but ubiquitous muscovite. Magmatic garnet, cordierite and andalusite occur locally (Clarke et al., 1976). Swarms of aplite, porphyry, and pegmatite dykes, locally containing garnet, fluorite, topaz, and tourmaline, cut or fringe the monzogranite bodies and pass into extensive zones of greisen and complex pegmatites mineralized with U, Sn, Mo, W, and Be (Kontak, 1987). Contacts of the batholith with surrounding supracrustal rocks are marked by well-developed contact aureoles of andalusite-cordierite type. Ghost stratigraphy defined by persistent trains of inclusions of Meguma Supergroup lithology can be traced into the pluton, in some cases for kilometres, and appear to be an important control of mineralization. Rb-Sr measurements on units of the South Mountain batholith suggest a narrow range of emplacement ages from 374 Ma for the granodiorite to 361 Ma for the monzogranite (Clarke and Halliday, 1980), with moderately high initial Sr ratios of 0.7079 to 0.7102. Younger ages with high initial Sr ratios have been reported at the southwest end of the batholith (Davis Lake, E22, Richardson et al., 1988), but these date the extensive hydrothermal activity documented by Dallmeyer and Keppie (1987), as the chemistry of the Davis Lake pluton resembles other monzogranites of the batholith (Kontak, 1987), and nearby Ar-Ar ages give "normal" values (Zentilli and Reynolds, 1985).

Whole rock ^{18}O values for the South Mountain batholith range from 10.1-12.0 mL^{-1} (Longstaffe et al., 1980) suggesting anatexis of ^{18}O-rich clastic sedimentary rocks, as indicated by petrographic observations and disturbed Rb-Sr systematics, played a part in the origin of the South Mountain batholith, but REE data (Muecke and Clarke, 1981) and isotope systematics indicate that the surrounding Meguma Supergroup cannot be the only source. The South Mountain batholith forms a classic S-type granite suite (Table 8.5). The plutons lithologically resemble those of the slightly younger Hercynian granite province of western Europe.

Plutons lithologically similar to the South Mountain batholith occur to the south and east. To the east, a chain of plutons, some of batholithic dimensions, stretches 200 km to the eastern tip of Nova Scotia. Although most plutons resemble the South Mountain, Liscomb contains gneissic material, possibly relicts of reworked basement (Giles and Chatterjee, 1987a). The most easterly plutons contain a tectonic foliation related to movements on the Cobequid Fault (Hill, 1991). ^{40}Ar-^{39}Ar dating (Dallmeyer and Keppie, 1984) and zircon dating (Hill, 1991) suggest that emplacement and deformation took place essentially simultaneously at 360 to 370 Ma.

South of the South Mountain batholith, the Port Mouton pluton contains similar peraluminous granites, locally beryl-bearing, but also includes minor tonalitic and gabbroic phases (Hope et al., 1988). The Shelburne pluton resembles lithologically the South Mountain batholith, but is cut by isograds associated with the Barrington Passage pluton, suggesting an older age (Rogers, 1985). Unpublished age determinations are 455 Ma (Rb-Sr) and about 370 Ma (zircon). Highly disturbed and aberrant isotope

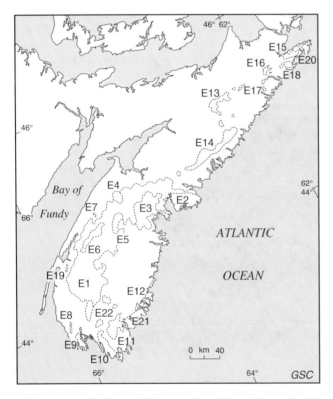

Figure 8.8. Index of plutons of southern Nova Scotia. Letter-number identifiers are keyed to Table 8.6.

patterns are common in this region. Permian Carboniferous metamorphism and hydrothermal activity has been documented by Dallmeyer and Keppie (1987) and Reynolds et al. (1987). This activity appears to be fault related. The only pluton for which there is strong evidence of late Carboniferous or younger emplacement is Wedgeport (E9, Cormier et al., 1988) for which U-Pb zircon dating suggests an emplacement age of 323 Ma.

Triassic tholeiitic volcanics outcrop along the Bay of Fundy (Greenough, Chapter 6, this volume). A large tholeiitic dyke or dykes of similar age trends northeast from Shelburne (Papezik and Barr, 1981) along the Atlantic coast. The rocks are poorly exposed, but magnetic anomalies suggest that the dyke(s) could be more than 100 m wide and persistent along strike for tens or even hundreds of kilometres. Observation and geophysical interpretation suggest that a single large body is more likely than a swarm of smaller dykes.

Dunnage Zone of Quebec

Gaspésie and the Eastern Townships of Quebec (Fig. 8.9) exhibit an intricate network of northwesterly transported nappes, including ophiolitic slices. Small slivers of Precambrian

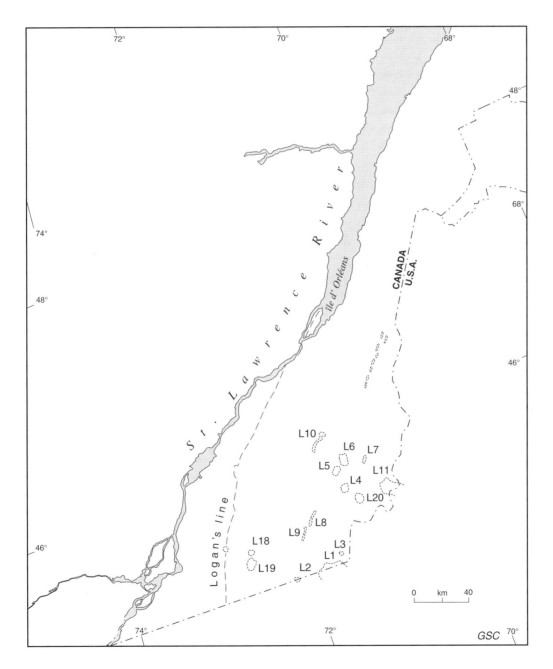

Figure 8.9. Index of plutons in the Eastern Townships region. Letter-number identifiers are keyed to Table 8.6.

(Grenville) crystalline basement occur in transported slices in this belt (Vallieres et al.,1978) and seismic basement thought to be Grenvillian is continuous under all but the southern edge (Stewart et al., 1986). A different type of Precambrian basement (Chain Lakes type) also occurs in a few slices (Cousineau, 1988).

The oldest known plutons comprise subvolcanic equivalents of calc-alkaline volcanics of the Ascot Formation which form elongate bodies of metatrondhjemite near Sherbrooke (Moulton Hill and Salmon River plutons, L8, L9; St-Julien, 1970). A Rb-Sr "errorchron" on the Moulton Hill body gave 505 Ma with an initial Sr ratio of 0.710 (Poole, 1980a).

Granitoid bodies are associated with ophiolitic rocks in the Thetford Mines region (Riordon, 1954). Sinuous pods of strongly deformed biotite granodiorite and biotite muscovite granite may be up to a kilometre in length. One body gave a K-Ar age of 485 Ma and a Rb-Sr isochron age of 466 Ma with an initial Sr ratio of 0.717 (Poole, 1980c). These bodies clearly did not originate in their ophiolitic surroundings, but have been tectonically interleaved and transported (Laurent et al., 1984). They resemble in age and petrography some granites from the Miramichi region of New Brunswick.

Early Silurian gabbroic rocks, apparently of island arc affinity, have been reported in the Cabano region (David and Gariepy, 1990). These rocks have some affinities with the suite of similar age reported by Currie and van Berkel (1989) from western Newfoundland.

Two clusters of Devonian plutonic rocks occur, one along the United States border in the Eastern Townships, and the other in eastern Gaspésie. Porphyry dykes and stocks occur in the Etchemin region south of Québec City, and some hydrothermal alteration zones in southern Gaspésie (for example Patapedia (Williams-Jones, 1982)) probably represent a surficial manifestation of plutons at depth.

A group of eight Devonian plutons, as well as numerous porphyry or rhyolite dykes, have recently been studied by Bourne (1984, 1986a, b) and Danis (1984, 1985). They show close petrographic and chemical similarities. The Stanstead (Averill) pluton on the international boundary consists of coarse biotite granite which is remarkably uniform and free from inclusions, despite local roof pendants of the Silurian-Devonian Saint Francis Group (Erdmer, 1982). The Graniteville and Hereford Mountian stocks (L2, L3, Fig. 8.9), a few kilometres away, appear petrographically identical, and may be connected at depth. All have been quarried for dimension stone. The roughly circular Scotstown pluton (L4) passes from minor quartz gabbro through tonalite to sphene-bearing biotite granite. Like most of these plutons, sparse late pegmatites contain muscovite. Northeastern plutons (Aylmer, L5, Winslow, L6, St. Cecile, L7) grade from diorite or tonalite to biotite granodiorite and granite, contain sphene and allanite and display a marked foliation parallel to tectonic grain. The Spider Lake pluton (L11) consists of several poorly exposed lobes stretching southward into Maine, each lobe displaying zonation from mafic rim to more salic core. All plutons show a narrow range of SiO_2 from about 68 to 75%, negative slope on REE patterns without a Eu anomaly, and low lithophile element contents. All exhibit contact aureoles up to 1.5 km wide, suggesting emplacement at about 3 kbars, and cut Acadian

structures. K-Ar dates fall in the range 380 to 360 Ma (Lowden, 1960), and U-Pb zircon ages in the range 384 to 374 Ma (Simonetti and Doig, 1990).

The cluster of plutons in Gaspésie consists of the large Mont McGerrigle pluton (L15) and numerous smaller high level plutons. The Mont McGerrigle pluton intrudes allochthonous Cambrian-Ordovician volcanic and sedimentary rocks. A hornfels aureole up to 1.5 km wide surrounds the pluton. The southern part of the pluton consists of reasonably homogeneous coarse amphibole granite whereas the northern part forms an intricate assemblage of hybridized rocks varying from syenite through granodiorite to diorite and gabbro (de Romer, 1977; Whalen and Gariepy, 1986). Dykes and small plugs of riebeckite-aegirine nepheline syenite also occur. The pluton exhibits superb examples of complex magma mixing processes probably involving three components (Whalen and Gariepy, 1986). A systematic K-Ar study by Whalen and Roddick (1987) which considered earlier work, suggests an emplacement age of about 390 Ma, but continuing high temperature interaction for 10 to 15 Ma.

Small stocks to the south of Mont McGerrigle consist of fine grained porphyries with oligoclase and corroded quartz phenocrysts in a matrix of quartz, alkali feldspar, plagioclase, biotite, and hornblende (Robert, 1966a, b) emplaced in the Lower Devonian Grande Grève Formation. Similar small stocks occur as far east as Murdochville. The Mont Breche igneous suite (L14), 15 km south of Mont McGerrigle, forms a high level composite porphyry pluton with salic phases intruding and agmatizing mafic phases. K-Ar dating reported by de Romer (1977) suggests that all these bodies have Middle to Late Devonian ages of 365 to 355 Ma.

Alkaline central complexes of the Monteregian alkaline province (Brome, L19, and Mount Shefford, L18) consist of two series of rocks, namely a dominant slightly undersaturated pyroxenite-gabbro-diorite series, and a subsidiary strongly undersaturated nepheline diorite to foyaite series. Nordmarkite of uncertain affinity occurs at Shefford. Eby (1984) suggested the dominant series was emplaced about 136 Ma, whereas the nepheline-bearing rocks were derived from a different protolith and emplaced about 120 Ma. The contemporary Mont Megantic igneous suite (L20, Fig. 8.9), comprising a central granite surrounded by slightly undersaturated mafic rocks forms part of the White Mountains magma series (Bedard, 1985). Both Monteregian and White Mountain provinces include extensive dyke swarms (Bedard, 1985; Greenough, Chapter 6, this volume).

Gander and Dunnage zones of New Brunswick

No Precambrian rocks have been identified in New Brunswick north of the Avalon Zone, but common lead studies by Bevier (1987) suggest that Precambrian basement of Avalonian isotopic characteristics is present at depth, and zircons with late Precambrian ages typical of Avalonian terranes have been recovered from younger granitoids (Roddick and Bevier, 1988). Figure 8.10 indicates plutons in this region.

The oldest known plutons comprise variably foliated to gneissic granites of Group 5a which commonly exhibit sinuous shapes attributed to postfoliation folding. Deformation varies from mild, with preserved granophyric textures, to intense foliation and mylonitization (Whalen, 1987). Compositions also vary from local biotite-hornblende granodiorite with tonalite to gabbro phases through biotite granite to predominant muscovite-biotite granite in which much of the muscovite is metamorphic. Megacrystic texture occurs locally. The plutons were emplaced in lower to middle Ordovician greenschist-grade mafic and salic volcanics and sediments of the Tetagouche Group. Rb-Sr isochron ages for two granite bodies gave 479 and 484 Ma with initial Sr ratios of 0.703, (Fyffe et al., 1977; Fyffe and Cormier, 1979; Poole, 1980d), but U-Pb zircon ages are 470 to 434 Ma (Bevier, 1988). Cobbles of gneissic, muscovite-bearing granite occur in Silurian conglomerates indicating late Ordovician or early Silurian metamorphism.

Late Silurian plutons have recently been identified. The undeformed Mount Elizabeth igneous suite (M12, Fig. 8.11; Whalen, 1987) consists of flow-banded troctolite (Portage Brook troctolite), a peraluminous granite suite (homogeneous fine grained biotite granite with traces of muscovite) and an alkaline granite suite (one feldspar

amphibole granite, probably with relicts of fayalite). It gave a U-Pb age on zircon of 414+11/-1 Ma (Bevier, 1988). The North Pole igneous suite (M9), one of the largest intrusive bodies in central New Brunswick, consists of biotite granite, locally megacrystic, with minor amphibole-bearing granodiorite-tonalite and late, muscovite-biotite granite with relict cordierite (Whalen, 1987). The suite has a U-Pb zircon age of 417 Ma (Bevier, 1988). These large, essentially bimodal complexes are here considered to be Group 8 suites like Mount Peyton in Newfoundland.

Peraluminous Group 7 granites south of these large bodies appear to be slightly younger than the composite Group 8 plutons. The foliated muscovite-bearing Lost Lake pluton (M25) gave a K-Ar age of 422 Ma. At the south end of the central batholithic belt, the Pokiok batholith consists of older, foliated muscovite-biotite granite cut by younger, massive fluorite-bearing megacrystic biotite granite, some of it exhibiting flow foliation (Lutes, 1987). Minor tonalite occurs on the west side of the batholith (Venugopal, 1981). A broad metamorphic aureole, ranging up to several kilometres in width, surrounds the batholith suggesting efficient

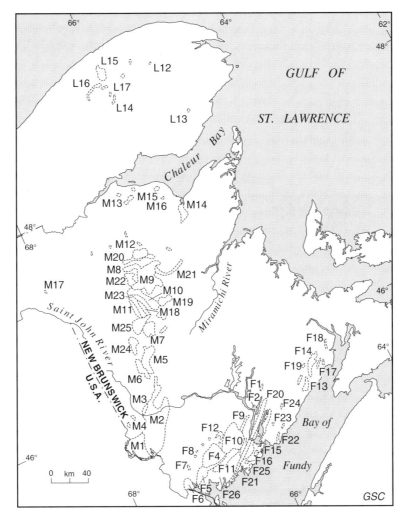

Figure 8.10. Index of plutons of New Brunswick and Gaspésie. Letter-number identifiers are keyed to Table 8.6. L = Gaspésie; M = northern New Brunswick; F = southern New Brunswick.

transport of heat, possibly due to convecting fluid. U-Pb zircon data show a range in time of emplacement from 415 Ma for the tonalite to 402 Ma for the latest granite (Bevier, 1988). Rb-Sr isotope data are highly disturbed (McCutcheon et al., 1981).

Two types of mid to late Devonian plutons occur in central New Brunswick, namely very evolved I-types transitional to A-type (Group 9a, Burnt Hill pluton and related bodies, Taylor et al., 1987; Whalen, 1986), and mid-Devonian, high-level composite suites (Charlo stocks).

Relatively equant, undeformed biotite granites with major rapakivi phases and minor aplitic to porphyritic phases occur at Burnt Hill, Dungarvon, and Trout Brook (M7, Fig. 8.11). These plutons exhibit narrow contact aureoles, and greisen-type mineralization with tin and tungsten. ^{40}Ar-^{39}Ar dating suggests an age of 381 Ma for Burnt Hill (Taylor et al., 1987). The mineralogy and chemistry resemble A-type granites (Whalen, 1986) but the rocks are more probably highly evolved I-types (compare Whalen et al., 1987).

The Charlo stocks (M13; Greiner, 1970; Stewart, 1979) of northern New Brunswick and the Mont Breche igneous suite (Robert, 1966a) of southern Gaspésie both consist of intimate mixtures of high-level salic and mafic rocks which intrude, brecciate, and hybridize each other in complex fashion. Both Mont Breche and Charlo exhibit sub-economic copper mineralization. The Benjamin River locality of the Charlo stocks returned a K-Ar age of 370 Ma (Stewart, 1979). These plutons are assigned to Group 9b.

A large north-northeast-trending Triassic tholeiitic dyke complex occurs in northeastern New Brunswick (Caraquet dyke, Burke et al., 1973; Greenough, Chapter 6, this volume). Outcrop is poor and most of the information comes from geophysics which suggests that en echelon

segments are up to 100 m wide and tens of kilometres long. Numerous other mafic dykes of Carboniferous or younger age occur in central New Brunswick.

Avalon Zone of New Brunswick

The Fundy coast of New Brunswick exhibits a Precambrian sequence of marble and quartzite (Green Head Group) generally in fault contact with tonalitic to granodioritic orthogneiss (Brookville Gneiss) emplaced at 605 Ma (Bevier et al., 1990), but containing "Grenvillian" inheritance (Olszewski et al., 1980; Currie et al., 1981). A major period of magmatism took place about 620 to 600 Ma (Watters, 1987; Bevier, 1988). These Group 2a plutons range from diorite and tonalite to granodiorite and granite, and form a co-magmatic differentiated I-type suite. Spectacular mixing and comingling textures occur between salic and mafic phases (Fig. 8.11). These plutons, typically elongate along the northeast tectonic grain, cut approximately coeval, presumably cogenetic volcanic rocks.

A slightly younger period of late Precambrian magmatism emplaced mafic to ultramafic plutons (Duck Lake, F23, Mechanics Settlement, F19), coarse granodiorite to granite, and distinctive red evolved granites with distinctive high U and Th contents. These Group 3a plutons, which exhibit transitions to associated bimodal volcanics and redbeds, give U-Pb zircon ages of 550-565 Ma (Currie, 1988; Bevier, 1988), and the redbeds are conformable with lowermost Cambrian strata.

Barr and White (1988) argued that the Brookville gneiss and Green Head Group formed a separate terrane (Brookville terrane) not related to the other rocks (Caledonia terrane). Currie et al. (1981) suggested that the "Brookville terrane" formed an infrastructure to the

Figure 8.11. Remobilized granitic rocks, Fairville pluton, New Brunswick. Note mafic dykes broken up and strewn through the matrix, as well as fragments of basement gneiss (medium grey) and an older aplitic granite. GSC 203034-Q

"Caledonia terrane". Recent geochemical and isotopic data (Whalen et al., 1994) show the Brookville gneiss to have an essentially identical, juvenile isotopic signature to the Group 2a plutons of the Caledonia terrane as well as similar age.

A major period of early Silurian magmatism northwest of Saint John emplaced a large, bimodal sheeted dyke swarm (Kingston dyke complex, F26; Currie, 1988; Eby and Currie, 1993) which stretches for more than 100 km with an average width of 4 km. The mafic dykes have tholeiitic composition, whereas the salic ones range from dacite to rhyolite. U-Pb zircon dating gave an age of 435 Ma (Doig et al., 1990), approximately the same as Ar-Ar dating on metamorphic muscovite on shear planes (Currie, unpublished data). The Saint George batholith and the contiguous Bocabec (F5) and Saint Stephen plutons (F6) (Fyffe et al., 1981) are now known to represent a complex and long-lived period of magmatism lasting from early Silurian to late Devonian (Currie, 1988). The plutons developed along northeast-trending, northwest-side down, normal faults. The oldest plutonic members include the late Silurian Welsford (F9; Payette and Martin, 1987) and Jake Lee Mountain (Currie, 1988) complexes, which probably originally formed part of a single linear body. Both are high level complexes, with volcanic equivalents. Welsford, includes peralkaline granite, mildly alkaline mafic material, mainly as dykes, and a variety of mixed phases. These plutons and the Kingston igneous suite belong to Group 6. The Saint Stephen pluton (Dunham, 1959; Paktunc, 1987) comprises peridotite, troctolite and gabbro which form a roughly equant intrusion about 2 km across. A K-Ar age of 401 Ma was obtained by Wanless et al. (1973). The Bocabec igneous suite (Pajari et al., 1974) forms a relatively thin sheet (Thomas and Willis, 1988) which displays spectacular abrupt transitions from gabbro to granodiorite. It gave a Rb-Sr isochron age of 394 Ma, with a relatively high initial ratio of 0.707 (Fyffe et al., 1981). The granitic part of the Saint George batholith is now known to comprise at least three plutons. The oldest phase, the equigranular biotite granite of the Magaguadavic pluton (F4), was emplaced at 395 Ma (U-Pb zircon, Bevier, 1988), while the youngest, the Mount Douglas (F10) was emplaced at 367 Ma (^{40}Ar-^{39}Ar, McLeod, 1990). The Saint Stephen, Bocabec, Magaguadavic and Utopia (F11) plutons are interpreted to form a suite of Group 8 plutons in southern New Brunswick, whereas the Mount Douglas pluton belongs to Group 9b. Rapakivi biotite granite of the Mount Douglas intrudes Late Precambrian granite (Currie, 1988) and Ordovician-Silurian sedimentary rocks. Rock types within it appear to be partially controlled by the Pendar Brook Fault, which it seals (McLeod, 1990). All the major plutons appear to be fault-controlled, but are undeformed. A number of minor satellite plutons to the northwest (Mount Pleasant, F12, Tower Hill-Pleasant Ridge, F7, Sorrell Hill-Beach Hill, F8) are significant because of associated mineralization.

Triassic mafic dykes (Group 10a) have been reported from southern New Brunswick, and Triassic basalts occur on Grand Manan Island (Greenough, Chapter 6, this volume). Re-examination of other reported dyke occurrences suggest that most are probably of pre-Late Devonian age.

NATURE AND ORIGIN OF MAGMATIC EVENTS

According to current petrological ideas, the origin and tectonic significance of plutonic rocks can be correlated (Pitcher, 1983; Pearce et al., 1984; Brown et al., 1984), but attempts to apply broad-scale classification to plutons of the Canadian Appalachians have not been very successful. The original definition of I- and S-type granite suites (White and Chappell, 1983) was based on narrow, persistent belts whose relations could be readily perceived by field mapping, petrographic, or chemical criteria. With the exception of the S-type suite of southern Nova Scotia and the A-type suite of the Topsails, such belts do not exist in the Canadian Appalachians where different types of plutons are intermingled. Attempts to classify by chemical discrimination diagrams fare even worse. A collection of 427 granite samples from a transect across the orogen in New Brunswick and Gaspésie plotted mainly in the within-plate field of Pearce et al. (1984) but close to the "triple junction" with the volcanic arc and collisional margin fields (Whalen, 1988). Such a result lacks utility.

Current research on origin of granites emphasizes isotopic data to "image" the source. Large volumes of such data are just beginning to appear for the Canadian Appalachians. Ayuso and Bevier (1991) argued on the basis of Pb isotope data that Pb in granites of New Brunswick and Gaspésie was derived from "Grenville-like" crust in the Miramichi and northwest, and from "Avalon-like" crust to the southeast with a relatively narrow zone of mixing in the middle. However, considering Pb, Nd, and O data for many of the same plutons, Whalen (1993), showed that plots of Pb versus O and Nd versus O tend to fall in linear arrays suggesting one component with light O, non-radiogenic Pb and heavy Nd, and another with heavy O, more radiogenic Pb, and light Nd. The first component clearly is of mantle origin, and appears to be predominant in most of the Humber, Dunnage and Avalon zone samples. The second component can plausibly be identified as re-melted sedimentary rock, or weathered material, and is predominant in the Gander Zone.

The youngest igneous rocks (Group 10, Triassic to Cretaceous) are not confined to any obvious terranes, and trend at low (Triassic) to high (Cretaceous) angles to older tectonic trends. Triassic igneous rocks are tholeiitic basalt, whereas younger rocks include alkali basalt and strongly alkaline mafic rocks, as well as nepheline diorite, syenite, and granite (Eby, 1984, 1985). Triassic basaltic rocks have been ascribed to rifting of the craton associated with opening of the Atlantic Ocean (Williams, 1979). Monteregian magmatism has been ascribed to "hot-spot" activity not directly related to ocean formation (Eby, 1985; Currie et al., 1986), but Bedard (1985) claimed a connection of White Mountains activity with opening of the Atlantic. Both Triassic basalts and plutons of the Monteregian and White Mountain series are clearly mantle derived, and contaminated only slightly by crustal material (Eby, 1985), with the possible exception of granite at Mont Megantic.

Post-Mississippian but pre-Mesozoic plutonism is rare in the Canadian Appalachian region, although dykes and flows occur in Pennsylvanian strata of southern New Brunswick. Rb-Sr, K-Ar and Ar-Ar dates in the range 320 to 220 Ma have been reported for granitic rocks from southwest Nova Scotia and the Cobequid Highlands. Many (all?) of these dates have been reset by hydrothermal activity along cataclastic zones (Kontak, 1987). Only the Wedgeport pluton (E9; 323 Ma, Cormier et al., 1988) may be equivalent to Permian-Carboniferous granitic plutonism of the northeastern United States. The widespread presence of thermal effects in the southern Bay of Fundy region (Dallmeyer and Keppie, 1987) suggests that other such plutons could be nearby, perhaps offshore.

Late Devonian to Mississippian granites (Group 9b) are connected with large scale wrench and normal faulting, which in some cases controls distribution of facies, even though the igneous rocks are massive and undeformed. The Ackley batholith (C23) seals the Dover-Hermitage fault in Newfoundland, fixing a minimum age for the juxtaposition of the Gander and Avalon zones in Newfoundland. Granites of similar age in southwest Newfoundland (Strawberry, S7, Isle aux Morts Brook, S7) also appear to postdate major faults. The South Mountain batholith may seal the "Tobeatic Fault" (Giles, 1985), and emplacement of related plutons to the east is controlled by the Cobequid Fault. The Saint George batholith is controlled by and seals a series of faults which separate the Avalon and Gander zones of southern New Brunswick (Currie, 1988). Group 9b plutons tend to be high-level, high-silica potash-rich bodies, associated with coeval volcanics. Rapakivi or megacrystic textures are common. They do not classify easily in standard schemes because they are transitional from evolved I- or S-types to A-types. Mineralization with lithophile elements (Sn, W, U) is virtually confined to groups 7b and 9. Devonian granites not obviously fault related, such as those of the Eastern Townships of Quebec, have low economic potential.

The source of the highly fractionated bodies of Group 9b remains debatable. Few of them contain significant mafic phases. Initial Sr ratios (0.705-0.707) suggest limited upper crustal involvement. Perhaps the contemporary Group 9a megacrystic plutons, which show observational evidence for metasomatism and anatexis (Pickerill et al., 1978) represent deeper equivalents of the high level Group 9b plutons. The peraluminous Devonian granites of southern Nova Scotia (Group 7b), which show many similarities with Group 9b, were derived in part from crustal peraluminous protoliths. The heat sources and heat transfer mechanisms required to sustain such large magmatic systems in the absence of major mafic magmatism remain unexplained. Accumulation of heat from crustal thickening during earlier deformation may be responsible (Pajari et al., 1981).

Early Devonian or Late Silurian composite plutons (Group 8) are common in Newfoundland and New Brunswick. Some marginal parts of Group 8 plutons such as Mount Peyton (N1), Mont McGerrigle (L15) and the older part of the Saint George batholith probably resulted from melting of deep-crustal material by mantle-derived mafic magma followed by emplacement of both mantle- and crustal derived components. The younger part of the Saint George batholith of New Brunswick suggests that such processes can trigger, or grade to, younger plutons with a crustal component. Contemporary two-mica granites on Cape Breton Island may represent a more indirect link to mantle activity where only heat has been supplied to crustal material. Deep seated composite early Devonian plutons appear to be devoid of mineralization, but high level examples in southeast Cape Breton Island and the New Brunswick-Gaspésie region exhibit Cu mineralization. The most characteristic ages for Late Devonian and

Early Devonian plutons are about 365 and 410 Ma respectively, but published determinations have a considerable range. It is presently unclear whether this range is real, or would be narrowed by improved age determinations.

Recognition of widespread and voluminous mid-Ordovician to Silurian magmatism (groups 5-7) has come with more precise dating methods. Plutons of this age occur throughout the Canadian Appalachians except in the Humber Zone but their character differs noticeably with position. In the Notre Dame Subzone of Newfoundland and the Gander Zone of New Brunswick, Silurian plutons cut deformed Ordovician plutons of orogenic character. In the Gander Zone of Newfoundland both Silurian and Ordovician plutons are variably deformed, and show no obvious distinction. In the Gander(?) Zone of northwestern Cape Breton Island, Silurian plutons in the central gneissic core are deformed whereas an Ordovician pluton in the less metamorphosed fringes is undeformed. In the Avalon Zone of southern New Brunswick, Silurian plutons are little deformed and of alkaline affinities (Chandler et al., 1987). Plutons of alkaline affinities (Group 6) probably result from mantle-driven deep crustal melting (Whalen et al., 1987), whereas plutons of Group 5a appear to be I- or M-type, and hence derived from the mantle, either directly or by relatively prompt remelting of mantle-derived material. The central deformed belt of groups 5b and 7a plutons contains a major crustal component, particularly in the case of the peraluminous plutons. The mechanism by which this component was incorporated into the plutons is uncertain. Field evidence shows that some plutons arose by anatectic melting of supracrustal rocks at relatively shallow depths (Currie and Pajari, 1981), and thermal modelling suggests that such anatectic melting could result from self-heating in collapsed and thickened sedimentary prisms (Pajari et al., 1981). Group 6 plutons seal normal faults along the periphery of the Avalon Zone in northern Nova Scotia and southern New Brunswick (Chandler et al., 1987; Currie, 1988) and show that penetrative late Silurian to Devonian deformation was insignificant in these regions.

Cambrian to early Ordovician plutonic rocks (Group 4) occur around Notre Dame Bay associated with relics of ocean floor, and are found in transported slices wherever similar rocks of the Dunnage Zone have been thrust over neighbouring zones. These dioritic to trondhjemitic plutons presumably represent differentiation and re-cycling of ocean floor and island arc material. Their limited distribution is noteworthy.

Late Proterozoic igneous rocks fall into three age divisions. The youngest (Group 3a, ages of 570-530 Ma) spans the Cambrian-Precambrian boundary, and basaltic volcanism locally continues into middle Cambrian time. Rock types range from leucogranite to melagabbro. In southern New Brunswick and Newfoundland magmatism is associated with transcurrent or transtensional faulting (Currie, 1988; O'Brien et al., 1988), and is bimodal with alkaline affinities, but in Cape Breton Island the rocks are more calc-alkaline (Barr and Raeside, 1986). Magmatic activity clearly postdates a slightly earlier subduction-related magmatic event, and appears to be related to crustal extension or transtension.

A-type plutons of late Precambrian age (Group 3b) occur in the Humber Zone of Newfoundland. These plutons are distinctly older than Group 3a (about 620-600 Ma). Alkaline plutons of similar age occur in the Canadian Shield adjacent to the Appalachian Orogen (Currie, 1976). Emplacement of these plutons probably accompanied an early stage of the opening of the Iapetus Ocean.

Plutons of Group 2 have been reported only from Avalonian terranes, and appear to be definitive of such terranes. Large plutons of Group 2a in the Avalon Peninsula and southern New Brunswick are classic I-type suites, and subduction related (Currie, 1988). These plutons give tightly constrained ages in the range 625-600 Ma. Group 2b plutons are, in general, not well dated, but dates range from 800 to 750 Ma, including a precise date on the Wandsworth gabbro (A18). High grade metamorphism and mafic dykes in New Brunswick and Cape Breton Island, and the presence of ocean-floor material in Newfoundland suggest Group 2b plutons may have been related to continental breakup. The three-fold division of Late Proterozoic magmatism into events at about 760, 620, and 560 Ma offers a distinctive criterion for recognizing Avalonian terranes which can be applied to terranes where the supracrustal rocks have been lost to erosion, or made unrecognizable by high-grade metamorphism (Jamieson et al., 1986). This situation is not unique to North America. Jenkins (1984) has shown that a similar division applies elsewhere.

Plutons of Group 1 are confined to Grenville inliers. However the extent of these inliers is unclear. Grenvillian ages have been obtained from the Cobequid Highlands, and Grenvillian affinities have been suggested for the Proterozoic marbles and quartzites found from Saint John to Cape Breton Island (Currie, 1983). Farrar (1984) has shown that Grenvillian basement underlies an Avalonian terrane in the United States, and Grenvillian basement beneath Avalonian terranes may be widespread.

The initial Sr ratio of most plutons falls in the range 0.703 to 0.709. Much of the plutonic material originated either in the mantle or in Rb-depleted parts of the deep crust such as granulite terranes. Except in local regions, contamination by sedimentary rocks or old high-level magmatic material must be limited. Along the eastern edge of the Gander Zone several plutons have high initial Sr ratios (Lockers Bay, C6, 0.715; Saint Lawrence, A4, 0.722; Tower Hill, N16, 0.727), and high ratios also occur in southwestern Nova Scotia (Davis Lake, E22, 0.729). Local anatexis of sediments may be responsible, but hydrothermal disturbance of isotope systematics seems more likely.

TECTONIC SIGNIFICANCE

Since the seminal synthesis of Williams (1979), tectonic analysis of the Canadian Appalachians has tended to be conducted in terms of tectonostratigraphic zones, although these zones have been variously redefined and subdivided (Williams et al., 1988). Analysis has concentrated on supracrustal sequences or structural arguments with little attention to the plutonic rocks. Data on the plutonic rocks offer new insights. In relation to the five major zones (Humber, Dunnage, Gander, Avalon and Meguma) plutonic rock distribution may be briefly summarized as follows. Group 1 plutons are confined to the Humber Zone. Group 2 plutons are essentially confined to the Avalon Zone, but examples may occur in probable Gander Zone rocks of northwest Cape Breton Island. Group 3 plutons occur mainly in the Avalon Zone, but examples occur in the Gander and Dunnage zones of Newfoundland and northwestern Cape Breton Island, and possibly in the Humber Zone of western Newfoundland.

Group 4 plutons of oceanic affinities occur only in the Dunnage Zone or nearby allochthons, but leucogranites of similar age (Group 4) occur in the Gander and/or Avalon zones in Cape Breton Island and southern Newfoundland. Plutons of Group 5a occur in the Notre Dame Subzone of the Dunnage Zone, and probably in the Gander Zone of Cape Breton Island. Group 5b is confined to the Gander Zone. Group 6 magmatism occurs in the Notre Dame Subzone of the Dunnage Zone, and along the western fringe of the Avalon Zone in Nova Scotia and New Brunswick. Group 7a magmatism is confined to the Gander Zone, and Group 7b to the Meguma Zone. Group 8 plutons are scattered across the orogen without consistent structural control. Groups 9 and 7b plutons occur in narrow belts along major, late terrane-bounding or dissecting faults. Group 10 magmatism is related to post-Appalachian events, possibly the opening of the Atlantic Ocean and/or passage of a mantle plume.

The Meguma Zone exhibits mid-Ordovician plutonism (Group 5b) of uncertain extent. This resembles plutonism in the Gander Zone (moderately peraluminous foliated plutons), and a late Ordovician or Silurian(?) bimodal volcanism (White Rock Formation) suggestive of contemporary Group 6 activity in the Avalon Zone and Notre Dame Subzone (Chandler et al., 1987). Metamorphic activity reached a peak about 385-400 Ma (Dallmeyer and Keppie, 1987). Relatively mafic plutons were emplaced near this peak. The major magmatism (Group 7b) is posttectonic, and resembles in age and character large plutons along the western periphery of the Avalon Zone (Ackley, C23, Saint George, F4, 5, 9, 10, 11) in its highly evolved and lithophile element-enriched character. The transcurrent fault-related character of this magmatism is unusually clear along the Cobequid fault system, and a case can be made for relation of the South Mountain batholith (E1) to the "Tobeatic fault" (Giles, 1985). Plutons of the Meguma Zone therefore suggest that it has been contiguous to the Avalon Zone at least since early Devonian time, and possibly since mid-Ordovician time. Further results on basement xenoliths and enclaves (Giles and Chatterjee, 1987a, b) may suggest even older connections (compare Krogh and Keppie, 1988).

The Avalon Zone contains relatively old (780-750 Ma) gabbro and diorite plutons (Group 2b), mafic dykes and high-grade metamorphism, but no granitic rocks of this age. Similar ages on mafic volcanics in the northeastern United States were ascribed by Rankin (1975) to breakup of the continental edge accompanying the opening of the Iapetus Ocean. Currie (1988) tentatively attributed events of this age in southern New Brunswick to rifting prior to the opening of Iapetus. Magmatism in the Avalon Zone (Group 2a) at 620 to 600 Ma comprises high level plutons exhibiting differentiation sequences from gabbro or diorite to granodiorite or granite, in many cases associated with cogenetic volcanic rocks, including extensive ignimbrite sheets. This association is a classic I-type suite suggesting subduction at a continental margin (Pitcher, 1983). A late Precambrian metamorphic belt possibly associated with this event outcrops in northwestern Cape Breton Island (Currie, 1986).

A slightly younger group of plutons in Avalonian terranes (Group 3a, 530-570 Ma) is more bimodal than the older suite. Locally they seem to be of arc-type character, but

667

distinctly alkaline plutons occur in Newfoundland. The double peak in ages (about 620 and 565 Ma) is similar to the younger Ordovician-Silurian and Early Devonian peaks.

The plutonic record suggests that the Avalon Zone underwent a complex late Precambrian history possibly commencing with rifting of a continent accompanied by dykes and formation of high grade metamorphic "core complexes" and eventually oceanic crust about 760 Ma. Subduction of uncertain polarity and emplacement of orogenic magmatic suites occurred about 620 Ma. The regime following subduction seems to have been dominated by transcurrent faulting in southern New Brunswick (Currie, 1988). Latest Precambrian magmatism about 550 Ma included both orogenic and alkaline components. The basement underlying Avalon terranes in Canada is little known, but where exposed has some similarities to Grenvillian rocks (Currie, 1983) although no plutons of Group 1 are known. One model which could accommodate these complexities envisages the Avalon Zone splitting off from cratonic America about 800 Ma, followed by re-accretion through subduction beneath the Avalonian terranes about 620 Ma, and a second breakup after 600 Ma. Configurations similar to the present Japan Sea and Japanese Islands (Currie, 1983, 1988) may be viable models for Avalon-Humber zone relations. The "Pan-African" ages of these igneous rocks can be explained by noting that Africa, North America and Europe were probably juxtaposed in a "Pangea" configuration at this time (Worsley et al., 1984). The complex tectonic sequence implied by plutonic rocks in the 620-550 Ma period may be the result of relative motions of various parts of this supercontinent.

The distinctive two-fold division of igneous activity at about 620 and 560 Ma "pins" the Avalon Zone to the Gander Zone in northwest Cape Breton Island, regardless of what subdivision of Cape Breton Island is adopted. The occurrence of late Avalonian ages (565 Ma) in southern and central Newfoundland suggests that Avalonian basement may be present beneath much of central Newfoundland.

Post-Precambrian magmatism in the Avalon Zone appears to be entirely related to normal or wrench faulting commencing in Silurian time and continuing until late Devonian. Early mafic phases (Kingston, F26, Saint Stephen, F6, Bocabec, F5) indicate influx of mantle material into the crust, and possibly crustal attenuation. The thermal anomaly created by influx of mantle material may have persisted to generate granites during Late Devonian transcurrent movements. The lack of mafic material associated with this later magmatism suggests it did not directly tap mantle sources, but it could have resulted from differentiation or recycling of an underplate.

The Gander Zone, according to the synthesis of Williams (1979), crosses Newfoundland, northern Cape Breton Island, and the Miramichi Highland of New Brunswick. These regions show similarities in plutonic record, but also significant differences. No Precambrian basement is definitely known in Newfoundland or New Brunswick, although small relicts may be present. Latest Precambrian (565 Ma) granite and volcano-sedimentary rocks in south Newfoundland (Dunning et al., 1988) are probably Avalonian. Parts of northwestern Cape Breton Island clearly have a Precambrian basement hosting late Precambrian magmatism. The sequence includes a marble-quartzite metasedimentary sequence, possibly correlative to the George River Group of southeastern Cape Breton Island,

and granitoid gneisses (Barr et al., 1987). The two-fold age division of late Precambrian igneous rocks implies an Avalonian provenance where present, but the central core of the highlands may be juvenile Silurian material.

Consideration of plutonic histories highlights some difficulties with current zonal schemes of Appalachian geology. Southeast-trending dextral transcurrent faulting in the Gulf of St. Lawrence (Stockmal et al., 1987) could line up central New Brunswick, with other parts of the Gander Zone, but the plutonic signature of the Miramichi region differs from that of Cape Breton Island and Newfoundland. Perhaps equivalents of the Miramichi region have been faulted away from these areas. Low angle, north over south translation of Dunnage subzones in Newfoundland has been demonstrated by kinematic analysis (Piasecki, 1988; van Berkel and Currie, 1988). The Dunnage and Gander zones of Newfoundland may form a stack of allochthonous sheets of uncertain thickness, rather than steep longitudinal areas.

ECONOMIC SIGNIFICANCE

Plutonic rocks of the Canadian Appalachians contain a variety of econmically valuable deposits. These deposits can be divided into five categories: namely (1) lithophile elements (Sn, W, U, Mo, Be); (2) Cu; (3) Ni and platinum group elements; (4) Au; and (5) industrial minerals (fluorite, pyrophyllite, and building stone).

Significant Sn-W mineralization occurs within high-silica granites of groups 9 (Mount Pleasant, F12, Kooiman et al., 1986; Burnt Hill, F8, Taylor et al., 1987) and 7b (East Kemptville, Kontak, 1987; Richardson et al., 1988). This mineralization is associated with late fracturing, hydrothermal alteration, quartz veining and greisenization near or at the rooves of the intrusions. Mineral occurrences of this type, locally associated with one or more of molybdenite, bismuthinite and uranium minerals are common in Group 9b plutons. Similar mineralization occurs in or around older plutons of Group 7a (in Newfoundland and New Brunswick, Tuach and Delany, 1987; Lake George, E4, Seal et al., 1988). U and W mineralization of quartz veins occurs within and around some Group 5b plutons in the Gander Zone of Newfoundland (see Dickson, 1984 for summary). Beryl is a common accessory in late phases of migmatitic Group 7a plutons in southwest Newfoundland, but has been little prospected. Minor mineralization by lithophile elements has also been reported from Group 6 plutons (Whalen and Currie, 1983; Payette and Martin, 1987). Alkali granite plutons of Group 3, and alkali phases of Group 8 have some potential for mineralization, particularly with Zr and U (Dickson, 1984) but the potential of other groups appears to be small.

Significant Cu mineralization in Appalachian granitoid rocks appears to be restricted to high-level Group 8 plutons in Gaspésie and New Brunswick (Murdochville, L12, Mont Breche, L14, Charlo stodes, M13, Mont McGerrigle, L15), most of which display mineralization of both pluton and aureole. At present the reason for the geographical concentration remains unknown, so that the potential of Group 8 intrusions elsewhere is uncertain.

Mafic phases of Group 8 intrusions in New Brunswick (Saint Stephen, F6, Portage Brook, M12) contain significant amounts of nickel, copper, and platinum group elements

(Paktunc, 1987, 1988; Dunham, 1959). These considerations suggest that mafic phases of Group 8 intrusions elsewhere, for example Mount Peyton and Twin Lakes in Newfoundland, may also have potential for similar mineralization. A mafic Group 3 pluton in southern New Brunswick (Mechanics Settlement, F19) has also proved to contain significant amounts of platinum group elements (Paktunc, 1988).

Recent exploration for Au in Atlantic Canada shows a complex relation between Au mineralization, large faults and granite plutonism. Gold mineralization follows large, complex fault zones with late (post-Devonian) movement. Mineralization, however, is of various ages, most of them older than the latest movements (for example see Tuach, 1987; Watters, 1987). Furthermore, significant gold mineralization invariably is spatially associated with a granitoid pluton, although this pluton may be much older than the structure, even Precambrian in age (Tuach, 1987; Watters, 1987). All the gold-bearing structures exhibit magmatism and/or hydrothermal activity of late Devonian or younger age. Since faults localize such activity, cause and effect relations are not clear. Significant gold concentrations may require several cycles of deformation and pluton-driven hydrothermal activity.

In addition to metallic ores, fluorite, pyrophyllite and building stone have been recovered from granitoid plutons in the Canadian Appalachians. Fluorite (+quartz+calcite) occurs in numerous southeast-trending veins 1 to 10 m wide cutting the Saint Lawrence pluton of high-level, high-silica peralkaline granite. Many plutons of groups 9 and 6 contain accessory fluorite, but the setting of the Saint Lawrence deposit is unique in the region. Pyrophyllite is mined from a contact zone of the Holyrood pluton (A1) near Foxtrap, Newfoundland. The altered zone stretches discontinuously for 7.6 km along the contact between granodiorite and volcanogenic sediments and breccia (Papezik et al., 1978). Building and ornamental stone have been recovered from quarries throughout the Canadian Appalachians, but current activity is centred west of Halifax in plutons of the South Mountain batholith, in the Saint George region of southern New Brunswick and in the Stanstead region of southern Quebec. The rocks in all cases are relatively leucocratic, massive posttectonic granites.

SUMMARY AND CONCLUSIONS

Plutonic rocks of the Canadian Appalachians fall into 10 groups. The oldest plutons (Group 1) occur only in Grenville basement, but the extent of this basement is unclear. More age dating, isotopic work on systems like Pb-Pb and Sm-Nd, and precise geochemical work is urgently needed to define the nature of basement across the orogen. Judging by initial results, this work is liable to result in radical revision of current ideas. Tholeiitic gabbro and diorite associated with high-temperature metamorphism were emplaced 800 to 750 Ma ago (Group 2b) possibly indicating continental breakup. Relics of this activity are found on both sides of the orogen suggesting linkage of cratonic North America and parts of "Avalonia" prior to 800 Ma. Late Proterozoic high-level diorite-granodiorite complexes with associated volcanics emplaced in the Avalon Zone 620 to 600 Ma (Group 2a) witness to subduction and major metamorphism. Slightly younger (570-530 Ma) plutons (Group 3) suggest a complex tectonic regime following subduction ending with breakup in latest

Precambrian to Cambrian time. Relicts of this breakup may be present in both Humber and Avalon zones, suggesting a linkage at around 600 Ma.

Late Cambrian to early Ordovician plutonism (510-470 Ma) appears to be mainly but not entirely of oceanic affinities (Group 4) and was terminated by ophiolite obduction in early to middle Ordovician. Major late Ordovician to Silurian plutonism clearly postdates emplacement of ophiolitic allochthons. In the Notre Dame Subzone of Newfoundland an older (465-450 Ma) pulse (Group 5a) reflects arc-related activity, whereas a younger phase (440-420 Ma) of high-level, bimodal, locally peralkaline plutonism (Group 6) reflects posttectonic magmatism into a cratonized substrate. Similar peralkaline magmatism associated with northwest-side down normal faulting occurs along the northwest edge of the Avalon Zone from the Antigonish Highlands to southern New Brunswick. In the Gander Zone plutonism was continuous from late Ordovician (mid-Ordovician in the Miramichi Highlands) to Early Devonian and associated with strong metamorphism and deformation (groups 5b and 7a). Many of these plutons appear to have developed by anatexis of deep-crustal material. Thermal modelling is required to explain the heat budget of this paroxysmal event.

Late Silurian to Early Devonian composite plutons (Group 8) were emplaced at the end of regional metamorphism and reflect an insurge of mantle material into hot crust. These plutons show no obvious zonal restrictions.

Late Devonian plutons (groups 9 and 7b) show association with major transcurrent faults, but are posttectonic both to the faults, and to earlier deformation and metamorphism. These high-silica, high level plutons commonly exhibit mineralization with U, Sn, W, and Mo. In southern Nova Scotia the late silicic plutons show a spatial relation to older, more mafic plutons. The thermal model required to generate these plutons along faults remains unexplained, but appears to be critical to development of both Au and lithophile element deposits. Thermal activity probably continued into Late Carboniferous-Permian time.

The Mesozoic breakup of Pangea is represented by sparse Triassic to Jurassic tholeiitic magmatism (Group 10). Strongly alkaline magmatism of Jurassic to Cretaceous age may represent passage of a plume.

ACKNOWLEDGMENTS

Any compilation on this scale necessitates collaboration with many colleagues. I am indebted to L.R. Fyffe, New Brunswick Department of Natural Resources, who supplied a compilation map of New Brunswick granites and drew my attention to numerous references, to Sandra Barr, Acadia University, who supplied a compilation of the plutonic rocks of Nova Scotia and invaluable guidance, and to J.P. Hayes, Mineral Development Division, Newfoundland Department of Mines and Energy who supplied a compilation map and various helpful reference material on the plutons of Newfoundland. Numerous colleagues have supplied reprints, references and unpublished data, as well as copious advice and admonition. An initial literature search was made by Jenifer Graves and Wanda Neely, and drafting assistance was provided by Normand Giguere and David Machin. The manuscript has benefitted from critical reading by Les Fyffe, Sandra Barr, John Hayes, Lawson Dickson, and Hank Williams. The inevitable errors and idiosyncracies that remain are my own.

REFERENCES

Abbott, D. and Barnett, D.E.
1969: Quartz monzonite in a drill hole at Bois Gagnon, Gloucester County, New Brunswick; New Brunswick Research and Productivity Council, Research Note 16, 10 p.

Allen, P.L. and Barr, S.M.
1983: The Ellison Lake pluton; a cordierite-bearing monzogranite intrusive body in southwestern Nova Scotia; Canadian Mineralogist, v. 21, p. 583-590.

Anderson, F.D. and Williams, H.
1970: Gander Lake, Newfoundland; Geological Survey of Canada, Map 1195A.

Ayuso, R.A. and Bevier, M.L.
1991: Regional differences in Pb isotopic compositions of feldspars in plutonic rocks of the northern Appalachian Mountains, U.S.A. and Canada: a geochemical method of terrane evaluation; Tectonics, v. 10, p. 191-212.

Barr, S.M. and MacDonald, A.S.
1983: Geology of the Mabou Highlands, western Cape Breton Island, Nova Scotia; Geological Association of Canada, Abstracts with Program, v. 8, p. A5.

Barr, S.M. and O'Beirne, A.M.
1981: Petrology of the Gillis Mountain pluton, Cape Breton Island, Nova Scotia; Canadian Journal of Earth Sciences, v. 18, p. 395-404.

Barr, S.M. and Raeside, R.P.
1986: Pre-Carboniferous tectono-stratigraphic subdivisions of Cape Breton Island; Maritime Sediments and Atlantic Geology, v. 22, p. 252-263.

Barr, S.M. and Setter, J.R.D.
1981: Petrology of granitoid rocks of the Boisdale Hills, central Cape Breton Island, Nova Scotia; Nova Scotia Department of Mines and Energy, Paper 84-1, 75 p.

Barr, S.M. and White, C.E.
1988: Petrochemistry of contrasting late Precambrian volcanic and plutonic associations, Caledonian Highlands, southern New Brunswick; Maritime Sediments and Atlantic Geology, v. 24, p. 353-372.

Barr, S.M., Dunning, G.R., Raeside, R.P., and Jamieson, R.A.
1990: Contrasting U-Pb ages from plutons in the Bras d'Or and Mira terranes of Cape Breton Island, Nova Scotia; Canadian Journal of Earth Sciences, v. 27, p. 1200-1208.

Barr, S.M., MacDonald, A.S., and Blenkinsop, J.
1986: The Cheticamp pluton: a Cambrian granodioritic pluton in the western Cape Breton Highlands, Nova Scotia; Canadian Journal of Earth Sciences, v. 23, p. 1686-1699.

Barr, S.M., O'Reilly, G.A., and O'Beirne, A.M.
1982: Geology and geochemistry of selected granitoid plutons of Cape Breton Island, Nova Scotia; Nova Scotia Department of Mines and Energy, Paper 82-1, 176 p.

Barr, S.M., Raeside, R.P., Dunning, G.R., and Jamieson, R.A.
1988: New U-Pb ages from the Cape Breton Highlands and correlations with southern Newfoundland; Geological Association of Canada, Abstracts with Program, v. 13, p. 11.

Barr, S.M., Raeside, R.P., and Jamieson, R.A.
1987: Geological map of the igneous and metamorphic rocks of northern Cape Breton Island, Nova Scotia; Geological Survey of Canada, Open File 1594.

Barr, S.M., Raeside, R.P., and van Breemen, O.
1986: Grenvillian basement in the northern Cape Breton Highlands, Nova Scotia; Canadian Journal of Earth Sciences, v. 24, p. 992-997.

Barr, S.M., Sangster, D.F., and Cormier, R.F.
1984: Petrology of early Cambrian and Devono-Carboniferous intrusions in the Loch Lomond complex, southeastern Cape Breton Island, Nova Scotia; in Current Research, Part A, Geological Survey of Canada, Paper 84-1A, p. 203-211.

Bedard, J.H.
1985: The opening of the Atlantic, the Mesozoic New England igneous province and the mechanisms of continental breakup; Tectonophysics, v. 113, p. 209-232.

Bell, K. and Blenkinsop, J.
1975: Geochronology of eastern Newfoundland; Nature, v. 254, p. 410-411.
1981: A geochronological study of the Buchans area, Newfoundland; Geological Association of Canada, Special Paper 22, p. 91-112.

Bell, K., Blenkinsop, J., Berger, A.R., and Jayasinghe, N.R.
1979: The Newport granite: its age, geological setting and implications for the geology of northeastern Newfoundland; Canadian Journal of Earth Sciences, v. 16, p. 264-269.

Bell, K., Blenkinsop, J., and Strong, D.F.
1977: The geochronology of some granitic bodies from eastern Newfoundland and its bearing on Appalachian evolution; Canadian Journal of Earth Sciences, v. 14, p. 456-476.

Bennett, G.
1965: The petrology of the Stewarton igneous complex; M.Sc. thesis, University of New Brunswick, Fredericton, New Brunswick.

Benson, D.G.
1974: Geology of the Antigonish Highlands, Nova Scotia; Geological Survey of Canada, Memoir 376.

Bevier, M.L.
1987: Isotopic ratios of Paleozoic granitoids from the Miramichi terrane, New Brunswick, and implications for the nature and age of basement rocks; Geological Survey of Canada, Paper 87-2, p. 43-50.
1988: U-Pb geochronologic studies of igneous rocks in New Brunswick; New Brunswick Department of Natural Resources, Mineral Resources Division, Information Circular.

Bevier, M.L., White, C.E., and Barr, S.M.
1990: Late Precambrian U-Pb ages for the Brookville gneiss, southern New Brunswick; Journal of Geology, v. 98, p. 935-965.

Boone, G.M. and Heisler, M.
1988: Penobscottian metamorphism in Maine: characteristics and preliminary isotopic ages; Geological Society of America, Abstracts with Programs, v. 20, p. 8.

Bostock, H.H.
1983: Precambrian rocks of the Strait of Belle Isle area; Geological Survey of Canada, Memoir 400, p. 1-73.

Bostock, H.H., Currie, K.L., and Wanless, R.K.
1979: The age of the Roberts Arm Group, north central Newfoundland; Canadian Journal of Earth Sciences, v. 16, p. 599-606.

Bourne, J.
1984: Géologie du complexe granitique du Lac aux Araignées; Ministère de l'Énergie et des Ressources du Québec, Rapport MB 84-22, 96 p.
1986a: Géologie du complexe granitique d'Aylmer; Ministère de l'Énergie et des Ressources du Québec, Rapport MB 86-40, 52 p.
1986b: Géologie du complexe granitique de Saint-Sébastien-Sainte-Cécile; Ministère de l'Énergie et des Ressources du Québec, Rapport MB 86-41, 49 p.

Bradley, D.A.
1962: Gisborne Lake and Terenceville map-areas; Geological Survey of Canada, Memoir 321.

Brown, G.C., Thorpe, R.S., and Webb, P.C.
1984: The geochemical characteristics of granitoids in contrasting arcs and comments on magma sources; Journal of the Geological Society of London, v. 141, p. 413-426.

Burke, K.B.S., Hamilton, J.B., and Gupta, V.
1973: The Caraquet dyke: its tectonic significance; Canadian Journal of Earth Sciences, v. 10, p. 1769-1768.

Butt, K.A.
1976: Genesis of granitic stocks in southwestern New Brunswick; Ph.D. thesis, University of New Brunswick, Fredericton, New Brunswick.

Campbell, R.M.
1980: Creignish Hills pluton; Nova Scotia Department of Mines and Energy, Mineral Resources Division, Report of Activities for 1980, p. 111-115.

Cawthorn, R.G.
1978: The petrology of the Tilting Harbour igneous complex, Fogo Island, Newfoundland; Canadian Journal of Earth Sciences, v. 15, p. 526-539.

Chandler, F.W. and Dunning, G.R.
1983: Four-fold significance of an Early Silurian U-Pb zircon age from rhyolite in redbeds, southwest Newfoundland; in Current Research, Part B, Geological Survey of Canada, Paper 83-1B, p. 419-421.

Chandler, F.W., Sullivan, R.W., and Currie, K.L.
1987: The age of the Springdale Group, western Newfoundland, and correlative rocks – evidence for a Llandovery overlap assemblage in the Canadian Appalachians; Transactions of the Royal Society of Edinburgh, v. 78, p. 41-49.

Chappell, B.W. and White, A.J.R.
1974: Two contrasting magma types; Pacific Geology, v. 8, p. 173-174.

Chatterjee, A.K. and Keppie, J.D.
1981: Polymetallic mineralization in the endo- and exo-contact of the Wedgeport pluton, Yarmouth County; Nova Scotia Department of Mines and Energy, Mineral Resources Division, Report of Activities for 1980, p. 43-46.

Chorlton, L.

1978: The geology of the La Poile map-area (11O/9), Newfoundland; Newfoundland and Labrador Department of Mines and Energy, Mineral Development Division, Report 78-5.

1980: The geology of the La Poile River area (11O/16), Newfoundland; Newfoundland and Labrador Department of Mines and Energy, Mineral Development Division, Report 80-3.

Chorlton, L. and Knight, I.

1983: Geology of the Grandy's Lake area (11O/15), Newfoundland; Newfoundland and Labrador Department of Mines and Energy, Mineral Development Division, Report 83-7.

Chorlton, L.B. and Dallmeyer, R.D.

1986: Geochronology of the early to middle Paleozoic tectonic development in the southwest Newfoundland Gander Zone; Journal of Geology, v. 94, p. 67-89.

Clarke, D.B. and Chatterjee, A.K.

1988: Sr-Nd isotopic study of the Liscomb complex and peraluminous granites in the Meguma Zone; Geological Association of Canada, Program with Abstracts, v. 13, p. A21.

Clarke, D.B. and Halliday, A.N.

1980: Strontium isotope geology of the South Mountain batholith, Nova Scotia; Geochimica et Cosmochimica Acta, v. 44, p. 1045-1058.

Clarke, D.B., Barr, S.M., and Donohoe, H.V.

1980: Granitoid and other plutonic rocks of Nova Scotia; in The Caledonides in the USA; (ed.) D.R. Wones; Virginia Polytechnic Institute and State University, Department of Geology, Memoir 2, p. 107-116.

Clarke, D.B., McKenzie, C.B., Muecke, G.K., and Richardson, S.W.

1976: Magmatic andalusite from the South Mountain batholith, Nova Scotia; Contributions to Mineralogy and Petrology, v. 56, p. 279-287.

Colman-Sadd, S.P.

1976: Geology of the Saint Albans map-area, Newfoundland (11M/13); Newfoundland and Labrador Department of Mines and Energy, Mineral Development Division, Report 76-4, p. 1-19.

1979: Geology of the Twillick Brook map-area (2D/4), Newfoundland; Newfoundland and Labrador Department of Mines and Energy, Mineral Development Division, Report 79-2, p. 1-23.

1980: Geology of parts of the Burnt Hill map-area (2D/5), Newfoundland; Newfoundland and Labrador Department of Mines and Energy, Mineral Development Division, Report 80-1, p. 79-88.

1984: Geology of the Cold Spring Pond map-area (12A/1), Newfoundland; Newfoundland and Labrador Department of Mines and Energy, Mineral Development Division, Report 84-1, p. 211-219.

Colman-Sadd, S.P. and Swinden, H.S.

1984: A tectonic window in central Newfoundland? Geological evidence that the Appalachian Dunnage Zone may be allochthonous; Canadian Journal of Earth Sciences, v. 21, p. 1349-1367.

Corey, M.C.

1987: A re-interpretation of U-Th ratios in the New Ross area: evidence for a zoned pluton; Nova Scotia Department of Mines and Energy, Mineral Resources Division, Report 87-5, p. 105-108.

Cormier, R.F.

1972: Radiometric ages of granitic rocks from Cape Breton Island, Nova Scotia; Canadian Journal of Earth Sciences, v. 9, p. 1074-1086.

1979: Rubidium-strontium isochron ages of Nova Scotian granitic rocks; Nova Scotia Department of Mines and Energy, Mineral Resources Division, 1978 Annual Report, p. 143-148.

Cormier, R.F. and Smith, T.E.

1973: Radiometric ages of granitic rocks from southwestern Nova Scotia; Canadian Journal of Earth Sciences, v. 10, p. 1201-1210.

Cormier, R.F., Keppie, J.D., and Odom, A.L.

1988: U-Pb and Rb-Sr geochronology of the Wedgeport granitoid pluton, southwestern Nova Scotia; Canadian Journal of Earth Sciences, v. 25, p. 255-261.

Cousineau, P.A.

1988: Le Groupe de Caldwell et le domaine océanique entre Saint-Joseph-de-Beauce et Sainte-Sabine-de-Bellechasse; Ministère de l'Énergie et des Ressources du Québec, Mémoire MM 87-02.

Coyle, M.L. and Strong, D.F.

1987: Geology of the Springdale Group: a newly recognized Silurian epicontinental-type caldera in Newfoundland; Canadian Journal of Earth Sciences, v. 24, p. 1135-1148.

Crouse, G.W.

1978: Geology of Head of Clearwater Brook (map-area K-13); New Brunswick Department of Natural Resources, Mineral Resources Division, Map Report 77-1.

Crouse, G.W. (cont.)

1979: Geology of Head of Little Southwest Nepisguit River, Head of North Pole Brook, Mitchell Lake region (map-areas L-9, L-10, and L-11); New Brunswick Department of Natural Resources, Mineral Resources Division, Map Report 70-4.

1981a: Geology of Napadogan and Miramichi lakes (map-area K-17) and Napadogan, Rocky, McLean and Ryan brooks (map-area K-18); New Brunswick Department of Natural Resources, Mineral Resources Division, Map Report 81-8.

1981b: Geology of parts of Burnthill, Clearwater and McKiel brooks (map-areas K-14, K-15 and K-16); New Brunswick Department of Natural Resources, Mineral Resources Division, Map Report 81-5.

Currie, K.L.

1976: The alkaline rocks of Canada; Geological Survey of Canada, Bulletin 239, 213 p.

1983: Repeated basement reactivation in the northeastern Appalachians; Geological Journal, v. 18, p. 223-239.

1984: A reconsideration of some geological relationships near St. John, New Brunswick; Geological Survey of Canada, Paper 84-1A, p. 193-201.

1986: Relations between metamorphism and magmatism near Cheticamp, Nova Scotia; Geological Survey of Canada, Paper 85-23.

1987a: A preliminary account of the geology of the Harry's River map area, southern Long Range of Newfoundland; in Current Research, Part A; Geological Survey of Canada, Paper 87-1A, p. 653-662.

1987b: Contrasting metamorphic terranes near Cheticamp Cape Breton Highlands, Nova Scotia; Canadian Journal of Earth Sciences, v. 24, p. 1121-1136.

1988: The end of the Avalon Zone in southern New Brunswick; Maritime Sediments and Atlantic Geology, v. 24, p. 171-190.

Currie, K.L. and Pajari, G.E.

1981: Anatectic peraluminous granites from the Carmanville area, northeastern Newfoundland; Canadian Mineralogist, v. 19, p. 147-161.

Currie, K.L., Pajari, G.E., and Pickerill, R.K.

1979: Tectono-stratigraphic problems in the Carmanville area, northeastern Newfoundland; Geological Survey of Canada, Paper 79-1A, p. 71-76.

Currie, K.L. and Piasecki, M.A.J.

1989: Kinematic model for southwestern Newfoundland based upon Silurian sinistral shearing; Geology, v. 17, p. 938-941.

Currie, K.L. and van Berkel, J.T.

1989: Tectonic significance of mafic intrusions in the Central Gneiss Terrane, southwestern Newfoundland; Atlantic Geology, v. 25, p. 181-190.

Currie, K.L., Eby, G.N., and Gittins, J.

1986: The petrology of the Mont Saint Hilaire complex, southern Quebec: an alkaline gabbro-peralkaline syenite association; Lithos, v. 19, p. 67-83.

Currie, K.L., Loveridge, W.D., and Sullivan, R.W.

1982a: The Deadmans Bay Pluton, northeastern Newfoundland: U-Pb study of zircon reveals a Grenvillian component; in Current Research, Part C, Geological Survey of Canada, Paper 82-1C, p. 119-124.

1982b: A U-Pb age on zircon from dykes feeding basal rhyolitic flows of the Jumping Brook complex, northwestern Cape Breton Island; in Current Research, Part C, Geological Survey of Canada, Paper 82-1C, p. 125-125.

Currie, K.L., Nance, R.D., Pajari, G.E., and Pickerill, R.K.

1981: A reconsideration of the pre-Carboniferous geology of St. John, New Brunswick; in Current Research, Part A, Geological Survey of Canada, Paper 81-1A, p. 23-30.

Currie, K.L., van Breemen, O., Hunt, P.A., and van Berkel, J.T.

1992: Age of granulitic gneisses south of Grand Lake, western Newfoundland; Atlantic Geology, v. 28, p. 153-161.

Dallmeyer, R.D.

1978: $^{40}Ar/^{39}Ar$ incremental-release ages of hornblende and biotite from Grenville basement rocks within the Indian Head Range complex, southwest Newfoundland; their bearing on Late Proterozoic-Early Paleozoic thermal history; Canadian Journal of Earth Sciences, v. 15, p. 1374-1379.

Dallmeyer, R.D. and Hibbard, J.

1984: Geochronology of the Baie Verte peninsula, Newfoundland: implications for the tectonic evolution of the Humber and Dunnage zones of the Appalachian orogen; Journal of Geology, v. 92, p. 489-512.

Dallmeyer, R.D. and Keppie, J.D.

1984: Geochronological constraints on the accretion of the Meguma terrane with North America; Geological Society of America, Abstracts with Program, v. 16, p. 11.

Dallmeyer, R.D. and Keppie, J.D. (cont.)

1987:   Polyphase late Paleozoic tectonothermal evolution of the south-western Meguma terrane, Nova Scotia: evidence from $^{40}Ar/^{39}Ar$ mineral ages; Canadian Journal of Earth Sciences, v. 24, p. 1242-1254.

Dallmeyer, R.D. and van Breemen, O.

1978:   The Hallowell quartz monzonite and Three Mile Pond granodiorite plutons, south central Maine; their bearing on post-Acadian cooling history; Contributions to Mineralogy and Petrology, v. 70, p. 61-73.

Dallmeyer, R.D., Blackwood, R.F., and Odom, A.L.

1981:   Age and origin of the Dover Fault: tectonic boundary between the Gander and Avalon zones of the northeastern Newfoundland Appalachians; Canadian Journal of Earth Sciences, v. 18, p. 1431-1442.

Dallmeyer, R.D., Hussey, E.M., O'Brien, S.J., and O'Driscoll, C.F.

1983a: Chronology of tectonothermal activity in the western Avalon Zone of the Newfoundland Appalachians; Canadian Journal of Earth Sciences, v. 20, p. 355-363.

Dallmeyer, R.D., Kean, B.F., Odom, A.L., and Jayasinghe, N.R.

1983b: Age and contact metamorphic effects of the Overflow Pond Granite, an undeformed pluton in the Dunnage Zone of the Canadian Appalachians; Canadian Journal of Earth Sciences, v. 20, p. 1639-1645.

Danis, D.

1984:   Géologie du complexe granitique du Scotstown; Ministère de l'Énergie et des Ressources du Québec, Rapport MB 84-21, 102 p.

1985:   Géologie du complexe granitique de Winslow; Ministère de l'Énergie et des Ressources du Québec Rapport MB 85-31.

David, J. and Gariepy, C.

1990:   Early Silurian orogenic andesites from the central Quebec Appalachians; Canadian Journal of Earth Sciences, v. 27, p. 632-643.

Davies, J.L., Tupper, W.M., Bachinski, D., Boyle, R.W., and Martin, R.F.

1969:   Geology and mineral deposits of the Nigadoo River-Millstream River area, Gloucester County, New Brunswick; Geological Survey of Canada, Paper 67-49.

de Albuquerque, C.A.R.

1977:   Geochemistry of the tonalitic and granitic rocks of the Nova Scotia southern plutons; Geochimica et Cosmochimica Acta, v. 41, p. 1-13.

de Grace, J.R., Kean, B.F., Hsu, E., and Green, T.

1976:   Geology of the Nippers Harbour map-area (2E/13), Newfoundland; Newfoundland and Labrador Department of Mines and Energy, Mineral Development Division, Report 76-3.

de Romer, H.S.

1977:   McGerrigle Mountains area; Quebec Department of Natural Resources, Mineral Resources Division, Geological Report 174, 233 p.

Dickson, W.L.

1982:   Geology of the Wolf Mountain (12A/2W) and Burnt Pond (12A/3E) map area, Newfoundland; Newfoundland and Labrador Department of Mines and Energy, Mineral Development Division, Report 82-5.

1983:   Geology, geochemistry and mineral potential of the Ackley Granite and parts of the Northwest Brook and Eastern Maelpaeg complexes, southeastern Newfoundland; Newfoundland and Labrador Department of Mines and Energy, Mineral Development Division, Report 83-6, 129 p.

1984:   Mineralised granites in southern Newfoundland; Newfoundland and Labrador Department of Mines and Energy, Mineral Development Division, Report 84-3, p. 149-153.

Dickson, W.L., Delaney, P.W., and Poole, J.C.

1985:   Geology of the Burgeo granite and associated rocks in the Ramea (11P/11) and LaHune (11P/10) map-areas, southern Newfoundland; Newfoundland and Labrador Department of Mines and Energy, Mineral Development Division, Report 85-1, p. 137-144.

Dickson, W.L., Hayes, J.P., O'Brien, S.J., and Tuach, J.

1988:   Late Devonian-early Carboniferous high-silica granites in southern Newfoundland; Geological Association of Canada, Program with Abstracts 13, p. A32.

Doig, R., Murphy, J.B., Nance, R.D., and Stokes, T.

1991:   Review of the geochronology of the Cobequid Highlands, Avalon Composite terrane, Nova Scotia; in Current Research, Part D, Geological Survey of Canada, Paper 91-1D, p. 71-78.

Doig, R., Nance, R.D., Murphy, J.B., and Casseday, R.P.

1990:   Evidence for Silurian sinistral accretion of Avalon composite terrane in Canada; Journal of the Geological Society of London, v. 147, p. 927-930.

Donohoe, H.V. and Wallace, P.I.

1979:   A preliminary map of the Cobequid Highlands of Nova Scotia; Nova Scotia Department of Mines and Energy, Mineral Resources Division, Report of Activities for 1978, p. 73-78.

Donohoe, H.V., Halliday, A.N., and Keppie, J.D.

1986:   Two whole-rock Rb-Sr isochrons from plutons in the Cobequid Highlands, Nova Scotia, Canada; Maritime Sediments and Atlantic Geology, v. 22, p. 148.

Dunham, K.C.

1959:   Petrography of the nickeliferous norite of Saint Stephen, New Brunswick; American Mineralogist, v. 35, p. 711-727.

Dunning, G.R.

1987a: U/Pb geochronology of the Coney Head complex, Newfoundland; Canadian Journal of Earth Sciences, v. 24, p. 1072-1075.

1987b: Geology of the Annieopsquotch Complex, southwest Newfoundland; Canadian Journal of Earth Sciences, v. 24, p. 1162-1174.

Dunning, G.R. and Chorlton, L.B.

1985:   The Annieopsquotch ophiolite belt of southwestern Newfoundland: geology and tectonic significance; Geological Society of America, Bulletin, v. 96, p. 1466-1476.

Dunning, G.R. and Krogh, T.E.

1985:   Geochronology of ophiolites in the Newfoundland Appalachians; Canadian Journal of Earth Sciences, v. 22, p. 1659-1670.

Dunning, G.R. and O'Brien, S.J.

1989:   Late Proterozoic-Early Paleozoic crust in the Hermitage flexure, Newfoundland Appalachians: U/Pb ages and tectonic significance; Geology, v. 17, p. 548-551.

Dunning, G.R., Barr, S.M., Raeside, R.P., and Jamieson, R.A.

1990:   U-Pb zircon, titanite and monozite ages in the Bras d'Or and Aspy terranes of Cape Breton Island, Nova Scotia; implications for igneous and metamorphic history; Geological Society of America, Bulletin, v. 102, p. 322-330.

Dunning, G.R., Carter, P.J., and Best, M.A.

1982:   Geology of Star Lake (west half); in Current Research, Part B, Geological Survey of Canada, Paper 82-1B, p. 21-26.

Dunning, G.R., Kean, B.F., Thurlow, J.G., and Swinden, H.S.

1987:   Geochronology of the Buchans, Roberts Arm and Victoria Lake groups, and Mansfield Cove complex, Newfoundland; Canadian Journal of Earth Sciences, v. 24, p. 1175-1184.

Dunning, G.R., Krogh, T.E., O'Brien, S.J., Colman-Sadd, S.P., and O'Neill, P.

1988:   Geochronologic framework for the central mobile belt in southern Newfoundland and the importance of Silurian orogeny; Geological Association of Canada/Mineralogical Association of Canada, Program with Abstracts, v. 13, p. A34.

Eby, G.N.

1984:   Geology of the Monteregian Hills alkaline province, Quebec; Geology, v. 12, p. 468-470.

1985:   Monteregian Hills II: Petrography, major and trace element geochemistry and strontium isotope chemistry of the eastern intrusions; Journal of Petrology, v. 26, p. 418-448.

Eby, G.N. and Currie, K.L.

1993:   Petrology and geochemistry of the Kingston complex, - a bimodal sheeted dyke suite in southern New Brunswick; Atlantic Geology, v. 29, p. 121-133.

Elias P.M. and Strong, D.F.

1982a: Timing of arrival of the Avalon Zone in the Newfoundland Appalachians, a new look at the Straddling Granite; Canadian Journal of Earth Sciences, v. 19, p. 1088-1094.

1982b: Paleozoic granitoid plutonism of southern Newfoundland: contrasts in timing, tectonic setting and level of emplacement; Transactions of the Royal Society of Edinburgh, v. 73, p. 43-57.

Emslie, R.F.

1978:   Anorthosite massifs, rapakivi granites and Late Proterozoic rifting of North America; Precambrian Research, v. 7, p. 61-98.

Erdmer, P.

1982:   Metamorphism at the northwest contact of the Stanhope pluton, Quebec Appalachians, Canada; Contributions to Mineralogy and Petrology, v. 76, p. 109-115.

1986:   Geology of the Long Range Inlier in Sandy Lake map-area, western Newfoundland; in Current Research, Part B, Geological Survey of Canada, Paper 86-1B, p. 19-27.

Ermanovics, I.F., Edgar, A.E., and Currie, K.L.

1967:   Evidence bearing on the origin of the Belleoram pluton, southern Newfoundland; Canadian Journal of Earth Sciences, v. 4, p. 413-430.

Evans, D.T.W., Kean, B.F., and Dunning, G.R.
1990: Geological studies, Victoria Lake Group, central Newfoundland; Newfoundland and Labrador Department of Mines and Energy, Mineral Development Division, Report 90-1, p. 131-144.

Fairbairn, H.W., Hurley, P.M., Pinson, W.H., and Cormier, R.F.
1960: Age of the granitic rocks of Nova Scotia; Geological Society of America, Bulletin 71, p. 399-414.

Farrar, S.S.
1984: The Goochland granulite terrane; remobilized Grenville basement in the eastern Virginia Piedmont; Geological Society of America, Special Paper 194, p. 215-277.

Fox, D. and van Berkel, J.T.
1988: Mafic-ultramafic occurrences in metasedimentary rocks of southwestern Newfoundland; in Current Research, Part B, Geological Survey of Canada, Paper 88-1B, p. 41-48.

French, V.A. and Barr, S.M.
1984: Age and petrology of the Gillanders Mountain intrusive complex, Lake Ainslie region, Cape Breton Island; Geological Association of Canada, Program with Abstracts, v. 9, p. 63.

Furey, D.J.
1985: Geology of the Belleoram pluton, southeastern Newfoundland; in Current Research, Part ?, Geological Survey of Canada, Paper 85-1, p. 151-156.

Fyffe, L.R. and Cormier, R.F.
1979: The significance of radiometric ages from the Gulquac Lake area of New Brunswick; Canadian Journal of Earth Sciences, v. 16, p. 2046-2052.

Fyffe, L.R., Irrinki, R.R., and Cormier, R.F.
1977: A radiometric age of deformed granitic rocks in north central New Brunswick; Canadian Journal of Earth Sciences, v. 14, p. 1687-1689.

Fyffe, L.R., Pajari, G.E., and Cherry, M.E.
1981: The Acadian plutonic rocks of New Brunswick; Maritime Sediments and Atlantic Geology, v. 17, p. 23-26.

Gaudette, H.E., Olszewski, W.J., and Donohoe, H.V.
1983: Age and origin of the basement rocks, Cobequid Highlands, Nova Scotia; Geological Society of America, Abstracts with Program, v. 15, p. 136.

Gaudette, H.E., Olszewski, W.J., and Jamieson, R.A.
1985: Rb-Sr ages of some basement rocks, Cape Breton Highlands, Nova Scotia; Geological Association of Canada, Program with Abstracts, v. 10, p. A20.

Giles, P.S.
1985: A major post-Visean sinistral shear zone; in new perspectives on Devonian and Carboniferous rocks of southwest Nova Scotia, (ed.) A.K. Chatterjee and D.B. Clarke; Nova Scotia Department of Mines and Energy, Mineral Resources Division, Preprint 85-3, p. 235-248.

Giles, P.S. and Chatterjee, A.K.
1987a: Peraluminous granites of the Liscomb complex; Nova Scotia Department of Mines and Energy, Mineral Resources Division, Report 87-1, p. 95-98.
1987b: Lower crustal xenocrysts and xenoliths in the Tangier dyke, eastern Meguma Zone, Nova Scotia; Nova Scotia Department of Mines and Energy, Mineral Resources Division, Report 87-5, p. 85-88.

Gray, C.M.
1984: An isotopic mixing model for the origin of granitic rocks in southeastern Australia; Earth and Planetary Science Letters, v. 70, p. 47-60.

Greiner, H.R.
1960: Pointe Verte, Gloucester and Restigouche counties, New Brunswick; New Brunswick Department of Natural Resources, Mineral Resources Division, Preliminary Map 60-2.
1970: Geology of the Charlo area (210/6), Restigouche County, New Brunswick; New Brunswick Department of Natural Resources, Mineral Resources Division, Map Series 70-2.

Ham, L.
1984: The Halfway Cove-Queensport pluton, Guysborough County, Nova Scotia; Nova Scotia Department of Mines and Energy, Mineral Resources Division, Report 84-1, 61 p.

Hayes, J.J.
1950: Preliminary map, Hodges Hill, Newfoundland; Geological Survey of Canada, Preliminary Map 51-5.

Hayes, J.P., Dickson, W.L., and Tuach, J.
1987: Preliminary map, Newfoundland granitoid rocks, nomenclature, distribution and possible classification; Newfoundland and Labrador Department of Mines and Energy, Mineral Development Division, Open File Map 87-15.

Herd, R.K. and Dunning, G.R.
1979: Geology of Puddle Pond map-area, southwestern Newfoundland; in Current Research, Part A, Geological Survey of Canada, Paper 79-1A, p. 305-310.

Hibbard, J.
1983: Geology of the Baie Verte peninsula, Newfoundland; Newfoundland and Labrador Department of Mines and Energy, Mineral Development Division, Memoir 2, 280 p.

Higgins, N.C. and Kerrich, R.
1982: Progressive O^{18} depletion during CO_2 separation from a carbon dioxide-rich hydrothermal fluid: evidence from the Gray River tungsten deposit, Newfoundland; Canadian Journal of Earth Sciences, v. 19, p. 2247-2257.

Hill, J.D.
1991: Petrology, tectonic setting and economic potential of Devonian peraluminous granitoid plutons in the Canso and Forest Hills area, eastern Meguma terrane, Nova Scotia; Geological Survey of Canada, Bulletin 383, 96 p.

Hogg, W.A.
1954: Geology of Red Island, Placentia Bay; M.Sc. thesis, Dalhousie University, Halifax, Nova Scotia, 50 p.

Hollister, V.F., Potter, R.R., and Barker, A.L.
1974: Porphyry-type deposits of the Appalachian orogen; Economic Geology, v. 69, p. 618-630.

Hope, T.L., Douma, S.L., and Raeside, R.P.
1988: Geology of the Port Mouton-Lockeport area, southwestern Nova Scotia; Geological Survey of Canada, Open File 1768.

Irrinki, R.R.
1978: Geology of Head of Dungarvon and Renous Rivers map-area L-13; New Brunswick Department of Natural Resources, Mineral Resources Division, Map Report 78-2.
1980: Geology of the Kennedy Lakes-Little Dungarvon and South Renous Rivers region, map-areas M-13, M-14, M-15 and part of M-16; New Brunswick Department of Natural Resources, Mineral Resources Division, Map Report 80-2.
1981: Geology of the Rocky, Sisters, and Clearwater Brooks-Todd Mountain region, map-areas L-14,L-15 and L-16, New Brunswick Department of Natural Resources and Energy, Mineral Resources Division, Map-Report 81-7.

Jamieson, R.A. and Craw, D.
1983: Reconnaissance mapping of the southern Cape Breton Highlands: a preliminary report; in Current Research, Part A, Geological Survey of Canada, Paper 83-1A, p. 263-268.

Jamieson, R.A. and Doucet, P.
1983: The Middle River-Crowdis Mountain area, southern Cape Breton Highlands; Geological Survey of Canada, Paper 83-1A, p. 269-275.

Jamieson, R.A., van Breemen, O., Sullivan, R.W., and Currie, K.L.
1986: The age of igneous and metamorphic events in the western Cape Breton Highlands; Canadian Journal of Earth Sciences, v. 22, p. 1891-1901.

Jayasinghe, N.R.
1978: Geology of the Wesleyville (2F/4) and the Musgrave Harbour East (2F/5) map areas, Newfoundland; Newfoundland and Labrador Department of Mines and Energy, Mineral Development Division, Report 78-8, p. 1-11.

Jayasinghe, N.R. and Berger, A.R.
1976: On the plutonic evolution of the Wesleyville area, Bonavista Bay, Newfoundland; Canadian Journal of Earth Sciences, v. 13, p. 1560-1570.

Jenkins, R.F.
1984: Ediacaran events; boundary relations and correlation of key sections, especially in Armorica; Geological Magazine, v. 121, p. 625-643.

Jenness, S.E.
1963: Terra Nova and Bonavista map-areas, Newfoundland; Geological Survey of Canada, Memoir 327.

Jones, B.E. and MacMichael, T.P.
1976: Observations on the geology of a portion of the Musquodoboit Harbour batholith; Nova Scotia Department of Mines and Energy, Mineral Resources Division, Report of Activities for 1975, p. 63-71.

Kamo, S.L., Gower, C.F., and Krogh, T.E.
1989: Birthdate for the Iapetus Ocean? A precise U-Pb zircon and baddeleyite age for the Long Range dykes, southeast Labrador; Geology, v. 17, p. 602-605.

673

Karlstrom, K.E., van der Pluijm, B.A., and Williams, P.F.
1982: Structural interpretation of the eastern Notre Dame Bay area, Newfoundland: regional post-Middle Silurian thrusting and asymmetrical folding; Canadian Journal of Earth Sciences, v. 19, p. 2325-2341.

Kean, B.F.
1977: Geology of the Victoria Lake map-area (12A/6), Newfoundland; Newfoundland and Labrador Department of Mines and Energy, Mineral Development Division, Report 77-4.
1983: Geology of the King George IV Lake map-area (12A/4), Newfoundland; Newfoundland and Labrador Department of Mines and Energy, Mineral Development Division, Report 83-4.
1984: Geology and mineral deposits of the Lushes Bight Group, Notre Dame Bay, Newfoundland; Newfoundland and Labrador Department of Mines and Energy, Mineral Development Division, Report 84-1, p. 141-156.

Kean, B.F. and Jayasinghe, N.R.
1980: Geology of the Lake Ambrose (12A/10)-Noel Pauls Brook (12A/9) map-areas, central Newfoundland; Newfoundland and Labrador Department of Mines and Energy, Mineral Development Division, Report 80-2.
1982: Geology of the Badger map-area (12A/16), Newfoundland; Newfoundland and Labrador Department of Mines and Energy, Mineral Development Division, Report 81-2.

Kelley, D.G.
1967: Baddeck and Whycocomagh map-areas, with emphasis on Mississippian stratigraphy in central Cape Breton Island, Nova Scotia; Geological Survey of Canada, Memoir 351.

Keppie, J.D. (comp.)
1979: Geological Map of the Province of Nova Scotia; Nova Scotia Department of Mines and Energy, Mineral Resources Division.

Keppie, J.D.
1984: The Appalachian collage; in The Caledonide Orogen, Scandanavia and related areas, (ed.) D.G. Gee and B. Sturt; J. Wiley and Sons, New York.

Keppie, J.D. and Halliday, A.N.
1986: Rb-Sr isotopic data from three suites of igneous rocks, Cape Breton Island, Nova Scotia; Maritime Sediments and Atlantic Geology, v. 22, p. 162-171.

Keppie, J.D. and Krogh, T.E.
1986: Detrital zircon ages indicating a North African provenance for the Goldenville Formation of Nova Scotia; Geological Association of Canada, Program with Abstracts, v. 11, p. 91.

Kirkham, R.V. and Soregaroli, A.E.
1975: Preliminary assessment of porphyry deposits in the Canadian Appalachians; in Report of Activities, Part A; Geological Survey of Canada, Paper 75-1A, p. 249-251.

Knapp, D., Kennedy, D., and Martineau, Y.
1979: Stratigraphy, structure and regional correlation of rocks at Grand Lake, western Newfoundland; in Current Research, Part A, Geological Survey of Canada, Paper 79-1A, p. 317-325.

Kontak, D.J.
1987: The East Kemptville leucogranite: a possible mid-Carboniferous topaz granite; Nova Scotia Department of Mines and Energy, Mineral Resources Division, Report 87-1, p. 87-94.

Kontak, D.J. and Strong, D.F.
1986: The volcano-plutonic Kings Point complex, Newfoundland; in Current Research, Part A, Geological Survey of Canada, Paper 86-1A, p. 465-470.

Kontak, D.J., Tuach, J., Strong, D.F., Archibald, D.A., and Farrar, E.
1988: Plutonic and hydrothermal events in the Ackley Granite, southeast Newfoundland, as indicated by total-fusion $^{40}Ar/^{39}Ar$ geochronology; Canadian Journal of Earth Sciences, v. 25, p. 1151-1160.

Kooiman, G.J.A., McLeod, M.J., and Sinclair, W.D.
1986: Porphyry tungsten-molybdenum ore bodies, polymetallic veins and replacement bodies, and tin-bearing greisen zones in the Fire Tower zone, Mount Pleasant, New Brunswick; Economic Geology, v. 81, p. 1356-1373.

Krogh, T.E. and Keppie, J.D.
1984: The U-Pb zircon geochronology of the eastern Meguma terrane of Nova Scotia; Geological Society of America, Programs with Abstracts, v. 16, p. 28.
1988: U-Pb ages of single zircon cores imply a Pan-African source for two Meguma granites; Geological Association of Canada, Program with Abstracts, v. 13, p. A69.

Krogh, T.E., Strong, D.F., O'Brien, S.J., and Papezik, V.S.
1988: Precise U-Pb zircon dates from the Avalon Terrane in Newfoundland; Canadian Journal of Earth Sciences, v. 25, p. 442-453.

Laurent, R., Taner, R.F., and Bertrand, J.
1984: Mise en place et pétrologie du granite associé au complexe ophiolitique de Thetford Mines, Québec; Canadian Journal of Earth Sciences, v. 21, p. 1114-1122.

Leech, G.B., Lowden, J.A., Stockwell, C.H., and Wanless, R.K.
1964: Age determinations and geological studies (including isotope ages report 4); Geological Survey of Canada, Paper 63-17.

Lentz, D.R., Lutes, G., and Hartree, R.
1988: Bi-Sn-Mo-W greisen mineralisation associated with the True Hill granite, south-western New Brunswick; Maritime Sediments and Atlantic Geology, v. 24, p. 321-338.

Loiselle, M.C. and Wones, D.R.
1979: Characteristics of anorogenic granites; Geological Society of America, Abstracts with Program, v. 11, p. 539.

Longstaffe, F.J., Smith, T.E., and Muehlenbachs, K.
1980: Oxygen isotope evidence for the genesis of Upper Paleozoic granitoids from southwestern Nova Scotia; Canadian Journal of Earth Sciences, v. 17, p. 132-141.

Lord, C.S.
1938: Megantic sheet, east half, Frontenac County, Quebec; Geological Survey of Canada, Map 379A.

Lorenz, B.E.
1982: Timing of igneous activity in the Dunnage mélange, Newfoundland; Geological Society of America, Abstracts with Program, v. 14, p. 35.

Lorenz, B.E. and Fountain, J.G.
1982: The South Lake igneous complex, Newfoundland; Canadian Journal of Earth Sciences, v. 19, p. 490-503.

Lowdon, J.A.
1960: Age determinations by the Geological Survey of Canada, Report 1: Isotopic ages; Geological Survey of Canada, Paper 60-7.

Lutes, G.
1987: Geology and geochemistry of the Pokiok batholith, New Brunswick; New Brunswick Department of Natural Resources, Mineral Resources Division, Report of Investigation 22, 55 p.

MacDonald, M.A. and Clarke, D.B.
1985: The petrology, geochemistry and economic potential of the Musquodoboit batholith, Nova Scotia; Canadian Journal of Earth Sciences, v. 22, p. 1633-1642.

MacDonald, M.A. and Horne, R.J.
1987: Petrological and geochemical aspects of a zoned pluton within the South Mountain batholith, central Nova Scotia; Nova Scotia Department of Mines and Energy, Mineral Resources Division, Report 87-5, p. 121-128.

MacDonald, M.A. and Smith, P.K.
1980: Geology of the Cape North area, northern Cape Breton Island, Nova Scotia; Nova Scotia Department of Mines and Energy, Paper 80-1, 11 p.

MacDonald, M.A., Corey, M.C., Ham, L.J., and Horne, R.J.
1987: South Mountain batholith project: progress report; Nova Scotia Department of Mines and Energy, Mineral Resources Division, Report 87-5, p. 99-104.

MacLellan, H.E., Taylor, R.P., and Lux, D.R.
1986: Geologic and geochronologic investigations of the Burnthill granite; New Brunswick Department of Natural Resources, Mineral Resources Division, Information Circular 86-2, p. 48-60.

Marillier, F., Keen, C.E., Stockmal, G.S., Quinlan, G., Williams, H., Colman-Sadd, S.P., and O'Brien, S.J.
1989: Crustal structure and surface zonation of the Canadian Appalachians: implications of deep seismic reflection data; Canadian Journal of Earth Sciences, v. 26, p. 305-321.

Martin, R.F.
1970: Petrogenetic and tectonic implications of two contrasting Devonian batholithic associations in New Brunswick, Canada; American Journal of Science, v. 268, p. 309-321.

Mattinson, J.M.
1975: Early Paleozoic ophiolite complexes of Newfoundland: isotopic ages of zircons; Geology, v. 3, p. 181-183.
1977: U-Pb ages of some crystalline rocks from the Burlington Peninsula, Newfoundland, and its implications for the age of Fleur de Lys metamorphism; Canadian Journal of Earth Sciences, v. 14, p. 2316-2324.

McCartney, W.D.
1967:   Whitbourne map-area, Newfoundland; Geological Survey of Canada, Memoir 341, 132 p.
1969:   Geology of the Avalon peninsula, southeastern Newfoundland; American Association of Petroleum Geology, Memoir 12, p. 115-129.

McCartney, W.D., Poole, W.H., Wanless, R.K., Williams, H., and Loveridge, W.D.
1966:   Rb/Sr age and geological setting of the Holyrood granite, southeast Newfoundland; Canadian Journal of Earth Sciences, v. 3, p. 947-958.

McCutcheon, S.R.
1986:   Mount Pleasant caldera project; New Brunswick Department of Natural Resources, Mineral Resources Division, Information Circular 86-2, p. 36.

McCutcheon, S.R., Lutes, G., Gauthier, G., and Brooks, C.
1981:   The Pokiok batholith: a contaminated Acadian intrusion with an anomalous Rb-Sr age; Canadian Journal of Earth Sciences, v. 18, p. 910-918.

McKenzie, C.B. and Clarke, D.B.
1975:   Petrology of the South Mountain batholith, Nova Scotia; Canadian Journal of Earth Sciences, v. 12, p. 1209-1218.

McLeod, M.J.
1990:   Geology, geochemistry and related mineral deposits of the Saint George batholith; New Brunswick Department of Natural Resources and Energy, Mineral Resources Report 5, 169 p.

McLeod, M.J., Taylor, R.P., and Lux, D.R.
in press: Geology, ^{40}Ar/^{39}Ar geochronology and Sn-W-Mo bearing sheeted veins of the Mount Douglas granite, SW New Brunswick; Canadian Institute of Mining and Metallurgy, Bulletin.

Muecke, G.K. and Clarke, D.B.
1981:   Geochemical evolution of the South Mountain batholith, Nova Scotia: rare earth element evidence; Canadian Mineralogist, v. 19, p. 133-145.

Murphy, J.B., Cameron, K., Dostal, J., Keppie, J.D., and Hynes, A.J.
1985:   Cambrian volcanism in Nova Scotia, Canada; Canadian Journal of Earth Sciences, v. 22, p. 599.

Murphy, J.B., Keppie, J.D., and Hynes, A.D.
1991:   Geology of the Antigonish Highlands, Nova Scotia; Geological Survey of Canada, Paper 89-10, 115 p.

Murphy, J.B., Pe-piper, G., Nance, R.D., and Turner, D.
1988:   A preliminary report on geology of the eastern Cobequid Highlands, Nova Scotia; in Current Research, Part B, Geological Survey of Canada, Paper 88-1B, p. 99-107.

Neale, E.R.W.
1963:   Pleasant Bay map-area, Nova Scotia; Geological Survey of Canada, Map 1119A.

Neale, E.R.W. and Nash, W.O.
1963:   Sandy Lake (east half), Newfoundland, 12 H (E 1/2); Geological Survey of Canada, Paper 62-28.

Neale, E.R.W. and Pajari, G.E.
1972:   Granitic intrusions; the Appalachian structural province; in Variations in Tectonic Style in Canada, (ed.) R.A. Price and R.J.W. Douglas; Geological Association of Canada, Special Paper 11, p. 223-243.

O'Beirne-Ryan, A.M., Barr, S.M., and Jamieson, R.A.
1986:   Contrasting petrology and age of two megacrystic granitoid plutons, Cape Breton Island, Nova Scotia; in Current Research, Part B; Geological Survey of Canada, Paper 86-1B, p. 179-189.

O'Brien, S.J.
1983:   Geology of the eastern half of the Peter Snout map-area (11P, 13E), Newfoundland; Newfoundland and Labrador Department of Mines and Energy, Mineral Development Division, Report 83-1, p. 57-67.

O'Brien, S.J. and Taylor, S.
1979:   Geology of Baine Harbour (1M/7) and Point Enragee (1M/6) map-areas, Newfoundland; Newfoundland and Labrador Department of Mines and Energy, Mineral Development Division, Report 79-1, p. 75-81.

O'Brien, S.J., Dickson, W.L., and Blackwood, R.F.
1986:   Geology of the central portion of the Hermitage Flexure area, Newfoundland; Newfoundland and Labrador Department of Mines and Energy, Mineral Development Division, Report 86-1, p. 189-208.

O'Brien, S.J., Nunn, G.A.G., Dickson, W.L., and Tuach, J.
1984:   Geology of the Terrenceville (11M/10) and Gisborne Lake (11M/15) map-areas, Newfoundland; Newfoundland and Labrador Department of Mines and Energy, Mineral Development Division, Report 84-4, 54 p.

O'Brien, S.J., O'Neill, P.P., King, A.F., and Blackwood, R.F.
1988:   Eastern margin of the Newfoundland Appalachians: a cross-section of the Avalon and Gander zones; Geological Association of Canada, Field Trip Guidebook B4.

O'Brien, S.J., Strong, P.G., and Evans, J.L.
1977:   The geology of Grand Bank (1M/4) and Lamaline (1L/13) map-areas, Burin Peninsula, Newfoundland; Newfoundland and Labrador Department of Mines and Energy, Mineral Development Division, Report 77-7.

Olszewski, W.J. and Gaudette, H.E.
1982:   Age of the Brookville gneiss and associated rocks, southeastern New Brunswick; Canadian Journal of Earth Sciences, v. 19, p. 2158-2166.

Olszewski, W.J., Gaudette, H.E., and Poole, W.H.
1980:   Rb-Sr whole rock and U-Pb zircon ages from the Greenhead Group, New Brunswick; Geological Society of America, Abstracts with Program, v. 12, p. 76.

O'Neill, P. and Lux, D.
1989:   Tectonothermal history and ^{40}Ar/^{39}Ar geochronology of the northeastern Gander Zone, Weirs Pond area; Newfoundland and Labrador Department of Mines and Energy, Mineral Development Division, Report 89-1, p. 131-139.

O'Reilly, G.A.
1977:   Field relations and mineral potential of the granitoid rocks of southeastern Cape Breton Island; Nova Scotia Department of Mines and Energy, Mineral Resources Division, Report 77-1.
1983:   Geology of the Sangster Lake, Larrys River and Forest Hill granitoid plutons; Nova Scotia Department of Mines and Energy, Mineral Resources Division, Seventh Annual Open House and Review, Program and Summaries, p. 57-60.

O'Reilly, G.A., Farley, E.J., and Charest, M.H.
1982:   Metasomatic-hydrothermal mineral deposits of the New Ross-Mahone Bay area, Nova Scotia; Nova Scotia Department of Mines and Energy, Mineral Resources Division, Paper 82-2.

Owen, J.V.
1986:   Geology of the Silver Mountain area, western Newfoundland; in Current Research, Part A, Geological Survey of Canada, Paper 86-1A, p. 515-522.

Owen, J.V. and Machin, D.C.
1987:   Petrography and geochemistry of some mafic dykes in the Long Range inlier, western Newfoundland; in Current Research, Part A, Geological Survey of Canada, Paper 87-1A, p. 305-316.

Owen, J.V., Campbell, J.E.M., and Dennis, F.A.R.
1987:   Geology of the Lake Michel area, Long Range inlier, western Newfoundland; in Current Research, Part A, Geological Survey of Canada, Paper 87-1A, p. 643-652.

Pajari, G.E.
1977:   Field guide to the geology and plutonic rocks of southeastern New Brunswick and the Penobscot Bay area of Maine; Department of Geology, University of New Brunswick, Fredericton, New Brunswick, 69 p.

Pajari, G.E., Currie, K.L., Cherry, M.E., and Pickerill, R.K.
1981:   Heat energy, metamorphism and plutonism: a northern Appalachian model involcing collapsed continental margins; Geological Society of America, Abstracts with Programs, v. 13, p. 169.

Pajari, G.E., Trembath, L.T., Cormier, R.F., and Fyffe, L.R.
1974:   The age of the Acadian deformation in southwestern New Brunswick; Canadian Journal of Earth Sciences, v. 11, p. 1309-1313.

Paktunc, A.D.
1987:   Nickel, copper, platinum and palladium relations in Ni-Cu deposits of the St. Stephen intrusion, New Brunswick; in Current Research, Part B, Geological Survey of Canada, Paper 87-1A, p. 543-553.
1988:   The Portage Brook troctolite: a layered intrusion in the New Brunswick Appalachians; Geological Survey of Canada, Paper 88-1B, p. 149-154.

Palmer, A.R.
1983:   The Decade of North American Geology: Geologic Time Scale; Geology, v. 11, p. 503-504.

Papezik, V.S. and Barr, S.M.
1981:   The Shelburne dyke, an early Mesozoic dyke in Nova Scotia; mineralogy, chemistry and regional significance; Canadian Journal of Earth Sciences, v. 18, p. 1346-1352.

Papezik, V.S. and Hodych, J.P.
1980:   Early Mesozoic diabase dykes of the Avalon Peninsula, Newfoundland; petrochemistry, mineralogy and origin; Canadian Journal of Earth Sciences, v. 17, p. 1417-1430.

Papezik, V.S., Keats, H.F., and Vahtra, J.
1978:   Geology of the Foxtrap pyrophyllite deposit, Avalon Peninsula, Newfoundland; Canadian Mining and Metallurgical Bulletin, v. 71, no. 799, p. 152-160.

Payette, C. and Martin, R.F.
1987:   The Welsford igneous complex, southern New Brunswick; rift related Acadian magmatism; in Current Research, Part A, Geological Survey of Canada, Paper 87-1A, p. 239-248.

Pearce, J.A., Harris, N.W.B., and Tindle, A.G.
1984:   Trace element determination diagrams for the tectonic interpretation of granitic rocks; Journal of Petrology, v. 25, p. 956-983.

Pe-Piper, G. and Piper, J.W.
1987:   The pre-Carboniferous rocks of the western Cobequid Hills, Avalon Zone, Nova Scotia; Maritime Sediments and Atlantic Geology, v. 23, p. 41-48.

Pe-Piper, G., Murphy, J.B., and Turner, D.S.
1989:   Petrology, geochemistry and tectonic setting of some Carboniferous plutons of the Eastern Cobequid Hills; Atlantic Geology, v. 25, p. 37-49.

Piasecki, M.A.J.
1988:   A major ductile shear zone in the Bay d'Espoir area, Gander terrane, southeastern Newfoundland; Newfoundland and Labrador Department of Mines and Energy, Mineral Development Division, Paper 88-1, p. 135-144.

Pickerill, R.K., Pajari, G.E., Currie, K.L., and Berger, A.R.
1978:   Carmanville map-area, Newfoundland; the northeastern end of the Appalachians; in Current Research, Part A, Geological Survey of Canada, Paper 78-1A, p. 209-216.

Pitcher, W.S.
1983:   Granite type and tectonic environment; in Mountain Building Processes, (ed.) K. Hsu; Academic Press, London, p. 19-40.

Poole, W.H.
1980a: Rb-Sr ages of some granitic rocks between Ludgate Lake and Negro Harbour, southwestern New Brunswick; in Current Research, Part C, Geological Survey of Canada, Paper 80-1C, p. 170-173.
1980b: Rb-Sr age study of the Moulton Hill granite, Sherbrooke area, Quebec; in Current Research, Part C, Geological Survey of Canada, Paper 80-1C, p. 185-189.
1980c: Rb-Sr age of granitic rocks in ophiolite, Thetford Mines-Black Lake area, Quebec; in Current Research, Part C, Geological Survey of Canada, Paper 80-1C, p. 181-184.
1980d: Rb-Sr ages of the "sugar" granite and Lost Lake granite, Miramichi anticlinorium, Hayesville map-area; in Current Research, Part C, Geological Survey of Canada, Paper 80-1C, p. 174-180.

Poole, W.H. and Lowden, W.D.
1980:   Rb-Sr age of Shunacadie pluton, central Cape Breton Island, Nova Scotia; in Current Research, Part C, Geological Survey of Canada, Paper 80-1C, p. 165-69.

Pringle, I.R.
1978:   Rb-Sr ages of silicic igneous rocks and deformation, Burlington Peninsula, Newfoundland; Canadian Journal of Earth Sciences, v. 15, p. 293-300.

Raeside, R.P., Barr, S.M., and Jong, W.
1984:   Geology of the Ingonish River-Wreck Cove area, Cape Breton Island; Nova Scotia Department of Mines and Energy, Mineral Resources Division, Report 84-1, p. 249-258.

Rankin, D.W.
1975:   The continental margin of eastern North America in the southern Appalachians: the opening and closing of the proto-Atlantic Ocean; American Journal of Science, v. 275A, p. 298-336.

Rast, N.
1979:   Precambrian meta-diabases of southern New Brunswick: the opening of the Iapetus Ocean?; Tectonophysics, v. 59, p. 127-137.

Reynolds, P.H., Elias, P.M., Muecke, G.K., and Grist, A.M.
1987:   Thermal history of the southwestern Meguma Zone, Nova Scotia from an $^{40}Ar/^{39}Ar$ and fission track dating study of intrusive rocks; Canadian Journal of Earth Sciences, v. 24, p. 1952-1965.

Reynolds, P.H., Taylor, K.A., and Morgan, W.R.
1981a: $^{40}Ar/^{39}Ar$ ages from the Botwood-Mount Peyton region, Newfoundland; possible paleomagnetic implications; Canadian Journal of Earth Sciences, v. 18, p. 1850-1855.

Reynolds, P.H., Zentilli, M., and Muecke, G.K.
1981b: K-Ar and $^{40}Ar/^{39}Ar$ geochronology of granitoid rocks from southern Nova Scotia: its bearing on the geological evolution of the Meguma Zone; Canadian Journal of Earth Sciences, v. 18, p. 386-394.

Richardson, J.M., Bell, K., Blenkinsop, J., and Watkinson, D.H.
1988:   Fluid saturation textures and Rb-Sr isotopic data from the East Kemptville tin deposit, southwestern Nova Scotia; in Current Research, Part A, Geological Survey of Canada, Paper 88-1A, p. 163-171.

Richardson, J.M., Spooner, E.T.C., and McAuslan, D.A.
1982:   The East Kemptville tin deposit, Nova Scotia; an example of a large tonnage-low grade, greisen-hosted deposit in the endocontact zone of a granite batholith; in Current Research, Part B, Geological Survey of Canada, Paper 82-1B, p. 27-32.

Riley, G.C.
1962:   Stephenville map-area, Newfoundland; Geological Survey of Canada, Memoir 325.

Riordon, P.H.
1954:   The Thetford Mines-Black Lake area, Megantic and Wolfe counties; Quebec Department of Natural Resources, Mineral Resources Division, Preliminary Report 295.

Robert, J.L.
1966a: Géologie de la région du Mont Hog's Back, Comte de Gaspé-Nord; Ministère de l'Énergie et des Ressources du Québec, Rapport RP 540, 30 p.
1966b: Géologie de la région du Mont Vallières-de-Saint-Réal, Comte de Gaspé-Nord; Ministère de l'Énergie et des Ressources du Québec, Rapport 549, 22 p.

Roddick, J.C. and Bevier, M.L.
1988:   Conventional and ion microprobe U-Pb ages of two Paleozoic granites, Miramichi terrane, New Brunswick; Geological Association of Canada; Programs with Abstracts, v. 13, p. A105.

Rogers, H.D.
1985:   Geology of the igneous-metamorphic complex of Shelburne and eastern Yarmouth counties, Nova Scotia; in Current Research, Part A, Geological Survey of Canada, Paper 85-1A, p. 773-777.

Ruitenberg, A.A.
1967:   Stratigraphy, structure and metallization of the Piskahegan-Rolling Dam area; Leidse Geologische Medelingen Deal, v. 40, p. 79-120.
1968:   Geology and mineral deposits, Passamaquoddy Bay area; New Brunswick Department of Natural Resources, Mineral Resources Division, Report of Investigation 9.
1970:   Mineralized structures in the Johnson Croft Annidale, Jordan River and Black River areas; New Brunswick Department of Natural Resources, Mineral Resources Division, Report of Investigation 13.

Ruitenberg, A.A. and Fyffe, L.R.
1982:   Mineral deposits associated with granitoid intrusions and related subvolcanic stocks in New Brunswick, and their relation to Appalachian tectonic evolution; Canada Institute of Mining Bulletin, June, p. 1-15.

Ruitenberg, A.A., Giles, P.S., Venugopal, D.V., Buttimer, S.M., McCutcheon, S.R., and Chandra, J.
1979:   Geology and mineral deposits, Caledonia area; New Brunswick Department of Natural Resources, Mineral Resources Division, Memoir 1.

St. Julien, P.
1970:   Geology of Disraeli area (eastern half), Frontenac, Wolfe and Megantic counties; Quebec Department of Natural Resources, Mineral Resources Division, Report 587.

St. Julien, P., Slivitsky, A.E., and Feininger, T.
1983:   A seismic reflection profile across the southern Quebec Appalachians; Geological Society of America, Memoir 158.

St. Peter, C.
1977:   Geology of parts of Restigouche, Victoria and Madawaska counties, northwestern New Brunswick; New Brunswick Department of Natural Resources, Mineral Resources Division, Report of Investigation 17.
1981:   Geology of North Branch Southwest Miramichi River, map-area J-14, J-15, J-16; New Brunswick Department of Natural Resources, Mineral Resources Division, Map Report 80-1.
1982:   Geology of Juniper-Knowlesville-Carlisle area map-areas I-16, I-17, I-18; New Brunswick Department of Natural Resources, Mineral Resources Division, Map Report 82-1.

Saunders, C.M. and Smyth, W.R.
1990:   Geochemical characteristics of the Gull Lake intrusive suite and Devils Room granite, western White Bay, Newfoundland; Newfoundland and Labrador Department of Mines and Energy, Mineral Development Division, Report 90-1, p. 183-199.

Saunders, C.M. and Tuach, J.
1989:   Zr-Nb-Y-REE mineralisation in the Cross Hills plutonic suite; Newfoundland and Labrador Department of Mines and Energy, Mineral Development Division, Report 89-1, p. 181-192.

Seal, R.R., Clark, A.H., and Morissy, C.J.
1988: Lake George, southwestern New Brunswick: a Silurian multi-stage polymetallic (Sb-W-Mo-Au-base metal) hydrothermal centre; Canadian Institute of Mining and Metallurgy, Special Volume 39.

Sexton, A.A. and Cotie, A.A.
1985: Petrogenesis and economic geology of the Sporting Mountain pluton, Cape Breton Island; in Current Research, Part A, Geological Survey of Canada, Paper 85-1A, p. 253-259.

Simonetti, A. and Doig, R.
1990: U-Pb and Rb-Sr geochronology of Acadian plutonism in the Dunnage Zone of the southeastern Quebec Appalachians; Canadian Journal of Earth Sciences, v. 27, p. 881-892.

Skinner, R.
1975: Geology of the Tetagouche Lakes, Bathurst and Nepisguit Falls map areas, New Brunswick, with emphasis on the Tetagouche Group; Geological Survey of Canada, Memoir 371, 168 p.

Smith, P.K.
1981: Geology of the Sherbrooke area, Guysborough County, Nova Scotia; Nova Scotia Department of Mines and Energy, Mineral Resources Division, Report 81-1, p. 77-94.

Smith, T.E.
1974: The geochemistry of the granitic rocks of Halifax County, Nova Scotia; Canadian Journal of Earth Sciences, v. 11, p. 650-657.
1975: Layered granitic rocks at Chebucto Head, Halifax County, Nova Scotia; Canadian Journal of Earth Sciences, v. 12, p. 456-463.

Smyth, W.R. and Schillereff, H.S.
1982: The pre-Carboniferous geology of southwest White Bay; Newfoundland and Labrador Department of Mines and Energy, Mineral Development Division, Report 82-1, p. 78-98.

Steiger, R.H. and Jager, E.
1977: Subcommission on geochronology: convention on the use of decay constants in geo- and cosmo-chronology; Earth and Planetary Science Letters, v. 36, p. 359-362.

Stevens, R.D., Delabio, R.N., and Lachance, G.R.
1982: Age determinations and geologic studies: K-Ar isotopic ages, Report 16; Geological Survey of Canada, Paper 82-2, 48 p.

Stewart, D.B., Unger, J.D., Phillips, J.D., Goldsmith, R., Poole, W.H., Spencer, C.P., Green, A.G., Loiselle, M.C., and St. Julien, P.
1986: The Quebec-western Maine seismic reflection profile; setting and first year results; in Reflection Seismology; The Continental Crust; (ed.) M. Barazangi and L. Brown; American Geophysical Union, Geodynamics Series, v. 14, p. 189-199.

Stewart, R.D.
1979: The geology of the Benjamin River intrusive complex, Restigouche County, New Brunswick; M.Sc. thesis, Carleton University, Ottawa, Ontario.

Stockmal, G.S., Colman-Sadd, S.P., Keen, C.E., O'Brien, S.J., and Quinlan, G.
1987: Collision along an irregular margin: a regional plate tectonic interpretation of the Canadian Appalachians; Canadian Journal of Earth Sciences, v. 24, p. 1098-1107.

Streckeisen, A.L.
1976: To each plutonic rock its proper name; Earth Science Reviews, v. 12, p. 1-33.

Strong, D.F.
1979: The Mount Peyton batholith, central Newfoundland: a bimodal calc-alkaline suite; Journal of Petrology, v. 20, p. 119-138.
1980: Granitoid rocks and associated mineral deposits of eastern Canada and western Europe; in The Continental Crust and its Mineral Deposits, (ed.) D.W. Strangway; Geological Association of Canada, Special Paper 20, p. 742-769.

Strong, D.F. and Dostal, J.
1980: Dynamic melting of Proterozoic upper mantle: evidence from rare earth elements in oceanic crust of eastern Newfoundland; Contributions to Mineralogy and Petrology, v. 72, p. 65-73.

Strong, D.F. and Dupuy, C.
1982: Rare earth elements in the bimodal Mount Peyton batholith; evidence of crustal anatexis by mantle derived magma; Canadian Journal of Earth Sciences, v. 19, p. 308-315.

Strong, D.F. and Harris, A.
1974: The petrology of the Mesozoic alkaline intrusives of central Newfoundland; Canadian Journal of Earth Sciences, v. 11, p. 1208-1220.

Strong, D.F. and Minatides, D.G.
1975: Geochemistry and tectonic setting of the Late Precambrian Holyrood plutonic series of eastern Newfoundland; Lithos, v. 8, p. 283-295.

Strong, D.F. and Payne, J.G.
1973: Early Paleozoic volcanism and metamorphism of the Moretons Harbour-Twillingate area, Newfoundland; Canadian Journal of Earth Sciences, v. 10, p. 1363-1379.

Strong, D.F., Dickson, W.L., O'Driscoll, C.F., and Kean, B.F.
1974: Geochemistry of eastern Newfoundland granitoid rocks; Newfoundland and Labrador Department of Mines and Energy, Mineral Development Division, Report 74-3.

Strong, D.F., O'Brien, S.J., Taylor, S.W., Strong, P.G., and Wilton, D.H.
1978: Geology of the Marystown (1M/3) and St. Lawrence (1M/4) map-areas, Newfoundland; Newfoundland and Labrador Department of Mines and Energy, Mineral Development Division, Report 77-8, 81 p.

Stukas, V. and Reynolds, P.H.
1974a: $^{40}Ar/^{39}Ar$ age dating of the Brighton gabbro complex, Lush's Bight terrane, Newfoundland; Canadian Journal of Earth Sciences, v. 11, p. 1485-1488.
1974b: $^{40}Ar/^{39}Ar$ dating of the Long Range dykes of Newfoundland; Earth and Planetary Science Letters, v. 22, p. 256-266.

Taylor, F.C.
1967: Reconnaissance geology of Shelburne map-area, Queens, Shelburne and Yarmouth counties, Nova Scotia; Geological Survey of Canada, Memoir 349.
1969: Geology of the Annapolis-St. Mary's Bay map-area, Nova Scotia; Geological Survey of Canada, Memoir 358.

Taylor, R.P., Lux, D.R., MacLellan, H.E., and Hubacher, F.
1987: Age and genesis of granite-related W-Sn-Mo mineral deposits, Burnthill, New Brunswick, Canada; Economic Geology, v. 82, p. 2187-2198.

Taylor, R.P., Strong, D.F., and Kean, B.F.
1980: The Topsails igneous complex: Silurian and Devonian peralkaline magmatism in western Newfoundland; Canadian Journal of Earth Sciences, v. 17, p. 425-439.

Taylor, S.W., O'Brien, S.J., and Swinden, H.S.
1979: Geology and mineral potential of the Avalon Zone and granitoid rocks of eastern Newfoundland; Newfoundland and Labrador Department of Mines and Energy, Mineral Development Division, Report 79-3, 50 p.

Teng, H.C. and Strong, D.F.
1976: Geology and geochemistry of the St. Lawrence peralkaline granite and associated fluorite deposits, southeast Newfoundland; Canadian Journal of Earth Sciences, v. 13, p. 1374-1385.

Thomas, M.D. and Willis, C.
1988: Gravity models of the Saint George batholith, New Brunswick Appalachians; Canadian Journal of Earth Sciences, v. 25, p. 1317-1330.

Thurlow, J.G.
1981: The Buchans Group: its stratigraphic and structural setting; Geological Association of Canada, Special Paper 22, p. 79-90.

Tuach, J.
1987: Mineralized environments, metallogenesis, and the Doucers Valley fault complex, western White Bay: a philosophy for gold exploration in Newfoundland; Newfoundland and Labrador Department of Mines and Energy, Mineral Development Division, Report 87-1, p. 129-144.

Tuach, J. and Delaney, P.W.
1987: Tungsten-molybdenum in the Granite Lake-Meelpaeg Lake area, Newfoundland; Newfoundland and Labrador Department of Mines and Energy, Mineral Development Division, Report 87-1, p. 113-127.

Tuach, J., Davenport, P.H., Dickson, W.L., and Strong, D.F.
1986: Geochemical trends in the Ackley granite, southeast Newfoundland: their relevance to magmatic-metallogenic processes in high silica granitoid systems; Canadian Journal of Earth Sciences, v. 23, p. 747-765.

Vallieres, A., Hubert, C., and Brooks, C.
1978: A slice of basement in the western margin of the Appalachian orogen, St. Malachie, Quebec; Canadian Journal of Earth Sciences, v. 15, p. 1242-1249.

van Berkel, J.T.
1987: Geology of the Dashwoods Pond, St. Fintans and Main Gut map areas, southwestern Newfoundland; in Current Research, Part A, Geological Survey of Canada, Paper 87-1A, p. 399-408.

van Berkel, J.T. and Currie, K.L.
1988: Geology of the Puddle Pond and Little Grand Lake map-area, southwestern Newfoundland; Newfoundland and Labrador Department of Mines and Energy, Mineral Development Division, Report 88-1.

van Staal, C.R.
1987: Tectonic setting of the Tetagouche Group in northern New Brunswick; implications for plate tectonic models of the northern Appalachians; Canadian Journal of Earth Sciences, v. 24, p. 1329-1344.

Venugopal, D.V.
1978: Geology of the Benton-Kirkland-Upper Eel River Bend map-area, G-22; New Brunswick Department of Natural Resources, Mineral Resources Division, Map Report 78-3.
1979: Geology of the Debec Junction-Gibson-Millstream-Temperance Vale-Meductic region, map-areas G-21, H-21, I-21 and H-22; Map Report 79-5.
1981: Geology of the Hartland-Woodstock-Nortondale region, map-areas H-19, H-20, I-20; New Brunswick Department of Natural Resources, Mineral Resources Division, Map Report 81-6.
1982: Geology of the upper parts Becauguimec, Keswick and Nashwaak rivers, map-areas I-19, J-19, J-20; New Brunswick Department of Natural Resources, Mineral Resources Division, Map Report 82-2.

Wanless, R.K., Stevens, R.D., Lachance, G.R., and Delabio, R.N.
1972: Age determinations and geological studies, K-Ar isotopic ages, Report 10; Geological Survey of Canada, Paper 71-2.
1973: Age determinations and geological studies, Report 11; Geological Survey of Canada, Paper 73-2.

Wanless, R.K., Stevens, R.D., Lachance, G.R., and Edmonds, C.M.
1965: Age determinations and geological studies, K-Ar isotopic ages; Geological Survey of Canada, Paper 65-17.
1966: Age determinations and geological studies, K-Ar isotopic ages; Geological Survey of Canada, Paper 66-17.

Watters, S.E.
1987: Gold-bearing rocks, Bay of Fundy coastal zone; New Brunswick Department of Natural Resources and Energy, Information Circular 87-2, p. 41-45.

Whalen, J.B.
1983: The Ackley City batholith, southeastern Newfoundland: evidence for crystal versus liquid-state fractionation; Geochimica et Cosmochimica Acta, v. 47, p. 1443-1457.
1986: A-type granites in New Brunswick; in Current Research, Part A, Geological Survey of Canada, Paper 86-1A, p. 297-300.
1987: Geology of a northern portion of the Central Plutonic Belt, New Brunswick; in Current Research, Part A, Geological Survey of Canada, Paper 87-1A, p. 209-217.
1988: Granitic rocks of New Brunswick and Gaspé, Quebec; a transect across the southern Canadian Appalachians; Geological Association of Canada, Program with Abstracts, v. 13, p. A133.
1993: Geology, petrography and geochemistry of Appalachian granites in New Brunswick and Gaspésie, Quebec; Geological Survey of Canada, Bulletin 436, 124 p.

Whalen, J.B. and Currie, K.L.
1983: The Topsails igneous terrane of western Newfoundland; in Current Research, Part A, Geological Survey of Canada, Paper 83-1A, p. 15-23.
1984a: A peralkaline granite near Hare Hill, western Newfoundland; in Current Research, Part A, Geological Survey of Canada, Paper 84-1A, p. 181-184.
1984b: The Topsails igneous terrane, western Newfoundland: evidence for magma mixing; Contributions to Mineralogy and Petrology, v. 87, p. 319-327.
1990: The Topsails suite, western Newfoundland: fractionation and magma-mixing in an "orogenic" A-type granite suite; in Ore-Bearing Granite Systems: Petrogenesis and Mineralising Processes, (ed.) H.J. Stein and J.L. Hannah; Geological Society of America, Special Paper 246, p. 287-299.

Whalen, J.B. and Gariepy, C.
1986: Petrogenesis of the McGerrigle plutonic complex, Gaspé, Quebec: a preliminary report; in Current Research, Part A, Geological Survey of Canada, Paper 86-1A, p. 265-274.

Whalen, J.B. and Roddick, J.C.M.
1987: K-Ar geochronology of the McGerrigle plutonic complex, Gaspésie Peninsula, Quebec; in Current Research, Part A, Geological Survey of Canada, Paper 87-1A, p. 375-380.

Whalen, J.B., Currie, K.L., and Chappell, B.W.
1987: A-type granites: geochemical characteristics and discrimination; Contributions to Mineralogy and Petrology, v. 95, p. 420-436.

Whalen, J.B., Currie, K.L., and van Breemen, O.
1985: Episodic Ordovician-Silurian magmatism in the Topsails igneous terrane, western Newfoundland; Transactions of the Royal Society of Edinburgh, v. 78, p. 17-28.

Whalen, J.B., Jenner, G.A., Currie, K.L., Barr, S.M., Longstaffe, F.J., and Hegner, E.
1994: Do geochemical and isotopic (Nd, O and Pb) data from granitoid rocks indicate repeated delamination in the evolution of the Avalon Zone of southern New Brunswick; Journal of Geology, v. 102, p. 138-156.

White, A.J.R. and Chappell, B.W.
1983: Granitoid types and their distribution in the Lachlan Fold Belt, southeastern Australia; in Circum Pacific plutonic terranes, (ed.) J.A. Roddick; Geological Society of America, Memoir 159, p. 21-30.

White, C.E., Barr, S.M., and Campbell, R.H.
1990: Petrology of the Creignish Hills pluton, Cape Breton Island, Nova Scotia; Atlantic Geology, v. 26, p. 109-124.

Wiebe, R.A.
1972: Igneous and tectonic events in northeastern Cape Breton Island, Nova Scotia; Canadian Journal of Earth Sciences, v. 9, p. 1262-1277.
1975: Origin and emplacement of Acadian granitic rocks, northern Cape Breton Island; Canadian Journal of Earth Sciences, v. 12, p. 252-262.

Williams, H.
1964: Botwood, Newfoundland (2E); Geological Survey of Canada, Map 60-1963.
1968: Geology, Wesleyville; Geological Survey of Canada, Map 1228A.
1977: The Coney Head complex: another Taconic allochthon in west Newfoundland; American Journal of Science, v. 277, p. 1279-1295.
1979: Appalachian Orogen in Canada; Canadian Journal of Earth Sciences, v. 16, p. 792-807.
1984: Miogeoclines and suspect terranes of the Canadian-Appalachian Orogen: tectonic patterns in the North Atlantic region; Canadian Journal of Earth Sciences, v. 21, p. 887-901.

Williams, H. and Payne, J.G.
1975: The Twillingate granite and nearby volcanic groups: an island arc complex in northeastern Newfoundland; Canadian Journal of Earth Sciences, v. 12, p. 982-995.

Williams, H. and Smyth, W.R.
1983: Geology of the Hare Bay allochthon; Geological Survey of Canada, Memoir 400, p. 109-132.

Williams, H. and St. Julien, P.
1982: The Baie Verte-Brompton line: early Paleozoic continent-ocean interface in the Canadian Appalachians; Geological Association of Canada, Special Paper 24, p. 177-207.

Williams, H., Colman-Sadd, S.P., and Swinden, H.S.
1988: Tectonic-stratigraphic subdivision of central Newfoundland; in Current Research, Part A, Geological Survey of Canada, Paper 88-1A, p. 91-98.

Williams, H., Dallmeyer, R.D., and Wanless, R.K.
1976: Geochronology of the Twillingate granite and Herring Neck Group, Notre Dame Bay, Newfoundland; Canadian Journal of Earth Sciences, v. 13, p. 1591-1601.

Williams, H., Gillespie, R.T., and Knapp, D.A.
1982: Geology of Pasadena map-area, Newfoundland; in Current Research, Part A, Geological Survey of Canada, Paper 82-1A, p. 282-288.

Williams, H., Gillespie, R.T., and van Breemen, O.
1985: A late Precambrian rift-related suite in western Newfoundland; Canadian Journal of Earth Sciences, v. 22, p. 1727-1735.

Williams-Jones, A.E.
1982: Patapedia; an Appalachian calc-silicate hosted copper prospect of porphyry affinity; Canadian Journal of Earth Sciences, v. 19, p. 438-455.

Wilton, D.H.C.
1985: Tectonic evolution of southwestern Newfoundland as indicated by granitoid petrogenesis; Canadian Journal of Earth Sciences, v. 22, p. 1080-1092.
1983: The geology and structural history of the Cape Ray fault zone in southwestern Newfoundland; Canadian Journal of Earth Sciences, v. 20, p. 1119.

Worsley, T.R., Nance, R.D., and Moody, J.B.
1984: Global tectonics and eustasy for the past 2 billion years; Marine Geology, v. 58, p. 373-400.

Wray, E.M.
1972: Petrography of the Tombstone stock, Nova Scotia; Canadian Journal of Earth Sciences, v. 9, p. 1455-1464.

Zentilli, M. and Reynolds, P.H.
1985: ^{40}Ar/^{39}Ar dating of micas from East Kemptville tin deposit, Yarmouth County, Nova Scotia; Canadian Journal of Earth Sciences, v. 22, p. 1546-1548.

ADDENDUM

As noted in the text, the information upon which subdivision and interpretation were based was current as of 1989. A large amount of new geochronological and isotopic data has appeared since that time. These data have generally supported, rather than undermined, the classification presented in the text and the conclusions drawn from it. Important conclusions to be drawn from this new data may be summarized as follows:

(1) Silurian plutonism marked the culmination of Paleozoic plutonism and was more widespread and important than suggested by the main text. The extent of this Silurian orogenic plutonism is complex. To the southeast, Silurian plutons are juxtaposed by a variety of processes with Avalonian plutons (O'Brien et al., 1991) and with Ordovician plutons (Colman-Sadd et al., 1992). To the north-west, some internal massifs contain mainly post-orgenic Silurian plutons, as noted in the main text, notably the Notre Dame and Dashwoods subzones of Newfoundland. However, orogenic Silurian plutonism is found west of these massifs (Cawood and Dunning, 1993). The reasons for this disposition are not presently known although magmatism along large transcurrent or transtensional faults appears to be involved.

(2) Plutonism of the Avalon Zone is of juvenile character. Combined Nd, Pb and O isotope studies (Whalen et al., 1994; Whalen, 1993; Kerr et al., 1993) show that the amount of recycled continental crust cannot exceed 25%, although a larger amount could come from young recycled underplate. These data suggest periodic delamination of a thick, mantle-derived underplate as the probable cause of cyclic volcanism in the Avalon Zone. By contrast, isotopic studies of the Gander Zone show that it is underlain by abundant old continental crust. Avalonian crust cannot be the source of the peraluminous plutons of the Gander Zone. Many current tectonic models extend Avalonian crust beneath the Gander Zone. If these models are correct, then the Avalonian crust must be composite with juvenile crust of the Avalon Zone proper melded with old continental crust sometime in the late Precambrian. Some indication that this type of model may be workable are provided by data from southern New Brunswick (Whalen et al., 1994). Thick continental crust is also present beneath the Meguma Zone (Eberz et al., 1991) If the speculations in the main text on relations between the Avalon and Meguma zones are correct, then some amalgamation of Avalonian crust to Meguma crust would be necessary prior to Ordovician time.

(3) Geophysical investigations (Quinlan et al., 1992) show that deep crust below the Canadian Appalachians comprises two, or possibly three blocks. Granitoid igneous rocks sample, or are derived from, deep crust. Contracts between distribution of granitoid plutons and the five-fold zonal distribution of supracrustal rocks must therefore be expected, as noted in the text, as least for certain time periods. Regional study of the deep crustal sources of plutons by means of isotope studies is now being actively pursued (for example Whalen et al., 1994), and should give further information on the tectonic history of the orogen.

A bibliography to support these points is appended. This bibliography is not an update of the main bibliography, but represents the author's idiosyncratic view of the most significant developments in the studies of Canadian Appalachian plutonic rocks in the past five years.

Selected bibliography

Barr, S.M. and Macdonald, A.B.
1989:   Geology of the Mabou Highlands, western Cape Breton Island, Nova Scotia; Nova Scotia Department of Mines, Paper 89-2, 65 p.

Bevier, M.L.
1990:   Preliminary geochronologic results for igneous and metamorphic rocks, New Brunswick; New Brunswick Department of Natural Resources, Information Circular 89-2, p. 208-212.

Cawood, P.A. and Dunning, G.R.
1993:   Silurian age for movement of the Baie Verte Line: implications for accretionary tectonics in the northern Appalachians; Geological Society of America, Abstracts with Programs, p. A422.

Colman-Sadd, S.P., Dunning, G.R., and Dec, T.
1992:   Dunnage-Gander relationships and Ordovician orogeny in central Newfoundland: a sediment provenance and U/Pb age study; American Journal of Science, v. 292, p. 317-355.

Dallmeyer, R.D., Doig, R., Nance, R.D., and Murphy, J.B.
1990: $^{40}Ar/^{39}Ar$ and U-Pb mineral ages from the Brookville Gneiss: implications for terrane analysis and evolution of Avalonian "basement" in southern New Brunswick; Atlantic Geology, v. 26, p. 247-257.

Dallmeyer, R.D. and Keppie, J.D.
1993: $^{40}Ar/^{39}Ar$ mineral ages from the southern Cape Breton Highlands and Creignish Hills, Cape Breton Island; evidence for a polyphase tectonothermal evolution; Jounral of Geology, v. 101, p. 467-482.

Dallmeyer, R.D. and Nance, R.D.
1990: $^{40}Ar/^{39}Ar$ ages of detrital muscovite within early Paleozoic overstep sequences, Avalon composite terrane, southern New Brunswick; implications for extent of late Precambrian tectonothermal overprint; Canadian Journal of Earth Sciences, v. 27, p. 1209-1214.
1992:   Tectonic implications of $^{40}Ar/^{39}Ar$ mineral ages from late Precambrian-Cambrian plutons, Avalon composite terrane, southern New Brunswick; Canadian Journal of Earth Sciences, v. 29, p. 2445-2462.

Dube, B., Dunning, G.R., Lauziere, K., and Roddick, C.
1993:   The Gondwanan-Laurentian suture: timing of deformation on the Cape Ray Fault, Newfoundland; Geological Society of America, Abstracts with Programs, p. A421.

Dunning, G.R., O'Brien, S.J., Colman-Sadd, S.P., and Blackwood, R.F.
1990:   Silurian orogeny in the Newfoundland Appalachians; Journal of Geology, v. 98, p. 895-913.

Dunning, G.R., O'Brien, B.H., Holdsworth, R.E., and Tucker, R.D.
1993:   Geochronology of Pan-African, Penobscot and Salinic shear zones on the Gondwanan margin, northern Appalachians; Geological Society of America, Abstracts with Programs, p. A421.

Eberz, G.W., Clarke, D.B., Chatterjee, A.K., and Giles, P.S.
1991:   Chemical and isotopic composition of the lower crust beneath the Meguma lithotectonic zone, Nova Scotia; evidence from granulite facies xenoliths; Contributions to Mineralogy and Petrology, v. 109, p. 69-88.

Elliot, C.G., Dunning, G.R., and Williams, P.F.

1991: New U-Pb zircon age constraints on the timing of deformation in north-central Newfoundland and implications for early Paleozoic Appalachian orogenesis; Geological Society of America, Bulletin 103, p. 125-135.

Greenough, J.D. and Fryer, B.J.

1991: Nd and Sr isotopic composition of the lamprophyric Malpeque Bay sill, Prince Edward Island; Candian Mineralogist, v. 29, p. 311-317.

Kempster, R.H.F., Clarke, D.B., Reynolds, P.H., and Chatterjee, A.K.

1989: Late Devonian lamprophyre dikes in the Meguma Zone of Nova Scotia; Canadian Journal of Earth Sciences, v. 26, p. 611-613.

Keppie, J.D., Dallmeyer, R.D., and Krogh, T.E.

1992: U-Pb and ^{40}Ar/^{39}Ar mineral ages from Cape North, northern Cape Breton Island: implications for accretion of the Avalon composite terrane; Canadian Journal of Earth Sciences, v. 29, p. 277-295.

Keppie, J.D. and Dallmeyer, R.D.

1989: ^{40}Ar/^{39}Ar mineral ages from Kellys Mountain, Cape Breton Island, Nova Scotia; implications for the tectonothermal evolution of the Avalon composite terrane; Canadian Journal of Earth Sciences, v. 26, p. 1509-1518.

Kerr, A., Hayes, J.P., Colman-Sadd, S.P., Dickson, W.L., and Butler, A.J.

1994: An integrated lithogeochemical database for the granitoid plutonic suites of Newfoundland; Newfoundland Geological Surveys Branch, Open File 2377.

Kontak, D.J.

1990: The East Kemptville topaz-muscovite leucogranite, Nova Scotia; I Geological setting and whole rock geochemistry; Canadian Mineralogist, v. 28, p. 787-825.

Kontak, D.J., Chaterjee, A.K., Reynolds, P.H., and Taylor, K.

1990: ^{40}Ar/^{39}Ar chronological study of the Liscomb complex, Meguma Terrane, southern Nova Scotia; Atlantic Geology, v. 26, p. 176-177.

MacDonald, M.A. and Horne, R.J.

1988: Petrology of the zoned peraluminous Halifax pluton; Atlantic Geology, v. 24, p. 33-45.

McLeod, M.J.

1990: Geology, geochemistry and related mineral deposits of the Saint George batholith, Charlotte, Queens and Kings counties, New Brunswick; New Brunswick Department of Natural Resources, Report 5, 169 p.

McLeod, M.J., Johnson, S.C., and Ruitenberg, A.A.

1994: Geological map of southeastern New Brunswick; New Brunswick Department of Natural Resources, Mineral Resource Map NR-6.

Miller, B.V., Dunning, G.R., Barr, S.M., and Raeside, R.B.

1993: Grenvillian basement in the northern Appalachian orogen; U-Pb age constraints on the origin and metamorphism of the Blair River complex, Cape Breton Island, Nova Scotia, Canada; Geological Society of America, Abstracts with Programs, p. A486.

Murphy, J.B., PePiper, G., Nance, R.D., Doig, R., and Miler, B.V.

1989: Geology of the eastern Coequid Highlands: U-Pb geochronological constraints; Nova Scotia Department of Mines and Energy, Mines and Minerals Branch, Report of Activities, p. 137-139.

Nance, R.D. and Dallmeyer, R.D.

1993: ^{40}Ar/^{39}Ar amphibole ages from the Kingston complex, New Brunswick; evidence for Siluro-Devonian tectonothermal activity and implications for the accretion of the Avalon composite terrane; Journal of Geology, v. 101, p. 375-388.

O'Brien, B.H., O'Brien, S.J., and Dunning, G.R.

1991: Silurian cover, late Precambrian-Early Ordovician basement, and the chronology of Silurian orogenesis in the Hermitage Flexure (Newfoundland Appalachians); American Journal of Science, v. 291, no. 8, p. 760-799.

Owen, J.V., Greenough, J.D., Fryer, B.J., and Longstaffe, F.J.

1992: Petrogenesis of the Potato Hill pluton, Newfoundland: transpression during the Grenvillian orogenic cycle; Journal of the Geological Society of London, v. 149, p. 923-935.

Quinlan, G.M., Hall, J., Williams, H., Wright, J.A., Colman-Sadd, S.P., O'Brien, S.J., Stockmal, G.S., and Marillier, F.

1992: Lithoprobe onshore seismic reflection transects across the Newfoundland Appalachians; Canadian Journal of Earth Sciences, v. 29, p. 1865-1877.

Reynolds, P.H., Jamieson, R.A., Barr, S.M., and Raeside, R.P.

1989: ^{40}Ar/^{39}Ar study of the Cape Breton Highlands, Nova Scotia: thermal histories and tectonic implications; Canadian Journal of Earth Sciences, v. 26, p. 2081-2091.

Rogers, H.D. and Barr, S.M.

1988: Petrology of the Shelburne and Barrington Passage plutons, southern Nova Scotia; Atlantic Geology, v. 24, p. 21-32.

Walker, J., Gower, S., and McCutcheon, S.R.

1991: Antinouri-Nicholas project, Restigouche and Gloucester counties, northern New Brunswick; New Brunswick Department of Natraul Resources, Information Circular 91-2, p. 87-100.

Whalen, J.B.

1993: Geology, petrography and geochemistry of Appalachian granites in New Brunswick and Gaspésie, Quebec; Geological Survey of Canada, Bulletin 436, 124 p.

Whalen, J.B., Jenner, G.A., Currie, K.L., Barr, S.M., Longstaffe, F.J., and Hegner, E.

1994: Do geochemical and isotopic (Nd, O and Pb) data from granitoid rocks indicate repeated delamination in the evolution of the Avalon Zone of southern New Brunswick; Journal of Geology, v. 102, p. 138-156.

Author's Address

K.L. Currie
Geological Survey of Canada
601 Booth Street
Ottawa, Ontario
K1A 0E8

Chapter 9

METALLOGENY

Chapter 9

METALLOGENY

H. Scott Swinden and S.M. Dunsworth

with contributions by:

G. Beaudoin, R.C. Boehner, J.L. Davies, W.L. Dickson, G. Duquette, D.T.W. Evans,
J.H. Fowler, L.R. Fyffe, M. Gauthier, A.F. Howse, B.F. Kean, T. Lane,
M. Malo, S.R. McCutcheon, C.F. O'Driscoll, A.A. Ruitenberg,
R.J. Ryan, A.L. Sangster, C.M. Saunders, C.R. van Staal

INTRODUCTION

H. Scott Swinden

Metallogeny in the context of the Appalachian Orogen

Metallogeny is the branch of geology that seeks to define the genetic relationships between the geological history of an area and its mineral deposits. Mineral deposits, in the broadest sense, form part of the same geological record as less-valuable rocks, and were deposited in response to processes in the same geological and tectonic environments. Mineral deposit studies both contribute to and benefit from understanding of the regional geological and tectonic development of an area. The deposits in some cases provide important data as to the geological processes operative at different times. On the other hand, regional geological models are essential to help constrain possible metallogenic models, when a large number of deposits having formed at several different times are present.

As recognized many years ago by McCartney and Potter (1962), the Canadian Appalachians provide a particularly good laboratory for the study of regional metallogeny. The regional geology is relatively well understood and interpreted in terms of well constrained tectonic models and a wide variety of mineral deposit types and ages provide a record of mineralization that spans the entire history of the orogen. This chapter considers the nature of the mineral deposits in the Canadian Appalachians and their place in the geological and tectonic framework of the orogen.

Swinden, H.S. and Dunsworth, S.M.
1995: Metallogeny; Chapter 9, in Geology of the Appalachian-Caledonian Orogen in Canada and Greenland, (ed.) H. Williams; Geological Survey of Canada, Geology of Canada, No. 6, p. 681-814 (also Geological Society of America, The Geology of North America, v. F-1).

The concept of metallogeny, as distinct from economic geology, was pioneered by de Launay (1900, 1913) who identified consistencies in the regional geographical variations in the occurrence of ores. Perhaps his major contribution was the introduction of the concepts of "Metallogenic Epoch" and "Metallogenic Province", which established time and place in the forefront of metallogenic analysis and encouraged subsequent workers to identify metallogenic epochs and provinces in virtually all parts of the world. However, the links between orogenic and metallogenic development, and the nature of the relationships between metallogenic cycles and orogenic cycles were first clearly propounded by Bilibin (1968) and colleagues in the Soviet Union. Using the framework of geosynclinal theory, in which geosynclines are considered as mobile belts that evolve through a number of stages into stable folded orogenic belts, Bilibin attempted to relate the occurrence of various types of ore deposits to three broad stages of orogenic development: (1) an initial phase of downwarping accompanied by marine volcanism and sedimentation; (2) a middle phase characterized by intense deformation and granitoid plutonism; and (3) a final stage of crustal adjustment characterized by sedimentation in successor basins. He suggested that the changes in geological processes that characterize these stages are accompanied by systematic changes in the nature of the associated mineralization through time and space, giving rise to metallogenic epochs and provinces. Bilibin's metallogenic concepts have been, and continue to be, enormously influential in metallogenic thinking, and exerted a major influence on early metallogenic interpretations of the Canadian Appalachians. McCartney and Potter (1962) were the first to interpret the metallogeny of the Canadian Appalachians in such a context and successfully placed all of its important mineral deposits within the metallogenic stages first proposed by Bilibin.

With the advent of plate tectonic theory, many workers were quick to translate Bilibin's metallogenic framework into a plate tectonic context. Although initial metallogenic attempts focussed mainly on correlation and reconstruction of pre-drift metallogenic provinces (e.g. Schuiling, 1967), attention soon shifted to consideration of the influence of plate tectonic processes on ore genesis (e.g. Guild, 1972; Mitchell and Bell, 1973; Mitchell and Garson, 1976; Sawkins, 1972; Zonenshain et al., 1972) and the interpretation of the distribution of various types of mineral deposits in ancient orogens in terms of the tectonic stages of the Wilson Cycle (e.g. Scheibner, 1974). Strong (1974) was the first to attempt such an analysis for the Canadian Appalachians. His model, which was influenced by the earlier work of McCartney and Potter (1962), showed that there was a regular and predictable relationship between certain ore deposits and their plate tectonic setting. Strong (1974) further attempted to apply this interpretation in a predictive sense by setting out some geological guidelines for recognizing potential mineralized environments.

Although many of the details of Appalachian geological and tectonic history upon which Strong's models were based have been superseded by more recent geological work, the metallogenic framework that he proposed has stood the test of time very well. His interpretation of the tectonic settings of many, if not most, important deposit types still forms the basis of regional metallogenic interpretations in the Canadian Appalachians. However, the wealth of more recent geological, geochemical, geochronological and isotopic information has resulted in substantial refinements to our understanding of both the geological history of the orogen and the settings of various mineral deposits. This, in turn, has lead to more sophisticated metallogenic models. The various types of mineral deposits can now be related much more precisely to specific stages in the geological development of the orogen than was previously possible. The improved geotectonic models for the accretionary history of the orogen permit a more complete understanding, not only of the place of mineral deposits in the geological history of the orogen, but also of the relationships between mineralizing processes and geological and tectonic processes.

Organization and scope

The purpose of this chapter is to describe the metallogeny of the Canadian Appalachians in such a way as to: (1) provide a comprehensive summary of the nature and distribution in time and space of all significant mineral deposits, and (2) to highlight the relationships between mineral deposit genesis and the overall geological development of the orogen. To this end, a strongly genetic approach to classification and description of deposits is adopted. The chapter is organized according to the stages of geological development of the orogen, as detailed elsewhere in this volume, and the descriptions of the various mineral deposits are given in the context of their place in this development. In this sense, the chapter provides a snapshot of our current understanding of the metallogenic history of the orogen and, perhaps, another milestone on the path originally pioneered by McCartney and Potter (1962) and Strong (1974).

The chapter is broadly organized into three geological and tectonic developmental stages of the Appalachian Orogen which, although here interpreted in a plate tectonic context, nonetheless broadly coincide with Bilibin's (1968) orogenic stages: (1) pre-accretion stages: mineral deposits include those from previous orogenic cycles that have been incorporated in the Appalachians as well as deposits formed at the continental margins and in the Iapetus Ocean during rift and drift phases of development; (2) synorogenic stages: mineral deposits are principally those formed as a result of faulting and granitoid plutonism that occurred during accretion of outboard terranes to the Laurentian margin; and (3) post-orogenic stages: mineral deposits are mainly those related to late- or post-orogenic faulting, granitoid intrusion, and development of successor basins across the entire width of the newly cratonized orogen.

The pre-accretion (Early Paleozoic and older) elements of the orogen define four distinct metallogenic entities, now represented by the Humber, Dunnage-Gander, Avalon and Meguma zones. Pre-accretion metallogeny proceeded independently in these zones and the contrasting suites of mineral deposits that formed in each zone reflect their diverse geological histories. Furthermore, the pre-accretion geological histories of these zones locally produced conditions for unique syn- or post-accretion mineralization (e.g. Mississippi Valley-type Zn deposits in carbonate rocks of the Humber Zone; metasomatic asbestos and talc deposits in ophiolitic ultramafic rocks). In such cases, the mineralization is described in the context of the zone in which it occurs.

Trans-zonal metallogeny in the Canadian Appalachians began at the Humber-Dunnage zone boundary during and immediately following initial accretion (Taconic Orogeny) with hydrothermal activity related to movements on major fault zones and granitoid intrusion. At this time, a pre-accretion metallogeny was still proceeding independently in outboard zones (eastern Dunnage, Gander, Avalon, Meguma). Deformation-related and granitoid-related mineralization increased in intensity and abundance as accretion and crustal thickening proceeded through the Silurian and Devonian. These styles of mineralization, which eventually encompassed all of the accreted outboard terranes, consist of two principal types: (1) structurally controlled mineralization with no apparent relationship to magmatic activity, and (2) granitoid-related mineralization. There is a spectrum of deposits between these end members. For example, mineralization related to major fault zones in northern New Brunswick and Gaspésie (e.g. Rocky Brook-Millstream Fault; Grand Pabos Fault) is associated with magmatism that is related to movement on these faults. For the purposes of this chapter, mineralization with a clear magmatic association, irrespective of a possible association to faulting, is discussed with granitoid-related mineralization.

The final stage in Appalachian metallogeny in Canada is mineralization related to the development of late Paleozoic overlap assemblages in Devonian to Permian successor basins. A wide variety of deposit types that reflect various stages of basin evolution are described.

Metallogenic discussions in this chapter are not strictly confined to "metals". Concentrations of useful non-metallic minerals (e.g. evaporites, fluorspar, asbestos, and talc) are locally of considerable importance to the mineral economy in eastern Canada, and the orebodies are readily integrated in the metallogenic history of the orogen. However, faced with the necessity of distinguishing between deposits that are appropriate to a discussion of metallogeny and those that are not, the decision has been made to exclude the following: (1) most industrial minerals which are used principally as stone (e.g. dimension stone, aggregate, slate, agricultural limestone); (2) fossil fuels; and (3) surficial materials.

NORTH AMERICAN MIOGEOCLINE: HUMBER ZONE

Introduction

S.M. Dunsworth and H. Scott Swinden

The northwestern margin of the Appalachian Orogen and adjacent St. Lawrence Platform represents an early Paleozoic passive Atlantic-type margin of the Laurentian continent (Williams, 1979) which originally lay northwest of the Iapetus Ocean (James et al., 1989). Remnants of this once extensive continental margin are recognized along the western edge of the orogen (Williams and Stevens, 1974; Williams, 1978, 1979) in southeastern Quebec, beneath the Gulf of St. Lawrence, in Cape Breton Island, and in western Newfoundland (Fig. 9.1).

Grenvillian basement granites and gneisses intruded by late Precambrian mafic dykes comprise the southeastern edge of the Canadian Shield and form basement to the Humber Zone. Late Proterozoic incipient rifting and eventual breakup of the Laurentian margin are recorded by siliciclastic and alkalic volcanic rocks. Following the rift-drift transition, siliciclastic sediments overlain by a dominantly carbonate succession record sedimentation on a passive continental margin until the Middle Ordovician when collision of outboard terranes during the Taconic Orogeny resulted in foundering of the margin and emplacement of allochthonous oceanic sedimentary and ophiolitic rocks. For the purposes of metallogenic discussions, oceanic successions of the Taconic allochthons are treated as part of the Dunnage Zone, from which they were derived.

The metallogeny of the Humber Zone includes at least six broad classes of deposits (Fig. 9.1): (1) iron-titanium deposits in western Newfoundland Grenville inliers; (2) Precambrian carbonate-hosted sulphides Cape Breton Island; (3) paleoplacer deposits in Cambrian rift successions; (4) copper deposits in Cambrian rift volcanic and sedimentary rocks; (5) Paleozoic carbonate-hosted deposits; and (6) Ba-Pb-Zn veins and disseminations in transported clastic rocks, Lower St. Lawrence region.

Iron-titanium deposits in western Newfoundland Grenville inliers

H. Scott Swinden

Amphibolite to granulite facies gneiss intruded by anorthosite is preserved in several Grenville inliers north and west of St. George's Bay, western Newfoundland. Lode deposits of titaniferous magnetite associated with the anorthosites occur in the Steel Mountain and Indian Head inliers (Heyl and Ronan, 1954; Colman-Sadd, 1974; Rose, 1969; Fig. 9.2). The deposits are generally considered to represent magmatic segregations and injections related to crystallization of anorthosites. They are geologically similar to, and probably correlative with, significant titaniferous deposits in the Grenville Province of Quebec and Ontario (e.g. Lac Allard, Magpie Mountain; Rose, 1969).

In the Steel Mountain Inlier, the main titaniferous occurrences (Bishop North, Bishop South, Hayes, several smaller occurrences; Fig. 9.2) form segregations and injections of magnetite and titanomagnetite in anorthosite. The Bishop North deposit, largest in the area, comprises a dyke-like body up to 15 m wide and 91 m long of coarse grained titanomagnetite with accessory plagioclase, pyroxene, and green spinel (Rose, 1969). Deposits in the Indian Head Inlier (Indian Head, Cliff, Upper and Lower Drill Brook, and Skindles; Fig. 9.2) were mined sporadically on a small scale from the early 1900s to the 1940s. They comprise mainly lenses of titaniferous magnetite, magnetite-ilmenite, and magnetite-ilmenite-hematite which parallel the gneissic banding in norite gneiss and pink soda-granite gneiss. The Indian Head occurrence consists of disseminated and conformable bands of titaniferous magnetite in gneissic gabbroic anorthosite. Magnetite and hematite occur together at the Cliff and Drill Brook deposits and pyrite, chalcopyrite, and molybdenite are accessories in the Upper Drill Brook deposit.

Precambrian carbonate-hosted sulphides, Cape Breton Island

A.L. Sangster

The Meat Cove zinc occurrence, hosted by a marble xenolith within the Lowland Brook syenite of the Precambrian Blair River Complex (Barr et al., 1987) (Fig. 9.3), is estimated to contain 3 990 000 tonnes grading 4% Zn (Energy, Mines and Resources Canada, 1989). It is the largest known sulphide mineral occurrence in central and northern Cape Breton Island and the only significant occurrence in the Blair River Complex. The mineralization consists of bands of disseminated, and less commonly massive, sphalerite with minor pyrrhotite, pyrite, and traces of germanium sulphides (germanite, renierite) and stannite (Chatterjee, 1979). The deposit is geologically similar to deposits and occurrences hosted by Grenville Supergroup marbles in Ontario, Quebec, and New York State (e.g. Balmat, New Calumet, Montauban; Sangster, 1970; Sangster and Bourne, 1982; Sangster et al., 1990a). Although originally interpreted as a skarn (Chatterjee, 1979), more recent studies indicate that the Meat Cove deposit is epigenetic or sedimentary exhalative in origin, formed from Mississippi Valley-type brines (Sangster et al., 1990a).

Paleoplacer deposits in Cambrian rift successions

M. Gauthier

Initial rifting of the Laurentian margin prior to and during opening of the Iapetus Ocean is recorded throughout the Canadian Appalachians by a basal terrestrial to shallow marine siliciclastic sedimentary sequence accompanied by mafic alkalic volcanism. Clastic sedimentation in shallow marine environments was accompanied by the concentration of heavy mineral sands in paleoplacers (Fig. 9.4).

Lower to Middle Cambrian quartzo-feldspathic clastic sedimentary rocks (greywacke, impure quartzite, microconglomerate) in southeastern Quebec contain numerous deposits of titaniferous rutile, zircon, and monazite which are potential sources of titanium and rare metals. The host

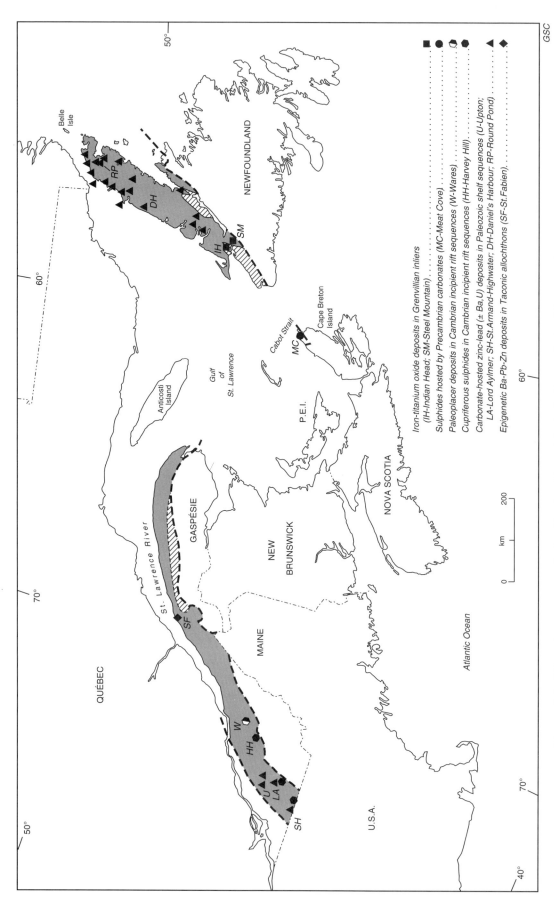

Figure 9.1. Mineral occurrences in the Humber Zone (shaded). Heavy dashed lines are boundaries of the Humber Zone. Post-middle Ordovician cover sequences are cross-hatched. Symbols represent either single deposits or groups of deposits.

Iron-titanium oxide deposits in Grenvillian inliers (IH-Indian Head; SM-Steel Mountain) . . . .
Sulphides hosted by Precambrian carbonates (MC-Meat Cove) . . . .
Paleoplacer deposits in Cambrian incipient rift sequences (W-Wares) . . . .
Cupriferous sulphides in Cambrian incipient rift sequences (HH-Harvey Hill) . . . . . .
Carbonate-hosted zinc-lead (± Ba,U) deposits in Paleozoic shelf sequences (U-Upton; LA-Lord Aylmer; SH-St.Armand-Highwater; DH-Daniel's Harbour; RP-Round Pond) . . . . . .
Epigenetic Ba-Pb-Zn deposits in Taconic allochthons (SF-St.Fabien) . . . .

GSC

sedimentary rocks occur at the base of the Pinnacle Formation and are underlain by alkalic volcanic rocks of the Tibbit Hill Formation (Fig. 9.5). In the Sutton area (Gauthier, 1985), the principal detrital minerals are titaniferous magnetite, ilmenite, zircon, and tourmaline, distributed in fine laminae comprising from 10% to 50% of the host quartzites. Hematite and rutile occur as secondary alteration products after titaniferous magnetite. Typical grades are in the range 51.2% Fe_2O_3, 27% TiO_2 and 3.2% Zr (Gauthier et al., 1989). Locally, all of the TiO_2 is contained in rutile rather than titaniferous magnetite. Deposits in the Bois-Francs area (e.g. Ware deposit; Gauthier et al., 1989) are hosted by impure sandstones and are dominated by zircon, monazite, and sphene.

The deposits are considered to be marine placers that accumulated on shoals along the southwestern margin of the Québec Reentrant (Dowling, 1988). There is still some controversy as to whether the heavy minerals were derived from disaggregation of the underlying Tibbit Hill volcanics

or from erosion of crystalline rocks in the adjacent Canadian Shield (particularly anorthosites in the Laurentian and Adirondack highlands; Dowling, 1988).

Copper deposits in Cambrian rift volcanic and sedimentary rocks

M. Gauthier

There are more than 200 copper showings in the Cambrian rift-related volcanic and sedimentary successions of southeastern Quebec between the Chaudière Valley and the Vermont border. Copper mineralization is widespread in chlorite-talc-sericite schist horizons in the Pinnacle Formation immediately above the Tibbit Hill metavolcanics and less commonly occurs in the overlying White Brook Dolostone and Gilman Quartzite (Fig. 9.4, 9.5). Chalcopyrite and bornite are the principal sulphides, with local subordinate chalcocite, pyrite, and sporadic molybdenite and gold. Copper grades locally attain 2% over metre widths.

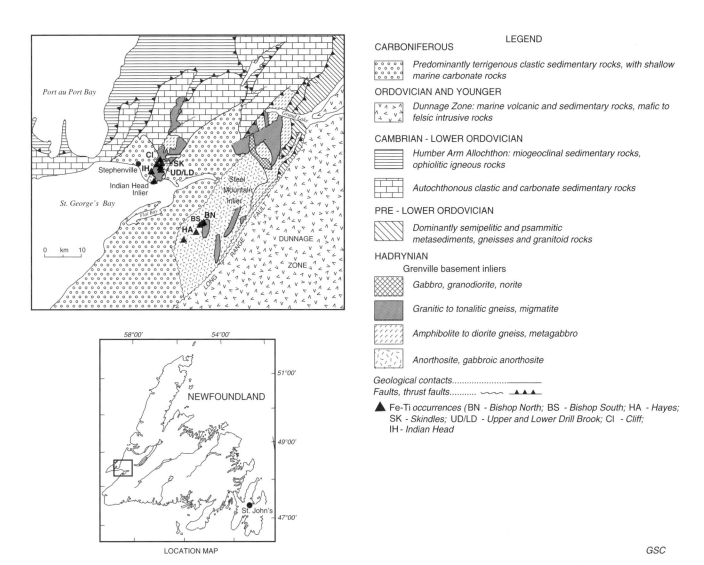

Figure 9.2. General geology of Grenville inliers of southwestern Newfoundland and location of the iron-titanium deposits.

Although a few of these showings were exploited during the American Civil War, only the Harvey Hill deposit (Fig. 9.4) has been a recent producer. This deposit yielded approximately 400 000 tonnes of ore grading 1.2% Cu and 3.6 g/t Ag between 1856 and 1976.

Harron (1976) interpreted these copper showings as metamorphically and tectonically modified sedimentary copper deposits similar to those in the Zairean and Zambian stratiform copper belts. In contrast, Gair and Slack (1980)

proposed that these copper showings are Besshi-type submarine exhalative deposits and can be correlated with important deposits in the southern Appalachian Blue Ridge (e.g. Ducktown, Great Gossan Lead). Gauthier (1985, 1986) favoured a hybrid syngenetic to syndiagenetic model, based in part on comparison with the Buena Esperanza deposit in Chile (Routhier, 1980), in which diagenetic epidotization of the underlying Tibbit Hill mafic volcanics provided copper to the nearby sedimentary environment. Later studies by Colpron (1987) suggested that at least some of these deposits formed as metasomatic replacements of paleoplacers during hydrothermal circulation along the Brome and Richardson faults, a refolded Taconic thrust zone (Charbonneau, 1975; Colpron, 1987) which forms the western boundary of the metamorphic core of the Sutton-Notre Dame Anticlinorium. According to this model, the paleoplacer beds would have acted as chemical traps for the mineralization during late Taconic or Acadian dynamic-metamorphism (Gauthier et al., 1989).

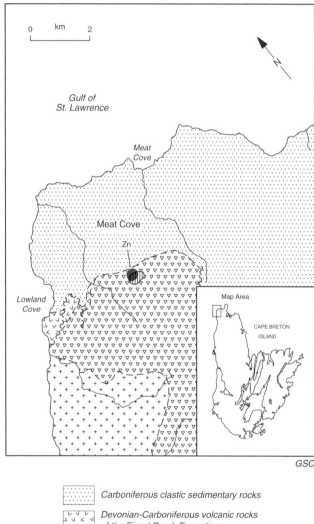
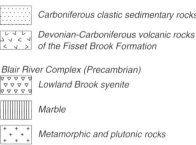

Figure 9.3. General geology in the area of the Meat Cove deposit (red dot), northeastern Cape Breton Island (after Sangster et al.,1990a).

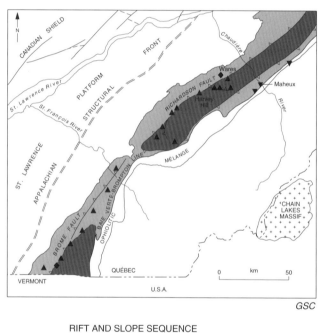
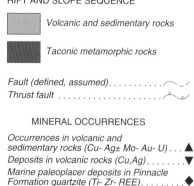

Figure 9.4. Principal paleoplacer and cupriferous sulphide occurrences in Cambrian rift successions of the Quebec Eastern Townships.

At the Maheux prospect (Fig. 9.4), hematized and epidotized basalt of the Lower Cambrian Caldwell Group hosts chalcocite and bornite veins. The mineralized volcanics are intercalated on a regional scale with red and green sandstone, conglomerate and shale and were erupted in a shallow marine to subaerial environment (Gariépy, 1978). Mineralized veinlets are nearly perpendicular to the bedding and zoned inwards from an epidote-rich wall through massive chalcocite and bornite to a quartz-bornite-specular hematite core (Gauthier et al., 1989; Trottier et al., 1989). The sulphides are notably rich in silver. The deposit has been interpreted to be a volcanic-hosted redbed copper deposit (cf. Kirkham, 1984; Bjørlekke and Sangster, 1981; Brown, 1989). Genetic models involve diagenetic mobilization and precipitation of copper in an arid, rift-dominated environment (Trottier et al., 1989).

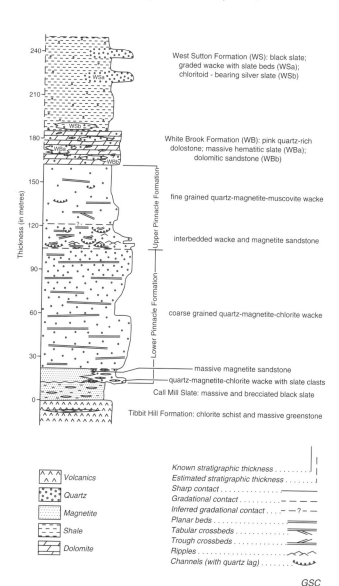

West Sutton Formation (WS): black slate; graded wacke with slate beds (WSa); chloritoid - bearing silver slate (WSb)

White Brook Formation (WB): pink quartz-rich dolostone; massive hematitic slate (WBa); dolomitic sandstone (WBb)

fine grained quartz-magnetite-muscovite wacke

interbedded wacke and magnetite sandstone

coarse grained quartz-magnetite-chlorite wacke

massive magnetite sandstone
quartz-magnetite-chlorite wacke with slate clasts
Call Mill Slate: massive and brecciated black slate

Tibbit Hill Formation: chlorite schist and massive greenstone

˄˄˄ Volcanics

Quartz

Magnetite

Shale

Dolomite

Known stratigraphic thickness . . . .
Estimated stratigraphic thickness . . . .
Sharp contact . . . .
Gradational contact . . . . – – –
Inferred gradational contact . . . . – —?– –
Planar beds . . . .
Tabular crossbeds . . . .
Trough crossbeds . . . .
Ripples . . . .
Channels (with quartz lag) . . . .

GSC

Figure 9.5. Stratigraphy of the basal Oak Hill Group (from Dowling, 1988) and location of the heavy mineral paleoplacer occurrences.

Paleozoic carbonate-hosted deposits

Overview

H. Scott Swinden

Cambrian-Ordovician shelf carbonate successions of the Laurentian margin provided geological environments for a variety of metallogenic events and mineral deposit types (Fig. 9.1), both during and after deposition of the host rocks. The earliest mineralizing event is probably represented by stratabound "Irish-type" polymetallic deposits of southeastern Quebec (e.g. Upton, Dunham), interpreted to be related to exhalative activity during shelf sedimentation. Possibly related epigenetic deposits in underlying volcanic rocks are found at Acton Vale. At about the same time, uraniferous phosphorites, now preserved in the Sainte-Armand-northern Vermont area, were forming in deeper water on the continental slope.

An epigenetic post-burial mineralizing event between the Middle Ordovician and Late Devonian produced numerous Mississippi Valley-type zinc deposits in the Cambrian-Ordovician carbonate succession west and south of the Long Range Inlier in western Newfoundland. The largest deposits (Newfoundland Zinc deposits at Daniel's Harbour) are associated with extensive karst features and breccia formation in Lower Ordovician strata.

A later (probably Carboniferous or younger) mineralizing event produced lead-rich deposits in Silurian carbonate strata on the eastern side of the Long Range Inlier (western White Bay) which may be metallogenically related to deposits in upper Paleozoic overlap successions.

"Mississippi Valley-type" Zn (± Pb) deposits in Cambrian-Ordovician carbonate rocks, western Newfoundland

S.M. Dunsworth, T. Lane, and H. Scott Swinden

"Mississippi Valley-type" sphalerite and galena occurrences are widespread both aerially and stratigraphically in the autochthonous Cambrian-Ordovician carbonate successions of the western Newfoundland Humber Zone (Fig. 9.6, 9.7). The Zn (± Pb) occurrences comprise two types (Saunders et al., 1992): (1) sphalerite±galena restricted to fractures, veinlets, stylolites, small clots and disseminations in predominantly Cambrian, low porosity/permeability dolostones; and (2) sphalerite±galena in extensive open spaces and in matrix to dolomite pseudobreccias and collapse breccias in predominantly Ordovician strata. Breccias are best developed in regressive marine strata in which dolomitization, karstification and fracturing has provided the necessary ground preparation and secondary porosity to allow ingress of the later mineralizing fluids. The two occurrence types differ mainly in the nature of the ground preparation leading to differences in the available secondary porosity. Lead isotope studies confirm that all of the occurrences formed during the same mineralizing event (Swinden et al., 1988a). Timing of mineralization is geologically constrained between the Middle Ordovician and Late Devonian, and the mineralizing event probably records movement of hydrothermal fluids out of the basement and

689

adjacent sedimentary basins during orogenesis related to accretion of outboard terranes to the Laurentian margin (Lane, 1990).

Zn±Pb occurrences in Cambrian strata

Small occurrences, comprising minor clots, disseminations, veinlets and fracture fillings of sphalerite and galena with associated white dolomite spar, are hosted by typically massive Cambrian dolostones of the March Point and Petit Jardin formations (Saunders et al., 1992). At Green Island Brook and Eddies Cove East (Fig. 9.6), brown and yellow

sphalerite occurs as disseminations within dolostone, coating joint surfaces and filling primary and secondary porosity. Galena also occurs along joint surfaces and forms massive clots with some sphalerite in open spaces in stromatolitic dolostone (Knight, 1984a). In the Pikes Feeder Pond area, sphalerite, galena, pyrite, and chalcopyrite occurs with dolomite spar as clots, disseminations and fracture cements in brecciated zones and replacements within dolostone. Sphalerite, galena and pyrite occur in stromatolitic microcrystalline dolostone at Watts Point and in veins at Wolf Brook, Beaver Pond, and Goose Arm in the Goose Arm area.

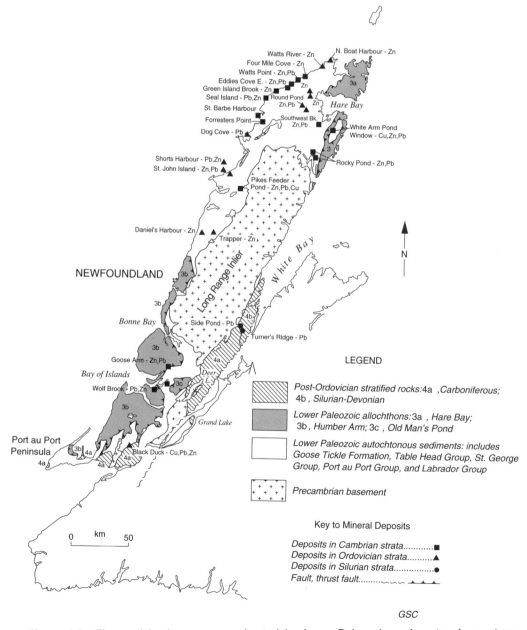

Figure 9.6. Zinc and lead occurrences hosted by lower Paleozoic carbonates in western Newfoundland (after Knight, 1984a; Saunders and Strong, 1986).

Zn±Pb occurrences in Ordovician strata

These deposits comprise mainly sphalerite (± pyrite and/or galena) hosted by dolomite pseudobreccia, a distinctive rock comprising up to 80% white dolomite spar surrounding rounded domains (pseudo-fragments) of dark dolomite (Cumming, 1968; Knight, 1980; Lane, 1984; Saunders et al., 1992). The largest and best known deposits of this type are found at the former Newfoundland Zinc mine at Daniel's Harbour (Fig. 9.6) where total ore reserves of about 6.6 million tonnes grading 7.9% Zn contained in more than a dozen mineralized zones were mined between 1975 and 1990. The Newfoundland Zinc deposits comprise numerous, northeast-trending, narrow lenticular to ribbon-like orebodies occurring along the margins of faults and several karst collapse breccia bodies with grey matrix (Crossley and Lane, 1984; Lane, 1990). The sphalerite mineralization is stratabound in pervasively dolomitized, veined, and brecciated carbonates. Crossley and Lane (1984) reported that the sphalerite in pseudobreccia occurs at three or more successive stratigraphic levels interconnected by veins. Ore lenses are stacked en echelon both along and across strike with abrupt contacts between ore and barren wall rock. Sphalerite distribution is controlled by anastomosing subvertical vein sets, which are stratabound within 10 to 30 m of the section.

The mineralization is related to northeast-trending basement faults which were active during sedimentation, burial, and uplift. A system of breccia bodies developed during Late Arenig subaerial exposure and synsedimentary faulting of the platform (Collins and Smith, 1975; Lane, 1984, 1990) during major sea level regression and convergent tectonism (Knight et al., 1991). Reactivation of these faults and intensive fracturing around karst breccia bodies coincided with the regional migration of hydrothermal, zinc-bearing brines during later burial. The corrosive brines developed extensive secondary porosity as psuedobreccia which became cemented by sulphides and dolomite.

The only other substantial zinc occurrence in Ordovician host rocks is the Round Pond deposit west of Hare Bay (Fig. 9.6), which contains approximately 400 000 tonnes grading 2% Zn (Born, 1983). The red-brown to green sphalerite occurs as disseminations, clots, and large crystals with white dolomite spar within irregular bodies of epigenetic dolostone, pseudobreccia, and salt-and-pepper textured dolomite.

Pb mineralization in Silurian carbonates, White Bay, western Newfoundland

C.M. Saunders

In western White Bay, a Silurian volcanic and sedimentary succession termed the Sops Arm Group is unconformable upon the Ordovician Southern White Bay Allochthon on the eastern side of the Long Range Inlier. This succession includes Silurian dolostone and limestone interbedded with felsic volcanic rocks and overlain by a dominantly fine grained siliciclastic succession. Carbonates in the lower part of the succession contain numerous lead occurrences, the largest of which are the Turner's Ridge and Side Pond deposits (Fig. 9.6). These occurrences typically comprise coarse- to fine-grained galena accompanied by calcite and minor barite and sphalerite in fractures and fracture stockworks in variably, locally intensely, brecciated dolostone. Brecciation, mineralization, and calcite alteration are most intense at the Turner's Ridge deposit, which contains approximately 200 000 tonnes grading 3 to 4% Pb (Saunders, 1991). The timing of mineralization is uncertain, but at Turner's Ridge, the dolostones are structurally overlain by rhyolite along a probable Carboniferous thrust fault. Galena occurs in brecciated rhyolite near the fault indicating that mineralization may be Carboniferous or younger.

Early Cambrian-late Middle Ordovician Ba-Zn-Pb, Cu-Ba and P-U mineralization, southeast Quebec

M. Gauthier

Cambrian-Ordovician transported-shelf successions in southern Quebec host vein- and breccia-type copper deposits, stratabound Ba-Zn-Pb-Cu mineralization (Gauthier, 1986; Paradis et al., 1990), and uraniferous phosphorites that record various metallogenic environments on the developing passive margin (Fig. 9.8).

The stratiform Upton deposit, hosted by impure brecciated carbonates of the Upton Group, contains approximately 950 000 tonnes grading 1.9% zinc, 0.6% lead, 0.15% Cu, 0.11% Cd, 13.5 g/t Ag, and 46.5% $BaSO_4$ (Paradis et al., 1990). Barite is the principal mineral of economic interest.

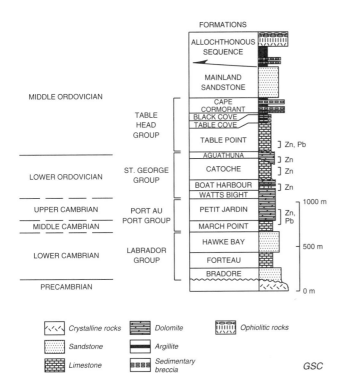

FORMATIONS

	ALLOCHTHONOUS SEQUENCE
	MAINLAND SANDSTONE
	CAPE CORMORANT
	BLACK COVE
	TABLE COVE
	TABLE POINT
	AGUATHUNA
	CATOCHE
	BOAT HARBOUR
	WATTS BIGHT
	PETIT JARDIN
	MARCH POINT
	HAWKE BAY
	FORTEAU
	BRADORE

MIDDLE ORDOVICIAN

LOWER ORDOVICIAN

UPPER CAMBRIAN

MIDDLE CAMBRIAN

LOWER CAMBRIAN

PRECAMBRIAN

TABLE HEAD GROUP

ST. GEORGE GROUP

PORT AU PORT GROUP

LABRADOR GROUP

] Zn, Pb
] Zn
] Zn
] Zn
Zn, Pb

1000 m
500 m
0 m

Legend:
- Crystalline rocks
- Sandstone
- Limestone
- Dolomite
- Argillite
- Sedimentary breccia
- Ophiolitic rocks

GSC

Figure 9.7. Stratigraphy of autochthonous shelf-facies strata and contained base metal occurrences of western Newfoundland (after Swinden et al., 1988a).

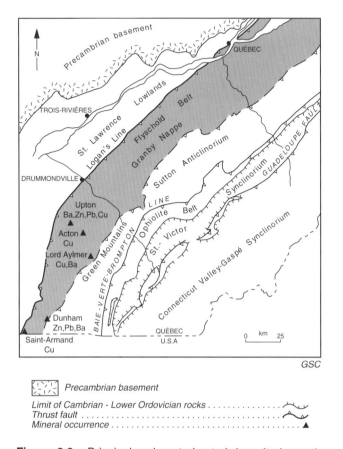

Figure 9.8. Principal carbonate-hosted deposits in southeastern Quebec (after Paradis et al., 1990).

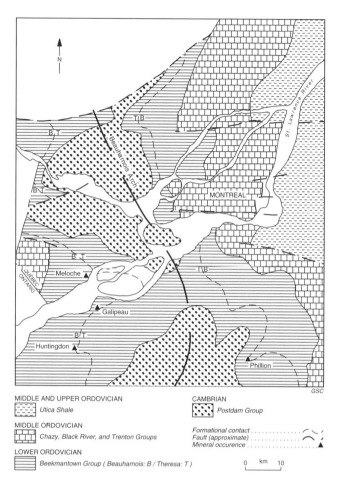

MIDDLE AND UPPER ORDOVICIAN
:::::: *Utica Shale*

MIDDLE ORDOVICIAN
▦ *Chazy, Black River, and Trenton Groups*

LOWER ORDOVICIAN
▤ *Beekmantown Group (Beauharnois: B / Theresa: T)*

CAMBRIAN
▨ *Postdam Group*

Formational contact . . . .
Fault (approximate) . . . .
Mineral occurence . . . . ▲

0   km   10

Figure 9.9. Carbonate-hosted deposits in the St. Lawrence Lowlands, southern Quebec (after Héroux and Tassé, 1990).

The deposit is interpreted to be syngenetic and probably represents an exhalative process like that postulated for the "Irish-type" stratiform sulphides (Hitzman and Large, 1986). A number of small sphalerite and galena showings that are probably similar in origin occur in Lower Cambrian dolostone assigned to the Dunham and White Brook formations of the Oak Hill Group. The Dunham sulphides, which occur in the matrix of breccia lenses, are stratabound and can be traced for over 25 km to the Vermont border. The small lead-zinc deposit at Brandon, Vermont, more than 200 km to the south, occurs in the same Cambrian formation and is also considered to be of "Irish-type" (Slack, 1991).

A predominantly cupriferous, broadly stratabound, deposit occurs in mineralized limestone breccia at Acton Vale. This deposit, which yielded about 16 300 tonnes grading 12% copper prior to closing in 1864, may be coeval and broadly related to the polymetallic mineralization in the overlying carbonates (e.g. Upton).

A horizon of brecciated uraniferous phosphatic dolostone, situated at Saint-Armand (Fig. 9.8), has been traced for more than 1 km along strike and continues southwards into Vermont (Prudhomme, 1985, 1986). The mineralization is associated with black phosphatic fragments and sphalerite is locally present. There is a considerable tonnage of uraniferous phosphatic dolostone, but the grade is low.

An unusually copper-rich, stratabound Cu-Ba occurrence, the Lord Aylmer deposit, occurs in a window of Cambrian-Ordovician platformal carbonates within the Granby Nappe (Fig. 9.8). The main zone consists of chalcopyrite and bornite (Chartrand et al., 1987) in extensively karst brecciated limestone, which formed prior to Taconic deformation (Trottier et al., 1989). Three types of mineralization are recognized in the deposit: (1) chalcopyrite, bornite, and calcite in a matrix that encloses fragments in limestone breccia; (2) lesser chalcopyrite-bornite-quartz-calcite veins emanating upward from the breccia zone; and (3) minor chalcopyrite and pyrite disseminations in intrakarstic sediments (Chartrand et al., 1987).

Stratabound Ba-Fe-(Pb-Zn) occurrences in Lower Ordovician carbonate rocks, St. Lawrence Lowlands

G. Beaudoin

There are two broad types of mineral occurrences in Lower and Middle Ordovician carbonate strata in the St. Lawrence Lowlands: (1) stratabound Ba-Fe-(Pb-Zn) occurrences, and 2) accessory sphalerite disseminated as millimetre-sized granules.

Three stratabound Ba-Fe-(Pb-Zn) examples (Huntingdon, Galipeau, Meloche, Fig 9.9) occur in the Lower Ordovician Beauharnois Formation of the Beekmantown Group west of Beauharnois Arch (Tassé and Schrijver, 1989, Fig. 9.9). A fourth occurrence (Phillion), located east of the Beauharnois Arch and hosted by the Theresa Formation of the Beekmantown Group, contains sphalerite as cement to a carbonate fault-breccia (Héroux and Tassé, 1990). Mineralization occurs in a stratabound dissolution porosity of matrix and lithoclasts (Tassé et al., 1987). Dissolution was accompanied by extensive silicification at the Huntingdon occurrence where barite is associated with the most intensely silicified zones. At Galipeau, precipitation of barite was accompanied by large amounts of iron sulphides with little silicification. Minerals precipitated from highly saline, relatively low temperature brines (Tassé et al., 1987).

Numerous minor occurrences of accessory sphalerite were found in Lower and Middle Ordovician carbonate rocks throughout the St. Lawrence Lowlands. Tassé and Schrijver (1989) suggested that accessory sphalerite precipitated as a result of thermochemical reduction of dissolved evaporitic sulphate from the Beekmantown Group.

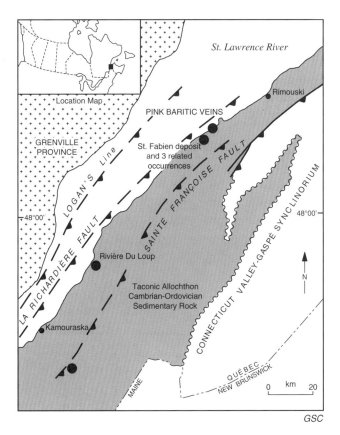

Figure 9.10. Location of epigenetic Ba-Pb-Zn occurrences (black dots) in the Taconic Allochthon of the Lower St. Lawrence region (after Williams-Jones et al., 1992). The Des Seigneuries thrust sheet, host to the deposits, is bounded by the La Richardière and Sainte Françoise thrust faults.

Ba-Pb-Zn veins and disseminations in transported clastic rocks, Lower St. Lawrence region

G. Beaudoin

Barite-galena-sphalerite veins are found in three areas within 125 km of each other in the Lower St. Lawrence region (Fig. 9.10). The veins are in the Lower Cambrian Grès verts unit or the Upper Cambrian Saint-Damase Formation and occur within the Taconic Des Seigneuries thrust sheet.

The Saint-Fabien deposit, the largest and only one to have been exploited, and three smaller vein occurrences are found in the Saint-Fabien area (Beaudoin et al., 1989; Williams-Jones et al., 1992). The Saint-Fabien deposit consists of barite-galena-sphalerite veins, and galena or barite disseminated in cavities associated with dolomitization of limestone clasts and carbonate cements in limestone conglomerates and sandstones of the Saint-Damase Formation. Mineralization is stratabound in the Saint-Damase Formation and is commonly contained within dolomitized areas. Vein and dissemination paragenesis can be generalized from early to late: white dolomite-ferroan dolomite-galena, barite, sphalerite-calcite (Beaudoin et al., 1989). Fluid inclusion data indicate that the mineralizing fluids had a temperature below 200°C with salinities ranging from 15 to 38 weight per cent NaCl equivalent (Williams-Jones et al., 1992).

Beaudoin et al. (1989) and Williams-Jones et al. (1992) suggested that fracturing resulted from fluid overpressuring and that mineralization occurred during mixing of a basinal brine expelled during or late in the Taconic Orogeny with oxidizing, near-surface, sulphate-bearing fluids.

DUNNAGE AND GANDER ZONES

Introduction

H. Scott Swinden

Stratified rocks of the Dunnage Zone comprise variably disrupted ophiolite suites and volcanic-volcaniclastic sequences that were formed in a complex series of island arcs, back-arc basins and transtensional basins, probably located near the margins of Iapetus (Dunning et al., 1991). Quartzite-dominated sedimentary sequences overlain by volcanic-volcaniclastic successions in the Gander Zone (e.g. Bathurst area, northern New Brunswick; Hermitage Flexure area, southern Newfoundland) may record sedimentation followed by volcanism at a tectonically active continental margin (Colman-Sadd, 1980; Colman-Sadd and Swinden, 1984; van Staal, 1987).

The Iapetus rift-drift transition has been dated stratigraphically as Early Cambrian (Williams and Hiscott, 1987), although no oceanic crust from this period is preserved. The earliest Iapetan magmatism is Late Cambrian, and recorded at three localities in Newfoundland: (1) Tally Pond island arc volcanics in the Victoria Lake Group, south-central Newfoundland (Dunning et al., 1991); (2) Sleepy Cove Group mafic volcanics in eastern Notre Dame Bay; and (3) Little Port Complex in the Humber Arm Allochthon. Dominantly ensimatic island arc volcanic

693

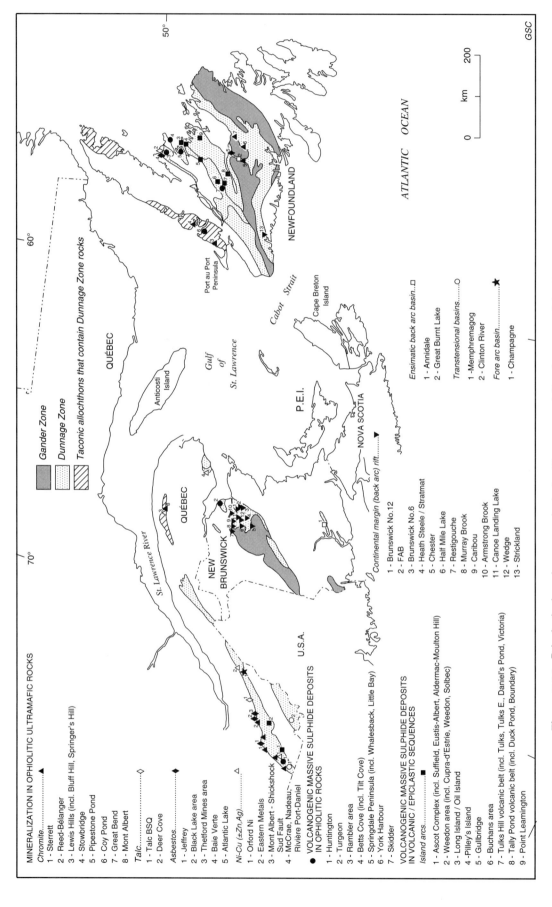

Figure 9.11. Principal mineral occurrences in the Dunnage and Gander zones of the Canadian Appalachians. Symbols may represent either single deposits or groups of deposits.

activity continued sporadically until the Middle Ordovician including at least five episodes preserved in central Newfoundland (Swinden, 1991) and the Ascot-Weedon volcanics of southeastern Quebec (Tremblay et al., 1989; Trottier and Gauthier, 1991). Lower Ordovician ophiolitic rocks are preserved throughout the Newfoundland Dunnage Zone as well as in the Taconic allochthons of western Newfoundland, the Elmtree Subzone of northern New Brunswick, and the ophiolite belt of southeastern Quebec. All are interpreted to record crustal extension in rifting arc or back-arc environments (Dunning and Chorlton, 1985; Jenner et al., 1988, 1991; van Staal et al., 1991; Trottier and Gauthier, 1991). Marine transtensional basins formed in the Late Ordovician in response to continuing accretion of outboard terranes to the Laurentian margins and marine environments persisted into the Silurian in some areas.

At least some of the Early Ordovician volcanic activity was approximately coeval with quartzose sedimentation at a continental margin east of Laurentia, now preserved mainly in the Gander Zone in eastern Newfoundland and northern New Brunswick. Subsequent ensialic back arc rifting environments at this continental margin are recorded in the Tetagouche Group of northern New Brunswick and its probable equivalents in the Hermitage Flexure of southern Newfoundland (Fyffe et al., 1990; van Staal et al., 1991; Swinden, 1991).

The Dunnage Zone is particularly rich in mineral deposits (Fig. 9.11) and has been a major contributor to the mineral economy of the Canadian Appalachian region for more than 150 years. Mineral deposits unique to the Dunnage-Gander zones comprise two very broad types: (1) mineralization in the ultramafic and mafic plutonic parts of ophiolite complexes (i.e. magmatic chromite and sulphides, epigenetic Ni arsenides and sulphides, and metasomatic products of the ultramafic rocks such as asbestos, talc, and magnesite) in central and western Newfoundland, Gaspésie, and southeastern Quebec; and (2) volcanic- and sediment-hosted massive sulphide deposits including various types of volcanogenic sulphide deposits in central and western Newfoundland, northern New Brunswick, and southeastern Quebec.

Mineralization in ophiolitic ultramafic rocks

S.M. Dunsworth, M. Gauthier and H. Scott Swinden

Significant occurrences of chromite are found in Newfoundland in the Bay of Islands (Snelgrove, 1934; Smith, 1958; Fogwill, 1964a, b; Malpas and Strong, 1975; Dunsworth et al., 1986) Pipestone Pond, Coy Pond, and Great Bend complexes (Snelgrove, 1934; Fogwill, 1964a, b; Colman-Sadd and Swinden, 1982), and in ophiolite suites of southeastern Quebec (Gauthier et al., 1990; Laurent and Kacira, 1987; Fig. 9.11). Minor occurrences of nickeliferous, locally platinum group element (PGE)-rich sulphides have been reported from the Bay of Islands Complex and numerous Ni occurrences are associated with dismembered ophiolitic rocks in Gaspésie. Nickeliferous deposits in southeastern Quebec, possibly of Outokumpu type, are apparently related to hydrothermal activity in the early

stages of the Taconic Orogeny (Gauthier et al., 1988, 1989; Auclair, 1990). Metasomatism of ophiolitic ultramafic rocks during accretion to the Laurentian margin resulted in the formation of talc and asbestos.

Chromite

The lower stratigraphic levels of the Bay of Islands Complex in the Humber Arm Allochthon, western Newfoundland, contain lenses, layers and dense disseminations of chromite associated with dunite. The principal occurrences include the Springers Hill and Bluff Head deposits in the Lewis Hills massif (Fig. 9.12) and the Stowbridge deposit in the North Arm Mountain massif. Springers Hill deposit, the largest of these, contains an estimated 9100 tonnes of massive and heavily disseminated refractory and chemical grade chromite occurring as discontinuous, massive to disseminated layers-schlieren enveloped by dunite bands and hosted by harzburgite. The main zone is truncated at depth by low angle mylonitic shear zones (Dunsworth et al., 1986) and the chromitiferous bands have been folded and transposed during ductile deformation within the oceanic domain. The main chromite occurrence at Bluff Head consists of thin layers of disseminated to massive metallurgical grade chromite hosted by massive dunite (Dahl and Watkinson, 1986; Dunsworth et al., 1986) whereas the Stowbridge showing consists of a 4 to 10 m wide zone of discontinuous chromite layers in a penetratively deformed dunite-clinopyroxene-dunite and wehrlite. Page and Talkington (1984) reported traces of PGEs in chromitite from the main Springers Hill and Stowbridge deposits.

Chromite is commonly widely disseminated in harzburgite and dunite in the Great Bend, Pipestone Pond, and Coy Pond complexes (Snelgrove, 1934; Fogwill, 1964a; Kean, 1974; Colman-Sadd and Swinden, 1982; Swinden, 1988a) in the Exploits Subzone of east-central Newfoundland (Fig. 9.11). Locally, chemical grade chromite is concentrated in pods up to 2 m thick and 6 m in length (e.g. Chrome Pond showing in the Pipestone Pond Complex; Swinden, 1988a).

Chromite deposits in the southeastern Quebec ophiolite belt (Fig. 9.13) comprise podiform and lenticular deposits in disrupted ophiolite sequences (Thetford area) and megablocks of ultramafic rocks within the Saint Daniel Mélange (Asbestos, Mount Orford areas). Chromite was first mined in Ham-Sud and Leeds townships of Estrie in 1861 and 1877 with maximum production during the First and Second World Wars. The Cr:Fe ratio of some of the chromite orebodies ranged up to 3:1.

The Reed-Bélanger and Hall mines in the Thetford Mines region are good examples of massive stratiform ophiolite-hosted chromite orebodies contained within the cumulate sequence. They consist of disseminated to semi-massive beds within basal dunite cumulate (Laurent and Kacira, 1987; Gauthier et al., 1990). The Reed-Bélanger mine produced 440 000 tonnes of 7% Cr_2O_3 and has reserves of at least 300 000 tonnes at 10% Cr_2O_3 (Marcotte, 1980). Podiform orebodies in southeastern Quebec are irregular in shape and locally high grade. They may occur in groups, as at the Sterrett mine in the Asbestos ophiolitic complex and at the Fletcher No. 4 mine (Mount Orford Complex), or in isolated bodies as at Caribou Mountain in the Thetford Mines region.

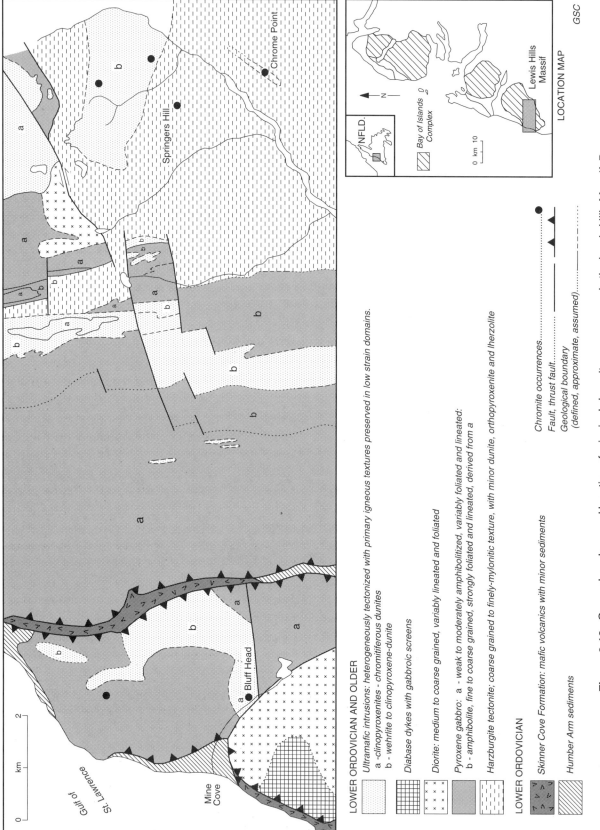

Figure 9.12. General geology and location of principal chromite occurrences in the Lewis Hills Massif, Bay of Islands Complex. Geology compiled from Dunsworth (1991) and Karson (1979).

LOWER ORDOVICIAN AND OLDER

Ultramafic intrusions: heterogeneously tectonized with primary igneous textures preserved in low strain domains.
a -clinopyroxenites - chromitiferous dunites
b - wehrlite to clinopyroxene-dunite

Diabase dykes with gabbroic screens

Diorite: medium to coarse grained, variably lineated and foliated

Pyroxene gabbro: a - weak to moderately amphibolitized, variably foliated and lineated:
b - amphibolite, fine to coarse grained, strongly foliated and lineated, derived from a

Harzburgite tectonite: coarse grained to finely-mylonitic texture, with minor dunite, orthopyroxenite and lherzolite

LOWER ORDOVICIAN

Skinner Cove Formation: mafic volcanics with minor sediments

Humber Arm sediments

Chromite occurrences............
Fault, thrust fault....................
Geological boundary
(defined, approximate, assumed)............

LOCATION MAP

GSC

696

Chromite occurs as disseminations and small lenses in a number of serpentinite bodies in Gaspésie, including the Weir, Mt. Serpentine, Mt. Albert bodies and ultramafic rocks near the Shickshock-Sud Fault. None of these deposits is of economic proportions.

Ni-PGE-base/precious metal occurrences

Pyrrhotite, pyrite, and pentlandite form disseminations and patches in ophiolitic gabbro at Rope Cove Canyon in the Bay of Islands Complex. The sulphides contain anomalous concentrations of PGEs and were interpreted by Lydon et al. (1990) to result from magmatic-hydrothermal activity in the plutonic part of the ophiolite suite. Minor nickeliferous occurrences elsewhere in the Bay of Islands Complex and in the Springers Hill area of the Lewis Hills massif (S.J. Edwards, unpub. report) have revealed slightly anomalous concentrations of PGEs in coarse grained to pegmatitic ortho- and clinopyroxenite veins/pods. The PGEs in this setting are not associated with sulphides.

Platiniferous chromitites have also been reported from the ophiolite suites of Estrie and Beauce, southeastern Quebec (Gauthier et al.,1990). PGEs occur both as euhedral alloys and sulphides or anhedral Pt-Pd alloys and Rh arsenides.

In Gaspésie, numerous nickeliferous sulphide occurrences have been reported from ultramafic rocks near the Shickshock-Sud Fault (Dugas et al., 1969) and in the McCrea, Nadeau, and Riviere Port Daniel ophiolitic mélanges south of the Grand Pabos Fault (Malo and Moritz, 1991). Mineralization occurs principally in serpentinite bodies within the mélanges and is associated with intense carbonatization, pyritization, and minor chromite. Millerite is the main primary nickel mineral and is commonly replaced by gaspéite (nickel-magnesium carbonate).

The Saint Daniel Mélange in the ophiolite belt of southeastern Quebec contains at least three pyrite-millerite-gersdorfite-chalcopyrite stockworks in metasomatized ultramafic rock, a deposit type which is not known elsewhere in the Canadian Appalachians. The Eastern Metals deposit (Auclair, 1990; Auclair et al., 1993), the most important example, comprises a Ni>Zn-rich North zone and a Cu>Zn>Ni-rich, locally auriferous, South zone totaling 1.09 million tonnes. The mineralization occurs in highly altered serpentinite at its contact with the enclosing Saint-Daniel Mélange clastic sediments (Fig. 9.13). The alteration sequence, from least- to most-altered rock consists of serpentinite, silicified serpentinite, talc-carbonate schist and silica-carbonate rock (listwaenite, birbirite; Gauthier et al., 1989; Auclair, 1990). Nickeliferous pyrite and chromite are locally reworked by sedimentary processes forming a lens of nickel-bearing pyrite with bedded and graded pyrite in the North zone (Trottier et al., 1989).

A second significant deposit of this type, the Orford Nickel deposit northeast of Sherbrooke (Fig. 9.13), is hosted by calcite-diopside listwaenite developed on the fringe of a serpentinized peridotite. Fine grained chromite disseminations at the Orford Nickel showing are the only evidence of the origin of the host-rocks. The chromite is surrounded by uvarovite, a chromitiferous garnet.

The resedimented pyrite and chromite at the Eastern Metals deposit is interpreted to occur at the base of the Middle Ordovician and younger Magog Group. This tightly constrains the timing of mineralization to the Middle Ordovician, post-dating ophiolite formation, but predating deposition of the Magog Group. The mineralization has been compared to the Hg-Au-Sb deposits of the California coast range, the Mother Lode Au-bearing listwaenites, the large Outokumpu Cu-Zn-Co-Ni sulphide lenses, and the Ni-Au deposit of Mount Vernon in Washington state (Trottier et al., 1989; Auclair et al., 1993). It is interpreted to have formed through hydrothermal activity related to obduction of the ophiolite suites during the Taconic Orogeny (Gauthier et al., 1989; Trottier et al., 1989; Auclair, 1990).

Talc and asbestos

Talc mining has been carried out in the Bolton region of southeastern Quebec since 1870 with modern operations commencing in 1938. The Talc BSQ deposits (Fig. 9.13) are associated with a thin sheet of serpentinized peridotite (the Pennington Sheet) which varies in thickness from several metres to 100 m, with the narrow portions having talc (steatite) and the thicker portions asbestos deposits. The thinnest zones of the sheet contain completely serpentinized peridotite, and in certain areas the peridotite has been pervasively transformed to steatite and talc-carbonate schists through metasomatism between magnesium-rich serpentinized peridotite and quartz-rich schists.

Several talc deposits are known on the Baie Verte Peninsula, western Newfoundland, although none have achieved production. All are contained within extensively metasomatized ultramafic rocks in fault zones. The largest deposit is near Deer Cove (Fig. 9.11), sited along the trace of the Deer Cove thrust fault.

The principal asbestos mining areas in the Quebec Appalachians are Asbestos, Black Lake, and Thetford Mines (Fig. 9.13-9.15). The asbestos deposits are commonly hosted by variably serpentinized ultramafic rocks, primarily harzburgite tectonite; a rare exception, the "C" deposit of the Lake Asbestos mine, occurs in a serpentinized dunite. The asbestos deposits occur in three distinct zones (Riordon, 1975), a southwestern zone (Jeffrey, Nicolet deposits), a central zone comprising the Vimy Ridge (Normandie, Penhale, Vimy Ridge), Black Lake (British Canadian, Asbestos Lake) and Thetford Mines (Bell-King, Beaver, Bennett-Martin) deposits, and a northeastern zone corresponding to the Pennington Sheet (Carey-Canada, National).

The Jeffrey Mine of JM Asbestos is the largest known asbestos deposit in the western world. Exploited since 1881, the orebody comprises 7% to 10% cross-fibre veins hosted by serpentinized harzburgite with numerous lenses of dunite and felsic rock. Reserves as of 1991 were 300 million tonnes. The Nicolet mine, which closed in 1968, is a lenticular orebody with particularly narrow asbestos veins.

The Vimy Ridge group of deposits has been mined since 1917. There are still reserves in the Normandie and Penhale deposits but the Vimy Ridge deposit is exhausted. The deposits are arranged en echelon within a partly serpentinized peridotite. The British Canadian orebody at Black Lake, mined since 1882, is hosted by a harzburgite with lesser dunite and felsic rock. The exceptionally long asbestos fibres comprise 2% of the rock. Reserves at the end of 1991 stood at approximately 29 million tonnes. The Lake Asbestos deposits (Fig. 9.14) were discovered under Black

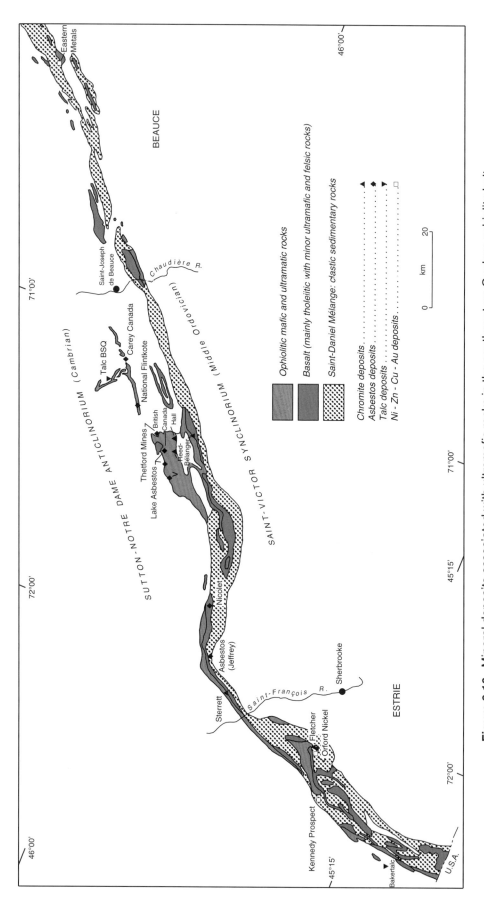

Figure 9.13. Mineral deposits associated with ultramafic rocks in the southeastern Quebec ophiolite belt. V – Vimy Ridge group of asbestos deposits including Vimy Ridge, Normandie, and Penhale deposits; Thetford Mines area includes Beaver, Bell-King, and Bennett-Martin deposits.

GSC

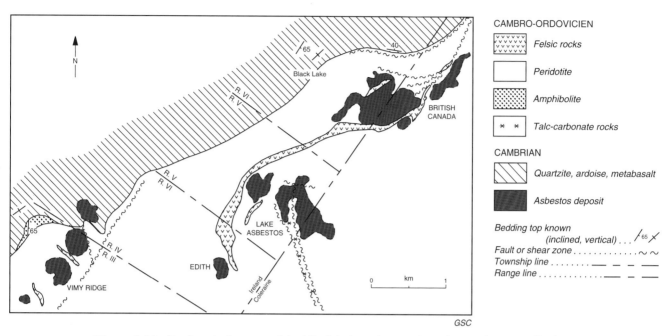

Figure 9.14. Geology in the area of the Black Lake asbestos deposits, southeastern Quebec (after Lamarche, 1984).

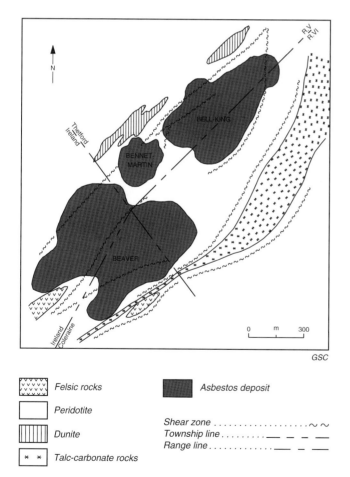

Figure 9.15. Geology in the area of the Thetford Mines asbestos deposits, southeastern Quebec (after Lamarche, 1984).

Lake in the 1950s. The "A" and "B" orebodies are hosted by serpentinized peridotite whereas the "C" orebody is hosted by serpentinized dunite. Reserves at the end of 1991 were 71 million tonnes. The Thetford Mines orebodies (Fig. 9.15) are generally oval-shaped, steeply dipping, and hosted by serpentinized peridotite. An important zone of talc-carbonate rocks borders these deposits to the east. Asbestos grades vary from 3 to 7%, with 25% asbestos observed locally. In 1991, reserves of the Bell Mines (Bell Asbestos Mines Ltd) were 6.4 million tonnes.

The Baie Verte asbestos deposit in west-central Newfoundland (Collins, 1984; Fig. 9.13), occurs in altered ultramafic rocks of the dismembered Advocate ophiolite complex. The tectonostratigraphic and structural setting is similar to that of the southeastern Quebec deposits. The mineralization comprises stockworks of crysotile cross-fibre, most of which are concentrated between brucitic dykes and sills. The orebody is elongate and canoe-shaped and yielded approximately 49 million tonnes of asbestos ore between 1963 and 1991. Substantial reserves remain below the mine workings.

Volcanogenic sulphide deposits in Ordovician ophiolite suites

Introduction

H. Scott Swinden and S.M. Dunsworth

Lower to Middle Ordovician ophiolite suites in the Dunnage Zone and Taconic allochthons contain widespread volcanogenic massive sulphide (VMS) occurrences and deposits (Fig. 9.11). Seven ophiolitic sequences in Newfoundland contain significant volcanogenic massive sulphide deposits. Six are in the Notre Dame Subzone: Point Rousse Complex, Betts Cove Complex, Advocate

Complex, Pacquet Harbour Group on Baie Verte Peninsula, Lushs Bight Group in western Notre Dame Bay, and Skidder Basalt in south-central Newfoundland. Another is the Bay of Islands Complex of the Humber Arm Allochthon. Volcanogenic massive sulphide deposits are also present in the ophiolitic Devereaux Formation in northern New Brunswick and in ophiolite suites of southeastern Quebec (Fig. 9.11).

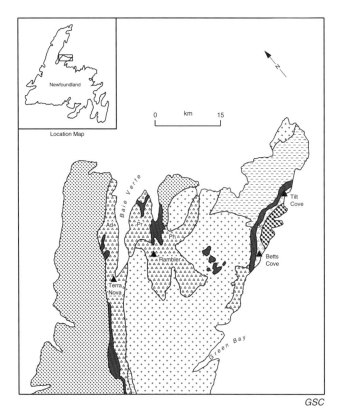

SILURIAN TO DEVONIAN

Undivided granitoid rocks; includes Silurian silicic volcanic porphyries

SILURIAN

Cape St. John and Micmac Lake groups: Mixed silisic and mafic volcanic rocks and associated sediments

LOWER ORDOVICIAN

Snooks Arm Group; mixed mafic volcanic rocks and epiclastic turbidites

OPHIOLITIC ROCKS (Ad-Advocate Complex; Pr-Pointe Rousse Complex; Ph-Pacquet Harbour Group; Bc-Betts Cove Complex)

Mafic volcanic and intrusive rocks

Ultramafic and associated mafic rocks

PRE-LOWER ORDOVICIAN

Fleur de Lys belt: dominantly semipelitic and psammitic metasediments, gneisses and granitoid rocks

Volcanogenic sulphide deposits . . . . ▲

Figure 9.16. General geology of Baie Verte Peninsula with principal ophiolite-hosted VMS deposits. Symbol for Rambler encompasses all deposits shown in Figure 9.17.

Ophiolitic volcanogenic sulphide deposits range from small occurrences to deposits up to approximately 10 000 000 tonnes (Tilt Cove). They are typically pyritic and rich in copper, zinc, and locally gold. Massive sulphides occur throughout the ophiolitic pillow lava unit and volcanogenic stockworks can be traced locally into the underlying sheeted dykes. The sulphides occur both as massive, stratiform lenticular bodies and as disseminations, stringers, and veins commonly associated with quartz and chlorite. The massive sulphides are generally deformed, but in places contain well developed primary banding and laminations, colloform and framboidal textures, and slump features.

The grades of the deposits vary from <1% to >10% copper, with high grade zones ranging up to 29% copper; most have significant gold values. Although zinc was not recovered or reported from many of the deposits, significant concentrations (from 2-5%) are locally present. Silver commonly ranges from <1 ppm to 25 ppm and lead, if present, typically occurs only in trace amounts.

All ophiolites that contain significant volcanogenic sulphide mineralization exhibit geochemical and/or isotopic evidence of eruption in supra-subduction settings (Fyffe et al., 1990; Swinden, 1991; Trottier and Gauthier, 1991). The pillow lavas and associated sheeted dykes commonly include island arc tholeiites (e.g. Betts Cove Complex, Lushs Bight Group, Devereaux Formation) and rocks that are transitional between arc tholeiites and Mid-Ocean Ridge Basalts (MORB) (e.g. Bay of Islands Complex). Ophiolites with dominantly MORB-like geochemistry (e.g. Annieopsquotch Complex; Dunning, 1987) are not known to contain significant volcanogenic mineralization. In the most heavily mineralized sequences (e.g. Betts Cove Complex, Pacquet Harbour Group, Lushs Bight Group, Huntington area of southeastern Quebec), there are significant volumes of refractory mafic volcanic rocks with boninitic affinity and these rocks show a spatial relationship to the volcanogenic mineralization (Swinden et al., 1989a, b; Trottier and Gauthier, 1991).

Despite the similarities of geological and tectonic settings of the Appalachian ophiolite-hosted VMS deposits, recent geochronological work confirms that they are not all of the same age. Mineralized ophiolite suites in western Newfoundland formed in the early Arenig (ca. 488 Ma; Dunning and Krogh, 1985), those in southeastern Quebec in the late Arenig (ca. 479 Ma; Dunning and Pederson, 1988) and those in northern New Brunswick in the Llanvirn (ca. 461 Ma; Sullivan et al., 1990) suggesting repetition of back-arc environments favourable for VMS formation.

Baie Verte Peninsula ophiolite complexes, west-central Newfoundland

H. Scott Swinden

Four separate ophiolitic successions are recognized on the Baie Verte Peninsula: the Advocate, Point Rousse, and Betts Cove complexes, and the Pacquet Harbour Group (Fig. 9.16). All exhibit a variably disrupted ophiolitic stratigraphy. The Betts Cove Complex is overlain by a volcanic-epiclastic cover sequence termed the Snooks Arm Group, interpreted to have been deposited in a back arc basin (Jenner and Fryer, 1980). Mafic volcanic rocks that form the top of the Baie Verte Peninsula ophiolite commonly

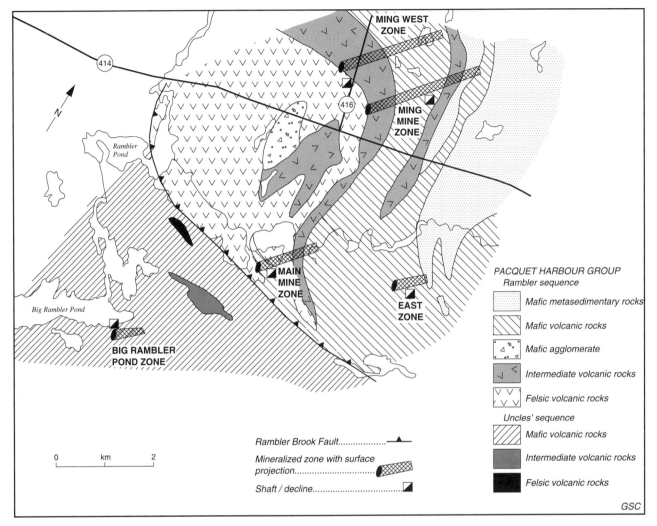

Figure 9.17. Geology of the Rambler mining camp (after Coates, 1990). East Zone and Big Rambler Pond deposits are volcanogenic stockworks. All others are massive sulphides with a mineralized stockwork in the footwall. See Figure 9.16 for location.

include island arc tholeiites and refractory volcanic rocks of boninitic affinity in association with the VMS deposits, suggesting that all formed in a similar tectonic environment. Although the presence of a rhyolite dome associated with ore in the Rambler area presents a significant geological contrast with ore environments elsewhere on the peninsula (e.g. Betts Cove Complex, where felsic volcanics are essentially absent), the significance of this rhyolite in the context of the tectonic environment of ore deposition is still a matter of debate (Hibbard, 1983; Tuach, 1988; Swinden et al., 1989b). Coish et al. (1982) and Swinden et al. (1989b) interpreted the Betts Cove Complex to have formed during a period of arc rifting in the earliest stages of back-arc basin opening and Swinden et al. (1989b) extended this interpretation to include the ore sequence in the Rambler area. Dunning and Krogh (1985) reported a U-Pb zircon date of 488 (+3/-2) Ma on gabbro from Betts Cove Complex, suggesting that the arc-rifting occurred during the earliest Arenig.

Cupriferous massive sulphides were first mined in Newfoundland in the 1860s on Baie Verte Peninsula (the Terra Nova and Tilt Cove deposits). The Tilt Cove deposit, in the Betts Cove Complex, eventually became the longest-lived massive sulphide producer in Newfoundland, with more than 8 million tonnes of ore produced from two main orebodies between 1864 to 1917 and 1957 to 1967. Mineralization consisted of massive pyrite-chalcopyrite-sphalerite (± Au, Ag, magnetite) underlain by an extensive alteration stockwork. The deposit contained significant amounts of zinc and minor nickel (Snelgrove, 1931; Papezik,1964). The other principal VMS deposit in the Betts Cove Complex, the Betts Cove mine, was a similar, although much smaller, massive sulphide body, underlain by a small, localized quartz-chlorite-sulphide alteration pipe and partially remobilized into later fault structures (Upadhyay and Strong, 1973; Saunders and Strong, 1986). The mine was in operation from 1875 to 1885 and produced about 118 000 tonnes of ore averaging 10% Cu (Snelgrove and Baird, 1953).

701

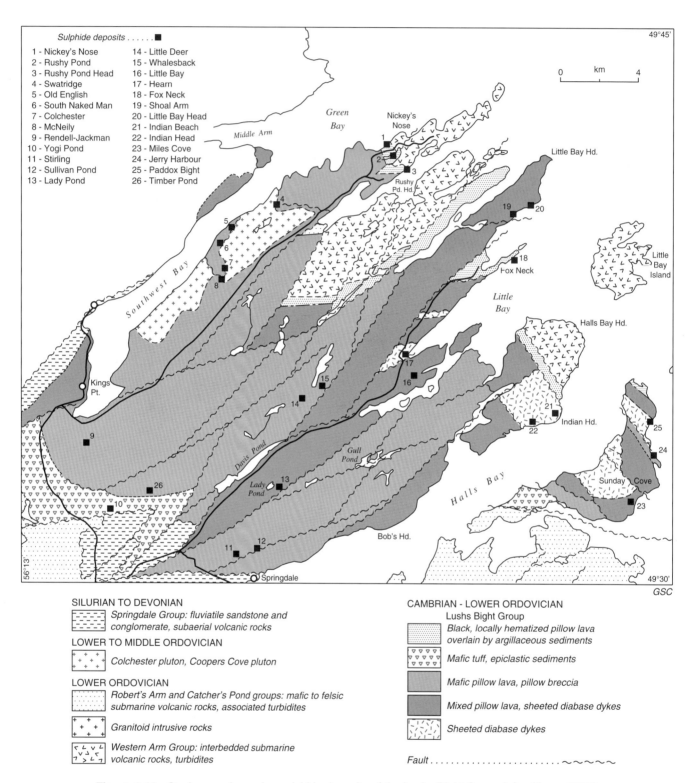

Figure 9.18. Geology and massive sulphide deposits of the Lushs Bight Group (after Kean, 1984).

Elsewhere on the Baie Verte Peninsula, mafic and felsic volcanic rocks of the Pacquet Harbour Group, probable correlatives of the Betts Cove Complex (Swinden et al., 1988b, 1989b; Swinden, 1991), host several massive and disseminated volcanogenic sulphide bodies (Fig. 9.17) that collectively comprise the "Rambler camp" (Gale, 1971; Heenan, 1973; Tuach, 1976, 1988; Tuach and Kennedy, 1978; Coates, 1990). The Cu-(Zn)-Au ore deposits (Rambler Main mine, Ming mine, East mine, Big Rambler Pond mine) produced about 4 500 000 tonnes between 1964 and 1982. The orebodies include both massive sulphides with footwall alteration pipes (Main, Ming, Ming West orebodies) and stockworks (East mine, Big Rambler Pond) and are stacked across a stratigraphic thickness of about 2500 m in and around the Rambler rhyolite, interpreted to be a felsic volcanic dome (Coates, 1990). Massive sulphides comprise conformable lenses of pyrite, chalcopyrite, and pyrrhotite with local sphalerite, minor galena and significant Au and Ag. Stockworks comprise cupriferous, pyritic sulphides in silica, sericite and chlorite alteration zones. All are deformed and metamorphosed, their plunge being concordant to mineral lineations and fold axes throughout the area.

The principal deposit in the Advocate Complex is the Terra Nova (Fig. 9.16), comprising massive blocks of pyrite, pyrrhotite, chalcopyrite with lesser sphalerite, arsenopyrite, and minor gold and silver. The sulphide blocks occur in a mélange with blocks of chlorite schist, mafic volcaniclastic rocks, graphitic slate, massive greenstone and nodular serpentinite and diorite in a matrix of graphitic slate (Hibbard, 1983).

Lushs Bight Group, central Newfoundland

B.F. Kean and D.T.W. Evans

The Lushs Bight Group in western Notre Dame Bay comprises structurally interleaved pillow lavas (dominantly island arc tholeiites and refractory lavas of boninitic affinity and sheeted dykes, Jenner et al., 1988) and is interpreted as a fragment of a supra-subduction ophiolite suite. It is the most productive of ophiolite suites in Newfoundland with respect to VMS deposits (Fig. 9.18). It was the site of a number of small high grade mining operations in the late 1800s and early 1900s (e.g. Little Bay, Swatridge, Colchester, McNeilly, Miles Cove) and experienced a revival in the 1960s and 70s, when mining resumed at the Little Bay deposit (total combined production in excess of 3 000 000 tonnes of 0.8-2.5% copper, 1878-1894 and 1961-1969), Whalesback deposit (3 800 000 tonnes of approximately 1% Cu between 1965 and 1972) and Little Deer deposit (about 75 000 tonnes grading 1.3% Cu in 1973-1974). VMS deposits in the Lushs Bight Group comprise mainly massive sulphides and/or stockworks within mafic volcanic rocks although alteration and stockwork mineralization locally penetrates the underlying sheeted diabase dykes and gabbro. Massive sulphides occur at different stratigraphic positions within the pillow sequence, some deposits near the pillow lava sheeted dyke contact (e.g. Miles Cove) and others apparently higher in the stratigraphy (e.g. Rendell-Jackman, Little Deer, Whalesback, McNeilly, and Colchester). Mafic volcanic rocks and sheeted dykes in the Lushs Bight Group are predominantly island arc tholeiites and there is a common spatial association between volcanogenic sulphides and refractory (boninitic) lavas and/or dykes (Swinden et al., 1989b).

Massive sulphide bodies are dominated by pyrite, chalcopyrite, pyrrhotite, magnetite, and lesser sphalerite, with local concentrations of gold and silver. Stockworks comprise cupriferous disseminations, sulphide veins and stringers, and quartz-sulphide veins in chloritized mafic volcanic rocks and dykes, commonly transformed to chlorite schist. The stockwork zones are generally considered to be feeders to the massive ores, although in many cases they are spatially removed or occur laterally to the massive lenses as a result of faulting and/or transposition.

Well developed primary banding and laminations, colloform and framboidal textures, and slump features are locally preserved (e.g. Rendell-Jackman, Stirling, and Lady Pond). However, the massive sulphides are generally deformed and contain a strongly developed tectonic banding which has obliterated all primary structures in most prospects. High grade thickened concentrations locally occur in fold noses.

The Lushs Bight Group is overlain by ferruginous chert, magnetite iron-formation, and siliceous chocolate-brown and black argillite which have been variously interpreted either as part of the overlying Western Arm Group (Marten, 1971; Upadhyay et al., 1971) or the top of the ophiolite complex (Kean, 1983, 1984). The sediments contain thin beds of pyrite, pyrrhotite, magnetite and rare chalcopyrite at the Nickey's Nose, Rushy Pond (Wheeler's Shaft), and Rushy Pond Head showings. The deposits are clearly exhalative, but lack base or precious metals and are best considered as sulphide facies iron-formations.

Locally, volcanogenic stockworks in the Lushs Bight Group occur within the sheeted dyke layer. An example of this type of mineralization is the Hearn prospect (Fig. 9.18), consisting of sulphide (pyrite, chalcopyrite, sphalerite, minor galena) and gold-bearing quartz veins (≤ 2 cm thick) and disseminated pyrite hosted by sheared chlorite schist in the sheeted dyke-pillow lava transition zone. The chloritized zones range up to 2 m in width and are intermittently developed over a distance of 1 km.

Bay of Islands Complex, Humber Arm Allochthon, western Newfoundland

B.F. Kean, H. Scott Swinden, and S.M. Dunsworth

The Bay of Islands Complex is one of the best exposed and most extensively studied ophiolite complexes in Newfoundland. Although widely interpreted as MORB-like recent geochemical data indicate that pillow lavas in the Bay of Islands Complex include island arc tholeiites, and ultramafic rocks show evidence of derivation from magmas of boninitic affinity. Recent workers have suggested that this ophiolite complex represents a transitional back-arc setting, with the subducting slab influencing magma generation (Elthon, 1991; Jenner et al., 1991). Trondhjemite from the Bay of Islands Complex has been dated as 486 +2/-1 Ma (Dunning and Krogh, 1985), suggesting that it is approximately coeval with the Betts Cove Complex.

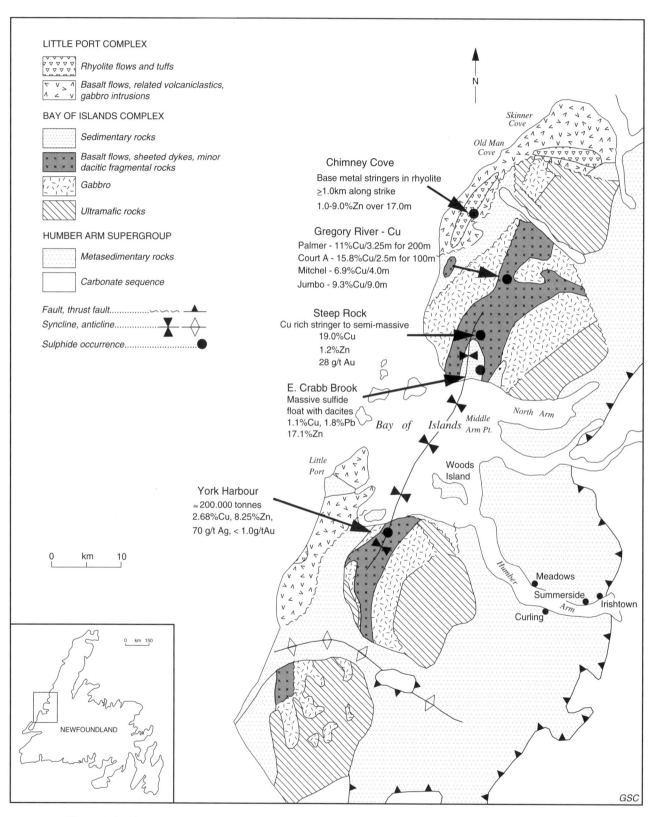

Figure 9.19. General geology and volcanogenic sulphide deposits in the Bay of Islands Complex, western Newfoundland (after MacDougall et al., 1991).

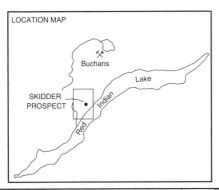

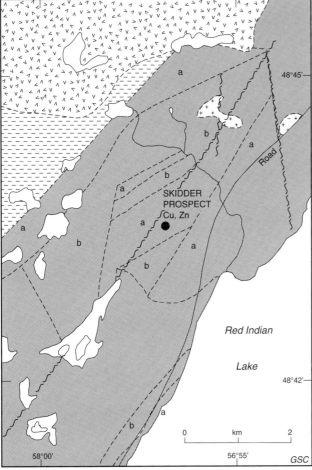

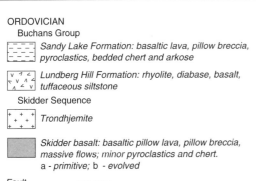

ORDOVICIAN

Buchans Group

Sandy Lake Formation: basaltic lava, pillow breccia, pyroclastics, bedded chert and arkose

Lundberg Hill Formation: rhyolite, diabase, basalt, tuffaceous siltstone

Skidder Sequence

Trondhjemite

Skidder basalt: basaltic pillow lava, pillow breccia, massive flows; minor pyroclastics and chert.
a - primitive; b - evolved

Fault....................

Figure 9.20. General geology in the area of the Skidder prospect (after Pickett, 1987).

The Bay of Islands Complex is host to several volcanogenic sulphide occurrences (Fig. 9.19). The largest of these, the York Harbour deposit, consists of numerous small (up to approximately 60 000 tonnes) lenses of brecciated massive chalcopyrite, pyrite, and sphalerite underlain by chalcopyrite-rich stringer stockwork zones (Graham, 1969; Duke and Hutchinson, 1974; MacDougall et al., 1991), which occur along a 350 to 400 m strike length at the contact between two basalt units. The deposits produced approximately 91 000 tonnes of ore prior to 1913 and are estimated to contain reserves of approximately 200 000 tonnes grading 2.68% Cu, 8.25% Zn, 35 to 70 g/t Ag and less than 1.0 g/t Au (MacDougall et al., 1991).

Skidder basalt, central Newfoundland

H. Scott Swinden

The Skidder basalt, comprises a sequence of tholeiitic pillow basalt, pillow breccia, and massive basalt, with lesser aquagene tuff and chert. Pickett (1987) interpreted the volcanic rocks to be MORB-like on the basis of whole rock, clinopyroxene and spinel geochemistry and suggested that the Skidder basalt is part of a dismembered ophiolite suite. The sequence occupies the same structural position with respect to the Buchans-Robert's Arm volcanic belt as the Annieopsquotch Complex to the south and the Hall Hill/Mansfield Cove ophiolitic complexes to the north.

The Skidder basalt hosts the Skidder prospect (Fig. 9.20), the only VMS deposit in this ophiolite belt. The deposit is composed of two semi-massive to massive lenses of variably layered pyrite with lesser amounts of chalcopyrite and low-Fe sphalerite, minor galena, hematite, and magnetite in a gangue of quartz, chlorite, and lesser calcite (Pickett, 1987). Disseminated pyrite occurs in a quartz-chlorite-talc-rich zone underlying the massive sulphides. Reserves are estimated at 200 000 tonnes grading 2% Cu, 2% Zn, and minor lead. Jasper and jasper-rich siltstone are spatially associated with the massive sulphides.

Devereaux Formation, northern New Brunswick

C.R. van Staal

The Devereaux Formation of the Fournier Group in the Elmtree Inlier (Belledune Subzone, northern New Brunswick) comprises a fragmented ophiolite suite containing gabbro, mylonitic amphibolite, plagiogranite, diabase dykes, and pillow lava (Pajari et al., 1977; Rast and Stringer, 1980; Langton and van Staal, 1989).

Kilometre-scale lenses or blocks of serpentinized harzburgite and pyroxenite together with lenses of pillow lava occur in a tectonic mélange that separates the Fournier Group from the Tetagouche Group (Winchester et al., 1992). Their presence suggests that despite the complex tectonic fragmentation during accretion, nearly all the geological elements typical of an ophiolite suite have been preserved to some extent. Pillow lavas have geochemical signatures transitional between island arc and ocean floor tholeiites, suggesting an ensimatic back-arc setting (Fyffe et al., 1990; Winchester et al., 1992). Coarse grained gabbro and plagiogranite from this ophiolite suite have yielded

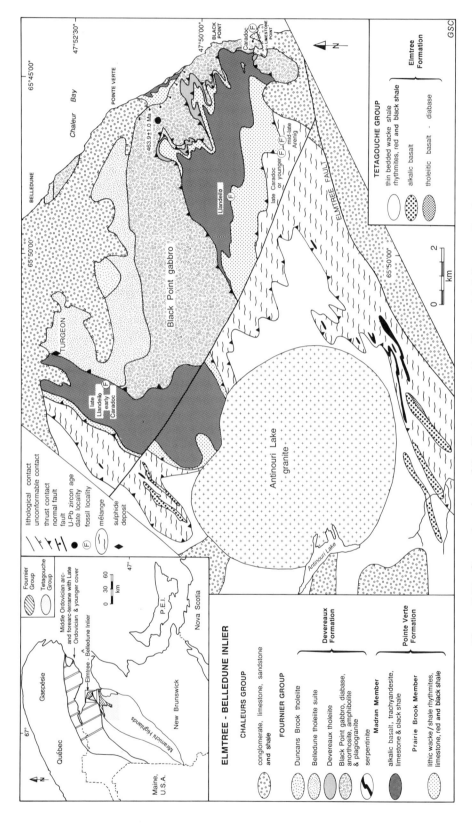

Figure 9.21. General geology of the Elmtree Inlier, New Brunswick (after van Staal and Langdon, 1990) and location of the Turgeon deposit (Zn-Pb-Cu).

U-Pb zircon dates of 464 ± 1 Ma and 460 ± 1 Ma, respectively (Sullivan et al., 1990), suggesting that it is younger than mineralized ophiolites in central Newfoundland.

Pillowed ferrobasalts of the Devereaux tholeiitic suite (Winchester et al., 1992) host the Turgeon VMS deposit (Davies et al., 1983; Kettles, 1987; Fyffe et al., 1990) which consists of two small pyritic zinc-lead-copper massive sulphide lenses (Fig. 9.21). The Powerline zone has been estimated to contain 1 to 2 million tonnes grading 4.0% zinc and 1.5% copper whereas the Beaver Pond zone is much smaller (Davies et al., 1983). Massive sulphides are associated with an extensive silica-chlorite-sulphide alteration zone (Fyffe et al., 1990).

Southeastern Quebec

M. Gauthier

Mafic volcanic rocks and underlying gabbros forming part of a dismembered ophiolite sheet approximately 5 km long and up to 300 m wide south of Eastman host a number of pyritic copper (± zinc) deposits, including the Huntington, Bolton, Ferrier and Ives mines (Fig. 9.22). The Bolton,

Ferrier, and Ives deposits comprise pyritic-cupriferous stockworks in the mafic volcanic rocks. The Huntingdon mine, however, comprises a massive sulphide-stockwork system consisting of four stratabound mineralized zones. The deposit produced 1.1 million tonnes of ore grading 0.9% copper, 0.062 g/t Au and 0.62 g/t Ag between 1950 and 1958 (Trottier et al., 1989). Remaining reserves are estimated at 800 000 tonnes grading 0.88% Cu (Carriere, 1957).

Trottier et al. (1989) recognized three important ore types in the Huntington deposit: (1) small, intensely deformed lenses of massive sulphide (pyrrhotite, lesser chalcopyrite, trace pyrite); (2) quartz and lesser carbonate veins and veinlets containing pyrrhotite, chalcopyrite and trace sphalerite, and widely disseminated pyrrhotite, associated with dominantly quartz-chlorite alteration; and (3) mineralized breccia, comprising volcanic fragments in a quartz-pyrrhotite-chalcopyrite matrix. The first ore type is interpreted as exhalative, the last two as part of the footwall stockwork (Trottier et al., 1989).

The mineralization at Huntington is hosted by refractory mafic volcanic rocks of boninitic affinity suggesting a tectonic environment comparable to the Baie Verte

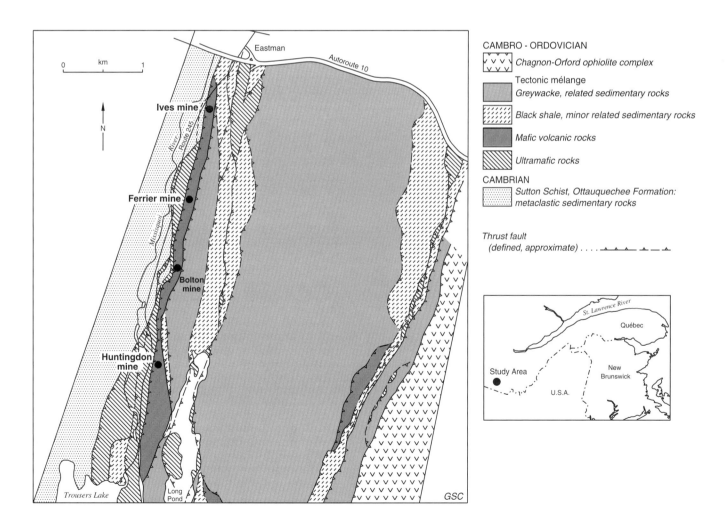

Figure 9.22. General geology and location of volcanogenic sulphide occurrences in the area of Huntingdon mine, southeastern Quebec (after Trottier et al., 1987).

Peninsula and Lushs Bight Group ophiolitic deposits. A recent U-Pb zircon age of 479 +3/-2 Ma from plagiogranite in the Thetford complex (Dunning and Pederson, 1988), indicates that this complex may be as much as 10 Ma younger than the Betts Cove Complex.

Volcanogenic sulphide deposits in Cambrian-Ordovician volcanic/epiclastic sequences

Introduction

H. Scott Swinden

Submarine volcanic and epiclastic sedimentary sequences in the Dunnage Zone record volcanism, sedimentation, and mineralization in a long-lived and complex series of Cambrian to Middle Ordovician island arcs and back-arc basins around the margins of Iapetus. These sequences contain many volcanic-hosted sulphide deposits, ranging in size from small to supergiant (Brunswick No. 12) and from relatively lean to extremely rich (Buchans). On both regional and deposit-level scales, they exhibit a wide spectrum of geological characteristics and tectonic settings typical of VMS deposits worldwide. In some sequences, the VMS deposits are cupriferous with varying amounts of Zn and local enrichments in Au (e.g. Lower-Middle Ordovician primitive arc sequences of north-central Newfoundland, some deposits in southeastern Quebec). Elsewhere, the deposits are polymetallic (e.g. Buchans). Even within single mining camps, there can be considerable variety in metal contents and ratios in the different deposits (e.g. Bathurst).

Recent tectonic models for various parts of the northern Appalachians emphasize the tectonic complexity of the remnants of Iapetus (e.g. Dunning and Pederson, 1988; Swinden et al., 1990a; Swinden, 1991; Dunning et al., 1991; van Staal et al., 1991). It is now recognized that VMS mineralization occurred in a great variety of geological and tectonic settings and at several different times during the Cambrian-Ordovician history of Iapetus (Fyffe et al., 1990; Trottier and Gauthier, 1991; Swinden, 1991). Intra-oceanic (ensimatic) island arc volcanism and associated VMS deposition in the Newfoundland Dunnage Zone occurred in primitive arc settings characterized by arc tholeiite and refractory magmatism (Victoria Lake Group, Wild Bight Group) as well as in more mature arc settings characterized by strongly calc alkalic volcanism (Buchans-Robert's Arm groups). Lower-Middle Ordovician arc volcanism in southeastern Quebec has a broadly similar character to the primitive volcanic arc sequences of central Newfoundland (Swinden and Thorpe, 1984), but the age of mineralization is not sufficiently well constrained to attempt a detailed correlation with specific volcanic episodes in Newfoundland.

Ensialic volcanism and VMS deposition occurred late in the Lower Ordovician during back-arc rifting and accompanying volcanism at or near a continental margin to the east of Laurentia. Eruption of dominantly rhyolitic volcanic rocks was accompanied by the formation of a large number of VMS deposits in the volcanic and closely associated sedimentary rocks of the Bathurst Subzone of northern New Brunswick and the Hermitage Flexure region of southern Newfoundland.

Deposits with broad Besshi-type characteristics (e.g. sediment-dominated settings with minor amounts of non-arc tholeiitic or alkalic mafic volcanism) formed in at least two Cambrian-Ordovician environments in the Canadian Appalachians: (1) Early Ordovician, intra-oceanic back arc rifting produced a cupriferous massive sulphide deposit in the Cold Spring Pond Formation of central Newfoundland and mineralization of a similar age and geological setting is found in the Annidale volcanic belt of southern New Brunswick; and (2) Late Ordovician-Early Silurian, transtensional basins developed in southeastern Quebec in response to arc continent collision. These contain both volcanic and clastic sediment-hosted massive sulphides of widely varying characteristics.

Central Newfoundland

H. Scott Swinden, B.F. Kean, D.T.W. Evans, and S.M. Dunsworth

Upper Cambrian to Middle Ordovician (about 513-460 Ma) volcanic and epiclastic sedimentary sequences in the central Newfoundland Dunnage Zone are rich in volcanogenic sulphides (Fig. 9.11). Five principal tectonic environments of massive sulphide formation in volcanic/epiclastic sequences are recognized (Swinden, 1991): (1) Late Cambrian to Early Ordovician primitive island arcs: dominantly polymetallic deposits associated with island arc tholeiites and rhyolites in the Tally Pond and Tulks Hill volcanic belts (e.g. Duck Pond, Tulks Hill); (2) probable Early Ordovician back-arcs: pyrrhotitic copper deposits related to volcanism of within plate, oceanic island affinity in the Cold Spring Pond Formation (e.g. Great Burnt Lake); (3) Early Ordovician mature island arcs: polymetallic deposits hosted by calc alkalic island arc volcanic rocks of the Buchans, Roberts Arm, and Cutwell groups (e.g. Buchans); (4) Middle Ordovician (Llanvirn or older) primitive island arcs: pyritic copper-zinc deposits related to rifting of a primitive, island arc in the Wild Bight and Exploits groups (e.g. Point Leamington); (5) Middle Ordovician (Llanvirn or younger) continental margin rift environments: Pb-Zn-Ag-rich deposits hosted by dominantly felsic volcanics in the Hermitage Flexure region (e.g. Strickland).

Massive sulphides in the volcanic/epiclastic sequences of central Newfoundland exhibit a considerable range in size and metal content. Although the majority are relatively small, more than 40 have quoted reserves and/or production of more than 200 000 tonnes and there are two deposits (Point Leamington and Duck Pond) with more than 10 000 000 tonnes of contained massive sulphides. In general, cupriferous deposits tend to be dominant in sequences comprising mainly mafic volcanic rocks whereas polymetallic deposits predominate in sequences with a more balanced proportion of mafic and felsic volcanics. Polymetallic deposits tend to be the richest; the massive sulphides in the Buchans mines graded about 24% combined base metals and similar grades are locally encountered in polymetallic deposits in the Victoria Lake Group. VMS deposits in back-arc volcanic sequences tend to be metal-poor (e.g. Great Burnt Lake).

Late Cambrian to Early Ordovician primitive island arcs

The Victoria Lake Group occupies the western side of the south-central part of the Exploits Subzone in south-central Newfoundland. It contains Upper Cambrian (about 513 Ma) arc tholeiites and rhyolites assigned to the Tally Pond volcanics (Dunning et al., 1991) and similar Lower Ordovician (about 498 Ma) rocks of the nearby Tulks Hill volcanics, which are characterized by predominantly felsic pyroclastic rocks with intercalated mafic flows, pillow lava, tuff, agglomerate and breccia (Fig. 9.23). Mafic flows are prevalent in the Tally Pond volcanics; mafic pyroclastic rocks in the Tulks Hill volcanics. Both volcanic sequences are structurally disrupted and imbricated; penetrative deformation within the Tulks Hill volcanics is relatively intense and has largely obliterated primary structures.

Geochemical data suggest mafic volcanics in both volcanic sequences are dominated by arc tholeiites and lesser calc alkalic basalts (Kean and Evans, 1988; Swinden et al., 1989b, Dunning et al., 1991; Swinden, 1991). The tectonic environments are best interpreted as primitive intra-oceanic island arcs.

Although the Tally Pond and Tulks Hill volcanic sequences (and their contained massive sulphide deposits) are geologically similar, their age difference renders mutual tectonic relationships uncertain. They may represent fragments of different island arcs; alternatively they may represent different volcanic and metallogenic episodes in a single long-lived arc system (Dunning et al., 1991).

Both belts contain significant accumulations of polymetallic (Cu-Zn-Pb-Au-Ag) volcanogenic massive sulphides. The deposits are commonly hosted by variably altered felsic volcanic rocks and include some or all of exhalative massive sulphides, transported massive sulphides, and disseminated and massive stockwork. Major volcanogenic massive sulphide deposits in the Tally Pond volcanics include the Duck Pond deposit and the geologically similar but smaller Boundary deposit (Fig. 9.23), as well as a number of minor showings. The Duck Pond deposit comprises a series of structurally disrupted ore lenses (the large Upper Duck, the smaller Lower Duck and Sleeper zones; Squires et al., 1990) with total geological reserves of about 4.3 million tonnes grading 3.58% Cu, 1.05% Pb, 6.63% Zn, 68.31 g/t Ag and 1.00 g/t Au. In addition, there is more than 10 million tonnes of barren pyrite. Widespread intense footwall alteration characterized by chlorite, carbonate, and sericite extends well beyond the limits of the massive sulphide bodies.

The five major volcanogenic sulphide deposits in the Tulks Hill volcanics (McKenzie et al., 1993) occur within a laterally extensive horizon of felsic volcanic rocks. Four of these (Tulks Hill and Tulks East in the south, Daniel's Pond and Bobby's Pond in the north) are massive sulphide-stockwork systems whereas the fifth (Jacks Pond) contains only stockwork. Massive sulphides in the southern part of the belt are typically hosted by felsic pyroclastic rocks characterized by siliceous, sericitic and in places chloritic alteration. Pyrite is the dominant sulphide (70%) with lesser sphalerite, galena, chalcopyrite and accessory arsenopyrite, pyrrhotite, and tetrahedrite-tennantite (Kean and Evans, 1988). The richest of these, the Tulks Hill deposit, contains 750 000 tonnes in four stratiform massive sulphide lenses grading 5-6% Zn, 2% Pb, 1.3% Cu, 41 g/t Ag, and 0.4 g/t Au (Kean and Thurlow, 1976; Jambor, 1984; Jambor and Barbour, 1987). Massive sulphides in the northern part of the belt are likewise associated with felsic volcanics and tend to be richer in Pb and Ag and poorer in Cu than their southern counterparts. Typical high grade intersections at Daniel's Pond assay generally less than 1% Cu, more than 11% combined lead and zinc, and more than 200 g/t Ag.

The polymetallic Victoria Mine deposit at the northern end of the Tulks Hill sequence is somewhat enigmatic in that although it has many geological characteristics of a VMS deposit (Kean and Evans, 1988), it occurs within a major fault zone and has probably been substantially remobilized (Desnoyers, 1990; McKenzie et al., 1993).

Probable Early Ordovician back-arcs

Pyrrhotitic, cupriferous massive sulphides hosted by within-plate tholeiites of ocean island affinity occur in the Cold Spring Pond Formation in the Exploits Subzone of south-central Newfoundland. The geological succession consists dominantly of epiclastic turbidites and argillites with lesser felsic and mafic volcanic rocks (Fig. 9.24). Mafic volcanic rocks, assigned to the North Salmon Dam Basalt (Swinden, 1988a) are geochemically of ocean island affinity and host several small, volcanogenic stockworks and one large VMS deposit, the Great Burnt Lake deposit. This deposit consists of a massive sulphide body (pyrrhotite, lesser pyrite, chalcopyrite, minor sphalerite) and a footwall chlorite-quartz stockwork with disseminated and stringer sulphides. The massive sulphide body, which is estimated to contain 680 000 tonnes of massive sulphide grading 2-3% Cu, occurs at the contact between tholeiitic basalt of non-arc, ocean island basalt affinity and overlying greywacke (Swinden, 1988a). The predominance of volcaniclastic sediments in the Cold Spring Pond Formation coupled with the affinity of the host basalts suggests that this deposit is best considered as Besshi-type (cf. Fox, 1984; Slack, 1990)

Early Ordovician mature island arcs

The Buchans-Robert's Arm volcanic belt (Dean, 1978) is a sequence of Lower Ordovician volcanic, subvolcanic and epiclastic rocks that can be traced in a sinuous outcrop pattern for more than 200 km along the eastern side of the Notre Dame Subzone in central Newfoundland (Fig. 9.25). The belt, as presently defined, includes rocks assigned to the Buchans, Robert's Arm, Cottrell's Cove, and Chanceport groups. Nearby rocks, which are coeval, lithologically and geochemically similar, and probably related, include the Cutwell Group between Triton Island and Halls Bay Head, and possibly parts of the Catchers Pond Group on the Springdale Peninsula (Swinden et al., 1988b; Szybinski et al., 1990).

Mafic volcanic rocks in the Buchans-Robert's Arm belt are geochemically bipartite (Swinden, 1992). Rocks in its eastern and central parts comprise a tholeiitic association consisting entirely of mafic volcanic rocks with geochemical affinities ranging from mid-ocean ridge basalts and non-arc alkalic basalts to transitional island arc tholeiites. The calc alkalic association occupies the western side of the belt and comprises a bimodal assemblage of rhyolite and calc alkalic basalt and andesite. The two associations are always in fault contact with each other. Swinden (1992) suggested

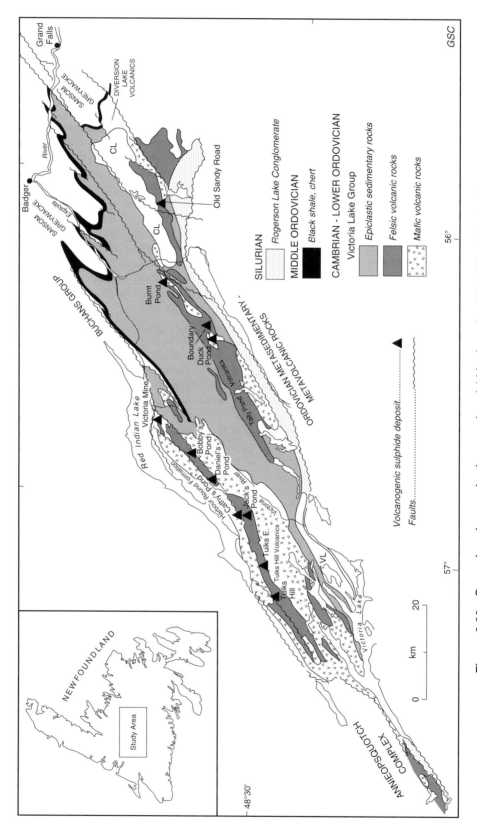

Figure 9.23. General geology and volcanogenic sulphide deposits of the Victoria Lake Group, south-central Newfoundland. CL – Crippleback Lake Quartz Monzonite (Precambrian); VL – Valentine Lake Quartz Monzonite (Precambrian).

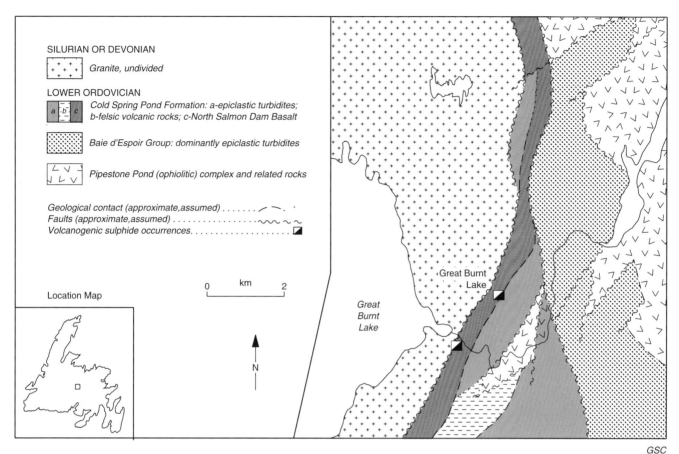

Figure 9.24. Simplified geology of the area of the Great Burnt Lake massive sulphide deposit, central Newfoundland (after Swinden, 1988a).

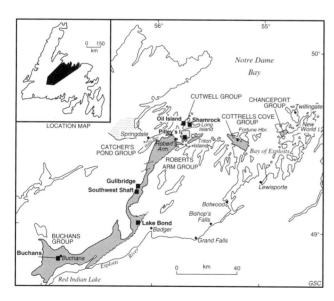

Figure 9.25. Location of the Buchans-Robert's Arm volcanic sequences (shaded) and related volcanic units (Cutwell and Catcher's Pond groups, stippled). See text for details. Volcanogenic sulphide deposits indicated by solid squares.

that this association records interleaving of rocks from disparate back-arc (tholeiitic association) and island arc (calc alkalic association) settings.

Most of the VMS deposits are hosted by the calc alkalic volcanic association. The bimodal nature of this volcanic association, particularly the presence of significant volumes of felsic volcanics, suggests that this sequence comprises the remnants of a relatively mature ensimatic island arc (Stephens et al., 1984; Swinden, 1991). Felsic volcanic rocks have yielded U-Pb zircon ages of 473 ± 2 Ma at two localities (Dunning et al., 1987).

Although the belt contains numerous minor occurrences of volcanogenic sulphides (Swinden and Sacks, 1986; Swinden, 1987), economically significant VMS deposits in the belt are concentrated mainly in the Buchans, Robert's Arm, and Cutwell groups in three areas: the Buchans mining camp, the Gullbridge-Lake Bond area, and the Pilley's Island-Long Island area. Deposits at the northern and southern ends of the belt (e.g. Buchans in the south and Pilley's Island in the north) include significant amounts of polymetallic massive sulphide, whereas central examples (e.g. Gullbridge, Lake Bond) comprise cupriferous stockworks.

The Buchans Group hosts the baritic, polymetallic orebodies of the Buchans camp (Fig. 9.26) which comprise Canada's highest grade volcanogenic massive sulphide camp. ASARCO Incorporated mined 16 196 876 tonnes of ore from five major orebodies between 1928 and 1984. The milling head grade averaged 14.51% Zn, 7.56% Pb, 1.33% Cu, 126 g/t Ag, and 1.37 g/t Au (Thurlow, 1990). The geology and history of these deposits are synthesized in two volumes dedicated to the Buchans camp (Swanson et al., 1981; Kirkham, 1987a) and numerous journal articles and guidebook descriptions.

The Buchans orebodies are hosted by felsic pyroclastic rocks and breccias assigned to the Buchans River Formation (Fig. 9.26); mafic volcanic and related sedimentary rocks occur stratigraphically above and below the ore. The stratigraphy of the Buchans Group is disrupted by a complex thrust system dominated by duplex and antiformal stack geometries (Calon and Greene, 1987; Thurlow, 1990). The relatively incompetent mineralized and altered pyroclastic rocks localized strain, and all of the major orebodies are bounded on at least one side by thrust faults.

Three distinct ore types occur at Buchans: (1) stockwork, (2) in-situ, and (3) transported. The last two types were of equal economic importance and accounted for 98% of production. The stockwork mineralization consists of networks of disseminated and veinlet pyrite with minor base metal sulphide and barite in a strongly silicified and chloritized country rock. The in-situ ores structurally overlie most of the stockwork mineralization, and are characterized by massive, brecciated, and streaky-banded textures (Jambor, 1987), which are now generally regarded as a product of ductile shear of the high grade ores (Thurlow, 1990). The transported ores consist of unsorted, matrix-supported resedimented breccia deposits hosted within felsic pyroclastics, thought to have been transported downslope as a series of gravity-driven debris flows (Thurlow, 1976; Thurlow and Swanson, 1981; Binney, 1987).

The Buchans ores are interpreted to have formed during a period of extension (Kirkham, 1987b) in an oceanic island arc. There is some evidence that caldera resurgence was a factor in ore genesis (Kirkham and Thurlow, 1987). High grade in situ ores are thought to have formed at the site of hydrothermal discharge above altered and mineralized footwall rocks. Evidence of ore deposition during extensional tectonics suggests channeling of transported ores into a graben (Thurlow and Swanson, 1981; Binney, 1987; McClay, 1987).

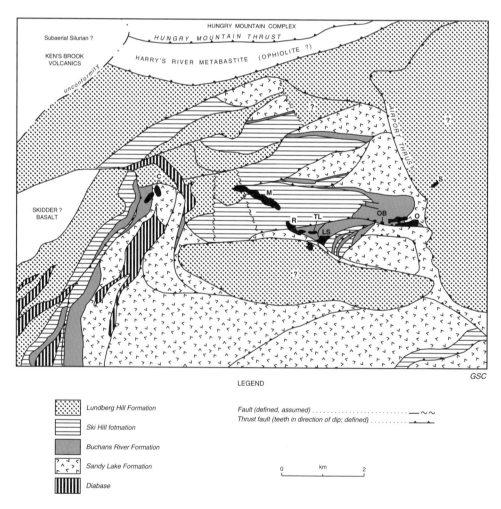

LEGEND

Lundberg Hill Formation

Ski Hill fotmation

Buchans River Formation

Sandy Lake Formation

Diabase

Fault (defined, assumed) . . . .

Thrust fault (teeth in direction of dip; defined) . . . .

0   km   2

GSC

Figure 9.26. Generalized geological map of the Buchans mining camp after Thurlow (1990). Black areas are orebody outlines projected to surface. C – Clementine; LS – Lucky Strike; M – MacLean; O – Oriental; OB – Old Buchans; R – Rothermere; S – Sandfill; TL – Two-Level.

At the northern end of the Buchans-Roberts Arm belt, the Pilley's Island deposits (Fig. 9.25) are hosted by calc alkalic mafic and felsic volcanic rocks of the Roberts Arm Group which are geochemically similar to those at Buchans. The main deposit consists of an upper zone of massive pyrite, chalcopyrite and minor sphalerite, barite in altered breccia, and a lower massive and stringer pyrite zone in an altered dacite flow. The sulphides are underlain and surrounded by sericitic alteration which contains stockwork pyrite±chalcopyrite, sphalerite, and galena mineralization. Other showings in the immediate vicinity include polymetallic massive Zn-Pb-rich sulphides (the Bull Road showing).

Between Pilley's Island and Buchans, a number of cupriferous volcanogenic stockworks locally attain sufficient size and grade to have supported mining. Best known of these is the Gullbridge deposit, which produced 3 million tonnes averaging 1.1% Cu (Pope et al., 1990) between 1967 and 1972. An interesting feature of this deposit is a thermal metamorphic overprint, attributed to nearby granite intrusion, that has produced a chlorite-cordierite-andalusite alteration assemblage within host metabasalt (Upadhyay and Smitheringale, 1972). The principal sulphides (i.e. pyrite, pyrrhotite, and chalcopyrite with traces of Zn, Pb and Ag) occur as disseminations, ribbons and rare lenses up to 10 by 1 m (Pope et al., 1990).

Middle Ordovician primitive island arcs

Thick volcanic and epiclastic sequences of Early to Middle Ordovician age assigned to the Wild Bight and Exploits groups outcrop in the Exploits Subzone of central and eastern Notre Dame Bay (Fig. 9.27). The Wild Bight Group hosts four volcanogenic sulphide deposits: Point Leamington, Lockport, Indian Cove, and Long Pond (Swinden, 1988b); the Exploits Group contains one principal volcanogenic sulphide deposit at Tea Arm and a number of minor stockwork-type sulphide occurrences.

Massive sulphides in these sequences are associated with a volcanic assemblage of island arc tholeiites, refractory mafic volcanic rocks, and rhyolites, which are interpreted to record the rifting of an early or middle Ordovician island arc (Swinden et al., 1990a). The massive sulphides are associated in space with the refractory volcanic rocks. Swinden et al. (1989a, b) suggested that massive sulphide deposition in this environment was promoted by the increased bulk permeability and heat flow inherent in the rifting environment and by the additional heat brought into the subsurface environment by the refractory magmas.

The largest VMS deposit in these sequences, Point Leamington, comprises a massive sulphide body underlain by an extensive stockwork system. It contains an estimated 13.8 million tonnes of stratabound, massive sulphide, grading 0.48% Cu, 1.92% Zn, 18.1 g/t Ag, and 0.9 g/t Au (Walker and Collins, 1988). The deposit occurs within a felsic volcanic lens, which overlies pillow basalts and is in structural and stratigraphic contact with a hanging wall of mafic hyaloclastic, pyroclastic, and volcaniclastic rocks. The massive sulphides consist of pyrite with accessory chalcopyrite, sphalerite, pyrrhotite, and arsenopyrite. Heavily silicified, chloritized, and sericitized rhyolite with stringer sulphides forms the footwall alteration zone while stringer and disseminated pyrite (≤10%) occurs in the pervasively silicified hanging wall sediments.

Middle Ordovician continental margin rifts

Dominantly rhyolitic volcanic rocks and associated sedimentary rocks in the Hermitage Flexure region of southern Newfoundland (Fig. 9.28) occupy the southern and eastern margins of the Exploits Subzone, a similar tectonostratigraphic position to that occupied by the Tetagouche Group in northern New Brunswick (Swinden and Thorpe, 1984; Stephens et al., 1984). The sequences are generally interpreted to reflect volcanism and sedimentation at or near the western margin of a continent or continental fragment to the east of Laurentia (Colman-Sadd, 1980; Swinden, 1991; Colman-Sadd et al., 1992). Rhyolite in these sequences has been dated (U-Pb zircon) at 466 ± 3 Ma (Dunning et al., 1990) and carbonaceous shales in the Bay d'Espoir area have yielded middle Ordovician fossils (Williams, 1991).

Massive sulphides occur in both the eastern (Barasway de Cerf) and western (Strickland) parts of the Hermitage Flexure (Fig. 9.28). The deposits are typically small, rich in Pb, Zn, Ag, and locally Sb and As, and contain little Au or Cu. The Strickland prospect, hosted by the Bay du Nord Group in southwestern Newfoundland, is the largest, comprising approximately 260 000 tonnes grading 5.25%

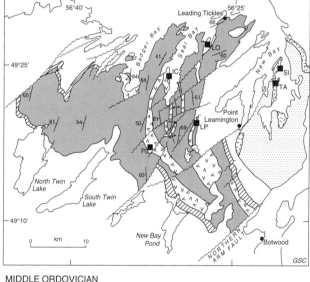

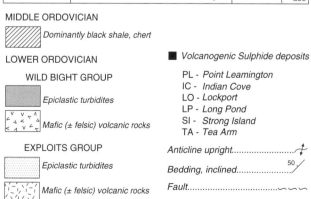

Figure 9.27. General geology and volcanogenic sulphide deposits of the Wild Bight and Exploits groups, central Notre Dame Bay.

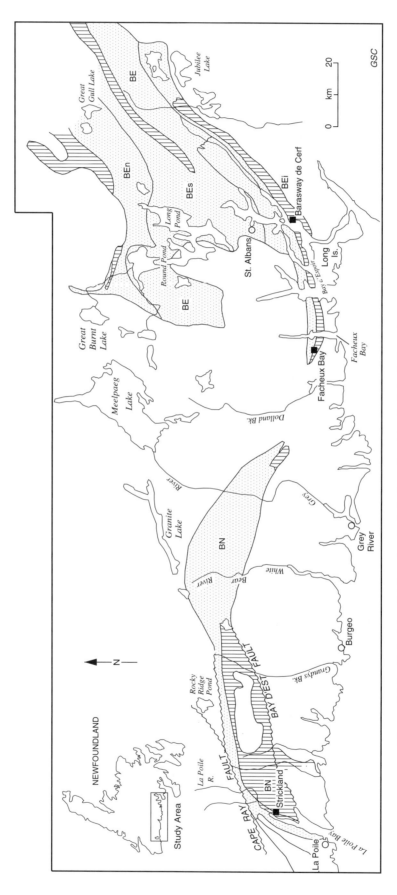

Figure 9.28. Ordovician felsic volcanic (ruled) and epiclastic (stippled) rocks and associated volcanogenic sulphide deposits (solid squares) in the Hermitage Flexure region, southern Newfoundland (after Swinden and Thorpe, 1984). BN – Bay du Nord Group; BE – Baie d'Espoir Group: BEi – Isle Galet Formation; BEs – St. Joseph's Cove Formation; BEn – North Steady Pond Formation.

combined Pb and Zn and 195 g/t Ag (D.R. Prince, unpublished report for Falconbridge, 1981). The mineralization comprises two exhalative massive sulphide zones (the Main and Silver Hill zones) and five stockwork zones (Wynne and Strong, 1984; Swinden, 1988c). The massive sulphide mineralization (sphalerite and galena) is locally associated with silver-rich lenses of carbonate-tremolite gangue. The stockwork zones occur in the hydrothermally altered stratigraphic footwall of the Main zone (Swinden, 1988c), and comprise heavily disseminated and veinlet chalcopyrite, pyrite, sphalerite, and galena in strongly chloritized mafic tuffs (Wynne and Strong, 1984).

New Brunswick
Early Ordovician back-arc basin, southern New Brunswick
A.A. Ruitenberg

Recent geological and geochronological work (McLeod et al., 1992a) has identified a Lower Ordovician (493 ± 2 Ma; 497 ± 10 Ma, U-Pb zircon) bimodal volcano-sedimentary belt (the Annidale belt) in southern New Brunswick. This belt, which is situated immediately north of the Avalon Zone, constitutes the northeastern extension of the St. Croix Subzone. The belt comprises intensely deformed pillow lavas, felsic flows, pyroclastic rocks, and clastic sedimentary rocks. Preliminary chemical data indicate an island arc or back arc origin (Ruitenberg and McLeod, 1991; McLeod et al., 1992a).

The Annidale belt contains a number of intensely deformed cupriferous, pyritic sulphide deposits, the largest of which is the Annidale deposit (Ruitenberg, 1970; Fig. 9.11). The mineralization is associated with thin basaltic to andesitic tuff units. The deposits are interpreted as volcanogenic with affinities to Besshi-type deposits (Ruitenberg and McLeod, 1991). In terms of age, geological setting, and tectonostratigraphy, they appear to resemble the Great Burnt Lake deposit in central Newfoundland.

Middle Ordovician continental margin (back-arc) rift, northern New Brunswick
C.R. van Staal and J.L. Davies

Massive sulphide deposits of the Bathurst camp in northern New Brunswick are hosted by the Lower to Middle Ordovician Tetagouche Group of the Bathurst Subzone (Fig. 9.29). They represent the largest concentration of volcanogenic massive sulphide (VMS) deposits in the Appalachian-Caledonian Orogen, with past production and published ore reserves of almost 300 million tonnes (Sangster, 1980). There are at least 94 deposits/occurrences in the Bathurst-Newcastle area, of which 56 are wholly or partly massive sulphide, and 74 are wholly or partly stockwork (McCutcheon, 1990a). The largest deposit is the supergiant Brunswick No. 12 deposit, estimated as of June 1990, to contain 97 million tonnes grading 8.99% Zn, 3.65% Pb, 0.32% Cu, and 9.68 g/t Ag (van Staal and Luff, 1990). Approximately 60 million tonnes of similar grade have been mined and there are separate copper reserves of 12.8 million

tonnes of 1.27% Zn, 0.45% Pb, 1.11% Cu, and 35.3 g/t Ag. Other deposits with estimated reserves and/or production of more than 10 million tonnes include Brunswick No. 6, Canoe Landing Lake, Caribou, Chester, FAB, and Heath Steele (Fig. 9.29). At present, ore is produced from the Brunswick No. 12, Caribou, Stratmat, and Heath Steele mines. The Brunswick No. 6 deposit produced 12 million tonnes of Zn-Pb-Ag ore from 1966 to 1983; and the Wedge deposit produced 1.36 million tonnes of Cu-Ag ore from 1962 to 1968. The Heath Steele deposit, mined since 1957, was forced to close in 1983 because of poor base metal prices but reopened in 1989. The Caribou mine produced 910 000 tonnes of secondarily enriched cupriferous sulphide ore between 1970 and 1974, and is now producing Zn-Pb-Ag-rich sulphide ore. Gold and silver concentrations in gossan were exploited in the Caribou and Heath Steele deposits in the 1980s and the voluminous gossan at the Murray Brook deposit has recently been mined for its gold content.

Recent structural and geochronological studies have resulted in a redefinition of the Tetagouche Group (van Staal, 1986, 1987; van Staal et al., 1988; van Staal and Fyffe, Chapter 3). Near the Tetagouche Falls, Bathurst and the Key Aracon Mine, the basal unit of the Tetagouche Group comprises fossiliferous Upper Arenig to lowermost Llanvirn shallow water, conglomerate, calcarenites and calcisiltites (Fig. 9.30). These pass conformably upwards into a felsic volcanic complex, dated (U-Pb zircon) at 466 ± 5 Ma. (Llanvirn to early Llandeilo, Sullivan et al., 1990; van Staal et al., 1991). Two formations are recognized in the felsic complex: (1) the Nepisiquit Falls Formation of mixed epiclastic/pyroclastic sediments (tuffites), generally characterized by quartz eyes and quartz- and feldspar-phyric rhyolite intrusions and/or flows; and (2) the Flat Landing Brook Formation of aphyric or feldspar-phyric rhyolite and pyroclastic flow and fall deposits with minor bodies of tholeiitic basalt (the Otter Brook tholeiites), which have geochemical characteristics of continental tholeiites (van Staal et al., 1991). The overlying Boucher Brook Formation comprises mainly volcaniclastic greywacke, fossiliferous Caradoc black phyllite with minor chert, and geochemically-distinct alkalic basalt (van Staal et al., 1991).

Most of the massive sulphide deposits in the Bathurst Subzone are hosted by the metamorphosed equivalents of fine grained clastic rocks and felsic tuffites of the Nepisiquit Falls and Boucher Brook formations (Fig. 9.30). The massive sulphide deposits are fine grained layered stratiform lenses that are pyrite-rich, barite-poor and texturally complex, comprising pyrite, sphalerite, galena, chalcopyrite, and pyrrhotite with minor arsenopyrite, marcasite, and tetrahedrite. Disseminated mineralization, mainly pyrite, underlies many of the lenses and many of the stratiform deposits have a laterally equivalent oxide (±carbonate and silicate) iron-formation. At least two subtypes can be recognized, based on stratigraphic setting and deposit characteristics (van Staal and Williams, 1984; van Staal, 1986; Fig. 9.29): (1) Brunswick-type deposits are facies equivalents of a laterally extensive Algoma-type, iron-formation (McAllister, 1960; Gross, 1965; Davies, 1972). They include the Brunswick No. 12 (Luff, 1975), No. 6 (Boyle and Davies, 1964), Austin Brook (Boyle and Davies, 1964; Davies, 1972), Flat Landing Brook (Troop, 1984),

715

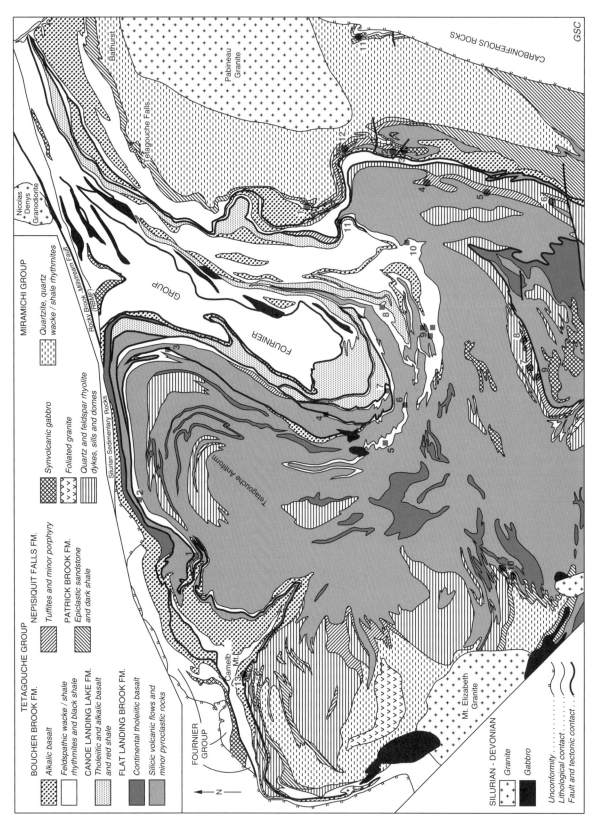

Figure 9.29. General geology of the Bathurst mining camp, northern New Brunswick, with locations of principal massive sulphide deposits. Red dots are Brunswick-type deposits: 1 – Brunswick No. 12; 2 – Brunswick No. 6; 3 – Austin Brook; 4 – Flat Landing Brook; 5 – Portage Brook; 6 – Captain; 7 – Heath Steele A,C; 8 – Heath Steele B; 9 – Heath Steele D; 10 – Half Mile Lake; 11 – Key Anacon; 12 – FAB; 13 – Murray Brook. Red squares are Caribou-type deposits: 1 – Caribou; 2 – McMaster; 3 – Armstrong Brook; 4 – California Lake; 5 – West Wedge; 6 – Wedge; 7 – Canoe Landing Lake; 8 – Willet; 9 – Nepisiquit A, B, C; 10 – Louvecourt; 11 – Headway.

and Heath Steele (Whitehead, 1973) deposits. The Brunswick-type deposits occur near the top of the Nepisiquit Falls Formation, and apparently formed during the earliest stages of felsic volcanism; (2) Caribou-type deposits are generally hosted by volcanic greywackes and black phyllites of the Boucher Brook Formation at the contact with the underlying felsic volcanics of the Flat Landing Brook or Nepisiquit Falls formations (McAllister, 1960; Helmstaedt, 1973; van Staal, 1986). They include the Nepisiquit A, B, C, Nine Mile Brook, Canoe Landing Lake, Orvan Brook, Murray Brook, and Caribou deposits. These deposits occupy a higher stratigraphic level than the Brunswick-type and formed during the end stages of felsic volcanism, probably close to the Llanvirn-Llandeilo boundary

(van Staal et al., 1992; van Staal and Sullivan, 1992). They are not associated with laterally extensive Algoma-type iron-formation.

The Tetagouche Group has been affected by at least five recognizable generations of folding. The geometry of the massive ores is defined by interference structures between F_1 and F_2 folds (van Staal and Williams, 1984; de Roo et al., 1990; de Roo and van Staal, 1990). The importance of structure with respect to the disposition of massive sulphides in the Bathurst camp is well illustrated by the recent discovery of an extension to the No. 12 orebody at a depth of more than 1000 m by drilling geological targets based on structural extrapolations (Hussey, 1992).

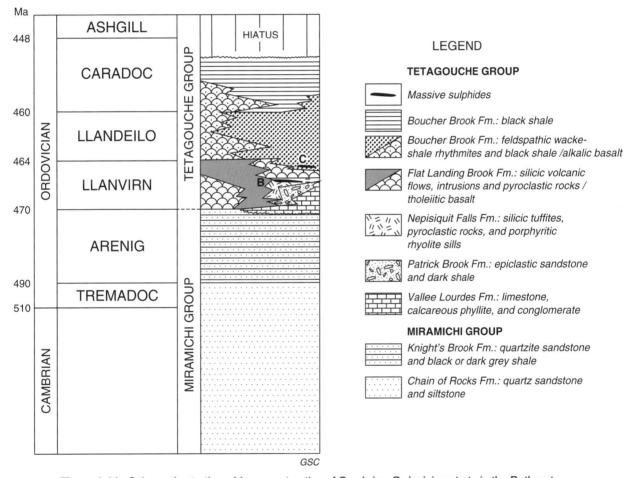

Figure 9.30. Schematic stratigraphic reconstruction of Cambrian-Ordovician strata in the Bathurst area showing stratigraphic setting of massive sulphide deposits. B – Brunswick-type deposits; C – Caribou-type deposits.

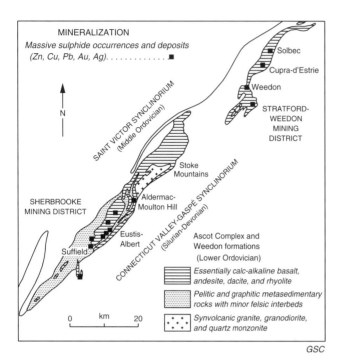

Figure 9.31. Geological setting of volcanogenic massive sulphide deposits in the Ascot-Weedon area, southeastern Quebec.

MINERALIZATION
Massive sulphide occurrences and deposits (Zn, Cu, Pb, Au, Ag) . . . . ■

GSC

The northern part of the Bathurst Subzone contains abundant pillow basalts structurally above the ore horizons that include distinct basaltic suites comprising various types of alkalic basalt with differentiates to trachyandesite and comendite, and tholeiitic basalt separated by major thrusts (van Staal et al., 1991). The bimodal rock association, geochemistry, and geochronology of the volcanics suggest that the Tetagouche Group volcanics represent a rift-related suite (van Staal et al., 1991) that formed during the early stages of the opening of a back-arc basin (van Staal, 1987). The abundance of felsic volcanic rocks indicates that this basin opened, at least in part, in continental crust like the modern Japan Sea or the basin between Okinawa and the mainland of China (Letouzey and Kimura, 1985).

Southeastern Quebec

M. Gauthier

Massive sulphide deposits are recognized in three principal geological environments in southeastern Quebec (Fig. 9.11): (1) Early to Middle Ordovician island arcs; (2) Middle Ordovician fore-arc basin; and (3) late Ordovician to early Silurian transtensional basins, formed during accretionary events of the Taconic Orogeny. In the first environment, cupriferous and polymetallic massive sulphides are similar to those in many parts of central Newfoundland. In the fore-arc basins, massive sulphide deposits are hosted by euxinic black slates and bear many similarities to shale-hosted deposits in the Selwyn Basin (Bossé et al., 1992), although they represent a different tectonic environment. Deposits formed in the transtensional

basins are more varied and include Besshi-type deposits associated with mafic volcanics (e.g. Clinton River), locally containing an unusual suite of metals (e.g. Memphremagog).

Early to Middle Ordovician island arcs

The Ascot Complex (Tremblay et al., 1989) and the Weedon formation in the Estrie-Beauce Subzone of southeastern Quebec (Fig. 9.31) comprise a structurally disrupted Lower to Middle Ordovician island arc volcanic and epiclastic complex (St-Julien and Hubert, 1975; Tremblay, 1992), generally termed the Ascot-Weedon belt, that hosts a number of volcanogenic base metal occurrences and former mines.

The Ascot Complex been studied in detail and subdivided by Tremblay et al. (1989) into the Sherbrooke, Eustis, and Stokes domains based on contrasts in the type and relative proportions of mafic and felsic volcanics. These authors identified two types of basaltic rocks, light rare earth element (LREE)-depleted island arc tholeiites and refractory mafic volcanics of boninitic affinity, and two types of felsic volcanic rocks having LREE-enriched and LREE-depleted signatures, respectively. They suggested that the Ascot Complex record successive stages of arc volcanism from an immature island arc (Sherbrooke Domain) to successively more mature settings near a continental margin (the Eustis and Stokes domains).

The principal deposits in the Ascot Complex include the Eustis-Albert, Suffield, and Aldermac-Moulton Hill deposits in the Eustis and Sherbrooke domains. All of these deposits occur in sericite schists at or near the contact with chloritic schists and slates. At the Eustis-Albert mines, the schist protolith is a coarse volcanic conglomerate which has been highly deformed along the Guadeloupe Fault (Dupuis, 1990). Suffield (7.0% Zn, 0.52% Pb, 0.91% Cu) and Aldermac-Moulton Hill (5.32% Zn, 1.89% Pb, 1.17% Cu) are zinc and lead-rich whereas the others are essentially cupriferous. With the exception of the Eustis, all deposits are vertically zoned, from a copper-rich base (stratigraphic footwall) to a zinc-rich top and lack footwall stockwork zones. Sulphide mineralogy is characterized by pyrite, chalcopyrite, sphalerite, some galena, and minor barite, arsenopyrite, sulphosalts, pyrrhotite, and tellurides in massive "banded" strata. The massive sulphides are characteristically concordant, localized in hinge zones of post-ore folds (Lamarche, 1965), and associated with oxide facies exhalite (Harron, 1976). The Aldermac-Moulton Hill mine produced 300 000 tonnes of ore between 1944 and 1953, the Eustis mine 1.6 million tonnes of ore grading 2.7% Cu between 1865 and 1939 (St-Julien and Lamarche, 1965), and the Suffield mine about 508 000 tonnes of ore intermittently between 1865 and 1958.

Deposits in the Weedon Formation (Cupra-d'Estrie, Weedon, Solbec), as in the Ascot Complex, are lenticular to tabular and enclosed by sericite schists in contact with chlorite schists. Iron- and manganese-rich horizons are generally associated with these deposits. At the Solbec mine, for example, a small siliceous iron-formation occurs along the massive sulphide lens (Sauvé et al., 1972). The Cupra-d'Estrie mine, the largest VMS orebody in this area, produced 2.18 million tonnes of ore grading 1.9% Cu, 2.4% Zn, 0.2% Pb, 0.01% Cd, 0.5 g/t Au, and 37.7 g/t Ag from 1965 to 1977. The ore consists of stratiform massive sulphides which are concordant with the foliation of the enclosing schists. Mafic metavolcanic rocks altered to quartz-chlorite and quartz-sericite schists occur in both the hangingwall and footwall (Sauvé et al., 1972). Up to 30% pyrite occurs disseminated and interlayered with the chloritic schist. The sulphides are mainly pyrite, chalcopyrite, and sphalerite with minor galena, tennantite, and bornite. The Weedon mine, which produced 1.45 million tonnes of ore grading 2.21% Cu and 0.35% Zn during the years 1910-1921, 1952-1959, and 1969-1973, is geologically similar to the Cupra-d'Estrie ore body, but is cut by a Devonian granite, and the schists have been transformed into cordierite-anthophyllite rocks by contact metamorphism.

Middle Ordovician fore-arc basin

The Saint Victor Synclinorium (Fig. 9.32) in the Estrie-Beauce Subzone is underlain by a Middle Ordovician fore-arc flysch sequence known as the Magog Group. This sequence discordantly overlies rocks of the ophiolite belt to the northwest and the island arc volcanics of the Ascot-Weedon belt to the southeast. Black shales and felsic tuffs in the Beauceville Formation of the Magog Group contain volcano-sedimentary exhalative massive sulphide occurrences. The largest of these is the Champagne deposit (Fig. 9.32) which consists of massive banded pyrite, pyrrhotite, and sphalerite (Gauthier et al., 1989; Bossé et al., 1992). Reserves are estimated at 172 000 tonnes grading 2.68% Zn, 0.45% Pb, 0.40% Cu, 19.7 g/t Ag, and 2.1 g/t Au. There is a barium-rich black shale horizon above the massive sulphides and an enrichment in gold (>2 g/t Au) in the upper part of the massive sulphide. Gold-bearing (>200 ppb) black shales also occur elsewhere in the Saint Victor Synclinorium (Godue, 1988).

Late Ordovician to Early Silurian transtensional basins

The Saint-Daniel Mélange in the Estrie-Beauce Subzone, although characterized by ophiolitic rocks, also preserves remnants of Ordovician volcanic/volcaniclastic sequences from other stratigraphic settings. The Bolton volcanics

(Ambrose, 1942) near Lake Memphremagog contain a clastic sedimentary unit that hosts the polymetallic (Zn-Cu-Pb-Sn-Sb-Ag) Memphremagog massive sulphide deposit (Fig. 9.33). This deposit, which was discovered in the late

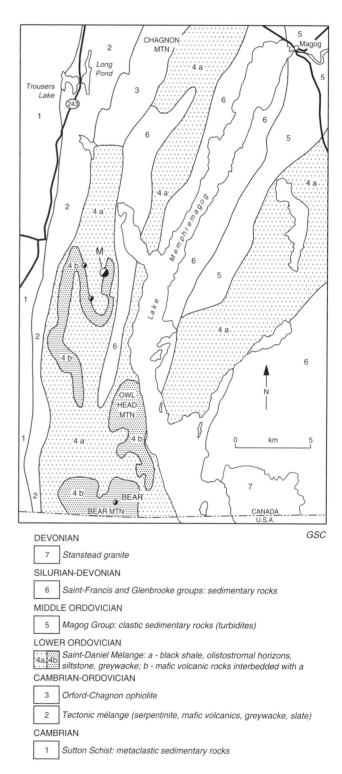

GSC

DEVONIAN

7 Stanstead granite

SILURIAN-DEVONIAN

6 Saint-Francis and Glenbrooke groups: sedimentary rocks

MIDDLE ORDOVICIAN

5 Magog Group: clastic sedimentary rocks (turbidites)

LOWER ORDOVICIAN

4a 4b Saint-Daniel Mélange: a - black shale, olistostromal horizons, siltstone, greywacke; b - mafic volcanic rocks interbedded with a

CAMBRIAN-ORDOVICIAN

3 Orford-Chagnon ophiolite

2 Tectonic mélange (serpentinite, mafic volcanics, greywacke, slate)

CAMBRIAN

1 Sutton Schist: metaclastic sedimentary rocks

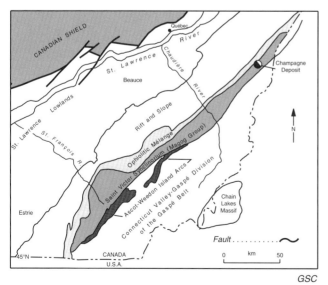

GSC

Figure 9.32. Location of the Magog Group and the Champagne sediment-hosted massive sulphide deposit.

Figure 9.33. Geological setting of the Memphremagog deposit (M) and related massive sulphide occurrences.

719

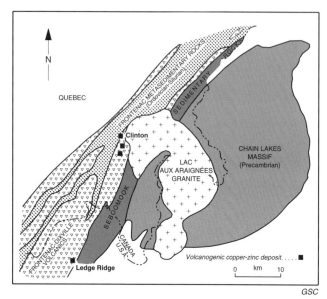

Figure 9.34. Geological setting of massive sulphides in the Frontenac belt, Clinton River area, southeastern Quebec and adjacent Maine.

1800s and from which about 2000 tonnes of ore was extracted between 1888 and 1924, is distinct in geological setting, tectonic environment and deposit characteristics from the cupriferous deposits that occur within the ophiolite suites.

The Memphremagog deposit comprises a massive sulphide lens grading 2.5% Zn, 1.5% Cu, 0.8% Pb, 0.4% Sn, up to 0.24% Sb, and 50 g/t Ag hosted by fine grained clastic sedimentary rocks at the contact with underlying pillow lavas (Trottier, 1987). The massive ore is brecciated, consisting mainly of pyrrhotite, pyrite (chalcopyrite, calcite) and sphalerite, chalcopyrite, galena, freibergite fragments in a pyrrhotite, calcite (pyrite, arsenopyrite, kersteritic stannite, meneghenite, gudmundite) matrix. The footwall mafic volcanics have been altered by seafloor processes but there is no well developed stockwork zone (Trottier and Gauthier, 1991).

Recent geochemical studies (Trottier et al.,1988, 1991) indicate that the mafic volcanics have geochemical signatures of transitional tholeiitic-alkalic affinity and are not of island arc derivation suggesting that the volcanics and associated massive sulphides may have formed in a transtensional basin during early stages of the Taconic Orogeny (Trottier et al., 1988, 1989). In this model, the mineralization is interpreted to be broadly Besshi-type (Trottier and Gauthier, 1991; Trottier et al., 1991) formed in an environment similar to the modern Guaymas Basin, with leaching of continentally derived sediments accounting for the unusual metal assemblage in the deposit.

The Frontenac belt on the southeastern side of the Connecticut Valley-Gaspé Synclinorium (Fig. 9.34) comprises two Upper Ordovician to Lower Silurian dominantly mafic volcanic units in a flysch sequence. In the Clinton River (Mégantic) area, five small stratiform massive sulphide bodies (the A, C, E, F, and O deposits) contain 1 490 000 tonnes grading 2.02% copper and 1.54% Zn. The "O" deposit, the most important, was mined from 1973 to 1975. The massive sulphides comprise dominantly pyrite

and lesser, chalcopyrite, sphalerite and galena, and occur at the contact between mafic porphyritic lavas and mafic to felsic volcaniclastics with lenses of pelitic sediment. Higher in the sequence, other mafic lavas are overlain by a horizon of crimson pelite that is rich in iron and manganese. The associated mafic volcanics are tholeiitic, and were interpreted by Chevé et al. (1983) to have been emplaced in a "pull-apart basin" environment similar to that of the modern Gulf of California.

AVALON ZONE
Introduction
H. Scott Swinden

The Avalon Zone is widely exposed in eastern Newfoundland, southeastern Cape Breton Island (Mira Subzone) and southern New Brunswick. It is generally viewed as an exotic terrane (or composite terrane) of Pan-African affinity (O'Brien et al., 1983), whose Precambrian rocks record events in the Grenvillian-Appalachian time gap. For the purposes of the present discussion and following Barr et al. (1990) and S.J. O'Brien et al. (1991), the Bras d'Or Subzone of Cape Breton Island (cf. Raeside and Barr, 1990), and the Burgeo Subzone of southern Newfoundland that contains rocks of Avalonian age south of the Cinq Cerf Fault Zone and in the Grey River enclave of southern Newfoundland (Dunning and O'Brien, 1989) are herein included in the Avalon Zone sensu lato. The earliest geological history in the Avalon Zone is recorded in the ca. 760 Ma Burin Group, southeastern Newfoundland, a fragment of oceanic crust and related rocks formed during a Proterozoic rifting event (Strong et al., 1978). Bimodal, ensialic volcanism with accompanying sedimentation and plutonism was initiated at about 680 Ma; the first of three main pulses of volcanism (ca. 680 Ma; 635-600 Ma; 575-550 Ma; S.J. O'Brien et al., 1991, 1992a, b; Barr et al., 1990, 1991; Bevier and Barr, 1990; Swinden and Hunt, 1991). The time separation of these magmatic events indicates that they cannot be related to a single tectono-magmatic episode (S.J. O'Brien et al., 1991). The 680 Ma and 635-600 Ma events are generally interpreted to record compressional regimes at one or more Andean-type continental margins (O'Driscoll and Strong, 1979; Barr and White, 1988; Dostal et al., 1990; O'Brien and Knight, 1988; Knight and O'Brien, 1988). The rocks were inhomogeneously deformed during one or more relatively mild events, and served as a basement to the products of the 570-550 Ma events which include both extensional (e.g. eastern Newfoundland, New Brunswick) and compressional (eastern Cape Breton Island) environments broadly synchronous with the rift-drift transition recorded on the Laurentian margin (Williams and Hiscott, 1987).

Latest Precambrian to early Paleozoic sedimentation occurred in a series of shallow marine basins across the Avalon Zone. It is now generally thought that docking of the Avalon Zone with the accreted terranes to the west was complete by the Silurian (Doig et al., 1990; S.J. O'Brien et al., 1991)

The Avalon Zone contains several types of mineral deposits formed mainly during late Proterozoic magmatism and early Paleozoic sedimentation (Fig. 9.35). Submergent facies of pre-600 Ma Andean-type arc volcanic sequences contain volcanic-hosted massive sulphide deposits in

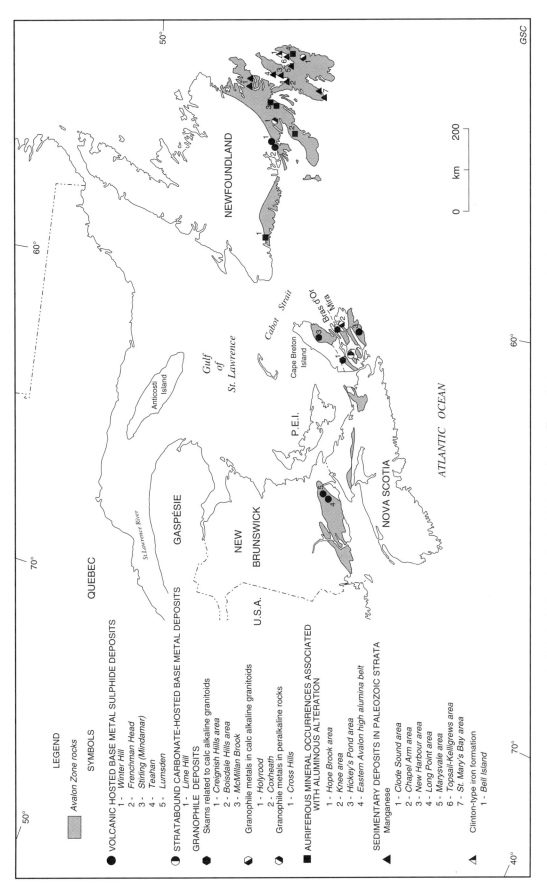

Figure 9.35. Principal mineral occurrences in the Avalon Zone. Symbols may represent either single deposits or groups of deposits.

Newfoundland, Cape Breton Island, and southern New Brunswick. An epithermal style of alteration and mineralization possibly related to the two younger episodes of Proterozoic magmatism (Stewart and Dunning, 1990; Huard, 1989; Hayes and O'Driscoll, 1990) is developed throughout the Newfoundland Avalon Zone and there is at least one possible upper Precambrian carbonate-hosted base metal sulphide deposit in central Cape Breton Island. Calc alkalic plutons related to the latest magmatic episode contain minor occurrences of granophile elements in Newfoundland and porphyry-style mineralization in Cape Breton Island. Peralkaline plutons of the same age in Newfoundland contain concentrations of rare metals. Early Paleozoic stable shelf sedimentation in the Newfoundland Avalon Zone resulted in the formation of extensive sedimentary manganese and iron deposits.

Upper Proterozoic volcanic-hosted massive sulphides

H. Scott Swinden

Volcanic-hosted massive sulphide deposits in the Avalon Zone are found in upper Proterozoic felsic and mafic volcanic sequences in southeastern Newfoundland, southeastern Cape Breton Island, and southern New Brunswick (Fig. 9.35).

The Tickle Point Formation and the coeval Furby's Cove Intrusive Suite, southeastern Newfoundland is an upper Proterozoic (ca. 680 Ma; Swinden and Hunt, 1991; O'Brien et al., 1992a, b) sequence of felsic volcanic rocks with interstratified basalt and andesite. The Tickle Point Formation with related intrusions is the oldest bimodal volcanic sequence in the Newfoundland Avalon Zone. Its tectonic environment has been variously interpreted as an extensional basin and range (Krogh et al., 1988), a continental margin island arc (e.g. Sears, 1990), and a transitional environment between the two (O'Driscoll and Strong, 1979). The Tickle Point Formation hosts two significant volcanogenic massive sulphide showings (Winter Hill and Frenchman Head deposits) and several other minor showings (Sears and O'Driscoll, 1989; Fig. 9.36). The Frenchman Head deposit comprises mainly disseminations and stringers of pyrite, chalcopyrite, sphalerite, and rare galena in felsic volcanics at the top of the Tickle Point Formation. The Winter Hill deposits may occupy a slightly higher stratigraphic interval. The main zone comprises a lower discordant stringer zone of chalcopyrite and pyrite associated with quartz, cordierite, chlorite, and rare andalusite, and an upper concordant lens of disseminated to massive sphalerite, galena, and pyrite (±chalcopyrite) in a carbonate and Ca-Mg-silicate gangue (Sears, 1990). The base metal mineralization passes laterally into a pyritic, base metal-poor facies that outcrops sporadically over more than 3 km of strike length.

The stratabound Mindamar (Stirling) Zn-Pb-Cu-Ag-Au deposit in southeastern Cape Breton Island is the largest volcanic-hosted massive sulphide in the Canadian Avalon Zone. It is hosted by the Stirling volcanics (Fig. 9.37), an apparently subduction-related (Dostal et al., 1990)

sequence of intercalated submarine felsic and mafic volcanics and derived sedimentary rocks. Felsic volcanic rocks have yielded a U-Pb zircon age of 681 +6/-2 Ma (Bevier et al., 1993), indicating that the Mindamar deposit is approximately coeval with the Winter Hill/Frenchman Head deposits in southeastern Newfoundland. The stratigraphic section at the Mindamar deposit comprises: (1) a footwall succession of dominantly felsic volcanic rocks; (2) a "mine series" comprising siliceous siltstones, three large lenticular quartz-carbonate bodies containing dolomite, magnesite, quartz, sericite, chlorite, barite, albite, alunite, and massive sulphide lenses; and (3) a hangingwall succession dominated by intermediate tuffs and tuff breccias with minor felsic flows and sedimentary rocks (Miller, 1980). The massive sulphide deposit comprises intensely deformed, conformable lenses of massive pyrite, sphalerite, galena, chalcopyrite, and tennantite which commonly occur within the siliceous siltstones between and along the contact between siltstone and quartz-carbonate rock. There is no recognizable footwall stockwork zone. Total production was approximately 1.1 million tonnes averaging 6.4% Zn, 1.5% Pb, 0.74% Cu, 75.2 g/t Ag, and 1.03 g/t Au (Miller, 1980).

There are several small volcanogenic sulphide deposits in the Caledonia Highlands of southern New Brunswick, hosted by the Eastern volcanic belt of the Coldbrook Group (Giles and Ruitenberg, 1977; a.k.a. Sequence A, Barr and White, 1988; Fig. 9.38). Volcanic rocks in this sequence, dated at 626-600 Ma (Bevier and Barr, 1990) comprise a partially submergent sequence of dominantly volcanogenic sediments and tuff and calc alkalic mafic and felsic volcanic flows, interpreted by Barr and White (1988) to record subduction at an Andean-type margin.

The Teahan deposit, which was mined for copper in the late 1800s, and the smaller Lumsden deposit, occur in intensely deformed, intercalated mafic flows, rhyolitic tuff and associated volcanogenic sediments (Ruitenberg et al., 1979). They comprise originally lenticular, irregular-shaped, roughly conformable masses of chalcopyrite, pyrite, sphalerite, and galena with minor tennantite forming fracture-fillings and cement to brecciated massive pyrite in a quartz-carbonate-talc gangue. The main ore shoots are structurally controlled in the axial zones of plunging folds. The mineralization is enveloped by chloritic alteration zones which, in turn, are enveloped by broad siliceous-talcose-micaceous alteration zones. Average assay results of drill core indicate 1% combined Zn and Cu with minor Ag and Au (Ruitenberg et al., 1979).

In summary, volcanogenic mineralization in the Canadian Avalon Zone apparently comprises at least two events, an older (ca. 680 Ma) event recorded in southeastern Newfoundland and Cape Breton Island, and a younger event in southern New Brunswick. Although the tectonic settings of the various sequences remain controversial, the results of recent workers indicate that these mineralizing events may have occurred at one or more late Proterozoic Andean-type continental margins. The occurrence of volcanic-hosted mineralization in these environments appears to be in part a function of the local presence of submergent environments, which are required for the deposition and preservation of such deposits.

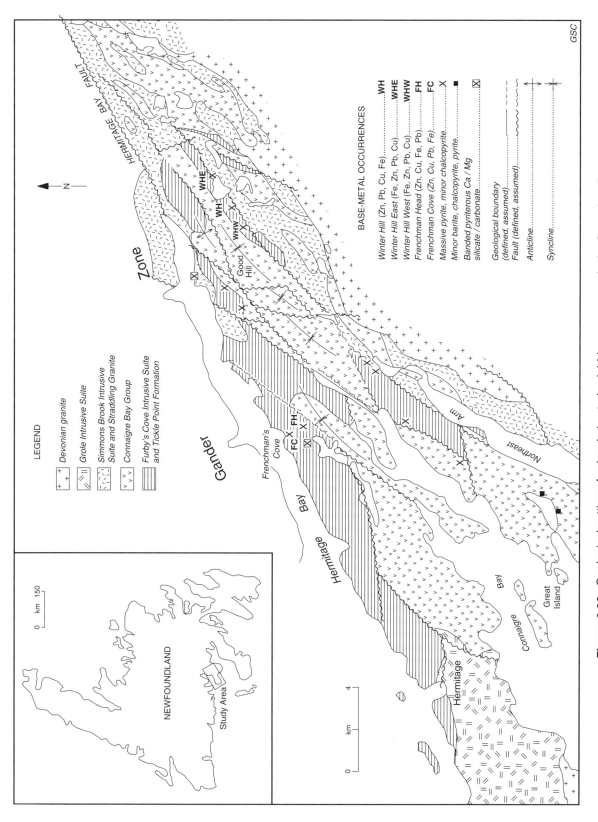

Figure 9.36. Geological setting of volcanic-hosted sulphide occurrences in the Connaigre Bay Group, Hermitage Peninsula, southeastern Newfoundland (after Sears and O'Driscoll, 1989, C.F. O'Driscoll, S. O'Brien, pers. comm., 1994).

Carbonate-hosted stratabound sulphide deposits

A.L. Sangster

The Lime Hill occurrence in central Cape Breton Island (Fig. 9.39), hosted by marble of the Precambrian Lime Hill gneissic complex (Justino and Sangster, 1987) consists of bands of massive and disseminated sphalerite, and minor pyrrhotite and pyrite, in serpentine dolomite and pure dolomite marble. It contains an estimated reserve of 136 000 tonnes grading 9% zinc or 1 800 000 tonnes grading

2% zinc (Energy, Mines and Resources, 1989). Wollastonite is present in calcitic marble at the contacts of granitic dykes (Sangster et al., 1990a) and traces of scheelite have been reported (Chatterjee, 1977a). Structural concentration of sphalerite in some fold hinges is evident in trench exposures. The sulphides are locally well layered and concordant with lithological variations in the host rocks.

Although historically considered a skarn (e.g. Milligan, 1970; Chatterjee, 1980), the deposit has recently been reinterpreted as a metamorphosed carbonate-hosted massive sulphide deposit by Sangster et al. (1990a), who

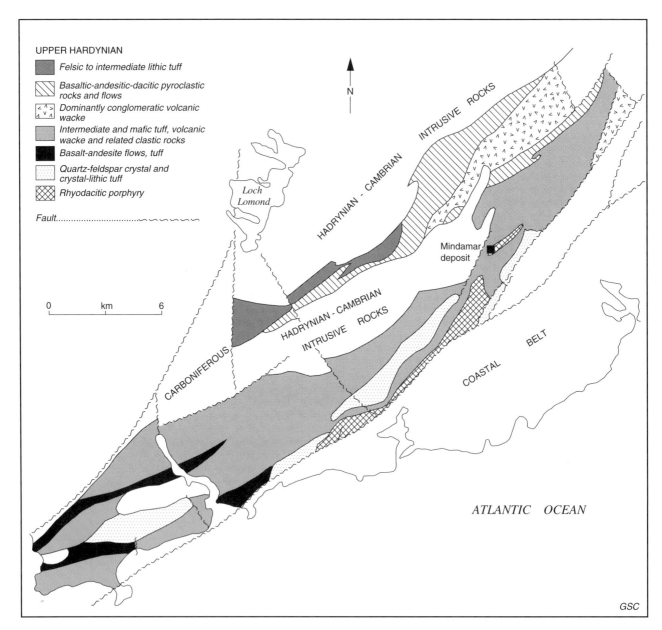

Figure 9.37. Geology of the Stirling area, southeastern Cape Breton Island (after Bevier et al., 1993) and location of the Mindamar (Stirling) volcanogenic sulphide deposit.

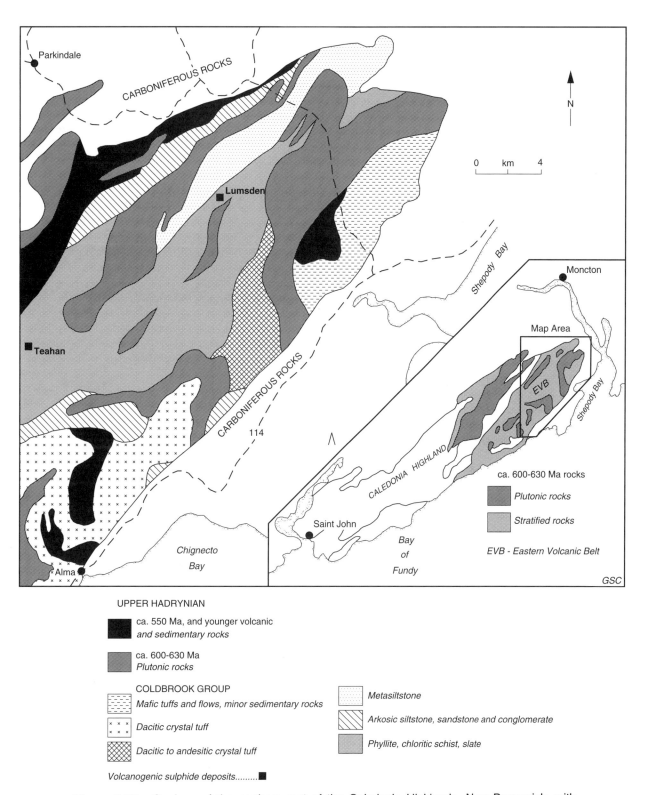

UPPER HADRYNIAN

■ ca. 550 Ma, and younger volcanic
and sedimentary rocks

▨ ca. 600-630 Ma
Plutonic rocks

COLDBROOK GROUP

Mafic tuffs and flows, minor sedimentary rocks

Dacitic crystal tuff

Dacitic to andesitic crystal tuff

Metasiltstone

Arkosic siltstone, sandstone and conglomerate

Phyllite, chloritic schist, slate

Volcanogenic sulphide deposits.........■

Figure 9.38. Geology of the northern part of the Caledonia Highlands, New Brunswick, with emphasis on the ca. 630 to 600 Ma plutonic and stratified rocks (after Barr and White, 1991). Location of the Teahan and Lumsden deposits is indicated.

suggested on the basis of lead isotopes that it might be in a fault block of Grenville basement similar to the Blair River Complex in northwestern Cape Breton Island. At present, there are no other geological criteria supporting a correlation of the Blair River Complex and Lime Hill gneiss. Raeside (1989, 1990) considered that the metamorphic history of the Lime Hill gneiss is more compatible with that prevalent elsewhere in the Aspy Subzone rather than in the Blair River Complex.

Mineralization related to late Precambrian-Cambrian granitoids

H. Scott Swinden and A.L. Sangster

Late Precambrian peralkaline intrusive rocks in the western part of the Newfoundland Avalon Zone contain occurrences of rare earth, base metal, and other granophile elements. The Cross Hills Plutonic Suite (Bradley, 1962; Tuach, 1984a; O'Brien et al., 1984; Tuach, 1991) consists of

a group of spatially and genetically related gabbro to diabase, granodiorite, biotite granite, peralkaline granite and minor syenite intrusions which outcrop over an area of 260 km² near Terrenceville, southeast Newfoundland (Fig. 9.40). These intrude upper Precambrian, subaerial, low grade, metavolcanic rocks (Belle Bay Formation) and metasedimentary rocks (Anderson's Cove Formation) of the Long Harbour Group (Williams, 1971; O'Brien and Nunn, 1980; O'Brien et al., 1984). Imprecise Rb-Sr ages of around 560 ± 30 Ma (Tuach, 1991) suggest an early Cambrian age.

Peralkaline granite in the Cross Hills Plutonic Suite is purple-brown, miarolitic, fine- to medium-grained and characterized by Na-amphibole, which is partly to

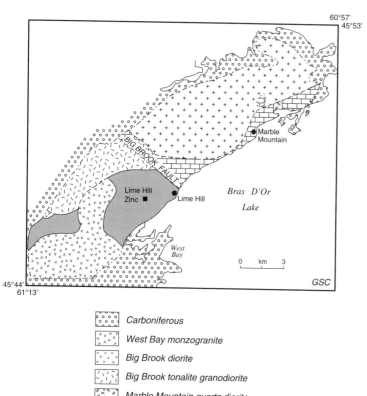

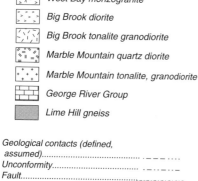

Figure 9.39. Geological setting of carbonate-hosted occurrences in the Lime Hill area, Cape Breton Island (after Sangster et al., 1990a).

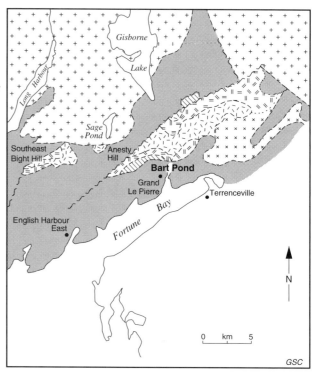

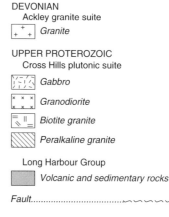

Figure 9.40. Geology of the Cross Hills Plutonic Suite and location of the mineralized peralkaline granite in the Bart Pond area (after Tuach, 1991).

completely replaced by chlorite. Granophyric intergrowths can commonly be recognized in hand-specimen, and small pegmatite patches occur locally. Tuffisite, indicative of gas streaming, is common as are small peralkaline aplite veins.

The peralkaline granite in the Bart Pond area and immediately adjacent rocks are host to numerous small Zr-Nb-Y-rare earth element (REE) occurrences (Tuach, 1991). The highest assay results were obtained from chips of small aplite veins which contain abundant fine grained (less than 0.1 mm), euhedral zircon disseminated in the matrix and concentrated in tuffisite and fractures within the veins. Various Zr-, Y-, Nb- and REE-bearing minerals have been identified including baddelyite (ZrO_2) and possibly fergusonite ($YNbO_4$).

Late Precambrian, calc-alkaline granitoids in the Avalon Zone of eastern Newfoundland contain minor disseminated Cu and Mo sulphides (e.g. Holyrood Granite; about 620 Ma; Krogh et al., 1988; Fig. 9.35). The Coxheath Hills Pluton in central Cape Breton Island (Fig. 9.41), a calc

alkaline pluton of similar age (626-616 Ma, Keppie et al., 1990), hosts a porphyry-type Cu-Mo-Ag deposit (Kirkham and Soregaroli, 1975; Chatterjee and Oldale, 1980). The pluton comprises an expanded calc alkalic I-type suite ranging from diorite to monzogranite (Thicke, 1987). It intrudes upper Precambrian volcanic rocks and is intruded by felsite and diabase dykes. Mineralization, hosted by a fine- to medium-grained quartz monzodiorite phase (Kirkham and Soregaroli, 1975; Thicke, 1987) comprises widespread chalcopyrite, bornite, and molybdenite in closely spaced quartz veinlets and hairline fractures. Mineralized veins are associated with potassic (K-feldspar) alteration accompanied by chlorite, epidote, and sericite. Gold values are associated with higher copper grades (Chatterjee and Oldale, 1980). Narrow shear zones with massive chalcopyrite associated with an albite-chlorite-sericite-K-feldspar alteration assemblage occur in the country rocks near the intrusive contact. The relationship of this mineralization to the porphyry-type copper mineralization is uncertain.

Slightly younger (about 560 Ma) calc alkaline plutons in the Bras d'Or Subzone of Cape Breton Island are associated with a number of polymetallic skarns (Fig. 9.42). Magnetite skarns are the most common, consisting of various mineral assemblages, including magnetite-pyrite-chalcopyrite± scheelite, sphalerite, molybdenite, and pyrrhotite (Hill, 1989, 1990). They include small magnetite skarns such as Glencoe Road in the Creignish Hills (magnetite-pyrite-chalcopyrite) and Scotch Lake in the Boisdale Hills (magnetite-pyrite-sphalerite; Hill, 1989, 1990; Wright, 1975). The best known sulphide skarn is the McMillan Brook occurrence which contains four types of mineralization: 1) massive sulphide replacements containing pyrite, chalcopyrite, pyrrhotite, bornite and sphalerite in forsterite-monticellite-diopside skarn; 2) copper-rich sulphides in fractures in siliceous limestone; 3) disseminated pyrite, chalcopyrite, bornite, molybdenite and fluorite in an adamellite dyke; and 4) disseminated cassiterite, scheelite, and molybdenite in contact metasomatic rocks (Chatterjee, 1977b). A second sulphide skarn comprising about 12 m of andradite rock with accessory diopside, abundant disseminated pyrrhotite, as much as 0.75% Cu and traces of wolframite was intersected in a drillhole at Glencoe (Chatterjee, 1979; Hill, 1989, 1990).

Epithermal(?) gold associated with aluminous alteration

H. Scott Swinden and C.F. O'Driscoll

The principal gold-mineralized areas in the Avalon Zone of Newfoundland are the "Eastern Avalon high-alumina belt" (Hayes and O'Driscoll, 1990) on the eastern Avalon Peninsula, the Cinq Cerf Bay area of southwestern Newfoundland (including the Hope Brook deposit), and the northern part of the Burin Peninsula (Hickey's Pond and related deposits; Fig. 9.35). The deposits vary considerably in geological setting, structural style and tenor, but all are associated with intense aluminous alteration.

The oldest auriferous mineralization associated with aluminous alteration is probably that in the Eastern Avalon high-alumina belt (Hayes and O'Driscoll, 1990). Extensive silicification, pyritization, and highly aluminous alteration occur within a shear zone in volcanic rocks of the

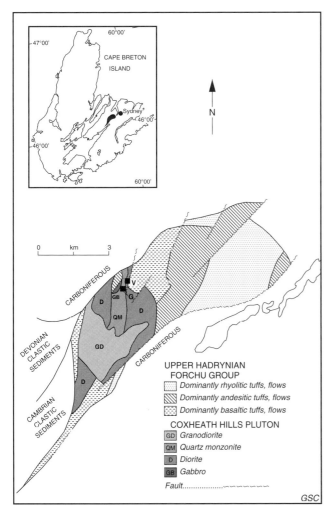

Figure 9.41. Geology in the area of the Coxheath deposits (black squares), Cape Breton Island. G – granite-hosted mineralization; V – volcanic-hosted mineralization (after Thicke, 1987).

Precambrian Harbour Main Group near the eastern margin of the Holyrood Intrusive Suite (Fig. 9.43). Pyrophyllite-rich rocks from the alteration zone in the Foxtrap area (Papezik and Keats, 1976; Papezik et al., 1978) are used in the manufacture of ceramics. Locally, the alteration contains anomalous concentrations of gold (Hayes and O'Driscoll, 1990) although no occurrences of economic size and grade have been identified. The age of the mineralization is constrained to about 620 Ma by the age of the host rocks

(about 623 ± 2 Ma, Krogh et al., 1988) and the age of the Holyrood Granite (620 ± 2 Ma; Krogh et al., 1988) which intrudes the alteration zone. Altered clasts occur in conglomerates at the base of the overlying Cambrian succession. Alteration and mineralization are interpreted to have occurred during hydrothermal activity related to intrusion of the Holyrood Granite (Papezik et al., 1978) or during shearing (Hayes and O'Driscoll, 1990).

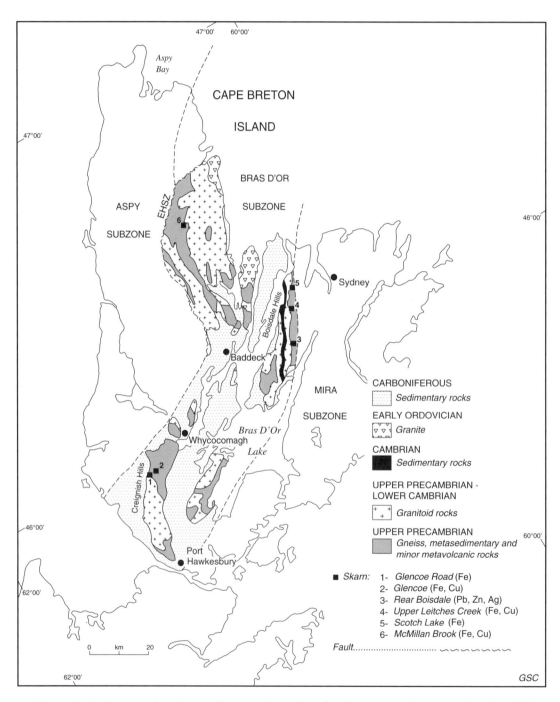

Figure 9.42. Skarns related to late Precambrian – Early Cambrian granitoid rocks in the Bras D'Or Subzone, Cape Breton Island. EHSZ – Eastern Highlands shear zone. Geology after Barr et al. (1990).

The most economically important gold deposit associated with aluminous alteration is the Hope Brook deposit in southwestern Newfoundland, discovered in 1983. Mining operations began in 1987 from reserves of 11.2 million tonnes grading 4.54 g/t Au. The deposit occurs near the northern end of an approximately 6 km long linear zone of intensely altered late Precambrian granitoid intrusions and tuffaceous and/or sedimentary rocks in the trace of the Cinq Cerf Fault Zone (B.H. O'Brien, 1989; Fig. 9.44). The alteration zone is characterized by pervasive silicification and a characteristic high-alumina assemblage which includes, andalusite, pyrophyllite, kaolinite, and sericite (muscovite and paragonite) with quartz, pyrite, and rutile. Pyrite, alunite, specularite, topaz, and fluorite are locally abundant. The mineralized rocks are variably, locally intensely, deformed in the Cinq Cerf Fault Zone and cut off and thermally metamorphosed by the Devonian Chetwynd Granite (390 ± 3 Ma; B.H. O'Brien et al., 1991) immediately north of the Hope Brook mine. Anomalous gold is present throughout the alteration zone, but ore grades are generally

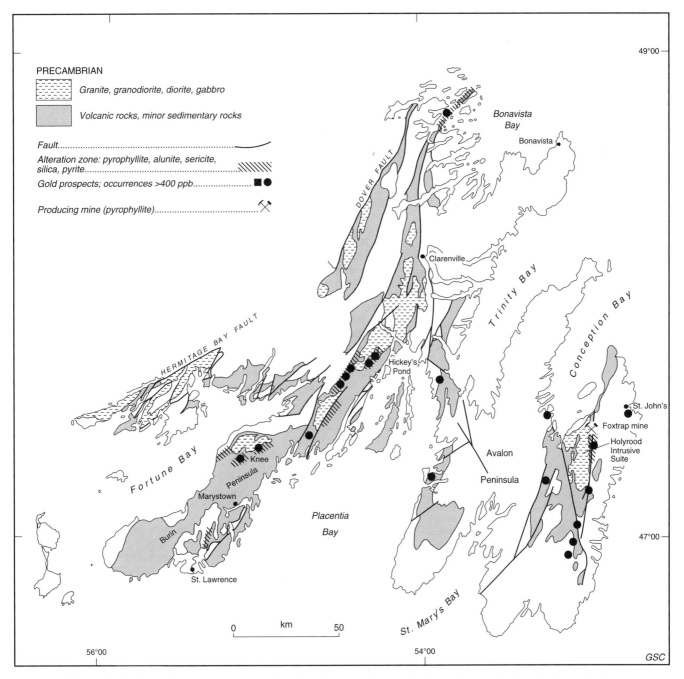

Figure 9.43. Upper Precambrian volcanic and related granitic rocks in the eastern part of the Newfoundland Avalon Zone showing epithermal aluminous, locally auriferous alteration zones.

higher near the Chetwynd Granite. Gold is disseminated within the alteration zone, associated with pyrite and the most intense silicification. Elevated copper values locally accompany the gold.

The age and genesis of the deposit are still a matter of debate. The deposit is intensely deformed and the broadly stratabound nature of the alteration probably reflects transposition of the altered rocks within the Cinq Cerf Fault Zone (Dubé, 1990). The aluminous style of alteration has led many authors to favour a pre-deformation epithermal (McKenzie, 1986) or mesothermal (Yule et al., 1990) origin for the mineralization whereas others have suggested that the mineralization and alteration are fault controlled and related to the Cinq Cerf Fault Zone (e.g. Stewart and Stewart, 1988). Whatever the case, geochronological evidence suggests that the earliest alteration, and possibly the mineralization, occurred in the late Proterozoic, bracketed

by U-Pb zircon dates of 576 ± 2 Ma on altered porphyry (the age of the protolith) and 567 ± 2 Ma on unaltered porphyry within the alteration zone (Stewart and Dunning, 1990; Stewart, 1992). Ore grades are principally developed in the contact aureole of the Chetwynd Granite suggesting that thermal metamorphism may have upgraded Au concentrations (Stewart and Dunning, 1990; Stewart, 1992).

On the Burin Peninsula in southeastern Newfoundland, a series of gold occurrences are associated with linear belts of aluminous alteration and silicification that are spatially associated with coeval granite intrusions (Fig. 9.43). Alteration is characterized by quartz-specularite-alunite ± pyrophyllite, pyrite, rutile, and lazulite. The best known occurrence at Hickey's Pond comprises a zone of advanced argillic alteration in approximately 590 ± 30 Ma (Dallmeyer et al., 1983) felsic tuffs and flows. Locally, a breccia of quartz-alunite-specularite clasts in a specularite-rich

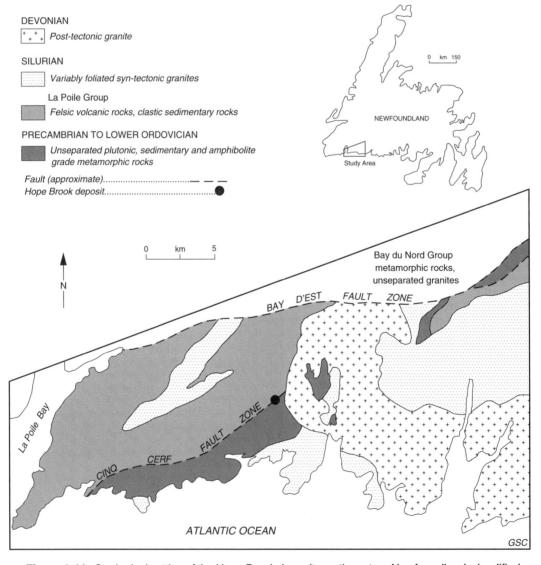

Figure 9.44. Geological setting of the Hope Brook deposit, southwestern Newfoundland, simplified after O'Brien and O'Brien (1990).

matrix has yielded assays of up to 6 g/t Au. Huard (1989) interpreted the mineralization to be epithermal, related to the felsic volcanism. This being the case, geochronological data permit the mineralization to be approximately coeval with that at Hope Brook.

Lower Paleozoic sedimentary deposits

C.F. O'Driscoll

Sedimentary manganese deposits

Manganese-rich sedimentary rocks occur throughout the Avalon, Burin and Bonavista peninsulas of eastern Newfoundland (Fig. 9.45) in the basal part of the lower Middle Cambrian Chamberlain's Brook Formation of the Adeyton Group (Jenness, 1963; Fletcher, 1972; Douglas, 1983). The Adeyton Group comprises Lower to lower Middle Cambrian red and green mudrocks and lesser intercalated limestone representing shallow marine sedimentation on a stable shelf. In the Conception Bay region of the Avalon Peninsula, these rocks unconformably overlie upper Precambrian volcanic, sedimentary, and plutonic rocks; elsewhere, they unconformably overlie quartzites of the Lower Cambrian Random Formation (Fletcher, 1972).

The manganese-rich rocks were deposited during an interruption in normal conditions of sedimentation. The manganese, derived from the weathering of the upper Precambrian rocks exposed around the Middle Cambrian basin, was originally precipitated in oxide form, with subsequent alteration to manganese carbonate during early diagenesis. Paleontological and stratigraphic thickness relationships across the Avalon Peninsula indicate that the manganese horizon was formed during a period of slow clastic sedimentation.

The manganese-rich horizon is progressively thicker from north to south, ranging from 1 m thick in Smith Sound, Trinity Bay, to 25 m in its southernmost exposures in St. Mary's Bay area. This is paralleled by a gradational lithofacies change from carbonate to mudrock facies. In the northernmost Trinity Bay exposures, the manganese horizon consists of a single massive, albeit impure, Mn-Ca carbonate bed, whereas in St. Mary's Bay, it consists primarily of red and green mudrock. In geographically intermediate areas near the heads of Trinity Bay and Conception Bay, it consists of interbedded carbonate and argillaceous beds.

Most of the manganese is in carbonate minerals (rhodochrosite and manganiferous calcite), although secondary manganese oxide minerals also occur. Diagenetic carbonate nodules are abundant and the manganese horizon locally contains anomalous concentrations of chamosite and/or phosphatic nodules and euhedral crystals of barite and pyrite. A limited number of bulk chemical analyses of material from the Conception Bay exposures returned a maximum of 33.35% Mn over approximately 0.5 m near Brigus (McCartney, 1967). Bulk chemical analyses of material from several other Conception Bay exposures yielded lower grades (e.g. 16.1% Mn over about 1.2 m at Kelligrews Brook, 10.69% Mn over 1.3 m at Manuels River). The potential reserves at Brigus, where the manganese content of the horizon is greatest, are probably not large, because of truncation of the manganese horizon by faulting

(McCartney, 1967). If it is assumed that the grade of the horizon along the southeastern margin of Conception Bay remains equal to or better than the grades determined by bulk analysis at Manuels River and Kelligrews Brook, then very large reserves grading 10% Mn may be available.

Sedimentary iron deposits

Bell Island, located in Conception Bay, eastern Newfoundland (Fig. 9.35), is underlain by iron deposits which supported the longest-lived mining operation in Newfoundland and still represent a substantial reserve of iron ore. More than 80 million tonnes of ore were produced between 1895 and 1966, and more than 50 million tonnes remain in roof pillars. Potential reserves beneath Conception Bay probably exceed 2 billion tonnes (Miller, 1983a). The cessation of mining was the result of changes in steel-making technology which resulted in a market preferance for low-phosphorous ores as well as increased competition from new sources of cheaper, higher grade ore.

The iron deposits consist mainly of oolitic hematite of Clinton-type. They are hosted by unmetamorphosed, gently dipping Ordovician sediments of the Bell Island and Wabana groups (Fig. 9.46), a shallow marine sequence of sandstones, siltstones, and shales preserved in a fault-bounded, north-northwest-trending basin centred 3 km north of Bell Island (Rose, 1952; Miller, 1983a, b). The Bell Island Group consists of approximately 1390 m of reddish-brown to grey micaceous sandstone, siltstone, and oolitic hematite beds (Ranger et al., 1984). The disconformably overlying Wabana Group consists of graptolitic black shale and sandstone, phosphatic pebble beds, and oolitic hematite and pyrite. Ranger (1979) and Ranger et al. (1984) suggested that the Bell Island and Wabana groups present a mixture of tidal flat and offshore bar facies developed in a broadly deltaic environment with a wide tidal zone. They suggested that iron was precipitated as chamositic oolites during periods of sediment starvation in subtidal and intertidal environments. Oxidation to ferric hydroxide took place during exposure to the atmosphere. The thicker ore beds are believed to represent the shoreward migration of the offshore bar facies. Ferric hydroxide was subsequently converted to hematite during diagenesis.

Iron ore at Bell Island occurs at several different stratigraphic levels within the Bell Island and Wabana groups (Fig. 9.46). In addition to well-defined ironstone beds, ferruginous sandstone and siltstone are common throughout the sequence. There are three hematite beds of economic importance, the Dominion (lower), Scotia (middle) and Gull Island (upper) formations (Ranger et al., 1984). The Scotia and Gull Island formations are 15 m apart vertically, and the Dominion Formation is 75 m vertically below the Scotia Formation. The Dominion Formation is the most economically important, ranging from 4 m thick on surface to 13 m thick in the underground workings and apparently continuous laterally in all directions. Surface exposures are dominantly oolitic ironstone, with up to 10% detrital quartz fragments and interstitial siderite, calcium phosphate, and pyrite (Ranger,1979; Ranger et al., 1984). Oolites are composed of hematite or alternating layers of hematite and chamosite. Thin lenticular shale beds are common, and in places, dilute the grade below economic levels.

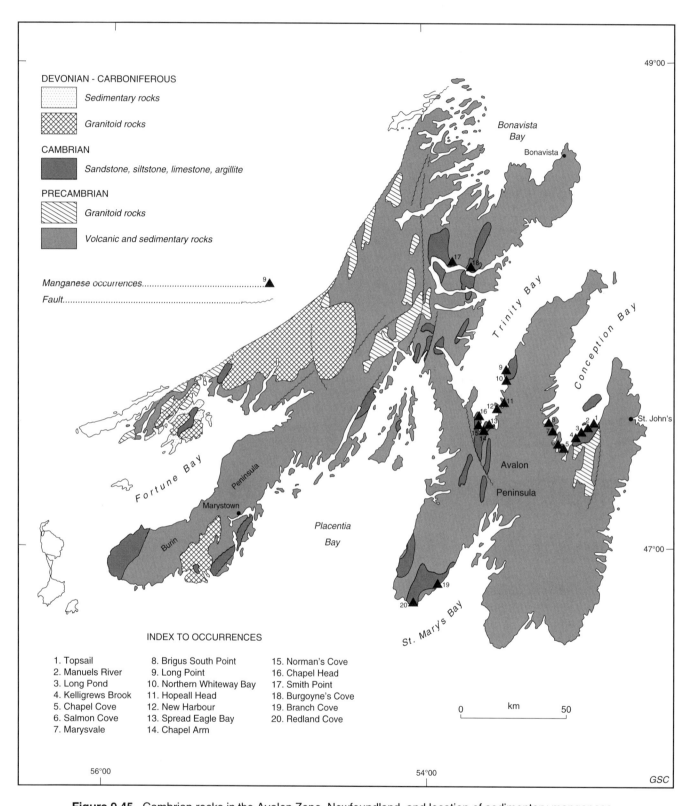

Figure 9.45. Cambrian rocks in the Avalon Zone, Newfoundland, and location of sedimentary manganese deposits (after Douglas, 1983)

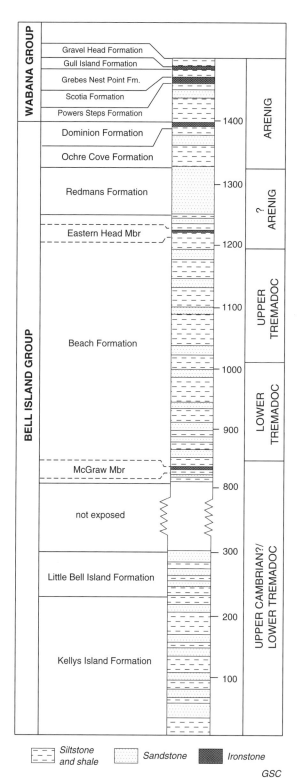

Figure 9.46. Stratigraphic section of the Bell Island and Wabana groups, and stratigraphic setting of the iron deposits (after Ranger et al., 1984). Vertical scale in metres.

MEGUMA ZONE

A.L. Sangster

The Meguma Zone, which outcrops only in the southwestern part of Nova Scotia (Fig. 9.47), is believed to have docked with the eastern side of the Avalon Zone during the Middle Devonian (Keppie, 1985). The Meguma Zone and overlying Annapolis Belt contain a more or less complete Cambrian to Lower Devonian depositional record. The Cambrian-Ordovician Meguma Supergroup (Schenk, Chapter 3), constituting approximately 90 per cent of the exposed sedimentary sequence, is subdivided into the basal Goldenville Group, comprising up to 6700 m of massive, thick-bedded, sandy flysch with thin interbeds of chloritic slate and siltstone, and the overlying Halifax Group, comprising between 7000 and 11 800 m of commonly carbonaceous, pyrrhotitic and pyritic, grey to black slate with thin interbeds of cross-laminated silty flysch (Schenk, Chapter 3). The age of the Meguma Supergroup is constrained by both paleontological and radiometric data. Shelly and trace fossils in the lower part of the Goldenville Group indicate a Cambrian age, consistent with a population of 650 to 560 Ma detrital zircons in the Goldenville Group (Krogh and Keppie, 1988). The Halifax Group contains Tremadoc graptolites (Cumming, 1985) and achritarchs (Schenk, Chapter 3) providing a younger limit for the age of the top of the Meguma Supergroup.

Regional work by Schenk (1970) suggests that the sandy flysch of the Goldenville Group may represent an abyssal plain deposit and the overlying Halifax Group shaly flysch may represent the lower part of a continental rise prism. Schenk (Chapter 3) has presented a depositional model for the Meguma Supergroup at a passive continental margin involving overall shoaling upward from submarine fans, through a channel-levee complex, to a prograding wedge, and finally shelf and near-shore lithologies.

The Halifax Group is overlain in the west and northeast by infolded keels of Ordovician to Lower Devonian clastic sedimentary and volcanic rocks of the Annapolis Belt. These rocks are assigned to the Annapolis Supergroup (Schenk, Chapter 4) made up, in ascending order, of the White Rock, Kentville, New Canaan, and Torbrook groups. The basal White Rock Group is defined in most areas by the presence of white quartzite (Taylor, 1965) and contains bimodal, rift related, alkalic-tholeiitic volcanic rocks (Keppie and Dostal, 1980). The overlying Kentville, New Canaan, and Torbrook groups represent a concordant Upper Silurian to Lower Devonian sedimentary sequence.

Two classes of mineral deposits formed before accretion of the Meguma Zone/Annapolis Belt sequence: (1) concordant base metal occurrences related to the Goldenville-Halifax transition zone; and (2) Clinton-type iron occurrences within the Silurian Torbrook Group of the Annapolis Belt. In addition, a suite of structurally controlled hydrothermal deposits is unique to the Meguma Zone, comprising concordant gold-bearing quartz veins hosted by the Goldenville Group and other small stratabound and cross-cutting veins (Fig. 9.47).

Goldenville-Halifax transition zone

The Goldenville-Halifax transition zone is a poorly exposed package of distinctive lithologies at the transition from sandy flysch in the uppermost part of the Goldenville

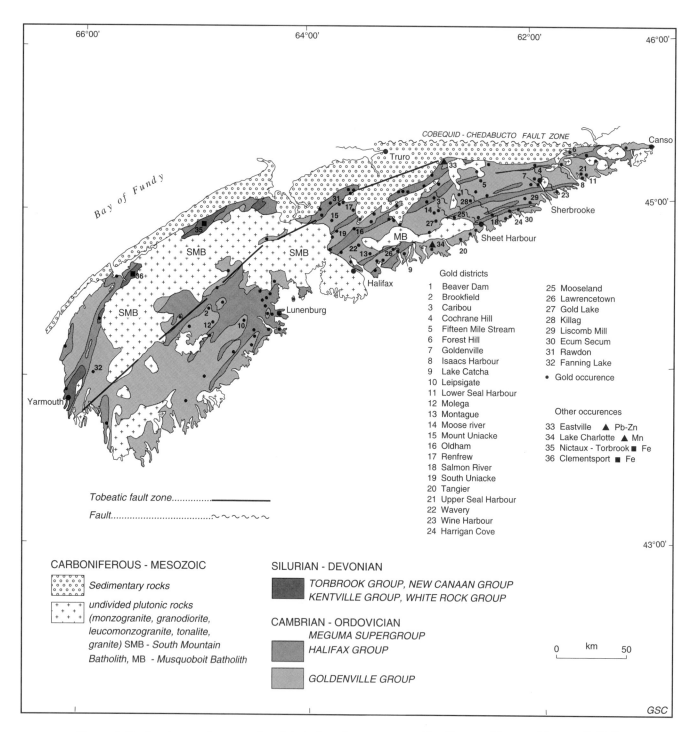

Figure 9.47. Geology and non-granite related mineral occurrences of the Meguma Zone and Annapolis Belt (modified after Sangster, 1990). Not all occurrences are named.

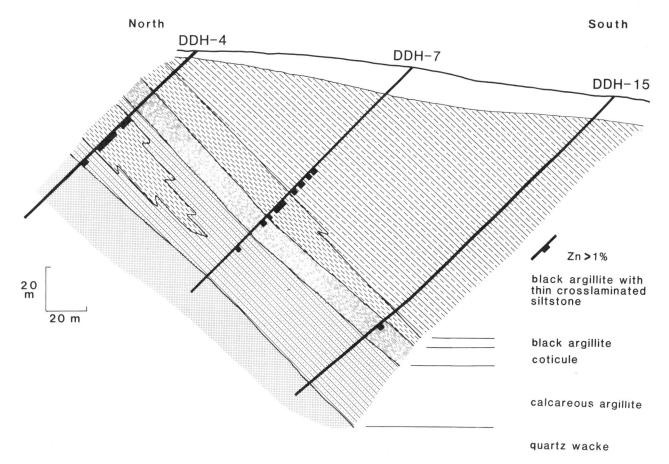

North

DDH-4

DDH-7

South

DDH-15

20 m

20 m

Zn > 1%

black argillite with
thin crosslaminated
siltstone

black argillite
coticule

calcareous argillite

quartz wacke

Figure 9.48. Vertical diamond drill section through the Goldenville-Halifax transition at Eastville illustrating the lithological sequence and relationship of elevated base metal values to coticules and black argillites (from Sangster, 1990).

Group to shaly flysch of the Halifax Group. The upper portion of the transition zone includes a manganese- and carbonate-rich unit that contains spessartine-rich garnet and Mn oxides, and is anomalously enriched in a spectrum of elements (including Mn, Ba, Cu, Pb, and Zn) that are commonly related to shale-hosted metal deposits. Regionally, the transition zone also contains significant occurrences of manganese (Hingston, 1985), tungsten (Shaw, 1983; Fisher, 1984), and tin (Wolfson, 1983).

Base metal mineralization representative of the Goldenville-Halifax transition zone is best developed at Eastville (Binney et al., 1986; MacInnis, 1986; Fig. 9.48). Mineralization can be traced along a 10 km strike length, mainly within the basal beds of the Halifax Group and to a lesser extent in manganese-rich lithologies of the Goldenville-Halifax transition zone. Combined Pb-Zn values tend to average 0.5 to 1.5 per cent over thicknesses up to 100 m though average thicknesses of 2 to 20 m are common. Pyrrhotite (up to 10 per cent) with minor pyrite are the only sulphides normally visible at combined metal values below 1.5 per cent. Above this grade, galena with lesser sphalerite occurs with pyrrhotite in small, 1 to 2 mm by 2 to 5 mm ovoid sulphide blebs within slate.

Zentilli et al. (1986) and Graves and Zentilli (1988a, b) suggested that the geochemistry of the Goldenville-Halifax transition zone resembles some Pacific metalliferous sediments; the whole-rock compositions approach average shale and average deep sea sediment of Lalou (1983), suggesting a pelagic depositional environment. Zentilli et al. (1986) concluded that the manganiferous sedimentary rocks had a sedimentary protolith and are not related to volcanic or hydrothermal processes.

Torbrook iron-formation

Sedimentary iron-formation occurs within the Lower Devonian Torbrook Group in the Clementsport and Nictaux-Torbrook basins (Gross, 1967; Smitheringale, 1973; Wright, 1975; Fig. 9.49). The deposits produced 317 000 tonnes of iron ore grading between 35 and 50 per cent Fe between 1825 and 1913. The iron ores consist of stratified hematite and/or magnetite intercalated with quartzite and slate. Individual units vary up to 3 m thick and occur at several distinct stratigraphic horizons. The ore is locally oolitic and, in addition to magnetite and hematite, contains iron silicate, detrital quartz and argillaceous material (Smitheringale, 1973). Gross (1967) classified the deposits as "oolitic-textured hematite-chamosite-siderite iron formation" similar to that mined at Wabana in Newfoundland. The magnetite content has resulted from the thermal metamorphism of the iron-bearing beds by the South Mountain Batholith.

Meguma Supergroup gold veins

Non-placer gold was first discovered in the Meguma Zone in 1858 and by 1900, the government of Nova Scotia had declared more than 60 formal gold districts. Production began (226 250 g) in 1862 and continued at a level of 310 000 to 930 000 g per year until 1910. Between that time and the late 1970s, there was only minor production in the Caribou and Goldenville districts (1937 to 1942). Total recorded production to 1985 was 43 540 000 g of gold with 15 per cent of that coming from the Goldenville district. A resurgence of exploration activity between 1983 and 1987 resulted in production of 368 810 g Au during 1987 and 1988.

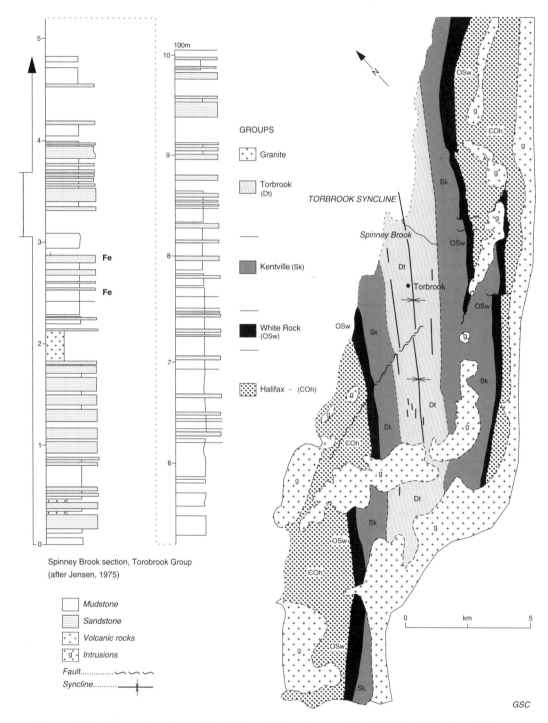

Figure 9.49. General geology of the Nictaux-Torbrook area (Fig. 9.47, No. 35 for location) and a stratigraphic section at Spinney Brook on the east limb of the Torbrook Syncline (simplified after Jenson, 1975) showing the stratigraphic setting of the oolitic ironstones (Fe); figure courtesy of G. Yeo.

Stratabound, gold-bearing vein arrays occur in the Meguma Zone mainly east of Halifax, and throughout the total exposed stratigraphy of the Goldenville Group (Fig. 9.47). The veins occur mostly on the flanks and crests of domes on regional anticlines. Most of the districts occur within greenschist facies rocks (Taylor and Schiller, 1966) and east of the South Mountain Batholith. However, some major occurrences (e.g. Cochrane Hill, Forest Hill, Upper Seal Harbour) are within amphibolite grade metamorphic rocks that are spatially associated with numerous Devonian-Carboniferous granitic intrusions.

Most gold occurrences are associated with thicker than normal interstratified argillite within the Goldenville Group. In the gold districts, the slate is commonly argillaceous rather than silty, carbon-rich and sulphidic with abundant pyrrhotite and arsenopyrite (Malcolm, 1929; Brunton, 1928; Smith, 1984).

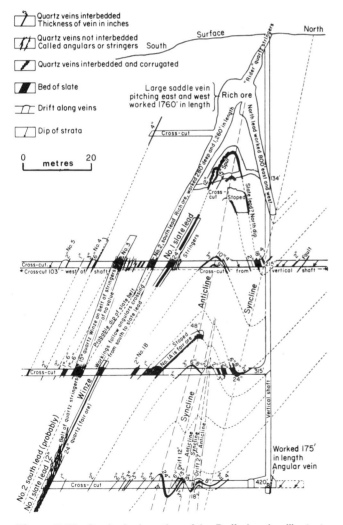

Figure 9.50. Geological section of the Dufferin mine illustrating concordant and angular gold-bearing veins. Productive veins were found on the crest of a secondary fold on the southern flank of the main anticline (modified by Guilbert and Park, 1986 after Malcolm and Faribault, 1912).

Meguma gold-bearing, concordant quartz veins (also referred to as bedding-parallel or stratiform for tabular veins and stratabound for less organized bedding parallel quartz) are located within or immediately below the upper margins of incompetent, impermeable argillite horizons in the Goldenville Group. Many of the districts are located on the steeper limb of a fold (e.g. Beaver Dam, Goldenville, Tangier; Smith and Kontak, 1986) or in a satellitic structure on the limb of a larger fold (e.g. Dufferin mine, Fig. 9.50). Henderson and Henderson (1986) pointed out that the Meguma deposits are mistakenly referred to as saddle reefs by many workers. True saddle reefs (Herman, 1914 in Boyle, 1986, Fig. 2; Ramsay, 1967) are restricted to fold hinges and are considerably thickened in the axial zones with only minor down-limb extension. Meguma veins are normally of constant thickness around the fold hinge and have significant down-dip extent beyond the fold hinge (e.g. Goldenville, Fig. 9.51) although significant exceptions are found where true saddle reefs do occur (e.g. Dufferin mine and Richardson vein, Upper Seal Harbour district).

Gold in Meguma Zone deposits occurs along vein-wallrock contacts, as medium- to coarse-grained aggregates in white quartz and less commonly as fine grained films between crack-seal laminae. Where gold occurs with arsenopyrite, it is most commonly present as fracture fillings, coatings on arsenopyrite crystals, or as small blebs within crystals. Coarse grained gold occurs with coarse grained quartz in concordant veins or at intersections of later structures (e.g. angulars, en echelon veins, northwest veins) and early veins. Gold assays approaching ore grades are reported in arsenopyritic slate or quartzite associated with quartz veins.

Quartz is the most abundant vein mineral and is commonly associated with calcite, ankerite, and ferroan dolomite (Kontak and Smith, 1988a). Arsenopyrite is variably present in all gold districts up to a maximum of about 5 to 10 per cent in wallrocks and less abundantly within the veins. The gold distribution within the veins commonly varies directly with the sulphide mineral content (pyrrhotite, chalcopyrite and galena). Scheelite is common in the Moose River veins and is an accessory in a few occurrences, locally with accompanying traces of stibnite (Smith and Kontak, 1986).

Several classifications of veins have been proposed within individual districts (Henderson, 1983; Smith, 1983; Haynes 1986; Henderson et al., 1986a). The following summary combines elements of these classifications.

Stratiform veins ("leads"). The earliest vein structures are concordant, laminated veins with crack-seal texture (Ramsay, 1980) that are interpreted to have formed during episodic hydraulic overpressure (Faribault, 1899; Graves, 1976; Graves and Zentilli, 1982; Henderson, 1983; Mawer, unpublished report, 1985, 1986, and 1987; Henderson et al., 1986b; O'Brien, 1985; Henderson and Henderson, 1990). These so-called "laminated" or "interbedded" veins are regionally concordant but crosscut bedding in detail and rarely cross from one stratigraphic horizon to another.

En echelon veins. En echelon veins form in slate or siltstone beds between massive wacke on the limbs of major folds. The veins fill extensional fractures which formed as a result of bedding-parallel shear during folding (Henderson et al., 1986a).

Angular veins. Some laterally extensive veins, which rarely exceed a few centimetres in thickness, resemble en echelon veins except that they crosscut several stratigraphic horizons and are continuous both above and below concordant veins.

Pegmatitic veins. Pegmatitic veins are restricted to areas around granitic intrusions and areas of higher grade metamorphism. At Cochrane Hill they occur both within and outside the pit area. The veins contain quartz, orthoclase, garnet, plagioclase, muscovite, biotite, andalusite, and traces of gold (Smith, 1983).

ac veins. Late extension parallel to the regional fold axes has produced quartz filled *ac* joints in the metawacke (Henderson, 1983; Williams and Hy, 1990). These are regional features and occur irrespective of the presence of gold-bearing veins but may contain traces of gold where they occur within gold districts.

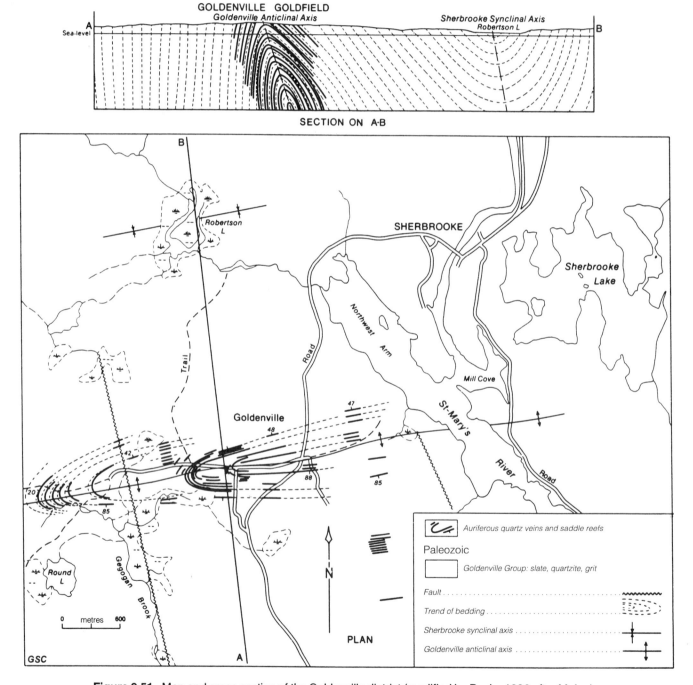

Figure 9.51. Map and cross section of the Goldenville district (modified by Boyle, 1986 after Malcolm and Faribault, 1912).

Northwest faults and veins. Subvertical quartz-calcite veins, vein breccias and sinistral kink bands are parallel to regional northwest sinistral faults and cut all structures in the Meguma Zone. They have a significant effect on gold distribution in some districts. Gold grades in some concordant veins at the Tangier deposit increase near intersections with northwest faults (D. Forgeron, pers. comm.). In the Brookfield, Caribou, and Rawdon districts (among others) the northwest faults host past-producing, gold-bearing veins where they cut slate which hosts mineralized concordant veins.

Miscellaneous structures. Some important gold mineralization is associated with structures which have developed locally. At the Mount Uniacke gold district, gold ore was found in intersections of concordant veins and barren veins developed along a normal fault associated with a secondary fold. At the Caribou mine, significant gold was mined from concordant veins and a stockwork zone on the flank of the major fold. In both the Leipsigate and Ecum Secum gold districts, gold mineralization occurs in faults subparallel to the fold axes.

The Goldenville district (Fig. 9.51) was the largest producer in the Meguma Zone with historical production of 6 535 730 g of gold (Sangster, 1990). The district included 105 veins, principally as concordant veins in an asymmetrical, east-trending anticline. The veins were exposed over a 518 m section across the anticline (Malcolm, 1976), occurring principally within and at the boundaries of slate beds, but locally within the adjacent greywacke as well. Both bedding-parallel ("leads") and crosscutting ("angulars") veins are found. In addition to gold, the veins contain galena, sphalerite, pyrrhotite, and arsenopyrite.

The Caribou mine produced 2 841 950 g of gold between 1869 and 1845. Gold occurs in three settings. Like the Goldenville district, there are high-grade concordant leads within slaty beds or at bedding contacts within quartzites as well as crosscutting lenses (angulars) branching off from the concordant veins. In addition, there is a large zone of network quartz veins in greywacke confined to a flexure in the southern limb of the northeast-trending Caribou anticline (Kontak and Smith, 1989), which produced substantial amounts of gold.

The genesis of the Meguma gold veins has been and remains controversial. Most researchers contend that several generations of veins are present (Graves, 1976; Henderson et al., 1986a, b; Haynes and Smith, 1983; Mawer, 1986) representing semi-continuous vein development throughout an extended period. However, some recent workers (e.g. Smith and Kontak, 1988a, b) have proposed that all Au-bearing veins formed simultaneously.

Three general mechanisms have been proposed for the origin of the gold veins: 1) syngenetic, hydrothermal deposition on the seafloor, 2) early syntectonic deposition from hydrothermal fluids of diverse origins, and 3) late syntectonic deposition from magmatic or deep crustal hydrothermal fluids.

Hunt (1868) was the first to propose a syngenetic origin for the quartz vein materials within the enclosing sediment. Hind (1869, 1870, 1872) supported Hunt's proposal for the laminated veins, but attributed other concordant veins to later processes. McBride (1978) reported a pyrite- and carbonate-rich stratum 1300 m below the Goldenville-Halifax transition zone that contains several small gold occurrences and proposed a syngenetic origin for the gold. Haynes and Smith (1983) and Haynes (1983, 1986, 1987) proposed a complex, hydrothermal, hot spring model. Haynes (1987) interpreted the laminated, concordant veins as fumarole related geyserite, and the crosscutting and semi-concordant mineralized veins as sub-surface feeders and parasites of the hydrothermal system. Documentation of structural features (Henderson and Henderson, 1986; Henderson et al., 1986a; Mawer, 1986, 1987; Williams and Hy, 1990) does not support a synsedimentary origin for these veins though it has been argued that synsedimentary components have been mobilized to form the veins.

Early syntectonic theories have been commonly used to explain the origin of the veins. Early workers, including Campbell and Poole (1862), Campbell (1863), Selwyn (1872), Gilpin (1888), Faribault (1899), Malcolm and Faribault (1912) and Malcolm (1929) stressed the importance of structure as a controlling factor and considered derivation of vein materials from enclosing sedimentary rocks, and "ascending" or "descending" hydrothermal fluids as transporting agents. Boyle (1966, 1979) supported a lateral secretion model. Modern variants of these theories support an early syntectonic, pre-granite, and pre-folding origin for the veins, with a source for the vein materials in the enclosing sedimentary rocks and material movement and deposition from fluids derived during the initial phases of prograde metamorphism (Graves, 1976; Graves and Zentilli, 1982; Henderson, 1983; Henderson and Henderson, 1986; Henderson et al., 1986b; Sangster et al., 1989; Sangster, 1990, 1992).

Late syntectonic theories of vein origin include magmatic hydrothermal processes related to Acadian granites and vein formation from deep, crustally derived metamorphic fluids at the time of Acadian plutonism but not directly from a magmatic source. A simple magmatic hydrothermal source for the vein materials and gold was proposed by Rickard (1927), Newhouse (1936), Emmons (1937), Douglas (1948), and Bell (1948). Recent proposals allow for a source of vein materials either directly from a granitic source, from the lower crust or mantle, or from the enclosing sedimentary rocks, mobilized by magmatic or deep crustal metamorphic fluids driven by magmatic heat (Smith and Kontak, 1986; Kontak and Smith, 1987, 1988b; Mawer, unpublished report, 1985, 1986, and 1987; Hy and Williams, 1986; Kontak et al., 1990a, b). In support of this argument, $^{40}Ar/^{39}Ar$ dates between 380-362 Ma have been recovered from vein-fill amphibole and mica from several deposits (Kontak et al., 1990b), consistent with models invoking protracted mineralization from metamorphogenic fluids broadly related to Acadian orogenesis and magmatism.

Some recent workers feel that elements of more than one model may explain the range of features present in the Meguma gold districts. In particular, Sangster (1992) has argued, based on stable isotopic evidence, that gold was preconcentrated in Goldenville Group carbonaceous sediments in an anoxic oceanic environment and later remobilized and reconcentrated to form the gold deposits. He suggested that the first gold veins to form would have been concordant veins during initial deformation and metamorphism. Vein formation and mineralization may have then proceeded episodically (perhaps including the period of 380 and 360 Ma suggested by Kontak et al., 1990b) in response to continuing deformation and intrusion of granite batholiths, resulting in later generations of veining.

SYN-ACCRETION STRATABOUND VOLCANIC-HOSTED MINERALIZATION

A.L. Sangster and A.A. Ruitenberg

Subaerial and locally subaqueous volcanic activity and associated sedimentation was widespread across the newly cratonized parts of the orogen during the Silurian and Early Devonian. The volcanic activity may have occurred partly in response to crustal thickening and extensional tectonics related to successive accretion of outboard terranes to the Laurentian margin. Volcanic-hosted, broadly stratabound base metal sulphide mineralization related to this stage of orogenic development occurs in Silurian submarine volcanic rocks in Cape Breton Island and in Early Devonian bimodal volcanic sequences in northern New Brunswick. The mineralization is interpreted to be volcanogenic, formed from hydrothermal activity.

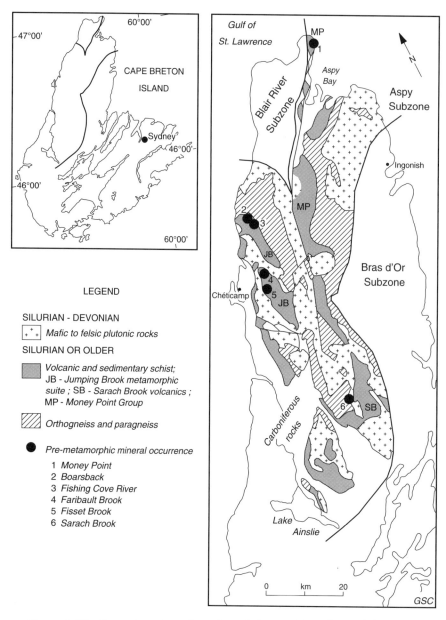

Figure 9.52. General geology of the Aspy Subzone, Cape Breton Island, and location of principal pre-metamorphic mineral occurrences (after Jamieson et al., 1990; Sangster et al., 1990b). Faribault Brook symbol includes several occurrences (see Sangster et al., 1990b).

LEGEND

SILURIAN - DEVONIAN

Mafic to felsic plutonic rocks

SILURIAN OR OLDER

Volcanic and sedimentary schist; JB - Jumping Brook metamorphic suite ; SB - Sarach Brook volcanics ; MP - Money Point Group

Orthogneiss and paragneiss

Pre-metamorphic mineral occurrence

1 Money Point
2 Boarsback
3 Fishing Cove River
4 Faribault Brook
5 Fisset Brook
6 Sarach Brook

Cape Breton Island

In the Aspy Subzone, Cape Breton Island, Silurian meta-volcanic and metasedimentary rocks of the Jumping Brook metamorphic suite near Chéticamp (~430 Ma; Currie et al., 1982; Jamieson et al., 1986, 1990) contain a number of locally stratiform arsenopyrite-sphalerite-galena-chalcopyrite-pyrrhotite-pyrite-gold occurrences (e.g. Faribault Brook, Fig. 9.52). The mineralization appears to be related to a regional stratigraphic interfingering of mafic volcanics and metasedimentary rocks with the distal part of a felsic volcanic pile and predates the regional metamorphism (~400 Ma; Jamieson et al., 1990; Reynolds et al., 1989). Sulphur and lead isotopic data are consistent with a genetic model involving hydrothermal circulation and exhalation related to the volcanic activity at ca 430 Ma (Sangster et al., 1990b). Some remobilization of the metals into veins occurred during the early stages of regional metamorphism circa 400 Ma (Sangster et al., 1990b). Stratiform occurrences elsewhere in Cape Breton Island including the Sarach Brook pyrite-sphalerite-(galena) occurrence in the Middle River gneiss and the Money Point pyrite occurrence (Fig. 9.52) are believed to be correlative with deposits in the Jumping Brook metamorphic suite (Jamieson et al., 1990).

Geochemical evidence suggests that although the older mafic volcanic rocks may be arc-related, the younger felsic volcanism with which the mineralization is associated may have been related to impingement of outboard island arc terranes on a continental margin (Barr and Jamieson, 1991). This implies that the mineralizing event was syn-accretion.

Northern New Brunswick

Early Devonian massive to pillowed basalt and subaqueous rhyolitic flows assigned to the Tobique Group in the Gaspé Belt, central and northeastern New Brunswick, contain stratabound base metal deposits. Mineralization is associated with rhyolite domes which interfinger with mafic volcanic and marine sedimentary rocks. Stratigraphic evidence indicates that this marine sequence was deposited in a foredeep during westward-directed thrusting in the vicinity of the Mirimichi Highlands (Ruitenberg, 1993a; Ruitenberg and McCutcheon, 1993). The mineralization is, therefore, interpreted to be volcanogenic and related to accretionary tectonics.

The Sewell Brook deposit (Wilson, 1991) consists of stratabound replacement bodies of Pb-Zn sulphides, with local enrichments of Cu and significant gold and silver. The mineralization is hosted by rhyolitic hyaloclastic breccias and is associated with extensive footwall sericite-chlorite alteration (Wilson, 1992). Although probably not strictly exhalative, the deposit appears to represent very high level sulphide deposition from sub seafloor hydrothermal fluids in favourable horizons (breccias) immediately beneath the sea floor (Wilson, 1991, 1992). The Gravel Hill deposit to the southwest (Ruitenberg, 1993) consists of disseminations, small pods and fracture-fillings of base metal sulphides, pyrite and silver minerals accompanied by green argillic and/or micaceous alteration and silicification. Mineralization is hosted by felsic hyaloclastites and adjacent sedimentary rocks.

SYN- AND POST-ACCRETION STRUCTURALLY CONTROLLED MESOTHERMAL/EPITHERMAL MINERALIZATION

Overview

H. Scott Swinden

Structurally controlled precious and base metal deposits with mesothermal and epithermal characteristics formed during the accretionary and post-accretionary history of the Canadian Appalachians. Some of these deposits have been known for some time. However, during the gold exploration boom of the 1980s, a significant number of new deposits were discovered and this renewed interest spurred considerable new research into the nature and origin of this class of deposit in the context of the geological and tectonic history of the orogen.

Deposits associated with aluminous alteration in southern and eastern Newfoundland (e.g. Hope Brook) are restricted to the Avalon Zone. Their original deposition is probably part of Avalon Zone history rather than accretionary Appalachian history. Deposits hosted by Meguma Supergroup quartz veins in the Meguma Zone are probably related to Appalachian tectonism (Kontak et al., 1990b). However this class of deposit is unique to the Meguma Zone and therefore was described previously. Some fault-related mineralization, particularly in northern New Brunswick and Gaspésie, comprises skarns and other mineral deposits related to magmatism along active major fault structures. These deposits are discussed later, because of their magmatic association.

Mesothermal and lesser epithermal mineralization accompanied orogenic events at the accreting Laurentian margin and throughout the maturing post-accretion orogen. There are a few characteristics that are common to most deposits. Mineralization is typically related to major structures but is commonly sited in secondary or higher order subsidiary structures. Alteration is typically propylitic and of relatively local extent. Carbonate alteration is ubiquitous and there is commonly a strong association of mineralization with rocks of contrasting competency (producing dilation during tectonism) and iron-rich lithologies (a chemical trap for gold). There is usually a marked association between gold and pyrite. However, comparison of characteristics of mineralization and alteration in different parts of the orogen show considerable complexities. In some areas, gold occurs in intensely altered country rocks whereas in others, it is in quartz veins associated with minimal alteration. Likewise, metal associations are not homogeneous. In some areas, there is an association of gold with antimony and arsenic whereas in others, the gold is associated with base metals. Clearly, local geological features have exercised considerable influence on the form and character of this class of deposits.

Although the timing of mineralization is not well constrained, such constraints as do exist indicate that structurally controlled mineralization was a protracted event that lasted (or occurred sporadically) from the Silurian to at least the Carboniferous. Ages of crosscutting plutons (Hudson and Swinden, 1989), host rocks (Ritcey, 1993), direct dating of alteration minerals (Ramezani, 1992) and geological evidence show that at least some gold mineralization in central Newfoundland was probably associated with Silurian orogenic events (Dunning et al., 1990). Mineralization occurs in Carboniferous rocks in southern New Brunswick, associated with local structures (Ruitenberg et al., 1990), demonstrating that mineralizing events continued until at least this time. Major, possibly transcrustal structures, which were subject to reactivation, were clearly a controlling factor in focussing hydrothermal fluids and mineralization.

Central and southwestern Newfoundland

H. Scott Swinden

Major gold discoveries in Newfoundland are a relatively recent phenomenon and the widespread occurrence of epigenetic gold deposits in Newfoundland has only been appreciated since the late 1970s. Documentation of geological settings and deposit characteristics is still at a relatively early stage. Most have characteristics typical of mesothermal deposits (Dubé, 1990; Evans, 1991), although some relatively low grade occurrences with epithermal characteristics have been identified in the Gander Lake area of the eastern Dunnage Zone (Evans, 1991).

There is typically an association between gold and secondary or tertiary structures related to long-lived, periodically reactivated structures (e.g. Cape Ray Fault, Doucers Valley Fault Complex, Baie Verte-Brompton Line, Green Bay Fault, and the Dunnage-Gander zone boundary; Fig. 9.53). There are few constraints on the actual age of mineralization, and little basis for postulating whether the deposits result from one or many mineralizing events, although available evidence suggests that many formed during the Silurian (Ritcey, 1993; Ramazani, 1993; Swinden, 1990), perhaps related to Silurian orogeny documented in Newfoundland (Dunning et al., 1990). As noted by numerous workers, extensive reactivation of major faults throughout central Newfoundland occurred at this time (e.g. Karlstrom et al., 1982; Tuach, 1987a; Szybinski, 1988).

Dubé (1990) has subdivided mesothermal deposits in western Newfoundland into two main types: "quartz-vein" type and "altered wallrock" type. The "quartz-vein type", the most common mode of mesothermal mineralization in Newfoundland, includes significant deposits in southwestern Newfoundland (e.g. Cape Ray), Baie Verte Peninsula (e.g. Deer Cove, Pine Cove), and Springdale Peninsula (e.g. Hammer Down), and may also encom- pass a number of deposits in the Victoria Lake Group (e.g. Valentine Lake, Midas Pond) and the eastern Dunnage Zone Au-Sb district (e.g. Moreton's Harbour, Hunan, Little River). The mineralization is hosted by a diverse variety of rocks including mafic and felsic volcanics, gabbro, ultramafic rocks, and sedimentary rocks, and occupies extension veins, stockwork breccias, or sheared quartz veins enclosed in high angle shear zones, all of which generally show brittle to brittle-ductile behavior. The gold is commonly accompanied by minor sulphides, especially pyrite but including some or all of galena, sphalerite, chalcopyrite, arsenopyrite, and stibnite. Gangue minerals are principally quartz with lesser carbonate and feldspar. Wallrock alteration is variable, but generally weak, and characterized by chlorite and carbonate, with minor sericite, pyrite, and green

mica. The "altered wallrock" subtype of mesothermal deposits is hosted by strongly altered rocks; mineralization is commonly distributed throughout the host rock and not located within cross-cutting mineralized veins (e.g. Stog'er Tight). The distinction between the two deposit types is usually clear in western Newfoundland, where mineralized quartz veins are generally not accompanied by extensively altered wallrock; however, it is less so in the south-central and eastern Dunnage Zone, where mineralized veins are typically associated with significant country rock alteration and mineralization (Evans et al., 1990; Evans, 1991, 1992).

The Cape Ray deposits, associated with the Cape Ray Fault Zone in southwestern Newfoundland (Fig. 9.54), are a key example of the quartz-vein type of mesothermal gold deposit. The main mineralized zones comprise electrum-bearing quartz veins with sphalerite, galena, chalcopyrite, pyrite, and rare arsenopyrite in sheared graphitic sedimentary rocks in an oblique splay of the Cape Ray Fault Zone (Wilton and Strong, 1986). Mineralization is localized in ductile reverse-oblique shear zones at the boundary between hangingwall chloritized mafic rocks and graphitic and footwall chloritic schist (Dubé, 1990). Related fracture-fillings and stockwork mineralization in the nearby Devonian (369 ± 13 Ma, Wilton, 1983) Windowglass Hill Granite (Fig. 9.54) comprises auriferous pyrite-rich (chalcopyrite, galena) extension veins. Dubé et al. (1991a) suggested that mineralization in both areas occurred during Late Devonian or later syn- to late-ductile shearing accompanying reactivation of the Cape Ray Fault Zone.

In the western White Bay area, structurally controlled gold mineralization occurs in Silurian volcanic and sedimentary rocks (Browning, Sops Arm mines) as well as in late Proterozoic granitoid rocks and the unconformably overlying Cambrian sedimentary rocks (Rattling Brook deposit; Fig. 9.55). Mineralization is spatially associated with the Doucers Valley Fault Complex, a probable splay of the Cabot fault system, which may have served as a conduit for mineralizing fluids (Tuach, 1987a). Lake sediment samples over the fault complex are enriched in Sb and As, indicating that hydrothermal systems were operative on a regional scale (Davenport and McConnell, 1989).

Gold occurrences in Silurian rocks are generally within shear zones. At the Browning mine, deformed syn-deformation, or possibly pre-deformation, quartz veins are hosted by sericitized shales and altered carbonate beds. The mineralization overlies a thrust plane defined by a chlorite schist horizon; anomalous gold is present both in boudinaged and broken veins oriented parallel to foliation, and in crosscutting veins.

In contrast, the Rattling Brook prospect (Tuach and Saunders, 1990; Saunders and Tuach, 1991; Poole, 1991) comprises auriferous fracture stockworks, veins and shear zones in the late Proterozoic Apsy pluton (megacrystic granodiorite), part of the Long Range Inlier, and in the unconformably overlying Lower Cambrian quartzites and limestones of the Coney Arm Group. Gold mineralization occurs in wide-spaced fractures and veins associated with an albite, quartz, ankerite, siderite, sericite, pyrite, and arsenopyrite assemblage that overprints an earlier potassic alteration. The absence of deformation features in the alteration and ore assemblages in veins and fractures suggests that the mineralization post-dates deformation in the area, and is Late Silurian or younger (Saunders and Tuach, 1991).

Quartz-vein-type mesothermal deposits on the Baie Verte and Springdale peninsulas, typified by the Deer Cove and Hammer Down deposits (Fig. 9.56), comprise gold-bearing quartz veins mainly hosted by Lower Ordovician mafic volcanic and intrusive rocks but also occurring in ophiolitic ultramafic and younger Silurian felsic volcanic rocks.

The Deer Cove deposit, hosted by mafic volcanic rocks of the Point Rousse Complex comprises auriferous quartz veins in volcanic rocks, sheeted dykes and gabbro immediately above their thrust contact with underlying ultramafic rocks (Gower et al., 1988; Patey, 1990). The Main zone comprises brecciated quartz veins in an envelope of silicified wall rock, locally accompanied by sericitization and serpentinization of the country rock. The gold occurs both in the quartz veins and in silicified wall rock, as free gold and associated with pyrite, minor chalcopyrite and arsenopyrite. The auriferous quartz-breccia lenses are aligned parallel to the shear fabric and plunge approximately along the line of intersection between the structure hosting the Main zone and the east-west cleavage associated with the underlying thrust fault (Gower et al., 1988). The zone of brecciated quartz veins and silicification is enclosed in a broad halo of carbonate alteration.

The Hammer Down gold deposit (Andrews, 1990; Andrews and Huard, 1991; Dubé et al., 1992) occurs in sulphide-rich quartz and quartz carbonate veins hosted by sheared mafic flows, tuffs, microgabbros and felsic porphyry dykes. The laminated, shear-vein type quartz veins tend to be closely associated with contact zones, faults and shear zones. Patches of fine grained sericite are variably distributed throughout the veins and minor chlorite is observed associated with pyrite and sericite. Carbonate occurs in varying proportions commonly associated with the sulphides. Pyrite is the dominant sulphide mineral with minor chalcopyrite, sphalerite, and galena; total sulphides range from <5% to >75% and gold is associated with pyrite (Andrews, 1990). The mineralized veins are localized in second-generation brittle-ductile high strain zones controlled by parasitic folds and the strong layer anisotropy induced by the presence of competent felsic dykes in the less competent mafic volcanic sequence (Dubé et al., 1992). These structures may be related to movement on the Green Bay Fault immediately to the west (Andrews, 1990).

The Stog'er Tight deposit on the Baie Verte Peninsula (Huard, 1990; Kirkwood and Dubé, 1992) is the best example of an "altered wallrock-type" gold deposit in western Newfoundland. Mineralization is hosted by an intensely and pervasively altered gabbro sill that cuts mafic volcanics of the Point Rousse Complex. Several zones have been outlined which conform in attitude to nearby thrust faults. The mineralization consists of bright pink (hematite-stained) strongly albitized, leucoxene-bearing gabbro that carries up to 10% coarse grained pyrite. Gold occurs as microveinlets and disseminated blebs within the pyrite. Quartz veins are locally present but they lack pyrite and gold, and in places crosscut the fabric. There are no anomalous concentrations of other elements commonly associated with mesothermal gold mineralization (e.g. Ag, Cu, As, Sb). With respect to alteration and its relationship with the gabbroic host rocks, the Stog'er Tight prospect is analogous to many Archean gold deposits, described, for example, by Phillips (1986) and Dubé et al. (1987). Gold is thought to have been localized in the gabbro by a combination of rheological (i.e. competency contrast resulting in dilation during tectonism) and chemical (iron acting as a chemical trap) factors (Dubé, 1990).

A third type of gold occurrence on the Baie Verte Peninsula comprises stratabound disseminated gold-pyrite mineralization hosted by chloritic, graphitic, argillaceous sedimentary rocks of the Betts Cove Complex (Nugget Pond deposit, Fig. 9.56). Unlike other gold occurrences in the area, the Nugget Pond deposit exhibits little evidence of deformation. Gold is intimately associated with coarsely recrystallized pyrite, stilpnomelane, and minor chalcopyrite, galena, and pyrrhotite (Swinden et al., 1990b).

Mesothermal gold occurrences in the Victoria Lake Group (Fig. 9.57) are geologically diverse; most are interpreted to be located in secondary or tertiary shear zones formed during late Silurian or early Devonian sinistral movement along the faults which bound the group (Evans et al., 1990). Some of the range of their characteristics are represented by the Valentine Lake and Midas Pond deposits. At the Valentine Lake prospect (Barbour, 1990), located along the southeast margin of the Victoria Lake Group near Victoria Lake, gold is contained mainly in quartz-tourmaline veins hosted by the late Precambrian (563 ± 2 Ma; Evans et al., 1990) Valentine Lake quartz monzonite and by the adjacent Silurian Rogerson Lake Conglomerate (Fig. 9.57). Shear deformation was focused along the nonconformity between the intrusive rocks and the overlying conglomerate and most of the significant auriferous veining occurs within 500 m of the nonconformity. Auriferous veins, which typically appear to occupy extensional fractures related to the shearing (Barbour, 1990), carry minor calcite, chlorite, and pyrite accompanied by accessory scheelite, tungstite, and traces of pyrrhotite, arsenopyrite, galena, chalcopyrite, sphalerite, and bornite. Wallrock alteration (sericite, albite, and variable silicification with lesser tourmaline and pyrite) is found only within 20 cm of the vein margins. Gold occurs within the quartz veins related to pyrite.

In contrast, the Midas Pond gold prospect (Evans, 1990) is hosted by extensively sheared and altered Lower Ordovician felsic and mafic pyroclastic rocks of the Tulks Hill volcanic belt. The mineralized zone consists of a subvertical zone of crosscutting quartz-pyrite-tourmaline-ankerite-paragonite veins developed near the contact between a ductilely deformed mafic breccia and the structurally overlying felsic pyroclastic rocks. It is structurally overlain by a zone of intense pyrophyllite-kaolinite-paragonite-silica alteration. The gold is associated with pyrite and variations in the pyrite content of the quartz veins are reflected in the gold grades.

The eastern Dunnage Zone is characterized by an extensive belt of Au ± Sb occurrences that extend from Bay d'Espoir in the south to Notre Dame Bay in the north (Fig. 9.58). Some of these have been known for many years (e.g. Kim Lake #2 prospect in the Bay d'Espoir area, Colman-Sadd and Swinden, 1982; Moreton's Harbour deposits in Notre Dame Bay; Gibbons, 1969; Kay and Strong, 1983) but most were discovered in the late 1980s. Mineralization occurs in a variety of host rocks including

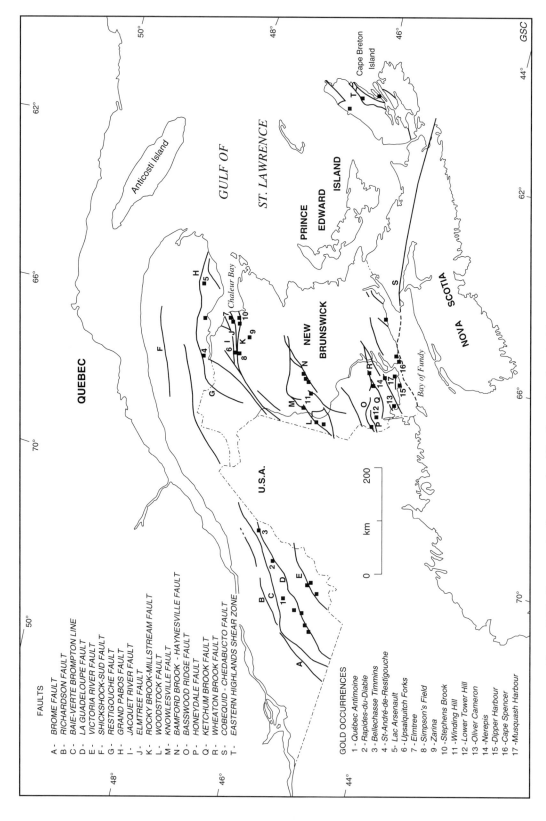

FAULTS

A -   BROME FAULT
B -   RICHARDSON FAULT
C -   BAIE-VERTE BROMPTON LINE
D -   LA GUADELOUPE FAULT
E -   VICTORIA RIVER FAULT
F -   SHICKSHOCK-SUD FAULT
G -   RESTIGOUCHE FAULT
H -   GRAND PABOS FAULT
I -   JACQUET RIVER FAULT
J -   ELMTREE FAULT
K -   ROCKY BROOK-MILLSTREAM FAULT
L -   WOODSTOCK FAULT
M -   KNOWLESVILLE FAULT
N -   BAMFORD BROOK - HAYNESVILLE FAULT
O -   BASSWOOD RIDGE FAULT
P -   HONEYDALE FAULT
Q -   KETCHUM BROOK FAULT
R -   WHEATON BROOK FAULT
S -   COBEQUID - CHEDABUCTO FAULT
T -   EASTERN HIGHLANDS SHEAR ZONE

GOLD OCCURRENCES

1 - Québec Antimoine
2 - Rapides-du-Diable
3 - Bellechasse Timmins
4 - St-André-de-Restigouche
5 - Lac Arsenault
6 - Upsalquitch Forks
7 - Elmtree
8 - Simpson's Field
9 - Zarina
10 - Stephens Brook
11 - Winding Hill
12 - Lower Tower Hill
13 - Oliver Cameron
14 - Nerepis
15 - Dipper Harbour
16 - Cape Spencer
17 - Musquash Harbour

Figure 9.53. Some major lineaments of the Canadian Appalachians and related gold occurrences.

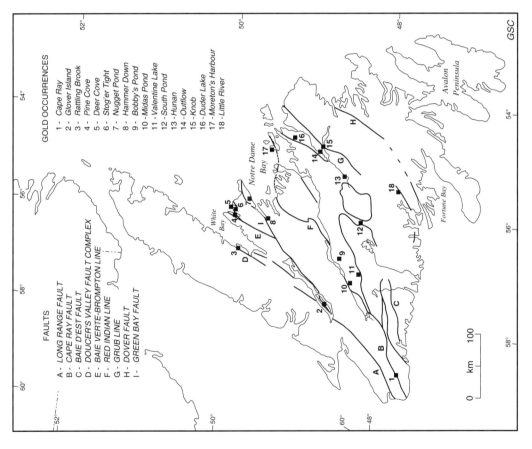

Figure 9.53. (cont.)

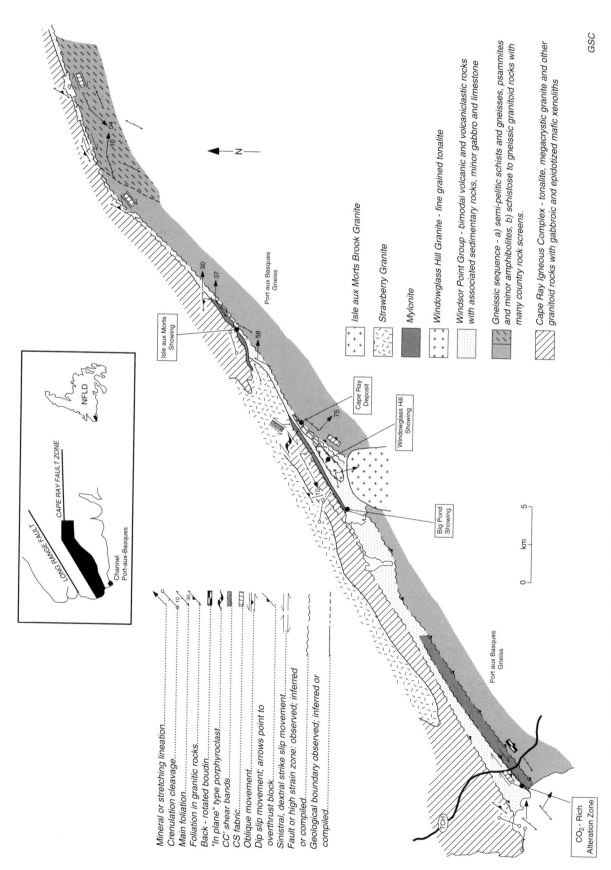

Figure 9.54. General geological relationships and gold occurrences along the Cape Ray Fault Zone, southwestern Newfoundland (figure courtesy of B. Dubé and K. Lauzière).

GSC

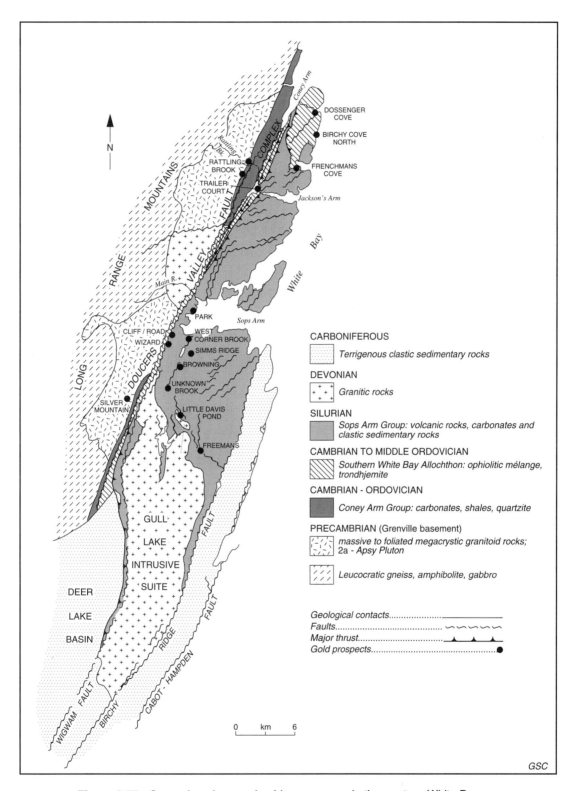

Figure 9.55. General geology and gold occurrences in the western White Bay area, north-central Newfoundland (after Tuach, 1987a).

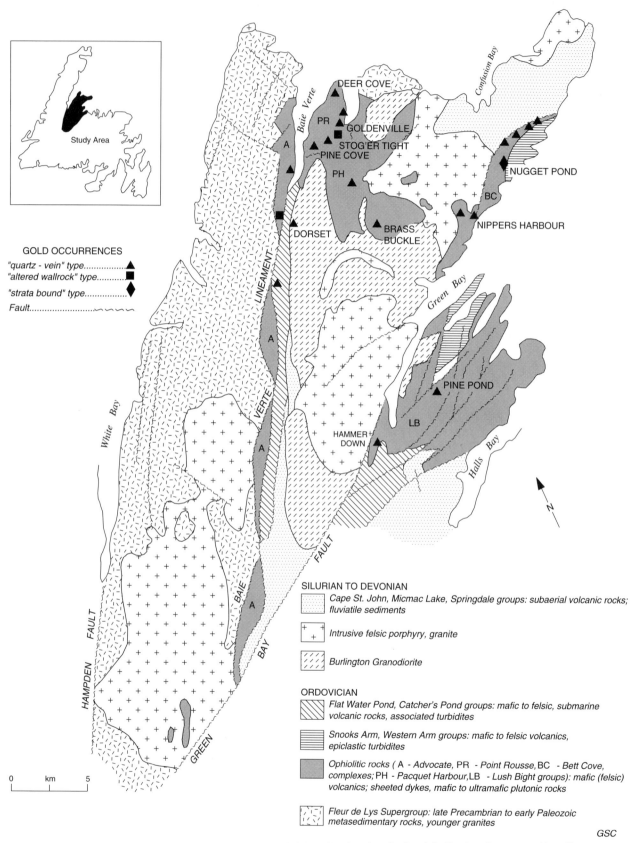

GOLD OCCURRENCES

"quartz - vein" type................ ▲
"altered wallrock" type........... ■
"strata bound" type................ ◆
Fault........................ ∼∼∼∼

SILURIAN TO DEVONIAN

Cape St. John, Micmac Lake, Springdale groups: subaerial volcanic rocks; fluviatile sediments

Intrusive felsic porphyry, granite

Burlington Granodiorite

ORDOVICIAN

Flat Water Pond, Catcher's Pond groups: mafic to felsic, submarine volcanic rocks, associated turbidites

Snooks Arm, Western Arm groups: mafic to felsic volcanics, epiclastic turbidites

Ophiolitic rocks (A - Advocate, PR - Point Rousse, BC - Bett Cove, complexes; PH - Pacquet Harbour, LB - Lush Bight groups): mafic (felsic) volcanics; sheeted dykes, mafic to ultramafic plutonic rocks

Fleur de Lys Supergroup: late Precambrian to early Paleozoic metasedimentary rocks, younger granites

GSC

Figure 9.56. General geology of the Baie Verte Peninsula – Springdale Peninsula area and location of significant gold occurrences.

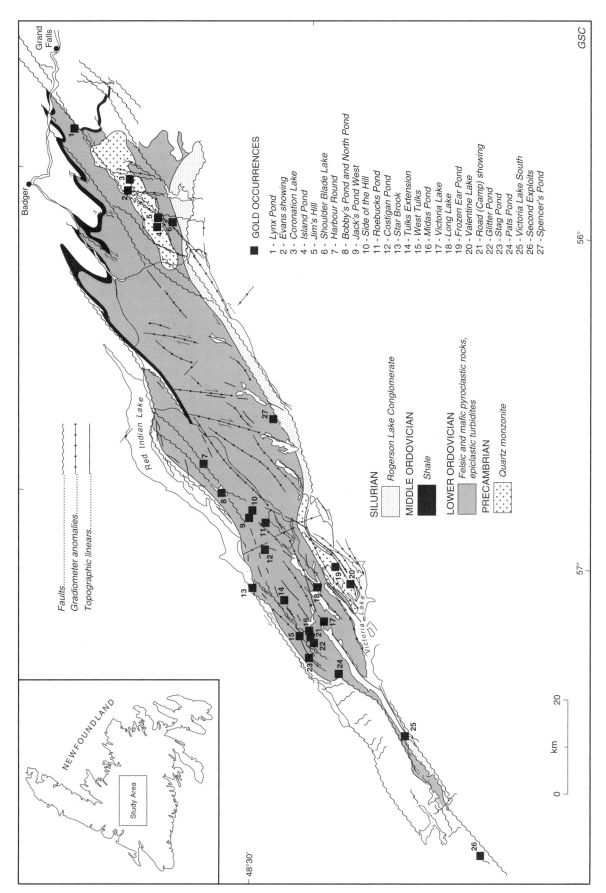

GSC

GOLD OCCURRENCES

1 - Lynx Pond
2 - Evans showing
3 - Coronation Lake
4 - Island Pond
5 - Jim's Hill
6 - Shoulder Blade Lake
7 - Harbour Round
8 - Bobby's Pond and North Pond
9 - Jack's Pond West
10 - Side of the Hill
11 - Roebucks Pond
12 - Costigan Pond
13 - Star Brook
14 - Tulks Extension
15 - West Tulks
16 - Midas Pond
17 - Victoria Lake
18 - Long Lake
19 - Frozen Ear Pond
20 - Valentine Lake
21 - Road (Camp) showing
22 - Glitter Pond
23 - Stag Pond
24 - Pats Pond
25 - Victoria Lake South
26 - Second Exploits
27 - Spencer's Pond

SILURIAN

Rogerson Lake Conglomerate

MIDDLE ORDOVICIAN

Shale

LOWER ORDOVICIAN

Felsic and mafic pyroclastic rocks, epiclastic turbidites

PRECAMBRIAN

Quartz monzonite

Faults..............
Gradiometer anomalies.........
Topographic linears..........

NEWFOUNDLAND

Study Area

km

0   20

Figure 9.57. Principal faults and linears interpreted from topographic and geophysical data in the Victoria Lake area. Gold occurrences are generally associated with second and third order structures.

749

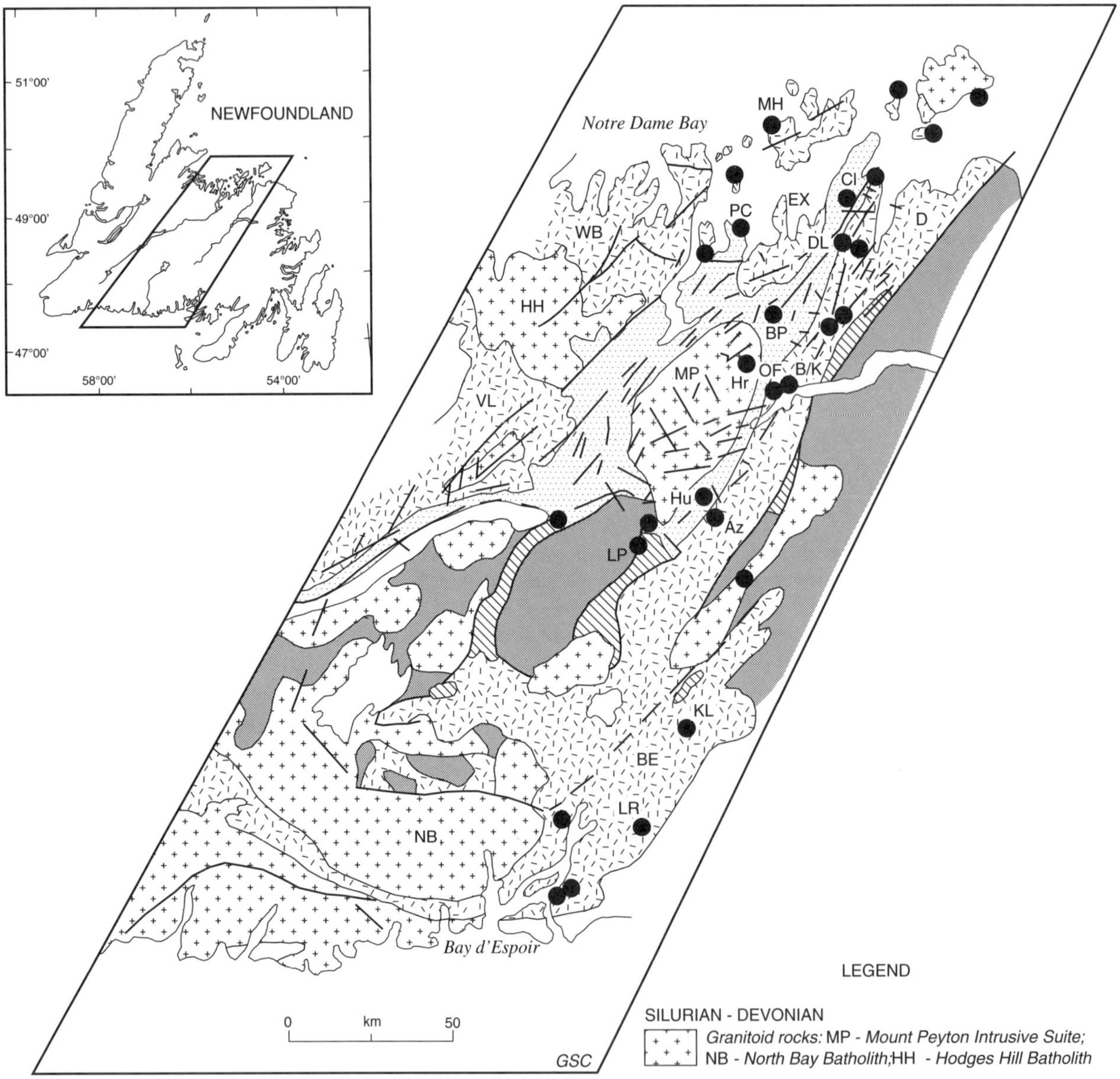

51°00'
49°00'
47°00'
58°00'
54°00'
NEWFOUNDLAND
Notre Dame Bay
MH
PC
EX
CI
D
DL
WB
HH
BP
MP
OF
B/K
Hr
VL
Hu
Az
LP
KL
BE
LR
NB
Bay d'Espoir
0
km
50
GSC

LEGEND

SILURIAN - DEVONIAN

Granitoid rocks: MP - *Mount Peyton Intrusive Suite;* NB - *North Bay Batholith;* HH - *Hodges Hill Batholith*

SILURIAN

Botwood, Indian Islands groups and equivalents: shallow marine to terrigenous sedimentary rocks, felsic to mafic volcanic rocks

ORDOVICIAN

Marine mafic to felsic volcanic rocks, epiclastic rocks: VL - *Victoria Lake Group;* WB - *Wild Bight Group;* D - *Davidsville Group;* BE - *Baie d'Espoir Group;* EX - *Exploits Group*

LOWER ORDOVICIAN

Ophiolitic rocks: Pipestone Pond, Coy Pond, Great Bend, Gander River complexes and equivalents

Gander Group, Spruce Brook Formation and equivalents: quartzose clastic sedimentary rocks, lesser pelitic rocks

Geological contacts

Fault or topographic / geophysical lineament

Mineral occurrences (Au *and/or* Sb *)*

Mineral occurrences (Au *and/or* Sb *):* Az - *Aztec;* B/K - *Bullet, Bowater, and Knob;* BP - *Big Pond;* CI - *Clutha;* DL - *Duder Lake (Stinger, Goldstash, Corvette)* Hu - *Hunan;* Hr - *Hurricane;* KL - *Kim Lake #2;* LP - *Lizard Pond;* LR - *Little River / Le Pouvoir;* MH - *Moreton's Harbour;* OF - *Outflow;* PC - *Powderhouse Cove*

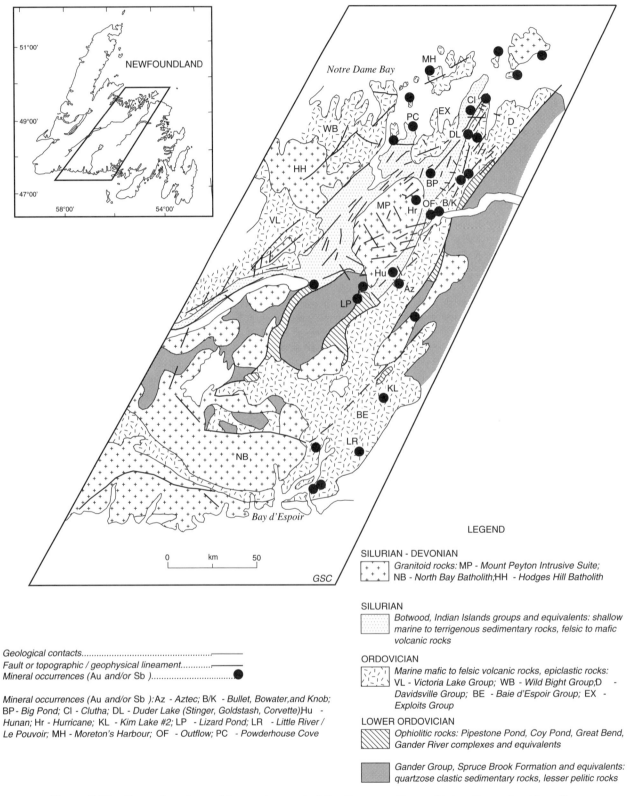

Figure 9.58. General geology of the eastern part of the Dunnage Zone with locations of gold and/or antimony occurrences.

sedimentary rocks of the Davidsville (Ordovician) and Botwood (Silurian) groups, mafic intrusions, and mafic volcanic rocks (Fig. 9.58). Evans (1991, 1992) has shown that a number of distinct subtypes, based on host rocks and accompanying alteration, can be identified: (1) deposits in quartz-carbonate veins in gabbro intrusions accompanied by intense Fe-carbonate alteration (e.g. Clutha, Big Pond); (2) quartz-carbonate veins in sediments accompanied by silicification and carbonatization (e.g. Stinger in the Duder Lake area); (3) quartz-carbonate veins in shear zones in greywackes and shales accompanied by minor carbonate alteration (e.g. Bullet); and (4) dilational quartz veins in deformed graphitic greywacke and shale (e.g. Bowater). The mineralized veins are typically associated with faults and shear zones which mark secondary and later structures. In this sense, the siting of the mineralization is similar to that in the Victoria Lake Group and may reflect hydrothermal activity during the same tectonic regime (Evans, 1991).

Unlike most deposits to the west, the eastern Dunnage Zone deposits are commonly rich in antimony. In fact, some of the occurrences in this belt are primarily stibnite showings (e.g. Kim Lake #2, Le Pouvoir, Hunan, Moreton's Harbour) and there is a regionally extensive Sb lake sediment anomaly over this area interpreted to be a relict of widespread Sb-rich hydrothermal activity related to tectonism (Davenport and Nolan, 1989). If all mineralization in the eastern Dunnage Zone is approximately coeval, its maximum age is constrained by the age of the Mount Peyton Intrusive Suite (maximum 423 ± 5 Ma; Reynolds et al., 1981) which hosts several $Au\pm Sb$ occurrences (e.g. Hurricane).

Cape Breton Island

A.L. Sangster

There are a number of structurally controlled mesothermal vein-type gold occurrences in the Bras d'Or and Aspy subzones in Cape Breton Island (Sangster et al., 1990b; Mengel et al. 1991; Fig. 9.59).

The largest concentration of veins in the Bras d'Or Subzone is in the St. Ann's area, where a north- to northwest-trending swarm of narrow shear veins and extension gash veins associated with north-trending fault zones cut Cambrian granitoid intrusions and upper Proterozoic dacite crystal tuff (MacDonald and Barr, 1985). White mica from the alteration zones yielded K-Ar ages of 517 ± 18 Ma (MacDonald and Barr, 1985). To the northwest near the Eastern Highlands Shear Zone, the INCO showing comprises pyritic, auriferous quartz veins in the late Precambrian (560 ± 2 Ma) Kathy Road diorite suite. Most are shear veins emplaced under brittle-ductile conditions during northnortheast thrusting. They have marginal hematized alteration zones which carry the best gold values. Sangster et al. (1990b) suggested on the basis of lead isotope studies that this deposit may be related to Silurian-Devonian movements on the Eastern Highlands Shear Zone.

There are a few small auriferous occurrences in the Aspy Subzone which are interpreted as mesothermal quartz vein type (Mengel et al., 1991). The best of these is the Second Gold Brook occurrence, comprising quartz veins carrying gold with pyrite, arsenopyrite, chalcopyrite, sphalerite, galena, and bismuth.

New Brunswick

A.A. Ruitenberg

Epigenetic gold mineralization in the New Brunswick Dunnage and Gander zones (Fig. 9.53) is interpreted to have been controlled by an interplay between structural and magmatic events during the late stages of the Devonian Acadian Orogeny (Ruitenberg et al., 1989, 1990). Quartz-carbonate and polymetallic veins generally occur in ductile shear and brittle fracture zones associated with steep-dipping terrane-bounding dextral wrench faults. Emplacement of mafic-felsic intrusions was influenced by these major fault systems and mineralization in the intrusions and associated contact metasomatic zones is concentrated in fracture zones that were at least in part generated by the intrusive activity (Ruitenberg and Fyffe, 1982; Fyffe and Pronk, 1985; Fyffe and MacLellan, 1988; MacLellan and Taylor, 1989; McLeod, 1990).

Numerous gold deposits occur in shear zones along the Elmtree and Rocky Brook-Millstream wrench-fault system (Fig. 9.53). The largest of these is the Elmtree deposit, a quartz-carbonate auriferous vein system hosted by hydrothermally altered and mineralized gabbro. Drill indicated reserves are about 350 000 tonnes grading 4.46 g/t Au (Tremblay and Dubé, 1991). Mineralization is intimately associated with the Elmtree Fault, a broad zone of intense shearing, fracturing, and deformation, which separates graphitic argillites of the Elmtree Formation (Tetagouche Group) from calcareous siltstones of the Chaleurs Group (Fig. 9.60), or mineralization is associated with parallel and en echelon splays from this fault. The West Gabbro zone of the Elmtree deposit consists of a hydrothermally altered and strongly deformed Ordovician(?) gabbroic sill intruding Elmtree Formation graphitic argillites and hornfels. Ductile and brittle deformation are locally very intense within the sill (Paktunc, 1987a, b; Murck, 1986). Plagioclase in the gabbro is saussuritized and ferromagnesian minerals are chloritized. Gold is associated with sulphide-rich zones cutting the coarse grained gabbro. Associated opaque minerals consist of varying amounts of arsenopyrite, pyrrhotite, and pyrite with minor chalcopyrite, stibnite, and sphalerite. Tremblay and Dubé (1991) compared the style of mineralization to some Archean gabbro-hosted deposits (e.g. Norbeau), and suggested that the host gabbro exerted both rheological and chemical controls in localizing the gold mineralization.

Numerous auriferous quartz veins occur in the Silurian and/or Devonian Chaleurs Group sedimentary and volcanic rocks along the Rocky Brook-Millstream Fault, particularly south of Elmtree and northwest of Bathurst (Fig. 9.60). These deposits (e.g. Zarina, Stephens Brook) consist of quartz-carbonate veins containing varying amounts of pyrite, sphalerite, galena, arsenopyrite and gold. Tremblay and Dubé (1991) noted that not all deposits formed from fluids of the same composition, nor do they all occur in the same structural settings, perhaps suggesting repeated mineralization through a protracted movement history on the fault and related splays.

Several gold-bearing quartz vein systems in the Upsalquitch-Jacquet River area occur in the vicinity of the intersection of the Rocky Brook-Millstream and Jacquet River faults (e.g. Simpsons Field; Fig. 9.61). Most of the discoveries are proximal to intrusions that may have been emplaced along splays of the Rocky Brook-Millstream Fault (Pronk and Burton, 1988).

Several auriferous quartz-carbonate veins occur along the Woodstock and Knowlesville faults that separate the Mirimachi Subzone from the Gaspé Belt, and along the Bamford Brook-Hainesville Fault, the boundary between the Miramichi Subzone and the Fredericton Belt (Fig. 9.53). At the Winding Hill prospect, Au and Ag are hosted by irregular quartz veins in shear zones that cut the lower Tetagouche Group.

In southwestern New Brunswick, gold-bearing quartz-carbonate veins occur in a highly deformed zone associated with the Basswood Ridge, Honeydale, and Ketchum Brook fault systems along the margins of and within the St. Croix Subzone (Fig. 9.53). Typical occurrences in the western part of this subzone occur at Lower Tower Hill, where quartz-carbonate veins carrying Au and Ag are hosted by polydeformed graphitic slate and quartzite of the Ordovician Cookson Group, immediately south of the Honeydale Fault. Farther to the northeast, auriferous quartz-carbonate

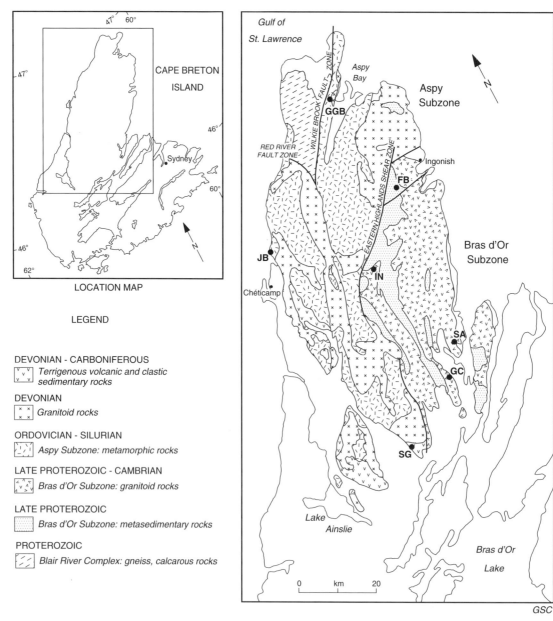

Figure 9.59. Generalized pre-Carboniferous geology of northwestern Cape Breton Island with vein-hosted gold and base metal occurrences (after Mengel et al., 1991): GGB – Grey Glen Brook; FB – Franey Brook; JB – Jerome Brook; IN – INCO showing; SA – St. Ann's area; GC – Goose Cove Brook; SG – Second Gold Brook.

veins occur in northeast-striking shear and brittle fracture zones associated with the Wheaton Brook and Ketchum Brook (e.g. Annidale) faults.

Southern New Brunswick contains numerous auriferous quartz-carbonate veins and stockworks associated with the intensely deformed southern margin of the Avalon Zone (Ruitenberg 1984; Ruitenberg et al., 1989). Deformation resulted from the Carboniferous juxtaposition of the Avalon and Meguma zones along the Cobequid-Chedabucto Fault (Rast and Grant, 1973; Ruitenberg et al., 1973; Ruitenberg and McCutcheon, 1982), which extends from the Cobequid Highlands of Nova Scotia westward under the Bay of Fundy. The structure of the gold-bearing zone north of the Cobequid-Chedabucto Fault Zone is dominated by northward- to northwestward-directed thrusts that were active during the Carboniferous (Rast and Grant, 1973; Ruitenberg et al., 1973). A penetrative fabric roughly parallel to layering is associated with this early deformation. Gold deposits are associated with a later thrusting phase that deformed the early penetrative fabric (Watters, 1985).

The best known gold deposits in the Bay of Fundy coastal area are the Cape Spencer-Millican Lake deposits (Fig. 9.62), discovered in the 1980s and mined as a low grade, open pit operation during the late 1980s. Reported proven reserves in 1988 for the Cape Spencer deposit included 550 360 tonnes grading 2.50 g/t gold (Gardiner, 1988). Gold deposits at Cape Spencer occur in and adjacent to quartz-carbonate veins hosted by intensely altered and sheared granite and altered sedimentary rocks of the Cape Spencer Formation. Pyrite and chalcopyrite are the most abundant sulphides, with minor galena and sphalerite. Two main alteration types can be distinguished: (1) a propylitic alteration (chlorite, albite, carbonate, epidote, and quartz±pyrite) that is pervasive in the granite, mafic volcanics, and porphyries; and (2) local chloritic alteration of Cape Spencer Formation sedimentary rocks at shear contacts. Generally, gold occurs with pyrite, mainly in illitic (illite, quartz, carbonate, pyrite) alteration haloes locally that overprint propylitic alteration adjacent to quartz carbonate veins. There is a broad halo of carbonate alteration around many of the illite-altered zones. Hematite-rich rocks appear to have been particularly favourable hosts for mineralization. The age of the gold mineralization is bracketed by the age of the youngest host strata (the fossiliferous Upper Carboniferous or Westphalian B Lancaster Formation) and by ^{40}Ar-^{39}Ar ages of earliest Permian on illite associated with the gold (Watters, 1987).

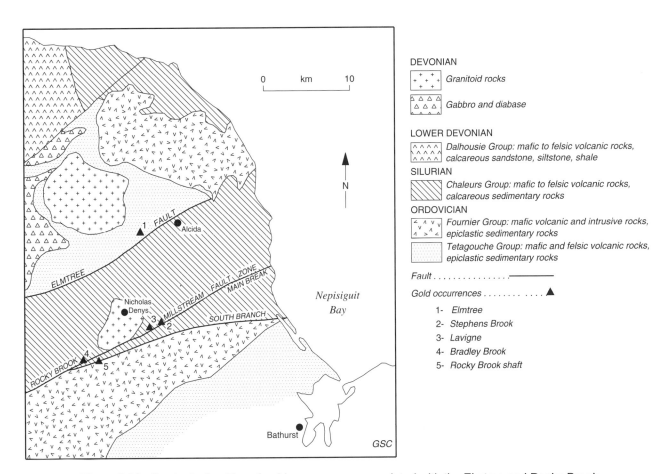

Figure 9.60. Geological setting of gold occurrences associated with the Elmtree and Rocky Brook - Millstream faults in the Bathurst area, northeastern new Brunswick (after Tremblay and Dubé, 1991).

Elsewhere in southern New Brunswick, gold-bearing quartz veins in the Musquash Harbour area (Sidwell, 1954) include a gold-bearing quartz stockwork in fine- to medium-grained intensely fractured, sericitized, and silicified porphyritic leucogranite along the west side of Musquash Harbour. The gold in this occurrence is associated with pyrite ±chalcopyrite±galena±sphalerite. At Little Dipper Harbour, quartz-carbonate veins with chalcopyrite, chalcocite, galena, and gold cut Carboniferous or older sandstone and conglomerate. The most prominent veins are situated immediately north of the thrust contact with upper Precambrian rhyolite of the Coldbrook Group. At Dipper Harbour, gold-bearing, quartz-hematite veins, up to 16 cm thick, cut intensely altered, upper Precambrian rhyolite of the Coldbrook Group.

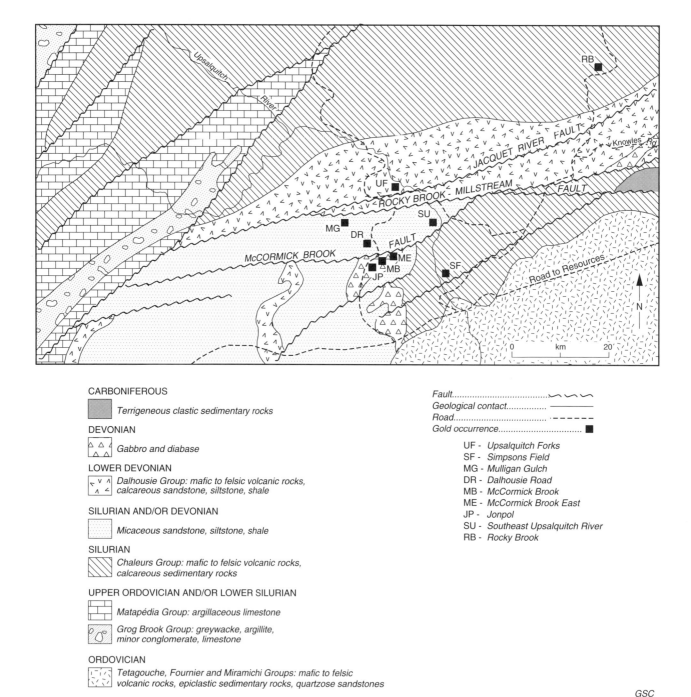

CARBONIFEROUS

Terrigeneous clastic sedimentary rocks

DEVONIAN

Gabbro and diabase

LOWER DEVONIAN

Dalhousie Group: mafic to felsic volcanic rocks, calcareous sandstone, siltstone, shale

SILURIAN AND/OR DEVONIAN

Micaceous sandstone, siltstone, shale

SILURIAN

Chaleurs Group: mafic to felsic volcanic rocks, calcareous sedimentary rocks

UPPER ORDOVICIAN AND/OR LOWER SILURIAN

Matapédia Group: argillaceous limestone

Grog Brook Group: greywacke, argillite, minor conglomerate, limestone

ORDOVICIAN

Tetagouche, Fournier and Miramichi Groups: mafic to felsic volcanic rocks, epiclastic sedimentary rocks, quartzose sandstones

Fault..
Geological contact................
Road...
Gold occurrence................................ ■

UF - Upsalquitch Forks
SF - Simpsons Field
MG - Mulligan Gulch
DR - Dalhousie Road
MB - McCormick Brook
ME - McCormick Brook East
JP - Jonpol
SU - Southeast Upsalquitch River
RB - Rocky Brook

GSC

Figure 9.61. General geology and gold occurrences in the Upsalquitch area, northern New Brunswick (after Ruitenberg et al., 1989).

Gaspésie

G. Duquette and M. Malo

The Grand Pabos Fault in southern Gaspésie (Fig. 9.53) is an east-trending, dextral strike slip fault that transects the Aroostook-Percé division of the Gaspé Belt in the east and forms the boundary between the Aroostook-Percé and Connecticut Valley-Gaspé divisions in the west where it passes into the oblique slip Restigouche Fault. In the Aroostook-Percé division, it cuts Upper Ordovician to Lower Silurian terriginous sedimentary rocks of the Honorat Group and calcareous sedimentary rocks of the Matapédia Group (Moritz et al., 1993; Fig. 9.63). Mineraliza-

tion related to the Grand Pabos Fault includes skarns in calcareous rocks of the Matapédia Group, and gold-bearing mesothermal quartz (±carbonate) veins with varying amounts of pyrite, base metals, stibnite, and arsenopyrite in subsidiary faults.

The most important mesothermal veins are the Lac Arseneault and Saint-André-de-Restigouche deposits (Fig. 9.63). The Lac Arseneault prospect comprises a series of well-mineralized, parallel quartz-epidote-carbonate veins cutting slightly mineralized, chloritized, and carbonatized greywacke of the Honorat Group. The largest vein (Baker) contains about 40 000 tonnes grading 15.42 g/t Au, 197 g/t Ag, 6.6% Pb, and 3.5% Zn to a depth of 60 m (Ministère de

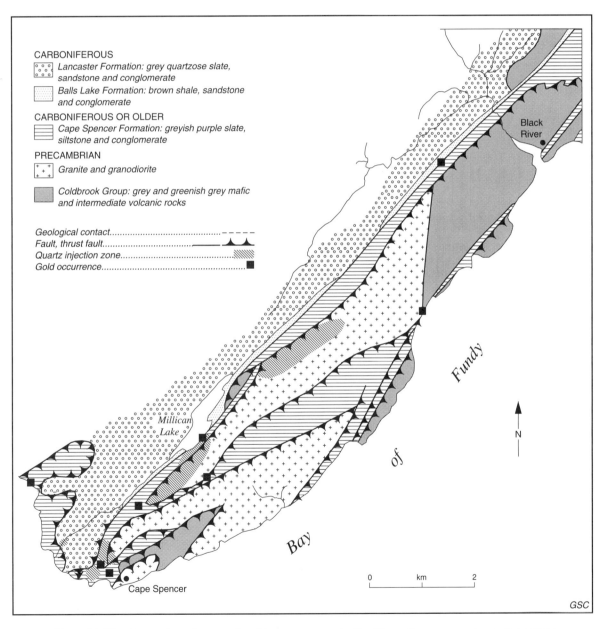

Figure 9.62. General geology and gold occurrences in the Cape Spencer area, southwestern New Brunswick (after Ruitenberg et al., 1989)

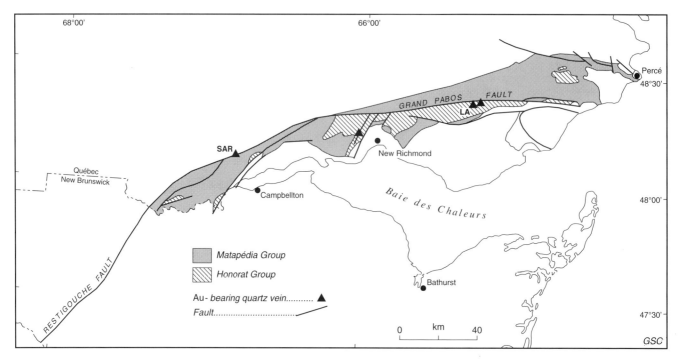

Figure 9.63. Mesothermal quartz-vein hosted gold occurrences associated with the Grand Pabos-Restagouche Fault system in Gaspésie (after Moritz et al., 1993). LA – Lac Arsenault prospect; SAR – Saint-André-de-Restigouche prospect.

l'Énergie et des Ressources, 1979). At the Saint-André-de-Restigouche prospect, Matapédia Group limestone hosts vertical stibnite-rich quartz veins that carry gold. Structural observations suggest compatibility between the orientation of structures in this prospect and movement on the nearby Restigouche Fault (Dubé et al., 1991b).

Three small antimony prospects occur near New Richmond, along the north shore of Chaleur Bay. Mineralization is confined to cross-faults in terrigenous rocks of the Honorat and Matapédia groups. Characteristically the sedimentary host rocks have been successively sheared, metamorphosed to hornfels, locally intruded by felsic or mafic dykes of assumed Devonian age, and mineralized by quartz-carbonate-pyrite-stibnite stringers which locally carry minor amounts of galena, gold, sphalerite, and chalcopyrite. Where dykes occur, mineralization is generally concentrated at the dyke-sediment interface.

Southeastern Quebec

M. Gauthier

Mesothermal and epithermal mineralization is associated with four major fault zones in southeastern Quebec, from west to east, the Taconic Brome-Richardson and Baie Verte-Brompton fault zones and the Acadian La Guadeloupe and Victoria River fault zones (Fig. 9.53).

The Brome and Richardson faults, have associated occurrences of disseminated chalcopyrite and bornite with minor chalcocite and sporadic gold and molybdenite. This style of occurrence is unique to the Humber Zone and discussed already under that heading.

The Baie Verte-Brompton Line which separates the Humber and Dunnage zones, hosts copper-bearing quartz stockworks in sheared gabbro, pyroxenite, and peridotite and pyrite-antimony-rich stockworks in sheared greywackes (e.g. Québec Antimoine mine, Gauthier et al., 1989; Normand et al., 1991). The Québec Antimoine (Ham-Sud) antimony deposit (Fig. 9.53), known since 1865, has been traced for about a kilometre and yielded 180 tonnes of concentrate. The mineralization consists of native antimony, stibnite, kermesite, senarmontite, and valentinite with pyrite (Normand et al., 1991) in a stockwork that is concordant to slaty cleavage in sheared sediments, and in a quartz vein occurring in a shear zone along the contact between the metasediments and a serpentinized peridotite.

Several types of mesothermal/epithermal mineralization are associated with the La Guadeloupe Fault in the Ordovician Ascot Complex (Fig. 9.64): 1) pyritic gold-bearing aluminous alteration in felsic volcanic rocks; 2) pyritiferous replacement bodies in banded iron-formation, and 3) gold-galena-pyrite in quartz stockwork affecting granite and porphyries. A gold-bearing stockwork also occurs in Mid-Ordovician black shales and gabbroic intrusions of the St. Victor Synclinorium.

The Timmins-Bellechasse prospect (Fig. 9.64) contains an estimated reserve of approximately 12 million tonnes grading 2.0 g/t Au (Trottier et al., 1989). It consists of a gold-bearing stockwork consisting of quartz-carbonate veins hosted by diorite sills intruding volcanic and clastic sedimentary rocks of the Magog Group. The veins are oriented parallel to the regional foliation and locally grade into breccia zones and carry free native gold intimately associated with galena. Minor pyrrhotite, pyrite, sphalerite, chalcopyrite, and arsenopyrite are also present (Gauthier et al., 1989). Geochemical alteration associated with the mineralization is restricted to the immediate vicinity of the veins where ilmenite has been converted to leucoxene in association with pyrrhotite (Trottier et al., 1989).

The Rapides-du-Diable deposit (Fig. 9.64), mined for gold around 1865 (Obalski, 1890), occurs in a quartz stockwork hosted by Beauceville Formation black shales and sandstones. The black shales are auriferous (>200 ppb Au; Godue, 1988) and are associated with important gold placers, the most extensive of which are found at Rivière-Gilbert 5 km northeast of Rapides-du-Diable. These deposits yielded several hundred kilograms of gold during the 19th century. They were worked again between 1959 and 1965 and 136 kg of gold was recovered in 1961 (Lasalle, 1980).

Silurian-Devonian limestones and conglomerates cut by the La Guadeloupe Fault also host epithermal mineralization. Visible gold has been reported in a conglomerate boulder found in the Hall Stream gold placers (Obalski, 1898) and present day mining at a nearby limestone quarry uncovered realgar and orpiment hosted by a calcite stockwork.

In the Lake Mégantic area, sheared porphyry dykes located close to the Victoria River Fault (Fig. 9.64) host native gold showings (Dresser, 1908). Silver mineralization is also known from the Victoria River Fault near La Patrie (Gauthier et al., 1989). Farther southwest, the Victoria River Fault is hidden by thick overburden which hosts

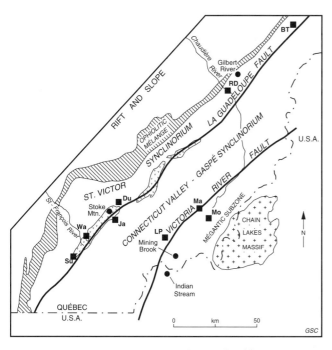

Figure 9.64. Mesothermal and epithermal gold occurrences (solid squares) associated with the La Guadeloupe and Victoria River faults, southeastern Quebec. BT – Bellechasse-Timmins; Du – Dudswell; Ja – Jackson; LP – La Patrie; Ma – Marston; Mo – Morrison; RD – Rapides-du-Diable; Su – Suffield; Wa – The Wall. Major gold placers shown by solid circles.

the important Mining Brook gold placers in Quebec (McGerrigle, 1935) and the Indian Stream gold placers in adjacent New Hampshire (Hitchcock, 1878).

SYN- AND POST-ACCRETION GRANITOID-RELATED MINERALIZATION

Overview

H. Scott Swinden

Granitoid intrusion occurred at many times and in response to various tectonic events during the accretionary and post-accretionary history of the Appalachian Orogen. Intrusive rocks range in composition from gabbro to high-silica granite, and include subalkaline, alkaline, and peralkaline types. Granitic rocks occur in all pre-accretion zones and a variety of tectonic settings. Granites with I-type, S-type and A-type characteristics can be recognized, as can large batholiths with mixed affinities (Currie, Chapter 8).

The wide variety of granites and magmatic episodes has inevitably produced a broad spectrum of mineral deposits (Fig. 9.65). These include important concentrations of granophile elements (e.g. East Kemptville, Nova Scotia, Mount Pleasant, New Brunswick), base and precious metals (e.g. Gaspé Copper, Madeleine mines, Quebec; Nigadoo River, Lake George, New Brunswick), Ni and platinum group elements (e.g. St. Stephen, New Brunswick), and industrial minerals (e.g. St. Lawrence fluorspar, Newfoundland). Although most of the economically significant granite-related mineralization in the Canadian Appalachians occurred during the Devonian, significant mineralization is also associated with older peraluminous magmatism (e.g. central Newfoundland, New Brunswick) and with Mesozoic rocks (alkaline granites, syenites, and carbonatites of southeastern Quebec and adjoining New England).

Newfoundland

W.L. Dickson

A wide variety of syn- and post-kinematic granitoid rocks is related to protracted Appalachian tectonism and magmatism in Newfoundland (Williams et al., 1989). Significant mineral occurrences are associated with three main types of plutonic rocks: (1) Middle to Late Silurian, syn-kinematic, peraluminous, biotite-muscovite granites (e.g. North Bay, Middle Ridge); 2) Devonian post-kinematic, alaskitic, high-level, biotite granites (e.g. Ackley, Granite Lake); and (3) Late Devonian peralkaline granites (e.g. St. Lawrence).

Deposits related to Silurian peraluminous granites

Synkinematic biotite-muscovite leucogranites are widespread in east-central Newfoundland. Radiometric ages indicate that most were emplaced in the Silurian and Early Devonian (Dunning et al., 1990). Minor mineralization is associated with several of these granites. In southeastern Newfoundland, extensive granite and pegmatite veins

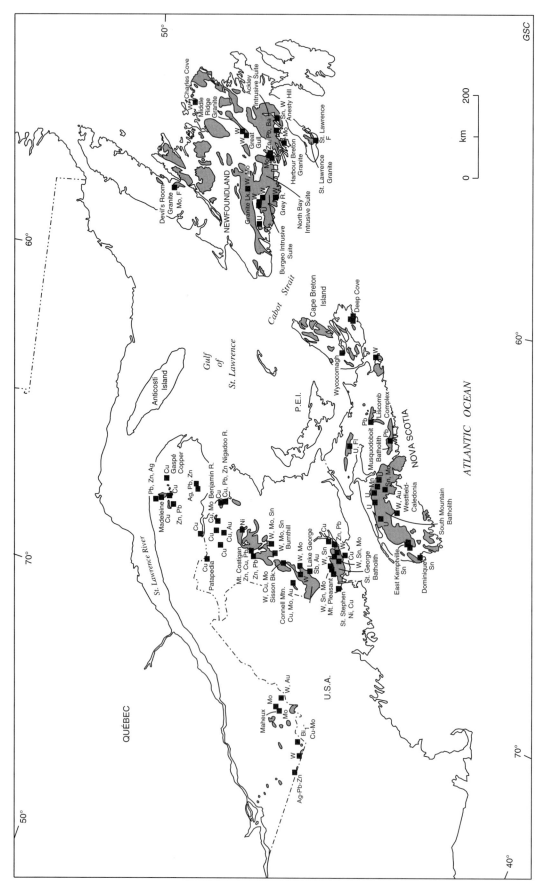

Figure 9.65. Granitoid plutonic rocks in the Canadian Appalachians (shaded) and principal granitoid-related mineral deposits.

occur in Ordovician sedimentary rocks at the contact with the East Bay Granite, a marginal pluton of the North Bay Granite Suite (Dickson, 1990). Near this contact, numerous minor occurrences of molybdenite in pegmatites grade outwards into a zone of barite-galena-sphalerite veining (Swinden, 1982; Fig. 9.66).

Farther north, extensive pegmatites with scheelite mineralization (the Great Gull prospect; Robertson and MacLean, 1991) occur in Middle Ordovician Davidsville Group greywackes near the contact with the Middle Ridge Granite (Fig. 9.65), which cuts the Gander-Dunnage zone boundary. In the northern Gander Zone, a minor tungsten occurrence (Charles Cove) is associated with a small granite pluton at Gander Bay.

Deposits related to Devonian high-silica metaluminous granites

There are two areas in southern Newfoundland where significant granophile mineralization is associated with Devonian high-silica granites: (1) the southern part of the Ackley Intrusive Suite, north of Fortune Bay; and (2) the western part of the North Bay Intrusive Suite, near Granite Lake (Fig. 9.65). A significant tungsten deposit near Grey River may also be related to granitoid intrusions of this type. Similar minor mineralization is associated with the Harbour Breton Granite (Mo) on the Hermitage Peninsula (Furey and Strong, 1986) and the Devils Room Granite (Mo, Fl) in White Bay (Saunders and Smyth, 1990; Saunders, 1991).

Ackley Intrusive Suite (Mo, Sn, W)

The Ackley Intrusive Suite, which underlies an area of approximately 2700 km² in southeastern Newfoundland, is dominated by K-feldspar megacrystic and equigranular biotite granite (Tuach, 1987b). It comprises seven granite plutons of which the Rencontre Lake and Sage Pond plutons at the southern margins are the highest-level and most evolved phases, and of most economic interest (Fig. 9.67). This post-kinematic Ackley suite has been dated at about 376-355 Ma (Dallmeyer et al., 1983; Tuach, 1987b; Kontak et al., 1988).

There are numerous endocontact, aplite-pegmatite-type molybdenite showings along the southern margin of the Rencontre Lake Granite (Fig. 9.67) at the contact with late Precambrian rhyolites of the Long Harbour Group (e.g. Motu, Ackley City, Crow Cliff-Dunphey Brook, and Wylie Hill). A maximum of 125 000 tonnes grading 0.3% MoS_2 (Smith, 1936) and possibly 4 400 000 tonnes grading about 0.1% MoS_2 have been indicated by exploration drilling of the Ackley City and Wylie Hill deposits, respectively.

The Ackley City and Wylie Hill showings occur in protrusions of the Ackley Intrusive Suite into the adjacent rhyolite. They comprise molybdenite as disseminations, pods, and fracture filings; associated minerals include quartz, chlorite, barite, calcite and fluorspar, pyrite, sphalerite, chalcopyrite and pyrrhotite (White, 1939). Associated alteration includes segregations of milky white quartz, patches of replacement muscovite, and local greisenization. At the Motu showing, located at the contact between a porphyritic aplite marginal phase and the adjacent rhyolites, molybdenite occurs as erratically distributed coarse rosettes and patches up to 20 cm in diameter,

as linings in miarolitic cavities, and as disseminated flakes mostly within the aplite. At the Crow Cliff-Dunphey Brook showing, molybdenite occurs as coarse grains within large quartz crystals and underlying massive vein quartz which forms irregular, more or less flat pods, with molybdenite crystals parallel to the long direction of the pods.

Endocontact, cassiterite-bearing, quartz-topaz greisen veins and tungsten-rich greisens are present in the Sage Pond Granite to the east (Dickson, 1982a; Tuach, 1984b; Fig. 9.67). The quartz-topaz veins outcrop along the southern contact and within the granite up to 3 km from the contact. The largest concentration of greisen veins lies within 1 km of the contact and is associated with protrusions of the granite into the country rocks. This granite host is typically biotite-bearing, with miarolitic cavities and some tuffisites. Veins southeast of Moulting Pond are typical, comprising sub-parallel quartz-topaz (±pyrite, hematite, fluorspar) vein swarms in sericitized and bleached alaskitic, quartz-feldspar-porphyritic, biotite granite below a massive quartz-topaz outcrop. Southeast of Anesty Hill, a large, layered greisen vein consisting of massive quartz-topaz±pyrite±fluorspar, with sparse grains of molybdenite within a lobe of the granite was interpreted to have resulted from ponding of fluorine-rich fluids near the granite roof (Tuach, 1984a) .

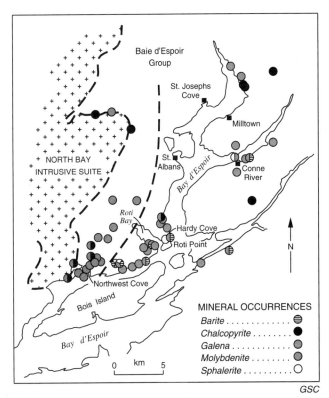

Figure 9.66. Granophile mineral occurrences around the eastern edge of the North Bay Intrusive Suite, south-central Newfoundland (after Swinden, 1982). Mineralization is zoned outward with Mo and Cu dominant near the intrusive margin of the batholith passing outwards (across heavy dashed line) into dominantly Pb-Zn-Ba showings.

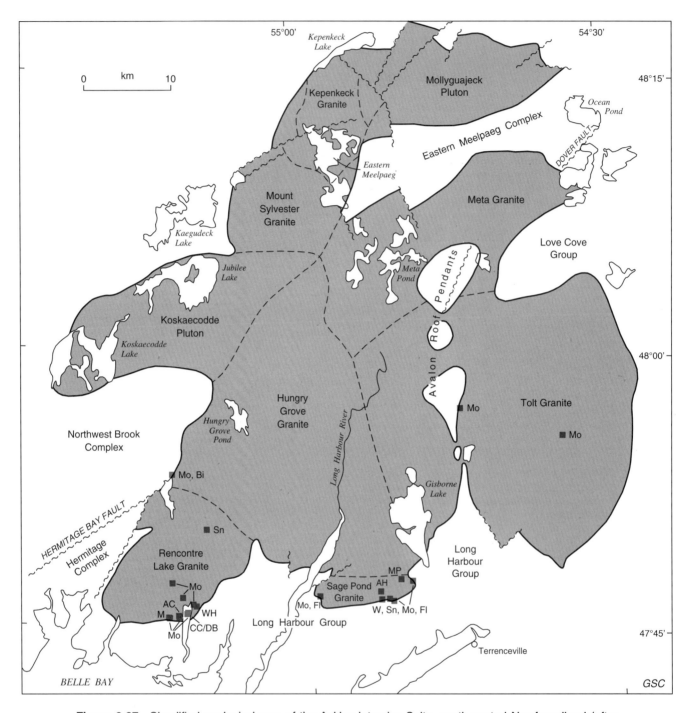

Figure 9.67. Simplified geological map of the Ackley Intrusive Suite, south-central Newfoundland (after Tuach, 1987a). Solid lines are external boundaries of the Ackley Intrusive Suite, dashed lines are internal boundaries of individual plutons. Red squares are mineral occurrences related to late intrusive phases (mainly Rencontre Lake and Sage Pond granites). M – Motu showing; AC – Ackley City showing; WH – Wylie Hill showing; CC/DB – Crow Cliff-Dunphey Brook showings; MP – Moulting Pond; AH – Anesty Hill; Mo – molybdenite; W – tungsten (wolframite-scheelite); Fl – fluorite; Bi – bismuth; Sn – tin.

Whalen (1980) and Dickson (1983) suggested that the increase in degree of fractionation towards the southern contact of the Rencontre Lake Granite resulted from in situ differentiation of a granitoid magma. Tuach et al. (1987) suggested that the northern parts of the Ackley Intrusive Suite were intruded at greater depths, or that they represent deeper levels of a larger magma chamber than the mineralized southern phases. The element enrichment-depletion trends, and the size of the Ackley Intrusive Suite, suggest that it may represent a now-frozen, geochemically stratified, magma chamber analogous to those modelled by Hildreth (1979, 1981) as a precursor to large-volume high-silica ash flows. Magmatic evolution of this large system ultimately resulted in Mo and Sn deposits in the Rencontre Lake and Sage Pond granites, respectively (Tuach et al., 1987; Tuach, 1987b).

Granite Lake area (W, Sn, Mo)

Molybdenum and tungsten occurrences are widespread in the Granite Lake area (Dickson, 1982b; Tuach and Delaney, 1986). The mineralization is associated with the Wolf Mountain Granite, an Early Devonian (396 +5/-3 Ma, Dunning et al., 1990) equigranular to feldspar-porphyritic biotite granite. It contains hydrothermal muscovite and occurs at the northwest end of the North Bay Intrusive Suite (Fig. 9.68). Wolframite and molybdenite (±scheelite and bismuthinite) occur in steeply dipping sheeted quartz veins, quartz-feldspar (pegmatite) veins, and quartz-muscovite greisen veins within the Wolf Mountain Granite and adjacent metasedimentary rocks and foliated tonalite. Traces of pyrite, fluorite, chalcopyrite, sphalerite, galena, and beryl are locally present (Tuach and Delaney, 1986). The mineralization is accompanied by minor potassium

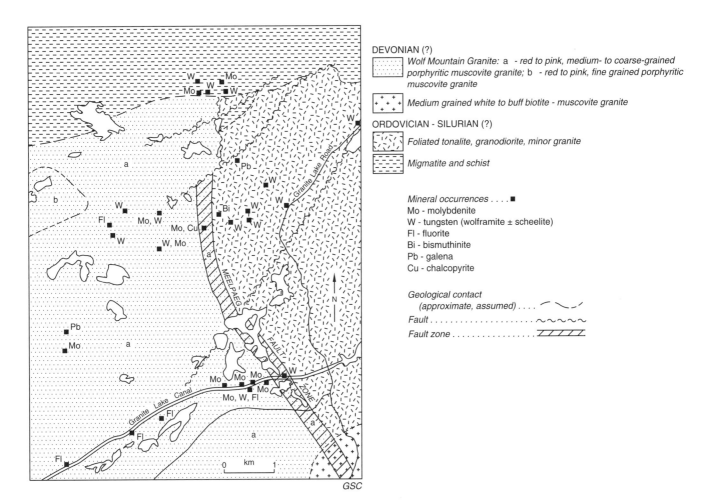

Figure 9.68. Simplified geological map of the Granite Lake area, central Newfoundland, with granophile mineral occurrences (after Tuach and Delaney, 1986).

761

metasomatism and was interpreted to have resulted from hydrothermal activity during and following emplacement of the Wolf Mountain Granite.

Grey River area (W)

The Grey River prospect (Higgins, 1980, 1985) on the south coast of Newfoundland (Fig. 9.65) comprises more than 600 veins delineated between 1956 and 1958. Extensive drilling and underground work indicated a probable 360 000 tonnes of ore, grading at 1.09% WO_3, and a further 160 000 tonnes possible.

Tungsten mineralization is contained within a series of quartz veins mainly hosted by amphibolite facies metamorphic rocks, but extending across a mylonitic contact into the Burgeo Intrusive Suite. Mineralization occurs in three forms: (1) as fractures coated with pyrite+scheelite±quartz; (2) tensional quartz veins with no displacement of the wall rocks; and (3) quartz lodes showing much displacement of the wall rocks. The lodes are the most important and are characterized by repeated reopening and injection of new quartz and mineralization into the open spaces created by the fault movements. The tungsten-bearing hydrothermal veins resulted from four pulses of quartz deposition (Higgins, 1980): (1) early stage of quartz-feldspar-molybdenite veins; (2) a composite stage of five vein types in the order quartz-bismuthinite, quartz-wolframite (Fe-rich), greisen, quartz-sulphides, and quartz-wolframite (Mn-rich); (3) a sulphide stage of silver-bearing quartz-galena-sphalerite veins; and (4) the late stage of zoned fluorspar-calcite-barite veins. The veins are believed to be related to nearby post-tectonic aplites, pegmatites, and leucogranites, which have a characteristic hydrothermal alteration. The intrusions are geochemically specialized and were interpreted by Higgins (1980) to be the apical expression of a buried post-kinematic specialized granite.

Deposits related to Late Devonian peralkaline granites

Fluorspar deposits are intimately associated with the Late Devonian (374 ± 2 Ma; Kerr et al.,1993) alkaline to peralkaline St. Lawrence Granite at the southern end of the Burin Peninsula (Fig. 9.69). This high level intrusion cuts Upper Precambrian to Cambrian sedimentary and volcanic rocks as well as its own volcanic carapace (Strong et al., 1978). The St. Lawrence mining district contains more than 40 individual fluorite veins (Howse et al., 1983) from which more than 3.4 million tonnes of ore have been mined since 1933. A further 9 million tonnes of reserves remain in the three main veins. Most fluorspar veins occur within the St. Lawrence Granite or its related porphyry dykes, as open-space fillings in tension fractures developed as a result of regional stress and contraction during cooling of the granite. Two major groups of veins are recognized: (1) low grade veins (e.g. Director, Tarefare, Blue Beach), containing about 70% CaF_2, are generally widest, averaging about 7 m but locally reaching widths of 30 m; (2) high grade veins (e.g. Lord and Lady Gulch, Iron Springs, and Mt. Margaret) average one metre or less wide, and are characterized by acid grade ore averaging 95% CaF_2. The fluorspar is mostly massive and coarsely crystalline; quartz is the most abundant gangue mineral, calcite is found in nearly all of the fluorspar veins and barite is locally common. Sulphides (galena, sphalerite, lesser pyrite, chalcopyrite) are fairly common, especially near the granite contact or in country-rock veins. Pitchblende is locally present and anomalous Mo and Sn values have been reported in overlying soils.

Nova Scotia
A.L. Sangster
Cape Breton Island contact metasomatic occurrences

In southeastern Cape Breton Island, a belt of skarn and related vein-type mineralization is associated with Devonian metaluminous granitoid intrusions (MacDonald, 1989). In the Gabarus Bay area, upper Proterozoic volcanic rocks in the coastal volcanic belt are intruded by the small, elliptical Deep Cove Granite stock (Fig. 9.70). Molybdenite veins and disseminations are developed along the contacts of the stock whereas polymetallic (Cu, Zn, Mo, Bi, Ag) mineralization occurs in the interior of the stock. The mineralization is generally associated with pervasive sericite-chlorite alteration and specifically associated with localized tabular zones of greisenization or K-feldspathization (MacDonald, 1989).

Farther north in the French Road-Blue Mountain area, Lower to Middle Cambrian calcareous argillite and siltstone have been contact metamorphosed to skarn and mineralized with Fe-Cu-Zn-Mo-Bi and overlapping Fe-Pb-Zn veins and disseminations (Fig. 9.70). The mineralization is thought to be related to Devonian-Carboniferous feldspar porphyry dykes similar to phases of the Deep Cove stock (MacDonald, 1989). Similar mineralization carrying Cu and Mo is found in hornfels surrounding the Gillis Mountain granite to the southwest.

A similar skarn in the Bras d'Or Subzone, the Whycocomagh Mountain diopside-andradite-forsterite skarn (Fig. 9.65), is hosted by Upper Precambrian marble adjacent to a composite Devonian granite porphyry, granodiorite, and diorite intrusion (Chatterjee, 1977a; Barr et al., 1984). The deposit contains magnetite with accessory scheelite, chalcopyrite, molybdenite, pyrite, pyrrhotite, and hematite.

Southern Nova Scotia-South Mountain Batholith and related intrusions

The Meguma Zone of Nova Scotia is intruded by Devonian and Carboniferous peraluminous granitoids (Clarke et al., 1980, 1985) consisting of granite, granodiorite, granodiorite porphyry and two-mica monzogranite with local tonalite and trondhjemite. The largest intrusion, the South Mountain Batholith (SMB) (about 370 Ma; Clarke and Halliday, 1980; Clarke et al., 1985), is composite and contains biotite granodiorite and biotite monzogranite plutons cut by plutons of two-mica monzogranite and leucomonzogranite (McKenzie and Clarke, 1976; MacDonald et al., 1989, 1990; Fig. 9.71). Prominent sets of northwest- and northeast-trending faults within the batholith apparently influenced magmatism during emplacement and were the focus for hydrothermal activity during subsequent tectono-thermal events (Horne et al., 1989; Corey and Horne, 1989). Smaller plutons of similar composition occur east of the South Mountain Batholith (Clarke et al., 1980; Hill, 1986) including the Musquodoboit Batholith (MacDonald and Clarke, 1985), the composite mafic/felsic Liscomb Complex (Giles and Chatterjee, 1987), and a number of smaller intrusions scattered throughout the eastern Meguma Zone, particularly in the Canso area (Fig. 9.71). A second group of smaller intrusions of broadly similar composition and with ages of ca. 315 Ma (e.g. Wedgeport pluton; Cormier et al., 1988) occur south and west of the batholith.

Granite-related mineral deposits and occurrences in the Meguma Zone are typically located in and peripheral to highly evolved, geochemically "specialized" (Tischendorf, 1977) intrusions of biotite-muscovite-bearing monzogranite, leucomonzogranite, and leucogranite. According to some workers (e.g. Zentilli and Reynolds, 1985; O'Reilly et al., 1985; Richardson et al., 1989), evidence from ^{40}Ar-^{39}Ar and Rb-Sr dating indicates that at least some specialized plutons within the South Mountain Batholith are Carboniferous and postdate the main phase of magmatism. This interpretation, however, remains controversial. Recent Pb-Pb, Rb-Sr and ^{40}Ar-^{39}Ar studies (e.g. Kontak and Cormier, 1991; Kontak and Chatterjee, 1992) indicate that

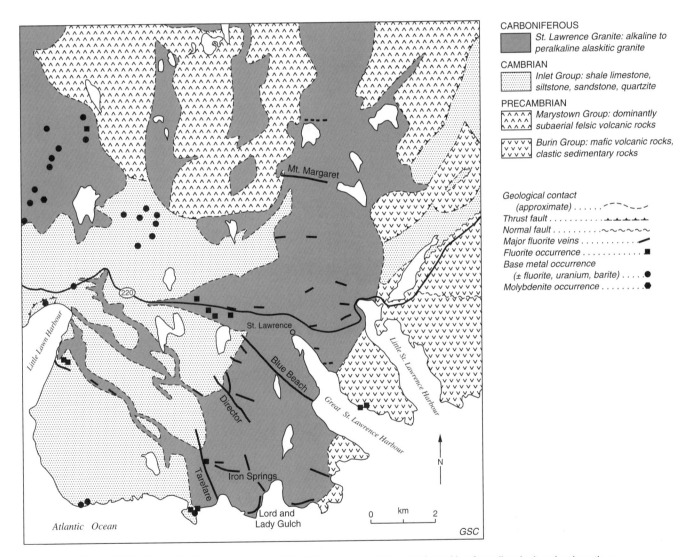

Figure 9.69. Simplified geological map of the St. Lawrence area, southern Newfoundland, showing locations of fluorite, base metal and Mo deposits and occurrences (after Howse et al., 1983).

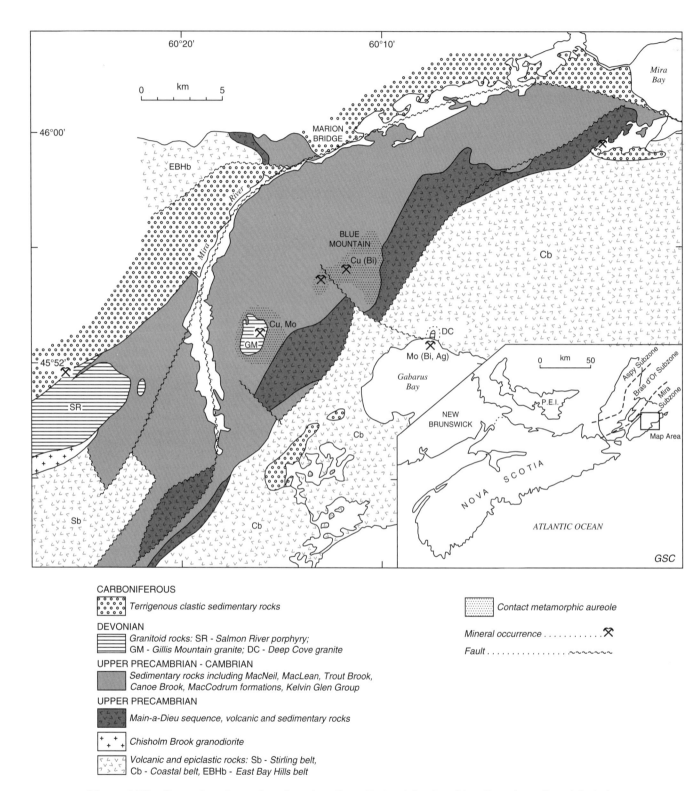

Figure 9.70. General geology of southeastern Cape Breton Island and location of granite-related skarn occurrences in the Gabarus Bay area, simplified after Barr et al. (1992).

younger Rb-Sr and ^{40}Ar-^{39}Ar ages reflect tectono-thermal resetting of the granite ages and that the specialized plutons represent evolved phases of the magmas emplaced at ca. 370 Ma.

Southwest Nova Scotia tin domain

The southwest Nova Scotia tin domain (Fig. 9.72) encompasses tin-bearing deposits in the Wedgeport and Davis Lake plutons and in the intervening Meguma Supergroup metasedimentary rocks (Chatterjee et al., 1985a; Wolfson, 1983; Richardson, 1985a, b, 1988a, b; Kontak et al., 1990a).

The East Kemptville tin deposit, the largest granophile element mineral deposit in the domain, is hosted by a topaz-muscovite leucogranite phase of the Davis Lake pluton (Kontak, 1990), a biotite monzogranite/leucomonzogranite body at the southwestern apex of the South Mountain Batholith (Fig. 9.71). Production from the deposit

began in October 1985 from a pre-mining reserve of 56 million tonnes grading 0.165% Sn at a cut-off grade of 0.05% Sn (Moyle, 1985). The deposit is a horizontal, tabular zone consisting of massive greisen and greisen-bordered veins and stockworks within variably-altered leucogranite. Mineralization occurs immediately below a horizontal roll in the steeply dipping contact between the intrusive rocks and enclosing Meguma Supergroup metagreywacke. The greisen-bordered vein and microfracture system is steeply dipping, with components oriented both subparallel and perpendicular to the intrusive contact (Richardson et al., 1982; Richardson, 1985a). Greisen-bordered veins occur over an area of 15 km^2, but only at the East Kemptville site do they occur in sufficient density to constitute a commercial deposit (Chatterjee et al., 1985a).

The greisens that host the tin ores are the end product of hydrothermal alteration that has affected much of the Davis Lake pluton. Essential minerals of the greisens are

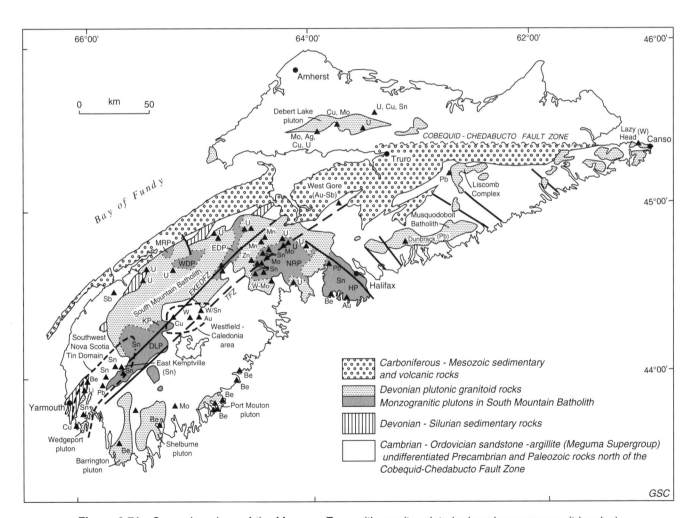

Figure 9.71. General geology of the Meguma Zone with granite-related mineral occurrences (triangles) (after MacDonald et al., 1990). DLP – Davis Lake Pluton; KP – Kejimkujik Pluton; MRP – Mores Road Pluton; WDP – West Dalhousie Pluton; EDP – East Dalhousie Pluton; NRP – New Ross Pluton; HP – Halifax Pluton. EKEDFZ – East Kemptville – East Dalhousie Fault Zone; TFZ – Tobeatic Fault Zone. Heavy dashed lines show extent of Southwest Nova Scotia tin domain and Westfield-Caledonia area.

fine- to very fine-grained bluish quartz, topaz, and greenish to silver mica (muscovite and zinnwaldite). Other minerals include cassiterite, pyrite, sphalerite, chalcopyrite, pyrrhotite, arsenopyrite, molybdenite, galena, wolframite, and fluorite. Richardson (1988a) described two types of alteration outside the greisen envelopes. The greisen is bordered by incompletely greisenized leucomonzogranite in which approximately 75 per cent of all feldspars have been replaced. Peripheral to this, the leucomonzogranite is white or blue and shows pervasive albite-white-mica alteration. Tin grades are high in the core veins and decrease with decreasing intensity of alteration away from these veins.

The southeast part of the Wedgeport pluton (about 315 Ma, Cormier et al., 1988) contains greisen-bordered sulphide-cassiterite veinlets and small zones of massive greisen with very different compositions. Chatterjee et al. (1985b) listed the major gangue minerals as quartz, muscovite, topaz, fluorite, beryl, tourmaline, and carbonates with minor metallic phases, including sphalerite, chalcopyrite, arsenopyrite, cassiterite, wolframite, scheelite, native bismuth, argentite, pyrite, and pyrrhotite.

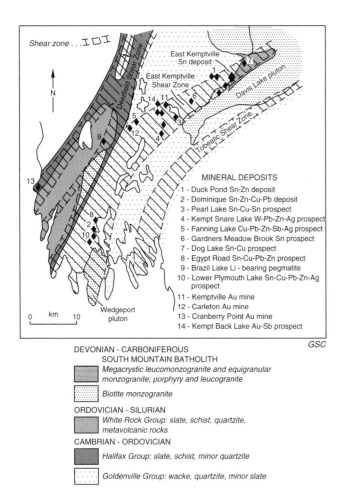

MINERAL DEPOSITS
1 - Duck Pond Sn-Zn deposit
2 - Dominique Sn-Zn-Cu-Pb deposit
3 - Pearl Lake Sn-Cu-Sn prospect
4 - Kempt Snare Lake W-Pb-Zn-Ag prospect
5 - Fanning Lake Cu-Pb-Zn-Sb-Ag prospect
6 - Gardners Meadow Brook Sn prospect
7 - Dog Lake Sn-Cu prospect
8 - Egypt Road Sn-Cu-Pb-Zn prospect
9 - Brazil Lake Li - bearing pegmatite
10 - Lower Plymouth Lake Sn-Cu-Pb-Zn-Ag prospect
11 - Kemptville Au mine
12 - Carleton Au mine
13 - Cranberry Point Au mine
14 - Kempt Back Lake Au-Sb prospect

GSC

DEVONIAN - CARBONIFEROUS
SOUTH MOUNTAIN BATHOLITH
Megacrystic leucomonzogranite and equigranular monzogranite; porphyry and leucogranite

Biotite monzogranite

ORDOVICIAN - SILURIAN
White Rock Group: slate, schist, quartzite, metavolcanic rocks

CAMBRIAN - ORDOVICIAN
Halifax Group: slate, schist, minor quartzite

Goldenville Group: wacke, quartzite, minor slate

Figure 9.72. Geology and mineral occurrences of the southwest Nova Scotia tin domain (after O'Reilly and Kontak, 1991).

There are a number of polymetallic, tin-base metal occurrences in sedimentary rocks of the Goldenville-Halifax transition zone (Kontak et al., 1990a) between the Duck Pond deposit, 1 km northwest of the East Kemptville deposit (Pitre and Richardson, 1989), and the Dominique-Plymouth area near the north end of the Wedgeport pluton (Wolfson, 1983; Fig. 9.72). The zone of mineralization is spatially related to the East Kemptville Shear Zone (Kontak et al., 1990a). Tin typically occurs in sulphide-cassiterite veinlets, crosscutting foliation, with associated chloritic and/or biotitic alteration. Wolfson (1983) also described thin layers of stratiform sulphide-cassiterite (up to 30%) in silicified and calcareous layers within the sedimentary rocks. Pyrite, rather than the regionally present pyrrhotite, is dominant in the cassiterite bands and is accompanied by sphalerite, pyrrhotite, arsenopyrite, and chalcopyrite. Alteration associated with the stratabound cassiterite bodies includes some silicification and development of chlorite, chloritoid, and hydrothermal garnet (Wolfson, 1983; Pitre, 1987; Kontak et al., 1990a). Kontak et al. (1990a) recently suggested that these deposits may be a surface manifestation of buried granite cupolas similar to the topaz leucogranite at East Kemptville. Hydrothermal fluids related to magmatic activity may have been focussed in the East Kemptville shear zone.

There is a distinctive lithium-bearing pegmatite cutting Silurian quartzite at Brazil Lake northeast of Yarmouth (Taylor, 1967; Fig. 9.72). The pegmatite consists of quartz, feldspar (clevelandite), and spodumene, with accessory hornblende, apatite, beryl, columbite-tantalite, cassiterite and holmquistite (Hutchinson, 1982). The pegmatite is poorly exposed and is not known to contain commercial quantities of lithium.

New Ross Pluton

The New Ross Pluton, located approximately 75 km west of Halifax, consists of muscovite-biotite monzogranite and leucomonzogranite and muscovite-topaz granite which cut biotite-granodiorite to monzogranite phases of the South Mountain Batholith (Charest et al., 1985; MacDonald et al., 1989). The highly differentiated northwestern part of the pluton is associated with metallic mineralization in greisen, pegmatite, and fissure veins and shear zones (O'Reilly et al., 1982; Logothetis, 1985; Fig. 9.73). The Millet Brook uranium deposit is the largest of these occurrences.

Polymetallic quartz-sericite and quartz-sericite-pyrite-fluorite greisens form selvages bordering quartz-sulphide veins or irregular zones of pervasive alteration (Logothetis, 1985; O'Reilly et al., 1982). The occurrences mainly contain Sn, W, Mo, Cu, and Zn and lesser U, Be, As, Nb, and Ta. The Long Lake and Turner tin occurrences, two larger greisens, are associated with apogranite (O'Reilly et al., 1982) and are the result of mineralization processes similar to those that occurred at the East Kemptville deposit.

Pegmatites of the New Ross Pluton are small, lensoid bodies that are spatially associated with the more evolved leucomonzogranite phases of the South Mountain Batholith (Charest, 1976). Where a spatial association with leucomonzonite is less obvious, the pegmatites contain leucogranitic or aplitic phases indicating they are probably related to the same period of intrusive activity.

Mineralization in the pegmatites is similar to that found in greisen deposits with: 1) Mo-bearing pegmatites and 2) Sn-W-F-bearing pegmatites with accessory Li, U, Ta, Nb, and Cu.

Late, hydrothermal mineralization occurs in veins, shears, joints and fractures within intrusive and, less commonly, Meguma Supergroup metasedimentary rocks. The common mineral assemblages are quartz-fluorite and quartz-calcite-molybdenite veins likely formed from late magmatic fluids and probably related to the period of pegmatite intrusion (Logothetis, 1985) and manganese-rich veins with massive pyrolusite, manganite and psilomelane, previously attributed to either meteoric or hypogene hydrothermal processes (O'Reilly et al., 1982; Logothetis, 1985). The manganese-rich veins cut sheared, hematite-stained and altered (argillic) megacrystic biotite monzogranite (e.g. Dean and Chapter mine). O'Reilly (1990) has reported leucomonzogranite in fault and/or shear zones adjacent to and beneath the Dean and Chapter Mine, suggesting a control by hydrothermal activity related to late magmatic fluids.

The Millet Brook uranium deposits, discovered in the late 1970s, consist of a number of small pitchblende-bearing fracture zones, the most important of which is the Upper Salter Lake zone (Robertson and Duncan, 1982). Reserves are reported to be close to 450 000 kg of contained U_3O_8 (Chatterjee et al., 1985c). At Millet Brook the host rock is massive to faintly foliated, medium- to coarse-grained, megacrystic, biotite granodiorite (Chatterjee et al., 1985c). Sooty pitchblende is the most important uranium mineral in the veins accompanied within 50 m of the surface by accessory coffinite, carnotite, Pb-autunite, and supergene metatorbernite and autunite. Gangue minerals include vein carbonate and minor pyrite and chalcopyrite (Robertson and Duncan, 1982), as well as sphalerite, proustite, quartz, phyllosilicates, and tourmaline (Chatterjee et al., 1985c). Hydrothermal alteration minerals in the host rocks adjacent to the veins include hematite, albite, kaolin, and minor muscovite, chlorite, and quartz. Cataclastic textures are present in the host rocks immediately adjacent to the veins.

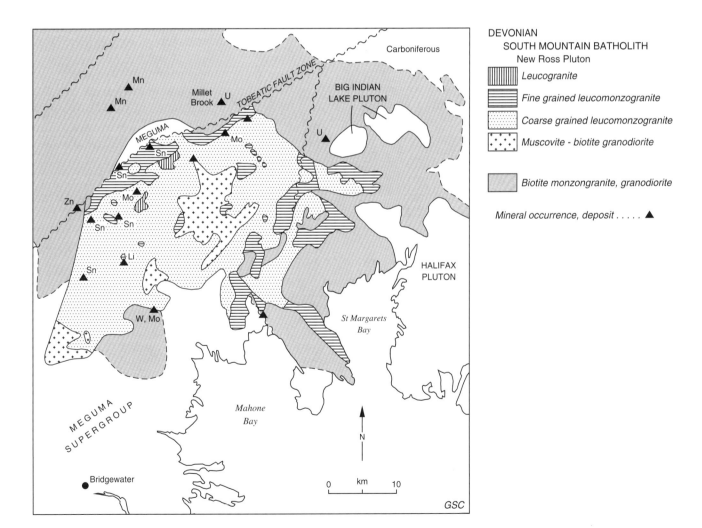

Figure 9.73. Geology of the New Ross Pluton (after MacDonald et al., 1990) and related mineral occurrences.

The Millet Brook uranium deposit is similar in character and probable origin to known granite-hosted intrabatholithic and peribatholithic uranium veins of the Bohemian, Central, Armorican, and Iberian massifs of Europe (Robertson and Duncan, 1982). Chatterjee et al. (1982, 1985c) related the origin of the veins to late-stage magmatic fluids derived from a two-mica leucomonzogranite concealed beneath the deposit but similar in character to the Lewis Lake muscovite-topaz leucogranite, located 10 km southeast of the Millet Brook deposit.

Westfield-Caledonia area

In the Westfield-Caledonia area (Fig. 9.71), K-feldspathized albite-bearing porphyry, albitized porphyry, and greisenized granitoids are associated with a granodiorite-monzogranite prominence of the South Mountain Batholith that intrudes slate and greywacke of the Meguma Supergroup (O'Reilly, 1985). The altered rocks locally host greisen zones, quartz veins, and quartz vein stockworks containing scheelite, cassiterite, molybdenite, arsenopyrite, and pyrite. The altered rocks also contain as much as 50 ppb gold, compared to a maximum of 15 ppb in unaltered granitoid rocks (O'Reilly, 1985). Meguma Supergroup slates south of the granodiorite-monzogranite prominence contain several quartz veins with associated greisen envelopes and quartz vein stockworks which contain scheelite-cassiterite mineralization with associated base metal sulphides, arsenopyrite, pyrite, marcasite, gold, and silver. This occurrence is characterized by the presence of anomalous gold with tin and a lack of geochemical specialization of the host granitic rocks. O'Reilly (1985) attributed this to the early loss of a fluid phase from the granodioritic magma and subsequent metasomatism of crystallized granodiorite by the Rb- and U-poor residual volatiles. Gold enrichment has been documented in several other metasomatized zones in proximity to sediment contacts throughout the South Mountain Batholith (MacDonald and O'Reilly, 1988).

West Dalhousie and East Dalhousie plutons

There are a number of small uranium prospects characterized by U-Cu (±Ag±P) associated with the West Dalhousie and East Dalhousie plutons (Fig. 9.74). These occurrences are similar to the Millet Brook deposits, having formed as en echelon pods within shear and fracture zones in the South Mountain Batholith. However, the host rocks range in composition from biotite granodiorite to leucomonzogranite and muscovite±topaz leucogranite (MacDonald and Ham, 1988). The uranium mineralization is dominated by pitchblende, and accessory amounts of torbernite and autunite. Associated minerals include pyrite, chalcopyrite, bornite, covellite, galena, chalcocite, sphalerite, and wolframite. Alteration minerals include albite, muscovite, biotite, chlorite, orthoclase, carbonates, and tourmaline (Chatterjee, 1983).

Miscellaneous occurrences, central and northern Nova Scotia

Numerous small occurrences and geochemically anomalous areas are associated with granites other than the South Mountain Batholith (Fig. 9.71).

(1) Minor occurrences of beryl are associated with pegmatites and quartz veins in spatial association with the Port Mouton, Barrington, and Shelburne plutons (Taylor, 1967).

(2) West Gore Sb-Au occurrence: auriferous stibnite-quartz veins fill fissures and constitute the matrix of slate breccia within northwest-striking fracture zones which cut Halifax Group slate. Sb concentrates shipped to England between 1914 and 1917 averaged 43.5 g/t Au (Douglas, 1940), as fine grained Au alloys and free gold intergrown with the stibnite. Kontak and Smith (1991) have shown that the West Gore deposit contrasts with typical Meguma Supergroup gold occurrences in that it apparently formed at higher pressure and from CO_2-rich fluids. The veins are similar in style and mineralogy to the Lake George antimony deposit in New Brunswick. Lead isotope analyses show a strong similarity between West Gore and East Kemptville galena, suggesting a late magmatic source for the mineralization (Sangster, 1990).

(3) Liscomb area: the Liscomb Complex consists of a suite of Devonian-Carboniferous granitoid rocks which have intruded Meguma Supergroup metasedimentary rocks and gabbroic and gneissic rocks of uncertain origin (Giles and Chatterjee, 1986). Two Pb-Zn-Ag± Sb occurrences are present in gneisses near the western margin of the complex.

(4) Eastern Meguma Zone: several mineral occurrences of possible igneous-related metasomatic origin are associated with small monzogranite plutons between

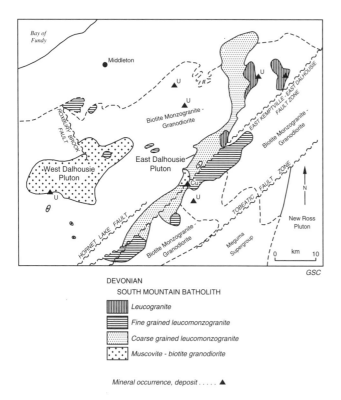

DEVONIAN
SOUTH MOUNTAIN BATHOLITH

Leucogranite
Fine grained leucomonzogranite
Coarse grained leucomonzogranite
Muscovite - biotite granodiorite

Mineral occurrence, deposit . . . . . ▲

Figure 9.74. Geology and principal U (Cu) occurrences in the area of the West Dalhousie and East Dalhousie plutons (patterned), southwestern Nova Scotia (after MacDonald et al., 1990).

Sherbrooke and Canso. At Lazy Head, disseminated scheelite is associated with pyrite, pyrrhotite, and chalcopyrite within spessartine-rich rocks.

(5) Musquodoboit Pluton: several small quartz-scheelite showings have been reported in the west central part of this intrusion. A small hydrothermal Pb-Zn-Ag-Cu-bearing quartz vein, the Dunbrack occurrence, cuts cordierite-bearing monzogranite within the intrusion (MacMichail, 1975; Dickie, 1978).

(6) Debert Lake Pluton: high level intrusive stockworks and collapse breccias in the Debert Lake composite pluton in northern Nova Scotia contain epithermal occurrences of U, Th, Mo± F, Ag, and Cu. Mineralization is characterized by pitchblende and high concentrations of Th and occurs in the high level intrusive rocks and their volcanic carapace associated with extensive alkali metasomatism (Chatterjee, 1984). These occurrences are associated with high level phases of the pluton which, in contrast to the dominantly peraluminous association of other granophile mineral occurrences in Nova Scotia, are acmite-normative (peralkaline) granites (A.K. Chatterjee, pers. comm., 1992).

New Brunswick

A.A. Ruitenberg and L.R. Fyffe

Introduction

Different families of granitoid rocks in New Brunswick were emplaced during successive episodes of middle Paleozoic terrane accretion and produced distinct suites of mineral deposits (Fig. 9.75): (1) subvolcanic Ordovician and Silurian-Devonian calc-alkaline intrusions with I-type characteristics, predominantly in the Belledune and Elmtree subzones of the Dunnage Zone and the Chaleurs Bay division of the Gaspé Belt, produced contact metasomatic and vein-type base metal sulphide and porphyry copper-molybdenum deposits with minor scheelite and gold; (2) Late Silurian to Early Devonian bimodal batholiths with mixed S- and I-type affinities, generally emplaced at deeper levels in the Miramichi and St. Croix subzones, produced antimony-gold, tungsten-molybdenum, and base metal vein deposits; (3) Middle to Late Devonian silica-rich granites with A-type features, emplaced at shallow depth and locally associated with terrestrial subvolcanic complexes, produced minor base metal skarns and veins and extensive exogranitic and endogranitic greisens rich in fluorine.

Deposits related to subvolcanic calc-alkaline intrusions

Numerous lead, zinc, and copper sulphide deposits are associated with high-level, mafic to felsic calc alkaline intrusions within Upper Ordovician-Devonian rocks of the Gaspé Belt (Chaleurs Bay division) in northwestern New Brunswick, and the Mascarene Belt in southern New Brunswick. Comagmatic volcanic rocks include basalt, andesite, and rhyolite. Three main types of deposits are recognized: 1) contact-metasomatic base metal sulphide; 2) vein-type base metal sulphide; and 3) porphyry copper-molybdenum-gold.

Contact metasomatic base metal deposits

A large number of contact metasomatic deposits are associated with small stocks of mainly intermediate composition that cut calcareous sedimentary rocks in the Aroostook-Percé division of the Gaspé Belt (Fig. 9.76). Typical of these is the Patapédia deposit, which occupies a brecciated and fractured skarn zone immediately east of the north-trending Restigouche Fault that cuts Upper Ordovician to Lower Silurian calcareous rocks of the Matapédia Group. The principal sulphides in the deposit are pyrrhotite, pyrite, and chalcopyrite with minor sphalerite and galena (Williams-Jones, 1982). Arsenopyrite is the main mineral in a breccia zone in the southern part of the deposit, but elsewhere it is fairly minor. The highest concentrations of sulphides coincide with the most intense metasomatism and with potash alteration in associated dykes. The Patapédia deposit is part of a large skarn zone that straddles the New Brunswick-Quebec border and includes the Mid-Patapedia deposit in Quebec.

Other skarns in the Matapédia Group (e.g. Burntland Brook, Popelogan, Fig. 9.76) comprise similar siliceous skarns carrying pyrite, pyrrhotite, chalcopyrite (±sphalerite, galena, molybdenite, scheelite, magnetite, arsenopyrite) associated with felsic stocks, dykes, and sills. The Madran deposit, 2 km east of the Antinouri Lake Granite (Fig. 9.77), comprises conformable lenses of massive to disseminated pyrite, pyrrhotite, chalcopyrite, sphalerite, silver minerals, and minor gold in a skarn zone within the Ordovician Elmtree Formation (Tetagouche Group) sedimentary rocks of the Dunnage Zone.

Base metal sulphide veins

The most prominent base metal sulphide vein system in the Chaleurs Bay division of the Gaspé Belt, the Nigadoo River deposit (Fig. 9.77), is located 17 km northwest of the city of Bathurst. Total tonnage was estimated to be 3 000 000 tonnes grading 0.23% Cu, 2.88% Pb, 2.84% Zn, and 125.84 g/t Ag (McAllister and LaMarche, 1972). The metallic minerals include pyrite, pyrrhotite, galena, sphalerite, chalcopyrite, and minor cassiterite, stannite, and gold, hosted by northwest-trending quartz-carbonate veins in Silurian calcareous metasedimentary rocks and a quartz-feldspar porphyry stock (Tupper et al., 1969). Sericite and kaolinite are ubiquitous alteration minerals. Structures in the mineralized zone are consistent with dextral movement along the Rocky Brook-Millstream fault system to the south.

Northwest-striking base metal sulphide veins (Halfmile, Pine Tree, Hachey, and Shaft deposits) associated with granodiorite porphyry intrusions are exposed in old prospecting trenches south of the Nicholas Denys stock along the trace of the Rocky Brook-Millstream Fault (Davies et al., 1969).

Porphyry copper-molybdenum-gold deposits

The largest known porphyry copper-molybdenum-gold deposit in New Brunswick occurs in the Connell Mountain granodiorite stock southeast of Woodstock (Venugopal, 1979; Thomas and Gleeson, 1988; Fig. 9.75). This stock is probably a cupola of the Ordovician Gibson Granodiorite (Bevier, 1989) exposed farther south. Pyrite, pyrrhotite, chalcopyrite, and minor molybdenite and gold occur as

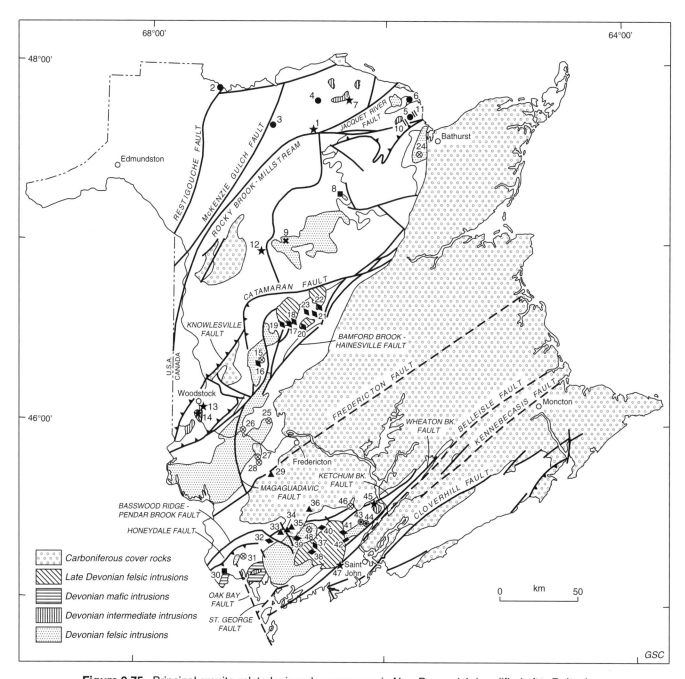

Figure 9.75. Principal granite-related mineral occurrences in New Brunswick (modified after Ruitenberg and Fyffe, 1982).

DEPOSITS RELATED TO SUBVOLCANIC CALC - ALKALINE INTRUSIONS

Contact metasomatic base metal deposit . . . . ●
- 2 - Patapedia
- 3 - Burntland Brook
- 4 - Popelogan
- 5 - Beresford
- 6 - Madran

Base metal sulphide vein . . . . \\
- 10 - Nicholas Denys
- 11 - Nigadoo River

Porphyry copper-molybdenum-gold deposit . . . . . .★
- 1 - Upsalquitch Forks
- 7 - Benjamin River
- 12 - Mount Costigan
- 13 - Connell Mountain
- 14 - Bulls Creek / Cobbler Sexton
- 45 - Evandale
- 47 - Eagle Lake

DEPOSITS RELATED TO BIMODAL BATHOLITHIC INTRUSIONS

Nickel-copper sulphide deposit . . . . ■
- 8 - Goodwin Lake
- 30 - St. Stephen

Base metal sulphide vein . . . . ✖
- 9 - Long Lake

Antimony-gold and tungsten-molybdenum vein . . . ⊗
- 15 - Sisson Brook
- 24 - Pabineau Lake
- 25 - Zealand Station
- 26 - Coac Stream
- 27 - Lake George W
- 28 - Lake George Sb-Au
- 31 - Waweig
- 46 - Pendar Brook
- 48 - Jimmy Hill / Otter Lake South

DEPOSITS RELATED TO HIGH SILICA GRANITES

Contact metasomatic base metal deposit . . . . ◆
- 42 - Nerepis

Base metal sulphide vein . . . . ✳
- 43 - Reserve Brook
- 44 - Johnson Croft

Tungsten-molybdenum-tin deposits
greisen and related endogranitic vein . . . . ◆
- 16 - Sisson Brook
- 17 - Burnthill
- 18 - Tin Hill
- 19 - Two and a Half Mile Brook
- 20 - Todd Mountain
- 21 - Falls Brook and Rocky Brook
- 22 - Little Dungarvon
- 23 - Sisters Brook
- 32 - Pleasant Ridge
- 37 - Mistake Lake
- 38 - Victoria Lake
- 39 - Deer Lake
- 40 - Howard Mountain
- 41 - Square Lake

Granophile / base metals in
subvolcanic stockwork and vein . . . . ▲
- 29 - Harvey (U)
- 33 - True Hill
- 34 - Mount Pleasant
- 35 - Mud Lake
- 36 - McNamara Prospect

Legend to Figure 9.75

disseminations and veinlets in intensely fractured and brecciated parts of the intrusion. Argillic and phyllic alteration are associated with the mineralized zone which averages 0.26% Cu (Lockhart, 1970). Other similar occurrences, which locally contain galena and sphalerite, have been found along the margin of the Gibson Granodiorite (e.g. Bulls Creek, Cobbler-Sexton deposits; Anderson 1968).

The Benjamin River porphyry copper-molybdenum deposit in northeastern New Brunswick (Fig. 9.76) occurs in the Benjamin River intrusive suite, a differentiated sequence of gabbro, granodiorite, and monzogranite that intrudes Silurian mafic and felsic volcanic rocks in the Gaspé Belt (Stewart, 1978). Pyrite, pyrrhotite, chalcopyrite, bornite, tetrahedrite, sphalerite, and minor molybde-

nite, associated with potassic, phyllic, argillic, and propylitic alteration occur within and along the contact of the granodioritic phase. The deposit is estimated to contain about 23 000 000 tonnes grading 0.18% Cu and 0.01% Mo.

The Upsalquitch Forks gold prospect, farther to the southwest, is associated with altered Upper Silurian-Lower Devonian (McCutcheon and Bevier, 1990) diorite and minor felsic porphyry intrusions near the intersection of the Rocky Brook-Millstream and Jacquet River faults. Disseminated pyrite, chalcopyrite, sphalerite, galena, and gold occur in quartz-carbonate veins in the porphyry.

A subvolcanic porphyry-type deposit, located at Mount Costigan on the southeast margin of the Gaspé Belt (Fig. 9.75), is hosted by Lower Devonian felsic volcanic rocks intercalated with shallow marine sedimentary rocks.

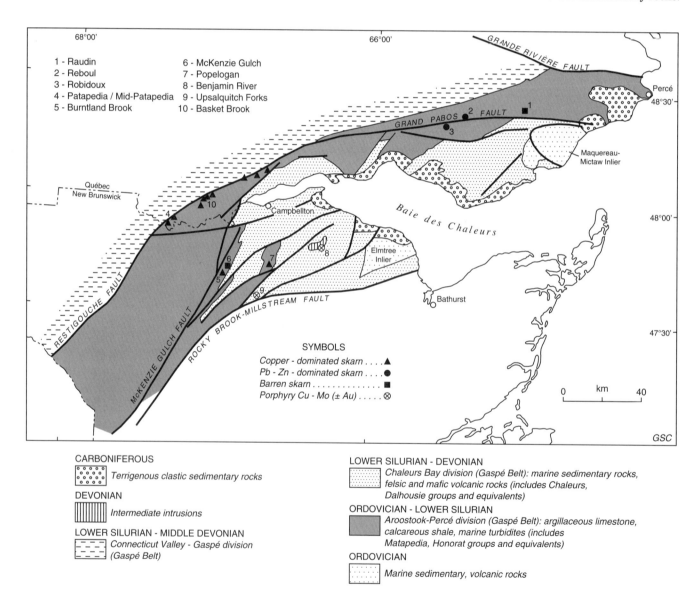

Figure 9.76. Simplified geology of northwestern New Brunswick and southern Gaspésie showing principal skarn deposits related to granitoid intrusion along major faults and porphyry-type deposits related to intermediate intrusive suites.

The Mount Costigan deposit forms a shallow, east-plunging breccia pipe about 200 m by 250 m in cross-section (Crevier, 1984; Fyffe and Pronk, 1985). Pyrite, sphalerite, galena, and minor chalcopyrite occur in a stockwork of veinlets, disseminations, and breccia fillings; the metallic minerals are coarse grained in the cavities of the breccias. Felsic flows and tuff, which surround the breccia pipe, are chloritized, sericitized, and silicified up to several hundred metres from the breccia pipe. Fluorite is common throughout the altered zone, but diminishes rapidly with increasing distance from the pipe.

Economically interesting copper occurrences carrying minor gold are associated with the Eagle Lake biotite granite porphyry stock in the Avalon Zone (Ruitenberg, 1969; Shewman, 1968; Fig. 9.75; 9.78). The stock is composed of an equigranular core and a porphyritic marginal phase that is intruded by numerous aplite dykes (Butt, 1976). Sulphide-bearing quartz veins and mineralized stringers with associated phyllic and argillic alteration are confined to highly fractured parts of the intrusion. Pyrite and chalcopyrite are the most abundant metallic minerals; tetrahedrite and galena occur mostly as veinlets cutting chalcopyrite.

An extensive copper-molybdenum-gold zone is located in the northwestern part of the Devonian Evandale Stock (Fig. 9.78), which intrudes Silurian sedimentary and vol-canic rocks in the southeastern Mascarene Belt (Ruitenberg, 1970; McCutcheon and Ruitenberg, 1987). The stock is composed of hornblende-bearing, biotite granodiorite and granite and carries mineralized fractures and quartz stringers that contain pyrite, chalcopyrite, molybdenite, and locally gold. Quartz veins with chalcopyrite, sphalerite, galena, and molybdenite occur in some fractures and locally contain varying amounts of scheelite.

Deposits related to bimodal batholiths

Three types of deposits are associated with gabbroic and granitic batholiths with mixed I- and S-type affinities in New Brunswick: (1) nickel-copper sulphide deposits; (2) base metal sulphide veins; and (3) antimony-gold and tungsten-molybdenum veins.

Nickel-copper sulphide deposits

Nickel-copper sulphides are associated with layered mafic intrusions (peridotite, troctolite, olivine gabbro, anorthosite, and norite; Paktunc, 1986, 1987a, b, 1988) in the St. Stephen area of the southwestern St. Croix Subzone (Fig. 9.78), and the Goodwin Lake area of the northern Miramichi Subzone (Fig. 9.75). The post-tectonic St. Stephen intrusion was emplaced at about 415 ± 19 Ma

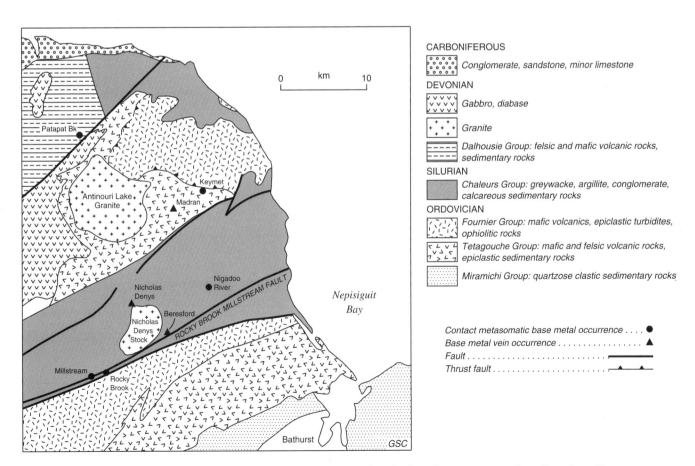

Figure 9.77. General geology and principal granite-related mineral occurrences of northeastern New Brunswick (simplified after Davies, 1979).

773

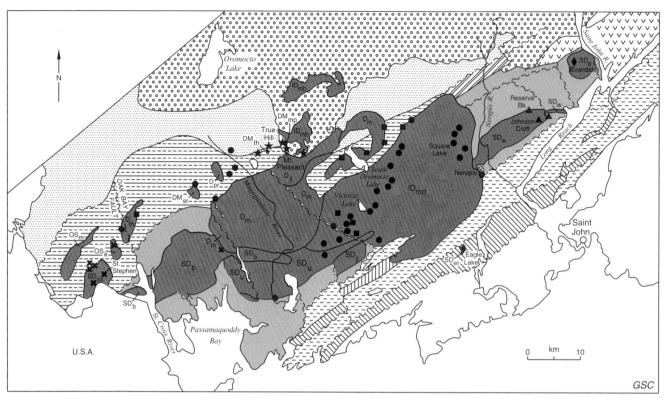

STRATIFIED ROCKS

LOWER DEVONIAN TO CARBONIFEROUS

Carboniferous

Hopewell, Petitcodiac groups: red and grey sedimentary rocks; Piskahegan Group: terrestrial volcanic rocks

SILURIAN TO LOWER DEVONIAN

Mascarene Group: shallow marine to terrestrial volcanic and sedimentary rocks

Fredericton Group: mainly turbidites, minor conglomerate, limestone

Kingston Dyke Complex

ORDOVICIAN ± SILURIAN

Queensbrook Formation: marine sedimentary rocks, minor volcanics

St. Croix Subzone

Digdeguash Formation; marine sediments, minor volcanic rocks

Mafic and felsic volcanics, lesser sediments

HADRYNIAN - CAMBRIAN

Avalon Zone

Matthews Lake beds: marine sediments and felsic volcanics

Granite, granodiorite and gabbro; volcanic and sedimentary rocks

INTRUSIVE ROCKS

LATE DEVONIAN TO MISSISSIPPIAN

DM | *Biotite granite (mp - Mount Pleasant porphyry; th - True Hill granite; bh - Beech Hill Granite; sr - Sorrel Ridge Granite)*

LATE DEVONIAN

ID | *Biotite granite (md - Mount Douglas granite; mb - MacDougall Brook porphyritic microgranite)*

EARLY - MIDDLE DEVONIAN

D | *Biotite granite (m - Magaguadavic granite); biotite - muscovite granite (jl - John Lee Brook granite; pr - Pleasant Ridge Granite; th - Tower Hill Granite)*

SILURIAN - DEVONIAN

SD | *Syenite (w - Weslford alkali granite, jl - Jake Lee Mountain Complex); biotite granite (er - Eagle Lake Granite), biotite - muscovite granite(u - Utopia Granite); granodiorite - quartz diorite (e - Evandale granodiorite); gabbro - (b - Bocabec Gabbro, ss - St. Stephen Gabbro)*

ORDOVICIAN - SILURIAN

OS | *m - Mohannas granodiorite; a - Allen Brook Gabbro*

MINERAL OCCURRENCES

Polymetallic sulphide veins
(Pb, Cu, Ag) . . . . ▲

Subvolcanic breccias and stockwork veins
(W, Mo, Sn, Bi, Zn, Cu, Pb, In, Nb, Ag) . . . ★

Greisen veins and pods
(Sn, Mo, W, Cu) . . . . ●

Contact metasomatic deposits
(W, Au, Mo, Cu, Zn, Pb, Ag, Bi) . . . . ⊗

Porphyry deposits
(Cu, Mo, W, Zn, Pb, Ag, Au, Bi) . . . . ◆

Intramagmatic deposits in felsic intrusions
(Mo, Cu) . . . . ■

Intramagmatic deposits in mafic intrusions
(Ni, Cu) . . . . ✖

Figure 9.78. Regional geology of southwestern New Brunswick and granite-related mineral occurrences (after McLeod et al., 1994a, b).

(Wanless et al., 1973) in a dilatant zone west of the Oak Bay sinistral wrench fault (Ruitenberg, 1967; Ruitenberg and Ludman, 1978). The associated deposits (Hale, 1950; Dunham, 1950; Paktunc, 1989) are composed of nickeliferous pyrrhotite, pentlandite, and chalcopyrite in massive bodies, disseminated blebs, matrix between olivine crystals, and remobilized sulphides hosted by olivine gabbro and gabbro. About 907 000 tonnes averaging 1.03% Ni, 0.47% Cu, and 0.10% Co have been delineated (Potter, 1967).

A large Devonian gabbroic sill in the Goodwin Lake area of the northern Miramichi Subzone (McNutt, 1962; Paktunc, 1988) contains pods of pyrrhotite, pentlandite, chalcopyrite, and minor cobaltite and bismuthinite. One drillhole is reported to have intersected 2.3 m grading 0.88% Ni and 0.53% Cu.

Base metal sulphide veins

The Long Lake deposit of central New Brunswick (Fig. 9.75) occurs in several northwest-striking fracture zones that cut the western lobe of the North Pole Stream pluton, comprising peraluminous biotite granite with less abundant biotite-muscovite granite and minor cordierite-bearing granite and porphyritic biotite granite dykes. None of these phases exhibits the geochemical specialization characteristic of the highly differentiated leucogranites associated with tin, tungsten, and uranium deposits (Strong, 1980). Base metal sulphides are found in the granite and country rocks as disseminations and stringers of pyrite with minor sphalerite, chalcopyrite and galena in fractured and silicified host rocks (Hauseux, 1980). Base metal sulphides within the granite are associated with quartz veins and quartz-feldspar porphyry dykes along west-northwest-trending fracture zones, suggesting that hydrothermal activity related to sulphide deposition was at least in part coeval with dyke emplacement. Uranium and molybdenum minerals occur within poorly exposed silicified granite breccia hosted by biotite-muscovite granite along a west-northwest-trending fault zone.

Antimony-gold and tungsten-molybdenum veins

The Lake George antimony-gold deposit is situated southeast of the Upper Silurian to Lower Devonian Pokiok Batholith, which intrudes the faulted boundary between Cambrian-Ordovician sedimentary rocks of the Miramichi Subzone and Silurian turbidites of the Fredericton Belt. The Pokiok Batholith contains four major units of variably textured hornblende-biotite granodiorite, and biotite and biotite-muscovite granite. Major and trace element analyses of these rocks (Lutes, 1987; Ruitenberg and Fyffe, 1982) show that the Pokiok Batholith lacks extreme differentiation and enrichment in lithophile elements.

The Lake George deposit comprises a system of stibnite, native antimony, and gold-bearing quartz veins that cut Silurian wacke and slate. Early scheelite-gold-bearing quartz-chlorite veins occur in the contact aureole of a small granodiorite stock adjacent to the Pokiok Batholith (Fig. 9.79), whereas the younger antimony-gold deposits cut across the stock. Stibnite occurs as irregular-shaped lenses and veins that are mainly concentrated in gently northeast-striking flexures in the north-dipping Hibbard

vein. The flexures occur where the vein cuts competent beds in the vicinity of regional fold hinges. The stibnite-bearing veins and bordering thin siliceous rinds transect a broad argillic alteration zone with pyrite, pyrrhotite, arsenopyrite, and bismuthinite. A separate extensive magnetite-quartz oxidation zone also postdates the argillic alteration. The stibnite-bearing quartz veins locally cut quartz-carbonate veins enveloped by greenish grey arsenopyrite-rich alteration halos; gold, and locally bismuthinite are commonly associated with the arsenopyrite. Gold is also associated with low-grade scheelite in early quartz-chlorite veins in parts of the contact aureole of the granodiorite stock. A general consensus from recent studies (e.g. Watson,

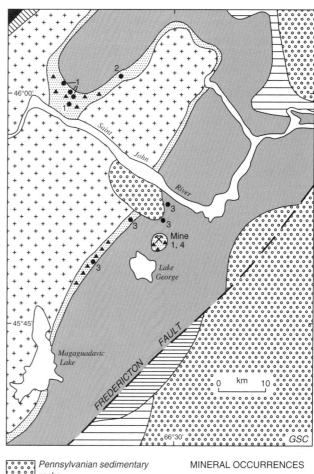

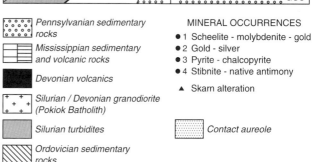

Figure 9.79. General geology and mineral occurrences in the area of the Lake George mine, southwestern New Brunswick (after Caron, 1991).

1981; Scratch et al., 1984; Seal, 1984) is that the scheelite mineralization is related to the emplacement of the granodiorite stock. However, there is still controversy as to whether Sb mineralization resulted from leaching of antimony- and gold-enriched shales through convection of connate seawater or from hydrothermal fluids generated during crystallization of a granodioritic magma that had been enriched in Sb-Au by assimilation of country rock (Scratch et al., 1984).

Scheelite-bearing stringers similar to those found at Lake George also occur in the Coac Stream area north of Saint John River and along a prominent fault in the Pendar Brook area in the northeastern St. Croix Subzone (Fig. 9.75). A rather uncommon occurrence along the northeast contact of the Pokiok Batholith at Zealand Station has disseminated molybdenite with abundant beryl in quartz-orthoclase pegmatite.

Gold, with or without arsenopyrite, also occurs in the contact aureoles of Devonian mafic and felsic plutons that intrude Lower Ordovician pelitic sedimentary and mafic volcanic rocks in the Waweig Stream area in the St. Croix Subzone.

The Jimmy Hill and Otter Lake South gold-bearing quartz-carbonate vein systems (McLeod, 1990) were discovered recently in the northeastern St. Croix Subzone. The Jimmy Hill veins, dark grey quartz with pyrite, local wolframite and bismuthinite, and fine grained visible gold with traces of molybdenite, cut a hornfelsed greywacke and slate roof pendant in the Lower Devonian Magaguadavic Granite. At Otter Lake South, the quartz-carbonate veins cut similar metasedimentary rocks intruded by quartz-feldspar porphyry dykes. They contain pyrite, pyrrhotite, chalcopyrite, bismuthinite, stibnite, arsenopyrite, silver minerals, and gold (Ruitenberg et al., 1989, 1990).

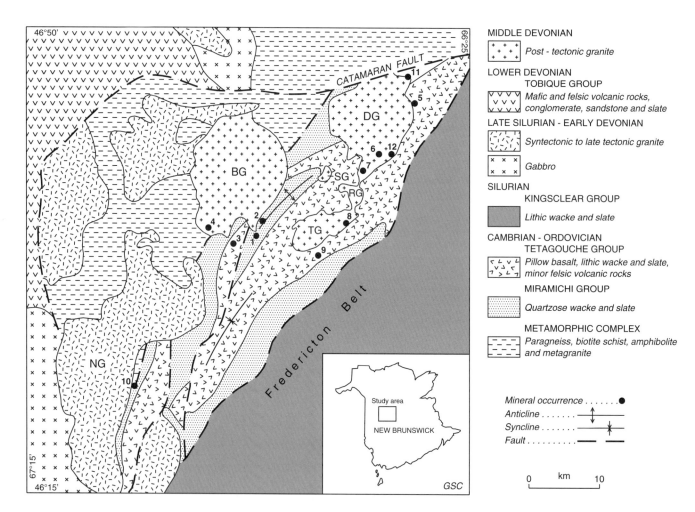

Figure 9.80. Regional geological setting of granites and related mineral occurrences in the Burnthill Brook area (after MacLellan et al., 1990). BG – Burnthill Granite; DG – Dungarvon Granite; NG – Nashwaak Granite; RG – Rocky Brook Granite; SG – Sisters Brook Granite; TG – Trout Lake Granite; Mineral occurrences: 1 – Burnthill tungsten (W-Mo-Sn); 2 – Tin Hill (Sn, W); 3 – Two and a Half Mile Brook (Sn, W); 4 – McLean Brook (W, Mo); 5 – Peaked Mountain (W, Mo); 6 – Dungarvon (Sn); 7 – Four Mile Brook (Sn); 8 – Trout Lake (Sn, W); 9 – Todd Mountain (Sn); 10 – Sisson Brook (W, Cu, Mo); 11 – Fall Brook (Sn,W); 12 – Cleveland (Sn).

The Sisson Brook W-Cu-Mo deposit (Fig. 9.80) occurs in the thermal aureole of the Devonian Nashwaak pluton along its contact with lower Paleozoic metagabbro, mafic and felsic metavolcanics, and metasediments (Lutes, 1981; Ruitenberg and McCutcheon, 1985; Fyffe et al., 1988; Nast and Williams-Jones, 1991). The mineralization comprises four steeply dipping vein sets contained in two mineralized zones: Zone "A" to the north consists of two steep-dipping vein- and quartz-stockwork-filled fracture zones in lower Paleozoic felsic and mafic metavolcanic and metasedimentary rocks, Zone "B" to the south consists of several north-northeast-trending tungsten-bearing zones and is transected by a molybdenum-rich zone and hosted mainly by gabbro. A multistage mineralization and alteration history has been documented by Nast and Williams-Jones (1991) in which early disseminated molybdoscheelite associated with amphibole alteration was followed by scheelite and molybdenite in quartz veins associated with biotite alteration and, finally, wolframite and chalcopyrite in quartz vein stockworks associated with phyllic alteration.

Deposits related to high-silica granites

Introduction

Peralkaline and silica-rich subalkaline granites occur as a cluster of plutons in the Burnthill area of central New Brunswick, as part of the Saint George Batholith in southern New Brunswick, and as satellite stocks along the northern boundary of the Saint George Batholith. These Upper Silurian to Upper Devonian granites have features similar to A-type granites (Loiselle and Wones, 1979; Rogers and Greenberg, 1981; Collins et al., 1982; Whalen, 1986). Three distinct types of deposits are associated with these plutons: (1) contact metasomatic base metal deposits in the Mascarene Belt; (2) base metal sulphide veins associated with Upper Silurian peralkaline granitic rocks in the Mascarene Belt; and (3) tungsten-molybdenum-tin-indium-bismuth deposits associated with Middle to Upper Devonian silica-rich and subalkaline granites in the central Miramichi and St. Croix subzones.

Contact metasomatic base metal deposits

The only known contact metasomatic base metal sulphide occurrence in southern New Brunswick is associated with the Upper Devonian Mount Douglas Granite at Nerepis in the Mascarene Belt (Ruitenberg, 1969 (Fig. 9.78)). Pods and small lenses of pyrite, pyrrhotite, sphalerite, galena, chalcopyrite, and tin minerals occur in a siliceous garnetiferous skarn zone in Upper Silurian calcareous sedimentary rocks.

Base metal sulphide veins

Typical base metal sulphide veins and breccia fillings are associated with northwest-trending faults that cut the peralkaline Welsford Complex at Reserve Brook and Johnson Croft in the southeastern Mascarene Belt (Fig. 9.78). Sphalerite, argentiferous galena, minor chalcopyrite, tetrahedrite, siderite, and quartz occur as irregular veins and pods in shear zones and breccias along the contact of the complex (Ruitenberg, 1970; McCutcheon and Ruitenberg, 1987).

Tungsten-molybdenum-tin deposits

One of the most extensively explored tungsten-molybdenum-tin deposits in the Miramichi Subzone is the Burnthill deposit (Fig. 9.80). It consists of numerous tungsten-, molybdenum-, and tin-bearing greisen veins associated with the Middle Devonian Burnthill Granite (Victor, 1957; Potter, 1969; Ruitenberg and McCutcheon, 1985; MacLellan et al., 1990). The Burnthill pluton and several others to the east (Rocky Brook, Sisters Brook, Trout Brook, and Dungarvon) occur just to the south of the Catamaran Fault.

The numerous W-Sn-Mo-Be-Fe-bearing quartz veins of the Burnthill deposit occur within the apical portion of a cupola on the southern margin of the Burnthill pluton, and in the quartz wacke and argillaceous metasedimentary rocks of the Miramichi Group. The steeply dipping vein swarms pinch and swell both vertically and horizontally, and are commonly branching and anastomosing. Potter (1969) distinguished two types of ore-bearing veins at Burnthill: (a) those with quartz, muscovite, minor beryl, molybdenite, wolframite, fluorite, scheelite, and cassiterite; and (b) those with quartz, topaz, minor wolframite, pyrrhotite, chalcopyrite, pyrite, arsenopyrite, fluorite, sphalerite, molybdenite, beryl, and cassiterite.

At Tin Hill, coarse cassiterite crystals (up to 2 cm in diameter), wolframite, magnetite, and pyrite occur in quartz veins along the western contact of a coarsely porphyritic granite cupola on the southeastern margin of the the Burnthill pluton. At Two And A Half Mile Brook, cassiterite- and wolframite-bearing quartz veins occur in quartz wacke and siltstone in the thermal aureole of the Burnthill pluton (Crosby, 1984). Similar W-Sn-bearing veins occur at Todd Mountain in the thermal aureole of the Trout Brook Granite (Wright, 1940; e.g. Sisters Brook, Falls Brook, Rocky Brook).

The endogranitic Little Dungarvon River prospect comprises an extensive zone of disseminated cassiterite in the Dungarvon pluton; some coarser cassiterite crystals line miarolitic cavities in the granite (Gardiner, 1985). High grade, narrow, tin-bearing veins cut medium grained biotite granite southwest of the Little Dunvargon River prospect. The veins consist of feldspar, quartz, cassiterite, fluorite, and beryl, and are associated with hematitic and potassic alteration. The cassiterite crystals exhibit dark brown unzoned cores enveloped by honey-coloured zoned rims.

In southwestern New Brunswick, tin-, tungsten-, and molybdenum-bearing greisen deposits are associated with the Upper Devonian Mount Douglas Granite, the eastern portion of the Saint George Batholith, and satellite plutons along the northern contact of the batholith (Fig. 9.78). Sheeted greisen vein systems are mainly concentrated in the Victoria Lake and Square Lake areas.

A number of tin, tungsten, and molybdenum deposits (e.g. True Hill, Mount Pleasant deposits) are associated with several satellite intrusions of Late Devonian to Early Carboniferous age, along the northern margin of the Saint George Batholith (Fig. 9.78).

The Mount Pleasant deposits (the North and Fire Tower zones; Kooiman et al., 1986; Sinclair et al., 1988) are situated along the western margin of the Mount Pleasant caldera (Ruitenberg, 1967; McCutcheon, 1984, 1990b).

Tungsten-molybdenum and tin (±base metal sulphides, indium and bismuth) are spatially associated with early fine grained granite and later granite porphyry intrusions emplaced in the Upper Devonian volcanic sequences. In the Fire Tower orebody, greisenized breccias that host the early porphyry tungsten-molybdenite deposits grade downward through microgranite breccias into a stockwork-veined granite porphyry (Kooiman et al., 1986). Two distinct types of later tin deposits are related to the granite porphyry: (1) tin-base metal veins and replacement bodies that are superimposed on the tungsten-molybdenum deposits of the granite porphyry; and (2) endogranitic tin and relatively minor base metal-sulphide-bearing greisen that forms irregular bodies at lower levels within the granite porphyry. The mineralogy of the tungsten-molybdenum deposits is characterized by wolframite, molybdenite, native bismuth, bismuthinite, arsenopyrite, and loellingite. The tin-base metal ores comprise an assemblage of cassiterite, sphalerite, galena, arsenopyrite-loellingite, pyrite, stannite, tennantite, tetrahedrite, and bornite (Ruitenberg, 1963, 1967; Boorman, 1968; Petruk, 1964, 1973). Varying amounts of indium are associated mainly with sphalerite. The main alteration minerals are quartz, fluorite, topaz, and chlorite.

Tungsten-molybdenum deposits in the North Zone comprise extensive mineralized fracture zones in breccia and granite. Wolframite, molybdenite, bismuth, bismuthinite, arsenopyrite, and loellingite are the most abundant minerals. A variety of tin deposits at different levels within the North Zone post-date the tungsten-molybdenum mineralization. Total tin reserves in the North Zone are 5 900 000 tonnes averaging 0.79% Sn; tungsten-molybdenum reserves in the North Zone are approximately 11 500 000 tonnes grading 0.20% W, 0.06% Mo, and 0.09% Bi (The Northern Miner, October 14, 1985). The True Hill deposit is associated with three granite porphyry cupolas within wacke and slate. Moderate to intense greisenization affected the apical parts of the cupolas, and hydrothermal pebble breccias and breccia dykes are developed locally along their flanks. In places, near the top of the cupolas, a contact breccia is developed between the granite porphyry and the sedimentary host rocks. The breccias at True Hill resemble those of the subvolcanic Mount Pleasant deposit (Ruitenberg, 1967; Lentz et al., 1988). Two types of mineral assemblages are present in the True Hill deposit: disseminated tungsten, molybdenum, and bismuth minerals hosted by greisen in both granite and adjacent metasedimentary rocks; and tin-sulphide lodes following faults, joints, and porous greisenized granite mainly in the granite porphyry. The principal minerals are cassiterite, sphalerite, chalcopyrite, galena, stannite, and native silver together with pyrite, iron-manganite, chlorite, quartz, and fluorite.

Gaspésie

G. Duquette

Northern Gaspésie

Polymetallic base metal deposits in northern Gaspésie are related to a cluster of intrusions that cut both volcanic and sedimentary rocks of the Taconic allochthons in the Humber Zone and the younger cover rocks of the Gaspé Belt

(Fig. 9.81). These include the large McGerrigle Mountains Pluton and numerous smaller high level plutons between the McGerrigle Mountains and Murdochville areas. All of the granites related to mineralization are hornblende-bearing with "I" type characteristics. Two geochemically distinct suites are present: the McGerrigle Mountains plutonic complex comprises a suite of Lower Devonian (391 ± 3 Ma; Whalen et al., 1991) intrusive rocks ranging from gabbro to granite and syenite (de Römer, 1977; Whalen and Gariepy, 1986; Whalen and Roddick, 1987) whereas smaller, probably younger (J. Whalen, pers. comm., 1991) intrusions in the Murdochville area are dominantly granitic.

Mineralization includes copper-rich skarns (e.g. Needle Mountain at the Gaspé Mine and Sullipek) and mesothermal veins and disseminations (e.g. Madeleine) in the metamorphosed country rocks as well as porphyry-type mineralization within the plutons (e.g. Copper mountain at the Gaspé Mine).

The Gaspé Mine, located at Murdochville, produced approximately 124 million tonnes of ore with 0.64% Cu and recoverable molybdenite, silver, bismuth, selenium, tellurium, and gold between 1955 and 1991 with interruptions in 1983-1984 and 1987-1989. At the end of 1990, reserves were reported to be 26.62 million tonnes (0.99% Cu) distributed as follows: 6.44 million tonnes (2.72% Cu) at Needle Mountain and 20.18 million tonnes of oxidized (malachite) and stockpiled ore (0.44% Cu) at Copper Mountain.

In the vicinity of the Gaspé Mine, calc-silicate hornfels and skarn of the Lower Devonian Grande Grève and Cap Bon Ami formations lie on the southern flank of the Champou syncline and surround the Copper Mountain plug, a Devonian (Early?) quartz porphyry intrusion. Progressive metamorphism toward the plug is recorded by the successive appearance of phlogopite, tremolite, diopside, wollastonite, and grossularite in the bleached limestones. After metamorphism, a few limestone units near the plug were converted into andradite-rich skarns (Alcock, 1982). Thereafter, two episodes of copper mineralization affected, to various degrees, all metamorphosed units. The earlier episode led to the formation of skarn-type orebodies and the later episode to porphyry copper-type deposits.

The skarn-type orebodies, known as the Needle Mountain orebodies, are characterized by disseminations and lesser fracture fillings and massive replacements of chalcopyrite, bornite and pyrrhotite. The host limestone shows widespread sodic alteration. The skarn-type orebodies are essentially stratiform and occur up to 2.5 km from the Copper Mountain plug. The limestone beds carrying the skarn-type orebodies grade laterally, toward the plug, into the previously mentioned andradite-rich units.

The porphyry copper-type mineralization is mainly represented by the Copper Mountain orebody, a stockwork, centred on the Copper Mountain plug, that contains more than 200 million tonnes grading a little more than 0.4% Cu and 0.02% Mo. Within the stockwork, both the metamorphosed limestones and the porphyry are enriched in K_2O (biotite and adularia) and contain fractures filled by pyrite, chalcopyrite, molybdenite, quartz, and anhydrite. Trace amounts of scheelite, fluorite, sphalerite, and galena occur throughout the mine and retrograde metamorphic minerals such as epidote and chlorite are common.

The Sullipek prospect, located approximately 30 km west of the Gaspé Mine, occupies a melanoskarn derived from uppermost Silurian marble. The skarn is intruded by abundant Devonian (Early?) felsic dykes and sills and is heavily mineralized with pyrite, pyrrhotite, magnetite, hematite, andradite, and chalcopyrite. The dykes, and to a lesser extent the melanoskarn, are locally cut by swarms of quartz veins that carry molybdenite and a little chalcopyrite. Retrograde epidote and chlorite are abundant in the wallrocks of the melanoskarn.

As at the Gaspé Mine, the host rocks to the Sullipek prospect have been regionally metamorphosed, intruded by felsic dykes, metasomatized (skarn) and affected by two periods of mineralization: an earlier event that formed a chalcopyrite-rich lode and a later event that formed molybdenite-bearing quartz veins (Duquette, 1983). Potassic and sodic alteration in association with mineralization have also been noted (Wares, 1983). Four ore shoots have been delineated totalling 540 000 tonnes and averaging 1.4% Cu, .025% MoS_2, and 7 g/t silver (Wares, 1983).

The former Madeleine Copper Mine lies within the northwestern segment of the metamorphic aureole of the McGerrigle Mountains Pluton. The Madeleine deposit yielded some 7.3 million tonnes grading a little more than 1.0% Cu and 7.88 g/t Ag between 1969 and 1982. The host rocks are Lower Ordovician allochthonous pelitic hornfels and magnetite-bearing metagreywacke. Copper occurs

where the metasedimentary rocks have been highly fractured and faulted, presumably during the emplacement of the McGerrigle Mountains Pluton. Ore came largely from the Main and South Footwall No. 1 zones. The Main zone is a chimney-like stockwork with a roughly elliptical shape up to 240 m in length that extends a vertical distance of about 400 m. Mineralization consists of minute veinlets of pyrrhotite, chalcopyrite, bornite, and argentiferous covellite; the latter two minerals clearly predominate in the core of the zone. The South Footwall No. 1 zone, which is approximately 165 m long and 480 m in depth, constitutes a near vertical, tabular orebody. It carries abundant crosscutting veinlets as well as sub-parallel veins of pyrrhotite and chalcopyrite.

Lead-zinc-silver±gold deposits (e.g. the Candego and Federal deposits) occur farther from the McGerrigle Mountains Pluton than the Madeleine deposit and may represent a distal lead-zinc or lower temperature zone to the Madeleine and Sullipek copper deposit areas.

The former Candego Mine northwest of the McGerrigle Mountains Pluton is reported to have yielded 68 497 tonnes grading 6.35% Pb, 4.28% Zn, 178 g/t Ag, and 0.5 g/t Au between 1948 and 1954. Ore came largely from two east-trending vertical veins within a longitudinal fault in Middle Ordovician allochthonous phyllite. Mineralization comprises pyrite, galena, sphalerite and minor pyrrhotite, chalcopyrite, tetrahedrite, and bournonite in a carbonate-

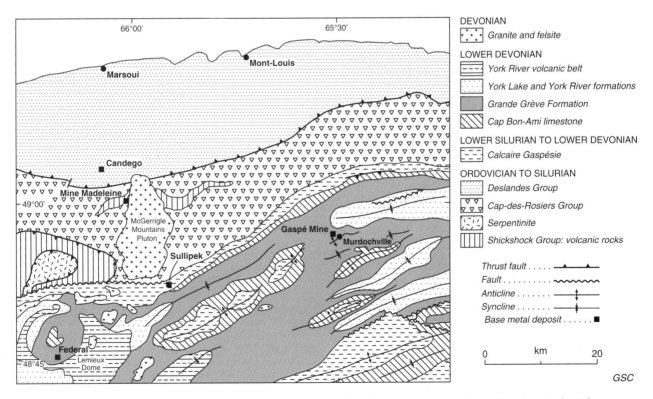

Figure 9.81. Regional geology of the central Gaspésie region and locations of granite-related mineral deposits (after Procyshyn et al., 1989).

779

quartz gangue. Gold follows pyrite whereas silver follows galena and, less commonly, tetrahedrite. A similar deposit, the Cromar vein, occurs some 2.4 km southeast of the Candego Mine. All are believed to have resulted from the filling of fault-related openings by hydrothermal solutions generated during emplacement of the McGerrigle Mountains Pluton.

The Federal prospect, to the southwest of the McGerrigle Mountains Pluton, comprises a series of near vertical quartz-carbonate-sphalerite-galena veins that intrude a block-faulted dome of Silurian and Lower Devonian limestones. Most veins are adjacent to felsic dykes which are believed to be offshoots of a deep-seated pluton. The vein system carries about 600 000 tonnes grading 3.95% Zn and 1.31% Pb

On surface, the mineral occurrences related to the McGerrigle Mountains Pluton are centred on the Shickshock-Sud Fault, in accord with the view (Lachance and Duquette, 1977) that this fault channelled both magmas that eventually formed the local granitic plutons and hydrothermal solutions that formed the metalliferous deposits. The same fault, or its easterly equivalent, might have played a similar role with regard to the Gaspé Mine.

Southern Gaspésie

A number of skarn localities have been identified along the Grand Pabos-Restigouche fault system in southern Gaspésie (Fig. 9.76). Skarns at the western end of the fault are Cu-rich, locally with silver, pyrrhotite, chalcopyrite, and pyrite (Williams-Jones, 1982; Savard, 1985), whereas those in the east are polymetallic, enriched in Zn, Pb, Ag, and Au along with Cu and contain abundant sphalerite, galena, and arsenopyrite (Savard, 1985; Malo et al., 1990). Cu skarns in southern Gaspésie are granite-related (e.g. Mid-Patapédia prospect, Williams-Jones, 1982) but recent study of the polymetallic style of mineralization at the Reboul prospect indicates that this mineralization is structurally controlled (Malo et al., 1990).

Copper-rich skarns in western Gaspésie include the Mid-Patapédia and Basket Brook occurrences. The Mid-Patapédia prospect is part of a continuous skarn zone that straddles the Quebec-New Brunswick border and includes the previously-described Patapédia deposit in New Brunswick. It consists of calc-silicate hornfels mineralized with pyrrhotite, pyrite, chalcopyrite and a little galena, sphalerite and younger arsenopyrite. The deposit is pipe-shaped and carries 4 million tonnes grading approximately 0.25% Cu and 5 g/t Ag (Ministére de l'Energie et des Ressources, 1983). The hornfels is derived from calcareous siltstones and sandstones and shaly limestones of the Upper Ordovician-Lower Silurian Matapédia Group. The mineralized rocks are associated with calc-alkaline dykes that are enriched in K_2O (Williams-Jones, 1982) and centrally located with respect to metamorphism. Metamorphism, dyke emplacement, and mineralization were all largely controlled by regional faulting. The Basket Brook prospect comprises a near-horizontal downward pointing wedge-shaped stockwork carrying veinlets of pyrrhotite, pyrite, chalcopyrite, and sphalerite with minor amounts of bornite and galena in an epidote, chlorite, sericite, quartz,

calcite, and garnet gangue. Felsic dykes everywhere border the mineralized hornfelses. More than 170 000 tonnes grading 0.57% Cu and 8.57 g/t Ag have been outlined by drilling (SOQUEM, 1975).

Silver-gold-lead-zinc-copper polymetallic skarn-type deposits related to the eastern part of the fault are represented by the Robidoux and Reboul prospects. The Robidoux prospect consists of pyrite, quartz, calcite, sphalerite, and galena hosted by metamorphosed siltstones and sandstones of the Honorat Group and limestones of the Matapédia Group that have been intruded by felsic dykes. To date only sub-economic values in silver, gold, lead, and zinc have been encountered (SOQUEM, 1972). At the Reboul prospect farther to the east, mineralization was preceded by a local metamorphic event producing marble and calc silicate hornfels and then a metasomatic event producing skarn. The mineralization comprises an early Cu-Zn-Ag event limited to areas of skarn, followed by intermediate Cu-Zn-Pb-Ag (Au)-bearing carbonate veins cutting all lithologies, and late Ag-Au-Zn-Pb quartz veins, hosted by marble, skarn and unaltered calcareous rock (Moritz et al., 1993).

The western and eastern skarn types share many similarities including the presence of an early high temperature Cu-mineralization (Williams-Jones, 1982) and formation from magmatic-meteoric fluids. Felsic dykes are commonly associated with the western skarns but rarely with the eastern skarns. Moritz et al. (1993) suggested that the western skarns were probably developed closer to an intrusion whereas the eastern skarns were more distal.

Southeastern Quebec

M. Gauthier

Introduction

Mineralization related to granite intrusions is present throughout the southeastern Quebec Appalachians and deposits of this type are also common in adjacent New England. Granophile mineralization is associated with three phases of granitoid intrusion: (1) possible porphyry-type occurrences are found in Ordovician intrusions that may have formed in the back-arc region to the Ascot-Weedon island arc; (2) granophile stockworks and veins are associated with late Paleozoic (Acadian) granitoids; and (3) auriferous occurrences with a Cu-Mo-Bi association that may be related to Mesozoic alkalic intrusions.

Back-arc felsic intrusions

There is a single occurrence of this type in the Estrie-Beauce Subzone. The Standard Asbestos occurrence comprises a siliceous stockwork in a muscovite granite which forms part of a fragment in the Saint-Daniel Mélange. The mineralization comprises a quartz-actinolite stockwork that contains molybdenite, chalcopyrite, and pyrite. Grab samples have yielded up to 0.68% MoS_2.

The granitoid that hosts the Standard Asbestos deposit is interpreted to be derived from the Chain Lakes terrane and the mineralization broadly resembles that in molybdenum-copper porphyry deposits at Catheart Mountain and Sally Mountain in Maine (Young, 1970; Harron, 1976; Atkinson, 1977). These deposits, in the Ordovician Attean quartz monzonite, display a typical porphyry-type metal zonation from molybdenite-rich core to chalcopyrite-rich rim, and extensive potassic and argillic alteration. Alteration at the Standard Asbestos occurrence is less well-developed, consisting mainly of silicification and sericitization.

Although there is little conclusive evidence as to the genesis of the Standard Asbestos occurrence, the most likely genetic model is that it represents porphyry-type mineralization, formed in a Colorado type environment behind the Ascot-Weedon arc (Gauthier et al., 1989)

Late Paleozoic granitoids

Two generations of late Paleozoic granitoids occur in southeastern Quebec. The first generation, emplaced along Acadian tectonic zones, includes the Saint-Sébastien and Sainte-Cécile granitic stocks as well as the Saint-Robert porphyry dykes. The Saint-Sébastien and Sainte-Cécile intrusions and the metamorphic aureole centred in the Saint-Robert dyke cluster are oval-shaped, with the long axis parallel to the faults.

Hydrothermal molybdenum deposits occur in the contact metamorphic aureole of the Saint-Sébastien and Sainte-Cécile stocks. The Copperstream-Frontenac deposit in Gayhurst township comprises a quartz stockwork, mineralized with molybdenite and minor pyrrhotite and chalcopyrite, that cuts hornfels zones. Proven reserves in two

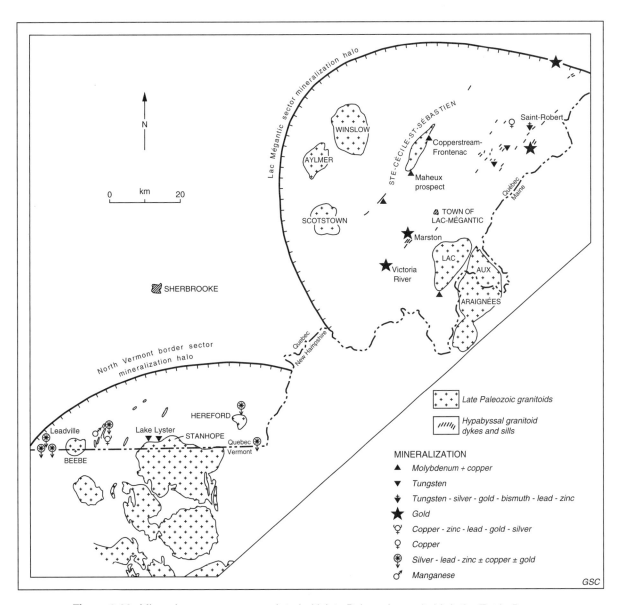

Figure 9.82. Mineral occurrences associated with late Paleozoic granitoids in the Estrie-Beauce region, southeastern Quebec.

zones total 2.7 million tonnes grading 0.52% MoS_2. The Maheux prospect in Whitton township comprises weak but widespread molybdenum, bismuth, silver, lead, and zinc mineralization in veins related to porphyritic dykes and a stockwork of quartz-molybdenite veins that cut Silurian-Devonian hornfels.

The Saint-Robert deposit consists of mineralized quartz veins that cut hornfels adjacent to porphyritic dykes. These veins have a complex paragenesis, composed of scheelite, chalcopyrite, pyrrhotite, pyrite, argentiferous galena, sphalerite, cosalite, gold, and molybdenite (Cattalani, 1987).

The second generation of Acadian plutons comprises two clusters of circular intrusions. The first, centred close to Lac-Mégantic, includes the Winslow, Aylmer, Scotstown, and Lac-aux-Araignées stocks; the second is an extension of Vermont's Northeastern Kingdom plutons and comprises the Beebe, Stanhope, and Hereford Mountains stocks (Fig. 9.82). These intrusions disappear toward the Connecticut Valley-Gaspé Synclinorium where small apophyses, dykes and sills represent their apices. Lead-, zinc-, and silver-bearing veins occur in the vicinities of these granitic cupolas. The Leadville mine on the west shore of Lake Memphremagog, carrying argentiferous galena with coarse pyrite in a quartz stockwork, is an example of such a vein. At Lake Lyster, garnet-diopside skarns carrying disseminated scheelite developed around the granitic stocks (Gauthier et al., 1989).

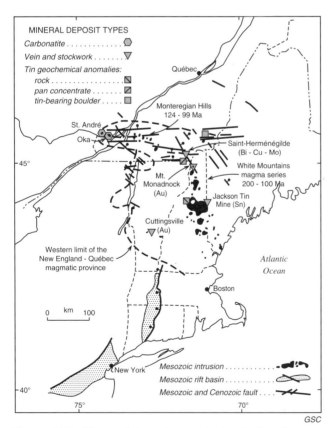

Figure 9.83. Mesozoic intrusions and related mineral occurrences in northern New England and southeastern Quebec (after McHone, 1984).

Mesozoic alkaline plutons

Mesozoic, rift-related, alkalic magmatism in southern Quebec and New England is represented by the Permian to Early Cretaceous White Mountains intrusions in northern New Hampshire and adjacent Massachusetts and the Cretaceous Monteregian Hills alkalic intrusions in southeastern Quebec (Fig. 9.83). The Monteregian Hills form a series of monadnocks rising above the St. Lawrence Lowlands and the alkaline intrusions represent the youngest thermal event.

Mineralization associated with Mesozoic alkalic intrusions in the northern Appalachians is best represented in the Cuttingsville area, Vermont, where gold, molybdenite, and bismuth are associated with a hydrothermally altered alkaline syenite stock (Robinson, 1986). Mineralization is found in a similar setting at the Saint-Herménégilde occurrence, just north of Mt. Monadnock in Quebec, where mineralization enriched in copper-bismuth-molybdenum is disseminated in fractures and greisens which occupy the hornfelsed pelites of the Saint Francis Group. Gauthier et al. (1989) suggested that the greisen dykes might be situated in the apex of a sub-surface syenitic stock. The Saint-Herménégilde occurrence contains anomalous copper (up to 0.2%), bismuth (up to 0.17%), gold (up to 0.18 g/t), molybdenum (<600 ppm), silver (<3.5 g/t), zinc (around 0.1%), lead (<0.1%), mercury (390 ppb), fluorine (610 ppm), and tellurium (32 ppm) and traces of Sn, W, and As. Metallic minerals include chalcopyrite, pyrrhotite, pyrite, molybdenite, marcasite, sphalerite, galena, native copper, native bismuth, and bismuth tellurides in a gangue of quartz, feldspar, muscovite, apatite, zoisite, and zircon (Gauthier et al., 1989)

UPPER PALEOZOIC OVERLAP ASSEMBLAGES
Geological overview
R.J. Ryan and R.C. Boehner

Upper Paleozoic and Mesozoic rocks in Atlantic Canada record molasse sedimentation, volcanism, and tectonism in successor basins. Sedimentation took place during the waning stages of the Acadian Orogeny and during the subsequent uplift of the orogen following the docking of the Avalon and Meguma zones (Fralick and Schenk, 1981; McCutcheon and Robinson, 1987).

Initial Late Devonian to early Viséan molasse deposition comprised alluvial, fluvial, and lacustrine sedimentation and local rift related volcanism in intermontane basins (Piskahegan, Fountain Lake, Horton and Anguille groups). The basin complex was flooded everywhere, except the Deer Lake Basin of western Newfoundland, by a major Viséan marine incursion (Kirkham, 1978). The Windsor/Codroy groups evaporitic basin received restricted marine to evaporitic marine and continental redbed deposits through middle to late Viséan with relatively little coincident tectonic activity. Continental sedimentation resumed in the Late Viséan-Namurian; the arid Viséan evolved into a more moderate pluvial to seasonal climatic regime in the Namurian to early Westphalian. Extensive shallow (rarely deep) lacustrine to brackish marginal marine(?) and wetland deposition produced major economic coal resources, especially in the late Westphalian. The late Carboniferous to early Permian records an evolution to more arid depositional conditions with extensive redbeds and local eolian deposits in the Permian. These conditions are generally similar to the pre-Zechstein evaporite basin conditions in northern Europe and Great Britain. During the late Permian a major erosion event (Ryan et al., 1990) was accompanied by reddening of the Upper Carboniferous strata.

Major lateral (dextral transpressive) motion along the Cobequid-Chedabucto Fault Zone culminated in the mid Westphalian to Stephanian with locally intense deformation and thrusting, especially in southern New Brunswick (Nance, 1987). In western Newfoundland, deformation occurred throughout the Carboniferous (with the main deformation occurring in post-Westphalian times) as a result of protracted strike slip movements on the Long Range and related faults.

The metallogeny of the Carboniferous basins reflects their sedimentary, tectonic and climatic histories. Several mineralizing episodes can be recognized: (1) Late Devonian to early Viséan: (a) minor occurrences of U are related to Upper Devonian rift-related felsic volcanics; (b) a few significant base metal occurrences are found near the contact with the overlying marine evaporitic rocks of the Viséan Windsor Group; (c) rare paleoplacer deposits occur at the unconformity between Horton Group conglomerate and Ordovician Meguma Supergroup slates; (2) Viséan: (a) abundant evaporitic deposits (e.g. anhydrite, salt, secondary gypsum hydrated from anhydrite, potash) formed at several stratigraphic levels in the Windsor/Codroy groups; (b) syngenetic low grade enrichments in Cu, Pb, Zn sulphides, pedogenic-diagenetic Fe and Mn and local barite and celestite replacements formed, especially near disconformity redox boundaries between marine carbonates and underlying redbeds; (3) Late Pennsylvanian: climate-controlled lead deposits formed in lower Westphalian sandstones (e.g., Yava deposit, Cape Breton Island); (4) Late Westphalian to Stephanian tectonism: numerous epigenetic mineral deposits formed, apparently from metalliferous brines developed deep in the basins and driven toward the basin margins where they interacted with reactive host rocks both along escape paths (faults and fractures) and reservoir-stratigraphic traps. The basal Windsor Group carbonate rocks were a primary host; (5) Late Permian: reddening of upper Carboniferous strata, especially in the Cumberland Basin, where low water tables generated numerous solution-roll front type Cu, Ag deposits.

The metallogeny of the Carboniferous basins is most easily described in terms of a number of distinct deposit types. Most of these are best developed in Nova Scotia, and so descriptions of type deposits are mostly drawn from there. Base metal deposits are subdivided into two major classes, those hosted by carbonate rocks (mainly marine carbonates of Windsor Group age) and those hosted by clastic sediments, including redbed Cu-Ag, U in sandstones and carbonaceous shales, shale-hosted (Kupferschiefer type?) base metals, and sandstone-hosted lead deposits. Non-metallic mineral deposits are associated with evaporites and include significant potash, salt, and gypsum/anhydrite. Other deposits of interest include epigenetic barite, celestite and uranium.

Carbonate-hosted base metal deposits

R.J. Ryan, R.C. Boehner, S.R. McCutcheon, and H. Scott Swinden

Carbonate hosted metal deposits are associated exclusively with the cyclic marine evaporitic marine and continental redbeds of the Viséan Windsor and Codroy groups. Although mineralization is found at several stratigraphic intervals, it is most persistent at the base of the Windsor/Codroy groups, in the Macumber Formation and the laterally equivalent Gays River Formation in Nova Scotia and New Brunswick, and the Ship Cove Formation in western Newfoundland. The Macumber and Ship Cove formations record basinal deposition on older Horton Group redbeds, and the Gays River Formation records carbonate buildups on pre-Carboniferous basement highs beyond the Horton Group pinchout.

The Horton Group siliciclastics immediately underlying the Macumber Formation are dominantly redbeds and appear to be re-reduced to green-grey to a depth of several metres or more adjacent to the contact. This marine/terrestrial contact zone, referred to as the Horton-Windsor contact, is a locus for numerous base metal, pyrite, siderite, and barite occurrences (Kirkham, 1978, 1985). Mineralization styles include both concordant and discordant deposits.

Concordant style

Concordant Cu, Pb, Zn occurrences lacking direct structural control are found in northern mainland Nova Scotia (e.g. Sylvan Valley Cu-Ba), Cape Breton Island (Yankee Line Road Pb-Zn-Cu, Frenchvale Cu-Zn-Pb), southern New Brunswick (e.g. Peekaboo Corner, Cedar Camp), and the Bay St. George Basin of western Newfoundland (e.g. Ship Cove, Ryan's Brook; Fig. 9.84).

At Yankee Line, Frenchvale, and Sylvan Valley, Cu-Pb-Zn sulphides, carbonates and oxides are concentrated along the Horton-Windsor contact at the redox boundary between the organic-rich Windsor carbonates and the underlying redbeds of the Horton Group. Additional occurrences of similarly-mineralized middle and upper Windsor carbonates also occur in areas of stratigraphic onlap (e.g. northeast mainland and northeast Cape Breton Island). In the Loch Lomond area where the Macumber and Gays River formations and overlying evaporites were not deposited, stratigraphically higher marine carbonate members and adjacent reduced redbed siliciclastics contain numerous chalcopyrite, sphalerite, pyrite, and galena occurrences.

There are several metallic mineral deposits in the lower Windsor strata in southern New Brunswick (Binney, 1975; Ruitenberg et al., 1977). Galena, sphalerite, chalcopyrite, and copper oxides occur in stromatolitic and/or micritic limestone of the Gays River Formation and the underlying Moncton redbeds (Horton equivalents). The largest of these is the Peekaboo Corner deposit, comprising galena and sphalerite occupying primary intercrystal, growth framework, and shelter pores as well as fractures, secondary intercrystal pores, and styolites in a substrate fringing algal buildups (McCutcheon, 1981).

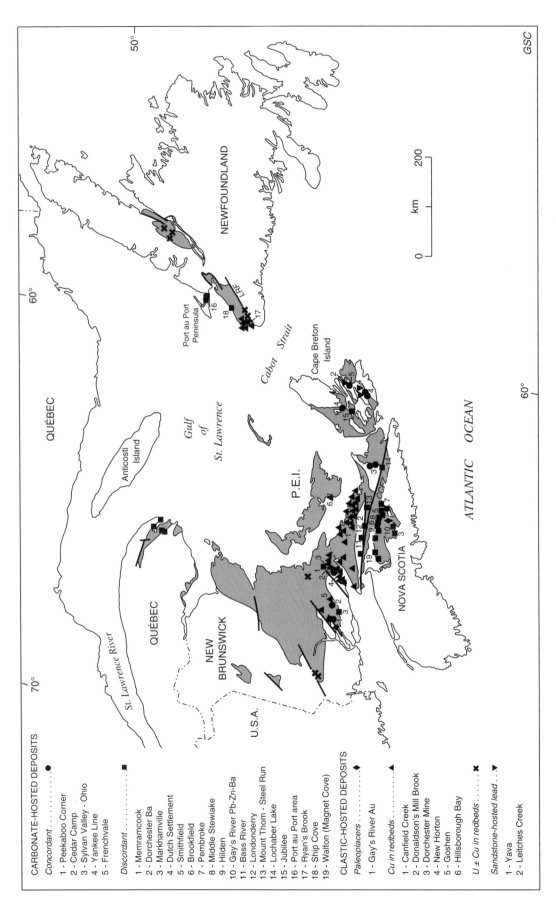

Figure 9.84. Principal metallic mineral deposits and/or districts in Carboniferous rocks of the Canadian Appalachians. Mineral deposits compiled after Ruitenberg et al. (1977), van der Poll (1978), Boehner and Ryan (1989), Knight (1984b), Hyde (1984).

CARBONATE-HOSTED DEPOSITS

Concordant . . . . ●

1 - Peekaboo Corner
2 - Cedar Camp
3 - Sylvan Valley - Ohio
4 - Yankee Line
5 - Frenchvale

Discordant . . . . ■

1 - Memramcook
2 - Dorchester Ba
3 - Markhamville
4 - Dutch Settlement
5 - Smithfield
6 - Brookfield
7 - Pembroke
8 - Middle Stewiake
9 - Hilden
10 - Gay's River Pb-Zn-Ba
11 - Bass River
12 - Londonderry
13 - Mount Thom - Steel Run
14 - Lochaber Lake
15 - Jubilee
16 - Port au Port area
17 - Ryan's Brook
18 - Ship Cove
19 - Walton (Magnet Cove)

CLASTIC-HOSTED DEPOSITS

Paleoplacers . . . . ◆

1 - Gay's River Au

Cu in redbeds . . . . ▲

1 - Canfield Creek
2 - Donaldson's Mill Brook
3 - Dorchester Mine
4 - New Horton
5 - Goshen
6 - Hillsborough Bay

U ± Cu in redbeds . . . ✕

Sandstone-hosted lead . . . ▼

1 - Yava
2 - Leitches Creek

Discordant (fault-related) style

Numerous base metal, silver, barite, manganese, and iron deposits occur near the Horton-Windsor group contact between Cheverie and Tennycape near the Bay of Fundy shoreline (Fig. 9.85). This area is extensively faulted and folded and is situated within 10-20 km south of the Cobequid-Chedabucto Fault Zone and immediately south of the superimposed early Mesozoic Fundy Graben.

Five types of mineralization (dominantly structurally controlled) were described by Boyle et al. (1976): (1) massive barite and massive sulphides in Macumber (Pembroke) limestone at Walton (Magnet Cove deposit); (2) barite-siderite without associated sulphides in Macumber (Pembroke)

limestone and locally Horton sandstone; (3) manganese and manganite-pyrolusite in Macumber/Horton limestone/sandstone; (4) hematite-limonite in Macumber/Horton limestone/sandstone (setting similar to 3); and (5) chalcopyrite, pyrite, malachite-atacamite in Horton sandstone.

The Magnet Cove deposit near Walton (Boyle, 1972; Boyle et al., 1976; Fig. 9.85) is hosted by basal Windsor Group Macumber Formation limestone and limestone breccia (Pembroke) and overlying anhydrite/gypsum at the intersection of west- and northwest-trending fault sets. The deposit comprises adjacent, but discrete, steeply plunging pipe-like massive barite bodies and a down-dip massive sulphide body. The barite and underlying sulphide bodies

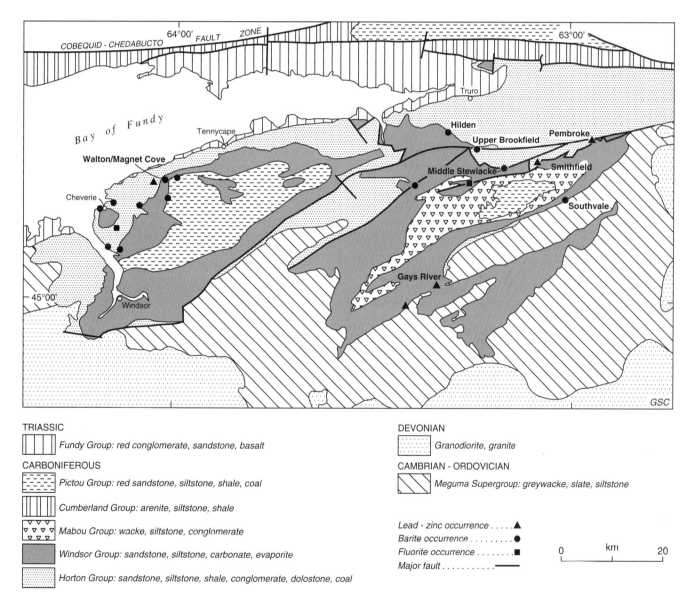

TRIASSIC

▦ Fundy Group: red conglomerate, sandstone, basalt

CARBONIFEROUS

▦ Pictou Group: red sandstone, siltstone, shale, coal

▦ Cumberland Group: arenite, siltstone, shale

▦ Mabou Group: wacke, siltstone, conglomerate

▦ Windsor Group: sandstone, siltstone, carbonate, evaporite

▦ Horton Group: sandstone, siltstone, shale, conglomerate, dolostone, coal

DEVONIAN

▦ Granodiorite, granite

CAMBRIAN - ORDOVICIAN

▦ Meguma Supergroup: greywacke, slate, siltstone

Lead - zinc occurrence . . . . . ▲
Barite occurrence . . . . ●
Fluorite occurrence . . . . ■
Major fault . . . . ▬

0   km   20

Figure 9.85. Geology and mineral occurrences of the Minas Basin (after Ravenhurst et al., 1990). 1 – Gays River; 2 – Southvale; 3 – Pembroke; 4 – Smithfield; 5 – Middle Stewiacke; 6 – Upper Brookfield; 7 – Hilden; 8 – Walton/ Magnet Cove.

extend together to a depth of approximately 330 m with Viséan(?) redbeds in the hangingwall. Below this point, only the sulphide body continues with anhydrite/gypsum in the hangingwall. The mineralogy is complex comprising barite, siderite, pyrite-marcasite (Ni enriched), galena, sphalerite, chalcopyrite, and silver minerals, various manganese oxides and hematite, and locally abundant hydrocarbons (liquid petroleum and carbonaceous material). The deposit has produced approximately 4.5 million tonnes of barite grading more than 90% $BaSO_4$ and 360 000 tonnes of sulphide ore grading 0.56% Cu, 4.33% Pb, 1.44% Zn, and 342 g/t Ag.

The complex structural zone in the Walton-Cheverie area extends en echelon to the northeast where several mineral occurrences are associated with faulted Macumber Formation limestone on the north side of Minas Basin (e.g. Smithfield, Middle Stewiacke, Hilden, Upper Brookfield; Fig. 9.85). In this area, there are also occurrences in younger Windsor Group limestones which are in fault contact with the Horton Group (e.g. Pembroke galena and calcite). The Smithfield occurrence, located in moderate- to steep-dipping, faulted Macumber limestone-breccia, contains 500 000 tonnes of abundant pyrite with sphalerite, galena, and minor barite and siderite. At Middle Stewiacke, west of Smithfield in the same fault system, a complex, structurally controlled, generally fine grained barite deposit forms irregular lenses and pods, with coarser (later) fracture fill barite and associated greenish calcite in fault brecciated Macumber Formation. Siderite and sulphides are a significant component in the Upper Brookfield and Hilden occurrences farther west in the same en echelon fault system. The Upper Brookfield barite-siderite deposit is unusual in that it is located in ferruginous red and grey sandstone-shale of the Horton Group within 30 m of the base of the adjacent Macumber Formation.

The largest carbonate-hosted deposit is the Gays River Pb, Zn deposit (Fig. 9.85) located in the Gays River bank complex, a fringing island ridge that lies on a Meguma basement high on the southern side of the Minas Basin. The bank complex extends for approximately 10 km along strike and is variably mineralized with several major galena and sphalerite zones at Gays River and a smaller zone near Dutch Settlement to the southwest. The area is structurally less disturbed than the Walton-Smithfield belt. The bank is flanked, and was formerly overlain, by a thick anhydrite-dominated evaporite sequence. The Gays River deposits contain approximately 12 million tonnes of 7% combined Pb, Zn and have produced 0.9 million tonnes (1978-1981). Abundant galena and sphalerite accompanied by trace and accessory chalcopyrite, pyrite, marcasite, calcite, barite, fluorite, and bituminous organic material (petroleum) occur dominantly as disseminations and concordant-stratiform bodies. Subordinate high grade mineralization includes concordant massive bodies at the evaporite-carbonate contact (15-25% Pb + Zn) and discordant-vein bodies (50% Zn + 10% Pb) within the carbonate and locally penetrating the underlying basement (Akande and Zentilli, 1984).

There are a number of structurally controlled Pb-Zn deposits in the Macumber Formation in central Cape Breton Island (Fig. 9.84), the largest of which is the Jubilee deposit (Hein et al., 1988). Galena and sphalerite with trace to accessory chalcopyrite, pyrite, calcite, barite, and fluorite are localized in a narrow northwest-trending (transverse) fault near the flanks of a broad anticline. There is abundant liquid hydrocarbon within the mineralized zone, as well as in adjacent evaporites. Reserve estimates for Jubilee are 500 000 tonnes of 6% Pb + Zn.

There are structurally controlled vein-replacement occurrences in several locations along the Cobequid-Chedabucto Fault Zone including Bass River (Ba, Fe), Londonderry (Fe, Ba), Mount Thom-Steele Run (Cu, Fe, Co, Au), and Lochaber Lake (Cu, Zn, Pb; Fig. 9.84). The occurrences are hosted by grey fine grained siliciclastics and limestone of Late Carboniferous Namurian to Westphalian and locally (Lochaber Lake?) Late Devonian age. Deposits in the Londonderry area comprise vein type ankerite, siderite, and barite in Silurian and Upper Carboniferous rocks adjacent to the Cobequid-Chedabucto Fault Zone which produced 3.5 million tonnes between 1849 and 1908. Similar fault-controlled vein replacement barite occurs in Upper Carboniferous rocks at Bass River. The Mount Thom-Steele Run occurrences comprise pyrite, chalcopyrite with siderite, and iron oxides enriched in Co and Au hosted by brecciated grey mudrocks of the Mabou Group (Namurian). The Lochaber Lake deposit on the structurally complex south side of the Antigonish Highlands comprises stratabound disseminations and stringers of chalcopyrite, malachite, and pyrite with minor barite, bornite, sphalerite, and galena along bedding planes and as fracture fill veins in a shaly limestone. Inferred reserves have been estimated at about 5 million tonnes of 0.33% Cu.

Along the western margin of the Cape Breton Highlands and in the area west of Lake Ainslie, fault slices of relatively unmetamorphosed mafic and felsic volcanic rocks of the Middle Devonian Fisset Brook Formation host a number of structurally controlled barite-fluorite veins. Although not hosted by carbonate rocks, they probably formed through similar processes related to Carboniferous basin evolution and so are included here. The Trout River vein group near the south end of the district includes four large veins containing an aggregate drill indicated and probable reserve of 4.3 million tonnes of ore grading 36.7% $BaSO_4$ and 18.5% CaF_2 (Felderhof, 1978). Most of the Trout River veins cut the Fisset Brook Formation rhyolite; the Campbell vein, however, occurs along a fault contact between Mississippian Horton Group clastic rocks and the Fisset Brook Formation (Felderhof, 1978), suggesting that the veins may be substantially younger than their host rocks. A cluster of similar but smaller veins in the Scotsville inlier at the north end of the district (Felderhof, 1978), cut mainly metamorphic rocks of late Proterozoic to Devonian age adjacent to basalt and rhyolite tuff of the Fisset Brook Formation.

Stratiform and vein-type barite, chalcopyrite, chalcocite, galena, sphalerite, and pyrite are also found at the contact between the Codroy and Anguille groups in the Bay St. George Basin (Knight, 1983, 1984b; Fig. 9.86). The mineralization generally occurs as centimetre sized clots, and as disseminations and joint surface coatings in limestone of the Ship Cove Formation and extends several metres downward into the underlying sandstone. The best mineralization occurs at the Ship Cove, Ryan's Brook, and North Branch River.

On Port au Port Peninsula in western Newfoundland, stratabound galena, sphalerite, marcasite, barite, and celestite occur at the base of lower Codroy Group limestone where it unconformably overlies Ordovician carbonates (e.g. Lead Cove; Fig. 9.84). Galena, sphalerite, pyrite, and marcasite occur as lenses, veins, and disseminations in limestone breccias (Knight, 1984b; Johnson, 1954) that were deposited in north trending, steep-sided, fault-controlled, Carboniferous paleokarst valleys (Knight, 1984b).

There are several small galena and sphalerite occurrences in calcite veins or the calcitic to dolomitic matrix of breccias in limestones of the Lower Devonian Grande Grève and Indian Cove formations at the east end of Gaspésie (Schrijver et al., 1988; Fig. 9.84). The fracture fillings are characteristically low grade, trend at a large angle to bedding, and lie close to regional, northwest-trending faults. Mineralization comprises galena with sphalerite, marcasite, and pyrite in a calcite-dolomite gangue. The age of the deposits is not well constrained. They may be related to the development of late Paleozoic successor basins or to unrelated events.

Genetic models

Current metallogenic interpetations suggest that carbonate-hosted mineralization in Carboniferous rocks may record two distinct mineralizing episodes: (1) an early diagenetic episode during which many if not most of the concordant deposits formed; and (2) a younger (Westphalian or later) epigenetic episode (or episodes) in which most of the discordant and perhaps some of the concordant, deposits formed.

The diagenetic origin, distribution, and zonation of the widespread dispersed concentrations of Cu, Pb, and Zn at the base of the Macumber and Gays River formations in central Cape Breton Island were interpreted by Kirkham (1978, 1985) to be genetically linked to the redox (reduction-oxidation) boundary at the base of the marine organic-rich Windsor Group (anoxic) where it rests concordantly, conformably to disconformably on red and grey green siliciclastics of the Horton Group. Concentration and deposition

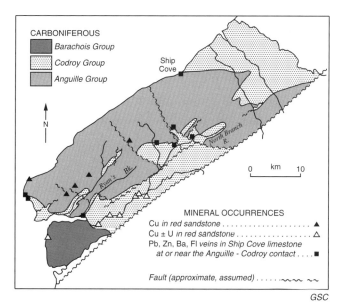

Figure 9.86. General geology and mineral occurrences in the Bay St. George Basin, western Newfoundland (after Knight, 1983).

occurred near the redox contact where anoxic carbonaceous pyritic strata were infiltrated by upward-migrating metalliferous brines derived from underlying redbeds. Although most common at the base of the Macumber Formation, these early diagenetic redox boundary metal concentrations occur at similar contacts elsewhere in the Windsor Group and at similar boundaries in other Carboniferous fluvial and lacustrine strata.

A number of arguments suggest that most of the discordant deposits, and perhaps some of the concordant deposits, are related to a later epigenetic mineralizing event. Boyle (1972) and Boyle et al. (1976) proposed a mineralization model for the diverse mineral occurrences and deposits in the Walton-Cheverie area which involved late Carboniferous tectonic activity. Faulting and fracturing related to the Cobequid-Chedabucto Fault Zone allowed penetration and circulation of brines and connate water which had been enriched in metals during residence in the Horton redbeds. Movement of these fluids (perhaps in the Triassic) accelerated or triggered by elevated geothermal gradient driven circulation and tectonism (rifting? and volcanism) resulted in metal deposition through reaction with chemically replaceable rocks, which were especially abundant at the Horton-Windsor contact.

A model for epigenetic mineralization in the central Nova Scotia area capable of relating the Gays River deposit to the nearby, although geologically dissimilar, Walton, Smithfield, and Pembroke deposits was developed by Akande and Zentilli (1984) and Ravenhust and Zentilli (1987). The conceptual model invokes the migration of hot mineralizing fluids from a deep part of the Carboniferous basin (specifically the deep Magdalen Basin) during a mid-Westphalian (about 300 Ma) tectonic episode. Deposits formed in the equivalent of hydrocarbon structural and stratigraphic traps (e.g. Gays River) and in faulted (escape path) areas where fluid mixing, chemistry, and host were appropriate for mineral precipitation, replacement, and/or potential sedimentary exhalative type deposition. This model accommodates the proposal for Walton by Boyle et al. (1976) and links the mineralogically diverse deposits of the area to the tectonic development of the late Paleozoic basins.

Hein et al. (1988) proposed an innovative carbonate deposition-mineral deposit model for the Jubilee deposit, in which enhanced carbonate deposition with thick buildup in the vicinity of a growth fault was followed by deposition of sulphides in locally hydrocarbon-saturated breccia porosity. Sulphide mineralization was interpreted to be a later phase of continuing early Viséan sedimentary exhalative processes. Timing of mineralization in this model is not compatible with the Ravenhurst and Zentilli (1987) post-Windsor model.

Clastic-hosted deposits

R.J. Ryan, R.C. Boehner, S.R. McCutcheon, and H. Scott Swinden

Paleoplacers

In Nova Scotia, paleoplacer Au occurs as flakes, grain coatings, and small sand size particles in Lower Carboniferous Horton Group sandstone and conglomerate. As well, cassiterite and wolframite occur as silt size detrital grains

787

in Upper Carboniferous sandstones (Ryan et al., 1988). The gold placers are concordant, stratabound, syn-depositional concentrations of reworked Au derived from early Paleozoic Au deposits in the Meguma Supergroup basement. Irregular unconformity surfaces on the basement acted as traps for the heavy minerals, which were also concentrated at stream gradient breaks between steep braided streams or alluvial fans and lower gradient meandering or anastomosing streams. The most noteworthy of these, the Gays River gold mine (Fig. 9.84), produced more than 93 300 g of gold.

Clastic-hosted Cu-Ag deposits (redbed deposits)

The sandstone-hosted deposits of the late Paleozoic basins include Cu-Ag deposits in Upper Carboniferous strata and U deposits in Lower Carboniferous to Triassic red or grey sandstone beds in Nova Scotia, southern New Brunswick, and western Newfoundland. The occurrences are concordant and stratabound. Cu-Ag and U are commonly found

together in these rocks, but are thought to represent separate mineralizing episodes with Cu-Ag mineralization representing an early (diagenetic?) enrichment by a solution front mechanism and U mineralization a later enrichment via a roll front mechanism.

Three forms of sandstone-hosted Cu-Ag deposits are recognized in northern Nova Scotia (cf. Papenfus, 1931): (1) nodules and concretions of chalcocite, bornite, and pyrite; (2) chalcocite and bornite replacing the pre-existing cements of the sandstone host rocks; and (3) chalcocite and pyrite replacing the coalified plant material contained within the strata. MacKay and Zentilli (1976) found the pitchblende in at least one of the deposits to occur as encrusting layers surrounding the Cu minerals and pyrite. Ryan et al. (1989) suggested that the textural relationships of the minerals indicate the following order of appearance, from oldest to youngest: (1) pyrite; (2) pyrite, sphalerite, and minor galena; (3) bornite and a trace of chalcopyrite; (4) chalcocite, digenite, bornite, barite, and silver; and (5) uranium. Stea et al. (1986) and Ryan et al. (1989)

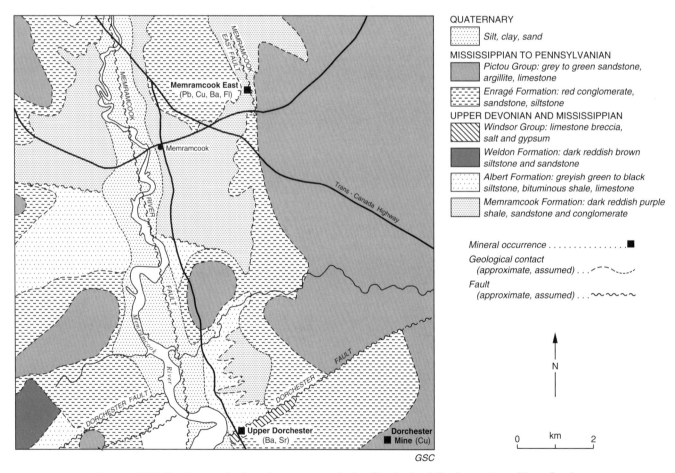

Figure 9.87. Geology and mineral occurrences in the Cumberland Basin, northern Nova Scotia (after Ryan et al., 1989).

showed that there is a depletion of Cu in the red sandstones adjacent to the Cu-Ag occurrences and suggested that the Cu and associated elements were scavenged from the sandstones and reconcentrated at redox boundaries.

Deposits of this nature are particularly common in the Cumberland Basin, northern Nova Scotia (Fig. 9.87). The largest of these occurs at Canfield Creek, where disseminated chalcocite, chalcocite, and bornite nodules up to 3 cm in diameter, and wispy infillings of chalcocite occur along the parting planes in grey sandstone. The deposit contains about 272 000 tonnes grading 1.2% Cu (O'Sullivan, 1981) and is open ended. The mineralization has been traced to a depth of approximately 110 m, and the ore grade zone is up to 5.2 m thick. The mineralization is most common within the basal 5 m of the grey channel sand bodies which coalesce to form sheet sandstones.

Some clastic-hosted Cu-Ag deposits in New Brunswick include the Dorchester, New Horton, and Goshen deposits (Ruitenberg et al., 1977; Fig. 9.84). The best known is the old Dorchester mine (Fig. 9.88) where chalcocite and chalcopyrite occur in fluvial-channel sandstone at the base of the Boss Point Formation. The sandstone is grey, rich in organic material, and overlies redbeds of the Enragé Formation. Some sections contain between 2% and 10% Cu but the average grade is less than 1% in a northeast-trending zone about 1000 m long and up to 6 m thick (McLeod and Ruitenberg, 1978; McLeod, 1979).

Low grade, generally minor Cu (up to 3000 g/t) and U (11 to 107 g/t) occurrences are found in sediments of the upper Codroy Group, western Newfoundland (Knight, 1976, 1983; Dunsmore, 1977a; Fig. 9.86) in fine grained, micaceous, carbonaceous sandstone beds interbedded with

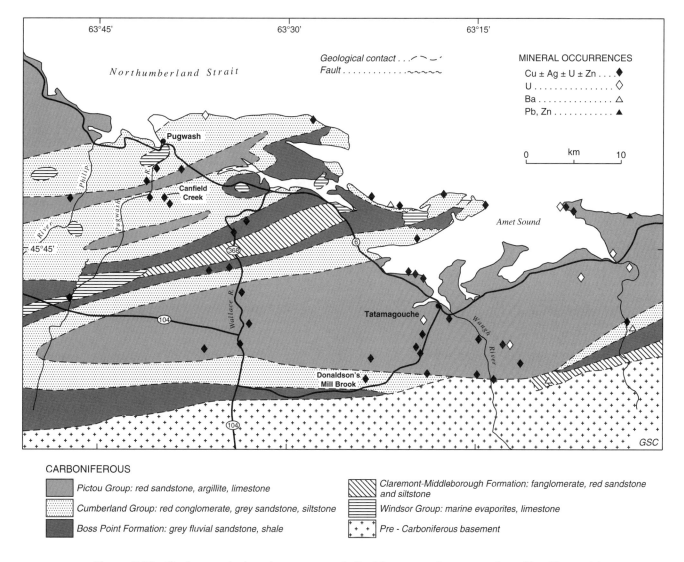

Figure 9.88. Geology and mineral occurrences in the Memramcook area, southern New Brunswick (after McLeod and Ruitenberg, 1978).

red siltstone and grey mudstone and in pebbles, mud clasts, and carbonaceous plant fossil lenses in basal sandstone units of meandering, fluvial sequences (Knight, 1984b).

Regionally, the Cu mineralization commonly occurs at or near the boundary between red and grey sandstones, implying a genetic link between colouration and the concentration of the Cu. Ryan (1988) suggested that arid conditions during the Late Permian lowered the water table within the basin resulting in secondary reddening of

the sandstones. Under these circumstances, in a basin margin onlap setting, additional oxygenated surface water runoff from the highland areas may have flowed into the permeable sandstone aquifers and continued basinward until they encountered the basinal waters. The Cu-Ag solution front model proposed here is similar to that proposed by Shockney et al. (1974) for the Naciemento and Paoli deposits in the southwestern United States. Cu and Ag were subject to selective leaching and transport from

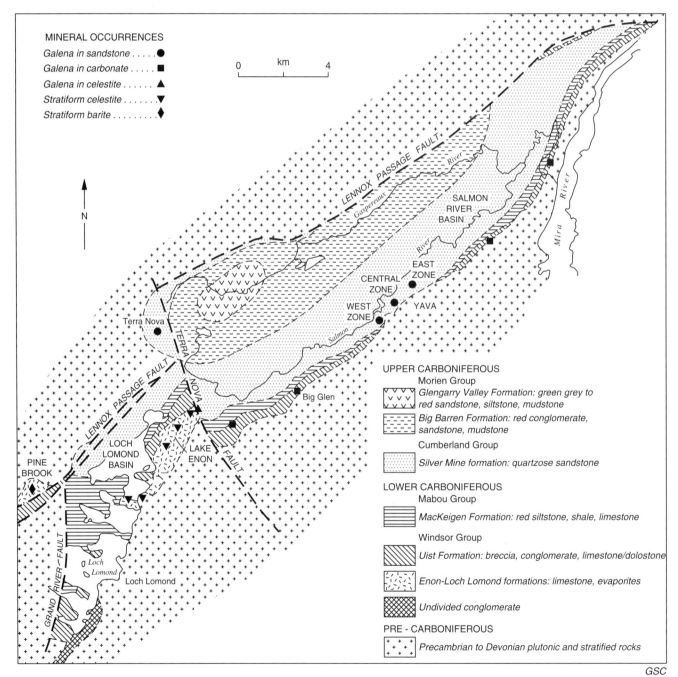

Figure 9.89. Geology and mineral deposits in the Salmon River Basin – Loch Lomond area, Cape Breton Island (after Vaillancourt and Sangster, 1986).

the strata during diagenetic reddening and the solutions migrated until they encountered reducing conditions. This occurred where sufficient coalified plant material and pyrite were preserved in the channel lags, at the interface with the reducing basinal fluids, near hydrocarbons associated with the evaporite diapirs, and at the boundary with overlying pyrite and coal bearing grey mudrocks.

Another version of the solution front Cu-Ag model is demonstrated by the Donaldson's Mill Brook occurrence in Nova Scotia (Fig. 9.84 and 9.88). The mineralized grey shale above a redbed sequence contains a 1 m thick zone of 1% Cu with high grade grab samples containing up to 583 g/t Ag. This is interpreted to be a Kuperschiefer-type occurrence in which the oxidizing groundwater depleted the underlying redbed sequence, either during or after reddening, and the Cu and associated metals were precipitated at the redox boundary with the overlying grey mudrocks. The mineralized mudrocks usually contain abundant coalified plant material and pyrite which has been replaced by the Cu mineralization.

U roll fronts and unconformity solution fronts

Uranium in the late Paleozoic basins may occur with or without associated Cu. Dunsmore (1977b) pointed out that most mineralization occurs near the base of fining-upward cycles in the fluvial strata where the uranium precipitated as a result of reducing conditions created by the carbon- and pyrite-rich lags. Uranium enrichment also occurs at the interface with underlying and overlying grey carbon-rich mudrocks (e.g. deposits in the North Brook Formation, Deer Lake Basin; Horton Group, Moncton Basin). The primary uranium mineralization is believed to be the result of uranium-rich solutions (probably groundwater) passing through the permeable sandstone beds and precipitating uranium at the redox boundary. As the movement of the redox boundary continues down-dip, only the uranium bound by the carbon- and pyrite-rich lags remains. Uranium with and without minor Cu occurs in the Deer Lake Basin in three main settings: (1) in paleokarstic breccias and bituminous carbonate rocks at the unconformity between the basal North Brook Formation and the Cambrian-Ordovician carbonate basement; (2) as stratiform bodies hosted by lacustrine grey mudstone and carbonate layers of the Rocky Brook Formation; and (3) in uraniferous sandstone till boulders thought to be derived from near the top of the Deer Lake Group (Gall, 1984). Two models of uranium concentration have been proposed for occurrences in the Deer Lake Basin (Vanderveer and Tuach, 1982; Hyde, 1984): (a) a model involving depositional/ stratigraphic control in which localized accumulations of organic debris in paleochannels were sites of later uranium deposition; and (b) a fault control model in which H_2S-bearing fluids from the fault produced chemical reduction in adjacent strata which provided a site for precipitation of uranium and other metals. Nova Scotia examples are compatible with mechanism (a).

Sandstone-hosted Pb deposits

This type of occurrence is best represented by the Yava (Salmon River) deposit, in southeastern Cape Breton Island (Bonham, 1983; Vaillancourt and Sangster, 1986;

Sangster and Vaillancourt, 1990a, b; Fig. 9.89). The ore is hosted by grey fluviatile quartzose sandstones of the Cumberland Group (Westphalian A) which overstep onto the basement rocks and Lower Carboniferous strata at the margin of the basin. The deposit consists of three zones of disseminated galena each overlying a basement depression and separated by intervening basement highs (Vaillancourt and Sangster, 1986). The Yava deposit was discovered in 1961 and three zones containing 11 million tonnes grading 3.5% Pb, an additional 5.9 million tonnes grading 3.3% Pb, and up to 70.0 million tonnes grading 2.0% Pb were delineated between 1962 and 1969. Mining between 1979 and 1981 produced approximately 0.4 million tonnes of 4.69% Pb.

The Yava deposit is a good example of a class of deposit termed "sandstone lead" deposits by Bjørlykke and Sangster (1981), characterized by disseminated galena in basal quartzose sandstones. Type examples of these deposits include the Laisvall and related deposits in Cambrian sandstones on the western edge of the Baltic Shield (Grip, 1967; Bjørlykke, 1980). At Yava, the sulphides, comprising primarily galena, pyrite, sphalerite with minor chalcopyrite, marcasite, and barite, occur as disseminations, patches, or aggregates in the matrix (pore space) of the sandstone and as replacement of coalified plant material. The mineralization is broadly stratabound, although in detail it transgresses beds and follows the trends of the underlying unconformity with the older strata and basement rocks. Bjørlykke and Sangster (1981), Vaillancourt and Sangster (1986), and Sangster and Vaillancourt (1990b) suggested a model involving groundwater transport of metals derived from the weathered basement rocks and from the destruction of the feldspars within the host strata into environments where sufficiently high H_2S concentrations triggered precipitation of the lead.

Sandstone hosted, unconformity-related galena and sphalerite also occur in Westphalian C Morien Group sandstones lithologically similar to those at Yava at Leitches Creek in the Sydney area.

Evaporites

R.C. Boehner, S.R. McCutcheon, J.H. Fowler, and A.F. Howse

The Windsor Group in Nova Scotia and New Brunswick, and the equivalent Codroy Group in Newfoundland comprise interstratified cyclical evaporites with subordinate redbeds, mudstone to fine sandstone with rare conglomerate, and numerous laterally persistent carbonate members. Evaporitic parts of this sequence contain deposits of gypsum, potash, and salt (Fig. 9.90).

Gypsum and anhydrite

Gypsum and minor anhydrite have been quarried from the Windsor Group in Nova Scotia for more than 200 years (Lewis and Holleman, 1983; Adams, 1991); current annual production is more than seven million tonnes. The largest single open pit gypsum quarry in the western world is operated by National Gypsum (Canada) Ltd. from which more than three million tonnes was produced in 1989. Gypsum was quarried from the lower Windsor Group in New Brunswick as early as 1840 and major deposits near Hillsborough produced gypsum until the early 1980s

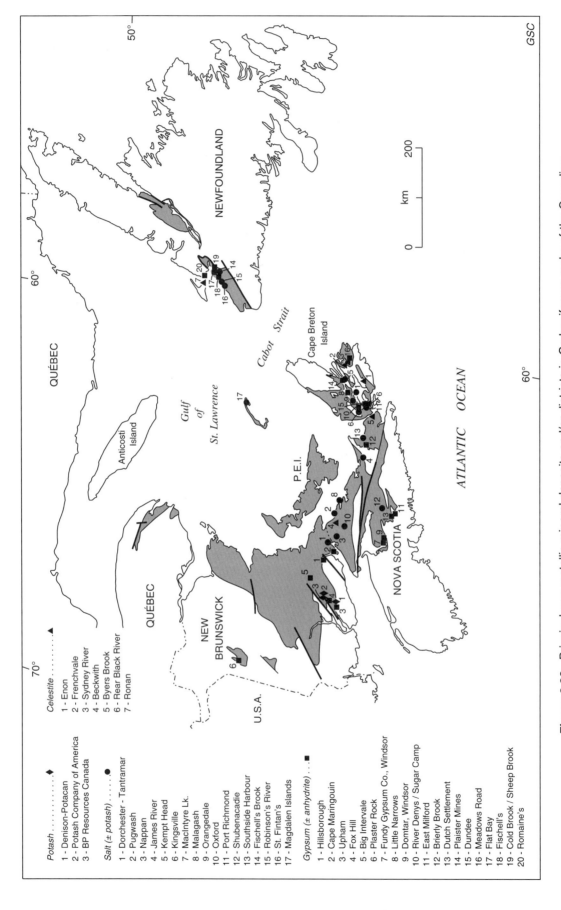

Figure 9.90. Principal non-metallic mineral deposits and/or districts in Carboniferous rocks of the Canadian Appalachians. Deposits compiled after Ruitenberg et al. (1977), Boehner and Ryan (1989), Knight (1984b), Webb (1988).

(Webb, 1988). Four major deposits of gypsum-anhydrite occur in the Codroy Group in the northern portion of the Bay St. George Basin. The largest of these at Flat Bay has been quarried since 1952.

Gypsum deposits in Nova Scotia resulted from hydration of anhydrite which forms a major proportion of the Windsor Group. The anhydrite occurs in two principal stratigraphic settings: (1) as thick (130-300 m) sections with thin interbeds of carbonate and mudstone at the base of the lower Windsor, Subzone A, Major Cycle 1, and (2) as thin (up to 10 m) interbeds with marine carbonates, redbeds and salt in the B Subzone, Major Cycle 2, and in the upper Windsor, Subzones C to E, Major Cycles 3 to 5.

Although anhydrite predominates in the subsurface away from the influence of groundwater, in near surface areas, it is replaced by gypsum. This process is variable in extent and depth (0-200 m) and is controlled by the movement of hydrating groundwater into permeable zones such as interbedded carbonates and redbeds. In the larger economic deposits, the hydration process has been enhanced by large scale structures particularly faults, folds, and solution collapse (salt removal), which have generated extensive fracture permeability. Hydration events are related to tectonics and unconformities which may include Early Carboniferous, Late Carboniferous (unconformity related), Early Mesozoic (regional unconformity), Early Cenozoic (regional unconformity), Late Cenozoic? and Recent events.

Salt

The first underground salt mine in Canada was operated at Malagash, Nova Scotia, in the Cumberland Basin, between 1919 and 1959. Salt is produced in a conventional underground mine at Pugwash, Nova Scotia, operated by The Canadian Salt Company Ltd. (approximately 800 000 tonnes annually) and in a modern solution mining vapour recompression operation by Sifto Canada Inc., at Nappan near Amherst, Nova Scotia (approximately 65 000 tonnes annually).

Twenty salt deposits of economic significance and 12 occurrences with potential have been identified in Windsor Group rocks in Nova Scotia (Boehner, 1986). Potash, sylvite and/or carnallite are present in trace to economic quantities in ten of the deposits. The two current producers at Pugwash and Nappan, as well as the abandoned mine at Malagash operate in Windsor Group diapirs. The salt sections in the Windsor Group are heterogeneous with numerous interbeds of anhydrite and mudrocks. Salt grade is highly variable depending upon the thickness of the zone of interest. In the Shubenacadie area of Minas Basin for example, the salt section in Major Cycle 1 is up to 300 m thick with an overall average section grade of approximately 85% NaCl. Thinner intervals of 10-30 m exceed 95% NaCl. Thin insoluble interbeds and inclusions of anhydrite and mudrocks are the dominant impurity.

In southern New Brunswick, salt occurs in both the Horton and Windsor groups. There is salt near the top of the Albert Formation (Gaultreau Member) in at least two places (Weldon, Corn Hill) and salt constitutes the major part of the Cassidy Lake Formation near Clover Hill, Millstream, Penobsquis, and Sackville. The only salt production is at Penobsquis, Potash Company of America's

potash deposit, where about 500 000 tonnes are mined annually in order to make room for underground disposal of tailings.

In western Newfoundland, major salt deposits with associated potash occur within the Codroy Group at St. Fintan's, Robinsons River, and Fischell's Brook (Knight, 1983). Anhydrite and gypsum, grey shale and mudstone, brown and red siltstone and mudstone, and some limestone and dolomite are associated with the salt sections. The salt is generally of good quality, low in impurities (<1% insolubles), banded, massive and fine- to coarse-granular and varies through white, grey, red, pink, and orange (Knight, 1983).

Potash

Deposits of potash have been identified in Nova Scotia, New Brunswick, and Newfoundland. However, only deposits in southern New Brunswick have been mined.

The three known New Brunswick potash deposits occur in the Mississippian (Viséan) Windsor Group of the Moncton Basin. The Penobsquis deposit, the first to be found, has been mined since 1983 by the Potash Company of America while production began at the Clover Hill deposit, owned by Potacan Mining Company, in 1985. The Millstream deposit, the last discovery, was found by BP Resources Canada Ltd. and has not been developed.

In the Moncton Basin, the Windsor Group consists, in ascending order, of the Gays River (nearshore carbonate algal buildups), Parleeville (nearshore carbonate algal buildups with interbedded siliciclastics), Macumber (basinal thin bedded carbonates), Upperton (basinal carbonates with interbedded gypsum and anhydrite) and Cassidy Lake (salt and potash) formations (McCutcheon, 1983). Potash deposits are hosted by the Cassidy Lake Formation (Anderle et al., 1979) which consists of a homogeneous halite unit (basal and middle halite members of Anderle et al., 1979) overlain by a sylvinite unit (potash member) and a heterogeneous halite unit (upper halite member). This three-fold subdivision, with local variations, is present in all three potash deposits. Cassidy Lake Formation rocks are distal to pre-Carboniferous uplands and lie near the axes of regional synclines that are tectonically modified syndepositional troughs (McCutcheon, 1981). They are flanked by anhydrite of the Upperton Formation resulting in a "bulls-eye" map distribution of Windsor Group rocks with Cassidy Lake evaporites in the centre.

A possible depositional model for Windsor Group evaporites in the Moncton Basin (McCutcheon, 1981) involves an "instantaneous" marine transgression followed by the start of carbonate deposition. Near shore deposition was in the form of algal carbonate buildups, locally with intercalated siliciclastics (Parleeville Formation). Basinward, thinly bedded carbonates were laid down (carbonate stage). High surface evaporation produced brines that migrated to local "deeps" in the basin floor where they became saturated with respect to sulphate, and nodular and mosaic gypsum (Upperton Formation) precipitated diagenetically just beneath the sediment-water interface. With increasing concentration, the locus of nodular sulphate precipitation shifted toward the basin margins and bedded sulphate began accumulating in the "deeps" (bedded sulphate stage). Bedded sulphate was followed by the homogeneous halite and sylvinite units (Cassidy Lake Formation) reflecting increasing brine concentration and a static environment (lower halite to sylvinite stage). Interbedded halite, anhydrite, and potash (Cassidy Lake Formation) overlie the sylvinite unit indicating numerous salinity changes related to a much shallowed basin. Mudstone-hosted, nodular to mosaic sulphate and slate of the Clover Hill Formation disconformably overlies the Cassidy Lake Formation and coarsening-upward clastic sedimentary rocks of the Hopewell Group overlie the Clover Hill Formation.

Potash, although volumetrically minor, is an important and potentially economically significant component of the saline evaporites in Nova Scotia. Of the ten potash salt localities, the most significant drilled prospect is at Malagawatch where three potash zones were intersected. Structural complexity, characterized by multiple repetition in large folds, and a restricted area for exploration and development limit the potential of the Malagawatch prospect. Similar geological environments, however, may be present in central Cape Breton Island as well as in other basins in Nova Scotia.

Salt deposits in Windsor Subzone A and Lower Subzone B strata of the Bay St. George Basin, southwestern Newfoundland (e.g. Fischells Brook, Robinsons and St. Fintans) contain significant amounts of potash. Low grade potash is distributed throughout the salt deposits where it is commonly zoned, with sylvite overlying carnallite.

Celestite-barite

R.C. Boehner, J.H. Fowler, and H. Scott Swinden

Epigenetic barite-celestite deposits occur in the Windsor Group in the Loch Lomond area of the Central Cape Breton Basin (Beckwith, Rear Black River, Frenchvale, and Sydney River deposits Fig. 9.90) and the Moncton Basin (Uppper Dorchester, Memramcook East, Fig. 9.88) and in the Codroy Group on the Port au Port Peninsula of western Newfoundland (Ronan, Fig. 9.90).

The Loch Lomond celestite deposits (Forgeron, 1977) are associated with Major Cycle 2 and the base of Major Cycle 3 of the Windsor Group (Fig. 9.89). The depositional setting was strongly influenced by a basin margin basement onlap unconformity. There are significant mineralized zones at four stratigraphic levels within a 150 m thick package in the Loch Lomond area. The stratabound mineralization occurs as tabular mantoes; most commonly in redbed-gypsum transitions at or near the tops or bases of carbonate members, as replacement of siliciclastics or gypsum, and locally carbonate. Hematite- enriched haloes are locally present in the redbed replacements. Minor Cu, Pb, and Zn sulphides and oxides are locally associated within and/or adjacent to the mineralization.

Depositional models for the Central Cape Breton Basin deposits are diverse and range from fault-related epigenetic, through syngenetic primary evaporitic precipitation, to early diagenetic recycling from a pre-enriched source, as well as late stage down-dip remobilization. Interpretations are hindered by post-Windsor Group faults and karstification solution collapse which may in part be Windsor age (i.e. recycled/cannibalized from dissolved evaporites) as well as Late Carboniferous to Recent. Anhydrite in the Windsor Group may contain Sr in the range of 100 to 1000 ppm. Gypsum produced in the near surface, by hydration of the

anhydrite, is unable to absorb all the Sr. Consequently, substantial Sr may have been liberated to migrate in the hydration fluid system. Diagenetic aragonite-calcite conversion also liberates significant Sr. Precipitation at sites away from dissolution and hydration could explain the origin of the Loch Lomond celestite. This process may have been active both during (syngenetic) and after (diagenetic/epigenetic) the deposition of the Windsor Group. It may have been facilitated and accentuated by the unique interaction of the locally steep paleotopography and juxtaposition of alluvial fan, evaporite and marine carbonate depositional environments as well as tectonic activity.

In New Brunswick, there are fault controlled vein-replacement occurrences of barite-celestite at Upper Dorchester and Memramcook East (Hamilton, 1968; Ruitenberg et al., 1977; Fig. 9.88). The occurrence near Upper Dorchester is hosted by grey siliciclastics and carbonates of the Horton and Windsor groups, respectively, whereas the occurrence near Memramcook East is hosted by Horton Group redbeds (Rose and Johnson, 1990). There are also replacement-type manganese occurrences in algal banks of the lower Windsor Group, in which manganese oxides fill primary and/or secondary porosity in the Gays River limestone. The best known deposit is at Markhamville (Fig. 9.84), where about 20 000 tonnes of high-grade manganese ore was mined prior to the turn of the century (Rose and Johnson, 1990).

The Ronan deposit is the largest of several barite-celestite occurrences on the Port au Port Peninsula, western Newfoundland (Johnson, 1954; McArthur and Knight, 1974). The deposit, consisting of granular and crystalline barite and celestite interbedded with sandy and shaly limestones of the lower Codroy Group, occurs in a north-trending Carboniferous paleovalley within Ordovician limestones. The deposit is estimated to contain more than 150 000 tonnes of mixed sulphate material in which celestite is subordinate to barite (Knight, 1983). Dix (1981) suggested that the barite/celestite formed during diagenesis closely following burial of the limestone beneath younger Carboniferous strata.

SUMMARY AND CONCLUSION

H. Scott Swinden

The metallogeny of the Canadian Appalachians is clearly a function of the geological and tectonic history of the orogen. It reflects a wide range of geological processes, in the mantle and at all crustal levels, that have led to the concentration of useful metals and/or minerals in economically significant quantities. The information contained in this chapter demonstrates that the Canadian Appalachian region is indeed, a good laboratory for the study of the metallogeny of a complex accretionary and collisional orogen. The accretion of outboard terranes to the Laurentian continental margin, intrusion, metamorphism, with concomitant uplift, preservation, and exposure of rocks representing a substantial part of the geological history of the orogen, provides an excellent opportunity to study and document the evolution of mineralization processes through the development of the orogen.

Formation of any mineral deposit requires: (1) a metal source; (2) a mechanism of mass transport; and (3) a favourable depositional environment. Within an evolving orogen, there are many opportunities for any or all of these requirements. During the various geotectonic stages of the Appalachian orogenic cycle, distinctly different metal sources, mechanisms of mass transport, and favourable depositional environments were available at different times and places as a result of variations in the accessibility of various mantle or crustal levels to metal-transporting media and the nature and intensity of tectonic activity. The resulting wide variety of mineral deposits were formed at different times, in distinct geological settings, and with different geological and geochemical characteristics.

The oldest mineral deposits in the Canadian Appalachians are the inherited products of earlier orogenic cycles, preserved in older continental crust that has been incorporated in the orogen. Examples of and analogues for these deposits are much more widespread in the ancestral orogens from which they were derived. Magmatic iron-titanium deposits in Humber Zone are geologically part of the Grenville Structural Province, whereas upper Proterozoic volcanogenic massive sulphides, carbonate-hosted, and granite-related deposits in Avalon Zone are geologically part of the Pan-African orogenic belt. Although unrelated to the Appalachian orogenic cycle, these deposits are important to Appalachian metallogeny as pre-concentrated metal sources. In some cases, these constitute stand-alone ore deposits (e.g. Mindamar). However, perhaps more significantly, they also provided metal-enriched sources which could be remobilized and/or upgraded by Appalachian orogenic processes. A good example of this is the Hope Brook deposit, where a late Proterozoic auriferous alteration zone was apparently upgraded during hydrothermal activity related to reactivation of the Cinq Cerf Fault Zone and ultimately during post-tectonic intrusion of the Chetwynd Granite.

At the start of rifting, ore-processes were still acting mainly at shallow levels on sialic crust. The principal mineral deposits preserved from this tectonic stage are beach placers in southeastern Quebec (e.g. Wares), formed through physical transport of heavy minerals from older upper crust near the margins of the orogen. Because incipient rifting in the Canadian Appalachians was apparently not accompanied by extensive hydrothermal activity, there does not seem to have been extensive remobilization of metals within the crust, or transfer of metals between crustal levels. However, it is possible that some of the sediment-hosted copper deposits of southeastern Quebec formed at this time, as a result of rift-related hydrothermal activity in the miogeocline. If so, this may represent the earliest Appalachian example of the concentration of juvenile metals from mantle sources.

The drift stage of orogen development provided the main opportunities for mass transfer from the mantle to the crust and mineral deposits associated with active tectonism in this stage were dominantly formed from juvenile metal sources. For example, podiform chromite deposits were formed at the base of ophiolitic sequences, from where they were subsequently transported tectonically into the upper crust. In active tectonic areas in the oceanic domain (e.g. spreading ridges, subduction zones), chalcophile and precious metals were transferred from the mantle to the oceanic crust by magmatic activity, then reconcentrated at and immediately below the seafloor as volcanogenic sulphide deposits (e.g. all ophiolite-hosted volcanogenic massive sulphide deposits), probably as a result of

hydro-thermal circulation driven by the magmatic heat. Locally, subduction near continental margins resulted in hydro-thermal circulation and ore formation on or near continental crust (e.g. Bathurst) and may have involved crustally derived metals as well. This ore-forming mechanism apparently operated mainly during extensional tectonic regimes, probably enhanced by the high heat flow and the focussing of hydrothermal fluids by extensional fracture systems. Geological and geochronological data show that this process was widespread both temporally and spatially in the pre-accretion orogen. It would appear that most of the observed deposits of this type formed in back arc basins near the Laurentian and Gondwanan margins of Iapetus; no remnants of Iapetan oceanic crust and no metal deposits likely to have formed on the Iapetan major ocean spreading ridge have been recognized to date.

During the drift phase, conditions were also favourable on the continental margins adjacent to Iapetus for the formation of certain mineral deposit types, involving dominantly crustal metal sources and high-level transport processes. On the Laurentian margin, rifting and hydrothermal activity resulted in formation of "Irish-type" Pb-Zn deposits in shelf carbonate sequences (e.g. Upton). Basinward on the passive margin, uraniferous phosphorites were also formed (e.g. St.-Armand). At about the same time, shallow seas spread across the stable Avalonian (Gondwanan) margin, from which sedimentary manganese and iron (e.g. Wabana) deposits were locally formed from metals liberated by weathering of the adjacent continental landmass. In terms of metal sources, all of these deposits represent mainly a redistribution of metals in the upper crust, with hydrothermal activity on the Laurentian shelf providing the main opportunity for transport of metals for significant vertical distances at elevated temperatures.

The beginning of accretion of Iapetan terranes to the Laurentian margin in the Early Ordovician provided opportunities for preservation of drift-related mineral deposits and for generation of new types of deposits. Because accretion apparently spanned a considerable period of time, the drift and accretionary stages of the orogen actually overlap. Between the Middle Ordovician to Early Silurian, the syn-accretion phase of metallogeny had begun in the accreted inboard terranes while the outboard terranes were still experiencing drift-phase metallogeny.

The processes of accretion included emplacement of outboard terranes upon the Laurentian margin with concomitant magmatic (both plutonic and volcanic) activity resulting from crustal thickening as well as the reactivation of pre-existing trans-crustal faults and formation of new faults. The widespread volcanic activity that accompanied and closely followed cratonization of outboard terranes at the Laurentian margin locally produced volcanogenic, high-level stockwork and exhalative base metal deposits (e.g. Sewell Brook). This mineralizing process probably is a consequence of the availability of both sea water in remnant marine basins at the continental margin coupled with the presence of heat sources supplied by the rising magmas. The resulting hydrothermal systems produced deposits that are similar in many respects to the Iapetan volcanogenic massive sulphide deposits formed during the pre-accretion stage. If the mineralizing process mainly occurred through the action of sea water-dominated hydrothermal systems, it would probably have involved mainly redistribution and concentration of metals in the upper crust rather than introduction of juvenile metals from the deep crust or mantle.

More or less contemporaneously with the formation of volcanogenic base metal deposits, movements on major faults and their subsidiary second and third order structures produced channelways and thermal gradients through which deep-seated hydrothermal fluids, perhaps formed in part through metamorphic dewatering at deep crustal levels, could move. Heat was locally provided by granitoid magma, utilizing the same structures for ascent. The resulting mineralizing process involved transfer of metals such as Au, As and base metals from the lower crust to the upper crust. Depositional environments were provided by medium- and high-level dilatant structures accompanied by chemical traps (forming mesothermal and epithermal deposits, respectively).

As crustal thickening related to accretion proceeded, deep partial melting and the ascent of granitic magmas provided further opportunities for the transfer of metals from the lower to upper crust. Granophile and base metals derived from the magma source regions were concentrated in the differentiating magmas and locally concentrated in orthomagmatic or magmatic/hydrothermal (e.g. pegmatite, vein) deposits. The ascending granitic magmas may also have been responsible for promoting hydrothermal circulation in the upper crust leading to remobilization of metals from pre-enriched sources and to the formation of new hydrothermal mineral deposits. The local presence of syn-accretion mafic intrusions locally containing concentrations of Ni (e.g. St. Stephen) and intermediate intrusions with porphyry-type base metal concentrations indicate that transfer of economic metals from the mantle to the crust was also occurring at this time, albeit at a considerably reduced scale relative to the drift phase. This is in accord with isotopic data for some syn- and post-orogenic granites that indicate a juvenile source (Fryer et al., 1992).

Accretionary processes were accompanied by dewatering of nearby sedimentary basins and where fluids encountered favourable depositional environments at higher structural levels on the Laurentian margin, they locally produced base metal and barite veins (e.g. St. Fabien). Similar processes may have been partially responsible for the formation of epigenetic carbonate-hosted Zn and Pb deposits in southeastern Quebec and western Newfoundland, although the timing and source of metals is not well constrained in these cases. These mineralizing processes resulted mainly in the redistribution of metals in the upper crust, probably not involving crustal-scale vertical mass transfer of metals.

Post-accretion mineralizing processes were accompanied by a dramatic decrease in the intensity of tectonism, with the result that mass transfer of economic minerals occurred principally within the upper crustal levels. Most sediment-hosted mineralization is believed to be related to movement on large faults within and around successor basins, with local fluid circulation in the immediate basement. Relief on basin-bounding faults provided the local opportunity for the formation of placer deposits derived from the immediately adjacent landmass (e.g. Gays River gold). Evaporitic basins produced thick deposits of sulphates which are, in themselves, of considerable economic significance, and which contributed soluble salts to basinal fluids, enhancing their metal-carrying capacity. Metals were probably derived from the sediments in the basins and from the immediately adjacent basement. Depositional environments were provided by porous units with available reduced sulphur, either sandstones (e.g. Dorchester, Yava) or carbonates (e.g. Gays River) or by dilatant zones in competent units (e.g. Lake Ainslie area).

In summary, the Canadian Appalachians provide a relatively detailed record of ore-forming processes that were evolving in concert with the developing orogen. The initiation of the Appalachian cycle involved mainly crustal processes with minimal tectonism and this is reflected in the high-level, generally non-tectonic nature of the associated deposits. As the cycle proceeded, the transfer of mantle material to the crust via magmatism in active tectonic environments was reflected in the widespread formation of mineral deposits from juvenile metals. The accretion of this juvenile material to the Laurentian margin provided both new mineral deposits and also sources of moderately concentrated juvenile metals that were available for mobilization and reconcentration during accretion-related tectonism and magmatism. The tectonic conditions prevalent during accretionary stages encouraged transfer of metals vertically from lower to upper crust (and from newly accreted juvenile crust to upper crust), and was accompanied by minor addition of new mantle-derived metals via the ascent of mafic intrusions. The increased involvement of the continental crust in ore-forming processes in these stages is reflected in the increasing involvement of granophile elements (e.g. Mo, Sn, W, U) in the ore-forming processes. As active tectonism waned, so too did the opportunities for substantial vertical transport of metals and post-accretion metallogeny involved mainly high-level, relatively low temperature processes, completing the cycle.

It is no accident that this metallogenic model is similar to that advanced by Bilibin (1968) many years ago. The empirical observations that led to these models are as valid today as they were half a century ago. The principal difference is that today, the plate tectonic framework provides actualistic and readily observable analogues for ancient processes. These analogues permit a more rigorous interpretation of the tectonic environments and processes in all stages of orogenic development and, in particular, allow many of our interpretations of ore-forming environments and processes to be confirmed by observation. A particularly striking example of this has been the discovery of massive sulphide deposits in a number of tectonic environments on the modern sea floor. Undoubtedly, as our understanding of orogenic processes are refined by studies of ancient and modern environments, our ability to interpret the contained mineral deposits in terms of regional metallogenic models and our ability to resolve individual metallogenic processes within ancient complex orogens will continue to be substantially enhanced.

ACKNOWLEDGMENTS

A great number of friends and colleagues have contributed to this chapter in many important ways. We would first like to thank all of the authors who contributed material for this chapter for their patience and willingness to rewrite and update information through many versions of the manuscript. The project was started by Paul Dean, who made initial contact with authors and left us a wealth of information with which to work.

We approached many colleagues in the Atlantic Provinces and Quebec during the writing of this material for new information, for reviews of parts of sections, for advice and for encouragement. Some of these were authors of sections whom we asked for additional help with material for which they were not directly responsible. These include, in alphabetical order: Sandra Barr, A.K. Chatterjee, Lawson Dickson, Martin Doyon, Benoit Dubé, Greg Dunning, Dave Evans, Becky Jamieson, George Jenner, Baxter Kean, Dan Kontak, Mike MacDonald, Michel Malo, Steve McCutcheon, Malcolm McLeod, Sean O'Brien, Peter Stewart, Joe Whalen, and Garry Yeo. H. Williams is thanked for reviews and comments.

REFERENCES

Adams, G.D.
1991: Gypsum and anhydrite resources in Nova Scotia; Nova Scotia Department of Natural Resources, Economic Geology Series 91-1, 293 p.

Akande, S.O. and Zentilli, M.
1984: Geologic, fluid inclusion, and stable isotope studies of the Gays River lead-zinc deposit, Nova Scotia, Canada; Economic Geology, v. 79, p. 1187-1211.

Alcock, J.B.
1982: Skarn and porphyry popper mineralization at Mines Gaspé, Murdochville, Québec; Economic Geology, v. 77, p. 971-999.

Ambrose, J.W.
1942: Preliminary map of Mansonville area, Quebec; Geological Survey of Canada, Paper 42-1.

Anderle, J.P., Crosby, K.S., and Waugh, D.C.E.
1979: Potash at Salt Springs, New Brunswick; Economic Geology, v. 74, p. 389-396.

Anderson, F.D.
1968: Woodstock, Millville and Coldstream map areas, Carleton and York counties, New Brunswick; Geological Survey of Canada, Memoir 353, 69 p.

Andrews, P.
1990: A summary of the geology and exploration history of the Hammer Down gold deposit, Springdale area, central Newfoundland; in Metallogenic Framework of Base and Precious Metal Deposits, Central and Western Newfoundland, (ed.) H.S. Swinden, D.T.W. Evans, and B.F. Kean; 8th IAGOD Symposium, Field Trip No. 1, Guidebook, Geological Survey of Canada, Open File 2156, p. 146-151.

Andrews, P. and Huard, A.A.
1991: Geology of the Hammer Down deposit, Green Bay, north-central Newfoundland; in Ore Horizons, Volume 1, (ed.) S. Swinden and A. Hogan; Newfoundland Department of Mines and Energy, Geological Survey Branch, p. 63-74.

Atkinson, D.
1977: Catheart Mountain, an Ordovician porphyry copper-molybdenum occurrence in northern Appalachia; Ph.D. thesis, University of Western Ontario, London, Ontario, 149 p.

Auclair, M.
1990: Eastern metals-une listwaenite cambro-ordovicienne gisaileé et minéralisée de Ni-Cu-Zn-Co-Au dans les Appalaches du Québec; thèse de matrise, Université du Québec à Montréal, Montréal, Québec, 152 p.

Auclair, M., Gauthier, M., Trottier, J., Jébrak, M. and Chartrand, F.
1993: Mineralogy, geochemistry and paragenesis of the Eastern Metals serpentinite-associated Ni-Cu-Zn deposit, Quebec Appalachians; Economic Geology, v. 88, p. 123-138.

Barbour, D.
1990: Valentine Lake gold prospect; in Metallogenic Framework of Base and Precious Metal Deposits, Central and Western Newfoundland, (ed.) H.S. Swinden, D.T.W. Evans, and B.F. Kean; 8th IAGOD Symposium, Field Trip No. 1, Guidebook, Geological Survey of Canada, Open File 2156, p. 73-76.

Barr, S.M. and Jamieson, R.A.
1991: Tectonic setting and regional correlation of Ordovician-Silurian rocks of the Aspy Terrane, Cape Breton Island, Nova Scotia; Canadian Journal of Earth Sciences, v. 28, p. 1769-1779.

Barr, S.M. and White, C.E.
1988: Petrochemistry of contrasting Late Precambrian volcanic and plutonic associations, Caledonian Highlands, southern New Brunswick; Maritime Sediments and Atlantic Geology, v. 24, p. 353-372.
1991: Geological mapping of the eastern Caledonian Highlands, Kings and Saint John counties, southern New Brunswick (C-NBCAMD); in Project Summaries for 1991, 16th Annual Review of Activities; New Brunswick Department of Natural Resources and Energy, Mineral Resources Branch, Information Circular 91-2, p. 190-194.

Barr, S.M., Raeside, R.P., and van Breeman, O.
1987: Grenvillian basement in the northern Cape Breton Highlands; Canadian Journal of Earth Sciences, v. 24, p. 992-997.

Barr, S.M., White, C.E., and MacDonald, A.S.
1992: Revision of upper Precambrian - Cambrian stratigraphy, southeastern Cape Breton Island, Nova Scotia; in Current Research, Part D; Geological Survey of Canada, Paper 92-1D, p. 21-26.

Barr, S.M., Bevier, M.L., White, C.E., and MacDonald, A.S.
1991: Geochronology and petrochemistry of volcanic and plutonic rocks, SE Cape Breton Island (Avalon Terrane), Canada; Geological Association of Canada, Programs with Abstracts, v. 16, p. A7.

Barr, S.M., Dunning, G.R., Raeside, R.P., and Jamieson, R.A.
1990: Contrasting U-Pb ages from plutons in the Bras D'Or and Mira Terranes of Cape Breton Island, Canada; Canadian Journal of Earth Sciences., v. 27, p. 1200-1208.

Barr, S.M., Yip Choy, R., Colwell, J.A., and Oldale, H.S.
1984: Granitoid rocks and associated copper skarn, Whycocomagh Mountain, Cape Breton Island, Nova Scotia; Maritime Sediments and Atlantic Geology, v. 20, p. 43-56.

Beaudoin, G., Schrijver, K., Marcoux, F., Calvez, J.Y.
1989: A vein and disseminated Ba-Pb-Zn deposit in the Appalachian thrust belt, St-Fabien, Quebec; Economic Geology, v. 84, p. 799-816.

Bell, K. and Blenkinsop, J.
1975: Geochronology of eastern Newfoundland; Nature, v. 254, p. 410-411.

Bell, L.V.
1948: Caribou mine; in Structural Geology of Canadian Ore Deposits; Canadian Institute of Mining and Metallurgy, Jubilee Volume, p. 927-936.

Bevier, M.L.
1989: Preliminary U-Pb geochronological results for igneous and metamorphic rocks, New Brunswick; in Project Summaries for 1989, 14th Annual Review of Activities, (ed.) S.A. Abbott; New Brunswick Department of Natural Resources and Energy, Minerals and Energy Division, Information Circular 89-2, p. 190-194.

Bevier, M.L. and Barr, S.M.
1990: U-Pb constraints on the stratigraphy and tectonic history of the Avalon Terrane, New Brunswick, Canada; Journal of Geology, v. 98, p. 53-63.

Bevier, M.L., Barr,S.M., White, C.E., Macdonald, A.S.
1993: U-Pb geochronologic constraints on the volcanic evolution of the Mira (Avalon) Terrane, southeastern Cape Breton Island, Nova Scotia; Canadian Journal of Earth Sciences, v. 30, p. 1-10.

Bilibin, Y.A.
1968: Metallogenic Provinces and Metallogenic Epochs; E.A. Alexandrov, (translation); Queen's College Press, Flushing, New York.

Binney, W.P.
1975: Copper occurrences in Lower Carboniferous sedimentary rocks of the Maritime Provinces; Geological Survey of Canada, Open File 281, 156 p.
1987: A sedimentological investigation of MacLean channel transported sulphide ores; in Buchans Geology, Newfoundland, (ed.) R.V. Kirkham; Geological Survey of Canada., Paper 86-24, p. 107-148.

Binney, W.P., Jenner, K.A., Sangster, A.L., and Zentilli, M.
1986: A stratabound lead-zinc deposit in Meguma Group metasediments at Eastville, Nova Scotia; Maritime Sediments and Atlantic Geology, v. 22, p. 65- 88.

Bjørlykke, A.
1980: Galena occurrences in Late Precambrian-Cambrian sandstone in Norway; Norges Geologisk Unders-kelse, no. 360, p. 95-97.

Bjørlykke, A. and Sangster, D.F.
1981: An overview of sandstone lead deposits and their relation to redbed copper and carbonate hosted lead-zinc deposits; Economic Geology, 75th Anniversary Volume, p. 179-213.

Boehner, R.C.
1986: Salt and potash resources in Nova Scotia; Nova Scotia Department of Mines and Energy, Bulletin, v. 5, 346 p.

Boehner, R.C. and Ryan, R.J.
1989:   Economic geology of Late Paleozoic-Early Mesozoic basins in Nova Scotia: Mid-Westphalian tectonism and metallogenesis; in Report of Activities, 1989, Part A; Nova Scotia Department of Mines and Energy, Mineral Resources Branch, Report 89-3, p. 103-108.

Bonham, O.J.H.
1983:   Mineralization controls at the Yava lead deposit, Salmon River, Cape Breton County, Nova Scotia; M.Sc. thesis, Dalhousie University, Halifax, Nova Scotia, 251 p.

Boorman, R.S.
1968:   Silver in some New Brunswick galenas; New Brunswick Research and Productivity Council, Research Note 11, 11 p.

Born, P.
1983:   Report on diamond drilling program of 1983 on the Round Pond property, Great Northern Peninsula, Newfoundland; Narex Ore Search Consultants Inc., Newfoundland Department of Mines, Open File 12P(51).

Bossé, J., Paradis, S., and Gauthier, M.
1992:   L'amas sulfuré de Champagne: un gite exhalatif dans les argilites ordoviciennes du Groupe de Magog, Appalaches du Québec; dans Recherches en cours, Partie D; Commission géologique du Canada, Étude 92-1D, p. 129-134.

Boyle, R.W.
1966:   Origin of gold and silver deposits of the Meguma Series, Nova Scotia; Geological Association of Canada/Mineralogical Association of Canada, Program, Annual Meeting.
1972:   The geology, geochemistry and origin of the barite, manganese and lead-zinc-copper-siver deposits of Walton-Cheverie areas, Nova Scotia; Geological Survey of Canada, Bulletin 166, 181 p.
1979:   The geochemistry of gold and its deposits; Geological Survey of Canada, Bulletin 280, 584 p.
1986:   Gold deposits in turbidite sequences: their geology, geochemistry, and history of the theories of their origin; in Turbidite-hosted Gold Deposits, (ed.) J.D. Keppie, R.W. Boyle, and S.J. Haynes; Geological Association of Canada, Special Paper 32, p. 1-14.

Boyle, R.W. and Davies, J.L.
1964:   Geology of the Austin Brook and Brunswick No. 6 sulphide deposits, Gloucester County, New Brunswick; Geological Survey of Canada, Paper 63-24, 23 p.

Boyle, R.W., Wanless, R.K., and Stevens, R.D.
1976:   Sulphur isotope investigation of the barite, manganese, and lead-zinc-copper-silver deposits of the Walton-Cheverie area, Nova Scotia, Canada; Economic Geology, v. 71, p. 749-762.

Bradley, D.A.
1962:   Gisborne Lake and Terrenceville map areas, Newfoundland (1M/15 and 10); Geological Survey of Canada, Memoir 321, 56 p.

Brown, A.C.
1989:   Sediment-hosted stratiform copper deposits: deposit-type, name, and related terminology; in Sediment-Hosted Stratiform Copper deposits, (ed.) E.C. Jowett and R.V. Kirkham; Geological Association of Canada, Special Paper 36, p. 39-51.

Brunton, S. Sir
1928:   Some observations on a so-called interbedded vein at Tangier, Nova Scotia; Mining Society of Nova Scotia, Annual Meeting, p. 784-803.

Butt, K.A.
1976:   Genesis of granitic rocks in southwestern New Brunswick; Ph.D. thesis, University of New Brunswick, Fredericton, New Brunswick, 234 p.

Calon, T.J. and Green, F.K.
1987:   Preliminary results of detailed structural analysis at Buchans; in Buchans Geology, Newfoundland, (ed.) R.V. Kirkham; Geological Survey of Canada, Paper 86-24, p. 273-288.

Campbell, J.S.
1863:   Nova Scotia Gold fields: A report to the Honourable Joseph Howe, Provincial Secretary; Halifax, 12 p.

Campbell, J.S. and Poole, H.
1862:   Report on the goldfields: a report to the commissioner of the Crown Lands; Halifax, 36 p.

Caron, A.
1991:   Lake George antimony-gold project, New Brunswick (C-NBCAMD); in Project Summaries for 1991; New Brunswick Department of Natural Resources and Energy, Mineral Resources Branch, Information Circular 91-2, p. 134-136.

Carriere, G.
1957:   Huntingdon Mine; in Structural Geology of Canadian Ore Deposits; Commonwealth C.I.M.M. Congress, v. 3, p. 462-466.

Cattalani, S.
1987:   A fluid inclusion and stable isotopic study of the St. Robert W-Ag-Bi vein deposit, Eastern Townships, Québec; M.Sc. thesis, McGill University, Montreal, Quebec, 107 p.

Charbonneau, J.M.
1975:   Analyse structurale des tectonites métamorphiques du Groupe de Oak Hill dans la region de Saint-Sylvestre, Appalaches du Quebec; thèse de matrise, Université Laval, Québec (Québec),61 p.

Charest, M.H.
1976:   Petrology, geochemistry and mineralization of the New Ross area, Lunenburg County, Nova Scotia; M.Sc. thesis, Dalhousie University, Halifax, Nova Scotia, 192 p.

Charest, M.H., Farley, E.J., and Clarke, D.B.
1985:   The northwestern part of the New Ross-Vaughan complex: petrology, geochemistry and mineral deposits; in Guide to the Granites and Mineral Deposits of Southwestern Nova Scotia, (ed.) A.K. Chatterjee and D.B. Clarke; Nova Scotia Department of Mines and Energy, Paper 85-3, p. 29-40.

Chartrand, F., Trottier, J., and Gauthier, M.
1987:   Epigenetic copper-barite mineralization in paleokarst at the Lord Aylmer deposit, southern Quebec Appalachians; Economic Geology, v. 82, p. 735-739.

Chatterjee, A.K.
1977a: Tungsten mineralization in the carbonate rocks of the George River Group, Cape Breton Island, Nova Scotia; Nova Scotia Department of Mines and Energy, Paper 77-7, 26 p.
1977b: Copper-zinc deposit at McMillan Brook, Victoria County, Cape Breton Island; in Report of Activities, 1976; Nova Scotia Department of Mines and Energy, Mines and Minerals Branch, Paper 77-1, p. 55-66.
1979:   Geology of the Meat Cove zinc deposit, Cape Breton Island, Nova Scotia; Nova Scotia Department of Mines and Energy, Report 79-3, 27 p.
1980:   Mineralization and associated wall rock alteration in the George River Group, Cape Breton Island, Nova Scotia; Ph.D. thesis, Dalhousie University, Halifax, Nova Scotia, 197 p.
1983:   Mineral deposit studies: contrasting granophile (Sn, W, Mo, Cu, U) deposits of Nova Scotia; in Report of Activities, 1982; Nova Scotia Department of Mines and Energy, Mines and Minerals Branch, Report 83-1, p. 49-51.
1984:   Devono-Carboniferous magmatism and epithermal U-Th-Mo-Ag-F mineralization in the Debert Lake area, eastern Cobequid Highlands; in Report of Activities, 1983; Nova Scotia Department of Mines and Energy, Mines and Minerals Branch, Report 84-1, p. 239-240.

Chatterjee, A.K. and Oldale, H.R.
1980:   Porphyry copper - molybdenum gold mineralization at Coxheath; in Mineral Deposits and Mineralogenic Provinces of Nova Scotia; Geological Association of Canada, Annual Meeting, Halifax, Nova Scotia, Field Trip No. 6, Guidebook, p. 3-10.

Chatterjee, A.K., Robertson, J., and Pollock, D.
1982:   A summary of the petrometallogenesis of the uranium mineralization at Millet Brook, South Mountain Batholith, Nova Scotia; in Report of Activities, 1981; Nova Scotia Department of Mines and Energy, Mines and Minerals Branch, Report 82-1, p. 57- 66.

Chatterjee, A.K., Strong, D.F., and Clarke, D.B.
1985a: Petrology of the polymetallic quartz-topaz greisen at East Kemptville; in Guide to the Granites and Mineral Deposits of Southwestern Nova Scotia, (ed.) A.K. Chatterjee and D.B. Clarke; Nova Scotia Department of Mines and Energy, Paper 85-3, preprint, p. 149-196.

Chatterjee, A.K., Strong, D.F., Clarke, D.B., and Keppie, J.D.
1985b: Geology and geochemistry of the Carboniferous Wedgeport pluton hosting Sn-W mineralization; in Guide to the Granites and Mineral Deposits of Southwestern Nova Scotia, (ed.) A.K. Chatterjee and D.B. Clarke; Nova Scotia Department of Mines and Energy, Paper 85-3, preprint, p. 201-216.

Chatterjee, A.K., Strong, D.F., Clarke, D.B., Robertson, J., Pollock, D., and Muecke, G.K.
1985c: Geochemistry of the granodiorite hosting uranium mineralization at Millet Brook; in Guide to the Granites and Mineral Deposits of Southwestern Nova Scotia, (ed.) A.K. Chatterjee and D.B. Clarke; Nova Scotia Department of Mines and Energy, Paper 85-3, preprint, p. 63-114.

Chevé, S.R., Brown, A.C., and Trzcienski, W.E., Jr.
1983:   Proto-rift related volcanogenic sulfide deposits, Clinton River area, Eastern Townships, Quebec; International Geological Correlation Program, Project 60, Symposium on Stratabound Sulphides of the Appalachian-Caledonian Orogen, Ottawa, Program with Abstracts, p. 9-10.

Clarke, D.B. and Halliday, A.N.
1980:   Strontium isotope geology of the South Mountain batholith; Geochimica et Cosmochimica Acta, v. 44, p. 1045-1058.

Clarke, D.B., Barr, S.M., and Donohoe, H.V.
1980:   Granitoid and other plutonic rocks of Nova Scotia; in Proceedings: The Caledonides in the U.S.A., (ed.) D. R. Wones; I.G.C.P. Project 27: Caledonide Orogen, Department of Geological Science, Virginia Polytechnic Institute and State University, Memoir 2, p. 107-116.

Clarke, D.B., Muecke, G.K., and Chatterjee, A.K.
1985:   The South Mountain batholith: geology, petrology and geochemistry; in Guide to the Granites and Mineral Deposits of Southwestern Nova Scotia, (ed.) A.K. Chatterjee and D.B. Clarke; Nova Scotia Department of Mines and Energy, Paper 85-3, Preprint, p. 1-14.

Coates, H.
1990:   Geology and mineral deposits of the Rambler Property; in Metallogenic Framework of Base and Precious Metal Deposits, Central and Western Newfoundland, (ed.) H.S. Swinden, D.T.W. Evans, and B.F. Kean; 8th IAGOD Symposium, Field Trip No. 1, Guidebook, Geological Survey of Canada, Open File 2156, p. 184-193.

Coish, R.A., Hickey, R., and Frey, F.A.
1982:   Rare earth element geochemistry of the Betts Cove ophiolite, Newfoundland: complexities in ophiolite formation; Geochimica et Cosmochimica Acta, v. 46, p. 2117-2134.

Collins, J.A. and Smith, L.
1975:   Zinc deposits related to diagenesis and intrakarstic sedimentation in the Lower Ordovician St. George Formation, western Newfoundland; Canadian Petroleum Geology, Bulletin, v. 23, p. 393-427.

Collins, M.J.
1984:   Baie Verte Mines Asbestos deposit; in Mineral Deposits of Newfoundland, A 1984 Perspective; Newfoundland Department of Mines and Energy, Mineral Development Division, Report 84-3, p. 98-104.

Collins, W.J., Beams, S.D., White, A.J.R., and Chappell, B.W.
1982:   Nature and origin of A-type granites with particular reference to southeastern Australia; Contributions to Mineralogy and Petrology, v. 80, p. 189-200.

Colman-Sadd, S.P.
1974:   Geology of the iron deposits near Stephenville, Newfoundland; M.Sc. Thesis, Memorial University of Newfoundland, St. John's, Newfoundland, 97 p.
1980:   Geology of south-central Newfoundland and evolution of the eastern margin of Iapetus; American Journal of Science, v. 280, p. 991-1017.

Colman-Sadd, S.P. and Swinden, H.S.
1982:   Geology and mineral potential of south-central Newfoundland; Newfoundland Department of Mines and Energy, Mineral Development Division, Report 82-8, 102 p.
1984:   A tectonic window in Central Newfoundland? Geological evidence that the Appalachian Dunnage Zone may be allochthonous; Canadian Journal of Earth Sciences, v. 21, p. 1349-1367.

Colman-Sadd, S.P., Dunning, G.R. and Dec, T.
1992:   Dunnage-Gander relationships and Ordovician orogeny in central Newfoundland: a sediment provenance and U/Pb age study; American Journal of Science, v. 292, p. 317-355.

Colpron, M.
1987:   Géologie de la Région de Sutton; Ministère de l'Energie et des Ressources du Quebec, MB-87-29, 60 p.

Corey, N.C. and Horne, R.J.
1989:   Polymetallic-precious metal potential of the Tobeatic Fault Zone in southwestern Nova Scotia; in Report of Activities, 1988, Part B; Nova Scotia Department of Mines and Energy, Mines and Minerals Branch, Report 89-1, p. 27-36.

Cormier, R.F., Keppie, J.D., and Odom, A.L.
1988:   U-Pb and Rb-Sr geochronology of the Wedgeport granitoid pluton, southwestern Nova Scotia; Canadian Journal of Earth Sciences, v. 25, p. 255-261.

Crevier, M.
1984:   Mount Costigan property (NTS 21 J/14E); unpublished report for Lac Minerals Ltd., New Brunswick Department of Natural Resources and Energy, Mineral Resources Division, Assessment File 473035, 20 maps.

Crosby, R.M.
1984:   McLean Brook area (NTS 21 J/10); unpublished report for Miramichi Lumber Company, New Brunswick Department of Natural Resources and Energy, Mineral Resources Division, Assessment File 473019, 12 p., 1 map.

Crossley, R.V. and Lane, T.E.
1984:   A guide to the Newfoundland Zinc Mines Limited ore bodies, Daniels Harbour; in Mineral Deposits of Newfoundland, 1984 Perspective, (ed.) H.S. Swinden; Newfoundland Department of Mines and Energy, Report 84-3 p. 45-51.

Cumming, L.M.
1968:   St. George-Table Head disconformity and zinc mineralization, western Newfoundland; Canadian Institute of Mining and Metallurgy, Bulletin, v. 61, p. 721-725.
1985:   A Halifax slate graptolite locality, Nova Scotia; in Current Research, Part A; Geological Survey of Canada, Paper 85-1A, p. 215-221.

Currie, K.L., Loveridge, W.D., and Sullivan, R.W.
1982:   A U-Pb age on zircon from dykes feeding basal rhyolitic flows of the Jumping Brook complex, northwestern Cape Breton Island, Nova Scotia; in Current Research, Part C; Geological Survey of Canada, Paper 82-1C, p. 125-128.

Dahl, R. and Watkinson, D.H.
1986:   Structural control on podiform chromite in Bay of Islands ophiolite, Springers Hill area, Newfoundland; in Current Research, Part B; Geological Survey of Canada, Paper 86-1B, p. 757-766.

Dallmeyer, R.D., Hussey, E.M., O'Brien, S.J., and O'Driscoll, C.F.
1983:   Chronology of tectonothermal activity in the western Avalon Zone of the Newfoundland Appalachians; Canadian Journal of Earth Sciences, v. 20, p. 355-363.

Davenport, P.H. and McConnell, J.W.
1989:   Lake sediment geochemistry in regional exploration for gold; in Prospecting in Areas of Glaciated Terrain, 1988, (ed.) D.R. MacDonald and K.A. Mills; Canadian Institute of Mining and Metallurgy, Geology Division, Halifax, Nova Scotia, p. 333-356.

Davenport, P.H. and Nolan, L.W.
1989:   Mapping the regional distribution of gold in Newfoundland using lake sediment geochemistry; in Current Research; Newfoundland Department of Mines, Geological Survey Branch, Report 89-1, p. 259-266.

Davies, J.L.
1972:   The geology and geochemistry of the Austin Brook area, Gloucester County, New Brunswick, with special emphasis on the Austin Brook iron formation; Ph.D. thesis, Carleton University, Ottawa, Ontario, 254 p.
1979:   Geological map of northern New Brunswick; New Brunswick Department of Natural Resources, Mineral Resources Branch, Map NR-3.

Davies, J.L., Fyffe, L.R., and McAllister, A.L.
1983:   Geology and massive sulphides of the Bathurst area, New Brunswick; in Field Trip Guidebook to Stratabound Sulphide Deposits, Bathurst Area, New Brunswick, Canada and West-Central New England, U.S.A., (ed.) D.F. Sangster; Geological Survey of Canada, Miscellaneous Report 36, p. 1-30.

Davies, J.L., Tupper,W.M., Bachinski, D., Boyle, R.W., and Martin, R.
1969:   Geology and mineral deposits of the Nigadoo River-Millstream River area, Gloucester County, New Brunswick; Geological Survey of Canada, Paper 67-49, 70 p.

Dean, P.L.
1978:   The volcanic stratigraphy and metallogeny of Notre Dame Bay; Memorial University of Newfoundland, St. John's, Geology Report 7, 204 p.

de Launay, L.
1900:   Sur les types régionaux des gites metallifères; Academie des Sciences, Comptes Rendus, Paris, v. 130, p. 743-746.

De Römer, H.S.
1977:   Region des Monts McGerrigle; Ministère de l'Énergie et des Ressources du Quebec, R.G.-174.

de Roo, J.A. and van Staal, C.R.
1990:   The structure of the Half Mine Lake region, Bathurst camp, New Brunswick; in Current Research, Part D; Geological Survey of Canada, Paper 91-1D, p. 179-186.

de Roo, J.A., Moreton, C., Williams, P.F., and van Staal, C.R.
1990:   The structure of the Heath Steele Mines region, Bathurst camp, New Brunswick; Atlantic Geology, v. 26, p. 27-41.

Desnoyers, D.
1990:   Victoria Mine prospect; in Metallogenic Framework of Base and Precious Metal Deposits, Central and Western Newfoundland; 8th IAGOD symposium, Field Trip No. 1, Guidebook, Geological Survey of Canada, Open File 2156, p. 65-67.

Dickie, J.R.
1978:   Geological, mineralogical and fluid inclusion studies at the Dunbrack lead-silver deposit, Musquodoboit Harbour, Halifax County, Nova Scotia; B.Sc. thesis, Dalhousie University, Halifax, Nova Scotia.

Dickson, W.L.
1982a: The southern contact of the Ackley Granite: location and mineralization; in Current Research; Newfoundland Department of Mines and Energy, Mineral Development Division, Report 82-1, p. 99-108.

Dickson, W.L. (cont.)

1982b: Geology of the Wolf Mountain (12A2/W) and Burnt Pond (12A3/E) map areas; Newfoundland Department of Mines and Energy, Mineral Development Division Report 82-5, 43 p.

1983: Geology, geochemistry and mineral potential of the Ackley Granite and parts of the Northwest Brook and Eastern Meelpaeg complexes, southeast Newfoundland (parts of map areas 1M/10, 11, 14, 15, 16; 2D/1, 2, 3 and 7); Newfoundland Department of Mines and Energy, Mineral Development Division, Report 83-6, 129 p.

1990: Geology of the North Bay Granite Suite and metasedimentary rocks in southern Newfoundland (NTS 11P/15E, 11P/16 and 12A/2E); Newfoundland Department of Mines and Energy, Geological Survey Branch, Report 90-3, 101 p.

Dix, G.R.

1981: The Codroy Group (Upper Mississippian), on the Port au Port Peninsula, Western Newfoundland: stratigraphy, paleontology, sedimentology and diagenesis; M.Sc. thesis, Memorial University of Newfoundland, St. John's, Newfoundland.

Doig, R., Nance, R.D., Murphy, J.B., and Cassidy, R.P.

1990: Evidence for Silurian sinistral accretion of Avalon composite terrane in Canada; Journal of the Geological Society of London, v. 147, p. 927-930.

Dostal, J., Keppie, J.D., and Murphy, J.B.

1990: Geochemistry of Late Proterozoic basaltic rocks from southeastern Cape Breton, Nova Scotia; Canadian Journal of Earth Sciences, v. 27, p. 619-631.

Douglas, G.V.

1940: Antimony at West Gore, Province of Nova Scotia; Department of Mines, Annual Report on Mines, 1939, Part 2, p. 37-49.

1948: Structure of the gold veins of Nova Scotia; in Structural Geology of Canadian Ore Deposits; Canadian Institute of Mining and Metallurgy, Jubilee Volume, p. 919-926.

Douglas, J.L.

1983: Geochemistry of the Cambrian manganese deposits of eastern Newfoundland; Ph.D. thesis, Memorial University of Newfoundland, St. John's, Newfoundland, 305 p.

Dowling, W.M.

1988: Depositional environment of the Lower Oak Hill Group, Southern Québec: implications for the Late Precambrian breakup of North America in the Québec reentrant; M.Sc. thesis, University of Vermont, U.S.A., 186 p.

Dresser, J.A.

1908: Rapport sur une dicouvertre récente d'or pres du Lac Mégantic, Québec; Commission géologique du Canada, publication 1032, 22 p.

Dubé, B.

1990: Contrasting styles of gold: only deposits in western Newfoundland (a preliminary report); in Current Research, Part B; Geological Survey of Canada, Paper 90-1B, p. 77-90.

Dubé, B., Guha, J., and Rocheleau, M.

1987: Alteration patterns related to gold mineralization and their relation to CO_2-H_2O ratios; Contributions to Mineralogy and Petrology, v 37, p. 267-291.

Dubé, B., Lauzière, K., and Gaboury, D.

1992: Preliminary report on the structural control of the Rendall-Jackman gold deposit, Springdale Peninsula, Newfoundland; in Current Research, Part D; Geological Survey of Canada, Paper 92-1D, p. 1-10.

Dubé, B., Lauzière, K., and Tremblay, A.

1991a: Observations on the structural control and tectonic setting of gold mineralization in the Cape Ray Fault Zone, southwestern Newfoundland; in Current Research, Part D; Geological Survey of Canada, Paper 90-1D, p. 135-145.

Dubé, B., Tremblay, A., Malo, M., Mengel, F., Lynch, G., Lauzière, K., and Godue, R.

1991b: Observations on the structural controls of gold-only deposits in the Canadian Appalachians with an exploration perspective; Paper presented at the Prospectors and Developers Annual Meeting, Toronto, Canada, March 1991.

Dugas, J., Assad, R., and Marleau, R.

1969: Metallogenic concepts in Gaspé; The Canadian Mining and Metallurgical Bulletin, v. 62, no. 688, p. 846-853.

Duke, N.A. and Hutchinson, R.W.

1974: Geologic relationships between massive sulfide bodies and ophiolitic volcanic rocks near York Harbour, Newfoundland; Canadian Journal of Earth Sciences, v. 11, p. 53-69.

Dunham, K.C.

1950: Petrography of the nickeliferous norite of St. Stephen, New Brunswick; The American Mineralogist, v. 35, p. 711-727.

Dunning, G.R.

1987: Geology of the Annieopsquotch Complex, southwest Newfoundland; Canadian Journal of Earth Sciences, v. 24, p. 1162-1174.

Dunning, G.R. and Chorlton, L.B.

1985: The Annieopsquotch ophiolite belt of southwest Newfoundland: geology and tectonic significance; Geological Society of America, Bulletin, v. 96, p. 1466-1476.

Dunning, G.R. and Krogh, T.E.

1985: Geochronology of ophiolites of the Newfoundland Appalachians; Canadian Journal of Earth Sciences, v. 22, p. 1659-1670.

Dunning, G.R. and O'Brien, S.J.

1989: Late Proterozoic-early Paleozoic crust in the Hermitage Flexure, Newfoundland Appalachians: U/Pb ages and tectonic significance; Geology, v. 17, p. 548-551.

Dunning, G.R. and Pederson, R.B.

1988: U/Pb ages of ophiolites and arc-related plutons of the Norwegian Caledonides: implications for the development of Iapetus; Contributions to Mineralogy and Petrology, v. 98, p. 13-23.

Dunning, G.R., Kean, B.F., Thurlow, J.G., and Swinden, H.S.

1987: Geochronology of the Buchans, Roberts Arm, and Victoria Lake groups and Mansfield Cove Complex, Newfoundland; Canadian Journal of Earth Sciences, v. 24, p. 1175-1184.

Dunning, G.R., O'Brien, S.J., Colman-Sadd, S.P., Blackwood, R.F., Dickson, W.L., O'Neill, P.P., and Krogh, T.E.

1990: Silurian orogeny in the Newfoundland Appalachians; Journal of Geology, v. 98, p. 895-913.

Dunning, G.R., Swinden, H.S., Kean, B.F., Evans, D.T.W., and Jenner, G.A.

1991: A Cambrian island arc in Iapetus: geochronology and geochemistry of the Lake Ambrose volcanic belt, Newfoundland Appalachians; Geological Magazine, v. 128, p. 1-17.

Dunsmore, H.E.

1977a: Uranium resources of the Permo-Carboniferous basin, Atlantic Canada; in Report of Activities, Part B; Geological Survey of Canada, Paper 77-1B, p. 247-253.

1977b: A new genetic model for uranium-copper mineralization, Permo-Carboniferous basin, northern Nova Scotia; in Report of Activities, Part B; Geological Survey of Canada, Paper 77-1B, p. 247-253.

Dunsworth, S.M.

1991: Geological map and descriptive notes of the central Lewis Hills massif, Newfoundland; Geological Survey of Canada, Open File 2332.

Dunsworth, S.M., Calon, T.J., and Malpas, J.

1986: Structural and magmatic controls on the internal geometry of the plutonic complex and its chromite occurrences in the Bay of Islands Ophiolite, Newfoundland; in Current Research, Part A; Geological Survey of Canada, Paper 86-1B, p. 471-482.

Dupuis, L.

1990: Controles géologique des amas sulfurés ordoviciens d'Eustis, Albert, Wheal Betsy, Capel et Victoria de l'arc insulare d'Ascot-Weedon, Appalaches du Sud du Québec; M.Sc. thesis, Université de Québec a Montral, Montréal, Québec, 154 p.

Duquette, G.

1983: Excursion géologique autour de monts McGerrigle; Ministère de l'Énergie et des Ressources du Québec, DV-86-06.

Elthon, D.

1991: Geochemical evidence for formation of the Bay of Islands ophiolite above a subduction zone; Nature, v. 354, p. 140-143.

Emmons, E.R.

1937: Gold deposits of the world; McGraw-Hill, New York, 562 p.

Energy, Mines and Resources

1989: Canadian mineral deposits not being mined in 1989; Mineral Policy Sector, Energy, Mines and Resources Canada, Mineral Bulletin 223, 308 p.

Evans, D.T.W.

1990: Midas Pond gold prospect; in Metallogenic Framework of Base and Precious Metal Deposits, Central and Western Newfoundland, (ed.) H.S. Swinden, D.T.W. Evans, and B.F. Kean; 8th IAGOD Symposium, Field Trip No. 1, Guidebook; Geological Survey of Canada, Open File 2156, p. 65-67.

1991: Gold metallogeny, eastern Dunnage Zone, central Newfoundland; in Current Research; Newfoundland Department of Mines and Energy, Geological Survey Branch, Report 91-1, p. 301-318.

1992: Gold metallogeny of the eastern Dunnage Zone, central Newfoundland; in Current Research; Newfoundland Department of Mines and Energy, Geological Survey Branch, Report 92-1, p. 231-244.

Evans, D.T.W., Kean, B.F., and Dunning, G.R.

1990: Geological studies, Victoria Lake Group, central Newfoundland; in Current Research; Newfoundland Department of Mines and Energy, Geological Survey Branch, Report 90-1, p. 144.

Faribault, E.R.
1899: The gold measures of Nova Scotia and deep mining; Canadian Mining Institute, Journal, v. 2, p. 119-129.

Felderhof, G.W.
1978: Barite, celestite and fluorite in Nova Scotia; Nova Scotia Department of Mines and Energy, Bulletin, v. 4, 464 p.

Fisher, B.E.
1984: A regional investigation of scheelite occurrences in the Meguma Group of Nova Scotia; B.Sc. thesis, Dalhousie University, Halifax, Nova Scotia, 210 p.

Fletcher, T.P.
1972: Geology and Lower to Middle Cambrian trilobite faunas of southwest Avalon, Newfoundland; Ph.D. thesis, Cambridge University, England.

Fogwill, W.D.
1964a: Chromite exploration, 1963 and 1964; Newfoundland Department of Mines, Agriculture and Resources, Mineral Resources Division, unpublished report, 12 p.
1964b: Chromite in Newfoundland; unpublished manuscript of paper presented to annual meeting, Canadian Institute of Mining and Metallurgy, Newfoundland Branch, November, 1964.

Forgeron, S.
1977: The Kaiser Celestite mining operation and mineral potential of the Loch Lomond Basin, Cape Breton, Nova Scotia; Nova Scotia Department of Mines and Energy, Open File Report, OFR 328, 100 p.

Fox, J.S.
1984: Besshi-type volcanogenic sulphide deposits: a review; Canadian Institute of Mining and Metallurgy, Bulletin, v. 77, no. 864, p. 57-68.

Fralick, P.W. and Schenk, P.E.
1981: Molasse deposition and basin evolution in a wrench tectonic setting: the late Paleozoic, eastern Cumberland Basin, Maritime Canada; in Sedimentation and Tectonics in Alluvial Basins, (ed.) A. Miall; Geological Society of Canada, Special Paper 23, p. 77-97

Fryer, B.J., Kerr, A., Jenner, G.A., and Longstaffe, F.J.
1992: Probing the crust with plutons: regional isotopic geochemistry of granitoid intrusions across insular Newfoundland; in Current Research; Newfoundland Department of Mines and Energy, Geological Survey Branch, Report 92-1, p. 119-139.

Furey, D.J. and Strong, D.F.
1986: Geology of the Harbour Breton Complex, Newfoundland; in Current Research, Part A; Geological Survey of Canada, Paper 86-1A, p 461-464.

Fyffe, L.R. and MacLellan, H.E.
1988: Lithogeochemistry (including gold) of altered and mineralized samples from the Burnthill Brook area (NTS 21 J/10) of central New Brunswick; New Brunswick Department of Natural Resources and Energy, Minerals and Energy Division, Geoscience Report 88-1, 92 p.

Fyffe, L.R. and Pronk, A.G.
1985: Bedrock and surficial geology: rock and till geochemistry in the Trousers Lake area, Victoria County, New Brunswick; New Brunswick Department of Natural Resources, Mineral Resources Division, Report of Investigation 20, 74 p.

Fyffe, L.R., Barr, S.M., and Bevier, M.L.
1988: Origin and U-Pb geochronology of amphibolite-facies metamorphic rocks, Miramichi Highlands, New Brunswick; Canadian Journal of Earth Sciences, v. 25, p. 1674-1686.

Fyffe, L.R., van Staal, C.R., and Winchester, J.H.
1990: Late Precambrian-Early Paleozoic volcanic regimes and associated massive sulphide deposits in the northeastern mainland Appalachians; Canadian Institute of Mining and Metallurgy, Bulletin, v. 83, no. 938, p. 70-78.

Gair, J.E. and Slack, J.F.
1980: Stratabound massive sulphide deposits of the U.S. Appalachians; in Review of Caledonian-Appalachian Stratabound Sulphides; Geological Survey of Ireland, Special Paper 5, p. 67-81.

Gale, G.H.
1971: An investigation of some sulphide deposits in the Rambler area, Newfoundland; Ph.D. thesis, University of Durham, 137 p.

Gall, Q.
1984: Petrography and diagenesis of the Carboniferous Deer Lake Group and Howley Formation, Deer Lake Subbasin, western Newfoundland; M. Sc. thesis, Memorial University of Newfoundland, St. John's, Newfoundland, 250 p.

Gardiner, C.D. (ed.)
1988: Canadian Mines Handbook 1988-89; Northern Miner Press Limited, p. 208.

Gardiner, W.W.
1985: Dungarvon claims (NTS 21 J/10E); unpublished report for Kidd Creek Mines; New Brunswick Department of Natural Resources and Energy, Mineral Resources, Assessment File 473103, 67 p., 1 map.

Gariepy, C.
1978: Stratigraphie et géochimie des laves cambriennes du Groupe de Caldwell dans la région du Lac Etchemin, Appalaches du Québec; thèse de matrise, Université de Montréal, Montréal, Québec, 161 p.

Gauthier, M.
1985: Synthèse métallogénique de l'Estrie et de la Beauce (secteur sud); Ministère l'Énergie et des Ressources du Québec, MB-85-20, 74 p.
1986: Synthese metallogenique de l'Estrie et de la Beauce (Secteur Centre-Ouest); Ministère l'Énergie et des Ressources du Québec, MB-86-55, 101 p.

Gauthier, M., Chartrand, F., Dupuis, L, and Trottier, J.
1988: Massive sulphide deposits of the southeastern Quebec Appalachians; in Proceedings of the 7th quadrennial IAGOD Symposium, Schweizerbart'sche Verlagsbuchhandlung, Stuttgart, Germany, p. 461-469.

Gauthier, M., Corrivaux, L., Trottier, J., Laflamme, G., Cabri, L.J., and Bergeron, M.
1990: Les chromitites platinifères de l'Estrie-Beauce, Appalaches du Sud du Québec; Mineralium Deposita, v. 25, p. 169-178.

Gauthier, M., Auclair, M., Bardoux, M., Blain, M., Boivert, D., Brassard, B., Chartrand, F., Darimont, A., Dupuis, L., Durocher, M., Gariepy, C., Godue, R., Jebrak, M., and Trottier, J.
1989: Synthèse métallogenique de l'Estrie et de la Beauce; Ministère de l'Énergie et des Ressources du Québec, MB-89-20, 631 p.

Gibbons, R.V.
1969: Geology of the Moreton's Harbour area, Newfoundland, with emphasis on the environment and mode of formation of the arsenopyrite veins; M.Sc. thesis, Memorial University of Newfoundland, St. John's, Newfoundland, 165 p.

Giles, P.S. and Chatterjee, A.K.
1986: Peraluminous granites of the Liscomb complex; in 10th Annual Open House and Review of Activities, Program and Summaries; Nova Scotia Department of Mines and Energy, Information Series No. 12, p. 83-90.
1987: Peraluminous granites of the Liscomb complex; in Report of Activities, 1986; Nova Scotia Department of Mines and Energy, Report 87-1, p. 95-98.

Giles, P.S. and Ruitenberg, A.A.
1977: Stratigraphy, paleogeography and tectonic setting of the Coldbrook Group in the Caledonia Highlands of southern New Brunswick; Canadian Journal of Earth Sciences, v. 14, p. 1263-1275.

Gilpin, E.
1888: Notes on the Nova Scotia gold veins; Royal Society of Canada, Transactions, v. VI, sec. IV, p. 63-70.

Godue, R.
1988: Étude métallogénique et lithogéochimique du Groupe de Magog, Estrie et Beauce, Québec; thèse de maîtrise, Université du Québec à Montréal, Québec, 70 p.

Gower, D., Graves, G., Walker, S., and MacInnis, D.
1988: Lode gold mineralization at Deer Cove, Point Rousse Complex, Baie Verte Peninsula; in The Volcanogenic Sulphide Districts of Central Newfoundland, (ed.) H.S. Swinden and B.F. Kean; Geological Association of Canada, Mineral Deposits Division, St. John's, Newfoundland, p. 43-48.

Graham, R.A.F.
1969: York Harbour Copper Mine: geology and geophysical surveys; unpublished report for Long Lac Mineral Exploration Limited, Newfoundland Department of Mines and Energy, Mineral Development Division, File 12G/01 (33).

Graves, M.C.
1976: The formation of gold-bearing quartz veins in Nova Scotia; M.Sc. Thesis, Dalhousie University, Halifax, Nova Scotia, 166 p.

Graves, M.C. and Zentilli, M.
1982: A review of the geology of gold in Nova Scotia; in Geology of Canadian Gold Deposits, (ed.) R.W. Hodder and W. Petruk; Canadian Institute of Mining and Metallurgy, Special Volume 24, p. 233-242.
1988a: The lithochemistry of metal enriched coticules in the Goldenville-Halifax transition zone of the Meguma Group, Nova Scotia; in Current Research, Part A; Geological Survey of Canada, Paper 88-1A, p. 251-262.
1988b: Geochemical characterization of the Goldenville Formation-Halifax Formation Transition of the Meguma Group, Nova Scotia: Data tables and sample location maps; Geological Survey of Canada, Open File 1829, 110 p.

Grip, E.
1967:   On the genesis of lead ores on the eastern border of the Caledonides in Scandinavia; Economic Geology, Monograph 3, p. 208-218.

Gross, G.A.
1965:   Geology of iron deposits in Canada; in General Geology and Evaluation of Iron Deposits; Geological Survey of Canada, Economic Geology Report 22.
1967:   Iron deposits in the Appalachian and Grenville regions of Canada; in Geology and Iron Deposits of Canada, v. 2; Geological Survey of Canada, Economic Geology Report 22, 111 p.

Guilbert, J.M. and Park, C.F.
1986:   Geology of Ore Deposits; W.H. Freeman, New York, 985 p.

Guild, P.W.
1972:   Metallogeny and the new global tectonics; 24th International Geological Congress, Proceedings, v. 4, p. 17-24.

Hale, W.E.
1950:   Variation in the gabbroic rocks of the St. Stephen area, Charlotte County, New Brunswick; M.Sc. thesis, University of New Brunswick, Fredericton, New Brunswick, 58 p.

Hamilton, J.B.
1968:   Barite occurrences in New Brunswick; Mineral Resources Branch, Department of Natural Resources, Fredericton, New Brunswick, Investigations Report 5, 23 p.

Harron, G.A.
1976:   Métallogénese des gites de sulfures des Cantons de l'Est; Ministère des Richesses Naturelle du Québec, Étude Speciale No. 27, 42 p.

Hauseux, M.
1980:   Long Lake claim group (21 O/2W); unpublished report for Canadian Occidental Petroleum Ltd., New Brunswick Department of Natural Resources and Energy, Mineral Resources, Assessment File 473155, 298 p.

Hayes, J.P. and O'Driscoll, C.F.
1990:   Regional setting and alteration within the eastern Avalon high-alumina belt, Avalon Peninsula, Newfoundland; in Current Research; Newfoundland Department of Mines and Energy, Geological Survey Branch, Report 90-1, p. 145-156.

Haynes, S.J.
1983:   Typomorphism of turbidite-hosted auriferous quartz veins, southern Guysborough County; in Report of Activities, 1982; Nova Scotia Department of Mines and Energy, Report 83-1, p. 183-224.
1986:   Geology and geochemistry of turbidite-hosted gold deposits, greenschist facies, eastern Nova Scotia; in Turbidite Hosted Gold Deposits, (ed.) J. Duncan Keppie, R.W. Boyle, and S.J. Haynes; Geological Association of Canada, Special Paper 32, p. 161-178.
1987:   Classification of quartz veins in turbidite-hosted gold deposits, greenschist facies, eastern Nova Scotia; Canadian Institute of Mining and Metallurgy, Bulletin, v. 80, no. 898, p. 37-51.

Haynes, S.J. and Smith, P.K.
1983:   Gold potential of the Meguma Group: new concepts; Maritime Sediments and Atlantic Geology, v. 19, p. 96.

Heenan, P.R.
1973:   The discovery of the Ming zone, Consolidated Rambler Mines Limited, Baie Verte, Newfoundland; Canadian Mining and Metallurgical Bulletin, v. 66, no. 729, p. 78-88.

Hein, F.J., Graves, M.C., and Ruffman, A.
1988:   The geology of the Jubilee Zinc-Lead Deposit, Victoria County, Cape Breton Island, Nova Scotia; Geological Survey of Canada, Open File 1891, 135 p.

Helmstaedt, H.
1973:   Structural geology of the Bathurst-Newcastle district; in Geology of New Brunswick, (ed.) N. Rast; New England Intercollegiate Geological Conference, Field Guide to Excursions, p. 34-46.

Henderson, J.R.
1983:   Analysis of structure as a factor controlling gold mineralization in Nova Scotia; in Report of Activities, 1982; Nova Scotia Department of Mines and Energy, Report 83-1, p. 265-282.

Henderson, J.R. and Henderson, M.N.
1990:   Crack-seal texture in bedding-parallel, gold-bearing, columnar-quartz veins: evidence of fossil water sills; in Mineral Deposit Studies in Nova Scotia, v. 1, (ed.) A.L. Sangster; Geological Survey of Canada, Paper 90-1, p. 163-168.

Henderson, J.R., Wright, T.O., and Henderson, M.N.
1986a: A history of cleavage and folding: an example from the Goldenville Formation, Nova Scotia; Geological Society of America Bulletin, v. 97, p. 1354-1366.

Henderson, J.R., Henderson, M.N., Mawer, C.K., Crocket, J.H., Clifford, P.M., and Fueten, F.
1986b: Timing and origin of auriferous quartz veins, Meguma Terrane, Nova Scotia; in Poster Volume, Gold '86; Geological Association of Canada, Mineral Deposits Division, p. 70-71.

Henderson, M.N. and Henderson, J.R.
1986:   Constraints on the origin of gold in the Meguma zone, Ecum Secum area, Nova Scotia; Maritime Sediments and Atlantic Geology, v. 22, p. 1-14.

Herman, H.
1914:   Economic Geology and Mineral Deposits of Victoria; Victoria (Australia) Geological Survey, Bulletin, v. 34, 36 p.

Héroux, Y. and Tassé, N.
1990:   Organic-matter alteration in an early Paleozoic basin: zonation around mineral showings compared to that around intrusions, St. Lawrence Lowlands, Quebec; Geological Society of America Bulletin, v. 102, p. 877-888.

Heyl, A.V. and Ronan, J.J.
1954:   The iron deposits of Indian Head area; in Contributions to the Economic Geology of Western Newfoundland; Geological Survey of Canada, Bulletin 27, p. 42-62.

Hibbard, J.P.
1983:   Geology of the Baie Verte Peninsula, Newfoundland; Newfoundland Department of Mines and Energy, Mineral Development Division, Memoir 2, 279 p.

Higgins, N.C.
1980:   The genesis of the Grey River tungsten prospect: A fluid inclusion, geochemical, and isotopic study; Ph.D. thesis, Memorial University of Newfoundland, St. John's, Newfoundland, 539 p.
1985:   Wolframite deposition in a hydrothermal vein system: the Grey River tungsten prospect, Newfoundland, Canada; Economic Geology, v. 80, p. 1297-1327.

Hildreth, E.W.
1979:   The Bishop Tuff: evidence for the origin of compositional zonation in silicic magma chambers; in Ash-Flow Tuffs, (ed.) C.E. Chapin and E.W. Elston; Geological Society of America, Special Paper 180, p. 43-76.
1981:   Gradients in silicic magma chambers: implications for lithostatic magmatism; Journal of Geophysical Research, v. 86, p. 10153-10192.

Hill, J.D.
1986:   Granitoid plutons of the Canso area, Nova Scotia; in Current Research, Part A; Geological Survey of Canada, Paper 86-1A, p. 185-192.

Hill, J.R.
1989:   Follow-up geology and geochemistry of metacarbonate formations and their contained mineral occurrences, Cape Breton Island, Nova Scotia; Geological Survey of Canada, Open File 2077, 43 p.
1990:   A geological and geochemical study of metacarbonate-rocks and contained mineral deposits, Cape Breton Island, Nova Scotia; in Mineral Deposit Studies in Nova Scotia, v. 1; Geological Survey of Canada, Paper 90-8, p. 3-30.

Hind, H.Y.
1869:   Notes on the structure of the Nova Scotia gold district; Nova Scotia Institute of National Science, Transactions, v. II, pt. 3, p. 102-109.
1870:   On two gneissoid series in Nova Scotia and New Brunswick, supposed to be the equivalents of the Huronian(Cambrian) and Laurentian; Quarterly Journal, Geological Society of London, v. XXVI, p. 468-479.
1872:   Report on the Mt. Uniacke, Oldham, and Renfrew gold mining districts; Halifax, 136 p.

Hingston, R.W.
1985:   The manganiferous slates of the Cambro-Ordovician Meguma Group at Lake Charlotte, Halifax County, Nova Scotia; B.Sc. thesis, Dalhousie University, Halifax, Nova Scotia, 67 p.

Hitchcock, C.H.
1878:   The Geology of New Hampshire, v. 5: Economic Geology; State of New Hampshire, Concord (N.H.), p. 1-69.

Hitzman, M.W. and Large, D.
1986:   A review and classification of the Irish carbonate-hosted base metal deposits; in Geology and Genesis of Mineral Deposits in Ireland, (ed.) C.J. Andrew, R.W.A. Crowe, S. Finlay, W.M. Pennell, and J.F. Payne; Irish Association For Economic Geology, Dublin, p. 217-238.

Horne, R.J., Corey, M.C., Ham, L.J., and MacDonald, M.A.
1989:   Lithogeochemical variation within the biotite rich envelope rocks of the eastern South Mountain Batholith: implications for its intrusive and post-intrusive history; in Report of Activities, 1988, Part B; Nova Scotia Department of Mines and Energy, Mines and Minerals Branch, Report 89-1, p. 37-50.

Howse, A.F., Dean, P.L., Swinden, H.S., Kean, B.F., and Morrissey, F.
1983:   Fluorspar deposits of the St. Lawrence area, Newfoundland: Geology and economic potential; Newfoundland Department of Mines and Energy, Mineral Development Division, Report 83-9, 21 p.

Huard, A.A.
1989: Epithermal alteration and gold mineralization in late Precambrian volcanic rocks on the northern Burin Peninsula, southeastern Newfoundland, Canada; M.Sc. thesis, Memorial University of Newfoundland, St. John's, Newfoundland, 273 p.
1990: The Noranda/Impala Stog'er Tight deposit; in Metallogenic Framework of Base and Precious Metal Deposits, Central and Western Newfoundland, (ed.) H.S. Swinden, D.T.W. Evans, and B.F. Kean; 8th IAGOD Symposium, Field Trip No. 1, Guidebook, Geological Survey of Canada, Open File 2156, p. 173-177.

Hudson, K.A. and Swinden, H.S.
1989: Geology and petrology of the Handcamp gold prospect, Robert's Arm Group, Newfoundland; in Current Research, Part B; Geological Survey of Canada, Paper 89-1B, p. 93-105.

Hunt, T.S.
1868: Report on the gold region of Nova Scotia; Geological Survey of Canada, 48 p.

Hussey, J.J.
1992: Deep exploration of the "Brunswick Horizon", Bathurst District, northern New Brunswick; Exploration and Mining Geology, v. 1, p. 187-195.

Hutchinson, H.E.
1982: Geology, geochemistry and genesis of the Brazil Lake pegmatites, Yarmouth County, Nova Scotia; B.Sc. thesis, Dalhousie University, Halifax, Nova Scotia, 105 p.

Hy, C. and Williams, P.F.
1986: Timing of quartz emplacement in the Meguma Terrane, Nova Scotia; Toronto International Conference Gold '86, Program with Abstracts, p. 75-76.

Hyde, R.S.
1984: Geology and mineralization of the Carboniferous Deer Lake Basin, western Newfoundland; in Mineral Deposits of Newfoundland, A 1984 Perspective; Newfoundland Department of Mines and Energy, Mineral Development Division, Report 84-3, p. 19-26.

Jambor, J.L.
1984: Mineralogy of the Tulks Zn-Pb-Cu massive sulphide deposit, Buchans area, Newfoundland; CANMET, Energy, Mines and Resources Canada, Division Report MRP/MSL 84-22(IR), 49 p.
1987: Geology and origin of orebodies in the Lucky Strike area; in Buchans Geology, Newfoundland, (ed.) R.V. Kirkham; Geological Survey of Canada, Paper 86-24, p. 75-106.

Jambor, J.L. and Barbour, D.M.
1987: Geology and mineralogy of the Tulks deposit; in Buchans Geology, Newfoundland, (ed.) R.V. Kirkham; Geological Survey of Canada, Paper 86-24, p. 219-226.

James, N.P., Barnes, C.R., Stevens, R.K., and Knight, I.
1989: A Lower Paleozoic continental margin carbonate platform, northern Canadian Appalachians; in Controls on Carbonate Platforms and Basin Development, (ed.) T. Crevello, R. Sarg, J.F. Read, and J.L. Wilson; Society of Economic Paleontologists and Mineralogists, Special Publication 44, p. 123-146.

Jamieson, R.A., Tallman, P.C., Plint, H.E., and Connors, K.A.
1990: Regional geological setting of pre-Carboniferous mineral deposits in the western Cape Breton Highlands; in Mineral Deposit Studies in Nova Scotia, (ed.) A.L. Sangster; Geological Survey of Canada, Paper 90-8, p. 77-100.

Jamieson, R.A., van Breeman, O., Sullivan, R.W., and Currie, K.L.
1986: The age of igneous and metamorphic events in the western Cape Breton Highlands, Nova Scotia; Canadian Journal of Earth Sciences, v. 23, p. 1891-1901.

Jenner, G.A. and Fryer, B.J.
1980: Geochemistry of the upper Snooks Arm Group basalts, Burlington Peninsula, Newfoundland: evidence against formation in an island arc; Canadian Journal of Earth Sciences, v. 17, p. 880-900.

Jenner, G.A., Evans, D.T.W., and Kean, B.F.
1988: The Lushs Bight Group revisited: new trace element and Sm/Nd isotopic evidence for its tectonic environment of formation (abstract); Geological Association of Canada, Program with Abstracts, v. 13, p. A61.

Jenner, G.A., Dunning, G.R., Malpas, J., Brown, M., and Brace, T.
1991: Bay of Islands and Little Port Complexes revisited: age, geochemical and isotopic evidence confirm supra-subduction zone origin; Canadian Journal of Earth Sciences, v. 28, p. 1635-1652.

Jenness, S.E.
1963: Terra Nova and Bonavista map-areas, Newfoundland; Geological Survey of Canada, Memoir 327, 184 p.

Jensen, L.R.
1975: The Torbrook Formation; Maritime Sediments, v. 11, p. 107-118.

Johnson, H.
1954: The strontium deposits of Port au Port Peninsula; in Contributions to the Economic Geology of Western Newfoundland; Geological Survey of Canada, Bulletin, v. 27, p. 1-19.

Justino, M.F. and Sangster, A.L.
1987: Geology in the vicinity of the Lime Hill zinc occurrence, southwestern Cape Breton Island, Nova Scotia; in Current Research, Part A; Geological Survey of Canada, Paper 87-1A, p. 555-561.

Karlstrom, K.E., van der Pluijm, B.A., and Williams, P.F.
1982: Structural interpretation of the eastern Notre Dame Bay area, Newfoundland: thrusting and asymmetrical folding; Canadian Journal of Earth Sciences, v. 19, p. 2325-2341.

Karson, J.A.
1979: Geological map and descriptive notes of the Lewis Hills massif, western Newfoundland; Geological Survey of Canada, Open File 628.

Kay, A. and Strong, D.F.
1983: Geologic and fluid controls in As-Sb-Au mineralization in the Moreton's Harbour area, Newfoundland; Economic Geology, v. 78, p. 1590-1604.

Kean, B.F.
1974: Notes on the geology of the Great Bend and Pipestone Pond ultramafic bodies; in Report of Activities, 1973; Newfoundland Department of Mines and Energy, Mineral Development Division, Report 74-1, p. 33-42.
1983: Geology and mineral deposits of the Lushs Bight Group in the Little Bay Head-Halls Bay Head Area; in Current Research; Newfoundland Department of Mines and Energy, Mineral Development Division, Report 83-1, p. 157-174.
1984: Geology and mineral deposits of the Lushs Bight Group, Notre Dame Bay, Newfoundland; in Current Research, Newfoundland Department of Mines and Energy, Mineral Development Division, Report 84-1, p. 141-156.

Kean, B.F. and Evans, D.T.W.
1988: Regional Metallogeny of the Victoria Lake Group, central Newfoundland; Newfoundland Department of Mines, Mineral Development Division, Report 88-1, p. 319-330.

Kean, B.F. and Thurlow, J.G.
1976: Geology, mineral deposits and mineral potential of the Buchans volcanic belt; unpublished Mining Subcommittee Report to the Buchans Task Force, Government of Newfoundland and Labrador, p. A4-1, A4-63.

Keppie J.D.
1985: Geology and tectonics of Nova Scotia; in Excursion 1: Appalachian geotraverse, (ed.) J.D. Keppie, K.L. Currie, J.B. Murphy, R.K. Pickerell, L.R. Fyffe, and P. St. Julien; Geological Association of Canada, Annual Meeting, Fredericton, New Brunswick, p. 181.

Keppie, J.D. and Dostal, J.
1980: Palaeozoic volcanic rocks of Nova Scotia; in International Geological Correlation Project 27; Virginia Polytechnical Institute and State University, Memoir 2, p. 249-256.

Keppie, J.D., Dallmeyer, R.D., and Murphy, J.B.
1990: Tectonic implications of $^{40}Ar/^{39}Ar$ hornblende ages from late Proterozoic-Cambrian plutons in the Avalon Composite Terrane, Nova Scotia; Geological Society of America, Bulletin, v. 102, p. 516-528.

Kerr, A., Dunning, G.R. and Tucker, R.D.
1993: The youngest Paleozoic plutonism of the Newfoundland Appalachians; U-Pb ages from the St. Lawrence and Francois granites, and their regional implications; Canadian Journal of Earth Sciences, v. 30, p. 2328-2333.

Kettles, K.R.
1987: The Turgeon mafic volcanic associated Fe-Cu-Zn sulphide deposit in the ophiolitic Fournier Group, northern New Brunswick; M.Sc. thesis, University of New Brunswick, Fredericton, New Brunswick, 202 p.

Kirkham R.V.
1978: Base metal and uranium distribution along the Windsor-Horton contact, central Cape Breton Island, Nova Scotia; in Current Research, Part B; Geological Survey of Canada, Paper 78-1B, p. 121-135.
1984: Volcanic redbed copper; in Canadian Mineral Deposit Types: A Geological Synopsis, (ed.) O.R. Eckstrand; Geological Survey of Canada, Economic Geology Report 36, 85 p.
1985: Base metals in Upper Windsor (Codroy) Group oolitic and stromatolitic limestones in the Atlantic provinces; in Current Research, Part A; Geological Survey of Canada, Paper 85-1A, p. 573-585.
1987a: Buchans Geology, Newfoundland; Geological Survey of Canada, Paper 86-24, 288 p. (editor).

Kirkham R.V. (cont.)
1987b: Tectonic setting of the Buchans Group; in Buchans Geology, Newfoundland; (ed.) R.V. Kirkham; Geological Survey of Canada, Paper 86-24, p. 23-34.

Kirkham, R.V. and Soregaroli, A.E.
1975: Preliminary assessment of porphyry deposits in the Canadian Appalachians; in Current Research, Part A; Geological Survey of Canada, Paper 75-1A, p. 249-252.

Kirkham, R.V. and Thurlow, J.G.
1987: Evaluation of a resurgent caldera and aspects of ore deposition and deformation at Buchans; in Buchans Geology, Newfoundland, (ed.) R.V. Kirkham; Geological Survey of Canada, Paper 86-24, p. 177-194.

Kirkwood, D. and Dubé, B.
1992: Structural control of sill-hosted gold mineralization: the Stog'er Tight gold deposit, Baie Verte Peninsula, northwestern Newfoundland; in Current Research, Part D; Geological Survey of Canada, Paper 92-1D, p. 211-222.

Knight, I.
1976: Geology of the Carboniferous of the Codroy Valley and northern Anguille Mountains; Newfoundland Department of Mines and Energy, Mineral Development Division, Report 76-1, p. 10-18.
1980: Cambro-Ordovician carbonate stratigraphy of western Newfoundland: sedimentation, diagenesis and zinc-lead mineralization; Summary of a paper given at the 82nd Annual General Meeting of the Canadian Institute of Mining and Metallurgy, Toronto, Ontario, April 20-24, 1980, Newfoundland Department of Mines and Energy, Open File NFLD, 1154, 34 p.
1983: Geology of the Carboniferous Bay St. George Subbasin, western Newfoundland; Newfoundland Department of Mines and Energy, Mineral Development Division, Memoir 1, 358 p.
1984a: Mineralization in Cambro-Ordovician rocks, western Newfoundland; in Mineral Deposits of Newfoundland, A 1984 Perspective; Newfoundland Department of Mines and Energy, Mineral Development Division, Report 84-3, p. 37-44.
1984b: Mineral deposits of the Carboniferous Bay St. George Subbasin; in Mineral Deposits of Newfoundland, A 1984 Perspective; Newfoundland Department of Mines and Energy, Mineral Development Division, Report 84-3, p. 1-13.

Knight, I. and O'Brien, S.J.
1988: Stratigraphy and sedimentology of the Connecting Point Group and related rocks, Bonavista Bay, Newfoundland: an example of a Late Precambrian Avalonian Basin; in Current Research; Newfoundland Department of Mines and Energy, Mineral Development Division, Report 88-1, p. 207-228.

Knight, I., James, N.P., and Lane, T.E.
1991: The Ordovician St, George unconformity, northern Appalachians: the relationship of plate convergence at the St. Lawrence promontory to the Sauk/Tippecanoe sequence boundary; Geological Society of America Bulletin, v. 103, p. 1200-1225.

Kontak, D.J.
1990: The East Kemptville topaz-muscovite leucogranite, Nova Scotia I: Geological setting and whole-rock geochemistry; Canadian Mineralogist, v. 28, p. 787-825.

Kontak, D.J. and Chatterjee, A.K.
1992: The East Kemptville tin deposit, Yarmouth county, Nova Scotia III: A lead isotope study of the leucogranite and the mineralized greisen: evidence for a 366 Ma metallogenic event; Canadian Journal of Earth Sciences, v. 29, p. 1180-1196.

Kontak, D.J. and Cormier, R.F.
1991: Geochronological evidence for multiple tectono-thermal overprinting events in the East Kemptville muscovite-topaz leucogranite, Yarmouth County Nova Scotia, Canada; Canadian Journal of Earth Sciences, v. 28, p. 209-224.

Kontak, D.J. and Smith, P.K.
1987: Meguma gold: the best kept secret in the Canadian mining industry; unpublished report presented at the Prospectors and Developers Association of Canada, Annual Meeting, March, 1987, 14 p.
1988a: Meguma gold studies IV: Chemistry of vein mineralogy; in Report of Activities, 1987, Part B; Nova Scotia Department of Mines and Energy, Mines and Minerals Branch, Report 88-1, p. 85-101.
1988b: Meguma gold studies VI: Integrated model for the genesis of Meguma-hosted lode gold deposits; in Report of Activities, 1987, Part B; Nova Scotia Department of Mines and Energy, Mines and Minerals Branch, Report 88-1, p. 111-119.
1989: Preliminary results of a fluid inclusion study of the stockwork zone vein mineralization, Caribou deposit, Meguma Terrane, Nova Scotia; in Mines and Minerals Branch Report of Activities, 1989, Part a; Nova Scotia Department of Mines and Energy, Mines and Minerals Branch, Report 89-3, p. 65-70.

1991: A mineralogical and fluid inclusion study of the West Gore Sb-Au deposit, Meguma Terrane, southern Nova Scotia; in Report of Activities, 1990; Nova Scotia Department of Mines and Energy, Mines and Minerals Branch, Report 91-1, pages 61-73.

Kontak, D.J., O'Reilly, G.A., and Chatterjee, A.K.
1990a: The southwest Nova Scotia tin domain, Yarmouth county, Nova Scotia: implications for tin metallogeny in the Meguma Terrane, Nova Scotia; in Report of Activities, 1989, Part B; Nova Scotia Department of Mines and Energy, Mines and Minerals Branch, Report 90-1, p. 13-32.

Kontak, D.J., Smith, P.K., Reynolds, P., and Taylor, K.
1990b: Geological and ^{40}Ar/^{39}Ar geochronological constraints on the timing of quartz vein formation in Meguma Group lode-gold deposits, Nova Scotia; Atlantic Geology, v. 26, p. 201-228.

Kontak, D.J., Tuach, J., Strong, D.F., Archibald, D.A., and Farrar, E.
1988: Plutonic and hydrothermal events in the Ackley Granite, southeast Newfoundland, as indicated by total-fusion ^{39}Ar/^{40}Ar geochronology; Canadian Journal of Earth Sciences, v. 25, p. 1151-1160.

Kooiman, G.J.A., McLeod, M.J., and Sinclair, W.D.
1986: Porphyry tungsten-molybdenum orebodies, polymetallic veins and replacement bodies, and tin-bearing greisen zones in the Fire Tower zone, Mount Pleasant, New Brunswick; Economic Geology, v. 81, p. 1356-1373.

Krogh, T.E. and Keppie, J.D.
1988: U-Pb ages of single zircon cores imply a pan African source for two Meguma granites (abstract); in Geological Association of Canada/Mineralogical Association of Canada, Program with Abstracts, v. 13, p. A69.

Krogh T.E., Strong, D.F., O'Brien, S.J., and Papezik., V.S.
1988: Precise U/Pb dates from the Avalon Terrane in Newfoundland; Canadian Journal of Earth Sciences, v. 25, p. 442-453.

LaChance, S. and Duquette, G.
1977: Region de Boisbuisson (NW); Ministère de l'Énergie et des Ressources du Québec, RG-187.

Lalou, C.
1983: Genesis of ferromanganese deposits: hydrothermal origin; in Hydrothermal Processes at Seafloor Spreading Centres, (ed.) P.A. Rona, K. Bostrom, L. Laubier, and K.L. Smith, Jr.; NATO Conference Series, Plenum Press, New York, p. 503-534.

Lamarche, R.Y.
1965: Géologie de la région de Sherbrooke, comté de Sherbrooke; Ph.D. thesis, Université Laval, Québec (Québec), 291 p.
1984: Contexte géologique et gènis des gîtes d'amiante du Sud du Québec, in The Geology of Industrial Minerals in Canada; The Canadian Institute of Mining and Metallurgy, Special Volume 29, p. 61-69.

Lane, T.E.
1984: Preliminary classification of carbonate breccias, Newfoundland Zinc Mines, Daniels Harbour, Newfoundland; in Current Research, Part A; Geological Survey of Canada, Paper 84-1A, p. 505-512.
1990: Dolomitization, brecciation and zinc mineralization and their paragenetic, stratigraphic and structural relationships in the Upper St. George Group (Ordovician) at Daniel's Harbour, western Newfoundland; Ph.D. thesis, Memorial University of Newfoundland, St. John's, Newfoundland, 565 p.

Langton, J.P. and van Staal, C.R.
1989: Geology of the Ordovician Elmtree Terrane (abstract); Geological Association of Canada, Program with Abstracts, v. 13, p. A11.

Lasalle, P.
1980: L'or dans les sédiments marbles: Formation des placers, extraction et occurrences dans le Sud-Est du Québec; Ministère de l'Énergie et des Ressources du Québec, DPV-745, 26 p.

Laurent, R. and Kacira, N.
1987: Chromite deposits in the Appalachian ophiolites; in Evolution of Chromium Ore Fields, (ed.) C.W. Stowe; Van Nostrand Reinhold Co., New York; p. 169-193.

Lentz, D.R., Lutes, G., and Hartree, R.
1988: Bi-Sn-Mo-W greisen mineralization associated with the True Hill granite, southwestern New Brunswick; Maritime Sediments and Atlantic Geology, v. 24, p. 321-338.

Letouzy, J. and Kimura, M.
1985: Okinawa Trough genesis: structure and evolution of a back-arc basin developed in a continent; Marine and Petroleum Geology, v. 2, p. 11-130.

Lewis, W.L. and Holleman, M.
1983: Gypsum in Atlantic Canada; paper presented at the 19th Forum on the Geology of Industrial Minerals, Toronto, Ontario Geological Survey, Miscellaneous Paper 114, p. 79-95.

Lockhart, A.W.
1970: Woodstock claim group (21 J/4E); unpublished report for Falconbridge Nickel Mines Ltd., New Brunswick Department of Natural Resources and Energy, Mineral Resources, Assessment Files 470232 and 470233.

Logothetis, J.
1985: Economic geology of the New Ross-Vaughan complex; in Guide to the Granites and Mineral Deposits of Southwestern Nova Scotia, (ed.) A.K. Chatterjee and D.B. Clarke; Nova Scotia Department of Mines and Energy, Paper 85-3, preprint, p. 41-62.

Loiselle, M.C. and Wones, D.R.
1979: Characteristics and origin of anorogenic granites (abstract); Geological Society of America, Abstracts with Programs, v. 11, p. 468

Luff, W.M.
1975: Structural geology of the Brunswick No. 12 open pit; Canadian Mining and Metallurgical Bulletin, v. 70, no. 782, p. 109-119.

Lutes, G.
1981: Geology of Deersdale-Head of Nashwaak River, map area J-17, and upper parts of Nashwaak River-McBean and Sisters brooks, map area J-18; New Brunswick Department of Natural Resources, Mineral Resources Division, Map Report 81-4, 26 p.
1987: Geology and geochemistry of the Pokiok Batholith, New Brunswick; New Brunswick Department of Natural Resources and Energy, Minerals and Energy Division, Report of Investigation 22, 55 p.

Lydon, J.W., Lavigne, J.G., and Roddick, J.C.M.
1990: The relationships of gold mineralization to the thermal and tectonic history of the Baie Verte Peninsula, Newfoundland (abstract); Geological Survey of Canada, Minerals Colloquium, Program with Abstracts, January, 1990, p. 26.

MacDonald, A.S.
1989: Metallogenic studies, southeastern Cape Breton Island; Nova Scotia Department of Mines and Energy, Paper 89-1, 99 p.

MacDonald, A.S. and Barr, S.M.
1985: Geology and age of polymetallic occurrences in volcanic and granitoid rocks, St. Ann's area, Cape Breton Island, Nova Scotia; in Current Research, Part B; Geological Survey of Canada, Paper 85-1B, p. 117-124.

MacDonald, M.A. and Clarke, D.B.
1985: The petrology, geochemistry and economic potential of the Musquodoboit batholith, Nova Scotia; Canadian Journal of Earth Sciences, v. 22, p. 1633-1642.

MacDonald, M.A. and Ham, L.J.
1988: Preliminary geological map of Gaspereau Lake, NTS Sheet 21A/15 and part of 21H/02; Nova Scotia Department of Mines and Energy, Open File map 88-016.

MacDonald, M.A. and O'Reilly, G.A.
1988: Gold enrichment associated with post-magmatic processes in the South Mountain Batholith, southwestern Nova Scotia; in Report of Activities, 1988, Part B; Nova Scotia Department of Mines and Energy, Mines and Minerals Branch, Report 89-1, p. 13-26.

MacDonald, M.A., Corey, M.C., Ham, L.J., and Horne, R.J.
1989: Petrographic and geochemical aspects of the South Mountain Batholith; in Report of Activities, 1989, Part A; Nova Scotia Department of Mines and Energy, Mines and Minerals Branch, Report 89-3, p. 75-80.

MacDonald, M.A., Corey, M.C., Ham, L.J., Horne, R.J., and Chatterjee, A.K.
1990: Recent advances in the geology of the South Mountain Batholith: anatomy and origin of a batholith; Atlantic Geology, v. 26, p. 180.

MacDougall, C., Perry, I., and MacInnis, D.
1991: Exploration potential of the York Harbour deposit: a high grade Cu-Zn, Cyprus-type volcanogenic massive sulphide deposit; in Ore Horizons, Volume 1, (ed.) S. Swinden and A. Hogan; Newfoundland Department of Mines and Energy, Geological Survey Branch, p. 99-118.

MacInnis, I.N.
1986: Lithogeochemistry of the Goldenville-Halifax Transition (GHT) of the Meguma Group in the manganiferous zinc-lead deposit at Eastville, Nova Scotia; B.Sc. thesis, Dalhousie University, Halifax, Nova Scotia, 138 p.

MacKay, R.M. and Zentilli, M.
1976: Mineralogical observation on the copper-uranium mineralization at Black Brook, Nova Scotia; in Report of Activities, Part B; Geological Survey of Canada, Paper 76-1B, p 343-344.

MacLellan, H.E. and Taylor, R.P.
1989: Geology and geochemistry of the Burnthill Granite and related W-Sn-Mo-F mineral deposits, central New Brunswick; Canadian Journal of Earth Sciences, v. 26, p. 499-514.

MacLellan, H.E., Taylor, R.P., and Gardner, W.W.
1990: Geology and geochemistry of Middle Devonian Burnthill Brook Granites and related tin-tungsten deposits, York and Northumberland Counties, New Brunswick; New Brunswick Department of Natural Resources and Energy, Mineral Resource Report 4, 95 p.

MacMichail, T.F.
1975: The origin of the lead-zinc-silver ores and alteration of the surrounding granite at the Dunbrack mine, Musquodoboit Harbour, Nova Scotia; B.Sc. thesis, Dalhousie University, Halifax, Nova Scotia.

Malcolm, W.
1929: Gold fields of Nova Scotia; Geological Survey of Canada, Memoir 156, 253 p.
1976: Gold fields of Nova Scotia; Geological Survey of Canada, Memoir 385 (reprint of Memoir 129 without plates or maps), 253 p.

Malcolm, W. and Faribault, E.R.
1912: Gold fields of Nova Scotia; Geological Survey of Canada, Memoir 20-E, 331 p.

Malo, M and Moritz, R.
1991: Géologie et métallogenie de la faille du Grand Pabos, régime de Raudinn-Weir, Gaspésie; Ministère de L'Énergie de des Ressources du Québec, MB 91-03, 48 p.

Malo, M., Moritz, R., Roy, F., Chagnon, A., and Bertrand, R.
1990: Géologie et métallogenie de la faille du Grand Pabos, région de Robadoux-Reboul, Gaspésie; Ministère de l'Énergie et des Ressources du Québec, MB-90-09, 76 p.

Malpas, J. and Strong, D. F.
1975: A comparison of chrome-spinels in ophiolites and mantle diapirs of Newfoundland; Geochimica et Cosmochimica Acta, v. 39, p. 1045-1060.

Marcotte, R.
1980: Gites et indices de chromite du Québec; Ministère de l'Énergie et des Ressources du Québec, DPV-724, 58 p.

Marten, B.E.
1971: Stratigraphy of volcanic rocks in the Western Arm area of the central Newfoundland Appalachians; Geological Association of Canada, Proceedings, v. 24, no. 1, p. 73-84.

Mawer, C.K.
1986: The bedding-concordant gold-quartz veins of the Meguma Group, Nova Scotia; in Turbidite Hosted Gold deposits, (ed.) J.D. Keppie, R.W. Boyle, and S.J. Haynes; Geological Association of Canada, Special Paper 32, p. 135-148.
1987: Mechanics of formation of gold-bearing quartz veins, Nova Scotia, Canada; Tectonophysics, v. 135, p. 99-119.

McAllister, A.L.
1960: Massive sulphide deposits in New Brunswick; Canadian Institute of Mining and Metallurgy Bulletin, v. 53, no. 573, p. 88-98.

McAllister, A.L. and Lamarche, R.Y.
1972: Mineral Deposits of Southern Quebec and Northern New Brunswick; 24th International Geological Congress, Montreal, Excursion A58-C58, Guidebook, 95 p.

McArthur, J.G. and Knight, I.
1974: Geology and industrial minerals of the Newfoundland Carboniferous; Geological Association of Canada, Annual Meeting, St. John's, Newfoundland, Fieldtrip Guidebook B-10, 43 p.

McBride, D.E.
1978: Geology of the Ecum Secum area, Halifax and Guysborough Counties, Nova Scotia; Nova Scotia Department of Mines and Energy, Report 78-1, 12 p.

McCartney, W.D.
1967: Whitbourne map area, Newfoundland; Geological Survey of Canada, Memoir 341, 133 p.

McCartney, W.D. and Potter R.R.
1962: Mineralization as related to structural deformation, igneous activity and sedimentation in folded geosynclines; Canadian Mining Journal, v. 83, no. 4, p. 83-87.

McClay, K.R.
1987: Aspects of the structural geology of the Buchans area; in Buchans Geology, Newfoundland, (ed.) R.V. Kirkham; Geological Survey of Canada, Paper 86-24, p. 47-58.

McCutcheon, S.R.
1981: Stratigraphy and paleogeography of the Windsor Group; New Brunswick Department of Natural Resources, Geological Surveys Branch, Open File Report 81-31, 210 p.
1983: Potash in the Central New Brunswick Platform?; Canadian Mining and Metallurgical Bulletin, v. 76, p. 70-76.
1984: Geology of the Mount Pleasant caldera; in 8th Annual Review of Activities, Project Resumes, 1983; New Brunswick Department of Natural Resources and Energy, Mineral Resources Division, Information Circular 83-3, p. 6-11.

McCutcheon, S.R. (cont.)

1990a: Base metal deposits of the Bathurst - Newcastle district; in Field Guide to Massive Sulphide Deposits in Northern New Brunswick, (ed.) L.R. Fyffe; Department of Natural Resources and Energy, Minerals and Energy Division, Fredericton, New Brunswick, p. 42-71.

1990b: The Late Devonian Mount Pleasant caldera complex: stratigraphy, mineralogy, geochemistry and geologic setting of a Sn-W deposit in southwestern New Brunswick; Ph.D. thesis, Dalhousie University, Halifax, Nova Scotia, 609 p.

McCutcheon, S.R. and Bevier, M.L.

1990: Implications of field relations and U-Pb geochronology for the age of gold mineralization and timing of Acadian deformation in northern New Brunswick; Atlantic Geology, v. 26, p. 237-246.

McCutcheon, S.R. and Robinson, P.T.

1987: Geological constraints on the genesis of the Maritimes Basin, Atlantic Canada; in Sedimentary Basins and Basin-Forming Mechanisms, (ed.) C. Beaumont and A.J. Tankard; Canadian Society of Petroleum Geologists, Memoir 12, p. 287-297.

McCutcheon, S.R. and Ruitenberg, A.A.

1987: Geology and mineral deposits of the Annidale-Nerepis area; New Brunswick Department of Natural Resources and Energy, Mineral Resources Division, Memoir 2, 141 p.

McGerrigle, H.W.

1935: Région du Mont Megantic, Sud-Est du Québec et ses placers auriferes; dans Rapport annuel du Service des Mines du Québec pour l'année 1935 (partre D); Ministère des Mines et des Pecheries du Québec.

McHone, J.G.

1984: Mesozoic igneous rocks of northern New England and adjacent Québec; Geological Society of America, Map and Chart Series MC-49.

McKenzie, C.B.

1986: Geology and mineralization of the Chetwynd deposit, southwestern Newfoundland, Canada; in Proceedings of Gold '86, an International Symposium on the Geology of Gold, (ed.) A.J. MacDonald; Toronto, Ont., p. 137-148.

McKenzie, C.B. and Clarke, D.B.

1976: Petrology of the South Mountain Batholith, Nova Scotia; Canadian Journal of Earth Sciences, v. 12, p. 1209-1218.

McKenzie, C.B., Desnoyers, D.W., Barbour, D. and Graves, R.M.

1993: Contrasting volcanic-hosted massive sulfide styles in the Tulks Belt, central Newfoundland; Exploration and Mining Geology, v. 2, p. 73-84.

McLeod, M.J.

1979: Geology and Mineral Deposits of the Hillsborough area, map area V-22 and V-23 (parts of 21H/15E, 21H/15W); New Brunswick Department of Natural Resources, Mineral Resources Branch, Map Report 79-6, 35 p.

1990: Geology, geochemistry and related mineral deposits of the Saint George Batholith, Charlotte, Queens and Kings Counties, New Brunswick; New Brunswick Department of Natural Resources and Energy, Mineral Resources Division, Mineral Resources Report 5, 169 p.

McLeod, M.J. and Ruitenberg, A.A.

1978: Geology and Mineral Deposits of the Dorchester area, Map area W-22, W-23 (21H/15E, 21H/16W); New Brunswick Department of Natural Resources, Mineral Resources Branch, Map Report 78-4, 27 p.

McLeod, M.J., Johnson, S.C., and Ruitenberg, A.A.

1994a: Geological map of Southwestern New Brunswick, New Brunswick Department of Natural Resources and Energy, Mineral Resources Division, Map NR-5.

1994b: Geological map of Southeastern New Brunswick; New Brunswick Department of Natural Resources and Energy, Mineral Resources Division, Map NR-6.

McLeod, M.J., Ruitenberg, A.A., and Krogh, T.E.

1992a: Geology and U-Pb geochronology of the Annidale Group, southern New Brunswick: Lower Ordovician volcanic and sedimentary rocks formed near the southeastern margin of Iapetus; Atlantic Geology, v. 28, p. 181-192.

McNutt, J.R.A.

1962: The petrology of the Goodwin Lake gabbro; M.Sc. thesis, University of New Brunswick, Fredericton, New Brunswick, 104 p.

Mengel, F., Godue, R., Sangster, A.L., Dubé, B., and Lynch, G.

1991: A progress report on the structural control of gold mineralization in the Cape Breton Highlands; in Current Research, Part D; Geological Survey of Canada, Paper 90-1D, p. 117-127.

Miller, C.

1980: Zinc-lead-copper-silver mineralization at Stirling; in Mineral Deposits and Mineralogenic Provinces of Nova Scotia; Geological Association of Canada, Annual Meeting, Halifax, Nova Scotia, Field Trip No. 6, Guidebook, p. 19-35.

Miller, H.G.

1983a: Geophysical constraints on the size and extent of the Wabana hematite deposit; Economic Geology, v. 78, p. 1017-1021.

1983b: A Geophysical interpretation of the geology of Conception Bay, Newfoundland; Canadian Journal of Earth Sciences, v. 20, p. 1421-1433.

Milligan, G.C.

1970: Geology of the George River Series, Cape Breton; Nova Scotia Department of Mines and Energy, Memoir 7, 111 p.

Ministere de l'Energie et des Ressources

1979: Rapport des geologues residents, 1978; Ministère des Richesses Naturelles du Québec, DPV-652, p. 46.

1983: Repertoire des fiches de gite mineral; Ministère de l'Énergie et des Ressources du Québec, DPV-845.

Mitchell, A.H.G. and Bell, J.D.

1973: Island-arc evolution and related mineral deposits; Journal of Geology, v. 81, p. 381-405.

Mitchell, A.H.G. and Garson, M.S.

1976: Mineralization at plate boundaries; Minerals Science and Engineering, v. 8, p. 129-169.

Moritz, R., Malo, M., Roy, F., and Chagnon, A.

1993: Skarn mineralization associated with the Grand Pabos- Restigouche Fault southern Gaspé Peninsula, Québec, Canada, in Proceedings of the 8th Quadrenniel IAGOD Symposium, Ottawa, Canada, (ed.) Y.T. Maurice, Schweizerbart'sche, Stuttgart, p. 271-284.

Moyle, J.E.

1985: East Kemptville tin project; in Guide to the Granites and Mineral Deposits of Southwestern Nova Scotia, (ed.) A.K. Chatterjee and D.B. Clarke; Nova Scotia Department of Mines and Energy, Paper 85-3, p. 199-200.

Murck, B.

1986: Petrographic Report on samples from the West Gabbro Zone; unpublished Company Report, Department of Earth and Planetary Sciences, University of Toronto.

Nance, R.D.

1987: Dextral transpression and late Carboniferous sedimentation in the Fundy coastal zone of southern New Brunswick; in Sedimentary Basins and Basin Forming Mechanisms,(ed.) C. Beaumont and A. J. Tankard; Canadian Society of Petroleum Geologists, Memoir 12, p. 363-377.

Nast, H.J. and Williams-Jones, A.E.

1991: The role of water-rock interaction and fluid evolution in forming the porphyry-related Sisson Brook W-Cu-Mo deposit, New Brunswick; Economic Geology, v. 86, p. 302-317.

Newhouse, W.H.

1936: A zonal gold distribution in Nova Scotia; Economic Geology, v. 31, p. 805-831.

Normand, C., Gauthier, M., and Jébrak, M.

1991: Antimony mineralization in the Brompton-Baie Verte structural zone (Quebec) (abstract); Geological Association of Canada, Program with Abstracts, v. 16, p. A91.

Obalski, J.

1890: Mines et minéraux de la province de Québec; Département de la Colonisation et des Mines du Québec, 174 p.

1898: Or dans la Province de Québec, Canada; Département de la Colonisation et des Mines du Québec, 85 p.

O'Brien, B.H.

1985: The formation of veins in greenschist facies rocks and the early deformation of the Meguma Group, eastern Nova Scotia; Nova Scotia Department of Mines and Energy, Paper 85-2, 35 p.

1989: Summary of the geology between La Poile Bay and Couteau Bay (11)/9 and 11O/16), southwestern Newfoundland; in Current Research; Newfoundland Department of Mines and Energy, Geological Survey Branch, Report 89-1, p. 105-120.

O'Brien, B.H. and O'Brien, S.J.

1990: Re-investigation and re-interpretation of the Silurian La Poile Group of southwestern Newfoundland; in Current Research; Newfoundland Department of Mines and Energy, Geological Survey Branch, Report 90-1. p. 317-326.

O'Brien, B.H., O'Brien, S.J., and Dunning, G.R.

1991: Silurian cover, Late Precambrian-Early Ordovician basement, and the chronology of Silurian orogenesis in the Hermitage Flexure (Newfoundland Appalachians); American Journal of Science, v. 291, p. 760-800.

O'Brien, S.J. and Knight, I.
1988:   The Avalonian geology of southwest Bonavista Bay: portions of the St. Brendan's (2C/13) and Eastport (2C/12) map areas; in Current Research; Newfoundland Department of Mines and Energy, Mineral Development Division, Report 88-1, p. 193-206.

O'Brien, S.J. and Nunn, G.A.G.
1980:   Terrenceville (1M/10) and Gisborne Lake (1M/15) map areas, Newfoundland; in Current Research; Newfoundland Department of Mines and Energy, Mineral Development Division, Report 80-1, p. 120-133.

O'Brien, S.J., Wardle, R.J., and King, A.F.
1983:   The Avalon Zone: a Pan-African terrane in the Appalachian Orogen of Canada; Geological Journal, v. 18, p. 195-222.

O'Brien, S.J., Nunn, G.A.G., Dickson, W.L., and Tuach, J.
1984:   Geology of the Terrenceville (1M/10) and Gisborne Lake (1M/15) map area, southeast Newfoundland; Newfoundland Department of Mines and Energy, Mineral Development Division, Report 84-4, p. 1-54.

O'Brien, S.J., Tucker, R.D., Dunning, G.R., and O'Driscoll, C.F.
1992a: Four-fold subdivision of the Late Precambrian magmatic record of the Avalon Zone type area (East Newfoundland): nature and significance (abstract); Geological Association of Canada, Program with Abstracts, v. 17.

O'Brien, S.J., O'Driscoll, C.F., and Tucker, R.D.
1992b: Reinterpretation of the geology of parts of the Hermitage Peninsula, southwestern Avalon Zone, Newfoundland; in Current Research, Newfoundland Department of Mines and Energy, Geological Survey Branch, Report 92-1, p. 185-194.

O'Brien, S.J., O'Brien, B.H., O'Driscoll, C.F., Dunning, G.R., Holdsworth, R.E., and Tucker, R.
1991:   Silurian orogenesis and the northwestern limit of Avalonian rocks in the Hermitage Flexure Newfoundland Appalachians (abstract); Geological Society of America, Program with Abstracts, v. 23, p. 109.

O'Driscoll, C.F. and Strong, D.F.
1979:   Geology and geochemistry of late Precambrian volcanic and intrusive rocks of south western Avalon Zone in Newfoundland; Precambrian Research, v. 8, p. 19-48.

O'Reilly, G.A.
1985:   Alkali metasomatism associated with W-Sn-Ag-Au mineralization in the granitoid rocks of the Westfield-Caledonia area, Queens County, Nova Scotia; in Guide to the Granites and Mineral Deposits of Southwestern Nova Scotia, (ed.) A.K. Chatterjee and D.B. Clarke; Nova Scotia Department of Mines and Energy, Paper 85-3, preprint, p. 217-232.
1990:   Petrographic and geochemical evidence for a hypogene origin of granite- hosted, vein type Mn mineralization at the New Ross Mines, Lunenburg County, Nova Scotia; Atlantic Geology, v. 26, p. 183.

O'Reilly, G.A. and Kontak, D.J.
1991:   The East Kemptville tin deposit and the Southwest Nova Scotia tin domain; Paper presented to the Mines Ministers Meeting, Halifax, Nova Scotia, 9 p.

O'Reilly, G.A., Farley, E.J., and Charest, M.H.
1982:   Metasomatic-hydrothermal mineral deposits of the New Ross-Mahone Bay area, Nova Scotia; Nova Scotia Department of Mines and Energy, Paper 82-2, 96 p.

O'Reilly, G.A., Gauthier, G., and Brooks, C.
1985:   Three Permo-Carboniferous Rb-Sr age determinations from the South Mountain Batholith, southwestern Nova Scotia; in Mines and Minerals Branch Report of Activities, 1984; Nova Scotia Department of Mines and Energy, Report 85-1, p. 143-152.

O'Sullivan, J.R.
1981:   Report on Geological, Geochemical, Geophysical and Diamond-Drilling Surveys on the Pugwash Claim Group; unpublished report, Nova Scotia Department of Mines and Energy, Mineral Assessment Report 11E/03A 13-E-33(03).

Page, N.J. and Talkington, R.W.
1984:   Palladium, platinum, rhodium, ruthenium and iridium in peridotites and chromitite from ophiolite complexes in Newfoundland; Canadian Mineralogist, v. 22, p. 137-149.

Pajari, G.E., Jr., Rast, N., and Stringer, P.
1977:   Paleozoic volcanicity along the Bathurst-Dalhousie geotraverse, New Brunswick, and its relations to structure; in Volcanic Regimes in Canada, (ed.) W.R.A. Baragar, L.C. Coleman, and J.M. Hall, Geological Association of Canada, Special Paper 16, p. 111-124.

Paktunc, A.D.
1986:   St. Stephen mafic-ultramafic intrusion and related nickel-copper deposits; in Current Research, Part A; Geological Survey of Canada, Paper 86-1A, p. 327-331.

1987a: Nickel, copper, platinum, and palladium relations in Ni-Cu deposits of the St. Stephen intrusion, New Brunswick; in Current Research, Part A; Geological Survey of Canada, Paper 87-1A, p. 543-553.
1987b: Platinum-group-element environments in New Brunswick; in 12th Annual Review of Activities, Project Resumes, 1987, (ed.) S.A. Abbott; New Brunswick Department of Natural Resources and Energy, Minerals and Energy Division, Information Circular 87-2, p. 132-136.
1988:   Nickel-copper sulphide mineralization associated with the Goodwin Lake intrusions, northern New Brunswick; in Current Research, Part B; Geological Survey of Canada, Paper 88-1B, p. 155-162.
1989:   Petrology of the St. Stephen intrusion and the genesis of related nickel-copper sulfide deposits; Economic Geology, v. 84, p 817-840.

Papenfus, E.B.
1931:   "Redbed" Copper Deposits in Nova Scotia and New Brunswick; Economic Geology, v. 26, p. 314-330.

Papezik, V.S.
1964:   Nickel mineralization at Tilt Cove, Notre Dame Bay, Newfoundland; Geological Association of Canada, Proceedings, v. 15, pt. 2, p. 224-225.

Papezik, V.S. and Keats, H.F.
1976:   Diaspore in a pyrophyllite deposit on the Avalon Peninsula, Newfoundland; Canadian Mineralogist, v. 14, p. 442-449.

Papezik, V.S., Keats, H.F., and Vahtra, J.
1978:   Geology of the Foxtrap pyrophyllite deposit, Avalon Peninsula, Newfoundland; Canadian Institute of Mining and Metallurgy, Bulletin, v. 71, no. 799, p. 152-160.

Paradis, S., Birkett, T.C., and Godue, R.
1990:   Preliminary investigation of the Upton sediment-hosted barite deposit, southern Québec Appalachians; in Current Research, Part B; Geological Survey of Canada, Paper 90-1B, p. 1-8.

Patey, K.S.
1990:   Lode gold mineralization at Deer Cove, Baie Verte Peninsula, Newfoundland; B.Sc. thesis, Memorial University of Newfoundland, St. John's, Newfounland, 97 p.

Petruk, W.
1964:   Mineralogy of the Mount Pleasant tin deposit in New Brunswick; Mines Branch, Ottawa, Technical Bulletin, v. 56, 37 p.
1973:   The tungsten-molybdenum-bismuth deposit of Brunswick Tin Mines Ltd.: its mode of occurrence, mineralogy and amenability to mineral beneficiation; Canadian Mining And Metallurgical Bulletin, v. 66, p. 113-130.

Phillips, G.N.
1986:   Geology and alteration in the Golden Mine, Kalgoorlie; Economic Geology, v. 81, p. 779-808.

Pickett, J.W.
1987:   Geology and geochemistry of the Skidder basalt; in Buchans Geology, Newfoundland, (ed.) R.V. Kirkham; Geological Survey of Canada, Paper 86-24, p. 195-218.

Pitre, C.V.
1987:   Paragenesis of sediment-hosted tin mineralization at the Duck Pond prospect, East Kemptville, Yarmouth County, Nova Scotia; B.Sc. thesis, Carleton University, Ottawa, Ontario, 55 p.

Pitre, C.V. and Richardson, J.M.
1989:   Paragenesis of veins of the Duck Pond tin prospect, Meguma Group, East Kemptville, Nova Scotia; Canadian Journal of Earth Sciences, v. 26, p. 2032-2043.

Poole, J.C.
1991:   Gold mineralization on the Rattling Brook property, Jackson's Arm area, White Bay, Newfoundland; in Ore Horizons, v. 1, (ed.) S. Swinden and A. Hogan; Newfoundland Department of Mines and Energy, Geological Survey Branch, p. 119-125.

Pope, A.J., Calon, T.J., and Swinden, H.S.
1990:   Stratigraphy, structural geology and mineralization in the Gullbridge area, central Newfoundland; in Metallogenic Framework of Base and Precious Metal Deposits, Central and Western Newfoundland, (ed.) H.S. Swinden, D.T.W. Evans, and B.F. Kean; 8th IAGOD Symposium, Field Trip No. 1, Guidebook, Geological Survey of Canada, Open File 2156, p. 93-100.

Potter, R.R.
1967:   Metallogenic investigations, Kennebecasis Zone; in Geological Investigations in New Brunswick, (ed.) R.R. Potter; New Brunswick Department of Lands and Mines, Mines Branch, Information Circular 67-1, p. 5-12.
1969:   The geology of Burnt Hill area and ore controls of the Burnt Hill tungsten deposit; Ph.D. thesis, Carleton University, Ottawa, Ontario, 199 p.

Procyshyn, E.L., Sinclair, W.D., and Williams-Jones, A.E.
1989:   Mineral deposits associated with Late Devonian plutons in the New Brunswick-Gaspé sector; Geological Association of Canada, Annual Meeting, Montreal, Field Trip Guidebook B-5.

Pronk, A.G. and Burton, D.M.
1988:   Till geochemistry as a technique for gold exploration in northern New Brunswick; Canadian Institute of Mining and Metallurgy Bulletin, v. 81, no. 915, p. 90-98.

Prudhomme, S.
1985:   Brèche uranifère de Saint-Armand; Ministère de l'Énergie et des Ressources du Québec, MB-85-19, 13 p.
1986:   Indice urano-zincifère de Saint-Armand; Ministère de l'Énergie et des Ressources du Québec, MB-86-16, 11 p.

Raeside, R.P.
1989:   The Lime Hill gneiss: a basement fragment in the Bras d'Or Zone of Cape Breton Island; Geological Survey of Canada, Open File 1998, 45 p.
1990:   Low-pressure metamorphism of the Lime Hill gneissic complex, Bras d'Or Terrane, Cape Breton Island, Nova Scotia; in Mineral Deposit Studies in Nova Scotia, v. 1, (ed.) A.L. Sangster; Geological Survey of Canada, Paper 90-8, p. 67-76.

Raeside, R.P. and Barr, S.M.
1990:   Geology and tectonic development of the Bras d'Or suspect terrane, Cape Breton Island, Nova Scotia; Canadian Journal of Earth Sciences, v. 27, p. 1371-1381.

Raeside, R.P., Barr, S.M, White, C.E. and Dennis, F.A.R.
1986:   Geology of the northernmost Cape Breton Highlands, Nova Scotia; in Current Research, Part A; Geological Survey of Canada, Paper 86-1A, p. 291-296.

Ramazani, J.
1992:   The geology, geochemistry and U-Pb geochronology of the Stog'er Tight gold prospect, Baie Verte Peninsula, Newfoundland; M.Sc. thesis, Memorial University of Newfoundland, St. John's, Newfoundland, 256 p.

Ramsay, J.G.
1967:   Folding and fracturing in rocks; McGraw Hill, London, 568 p.
1980:   The crack-seal mechanism of rock deformation; Nature, v. 284, p. 135-139.

Ranger, M.J.
1979:   The stratigraphy and depositional environment of the lower Ordovician Bell Island and Wabana Groups, Conception Bay, Newfoundland; M.Sc. thesis, Memorial University of Newfoundland, St. John's, Newfoundland, 216 p.

Ranger, M.J., Pickerill, R.K., and Fillion, D.
1984:   Lithostratigraphy of the Cambrian?-Lower Ordovician Bell Island and Wabana groups of Bell, Little Bell and Kellys islands, Conception Bay, eastern Newfoundland; Canadian Journal of Earth Sciences, v. 21, p. 1245-1261.

Rast, N. and Grant, R.
1973:   Transatlantic correlation of the Variscan-Appalachian Orogeny; American Journal of Science, v. 273, p. 572-579.

Rast, N. and Stringer, P.
1980:   A geotraverse across a deformed Ordovician ophiolite and its Silurian cover, Northern New Brunswick, Canada; Tectonophysics, v. 69, p. 221-245.

Ravehurst, C.E. and Zentilli, M.
1987:   A model for the evolution of hot (200°C) overpressured brines under an evaporite sea: the Fundy/Magdalen Carboniferous Basin of Atlantic Canada and its associated Pb-Zn-Ba deposits; in Sedimentary Basins and Basin Forming Mechanisms, (ed.) C. Beaumont and A.J. Tankard; Canadian Society of Petroleum Geologists, Memoir 12, p. 335-349.

Ravenhurst, C.E., Reynolds, P.H., Zentilli, M., Krueger, H.W., and Blenkinsop, J.
1990:   Strontium isotopic geochemistry of lead-zinc/barite deposits and host rocks of the Carboniferous Minas Sub-basin, Nova Scotia; in Mineral Deposit Studies in Nova Scotia, v. 1, (ed.) A.L. Sangster; Geological Survey of Canada, Paper 90-1, p. 115-162.

Reynolds, P.H., Taylor, K.A., and Morgan, W.R.
1981:   40Ar/39Ar ages from the Botwood-Mount Peyton region, Newfoundland: possible paleomagnetic implications; Canadian Journal of Earth Sciences, v. 18, p. 1850-1855.

Reynolds, P.H., Jamieson, R.A., Barr, S.M., and Raeside, R.P.
1989:   A 40Ar/39Ar dating study in the Cape Breton Highlands, Nova Scotia: thermal histories and tectonic implications; Canadian Journal of Earth Sciences, v. 26, p. 2081-2091.

Richardson, J.M.
1985a: Magmatic fractionation and metasomatic alteration of the Davis Lake-East Kemptville complex and its relationship to the rest of the South Mountain Batholith; in Guide to the Granites and Mineral Deposits of Southwestern Nova Scotia, (ed.) A.K. Chatterjee and D.B. Clarke; Nova Scotia Department of Mines and Energy, Paper 85-3, preprint, p. 131-148.

1985b: The East Kemptville greisen-hosted tin deposit, Nova Scotia: overview of the geological relationships and geochemistry; in Granite-related Mineral Deposits: Geology, Petrogenesis and Tectonic Setting; Canadian Institute of Mining and Metallurgy, Geology Division, Extended Abstracts, Conference on Granite related Mineral Deposits, September 15-18, Halifax, Canada, p. 203-204.
1988a: Field and textural relationships of alteration and greisen-hosted mineralization at the East Kemptville tin deposit, Davis Lake monzogranite, southwest Nova Scotia; in Recent Advances in the Geology of Granite-Related Mineral Deposits; Canadian Institute of Mining and Metallurgy, Special Volume 39, p. 265- 279.
1988b: Genesis of the East Kemptville Greisen-hosted tin deposit, Davis Lake Complex, southwestern Nova Scotia, Canada; Ph.D. thesis, Carleton University, Ottawa, Ontario, 293 p.

Richardson, J.M., Spooner, E.T.C., and McAuslin, D.A.
1982:   The East Kemptville tin deposit, Nova Scotia: an example of a large tonnage, low grade, greisen-hosted tin deposit in the endocontact zone of a granite batholith; in Current Research, Part B; Geological Survey of Canada, Paper 82-1B, p. 27-32.

Richardson, J.M., Bell, K., Blenkinsop, J., and Watkinson, D.H.
1989:   Rb-Sr age and geochemical distinctions between the Carboniferous tin-bearing Davis Lake complex and the Devonian South Mountain Batholith, Meguma Terrane, Nova Scotia; Canadian Journal of Earth Sciences, v. 26, p. 2044-2061.

Rickard, T.A.
1927:   Gold mining in Nova Scotia; Nova Scotia Department of Public Works and Mines, Report on the Mines, 1926, p. 163-177.

Riordon, P.H.
1975:   Geologie des gites d'amiante du sud-est Quebecois; Ministère des Richesses naturelles du Québec, ES 18, 100 p.

Ritcey, D.H.
1993:   Geology, U-Pb geochronology, and stable isotope geochemistry of the Hammer Down gold prospect, Green Bay District, Newfoundland; M.Sc. thesis, Memorial University of Newfoundland, St. John's, Newfoundland, 229 p.

Robertson, D.J. and Duncan, D.R.
1982:   Geology and exploration history of the Millet Brook uranium deposit (abstract and reprint of presentation); Canadian Institute of Mining and Metallurgy Bulletin, v. 75, no. 839, p. 130.

Robertson, D.J. and MacLean, S.A.
1991:   Geology of the Great Gull scheelite prospects, central Newfoundland; in Ore Horizons, v. 1, (ed.) S. Swinden and A. Hogan; Newfoundland Department of Mines and Energy, Geological Survey Branch, p. 33-42.

Robinson, G.R.
1986:   Gold and associated metals at Cuttingsville, Vermont (abstract); Geological Society of America, Programs with Abstracts, v. 18, no. 1, p. 62-63.

Rogers, J.W. and Greenberg, J.K.
1981:   Trace elements in continental-margin magmatism, Part III: Alkali granites and their relationship to cratonization; Geological Society of America Bulletin, v. 92, p. 6-9.

Rose, D.G. and Johnson, S.C.
1990:   New Brunswick computerized mineral occurrence database; New Brunswick Department of Natural Resources and Energy, Mineral Resources Division, Mineral Resource Report 3, 69 p. and 7 diskettes.

Rose, E.R.
1952:   Torbay map area, Newfoundland; Geological Survey of Canada, Memoir 165, 64 p.
1969:   Geology of titanium and titaniferous deposits of Canada; Geological Survey of Canada, Economic Geology Report 25, 177 p.

Routhier, P.
1980:   Ou sont les métaux pour l'avenir? Les provinces métalliques-essai de métallogenie globale; Bureau de Recherche Geokogique et Minieres, Orleans, France, Memoir 105, 410 p.

Ruitenberg, A.A.
1963:   Tin mineralization and associated rock alteration at Mount Pleasant, Charlotte County, New Brunswick; M.Sc. thesis, University of New Brunswick, Fredericton, New Brunswick, 172 p.
1967:   Stratigraphy, structure and metallization Piskahegan-Rolling Dam area, northern Appalachians, New Brunswick, Canada; Leidse Geologische Mededelingen, v. 40, p. 79-120.
1969:   Mineral deposits in granitic intrusions and related metamorphic aureoles in parts of the Welsford, Loch Alva, Musquash and Pennfield areas; New Brunswick Department of Natural Resources, Mineral Resources Branch, Report of Investigation 9, 24 p.

1970: Mineralized structures in the Johnson Croft, Annidale, Jordan Mountain, and Black River areas; New Brunswick Department of Natural Resources, Mineral Resources Branch, Report of Investigation 13, 28 p.

1984: Geology and mineralogy of gold deposits in the Cape Spencer-Black River area; in 9th Annual Review of Activities; New Brunswick Department of Natural Resources, Mineral Resources Division, Information Circular 84-2, p. 7-15.

Ruitenberg, A.A.

1993: Geology of the Gravel Hill volcanogenic base metal sulphide deposits; in Current Research, New Brunswick Department of Natural Resources, Mineral Resources, Information Circular 93-1, p. 104-106.

Ruitenberg, A.A. and Fyffe, L.R.

1982: Mineral deposits associated with granitoid intrusions and related subvolcanic stocks in New Brunswick and their relationship to Appalachian tectonic evolution; Canadian Institute of Mining and Metallurgy Bulletin, v. 75, no. 842, p. 83-97.

Ruitenberg, A.A. and Ludman, A.

1978: Stratigraphy and tectonic setting of Early Paleozoic sedimentary rocks of the Wirral-Big Lake area, southwestern New Brunswick and Southeastern Maine; Canadian Journal of Earth Sciences, v. 15, p. 22-32.

Ruitenberg, A.A. and McCutcheon, S.R.

1982: Acadian and Hercynian structural evolution of southern New Brunswick; in Major Structural Zones and Faults of the Northern Appalachians, (ed.) P. St-Julien and J. Béland; Geological Association of Canada, Special Paper 24, p. 131-148.

1985: Tungsten, molybdenum and tin deposits in New Brunswick; Geological Association of Canada/Mineralogical Association of Canada, Joint Annual Meeting, Fredericton, New Brunswick, Excursion 13, Guidebook, 33 p.

1993: Syn- and post-collisional mineral deposits in the Canadian Appalachians; Exploration and Mining Geology, v. 2, p. 390-392.

Ruitenberg, A.A. and McLeod, M.J.

1991: Metallogenic implications of a newly discovered Ordovician island-arc sequence in southern New Brunswick (abstract); Geological Association of Canada, Programs with Abstracts, v. 16, p. A108.

Ruitenberg, A.A., Johnson, S.C., and Fyffe, L.R.

1990: Epigenetic gold deposits and their tectonic setting in the New Brunswick Appalachians; Canadian Institute of Mining and Metallurgy Bulletin, v. 83, no. 934, p. 43-55.

Ruitenberg, A.A., Venugopal, D.V., and Giles, P.S.

1973: "Fundy Cataclastic Zone", New Brunswick: evidence for post-Acadian penetrative deformation; Geological Society of America Bulletin, v. 84, p. 3029-3044.

Ruitenberg, A.A., McCutcheon, S.R., Venugopal, V., and Pierce, G.A.

1977: Mineralization related to post-Acadian tectonism in southern New Brunswick; Geoscience Canada, v. 4, p. 13-22.

Ruitenberg, A.A., Giles, P.S., Venugopal, D.V., Buttimer, S.M., McCutcheon, S.R., and Chandra, J.

1979: Geology and mineral deposits, Caledonia area; New Brunswick Department of Natural Resources, Mineral Resources Branch, Memoir 1, 213 p.

Ruitenberg, A.A., McCutcheon, S.R., Watters, S.E., McLeod, M.J., Burton, D.J., and Hoy, D.

1989: Field guide to gold occurrences in New Brunswick; New Brunswick Department of Natural Resources and Energy, Minerals and Energy Division, Field Guidebook No. 1, 63 p.

Ryan, R.J.

1988: Nova Scotia's copper silver domain: the Cumberland Basin; Information pamphlet released at the annual meeting, Prospectors and Developers Association, March, 1988, 15 p.

Ryan, R.J., Grist, Z., and Zentilli, M.

1990: Thermal history of the Maritimes Basin: evidence from apatite fission track analysis; in Report of Activities, 1990; Nova Scotia Department of Mines and Energy, Mines and Minerals Branch, Report 91-1, p. 27-32.

Ryan R.J., Boehner, R.C., Stea, R.R., and Rogers, P.J.

1989: Geology, geochemistry and exploration applications for the Permo-Carboniferous redbed copper deposits of the Cumberland Basin, Nova Scotia, Canada; in Sediment-Hosted Stratiform Copper Deposits, (ed.) R.W. Boyle, C.J. Jefferson, E.C. Jowett, and R.V. Kirkham; Geological Association of Canada, Special Paper 36, p. 245-256.

Ryan, R.J., Turner, R., Stea, R.R., and Rogers, P.J.

1988: Heavy minerals and sediment dispersal trends as exploration tools for potential tin and gold paleoplacers in Carboniferous rocks, Nova Scotia; in Prospecting in Areas of Glaciated Terrane, 1988, (ed.) D. MacDonald and K. Mills; Canadian Institute of Mining and Metallurgy, p. 41-56.

St-Julien, P. and Hubert, C.

1975: Evolution of the Taconian Orogen in the Quebec Appalachians; in Tectonics and Mountain Ranges; American Journal of Science, v. 275-A, p. 337-362.

St-Julien, P. and Lamarche, R.Y.

1965: Geology of Sherbrooke area, Sherbrooke County, Quebec; Quebec Department of Natural Resources, Report 530, 34 p.

Sangster, A.L.

1970: Metallogeny of base metal, gold and iron deposits of the Grenville Province of southeastern Ontario; Ph.D. thesis, Queen's University, Kingston, Ontario, 356 p.

1990: Metallogeny of the Meguma Terrane, Nova Scotia; in Mineral Deposit Studies in Nova Scotia, Volume 1, (ed.) A.L. Sangster; Geological Survey of Canada, Paper 90-8, p. 115-162.

1992: Light stable isotope evidence for a metamorphogenic origin for bedding- parallel: gold-bearing veins in Cambrian flysch, Meguma Group, Nova Scotia; Exploration and Mining Geology, v. 1, p. 69-79.

Sangster, A.L. and Bourne, J.

1982: Geology of the Grenville Province and regional metallogenesis of the Grenville Supergroup; in Precambrian Sulphide Deposits, (ed.) R.W. Hutchinson, C.D. Spence, and J.M. Franklin, Geological Association of Canada, Special Paper 25, p. 91-126.

Sangster, A.L., Justino, M.F., and Thorpe, R.I.

1990a: Metallogeny of Proterozoic marble-hosted zinc occurrences at Lime Hill and Meat Cove, Cape Breton Island, Nova Scotia; in Mineral Deposit Studies in Nova Scotia, v. 1, (ed.) A.L. Sangster; Geological Survey of Canada, Paper 90-8, p.31-66.

Sangster, A.L., Thorpe, R.I., and Chatterjee, A.K.

1990b: A reconnaissance lead isotopic study of mineral occurrences in pre-Carboniferous basement rocks of northern and central Cape Breton Island, Nova Scotia; in Mineral Deposit Studies in Nova Scotia, v. 1, (ed.) A.L. Sangster; Geological Survey of Canada, Paper 90-8, p. 101-114

Sangster, A.L., Bretzlaff, R., Graves, M.C., and Zentilli, M.

1989: Stratigraphic variation of ^{34}S compositions in the Meguma Group: implications for paleoenvironment and mineralization; Atlantic Geology, v. 25, p. 168.

Sangster, D.F.

1980: A review of Appalachian stratabound sulphides in Canada; in Review of Caledonian-Appalachian Stratabound Sulphides, (ed.) F.M. Vokes and E. Zachrisson; Geological Survey of Ireland, Special Paper 5, p. 7-18.

Sangster, D.F. and Vaillancourt, P.D.

1990a: Geomorphology in the exploration for undiscovered sandstone-lead deposits, Salmon River Basin, Nova Scotia; Canadian Institute of Mining and Metallurgy Bulletin, v. 83, p. 62-68.

1990b: Geology of the Yava sandstone-lead deposit, Cape Breton Island, Nova Scotia; in Mineral Deposit Studies in Nova Scotia, v. 1, (ed.) A.L. Sangster; Geological Survey of Canada, Paper 90-8, p. 203-244

Saunders, C.M.

1991: Mineralization in western White Bay; in Current Research; Newfoundland Department of Mines and Energy, Geological Survey Branch, Report 91-1, p. 335-348.

Saunders, C.M. and Smyth, W.R.

1990: Geochemical characteristics of the Gull Lake intrusive suite and Devils Room Granite, western White Bay, Newfoundland; in Current Research; Newfoundland Department of Mines and Energy, Geological Survey Branch, Report 90-1, p. 183-200.

Saunders, C.M. and Strong, D.F.

1986: Alteration-zonation related to variations in water/rock ratio at the Betts Cove ophiolite massive sulphide deposit, Newfoundland, Canada; Metallogeny of Basic and Ultramafic Rocks, (ed.) M.J. Gallagher, R.A. Ixer, C.R. Neary, and H.M. Prichard; Institute of Mining and Metallurgy, London, p. 161-175.

Saunders, C.M. and Tuach, J.

1991: Potassic and sodic alteration accompanying gold mineralization at the Rattling Brook deposit, western White Bay, Newfoundland; Economic Geology, v. 86, p. 555-569.

Saunders, C.M., Strong, D.F., and Sangster, D.F.

1992: Carbonate-hosted lead-zinc deposits of western Newfoundland; Geological Survey of Canada, Bulletin 219, 78 p.

Sauvé, P., Cloutier, J.P., and Genois, G.

1972: Gisements de métaux usuels du Sud-Est du Québec; 24th International Geological Congress, Montreal; Field Trip Guidebook B-07, 24 p.

Savard, M.
1985: Indices minéralisés du sud de la Gaspésie; Ministère de l'Energie et des Ressources du Québec, ET 83-08, 92 p.

Sawkins, F.J.
1972: Sulfide ore deposits in relation to plate tectonics; Journal of Geology, v. 80, p. 377-397.

Scheibner, E.
1974: A plate tectonic model of the Palaeozoic tectonic history of New South Wales; Journal of the Geological Society of Australia, v. 20, p. 405-426.

Schenk, P.E.
1970: Regional variation of the flysch-like Meguma Group (Lower Paleozoic) of Nova Scotia compared to recent sedimentation of the Scotia Shelf; Geological Association of Canada, Special Paper 7, p. 127-153.

Schrijver, K., Marcoux, E., Beaudoin, G., and Calvez, J.Y.
1988: Pb-Zn occurrences and their Pb-isotopic signatures bearing on metallogeny and mineral exploration-Paleozoic sedimentary rocks, northern Appalachians, Quebec; Canadian Journal of Earth Sciences, v. 25, p. 1777-1790.

Schuiling, R.D.
1967: Tin belts on the continents around the Atlantic Ocean; Economic Geology, v. 62, p. 540-550.

Scratch, R.B., Watson, G.P., Kerrich, R., and Hutchinson, R.W.
1984: Fracture-controlled antimony-quartz mineralization, Lake George deposit, New Brunswick; Economic Geology, v. 79, p. 1159-1186.

Seal, R.R.
1984: The Lake George W-Mo stockwork deposit, southwestern New Brunswick; M.Sc. thesis, Queen's University, Kingston, Ontario, 202 p.

Sears, W.A.
1990: A geochemical, petrographic and metallogenic analysis of volcanogenic sulphide deposition within the Connaigre Bay Group, Hermitage Peninsula, southern Newfoundland; M.Sc. thesis, Memorial University of Newfoundland, St. John's, Newfoundland, 282 p.

Sears, W.A. and O'Driscoll, C.F.
1989: Metallogeny of the Connaigre Bay Group, southern Newfoundland; in Current Research; Newfoundland Department of Mines and Energy, Geological Survey Branch, Report 89-1, p. 193-200.

Selwyn, A.R.C.
1872: Notes and observations on the goldfields of Quebec and Nova Scotia; Geological Survey of Canada, Report of Progress, 1870-71, p. 252-282.

Shaw, W.G.
1983: A geological investigation of the Lazy Head tungsten deposit, Guysborough County, and a comparison with the Moose River deposit, Halifax County, Nova Scotia; M.Sc. thesis, Dalhousie University, Halifax, Nova Scotia, 229 p.

Shewman, R.W.
1968: Geological report on the Evandale stock, New Brunswick; unpublished report for Rio Tinto Canadian Exploration, New Brunswick Department of Natural Resources and Energy, Mineral Resources Division, Assessment File 470171.

Shockney, P.N., Renfro, A.R., and Peterson, R.J.
1974: Copper-silver solution fronts at Paoli, Oklahoma; Economic Geology, v. 69, p. 266-268.

Sidwell, K.O.J.
1954: Geology and geochemistry of Group 42A, National Management Limited; New Brunswick Department of Natural Resources and Energy, Minerals and Energy Division, Assessment File 470025.

Sinclair, W.D., Kooiman, G.J.A., and Martin, D.A.
1988: Geological setting of granites and related tin deposits in the North Zone, Mount Pleasant, New Brunswick; in Current Research, Part B; Geological Survey of Canada, Paper 88-1B, p. 201-208.

Slack, J.F.
1990: Geologic features of Besshi-type massive sulphide deposits; 8th IAGOD symposium, Ottawa, Ont., Canada, Program with Abstracts, p. A191-A192.
1991: Preliminary Assessment of Metallic Mineral Resources in the Glens Falls 1º X 2º Quadrangle, New York, Vermont and New Hampshire; U.S. Geological Survey, Bulletin 1887, p. R1-R26.

Smith, C.H.
1958: The Bay of Islands igneous complex, western Newfoundland; Geological Survey of Canada, Memoir 290, 132 p.

Smith, P.K.
1983: Geology of the Cochrane Hill gold deposit, Guysborough County, Nova Scotia; in Report of Activities, 1982; Nova Scotia Department of Mines and Energy, Mines and Minerals Branch, Report 83-1, p. 225-256.
1984: Geology and lithogeochemistry of the Cochrane Hill gold deposit-an indication of metalliferous source beds; Nova Scotia Department of Mines and Energy, Report 84-1, p. 203-214.

Smith, P.K. and Kontak, D.J.
1986: Meguma gold studies: Advances in geological insight as an aid to gold exploration; in 10th Annual Open House and Review of Activities, Program and Summaries; Nova Scotia Department of Mines and Energy, Information Series 12, p. 105-114.
1988a: Meguma gold studies I: Generalized geological aspects of the Beaver Dam gold deposit, eastern Meguma zone, Nova Scotia; in Mines and Minerals Branch Report of Activities, 1987, Part B; Nova Scotia Department of Mines And Energy, Report 88-1, p. 45-60.
1988b: Meguma gold studies II: Vein morphology, classification and information, a new interpretation of "crack seal" quartz veins; in Mines and Minerals Branch Report of Activities, 1987, Part B; Nova Scotia Department of Mines And Energy, Report 88-1, p. 61-76.

Smith, W.S.
1936: Report on Rencontre East molybdenite; unpublished report, Geological Survey of Newfoundland, 18 p.

Smitheringale, W.G.
1973: Geology of parts of Digby, Bridgetown, and Gaspereau Lake map areas, Nova Scotia; Geological Survey of Canada, Memoir 375, 78 p.

Snelgrove, A.K.
1931: Geology and ore deposits of the Betts Cove-Tilt Cove area, Notre Dame Bay, Newfoundland; Canadian Mining and Metallurgical Bulletin, v. 228, p. 477-519.
1934: Chromite deposits of Newfoundland; Newfoundland Geological Survey, Bulletin No. 1.

Snelgrove, A.K. and Baird, D.M.
1953: Mines and mineral occurrences of Newfoundland; Newfoundland Department of Mines and Energy, Information Circular 4, 149 p.

SOQUEM
1972: Report on the geology and geochemistry of DDH's 7104 to 7109, Anomaly No. 21, Project 1-343; Ministère de l'Énergie et des Ressources du Québec, GM-28012.
1975: Propriété de Basket Brook, cantons de Mann et de Restigouche, Project 11-349; Ministère de l'Énergie et des Ressources du Québec, GM-31282, p. 2.

Squires, G.C., MacKenzie, A. Colin, and McInnis, D.
1990: Geology and genesis of the Duck Pond volcanigenic massive sulphide deposit; in Metallogenic Framework of Base and Precious Metal Deposits, Central and Western Newfoundland, (ed.) H.S. Swinden, D.T.W. Evans, and B.F. Kean; 8th IAGOD Symposium, Field Trip No. 1, Guidebook, Geological Survey of Canada, Open File 2156, p. 56-64.

Stea, R.R., Day, T.E., and Ryan, R.J.
1986: The relationship of till and bedrock geochemistry in northern Nova Scotia and its metallogenic implications; in Prospecting in Areas of Glaciated Terrain, 1986; Institute of Mining and Metallurgy, p. 241-260

Stephens, M.B., Swinden, H.S., and Slack, J.
1984: Correlation of massive sulphide deposits in the Appalachian-Caledonian Orogen on the basis of paleo-tectonic setting; Economic Geology, v. 76, p. 1442-1478.

Stewart, P.W.
1992: The origin of the Hope Brook mine, Newfoundland: a shear zone-hosted acid-sulphate gold deposit; Ph.D. thesis, University of Western Ontario, London, Ontario, 780 p.

Stewart, P.W. and Dunning, G.R.
1990: The LaPoile Group, the Hope Brook Mine and geochronology (abstract); Program with Abstracts; Atlantic Geoscience Society, Symposium, 1990, p. 30.

Stewart, P.W. and Stewart, J.W.
1988: The relative timing of gold mineralization at Hope Brook, Newfoundland (abstract); Geological Association of Canada, Program with Abstracts, v. 13, p. A118.

Stewart, R.D.
1978: The geology of the Benjamin River intrusive complex; M.Sc. thesis, Carleton University, Ottawa, Ontario, 126 p.

Strong, D.F.
1974: Plate tectonic setting of Appalachian-Caledonian mineral deposits as indicated my Newfoundland examples; Society of Mining Engineers AIME, Transactions, v. 256, p. 121-128.
1980: Granitoid rocks and associated mineral deposits of eastern Canada and western Europe; in The Continental Crust and its Mineral Deposits, (ed.) D.W. Strangway; Geological Association of Canada, Special Paper 20, p. 741-769.

Strong, D.F., O'Brien, S.J., Taylor, S.W., Strong, P.G., and Wilton, D.H.C.
1978: Geology of the Marystown (1M/3) and St. Lawrence (1L/14) map areas, Newfoundland; Newfoundland Department of Mines and Energy, Mineral Development Division, Report 77-8, 81 p.

Sullivan, R.W., van Staal, C.R., and Langton, J.P.
1990: U-Pb zircon ages of plagiogranite and gabbro from the ophiolitic Deveraux Formation, Fournier Group, northeastern New Brunswick; in Radiogenic age and isotope studies; Geological Survey of Canada, Paper 89-2, p. 119-122.

Swanson, E.A., Strong, D.F., and Thurlow, J.G. (ed.)
1981: The Buchans Orebodies: Fifty Years of Geology and Mining; Geological Association of Canada, Special Paper 22, 350 p.

Swinden, H.S
1982: Metallogenesis in the Bay d'Espoir area, southern Newfoundland: summary and implications for regional metallogeny in the Newfoundland Central Mobile Belt; Canadian Mining and Metallurgical Bulletin, v.75, no. 839, p. 172-183.
1987: Geology and mineral occurrences in the central and northern parts of the Robert's Arm Group, central Newfoundland; in Current Research, Part A; Geological Survey of Canada, Paper 87-1A, p. 381-390.
1988a: Geology and Economic Potential of the Pipestone Pond area (12A/1 NE; 12A/8E), central Newfoundland; Newfoundland Department of Mines, Mineral Development Division, Report 88-2, 88 p.
1988b: Volcanogenic sulphide deposits of the Wild Bight Group, Notre Dame Bay; in The Volcanogenic Sulphide Districts of Central Newfoundland, (ed.) H.S. Swinden and B.F. Kean; Geological Association of Canada, Mineral Deposits Division, p. 178-193.
1988c: Volcanogenic sulphide mineralization in the Hermitage Flexure; in The Volcanogenic Sulphide Districts of Central Newfoundland, (ed.) H.S. Swinden and B.F. Kean; Geological Association of Canada, Mineral Deposits Division, p. 165-173.
1990: Regional geology and metallogeny of central Newfoundland; in Metallogenic Framework of Base and Precious Metal Deposits, Central and Western Newfoundland, (ed.) H.S. Swinden, D.T.W. Evans, and B.F. Kean; 8th IAGOD Symposium, Field Trip No. 1, Guidebook, Geological Survey of Canada, Open File 2156, p. 1-27.
1991: Paleotectonic settings of volcanogenic massive sulphide deposits in the Dunnage Zone, Newfoundland Appalachians; Canadian Institute of Mining and Metallurgy Bulletin, v. 84, no. 946, p. 59-69.
1992: Bipartite massive sulphide metallogeny in the Buchans-Robert's Arm belt, central Newfoundland indicated by volcanic whole rock geochemistry (abstract); Geological Association of Canada, Program with Abstracts, v. 17, p. A108.

Swinden, H.S. and Hunt, P.A.
1991: A U-Pb zircon age from the Connaigre Bay Group, southwestern Avalon Zone, Newfoundland: implications for regional correlations and metallogenesis; in Radiogenic Age and Isotope Studies, Report 4; Geological Survey of Canada, Paper 90-2, p. 3-10.

Swinden, H.S. and Sacks, P.E.
1986: Stratigraphy and economic geology of the southern part of the Roberts Arm Group, central Newfoundland; in Current Research, Part A; Geological Survey of Canada, Paper 86-1A, p. 213-220.

Swinden, H.S. and Thorpe, R.I.
1984: Variations in style of volcanism and massive sulphide deposition in the Newfoundland Central Mobile Belt; Economic Geology, v. 76, p. 1596-1619.

Swinden, H.S., Jenner, G.A., and Kean, B.F.
1989a: The significance of refractory source melting in volcanogenic sulphide mineralization: examples from central Newfoundland (abstract); Geological Association of Canada, Program with Abstracts, v. 14, p. A22.

Swinden, H.S., Kean, B.F., and Dunning, G.R.
1988b: Geological and paleotectonic settings of volcanogenic sulphide mineralization in Central Newfoundland; in The Volcanogenic Sulphide Districts of Central Newfoundland, (ed.) H.S. Swinden and B.F. Kean; Geological Association of Canada, Mineral Deposits Division, St. John's, Newfoundland, p. 5-27.

Swinden, H.S., Lane, T.E., and Thorpe, R.I.
1988a: Lead-isotopic compositions of galena in carbonate-hosted deposits of western Newfoundland: evidence for diverse lead sources; Canadian Journal of Earth Sciences, v. 25, p. 593-602.

Swinden, H.S., McBride, D., and Dubé, B.
1990b: Preliminary geological and mineralogical notes on the Nugget Pond gold deposit, Baie Verte Peninsula, Newfoundland; in Current Research; Newfoundland Department of Mines and Energy, Geological Survey Branch, Report 90-1, p. 201-215.

Swinden, H.S., Jenner, G.A., Kean, B.F., and Evans, D.T.W.
1989b: Volcanic rock geochemistry as a guide for massive sulphide exploration in central Newfoundland; in Current Research; Newfoundland Department of Mines and Energy, Mineral Development Division, Report 89-1, p. 201-219.

Swinden, H.S., Jenner, G.A., Fryer, B.J., Hertogen, J., and Roddick, J.C.
1990a: Petrogenesis and paleotectonic history of the Wild Bight Group, an Ordovician rifted island arc in central Newfoundland; Contributions to Mineralogy and Petrology, v. 105, p. 219-241.

Szybinski, Z.A.
1988: New interpretation of the structural and stratigraphic setting of the Cutwell Group, Notre Dame Bay, Newfoundland; in Current Research, Part B; Geological Survey of Canada, Paper 88-1B, p. 263-270.

Szybinski, Z.A., Swinden, H.S., O'Brien, F.H.C., Jenner, G.A., and Dunning, G.R.
1990: Correlation of Ordovician volcanic terranes in the Newfoundland Appalachians: lithological, geochemical and age constraints (abstract); Geological Association of Canada, Program with Abstracts, v. 15, p. A128.

Tassé, N. and Schrijver, K.
1989: Formation of accessory sphalerite by thermochemical sulphate reduction in Lower Paleozoic carbonate rocks (St. Lawrence Lowlands, Québec, Canada); Chemical Geology, v. 80, p. 55-70.

Tassé, N., Schrijver, K., Héroux, Y., and Chagnon, A.
1987: Étude gitologique et évaluation du potentiel minéral des Basse-Terres du Saint-Laurent; Ministère de l'Énergie et des Ressources du Québec, MB 87-46, 268 p.

Taylor, F.C.
1965: Silurian-Ordovician relationships and Silurian stratigraphy in southwestern Nova Scotia; Geological Survey of Canada, Paper 64-13, 24 p.
1967: Reconnaissance geology of Shelburne map area, Queens, Shelburne, and Yarmouth Counties, Nova Scotia; Geological Survey of Canada, Memoir 349, 83 p.

Taylor, F.C. and Schiller, E.A.
1966: Metamorphism of the Meguma Group of Nova Scotia; Canadian Journal of Earth Sciences, v. 3, p. 959-974.

Thicke, M.J.
1987: The Geology of Late Hadrynian metavolcanic and granitoid rocks of the Coxheath-northern East Hills area, Cape Breton Island, Nova Scotia; M.Sc. thesis, Acadia University, Wolfville, Nova Scotia, 300 p.

Thomas, R.D. and Gleeson, C.F.
1988: Metallogeny of the Woodstock area, New Brunswick; Geological Survey of Canada, Open File 1726, 92 p.

Thurlow, J.G.
1976: Occurrence, origin and significance of mechanically transported sulphide ores at Buchans, Newfoundland; in Volcanic Processes in Ore Genesis; Geological Society of London, Special Publication 7, p. 127.
1990: Geology of the Buchans orebodies-A 1990 summary; in Metallogenic Framework of Base and Precious Metal Deposits, Central and Western Newfoundland, (ed.) H.S. Swinden, D.T.W. Evans, and B.F. Kean; 8th IAGOD Symposium, Field Trip No. 1, Guidebook, Geological Survey of Canada, Open File 2156, p. 84-91.

Thurlow, J.G. and Swanson, E.A.
1981: Geology and ore deposits of the Buchans area, central Newfoundland; in The Buchans Orebodies: Fifty Years of Geology and Mining, (e.d.) E.A. Swanson, D.F. Strong, and J.G. Thurlow; Geological Association of Canada, Special Paper 22, p. 114-142.

Tischendorf, G.
1977: Geochemical and petrographic characters of silicic magmatic rocks associated with rare-element mineralization; in Metallization Associated with Acid Magmatism, (ed.) M. Stemprok and others; Czechoslovakia Geological Survey, Prague, v. 2, p. 41-96.

Tremblay, A.
1992: Tectonic and accretionary history of Taconian oceanic rocks of the Quebec Appalachians; American Journal of Science, v. 292, p. 229-252.

Tremblay, A. and Dubé, B.
1991: Structural relationships between some gold occurrences and fault zones in the Bathurst area, northern New Brunswick; in Current Research, Part D; Geological Survey of Canada, Paper 90-1D, p. 89-100.

Tremblay, A., Hébert, R., and Bergeron, M.
1989: Le complexe d'Ascot des Appalaches du sud du Québec: pétrologie et géochimie; Canadian Journal of Earth Sciences, v. 26, p. 2407-2420.

Troop, D.G.
1984:   The petrology and geochemistry of Ordovician banded iron formations and associated rocks at the Flat Landing Brook massive sulphide deposit, northern New Brunswick; M.Sc. thesis, University of Toronto, Toronto, Ontario.

Trottier, J.
1987:   Synthese metallogenique des depots sulfureux volcanogenes de la ceinture ophiolitique des Appalaches du Quebec, région de l'Estrie; These de Doctorat, École Polytechnique de Montréal, 201 p.

Trottier, J. and Gauthier, M.
1991:   Regional tectonic setting and chemistry of base metal deposits in the Eastern Townships, Quebec; Canadian Institute of Mining and Metallurgy Bulletin, v. 84, no. 946, p. 59-69.

Trottier, J., Brown. A.C., and Gauthier, M.
1991:   An Ordovician rift environment for the Memphremagog polymetallic massive sulphide deposit, Appalachian Ophiolite Belt, Quebec; Canadian Journal of Earth Sciences, v. 28, p. 1887-1904.

Trottier, J., Chartrand, F., and Gauthier, M.
1989:   Metallogeny of the southeastern Quebec Appalachians; Geological Association of Canada, Annual meeting, Field trip guidebook B-4, 71 p.

Trottier, J., Gauthier, M., and Brown, A.C.
1987:   Geology and lithogeochemistry of the Huntingdon deposit, Cyprus-type mineralization in the ophiolite belt of the southeastern Quebec Appalachians; Economic Geology, v. 82, p. 1483-1504.
1988:   The Memphremagog deposit; an unusual Sn- and Sb-bearing massive sulphide deposit in the ophiolite belt of the southeastern Quebec Appalachians (abstract); Geological Association of Canada, Program with Abstracts, v. 13, p. A12.

Tuach, J.
1976:   Structural and stratigraphic setting of the Ming and other sulphide deposits in the Rambler area, Newfoundland; M.Sc. thesis, Memorial University of Newfoundland, St. John's, Newfoundland, 280 p.
1984a: Metallogenic studies of granite-associated mineralization in the Ackley Granite and Cross Hills Plutonic Complex, Fortune Bay area, Newfoundland; in Current Research, Part A; Geological Survey of Canada, Paper 84-1A, p. 499-504.
1984b: Tungsten analyses from quartz topaz greisen veins, and the economic significance of lithogeochemical trends in the Ackley Granite; Newfoundland Department of Mines and Energy, Mineral Development Division, Open File 1M (214).
1987a: Mineralized environments, metallogenesis, and the Doucers Valley Fault complex, western white Bay: a philosophy for gold exploration in Newfoundland; in Current Research; Newfoundland Department of Mines and Energy, Mineral Development Division, Report 87-1, p. 129-144.
1987b: The Ackley high-silica magmatic/metallogenic system and associated post-tectonic granites, southeast Newfoundland; Ph.D. thesis, Memorial University of Newfoundland, St. John's, Newfoundland, 455 p.
1988:   Geology and sulphide mineralization in the Pacquet Harbour Group; in The Volcanogenic Sulphide Districts of Central Newfoundland, (ed.) H.S. Swinden and B.F. Kean; Geological Association of Canada, Mineral Deposit Division, St. John's, Newfoundland, p. 49-53.
1991:   The geology and geochemistry of the Cross Hills Plutonic Suite, Fortune Bay, Newfoundland (NTS 1M/10); Newfoundland Department of Mines and Energy, Geological Survey Branch, Report 91-2, 73 p.

Tuach, J. and Delaney, P.W.
1986:   Tungsten-molybdenum in the Granite Lake-Meelpaeg Lake area, Newfoundland; in Current Research; Newfoundland Department of Mines and Energy, Mineral Development Division, Report 86-1, p. 113-128.

Tuach, J. and Kennedy, M. J.
1978:   The geologic setting of the Ming and other sulphide deposits, Consolidated Rambler Mines, Northwest Newfoundland; Economic Geology, v. 73, p. 192-206.

Tuach, J. and Saunders. C.M.
1990:   A field guide to the geology and mineralization of western White Bay; in Metallogenic Framework of Base and Precious Metal Deposits, Central and Western Newfoundland, (ed.) H.S. Swinden, D.T.W. Evans, and B.F. Kean; 8th IAGOD Symposium, Field Trip No. 1, Guidebook, Geological Survey of Canada, Open File 2156, p. 205-211.

Tuach, J., Davenport, P.H., Dickson, W.L., and Strong, D.F.
1987:   Geochemical trends in the Ackley Granite, southeast Newfoundland: their relevance to magmatic-metallogenic processes in high-silica granitoid systems; Canadian Journal of Earth Sciences, v. 23, p. 747-765.

Tupper, W.M., Boyle, R.W., and Martin, R.
1969:   The geology and geochemistry of the Nigadoo sulphide deposits, Gloucester County, New Brunswick; in Geology and Mineral Deposits of the Nigadoo River-Millstream River Area, Gloucester County, New Brunswick; Geological Survey of Canada, Paper 67-49, p. 59-64.

Upadhyay, H.D. and Smitheringale, W.G.
1972:   Geology of the Gullbridge copper deposit, Newfoundland: volcanogenic sulphides in cordierite-anthophyllite rocks; Canadian Journal of Earth Sciences, v. 9, p. 1061-1073.

Upadhyay, H.D. and Strong, D.F.
1973:   Geological setting of the Betts Cove copper deposits, Newfoundland: an example of ophiolitic mineralization; Economic Geology, v. 68, p. 161-167.

Upadhyay, H.D., Dewey, J.F., and Neale, E.R.W.
1971:   The Betts Cove Ophiolitic Complex, Newfoundland: Appalachian oceanic crust and mantle; Geological Association of Canada, Proceedings, v. 24, p. 27-34.

Vaillancourt, P.D. and Sangster, D.F.
1986:   Isotope and hydrocarbon studies of the Yava sandstone-lead deposit, Cape Breton Island, Nova Scotia: a progress report; in Current Research, Part A; Geological Survey of Canada, Report 86-1A, p. 133-140.

van de Poll, H.W.
1978:   Paleoclimatic control and stratigraphic limits of synsedimentary mineral occurrences in Mississippian-Early Pennsylvanian strata of eastern Canada; Economic Geology, v. 73, p. 1069-1081.

van Staal, C.R.
1986:   Preliminary results of structural investigations in the Bathurst Camp of northern New Brunswick; in Current Research, Part A; Geological Survey of Canada, Paper 86-1A, p. 193-204.
1987:   Tectonic setting of the Tetagouche Group in northern New Brunswick: implications for plate tectonic models of the northern Appalachians; Canadian Journal of Earth Sciences, v. 24, p. 1329-1351.

van Staal, C.R. and Langdon, J.P.
1990:   Geology of Ordovician massive sulphide deposits and their host rocks in northern New Brunswick; in Field Guide to Massive Sulphide Deposits in Northern New Brunswick, (ed.) L.R. Fyffe; Department of Natural Resources and Energy, Minerals and Energy Division, Fredericton, New Brunswick, p. 15-27.

van Staal, C.R. and Luff, W.M.
1990:   The Brunswick No. 12 and No. 6 mines, Brunswick Mining and Smelting Corporation Ltd.; in Field Guide to Massive Sulphide Deposits in Northern New Brunswick, (ed.) L.R. Fyffe; Department of Natural Resources and Energy, Minerals and Energy Division, Fredericton, New Brunswick, p. 42-71.

van Staal, C.R. and Sullivan, R.W.
1992:   Significance of U-Pb zircon dating of silicic volcanic rocks associated with the Brunswick-type massive sulphide deposits of the Tetagouche Group (abstract); Geological Association of Canada, Program with Abstracts, v. 17, p. A112.

van Staal, C.R. and Williams, P.F.
1984:   Structure, origin and concentration of the Brunswick 12 and 6 orebodies; Economic Geology, v. 79, p. 1669-1692.

van Staal, C.R., Fyffe, L.R., Langton, J.P. and McCutcheon, S.R.
1992:   The Ordovician Tetagouche Group, Bathurst Camp, northern New Brunswick, Canada: history, tectonic setting, and distribution of massive-sulfide deposits; Exploration and Mining Geology, v. 1, p. 93-103.

van Staal., C.R., Winchester, J.A., and Bédard, J.H.
1991:   Geochemical variations in Middle Ordovician volcanic rocks of the northern Miramichi Highlands and their tectonic significance; Canadian Journal of Earth Sciences, v. 28, p. 1031-1049.

van Staal, C.R., Winchester, J.A., and Cullen, R.
1988:   Evidence for D1-related thrusting and folding in the Bathurst-Millstream River area, New Brunswick; in Current Research, Part B; Geological Survey of Canada, Paper 88-1B, p. 135-148.

Vanderveer, D.G. and Tuach, J.
1982:   Field trip guide, Deer lake Basin; in Prospecting in Areas of Glaciated Terrain, 1982, (ed.) P.H. Davenport; Canadian Institute of Mining and Metallurgy, St. John's, Newfoundland.

Venugopal, D.V.
1979:   Geology of Debec Junction-Gibson Millstream-Temperance Vale-Meductic region, map-areas G-21, H-21, I-21, and H-22 (Parts of 21 J/3, 21 J/4, 21 G/13, 21 G/14); New Brunswick Department of Natural Resources, Mineral Resources Branch, Map Report 79-5, 36 p.

Victor, I.
1957:   The Burnthill wolframite deposit; Economic Geology, v. 52, p. 149-169.

Walker, S.D. and Collins, C.
1988: The Point Leamington massive sulphide deposit; in The Volcanogenic Sulphide Districts of Central Newfoundland, (ed.) H.S. Swinden and B.F. Kean; Geological Association of Canada, Mineral Deposits Division, p. 193-198.

Wanless, R.K., Stevens, R.D., Lachance, G.R., and Delabio, R.N.
1973: K-Ar age determinations of samples from various areas of New Brunswick; in Age Determinations and Geological Studies, K-Ar isotopic ages, Report 11; Geological Survey of Canada, Paper 73-2, p. 78-90.

Wares, R.
1983: Synthese metallogenique du gite Sullipek et de ses environs; Ministère de l'Énergie et des Ressources du Québec, DP-83-02.

Watson, G.P.
1981: Geology and geochemistry of the alteration zone surrounding quartz-stibnite mineralization, Lake George antimony mine deposit, New Brunswick; M.Sc. thesis, University of New Brunswick, Fredericton, New Brunswick, 208 p.

Watters, S.E.
1985: Bay of Fundy gold-bearing rocks; in 10th Annual Review of Activities, Project Resumes, 1985, (ed.) S.A. Abbott; New Brunswick Department of Forests, Mines and Energy, Mineral Resources Division, Information Circular 85-3, p. 3-5.
1987: Gold-bearing rcks, Bay of Fundy Coastal Zone; in 12th Annual Review of Activities, Project Resumes, 1987, (ed.) S.A. Abbott; New Brunswick Department of Forests, Mines and Energy, Mineral Resources Division, Information Circular 87-2, p. 41-44.

Webb, T.C.
1988: Industrial mineral resources of New Brunswick, 1987; Minerals and Energy Division, New Brunswick Department of Natural Resources and Energy, Fredericton, New Brunswick, unpublished report, 85 p.

Whalen J.B.
1980: Geology and geochemistry of the molybdenite showings of the Ackley City Batholith, southeast Newfoundland; Canadian Journal of Earth Sciences, v. 17, p. 1246-1258.
1986: A-type granites in New Brunswick; in Current Research, Part A; Geological Survey of Canada, Paper 86-1A, p. 297-300.

Whalen, J.B. and Gariepy, C.
1986: Petrogenesis of the McGerrigle plutonic complex, Gaspe, Quebec: a preliminary report; in Current Research, Part A; Geological Survey of Canada, Paper 86-1A, p. 265-274.

Whalen, J.B. and Roddick, J.C.M.
1987: K-Ar geochronology of the McGerrigle plutonic complex, Gaspésie Peninsula, Quebec; in Current Research, Part A; Geological Survey of Canada, Paper 87-1A, p. 375-380.

Whalen, J.B., Mortenson, J.K., and Roddick, J.C.
1991: Implications of U-Pb and K-Ar geochronology for petrogenesis and cooling history of the McGerrigle plutonic complex, Gaspé, Quebec; Canadian Journal of Earth Sciences, v. 28, p. 754-761.

White, D.E.
1939: Geology and molybdenite deposits of the Rencontre East area, Fortune Bay, Newfoundland; Ph.D. thesis, Princeton University, Princeton, 118 p.

Whitehead, R.
1973: Environment of stratiform sulphide deposition; variation in Fe/Mn ratio in host rocks at Heath Steele Mine, New Brunswick, Canada; Mineralium Deposita, v. 8.

Williams, H.
1971: Geology of Belleoram map area (1M/11), Newfoundland; Geological Survey of Canada, Paper 70-65, 39 p.
1978: Geological development of the northern Appalachians: its bearing on the evolution of the British Isles; in Crustal Evolution in Northwestern Britain and Adjacent Regions, (ed.) D.R. Bowes and B.E. Leake; Seal House Press, Liverpool, England, p. 1-22.
1979: Appalachian Orogen in Canada; Canadian Journal of Earth Sciences, v. 16, p. 792-807.

Williams, H. and Hiscott, R.N.
1987: Definition of the Iapetus rift-drift transition in western Newfoundland; Geology, v. 15, p. 1044-1047.

Williams, H. and Stevens, R.K.
1974: The ancient continental margin of eastern North America; in The Geology of Continental Margins, (ed.) C.A. Burke and C.L. Drake; Springer Verlag, New York, p. 781-796.

Williams, H., Dickson, W.L., Hayes, J.P., Currie, K.L., and Tuach, J.
1989: Preliminary report on a classification of Newfoundland granitic rocks and their relations to tectonostratigraphic zones and lower crustal blocks; in Current Research, Part B; Geological Survey of Canada, Paper 89-1B, p. 47-54.

Williams, P.F. and Hy, C.
1990: Origin and deformational and metamorphic history of gold-bearing quartz veins on the Eastern Shore of Nova Scotia; in Mineral Deposit Studies in Nova Scotia, v. 1, (ed.) A.L. Sangster; Geological Survey of Canada, Paper 90-1, p. 169-194.

Williams, S.H.
1991: Graptolites from the Baie D'Espoir Group, south-central Newfoundland; in Current Research; Newfoundland Department of Mines and Energy, Geological Survey Branch, Report 91-1, p. 175-178.

Williams-Jones, A.E.
1982: Patapedia: an Appalachian calc-silicate-hosted copper prospect of porphyry affinity; Canadian Journal of Earth Sciences, v. 19, p. 438-455.

Williams-Jones, A.E., Schrijver, K., Doig, R., and Sangster, D.F.
1992: A model for epigenetic Ba-Pb-Zn mineralization in the Appalachian thrust belt, Québec; Economic Geology, v. 87, p. 154-174.

Wilson, R.A.
1991: A new base metal discovery in the Siluro-Devonian Tobique belt of New Brunswick; Prospectors and Developers Annual Meeting, Abstract T-15.
1992: Petrographic features of Siluro-Devonian felsic volcanic rocks in the Riley Brook area, Tobique Zone, New Brunswick: implications for base metal mineralization at Sewell Brook; Atlantic Geology, v. 28, p. 115-135.

Wilton, D.H.C.
1983: The geology and structural history of the Cape Ray Fault Zone in southwestern Newfoundland, Canada; Canadian Journal of Earth Sciences, v. 20, p. 1119-1133.

Wilton, D.H.C. and Strong, D.F.
1986: Granite-related gold mineralization in the Cape Ray Fault Zone of Southwestern Newfoundland; Economic Geology, v. 81, p. 281-295.

Winchester, J.A., van Staal, C.R., and Langton, J.P.
1992: The Ordovician volcanics of the Elmtree-Belledune inlier and their relationship to volcanics of the northern Miramichi Highlands, New Brunswick; Canadian Journal of Earth Sciences, v. 29, p. 1430-1447.

Wolfson, I.
1983: Geology, mineralogy and geochemistry of tin mineralization in the Dominique-Plymouth area, Yarmouth County, Nova Scotia; M.Sc. thesis, Dalhousie University, Halifax, Nova Scotia, 411 p.

Wright, J.D.
1975: Iron deposits of Nova Scotia; Nova Scotia Department of Mines, Economic Geology Series 75-1, 154 p.

Wright, W.J.
1940: Molybdenite prospect at Pabineau Lake, Gloucester County, New Brunswick; New Brunswick Department of Lands and Mines, Mines Branch, Paper 40-1, 8 p.

Wynne, P.J. and Strong, D.F.
1984: The Strickland prospect of southwestern Newfoundland: a lithogeochemical study of metamorphosed and deformed volcanogenic massive sulphides; Economic Geology, v. 79, p. 1620-1642.

Young, R.S.
1970: Mineral exploration and development in Maine; in Ore Deposits in the United States, 1933/1967 (Graton Sales Volume); American Institute of Mining Engineering, v. 1, p. 125-139.

Yule, A., McKenzie, C., and Zentilli, M.
1990: Hope Brook: a new Appalachian gold deposit in Newfoundland, Canada; Chronique de la Recherche Minière, no. 498, p. 29-42.

Zentilli, M. and Reynolds, P.H.
1985: $^{40}Ar/^{39}Ar$ dating of micas from the East Kemptville tin deposit, Yarmouth County, Nova Scotia; Canadian Journal of Earth Sciences, v. 22, p. 1546-1548.

Zentilli, M., Graves, M.C., Mulja, T., and MacInnes, I.
1986: Geochemical characterization of the Halifax-Goldenville transition of the Meguma Group of Nova Scotia: preliminary report; in Current Research, Part A; Geological Survey of Canada, Paper 86-1A, p. 423-428.

Zonenshain, L.P., Kuzmin, V.I., Kovalenko, V.I., and Saltykovsky, A.J.
1972: Mesozoic structural-magmatic pattern and metallogeny of the western part of the Pacific belt; Earth and Planetary Science Letters, v. 22, p. 96-109.

Authors' Addresses

G. Beaudoin
Faculté des Sciences de Génie
Départment de Géologie et de Génie Géologique
Université Lavel, Québec (Québec)
G1K 7B4

R.C. Boehner
Box 698
Nova Scotia Department of Natural Resources
Halifax, Nova Scotia
B3J 2T9

J.L. Davies
New Brunswick Department of Natural Resources and Energy
P.O. Box 6000
Fredericton, New Brunswick
E3B 5H1

S.M. Dunsworth
P.O. Box 621
Portugal Cove
Newfoundland
A0A 3K0

G. Duquette
5 Place des Lynx
C.P. 715
Sainte-Anne-des Monts, Québec
G0E 2G0

W.L. Dickson
Geological Survey Branch
Newfoundland Department of Natural Resources
P.O. 8700
St. John's, Newfoundland
A1B 4J6

D.T.W. Evans
Geological Survey Branch
Newfoundland Department of Natural Resources
P.O. 8700
St. John's, Newfoundland
A1B 4J6

J.H. Fowler
58 Arlington Avenue
Halifax, Nova Scotia
B3N 2A1

L.R. Fyffe
New Brunswick Department of Natural Resources and Energy
P.O. Box 6000
Fredericton, New Brunswick
E3B 5H1

M. Gauthier
Département des sciences de la Terre
Université du Québec à Montréal
C.P. 8 888
succursale «A»
Montréal (Québec)
H3C 3P8

A.F. Howse
Geological Survey Branch
Newfoundland Department of Natural Resources

P.O. 8700
St. John's, Newfoundland
A1B 4J6

B.F. Kean
Geological Survey Branch
Newfoundland Department of Natural Resources
P.O. 8700
St. John's, Newfoundland
A1B 4J6

T.E. Lane
Teck Exploration Ltd.
Suite 7000
1 First Canadian Place
P.O. Box 170
Toronto, Ontario
M5X 1G9

M. Malo
Quebec Geoscience Centre
INRS - Géoressources
C.P. 7 500
Sainte-Foy (Québec)
G1V 4C7

S.R. McCutcheon
Department of Natural Resources and Energy
P.O. Box 50
Bathurst, New Brunswick
E2A 3Z1

C.F. O'Driscoll
Geological Survey Branch
Newfoundland Department of Natural Resources
P.O. 8700
St. John's, Newfoundland
A1B 4J6

R.J. Ryan
Nova Scotia Department of Natural Resources
Box 698
Halifax, Nova Scotia
B3J 2T9

A.A. Ruitenberg
Minerals and Energy Division
Department of Natural Resources and Energy
P.O. Box 1547
Sussex, New Brunswick
E3B 5H1

A.L. Sangster
Geological Survey of Canada
601 Booth St.
Ottawa, Ontario
K1A 0E8

H.S. Swinden
Geological Survey Branch
Newfoundland Department of Natural Resources
P.O. 8700
St. John's, Newfoundland
A1B 4J6

C.R. van Staal
Geological Survey of Canada
601 Booth Street
Ottawa, Ontario
K1A 0E8

Printed in Canada

Chapter 10

PALEONTOLOGICAL CONTRIBUTIONS TO PALEOZOIC PALEOGEOGRAPHIC AND TECTONIC RECONSTRUCTIONS

Chapter 10

PALEONTOLOGICAL CONTRIBUTIONS TO PALEOZOIC PALEOGEOGRAPHIC AND TECTONIC RECONSTRUCTIONS

G.S. Nowlan and R.B. Neuman

INTRODUCTION

Fossils provide essential information for the determination of prior locations of the components of ancient orogenic belts such as the Appalachians. In this chapter we review the record of fossils from the Canadian Appalachians that, together with those from the Appalachians in the United States (Neuman et al., 1989), assist in determining the paleogeographic and tectonic evolution of this orogenic system. The rocks of the Canadian Appalachians are more fossiliferous than their counterparts in the United States because they are generally less deformed and metamorphosed, and because there are important differences in the geology of the orogenic belt in the two countries.

Paleontological studies in Canada have long contributed to the development of ideas on the history of the orogen. The Cambrian faunas of the Avalon Zone in Newfoundland and southern New Brunswick, and correlatives in northwestern Europe were assigned by Walcott (1891) to an "Atlantic Coast Province", considered by him to be different from those of the "Appalachian Province" elsewhere in the Canadian and U.S. Appalachians. Following the introduction of the idea of continental drift (Wegener, 1928), Grabau (1936) explained the similarity of the stratigraphy and faunas of eastern North America and northwestern Europe by deposition in contiguous synclines that were parts of his hypothetical "Pangea." After the general acceptance of continental drift, Wilson (1966) proposed that a Paleozoic "proto-Atlantic Ocean" preceded the present Atlantic. In his view the Atlantic Cambrian faunas populated the eastern margin of the "proto-Atlantic Ocean", contemporaneous Pacific faunas populated its western margin, and the complex area of the northern Appalachians between them resulted from the post-Cambrian convergence of the opposite margins of that old ocean basin. With the appreciation that the orogen consists of a collage of suspect terranes (discrete crustal blocks whose paleogeography cannot be

definitely ascertained; Williams and Hatcher, 1983), paleontological information becomes increasingly important in the identification of such terranes as has been demonstrated in the Cordillera of western North America (Jones et al., 1977).

Paleontology can be used for such purposes because fossils have the potential to yield information concerning ancient habitats as well as ages. Of the several kinds of discriminations that can be made, those that indicate paleoclimates and former geographic isolation are particularly important for determination of paleogeography. Tropical biotas and their physical environments can be distinguished from boreal biotas of high latitudes (realms); within these realms are areas that contain variously distinctive biotic associations, the products of isolation by barriers to biotic exchanges; the term "provinces" is commonly applied to the more important of these.

Comparison of modern and ancient realms and provinces requires the comparison of environmentally similar biotas; in aquatic settings, for example, variable factors include water depth, sediment supply, and salinity. Different organisms reflect different aspects of their environments; those that live out their adult stages in shallow waters attached on or near the seafloor (such as brachiopods and corals) are more likely to reflect the temperatures of that water and its neighbouring landmasses than are mobile or free-swimming organisms such as many molluscs and trilobites, pelagic (floating) organisms such as many graptolites, and eurybathic organisms such as conodonts.

The literature of paleobiogeography consists largely of works concerned with the distribution of constituent taxa of specific groups of organisms; many of these papers are accompanied by small-scale maps. This contribution is based on that literature, on the current knowledge of the authors, and discussions with their colleagues. The main outlines of the occurrence and distribution of Phanerozoic fossils on the larger continental blocks is reasonably well known, but those of the Precambrian are poorly known, as are those of crustal fragments, or the suspect terranes that constitute the Appalachians. Relevant data are presented here chronologically in the context of the lithotectonic zones, belts, and basins adopted for this volume. Interpretations that follow are largely consistent with commonly accepted reconstructions, as the reconstructions incorporate many of these interpretations.

Nowlan, G.S. and Neuman, R.B.
1991: Paleontological contributions to Paleozoic paleogeographic and tectonic reconstructions; Chapter 10 in Geology of the Appalachian/Caledonian Orogen in Canada and Greenland, (ed.) H. Williams; Geological Survey of Canada, Geology of Canada, no. 6, p. 815-842 (also Geological Society of America, The Geology of North America, v. F-1).

Fossils that occur throughout the Appalachian Orogen have useful application in the reconstruction of its development. Stromatolites in carbonate rocks provide information on the geographic setting of some of the oldest (middle Proterozoic Helikian, >800 Ma) rocks of the Avalon Zone. Late Proterozoic Hadrynian (900-570 Ma) imprints of soft-bodied organisms, trace fossils, and enigmatic shelly fossils, also in the Avalon Zone, suggest the outlines of a later ancient geography. The "Atlantic realm" Cambrian fauna is now known to characterize the Avalon Zone, distinctive from the "Pacific realm" of the Humber miogeocline, but the few Cambrian fossils from the intervening terranes have yielded little information. Increased organic diversity and abundance beginning with the Ordovician, together with changing geological patterns and diverse marine habitats provide abundant materials for paleogeographic assessments throughout the orogen. Marine deposition was considerably reduced after the Acadian Orogeny, and the Middle Devonian and Carboniferous record is based largely on nonmarine faunas and floras in isolated successor basins.

This paper summarizes the occurrence of fossils of inferred or potential biogeographic significance using successively younger intervals of geological time. For the Precambrian-Cambrian interval the discussion treats different kinds of fossils separately; for later intervals (Tremadoc, Arenig-Llandeilo, Caradoc-Ashgill, Silurian-Early Devonian, and Middle Devonian-Carboniferous), fosssils and rocks are treated from northwest to southeast according to early Paleozoic tectonostratigraphic zones or middle Paleozoic belts and late Paleozoic basins appropriate for those time intervals. Individual occurrences of fossils referred to in the text are shown in Figures 10.1, 10.6, and 10.8. Middle Ordovician and older occurrences are shown in Figure 10.1 and coded with respect to age and zone in which they occur (see caption for explanation). Late Ordovician to Late Devonian occurrences are shown in Figure 10.6 and coded with respect to the belt in which they occur (see caption for explanation). Carboniferous and Permian occurrences are shown similarly in Figure 10.8.

PRECAMBRIAN-CAMBRIAN
Stromatolites

The oldest fossils in the Appalachians are stromatolites in rocks of the Avalon Zone in New Brunswick and Newfoundland. In New Brunswick, the stromatolite, *Archaeozoon acadiense* Matthew 1890 occurs in the Middle Proterozoic (Helikian) Green Head Group near Saint John (AP-1; Hofmann, 1974). In addition to stromatolitic limestone, the lower and middle parts of the Green Head Group consist of quartzite and other rocks characteristic of deposition in shallow water. The age of the formation is poorly constrained in the 900-1200 Ma range. The stromatolites and other evidence of shallow-water deposition suggest correlation of the Green Head with the shallow-water sequence of the Grenville Group of southern Ontario and Quebec that contains similar stromatolitic limestone (Currie, 1987), a correlation that leads to the inference that the basement of the Avalon Zone is a rifted fragment of Grenville crust.

Large blocks of stromatolitic limestone occur in the Burin Group on the Burin Peninsula, Newfoundland (AP-2; Strong et al. 1978; O'Brien et al., 1983). The structure of these stromatolites differs from that of *Archaeozoon* (Hofmann in O'Brien and King, 1982), but they are alike in indicating Middle Proterozoic shallow-water environments.

Ediacaran fossils

About 30 species of soft-bodied organisms similar to those of the Ediacaran fauna of South Australia, are known from the upper Conception Group and the lower St. John's Group (Upper Proterozoic, Hadrynian) in the Avalon Zone of Newfoundland (Fig. 10.1, 10.2; AP-3; Misra, 1969; Anderson, 1978; Anderson and Conway Morris, 1982). Most species are unique to this area, perhaps because they represent a deeper water environment than that of the more common Ediacaran faunas in South Australia (Conway Morris and Rushton, 1988).

Such fossils are known elsewhere in the Appalachians only in the Avalon Zone and related Carolina Terrane (Neuman et al., 1989). The latter, like the Avalon Zone of Newfoundland, contains the Cambrian "Atlantic realm" fauna. The Cambrian sequences of the Newfoundland Avalon Zone contain far less volcanic rocks than the middle Cambrian sequence of the Carolina Terrane. The fossil occurrences suggest a similar paleogeography. Ediacaran fossils have not been found in upper Proterozoic rocks (Hadrynian equivalents) in the Humber Zone.

Small shelly fossils

Small calcareous and phosphatic fossils from rocks in the Avalon Zone of Newfoundland, Nova Scotia, and Massachusetts range in age from latest Precambrian (uppermost Vendian) into the Middle Cambrian. Microfaunas have been described from Newfoundland (Bonavista, Avalon, and Burin peninsulas; AC-1, AC-2, AC-3) by Bengston and Fletcher (1983); from the northern Antigonish Highlands of Nova Scotia (AC-4) by Landing et al. (1980); and from Cape Breton Island (AC-5) by Matthew (1903); for Massachusetts occurrences see Landing (1988). These fossils include the remains of several different kinds of organisms, including conodont-like forms, molluscs, hyoliths, and other less well-known groups, the first appearances of which precede the first appearance of trilobites. Such fossils occur in shallow-water sedimentary rocks of several kinds, most abundantly and most diversified in massively bedded peritidal limestone, and decreasingly abundant and diverse in nodular limestone interbedded with wackestone, and variously burrowed red shale, with and without calcareous nodules (Landing, 1988).

Small shelly fossils of the same age range are well known from such distant areas as Siberia and southern China, and although their distribution is far from cosmopolitan (Rozanov, 1984), provincialism within these faunas has yet to be demonstrated. Most closely related to the Avalon occurrences are those in the British Midlands (Brasier, 1984) and in Shropshire (Cobbold, 1921) that further confirm the prior links between those parts of the British Isles and the Avalon Zone.

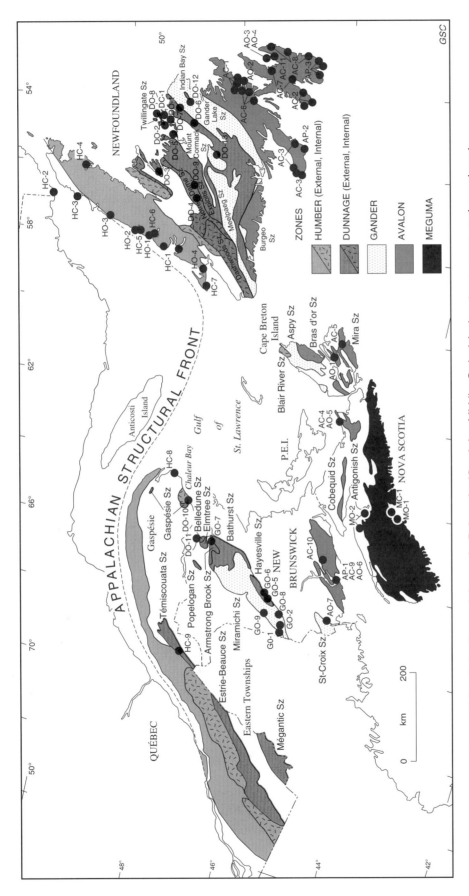

Figure 10.1. Late Precambrian-early Paleozoic (Cambrian-Middle Ordovician) outcrop areas and zonal and subzonal subdivision of the Canadian Appalachian region with locations of fossil localities discussed in text. Localities coded as to zone as follows: A – Avalon Zone; D – Dunnage Zone; G – Gander Zone; H – Humber Zone; M – Meguma Zone; and as to age as follows P – Precambrian, C – Cambrian, O – Ordovician. A combination of two letter codes followed by a number is an unique identifier of zone, age, and location.

Palynomorphs and trace fossils

These very different kinds of fossils have proved useful in establishing the ages of the few rock units in which they have been found. The occurrences of each appear to reflect particular depositional facies, the Canadian occurrences having counterparts elsewhere that suggest ancient geographic links. In composition and type of preservation, the best known assemblages of palynomorphs from the Appalachians are acritarchs and nonseptate organic filaments from upper Proterozoic (Hadrynian) rocks of the Avalon Zone in Newfoundland. They are known from eleven different stratigraphic units (Hofmann et al., 1979) including the Musgravetown and Connecting Point groups of the Placentia Bay-Trinity Bay area (AC-6), the Hodgewater Group in Conception Bay-St. Mary's Bay area (AC-7), and the Conception, St. John's and Signal Hill groups in the easternmost Avalon Peninsula (AC-8). Despite their poor preservation, they resemble those in the lower Dalradian succession of Scotland and the Brioverian of the Armorican Massif of France (Hofmann et al., 1979).

Sequences of late Precambrian and Early Cambrian ichnofossil assemblages are similarly rich and diversified in siliciclastic rocks of the Avalon Zone of Newfoundland and many other places worldwide (Crimes and Anderson, 1985; Narbonne and Myrow, 1988). Comparable assemblages are not known elsewhere in the Appalachians, but their widespread distribution across Eurasia and in northwestern Canada suggests their lack of provincialism (Crimes, 1987; Narbonne and Myrow, 1988).

The trace fossil *Oldhamia* occurs throughout the world in rocks of Cambrian and possibly late Proterozoic (Hadrynian) age that were probably deposited in deep water (Hofmann and Cecile, 1981). It has long been known to occur at several places in the northern Appalachians of the United States, but there were no confirmed occurrences in the Canadian Appalachians until its discovery in abundance at several places in the Blow Me Down Brook Formation of western Newfoundland. Its presence here led to interpretation of those rocks as having been derived from the rifted margin of North America (HC-1; Lindholm and Casey, 1989, 1990).

Trilobites

Trilobites, the most common, abundant and widespread macrofossil in Cambrian rocks, were major factors in Wilson's (1966) prescient identification of exotic terranes

Figure 10.2. Ediacaran fossils from eastern Newfoundland (locality AR-3 of Fig. 10.1). **A** – Several spindle-shaped forms, Mistaken Point Formation, Mistaken Point, Newfoundland (lens cap, diameter 5 cm); **B** – Bush-like form, Mistaken Point Formation, Mistaken Point; **C** – Large discoidal form, incomplete, 20 cm in diameter, Trepassey Formation, near Green Point, Conception Bay. See Anderson and Conway Morris (1982). Specimen A is still in place in outcrop; specimens B and C are stored in the Royal Ontario Museum.

in the Appalachians. Continuing investigations amply confirm Walcott's (1891) distinction of "Atlantic" and "Pacific" faunal realms, and the occurrences of assemblages of Cambrian trilobites and other fossils diagnostic of these faunal realms can now be described in terms of the currently accepted lithotectonic zones of the Appalachians.

Pacific realm Early Cambrian trilobite assemblages characterized by *Olenellus* are confined to upper Lower Cambrian rocks of the Laurentian margin and its miogeocline (Fritz, 1972), i.e. the Humber Zone. Archeocyathid reefs are associated with these trilobites in the Forteau Formation of western Newfoundland and adjacent Labrador (HC-2; Whittington and Kindle, 1969; Debrenne and James, 1981; James and Kobluk, 1978). Older Cambrian fossils are not known from either the St. Lawrence Platform or the Humber miogeocline, probably because of inhospitable environments or lack of deposition in those areas. Early and Middle Cambrian trilobites in the allochthonous Taconic sequence of the Humber Zone in New York and Quebec have elements of both Atlantic and Pacific realms (Palmer, 1971); similar mixtures will probably be found in allochthonous rocks within the Humber Zone of Newfoundland.

The Atlantic realm Early Cambrian assemblages in the Avalon zone in Newfoundland and Nova Scotia include some elements older than those of the Pacific realm in Canada. Here they are preceded by the small shelly fossils, mentioned above, followed by the early trilobites (*Callavia*, etc.), and by protolenids, the latter contemporaneous with *Olenellus* of the Humber Zone (North, 1971).

The Middle and Upper Cambrian rocks of the Humber Zone lie at the eastern edge of a blanket of shallow-water carbonate rocks. These rocks (Fig. 10.1, HC-3 to 7) contain a sequence of polymerid trilobites and other fossils of essentially continent-wide distribution (Lochman-Balk and Wilson, 1958; Palmer, 1969). The eastern limit of the Humber Zone carbonate sequence is approximately the boundary between the Cambrian and Early Ordovician shallow-water shelf and the adjacent deep waters, most clearly seen in the allochthonous Cow Head Group of western Newfoundland (James and Stevens, 1986). Here limestone clasts in breccia and conglomerate beds contain trilobites that indicate their deposition at different water depths. Trilobites resembling those of the North American shelf indicate shallow-water deposition, whereas agnostid trilobites, best known from the Acado-Baltic region, indicate contemporaneous deposition in cooler or deeper water. The edge of the Cambrian-Early Ordovician carbonate shelf is marked by similar but less well exposed conglomerates at several places along the southern shore of the St. Lawrence River, and in the United States at Highgate Falls, Vermont (Landing, 1983).

At the southeastern edge of the Humber Zone in Quebec, trilobites from the Corner-of-the-Beach and Murphy Creek formations near Percé (HC-8) suggest a carbonate bank setting, perhaps as a large block that was derived from the bank margin, or as an island offshore from the bank margin. Thirteen North American late Middle and early Late Cambrian genera in these rocks are congeneric with trilobites from the carbonate bank and platform, but the species of the 10 genera that were specifically determinable are endemic (Fritz et al., 1970), suggesting that they lived at a

distance great enough to preclude genetic interchange with their congeners on the bank. Boulders and cobbles of oolite and other shallow-water limestones occur in conglomerate of Ashgill-Llandovery age in a comparable setting near Cabano, Quebec (HC-9; Lespérance and Greiner, 1969; Lajoie et al., 1968; David et al., 1985). Study of a few samples that yielded the Late Cambrian-Early Ordovician brachiopod *Schizambon* and indeterminate trilobites (Palmer in Lespérance and Greiner, 1969) suggests that further studies may infer a nearby wide tract of outlying carbonate rocks of this age range, now largely buried.

The only Cambrian fossils from the Dunnage Zone are trilobites of the genera *Kootenia* and *Bailiella* from limestone within a volcanic olistolith in the Dunnage Mélange on Dunnage Island, Newfoundland (DC-1; Kay and Eldredge, 1968). Although first interpreted as a mixture of Atlantic and Pacific forms (Kay and Eldredge, 1968), reassessment of these fossils indicates that the two genera are of dubious value as provincial indicators, and deformation prohibits specific identification (Dean, 1985).

Most of the Middle and Late Cambrian trilobites in the shales and siltstones of the Avalon Zone belong to Walcott's (1891) Atlantic Realm, now widely referred to as the Acado-Baltic Province. These and other fossils have long been known from important stratigraphic sequences most of which were discovered by pioneer Canadian geological investigators. In New Brunswick such fossils occur in the environs of the city of Saint John (AC-9; Hayes and Howell, 1937; Currie, 1987) and along Hanford Brook about 40 km northeast of Saint John (AC-10; Matthew, 1895; McCutcheon, 1987). In Nova Scotia they are best known from Cape Breton Island (AC-5; Hutchinson 1952), and they are now known from a small area in the northern Antigonish Highlands of the mainland (AC-4; Landing at al., 1980; Keppie and Schenk in King, 1982). These fossils occur in rocks that occupy several synclines in the Avalon and Burin peninsulas, Newfoundland, each with its special features, the most accessible being the exposures along Manuels River (AC-11), southeastern Conception Bay (Howell, 1925; Anderson, 1987).

Genera and species of trilobites from rocks of the Avalon Zone, notably species of *Paradoxides*, are congeneric, and many are conspecific with those from contemporaneous rocks of similar kinds in the Atlantic borderlands from Norway to Morocco (Dean, 1985), but these taxa are unknown in the carbonate rocks of the St. Lawrence Platform or from the Humber Zone.

Trilobites, including *Paradoxides* and other Middle Cambrian Acado-Baltic genera, were recently reported from the Meguma Zone in the uppermost part of the Goldenville Group, the lower part of the Meguma Supergroup, Nova Scotia (MC-1; Pratt and Waldron, 1991). The trilobites and associated pelmatozoan debris occur in a 44 cm thick bed of fine grained sandstone that is interbedded with massive, fine grained sandstone typical of the unit. This is the first published report of Cambrian fossils from the Meguma Supergroup and their paleogeographic and biostratigraphic significance confirm the interpretation, based on sedimentology, that these rocks were deposited on the eastern side of Iapetus over a long period encompassing much of the Cambrian and Early Ordovician.

ORDOVICIAN

Tremadoc

Fossils in the earliest Ordovician rocks of the Humber and Avalon zones reflect continuity with those of Cambrian age; occurrences in the Dunnage and Gander zones suggest early manifestations of later patterns, and the few in the Meguma Zone remain insufficient for paleogeographic interpretation.

In the Humber Zone, the persistence of the carbonate bank margin through Tremadoc time is shown by the alternation of trilobite-bearing limestones and graptolite-bearing shales in the allochthonous Cow Head Group. Trilobite genera in bedded limestone of the Broom Point Member of the Green Point Formation (HO-1) and in clasts in the Stearing Island Member of the Shallow Bay Formation (HO-2) are characteristic of the carbonate-shelf environment, whereas shales interbedded with Broom Point limestone contain deeper water fossils such as *Rhabdinopora flabelliformis* (formerly *Dictyonema flabelliforme*) and other graptolites of the Pacific faunal realm (James and Stevens, 1986). In the Avalon Zone coeval graptolites including *R. flabelliformis* occur in carbonate-free siliciclastic rocks that contain assemblages more readily correlated with those in Wales that are typical of the Atlantic faunal realm.

Tremadoc fossils are known from only two localities in the Dunnage Zone of Newfoundland: graptolites, dominated by *R. flabelliformis* in the matrix of the Dunnage Mélange (DO-1) (Hibbard et al., 1977) and a cephalopod-rich shelly fauna (ellesmeroceratids according to Flower in Kay, 1967) with an unidentified pliomerid trilobite (Dean, 1976) in limestone of the basal unit of the Summerford Group on nearby New World Island (DO-2; Horne, 1970). The scarcity of rocks of this age prohibits their interpretation with confidence, but a warm, shallow-water environment is suggested by the limestone and its fossils, perhaps confined to a small peri-insular area adjacent to a deep-water trench in which the Dunnage Mélange was accumulating.

Tremadoc graptolites are also known from black shale in the Miramichi Group at two places (Benton, GO-1, and Meductic, GO-2; Fyffe et al., 1983) in the Miramichi Subzone of the Gander Zone of New Brunswick, but their small numbers and little known stratigraphic relationships prohibit meaningful paleogeographic interpretation.

Faunas from the Tremadoc of the Avalon Zone are similar to those from Wales and the Welsh Borderland. For example, the McLeod Brook Formation of Cape Breton Island, Nova Scotia (AO-1) is lithologically similar to typical Tremadoc shale in Shropshire and contains similar trilobites (Hutchinson, 1952). Trilobites from the Random Formation of eastern Newfoundland (AO-2) also show strong affinities to those of Wales (Dean, 1985). Tentative identification of *Adelograptus tenellus* (Linnarsson) in the earliest Tremadoc part of the Cookson Group in the St. Croix Subzone in southwestern New Brunswick (AO-7; Fyffe and Riva, 1990) links the St. Croix Subzone with the Baltic area and other parts of Europe.

The few identifiable fossils from the upper part of the Meguma Supergroup (Halifax Group), Meguma Zone are acritarchs (MO-1; Schenk, 1982) and graptolites (MO-2; Crosby, 1962) of Tremadoc age. These, together with trace fossils, suggest a deep-water environment but permit no paleogeographic interpretation. However, the recent discovery of Cambrian fossils of Acado-Baltic affinity (Pratt and Waldron, 1991) supports the Gondwanan affinity inferred for these rocks by Schenk (1982) based on sedimentological and other data. The Halifax Group has yielded one trilobite specimen (Fig. 10.3) from Spindler Cove, north of The Ovens Park (MO-1), but poor preservation prohibits its identification. This specimen demonstrates that the Halifax Group is not barren, and that fossils sufficiently well preserved to permit generic identification probably remain to be found.

Arenig-Llandeilo

Most fossil groups exhibit increased provincialism during Arenig and early Llanvirn time, perhaps because of the great burst of biological diversity that took place together with the paleogeographic patterns that are inferred to have prevailed, particularly a wide Iapetus Ocean containing islands of several kinds. Attention has been called to the contrasts between warm- and cold-water pelagic organisms at this time such as conodonts (Barnes and Fåhraeus, 1975), graptolites (Skevington, 1974), and acritarchs (Achab, 1988), and to the more complex provincialism of shallow-water benthic organisms such as trilobites (Whittington and Hughes, 1972) and brachiopods (Williams, 1973; Neuman, 1984). Manifestations of this differentiation throughout the Canadian Appalachians have been particularly significant in discriminating its constituent zones and terranes. Provincialism was considerably reduced in late Llanvirn/early Llandeilo time, probably as a result of tectonic events ascribed to Taconic Orogeny that reduced the distances between the opposite shores of Iapetus and the remaining islands in it.

Figure 10.3. Indeterminate trilobite specimen (GSC 96770) from the Meguma Supergroup, Halifax Group, (locality MO-1 of Fig. 10.1) X3.7, GSC loc. 91960, at Spindler Cove north of the Ovens Park (see Schenk, 1982).

A major breakthrough in the understanding of the Humber Zone came with the interpretation of deep-water Cambro-Ordovician rocks in western Newfoundland as Taconic klippen that overlie shallow-water carbonates of the same age (Rodgers and Neale, 1963). Studies of the Newfoundland Humber Zone fossils of Arenig-Llanvirn age, together with sedimentology of the rocks that contain them, defined final stages of the continental shelf (St. George Group; Ulrich and Cooper, 1938; Whittington and Kindle, 1969; Barnes and Tuke, 1970; Flower, 1978; Stouge, 1982; Boyce et al., 1988), its margin, and its penecontemporaneous erosion/deposition products (HO-2; Cow Head Group; James and Stevens, 1986; Pohler et al., 1987; Ross and James, 1987). The foundering and fragmentation of the shelf is recorded in the rocks of the Table Head Group (Klappa et al., 1980) that contain a more diversified assemblage of benthic fossils (HO-3; Whittington and Kindle, 1963; Whittington, 1965; Ross and Ingham, 1970; Ross and James, 1987) that together with their conodonts of both North American and North Atlantic provinces (Stouge, 1984) record deposition in progressively deepening water prior to emplacement of the Humber Arm Allochthon and other manifestations of Taconic Orogeny. Trilobites of Avalon Zone affinity have been reported from continental slope deposits where they occur with brachiopods of eastern North American affinity (Dean, 1985).

Division of the Dunnage Zone of Newfoundland into Notre Dame Subzone and Exploits Subzone was prompted by several factors including their contrasting fossils of Arenig age (Williams et al., 1988). Those in the Notre Dame Subzone are known from only two places. At one of them (South Catcher Pond; DO-3) limestone associated with volcanic rocks yielded early Arenig cool-water, North Atlantic province conodonts (Bergström et al., 1972), together with a unique assemblage of trilobites (Dean, 1970) and brachiopods of the genus *Syntrophia*, typical of the North American platform (Boucot, 1973). From the second place, near Buchans (DO-4), late Arenig-early Llanvirn warm-water, North American province conodonts were extracted from limestone clasts in a volcanic breccia (Nowlan and Thurlow, 1984). Although conodont assemblages of this kind are best known from rocks of the North American platform margin such as the Table Head Group, they occur in Norway in an oceanic setting (Bergström, 1979) where the fossiliferous Hølonda Limestone, part of the lower Hovin Group, overlies an ophiolite fragment (Furnes et al., 1985). A similar conodont fauna also occurs in marine chert in mélange beneath an ophiolite klippe on the Port au Port Peninsula, Newfoundland (HO-4; Nowlan and Thurlow, 1984) suggesting that it was quite widespread in the western Iapetus Ocean. The Toquima-Table Head affinities of the Hølonda Limestone macrofossils suggest North American affinities (Bruton and Bockelie, 1980) or an offshore volcanic edifice within the North American climatic realm (Neuman and Bruton, 1989). This fauna, however, may have had a trans-Iapetan distribution with the Newfoundland occurrences representing staging sites for migration across the ocean (Bergström, 1979; Nowlan and Thurlow, 1984).

Arenig-Llandeilo fossils have been found at many places in the Exploits Subzone. Only one occurrence is of early Arenig age – an assemblage of North Atlantic province conodonts from a limestone boulder in the Dunnage Mélange (DO-5; Hibbard et al., 1977). Those of late Arenig age are dominantly brachiopods associated with smaller numbers of trilobites that occur in shallow-water volcaniclastic rocks in the Summerford Group of New World Island, Newfoundland (DO-2; Neuman, 1976, 1984; Dean, 1974) and siliciclastic rocks of the Davidsville Group southwest of Gander (DO-6; McKerrow and Cocks, 1977; 1986) and near Weir's Pond (DO-7; Boyce in O'Neill, 1987; see also conodonts of North Atlantic province affinity in Fig. 10.5). These assemblages, characterize the Celtic biogeographic province that probably represents peri-insular settings and water temperatures intermediate between the warm waters of the North American continental margin and the cold waters of the Armorican and Baltic platform margins (Neuman, 1984).

Fossils of late Llanvirn-early Llandeilo age, largely conodonts of North Atlantic provincial affinities, occur in limestone at many localities in the Exploits Subzone (e.g. Cobbs Arm Limestone, DO-8, Fåhraeus and Hunter, 1981; Victoria Lake Group, DO-9, Kean and Jayasinge, 1980; Stouge, 1980b), recording a transition into warmer water environments. Associated trilobites and brachiopods in the Summerford Group belong to genera that characterize the Scoto-Appalachian biogeographic province (Dean, 1971; McKerrow and Cocks, 1977). They confirm this warming trend, as do brachiopods in limestone at the base of the Davidsville Group at Weir's Pond (DO-7; Neuman, written communication to Colman-Sadd, 1987). The time of the eradication of the Celtic assemblage and its replacement by Scoto-Appalachian brachiopods corresponds to the closing of a wide segment of the Iapetus Ocean basin recorded by Llanvirn/Llandeilo obduction of ophiolites in western Newfoundland and juxtapositioning of the Exploits Subzone and the Gander Zone (Williams and Piasecki, 1990).

The Gander Zone in Newfoundland lies east of the Dunnage Zone (Gander Lake Subzone of Williams et al., 1988) and in inliers surrounded by rocks of the Dunnage Zone (Mount Cormack and Meelpaeg subzones of Williams et al., 1988). The zone is characterized by its monotonous, unfossiliferous, clastic sedimentary rocks largely lacking in volcanic rocks, overlain in several places by rocks containing Celtic province shelly fossils of Arenig-Llanvirn age. Fossils of probable late Arenig age occur above the Gander Lake Subzone, in tuffaceous rocks associated with pillow lavas and other volcanic rocks of the Indian Bay Formation, now assigned to the Indian Bay Subzone (Williams, Chapter 2) (Fig. 10.4; DO-12; Wonderley and Neuman, 1984; O'Neill and Knight, 1988). No younger fossiliferous rocks are known from this area. Above the Mount Cormack Subzone, limestone clasts and siltstone matrix of probable debris-flow origin have brachiopods of probable Llanvirn age and Celtic affinities (DO-13; Neuman in Colman-Sadd and Swinden, 1984), and similar rocks nearby have trilobites and brachiopods of Llanvirn-Llandeilo age and Scoto-Appalachian affinities like those of the limestone at the base of the Davidsville Group at Weir's Pond (Boyce, 1987; Neuman, written communication to Colman-Sadd, 1987).

The Dunnage Zone in Quebec and New Brunswick is narrower, less well defined, and less fossiliferous than in Newfoundland. The only Middle Ordovician fossils are late Llanvirn-early Llandeilo conodonts at two places: (1) calcareous turbidite of the Mictaw Group in the Port-Daniel area, southern Gaspésie (DO-10; De Broucker, 1984; Nowlan, unpublished data); and (2) interpillow limestone of the

Fournier Group in the Elmtree inlier northwest of Bathurst, New Brunswick (DO-11; Nowlan, unpublished data; Fyffe, 1987).

The Miramichi Anticlinorium exposes the Gander Zone in New Brunswick (Miramichi Subzone, Fig. 10.1). The oldest rocks here are slate and quartzite of the Miramichi Group that have yielded Tremadoc graptolites at the two places cited previously. Similar but totally unfossiliferous rocks comprise the Gander Group in Newfoundland. Late Arenig brachiopods of Celtic province affinities occur in tuffaceous siltstone that overlies the slate and quartzite at two places in New Brunswick (Rocky Brook, GO-5, and Middle Hayden Brook, GO-6); tuff-free siltstone with the same brachiopods at Tetagouche Falls (GO-7) is an apparent lateral equivalent of nearby sandy limestone that contains North Atlantic province conodonts (Nowlan, 1981a; Neuman, 1984; Fyffe, 1987). Carbonate rocks bearing conodonts of late Llanvirn-early Llandeilo age occur in the Belle Lake Slate (GO-8; Nowlan, 1981a). A similar fauna is present in the Craig Brook Limestone (GO-9; Nowlan in St. Peter, 1982) where the conodonts are accompanied by ostracodes of mixed European and North American aspect (Copeland in St. Peter, 1982) and brachiopods that have not been examined since they were discovered a century ago (St. Peter, 1982).

Fossils of Arenig age in the Avalon Zone are sparse but distinctive. Inarticulate brachiopods from Bell Island, Newfoundland (AO-3) are similar to those from the Armorican Quartzite, France (Van Ingen, 1914). The Wabana sedimentary iron ores (AO-4) contain distinctive European trilobites (Dean and Martin, 1978) and abundant specimens of the arthropod trace fossil *Cruziana*. The linkage between the Avalon Zone and parts of Europe is supported by the finding that the graptolite *Didymograptus* (sensu lato) *simulans* Elles and Wood, from near the base

of the Wabana Group is conspecific with the type specimens from the Skiddaw Group in the English Lake District (Williams, 1990). Small amounts of oolitic hematite containing inarticulate brachiopods also occur in the northern Antigonish Highlands of mainland Nova Scotia (AO-5; Keppie and Schenk in King, 1982). This iron-rich sedimentary facies and the fossils associated with it probably represent a circum-Gondwana nearshore environment (Ranger in King, 1982; Dean, 1985). Elsewhere in the Avalon Zone, fossils of Arenig age are limited to graptolites of uncertain provincial affinities in the Suspension Bridge Formation, the uppermost unit of the Saint John Group in New Brunswick (AO-6). Younger Ordovician rocks are not known from the Avalon Zone in Canada.

Caradoc-Ashgill

Taconic orogenic events profoundly altered Appalachian depositional and paleontological patterns. Fossiliferous Ordovician post-Taconic rocks are known from only one place in western Newfoundland (Long Point Group of the Clam Bank Belt). They are more abundant in successor basins of the Gaspé Belt of Quebec and New Brunswick, and in the Exploits Subzone and Badger Belt of central Newfoundland. No rocks of this age are known from eastern Newfoundland, and there are only a few occurrences in central and southern New Brunswick and Nova Scotia.

The Long Point Group on Port au Port Peninsula, western Newfoundland (Fig. 10.1; CbO-1, Fig. 10.6) is interpreted to post-date the emplacement of the Humber Arm Allochthon. Its basal contact was interpreted as an unconformity (Rodgers, 1965) but it may be a fault (Stockmal and Waldron, 1990; Williams, Chapter 1). In its lower part, richly fossiliferous nodular argillaceous limestone contains brachiopods (Cooper, 1956), trilobites (Dean, 1979), the

Figure 10.4. Brachiopods from the Indian Bay Formation, Dunnage Zone, Newfoundland (locality DO-12 of Fig. 10.1). Opposite faces of a bedding surface of tuffaceous siltstone showing specimens of three genera. 1 – *Orthambonites* sp., ventral valve, internal mould on A, external mould on B; 2 – *Tritoechia* sp., dorsal valve, internal mould on A, external mould on B; 3 – new genus related to *Strophomena*, dorsal valve, external mould on A, internal mould on B, GSC 69247, X1; United States Geological Survey photograph from Wonderley and Neuman (1984).

Figure 10.5. Ordovician conodonts from the Appalachian orogen representing the North Atlantic Province (A-E), the North American Midcontinent Province (G-I), a cosmopolitan form (F), and a cold water form present in uppermost Ordovician strata (J-M). **A,B** – *Pygodus* cf. *P. anserinus* (Hadding), Davidsville Group, Weir's Pond, Newfoundland (locality DO-7 of Fig. 10.1), GSC loc. 96217, pygodiform (*g*) and haddingodiform (*f*) elements, X 110 and X125, GSC 96286, 96287. **C** – Microzarkodina flabellum Lindström), Tetagouche Group, Middle Hayden Brook, New Brunswick (locality GO-6, Fig. 10.1), GSC loc. 96067, ozarkodiniform (*f*) element, X120, GSC 64332; **D** – *Prioniodus alobatus* Bergström, Tetagouche Group, Camel Back Mountain, New Brunswick (locality ChO-1, Fig. 10.6), GSC loc. 96064, amorphognathiform (*g*) element, X42, GSC 64407; **E** – *Protopanderodus liripipus* Kennedy, Barnes and Uyeno, Tetagouche Group, Camel Back Mountain, New Brunswick (locality ChO-1, Fig. 10.6), GSC loc. 96064, costate element, X66, GSC 66401; **F** – *Panderodus gracilis* (Branson and Mehl), White Head Formation, Amphitheatre at Percé, Quebec (locality APO-1, Fig. 10.6 for this and specimens from all White Head Formation localities), compressiform (*e*) element, X60, GSC 66448; **G** – *Oulodus rohneri* Ethington and Furnish, White Head Formation, type section at Cap Blanc, Percé, Quebec, GSC loc. 96573, oulodiform (*g*) element, X65, GSC 66414; **H** – *Belodina confluens* Sweet, White Head Formation, type section at Cap Blanc, Percé, Quebec, GSC loc. 96572, grandiform (*p*) element, X85, GSC 66423; **I** – *Phragmodus undatus* Branson and Mehl, White Head Formation, Amphitheatre section, Percé, Quebec, GSC loc. 96557, phragmodiform (*a*) element, X60, GSC 66429; **J-M** – *Gamachignathus ensifer* McCracken, Nowlan and Barnes, White Head Formation, type section, Cap Blanc (J,M) and Flynn Road section (K,L), Percé, Québec; J, gothodiform (*a*-2) element, GSC loc. 96572, X90, GSC 66435; K, falodiform (*e*-2) element, GSC loc. 96601, X100, GSC 66438; L, prioniodiform (*f*) element, GSC loc. 96605, X95, GSC 66441; M, modified prioniodiform (*g*) element, GSC loc. 96572, X65, GSC 66440.

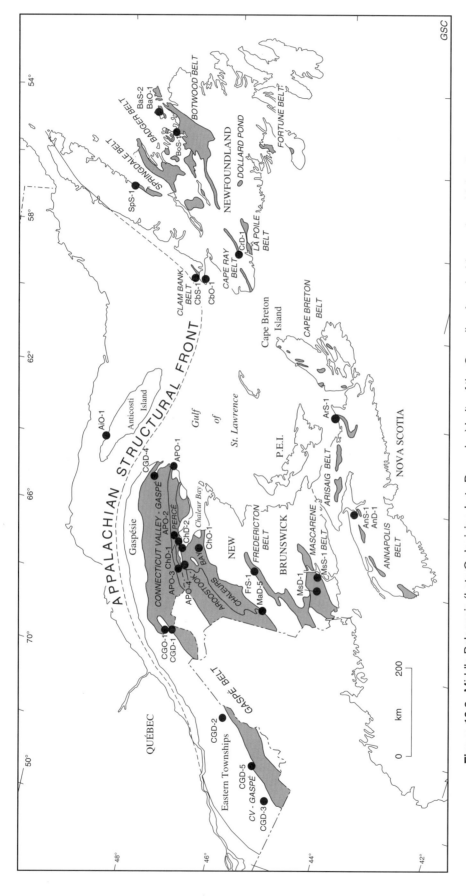

Figure 10.6. Middle Paleozoic (Late Ordovician-Late Devonian) belts of the Canadian Appalachian region with fossil localities discussed in text. Localities are coded as follows: Ai – Anticosti Island; An – Annapolis Belt; Ar – Arisaig Belt; Ba – Badger Belt; Bo – Botwood Belt; Cb – Clam Bank Belt; Ch – Chaleurs Bay Division of the Gaspé Belt; Fr – Fredericton Belt; CG – Connecticut Valley-Gaspé Division of the Gaspé Belt; AP – Aroostook-Percé Division of the Gaspé Belt; Cr – Cape Ray Belt; Ms – Mascarene Belt; O – late Ordovician; S – Silurian; D – Devonian. A combination of two letter codes followed by a number is an unique identifier of belt, age, and locality.

cephalopod *Gonioceras*, ostracodes, tabulate corals and bryozoans (Bolton, 1965; Copeland and Bolton, 1977) of middle Caradoc age that strongly resemble those in the central and southern Appalachians and Scotland. The Long Point Group contains North Atlantic province conodonts in the lower part of the sequence and North American province conodont assemblages in its upper part (Fåhraeus, 1973).

Faunas of Ashgill age occur in shallow-water carbonate rocks on Anticosti Island (AiO-1, Fig. 10.6; Lespérance, 1981) immediately to the west of the orogen, and in a wide variety of rocks in the Gaspé Belt (Aroostook-Percé Division; Bourque et al., Chapter 3) including shallow- to deep-water carbonates and siliciclastic rocks ranging from shale to conglomerate. Fossils of this age from Anticosti Island include brachiopods that belong to genera that link it to the North American brachiopod biogeographic province of Williams (1973) and rugose and tabulate corals that, with some exceptions, have similar affinities (Bolton in Lespérance, 1981; Elias, 1982). Latest Ordovician ostracodes here include several taxa of European aspect (Copeland in Lespérance, 1981).

Most of the carbonate rocks of the Gaspé Belt (Aroostook-Percé Division) were deposited in water that was deeper than that of Anticosti Basin. The rocks and fossils at its northeastern end near Percé, show evidence of shallower-water deposition compared to that of equivalent rocks to the southwest. The European affinities of the brachiopods near Percé (APO-1; Fig. 10.6), known since the work of Schuchert and Cooper (1930), was confirmed by Williams (1973) who classed them with his North European province. Associated trilobites belong to genera of the Remopleurid trilobite province of Whittington and Hughes (1972). More recent work has led to the assignment of specific generic suites to depth-controlled benthic assemblage communities (Lespérance et al., 1987). Conodonts of the Anticosti and Percé sequences are similar in that both are dominated by representatives of the North American Midcontinent Province (Nowlan, 1981b; Nowlan and Barnes, 1981; McCracken and Barnes, 1981).

Of special interest is the occurrence in the uppermost Ordovician beds at Anticosti and at Percé of assemblages of brachiopods known widely as the Hirnantian fauna (Lespérance and Sheehan, 1976; Cocks and Copper, 1981). This assemblage is characterized by a significant reduction in the number of brachiopod genera compared to the assemblages that preceded it, presumably in response to rapid cooling of the world's oceans coincident with the latest Ordovician glaciation that is demonstrated by extensive glacial deposits of late Ashgill age in sub-Saharan Africa. Conodonts in the highest Ordovician beds on Anticosti Island and at Percé include forms (e.g. *Gamachignathus*; Fig. 10.5) whose ancestors are found in cool-water North Atlantic province assemblages (Nowlan, 1981b; McCracken and Nowlan, 1988). The relative completeness of strata across the Ordovician-Silurian boundary in the Appalachians is discussed at the end of this section.

Fossils are rare in the deep-water sedimentary rocks of the western part of the Gaspé Belt (Aroostook-Percé Division) of Quebec and New Brunswick, but tabulate corals, apparently resedimented from a carbonate bank, occur in siliciclastic rocks of the Honorat Group near Carleton, Quebec (APO-2, Fig. 10.6; Bolton, 1980). Conodonts in some distal debris flows include North Atlantic province forms together with resedimented taxa that represent a spectrum of

biofacies in the North American midcontinent province (Fig. 10.5; APO-3, 4, Fig. 10.6; Nowlan, 1981b, 1983a). Graptolites assemblages, reported recently by Riva and Malo (1988), consist of cosmopolitan species of Middle and early Late Ordovician age.

Fossils of Late Ordovician age are rare in rocks that border the Aroostook-Percé Division of the Gaspé Belt. Conodonts of probable Ashgill age occur in the matrix of an otherwise unfossiliferous carbonate breccia in the Becaguimec area, New Brunswick (Nowlan in St. Peter, 1982), in the Connecticut Valley-Gaspé Division of the Gaspé Belt. Graptolites and cyclopygid trilobites occur at several places in siliciclastic rocks of the Cabano Formation in the northwestern Connecticut Valley-Gaspé Division in the Lake Témiscouata area, Quebec (CGO-1, Fig. 10.6; David et al., 1985; Lespérance, pers. comm., 1987), as does a more varied assemblage including the bivalves *Ambonychia* sp. and *Modiolopsis*? and the trilobite *Encrinurus* sp. in similar rocks in Maine near the Quebec border (Roy, 1989) that are probably laterally contiguous with the Cabano Formation. The absence of limestone from the sequences containing these fossils, and the presence of forms that suggest both deep- and shallow-water environments contrast sharply with Aroostook-Percé Division occurrences. This difference may be explained by northwestward thrusting of the Aroostook-Percé rocks (Lespérance, pers. comm., 1987).

Caradoc graptolites are known from black shale at several places in the Chaleurs Bay Division of the Gaspé Belt, but the only known shelly fossils are those from limestone interbedded with basalt at Camel Back Mountain, New Brunswick (ChO-1; Skinner, 1974). Trilobites here have southern Appalachian affinities (Dean, 1976), and associated conodonts are of North Atlantic province affinity (Nowlan, 1981a; Fig. 10.5 herein). The volcanic rocks here are probably of mid-ocean-ridge origin, faulted southward over the sequence that contains Celtic-province fossils of Arenig age (van Staal, 1987), relations that provide evidence for telescoping of the subdivisions recognized in the Newfoundland part of the Gander Zone mentioned earlier.

Caradoc and Ashgill-age rocks in central Newfoundland are confined to the Exploits Subzone and Badger Belt, mostly graptolite-bearing black shale of Caradoc age that is interpreted to record tectonic quiescence (Bergström et al., 1974; van der Pluijm et al., 1987). Shelly fossils appear to be concentrated in the eastern Notre Dame Bay area, on and near New World Island of the Badger Belt (BaO-1; Fig. 10.6). Most occurrences are of Ashgill age in coarse grained epiclastic rocks, but notable exceptions are the large blocks of coral-algal reefal limestone in conglomerate assigned to both the Sansom Formation (Horne and Johnson, 1970; McKerrow and Cocks, 1981) and the Goldson Formation (Elliott et al., 1989). The Caradoc age of one of these blocks, determined from study of its algae and other fossils (Elliott et al., 1989), may apply to all of them, but reefs of this age are not known elsewhere in the Badger Belt.

Ashgill-age fossils, largely brachiopods, occur in clastic rocks ranging from mudstone to conglomerate at several places in the Badger Belt on New World Island (McKerrow and Cocks, 1978; Neuman, 1984; van der Pluijm et al., 1987; Boyce, 1987; Elliott et al., 1989) in both coherent bedded sequences (Sansom Formation) and in olistoliths in Silurian mélange. Assemblages of brachiopods differ from place to place, their generic composition apparently governed by local environments. The most common and diversified assemblages consist of genera (e.g. *Christiania, Dolerorthis, Plectatrypa, Sampo*) that characterize the North-European brachiopod biogeographic province of Williams (1973). Associated fossils include smaller numbers of gastropods, bivalves, trilobites, and tabulate and rugose corals. The rugose corals, under study by R.J. Elias, are of special interest in view of his finding (Elias, 1982) that Ashgill-age solitary rugose corals from Maine, Gaspésie, and Anticosti Island permit recognition of a Maritime subprovince of his Red River-Stony Mountain biogeographic province, thus linking the faunas of the interior of the North American craton and the faunas of the Iapetan continental margin.

Glaciation and attendant eustatic lowering of sea level at or near the end of the Ordovician have been identified by many workers as the cause of stratigraphic discontinuities and worldwide faunal changes (e.g. Sheehan, 1975; Spjeldnaes, 1981; Hambrey, 1985). Lowermost Silurian (lower Llandovery) rocks were not deposited on the North American craton where contrasts between Ordovician and Silurian faunas reflect a lengthy gap in the sedimentary and paleobiological record. Deposition across the Ordovician-Silurian boundary interval was nearly continuous on Anticosti Island and in the Aroostook-Percé Division of the Gaspé Belt. Facies changes in both places reflect significant regression (shoaling) at or near the system boundary followed by Silurian transgression (deepening) accompanied by the introduction of Early Silurian faunal elements (Petryk, 1981; Brenchley, 1988; Lespérance, 1988). Latest Ordovician and earliest Silurian faunas are absent from the Badger Belt where evidence of crustal instability suggests that structural events were more significant than eustatic sea level changes in controlling the nature and distribution of the rocks and fossils at the systemic boundary.

SILURIAN-EARLY DEVONIAN

The depositional pattern of Silurian sediments in the Appalachian Orogen is extremely heterogeneous, with basinal sedimentation continuing in troughs like the Aroostook-Percé Division of the Gaspé Belt, and nonmarine to shallow marine sedimentation developing locally on a rugged landscape that resulted from pre-Silurian tectonism. The age of lowermost Silurian strata deposited in the region varies considerably (Nowlan, 1983b).

The paleontological record shows that biogeographic patterns in the Appalachians during the Silurian and Early Devonian were essentially continuations from those of the Late Ordovician, although workers on each of several different fossil groups have used different names for similar biogeographic subdivisions. For example, the Late Ordovician faunas of the Appalachians were classed from their brachiopods as North American and North European provinces of a cosmopolitan realm (Williams, 1973). By contrast, Silurian and Lower Devonian brachiopod-based biogeographic units place most of the Appalachians in the North Atlantic region of the North Silurian biogeographic realm, but outboard belts (Mascarene, Arisaig, and Annapolis in Fig. 10.6) are assigned to the Old World realm (Boucot, 1975); the Lower Devonian term for the larger part of the Appalachians is the Appalachian Brachiopod province (Boucot et al., 1969; Johnson and Boucot, 1973; Lespérance and Sheehan, 1988). The term for the same

area in the Early Devonian based on rugose corals is Eastern Americas realm (Oliver and Pedder, 1976). Of these schemes, the one based on brachiopods best reflects the proposition that a major part of the Iapetus Ocean had closed during the Middle Ordovician and the remaining oceanic separation isolated only the outermost Appalachian belts.

Continuity with the Late Ordovician is particularly notable in the stratigraphic and faunal successions of Anticosti Island and Gaspé Belt. For example, on Anticosti Island the earliest Silurian trilobites "consist of holdovers from the Ordovician, and show little change from their ancestors" (Lespérance, 1988, p. 359). The rocks of the Gaspé Belt, particularly those at Percé are deeper water deposits than those of Anticosti Island, differences that are reflected in the contrasting composition of the Early Silurian faunas of the two areas (Lespérance et al., 1987). Lowermost Silurian strata in the main part of the Aroostook-Percé Division to the southwest are virtually devoid of fossils, perhaps because of the major anoxic event recognized near the boundary elsewhere (Nowlan et al., 1988). Faunas tend to recover in the late Llandovery.

Local depositional environments are probably responsible for differences in Early Silurian faunas in the Gaspé Belt, and in those of Late Silurian and Early Devonian age in its Connecticut Valley-Gaspé and Chaleurs Bay divisions that contain mainly siliciclastic rocks and abundant carbonate complexes that suggest low paleolatitudes (Bourque et al., 1986; Cocks and Fortey, 1982). The influence of environments on the occurrence and distribution of Early Devonian brachiopods and trilobites is well shown in eastern Gaspésie (Lespérance and Sheehan, 1988).

Southeast of the Gaspé Belt, the thick accumulation of shale and turbiditic sandstone of the Fredericton Belt (Fig. 10.6) contains graptolites of Wenlock and Ludlow age (FrS-1). These rocks are of special significance because they appear to be the remnants of a deep-water trough separating the Chaleurs Bay Division of the Gaspé Belt, bearing faunas of Boucot's (1974) North Silurian realm, from the Mascarene Belt with its Old World realm fossils. The Fredericton Belt may have been the site of the Iapetus Ocean during Silurian time (McKerrow and Ziegler, 1971; Rast and Stringer, 1980). Evidence of oceanic crust is lacking but there is no doubt that the trough occupies an axis that separates markedly different shelly and ostracode faunas.

Silurian rocks of the Mascarene Belt in southwestern New Brunswick, like those of the Eastport area in adjacent southeastern Maine, are largely volcanogenic with interbedded sedimentary rocks that contain scattered occurrences of brachiopods and other fossils (Boucot et al., 1966). All but the latest Silurian fossils in these areas belong to cosmopolitan taxa, but some ostracodes of late Silurian (Pridoli) age in the Jones Creek Formation of southwestern New Brunswick (MsS-1), are distinctive of the north-European (Old World) realm (Copeland and Berdan, 1977; Berdan, 1983; Schallreuter and Siveter, 1985). The same ostracodes occur in the latest Silurian part (Stonehouse Formation; Fig. 10.7) of the sequence of marine and nonmarine fine grained epiclastic rocks of the Arisaig Belt, Nova Scotia (ArS-1; Copeland and Berdan, 1977). The brood pouches of these ostracodes suggest that they had restricted dispersal capacity because their life cycle did not include a free-swimming larval stage, and they began life as relatively immobile young individuals (Cocks and Fortey, 1982).

Late Silurian vertebrate faunas were highly provincial (Halstead and Turner, 1973; Dineley and Loeffler, 1979, Young, 1981). The Late Silurian North American and European subprovinces identified by Dineley and Loeffler (1979) are based on agnathans with abundant diverse heterostracans and rare osteostracans in North America and abundant, diverse osteostracans and rare heterostracans in Europe. These two provinces have been termed the Cephalaspid Province by Young (1981) who attributed their breakdown in the Devonian to the closing of Iapetus. Distribution of thelodonts and other agnathans were used to evaluate proposed continental reconstructions (Turner and Tarling, 1982). A report of thelodonts of Late Silurian age from the White Rock Formation in the Annapolis Belt of southern Nova Scotia (AnS-1; Bouyx and Goujet, 1985) suggests that the Avalon and Meguma zones may have been in close proximity in the Late Silurian. Their fauna suggests that the Annapolis Belt was part of a biogeographic region encompassing the Anglo-Welsh region, eastern North America and Greenland. The Old World affinities of the Annapolis Valley sequence are confirmed by the Rhenish assemblage of Early Devonian brachiopods in the Torbrook Formation (AnD-1; Boucot, 1960).

No Silurian-Early Devonian fossils are known from eastern Newfoundland, and the few occurrences of Silurian age suggest that the central and western part of the island lay within Boucot's (1975) North Atlantic region of the North Silurian realm. The rarity of marine Silurian fossils in western and central Newfoundland as compared to other areas of the Appalachian Orogen indicates a major difference in lithofacies, and limits paleobiogeographic assessment of the area. This part of the orogen is the site of convergence of the Notre Dame and Exploits subzones that probably began in the Late Ordovician and Early Silurian.

The Upper Silurian (Pridoli) Clam Bank Group on Port au Port Peninsula, southwestern Newfoundland (CbS-1) may be an outlying part of the Anticosti Basin, but its fossils (Fåhraeus, 1974) are inadequate for biogeographic placement. Although Silurian-Early Devonian fossils of central Newfoundland remain poorly known, they are probably amenable to community analysis to obtain some indication of the post-Taconic-pre-Acadian history of the region. Volcanic rocks of the Springdale Group (Springdale Belt, Fig. 10.6) are unfossiliferous, but their Early Silurian (Llandovery) age (429+6/-5 Ma) was determined isotopically (Chandler et al., 1987). Shallow-water marine sedimentary rocks (Spruce Ridge and Natlins Cove formations of Lock, 1969) overlie volcanic rocks of the Sops Arm Group (SpS-1). These formations have yielded poorly preserved

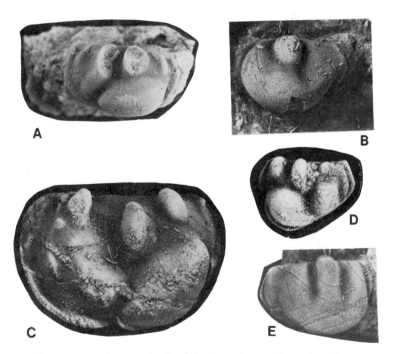

Figure 10.7. Ostracodes from the Stonehouse Formation, Nova Scotia (locality ArS-1, Fig. 10.6). **A** – *Londinia arisaigensis* Copeland, right valve, X12, heteromorph, GSC 14563a; **B** – *Frostiella* sp. cf. *F. plicata* (Jones), left valve, X9, heteromorph, GSC 14513; **C** – *Nodibeyrichia pustulosa* (Hall), right valve, X16, heteromorph, GSC 14503; **D** – *Hemsiella maccoyiana sulcata* (Reuter), left valve, X17, heteromorph, GSC 14512; **E**, *Kloedenia wilkensiana* (Jones), right valve, X17, tecnomorph, GSC 14514.

fossils including corals and brachiopods possibly as old as Llandovery, conodonts of probable Ludlow age, and others of doubtful latest Silurian (Pridoli) age (Berry and Boucot, 1970; Chandler et al., 1987). The Botwood Group (BoS-1; Botwood Belt, Fig. 10.6) also consists of a volcanic unit (Lawrenceton Formation) overlain by sandstone and silt-stone (Wigwam Formation; Chandler et al., 1987) that contains Early and possibly Late Silurian (Llandovery, Wenlock, and Ludlow?) brachiopods and other shelly fossils and a local graptolite occurrence (Berry and Boucot, 1970; Blackwood, 1982). Lower Silurian (Llandovery) fossilifer-ous rocks on New World Island (BaS-2, Fig. 10.6), sepa-rated from the Botwood Group by a major fault, contain brachiopod genera of the North Atlantic region, but the rocks containing them and their distribution reflect the persistence of crustal instability that began in the Ordovi-cian (McKerrow and Cocks, 1978). Rocks younger than Llandovery are not known from this area.

Early Devonian plants from the Cape Ray Belt of south-western Newfoundland (CrD-1; Dorf and Cooper, 1943), Gaspésie (CGD-4; Gensel and Andrews, 1984), northern New Brunswick (ChD-2, Gensel and Andrews, 1984), southern Quebec (CGD-5) and northern New Hampshire (Hueber et al., 1990), and north-central Maine (Andrews et al., 1977) indicate that terrestrial environments pre-vailed through much of the central part of the northern Appalachians at this time. These fossils are preserved in rocks that range from very fine grained siltstone to coarse grained sandstone; in Maine and northern New Brunswick they indicate intermittent river flooding along channels and in deltas, some having volcanic highlands in their headwaters, and in Gaspésie they record the deposition of a complete clastic wedge (Lawrence and Rust, 1988). In the world paleophytogeographical scheme of Ziegler et al. (1981), these floras, together with those from Ontario, Scotland, Portugal and Libya constitute the Appalachian subunit of the Equatorial floral subprovince.

MIDDLE DEVONIAN-CARBONIFEROUS

The onset of major paleogeographic realignments resulting from the Acadian Orogeny are recorded in Middle Devonian strata and their fossils, followed by the major Carboniferous post-orogenic effects. In many areas (e.g. Newfoundland and the Mascarene Belt) nonmarine sedimentation began in the Late Silurian or Early Devonian. The firm estab-lishment of a redbed facies across much of the North Atlantic region by the Middle Devonian eliminated most marine faunas.

Middle Devonian (Eifelian-Givetian) marine fossiliferous rocks occur in the northwestern Connecticut Valley-Gaspé Division of the Gaspé Belt and the adjacent foreland. Those of the foreland have been eroded and are preserved only as blocks in the Mesozoic diatreme breccia at Saint-Hélène Island near Montreal (Boucot et al., 1986). The Lower and Middle Devonian sequence reconstructed from these blocks is similar in part to that of western New York, and through-out the fossils can be referred to the "Appohimchi Sub-province" (equivalent to the North American platform) of the Eastern Americas Realm (Boucot et al., 1986).

In the northwestern Connecticut Valley-Gaspé Division of the Gaspé Belt of southern Quebec several limestone units including the Touladi Limestone (CGD-1), the Famine Lime-stone (CGD-2) and the Mountain House Wharf Limestone

(CGD-3) contain brachiopods of mixed provincial affinity (Boucot and Drapeau, 1968; Boucot and Johnson, 1967; Boucot, 1975), indicating that Old World forms had almost reached the main part of the North American continent by the Middle Devonian. Corals from the Famine Limestone are described as being of eastern North American aspect (Oliver, 1971). Coeval sedimentary rocks in the adjacent Appalachians of northeastern Maine are terrestrial sequences that contain plant fossils (Neuman et al., 1989).

The virtual absence of marine Middle and Upper Devonian from the Canadian Appalachians is a result of the Acadian Orogeny which was mainly responsible for what has been termed the Old Red Sandstone Continent (see Goldring and Langenstrassen, 1979). Biogeographic assessments of the Middle and Upper Devonian of the Appalachian region must rely upon nonmarine faunas and floras. The oldest post-Acadian rocks in Canada have recently been shown to be Late Devonian, based on spores. Some of these represent species that were restricted to the southern parts of the Old Red Sandstone continent (MsD-1; McGregor and McCutcheon, 1988). The distribution of ver-tebrate fossils in the Devonian has been used to identify faunal provinces (Westoll, 1979; Young, 1981). The Euramerican province (Young, 1981) embraces all of the Canadian Appalachians until the Late Devonian when vertebrate fossil distribution suggests that Euramerica and Gondwana became contiguous. A continuous land con-nection is postulated to have developed at or near the Frasnian-Famennian boundary (Young, 1981). Devonian vertebrate faunas in the Canadian Appalachians, particu-larly those from the Late Devonian Escuminac Formation (ChD-1), are extremely well preserved but a taxonomic and paleobiogeographic synthesis remains to be completed. These exquisite vertebrate faunas are associated with equally well preserved plant fossils that also require modern taxonomic and paleobiogeographic treatment.

Marine deposition in the Canadian Appalachians resumed in the Early Carboniferous (Viséan) with the overlap of the southern margin of the Old Red Sandstone continent by an extensive North Hercynian Ocean. Marine strata repre-senting shallow-water deposition are preserved in the Windsor Group, recognized widely in the Maritime Prov-inces, and the Codroy Group of Newfoundland (Fig. 10.8). The faunas present in these marine strata are closely related to those of Western Europe as was shown by Bell (1929, 1948, 1959) in his extensive pioneering work on the Carboniferous strata of eastern Canada. These faunas dif-fer markedly from the endemic early Carboniferous (Mississippian) assemblages of the eastern interior and Mississippi Valley regions of North America.

Ostracodes (BGC-1; Dewey, 1983) and conodonts (CBC-1, von Bitter, 1976; BGC-2, von Bitter and Plint-Geberl, 1982) confirm the affinity of these faunas with the widespread Viséan faunas of western Europe. The affinity is so strong that it is possible to define and apply conodont zones between eastern Canada and western Europe (e.g. the *Taphrognathus transatlanticus* Zone of von Bitter and Austin, 1984).

Early Carboniferous spore assemblages from the Canadian Appalachians resemble those from western Europe (Utting, 1987a). Spores known from the Windsor Group (Utting, 1987a) of eastern Canada and the Albert Formation of New Brunswick (Utting, 1987b) are assigned

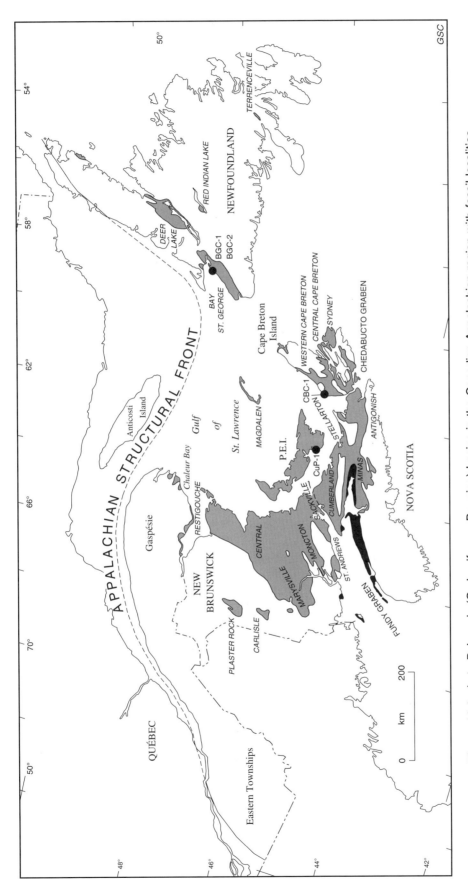

Figure 10.8. Late Paleozoic (Carboniferous-Permian) basins in the Canadian Appalachian region with fossil localities discussed in text. Localities coded as to basins as follows: CB – Western Cape Breton; Cu – Cumberland; BG – Bay St. George; and to age as follows: C – Carboniferous; P – Permian. A combination of two codes followed by a number is an unique identifier of basin, age, and locality.

to the *Vallatisporites* Region of Van der Zwan (1981), embracing a subtropical dry belt that extended from eastern Canada through the British Isles and Denmark.

The nonmarine Pennsylvanian sedimentary rocks of the Canadian Appalachian region were deposited in a complex of basins, the most important of which are the Cumberland, Minas and Sydney (Bell, 1944 and Fig. 10.8 (updated). These basins include several fossiliferous coal deposits and the stratigraphy has been based on rock units with distinctive flora and fauna (Bell, 1944; Copeland, 1957). The Pennsylvanian fauna (mostly arthropods and bivalves) and flora are similar to Namurian and Westphalian fossils of western Europe (Bell, 1944; Copeland, 1957). Fossil spores occur in strata above those bearing macrofossils and suggest that the highest beds are Stephanian to early Permian (Hacquebard and Donaldson, 1963). The youngest Paleozoic macrofossils known from the Appalachian region are Permian vertebrates from Prince Edward Island (CuP-1, Fig. 10.8; Langston, 1963). Paleobiogeographic assessments of youngest Paleozoic strata in the Canadian Appalachians are not available.

THERMAL MATURATION PATTERNS DEFINED BY FOSSILS

Fossils can also be used as indicators of temperature. Patterns of thermal development based on fossils such as conodonts and palynomorphs contribute to tectonic interpretations. For example, Nowlan (1983a) demonstrated that Upper Ordovician strata of the Grog Brook Group in New Brunswick contained conodonts with low thermal alteration indices in an area characterized by much higher levels. This anomaly indicated that the strata had never been deeply buried and permitted development of a hypothesis that thrust faulting of Late Ordovician-Early Silurian age took place in the area thus demonstrating tectonic activity in post-Taconic and pre-Acadian time.

A comprehensive analysis of conodont colour alteration indices (CAI) in the Appalachian region (Nowlan and Barnes, 1987) shows that autochthonous strata in the Humber Zone of western Newfoundland range from a low of CAI 1 in the south to 3.5 in the north. Within the Taconic allochthons, CAI values range from 1 in the ancient slope sediments preserved in the frontal thrusts to 5 near the upper obducted ophiolites, where temperatures probably exceeded 300°C. These regional variations suggest that the two main Taconic allochthons in the Newfoundland Humber Zone were discrete bodies each with a regional extent similar to their present outcrop areas. CAI values in the orogenic belt itself show a wide variation: those in the Dunnage and Gander zones are typically 4 to 6 (suggesting temperatures of about 190-400°C) although locally there are anomalously low and high values that require explanation. Nowlan and Barnes (1987) attributed much of the pattern of CAI isopleths in the Appalachian region to intrusive activity that locally produced CAI values of 8 (the maximum value, representing about 600°C). They also suggested that thermal maturation patterns in parts of the Canadian Appalachian region may be attributed to the passage of Mesozoic hotspot tracks, particularly in the region of Montreal, which may have overprinted much of the Paleozoic thermal history of the region.

Palynomorphs can also be used for estimating thermal maturation patterns, but a comprehensive assessment of Appalachian data remains to be done.

DISCUSSION AND CONCLUSIONS

Paleontologists have been important contributors to the present knowledge of the geology of the Appalachian Orogen and to evolution of ideas concerning its development because fossils provide information that can be used in assessing the paleogeography and tectonics of an orogenic belt. Fossil assemblages provide information on the age of the enclosing sediments, past climates, water depths, and paleoenvironments. Global patterns of fossil distribution play a leading role in the recognition of paleocontinental masses and their boundaries, as well as the proximity of continents through time. Thus paleontology is one of several disciplines drawn upon for the construction of paleogeographic maps (Ziegler et al., 1979). At a more local scale the juxtaposition of different biofacies may permit the estimation of horizontal displacement. Some fossils, particularly microfossils such as conodonts and palynomorphs, may be treated as inanimate objects that respond to physical parameters such as temperature and thus provide information on depth of burial, proximity of plutons, and other important aspects related to tectonic development.

Data derived from fossils and their distribution have several advantages that are important for their use in interpretation of paleogeographic, structural or tectonostratigraphic problems. Foremost among these is the fact that fossil data are independent of tectonic models (Fortey and Cocks, 1986). They can therefore provide impartial tests for tectonic hypotheses, and may themselves lead to the formulation of tectonic hypotheses. Secondly, the amount of detail available in paleontological information is considerable because of the complexity of the organisms that provide the material. Identification of a trilobite or conodont is less likely to be in error than an analytical test performed on the enclosing rock, because the amount of information available from their complex morphology provides many constraints on their interpretation. Furthermore, errors made are more likely to be recognized because the fossils are illustrated and additional material can be collected. Thirdly, the fossil record provides a history of biological evolution, the nature of which is considerably different from the physical parameters employed in other aspects of tectonic analysis, mainly because evolution has proceeded in certain identifiable directions. The importance of each fossil specimen can therefore be related to the evolution of successive assemblages, the development of which is related to biological and environmental, rather than physical factors. It is therefore not surprising that paleontology has served as a catalyst to new interpretations of the Appalachian Orogen and also as a corroborator of well established hypotheses.

It is the value of fossils as time indicators that allows for the fundamental observation that the Appalachian Orogen developed through Paleozoic accretion to the continental nucleus of North America. Knowledge of the relative age of strata in the component parts of the orogen has been important for initial recognition and characterization of those parts and has elucidated the timing of events responsible for their assembly. For example, recognition that the

Cow Head Group contains Middle Cambrian to Middle Ordovician fossils and that it overlies a sequence of strata of similar age permitted the revolutionary recognition of Taconic klippen in western Newfoundland (Rodgers and Neale, 1963), an idea that had a profound effect on the development of subsequent tectonic models. The value of biostratigraphy for determining temporal relationships has been critical to understanding of the timing of Appalachian events. For example, the end of the Taconic Orogeny in Newfoundland is known from the Caradoc age of the Long Point Group which directly overlies allochthonous strata. Similarly post-Acadian units have been dated as Late Devonian.

Perhaps the earliest application of paleontology to understanding of the Appalachian Orogen is the use of provincialism in Cambrian trilobites to demonstrate the distinctiveness of the Humber and Avalon zones. Faunal evidence played a role in Schuchert's (1923) recognition of geanticlines separating the St. Lawrence (Humber) and Acadian (Avalon) geosynclines, a model that was influential for many years in discussions of the Appalachian Orogen. The development of models of geosynclinal development (e.g. Kay, 1951) relied heavily on the paleobiogeographic and paleoenvironmental implications of fossils. For many years, paleontologists pointed out the similarity of Avalon Zone faunas to those of Europe (e.g. Grabau, 1936; Hutchinson, 1962). Faunal differentiation across the orogen also contributed to Williams' (1964) idea of the two-sided symmetry of the Appalachian Orogen, which in turn, led to the definition of the proto-Atlantic Ocean by Wilson (1966). In Wilson's model the Cambrian sequences of the Atlantic province were remnants of the land that lay on the eastern margin of the proto-Atlantic Ocean (Avalon Zone) and the complex area between Atlantic province faunas (Dunnage and Gander zones) and those of the eastern part of North America (Humber Zone) resulted from a post-Cambrian convergence of the two sides of an ancient ocean. In fact, Wilson's (1966, p. 676) opening paragraph draws attention to the importance of faunal realms by quoting a passage from Hutchinson (1962) in which he noted the difficulty of correlation of strata in the Avalon Peninsula of Newfoundland with those in the rest of North America based on trilobites.

It is now clearly established that Cambrian fossils show strong biogeographic control in the Appalachian Orogen with trilobites of one faunal realm (Atlantic) characterizing the Avalon Zone on the eastern margin of Iapetus and another (Pacific) characterizing the Humber Zone on the western margin of Iapetus. Species characteristic of the Avalon Zone were also capable of trans-Iapetus migration from time to time. Faunas from the Dunnage Zone are sparse but provide useful information on the extent of the Laurentian (North American) margin in the Cambrian and in some cases exhibit endemism that may be the result of development on offshore shoals or oceanic islands developed in the same climatic belt.

The faunal differentiation between the two sides of the Iapetus Ocean diminishes from the Cambrian to the Ordovician, although provincially distinct faunas continue because the ocean remained a formidable barrier to migration for many groups. Early to Middle Ordovician benthic trilobite and brachiopod faunas occurring around the margins of North America (Humber Zone) are quite distinctive from those in the Iapetus Ocean. In the early Middle Ordovician, for example, this distinctive generic assemblage is referred to the Toquima-Table Head faunal realm (Ross and Ingham, 1970), which occupied a belt peripheral to the platform. Elements of faunas typical of the Iapetus Ocean and its southeastern margin (Gondwanaland) occur rarely in these slope faunas. Coeval assemblages of largely different brachiopods and trilobites occur in rocks that were deposited on the fringes of islands within Iapetus and on the margins of the Armorican Platform (Celtic Province of Neuman, 1984). Ocean-closing events associated with the Taconic Orogeny are recorded in the Exploits Subzone of the Dunnage Zone, Newfoundland where volcaniclastic rocks containing Arenig-Llanvirn age fossils of the Celtic Province are overlain by Llandeilo age carbonate rocks containing fossils of North American affinities (McKerrow and Cocks, 1986).

Conodonts of the North Atlantic province and graptolites provide important links across Iapetus because both groups occurred in open ocean waters on both sides of the ocean. Distinct conodont assemblages, however, occurred within North America in the warmer, presumably more saline waters (North American Midcontinent province). Rarely, these North American faunas occurred within the Iapetus Ocean, their distribution being attributed to islands that served as staging sites for migration.

Provincialism has been used to delineate the physical axis of the Iapetus Ocean. For example, McKerrow and Cocks (1977) suggested that the provincial affinity of benthic faunas could be used to identify the suture at the centre of the Iapetus Ocean during the Ordovician. Citing the North American aspect of Llandeilo benthic faunas as far east as New World Island, Newfoundland, and the apparent European early Paleozoic faunas farther to the east, they concluded that the Reach Fault marks the suture at which the Iapetus Ocean closed at the end of the Early Devonian. Stouge (1980a) observed that Llandeilo conodonts were the same on both sides of the Reach Fault and cast doubt on the validity of the Reach Fault as the suture. Conodonts of the North Atlantic Province have a trans-Iapetan distribution and it is therefore not surprising that the same conodonts occur on both sides. It is clear that owing to the complexity of real distributions and the relative crudity of faunal comparisons at the generic level, identification of a single line of suture is likely to be approximate at best.

Large differences in the Early Ordovician paleolatitudes of the Laurentian and Gondwanan-Iberian-Armorican continental cratons determined from paleomagnetics indicate a width of about 3000 km for the Iapetus Ocean (Fig. 10.9; Smith et al., 1973; Ziegler et al., 1979; van der Voo, 1988). Such a width is largely consistent with the paleontological record (Cocks and Fortey, 1982; Neuman, 1984; McKerrow and Cocks, 1986) and the estimates of Williams (1980). Fossils assist in the estimation of the former location of terranes in the mobile belts of the Appalachians and Caledonides. Comparison of Ordovician rocks and fossils of the Appalachian lithotectonic zones with those of the British-Irish Caledonides strongly supports the existence, and helps define the extent, of the Iapetus Ocean (Williams, 1978; Harper and Parkes, 1989). This is apparent from comparison of the distribution of rocks and fossils in the North American Appalachian and British-Irish Caledonian orogens. In the Appalachians the east-to-west progression from the Laurentian continental margin (Humber Zone/Toquima-Table Head Province) through those of mid-oceanic settings (Dunnage Zone/Celtic Province) to peri-Gondwanan sites (Gander Zone/Celtic Province) compares well with the Caledonian west-to-east progression from the Gondwanan continental margin (Anglesey Zone-Rosslare terrane/Celtic Province), across a continental margin volcanic arc (Bellewstown and Grangegeeth terranes/Celtic Province) (Murphy, 1987; Harper et al., 1990), to a fragment of the Laurentian continental margin (northern Scotland) and a marginal volcanic arc (County Mayo, Ireland) (Hebrides and Northwestern terranes/Toquima-Table Head Province). Although these and other comparisons are consistent in their support of the existence of the Iapetus Ocean, the outline of its shores and the position of islands within it can be only vaguely estimated, and its bathymetry remains largely unknown. These uncertainties remain in part because there are unanswered questions concerning biological aspects of oceanic dispersal, as well as questions concerning the systems of water currents that prevailed in the ocean and their effects on dispersal and migration. Despite these uncertainties, a schematic map of the Early Ordovician Iapetus Ocean basin showing estimated locations of its opposite shores and some of the fossiliferous parts of these terranes resembles a map of a modern tectonically active ocean basin (Neuman and Max, 1989).

The breakdown of strong provinciality in the Caradoc can be interpreted as evidence of the increasing proximity of the margins of the Iapetus Ocean during the Middle Ordovician. Alternatively, it could signal a radical change in ocean current patterns as a result of changing continental configurations in the region that permitted trans-oceanic migration, but this is considered less likely. Provincialism was reduced still further in the Late Ordovician with the progressive narrowing of Iapetus and many Silurian forms are considered to be cosmopolitan.

In the context of a continually contracting Iapetus Ocean, it must be explained why early Silurian faunas seem to be cosmopolitan whereas later Silurian and Early Devonian faunas are provincial at a time when the ocean was presumably narrower. Perhaps the massive terminal Ordovician extinction event reduced diversity to the extent that recognition of provincialism in the Early Silurian is precluded by the sheer lack of diversity in many fossil groups. The provincialism demonstrated by ostracodes, brachiopods and vertebrates during the late Silurian supports the idea that the Fredericton Trough was the Silurian site of the axis of the remnant Iapetus Ocean. Alternatively, it is possible that the Silurian outcrop areas outboard of the Fredericton axis are suspect terranes or complexes of terranes derived from a different latitudinal or paleo-oceanographic belt. However, it is probable that geographic distance played a less important role in provincialism at this time. Boucot (1974, 1975) has suggested that provincialism developed in the Late Silurian because of regional reef development and the presence of hypersaline waters on the large carbonate platforms. He attributed the pronounced provincialism of the early Devonian to regression of seas, persistence of areas of hypersalinity and the presence of a strong climatic gradient. Whatever the cause, provincialism in late Silurian and Devonian organisms has demonstrated a progressive increase in similarity of faunas during the period until forms characteristic of the eastern margin of Iapetus invade the western areas by the Middle Devonian. After that time, faunas and floras of the region are similar to those of western Europe, confirming the idea that Iapetus had closed.

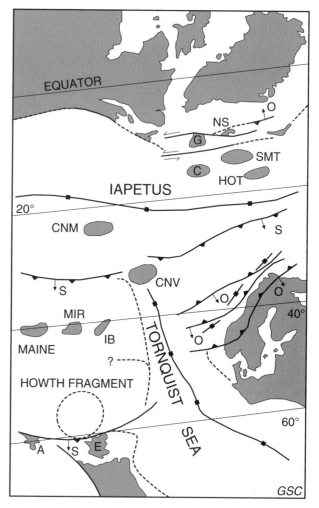

Figure 10.9. Paleogeographic map of the Early Ordovician Iapetus Ocean and bordering continents and continental fragments. Terranes destined for the northern Appalachians include parts of the Armorican Continental margin (lower left): Avalon terrane (A), northern Maine, Miramichi (MIR), Indian Bay (IB), and the disjunct Central Newfoundland Microcontinent (CNM=Mount Cormack Subzone); the Central Newfoundland Volcanics (CNV=Summerford Group and related rocks) lie astride a spreading ridge. Southern England (E) and the Howth fragment remain unrifted from Armorica/Gondwana represented by the outline of the west coast of France. Terranes destined for the Irish, British and Scandinavian Caledonides lie seaward of Laurentia (top): Northern Scotland (NS), Connemara (C), Grampia (G), South Mayo Trough (SMT), and the Hølonda Terrane (HOT). Details off the Baltic shore (lower right) omitted. Spreading ridges shown by lines with blocks; polarity of obduction (O) and subduction (S) shown with arrows. Modified from Neuman and Max (1989, fig. 4f).

Fossils are one of the main contributors to the zonal subdivisions used as the framework for this volume. In more recent tectonic interpretations of the Canadian Appalachians the concept of suspect terranes has been applied. A suspect terrane is an internally homogeneous belt of strata characterized by similar stratigraphy, structure, tectonic history, mineral deposits, paleomagnetic signatures and, above all, faunas. The faunas not only serve to distinguish the terrane but they are the keys in assessing its origin from the points of view of age, environment of deposition, and possible paleogeography.

Many elements of paleogeography and tectonostratigraphy in the Canadian Appalachians remain enigmatic. The finding of a single new fossil locality in an orogenic setting can commonly provide the impetus for a breakthrough in tectonic modelling. The new fossils can provide information not only on the age of the strata, but also on past climates, water depths, environments and paleogeography and, in some cases, thermal history. Future applications of fossils will involve geochemical evaluation of fossil materials as an aid to understanding paleo-oceanography (e.g. Brand and Morrison, 1987). The value of paleontological contributions to paleogeographic and tectonic reconstructions of the Appalachian orogen is immense. Many exciting new discoveries remain to be made.

ACKNOWLEDGMENTS

The preparation of this chapter has benefitted greatly from discussions with many colleagues including T.E. Bolton, M.J. Copeland, W.H. Fritz, D.C. McGregor, A.E.H. Pedder and J. Utting (Geological Survey of Canada), R.J. Elias (University of Manitoba), H.J. Hofmann and P.J. Lespérance (University of Montréal), and F.M. Hueber (Smithsonian Institution). M.M. Anderson (Memorial University of Newfoundland) graciously provided the photographs of Ediacaran material in Figure 10.2; W.H. Fritz arranged for the photograph of the trilobite specimen in Figure 10.3; and M.J. Copeland kindly supplied the photographs of ostracodes for Figure 10.7.

REFERENCES

Achab, A.
1988: Mise en évidence d'un provincialisme chez les chitinozoaires ordoviciens; Canadian Journal of Earth Sciences, v. 25, p. 635-638.

Anderson, M.M.
1978: Ediacaran fauna; in Yearbook of Science and Technology, D.N. Lapedes (ed.); McGraw-Hill, New York, p. 146-149.
1987: Stratigraphy of Cambrian rocks at Bacon Cove, Duffs, and Manuels River, Conception Bay, Avalon Peninsula, eastern Newfoundland; in northeastern section of the Geological Society of America, (ed.) D.C. Roy; Geological Society of America Centennial Field Guide, v. 5, p. 467-472.

Anderson, M.M. and Conway Morris, S.
1982: A review, with descriptions of four unusual forms of the soft-bodied fauna of the Conception and St. John's groups (Late Precambrian), Avalon Peninsula, Newfoundland, in (ed.) B. Mamet and M.J. Copeland; Third North American Paleontological Convention, Proceedings, v. 1, p. 1-12: Business and Economic Service Ltd., Toronto.

Andrews, H.N., Kasper, A.E., Forbes, W.H., Gensel, P.G., and Chaloner, W.G.
1977: Early Devonian flora of the Trout Valley Formation of northern Maine; Review of Palaeobotany and Palynology, v. 23, p. 255-285.

Barnes, C.R. and Fåhraeus, L.E.
1975: Provinces, communities and the proposed nektobenthic habit of Ordovician conodontophorids; Lethaia, v. 8, p. 133-149.

Barnes, C.R. and Tuke, M.F.
1970: Conodonts from the St.George Formation (Ordovician), northern Newfoundland; Geological Survey of Canada, Bulletin 187, p. 79-97.

Bell, W.A.
1929: Horton-Windsor district, Nova Scotia; Geological Survey of Canada, Memoir 155, 268 p.
1944: Carboniferous rocks and fossil floras of northern Nova Scotia; Geological Survey of Canada, Memoir 328, 276 p.
1948: Early Carboniferous strata of St.Georges Bay area, Newfoundland; Geological Survey of Canada, Bulletin 10, 45 p.
1959: Mississippian Horton Group of type Windsor-Horton district, Nova Scotia; Geological Survey of Canada, Memoir 314, 112 p.

Bengston, S. and Fletcher, T.P.
1983: The oldest sequence of skeletal fossils in the Lower Cambrian of southeastern Newfoundland; Canadian Journal of Earth Sciences, v. 20, p. 525-536.

Berdan, J.M.
1983: Biostratigraphy of Upper Silurian and Lower Devonian ostracodes in the United States; in Applications of Ostracoda-Proceedings of the 8th International Symposium on Ostracoda, July 26-29, Houston, Texas, (ed.) R.F.Maddox; University of Houston, Department of Geosciences, p. 313-337.

Bergström, S.M.
1979: Whiterockian (Ordovician) conodonts from the Hølonda Limestone of the Trondheim region, Norwegian Caledonides; Norsk Geologisk Tidsskrift, v. 59,p. 295-307.

Bergström, S.M., Epstein, A.G., and Epstein, J.B.
1972: Early Ordovician North Atlantic province conodonts in eastern Pennsylvania; United States Geological Survey, Professional Paper 800-D, p. D37-D44.

Bergström, S.M., Riva, J., and Kay, M.
1974: Significance of conodonts, graptolites, and shelly faunas from the Ordovician of western and north-central Newfoundland; Canadian Journal of Earth Sciences, v. 11, p. 1625-1660.

Berry, W.B.N. and Boucot, A.J. (ed.)
1970: Correlation of the North American Silurian rocks; Geological Society of America, Special Paper 102, 289 p.

Blackwood, R.F.
1982: Geology of the Gander Lake (2D/15) and Gander River (2E/2) area; Newfoundland and Labrador Department of Mines and Energy, Mineral Development Division, Report 82-4, 56 p.

Bolton, T.E.
1965: Ordovician and Silurian tabulate corals Labyrinthites, Arcturia, Troedssonites, Multisolenia, and Boreaster; Geological Survey of Canada, Bulletin 134, p. 15-39.
1980: Colonial coral assemblages and associated fossils from the Late Ordovician Honorat Group and White Head Formation, Gaspé Peninsula, Quebec; in Current Research, Part C; Geological Survey of Canada, Paper 80-1C, p. 13-28.

Boucot, A.J.
1960: Implications of Rhenish Lower Devonian brachiopods from Nova Scotia; 21st International Geological Congress, Norden, Part 12, Regional Palaeogeography, p. 129-137.
1973: The Lower Ordovician brachiopod Syntrophia cf. S. arethusa (Billings, 1862) from South Catcher Pond, northeastern Newfoundland; Canadian Journal of Earth Sciences, v. 10, p. 427-430.
1974: Silurian and Devonian biogeography; in Paleogeographic Provinces and Provinciality, (ed.) C.A. Ross; Society of Economic Paleontologists and Mineralogists, Special Publication 21, Tulsa, Oklahoma, p. 165-176.
1975: Evolution and extinction rate controls; Developments in Palaeontology and Stratigraphy No. 1, Elsevier, 427 p.

Boucot, A.J. and Drapeau, G.
1968: Siluro-Devonian rocks of Lake Memphremagog, and their correlatives in the Eastern Townships; Quebec Department of Natural Resources, Special Paper 1.

Boucot, A.J. and Johnson, J.G.
1967: Paleogeography and correlation of the Appalachian Province Lower Devonian sedimentary rocks; Tulsa Geological Society Digest, v. 35, p. 35-87.

Boucot, A.J., Brett, C.E., Oliver, W.A., Jr., and Blodgett, R.B.
1986: Devonian faunas of the Sainte-Hélène Island breccia, Montreal, Quebec; Canadian Journal of Earth Sciences, v. 23, p. 2047-2056.

Boucot, A.J., Johnson, J.G. and Talent, J.A.
1969: Early Devonian brachiopod zoogeography; Geological Society of America, Special Paper 119, 106 p.

Boucot, A.J., Johnson, J.G., Harper, C.W. and Walmsley, V.G.
1966:   Silurian brachiopods and gastropods of southern New Brunswick; Geological Survey of Canada, Bulletin 140, 45 p.

Bourque, P.-A., Amyot, G., Desrochers, A., Gignac, H., Gosselin, C., Lachambre, G., and Laliberté, J.-Y.
1986:   Silurian and Lower Devonian reef and carbonate complexes of the Gaspé Basin, Quebec: a summary; Bulletin of Canadian Petroleum Geology, v. 34, p. 452-489.

Bouyx, E. and Goujet, D.
1985:   Découverte de vertébrés dans le Silurien supérieur de la zone de Meguma (Nouvelle-Écosse, Canada): implications paléogéographiques; Comptes Rendus de l'Académie des Sciences, v. 301, series II, No. 9, p. 711-714.

Boyce, W.D.
1987:   Cambro-Ordovician trilobite biostratigraphy in central Newfoundland; Newfoundland Department of Mines and Energy, Mineral Development Division, Report 87-1, p. 335-341.

Boyce, W.D., Ash, J.S., and Knight, I.
1988:   Biostratigraphic studies of Ordovician carbonate rocks in western Newfoundland; Newfoundland Department of Mines and Energy, Mineral Development Division, Report 88-1, p. 75-83.

Brand, U. and Morrison, J.
1987:   Biogeochemistry of fossil marine invertebrates; Geoscience Canada, v. 14, p. 85-107.

Brasier, M.D.
1984:   Microfossils and small shelly fossils from the Lower Cambrian *Hyolithes* Limestone at Nuneaton, English Midlands; Geological Magazine, v. 121, p. 229-253.

Brenchley, P. J.
1988:   Environmental changes close to the Ordovician Silurian boundary; in A global analysis of the Ordovician-Silurian boundary, (ed.) L.R.M. Cocks and R.B. Rickards; British Museum (Natural History), Geology Series, v. 43, p. 377-385.

Bruton, D.L. and Bockelie, J.F.
1980:   Geology and paleontology of the Hølonda area, western Norway – A fragment of North America?; in The Caledonides in the U.S.A., (ed.) D.R. Wones; Virginia Polytechnic Institute and State University, Department of Geological Sciences, Memoir 2, p. 41-47.

Chandler, F.W., Sullivan, R.W., and Currie, K.L.
1987:   The age of the Springdale Group, western Newfoundland, and correlative rocks: evidence for a Llandovery overlap assemblage in the Canadian Appalachians; Royal Society of Edinburgh Transactions, Earth Sciences, v. 78, p. 41-49.

Cobbold, E.S.
1921:   The Cambrian horizons of Comley (Shropshire) and their Brachiopoda, Pteropoda and Gasteropoda, etc.; Quarterly Journal of the Geological Society of London, v. 76, p. 325-386.

Cocks, L.R.M. and Copper, P.
1981:   The Ordovician-Silurian boundary at the eastern end of Anticosti Island, Quebec; Canadian Journal of Earth Sciences, v. 18, p. 1029-1034.

Cocks, L.R.M. and Fortey, R.A.
1982:   Faunal evidence for oceanic separation in the Palaeozoic of Britain; Journal of the Geological Society of London, v. 139, p. 465-478.

Colman-Sadd, S.P. and Swinden, H.S.
1984:   A tectonic window in central Newfoundland? Evidence that the Appalachian Dunnage zone may be allochthonous; Canadian Journal of Earth Sciences, v. 21, p. 1349-1367.

Conway Morris, S. and Rushton, A.W.A.
1988:   Precambrian to Tremadoc biotas in the Caledonides; in The Caledonian-Appalachian orogen, (ed.) A.L.Harris and D.J. Fettes; Geological Society, Special Publication 38, p. 93-109.

Cooper, G.A.
1956:   Chazyan and related brachiopods; Smithsonian Miscellaneous Collections, v. 127, pt. 1, 1024 p.

Copeland, M.J.
1957:   The arthropod fauna of the Upper Carboniferous rocks of the Maritime Provinces; Geological Survey of Canada, Memoir 286, 110 p.

Copeland, M.J. and Berdan, J.M.
1977:   Silurian and Early Devonian beyrichiacean ostracode provincialism in northeastern North America; in Report of Activities, Part B; Geological Survey of Canada, Paper 77-1B, p. 15-24.

Copeland, M. J. and Bolton, T. E.
1977:   Additional paleontological observations bearing on the age of the Lourdes Formation (Ordovician), Port au Port Peninsula, western Newfoundland; in Report of Activities, Part B; Geological Survey of Canada, Paper 77-1B, p. 1-5.

Crimes, T.P.
1987:   Trace fossils and correlation of late Precambrian and early Cambrian strata; Geological Magazine, v. 124, p. 97-119.

Crimes, T.P. and Anderson, M.M.
1985:   Trace fossils from Late Precambrian-Early Cambrian strata of southeastern Newfoundland (Canada): Temporal and environmental implications; Journal of Paleontology, v. 59, p. 310-343.

Crosby, D.G.
1962:   Wolfville map-area, Nova Scotia (21 H/1); Geological Survey of Canada, Memoir 325, 67 p.

Currie, K.L.
1987:   The Avalonian terrane around Saint John, New Brunswick, and its deformed Carboniferous cover; in northeastern section of the Geological Society of America, (ed.) D.C. Roy; Geological Society of America Centennial Field Guide, v. 5, p. 403-408.

David, J., Chabot, N., Marcotte, C., Lajoie, J., and Lespérance, P.J.
1985:   Stratigraphy and sedimentology of the Cabano, Pointe aux Trembles, and Lac Raymond formations, Témiscouata and Rimouski counties, Quebec; in Current Research, Part B; Geological Survey of Canada, Paper 85-1B, p. 491-497.

Dean, W.T.
1970:   Lower Ordovician trilobites from the vicinity of South Catcher Pond, northeastern Newfoundland; Geological Survey of Canada, Paper 70-44, 15 p.
1971:   Ordovician trilobites from the central volcanic mobile belt at New World Island, northeastern Newfoundland; Geological Survey of Canada, Bulletin 210, 37 p.
1974:   Lower Ordovician trilobites from the Summerford Group at Virgin Arm, New World Island, northeastern Newfoundland; Geological Survey of Canada, Bulletin 240, 43 p.
1976:   Some aspects of Ordovician correlation and trilobite distribution in the Canadian Appalachians; in The Ordovician System, (ed.) M.G. Bassett; Cardiff, University of Wales Press and National Museum of Wales, p. 227-250.
1979:   Trilobites from the Long Point Group (Ordovician), Port au Port Peninsula, south-western Newfoundland; Geological Survey of Canada, Bulletin 290, p. 1-53.
1985:   Relationships of Cambrian-Ordovician faunas in the Caledonide-Appalachian region, with particular reference to trilobites; in The Tectonic Evolution of the Caledonide-Appalachian Orogen, (ed.) R.A. Gayer; Braunschweig/Wiesbaden, Friedrik Vieweg & Sohn, p. 17-47.

Dean, W.T. and Martin, F.
1978:   Lower Ordovician acritarchs and trilobites from Bell Island, eastern Newfoundland; Geological Survey of Canada, Bulletin 284, p. 1-35.

Debrenne, F. and James, N.P.
1981:   Reef-associated archaeocyathans from the Lower Cambrian of Labrador and Newfoundland; Palaeontology, v. 24, p. 343-378.

De Broucker, G.
1984:   Stratigraphie et structure des groupes de Mictaw et de Maquereau, région de Port-Daniel (Gaspésie); Ministère de l'Énergie et des Ressources, Québec; Direction Générale de l'Exploration Géologique at Minérale, Séminaire d'Information, DV 84-18, p. 3-16.

Dewey, C.P.
1983:   Ostracode paleoecology of the lower Carboniferous of western Newfoundland; in Applications of Ostracoda-Proceedings of the 8th International Symposium on Ostracoda, July 26-29, 1982, Houston, Texas, (ed.) R.F. Maddocks; University of Houston, Department of Geosciences, p. 104-111.

Dineley, D.L. and Loeffler, E.J.
1979:   Early vertebrates and the Caledonian earth movements, in The Caledonides of the British Isles Reviewed; Geological Society of London, Special Publication 8, p. 411-414.

Dorf, E. and Cooper, J.R.
1943:   Early Devonian plants from Newfoundland; Journal of Paleontology, v. 17, p. 264-270.

Elias, R.J.
1982:   Latest Ordovician solitary rugose coral; Bulletins of American Paleontology, v. 81, 116 p.

Elliott, C. G., Barnes, C. R. and Williams, P. F.
1989:   Southwest New World Island stratigraphy: new fossil data, new implications for the history of the Central Mobile Belt, Newfoundland; Canadian Journal of Earth Sciences, v. 26, p. 2062-2074.

Fåhraeus, L.E.
1973: Depositional environments and conodont-based correlation of the Long Point Formation (Middle Ordovician), western Newfoundland; Canadian Journal of Earth Sciences, v. 10, p. 1822-1833.
1974: Lower Paleozoic stratigraphy of the Port au Port area, west Newfoundland; Geological Association of Canada-Mineralogical Association of Canada Annual Meeting, St. John's, Newfoundland, Field Trip Guidebook B-4, 16 p.

Fåhraeus, L.E. and Hunter, D.R.
1981: Paleoecology of selected conodontophorid species from the Cobbs Arm Formation (Middle Ordovician), New World Island, north-central Newfoundland; Canadian Journal of Earth Sciences, v. 18, p. 1653-1665.

Flower, R.H.
1978: St. George and Table Head cephalopod zonation in western Newfoundland; in Current Research, Part A; Geological Survey of Canada, Paper 78-1A, p. 217-224.

Fortey, R.A. and Cocks, L.R.M.
1986: Fossils and tectonics: Journal of the Geological Society, v. 143, p. 149-150.

Fritz, W.H.
1972: Lower Cambrian trilobites from the Sekwi Formation type section, Mackenzie Mountains, northwestern Canada; Geological Survey of Canada, Bulletin 212, p. 1-90.

Fritz, W.H., Kindle, C.H., and Lespérance, P.J.
1970: Trilobites and stratigraphy of the Middle Cambrian Corner-of-the-Beach Formation, eastern Gaspé Peninsula, Quebec; Geological Survey of Canada, Bulletin 187, p. 43-58.

Furnes, H., Ryan, P.D., Grenne, T., Roberts, D., Sturt, B.A., and Prestvik, T.
1985: Geological and geochemical classification of the ophiolitic fragments in the Scandinavian Caledonides; in The Caledonide Orogen-Scandinavia and related areas, (ed.) D.G. Gee and B.A. Sturt; Chichester, John Wiley, v. 2, p. 657-669.

Fyffe, L.R.
1987: Stratigraphy and tectonics of Miramichi and Elmtree terranes in the Bathurst area, northeastern New Brunswick; in northeastern section of the Geological Society of America, (ed.) D.C. Roy; Geological Society of America Centennial Field Guide, v. 5, p. 389-393.

Fyffe, L.R. and Riva, J.
1990: Revised stratigraphy of the Cookson Group of southwestern New Brunswick in and adjacent Maine; Atlantic Geology, v. 26, p. 271-275.

Fyffe, L.R., Forbes, W.H., and Riva, J.
1983: Graptolites from the Benton area of west-central New Brunswick and their regional significance; Maritime Sediments and Atlantic Geology, v. 19, p. 117-125.

Gensel, P.G. and Andrews, H.N.
1984: Plant life in the Devonian; Praeger Publishers, New York, 380 p.

Goldring, R. and Langenstrassen, F.
1979: Open shelf and near-shore clastic facies in the Devonian, in The Devonian System, (ed.) M.R. House, C.T. Scrutton, and M.G. Bassett; Palaeontological Association, Special Papers in Palaeontology, v. 23, p. 81-97.

Grabau, A.W.
1936: Paleozoic formations in the light of the pulsation theory, Volume II, Part 1, Caledonian and St. Lawrence geosynclines; Peking University Press, National University of Peking, 751 p.

Hacquebard, P.A., and Donaldson, J.R.
1963: Stratigraphy and palynology of the Upper Carboniferous coal measures in the Cumberland Basin of Nova Scotia; Cinquième Congres International Stratigraphique Géologie du Carbonifère, Paris, p. 1157-1169.

Halstead, L.B. and Turner, S.
1973: Silurian and Devonian ostracoderms; in Atlas of Palaeobiogeography, (ed.) A. Hallam; Elsevier, Amsterdam, p. 67-79.

Hambrey, M.J.
1985: The Late Ordovician-Early Silurian glacial period; Palaeogeography, Palaeoclimatology, and Palaeoecology, v. 51, p. 273-289.

Harper, D.A.T. and Parkes, M.A.
1989: Palaeontological constraints on the definition and development of Irish Caledonide terranes; Journal of the Geological Society of London, v. 146, p. 413-415.

Harper, D.A.T., Parkes, M.A., Hoey, A.N., and Murphy, F.C.
1990: Intra-Iapetus brachiopods from the Ordovician of eastern Ireland: Implications for Caledonide correlation; Canadian Journal of Earth Sciences, v. 27, p. 1757-1761.

Hayes, A.O. and Howell, B.F.
1937: Geology of Saint John, New Brunswick; Geological Society of America, Special Paper 5, 146 p.

Hibbard, J.P., Stouge, S., and Skevington, D.
1977: Fossils from the Dunnage Melange, north-central Newfoundland; Canadian Journal of Earth Sciences, v.14, p. 1176-1178.

Hofmann, H.J.
1974: The stromatolite Archaeozoon acadiense from the Proterozoic Green Head Group of Saint John, New Brunswick; Canadian Journal of Earth Sciences, v. 11, p. 1098-1115.

Hofmann, H.J. and Cecile, M.P.
1981: Occurrences of Oldhamia and other trace fossils in Lower Cambrian (?) argillites, Niddery Lake Map Area, Selwyn Mountains, Yukon Territory; in Current Research, Part A; Geological Survey of Canada, Paper 81-1A, p. 281-290.

Hofmann, H.J., Hill, J., and King, A.F.
1979: Late Precambrian microfossils, southeastern Newfoundland; in Current Research, Part B; Geological Survey of Canada, Paper 79-1B, p. 83-92.

Horne, G.S.
1970: Complex volcanic-sedimentary patterns in the Magog belt of northeastern Newfoundland; Geological Society of America Bulletin, v. 81, p. 1767-1788.

Horne, G.S. and Johnson, J.H.
1970: Ordovician algae from boulders in Silurian deposits of New World Island, Newfoundland; Journal of Paleontology, v. 44, p. 1055-1059.

Howell, B.F.
1925: The faunas of the Cambrian Paradoxides beds at Manuels, Newfoundland; Bulletins of American Paleontology, v. 11, no. 43, 140 p.

Hueber, F.M., Bothner, W.A., Hatch, N.L., Finney, S.C., and Aleinikoff, J.N.
1990: Devonian plants in southern Quebec and northern New Hampshire, and the age of the Connecticut Valley trough; American Journal of Science, v. 290, p. 360-395.

Hutchinson, R.D.
1952: The stratigraphy and trilobite faunas of the Cambrian sedimentary rocks of Cape Breton Island, Nova Scotia; Geological Survey of Canada, Memoir 263, 124 p.
1962: Cambrian stratigraphy and trilobite faunas of southeastern Newfoundland; Geological Survey of Canada, Bulletin 88, 156 p.

James, N.P. and Kobluk, D.R.
1978: Lower Cambrian patch reefs and associated sediments, southern Labrador; Sedimentology, v. 25, p. 1-32.

James, N.P. and Stevens, R.K.
1986: Stratigraphy and correlation of the Cambro-Ordovician Cow Head Group, western Newfoundland; Geological Survey of Canada, Bulletin 366, 143 p.

Johnson, J.G. and Boucot, A.J.
1973: Devonian brachiopods; in Atlas of Palaeobiogeography, (ed.) A. Hallam; Elsevier, Amsterdam, p. 89-96.

Jones, D.L., Silberling, N.J., and Hillhouse, J.
1977: Wrangellia: a displaced terrane in northwestern North America; Canadian Journal of Earth Sciences, v. 14, p. 2565-2577.

Kay, M.
1951: North American geosynclines; Geological Society of America, Memoir 48, 143 p.
1967: Stratigraphy and structure of northeastern Newfoundland bearing on drift in north Atlantic; American Association of Petroleum Geologists, Bulletin, v. 51, p. 579-600.

Kay, M. and Eldredge, N.
1968: Cambrian trilobites in central Newfoundland volcanic belt; Geological Magazine, v. 105, p. 372-377.

Kean, B.F. and Jayasinghe, N.R.
1980: Badger map area (12A/16), Newfoundland; Newfoundland Department of mines and Energy, Mineral Development Division, Report 80-1, p. 37-43.

King, A.F.
1982: Guidebook for Avalon and Meguma zones; Memorial University of Newfoundland, Department of Earth Sciences, Report 9, 308 p.

Klappa, C.F., Opalinski, P.R., and James, N.P.
1980: Middle Ordovician Table Head Group of western Newfoundland: a revised stratigraphy; Canadian Journal of Earth Sciences, v. 17, p. 1007-1019.

Lajoie, J., Lespérance, P.J., and Béland, J.
1968: Silurian stratigraphy and paleogeography of Matapédia-Témiscouata region, Quebec; American Association of Petroleum Geologists, Bulletin, v. 52, p. 615-640.

Landing, E.

1983:   Highgate Gorge: Upper Cambrian and Lower Ordovician continental slope deposition and biostratigraphy, northwestern Vermont; Journal of Paleontology, v. 576, p. 1149-1187.

1988:   Lower Cambrian of eastern Massachusetts: stratigraphy and small shelly fossils; Journal of Paleontology, v. 62, p. 661-695.

Landing, E., Nowlan, G.S., and Fletcher, T.P.

1980:   A microfauna associated with Early Cambrian trilobites of the *Callavia* Zone, northern Antigonish Highlands, Nova Scotia; Canadian Journal of Earth Sciences, v. 17,p. 400-418.

Langston, W., Jr.

1963:   Fossil vertebrates and the Late Palaeozoic red beds of Prince Edward Island; National Museum of Canada, Bulletin 187.

Lawrence, D.A. and Rust, B.R.

1988:   The Devonian clastic wedge of eastern Gaspé and the Acadian orogeny; in Devonian of the World, v. 1, (ed.) N.J. McMillan, A.F. Embry, and D.J. Glass, Canadian Society of Petroleum Geologists, Memoir 14, p. 53-64.

Lee, H.J. and Noble, J.P.A.

1977:   Silurian stratigraphy and depositional environments: Charlo-Upsalquitch Forks area, New Brunswick; Canadian Journal of Earth Sciences, v. 14, p. 2533-2542.

Lespérance, P.J., (ed.)

1981:   IUGS Subcommission on Silurian Stratigraphy, Ordovician-Silurian Boundary Working Group, Field Meeting, Anticosti-Gaspé, Quebec; in Volume II: Stratigraphy and Paleontology; Department of Geology, University of Montreal, 321 p.

1988:   Percé, Quebec; in A global analysis of the Ordovician-Silurian boundary, (ed.) L.R.M. Cocks, and R.B. Rickards; British Museum (Natural History), Geology Series, v. 43, p. 239-245.

Lespérance, P.J. and Greiner, H.R.

1969:   Squatec-Cabano area, Rimouski, Rivière-du-Loup and Témiscouata counties; Quebec Department of Natural Resources, Mines Branch, Geological Report 128, 111 p.

Lespérance, P.J., Malo, M., Sheehan, P.M., and Skidmore, W.B.

1987:   A stratigraphical and faunal revision of the Ordovician-Silurian strata of the Percé area, Quebec; Canadian Journal of Earth Sciences, v. 24, p. 117-134.

Lespérance, P.J. and Sheehan, P.M.

1976:   Brachiopods from the Hirnantian Stage (Ordovician-Silurian) at Percé, Quebec; Palaeontology, v. 19, p. 719-731.

1988:   Faunal assemblages of the upper Gaspé limestones, Early Devonian of eastern Gaspé, Quebec; Canadian Journal of Earth Sciences, v. 25, p. 1432-1439.

Lindholm, R.M. and Casey, J.F.

1989:   Regional significance of the Blow Me Down Brook Formation, western Newfoundland: new fossil evidence for an Early Cambrian age; Geological Society of America Bulletin, v. 101, p. 1-13.

1990:   The distribution and possible biostratigraphic significance of the ichnogenus *Oldhamia* in the shales of the Blow Me Down Brook Formation, western Newfoundland; Canadian Journal of Earth Sciences, v. 27, p. 1270-1287.

Lochman-Balk, C. and Wilson, J.L.

1958:   Cambrian biostratigraphy in North America; Journal of Paleontology, v. 32, p. 313-350.

Lock, B.E.

1969:   Silurian rocks of West White Bay area, Newfoundland; in North Atlantic-geology and continental drift, a symposium, (ed.) M. Kay ; American Association of Petroleum Geologists, Memoir 12, p. 433-442.

Matthew, G.F.

1895:   The *Protolenus* fauna; New York Academy of Sciences, Transactions, v. 14, p. 101-153.

1903:   Report on the Cambrian rocks at Cape Breton; Geological Survey of Canada, Publication No. 797.

McCracken, A.D. and Barnes, C.R.

1981:   Conodont biostratigraphy and paleoecology of the Ellis Bay Formation, Anticosti Island, Quebec; with special reference to Late Ordovician-Early Silurian chronostratigraphy and the systemic boundary; Geological Survey of Canada, Bulletin 329, p. 51-134.

McCracken, A.D. and Nowlan, G.S.

1988:   The Gamachian Stage and Fauna 13; in Proceedings of the Canadian Paleontology and Biostratigraphy Seminar, Albany, New York, September 26-29, 1986; New York State Museum, Bulletin 462, p. 71-79.

McCutcheon, S.R.

1987:   Cambrian stratigraphy in the Hanford Brook area, southern New Brunswick; in northeastern section of the Geological Society of America, (ed.) D.C. Roy; Geological Society of America Centennial Field Guide, v. 5, p. 399-402.

McGregor, D.C. and McCutcheon, S.R.

1988:   Implications of spore evidence for Late Devonian age of the Piskahegan Group, southwestern New Brunswick; Canadian Journal of Earth Sciences, v. 25, p. 1349-1364.

McKerrow, W.S. and Cocks, L.R.M.

1976:   Progressive faunal migration across the Iapetus Ocean; Nature, v. 263, p. 304-305.

1977:   The location of the Iapetus suture in Newfoundland; Canadian Journal of Earth Sciences, v. 14, p. 488-495.

1978:   A lower Paleozoic trench-fill sequence, New World Island, Newfoundland; Geological Society of America Bulletin, v. 89, p. 1121-1132.

1981:   Stratigraphy of eastern Bay of Exploits, Newfoundland; Canadian Journal of Earth Sciences, v. 18, p. 751-764.

1986:   Oceans, island arcs and olistostromes: the use of fossils in distinguishing sutures, terranes and environments around the Iapetus Ocean; Journal of the Geological Society, v. 143, p. 185-191.

McKerrow, W.S. and Ziegler, A.M.

1971:   The Lower Silurian paleogeography of New Brunswick and adjacent areas; Journal of Geology, v. 71, p. 635-646.

Misra, S.B.

1969:   Late Precambrian(?) fossils from southeastern Newfoundland; Geological Society of America Bulletin, v. 80, p. 2133-2140.

Murphy, F.C.

1987:   Evidence for Late Ordovician amalgamation of volcanogenic terranes in the Iapetus suture zone, eastern Ireland; Royal Society of Edinburgh, Earth Sciences, Transactions, v. 78, p. 153-167.

Narbonne, G.M. and Myrow, P.

1988:   Trace fossil biostratigraphy in the Precambrian-Cambrian boundary interval; New York State Museum, Bulletin 463, p. 72-76.

Neuman, R.B.

1976:   Early Ordovician (early Arenig) brachiopods from Virgin Arm, New World Island, Newfoundland; Geological Survey of Canada, Bulletin 261, p. 11-61.

1984:   Geology and paleobiology of islands in the Ordovician Iapetus Ocean: Review and implications; Geological Society of America Bulletin, v. 95, p. 1188-1201.

Neuman, R.B. and Bruton, D.L.

1989:   Brachioods and trilobites from the Ordovician Lower Hovin Group (Arenig/Llanvirn), Hølonda area, Trondheim region, Norway; new and revised taxa, and paleogeographic interpretation; Norges Geologiske Undersøkelse Bulletin 414, p. 49-89.

Neuman, R.B. and Max, M.D.

1989:   Penobscottian-Grampian-Finnmarkian orogenies as indicators of terrane linkages; in Terranes in the Circum-Atlantic Paleozoic Orogens, (ed.) R.D. Dallmeyer; Geological Society of America, Special Paper 230, p. 31-45.

Neuman, R.B., Palmer, A.R., and Dutro, J.T., Jr.

1989:   Paleontological contributions to Paleozoic paleogeographic reconstructions of the Appalachians; in The Appalachian-Ouachita Orogen in the United States, (ed.) R.D. Hatcher, Jr., W.A. Thomas, and G.W. Viele; Geological Society of America, The Geology of North America, v. F-2, p. 375-384.

North, F.K.

1971:   The Cambrian of Canada and Alaska; in Cambrian of the New World, (ed.) C.H. Holland; London, Wiley-Interscience, p. 219-324.

Nowlan, G.S.

1981a: Some Ordovician conodont faunules from the Miramichi Anticlinorium, New Brunswick; Geological Survey of Canada, Bulletin 345, 34 p.

1981b: Late Ordovician-Early Silurian conodont biostratigraphy of the Gaspé Peninsula-a preliminary report; in Subcommission on Silurian Stratigraphy, Ordovician-Silurian Boundary Working Group, Field Meeting, Anticosti-Gaspé, Quebec, Volume 2, (ed.) P.J. Lespérance; Stratigraphy and Paleontology; Université de Montréal, Département de géologie, p. 257-291.

1983a: Biostratigraphic, paleogeographic, and tectonic implications of Late Ordovician conodonts from the Grog Brook Group, northwestern New Brunswick; Canadian Journal of Earth Sciences, v. 20, p. 651-671.

1983b: Early Silurian conodonts of eastern Canada; Fossils and Strata, No. 15, p. 95-110.

Nowlan, G.S. and Barnes, C.R.

1981: Late Ordovician conodonts from the Vauréal Formation, Anticosti Island, Quebec; Geological Survey of Canada, Bulletin 329, p. 1-49.

1987: Thermal maturation of Paleozoic strata in eastern Canada from conodont colour alteration index (CAI) data with implications for burial history, tectonic evolution, hotspot tracks and mineral and hydrocarbon exploration; Geological Survey of Canada, Bulletin 367, 47 p.

Nowlan, G.S., Goodfellow, W.D., McCracken, A.D., and Lenz, A.C.

1988: Geochemical evidence for sudden biomass reduction near the Ordovician-Silurian boundary, in northwestern Canada; Fifth International Symposium on the Ordovician System, Abstracts, p. 66.

Nowlan, G.S. and Thurlow, J.G.

1984: Middle Ordovician conodonts from the Buchans Group, central Newfoundland, and their significance for regional stratigraphy of the Central Volcanic Belt; Canadian Journal of Earth Sciences, v. 21, p. 284-296.

O'Brien, S.J. and King, A.F.

1982: The Avalon zone in Newfoundland, in Field guide for Avalon and Meguma zones, (ed.) A.F. King; Memorial University of Newfoundland, Report 9, p. 1-64.

O'Brien, S.J., Wardle, R.J., and King, A.F.

1983: The Avalon zone: a pan-African terrane in the Appalachian orogen of Canada; Geological Journal, v. 18, p. 195-222.

Oliver, W.A., Jr.

1971: The coral fauna and age of the Famine Limestone in Quebec; in Paleozoic perspectives: a paleontological tribute to (ed.) G. Arthur Cooper, J.T. Dutro; Smithsonian Contributions to Paleo-biology, No. 3, p. 193-202.

Oliver, W.A., Jr. and Pedder, A.E.H.

1976: Biogeography of Late Silurian and Devonian rugose corals in North America; in Historical biogeography and the changing environment, (ed.) J. Gray and A.J. Boucot; Corvallis, Oregon State University Press, p. 131-145.

O'Neill, P.

1987: Geology of the west half of the Weir's Pond (2E/1) map area; Newfoundland Department of Mines and Energy, Mineral Development Division, Report 87-1, p. 271-281.

O'Neill, P. and Knight, I.

1988: Geology of the east half of the Weir's Pond (2E/1) map area and its regional significance; Newfoundland Department of Mines and Energy, Mineral Development Division, Report 88-1, p. 165-176.

Palmer, A.R.

1969: Cambrian trilobite distributions in North America and their bearing on Cambrian paleogeography of Newfoundland; American Association of Petroleum Geologists, Memoir 12, p. 139-144.

1971: The Cambrian of the Appalachian and eastern New England regions, eastern United States; in Cambrian of the New World, (ed.) C.H. Holland; John Wiley and Sons, p. 169-217.

Petryk, A.A.

1981: Stratigraphy, sedimentology, and paleogeography of the Upper Ordovician-Lower Silurian of Anticosti Island, Quebec; in (ed.) P.J. Lespérance; Field meeting, Anticosti-Gaspé, Quebec; Département de géologie, Université de Montréal, v. 2, p. 11-39.

Pohler, S.L., Barnes, C.R., and James, N.P.

1987: Reconstructing a lost faunal realm: conodonts from mega-conglomerates of the Ordovician Cow Head Group, western Newfoundland; in Conodonts-Investigative Techniques and Applications, (ed.) R.L. Austin; Ellis Horwood, Chichester, p. 341-360.

Pratt, B.R. and Waldron, J.W.F.

1991: A Middle Cambrian trilobite faunule from the Meguma Group of Nova Scotia; Canadian Journal of Earth Sciences, v. 28, p. 1843-1853.

Rast, N. and Stringer, P.

1980: A geotraverse across a deformed Ordovician ophiolite and its Silurian cover, northern New Brunswick; Tectonophysics, v. 69, p. 221-245.

Riva, J. and Malo, M.

1988: Age and correlation of the Honorat Group, southern Gaspé Peninsula; Canadian Journal of Earth Sciences, v. 25, p. 1618-1628.

Rodgers, J.

1965: Long Point and Clam Bank formations, Western Newfoundland; Proceedings of the Geological Association of Canada, v. 16, p. 83-94.

Rodgers, J. and Neale, E.R.W.

1963: Possible "Taconic" klippen in western Newfoundland; American Journal of Science, v. 261, p. 713-730.

Ross, R.J., Jr. and Ingham, J.K.

1970: Distribution of the Toquima-Table Head (Middle Ordovician, Whiterock) Faunal Realm in the northern hemisphere; Geological Society of America Bulletin, v. 81, p. 393-408.

Ross, R.J., Jr. and James, N.P.

1987: Brachiopod biostratigraphy of the Middle Ordovician Cow Head and Table Head groups, western Newfoundland; Canadian Journal of Earth Sciences, v. 24, p. 70-95.

Roy, D.C.

1989: The Depot Mountain Formation: transition from syn- to post-Taconian basin along the Baie Verte-Brompton Line in northwestern Maine; in Studies in Maine Geology, Structure and Stratigraphy, v. 2, (ed.) R.D. Tucker and R.G. Marvinney; Maine Geological Survey, p. 85-99.

Rozanov, A. Yu.

1984: The Precambrian-Cambrian boundary in Siberia; Episodes, v. 7, p. 20-24.

St. Peter, C.

1982: Geology of Juniper-Knowlesville-Carlisle area, map areas I-16, I-17, I-18 (parts of 21J/11 and 21 J/06); New Brunswick Department of Natural Resources, Mineral Resources Division, Map Report 82-1, 82 p.

Schallreuter, R.E.L. and Siveter, D.J.

1985: Ostracodes across the Iapetus Ocean; Palaeontology, v. 28, p. 577-598.

Schenk, P.E.

1982: Stratigraphy and sedimentology of the Meguma Zone and part of the Avalon Zone; in Field guide for Avalon and Meguma zones, (ed.) A.F. King; Memorial University of Newfoundland, Department of Earth Sciences, Report 9, p. 189-247.

Schuchert, C.

1923: Sites and nature of the North American geosynclines; Geological Society of America Bulletin, v. 34, p. 151-230.

Schuchert, C. and Cooper, G.A.

1930: Upper Ordovician and Lower Devonian stratigraphy and paleontology of Percé, Quebec; American Journal of Science, series 5, v. 220, p. 161-176, 265-288, 365-392.

Sheehan, P.M.

1975: Brachiopod synecology in a time of crisis (Late Ordovician-Early Silurian); Paleobiology, v. 1, p. 205-212.

Skevington, D.

1974: Controls influencing the composition and distribution of Ordovician graptolite faunal provinces; in Graptolite studies in honour of (ed.) O.M.B. Bulman, R.B. Rickards, D.E. Jackson, and C.P. Hughes; Palaeontological Association, Special Papers in Palaeontology, 13, p. 59-73.

Skinner, R.

1974: Geology of Tetagouche Lakes, Bathurst, and Nepisiguit Falls map-areas, New Brunswick; Geological Survey of Canada, Memoir 371, 133 p.

Smith, A.G., Briden, J.C., and Drewry, G.E.

1973: Phanerozoic world maps; in Organisms and Continents Through Time, (ed.) N.F. Hughes; Palaeontological Association, Special Papers in Palaeontology, 12, p. 1-42.

Spjeldnaes, N.

1981: Lower Palaeozoic climatology; in Lower Palaeozoic of the Middle East, eastern and southern Africa, and Antarctica, (ed.) C.H. Holland; John Wiley and Sons, p. 199-256.

Stockmal, G.S., and Waldron, J.W.F.

1990: Structure of the Appalachian deformation front in western Newfoundland: implications of multichannel seismic reflection data; Geology, v. 18, p. 765-768.

Stouge, S.

1980a: Conodonts from the Davidsville Group, northeastern Newfoundland; Canadian Journal of Earth Sciences, v. 17, p. 268-272.

1980b: Lower and Middle Ordovician conodonts from central Newfoundland and their correlatives in western Newfoundland; Newfoundland Department of Mines and Energy, Mineral Development Division, Report 80-1, p. 134-142.

1982: Preliminary conodont biostratigraphy and correlation of Lower to Middle Ordovician carbonates of the St. George Group, Great Northern Peninsula, Newfoundland; Newfoundland Department of Mines and Energy, Mineral Development Division, Report 82-3, 59 p.

1984: Conodonts of the Middle Ordovician Table Head Formation, western Newfoundland; Fossils and Strata, No. 16, 145 p.

Strong, D.F., O'Brien, S.J., Taylor, S.W., Strong, P.G., and Wilton, D.H.

1978: Aborted Proterozoic rifting in eastern Newfoundland; Canadian Journal of Earth Sciences, v. 15, p. 117-131.

Turner, S. and Tarling, D.H.

1982: Thelodont and other agnathan distributions as tests of Lower Paleozoic continental reconstructions; Palaeogeography, Palaeoclimatology, Palaeoecology, v. 39, p. 295-311.

Ulrich, E.O. and Cooper, G.A.

1938: Ozarkian and Canadian Brachiopoda; Geological Society of America, Special Paper 13, 323 p.

Utting, J.

1987a: Palynology of the Lower Carboniferous Windsor Group and Windsor-Canso boundary beds of Nova Scotia, and their equivalents in Quebec, New Brunswick and Newfoundland; Geological Survey of Canada, Bulletin 374, 93 p.

1987b: Palynostratigraphic investigation of the Albert Formation (Lower Carboniferous) of New Brunswick; Palynology, v. 11, p. 73-96.

van der Pluijm, B.A., Karlstrom, K.E., and Williams, P.F.

1987: Fossil evidence for fault-derived stratigraphic repetition in the northeastern Newfoundland Appalachians; Canadian Journal of Earth Sciences, v. 24, p. 2337-2350.

van der Voo, R.

1988: Paleozoic paleogeography of North America, Gondwana, and intervening displaced terranes: comparisons of paleomagnetism with paleoclimatology and biogeographical patterns; Geological Society of America Bulletin, v. 100, p. 311-324.

Van der Zwan, C.J.

1981: Palynology, phytogeography and climate of the Lower Carboniferous; Review of Palaeobotany and Palynology, v. 33, p. 279-310.

Van Ingen, G.

1914: Cambrian and Ordovician faunas of southeastern Newfoundland; Geological Society of America Bulletin, v. 25, p. 138.

van Staal, C.R.

1987: Tectonic setting of the Tetagouche Group in northern New Brunswick: implications for plate tectonic models of the northern Appalachians; Canadian Journal of Earth Sciences, v. 24, p. 1329-1351.

von Bitter, P.H.

1976: Paleoecology and distribution of Windsor Group (Visean-?early Namurian) conodonts, Port Hood Island, Nova Scotia; in Conodont Paleoecology, (ed.) C.R. Barnes; Geological Association of Canada, Special Paper 15, p. 225-241.

von Bitter, P.H. and Austin, R.L.

1984: The Dinantian *Taphrognathus* transatlanticus conodont Range Zone of Great Britain and Atlantic Canada; Palaeontology, v. 27, p. 95-111.

Von Bitter, P.H. and Plint-Geberl, H.A.

1982: Conodont biostratigraphy of the Codroy Group (Lower Carboniferous), southwestern Newfoundland; Canadian Journal of Earth Sciences, v. 19, p. 193-221.

Walcott, C.D.

1891: Correlation papers-Cambrian; United States Geological Survey, Bulletin 81, 447 p.

Wegener, A.L.

1928: The origins of continents and oceans, 4th edition, revised 1966, English translation by John Biram; New York, Dover Publications, 246 p.

Westoll, T.S.

1979: Devonian fish biostratigraphy; in The Devonian System, (ed.) M.R. House, C.T. Scrutton, and M.G. Bassett; Palaeontological Association, Special Papers in Palaeontology, 23, p. 341-353.

Whittington, H.B.

1965: Trilobites of the Ordovician Table Head Formation, western Newfoundland; Harvard University Museum of Comparative Zoology, Bulletin, v. 132, p. 275-441.

Whittington, H.B. and Hughes, C.P.

1972: Ordovician geography and faunal provinces deduced from trilobite distribution; Royal Society of London Philosophical Transactions, series B, v. 263, p. 235-278.

Whittington, H.B. and Kindle, C.H.

1963: Middle Ordovician Table Head Formation, western Newfoundland; Geological Society of America Bulletin, v. 74, p. 745-758.

1969: Cambrian and Ordovician stratigraphy of western Newfoundland; in North Atlantic- Geology and Continental Drift, (ed.) M. Kay; American Association of Petroleum Geologists, Memoir 12, p. 655-664.

Williams, A.

1973: Distribution of brachiopod assemblages in relation to Ordovician paleogeography; in Organisms and Continents Through Time, (ed.) N.F. Hughes, Palaeontological Association, Special Papers in Palaeontology, 12, p. 241-269.

Williams, H.

1964: The Appalachians in northeastern Newfoundland: a two-sided symmetrical system; American Journal of Science, v. 262, p. 1137-1158.

1978: Geological development of the northern Appalachians; Its bearing on the evolution of the British Isles; in Crustal Evolution in Northwestern Britain and Adjacent Regions, (ed.) D.R. Bowes and B.E. Leake, Geological Society of London, Special Publication 10, p. 1-22.

1980: Structural telescoping across the Appalachian Orogen and the minimum width of the Iapetus Ocean; in The Continental Crust and its Mineral Deposits, (ed.) D.W. Strangway; Geological Association of Canada, Special Paper Number 20, p. 421-440.

Williams, H. and Hatcher, R.D. Jr.

1983: Appalachian suspect terranes; in Contributions to the Tectonics and Geophysics of Mountain Chains, (ed.) R.D. Hatcher, Jr., H. Williams, and I. Zietz; Geological Society of America, Memoir 158, p. 33-53.

Williams, H. and Piasecki, M.A.J.

1990: The Cold Spring Melange and a possible model for Dunnage-Gander zone interaction in central Newfoundland; Canadian Journal of Earth Sciences, v. 27, p. 1126-1134.

Williams, H., Colman-Sadd, S.P., and Swinden, H.S.

1988: Tectonic-stratigraphic subdivisions of central Newfoundland; in Current Research, Part B; Geological Survey of Canada, Paper 88-1B, p. 91-98.

Williams, S.H.

1990: An Arenig graptolite from Bell Island, eastern Newfoundland - its biostratigraphic and paleogeographic significance; Atlantic Geology, v. 26, p. 43-55.

Wilson, J.T.

1966: Did the Atlantic close and then re-open?; Nature, v. 211, p. 676-681.

Wonderley, P.F. and Neuman, R.B.

1984: The Indian Bay Formation: Fossiliferous Early Ordovician volcanogenic rocks in the northern Gander Terrane, Newfoundland; Canadian Journal of Earth Sciences, v. 21, p. 525-532.

Young, G.C.

1981: Biogeography of Devonian vertebrates; Alcheringa, v. 5, p. 225-243.

Ziegler, A.M., Bambach, R.K., Parrish, J.T., Barrett, S.F., Gierlowski, E.H., Parker, W.C., Raymond, A., and Sepkoski, J.J., Jr.

1981: Paleozoic biogeography and climatology; in Paleobotany, Paleoecology and Evolution, (ed.) K.J. Niklas; Praeger Publishers, New York, v. 2, p. 231-266.

Ziegler, A.M., Scotese, C.R., McKerrow, W.S., Johnson, M.E., and Bambach, R.K.

1979: Paleozoic paleogeography; Annual Review of Earth and Planetary Sciences, v. 7, p. 473-502.

ADDENDUM

New fossils finds and continuing paleontological research provide information that complements the foregoing summary. Of special interest in the Precambrian - Cambrian interval is the international ratification of the base of the Cambrian at the base of the *Phycodes pedum* Trace Fossil Zone at Fortune Head, of the Avalon Zone in southeastern Newfoundland (Narbonne et al., 1987; Landing, 1994). As a consequence, the earliest Cambrian is now marked by the appearance of trace fossils exhibiting deep complex burrows and is succeeded by the appearance of diversified skeletal metazoans commonly considered to represent the base of the Cambrian in the past. Earliest Cambrian trace fossils lack the strong depth-control and environmental restriction of younger trace fossils and there are species with worldwide distribution and a relatively short stratigraphic range. Small shelly fossils are markedly provincial and the only really diverse faunas are restricted to shallow water facies (Landing, 1991; Landing and Murphy, 1991).

New fossils discoveries confirm the contrasting paleontological composition of the Notre Dame and Exploits subzones of the Dunnage Zone in Newfoundland (Williams et al., 1988). Early Ordovician (lower to middle Arenig) graptolites from several places in the Notre Dame Bay Subzone are like those of the Cow Head Group in western Newfoundland (Williams, 1992), whereas trilobites and graptolites (*Cyclopyge* and *Undulograptus*) of late Arenig age in the Exploits Subzone of Newfoundland confirm its Gondwanan affinities (Williams et al., 1992). Statistical tests have demonstrated that the Early Ordovician brachiopods from the Exploits Subzone are significantly different from coeval assemblages of western Newfoundland (Celtic and Toquima-Table Head provinces, respectively) (Harper, 1992; Neuman and Harper, 1992; Fortey and Cocks, 1992; Fortey and Mellish, 1992). The schematic paleogeographic map (Fig. 10.9) takes these differences into account, whereas the hypothetical volcanic arc shown on maps by Scotese and McKerrow (1990) and Cocks and McKerrow (1993) does not.

A distinctive assemblage of Ashgill-age trilobites and micromorphic brachiopods (*Foliomena* fauna) occur in the Pyle Mountain Argillite (Neuman, in press) in the southward continuation of the Aroostook-Percé Division of the Gaspé Belt (Fig. 10.6) into northeastern Maine, 32 km west of the international boundary at the latitude of Presque Isle, Maine (46°41′N). This remains the only North American occurrence of this assemblage following the re-identification by Cocks and Rong (1988) of its eponymous brachiopod from the White Head Formation at Percé, Quebec (APO-1, Fig. 10.6) (Sheehan and Lespérance, 1978). The *Foliomena* fauna is considered indicative of cool, deep waters, and the Maine occurrence confirms this environment in the Late Ordovician.

The presence of a Late Ordovician shoreline within the Gaspé Belt was inferred from a recently discovered coquinoid concentration of the inarticulate brachiopod *Eodinobolus* and other fossils in sandstone in northernmost Maine, about 7 km southeast of the international boundary, midway between areas assigned to that belt in Figure 10.6 (Pollock, 1993). The sequence to which these rocks were assigned was interpreted to be related to coeval rocks of the central Appalachians (Pollock et al., in press).

Also newly reported are late Middle to Late Ordovician conodonts from limestone of the Goss Point beds in the Mascarene Belt on the Letang Peninsula of southwestern New Brunswick (McLeod et al., 1994). The recovery of conodonts of both the Midcontinent and Atlantic faunal realms is unexpected for Ordovician rocks resting on Avalon basement. It may be that this part of the Letang Peninsula is a dislocated part of the Avalon Terrane that was located farther west in the Iapetus Ocean in Late Ordovician time, or else the whole ocean was smaller and allowed trans-Iapetan migration.

Cocks, L.R.M. and McKerrow, W.S.
1993: A reassessment of the early Ordovician 'Celtic' brachiopod province; Journal of the Geological Society of London, v. 150, p. 1039-1042.

Cocks, L.R.M. and Rong, Jia-yu
1988: A review of the late Ordovician *Foliomena* brachiopod fauna with new data from China, Wales, and Poland; Palaeontology, v. 31, p. 53-67.

Fortey, R.A. and Cocks, L.R.M.
1992: The early Palaeozoic of the North Atlantic region as a test case for the use of fossils in continental reconstruction; Tectonophysics, v. 206, p. 147-158.

Fortey, R.A. and Mellish, C.J.T.
1992: Are some fossils better than others for inferring palaeogeography? The early Ordovician of the North Atlantic region as an example; Terra Nova, v. 4, p. 210-216.

Harper, D.A.T.
1992: Ordovician provincial signals from the Appalachian - Caledonian terranes; Terra Nova, v. 4, p. 204-209.

Landing, E.
1991: Upper Precambrian through Lower Cambrian of Cape Breton Island: Faunas, paleoenvironments, and stratigraphic revision; Journal of Paleontology, v. 65, p. 570-595.

1994: Precambrian-Cambrian boundary global stratotype ratified and a new perspective of Cambrian time; Geology, v. 22, p. 179-182.

Landing, E. and Murphy, J.B.
1991: Uppermost Precambrian(?) - Lower Cambrian of mainland Nova Scotia: Faunas, depositional environments, and stratigraphic revision; Journal of Paleontology, v. 65, p. 382-396.

McLeod, M.J., Nowlan, G.S., and McCracken, A.D.
1994: Tectonic and economic significance of an Ordovician conodont discovery in southwestern New Brunswick; in Current Research 1993, (ed.) S.A.A. Merlini; New Brunswick Department of Natural Resources and Energy, Minerals and Energy Division, Miscellaneous Report No. 12, p. 121-128.

Narbonne, G.M., Myrow, P.M., Landing, E., and Anderson, M.M.
1987:   A candidate stratotype for the Precambrian - Cambrian boundary, Fortune Head, Burin Peninsula, southeastern Newfoundland; Canadian Journal of Earth Sciences, v. 24, p. 1277-1293.

Neuman, R.B.
in press: Late Ordovician (Ashgillian) *Foliomena* fauna brachiopods from northeastern Maine, U.S.A.; Journal of Paleontology, v. 68

Neuman, R.B. and Harper, D.A.T.
1992:   Paleogeographic significance of Arenig-Llanvirn Toquima-Table Head and Celtic brachiopod assemblages; in Global Perspectives on Ordovician Geology, (ed.) B.D. Webby and J.R. Laurie; A.A. Balkema, Rotterdam, p. 241-254.

Pollock, S.G.
1993:   Terrane sutures in the Maine Appalachians, U.S.A. and adjacent areas; Geological Journal, v. 28, p. 45-67.

Pollock, S.G., Harper, D.A.T., and Rohr, D.
in press: Late Ordovician nearshore faunas and depositional environments, northwestern Maine; Journal of Paleontology, v. 68.

Scotese, C.R. and McKerrow, W.S.
1990:   Revised World maps and introduction; in Palaeozoic Palaeogeography and Biogeography, (ed.) W.S. McKerrow and C.R. Scotese; Geological Society of London, Memoir 12, p. 1-21.

Sheehan, P.M. and Lespérance, P.J.
1978:   The occurrence of the Ordovician brachiopod *Foliomena* at Percé, Quebec; Canadian Journal of Earth Sciences, v. 15, p. 454-458.

Williams, H., Colman-Sadd, S.P., and Swinden, H.S.
1988:   Tectonic - stratigraphic subdivisions of central Newfoundland; in Current Research, Part B; Geological Survey of Canada, Paper 88-1B, p. 91-98.

Williams, S.H.
1992:   Lower Ordovician (Arenig-Llanvirn) graptolites from the Notre Dame Subzone, central Newfoundland; Canadian Journal of Earth Sciences, v. 29, p. 1717-1733.

Williams, S.H., Boyce, W.D., and Colman-Sadd, S.P.
1992:   A new Lower Ordovician (Arenig) faunule from the Coy Pond Complex, central Newfoundland, and a refined understanding of the closure of the Iapetus Ocean; Canadian Journal of Earth Sciences, v. 29, p. 2046-2057.

Authors' Addresses

R.B. Neuman
E-311
Natural History Museum
Smithsonian Institute
Washington, D.C. 20560
U.S.A.

G.S. Nowlan
Geological Survey of Canada
Institute of Sedimentary and Petroleum Geology
3303- 33rd St. N.W.
Calgary, Alberta
T2L 2A7

Printed in Canada

Chapter 11

SUMMARY AND OVERVIEW

Chapter 11

SUMMARY AND OVERVIEW

Harold Williams

PREAMBLE

This chapter summarizes, and repeats without referencing, information presented in preceding chapters. Some of these chapters were written more than 5 years ago and there have been important subsequent changes. A few new references are therefore cited. This chapter was written also with an eye toward suitability for a Canadian overview volume.

The summary follows the systematics introduced earlier, treating all rocks according to the four broad temporal divisions; lower Paleozoic and older rocks, middle Paleozoic rocks, upper Paleozoic rocks, and Mesozoic rocks. The rocks of each temporal division are subdivided into spatial divisions. Thus, the lower Paleozoic and older rocks are separated into the Humber, Dunnage, Gander, Avalon, and Meguma zones and subzones as depicted in Figure 11.1. The middle Paleozoic rocks are separated into six belts: Gaspé, Fredericton, Mascarene, Arisaig, Cape Breton, and Annapolis for the mainland; and Clam Bank, Springdale, Cape Ray, Badger, La Poile, Botwood, and Fortune for Newfoundland (Fig. 11.2). The upper Paleozoic rocks define a number of basins, and Mesozoic rocks define graben (Fig. 11.3). A compilation and classification of volcanic rocks for the Canadian Appalachian region is provided for comparisons with other divisions (Fig. 11.4).

The Canadian Appalachians provide an excellent example of an orogen that built up through accretion and eventual continental collision. In this model of a typical Wilson cycle, the Humber Zone is the Appalachian miogeocline or continental margin of Laurentia, and outboard zones are accreted parts of the orogen or suspect terranes. These zones are the fundamental divisions upon which all younger rocks and events are superposed. Rocks and relationships are described first for the Humber Zone, then successively outboard zones. The earliest interaction among zones occurred in the Ordovician with accretion of western parts of the Dunnage Zone to the Humber Zone and amalgamation of eastern parts of the Dunnage Zone to the Gander Zone. Middle Paleozoic Belts comprise mainly cover sequences, but some belts are confined to zones

Williams, H.
1995: Summary and overview; Chapter 11 in Geology of the Appalachian-Caledonian Orogen in Canada and Greenland, (ed.) H. Williams; Geological Survey of Canada, Geology of Canada, no. 6, p. 843-890 (also Geological Society of America, The Geology of North America, v. F-1).

(Annapolis Belt above Meguma Zone) or parts of zones (Badger Belt above eastern Dunnage Zone), and others occur at zone boundaries. There is no evidence that the Avalon and Meguma zones were incorporated into the orogen before the Silurian-Devonian, and open marine tracts probably existed in the Newfoundland Dunnage Zone after Ordovician deformation of bordering areas. The arrival of the Avalon and Meguma zones apparently coincided with important middle Paleozoic tectonism. Upper Paleozoic basins developed upon the accreted orogen. Most of their rocks are terrestrial and undeformed, but deformation is important locally and a few granite batholiths are dated isotopically as late Paleozoic. Graben are treated last as their rocks are unconformable upon deformed upper Paleozoic rocks. The graben are related to rifting that initiated the Atlantic Ocean.

These accounts are followed by geophysical characteristics of the orogen and its offshore extensions, paleontology, metallogeny, and a summary of zonal linkages and accretionary history, in an attempt to explain orogenic development.

ANCESTRAL NORTH AMERICAN MARGIN: HUMBER ZONE

The lower Paleozoic rocks of the Humber Zone are coextensive with the cover rocks of the St. Lawrence Platform. Its western limit is the Appalachian structural front which separates the deformed rocks of the orogen from equivalent undeformed rocks of the St. Lawrence Platform. Its eastern boundary with the Dunnage Zone is the Baie Verte-Brompton Line, a steep structural zone marked by discontinuous ophiolite occurrences. An external division and internal division of the Humber Zone are defined on structural and metamorphic contrasts (see Fig. 3.1).

External Humber Zone

The stratigraphic and structural elements of the external Humber Zone fit the model of an evolving continental margin. It began with: (a) rifting of a Grenville crystalline basement and deposition of upper Precambrian to Lower Cambrian clastic sedimentary and volcanic rocks with coeval dyke swarms and carbonatite intrusions, the rift stage, (b) deposition of a Cambrian-Ordovician mainly carbonate sequence, the passive margin stage, (c) deposition of Middle Ordovician clastic rocks of outboard derivation that transgress the carbonate sequence and are the first intimation of offshore disturbance, the foreland basin stage, and (d) emplacement of allochthons in the Middle

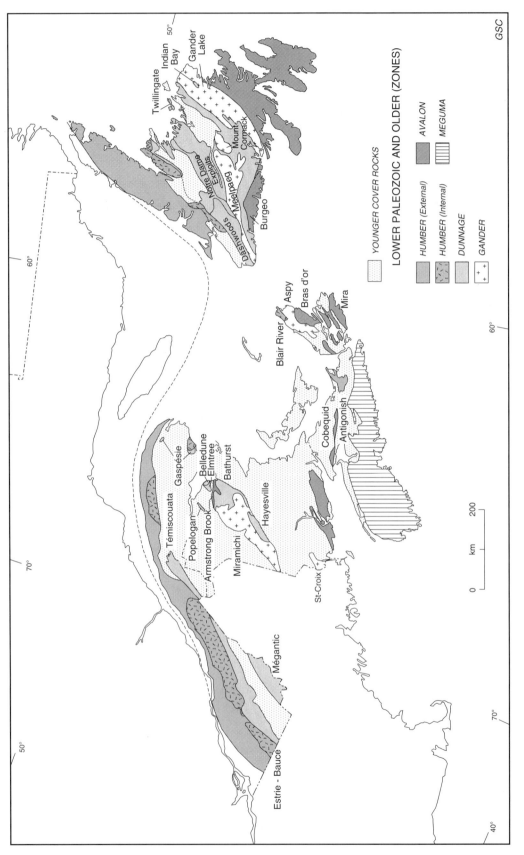

Figure 11.1. Zones and subzones of the Canadian Appalachian region.

YOUNGER COVER ROCKS

LOWER PALEOZOIC AND OLDER (ZONES)

HUMBER (External)

HUMBER (Internal)

DUNNAGE

GANDER

AVALON

MEGUMA

GSC

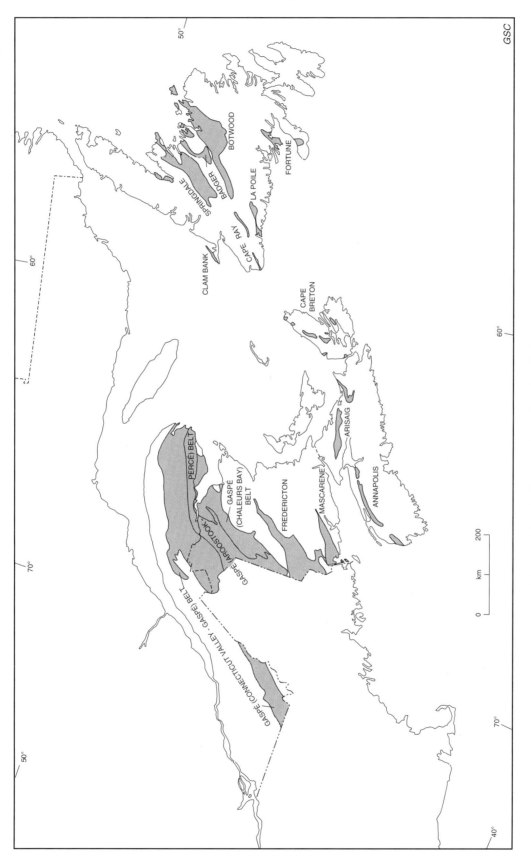

Figure 11.2. Belts of the Canadian Appalachian region.

847

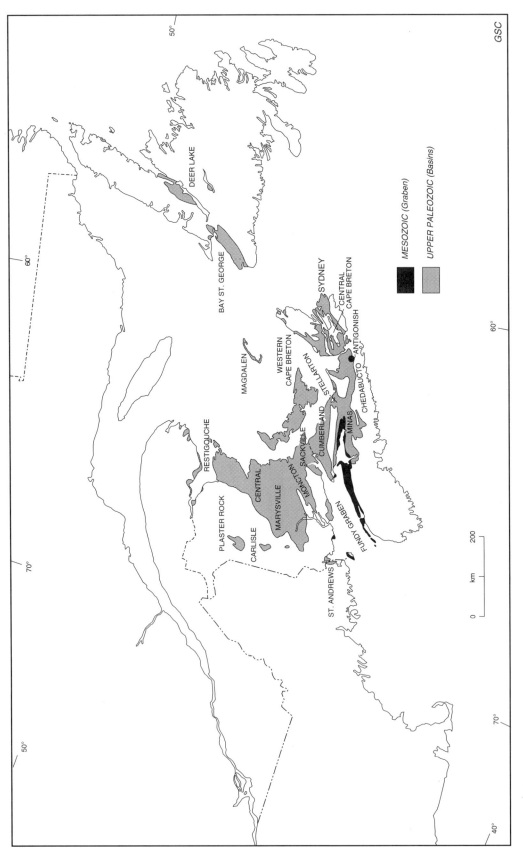

Figure 11.3. Basins and graben of the Canadian Appalachian region.

GSC

Ordovician that are a sampling of rocks from the continental slope and rise and adjacent oceanic crust, the destructive stage. Any of these rocks can be used to define the Humber Zone. Naturally, the more elements recognized, the sharper the definition. Rocks of the internal Humber Zone are polydeformed and metamorphosed so that their stratigraphic and sedimentological records are less informative. The available data suggest deposition on the continental slope and rise, mainly upon a continental Grenville basement, but possibly linked to an oceanic basement in Newfoundland. The internal Humber Zone is important for its structural and metamorphic records of the destruction of the continental margin. These are in general agreement with the stratigraphic and sedimentological records of the external Humber Zone.

Rift stage

Basement rocks of the Humber Zone are mainly gneisses, schists, granitoids, and metabasic rocks that are part of the Grenville Structural Province of the adjacent Canadian Shield (Fig. 3.2). Most metamorphic and plutonic ages are between 1200-1000 Ma with some about 1500 Ma. The basement rocks occur in both external and internal divisions of the Humber Zone. Occurrences in the external Humber Zone are restricted to western Newfoundland. These are the Long Range Inlier, the largest of the entire Appalachian Orogen, and the Belle Isle and Indian Head inliers to its north and south, respectively. Other small faulted examples occur nearby at Ten Mile Lake and Castors River. Internal Humber Zone occurrences in Newfoundland are the East Pond Metamorphic Suite at Baie Verte Peninsula, the Cobble Cove Gneiss at Glover Island of Grand Lake, the Steel Mountain Inlier south of Grand Lake, and possible basement rocks of the Cormacks Lake Complex of the Dashwoods Subzone in southwestern Newfoundland.

Metamorphic and plutonic rocks of the Blair River Complex that occur at the northwest corner of Cape Breton Island, Nova Scotia, are also Grenville basement. Other Precambrian crystalline rocks occur in three widely separated areas of the Quebec internal Humber Zone. These are the Port Daniel Inlier, the Sainte-Marguerite Fault Slice, and the Sainte-Helène-de-Chester Fault Slice, from east to west.

Occurrences in the external Humber Zone, such as the Belle Isle and Indian Head inliers, form the cores of northeast-trending anticlines. The Long Range Inlier and the smaller Ten Mile Lake and Castors River inliers were brought to the surface by steep reverse faults or west-directed thrusts. The basement rocks are overlain by upper Precambrian or Lower Cambrian cover rocks with preserved unconformable relationships in most places. Occurrences in the internal Humber Zone of Newfoundland are thrust slices or structural culminations in high-grade metamorphic areas. These basement rocks were deformed with their cover and a basement-cover distinction is not everywhere clear. Thus, the boundaries of the East Pond Metamorphic Suite and Cormacks Lake Complex are poorly defined.

The Blair River Complex of Cape Breton Island is separated from metamorphic rocks of the Aspy Subzone to the east by a wide zone of mylonite developed along the Wilkie Brook Fault. The Port Daniel Inlier occurs in a major fault zone bordering the Maquereau Dome of Gaspésie. The Sainte-Marquerite Complex forms three faulted slivers up to 3 km long within the Richardson Fault Zone, which forms the boundary between the internal and external divisions of the Humber Zone in Quebec. The Sainte-Helène-de-Chester occurrence is a fault slice within the Bennett schists, a few kilometres northwest of the Baie Verte-Brompton Line.

Rocks of basement inliers are mainly of igneous protolith with minor marble, quartzite, and metaclastic rocks. The extensive Long Range Inlier has mainly quartz-feldspar gneisses and granites. In the north, the rocks are biotite-quartz-feldspar gneiss and hornblende-biotite-quartz-feldspar gneiss with lesser quartz-rich gneiss, pelitic schist, amphibolite, and calc-silicate gneiss. In the south, they are pink to grey quartz-feldspar gneiss, quartz-rich gneiss, pelitic to psammitic schist, amphibolite, and minor marble and calc-silicate rock. Small mafic plutons, now amphibolite, hypersthene amphibolite, metaperidotite, and metadiorite cut the leucocratic gneisses. The mafic plutons are cut by distinctive megacrystic foliated to massive granites of at least two suites. Pegmatites that occur in most outcrops probably relate to the granites. Northeast-trending diabase dykes of the Long Range Swarm cut all other rocks.

Coarse grained white anorthosite and banded hypersthene gabbro of the Indian Head Inlier are conspicuous in coastal headlands and these are intruded by massive to foliated pink granites and coarse grained pegmatites.

The Steel Mountain Inlier consists of coarse grained grey to pale purple anorthosite, gabbroic anorthosite, norite, and minor pyroxenite in the south and granitic, granodioritic, and amphibolitic gneisses with crosscutting bodies of granite and syenite in the north. Some small occurrences of marble and quartzite near Grand Lake are part of the basement complex, others may be Paleozoic structural enclaves.

The dominant lithology of the Blair River Complex is quartzo-feldspathic gneiss with associated amphibolite, granitic gneiss, and minor calcareous rocks. Small bodies of gabbroic to granitic rocks, syenite, and anorthosite are intimately mixed with the gneisses. In other places, anorthosite cuts the gneisses and is in turn cut by syenite.

Foliation patterns in the Precambrian inliers are extremely variable. In the northern part of the Long Range Inlier broad areas of shallowly dipping foliation are separated by steep structural zones or by granitic plutons and faults.

Banding and gneissic foliation in rocks of the southern massif of the Indian Head Inlier trend northwest, perpendicular to the length of the inlier and to Paleozoic structures. Similarly, anorthosites of the Steel Mountain Inlier have a mild northwest-trending foliation in the west that is overprinted by a Paleozoic northeast fabric toward the east.

The dominant trend of fold axes in the Cormacks Lake Complex is northwest, and northwest trends also occur in the East Pond Metamorphic Suite. These are like Grenville structural trends in the Steel Mountain and Indian Head inliers, and suggest Precambrian ages for these complexes.

Precambrian regional metamorphism is uniformly high grade. Amphibolite to granulite facies in the northern part of the Long Range Inlier was accompanied by the emplacement of mafic plutons. Pelitic rocks contain microcline-sillimanite- cordierite (or garnet) assemblages, and remnants of hyper- sthene are preserved in some localities.

In central and southern portions of the Long Range Inlier, amphibolite and locally granulite facies mineral assemblages (two-pyroxene gneisses) are recorded, with metamorphic facies increasing from amphibolite in the northeast to granulite in the southwest. The metamorphism predates most of the large granite plutons and possibly the gabbro-anorthosite suite.

Metamorphism in mafic gneisses of the Indian Head Inlier reaches amphibolite and granulite facies. Biotite-hornblende-plagioclase gneisses are common and two-pyroxene plagioclase gneisses occur locally. Sapherine occurs at one locality.

Granitic gneisses and schists of the Steel Mountain Inlier contain amphibolite- to granulite-facies mineral assemblages. This high-grade metamorphism may be older than nearby anorthosites.

In the external Humber Zone, retrograde metamorphism affected the Long Range, Belle Isle, and Indian Head inliers. It occurred in at least two phases in the Long Range Inlier. The first preceded the emplacement of the Long Range mafic dykes. The second postdated the dykes and related volcanic rocks and increases in intensity from west to east. Relationships in the Belle Isle Inlier indicate that late Precambrian faulting affected coarse clastic cover rocks before the emplacement of the Long Range dyke swarm. This implies rift-related deformation and Precambrian retrogression related to extensional tectonism. The second metamorphic event is Paleozoic.

Upper Precambrian-Lower Cambrian clastic sedimentary and volcanic rocks record the rifting stage of the Laurentian margin and initiation of the Appalachian cycle (Fig. 3.11). They are mainly terrestrial red arkosic sandstones and conglomerates, bimodal volcanic rocks, marine quartzites, and greywackes. Abrupt changes in thickness and stratigraphic order are characteristic and these features contrast with uniform stratigraphy, constant thickness and broad lateral continuity of overlying Lower Cambrian marine formations (Fig. 3.14). Extensive swarms of mafic dykes are in places coeval with mafic flows. Isotopic ages of volcanic rocks, mafic dykes, and related small anorogenic plutons range between 620 and 550 Ma.

Rocks and relations are well-preserved in the external Humber Zone of Newfoundland and locally in the internal Humber Zone of Newfoundland and Quebec. Correlatives are represented in the basal parts of stratigraphic sections that make up the sedimentary slices of Taconic allochthons. Metamorphosed clastic rocks and mafic volcanic rocks and dykes, now amphibolites, that occur in basal parts of stratigraphic sections in the internal Humber Zone are also correlatives.

Examples in southeast Labrador, just outside the Appalachian deformed zone, are virtually continuous with mildly deformed rocks of the nearby orogen in Newfoundland. Other discrete occurrences at Lake Melville and Sandwich Bay of Labrador are 300 km farther north and lie at the northwest limit of the Grenville Structural Province. In Quebec, the Grenville dyke swarm dated at 575 Ma extends 700 km inland beyond the Appalachian

structural front along the Ottawa Valley. Coeval anorogenic intrusions occur along the Saguenay, St. Lawrence, and Ottawa valleys.

Three formations are recognized among the upper Precambrian-Lower Cambrian clastic sedimentary and volcanic rocks in the external Humber Zone of western Newfoundland. A basal clastic unit (Bateau Formation) is overlain by mafic volcanic rocks (Lighthouse Cove Formation), in turn overlain by red arkosic sandstones (Bradore Formation). The formations are transgressive northwestward with all units locally resting on Grenville basement. The Bradore Formation is overlain by fossiliferous shale, siltstone, and limestone of the Lower Cambrian Forteau Formation, the basal formation of the succeeding carbonate sequence.

Mafic dykes of the Long Range swarm are coeval with the Lighthouse Cove Formation and they cut the Bateau Formation. Chemical and petrographic data show that they are typical tholeiites and there is no detectable difference between the Long Range dykes and Lighthouse Cove volcanic rocks of southeast Labrador, Belle Isle, and northeast Newfoundland.

Some northeast-trending faults in the basement coincide with thickness changes in the Bradore Formation. Other contemporary faults at Belle Isle affected the Bateau Formation but not nearby mafic dykes. The surface on which the Bateau, Lighthouse Cove, and Bradore formations accumulated was therefore irregular with sharp local relief that led to significant thickness variations.

At Hughes Lake in the internal Humber Zone of western Newfoundland, a deformed and metamorphosed bimodal volcanic suite and associated granite are overlain by psammitic and pelitic schists. The granite is dated at 602 ± 10 Ma and interpreted as a high-level anorogenic pluton. Silicic volcanic rocks have trace-element characteristics typical of rhyolitic end members of alkali igneous suites. Mafic flows and dykes are tholeiites, similar to the Lighthouse Cove Formation and dykes of the Long Range Swarm. Basal metaclastic rocks are correlated with the Bradore Formation although they are much thicker.

The age of the granite, its hypersolvus textural features, its association with a bimodal volcanic suite, the alkaline affinities of associated silicic volcanic rocks, and the presence of mafic dykes and flows all suggest a consanguineous igneous suite that formed in a rift setting.

The Tibbit Hill volcanics in the internal Humber Zone of Quebec are equated with occurrences in Newfoundland and nearby examples in Vermont. They are predominantly altered basalts that are polydeformed and metamorphosed. Abundances of immobile elements and enrichment patterns indicate transitional or alkalic basalts of "Within Plate" affinity, contaminated by continental crust. Comenditic metafelsites that form a minor unit in the volcanic sequence are dated at 554 +4/-2 Ma.

Outside the Appalachian deformed zone, the Double Mer Formation is a sequence of red arkosic sandstones, conglomerates, siltstones, and shales that occurs in graben at Lake Melville and Double Mer. Bedding attitudes suggest broad open folding with thicknesses of sediment possibly reaching 5 km. Conglomerates at the northern margin of the Double Mer graben are interpreted as fanglomerates.

Coupled with northerly paleocurrents elsewhere in the graben, the pattern suggests a half graben with a faulted northern margin and a north-sloping floor. Relationships to the south suggest that the Mealy Mountains formed a contemporary upland at the southern edge of the Lake Melville Graben.

The Double Mer Formation is traditionally correlated with the Bradore Formation of southeast Labrador. Its depositional graben are parallel to the northeast direction of late Precambrian-Cambrian mafic dykes. A northeast-trending dyke, which crosses a small graben at Sandwich Bay, yields a U-Pb zircon and baddeleyite age of 615 ± 2 Ma.

The Grenville dyke swarm and Ottawa Graben are probably linked with the Tibbit Hill volcanic rocks, although the Tibbit Hill volcanic rocks are alkalic to transitional, whereas the dyke rocks are tholeiitic. The mafic dyke-volcanic episode is roughly coeval with the emplacement of alkali complexes, including several carbonatite complexes dated at 565 Ma.

Faults, contemporaneous with sedimentation and locally predating mafic dyke intrusion, may have been listric normal faults like those so common beneath breakup unconformities at modern continental margins. Overlying marine shales and limestones mark an abrupt termination of coarse clastic deposition and a broad submergence and permanent change in the tectonic regime. This is interpreted to mark the transition from continental rifting to ocean spreading. The marine transgression and onset of carbonate deposition is regarded as the result of cooling and thermal subsidence, rather than a eustatic rise of sealevel, as it is synchronous with the cessation of rifting.

Widths of the Ottawa Graben and Lake Melville-Double Mer graben are in the order of 50 to 100 km, typical of modern rift valleys.

The Ottawa Graben has the geometry of an aulacogen or failed arm of a triple rift system. Its setting at the Sutton Mountains structural salient is interpreted as an original deep embayment in the rifted margin. The Long Range Swarm, and Lake Melville and Double Mer graben are parallel to the ancient Iapetus continental margin, much as Triassic-Jurassic dyke swarms and graben of the Appalachian Orogen are parallel to the modern Atlantic margin.

In southeast Labrador, the trends of the late Precambrian-Early Cambrian Iapetus margin and Mesozoic Atlantic margin are almost perpendicular. The distribution of Iapetan clastic dykes and outliers of horizontal cover rocks indicate that the present exposed surface of the Canadian Shield was the surface at the time of Iapetan rifting. This provides the possibility of studying listric normal faults and transfer faults related to Iapetan rifting, and also similar features related to the opening of the Atlantic Ocean that should be superimposed on Iapetan effects at high angles. The Cartwright Fracture Zone at the modern Atlantic margin off Labrador may reflect the Lake Melville rift system. Other examples of ancestral controls and micmicry abound.

Passive margin stage

A Cambrian-Ordovician carbonate sequence occurs above the rift related rocks of the Humber Zone or it directly overlies Grenville basement. These rocks also extend across the St. Lawrence Platform. They are mainly limestones and dolomites with lesser siliciclastic rocks that range in age from Early Cambrian to Middle Ordovician. They record the development of a passive continental margin, and the beginning of Taconic Orogeny at the margin. The Cambrian-Ordovician carbonate sequence is the most distinctive and characteristic element of the Humber Zone. The stratigraphic analysis of the platformal rocks and equivalent continental slope/rise facies in Newfoundland is as sophisticated as that of any continental margin analysis, regardless of age.

The rocks are preserved best in western Newfoundland throughout the external Humber Zone and nearby (Fig. 3.15). They are mainly hidden by Taconic allochthons along the Quebec segment of the Humber Zone. Small examples occur at Gaspésie and more deformed examples occur in the Sutton Mountains and Lake Champlain area to the west. The rocks occur in the St. Lawrence Lowlands west of Québec City, and they occur at Mingan Islands on the north shore of the St. Lawrence River.

In Newfoundland, the carbonate sequence is divided into the Labrador, Port au Port, St. George, and Table Head groups. In the St. Lawrence Lowlands of Quebec, the succession includes the Potsdam, Beekmantown, Chazy, Black River and Trenton groups; and the Romaine and Mingan formations on the north shore of the St. Lawrence River.

Where the rocks occur in thrust slices of the Sutton Mountains and Lake Champlain areas they include the Oakhill Group, Milton Dolomite, Rock River Formation, and Phillipsburg Group. Cambrian carbonates in Gaspésie are the Murphy Creek and Corner-of-the-Beach formations.

Present exposures of the carbonate sequence were part of a continental margin platform, the landward portion of which is preserved only in Quebec. There, a rolling and faulted paleotopography is onlapped by progressively younger rock units northwards across the St. Lawrence Lowlands. At Montmorency Falls near Québec City, the Middle Ordovician Trenton Group lies directly on Grenville basement. At Mingan Islands and in the subsurface of Anticosti Island, upper Lower Ordovician carbonates overlie Precambrian basement. The landward edge of Cambrian carbonates was between Gaspésie and Anticosti Island and southeast of the St Lawrence Lowlands. This implies that a relatively narrow Cambrian shelf was overstepped by Ordovician rocks. The shelf edge is nowhere preserved. Coeval rocks of the continental slope are preserved in Taconic allochthons.

Carbonate sedimentation occurred in two tectonic settings: (1) trailing passive margin sedimentation from the late Early Cambrian to the late Early Ordovician, and (2) Middle Ordovician carbonate sedimentation along the cratonic edge of the Taconic foreland basin. The Early Cambrian passive margin history in western Newfoundland is preserved in the upper part of the Labrador Group (Forteau and Hawke Bay formations) and comprises mixed siliciclastic and carbonate marine strata laid down above nonmarine to nearshore sandstones (Bradore Formation, early rift facies). Equivalent strata of the Oakhill Group and Milton Dolomite occur only in thrust slices in Quebec. Carbonate deposition was established throughout western Newfoundland during the Middle Cambrian and continued to the Early Ordovician. During this period of about 70 Ma, strata belonging to the Port au Port and St. George groups formed the major part of the 1500 m sequence. In Quebec, carbonate sedimentation was not widespread until the Early Ordovician when the Beekmantown and Phillipsburg groups were deposited above sandstones of the Potsdam Group. Cambrian carbonate sedimentation is recorded near Percé of Gaspésie by Middle and Upper Cambrian carbonates, and shallow water, Cambrian carbonate clasts in allochthonous Ordovician deepwater sediments.

The St. George Unconformity is a major break between the St. George and Table Head groups in Newfoundland. Dramatic variations in thickness reflect a combination of faulting, uplift, and erosion. On Mingan Islands, the Romaine Formation was tilted before erosion so that the unconformity is a planar karst surface which bevels several units.

Foreland basin stage

Clastic rocks above the carbonate sequence represent a major reversal in the direction of sediment supply to the Humber Zone, from the North American continent to outboard elements of the Dunnage Zone. Autochthonous and allochthonous examples occur in the Humber Zone, and the rocks extend beyond the western limit of Appalachian deformation to the St. Lawrence Platform (Fig. 3.27). The clastics contain recycled quartz grains, fresh and weathered feldspars, a variety of sedimentary rock fragments, mafic to felsic volcanic rock fragments, local serpentinite grains, and a heavy mineral suite containing green and green-brown hornblende, hypersthene, and chromite.

The autochthonous flysch lies stratigraphically above the Cambrian-Ordovician carbonate sequence, except where original relationships have been obscured by thrusting. Thicknesses range from about 250-4000 m, and ages range from Llanvirn in Newfoundland to Caradoc and Ashgill in Quebec.

Allochthonous flysch lies stratigraphically above Lower Ordovician shales and limestone breccias, interpreted as continental slope and rise deposits. These occurrences range in thickness from about 200-2500 m, and in age from Arenig to Caradoc.

In western Newfoundland, autochthonous flysch overlies the Table Head Group, a predominantly shallow-water limestone unit. In Quebec, parautochthonous flysch that occurs along the south side of the St. Lawrence River was deposited along the axis of a foreland basin, with paleocurrents predominantly to the west in the lower to middle part of the sequence, and to the east at its top. In the intervening Gulf of St. Lawrence, shaly flysch occurs in the subsurface beneath Anticosti Island.

Autochthonous flysch also extends about 100 km east of Quebec City to the Canada-United States border, and across strike to the northwest for about 50 km onto the St. Lawrence Platform. In the core of the Chambly-Fortierville Syncline, the flysch is overlain conformably by redbeds of the Rivière Bécancour Formation. This is the only place in the Canadian Appalachians where terrestrial molasse sediments are preserved directly above Ordovician flysch. A deep seismic reflection profile across the eastern end of the Chambly-Fortierville Syncline shows that the flysch is in part buried by thrust sheets, and is itself somewhat telescoped by folding and thrusting.

Thrust slices that overlie the autochthon locally contain units of sandy flysch at the top of their stratigraphic sections. These allochthonous units are older than those of the adjacent autochthon. They overlie Lower Ordovician shales, limestone breccias, and older rocks that are interpreted as continental slope and rise deposits.

In western Newfoundland, allochthonous flysch is found in lower structural slices of the Humber Arm Allochthon in the Port-au-Port area, in the Bay of Islands area, and north of Bonne Bay where it overlies limestone conglomerates at the top of the Cow Head Group. The age of the flysch is Arenig to Llanvirn. Sole markings indicate derivation from the north to northeast.

In Quebec, the oldest allochthonous flysch units are of relatively proximal character and late Arenig to early Llanvirn age.

The St. George Unconformity in Newfoundland developed near the carbonate platform edge. An equivalent unconformity is only weakly developed or absent landward in the St. Lawrence Lowlands. This Middle Ordovician unconformity is interpreted to represent erosion along the crest of a migrating peripheral bulge associated with loading and landward migration of Taconic thrust sheets. Flysch was deposited in an evolving foreland basin developed upon the shallow marine carbonate platform, with rapid subsidence in the order of kilometres.

The transition from carbonate platform to deep foreland basin in the autochthon of western Newfoundland includes a complicated record of platform collapse, fault movements, and derivation of limestone boulder conglomerates from the crests of fault blocks. In particular, the Cape Cormorant boulder conglomerates contain a mixture of Cambrian and Ordovician carbonate clasts shed into the developing black shale and flysch environment from the tops of uplifted blocks. Elsewhere in western Newfoundland and also in the Quebec City area, deepening and tilting of the foundering platform edge generated wet-sediment slides in argillaceous facies.

Black shale deposits probably represent local sites of low siliciclastic sediment supply during the early phases of basin filling. There is no obvious relationship between times of early black shale deposition and times of Ordovician eustatic sea-level rise, as has been suggested for black shales of the British Isles. Nevertheless, unusually fine grained units within thick flysch formations may have been deposited during times of high sea level.

Some autochthonous flysch units in the Québec City area and Gaspésie have been interpreted as small submarine fans. Other flysch units between Montréal and Québec City appear to have been deposited on a basin floor controlled by differential subsidence, and characterized by active intrabasinal horsts. Because of strong fault control, morphological submarine fans were apparently not formed.

Intense deformation and production of mélange below the allochthonous flysch units may partly pre-date the end of flysch deposition. Abrupt basal contacts are incompatible with fan progradation, and suggest deposition in small basins, probably on slopes formed by stacking of Taconic thrust sheets.

Paleoflow data and stratigraphic analyses support a model for earliest flysch deposition near promontories of the ancestral North American margin and latest deposition in reentrants. The earliest deposition in southwestern Newfoundland and eastern Gaspésie occurred close to the St. Lawrence Promontory. The latest deposition of Quebec occurred in the Quebec Reentrant, an area that was removed from early collisional effects. The ages of other allochthonous and autochthonous flysch units can be explained by progressive collision of an island arc or arcs with an irregular continental margin.

The difference in age between allochthonous and autochthonous flysch in the same part of the orogen is interpreted as a function of the distance that the allochthons were transported toward the continental margin.

Destructive stage

Taconic allochthons, above the Middle Ordovician clastic unit of the autochthonous section, are a spectacular structural feature of the external Humber Zone. Contorted marine shales, sandstones, and mélanges in lower structural slices contrast sharply with the underlying mildly deformed carbonate sequence. Volcanic rocks and ophiolite suites in higher slices are particularly out-of-context with respect to Grenville basement and its cover rocks. Allochthons are absent or poorly defined in internal parts of the Humber Zone, except for occurrences of small disrupted ophiolitic rocks and mélanges.

The best known examples in western Newfoundland are the Humber Arm and Hare Bay allochthons. The Old Mans Pond Allochthon is an eastern outlier of the Humber Arm Allochthon. The Southern White Bay Allochthon is a small occurrence on the east side of the Great Northern Peninsula (Fig. 3.38).

Transported rocks in Quebec form a series of structural slices, the Lower St. Lawrence River nappes, that make up most of the exposed external Humber Zone. The largest nappes from north to south are the Sainte Anne River, Marsoui River, Shickshock, Lower St. Lawrence Valley, Chaudière, St. Henedine, Bacchus, Point-de-Levy, Quebec Promontory, Grandby, and Stanbridge (Fig. 3.38).

The western edge of the transported rocks is Logan's Line, first defined at Québec City and traceable southward to Vermont. Its northern course is covered by the St. Lawrence River, but it is exposed near the tip of Gaspésie. In Newfoundland, Logan's Line is discontinuous and defined by the leading edges of the Humber Arm and Hare Bay allochthons (Fig. 3.38).

The Humber Arm and Hare Bay allochthons of western Newfoundland are described as they exhibit a variety of igneous and metamorphic rocks in upper structural slices that are rare or absent in most Appalachian examples. Especially well-known are the ophiolite suites and metamorphic soles of the Bay of Islands Complex in the Humber Arm Allochthon and the St. Anthony Complex of the Hare Bay Allochthon. A complete stratigraphy from late Precambrian to Early Ordovician has been deciphered among sedimentary rocks of the Humber Arm Allochthon and some facies, such as the Cambrian-Ordovician Cow Head breccias, are renowned for their sedimentologic and paleontologic characteristics. Gros Morne National Park, situated in the Humber Zone of western Newfoundland, was declared a World Heritage Site in 1987, mainly because of its rocks and geological relationships.

The Humber Arm (Fig. 3.39) and Hare Bay (Fig. 3.43) allochthons occupy structural depressions. The Humber Arm Allochthon is surrounded to the east by the Cambrian-Ordovician carbonate sequence and separated from the Hare Bay Allochthon by a major structural culmination that exposes Grenville basement of the Long Range Inlier. These structural depressions and culminations are later than the emplacement of the allochthons. However, conodont colouration indices in underlying and surrounding rocks indicate that the allochthons were never much larger then their present areas.

Conspicuous carbonate slivers occur beneath the Bay of Islands Complex and above mélange of lower structural levels. They are gently dipping, upright sections of thick bedded, white and grey limestones and dolomites of the St. George and Table Head groups. These are interpreted either as integral parts of the allochthon, entrained during Middle Ordovician assembly and transport, or features of later gravity collapse and transport of ophiolites and underlying carbonates across lower levels of the already assembled allochthon.

Mélanges are extensively developed among sedimentary rocks of the Humber Arm Allochthon (Fig. 3.39). They consist mainly of greywacke, quartzite, dolomitic shale, chert, and limestone blocks in a black, green, and red scaly shale matrix. Volcanic blocks are most common in structurally higher mélanges and some contain a sampling of ophiolitic blocks in a greasy green serpentinite matrix. Volcanic blocks range in diameter from a few metres to a few kilometres. Blocks beyond a kilometre across are oblate to discoidal and the largest are the flat ophiolite massifs of the Bay of Islands Complex. There is every gradation between the smallest equidimensional blocks and the largest structural slices of the allochthon.

Rocks of lower sedimentary slices of the Humber Arm Allochthon are mainly eastern deeper water correlatives of the passive margin carbonate sequence, but relationships at deposition are nowhere preserved. The oldest allochthonous clastic rocks are correlated with autochthonous rift-related clastics. They were derived from a crystalline Grenville source before and during continental breakup and before the development of the carbonate sequence. Volcanic rocks, such as those at the base of the Blow-Me-Down Brook Formation in the Humber Arm Allochthon and within the Maiden Point Formation of the Hare Bay Allochthon, are possible Lighthouse Cove correlatives. Overlying shales, quartzites, and polymictic conglomerates with Lower to Middle Cambrian carbonate clasts are coeval with breakup and the initiation of the carbonate sequence. The condensed limestone breccia, thin platy limestone, shale units of the Cow Head Group are the continental slope/rise equivalents of the carbonate platform sequence. At Cow Head, repeated belts are interpreted as thrust slices developed during transport and emplacement. The pattern of abundant coarse limestone breccias in western slices and finer, thinner and fewer breccias in eastern slices implies a proximal (west) to distal (east) facies arrangement that is still discernible across the thrust slices (Fig. 3.40). Overlying sandstones with chromite and other ophiolite detritus are part of the foreland basin flysch units that appear first in the allochthonous sequences.

Volcanic rocks and ophiolite suites of higher structural slices are of unknown paleogeography with respect to other rocks of the Humber Zone. The Skinner Cove volcanic rocks of the Humber Arm Allochthon have been interpreted as oceanic islands, possibly seamounts. Alternatively, they could represent rift-facies volcanic rocks like those at the base of the Blow-Me-Down Brook Formation. The Bay of Islands Complex is interpreted as oceanic crust and mantle. Metamorphism in volcanic rocks and dykes is interpreted as a depth controlled static seafloor hydration. A lack of metamorphism in deeper gabbros possibly reflects an absence of surficial fluids necessary to accomplish the hydration.

The metamorphic soles of the ophiolite suites, now subhorizontal or steeply dipping surfaces frozen into the ophiolite sequences, are interpreted as high temperature shear zones resulting from transport of hot mantle and oceanic crust. The best example occurs beneath the White Hills Peridotite of the Hare Bay Allochthon where the Ireland Point Volcanics, Goose Cove Schist, and Green Ridge Amphibolite were accreted to the base of the peridotite. Contacts between rock units within the accreted metamorphic sole that were first interpreted as gradational are now interpreted as ductile shears. Thus the Green Ridge Amphibolite is retrograded toward its base; the Goose Cove Schist is prograded toward its top, and the overall inverted metamorphic gradient is fortuitous. The juxtaposition of oceanic and continental margin rocks along this shear zone suggests that it represents the interface between down-going and overriding plates in a subduction zone. The lithologies and pressure/temperature paths suggest assembly at a depth of 10 km where the geothermal gradient was abnormally high because the overriding plate was hot. Rocks of the upper plate were metamorphosed somewhere else and they were cooling when juxtaposed with the continental plate.

An integrated geochronological, isotopic, and geochemical study of the Little Port and Bay of Islands complexes of the Humber Arm Allochthon shows: (1) a significant age difference between a U-Pb zircon age of 505 +3/-2 Ma for the Little Port Complex and a U-Pb zircon and baddaleyite age of 484 ± 5 Ma for the Bay of Islands Complex, (2) Little Port trondhjemites are characterized by initial E_{Nd} values of -1 to +1 whereas those in the Bay of Islands Complex are +6.5, and (3) geochemical signatures in mafic and felsic volcanic rocks of the complexes are diverse and show a complete gradation between volcanic arc and non-volcanic arc patterns. These data contradict an earlier interpretation that the complexes were coeval parts of the same ophiolite suite connected by a mid-ocean ridge transform fault. An alternative interpretation relates the Little Port Complex to a volcanic arc and the Bay of Islands Complex to a supra-subduction zone setting.

Structural stacking within the allochthons indicates that the highest volcanic and ophiolitic slices are the farthest travelled. The occurrence of ophiolitic detritus in allochthonous Lower Ordovician sandstones and the local presence of volcanic and ophiolitic blocks in basal mélanges, indicate assembly from east to west and emplacement as already assembled allochthons.

The extensive mélanges developed among the lower slices of the allochthons are interpreted as the result of surficial mass wastage with later structural overriding. The broad flat Maiden Point slice of the Hare Bay Allochton has internal recumbent folds throughout that pre-date the formation of underlying mélange, and blocks of ampibolites

and greenschists of the Bay of Island metamorphic sole occur in underlying low grade mélange. Thus many of the structural styles exhibited within the allochthons predated assembly and mélange formation. The leading edges of the higher structural slices in both the Humber Arm and Hare Bay allochthons have a tendency to disrupt, repeat, and disintegrate.

Complex internal structures of the lower sedimentary slices were developed during transport and emplacement. As a generality, the lowest sedimentary slices contain the stratigraphically youngest rocks. The slices have little morphological expression and their outlines and external geometries are in places poorly known. Higher igneous slices have clearer morphological expression and sharper boundaries. Some of their internal structures predate transport. Others relate to transport and emplacement.

The earliest indication of assembly of the allochthons is the reversal in sedimentary provenance as recorded in the stratigraphy of lower structural slices. This and other features indicate diachroneity of assembly and emplacement along the length of the Canadian Appalachians with earlier events in Newfoundland compared to Quebec. An unconformable Llandeilo cover on the Bay of Islands Complex is interpreted as coeval with transport, and the Caradoc Long Point Group was deposited after emplacement of the Humber Arm Allochthon.

Ordovician structures are confused in places where middle Paleozoic thrusts bring Grenville basement rocks and the Cambrian-Ordovician carbonate sequence above the Taconic allochthons. Final emplacement of uppermost slices of ophiolitic and/or volcanic rocks and carbonate slivers may relate to gravity collapse of an overthickened allochthon.

Preservation of allochthons such as those in western Newfoundland suggests burial soon after emplacement. A lack of ophiolitic and other detritus in Silurian, Devonian, and Carboniferous rocks suggests prolonged burial. Present exposure is the result of Mesozoic and Tertiary uplift.

Internal Humber Zone

The internal Humber Zone consists of intensely deformed and regionally metamorphosed rocks of greenschist to amphibolite facies that contrast with less deformed and relatively unmetamorphosed rocks of the external Humber Zone and adjacent Dunnage Zone (Fig. 3.45). It consists of an infrastructure of gneisses and subordinate schists, and a cover sequence of metaclastic rocks with minor metavolcanic rocks and marble, and tectonic slivers of mafic-ultramafic rocks. The infrastructure is mainly equivalent to Grenville basement, now remoulded with the cover sequence. The cover rocks are equivalent to allochthonous and authochthonous sequences of the external Humber Zone. The infrastructure is most extensive and varied in Newfoundland, with only small structural enclaves recognized in Quebec. In both areas, the contact with the cover sequence is generally tectonic, although locally it has been interpreted as an unconformity.

The Dashwoods Subzone of Newfoundland contains large tracts of paragneiss that may be equivalent to similar rocks in the internal Humber Zone.

Westward-directed thrusts bring metaclastic rocks of the internal Humber Zone against the external Humber Zone or the two are separated by steep faults. The continuity of the internal Humber Zone in Newfoundland is disrupted by Carboniferous cover rocks and in Quebec, Silurian-Devonian rocks locally overlie the older deformed rocks.

The infrastructure records the deepest tectonism in the Humber Zone, as it occupied the lowest stratigraphic levels. The presence of Grenville basement this far east in the Humber Zone indicates that protoliths of the internal Humber Zone were deposited near the less deformed miogeoclinal rocks to the west. In Newfoundland, the cover sequence is interpreted to overlap both continental and oceanic substrates.

The lower portions of the cover sequence are remarkably similar; in general they are characterized by metabasalts and coarse metasedimentary rocks that are locally of shallow water origin, and in some places they include felsic metavolcanic rocks. In Newfoundland, these units include lithic correlatives of the lower Labrador Group of the external Humber Zone. In Quebec, they include the Tibbit Hill, the Montagne Saint-Anselme, and Shickshock metavolcanics, as well as lower parts of the Shickshock and Maquereau groups. All of the metabasalts in these units have tholeiitic rift-related petrochemical signatures. Thus, regional correlation and petrochemical analyses indicate that the cover sequence was deposited in a late Precambrian rift setting. Younger strata of the cover sequence most reasonably represent deposits of a contin-ental slope and rise.

ACCRETED TERRANES

Dunnage Zone

The Dunnage Zone is recognized by its abundant volcanic assemblages, ophiolite suites, and mélanges (Fig. 3.61). Sedimentary rocks include slates, greywackes, epiclastic volcanic rocks, cherts, and minor limestones all of marine deposition. Stratigraphic sequences are variable and formations are commonly discontinuous. Most rocks are of Late Cambrian to Middle Ordovician age.

An ophiolitic basement to its volcanic-sedimentary sequences was part of the original Dunnage Zone definition. Accordingly, in plate tectonic models the zone was described as vestiges of the Iapetus Ocean, preserved between the continental Humber and Gander zones. It is now clear that whereas some volcanic-sedimentary sequences of the Dunnage Zone are conformable above the ophiolite suite, others of Early and Middle Ordovician age are unconformable upon disturbed and eroded ophiolite suites and ophiolitic mélanges, and in eastern Newfoundland and northern New Brunswick Lower and Middle Ordovician volcanic-dominated sequences are stratigraphically above Gander Zone clastic sedimentary rocks. Rocks that link a Dunnage oceanic basement in one place with a Gander continental clastic sequence in another are overstep sequences. Regardless of basement relationships, the Middle Ordovician or older volcanic-sedimentary sequences between the Humber and Gander zones are included as part of the Dunnage Zone. Locally in central Newfoundland and Quebec, Middle Ordovician clastic rocks on a mélange substrate are conformable below Lower Silurian rocks. The Silurian rocks are assigned to the Badger Belt in central Newfoundland, but in Quebec they are viewed either as part of the Dunnage Zone or part of the Gaspé Belt.

The Dunnage Zone is widest in northeast Newfoundland, although it is narrow and disappears southwestward at Cape Ray. It is absent on the opposite side of Cabot Strait in Cape Breton Island. Ophiolitic rocks, mélanges, and volcanic sequences reappear in northern New Brunswick. The Dunnage Zone is hidden by the Gaspé Belt in much of eastern Quebec. In eastern Gaspésie, the Humber-Dunnage boundary is offset by dextral transcurrent faults and it follows the Shickshock-Sud Fault farther southwest. Dunnage Zone rocks reappear in the Témiscouata inlier and they form a continuous belt of ophiolitic rocks, mélanges, and marine sedimentary and volcanic rocks throughout the eastern townships of Quebec.

The Humber-Dunnage boundary, the Baie Verte-Brompton Line, is everywhere a tectonic junction, and there are no Middle Ordovician cover rocks that link the Humber and Dunnage zones. The Gander-Dunnage boundary has always been problematic. In northeast Newfoundland, it is a tectonic boundary marked by mafic-ultramafic rocks of the Gander River Complex. In central and southern Newfoundland the boundaries are structural with allochthonous Dunnage Zone rocks above Gander Zone rocks. The Dunnage-Gander boundary beneath the Indian Bay Subzone of northeast Newfoundland is interpreted as stratigraphic.

In New Brunswick, Dunnage Zone volcanic rocks overlie an ophiolitic substrate on both sides of the Rocky Brook-Millstream Fault. Farther south in the Bathurst area, the volcanic rocks overlie quartzites of the Gander Zone with a conglomerate unit marking the stratigraphic base. Tectonic emplacement of northern belts above southern belts occurred in the Late Ordovician-Early Silurian and produced the largest known blueschist belt in the Appalachian Orogen.

Newfoundland

The Newfoundland Dunnage Zone is separated into two large subzones and a few smaller ones. The two large divisions are the Notre Dame Subzone in the west and the Exploits Subzone in the east, separated by the Red Indian Line, a major tectonic boundary traceable across Newfoundland (Fig. 3.62). Two small areas of volcanic and sedimentary rocks are assigned to the Twillingate and Indian Bay subzones in northeast Newfoundland. A larger area of metamorphic rocks, tonalites and mafic-ultramafic complexes, the Dashwoods Subzone, extends from Grand Lake to Port aux Basques of southwest Newfoundland.

The Notre Dame and Exploits subzones have numerous contrasts as follow: (a) Early to early Middle Ordovician faunas of the Notre Dame Subzone have North American affinities whereas those of the Exploits Subzone have Celtic affinities, (b) the Notre Dame Subzone was affected by Taconic Orogeny whereas nearby parts of the Exploits Subzone are unaffected, (c) the subzones have different plutonic histories, (d) the Exploits Subzone contains more sedimentary rocks and mélanges, (e) lead isotopic signatures in volcanogenic sulphide deposits are different in each zone, and (f) structural inliers of the Gander Zone (Mount Cormack and Meelpaeg subzones) are confined to the Exploits Subzone.

The Notre Dame Subzone has one of the widest ophiolite recurrences in the Appalachian Orogen (the Birchy-Advocate-Point Rousse-Betts Cove corridor). The stratigraphic order of ophiolitic units is toward the east or southeast. The overall structural arrangement indicates westward verging imbricate slices with eastward-facing stratigraphy. Intensity of deformation and metamorphism decreases progressively from west to east, or upward through this structural pile.

Recent U-Pb zircon geochronological studies indicate that the Notre Dame Subzone ophiolites are Early Ordovician and about the same age as the Bay of Islands Complex. Ophiolitic components of the Birchy Complex at the Baie Verte-Brompton Line are structurally commingled with metaclastic rocks of the internal Humber Zone, and some may have been oceanic basement to the eastern depositional edge of the metaclastic rocks.

The Exploits Subzone is a composite and structurally complex collection of rocks of varying ages, geochemical groupings and tectonic environments. U-Pb zircon dating of the volcanic sequences within the Victoria Lake Group has identified three age groupings; 513 ± 2 Ma for the Tally Pond volcanics, $498 +6/-4$ Ma for the Tulks Hill volcanics, and $462 +4/-2$ Ma for the Victoria Mine sequence. The Roebucks quartz monzonite, intrusive into the Tulks Hill volcanic rocks, is dated at 495 ± 4 Ma and is probably coeval with the volcanism. Two large plutons, the Valentine Lake and Cripple Back Lake quartz monzonites, dated as 563 ± 2 Ma and $565 +4/-2$ Ma, are late Precambrian, and interpreted to be structurally emplaced within the Victoria Lake Group.

Mélanges occur all across the northeast Exploits Subzone. These include the Dunnage, Dog Bay Point, and Carmanville mélanges from west to east. The mélanges have been interpreted mainly as major slumps. However, some features are difficult to explain by simple surficial slumping; such as structural and metamorphic variations in the Carmanville Mélange that imply tectonic recycling, injection of resistive volcanic blocks by dykes of pebbly shale, and the Dunnage Mélange is the locus for contemporary intrusions.

The Twillingate Subzone contains a large tonalite batholith dated at 510 Ma. It cuts mafic volcanic rocks and it is older than ophiolite suites of the Notre Dame and Exploits subzones. Amphibolite facies metamorphism, intense deformation, and mylonitization along the southern margin of the Twillingate Subzone are out of context with adjacent, younger, less metamorphosed and less deformed rocks of the Notre Dame Subzone.

The Indian Bay Subzone has volcanic rocks and fossiliferous sedimentary rocks typical of the Dunnage Zone, but they occur within the northeast Gander Zone.

The Dashwoods Subzone has metaclastic rocks that are possible Humber Zone correlatives. These may represent Humber inliers in the western Dunnage Zone. Where Humber metaclastic rocks and Dunnage ophiolitic rocks are structurally commingled, metamorphosed, and intruded together, a Humber-Dunnage distinction is impractical.

New Brunswick

Six subzones are recognized in the New Brunswick Dunnage Zone. North of the Rocky Brook-Millstream Fault, the Belledune, Elmtree, and Popelogan subzones occur in two discrete Ordovician inliers that are surrounded by middle Paleozoic rocks of the Chaleurs Uplands. South of the Rocky Brook-Millstream Fault, the Armstrong Brook, Bathurst, and Hayesville subzones occur in the Miramichi Highlands (Fig. 3.80).

The Belledune Subzone contains the Devereaux and Point Verte formations. The Devereaux Formation has ophiolitic components of mainly gabbros, dykes, and basalts. Its isotopic age of 463 ± 1 Ma is significantly younger than the ophiolite suites of the Newfoundland Dunnage Zone. The Point Verte Formation consists of basalt, greywacke, shale, limestone, and chert. Most of its basalts are alkalic, atypical of ophiolitic suites.

Sedimentary and volcanic rocks of the Elmtree Subzone (Elmtree Formation) are compositionally similar to rocks in the Tetagouche Group of the Bathurst Subzone. Their age is probably Llanvirn-early Caradoc, like other Ordovician volcanic rocks in northern New Brunswick. A 1-2 km thick mélange, characterized by chaotically deformed sedimentary rocks, separates the Elmtree Formation and overlying Pointe Verte Formation. The mélange is interpreted as a major thrust zone that superposes older rocks above younger rocks.

Volcanic rocks of the Popelogan Subzone (Balmoral Group) have a chemistry unlike those of the Belledune Subzone and subzones south of the Rocky Brook-Millstream Fault.

The Armstrong Brook Subzone has basalts that are primitive tholeiites with Mid-Ocean Ridge Basalt (MORB) characteristics, including flat rare earth element (REE) patterns. These basalts have a thin, discontinuous layer of mylonitic gabbro at their base along the contact with underlying, very highly strained blueschists. The blueschists are part of an alkalic basalt suite, which is chemically similar to the Llandeilo high-Cr alkalic basalts of the Pointe Verte Formation in the Belledune Subzone.

The Tetagouche Group of the Bathurst Subzone is dominated by a large complex of felsic metavolcanic rocks that have compositions mainly ranging from dacite to rhyolite. The lowest parts of the Tetagouche Group include shallow-water deposits that disconformably overlie relatively deep-water turbidites of the Miramichi Group (Gander Zone). Quartzite and phyllite pebbles derived from the Miramichi Group are locally abundant in felsic agglomerates of the Bathurst Subzone, suggesting that the felsic volcanic magma came up through crust underlain by the Miramichi Group. This supports other indications that both the Miramichi and Tetagouche groups are underlain by a common continental crust.

A conglomerate similar to the basal Tetagouche Group marks the Penobscot unconformity in Maine. This phase of development in nearby New Brunswick was apparently unrelated to pentrative deformation and immediately preceded extrusion of volcanic rocks.

Although some of the felsic volcanic rocks in the Bathurst Subzone are autochthonous or parautochthonous with respect to the underlying clastic sedimentary rocks of the Miramichi Group, most volcanic rocks are allochthonous.

Ordovician stratigraphy of the Hayesville Subzone is similar to that in the Bathurst Subzone to the north.

There are important differences between the Armstrong Brook and Bathurst subzones: (1) north of their tectonic boundary, basalts are compositionally like those found at mid-ocean ridges or are intermediate between mid-ocean ridge and island-arc tholeiites whereas to the south the volcanic rocks are typical of a continental margin rift sequence; (2) no silicic volcanic rocks are interlayered with basalts of the Armstrong Brook Subzone; (3) the sedimentary rocks overlying basalts in the Armstrong Brook Subzone contain numerous limestone lenses and coarse, thick-bedded quartz- and feldspar-rich lithic wackes as in the Pointe Verte Formation, which are rare or absent in the Bathurst Subzone; and (4) red phyllite, chert, jasper, and Zn-Pb-Cu-Ag type massive sulphides, typical of metalliferous sediments found in the Bathurst Subzone, are absent in the Armstrong Brook Subzone, which contains only one Cyprus type Cu-pyrite sulphide prospect, i.e., Middle River Copper.

Lower to Middle Ordovician rocks in the Bathurst and Hayesville subzones are similar to rocks of the Exploits and Indian Bay subzones in Newfoundland, and all contain the same faunas. The correlation with the Exploits Subzone is further strengthened by isotopic dates of volcanic units and similarities between lead isotopes from massive sulphides of the Bathurst and Exploits subzones.

Quebec

The Dunnage Zone of Quebec is divided into the Estrie-Beauce, Mégantic, Témiscouata, and Gaspésie subzones from southwest to northeast. Marine volcanic and sedimentary rocks range in age from Early Ordovician to Early Silurian. Rocks of the Estrie-Beauce, Témiscouata, and Gaspésie subzones are interpreted as laterally equivalent. The Mégantic Subzone has different rocks and stratigraphy.

The Estrie-Beauce Subzone comprises three tectonostratigraphic units: (1) the Saint-Daniel Mélange, which includes the ophiolite complexes at Thetford-Mines, Asbestos, and Mount Orford, (2) the Ascot Complex, and (3) the Magog Group. High-grade metamorphic rock slices in the Saint-Daniel Mélange are correlated with the Chain Lakes Massif and the mélange contains olistoliths of Caldwell sandstones of the internal Humber Zone. The Ascot Complex has three distinctive lithotectonic domains that are dominated by felsic volcanic rocks. These are separated by chaotic phyllites correlated with the Saint-Daniel Mélange. The Magog Group is a 10 km-thick flysch-dominated sequence of Late Llandeilo to Early Caradoc age that unconformably overlies the Saint-Daniel Mélange.

The Témiscouata Subzone contains mainly mélange-type rocks that are separated from the allochthonous Québec Supergroup of the external Humber Zone by the Lac-des-Aigles Fault. Southeast of the fault, the chaotic Trinité Group is unconformably overlain by the Middle Ordovician to Lower Silurian Cabano Formation. The Cabano Formation is conformably overlain by the

Llandovery Pointe-aux-Trembles and Lac Raymond formations which are in turn conformably overlain by the Silurian-Devonian cover sequence of the Gaspé Belt. Chaotic rocks of Trinité Group are correlated with the Saint-Daniel Mélange of the Estrie-Beauce Subzone and the Cabano Formation is correlated with the Magog Group. The Pointe-aux-Trembles and Lac Raymond formations are interpreted as arc-related volcanic and sedimentary suites, much like the Ascot Complex and Magog Group, but somewhat younger. Based on this interpretation, the Cabano/Pointe-aux-Trembles/Lac Raymond formations are included in the Témiscouata Subzone of the Dunnage Zone.

The Gaspésie Subzone is defined mainly on ophiolitic mélanges along the Shickshock-Sud Fault and at the periphyry of the Maquereau Inlier. Clastic sequences of Llanvirn to Caradoc age unconformably overlie mélanges as follow: (1) the Llanvirn Neckwick Formation of the basal Mictaw Group is unconformable upon the Rivière Port-Daniel Mélange, (2) the Llandeilo Dubuc Formation of the upper Mictaw Group overlies the Rivière du Milieu mélange, and (3) the Llanvirn Arsenault Formation rests unconformably on the McCrea mélange. The Silurian Chaleurs Group unconformably overlies the Mictaw Group.

A complete sequence from the basal Arsenault Formation to Silurian rocks of the Gaspé Belt may exist in the western part of the Aroostook-Percé Anticlinorium. This introduces difficulties in defining a Dunnage Zone-Gaspé Belt boundary. If the sequence is complete and gradational, the boundary is arbitrary as it is in the Témiscouata Subzone. If there is a hiatus between the Arsenault Formation and overlying rocks (Garin Formation), the boundary is most reasonably placed at the hiatus. Nevertheless, a structural unconformity that is present in many other places is missing at this locality. A parallel situation exists in the Exploits Subzone of northeast Newfoundland where the Dunnage Mélange is overlain by Caradoc shales that are gradational with a greywacke unit that is Llandovery at its top.

The Mégantic Subzone includes volcanic and sedimentary rocks of the Clinton Formation and the ophiolitic Chesham Mélange, which is correlated with the Hurricane Mountain Mélange of Maine. The Frontenac Formation is removed from the Mégantic Subzone as its age is Silurian-Devonian (Tremblay et al., addenda in Chapter 3).

Summary and conclusions

In Newfoundland, most features of the Ordovician volcanic rocks of the Notre Dame Subzone suggest an ancient island arc built upon an ophiolitic substrate. Other geochemical studies suggest marine volcanism other than that of island arcs (Fig. 11.4). Eastward subduction is favoured, as there is no evidence for a proximal island arc in the stratigraphic record of the Humber Zone.

Similar chemical trends suggesting island arc as well as other volcanic regimes are established for rocks of the Exploits Subzone.

Lower Ordovician (Arenig-Llanvirn) black shale and limestone occurrences throughout both the Notre Dame and Exploits subzones are sporadic, localized, widely separated, and all of somewhat different ages. The situation suggests evolving, multiple Early Ordovician tectonic

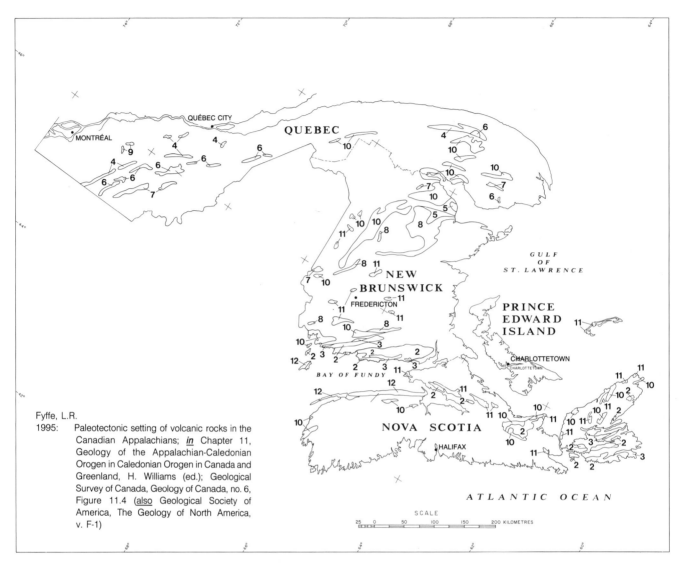

Fyffe, L.R.
1995: Paleotectonic setting of volcanic rocks in the Canadian Appalachians; *in* Chapter 11, Geology of the Appalachian-Caledonian Orogen in Caledonian Orogen in Canada and Greenland, H. Williams (ed.); Geological Survey of Canada, Geology of Canada, no. 6, Figure 11.4 (<u>also</u> Geological Society of America, The Geology of North America, v. F-1)

Figure 11.4. Paleotectonic setting of volcanic rocks in the Canadian Appalachians (compiled by L.R. Fyffe, 1991).

elements without significant continuity, even within a particular subzone. The idea of a Dunnage collage in Newfoundland is also supported by: (a) the transitional chemical affinities of most Dunnage ophiolite suites, implying that none represents the crust of a major ocean, (b) some volcanic rocks overlie an ophiolitic substrate but others contain zircons with inheritance from older crust, (c) volcanic rocks and tonalites of the Twillingate Subzone are older than surrounding ophiolite suites, and (d) late Precambrian ages for quartz monzonites (Valentine Lake and Cripple Back Lake) of the Exploits Subzone.

In New Brunswick, volcanic rocks of the Tetagouche Group are interpreted as a continental margin, rift sequence, produced in late Arenig/Llanvirn times. They formed during the opening of an ensialic back-arc basin, which subsequently developed into a marginal sea floored by upper Landeilo-lower Caradoc ophiolitic rocks of the Belledune and Armstrong Brook subzones. Other volcanic rocks of the Popelogan Subzone are interpreted as part of a Middle Ordovician volcanic arc.

In Quebec, the Ascot Complex of the Estrie-Beauce Subzone and correlatives may represent a dismembered volcanic arc complex. The Saint-Daniel, Trinité, and Gaspésie mélanges are probably all remnants of a single accretionary prism, and the Magog, Cabano-Pointe aux Trembles-Lac Raymond, and Mictaw-Arsenault units are overlying deposits of a forearc basin. Relationships suggest diachronous arc volcanism and forearc sedimentation; older in the southwest (Estrie-Beauce Subzone), and younger in the northeast (Gaspésie Subzone).

The Dunnage Zone is therefore a complex composite entity. Its definition is clearest where its volcanic sequences and mélanges are developed upon an ophiolitic substrate. However, other volcanic assemblages overlie continental rocks of the Gander Zone.

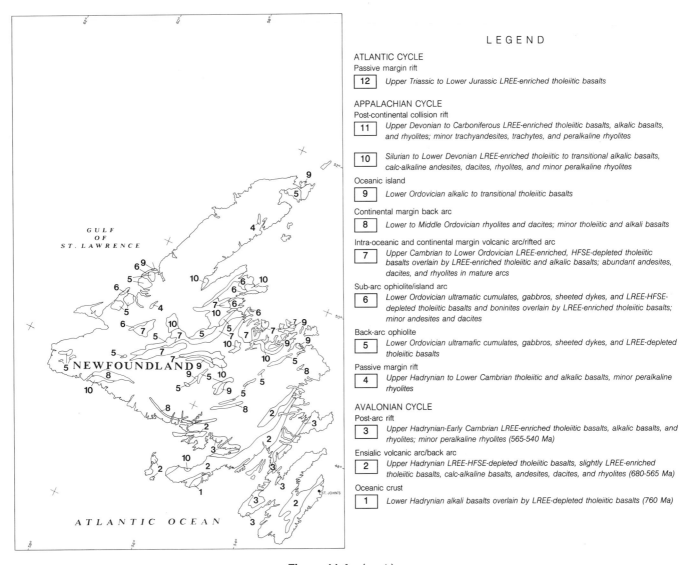

ATLANTIC CYCLE
Passive margin rift

12 Upper Triassic to Lower Jurassic LREE-enriched tholeiitic basalts

APPALACHIAN CYCLE
Post-continental collision rift

11 Upper Devonian to Carboniferous LREE-enriched tholeiitic basalts, alkalic basalts, and rhyolites; minor trachyandesites, trachytes, and peralkaline rhyolites

10 Silurian to Lower Devonian LREE-enriched tholeiitic to transitional alkalic basalts, calc-alkaline andesites, dacites, rhyolites, and minor peralkaline rhyolites

Oceanic island

9 Lower Ordovician alkalic to transitional tholeiitic basalts

Continental margin back arc

8 Lower to Middle Ordovician rhyolites and dacites; minor tholeiitic and alkali basalts

Intra-oceanic and continental margin volcanic arc/rifted arc

7 Upper Cambrian to Lower Ordovician LREE-enriched, HFSE-depleted tholeiitic basalts overlain by LREE-enriched tholeiitic and alkalic basalts; abundant andesites, dacites, and rhyolites in mature arcs

Sub-arc ophiolite/island arc

6 Lower Ordovician ultramafic cumulates, gabbros, sheeted dykes, and LREE-HFSE-depleted tholeiitic basalts and boninites overlain by LREE-enriched tholeiitic basalts; minor andesites and dacites

Back-arc ophiolite

5 Lower Ordovician ultramafic cumulates, gabbros, sheeted dykes, and LREE-depleted tholeiitic basalts

Passive margin rift

4 Upper Hadrynian to Lower Cambrian tholeiitic and alkalic basalts, minor peralkaline rhyolites

AVALONIAN CYCLE
Post-arc rift

3 Upper Hadrynian-Early Cambrian LREE-enriched tholeiitic basalts, alkalic basalts, and rhyolites; minor peralkaline rhyolites (565-540 Ma)

Ensialic volcanic arc/back arc

2 Upper Hadrynian LREE-HFSE-depleted tholeiitic basalts, slightly LREE-enriched tholeiitic basalts, calc-alkaline basalts, andesites, dacites, and rhyolites (680-565 Ma)

Oceanic crust

1 Lower Hadrynian alkali basalts overlain by LREE-depleted tholeiitic basalts (760 Ma)

Figure 11.4. (cont.)

There is little evidence for Ordovician penetrative deformation in most of the Dunnage Zone, more than that its Ordovician mélanges indicate mobility, and that the makeup of clastic sedimentary sequences imply uplift and erosion. In many other places, even where Ordovician deformation is demonstrated by unconformable relationships, the deformation is mild compared to middle Paleozoic deformation.

Gander Zone

The type area of the Gander Zone is northeast Newfoundland where a thick monotonous sequence of poly-deformed quartz greywacke, quartzite, siltstone and shale grade eastward into psammitic schist, gneiss and migmatite. These sedimentary rocks contrast with the mixed volcanic-sedimentary assemblages of mainly lower grade rocks of the bordering Dunnage and Avalon zones (Fig. 3.111a). The age of the Gander Zone sedimentary rocks in Newfoundland is constrained mainly by the age of its oldest cover rocks. These are Arenig to Llandeilo, and if the fossiliferous rocks are in stratigraphic contact with the monotonous clastic sequence, the Gander Zone clastics are mainly Early Ordovician and earlier. Tremadoc to middle-late Arenig graptolites occur in the clastic sequence of the Gander Zone in New Brunswick. An occurrence of *Oldhamia* in Gander Zone correlatives in Maine indicates a Cambrian age.

Clastic rocks of the type area in Newfoundland extend southward to Bay d'Espoir and crystalline rocks extend around the Hermitage Flexure to southwestern Newfoundland. However, the nature, or even the existence, of the Gander Zone in southwestern Newfoundland is debatable and most rocks there are now assigned to the Burgeo Subzone, a possible equivalent of the Bras d'Or Subzone of Cape Breton Island. The Aspy Subzone of Cape Breton Island is included with the Gander Zone because of plutonic and metamorphic styles, but most of its dated rocks are younger and atypical. In New Brunswick, the Gander Zone includes the clastic sequence of the Miramichi Group in the Miramichi Subzone, and clastic rocks of the Cookson Group in the St. Croix Subzone.

The Gander-Avalon boundary is a sharp tectonic junction in Newfoundland: the Dover and Hermitage Bay faults. An unconformity between a basal conglomerate on the Cheticamp Lake Gneiss of the Aspy Subzone and deformed diorite of the Bras d'Or Subzone in Nova Scotia indicates a depositional contact (Lin, 1993). If the Aspy Subzone is a Gander Zone equivalent, and the Bras d'Or Subzone an Avalon Zone equivalent, Gander and Avalon zones in Nova Scotia locally have the same late Precambrian plutons and Ordovician-Silurian cover rocks. In New Brunswick, the Gander-Avalon boundary is hidden by Silurian rocks.

Even before the wide acceptance of plate tectonics, the Gander Zone clastic rocks were viewed as a prism of sediment built up parallel to an Avalon shoreline on the eastern side of a Paleozoic ocean. The idea persisted chiefly because the Gander clastic rocks are of continental affinity and Gander Zone gneisses and migmatites resemble continental basement. Accordingly, the Gander Zone is commonly termed the eastern margin of Iapetus. However, there are no confirmed basement relationships or stratigraphic analyses comparable to those for the Humber miogeocline. Thus, the Gander Zone has also been viewed as a suspect terrane.

Almost one half of the area of the Gander Zone consists of granitic intrusions and one half of the remainder comprises metamorphic rocks ranging from greenschist to upper amphibolite facies. Foliated biotite granite, foliated to massive garnetiferous muscovite-biotite leucogranite, and potassic megacrystic biotite granite are characteristic. Since granites of this type and size are atypical of oceans or ophiolitic basement, it is concluded that the Gander Zone lies above continental crust. Regional gravity, magnetic, and seismic signatures support this conclusion.

Rocks of the Gander Zone exhibit folded cleavages or schistosities and are almost everywhere more deformed than rocks in the adjacent Dunnage and Avalon zones. Coupled with evidence of Early to Middle Ordovician juxtapositioning with ophiolitic rocks of the Dunnage Zone, relationships hint at pre-Middle Ordovician deformation. The latest isotopic data suggest that major deformation and regional metamorphism are Silurian.

Newfoundland

The Newfoundland Gander Zone comprises three discrete subzones (Fig. 3.111a). The type area and its southward continuation is the Gander Lake Subzone. The inland areas are the Mount Cormack and Meelpaeg subzones. These inland subzones are structural windows or core complexes and relationships indicate a two-layer crust, with Dunnage Zone rocks above Gander Zone rocks. The Gander Zone was extended previously to include rocks between La Poile Bay and Grey River, Burgeo Subzone, chiefly because of metamorphic and plutonic styles, and because these rocks occur along strike with Gander Zone metamorphic and foliated plutonic rocks in the Hermitage Flexure. The Burgeo Subzone is now assigned to the Avalon Zone and correlated with rocks of the Bras d'Or and Aspy subzones of Cape Breton Island.

Three lithic units are recognized in the Gander Lake Subzone in the vicinity of Gander Lake and eastward (Fig. 3.11b). From west to east, these are the low grade clastic rocks of the Gander Group, low- to medium-grade metamorphic rocks of the Gander Group (Square Pond Gneiss), and medium- to high-grade metamorphic rocks of the Hare Bay Gneiss.

The Gander Group consists of interbedded psammite and pelite with quartz-granule sandstone and quartzite. Its depositional setting is uncertain. Sharp interbeds of quartz sandstone within pelite suggest distinct pulses of sand transport into a mud-dominated shallow marine or shelf environment. Thick, graded sandstone beds suggest turbidity currents and a deeper basinal setting. A major lithic change between the monotonous clastic sedimentary rocks and overlying mixed lithologies of the Indian Bay Subzone coincides with the unconformity between the Davidsville Group and Gander River Complex of the nearby Dunnage Zone.

The Square Pond Gneiss consists of psammitic and semipelitic metasedimentary rocks with zones of schist and migmatite. Psammite with a "pinstripe" or "herring bone" banding is characteristic. The Hare Bay Gneiss consists of migmatitic and tonalitic gneiss containing xenoliths and rafts of paragneiss and amphibolite. All contacts are now regarded as gradational.

Metamorphic rocks extend along the eastern margin of the Gander Lake Subzone to Bay d'Espoir on the south coast. There, the Little Passage Gneiss is correlated with the Square Pond and Hare Bay gneisses to the north. The boundary between the Little Passage Gneiss and rocks of the Dunnage Zone is a major ductile shear with left lateral sense of displacement. Isotopic ages indicate Silurian tectonism.

The Mount Cormack Subzone is an oval area of clastic rocks and metamorphic equivalents correlated with the Gander Group. It is surrounded, or almost so, by ophiolite complexes whose sequences of stratigraphic units everywhere face outwards from the Mount Cormack Subzone. Accordingly, the Mount Cormack Subzone is interpreted as a window of Gander Zone rocks exposed through an allochthonous cover of ophiolitic and volcanic-sedimentary rocks of the Dunnage Zone.

The Meelpaeg Subzone consists of psammitic rocks and metamorphic or migmatitic equivalents. These rocks are correlated with clastic rocks of the Mount Cormack Subzone and the Gander Group. Boundaries of the Meelpaeg Subzone are tectonic, where not truncated by intrusions.

In all parts of the Gander Lake Subzone there is a metamorphic progression southeastward that culminates in migmatitic gneisses intruded by syntectonic granitoid plutons. The juxtaposition of migmatites of the Gander Lake Subzone and greenschist or lower grade Precambrian rocks of the Avalon Zone is one of the sharpest metamorphic breaks at any zone boundary.

The Mount Cormack Subzone has a systematic progression of concentric metamorphic isograds that are roughly conformable with its oval outline (Fig. 3.114). From the periphery inward, they indicate a rapid increase in grade from greenschist to migmatitic upper amphibolite facies. The Through Hill Granite in the centre of the subzone has a variety of features that suggest an origin through anatexis of supracrustal rocks.

Structural styles, metamorphism, and plutonism are all consistent with the model of a major Dunnage allochthon above the Gander Zone. A two layer crust explains the presence of widespread granitic plutons throughout the ophiolitic Dunnage Zone and more intense metamorphism, migmatization and plutonism in the Gander Zone. Broad zones of ductile shearing were later. Where shearing affected rocks of both the Dunnage and Gander zones, structural, metamorphic and plutonic styles have no regard for the zone boundary.

Cape Breton Island, Nova Scotia

The Aspy Subzone is characterized by a variety of low- to high-grade metamorphic rocks intruded by Silurian-Devonian granitic rocks. It is included with the Gander Zone mainly because: (1) the Aspy Subzone lies between rocks of Humber affinity to the northwest and Avalon affinity to the southeast, (2) structural, metamorphic and plutonic styles of the Aspy Subzone are more akin to those of the Gander Zone than to any other zone, and (3) rocks of the Aspy Subzone were correlated with those of the Burgeo Subzone and the La Poile Belt of southwest Newfoundland. However, no monotonous psammitic sequence like that of the Newfoundland and New Brunswick Gander zones is present and there are no indigenous layered rocks of definite Ordovician or older age. Correlation with the Burgeo Subzone also suggests Avalon Zone affinity.

The Eastern Highlands shear zone and associated faults overprint the Aspy-Bras d'Or subzone boundary, that was originally stratigraphic (Lin, 1993). Seismic data suggest that the boundary dips to the south, and that the Bras d'Or Subzone is thrust against the Aspy Subzone. This is consistent with the presence of southerly dipping thrusts within the Eastern Highlands shear zone. The Chéticamp Pluton and associated diorites and gneisses in the western part of the Aspy Subzone suggest Bras d'Or Subzone affinities.

The stratified rocks of the Aspy Subzone are greenschist to amphibolite facies phyllites and schists, and upper amphibolite facies gneisses of volcanic and sedimentary protoliths. A diverse suite of mainly Silurian and Devonian plutons, which range from strongly foliated to massive, is a distinctive feature.

Complex deformation and metamorphism, combined with poor exposure, make it difficult to determine whether tectonic or stratigraphic breaks exist within the metavolcanic-metasedimentary sequences, and the relations among phyllites, schists, and gneisses are controversial. U-Pb zircon ages of felsic volcanic rocks indicate that they are early Silurian. Mafic volcanic units have petrological characteristics like those of volcanic-arc tholeiites, and are interpreted to have formed in a subduction zone. Metamorphism and protracted deformation occurred during Silurian to Devonian. The latest model has the Aspy Subzone as an island arc separated from the Bras d'Or Subzone by a small back-arc basin (Lin, 1993).

New Brunswick

The New Brunswick Gander Zone is characterized by a thick, complexly deformed sequence of Cambrian-Ordovician quartz sandstones and pelite that constitute the Miramichi and Cookson groups in the Miramichi and St. Croix subzones, respectively. Inclusion of the St. Croix Subzone extends the Gander Zone of New Brunswick southward to the Avalon Zone boundary, corresponding with its limits in Newfoundland. This boundary in New Brunswick is mainly hidden by Silurian to Lower Devonian volcanic and sedimentary rocks of the Mascarene Belt. The St. Croix Subzone is separated from the Miramichi Subzone by Silurian turbidites of the Fredericton Belt (Fig. 3.80). A wedge of complexly deformed pelite and psammite of Gander aspect has recently been recognized along the Fredericton Fault, which transects the central portion of the Fredericton Belt.

The Miramichi Group in the northern part of the Miramichi Highlands consists of dark grey to black, locally graphitic shales, rhythmically interbedded with medium- to thick-bedded, light grey to olive-green, graded quartz sandstones. Greenschist-facies quartz sandstones and shales in the central Miramichi Highlands pass westward through a steep metamorphic gradient into amphibolite-facies rocks. These are thin banded, fine grained amphibolites which contain dark green hornblende-rich bands alternating with light grey plagioclase-rich bands. The amphibolites are interlayered with cordierite-andalusite-bearing pelites and psammites, and with granitic gneiss containing microcline augen. The granitic gneisses were originally interpreted as Precambrian basement to the Miramichi Group. U-Pb ages are Ordovician.

The St. Croix Subzone contains black carbonaceous slate and minor thin-bedded wacke overlain by a pillowed basalt member and a thick sandstone-rich sequence at the top. Graptolites from the base of the sequence are Early Ordovician (Tremadoc), and Caradoc near its top. Northeast of the Saint John River, a thick steeply dipping volcanic succession is included in the subzone.

The tectonometamorphic history of the Gander and Dunnage zones of New Brunswick are remarkably similar. There is no correspondence between structural styles and stratigraphic divisions. At least five generations of structures have been recognized by overprinting relationships. These are interpreted as part of one continous orogenic event that is constrained between the Late Ordovician and Middle Devonian. The earliest structures are interpreted to have formed by progressive thrusting in an accretionary wedge (Fig. 3.97). Exhumation of parts of the accretionary wedge started at least in the Early Silurian.

Avalon Zone

The Avalon Zone is defined in most places by its well preserved upper Precambrian sedimentary and volcanic rocks and overlying Cambrian-Ordovician shales and sandstones. In a few places it is defined by isotopic ages and intrusive relationships. The Avalon Terrane or Avalon Composite Terrane are other names for the Avalon Zone.

The Avalon Zone (Maps 1 and 2, and Fig. 3.126) is the broadest of Canadian Appalachian zones and is more than twice the combined width of zones to the west in its type area of Newfoundland. It is much narrower in Nova Scotia and southeast New Brunswick.

Boundaries of the Avalon Zone are major faults. In Newfoundland, its western boundary is the Dover-Hermitage Bay Fault. In New Brunswick, its western boundary is the Belleisle Fault or the Taylor Brook Fault farther west. The

Burgeo Subzone is a recent extension of the Avalon Zone in Newfoundland. Ambiguity exists regarding the full extent of the Avalon Zone in Cape Breton Island, Nova Scotia. One interpretation places the northwestern Avalon boundary between the Bras d'Or and Mira subzones; another suggests that all of Cape Breton Island, including Grenvillian basement at its northwest extremity, belongs to the Avalon Zone. The southern boundary of the Avalon Zone with the Meguma Zone is the Cobequid Fault of Nova Scotia. These faults disrupt the continuity of lithic belts within the Avalon Zone and account for its extreme variability in width from Newfoundland to New Brunswick.

The upper Precambrian sedimentary and volcanic rocks were affected by late Precambrian Avalonian Orogeny, expressed by granite plutonism and generally mild deformation. More intense Precambrian deformation occurs locally. Middle Paleozoic orogeny was intense, especially where the zone is narrow in Nova Scotia.

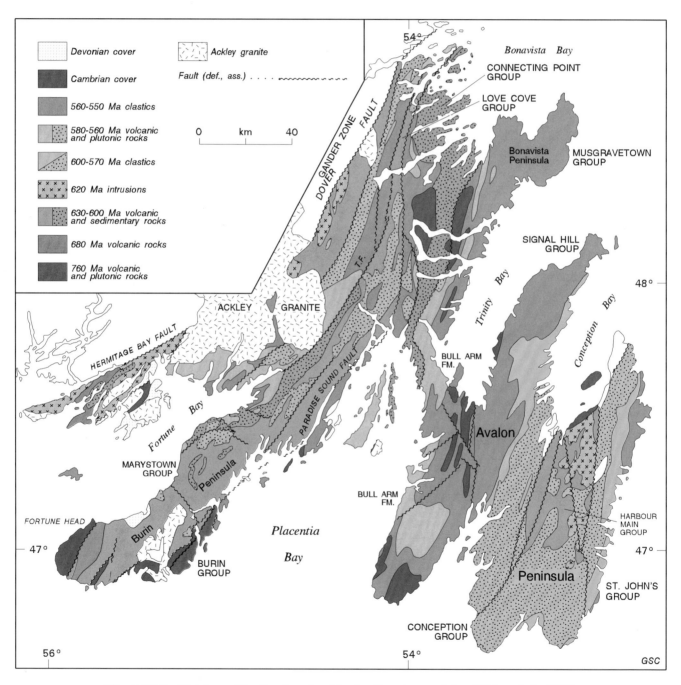

Figure 11.5. Divisions of the Newfoundland Avalon Zone by age (after O'Brien et al., 1990).

The Avalon Zone differs from other zones of the Appalachian Orogen in the following ways: its upper Precambrian rocks evolved during the Grenville-Appalachian time gap, although they are in part coeval with the early stages of the Appalachian cycle; its full complement of Cambrian strata, mainly shales, contain Acado-Baltic trilobite faunas distinctive from those of the Humber miogeocline; the widest expanse of the Avalon Zone across the Grand Banks of Newfoundland is virtualy unaffected by Paleozoic deformation; and the zone has distinctive mineral deposits.

Newfoundland

The oldest rocks of the Newfoundland Avalon Zone are those of the Precambrian Burin Group, a mixed assemblage of sedimentary and volcanic rocks with a stromatolite-bearing carbonate olistostrome near its base. A comagmatic gabbro sill is dated at 763 +2.2/-1.8 Ma. The rocks occur in a fault-bounded area.

Later Precambrian rocks underlie most of the type area of the Avalon Peninsula and similar rocks occur west of the Avalon Peninsula (Fig. 3.128). They are separable into three lithological units that have at least local stratigraphic significance: a basal volcanic unit dominated by ignimbrites and volcanic breccias; a middle marine sedimentary unit of grey-green sandstones, siltstones, and siliceous argillites overlain by shales and sandstones; and an upper unit of terrestrial sedimentary and volcanic rocks dominated by red and grey sandstones and conglomerates.

The basal volcanic unit has several kilometres of felsic to mafic, subaerial to submarine volcanic rocks. Original textures of ignimbrites, welded tuffs, and flow-banded rhyolites are remarkably well-preserved.

The middle marine unit is dominated by green to grey siliceous fine grained sedimentary rocks. It also includes conglomerate, mixtite of glacial origin, tuff, agglomerate, minor pillow lava, and mafic dykes. Its thickness in the southern Avalon Peninsula is about 4.5 km. Some clasts in Precambrian glacial deposits, such as quartzite, foliated granite, and detrital garnets are exotic to the Newfoundland Avalon Zone. Casts and impressions of soft-bodied metazoans of *Ediacaran* type are known from the top of the unit.

The upper unit of terrestrial sedimentary and volcanic rocks has a thickness in excess of 5 km. Volcanic rocks occur locally at or near the base of the section. In other places, grey sandstones at the base represent a transition from shallow marine to subaerial environments. Thick bedded red arkosic sandstones and red conglomerates higher in the succession indicate a major shoaling- and coarsening-upward cycle.

Precambrian-Cambrian relationships are variable, though well known, throughout most of the Newfoundland Avalon Zone. Preserved sections of upper Precambrian rocks in places pass upwards into Cambrian strata without significant disconformity. This is exemplified by the section at Fortune Head that is the Precambrian-Cambrian boundary stratotype. In other places, unconformities are common both within the upper Precambrian successions and below the Cambrian cover.

Late Precambrian intrusive rocks are common. One suite, dated isotopically from 620 to 580 Ma, consists of hornblende and biotite granite, granodiorite, and diorite with calc-alkaline affinities. Its isotopic ages and field relations demonstrate a coeval relationship with the basal volcanic unit. Another suite of smaller plutons, confined to the western part of the Avalon Zone, consists of gabbro, alkali granite, riebeckite peralkaline granite, and related hybrid rocks. Preliminary Rb-Sr ages for this suite are from 560 to 540 Ma. These intrusions are peralkaline and probably related to volcanic rocks at the base of the upper terrestrial unit.

Evidence for late Precambrian Avalonian Orogeny is provided by granite intrusion, block faulting, gentle folding, and low grade regional metamorphism. High strain zones and penetrative fabrics indicate local intense Precambrian deformation. The effects of Paleozoic deformation, plutonism, and regional metamorphism decrease eastward and locally Cambrian-Ordovician rocks of the eastern Avalon Peninsula are subhorizontal.

In the western Avalon Zone of the Hermitage Bay Peninsula, at least four prominent orogenic events are recognized: ca. 760 Ma, ca. 680 Ma, ca. 630-600, and ca. 575-550 (O'Brien et al., 1990, 1992, 1994). Intrusive relations and ductile shears in Precambrian rocks that are unconformably overlain by other upper Precambrian rocks demonstrate distinctive tectonic events that occurred at least twice prior to the deposition of a Cambrian platformal cover. Coeval Precambrian tectonic events are recorded in similar Precambrian rocks in the Burgeo Subzone. There, the Precambian rocks have an unconformable Silurian cover and the northern boundary of the Burgeo Subzone was the locus of a widespread and complex Silurian tectonothermal event which reactivated funadamental Precambrian structures (O'Brien et al., 1992).

A recent subdivision of Precambrian rocks of the Newfoundland Avalon Zone based on isotopic ages is depicted in Figure 11.5. Differences between this subdivision and the broad lithic correlations (Fig. 3.128) imply major facies changes and/or separate terranes within the Newfoundland Avalon Zone.

Nova Scotia

Rocks of the Avalon Zone in Nova Scotia occur in Cape Breton Island, the Antigonish Highlands and Cobequid Highlands. The Mira Subzone of southeastern Cape Breton Island has upper Precambrian volcanic and sedimentary rocks overlain by Cambrian-Ordovician rocks typical of the Avalon Zone in its type area. The adjoining Bras d'Or Subzone to the north is regarded as Avalonian in some analyses because: (1) its Precambrian marbles and quartzites of the George River Group are correlated with the Green Head Group of the Avalon Zone in New Brunswick, and (2) it has Cambrian shales with a fauna typical of the Avalon Zone, although the Cambrian rocks may be allochthonous. Another view defines the Avalon Zone on its Cambrian-Ordovician overstep sequence that contains an Acado-Baltic fauna, and/or its Silurian-Lochkovian cover that contains a Rhenish-Bohemian fauna, and all of Cape Breton Island is assigned to the Avalon Zone. Whether or not Cape Breton Island represents a narrow cross-section of the entire Appalachian Orogen, or an oblique section of the Avalon Zone with progressively deeper levels from southeast to northwest, is debatable. Expressed in another way, does the Iapetus suture cross Cape Breton Island or does it lie offshore to the north?

The oldest rocks of Cape Breton Island are ortho- and paragneisses and amphibolites in the Blair River Complex, with metamorphic ages of about 1000 Ma. Strongly deformed metasedimentary rocks of the George River Group and correlatives consist of quartzite, carbonate, greywacke, slate and minor metavolcanic rocks. These may also be affected by events in the order of 1000 Ma. Contacts between gneisses and metasediments are tectonic.

Upper Precambrian rocks in the Mira Subzone (Fig. 3.126) are subaerial and shallow marine pyroclastic rocks, flows, and small intrusions. In the Antigonish Highlands they consist of a bimodal sequence of basalt, basaltic andesite, and rhyolite overlain by a thick succession of turbidites with volcanic clasts, and minor basalts. Overlying turbidites are tuffaceous, green and siliceous. In the northern and western Cobequid Highlands, the upper Precambrian rocks have a similar stratigraphy and volcanic geochemistry to those of the Antigonish Highlands.

In the Mira Subzone, the upper Precambrian rocks are overlain by Cambrian-Ordovician sediments with an Acado-Baltic fauna. In the Boisdale Hills of the Bras d'Or Subzone, a Middle Cambrian section contains a thick bimodal volcanic suite. In the Antigonish Highlands, Cambrian-Ordovician sediments rest unconformably upon deformed upper Precambrian rocks. Bimodal volcanic rocks are interbedded with the Lower Cambrian sedimentary rocks.

Plutonic rocks are common throughout the Avalon Zone of Nova Scotia and they fall into two age groups: about 635-600 Ma and about 565-490 Ma. The calc-alkaline geochemistry of the older group of plutons suggests that they are genetically related to coeval volcanic rocks. Unlike southern Cape Breton Island, the plutons in the Antigonish and Cobequid highlands cut polydeformed volcanic-sedimentary rocks. However, ages indicate that sedimentation, deformation, and intrusion were penecontemporaneous.

The younger group of plutons may be the subvolcanic equivalents of upper Precambrian-Cambrian volcanic rocks.

New Brunswick

The Avalon Zone of New Brunswick has mainly upper Precambrian to Lower Cambrian volcanic, sedimentary, and intrusive rocks. Most of the upper Precambrian volcanic rocks are assigned to the Coldbrook Group. It is inferred to overlie carbonates and clastic rocks of the Green Head Group, including the spatially associated Brookville Gneiss. It is overlain by redbeds of the Ratcliffe Brook Formation and shales of the Cambrian-Ordovician Saint John Group. Isotopic dates define two groups of intrusions at about 600-550 Ma and 547-510 Ma.

Metacarbonates and clastic sedimentary rocks of the Green Head Group and Brookville Gneiss are either basement to the Coldbrook Group or a separate unrelated terrane (Fig. 3.144). The Green Head Group consists of a lower mainly carbonate unit (Ashburn Formation) and an upper mainly clastic unit (Martinon Formation). Stromatolites in the Ashburn Formation suggest a mid Proterozoic age (1400-1000 Ma). The Brookville Gneiss consists of a diapiric core of orthogneiss surrounded by a mantle of paragneiss, largely formed from the Green Head Group. One isotopic age of about 605 Ma and another that defines

a maximum age of 641 Ma suggest that the Brookville Gneiss is much younger than previously considered. Its contact with the Green Head Group is tectonic.

The Martinon Formation may be equivalent to the lower part of the Coldbrook Group as: (1) it unconformably overlies other rocks with the contact marked by a distinctive marble conglomerate, (2) it is of turbiditic origin and therefore sedimentologically different from Ashburn carbonates, and (3) it contains numerous basalt sills suggesting that it is akin to volcanic rocks of the Coldbrook Group.

The contact between the Green Head Group and Coldbrook Group was considered to be an unconformity because of the more deformed and metamorphosed nature of the Green Head Group. If the Martinon Formation is equivalent to the lower part of the Coldbrook Group, an unconformity is expected.

Three lithological divisions are recognized in the Coldbrook Group: (1) a subaqueous sequence, at least 5000 m thick, that is dominated by fine grained mafic and felsic tuffaceous rocks, massive and pillowed mafic flows, greenish grey and purple siltstone, and greenish grey sandstone; (2) a subaerial sequence, several kilometres thick, composed of felsic and lesser mafic volcanic rocks with minor sedimentary rocks; and (3) a turbiditic sequence of greenish, thin bedded, siliceous siltstones and sandstones (Fig. 3.145).

Volcanic rocks of the first division are the oldest and approximately coeval with granite and diorite dated at 600-550 Ma. Those of the second division are terrestrial with ages of about 547-510 Ma and in part co-magmatic with syenogranitic and gabbroic plutons. Volcanic rocks of both divisions are mainly bimodal. Other volcanics of the Coldbrook Group are calc-alkalic.

The Ratcliffe Brook Formation is predominantly a redbed sequence with minor limestone (Fig. 3.146). The lithology and stratigraphy of these rocks are distinct from Cambrian shales of the Saint John Group. They unconformably overlie pillow basalt of the Coldbrook Group and a 625 Ma granite east of Saint John. The redbeds postdate the main Coldbrook volcanism and they are overlain by trilobite- bearing Cambrian rocks.

The first episode of deformation is related to the emplacement of Brookville orthogneiss diapirs into the Green Head Group. U-Pb zircon ages indicate that this deformation is much younger than previously considered. The fact that amphibolite grade metamorphism only occurs in Green Head rocks that are spatially associated with Brookville gneisses, suggests that metamorphism and orthogneiss emplacement were coeval. This style of deformation does not affect the Coldbrook Group or younger rocks.

Correlations and interpretation

Marbles and quartzites of the George River and Green Head groups were interpreted as a cover sequence above crystalline basement, but contacts are tectonic and this interpretation is not supported by the latest isotopic ages. If stromatolitic blocks in the Newfoundland Burin Group are correlatives of the stromatolitic Green Head Group, the marbles and quartzites are older than the 763 Ma age of the Burin Group. Since the Burin Group has ophiolitic chemical affinities, implying the existence of oceanic crust, possibly its stromatolitic blocks were derived from a contemporary carbonate platform. Correspondence in age, composition, and regional setting of the Burin Group with Pan African ophiolitic rocks suggest African links.

Upper Precambrian volcanic and sedimentary rocks, dated at about 600 Ma, are the products of extensive subaerial and submarine eruptions and they occur in Newfoundland, Nova Scotia, and New Brunswick. In the type area of the Avalon Peninsula in Newfoundland, the volcanic rocks are high alumina, low titanium basalts of transitional to mildly alkaline chemical affinity. West of the Avalon Peninsula, they are calc-alkaline and tholeiitic. In Nova Scotia, correlative basalts are tholeiites of continental rift affinity, others are calc-alkalic basaltic andesites, and associated rhyolites have volcanic arc affinities. In New Brunswick, lower parts of the volcanic sequence are calc-alkaline. Almost everywhere, these volcanic rocks are coeval with voluminous calc-alkaline, diorite-granodiorite-granite intrusions without significant metamorphism or structural complications. The geochemistry of the volcanic rocks of southern Cape Breton Island and southern New Brunswick indicates that they were erupted in a volcanic arc environment. In both regions, the mafic rocks display a remarkably similar compositional zonation which resembles the across-arc variations observed in modern volcanic arcs. The progressive compositional changes that include a transition from island arc tholeiites along the southeastern coasts to calc-alkaline rocks inland to the northwest, suggest a northwest-dipping subduction zone with a trench located to the southeast.

This volcanic-plutonic activity was at least in part contemporaneous with sedimentation in deep marine basins. The extremely thick basin-fills in Newfoundland include complex successions of turbidites with tillites and olistostromes. Similar marine successions of mainly greenish siliceous rocks occur in Nova Scotia and New Brunswick. This phase of development may have lasted for about 50 Ma.

The discrete occurrences of upper Precambrian rocks in Cape Breton Island, Antigonish Highlands, and Cobequid Highlands of Nova Scotia are viewed as part of a single magmatic-volcanic complex. A model of a rifted ensialic magmatic arc floored by continental basement is favoured. Diachronous deformation associated with opening and closing of intra-arc rifts may have controlled Avalonian Orogeny.

The mild effects of Paleozoic and late Precambrian deformation on the wide Newfoundland Avalon Zone suggest that the present configuration of alternating upper Precambrian volcanic and sedimentary belts is a relict of late Precambrian evolution. The pattern indicates volcanic ridges flanked and separated by marine sedimentary basins. A modern analogue for the alternating volcanic and sedimentary belts in Newfoundland may be the Marianas region of the Pacific Ocean. This model is supported by a comparison of the upper Precambrian marine sedimentary rocks and recent counterparts in basins adjacent to existing island arcs.

Comparisons between the late Precambrian evolution of the Avalon Zone and Pan-African terranes suggest similar tectonic controls and support a late Precambrian accretionary model for the Canadian Avalon Zone.

Upper Precambrian to Cambrian terrestrial clastic rocks were deposited as shoaling-upward fan-deltas and alluvial fans within fault-bounded basins. Associated volcanic rocks are terrestrial bimodal suites. Comagmatic plutons are strongly alkaline or peralkaline and dated at about 550 Ma. These volcanic and sedimentary rocks are completely different compared to the older volcanic and sedimentary rocks. This phase of activity followed Avalonian Orogeny. It may be explained by a model of pull-apart basins or continental rift basins, like the Basin and Range Province of the western United States. If the isotopic ages of about 550 Ma are reliable, this phase of activity was penecontemporaneous with the rift-drift transition as defined in the stratigraphy of the Humber miogeocline, and a major episode of worldwide rifting. Other models relate this phase of activity to the final stages of an accretionary event. The choice here is between interpreting the upper Precambrian-Cambrian rocks as volcanics and molasse related to collisional tectonics, or the fill of rift basins formed in advance of imminent Cambrian subsidence and marine transgression.

The synchroneity of late Precambrian subduction beneath the Avalon Zone and rifting in the Humber Zone implies that the Avalon Zone was not located within the Iapetus Ocean. Paleomagnetic data suggest affinities with Gondwana and Armorica. Preservation of the mildly deformed Avalonian magmatic arc sequences contrasts with their structural styles and general absence in collisional orogens. This suggests that termination of late Precambrian subduction was not the result of continent-continent collision, but instead may reflect global plate reorganization associated with the breakup of a late Proterozoic supercontinent.

Meguma Zone

The Meguma Zone is defined by a thick siliciclastic sequence, the Meguma Supergroup, that ranges in age from Late Cambrian or older to Early Ordovician. It occupies all of the southern mainland of Nova Scotia and extends offshore (Fig. 3.149, Map 2). If its strata were unfolded, the restored width would exceed the present width of the Canadian Appalachians. The base of the Meguma Supergroup is not exposed. The boundary between the Meguma Zone and Avalon Zone is the Cobequid Fault (also called Glooscap Fault and Minas Geofracture).

The Meguma Supergroup is overlain by another thick siliciclastic succession, the Annapolis Supergroup of Silurian and Early Devonian age. The Annapolis Supergroup defines the Annapolis Belt. The separation of a Meguma Zone and Annapolis Belt is artificial as: (1) both zone and belt define the same geographic area; (2) some rocks of the Annapolis Supergroup are possibly Early Ordovician and therefore older than those normally found in middle Paleozoic belts; (3) Acadian Orogeny was the first major event to affect rocks of both the Meguma Zone and Annapolis Belt; and (4) the Meguma Supergroup and Annapolis Supergroup are viewed as integral parts of the same stratigraphic section and interpreted in the same model of an evolving continental margin. The Meguma Zone and Annapolis Belt are defined only to maintain the systematics of the time-slice subdivision used here. Both are integral parts of the Meguma terrane, since there are no established linkages until Carboniferous time.

Stratigraphy

The Meguma Supergroup is divided into the Goldenville and Halifax groups (Fig. 3.151). Contacts between all formations within these groups are gradational and conformable. The maximum measured thickness of the Goldenville Group is 6.7 km. The Halifax Group is 11.8 km. The overall minimal thickness of each group is about 7 km.

The contact with the overlying Annapolis Supergroup is a paraconformity, but locally it may be an angular unconformity, a disconformity, and a conformable contact. The contact has also been interpreted locally as a thrust. In general, uppermost strata of the Meguma Supergroup are marine whereas basal strata of the overlying Annapolis Supergroup are subaerial volcaniclastic rocks. Because fossils below the boundary are Tremadoc and those above are Caradoc or younger, a paraconformity is implied.

Provenance and depositional model

The dispersal pattern in the lower portion of the Meguma Supergroup is remarkably constant both regionally and stratigraphically. In general, the sedimentary transport direction is northward in the southwest exposure area, changing gradually through 90 degrees in the central area, to eastward in the far eastern extremity of the exposure area (Fig. 3.152). This dispersal pattern is very similar to recent depositional patterns at continental margins of the western North Atlantic.

Conservative palinspastic calculations indicate that the volume of the Meguma Supergroup in Nova Scotia is at least equal to a block with an area of Ontario and a height of 5 km.

Petrography of the Meguma Supergroup indicates that the source area was rich in quartz, biotite, and plagioclase but relatively poor in potash feldspar. The source area was vast, deeply eroded, cratonic, quartz-rich, granodioritic, and presumably of Precambrian age.

The uppermost part of the Meguma Supergroup along the northwest rim of the zone contains a diamictite that may be a glaciomarine tillite or drift-ice deposit. A relative drop in sea level, suggested by shoaling in the upper part of the Halifax Group, may reflect significant global glacioeustatic sea-level changes.

The model for deposition of the Meguma Supergroup is a continental margin. The Goldenville Group is interpreted as an ancient abyssal-plain fan. Thicker sandstone packages in the Goldenville Group are submarine-fan channel complexes flanked by raised levees; currents on the levees diverged from those in the channel axis. The Halifax Group is interpreted to be the mid- or upper-fan area of a muddy deep-sea fan, passing upwards into a prograding continental slope and shelf. The lower part of the group resembles modern deep-sea channel-levee complexes; the upper may have accumulated principally from turbidity currents on a rapidly prograding continental slope. The decrease in grain size from the Goldenville to Halifax groups probably results from submergence of the source-continent.

Summary and North Atlantic connections

Since the Meguma Zone was the last to accrete against eastern North America, its derivation should be the easiest to solve by trans-Atlantic correlation.

In terms of sequence stratigraphy, the Meguma Supergroup is a type 1 sequence, equivalent to a supercycle. The Goldenville and basal part of the Halifax groups comprise a low-stand systems tract of basin-floor fan overlain by a slope fan and prograding wedge. The upper part of the Halifax Group consists of transgressive systems and highstand systems tracts. Although the Meguma Supergroup was deposited on a continental margin, its cycle corresponds in time with the Sauk cratonic sequence. Vectoral and scalar sedimentary structures indicate that the source-area lay to the present southeast.

Stratigraphic sections, provenance, sequence stratigraphy, paleontology, igneous petrology, and geophysics suggest that the Meguma Zone is a part of the continental margin of Gondwana, stranded against North America following Jurassic rifting. Other remnants are postulated in Mali, Mauritania, and southern Morocco. Rifting of the Gondwana margin created a plethora of microcontinents or microplates, including present parts of North Africa, the Middle East, central and southwestern Europe, eastern and western Avalonia, Florida, and the Meguma Zone.

There were several previous suggestions of correlation between the Meguma Supergroup of the Meguma Zone and the Gander Group of the Gander Zone, as both are thick siliciclastic sequences of about the same age. The correlation was always hampered because the two sequences are separated by the Avalon Zone. However, if the Avalon Zone was not part of the Iapetus Ocean, then both the Gander Group and Meguma Supergroup could have built up as a single unit at the eastern margin of Iapetus. The different accretionary histories of the Gander Group and Meguma Supergroup and the present arrangement of zones are circumstances of the complex closing history of Iapetus.

MIDDLE PALEOZOIC BELTS

Rocks of middle Paleozoic belts are less distinctive compared to those of early Paleozoic zones. There are no middle Paleozoic ophiolite suites, few mélanges, and other rocks that can be related to plate boundaries and plate interactions. Most of the rocks occur in successor basins with terrestrial rocks predominant. The belts are most extensive in Quebec, New Brunswick and Nova Scotia. From west to east, these are the Gaspé, Fredericton, Mascarene, Arisaig, Cape Breton, and Annapolis belts (Map 2 and Fig. 11.2). The middle Paleozoic record in Newfoundland is fragmentary and there is no correspondence between Newfoundland and mainland belts. From west to east the Newfoundland belts are Clam Bank, Springdale, Cape Ray, Badger, Botwood, La Poile, and Fortune. The broad offshore area of the Newfoundland Grand Banks has middle Paleozoic rocks unlike those of the onland orogen.

The belts are defined on lithological and stratigraphic contrasts, but some eastern mainland belts also have faunal, plutonic, and metallogenic distinctiveness. In areas affected by Ordovician deformation, the rocks of middle Paleozoic belts are unconformable on those of early Paleozoic zones. Thus in Newfoundland, the Springdale Belt straddles the Humber-Dunnage zone boundary. In areas unaffected by Ordovician deformation, lower and middle Paleozoic rocks are conformable and the oldest deposits of middle Paleozoic belts coincide with early Paleozoic zones.

Thus the Annapolis Belt and Meguma Zone of Nova Scotia define the same area, and the Badger Belt of Newfoundland lies within the Exploits Subzone of the Dunnage Zone.

Where stratigrahic sections are conformable, Caradoc and older rocks are assigned to zones or subzones in most cases. Thus in central Newfoundland, Caradoc shales are assigned to the Exploits Subzone and conformable grey-wackes, above the shales, to the Badger Belt. However, Caradoc rocks are included in the Clam Bank Belt where they are part of the unconformable cover to rocks of the Humber Zone. Other problems arise in the Quebec Dunnage Zone where a mélange substrate has a Llanvirn or Llandeilo unconformable cover that is structurally conformable with Caradoc and Silurian rocks of the Gaspé Belt. One interpretation places a boundary at a hiatus between Llanvirn rocks and overlying conformable Caradoc rocks (Bourque et al., Chapter 4); another includes all of the Ordovician rocks and some Lower Silurian rocks in the Dunnage Zone (Tremblay et al., Chapter 3).

Stratigraphic sections of most middle Paleozoic belts show an upward change from marine to terrestrial rocks, with all rocks deformed together and cut by plutons. This change from marine to terrestrial conditions preceded middle Paleozoic orogenesis.

Quebec, New Brunswick, and Nova Scotia
Gaspé Belt

Middle Paleozoic rocks of the Gaspé Belt belong to three major structural divisions, from north to south, the Connecticut Valley-Gaspé Synclinorium, the Aroostook-Percé Anticlinorium and the Chaleurs Bay Synclinorium (Fig. 4.1 and 11.2).

The middle Paleozoic rocks are separable into four broad temporal and lithological packages (Fig. 4.2): (1) Middle Ordovician-lowermost Silurian (Llanvirn? to Llandovery) deep water fine grained siliciclastic and carbonate facies; (2) Silurian-lowermost Devonian (Llandovery to Lochkovian) shallow to deep shelf facies; (3) Lower Devonian (Pragian-Emsian) mixed siliciclastic and carbonate fine grained deep shelf facies; and (4) upper Lower to Upper Devonian (Emsian to Frasnian) nearshore to terrestrial coarse grained facies. Rocks of the Connecticut Valley-Gaspé and the Chaleurs Bay synclinoria are either in stratigraphic continuity with those of the Aroostook-Percé Anticlinorium, or contacts are faults. Along the northern margin of the Connecticut Valley-Gaspé Synclinorium, the middle Paleozoic sequence is unconformable above pre-Taconic rocks of the Humber Zone. In the Chaleurs Bay Synclinorium, the Silurian sequence is unconformable above the pre-Taconic Maquereau-Mictaw groups of the Humber Zone and rocks of the Elmtree Inlier in the Dunnage Zone. In southern Gaspésie and northern New Brunswick, the middle Paleozoic sequences are overlain unconformably by Carboniferous rocks.

Upper Ordovician-lowermost Silurian rocks occupied a relatively deep marine basin. Sedimentological studies of siliciclastic rocks suggest a source containing volcanic, metamorphic, sedimentary, plutonic and ultrabasic rocks, such as the Elmtree inlier to the south. A progressive change from deeper terrigenous deposition to shallower

carbonate deposition reflects basin infilling with carbonate detritus derived from the contemporary Anticosti Platform to the northeast.

Palinspastic restoration of Gaspésie indicates that rocks of the northern part of the Gaspé Belt developed in an unstable shelf setting that followed the irregular course of the Humber Zone margin along the Gaspésie . Repetitions of facies belts in map pattern, implying repeated southward-facing shorelines, are explained by dextral offsets of a single shoreline on major transcurrent faults. Shallow marine shales on the south side of Chaleurs Bay are interpreted as part of the same shelf rather than an opposing shelf on the opposite side of the Gaspé depositional basin. Structural analysis supports a wrench tectonics model for Acadian deformation.

Three distinctive unconformities occur in the middle Paleozoic sequence of Quebec and adjacent New Brunswick (Fig. 4.3). The oldest unconformity corresponds to Taconic Orogeny. The second unconformity is dated as late Ludlow-early Pridoli and corresponds to the Salinic Disturbance recognized in Maine. It is either an angular unconformity or an erosional disconformity. The third unconformity is angular and separates Middle or Upper Devonian rocks from Carboniferous rocks. It corresponds to the Acadian Orogeny.

The Gaspé Belt is a successor basin located in the embayment at the junction between the Québec Reentrant and St. Lawrence Promontory of the Humber Zone. This Upper Ordovician to Middle Devonian basin was mostly sited above the Dunnage Zone between the already destroyed Taconic margin (Humber Zone, Fig. 4.14), and the Miramichi Subzone of the Gander Zone. Coeval volcanism was generated in an intraplate tectonic environment.

Fredericton Belt

The Fredericton Belt comprises a thick sequence of Silurian turbidites lying between the Miramichi and St. Croix subzones of New Brunswick. Basement relationships of the Fredericton Belt are unknown, but clastics of Gander Zone aspect are brought to the surface along the Fredericton Fault that transects the central portion of the belt (Fig. 3.80).

The Fredericton Belt has been interpreted as a remnant of a middle Paleozoic ocean that closed by subduction during Acadian orogenesis. This model is supported by the presence of Early Devonian Eastern Americas brachiopod assemblages in the Aroostook-Percé division of the Gaspé Belt to the northwest and Rhenish-Bohemian brachiopod assemblages in Silurian-Devonian rocks of the Mascarene Belt to the southeast. The absence of contemporary arc-volcanic rocks do not support the model and Nd-isotope studies on granites suggest that the crustal basement to the Fredericton Belt is similar to that of the St. Croix and Miramichi subzones. The Fredericton Belt has also been interpreted as a foredeep formed by tectonic loading in front of southward-advancing nappes. In this model, its Wenlock to Ludlow lithic wackes represent a clastic wedge that prograded over more mature Llandovery quartz wackes as the tectonic landmass of the Miramichi Subzone emerged to the northwest. Finer grained, calcareous beds to the southeast represent a distal facies.

Mascarene Belt

The Mascarene Belt of southern New Brunswick consists of a lithologically and structurally diverse sequence of Silurian to Lower Devonian volcanic and sedimentary rocks intruded by locally voluminous, Silurian-Devonian mafic intrusions, granitoid complexes, and felsic dykes. Five subbelts are recognized on facies and stratigraphy but faunas and depositional environments are everywhere the same. The Silurian-Devonian sequences reflect filling of a basin in which fairly deep-water deposition, in the Early Silurian, gave way to shallow water and eventually subaerial deposition in the Early Devonian. Most of its rocks were deposited above an Avalon Zone basement.

Arisaig Belt

Upper Ordovician to Lower Devonian rocks of the Arisaig Belt (Fig. 4.25) developed upon the Avalon Zone. A Silurian coastal section is complete and consists of shallow marine to brackish water fossiliferous shales and siltstones. In the Antigonish Highlands, underlying volcanic rocks are predominantly continental, bimodal, within-plate, rift varieties. In the Cobequid Highlands, volcanic rocks are scarce and the sequence is dominated by sedimentary rocks similar to those of the Antigonish Highlands.

Cape Breton Belt

Most middle Paleozoic rocks in Cape Breton Island are within the Aspy Subzone and they were described with the Gander Zone. However, other discrete occurrences of Silurian and Devonian units are present. These include a sedimentary unit near the Mabou Highlands, the Ingonish Island Rhyolite, the Fisset Brook Formation, and the McAdam Lake Formation (Fig. 4.30).

The chemical characteristics of the volcanic rocks in the Fisset Brook Formation are consistent with a within-plate setting. The unit is generally considered to have formed in a post-Acadian pull-apart basin. Post-tectonic middle to late Devonian plutons, with typical "A-type" features, are inferred to be plutonic equivalents of Fisset Brook rhyolites.

A model for the Late Ordovician to Early Devonian development of the Aspy Subzone depends on whether all of Cape Breton Island is part of the Avalon Zone or if its subzones are discrete entities. The geochemistry of the volcanic rocks implies subduction. One model suggests a Silurian arc with subduction beneath the Laurentian margin. Another model suggests local Silurian subduction within the Avalon Zone. Regardless of models, the volcanic and plutonic rocks of Cape Breton Island record a change from Silurian-Devonian compression to local Devonian-Carboniferous extension.

Annapolis Belt

The Annapolis Belt is defined by thick sequences of fine grained shallow marine siliciclastic sedimentary rocks and volcaniclastic rocks of the Annapolis Supergroup. It ranges in age from Early Ordovician to Early Devonian.

Separation of the Annapolis Belt and Meguma Zone is artificial as both are integral parts of the same terrane. This terrane has well-defined metallogenic, plutonic, and tectonic characteristics that contrasts with those of the adjacent Avalon Zone. Intense deformation in the vicinity of the Glooscap Fault supports major transport to juxtapose these very different terranes.

The model for deposition of the Annapolis Supergroup is one of relatively shallow-water sedimentation on a Gondwanan continental shelf. Ordovician volcanic rocks at the base of the Annapolis Supergroup signal the first volcanic event and subaerial exposure. Sedimentary rocks are part of a shelf margin system. Some are related to Saharan glaciation that was of late Caradoc and younger age. Silurian volcanic rocks indicate further subaerial exposure and overlying Ludlow black slates suggest post-glacial submergence. Lower Devonian rocks contain the first abundant shelly fossils. These belong to the Rhenish-Bohemian realm that includes middle and southern Europe and North Africa.

Arguments for a West African source for the Meguma Supergroup are applicable also to the Annapolis Supergroup. Both supergroups are viewed as integral parts of the same stratigraphic section and interpreted in the same model of an evolving continental margin.

Newfoundland

Rocks included in this temporal category in Newfoundland range in age from Middle Ordovician (Caradoc) to Middle Devonian, although the majority are Silurian. Stratigraphic sections include red sandstones and conglomerates toward their tops, and terrestrial volcanic rocks are abundant in central Newfoundland belts. Middle Ordovician to Lower Silurian rocks, where present, display sharp contrasts from one belt to another, e.g. limestones of the Clam Bank Belt compared to coarse turbidites of the Badger Belt. Net stratigraphic thicknesses in places exceed several kilometres.

Clam Bank Belt

The Clam Bank Belt is defined on a homoclinal sequence of Middle Ordovician to Devonian rocks along the western shore of the Port au Port Peninsula (Fig. 4.39, 4.40). The Long Point Group at the base is a carbonate sequence of Caradoc age. The conformably overlying Clam Bank Group is a redbed sequence that contains a Silurian-Devonian fauna. These rocks provide a record of deposition and deformation between the Middle Ordovician emplacement of the Humber Arm Allochthon and deposition of Carboniferous cover rocks, in an area where relationships are otherwise hidden by the Gulf of St. Lawrence.

Rocks of the Clam Bank Belt were viewed traditionally as unconformable upon the Humber Arm Allochthon, linking the allochthon with platformal rocks farther west. However, recent interpretations favour a faulted contact, with rocks of the Clam Bank Belt thrust eastward as the upper level of a structural triangle zone.

Springdale Belt

The Springdale Belt is defined on its extensive terrestrial volcanic rocks and associated terrestrial to shallow marine clastic sedimentary rocks (Fig. 4.41). Middle Silurian fossils occur at White Bay and volcanic rocks at Springdale are dated isotopically as Silurian. Unconformities between rocks of the Springdale Belt and underlying Ordovician or older rocks are recorded wherever basal relationships are preserved. Granites of Silurian and Devonian age cut rocks of the Springdale Belt and all are locally overlain by undeformed Carboniferous rocks.

The latest studies indicate that depositional patterns within the Springdale Belt were controlled by volcanic centres that terminated as cauldron subsidence features. The most extensive areas of volcanic rocks and the thickest volcanic sections occur side by side with consanguineous high level plutons. Four or more nested volcanic centres are suggested, all partly interlocking and aligned northeasterly (Fig. 4.45). In the east, unconformities with ophiolitic rocks of the Dunnage Zone are exposed locally and an ophiolitic substrate is indicated by serpentinite inclusions in volcanic and intrusive rocks. In the west at White Bay, volcanic and sedimentary rocks overlie the Humber Zone.

The numerous, long, linear, steep, northeast-trending faults that transect the Springdale Belt may follow middle Paleozoic structures that controlled volcanism, plutonism, and sedimentation at or near the Humber-Dunnage boundary. The lithologies and chemistry of the volcanic rocks seem to refute a subduction model and it is difficult to define a contemporary oceanic tract.

Cape Ray Belt

The Cape Ray Belt is defined by discontinuous occurrences of Silurian and Devonian terrestrial sedimentary and bimodal volcanic rocks that are traceable from Cape Ray 150 km northeastward (Fig. 4.46). Silurian redbeds are unconformable above the Ordovician Annieopsquotch ophiolite complex of the Notre Dame Subzone toward the northeast, and Devonian volcanic rocks and conglomerates are unconformable on Ordovician tonalites of the Dashwoods Subzone near Cape Ray.

The rocks occur along the Cape Ray fault zone that was viewed as a suture between continental rocks of the Humber Zone and Gander Zone. The latest studies suggest that the Dashwoods Subzone to the northwest includes rocks typical of both the Humber and Dunnage zones, all structurally commingled, intruded, and metamorphosed together. Southeast of the fault zone, the Port aux Basque Gneiss is correlated with rocks of the Meelpaeg Subzone, implying Gander affinities. The Springdale and Botwood belts appear to converge southwestward into the Cape Ray Belt, implying a paleogeography unrelated to Ordovician tectonic elements.

The narrow Cape Ray Belt coincides, at least in part, with an ancestral zone of intense deformation and mylonitization. An intracratonic ductile shear zone may have localized Devonian deposition. After deposition, the Devonian rocks were deformed and mylonitized (Dubé and Lauzière, 1994). Deformation is much more intense in the southern Cape Ray Belt than elsewhere in the Springdale and Botwood belts.

Badger Belt

The Badger Belt (Fig. 4.47) is distinguished by its preponderance of greywackes and overlying polymictic conglomerates of the Badger group that exceed 3 km in thickness. The rocks are dated by shelly faunas of Middle to Late Llandovery age. The thick marine clastic sequences are conformable above Middle Ordovician (Caradoc) black graptolitic shales of the Exploits Subzone. Silurian mélanges, locally with fossiliferous shale matrices and containing huge Ordovician volcanic, limestone, and black argillite blocks are a distinctive lithology in upper parts of stratigraphic sections. The rocks range from deep to shallow marine and they contrast with redbeds and terrestrial volcanic rocks of other Newfoundland belts.

All of the sections exhibit the same pattern of large-scale coarsening upward, which is interpreted as a shoaling and infilling of one or several marine basins. A variety of sedimentary features indicate a mainly northern provenance. This is supported by a local unconformity between Ordovician volcanic rocks and Silurian conglomerate at northern New World Island. Coupled with sub-Silurian unconformities throughout the adjacent Springdale Belt, an emergent landmass to the west and northwest is indicated.

Whereas the greywackes and conglomerates of the Badger Belt almost everywhere overlie Caradoc shales, the shales overlie various older assemblages such as the Victoria Lake Group, Wild Bight Group, Exploits Group, Summerford Group, Dunnage Mélange, and deformed mafic-ultramafic rocks of the Gander River Complex, from west to east.

Structural styles of the Badger Belt with southeast thrusting, or northwestward underplating of outboard elements from the southeast, coupled with mélange formation, resemble structural styles of accretionary prisms either in forearc or foreland basin settings. Some Silurian mélanges are olistostromes that mark the leading edges of contemporary thrust faults. The significance of some other mélanges is debatable.

The Silurian structural style was possibly controlled by an encroaching Gander terrane that underthrust rocks of the Exploits Subzone and led to southeast thrusting as the heretofore undeformed Paleozoic rocks of the Exploits Subzone were compressed against the stabilized Taconic deformed parts of the Notre Dame Subzone. Geochronological evidence of an important Silurian metamorphic and plutonic event along the eastern Gander Zone also fits this model.

Botwood Belt

Rocks of the Botwood Belt (Fig. 4.53) are mainly terrestrial volcanic rocks overlain by fluviatile red and grey cross-bedded sandstones. Polymictic conglomerates occur in places along its western margin and greywackes like those of the Badger Belt occur beneath its volcanic rocks in the northeast and along its eastern margin.

Correlation between greywackes and conglomerates of the Badger Belt with those of the Botwood Belt implies continuity of these rocks across the Exploits Subzone. Stratigraphic relationships between greywackes of Badger type and underlying Ordovician rocks, of the Davidsville Group to the east indicate that the Davidsville Group was part of the conformable substrate beneath the Silurian deposits of eastern Notre Dame Bay. An unconformity between the Davidsville Group and ophiolitic rocks of the Gander River Complex at the Dunnage Zone-Gander zone boundary, and an absence of Silurian rocks above the Gander Zone suggest an eastern limit to the depositional basin.

Recognition of the Dog Bay Line as a major structural junction that separates different Silurian rock groups requires subdivision of the Botwood Belt into western and eastern parts (Williams et al., 1993). A new belt, the Indian Islands Belt, is introduced for the narrow part of the former Botwood Belt southeast of the Dog Bay Line. Its rocks are Silurian calciferous shales, sandstones, conglomerates, and redbeds assigned to the Indian Islands Group. They are disconformable upon upon Ordovician shales and mélange exposed along the west side of Gander Bay.

The Dog Bay Line is marked by a wide zone of disrupted Ordovician rocks and it is an important Silurian tectonic boundary that may (see addendum to Chapter 4) mark the terminal Iapetus Suture.

La Poile Belt

The La Poile Belt forms an integral part of the Hermitage Flexure and its rocks occur above the Burgeo Subzone in southwestern Newfoundland (Fig. 4.56). Two separate outcrop areas consist of subaerial felsic volcanic and associated epiclastic rocks interdigitated with crossbedded quartz sandstones of nonmarine shallow-water origin. The rocks are assigned to the La Poile Group. Some of its felsic volcanic rocks are dated isotopically as Silurian and the group is 3.5 to 5 km thick.

Conglomerates at the base of the La Poile Group lie nonconformably on the Roti Granite dated at 563 ± 4 Ma in the northeast part of the belt. Rocks that are cut by the Roti Granite include amphibolite (Cinq Cerf gneiss) and less metamorphosed green siltstones, argillites, and volcanic rocks of the Burgeo Subzone.

The structural styles and basal relationships of the La Poile Group contrast with those in other Newfoundland belts. Northwest thrusting of Silurian rocks above the Ordovician Bay du Nord Group, northwest thrusting of southerly belts of mainly upper Precambrian rocks above the La Poile Group, and an unconformity between Silurian and upper Precambrian-Cambrian rocks are unique to southwestern Newfoundland.

The La Poile Belt separates rocks assigned to the Dunnage Zone to the north from upper Precambrian rocks of the Burgeo Subzone to the south. Gander Group correlatives are absent in this area, although syntectonic plutons and Paleozoic structures typical of the Gander Zone extend around the Hermitage Flexure and into the La Poile area.

Fortune Belt

Several middle Paleozoic units occur along the north shore of Fortune Bay and another occurs on its southeast side at Grand Beach. These overlie rocks of the Avalon Zone. The oldest rocks of northern Fortune Bay are mainly red arkosic sandstones and conglomerates with clasts of metamorphic and plutonic rocks derived from the adjacent Gander Zone. They are nonconformable upon the late Precambrian

Simmons Brook Batholith. A younger and less deformed conglomerate unit nearby contains Late Devonian plants. All of these rocks are cut by Devonian granite.

Volcanic rocks at Grand Beach of the Burin Peninsula are dated isotopically as Devonian and they are interpreted as cover to upper Precambrian rocks.

Sedimentology and intrusive relationships among middle Paleozoic rocks of the Fortune Belt indicate fluviatile and alluvial deposition between periods of granite intrusion and contemporary deformation. This is supported by the synchronous ages of silicic volcanic rocks at Grand Beach and Acadian plutonism. Possibly, the synorogenic deposits of the Fortune Belt reflect transcurrent faulting and uplift associated with the accretion of the Avalon Zone against the Gander Zone.

Grand Banks

Seismic stratigraphy, drill cores, and palynological analyses indicate that 8000 m of Cambrian to Devonian rocks cover some 50 000 km^2 east of the Avalon Peninsula on the Grand Banks of Newfoundland. These rocks are mainly marine and they contrast with mainly terrestrial onshore sections. Formline structural mapping revealed 4000 m of Ordovician-Silurian strata that are gently folded about north-northwest axes. These are overlain unconformably by a synclinal outlier of Devonian strata 700 m thick.

Upper Ordovician and Silurian rocks are grey laminated fissile siltstones with burrows and fragments of brachiopods and bryozoa, light grey calcareous siltstones which are laminated and bioturbated with well-sorted beds of fossil fragments, and non-fossiliferous, bioturbated siltstones. The rock types and fauna indicate deposition in a variety of shallow marine environments. Devonian rocks are redbeds of continental affinity.

Mild deformation in the Paleozoic rocks offshore supports the onland observation of decreasing intensity of Paleozoic deformation from west to east across the Newfoundland Avalon Zone. Furthermore, there is no middle Paleozoic plutonism or regional metamorphism recorded in the offshore sections. The presence of an offshore Devonian basin of terrestrial rocks records a change to subaerial conditions as noted in the onland successions, although this change is later in the case of the offshore.

Overview of middle Paleozoic belts

Lithologies and stratigraphies of Middle Ordovician to lowermost Silurian rocks of middle Paleozoic belts are different, and they imply deposition in discrete basins, separated by emergent lands. The rocks of middle Paleozoic belts everywhere overlie rocks of older zones, indicating that all the important elements of the orogen were established by middle Paleozoic time. Nowhere are middle Paleozoic rocks conformable on a contemporary ophiolitic basement, indicating the existence of an important middle Paleozoic ocean.

The oldest rocks of the Gaspé Belt are Caradoc and these are conformable on Llanvirn strata. The latter may be part of the same sequence or an important hiatus may be present. Regardless, there is no structural discordance and the section spans the interval of Taconic orogeny that affected the Humber Zone to the north. Since the Gaspé sequence in question lies on an ophiolitic mélange substrate, the formation of the mélange may represent disruption and an early phase of Taconic Orogeny in this area. A similar situation exists in the Badger Belt of Newfoundland where Silurian marine clastics overlie Caradoc shales, in turn deposited on the Dunnage Mélange.

The geographic and tectonic setting of the Clam Bank Belt in Newfoundland suggests a connection with the northern portion of the Connecticut Valley-Gaspé division of the Gaspé Belt, and rocks of the Clam Bank Group may correlate with the Griffon Cove River Formation of northeastern Gaspésie. Southern parts of the Connecticut Valley-Gaspé division and the Aroostook-Percé division have no Newfoundland counterparts.

Silurian greywackes of the Fredericton Belt resemble those of the Badger Belt. However, the Llandovery shelly faunas of the Badger Belt are absent in the Fredericton Belt, and the Badger and Fredericton belts have very different positions compared to the early Paleozoic zonation of Newfoundland and New Brunswick.

Silurian and Devonian rocks are conformable in most places. Local stratigraphic sections record upward shoaling with marine sections capped by terrestrial redbeds. This change is rarely contemporaneous across the belts and in some cases may be diachronous along the course of a single belt. The change is recorded earliest in the Lower to Middle Silurian rocks of the Newfoundland Botwood Belt. It appears to be latest in the Gaspé and Annapolis belts where marine Lower Devonian rocks are present.

Middle Paleozoic deformation affected the rocks of all belts, although it is more intense in some areas than in others. Silurian plutonism and metamorphism accompanied deposition and/or deformation in some places, but there are few well-documented examples of Silurian or Devonian rocks that make up high grade regionally metamorphosed parts of the orogen. Granites cut middle Paleozoic rocks of all belts, with the exception of the Clam Bank Belt and eastern parts of the Avalon Peninsula and offshore. Most are post tectonic with respect to deformation in the middle Paleozoic rocks; but some are syntectonic and others are spatially and genetically related to midddle Paleozoic volcanic rocks. The onset of middle Paleozoic deformation is not everywhere sharply defined. Isotopic ages indicate Silurian onset and that rocks as young as Late Devonian are locally involved. The shoaling-upward trend occurred everywhere in advance of deformation and its age variations imply diachronous onset of Acadian Orogeny.

There is no obvious relationship between middle Paleozoic belts and the distribution of middle Paleozoic intrusive and metamorphic rocks, except in places such as the Springdale Belt where plutons are subvolcanic, and the Annapolis Belt where its plutonic history is distinctive.

Middle Paleozoic volcanism, deformation, plutonism and metamorphism are all more important in interior parts of the Newfoundland Appalachians and decrease westward across the Humber Zone and eastward across the Avalon Zone. This applies also to Cape Breton Island and Quebec-New Brunswick cross sections. It does not apply to the outboard Meguma Zone and Annapolis Belt where plutonism is intense with local high grade metamorphism.

Major strike-slip displacements, megashears and transpression zones are words of increasingly common usage in tectonic models for the middle Paleozoic development of the Canadian Appalachians. Transcurrent motions may have provided the controls for middle Paleozoic volcanism and deposition in tectonically active basins. Continued movements possibly led to deformation in the same areas.

The significance of Silurian-Gedinnian, Rhenish-Bohemian fauna restricted to mainland belts developed upon the Avalon and Meguma zones and subsequent appearance on the Laurentian margin during the Siegenian and Emsian is difficult to assess. The analysis seems the same as that for Cambrian and Early Ordovician faunas on opposite sides of the orogen that lost their distinctiveness in the Middle Ordovician. Most evidence suggests Avalonian accretion during the Silurian, although the Meguma terrane was accreted later. The Miramichi Highlands may have been an effective land barrier between the mainland belts, or possibly the Fredericton Belt was an effective marine barrier. No such middle Paleozoic faunal distinction is recognized in Newfoundland, or at least no equivalent analysis exists.

LATE PALEOZOIC BASINS

Upper Paleozoic rocks of the Canadian Appalachian region are mainly of Carboniferous age but they include Upper Devonian beds at the base of some sections and Permian strata at the top of some others. The rocks are mainly coarse-to fine-grained continental red and grey sedimentary rocks that include fluvial and fluvio-lacustrine strata, coal measures, marine limestone, and evaporites. Volcanic rocks of bimodal aspect occur locally, most commonly at the bases of mainland stratigraphic sections. The rocks extend across the exposed orogen as an unconformable cover on lower Paleozoic zones and middle Paleozoic belts and extend offshore and underlie much of the Gulf of St. Lawrence, the southern Grand Banks, and the northeast Newfoundland Shelf. They occur in discrete depocentres, some of which are connected by higher stratigraphic units. The depocentres are referred to as basins. They generally trend northeast, parallel to older structures, and they are in places bounded by faults that partly controlled their initiation and subsequent evolution. The basins are best developed and best preserved from Gaspésie to Cape Breton Island. From northwest to southeast these are the Restigouche, Plaster Rock, Carlisle, Central, Marysville, Moncton, Sackville, Cumberland, Minas, Stellarton, Antigonish, Western Cape Breton, Central Cape Breton, and Sydney basins (Fig. 11.3). Two depocentres in Newfoundland are the Bay St. George and Deer Lake basins; with small redbed outliers at Conche, Red Indian Lake, Terrenceville, and Spanish Room. The Magdalen Basin in the Gulf of St. Lawrence is the largest and deepest structure, and its rocks are continuous with adjacent mainland basins and the Bay St. George Basin in Newfoundland. The name Maritimes Basin is a general term for all upper Paleozoic rocks in Atlantic Canada.

A relatively narrow, northeast-trending, fault-bounded and fault-dominated central region from the Bay of Fundy to western Newfoundland has thicknesses of upper Paleozoic rocks in excess of 12 km. These rocks are extensively deformed and they contrast with thinner, undeformed rocks of bordering areas. The central region is referred to as the Maritimes Rift and it contains the Moncton, Sackville, Cumberland, Magdalen, and Bay St. George basins. The bordering area to the north is the New

Brunswick Platform, although the cover rocks are not conventional platformal deposits. It contains the Plaster Rock, Carlisle, Marysville, and Central basins. The Maritimes Rift has a "horst and graben" style of basement morphology. Local activity on normal, thrust, and strike-slip faults resulted in a complex temporal and spatial pattern of stratigraphic units. The upper Paleozoic rocks at Chaleur Bay in the north and Passamaquoddy Bay in the south are isolated occurrences, the Restigouche and St. Andrews basins, respectively.

The oldest rocks are coarse fanglomerates (Late Devonian) largely confined to the Maritimes Rift and associated with boundary faults. Deformation affected the thickest sections. Middle and Upper Carboniferous rocks are in places unconformable upon deformed rocks of the Maritimes Rift and locally they overlap boundary faults onto adjacent platformal areas. Other evidence of unconformable overlap and local episodic deformation is common within both lower and upper parts of Carboniferous sections. The youngest rocks are redbeds (Early Permian) of Prince Edward Island.

The former extent of upper Paleozoic rocks in Atlantic Canada is debatable. Some estimates imply that up to 3000 m of sedimentary cover has been eroded since the Early Permian. This suggests that most, if not all of the Atlantic region, was once covered by Upper Paleozoic strata.

A change from Lower Carboniferous thick redbed accumulations to Middle and Upper Carboniferous grey and red fluvioclastic deposits with coal measures is equated with a major paleoclimatic change from arid to humid tropical.

Late Paleozoic deformation is most intense along the northwest shoreline of the Bay of Fundy in southern New Brunswick. There the Carboniferous rocks are involved in thrusts with polyphase deformation and subhorizontal penetrative cleavage. Granitic intrusions of Early Carboniferous age occur locally at Mount Pleasant and the Saint John area of southern New Brunswick, in the Cobequid Highlands of northwestern Nova Scotia, and in the Meguma Zone of Nova Scotia.

Upper Palaeozoic strata have played a significant role in the economic development of the Atlantic region. Oil, natural gas, oil shale, coal, salt, gypsum, potash, lime, building stone and base metals have been produced intermittently for over a century and new discoveries are possible. Tin, tungsten, and molybdenum occur in Upper Devonian/Lower Carboniferous felsic volcanic and intrusive rocks at Mount Pleasant, southern New Brunswick.

Stratigraphy

The original stratigraphic subdivision of the Maritimes Basin Carboniferous sequence into six major units is to a large extent still in wide usage. The group names and their biostratigraphic ages based on European stages are listed below.

The Horton Group of Late Devonian-Early Carboniferous (Late Famennian-Tournaisian) age consists of red and grey-green polymictic conglomerates, arkosic sandstones, mudstones, oil shales, and minor non-marine evaporites. Horton Group strata overlie pre-Acadian basement with angular unconformity. The Horton equivalent in Newfoundland is the Anguille Group.

The Windsor Group of Early Carboniferous (Viséan-early Namurian) age consists of marine limestones, evaporites, and intercalated redbeds. It overlies the Horton Group and represents the only known marine incursion into the Maritimes Basin prior to the Late Carboniferous, at a time when the paleoclimate was arid. The most extensive and thickest Windsor sequence occurs in the Magdalen Basin. The Windsor equivalent in Newfoundland is the Codroy Group.

The Canso-Riversdale Group (Mabou Group in Nova Scotia) is of Early-Middle Carboniferous (mainly Namurian) age. It consists of red and grey terrestrial strata with a variety of facies that range from evaporitic to lacustrine. It is gradational and conformable above the Windsor Group.

The Cumberland Group of Westphalian age consists of approximately 2700 m of red and grey fluvial conglomerates, sandstones, and mineable coal measures. It includes the classical coal measures and plant-bearing beds in the coastal cliffs of the Joggins area, northwestern Nova Scotia. The sequence is gradational with the Canso-Riversdale Group below and locally oversteps older Carboniferous strata onto basement. Equivalent strata in the Sydney Basin of Nova Scotia are the Morien Group of Westphalian C to Stephanian age. Newfoundland equivalents are the Barachois Group.

The Northumberland Strait Supergroup includes the Pictou Group and overlying Prince Edward Island Group. The age of the supergroup extends from Middle Carboniferous to late Early Permian (early Westphalian C to Artinskian). The Pictou Group is conformable above the Cumberland Group. In Nova Scotia it has two major facies: a mainly grey, coal bearing facies and a predominantly redbed facies. Pictou strata of the New Brunswick Platform and the Moncton Basin consist of alternating grey and red facies. The Prince Edward Island Group is similar to the Pictou Group of New Brunswick except that its rocks are red, rather than grey and red, and they lack coal.

Tectonic history

Tournaisian to lower Viséan strata of the Horton Group and equivalents were deposited in a series of rift basins, probably half-graben, as indicated by basin-margin facies and patterns of stratal thickening and paleoflow. Marine seismic sections show fault-bounded structures filled with probable Horton strata beneath much of the Gulf of St. Lawrence. The Deer Lake Basin in Newfoundland has evidence for Tournaisian to Viséan movements on the Cabot Fault Zone. The widespread and apparently coeval occurrences of organic-rich lacustrine strata in the Horton Group imply some interconnection between the depocentres and suggest a period of accelerated regional subsidence. The earliest phase of Carboniferous deformation, involving folding, faulting, and uplift affected Horton strata in some basins prior to the deposition of the Windsor Group.

Strong topographic relief, partly inherited, persisted throughout deposition of the mid to upper Viséan Windsor Group, as indicated by facies relations and basin-margin facies. The unusually thick and extensive evaporites of Nova Scotia and offshore probably accumulated in an embayment or semi-enclosed basin, remote from a major Viséan seaway. The onset of Windsor sedimentation was sudden, with the Viséan sea inundating a broad, complex area. Depositional cycles identified in places in the Windsor Group may reflect glacioeustatic events.

The conformable change from the marine Windsor Group to evaporitic lacustrine Canso-Riversdale and Mabou groups indicates a return to non-marine conditions. This is roughly coeval with a major phase of Gondwanan glaciation and lowering of sea level.

An important tectonic phase, accompanied by widespread faulting, local uplift and concomitant basin development affected much of Nova Scotia and adjacent areas in the mid Carboniferous. The effects of this activity include: (1) a change from lacustrine and fine grained alluvial strata of the Mabou Group to the coarser clastic facies of the predominantly alluvial Cumberland (Riversdale) and Pictou groups; (2) basin inversion and faulting in southern New Brunswick; (3) faulting and mylonitization of the Early Carboniferous Cape Chignecto pluton at the western end of the Cobequid Highlands; (4) intensive deformation, including overturning of Lower Carboniferous strata in the Minas Basin and areas adjacent to the Cobequid Fault; (5) formation of the Stellarton Basin as an extensional structure in association with the Cobequid-Hollow fault system; (6) unconformities, commonly angular, between the Mabou and Cumberland-Pictou groups at numerous locations, with Namurian and Westphalian A-C strata missing; and (7) a widespread thermal event within the Meguma Zone dated at 320-300 Ma, approximately mid-Namurian to Stephanian.

Westphalian to Early Permian sedimentation was predominantly alluvial. Upper Carboniferous sediments encroached on most adjacent basement areas suggesting a progressive diminution of tectonic activity. In the Sydney Basin and elsewhere, faults that cut Lower Carboniferous strata cannot be traced into Upper Carboniferous strata, and paleoflow data suggest progressively reduced intrabasinal relief.

Upper Lower Permian redbeds of Prince Edward Island exhibit gentle folds indicating continued tectonic activity. The presence of coals of bituminous rank in many Westphalian sections implies substantial burial, possibly to a depth of several kilometres. This suggests that most upland areas were buried during the Permian, at which time the Atlantic area formed an extensive alluvial plain.

Subsidence in the Atlantic region probably terminated by the early Triassic. Fission-track analysis of apatites in a variety of rocks in Nova Scotia indicate Permian to early Mesozoic cooling ages below about 100°C, probably because of uplift. This uplift was coeval with the onset of Atlantic seafloor spreading and the breakup of Pangea.

Structural models and correlations

There are two main interpretations for the structural controls of the Maritimes Basin. One favours a stress regime that was largely extensional (continental rifting) punctuated by periods of transverse compression; the other favours a dextral wrench tectonic setting. Evidence cited in support of rifting and transverse compression includes: (1) texture, chemistry and stratigraphic relations of Mid Devonian to Lower Carboniferous volcanic rocks, including those of the Devonian Fisset Brook Formation and Fountain Lake Group of Nova Scotia and Viséan volcanics

on the Îles de la Madeleine, that suggest eruption in a terrestrial setting related to intraplate continental rifting; (2) the apparent "basin and range", block-faulted topography of the basement floor; (3) the presence of thick alluvial fanglomerates adjacent to upland source areas; (4) the presence of northeast-trending normal and high angle reverse faults in southern New Brunswick and locally in Nova Scotia and Newfoundland; and (5) the lack of evidence for large-scale lateral offset of Windsor Group strata on opposite sides of the Belleisle Fault in New Brunswick.

A wrench tectonic model is based on dextral strike-slip motion documented on major faults that transect upper Paleozoic strata, and the following features and observations: (1) basins bordered by demonstrable steep dextral faults fit the pull-apart model; (2) the geometry of folds within the basins with axes oblique to boundary faults is a well known pattern in wrench tectonic zones; (3) exceptionally thick homoclinal sections with consistent dips in a direction parallel to the long dimension of the basins imply a shifting depocentre in unison with lateral fault movements; some of these thicknesses are much greater than depth to basement; (4) continuity and apparent steep dips of faults associated with basin development and their tendency to splay and enclose elongate blocks of basement rocks, similar to piercement structures along the Alpine Fault Zone of New Zealand; (5) reverse movements noted on some boundary faults are also consistent with transpression in a zone of wrench tectonics; and (6) a major zone of thrusting that involves basement rocks, coeval plutons, recumbent folds, and penetrative fabrics in southeast New Brunswick can be explained as the compressional zone resulting from dextral offset on the major Cobequid Fault and westward advancement of the Meguma Zone against the Avalon Zone.

The age of faulting is generally unknown, apart from a few cases where unconformable overstep is preserved or provenance relations are known. Most faults had important post-Carboniferous movement, and others followed ancestral basement structures, which complicates structural analyses. Possibly, the Maritimes Basin began as a rift system, then evolved partially or entirely as a wrench-tectonic system.

Upper Paleozoic rocks of Atlantic Canada are very different in structural style and tectonic setting compared to correlatives in the southern Appalachians of the United States. In Atlantic Canada the rocks occur in broad undeformed areas of little relief that are mainly beneath sealevel. This contrasts with the deformed, elevated upper Paleozoic rocks in the Valley and Ridge province of the U.S. Appalachians, and with the Piedmont Province in the south that contains large Carboniferous plutons in a contemporary regional metamorphic terrane that lacks identifiable upper Paleozoic layered rocks. Even the contrasts with nearby New England are marked in that the upper Paleozoic rocks of New England are polydeformed, intruded, and affected by regional metamorphism in amphibolite facies. Atlantic Canada has no equivalent to the Alleghanian or Hercynian structural fronts as deformation in Canada is confined to narrow fault-bounded areas bordered by broad areas of little or no deformation. The control of Alleghanian Orogeny in the U.S. Appalachians was the head-on collision of Laurentia with northwest Africa as the indentor. Possibly the Alleghanian thrusting in the central and

MESOZOIC GRABEN

Mesozoic rocks of Atlantic Canada are mainly Triassic and Lower Jurassic continental redbeds, tholeiitic basalts, and related mafic dykes and small intrusions. The sedimentary and volcanic rocks up to 3500 m thick occur in the Fundy Graben of the Bay of Fundy area with a few small outcrops in the Chedabucto Graben to the east (Fig. 11.3, 6.1). Geophysical surveys indicate that Mesozoic strata underlie much of the Bay of Fundy and extend offshore from Chedabucto Bay to the Orpheus Graben. Mafic dykes and small intrusions that are Triassic to Early Cretaceous occur beyond the limits of the Fundy and Chedabucto graben (Fig. 6.2). Small isolated outcrops of Cretaceous clays and sands also occur outside the Fundy and Chedabucto graben in central Nova Scotia and southwestern Cape Breton Island.

The Mesozoic rocks are related to the early stages of rifting and drifting that led to opening of the Atlantic Ocean. The Fundy Graben is typical of Mesozoic graben present along the length of the Appalachian Orogen and Atlantic continental shelf. It is bounded by faults and sited, in part, above the Carboniferous Minas Basin and the Avalon-Meguma zone boundary.

Stratigraphy, age, and depositional environment

The Triassic-Lower Jurassic rocks of the Fundy Graben are assigned to the Fundy Group. The rocks are best exposed in Nova Scotia where they consist of the Wolfville Formation, Blomidon Formation, North Mountain Basalt, and Scots Bay Formation, from bottom to top. Table 6.1 summarizes correlations of Mesozoic formations in Atlantic Canada.

The Wolfville Formation lies unconformably upon Carboniferous rocks of the Minas Basin and pre-Carboniferous metamorphic rocks and Devonian granites. Exposed thicknesses range from 60 m to 750 m and the formation is apparently thicker in the centre of the Bay of Fundy (Fig. 6.3d). The rocks comprise a red sequence of coarse breccias, conglomerates, sandstones, and mudstones. The Wolfville Formation grades laterally and vertically into the Blomidon Formation which has exposed thicknesses from 7 m to 370 m (Fig. 6.3c). The Blomidon Formation has planar, crossbedded and cross-laminated sandstones together with horizontal and cross-laminated siltstones, mudstones, and claystones.

The Lower Jurassic North Mountain Basalt, locally up to 400 m thick, overlies the Blomidon Formation and extends throughout most of the Bay of Fundy.

The basalts have undergone zeolite to greenschist facies metamorphism and a middle unit at North Mountain contains a variety of zeolite minerals.

The Lower Jurassic Scots Bay Formation (Fig. 6.1b, a) consists of calcareous sandstones, calcareous siltstones, limestones, and silicified nodules of limestone and tree trunks. It is 2 m to 7 m thick and overlies the North Mountain Basalt.

Coarse breccias of the Chedabucto Formation at Chedabucto Bay (Fig. 6.1b, f) exceed 61 m in thickness and the rocks resemble those of the Wolfville Formation. In New Brunswick, Mesozoic sedimentary rocks occur on Grand Manan Island (Fig. 6.1b, d) and close to the northwest faulted margin of the Fundy Graben.

Lower Cretaceous deposits of central Nova Scotia are unconsolidated, red, grey and white kaolinitic clays and white, unconsolidated, almost pure silica sands (Fig. 6.1a, g). The clay deposits are economically important sources of pottery clay and the silica sands have various industrial uses. Similar silica sands occur in southern New Brunswick. The sedimentary rocks occur in depressions and sinkholes that have escaped erosion. Compaction studies on coals indicate that over 700 m of Cretaceous sedimentary rocks may have covered Nova Scotia.

Mesozoic ages for rocks of the Fundy Graben are based on pelecypods, amphibian, and reptilian remains and footprints as well as fish scales and bone fragments. A variety of late Carnian to early Norian trace fossils occur in the upper Wolfville Formation and its dinosaur footprints may be the oldest in Canada. The oldest rocks in the Fundy Graben are Anisian, or early Middle Triassic.

The McCoy Brook Formation along the northern shores of Minas Basin is well-dated by an Early Jurassic (Hettangian) tetrapod fauna. Its much-publicized fossil discoveries may help explain world-wide faunal extinctions at the Triassic-Jurassic boundary. One theory equates faunal extinction with meteorite impact at Manicouagan, Quebec.

Clastic rocks of the Wolfville Formation have been interpreted as deposits of a proximal alluvial fan together with braided river and eolian sand dune settings. Paleoflow directions (summarized in Fig. 6.3e) show that alluvial fan detritus was derived from highlands bordering the Fundy Graben and generally moved south on the north side of the graben and north on its south side.

Facies relationships and sedimentary structures indicate that cross-laminated and crossbedded clastic rocks of the Blomidon Formation represent the distal sheet-flood deposits of alluvial fans as well as sand flats, playa mudflats, and lakes. Periodic movement on the Cobequid Fault produced numerous depositional cycles of sand-flat sandstones followed by playa mudstones and/or lacustrine claystones. Flow directions appear similar to those for the Wolfville Formation.

Calcareous clastic rocks, limestones and silicified organic material of the Scots Bay Formation represent carbonate sedimentation in a near-shore lacustrine environment.

The Mesozoic paleoclimate during deposition in the Fundy and Chedabucto graben varied from semiarid, as indicated by the paucity of weathered grains, scarcity of carbon, and the oxidized state of iron in the sediments, to hot and humid with high precipitation and an annual dry season. Eolian sandstones, caliche paleosols, alluvial fan conglomerates and playa gypsiferous mudstones confirm arid to semiarid conditions with seasonal precipitation. Lycopsid megaspores in the Blomidon Formation support a desert-like climate. Southwest to northwest paleowind directions inferred from eolian sands of the Fundy Graben reflect the direction and effect of prevailing subtropical trade winds that led to an arid environment.

Primary sedimentary structures show that Cretaceous deposits are deltaic. The purity of the clay and silica sand may be the result of intense weathering during the Early Tertiary in a humid subtropical climate with high rainfall.

Igneous rocks (Fig. 6.2) fall into three age categories: Late Permian-Early Triassic, Late Triassic-Early Jurassic, and Middle Jurassic-Early Cretaceous. Upper Permian-Lower Triassic rocks are represented by the Malpeque Bay Sill in Prince Edward Island. Upper Triassic-Lower Jurassic rocks include the North Mountain Basalt and Shelburne Dyke of Nova Scotia, the Minister Island and Caraquet dykes in New Brunswick, two dykes on Anticosti Island, and the Avalon Dyke in Newfoundland. Middle Jurassic-Lower Cretaceous rocks include the alkaline Budgells Harbour pluton and associated lamprophyre dykes of the Notre Dame Swarm in Newfoundland.

Igneous activity reached a maximum during the Early Jurassic with extrusion of the voluminous North Mountain Basalt and emplacement of mafic dykes.

High-Ti phlogopite in the Malpeque Bay Sill indicates an affinity with alkaline igneous rocks such as lamprophyres. The sill is chemically similar to nephelinites. The North Mountain basalts are high-Ti quartz normative tholeiites. The Avalon, Shelburne, Caraquet, and Anticosti Island dykes are high-Ti quartz normative tholeiites, but the Caraquet Dyke is transitional between high-Ti and low-Ti types (Table 6.3). Lamprophyric rocks from the Notre Dame Swarm have evolved compositions and their alkaline nature may reflect extreme source-region metasomatism.

Structure and interpretation

The Fundy Graben is a major asymmetrical syncline with a steeper north limb and gentle plunge to the southwest (Fig. 6.1). The graben is bounded to the north by east-west faults of the extensive Cobequid system (Fig. 6.2) and it is cut by north-south faults that offset Lower Jurassic strata and all other faults. Total vertical movement on the east-west faults was at least 3500 m as given by the thickness of Mesozoic strata in the Bay of Fundy. Greater movement may have occurred at the southern end of the graben where seismic studies indicate 9000 m of Mesozoic strata. The detailed stratigraphy of alluvial fan deposits adjacent to the boundary faults reflects, at least in part, the timing and nature of fault movement. It began in the Triassic. A seismic reflection study of the offshore extension of the Cobequid Fault along the Orpheus Graben indicates that movement continued into the Early Cretaceous. Seismic evidence indicates that the Cobequid Fault dips eastward below the Bay of Fundy, apparently representing a major detachment surface formed through reactivation of a Paleozoic basement thrust. This suggests a detachment model for graben formation.

The northeast-southwest orientation of the Caraquet, Minister Island, Shelburne, and Avalon dykes (Fig. 6.4a) indicates a northwest-southeast direction of maximum extension during the Early Jurassic. The orientation of Early Cretaceous lamprophyre dykes in Notre Dame Bay is north-northwest-south-southeast (Fig. 6.4b) indicating a local east-northeast-west-southwest maximum extension direction.

North-south to northwest-southeast wrench or transverse faults displaced the vertical Caraquet, Minister Island and Shelburne dykes. Brittle transcurrent faults of

similar orientation influenced topography and controlled drainage on the eastern shore of Nova Scotia. They overprint all earlier structures and follow older, Devonian joint systems. These are also parallel to oceanic fracture zones and may represent transfer faults.

The North Mountain Basalt and coeval dykes of Atlantic Canada are the most northerly occurrence of igneous rocks of the Eastern North American dolerite province. This province extends all along the eastern seaboard and is the most extensive magmatic province related to opening of the Atlantic Ocean. Ages for the Eastern North American dolerites range from 205-165 Ma with a peak at 190 Ma. The province's western boundary (Fig. 6.7) is almost coincident with the eastern boundary of the Appalachian miogeocline. Dykes tend to decrease in number to the north so that only five major dykes are identified in Atlantic Canada. Early Jurassic igneous activity and onland basin subsidence ceased abruptly by the Middle Jurassic when North Africa separated from Eastern North America, with the rifted margin parallel to structural trends of the Appalachian Orogen.

Middle Jurassic to Cretaceous intrusions define the New England-Quebec magmatic province (130-90 Ma). Igneous activity was largely restricted to more northerly areas around the Atlantic Ocean and produced small, localized eruptions of predominantly alkaline to strongly alkaline rocks. This province includes the lamprophyres at Notre Dame Bay, Newfoundland and alkaline intrusions of the Monteregian Hills in Quebec. It affected a narrower band of continental crust, and resulted in a rift axis that crossed Paleozoic structural trends when the Grand Banks of Newfoundland separated from the Iberian Peninsula with subsequent opening of the Labrador Sea.

GEOPHYSICAL CHARACTERISTICS

Gravity, magnetic, and seismic studies are the most important that allow comparisons between surface geology and geophysical expression. Magnetic anomalies are the best indicators of near surface boundaries. The gravity data provide information on the deeper structure. Magnetic and gravity maps also allow extension of geological features where exposure is poor and beneath marine areas. Deep seismic reflection data allow comparisons between lower crustal blocks and surface geological features.

Gravity signatures

The gravity field of the Appalachian Orogen is significantly higher than the adjoining Grenville Province where typical Bouguer anomalies are between -60 mGal and 0 mGal (Map 3). A strong eastward gravity gradient from predominantly negative to predominantly positive values is located in the eastern part of the Humber Zone, especially in Newfoundland and offshore. The gradient is typical of the paired anomaly signature attributed to the transition from an older to a younger crust with the positive values over the younger region. The regional negative gravity field extends well south of the Humber Zone in Quebec, and a marked gradient in Newfoundland crosses acutely the exposed Long Range Grenville inlier. There is no clear relationship between the gravity gradient and the Baie Verte-Brompton Line or between the gradient and the edge of the Grenville basement as defined by deep seismic reflection experiments. Whether the gradient represents the edge of a Paleozoic passive continental margin, a Paleozoic collisional zone, or Mesozoic extensional effects related to the modern Atlantic, is unclear.

The trend of the Long Range dyke swarm is parallel to the prominent gravity gradient that crosses the Long Range Inlier and the dykes are most abundant at the change from negative to positive values. A more northerly trend of the gravity gradient at the Strait of Belle Isle follows a change in the direction of the dykes. Possibly the locus of dyke injection and the gravity gradient have the same late Precambrian control in the deeper crust.

Gravity and magnetic anomalies associated with the exposed Tibbit Hill volcanic rocks in Quebec suggest a much more voluminous volcanic mass in subsurface, up to 250 km long, 45 km wide, and 8 km thick. It is convex toward the northwest and marks a triple rift junction involving the Ottawa Graben and Quebec Reentrant in the Appalachian Orogen.

Strong positive Bouguer anomalies east of the gravity gradient coincide with the Dunnage Zone in Newfoundland. A positive Bouguer anomaly of 50 mGal in northeast Newfoundland is the broadest positive Bouguer anomaly in the interior part of the Appalachian Orogen. However, the anomaly is less than that predicted if mafic volcanic and ophiolitic rocks extended to mantle depths. The region of highest values is coincident with the Notre Dame Subzone. A linear belt of anomalies in excess of +30 mGal extends from the northeast coast southwest to Cape Ray. The eastern edge of this belt generally follows the 0 mGal contour. At the northeast coast of Newfoundland it approximately coincides with the Dunnage-Gander zone boundary. Where the Dunnage Zone is narrow in southwest Newfoundland, the positive Bouguer anomaly field is also narrow.

The presence of the Dunnage Zone beneath the Gulf of St. Lawrence is difficult to document using gravity data (Map 3), as the lower Paleozoic rocks are overlain by upwards of 9 km of Carboniferous cover rocks. Similarly in New Brunswick, large amplitude, long wavelength Bouguer gravity anomalies are basement anomalies suppressed by the thick middle Paleozoic cover.

The Gander Zone in Newfoundland is typified by negative, long wavelength, Bouguer gravity anomalies. Local positive anomalies may be indicative of ophiolite complexes or mafic intrusions. The eastern boundary of the Gander Zone with the Avalon Zone is delineated in Newfoundland and offshore to the north and south by a transition from the predominantly negative gravity anomalies of the Gander Zone to the alternating positive and negative anomalies of the Avalon Zone. Tracing the boundary between southern Newfoundland and Cape Breton Island and between the Gander and Avalon zones in New Brunswick and in the Gulf of St. Lawrence is difficult.

In New Brunswick, the Gander Zone gravity anomalies are mainly negative. The anomalies are negative relative to the Avalon Zone to the south and east but are more positive than those in the Humber Zone in the vicinity of the Gaspé Belt.

The Avalon Zone has arcuate belts of alternating positive and negative gravity and magnetic anomalies.

The Meguma Zone has mainly negative anomalies centred on large granite intrusions.

Magnetic signatures

The magnetic signatures are more complicated and variable than the gravity field (Map 4). This arises from the nature of the magnetic anomalies, the greater density of observations, and their greater dependence on depth to source. The magnetic signatures are helpful in tracing zone boundaries and establishing the structural fabric within particular zones.

In the Humber Zone, magnetic anomalies tend to be broad and of low amplitude indicating deep magnetic sources beneath the Cambrian-Ordovician carbonate sequence. Pronounced high amplitude, short wavelength magnetic anomalies occur throughout the Long Range Inlier of Newfoundland (Map 4) and over the Blair River Complex of Cape Breton Island. Moderate amplitude magnetic anomalies occur over major obducted ophiolitic bodies in Quebec and Newfoundland. The boundary between the Humber and Dunnage zones in Quebec and New Brunswick is expressed by the arcuate pattern of high magnetic anomalies which can be traced from the Brompton area of Quebec to Chaleur Bay (Map 4). The boundary is difficult to follow beneath sediments of the Anticosti Basin because of the depth to source.

Dunnage Zone magnetic data exhibit several horizontal wavelengths and variable intensities. Anomalies have short wavelengths, and high amplitudes that reflect its local volcanic belts. In general the magnetic anomalies are positive. The highest amplitude magnetic anomalies, in excess of 6000nT coincide with Jurassic plutons in central Newfoundland. Anomalies expressing mafic and ultramafic plutons along both sides of the Dunnage Zone are less intense, typically having amplitudes of a few hundred nanoTeslas and wavelengths of a few kilometres to tens of kilometres. These patterns can be used to constrain the geometry of various features and to trace them offshore northeast of Newfoundland. The fundamental differences in the magnetic character between the western and eastern parts of the Dunnage Zone in Newfoundland supports its subdivision into Notre Dame and Exploits subzones.

Several major plutons, such as the Mount Peyton in the Newfoundland Dunnage Zone, are recognized by their elliptical signatures of a perpheral high surrounding a magnetic low over the pluton itself. This contrasts with more typical magnetic highs over the plutons in New Brunswick, e.g. the St. George Batholith.

The Dunnage Zone in New Brunswick is typified by positive anomalies which begin at Chaleur Bay and continue inland to the Maine border. Extrapolation of boundaries depicted in Figure 7.6 are a compromise between gravity, magnetic, and geological information.

The Gander Zone has moderate amplitude, long wavelength, negative magnetic anomalies which can be traced throughout Newfoundland and offshore. The overall negative anomalies are interrupted by the narrow belt of positive anomalies associated with the Gander River Complex along the eastern margin of the Dunnage Zone. Subsequent detailed studies in northeast Newfoundland reveal two subtle positive anomaly belts east of the Dunnage zone boundary. These have been ascribed to the presence of mafic and ultramafic material in local thrusts or intrusions.

In New Brunswick, the Gander Zone is also typified by negative magnetic anomalies with minor positive areas in its northeastern part.

The arcuate belts of alternating positive and negative magnetic anomalies of the Avalon Zone are 20-100 km wide with amplitudes of 200-400 nT. Onshore in Newfoundland the positive anomalies correlate with belts of Precambrian volcanic rocks. This signature may be traced northeast of Newfoundland to the continental shelf in the vicinity of the Charlie-Gibbs Fracture Zone. South of Newfoundland the arcuate anomaly patterns are terminated by the Collector Anomaly which expresses the Avalon-Meguma zone boundary (Map 4). Recent deep reflection data, refraction data, and additional magnetic data have corroborated the placement of this boundary offshore in the Tail of the Banks area. More recent studies south of Newfoundland have led to the recognition of distinctive Avalon magnetic signatures which can be traced southward into a major northwest-striking fault. Several other northwest-trending features are interpreted as Carboniferous or younger faults. Similar features have been recognized in both the gravity and magnetic data in the southern Avalon Zone of Newfoundland and across the Avalon-Gander zone boundary in Bonavista Bay.

A magnetic signature adjacent to the Fredericton-Norembega Fault that crosses the Fredericton Belt and Carboniferous basins may mark the Gander-Avalon zone boundary in New Brunswick. This implies that the St. Croix Subzone lies on the Avalon side of this boundary. An extension of the Collector Anomaly passes through Chedabucto Bay and Minas Basin along the trace of the Cobequid Fault. These broad positive anomalies mark the southeastern extent of the Avalon Zone on land and are interpreted as the southern boundary of the zone offshore.

The Meguma Zone has distinctive northeast-trending linear magnetic anomalies of moderate amplitude. These correlate with the general structural trends of the folded Meguma Supergroup and they are truncated by Devonian plutons. The characteristic linear patterns can be used to extend the Meguma Zone offshore, especially near the Nova Scotian coast. The zone is not found in Newfoundland but is interpreted to underlie the southeastern parts of the Grand Banks (Fig. 7.6).

In addition to magnetic expressions of structural trends in the major Appalachian zones, there are other features which appear on large scale magnetic maps and on individual track records offshore. Linear, short wavelength, low amplitude anomalies northeast of Newfoundland may reflect Mesozoic dykes, based on the presence of similar dykes and anomalies on land. A magnetic signature of the Avalon Dyke extends offshore, but drilling attempts failed to prove its presence. A similar offshore feature in the Avalon Zone south of Newfoundland may express another dyke. There are similar magnetic signatures over dykes in Nova Scotia and New Brunswick. All of these features along with the strongly magnetic Mesozoic plutons of central Newfoundland provide evidence of Mesozoic tectonic activity.

Also evident on magnetic maps, especially magnetic shadowgrams, and to a lesser extent on gravity maps, are regional northwest-trending lineations which appear to truncate major gravity and magnetic anomaly patterns (Maps 3 and 4). High resolution aeromagnetic data from the Gulf of St. Lawrence support the notion of northwest-trending right-lateral faults that offset zones and their boundaries.

Northwest-trending magnetic patterns are also recognized in Newfoundland and throughout the Grand Banks and extreme offshore areas. In Newfoundland, the trends are interpreted as two sets of linears that overprint the Dover Fault and the Ackley Granite suite that stitches the Gander and Avalon zones. Using gravity, magnetic, topographic, and geological maps it is possible to identify a series of such features which occur almost regularly across Newfoundland (Fig. 7.6). Similar features can be recognized in Nova Scotia, New Brunswick, and in the adjacent offshore regions (Fig. 7.6). At present there is no structural analysis of fracture patterns or their timing. The association of the magnetic anomalies with bathymetric and topographic features imply late structural features unrelated to Appalachian trends, and they are parallel to the Newfoundland Fracture Zone on the southern nose of the Grand Banks. The features are also perpendicular to the general trends of known Mesozoic dykes. On the Grand Banks, transfer faults related to Cretaceous and younger basin formation have similar orientation, suggesting that the magnetic anomalies reflect transfer faults related to the Mesozoic opening of the Atlantic Ocean.

A prominent positive magnetic anomaly, the East Coast Magnetic Anomaly, occurs at the morphological shelf edge east of Nova Scotia and southward, but it is absent along the torturous rifted margin of the Grand Banks in Newfoundland. One suggestion is that the East Coast Magnetic Anomaly is a Paleozoic collisional zone that was the locus for Mesozoic opening of the North Atlantic. Part of the collisional zone occurs inland in the southeastern United States (Brunswick Magnetic Anomaly) and it may be truncated off Nova Scotia by the axis of Atlantic spreading, and therefore displaced to the African continental margin.

Seismic reflection

Recent information on the crustal configuration of the Canadian Appalachian region has come from deep seismic reflection experiments. These data from Newfoundland and offshore (Fig. 7.1) have dramatically changed our interpretation of the subsurface of the orogen.

The initial survey northeast of Newfoundland completed in 1984 demonstrated three lower crustal blocks. A Grenville lower crustal block was interpreted as the ancient North American continent, extending in subsurface beneath an allochthonous Dunnage Zone. It abuts a Central lower crustal block at mid crustal to mantle depths. The central block is overlain by rocks of the eastern Dunnage Zone and Gander Zone. The boundary between the Gander and Avalon zones, the Dover Fault, is seen to be a major, steep feature having distinctly different seismic character on either side. It is interpreted as the boundary between the Central and Avalon lower crustal blocks. Subsequent surveys in 1986 in the Gulf of St. Lawrence confirmed this general interpretation of lower crustal blocks. The reflection data also indicate that the Grand Banks are underlain by rocks typical of the Avalon Zone. Both refraction and deep reflection data on the southern Grand Banks indicate that the Avalon-Meguma zone boundary is vertical and cuts the deep crust. Other reflection data from the Bay of Fundy and Gulf of Maine further imply that the Meguma Zone has a distinctive basement that defines the Sable lower crustal block (Keen et al., 1990).

The Grenville lower crustal block is wedge-shaped and its subsurface edge follows the form of the Appalachian structural front. It corresponds with the Humber tectonostratigraphic zone. The Dunnage Zone is allochthonous above the opposing Grenville and Central lower crustal blocks. Their boundary is interpreted as a deep collisional zone, the Iapetus Suture. The Gander Zone may be the surface expression of the Central lower crustal block, or it too may be allochthonous. The Avalon Zone corresponds to the Avalon lower crustal block. The Meguma Zone corresponds to the Sable lower crustal block. Results of the 1989 onland Lithoprobe line in Newfoundland failed to show a significant distinction between the Central and Avalon lower crustal blocks, so that the situation beneath the Dunnage Zone is simpler with a Grenville block or Laurentian margin juxtaposed with a combined Central-Avalon block or Gondwana margin (Quinlan et al., 1992). The spatial coherence, if not seismic continuity, between lower crustal blocks and surface zones implies genetic links and common controls during the early Paleozoic development of the Canadian Appalachians.

PALEONTOLOGY

The paleontology of the Canadian Appalachian region contributes to its zonal division. Losses of provincialism coincide with times of accretion.

Precambrian-Cambrian

The oldest fossils are stromatolites, *Archaeozoon acadiense*, in the Middle Proterozoic Green Head Group of the New Brunswick Avalon Zone. The fossils and other evidence of shallow-water deposition suggest correlation with the Grenville Group of southern Ontario and Quebec that contains similar stromatolitic limestone. This implies that Grenville rocks occur in the Avalon Zone.

Soft-bodied organisms of Ediacaran type are known from upper Precambrian rocks in the Newfoundland Avalon Zone, and small calcareous and phosphatic fossils of latest Precambrian to Middle Cambrian age precede the first appearance of trilobites.

The best known assemblages of late Precambrian palynomorphs are acritarchs and nonseptate organic filaments from the Newfoundland Avalon Zone. They resemble those in the lower Dalradian succession of Scotland and the Brioverian of the Armorican Massif of France.

Late Precambrian and Early Cambrian ichnofossil assemblages are similarly rich and diversified in siliciclastic rocks of the Avalon Zone. Comparable assemblages are not known elsewhere in the Appalachians, but their widespread distribution across Eurasia and in northwestern Canada suggests a lack of provincialism.

Continuing investigations of trilobites confirm the "Atlantic" and "Pacific" faunal realms known for about 100 years. Middle and Upper Cambrian rocks of the Humber Zone contain a sequence of polymerid trilobites and other fossils of essentially continent-wide distribution. Early and Middle Cambrian trilobites in the Taconic allochthons of the Humber Zone have elements of both Atlantic and Pacific realms. Trilobites resembling those of the North American shelf indicate shallow-water deposition, whereas

agnostid trilobites, best known from the Acado-Baltic region, indicate contemporaneous deposition in cooler or deeper water.

Genera and species of trilobites from rocks of the Avalon Zone, notably species of *Paradoxides*, are congeneric, and many are conspecific with those from contemporaneous rocks of similar kinds in the Atlantic borderlands from Norway to Morocco, but these taxa are unknown in the carbonate rocks of the St. Lawrence Platform or from the Humber Zone.

Trilobites, including *Paradoxides* and other Middle Cambrian Acado-Baltic genera, were recently reported from the Meguma Zone in the uppermost part of the Goldenville Group. Their paleogeographic and biostratigraphic significance confirm that these rocks were deposited on the eastern side of Iapetus over a long period encompassing much of the Cambrian and Early Ordovician.

Ordovician

Earliest Ordovician fossils of the Humber and Avalon zones reflect continuity with those of Cambrian age. Tentative identification of *Adelograptus tenellus* (Linnarsson) in the earliest Tremadoc part of the Cookson Group in the St. Croix Subzone in southwestern New Brunswick links that part of the sequence with the Baltic area and other parts of Europe.

Most fossil groups exhibit increased provincialism during Arenig and early Llanvirn time. Provincialism was considerably reduced in late Llanvirn-early Llandeilo time.

Division of the Dunnage Zone of Newfoundland into the Notre Dame and Exploits subzones was prompted by several factors including contrasting fossils of Arenig age. Those in the Notre Dame Subzone are conodonts, trilobites, and brachiopods of North American affinity. Arenig-Llandeilo fossils in the Exploits Subzone are dominantly brachiopods associated with smaller numbers of trilobites. These assemblages, and similar ones in the Gander Zone, characterize the Celtic biogeographic province that probably represents peri-insular settings and water temperatures intermediate between the warm waters of the North American continental margin and the cold waters of the Armorican and Baltic platform margins.

Fossils of late Llanvirn-early Llandeilo age, largely conodonts of North Atlantic provincial affinities, occur in limestone at many localities in the Exploits Subzone, recording a transition into warmer water environments. Associated trilobites and brachiopods of the Scoto-Appalachian biogeographic province, confirm this warming trend.

The Dunnage Zone in Quebec and New Brunswick is narrower, less well defined, and has fewer fossils. Late Arenig brachiopods of Celtic province affinities occur in tuffaceous siltstones that overlie slate and quartzite of the Miramichi Subzone.

Fossils of Arenig age in the Avalon Zone are sparse but distinctive. Inarticulate brachiopods from Bell Island, Newfoundland are similar to those from the Armorican Quartzite, France. The Wabana sedimentary iron ores contain distinctive European trilobites and abundant specimens of the arthropod trace fossil *Cruziana*. The linkage between the Avalon Zone and parts of Europe is supported by the graptolite *Didymograptus* (sensu lato) *simulans*, from near the base of the Wabana Group that is conspecific with the type specimens from the Skiddaw Group in the English Lake District. Oolitic hematite containing inarticulate brachiopods also occurs in the northern Antigonish Highlands of Nova Scotia. This sedimentary facies and associated fossils probably represent a circum-Gondwana nearshore environment.

Taconic Orogeny profoundly altered Appalachian depositional and paleontological patterns. The Long Point Group of western Newfoundland is richly fossiliferous and its middle Caradoc forms resemble those in the central and southern Appalachians and Scotland. Ashgill faunas occur in shallow-water carbonate rocks on Anticosti Island, and in a wide variety of rocks in the Gaspé Belt (Aroostook-Percé division). The Anticosti examples include brachiopods that link it to the North American brachiopod biogeographic province, and rugose and tabulate corals that, with some exceptions, have similar affinities. Most of the carbonate rocks of the Gaspé Belt (Aroostook-Percé division) were deposited in deeper water than that of the Anticosti Basin. Recent work has led to the assignment of specific generic suites to depth-controlled benthic assemblage communities. Conodonts of the Anticosti and Percé sequences are similar in that both are dominated by representatives of the North American Midcontinent Province.

The Hirnantian assemblages of brachiopods in the uppermost Ordovician beds at Anticosti Island and Percé are of special interest. This assemblage is characterized by a significant reduction in the number of brachiopod genera, presumably in response to rapid cooling of the world's oceans coincident with the latest Ordovician Saharan glaciation. Late Ordovician trilobites have southern Appalachian affinities, and associated conodonts are of North Atlantic province affinity.

Ashgill fossils, largely brachiopods, occur in clastic rocks of the Badger Belt at New World Island. Assemblages of brachiopods differ from place to place, but their generic composition was apparently governed by local environments. The most common and diversified assemblages consist of genera (e.g. *Christiania, Dolerorthis, Plectatrypa, Sampo*) that characterize the North-European brachiopod biogeographic province.

Silurian-Early Devonian

Silurian and Early Devonian biogeographic patterns were essentially continuations from the Late Ordovician, although there are different names for similar biogeographic subdivisions. For example, the Late Ordovician faunas of the Appalachians were classed from their brachiopods as North American and North European provinces of a cosmopolitan realm. By contrast, Silurian and Early Devonian brachiopod-based biogeographic units place most of the Appalachians in the North Atlantic region of the North Silurian biogeographic realm, with outboard belts (Mascarene, Arisaig, and Annapolis) assigned to the Old World realm. The Early Devonian name for the larger part of the Appalachians is the Appalachian Brachiopod Province. The term for the same area in the Early Devonian, based on rugose corals, is the Eastern Americas Realm. Local depositional environments may be partly responsible for differences in Early Silurian, Late Silurian, and Early Devonian faunas in the Gaspé Belt.

The Fredericton Belt was a marine trough separating the North Silurian faunas of the Chaleurs Bay division of the Gaspé Belt from the Old World faunas of the Mascarene Belt. Evidence of oceanic crust is lacking but the Fredericton Belt separates markedly different shelly and ostracode faunas. Possibly the Miramichi Highlands was a Silurian land barrier.

All but the latest Silurian fossils of the Mascarene Belt belong to cosmopolitan taxa, but some ostracodes of Late Silurian (Pridoli) age are north-European (Old World) realm. The same ostracodes occur in uppermost Silurian strata of the Arisaig Belt.

A report of thelodonts of Late Silurian age from the White Rock Formation in the Annapolis Belt suggests that the Avalon and Meguma zones were in proximity in the Late Silurian. The Old World affinities of the Annapolis Valley sequence are also confirmed by the Rhenish assemblage of Early Devonian brachiopods in the Torbrook Formation.

Middle Devonian-Carboniferous

In the northwestern Connecticut Valley-Gaspé division of the Gaspé Belt several limestone units contain brachiopods of mixed provincial affinity, indicating that Old World forms had almost reached the main part of the North American continent by the Middle Devonian.

The Lower Carboniferous (Viséan) Windsor and Codroy groups contain marine fauna related to those of Western Europe. These faunas differ markedly from the endemic early Carboniferous (Mississippian) assemblages of the eastern interior and Mississippi Valley regions of North America. Ostracodes and conodonts confirm the affinity of these faunas with the widespread Viséan faunas of western Europe. Early Carboniferous spore assemblages from the Canadian Appalachians also resemble those from western Europe. Pennsylvanian faunas (mostly arthropods and bivalves) and flora are similar to Namurian and Westphalian fossils of western Europe.

Discussion and conclusions

Perhaps the earliest application of paleontology to subdividing the Appalachian Orogen was the recognition of provincialism in Cambrian trilobites in what are now the Humber and Avalon zones. This differentiation also contributed to the idea of a two-sided symmetrical orogen, which in turn, led to the definition of the Iapetus Ocean.

The faunal differentiation between the two sides of the Iapetus Ocean diminishes from the Cambrian to the Ordovician, although provinciality continues because the ocean remained a formidable barrier to migration for many groups. Early to Middle Ordovician benthic trilobite and brachiopod faunas occurring around the margins of North America (Humber Zone) are quite distinctive from those in the Iapetus Ocean. For example, in the early Middle Ordovician, the Toquima-Table Head faunal realm occupied a belt peripheral to the platform. Coeval assemblages of largely different brachiopods and trilobites occur in rocks that were deposited on the fringes of islands within Iapetus and on the margins of the Armorican Platform (Celtic Province). Ocean-closing events associated with Ordovician orogeny are recorded in the Exploits Subzone of the

Dunnage Zone where volcaniclastic rocks containing Arenig/Llanvirn fossils of the Celtic Province are overlain by Llandeilo carbonate rocks containing fossils of North American affinities. This corresponds with the time of ophiolite obduction in western Newfoundland and it followed juxtapositioning of the Exploits Subzone and the Gander Zone.

Conodonts of the North Atlantic province and graptolites provide important links across Iapetus because both groups occurred in open ocean waters. Distinct conodont assemblages occurred within North America in the warmer, presumably more saline waters (North American Midcontinent province).

The breakdown of strong provinciality in the Caradoc can be interpreted as evidence of the increasing proximity of the margins of the Iapetus Ocean during the Middle Ordovician. Provincialism was reduced further in the Late Ordovician with the progressive narrowing of Iapetus, and many Silurian forms are considered to be cosmopolitan.

In the context of a continually contracting Iapetus Ocean, it must be explained why early Silurian faunas seem to be cosmopolitan whereas later Silurian and Early Devonian faunas are provincial at a time when the ocean was presumably narrower. Perhaps the massive terminal Ordovician extinction event reduced diversity to the extent that recognition of provincialism in the Early Silurian is precluded by the lack of diversity in many fossil groups. The provincialism demonstrated by ostracodes, brachiopods and vertebrates during the Late Silurian supports the idea that the Fredericton Belt was the site of a Silurian seaway. However, it is probable that environmental factors and not geographic distance played an important role in provincialism at this time. Whatever the controls, provincialism in late Silurian and Devonian organisms change progressively until European or Old World forms invaded North America by the Middle Devonian. After that time, Carboniferous faunas and floras of the Canadian Appalachian region are similar to those of western Europe, confirming that accretion was complete.

METALLOGENY

Mineral deposits are related in time and space to the Appalachian orogenic cycle. Just as the rocks of any orogen contrast with coeval rocks of adjacent platforms, so too are mineral deposits as characteristic of orogens as the rocks and processes themselves. There is a regular and somewhat predictable relationship between kinds and ages of mineral deposits and orogenic development. Improved plate tectonic models for the Canadian Appalachians and improved accretionary analyses allow a more complete understanding of mineral deposits and their relationships to constructional and destructional stages of orogenic development than was heretofore possible. The Canadian Appalachian region is therefore an excellent laboratory for metallogenic studies in a complex accretionary and collisional orogen.

The mineral deposits are grouped according to three main stages of orogenic development: (1) pre-orogenic — includes mineral deposits from previous orogenic cycles and deposits formed at continental margins and in the intervening ocean during rift and drift phases of development; the deposits are confined to tectonostratigraphic zones; (2) syn-orogenic — includes mineral deposits

controlled by faulting and granitoid plutonism that occurred during accretion of outboard terranes; these deposits occur in both zones and belts; and (3) post-orogenic — includes mineral deposits related to late- or post-orogenic faulting, granitoid intrusion, and development of successor basins; these deposits occur across the entire orogen.

The pre-orogenic late Precambrian to Middle Ordovician-Lower Silurian elements of the orogen define four metallogenic entities, now represented by the Humber, Dunnage-Gander, Avalon and Meguma zones. Pre-accretion metallogeny proceeded independently in these entities and their contrasting deposits reflect their diverse geological histories.

Syn-orogenic mineral deposits formed during and immediately following initial accretion (Taconic Orogeny) at the Humber-Dunnage zone boundary with hydrothermal activity related to major faults and granitoid intrusions. At this time, a pre-accretion metallogeny was still proceeding independently in outboard zones (eastern Dunnage, Gander, Avalon, Meguma). Deformation-related and granitoid-related mineralization increased in intensity and abundance as accretion and crustal thickening proceeded through the Silurian and Devonian. Syn-orogenic mineralization, which eventually encompassed all of the accreted outboard terranes, consists of two principal types: (1) structurally controlled and (2) granitoid-related.

The post-orogenic stage of mineralization is related to the development of late Paleozoic overlap assemblages in Devonian to Permian successor basins.

Pre-orogenic deposits
Humber Zone

The development of the Humber Zone began with late Precambrian-Early Cambrian rifting of Grenville basement and concomitant dyke intrusion, volcanism, and clastic sedimentation. This was followed by a passive margin stage with a Cambrian-Ordovican carbonate sequence, and destruction of the margin by ophiolite obduction in the Middle Ordovician.

The metallogeny of the Humber Zone includes at least six broad classes of deposits (Fig. 9.1): (1) iron-titanium oxide deposits in Grenville anorthosite inliers of western Newfoundland; (2) sulphides hosted by Precambrian carbonates of Cape Breton Island; (3) paleoplacer heavy mineral occurrences in Cambrian incipient rift sequences of southeastern Quebec; (4) cupriferous sulphide occurrences in Cambrian incipient rift sequences of southeastern Quebec; (5) zinc and lead occurrences, locally accompanied by Ba, U, and other metals, in the lower Paleozoic carbonate sequence of southeastern Quebec and western Newfoundland; and (6) epigenetic Ba-Pb-Zn deposits in Cambrian clastic rocks in the Lower St. Lawrence Lowlands. All of these deposit types are related to rocks that define the Humber Zone.

Dunnage and Gander zones

Cambrian and Ordovician volcanic rocks of the Dunnage Zone are particularly rich in mineral deposits (Fig. 9.11). Ensimatic island arc volcanic activity continued sporadically from the Cambrian to the Middle Ordovican. Lower Ordovician ophiolitic rocks are preserved throughout the

Newfoundland Dunnage Zone as well as in Taconic allochthons of western Newfoundland, the Elmtree Subzone of New Brunswick, and the ophiolite belt of southeastern Quebec. All are interpreted to record rifting in arc or back-arc environments. At least some of the volcanic activity was approximately coeval with clastic sedimentation of the Gander Zone. Subsequent ensialic back-arc rifting environments are recorded in northern New Brunswick and probable equivalents in the Hermitage Flexure region of southern Newfoundland.

Mineral deposits unique to the Dunnage-Gander zones comprise two broad types: (1) mineralization in the ultramafic and mafic plutonic parts of ophiolite complexes (i.e. magmatic chromite and sulphides, epigenetic Ni arsenides and sulphides, and metasomatic products of the ultramafic rocks such as asbestos, talc, and magnesite) in central and western Newfoundland, Gaspésie, and southeastern Quebec; (2) volcanic- and sediment-hosted massive sulphide deposits including various types of volcanogenic sulphide deposits in central and western Newfoundland, northern New Brunswick, and southeastern Quebec.

Significant occurrences of chromite are found in Newfoundland in the Bay of Islands, Pipestone Pond, Coy Pond, and Great Bend complexes, and in ophiolite suites of southeastern Quebec (Fig. 9.11). Minor occurrences of nickeliferous, locally platinum group elements (PGE)-rich sulphides have been reported from the Bay of Islands Complex and there are numerous Ni occurrences associated with dismembered ophiolitic rocks in Gaspésie. Nickeliferous deposits in southeastern Quebec, possibly of Outokumpu type, are apparently related to hydrothermal activity in the early stages of Taconic Orogeny. Metasomatism of ophiolitic ultramafic rocks during accretion to the Laurentian margin resulted in the formation of talc and asbestos.

Despite the similarities of geological and tectonic settings of the Appalachian ophiolite-hosted volcanogenic massive sulphide (VMS) deposits, recent geochronological work confirms that not all are of the same age. Mineralized ophiolites in western Newfoundland formed in the early Arenig (ca. 488 Ma), those in southeastern Quebec in the late Arenig (ca. 479 Ma) and those in northern New Brunswick in the Llanvirn (ca. 461 Ma) suggesting repetition of back-arc environments favourable for VMS formation.

Submarine volcanic and epiclastic sedimentary sequences in the Dunnage Zone record volcanism, sedimentation, and mineralization in a long-lived and complex series of Cambrian to Middle Ordovician island arcs and back-arc basins around the margins of Iapetus. These sequences are rich in volcanic-hosted massive sulphides, both as massive exhalative and stockwork deposits. The deposits range in size from small to supergiant (Brunswick No. 12) and from relatively lean to extremely rich (Buchans).

Recent tectonic models for various parts of the northern Appalachians emphasize the complexity of the tectonic development of the remnants of Iapetus. All ophiolite suites that contain significant volcanogenic sulphide mineralization exhibit geochemical and/or isotopic evidence of development in supra-subduction settings. It is now recognized that VMS mineralization occurred in a variety of geological and tectonic settings and at several different times during the Cambrian-Ordovician history of Iapetus.

Avalon Zone

The Avalon Zone contains several types of mineral deposits formed mainly during late Precambrian magmatism and early Paleozoic sedimentation (Fig. 9.35). Submergent facies of pre-600 Ma Andean-type arc volcanic sequences contain volcanic-hosted massive sulphide deposits in Newfoundland, Cape Breton Island, and southern New Brunswick. An epithermal style of alteration and mineralization possibly related to two younger episodes of Precambrian magmatism is developed throughout the Newfoundland Avalon Zone and there is at least one possible late Precambrian carbonate-hosted base metal sulphide deposit in central Cape Breton Island. Calc-alkalic plutons related to a late magmatic episode contain minor occurrences of granophile elements in Newfoundland and porphyry-style mineralization in Cape Breton Island. Peralkaline plutons of the same age in Newfoundland contain concentrations of rare metals. Early Paleozoic stable shelf sedimentation in the Newfoundland Avalon Zone resulted in the formation of extensive sedimentary manganese and iron deposits.

Styles and ages of mineralization in the Bras d'Or and Burgeo subzones support other evidence for assignment to the Avalon Zone.

Meguma Zone

Two classes of mineral deposits formed during the pre-accretion history of the Meguma Zone/Annapolis Belt: (1) Concordant base metal occurrences related to the Goldenville-Halifax group transition zone; and (2) Clinton-type iron occurrences within the Devonian Torbrook Group of the Annapolis Belt. In addition, there is a suite of structurally controlled hydrothermal deposits that are unique to the Meguma Zone, including concordant gold-bearing quartz veins hosted by the Goldenville Group and other small stratabound and crosscutting veins (Fig. 9.47).

Three general mechanisms have been proposed for the origin of the Meguma gold veins: 1) syngenetic, hydrothermal deposition on the seafloor, 2) early syntectonic deposition from hydrothermal fluids of diverse origins, and 3) late syntectonic deposition from magmatic or deep crustal hydrothermal fluids.

Syn-orogenic deposits

Structurally controlled mesothermal/epithermal mineralization

Structurally controlled precious and base metal deposits with mesothermal and epithermal characteristics formed throughout the accretionary and post-accretionary history of the Canadian Appalachians. During the gold exploration boom of the 1980s, a significant number of new deposits were discovered and renewed interest spurred research into the nature and origin of this class of deposit.

Mesothermal and lesser epithermal mineralization accompanied orogenic events at the accreting Laurentian margin and throughout the maturing post-accretion orogen. There are a few characteristics that are common to most deposits. Mineralization is typically related to major structures but it is commonly sited in second or third order subsidiary structures. Alteration is typically propylitic and of relatively local extent. Carbonate alteration is ubiquitous and there is commonly a strong association with rocks of contrasting competency (producing dilation during tectonism) and iron-rich lithologies (a chemical trap for gold). There is usually a marked association between gold and pyrite. However, comparison of characteristics of mineralization and alteration in different parts of the orogen show considerable complexities. In some areas, gold occurs in intensely altered country rocks whereas in others, it is in quartz veins associated with minimal alteration. Likewise, metal associations are not homogeneous. In some areas, gold is associated with antimony and arsenic whereas in others, it is associated with base metals. Clearly, local geological features have exercised considerable influence on the form and character of this class of deposits.

Although the timing of mineralization is not well constrained, such constraints as do exist indicate that structurally controlled mineralization was a protracted event that lasted from the Silurian to the Carboniferous. Ages of cross-cutting plutons and geological evidence show that at least some gold mineralization in central Newfoundland was probably associated with Silurian orogenic events. Mineralization occurs in Carboniferous rocks in southern New Brunswick. Major, possibly transcrustal, structures which were subject to reactivation were clearly a controlling factor in focussing hydrothermal fluids and mineralization.

Granitoid-related mineralization

Granitoid intrusion occurred at many times and in response to various tectonic events during the accretionary and post-accretionary history of the Appalachian Orogen. Intrusive rocks range in composition from gabbro to high-silica granite, and include subalkaline, alkaline, and peralkaline types. Granitic rocks occur in all zones and a variety of tectonic settings. Granites with I-type, S-type and A-type characteristics can be recognized, as well as large batholiths with mixed affinities.

The wide variety of granites and magmatic episodes has produced a broad spectrum of mineral deposits (Fig. 9.65). These include important concentrations of granophile elements (e.g. East Kemptville, Nova Scotia; Mount Pleasant, New Brunswick), base and precious metals (e.g. Gaspé Copper and Madeleine Mines, Quebec; Nigadoo River and Lake George, New Brunswick), Ni and platinum group elements (e.g. St. Stephen, New Brunswick), and industrial minerals (e.g. St. Lawrence fluorspar, Newfoundland). Although most of the economically significant granite-related mineralization in the Canadian Appalachians occurred during the Devonian, significant mineralization is also associated with older peraluminous plutons (e.g. central Newfoundland, New Brunswick) and with younger Carboniferous plutons (notably the peralkaline granites of southern Newfoundland), and Mesozoic plutons (alkaline granites, syenites, and carbonitites of southeastern Quebec and adjoining New England).

Post-orogenic deposits

Several mineralizing episodes can be recognized in late Paleozoic overlap assemblages: (1) Late Devonian to early Viséan: (a) minor occurrences of U are related to late Devonian rift-related felsic volcanics; (b) base mineral occurrences are found near the contact with the overlying marine evaporitic rocks of the Viséan Windsor Group; (c) rare paleoplacer deposits occur at the unconformity between the Meguma Supergroup slates and Horton Group conglomerates; (2) Viséan: (a) abundant evaporitic deposits (e.g.. anhydrite, salt, secondary gypsum hydrated from anhydrite, potash) formed at several stratigraphic levels in the Windsor/Codroy groups; (b) syngenetic low grade enrichments in Cu, Pb, Zn sulphides, pedogenic-diagenetic Fe and Mn and local barite and celestite replacements formed, especially near disconformity redox boundaries between marine carbonates and underlying redbeds; (3) Late Pennsylvanian: climate- controlled lead deposits formed in lower Westphalian sandstones (e.g., Yava Deposit, Cape Breton Island); (4) Late Westphalian to Stephanian: numerous epigenetic mineral deposits formed, apparently from metalliferous brines developed deep in the basins and driven toward the basin margins where they interacted with reactive host rocks both along escape paths (faults and fractures) and reservoir-stratigraphic traps. The basal Windsor Group carbonate rocks were a primary host; (5) Late Permian: reddening of upper Carboniferous strata, especially in the Cumberland Basin, where low water tables generated numerous solution-roll front type Cu, Ag deposits.

Summary and conclusions

The present plate tectonic model for mineral deposits used here is similar to models advanced many years ago, based on geosynclinal development and the orogenic cycle. The principal difference is that the plate tectonic framework provides actualistic and readily observable analogues for ancient processes. The empirical observations that led to former models are as valid today as they were half a century ago. As the geological database expands there will be added refinements to tectonic and metallogenic models leading to an integrated understanding of the orogen.

The oldest mineral deposits are the products of earlier orogenic cycles, preserved in continental crust that was incorporated in the Appalachian cycle. Thus, magmatic iron-titanium deposits in the Humber Zone are geologically part of the Grenville orogenic belt, whereas late Proterozoic volcanogenic massive sulphides, carbonate-hosted, and granite-related deposits in the Avalon Zone are geologically part of the Pan-African orogenic belt. Some of these occurrences provided metals which were remobilized and/or upgraded by Appalachian orogenic processes.

During the rift stage, the principal mineral deposits were beach placers in southeastern Quebec. Because incipient rifting in the Canadian Appalachians was apparently not accompanied by extensive hydrothermal activity, there does not seem to have been extensive remobilization of metals within the crust, or transfer of metals between crustal levels.

During the drift stage, conditions were favourable on the continental margins adjacent to Iapetus for the formation of certain mineral deposit types such as Pb-Zn in shelf carbonate sequences. Basinward on the passive margin, uraniferous phosphorites were also formed. At about the same time, shallow seas spread across the stable Avalon Zone with sedimentary manganese and iron deposits. This stage of development provided the main opportunities for mass transfer from the mantle to the crust and mineral deposits such as chromite and base metals in ophiolitic suites formed from juvenile metal sources. Local subduction near continental margins resulted in hydrothermal circulation and ore formation on or near continental crust (e.g. Bathurst) and may have involved crustally derived metals as well. This ore-forming episode apparently operated mainly during extensional tectonic regimes, probably enhanced by the high heat flow and the focussing of hydrothermal fluids by extensional fracture systems. Most of the deposits of this type formed in back arc basins as no remnants of true oceanic crust and no metal deposits of a major spreading ridge are recognized.

The beginning of accretion at the Laurentian margin in the Early Ordovician led to transport of ophiolite suites and their mineral deposits and led to the generation of new types of deposits. Because accretion apparently spanned a considerable period, the drift and accretionary stages of the orogen overlap. Between the Middle Ordovician to Early Silurian, syn-accretion metallogeny began in the accreted inboard terranes while the outboard terranes were still experiencing drift-phase metallogeny. Thus evidence for Silurian subduction in the Aspy Subzone is supported by occurrences of volcanogenic sulphide deposits.

The processes of accretion included reactivation of pre-existing trans-crustal faults, formation of new faults, and crustal thickening that produced granitoid magmas. Heat was locally provided by the magmas that utilized coeval structures for ascent. As crustal thickening proceeded deeper partial melting and granitic magma generation provided further opportunities for the transfer of metals from the lower to upper crust.

Accretionary processes were accompanied by dewatering of nearby sedimentary basins liberating fluids that produced base metal and barite veins. Similar processes may have been partially responsible for the formation of epigenetic carbonate-hosted Zn and Pb deposits, although the timing and source of metals is not well constrained.

Post-accretion mineralizing processes were accompanied by a dramatic decrease in the intensity of tectonism, with the result that mass transfer of minerals occurred principally within the upper crustal levels. Most sediment hosted mineralization is believed to be related to movement on large faults within and around successor basins, with local fluid circulation in the immediate basement. Relief on boundary faults provided opportunities for placer deposits derived from the immediately adjacent highlands. Evaporitic basins produced thick deposits of sulphates which contributed soluble salts to basinal fluids, enhancing their metal-carrying capacity. Metals were probably derived from the sediments in the basins and from the immediately adjacent basement. Depositional environments were provided by porous units: sandstones, carbonates, or dilatant zones in competent units.

Mesozoic rocks, structures, and intrusions are related to the Atlantic spreading episode and another cycle of metallogeny.

ZONAL LINKAGES AND ACCRETIONARY HISTORY

Stratigraphic and sedimentological analyses of the Canadian Appalachians indicate that its elements were assembled during two major accretionary events. Emplacement of allochthons across the Humber Zone and interaction of the Dunnage Zone and Humber Zone in the Early and Middle Ordovician was the first event. It is attributed to northwestward obduction of oceanic crust and mantle and head-on collision between a continental margin and an island arc. Its structural effects and stratigraphic expression are recognized in the Humber and adjacent Dunnage zones and they are attributed to Taconic Orogeny (Map 5). The Gander Zone and Dunnage Zone in Newfoundland also interacted at this time with southeastward ophiolite obduction, but gaps probably existed in the central Dunnage Zone until the Silurian. Accretion of the Avalon Zone to the Gander Zone was later, in the Silurian-Devonian, and accretion of the Meguma Zone to the Avalon Zone occurred in the Devonian. Structural, plutonic, and metamorphic effects of these later events are attributed to Acadian Orogeny in its broadest sense (Map 6). Its surface effects were probably controlled by compression and collision between deep crustal blocks. Carboniferous (Alleghanian) deformation is recorded by major transcurrent faults and attendant thrust zones. These effects were superposed on the already assembled orogen.

Ordovician effects of the first event are absent in the Avalon and Meguma zones (Map 5). All zones were affected by middle Paleozoic deformation, except for eastern parts of the Newfoundland Avalon Zone (Map 6).

Precambrian orogenic effects unrelated to the Appalachian cycle are restricted to the Humber and Avalon zones. In the Humber Zone, the present positions of Grenville inliers are controlled by Paleozoic structures so that their dimensional orientations are parallel to Appalachian structural trends and facies belts. Internal fabrics and fold axes trend northwest in some examples, perpendicular to Appalachian structures. As northwesterly trends are common in the Grenville Structural Province of the Canadian Shield, this suggests minimal or no rotation during Paleozoic deformations. The Avalon Zone encompasses a variety of diverse Precambrian elements, implying a composite makeup. These elements were assembled in the late Precambrian as Cambrian rocks have similar stratigraphies and distinctive Atlantic realm faunas.

Humber-Dunnage interaction (Taconic Orogeny)

Stratigraphic expression and sedimentological linkages

The first indication of major instability in the external Humber Zone is the sub-Middle Ordovician St. George Unconformity. It is interpreted to represent erosion along a migrating peripheral bulge associated with assembly and transport of Taconic allochthons.

The transition from carbonate platform to deep foreland basin includes a complicated record of platform collapse, fault movements, and deposition of limestone boulder conglomerates. Subaerial erosion of rocks that lay mainly east of the Humber Zone produced polymictic flysch units of chiefly Middle Ordovician age that are found now in both autochthonous and allochthonous successions, and they extend westward beyond the limit of Appalachian deformation. The deposition of deep marine clastics above shallow marine carbonates indicates rapid subsidence under the load of Taconic thrust sheets.

The earliest indication of assembly of Taconic allochthons is the reversal in sedimentary provenance recorded in the stratigraphy of their lower structural slices. In Newfoundland, the easterly derived clastic rocks are as old as Arenig. The youngest rocks of the underlying autochthon are Llanvirn. The North Arm Mountain massif of the Bay of Islands Complex has an unconformable cover of Llandeilo age, interpreted as coeval with transport. The first post-orogenic phase of sedimentation is represented by the Caradoc Long Point Group. These and other features bracket the timing of Taconic Orogeny and indicate that Taconic events in Newfoundland are older than those in Quebec.

Discontinuous ophiolite complexes along the Baie Verte-Brompton Line are associated with mélanges, either occurring as huge blocks within mélange (Quebec) or as basement to a mélange cover (Quebec and Newfoundland). Clastic sequences, such as the Magog Group and correlatives, occur above mélanges of the Estrie-Beauce, Témiscouata, and Gaspésie subzones. The clastic sequences are as old as Llanvirn and some have sedimentary links with Taconic allochthons of the Humber Zone. The mélanges in Quebec also contain deformed clasts from the adjacent internal Humber Zone. Similarly in Newfoundland, the Flat Water Pond Group of probable early Middle Ordovician age contains metamorphic clasts from the internal Humber Zone. These relationships suggest that mélange formation on the Baie Verte-Brompton Line was coeval with early assembly of Taconic allochthons and that the allochthons were in a sediment shedding position when clastic aprons like the Magog Group were deposited. The sedimentary evidence also indicates that parts of the internal Humber Zone were deformed and metamorphosed by early Middle Ordovician time.

Structural expression

In the external Humber Zone, different structural slices of Taconic allochthons exhibit different deformational styles, and most structures were imprinted prior to or during the assembly and transport of the allochthons. These vary from intense foliations, tectonic banding, and folded schistosities in ophiolitic rocks and their metamorphic soles, to scaly cleavages, rootless folds, and overturned beds in sedimentary rocks. Lower slices have internally complex geometries of rock units and the slices are rarely morphologically distinct. Higher structural slices are of simpler internal makeup and some have marked morphological expression. Stratigraphic relationships, palinspastic restorations, and structural considerations all indicate that the allochthons were assembled from east to west and that the structurally highest slices travelled the farthest. Assembly began in the Early Ordovician and final emplacement was Middle to Late Ordovician.

Limestone breccias of the Cow Head Group and overlying easterly derived clastic rocks in western Newfoundland occur in repeated east-dipping, east-facing stratigraphic sections. The coarsest limestone breccias are in western sections with finer, thinner, and fewer breccias in eastern sections. The overlying clastic rocks are everywhere in thrust contact with an allochthonous slice, indicating entrainment in a west-directed thrust complex soon after deposition.

Early recumbent folds with penetrative cleavage in some higher sedimentary slices have subhorizontal axes and face westward, in the general direction of tectonic transport. These structures are coeval with assembly and transport.

The metamorphic soles of allochthonous ophiolite suites are interpreted as high temperature shear zones resulting from transport of hot mantle and oceanic crust. The juxtaposition of oceanic and continental rocks within the dynamothermal soles indicate an interface between underlying and overriding plates. Hornblende cooling ages are Early and Middle Ordovician.

Rocks of the internal Humber Zone are characterized by multiphase deformation and metamorphism that began with the interaction of the continental Humber Zone and the oceanic Dunnage Zone. Relationships in Newfoundland suggest that the multideformed cover sequence overlaps the ancient continent-ocean transition. Associated mélanges display all structures evident in surrounding rocks. Hence their formation preceded metamorphism and penetrative deformation.

The earliest deformation is characterized by ductile shear zones and thrusts, that in places contain ultramafic rocks. The ultramafic rocks were incorporated before the earliest fabrics developed, and some shear zones may represent extensions of ophiolitic mélange zones. This suggests that the earliest deformation involved tectonic transport from the oceanic Dunnage Zone onto the Humber miogeocline.

Peak metamorphic conditions of the internal Humber Zone are greenschist to lower amphibolite facies in most places. Of particular note are blueschist facies in the Tibbit Hill mafic volcanic rocks of Quebec and eclogite facies in the infrastructure of Newfoundland. The Dunamagon Granite in Newfoundland, dated isotopically at 460 ± 12 Ma, cuts the metamorphic rocks. Thus an Early Ordovician age is inferred for the tectonothermal events. Cooling ages from Newfoundland indicate that regional metamorphism had subsided by the Middle Silurian. Cooling ages in Quebec are Late Ordovician in most areas.

In Gaspésie, early deformation of the Maquereau Group of the internal Humber Zone predated tectonic accretion to adjacent ophiolitic mélange, that in turn predated deposition of the Llanvirn Mictaw Group.

Ordovician deformation in the Dunnage Zone is mild compared to that in adjacent parts of the Humber Zone. Sub-Silurian unconformities are everywhere present across the Notre Dame Subzone, and locally present in the Estrie-Beauce, Témiscouata, and Gaspésie subzones. In New Brunswick, Ordovician rocks north of the Rocky Brook-Millstream Fault are mildly deformed and overlain unconformbly by Silurian strata.

The Notre Dame Subzone has one of the widest ophiolite belts in the Appalachian Orogen exposed along its coastal section. The overall structural arrangement indicates westward verging imbricate slices with eastward-facing stratigraphy. Intensity of deformation and metamorphism decreases progressively from west to east, or upward through the structural pile. An unconformable Silurian cover indicates that the imbrication is Ordovician.

Deformation and metamorphism in the Dashwoods Subzone are interpreted as Ordovician, based on isotopic dates and crosscutting relationships. Along the southeast margin of the Dashwoods Subzone, Devonian rocks of the Windsor Point Group are nonconformable on foliated tonalites and granites.

Older rocks in the Dunnage Zone

The Twillingate Subzone of Newfoundland is a small area of mafic volcanic rocks cut by tonalite dated at 510 ± 16 Ma (Fig. 3.76). Amphibolite facies metamorphism, intense deformation, and mylonitization along its southern margin are out of context with low grade rocks of the Notre Dame Subzone. Mafic dykes of the Notre Dame Subzone cut mylonitic tonalite and have cooling ages up to 470 Ma. The Twillingate tonalite is significantly older than ophiolite complexes of the Dunnage Zone.

Crystalline rocks occur in the Saint-Daniel Mélange and in ophiolitic mélange northwest of the Maquereau Dome of Gaspésie. Some of these are correlated with the Precambrian Chain Lakes massif of Maine. Relationships in the Mégantic Subzone also imply early to middle Ordovician interaction with the Chain Lakes massif. Two large plutons in the Exploits Subzone, the Valentine Lake and Cripple Back Lake quartz monzonites, dated at 563 ± 2 Ma and 565 +4/-2 Ma, are structurally emplaced and of uncertain significance.

Dunnage-Gander amalgamation

Ordovician sedimentary links between the Exploits Subzone and Gander Lake Subzone in Newfoundland are provided by Upper Arenig-Lower Llanvirn and Upper Llanvirn-Lower Llandeilo conglomerates that contain a sampling of ophiolitic clasts (Dunnage rocks) and quartzose sedimentary clasts (Gander rocks). Some of the ophiolitic clasts have fabrics that predate incorporation in the conglomerates, but whether or not the same is true for Gander clasts is debatable. An occurrence of ophiolitic olistostromal mélange at the Mount Cormack-Exploits subzone boundary contains ultramafic and quartzite blocks in a black shaly matrix, implying an Ordovician age. There is no evidence of fabrics in quartzite blocks before incorporation in the mélange.

The structural evidence for the Exploits Subzone (Dunnage) above the Mount Cormack Subzone (Gander) in Newfoundland is especially clear. The age of the Pipestone Pond Complex at 494 Ma sets a lower limit to the time of Dunnage transport. Metamorphic isograds in the Mount Cormack Subzone have a concentric pattern and the synmetamorphic Through Hill Granite, associated with the highest grade metamorphic rocks, is dated at 464 Ma (Colman-Sadd et al., 1992). These features suggest that the Mount Cormack Subzone is not a simple structural window, but rather a core complex with attendant metamorphism and plutonism. The Partridge Berry Hills Granite dated at 474 Ma stitches the Mount Cormack-Exploits boundary (Colman-Sadd et al., 1992). This and sedimentological evidence all support an Early Ordovician age for Dunnage-Gander interaction.

Rocks of the Gander Zone almost everywhere exhibit folded cleavages or schistosities and appear to be more deformed than rocks in the adjacent Dunnage and Avalon zones. Coupled with the presence of unconformities between Lower and/or lower Middle Ordovician conglomerates and mafic-ultramafic rocks of the Gander River

Complex, relationships hint at pre-Middle Ordovician deformation in the Gander Zone. Relations in the Mount Cormack Subzone confirm Ordovician deformation, metamorphism and intrusion. However, the major deformation and regional metamorphism along the eastern margin of the Newfoundland Gander Zone is Silurian. This roughly coincides with imbrication and development of Silurian mélanges throughout the northeast Exploits Subzone.

The Baie Verte-Brompton Line and the northern Exploits-Gander Lake subzone boundary on opposite sides of the Newfoundland Dunnage Zone have several important similarities: (a) both are marked by discontinuous mafic-ultramafic complexes; (b) there is a sub-Middle Ordovician unconformity developed above mafic-ultramafic rocks along the Exploits-Gander boundary and an inferred sub-Middle Ordovician unconformity above mafic-ultramafic rocks of the Baie Verte-Brompton Line; (c) Ordovician conglomerates above the unconformities are of local derivation, unsorted, immature, and typical of rapidly evolving sources; (d) the conglomerates contain sedimentary clasts, deformed in some examples, that indicate linkages with the deformed Humber Zone to the west and the Gander Zone to the east; (e) mélanges occur locally at both boundaries; (f) in a general way, regional metamorphism increases in adjacent rocks outward and away from the Dunnage zone, although the times of metamorphism may be different (recent chronological studies of the internal Humber Zone in Newfoundland indicate important Silurian plutonism and high grade regional metamorphism (Cawood et al., 1994); (g) the present steep to overturned structures change outwards from the Dunnage Zone into flatter structures in metaclastic rocks; and (h) polarity of first structural transport is away from the Dunnage Zone with later transcurrent movements.

In contrast to the mainly tectonic Dunnage-Gander contacts in Newfoundland, most of the Tetagouche Group in the New Brunswick Dunnage Zone represents a conformable to disconformable stratigraphic cover to the Miramichi Group of the Gander Zone. Although contacts are commonly tectonized, field relationships do not support the idea that Middle Ordovician volcanic and sedimentary rocks represent the remnants of an enormous klippe of Dunnage Zone rocks above Gander Zone rocks. A disconformity between the basal Tetagouche Group and the Miramichi Group is marked by a thin layer of conglomerate. A similar conglomerate in Maine at the contact between the Grand Pitch and Shin Brook formations marks the sub-Middle Ordovician Penobscot unconformity. However in New Brunswick, the Miramichi and Tetagouche groups are deformed together and there is no correspondence between structural styles and stratigraphic divisions.

Faunal and other distinctions between the Notre Dame and Exploits subzones in Newfoundland suggest separation during their Early and early Middle Ordovician development. A sub-Silurian unconformity and absence of Caradoc shales and younger Ordovician rocks throughout the Notre Dame Subzone contrast with continuous deposition in north-central parts of the Exploits Subzone. The earliest subzone linkage is provided by volcanic clasts from the Notre Dame Subzone found in Upper Ordovician-Lower Silurian conglomerates and olistostrome of the Exploits Subzone. Structural evidence for a late Ordovician-early Silurian collisional event is scarce, except for eastward thrusting in the Exploits Subzone and local development of lower Silurian mélanges. Paleontological evidence suggests links between Laurentia and the Exploits Subzone by Llandeilo time.

The St. Croix Subzone-Avalon Zone boundary in New Brunswick is mainly hidden by Silurian to Lower Devonian volcanic and sedimentary rocks of the Mascarene Belt. The Silurian rocks are unconformable upon Ordovician rocks at Cookson Island.

Although Dunnage-Gander relations resemble Dunnage-Humber relations on the opposite side of the orogen, some important differences in the east are: (1) no stratigraphic analysis exists for a Gander Zone continental margin comparable to that for the Humber Zone; (2) metaclastic rocks that define the Gander Zone have no defined basement and their maximum age is unknown; (3) there is no stratigraphic or sedimentological expression of ophiolite emplacement in rocks of the Gander Zone; (4) ophiolite complexes of the Exploits Subzone lack dynamothermal soles of Bay of Islands type; (5) extensive mélanges and allochthons of Taconic style are absent in the east; (6) in both northeast Newfoundland and New Brunswick lower and middle Ordovician rocks are overstep sequences either stratigraphically above the Gander Zone or conformable and unconformable above rocks of the Dunnage Zone; (7) penetrative deformation related to Early Ordovician Dunnage-Gander interaction is poorly defined or absent, and rocks of the Miramichi and Bathurst subzones in New Brunswick have the same stuctural history; and (8) rocks of the Gander and Dunnage zones are cut by large middle Paleozoic plutons that confuse Ordovician relationships.

Gander-Avalon-Meguma accretion (Acadian Orogeny)

Accretion of the Avalon Zone to the Gander Zone is considered to be Silurian as there are no confirmed earlier linkages. The Meguma Zone and Annapolis Belt maintained a distinctiveness into the Devonian. The earliest linkage with the Avalon Zone is Carboniferous, so that the time of Avalon-Meguma accretion is considered Devonian.

The expressions of Silurian-Devonian accretion and the styles of Acadian Orogeny contrast sharply with expressions of Ordovician accretion and styles of Taconic Orogeny. The boundaries of Ordovician accreted zones are marked by ophiolite complexes and mélanges. In contrast, the Gander-Avalon boundary and the Avalon-Meguma boundary are steep faults with local wide zones of ductile shearing. Furthermore, whereas the evidence of an oceanic tract between the Humber and Gander zones is virtually unassailable, oceanic vestiges are absent at the boundaries of more outboard Zones. The mechanism of earliest accretion was by obduction of the Dunnage Zone across the opposing Humber and Gander zones. Later accretion of the Avalon and Meguma zones was controlled by transcurrent movements.

The structural effects of Acadian Orogeny are more widespread than those of earlier events. Whereas Taconic Orogeny is confined to the eastern Humber Zone and adjacent Dunnage Zone, with coeval effects at the Dunnage-Gander zone boundary (Map 5), Acadian Orogeny affected the entire exposed orogen, except for parts of the eastern Avalon Zone in Newfoundland (Map 6). This phase of orogenesis also coincides with the plutonic and metamorphic peaks in the development of the orogen.

Taconic Orogeny affected Cambrian-Ordovician marine rocks and was followed by marine deposition. Acadian Orogeny affected Silurian-Devonian rocks that were terrestrial in upper parts of stratigraphic sections, and the orogenic event was followed by terrestrial deposition. Acadian Orogeny marked a complete and permanent change in the development of the orogen.

Where middle Paleozoic stratigraphic sections are complete, such as in the Gaspé Belt, the Acadian event is dated as middle to late Devonian. In the absence of Devonian rocks, the stratigraphic definition of Acadian Orogeny is less certain. Lower and Middle Devonian rocks of the Canadian Appalachians are everywhere deformed, albeit mildly in some places. Large Devonian plutons cut most zones and belts of the orogen. Recent isotopic ages indicate a major Silurian orogenic event in central Newfoundland and there is evidence for a similar event in the Aspy Subzone and elsewhere. It affected Ordovician and Silurian rocks and it predated the emplacement of granites dated at about 400 Ma.

Sedimentary and stratigraphic linkages

The earliest linkage between the Gander and Avalon zones in Newfoundland is Silurian or Devonian, dependent on the age of the Cinq Isles and Pools Cove formations of the Fortune Belt that contain Gander Zone plutonic and metamorphic detritus and that lie unconformably on rocks of the Avalon Zone. The middle Paleozoic rocks of the Gaspé Belt overlap the Humber, Dunnage, and Gander zones, indicating that these zones were accreted before the end of the Ordovician. In New Brunswick, Lower Silurian rocks of the Fredericton Belt link the Miramichi and St. Croix subzones and Silurian rocks of the Mascarene Belt link the St. Croix Subzone and Avalon Zone.

The Meguma Zone has a dintinct Devonian stratigraphy and plutonic expression. Its first link with the Avalon Zone is established by the age of Carboniferous cover rocks.

The spread of the Late Silurian-Lochkovian Rhenish-Bohemian fauna onto Laurentia during the Pragian and Emsian agrees with other evidence for middle Paleozoic Avalon and Meguma accretion.

Structural expression

The effects of Acadian deformation extend all the way across the Canadian Appalachians from the Humber Zone in the west to the Meguma Zone in the east. At the surface, the Acadian foreland structural front of the Humber Zone coincides roughly with Logan's Line or the foreland limit of Taconic allochthons.

In Quebec, rocks of the northwestern portion of the Connecticut Valley-Gaspé division of the Gaspé Belt have a structural style like that of the Valley and Ridge Province of the U.S. Appalachians. Folds have wavelengths and amplitudes measured in kilometres and thrust faults occur on the north flanks of some anticlines. Across a zone of steep faults, rocks of the southeastern Connecticut Valley-Gaspé division are tightly folded with slaty cleavage. Tight to isoclinal, nearly upright folds on all scales are also a characteristic feature of the Aroostook-Percé division of the Gaspé Belt. Rocks of the Chaleurs Bay division are less deformed compared to the Aroostook-Percé division.

Major dextral faults in the Gaspé Belt are one of the most important features of Acadian Orogeny. Repetitions of the Matapédia Group in map pattern, implying several southward-facing shorelines, are explained by dextral offsets of a single shoreline on major transcurrent faults. Shallow marine shales on the south side of Chaleur Bay are interpreted as part of the same shelf rather than an opposing shelf on the opposite side of the Gaspé Belt.

In New Brunswick, Silurian rocks of the Fredericton Belt have polyphase deformation in the south, whereas only one phase of folding is recognized in most of its northern part.

The Arisaig Belt of Nova Scotia has shallow southeast-plunging upright folds. The Middle Devonian McAras Brook Formation postdates these structures suggesting a mid-Devonian age of deformation. The counter-clockwise orientation of the folds suggests that they were produced by dextral motions on the Hollow Fault. In the northern Cobequid Highlands, folding of Silurian-lower Devonian sedimentary rocks also pre-dates deposition of Upper Devonian redbeds and volcanic rocks. In the southern Antigonish Highlands, deformation is heterogenous and is restricted to shear zones, with local evidence of thrusting and development of positive flower structures. At least some of this deformation affected Upper Devonian rocks and is therefore younger than that in the northern Antigonish Highlands.

The Annapolis Belt has broad, shallowly plunging, upright folds with sub-vertical cleavage. Acadian Orogeny is the first major event to affect the Meguma Zone/Annapolis Belt.

In Newfoundland, the morphological Acadian front is a west directed thrust or steep reverse fault. Offshore seismic reflection profiles suggest a triangle zone similar to those developed at the foreland edges of other thrust belts. Devonian deformation is defined at Port au Port where the Silurian-Devonian Clam Bank Group is deformed and nearby Carboniferous rocks are undeformed. At Baie Verte Peninsula, structural styles of Silurian rocks vary from mild open folding to intense polyphase deformation, and the latest structures are east directed. Much of the plutonism and regional metamorphism in the internal Newfoundland Humber Zone is locally Silurian (Cawood et al., 1994).

Silurian rocks of the Springdale Belt are mildly deformed in most places where they lie unconformably upon deformed Ordovician or older rocks. Deformation is more intense throughout the Badger and Botwood belts. There, Ordovician and Silurian rocks are in places conformable and deformed together. In the Botwood Belt, a Silurian isotopic age for volcanic rocks that are unconformable above steeply dipping folded rocks of the Silurian Botwood Group is further evidence for Silurian deformation. Also, a 408 ± 2 Ma age for the Loon Bay intrusive suite places a Late Silurian upper limit on the timing of thrust faulting and folding in the Badger Belt.

Intensity of deformation increases southwestward along the Exploits Subzone and Botwood Belt toward the narrow Cape Ray Belt. Devonian rocks of the Windsor Point Group are polydeformed at Cape Ray. East over west oblique thrusting produced mylonites in the Windsor Point Group and this polarity of structure corresponds to northwest thrusting in the Silurian La Poile Belt.

Isotopic dates suggest a Silurian age for the intense deformation that increases eastward and southeastward across the Gander Lake Subzone and extends westward around the Hermitage Flexure.

The Gander-Avalon zone boundary along the Dover Fault is a wide zone of mylonitization that affects granite and metamorphic rocks on the Gander side and less deformed upper Precambrian sedimentary and volcanic rocks on the Avalon side. Deformation within the Avalon Zone is mild compared to that of the adjacent Gander Zone. Tight upright folds affect parts of the western Avalon Zone, whereas eastern parts have open folds. Some Cambrian-Ordovician rocks in the eastern Avalon Zone, such as those at Conception Bay, are subhorizontal. Offshore, Paleozoic deformation is uniformly mild or absent across the Grand Banks. A mild erosional unconformity separates marine Silurian and terrestrial Devonian rocks.

Metamorphic expression

Where tectonostratigraphic zones are wide, such as in northeast Newfoundland and Quebec-New Brunswick, middle Paleozoic metamorphism occurs in relatively narrow belts. Where zones are narrow, as in southwest Newfoundland and Cape Breton Island, regional metamorphism affects most rocks and zones. Intensity of regional metamorphism, mainly of Silurian age, increases eastward across the Newfoundland Gander Lake Subzone. Metamorphism is low grade throughout most of the Avalon Zone and there is a sharp metamorphic contrast in most places across the Gander-Avalon zone boundary, especially across the Dover Fault in northeast Newfoundland. Silurian metamorphism is high grade in the Aspy Subzone. The metamorphic facies of the Meguma Zone is greenschist, with an increase to amphibolite facies in the west. Hornblende-hornfels facies is developed close to granitic batholiths.

Plutonic expression

Middle Paleozoic plutons extend across the orogen from eastern parts of the Humber Zone to western parts of the Avalon Zone in Newfoundland and across the Avalon and Meguma zones in Nova Scotia. There are few middle Paleozoic plutons in the exposed Humber Zone, except for its internal parts in Newfoundland. One exception is the Devonian Mount McGerrigle pluton of Quebec, which cuts transported rocks near the Appalachian structural front. Another is the Devonian Devils Room granite of White Bay, Newfoundland that cuts all fabrics in adjacent gneisses of the Long Range Inlier.

Some post-tectonic middle Paleozoic plutons have a zonal preference, others cut zone boundaries or follow coeval volcanic belts that transgress zone boundaries. Alkali plutons that span the eastern Humber Zone and Notre Dame Subzone are coeval with Silurian volcanism and a series of nested calderas that cross the Humber-Dunnage boundary. Acadian composite plutons with early peripheral mafic phases and later granitic phases are common in the Exploits Subzone, but also cut the Red Indian Line. Foliated to massive middle Paleozoic biotite granites and garnetiferous muscovite leucogranites are typical of the Gander Zone and occur in all subzones. Most of these are of Silurian-Devonian age but some are Ordovician. Lineations and foliations in deformed examples are everywhere

887

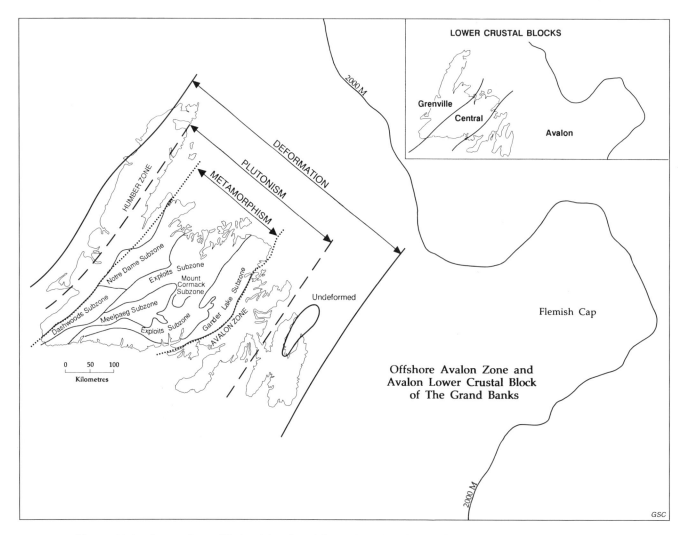

Figure 11.6. Comparison of limits of Acadian deformation, plutonism and upper greenschist to amphibolite facies regional metamorphism across the Newfoundland Appalachians and Atlantic continental margin.

parallel to those of host or nearby migmatites and mylonites, suggesting emplacement during shearing. Undeformed examples indicate that intrusive activity outlasted ductile shearing. A variety of features of the leucogranites suggest an origin through anatexis of supracrustal rocks, such as numerous xenoliths of partially assimilated country rock, a peraluminous chemistry, and high strontium isotope initial ratios. This underscores their localization in high grade areas of the Gander Zone and nearby anomalously high grade areas of the Exploits Subzone.

Sm-Nd isotopic characeristics of the Gander Group sedimentary rocks and the Hare Bay Gneiss are consistent with derivation from the same protolith sedimentary sequence. These rocks and Silurian-Devonian intrusions contrast with those of the Dunnage, Humber, and Avalon zones confirming fundamental differences in the lower crust (D'Lemos et al., 1994).

Coarse grained porphyritic biotite granites cut all other plutons and zone boundaries. Petrological and age differences among intrusions of the Ackley suite on opposite sides of the Gander-Avalon boundary may reflect underlying crustal contrasts, in accord with the seismic data for a vertical boundary between the Central and Avalon lower crustal blocks. Rocks as young as Late Devonian are cut by the Belleoram Granite in the western Avalon Zone of Fortune Bay.

In New Brunswick, an Acadian magmatic suite of bimodal plutonic and extrusive rocks crosses all boundaries between the Dunnage, Gander and Avalon zones, including a late Ordovician-early Silurian blueschist suture that links the Dunnage and Gander zones.

The Meguma Zone has a distinct Devonian plutonic history and its middle Paleozoic intrusions do not cross the Meguma-Avalon zone boundary.

Discussion

The lack of complete Silurian and Devonian sections in the Humber Zone and St. Lawrence Platform precludes a stratigraphic analysis of the sedimentological effects of Acadian Orogeny. Possibly, easterly derived Silurian-Devonian redbeds of the Clam Bank Belt are a molasse related to uplift in interior parts of the orogen. Likewise, Silurian-Devonian redbeds and conglomerates of the Fortune Belt and the Devonian redbeds that overlie Silurian rocks with mild unconformity on the Grand Banks may represent molasse deposits on the opposing side of the orogen.

The shape of the ancient continental margin of eastern North America had an important control on the local extent and intensity of Acadian deformation. At the St. Lawrence Promontory, deformation in the Devonian Windsor Point Group is more intense than that in Silurian-Devonian rocks elsewhere. Furthermore, Paleozoic rocks of the Newfoundland Reentrant at the Strait of Belle Isle are locally outside the Appalachian deformed zone, although the rocks are part of the Appalachian system by any other definition.

Juxtaposed lower crustal blocks beneath the Dunnage Zone explain the presence and abundance of large granitic plutons that cut its oceanic rocks. The localization of deformed middle Paleozoic plutons to the Gander Zone may reflect diapiric emplacement during doming of a lighter Gander crust through a heavier Dunnage cover. Ductile shearing may also localize plutonism here.

In Newfoundland, the area affected by Acadian Orogeny is symmetrically disposed between the edges of the wide opposing Grenville and Avalon lower crustal blocks (Fig. 11.6). Deformation at the surface spans the eastern portion of the Grenville lower crustal block, all of the Central lower crustal block, and western parts of the Avalon lower crustal block. The limits of plutonism also span these regions but within narrower limits. High grade regional metamorphism is more restrictive, mainly above the Central lower crustal block.

In Nova Scotia, Acadian Orogeny affected the Mira Subzone (Avalon Zone and Avalon lower crustal block) but intensity of deformation, regional metamorphism and middle Paleozoic plutonism are less important than in the Bras d'Or and Aspy Subzones to the northwest (Central lower crustal block). This resembles the situation in Newfoundland. Similarly in New Brunswick, middle Paleozoic plutons are less abundant in the Avalon Zone compared to the adjacent Gander Zone. However, the Meguma Zone of Nova Scotia, outboard of the Avalon, exhibits the full effects of Acadian deformation, metamorphism, and plutonism. This supports other data for a Sable lower crustal block that had its own unique controls on surface orogenic effects in the Meguma Zone.

These spatial relations imply that Acadian Orogeny resulted from the middle Paleozoic collisional interaction between lower crustal blocks. The main zone of Acadian deformation, regional metamorphism and plutonism expectedly coincides with the narrow, intervening Central lower crustal block and the edges of the Grenville and Avalon lower crustal blocks. Where the Avalon Zone and corresponding lower crustal block are especially wide in Newfoundland, the basement behaved rigidly and cover rocks are undeformed across its eastern parts.

Alleghanian Orogeny

Late Paleozoic deformation in the Canadian Appalachians is mild or non existent (Map 6). Late Paleozoic deformation affected the thickest sections of the Maritimes Rift from New Brunswick to western Newfoundland, and it is most intense along the northwest shoreline of the Bay of Fundy and along the Avalon-Meguma zone boundary. Middle and Upper Caboniferous rocks of the Maritimes Rift are in places unconformable on deformed Lower Carboniferous rocks and locally they overlap boundary faults onto adjacent stable areas. Along the northwest shore of the Bay of Fundy, Carboniferous rocks are involved in thrusts with polyphase deformation and subhorizontal penetrative cleavage. A few small plutons, dated isotopically as Carboniferous, occur outside the late Paleozoic basins in older rocks of the Avalon and Meguma zones.

The localization of late Paleozoic deformation along narrow faulted belts and along the Avalon-Meguma boundary favours a model of wrench tectonics. Dextral movement on the Cobequid Fault explains local deformation in the fault zone, and westward movement of the Meguma Zone (Sable lower crustal block) against the Avalon Zone (Avalon lower crustal block) is a simple and elegant explanation for the siting and polarity of thrusting along the northwest shore of the Bay of Fundy. A wrench tectonic regime in Atlantic Canada complements a compressional regime in the U.S. Appalachians.

ACKNOWLEDGMENTS

This is another opportunity to thank all those who contributed to this volume and whose contributions are summarized in this chapter. Their names are listed elsewhere.

REFERENCES

Cawood, P.A., Dunning, G.R., van Gool, J.A.M., and Lux, D.
1994: Collisional tectonics along the Appalachian margin of Laurentia: constraints from Corner Brook Lake and Baie Verte; in New Perspectives in the Appalachian-Caledonian Orogen, Geological Association of Canada Nuna Conference in honor of Dr. Harold Williams, Program and abstracts, p. 7-8.

Colman-Sadd, D.P., Dunning, G.R., and Dec, T.
1992: Dunnage-Gander relationships and Ordovician orogeny in central Newfoundland: A sediment provenance and U/Pb age study; American Journal of Science, v. 292, p. 317-355.

D'Lemos, R.K., King, T., and Holdsworth, R.
1994: The relationship between migmatization and granite magmatism in the northeastern Gander Zone: evidence from Sm-Nd isotopic characteristics; in New Perspectives in the Appalachian-Caledonian Orogen, Geological Association of Canada Nuna Conference in honor of Dr. Harold Williams, Program and abstracts, p. 12.

Dubé, B. and Lauziere, K.
1994: Deformational events and their timing along the Cape Ray Fault, Newfoundland Appalachians; in New Perspectives in the Appalachian-Caledonian Orogen, Geological Association of Canada Nuna Conference in honor of Dr. Harold Williams, Program and abstracts, p. 13.

Fyffe, L.R. and Swinden, H.S.
1992: Paleotectonic setting of Cambro-Ordovician volcanic rocks in the Canadian Appalachians; Geoscience Canada, v. 18, no. 4, p. 145-157.

Lin, S.
1993: Relationship between the Aspy and Bras d'Or "terranes" in the northern Cape Breton Highlands, Nova Scotia; Canadian Journal of Earth Sciences, v. 30, p. 1773-1781.

Keen, C.E., Kay, W.A., Keppie, D., Marillier, F., Pe-Piper, G, and Waldron, J.W.F.
1990:   Deep seismic reflection data from the Bay of Fundy and Gulf of Maine; Lithoprobe East Transect, Report No. 13, Memorial University of Newfoundland, p. 111-116.

Nance, R.D. and Murphy, J.B.
1990:   Kinematic history of the Bass River Complex, Nova Scotia: Cadomian tectonostratigraphic relations in the Avalon terrane of the Canadian Appalachians; in The Cadomian Orogeny, (ed.) R.S. D'Lemos, R.A. Strachan, and C.G. Topley; Geological Society of London, Special Publication No. 51, p. 395-406.

O'Brien, S.J., Strong, D.F., and King, A.F.
1990:   The Avalon Zone type area: southeastern Newfoundland Appalachians; in Avalon and Cadomian Geology of the North Atlantic, (ed.) R.A. Strachan and G.K. Taylor, Blackie, Glasgow, p. 166-194.

O'Brien, S.J., Dunning, G.G., Tucker, R.D., O'Driscoll, C.F., and O'Brien, B.H.
1992:   On the nature, timing and relationships of late Precambrian tectonic events on the southeastern (Gondwanan) margin of the Newfoundland Appalachians (abstract); Geological Association of Canada (Newfoundland), The Tuzo Wilson Cycle: a 25th anniversary symposium, Program and Abstracts volume, p. 21.

O'Brien, S.J., Tucker, R.D., O'Driscoll, C.F.
!994:   Neoproterozoic basement-cover relationships and the tectono-magnetic recordof the Avalon Zone on the Hermitage Peninsua and environs, Newfoundland; in New Perspectives in the Appalachian-Caledonian Orogen, Geological Association of Canada Nuna Conference in honor of Dr. Harold Williams, Program and abstracts, p. 21-22.

Quinlan, G.M., Hall, J., Williams, H., Wright, J.A., Colman-Sadd, S.P., O'Brien, S.J., Stockmal, G.S., and Marillier, F.
1992:   Lithoprobe onshore seismic reflection transects across the Newfoundland Appalachians; Canadian Journal of Earth Sciences, v. 29, p. 1865-1877.

Williams, H., Currie, K.L., and Piasecki, M.A.J.
1993:   The Dog Bay Line: a major Silurian Tectonic boundary in northeast Newfoundland; Canadian Journal of Earth Sciences, v. 30, p. 2481-2494.

Author's Address

H. Williams
Department of Earth Sciences
Memorial University of Newfoundland
St. John's
Newfoundland
A1B 3X5

Printed in Canada

Chapter 12

CALEDONIDES OF EAST GREENLAND

Chapter 12

CALEDONIDES OF EAST GREENLAND

A.K. Higgins

INTRODUCTION

Greenland, the largest island in the world, is situated at the northeast corner of the North American plate. Nares Strait, a channel in places as narrow as 20 km, separates western North Greenland from Ellesmere Island. Geologically the areas on the two sides of the strait have much in common, and there is little evidence to support the large sinistral strike-slip movements proposed along the strait in response to seafloor spreading farther south (see Dawes and Kerr, 1982; Okulitch et al., 1990). Baffin Bay and Labrador Sea, which separate the coast of West Greenland from the coasts of Baffin Island and Labrador, developed by seafloor spreading that began at about the Cretaceous-Tertiary boundary and terminated by the Early Oligocene. East of Greenland the North Atlantic Ocean opened during the Tertiary, and fragmented the Caledonide Orogen (Fig. 12.1). In East Greenland the Caledonides can be traced from latitude 70° to 82°N. Predrift reconstructions produce a variety of configurations for the reassembled Caledonide Orogen, but there is a general consensus that parts of Spitsbergen represent the northern extension of the East Greenland Caledonides, whereas the southern extension is to be seen in the Caledonides of the British Isles and the Appalachians of eastern North America (Harland and Gayer, 1972; Harland, 1985; Ziegler, 1985). General reviews of the East Greenland Caledonides are given by Haller (1971), Henriksen and Higgins (1976), Higgins and Phillips (1979), and Henriksen (1985).

The East Greenland Caledonides form a coastal belt 1200 km long and up to 250 km wide (Fig. 12.1). The greater part comprises crystalline complexes made up of a variety of Archean and Proterozoic gneissic and granitic rocks together with metasedimentary rocks, variably reworked during the Caledonian Orogeny. Upper Proterozoic (Eleonore Bay Group) and lower Paleozoic sedimentary sequences outcrop mainly between latitudes 72° and 74°N, with northern outcrops in Kronprins Christian Land (79°-82°N) (Fig. 12.1). A broad zone of post-Caledonian (Devonian-Cretaceous) sedimentary rocks covers eastern coastal areas (70°-75°N), and small areas of post-Caledonian

sedimentary rocks occur in east and north Kronprins Christian Land. Tertiary basalts hide the continuation of the Caledonides south of 70°N.

Some of the first geologists to visit the crystalline complexes of the inner fiords had considered the gneissic rocks to be Archean in age, only superficially disturbed by Caledonian events (Parkinson and Whittard, 1931; Odell, 1939, 1944). However, members of Lauge Koch's geological expeditions at about the same time introduced the concept of a deep-seated Caledonian Orogeny, in which the infracrustal rock units were viewed as essentially "transformed" by Caledonian orogenesis, and the supracrustal rocks were all viewed as metamorphic representatives of the Upper Proterozoic Eleonore Bay Group (Backlund, 1930; Wegmann, 1935). Subsequent detailed fieldwork, mainly in the southern half of the Greenland Caledonides, appeared to confirm the basic concept (Haller, 1955, 1958; Wenk and Haller, 1953), most clearly described in terms of "stockwerke" tectonics with a complex deep-seated infrastructure and an overlying simpler suprastructure (Haller, 1953, 1971). Former basement rocks within the infrastructure were recognized in southern areas (Fig. 15 of Haller, 1971), although they were considered petrogenetically rejuvenated during the Caledonian Orogeny. In the northern half of the Greenland Caledonides (north of 76°N), substantial parts of the crystalline complexes were interpreted to have been affected by a Precambrian Carolinidian orogenic event prior to that of the Caledonian (Haller, 1961, 1971). These viewpoints were largely based on geological arguments, as the first K-Ar isotopic ages became available only after the termination of Lauge Koch's expeditions in 1958 (Haller and Kulp, 1962).

The Geological Survey of Greenland (GGU) carried out systematic geological mapping in the Scoresby Sund region (70°-72°N) from 1968-72, and a variety of special studies farther north (72°-75°N) from 1973-78. During the Scoresby Sund expeditions it soon became clear that many of the infracrustal rock units within the Caledonian fold belt exhibited the same geological characteristics as the Archean and Proterozoic basement complexes that make up the crystalline shield of Greenland. Rb-Sr whole rock and zircon isotopic studies have subsequently confirmed their original Archean or Proterozoic age of formation, although K-Ar and Rb-Sr mineral ages are everywhere witness to a Caledonian metamorphic overprint. North of 72°N, in the classic areas studied by John Haller (Haller, 1953, 1955, 1958), Rb-Sr isotopic studies have demonstrated an important Early Proterozoic episode of granite intrusion. A variety of isotopic studies in a broad migmatite

Higgins, A.K.
1995: Caledonides of East Greenland; Chapter 12 in Geology of the Appalachian-Caledonian Orogen in Canada and Greenland, (ed.) H. Williams; Geological Survey of Canada, Geology of Canada, no. 6, p. 891-921 (also Geological Society of America, The Geology of North America, v. F-1).

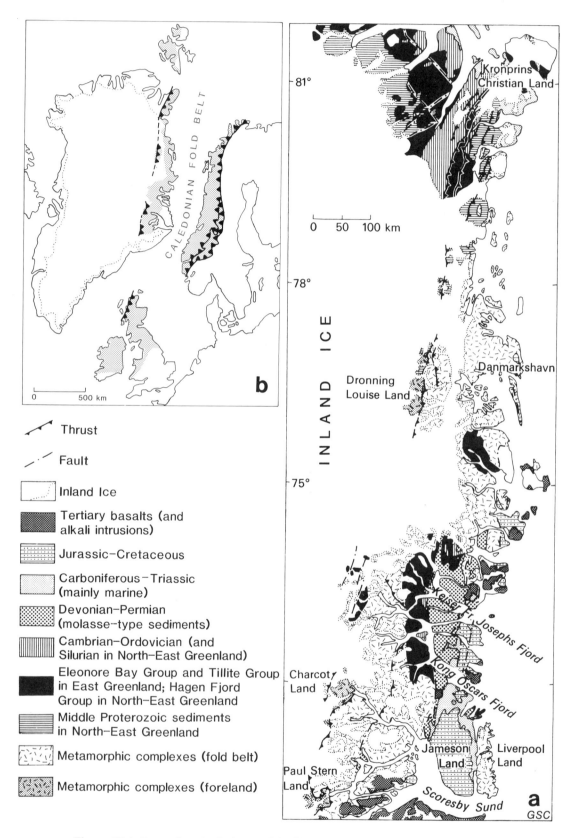

Figure 12.1. General geological map of the East Greenland Caledonian fold belt **(a)**, and one of the predrift reconstruction of the circum-Atlantic Caledonides **(b)**.

and granite zone suggest that both Middle Proterozoic and Caledonian phases of migmatite formation and granite emplacement were important.

The age of the supracrustal elements of the crystalline complexes, formerly assumed to be part of the Eleonore Bay Group, was called into question by ca. 1000 Ma Rb-Sr whole rock ages on some of the metasedimentary rocks. Subsequently, 1000 Ma ages were determined by Rb-Sr and zircon isotopic studies on granites that intruded migmatized supracrustal rocks. Other supracrustal sequences have proved to be Early Proterozoic in age. One of the most widespread of the metasedimentary sequences is known in the Scoresby Sund region as the Krummedal supracrustal sequence.

Much Geological Survey of Greenland fieldwork in central and North-East Greenland since 1978 has been concerned with the oil prospects of the post-Caledonian basins. However, the northern extremity of the Greenland Caledonides was reached in 1980 during the Geological Survey of Greenland North Greenland expeditions (Hurst et al., 1985), and a large area of North-East Greenland between latitudes 76° and 78°N was the focus of expedition activity from 1988-90.

The East Greenland Caledonides are described here in terms of a series of north-trending zones. These zones are assumed to be the present-day surface reflection of a series of thick thrust wedges of infracrustal and supracrustal rocks. It is envisaged that during the Caledonian Orogeny, complexes of migmatites and granites strongly affected by Caledonian events were thrust westward over regions less disturbed by Caledonian metamorphism and deformation, which in turn were thrust westward over the Caledonian foreland.

FORELAND AREAS

The Caledonian foreland is exposed in two major tectonic windows along the border of the Inland Ice in the Scoresby Sund region: the Gåseland-Paul Stern Land and Charcot Land windows. Farther north, between latitudes 73° and 74°N, low-grade metasedimentary and volcanic rocks in a fault-bounded region centred on Eleonore Sø may occupy a tectonic window. Dronning Louise Land, a large nunatak area extending from 76° to 77°30′N, is divided centrally by a north-trending thrust system; the western half is viewed as part of the autochthonous Precambrian shield. In Kronprins Christian Land the marginal Caledonian thrust belt can be traced over a distance of about 200 km; Caledonian thrust sheets here are displaced westward over a foreland of mainly lower Paleozoic sedimentary rocks.

Gåseland-Paul Stern Land window

Foreland rocks outcrop in westernmost Gåseland, Paul Stern Land, and adjacent areas. The main geological relationships were established by E. Wenk and P. Stern (Wenk, 1961), while additional investigations and different interpretations were given by Phillips et al. (1973) and Moncrieff (1989). Three main rock divisions are recognized: a gneissic basement, tillites preserved in lens-like pockets in the eroded basement, and an overlying marble sequence. Caledonian thrust units overlie and surround the window (Fig. 12.1, 12.4).

The infracrustal basement rocks comprise hornblende-biotite gneisses, strongly folded, with migmatic developments in places, and a later phase of augen granite. Discordant basic dykes cut this complex, and have been metamorphosed under amphibolite facies conditions. K-Ar mineral ages of 1890 and 2290 Ma suggest that this basement complex is Archean or Early Proterozoic in age (Haller and Kulp, 1962).

Tillitic developments are preserved at nine localities in Gåseland and Paul Stern Land, in lens-like pockets in the eroded peneplain of the basement surface, and are overlain by the marble sequence. The best tillite sequence is in western Paul Stern Land, a lens-like development 100 m across and said to be up to 40 m thick (Phillips and Friderichsen, 1981); it is divided into 21 members, 10 of which are tillites, while several of the intervening mudstone and sandstone units contain dropstones. Clasts are mainly of granite or quartzite, and are up to 50 cm across. Moncrieff (1989) has presented a 21.5 m thick sedimentary log from the same locality, Støvfanget, the type section for his Støvfanget Formation. Wenk (1961) had originally suggested the tillitic developments were part of the "Basal Series" of the Eleonore Bay Group. Phillips and Friderichsen (1981) favoured a late Precambrian (Vendian) age, an interpretation confirmed by Moncrieff (1989), who correlated the Støvfanget Formation (and the Tillit Nunatak Formation of Charcot Land - see below) with the top of the upper tillite formation of the Tillite Group in the central fiord zone.

The marble sequence overlying the basement gneisses and tillites is 0 to 300 m thick, but is truncated and much deformed by the overlying Caledonian thrust units. The marbles are pale yellow and strongly laminated with streaks of chlorite and quartz. Wedge-shaped sheets of quartz-sericite schist and chlorite-carbonate schist occur within the marble sequence, the former probably representing mylonitic gneiss. An early Paleozoic age has been suggested for the sequence (Phillips et al., 1973; Phillips and Friderichsen, 1981), but in the absence of fossils this interpretation is speculative.

Charcot Land window

The Charcot Land window is a broad region of distinctive rock types bounded by Caledonian thrusts to the east and west. It encompasses the greater part of Charcot Land itself, and a narrow strip of western Hinks Land (Fig. 12.1, 12.4). Four rock divisions are distinguished: a gneissic basement, a supracrustal sequence, an intrusive suite, and a small outcrop of tillite.

The basement complex comprises biotite and hornblende gneisses, with lenses, layers, and thick bands of amphibolite. K-Ar hornblende ages of 2100 Ma (Hansen et al., 1981) and U-Pb zircon ages on quartzofeldspathic gneiss of 2800 Ma (Hansen et al., 1987a) indicate an Archean to Early Proterozoic age of formation.

The Charcot Land supracrustal sequence rests directly on a peneplained surface of gneissic basement (Fig. 12.2). It comprises a succession at least 2000 m thick of white, yellow, or orange limestone and dolomite, mica quartzite, black and grey shale, tuffs, gabbros, and basic lavas. The sequence exhibits a pronounced increase in metamorphic

Figure 12.2. Yellow-white carbonates and basic lavas of the Lower Proterozoic Charcot Land supracrustal sequence resting on peneplained gneissic basement. Photo: Lauge Koch's aerial photograph collection, Geological Museum, Copenhagen.

grade from low greenschist facies in the south to high amphibolite facies in the north (Steck, 1971). The isograds trend approximately east, perpendicular to the Caledonian trends. Hansen et al. (1981) deduced an age of 1850 to 1900 Ma as most likely for the time of metamorphism. This supracrustal sequence is the "Grüngestein-Marmor-Serie" of Wenk (1961) and Vogt (1965), who had compared it with the sequences in Gåseland to the south and at Eleonore Sø in the north.

Two major intrusions occur within the Charcot Land window. One is a hornblende-biotite quartz diorite or granodiorite which makes up most of Tillit Nunatak; offshoots from it penetrate the Charcot Land supracrustal sequence, and a tillite overlies it unconformably. It has yielded a U-Pb zircon age of 1800 to 1900 Ma (Hansen et al., 1981). The second intrusion is a large body of pegmatitic muscovite granite outcropping over an area of 20 km by 20 km, and notable for its spectacular trails of gneiss and amphibolite inclusions. Rb-Sr whole rock analyses suggest an age of intrusion of about 1850 Ma (Hansen et al., 1981).

An outcrop of tillite about 1500 m by 500 m in area occupies the south summit of Tillit Nunatak, where a sequence 50 m thick rests unconformably on the quartz diorite intrusion. Boulders and clasts range up to 1 to 2 m in size, about 80 per cent of them comprising quartz diorite or gneiss, and the remaining 20 per cent mainly shale, greenstone, or limestone derived from the Charcot Land

supracrustal sequence (Steck, 1971; Henriksen, 1981). Moncrieff (1989) presented a 50 m sedimentary log of the sequence, his Tillit Nunatak Formation, and correlated it with the upper tillite of the Vendian Tillite Group.

On the grounds that the tillites exposed in both the Gåseland-Paul Stern Land and Charcot Land windows are affected by westward-verging Caledonian folds and thrusts, Manby and Hambrey (1989) have claimed that the rocks exposed are allochthonous and not true foreland. However, this author would argue that the Caledonian deformation and very low-grade metamorphism of the foreland sequences is what might be expected in the immediate proximity of major Caledonian thrusts.

Eleonore Sø

A fault-bounded region of low-grade metasedimentary and volcanic rocks occurs in the nunatak region between 73°30′ and 74°N (Fig. 12.4). The main outcrops and rock divisions were described by Katz (1952a), whereas John Haller recognized southern outcrops using aerial observations (Koch and Haller, 1971). Katz had correlated the main part of the sequence with the upper Eleonore Bay Group, while Haller (1971) referred the sequence to the "Basal Series" of the same group on the basis of comparisons with similar sequences in Charcot Land and Gåseland. Haller (1971) also raised the possibility that these rocks occupy a window overridden by parts of the Caledonian mountain chain.

The main sequence, known as the "Eleonore Sø Series", comprises more than 1000 m of sandstone, mudstone, calcareous shale, limestone, and dolomite. This is overlain by at least 1000 m of low-grade volcanic greenstone, intruded by small porphyry bodies. A thrust unit of about 1000 m of white quartzite, the "Slottet quartzite", overlies the other two sequences.

Katz (1952a) interpreted the volcanic rocks as ophiolitic, an interpretation upheld by Haller (1971, Fig. 32) who interpreted most of the basic rocks in the Caledonides as ophiolitic and related to the first depositional cycle of the Caledonian geosyncline. These various basic rocks proved subsequently to have a variety of ages and origins (Henriksen and Higgins, 1976), although the age of the Eleonore Sø greenstones remains enigmatic.

Dronning Louise Land

Dronning Louise Land is an extensive nunatak complex extending between latitudes 76° and 77°30′N, and is divided into two parts by a north-trending imbricate thrust zone (Fig. 12.3). The western part comprises a gneiss complex and overlying sedimentary rocks, and is viewed as part of the autochthonous Greenland shield. The eastern part (eastern hinterland) comprises a group of schists and gneisses making up Caledonian thrust sheets displaced westward over the foreland. Dronning Louise Land is best known from the studies of Peacock (1956, 1958), whose principal conclusions were confirmed and supplemented by Geological Survey of Greenland fieldwork in 1989 (Friderichsen et al., 1990).

The basement gneiss complex of the western foreland, probably of Late Archean to Early Proterozoic age, is dominated by orthogneisses which are unconformably overlain by sedimentary rocks of the Trekant Series and the Zebra Series. Both gneisses and the Trekant Series are intruded by a suite of dolerites. Locally the Zebra Series rests unconformably on basement gneisses, Trekant Series, and dolerites (Friderichsen et al., 1990). Rocks of the western foreland become increasingly disturbed by Caledonian deformation toward the imbricate thrust zone.

The Trekant Series (sandstone, siltstone) is correlated with the Independence Fjord Group of North Greenland. The Zebra Series (quartzite, mudstone, siltstone, limestone) locally contains *Skolithos* trace fossils which demonstrate an age no older than Late Proterozoic. *Skolithos* has long been known in glacially transported quartzite erratics in East and North-East Greenland (Fig. 48 of Haller, 1971)

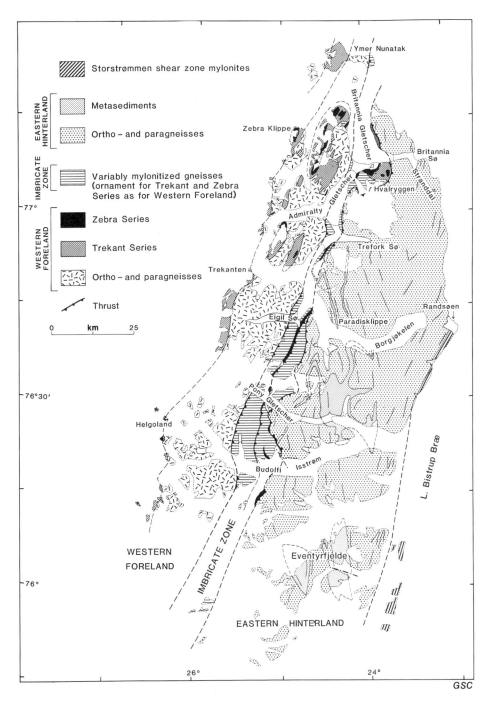

Figure 12.3. Dronning Louise Land, divided by Caledonian thrusts into western foreland, imbricate zone, and eastern hinterland. After an original figure by J.D. Friderichsen (pers. comm., 1991).

897

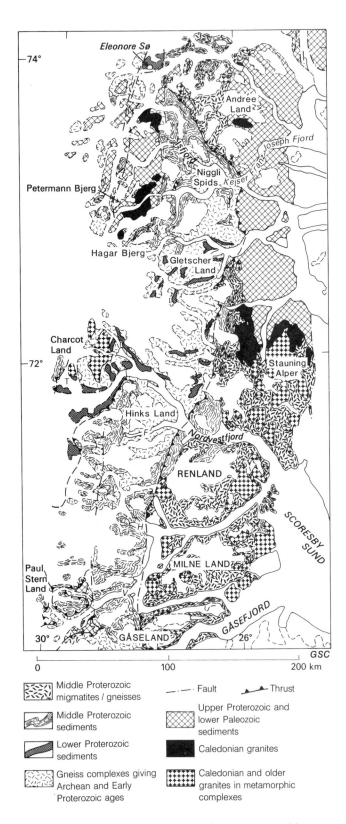

Figure 12.4. Geological divisions of the metamorphic complexes in the inner fjord zone 70° to 74°N. Distinction is made between Caledonian granites, the age of which is stratigraphically determined, and other granites in the metamorphic complexes. F = Flyverfjord, H = Harefjord, T = Tillit Nunatak, V = Vestfjord..

and it is possible these erratics originated from formerly widespread outcrops of the Zebra Series beneath the Inland Ice.

Kronprins Christian Land

The eastern half of Kronprins Christian Land comprises Caledonian thrust units of gneisses and sedimentary rocks that are displaced westward over a foreland composed largely of lower Paleozoic platform carbonates (Fig. 12.1, 12.16). The latter occupy the western half of Kronprins Christian Land, and form part of a continuous belt of Lower Paleozoic rocks traceable to the northwest and westward across North Greenland.

The lower Paleozoic rocks are today correlated with the better known succession of North Greenland (Hurst et al., 1985), and some of the ambiguities surrounding the earlier stratigraphic usage, described by Peel (1980), have been resolved. The stratigraphically highest unit exposed in the foreland sequence is the Profilfjeldet Formation, formed of turbiditic shales of Llandovery to Middle Wenlock age overlying shallow water platform carbonates. These rocks indicate a maximum age for the emplacement of the Caledonian thrust sheets.

CRYSTALLINE COMPLEXES

The greater part of the East Greenland Caledonides comprises crystalline complexes (Fig. 12.1). The best known areas are in the southern part of the fold belt (70°-74°N) where the activities of both Lauge Koch's expeditions and the Geological Survey of Greenland have been concentrated. The crystalline areas are treated here from west to east in the southern region, and from south to north along the outer coast, largely following the geographic divisions of Henriksen and Higgins (1976).

Vestfjord-Hinks Land gneiss and schist zone

This zone extends between latitudes 70° and 72°N in the inner part of the Scoresby Sund region (Fig. 12.4). It is bounded to the west by the Caledonian marginal thrust system, which outlines the Gåseland-Paul Stern Land and Charcot Land windows, and to the east by a major north-south-trending lineament which was apparently a major thrust in Caledonian time, and subsequently has been reactivated as a normal fault. Internally the zone comprises two principal lithostratigraphic units, The Flyverfjord infrastructural complex and the Krummedal supracrustal sequence, the latter in some areas built up of several thrust sheets. The two units are folded together with variable intensity, the original basement-cover relationships being most clearly seen in the Hinks Land region.

Flyverfjord infracrustal complex

The basement rocks comprise a variety of biotite- and hornblende-bearing gneisses (Fig. 12.5), amphibolite and ultrabasic bands and lenses, and several major granite bodies. These are dissected by folded, discordant amphibolite dykes, commonly 5 to 10 m wide, and in some places up to 50 m wide. The dykes postdate at least two phases of deformation, and were themselves affected by at least one phase of deformation and amphibolite facies metamorphism. These relationships are best exposed in Hinks Land, the walls of Flyverfjord, and north of innermost Nordvestfjord.

Isotopic age determinations indicate a general Archean age for the Flyverfjord infracrustal complex. Gneisses in Flyverfjord have yielded a Rb-Sr whole rock age of 2935 Ma (Rex and Gledhill, 1974), although with a large uncertainty. A zircon age on a foliated biotite granite containing partly digested folded amphibolites north of inner Nordvestfjord is 2300 Ma, while another zircon age on augen gneiss near Harefjord suggests an age of crystallization of 2520 Ma (Steiger et al., 1979).

Krummedal supracrustal sequence

The supracrustal rocks overlying and interfolded with the Flyverfjord infracrustal sequence are distinguished informally as the Krummedal supracrustal sequence (Higgins, 1974, 1988). The sequence is widely exposed, and comprises several thousand metres of mainly psammite and mica schist, commonly developed as regular alternations in beds 10 to 50 cm thick (Fig. 12.6). Original sedimentary structures are rarely preserved, but the sequence may represent deep-water turbidites.

Higher levels of the supracrustal sequence are locally of higher grade and coarser grained than lower levels, in some cases gneissic with quartzofeldspathic veins. This apparent reverse metamorphic zoning is interpreted as a consequence of thrusting of higher grade rocks over lower grade rocks (Higgins, 1974; Phillips et al., 1973).

The basal levels of the sequence on Hinks Land comprise about 100 m of white marble associated with amphibolite. Farther south, up to 200 m of amphibolite associated with impure quartzite are developed, and here the amphibolites are interpreted to be metavolcanic rocks (Phillips et al., 1973).

The Middle Proterozoic age of the Krummedal supracrustal sequence is based on geological arguments combined with Rb-Sr whole rock determinations suggesting a major metamorphism at 1122 Ma (see discussion in Higgins, 1988); the latter age is supported by U-Pb studies on monazite (Hansen et al., 1978).

Figure 12.5. Fold interference patterns in banded leucocratic and hornblende gneisses of the Flyverfjord infracrustal complex, Hinks Land. Photo by J.D. Friderichsen.

Central metamorphic complex

The classic areas of the central metamorphic complex occupy the inner fiord region between latitudes 72° and 74°N (Fig. 12.7), and the "stockwerke" model of the East Greenland Caledonides most clearly expounded by Haller (1955, 1971) derives from fieldwork here. The three main infracrustal units (Gletscherland complex, Hagar sheet, Niggli Spids dome) with their spectacular involuted nappe, dome, and mushroom shapes, were viewed by Haller as products of highly mobile Caledonian fronts of migmatization, essentially younger than their cover of metasedimentary rocks. A thermal convection model that could explain the intricate forms of the infracrustal complexes has been presented by Talbot (1979). Recent fieldwork and isotopic studies (Higgins et al., 1981; Rex and Gledhill, 1981) indicated that the genetic features of the infracrustal units are essentially Archean or Early Proterozoic and the enveloping metasedimentary rocks are Middle Proterozoic in age, although both levels have been strongly affected by Caledonian overprinting.

The region of the central metamorphic complex thus has much in common with the Vestfjord-Hinks Land gneiss and schist zone, and can be viewed as its northward extension (Fig. 12.4, 12.7). Its western border is the Inland Ice, with a strip of Eleonore Bay Group sedimentary rocks (Petermann Series) overlying it in the nunatak zone between 73° and 74°N; the eastern boundary is a north-trending thrust or fault line traceable throughout the region.

Gletscherland complex

The Gletscherland complex is the southernmost of the three infracrustal units. It extends over most of Gletscherland, the southern half of Suess Land, and extends southward into Nathorst Land. It comprises a great variety of biotite and

Figure 12.6. Alternations of massive sandstone and thin mica schist beds, Krummedal supracrustal sequence, south Hinks Land. Photo by A.K. Higgins.

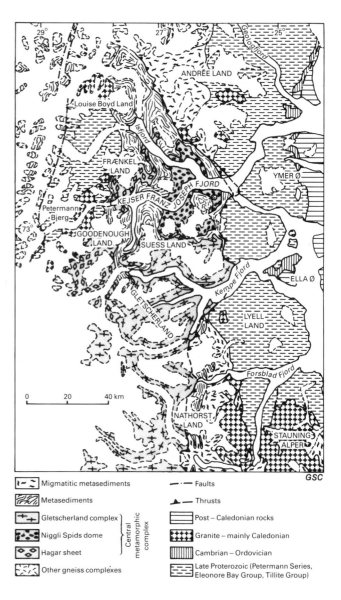

Figure 12.7. Geological setting and distribution of three subdivisions (Gletscherland complex, Hagar sheet, Niggli Spids dome) of the central metamorphic complex.

hornblende gneisses, amphibolite, dioritic and granitic bodies, and massive units of augen gneiss (Haller, 1955; Zweifel, 1959; Higgins et al., 1981). Discordant amphibolite dykes are conspicuous locally. The gneisses in the south have yielded a Rb-Sr "errorchron" age of 2450 Ma (Rex et al., 1977).

East-west-trending fold structures dominate the Gletscherland complex, and form the transverse trends of Wegmann (1935) and Haller (1955, 1970, 1971), which because they conflicted with the regional north Caledonian trends were interpreted as "inherited" from a Caledonian transformed older basement. They are today viewed as a Proterozoic deformation pattern (Higgins et al., 1981). Wegmann (1935, p. 26) had admitted the latter possibility, but with the proviso that this could hardly be the case since relics of sedimentary rocks within the complex were believed to be parts of the Eleonore Bay Group. These sedimentary relics, which include bands of marble and mica schists, are now considered to be Middle Proterozoic or older.

Hagar sheet

The Hagar sheet extends from western Gletscherland, where it adjoins the Gletscherland complex, northward through Goodenough Land, part of Suess Land, and into Frænkel Land. It has the form of an extensive nappe of infracrustal rocks, overturned eastward and enveloped by supracrustal rocks. The infracrustal rocks comprise banded biotite and hornblende gneisses, nebulitic veined gneisses, granitic gneisses, and foliated granites, with locally discordant amphibolite dykes (Fig. 12.8). Five Rb-Sr isochron ages of between 1980 and 1725 Ma have been obtained from bodies of foliated granite and leucocratic gneiss (Higgins et al., 1978; Rex and Gledhill, 1981), and suggest that significant Early Proterozoic orogenic activity contributed to the genesis of the Hagar sheet. Haller (1953) and Wenk and Haller (1953) provided an account of structural and petrological features.

Niggli Spids dome

A dome-shaped mass of mainly granitic gneisses outcrops in northern Suess Land, eastern Frænkel Land, and part of southwest Andrée Land. Its structure and petrology have been described by Haller (1953, 1955, 1971). The dominant rock types are homogeneous, well-foliated granitic gneiss, augen gneiss, and porphyritic granodioritic gneiss, which form spectacular exposures on the steep walls of Kejser Franz Joseph Fjord (Fig. 12.9). All rock types appear to have been derived from plutonic rocks. The general structure is superficially simple, with overlying metasedimentary rocks dipping off the dome to the north, east, and west.

Figure 12.8. Leucocratic granitic gneiss unit of Hagar sheet in southwest Suess Land, which has Rb-Sr isochron ages of ca. 1900 Ma (Rex and Gledhill, 1981). The gneisses are cut by several dark amphibolite dykes. Mountain summits reach 2000 m. Photo by A.K. Higgins.

The Niggli Spids dome is assumed to contain an infracrustal basement unit, older than the overlying metasedimentary rocks, and possibly of the same age of formation as the Gletscherland complex or Hagar sheet. However, isotopic age studies have not as yet confirmed this interpretation.

Metasedimentary rocks

Thick metasedimentary sequences overlie and are interfolded with the three infracrustal units of the central metamorphic complex, and have been broadly equated with the Krummedal supracrustal sequence (Higgins et al., 1981; Higgins, 1988). These have been described by Haller (1953, 1955, 1971), although he identified them as metamorphic representatives of the Eleonore Bay Group.

Basal units of the supracrustal sequence include in many areas yellow to orange weathering marbles and calcareous schists, particularly conspicuous around the margin of the Niggli Spids dome. The major part of the sequence in most areas, several kilometres in thickness, comprises red-brown weathering alternations of psammitic, semipelitic, and pelitic beds (Fig. 12.9). Metamorphism varies between low and high amphibolite facies.

Rb-Sr whole rock isotopic studies on the metasedimentary rocks have met with little success (Rex and Gledhill, 1981). Mineral ages on the metasedimentary rocks, and on the various infracrustal units, are Caledonian, and testify to a regional Caledonian metamorphic overprinting (Rex and Higgins, 1985).

Gåsefjord-Stauning Alper migmatite and granite zone

A zone dominated by migmatitic metasedimentary rocks, with profuse intrusions of diorite to granite of pre-Caledonian and Caledonian age, extends from Gåsefjord to the Stauning Alper and into west Lyell Land (70°-72°40′N). The zone is up to 90 km wide, bounded to the west by a thrust or steep fault, and to the east in fault contact with post-Caledonian rocks, except in the northern Stauning Alper where the marginal Caledonian granites intrude the Late Proterozoic Eleonore Bay Group (Fig. 12.4, 12.7).

The migmatite and granite zone has a complex genesis. An early orogenic phase (Middle Proterozoic) led to emplacement of major sheets of augen granite into a thick supracrustal sequence, and was accompanied by migmatization, high-grade metamorphism and large-scale deformation. A second (Caledonian) orogenic phase led to the emplacement of a suite of late kinematic to postkinematic plutons, associated with further deformation and metamorphism.

Figure 12.9. View from Frænkel Land eastward across Kejser Franz Joseph Fjord to Suess Land, showing light coloured, well-foliated granitic gneisses and augen gneisses of the Niggli Spids dome overlain by dark coloured metasedimentary rocks. The highest peak, Payer Tinde, is 2320 m high. Photo by A.K. Higgins.

Figure 12.10. Migmatitic paragneisses cut by several generations of neosome veins in the northern part of the Gåsefjord-Stauning Alper migmatite and granite zone. Photo by N. Henriksen.

Migmatitic metasedimentary rocks

Migmatitic rocks dominate the zone, varying in appearance from migmatitic paragneisses to granitic migmatite with increasing proportions of neosome (Fig. 12.10). Finely-banded varieties dissected by thin, irregular neosome veins are most common. Granitic migmatites typically consist of abundant, rotated paleosome components enveloped by a structureless granitic neosome. The paleosome component comprises relicts of supracrustal rocks, mainly schists, siliceous gneiss, quartzite, and marble.

In a few places, the migmatite and granite zone contains large enclaves of only slightly migmatized meta-sedimentary rocks. The most extensive is at the west margin of the zone on both sides of Nordvestfjord, where a sequence up to 8 km thick is preserved. Another well-preserved sequence about 5 km thick occurs in eastern Milne Land. These sequences comprise pelitic to semi-pelitic schists and psammitic rocks, commonly developed as well-bedded alternations of psammite and mica schist. They resemble the development of the Krummedal supra-crustal sequence in the adjacent gneiss and schist zone to the west, and a correlation is assumed (Higgins, 1974, 1988; Henriksen et al., 1980).

Figure 12.11. Dome-shaped body of augen granite enveloped by migmatitic paragneisses, south Renland. The cliffs are up to 1800 m high. Photo: Lauge Koch's aerial photograph collection, Geological Museum, Copenhagen.

Early plutonic phase

The early plutonic phase is represented by extensive thick sheets and massive bodies of foliated garnetiferous augen granite, which are particularly conspicuous in Renland, Milne Land, and parts of the southern Stauning Alper (Chadwick, 1971, 1975). The augen consist of aggregates of microcline commonly up to 4 cm in diameter. The sheets are clearly intrusive, and while generally conformable to the banding and foliation in the enveloping migmatitic gneisses, discordant relationships are preserved locally. They have been deformed into a series of major, nappe-like, recumbent folds and domes (Fig. 12.11).

Isotopic studies have yielded a zircon age of 1053 Ma, and a Rb-Sr whole rock age of 987 Ma (Steiger et al., 1979). These ages are taken to indicate that emplacement of the granites, together with the associated deformation and migmatization, took place in the Middle Proterozoic.

Caledonian plutonic phase

Late kinematic to postkinematic plutonic rocks outcrop extensively in the Gåsefjord-Stauning Alper migmatite and granite zone, especially in the eastern half of the zone in Milne Land and the Stauning Alper. They are mainly granite intrusions, but also include granodiorites, monzonites, and syenites (Fig. 12.12). The largest bodies exceed 10 km in diameter.

A 500 m thick sheet of hypersthene monzonite in Renland is a useful marker unit. It cuts strongly folded migmatitic gneiss and augen granites, but is itself affected by a migmatitic episode. A Rb-Sr whole rock isochron of 475 Ma indicates a Caledonian intrusive age (Steiger et al., 1979), and also demonstrates that the late migmatization event must be Caledonian.

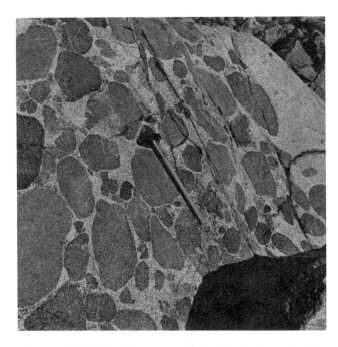

Figure 12.12. Caledonian granodiorite intrusion in east Milne Land containing a variety of quartz diorite inclusions. Photo by A.K. Higgins.

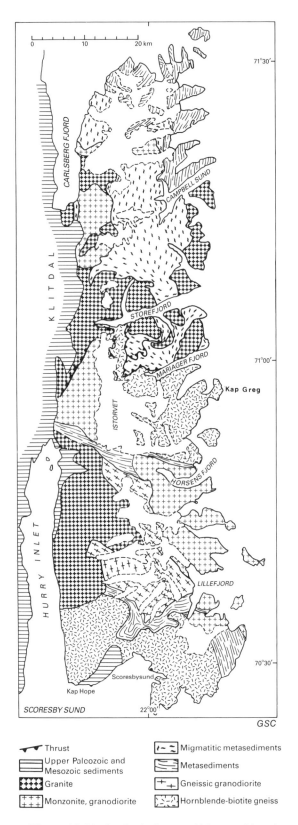

GSC

Thrust		Migmatitic metasediments
Upper Palœozoic and Mesozoic sediments		Metasediments
Granite		Gneissic granodiorite
Monzonite, granodiorite		Hornblende-biotite gneiss

Figure 12.13. Geological map of Liverpool Land.

CALEDONIDES OF EAST GREENLAND

Many of the late intrusions have now been dated by Rb-Sr whole rock and U-Pb zircon techniques, and have given ages mainly in the range 450 to 350 Ma (Rex et al., 1976; Hansen and Tembusch, 1979; Steiger et al., 1979; Rex and Gledhill, 1981; Hansen et al., 1987b). One of the suite of leucocratic granites that border the northeast part of the zone between the Stauning Alper and Forsblad Fjord, and which cuts the Late Proterozoic Eleonore Bay Group, has given a Rb-Sr isochron age of 445 Ma (Rex et al., 1976). The youngest ages recorded are from a suite of small plutons of pinkish granites in east Milne Land; they have yielded a Rb-Sr whole rock age of 373 Ma (Hansen and Tembusch, 1979) and a U-Pb zircon age of 352 Ma (Hansen et al., 1987b).

Mineral ages on all rock types, intrusive and country rocks, are Caledonian and witness to regional Caledonian metamorphism.

Andrée Land

A wedge-shaped region of metasedimentary rocks and granites occupies central Andrée Land and extends into the nunataks just north of latitude 74°N (Fig. 12.7). It is bounded by a thrust to the west and by the Late Proterozoic Eleonore Bay Group to the east. The eastern contact is interpreted as a reactivated fault.

The metasedimentary rocks form thick, well-bedded sequences that are dark or rusty weathering, and generally flat lying or gently inclined. They comprise pelitic, semipelitic, and psammitic rock types, and some indication of the variation in lithology is given by Haller (1953). Semiconcordant granitic or pegmatitic sheets, veins, and lenses are common. Outcrops can in many places be described as migmatitic, although the intensity of migmatization is much less than in the Gåsefjord-Stauning Alper granite and migmatite zone.

The granites are mostly sheet-formed, and include gneissic granites with irregular-veined textures or feldspathic augen, as well as normal, medium grained, leucocratic granites. Some of the sheets have been deformed by spectacular major folds (photo 104 of Haller, 1971). Two granite sheets have given Rb-Sr isochron ages of about 1000 Ma (Rex and Gledhill, 1981), which implies that the metasedimentary rocks that they invade are Middle Proterozoic or older, and probably comparable to the Krummedal supracrustal sequence of the Scoresby Sund region. Other granites from this zone have yielded Caledonian ages (Rex and Gledhill, 1981).

Potassium-argon mineral ages from the region have yielded Caledonian ages (407-431 Ma: Rex and Higgins, 1985), and testify to a regional Caledonian metamorphic event.

Liverpool Land

Liverpool Land comprises a complex of crystalline rocks 30 km wide from east to west and 120 km from north to south, situated east of the Mesozoic Jameson Land basin and totally isolated from other crystalline complexes (Fig. 12.13). A variety of gneisses, migmatites, and metasedimentary rocks are intruded by abundant plutonic bodies.

Gneisses and metasedimentary rocks

Liverpool Land is conveniently divided into three gneissic complexes. The structurally lowest, and presumed oldest, is sited centrally. It comprises about 2500 m of well-banded, flat-lying granitic gneisses (the Kap Greg gneisses of Kranck, 1935), which are overlain conformably by a sequence of marble, amphibolite, and siliceous gneiss. The north and south boundaries of this complex are tectonic.

A southern complex, centred on the town of Scoresbysund, comprises a structurally lower part of hornblende-biotite gneisses and metasedimentary rocks, and a higher part of gneissic granodiorites which grade laterally into their more homogeneous plutonic equivalents. Eclogitic inclusions are also found in this region (Kranck, 1935). It is envisaged that the granodiorites were emplaced into the metasedimentary sequence, and that both were deformed and metamorphosed to amphibolite or granulite facies, and juxtaposed with eclogitic rocks originating at great depth (Cheeney, 1985).

Northern Liverpool Land is dominated by metasedimentary rocks, several thousand metres thick, and their migmatized equivalents. The rock types include hornblende-biotite paragneisses, semipelitic and pelitic schists, yellow and white marbles, and quartzitic rocks. The metasedimentary sequence has been disturbed by two phases of isoclinal folding and amphibolite facies metamorphism, with migmatization increasing southward. To some extent these supracrustal rocks resemble the Krummedal supracrustal sequence of the inner fiord zone, except that marbles are more abundant in Liverpool Land. Comparison might also be made with the Eleonore Bay Group, of which unmetamorphosed representatives are known in Canning Land 20 km to the north (Caby, 1972); these two sequences, however, have very different structural and metamorphic histories.

The metamorphic rocks of Liverpool Land were initially thought to be Archean, until Koch (1929) suggested they might represent a deep-seated level of the Caledonian fold belt, a view supported by Kranck (1935). Current opinion is that much of the metamorphism is pre-Caledonian, with extensive reworking during the Caledonian Orogeny. Isotopic evidence for pre-Caledonian events is so far limited to K-Ar hornblende ages of 1134 and 1193 Ma (Hansen et al., 1973; Hansen and Friderichsen, 1987).

Plutonic rocks

Synkinematic and postkinematic igneous rocks are abundantly developed in Liverpool Land, and have been divided into an older suite of granodioritic and monzonitic types, and a younger granitic suite (Coe and Cheeney, 1972; Coe, 1975).

The older suite is represented by hornblende granodiorite (a somewhat inhomogeneous body in the southeast), quartz monzonite, and quartz monzodiorite. The younger plutons are mainly biotite granite, of which the largest body is the Hurry Inlet granite exposed over a region of 30 km by 15 km (Coe, 1975). However, Rb-Sr whole rock, U-Pb zircon, and K-Ar mineral ages suggest both suites are Caledonian and were separated by only a short time interval (Hansen and Friderichsen, 1987).

905

Grandjean Fjord metamorphic complex

The crystalline region extending between latitudes 74° and
76°N between Payer Land and Bessel Fjord has been
termed the Grandjean Fjord mountain belt (Haller, 1971)
or Grandjean Fjord metamorphic complex (Henriksen and
Higgins, 1976). It is bounded to the west by the Inland Ice,
and to the east by downfaulted post-Caledonian sedimentary rocks (Fig. 12.14). A broad region of sedimentary rocks
of the upper Proterozoic Eleonore Bay Group occupies a
graben-like structure centred on Ardencaple Fjord (see
below).

The Grandjean Fjord metamorphic complex comprises
infracrustal and supracrustal rock units. The former
mainly occur in the western inland areas, and the latter in
the eastern areas on both sides of Grandjean Fjord and
south of Bessel Fjord.

The infracrustal rocks between Payer Land and inner
Grandjean Fjord were distinguished by Mittelholzer (1941)
and Leedal (1952) as pre-Caledonian basement. They comprise mainly hornblende and hornblende-biotite gneisses
and amphibolites (Fig. 12.15); Larsen (1981) also distinguished hypersthene gneisses and enderbitic rocks, and
concurred that the infracrustal rock units are part of a
pre-Caledonian basement. Haller's (1956, 1971) viewpoint
that they are Caledonian synorogenic granites and migmatites was based largely on aerial observations.

The arcuate belt of infracrustal rocks between Grandjean Fjord and inner Bessel Fjord, west of the Eleonore Bay
Group outcrop (Fig. 12.14), comprises dominantly migmatitic gneisses and foliated orthogneisses. Throughout the
area they are interleaved with a variety of supracrustal
rocks, mainly brown weathering mica schists and black
amphibolites (Henriksen et al., 1989). Unpublished Sm-Nd
analyses on some of the orthogneisses indicate an initial
Early Proterozoic age of emplacement (F. Kalsbeek,
pers. comm., 1990).

Thick sequences of metasedimentary rocks of probable
Middle Proterozoic age overlie the infracrustal gneisses,
and are most extensive in C.H. Ostenfeld Land and south
of Bessel Fjord (Fig. 12.14). Contacts with the younger
Eleonore Bay Group are tectonic. The metasedimentary
sequences have been designated as the Smallefjord
sequence (Henriksen et al., 1989) and comprise dominantly
semipelites and psammites, variable migmatized.

Dronning Louise Land

The north-trending imbricate thrust zone in Dronning
Louise Land separates a western foreland region (described
above) from an eastern hinterland which is part of the
Caledonian fold belt (Fig. 12.3). The imbricate zone is
characterized by numerous thrust slices within a belt up to
15 km wide. Individual thrust sheets vary from a few
metres to several hundred metres in thickness, and involve
basement gneisses, metasedimentary cover rocks, and
metadolerites (Friderichsen et al., 1990). The cover rocks
(which include the Britannia Sø Group of Peacock, 1956,
1958) are considered to correspond to the Trekant and
Zebra series, and provide a link between the foreland and
the eastern hinterland.

The eastern hinterland can be divided into two assemblages: high-grade, migmatitic basement gneiss complexes,
and medium-grade, nonmigmatitic cover sequences. The

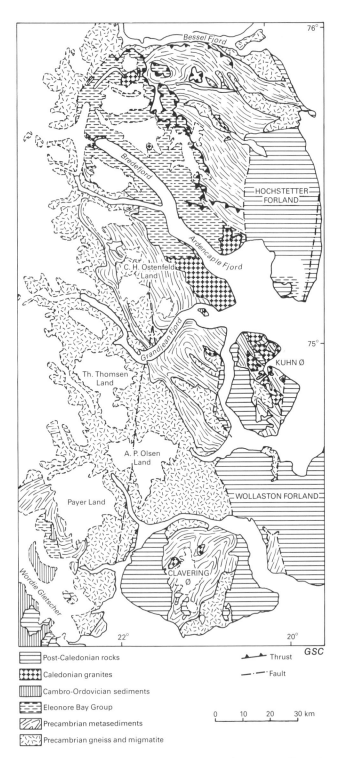

Figure 12.14. Geological map of the region between Clavering
Ø and Bessel Fjord showing the distribution of the Precambrian gneisses, migmatites, and metasedimentary rocks
that make up the Grandjean Fjord metamorphic complex.

basement gneiss complexes resemble those of the western foreland, but over wide areas seem to have been affected by intense Caledonian deformation. The cover sequences are generally preserved in the cores of what are interpreted as early Caledonian isoclinal synclines, and a correlation of some units with the Zebra Series of the foreland has been suggested (Friderichsen et al., 1990).

Bessel Fjord-Ingolf Fjord metamorphic complex

This general designation was used by Haller (1971) and Henriksen and Higgins (1976) for a broad zone of migmatites, gneisses, and local supracrustal rocks extending from Bessel Fjord (76°N) to Ingolf Fjord (80°40′N). The northern part of this region (Fig. 12.16) is that supposedly affected by the Late Proterozoic Carolinidian Orogeny, and subsequently reworked during the Caledonian Orogeny. Haller (1970, 1971) envisaged that the former was responsible for the development of older north-northwest fold systems, whereas younger Caledonian structures had north-northeast trends. However, Jepsen and Kalsbeek (1981, 1985), as a result of fieldwork in Kronprins Christian Land, reported that no evidence existed for a Late Proterozoic Carolinidian Orogeny.

The southern part of the region (76°-78°N) was extensively investigated in the years 1988-90 (Chadwick et al., 1990; Chadwick and Friend, 1991; Friderichsen et al., 1991). The oldest rocks represented in the Dove Bugt region (Fig. 12.16) appear to be supracrustal rocks, which form thin but persistant units within the quantitatively dominant orthogneisses. The most prominent supracrustal rocks are marbles and associated calc-silicates, whereas rusty weathering mica schists and quartzites are locally important. Concordant amphibolite layers within the supracrustal units are interpreted as former basic intrusions. Around Skærfjorden, ultramafites, eclogites, and metasedimentary rocks are preserved as inclusions or trails of enclaves within migmatitic orthogneisses.

The areas around Dove Bugt and Skærfjorden are dominated by a heterogeneous complex of high-grade, migmatitic grey gneisses. Later intrusive rocks include metaporphyries and metagabbro sheets, and metadolerite dykes. All the crystalline rocks were affected by thorough metamorphism, and up to three fabrics are recognized. Several major late shear zones are found, one of which follows the east side of Dronning Louise Land, and extends north-northeast into Jökelbugten. These hinder correlation of structures and lithologies between the eastern hinterland of Dronning Louise Land and the Dove Bugt region.

Figure 12.15. Isoclinal recumbent fold in banded leucocratic gneisses of the Grandjean Fjord metamorphic complex. North wall of inner Grandjean Fjord, upper half of 1300 m high cliff. Photo: Lauge Koch's aerial photograph collection, Geological Museum, Copenhagen.

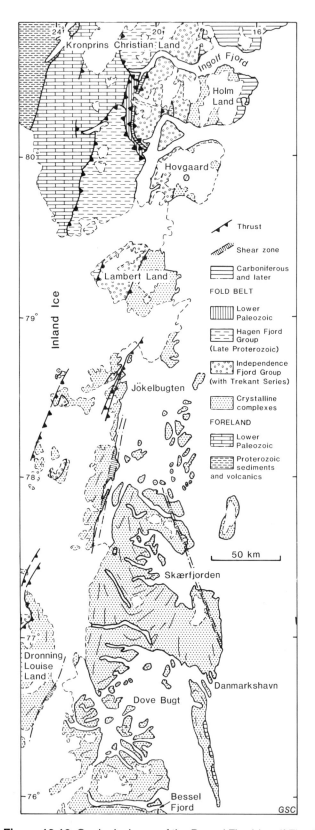

Figure 12.16. Geological map of the Bessel Fjord-Ingolf Fjord metamorphic complex.

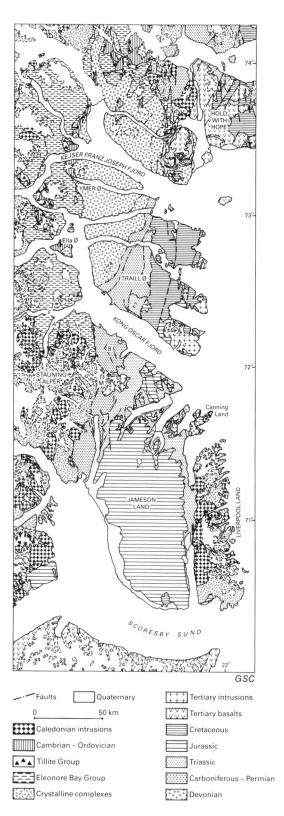

Figure 12.17. The outer fjord zone 70°-74°N, showing main outcrops of the Upper Proterozoic and lower Paleozoic sedimentary sequences, Caledonian granites, and the post-Caledonianmainly sedimentary sequences. A = Alpefjord, F = Forsblad Fjord, B = Blyklippen, M = Malmbjerg.

Although the crystalline rocks of the Bessel Fjord-Ingolf Fjord region clearly form part of the hinterland of the East Greenland Caledonides, the only published isotopic data on gneisses at Danmarkshavn (Steiger et al., 1976), as well as unpublished Sm-Nd analyses (F. Kalsbeek, pers. comm., 1990), suggest most of the crystalline rocks are Archean and Early Proterozoic. The degree of Caledonian reworking is uncertain.

UPPER PROTEROZOIC SEDIMENTARY ROCKS

Upper Proterozoic sedimentary sequences that have been affected only by Caledonian deformation and metamorphism occur in four main areas. The most extensive are the classic outcrops of the Eleonore Bay Group and Tillite Group in central East Greenland (Fig. 12.17). The Petermann Series occurs in the inner fjord zone, whereas the enigmatic Eleonore Sø Series is found in the nunataks to the northwest. Both have usually been compared with the Eleonore Bay Group, but the Eleonore Sø Series (described above) has a different development and is a possible autochthonous foreland sequence. Around Ardencaple Fjord, extensive outcrops of the Eleonore Bay Group are preserved in a graben-like structure. In Kronprins Christian Land, Upper Proterozoic clastic rocks and carbonates are referred to as the Hagen Fjord Group.

Eleonore Bay Group

The classic outcrops of the Eleonore Bay Group occur in a ca. 250 km long north-trending belt in the central fjord zone between latitudes 71°35' and 74°25'N. The sedimentary rocks are folded mainly in large open structures, and apart from a marginal zone adjacent to the metamorphic complexes, are nonmetamorphic. As stated in the introduction, the general assumption that all metamorphic supracrustal remnants within the metamorphic complexes were formerly Eleonore Bay Group has been negated by new geological observations and isotopic studies. However, some of the high-grade metasedimentary rocks bordering the Eleonore Bay Group in the Forsblad Fjord region, now sillimanite schists and migmatites, may have been formed by transformation of the Eleonore Bay Group during the Caledonian Orogeny (Peucat et al., 1985; Higgins, 1988).

The Eleonore Bay Group is divided into a lower sequence of monotonous psammitic and semipelitic sedimentary rocks up to 8000 m thick, and an upper more variable and colourful sequence of quartzite, mudstone, and carbonate up to 4000 m thick (Haller, 1971). Haller's (1971) terminology is used here, but following detailed sedimentological studies in the period 1988-90 the divisions into "series" and "beds" are currently being brought into line with modern lithostratigraphic practice, and the Eleonore Bay Group is to be upgraded to a Supergroup (M. Sønderholm, pers. comm., 1990).

The great thickness of the lower Eleonore Bay Group is only known from the Forsblad Fjord-Alpefjord region. In other areas, thicknesses of 2000 to 4000 m are present. Caby (1976) interpreted the sequence of dominantly quartzites and shales as very shallow water deposits of a subsiding and fluctuating, wide deltaic zone. Caby and Bertrand-Sarfati (1988) tentatively proposed a continental, mainly fluvial origin for the lower two divisions, and a shallow marine origin for the upper division. It is probable that the variation in the recorded thicknesses is a reflection of the mode of preservation. The boundary with the adjacent metamorphic complexes is considered to be a fault contact reworked during the Caledonian Orogeny (Henriksen et al., 1989).

The upper Eleonore Bay Group comprises three main divisions, which have been divided into many subdivisions (Katz, 1952b; Eha, 1953; Fränkl, 1953; Caby and Bertrand-Sarfati, 1988; Sønderholm and Tirsgaard, 1990). The sequence shows an exceptionally constant lithological development over its entire outcrop area. The Quartzite Series comprises about 2500 m of well-sorted sandstones, siltstone, and shale, with sedimentary structures indicating shallow water deposition. The Multicoloured Series comprises 1200 m of colourful, banded shales, mudstones, and carbonates, representing a transitional phase between siliciclastic shelf sedimentation of the Quartzite Series and the carbonate platform of the Limestone-Dolomite Series (Fig. 12.18). The Limestone-Dolomite Series is up to 1500 m thick, and exhibits sedimentary features suggesting deposition on an extensive carbonate platform. Stromatolitic biostromes are abundant at many levels and are interpreted as growing in the intertidal zone (Caby and Bertrand-Sarfati, 1988).

Acritarchs recovered from the upper Eleonore Bay Group indicate a Late Riphean age for the Quartzite Series and Multicoloured Series, and a very Late Riphean age for the Limestone-Dolomite Series (Vidal, 1976). Bertrand-Sarfati and Caby (1976) have suggested that stromatolites from the Limestone-Dolomite Series can be correlated with Vendian rather than Late Riphean forms.

Petermann Series

A strip of largely nonmetamorphic supracrustal rocks known as the Petermann Series occurs about 60 km west of the main Eleonore Bay Group exposures at the head of Kejser Franz Joseph Fjord. This series outcrops over a region about 100 km from north to south and up to 30 km wide, bounded by a fault on the west side, and disturbed by a simple pattern of north-trending folds.

The strikingly banded Petermann Series (Fig. 12.19) was originally compared in general terms with the Eleonore Bay Group (Wordie, 1930). All subsequent workers have made similar comparisons, and Wenk and Haller (1953), who established the first reasonably detailed lithological division, suggested correlation with the highest levels of the lower Eleonore Bay Group and lowest levels of the upper Eleonore Bay Group.

The sequence is about 6500 m thick, comprising dominantly sandstone and mudstone lithologies. Metamorphic grade is generally low, but increases eastward such that shaly lithologies pass eastward into spotted phyllites, and garnet biotite schists. Wenk and Haller (1953) claimed there was a gradual transition farther eastward into the gneisses of the metamorphic complex (Hagar sheet), supporting the observations of Huber (1950). Higgins et al. (1981) viewed the contact between the metamorphic Petermann Series and banded gneisses as a cover-basement contact overprinted by Caledonian metamorphism, an interpretation more in line with that of Odell (1939).

The Petermann Series is intruded by several large plutons of Caledonian granite.

Figure 12.18. Part of the upper Eleonore Bay Group succession in outer Forsblad Fjord. Cliffs are up to 1700 m high. Photo by N. Henriksen.

Ardencaple Fjord

Ardencaple Fjord and its northwest extension, Bredefjord, expose impressive profiles through a largely sandstone and shale sequence up to 3500 m thick (Fig. 12.14). The sequence occupies a northwest-trending zone about 80 km long by 40 km wide between latitudes 75°15′ and 76°N. Sommer (1957) correlated the sequence with the uppermost levels of the lower Eleonore Bay Group and lower part of the upper Eleonore Bay Group, a correlation confirmed by Sønderholm et al. (1989) who also resolved the problem of the stratigraphic position of Sommer's (1957) "Brædal-Quartzite". Studies of the contact zones of the Eleonore Bay Group rocks with adjacent metamorphic rocks, and the style of Caledonian deformation (Henriksen et al., 1989), suggest a two-stage tectonic model. Pre-Caledonian or early Caledonian downward displacement of the Eleonore Bay Group along extensional shear zones was succeeded by Caledonian compression, deforming the shear zones and in some cases reactivating them as thrust faults.

A number of light coloured Caledonian granites intrude the Eleonore Bay Group, and are the northernmost known examples of Caledonian plutonic rocks in East Greenland.

Figure 12.19. The Late Proterozoic Petermann Series. The snow-capped peak on the right is Petermann Bjerg, 2940 m high. Note general resemblance to the Eleonore Bay Group in Figure 12.18.

Tillite Group

The Tillite Group overlies the Eleonore Bay Group with slight angular unconformity, and outcrops as a narrow belt in the central fiord zone between latitudes 71°35′ and 74°25′N (Fig. 12.17). It comprises a 500 to 800 m thick sequence of distinctive sedimentary rocks with two diamictite (tillite) formations of regional extent. Five formations are distinguished, and on the basis of recent fieldwork have been described and redefined by Hambrey and Spencer (1987).

Both the lower and upper diamictite formations comprise a variety of tillite types, glaciolacustrine sedimentary rocks, and fluvioglacial deposits. The lower formation includes mainly locally derived blocks of Eleonore Bay Group lithologies. The upper formation includes numerous crystalline boulders in the tillite levels, of which red granites are conspicuous. Other sedimentary rocks of the Tillite Group were deposited in a dominantly shallow marine or lacustrine environment, with locally distinctive levels characterized by stromatolites, desiccation cracks, and halite pseudomorphs.

The diamictite levels are envisaged as deposited on an extensive shallow shelf, at the margin of a subpolar ice sheet resting on an extensive landmass (Hambrey, 1989; Moncrieff, 1989). Paleocurrent data suggest derivation from the southwest.

Acritarch studies by Vidal (1976) demonstrate that the Tillite Group is of Vendian age. Hambrey and Spencer (1987) confirmed the remarkable similarity of the East Greenland and northeast Svalbard Vendian successions, which suggests the two areas were juxtaposed in Late Proterozoic times.

Kronprins Christian Land

Upper Proterozoic clastic rocks and carbonates in the Caledonian nappes of Kronprins Christian Land (Fig. 12.1, 12.16) are referred to as the Hagen Fjord Group (Haller and Kulp, 1962; Haller, 1971). They overlie the Independence Fjord Group unconformably, and are in turn overlain by lower Paleozoic carbonates and clastic rocks (Hurst and McKerrow, 1981).

The oldest unit, the Rivieradal Sandstone of Fränkl (1954), is only known in the allochthonous Vandredalen nappe (Hurst and McKerrow, 1985). The Rivieradal Sandstone is a sequence of turbiditic sandstone and some conglomerate up to 2500 m thick, the upper part of which has yielded Late Riphean acritarchs. It is overlain by the Campanuladal Formation.

The Campanuladal Formation is known in the foreland shelf region and in two of the nappe units; the sequences 300 to 600 m thick in the nappe units, are thought to have been deposited in a marine deep water basin east of the present outer coast. In the parautochthonous region around Ingolf Fjord it is represented by 0 to 80 m of shallow water mudstone and sandstone.

The succeeding Fyns Sø Formation is a characteristic stromatolitic dolomite known in the foreland, in two of the nappes, and in the parautochthonous region around Ingolf Fjord. It is up to 320 m thick in its type area, but in parts of Kronprins Christian Land it may be reduced to a few metres because of tectonic complications.

LOWER PALEOZOIC ROCKS

Lower Paleozoic rocks outcrop in two regions within the East Greenland Caledonides; central East Greenland and Kronprins Christian Land. The central East Greenland sequence consists of up to 4000 m of Cambrian-Ordovician limestones and dolomites, which resemble the sequences in Newfoundland, northwest Scotland, and Spitsbergen (Swett and Smit, 1972). The East Greenland sequences are best known from the work of Cowie and Adams (1957), summarized by Haller (1971), with more recent work by Frykman (1979), Peel and Cowie (1979), and Peel (1982). The sequence in Kronprins Christian Land occurs in one of the nappe units, and comprises Ordovician and Silurian carbonates that can be correlated with the foreland carbonate shelf sequence to the west.

Central East Greenland

Cambrian and Ordovician platform sedimentary rocks form a number of small outcrops within the Caledonian fold belt between 71°45′ and 74°15′N (Fig. 12.17), but Silurian strata are not known. The sequence conformably overlies the Tillite Group, and is overlain with major unconformity by Devonian sandstones and conglomerates (Fig.12.20).

The Cambrian sequence is divided into five formations. It begins with sandstones and siltstones of the Kløftelv and Bastion formations, which are overlain by carbonates of the Ella Island and Hyolithus Creek formations of late Early Cambrian age. A Middle Cambrian age was assumed previously for the succeeding unit, the Dolomite Point Formation (Cowie, 1971), requiring a major unconformity below the Early Ordovician Antiklinalbugt Formation. However, upper beds of the Dolomite Point Formation have yielded conodonts of Early Ordovician age (Frykman, 1979; Miller and Kurtz, 1979). Faunas of Middle and Late Cambrian age

Figure 12.20. Antiklinalbugt on the southwest coast of Ella Ø. Folded lower Paleozoic sedimentary rocks are unconformably overlain by Devonian conglomerates (D)

are still unknown, and Swett (1981) suggested that their absence probably reflected slow sedimentation under schizohaline conditions.

Ordovician strata attain a thickness in excess of 3 km, three times greater than that known in northern Greenland, although the formations in the two regions show a strong resemblance in development (Peel, 1982). These formations include the lower Lower Ordovician Antiklinalbugt Formation (212-270 m), middle to upper Lower Ordovician Cape Weber Formation (1040-1165 m), upper Lower to lower Middle Ordovician Narwhale Sound Formation (460 m), and Middle to Upper Ordovician Heim Bjerge Formation (up to 1200 m). This succession of alternating limestones and dolomites has yielded diverse shelf-type Pacific faunas.

Kronprins Christian Land

Three nappe sheets are recognized in the Caledonian fold belt of Kronprins Christian Land (Hurst and McKerrow, 1981), of which two preserve lower Paleozoic rock units.

The Finderup Land nappe in northern Kronprins Christian Land contains an upper sequence of green bioturbated siltstones and white quartzites with *Skolithos*, assigned to the Kap Holbæk Formation. This has sometimes been viewed as Early Cambrian, although it has been argued by Peel (1985) that it is more probably of Late Proterozoic age.

The Sæfaxi Elv nappe, found in the area south of Ingolf Fjord (Fig. 12.16), comprises a sequence of dolomites, limestones, and shales of Ordovician and Silurian age. The lower dolomite unit contains sandstone fissures, which led Fränkl (1954, 1955) to infer the presence of the Kap Holbæk Formation, but the unit is now thought to be Early Ordovician (Peel et al., 1981). The overlying thin-bedded black limestones and shales are viewed as deeper marine equivalents of the platform carbonate strata in the foreland to the west.

CALEDONIAN OROGENY

Caledonian deformation, metamorphism, and plutonism have affected the entire length of the fold belt to varying degrees, although plutonic rocks are largely confined to the southern half of the region. Deformation and metamorphism overprint the features developed during Archean, Early Proterozoic, and Middle Proterozoic orogenic episodes, such that in the crystalline complexes that dominate the fold belt, it is difficult or impossible to distinguish pre-Caledonian and Caledonian events. Interpretations presented here are based largely on Geological Survey of Greenland fieldwork over the past 20 years, although parts of the fold belt have only been visited cursorily since the pioneer work of Lauge Koch's expeditions and other early workers (Haller, 1971; Henriksen and Higgins, 1976).

Structure

Caledonian structure in the Upper Proterozoic and lower Paleozoic successions of central East Greenland is characterized by rather open major folds and flexures (Fig. 12.20) with generally north trends, and associated faults (Eha, 1953; Fränkl, 1953; Wenk and Haller, 1953; Haller, 1971; Bengaard, 1989). Structures affecting the lower part of the

Upper Proterozoic sequence in Andrée Land include classic box folds, which imply thin-skinned displacement along a thrust or décollement surface located at the contact between the sedimentary rocks and older crystalline complexes (Higgins et al., 1981). On a regional scale, east-west shortening appears limited; Eha (1953) estimated 5.3 per cent for one area, Higgins et al. (1981) estimated 15 per cent for another area, while Manby and Hambrey (1989) suggested that thrusting and shortening on bedding plane detachments probably exceeded 30 per cent. North- and east-trending faults are important in many areas.

Within the crystalline complexes that dominate the fold belt at the present erosion levels, the age of the structures is debatable. Many of the spectacular nappe-like structures formerly viewed as Caledonian (Haller, 1970, 1971) are now thought to be pre-Caledonian. The transverse fold trends of parts of the region, which were such a puzzle to early workers (e.g. Wegmann, 1935), are likely to be Archean or Early Proterozoic in age. In most crystalline regions sets of late north-trending structures can be distinguished and are thought to be Caledonian, but because of the limited distribution of Late Proterozoic and Paleozoic cover sequences, the regional extent and intensity of Caledonian deformation is difficult to assess (Higgins and Phillips, 1979; Henriksen, 1985).

The crystalline complexes have been moved westward on major thrust planes, apparently as more or less intact thick thrust wedges or sheets in which Caledonian penetrative deformation was largely confined to the borders of the sheets. In the southern part of the fold belt the marginal thrust system follows an undulating course through the nunatak zone at the margin of the Inland Ice, the thrust planes being inclined at very low angles and associated with extensive mylonite developments. The thrust zone itself has a complex history, divisible in some areas into early deformation involving emplacement of numerous thin thrust sheets, and later deformation with thrusting that truncated and refolded structures of the earlier phase (Phillips et al., 1973). At different localities along the margin of the fold belt, displacements of at least 40 km can be demonstrated, and speculation as to the original position of the rock units has led to suggestions of 100 km displacements (Homewood, 1973; Higgins, 1974). Estimates of shear strain in Paul Stern Land (72°15′N) suggest a minimum of 130 km displacement to the west-southwest.

The marginal thrust system is traceable almost continuously from 70° to 72°N, is largely hidden beneath the Inland Ice from 72° to 76°N, reappears in Dronning Louise Land (76°-77°N) and after a short gap, continues in Kronprins Christian Land (79°-82°N; Fig. 12.1). In the extreme north the thrust system affects Proterozoic and lower Paleozoic sedimentary rocks in a series of thrust sheets and glide nappes (Fränkl, 1955; Hurst and McKerrow, 1981, 1985). Spectacular imbricate zones are locally developed above the main thrust planes. Facies variations within the three main nappe units of Kronprins Christian Land imply westward displacements of at least 150 km (Hurst et al., 1983). Studies of the sedimentological development of the lower Paleozoic Franklinian basin which extends east-west across North Greenland suggest that erosion of the rising mountains and advancing Caledonian nappes in North-East Greenland provided much of the source for clastic detritus in the Silurian turbidites (Surlyk, 1982). It is also suggested that depression of the crust

ahead of the advancing nappes led directly to foundering of the carbonate platform and expansion of basinal turbidite deposition in the latest Llandovery (Hurst et al., 1983).

Metamorphism

Caledonian metamorphism can only be distinguished from earlier metamorphic episodes in the Upper Proterozoic and lower Paleozoic rocks. Only the lower part of the Upper Proterozoic (Eleonore Bay Group) sequence shows significant metamorphism, and in southern parts of the fold belt metamorphic grade increases rapidly stratigraphically downward through greenschist facies to high amphibolite facies adjacent to the metamorphic complexes (Higgins et al., 1981).

Within the crystalline complexes the metamorphic zonation, like the deformation pattern, is a composite feature, and is the result of Caledonian metamorphism overprinting several pre-Caledonian metamorphic episodes. Amphibolite grade Caledonian metamorphism appears to have affected much of the southern crystalline regions, and where associated with migmatite developments and Caledonian monzonite intrusions as in Renland (Chadwick, 1975), may locally have reached granulite facies. Henriksen et al. (1980) described the metamorphic conditions in a 130 km wide cross-section of the crystalline complexes. A segment of the Gåseland-Hinks Land gneiss and schist zone is characterized by amphibolite facies conditions with widespread kyanite. The western part of the Gåsefjord-Stauning Alper migmatite and granite zone exhibits high-pressure amphibolite facies conditions with abundant sillimanite, whereas the eastern part was overprinted by low-pressure amphibolite facies conditions, and cordierite-bearing assemblages are common. Late-stage retrogressive alteration is widespread.

The influence of Caledonian regional metamorphism in the crystalline complexes is perhaps best appreciated from the pattern of Rb-Sr and K-Ar mineral ages, of which nearly all those from the region 70° to 74°N are Caledonian; the majority fall in the range 440 to 365 Ma (Higgins and Phillips, 1979; Rex and Higgins, 1985). However, Caledonian reactivation appears only very locally to have upset Rb-Sr whole rock isotopic systems.

In the northern half of the fold belt the intensity of Caledonian metamorphism is uncertain, but appears to have been less pervasive than in southern areas. The Upper Proterozoic and lower Paleozoic sedimentary rocks in Kronprins Christian Land show no evidence of significant Caledonian metamorphism, but the Eleonore Bay Group sequence around Ardencaple Fjord exhibits Caledonian amphibolite facies metamorphism in its deepest levels (Henriksen et al., 1989). Isotopic studies have been initiated on samples from the Dove Bugt and Skærfjorden areas collected in 1989-90 with the purpose of ascertaining the extent of Caledonian overprinting on the crystalline rocks.

Plutonic rocks

Caledonian plutonic rocks, apart from a few small bodies and pegmatites, are confined to the southern half of the fold belt. They can be divided into two groups: mainly leucocratic granites emplaced generally in the boundary zone between the crystalline complexes and the Upper Proterozoic sedimentary sequences, the crosscutting relationships

Figure 12.21. Gently folded and faulted Eleonore Bay Group cut by leucocratic Caledonian granites, Andrée Land.

of which demonstrate their Caledonian age (Fig. 12.21); and a much more varied plutonic suite within the crystalline complexes, the Caledonian age of which is indicated by Rb-Sr whole rock isotopic ages or U-Pb zircon ages. A number of bodies previously regarded as Caledonian (Haller, 1970, 1971) have Early Proterozoic and Middle Proterozoic isotopic ages (Rex and Gledhill, 1981).

The earliest Caledonian plutonic rocks form part of the varied suite within the crystalline complexes, of which the great majority occur within the Gåsefjord-Stauning Alper migmatite and granite zone or in Liverpool Land (see also above). These early bodies include sheets of diorite, monzonite, and granodiorite emplaced into Middle Proterozoic migmatitic rocks, and give isochron ages of up to 475 Ma. A Caledonian monzonite sheet in Renland predates a late phase of migmatization (Chadwick, 1975), implying that some of the migmatization in this zone is Caledonian. Other, slightly younger, plutonic rocks within the crystalline complexes include hornblende syenites, pink and grey fine grained granites, and leucocratic biotite-muscovite granites (Henriksen et al., 1980; Henriksen, 1986; Henriksen and Higgins, 1987). Ultramafic and alkaline rocks are uncommon. The youngest suite is pink granite in Milne Land (373-352 Ma; Hansen and Tembusch, 1979; Hansen et al., 1987a).

The Caledonian plutonic rocks emplaced in the contact zone between the crystalline rocks and the Eleonore Bay Group are almost exclusively leucocratic biotite-muscovite granites. They are conspicuous for their colour and the spectacular crosscutting relationships. They are common between 72° and 74°N, with other examples bordering the Eleonore Bay Group around Ardencaple Fjord. A few

examples have isotopic ages of 560 to 377 Ma (Rex and Gledhill, 1981), although the older ages have large uncertainties and are regarded as "errorchrons".

The initial strontium isotopic ratios of the Caledonian granites are generally high with a wide range (0.710-0.736), implying that they have been contaminated by material with a previous crustal history (Rex and Gledhill, 1981). They were probably derived by anatectic melting of the local Precambrian metamorphic and migmatitic rocks at depth during Caledonian metamorphism. The ^{18}O-rich character values (>9.5 ‰) reflect the crustal origin of these granitoids (Harmon, 1984).

POST-CALEDONIAN ROCKS

Upper Paleozoic rocks

After the main phase of the Caledonian Orogeny, fault-bounded Devonian continental basins were initiated; the present exposures are found between latitudes 70°30′ and 73°40′N (Fig. 12.17). The deposits comprise Middle to Upper Devonian nonmarine conglomerates, sandstones, and arkoses with an aggregate thickness of about 10 000 m, and continental Upper Carboniferous to Lower Permian deposits totalling 4000 to 5000 m. A new subdivision of the Devonian sequence supplemented by detailed sedimentological investigations has been presented by Friend et al. (1976), partly based on the earlier work of Bütler (1959, 1961a).

Friend et al. (1983) suggested that the Devonian basins were initiated by four north-trending fracture zones, probably largely generated by wrench stresses. Larsen and

Bengaard (1991) proposed that the Middle Devonian basin formation can be related to extensional collapse of an over-thickened Caledonian crustal welt, comparable with the scenario put forward by McClay et al. (1986) for the Devonian basins of northern Great Britain.

The Devonian sedimentary rocks were deposited in local basins, with drainage in places axial, and in other places marginal from both east and west sides reflecting the uplift of new source areas and local folding at different stages in the basin development.

Devonian acid and basic volcanism took place intermittently. Up to 1000 m of silicic tuffs and rhyolites occur in some sequences and are cut by several generations of lamprophyre dykes. Two intrusive bodies occur within the Devonian sequence and can be dated on stratigraphic criteria. One body cuts a slightly folded Devonian formation, and is overlain unconformably by a later Devonian formation; repeated uplift of this intrusive body is reflected in the structure and stratigraphy of the overlying formation (Alexander-Marrack and Friend, 1976; Friend et al., 1983).

The Devonian sequence was affected by late Caledonian open folding and minor thrusting and faulting, reflecting both compressional and tensional events (Haller, 1971; Friend et al., 1983).

Upper Carboniferous-Lower Permian sedimentary rocks overlie Upper Devonian rocks with angular unconformity in central East Greenland. They comprise continental conglomerates, arkoses, and sandstones with strong lateral facies variations and are divided into four formations (Bütler, 1961b; Perch-Nielsen et al., 1972). A regional marine transgression in the Upper Permian marks the beginning of the cratogenic evolution that characterizes the Mesozoic and Tertiary.

In Kronprins Christian Land (78°-81°N), small areas of marine Permo-Carboniferous strata occur along the outer coast (Håkansson and Stemmerik, 1984).

Mesozoic rocks

A major north-trending Mesozoic sedimentary basin occurs in the coastal zone of East Greenland. It extends onshore and offshore over a distance of 800 km from the southern tip of Jameson Land (70°25′N) to the little known areas north of Germania Land (77°45′N). The exposed width of the basin is 150 km in the Jameson Land region, where the succession is thickest and most complete (Fig. 12.17). The west margin of the basin is formed by the so-called post-Devonian main fault (Vischer, 1943), and the development of the basin can be related to episodes of faulting, and uplift, downthrow, and tilting of fault blocks (Vischer, 1943; Haller, 1971; Birkelund et al., 1974; Surlyk, 1978). The fault pattern has been related to rifting accompanying the initial opening of the North Atlantic. The development and structure of the basins on the East Greenland shelf have been summarized by Larsen (1990).

Most of the Triassic sequence consists of clastic deposits, with some gypsum and minor limestone intercalations. The Jurassic sequence is complete and consists of clastic rocks, the older part being nonmarine and the younger marine. The sequence is very fossiliferous, with boreal ammonite faunas giving way upward to faunas with

European or Volgian affinities (Callomon and Birkelund, 1982). The Cretaceous sedimentary rocks are marine, and predominantly clastic.

Small areas of Upper Cretaceous to Lower Jurassic strata occur in eastern Kronprins Christian Land at about 81°N (Håkansson and Stemmerik, 1984).

Tertiary rocks

The southern continuation of the East Greenland Caledonides is hidden beneath a cover of Tertiary basalts from 68° to 70°N (Deer, 1976; Watt et al., 1986; Fig. 12.1). Other Tertiary basalts occur in the coastal region from 74° to 75°30′N, and in the nunatak region centred around 74°N (Noe-Nygaard, 1976). Dykes and sills are widespread in the post-Caledonian rocks between latitudes 70° and 75°N, and farther south the celebrated coast-parallel dyke swarm cuts the basalt sequence (Deer, 1976).

The extensive Tertiary basalt lavas south of Scoresby Sund have an aggregate thickness of 3200 m; thicknesses decrease inland to 1500 m around Gåsefjord and 300 to 800 m on Gåseland and Milne Land, where the basalts lap onto Precambrian and Caledonian gneisses and Jurassic sedimentary rocks (Fig. 12.22). The lavas are divisible into five formations that form two separate sequences (Larsen and Watt, 1985; Watt et al., 1986; Larsen et al., 1989). The sequences are interpreted to be produced in two volcanic episodes related to failed rifting episodes during the opening of the North Atlantic Ocean. At the Atlantic coast the remains of a third separate lava sequence are found, apparently formed during active spreading.

MINERAL RESOURCES AND OIL EXPLORATION

Between 1952 and 1984 Nordisk Mineselskab A/S held exclusive rights for mineral exploration in East Greenland between latitudes 70° and 74°30′N, the southern part of the Caledonian fold belt (Harpøth et al., 1986). Lead and zinc were mined at Blyklippen near Mesters Vig from 1956 to 1963, and a molybdenum deposit was found at Malmbjerg in 1954, although it has not yet been exploited (Fig. 12.17). The old concessions were relinquished in 1984 in connection with negotiations for a petroleum exploration concession in the Jameson Land area by Atlantic Richfield Company (ARCO) in co-operation with Nordisk Mineselskab.

Mesters Vig (Blyklippen)

Several lead-zinc deposits occur in the Mesters Vig area, all genetically related to the Tertiary alkaline intrusions of the Werner Bjerge. The largest deposit, that at Blyklippen, was mined and is now depleted. It was situated in Carboniferous sandstones in a system of north-northwest-trending veins associated with faults; galena and sphalerite were present in the quartz veins.

The deposit was discovered in 1948, drilled in 1951 and 1953, and mined from 1956 to 1963; 58 000 tons (52 600) of galena concentrate and 75 000 (68 000 t) tons of sphalerite concentrate were recovered with mean concentrations of 12 per cent Pb and 10 per cent Zn (Harpøth et al., 1986).

Figure 12.22. Tertiary basalts filling a pre-basalt valley eroded in gneisses and migmatites in Gåseland. Photo by M. Watt.

Werner Bjerge (Malmbjerg)

In 1954 a molybdenite occurrence named Malmbjerg was discovered in the Werner Bjerge, associated with a complex of Tertiary alkaline intrusions (Bearth, 1959). Both intrusions and the surrounding Carboniferous sedimentary rocks were affected by pneumatolytic and hydrothermal activity, and the ore-impregnated zones are conspicuous for their red, yellow, and ochre shades caused by coatings of hematite and limonite. The deposit has been drilled extensively, and reserves have been estimated close to 200 million tons (181 million tonnes) with a grade of 0.25 per cent MoS_2. Exploration has been hindered by the large initial costs arising from its remote location.

Jameson Land (petroleum exploration)

The Jameson Land basin contains a very thick upper Paleozoic-Mesozoic sedimentary sequence. From a petroleum point of view, the Lower Permian is considered as "economic basement", while the Jurassic is nonprospective in most areas because it forms the topographic surface. The Upper Permian is the most promising target as it contains potential source rocks, possible reservoir rocks, and is moderately deeply buried (Surlyk et al., 1986).

In 1984 a consortium of ARCO-Nordisk Mineselskab was granted the rights to petroleum exploration in an area of 9800 km² onshore Jameson Land for a period of 12 years. Field operations were initiated in 1985, and more than 1500 km of high-quality seismic were completed by 1990, when the concession was given up without drilling.

ACKNOWLEDGMENTS

This paper is published with the permission of the Geological Survey of Greenland. N. Henriksen and J.D. Friderichsen are thanked for their helpful comments on the text; Ulla Johansen, Tove Buus, and Lena Blomgren for the typing of the manuscript; and Gurli Hansen for drafting the figures. Except where otherwise stated, photographs are by the author.

REFERENCES

Alexander-Marrack, P.D. and Friend, P.F.
1976:   Devonian sediments of East Greenland, III, The eastern sequence, Vilddal Supergroup and part of the Kap Kolthoff Supergroup; Meddelelser om Grønland, v. 206, no. 3, 122 p.

Backlund, H.G.
1930:   Contributions to the geology of Northeast Greenland; Meddelser om Grønland, v. 74, no. 11, p. 207-296.

Bearth, P.
1959:   On the alkali massif of the Werner Bjerge in East Greenland; Meddelelser om Grønland, v. 153, no. 4, 63 p.

Bengaard, H.-J.
1989:   Geometrical and geological analysis of photogrammetrically measured deformed sediments of the fjord zone, central East Greenland; Grønlands Geologiske Undersøgelse, Open-File Series no. 89/6, 101 p.

Bertrand-Sarfati, J. and Caby, R.
1976:   Carbonates et stromatolites du sommet du Groupe d'Eleonore Bay (Précambrian terminal) au Canning Land (Groenland oriental); Grønlands Geologiske Undersøgelse, Bulletin no. 119, 51 p.

Birkelund, T., Perch-Nielsen, K., Bridgwater, D., and Higgins, A.K.
1974:   An outline of the geology of the Atlantic coast of Greenland; in The Ocean Basins and Margins, v. 2, (ed.) A.E.M. Nairn and F.G. Stehli; Plenum Press, New York, p. 125-159.

Bütler, H.
1959:   Das Old Red-Gebiet am Moskusoksefjord, Attempt at a correlation of the series of various Devonian areas in central East Greenland; Meddelelser om Grønland, v. 160, no. 5, 188 p.
1961a: Devonian deposits of central East Greenland; in Geology of the Arctic v. 1, (ed.) G.O. Raasch; Toronto University Press, p. 188-196.
1961b: Continental Carboniferous and Lower Permian in central East Greenland; in Geology of the Arctic, v. 1, (ed.) G.O. Raasch; Toronto University Press, p. 205-213.

Caby, R.
1972:   Preliminary results of mapping in Caledonian rocks of Canning Land and Wegener Halvø, East Greenland; Grønlands Geologiske Undersøgelse, Rapport no. 48, p. 21-38.
1976:   Investigations on the Lower Eleonore Bay Group in the Alpefjord region, central East Greenland; Grønlands Geologiske Undersøgelse, Rapport no. 80, p. 102-106.

Caby, R. and Bertrand-Sarfati, J.
1988:   The Eleonore Bay Group, central East Greenland; in Later Proterozoic Stratigraphy of the Northern Atlantic Regions, (ed.) J.A. Winchester; Blackie and Son Ltd., Glasgow, p. 212-236.

Callomon, J.H. and Birkelund, T.
1982:   The ammonite zones of the Boreal Volgian (Upper Jurassic) in East Greenland; in Arctic Geology and Geophysics, (ed.) A.F. Embry and H.R. Balkwill; Canadian Society of Petroleum Geologists, Memoir no. 8, p. 349-369.

Chadwick, B.
1971:   Preliminary account of the geology of south-east Renland, Scoresby Sund, East Greenland; Grønlands Geologiske Undersøgelse, Rapport no. 34, 32 p.
1975:   The structure of south Renland, Scoresby Sund, with special reference to the tectonometamorphic evolution of a southern internal part of the Caledonides of East Greenland; Grønlands Geologiske Undersøgelser, Bulletin no. 112, 67 p.

Chadwick, B. and Friend, C.R.L.
1991:   The high-grade gneisses in the south-west of Dove Bugt: an old gneiss complex in a deep part of the Caledonides of North-East Greenland; Grønlands Geologiske Undersøgelse, Rapport no. 152, p. 103-111.

Chadwick, B., Friend, C.R.L., and Higgins, A.K.
1990:   The crystalline rocks of western and southern Dove Bugt, North-East Greenland; Grønlands Geologiske Undersøgelse, Rapport no. 148, p. 127-132.

Cheeney, R.F.
1985:   The plutonic igneous and high-grade metamorphic rocks of south Liverpool Land, East Greenland, part of a supposed Caledonian and Precambrian complex; Grønlands Geologiske Undersøgelse, Rapport no. 123, 39 p.

Coe, K.
1975:   The Hurry Inlet granite and related rocks of Liverpool Land; Grønlands Geologiske Undersøgelse, Bulletin no. 115, 34 p.

Coe, K. and Cheeney, R.F.
1972:   Preliminary results of mapping in Liverpool Land, East Greenland; Grønlands Geologiske Undersøgelse, Rapport no. 48, p. 7-20.

Collinson, J.D.
1980:   Stratigraphy of the Independence Fjord Group (Proterozoic) of eastern North Greenland; Grønlands Geologiske Undersøgelse, Rapport no. 99, p. 7-23.

Cowie, J. W.
1971:   The Cambrian of the North America Arctic regions; in Cambrian of the New World, (ed.) C.H. Holland; London, Interscience, p. 325-383.

Cowie, J.W. and Adams, P.J.
1957:   The geology of the Cambro-Ordovician rocks of central East Greenland, part 1; Meddelelser om Grønland, v. 153, no. 1, 193 p.

Dawes, P.R. and Kerr, J.W. (ed.)
1982:   Nares Strait and the drift of Greenland - a conflict in plate tectonics; Meddelelser om Grønland, Geoscience, no. 8, 392 p.

Deer, W.A.
1976:   Tertiary igneous rocks between Scoresby Sund and Kap Gustav Holm, East Greenland; in Geology of Greenland, (ed.) A. Escher and W.S. Watt; Geological Survey of Greenland, Copenhagen, p. 405-429.

Eha, S.
1953:   The Pre-Devonian sediments on Ymers Ø, Suess Land, and Ella Ø (East Greenland) and their tectonics; Meddelelser om Grønland, v. 111, no. 2, 105 p.

Fränkl, E.
1953:   Geologische Untersuchungen in Ost-Andrées Land (NE-Grønland); Meddelelser om Grønland, v. 113, no. 4, 160 p.
1954:   Vorläufige mitteilung über die Geologie von Kronprins Christians Land (NE-Grönland); Meddelelser om Grønland, v. 116, no. 2, 85 p.
1955:   Weitere Beiträge zur Geologie von Kronprins Christians Land (NE-Grönland); Meddelelser om Grønland, v. 103, no. 7, 35 p.

Friderichsen, J.D., Gilotti, J.A., Henriksen, N., Higgins, A.K., Hull, J.M., Jepsen, H.F., and Kalsbeek, F.
1991:   The crystalline rocks of Germania Land, Nordmarken and adjacent areas, North-East Greenland; Grønlands Geologiske Undersøgelse, Rapport no. 152, p. 85-94.

Friderichsen, J.D., Holdsworth, R.E., Jepsen, H.F., and Strachan, R.A.
1990:   Caledonian and pre-Caledonian geology of Dronning Louise Land, North-East Greenland; Grønlands Geologiske Undersøgelse, Rapport no. 148, p. 133-141.

Friend, P.F., Alexander-Marrack, P.D., Allen, K.C., Nicholson, J., and Yeats, A.K.
1983:   Devonian sediments of East Greenland, VI, Review of results; Meddelelser om Grønland, v. 206, no. 6, 96 p.

Friend, P.F., Alexander-Marrack, P.D., Nicholson, J., and Yeats, A.K.
1976:   Devonian sediments of East Greenland, II, Sedimentary structures and fossils; Meddelelser om Grønland, v. 206, no. 2, 91 p.

Frykman, P.
1979:   Cambro-Ordovician rocks of C.H. Ostenfeld Nunatak, northern East Greenland; Grønlands Geologiske Undersøgelse, Rapport no. 91, p. 125-132.

Håkansson, E. and Stemmerik, L.
1984:   Wandel Sea Basin - North Greenland equivalent to Svalbard and the Barents Shelf; in Norwegian Petroleum Society, Petroleum Geology of the North European Margin; Graham and Trotman, London, p. 97-107.

Haller, J.
1953:   Geologie und Petrographie von West-Andrées Land und Ost-Fraenkels Land (NE-Grönland); Meddelelser om Grønland, v. 113, no. 5, 196 p.
1955:   Der "Zentrale Metamorphe Komplex" von NE-Grönland, Teil I, Die geologische Karte von Suess Land, Gletscherland und Goodenoughs Land; Meddelelser om Grønland, v. 73, part 1, no. 3, 174 p.
1956:   Geologie der Nunatakker Region von Zentral-Ostgrönland; Meddelelser om Grønland, v. 154, no. 1, 172 p.
1958:   Der "Zentrale Metamorphe Komplex" von NE-Grönland, II, Die geologische Karte der Staunings Alper und des Forsblads Fjordes; Meddelelser om Grønland, v. 154, no. 3, 153 p.
1961:   The Carolinides: an orogenic belt of late Precambrian age in Northeast Greenland; in Geology of the Arctic, v. 1, (ed.) G.O. Raasch; Toronto University Press, p. 155-159.
1970:   Tectonic map of East Greenland (1:500 000), An account of tectonism, plutonism, and volcanism in East Greenland; Meddelelser om Grønland, v. 171, no. 5, 286 p.
1971:   Geology of the East Greenland Caledonides; Interscience Publishers, New York, 413 p.

Haller, J. and Kulp, J.L.
1962: Absolute age determinations in East Greenland; Meddelelser om Grønland, v. 171, no. 1, 77 p.

Hambrey, M.J.
1989: The Late Proterozoic sedimentary record of East Greenland: its place in understanding the evolution of the Caledonide orogen; in The Caledonide Geology of Scandinavia, (ed.) R.A. Gayer; Graham and Trotman, London, p. 257-262.

Hambrey, M.J. and Spencer, A.M.
1987: Late Precambrian glaciation of East Greenland; Meddelelser om Grønland, Geoscience, no. 19, 50 p.

Hansen, B.T. and Friderichsen, J.D.
1987: Isotopic age dating in Liverpool Land; Grønlands Geologiske Undersøgelse, Rapport no. 134, p. 25-37.

Hansen, B.T. and Tembusch, H.
1979: Rb-Sr isochron ages from east Milne Land, Scoresby Sund, East Greenland; Grønlands Geologiske Undersøgelse, Rapport no. 95, p. 96-101.

Hansen, B.T., Frick, U., and Steiger, R.H.
1973: The geochronology of the Scoresby Sund area, 5: K/Ar mineral ages; Grønlands Geologiske Undersøgelse, Rapport no. 58, p. 59-61.

Hansen, B.T., Higgins, A.K., and Bär, M.T.
1978: Rb-Sr and U-Pb age patterns in polymetamorphic sediments from the southern part of the East Greenland Caledonides; Geological Society of Denmark, Bulletin no. 27, p. 55-62.

Hansen, B.T., Higgins, A.K., and Borchardt, B.
1987a: Archaean U-Pb zircon ages from the Scoresby Sund region, East Greenland; Grønlands Geologiske Undersøgelse, Rapport no. 134, p. 19-24.

Hansen, B.T., Steiger, R.H., Henriksen, N., and Borchardt, B.
1987b: U-Pb and Rb-Sr age determinations on Caledonian plutonic rocks in the central part of the Scoresby Sund region, East Greenland; Grønlands Geologiske Undersøgelse, Rapport no. 134, p. 5-18.

Hansen, B.T., Steiger, R.H., and Higgins, A.K.
1981: Isotopic evidence for a Precambrian metamorphic event within the Charcot Land window, East Greenland Caledonian fold belt; Geological Society of Denmark, Bulletin no. 29, p. 151-160.

Harland, W.B.
1985: Caledonide Svalbard; in The Caledonide Orogen - Scandinavia and Related Areas, (ed.) D.G. Gee and B.A. Sturt; John Wiley, Chichester, p. 999-1016.

Harland, W.B. and Gayer, R.A.
1972: The Arctic Caledonides and earlier oceans; Geological Magazine, v. 109, p. 289-314.

Harmon, R.S.
1984: Stable isotope geochemistry of Caledonian granitoids from the British Isles and East Greenland; Physics of the Earth and Planetary Interiors, v. 35, p. 105-120.

Harpøth, O., Pedersen, J.L., Schønwandt, H.K., and Thomassen, B.
1986: The mineral occurrences of central East Greenland; Meddelelser om Grønland, Geoscience, no. 17, 139 p.

Henriksen, N.
1981: The Charcot Land tillite, Scoresby Sund, East Greenland; in Earth's Pre-Pleistocene Glacial Record, (ed.) M.J. Hambrey and W.B. Harland; Cambridge Science series, p. 776-777.
1985: The Caledonides of central East Greenland, 70°-76°N; in The Caledonide Orogen - Scandinavia and Related Areas, (ed.) D.G. Gee and B.A. Sturt; John Wiley, Chichester, p. 1095-1113.
1986: Scoresby Sund, 1:500 000 sheet 12, descriptive text; Grønlands Geologiske Undersøgelse, Copenhagen, 27 p.

Henriksen, N. and Higgins, A.K.
1976: East Greenland Caledonian fold belt; in Geology of Greenland, (ed.) A. Escher and W.S. Watt; Geological Survey of Greenland, Copenhagen, p. 182-246.
1987: Descriptive text to 1:100 000 map sheets Rødefjord 70 Ø.3 Nord and Kap Leslie 70 Ø.2 Nord; Grønlands Geologiske Undersøgelse, Copenhagen, 34 p.

Henriksen, N., Friderichsen, J.D., Strachan, R.A., Soper, N.J., and Higgins, A.K.
1989: Caledonian and pre-Caledonian geology of the region between Grandjean Fjord and Bessel Fjord (75°-76°N), North-East Greenland; Grønlands Geologiske Undersøgelse, Rapport no. 145, p. 90-97.

Henriksen, N., Perch-Nielsen, K., and Andersen, C.
1980: Descriptive text to 1:100 000 sheets Sydlige Steuning Alper 71 Ø.2 N and Frederiksdal 71 Ø.3 N; Grønlands Geologiske Undersøgelse, Copenhagen, 46 p.

Higgins, A.K.
1974: The Krummedal supracrustal sequence around inner Nordvestfjord, Scoresby Sund, East Greenland; Grønlands Geologiske Undersøgelse, Rapport no. 68, 34 p.
1988: The Krummedal supracrustal sequence in East Greenland; in Later Proterozoic Stratigraphy of the Northern Atlantic Regions, (ed.) L.A. Winchester; Blackie and Son Ltd., Glasgow, p. 86-96.

Higgins, A.K. and Phillips, W.E.A.
1979: East Greenland Caledonides - an extension of the British Caledonides; in The Caledonides of the British Isles - Reviewed, (ed.) A.L. Harris, C.H. Holland, and B.E. Leake; Geological Society of London, Special Publication, no. 8, p. 19-31.

Higgins, A.K., Friderichsen, J.D., Rex, D.C., and Gledhill, A.R.
1978: Early Proterozoic isotopic ages in the East Greenland Caledonian fold belt; Contributions to Mineralogy and Petrology, v. 67, p. 87-94.

Higgins, A.K., Friderichsen, J.D., and Thyrsted, T.
1981: Precambrian metamorphic complexes in the East Greenland Caledonides (72°-74°N) - their relationships to the Eleonore Bay Group, and Caledonian orogenesis; Grønlands Geologiske Undersøgelse, Rapport no. 104, p. 5-46.

Homewood, P.
1973: Structural and lithological divisions of the western border of the East Greenland Caledonides in the Scoresby Sund region; Grønlands Geologiske Undersøgelse, Rapport no. 57, 27 p.

Huber, W.
1950: Geologisch-Petrographische Untersuchungen in der innern Fjordregion des Kejser Franz Josephs Fjordsystems in Nordostgrönland; Meddelelser om Grønland, v. 151, no. 3, 83 p.

Hurst, J.M. and McKerrow, W.S.
1981: The Caledonian nappes of Kronprins Christian Land, eastern North Greenland; Grønlands Geologiske Undersøgelse, Rapport no. 106, p. 15-19.
1985: Origin of the Caledonian nappes of eastern North Greenland; in The Caledonide Orogen - Scandinavia and Related Areas, (ed.) D.G. Gee and B.A. Sturt; John Wiley, Chichester, p. 1065-1069.

Hurst, J.M., Jepsen, H.F., Kalsbeek, F., McKerrow, W.S., and Peel, J.S.
1985: The geology of the northern extremity of the East Greenland Caledonides; in The Caledonide Orogen - Scandinavia and Related Areas, (ed.) D.G. Gee and B.A. Sturt; John Wiley, Chichester, p. 1047-1069.

Hurst, J.M., McKerrow, W.S., Soper, N.J., and Surlyk, F.
1983: The relationship between Caledonian nappe tectonics and Silurian turbidite deposition in North Greenland; Journal of the Geological Society of London, v. 140, p. 123-132.

Jepsen, H.F. and Kalsbeek, F.
1981: Non-existence of the Carolinidian orogeny in the Prinsesse Caroline-Mathilde Alper of Kronprins Christian Land, eastern North Greenland; Grønlands Geologiske Undersøgelse, Rapport no. 106, p. 7-14.
1985: Evidence for non-existence of a Carolinidian fold belt in eastern North Greenland; in The Caledonide Orogen - Scandinavia and Related Areas, (ed.) D.G. Gee and B.A. Sturt; John Wiley, Chichester, p. 1071-1076.

Katz, H.R.
1952a: Ein Querschnitt durch die Nunatakzone Ostgrönlands; Meddelelser om Grønland, v. 144, no. 8, 65 p.
1952b: Zur Geologie von Strindbergs Land (NE-Grönland); Meddelelser om Grønland, v. 111, no. 1, 150 p.

Koch, L.
1929: The geology of East Greenland; Meddelelser om Grønland, v. 73, part 2, no, 1, 204 p.

Koch, L. and Haller, J.
1971: Geological map of East Greenland 72°-76°N. Lat. (1:250 000); Meddelelser om Grønland, v. 183, 26 p.

Kranck, E.H.
1935: On the crystalline complex of Liverpool Land; Meddelelser om Grønland, v. 95, no. 7, 122 p.

Larsen, H.C.
1981: A high-pressure granulite facies complex in northwest Payers Land, East Greenland fold belt; Geological Society of Denmark, Bulletin no. 29, p. 161-174.
1990: The East Greenland shelf; in The Arctic Ocean Region, (ed.) A. Grantz, (G.) L. Johnson, and J.F. Sweeney; Geological Society of America, Boulder, Colorado, The Geology of North America, v. L, p. 185-210.

Larsen, L.M. and Watt, W.S.
1985: Episodic volcanism during break-up of the North Atlantic: evidence from the East Greenland plateau basalts; Earth and Planetary Science Letters, v. 73, p. 105-116.

Larsen, L.M., Watt, W.S., and Watt, M.
1989: Geology and petrology of the Lower Tertiary plateau basalts of the Scoresby Sund region, East Greenland; Grønlands Geologiske Undersøgelse, Bulletin no. 157, 164 p.

Larsen, P.-H. and Bengaard, H.J.
1991: Devonian basin initiation in East Greenland: a result of sinistral wrench faulting and Caledonian extensional collapse; Journal of the Geological Society of London, v. 148, p. 355-368.

Leedal, G.P.
1952: The crystalline rocks of East Greenland between latitudes 74°30' and 75°N; Meddelelser om Grønland, v. 142, no. 6, 80 p.

Manby, G.M. and Hambrey, M.J.
1989: The structural setting of the Late Proterozoic tillites of East Greenland; in The Caledonide Geology of Scandinavia, (ed.) R.A. Gayer; Graham and Trotman, London, p. 299-312.

McClay, K.R., Norton, M.G., Coney, P., and Davis, G.H.
1986: Collapse of the Caledonian orogen and the Old Red Sandstone; Nature, v. 323, p. 147-149.

Miller, J.F. and Kurtz, V.E.
1979: Reassignment of the Dolomite Point Formation of East Greenland from the Middle Cambrian (?) to the Lower Ordovician based on conodonts; Geological Society of America, Abstracts with Programs, v. 11, no. 7, p. 480.

Mittelholzer, A.E.
1941: Die Kristallingebiete von Clavering-Ø und Payer Land (Ostgrönland); Meddelelser om Grønland, v. 114, no. 8, 42 p.

Moncrieff, A.C.M.
1989: The Tillite Group and related rocks of East Greenland: implications for Late Proterozoic palaeogeography; in The Caledonide Geology of Scandinavia, (ed.) R.A. Gayer; Graham and Trotman, London, p. 285-297.

Noe-Nygaard, A.
1976: Tertiary igneous rocks between Shannon and Scoresby Sund, East Greenland; in Geology of Greenland, (ed.) A. Escher and W.S. Watt; Geological Survey of Greenland, Copenhagen, p. 387-402.

Odell, N.
1939: The structure of the Kejser Franz Josephs Fjord region, north-east Greenland; Meddelelser om Grønland, v. 119, no. 6, 54 p.

Odell, N.E.
1944: The petrography of the Franz Josef Fjord region, North-East Greenland, in relation to its structures; Transactions of the Royal Society of Edinburgh, v. 61, no. 1, p. 221-246.

Okulitch, A.V., Dawes, P.R., Higgins, A.K., Soper, N.J., and Christie, R.L.
1990: Towards a Nares Strait solution: structural studies on south-eastern Ellesmere Island and northwestern Greenland; Marine Geology, v. 93, p. 369-384.

Parkinson, M.M.L. and Whittard, W.F.
1931: The geological work of the Cambridge expedition to East Greenland in 1929; Quarterly Journal of the Geological Society of London, v. 87, p. 650-674.

Peacock, J.D.
1956: The geology of Dronning Louise Land, N.E. Greenland; Meddelelser om Grønland, v. 137, no. 7, 38 p.
1958: Some investigations into the geology and petrography of Dronning Louise Land, N.E. Greenland; Meddelelser om Grønland, v. 157, no. 4, 139 p.

Peel, J.S.
1980: Geological reconnaissance in the Caledonian foreland of eastern North Greenland with comments on the Centrum Limestone; Grønlands Geologiske Undersøgelse, Rapport no. 99, p. 61-79.
1982: The Lower Palaeozoic of Greenland; in Proceedings of the Third Arctic Geology Symposium, (ed.) A.F. Embry and H.B. Balkwill; Canadian Society of Petroleum Geology, Memoir no. 8, p. 309-330.
1985: Cambrian-Silurian platform stratigraphy of eastern North Greenland; in The Caledonide Orogen - Scandinavia and Related Areas, (ed.) D.G. Gee and B.A. Sturt; John Wiley, Chichester, p. 1076-1094.

Peel, J.S. and Cowie, J.W.
1979: New names for Ordovician formations in Greenland; Grønlands Geologiske Undersøgelse, Rapport no. 91, p. 117-124.

Peel, J.S., Ineson, J.R., Lane, P.D., and Armstrong, H.A.
1981: Lower Palaeozoic stratigraphy around Danmarks Fjord, eastern North Greenland; Grønlands Geologiske Undersøgelse, Rapport no. 106, p. 21-27.

Perch-Nielsen, K., Bromley, R.G., Birkenmajer, K., and Aellen, M.
1972: Field observations in Palaeozoic and Mesozoic sediments of Scoresby Land and northern Jameson Land; Grønlands Geologiske Undersøgelse, Rapport no. 48, p. 39-59.

Peucat, J.J., Tisserant, D., Caby, R., and Clauer, N.
1985: Resistance of Zr to U-Pb resetting in a prograde metamorphic sequence of Caledonian age in East Greenland; Canadian Journal of Earth Sciences, v. 22, p. 330-338.

Phillips, W.E.A. and Friderichsen, J.D.
1981: The Late Precambrian Gåseland tillite, Scoresby Sund, East Greenland; in Earth's Pre-Pleistocene Glacial Record, (ed.) M.J. Hambrey and W.B. Harland; Cambridge University Press, p. 773-775.

Phillips, W.E.A., Stillman, C.J., Friderichsen, J.D., and Jemlin, L.
1973: Preliminary results of mapping in the western gneiss and schist zone around Vestfjord and inner Gåsefjord, south-west Scoresby Sund; Grønlands Geologiske Undersøgelse, Rapport no. 58, p. 17-32.

Rex, D.C. and Gledhill, A.
1974: Reconnaissance geochronology of the infracrustal rocks of Flyverfjord, Scoresby Sund, East Greenland; Geological Society of Denmark, v. 23, p. 49-54.

Rex, D.C. and Gledhill, A.R.
1981: Isotopic studies in the East Greenland Caledonides (72°-74° N) - Precambrian and Caledonian ages; Grønlands Geologiske Undersøgelse, Rapport no. 104, p. 47-72.

Rex, D.C. and Higgins, A.K.
1985: Potassium-argon mineral ages from the East Greenland Caledonides between 72° and 74°N; in The Caledonide Orogen - Scandinavia and Related Areas, (ed.) D.G. Gee and B.A. Sturt; John Wiley, Chichester, p. 1115-1123.

Rex, D.C., Gledhill, A.R., and Higgins, A.K.
1976: Progress report on geochronological investigations in the crystalline complexes of the East Greenland Caledonian fold belt between 72° and 74°N; Grønlands Geologiske Undersøgelse, Rapport no. 80, p. 127-133.
1977: Precambrian Rb-Sr isochron ages from the crystalline complexes of inner Forsblad Fjord, East Greenland fold belt; Grønlands Geologiske Undersøgelse, Rapport no. 85, p. 122-126.

Sonderholm, M. and Tirsgaard, H.
1990: Sedimentological investigation of the Multicoloured 'series' (Eleonore Bay Group, Late Precambrian) in the Scoresby Land - Andrée Land region, North-East Greenland; Grønlands Geologiske Undersøgelse, Rapport no. 148, p. 115-122.

Sønderholm, M., Collinson, J.D., and Tirsgaard, H.
1989: Stratigraphic and sedimentological studies of the Eleonore Bay Group (Precambrian) between 73°30 and 76°N in East Greenland; Grønlands Geologiske Undersøgelse, Rapport no. 145, p. 97-102.

Sommer, M.
1957: Geologische Untersuchungen in den praekambrischen Sedimenten zwischen Grandjeans Fjord und Bessels Fjord in NE-Grönland; Meddelelser om Grønland, v. 160, 56 p.

Steck, A.
1971: Kaledonische metamorphose der praekambrischen Charcot Land Serie, Scoresby Sund, Ost-Grönland; Grønlands Geologiske Undersøgelse, Bulletin no. 97, 69 p.

Steiger, R.H., Hansen, B.T., Schuler, C.H., Bär, M.T., and Henriksen, N.
1979: Polyorogenic nature of the southern Caledonian fold belt in East Greenland: an isotopic age study; Journal of Geology, v. 87, p. 475-495.

Steiger, R.H., Harnik-Soptrajanova, G., Zimmermann, E., and Henriksen, N.
1976: Isotopic age and metamorphic history of the banded gneisses at Danmarkshavn, East Greenland; Contributions to Mineralogy and Petrology, v. 57, p. 1-24.

Surlyk, F.
1978: Submarine fan sedimentation along fault scarps on tilted fault blocks (Jurassic-Cretaceous boundary, East Greenland); Grønlands Geologiske Undersøgelse, Bulletin no. 128, 134 p.

1982: Nares Strait and the down-current termination of the Silurian turbidite basin of North Greenland; in Nares Strait and the Drift of Greenland - A Conflict in Plate Tectonics, (ed.) P.R. Dawes and J.W. Kerr; Meddelelser om Grønland, Geoscience no. 8, p. 147-150.

Surlyk, F., Hurst, J.M., Piasecki, S., Rolle, F., Scholle, P.A., Stemmerik, L., and Thomsen, E.
1986: The Permian of the western margin of the Greenland Sea - a future exploration target; American Association of Petroleum Geologists, Memoir no. 40, p. 629-659.

Swett, K.
1981: Cambro-Ordovician strata in Ny Friesland, Spitsbergen and their paleotectonic significance; Geological Magazine, v. 118, p. 225-250.

Swett, K. and Smit, D.E.
1972: Cambro-Ordovician shelf sedimentation of western Newfoundland, northwest Scotland and central East Greenland; Proceedings of the 24th International Geological Congress, Canada, v. 6, p. 33-41.

Talbot, C.J.
1979: Infracrustal migmatitic upwellings in East Greenland interpreted as thermal convective structures; Precambrian Research, v. 8, p. 77-93.

Vidal, G.
1976: Late Precambrian acritarchs from the Eleonore Bay Group and Tillite Group in East Greenland; Grønlands Geologiske Undersøgelse, Rapport no. 78, 19 p.

Vischer, A.
1943: Die postdevonische Tectonic von Ostgrönland zwischen 74° und 75°N. Br., Kuhn Ø, Wollaston Forland, Clavering Ø und angrezende Gebiete; Meddelelser om Grønland, v. 133, no. 1, 195 p.

Vogt, P.
1965: Zur Geologie von Südwest-Hinks Land (Ostgrönland); Meddelelser om Grønland, v. 154, no. 5, 24 p.

Watt, W.S., Larsen, L.M., and Watt, M.
1986: Volcanic history of the Lower Tertiary plateau basalts in the Scoresby Sund region, East Greenland; Grønlands Geologiske Undersøgelse, Rapport no. 128, p. 145-156.

Wegmann, C.E.
1935: Preliminary report on the Caledonian orogeny in Christian X's Land (North-East Greenland); Meddelelser om Grønland, v. 103, no. 3, 59 p.

Wenk, E.
1961: On the crystalline basement and the basal part of the pre-Cambrian Eleonore Bay Group in the southwest part of Scoresby Sund; Meddelelser om Grønland, v. 168, no. 1, 54 p.

Wenk, E. and Haller, J.
1953: Geological explorations in the Petermann region, western part of Frænkels Land, East Greenland; Meddelelser om Gønland, v. 111, no. 3, 48 p.

Wordie, J.M.
1930: Cambridge East Greenland Expedition 1929, Ascent of Peterman Peak; Geographical Journal, v. 75, p. 481-498.

Ziegler, P.A.
1985: Late Caledonian framework of western and central Europe; in The Caledonide Orogen - Scandinavia and Related Areas, (ed.) D.G. Gee and B.A. Sturt; John Wiley, Chichester, p. 3-18.

Zweifel, H.
1959: Geologie und Petrographie von Nathorst Land (NE-Grönland); Meddelelser om Grønland, v. 160, no. 3, 94 p.

ADDENDUM

Rb-Sr and Sm-Nd model ages of basement gneiss lithologies in the southern half of the Bessel Fjord-Ingolf Fjord metamorphic complex show that most of the rocks are related to an Early Proterozoic event of crust formation around 2000 Ma (Kalsbeek et al., 1993). U-Pb zircon (SHRIMP) analyses on three samples, also reported by Kalsbeek et al., yielded ages between 1974 and 1739 Ma. Tucker et al. (1993) obtained similar results for a granitoid gneiss in southeast Dronning Louise Land, for which U-Pb analysis of a zircon suite gave a crystallization age of 1909 Ma.

The eclogites from the Bessel Fjord-Ingolf Fjord metamorphic complex around Skærfjorden have been shown to form part of a widespread medium-temperature eclogite province extending from latitude 76°40'N to 78°N (Gilotti, 1993). Eclogite, garnet-clinopyroxenite, garnet-websterite and websterite bodies have been recorded, and field relations and preliminary geochronology indicate that the eclogite facies metamorphism is Caledonian (Brueckner and Gilotti, 1993). Widespread regional Caledonian metamorphism is also indicated by 420-400 Ma U-Pb lower intercepts on zircons from the gneisses (Kalsbeek et al., 1993; Tucker et al., 1993), and by $^{40}Ar/^{39}Ar$ mineral cooling ages obtained from a variety of rock samples from the area (Dallmeyer et al., in press).

West of the Caledonian thrust zone in Dronning Louise Land, the foreland rock units include basement gneisses and two cover sequences (Trekant and Zebra series). Isotopic mineral ages have been obtained on detrital material from the sandstones of the younger Zebra series; U-Pb dates for detrital zircon grains define an age range of detritus between 3001 and 1700 Ma, and detrital muscovites record $^{40}Ar/^{39}Ar$ mineral cooling ages of ca. 1700-1600 Ma (Tucker et al., 1993). As the Zebra series contains trace fossils and *Skolithus* tubes, a correlation with the uppermost part of the Hagen Fjord Group of eastern North Greenland is considered most probable (Clemmensen and Jepsen, 1992). In Kronprins Christian Land the Hagen Fjord Group is represented in Caledonian thrust sheets as well as in the adjacent foreland areas. Clemmensen and Jepsen describe the lithostratigraphy and geological setting of the group, which they redefine to contain only Upper Proterozoic, mainly shallow marine shelf deposits.

The Late Proterozoic Eleonore Bay Group of central East and North-East Greenland has been raised to Supergroup status by Sønderholm and Tirsgaard (1993). Five new groups have been erected, corresponding to the former "series" divisions. The structural setting of the northernmost outcrops of the Eleonore Bay Supergroup centred on Ardencaple Fjord has been analyzed by Soper and Higgins (1993), who extend their model to the general problem of how extensive remnants of the succession came to be preserved as enclaves within the older metamorphic basement after Caledonian deformation and uplift. They infer that the Eleonore Bay Supergroup is preserved in pre-Caledonian graben formed during a Vendian rift episode, and that the extensional shear zones forming the contacts with the adjacent crystalline rocks were reactivated in compression during the Caledonian Orogeny.

On the basis of their detailed studies in Dronning Louise Land, Strachen et al. (1992) developed a Caledonian tectonic model, which involves early low-angle northward directed sinistral shear, followed by partitioned sinistral strike slip along the Storstrømmen shear zone and oblique ductile thrusting.

Recent work on the post-Caledonian continental Devonian sediments of North-East Grenland includes a new lithostratigraphy (Olsen and Larsen, 1993), and a sedimentary basin analysis (Olsen, 1993).

Brueckner, H.K. and Gilotti, J.A.
1993: Preliminary age constraints on the timing of eclogite facies metamorphism, North-East Greenland Caledonides. Geological Society of America Abstracts with Programs, v. 25, p. 340.

Clemmensen, L.B. and Jepsen, H.F.
1992: Lithostratigraphy and geological setting of Upper Proterozoic shoreline-shelf deposits, Hagen Fjord Group, eastern North Greenland. Grønlands Geologiske Undersøgelse, Report no. 157, 27 p.

Dallmeyer, R.D., Strachan, R.A., and Henriksen, N.
in press: $^{40}Ar/^{39}Ar$ mineral age record in NE Greenland - implications for tectonic evolution of the North Atlantic Caledonides. Journal of the Geological Society of London.

Gilotti, J.A.
1993: Discovery of a medium-temperature eclogite province in the Caledonides of North-East Greenland. Geology, v. 21, p. 523-526.

Kalsbeek, F., Nutman, A.P., and Taylor, P.N.
1993: Palaeoproterozoic basement province in the Caledonian fold belt of North-East Greenland. Precambrian Research, v. 63, p. 163-178.

Olsen, H.
1993: Sedimentary basin analsis of the continental Devonian basin in North-East Greenland. Grønlands Geologiske Undersøgelse, Bulletin no. 168, 80 p.

Olsen, H. and Larsen, P.-H.
1993: Lithostratigraphy of the continental Devonian sediments in North-East Greenland. Grønlands Geologiske Undersøgelse, Bulletin no. 165, 108 p.

Sønderholm, M. and Tirsgaard, H.
1993: Lithostratigraphic framework of the Upper Proterozoic Eleonore Bay Supergroup of East and North-East Greenland. Grønlands Geologiske Undersøgelse, Bulletin no. 167, 38 p.

Soper, N.J. and Higgins, A.K.
1993: Basement-cover relationships in the East Greenland Caledonides: evidence from the Eleonore Bay Supergroup at Ardencaple Fjord. Transactions of the Royal Society of Edinburgh: Earth Sciences, v. 84, p. 103-115.

Strachan, R.A., Holdsworth, R.E., Friderichsen, J.D., and Jepsen, H.F.
1992: Regional Caledonian structure within an oblique convergence zone, Dronning Louise Land, NE Greenland. Journal of the Geological Society of London, v. 149, p. 359-371.

Tucker, R.D., Dallmeyer, R.D., and Strachan, R.A.
1993: Age and tectonothermal record of Laurentian basement, Caledonides of NE Greenland. Journal of the Geological Socity of London, v. 150, p. 371-379.

Author's Address

A.K. Higgins
Geological Survey of Greenland
Oster Voldgade 10
DK-1350 Kobenhavn K
Denmark

Printed in Canada

AUTHOR INDEX

INDEX